WORLDMARK YEARBOOK 2000

ISSN 1527-6503

WORLDMARK YEARBOOK 2000

Volume 3

S–Z

International Organizations

Index

Mary Rose Bonk, Editor

GALE GROUP

Detroit
New York
San Francisco
London
Boston
Woodbridge, CT

Editor: Mary Rose Bonk
Associate Editors: Nancy Matuszak, Matthew May, John F. McCoy, Michael Reade
Assistant Editors: R. David Riddle, Chris Romig, Christy Wood
Permissions Manager: Maria Franklin
Permissions Specialist: Margaret Chamberlain
Permissions Associate: Shalice Shah-Caldwell
Image Cataloger: Mary Grimes

Composition Manager: Mary Beth Trimper
Assistant Production Manager: Evi Seoud
Manufacturing Manager: Dorothy Maki
Senior Buyer: Wendy Blurton
Product Design Manager: Cynthia Baldwin
Senior Art Director: Michelle DiMercurio
Graphic Specialist: Christine O'Bryan
Indexing Specialists: Susan Kelsch, Amy Suchowski, Cindy Tsiang

27500 Drake Rd.
Farmington Hills, MI 48331-3535

British Library Cataloguing in Publication Data. A Catalogue record of this book is available from the British Library.
ISBN 0–7876–4931–7 (2-Volume set)
ISBN 0–7876–4932–5 (Volume 1)
ISBN 0–7876–4933–3 (Volume 2)
ISBN 0–7876–5088–9 (Volume 3)
ISSN 1527–6503

Printed in the United States of America

10 9 8 7 6 5 4 3 2 1

CONTENTS

FOREWORD

A NEW, YEARLY REFERENCE FOR A NEW ERA

As a new addition to the Worldmark family, *Worldmark Yearbook* presents to users a comprehensive profile of 229 of the world's nations and terrritories and their current events. Information often scattered throughout books, articles, and various agencies is compiled here in an easy to use reference. It complements other Worldmark publications by presenting the most current information on countries around the world. The Worldmark line already provides you with an in-depth historical account for these countries in *Worldmark Encyclopedia of the Nations*. With the introduction of the *Worldmark Yearbook*, you can be kept up-to-date with the latest developments.

The Year in Review

Worldmark Yearbook begins with a look at the year in review. From Tokyo to Berlin, and from Moscow to Auckland and Pretoria, major events from the past year are highlighted and placed in an international perspective. The main focus of the *Yearbook* is on national events, so the Year in Review provides a broader look at how national issues can affect regional neighbors and the world.

Where in the World?

Each country is accompanied by a map that places it within its area of the world and which details major cities and landmarks. Geography often plays an important role in a nation's politics and economy. The expansive maps can help *Yearbook* users better understand how geography may affect what goes on within a nation. In addition, in a world increasingly referred to as a ''global village,'' national events often spill over into neighboring states. To easily identify neighbors, six color regional maps are on hand for reference.

National Symbols

An added feature of the *Worldmark Yearbook* is the color illustration of national symbols, including flags and official insignia. Each country's official flag is represented and a brief description is included in each entry. A country's official insignia, such as an emblem or seal, is also represented.

Profiles

A national profile highlights basic information for each country. The national capital, monetary unit, anthem, and climate are a few examples of the data available at the beginning of each entry.

Background Check

A yearbook of current events would be of little use without a context in which to put those events. *Worldmark Yearbook*'s introductory survey provides users with this context. Recent history from the mid-twentieth century onward is outlined to provide background in and support for the events of the past year. But history itself does not present a comprehensive picture of how a nation operates. Do you want to know how a country's government is set up? Who can participate in it? How the economy operates? What are the major industrial and economic developments of the past few years? The answers to these questions, and more, can be found in the introductory survey.

Analyzing the Year's Events

A timeline is included in each entry and lists chronologically the events of the past year. Selecting the most significant of these events, an analysis explains them in greater depth and pulls together the threads of politics, economy, and culture to create a cohesive picture of a distinct nation. Economic struggles, cultural revivals, political triumphs: the analysis paints a more personal picture of each country, which can not be as easily portrayed through the simple listing of facts or statistics.

Who's Who?

Who's been running the country while these events have taken place? To find out, users can reference the *Yearbook*'s directory, which, when possible, includes information on how to contact an individual or group. Main government functionaries, including the heads and ministers of state, are listed, as are political organizations active within the country and its major judicial courts. In this "global village," knowing one's neighbors is important, and maintaining official contact with them is a vital part of government operations. In this light, users will find a comprehensive listing of diplomatic representation for each country.

Just the Facts

If hard data is what you're looking for, you can find it in the statistical survey. Population size and growth, economic output and development, major industrial production, and more can be located in the statistical survey. Statistics complement the analysis by explaining numerically much of what's going on in a country's society and economy. It creates a numerative picture that can be easily compared with that of other nations.

Want More?

Check out the Further Reading section for sources of additional information on statistics, current events, and historical background. You can also refer to the International Organizations listing. It provides contact information and a brief description of over 1,800 international organizations that are involved in a variety of global concerns.

At Your Fingertips

Are you looking for something specific? A glance at the *Yearbook*'s comprehensive index will help you find your way. The user-friendly index covers personal names, subjects, and geographies and can refer you quickly to the information you need.

A COMPREHENSIVE TOOL TO TODAY'S WORLD

The *Worldmark Yearbook* pulls together the many components that make up a government and a society to give users a well-balanced, comprehensive, and illustrative source of information on the world's nations. Its many headers make quick reference points to easily locate information. *Worldmark Yearbook* is a proud new edition to the Worldmark line and will enhance the libraries of all reference users. This is just the beginning.

We encourage you to contact us with comments or suggestions. Tell us what you want to see in the *Yearbook* and how we can better meet your needs. Comments and suggestions can be sent to: The Editors, *Worldmark Yearbook*, The Gale Group, 27500 Drake Road, Farmington Hills, MI 48331. Or, call toll free at 1-800-877-4253.

ACKNOWLEDGEMENTS

For editorial and technical assistance that helped keep this project on track and on time, the editors are extremely grateful to the following Gale Group contributors:

Richard Antonowicz, programmer/analyst, **Pamela A. Dear**, associate editor, **Shelly Dickey**, senior editor, **Kathy Droste**, editor, **Anthony Gerring**, technical support, **Bernard Grunow**, editor, **Amanda Moran**, senior market analyst, **Rita Runchock**, managing editor, and **Phyllis Spinelli**, associate editor.

For editorial and textual contributions to the *Worldmark Yearbook*, the editors are indebted to the following:

Advanced Information Consultants, Canton, Michigan, **Kimberly Burton**, Ann Arbor, Michigan, **Michael Dawson**, Carlsbad, California, **Eastword Publications Development**, Cleveland, Ohio, **Editorial Code and Data, Inc. (ECDI)**, Southfield, Michigan, **GGS Information Services**, York, Pennsylvania, **Richard Clay Hanes**, Eugene, Oregon, **Margery Heffron**, Exeter, New Hampshire, **Paul Kobel**, Charlotte, North Carolina, **John Macaulay**, Miami, Florida, and **Vocabula Communications Company**, Lexington, Massachusetts.

For permission to take material from personal or published sources, use of images, and for other courtesies extended during the preparation of this edition, the editors are grateful to the following sources:

Eastword Publications Development, **The Flag Institute**, **International Monetary Fund**, Publications Services Unit, **Maryland Cartographics**, **UNESCO (United Nations Education, Scientific and Cultural Organization)**, **UNIDO (United Nations Industrial Development Organization)**, **United Nations Publications**, author of original material, **Oxford University Press, Inc.**, **Pascal Vagnat**, and the **World Bank**.

KEY TO ABBREVIATIONS

ABEDA: Arab Bank for Economic Development in Africa

ACC: Arab Cooperation Council

ACCT: Agence de Cooperation Culturelle et Technique; see Agency for Cultural and Technical Cooperation

ACP: African, Caribbean, and Pacific Countries

AfDB: African Development Bank

AFESD: Arab Fund for Economic and Social Development

AG: Andean Group

AL: Arab League

ALADI: Asociacion Latinamericana de Intergracion; see Latin American Integration Association (LAIA)

AMF: Arab Monetary Fund

AMU: Arab Maghreb Fund

ANZUS: Australia-New Zealand-United States Security Trust

APEC: Asia Pacific Economic Cooperation

AsDB: Asian Development Bank

ASEAN: Association of Southeast Asian Nations

BAD: Banque Africaine de Developpement; see African Development Bank (AfDB)

BADEA: Banque Arabe de Developpement Economique en Afrique; see Arab Bank for Economic Development in Africa (ABEDA)

BCIE: Banco Centroamericano de Integracion Economico; see Central American Bank for Economic Integration (BCIE)

BDEAC: Banque de Development des Etats de l'Afrique Centrale; see Central African States Development Bank

Benelux: Benelux Economic Union

BID: Banco Interamericano de Desarrollo; see Inter-American Development Bank (IADB)

BIS: Bank for International Settlements

BOAD: Banque Ouest-Africaine de Developpement; see West African Development Bank (WADB)

BSEC: Black Sea Economic Coorperation Zone

C: Commonwealth

CACM: Central American Common Market

CAEU: Council of Arab Economic Unity

CARICOM: Caribbean Community and Common Market

CBSS: Council of Baltic Sea States

CCC: Customs Cooperation Council

CDB: Caribbean Development Bank

CE: Council of Europe

CEAO: Communaute Economique de l'Afrique de l'Ouest; see West African Economic Community (CEAO)

CEEAC: Communaute Economique des Etats de l' Afrique Centrale; see Economic Community of Central African States (CEEAC)

CEI: Central European Initiative

CEMA: Council for Mutual Economic Assistance; also known as CMEA or Comecon

CEPGL: Communaute Economique de Pays des Grands Lacs; see Economic Community of the Great Lakes Countries (CEPGL)

CERN: Conseil European pour la Recherche Nucleaire; see European Organization for Nuclear Research (CERN)

CG: Contadora Group

CIS: Commonwealth of Independent States

CMEA: Council for Mutual Economic Assistance (CEMA); also known as Comeecon

COCOM: Coordinating Committee on Export Controls

Comecon: Council for Mutual Economic Assistance (CEMA); also known as CMEA

CP: Colombo Plan

CSCE: Conference on Security and Cooperation in Europe

DC: Developed country

EADB: East African Development Bank

EBRD: European Bank for Reconstruction and Development

EC: European Community; see European Union (EU)

ECA: Economic Commission for Africa

ECAFE: Economic Commission for Asia and the Far East; see Economic and Social Commission for Asia and the Pacific (ESCAP)

ECE: Economic Commission for Europe

ECLA: Economic Commission for Latin America; see Economic Commission for Latin America and the Caribbean (ECLAC)

ECLAC: Economic Commission for Latin America and the Caribbean

ECO: Economic Cooperation Organization

ECOSOC: Economic and Social Council

ECOWAS: Economic Community of West African States

ECSC: European Coal and Steel Community

ECWA: Economic Commission for Western Asia; see Economic and Social Council for Western Asia (ESCWA)

EEC: European Economic Community

EFTA: European Free Trade Association

EIB: European Investment Bank

Entente: Council of the Entente

ESA: European Space Agency

ESCAP: Economic and Social Commission for Asia and the Pacific

ESCWA: Economic and Social Commission for Western Asia

EU: European Union

Euratom: European Atomic Energy Community

FAO: Food and Agriculture Organization

FLS: Front Line States

FZ: Franc Zone

G-2: Group of 2

G-3: Group of 3

G-5: Group of 5

G-6: Group of 6

G-7: Group of 7

G-8: Group of 8

G-9: Group of 9

G-10: Group of 10

G-11: Group of 11

G-15: Group of 15

G-19: Group of 19

G-24: Group of 24

G-30: Group or 30

G-33: Group of 33

G-77: Group of 77

GATT: General Agreement on Tariff and Trade

Habitat: Commission on Human Settlements

IADB: Inter-American Development Bank

IAEA: International Atomic Energy Agency

IBEC: International Bank for Economic Cooperation

IBRD: International Bank for Reconstruction and Development

ICAO: International Civil Aviation Organization

ICC: International Chamber of Commerce

ICEM: Intergovernmental Committee for European Migration; see International Organization for Migration (IOM)

ICFTU: International Confederation of Free Trade Unions

ICJ: International Court of Justice

ICM: Intergovernmental Committee for Migration; see International Organization for Migration (IOM)

ICRC: International Committee of the Red Cross

ICRM: International Red Cross and Red Crescent Movement

IDA: International Development Association

IDB: Islamic Development Bank

IEA: International Energy Agency

IFAD: International Fund for Agriculture Development

IFC: International Finance Corporation

IFCTU: International Federation of Christian Trade Unions

IFRCS: International Federation of Red Cross and Red Crescent Societies

IGADD: Inter-Governmental Authority on Drought and Development

IIB: International Investment Bank

ILO: International Labor Organization

IMCO: Intergovernmental Maritime Consultative Organization; see International Maritime Organization (IMO)

IMF: International Monetary Fund

IMO: International Maritime Fund

INMARSAT: International Maritime Satelite Organization

INTELSAT: International Telecommunications Satellite Organization

INTERPOL: International Criminal Police Organization

IOC: International Olympic Committee

IOM: International Organization for Migration

ISO: International Organization for Standardization

ITU: International Telecommunications Union

LAES: Latin American Economic System

LAIA: Latin American Integration Association

LAS: League of Arab States; see Arab League (AL)

LDC: Less developed country

LLDC: Least developed country

LORCS: League of Red Cross and Red Crescent Societies

MERCOSUR: Mercado Commun del Cono Sur; see Southern Cone Common Market

MINURSO: United Nations Mission for the Referendum in Western Sahara

MTCR: Missile Technology Control Regime

NACC: North Atlantic Cooperation Council

NAM: Nonaligned Movement

NATO: North Atlantic Treaty Organization

NC: Nordic Council

NEA: Nuclear Energy Agency

NIB: Nordic Investment Bank

NIC: Newly industrializing country; see newly industrializing economy (NIE)

NIE: Newly industrializing economy

NSG: Nuclear Suppliers Group

OAPEC: Organization of Arab Petroleum Exporting Countries

OAS: Organization of American States

OAU: Organization of African Unity

OECD: Organization for Economic Cooperation and Development

OECS: Organization of Eastern Caribbean States

OIC: Organization of the Islamic Conference

ONUSAL: United Nations Observer Mission in El Salvador

OPANAL: Organismo para la Proscripcion de las Armas Nuclearea en la America Latina y el Caribe; see Agency for the Prohibition of Nuclear Weapons in Latin America and the Caribbean

OPEC: Organization of Petroleum Exporting Countries

OSCE: Organization on Security and Cooperation in Europe

PCA: Permanent Court of Arbitration

PPP: Partnership for Peace

RG: Rio Group

SAARC: South Asian Association for Regional Cooperation

SACU: South African Customs Union

SADC: South African Development Community

SADCC: South African Development Coordination Conference

SELA: Sistema Economico Latinamericana; see Latin American Economic System (LAES)

SPARTECA: South Pacific Regional Trade and Economic Cooperation Agreement

SPC: South Pacific Commission

SPF: South Pacific Forum

UDEAC: Union Douaniere et Economique de l'Afrique Centrale; see Central African Customs and Economic Union (UDEAC)

UN: United Nations

UNAVEM II: United Nations Angola Verification Mission

UNAMIR: United Nations Assistance Mission for Rwands

UNCTAD: United Nations Conference on Trade and Development

UNDOF: United Nations Disengagement Observer Force

UNDP: United Nations Development Program

UNEP: United Nations Environment Program

UNESCO: United Nations Educational, Scientific, and Cultural Organization

UNFICYP: United Nations Forces in Cyprus

UNFPA: United Nations Fund for Population Activities; see UN Population Fund (UNFPA)

UNHCR: United Nations Office of the High Commissioner for Refugees

UNICEF: United Nations Children's Fund

UNIDO: United Nations Industrial Development Organization

UNIFIL: United Nations Interim Force in Lebanon

UNIKOM: United Nations Iraq-Kuwait Observation Mission

UNITAR: United Nations Institute for Training and Research

UNMIH: United Nations Mission in Haiti

UNMOGIP: United Nations Military Observer Group in India and Pakistan

UNOMIG: United Nations Observer Mission in Georgia

UNOMIL: United Nations Observer Mission in Liberia

UNOMOZ: United Nations Operation in Mozambique

UNOMUR: United Nations Observer Mission Uganda-Rwanda

UNOSOM: United Nations Operation in Somalia

UNPROFOR: United Nations Protection Force

UNRISD: United Nations Research Institute for Social Development

UNRWA: United Nations Relief and Works Agency for Palestine Refugees in the Near East

UNTAC: United Nations Transitional Authority in Cambodia

UNTSO: United Nations Truce Supervision Organization

UNU: United Nations University

UPU: Universal Postal Union

USSR/EE: USSR/Eastern Europe

WADB: West African Development Bank

WCL: World Confederation of Labor

WEU: Western European Union

WFC: World Food Council

WFP: World Food Program

WFTU: World Federation of Trade Unions

WHO: World Health Organization

WIPO: World Intellectual Property Organization

WMO: World Meteorological Organization

WP: Warsaw Pact

WTO: World Trade Organization

WtoO: World Tourism Organization

ZC: Zangger Committee

IMPERIAL/METRIC CONVERSION KEY

WHEN YOU KNOW	MULTIPLY BY	TO FIND
Length		
Millimeters (mm)	0.04	inches (in)
Centimeters (cm)	0.4	inches (in)
Meters (m)	3.3	feet (ft)
Meters (m)	1.1	yards (yd)
Kilometers (km)	0.6	miles (mi)
Area		
sq. centimeters (cm^2)	0.155	sq. inches (in^2)
sq. meters (m^2)	10.76	sq. feet (ft^2)
sq. meters (m^2)	1.2	sq. yards (yd^2)
sq. kilometers (km^2)	0.4	sq. miles (mi^2)
hectares (ha)	2.5	acres
Weight		
grams (g)	0.035	ounces (oz)
kilograms (km)	2.2	pounds (lbs)
metric tons (t)	1.1	short tons (2,000 lbs)
Volume		
milliliters (ml)	0.03	fluid ounces (fl oz)
liters (L)	2.1	pints (pt)
liters (L)	1.06	quarts (qt)
liters (L)	0.26	gallons (gal)
cubic meters (m^3)	35.0	cubic feet (ft^3)
cubic meters (m^3)	1.3	cubic yards (yd^3)
Temperature		
Celsius (°C)	9/5 +32	Fahrenheit (°F)

WHEN YOU KNOW	MULTIPLY BY	TO FIND
Length		
inches (in)	25.4	millimeters
inches (in)	2.54	centimeters (cm)
feet (ft)	30.5	centimeters (cm)
yards (yd)	0.9	meters (m)
miles (m)	1.1	kilometers (km)
Area		
sq. inches (in^2)	6.45	sq. centimeters (cm^2)
sq. feet (ft^2)	0.09	sq. meters (m^2)
sq. yards (yd^2)	0.84	sq. meters (m^2)
sq. miles (mi^2)	0.4	sq. kilometers (km^2)
acres	0.4	hectares (ha)
Weight		
ounces (oz)	28.0	grams (g)
pounds (lbs)	0.45	kilograms (kg)
short tons (2,000 lbs)	0.9	metric tons (t)
Volume		
fluid ounces (fl oz)	30.0	milliliters (ml)
pints (pt)	0.47	liters (L)
quarts (qt)	.95	liters (L)
gallons (gal)	3.8	liters (L)
cubic feet (ft^3)	0.03	cubic meters (m^3)
cubic yards (yd^3)	0.76	cubic meters (m^3)
Temperature		
Fahrenheit (°F)	5/9 −32	Celsius (°C)

STATUS OF NATIONS

COUNTRY NAME: SYSTEM OF GOVERNMENT

Afghanistan: transitional government

Albania: emerging democracy

Algeria: republic

American Samoa: unincorporated territory of the United States

Andorra: parliamentary democracy

Angola: transitional government, nominally a multiparty democracy with a strong presidential system

Anguilla: British crown colony

Antigua and Barbuda: parliamentary democracy

Argentina: republic

Armenia: republic

Aruba: parliamentary

Australia: democratic, federal-state system recognizing the British monarch as sovereign

Austria: federal republic

Azerbaijan: republic

Bahamas, The: commonwealth

Bahrain: traditional monarchy

Bangladesh: republic

Barbados: parliamentary democracy

Belarus: republic

Belgium: federal parliamentary democracy under a constitutional monarch

Belize: parliamentary democracy

Benin: republic under multiparty democratic rule

Bermuda: British crown colony

Bhutan: monarchy

Bolivia: republic

Bosnia and Herzegovina: emerging democracy

Botswana: parliamentary republic

Brazil: federal republic

British Virgin Islands: British crown colony

Brunei: constitutional sultanate

Bulgaria: republic

Burkina Faso: parliamentary

Burundi: republic

Cambodia: multiparty liberal democracy under a constitutional monarchy

Cameroon: unitary republic; multiparty presidential regime (opposition parties legalized in 1990)

Canada: federation with parliamentary democracy

Cape Verde: republic

Cayman Islands: British crown colony

Central African Republic: republic

Chad: republic

Chile: republic

China: Communist state

Christmas Island: territory of Australia

Colombia: republic; executive branch dominates government structure

Comoros: independent republic

Congo, Democratic Republic of: dictatorship; presumably undergoing a transition to representative government

Congo, Republic of: republic

Cook Islands: self-governing parliamentary democracy

Costa Rica: democratic republic

Cote d'Ivoire: republic

Croatia: presidential/parliamentary democracy

Cuba: Communist state

Cyprus: republic

Czech Republic: parliamentary democracy

Denmark: constitutional monarchy

Djibouti: republic

Dominica: parliamentary democracy

Dominican Republic: republic

Ecuador: republic

Egypt: republic

El Salvador: republic

Equatorial Guinea: republic in transition to multiparty democracy

Eritrea: transitional government

Estonia: parliamentary democracy

Ethiopia: federal republic

Falkland Islands: British crown colony

Faroe Islands: part of the Kingdom of Denmark; self-governing overseas administrative division of Denmark since 1948

Fiji: republic

Finland: republic

France: republic

French Guiana: French overseas department

French Polynesia: territory of France

Gabon: republic; multiparty presidential regime

Gambia, The: republic under multiparty democratic rule

Georgia: republic

Germany: federal republic

Ghana: constitutional democracy

Gibraltar: British crown colony

Greece: parliamentary republic

Greenland: part of the Kingdom of Denmark; self-governing overseas administrative division of Denmark since 1979

Grenada: parliamentary democracy

Guadeloupe: overseas department and administrative region of France

Guam: unincorporated territory of the United States

Guatemala: republic

Guernsey: dependency of the British crown

Guinea: republic

Guinea-Bissau: republic

Guyana: republic

Haiti: republic

Honduras: republic

Hungary: republic

Iceland: constitutional republic

India: federal republic

Indonesia: republic

Iran: theocratic republic

Iraq: republic

Ireland: republic

Israel: republic

Italy: republic

Jamaica: parliamentary democracy

Japan: constitutional monarchy

Jersey: dependency of the British crown

Jordan: constitutional monarchy

Kazakstan: republic

Kenya: republic

Kiribati: republic

Korea, North: Communist state; one-man dictatorship

Korea, South: republic

Kuwait: nominal constitutional monarchy

Kyrgyzstan: republic

Laos: Communist state

Latvia: parliamentary democracy

Lebanon: republic

Lesotho: parliamentary constitutional monarchy

Liberia: republic

Libya: Jamahiriya (a state of the masses) in theory, governed by the populace through local councils; in fact, a military dictatorship

Liechtenstein: hereditary constitutional monarchy

Lithuania: parliamentary democracy

Luxembourg: constitutional monarchy

Macau: Chinese province

Macedonia: emerging democracy

Madagascar: republic

Malawi: multiparty democracy

Malaysia: constitutional monarchy

Maldives: republic

Mali: republic

Malta: parliamentary democracy

Man, Isle of: British crown dependency

Marshall Islands: constitutional government in free association with the US

Martinique: overseas department and administrative region of France

Mauritania: republic

Mauritius: parliamentary democracy

Mayotte: territory of France

Mexico: federal republic operating under a centralized government

Micronesia, Federated States of: constitutional government in free association with the US

Midway Islands: territory of the United States

Moldova: republic

Monaco: constitutional monarchy

Mongolia: republic

Montenegro: republic

Montserrat: British crown colony

Morocco: constitutional monarchy

Mozambique: republic

Myanmar: military regime

Namibia: republic

Nauru: republic

Nepal: parliamentary democracy

Netherlands: constitutional monarchy

Netherlands Antilles: parliamentary

New Caledonia: territory of France

New Zealand: parliamentary democracy

Nicaragua: republic

Niger: republic

Nigeria: republic transitioning from military to civilian rule

Niue: self-governing parliamentary democracy

Norfolk Island: territory of Australia

Northern Mariana Islands: commonwealth

Norway: constitutional monarchy

Oman: monarchy

Pakistan: federal republic

Palau: constitutional government in free association with the US

Panama: constitutional republic

Papua New Guinea: parliamentary democracy

Paraguay: republic

Peru: republic

Philippines: republic

Poland: democratic state

Portugal: parliamentary democracy

Puerto Rico: commonwealth

Qatar: traditional monarchy

Reunion: overseas department of France

Romania: republic

Russia: republic

Rwanda: republic

Saint Helena: British dependency

Saint Kitts and Nevis: constitutional monarchy

Saint Lucia: constitutional monarchy

Saint Pierre and Miquelon: French territorial collectivity

Saint Vincent and the Grenadines: constitutional monarchy

Samoa: constitutional monarchy under native chief

San Marino: republic

Sao Tome and Principe: republic

Saudi Arabia: monarchy

Senegal: republic under multiparty democratic rule

Serbia: republic

Seychelles: republic

Sierra Leone: constitutional democracy

Singapore: republic within Commonwealth

Slovakia: parliamentary democracy

Slovenia: parliamentary democratic republic

Solomon Islands: parliamentary democracy

Somalia: none

South Africa: republic

Spain: parliamentary monarchy

Sri Lanka: republic

Sudan: transitional

Suriname: republic

Swaziland: monarchy

Sweden: constitutional monarchy

Switzerland: federal republic

Syria: republic under military regime since March 1963

Taiwan: multiparty democratic regime headed by popularly elected president

Tajikistan: republic

Tanzania: republic

Thailand: constitutional monarchy

Togo: republic under transition to multiparty democratic rule

Tonga: hereditary constitutional monarchy

Trinidad and Tobago: parliamentary democracy

Tunisia: republic

Turkey: republican parliamentary democracy

Turkmenistan: republic

Turks and Caicos: British dependency

Tuvalu: constitutional monarchy with a parliamentary democracy

Uganda: republic

Ukraine: republic

United Arab Emirates: federation with specified powers delegated to the UAE federal government and other powers reserved to member emirates

United Kingdom: constitutional monarchy

United States: federal republic; strong democratic tradition

Uruguay: republic

Uzbekistan: republic; effectively authoritarian presidential rule, with little power outside the executive branch; executive power concentrated in the presidency

Vanuatu: republic

Vatican City: monarchical-sacerdotal state

Venezuela: republic

Vietnam: Communist state

Virgin Islands: territory of the United States

Wallis and Futuna: French overseas territory

Yemen: republic

Zambia: republic

Zimbabwe: parliamentary democracy

SOURCES OF STATISTICS

GEOGRAPHY—1

SOURCE. U.S. Central Intelligence Agency (CIA) 1998, *The World Factbook 1998* [Online]. Available: http://www.cia.gov/cia/publications/factbook/index.html [October 1999].

NOTES.

Comparative area—Based on total area equivalents. Most entities are compared with the entire United States or one of the 50 states. The smaller entities are compared with Washington, D.C. (178 square km, 69 square miles), or the Mall in Washington, D.C. (0.59 square km, 0.23 square miles, 146 acres).

km—Kilometers.

Land area—Aggregate of all surfaces delimited by international boundaries and/or coastlines, excluding inland water bodies (lakes, reservoirs, rivers).

Land use—Human use of the land surface is categorized as *arable land*—land cultivated for crops that are replanted after each harvest (wheat, maize, rice); *permanent crops*—land cultivated for crops that are not replanted after each harvest (citrus, coffee, rubber); *meadows and pastures*—land permanently used for herbaceous forage crops; *forest and woodland*—land under dense or open stands of trees; and *other*—any land type not specifically mentioned above (urban areas, roads, deserts).

mi—Miles.

NA—Data are not available.

Total area—Sum of all land and water area delimited by international boundaries and/or coastlines.

DEMOGRAPHICS—2A

SOURCE. U.S. Bureau of the Census (1998). *International Database 1998* [Online]. Available: http://www.census.gov:80/ipc/www/wp98.html [October 1999].

NOTES.

NA—Data are not available.

DEMOGRAPHICS—2B

SOURCE. U.S. Central Intelligence Agency (CIA) 1998, *The World Factbook 1998* [Online]. Available: http://www.cia.gov/cia/publications/factbook/index.html [October 1999].

NOTES.

NA—Data are not available.

HEALTH PERSONNEL—3

SOURCE. The World Bank, *World Development Indicators 1999* (March 1999), pages 90–92. Reprinted with permission.

United Nations Development Program (UNDP) and Oxford University Press, *Human Development Report 1999*, pages 172–175. Reprinted with permission.

NOTES.

Public Health Expenditure—This category consists of recurrent and capital spending from government (central and local) budgets, external borrowings and grants (including donations from international agencies and nongovernmental organizations), and social (or compulsory) health insurance funds.

Private Health Expenditure—This category includes direct household (out-of-pocket) spending, private insurance, charitable donations, and direct service payments by private corporations.

FOOTNOTES.

a—Data are for the most recent year available.

b—Data may not sum to totals because of rounding.

c—Data refer to 1993 or a year around 1993.

HEALTH CARE INDICATORS—4

SOURCE. The World Bank, *World Development Indicators 1999* (March 1999), pages 94–112. Reprinted with permission.

United Nations Development Program (UNDP) and Oxford University Press, *Human Development Report 1999*, pages 211–214. Reprinted with permission.

NOTES.

Percentage of Population with Access to Safe Water.—This is the share of the population with reasonable access to an adequate amount of safe water (including treated surface water and untreated but uncontaminated water, such as from springs, sanitary wells, and protected boreholes). In urban areas the source may be a public fountain or standpipe located not more than 200 meters away. In rural areas the definition implies that members of a household do not have to spend a disproportionate part of the day fetching water. An adequate amount of safe water is that needed to satisfy metabolic, hygienic, and domestic requirements—usually about 20 liters a person a day. The definition of safe water has changed over time.

Percentage of Population with Access to Sanitation.—This is the share of the population with at least adequate excreta disposal facilities that can effectively prevent human, animal, and insect contact with excreta. Suitable facilities range from simple but protected pit latrines to flush toilets with sewerage. To be effective, all facilities must be correctly constructed and properly maintained.

Adult HIV Prevalence—This is the percentage of people aged 15-49 who are infected with human immunodeficiency virus (HIV).

FOOTNOTES.

...—Data are not available.

a—Data are for most recent year available.

b—Official estimate.

c—UNICEF-WHO estimate based on statistical modeling.

d—Indirect estimate based on a sample survey.

e—Based on a survey covering 30 provinces.

f—Based on a sample survey.

INFANTS & MALNUTRITION—5

SOURCE. United Nations Children's Fund (UNICEF), *The State of the World's Children 1999*, pages 94–105.

The World Bank and Oxford University Press, *Entering the 21st Century: World Development Report 1999/2000* (August 1999), pages 232 and 233. Reprinted with permission.

The World Bank, *World Development Indicators 1999* (March 1999), pages 98–101. Reprinted with permission.

NOTES.

Under-five mortality rate—Probability of dying between birth and exactly five years of age expressed per 1,000 live births.

Low birthweight—Weights at birth that are less than 2,500 grams.

TB—Tuberculosis

DPT—Diphtheria, pertussis (whooping cough) and tetanus.

Prevalence of child malnutrition—Expressed in percentage of children under age 5.

FOOTNOTES.

NA—Data are not available.

x—Indicates data that refer to years other than those specified, differ from the standard definitions, or refer to only part of a country.

a—Data are for the most recent year available within the range listed.

b—Data are for the most recent year available within the range.

ETHNIC DIVISION—6

SOURCE. U.S. Central Intelligence Agency (CIA) 1998, *The World Factbook 1998* [Online]. Available: http://www.cia.gov/cia/publications/factbook/index.html [October 1999].

NOTES.

Tables show the major ethnic divisions of peoples in the given country for the most recent year available. When available, the distribution is shown in percent.

NA—Data are not available.

RELIGION—7

SOURCE. U.S. Central Intelligence Agency (CIA) 1998, *The World Factbook 1998* [Online]. Available: http://www.cia.gov/cia/publications/factbook/index.html [October 1999].

NOTES.

Tables show major religious denominations of the peoples of the given country for the most recent year available. When available, the distribution is shown in percent.

NA—Data are not available.

MAJOR LANGUAGES—8

SOURCE. U.S. Central Intelligence Agency (CIA) 1998, *The World Factbook 1998* [Online]. Available: http://www.cia.gov/cia/publications/factbook/index.html [October 1999].

NOTES.

Tables show major language(s) spoken by inhabitants of the given country for the most recent year available. When available, the distribution is shown in percent.

NA—Data are not available.

PUBLIC EDUCATION EXPENDITURES—9

SOURCE. The World Bank, *World Development Indicators 1999* (March 1999), pages 74–77. Reprinted with permission.

NOTES.

The data on education spending refer solely to public spending—that is, government spending on public education plus subsidies for private education. The data generally exclude foreign aid for education.

They also may exclude religious schools, which play a significant role in many developing countries.

The percentage of GNP devoted to education can be interpreted as reflecting a country's effort in education. Often it bears a weak relationship to measures of output of the education system, as reflected in educational attainment. The pattern suggests wide variations across countries in the efficiency with which the government's resources are translated into education outcomes.

Public Expenditures of Education.—This is the percentage of GNP accounted for by public spending on public education plus subsidies to private education at the primary, secondary, and tertiary levels.

Expenditure of Teaching Materials.—The public spending on teaching materials (textbooks, books, and other scholastic supplies) as a percentage of total public spending on primary or secondary education.

FOOTNOTES.

1—Data are for years or periods other than those specified.

EDUCATION ATTAINMENT—10

SOURCE. United Nations Education, Scientific, and Cultural Organization and Bernan Press, *UNESCO 1999 Statistical Yearbook*, pages 51–64. Reprinted with permission.

NOTES.

The percentage distribution of the population aged 25 years and over according to the highest level of education attained reflects both the outcomes of participation in education in the past and the educational composition of the population. These data have been collected mainly during national population censuses and sample surveys. The six levels of educational attainment presented here are based on the International Standard Classification of Education (ISCED) and are defined as follows:

No schooling—Refers to persons who have completed less than one year of primary education.

Primary education incomplete—Includes all persons who have completed at least one grade of primary education but who did not complete the final grade of this level of education.

Primary education completed—Refers to all persons who have completed the final grade of primary education but did not enter secondary education.

Attended lower secondary education—Comprises all persons who have attended lower secondary education but not (upper) secondary education.

Attended (upper) secondary education—Includes all persons who have attended (upper) secondary education but not post-secondary education.

Post-secondary education—Refers to all persons who have completed secondary education and attended post-secondary education.

FOOTNOTES.

1—Not including persons with no schooling or less than one year of primary education.

2—The category ''No Schooling'' comprises illiterates.

3—''Completed primary education'' refers to the last two years of primary education.

4—Persons who can read and write have been counted with ''incomplete primary.''

5—Not including population attending and never attended school.

6—Data refer only to persons who have attended school but left school.

7—Based on a sample survey of 35,502 persons.

8—Not including persons still in school.

9—Based on a sample survey of 51,372 persons.

10—Post-secondary education refers to universities only.

11—Not including transients and residents of former canal zone.

12—The category ''No schooling'' refers to those who have attended less than one grade of primary education.

13—Not including armed forces stationed in the area.

14—Lower secondary education refers to ''intermedio'' level of education. (Upper) secondary education refers to ''Media,'' ''Tecnica'' and ''Normal'' education.

15—Not including rural population of Northern Brazil.

16—Not including persons whose level of education is unknown.

17—Not including Jammu and Kashmir.

18—Not including persons still attending school for whom the level is unknown.

19—Household survey results based on a sample of 6,393 households. The category of ''No schooling'' includes illiterates.

20—(Upper) secondary education includes 'polytechnic'; post-secondary education refers to universities only.

21—Data are based on a sample of 8,619 households (5,563 urban and 3,056 rural) from the 1993 Demographic and Health Survey.

22—''Incomplete primary education'' refers to grades 1 to 4 and ''Complete primary education'' refers to grades 5 to 8.

23—Not including expatriate workers and their families.

24—The category ''No schooling'' includes persons who are still in school.

25—The category ''No schooling'' comprises persons who did not state their level of education.

26—Based on a 20% sample of census returns.

LITERACY RATES—11A

SOURCE. United Nations Education and Culture Organization (UNESCO), *Compendium of Statistics on Illiteracy* (1995 Edition), pages 40–49. Reprinted with permission.

NOTES.

Literacy statistics are concerned with the stock of persons who have successfully acquired the basic reading, writing and numerical skills essential for personal growth and cohesion within contemporary societies. Levels of literacy within a population constitute on the one hand a reflection of the level of development and accomplishments of the education systems, and on the other hand a pointer on the potential for human input into further economic, social and cultural development. Literacy rate has therefore been widely used as a key common indicator for monitoring and assessing progress in the current world thrusts of Education for All and Human Resources Development, and has been regularly incorporated into various reports and publications.

As the national statistics on literacy made available to UNESCO are collected during population censuses that usually take place once every ten years, estimations and projections are carried out to fill the data gaps for the years between two censuses, as well as to provide projections showing likely progress in literacy for the future.

Literacy continues to progress in the world. Adult literacy rate, or the percentage of literates within the adult population aged 15 years and over, has been steadily growing in all countries. Entering the 1990s, over three-quarters (75.3 percent) of the world's adult population have become literate—increasing from 69.5 percent in 1980. Based on the past trends, it is estimated that the overall literacy rate in the world has further improved to 77.4 percent in 1995, and is projected to reach 80 percent at the beginning of the 21st century.

The literate adult population in the world has undergone phenomenal expansion during the past fifteen years from 1980 to 1995, and is projected to further increase in the future. In absolute numbers, the adult literate population in the world rose from 2 billion in 1980 to an estimated 3 billion in 1995, i.e. by 1 billion persons. If the current rate of progress can be maintained, the number of adult literates in the world may reach 3.4 billions in the year 2000, and 4.2 billion in 2010.

Despite these signs of positive progress in both literacy rates and number of literates, one may notice that there remains a large illiterate population in the world of today—numbering some 885 million adults aged 15 years and over—and that this illiterate population increased from an estimated 877 million in 1980. The expansion of the world's illiterate population seems to have reached its turning point during the first half of the 1990s. The projections show that if the past trend were to continue, this world total would gradually decrease to some 881 million by the year 2000. But the huge mass of more than 880 million illiterates shall continue to constitute major challenges to education in the future.

Literate—A person is literate who can with understanding both read and write a short simple statement on his everyday life.

Illiterate—A person is illiterate who cannot with understanding both read and write a short simple statement on his everyday life.

Adult—Refers to persons aged 15 years or older.

LITERACY RATES—11B

SOURCE. United Nations Children's Fund (UNICEF), *The State of the World's Children 1999*, pages 106–109.

NOTES.

Adult Literacy Rate—Percentage of persons aged 15 years and over who can read and write.

-—Data not available.

X—Indicates data that refer to years or periods other than those specified, differ from the standard definitions, or refer to only part of a country.

POLITICAL PARTIES—12

SOURCE. U.S. Central Intelligence Agency (CIA) 1998, *The World Factbook 1998* [Online]. Available: http://www.cia.gov/cia/publications/factbook/index.html [October 1999].

NOTES.

When available, political party representation is shown for the lower house of the legislative branch of government. The lower house was chosen in order to present, in most cases, a picture of the electoral results of voting by the general public. The name of this legislative body is shown in the legend of the given table.

When available, election results are shown as a percent distribution of votes in the most recent election. Otherwise, percent distribution of seats, or number of seats, by political party is shown. If there are no political parties or there is one-party rule, this information is provided in place of tabular data.

Wherever possible, political party names have been presented in English translation.

NA—Data are not available.

GOVERNMENT BUDGETS—13A

SOURCE. International Monetary Fund (IMF), *Government Finance Statistics Yearbook 1998*, pages 18–421.

FOOTNOTES.

f—Forecast.

p—Preliminary / provisional.

....—Data not available.

——Zero or less than half a significant digit.

GOVERNMENT BUDGET—13B

SOURCE. U.S. Central Intelligence Agency (CIA) 1998, *The World Factbook 1998* [Online]. Available: http://www.cia.gov/cia/publications/factbook/index.html [October 1999].

NOTES.

IMF data were obtained primarily by means of a detailed questionnaire distribution to government finance statistics correspondents, who are usually located in each country's respective ministry of finance or central bank. Three of the six categories of central government expenditure shown in the IMF tables are comprised of subcategories, whose subtotals have been summed. Below is a list of these subcategories.

Education/Health—Also includes *Welfare* and *Social security.*

Industry—Includes *Fuel and energy; Agriculture, forestry, fishing, and hunting; Mining, manufacturing, and construction; Transportation and communication;* and *Other economic affairs and services.*

Other—Includes *Recreational, cultural, and religious affairs and other expenditures.*

Some of the subcategory data are incomplete for Guatemala, India, and Nepal, and consequently have been calculated as zero (0).

Minor differences between published totals and the sum of components are attributable to rounding.

Following are definitions of acronyms and terms pertinent to these tables.

Central government—All units representing the territorial jurisdiction of the central authority throughout a country.

CY—Calendar year: 12-month year beginning January 1 and ending the following December 31.

A dash (-)—Data are nil or negligible.

est.—Estimate.

Expenditure—All nonrepayable payments by government, including both capital and current expenditures and regardless of whether goods or services were received for such expenditures.

FY—Fiscal Year: presented within the calendar year containing the greatest number of months for that fiscal year. Fiscal years ending June 30 are presented within the same calendar year. For example, the fiscal year July 1, 1995–June 30, 1996 is shown within the calendar year 1996.

Government—All units that implement public policy by providing nonmarket services and transferring income; these units are financed mainly by compulsory levies on other sectors.

NA—Data are not available.

Revenue—All nonrepayable government receipts other than grants.

MILITARY AFFAIRS—14A

SOURCE. U.S. Central Intelligence Agency (CIA) 1998, *The World Factbook 1998* [Online]. Available: http://www.cia.gov/cia/publications/factbook/index.html [October 1999].

FOOTNOTES.

e—Estimate based on partial or uncertain data.

NA—Data not available.

0—Nil or negligible.

MILITARY AFFAIRS—14B

SOURCE. U.S. Arms Control and Disarmament Agency, *World Military Expenditures and Arms Transfers 1996* (WMEAT), (July 1997), pages 57–99 and 108–150.

NOTES.

Military Expenditures

For NATO countries, military expenditures are from NATO publications and are based on the NATO definition. In this definition, (a) civilian-type expenditures of the defense ministry are excluded and military-type expenditures of other ministries are included; (b) grant military assistance is included in the expenditures of the donor country; and (c) purchases of military equipment for credit are included at the time the debt is incurred, not at the time of payment.

For other non-communist countries, data are generally the expenditures of the ministry of defense. When these are known to include the costs of internal security, an attempt is made to remove these expenditures. A wide variety of data sources is used for these countries, including the publications and data resources of other U.S. government agencies, standardized reporting to the United Nations by country, and other international sources.

It should be recognized by users of the statistical tables that the military expenditure data are of uneven accuracy and completeness. For example, there are indications or reasons to believe that the military expenditures reported by some countries consist mainly or entirely of recurring or operating expenditures and omit all or most capital expenditures, including arms purchases.

In some cases it is believed that a better estimate of total military expenditures is obtained by adding to nominal military expenditures the value of arms imports. It must be cautioned, however, that this method may over- or underestimate the actual expenditures in a given year due to the fact that payment for arms may not coincide in time with deliveries. Also, arms acquisitions in some cases may be financed by, or consist of grants from, other countries.

For countries that have major clandestine nuclear or other military weapons development programs, such as Iraq, estimation of military expenditures is extremely difficult and especially subject to errors of underestimation.

Further information in the quality of the military expenditure data presented for countries throughout the world will be difficult to achieve without better reporting by the countries themselves. As has been noted elsewhere, "There is growing evidence that important amounts of security expenditures may not enter the accounts or the national budgets of many developing countries." Among the mechanisms commonly used to obscure such expenditures are: double-bookkeeping budget categories, military assistance, and manipulation or foreign exchange.

Particular problems arise in estimating the military expenditures of communist countries due to the exceptional scarcity and ambiguity of released information. As in past editions of this publication, data on the military expenditures of the Soviet Union are based on Central Intelligence Agency (CIA) estimates. For most of the series, these are estimates of what it would cost in the United States in dollars to develop, procure, staff, and operate a military force similar to that of the Soviet Union. Estimates of this type—that is, those based entirely on one country's price pattern—generally overstate the relative size of the second country's expenditures in intercountry comparisons. Also, such estimates are not consistent with the methods used here for converting other countries' expenditures into dollars.

Nevertheless, the basic CIA estimates are the best available for present purposes; in fact, there are no alternative estimates that can inspire confidence and have the capability to detect relatively small changes over time, such as the slowdown and decline in Soviet military spending that the CIA estimates have indicated.

For Russia, estimated military spending trends in rubles are used in conjunction with dollar estimates for earlier years to make rough estimates of spending in dollars.

For former Warsaw Pact countries other than the Soviet Union, the estimates of military expenditures through 1989 are from Thad P. Alton et al. These estimates cover the officially announced state budget expenditures on national defense and thus understate total military expenditures to the extent of possible defense outlays by non-defense agencics of the central government, local governments, and economic enterprises. Possible subsidization of military procurement may also cause understatement. The dollar estimates were derived by calculating pay and allowances at the current full U.S. average rates for officers and for lower ranks. After subtraction of pay and allowances, the remainder of the official defense budgets in national currencies was converted into dollars at overall rates based on comparisons of the various countries' GNPs expressed in dollars and in national currencies. The rates are based in part on the purchasing power parites (PPPs) estimated by the International Comparison Project of the United Nations, including there latest (Phase V) versions.

Estimates for these countries in 1990 and 1991 are based on total military spending in national currency as reported by the respective governments to the UN (in most cases) or the IMF. These expenditures in toto are converted to dollars at the Alton GNP conversion rates for 1989 as adjusted to 1991 by the respective U.S. and national GNP deflators (per the World Bank), without estimating personnel compensation separately at U.S. dollar rates, as was done for earlier years. The resulting military conversion rates (in national currency per dollar) are substantially lower than the 1991 market rate, and approximately the same as the implied rate for GNP.

Estimates for the newly independent states of the former Soviet Union, Yugoslavia, and Czechoslovakia and other former Warsaw Pact countries present difficulties due to scarcity of reliable data in national currencies and to problems in converting to dollars. The basic method employed for most of these countries was to establish the ratio of military expenditures to GNP in national currency and then to multiply this ratio by the World Bank's estimate of GNP in dollars as converted to international dollars by estimate PPPs and reported in the *World Bank Atlas 1997.* This method implicitly converts military spending at the GNP-wide PPP, which, as with conversion by exchange rates, preserves the same ME/GNP ratio in dollars as obtains in national currency.

Data for China are based on U.S. Government estimates of the yuan costs of Chinese forces, weapons, programs, and activities. Costs in yuan are here converted to dollars using the same estimated conversion rate as used for GNP. Due to the exceptional difficulties in both estimating yuan costs and converting them to dollars, comparisons of Chinese military spending with other data should be treated as having a wide margin of error.

Other published sources used include the *Government Finance Statistics Yearbook,* issued by the International Monetary Fund; *The World Factbook,* produced annually the Central Intelligence Agency; *The Military Balance,* issued by the International Institute for Strategic Studies (London); and the *SIPRI Yearbook: World Armaments and Disarmament,* issued by the Stockholm International Peach Research Institute.

Gross National Product (GNP)

GNP represents the total output of goods and services produced by residents of a country, valued at market prices. The source of GNP data for most noncommunist countries is the International Bank for Reconstruction and Development (World Bank).

For a number of countries whose GNP is dominated by oil exports (Bahrain, Kuwait, Libya, Oman, Qatar, Saudi Arabia, and the United Arab Emirates), the World Bank's estimate of deflated (or constant price) GNP in domestic currency tends to understate increases in the monetary value of oil exports, and thus of GNP, resulting from oil price increases. These World Bank estimates are designed to measure real (or physical) product. An alternative estimate of constant-price GNP was therefore obtained using the implicit price deflator for U.S. GNP (for lack of a better national deflator). This considered appropriate because a large share of the GNP of these countries is realized in U.S. dollars.

GNP estimates of the Soviet Union are by the CIA, as published in its *Handbook of Economic Statistics 1990* and updated. GNP estimates for other Warsaw Pact countries through 1989 are from ''East European Military Expenditures, 1965–1978,'' by Thad P. Alton and others, *op. cit.,* as updated and substantially revised by the authors. These estimates through 1989 have been extended to 1990 and 1991 on the basis of estimates for those years in the CIA's *Handbook of Economic Statistics, 1992.*

Estimates of GNP in 1992–1994 for successor states to the Soviet Union, Yugoslavia, and Czechoslovakia are based on World Bank estimates of GNP per capita employing PPPs and population, as published in the *World Bank Atlas 1997.*

GNP data for China are based on World Bank estimates in yuan. These are in line with estimates of GDP in Western accounting terms made by Chinese authorities. Converting estimates in yuan to dollars is highly problematic, however, due to the inappropriateness of the official exchange rate and lack of sufficient yuan price information by which to reliably estimate PPPs. (The ratio of the highest to the lowest estimates by various sources of China's GNP is on the order of 6 or 7 to 1, which would make the world rank of China's GNP vary between about third or fourth and twelfth). The conversion rate used here is based on a PPP estimated for 1981 and moved by respective U.S. and China implicit GNP deflators to 1994.

GNP estimates for a few non-communist countries are from the CIA's *Handbook of Economic Statistics* cited above. Estimates for the other communist countries are rough approximations.

Military Expenditures-to-GNP Ratio

It should be noted that the meaning of the ratio of military expenditures to GNP differs somewhat between most communist (or previously communist) and other countries. For non-communist countries, both military expenditures and GNP are converted from the national currency unit to dollars at the same exchange rate; consequently, the ratio of military expenditures to GNP is the same in dollars as in the national currency and reflects national relative prices. For communist countries, however, military expenditures and GNP are converted differently. Soviet military expenditures, as already noted, are estimated in a way designed to show the cost of the Soviet armed forces in U.S. prices, as if purchased in this country. On the other hand, the Soviet GNP estimates used here are designed to show average relative size when both U.S. and Soviet GNP are valued and compared at both dollar and ruble prices. The Soviet ratio of military expenditures to GNP in ruble terms, the preferred method of comparison, is estimated to have been 15-18% in that country's latest years.

The estimated ratio for Russia derived here in dollars is probably somewhat overstated since military spending in dollars is relative to earlier estimates for the Soviet Union, while GNP estimates (at PPPs) are by the World Bank. Russia's burden ratio in ruble term is preferably estimated to be under 10%.

For Eastern European countries before 1992, the ratios of military expenditures to GNP in dollars were about twice the ratios that would obtain in domestic currencies. However, since official military budgets in these countries probably substantially understated their actual military expenditures, the larger ratios on dollar estimates are believed to be the better approximations of the actual ratios.

Central Government Expenditures (CGE)

These expenditures include current and capital (developmental) expenditures plus net lending to the government enterprises by central (or federal) governments. A major source is the International Monetary Fund's *Government Finance Statistics Yearbook.* The category used here is ''Total Expenditures and Lending minus Repayment, Consolidated Central Government.''

Other sources for these data are the International Monetary Fund, *International Financial Statistics* (monthly); OECD, *Economics Surveys;* and CIA, *The World Factbook* (annual). Data for Warsaw Pact countries are from national publications and are supplied by Thad P. Alton and others. For all Warsaw Pact countries and China, conversion to dollars is at the implicit rates used for calculating dollar estimates of GNP.

For all countries, with the same exceptions as noted above for the military expenditures-to-GNP ratio, military expenditures and central government expenditures are converted to dollars at the same rate; the ratio of the two variables is thus the same in dollars as in national currency.

It should be noted that for the Soviet Union, China, Iran, Jordan, and possibly others, the ratio of military expenditures to central government expenditures may be overstated, inasmuch as the estimate for military expenditures is obtained at least in part independently of nominal budget or government expenditure data, and it is possible that not all estimated military expenditures pass through the nominal central government budget.

Population

Population estimates are for midyear and are made available to ACDA by the U.S. Bureau of the Census.

Armed Forces

Armed forces refer to active-duty military personnel, including paramilitary forces if those forces resemble regular units in their organization, equipment, training, or mission. Reserve forces are not included unless specifically noted.

Figures for the United States and all other North American Treaty Organization (NATO) countries are as reported by NATO. Estimates of the number of personnel under arms for other countries are provided by U.S. Government sources. The armed forces series for the Soviet Union includes all special forces judged to have national security missions (e.g., KGB border guards) and excludes uniformed forces primarily performing noncombatant services (construction, railroad, civil defense, and internal security troops).

Arms Transfers

Arms transfers (arms imports and exports) represent the international transfer (under terms of grant, credit, barter, or cash) of military equipment, usually referred to as ''conventional,'' including weapons of war, parts thereof, ammunition, support equipment, and other commodities designed for military use. Among the items included are tactical guided missiles use. Among the items included are tactical guided missiles and rockets, military aircraft, naval vessels, armored and nonarmored military vehicles, communications and electronic equipment, artillery, infantry weapons, small arms, ammunition, other ordinance, parachutes, and uniforms. Dual use equipment, which can have application in both military and civilian sectors, is included when its primary mission is identified as military. The building of defense production facilities and licensing fees paid as royalties for the production of military equipment are included when they are contained in military transfer agreements. There have been no international transfers of purely strategic weaponry. Military services such as training, supply operations, equipment repair, technical assistance, and construction are included where data are available. Excluded are foodstuffs, medical equipment, petroleum products and other supplies.

Redefinition of U.S. Arms Exports. The scope of U.S. arms exports data was modified in the *WMEAT 1995* edition. These exports include both government-to-government transfers under the Foreign Military Sales (FMS), Military Assistance Program (MAP), and other programs administered by the Department of Defense, and commercial (enterprise-to-government) transfers licensed by the Department of State under International Traffic in Arms Regulations. Under the previous practice, the material component (arms, equipment, and ''hardware'' items) of FMS and MAP sales was included, while the military services component was excluded (although the magnitude and general destination of the omitted services was reported in these Statistical Notes).

Beginning with the previous edition, both the material and the military services components of FMS and other government-to-government sales (such as the International Military Education and Training Program—IMET) are included in total U.S. arms exports as reported here. The commercial sales category, covering both material and military services, was included in its entirety.

The omission of FMS and other military services prior to the previous edition had been intended to improve comparability with available estimates of the arms exports of other countries, which tended to contain a much smaller services component and/or were subject to significant underestimation (services being less easily observed). The increasing importance of these services and the desire to present a full picture of U.S. arms exports consistent with other sources prompted the change to inclusion. Users should be aware, however, of both the lower true share of services in other countries' arms exports and the tendency to underestimate them. It should also be noted that a portion of the IMET program is devoted to programs that promote improved civil-military relations.

The change in scope of U.S. arms exports increased their overall volume by amounts ranging over the last decade from $2.3 billion (current dollars) to $3.7 billion for deliveries and $2.3 billion to $7.3 billion for agreements.

The statistics contained herein are estimates of the value of goods actually delivered during the reference year, in contrast both to payments and the value of programs, agreements, contracts, or orders concluded during the period, which are expected to result in future deliveries.

U.S. Arms Imports. U.S. arms import data in this and the previous four editions of WMEAT are revised upward substantially from earlier editions. The present series consist of data obtained from the Department of Commerce, Bureau of Economic Analysis (BEA), including (a) imports of military-type (formerly ''special category'') goods, as compiled by the Bureau of the Census, and (b) Department of Defense direct expenditures abroad for major equipment, as compiled from DOD data by BEA. The goods in (a) include: complete military aircraft, all types; engines and turbines for military (naval) ships and boats; tanks, artillery, missiles, guns, and ammunition; military apparel and footwear; and other military goods, equipment, and parts.

Data on countries other than the United States are estimates by U.S. Government sources. Arms transfer data for the Soviet Union and other former communist countries are approximations based on limited information.

It should be noted that the arms transfer estimates for the most recent year, and to a lesser extent for several preceding years, tend to a lesser extent for several preceding years, tend to be understated. This applies to both foreign and U.S. arms exports. In former case, information on transfers comes from a variety of sources and is sometimes acquired and processed with a considerable time lag. In the U.S. case, commercial arms transfer licenses are now valued for three years, causing a delay in the reporting of deliveries made on them to statistical agencies.

Close comparisons between the estimated values shown for arms transfers and for GNP and military expenditures are not warranted. Frequently, weapons prices do not reflect true production costs. Furthermore, much of the international arms trade involves offset or barter arrangements, multiyear loans, discounted prices, third party payments, and partial debt forgiveness. Acquisitions of armaments thus may not [necessarily] impose the burden on an economy, whether in the same or in other years, that is implied by the estimated equivalent U.S. dollar value of the shipment. Therefore, the value of arms imports should be compared to other categories of data with care.

Total Imports and Exports

The values for imports and exports cover merchandise transactions and come mainly from International Financial Statistics published by the IMF. The trade figures for presently and formerly communist countries and from the CIA's *Handbook of Economic Statistics, 1996* edition.

FOOTNOTES.

e—Estimate based on partial or uncertain data.

NA—Data not available.

p—Estimate based on purchasing power parities.

r—Rough estimate.

0—Nil or negligible.

1—Estimated by adding arms imports to data on military expenditures, which are believed to exclude arms purchases. However, it should be noted that the value of arms deliveries in a given year may differ significantly from actual expenditures on arms imports in that year.

2—This ratio is calculated from the dollar values shown in previous columns. In most cases it also is equal to the ratio that could be calculated from national currency values, since both numerator and denominator are usually converted into dollars by the same exchange rate or other conversion factor. In the case of this country, however, the two variables are converted at different rates, yielding a different ratio than would obtain in national currency. The ratio for Russia in rubles terms, for example, is believed to be less than 10 percent in 1995.

3—This series or entry is believed to omit a major share of total military expenditures, probably including most expenditures on arms procurement.

4—Germany, (The Federal Republic of), was known as West Germany through 1990. Thereafter, Germany refers to the unified Germany.

5—In order to reduce distortions in the trend for worked, region, and organization totals caused by data gaps for individual countries and years, rough approximations for all gaps are included in the totals.

6—To avoid the appearance of excessive accuracy, arms transfer data by country are rounded, with greater severity for larger amounts. All country group totals for arms exports and arms imports shown here are the sums of rounded country data. Consequently, world totals for arms imports and arms exports will not be equal.

7—Total imports and exports usually are as reported by individual countries and the extent to which arms transfers are included is often uncertain. Imports are reported ''cif'' (including cost of shipping, insurance, and freight) and exports are reported ''fob'' (excluding these costs). For these reasons and because of divergent sources, world totals for imports and exports are not equal.

8—Because some countries exclude arms imports or exports from their trade statistics and their ''total'' imports and exports are therefore understated, and because arms transfers may be estimated independently for trade data, the resulting ratios of arms to total imports or exports may be overstated and may even exceed 100 percent.

9—Some part of estimated total military expenditures may not be included in announced central budget expenditures. The ratio of ME to CGE therefore may be overstated.

10—Included major equipment purchased by the U.S. Army Corps of Engineers for use in military construction projects in Saudi Arabia, recorded in U.S. accounts as U.S. imports.

11—U.S. arms imports data shown here is revised upward substantially form reports before 1993.

12—Little data are available because of an ongoing civil war.

CRIME—15

SOURCE. Crime Prevention and Criminal Justice Division, United Nations Criminal Justice Information Network (UNCJIN), *The Fifth United Nations Survey of Crime Trends and Operations of Criminal Justice Systems* [Online], Available: http://www.uncjin.org/stats/wcs.html [October 1999]. Reprinted with permission.

NOTES.

The major goal of the Fifth United Nations Survey is to collect data on the incidence of reported crime and the operations of criminal justice systems with a view to improving the dissemination of that information globally. To that end, the Survey should facilitate an overview of trends and interrelationships between various parts of the criminal justice system so as to promote informed decision making in its administration, nationally and cross-nationally.

As with data collected for the Fourth United Nations Survey, these data demonstrate the difficulty of comparing crime internationally. One difficulty is that the vast majority of incidents that become known to the police come from reports by victims. Thus, credibility becomes a statistical determinant. Another difficulty is that comparison is severely undermined by differences in legal definitions and by administrative procedures regarding counting, classification, and disclosure. The researcher should be aware of these shortcomings when using these data.

NA—Data are not available.

TOTAL LABOR FORCE—16

SOURCE. U.S. Central Intelligence Agency (CIA) 1998, *The World Factbook 1998* [Online]. Available: http://www.cia.gov/cia/publications/factbook/index.html [October 1999].

NOTES.

Data show the number of persons in the labor force for the most recent year available.

NA—Data are not available.

UNEMPLOYMENT RATE—17

SOURCE. U.S. Central Intelligence Agency (CIA) 1998, *The World Factbook 1998* [Online]. Available: http://www.cia.gov/cia/publications/factbook/index.html [October 1999].

NOTES.

Data show the rate of unemployment in percent for the most recent year available.

NA—Data are not available.

ENERGY PRODUCTION—18

SOURCE. U.S. Central Intelligence Agency (CIA) 1998, *The World Factbook 1998* [Online]. Available: http://www.cia.gov/cia/publications/factbook/index.html [October 1999].

NOTES.

Btu—British thermal units.

TRANSPORTATION—19

SOURCE. U.S. Central Intelligence Agency (CIA) 1998, *The World Factbook 1998* [Online]. Available: http://www.cia.gov/cia/publications/factbook/index.html [October 1999].

NOTES.

Following are CIA definitions of terms used in these tables.

Airports—Only airports with usable runways are included in this listing. Not all airports have facilities for refueling, maintenance, or air traffic control. Paved runways have concrete or asphalt surfaces; unpaved runways have grass, dirt, sand, or gravel surfaces.

DWT—Deadweight tons.

GRT—Gross register tons.

km—Kilometers.

m—Meters.

Merchant marine—All ships engaged in the carriage of goods. All commercial vessels (as opposed to all nonmilitary ships), which excludes tugs, fishing vessels, offshore oil rigs, etc. Also, a grouping of merchant ships by nationality or register.

NA—Data are not available.

TOP AGRICULTURAL PRODUCTS—20

SOURCE. U.S. Central Intelligence Agency (CIA) 1998, *The World Factbook 1998* [Online]. Available: http://www.cia.gov/cia/publications/factbook/index.html [October 1999].

NOTES.

GDP—Gross Domestic Product: the value of all goods and services produced within a nation in a given year.

GDP & MANUFACTURING SUMMARY—21

SOURCE. United Nations Industrial Development Organization (UNIDO), *Industrial Development Global Report 1996*, pages 129—254. Reprinted with permission.

NOTES.

Gross domestic product (GDP)—All economic activity in a given country, including activity engaged in by foreign nationals. For example, assets of a General Motors plant in Mexico would contribute to Mexico's GDP. *Real GDP* measures economic activity in constant prices, that is, after adjustments for inflation.

Manufacturing value added (MVA)—The value of output minus the cost of raw materials and other inputs.

FOOTNOTES.

1—Value originating from the National Accounts Statistics.

2—In 1990 constant prices.

3—The data presented here are for activities in the former Federal Republic of Germany and do not include those of the former Democratic Republic of Germany, even after unification in 1990.

Numbers in *italics*—Estimated by UNIDO, Research and Publication Division, Research and Studies Branch.

NA—No value available.

——Value is less than half a unit.

ECONOMIC INDICATORS—22

SOURCE. U.S. Central Intelligence Agency (CIA) 1998, *The World Factbook 1998* [Online]. Available: http://www.cia.gov/cia/publications/factbook/index.html [October 1999].

NOTES.

Following are CIA definitions of acronyms and terms used in these tables.

est.—Estimate.

External debt—The amount of debt owed to foreign entities by the given country.

GDP—Gross domestic product: the value of all goods and services produced within a nation in a given year. Methodology: GDP dollar estimates for all countries are derived from purchasing power parity (PPP) calculations rather than from conversions at official currency exchange rates. The PPP method involves the use of international dollar price weights, which are applied to the quantities of goods and services produced in a given economy. The data derived from the PPP method provide a better comparison of economic well-being between countries. The division of a GDP estimate in domestic currency by the corresponding PPP estimate in dollars gives the PPP conversion rate. When priced in PPPs, $1,000 will buy the same market basket of goods in any country. Whereas PPP estimates for OECD countries are quite reliable, PPP estimates for developing countries are often rough approximations. Most of the GDP estimates are based on extrapolation of numbers published by the UN International Comparison Program and by Professors Robert Summers and Alan Heston of the University of Pennsylvania and their colleagues. Note: the numbers for GDP and other economic data can not be chained together from successive volumes of the *Factbook* because of changes in the U.S. dollar measuring rod, revisions of data by statistical agencies, use of new or different sources of information, and changes in national statistical methods and practices.

Inflation rate—An increase in prices unrelated to value.

NA—Data are not available.

National product—The total output of goods and services in a given country. (See gross domestic product).

BALANCE OF PAYMENTS SUMMARY—23

SOURCE. United Nations Conference on Trade and Development, *Handbook of International Trade and Development Statistics*, pages 214–241.

NOTES.

The following explanatory notes are intended to provide a brief description of the balance of payments categories presented. In actual practice, there are many exceptions to the definitions of categories, and for these the reader should refer to the country notes in the *Balance of Payments Yearbook*.

Exports of goods (fob)—The export figure here differs from that reported in the trade returns because of adjustments for coverage, valuation, timing, inland freight, etc. Such adjustments to the reported export and import figures are necessary in order to make the trade statistics compatible with the concepts employed in the balance of payments. In particular, valuation adjustments are required in those cases in which the market price at which goods have been sold differs from the price used for customs' purposes. This problem in valuation is probably more important for imports than for exports and is likely to be a factor whenever there is a long delay between the date of sale and the date at which an import duty becomes payable.

The coverage of goods in Balance of Payments Manual, 5th edition, has been expanded to include (a) the value of goods (on a gross basis) received/sent for processing and their subsequent export/import in the form of processed goods; (b) the value of repairs on goods; and (c) the value of goods procured in ports by carriers. In Balance of Payments Manual, 4th edition, the net value between goods imported for processing and subsequently re-exported was included in processing services; repairs of goods and goods procured in ports by carriers were also included under services.

Imports of goods (fob)—Adjustments for coverage, valuation, timing, etc., are made to imports reported in trade returns, as described in the notes above. In addition, an adjustment is made to convert imports from a cif to an fob basis for those countries reporting imports cif. The import figures reported here include imports of non-monetary gold.

Balance of goods—Measured on a fob/fob basis and including transactions in monetary gold.

Services and income-debit (total)—Total payments for services and income.

The Balance of Payments Manual, 5th edition, classifies income and services separately; in Balance of Payments Manual, 4th edition, income was a subcomponent of services. Balance of Payments Manual, 5th edition, also reclassifies certain income and services transactions. In Balance of Payments Manual, 4th edition, labor income included non-resident workers' expenditures as well as workers' earnings; in Balance of Payments Manual, 5th edition, workers' earnings are classified under compensation of employees in the income category, and their expenditures appear under travel services. In Balance of Payments Manual, 4th edition, compensation of resident staff of foreign embassies and military bases and of international organizations was included under government services; this compensation is classified as a credit item of compensation of employees in Balance of Payments Manual, 5th edition. Balance of Payments Manual, 4th edition, treated payments for the use of patents, copyrights, and similar non-financial intangible assets as property income; these are regarded as subcomponents of other services in Balance of Payments Manual, 5th edition. In general, the Balance of Payments Manual, 5th edition, concept of income covers investment income plus all forms of compensation of employees; whereas, in Balance of Payments Manual, 4th edition, the concept included investment income, most forms of labor income (including workers' expenditures abroad), and property income.

Services and income-credit (total)—Counterpart to service and income-debit (total).

Current transfers: government-net—Current transfers are classified, according to the sector of the compiling economy, into two main categories: general government and other sectors. General government transfers comprise current international cooperation, which covers current transfers—in cash or in kind—between governments of different economies or between governments and international organizations.

Current transfers: other sectors-net—Current transfers between other sectors of an economy and non-residents comprise those occurring between individuals, between non-governmental institutions or organizations (or between the two groups), or between non-resident governmental institutions and individuals or non-governmental institutions. The same basic items (described in paragraphs 298 to 300 of the IMF Manual) for the government sector are generally applicable to other sectors, although there are some differences within components. In addition, there is the category of workers' remittances.

Balance of current account—Covered in the current account are all transactions (other than those in financial items) that involve economic values and occur between resident and non-resident entities. Also covered are offsets to current economic values provided or acquired without a quid pro quo. Specifically, the major classifications are goods and services, income, and current transfers.

FOOTNOTES.

f.o.b—Free On Board, i.e., the value of goods does not include insurance and freight charges.

..—Data are not available.

——Data are nil or negligible.

EXCHANGE RATES—24

SOURCE. U.S. Central Intelligence Agency (CIA) 1998, *The World Factbook 1998* [Online]. Available: http://www.cia.gov/cia/publications/factbook/index.html [October 1999].

NOTES.

Following are CIA definitions of acronyms and terms used in these tables.

Exchange rate—The official value of a nation's monetary unit, at a given date or over a given period of time, as expressed in units of local currency per U.S. dollar and as determined by international market forces or official fiat. These often have little relation to domestic output. In developing countries with weak currencies, the exchange rate estimate in GDP (gross domestic product) in dollars is typically one-fourth to one-half the PPP (purchasing power parity) estimate. Although exchange rates may suddenly go up or down by 10% or more, real output may have remained unchanged. On January 12, 1994, for example, the 14 countries of the African Financial Community (whose currencies are tied to the French franc) devalued their currencies by 50%. This move, of course, did not cut the real output of their countries by half.

BMR—Black Market rate.

NA—Data are not available.

SOURCE. U.S. Central Intelligence Agency (CIA) 1998, *The World Factbook 1998* [Online]. Available: http://www.cia.gov/cia/publications/factbook/index.html [October 1999].

NOTES.

Top import origins are distributed in percent when data are available.

Following are CIA definitions of the acronyms and terms used here.

BLEU—Belgium-Luxembourg Economic Union.

Caricom—Caribbean Community and Common Market.

CEMA—Council for Mutual Economic Assistance; also known as CMEA or Comecon.

c.i.f.—Cost, insurance, freight.

CIS—Commonwealth of Independent States.

CMEA—Council for Mutual Economic Assistance; also known as CEMA or Comecon.

ECOWAS—Economic Community of West African States.

EFTA—European Free Trade Association.

est.—Estimate.

EU—European Union.

f.o.b.—Free on board.

FSU—Former Soviet Union.

NA—Data are not available.

OECD—Organization for Economic Cooperation and Development.

OECS—Organization of Eastern Caribbean States.

OPEC—Organization of Petroleum Exporting Countries.

SACU—South African Customs Union.

UAE—United Arab Emirates.

UK—United Kingdom.

U.S.—United States.

U.S.S.R.—Union of Soviet Socialist Republics (Soviet Union).

TOP EXPORT—26

SOURCE. U.S. Central Intelligence Agency (CIA) 1998, *The World Factbook 1998* [Online]. Available: http://www.cia.gov/cia/publications/factbook/index.html [October 1999].

NOTES.

Top export destinations are distributed in percent when data are available.

Following are CIA definitions of the acronyms and terms used in these tables.

BLEU—Belgium-Luxembourg Economic Union.

Caricom—Caribbean Community and Common Market.

CEMA—Council for Mutual Economic Assistance; also known as *CEMA* or *Comecon.*

c.i.f.—Cost, insurance, freight.

CIS—Commonwealth of Independent States.

CMEA—Council for Mutual Economic Assistance; also known as *CEMA* or *Comecon.*

ECOWAS—Economic Community of West African States.

EFTA—European Free Trade Association.

est.—Estimate.

EU—European Union.

f.o.b.—Free on board.

FSU—Former Soviet Union.

NA—Data are not available.

OECD—Organization for Economic Cooperation and Development.

OECS—Organization of Eastern Caribbean States.

OPEC—Organization of Petroleum Exporting Countries.

SACU—South African Customs Union.

UAE—United Arab Emirates.

UK—United Kingdom.

U.S.—United States.

U.S.S.R.—Union of Soviet Socialist Republics (Soviet Union).

FOOTNOTES.

f.o.b—Free on board, i.e., the value of goods does not include insurance and freight charges.

..—Data are not available.

——Data are nil or negligible.

FOREIGN AID—27

SOURCE. U.S. Central Intelligence Agency (CIA) 1998, *The World Factbook 1998* [Online]. Available: http://www.cia.gov/cia/publications/factbook/index.html [October 1999].

NOTES.

Following are CIA definitions of terms used in these tables.

Donor—Country that pledges official economic aid to another country.

NA—Data are not available.

ODA—Official development assistance. ODA refers to financial assistance which is concessional in characters, has the main objective of promoting economic development and welfare in less developed countries (LDCs), and contains a grant element of at least 25 percent.

OOF—Other official flows. OOF also refers to official government assistance, but with a main objective other than development and with a grant element less than 25 percent. Transactions include official export credits (such as Export-Import Bank credits), official equity and portfolio investment, and debt reorganization by the official sector that does not meet concessional terms. Aid is considered to have been committed when the parties involved initial agreements constituting a formal declaration of intent.

Recipient—Country that receives official economic aid from another country.

IMPORT AND EXPORT COMMODITIES—28

SOURCE. U.S. Central Intelligence Agency (CIA) 1998, *The World Factbook 1998* [Online]. Available: http://www.cia.gov/cia/publications/factbook/index.html [October 1999].

NOTES.

Category 39: *Top Import Origins* and Category 40: *Top Export Destinations* provide corresponding year of commodity imports/exports respectively.

When available, commodities are distributed in percent.

SAINT HELENA

CAPITAL: Jamestown.

FLAG: The flag of St. Helena is blue with the flag of the United Kingdom in the upper hoist-side quadrant and the Saint Helenian shield centered on the outer half of the flag; the shield features a rocky coastline and three-masted sailing ship.

ANTHEM: *God Save the Queen.*

MONETARY UNIT: 1 Saint Helenian pound (£s) = 100 pence.

WEIGHTS AND MEASURES: The imperial system is used.

HOLIDAYS: Celebration of the Birthday of the Queen, 10 June.

TIME: Noon = noon GMT.

LOCATION AND SIZE: St. Helena, a British colony 122 sq. km (47 sq. mi.) in area, is a mountainous island in the South Atlantic Ocean at approximately 16°S and 5°45′W, about 1,930 km (1,200 mi.) from the west coast of Africa. The maximum elevation, at Diana's Peak, is 828 m (2,717 ft.).

CLIMATE: Southeast trade winds give the island a pleasant climate, despite its tropical location. The temperature at Jamestown, the capital, on the north coast, ranges from 18° to 29°C (65–85°F); inland, as the elevation rises, temperatures are somewhat cooler. Rainfall ranges to an annual maximum of about 100 cm (40 in.).

INTRODUCTORY SURVEY

RECENT HISTORY

Uninhabited when first sighted by the Portuguese navigator Joio da Nova Castella in 1502, and claimed by the Dutch in 1633, St. Helena was garrisoned in 1659 by the British East India Company, captured by the Dutch in 1673, and retaken that same year by the English. It became famous as the place of Napoleon's exile, from 1815 until his death in 1821, and passed to the crown in 1834.

GOVERNMENT

The island is administered by a governor, with the aid of a Legislative Council that includes, in addition to the governor, 2 ex-officio and 12 elected members. General elections were last held in July 1993. Council committees, a majority of whose members belong to the Legislative Council, are appointed by the governor and charged with executive powers and general supervision of government departments. An island council consists of an administrator, three appointed members, and eight elected members.

Judiciary

The Supreme Court of St. Helena, headed by a chief justice, has full criminal and civil jurisdiction. Trial is by a jury of eight. Other judicial institutions include a magistrate's court, a small claims court, and a juvenile court.

ECONOMIC AFFAIRS

Public Finance

St. Helena coins of 1, 2, 5, 10, and 50 pence and 1 pound and notes of 5 and 10 pounds are legal tender; their value is on a par with their UK equivalents.

Industry

The domestic economy is based on agriculture. Considerable revenue is derived from the sale of stamps; however, the fishing industry provides the chief source of livelihood. The main crops are potatoes, sweet potatoes, corn, and vegetables. St. Helenians also are employed on Ascension and the

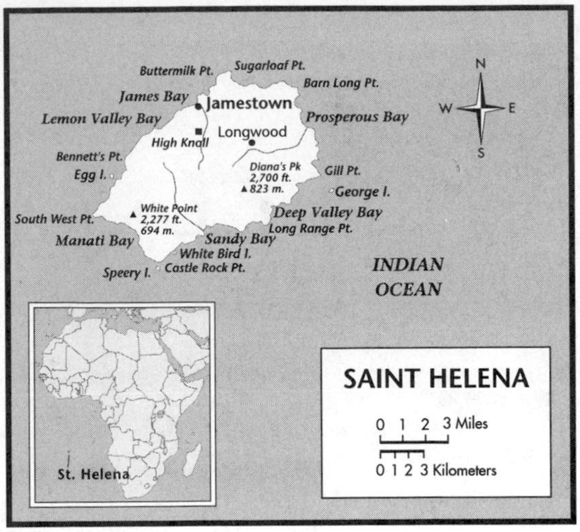

Falkland Islands. The fish catch in 1994 was 631 tons. There are no exploitable minerals, and virtually all timber is imported ($286,000 in 1995). St. Helena also imports all of its consumer and capital goods. The UK and South Africa are St. Helena's best customers and suppliers. Imports came to 4.7 million pounds in 1992–93; British aid amounted to £ 7.98 million in 1993–94.

Banking and Finance

The colony's revenues in 1992/93 amounted to US$11.2 million; expenditures totaled US$11 million. Domestic revenues include succession and death duties, an entertainment tax, a head tax, taxes on motor vehicles and on shops, and personal income and company taxes. The graduated personal income tax rate ranges from 10% to 30%. The company tax was 25% of net distributable profits.

SOCIAL WELFARE

There is an unemployment relief system and workers' compensation is paid for death or disablement. The sole labor union, the St. Helena General Workers' Union had about 472 members in 1992; approximately two-thirds of the labor force works for the government.

Healthcare

Health facilities include a hospital of 54 beds as well as six clinics and a mental hospital.

EDUCATION

The population is entirely literate. Education is free and compulsory between the ages of 5 and 15. There are 8 primary schools and one high school. A free public library is located in Jamestown, and there are branch libraries in several rural districts. Longwood House, Napoleon's home in exile, is now French property and a museum.

1999 KEY EVENTS TIMELINE

June

- A delegation of athletes travels to Gotland, Sweden, to participate with athletes from twenty-two island nations in the eighth Natwest Island Games. After being unable to participate in the games for several years due to lack of funding the Saint Helena Foreign and Commonwealth Office raised sufficient funding for the Island Games Association (formed in 1998), under which St. Helena's athletes could play.

October

- Radio St. Helena makes its final wide-distance broadcast on October 23. Cable and Wireless PLC in Briars, St. Helena, took the transmission equipment out of service and replaced it with a transmitter that does not have broadcast capability, but does transmit radio signals across the local area.

November

- A ship that set sail from Wales, carrying essential supplies to the tiny British dependency of Saint Helena, breaks down en-route and must dock to await repairs at Brest in France; some 5,000 islanders in the South Atlantic fear it won't reach them before Christmas.

ANALYSIS OF EVENTS: 1999

CULTURE AND SOCIETY

The mountainous terrain of Saint Helena makes air service impossible, so residents are dependent on sea transportation for receiving materials and supplies and for travel to and from the island. In October a cargo ship backtracked three hundred miles to pick up a child living on Saint Helena. Doctors believe the girl had a serious medical condition needing further diagnosis and treatment unavailable on the island.

DIRECTORY

CENTRAL GOVERNMENT

Head of State

Queen
Elizabeth II, Monarch, Buckingham Palace, London SWIA 1AA, England
PHONE: +44 (171) 9304832

Governor and Commander in Chief
David Hollamby, Office of the Governor, The Castle, Jamestown, St. Helena
PHONE: +290 2555
FAX: +290 2598

Ministers

Chairman of the Agriculture and Natural Resources Committee
Terrence Richards, Agriculture and Natural Resources Committee
PHONE: +290 2601

Education Committee Chairman
Bobby Robertson, Education Committee
PHONE: +290 4746

Employment and Social Services Committee Chairman
Bill Drabble, Employment and Social Services Committee
PHONE: +290 4224

Public Health Committee Chairman
Eric W. George, Public Health Committee
PHONE: +290 2758

Public Works and Services Committee Chairman
Stedson George, Public Works and Services Committee
PHONE: +290 4331

JUDICIAL SYSTEM

The Supreme Court

Magistrates' Courts

Court of Appeal

FURTHER READING

Articles

Cohen, Mike (Associated Press). ''Cargo Ship Turns Back to Rescue Sick Girl, 6.'' *The Plain Dealer* (October 27, 1999): 5A.

Internet

St. Helena. Available Online @ http://www.sthelena.se/ (November 17, 1999).

St. Helena Institute. Available Online @ http://website.lineone.net/~sthelena/ (November 17, 1999).

''St. Helena Radio Day 1999.'' Available Online http://www.sthelena.se/radiosth.htm (November 17, 1999).

SAINT HELENA: STATISTICAL DATA

For sources and notes see "Sources of Statistics" in the front of each volume.

GEOGRAPHY

Geography (1)

Area:

Total: 410 sq km.

Land: 410 sq km.

Water: 0 sq km.

Note: includes Ascension, Gough Island, Inaccessible Island, Nightingale Island, and Tristan da Cunha Island.

Area—comparative: slightly more than two times the size of Washington, DC.

Land boundaries: 0 km.

Coastline: 60 km.

Climate: Saint Helena—tropical; marine; mild, tempered by trade winds; Tristan da Cunha—temperate; marine, mild, tempered by trade winds (tends to be cooler than Saint Helena).

Terrain: Saint Helena—rugged, volcanic; small scattered plateaus and plains.

Natural resources: fish.

Land use:

Arable land: 6%

Permanent crops: NA%

Permanent pastures: 6%

Forests and woodland: 6%

Other: 82% (1993 est.).

HUMAN FACTORS

Demographics (2A)

	1990	1995	1998	2000	2010	2020	2030	2040	2050
Population	6.7	6.9	7.1	7.2	7.6	7.7	7.6	7.3	6.8
Net migration rate (per 1,000 population)	NA	NA	NA	NA	NA	NA	NA	NA	NA
Births	NA	NA	NA	NA	NA	NA	NA	NA	NA
Deaths	NA	NA	NA	NA	NA	NA	NA	NA	NA
Life expectancy - males	71.6	72.3	72.7	72.9	74.1	75.1	75.9	76.7	77.3
Life expectancy - females	77.5	78.5	79.0	79.3	80.8	82.1	83.1	83.9	84.5
Birth rate (per 1,000)	13.4	14.3	14.1	13.5	10.6	9.2	8.8	8.1	8.1
Death rate (per 1,000)	6.4	6.2	6.5	6.4	7.3	8.9	11.4	13.8	16.6
Women of reproductive age (15-49 yrs.)	1.8	2.0	2.0	2.0	2.0	1.8	1.5	1.4	1.3
of which are currently married	NA	NA	NA	NA	NA	NA	NA	NA	NA
Fertility rate	1.5	1.5	1.5	1.5	1.5	1.5	1.6	1.6	1.6

Except as noted, values for vital statistics are in thousands; life expectancy is in years.

Ethnic Division (6)

African descent, white.

Religions (7)

Anglican (majority), Baptist, Seventh-Day Adventist, Roman Catholic.

Languages (8)

English.

GOVERNMENT & LAW

Political Parties (12)

The legislative branch is a unicameral Legislative Council (15 seats, including the governor, 2 ex-officio and 12 elected members; members are elected by popular vote to serve four- year terms). All seats are held by independents.

Government Budget (13B)

Revenues	$11.2 million
Expenditures	$11 million
Capital expenditures	NA

Data for FY92/93. NA stands for not available.

Military Affairs (14A)

Defense is the responsibility of the UK.

LABOR FORCE

Labor Force (16)

Total	2,416
Professional, technical, and related workers	8.7%
Managerial, administrative, and clerical	12.8%
Sales people	8.1%
Farmer, fishermen, etc.	5.4%
Craftspersons, production process workers	14.7%
Others	50.3%

A large proportion of the work force has left to seek employment overseas. Data for 1991 est. Percent distribution for 1987.

Unemployment Rate (17)

Rate not available.

PRODUCTION SECTOR

Electric Energy (18)

Capacity	4,000 kW (1995)
Production	6 million kWh (1995)
Consumption per capita	887 kWh (1995)

Transportation (19)

Highways:

total: NA km (Saint Helena 118 km, Ascension NA km, Tristan da Cunha NA km)

Merchant marine: none

Airports: 1 (1997 est.)

Airports—with paved runways:

total: 1

over 3,047 m: 1 (1997 est.)

Top Agricultural Products (20)

Maize, potatoes, vegetables; timber production being developed; fishing, including crawfishing on Tristan da Cunha.

FINANCE, ECONOMICS, & TRADE

Economic Indicators (22)

No data available.

Exchange Rates (24)

Exchange rates:

Saint Helenian pounds (£) per US$1

January 1998	0.6115
1997	0.6047
1996	0.6403
1995	0.6335
1994	0.6529
1993	0.6658

The Saint Helenian pound is at par with the British pound.

Top Import Origins (25)

$14.434 million (c.i.f., 1995)

Origins	%
United Kingdom	NA
South Africa	NA

NA stands for not available.

Top Export Destinations (26)

$704,000 (f.o.b., 1995).

Destinations	%
South Africa	NA
United Kingdom	NA

NA stands for not available.

Economic Aid (27)

Recipient: $5.3 million from UK (1997).

Import Export Commodities (28)

Import Commodities	Export Commodities
Food	Fish (frozen canned and salt-dried skipjack Tuna)
Beverages	Handicrafts
Tobacco	
Fuel oils	
Animal feed	
Building materials	
Motor vehicles and parts	
Machinery and parts	

SAINT KITTS AND NEVIS

CAPITAL: Basseterre.

FLAG: Two thin diagonal yellow bands flanking a wide black diagonal band separate a green triangle at the hoist from a red triangle at the fly. On the black band are two white five-pointed stars.

ANTHEM: *National Anthem,* beginning "O land of beauty."

MONETARY UNIT: The East Caribbean dollar of 100 cents is the national currency. There are coins of 1, 2, 5, 10, and 25 cents and 1 East Caribbean dollar, and notes of 5, 10, 20, and 100 East Caribbean dollars. EC$1 = US$0.37037 (US$1 = EC$2.70).

WEIGHTS AND MEASURES: The imperial system is used.

HOLIDAYS: New Year's Day, 1 January; Labor Day, 1st Monday in May; Bank Holiday, 1st Monday in August; Independence Day, 19 September; Prince of Wales's Birthday, 14 November; Christmas, 25 December; Boxing Day, 26 December; Carnival, 30 December. Movable religious holidays include Good Friday and Whitmonday.

TIME: 8 AM = noon GMT.

LOCATION AND SIZE: Saint Kitts is situated in the Leeward Islands. It has a total area of 269 square kilometers (104 square miles), slightly more than 1.5 times the size of Washington, D.C. Nevis lies southeast of Saint Kitts, across a channel called the Narrows; it has a land area of 93 square kilometers (36 square miles). The capital city, Basseterre, is located on Saint Kitts.

CLIMATE: Temperatures range from 20°C (68°F) to 29°C (84°F) all year long. Northeast trade winds are constant. Rain usually falls between May and November, averaging 109 centimeters (43 inches) a year.

Federation of Saint Kitts and Nevis

INTRODUCTORY SURVEY

RECENT HISTORY

Saint Kitts, Nevis, and Anguilla (the most northerly island of the Leeward chain) were incorporated with the British Virgin Islands into a single colony in 1816. The territorial unit of Saint Kitts-Nevis-Anguilla became part of the Leeward Islands Federation in 1871 and belonged to the Federation of the West Indies from 1958 to 1962. In 1967, the three islands became an associated state with full internal autonomy under a new constitution. After the Anguilla islanders rebelled in 1969, British paratroopers intervened, and Anguilla was allowed to secede in 1971.

Saint Kitts and Nevis became an independent federated state within the Commonwealth on 19 September 1983. Under the arrangement, Nevis was given its own legislature and the power to secede from the federation. The People's Action Movement/Nevis Reformation Party coalition won a majority of seats in the 1984 and 1989 elections, but lost to the Labour Party in 1993.

Beginning in June 1996, officials on Nevis announced plans to secede from the federation. On 11 August 1998, the independence referendum for Nevis received only 62% of the vote, short of the two-thirds majority needed.

The islands received attention from American law enforcement officials in August 1998 when drug trafficker Charles Miller (also known as Cecil Connor) threatened to kill American college stu-

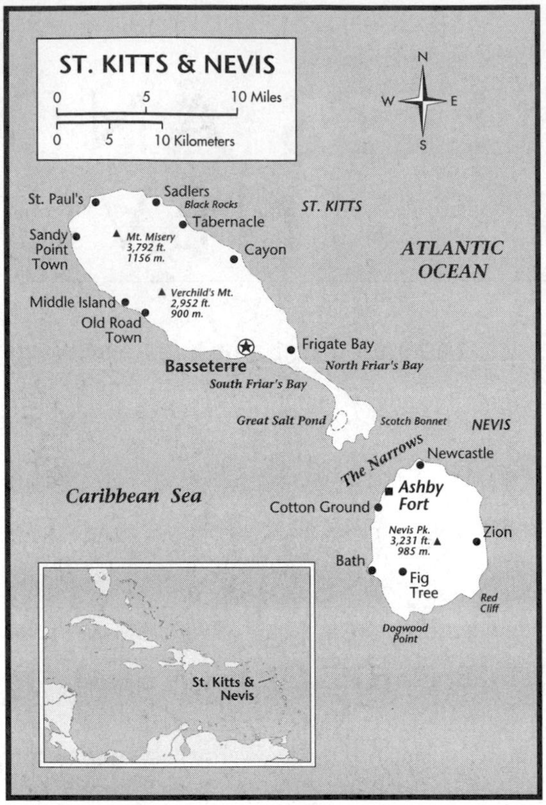

dents in Saint Kitts and Nevis if the government handed him over to the United States.

GOVERNMENT

Saint Kitts–Nevis is a federation of the two islands. Under the 1983 constitution, the British monarch is head of state and is represented by a governor-general. The nation is governed under a parliamentary system, with legislative power vested in the single-chamber House of Assembly, consisting of the speaker, three appointed senators, and eleven elected members. Nevis also has its own legislative assembly and the right to secede under certain conditions.

Judiciary

The Eastern Caribbean Supreme Court, established on Saint Lucia, administers the judicial system. Magistrates' courts deal with petty criminal and civil cases.

Political Parties

The four political parties holding seats in the House of Assembly are the Labour Party (also known as the Workers' League), the People's Action Movement, the Nevis Reformation Party, and the Concerned Citizen's Movement.

DEFENSE

Antigua and Barbuda, Dominica, Grenada, Saint Kitts and Nevis, Saint Lucia, and Saint Vincent and the Grenadines created a Regional Security System in 1985.

ECONOMIC AFFAIRS

The economy is based on tourism and agriculture, particularly on sugar, which generated some 55% of export revenues in 1994. The government has been making efforts to attract foreign investment, to diversify the economy industrially, to expand tourism, and to improve local food production.

Public Finance

The U.S. CIA estimates that, in 1996, government revenues totaled approximately $100.2 million and expenditures $100.1 million. External debt totaled $45.3 million.

Income

In 1998, Saint Kitts and Nevis's gross national product (GNP) was $250 million, or about $6,130 per person. For the period 1985–95, the average inflation rate was 5.5%, resulting in a real growth rate in per person GNP of 4.6%.

Industry

The principal manufacturing plant and largest industrial employer is the Saint Kitts Sugar Manufacturing Corp., which grinds and processes sugarcane for export. Saint Kitts and Nevis has transformed small electronic plants into the largest electronics assembly industry in the Eastern Caribbean. Its apparel assembly industry has also become very successful in recent years.

Banking and Finance

Saint Kitts and Nevis had a relatively simple system of public and private financial institutions in the 1980s. As a member of the Organization of Eastern Caribbean States (OECS), it had as its central monetary authority the Eastern Caribbean Central Bank (ECCB), headquartered in Basseterre.

Foreign reserves, excluding gold, amounted to US$36.2 million as of November 1996. Securities transactions on international exchanges are performed by the banks.

Economic Development

The government has attempted to halt the decline of the sugar industry by restructuring the sector, and has encouraged agricultural diversification and the establishment of small industrial enclaves linked to the international export market. Three industrial estates have been developed, two in Saint Kitts and one on Nevis. The tourist industry has received considerable government support. The Development and Finance Corporation is the principal development agency.

The country's immediate plans continue to be aimed at diversifying the economy. Construction projects in the private and public sectors are expected to contribute substantially to the moderate economic growth that Saint Kitts and Nevis should experience in the near future. A plan to double livestock production by the year 2000 with the development of lands for animal production was launched in 1996. In addition, the tourism industry is expected to keep growing as a result of successful promotions.

SOCIAL WELFARE

Effective 1978, a social security system replaced the existing fund as the provider of old-age, disability, survivor, sickness, and maternity benefits. The Ministry of Women's Affairs promotes women's rights and provides counseling for abused women.

Healthcare

The average life expectancy is 70 years. There were 2,200 inhabitants per doctor in 1992.

Housing

The Housing Authority began a program of low-cost home construction in 1977. In 1996, the National Housing Corporation and the Social Security Board signed a $30.6 million loan for the construction of new housing.

EDUCATION

The literacy rate is about 97%. Education is free and compulsory for twelve years. In 1991, there were 32 primary schools with 350 teachers and 7,236 students. At the secondary-level schools in 1992 there were 312 teachers and 4,402 students enrolled. At the postsecondary level, there were 51 teachers and 394 students enrolled.

1999 KEY EVENTS TIMELINE

January

- The National Conference on Telecommunications Reform convenes.

February

- The International Monetary Fund (IMF) announces a $2.3 million loan to Saint Kitts and Nevis to aid in recovery from Hurricane Georges.

May

- Prime Minister Denzil Douglas is reelected as leader of the ruling Saint Kitts and Nevis Labour Party at the party's annual conference.

June

- Taiwan's Prime Minister Vincent Siew makes a three-day official visit to Saint Kitts and Nevis.
- The annual Saint Kitts and Nevis music festival begins.

August

- China donates $1 million in foreign aid, allocated for the islands' social programs.
- A Constitutional Task Force delivers its report to the National Assembly.

September

- The first national disaster plan in the islands' history is unveiled.
- Prime Minister Douglas addresses the United Nations.

October

- Prime Minister Douglas meets with U.S. Secretary of State Madeline Albright.

ANALYSIS OF EVENTS: 1999

BUSINESS AND THE ECONOMY

In the last decades of the twentieth century Saint Kitts and Nevis, the last country in the Eastern Caribbean to have a "mono-culture" economy dominated by the sugar industry, began to focus on diversifying its agricultural sector and encouraging the growth of new industries through a variety of investment incentives. By 1987 tourism had become the islands' major source of foreign exchange earnings, and the tourist industry had also fostered a boom in construction. In the latter half of the 1990s economic growth remained strong, averaging between 3% and 5%, with inflation remaining under 3%.

A succession of natural disasters toward the end of 1998, however, presented serious economic challenges for the country as it recovered from the effects of Hurricane Georges in September 1998 and flash floods and landslides that November. An estimated 25% of the sugar crop was lost, resulting in a decrease in national revenue. In February 1999 Prime Minister Denzil Douglas announced that the government would be unable to pay sugar industry workers the annual bonus they had received in the preceding three years, and the Saint Kitts Sugar Manufacturing Corporation (SSMC) considered several cost-cutting measures, including early retirement for some employees. Growth for 1998 slowed to 3.5% and was expected to drop to 2% for 1999.

The government pledged to continue its efforts to promote tourism, including maintaining and upgrading tourist attractions and expanding promotional activities in North American and European markets.

GOVERNMENT AND POLITICS

Following the electoral defeat of Nevis's secession attempt in mid-1998 the islands of Saint Kitts and Nevis remained united in the federation that has existed since their joint independence from Great Britain in 1983. Early in 1999 Prime Minister Denzil Douglas announced that constitutional reform to satisfy the needs of both islands would be a government priority. A constitutional task force was created and delivered its report to the National Assembly at the end of August.

Douglas's seat as prime minister was confirmed at the annual conference of the ruling Saint Kitts and Nevis Labour Party in May, when he was reelected head of the party, a position he has held since 1995. The next general election was scheduled for 2000.

In September, as Saint Kitts and Nevis celebrated sixteen years of independence, Douglas addressed the United Nations. In October he was part of a group of foreign ministers from Caribbean nations meeting with U.S. Secretary of State Madeleine Albright in New York City. The group discussed issues of common concern in the region, such as the recent World Trade Organization ruling on the export of bananas and other fruit produced by Caribbean Community (CARICOM) nations, concerns over drug trafficking and gun-running in the region, and membership of the Organization of Eastern Caribbean States (OECS) in the Inter-American Development Bank (IDB).

A key economic development area targeted by Prime Minister Douglas was telecommunications reform, and a national conference on the issue was held in January.

CULTURE AND SOCIETY

Since coming to office in 1995 the Labour Party has built over five hundred affordable homes and planned to complete another five hundred in 1999, three hundred of which were to be allocated to victims of Hurricane Georges. By early 1999 the government had also approved 1,375 applications for emergency relief to families affected by the disaster, much of it designed to help in furnishing building materials and labor for hurricane-related repairs.

In February the International Monetary Fund approved emergency aid totaling US$2.3 million to support further relief efforts, as well as economic recovery measures to rehabilitate the country's infrastructure and tourism facilities. According to IMF estimates overall hurricane damage amounted to an estimated US$400 million. In September the prime minister and the minister of national security launched the country's first national disaster plan to improve readiness for future natural disasters.

Government social welfare initiatives in 1999 included a poverty survey and several programs concerning women, such as violence against women and children, women in poverty, women in leadership positions, and access of women to health services and treatments. The prime minister also announced his intention to establish a National Council on Drug Abuse Prevention to coordinate institutional programs for substance abuse prevention, treatment, and rehabilitation.

The fourth annual music festival of Saint Kitts and Nevis took place in June and featured four evenings of performances by artists including Dru Hill, Shanice, Beenie Man, and Square One. Sponsors of the festival claimed that over time it had met the desired goals of increasing international awareness of the islands, raising local cultural involvement, and improving the standards of local musical groups.

DIRECTORY

CENTRAL GOVERNMENT

Head of State

Prime Minister
Denzil Douglas, Office of the Prime Minister, Government Headquarters, Basseterre, Saint Kitts and Nevis
PHONE: +(809) 4652521
FAX: +(809) 4651001

Ministers

Minister of National Security, Foreign Affairs, Finance, Planning and Information
Denzil Douglas, Ministry of National Security, Foreign Affairs, Finance, Planning and Information, Government Headquarters, Basseterre, Saint Kitts and Nevis
PHONE: +(809) 4652521
FAX: +(809) 4651001

Minister of Trade, Industry and CARICOM Affairs
Sam Condor, Ministry of Trade, Industry and CARICOM Affairs, Church Street, PO Box 186, Basseterre, Saint Kitts and Nevis
PHONE: +(809) 4652521
FAX: +(809) 4651778

Minister of Youth, Sports and Community Affairs
Sam Condor, Ministry of Youth, Sports and Community Affairs, Church Street, PO Box 186, Basseterre, Saint Kitts and Nevis
PHONE: +(809) 4652521
FAX: +(809) 4651778

Minister of Agriculture, Lands and Housing
Timothy Harris, Ministry of Agriculture, Lands and Housing, Government Headquarters, PO Box 186, Basseterre, Saint Kitts and Nevis
PHONE: +(809) 4652521
FAX: +(809) 4652635

Minister of Communications, Works and Public Utilities
Cedric Liburd, Ministry of Communications, Works and Public Utilities, Government Headquarters, PO Box 186, Basseterre, Saint Kitts and Nevis
PHONE: +(809) 4652521
FAX: +(809) 4650604

Minister of Education, Labor and Social Security
Rupert Herbert, Ministry of Education, Labor and Social Security, Cayon Street, PO Box 333, Basseterre, Saint Kitts and Nevis
PHONE: +(809) 4652521
FAX: +(809) 4659069

Minister Tourism, Culture and Environment
G. A. Dwyer Astaphan, Ministry of Tourism, Culture and Environment, Pelican Mall, Bay Rd., PO Box 132, Basseterre, Saint Kitts and Nevis
PHONE: +(809) 4654040
FAX: +(809) 4658794

Minister of Health and Women's Affairs
Earl Asim Martin, Ministry of Health and Women's Affairs, Church Street, PO Box 186, Basseterre, Saint Kitts and Nevis
PHONE: +(809) 4652521
FAX: +(809) 4651316

POLITICAL ORGANIZATIONS

People's Action Movement (PAM)

PHONE: +(809) 4662303
FAX: +(809) 4657268
E-MAIL: pubrel@pamskb.com.com
TITLE: Leader
NAME: Kennedy Simmonds

Reformation Party (NRP)

TITLE: Leader
NAME: Joseph Parry

Saint Kitts-Nevis Labor Party (SKLP)

NAME: Denzil Douglas

Concerned Citizens' Movement (CCM)

DIPLOMATIC REPRESENTATION

Embassies in Saint Kitts and Nevis

Japan
Government Headquarters, Basseterre, Saint Kitts and Nevis
PHONE: +(869) 4652521
FAX: +(869) 4655202

JUDICIAL SYSTEM

Eastern Caribbean Supreme Court

FURTHER READING

Articles

''Four Seasons Resort Nevis Reopens Following Refurbishment.'' *Travel Weekly* (Jan. 7, 1999): 58.

Marcus, Frances Frank. ''Doing What the Islands Do Best'' *The New York Times* 24 Oct. 1999 p. TR10.

Myers, Gay Nagle. ''Resilient People, Island Shake off Georges' September Visit.'' *Travel Weekly* (Nov. 19, 1998): C8.

Books

Moll, Verna P. *Saint Kitts–Nevis*. Santa Barbara, Calif.: Clio Press, 1995.

Internet

''The 1999 New Year's Message'' (Prime Minister Denzil Douglas); ''Saint Kitts and Nevis Prime Minister Douglas Presents Caribbean Concerns to United States Secretary of State;'' ''Media Releases.'' Available Online @ http://www.stkittsnevis.net.

SAINT KITTS AND NEVIS: STATISTICAL DATA

For sources and notes see "Sources of Statistics" in the front of each volume.

GEOGRAPHY

Geography (1)

Area:

Total: 269 sq km.

Land: 269 sq km.

Water: 0 sq km.

Area—comparative: 1.5 times the size of Washington, DC.

Land boundaries: 0 km.

Coastline: 135 km.

Climate: subtropical tempered by constant sea breezes; little seasonal temperature variation; rainy season (May to November).

Terrain: volcanic with mountainous interiors.

Natural resources: NEGL.

Land use:

Arable land: 22%

Permanent crops: 17%

Permanent pastures: 3%

Forests and woodland: 17%

Other: 41% (1993 est.).

HUMAN FACTORS

Infants and Malnutrition (5)

Under-5 mortality rate (1997) 37

% of infants with low birthweight (1990-97) 9

Births attended by skilled health staff % of total[a] . . .NA

% fully immunized (1995-97)

TB 99

DPT 100

Polio 100

Measles 97

Prevalence of child malnutrition under age 5 (1992-97)[b] .... NA

Demographics (2A)

	1990	1995	1998	2000	2010	2020	2030	2040	2050
Population	39.9	41.0	42.3	43.4	50.2	56.8	62.5	66.7	68.8
Net migration rate (per 1,000 population)	NA	NA	NA	NA	NA	NA	NA	NA	NA
Births	NA	NA	NA	NA	NA	NA	NA	NA	NA
Deaths	NA	NA	NA	NA	NA	NA	NA	NA	NA
Life expectancy - males	61.7	63.5	64.5	65.2	68.3	71.0	73.1	74.9	76.3
Life expectancy - females	67.6	69.7	70.8	71.6	75.0	77.8	80.0	81.8	83.1
Birth rate (per 1,000)	24.6	23.5	22.9	22.4	19.3	15.9	14.1	12.5	11.5
Death rate (per 1,000)	11.7	9.6	8.5	7.8	5.5	5.1	5.8	7.7	9.7
Women of reproductive age (15-49 yrs.)	9.4	10.6	11.6	12.3	14.5	15.6	16.0	15.7	15.4
of which are currently married	NA	NA	NA	NA	NA	NA	NA	NA	NA
Fertility rate	2.8	2.6	2.5	2.4	2.1	2.0	1.9	1.9	1.8

Except as noted, values for vital statistics are in thousands; life expectancy is in years.

Ethnic Division (6)

Black.

Religions (7)

Anglican, other Protestant sects, Roman Catholic.

Languages (8)

English.

EDUCATION

Educational Attainment (10)

Age group (1980)	25+
Total population	16,771
Highest level attained (%)	
No schooling	1.1
First level	
Not completed	29.0
Completed	NA
Entered second level	
S-1	66.6
S-2	NA
Postsecondary	2.3

GOVERNMENT & LAW

Political Parties (12)

House of Assembly	% of seats
Nevis Labor Party (SKNLP)	58
People's Action Movement (PAM)	41

Government Budget (13B)

Revenues	$100.2 million
Expenditures	$100.1 million
Capital expenditures	$41.4 million

Data for 1996 est.

Military Affairs (14A)

Total expenditures	$NA
Expenditures as % of GDP	NA%

NA stands for not available.

LABOR FORCE

Labor Force (16)

Total	18,172
Services	69%
Manufacturing	31%

Total for June 1995.

Unemployment Rate (17)

4.3% (May 1995)

PRODUCTION SECTOR

Electric Energy (18)

Capacity	16,000 kW (1995)
Production	81 million kWh (1995)
Consumption per capita	1,976 kWh (1995)

Transportation (19)

Highways:

total: 320 km

paved: 136 km

unpaved: 184 km (1996 est.)

Merchant marine: none

Airports: 2 (1997 est.)

Airports—with paved runways:

total: 2

1,524 to 2,437 m: 1

under 914 m: 1 (1997 est.)

Top Agricultural Products (20)

Sugarcane, rice, yams, vegetables, bananas; fishing potential not fully exploited.

FINANCE, ECONOMICS, & TRADE

Economic Indicators (22)

National product: GDP—purchasing power parity—$235 million (1996 est.)

National product real growth rate: 5.8% (1996 est.)

National product per capita: $5,700 (1996 est.)

Inflation rate—consumer price index: 3.1% (1996)

Exchange Rates (24)

Exchange rates: East Caribbean dollars (EC$) per US$1—2.7000 (fixed rate since 1976)

Top Import Origins (25)

$131.5 million (f.o.b., 1996 est.) Data are for 1994.

Origins	%
United States	45
Caricom nations	18.8
United Kingdom	12.5
Canada	4.2
Japan	4.2

Top Export Destinations (26)

$39.1 million (f.o.b., 1996 est.) Data are for 1994.

Destinations	%
United States	46.6
United Kingdom	26.4
Caricom nations	9.8

Economic Aid (27)

Recipient: ODA, $NA. NA stands for not available.

Import Export Commodities (28)

Import Commodities	Export Commodities
Machinery	Machinery
Manufactures	Food
Food	Electronics
Fuels	Beverages and tobacco

SAINT LUCIA

CAPITAL: Castries.

FLAG: On a blue background is a yellow triangle surmounted by a black arrowhead whose outer edges are bordered in white.

ANTHEM: *Sons and Daughters of Saint Lucia.*

MONETARY UNIT: The East Caribbean dollar (EC$) of 100 cents is the national currency. There are coins of 1, 2, 5, 10, and 25 cents and 1 dollar, and notes of 5, 10, 20, and 100 East Caribbean dollars. EC$1 = US$0.37037 (or US$1 = EC$2.70).

WEIGHTS AND MEASURES: The metric system has been introduced, but imperial measures are still commonly employed.

HOLIDAYS: New Year's Day, 1 January; Carnival, 8–9 February; Independence Day, 22 February; Labor Day, 1 May; Queen's Official Birthday, 5 June; Bank Holiday, 1st Monday in August; Thanksgiving Day, 1st Monday in October; Saint Lucia Day, 13 December; Christmas Day, 25 December; Boxing Day, 26 December. Movable religious holidays include Good Friday, Easter Monday, Whitmonday, and Corpus Christi.

TIME: 8 AM = noon GMT.

LOCATION AND SIZE: The Caribbean island of Saint Lucia, part of the Windward Islands group of the Lesser Antilles, has a total area of 620 square kilometers (239 square miles), slightly less than 3.5 times the size of Washington, D.C. The capital city, Castries, is located on Saint Lucia's northwest coast.

CLIMATE: The average yearly temperature on Saint Lucia is 26°C (79°F). The average rainfall at sea level is 231 centimeters (91 inches) a year. Like the rest of the West Indies, Saint Lucia is vulnerable to hurricanes.

INTRODUCTORY SURVEY

RECENT HISTORY

Saint Lucia established full internal self-government in 1967 and on 22 February 1979 became an independent member of the Commonwealth of Nations.

The first three years of independence were marked by political turmoil and civil strife, as leaders of rival parties fought bitterly. In 1982, the conservative United Workers' Party (UWP) won fourteen of seventeen seats in the House of Assembly. Party leader and prime minister John Compton became prime minister at independence and has governed ever since. In 1992, the UWP controlled eleven parliamentary seats. After fifteen years out of office, the Saint Lucia Labour Party won control in 1997.

Saint Lucia suffered back-to-back tropical storms in 1994 and 1995 that caused losses of 65% and 20% of each of those years' banana crops, respectively.

GOVERNMENT

Under the 1979 constitution, the British monarch, as official head of government, is represented by a governor-general. Executive power is exercised by the prime minister and cabinet. There is a two-chamber parliament consisting of a Senate with eleven members and a House of Assembly with seventeen representatives.

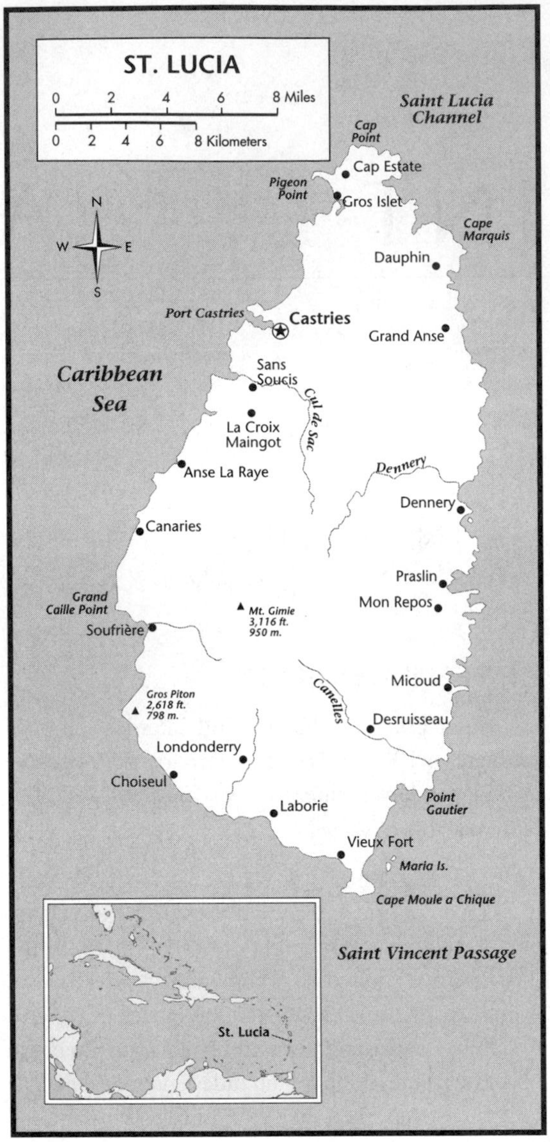

Judiciary

The lowest court is the district or magistrate's court, above which is the Court of Summary Jurisdiction. The Eastern Caribbean Supreme Court, with international jurisdiction, is seated in Castries.

Political Parties

The majority parties are United Workers' Party (UWP), led by John Compton, and the Saint Lucia Labour Party (SLP), led by Julian Hunte.

DEFENSE

There are no armed forces; the police department numbers 300. A regional defense pact provides for joint defense and disaster contingency plans.

ECONOMIC AFFAIRS

Agriculture has traditionally been the main economic activity on Saint Lucia, which is the leading producer of bananas in the Windward Islands group. Tourism has recently become an equally important economic activity.

Public Finance

The U.S. CIA estimates that, in 1995, government revenues totaled approximately $121 million and expenditures $127 million, including capital expenditures of $104 million. External debt totaled $222.7 million. Most of the income came from customs duties, income taxes, taxes on goods and services, and taxes on international trade and transactions.

Income

In 1998, the gross national product (GNP) was $546 million, or about $3,410 per person. For the period 1985–95, the average inflation rate was 3.2%, resulting in a real growth rate in per person GNP of 3.9%.

Industry

Saint Lucia's manufacturing industry is the largest in the Windward Islands, with many assembly plants producing apparel, electronic components, plastic products, and paper and cardboard boxes. Textile production fell in 1996 from the closing of some garment factories.

Banking and Finance

In early 1981, the government-owned Saint Lucia National Bank and the Saint Lucia Development Bank were opened. There were seven commercial banks as of 1993. Saint Lucia is a member of the Eastern Caribbean Central Bank, which is responsible for the administration of the country's monetary policies, the regulation of exchange control and supervision of commercial banks and other financial institutions for the islands belonging to the Organization of Eastern Caribbean States.

Foreign reserves, excluding gold, totaled $56.4 million in November 1996. Saint Lucia has no securities exchange.

Economic Development

Since the establishment the National Development Corp. in 1971, Saint Lucia has succeeded in diversifying its economy. Saint Lucia has the most highly developed infrastructure of all the Windward Islands, with an international airport, a highway system that connects the important coastal and agricultural areas with the political and commercial centers, and a fully automated telephone system with direct dialing to most parts of the world.

At the end of 1996, the Lewis government unveiled a job-creating budget aimed at boosting his party's flagging fortunes. A $242 million package devoted $136 million to current expenditure and $104 million to capital items.

SOCIAL WELFARE

The National Insurance Scheme provides all workers from age 16 to 60 with old-age, disability, survivor, sickness, and maternity coverage, as well as workers' compensation. Efforts have been made to improve the status of women, especially in employment.

Healthcare

There were 5 hospitals on Saint Lucia with 528 beds in 1988. In 1991, there were 59 doctors, or about one doctor per 2,306 inhabitants. Malnutrition and intestinal disorders are the main health problems. The average life expectancy is 70 years.

Housing

The demand for private ownership of homes far exceeds the supply. In 1980, the majority of housing (74%) was built of wood.

EDUCATION

In 1993, the literacy rate was estimated at over 72%, an improvement over the average of 67% from 1988–93. In 1992, there were 88 primary schools with 1,204 teaching staff and 32,545 students. There were 524 secondary teachers and 9,550 students. Institutions of higher learning include a branch of the University of the West Indies and the Sir Arthur Lewis Community College.

1999 KEY EVENTS TIMELINE

February

- China says it will build a $4 million sports stadium for Saint Lucia. The stadium is a reward for breaking diplomatic ties with Taiwan.

March

- American Airlines announces it will end direct service between Miami, Florida, and Saint Lucia on April 15. The company's decision is made after Saint Lucia's government says it will not continue to pay $1.5 million in marketing subsidies to the airline.

April

- The World Trade Organization (WTO) upholds the U.S. government's right to impose trade penalties on Europe over its banana import rules. At issue are Caribbean bananas from countries like Dominica that enjoy special tariffs in Europe.
- Saint Lucia residents are told they will have to pay as much as 100 percent more for water after the recent privatization of the public utility. Water rates had not increased in 10 years.

May

- Prime Minister Kenny Anthony says the Caribbean needs a press complaint office to ''redress inaccurate and dishonest reporting.''
- Drug enforcement officers from Saint Lucia and the United States intercept about fifty-seven pounds of cocaine, valued at $370,000, being flown from Saint Lucia to Puerto Rico.

June

- Prime Minister Anthony announces that the government would resume the practice of hanging death row prisoners who have exhausted all recourse to appeal. His announcement comes after neighboring islands also resumed the punishment.

ANALYSIS OF EVENTS: 1999

BUSINESS AND THE ECONOMY

Seeking to position itself as a major port of call for cruise ships, Saint Lucia completed a major reconstruction project at its main port. A new berth will handle ships up to 1,000 feet in length, while a second berth will accommodate ships up to 750 feet in length.

GOVERNMENT AND POLITICS

Prime Minister Kenny Anthony announced in August that he would introduce anti-corruption legislation in an attempt to make the government more accountable. The new corruption measures would be modeled on a similar proposal being discussed by the Jamaican legislature.

Anthony, who became prime minister after his Labor Party won a majority in the 1997 elections, considered bringing charges against members of the former government. Anthony did not publicly release details of alleged wrongdoing, but did say former prime ministers were not implicated.

Members of the Organization of American States (OAS) visited the island in April to assist the government with electoral reforms. Officials said new measures, including the verification of voters' lists and the clearing up electoral boundary issues, would strengthen the country's democracy.

In other political matters Saint Lucia and other Caribbean nations spent most of 1999 lobbying in Europe and the United States to protect their banana industries, which make up a considerable portion of their economies. Bananas are the top export for Saint Lucia, which privatized the industry to make it more efficient. Some officials believe a quota allocation is the best course of action in order to continue the preferential access Caribbean fruit producers have to the European market.

Several European nations have rules that favor bananas imported from their former colonies in the Caribbean and restricting bananas imported from places like Honduras, Ecuador, and Colombia. The United States, under lobbying pressure by U.S. fruit companies growing bananas in Central and South America, challenged the preferential treatment at the World Trade Organization (WTO). In February the WTO upheld the U.S. government's right to impose trade penalties on Europe over its banana import rules. Following this decision the United States announced it would impose 100% tariffs on many European products, including clothes, food, and jewelry. While Europe and the United States argued about fair trade practices, small Caribbean nations like Dominica, Saint Lucia, Saint Vincent and the Grenadines joined lobbying efforts in an attempt to be heard. No solution to the issue was expected in 1999.

CULTURE AND SOCIETY

During 1999 the Caribbean was shocked by brutal homicides on several islands, renewing calls for the hanging of convicted criminals on death row. Saint Lucia, which held seven inmates on death row, announced that hangings would resume for prisoners who had exhausted their appeals.

In June Saint Lucian officials objected to British criticism of Caribbean countries supporting the death penalty. Saint Lucia's Foreign Minister George Odlum said some Caribbean countries were facing large increases in violent crime and had no choice but to implement the death penalty. Saint Lucia and other Caribbean countries, upset at Britain's position, met in Grenada in June to discuss the creation of a Caribbean court to replace Britain's Privy Council, a colonial vestige that still has final say on legal appeals.

DIRECTORY

CENTRAL GOVERNMENT

Head of State

Governor-General
Pearlete Louisy, Office of the Governor General
PHONE: +(758) 4532481; 4532483
FAX: +(758) 4532731

Prime Minister
Kenny Anthony, Office of the Prime Minister
PHONE: +(758) 4537880
FAX: +(758) 4237352

Ministers

Deputy Prime Minister
Mario Michel, Office of the Deputy Prime Minister
PHONE: +(758) 4522476
FAX: +(758) 4532299

Minister of Agriculture, Fisheries, Forestry, and the Environment
Cassius Elias, Ministry of Agriculture, Fisheries, Forestry, and the Environment
PHONE: +(758) 4522526
FAX: +(758) 4536314

Minister of Commerce, Industry, and Consumer Affairs
Walter Francois, Ministry of Commerce, Industry, and Consumer Affairs
PHONE: +(758) 4519763
FAX: +(758) 4537347

Minister of Communications, Works, Transport and Public Utilities
Calixte George, Ministry of Communications, Works, Transport and Public Utilities
PHONE: +(758) 4524444
FAX: +(758) 4532769

Minister of Community Development, Culture, Local Government and Cooperatives
Damian Greaves, Ministry of Community Development, Culture, Local Government and Cooperatives
PHONE: +(758) 4531487
FAX: +(758) 4537921

Minister of Education, Human Resource Development, Youth and Sports
Mario Michel, Ministry of Education, Human Resource Development, Youth and Sports
PHONE: +(758) 4522476
FAX: +(758) 4532299

Minister of Finance Planning, Information Services and the Public Service
Kenny D. Anthony, Ministry of Finance Planning, Information Services and the Public Service
PHONE: +(758) 4537880
FAX: +(758) 4537352

Minister of Health, Human Services, Family Affairs and Gender Relations
Sarah Flood Beabrun, Ministry of Health, Human Services, Family Affairs and Gender Relations
PHONE: +(758) 4522859; 4531960
FAX: +(758) 4525655

Minister of Legal Affairs, Home Affairs and Labor
Velon John, Ministry of Legal Affairs, Home Affairs and Labor
PHONE: +(758) 4523622; 4523772
FAX: +(758) 4536315

Minister for Tourism, Civil Aviation and Offshore Financial Services
Philip J. Pierre, Ministry for Tourism, Civil Aviation and Offshore Financial Services
PHONE: +(758) 4516643
FAX: +(758) 4516986

Attorney General
Petrus Compton, Office of the Attorney General
PHONE: +(758) 4523622; 4523772
FAX: +(758) 4536315

POLITICAL ORGANIZATIONS

United Workers Party (UWP)

E-MAIL: uwp@iname.com
TITLE: Leader
NAME: John Compton

Saint Lucia Labor Party (SLP)

2nd Floor, Tom Walcott Building, Jeremie Street, PO Box 427, Castries, Saint Lucia
PHONE: +(758) 4518446; 4526934
FAX: +(758) 4519389
E-MAIL: slp@candw.lc
TITLE: Leader
NAME: Julian Hunte

DIPLOMATIC REPRESENTATION

Embassies in Saint Lucia

United Kingdom
British High Commission, Derek Walcott Square, PO Box 227, Castries, Saint Lucia
PHONE: +(809) 4522484
FAX: +(809) 4531543

JUDICIAL SYSTEM

Eastern Caribbean Supreme Court

Court of Summary Jurisdiction

FURTHER READING

Articles

''Another Decline in Banana Prices.'' *Caribbean Week* (July 31, 1999).

''Banana Dispute Sparks Caribbean Fear, Resentment of U.S.'' *The Associated Press*, 12 April 1999.

''China Agrees to Build New Stadium for Saint Lucia.'' *The Associated Press*, 6 February 1999.

''Clamor for Capital Punishment Sweeps Crime-Plagued Caribbean.'' *The Associated Press*, 22 June 1999.

''Minister Criticizes British Stance.'' *Caribbean Week* (June 28, 1999).

''St Lucia Anti-Corruption Law Coming.'' *Caribbean Week* (August 6, 1999).

''U.S. Ambassador Says Caribbean Too Eager to 'Cozy Up' with Cuba.'' *The Associated Press*, 20 August 1999.

SAINT LUCIA: STATISTICAL DATA

For sources and notes see "Sources of Statistics" in the front of each volume.

GEOGRAPHY

Geography (1)

Area:

Total: 620 sq km.

Land: 610 sq km.

Water: 10 sq km.

Area—comparative: 3.5 times the size of Washington, DC.

Land boundaries: 0 km.

Coastline: 158 km.

Climate: tropical, moderated by northeast trade winds; dry season from January to April, rainy season from May to August.

Terrain: volcanic and mountainous with some broad, fertile valleys.

Natural resources: forests, sandy beaches, minerals (pumice), mineral springs, geothermal potential.

Land use:

Arable land: 8%

Permanent crops: 21%

Permanent pastures: 5%

Forests and woodland: 13%

Other: 53% (1993 est.)

HUMAN FACTORS

Demographics (2A)

	1990	1995	1998	2000	2010	2020	2030	2040	2050
Population	139.6	147.2	152.3	155.7	174.3	194.0	210.0	220.3	223.7
Net migration rate (per 1,000 population)	NA	NA	NA	NA	NA	NA	NA	NA	NA
Births	NA	NA	NA	NA	NA	NA	NA	NA	NA
Deaths	NA	NA	NA	NA	NA	NA	NA	NA	NA
Life expectancy - males	66.5	67.4	67.9	68.3	70.2	71.8	73.2	74.5	75.5
Life expectancy - females	74.0	74.7	75.5	76.0	78.2	80.0	81.5	82.7	83.6
Birth rate (per 1,000)	25.1	24.9	22.5	20.8	17.2	14.7	12.8	11.4	10.6
Death rate (per 1,000)	6.1	5.8	5.6	5.5	5.2	5.4	6.3	8.3	10.6
Women of reproductive age (15-49 yrs.)	34.7	38.8	41.4	43.5	51.3	54.1	53.8	50.8	47.8
of which are currently married	NA	NA	NA	NA	NA	NA	NA	NA	NA
Fertility rate	2.7	2.6	2.3	2.2	1.9	1.8	1.8	1.8	1.8

Except as noted, values for vital statistics are in thousands; life expectancy is in years.

Infants and Malnutrition (5)

Under-5 mortality rate (1997) 29
% of infants with low birthweight (1990-97) 8
Births attended by skilled health staff % of total[a] . . .NA
% fully immunized (1995-97)
TB 100
DPT 98
Polio 98
Measles 95
Prevalence of child malnutrition under age 5 (1992-97)[b] . . . . .NA

Ethnic Division (6)

Black 90%
Mixed 6%
East Indian 3%
White 1%

Religions (7)

Roman Catholic 90%
Protestant 7%
Anglican 3%

Languages (8)

English (official), French patois.

EDUCATION

Educational Attainment (10)

Age group (1991) . . . . .25+
Total population 49,031
Highest level attained (%)
No schooling 0
First level
Not completed 75.5
Completed . . . . .NA
Entered second level
S-1 . . . . .21.2
S-2 . . . . .NA
Postsecondary 3.4

GOVERNMENT & LAW

Political Parties (12)

House of Assembly	No. of seats
Saint Lucia Labor Party (SLP)	16
United Workers' Party (UWP)	1

Government Budget (13B)

Revenues . . . .$155 million
Expenditures . . . .$169 million
Capital expenditures . . . .$48 million

Data for FY96/97 est.

Military Affairs (14A)

Total expenditures (1991) . . . .$5 million
Expenditures as % of GDP (1991) 2%

For police force.

LABOR FORCE

Labor Force (16)

Total 43,800
Agriculture 43.4%
Services 38.9%
Industry and commerce 17.7%

Data for 1983 est.

Unemployment Rate (17)

15% (1996 est.)

PRODUCTION SECTOR

Electric Energy (18)

Capacity . . . . .22,000 kW (1995)
Production . . . . .110 million kWh (1995)
Consumption per capita . . . . .705 kWh (1995)

Transportation (19)

Highways:
total: 1,210 km
paved: 63 km
unpaved: 1,147 km (1996 est.)
Merchant marine: none
Airports: 2 (1997 est.)
Airports—with paved runways:
total: 2
2,438 to 3,047 m: 1
1,524 to 2,437 m: 1 (1997 est.)

Top Agricultural Products (20)

Bananas, coconuts, vegetables, citrus, root crops, cocoa.

FINANCE, ECONOMICS, & TRADE

Economic Indicators (22)

National product: GDP—purchasing power parity—$600 million (1996 est.)

National product real growth rate: 0.8% (1996 est.)

National product per capita: $3,800 (1996 est.)

Inflation rate—consumer price index: -2.3% (1996 est.)

Exchange Rates (24)

Exchange rates: East Caribbean dollars (EC$) per US$1—2.7000 (fixed rate since 1976)

Top Import Origins (25)

$270.6 million (f.o.b., 1996 est.) Data are for 1995.

Origins	%
United States	36
Caricom countries	22
United Kingdom	11
Japan	5
Canada	4

Top Export Destinations (26)

$79.5 million (f.o.b., 1996 est.) Data are for 1995.

Destinations	%
United Kingdom	50
United States	24
Caricom countries	16

Economic Aid (27)

Recipient: ODA, $NA. NA stands for not available.

Import Export Commodities (28)

Import Commodities	Export Commodities
Food 23%	Bananas 41%
Manufactured goods 21%	Clothing
Machinery and transportation equipment 19%	Cocoa
Chemicals	Vegetables
Fuels	Fruits
	Coconut oil

Balance of Payments (23)

	1988	1989	1990	1991	1992
Exports of goods (f.o.b.)	119.1	112.0	127.3	110.3	122.8
Imports of goods (f.o.b.)	−194.5	−240.9	−238.7	−261.4	−275.4
Trade balance	−75.4	−128.9	−111.4	−151.1	−152.6
Services - debits	−83.4	−92.8	−121.0	−124.5	−129.9
Services - credits	126.7	146.4	160.2	186.0	204.8
Private transfers (net)	14.2	13.7	15.2	17.3	16.1
Government transfers (net)	5.4	5.3	0.3	4.5	6.4
Long-term capital (net)	25.0	35.3	49.8	62.0	67.3
Short-term capital (net)	−15.0	18.6	5.1	−0.2	2.3
Errors and omissions	5.0	8.2	8.3	13.7	−7.8
Overall balance	2.6	5.7	6.5	7.7	6.5

SAINT PIERRE AND MIQUELON

CAPITAL: Saint-Pierre.

FLAG: The flag of France is used.

MONETARY UNIT: 1 French franc (F) = 100 centimes.

WEIGHTS AND MEASURES: The metric system is used.

HOLIDAYS: National Day, 14 July.

TIME: 8 AM = noon GMT.

LOCATION AND SIZE: The French territorial collectivity of Saint Pierre and Miquelon is an archipelago in the North Atlantic Ocean, between 46°45′ and 47°10′N and 56°5′ and 56°25′W, located about 24 km (15 mi.) W of Burin Peninsula on the south coast of Newfoundland. It consists of three main islands, Saint Pierre, Miquelon, and Langlade—the two latter linked by a low, sandy isthmus—and several small ones. The length of the group is 43 km (27 mi.) N–S, and it measures 22 km (14 mi.) E–W at its widest extent. The total area is 242 sq. km (93 sq. mi.).

CLIMATE: The climate is mild and humid, with temperatures reaching −10°C (14°F) in the winter and 20°C (68°F) in the summer. Average annual rainfall is about 59 inches (150 cm).

INTRODUCTORY SURVEY

RECENT HISTORY

The first permanent French settlement dates from 1604, and, except for several periods of British rule, the islands have remained French ever since. They became a French overseas territory in 1946, an overseas department in 1976, and a territorial collectivity in 1985. The islands were the focus of a maritime boundary dispute between Canada and France, but in 1992 an arbitration panel awarded the islands an exclusive economic zone area of 12,348 sq km (4,768 sq mi) as a settlement.

Territorial Collectivity of Saint Pierre and Miquelon

ECONOMIC AFFAIRS

The economy has traditionally centered around fishing and by servicing the fishing fleets operating off the coast of Newfoundland, but the number of ships stopping at Saint Pierre has been declining in recent years. The annual cod catch declined from 13,000 tons in 1991 to 86 tons in 1994. Total exports in 1994 amounted to US$13.74 million, while imports totaled US$42 million, requiring heavy subsidies from France.

DIRECTORY

CENTRAL GOVERNMENT

Head of State

Prefect

Remi Thuau, Office of the Prefect, 97500 Saint Pierre and Miquelon

PHONE: +508 411010

FAX: +508 414738

Chief of the Cabinet

Francois Chauvin, Office of the Chief of the Cabinet

Secretary General

Jean-Pierre Tressard, Office of the Secretary General

Ministers

Minister of the Development Agency of Saint Pierre and Miquelon

Ministry of the Development Agency of Saint Pierre and Miquelon, Rue Borda-Palais Royal, 97500 Saint Pierre and Miquelon

PHONE: +508 417114

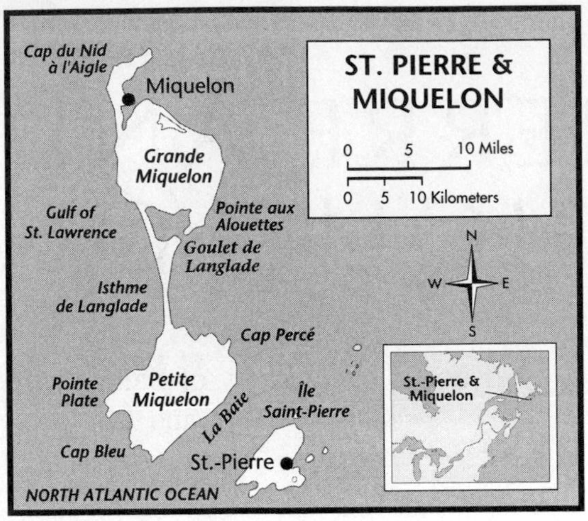

FAX: +508 417115
E-MAIL: sodepar@cancom.net

President of the General Council

Bernard Le Soavec, Office of the President of the General Council, Place de l'Eglise Place, 97500 Saint Pierre and Miquelon
PHONE: +508 414622
FAX: +508 412297
E-MAIL: conseilg@cancom.net

POLITICAL ORGANIZATIONS

Union pour la Democratie Francaise-UDF (Union for French Democracy)

TITLE: Leader
NAME: Gerard Grignon

Parti Socialiste-PS (Socialist Party)

TITLE: Leader
NAME: Marc Plantagenest

Rassemblement pour la Republique-RPR (Rally for the Republic)

TITLE: Leader
NAME: Victor Reux

SAINT PIERRE AND MIQUELON: STATISTICAL DATA

For sources and notes see "Sources of Statistics" in the front of each volume.

GEOGRAPHY

Geography (1)

Area:

Total: 242 sq km.

Land: 242 sq km.

Water: 0 sq km.

Note: includes eight small islands in the Saint Pierre and the Miquelon groups.

Area—comparative: 1.5 times the size of Washington, DC.

Land boundaries: 0 km.

Coastline: 120 km.

Climate: cold and wet, with much mist and fog; spring and autumn are windy.

Terrain: mostly barren rock.

Natural resources: fish, deepwater ports.

Land use:

Arable land: 13%

Permanent crops: NA%

Permanent pastures: NA%

Forests and woodland: 4%

Other: 83% (1993 est.).

HUMAN FACTORS

Demographics (2B)

Population (July 1998 est.)	6,914
Age structure:	
0-14 years	NA
15-64 years	NA
65 years and over	NA
Population growth rate (1998 est.)	0.76%
Birth rate, 1998 est. (births/1,000 population)	12.45
Death rate, 1998 est. (deaths/1,000 population)	5.49
Net migration rate, 1998 est. (migrant(s)/1,000 population)	0.59
Infant mortality rate, 1998 est. (deaths/1,000 live births)	8.62
Life expectancy at birth (years):	
Total population	76.91
Male	75.35
Female (1998 est.) *: 78.79	
Total fertility rate, 1998 est. (children born/woman)	1.6

Ethnic Division (6)

Basques and Bretons (French fishermen).

Religions (7)

Roman Catholic 99%.

Languages (8)

French.

EDUCATION

Educational Attainment (10)

Age group (1982)	16+
Total population	4,282
Highest level attained (%)	
No schooling	0.8
First level	
Not completed	53.2
Completed	NA

Entered second level	
S-1	40.8
S-2	NA
Postsecondary	4.9

GOVERNMENT & LAW

Political Parties (12)

General Council	No. of seats
Rassemblement pour la Republique (RPR)	15
Other	4

Government Budget (13B)

Revenues	$28 million
Expenditures	$28 million
Capital expenditures	$7.8 million

Data for 1992 est.

Military Affairs (14A)

Defense is the responsibility of France.

LABOR FORCE

Labor Force (16)

Total 2,971 (1995)

Unemployment Rate (17)

11% (1996)

PRODUCTION SECTOR

Electric Energy (18)

Capacity	27,000 kW (1995)
Production	42 million kWh (1995)
Consumption per capita	6,216 kWh (1995)

Transportation (19)

Highways:

total: 114 km

paved: 69 km

unpaved: 45 km (1994 est.)

Merchant marine: none

Airports: 2 (1997 est.)

Airports—with paved runways:

total: 2

914 to 1,523 m: 2 (1997 est.)

Top Agricultural Products (20)

Vegetables; cattle, sheep, pigs; fish catch of 14,800 metric tons (1994).

FINANCE, ECONOMICS, & TRADE

Economic Indicators (22)

National product: GDP—purchasing power parity—$74 million (1996 est.)

National product real growth rate: NA%

National product per capita: $11,000 (1996 est.)

Inflation rate—consumer price index: NA%

Exchange Rates (24)

Exchange rates:	
French francs (F) per US$1	
January 1998	6.0836
1997	5.8367
1996	5.1155
1995	4.9915
1994	5.5520
1993	5.6632

Top Import Origins (25)

$70.2 million (c.i.f., 1995)

Origins	%
United States	NA
European Union	NA
Kenya	NA
Tanzania	NA

NA stands for not available.

Top Export Destinations (26)

$5 million (f.o.b., 1995) Data are for 1990.

Destinations	%
United States	58
France	17
United Kingdom	11
Canada	NA
Portugal	NA

NA stands for not available.

Economic Aid (27)

Recipient: ODA, $NA. NA stands for not available.

Import Export Commodities (28)

Import Commodities	Export Commodities
Meat	Fish and fish products
Clothing	Fox and mink pelts
Fuel	
Electrical equipment	
Machinery	
Building materials	

SAINT VINCENT AND THE GRENADINES

CAPITAL: Kingstown.

FLAG: Three vertical bands of blue, yellow, and green; centered on the yellow band are three green diamonds arranged in a v-pattern.

ANTHEM: *National Anthem*, beginning "Saint Vincent! Land so beautiful."

MONETARY UNIT: The East Caribbean dollar (EC$) of 100 cents is the national currency. There are coins of 1, 2, 5, 10, and 25 cents and 1 dollar, and notes of 5, 10, 20, and 100 East Caribbean dollars. EC$1 = US$0.3704 (US$1 = EC$2.70).

WEIGHTS AND MEASURES: The imperial measures are used.

HOLIDAYS: New Year's Day, 1 January; Labor Day, 1 May; CARICOM Day, 5 July; Carnival, 6 July; Bank Holiday, 1st Monday in August; Independence Day, 27 October; Christmas Day, 25 December; Boxing Day, 26 December. Movable religious holidays include Good Friday, Easter Monday, and Whitmonday.

TIME: 8 AM = noon GMT.

LOCATION AND SIZE: Saint Vincent is located in the Windward Islands group of the Lesser Antilles. Scattered between Saint Vincent and Grenada are more than 100 small islands called the Grenadines, half of which belong to Saint Vincent and the other half to Grenada. The total land area of the country is 340 square kilometers (131 square miles), slightly less than twice the size of Washington, D.C. The capital city, Kingstown, is located on the southeast coast of the island of Saint Vincent.

CLIMATE: The islands enjoy a pleasant tropical climate all year round, with a yearly average temperature of 26°C (79°F). The warmest month is September, with an average temperature of 27°C (81°F); the coolest is January, with an average temperature of 25°C (77°F). The average yearly rainfall on Saint Vincent is 231 centimeters (91 inches).

INTRODUCTORY SURVEY

RECENT HISTORY

Saint Vincent was administered as a crown colony within the Windward Islands group from 1833 until 1960, when it joined the Federation of the West Indies. The federation fell apart in 1962, and Saint Vincent became a self-governing state in association with the United Kingdom seven years later. On 27 October 1979, Saint Vincent and the Grenadines achieved full independence as a member of the Commonwealth of Nations.

During the first months of independence, the young nation faced a rebellion by a group of Rastafarians (semi-political, semi-religious cult based in Jamaica) attempting to secede. The revolt was put down with help from neighboring Barbados. Otherwise, the political system has had few disruptions. The government at independence under the Saint Vincent Labor Party gave way to the New Democratic Party (NDP) in 1984, with the NDP renewing its electoral majority in 1989 and again in 1994.

GOVERNMENT

The British monarch, represented by a governor-general, is formally head of the government. Executive power is in the hands of the prime minister and cabinet, who are members of the majority party in the legislature. The single-chamber legislature is a twenty-one-seat House of Assembly consisting of fifteen elected representatives and six appointed senators. The nation is divided into eight local districts, two of which cover the Grenadines.

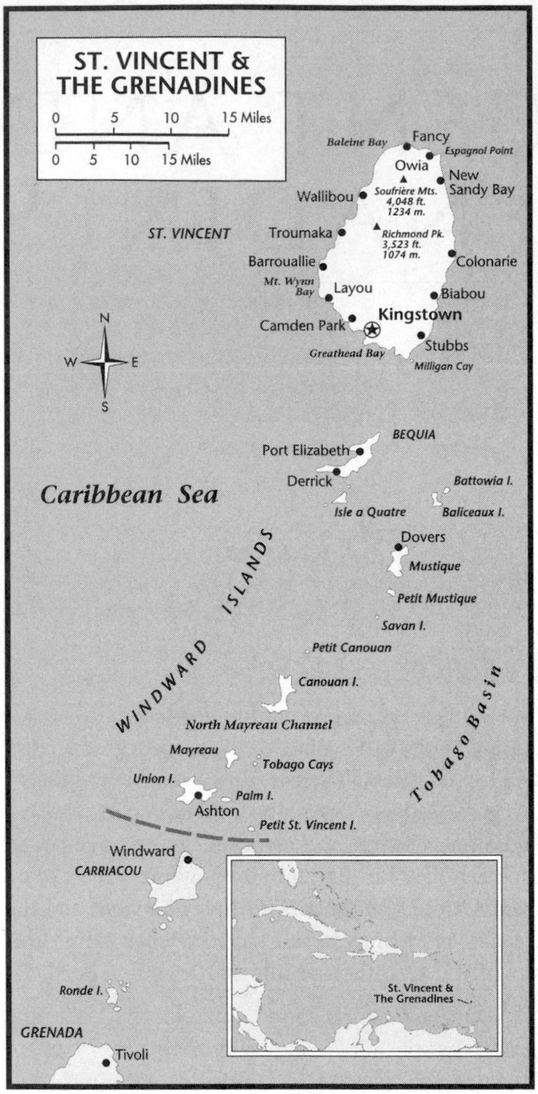

Judiciary

The islands are divided into three judicial districts, each with its own magistrate's court. Appeals may be carried to the East Caribbean Supreme Court, based in Saint Lucia.

Political Parties

There are two major parties and three minor parties on the islands. The majority party is the New Democratic Party (NDP), led by Prime Minister James FitzAllen Mitchell. The Saint Vincent Labour Party (SVLP) was in power at independence and governed until the 1984 elections.

DEFENSE

There are no armed forces except those of the police department. A regional defense pact provides for joint coast-guard operations, military exercises, and disaster contingency plans.

ECONOMIC AFFAIRS

In recent years, tourism and manufacturing have been expanding steadily. However Saint Vincent and the Grenadines continues to rely heavily on agriculture for its economic progress.

Public Finance

Most of the government's income comes from customs duties and taxes. The leading categories of expenditures are education, public works, and health. The U.S. CIA estimates that, in 1996, government revenues totaled approximately $80 million and expenditures $118 million, including capital expenditures of $39 million. External debt totaled $74.9 million, approximately 64% of which was financed abroad.

Income

In 1998, the gross national product (GNP) was $274 million, or about $2,420 per person. For the period 1985–95, the average inflation rate was 3.6%, resulting in a real growth rate in per person GNP of 3.9%.

Industry

A large amount of industrial activity centers on the processing of agricultural products. Nonagricultural industries include several garment factories, a furniture factory, an electronics plant, and a corrugated cardboard box plant.

Banking and Finance

In the late 1980s, the nation's largest financial institution was the state-run National Commercial Bank, and various branches of Canadian banks were prominent in Kingstown. Foreign reserves, excluding gold, totaled $23.2 million in November 1996. The government has established arrangements for offshore banking corporations, with direct connections to Swiss banking facilities. There is no securities exchange.

Economic Development

A 1982 loan from the Caribbean Development Bank was designed to stimulate and redirect agricultural production, support the tourist industry, and contribute to the creation of the infrastructure necessary for industrial development.

SOCIAL WELFARE

The social security system provides benefits for old age, disability, death, sickness, and maternity. Employers fund a compulsory workers' compensation program. Saint Vincent has an extensive program of community development, and a national family planning program. A new law requiring that women receive equal pay for equal work went into effect in 1990. In 1995, a domestic violence law was passed that established a family court to handle cases of spousal abuse.

Healthcare

As of 1988, Kingstown had a general hospital with 204 beds, and there were 3 rural hospitals. Approximately 35 outpatient clinics provide medical care throughout the nation. In 1990, there were an estimated 2.7 hospital beds per 1,000 inhabitants. In 1991, there were fifty-five doctors.

Gastrointestinal diseases continue to be a problem, although they are less so than in the past. Life expectancy is 73 years.

Housing

The government has undertaken housing renewal projects in both rural and urban areas and has sought to provide housing for workers on industrial estates. Another government program supplies building materials at low cost to working people.

EDUCATION

Primary education, which lasts for seven years, is free but not compulsory. In 1993, enrollment in primary schools was 21,386 with 1,080 teaching staff. In secondary schools the same year, there were 9,870 students and 395 teachers. The government-assisted School for Children with Special Needs serves handicapped students.

At the postsecondary level, there are a teachers' training college, affiliated with the University of the West Indies, and a technical college. Adult education classes are offered by the Ministry of Education. Literacy is estimated at 96%.

1999 KEY EVENTS TIMELINE

January

- The government eliminates export taxes.

April

- New legislation imposes mandatory imprisonment for illegal possession of firearms.

May

- Saint Vincent and the Grenadines defies whaling limits set by the International Whaling Commission.

- The Caribbean Development Bank predicts continued growth for Saint Vincent and the Grenadines.

June

- The government launches a three-month employment survey.
- Saint Vincent's finance minister declares that European currency fluctuations are costing the government millions in aid money.

July

- Carnival festivities take place on Saint Vincent and other islands.
- The Health Ministry reports a rise in the incidence of obesity and related medical conditions.

September

- Businessman Higman Peters announces the formation of new political party, the National Democratic Labour Party (NDLP).

- New Unity Labour Party (ULP) head Dr. Ralph Gonsalves replaces retiring legislator Vincent Beache as the opposition leader in parliament.

October

- Dead fish, including snapper, parrotfish, and angelfish among others, are washing up on beaches around the Caribbean, especially in Barbados, St. Vincent and the Grenadines, Grenada, and Trinidad and Tobago. Health inspectors and scientists are searching for an explanation.

- The St. Vincent and the Grenadines Teachers Union asks for help from cabinet members to help resolve their salary dispute with the Ministry of Finance. Teachers are seeking a 20 percent increase in pay, while the government offer includes a 12 percent increase over three years with a 5 percent increase the first year.

ANALYSIS OF EVENTS: 1999

BUSINESS AND THE ECONOMY

Although tourism plays an important role in the economy of Saint Vincent and the Grenadines, agriculture, including the vulnerable banana industry, which had experienced an overall decline beginning in the early 1990s, remained dominant. A three-year recovery program begun in 1998 with support from the European Union (EU), in addition to the elimination of the islands' export tax in 1999, was expected to raise banana production significantly. However, a U.S. initiative against preferential treatment by the EU toward banana imports from the Caribbean continued to pose a serious threat to the industry. A Caribbean Development Bank report released in May stated that gross domestic product for Saint Vincent and the Grenadines had risen from 2.2% in 1997 to an estimated 5.5% in 1998 and predicted an increase in economic growth for 1999.

In June Finance Minister Arnhim Eustace announced that fluctuations in the value of the euro on European financial markets were reducing by millions of dollars the actual value of foreign aid given to Saint Vincent and the Grenadines. This represents a loss of government revenue that the Finance Ministry believes will adversely affect the salaries of teachers and other public-sector employees.

Unemployment remained high in 1999, officially noted at 20%. Some estimates, however, claim the rate may actually be twice that amount. In June the government initiated an employment survey designed to provide an accurate picture of the country's employment levels and labor market and to gain other valuable information about its work force.

In an effort to stimulate the country's economy and boost employment the Saint Vincent and the Grenadines Chamber of Industry and Commerce launched a new-business start-up project in February in conjunction with the Caribbean Development Bank (CDB). A committee from the chamber of commerce will screen business ideas submitted by fledgling entrepreneurs. The most promising should receive funding by the CDB and receive expert advice by local business consultants as well as experts in the United States.

Representatives of Saint Vincent and the Grenadines rejected requests during the 51st annual general meeting of the International Whaling Commission (IWC) that the country modify the traditional whaling practices of its modest industry. Despite allegations that Saint Vincent and the Grenadines had violated its annual whale quota set by the IWC, the representatives claimed that their whale hunters were not at fault. The matter has not been resolved.

GOVERNMENT AND POLITICS

Following an early general election in 1998 the ruling New Democratic Party (NDP) was in the midst of a record fourth term in office, led by James Mitchell, who had served as prime minister since 1984. In September Dr. Ralph Gonsalves assumed leadership of the parliamentary opposition, having earlier replaced Vincent Beache as leader of the Unity Labour Party. The same month Higman Peters, a businessman and manager of the Caribbean Institute of Technology, announced plans to form a third political party, the National Democratic Labour Party (NDLP).

Internationally the top political issues involved the United States. The controversy continued over the U.S. response to Latin American banana exports to European Union (EU) countries. The United States launched an initiative through the World Trade Organization (WTO) to eliminate the preferences given to banana exports from former European colonies, such as Saint Vincent and the Grenadines. In a second controversy, government cooperation with U.S. anti-drug efforts drew protest by marijuana growers on the islands after over one million marijuana plants were destroyed the previous December during a cooperative U.S.–Saint Vincent military effort.

Domestically, a confrontation between the nation's teacher's union and the government made headlines throughout most of the year. Between January and September the teachers lowered their demands for a salary increase from 30% to 20%, but as of September their bargaining position was still far from that of the government's, which was offering only an 8% increase. Earlier in the year the teachers staged a two-day strike over the issue.

CULTURE AND SOCIETY

A major health problem in Saint Vincent and the Grenadines was highlighted in July in a statement by government nutritionist Andrea Robin.

She stated that obesity in both adults and children was on the rise, accompanied by an increase in the incidence of associated medical conditions including diabetes, hypertension, heart disease, stroke, and certain cancers.

The public disapproved of plans to remove a World War I monument, the ''Iron Man'' cenotaph, from the center of Kingstown. The government intended to relocate the monument to the Bay Street area to make way for a market expansion project in the center of Kingstown. Citizens claimed the move illustrated a lack of respect for the soldiers memorialized by the statue, who had fought with British troops against the Germans during the first world war. Critics also noted that the traditional location in the center of the capital was a more appropriate setting for the annual Memorial Day service held by the military, police, and other uniformed groups.

The traditional carnival festivities that are the country's major tourist attraction were held in July at several locations on Saint Vincent and in the Grenadines.

DIRECTORY

CENTRAL GOVERNMENT

Head of State

Queen
Elizabeth II, Monarch

Governor-General
David Jack, Office of the Governor-General
PHONE: +(809) 4561401
FAX: +(809) 4572157

Ministers

Prime Minister
James F. Mitchell, Office of the Prime Minister
PHONE: +(809) 4561703
FAX: +(809) 4572152

Minister of Agriculture and Labor
Jeremiah C. Scott, Ministry of Agriculture and Labor
PHONE: +(809) 4561111
FAX: +(809) 4571688

Minister of Communications and Works
Glenford Stewart, Ministry of Communications and Works
PHONE: +(809) 4561111; 4572039
FAX: +(809) 4537702

Minister of Education, Women's and Ecclesiastical Affairs
Alpian Rudolph Otway Allen, Ministry of Education, Women's and Ecclesiastical Affairs
PHONE: +(809) 4561111; 4572018
FAX: +(809) 4532299; 4571114

Minister of Finance and Public Service
Arnhim Eustace, Ministry of Finance and Public Service
PHONE: +(809) 4561111
FAX: +(809) 4531648

Minister of Foreign Affairs, Tourism and Information
Allan Cruickshank, Ministry of Foreign Affairs, Tourism and Information
PHONE: (809) 4561111; 4561502; 4562060
FAX: +(809) 4562610
E-MAIL: mail@svgtourism.com

Minister of Health and the Environment
Thomas Saint Clair, Ministry of Health and the Environment
PHONE: +(809) 4561111; 4571892
FAX: +(809) 4525655; 4572684

Minister of Housing and Community Development, Youth and Sports
Monty Roberts, Ministry of Housing and Community Development, Youth and Sports
PHONE: +(809) 4561111; 4572932
FAX: +(809) 4537921

Minister of Justice
Carl Joseph, Ministry of Justice
PHONE: +(809) 4561111
FAX: +(809) 4572898

Minister of Trade, Industry and Consumer Affairs
John C. A. Horne, Ministry of Trade, Industry and Consumer Affairs
PHONE: +(809) 4561111
FAX: +(809) 4572880

POLITICAL ORGANIZATIONS

New Democratic Party (NDP)

NAME: James F. Mitchell

United People's Movement (UPM)

NAME: Adrian Saunders

National Reform Party (NRP)

NAME: Joel Miguel

Unity Labor Party (ULP)

E-MAIL: ulpweb@aol.com
NAME: Ralph Gonsalves

DIPLOMATIC REPRESENTATION

Embassies in Saint Vincent

Taiwan
Chinese Embassy, Murray's Rd., PO Box 878, Kingstown, Saint Vincent
PHONE: +(809) 4562431
FAX: +(809) 4571934

United Kingdom
Granby Street, PO Box 182, Kingstown, Saint Vincent
PHONE: +(809) 4571701
FAX: +(809) 4562750

Venezuela
Granby Street, PO Box 852, Kingstown, Saint Vincent
PHONE: +(809) 4561374

JUDICIAL SYSTEM

Eastern Carribean Supreme Court

PO Box 1093, Castries, Saint Lucia
PHONE: +(758) 4522574

FURTHER READING

Articles

''Doing What the Islands Do Best'' *The New York Times*, Oct. 24, 1999: TR10.

Harrigan, Bill. ''Saint Vincent and the Grenadines: The Quintessential Caribbean.'' *Skin Diver* (September 1999): 88.

''An Outpost in the Banana and Marijuana Wars.'' *The New York Times* (March 4, 1999): A4.

Pickthall, Barry. ''The Grenadines: Sea of Tranquility.'' *Yachting* (September 1999): 88.

Books

Background Notes: Saint Vincent and the Grenadines. U.S. Department of State, Bureau of Public Affairs. Washington, D.C.: U.S. Government Printing Office, 1998.

Young, Virginia Heyer. *Becoming West Indian: Culture, Self, and Nation in Saint Vincent.* Washington, D.C.: Smithsonian Institution Press, 1993.

SAINT VINCENT AND THE GRENADINES: STATISTICAL DATA

For sources and notes see "Sources of Statistics" in the front of each volume.

GEOGRAPHY

Geography (1)

Area:

Total: 340 sq km.

Land: 340 sq km.

Water: 0 sq km.

Area—comparative: twice the size of Washington, DC.

Land boundaries: 0 km.

Coastline: 84 km.

Climate: tropical; little seasonal temperature variation; rainy season (May to November).

Terrain: volcanic, mountainous.

Natural resources: NEGL.

Land use:

Arable land: 10%

Permanent crops: 18%

Permanent pastures: 5%

Forests and woodland: 36%

Other: 31% (1993 est.)

HUMAN FACTORS

Ethnic Division (6)

Black, white, East Indian, Carib Amerindian.

Religions (7)

Anglican, Methodist, Roman Catholic, Seventh-Day Adventist.

Languages (8)

English, French patois.

Demographics (2A)

	1990	1995	1998	2000	2010	2020	2030	2040	2050
Population	112.8	117.6	119.8	121.2	131.9	145.6	155.9	162.4	163.4
Net migration rate (per 1,000 population)	NA	NA	NA	NA	NA	NA	NA	NA	NA
Births	NA	NA	NA	NA	NA	NA	NA	NA	NA
Deaths	NA	NA	NA	NA	NA	NA	NA	NA	NA
Life expectancy - males	69.1	71.2	72.0	72.6	74.9	76.7	77.9	78.9	79.5
Life expectancy - females	72.3	74.2	75.1	75.7	78.2	80.2	81.8	83.1	84.0
Birth rate (per 1,000)	24.0	19.6	18.7	17.9	16.2	13.4	11.8	11.0	10.2
Death rate (per 1,000)	6.1	5.5	5.3	5.2	4.8	5.1	6.3	8.6	11.2
Women of reproductive age (15-49 yrs.)	28.4	31.4	33.7	34.9	38.1	38.3	37.7	35.0	33.6
of which are currently married	NA	NA	NA	NA	NA	NA	NA	NA	NA
Fertility rate	2.7	2.1	2.0	1.9	1.8	1.8	1.8	1.8	1.8

Except as noted, values for vital statistics are in thousands; life expectancy is in years.

EDUCATION

Educational Attainment (10)

Age group (1980)	25+
Total population	32,444
Highest level attained (%)	
No schooling	2.4
First level	
Not completed	88.0
Completed	NA
Entered second level	
S-1	8.2
S-2	NA
Postsecondary	1.4

GOVERNMENT & LAW

Political Parties (12)

House of Assembly	No. of seats
New Democratic Party (NDP)	12
Unity Labor Party (ULP)	3

Government Budget (13B)

Revenues	$80 million
Expenditures	$118 million
Capital expenditures	$39 million

Data for 1996 est.

Military Affairs (14A)

Total expenditures	$NA
Expenditures as % of GDP	NA%

NA stands for not available.

Crime (15)

Crime rate (for 1994)	
Crimes reported	7,994
Total persons convicted	NA
Crimes per 100,000 population	7,202
Persons responsible for offenses	
Total number of suspects	4,288
Total number of female suspects	508
Total number of juvenile suspects	17

LABOR FORCE

Labor Force (16)

Total	67,000
Agriculture	26%
Industry	17%
Services	57%

Data for 1984 est. Percent distribution for 1980 est.

Unemployment Rate (17)

35%-40% (1994 est.)

PRODUCTION SECTOR

Electric Energy (18)

Capacity	14,000 kW (1995)
Production	64 million kWh (1995)
Consumption per capita	545 kWh (1995)

Transportation (19)

Highways:

total: 1,040 km

paved: 320 km

unpaved: 720 km (1996 est.)

Merchant marine:

total: 799 ships (1,000 GRT or over) totaling 8,063,755 GRT/12,629,612 DWT

Airports: 6 (1997 est.)

Airports—with paved runways:

total: 5

914 to 1,523 m: 2

under 914 m: 3 (1997 est.)

Airports—with unpaved runways:

total: 1

under 914 m: 1 (1997 est.)

Top Agricultural Products (20)

Bananas, coconuts, sweet potatoes, spices; small numbers of cattle, sheep, pigs, goats; small fish catch used locally.

FINANCE, ECONOMICS, & TRADE

Economic Indicators (22)

National product: GDP—purchasing power parity—$259 million (1996 est.)

National product real growth rate: 1% (1996 est.)

National product per capita: $2,200 (1996 est.)

Inflation rate—consumer price index: 3.6% (1996)

Exchange Rates (24)

Exchange rates: East Caribbean dollars (EC$) per US$1—2.7000 (fixed rate since 1976)

Top Import Origins (25)

$127 million (f.o.b., 1996) Data are for 1995.

Origins	%
United States	36
Caricom countries	28
United Kingdom	13

Top Export Destinations (26)

$46 million (f.o.b., 1996).

Destinations	%
Russia	NA
Ukraine	NA
Eastern Europe	NA
Western Europe	NA

NA stands for not available.

Economic Aid (27)

Recipient: ODA, $NA. NA stands for not available.

Import Export Commodities (28)

Import Commodities	Export Commodities
Foodstuffs	Bananas 39%
Machinery and equipment	Eddoes and dasheen (taro)
Chemicals and fertilizers	Arrowroot starch
Minerals and fuels	Tennis racquets

Balance of Payments (23)

	1992	1993	1994	1995	1996
Exports of goods (f.o.b.)	79	57	49	62	52
Imports of goods (f.o.b.)	−117	−118	−115	−119	−127
Trade balance	−38	−61	−67	−57	−75
Services - debits	−55	−55	−71	−71	−70
Services - credits	64	65	67	78	98
Private transfers (net)	−2	−2	1	−2	−1
Government transfers (net)	10	9	10	11	13
Overall balance	−21	−44	−58	−41	−35

SAMOA

CAPITAL: Apia.

FLAG: The upper left quarter of the flag is blue and bears five white, five-rayed stars representing the Southern Cross; the remainder of the flag is red.

ANTHEM: *The Flag of Freedom.*

MONETARY UNIT: The Samoa tala (WS$) is a paper currency of 100 sene. There are coins of 1, 2, 5, 10, 20, and 50 sene and 1 tala, and notes of 2, 5, 10, 20, and 100 talas. WS$1 = US$0.3969 (or US$1 = WS$2.5195).

WEIGHTS AND MEASURES: British weights and measures are used.

HOLIDAYS: New Year's, 1–2 January; Independence Holidays (first three workdays of June), Anzac Day, 25 April; Christmas Day, 25 December; Boxing Day, 26 December. Movable religious holidays are Good Friday, Easter Monday, and Whitmonday.

TIME: 1 AM = noon GMT.

LOCATION AND SIZE: Samoa consists of the islands of Savai'i and Upolu and several smaller islands, of which only two are inhabited. The group, situated in the Pacific Ocean among the South Sea islands, has a total land area of 2,860 square kilometers (1,104 square miles), slightly smaller than the state of Rhode Island. Its total boundary length is 371 kilometers (231 miles).

CLIMATE: The mean daily temperature is about 27°C (81°F). Rainfall averages 287 centimeters (113 inches) annually, and the average yearly relative humidity is 83%.

Independent State of Samoa
Malo Sa'oloto Tuto'atasi o Samoa i Sisifo

INTRODUCTORY SURVEY

RECENT HISTORY

In 1946, the UN General Assembly and New Zealand began a process leading toward ultimate self-government of Samoa. On 1 January 1962, Samoa became an independent nation. Upon independence, Fiame Faumuina Mataafa was Samoa's first prime minister (1962–70) and served again in that post from 1973 until his death in 1975.

During the late 1970s and early 1980s, Samoa suffered from a worsening economy and growing political and social unrest. A divisive public-sector strike from 6 April to 2 July 1981 cut many essential services to a critical level. Controversy erupted in 1982 over the signing by the HRPP (Human Rights Protection Party) government in August of a protocol with New Zealand that reduced the right of Samoans to New Zealand citizenship. Va'ai Kolone became prime minister in January 1986 as head of a new coalition government. In 1991, Samoa held its first elections in which all adult citizens were permitted to vote. The HRPP kept the majority.

GOVERNMENT

The powers and functions of the head of state are far-reaching. All legislation must have his assent before it becomes law. He also has power to grant pardons and reprieves and to suspend or commute any sentence by any court. The parliament consists of the head of state and the Fono, which is made up of one elected member from each of forty-five Samoan constituencies. Local government is carried out by the village *fono*, or council.

Judiciary

The Supreme Court has full civil and criminal jurisdiction for the administration of justice in Samoa. The Court of Appeal consists of three judges who may be judges of the Supreme Court. Magistrates' courts are subordinate courts with varying degrees of authority.

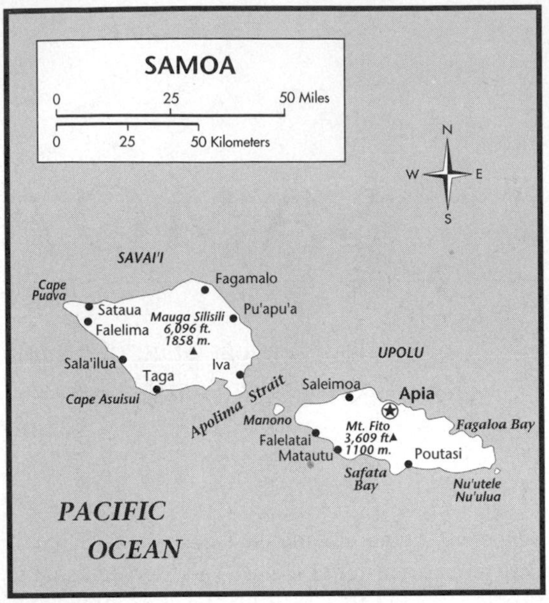

The Land and Titles Court has jurisdiction in disputes over Samoan land and succession to Samoan titles. Some civil and criminal matters are handled by village traditional courts, which apply a very different procedure than that used in the official western-style courts.

Political Parties

Political parties are becoming increasingly important. In a general election held in April 1991, the ruling HRPP (Human Rights Protection Party) won twenty-eight of forty-seven seats in the Parliament (Fono) and made Tofilau Eti prime minister again. Tofilau kept his post as prime minister after the 1996 elections.

DEFENSE

Samoa has no armed forces.

ECONOMIC AFFAIRS

The economy is based largely on agriculture. Tourist revenues and earnings by overseas workers are important sources of foreign exchange. Economic performance has suffered since 1990 due to the devastation to crops and tourism caused by two cyclones. Western Samoa has the highest unemployment rate and the lowest wages in Oceania, the south Pacific region.

Public Finance

Government budgets have commonly shown deficits in recent years. The U.S. CIA estimates that, in 1995, government revenues totaled approximately $78.6 million and expenditures $81.9 million. External debt totaled $141 million.

Income

In 1998, Samoa's gross national product (GNP) was $180 million, or about $1,020 per person. For the period 1985–95, the average inflation rate was 10.6%, resulting in a decline in per person GNP of 0.4%.

Industry

Industries include food- and timber-processing facilities, a brewery, cigarette and match factories, and small individual enterprises for processing coffee and for manufacturing curios, soap, carbonated drinks, light metal products, garments, footwear, and other consumer products.

Banking and Finance

Legislation in 1974 set up the Monetary Board to act as the central bank. The activities of the Monetary Board were taken over in May 1984 by the new Central Bank of Samoa. An Australian bank, ANZ, acquired the government's 25% stake in the Bank of Samoa (BWS), becoming its outright owner. The BWS is the largest bank in the country, with assets of about A $16 million (US $13 million).

Parliament passed legislation in early 1988 to allow the setting up of an offshore banking center. More the 1,000 companies have registered in Apia under the new tax haven legislation. The banking center's contribution to the national budget in 1994 is estimated to have been about WS 400,000.

Economic Development

The government has consistently stressed diversification of agriculture. It has also sought through a series of plans to promote growth in manufacturing, forestry, fishing, hydroelectric power, and tourism.

Investment increased significantly in 1990 and 1992, mainly due to increased public capital expenditure. External aid has been a major source of public investment financing, providing approximately 68% of capital expenditure in 1991 and 47% in 1992.

SOCIAL WELFARE

A social security system provides for employee retirement pensions, disability benefits, and death benefits. Workers' compensation is compulsory.

In Samoan society, obligations to the *aiga*, or extended family, are often given precedence over individual rights. While there is some discrimination against women, they can play an important role in society, especially female *matai*, or heads of families.

Healthcare

In 1990, there were fifty physicians in Samoa. District nurses are stationed at strategic points throughout the islands. Child health clinics are a regular feature of their work. Life expectancy was 70 years.

Housing

Many Samoans live in traditional houses called *fales*. A fale is usually round or oval, with pebble floors and a thatch roof. It has no walls, being supported on the sides by posts. Coconut-leaf blinds can be lowered to keep out wind and rain. A popular Samoan-European type of dwelling is an oblong concrete house with some walls, often with separate rooms in each corner; like the fale it is open at the sides.

EDUCATION

The adult literacy rate is estimated to be 97%. In 1991, there were 30 primary schools with 7,884 pupils and 524 teachers. At the secondary level, there were 3,643 pupils and 266 teachers.

The University of Samoa offers courses in both arts and sciences.

1999 KEY EVENTS TIMELINE

February

- A Utah company running a home for troubled teenagers in Samoa goes out of business, allegedly leaving five U.S. youths in the country with almost no supervision.

March

- Former Samoan Prime Minister Tofilau Eti Alesana dies at his home in Apia at age 74.
- The International Olympic Committee (IOC) expels six of its members for taking hundreds of thousands of dollars from officials while determining the site for the 2002 Winter Games. Paul Wallwork of Samoa is dismissed by a vote of 67 to 19.

July

- Samoa's Minister of Public Works Luagalau Levaula Samuelu Kamu is shot and killed at a political function.

August

- A Samoan legislator, a Cabinet minister's son, and a third man are charged with the assassination of Luagalau Levaula Samuelu Kamu.

September

- Samoa joins other Pacific island nations for a two-day conference discussing the problems that threaten their existence, including climate change, toxic waste, and trade.

October

- Samoa's national team advances to the quarterfinals of the Rugby World Cup, but loses to Scotland, 35 to 20.

ANALYSIS OF EVENTS: 1999

BUSINESS AND THE ECONOMY

In September Samoa and dozens of other small island states worked together in an unprecedented special session of the United Nations General Assembly to demand greater efforts to combat climate change, rising seas, and globalization. Islands like Samoa are concerned not only about their economic survival, but also about basic survival. Some of the smallest Pacific islands already have disappeared below rising waters.

"We are talking about very small nations that do not normally have access to the councils of the world," Tuiloma Neroni Slade, Samoa's envoy to the United Nations (U.N.) and chairman of the 43–strong Alliance of Small Island States, told the *Guardian* of London. Slade said the U.N. meeting would focus attention on the islands, thirty-six of which are full members of the United Nations. "These problems are not of the making of those countries and none have the capacity to deal with global issues," Slade said. "We want to tell the U.N. system to bear in mind the small size of our countries and the smallness of our resources," he said.

One of the most pressing problems for island states like Samoa are rising waters. While the larger islands should not be completely covered by rising ocean levels, many islands would lose valuable coastal land and force the relocation of millions of people. There are other environmental problems as well. Samoa has been hit by three cyclones in the past four years, an anomaly for the island.

GOVERNMENT AND POLITICS

In March Samoans were shocked to hear that the International Olympic Committee (IOC) had expelled the Samoan representative, Paul Wallwork, for accepting bribes. The IOC voted to expel six members for taking hundreds of thousands of dollars from officials while deciding the location of the 2002 Winter Games, to be held in Salt Lake City, Utah. Wallwork was dismissed by a vote of 67 to 19. The decision strained relations between the IOC and Samoa.

In October the new Samoa National Olympic Committee, which broke away from the national Sports Federation, held a press conference to announce it would continue to encourage sports on the island and would work in cooperation with the IOC. Leau Akeripa, the new president of the country's Olympic committee, acknowledged the work performed by Wallwork, saying he had achieved much success in the development of Samoa's sports programs.

Former Prime Minister Tofilau Eti Alesana died in his home in Apia four months after his resignation. He was 74. Tofilau Eti resigned because of poor health in November 1998, after sixteen years in office. He was suffering from cancer and announced his resignation in a speech to parliament. The speech was his first public appearance since returning from New Zealand, where he was receiving medical treatment. Parliament elected Tofilau Eti's deputy, Tuila'epa Sailele Malielegaoi, to succeed him.

Samoa's Minister of Public Works, Luagalau Levaula Samuelu Kamu, was shot and killed in July while at a political function hosted by the ruling Human Rights Protection Party, which was celebrating its twentieth anniversary. A Samoan legislator, a Cabinet minister's son, and a third man were charged with the assassination. Over two hundred people attended the three–hour government memorial service, which was held under police protection. Among those attending were political and traditional leaders from Samoa and the Pacific region.

CULTURE AND SOCIETY

Small nations rarely have a chance at winning world cup tournaments. Most don't even come close. But the island nation of Samoa impressed world rugby fans in October, when it advanced to the quarterfinals of the Rugby World Cup. Samoa's team triumphed over both Japan and Wales. Expections at home were high, and other Pacific island nations also began to cheer for Samoa. The team, however, was defeated by Scotland in the quarterfinals by a score of 35 to 20.

''Rugby has put us on the map,'' Samoan team captain Pat Lam told the *Guardian* of London. ''It is a source of national pride. We had been shaken by our defeat by Argentina and it was vital for the people back home that we should qualify for the quarterfinals. Anything less would have been a failure.''

DIRECTORY

CENTRAL GOVERNMENT

Head of State

Chief of State
Malietoa Tanumafili, Office of the Chief of State

Ministers

Prime Minister
Tuila'epa Sailele Malielegaoi, Office of the Prime Minister

Ministers

Minister of Agriculture, Forestry, Fisheries, and Meteorological Services
Teofilo Molio'o, Ministry of Agriculture, Forestry, Fisheries, and Meteorological Services

Minister of Commerce, Trade, and Industry
Tuila'epa Sailele Malielegaoi, Ministry of Commerce, Trade, and Industry, PO Box 862, Samoa
PHONE: +685 20471
FAX: +685 21646
E-MAIL: TIPU@samoa.net

Minister of Education
Fiame Naomi Mata'afa, Ministry of Education

Minister of Finance
Tuila'epa Sailele Malielegaoi, Ministry of Finance

Minister of Foreign Affairs
Tuila'epa Sailele Malielegaoi, Ministry of Foreign Affairs

Minister of Health
Misa Telefoni Retzlaff, Ministry of Health

Minister of Immigration
Tofilau Eti Alesana, Ministry of Immigration

Minister of Internal Affairs
Tofilau Eti Alesana, Ministry of Internal Affairs

Minister of Justice
Solia Papu Vaai, Ministry of Justice

Minister of Labor
Polataivao Fosi, Ministry of Labor

Minister of Lands, Survey, and Environment
Tuala Sale Tagaloa Kerslake, Ministry of Lands, Survey, and Environment

Minister of Police and Prisons
Tofilau Eti Alesana, Ministry of Police and Prisons

Minister of Public Service
Tofilau Eti Alesana, Ministry of Public Service

Minister of Public Works
Luagalau Levaula Kamu, Ministry of Public Works

Minister of Transportation and Civil Aviation
Hans Joachim Keil, Ministry of Transportation and Civil Aviation

Minister of Treasury, Inland Revenue, and Customs
Tuila'epa Sailele Malielegaoi, Ministry of Treasury, Inland Revenue, and Customs

Minister of Women's Affairs
Leniu Tofaeono Avamagalo, Ministry of Women's Affairs

POLITICAL ORGANIZATIONS

Human Rights Protection Party (HRPP)

TITLE: Chairman
NAME: Tofilau Eti Alesana

Samoan National Development Party (SNDP)

TITLE: Chairman
NAME: Tapua Tamasese Efi

JUDICIAL SYSTEM

Supreme Court

Court of Appeal

Magistrate Court

Land & Titles Court

FURTHER READING

Articles

''New Charge in Samoa Murder Case: Prime minister was Target.'' The Associated Press, 13 August 1999.

''Program Collapses, Leaving Five Teens Stranded in Samoa.'' The Associated Press, 18 February 1999.

''Prominent Samoans Face Murder Charges in Minister's Assassination.'' The Associated Press, 5 August 1999.

''Samoa Politician Shot Dead at Political Function.'' The Associated Press, 18 July 1999.

''Small Islands Get to Voice Concerns About Climate, Waste, Trade.'' The Associated Press Writer, 25 September 1999.

''Tiny Islands Say Big Nations Ignore Their Problems.'' The Associated Press, 4 February 1999.

SAMOA: STATISTICAL DATA

For sources and notes see "Sources of Statistics" in the front of each volume.

GEOGRAPHY

Geography (1)

Area:

Total: 2,860 sq km.

Land: 2,850 sq km.

Water: 10 sq km.

Area—comparative: slightly smaller than Rhode Island.

Land boundaries: 0 km.

Coastline: 403 km.

Climate: tropical; rainy season (October to March), dry season (May to October).

Terrain: narrow coastal plain with volcanic, rocky, rugged mountains in interior.

Natural resources: hardwood forests, fish.

Land use:

Arable land: 19%

Permanent crops: 24%

Permanent pastures: 0%

Forests and woodland: 47%

Other: 10%

HUMAN FACTORS

Infants and Malnutrition (5)

Under-5 mortality rate (1997) 52

% of infants with low birthweight (1990-97) 6

Births attended by skilled health staff % of total[a] . . .NA

% fully immunized (1995-97)

TB 99

DPT 99

Polio 99

Measles 99

Prevalence of child malnutrition under age 5 (1992-97)[b] . . . . .NA

Demographics (2A)

	1990	1995	1998	2000	2010	2020	2030	2040	2050
Population	186.2	209.4	224.7	235.3	288.1	341.1	391.7	435.5	471.0
Net migration rate (per 1,000 population)	NA	NA	NA	NA	NA	NA	NA	NA	NA
Births	NA	NA	NA	NA	NA	NA	NA	NA	NA
Deaths	NA	NA	NA	NA	NA	NA	NA	NA	NA
Life expectancy - males	64.0	66.0	67.1	67.8	71.0	73.5	75.5	77.0	78.1
Life expectancy - females	68.9	70.9	72.0	72.7	75.9	78.5	80.6	82.2	83.4
Birth rate (per 1,000)	34.2	31.7	29.6	28.0	22.9	19.9	17.0	15.0	13.9
Death rate (per 1,000)	6.6	5.9	5.5	5.3	4.6	4.4	4.8	5.9	7.3
Women of reproductive age (15-49 yrs.)	42.6	48.3	53.2	56.8	76.5	91.1	102.6	109.3	110.9
of which are currently married	NA	NA	NA	NA	NA	NA	NA	NA	NA
Fertility rate	4.7	4.0	3.7	3.5	2.7	2.3	2.2	2.1	2.0

Except as noted, values for vital statistics are in thousands; life expectancy is in years.

Ethnic Division (6)

Samoan	92.6%
Euronesians	7%
Europeans	0.4%

Euronesians are persons of European and Polynesian blood.

Religions (7)

Christian 99.7% (about one-half of population associated with the London Missionary Society; includes Congregational, Roman Catholic, Methodist, Latter-Day Saints, Seventh-Day Adventist).

Languages (8)

Samoan (Polynesian), English.

EDUCATION

Educational Attainment (10)

Age group (1991)	25+
Total population	59,902
Highest level attained (%)	
No schooling[2]	2.7
First level	
Not completed	31.6
Completed	NA
Entered second level	
S-1	31.4
S-2	26.0
Postsecondary	5.6

Literacy Rates (11B)

Adult literacy rate	
1980	
Male	-
Female	-
1995	
Male	-
Female	98%

GOVERNMENT & LAW

Political Parties (12)

Legislative Assembly	% of seats
Human Rights Protection Party (HRPP)	45.17
Samoan National Development Party (SNDP)	27.1
Independents	23.7

Government Budget (13B)

Revenues	$52 million
Expenditures	$99 million
Capital expenditures	$37 million

Data for FY96/97 est.

Military Affairs (14A)

Total expenditures	$NA
Expenditures as % of GDP	NA%

LABOR FORCE

Labor Force (16)

Total	82,500
Agriculture	65%
Services	30%
Industry	5%

Data for 1991 est. Percent distribution for 1995 est.

Unemployment Rate (17)

Rate not available.

PRODUCTION SECTOR

Electric Energy (18)

Capacity	21,700 kW (1996 est.)
Production	56.3 million kWh (1996 est.)
Consumption per capita	310 kWh (1995)

Transportation (19)

Highways:

total: 790 km

paved: 332 km

unpaved: 458 km (1996 est.)

Merchant marine:

total: 1 roll-on/roll-off cargo ship (1,000 GRT or over) totaling 3,838 GRT/5,536 DWT (1997 est.)

Airports: 3 (1997 est.)

Airports—with paved runways:

total: 1

2,438 to 3,047 m: 1 (1997 est.)

Airports—with unpaved runways:

total: 2

under 914 m: 2 (1997 est.)

Top Agricultural Products (20)

Coconuts, bananas, taro, yams.

FINANCE, ECONOMICS, & TRADE

Economic Indicators (22)

National product: GDP—purchasing power parity—$450 million (1996 est.)

National product real growth rate: 5.9% (1996 est.)

National product per capita: $2,100 (1996 est.)

Inflation rate—consumer price index: 7.5% (1996)

Exchange Rates (24)

Exchange rates:

Tala (WS$) per US$1

January 1998	2.7556
1997	2.5562
1996	2.4618
1995	2.4722
1994	2.5349
1993	2.5681

Top Import Origins (25)

$100 million (c.i.f., 1996)

Origins	%
New Zealand	37
Australia	22
Fiji	15
United States	13

Top Export Destinations (26)

$10 million (f.o.b., 1996) Data are for 1996.

Destinations	%
New Zealand	48
American Samoa	11
Australia	10
Germany	7
United States	3

Economic Aid (27)

Recipient: ODA; $8.7 million bilateral aid from Australia (FY96/ 97 est.); $5 million bilateral aid from NZ (FY95/96).

Import Export Commodities (28)

Import Commodities	Export Commodities
Intermediate goods 50%	Coconut oil and cream
Food 26%	Copra
Capital goods 12%	Fish
	Beer

MANUFACTURING SECTOR

GDP & Manufacturing Summary (21)

	1980	1985	1990	1992	1993	1994
Gross Domestic Product						
Millions of 1990 dollars	183	179	176	167	177	165
Growth rate in percent	3.00	5.96	−4.47	−3.26	6.00	7.10
Per capita (in 1990 dollars)	1,149.8	1,119.2	1,085.1	1,014.2	1,062.1	975.0
Manufacturing Value Added						
Millions of 1990 dollars	15	15	14	15	15	*15*
Growth rate in percent	3.02	4.53	−4.55	0.81	*1.28*	*0.61*
Manufacturing share in percent of current prices	4.6	13.8	7.9	NA	NA	NA

SAN MARINO

CAPITAL: San Marino.

FLAG: The flag is divided horizontally into two equal bands, sky blue below and white above, with the national coat of arms superimposed in the center.

ANTHEM: *Onore a te, onore, o antica repubblica (Honor to You, O Ancient Republic).*

MONETARY UNIT: San Marino principally uses the Italian lira (L) as currency; Vatican City State currency is also honored. The country issues its own coins in standard Italian denominations in limited numbers as well. Coins of San Marino may circulate in both the republic and in Italy. L1 = $0.0006 (or $1 = L1,611.3).

WEIGHTS AND MEASURES: The metric system is the legal standard.

HOLIDAYS: New Year's Day, 1 January; Epiphany, 6 January; Anniversary of St. Agatha, second patron saint of the republic, and of the liberation of San Marino (1740), 5 February; Anniversary of the Arengo, 25 March; Investiture of the Captains-Regent, 1 April and 1 October; Labor Day, 1 May; Fall of Fascism, 28 July; Assumption and August Bank Holiday, 14–16 August; Anniversary of the Foundation of San Marino, 3 September; All Saint's Day, 1 November; Commemoration of the Dead, 2 November; Immaculate Conception, 8 December; Christmas, 24–26 December; New Year's Eve, 31 December. Movable religious holidays include Easter Monday and Ascension.

TIME: 1 PM = noon GMT.

LOCATION AND SIZE: San Marino is the third-smallest country in Europe, with an area of 60 square kilometers (23.2 square miles), about one-third the size of Washington, D.C. It is a landlocked state completely surrounded by Italy, with a total boundary length of 39 kilometers (24 miles).

CLIMATE: The climate is that of northeastern Italy: rather mild in winter, but with temperatures frequently below freezing, and warm and pleasant in the summer, reaching a maximum of 26°C (79°F). Annual rainfall averages about 89 centimeters (35 inches).

The Most Serene Republic of San Marino
La Serenissima Repubblica di San Marino

INTRODUCTORY SURVEY

RECENT HISTORY

San Marino is the oldest republic in the world. It is the sole survivor of the independent states that existed in Italy at various times, from the downfall of the western Roman Empire to the proclamation of the Kingdom of Italy in 1861.

Because of the poverty of the region and the mountainous terrain, San Marino was rarely disturbed by outside powers. It was briefly held by Cesare Borgia (an Italian military and church leader) in 1503, but in 1549, its sovereignty was confirmed by Pope Paul III. In 1739, a military force under Cardinal Giulio Alberoni occupied San Marino. In the following year, Pope Clement II terminated the occupation and signed a treaty of friendship with the tiny republic. Napoleon allowed San Marino to retain its liberty.

In 1849, Giuseppe Garibaldi, the liberator of Italy, took refuge from the Austrians in San Marino. San Marino and Italy entered into a treaty of friendship in 1862, which is still in effect. In 1922–43, during the period of Benito Mussolini's rule in Italy, San Marino adopted a fascist type of government. Despite its neutrality in World War II (1939–45), San Marino was bombed by Allied planes on 26 June 1944. The raid caused heavy damage, especially to the railway line, and killed a number of persons.

Since 1945, government control has shifted between parties of the right and left, often in coali-

tions. In 1986, a Communist–Christian Democratic coalition came to power.

GOVERNMENT

Legislative power is exercised by the Grand and General Council of sixty members, regularly elected every five years by universal vote at age 18. The council elects from among its members a State Congress of ten members, which makes and carries out most administrative decisions. Two members of the council are named every six months to head the executive branch of the government; one represents the town of San Marino and the other the countryside.

Judiciary

There is a civil court, a criminal court, and a superior court. Most criminal cases are tried before Italian magistrates because, with the exception of minor civil suits, the judges in cases in San Marino are not allowed to be citizens of San Marino. The highest appellate court is the Council of Twelve.

Political Parties

The political parties in San Marino have close ties with the corresponding parties in Italy.

DEFENSE

The San Marino militia officially consists of all able-bodied citizens between the ages of 16 and 55, but the armed forces are principally for purposes of ceremonial display.

ECONOMIC AFFAIRS

Farming was formerly the principal occupation, but it has been replaced in importance by light manufacturing. However, the main sources of income are tourism and payments by citizens of San Marino (Sanmarinese) living abroad. Some government revenue comes from the sale of postage stamps and coins and from a subsidy by Italy.

Public Finance

The government derives its revenues mainly from the worldwide sale of postage stamps, direct and indirect taxes, and yearly subsidies by the Italian government. State budgets have increased sharply in recent years.

The U.S. CIA estimates that, in 1995, government revenues totaled approximately $320 million and expenditures $320 million.

Income

In 1995, San Marino's gross domestic product (GDP) was $380 million at current prices, or $15,800 per person. In 1995, the average inflation rate was 5.5%.

Industry

Manufacturing is limited to light industries such as textiles, bricks and tiles, leather goods, clothing, and metalwork.

Banking and Finance

The principal bank, the Cassa di Risparmio, was founded in 1882. Other banks include the Banca Agricola and the Cassa Rurale. There are no securities transactions in San Marino.

Economic Development

In addition to promoting tourism in San Marino, the government has encouraged the establishment of small-scale industries and service-oriented enterprises by offering tax exemptions for five to ten years.

SOCIAL WELFARE

The government maintains a comprehensive social insurance program, including disability, family supplement payments, and old-age pensions. In 1982, Sanmarinese women who married foreign citizens were given the right to keep their citizenship.

Healthcare

All citizens receive free, comprehensive medical care. Estimated average life expectancy in 1996 was 77 years.

Housing

In 1986, San Marino had 7,926 dwellings, nearly all with electricity and piped-in water. Most new construction is financed privately.

EDUCATION

Primary education is compulsory for all children between the ages of 6 and 14; the adult literacy rate is about 98%. The program of instruction is patterned after Italy's.

In 1993, there were 14 elementary schools, with 1,166 students and 219 teachers; middle and upper-secondary schools enrolled 1,157 pupils during the same year. San Marino students are able to enroll at Italian universities.

1999 KEY EVENTS TIMELINE

January

- San Marino joins the European Monetary Union as Italy cedes monetary responsibilities to the European Central Bank. Previously, San Marino's currency was pegged to the Italian lira.

March

- Pressure from the United Kingdom and the United States prompts a series of raids on San Marino retail outlets suspected of selling pirated music recordings. For several years San Marino has been recognized as a major European center for music piracy. The Italian music industry organization FPM allied with the U.K. and U.S. governments to put more pressure on San Marino's government to enforce its 1997 law against piracy.

May

- Michael Schumacher of Team Ferrari edges out rival David Coulthard of McLaren by 4.26 seconds in the San Marino Grand Prix. It was Ferrari's first victory in sixteen years at this stop on the Formula One circuit.

August

- A new law intended to protect Sanmarinese citizenship prohibits female workers under the age of fifty from working as domestic servants. The goal was to protect elderly male citizens from falling for young female servants looking to obtain citizenship via marriage.

September

- Two new Captains-Regent, San Marino's heads of state, are elected. Marino Bollini and Giuseppe Arzilli take office on the first day of October to begin six-month terms.

November

- San Marino, along with four other countries, is admitted into the United Nations Food and Agriculture Organization (FAO) at the Ministerial Conference in Rome. The FAO membership is now 180 countries plus one organization, the European Community.

ANALYSIS OF EVENTS: 1999

BUSINESS AND THE ECONOMY

San Marino joined the European Monetary Union in January and adopted the euro as its currency. This currency, planned to be phased in as a single currency used across western Europe, will likely boost tourism, already 60% of San Marino's economy, by making it easier for tourists to spend the money they have brought with them. A negative affect is that Sanmarinese banks are likely to see a drop in revenue from currency exchange fees, which will no longer be necessary.

GOVERNMENT AND POLITICS

San Marino is a small country, but it is among the world's oldest republics. The Sanmarinese have maintained their statehood in part because of a national passion for the democratic political process. From May 1998 through 1999 the ruling government coalition has been composed of the Partito Democratico Cristiano Sammarinese (PDCS) and the Partito Socialista Sammarinese (PSS).

The government has pursued policies to maintain strong public finances throughout the 1990s, but a widening deficit in recent years has led to reforms in 1999. Rising public sector wages have been reduced by a halt in recruitment and overall public expenditure received a 5% cut.

CULTURE AND SOCIETY

Citizenship in San Marino is a valuable commodity. Despite its small size, the standard of living is quite high. Consequently, obtaining Sammarinese citizenship is often an attractive goal for outsiders. The government passed a law in August that prohibits employing female household servants under the age of fifty. The measure was made to minimize the potential for elderly male citizens to fall for their young, foreign female help and providing citizenship via marriage. Since the only ways to become a citizen are to reside in San Marino for thirty years, to be born to Sammarinese parents, or to marry a male citizen, the government decided it was in the country's best interests to pass such a law.

DIRECTORY

CENTRAL GOVERNMENT

Head of State

Capitani Reggenti
Marino Bollini and Giuseppe Arzilli, Office of Capitani Reggenti

Departments

Secretary of Foreign and Political Affairs
Gabriele Gatti, Department of Foreign and Political Affairs, Palazzo Begni, Contrada Omerelli, San Marino
PHONE: +39 882293
FAX: +39 882814
E-MAIL: affaripolitic@omniway.sm

Secretary of Interior Affairs and Civil Protection
Antonio Lazzaro Volpinari, Department of Interior Affairs and Civil Protection, Parva Domus, Liberta, San Marino
PHONE: +39 882196
FAX: +39 882261
E-MAIL: affariinterni@omniway.sm

Secretary of Finance, Budget, Programming, and Information
Clelio Galassi, Department of Finance, Budget, Programming, and Information, Palazzo Begni, Contrada Omerelli, San Marino
PHONE: +39 882242
FAX: +39 882244
E-MAIL: segr.finanze@omniway.sm

Secretary of Industry, Handicraft, and Economic Cooperation
Florenzo Stolfi, Department of Industry, Handicraft, and Economic Cooperation, Palazzo Mercuri, Contrada del Collegio, San Marino
PHONE: +39 882528
FAX: +39 882529
E-MAIL: dic.industria@omniway.sm

Secretary of Environment, Territory and Agriculture
Augusto Casall, Department of Environment, Territory and Agriculture, Contrada Omerelli, San Marino
PHONE: +39 882470
FAX: +39 882474
E-MAIL: info_territ@omniway.sm

Secretary of Tourism, Commerce and Sport
Claudio Podeschi, Department of Tourism, Commerce and Sport, Palazzo del Turismo, Contrada Omagnano, San Marino
PHONE: +39 882420
FAX: +39 882998
E-MAIL: statoturismo@omniway.sm

Secretary of Public Health and Social Security
Luciano Clavatta, Department of Public Health and Social Security, Via V. Scialoja, San Marino
PHONE: +39 994438
FAX: +39 903967
E-MAIL: dic.sanita@omniway.sm

Secretary of Relations with the Administrative Castles
Cesare Antonio Gasperoni, Department of Relations with the Administrative Castles, Via A. di Superchio, San Marino
PHONE: +39 882522
FAX: +39 882524
E-MAIL: aass@omniway.sm

Secretary of Education, Culture, Justice and Social Affairs
Sante Canducci, Department of Education, Culture, Justice and Social Affairs, Palazzo Begni, Contrada Omeralli, San Marino
PHONE: +39 882250
FAX: +39 882301

Secretary of Labor and Cooperation
Romeo Morri, Department of Labor and Cooperation, Palazzo Mercuri, Contrada del Collegio, San Marino

PHONE: +39 882532
FAX: +39 882535

POLITICAL ORGANIZATIONS

Christian Democratic Party (PDCS)

TITLE: Secretary General
NAME: Piermarino Menicucci

Democratic Progressive Party (PPDS)

TITLE: Secretary General
NAME: Stefano Macina

Reformists Socialists (RS)

TITLE: Secretary General
NAME: Maurizio Rattini

Communist Renewal (RC)

NAME: Giuseppe Amichi

JUDICIAL SYSTEM

Council of Twelve

FURTHER READING

Articles

Dezzani, Mark. ''San Marino Acts Against Piracy–At Last.'' *Billboard.* (March 27, 1999): 68.

Books

Catling, Christopher. *Umbria, the Marches and San Marino.* Lincolnwood, Ill.: Passport Books, 1994.

Duursma, Jorri. *Self-Determination, Statehood, and International Relations of Micro-states: the Cases of Liechtenstein, San Marino, Monaco, Andorra, and the Vatican City.* New York: Cambridge University Press, 1996.

SAN MARINO: STATISTICAL DATA

For sources and notes see "Sources of Statistics" in the front of each volume.

GEOGRAPHY

Geography (1)

Area:

Total: 60 sq km.

Land: 60 sq km.

Water: 0 sq km.

Area—comparative: about 0.3 times the size of Washington, DC.

Land boundaries:

Total: 39 km.

Border countries: Italy 39 km.

Coastline: 0 km (landlocked).

Climate: Mediterranean; mild to cool winters; warm, sunny summers.

Terrain: rugged mountains.

Natural resources: building stone.

Land use:

Arable land: 17%

Permanent crops: NA%

Permanent pastures: NA%

Forests and woodland: NA%

Other: 83% (1993 est.).

HUMAN FACTORS

Infants and Malnutrition (5)

Under-5 mortality rate (1997) 6

% of infants with low birthweight (1990-97) . . . . . .NA

Births attended by skilled health staff % of total[a] . . .NA

% fully immunized (1995-97)

TB 97

DPT 98

Polio 100

Measles 96

Prevalence of child malnutrition under age 5 (1992-97)[b] . . . . .NA

Demographics (2A)

	1990	1995	1998	2000	2010	2020	2030	2040	2050
Population	23.1	24.3	24.9	25.2	26.3	26.8	27.1	26.9	26.5
Net migration rate (per 1,000 population)	NA	NA	NA	NA	NA	NA	NA	NA	NA
Births	NA	NA	NA	NA	NA	NA	NA	NA	NA
Deaths	NA	NA	NA	NA	NA	NA	NA	NA	NA
Life expectancy - males	76.8	77.3	77.5	77.7	78.4	78.9	79.4	79.7	80.0
Life expectancy - females	85.2	85.3	85.3	85.4	85.5	85.7	85.8	85.9	86.0
Birth rate (per 1,000)	11.5	11.0	10.5	10.3	8.6	9.0	8.9	8.7	9.2
Death rate (per 1,000)	6.7	7.6	8.1	8.4	9.8	11.0	12.4	13.7	14.6
Women of reproductive age (15-49 yrs.)	5.9	6.1	6.0	5.9	5.8	5.6	5.2	5.3	5.2
of which are currently married	NA	620.0	NA	NA	NA	NA	NA	NA	NA
Fertility rate	1.5	1.5	1.5	1.5	1.5	1.6	1.6	1.6	1.7

Except as noted, values for vital statistics are in thousands; life expectancy is in years.

Ethnic Division (6)

Sammarinese, Italian.

Religions (7)

Roman Catholic

Languages (8)

Italian.

GOVERNMENT & LAW

Political Parties (12)

Great and General Council	% of seats
Christian Democratic Party (PDCS)	41.4
San Marino Socialist Party (PSS)	23.7
Democratic Progressive Party (PDP)	18.6
Popular Alliance (AP)	7.7
Democratic Movement (MD)	5.3
Communist Refoundation (RC)	3.3

Government Budget (13B)

Revenues	$320 million
Expenditures	$320 million
Capital expenditures	$26 million

Data for 1995 est.

Military Affairs (14A)

Total expenditures (1995)	$3.7 million
Expenditures as % of GDP (1995)	1%

LABOR FORCE

Labor Force (16)

Total	15,600
Services	55%
Industry	43%
Agriculture	2%

Data for 1995. Percent distribution for 1995.

Unemployment Rate (17)

3.6% (April 1996)

PRODUCTION SECTOR

Electric Energy (18)

Electricity supplied by Italy.

Transportation (19)

Highways:
total: 220 km
paved: NA km
unpaved: NA km
Airports: none

Top Agricultural Products (20)

Wheat, grapes, maize, olives; cattle, pigs, horses, meat, cheese, hides.

FINANCE, ECONOMICS, & TRADE

Economic Indicators (22)

National product: GDP—purchasing power parity—$500 million (1997 est.)

National product real growth rate: 4.8% (1994 est.)

National product per capita: $20,000 (1997 est.)

Inflation rate—consumer price index: 5.3% (1995)

Exchange Rates (24)

Exchange rates:	
Italian lire (Lit) per US$1	
January 1998	1,787.7
1997	1,703.1
1996	1,542.9
1995	1,628.9
1994	1,612.4
1993	1,573.7

Economic Aid (27)

Recipient: ODA, $NA. NA stands for not available.

SÃO TOMÉ AND PRÍNCIPE

CAPITAL: São Tomé.

FLAG: The flag consists of three unequal horizontal stripes of green, yellow, and green; there is a red triangle at the hoist, and two black stars on the yellow stripe.

ANTHEM: *Independência Total (Total Independence).*

MONETARY UNIT: The dobra (Db) is equal to 100 centimos. There are coins of 50 centimos and 1, 2, 5, 10, and 20 dobras, and notes of 50, 100, 500, and 1,000 dobras. Db1 = $0.00914 (or $1 = Db516.70)

WEIGHTS AND MEASURES: The metric system is used.

HOLIDAYS: New Year's Day, 1 January; Martyrs' Day, 4 February; Labor Day, 1 May; Independence Day, 12 July; Armed Forces Day, first week in September; Farmers' Day, 30 September. The principal Christian holidays are also observed.

TIME: GMT.

LOCATION AND SIZE: São Tomé and Príncipe, the smallest country in Africa, lies in the Gulf of Guinea off the west coast of Gabon.

The nation has an area of 964 square kilometers (371 square miles), of which São Tomé comprises 855 square kilometers (330 square miles), and Príncipe 109 square kilometers (42 square miles). Comparatively, the nation's combined area is slightly less than 5.5 times the size of Washington, D.C.

São Tomé has a coastline of 141 kilometers (88 miles); Príncipe's shoreline is 209 kilometers (130 miles). The capital city, São Tomé, is located on the northeast coast of the island of São Tomé.

CLIMATE: Coastal temperatures average 27°C (81°F), but the mountain regions average only 20°C (68°F). From October to May, São Tomé and Príncipe receive between 380 and 510 centimeters (150–200 inches) of rain.

Democratic Republic of São Tomé and Príncipe
República Democrática de São Tomé e Príncipe

INTRODUCTORY SURVEY

RECENT HISTORY

The Committee for the Liberation of São Tomé and Príncipe (later renamed the Movement for the Liberation of São Tomé and Príncipe—MLSTP) was formed in 1960 and recognized by Portugal in 1974 as the sole legitimate representative of the people of São Tomé and Príncipe. On 12 July 1975, the islands achieved full independence. On the same day, Manuel Pinto da Costa, the secretary-general of the MLSTP, became the country's first president. In 1979, Prime Minister Miguel dos Anjos da Cunha Lisboa Trovoada was arrested and charged with attempting to seize power. By 1985, São Tomé and Príncipe had begun to establish closer ties with the West.

In 1990, a new policy of *abertura*, or political and economic "opening," led to the legalization of opposition parties and direct elections with secret balloting. A number of groups united as the Party of Democratic Convergence-Group of Reflection (PDC-GR), led by former prime minister Miguel Trovoada, who was elected president on 3 March 1991. In 1992, the government began a strict program to strengthen the economy. Gasoline prices increased, and the currency was devalued by 40%. The measures prompted massive demonstrations and calls for the dissolution of the government. The parliament then appointed Norberto Alegre prime minister, who then formed a new government.

The PDC-GR continued to dominate the central government in 1993, but opposition grew as the PDC-GR was increasingly seen as corrupt and

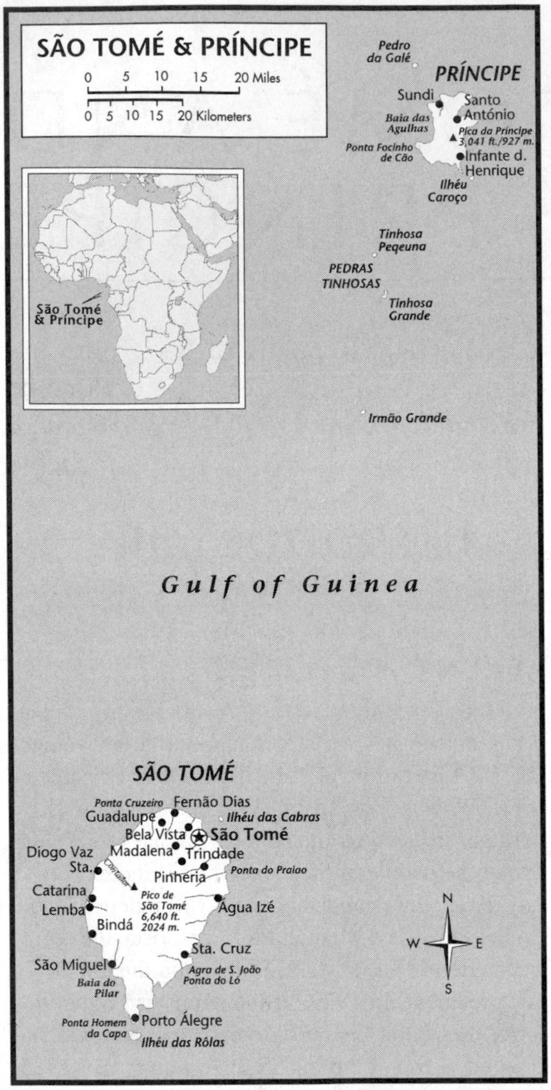

complacent. In August 1995, five army officers led a bloodless coup, but the elected government was put back into power with the promise to institute reforms and bring opposition members into the government.

GOVERNMENT

A new constitution was adopted by the People's Assembly in April 1990. The president is chosen for a maximum of two five-year terms. The People's Assembly, comprising fifty-five members, is elected to four-year terms. Voting is universal at age 18.

Judiciary

The highest court is the Supreme Tribunal, which is named by and responsible to the People's Assembly. The constitution affords parties in civil cases the right to a fair public trial and a right to appeal. The constitution also affords criminal defendants a public trial before a judge as well as legal representation.

Political Parties

On 15 October 1974, the government of Portugal recognized the Movement for the Liberation of São Tomé and Príncipe (Movimento de Libertação de São Tomé e Príncipe—MLSTP) as the sole legitimate representative for the islands. After independence, the MLSTP became the only political party. With the legalization of opposition party activity, Miguel Trovoada, an MLSTP founder who had been exiled, formed the Democratic Convergence Party-Group of Reflection (PDC-GR). Other parties include The Democratic Opposition Coalition (CODO) and the Christian Democratic Front (FDC).

DEFENSE

A small citizen's army was formed in 1975 by the Movement for the Liberation of São Tomé and Príncipe (MLSTP) government after Portuguese troops were withdrawn.

ECONOMIC AFFAIRS

São Tomé and Príncipe is one of the poorest countries in the world. The economy is based on cocoa-producing plantation agriculture, but the fall of cocoa prices since the early 1980s has created serious problems for the government, which abandoned socialist-style economic policies in favor of market-style policies.

Income

In 1998, gross domestic product (GDP) was estimated at $40 million, or $280 per person. During 1985–95, the average annual decline of per capita GNP was 2.1%, and the average annual inflation rate was 40.1% at that time.

Industry

São Tomé has very little industry. Soap, beverages, finished wood and furniture, bread, textiles, bricks and ceramics, garments, and palm oil are produced on the islands.

SOCIAL WELFARE

Before independence in 1975, social welfare was handled largely by the plantation corporations, missionaries, and private agencies. After independence, the government assumed responsibility for fostering community well-being.

Healthcare

Malnutrition continues to plague the country. An estimated 220 cases of tuberculosis per 100,000 people were reported in 1990. Life expectancy was 64 years. In 1983, there were sixteen hospitals and dispensaries, and, by 1989, sixty-one doctors.

Housing

Housing on the islands varies greatly, from the estate houses of the plantation headquarters to the thatch huts of the plantation laborers. Some town buildings are wooden; others are mud block with timber, as are plantation-labor dormitories.

EDUCATION

Schooling is compulsory for three years only. Primary education is for four years, and secondary has two stages: the first four years are followed by three years.

In 1989, there were 19,822 pupils in 64 primary schools with 559 teachers; in general secondary schools, there were 318 teachers and 7,446 pupils. In the mid-1990s, adult literacy was estimated at 75%.

1999 KEY EVENTS TIMELINE

January

- A new government led by Prime Minister Guilherme Posser da Costa is sworn in two months after elections.
- Several Central Bank officials are dismissed in an embezzlement scandal.

February

- Island Oil Exploration begins to develop an offshore logistics center.

March

- Corruption allegations lead to the dismissal of the Central Bank's governor and the resignation of the finance minister.

September

- World Bank and International Monetary Fund officials visit São Tomé to discuss the details of a structural adjustment agreement and financing program.

October

- Representatives from São Tomé participate in a media seminar in Mozambique.

ANALYSIS OF EVENTS: 1999

BUSINESS AND THE ECONOMY

São Tomé and Príncipe's economy is heavily dependent on cocoa exports for income. A chronic cocoa shortage brought on by drought and inefficient farming methods, as well as falling world cocoa prices, has hurt the nation's balance of payments. Most of the nation's food, fuel, and manufactured goods are supplied by imports. The government hopes that economic prospects will improve with the development of offshore petroleum deposits in the Gulf of Guinea and the development of the tourism industry.

Over the years São Tomé and Príncipe's external debt grew to approximately US $300 million, a sum fourteen times larger than its annual export income. It has fallen behind in servicing its foreign debt and relies primarily on debt rescheduling and foreign aid. In 1999 the country began to seek a classification of HIPC (Highly Indebted Poor Country) to qualify for debt relief. It also began to work with officials from the International Monetary Fund (IMF) and the World Bank to establish a structural adjustment program that would qualify it for financing from both organizations. IMF and World Bank representatives made a two-week visit to São Tomé and Príncipe in September to investigate conditions in the country and work on the details of the program. Once an agreement was reached, additional aid would be forthcoming from France and possibly other donor nations. In addition the European Union offered balance of payments support.

In late February Island Oil Exploration began work on an offshore logistics center on Sao Tome, having won exclusive financing and development rights for a period of six months. The facility was intended to serve as both a center for offshore oil and gas operations and as a commercial shipping port, complete with a repair station and dry dock. Initial development costs were estimated at $60 million.

Meanwhile Geco-Prakla proceeded with its seismic survey of twenty-two offshore blocks as specified in a contract between the government and Mobil Corporation, which had caused disagreement between São Tomé and Príncipe and Equatorial Guinea over the location of offshore boundaries between the two nations.

GOVERNMENT AND POLITICS

Legislative elections in November 1998 brought the Movement for the Liberation of São Tomé and Príncipe-Social Democratic Party (MLSTP-PSD) to power with 31 seats, and in December 1998 Guilherme Posser da Costa was appointed prime minister. There was a delay in the formation of a new government, however, when sitting President Trovoada vetoed Posser da Costa's initial cabinet nominations, an action criticized by the MLSTP-PSD. Posser da Costa submitted a revised list of nominations, and in January 1999 the new Council of Ministers was sworn in.

The Central Bank was at the center of several scandals early in 1999. In January several senior bank officials were dismissed in connection with the embezzlement of about US $1 million. The bank's governor was dismissed in March on suspicion of corruption, and a government investigation of the bank led to the resignation of a finance minister in that same month.

São Tomé and Príncipe and neighboring Equatorial Guinea have been involved in a border dispute over an offshore oil development. Equatorial Guinea challenged the legitimacy of twenty-two licenses awarded to Mobil by São Tomé and Príncipe as part of a technical cooperation agreement. The two countries have been working towards a resolution since 1998, and by mid-1999 the two were close to reaching an accord. Both did agree to accept the United Nations Convention of the Law of the Sea.

In January President Trovoada was one of several African heads of state to travel to Gabon to attend the inauguration of that nation's reelected president, Omar Bongo.

CULTURE AND SOCIETY

Poverty remained widespread in São Tomé and Príncipe, a legacy of the islands' plantation economy. Its cultural heritage as former Portuguese colonies was highlighted by an October conference on the state of the media in Africa's five Portuguese-speaking (Lusophone) countries. (The other Lusophone nations are Angola, Mozambique, Guinea-Bissau, and Cape Verde.) Delegates to the conference, held in Maputo, Mozambique, united in protesting the intimidation and persecution of journalists by the Angolan government, which had prevented some of its reporters from attending the conference and detained a newspaper editor the previous week.

The journalists at the conference drew up a declaration of goals for both their profession and their governments. It advocated respect for, and legislation guaranteeing, freedom of the press and editorial independence; state economic assistance to the media through lower telecommunications taxes and reduced customs duties on publications; and an exemption from visa requirements for journalists traveling between the Lusophone countries.

DIRECTORY

CENTRAL GOVERNMENT

Head of State

President
Miguel Trovoada, Office of the President

Ministers

Prime Minister
Guilherme Posser da Costa, Office of the Prime Minister, Praca Yon Gato, CP 302, São Tomé and Príncipe
PHONE: +239 (12) 22890
FAX: +239 (12) 21670

Minister of Labor and Civil Service
Emilio Guadalupe Fernandes Lima, Ministry of Labor and Civil Service

Minister of Equipment and Environment
Arlindo Afonso Carvalho, Ministry of Equipment and Environment

Minister of Commerce, Industry, and Tourism
Cosme Afonso Da Trindade Rita, Ministry of Commerce, Industry, and Tourism

Minister of Economics
Maria Das Neves Ceita Batista de Sousa, Ministry of Economics

Minister of Education and Culture
Peregrino Do Sacramento Da Costa, Ministry of Education and Culture

Secretary of Youth, Sports, and Skills Training
Luis Yaz de Sousa Bastos, Department of Youth, Sports, and Skills Training

Minister of Public Administration and Labor
Emilio Lima, Ministry of Public Administration and Labor

Minister of Internal and Territorial Administration
Manuel Da Cruz Margal Lima, Ministry of Internal and Territorial Administration

Minister of Health
Antonio Marques Lima, Ministry of Health

Minister of Infrastructure, Natural Resources, and Environment
Luis Alberto Dos Prazeres, Ministry of Infrastructure, Natural Resources, and Environment

Minister of Justice and Parliamentary Affairs
Alberto Paulino, Ministry of Justice and Parliamentary Affairs

Minister of Planning, Finance, and Cooperation
David Adelino Castelo, Ministry of Planning, Finance, and Cooperation, CP 168 STP, São Tomé and Príncipe
PHONE: +239 (12) 22372
FAX: +239 (12) 22182

Minister of Defense
Captain Joao Quaresma Bexigas, Ministry of Defense

Minister of Foreign Affairs and São Toméan Communities
Paulo Jorge de Espirito Santo, Ministry of Foreign Affairs and São Toméan Communities, CP 111, São Tomé and Príncipe
PHONE: +239 (12) 21166
FAX: +239 (12) 22597

Minister of Economy, Agriculture, Trade, Tourism, and Fisheries
Maria das Neves Baptista, Ministry of Economy, Agriculture, Trade, Tourism, and Fisheries

POLITICAL ORGANIZATIONS

Christian Democratic Front (FDC)

TITLE: Leader
NAME: Alphonse Dos Santos

Independent Democratic Action (ADI)

TITLE: Leader
NAME: Carlos Neves

Movimiento de Libertacao de São Tomé e Príncipe (Movement for the Liberation of São Tomé and Príncipe)

NAME: Manuel Pinto Da Costa

Partido da Convergencia Democratica-PCD (Party for Democratic Convergence)

TITLE: Secretary General
NAME: Armindo Aguiar

JUDICIAL SYSTEM

Supreme Tribunal

FURTHER READING

Articles

"Geco-Prakla Has Also Been Busy off São Tomé." *Offshore* (April 1999): 15.

"Regional Initiatives: Emphasis on Cooperation." *Institutional Investor International Edition* (August 1999): 2S9.

"São Tomé and Príncipe." *The Oil and Gas Journal* (April 12, 1999): 75.

Books

"São Tomé and Príncipe." *Africana: The Encyclopedia of the African and African American Experience*. Henry Louis Gates, Jr. and Kwame Anthony Appiah, ed. New York: Basic Books, 1999.

Garfield, Robert. *A History of São Tomé Island, 1470–1655: The Key to Guinea*. San Francisco: Mellen Research University Press, 1992.

Shaw, Caroline S. *São Tomé and Príncipe*. Santa Barbara, Calif.: Clio Press,1994.

Torp, Jens Erik, L.M. Denny, and Donald I. Ray. *Mozambique, São Tomé and Príncipe: Economics, Politics and Society*. London: Pinter Publishers, 1989.

Internet

Africaonline. Available Online @ http://www.africaonline.com. (October 28, 1999).

Africa News Online. Available Online @ http://www.africanews.org/west/mauritania/. (November 12, 1999).

CIA World Factbook. Available Online @ http://www.odci.gov/cia/publications/factbook/mr.html. (November 12, 1999).

Integrated Regional Information Network (IRIN). Available Online @ http://www.reliefweb.int/IRIN. (October 28, 1999).

SÃO TOMÉ AND PRÍNCIPE: STATISTICAL DATA

For sources and notes see "Sources of Statistics" in the front of each volume.

GEOGRAPHY

Geography (1)

Area:

Total: 960 sq km.

Land: 960 sq km.

Water: 0 sq km.

Area–comparative: more than five times the size of Washington, DC.

Land boundaries: 0 km.

Coastline: 209 km.

Climate: tropical; hot, humid; one rainy season (October to May).

Terrain: volcanic, mountainous.

Natural resources: fish.

Land use:

Arable land: 2%

Permanent crops: 36%

Permanent pastures: 1%

Forests and woodland: NA%

Other: 61% (1993 est.).

HUMAN FACTORS

Infants and Malnutrition (5)

Under-5 mortality rate (1997)	78
% of infants with low birthweight (1990-97)	7
Births attended by skilled health staff % of total[a]	NA
% fully immunized (1995-97)	
TB	70
DPT	73
Polio	73
Measles	60
Prevalence of child malnutrition under age 5 (1992-97)[b]	NA

Demographics (2A)

	1990	1995	1998	2000	2010	2020	2030	2040	2050
Population	119.5	137.1	150.1	159.8	219.1	291.9	370.1	446.9	518.3
Net migration rate (per 1,000 population)	NA	NA	NA	NA	NA	NA	NA	NA	NA
Births	NA	NA	NA	NA	NA	NA	NA	NA	NA
Deaths	NA	NA	NA	NA	NA	NA	NA	NA	NA
Life expectancy - males	60.8	62.0	62.9	63.5	66.3	68.7	70.9	72.7	74.3
Life expectancy - females	63.1	64.6	65.9	66.7	70.3	73.6	76.3	78.6	80.4
Birth rate (per 1,000)	42.1	43.5	43.5	43.1	38.3	32.0	26.0	21.4	18.2
Death rate (per 1,000)	9.9	9.0	8.3	7.8	5.9	4.5	3.9	4.0	4.5
Women of reproductive age (15-49 yrs.)	24.7	29.4	32.8	35.1	49.6	71.2	96.2	121.4	140.8
of which are currently married	NA	NA	NA	NA	NA	NA	NA	NA	NA
Fertility rate	6.3	6.3	6.2	6.1	5.2	4.1	3.0	2.5	2.2

Except as noted, values for vital statistics are in thousands; life expectancy is in years.

Ethnic Division (6)

Mestico, angolares (descendants of Angolan slaves), forros (descendants of freed slaves), servicais (contract laborers from Angola, Mozambique, and Cape Verde), tongas (children of servicais born on the islands), Europeans (primarily Portugese).

Religions (7)

Roman Catholic, Evangelical Protestant, Seventh-Day Adventist.

Languages (8)

Portuguese (official).

EDUCATION

Educational Attainment (10)

Age group (1981)	25+
Total population	33,308
Highest level attained (%)	
No schooling[2]	56.6
First level	
Not completed	18.0
Completed	19.3
Entered second level	
S-1	4.6
S-2	1.3
Postsecondary	0.3

GOVERNMENT & LAW

Political Parties (12)

National Assembly	No. of seats
Movement for the Liberation of Sao Tome and Principe (MLSTP)	27
Party for Democratic Convergence-Reflection Group (PCD-GR)	14
Independent Democratic Action (ADI)	14

Government Budget (13B)

Revenues	$58 million
Expenditures	$114 million
Capital expenditures	$54 million

Data for 1993 est.

Crime (15)

Crime rate (for 1994)	
Crimes reported	1,256
Total persons convicted	NA
Crimes per 100,000 population	1,005
Persons responsible for offenses	
Total number of suspects	1,339
Total number of female suspects	168
Total number of juvenile suspects	38

LABOR FORCE

Labor Force (16)

There are shortages of skilled workers.

Unemployment Rate (17)

28% (1996 est.)

PRODUCTION SECTOR

Electric Energy (18)

Capacity	6,000 kW (1995)
Production	16 million kWh (1995)
Consumption per capita	114 kWh (1995)

Transportation (19)

Highways:

total: 320 km

paved: 218 km

unpaved: 102 km (1996 est.)

Merchant marine:

total: 1 cargo ship (1,000 GRT or over) totaling 1,096 GRT/1,105 DWT (1997 est.)

Airports: 2 (1997 est.)

Airports—with paved runways:

total: 2

1,524 to 2,437 m: 1

914 to 1,523 m: 1 (1997 est.)

Top Agricultural Products (20)

Cocoa, coconuts, palm kernels, copra, cinnamon, pepper, coffee, bananas, papaya, beans; poultry, fish.

FINANCE, ECONOMICS, & TRADE

Economic Indicators (22)

National product: GDP–purchasing power parity–$154 million (1996 est.)

National product real growth rate: 1.5% (1996 est.)

National product per capita: $1,000 (1996 est.)

Inflation rate–consumer price index: 60% (1996 est.)

GOVERNMENT & LAW

Military Affairs (14B)

	1990	1991	1992	1993	1994	1995
Military expenditures						
Current dollars (mil.)	NA	NA	NA	NA	NA	NA
1995 constant dollars (mil.)	NA	NA	NA	NA	NA	NA
Armed forces (000)	1	1	1	1	3	NA
Gross national product (GNP)						
Current dollars (mil.)	28	29	30	30	32	33
1995 constant dollars (mil.)	32	32	32	32	33	33
Central government expenditures (CGE)						
1995 constant dollars (mil.)	NA	NA	NA	NA	NA	NA
People (mil.)	.1	.1	.1	.1	.1	.1
Military expenditure as % of GNP	NA	NA	NA	NA	NA	NA
Military expenditure as % of CGE	NA	NA	NA	NA	NA	NA
Military expenditure per capita (1995 $)	NA	NA	NA	NA	NA	NA
Armed forces per 1,000 people (soldiers)	8.1	7.9	7.7	7.5	21.9	NA
GNP per capita (1995 $)	258	251	246	240	239	238
Arms imports[6]						
Current dollars (mil.)	5	0	0	0	0	0
1995 constant dollars (mil.)	6	0	0	0	0	0
Arms exports[6]						
Current dollars (mil.)	0	0	0	0	0	0
1995 constant dollars (mil.)	0	0	0	0	0	0
Total imports[7]						
Current dollars (mil.)	21	25	25	22	24[e]	NA
1995 constant dollars (mil.)	24	28	27	23	24[e]	NA
Total exports[7]						
Current dollars (mil.)	4	6	5	5	7[e]	NA
1995 constant dollars (mil.)	5	7	5	5	7[e]	NA
Arms as percent of total imports[8]	23.8	0	0	0	0	NA
Arms as percent of total exports[8]	0	0	0	0	0	NA

MANUFACTURING SECTOR

GDP & Manufacturing Summary (21)

	1980	1985	1990	1992	1993	1994
Gross Domestic Product						
Millions of 1990 dollars	70	52	55	60	67	68
Growth rate in percent	2.59	−1.61	3.80	4.51	12.23	1.30
Per capita (in 1990 dollars)	744.0	486.7	463.5	481.2	527.2	521.8
Manufacturing Value Added						
Millions of 1990 dollars	4	3	3	3	3	*3*
Growth rate in percent	0.00	−8.68	5.18	4.34	*11.46*	*1.38*
Manufacturing share in percent of current prices	9.1	9.3	6.2	7.4	6.7	NA

FINANCE, ECONOMICS, & TRADE

Balance of Payments (23)

	1980	1985	1990
Exports of goods (f.o.b.)	17	7	4
Imports of goods (f.o.b.)	−15	−20	−13
Trade balance	2	−13	−9
Services - debits	−7	−10	−9
Services - credits	6	2	4
Private transfers (net)	−0	5	2
Government transfers (net)	1	−0	−
Overall balance	1	−16	−12

Exchange Rates (24)

Exchange rates:	
Dobras (Db) per US$1	
December 1997	7,003.9
1997	4,552.5
1996	2,203.2
1995	1,420.3
1994	732.6
1993	429.9

Top Import Origins (25)

$19.6 million (c.i.f., 1996 est.) Data are for 1995 est.

Origins	%
Ghana	17.1
China	13.3
France	12.5
Cameroon	6.0

Top Export Destinations (26)

$4.9 million (f.o.b., 1996 est.).

Destinations	%
Netherlands	75.7
Germany	1.2
Portugal	1.1

Economic Aid (27)

Recipient: ODA, $NA. NA stands for not available.

Import Export Commodities (28)

Import Commodities	Export Commodities
Machinery and electrical equipment	Cocoa 95%
Food products	Copra
Petroleum products	Coffee
	Palm oil

SAUDI ARABIA

CAPITAL: Riyadh (Ar-Riyad).

FLAG: The national flag bears in white on a green field the inscription, in Arabic, "There is no god but Allah, and Mohammad is the messenger of Allah." There is a long white sword beneath the inscription; the sword handle is toward the fly.

ANTHEM: The National Anthem is a short instrumental selection.

MONETARY UNIT: The Saudi riyal (SR) is divided into 20 qursh (piasters), in turn divided into 5 halalah. There are coins of 1, 5, 10, 25, 50, and 100 halalah and notes of 1, 5, 10, 50, 100, and 500 riyals. SR1 = $0.2670 (or $1 = SR3.7450).

WEIGHTS AND MEASURES: The metric system has been officially adopted.

HOLIDAYS: Muslim religious holidays include 1st of Muharram (Muslim New Year), 'Id al-Fitr, and 'Id al-'Adha'.

TIME: 3 PM = noon GMT.

LOCATION AND SIZE: Saudi Arabia, the third-largest country in Asia, constitutes about four-fifths of the Arabian Peninsula in Southwest Asia. Its precise area is difficult to specify because several of its borders are incompletely marked out. According to United Nations estimates, the nation has an area of 1,960,582 square kilometers (756,985 square miles), slightly less than one-fourth the size of the United States, with a total estimated boundary length of 7,055 kilometers (4,384 miles).

Saudi Arabia's capital city, Riyadh, is located in the east central part of the country.

CLIMATE: From May to September, the hottest period, daytime temperatures reach 54°C (129°F) in the interior and are among the highest recorded anywhere in the world. From October through April, the climate is more moderate, with evening temperatures between 16° and 21°C (61° and 70°F). Average annual rainfall is 9 centimeters (3.5 inches).

Kingdom of Saudi Arabia
Al-Mamlakah al-'Arabiyah as-Sa'udiyah

INTRODUCTORY SURVEY

RECENT HISTORY

With the discovery of oil in the 1930s, the history of Saudi Arabia was changed forever. Reserves have proved vast—about one-fourth of the world's total—and production, begun in earnest after World War II, has provided a huge income, much of it expended on construction and social services. Saudi Arabia's petroleum-derived wealth has considerably enhanced the country's influence in world economic and political forums.

Since the 1980s, the government has regulated its petroleum production to stabilize the international oil market and has used its influence as the most powerful moderate member of the Organization of Petroleum Exporting Countries (OPEC) to restrain the more radical members.

Political life in Saudi Arabia has remained basically stable in recent decades, despite several abrupt changes of leadership. Crown Prince Fahd ibn-'Abd al-'Aziz as-Sa'ud ascended the throne in 1982 after his half-brother, King Khaled, died of a heart attack. King Fahd has encouraged continuing modernization while seeking to preserve the nation's social stability and Islamic heritage.

Saudi Arabia's wealth and selective generosity have given it great political influence throughout the world and especially in the Middle East. It suspended aid to Egypt after that country's peace talks with Israel at Camp David, Maryland, but renewed relations in 1987. It secretly contributed

substantial funds to U.S. president Ronald Reagan's administration for combating communist regimes in Central America. It supported Iraq during the war with Iran and tried, in vain, to prevent Iraq's conflict with Kuwait.

When Iraq invaded Kuwait in 1990, Saudi Arabia, fearing Iraqi aggression, radically altered its traditional policy to permit the stationing of foreign troops on its soil. Riyadh made substantial contributions of arms, oil, and funds to the allied victory, cutting off subsidies to and expelling Palestinians and workers from Jordan, Yemen, and other countries that had supported Iraq in the period after the invasion.

Saudi Arabia and the United States consult closely on political, economic, commercial, and security matters. These supports became more visible following the Gulf War (1991), and continued Iraqi resistance to disarm.

GOVERNMENT

Saudi Arabia is a religiously based monarchy in which the sovereign's dominant powers are regulated according to Muslim law (*Shari`ah*), tribal law, and custom. There is no written constitution; laws must be compatible with Islamic law. The Council of Ministers is appointed by the king to advise on policy, originate legislation, and supervise the growing bureaucracy. In 1992, King Fahd announced the creation of the *Majlis al Shura*, an advisory body that would provide a forum for public debate.

Judiciary

The king acts as the highest court of appeal and has the power of pardon. The judiciary consists of lower courts that handle misdemeanors and minor civil cases; high courts of Islamic law (Shari'ah), and courts of appeal. An eleven-member Supreme Council of Justice reviews all sentences of execution, cutting, or stoning.

Political Parties

There are no political parties in Saudi Arabia.

DEFENSE

Saudi Arabia's armed forces totaled 105,500 personnel in 1995, having doubled in size to meet the Iraqi threat. The army had 73,000 personnel; the navy, 13,500 personnel; and the air force, 18,000 personnel. In 1995, Saudi Arabia spent $13.2 billion on its own forces.

ECONOMIC AFFAIRS

The economy depends heavily on oil production. Oil provided over 90% of export value and 75% of government revenues in 1995.

The government has tried to diversify the economy by developing industries using petroleum, including steel and petrochemical manufacture. The economy is open to private investors, but the government plays a significant role in the economy.

Public Finance

Deficits have been common since 1983, as oil revenues have declined. Deep budget cuts over the past two years have combined to reduce the deficit to approximately 3% of GDP. In 1996, government revenues totaled approximately $35.1 billion and expenditures $40 billion. External debt totaled $18.9 billion. The 1996 budget increased spending on education, transportation and telecommunications, and infrastructure in response to high population growth. To finance the deficit, the government borrows from domestic financial markets. As of 1996, domestic debt stood at $98 billion or 76% of GDP.

Income

Saudi Arabia's gross national product (GNP) was $133.5 billion, or about $6,800 per person. For the period 1985–95, the average inflation rate was 2.7%, resulting in a decline in per person GNP of 1.9%.

Industry

The country is attempting to diversify its manufacturing. Industrial products include cement, steel, glass, metal manufactures, automotive parts, building materials, and other industrial products, along with petroleum refinery products and petrochemicals. Total refinery capacity in 1995 was 1.6 million barrels per day. Saudi Arabia produced 588 million barrels of refined petroleum products in 1995.

Banking and Finance

The Saudi Arabian Monetary Agency (SAMA) was established by royal decree in 1952 to maintain the internal and external value of currency. The agency issues notes and coins with 100% cover in gold and convertible foreign exchange and regulates all banks and exchange dealers. Its foreign assets were once larger than those held by any other single banking institution, totaling an estimated $8,223 million in November 1996.

In 1997 there were 12 commercial banking houses. Credit and development agencies include the Agricultural Credit Bank, Saudi Credit Bank, and the Saudi Investment Banking Corp. Since the end of the Gulf conflict, commercial banks' credit exposure to the public and private sectors rose rapidly, by 54.4% and 88.8%, respectively, between the end of 1991 and the end of 1994.

There is a stock exchange in Saudi Arabia, created in 1990 as an over-the-counter market in which the commercial banks buy and sell shares by means of an electronic trading system.

Economic Development

Saudi Arabia's fifth economic development plan (1990–95) stressed economic diversification. This plan supported industry, agriculture, finance, and business services. An important goal of the sixth plan (1995–2000) is to reduce water consumption by 2% annually over the plan's period.

Saudi Arabia is a major supplier of economic aid to developing countries. It has provided two-thirds of OPEC assistance to some sixty nations since 1972; the cumulative aid total exceeds $30 billion. During the Iran-Iraq war, substantial foreign aid was given to Iraq to prevent a victory by Iran and possible exportation of the Ayatollah Khomeini's Shi'i revolution.

SOCIAL WELFARE

Social welfare in Saudi Arabia is traditionally provided through the family or tribe. Those with no family or tribal ties have recourse to the traditional Islamic religious foundations or may request government relief. Social insurance provides health care, disability, death, old-age pension, and survivor benefits for workers and their families. A large company with many employees in Saudi Arabia, ARAMCO, has a welfare plan for its employees that includes pension funds, accident compensation, and free medical care.

In 1993, only 5% of the labor force was female. Extreme modesty of dress is required, and women are not permitted to drive motor vehicles.

Healthcare

In 1990, hospital beds per 1,000 people equaled 3.35. Health personnel in 1990 included 21,110 physicians, 1,967 dentists, and 48,066 nurses (about 3.8 per 1,000).

In 1992, 97% of the population had access to health care services. Total health care expenditures in 1990 were $4,784 million.

Saudi Arabia still suffers from severe health problems. A major cause of disease is malnutrition, leading to widespread scurvy, rickets, night blindness, and anemia, as well as low resistance to tuberculosis. Dysentery attacks all ages and classes, trachoma is common, and typhoid is widespread.

In 1991, 95% of the population had access to safe water and 86% had adequate sanitation. In 1960, life expectancy at birth was 43 years; it averaged 70 years by 1995.

Housing

The continuing inflow of rural people to towns and cities, coupled with the rise in levels of expectation among the urban population, has created a serious housing problem; improvement in urban housing is one of Saudi Arabia's foremost economic needs. Some 506,800 dwelling units were built during 1974–85. In 1984, 78,884 building permits were issued, 84% of these for concrete dwellings and 8% for housing units of blocks and bricks.

EDUCATION

The literacy rate was 63% in 1995. In 1993, there were 11,244 primary schools, with 2.2 million pupils and 153,556 teachers. Secondary schools had 1.2 million students and 108,820 teachers. Higher education was pursued in 90 institutions, with 201,090 students and 14,394 instructors.

1999 KEY EVENTS TIMELINE

January

- Saudi Arabia calls for easing economic sanctions against Iraq to ease the plight of Iraqi people.
- Saudi officials are irritated with Iraqi leader Saddam Hussein, who has called on Arabs to overthrow leaders who do not support Iraq. The official Saudi Press Agency urges Iraqis to oust Hussein.

March

- The Pentagon offers Persian Gulf states access to intelligence on Iraqi and Iranian missile launches.
- OPEC members agree to cut crude-oil production by 2.1 million barrels a day and maintain lower levels of output for a full year starting April 1; Saudi Arabia agrees to slash output by 585,000 barrels of crude a day.
- Hundreds of thousands of Muslim pilgrims take part in the annual Hajj in Saudi Arabia; officials say economic problems around the world have decreased the number of international visitors who came for the Hajj, a sacred ritual.
- Iraq orders more than 18,000 Iraqi Hajj pilgrims to refuse to accept Saudi Arabia's "charity" offer to pay for their trip to Muslim holy sites and demands their immediate return.

May

- OPEC nations keep their pledge to trim oil production; Saudi Arabia raises its domestic gasoline prices by 50 percent to stem rising consumption.

July

- A tent catches fire at a pre-wedding party in eastern Saudi Arabia and collapses on guests, wounding 132 people.
- Four leading Muslim activists who call for the eviction of American troops from Saudi Arabia are released from jail.

August

- Saudi Arabia's royal family attends the funeral of King Fahd's oldest son, Prince Faisal, who at age 61 was minister of sports and youth.
- Women are not allowed to drive or travel alone, but the government announces they will be allowed to surf the World Wide Web.
- In the highest level visit by an Iraqi official to the Middle Eastern kingdom since the Gulf War, Iraq's justice minister travels to Saudi Arabia to discuss a regional anti-terrorism pact with Crown Prince Abdullah.

September

- Rumors circulate about King Fahd's health as he returns home after a two-month stay in Spain.
- Kobe Steel Ltd., a major Japanese steel and aluminum maker, announces it will build two chemical plants in Saudi Arabia worth a combined $160 million.

October

- The United States deports Hani Abdel Rahim el-Sayegh, a suspect in the 1996 bombing of a U.S. military-housing complex in Saudi Arabia.
- Amnesty International says it wants to observe the trial of Hani Abdel Rahim el-Sayegh.
- A leading member of the Saudi consultative council welcomes the first women to attend a council meeting in the kingdom.
- France's defense minister travels to Saudi Arabia to conclude an arms deal worth nearly $4 billion.
- Saudi Arabia says it is not seeking the extradition of Osama bin Laden, a former Saudi stripped of his citizenship in 1994, suspected of bombing two U.S. embassies in East Africa.
- Gen. Pervaiz Musharraf, the military leader who overthrew Pakistan's government in October, travels to Saudi Arabia for high-level talks.
- Three Pakistani men convicted of murder and armed robbery are beheaded, bringing the number of people beheaded in Saudi Arabia in 1999 to at least 96.

ANALYSIS OF EVENTS: 1999

BUSINESS AND THE ECONOMY

For most observers it was difficult to believe the Saudis, long known for their lack of frugality, were in a savings mode in 1999. Residents were asked to turn down their air conditioners to save electricity. They also saw the price of gasoline go up by 50 percent.

A drop in world oil prices is partly to blame. Oil contributes more than 40 percent to the country's GDP. With world oil prices down by 40 percent since 1997, the Saudis were trying to find new investments to finance their massive budget deficit, which was projected to grow between $12 billion to $15 billion in 1999, up from $5 billion in 1998.

Saudi Arabia earned $45 billion from oil exports in 1997, and saw revenues fall by one-third 1998. With the decrease in prices, the Saudis increased pressure on OPEC members to cut production.

Finally in March, OPEC members agreed to cut crude-oil production by 2.1 million barrels a day, and promised to maintain lower levels of output for a full year starting April 1. Saudi Arabia, OPEC's largest producer, said it would slash output by 585,000 barrels of crude oil a day. An economic collapse is not likely, but it may take some time for the Saudis to get rid of their large deficit. Crown Prince Abdullah, during a Gulf summit in Abu Dhabi in February, warned that the boom period was over, adding that Saudis "must get used to a new way of life that is not based on total dependence on the state."

GOVERNMENT AND POLITICS

Despite its economic woes, Saudi Arabia remained an important mediator in regional politics. *The Washington Post* reported that Saudi diplomats played a major role in convincing Libyan leader Moammar Gaddafi to hand over two suspects in the 1988 bombing of a Pan Am jetliner over Lockerbie, Scotland. Prince Bandar bin Sultan, Saudi Arabia's ambassador, told the press he wanted a solution that was "just" for Libya and for the relatives of the people who died in the bombing. In October, Crown Prince Abdullah traveled to Algiers to help ease tensions between Algeria and Morocco. The

Saudis also were trying to smooth relations with Iraq. In August, Iraq's justice minister traveled to Saudi Arabia to discuss a regional anti-terrorism pact with Crown Prince Abdullah. It was the highest level visit by an Iraqi official to the Saudi Arabia since the Gulf War.

CULTURE AND SOCIETY

Saudi Arabian women have few rights in their country. They must be veiled and segregated from men. They are not allowed to drive, or travel without a male guardian's permission. Which is why recent measures caught Saudis by surprise. First, women were allowed to surf the web, but more importantly, they were allowed to attend meetings of the consultative council, the closest institution that Saudi Arabia has to parliament.

Sheik Mohammed Bin Jubeir, head of the 90-member council, welcomed women to their meetings, saying the council could benefit from women's opinions on several issues. It was unclear whether the measure meant that women would get to play a larger role in their society.

For centuries, people who lived in the Saudi desert found natural ways to keep cool, but with so much money coming into the country through oil sales in the latter part of the 20th century, the Saudis became hooked on air conditioner. The Washington Post reported that the use of air conditioners has become so prevalent that it is hurting the nation economically. Saudi officials said citizens must conserve energy or the government will have to find an estimated $120 billion to triple the country's electrical capacity. Air conditioning accounts for as much as 70 percent of the country's electricity consumption during peak periods.

Tuwaijri's office has started an energy-saving campaign, telling people to turn lights off, and set the thermostat at higher temperatures to conserve power. Already, Saudi Arabia has asked industries to shift production schedules to hours of the day when demand for electricity is lighter. Even then, the government has cut power to industries and other heavy users during peak periods. Raising electricity rates to cover the cost of production has been suggested.

In March, about 2 million pilgrims made their way to the plain of Mina, six miles north of Mecca. Their final destination was Mount Arafat, where they prayed and reflected on their lives. The migration was a key component of the annual Hajj, a solemn ritual of prayer and reflection. Mount Arafat is where the Prophet Mohammed gave his last sermon centuries ago. In 1999, the ritual took place on Islam's holy day for the first time since 1994. It gave the Hajj the status of a ''greater pilgrimage.''

About 1.2 million pilgrims were Saudis. Officials said fewer foreigners came in 1999, mostly because of economic troubles in Southeast Asia, and oil-producing countries. Only 70,000 people came from Indonesia, compared to 230,000 in 1998, according to some estimates.

Iraqi pilgrims threatened to turn the Hajj into a political event. More than 18,000 of them congregated at the border demanding to be allowed to travel to Mount Arafat. The Saudis, who consider themselves custodians of the holy sites and have historically welcomed all pilgrims, said they would not stand in the way of the Iraqis, even though the two countries have had no diplomatic relations since 1990. The Iraqi pilgrims pressed for funds frozen by the United Nations to cover their expenses. The UN considered freeing up to $2,000 per pilgrim, but dropped the plan when Iraq demanded that the money be deposited directly in its central bank. Iraqi leader Saddam Hussein accused the Saudis of mistreating the pilgrims, and ordered them to come home. The pilgrims, he said, would not accept Saudi ''charity.''

The Saudis have modernized their facilities, and were better prepared for disasters during the Hajj. They provided phone, mail and telex services, and set up more than 27,000 fireproof tents. They wanted to prevent a repeat of the 1997 fire that spread through a tent city in Mina, killing 340 pilgrims. The Saudis mobilized more than 45,000 people to assist the pilgrims. Thirty-five bakeries prepared more than 5 million loaves of bread daily during the Hajj. Among the world's highest-ranking officials who performed the Hajj were Sudanese President Omar el-Bashir, Malaysian Prime Minister Mahathir Mohamad and Aslan Maskhadov, president of Russia's Chechnya region.

A large number of beheadings during 1999 continued to worry international human rights groups, which claimed convicts generally do not receive a fair trial in Saudi courts. In October, three Pakistani men convicted of murder and armed robbery were beheaded, bringing the number of executions to at least 96 for 1999. In 1998, there were at least 29 executions.

Saudi Arabia's courts impose death sentences for murder, rape, drug trafficking and armed robbery. Executions are usually carried out with a sword in public. Many of those who were executed in 1999 were foreign nationals accused of drug smuggling.

DIRECTORY

CENTRAL GOVERNMENT

Head of State

King, Prime Minister, and Custodian of the Two Holy Mosques
Fahd Bin Abdul Aziz al-Saud, Monarch
PHONE: +966 488222

Ministers

Crown Prince, Deputy Prime Minister, and Inspector General
Bin Abdul Aziz al-Saud, Office of the Deputy Prime Minister
PHONE: +966 4024600

Minister of Labor and Social Affairs
Musaid Bin Muhammad al-Sinani, Ministry of Labor and Social Affairs, Omar Bin al-Khatab Street, Riyadh 11157, Saudi Arabia
PHONE: +966 4771480
FAX: +966 4777336

Minister of Justice
Abdullah Bin Muhammad Bin Ibrahim al-Ashaikh, Ministry of Justice, University Street, Riyadh 11137, Saudi Arabia
PHONE: +966 4057777

Minister of Interior
Bin Abdul Aziz al-Saud, Ministry of Interior, PO Box 2933, Riyadh 11134, Saudi Arabia
PHONE: +966 4011944
FAX: +966 4031185

Minister of Information
Fouad Bin Abdul Salaam Bin Muhammad Farsi, Ministry of Information, Nasseriya Street, Riyadh 11161, Saudi Arabia
PHONE: +966 4014440
FAX: +966 4023570

Minister of Industry and Electricity
Hashim Bin Abdullah Bin Hashim al-Yamani, Ministry of Industry and Electricity, PO Box 5729, Omar Bin al-Khatab Road, Riyadh 11127, Saudi Arabia
PHONE: +966 4772722
FAX: +966 4775451

Minister of Higher Education
Khalid Bin Muhammad al-Angary, Ministry of Higher Education, King Faisal, Hospital Street, Riyadh 11153, Saudi Arabia
PHONE: +966 4644444
FAX: +966 4419004

Minister of Health
Osama Bin Abdul Majeed Shobokshi, Ministry of Health, Airport Road, Riyadh 11176, Saudi Arabia
PHONE: +966 4012220
FAX: +966 4029876

Minister of Foreign Affairs
Bin Abdul Aziz al-Saud, Ministry of Foreign Affairs, Nasseriya Street, Riyadh 11124, Saudi Arabia
PHONE: +966 4066836
FAX: +966 4030159

Minister of Finance and National Economy
Ibrahim al-Asaf, Ministry of Finance and National Economy, Airport Road, Riyadh 11177, Saudi Arabia
PHONE: +966 4050000
FAX: +966 4059202

Minister of Education
Muhammad Bin Ahmed al-Rasheed, Ministry of Education, Airport Road, Riyadh 11148, Saudi Arabia
PHONE: +966 4042888
FAX: +966 4042365

Minister of Defense and Aviation
Bin Abdul Aziz al-Saud, Ministry of Defense and Aviation, Airport Road, Riyadh 11195, Saudi Arabia
PHONE: +966 4785900
FAX: +966 4011336

Minister of Communication and Transportation
Nasir Bin Muhammad al-Salloum, Ministry of Communication and Transportation, Airport Road, Riyadh 11195, Saudi Arabia
PHONE: +966 4042928
FAX: +966 4031401

Minister of Commerce
Osama Bin Jafar Bin Ibrahim Faqih, Ministry of Commerce, PO Box 1774, Airport Road, Riyadh 11162, Saudi Arabia
PHONE: +966 4011110
FAX: +966 4038421

Minister of Agriculture and Water
Abdullah Bin Abdul Aziz Bin Mu'amar, Ministry of Agriculture and Water, Airport Road, Riyadh 11195, Saudi Arabia
PHONE: +966 401666
FAX: +966 4031415

Minister of Public Works and Housing
Bin Abdul Aziz al-Saud, Ministry of Public Works and Planning, Washem Street, Riyadh 11151, Saudi Arabia
PHONE: +966 4022268
FAX: +966 4022723; 4027373

Minister of Post, Telegraph and Telephone
Ali Bin Talal al-Juhani, Ministry of Post, Telegraph and Telephone, Intercontinental Road, Riyadh 11112, Saudi Arabia
PHONE: +966 4637225
FAX: +966 4052310

Minister of Planning
Abdul Wahab Bin Abdul Salam Attar, Ministry of Planning, PO Box 1358, Riyadh 11183, Saudi Arabia
PHONE: +966 4023562

Minister of Pilgrimage
Mahmoud Bin Muhammad Safar, Ministry of Pilgrimage, Omar Bin al-Khatab Street, Riyadh 11183, Saudi Arabia
PHONE: +966 4022200

Minister of Petroleum and Mineral Resources
Ali Bin Ibrahim al-Naimi, Ministry of Petroleum and Mineral Resources, PO Box 757, Airport Road, Riyadh 11189, Saudi Arabia
PHONE: +966 4781661
FAX: +966 4793596

Minister of Municipal and Rural Affairs
Muhammad Bin Ibrahim al-Jarallah, Ministry of Municipal and Rural Affairs, Nasseriya Street, Riyadh 11136, Saudi Arabia
PHONE: +966 4415434

DIPLOMATIC REPRESENTATION

Embassies in Saudi Arabia

Australia
Diplomatic Quarter, PO Box 94400, Riyadh 11693, Saudi Arabia
PHONE: +966 (1) 4887788
FAX: +966 (1) 4887973

Belgium
Quartier Diplomatique, Main Road 2, Lot A2, Riyadh 11693, Saudi Arabia
PHONE: +996 (1) 4882888
FAX: +996 (1) 4882033

Ireland
Diplomatic Quarter, PO Box 94349, Riyadh 11693, Saudi Arabia
PHONE: +966 (1) 4882300
FAX: +966 (1) 4880927

Italy
Diplomatic Quarter, PO Box 94389, Riyadh 11693, Saudi Arabia
PHONE: +966 (1) 4881212
FAX: +966 (1) 4881951

New Zealand
Diplomatic Quarter, PO Box 94397, Riyadh 11693, Saudi Arabia
PHONE: +966 (1) 4887988
FAX: +966 (1) 4887912

South Africa
PO Box 94006, Riyadh 11693, Saudi Arabia
PHONE: +966 (1) 4543723
FAX: +966 (1) 4543727

United Kingdom
PO Box 94351, Riyadh 11693, Saudi Arabia
PHONE: +966 (1) 4880077
FAX: +966 (1) 4882373

JUDICIAL SYSTEM

Department of Shari'ah

FURTHER READING

Articles

''Abdullah and the Ebbing Tide.'' *The Economist* (January 23, 1999): 41.

''Alwaleed's Kingdom, The Mystery of the World's Second-Richest Businessman.'' *The Economist* (February 27, 1999): 67.

''Does Big Oil Have Saudi Arabia Over a Barrel?'' *Business Week* (February 22, 1999): 35.

''Gas Prices Rise Quickly as OPEC Cuts Output.'' *The Seattle Times*, 23 March 1999.

''It's Cool to Be Hot in Saudi Arabia.'' *The Seattle Times*, 8 June 1999.

''Mega Mosques: Islam's Holiest Sites are Expanded to Accommodate the Growing Number of Pilgrims.'' *Time International*, (April 5, 1999): 72.

''Muslim Pilgrims Perform Key Part of Annual Hajj.'' *The Seattle Times*, 26 March 1999.

''Oil Ministers Appear Poised for More Cuts.'' *The Oil Daily*, 11 March 1999.

''OPEC Unity Means Gas Prices Won't Decline.'' *The Seattle Times*, 5 May 1999.

''Role of Saudi Women Reconsidered.'' The Associated Press, 3 October 1999.

''Saudi Arabia: Other Income Welcome.'' *Petroleum Economist*, (January 11, 1999): 36.

''Saudi Arabia Seeks Algeria-Morocco Rapprochement.'' Reuters, 11 October 1999.

''Saudi Arabia Sends Warning: 'Cut or Else.''' *The Oil Daily*, 12 March 1999.

''Saudi Arabia's Quiet Diplomacy Succeeds.'' *The Seattle Times*, 9 April 1999.

''Saudis Behead Convicted Killer.'' The Associated Press, 15 October 1999.

''Saudis Raise Price Terms for Sales to U.S.'' *The Oil Daily*, 6 January 1999.

''Saudi Women Attend Consultative Council Meeting for First Time.'' The Associated Press, 4 October 1999.

''They Can't Drive or Travel Alone, but Now Saudi Women Can Surf the Web.'' The Associated Press, 13 August 1999.

''Top Saudi Cleric Wield Political Cloud, Faces Modern Issues.'' The Associated Press, 4 August 1999.

SAUDI ARABIA: STATISTICAL DATA

For sources and notes see "Sources of Statistics" in the front of each volume.

GEOGRAPHY

Geography (1)

Area:

Total: 1,960,582 sq km.

Land: 1,960,582 sq km.

Water: 0 sq km.

Area—comparative: slightly more than one-fifth the size of the US.

Land boundaries:

Total: 4,415 km

Border countries: Iraq 814 km, Jordan 728 km, Kuwait 222 km, Oman 676 km, Qatar 60 km, UAE 457 km, Yemen 1,458 km.

Coastline: 2,640 km.

Climate: harsh, dry desert with great extremes of temperature.

Terrain: mostly uninhabited, sandy desert.

Natural resources: petroleum, natural gas, iron ore, gold, copper.

Land use:

Arable land: 2%

Permanent crops: 0%

Permanent pastures: 56%

Forests and woodland: 1%

Other: 41% (1993 est.).

HUMAN FACTORS

Demographics (2A)

	1990	1995	1998	2000	2010	2020	2030	2040	2050
Population	15,870.5	18,729.6	20,786.0	22,245.8	31,198.5	43,255.4	58,249.8	76,307.6	97,119.8
Net migration rate (per 1,000 population)	NA	NA	NA	NA	NA	NA	NA	NA	NA
Births	NA	NA	NA	NA	NA	NA	NA	NA	NA
Deaths	NA	NA	NA	NA	NA	NA	NA	NA	NA
Life expectancy - males	64.1	66.8	68.2	69.2	73.0	75.7	77.6	78.8	79.6
Life expectancy - females	67.2	70.3	72.0	73.1	77.7	80.9	83.1	84.6	85.5
Birth rate (per 1,000)	40.8	38.8	37.6	37.2	36.8	34.7	31.9	29.0	25.7
Death rate (per 1,000)	6.7	5.5	5.0	4.7	3.9	3.6	3.5	3.4	3.2
Women of reproductive age (15-49 yrs.)	3,027.5	3,591.3	4,034.1	4,371.7	6,432.7	9,235.1	13,275.1	18,104.5	23,718.0
of which are currently married	NA	NA	NA	NA	NA	NA	NA	NA	NA
Fertility rate	6.6	6.5	6.4	6.3	5.7	5.1	4.5	3.8	3.3

Except as noted, values for vital statistics are in thousands; life expectancy is in years.

Health Personnel (3)

Total health expenditure as a percentage of GDP, 1990-1997[a]

Public sector 6.4

Private sector 1.6

Total[b] 8.0

Health expenditure per capita in U.S. dollars, 1990-1997[a]

Purchasing power parity .221

Total 161

Availability of health care facilities per 100,000 people

Hospital beds 1990-1997[a] 250

Doctors 1993[c] 166

Nurses 1993[c] 348

Health Indicators (4)

Life expectancy at birth

1980 61

1997 71

Daily per capita supply of calories (1996) 2,735

Total fertility rate births per woman (1997) 5.9

Maternal mortality ratio per 100,000 live births (1990-97) . . . . .18[b]

Safe water % of population with access (1995) 93

Sanitation % of population with access (1995) 86

Consumption of iodized salt % of households (1992-98)[a] . . . .

Smoking prevalence

Male % of adults (1985-95)[a] 53

Female % of adults (1985-95)[a] . . . .

Tuberculosis incidence per 100,000 people (1997) 46

Adult HIV prevalence % of population ages 15-49 (1997) 0.01

Infants and Malnutrition (5)

Under-5 mortality rate (1997) 28

% of infants with low birthweight (1990-97) 7

Births attended by skilled health staff % of total[a] . . .90

% fully immunized (1995-97)

TB 99

DPT 92

Polio 92

Measles 87

Prevalence of child malnutrition under age 5 (1992-97)[b] . . . . .NA

Ethnic Division (6)

Arab 90%

Afro-Asian 10%

Religions (7)

Muslim 100%

Languages (8)

Arabic.

EDUCATION

Public Education Expenditures (9)

Public expenditure on education (% of GNP)

1980 . . . . .4.1

1996 . . . . .5.5[1]

Expenditure per student

Primary % of GNP per capita

1980 . . . . .18.9

1996 . . . . .36.6[1]

Secondary % of GNP per capita

1980 . . . .

1996 . . . .

Tertiary % of GNP per capita

1980 . . . . .109.3

1996 . . . . .70.6[1]

Expenditure on teaching materials

Primary % of total for level (1996) . . . .

Secondary % of total for level (1996) . . . .

Primary pupil-teacher ratio per teacher (1996) 13

Duration of primary education years (1995) 6

Literacy Rates (11A)

In thousands and percent[1]	1990	1995	2000	2010
Illiterate population (15+ yrs.)	3,821	3,871	4,146	4,461
Literacy rate - total adult pop. (%)	59.1	62.8	67.0	74.4
Literacy rate - males (%)	68.7	71.5	74.9	80.6
Literacy rate - females (%)	43.7	50.2	56.1	66.5

GOVERNMENT & LAW

Political Parties (12)

The legislative branch is a consultative council (90 members and a chairman appointed by the king for four-year terms). No political parties are allowed.

Government Budget (13B)

Revenues .... $47.5 billion
Expenditures .... $52.3 billion
Capital expenditures .... NA

Data for 1998 est. NA stands for not available.

LABOR FORCE

Labor Force (16)

Government 40%
Industry, construction, and oil 25%
Services 30%
Agriculture 5%

35% of the population in the 15-64 age group is non-national (July 1998 est.)

Unemployment Rate (17)

Rate not available.

PRODUCTION SECTOR

Electric Energy (18)

Capacity .... 20.9 million kW (1995)
Production .... 65 billion kWh (1995)
Consumption per capita .... 3,470 kWh (1995)

GOVERNMENT & LAW

Military Affairs (14B)

	1990	1991	1992	1993	1994	1995
Military expenditures						
Current dollars (mil.)	23,160[e]	35,510[e]	35,010[e]	20,480[e]	17,200	17,210
1995 constant dollars (mil.)	26,620[e]	39,240[e]	37,650[e]	21,470[e]	17,360	17,210
Armed forces (000)	146	191	172	172	164	175
Gross national product (GNP)						
Current dollars (mil.)	112,600	124,800	128,600	122,700	121,400	127,600
1995 constant dollars (mil.)	129,500	137,900	138,400	128,600	124,400	127,600
Central government expenditures (CGE)						
1995 constant dollars (mil.)	43,900[e]	NA	51,950[e]	52,420[e]	NA	NA
People (mil.)	15.9	16.1	16.7	17.4	18.0	18.7
Military expenditure as % of GNP	20.6	28.5	27.2	16.7	14.2	13.5
Military expenditure as % of CGE	60.6	NA	72.5	41.0	NA	NA
Military expenditure per capita (1995 $)	1,677	2,436	2,249	1,235	977	919
Armed forces per 1,000 people (soldiers)	9.2	11.9	10.3	9.9	9.1	9.3
GNP per capita (1995 $)	8,157	8,561	8,266	7,398	6,894	6,815
Arms imports[6]						
Current dollars (mil.)[10]	6,900	8,400	7,800	7,600	6,400	8,600
1995 constant dollars (mil.)[10]	7,930	9,282	8,390	7,968	6,560	8,600
Arms exports[6]						
Current dollars (mil.)	0	0	5	0	20	40
1995 constant dollars (mil.)	0	0	5	0	21	40
Total imports[7]						
Current dollars (mil.)	24,070	29,080	33,700	28,200	23,340	27,460
1995 constant dollars (mil.)	27,660	32,130	36,250	29,560	23,920	24,760
Total exports[7]						
Current dollars (mil.)	44,420	47,800	50,280	42,390	42,610	NA
1995 constant dollars (mil.)	51,050	52,820	54,080	44,450	43,680	NA
Arms as percent of total imports[8]	28.7	28.9	23.1	27.0	27.4	31.3
Arms as percent of total exports[8]	0	0	0	0	0	NA

Transportation (19)

Highways:

total: 162,000 km

paved: 69,174 km

unpaved: 92,826 km (1996 est.)

Pipelines: crude oil 6,400 km; petroleum products 150 km; natural gas 2,200 km (includes natural gas liquids 1,600 km)

Merchant marine:

total: 76 ships (1,000 GRT or over) totaling 1,009,059 GRT/1,329,377 DWT ships by type: bulk 1, cargo 13, chemical tanker 6, container 3, liquefied gas tanker 1, livestock carrier 5, oil tanker 22, passenger 1, refrigerated cargo 4, roll-on/roll-off cargo 12, short-sea passenger 8 (1997 est.)

Airports: 202 (1997 est.)

Airports—with paved runways:

total: 70

over 3,047 m: 30

2,438 to 3,047 m: 12

1,524 to 2,437 m: 23

914 to 1,523 m: 3

under 914 m: 2 (1997 est.)

Airports—with unpaved runways:

total: 132

over 3,047 m: 1

2,438 to 3,047 m: 5

1,524 to 2,437 m: 77

914 to 1,523 m: 36

under 914 m: 13 (1997 est.)

Top Agricultural Products (20)

Wheat, barley, tomatoes, melons, dates, citrus; mutton, chickens, eggs, milk.

MANUFACTURING SECTOR

GDP & Manufacturing Summary (21)

Detailed value added figures are listed by both International Standard Industry Code (ISIC) and product title.

	1980	1985	1990	1994
GDP ($-1990 mil.)[1]	116,723	84,895	82,997	89,857
Per capita ($-1990)[1]	12,154	6,712	5,172	5,149
Manufacturing share (%) (current prices)[1]	5.0	7.8	8.2	NA
Manufacturing				
Value added ($-1990 mil.)[1]	4,085	5,715	6,736	7,929
Industrial production index	100	168	230	254
Value added ($ mil.)	5,283	4,518	5,261	6,780
Gross output ($ mil.)	9,586	13,213	17,468	25,813
Employment (000)	80	132	122	183
Profitability (% of gross output)				
Intermediate input (%)	NA	NA	NA	NA
Wages and salaries inc. supplements (%)	NA	NA	NA	NA
Gross operating surplus	NA	NA	NA	NA
Productivity ($)				
Gross output per worker	119,681	96,701	143,509	139,927
Value added per worker	75,727	33,164	43,793	37,029
Average wage (inc. supplements)	NA	NA	NA	NA
Value added ($ mil.)				
311/2 Food products	267	290	300	398
313 Beverages	54	38	30	37
314 Tobacco products	40	28	23	22
321 Textiles	23	22	20	28
322 Wearing apparel	7	5	5	5
323 Leather and fur products	6	5	5	4
324 Footwear	2	1	1	1
331 Wood and wood products	11	9	9	11
332 Furniture and fixtures	45	35	35	39
341 Paper and paper products	68	79	110	157
342 Printing and publishing	48	52	56	69
351 Industrial chemicals	447	904	1,868	2,663
352 Other chemical products	159	101	158	131
353 Petroleum refineries	2,964	1,649	844	818
354 Miscellaneous petroleum and coal products	156	112	96	124
355 Rubber products	9	5	7	6
356 Plastic products	170	162	148	206
361 Pottery, china and earthenware	17	17	20	21

	1980	1985	1990	1994
362 Glass and glass products	*18*	*18*	*21*	*21*
369 Other non-metal mineral products	505	*578*	619	*833*
371 Iron and steel	*17*	*64*	*342*	*516*
372 Non-ferrous metals	*1*	*5*	*17*	*24*
381 Metal products	*128*	*181*	*289*	*358*
382 Non-electrical machinery	*30*	*40*	*63*	*71*
383 Electrical machinery	*47*	*68*	*105*	*132*
384 Transport equipment	*14*	*20*	*31*	*39*
385 Professional and scientific equipment	*1*	*2*	*2*	*2*
390 Other manufacturing industries	*29*	*29*	*38*	*44*

FINANCE, ECONOMICS, & TRADE

Economic Indicators (22)

National product: GDP—purchasing power parity—$206.5 billion (1997 est.)

National product real growth rate: 4% (1997 est.)

National product per capita: $10,300 (1997 est.)

Inflation rate—consumer price index: 0% (1997 est.)

Exchange Rates (24)

Exchange rates: Saudi riyals (SR) per US$1—3.7450 (fixed rate since June 1986)

Top Import Origins (25)

$25.4 billion (f.o.b., 1996)

Origins	%
United States	NA
European Union	NA
Kenya	NA
Tanzania	NA

NA stands for not available.

Top Export Destinations (26)

$56.7 billion (f.o.b., 1996) Data are for 1996 est.

Destinations	%
Japan	17
United States	15
South Korea	10
Singapore	8
France	5

Economic Aid (27)

Donor: pledged $100 million in 1993 to fund reconstruction of Lebanon.

Import Export Commodities (28)

Import Commodities	Export Commodities
Machinery and equipment	Petroleum and petroleum products 90%
Foodstuffs	
Chemicals	
Motor vehicles	
Textiles	

Balance of Payments (23)

	1992	1993	1994	1995	1996
Exports of goods (f.o.b.)	50,287	42,395	42,614	50,041	56,703
Imports of goods (f.o.b.)	−30,248	−25,873	−21,325	−25,650	−25,358
Trade balance	20,039	16,522	21,289	24,390	31,346
Services - debits	−34,226	−26,764	−20,453	−21,267	−24,414
Services - credits	10,844	9,491	7,379	8,467	9,097
Private transfers (net)	−1,000	−800	−600	−300	−300
Government transfers (net)	−13,397	−15,717	−18,102	−16,616	−15,513
Overall balance	−17,741	−17,268	−10,487	−5,326	215

SENEGAL

CAPITAL: Dakar.

FLAG: The flag is a tricolor of green, yellow, and red vertical stripes; at the center of the yellow stripe is a green star.

ANTHEM: Begins ''Pincez, tous, vos koras, frappez les balafons'' (''Pluck your koras, strike the balafons'').

MONETARY UNIT: The Communauté Financière Africaine franc (CFA Fr) is the national currency. There are coins of 1, 2, 5, 10, 25, 50, 100, and 500 CFA francs, and notes of 50, 100, 500, 1,000, 5,000, and 10,000 CFA francs. CFA Fr1 = $0.00196 (or $1 = CFA Fr510.65).

WEIGHTS AND MEASURES: The metric system is the legal standard.

HOLIDAYS: New Year's Day, 1 January; Independence Day, 4 April; Labor Day, 1 May; Day of Association, 14 July; Assumption, 15 August; All Saints' Day, 1 November; Christmas, 25 December. Movable religious holidays include 'Id al-Fitr, 'Id al-'Adha', Milad an-Nabi, Good Friday, Easter Monday, Ascension, and Pentecost Monday.

TIME: GMT.

LOCATION AND SIZE: Situated on the western bulge of Africa, Senegal has a land area of 196,190 square kilometers (75,749 square miles), slightly smaller than the state of South Dakota. The total boundary length of Senegal is 3,171 kilometers (1,970 miles). Senegal's capital city, Dakar, is located on the Atlantic coast.

CLIMATE: The average annual rainfall ranges from 34 centimeters (13 inches) in the extreme north to 155 centimeters (61 inches) in the southwest. Temperatures vary according to the season. At Dakar, during the cool season (December–April), the average daily maximum is 26°C (79°F) and the average minimum is 17°C (63°F). During the hot season (May–November), the averages are 30°C (86°F) and 20°C (68°F).

Republic of Senegal
République du Sénégal

INTRODUCTORY SURVEY

RECENT HISTORY

Under the constitution of 1946, Senegal was given two deputies in the French parliament, and a Territorial Assembly was established. The following year, Senegal accepted the new French constitution and became a self-governing republic within the French Community.

In June 1960, Senegal joined the Mali Federation together with Mali and French Sudan, but conflicting views soon led to its breakup; a month later the Legislative Assembly of Senegal proclaimed Senegal's national independence. A new constitution was adopted, and on 5 September 1960, Léopold-Sédar Senghor was elected president and Mamadou Dia became prime minister, retaining the position he had held since 1957 as head of the government. In 1962, the legislature overthrew Dia's government, and Senghor was elected by unanimous vote as head of government. Less than three months later, the electorate approved a new constitution that abolished the post of prime minister and made the president both chief of state and head of the executive branch.

Having been reelected in 1968, 1973, and 1978, Senghor resigned as president at the end of 1980 and was succeeded by Abdou Diouf. In February 1982, Senegal and The Gambia formed the Confederation of Senegambia with Diouf as president. The two countries pledged to integrate their armed and security forces, form an economic and monetary union, and coordinate foreign policy and

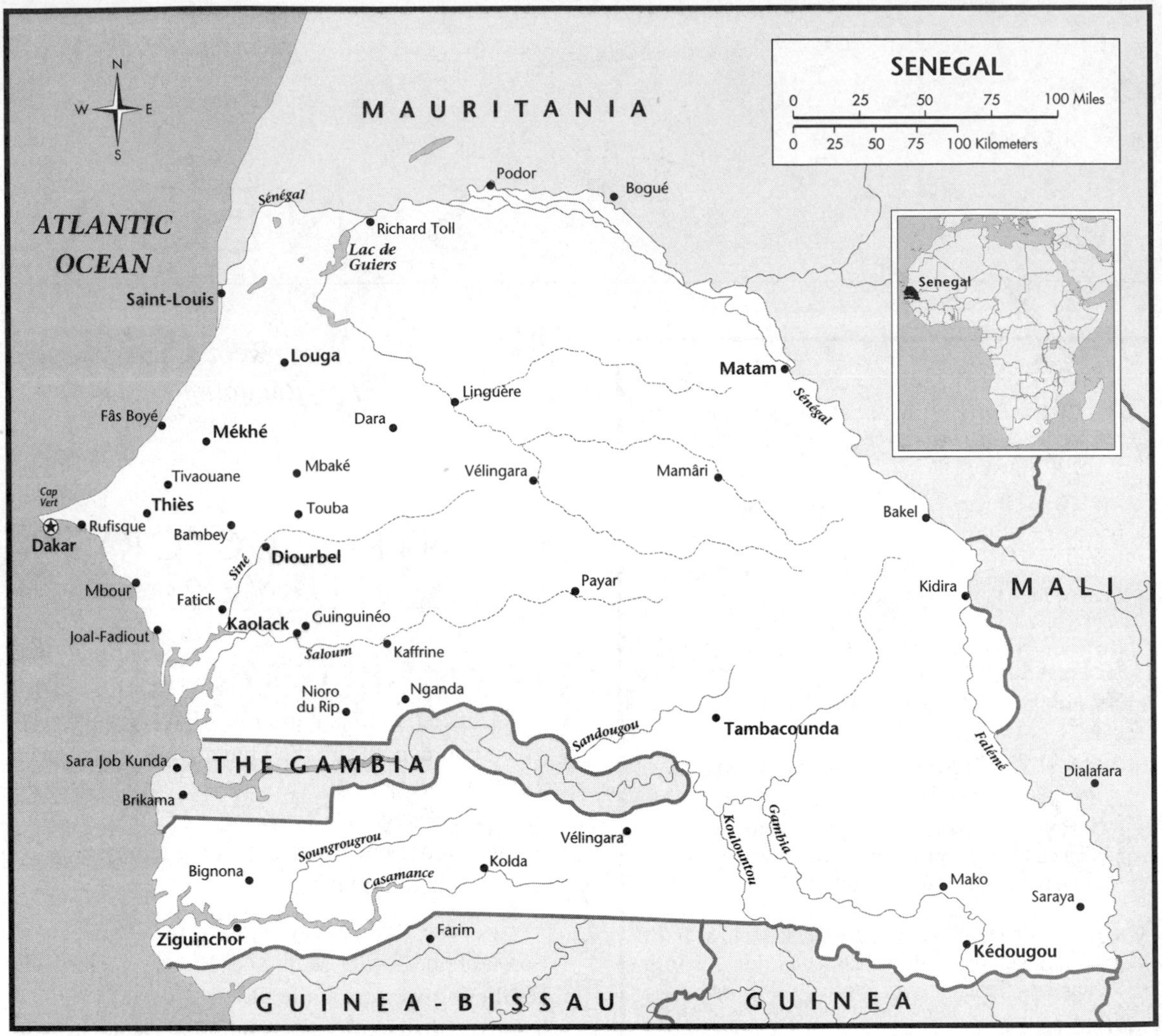

communications. However, the confederation dissolved in 1989.

Diouf was elected to a full term as president of Senegal on 27 February 1983. He reformed the government, making it less corrupt and more efficient. In the 1988 national elections, Diouf carried 77% of the vote. In April 1989, a nationwide state of emergency was declared and a curfew imposed in Dakar after rioters, enraged by reports of the killing of hundreds of Senegalese in Mauritania, killed dozens of Mauritanians. Relations with Mauritania were broken, and armed clashes along the border and internal rioting led to the expulsion of most Mauritanians residing in Senegal. Diplomatic relations were reestablished in April 1992, and the northern border along the Senegal River was reopened.

Diouf again won reelection in February 1993.

In the southernmost province of Casamance, a separatist group, the Movement of Democratic Forces of the Casamance (MFDC), has challenged the armed forces for years. A July 1993 cease-fire agreement appears to be holding, although there are numerous charges of human rights abuses by both sides, and hundreds have been killed. From 1990 to 1993, Senegalese armed forces played a major role in the peacekeeping effort in Liberia.

GOVERNMENT

Legislative power is exercised by a 120-member National Assembly, elected for five years simultaneously with the president. Under the 1963 constitution, as amended, the president of the republic determines national policy and has the power to dissolve the National Assembly. If the president asks the assembly to reconsider a measure it has enacted, the bill must be passed again by a three-fifths majority before it becomes law.

Judiciary

The High Council of the Magistrature, founded in 1960 and headed by the president, determines the constitutionality of laws and international commitments and decides when members of the legislature and the executive have exceeded their authority. A sixteen-member High Court of Justice, founded in 1962 and elected by the National Assembly from among its own members, presides over impeachment proceedings. The Supreme Court, founded in 1960, is made up of members appointed by the president of the republic on the advice of the High Council of the Magistrature.

Political Parties

Since independence in 1960, the UPS (Union Progressiste Sénégalaise—UPS) has been the dominant political party. In 1976, the UPS changed its name to the Senegalese Socialist Party (Parti Socialiste Sénégalais—PS), after joining the Socialist International. There was no legal opposition party from 1966 until 1974, when the Senegalese Democratic Party (Parti Démocratique Sénégalais—PDS) was formed in order to meet the constitutional requirement for a responsible opposition.

DEFENSE

Senegal's armed forces totaled about 13,400 men in 1995. The army had 12,000 men; the navy, 700; and the air force, 650. Military outlays in 1995 were about $74 million.

ECONOMIC AFFAIRS

Senegal's economy is based on its agricultural sector, primarily peanut production, and a modest industrial sector. Agriculture is highly vulnerable to declining rainfall, expansion of the desert onto farm land, and changes in world food prices. Senegal's resource-poor economy remains fragile and dependent upon foreign donors.

Public Finance

Although Senegal's finances are recorded as being in balance each year, in fact the country has run persistent deficits since 1976, generally covered by foreign aid, which represented 39.1% of the budget in 1995. From 1987 to 1990, Senegal's fiscal deficit grew from 2.6% of GDP to 4.3% of GDP. The U.S. CIA estimates that, in 1996, government revenues totaled approximately $876 million and expenditures $1.2 billion. External debt totaled $3.8 billion.

Income

In 1998, Senegal's gross national product (GNP) was $4.78 billion at current prices, or about $530 per person. For the period 1985–95, the average inflation rate was 3.7%, resulting in a decline in per person GNP of 1.2%.

Industry

Senegal's manufacturing sector plays a key role in the country's economy. Oil mills, sugar refineries, fish canneries, and flour mills are especially important. Textiles, leather goods, chemicals, paper, wood products, and building materials are also important manufactures.

Banking and Finance

In 1959, the Central Bank of the West African States (Banque Centrale des États de l'Afrique de l'Ouest—BCEAO) succeeded the Currency Board of French West Africa as the bank of issue for the former French West African territories. In 1962, it was reorganized as the joint note-issue bank of Benin, Côte d'Ivoire, Mauritania (which left in 1973), Niger, Senegal, Togo, and Upper Volta (now Burkina Faso), members of the West African Monetary Union. Foreign reserves, as of September 1995, stood at more than $277 million. Money supply, as measured by M2, amounted to CFA Fr463.8 billion at the end of 1995.

The most significant development bank is the government-controlled National Development Bank of Senegal, which participates in development projects and provides credit for government organizations, mixed societies, and cooperatives.

There are no securities exchanges in Senegal.

Economic Development

Senegal's development program addresses the basic problems encountered by Senegal's economy: lack of diversified output, the inefficiency of investments, the role of state in economic activity, and the excessive expansion of domestic consumer demand. These problems have been partly addressed by programs focusing on food self-sufficiency, fishing, and tourism, and by strengthening high-return activities. Since 1994, the government has made progress in privatizing state-owned enterprises, reducing labor costs to improve competitiveness in the manufacturing sector, and liberalizing trade by eliminating export subsidies and removing restrictions on certain strategic imports.

SOCIAL WELFARE

Since 1956, a system of family allowances for wage earners has provided small maternity and child benefits. Shared equally by employer and employee is a 6% contribution to a fund for general medical and hospital expenses. In addition, employers contribute 4.8% of gross salary to a retirement fund and employees 3.2%; the retirement age is 65.

Discrimination against women is widespread in education and employment. Non-Muslims also may face discrimination in civil, political, or economic matters, even though minority religions are protected under the law.

Healthcare

In 1992, there were 5 doctors per 100,000 people, and in 1990, there were 0.8 hospital beds per 1,000 people. In 1993, only 40% of the population had access to health care services.

Major health problems include measles and meningitis, along with water-related diseases such as malaria and schistosomiasis. In 1993, only 50% of the population had access to safe water and only 58% had adequate sanitation. Life expectancy was about 50 years in 1995.

Housing

Most housing in Dakar is like that of a European city. Elsewhere, housing ranges from European-type structures to the circular mud huts with thatched roofs common in villages.

EDUCATION

Education is compulsory at the primary level between ages 6 and 12; however, because of a lack of facilities, just over half the children in this age group attend school. In 1993, there were 2,559 primary schools in which 773,386 students were enrolled. At the secondary level, 182,140 students were attending schools the same year. Universities and equivalent institutions had 23,001 students in 1992. Illiteracy rates are high: in 1995, some 67% of adults were illiterate.

1999 KEY EVENTS TIMELINE

January

- CFA franc becomes directly tied to the European Union's currency at a fixed exchange rate of CFA 655.96: C1.
- Prospects for peace in Casamance brighten following a two-hour meeting between President Abdoul Diouf and Father Augustine Diamacoune Senghor, the political leader of the separatist Mouvement des Forces Démocratiques de Casamance (MFDC). Father Diamacoune is accompanied by Sidy Badji, leader of the moderate MFDC faction.
- Female circumcision is outlawed in Senegal.

March

- Senegal sends the annex protocol on women in the African Human Rights Charter to the African Commission on Human and People's Rights.
- The first International Congress on Traditional Medicine and HIV/AIDS takes place in Dakar. Senegal has one of the lowest HIV/AIDS seroprevalence rates in Africa.
- An information network is established to raise public and government awareness of the dangers arising from incineration as a method for disposing wastes.
- A business arbitration court is inaugurated.
- Government organizes a police crackdown on crime in Dakar and other cities.

April

- The government condemns Guinea-Bissau Prime Minister Fadul's suggestion that a referendum for self-rule be held in the Casamance.
- President Diouf appoints nine members to the National Elections Observatory (ONEL), including General Amadou Abdoulaye Dieng as head.
- The National Telecommunications Company of Senegal, SONATEL, announce its intention to invest 200 billion CFA over the next five years that will provide Senegal with 200,000 additional phone lines by the end of 2000.
- The army reports that 19 soldiers have died in clashes with guerrillas.

- The Civil Forum, a coalition of Senegalese civil society associations, organizes a conference on transparency in public affairs.

May

- The NGO, Wetlands International, opens a regional office in Dakar.
- Fighting resumes in the Casamance.
- Opposition parties form the Front pour la regularité et la transparence des élections (FRTE). They call for the resignation of General Dieng as head of ONEL when it is discovered that he was personally involved in establishing a re-election committee, Horizon 2000, for President Diouf.

June

- Dakar hosts first meeting of UEMOA's women ministers.
- The PS expels former foreign minister Moustapha Niasse following his castigation of the party's policies.
- Millions of pilgrims flock to the holy city of Touba for annual celebrations.
- Trade unions organize a country-wide strike mainly to press for higher wages.
- Casamance rebel leader denies plan to federate the Casamance with The Gambia and Guinea-Bissau. Separatists ask for a ceasefire.
- The FTRE holds a protest march in Dakar to remove General Dieng. The march turns violent as large numbers of police intervene.
- French and government police arrest prominent local leaders of the PS in the town of Louga on charges of laundering drug money.

July

- Kofi Annan makes a three-day visit to Senegal, the first stage of a five-country Africa tour. He welcomes the ceasefire agreement ending the fighting in Sierra Leone, signed in Lomé.
- Two mosques belonging to a fundamentalist cult are burnt down by disciples of the Mouride Brotherhood.
- Rebels end second round of peace talks in Banjul.
- General Dieng resigns as head of ONEL and is replaced by Conseil d'Etat, Louis Pereira de Carvalho.
- IMF disburses the second tranche of Senegal's ESAF worth $47.7 million.

August

- Tamaro Touré becomes the first African female head of the regional International Planned Parenthood Foundation (IPPF). The IPPF comprises 43 sub-Saharan countries.
- At least 16 fishermen die in storm.
- A $5 million loan from the BOAD is made for water project.
- Many are left homeless by heavy rains.
- Ten people are kidnapped in the Casamance.
- A truth and reconciliation body on torture is proposed.

September

- Protective measures for fisherman are instituted.

October

- Opposition leader Abdoulaye Wade is greeted by millions upon returning to Senegal.
- Senegal's wrestling champions are decorated.

November

- International Amateur Athletic Federation (IAAF) names Lamine Diack of Senegal as its new president.

December

- Casamance Movement of Democratic Forces, armed separatists fighting for independence in Casamance in the south, agree to a ceasefire with government forces.
- Armed separatists deny responsibility for attacks in the south at the end of 1999, stating that they continue to be committed to the ceasefire.

ANALYSIS OF EVENTS: 1999

BUSINESS AND THE ECONOMY

Senegal's macro-economic performance has been good, but the benefits of austerity are not reaching most people. The IMF approved Senegal's economic performance in June. The economy has been growing at a robust 5–6 percent yearly, but wages in both the public and private sector have remained flat while inflation has reduced the average family's purchasing power. A general strike in June indicated that workers were dissatisfied with government policies. In the eyes of wage-earning Senegalese, privatization has raised prices further, especially in utilities that probably were being held artificially low. More strikes are expected in the months leading into the elections.

Although flooding caused hardship for some, solid rains encouraged agricultural production, tripling the cotton harvest to about 45,000 tons and boosting peanut harvests as well. Industrial chemicals of Senegal (ICS), a major producer of fertilizers for the Indian market and Senegal's largest export, had a good year and has been promised $121 in investment over the next few years. Only fish exports were expected to decline as a result of depletion of stocks. GDP in 1999 was expected to reach 6.5% and then fall to 6% in 2000.

Inflation was expected to remain at around 2% into 2000, but poverty and undernourishment will have to be addressed. Senegal once enjoyed an enviable level of prosperity, but a third of households now fall below the poverty line, and one-quarter of all Senegalese suffer from chronic malnutrition. Unemployment, partial employment and low wages have contributed to a culture of labor export. Over the past five years, the number of health care workers has declined by 20 percent. Because of high fees, 50 percent of the children in five of the country's ten regions do not attend school. The government's announcement to invest $327 million in the Casamance region (in the extreme south of Senegal, just north of the border with Guinea-Bissau) is the scale of funding required for rehabilitation of public infrastructure generally.

GOVERNMENT AND POLITICS

The upcoming elections in 2000 and the peace initiatives in Casamance gained the most attention in 1999. Based on talks held in Banjul, Casamance separatists agreed to negotiate with the Senegalese government under the leadership of Father Augustine Diamacoune Senghor. He has favored peace talks, but more radical elements within the Mouvement des Forces Démocratiques de Casamance (MFDC) supported secession by military means if necessary. The government condemned, for example, Guinea-Bissau Prime Minister Fadul's suggestion that a referendum for self-rule be held in the Casamance. The conferees demanded the withdrawal of Senegalese military forces from Casamance, the closing of military bases established after 1979 there, and a ceasefire. The government and donors such as USAID have made significant commitments to rebuilding and developing the southern province.

As for elections, Moustapha Niasse, former foreign minister, was expelled from the ruling Socialist Party (PS) party after characterizing the PS as a small minority that "steals from the state on a grand scale." He also criticized the new law removing the limit on presidential terms, and the appointment of a Diouf colleague as head of the elections monitoring commission. Mr. Niasse had been a former political secretary of the PS, and was expected to run for election next year. He could be joined by Djibo Ka, who left the party in late 1997 to form the Union et le Renouveau Démocratique (URD). They will have to contend with Abdoulaye Wade of the Parti Démocratique Sénégalais (PDS), whose return to Dakar in October was greeted by an estimated 2 million supporters. He was given a similar reception in St. Louis.

Senegalese are losing faith in their political system. Many believe that change will not occur via legal, electoral means. The PS tightly controls the country via patronage, party machinery, and the state media. The dissension in its ranks has more to do with internal rivalries than with substantive policy issues. A widespread demand for change exists outside the party, but most Senegalese remain skeptical that the February elections will be its vehicle.

CULTURE AND SOCIETY

1999 marked a number of firsts for women. Dakar hosted the first meeting of UEMOA's women ministers, who concluded that the low schooling rate for girls and women's low representation in decision-making bodies accounted for two of the key constraints to building greater equality in social, economic, and political affairs. In a workshop held in Dakar to amend the African Human Rights Charter, women agreed that they should have the same rights and responsibilities as men in marriage and jointly run the home. They called on governments to eliminate polygamy, albeit gradually. Senegalese Tamaro Touré became the first African woman to head the executive committee of the African region of International Planned Parenthood. Touré is also the chairperson of the Senegalese Association for Family Welfare.

In January, Senegal became the first African nation to pass a law against excision (removal of the clitoris). UNICEF estimates that 700,000 girls and women in Senegal have been excised. The law carries a penalty of six months to five years in prison, mainly because circumcision is often performed under non-sterile conditions using blunt instruments that mutilate and sometimes cause death. Some legislators and religious leaders objected that the law interfered with religious and cultural practices. Women's rights activist Aya Ndiaye said that a grassroots information campaign was needed to stop the practice, especially in rural areas.

DIRECTORY

CENTRAL GOVERNMENT

Head of State

President
Abdou Diouf, Office of the President, Avenue Rourne, BP 4026, Dakar, Senegal
PHONE: +221 8231088

Prime Minister
Mamadou Lamine Loum, Office of the Prime Minister, Building Administratif, Avenue Rourne, Dakar, Senegal
PHONE: +221 8231148

Ministers

Minister of Agriculture
Robert Sagna, Ministry of Agriculture, BP 4055, Dakar, Senegal
PHONE: +221 8231000

Minister of Foreign Affairs and Expatriate Senegalese
Jacques Baudin, Ministry of Foreign Affairs, rue Pierre Millou, Dakar, Senegal
PHONE: +221 8234284

Minister of Economic Affairs, Finance, and Planning
Mouhamed el-Moustapha Diagne, Ministry of Economic Affairs, avenue Carde, BP 4017, Dakar, Senegal
PHONE: +221 8239699

Minister of Commerce and Handicrafts
Khalifa Sall, Ministry of Commerce, Building Adminstratif, Dakar, Senegal
PHONE: +221 8237429

Minister of Tourism and Air Transport
Tijane Sylla, Ministry of Tourism, 23. rue Dr Calmette, Dakar, Senegal
PHONE: +221 8210534
FAX: +221 8229413

Minister of Communications
Aissata Tall Sall, Ministry of Communication, blvd. de la Republique, Dakar, Senegal
PHONE: +221 8231065

Minister of Energy, Mines and Industry
Magued Diouf, Ministry of Energie, 122 bis, ave. Andre Peytavin, BP 5564, Dakar, Senegal
PHONE: +211 8229626
FAX: +221 8225594

Minister of Development and Water Resources
Mamadou Faye, Ministry of Development and Water Resources, Building Administratif, BP 4055, Dakar, Senegal
PHONE: +221 8233974

Minister of Housing and Town Planning
Abdourahmane Sow, Ministry of Housing and Town Planning, rue Dr Calmette, Dakar, Senegal
PHONE: +221 8239127
FAX: +221 8233278

Minister of Scientific Research and Technology
Balla Moussa Daffe, Ministry of Scientific Research and Technology, rue des Essarts, Dakar, Senegal
PHONE: +221 8222205

Minister of African Economic Integration
Abibatou Mbaye, Ministry of African Economic Integration, Building Administratif, BP 4077, Dakar, Senegal
PHONE: +221 8238716

Minister of National Education
Andre Sonko, Ministry of National Education, rue Dr Calmette, Dakar, Senegal
PHONE: +221 8214123
FAX: +221 8211258

POLITICAL ORGANIZATIONS

African Party for Democracy and Socialism or And-Jef (PADS)

TITLE: Secretary General
NAME: Landing Savane

Democratic League-Labor Party Movement

NAME: Abdoulaye Bathily

Democratic and Patriotic Convention

NAME: Iba Der Thiam

Socialist Party (PS)

TITLE: President
NAME: Abdou Diouf

National Democratic Rally (RND)

NAME: Madier Diouf

DIPLOMATIC REPRESENTATION

Embassies in Senegal

Canada
45, ave. de la Republique, BP 3373, Dakar, Senegal
PHONE: +221 8239290
FAX: +221 8238749

France
1, rue Assane Ndoye, Dakar, Senegal
PHONE: +221 239181

Germany
20, avenue Pasteur, Dakar, Senegal
PHONE: +221 222519

Israel
Hotel Sofitel Teranga, place de l'Independance, BP 3380, Dakar, Senegal
PHONE: +221 231044
FAX: +221 226153

Italy
rue Alpha Achamiyou Tall, Dakar, Senegal
PHONE: +221 220578
FAX: +221 217580

United Kingdom
20, rue du Docteur Guillet, BP 6025, Dakar, Senegal
PHONE: +221 237392
FAX: +221 232766

United States
avenue Jean XXIII, BP 49, Dakar, Senegal
PHONE: +221 234296
FAX: +221 222991

JUDICIAL SYSTEM

High Council of the Magistrature

High Court of Justice

Supreme Court

FURTHER READING

Articles

Economist Intelligence Unit Ltd. *EIU Country Reports*. London: *The Economist*, 1999.

Books

Africa South of the Sahara. London: Europa Publishers, 1998, s.v. ''Senegal.''

Colvin, Lucie Gallistel, ed. *Historical Dictionary of Senegal*. African Historical Dictionaries, No. 23. Metuchen, N.J. and London: The Scarecrow Press, 1981.

Dilly, Roy and Jerry Eades, eds. *Senegal*. World Bibliographical Series, Vol. 166. Oxford, England: Clio Press, 1994.

Gellar, Sheldon. *Senegal: An African Nation between Islam and the West*. Second Edition. Boulder, Colorado: Westview Press, 1995.

Senghor, Léopold Sédar. *The Collected Poetry*. Charlottesville: University of Virginia Press, 1991.

Sharp, Robin. *Senegal: A State of Change*. Oxford, UK: Oxfam, 1994.

Internet

Africaonline. Available Online @ http://www.africaonline.com (November 2, 1999).

Africa News Online. Available Online @ http://www.africanews.org/west/stories/1999_feat1.html (November 2, 1999).

Integrated Regional Information Network (IRIN). Available Online @ http://www.reliefweb.int/IRIN (November 2, 1999).

SENEGAL: STATISTICAL DATA

For sources and notes see "Sources of Statistics" in the front of each volume.

GEOGRAPHY

Geography (1)

Area:

Total: 196,190 sq km.

Land: 192,000 sq km.

Water: 4,190 sq km.

Area—comparative: slightly smaller than South Dakota.

Land boundaries:

Total: 2,640 km.

Border countries: The Gambia 740 km, Guinea 330 km, Guinea- Bissau 338 km, Mali 419 km, Mauritania 813 km.

Coastline: 531 km.

Climate: tropical; hot, humid; rainy season (May to November) has strong southeast winds; dry season (December to April) dominated by hot, dry, harmattan wind.

Terrain: generally low, rolling, plains rising to foothills in southeast.

Natural resources: fish, phosphates, iron ore.

Land use:

Arable land: 12%

Permanent crops: 0%

Permanent pastures: 16%

Forests and woodland: 54%

Other: 18% (1993 est.).

HUMAN FACTORS

Demographics (2A)

	1990	1995	1998	2000	2010	2020	2030	2040	2050
Population	7,408.2	8,790.4	9,723.1	10,390.3	14,362.2	19,497.5	25,620.8	32,448.4	39,689.5
Net migration rate (per 1,000 population)	NA	NA	NA	NA	NA	NA	NA	NA	NA
Births	NA	NA	NA	NA	NA	NA	NA	NA	NA
Deaths	NA	NA	NA	NA	NA	NA	NA	NA	NA
Life expectancy - males	51.4	53.4	54.6	55.4	59.3	62.9	66.3	69.1	71.6
Life expectancy - females	56.3	58.8	60.3	61.3	66.0	70.2	73.9	76.9	79.4
Birth rate (per 1,000)	48.3	46.0	44.4	43.4	39.5	35.2	30.4	26.4	22.9
Death rate (per 1,000)	14.1	12.1	11.0	10.4	7.8	6.1	5.0	4.6	4.5
Women of reproductive age (15-49 yrs.)	1,686.1	1,998.6	2,219.4	2,383.0	3,409.2	4,745.2	6,492.8	8,471.1	10,466.8
of which are currently married	NA	NA	NA	NA	NA	NA	NA	NA	NA
Fertility rate	6.6	6.4	6.2	6.0	5.3	4.5	3.8	3.2	2.8

Except as noted, values for vital statistics are in thousands; life expectancy is in years.

Health Personnel (3)

Total health expenditure as a percentage of GDP, 1990-1997[a]

Public sector	1.2
Private sector	NA
Total[b]	NA

Health expenditure per capita in U.S. dollars, 1990-1997[a]

Purchasing power parity	NA
Total	NA

Availability of health care facilities per 100,000 people

Hospital beds 1990-1997[a]	70
Doctors 1993[c]	7
Nurses 1993[c]	35

Health Indicators (4)

Life expectancy at birth

1980	45
1997	52

Daily per capita supply of calories (1996)	2,394
Total fertility rate births per woman (1997)	5.6
Maternal mortality ratio per 100,000 live births (1990-97)	510[d]
Safe water % of population with access (1995)	50
Sanitation % of population with access (1995)	....
Consumption of iodized salt % of households (1992-98)[a]	9

Smoking prevalence

Male % of adults (1985-95)[a]	....
Female % of adults (1985-95)[a]	....

Tuberculosis incidence per 100,000 people (1997)	223
Adult HIV prevalence % of population ages 15-49 (1997)	1.77

Infants and Malnutrition (5)

Under-5 mortality rate (1997)	124
% of infants with low birthweight (1990-97)	4
Births attended by skilled health staff % of total[a]	47

% fully immunized (1995-97)

TB	80
DPT	65
Polio	65
Measles	65

Prevalence of child malnutrition under age 5 (1992-97)[b]	22

Ethnic Division (6)

Wolof	36%
Fulani	17%
Serer	17%
Toucouleur	9%
Diola	9%
Mandingo	9%
European and Lebanese	1%
Other	2%

Religions (7)

Muslim	92%
Indigenous beliefs	6%
Christian	2%

Christians are mostly Roman Catholic.

Languages (8)

French (official), Wolof, Pulaar, Diola, Mandingo.

EDUCATION

Public Education Expenditures (9)

Public expenditure on education (% of GNP)

1980	....
1996	3.5

Expenditure per student

Primary % of GNP per capita

1980	24.6
1996	10.6

Secondary % of GNP per capita

1980	....
1996	....

Tertiary % of GNP per capita

1980	442.7
1996	....

Expenditure on teaching materials

Primary % of total for level (1996)	4.0
Secondary % of total for level (1996)	....
Primary pupil-teacher ratio per teacher (1996)	58
Duration of primary education years (1995)	6

Literacy Rates (11A)

In thousands and percent[1]	1990	1995	2000	2010
Illiterate population (15+ yrs.)	2,844	3,084	3,346	3,833
Literacy rate - total adult pop. (%)	28.8	33.1	37.5	46.7
Literacy rate - males (%)	38.8	43.0	47.4	55.9
Literacy rate - females (%)	19.0	23.2	27.7	37.6

GOVERNMENT & LAW

Political Parties (12)

National Assembly	% of seats
Socialist Party (PS)	50.19
Senegalese Democratic Party (PDS)	19
Senegalese Democratic Union-Renewal (UDS-R)	13
African Party for Democracy and Socialism (And-Jef/PADS)	5
Democratic League-Labor Party Movement (LD-MPT)	4
Democratic and Patriotic Convention (CDP/Garab-Gi)	2
Others	5

Government Budget (13B)

Revenues	$885 million
Expenditures	$885 million
Capital expenditures	$125 million

Data for 1996 est.

LABOR FORCE

Labor Force (16)

Agriculture 60%.

Unemployment Rate (17)

Rate not available; urban youth 40%

Military Affairs (14B)

	1990	1991	1992	1993	1994	1995
Military expenditures						
Current dollars (mil.)	81[e]	83[e]	120[e]	106[e]	72	76
1995 constant dollars (mil.)	94[e]	91[e]	129[e]	111[e]	73	76
Armed forces (000)	18	18	18	18	14	14
Gross national product (GNP)						
Current dollars (mil.)	3,951	4,057	4,279	4,295	4,341	4,672
1995 constant dollars (mil.)	4,540	4,483	4,603	4,503	4,450	4,672
Central government expenditures (CGE)						
1995 constant dollars (mil.)	NA	NA	937[e]	NA	NA	NA
People (mil.)	7.4	7.7	7.9	8.2	8.5	8.8
Military expenditure as % of GNP	2.1	2.0	2.8	2.5	1.7	1.6
Military expenditure as % of CGE	NA	NA	13.7	NA	NA	NA
Military expenditure per capita (1995 $)	13	12	16	14	9	9
Armed forces per 1,000 people (soldiers)	2.4	2.3	2.3	2.2	1.6	1.6
GNP per capita (1995 $)	613	585	580	548	524	531
Arms imports[6]						
Current dollars (mil.)	0	10	10	10	5	5
1995 constant dollars (mil.)	0	11	11	10	5	5
Arms exports[6]						
Current dollars (mil.)	0	0	0	0	0	0
1995 constant dollars (mil.)	0	0	0	0	0	0
Total imports[7]						
Current dollars (mil.)	1,314	1,173	1,034	967	704	1,385[e]
1995 constant dollars (mil.)	1,510	1,296	1,112	1,014	722	1,385[e]
Total exports[7]						
Current dollars (mil.)	762	702	673	457	340	593[e]
1995 constant dollars (mil.)	876	775	724	479	349	593[e]
Arms as percent of total imports[8]	.0	.9	1.0	1.0	.7	.4
Arms as percent of total exports[8]	0	0	0	0	0	0

PRODUCTION SECTOR

Electric Energy (18)

Capacity . . . . .303,440 kW (1997)

Production . . . . .1.027 billion kWh (1997 est.)

Consumption per capita . . . . .109 kWh (1997 est.)

Transportation (19)

Highways:

total: 14,576 km

paved: 4,271 km

unpaved: 10,305 km (1996 est.)

Waterways: 897 km total; 785 km on the Senegal river, and 112 km on the Saloum river

Merchant marine:

total: 1 bulk ship, 1,995 GRT/3,775 DWT (1997 est.)

Airports: 20 (1997 est.)

Airports—with paved runways:

total: 10

over 3,047 m: 1

1,524 to 2,437 m: 7

914 to 1,523 m: 2 (1997 est.)

Airports—with unpaved runways:

total: 10

1,524 to 2,437 m: 5

914 to 1,523 m: 4

under 914 m: 1 (1997 est.)

Top Agricultural Products (20)

Peanuts, millet, corn, sorghum, rice, cotton, tomatoes, green vegetables; cattle, poultry, pigs; fish.

MANUFACTURING SECTOR

GDP & Manufacturing Summary (21)

Detailed value added figures are listed by both International Standard Industry Code (ISIC) and product title.

	1980	1985	1990	1994
GDP ($-1990 mil.)[1]	4,326	5,019	5,703	5,922
Per capita ($-1990)[1]	781	787	778	731
Manufacturing share (%) (current prices)[1]	15.3	12.8	13.6	*21.7*
Manufacturing				
Value added ($-1990 mil.)[1]	493	631	775	*766*
Industrial production index	100	101	118	121
Value added ($ mil.)	266	268	421	*234*
Gross output ($ mil.)	1,070	926	1,583	*976*
Employment (000)	32	30	32	*23*
Profitability (% of gross output)				
Intermediate input (%)	75	71	73	*74*
Wages and salaries inc. supplements (%)	*10*	*11*	10	*9*
Gross operating surplus	*14*	*18*	16	*16*
Productivity ($)				
Gross output per worker	33,812	18,250	48,738	*42,416*
Value added per worker	8,400	3,695	12,956	*10,869*
Average wage (inc. supplements)	*3,508*	*3,240*	5,056	*4,053*
Value added ($ mil.)				
311/2 Food products	113	*100*	221	*124*
313 Beverages	12	*9*	17	*11*
314 Tobacco products	7	7	16	*10*
321 Textiles	33	*23*	13	2
322 Wearing apparel	10	*7*	—	*3*
323 Leather and fur products	5	*4*	—	*3*
324 Footwear	2	*1*	—	*1*
331 Wood and wood products	2	*7*	1	*1*
332 Furniture and fixtures	2	—	—	—
341 Paper and paper products	4	2	5	*4*
342 Printing and publishing	6	*7*	10	*6*
351 Industrial chemicals	16	*12*	13	*21*
352 Other chemical products	5	*15*	24	*13*
353 Petroleum refineries	18	*11*	27	*10*
354 Miscellaneous petroleum and coal products	—	—	—	—
355 Rubber products	—	—	—	—
356 Plastic products	—	*4*	11	*6*
361 Pottery, china and earthenware	—	—	—	—
362 Glass and glass products	—	—	—	—

	1980	1985	1990	1994
369 Other non-metal mineral products	12	*21*	31	*20*
371 Iron and steel	—	—	—	—
372 Non-ferrous metals	—	—	—	—
381 Metal products	10	*21*	11	*8*
382 Non-electrical machinery	3	*7*	2	*1*
383 Electrical machinery	1	*1*	3	*2*
384 Transport equipment	5	*7*	17	*6*
385 Professional and scientific equipment	—	—	—	—
390 Other manufacturing industries	—	—	—	—

FINANCE, ECONOMICS, & TRADE

Economic Indicators (22)

National product: GDP—purchasing power parity—$15.6 billion (1997 est.)

National product real growth rate: 4.7% (1997 est.)

National product per capita: $1,850 (1997 est.)

Inflation rate—consumer price index: 2.5% (1997 est.)

Exchange Rates (24)

Exchange rates:	
CFA francs (CFAF) per US$1	
January 1998	608.36
1997	583.67
1996	511.55
1995	499.15
1994	555.20
1993	283.16

Beginning 12 January 1994, the CFA franc was devalued to CFAF 100 per French franc from CFAF 50 at which it had been fixed since 1948.

Top Import Origins (25)

$1.4 billion (f.o.b., 1996)

Origins	%
Germany	NA
United States	NA
Japan	NA
United Kingdom	NA
Italy	NA

NA stands for not available.

Top Export Destinations (26)

$986 million (f.o.b., 1996) Data are for 1993.

Destinations	%
France	NA
United Kingdom	NA
China	NA
Germany	NA
Japan	NA

NA stands for not available.

Economic Aid (27)

Recipient: ODA, $439 million (1993).

Import Export Commodities (28)

Import Commodities	Export Commodities
Foods and beverages	Fish
Consumer goods	Ground nuts (peanuts)
Capital goods	Petroleum products
Petroleum products	Phosphates
	Cotton

Balance of Payments (23)

	1991	1992	1993	1994	1995
Exports of goods (f.o.b.)	848	861	737	819	993
Imports of goods (f.o.b.)	−1,114	−1,192	−1,087	−1,022	−1,243
Trade balance	−266	−331	−350	−203	−250
Services - debits	−798	−827	−744	−657	−790
Services - credits	534	578	497	476	600
Private transfers (net)	322	329	275	353	344
Government transfers (net)	9	33	43	35	38
Overall balance	−200	−218	−279	3	58

SERBIA

Srbija

CAPITAL: Belgrade.

FLAG: The flag is a tricolor of red (top), blue, and white (bottom).

MONETARY UNIT: The official currency is the Yogoslav New Dinar (YD) where 1 YD = 100 paras. Notes are in denominations of YD50, 20, 10, and 5. Coins are in denominations of 50, 20, 10, and 1 paras. As of November 1998, the official exchange rate was $1 = YD 9.90.

WEIGHTS AND MEASURES: The metric system is in use.

HOLIDAYS: New Year, 1–2 January; Statehood Day, 27 April; Labour Days, 1–2 May; Victory Day, 9 May; St. Vitus Day, 28 June; Public Holiday, 7 July; Republic Days, 29–30 November

TIME: 1 PM = noon GMT.

LOCATION AND SIZE: Serbia is a landlocked country located in the Balkans of Southeastern Europe, near the Adriatic Sea. Its total area is 88,412 square kilometers, which is slightly larger than Maine. It shares borders with Albania (114 kilometers), Bosnia–Herzegovina (312 kilometers), Bulgaria (318 kilometers), Croatia (241 kilometers), Hungary (151 kilometers), The Former Yugoslav Republic of Macedonia (221 kilometers), Montenegro (211 kilometers), and Romania (476 kilometers). The capital city, Belgrade, is located in the north–central part of the country.

CLIMATE: In the north, the climate is continental. Winters are cold and summers are hot and humid. Rainfall is well-distributed throughout the year. In the south, summers and autumns are hot and dry, while winters are relatively cold with heavy snowfall.

INTRODUCTORY SURVEY

RECENT HISTORY

With a population of over ten million augmented by another 700,000 Montenegrins, Serbia is a mountainous country whose capital, Belgrade, sits in a bend of the Danube River. Since the seventh century this region has been influenced by cultural differences and recurrent movements for independence to create a complex and fiercely nationalistic culture.

In the nineteenth century the center of Slavic revival was Serbia and one of its main demands was the release of Croatia and Slovenia from the Austro-Hungarian Empire. This movement for national self-determination culminated in the assassination of the Archduke Franz Ferdinand, heir to the Austro-Hungarian Empire, by a Serbian nationalist in 1914. The act precipitated World War I.

By the close of 1915 the Central Powers (Austria-Hungary, Germany, and the Ottoman Empire) had overrun the Serbian Army and the Serbian government withdrew to the Mediterranean island of Corfu. There, in July 1917 at the close of World War I, the representatives of the different South Slav nations issued a call for the creation of a joint Serbian state composed of Croats, Slovenes, and Serbs, with Macedonia and Montenegro merged into the new Serbia. Formed after the Central Powers' defeat, the name of the new state was the Kingdom of the Serbs, Croats, and Slovenes. The new head of state, King Alexander, turned out to be a harsh dictator who abolished the parliament, re-

drew internal boundaries with no regard to ethnicity, and changed the name of the new country to Serbia.

Revolutionary opposition from Croatian and Macedonian nationalists led to the assassination of the king, while the social crises caused by the world wide economic depression of the 1930s strengthened the hand of the non-democratic political forces of both the fascist and communist varieties. This tumultuous period ended when the armies of Germany, Italy, Hungary, and Bulgaria invaded Serbia in 1941 during World War II. In the battle hundreds of thousands of Serbs died, many in massacres carried out by the fascist occupying state of Croatia. Instead of accepting German occupation, as some western and central European politicians had done, a number of Yugoslav resistance movements engaged in campaigns of disruption, sabotage, and ambush against the fascist occupying powers. Gradually the different resistance groups came under the leadership of the Communist group led by Marshal Josip Tito.

At the close of World War II Serbia was already under effective control of Tito's Communist Party. The non-communist political parties boycotted the elections in November 1945 and the People's Front, led by the communists, won 90 percent of the vote. They took power, abolished the monarchy and declared the existence of the Federative People's Republic of Serbia. Serbia consisted of five nations—Slovenia, Croatia, Serbia, Macedonia, and Montenegro. Bosnia and Herzegovina served as a large buffer state on the West. Tito nationalized the economy and collectivized the farms.

In the period immediately after the end of World War II, American Cold War policymakers operated on the belief that, behind the ''Iron Curtain,'' eastern Europe had become a monolithic Communist bloc under the dictate of the U.S.S.R. In Serbia and its constituent republics, however, this description proved not to be the case. Although on questions of form Serbia might borrow from the Soviet Union (it drew its 1946 constitution, for instance, from the Soviet Union's 1936 constitution), but on issues of substance it set its own policies, reflecting the country's historic commitment to autonomy and independence in state-to-state relations. In 1948, the U.S.S.R. tried to pressure Tito to accept Soviet leadership in setting Cold War strategy against the capitalist West, but Tito refused to allow the Soviet Union to dictate Yugoslav foreign policy. As punishment, Serbia was expelled from the Cominform, the association of world communist parties. Responding to Stalin's threats, Tito declared that, if invaded, the Serbian partisans would take their guns back into the mountains to defend their homeland, just as they had during the German occupation. Adopting a non–aligned stance in foreign policy and taking money and development programs from whomever offered them (as long as there were no strings attached), Tito maintained Yugoslav autonomy. Tito died in 1980 and was replaced by a rotating presidency.

The harsh and bloody struggle to establish and maintain its independence transformed Serbia from an oppressed national minority into an oppressive leader of post-World War II Serbia. Ethnic Serbs dominated the armed forces and the leading offices of the central government. In 1986 the quasi-governmental Serbian Academy of Arts and Sciences officially adopted the goal of establishing in Serbia a ''Greater Serbia'' bringing all ethnic Serbs into a single state.

At the same time, however, the Soviet Union's collapse brought about major changes in the governance of Serbia. As its constituent republics grew openly restless under Serb domination and as the different political forces began to adapt to democratic elections, the issue of Serbian nationalism took on increased importance. The old Communist Party leadership learned how to stay in power by inciting Serbian nationalism. In 1987 Slobodan Milosevic, a master of this demagogic appeal to a greater Serbia, became the head of the Serbian Communist Party. In the early 1990s he served two terms as president of the constituent republic of Serbia. Prohibited by the Serbian constitution from serving a third term in that post, in July 1997 he was elected president of the federal Yugoslav state.

Serbia dissolved in the 1990s as its constituent states declared their independence. Slovenia and Croatia left the federation in 1991, though not without a war which leveled entire cities. Serbia and Montenegro proclaimed their own joint independence in April 1992 and made a failed attempt to aid the ethnic Serbs in Bosnia and Herzegovina.

GOVERNMENT

Although the United States refuses to recognize it, Serbia and Montenegro call themselves the ''Federal Republic of Serbia.'' [The official U.S. view is that the former Socialist Federal Republic of Yugoslavia (SFRY) has dissolved and has not been reconstituted. Most European states, however, have recognized the Federal Republic of Serbia.] The Federal Republic of Serbia is composed of the two republics, Serbia and Montenegro, plus two provinces-Kosovo and Vojvodna. In April 1992 Serbia and Montenegro adopted a new constitution based on a bi-cameral parliamentary form of government in which the two republics are each guaranteed a proportion of the seats.

Judiciary

The judicial system is composed of the Federal Court (the Savezni Sud) whose judges are elected by the Federal Assembly and serve nine-year terms, and the Constitutional Court whose judges are elected by the Federal Assembly and also serve nine-year terms.

Political Parties

The main political parties in Serbia and Montenegro are the Serbian Socialist Party (SPS—the continuation of the former Communist Party led by Slobodan Milosevic); the Serbian Radical Party (SRS); the Serbian Renewal Movement (SPO); the Democratic Party (DS); the Democratic Party of Serbia (DSS); the People's Party of Montenegro (NS); the Socialist People's Party of Montenegro (SNP); the Social Democratic Party of Montenegro (SDP); the Liberal Alliance of Montenegro; the Democratic Community of Vojvodina Hungarians (DZVM); the League of Social Democrats of Vojvodina (LSV); the Reformist Democratic Party of Vojvodina (RDSV); the Democratic Alliance of Vojvodina Croats (DSHV); the League of Communists–Movement for Serbia (SK-PJ); the Democratic Alliance of Kosovo (LDK); the Democratic League of Albanians (QOSJA); the Parliamentary Party of Kosovo (PPK); the Party of Democratic Action (JUL); the Civic Alliance of Serbia (GSS); the Yugoslav United Left (JUL); the New Democracy (ND); the Alliance of Vojvodina Hungarians (SVM); and the Serbian Democratic Movement (DEMOS, a coalition of parliamentary parties).

DEFENSE

Serbia and Montenegro have an army, a navy, and an air force. The age for military service is nineteen years-old. Serbia has 2.2 million men fit for military service; Montenegro has almost 150,000 men fit for military service. Serbia and Montenegro spend yearly $911 million on the military.

ECONOMIC AFFAIRS

The 1991 collapse of the Socialist Federal Republic of Yugoslavia plus the devastation from internecine warfare in Bosnia and in Kosovo have severely hampered trade and manufacture. During this upheaval the interdependent economic system of the old Yugoslav federation was thoroughly disrupted and the industrial sector suddenly lacked both suppliers and markets. The gross domestic product of Serbia and Montenegro fell by half between 1992 and 1993. Although United Nations sanctions were suspended in late 1995 the economy failed to revive.

Public Finance

In August 1999 the International Monetary Fund (IMF) approved $19 million in loans to the former Yugoslav Republic of Macedonia to help compensate for the shortfall in commercial activity and the social dislocation caused by the crush of refugees resulting from the Serbian occupation of Kosovo. Thus, the IMF quarantined Serbia while lending money to its former member republics.

Income

The gross domestic product of Serbia for 1999 stood at $25.4 billion, an increase of 3.5 percent from the previous year. Gross domestic product per capita was $2,300. The U.S. Central Intelligence Agency estimated the unemployment rate to be over 35 percent, while the consumer price measure of inflation stood at 48 percent.

Industry

Serbia and Montegnegro have an industrial base in the production on aircraft, trucks, automobiles, tanks and weapons, metallurgy and mining (steel, aluminum, copper, lead, zinc, chromium, antimony, bismuth, cadmium), consumer goods, electronics, agricultural machinery, petroleum products, chemicals, and pharmaceuticals. Industrial output, however, fell by 20 percent in 1991. In 1997 the industrial production growth rate was 8 percent.

Banking and Finance

Serbian external debt stands at $112.2 billion. The war in Kosovo precluded the Serbian government from negotiating with the main capitalist financial institutions—the International Monetary Fund and the World Bank (WB)—where free market reforms such as the privatization of government enterprises are routinely exchanged for capital to rebuild the economy. Instead, sanctions virtually ostracized Serbia from relations with international financial institutions. In 1998 the Western economies imposed an investment ban and a freeze in assets to punish Serbia for its repressive campaign in Kosovo.

Economic Development

As a result of these international sanctions the Serbian economy went into hibernation in the late 1990s. Western commentators blamed the Serbian political leadership for sacrificing economic reform for political control. But the real problem was the international quarantine of the Yugoslav economy. Serbia did what it could to improve the investment climate. By establishing a new currency in June 1993 Serbia succeeded in curbing the hyperinflation that plagued the economy during the 1991 dissolution of Yugoslav Federation. Prices remained fairly stable from 1995 to 1997, although inflationary pressures regained strength in 1998.

SOCIAL WELFARE

The social welfare system in Serbia, like every other feature of life in Serbia, has been hobbled by ethnic divisions, the disruptions of war, and the negative effect of the war on the economy. The unemployment rate in excess of 35 percent and the inflation rate which cuts 48 percent into consumer budgets mean that the level of social welfare in Serbia is under siege and may result in more political opposition to Milosevic.

Healthcare

Life expectancy at birth in Serbia is 73.45 years (71.03 for men and 76.05 for women); in Montenegro it is 76.32 years at birth (72.87 for men and 80.07 for women). The infant mortality rate is 16.49 deaths per thousand live births; in Montenegro it is 10.99 deaths per thousand live births. The fertility rate is 1.74 children born per woman in Serbia and 1.76 children born per woman in Montenegro.

Housing

Realiable information about housing in Serbia was unavailable.

EDUCATION

Reliable information about education in Serbia was unavailable.

1999 KEY EVENTS TIMELINE

January

- Forty-five ethnic Albanians are slain outside Racak, spurring international efforts for a peace settlement.
- Serb police kill 24 Kosovo Albanians in a raid on a suspected rebel hideout.
- Western allies demand that Kosovo Albanians and Serbs attend a peace conference or face NATO air strikes; both sides meet in early February without reaching an agreement.

February

- Serbian forces deploy along the Macedonian border, digging in across the frontier from thousands of NATO forces gathering for a possible peacekeeping mission.
- Serbian forces bombard KLA positions in the north, and rebels launch several attacks on army and police.

March

- Bombings kill seven and injure dozens of ethnic Albanians in the government-held towns of Kosovska Mitrovica and Podujevo. Both sides accuse the other of the attacks.
- Three days after peace talks resume in Paris, Kosovo Albanians unilaterally sign a peace deal calling for interim broad autonomy and for 28,000 NATO troops to implement it. Serb delegation refuses to sign accord, and the peace talks are suspended.
- International peace monitors leave Serbia, citing lack of security and possibility of NATO air strikes.
- As many as 40,000 Serbian forces in Kosovo launch an offensive against rebels in the north and in the central Drenica region, burning homes and sending thousands of people fleeing.
- U.S. President Bill Clinton says Serbian President Slobodan Milosevic has rejected diplomacy, and NATO bombings punish Serbia for refusing to make peace in Kosovo.
- NATO forces target downtown Belgrade, as NATO members agree to escalate the bombing campaign against Serbia.

April

- NATO bombs the headquarters of Serbia's state television, knocking the country's main source of news off the air.

May

- Serbian authorities hand over three captured U.S. soldiers to the Rev. Jesse Jackson, ending their 32 days in captivity.
- NATO warplanes drop bombs on Belgrade, mistakenly setting the Chinese embassy ablaze.
- In the first sign of growing resistance by Serbian citizens since NATO began bombing the country, hundreds of angry parents rally in two Serbian towns to demand the return of their sons from combat duty in Kosovo.
- The United Nations war-crimes tribunal announces it has indicted Milosevic and four top aides for atrocities in Kosovo; the United States calls on Milosevic to surrender to the tribunal in The Hague.

June

- Milosevic's government accepts an international peace plan after more than 2 months of NATO air strikes.
- NATO forces arrive in Kosovo, and peacekeeping troops enter only hours after a Russian armored column enters the province's capital; Russian officials call it a mistake.
- U.S. President Clinton mounts a campaign to remove Milosevic from power, offering a $5 million bounty for information leading to Milosevic's arrest

July

- In a direct challenge to Milosevic, Serbian opposition leader Zoran Djindjic returns from self-imposed exile, saying he prefers to risk imprisonment to the prospect of civil war.

August

- In a last-minute concession aimed at defusing Serbia's political crisis, Milosevic's party offers to hold early elections. The opposition says it is prepared to accept the offer only if Milosevic steps down and allows a transitional government before any elections.
- More than 150,000 protesters call for Milosevic's removal in one of the biggest anti-government demonstrations in the capital in several years.

September

- Nationwide anti-government rallies attract modest turnout; the low numbers are blamed on opposition infighting and divided leadership.

October

- In a major policy shift, Milosevic says he will allow pro-Western Montenegro to leave the Yugoslav federation without bloodshed if the republic decides to do so.
- Serbian police block an anti-government march of about 10,000 people in the 12th straight day of protests across Serbia.

November

- The Serbian parliament agrees to debate an opposition demand for early general elections.
- As U.S. President Bill Clinton tours south-eastern Europe, Serbia's state-controlled media scorns his stop in the Balkans as a ''return to the scene of the crime,'' a derogatory reference to where NATO dropped its bombs in the spring.
- Bishop Artemije, leader of Kosovo's Serbs, conducts what he describes as a very constructive meeting with U.S. President Bill Clinton, in which he detailed Kosovar Serb concerns: 400 people killed and 80 churches attacked since NATO took over in June.

December

- Russian Foreign Minister Igor Ivanov sharply criticizes peacekeeping efforts in Kosovo, saying international forces are ignoring atrocities against Serbs.
- Only a month after Yugoslav authorities reopen the railway from Belgrade to the Adriatic coast, passengers already face delays as, at the border of Serbia and Montenegro, the train is stopped so the engine can be changed and travelers are forced to wait while officials patrol the corridors.

ANALYSIS OF EVENTS: 1999

BUSINESS AND THE ECONOMY

War has devastated Serbia's economy. After 10 years of sanctions, and more than 70 days of NATO bombings, Serbian industry was essentially shut down. Power plants and refineries were knocked out of service by the bombings, and skilled workers left the republic by the thousands. Many of those who stayed joined the ranks of the unemployed, estimated at 60 percent in the Serbian republic.

While the rest of the Balkans received assistance from the World Bank and the International Monetary Fund, Serbia was not likely to get any reconstruction aid as long as Milosevic stayed in power. In the long run, that was more likely to hurt the Serbs, and the surrounding region, and do little to oust Milosevic.

Trade problems, loss of foreign investments, and Kosovo refugees could cost as much as $2 billion to the neighboring countries of Albania, Bosnia and Herzegovina, Bulgaria, Croatia, Macedonia and Romania during 1999 alone, according the IMF. By October, pressure built to end sanctions against Serbia. Some nations began to press the European Union for financial help to remove dropped bridges blocking the Danube River in Serbia, which was hurting trade in neighboring countries. United Nations Secretary General Kofi Annan said Serbia could face a humanitarian disaster without help for 700,000 refugees living in the republic. As the winter months approached, even Serb opposition leaders asked the United States to lift sanctions against Serbia. Fear of Serbians freezing to death led the EU to send heating oil to Serbian towns under opposition control.

In the meantime, Milosevic has spent millions of dollars on cosmetic repairs to convince Serbian voters that his government can rebuild the republic without foreign help. Economists said he was printing money to pay for those repairs, which could trigger hyperinflation.

GOVERNMENT AND POLITICS

Western nations began 1999 trying to bring Serbia and the Kosovo Liberation Army (KLA) back to the negotiating table. In January, NATO generals warned Serbian President Slobodan Milosevic to respect the October 1998 agreement to end the fighting. If he didn't halt the assaults on ethnic Albanians, Serbia would face possible military strikes. The Serbs continued to pound the KLA, which was seeking independence from Serbia.

By March, the Kosovo conflict had grown worse, with atrocities reported nearly every day. Among the worst incidents were the killing of 45 ethnic Albanians outside Racak, and the killing of 24 suspected KLA rebels.

Milosevic would not soften his position on Kosovo. By February, he deployed thousands of Serbian forces along the Macedonian border, and kept sending more troops into Kosovo. On the other side of the Macedonian border, thousands of NATO forces gathered for a possible peacekeeping mission.

Some political analysts believed Milosevic would not change his policy on Kosovo because he was confident that the United States and NATO would not attack Serbia. By early March, peace talks resumed in Paris as Kosovo spun into a deeper crisis. In Paris, Kosovo Albanians unilaterally signed a peace deal calling for interim broad autonomy and for 28,000 NATO troops to implement it. But the Serbs refused to sign the accord, and talks were suspended. Citing lack of security, international peace observers left Serbia.

In Kosovo, Yugoslav forces grew to as many as 40,000. The troops began a major offensive against KLA rebels in north and central Drenica, burning homes and forcing thousands of Kosovo Albanians to flee the area.

Finally on March 24, NATO began to bomb key Serbian military targets in Serbia and Kosovo. In a message to the American people, President Bill Clinton explained why the United States and NATO began a military offensive against Yugoslav forces.

''Now they've started moving from village to village, shelling civilians and torching their houses,'' Clinton said. ''We've seen innocent people taken from their homes, forced to kneel in the dirt and sprayed with bullets. Kosovar men dragged from their families, fathers and sons together, lined up, and shot in cold blood. This is not war in the traditional sense. It is an attack by tanks and artillery on a largely defenseless people, whose leaders already have agreed to peace. Ending this tragedy is a moral imperative. It is also important to America's national interests.''

As NATO bombs fell over Serbia, the United Nations war-crimes tribunal in The Hague indicted Milosevic and four top aides for atrocities in Kosovo. On June 3, Serbian lawmakers met in emergency session in Belgrade to approve a peace plan. That same day, Milosevic's government accepted international terms, which called for the removal of Serbian troops from Kosovo, and giving substantial autonomy to the province. On June 12, thousands of NATO troops entered Kosovo to enforce the peace agreement and ensure the safe return of 850,000 ethnic Albanians.

Despite 11 weeks of NATO bombings, Milosevic remained defiant, claiming victory even after capitulating to peace demands. He went on television to report military casualties, saying 462 soldiers and 114 police officers were killed during the bombings. NATO said as many as 5,000 died, with civilian casualties estimated at 2,000.

The United States continued to pressure Serbia's government, offering $5 million for information leading to Milosevic's arrest and conviction on war crimes charges. The bounty was part of a complex plan to force Milosevic from office.

According to some NATO officials, most Serbians will eventually leave Kosovo, allowing the ethnic Albanians to hold a referendum to break away from Serbia in three to five years. Montenegro could leave as well, but Western countries have tried to persuade Montenegro's pro-western president Milo Djukanovic to remain.

In Serbia, Milosevic remained in power despite widespread protests and a growing opposition. In August, he faced the stiffest challenge to his rule, when more than 100,000 Serbians called for his ousting during a march in Belgrade. Milosevic's party attempted to derail the protest by offering to hold early elections, but the opposition rejected his plan, and demanded his immediate resignation.

By late October, Milosevic's government tried to improve its chances for political survival by announcing that it would not stand in the way of Montenegro if its citizens voted to leave the Yugoslav federation. President Djukanovic said Montenegro was considering a referendum to seek its own course if Serbia would not agree to a major restructuring of the Yugoslav federation.

CULTURE AND SOCIETY

During 1999, the people of Serbia and Kosovo were the subject of numerous articles that attempted to dissect the effects of years of war and ethnic tensions that culminated with the tragedy of Kosovo.

The New Republic wrote about the complicity of the Serbs in ethnic cleansing. It was a mistake, the reporter wrote, to believe ethnic cleansing in Kosovo was simply a Milosevic creation that had no widespread support in Serbia. *Mother Jones* magazine offered a sympathetic portrayal of the thousands of young Serbs who left the country to avoid fighting for Milosevic. Many of these Serbs crossed the border into Hungary, where they were quickly sent to refugee camps. Draft dodgers face up to 20 years in prison in Serbia. *The Washington Post* wrote about soldiers haunted by the ''dirty war.'' In his article, reporter William Booth wrote about soldiers who believe they were misled into fighting a war that should never have occurred. These soldiers, Booth wrote, were ''angry, defensive, frustrated, defeated, proud, disillusioned, exhausted. They have returned home to rebuild their lives in a broken country. They go to funerals. They sit for hours in cafes. And they say it is strange, how they feel. Some have begun to call it the 'Kosovo Syndrome.' ''

DIRECTORY

CENTRAL GOVERNMENT

Serbia and Montenegro assert that together they compose a joint independent state, the Federal Republic of Yugoslavia. The United States does not formally recognize this entity as a state. Both Montenegro and Serbia have a significant amount of autonomy. Members of the governments of the Federal Republic of Yugoslavia and the Republic of Montenegro are noted here.

FEDERAL REPUBLIC OF YUGOSLAVIA

Head of State

President of the Federal Republic of Yugoslavia
Slobodan Milosevic, Office of the Federal President, Savezna Skupstina, 11000 Belgrade, Serbia, Yugoslavia

Prime Minister
Momir Bulatovic, Office of the Federal Prime Minister, Palace of Federation, Belgrade, Serbia, Yugoslavia
PHONE: +381 (11) 3117087

Ministers

Deputy Prime Minister
Vladan Kutlesic, Office of the Deputy Prime Minister

Deputy Prime Minister
Zoran Lilic, Office of the Deputy Prime Minister

Deputy Prime Minister
Nikola Sainovic, Office of the Deputy Prime Minister

Deputy Prime Minister
Danilo Vuksanovic, Office of the Deputy Prime Minister

Deputy Prime Minister
Jovan Zebic, Office of the Deputy Prime Minister

Minister of Agriculture
Nedeljko Sipovac, Ministry of Agriculture

Minister for Coordination of Relations with International Organizations
Nebojsa Maljkovic, Ministry for Coordination of Relations with International Organizations

Minister of Defense
Pavle Bulatovic, Ministry of Defense

Minister of Domestic Trade
Slobodan Nenadovic, Ministry of Domestic Trade

Minister for Economy and Industry
Rade Filipovic, Ministry for Economy and Industry

Minister of Finance
Bozidar Gazivoda, Ministry of Finance

Minister of Foreign Affairs
Zivadin Jovanovic, Ministry of Foreign Affairs

Minister of Foreign Trade
Borisa Vukovic, Ministry of Foreign Trade

Minister of Internal Affairs
Zoran Sokolovic, Ministry of Internal Affairs

Minister of Justice
Zoran Knezevic, Ministry of Justice

Minister of Labor, Health, and Social Policy
Miroslav Ivanisevic, Ministry of Labor, Health, and Social Policy

Minister of Science, Development, and Ecology
Jagos Zelenovic, Ministry of Science, Development, and Ecology

Minister of Sport
Velizar Djeric, Ministry of Sport

Minister of Telecommunication
Dojcilo Radojevic, Ministry of Telecommunication

Minister of Trade and Tourism
Djordje Siradovic, Ministry of Trade and Tourism

Minister of Transportation
Dejan Drobnjakovic, Ministry of Transportation

GOVERNMENT OF THE REPUBLIC OF SERBIA

Head of State

President of Serbia
Milan Milutinovic, Office of the President

Prime Minister
Mirko Marjanovic, Office of the Prime Minister

Ministers

Minister of Agriculture, Forestry and Water Management
Jovan Babovic, Ministry of Agriculture, Forestry and Water Management

Minister of Civil Engineering
Dejan Kovacevic, Ministry of Civil Engineering

Minister of Culture
Zeljko Simic, Ministry of Culture

Minister of Economic and Property Transformation
Jorgovanka Tabakovic, Ministry of Economic and Property Transformation

Minister of Education
Jovo Todorovic, Ministry of Education

Minister of Energy and Mining
Zivota Cosic, Ministry of Energy and Mining

Minister of Environment
Branislav Blazic, Ministry of Environment

Minister of Family Welfare
Miroslav Nedeljkovic, Ministry of Family Welfare

Minister of Finance
Borislav Milacic, Ministry of Finance

Minister of Health
Leposava Milicevic, Ministry of Health

Minister of Industry
Luka Mitrovic, Ministry of Industry

Minister of Information
Aleksandar Vucic, Ministry of Information

Minister of Internal Affairs
Vlajko Stojiljkovic, Ministry of Internal Affairs

Minister of Justice
Dragoljub Jankovic, Ministry of Justice

Minister of Labor, Veterans Affairs, and Social Welfare
Tomislav Milenkovic, Ministry of Labor, Veterans Affairs, and Social Welfare

Minister of Local Self-Administration
Gordana Pop-Lazie, Ministry of Local Self-Administration

Minister of Relations with Serbs Outside Serbia
Miroslav Mircic, Ministry of Relations with Serbs Outside Serbia

Minister of Religious Affairs
Milovan Radovanovic, Ministry of Religious Affairs

Minister of Science and Technology
Branislav Ivkovic, Ministry of Science and Technology

Minister of Sports and Youth
Zoran Andjelkovic, Ministry of Sports and Youth

Minister of Tourism
Slobodan Cerovic, Ministry of Tourism

Minister of Trade
Zoran Krasic, Ministry of Trade

Minister of Transport
Dragan Todorovic, Ministry of Transport

POLITICAL ORGANIZATIONS

Serbian Socialist Party (SPS)

NAME: Slobodan Milosevic

Serbian Radical Party (SRS)

NAME: Vojislav Seselj

Serbian Renewal Movement (SPO)

TITLE: President
NAME: Vuk Draskovic

Democratic Party (DS)

NAME: Zoran Djindjic

Democratic Alliance of Kosovo (LDK)

TITLE: President
NAME: Ibrahim Rugova

Democratic League of Albanians

NAME: Rexhep Qosja

DIPLOMATIC REPRESENTATION

Embassies in Serbia

Australia
13 Cika Ljubina, YU-11000 Belgrade, Serbia
PHONE: +381 (11) 624655
FAX: +381 (11) 624029

Italy
Bircaninova Ulica 11, YU-11000 Belgrade, Serbia
PHONE: +381 (11) 659722
FAX: +381 (11) 688116

United Kingdom
Generala Zdanova 46, YU-11000 Belgrade, Serbia
PHONE: +381 (11) 645055
FAX: +381 (11) 659651

JUDICIAL SYSTEM

Savezni Sud (Federal Court)

Constitutional Court

FURTHER READING

Articles

''Albania and Serbia Need Major Repairs to Infrastructure.'' *Newsweek International*, (June 14, 1999): 30.

''Analysts Say Lust for Power Makes Milosevic Tick.'' *The Seattle Times*, April 11, 1999.

''Downtown Belgrade Targeted as NATO Escalates Bombing.'' *The Seattle Times*, March 31, 1999.

''The Ground War Scenario.'' *The Economist* (May 29, 1999): 46.

''If You Rebuild It . . . A New Serbia.'' *The New Republic* (May 17, 1999): 19.

''Kosovo Resurgent.'' *The Economist* (September 25, 1999): 57.

''Kosovo Serb Leaders Form National Assembly.'' Reuters, October 25, 1999.

''Milosevic Ignores His Scars and Stubbornly Presses On.'' *The New York Times*, October 30, 1999.

''Milosevic Tells Serbs: We Won.'' *The Seattle Times*, June 11, 1999.

''Milosevic's Willing Executioners.'' *The New Republic*, (May 10, 1999): 26.

''Montenegro and Serbia Discuss Ties.'' Reuters, October 24, 1999.

''NATO Raid Knocks Serb TV Off the Air.'' *The Seattle Times*, April 23, 1999.

''On the Edge of the Knife.'' *The Economist* (March 20, 1999): 54.

''Peace Agreement Leaves Winners and Losers.'' *The Seattle Times*, June 10, 1999.

''Serb Opposition Asks U.S. Envoy to End Sanctions.'' Reuters, October 25, 1999.

''Serbia and Montenegro: Montenegro's Push for Autonomy May Hurt Republic.'' *The Economist* (July 17, 1999): 44.

''Serbia's Lost Generation.'' *Mother Jones* (September 1999): 48.

''Serbia's Opposition.'' *The Economist* (October 9, 1999): 61.

''The Serbian Kickback: With the Bombing Over, Strong Anti-Milosevic Sentiment is

Producing a Political Groundswell.'' *Time International*, (July 19, 1999): 20.

Text of President Clinton's Speech on Decision to Bomb, The Associated Press, March 25, 1999.

''Vengeance of a Victim Race.'' *Newsweek* (April 12, 1999): 42.

''A Widening Conflict.'' *The Economist* (April 10, 1999): 21.

''Serbia Releases Three U.S. Soldiers to Jackson.'' *The Seattle Times*, May 2, 1999.

''In Serbia, War Tales Cloud Opening Day.'' *The Seattle Times*, September 8, 1999.

SERBIA: STATISTICAL DATA

For sources and notes see "Sources of Statistics" in the front of each volume. The following information is for Serbia and Montenegro combined unless otherwise noted.

GEOGRAPHY

Geography (1)

Area:

Total: 102,350 sq km (Serbia 88,412 sq km; Montenegro 13,938 sq km).

Land: 102,136 sq km (Serbia 88,412 sq km; Montenegro 13,724 sq km).

Water: 214 sq km (Serbia 0 sq km; Montenegro 214 sq km).

Area—comparative: slightly smaller than Kentucky (Serbia is slightly larger than Maine; Montenegro is slightly smaller than Connecticut).

Land boundaries:

Total: 2,246 km.

Border countries: Albania 287 km (114 km with Serbia, 173 km with Montenegro), Bosnia and Herzegovina 527 km (312 km with Serbia, 215 km with Montenegro), Bulgaria 318 km (with Serbia), Croatia (north) 241 km (with Serbia), Croatia (south) 25 km (with Montenegro), Hungary 151 km (with Serbia), The Former Yugoslav Republic of Macedonia 221 km (with Serbia), Romania 476 km (with Serbia).

Note: the internal boundary between Montenegro and Serbia is 211 km.

Coastline: 199 km (Montenegro 199 km, Serbia 0 km).

Climate: in the north, continental climate (cold winter and hot, humid summers with well distributed rainfall); central portion, continental and Mediterranean climate; to the south, Adriatic climate along the coast, hot, dry summers and autumns and relatively cold winters with heavy snowfall inland.

Terrain: extremely varied; to the north, rich fertile plains; to the east, limestone ranges and basins; to the southeast, ancient mountains and hills; to the southwest, extremely high shoreline with no islands off the coast.

Natural resources: oil, gas, coal, antimony, copper, lead, zinc, nickel, gold, pyrite, chrome.

Land use:

Arable land: NA%

Permanent crops: NA%

Permanent pastures: NA%

Forests and woodland: NA%

Other: NA%

HUMAN FACTORS

Ethnic Division (6)

Serbs	63%
Albanians	14%
Montenegrins	6%
Hungarians	4%
Other	13%

Religions (7)

Orthodox	65%
Muslim	19%
Roman Catholic	4%
Protestant	1%
Other	11%

Languages (8)

Serbo-Croatian 95%, Albanian 5%.

GOVERNMENT & LAW

Political Parties (12)

Chamber of Citizens	No. of seats
Serbian Socialist Party (SPS) Yugoslav United Left (JUL), New Democracy (ND)	64
Zajedno	22
Democratic Party of Socialists of Montenegro (DPSCG)	20
Serbian Radical Party (SRS)	16
People's Party of Montenegro (NS)	8
Alliance of Vojvodina Hungarians (SVM)	3
Other	5

Government Budget (13B)

Revenues ....NA
Expenditures ....NA
Capital expenditures ....NA

NA stands for not available.

Military Affairs (14B)

	1992	1993	1994	1995
Military expenditures				
Current dollars (mil.)	NA	NA	2,900[P,E]	NA
1995 constant dollars (mil.)	NA	NA	2,943[P,E]	NA
Armed forces (000)	137[e]	100	130	130
Gross national product (GNP)				
Current dollars (mil.)	13,500	NA	NA	20,600
1995 constant dollars (mil.)	14,520	NA	NA	20,600
Central government expenditures (CGE)				
1995 constant dollars (mil.)	NA	NA	NA	NA
People (mil.)	10.5	10.5	10.5	10.6
Military expenditure as % of GNP	NA	NA	NA	NA
Military expenditure as % of CGE	NA	NA	NA	NA
Military expenditure per capita (1995 $)	NA	NA	282	NA
Armed forces per 1,000 people (soldiers)	13.1	9.5	12.3	12.3
GNP per capita (1995 $)	1,389	NA	NA	1,948
Arms imports[6]				
Current dollars (mil.)	0	0	0	0
1995 constant dollars (mil.)	0	0	0	0
Arms exports[6]				
Current dollars (mil.)	0	0	0	0
1995 constant dollars (mil.)	0	0	0	0
Total imports[7]				
Current dollars (mil.)	NA	NA	NA	NA
1995 constant dollars (mil.)	NA	NA	NA	NA
Total exports[7]				
Current dollars (mil.)	NA	NA	NA	NA
1995 constant dollars (mil.)	NA	NA	NA	NA
Arms as percent of total imports[8]	NA	NA	NA	NA
Arms as percent of total exports[8]	NA	NA	NA	NA

HUMAN FACTORS

Demographics (2A)

	1990	1995	1998	2000	2010	2020	2030	2040	2050
Population	NA	10,437.1	10,526.1	10,529.5	10,711.0	10,668.6	10,385.6	9,889.1	9,194.8
Net migration rate (per 1,000 population)	NA	NA	NA	NA	NA	NA	NA	NA	NA
Births	NA	NA	NA	NA	NA	NA	NA	NA	NA
Deaths	NA	NA	NA	NA	NA	NA	NA	NA	NA
Life expectancy - males	NA	70.0	70.8	71.3	73.5	75.4	76.9	78.1	79.0
Life expectancy - females	NA	74.9	75.8	76.3	78.9	81.0	82.6	83.9	84.9
Birth rate (per 1,000)	NA	13.0	12.6	12.5	11.0	9.1	8.1	7.1	6.3
Death rate (per 1,000)	NA	9.7	9.7	9.7	10.5	10.9	11.7	13.3	14.7
Women of reproductive age (15-49 yrs.)	NA	2,582.4	2,630.4	2,617.3	2,549.7	2,458.0	2,243.6	1,958.1	1,718.1
of which are currently married	NA	NA	NA	NA	NA	NA	NA	NA	NA
Fertility rate	NA	1.8	1.8	1.7	1.6	1.5	1.4	1.4	1.3

Except as noted, values for vital statistics are in thousands; life expectancy is in years.

LABOR FORCE

Labor Force (16)

Total (million)	2.178
Industry	41%
Services	35%
Trade and tourism	12%
Transportation and communication	7%
Agriculture	5%

Data for 1994.

Unemployment Rate (17)

More than 35% (1995 est.)

PRODUCTION SECTOR

Electric Energy (18)

Capacity	11.779 million kW (1995)
Production	33.4 billion kWh (1995)
Consumption per capita	3,009 kWh (1995)

Transportation (19)

Highways:

total: 49,525 km

paved: 28,873 km

unpaved: 20,652 km (1996 est.)

Waterways: NA km

Pipelines: crude oil 415 km; petroleum products 130 km; natural gas 2,110 km

Merchant marine:

total: 20 ships (1,000 GRT or over) totaling 322,391 GRT/533,935 DWT (owned by Montenegro) ships by type: bulk 6, cargo 11, container 3 note: Montenegrin ships operate under the flag of Malta (1997 est.)

Airports: 48 (Serbia 43, Montenegro 5) (1997 est.)

Airports—with paved runways:

total: 18

over 3,047 m: 2 (Serbia 2, Montenegro 0)

2,438 to 3,047 m: 5 (Serbia 3, Montenegro 2)

1,524 to 2,437 m: 5 (Serbia 4, Montenegro 1)

914 to 1,523 m: 2 (Serbia 2, Montenegro 0)

under 914 m: 4 (Serbia 4, Montenegro 0) (1997 est.)

Airports—with unpaved runways:

total: 30

1,524 to 2,437 m: 2 (Serbia 2, Montenegro 0)

914 to 1,523 m: 14 (Serbia 13, Montenegro 1)

under 914 m: 14 (Serbia 13, Montenego 1) (1997 est.)

Top Agricultural Products (20)

Cereals, fruits, vegetables, tobacco, olives; cattle, sheep, goats.

FINANCE, ECONOMICS, & TRADE

Economic Indicators (22)

National product: GDP—purchasing power parity—$24.3 billion (1997 est.)

National product real growth rate: 7% (1997 est.)

National product per capita: $2,280 (1997 est.)

Inflation rate—consumer price index: 7% (1997)

Exchange Rates (24)

Exchange rates:

Yugoslav New Dinars (YD) per US $1	
Official rate:	
December 1997	5.85
September 1996	5.02
early 1995	1.5
Black market rate:	
December 1997	8.9
early 1995	2 to 3

Top Import Origins (25)

$6.2 billion (1996 est.)

Origins	%
Germany	NA
Italy	NA
Russia	NA

NA stands for not available.

Top Export Destinations (26)

$2.8 billion (1996 est.).

Destinations	%
Russia	NA
Italy	NA
Germany	NA

NA stands for not available.

Economic Aid (27)

Recipient: ODA, $NA. NA stands for not available.

Import Export Commodities (28)

Import Commodities	Export Commodities
Machinery and transport equipment	Manufactured goods
Fuels and lubricants	Food and live animals
Manufactured goods	Raw materials
Chemicals	
Food and live animals	
Raw materials	

SEYCHELLES

Republic of Seychelles

CAPITAL: Victoria.

FLAG: The flag is made up of five oblique bands of (left to right) blue, yellow, red, white, and green.

ANTHEM: Begins "Seychellois both staunch and true."

MONETARY UNIT: The Seychelles rupee (R) is a paper currency of 100 cents. There are coins of 5, 10, and 25 cents and 1, 5, 10, 20, 25, 50, 100, 1,000, and 1,500 rupees and notes of 10, 25, 50, and 100 rupees. R1 = $0.20056 (or $1 = R4.986).

WEIGHTS AND MEASURES: The metric system is the legal standard.

HOLIDAYS: New Year's, 1–2 January; Labor Day, 1 May; National Day, 5 June; Independence Day, 29 June; Assumption, 15 August; All Saints' Day, 1 November; Immaculate Conception, 8 December; Christmas, 25 December. Movable religious holidays include Good Friday, Easter Monday, Corpus Christi, and Ascension.

TIME: 4 PM = noon GMT.

LOCATION AND SIZE: Seychelles, an archipelago in the Indian Ocean, consists of an estimated 115 islands, most of which are not permanently inhabited. The second-smallest country in Africa, Seychelles has an area of 455 square kilometers (176 square miles), slightly more than 2.5 times the size of Washington, D.C. The capital city of Seychelles, Victoria, is located on the island of Mahé.

CLIMATE: Coastal temperatures are fairly constant at about 27°C (81°F) throughout the year. At higher altitudes, temperatures are lower, especially at night. Average annual rainfall is 236 centimeters (93 inches) at sea level, as much as 356 centimeters (140 inches) in the mountains, and much lower on the southwestern coral islands, averaging about 50 centimeters (20 inches) a year on Aldabra.

INTRODUCTORY SURVEY

RECENT HISTORY

Seychelles achieved independence at 12:05 AM on 29 June 1976. Richard Marie Mancham, leader of the conservative Seychelles Democratic Party, became president on independence, heading a coalition government that included Seychelles People's United Party (SPUP) leader France Albert René as prime minister. Mancham was overthrown by a coup on 5 June 1977 and went into exile, and René became president. He suspended the constitution, dismissed the legislature, and ruled by decree.

The constitution of March 1979, adopted by referendum, established a one-party state. René was reelected president without opposition in June 1984. Since then, Seychelles has made progress economically and socially. Under rising pressure to democratize, in December 1991, René agreed to reform the system.

Multiparty elections were held in July 1992, and many dissidents, including Mancham, returned from exile. Finally, in June 1993, 73% of the voters approved a new constitution providing for multiparty government.

GOVERNMENT

In June 1993, 74% of the voters approved a new constitution drafted by a bipartisan commission. It called for multiparty elections of a president and a National Assembly of thirty-three members.

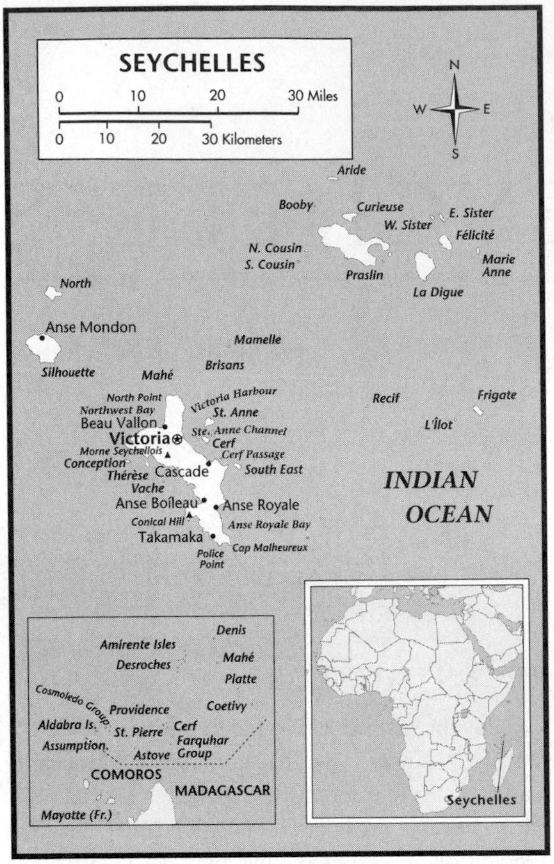

Judiciary

Cases are first tried in Magistrates' courts. The Supreme Court hears appeals and takes original jurisdiction of some cases, and the Court of Appeal hears appeals from the Supreme Court. Appointment to the post of chief justice is made by the president of Seychelles.

Political Parties

The Seychelles People's Progressive Front (SPPF), was established in 1979 as the sole legal party, with the avowed objective of creating a socialist state. The Seychelles Democratic Party (SDP) was declared to have ''disappeared,'' and at least three opposition groups were in exile.

After President René's 1991 announcement of a return to multiparty democracy, many dissidents returned from exile and the Democratic Party (DP) was reestablished, as well as the Seychelles Party (PS), the Seychelles Democratic Movement (MSPD), and the Seychelles Liberal Party (SLP).

DEFENSE

The Seychelles People's Liberation Army (SPLA) was merged with a People's Militia in 1981 to form the Seychelles People's Defense Force (SPDF) of 1,300 men. In 1995, the SPLA spent $14 million on defense.

ECONOMIC AFFAIRS

With the opening of an international airport in 1971, the Seychelles economy began to move away from cash crops to the development of tourism. In 1996, tourism accounted for over 20% of the gross domestic product (GDP), employed 30% of the labor force, and provided over 70% of foreign exchange earnings. Seychelles depends heavily on imports and financial aid.

Public Finance

Annual budgets of increasing deficits were common in the 1980s. The public sector is responsible for two-thirds of Seychelles' employment, and the budget amounts to about 50% of GDP. Public investment focuses on social and physical infrastructure, tourism, and export activities. The U.S. CIA estimates that, in 1993, government revenues totaled approximately $227.4 million and expenditures $263 million, including capital expenditures of $54 million. External debt totaled $181 million.

Income

In 1998, Seychelles' gross national product (GNP) was $507 million, or about $6,450 per person. For the period 1985–95, the average inflation rate was 3.3%, resulting in a real growth rate in per person GNP of 4.2%.

Industry

The largest plant is the tuna cannery, opened in 1987 and privatized in 1995 with a 60% purchase by Heinz Inc. of the United States. The rest are small and process local agricultural products, including tea, copra, and vanilla. There is a plastics factory, a brewery and soft drink bottler, and a cinnamon distiller.

Banking and Finance

The Seychelles Monetary Authority, established in 1978 as the bank of issue, became the Central Bank of Seychelles in 1983.

Money supply, as measured by M2, totaled R1,210.1 million at the end of 1995. There is no stock exchange in Seychelles.

Economic Development

The 1985–89 plan sought to create jobs and emphasized developing cash crops, tourism, and the fishing industry. The 1990–94 plan emphasized the need to attract foreign investment. Of considerable interest to donors in the 1990s is the ten-year plan to improve the Seychelles environment.

SOCIAL WELFARE

The National Provident Fund makes payments for marriage, emigration, disability, survivors, and old age. There is also a workers' compensation scheme. Health services are free for all residents.

Healthcare

There are five hospitals, and, in 1990, there were ninety doctors. The average life expectancy is 71 years. In 1994, 3.9% of the gross national product (GNP) went to health expenditures.

Housing

Most homes are of wood or stone with corrugated iron roofs; many rural houses are thatched.

EDUCATION

Public education is free and compulsory for children between the ages of 6 and 15. In 1994, there were 9,911 students in primary schools and 9,280 students in secondary schools. Seychelles does not provide education at university level, but many students study abroad, especially in the United Kingdom. Adult literacy was estimated at 79%.

1999 KEY EVENTS TIMELINE

January

- German businessman Claus Mittermeyer criticizes British Prime Minister Tony Blair while Blair vacations in Seychelles over the New Year holiday. Mittermeyer is deported and his personal property confiscated without payment.
- The Eden Project begins its work to preserve rare botanical species throughout the world, including Seychelles.
- The government announces efforts to increase the attractiveness of Seychelles as an offshore banking center through taxation treaties with numerous countries.

February

- President France Albert Rene's communications advisor travels abroad to promote the benefits of a new $100 per capita tax associated with the Seychelles ''Goldcard'' campaign, designed to attract tourists. The program is later dropped.

March

- The U.S. State Department releases a report for 1998 stating that rising ocean temperature in the Indian Ocean is detrimental to the survival of coral reefs.
- The Organization for Economic Cooperation and Development (OECD) reported that Official Development Assistance (ODA) to Seychelles for 1997 was only $8 million, the lowest of all nations receiving such aid in sub-Saharan Africa.
- Seychelles external debt reaches a critical point, with arrears of payments exceeding annual exports.
- The government reports that the number of foreigners who obtained Seychelles nationality diminished after the country suspended the Economic Citizenship Program, which sold citizenship for $25,000.

April

- The Chamber of Commerce complains about government regulations and management of the country's economy.
- Vice President James Michel visits Mauritius and India to discuss cooperation in the Indian Ocean.
- The government proposes two measures to curb black market hard currency transactions—allowing tourists to prepay hotel and other service bills and freezing up to 20% of foreign exchange earnings.
- Bertrand Rassool, an economist and former government minister, is named ambassador to the United Kingdom.

July

- The Ministry of Environment meets with major industries and sponsors a workshop on governmental policies and regulations on environmental assessment and measurement standards.

August

- The Second Games of the Youth and Sport Commission of the Indian Ocean are held in Mahe, Seychelles. Over seven hundred of the participating youths are under age eighteen.

- Seychelles delegates attend a Southern Africa Development Community (SADC) meeting held in Mozambique.

September

- Alexander Resources International, Ltd., a satellite technology and communications firm, announces an agreement to create Tele-Trade Centers throughout the region of southern Africa, including Seychelles, to facilitate satellite-based global communications.

October

- The U.S. Department of Agriculture extends $83 million in export credit guarantees to the Southern African Development Community (SADC) and certifies Seychelles' eligibility.
- Seychelles becomes the first nation to ratify the International Labor Organization's (ILO) treaty on international standards for child labor.
- Seychelles hosts a major health conference, with health ministers from east, central, and southern Africa in attendance.
- Air Seychelles signs a credit guarantee to purchase Boeing 767 aircraft with International Leasing Finance Corporation.

ANALYSIS OF EVENTS: 1999

BUSINESS AND THE ECONOMY

The Seychellois economy remained in flux during 1999, with some major initiatives in tourism counterbalanced by macroeconomic problems and policy errors. After the failure of the Seychelles ''Goldcard'' campaign, the Chamber of Commerce voiced opposition about government regulations and overall management of the economy. It was announced in March that Seychelles' external debt had reached an unsustainable level, with payment arrears equal to one year's exports. The weakening of the national currency, the rupee, resulted in the growth of a black market among business operators and foreign exchange speculators. The government responded in April by proposing measures to halt the currency black market and end the expatriation of hard currency by calling for a partial freeze on foreign exchange earnings.

Despite these negative events a number of positive achievements occurred in business and the economy. In January the government announced that it was going to actively pursue the development of Seychelles as an offshore banking center. Alexander Resources International, Ltd., an international satellite communication firm, announced in September its plans to create a region-wide global telecommunications and video conferencing network for southern Africa.

In a move intended to boost the tourism business Air Seychelles signed a contract for the delivery of new Boeing 767-300 aircraft, a sign that the financially troubled airline might have turned an important corner. In addition economic diversification made a positive advance when the Seychelles Marketing Board announced the purchase of a second-hand aircraft for its prawn industry operating in outlying islands. Meanwhile Vice President James Michel assessed the upcoming reopening of The Beau Vallon promenade, which had been closed for ten years due to de-vegetation and related environmental concerns. The government participated in a contingent of fourteen African nations in September. The group signed an agreement with various U.S. agencies for cooperation in the improvement of transportation infrastructure and the expansion of commercial relations and tourism promotion.

GOVERNMENT AND POLITICS

The government continued to send mixed signals to the business sector during 1999. It made efforts to manage Seychelles' increasing external debt problems, stagnant tourism, and business dissatisfaction with macroeconomic policies. In early 1999 the government unveiled a tourism campaign called the Seychelles ''Goldcard'' program, intended to attract tourists to Seychelles and increase the islands' stagnant tourism industry. The program was criticized by the Seychelles Chamber of Commerce for its taxation of incoming tourists and was abandoned.

Despite these problems the government achieved new milestones in conservation, public housing, and in promoting Seychelles as the place for international meetings. The Baba Housing Estate and the Grand Anse Praslin housing village, inaugurated in 1999, reflected what Dolor Ernesta, the Minister for Land Use and Habitat, called the government's continued progress toward increasing homeownership for the islands' less fortunate families. The government also announced the need for the creation of a National Cancer Center, with Health Minister Jacquelin Dugasse pointing out a 66% increase in cancer during the 1990s.

Other positive developments included the ratification of the ILO Convention on Child Labor in October and the hosting of a meeting of the health ministers from east, central, and southern Africa to discuss pressing health issues for the next millenium. In late October the country also won praise from Ms. Kuki Daniel of the United Nations International Drug Control Program for the country's comprehensive anti-drug strategy, which emphasized a balanced approach to controlling supply and demand through interdiction and rehabilitation.

CULTURE AND SOCIETY

Seychelles hosted a variety of cultural and social events in 1999. In August it was host to the 2nd Games of the Youth and Sports Commission of the Indian Ocean, bringing together over seven hundred young athletes from Indian Ocean nations. In addition to athletic events a cultural component was held for six days, reflecting the participating nations of Comoros, Madagascar, Mauritius, Reunion, and Seychelles.

Agriculture and Marine Resources Minister Ronny Jumeau presented the Minister's trophy for excellence in hospitality to Kathleen Fonseka, the proprietor of Marie Antoinette Restaurant, during the U-First Campaign award ceremonies. Next year's theme, ''A National Commitment to Seychellois Hospitality,'' was also unveiled.

In September a poppy ball featuring local and international celebrities provided an evening of dance, dinner, and entertainment. Featured guests included the world famous magician Paul Daniels and his wife and TV personality Debbie Daniels.

DIRECTORY

CENTRAL GOVERNMENT

Head of State

President
France Albert Rene, Office of the President

Vice President
James Michel, Office of the Vice President

Ministers

Ministry of Agriculture and Marine Resource
Ronny Jumeau, Ministry of Agriculture and Marine Resources

Ministry of Education
Danny Faure, Ministry of Education

Ministry of Health
Jacqueline Dugasse, Ministry of Health

Ministry of Local Government and Sports
Sylvette Pool, Ministry of Local Government and Sports

Ministry of Youth and Culture
Patrick Pillay, Ministry of Youth and Culture

POLITICAL ORGANIZATIONS

Seychelles People's Progressive Front

NAME: France Albert Rene

Democratic Party

United Opposition

NAME: Wavel Ramkalawan

DIPLOMATIC REPRESENTATION

Embassies in Seychelles

United States
PO Box 231, Victoria House, Victoria, Mahe Island, Seychelles
PHONE: +248 25256
FAX: +248 25189

United Kingdom
PO Box 161, 3rd Floor, Victoria House, Victoria, Mahe Island, Seychelles
PHONE: +248 225225
FAX: +248 225127

JUDICIAL SYSTEM

Supreme Court

Court of Appeal

FURTHER READING

Articles

''Alexander to Implement Communications Network in 21 African Countries.'' *Business Wire*, August 4, 1999.

''USDA Extends Export Credits to Africa Nations.'' Reuters, October 5, 1999.

Books

Vine, Peter. *Seychelles.* 2nd ed. London: Immel Publishing, 1992.

Internet

Islands of the Indian Ocean. Available Online @ http://www.runisland.com/index1.html (November 12, 1999).

SEYCHELLES: STATISTICAL DATA

For sources and notes see "Sources of Statistics" in the front of each volume.

GEOGRAPHY

Geography (1)

Area:

Total: 455 sq km.

Land: 455 sq km.

Water: 0 sq km.

Area—comparative: 2.5 times the size of Washington, DC.

Land boundaries: 0 km.

Coastline: 491 km.

Climate: tropical marine; humid; cooler season during southeast monsoon (late May to September); warmer season during northwest monsoon (March to May).

Terrain: Mahe Group is granitic, narrow coastal strip, rocky, hilly; others are coral, flat, elevated reefs.

Natural resources: fish, copra, cinnamon trees.

Land use:

Arable land: 2%

Permanent crops: 13%

Permanent pastures: NA%

Forests and woodland: 11%

Other: 74% (1993 est.)

HUMAN FACTORS

Demographics (2A)

	1990	1995	1998	2000	2010	2020	2030	2040	2050
Population	73.4	77.0	78.6	79.7	84.3	88.8	92.8	95.2	95.4
Net migration rate (per 1,000 population)	NA	NA	NA	NA	NA	NA	NA	NA	NA
Births	NA	NA	NA	NA	NA	NA	NA	NA	NA
Deaths	NA	NA	NA	NA	NA	NA	NA	NA	NA
Life expectancy - males	64.0	64.7	66.1	67.1	70.8	73.4	75.4	76.9	78.0
Life expectancy - females	74.8	75.8	75.5	75.3	77.1	79.6	81.5	82.9	84.0
Birth rate (per 1,000)	22.0	20.6	19.7	19.0	16.0	14.1	12.3	11.2	10.7
Death rate (per 1,000)	7.4	6.8	6.6	6.5	5.8	5.7	6.6	8.8	11.3
Women of reproductive age (15-49 yrs.)	19.2	21.4	22.5	23.2	25.2	23.8	22.5	21.2	20.1
of which are currently married	NA	NA	NA	NA	NA	NA	NA	NA	NA
Fertility rate	2.3	2.0	2.0	2.0	1.8	1.8	1.8	1.8	1.8

Except as noted, values for vital statistics are in thousands; life expectancy is in years.

Infants and Malnutrition (5)

Under-5 mortality rate (1997)	18
% of infants with low birthweight (1990-97)	10
Births attended by skilled health staff % of total[a]	NA
% fully immunized (1995-97)	
TB	100
DPT	98
Polio	98
Measles	100
Prevalence of child malnutrition under age 5 (1992-97)[b]	NA

Ethnic Division (6)

Seychellois (mixture of Asians, Africans, Europeans).

Religions (7)

Roman Catholic	90%
Anglican	8%
Other	2%

Languages (8)

English (official), French (official), Creole.

EDUCATION

Educational Attainment (10)

Age group (1987)	25+
Total population	30,912
Highest level attained (%)	
No schooling	12.1
First level	
Not completed	44.9
Completed	NA
Entered second level	
S-1	35.7
S-2	NA
Postsecondary	4.6

Literacy Rates (11B)

Adult literacy rate	
1980	
Male	-
Female	-
1995	
Male	83%
Female	86%

GOVERNMENT & LAW

Political Parties (12)

National Assembly	No. of seats
Elected	
Seychelles People's Progressive Front (SPPF)	24
Democratic Party (DP)	1
Awarded	
Seychelles People's Progressive Front (SPPF)	6
Democratic Party (DP)	1
United Opposition (UO)	3

The 10 awarded seats are apportioned according to the share of each party in the total vote.

Government Budget (13A)

Year: 1995

Total Expenditures: 1,276.5 Millions of Rupees

Expenditures as a percentage of the total by function:

General public services and public order	15.69
Defense	4.32
Education	12.20
Health	8.01
Social Security and Welfare	14.41
Housing and community amenities	.31
Recreational, cultural, and religious affairs	.96
Fuel and energy	-
Agriculture, forestry, fishing, and hunting	1.99
Mining, manufacturing, and construction	.38
Transportation and communication	4.60
Other economic affairs and services	4.70

Military Affairs (14A)

Availability of manpower	
Males age 15-49 (1998 est.)	22,107
Fit for military service	
Males (1998 est.)	11,111
Total expenditures (1995)	$13.7 million
Expenditures as % of GDP	NA%

NA stands for not available.

LABOR FORCE

Labor Force (16)

Total	26,000
Industry	19%
Services	57%

Government	14%
Fishing, agriculture, and forestry	10%

Data for 1996. Percent distribution for 1989.

Unemployment Rate (17)

Rate not available.

PRODUCTION SECTOR

Electric Energy (18)

Capacity	28,000 kW (1995)
Production	125 million kWh (1995)
Consumption per capita	1,719 kWh (1995)

Transportation (19)

Highways:
total: 280 km
paved: 176 km
unpaved: 104 km (1996 est.)
Merchant marine: none
Airports: 14 (1997 est.)
Airports—with paved runways:
total: 8
2,438 to 3,047 m: 1
914 to 1,523 m: 5
under 914 m: 2 (1997 est.)
Airports—with unpaved runways:
total: 6
914 to 1,523 m: 2
under 914 m: 4 (1997 est.)

Top Agricultural Products (20)

Coconuts, cinnamon, vanilla, sweet potatoes, cassava (tapioca), bananas; broiler chickens; tuna fishing (expansion under way).

FINANCE, ECONOMICS, & TRADE

Economic Indicators (22)

National product: GDP—purchasing power parity—$550 million (1997 est.)

National product real growth rate: NA%

National product per capita: $7,000 (1997 est.)

Inflation rate—consumer price index: -0.3% (1995 est.)

Exchange Rates (24)

Exchange rates:

Seychelles rupees (SRe) per US$1

January 1998	5.1901
1997	5.0263
1996	4.9700
1995	4.7620
1994	5.0559
1993	5.1815

Top Import Origins (25)

$238 million (c.i.f., 1995) Data are for 1993.

Origins	%
China	NA
Singapore	NA
South Africa	NA
United Kingdom	NA

NA stands for not available.

MANUFACTURING SECTOR

GDP & Manufacturing Summary (21)

	1980	1985	1990	1992	1993	1994
Gross Domestic Product						
Millions of 1990 dollars	256	275	369	376	390	379
Growth rate in percent	−2.55	10.29	7.56	1.95	3.90	−3.00
Per capita (in 1990 dollars)	4,060.7	4,237.8	5,265.5	5,292.6	5,422.5	5,259.9
Manufacturing Value Added						
Millions of 1990 dollars	19	20	34	35	39	*41*
Growth rate in percent	18.21	8.44	14.91	2.39	12.37	*5.60*
Manufacturing share in percent of current prices	8.0	10.6	9.9	10.4	11.7 NA	

Top Export Destinations (26)

$56.1 million (f.o.b., 1995) Data are for 1993.

Destinations	%
France	NA
United Kingdom	NA
China	NA
Germany	NA
Japan	NA

NA stands for not available.

Economic Aid (27)

Recipient: ODA, $NA. NA stands for not available.

Import Export Commodities (28)

Import Commodities	Export Commodities
Manufactured goods	Fish
Food	Cinnamon bark
Petroleum products	Copra
Tobacco	Petroleum products (re-exports)
Beverages	
Machinery and transportation equipment	

Balance of Payments (23)

	1992	1993	1994	1995	1996
Exports of goods (f.o.b.)	48	51	52	53	92
Imports of goods (f.o.b.)	−181	−216	−189	−214	−251
Trade balance	−132	−165	−137	−161	−159
Services - debits	−91	−109	−99	−125	−131
Services - credits	199	218	202	224	254
Private transfers (net)	20	22	15	13	22
Government transfers (net)	−3	−6	−7	−5	−6
Overall balance	−7	−39	−26	−54	−20

SIERRA LEONE

CAPITAL: Freetown.

FLAG: The national flag is a tricolor of green, white, and blue horizontal stripes.

ANTHEM: Begins ''High we exalt thee, realm of the free, Great is the love we have for thee.''

MONETARY UNIT: The leone (Le) is a paper currency of 100 cents. There are coins of 1/2, 1, 5, 10, 20, and 50 cents, and notes of 1, 2, 5, 10, 20, 50, 100, and 500 leones. Le1 = $0.00108 (or $1 = Le928.488).

WEIGHTS AND MEASURES: The metric system is employed.

HOLIDAYS: New Year's Day, 1 January; Independence Day, 27 April; Bank Holiday, August; Christmas, 24–25 December; Boxing Day, 26 December. Movable religious holidays include Good Friday, Easter Monday, Whitmonday, 'Id al-Fitr, 'Id al-'Adha', and Milad an-Nabi.

TIME: GMT.

LOCATION AND SIZE: Situated on the west coast of Africa, Sierra Leone has an area of 71,740 square kilometers (27,699 square miles), slightly smaller than the state of South Carolina. It has a total boundary length of 1,364 kilometers (847 miles). In addition to the mainland proper, Sierra Leone also includes the offshore Banana Turtle islands and Sherbro Island, and other small islands. Sierra Leone's capital city, Freetown, is located on the Atlantic Coast.

CLIMATE: Temperatures and humidity are high, and rainfall is heavy. The average temperature is about 27°C (81°F) on the coast and almost as high on the eastern plateau. There are distinct wet and dry seasons, with rainfall averaging more than 315 centimeters (125 inches) a year for the country as a whole.

Republic of Sierra Leone

INTRODUCTORY SURVEY

RECENT HISTORY

In 1958, Milton Margai became Sierra Leone's first prime minister; in 1960, he led a delegation to London, England, to establish conditions for full independence. Sierra Leone became an independent country within the British Commonwealth of Nations on 27 April 1961. After the 1967 national elections, there were two successive military coups, and a state of emergency was declared in 1970. In 1971, a new constitution was adopted, and the country was declared a republic on 19 April 1971. Siaka Stevens, then prime minister, became the nation's first president. An alleged plot to overthrow Stevens failed in 1974, and in March 1976, he was elected without opposition for a second five-year term as president. In 1978, a new constitution was adopted, making the country a one-party state.

Stevens did not run for reelection as president in 1985, yielding power to his handpicked successor, Major General Joseph Saidu Momoh, the armed forces commander. By 29 April 1992, Momoh was overthrown in a military coup and fled to Guinea. A National Provisional Ruling Council (NPRC) was created but, shortly afterward, the head of the five-member junta, Lieutenant Colonel Yahya, was arrested by his colleagues and replaced by Captain Valentine Strasser, who was formally designated head of state.

The Strasser government soon limited the status of the 1991 constitution by a series of decrees and public notices. The NPRC (National Provi-

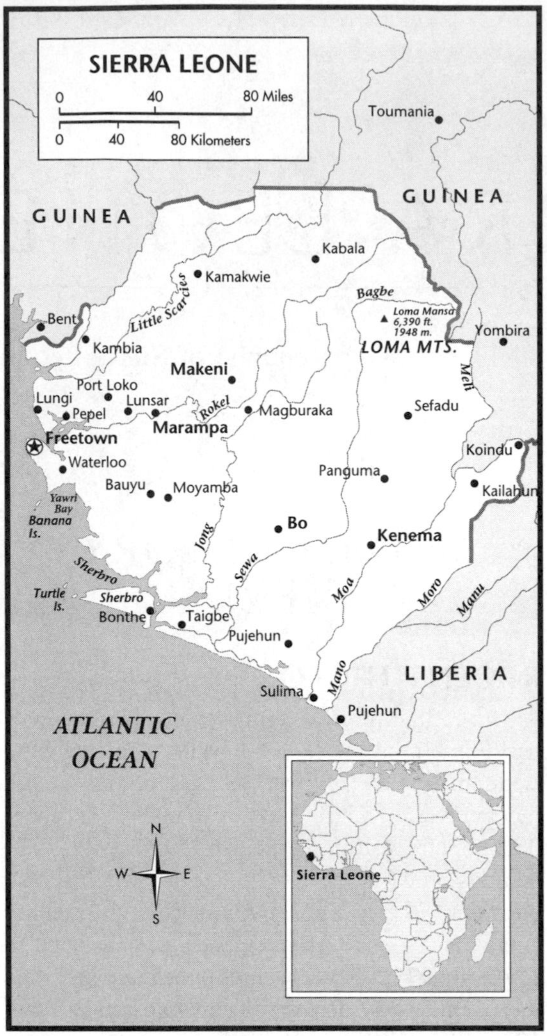

sional Ruling Council) dissolved parliament and political parties and ruled by decree. There was fighting in the southeast, where the forces of the National Patriot Front of Liberia and Sierra Leone dissidents were fighting with Sierra Leone armed forces. Forces from the Economic Community of West African States (ECOWAS) Monitoring Group sought to create a cease-fire zone along the boundary between the two countries. In November 1993, Strasser announced a unilateral cease-fire and an amnesty for rebels. In November 1993, Strasser issued a timetable for a transition to democracy to culminate in general elections in late 1995. A month later, the NPRC released a ''Working Document on the Constitution'' to serve as the basis for public debates leading to a constitutional referendum in May 1995.

Strasser was overthrown in 1996 by his deputy Brigadier Julius Maada Brio and given safe passage out of the country. Presidential and parliamentary elections took place in February 1996 and were met with violent opposition by rebel forces resulting in 27 deaths. Ahmad Tejan Kabbah won the presidency in a run-off on 15 March 1996. In May 1997, however, Major Johnny Paul Koromah led a coup that overthrew Kabbah. Civil war broke out and armed gangs fought. In February 1998, Nigeria led a force of peacekeeping troops into Sierra Leone that ousted Koromah's ruling military council and restored President Kabbah to power. After the fighting was over, Freetown was heavily damaged and still had to deal with looters, vigilante gangs, disease, and food shortages. Approximately 250,000 people fled the country.

The Armed Forces Revolutionary Council and the Revolutionary United Front, however, started a campaign of terror against civilians called ''Operation No Living Thing.'' The rebels have killed civilians and looted and destroyed villages. Victims have their hands or feet amputated as ''messages'' to the Kabbah government.

In October 1998, the restored government executed some two dozen people who had been convicted for taking part in the 1997 coup.

GOVERNMENT

A new constitution came into force on 1 October 1991, but since that time it has been superseded by a number of coups led by the military. In May 1997, a military coup, led by Major Johnny Paul Koromah, overthrew the most recently elected government. However, in February 1998, Nigerian troops entered the country to restore the elected government.

Judiciary

Magistrates hold court in the various districts and in Freetown, administering the English-based code of law. Appeals from magistrates' courts are heard by the High Court, which also has unlimited original civil and criminal jurisdiction. Appeals from High Court decisions may be made to the Court of Appeal and finally to the Supreme Court, consisting of a chief justice and not fewer than three other justices.

Political Parties

A multiparty presidential election was held in February 1996, and Ahmad Tejan Kabbah of the National Peoples Party won in a run-off. There were fifteen parties registered for the 1996 elections. In 1997, the elected government was overthrown by Major Johnny Paul Koromah, a member of the newly formed Armed Forces Revolutionary Council, but in February 1998, he fled the country after Nigerian troops entered the country to restore the elected government.

DEFENSE

In 1993, the Sierra Leone armed services had 14,200 members, in addition to an 800-member state security force. Military service is voluntary.

ECONOMIC AFFAIRS

Although Sierra Leone is a potentially rich country with diverse resources, the economy has been severely depressed over the past two decades. Currently, agriculture employs 70% of the labor force.

Public Finance

The U.S. CIA estimates that, in 1995, government revenues totaled approximately $75 million and expenditures $128 million. External debt totaled $1.4 billion, approximately 96% of which was financed abroad.

Income

In 1998, Sierra Leone's gross national product (GNP) was $680 million at current prices, or about $140 per person. For the period 1985–95, the average annual inflation rate was 61.5%, resulting in a decline in per person GNP of 3.4%.

Industry

The Wellington Industrial Estate, covering 46 hectares (113 acres) just east of Freetown, was developed in the 1960s by the government to encourage investments. Its factories produce a variety of products, including cement, nails, shoes, oxygen, cigarettes, beer and soft drinks, paint, and knitted goods.

Banking and Finance

The Bank of Sierra Leone, established in 1963, is the central bank and bank of issue. The Banking Act of 1964 provides for the regulation of commercial banks by the central bank, including the control of money supply. Poor revenue collection, failure to control expenditure and heavy debt servicing requirements as a result of past borrowing characterized government finances in the 1980s and early 1990s. Money supply, as measured by M2, totaled Le31,091 million at the end of 1991. Total foreign reserves at the end of 1991 were $9.6 million, excluding gold.

There is no securities exchange in Sierra Leone.

Economic Development

The Sierra Leone government, in addition to stabilizing its balance-of-payment and budgetary deficits and meeting its debt obligations, seeks investors in its mining sector. In 1997, a parallel economy, lawless conditions, a crumbling infrastructure, and a military coup continued to constrain economic growth.

SOCIAL WELFARE

Since 1946, the Welfare Department has sponsored child welfare and domestic affairs programs, promoted youth groups, and set up programs for the care of the aged, the blind, and the mentally handicapped. In 1955, these services were reorganized into the Ministry of Social Welfare. The constitution guarantees equal rights for females, and women have held prominent positions in government. However, discrimination and violence against women are frequent.

Healthcare

In 1992, only 38% of the population had access to health care services. That year, there were 7 doctors per 100,000 people. In 1990, there was one hospital bed per 1,000 inhabitants.

With technical assistance from World Health Organization (WHO) and United Nations Children's Fund (UNICEF), a disease-control unit reduced the incidence of sleeping sickness and yaws and began a leprosy-control campaign. Malaria, tuberculosis, and schistosomiasis remain serious health hazards, however, as is malnutrition, with the calorie supply meeting only 83% of minimum requirements in 1992. Life expectancy is only 37 years, one of the lowest in the world. UNICEF estimated that Sierra Leone had the world's highest mortality rate in 1994.

Housing

Many of the older two-story wooden houses in Freetown are being replaced by structures built largely of concrete blocks, with corrugated iron or cement-asbestos roofs. Village houses in the provinces are traditionally made of sticks, with mud walls and thatch or grass roofs, and they may be circular or rectangular in shape.

EDUCATION

In 1990, Sierra Leone's 1,795 primary schools had 10,850 teachers and a total enrollment of

367,426 pupils, and secondary schools had 102,474 pupils and 5,969 teachers. In 1995, the adult illiteracy rate was estimated to be 69% (males, 54.6%; females, 81.8%).

In 1990, all higher level institutions were reported to have 4,742 pupils and 600 teaching personnel.

1999 KEY EVENTS TIMELINE

January

- Rebels attack President Ahmed Tejan Kabbah's government forces in and around the capital of Freetown. Thousands of people are killed or injured and over 150,000 are dislocated; the city is destroyed.
- Diplomatic and foreign aid personnel are evacuated to Conakry, Guinea, where the United Nations (UN) holds weekly meetings to monitor the situation.
- The West African States Monitoring Observer Group (ECOMOG), led by Nigerian forces, battles the rebels and retakes Freetown.

February

- Charges made by a U.N. human rights commission accuse ECOMOG forces of torturing, raping, and executing rebel suspects. The report also cites amputations, rapes, and decapitations committed by the rebels. A Sierra Leonean representative to the U.N. characterizes the ECOMOG allegations in the report as inaccurate and unfair.

March

- A British warship fires upon a local passenger boat, killing all 137 passengers aboard.
- President Kabbah bows to pressure to grant temporary amnesty to Corporal Foday Sankoh, leader of the Revolutionary United Front (RUF), who is accused of treason. The conciliatory gesture is a step toward peace, but is also meant to obtain the release of hundreds of abducted children held hostage by the rebels.
- The Sierra Leone People's Democratic League (PDL) calls for a sovereign national conference to decide the direction of the country.

April

- Rebels gain control of two-thirds of the country. Although most evacuated international relief workers return to Freetown, government forces and peacekeeping units control only the capital and the surrounding area.
- Nigeria accuses neighbors Liberia and Burkina Faso of shipping arms and ammunition to the rebels, and threatens air strikes unless the shipments are stopped.
- ECOMOG launches a major military offensive in the north of the country.

May

- President Ahmad Tejan Kabbah and detained rebel leader Foday Sankoh sign a ceasefire agreement in Lomé, Togo, aimed at obtaining conditions leading to a peace accord.

June

- Civilians in the northern provincial headquarter town of Kambia report that AFRC/RUF rebels have burned and looted Kambia and several towns and villages including Masungbala, Madina, Kukuna, Sentai, Kamba, Rokupr, Upper Dixing, Rosinor and Royama.
- In a sixty page document, Human Rights Watch reports that rebels intentionally and systematically murdered, mutilated, and raped civilians during the January offensive.
- Foday Sankoh accepts the government's offer of four ministerial positions and three deputies, bringing the rebel presence in government to seven. Donors gathered at U.N. headquarters in New York to pledge aid for rebuilding Sierra Leone.

July

- The Sierra Leonean government and the RUF sign a peace accord in Lomé, Togo.
- Gambia's airline, Air Dabia, is banned in Sierra Leone after the airline fails to convey President Kabbah and his entourage for the signing of the Togo peace accord.
- Ninety-eight death row prisoners, including senior former government officials accused of treason, are freed in line with a presidential amnesty.

August

- Fighting continues between the RUF and pro-government troops in Lunsar.

- An escalation of violent crime, including armed robbery, leads ECOMOG and government officials to introduce mobile patrols.
- About thirty foreigners, including Nigerian and British soldiers, U.N. monitors, aid workers, and journalists, are abducted.
- A high-level U.S. delegation visits Sierra Leone to determine where and how much foreign recovery and development assistance should be allocated over the coming months.

September

- Sierra Leonean refugees in Guinea-Bissau find themselves victimized by the aftermath of Bissau's civil strife, caused by a military revolt in May.
- Foday Sankoh begins his return from exile via Ouagadougou, Burkina Faso.

October

- Foday Sankoh and former military junta leader Johnny Paul Koroma return to Sierra Leone on the same flight from Liberia.
- Sankoh criticizes President Kabbah for not offering former rebels the ministries of justice, finance, or foreign affairs in the new unity government.
- The Commonwealth Ministerial Action Group (CMAG) calls on the international community to help Sierra Leone rebuild after the devastating civil war. USAID plans to develop a strategy for sustainable activities in the country.

December

- Sam Bockarie, a rebel leader known as Mosquite, is reported to have left the country. The rebel leader has been illegally mining diamonds, and fled the country to escape conflict with other rebels.
- The United Nations Security Council is considering a reommendation to increase the number of peacekeeping forces from 6,000 to 10,000 to compensate for the withdrawal of Nigerian troops planned for February.
- The International Federation of Journalists declares Sierra Leone the most dangerous country for journalists. Ten of the 19 journalists who were killed in Africa in 1999 died in Sierra Leone.

ANALYSIS OF EVENTS: 1999

BUSINESS AND THE ECONOMY

Sierra Leone's economy is moribund and will require fundamental restructuring and millions of dollars in foreign assistance to become productive again. Most investors, however, are staying away until they can be convinced the civil war that has ravaged the country is truly over. This presents a problem because a revival of the economy is a prerequisite to maintaining peace and moving Sierra Leone from the ranks of the world's most impoverished nations. Parts of the country remain under rebel control, and citizens face uncertain economic circumstances for the foreseeable future.

The best hope for the economy lies in a revival of the mineral sector. Diamonds continued to be exported during the war, and a great potential exists for the mining of rutile (titanium oxide), gold, and platinum. In June Sierra Leone shipped 7,000 tons of rutile to the United States, the first cargo since the mine was closed by rebels in early 1995.

Agriculture must also be reinvigorated. The government's policy of subsidizing food supply to the mining sector while exploiting farmers has not helped the country's agricultural industry. International assistance will be a major factor if Sierra Leone's economy is to turn itself around.

GOVERNMENT AND POLITICS

The year 1999 began with a bloody struggle between Nigerian-led ECOMOG troops and rebel fighters of the Armed Forces Revolutionary Council (AFRC) of Johnny Paul Koroma, and the Revolutionary United Front (RUF) of Foday Sankoh. Sadly, the principal victims of the conflict have been civilians caught between the two sides.

The January offensive marked the beginning of the end of the civil war in Sierra Leone. The war began in 1991, a spillover from the war in Liberia. Fighting was fueled by competition to control the country's lucrative diamond trade and other mineral wealth. ECOMOG and private security companies hired professional mercenaries and intervened with some success, but they did not drive out the rebels. The United Nations Office of the High Commission for Refugees (UNHCR) estimates that between 1991 and 1996, 15,000 people were killed, 450,000 made homeless, and one million people were displaced as a result of the fighting. Thousands of civilians were raped, tortured, and mutilated. Estimates place the total lives lost since 1991 between 20,000 to 50,000.

In January 1999 the rebel offensive convinced the Kabbah government it needed to negotiate a settlement. During the January offensive the civilian predicament became abundantly clear. Civilians in Freetown were told by the rebels either to join the rebel side or be burnt in their homes. The government warned civilians not to stay in their homes on penalty of death.

A peace deal was reached in July. It gave full amnesty to the rebels and allowed for power-sharing in a new government. Based on the accords rebel leaders will be allowed four cabinet posts and three deputy ministerial positions. But dissension has already broken out over whether the posts are "senior" enough to satisfy the rebels. Kabbah maintains that classification of the posts is his prerogative. In the meantime a new rebel group has emerged, the Sierra Leone Popular Army (SLPA), which has mounted cross-border attacks into Guinea. Concern exists that this group, loosely connected to the RUF, may provide Guinean dissidents with a platform from which to stage attacks on Guinea's government.

CULTURE AND SOCIETY

One of the worst crimes of the war has been the forcible induction of children as soldiers. Children were conscripted with promises of wealth or they were made participants in secret societies. In many cases the children were removed from their families and cut off from values by having to kill parents or local authorities. This practice effectively created a generation of war-scarred children, trained as brutal killers.

Stability in Sierra Leonean requires a balance among elite forces as well as redress of ethnic-regional divides within the national society. Tension also exists between the elite, rural land-owning families and the impoverished descendants of the farm labor class, released from domestic slavery in 1927. Many of these people, especially youth without career prospects, were attracted to the RUF. It remains to be seen whether Foday Sankoh, serving as a member of government, can bring about improvement in their material conditions.

If the peace in Sierra Leone holds, people will leave their refugee shelters in Guinea and Liberia and return to their homeland. Homes and villages will have to be rebuilt, and humanitarian organizations have found that in some isolated areas severe malnutrition has reached 30% of children under the age of five, a catastrophic situation. Water sources have been contaminated and fields have gone uncultivated. In northern and eastern provinces, 75% of farmers were unable to plant this year.

The health of Sierra Leone's population is a critical social issue. Non-governmental organizations (NGOs) reported outbreaks of malaria, including chloroquine-resistant strains. Measles was listed as one of the leading causes of death this past year. The second highest health risk was cholera. Cholera cases increased in the Freetown peninsula from 633 to 863 by the end of September. Diarrhea and dysentery were also reported as major problems in some areas. With food and medicine in short supply a return to normalcy will be slow and difficult.

DIRECTORY

CENTRAL GOVERNMENT

Head of State

President and Minister of Defense
Alhaji Ahmad Tejan Kabbah, Office of the President, State House, Independence Avenue, Freetown, Sierra Leone
FAX: +232 (22) 231404

Vice President
Albert Joe Demby, Office of the Vice President, State House, Independence Avenue, Freetown, Sierra Leone

Ministers

Minister of Finance
James O.D. Jonah, Ministry of Finance, Development and Economic Planning, NDB Building, 3rd Floor, Siaka Stevens Street, Freetown, Sierra Leone

Minister of Agriculture, Forestry and Marine
Okere Adams, Ministry of Agriculture, Forestry and Marine Resources

Minister of Tourism and Culture
A.B.S. Jomo-Jalloh, Ministry of Tourism and Culture

Minister of Transport and Communications
Momah Pujeh, Ministry of Transport and Communications

Minister of Political and Parliamentary Affairs
Abu A. Koroma, Ministry of Political and Parliamentary Affairs

Minister of Labor and Industrial Relations
Alpha Timbo, Ministry of Labor and Industrial Relations

Minister of Rural Development and Local Government
J.B. Dauda, Ministry of Rural Development and Local Government

Minister of Justice and Attorney-General
Solomon Berewa, Ministry of Justice and Attorney-General

Minister of Lands, Housing, Country Planning and the Environment
Peter Vandy, Ministry of Lands, Housing, Country Planning and the Environment

Minister of Health and Sanitation
Ibrahim I. Tejan-Jalloh, Ministry of Health and Sanitation

Minister of Social Welfare, Gender and Child Affairs
Shirley Gbujama, Ministry of Social Welfare, Gender and Child Affairs

Minister of Development and Economic Planning
Kadie Sesay, Ministry of Development and Economic Planning

Minister of Foreign Affairs and International Cooperation
Sama S. Banya, Ministry of Foreign Affairs and International Cooperation

Minister of Trade and Industry
Mike Lamin, Ministry of Trade and Industry

Minister of Energy and Power
Alimamy Pallo Bangura, Ministry of Energy and Power

Minister of Education, Youth and Sport
Alpha T. Wurie, Ministry of Education, Youth and Sport

Minister of Mineral Resources
Alhaji Mohamed Swarry Deen, Ministry of Mineral Resources

Minister of Information and Broadcasting
Julius Spencer, Ministry of Information and Broadcasting

POLITICAL ORGANIZATIONS

National Republican Party (NRP)

TITLE: Leader
NAME: Sahr Stephen Mambu

Coalition for Progress Party (CPP)

TITLE: Leader
NAME: Geredine Williams-Sarho

National Democratic Alliance (NDA)

TITLE: Leader
NAME: Amadu M. B. Jalloh

People's Progressive Party

TITLE: Chairman
NAME: Abass Chernok Bundu

People's National Convention (PNC)

TITLE: Chairman
NAME: Edward John Kargbo

Social Democratic Party (SDP)

TITLE: Leader
NAME: Andrew Victor Lungay

National Unity Movement

TITLE: Leader
NAME: John Desmond Fashole Luke

National Alliance for Democracy Party (NADP)

TITLE: Leader
NAME: Mohamed Yahya Sillah

National People's Party

TITLE: Leader
NAME: Andrew Turay

Democratic Center Party (DCP)

TITLE: Chairman
NAME: Abu Koroma

National Union Party (NUP)

TITLE: Chairman
NAME: John Karimu

All People's Congress

TITLE: Chairman
NAME: Edward Mohammed Turay

People's Democratic Party (PDP)

TITLE: Chairman
NAME: Thaimu Bangura

United National Peoples Party (UNPP)

TITLE: Chairman
NAME: John Karifa-Smart

Sierra Leone People's Party (SLPP)

TITLE: Chairman
NAME: Alhaji Ahmad Tejan Kabbah

DIPLOMATIC REPRESENTATION

Embassies in Sierra Leone

United Kingdom
Spur Road, Freetown, Sierra Leone
TITLE: High Commissioner
NAME: Peter A. Penfold

JUDICIAL SYSTEM

Supreme Court

FURTHER READING

Books

Alie, Joe A.D. *A New History of Sierra Leone.* New York: St. Martin's Press, 1990.

Davies, Clarice et al., eds. *Women of Sierra Leone: Traditional Voices.* Freetown: Partners in Adult Education, Women's Commission, 1992.

Internet

Africaonline. Available Online @ http://www.africaonline.com (November 1, 1999).

Africa News Online. Available Online @ http://www.africanews.org/west/stories/1999_feat1.html (November 1, 1999).

Integrated Regional Information Network (IRIN). Available Online @ http://www.reliefweb.int/IRIN (November 1, 1999).

SIERRA LEONE: STATISTICAL DATA

For sources and notes see "Sources of Statistics" in the front of each volume.

GEOGRAPHY

Geography (1)

Area:

Total: 71,740 sq km.

Land: 71,620 sq km.

Water: 120 sq km.

Area—comparative: slightly smaller than South Carolina.

Land boundaries:

Total: 958 km.

Border countries: Guinea 652 km, Liberia 306 km.

Coastline: 402 km.

Climate: tropical; hot, humid; summer rainy season (May to December); winter dry season (December to April).

Terrain: coastal belt of mangrove swamps, wooded hill country, upland plateau, mountains in east.

Natural resources: diamonds, titanium ore, bauxite, iron ore, gold, chromite.

Land use:

Arable land: 7%

Permanent crops: 1%

Permanent pastures: 31%

Forests and woodland: 28%

Other: 33% (1993 est.)

HUMAN FACTORS

Demographics (2A)

	1990	1995	1998	2000	2010	2020	2030	2040	2050
Population	4,283.2	4,589.3	5,080.0	5,509.3	7,379.6	9,689.6	12,406.2	15,344.8	18,369.5
Net migration rate (per 1,000 population)	NA	NA	NA	NA	NA	NA	NA	NA	NA
Births	NA	NA	NA	NA	NA	NA	NA	NA	NA
Deaths	NA	NA	NA	NA	NA	NA	NA	NA	NA
Life expectancy - males	41.7	44.1	45.6	46.6	51.9	57.4	62.5	67.0	70.7
Life expectancy - females	47.0	49.9	51.7	52.9	59.0	64.9	70.1	74.6	78.0
Birth rate (per 1,000)	47.9	47.6	46.2	45.1	39.9	35.2	29.6	24.9	21.3
Death rate (per 1,000)	20.9	18.7	17.3	16.3	12.1	8.9	6.6	5.3	5.0
Women of reproductive age (15-49 yrs.)	1,007.7	1,056.3	1,155.8	1,248.7	1,752.3	2,411.1	3,218.1	4,146.5	4,992.7
of which are currently married	NA	NA	NA	NA	NA	NA	NA	NA	NA
Fertility rate	6.5	6.4	6.2	6.1	5.3	4.3	3.5	2.9	2.5

Except as noted, values for vital statistics are in thousands; life expectancy is in years.

Health Indicators (4)

Life expectancy at birth

1980	35
1997	37
Daily per capita supply of calories (1996)	2,002
Total fertility rate births per woman (1997)	6.1
Maternal mortality ratio per 100,000 live births (1990-97)	....
Safe water % of population with access (1995)	34
Sanitation % of population with access (1995)	....
Consumption of iodized salt % of households (1992-98)[a]	75
Smoking prevalence	
Male % of adults (1985-95)[a]	....
Female % of adults (1985-95)[a]	....
Tuberculosis incidence per 100,000 people (1997)	315
Adult HIV prevalence % of population ages 15-49 (1997)	3.17

Infants and Malnutrition (5)

Under-5 mortality rate (1997)	316
% of infants with low birthweight (1990-97)	11
Births attended by skilled health staff % of total[a]	25
% fully immunized (1995-97)	
TB	38
DPT	26
Polio	28
Measles	26
Prevalence of child malnutrition under age 5 (1992-97)[b]	NA

Ethnic Division (6)

20 native African tribes	90%
Temne	30%
Mende	30%
Other	30%
Creole	10%

Creole are descendents of freed Jamaican slaves who were settled in the Freetown area in the late-eighteenth century.

Religions (7)

Muslim	60%
Indigenous beliefs	30%
Christian	10%

Languages (8)

English (official, regular use limited to literate minority), Mende (principal vernacular in the south), Temne (principal vernacular in the north), Krio (English-based Creole, spoken by the descendents of freed Jamaican slaves who were settled in the Freetown area, a lingua franca and a first language for 10% of the population but understood by 95%).

EDUCATION

Public Education Expenditures (9)

Public expenditure on education (% of GNP)

1980	3.5
1996	....
Expenditure per student	
Primary % of GNP per capita	
1980	....
1996	....
Secondary % of GNP per capita	
1980	....
1996	....
Tertiary % of GNP per capita	
1980	....
1996	....
Expenditure on teaching materials	
Primary % of total for level (1996)	....
Secondary % of total for level (1996)	....
Primary pupil-teacher ratio per teacher (1996)	....
Duration of primary education years (1995)	7

Educational Attainment (10)

Age group (1985)	5+
Total population	1,315,897
Highest level attained (%)	
No schooling	64.5
First level	
Not completed	18.7
Completed	1.8
Entered second level	
S-1	9.7
S-2	3.8
Postsecondary	1.5

Literacy Rates (11A)

In thousands and percent[1]	1990	1995	2000	2010
Illiterate population (15+ yrs.)	1,649	1,727	1,805	1,960

In thousands and percent[1]	1990	1995	2000	2010
Literacy rate - total adult pop. (%)	26.9	31.4	36.2	46.6
Literacy rate - males (%)	40.1	45.4	50.6	60.8
Literacy rate - females (%)	14.4	18.2	22.6	33.1

GOVERNMENT & LAW

Political Parties (12)

House of Representatives	No. of seats
Sierra Leone Peoples Party (SLPP)	27
United National Peoples Party (UNPP)	17
Peoples Democratic Party (PDP)	12
All Peoples Congress (APC)	5
National Unity Party (NUP)	4
Democratic Center Party (DCP)	3

First elections since the former House of Representatives was shut down by the military coup of 29 April 1992.

Government Budget (13B)

Revenues	$96 million
Expenditures	$150 million
Capital expenditures	NA

Data for 1996 est. NA stands for not available.

Military Affairs (14B)

	1990	1991	1992	1993	1994	1995
Military expenditures						
Current dollars (mil.)	6	14[e]	15[e]	14	44[e]	41[e]
1995 constant dollars (mil.)	6	16[e]	16[e]	15	45[e]	41[e]
Armed forces (000)	5	5	6	6	13	14[e]
Gross national product (GNP)						
Current dollars (mil.)	520	535	596	618	652	671
1995 constant dollars (mil.)	597	592	641	648	669	671
Central government expenditures (CGE)						
1995 constant dollars (mil.)	66	141	143	148	163	142
People (mil.)	4.3	4.4	4.3	4.3	4.4	4.6
Military expenditure as % of GNP	1.1	2.7	2.5	2.3	6.8	6.1
Military expenditure as % of CGE	9.8	11.3	11.4	10.2	27.7	28.9
Military expenditure per capita (1995 $)	2	4	4	3	10	9
Armed forces per 1,000 people (soldiers)	1.2	1.1	1.4	1.4	2.9	3.1
GNP per capita (1995 $)	139	134	148	150	151	145
Arms imports[6]						
Current dollars (mil.)	0	0	10	0	0	0
1995 constant dollars (mil.)	0	0	11	0	0	0
Arms exports[6]						
Current dollars (mil.)	0	0	0	0	0	0
1995 constant dollars (mil.)	0	0	0	0	0	0
Total imports[7]						
Current dollars (mil.)	149	163	146	147	151	135
1995 constant dollars (mil.)	171	180	157	154	155	135
Total exports[7]						
Current dollars (mil.)	138	145	149	118	115	25
1995 constant dollars (mil.)	159	160	160	124	118	25
Arms as percent of total imports[8]	0	0	6.8	0	0	0
Arms as percent of total exports[8]	0	0	0	0	0	0

LABOR FORCE

Labor Force (16)

Total (million)	1.369
Agriculture	65%
Industry	19%
Services	16%

Only about 65,000 wage earners. Data for 1981 est. Percent distribution for 1981-1985.

Unemployment Rate (17)

Rate not available.

PRODUCTION SECTOR

Electric Energy (18)

Capacity	126,000 kW (1995)
Production	230 million kWh (1995)
Consumption per capita	48 kWh (1995)

Transportation (19)

Highways:

total: 11,700 km

paved: 1,287 km

unpaved: 10,413 km (1996 est.)

Waterways: 800 km; 600 km navigable year round

Merchant marine: none

Airports: 10 (1997 est.)

Airports—with paved runways:

total: 3

over 3,047 m: 1

914 to 1,523 m: 2 (1997 est.)

Airports—with unpaved runways:

total: 7

914 to 1,523 m: 5

under 914 m: 2 (1997 est.)

Top Agricultural Products (20)

Rice, coffee, cocoa, palm kernels, palm oil, peanuts; poultry, cattle, sheep, pigs; fish.

FINANCE, ECONOMICS, & TRADE

Economic Indicators (22)

National product: GDP—purchasing power parity—$2.65 billion (1997 est.)

National product real growth rate: -27% (1997 est.)

National product per capita: $540 (1997 est.)

Inflation rate—consumer price index: 40% (1997 est.)

Exchange Rates (24)

Exchange rates:	
Leones (Le) per US$1	
December 1997	1,312.37
1997	967.72
1996	920.73
1995	755.22
1994	586.74
1993	567.46

Top Import Origins (25)

$211 million (c.i.f., 1996)

Origins	%
Cote d'Ivoire	NA
European Union countries	NA
India	NA

NA stands for not available.

MANUFACTURING SECTOR

GDP & Manufacturing Summary (21)

	1980	1985	1990	1992	1993	1994
Gross Domestic Product						
Millions of 1990 dollars	426	494	547	460	432	432
Growth rate in percent	2.91	7.46	2.50	−13.62	−6.15	0.15
Per capita (in 1990 dollars)	131.8	137.9	136.8	109.7	100.5	98.2
Manufacturing Value Added						
Millions of 1990 dollars	57	55	38	72	*81*	*85*
Growth rate in percent	−4.88	−15.55	−7.06	20.02	*12.16*	*5.53*
Manufacturing share in percent of current prices	7.5	4.8	7.1	8.7	NA	NA

Top Export Destinations (26)

$47 million (f.o.b., 1996); note—much reduced in 1997 by civil warfare Data are for 1996.

Destinations	%
Germany	31
former Yugoslavia	16.5
Italy	13
Croatia	10
France	7
Austria	7
United States	5

Economic Aid (27)

Recipient: ODA, $NA. NA stands for not available.

Import Export Commodities (28)

Import Commodities	Export Commodities
Foodstuffs	Manufactured goods 50.7%
Machinery and equipment	Machinery and transport equipment 31.4%
Fuels and lubricants	Chemicals 10.5%
	Food 3.8%

Balance of Payments (23)

	1990	1991	1992	1993	1994
Exports of goods (f.o.b.)	149	150	150	118	116
Imports of goods (f.o.b.)	−140	−139	−139	−187	−189
Trade balance	8	11	11	−69	−73
Services - debits	−146	−81	−78	−67	−165
Services - credits	62	76	54	61	102
Private transfers (net)	7	7	3	14	46
Government transfers (net)	—	3	4	3	1
Overall balance	−70	15	−5	−58	−89

SINGAPORE

Republic of Singapore

CAPITAL: Singapore.

FLAG: The flag consists of a red stripe at the top and a white stripe on the bottom. On the red stripe, at the hoist, are a white crescent opening to the fly and five white stars.

ANTHEM: *Long Live Singapore.*

MONETARY UNIT: The Singapore dollar (s$) of 100 cents is a freely convertible currency. There are coins of 1, 5, 10, 20, and 50 cents and 1 dollar and notes of 2, 5, 10, 20, 50, 100, 500, 1,000, and 10,000 dollars. s$1 = US$0.70972 (or US$1 = s$1.409).

WEIGHTS AND MEASURES: The metric system is in force, but some local measures are used.

HOLIDAYS: Major Western, Chinese, Malay, and Muslim holidays are celebrated, some of which fall on annually variable dates because of the calendars used. Major holidays include New Year's Day, 1 January; Chinese New Year; Good Friday; Vesak Day (Buddhist festival); Labor Day, 1 May; Hari Raya Puasa (Muslim festival); National Day, 9 August; Hari Raya Haji (Malay Muslim festival); Dewali; Christmas, 25 December.

TIME: 8 PM = noon GMT.

LOCATION AND SIZE: The Republic of Singapore, the second smallest country in Asia, consists of Singapore Island and several smaller adjacent islets. Situated in the Indian Ocean off the southern tip of the Malay Peninsula, Singapore has an area of 632.6 square kilometers (244.2 square miles), slightly less than 3.5 times the size of Washington, D.C. Singapore is connected to the nearby western portion of Malaysia by a causeway across the narrow Johore Strait. Singapore's capital city, Singapore, is located on the country's southern coast.

CLIMATE: The climate is tropical, with heavy rainfall and high humidity. The range of temperature is slight; the average annual maximum is 31°C (88°F), and the average minimum 24°C (75°F). The annual rainfall of 228 centimeters (90 inches) is distributed fairly evenly throughout the year, ranging from 39 centimeters (15 inches) in December to 28 centimeters (11 inches) in May. It rains about every other day.

INTRODUCTORY SURVEY

RECENT HISTORY

In 1959, Singapore became a self-governing state, and on 16 September 1963, it joined the new Federation of Malaysia (formed by bringing together the previously independent Malaya and Singapore and the formerly British-ruled northern Borneo territories of Sarawak and Sabah).

However, Singapore, with its mostly urban Chinese population and highly commercial economy, found itself at odds with the Malay-dominated central government of Malaysia. Frictions mounted, and on 9 August 1965, Singapore separated from Malaysia to become wholly independent in its own right as the Republic of Singapore. Singapore, Indonesia, Malaysia, the Philippines and Thailand formed the Association of South-East Asian Nations (ASEAN) in 1967. Brunei became a member of ASEAN in 1984. Harry Lee Kuan Yew, a major figure in the move toward independence, served as Singapore's first prime minister.

The People's Action Party (PAP) founded in 1954 has been the dominant political party, winning every general election since 1959. The PAP's popular support has rested on economic growth and improved standards of living along with unrelenting repression of opposition leaders. The PAP won all parliamentary seats in the general elections from 1968 to 1980.

On 28 November 1990, Lee Kuan Yew, Prime Minister of Singapore for over thirty-one years, transferred the prime ministership to Goh Chock

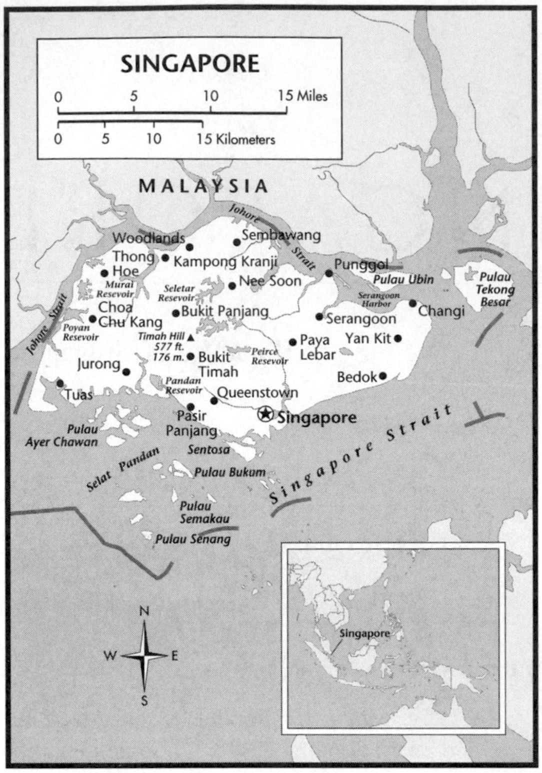

Tong, the former first deputy prime minister. Singapore's first direct presidential elections were held on 28 August 1993, and Ong Teng Cheong became the first elected president.

GOVERNMENT

The constitution of the Republic of Singapore provides for a single-chamber parliamentary form of government. Singapore practices universal suffrage, and voting has been compulsory for all citizens over 21 since 1959. In 1993, the unicameral legislature consisted of an eighty-one elected-member parliament and six nominated members (NMPs) appointed by the president.

The prime minister, who commands the confidence of a majority of parliament, acts as effective head of government, and appoints the cabinet. The president is elected for a term of six years.

Judiciary

The judiciary includes the Supreme Court as well as district, magistrate, and special courts. Minor cases are heard in the country's ten magistrate courts and in district courts (two civil and four criminal), each presided over by a district judge. The Supreme Court is headed by a chief justice and is divided into the High Court and the Court of Appeal. In its appeals jurisdiction, the High Court hears criminal and civil appeals from the magistrate and district courts.

Political Parties

There were twenty-two registered political parties at the beginning of 1993. The ruling People's Action Party (PAP) of former Prime Minister Lee Kuan Yew has dominated the country since 1959. The main opposition parties are the Singapore Democratic Party (SDP) and the Workers' Party (WP). Smaller minority parties are the United People's Front, the Singapore Malays' National Organization, and the Singapore Solidarity Party.

In the 1997 parliamentary elections, the PAP maintained its control by winning eighty-one of eighty-three seats up for election.

DEFENSE

Required national military service has been in effect since 1967. Male citizens are called up for twenty-four months' full-time military service at age 18. Singapore's armed forces numbered 53,900 (with 33,800 draftees) in 1995.

In 1995, the army had an estimated 45,000 personnel; the navy had 4,900 personnel. The 1995 defense budget was US$4 billion, or 6.5% of the gross domestic product (GDP).

ECONOMIC AFFAIRS

Singapore's most significant natural resource is a deep-water harbor. By the early 1980s, Singapore had built a strong, diversified economy, giving it an economic importance in Southeast Asia out of proportion to its small size.

In the late 1980s, Singapore began to further diversify its economy, making it capable of providing manufacturing, financial, and communications facilities for multinational firms. One of the fastest-growing sectors of Singapore's economy was international banking and finance.

The inflation rate based on consumer prices in 1996 was 1.3%. Since 1992, property prices have doubled and residential property prices were still climbing in mid-1995. The main constraints on Singapore's economic performance are labor shortages, rising labor costs, and erosion of productivity. The gross domestic product (GDP) grew by over 10% in 1993 and 1994, but growth cooled to 8.8% in 1995 and 6.9% in 1996.

Public Finance

Budget surpluses have been registered since 1987. In 1996, government revenues totaled approximately US$17.3 billion and expenditures $12.9 billion. External debt totaled US$3.2 million.

Income

In 1998, Singapore's gross national product (GNP) was US$95.1 billion, or approximately US$30,060 per person. For the period 1985–95, the average annual inflation rate was 3.9%, resulting in a real growth rate in per person GNP of 6.2%.

Industry

Industry in 1995 accounted for 28% of the gross domestic product (GDP). The electronics industry is the most important sector of manufacturing. Petroleum refining is a well-established industry in Singapore. Other major industries are oil drilling equipment, rubber processing and rubber products, processed food and beverages, ship repair, entrepôt (intermediary) trade, financial services, and biotechnology.

Banking and Finance

In 1988, Singapore had foreign reserves worth about $533 billion. Many sources of finance are available to organizations doing business in Singapore. The Monetary Authority of Singapore (MAS) requires banks to observe its policy of discouraging the internalization of the Singapore dollar. The MAS performs the functions of a central bank, except for the issuing of currency.

As of 1996, Singapore had 140 commercial banks. Since 1971, the government has sought to attract representation by a variety of foreign banks in terms of countries and geographical regions. Most of the new foreign banks allowed into Singapore have been offshore banks that have concentrated on foreign-exchange transactions. The Post Office Savings Bank (POSBank) is the national savings bank (est. 1877).

Currency is issued by the Singapore Board of Commissioners of Currency.

Over 300 companies are listed on the Stock Exchange of Singapore. The Singapore International Monetary Exchange (SIMEX) opened in 1984.

Economic Development

In the 1990s, Singapore's economic development strategy emphasized both the manufacturing and service sectors. The Economic Development Board (EDB), formed in 1961, has guided Singapore's industrialization. Early emphasis was placed on promoting investment in manufacturing. The Strategic Economic Plan (SEP) announced in 1991 focused on education and human resources to enhance export competitiveness. Emphasis on developing the service sector has been supported and enhanced by the Operational Headquarters (OHQ) program, encouraging companies to use Singapore as regional headquarters or as a central distribution center. The Creative Business Program promotes investment in the film, media, and publishing, arts and entertainment, textile, fashion and design sectors. Singapore's globalization strategy hinges on making a transformation from a production-driven economy to an innovation-driven one.

SOCIAL WELFARE

Besides institutionalized care, the Ministry of Community Development administers foster and homemaker service programs for needy young persons. The government operates child care centers and welfare homes for aged and destitute persons. Social welfare assistance is also provided by mutual-benefit organizations and voluntary services.

The Central Provident Fund, a public pension and retirement program, provides lump-sum benefits for old age, disability, death, sickness, and maternity. Employers fund workers' compensation benefits for job-related injuries.

Women's legal rights are equal to those of men in most areas, including civil liberties, employment, business, and education. Women comprise over half the labor force and are well-represented in the professions. However, they still fill most low-paying clerical positions, and their average salary is only 70% that of men.

Prison conditions are considered good, but there are reports of mistreatment of prisoners. Caning is a common form of punishment.

Healthcare

Singapore's population enjoys one of the highest health levels in all of Southeast Asia due largely to good housing, sanitation, and water supply, as well as the best hospitals and other medical facilities in the region. Fully 100% of the population has access to safe drinking water, and 99% has adequate sanitation. Nutritional standards are among the highest in Asia. Life expectancy is 76 years.

In 1989, there were 9,801 hospital beds in Singapore (3.5 per 1,000 people). In 1991, there were 3,779 doctors and 10,240 nurses. There was 1 doctor per 837 people in 1992. In the same year, 100% of the population had access to health care services.

The principal causes of death are heart disease and cancer.

Housing

In 1985, as a result of government-sponsored efforts, 2.1 million persons—or 84% of the total population of Singapore—lived in 551,767 apartments under the management of the Housing and Development Board. Some 397,180 units had been sold to the public. In 1990, 84% of all housing units were apartments, 7% were bungalows and terrace houses, 5% were condominiums, and 1% were dwellings with attap or zinc roofs. The total number of housing units in 1992 was 758,000.

EDUCATION

Literacy is about 91%. All children who are citizens are entitled to free primary education. Primary schooling is available in all four official languages. Upon completion of primary school, students can receive vocational training, or if they qualify, they can take four or five years of secondary schooling leading to two-year courses in junior colleges or three-year courses in school centers at the pre-university level.

In 1989, there were 203 primary schools in Singapore and 145 secondary schools and junior colleges. The total school population in 1991 was 260,286 primary and 185,693 secondary and pre-university students. The National University of Singapore was established on 8 August 1980.

1999 KEY EVENTS TIMELINE

January

- Chee Soon Juan, secretary-general of the Singapore Democratic Party, pleads not guilty after being charged for speaking in public without a permit.
- SingTel, under pressure from competition at home, expands its regional presence with the purchase of an 18.6% stake in Thailand's largest mobile phone company, Advanced Info Service.

February

- In collaboration with IGA, Singapore opens a U.S. style supermarket; the new supermarket is targeted at Singapore's American community.
- Chee Soon Juan is fined $830 for a speech he gave in the financial district without a permit. He refuses to pay the fine and is jailed for seven days.
- Singapore's Senior Minister Lee Kuan Yew speaks out on regional issues.

March

- Singapore ONE, a government funded national network offering interactive services, is inaugurated as part of Singapore's effort to be a "wired" economy.
- The eleventh execution of the year is carried out.
- Michael Yap, known for his blueprint of Singapore as an "intelligent island," heads the National Computer Board.

April

- Singapore Telecom Ltd., the government owned Internet carrier, has competition from a new carrier, Starhub Pte. Ltd. Starhub is backed by NTT Communications Corp and British Telecommunications PLC.
- Technopreneurship Investment Fund, a government program aimed at encouraging creativity and local entrepreneurship, is established with a $1 billion fund to finance homegrown high-tech entrepreneurs and venture capitalists.

May

- A public apology is made by SingNet, the Singaporean internet service provider, for secretly scanning about 200,000 computers.
- Singapore's government decides to open its banking sectors to outside players. Led by Deputy Prime Minister Lee Hsien Loong, also known as B.G. Lee, the restructuring removes some of the limits on foreign ownership in the banking sector.
- The eight-deck Sun Vista cruise ship, owned by Singapore's Sun Cruises Co., sinks off of Malaysia on its last leg of a six-day cruise due to a fire in the main engine room; all 1,093 passengers and crew are rescued.
- Chee Soon Juan, who was tried for speaking in public without a permit, has his fine reduced, and he will not be barred from elections.

July

- Nick Leeson, convicted for unauthorized futures trades in the Singapore office of Baring's Bank, is released from prison three years early for good behavior and returns to Britain.
- With access to the Internet and foreign internet service providers, Singapore loosens some of its societal restraints: literary and theater productions are allowed greater freedom. Censorship, however, is still prevalent.
- After six years in office Singapore's first elected president, Ong Teng Cheong, announces his retirement.

August

- The government announces a new plan to create a "Boston of Asia," a vibrant university scene that is hoped to heighten Singapore's economic competitiveness.
- Singapore taxis are equipped with Navstar satellite guided systems, a global positioning system.
- An American man is jailed for six months and is fined $590 for a drunken rampage on a Singapore Airlines flight, during which he threatened to open an exit door in mid-flight.
- A drunken, unemployed British national is sentenced to a 12-month jail term in Singapore and fined $590 for punching a crew member on a flight from Australia to London, England.
- Western Digital Corp announces that it will lay off 2,500 Singaporean employees and hire 1,700 new workers in Malaysia; about 10,000 workers in Singapore will be laid off in 1999.
- S.R. Nathan, former ambassador to the United States and former head of Singapore's Internal Security Department, is certified by the Presidential Elections Committee as the only eligible candidate to contest elections.
- Goh Chock Tong gives an annual National Day Rally speech, outlining a blueprint for creating a "world-class economy" and "world class home" in Singapore.
- Paddy Chew, Singapore AIDS activist, dies at age thirty-nine.
- George Yeo, Minister for Trade and Industry, leads a business delegation representing twenty-six companies from various industries to explore business prospects in Beijing and the Chinese provinces of Henan and Fujian.
- Indonesian investigators cite "unlawful interference" in Singapore's Silk Air crash probe.
- Singapore's presidential election is called off when the only candidate approved for the presidency, S.R. Nathan, is appointed to the position.
- The World Economic Forum's 1999 Global Competitiveness Report ranks Singapore as the world's most competitive economy.

September

- Two record labels, Kitho Lab and Ambient Asia, are launched in Singapore.
- SingTel and Commerce One, Inc. launch sites for electronic procurement.
- The Minister of National Development, Mah Bow Tan, announces the release of land for sale: 5,000 to 6,000 private housing units, 2,000 to 3,000 executive condominium units (partially government subsidized), and 75,000 square meters of net commercial space.
- LycosAsia, a 50–50 joint internet venture between U.S. web media company Lycos and Singapore Telecom, is announced.
- Prime Minister Goh Chok Tong encourages the Association of South East Asian Nations (ASEAN) to move more quickly toward trade liberalization. ASEAN's goal is to create a Free Trade Area (FTA) by 2015.
- Singapore contributes 250 military personnel including a medical detachment and two landing ships to the East Timor peacekeeping force.
- A first stage free trade agreement between New Zealand and Singapore is intended to serve as a catalyst for a wider free trade movement.
- A devastating earthquake in Taiwan results in a shortage of electronic components for Singapore's manufacturers.
- Crude oil prices soar as OPEC producers curb output until March 2000; the profit margin for refiners in Singapore slips as these prices climb, and producers reduce capacity as the market remains soft.
- For the eighth consecutive year Changi airport is named "Best Airport in the World" by *Business Traveler* (Asia-Pacific) and "Best Airport" by *Conde Nast Traveler* (UK).
- Yahoo! Inc. introduces Yahoo! Singapore.

October

- U.S. Defense Secretary William Cohen discusses regional issues, East Timor, and bilateral military ties with Singapore's leaders.
- In a project funded by Microsoft Corp., ''Project I-Campus,'' the Massachusetts Institute of Technology and the University of Singapore (NUS) cooperate in distance learning.

November

- At an informal meeting in Manila, Philippines, six members of the Association of Southeast Asian Nations (ASEAN)—Brunei, Indonesia, Malaysia, Philippines, Singapore, and Thailand—agree to establish a free-trade zone by eliminating duties on most goods traded in the region by 2010. The remaining four newer and less-developed nation members—Cambodia, Laos, Myanmar (Burma), and Vietnam—will eliminate duties by 2015. Rice will be excluded from trade agreements, however.

ANALYSIS OF EVENTS: 1999

BUSINESS AND THE ECONOMY

Singapore has not escaped the economic crisis that swept across much of Asia in 1997. Gross domestic product (GDP) growth in 1997 was at 8%. In 1998 growth fell to 1.5%. At the end of 1998 Singapore was falling into its deepest recession since gaining independence. About 10,000 Singaporean workers will be laid off in 1999. The failure of Malaysia and Indonesia to recover from the financial crisis as quickly as Singapore has done will continue to affect the small nation.

Signs of Singapore's recovery, however, are evident. The government forecasts growth of the GDP by 4% to 5%. In addition the economy expanded 6.7% in the second quarter of 1999, its fastest growth rate since the end of 1997. At midyear Singapore's Big Four banks generated profits of over 90% for the same period in 1998. Singapore and Hong Kong continue to be regional rivals for multinational headquarters and international business centers.

With the goal of increasing economic competitiveness Singapore responded to the regional economic crisis with a gradualist approach to reform. Reforms in progress prior to the financial crisis were kept in place and made primarily in the banking, legal, and technology sectors. Foreign institutions would be allowed to acquire Singaporean banks, and foreign Internet service providers began to compete with the nationally owned Singapore Telecom Ltd. The costs of doing business were made cheaper, while the costs of labor were cut by reducing employers' contributions to the Central Provident Fund (CPF). Ministerial salaries and transportation fees were also cut.

Singapore's leaders believe that future growth will come from the software and high-tech manufacturing services. They have begun to emphasize creativity, entrepreneurship, and innovation. The government is backing this new position with a $1 billion fund called the Technopreneurship Investment Fund. The effect of this new emphasis on business development is filtering down to all levels of the educational system and is loosening the political system.

Other infrastructure investments include extension of the Mass Rapid Transit (MRT) system, an electronic road pricing system, and the completion of a petrochemical complex on an outer island. Although land sales were halted in June 1998, the economy continues to show signs of recovery and the government plans to release land for commercial and private housing developments in 2000. Project completion times for developers will be cut back from eight to six years.

GOVERNMENT AND POLITICS

Singapore's President Ong Teng Cheong retired, allegedly for heath reasons. Only one candidate, S.R. Nathan, succeeds in attaining certification from the Presidential Elections Committee; two other hopeful candidates were rejected. As the only candidate approved to run for the office, Nathan, a former ambassador to the United States and former head of the Internal Security Department, is automatically appointed president, and the election is canceled.

Although the government has relaxed some of its restrictions because of the new emphasis on entrepreneurship and innovation, it still maintains a tight hold on its citizenry. Opposition politician Chee Joon Suan was sentenced to jail twice this year. Chee made public speeches without a permit. Free speech, though allowed by Singapore's constitution, can only be practiced if it does not threaten public order or national security. The Internal Security Act, allowing detention without trial, remains in force.

Former Indonesian president B.J. Habibie criticized Singapore for maintaining a ''racist'' policy. The claim, made while Habibie was still in office, was made in regards to Singapore's stipulation that Malays can never be made senior officers in the Singaporean army. Singapore responded by publishing the photographs of Malay officers in the newspaper.

CULTURE AND SOCIETY

Singapore is one of the most densely populated countries in the world and it has a literacy rate of 91%. Reform of the education system is proceeding on several fronts. Classrooms are shifting from being a place for rote memorization and rigid behavior to the source for increasing interaction. Computers have been added to the classroom environment, while the country's computer-to-household ratio is 45 to 100. By the end of 1999 every household is intended to have the ability to hook up to Singapore One service.

The arts are benefiting from the relaxation in Singapore's traditionally strict society. Censors were overruled when they attempted to change the title of the James Bond spoof, ''Austin Powers: The Spy Who Shagged Me.'' In addition the government continues to emphasize the use of English, knowledge of which it sees as a competitive edge within the region and globally. The use of the local patois ''Singlish,'' which combines English with Malay and Chinese words and sentence structure, is also spreading.

DIRECTORY

CENTRAL GOVERNMENT

Head of State

President
S.R. Nathan, Office of the President, Istana Singapore, Orchard Road, Singapore 238823

Ministers

Prime Minister
Goh Chok Tong, Office of the Prime Minister

Senior Minister
Lee Kuan Yew, Office of the Senior Minister

Deputy Prime Minister
Le Hsien Loong, Office of the Deputy Prime Minister

Deputy Prime Minister and Minister for Defence
Tony Tan Keng Yam, Ministry of Defence

Minister for Law and Minister for Foreign Affairs
S. Jayakumar, Ministry of Law

Minister for Finance
Richard Hu Tsu Tau, Ministry of Finance Headquarters, 100 High Street, #10-01, The Treasury, Singapore 179434

Minister for Information and the Arts, and Minister for the Environment
Lee Yock Suan, Ministry of Information and the Arts

Minister for Home Affairs
Wong Kan Seng, Ministry of Home Affairs

Minister for Communications and Information Technology
Yeo Cheow Tong, Ministry of Communications and Information Technology, #39-00 PSA Building, 460 Alexandra Road, Singapore 119963
E-MAIL: Feedback@mcit.gov.sg

Minister for Trade and Industry
George Yong-Boon Yeo, Ministry of Trade and Industry

Minister for Manpower
Lee Boon Yang, Ministry of Manpower, 18 Havelock Road, Singapore 059764
PHONE: +65 5341511
FAX: +65 5344840

Minister for National Development
Mah Bow Tan, Ministry of National Development

Minister without Portfolio, Prime Minister's Office
Lim Boon Heng, Office of the Minister without Portfolio

Minister for Health and Second Minister for Finance
Lim Hng Kiang, Ministry of Health

Minister for Community Development and Minister-in-charge of Muslim Affairs
Abdullah Tarmugi, Ministry of Community Development

Minister for Education and Second Minister for Defence
Teo Chee Hean, Ministry of Education, Kay Siang Road, Singapore 248922
PHONE: +65 4739111
FAX: +65 4756128
E-MAIL: contact@moe.edu.sg

POLITICAL ORGANIZATIONS

People's Action Party (PAP)

TITLE: Secretary General
NAME: Goh Chok Tong

Singapore Democratic Party (SDP)

NAME: Chee Soon Juan

Workers' Party (WP)

NAME: J.B. Jeyaretnam

National Solidarity Party (NSP)

NAME: C.K. Tan

Singapore People's Party (SPP)

NAME: Chiam See Tong

Democratic Action Party

TITLE: Secretary General
NAME: Lim Kit Siang

DIPLOMATIC REPRESENTATION

Embassies in Singapore

Belgium
8 Shenton Way, Temasek Tower #14–10, Singapore 068811
PHONE: +65 2207677
FAX: +65 2226976
E-MAIL: Singapore@diplobel.org
TITLE: Ambassador
NAME: Patrick van Haute

Brazil
101 Thomson Road, #10–05 United Square, Singapore 307591
PHONE: +65 2566001
FAX: +65 2566619
E-MAIL: cinbrem@singnet.com.sg

Germany
Far East Shopping Centre, 14th Floor, 545 Orchard Road, Singapore 238882
PHONE: +65 7371355
FAX: +65 7372653
E-MAIL: german@singnet.com.sg
TITLE: Ambassador
NAME: Jurgen Ostreich

South Africa
15th Floor, Odeon Towers, 331 North Bridge Road, Singapore 188720
PHONE: +65 3393319
FAX: +65 3370196
TITLE: High Commissioner
NAME: C. H. Werth

United States
27 Napier Road, Singapore 258508
PHONE: +65 4769100
FAX: +65 4769340
TITLE: Ambassador
NAME: Steven J. Green

JUDICIAL SYSTEM

Supreme Court

Court of Appeals

District Court

Magistrate Court

Special Courts

FURTHER READING

Articles

''Future Trends: On the Loose in Singapore.'' *The Economist* 351 (May 29, 1999): 38.

Juan, Chee Soon. ''Backwardness in Singapore.'' *The Asian Wall Street Journal Weekly* 21, 28 (July 12, 1999).

''Lee Hsien Loong, Deputy Prime Minister, Singapore.'' *Business Week*, 14 June 1999, p. 75.

''Remaking Singapore Inc.'' *Business Week*, 5 April 1999, p. 19.

''Singapore Online: This Small Island Nation Is Wiring Itself Up.'' *Time International* 154 (October 11, 1999): 60.

''Singapore's Speech-Maker.'' *The Economist* 350 (February 6, 1999): 42.

''Speaking Out in Singapore.'' *The Economist* 350 (January 9, 1999): 38.

Books

Chee, Soon Juan. *To Be Free: Stories from Asia's Struggle Against Oppression.* Australia: Monash University, 1998.

Milne, R. S. *Singapore: The Legacy of Lee Kuan Yew.* Boulder, Colorado: Westview Press, 1990.

Minchin, James. *No Man Is an Island: A Study of Singapore's Lee Kuan Yew.* Sydney: Allen and Unwin, 1986.

Seow, Francis T. *To Catch a Tartar.* New Haven, Connecticut: Yale University Southeast Asian Studies, 1994.

Internet

Simply Singapore. Available Online @ .com/TBV/sguide.html (November 2, 1999).

SINGAPORE: STATISTICAL DATA

For sources and notes see "Sources of Statistics" in the front of each volume.

GEOGRAPHY

Geography (1)

Area:

Total: 647.5 sq km.

Land: 637.5 sq km.

Water: 10 sq km.

Area—comparative: slightly more than 3.5 times the size of Washington, DC.

Land boundaries: 0 km.

Coastline: 193 km.

Climate: tropical; hot, humid, rainy; no pronounced rainy or dry seasons; thunderstorms occur on 40% of all days (67% of days in April).

Terrain: lowland; gently undulating central plateau contains water catchment area and nature preserve.

Natural resources: fish, deepwater ports.

Land use:

Arable land: 2%

Permanent crops: 6%

Permanent pastures: NA%

Forests and woodland: 5%

Other: 87% (1993 est.).

HUMAN FACTORS

Demographics (2A)

	1990	1995	1998	2000	2010	2020	2030	2040	2050
Population	3,038.7	3,325.8	3,490.4	3,571.7	3,895.5	4,136.6	4,290.1	4,276.6	4,160.7
Net migration rate (per 1,000 population)	NA	NA	NA	NA	NA	NA	NA	NA	NA
Births	NA	NA	NA	NA	NA	NA	NA	NA	NA
Deaths	13.9	NA	NA	NA	NA	NA	NA	NA	NA
Life expectancy - males	73.6	74.5	75.5	76.1	78.0	79.2	79.9	80.4	80.6
Life expectancy - females	78.8	80.7	81.8	82.5	84.4	85.6	86.2	86.6	86.8
Birth rate (per 1,000)	16.8	14.6	13.8	13.0	9.8	10.0	9.2	8.5	9.1
Death rate (per 1,000)	4.6	4.7	4.7	4.7	5.6	7.2	9.8	12.9	14.5
Women of reproductive age (15-49 yrs.)	934.2	1,011.1	1,044.0	1,048.5	1,015.7	904.1	853.9	822.7	777.2
of which are currently married	454.4	NA	NA	NA	NA	NA	NA	NA	NA
Fertility rate	1.6	1.5	1.5	1.5	1.5	1.6	1.6	1.6	1.7

Except as noted, values for vital statistics are in thousands; life expectancy is in years.

Health Personnel (3)

Total health expenditure as a percentage of GDP, 1990-1997[a]

Public sector	1.5
Private sector	1.9
Total[b]	3.3

Health expenditure per capita in U.S. dollars, 1990-1997[a]

Purchasing power parity	829
Total	943

Availability of health care facilities per 100,000 people

Hospital beds 1990-1997[a]	360
Doctors 1993[c]	147
Nurses 1993[c]	416

Health Indicators (4)

Life expectancy at birth

1980	71
1997	76
Daily per capita supply of calories (1996)	....
Total fertility rate births per woman (1997)	1.7
Maternal mortality ratio per 100,000 live births (1990-97)	10[c]
Safe water % of population with access (1995)	100
Sanitation % of population with access (1995)	100
Consumption of iodized salt % of households (1992-98)[a]	....

Smoking prevalence

Male % of adults (1985-95)[a]	32
Female % of adults (1985-95)[a]	3
Tuberculosis incidence per 100,000 people (1997)	48
Adult HIV prevalence % of population ages 15-49 (1997)	0.15

Infants and Malnutrition (5)

Under-5 mortality rate (1997)	4
% of infants with low birthweight (1990-97)	7
Births attended by skilled health staff % of total[a]	100

% fully immunized (1995-97)

TB	98
DPT	93
Polio	94
Measles	89
Prevalence of child malnutrition under age 5 (1992-97)[b]	NA

Ethnic Division (6)

Chinese	76.4%
Malay	14.9%
Indian	6.4%
Other	2.3%

Religions (7)

Buddhist (Chinese), Muslim (Malays), Christian, Hindu, Sikh, Taoist, Confucianist.

Languages (8)

Chinese (official), Malay (official and national), Tamil (official), English (official).

EDUCATION

Public Education Expenditures (9)

Public expenditure on education (% of GNP)

1980	2.8
1996	3.0[1]

Expenditure per student

Primary % of GNP per capita

1980	6.8
1996	7.8[1]

Secondary % of GNP per capita

1980	....
1996	....

Tertiary % of GNP per capita

1980	40.6
1996	28.2[1]

Expenditure on teaching materials

Primary % of total for level (1996)	0.0
Secondary % of total for level (1996)	....
Primary pupil-teacher ratio per teacher (1996)	22[1]
Duration of primary education years (1995)	6

Educational Attainment (10)

Age group (1990)[20]	25+
Total population	1,860,878

Highest level attained (%)

No schooling	14.3
First level	
Not completed	11.2
Completed	16.5
Entered second level	
S-1	36.9
S-2	13.7
Postsecondary	7.6

Literacy Rates (11A)

In thousands and percent[1]	1990	1995	2000	2010
Illiterate population (15+ yrs.)	227	196	165	98
Literacy rate - total adult pop. (%)	89.1	91.1	92.9	96.2
Literacy rate - males (%)	95.1	95.9	96.6	97.9
Literacy rate - females (%)	83.0	86.3	89.1	94.5

GOVERNMENT & LAW

Political Parties (12)

Parliament	% of seats
People's Action Party (PAP) (in contested constituencies)	65
Other	35

Military Affairs (14B)

	1990	1991	1992	1993	1994	1995
Military expenditures						
Current dollars (mil.)	2,106	2,680	2,987	3,123	3,304	3,970
1995 constant dollars (mil.)	2,421	2,961	3,213	3,274	3,386	3,970
Armed forces (000)	56	56	56	56	56	60
Gross national product (GNP)						
Current dollars (mil.)	48,830	54,440	59,630	66,600	74,960	84,300
1995 constant dollars (mil.)	56,120	60,160	64,140	69,820	76,840	84,300
Central government expenditures (CGE)						
1995 constant dollars (mil.)	12,150	13,650	12,890	14,220	14,930	16,550
People (mil.)	3.0	3.1	3.2	3.2	3.3	3.3
Military expenditure as % of GNP	4.3	4.9	5.0	4.7	4.4	4.7
Military expenditure as % of CGE	19.9	21.7	24.9	23.0	22.7	24.0
Military expenditure per capita (1995 $)	797	957	1,019	1,020	1,036	1,191
Armed forces per 1,000 people (soldiers)	18.4	18.1	17.8	17.1	17.1	18.0
GNP per capita (1995 $)	18,470	19,430	20,350	23,500	23,500	25,290
Arms imports[6]						
Current dollars (mil.)	250	380	220	130	230	200
1995 constant dollars (mil.)	287	420	237	136	236	200
Arms exports[6]						
Current dollars (mil.)	30	50	30	20	20	30
1995 constant dollars (mil.)	34	55	32	21	21	30
Total imports[7]						
Current dollars (mil.)	60,900	66,290	72,180	85,230	102,700	124,500
1995 constant dollars (mil.)	69,990	73,250	77,640	89,360	105,200	124,500
Total exports[7]						
Current dollars (mil.)	52,750	59,020	63,480	74,010	96,830	118,300
1995 constant dollars (mil.)	60,630	65,220	68,280	77,590	99,250	118,300
Arms as percent of total imports[8]	.4	.6	.3	.2	.2	.2
Arms as percent of total exports[8]	.1	.1	0	0	0	0

Government Budget (13A)

Year: 1997

Total Expenditures: 23,757 Millions of Dollars

Expenditures as a percentage of the total by function:

General public services and public order	13.32
Defense	28.90
Education	18.82
Health	6.72
Social Security and Welfare	1.79
Housing and community amenities	9.04
Recreational, cultural, and religious affairs	.07
Fuel and energy	--
Agriculture, forestry, fishing, and hunting	.25
Mining, manufacturing, and construction	.03
Transportation and communication	6.90
Other economic affairs and services	9.99

Crime (15)

Crime rate (for 1997)

Crimes reported	36,200
Total persons convicted	10,100
Crimes per 100,000 population	975

Persons responsible for offenses

Total number of suspects	15,800
Total number of female suspects	3,000
Total number of juvenile suspects	2,150

LABOR FORCE

Labor Force (16)

Total (million)	1.856
Financial, business, and other services	33.5%
Manufacturing	25.6%
Commerce	22.9%
Construction	6.6%
Other	11.4%

Data for 1997 est. Percent distribution for 1994.

Unemployment Rate (17)

3% (1997 est.)

PRODUCTION SECTOR

Electric Energy (18)

Capacity	4.513 million kW (1995)
Production	21 billion kWh (1995)
Consumption per capita	7,234 kWh (1995)

Transportation (19)

Highways:

total: 3,010 km

paved: 2,932 km (including 150 km of expressways)

unpaved: 78 km (1995 est.)

Merchant marine:

total: 856 ships (1,000 GRT or over) totaling 18,463,338 GRT/29,322,743 DWT

Airports: 9 (1997 est.)

Airports—with paved runways:

total: 9

over 3,047 m: 2

2,438 to 3,047 m: 1

1,524 to 2,437 m: 4

914 to 1,523 m: 1

under 914 m: 1 (1997 est.)

Top Agricultural Products (20)

Rubber, copra, fruit, vegetables; poultry.

MANUFACTURING SECTOR

GDP & Manufacturing Summary (21)

Detailed value added figures are listed by both International Standard Industry Code (ISIC) and product title.

	1980	1985	1990	1994
GDP ($-1990 mil.)[1]	17,677	23,864	34,991	48,175
Per capita ($-1990)[1]	7,320	9,329	12,936	17,077
Manufacturing share (%) (current prices)[1]	28.0	22.0	28.0	*24.6*
Manufacturing				
Value added ($-1990 mil.)[1]	5,309	5,736	10,343	13,842
Industrial production index	100	104	165	197
Value added ($ mil.)	4,004	4,861	11,918	20,593
Gross output ($ mil.)	15,278	17,575	39,345	65,878
Employment (000)	287	254	352	366
Profitability (% of gross output)				
Intermediate input (%)	75	73	70	69
Wages and salaries inc. supplements (%)	8	10	10	10
Gross operating surplus	17	17	20	21
Productivity ($)				

	1980	1985	1990	1994
Gross output per worker	53,564	69,711	112,432	181,199
Value added per worker	13,942	19,137	33,888	56,329
Average wage (inc. supplements)	4,168	7,290	10,839	17,794
Value added ($ mil.)				
311/2 Food products	121	180	322	512
313 Beverages	52	76	139	200
314 Tobacco products	25	35	64	138
321 Textiles	70	28	72	72
322 Wearing apparel	127	157	294	241
323 Leather and fur products	7	5	11	22
324 Footwear	9	5	9	10
331 Wood and wood products	84	43	55	52
332 Furniture and fixtures	40	61	89	134
341 Paper and paper products	45	82	189	303
342 Printing and publishing	128	229	514	974
351 Industrial chemicals	52	138	584	754
352 Other chemical products	143	267	600	1,028
353 Petroleum refineries	*513*	*315*	*725*	*987***
354 Miscellaneous petroleum and coal products	*173*	*82*	*192*	*274*
355 Rubber products	44	21	35	61
356 Plastic products	84	102	327	574
361 Pottery, china and earthenware	*1*	—	*2*	*6*
362 Glass and glass products	*10*	*5*	*31*	*74*
369 Other non-metal mineral products	82	140	149	332
371 Iron and steel	62	48	97	99
372 Non-ferrous metals	9	17	41	52
381 Metal products	206	298	730	1,331
382 Non-electrical machinery	319	370	2,737	5,417
383 Electrical machinery	950	1,538	2,707	4,962
384 Transport equipment	500	470	890	1,492
385 Professional and scientific equipment	80	89	200	381
390 Other manufacturing industries	69	58	114	111

FINANCE, ECONOMICS, & TRADE

Economic Indicators (22)

National product: GDP—purchasing power parity—$84.6 billion (1997 est.)

National product real growth rate: 6% (1997 est.)

National product per capita: $24,600 (1997 est.)

Inflation rate—consumer price index: 1.8% (1997 est.)

Exchange Rates (24)

Exchange rates:

Singapore dollars (S$) per US$1

January 1998	1.7533
1997	1.4848
1996	1.4100
1995	1.4174
1994	1.5274
1993	1.6158

Balance of Payments (23)

	1992	1993	1994	1995	1996
Exports of goods (f.o.b.)	66,565	77,858	97,919	118,456	126,012
Imports of goods (f.o.b.)	−68,388	−80,582	−96,565	−117,391	−123,731
Trade balance	−1,823	−2,724	1,354	1,065	2,281
Services - debits	−16,176	−18,930	−21,862	−26,486	−29,088
Services - credits	24,153	26,599	33,382	40,739	42,100
Private transfers (net)	−134	−115	−133	−160	−173
Government transfers (net)	−367	−413	−515	−658	−837
Overall balance	5,652	4,418	12,226	14,499	14,284

Top Import Origins (25)

$133.9 billion (1997 est.) Data are for 1995.

Origins	%
Japan	21
Malaysia	15
United States	15
Thailand	5
Taiwan	4
South Korea	4

Top Export Destinations (26)

$125.6 billion (1997 est.) Data are for 1995.

Destinations	%
Malaysia	19
United States	18
Hong Kong	9
Japan	8
Thailand	6

Economic Aid (27)

$NA. NA stands for not available.

Import Export Commodities (28)

Import Commodities	Export Commodities
Aircraft	Computer equipment
Petroleum	Rubber and rubber products
Chemicals	Petroleum products
Foodstuffs 4%	Telecommunications equipment

SLOVAKIA

CAPITAL: Bratislava.

FLAG: Horizontal bands of white (top), blue, and red superimposed with a crest of a white double cross on three blue mountains.

ANTHEM: *Nad Tatru sa blyska (Over Tatra it lightens).*

MONETARY UNIT: The currency of the Slovak Republic is the Slovak koruna (Sk) consisting of 100 hellers, which replaced the Czechoslovak Koruna (Kcs) on 8 February 1993. There are coins of 10, 20, and 50 hellers and 1, 2, 5, and 10 korun, and notes of 20, 50, 100, 500, 1,000 and 5,000 korun. Sk1 = $0.03215 (or $1 = Sk31.1).

WEIGHTS AND MEASURES: The metric system is the legal standard.

HOLIDAYS: New Year's Day, 1 January; May Day, 1 May; Anniversary of Liberation, 8 May; Day of the Slav Apostles, 5 July; Anniversary of the Slovak National Uprising, 29 August; Reconciliation Day, 1 November; Christmas, 24–26 December. Movable holiday is Easter Monday.

TIME: 1 PM = noon GMT.

LOCATION AND SIZE: Slovakia, a landlocked country located in Eastern Europe, is about twice the size of the state of New Hampshire with a total area of 48,845 square kilometers (18,859 square miles). It has a total boundary length of 1,355 kilometers (842 miles). Slovakia's capital city, Bratislava, is located on the southwestern border of the country.

CLIMATE: In July the average temperature is 21°C (70°F). The average temperature in January is −1°C (30°F). Rainfall averages roughly 49 centimeters (19.3 inches) a year, and can exceed 200 centimeters (80 inches) annually in the Tatry (Tatras) mountains.

Slovak Republic
Slovenska Republika

INTRODUCTORY SURVEY

RECENT HISTORY

Slovakia grew out of the Eastern European communist nation of Czechoslovakia. The National Front government ran Czechoslovakia as a democracy until 1948, when a military coup with backing from the Soviet Union forced President Benes to accept a government headed by Klement Gottwald, a communist. A wave of purges and arrests rolled over the country from 1949 to 1954. Gottwald died a few days after Stalin, in March 1953. His successors clung to harsh Stalinist methods of control, holding Czechoslovakia in a tight grip until well into the 1960s.

Soviet leader Nikita Khrushchev led a movement of liberalization in the Soviet Union. This atmosphere encouraged liberals within the Czechoslovak party to try to emulate his leadership style. In January 1968, Alexander Dubček was named head of the Czechoslovak Communist Party, the first Slovak ever to hold the post. Under Dubček, Czechoslovakia embarked on a radical liberalization, termed "socialism with a human face." The leaders of the Soviet Union and other eastern bloc (communist) nations viewed these developments—termed the "Prague Spring"—with alarm. Communist leaders issued warnings to Dubček.

Finally, on the night of 20–21 August 1968, military units from almost all the Warsaw Pact (communist) nations invaded Czechoslovakia, to "save it from counter-revolution." Dubček and other officials were arrested, and the country was placed under Soviet control. A purge of liberals

followed, and Dubček was expelled from the Communist Party. Between 1970 and 1975, nearly one-third of the party was dismissed, as the new Communist Party leader, Gustav Husak, consolidated his power, reuniting the titles of party head and republic president.

In the 1980s, liberalization in the Soviet Union, under the concepts of *perestroika* (restructuring) and *glasnost* (openness), once again set off political change in Czechoslovakia. After ignoring Soviet leader Mikhail Gorbachev's calls for Communist Party reform, in 1987 Husak announced his retirement. His replacement was Milos Jakes.

In November 1989, thousands gathered in Prague's Wenceslas Square, demanding free elections. This ''velvet revolution,'' so called because it was not violent, ended on 24 November, when Jakes and all of his government resigned. Vaclav Havel, a Czech playwright and dissident, was named president on 29 December 1989, while Dubček was named leader of the National Assembly.

There was less enthusiasm for returning to an economic system of private ownership among the Slovaks than the Czechs who made up the two main ethnic groups within the country. Vladimir Meciar, the Slovak premier, was a persuasive voice for growing Slovak separatism from the Czechs. In July 1992, the new Slovak legislature issued a declaration of sovereignty and adopted a new constitution as an independent state, to take effect 1 January 1993. By the end of 1992, it was obvious that separation was inevitable. The two prime ministers, Klaus and Meciar, agreed to the peaceful separation—the so-called velvet divorce, which took effect 1 January 1993.

The new Slovak constitution created a 150-seat National Assembly, which elects the head of state, the president. The Meciar government rejected the political and economic liberalization that the Czechs were pursuing, attempting instead to retain a socialist-style government. Swift economic decline, especially compared to the Czechs' growing prosperity, caused him to lose a vote of no-confidence in March 1994. Michal Kovac was elected to a five-year term in February 1993 and appointed Meciar prime minister in December 1994. However, Meciar again was slow to implement economic reforms, and Kovac stopped much of Meciar's more authoritarian legislation. Kovac's term, however, ended in March 1998 and Meciar abstained from voting in parliament for a new president, thus keeping the position vacant. Meciar assumed presidential duties and began building up his base of power. Tens of thousands of protesters marched in Bratislava against Meciar's administration.

GOVERNMENT

The constitution that the Slovak National Assembly adopted in July 1992 created a single-chamber legislature of 150 members. The government is formed by the leading party, or coalition of parties, and headed by a prime minister. The president is head of state.

Judiciary

The judicial system consists of a republic-level Supreme Court; several regional courts; and thirty-eight local courts responsible for individual districts. The highest judicial body, the ten-judge Constitutional Court in Koice, rules on the constitutionality of laws as well as the decisions of lower level courts.

Political Parties

The single most popular party is Meciar's Movement for a Democratic Slovakia. The Party of the Democratic Left, the Christian Democratic Movement, and the Association of Slovak Workers are other leading parties.

DEFENSE

The total active armed forces of Slovakia number 42,600, including 25,000 army personnel and 12,200 air force personnel. Additionally, there is an estimated reserve national guard force of 20,000. The defense budget for 1996 was $450 million.

ECONOMIC AFFAIRS

The economy of Slovakia is highly industrialized, although the industrial structure is less developed than that of the Czech Republic. Agriculture, which had been the most important area of the economy before the communist era, now plays a smaller role than industry. The real growth rate of the gross domestic product in 1996 was 6.7%.

Public Finance

Since the dissolution of Czechoslovakia, the Slovak government has implemented several measures to compensate for the large loss of fiscal transfers it received from the Federation, which were equivalent to between Sk20-25 billion in 1992.

Economic growth of 3% in 1993, 9% in 1994, and 2% in 1995 led to an increase in state revenues of 17% in 1994, to about $6.1 billion. This, in turn, led to renewed budget increases of some 6% of GDP in 1995, to approximately $6.4 billion. External debt totaled $4.6 billion.

Income

In 1998, Slovakia's gross national product (GNP) was $19.95 billion, or about $3,700 per person. For the period 1985–95, the average annual inflation rate was 10.4%, resulting in a decline in per person GNP of 2.6%.

Industry

Major industries include heavy engineering, armaments, iron and steel production, nonferrous metals, and chemicals.

Banking and Finance

Four years after the Soviet system relinquished control over the eastern bloc, Slovakia formed a national bank. In January 1992, the banking system of Czechoslovakia was split. From that point on the National Bank of Slovakia was charged with the responsibility of circulating currency and regulating the banking sector.

The Bratislava Stock Exchange opened on 8 July 1990. The Bratislava Option and Futures Exchange opened in 1994.

Economic Development

The government has slowed economic reforms due to the social burden imposed by the transformation to a market economy. Anticipated measures include stimulation of demand through price subsidies and public spending. Slovakia's most successful structural reform has been privatization.

SOCIAL WELFARE

Slovak law guarantees the equality of all citizens and prohibits discrimination. Health care, retirement benefits, and other social services are provided regardless of race, sex, religion, or disability. Women and men are equal under the law, enjoying the same property, inheritance, and other rights, and receiving equal pay for equal work. Slovakia's Roma (Gypsy) population experiences higher levels of unemployment and housing discrimination than ethnic Slovaks.

Healthcare

Life expectancy is 73 years, and infant mortality is 12.4 per 1,000 live births. The country had 17,419 physicians in 1992 and 61,573 hospital beds. Tuberculosis has been on the rise in Slovakia; there were 1,760 diagnosed cases in 1994.

Housing

With a shortage of 500,000 apartments, the Slovak Republic planned to build 200,000 new ones by the year 2000. As of 1992, 80,000 people were on waiting lists.

EDUCATION

Slovakia has an estimated adult literacy rate of more than 99%. Education is compulsory for ten years, approximately up to the age of 18.

In 1994, 338,291 children attended the elementary schools, and 251,221 students attended general secondary schools. Another 237,130 pupils attended the 493 specialized and technical secondary schools.

In 1994, 82,223 students were enrolled at the universities.

1999 KEY EVENTS TIMELINE

January

- The former Minister of Industry and head of the Slovak Gas Company (SPP), Jan Ducky, is shot dead at the front of his apartment building in Bratislava.
- A U.S. delegation lead by representative Ben Gillman visits Slovakia.
- The parliament passes a bill legalizing bilingual report cards.
- The chairman of the parliament introduces the Infoage project, proposing the full-scale introduction of the Internet into the Slovak school system.
- The Slovak Minister of Foreign Affairs meets with Madeline Albright in New York.

February

- The premiers of Hungary and Slovakia sign a protocol on implementation of the Slovak-Hungarian Agreement.
- The one hundredth HIV positive patient is discovered in Slovakia.
- A spaceship carrying the first Slovak astronaut, Ivan Bella, is launched from Bajkonur.
- Parliamentary immunity for the former Minister of Interior, Gustav Krajci, is revoked.

March

- Chairman of the Slovak National Party (SNS), Jan Slota urges the participants of a meeting to attack Budapest with tanks.
- The government announces that Slovakia is not buying the Russian S-300 antiaircraft systems.
- The chairman of the parliament announces that presidential elections will be held on May 15, 1999.
- The government continues to support NATO operations in Kosovo.
- Nine men, all members of an organized crime group, are shot dead in Dunajska Streda.
- Protest demonstrations are held in Bratislava against NATO operations in Serbia.

April

- The government opens Slovakia's air space for NATO planes.
- The parliament agrees that the former head of the secret service (SIS) should have his immunity revoked and be investigated for abuse of office.
- The police take four former high officers of the secret service into custody.
- Eleven candidates, including the former premier, Vladimir Meciar, qualify to run in the presidential elections.
- The former director of SIS is taken into custody.
- Slovakia's territory is opened for ground transport of NATO technology and personnel.
- Minister of Foreign Affairs Eduard Kukan becomes one of two UN Commissioners for Kosovo.
- Austrian Chancellor Viktor Klima is disappointed with the Slovak government's decision concerning delays in the closure of the nuclear power plant in Jaslovske Bohunice.

May

- The mayor of Kosice, Rudolf Schuster' wins the first round of presidential elections with 34.58% of the vote, over Vladimir Meciar with 27.18%. Schuster wins in the runoff.
- More than 3,000 unionized construction workers hold a protest demonstration in Bratislava.
- A 29 year Ukrainian man, Oleg T., is formally charged with the murder of Jan Ducky.

June

- Ambassador Jan Kubis becomes the chairman of OSCE.
- The government raises the bottom level of value added tax from 6 to 10%.
- The Minister of Transportation, Post and Telecommunication is accused of corruption in the case of the GSM 1800 mobile phone network privatization.
- Special investigators on economic crime start to collect evidence in East Slovakian Steel Plants in Kosice, a stronghold of the former premier, Vladimir Meciar.
- The Confederation of Trade Unions claims that working conditions in Slovakia are critical and asks for adequate compensation.
- The director of the Department for European Integration, Roman Filistein, is fired for being suspected of corruption.
- Three hundred Slovak Roma asks for asylum in Finland and hundreds more are leaving the country.

July

- Slovak Roma leaders call for the resignation of Pal Csaky, the Vice Premier for Minorities.
- Responding to the uncontrolled influx of Roma, Finland and Norway install a visa regime with Slovakia.
- UN Secretary General Kofi Annan arrives for a three-day visit to Slovakia.
- Ivan Lexa, the former head of the secret service, is released from prison.
- Vladimir Meciar's Movement for Democratic Slovakia submits to President Schuster a petition asking for a referendum on issues related to the use of minority languages.
- NOKIA criticizes Minister of Transportation, Post and Telecommunication Gabriel Palacka for the lack of transparency in the competition for GSM 1800 mobile network frequency.

August

- IMF criticizes the lax approach of the government to fiscal policy.
- A well-known organized crime figure, Peter Steinhubel, is gunned down in Bratislava.
- President Schuster accepts the resignation of the Transportation, Post and Telecommunications Minister.
- Hundreds of Slovak Roma are waiting for asylum in Finland, Norway, and Belgium.
- President Schuster refuses to call a referendum on issues related to the use of minority languages.
- A delegation from the U.S. House of Representatives arrives for an official visit to Slovakia.

September

- In Bratislava and Kosice, teachers demonstrate against declining conditions in Slovak schools.
- Slovakia joins the International Agreement Against Money Laundering.
- The state loses control over Priemyselna Bank. A group around the former boss of East Slovak Steel Plants, Alexender Rezes, is suspected to be behind the transaction.
- The premiers of Hungary and Slovakia, Viktor Orban and Mikulas Dzurinda respectively, sign an international agreement on the construction of a bridge between Esztergom and Sturovo.
- The president of the Czech Republic, Vaclav Havel, arrives for an official visit to Slovakia.
- President Clinton declares that the U.S. will maintain an ''open door policy'' toward Slovakia.
- The Confederation of Trade Unions holds a 35,000-man rally in Bratislava.

October

- The prime ministers of the Czech Republic, Poland, Hungary and Slovakia meet in a Slovak mountain resort, Tatranska Javorina. They agree to coordinate efforts to combat organized crime and international money laundering.
- The Minister of Economy, Ludovit Cernak, who is implicated in a number of corruption scandals, resigns.
- The European Commission recommends that membership negotiations begin with Malta and five East European countries, including Slovakia.
- The Slovak government's decision to postpone the shutdown of the Jaslovske Bohunice nuclear power plant outrages Austria.
- Another well known organized crime boss is gunned down in Bratislava.

- President Schuster meets with the Pope during his official visit to the Vatican.

November

- The Czech Republic and the Slovak Republic reach an agreement over financial matters related to their split in 1995. The Czechs will return 4.5 tons of gold to the Slovaks, and will write off an estimated one billion in debt. Both countries will cancel claims against each other's banks. The agreement must be ratified by the two countries' parliaments.

ANALYSIS OF EVENTS: 1999

BUSINESS AND THE ECONOMY

The Slovak economy is expected to maintain a 3 percent growth of GDP in 1999, in spite of the chronic shortage of domestic capital and foreign direct investment. Conditions in the country's largest industrial plant, Vychodo Slovenske Zeleziarne (East Slovak Steel Plants) were stabilized, and the firm started to pay its creditors. Due mainly to a large proportion of bad loans, the banking sector is loosing more than 10 billion Slovak Crown (Sk) every month (US$2–3 million), and with interest rates exceeding 20 percent Slovak banks have a hard time competing with foreign financial institutions.

The level of annual inflation is at 19 percent, due mainly to partial liberalization of energy prices. More than 500,000 workers, or close to 18 percent of the active workforce, are out of work. Direct foreign investment dropped to one-third of the 1998 level, and it is continuing to be problematic since legal and political conditions in Slovakia do not favor foreign investors in comparison with Hungary or Poland. The budget deficit for year 2000 is Sk18 billion, or close to 10 percent of the total budget. Slovakia's debt burden exceeded 60 percent of GDP, and the country's external debt is almost 30 percent of GDP.

Privatization of state firms and the revision of some problematic cases of privatization from the Meciar government were plagued by a lack of professionalism and corruption scandals. Two cabinet ministers and their deputies resigned. The government still holds onto a majority stake in the Gas Industry, Electric Energy Plants, Telecommunications, Slovak Saving Institute, and Slovak Insurance Agency.

The Ministry of Finances was sharply criticized for the imposed restrictive measures and for moving slowly on legislation concerning foreign direct investment, bankruptcy, special economic zones, and the reconstruction of the banking sector. Entrepreneurs are especially critical of the increases in value added tax, import tariffs, restriction on tax write-offs, the proposed transport tax, and the shortage of domestic financial resources.

GOVERNMENT AND POLITICS

Rudolf Schuster became the first directly elected president of Slovakia, yet the governing coalition remains fragmented, lacks professionalism, and is unable to agree on solutions for social and economic problems of the country. Most leading positions in state owned firms, state administration, and civil service were filled according to a formula agreed upon during the coalition talks in 1998. More than 2000 high-seniority civil servants were replaced by May 1999. Party membership remains the most relevant condition in the selection process, while the professional background is secondary. Adding more instability to the coalition government, the leader of the Christian Democratic Movement (KDH), Jan Carnogursky, succeeded in splintering the Slovak Democratic Coalition (SDK). After one year in power, the new legislature was able to pass 71 bills. Until the end of 2002, they need to pass more than 250 bills only concerning requirements for EU integration.

The most successful resort of the new government appears to be the Ministry of Foreign Affairs. Minister Eduard Kukan and the Vice Premier for European Integration, Pavol Hamzik, were prized for the pro-active foreign policy of Slovakia, namely for Slovakia's constructive role in the Kosovo crisis, and for the recommendation of the European Commission that Slovakia be formally invited to negotiations.

CULTURE AND SOCIETY

Two main sets of issues captured the attention of the Slovak society in 1999: (1) corruption scandals involving high government officials, and (2) organized crime. The corruption scandals brought down two ministers, their deputies, the department director of the Vice Premier for European Integration, the government director of informatics, and others. The first scandal erupted when the Slovak

Telecom (ST) offered a public tender for the GSM 1800 mobile phone network, and a couple of high government officials asked for a bribe from a Telenor Slovakia CEO. As a result of this, ST was forced to sell the network for US$10 million to two local mobile phone networks, a price that was 25 times lower then the one paid for a similar network in Hungary. The Minister of Economy, on the other hand, was involved in several scandals before his resignation, and these are still waiting for the decision of the courts. The official reason given for the minister's resignation was mis-management of the Nafta Gbely affair, where a former Meciar privatizer of the largest gas storage facility in Slovakia initially agreed to transfer his shares to the FNM, but ended up selling them to a U.S. firm.

Organized crime activities in Slovakia were big news because of the rate at which the heads of organized crime were killing each other. The year started with the murder of Jan Ducky, the former Minister for Industry and the director of the state gas company. Then, nine members of the so-called Papay Gang were executed by a commando in a café in Dunajska Streda. Several organized crime bosses were assassinated in Kosice, Zilina, and Bratislava. A corruption scandal related to organized crime rings in the country rocked the police force in the northern part of the country. The Minister of Interior placed the blame on the ''Syndicate,'' a secret Ukrainian organized crime ring directed by former KGB officers operating throughout Europe. In short, according to opinion polls, crime remains the greatest concern of Slovak citizens. Less than 50 percent of all cases, and only approximately 70 percent of murders are solved. The premiers of the Visegrad Four agreed to take a coordinated effort to solve this problem.

DIRECTORY

CENTRAL GOVERNMENT

Head of State

President
Rudolf Schuster, Office of the President, Stefanikova ul. 1, PO Box 128, 811 04 Bratislava, Slovakia
PHONE: +421 (7) 3598246
FAX: +421 (7) 394239

Ministers

Prime Minister
Mikulas Dzurinda, Office of the Prime Minister, nám. Slobody 1, 813 70 Bratislava, Slovakia
PHONE: +421 (7) 3595111
FAX: +421 (7) 315484

Ministry of Privatization
Maria Machova, Ministry of Privatization, Drienova ul. 24, 820 09 Bratislava, Slovakia
PHONE: +421 (7) 5234332
FAX: +421 (7) 5233335

Ministry of Agriculture
Pavel Koncos, Ministry of Agriculture, Dobrovicova 12, 813 31 Bratislava, Slovakia
PHONE: +421 (7) 5490581
FAX: (+421 7 5322150

Ministry of Construction and Public Works
Istvan Harna, Ministry of Construction and Public Works, Spitálska 8, 816 44 Bratislava, Slovakia
PHONE: +421 (7) 544111; 5321316
FAX: +421 (7) 5322150

Ministry of Culture
Milan Knazko, Ministry of Culture, Dobrovicova ul. 12, 813 31 Bratislava, Slovakia
PHONE: +421 (7) 5490781; 5367726
FAX: +421 (7) 5323528

Ministry of Defense
Pavol Kanis, Ministry of Defense, Kutuzovova ul. 8, 832 47 Bratislava, Slovakia
PHONE: +421 (7) 5254500
FAX: +421 (7) 5258904

Ministry of Economy
Ludovit Cernak, Ministry of Economy, Mierov 19, 827 15 Bratislava, Slovakia
PHONE: +421 (7) 5741111
FAX: +421 (7) 5237827

Ministry of Education and Science
Milan Ftacnik, Ministry of Education and Science, Hlbok 2, 813 30 Bratislava, Slovakia
PHONE: +421 (7) 5495772
FAX: +421 (7) 5497098

Ministry of Environment
Laszlo Miklos, Ministry of Environment, Hlbok 2, 812 35 Bratislava, Slovakia
PHONE: +421 (7) 5492451
FAX: +421 (7) 5497267

Ministry of Finance
Brigita Schmognerova, Ministry of Finance, Stefanovicova ul 5, 813 08 Bratislava, Slovakia

PHONE: +421 (7) 5497541
FAX: +421 (7) 5498042

Ministry of Foreign Affairs
Eduard Kukan, Ministry of Foreign Affairs
PHONE: +421 (7) 801902
FAX: +421 (7) 802586

Ministry of Health Care
Tibor Sagat, Ministry of Health Care, Limbov 2, 833 41 Bratislava, Slovakia
PHONE: +421 (7) 5377943
FAX: +421 (7) 5377934

Ministry of Interior
Ladislav Pittner, Ministry of Interior, Pribinova ul 2, 812 72 Bratislava, Slovakia
PHONE: +421 (7) 5323659
FAX: +421 (7) 5367746

Ministry of Justice
Jan Carnogursky, Ministry of Justice, nám. Zupné 13, 811 Bratislava, Slovakia
PHONE: +421 (7) 5353111; 5353204
FAX: +421 (7) 5315952; 5330732

Ministry of Labor, Social Affairs and Family
Peter Magvasi, Ministry of Labor, Social Affairs and Family, Spitálska 4-6, 816 43 Bratislava, Slovakia
PHONE: +421 (7) 5321245; 5441714; 5442415
FAX: +421 (7) 5321258; 5362150

Ministry of Transportation, Posts and Telecommunications
Gabriel Palacka, Ministry of Transportation, Posts and Telecommunications, Mileticova 19, 820 06 Bratislava 26, Slovakia
PHONE: +421 (7) 5254753
FAX: +421 (7) 5254800

POLITICAL ORGANIZATIONS

Association of Slovak Workers (ZRS)

TITLE: Chairman
NAME: Jan Luptak

Hungarian Coalition Party (SMK)

NAME: Bela Bugar

Movement for a Democratic Slovakia (MKDH)

TITLE: Chairman
NAME: Vladimir Meciar

Party of Civic Understanding (SOP)

TITLE: Chairman
NAME: Rudolf Schuster

Party of the Democratic Left (SDL)

TITLE: Chairman
NAME: Jozef Migas

Slovak Democratic Coalition (SDK)

NAME: Mikulas Dzurinda

Slovak National Party (SNS)

TITLE: Chairman
NAME: Jan Slota

Social Democratic Party of Slovakia (SSDS)

TITLE: Chairman
NAME: Jaroslav Volf

DIPLOMATIC REPRESENTATION

Embassies in Slovakia

Austria
Holubyho 11, 811 03 Bratislava, Slovakia
PHONE: +421 (7) 5311103
FAX: +421 (7) 5313145

Belgium
Frana Krála 5, 811 05 Bratislava, Slovakia
PHONE: +421 (7) 5391338
FAX: +421 (7) 5394296

Bulgaria
Kuzmanyho 1A, 811 06 Bratislava, Slovakia
PHONE: +421 (7) 5315308

China
Jancova 8, 811 06 Bratislava, Slovakia
PHONE: +421 (7) 5314577
FAX: +421 (7) 5316551

Croatia
Grösslingova 47, 811 09 Bratislava, Slovakia
PHONE: +421 (7) 5361413
FAX: +421 (7) 5361403

Cuba
Matuskova 10, 831 01 Bratislava, Slovakia
PHONE: +421 (7) 5377960

Czech Republic
Panenská 33, 810 00 Bratislava, Slovakia
PHONE: +421 (7) 5334361
FAX: +421 (7) 5333410

France
Hlavné 7, 811 01 Bratislava, Slovakia
PHONE: +421 (7) 5335725

FAX: +421 (7) 5335719

Germany
Hviezdoslavovo 10, 811 02 Bratislava, Slovakia
PHONE: +421 (7) 5319640
FAX: +421 (7) 5319634

Hungary
Sedlarska ul. 62, 814 25 Bratislava, Slovakia
PHONE: +421 (7) 5330541
FAX: +421 (7) 5335484

Italy
Cervenova 19, 811 03 Bratislava, Slovakia
PHONE: +421 (7) 5313195
FAX: +421 (7) 5313202

Netherlands
Frana Krala 5, 811 05 Bratislava, Slovakia
PHONE: +421 (7) 5391577
FAX: +421 (7) 5391075

Poland
Hummelova 4, 814 91 Bratislava, Slovakia

Romania
Frana Krála 11, 811 05 Bratislava, Slovakia
PHONE: +421 (7) 5491665

Russia
Godrova 4, 811 02 Bratislava, Slovakia
PHONE: +421 (7) 5313468
FAX: +421 (7) 5334910

Serbia
Palkovicova 16, 821 08 Bratislava, Slovakia
PHONE: +421 (7) 5499422
FAX: +421 (7) 5499477

South Africa
Jancova 8, 811 02 Bratislava, Slovakia
PHONE: +421 (7) 5311582
FAX: +421 (7) 5312581

Spain
Grosslingova 35, 811 09 Bratislava, Slovakia
PHONE: +421 (7) 5362294
FAX: +421 (7) 5362320

Turkey
Holubyno 11, 811 03 Bratislava, Slovakia
PHONE: +421 (7) 5315504
FAX: +421 (7) 5315606

Ukraine
Radvanská 35, 811 01 Bratislava, Slovakia
PHONE: +421 (7) 5331672
FAX: +421 (7) 5312561

United Kingdom
Panská 16, 811 01 Bratislava, Slovakia
PHONE: +421 (7) 5319632
FAX: +421 (7) 5310002

United States
Hviezdoslavovo 4, 811 02 Bratislava, Slovakia
PHONE: +421 (7) 5330861
FAX: +421 (7) 5335439

JUDICIAL SYSTEM

Constitutional Court

Supreme Court

Regional Court

Local Courts

FURTHER READING

Articles

Chance, David. ''Emerging Markets: Fast Track to EU Wider but Slower.'' Reuters, October 14, 1999.

Hajosi, Milan. ''Good News For Investors.'' *Euroforum* (January 1999): 34–35.

''IMF Warns on Slovak Budget, Current Account, Banks.'' *The Economist* (May 22, 1999): 56.

Matas, Michal. ''Industrial Policy.'' *Euroforum*, (February 1999): 30–34.

''A New Day in Slovakia.'' *The Economist*, 350 (March 6, 1999): 100.

Shepherd, Robin. ''Focus: Slovak Trade Deficit Narrows Further.'' Reuters, September 27, 1999.

''Slovakia.'' *Newsweek International* (April 26, 1999): 31.

''The Unsecret Service: Slovakia's Intelligence Chief Lists the Agency's Alleged Dirty Tricks.'' *Time International* 153 (March 15, 1999): 19.

Books

Goldman, Minton F. *Slovakia Since Independence: a Struggle for Democracy.* Westport, Conn.: Praeger, 1999.

Jacobs, Michael. *Czech and Slovak Republics.* 2nd ed. London: A. and C. Black, 1999.

Kirschbaum, Stanislav J. *Historical Dictionary of Slovakia.* Lanham, Md.: Scarecrow Press, 1999.

Slovak Republic: A Strategy for Growth and European Integration. Washington, D.C.: World Bank, 1998.

Internet

EUnet Slovakia. Available Online @ http://slovakia.eunet.sk/slovakia/news.asp (November 9, 1999).

Slovakia Daily Surveyor. Available Online @ http://www.slovensko.com (November 9, 1999).

SLOVAKIA: STATISTICAL DATA

For sources and notes see "Sources of Statistics" in the front of each volume.

GEOGRAPHY

Geography (1)

Area:

Total: 48,845 sq km.

Land: 48,800 sq km.

Water: 45 sq km.

Area—comparative: about twice the size of New Hampshire.

Land boundaries:

Total: 1,355 km.

Border countries: Austria 91 km, Czech Republic 215 km, Hungary 515 km, Poland 444 km, Ukraine 90 km.

Coastline: 0 km (landlocked).

Climate: temperate; cool summers; cold, cloudy, humid winters.

Terrain: rugged mountains in the central and northern part and lowlands in the south.

Natural resources: brown coal and lignite; small amounts of iron ore, copper and manganese ore; salt.

Land use:

Arable land: 31%

Permanent crops: 3%

Permanent pastures: 17%

Forests and woodland: 41%

Other: 8% (1993 est.).

HUMAN FACTORS

Demographics (2A)

	1990	1995	1998	2000	2010	2020	2030	2040	2050
Population	NA	5,368.0	5,393.0	5,401.1	5,636.3	5,738.8	5,681.9	5,520.9	5,214.9
Net migration rate (per 1,000 population)	NA	NA	NA	NA	NA	NA	NA	NA	NA
Births	NA	NA	NA	NA	NA	NA	NA	NA	NA
Deaths	NA	NA	NA	NA	NA	NA	NA	NA	NA
Life expectancy - males	NA	68.4	69.4	70.0	72.7	74.8	76.5	77.9	78.9
Life expectancy - females	NA	76.5	77.1	77.6	79.8	81.6	83.0	84.1	85.0
Birth rate (per 1,000)	NA	11.5	10.0	10.6	12.8	9.4	9.3	8.2	7.2
Death rate (per 1,000)	NA	9.9	9.5	9.4	9.3	9.6	10.9	12.6	13.9
Women of reproductive age (15-49 yrs.)	NA	1,406.6	1,438.3	1,448.5	1,406.4	1,337.5	1,242.8	1,095.3	1,037.1
of which are currently married	NA	NA	NA	NA	NA	NA	NA	NA	NA
Fertility rate	NA	1.5	1.3	1.3	1.7	1.6	1.5	1.5	1.4

Except as noted, values for vital statistics are in thousands; life expectancy is in years.

Health Indicators (4)

Life expectancy at birth
- 1980 70
- 1997 73

Daily per capita supply of calories (1996) 3,030
Total fertility rate births per woman (1997) 1.4
Maternal mortality ratio per 100,000 live births (1990-97) . . . . 8[b]
Safe water % of population with access (1995) . . . .
Sanitation % of population with access (1995) 51
Consumption of iodized salt % of households (1992-98)[a] . . . .
Smoking prevalence
- Male % of adults (1985-95)[a] 43
- Female % of adults (1985-95)[a] 26

Tuberculosis incidence per 100,000 people (1997) 35
Adult HIV prevalence % of population ages 15-49 (1997) 0.01

Infants and Malnutrition (5)

Under-5 mortality rate (1997) 11
% of infants with low birthweight (1990-97) . . . . NA
Births attended by skilled health staff % of total[a] . . 100
% fully immunized (1995-97)
- TB 90
- DPT 98
- Polio 98
- Measles 98

Prevalence of child malnutrition under age 5 (1992-97)[b] . . . . NA

Ethnic Division (6)

Slovak 85.7%
Hungarian 10.7%
Gypsy 1.5%
Czech 1%
Ruthenian 0.3%
Ukrainian 0.3%
German 0.1%
Polish 0.1%
Other 0.3%

The 1992 census figures underreport the Gypsy/Romany community which could reach 500,000 or more).

Religions (7)

Roman Catholic 60.3%
Atheist 9.7%
Protestant 8.4%
Orthodox 4.1%
Other 17.5%

Languages (8)

Slovak (official), Hungarian.

EDUCATION

Public Education Expenditures (9)

Public expenditure on education (% of GNP)
- 1980 . . . .
- 1996 . . . . 4.9

Expenditure per student
- Primary % of GNP per capita
  - 1980 . . . .
  - 1996 . . . . 23.4
- Secondary % of GNP per capita
  - 1980 . . . .
  - 1996 . . . . 4.1[1]
- Tertiary % of GNP per capita
  - 1980 . . . .
  - 1996 . . . . 30.5

Expenditure on teaching materials
- Primary % of total for level (1996) . . . .
- Secondary % of total for level (1996) 0.5

Primary pupil-teacher ratio per teacher (1996) 19
Duration of primary education years (1995) 4

Educational Attainment (10)

Age group (1991) . . . . 25+
Total population 3,144,143
Highest level attained (%)
- No schooling 0.7
- First level
  - Not completed 37.9
  - Completed . . . . NA
- Entered second level
  - S-1 . . . . 50.9
  - S-2 . . . . NA
- Postsecondary 9.5

GOVERNMENT & LAW

Political Parties (12)

National Council	% of seats
Movement for a Democratic Slovakia (HZDS)	35
Party of the Democratic Left (SDL)	10.4
Hungarian coalition	10.2
Christian Democratic Movement (KDH)	10.1
Democratic Union (DU)	8.6
Association of Slovak Workers (ZRS)	7.3
Slovak National Party (SNS)	5.4

Government Budget (13B)

Revenues . . . .$5.7 billion

Expenditures . . . .$6.4 billion

Capital expenditures . . . . .NA

Data for 1996. NA stands for not available.

Military Affairs (14B)

	1993	1994	1995
Military expenditures			
Current dollars (mil.)	405	440	577
1995 constant dollars (mil.)	425	451	577
Armed forces (000)	33	47	52
Gross national product (GNP)			
Current dollars (mil.)	16,740	18,510	19,380
1995 constant dollars (mil.)	17,550	18,510	19,380
Central government expenditures (CGE)			
1995 constant dollars (mil.)	8,266	7,204	8,489
People (mil.)	5.3	5.3	5.4
Military expenditure as % of GNP	2.4	2.4	3.0
Military expenditure as % of CGE	5.1	6.3	6.8
Military expenditure per capita (1995 $)	80	84	108
Armed forces per 1,000 people (soldiers)	6.2	8.8	9.7
GNP per capita (1995 $)	3,300	3,467	3,619
Arms imports[6]			
Current dollars (mil.)	230	30	290
1995 constant dollars (mil.)	241	31	290
Arms exports[6]			
Current dollars (mil.)	40	30	70
1995 constant dollars (mil.)	42	31	70
Total imports[7]			
Current dollars (mil.)	6,655	6,823	9,216
1995 constant dollars (mil.)	6,977	6,994	9,216
Total exports[7]			
Current dollars (mil.)	5,451	6,587	8,585
1995 constant dollars (mil.)	5,715	6,752	8,585
Arms as percent of total imports[8]	3.5	.4	3.1
Arms as percent of total exports[8]	0.7	.5	.8

Crime (15)

Crime rate (for 1997)

Crimes reported 92,400

Total persons convicted 43,600

Crimes per 100,000 population 1,700

Persons responsible for offenses

Total number of suspects 43,800

Total number of female suspects 3,000

Total number of juvenile suspects 4,650

LABOR FORCE

Labor Force (16)

Total (million) 2.352

Industry 29.3%

Agriculture 8.9%

Construction 8.0%

Transport and communication 8.2%

Services 45.6%

Data for 1994.

Unemployment Rate (17)

12.8% (1997 est.)

PRODUCTION SECTOR

Electric Energy (18)

Capacity . . . . .7.115 million kW (1995)

Production . . . . .23.223 billion kWh (1995)

Consumption per capita . . . . .4,698 kWh (1995)

Transportation (19)

Highways:

total: 36,608 km

paved: 36,059 km (including 215 km of expressways)

unpaved: 549 km (1996 est.)

Waterways: 172 km on the Danube

Pipelines: petroleum products NA km; natural gas 2,700 km

Transportation (19)

Merchant marine:

total: 3 cargo ships (1,000 GRT or over) totaling 15,041 GRT/19,517 DWT (1997 est.)

Airports: 13 (1997 est.)

Airports—with paved runways:

total: 8

over 3,047 m: 1

2,438 to 3,047 m: 3

1,524 to 2,437 m: 2

914 to 1,523 m: 1

under 914 m: 1 (1997 est.)

Airports—with unpaved runways:

total: 5

914 to 1,523 m: 2

under 914 m: 3 (1997 est.)

Top Agricultural Products (20)

Grains, potatoes, sugar beets, hops, fruit; hogs, cattle, poultry; forest products.

MANUFACTURING SECTOR

GDP & Manufacturing Summary (21)

Detailed value added figures are listed by both International Standard Industry Code (ISIC) and product title.

	1980	1985	1990	1994
GDP ($-1990 mil.)[1]	NA	13,370	14,323	11,441
Per capita ($-1990)[1]	NA	2,601	2,725	2,145
Manufacturing share (%) (current prices)[1]	NA	NA	NA	NA
Manufacturing				
Value added ($-1990 mil.)[1]	NA	NA	NA	NA
Industrial production index	NA	NA	NA	NA
Value added ($ mil.)	NA	NA	NA	2,690
Gross output ($ mil.)	NA	*10,608*	*12,472*	9,139
Employment (000)	NA	*588*	*617*	445
Profitability (% of gross output)				
Intermediate input (%)	NA	NA	NA	71
Wages and salaries inc. supplements (%)	NA	NA	NA	11
Gross operating surplus	NA	NA	NA	18
Productivity ($)				
Gross output per worker	NA	*18,043*	*20,219*	20,541
Value added per worker	NA	NA	NA	6,047
Average wage (inc. supplements)	NA	*1,534*	*1,677*	2,327
Value added ($ mil.)				
311/2 Food products	NA	NA	NA	285
313 Beverages	NA	NA	NA	81
314 Tobacco products	NA	NA	NA	NA
321 Textiles	NA	NA	NA	107
322 Wearing apparel	NA	NA	NA	94
323 Leather and fur products	NA	NA	NA	18
324 Footwear	NA	NA	NA	41
331 Wood and wood products	NA	NA	NA	62
332 Furniture and fixtures	NA	NA	NA	NA
341 Paper and paper products	NA	NA	NA	123
342 Printing and publishing	NA	NA	NA	69
351 Industrial chemicals	NA	NA	NA	167
352 Other chemical products	NA	NA	NA	98
353 Petroleum refineries	NA	NA	NA	183
354 Miscellaneous petroleum and coal products	NA	NA	NA	NA
355 Rubber products	NA	NA	NA	57
356 Plastic products	NA	NA	NA	54
361 Pottery, china and earthenware	NA	NA	NA	2
362 Glass and glass products	NA	NA	NA	64
369 Other non-metal mineral products	NA	NA	NA	107
371 Iron and steel	NA	NA	NA	259
372 Non-ferrous metals	NA	NA	NA	40
381 Metal products	NA	NA	NA	183
382 Non-electrical machinery	NA	NA	NA	235
383 Electrical machinery	NA	NA	NA	151
384 Transport equipment	NA	NA	NA	147
385 Professional and scientific equipment	NA	NA	NA	49
390 Other manufacturing industries	NA	NA	NA	13

FINANCE, ECONOMICS, & TRADE

Economic Indicators (22)

National product: GDP—purchasing power parity—$46.3 billion (1997 est.)

National product real growth rate: 5.9% (1997 est.)

National product per capita: $8,600 (1997 est.)

Inflation rate—consumer price index: 6% (1997)

Exchange Rates (24)

Exchange rates:	
Koruny (Sk) per US$1	
January 1998	35.50
1997	33.616
1996	30.654
1995	29.713
1994	32.045
1993	30.770

Top Import Origins (25)

$11.1 billion (f.o.b., 1996) Data are for 1996.

Origins	%
European Union	36.9
Germany	14.7
Italy	6.0
Czech Republic	24.8
FSU	17.7

Top Export Destinations (26)

$8.8 billion (f.o.b., 1996) Data are for 1996.

Destinations	%
European Union	41.3
Germany	20.9
Austria	6.0
Czech Republic	30.6
FSU	7.1

Economic Aid (27)

$NA. NA stands for not available.

Import Export Commodities (28)

Import Commodities	Export Commodities
Machinery and transport equipment 35.4%; fuels 17.0%; intermediate manufactured goods 15.5%; miscellaneous manufactured goods 9.0%	Machinery and transport equipment 22.8%; chemicals 12.2%; miscellaneous manufactured goods 11.9%; raw materials 4.4%

Balance of Payments (23)

	1993	1994	1995	1996
Exports of goods (f.o.b.)	5,452	6,743	8,591	8,824
Imports of goods (f.o.b.)	−6,365	−6,634	−8,820	−11,106
Trade balance	−912	109	−229	−2,283
Services - debits	−1,890	−1,875	−2,101	−2,297
Services - credits	2,124	2,416	2,628	2,290
Private transfers (net)	4	6	17	9
Government transfers (net)	95	62	76	192
Overall balance	−579	719	390	−2,091

SLOVENIA

CAPITAL: Ljubljana

FLAG: Equal horizontal bands of white (top), blue, and red with seal superimposed on upper hoist side.

ANTHEM: *Zive naj vsi narodi.* (The national anthem begins, "Let all nations live . . .")

MONETARY UNIT: The currency of Slovenia is the tolar (Slt), which consists of 100 stotinov. There are coins of 50 stotinov and 1, 2, and 5 tolars, and notes of 10, 20, 50, and 200 tolars. Slt1 = $0.00752 (or $1 = Slt133.0).

WEIGHTS AND MEASURES: The metric system is in force.

HOLIDAYS: New Year, 1–2 January; Preeren Day, Day of Culture, 8 February; Resistance Day, 27 April; Labor Days, 1–2 May; National Statehood Day, 25 June; Assumption, 15 August; Reformation Day, 31 October; All Saints' Day, 1 November; Christmas Day, 25 December; Independence Day, 26 December. Movable holidays are Easter Sunday and Monday.

TIME: 1 PM = noon GMT.

LOCATION AND SIZE: Located in central Europe, Slovenia is slightly larger than the state of New Jersey with a total area of 20,296 square kilometers (7,836 square miles). Slovenia has a total boundary length of 999 kilometers (621 miles). The capital city, Ljubljana, is located near the center of the country.

CLIMATE: Slovenia's climate is Mediterranean on the coast, and continental in the plateaus and valleys to the east. July's mean temperature is 22°C (72°F). The mean temperature in January is 1°C (31°F). Rainfall averages 62 centimeters (24 inches) a year.

Republic of Slovenia
Republika Slovenije

INTRODUCTORY SURVEY

RECENT HISTORY

Slovenia was divided in 1941 among Germany, Italy, and Hungary. Resistance movements were initiated by nationalist groups and by communist-dominated Partisans. Spontaneous resistance to the Partisans by the non-communist peasantry led to a bloody civil war in Slovenia. The other Yugoslav states also suffered civil war. All were largely under foreign occupiers, who encouraged the bloodshed.

With the entry of the Soviet Union's armies into Yugoslav territory in October 1944, the Partisans swept over Yugoslavia in pursuit of the retreating German forces. The Partisans took over Croatia, launching a campaign of executions and large-scale massacres.

All the republics of the former Federal Socialist Republic of Yugoslavia share a common history between 1945 and 1991, the year of Yugoslavia's break-up. The World War II Partisan resistance movement, controlled by the Communist Party of Yugoslavia and led by Marshal Josip Broz Tito, won a civil war waged against nationalist groups.

A conflict erupted between Tito and the Russian leader Josef Stalin in 1948, and Tito was expelled from the Soviet Bloc. Yugoslavia then developed its own brand of communism based on workers' councils and self-management of enterprises and institutions. Yugoslavia became the leader of the nonaligned group of nations (those

countries that were allies neither of the United States nor the Soviet Union).

The Yugoslav communist regime relaxed its central controls somewhat. This allowed for the development of more liberal wings of communist parties, especially in Croatia and Slovenia. Also, nationalism reappeared, with tensions especially strong between Serbs and Croats in the Croatian republic. This led Tito to repress the Croatian and Slovenian ''springs'' (freedom movements like the one in Czechoslovakia in 1968) in 1970–71.

The 1974 constitution shifted much of the decision-making power from the federal level to the republics, further decentralizing the political process. Following Tito's death in 1980, there was an economic crisis. Severe inflation and inability to pay the nation's foreign debts led to tensions between the different republics and demands for a reorganization of the Yugoslav federation into a confederation of sovereign states.

Pressure toward individual autonomy for the regions, as well as a market economy, grew stronger, leading to the formation of non-communist political parties. By 1990, these parties were able to win majorities in multiparty elections in Slovenia and then in Croatia, ending the era of Communist Party monopoly of power.

Slovenia and Croatia declared their independence on 25 June 1991. On 27 June 1991, the Yugoslav army tried to seize control of Slovenia but was met by heavy resistance from Slovenian ''territorial guards.'' The ''guards'' surrounded Yugoslav army tank units, isolated them, and engaged in close combat. In most cases, the Yugoslav units surrendered to the Slovenian forces. Over 3,200 Yugoslav army soldiers surrendered and were well treated by the Slovenes, who gained favorable publicity by having the prisoners call their parents all over Yugoslavia to come to Slovenia and take their sons back home.

The war in Slovenia ended in ten days due to the intervention of the European Community, which negotiated a cease-fire. Thus Slovenia was able to remove itself from Yugoslavia with a minimum of casualties, although the military operations caused considerable damage to property estimated at almost $3 billion.

On 23 December 1991, a new constitution was adopted by Slovenia establishing a parliamentary democracy with a two-chamber legislature. International recognition came first from Germany on 18 December 1991, from the European Community (EC) on 15 January 1992, and finally from the United States on 7 April 1992. Slovenia was accepted as a member of the United Nations on 23 April 1992 and has since become a member of many other international organizations.

In December 1992, a coalition government was formed by the Liberal Democrats, Christian Democrats, and the United List Group of Leftist Parties. Dr. Milan Kucan was elected president and Dr. Janez Drnovek became prime minister.

GOVERNMENT

Slovenia is a republic based on a constitution adopted on 23 December 1991. The constitution provides for a National Assembly as the highest legislative authority with ninety seats. Deputies are elected to four-year terms of office. The National Council, with forty seats, has an advisory role. Council members are elected to five-year terms of office and may propose laws to the National Assembly, request it to review its decisions, and may demand the calling of a constitutional referendum.

The executive branch consists of a president of the republic who is also supreme commander of the armed forces, and is elected to a five-year term of office, limited to two consecutive terms. The president calls for elections to the National Assembly, proclaims the adopted laws, and proposes candidates for prime minister to the National Assembly. Since 1993, the government has consisted of fifteen ministries.

Judiciary

The judicial system consists of local and district courts and a Supreme Court that hears appeals. A nine-member Constitutional Court resolves jurisdictional disputes and rules on the constitutionality of legislation and regulations. The Constitutional Court also acts as a final court of appeal in cases requiring constitutional interpretation.

The constitution guarantees the independence of judges. Judges are appointed to permanent positions subject to an age limit. The constitution affords criminal defendants a presumption of innocence, open court proceedings, the right to an appeal, a prohibition against double jeopardy, and a number of other due process protections.

Political Parties

The last parliamentary elections were held on 10 November 1996, with seven parties receiving enough votes to gain representation in the National Assembly. The Liberal Democratic Party held twenty-five seats; Slovene People's Party, nineteen; Social Democrats of Slovenia, sixteen; Christian Democrats, ten; and United List (former communists and allies), ten.

DEFENSE

The Slovenian armed forces number 9,550 active duty soldiers and 53,000 reservists who are required to give seven months of service. The paramilitary police has 4,500 actives and 5,000 reserves. Defense spending amounted to 3.6% of the gross domestic product (GDP) in 1995.

ECONOMIC AFFAIRS

Before its independence, Slovenia was the most highly developed and wealthiest republic of the former Yugoslav Socialist Federal Republic (which broke apart in 1991), with a per person income more than double that of the Yugoslav average, and nearly comparable to levels in neighboring Austria and Italy. The painful transition to a market-based economy has been aggravated by the disruption of intra-Yugoslav trade.

Public Finance

Fiscal policy supported a cautious monetary position maintained by the Central Bank in the early 1990s. In 1995, general government revenue was equivalent to 29% of GDP. General government expenditures in 1995 also came to 29% of GDP, with budgets and pension funds accounting for over four-fifths of expenditures. The government has kept its budget deficit within 1% of GDP every year since 1991, which has attracted foreign investors and financial market activity. The CIA estimates that, in 1995, government revenues totaled approximately $6.6 billion and expenditures $6.6 billion. External debt totaled $2.9 billion.

Income

In 1998, Slovenia's gross national product (GNP) was about $19.4 billion, or about $9,760 per person.

Industry

Manufacturing is the most prominent economic activity and is widely diversified. Important manufacturing areas include electrical and nonelectrical machinery, metal processing, chemicals, textiles and clothing, wood processing and furniture, transport equipment, and food processing.

Banking and Finance

The Bank of Slovenia is the country's central bank, and it is independent of the government. It has pursued a tight monetary and credit policy, aimed at the gradual reduction of inflation, since the introduction of the tolar in October 1991. Foreign reserves, at the end of 1996, amounted to $2.4 billion. Time commercial bank reserves were $1.9 billion.

At the end of 1996, the Bank of Slovenia changed its method of calculating the revalorization rate for other banks. From January 1997, the rate is derived from price increases over the preceding six months, instead of four months as before.

The Ljubljana Stock Exchange, abolished in 1953, was reopened in December 1989.

Economic Development

Slovenia has become a member of the IMF as well as the World Bank, from which it has obtained an $80 million loan for financial rehabilitation. The EBRD has loaned Slovenia $50 million for the improvement of the railway sector.

SOCIAL WELFARE

The constitution provides for special protection against economic, social, physical, or mental exploitation or abuse of children. Women and men have equal status under the law. Discrimination against women or minorities in housing, jobs, or other areas, is illegal. Officially, both spouses are equal in marriage, and the constitution asserts the state's responsibility to protect the family. Women are well represented in business, academia, and government, although they still hold a disproportionate share of lower-paying jobs.

Healthcare

The life expectancy at birth is 75 years. The infant mortality rate dropped from 15 to 6 infant deaths per 1,000 live births during 1980–94.

Housing

No recent information is available.

EDUCATION

Slovenia has a high literacy rate. There were 842 primary schools with 102,184 students in 1994. At the secondary level, 13,919 teachers taught 214,042 students.

In 1994, there were 43,249 enrolled university students.

1999 KEY EVENTS TIMELINE

March

- Slovenia announces that, if necessary, NATO may use its airspace for military action against Yugoslavia.
- The government's only official borrowing to finance its budget is issued in the form of a 10-year bond amounting to $408 million; the bond is issued on the best terms of any so far in post-communist Eastern Europe.

June

- In the first ever visit by a U.S. President, Bill Clinton meets with Slovene President Milan Kucan and Prime Minister Janez Drnovsek in Ljubljana.

August

- The government institutes a value-added tax (VAT). Following the model exercised by the European Union (EU), Slovenia institutes a tax paid at various stages of economic production as an item has value ''added.'' The tax is intended to allay Slovenia's inflation rate in preparation for accession to the EU.

September

- Pope John Paul II celebrates mass before an audience of over 100,000 near Maribor. The ceremony is centered on the beatification of a 19th century Slovene bishop, Anton Martin Slomsek. The Pope urges Slovenes to avoid nationalist extremism.

October

- The annual summit of the Central European Free Trade Association meets in Budapest. Slovenia opposes further liberalization in trade of agricultural products, citing differing levels of farm subsidies among the CEFTA members. Members agree on the abolition of tariffs on industrial goods by the beginning of 2000.

November

- Slovenia and three other former Yugoslav states (Macedonia, Croatia, Bosnia-Herzegovina) call on the UN to force the current Yugoslavia to reapply for UN membership. The Federal Republic of Yugoslavia (FRY), now composed of Serbia and Montenegro, holds the seat of the

Socialist FRY, which fell apart after Slovenia seceded in 1991.

- Irish Prime Minister Bertie Ahern makes an official visit to Slovenia, intended to build Irish-Slovenian relations in anticipation of Slovenia's eventual accession to the EU.

ANALYSIS OF EVENTS: 1999

BUSINESS AND THE ECONOMY

Spurred by the goal of accession to the European Union, the Slovenian government has begun to implement a reform strategy in the banking sector. The strategy, developed by Liberal Democrat Minister for European Affairs Igor Bavcar and the government's chief EU negotiator, Janez Popoonik, is composed of three main components. By privatizing the two biggest state-owned banks, opening the market to foreign competition, and encouraging consolidation, the goal is to reduce the amount of GDP produced in the public sector (55 percent in 1998) and to reduce the cost of credit.

Bank privatization has begun, albeit slowly and after several delays. The banks being privatized are Nova Ljubljanska Banka and Nova Kreditna Banka Maribor the first and third largest banks, respectively, in Slovenia. Together they hold 40 percent of the banking market. Likewise beginning slowly is the consolidation of some of the twenty-four commercial and six savings banks (which serve a population of only two million).

The major accomplishment of the strategy so far in 1999 came in August, when the legislature passed a law allowing foreign banks to open branches in Slovenia. Previously, the government had required foreign banks to establish Slovenian-incorporated subsidiaries with $30 million of reserve capital, a high hurdle for such a small market.

The August banking legislation will bring Slovenian rules of regulation and supervision in line with those of EU, although Slovenian banking is already in very good condition. The capital strength of virtually all banks is above the 8 percent minimum reserve ratio. In other economic legislation, the foreign ownership of land has been permitted.

Inflation is running at an annualized rate of six to seven percent at the end of August, very close to the EU guidelines for applicant states. Planning for the future, the governing coalition is beginning to reform the Slovenian pension system. This system is projected to draw 12 percent of GDP in a decade (compared to 3.9 percent in 1999).

Worries about the small size and fragmented nature of the Slovenian economy have pushed the government to advance the sales of stakes in up to thirty state companies in the next two years, including railways, steel, and telecommunications. Trade with countries in former Yugoslav federation is down to 15 percent (from nearly two-thirds in 1991), and most of that is with neighboring Croatia. Reform in Slovenia is moving ahead in some cases slowly, but it is being established through the democratic political process. Despite the broad privatization since 1991, 17 percent of consumer prices are still government controlled.

GOVERNMENT AND POLITICS

The coalition government of Liberal Democrats, the People's Party, and the Democratic Pensioners remains stable and is expected to retain power in its present form until the next national elections in 2000. The talks with the European Union have been successful, with observers placing Slovenia with the top candidates for inclusion. To this end, Slovenia has completed eight of thirty-one "chapters" in the Entry Negotiations that were started by the EU in March 1998. Despite the progress towards membership, turmoil in the EU has slowed the process for all of the applicant states. Germany, the largest contributor to the EU budget, is attempting to distribute the burden somewhat and to promote a reduction in support for the beneficiary states, Ireland, Greece, Spain and Portugal. While this debate continues, Slovenia must wait. As a direct result of the delay, Slovenia and five others negotiating to join the EU have called upon Brussels to set a date for the conclusion of the talks. In the Estonia meeting in October, the ministers from Slovenia, the Czech Republic, Hungary, Poland, Cyprus, and Estonia declared in a joint statement that talks should be completed by the end of next year, or 2001 at the very latest.

CULTURE AND SOCIETY

At an October Meeting with EU officials at the WTO, Slovenia negotiated a series of bilateral talks with Austria over the issue of Lipizzaner horses. The breed was founded in 1580 in Lipica, then part of the Habsburg Empire, now in Slovenia. They are a major tourist attraction in Austria, where some of the horses were moved after World War One. Earlier this year Slovenia notified the WTO that it had issued a decree protecting the geographical origin of the horses. Slovenian officials argue that the breed is a part of their national heritage. The decree is an attempt by the Slovenian nation to recover its standing as contributing member to European culture.

DIRECTORY

CENTRAL GOVERNMENT

Head of State

President
Milan Kucan, Office of the President, Erjavceva 17, 1000 Ljubljana, Slovenia
PHONE: +386 (61) 1781222
FAX: +386 (61) 1781357

Prime Minister
Janez Drnovsek

Ministers

Minister of Foreign Affairs
Boris Frlec, Ministry of Foreign Affairs, Gregorciceva 25, 1000 Ljubljana, Slovenia
PHONE: +386 (61) 1782000
FAX: +386 (61) 1782340

Minister of Interior
Borut Suklje, Ministry of Interior, Stefanova 2, 1000 Ljubljana, Slovenia
PHONE: +386 (61) 1724217
FAX: +386 (61) 214330

Minister of Finance
Mitja Gaspari, Ministry of Finance, Zupanciceva 3, 1000 Ljubljana, Slovenia
PHONE: +386 (61) 1763400
FAX: +386 (61) 1763417

Minister of Defense
Franci Demsar, Ministry of Defense, Kardeljeva Ploscad 25, 1000 Ljubljana, Slovenia
PHONE: +386 (61) 1319100
FAX: +386 (61) 1319145

Minister of Education and Sport
Pavel Zgaga, Ministry of Education and Sport, Zupanciceva 6, 1000 Ljubljana, Slovenia
PHONE: +386 (61) 1785437
FAX: +386 (61) 1785669

Minister of Science and Technology
Lojze Marincek, Ministry of Science and Technology, Trg Osvobodilne Fronte 13, 1000 Ljubljana, Slovenia
PHONE: +386 (61) 1784600
FAX: +386 (61) 1784719

Minister of Culture
Jozef Skolc, Ministry of Culture, Cankarjeva 5, 1000 Ljubljana, Slovenia
PHONE: +386 (61) 1785900
FAX: +386 (61) 1785901

Minister of Transport and Communications
Anton Bergauer, Ministry of Transport and Communication, Langusova 4, 1000 Ljubljana, Slovenia
PHONE: +386 (61) 1788000
FAX: +386 (61) 1788139

Minister of Agriculture and Forestry
Ciril Smrkolj, Ministry of Agriculture and Forestry, Dunajska 56-8, 1000 Ljubljana, Slovenia
PHONE: +386 (61) 1789000
FAX: +386 (61) 1789021

Ministry of Justice
Tomaz Marusic, Ministry of Justice, Zupanciceva 3, 1000 Ljubljana, Slovenia
PHONE: +386 (61) 1785211
FAX: +386 (61) 210200

Minister of Health
Marjan Jereb, Ministry of Health, Stefanova 5, 1000 Ljubljana, Slovenia
PHONE: +386 (61) 1786001
FAX: +386 (61) 1786058

Minister of Labor, Family and Social Affairs
Anton Rop, Ministry of Labor, Family and Social Affairs, Kotnikova 5, 1000 Ljubljana, Slovenia

POLITICAL ORGANIZATIONS

Liberal Democratic Party

TITLE: Chairman
NAME: Janez Drnovsek

Slovene Christian Democrats

TITLE: Chairman
NAME: Lozje Peterle

Slovene People's Party

TITLE: Chairman
NAME: Marjan Podobnik

DIPLOMATIC REPRESENTATION

Embassies in Slovenia

France
Barjanska 1, 1000 Ljubljana, Slovenia
PHONE: +386 (61) 1264525
FAX: +386 (61) 1250465

Germany
Presernova 27, 1000 Ljubljana, Slovenia
PHONE: +386 (61) 1790300
FAX: +386 (61) 1254210

Italy
Snezniska 8, 1000 Ljubljana, Slovenia
PHONE: +386 (61) 1262194
FAX: +386 (61) 1253302

United Kingdom
Trg Republike 3/IV, 1000 Ljubljana, Slovenia
PHONE: +386 (61) 1257191
FAX: +386 (61) 1250174

United States
Prazakova 4, 1000 Ljubljana, Slovenia
PHONE: +386 (61) 301427
FAX: +386 (61) 301401

JUDICIAL SYSTEM

Supreme Court

Tavcarjeva 9, 1000 Ljubljana, Slovenia
PHONE: +386 (61) 1323133
FAX: +386 (61) 1344807

Constitutional Court

Beethovnova 10, 1000 Ljubljana, Slovenia
PHONE: +386 (61) 210448
FAX: +386 (61) 210451

District Courts

Local Courts

FURTHER READING

Articles

Choudhry, Taufiq. ''Inder Cointegration Tests.'' *Journal of Macroeconomics* 21 (Spring 1999): 293.

Jones, Colin. ''Independent Stance.'' *The Banker* 149 (January 1999): 39.

———. ''Looking to the Future.'' *The Banker* 149 (August 1999) 37.

Koenig, Robert. ''Money and Mentors Hold on to Young Researchers.'' *Science* 283 (January 1, 1999): 25.

''Slovenia: The View from the Outside Looking In.'' *Time International* 153 (March 22, 1999): 31.

Vodopivec, Milan. ''Does the Slovenian Public Work Program Increase Participants' Chances to Find a Job?'' *Journal of Comparative Economics* 27 (March 1999): 113.

Internet

Central Europe Online-Slovenia. Available Online @ http://www.centraleurope.com/country/sloveniatoday/ (November 16, 1999).

Slovenia Quarterly Magazine. Available Online @ http://www.arctur.si/slovenia/ (November 16, 1999).

SLOVENIA: STATISTICAL DATA

For sources and notes see "Sources of Statistics" in the front of each volume.

GEOGRAPHY

Geography (1)

Area:

Total: 20,256 sq km.

Land: 20,256 sq km.

Water: 0 sq km.

Area—comparative: slightly smaller than New Jersey.

Land boundaries:

Total: 1,334 km.

Border countries: Austria 330 km, Croatia 670 km, Italy 232 km, Hungary 102 km.

Coastline: 46.6 km.

Climate: Mediterranean climate on the coast, continental climate with mild to hot summers and cold winters in the plateaus and valleys to the east.

Terrain: a short coastal strip on the Adriatic, an alpine mountain region adjacent to Italy, mixed mountain and valleys with numerous rivers to the east.

Natural resources: lignite coal, lead, zinc, mercury, uranium, silver.

Land use:

Arable land: 12%

Permanent crops: 3%

Permanent pastures: 28%

Forests and woodland: 51%

Other: 6% (1993 est.)

HUMAN FACTORS

Demographics (2A)

	1990	1995	1998	2000	2010	2020	2030	2040	2050
Population	NA	1,970.5	1,971.7	1,970.1	1,975.9	1,916.7	1,804.2	1,660.3	1,484.0
Net migration rate (per 1,000 population)	NA	NA	NA	NA	NA	NA	NA	NA	NA
Births	NA	NA	NA	NA	NA	NA	NA	NA	NA
Deaths	NA	NA	NA	NA	NA	NA	NA	NA	NA
Life expectancy - males	NA	70.8	71.5	71.9	74.0	75.7	77.1	78.2	79.0
Life expectancy - females	NA	78.5	79.0	79.4	81.0	82.4	83.5	84.4	85.2
Birth rate (per 1,000)	NA	9.5	8.6	9.4	9.1	7.0	6.7	6.1	5.4
Death rate (per 1,000)	NA	9.6	9.6	9.7	10.8	12.0	13.7	16.0	17.8
Women of reproductive age (15-49 yrs.)	NA	511.8	516.8	513.1	466.4	410.5	355.3	293.9	259.1
of which are currently married	NA	NA	NA	NA	NA	NA	NA	NA	NA
Fertility rate	NA	1.3	1.2	1.3	1.4	1.3	1.3	1.3	1.3

Except as noted, values for vital statistics are in thousands; life expectancy is in years.

Health Personnel (3)

Total health expenditure as a percentage of GDP, 1990-1997[a]

Public sector	7.1
Private sector	NA
Total[b]	NA

Health expenditure per capita in U.S. dollars, 1990-1997[a]

Purchasing power parity	NA
Total	NA

Availability of health care facilities per 100,000 people

Hospital beds 1990-1997[a]	570
Doctors 1993[c]	219
Nurses 1993[c]	686

Health Indicators (4)

Life expectancy at birth	
1980	70
1997	75
Daily per capita supply of calories (1996)	3,117
Total fertility rate births per woman (1997)	1.3
Maternal mortality ratio per 100,000 live births (1990-97)	5[b]
Safe water % of population with access (1995)	98
Sanitation % of population with access (1995)	98
Consumption of iodized salt % of households (1992-98)[a]	....
Smoking prevalence	
Male % of adults (1985-95)[a]	35
Female % of adults (1985-95)[a]	23
Tuberculosis incidence per 100,000 people (1997)	30
Adult HIV prevalence % of population ages 15-49 (1997)	<0.005

Infants and Malnutrition (5)

Under-5 mortality rate (1997)	6
% of infants with low birthweight (1990-97)	NA
Births attended by skilled health staff % of total[a]	100
% fully immunized (1995-97)	
TB	98
DPT	91
Polio	98x
Measles	92
Prevalence of child malnutrition under age 5 (1992-97)[b]	NA

Ethnic Division (6)

Slovene	91%
Croat	3%
Serb	2%
Muslim	1%
Other	3%

Religions (7)

Roman Catholic	70.8%
Lutheran	1%
Muslim	1%
Atheist	4.3%
Other	22.9%

Roman Catholic includes 2% Uniate.

Languages (8)

Slovenian 91%, Serbo-Croatian 6%, other 3%.

EDUCATION

Public Education Expenditures (9)

Public expenditure on education (% of GNP)	
1980	....
1996	5.8[1]
Expenditure per student	
Primary % of GNP per capita	
1980	....
1996	20.5[1]
Secondary % of GNP per capita	
1980	....
1996	24.2[1]
Tertiary % of GNP per capita	
1980	....
1996	37.6[1]
Expenditure on teaching materials	
Primary % of total for level (1996)	....
Secondary % of total for level (1996)	6.1[1]
Primary pupil-teacher ratio per teacher (1996)	14
Duration of primary education years (1995)	8

Educational Attainment (10)

Age group (1991)	25+
Total population	1,272,409
Highest level attained (%)	
No schooling	0.7
First level	

Not completed	45.1
Completed	NA
Entered second level	
S-1	42.4
S-2	NA
Postsecondary	10.4

Literacy Rates (11B)

Adult literacy rate	
1980	
Male	-
Female	-
1995	
Male	100%
Female	99%

GOVERNMENT & LAW

Political Parties (12)

National Assembly	% of seats
Liberal Democratic (LDS)	27.01
Slovene People's Party (SLS)	19.38
Social Democratic Party of Slovenia (SDS)	16.13
Slovene Christian Democrats (SKD)	9.62
United List (former Communists and allies) (ZLSD)	9.03
Democratic Party of Retired (Persons) of Slovenia (DeSUS)	4.32
Slovene National Party (SNS)	3.22

Government Budget (13B)

Revenues	$8.48 billion
Expenditures	$8.53 billion
Capital expenditures	$455 million

Data for 1996 est.

Military Affairs (14B)

	1992	1993	1994	1995
Military expenditures				
Current dollars (mil.)	341	195	438	344
1995 constant dollars (mil.)	367	204	449	344
Armed forces (000)	15	12	17	10
Gross national product (GNP)				
Current dollars (mil.)	18,670	19,470	21,040	22,600
1995 constant dollars (mil.)	20,080	20,420	21,570	22,600
Central government expenditures (CGE)				
1995 constant dollars (mil.)	NA	8,654	9,295	9,904
People (mil.)	2.0	2.0	2.0	2.0
Military expenditure as % of GNP	1.8	1.0	2.1	1.5
Military expenditure as % of CGE	NA	2.4	4.8	3.5
Military expenditure per capita (1995 $)	187	104	229	1.7
Armed forces per 1,000 people (soldiers)	7.6	6.1	8.7	5.1
GNP per capita (1995 $)	10,220	10,400	11,000	11,550
Arms imports[6]				
Current dollars (mil.)	0	0	10	30
1995 constant dollars (mil.)	0	0	10	30
Arms exports[6]				
Current dollars (mil.)	10	0	5	5
1995 constant dollars (mil.)	11	0	5	5
Total imports[7]				
Current dollars (mil.)	6,142	6,529	7,304	9,452
1995 constant dollars (mil.)	6,607	6,845	7,487	9,452
Total exports[7]				
Current dollars (mil.)	6,142	6,529	7,304	8,286
1995 constant dollars (mil.)	6,607	6,845	7,487	8,286
Arms as percent of total imports[8]	0.0	0	.1	.3
Arms as percent of total exports[8]	0.2	0	.1	.1

Crime (15)

Crime rate (for 1997)	
Crimes reported	37,200
Total persons convicted	24,000
Crimes per 100,000 population	1,900
Persons responsible for offenses	
Total number of suspects	30,200
Total number of female suspects	3,850
Total number of juvenile suspects	4,600

LABOR FORCE

Labor Force (16)

Total 857,400

Services 62%

Industry 36%

Agriculture 2%

Data for 1995.

Unemployment Rate (17)

7.1% (1997 est.)

PRODUCTION SECTOR

Electric Energy (18)

Capacity . . . . .2.524 million kW (1995)

Production . . . . .11.615 billion kWh (1995)

Consumption per capita . . . . .5,759 kWh (1995)

Transportation (19)

Highways:

total: 14,910 km

paved: 12,226 km (including 231 km of expressways)

unpaved: 2,684 km (1996 est.)

Waterways: NA

Pipelines: crude oil 290 km; natural gas 305 km

Merchant marine:

total: 13 ships (1,000 GRT or over) totaling 223,976 GRT/373,462 DWT (controlled by Slovenian owners) ships by type: bulk 8, cargo 5 note: ships operate under the flags of Antigua and Barbuda, Liberia, Saint Vincent and the Grenadines, and Singapore; no ships remain under the Slovenian flag (1997 est.)

Airports: 14 (1997 est.)

Airports—with paved runways:

total: 6

over 3,047 m: 1

2,438 to 3,047 m: 1

1,524 to 2,437 m: 1

914 to 1,523 m: 2

under 914 m: 1 (1997 est.)

Airports—with unpaved runways:

total: 8

1,524 to 2,437 m: 2

914 to 1,523 m: 2

under 914 m: 4 (1997 est.)

Top Agricultural Products (20)

Potatoes, hops, wheat, sugar beets, corn, grapes; cattle, sheep, poultry.

MANUFACTURING SECTOR

GDP & Manufacturing Summary (21)

Detailed value added figures are listed by both International Standard Industry Code (ISIC) and product title.

	1980	1985	1990	1994
GDP ($-1990 mil.)[1]	8,957	9,240	8,679	7,796
Per capita ($-1990)[1]	4,889	4,913	4,525	4,014
Manufacturing share (%) (current prices)[1]	40.8	40.8	32.5	*29.6*
Manufacturing				
Value added ($-1990 mil.)[1]	NA	NA	2,582	NA
Industrial production index	NA	NA	NA	NA
Value added ($ mil.)	*3,390*	*2,219*	2,509	4,837
Gross output ($ mil.)	*17,050*	*9,380*	7,900	13,913
Employment (000)	*506*	*399*	356	265
Profitability (% of gross output)				
Intermediate input (%)	*80*	*76*	66	*65*
Wages and salaries inc. supplements (%)	*18*	*18*	22	*15*
Gross operating surplus	*2*	*6*	*12*	*20*
Productivity ($)				
Gross output per worker	*31,827*	*21,926*	21,226	48,679
Value added per worker	*6,649*	*5,386*	7,239	16,924
Average wage (inc. supplements)	*6,049*	*4,102*	*4,788*	*7,919*
Value added ($ mil.)				
311/2 Food products	*45*	*82*	*214*	608
313 Beverages	*53*	*39*	*64*	102
314 Tobacco products	*1*	*3*	*26*	25
321 Textiles	*668*	379	*291*	176
322 Wearing apparel	*2*	*4*	*24*	194
323 Leather and fur products	NA	NA	153[d]	121
324 Footwear	NA	NA	[d]	65

	1980	1985	1990	1994
331 Wood and wood products	*87*	*93*	97	147
332 Furniture and fixtures	*112*	*79*	*78*	246
341 Paper and paper products	*52*	*68*	*86*	299
342 Printing and publishing	*31*	*45*	*80*	190
351 Industrial chemicals	*103*	*80*	*117*	389
352 Other chemical products	*95*	*72*	*126*	216
353 Petroleum refineries	*9*	*5*	5	10
354 Miscellaneous petroleum and coal products	NA	NA		
355 Rubber products	*71*	*46*	*55*	*79*
356 Plastic products	*85*	*50*	*50*	80
361 Pottery, china and earthenware	*7*	*6*	*11*	*15*
362 Glass and glass products	*29*	*18*	*22*	*34*
369 Other non-metal mineral products	*126*	*79*	*92*	*147*
371 Iron and steel	*204*	*119*	*98*	469
372 Non-ferrous metals	*135*	*44*	*66*	14
381 Metal products	*247*	*209*	*253*	230
382 Non-electrical machinery	*542*	*263*	*175*	290
383 Electrical machinery	*460*	272	273	397
384 Transport equipment	*134*	*103*	144	187
385 Professional and scientific equipment	*3*	*7*	*20*	85
390 Other manufacturing industries	*90*	*54*	*43*	20

FINANCE, ECONOMICS, & TRADE

Economic Indicators (22)

National product: GDP—purchasing power parity—$19.5 billion (1997 est.)

National product real growth rate: 3.25% (1997 est.)

National product per capita: $10,000 (1997 est.)

Inflation rate—consumer price index: 9.7% (1996)

Exchange Rates (24)

Exchange rates:

Tolars (SIT) per US$1

January 1998	171.30
1997	159.69
1996	135.36
1995	118.52
1994	128.81
1993	113.24

Top Import Origins (25)

$9.5 billion (f.o.b., 1996) Data are for 1996.

Origins	%
Germany	22
Italy	17
France	10
Austria	10
Croatia	6
United States	3

Balance of Payments (23)

	1992	1993	1994	1995	1996
Exports of goods (f.o.b.)	6,681	6,083	6,830	8,350	8,370
Imports of goods (f.o.b.)	−5,892	−6,237	−7,168	−9,305	−9,252
Trade balance	789	−154	−338	−954	−882
Services - debits	−1,188	−1,184	−1,294	−1,622	−1,688
Services - credits	1,331	1,507	2,139	2,462	2,547
Private transfers (net)	−31	−59	−109	−82	−79
Government transfers (net)	77	81	201	173	141
Overall balance	978	191	600	−23	39

Top Export Destinations (26)

$8.3 billion (f.o.b., 1996) Data are for 1996.

Destinations	%
Germany	31
former Yugoslavia	16.5
Italy	13
Croatia	10
France	7
Austria	7
United States	5

Economic Aid (27)

Recipient: ODA, $5 million (1993).

Import Export Commodities (28)

Import Commodities	Export Commodities
Machinery and transport equipment 33.8%	Manufactured goods 50.7%
Manufactured goods 30.4%	Machinery and transport equipment 31.4%
Chemicals 12.1%	Chemicals 10.5%
Fuels and lubricants 6.6%	Food 3.8%
Food 8.4%	

SOLOMON ISLANDS

CAPITAL: Honiara.

FLAG: The flag consists of two triangles, the upper one blue, the lower one green, separated by a diagonal gold stripe; on the blue triangle are five white five-pointed stars.

ANTHEM: *God Save the Queen.*

MONETARY UNIT: The Solomon Islands dollar (SI$), a paper currency of 100 cents, was introduced in 1977, replacing the Australian dollar, and became the sole legal tender in 1978. There are coins of 1, 2, 5, 10, 20, and 50 cents and 1 dollar, and notes of 2, 5, 10, 20, and 50 dollars. SI$1 = US$0.277 (or US$1 = SI$3.61).

WEIGHTS AND MEASURES: The metric system is in force.

HOLIDAYS: New Year's Day, 1 January; Queen's Birthday, June; Independence Day, 7 July; Christmas, 25 December; Boxing Day, 26 December. Movable religious holidays include Good Friday, Easter Monday, and Whitmonday.

TIME: 11 PM = noon GMT.

LOCATION AND SIZE: The Solomon Islands consist of a chain of six large and numerous small islands situated in the South Pacific. The Solomon Islands have an area of 28,450 square kilometers (10,985 square miles), slightly larger than the state of Maryland.

The largest island is Guadalcanal, covering 5,302 square kilometers (2,047 square miles). The total coastline of the Solomon Islands is 5,313 kilometers (3,301 miles).

The capital city, Honiara, is located on the island of Guadalcanal.

CLIMATE: The average daily temperature is about 26–27°C (79–81°F); annual rainfall averages 210 centimeters (83 inches); humidity is about 80%. Damaging cyclones occur periodically.

INTRODUCTORY SURVEY

RECENT HISTORY

In the decades following World War II, the Solomons moved gradually toward independence. The islands achieved internal self-government in 1976 and became an independent member of the Commonwealth of Nations on 7 July 1978.

Francis Billy Hilly became the Solomon Islands' new prime minister in June 1993. Hilly has worked with the Melanesian Spearhead Conference to ease tension between the Solomon Islands and Papua New Guinea. In 1994, the parliament voted to replace Hilly with Solomon Mamaloni, leader of the National Unity and Reconciliation Group, the largest political party in the parliament.

GOVERNMENT

The Solomon Islands are a parliamentary democracy with a prime minister and a single-chamber forty-seven-member National Parliament.

Judiciary

The judicial system consists of the High Court, magistrate courts, and local courts. Appeals from magistrate courts go to the High Court; customary land appeals courts hear appeals from the local courts.

Political Parties

Parties have included the People's Alliance Party (PAP), the National Democratic Party (NDP), and the Nationalist Front for Progress. The Group for National Unity and Reconciliation (GNUR), led by Solomon Mamaloni, gained the

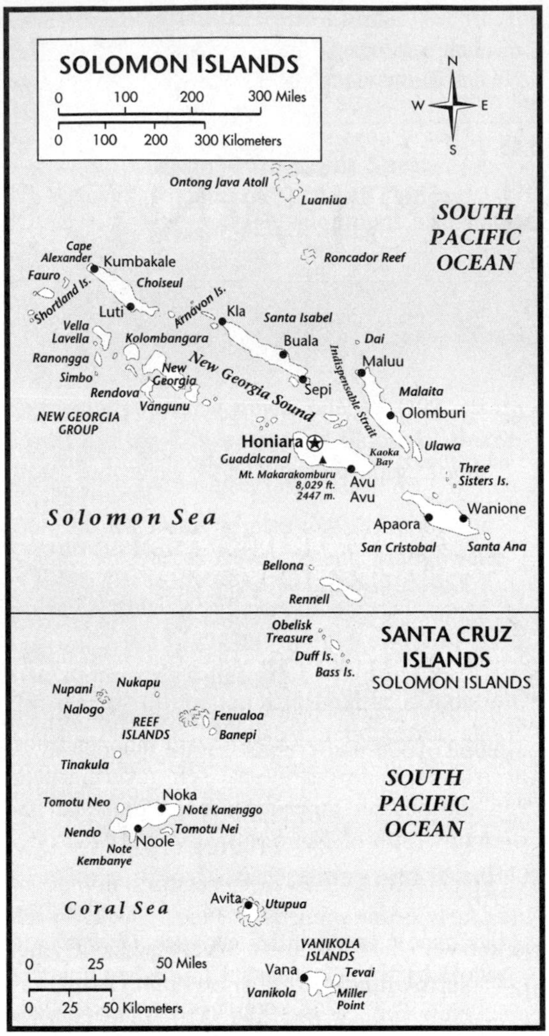

most seats in the 1993 election. Other parties receiving seats were the National Action Party, the Labour Party, and the Christian Fellowship Group.

DEFENSE

The Solomon Islands have no military forces.

ECONOMIC AFFAIRS

At least 80% of the population is tied to farming. The economy depends on the export of copra (dried coconut meat), timber, and fish. Production of other cash commodities—particularly cocoa, spices, and palm oil—has grown in recent years.

Public Finance

In 1995, the government had revenues of $83 million and expenditures of $101.9 million, which have edged it to the brink of insolvency. The 1996 budget was expected to show a deficit in excess of $20 million despite government attempts to restrain spending.

Income

In 1998, Solomon Islands' gross national product (GNP) was $311 million at current prices, or about $750 per person. For the period 1985–95, the average inflation rate was 11.7%, resulting in a real growth rate in per person GNP of 2.2%.

Industry

The leading industries are fish processing and timber milling; soaps are made from palm oil and coconut oil. Small firms produce goods for local consumption, including biscuits, tobacco products, rattan furniture, and baskets and mats.

Banking and Finance

The Solomon Islands Monetary Authority became the Central Bank of the Solomon Islands (CBSI) in January 1983. Three commercial banks also operate on the islands.

The government participates in private investment projects through a holding company, the Investment Corp. of Solomon Islands (ICSI), the successor to the Government Shareholding Agency.

Money supply, as measured by M2, totaled SI$312.58 at the end of 1994. There is no securities market.

Economic Development

The government has attempted to diversify agricultural production in order to make the economy less vulnerable to world price fluctuations of such key cash crops as copra. Important development projects during the 1980s included new sawmills, a fish cannery, a spice industry, and the Lungga hydroelectric plant. Fisheries receive significant portions of development funds. A rubber industry is being developed, and plans are under way to export the indigenous ngali nut as an upscale confectionery product under the name "Solomons nut." In 1991, plans were announced for a SI$60 million hydroelectric plant on Guadalcanal, financed by the Asian Development Bank.

Foreign assistance plays an essential role in the nation's development strategy, with Australia and Japan the largest donors.

SOCIAL WELFARE

A National Provident Fund provides old-age, disability, and survivor benefits. Most organized welfare services are provided by church missions.

Much assistance is traditionally provided through the extended family.

Healthcare

Malaria and tuberculosis are widespread. In 1985, there were 38 doctors, 6 government hospitals, a mission hospital, and 124 clinics. Average life expectancy was 63 years.

Housing

The government has built low-cost housing projects in Honiara to help ease congestion. Outside Honiara, housing is primitive, with overcrowding a problem. As of 1990, 82% of urban and 58% of rural dwellers had access to a public water supply, while 73% of the urban population had access to sanitation services.

EDUCATION

About 60% of the adult population is estimated to be literate. In 1994, there were 60,493 students and 2,514 teachers in primary schools. Secondary schools had 7,811 pupils that year.

Higher education is provided by the Solomon Islands Teachers College, the Honiara Technical Institute, and the University of the South Pacific.

1999 KEY EVENTS TIMELINE

June

- The Solomon Islands joins in a South Pacific Forum call for the establishment of a regional free trade zone.
- Thousands are displaced as tensions flare between natives and ethnic Malaitans on Guadalcanal.
- The Solomon Islands government passes the Emergency Powers Act restricting journalistic activity.
- The prime minister of Fiji holds talks with Guadalcanalese militants in the hopes of mediating a peace settlement.

September

- The Solomon Islands supports a request that Taiwanese membership be included on the agenda of the United Nations General Assembly.

ANALYSIS OF EVENTS: 1999

BUSINESS AND THE ECONOMY

At the South Pacific Forum in June, the Solomon Islands supported a Pacific Free Trade Area (FTA) that would initially include 14 countries in the region, with the possibility of future expansion. The proposed FTA would offset ramifications of the imminent expiration of the Lome Convention agreement between the European Union (EU) and 71 African, Caribbean, and Pacific (ACP) countries. The Lome Convention agreement is subject to a World Trade Organization (WTO) waiver set to expire in February 2000, though the EU has announced plans to petition for an extension to 2006. The South Pacific FTA would also buffer its members against the expiration in 2006 of the WTO waiver for the U.S. Compact of Free Association, which selectively offers trade preferences to member countries. The Solomon Islands tuna industry, one of the nation's only abundant sources of income, will lose its tariff-free access to the EU market and be forced to compete with tuna producers like the Philippines and Thailand, who pay a 24 percent import duty.

The Solomon Islands were particularly hard hit by the recent economic downturn in Asia. Unlike other South Pacific nations, which were largely unaffected by problems in Asia, countries in Asia receive two-thirds of the Solomon Islands' exports.

The Solomon Islands were among the Asian Pacific countries named in an October report by the United Nations' International Labor Organization (ILO) that could benefit economically from a commitment of resources to repairing and expanding of national infrastructure. Road construction and repair were among the projects specifically outlined in the ILO report. In the Solomon Islands, where both unemployment and population growth are high, benefits would come in the form of an increase in jobs and personal income and cheaper and more efficient transportation.

Airlines in the Solomon Islands issued assurances that their equipment and procedures were in conformity with international standards for larger airlines and evinced no problems with a satellite switchover that could have affected users of the satellite-based Global Positioning System (GPS). In August, the 24 satellites of the GPS began a new cycle of 1,024 weeks after completing the original cycle begun at the satellites' launch in 1980.

Operations at the Gold Ridge gold mine on Guadalcanal were interrupted by the ethnic violence that broke out on the island in June and July. Australian-based Ross Mining, which owns the gold mine, has also been subjected to lawsuits filed by local inhabitants, represented by an Australian law firm. The lawsuits charge environmental irresponsibility on the part of Ross Mining.

GOVERNMENT AND POLITICS

In response to attacks on Malaitans living on Guadalcanal, the government of the Solomon Islands passed the Emergency Powers Act, which prohibits journalists from reporting that may promote violence or unrest or jeopardize the stability of the state. The Committee to Protect Journalists (CPJ) in New York denounced the legislation as a violation of the nineteenth article of the Universal Declaration of Human Rights.

In September, the Solomon Islands joined other island nations in requesting that United Nations membership for Taiwan be included on the agenda of the General Assembly. The request was opposed by the U.S., which supports Beijing's "one China" policy, whereby Taiwan is considered part of mainland China. The Assembly expelled Taiwan from the U.N. in 1971 and gave its seat to Beijing.

CULTURE AND SOCIETY

Tensions erupted between natives of Guadalcanal and ethnic Malaitans living on Guadalcanal, where the Solomon Islands capital of Honiara is situated. Armed Guadalcanalese drove some 20,000 Malaitans from their homes, 6,000 of whom fled back to Malaita. Some Malaitans offered resistance to the Guadalcanalese, who regard the Malaitans as squatters and blame them for an increase in crime around the capital. Since there is only a small police force and no standing army in the Solomon Islands, such violence is difficult to check. The national government appropriated half a million dollars toward the repatriation of Malaitans displaced by the violence, though officials in Honiara announced in July that about $200,000 of the apportioned funds had been misplaced. The prime minister of Fiji traveled to the Solomon Islands to hold talks with the Guadalcanalese militants, who agreed to dissolve their militia groups and surrender their weapons to local chiefs.

In September, The United Nations Environment Program cited the Solomon Islands as one of the small island developing states (SIDS) whose environment has become increasingly threatened by increases in population, consumption, and urbanization, all of which increase pollution factors. Tourism, a mainstay of small island economies, also has put more pressure on the environment. A severe lack of resources compounds the environmental problems faced by the Solomon Islands and other countries.

DIRECTORY

CENTRAL GOVERNMENT

Head of State

Queen
Elizabeth II, Monarch

Governor-General
Moses Pitakaka, Office of the Governor-General

Prime Minister
Bartholomew Ulufa'alu, Office of the Prime Minister, PO Box G1, Honiara, Guadalcanal, Solomon Islands
PHONE: +677 21863
FAX: +677 26088

Ministers

Minister of Foreign Affairs and Trade Relations
Patteson Oti, Ministry of Foreign Affairs and Trade Relations, PO Box G-10, Honiara, Guadalcanal, Solomon Islands
PHONE: +677 21250; 22223
FAX: +677 20351

Minister of Commerce, Employment and Tourism
Moses K. Mose, Ministry of Commerce, Employment and Tourism, PO Box G26, Honiara, Guadalcanal, Solomon Islands
PHONE: +677 21849
FAX: +677 25084; 22808
E-MAIL: ps@commerce.gov.sb

Minister of Agriculture and Fisheries
S. Steve Aumanu, Ministry of Agriculture and Fisheries, PO Box G13, Honiara, Guadalcanal, Solomon Islands
PHONE: +677 21326; 21327; 21328
FAX: +677 21955
E-MAIL: drsteve@welkam.solomon.com.sb

Minister of Education and Human Resource Development
Ronnie Mannie, Ministry of Education and Human Resource Development, PO Box G28, Honiara, Guadalcanal, Solomon Islands
PHONE: +677 23900
FAX: +677 20485

Minister of Energy, Mines and Water Resources
Walton Naeson, Ministry of Energy, Mines and Water Resources, PO Box G37, Honiara, Guadalcanal, Solomon Islands
PHONE: +677 21521
FAX: +677 20392; 25811

Minister of Finance
Alpha Kimata, Ministry of Finance, PO Box G26, Honiara, Guadalcanal, Solomon Islands
PHONE: +677 23700
FAX: +677 20392
E-MAIL: finance@welkam.solomon.com.sb

Minister of Forestry, Environment and Conservation
Hilda Kari, Ministry of Forestry, Environment and Conservation, PO Box G24, Honiara, Guadalcanal, Solomon Islands
PHONE: +677 25849
FAX: +677 21245

Minister of Health and Medical Services
Dick Warakohia, Ministry of Health and Medical Services, PO Box 349, Honiara, Guadalcanal, Solomon Islands
PHONE: +677 20830
FAX: +677 20085

Minister of Home Affairs
Leslie Boseto, Ministry of Home Affairs, PO Box G11, Honiara, Guadalcanal, Solomon Islands
PHONE: +677 21621
FAX: +677 22606

Minister of Indigenous Business
Ministry of Indigenous Business, PO Box G19, Honiara, Guadalcanal, Solomon Islands
PHONE: +677 26851
FAX: +677 26861
E-MAIL: Caroline@welkam.solomon.com.sb

Minister of Justice and Legal Affairs
Edmond Andresen Karaer, Ministry of Justice and Legal Affairs, PO Box 404, Honiara, Guadalcanal, Solomon Islands
PHONE: +677 21181
FAX: +677 25610

Minister of Lands and Housing
Jackson Piasi, Ministry of Lands and Housing, PO Box G38, Honiara, Guadalcanal, Solomon Islands
PHONE: +677 21430
FAX: +677 26134

Minister of National Planning and Development
Fred Fono, Ministry of National Planning and Development, PO Box G30, Honiara, Guadalcanal, Solomon Islands
PHONE: +677 38255
FAX: +677 38259
E-MAIL: sekera@welkam.solomon.com.sb

Minister of Provincial Government and Rural Development
Japhet Waipora, Ministry of Provincial Government and Rural Development, PO Box G35, Honiara, Guadalcanal, Solomon Islands
PHONE: +677 21140
FAX: +677 21289

Minister of Police and National Security
R. Mesepitu, Ministry of Police and National Security, PO Box G1723, Honiara, Guadalcanal, Solomon Islands
PHONE: +677 22208
FAX: +677 25949
E-MAIL: compol@welkam.solomon.com.sb

Minister of Transport, Works, Communications, and Aviation
Baddeley Devesi, Ministry of Transport, Works, Communications, and Aviation, PO Box G8, Honiara, Guadalcanal, Solomon Islands
PHONE: +677 21141
FAX: +677 20182

Minister of Youth, Women, Sports, and Recreation
Gordon Mara, Ministry of Youth, Women, Sports, and Recreation, PO Box G39, Honiara, Guadalcanal, Solomon Islands
PHONE: +677 25490
FAX: +677 21689
E-MAIL: wdd@welkam.solomon.com.sb

POLITICAL ORGANIZATIONS

Alliance for Change (AC)

Solomon Islands Liberal Party (SILP)

NAME: Bartholomew Ulufa'alu

National Action Party (NAP)

Group for National Unity and Reconciliation Progressive Party

NAME: Solomon Mamaloni

People's Alliance Party (PAP)

Solomon Islands Labor Party

NAME: Joses Tuhanuku

United Party (SIUPA)

DIPLOMATIC REPRESENTATION

Embassies in Solomon Islands

Australia
Cnr Hibiscus Ave. and Mud Alley, Honiara, Guadalcanal, Solomon Islands
PHONE: +677 21561
FAX: +677 23691

China
Panatina Plaza, 1st Fl., Honiara, Guadalcanal, Solomon Islands
PHONE: +677 38050
FAX: +677 38060

Japan
National Provident Fund Building, Mendana Avenue, Honiara, Guadalcanal, Solomon Islands
PHONE: +677 22953
FAX: +677 21006

United Kingdom
Telekon House, Mendana Ave., Honiara, Guadalcanal, Solomon Islands
PHONE: +677 21705; 21706
FAX: +677 21549
TITLE: High Commissioner
NAME: B. N. Connelly

JUDICIAL SYSTEM

High Court and Court of Appeal

Honiara, Guadalcanal, Solomon Islands
PHONE: +677 21362
FAX: +677 22702

FURTHER READING

Articles

Clausen, Lisa. ''Strife in the Happy Isles: Decades of Tension Take a Violent Turn as Militants in the Solomon Islands Vent Ethnic Grievances.'' *Time International* 153 (June 28, 1999): 52.

Giarelli, Andrew. ''Ethnic cleansing.'' *World Press Review* 46 (September 1999): 22.

Books

Trade Policy Review: Solomon Islands. Geneva, Switzerland: World Trade Organization, 1999.

Internet

Ethnic Tension on Guadalcanal. Available Online @ http://www.geocities.com/TheTropics/Harbor/2946/ (November 13, 1999).

Solomon Islands Department Ministry of Commerce Available Online @ http://www.commerce.gov.sb/ (November 13, 1999).

SOLOMON ISLANDS: STATISTICAL DATA

For sources and notes see "Sources of Statistics" in the front of each volume.

GEOGRAPHY

Geography (1)

Area:

Total: 28,450 sq km.

Land: 27,540 sq km.

Water: 910 sq km.

Area—comparative: slightly smaller than Maryland.

Land boundaries: 0 km.

Coastline: 5,313 km.

Climate: tropical monsoon; few extremes of temperature and weather.

Terrain: mostly rugged mountains with some low coral atolls.

Natural resources: fish, forests, gold, bauxite, phosphates, lead, zinc, nickel.

Land use:

Arable land: 1%

Permanent crops: 1%

Permanent pastures: 1%

Forests and woodland: 88%

Other: 9% (1993 est.)

HUMAN FACTORS

Infants and Malnutrition (5)

Under-5 mortality rate (1997) .... .28

% of infants with low birthweight (1990-97) .... .20

Births attended by skilled health staff % of total[a] . . .NA

% fully immunized (1995-97)

- TB .... .73
- DPT .... .72
- Polio .... .70
- Measles .... .68

Prevalence of child malnutrition under age 5 (1992-97)[b] .... .NA

Demographics (2A)

	1990	1995	1998	2000	2010	2020	2030	2040	2050
Population	335.5	399.2	441.0	470.0	619.7	767.3	911.2	1,043.5	1,157.8
Net migration rate (per 1,000 population)	NA	NA	NA	NA	NA	NA	NA	NA	NA
Births	NA	NA	NA	NA	NA	NA	NA	NA	NA
Deaths	NA	NA	NA	NA	NA	NA	NA	NA	NA
Life expectancy - males	66.8	68.4	69.3	69.8	72.4	74.5	76.1	77.3	78.3
Life expectancy - females	71.6	73.4	74.4	75.1	78.0	80.3	82.0	83.3	84.3
Birth rate (per 1,000)	40.7	38.5	36.6	35.2	27.5	22.4	19.0	16.2	14.6
Death rate (per 1,000)	5.2	4.5	4.2	4.0	3.4	3.4	3.7	4.4	5.6
Women of reproductive age (15-49 yrs.)	71.1	88.0	99.3	107.4	154.6	207.4	248.0	275.4	287.1
of which are currently married	NA	NA	NA	NA	NA	NA	NA	NA	NA
Fertility rate	6.3	5.6	5.1	4.8	3.4	2.6	2.2	2.1	2.0

Except as noted, values for vital statistics are in thousands; life expectancy is in years.

Ethnic Division (6)

Melanesian	93%
Polynesian	4%
Micronesian	1.5%
European	0.8%
Chinese	0.3%
Other	0.4%

Religions (7)

Anglican	34%
Roman Catholic	19%
Baptist	17%
United (Methodist/Presbyterian)	11%
Seventh-Day Adventist	10%
Other Protestant	5%
Traditional beliefs	4%

Languages (8)

Melanesian pidgin in much of the country is lingua franca, English spoken by 1%-2% of population. 120 indigenous languages.

EDUCATION

Literacy Rates (11B)

Adult literacy rate	
1980	
Male	-
Female	-
1995	
Male	-
Female	56%

GOVERNMENT & LAW

Political Parties (12)

National Parliament	No. of seats
Group for National Unity and Reconciliation (GNUR)	21
People's Alliance Party (PAP)	7
National Action Party of Solomon Islands (NAPSI)	5
Solomon Islands Labor Party (SILP)	4
United Party (UP)	4
Independents	6
Other	3

Government Budget (13B)

Revenues	$147 million
Expenditures	$168 million
Capital expenditures	NA

Data for 1997 est. NA stands for not available.

Military Affairs (14A)

Total expenditures	$NA
Expenditures as % of GDP	NA%

NA stands for not available.

LABOR FORCE

Labor Force (16)

Total	26,842
Services	41.5%
Agriculture, forestry, and fishing	23.7%
Commerce, transport, and finance	21.7%
Construction, manufacturing, and mining	13.1%

Data for 1992 est.

Unemployment Rate (17)

Rate not available.

PRODUCTION SECTOR

Electric Energy (18)

Capacity	12,000 kW (1995)
Production	30 million kWh (1995)
Consumption per capita	75 kWh (1995)

Transportation (19)

Highways:

total: 1,360 km

paved: 34 km

unpaved: 1,326 km (includes about 800 km of private plantation roads) (1996 est.)

Merchant marine: none

Airports: 32 (1997 est.)

Airports—with paved runways:

total: 2

1,524 to 2,437 m: 1

914 to 1,523 m: 1 (1997 est.)

Airports—with unpaved runways:

total: 30

1,524 to 2,437 m: 1

914 to 1,523 m: 9

under 914 m: 20 (1997 est.)

Top Agricultural Products (20)

Cocoa, beans, coconuts, palm kernels, rice, potatoes, vegetables, fruit; cattle, pigs; timber; fish.

FINANCE, ECONOMICS, & TRADE

Economic Indicators (22)

National product: GDP—purchasing power parity—$1.27 billion (1997 est.)

National product real growth rate: 3.5% (1997 est.)

National product per capita: $3,000 (1997 est.)

Inflation rate—consumer price index: 11.8% (1996)

Exchange Rates (24)

Exchange rates:

Solomon Islands dollars (SI$) per US$1

November 1997	3.7580
1997	3.5664
1995	3.4059
1994	3.2914
1993	3.1877

Top Import Origins (25)

$152 million (c.i.f., 1995 est.)

Origins	%
Russia	NA
Ukraine	NA
Poland	NA
Germany	NA

NA stands for not available.

Top Export Destinations (26)

$168 million (f.o.b., 1995) Data are for 1991.

Destinations	%
Japan	39
United Kingdom	23
Thailand	9
Australia	5
United States	2

Economic Aid (27)

Recipient: ODA, $8.625 million from Australia (FY96/97 est.); $3.3 million from. NZ (FY95/96).

Import Export Commodities (28)

Import Commodities	Export Commodities
Plant and machinery	Timber
Manufactured goods	Fish
Food and live animals	Palm oil
Fuel	Cocoa
	Copra

SOMALIA

CAPITAL: Mogadishu (Muqdisho).

FLAG: The national flag is light blue with a five-pointed white star in the center.

ANTHEM: *Somalia Hanolato (Long Live Somalia).*

MONETARY UNIT: The Somali shilling (SH) of 100 cents is a paper currency. There are coins of 1, 5, 10, and 50 cents and 1 shilling, and notes of 5, 10, 20, 100, 500, and 1,000 shillings. SH1 = $0.01 (or $1 = SH100).

WEIGHTS AND MEASURES: The metric system is in use.

HOLIDAYS: New Year's Day, 1 January; Labor Day, 1 May; National Independence Day, 26 June; Foundation of the Republic, 1 July. Muslim religious holidays include 'Id al-Fitr, 'Id al-'Adha', 'Ashura, and Milad an-Nabi.

TIME: 3 PM = noon GMT.

LOCATION AND SIZE: Situated on the horn of East Africa, Somalia has an area of 637,660 square kilometers (246,202 square miles), slightly smaller than the state of Texas. It has a total boundary length of 5,391 kilometers (3,350 miles). Somalia's capital city, Mogadishu, is located on the Indian Ocean coast.

CLIMATE: Somalia has a tropical climate, and there is little seasonal change in temperature. In the low areas, the mean temperature ranges from about 24°C to 31°C (75°F to 88°F). Rain falls in two seasons of the year.

INTRODUCTORY SURVEY

RECENT HISTORY

In 1940–41, the British conquered Italian Somaliland from Italian troops, who were allied with Nazi Germany. British administration over the rest of Italian Somaliland continued until 1950, when Italy, through the UN, gained administrative control again. However, in 1949, the UN General Assembly resolved that Italian Somaliland would receive its independence in 1960.

By the end of 1956, Somalis were in almost complete charge of domestic affairs. Meanwhile, Somalis in British Somaliland were demanding self-government. As Italy agreed to grant independence on 1 July 1960 to its trust territory, the United Kingdom gave its protectorate independence on 26 June 1960, thus enabling the two Somali territories to join in a united Somali Republic on 1 July 1960. On 20 July 1961, the Somali people ratified a new constitution, drafted in 1960, and one month later confirmed Aden 'Abdullah Osman Daar as the nation's first president.

Somalia was involved in many border clashes with Ethiopia and Kenya. Soviet influence in Somalia grew after Moscow agreed in 1962 to provide substantial military aid. Abdirashid 'Ali Shermarke, who was elected president in 1967, was assassinated on 15 October 1969. Six days later, army commanders seized power with the support of the police. The military leaders dissolved parliament, suspended the constitution, arrested members of the cabinet, and changed the name of the country to the Somali Democratic Republic. Major General Jalle Mohamed Siad Barre, commander of the army, was named chairman of a twenty-five-member Supreme Revolutionary Council (SRC) that assumed the powers of the president, the Supreme Court, and the National Assembly. Siad Barre was later named president.

In 1970, President Siad Barre proclaimed "scientific socialism" as the republic's guiding ideology. Controversy arose in the mid-1970s over Somalia's links to the Soviet Union and its support

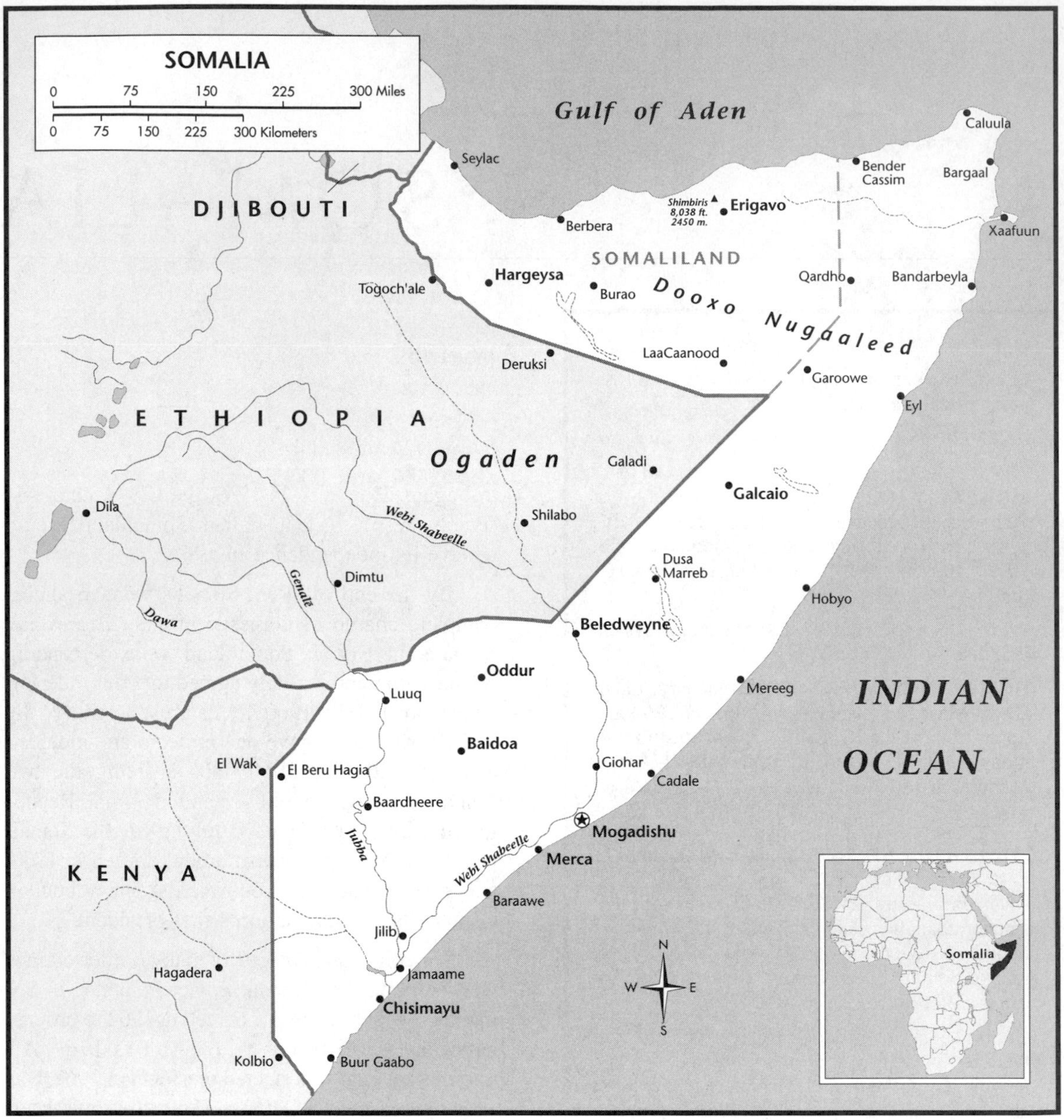

of the Western Somali Liberation Front in Ethiopia's Ogaden region.

In January 1986, Siad Barre met three times with Ethiopia's head of state in an effort to improve relations between the two countries, but no agreement was reached. In addition, internal dissent continued to mount.

In February 1987, relations between Somalia and Ethiopia worsened following an Ethiopian attack. By 1990, the Somali regime was losing control. Armed resistance from guerrilla groups was turning the Somali territory into a death trap.

In 1990, Barre was ousted and, in January 1991, he fled Mogadishu. The United Somali Congress (USC) seized the capital. The economy broke down, and the country turned into chaos as armed groups terrorized the population and disrupted shipments of food. Several hundred thousand people were killed, and far more were threatened by starvation. Over a half million fled to Kenya. As the starvation and total breakdown of public services was publicized in the western media, calls for the UN to intervene mounted.

Late on 3 December 1992, the UN Security Council passed a resolution to deploy a massive

U.S.-led international military intervention (UNITAF-United Task Force) to safeguard relief operations. By the end of December, faction leaders Muhammad Farah Aideed and Ali Maludi Muhammad had pledged to stop fighting. The UNITAF spread throughout the country, and violence decreased dramatically.

Although the problem of relief distribution had largely been solved, there was no central government, few public institutions, and local warlords and their forces became increasingly bold. By early 1993, over 34,000 troops from 24 United Nations members—75% from the United States—were deployed. Starvation was virtually ended, and some order had been restored. Yet, little was done to achieve a political solution or to disarm the factions. From January 1993 until 27 March, fifteen armed factions met in Addis Ababa, Ethiopia, to end hostilities and form a transitional National Council for a two-year period to serve as the political authority in Somalia.

On 4 May 1993, Operation Restore Hope, as the relief effort was labeled, was declared successful, and U.S. force levels were sharply reduced. A second relief effort, UNOSOM II, featured Pakistani, American, Belgian, Italian, Moroccan, and French troops, commanded by a Turkish general. On 23 June 1993, twenty-three Pakistani solders were killed in an ambush. General Aideed's forces were blamed and a $25,000 bounty was placed on Aideed's head. Mogadishu became a war zone.

In early October 1993, eighteen U.S. Army Rangers were killed and seventy-five were wounded in a firefight. American public opinion and politicians pressured President Bill Clinton to withdraw American troops. New discussions in Kenya and in Mogadishu reached agreements that teetered on collapse. After the withdrawal of the foreign peacekeeping troops, General Muhammad Farah Aideed became Somalia's ''self-declared'' president. Fighting continued in 1995 and 1996.

The hope for an end to the violence came with the death of General Aideed on 1 August 1996. Aideed's rivals declared a cease-fire, although his son and successor, Hussein Muhammad Aideed, promised to renew the fight. Despite the lack of a functioning central government, Somalia's economy was still functioning at the end of 1996, and harvests have even improved since the early 1990s. The factional splits between warlord rivals are not based on ideology, religion, or political issues. Their ongoing struggle for power and riches provides little chance for national unity and the restoration of a central government.

In December 1997 the leaders of the main rival factions met in Cairo and agreed on a plan that would end the civil war and restore the national government. The plan was supposed to establish a conference on national reconciliation by 15 May 1998, but there were delays. In April 1998, the Red Cross pulled its personnel out of Somalia after ten of its workers were abducted in Mogadishu.

GOVERNMENT

Since the overthrow of President Siad Barre in June 1991, Somalia has had no viable central government. Some fifteen armed factions have been fighting for control.

Although it is unrecognized as an independent nation, Somalia's northern province declared its independence on 18 May 1993. The state of ''Somaliland'' (its name during British colonial rule) has its own army, police force, currency, and judicial system. Rebels and armed gangs also battle throughout the region, however, and the self-declared government controls only half of the ''republic.''

Judiciary

As a result of the civil disorder in recent years, most of the structure for the administration of justice has collapsed. Islamic law and traditional courts continue to be applied to settle disputes over property and criminal offenses. In the northwest, the self-declared Republic of Somaliland uses the pre-1991 penal code.

Political Parties

President Siad Barre's SRSP (Somali Revolutionary Socialist Party) was the sole legal party at the time of his overthrow in January 1991. The Somali National Movement (SNM) has seized control of the north. Armed factions have divided up the territory as they fight and negotiate to expand their influence. Although many of them bear the titles of political parties, such as the Somali Democratic Movement, the Somali National Union, and the United Somali Congress (USC), they do not have national bases of support. The USC controlled Mogadishu and much of central Somalia until late in 1991 when it split into two major factions, Aideed's Somali National Alliance (SNA) and Ali Mahdi's Somali Salvation Alliance.

DEFENSE

The regular armed forces disintegrated in the revolution of 1991. Clan gangs armed with imported weapons terrorized the country and continue to fight among themselves. UN troops withdrew in 1995.

ECONOMIC AFFAIRS

Since 1990, Somalia's primarily agricultural economy has fallen apart owing to drought and a drawn-out civil war, which has left the country without central authority. By early 1992, virtually all trade, industry, and agriculture had stopped, large numbers of people were forced from their homes, and more than six million were at risk of starvation. In 1993, donors pledged $130 million toward Somalia's reconstruction, and good rains and increased stability eased the food situation. The aid, together with good harvests and increased stability, helped ease the food situation so that few communities were at risk of widespread famine in 1997.

Public Finance

The Somali budget has been in deficit since the early 1970s. Disintegration of the national economy since 1991 has led to relief and military intervention by the UN. No central government existed as of 1997, so there was no functioning system of civil administration to collect and disburse public finances.

Income

In 1995, Somalia's gross national product (GNP) per person was less than $765.

Industry

Industries mainly serve the domestic market and, to a lesser extent, provide some of the needs of Somalia's agricultural exports. The most important industries are the petroleum refinery, the state-owned sugar plants, an oilseed-crushing mill, and a soap factory. Newer industries manufacture corrugated iron, paint, cigarettes and matches, aluminum utensils, cardboard boxes and polyethylene bags, and textiles.

Banking and Finance

The Central Bank of Somalia, a government institution with branches in every region, controls the issue of currency and performs the central banking functions of the state.

Money supply, as measured by M2, totaled SH161.63 million at the end of 1990. Total reserves at that time, excluding gold, amounted to $11.4 million.

A new bank, the Barakat Bank of Somalia, was established in Mogadishu at the end of October 1996. Initially capitalized at $2 million, the bank will use the dollar as its working currency and will specialize in small loans to Somali traders, foreign currency exchange, and currency transactions abroad.

There are no securities exchanges in Somalia.

Economic Development

Successive Somali governments have sought to stimulate production in all sectors of agriculture, commerce, and industry. However, drought, inflation, civil strife, and the rise of oil prices have severely hampered these programs.

Economic development in the latter part of the 1990s will be devoted in large part to the rebuilding of the Somali civil administration.

SOCIAL WELFARE

The internal fighting and widespread drought conditions between 1989 and late 1992 have totally destroyed the government's provision of social services. Private humanitarian agencies tried to fill the needs but fighting, extortion, and the activities of armed factions and looters chased many of them away. Civilians are often the victims of indiscriminate attacks.

Healthcare

Because of the ongoing civil strife, hospitals are without drugs and illnesses are on the rise. Malaria and intestinal parasites are widespread. Water has been cut off to the capital city of Mogadishu leaving the people to rely on well water, which is scarce and often contaminated. Major operations are often performed without anesthetic.

Housing

Development schemes aided by UN and foreign assistance programs have helped alleviate housing shortages in Mogadishu and Hargeysa. The typical Somali house is either a round or a rectangular hut with a thatched or metal roof.

EDUCATION

Somalia's educational system collapsed with the government in 1992. Few schools kept operating, and even the Somali National University was closed in 1991. Some schools began reopening in 1996. In 1990, the UN Educational, Scientific and

Cultural Organization (UNESCO) estimated the adult literacy rate to be 24% (males, 36%; and females, 14%).

1999 KEY EVENTS TIMELINE

March

- Thousands flee devastating drought in the southern portion of country as fighting blocks supply route from Mogadishu.

April

- Fighting flares between groups led by rival warlords in Mogadishu.

June

- Ethiopia-Eritrea war spreads to Somalia as Ethiopian forces seize Baidoa and create a buffer region to keep Eritrea from using Somalia as a staging area for attacks.

July

- Humanitarian coalition warns that 1 million Somalis are at risk of starvation in drought-ridden southern Somalia and asks for $17.5 million from foreign donors.

August

- UN issues report describing Somalia as a country without a national government.
- Kenya closes border with Somalia following incursions by Somali militiamen.

September

- Chief UNICEF health officer is killed in armed attack on vehicle of aid workers; five others are wounded.
- International agencies suspend aid to Somalia in wake of attack on UNICEF workers.

November

- At the regional meeting of the inter-Governmental Authority on Development in Djibouti, Djibouti president Ismael Omar Guelleh puts forth a plan for peace in Somalia.

December

- Leaders of five warloads who control parts of the capital, Mogadishu, agree to cooperate on the running of the port and the airport, both of which have been closed since the United Nations peacekeepers left in 1995.
- The United Nations Security Council condemns violations of the arms embargo on Somalia, and estimates that about 600,000 Somalians are in urgent need of food aid.

ANALYSIS OF EVENTS: 1999

BUSINESS AND THE ECONOMY

Despite the state of virtual anarchy in Somalia, economic conditions in some parts of the country were no worse, and in some cases even better, than those in African countries with functioning governments. Some observers cited the freedom enjoyed by entrepreneurs not subject to bureaucratic red tape, official corruption, and state monopolies.

Others noted that the intricate pattern of clan relationships that had destroyed the country's political unity had actually constituted a positive economic force in some instances (such as the international clan-supported financial network that facilitated the transfer of remittances from Somalis abroad to their families at home, a major financial resource that amounted to hundreds of million of dollars annually). An important example of entrepreneurial success was the re-establishment of telephone systems throughout the country, facilitating trade and keeping Somalis in touch with family members abroad.

However, the difficulty of establishing economic normalcy amidst political chaos was highlighted by a currency crisis, the result of a mass distribution of counterfeit Somali shillings that reduced the value of the shilling against the U.S. dollar from 7.5 to 10,000. As of August, four competing versions of the national currency were reported to be in circulation.

Somalia's most severe economic problem was the drought that had plagued the southern portion of the country for three years and now placed as many as one million Somalis at risk of starvation.

GOVERNMENT AND POLITICS

In 1999 Somalia remained in a state of lawlessness and anarchy described as a ''black hole'' in a United Nations report issued in August—a situation that had existed since 1991, when dictator Mohammed Siad Barre had been ousted and rival warlords began fighting for control of the country. Since that time, no effective central government had ever been established, and the August report noted that Somalians had no governing authority to perform many of the basic functions of a state, such as providing social services and regulating commerce and transport. The absence of law enforcement had encouraged a proliferation of criminal activity, which now accounted for much of the violence that still plagued the country.

Clan warfare and criminal violence had exacerbated the effects of a three-year-old drought by blocking the transport of relief supplies to drought-stricken southern areas, and Somalia's neighbors were also affected by the prevailing lawlessness. In August Kenyan president Daniel arap Moi closed the border between his country and Somalia due to incursions of armed Somali militiamen into Kenya and the smuggling of firearms and other illegal items into the country. A situation that posed an even greater threat to the stability of the Horn of Africa region was the involvement of Somalia in a new round of hostilities between Ethiopia and Eritrea, which were reported to be conducting a "proxy war" by enlisting rival Somali groups to fight for their respective sides. By the summer of 1999, Ethiopia had taken control of key areas in southern Somalia in order to establish a buffer zone against Eritrean-sponsored attacks.

In the northwest, Somaliland, which had broken away from Somalia and declared independence in 1991, continued to struggle toward internal democracy and international recognition, without which it was unable to obtain Western aid or take other important steps toward economic stability. Nevertheless, it had achieved a degree of political stability absent elsewhere in the country except for Puntland, in the northeast, which had also established a relatively organized local administration, although unlike Somaliland it had not declared independence from Somalia.

In September, the top health officer of the United Nations Children's Fund (UNICEF) was killed when a vehicle he was riding in was attacked in central Somalia, in what was described as a robbery attempt. Five other UNICEF aid workers were injured, and international agencies suspended relief operations in the country, fearing for the safety of their personnel.

CULTURE AND SOCIETY

In the absence of a functioning national government, Somali society was held together by the clan relationships that had long been its traditional source of cohesion. Clan leaders in the northern areas of Somaliland and Puntland had demonstrated cooperation in bringing a measure of administrative organization to these regions, and in the south the most warlike leaders were losing support. After a decade of death and destruction, the nation's poets, who traditionally praised the glory of war, had begun to condemn it. According to the United Nations report issued in August about half the country enjoyed peace.

However, as many as one million Somalis in the south of the country faced a serious humanitarian disaster from the continuing drought in that region, from which tens of thousands of refugees had fled by the end of March, many seeking food and shelter in camps set up by UNICEF. In July a coalition of donor nations, relief groups, and international agencies issued new warnings about the magnitude of the crisis and asked for $17.5 million in aid for the region.

DIRECTORY

CENTRAL GOVERNMENT

Head of State

In January 1991 the government of President Muhammad Siad Barre was overthrown, and Somalia fell into a civil war marked by inter-clan fighting. The country is presently in a state of anarchy, with no functioning central government. Like the central leadership, other government institutions, such as the legislature and judiciary, are not currently in operation.

Unrest and clan fighting continue throughout most of Somalia, but some orderly government has been established in the northern part of the country. In May 1991 the clan elders in former British Somaliland established the Independent Republic of Somaliland. This region, though unrecognized by any government, has managed to maintain a stable existence through the dominance of the ruling clan and the economic infrastructure left behind by British, Russian, and American military assistance programs. The economy in this region has grown, and in February 1996 the European Union (EU) agreed to finance the reconstruction of the port of Berbera.

POLITICAL ORGANIZATIONS

Numerous clans are currently fighting for power within the country.

DIPLOMATIC REPRESENTATION

Embassies in Somalia

United States

U.S. interests in Somalia are represented by the U. S. embassy in Kenya.
PO Box 30137, Unit 64100, Nairobi APO AE 09831, Kenya
PHONE: +254 (2) 334141
FAX: +254 (2) 340838

JUDICIAL SYSTEM

Since civil unrest took over the nation and its infrastructure collapsed, many regions in Somalia have reverted to the rule of Islamic law.

FURTHER READING

Articles

''A Failed State That is Succeeding in Parts.'' *The Economist* (August 28, 1999): 33.

''A Nomad's Life is Hard.'' *The Economist* (August 7, 1999): 35.

Santoro, Lara. ''Why War is Spreading in Horn of Africa.'' *The Christian Science Monitor* (July 22, 1999): 6.

''U.N. Report Describes Somalia's Swift Descent into Anarchy.'' *The New York Times*, August 19, 1999, p. A7.

Books

Besteman, Catherine and Lee V. Cassanelli, eds. *The Struggle for Land in Southern Somalia: The War Behind the War*. Boulder, Colorado: Westview Press, 1996.

Mubarak, Jamil Abdalla. *From Bad Policy to Chaos in Somalia: How an Economy Fell Apart*. Westport, Connecticutt: Praeger, 1996.

Internet

Somali Electronic Journal. Available Online @ http://www.angelfire.com/ms/sej/ (November 9, 1999).

Somali News Page. Available Online @ http://www.etek chalmers.se/e3hassan/news.html (November 9, 1999).

SOMALIA: STATISTICAL DATA

For sources and notes see "Sources of Statistics" in the front of each volume.

GEOGRAPHY

Geography (1)

Area:

Total: 637,660 sq km.

Land: 627,340 sq km.

Water: 10,320 sq km.

Area—comparative: slightly smaller than Texas.

Land boundaries:

Total: 2,366 km.

Border countries: Djibouti 58 km, Ethiopia 1,626 km, Kenya 682 km.

Coastline: 3,025 km.

Climate: principally desert; December to February—northeast monsoon, moderate temperatures in north and very hot in south; May to October—southwest monsoon, torrid in the north and hot in the south, irregular rainfall, hot and humid periods (tangambili) between monsoons.

Terrain: mostly flat to undulating plateau rising to hills in north.

Natural resources: uranium and largely unexploited reserves of iron ore, tin, gypsum, bauxite, copper, salt.

Land use:

Arable land: 2%

Permanent crops: 0%

Permanent pastures: 69%

Forests and woodland: 26%

Other: 3% (1993 est.)

HUMAN FACTORS

Demographics (2A)

	1990	1995	1998	2000	2010	2020	2030	2040	2050
Population	6,675.1	6,256.4	6,841.7	7,433.9	10,131.6	13,312.1	17,241.9	21,628.6	26,242.6
Net migration rate (per 1,000 population)	NA	NA	NA	NA	NA	NA	NA	NA	NA
Births	NA	NA	NA	NA	NA	NA	NA	NA	NA
Deaths	NA	NA	NA	NA	NA	NA	NA	NA	NA
Life expectancy - males	44.7	44.7	44.7	44.7	48.2	51.8	55.5	59.1	62.6
Life expectancy - females	47.8	47.8	47.8	47.8	51.8	56.0	60.1	64.1	67.9
Birth rate (per 1,000)	47.9	43.0	46.8	47.5	43.5	39.2	34.8	29.7	25.5
Death rate (per 1,000)	19.3	18.1	18.5	18.7	15.3	12.5	10.4	8.8	7.6
Women of reproductive age (15-49 yrs.)	1,537.1	1,415.8	1,564.0	1,708.3	2,336.1	3,121.6	4,185.1	5,466.7	6,936.6
of which are currently married	NA	NA	NA	NA	NA	NA	NA	NA	NA
Fertility rate	7.3	6.3	7.0	7.2	6.4	5.5	4.5	3.7	3.0

Except as noted, values for vital statistics are in thousands; life expectancy is in years.

Infants and Malnutrition (5)

Under-5 mortality rate (1997) 211
% of infants with low birthweight (1990-97) 16
Births attended by skilled health staff % of total[a] . . .NA
% fully immunized (1995-97)
TB 37
DPT 19
Polio 19
Measles 25
Prevalence of child malnutrition under age 5 (1992-97)[b] . . . . .NA

Ethnic Division (6)

Somali 85%
Bantu, Arabs 30,000

Religions (7)

Sunni Muslim.

Languages (8)

Somali (official), Arabic, Italian, English.

GOVERNMENT & LAW

Military Affairs (14B)

	1990	1991[12]	1992	1993	1994	1995
Military expenditures						
Current dollars (mil.)	8	NA	NA	NA	NA	NA
1995 constant dollars (mil.)	9	NA	NA	NA	NA	NA
Armed forces (000)	47	NA	NA	NA	NA	NA
Gross national product (GNP)						
Current dollars (mil.)	904	917[e]	790[e]	806[e]	775[e]	810[e]
1995 constant dollars (mil.)	1,034	1,013[e]	850[e]	845[e]	794[e]	810[e]
Central government expenditures (CGE)						
1995 constant dollars (mil.)	NA	NA	NA	NA	NA	NA
People (mil.)	8.3	8.3	8.0	8.1	8.6	9.2
Military expenditure as % of GNP	.9	NA	NA	NA	NA	NA
Military expenditure as % of CGE	NA	NA	NA	NA	NA	NA
Military expenditure per capita (1995 $)	1	NA	NA	NA	NA	NA
Armed forces per 1,000 people (soldiers)	5.6	NA	NA	NA	NA	NA
GNP per capita (1995 $)	125	122	106	104	93	88
Arms imports[6]						
Current dollars (mil.)	30	30	0	5	0	0
1995 constant dollars (mil.)	34	33	0	5	0	0
Arms exports[6]						
Current dollars (mil.)	0	0	0	0	0	0
1995 constant dollars (mil.)	0	0	0	0	0	0
Total imports[7]						
Current dollars (mil.)	249[e]	151[e]	184[e]	205[e]	NA	NA
1995 constant dollars (mil.)	286[e]	167[e]	198[e]	215[e]	NA	NA
Total exports[7]						
Current dollars (mil.)	150[e]	91[e]	118[e]	117[e]	NA	100[e]
1995 constant dollars (mil.)	172[e]	101[e]	127[e]	123[e]	NA	100[e]
Arms as percent of total imports[8]	12.0	19.9	0	2.4	NA	NA
Arms as percent of total exports[8]	0	0	0	0	NA	0

EDUCATION

Literacy Rates (11B)

Adult literacy rate

1980

Male 8%

Female 1%

1995

Male 36%

Female 14%

GOVERNMENT & LAW

Political Parties (12)

Legislative branch: unicameral People's Assembly or Golaha Shacbiga. The Golaha Shacbiga is not functioning. The United Somali Congress or USC ousted the former regime on 27 January 1991; formerly the only party was the Somali Revolutionary Socialist Party or SRSP, headed by former President and Commander in Chief of the Army Major General Mohamed Siad Barre.

Government Budget (13B)

Revenues . . . . .NA

Expenditures . . . . .NA

Capital expenditures . . . . .NA

NA stands for not available.

LABOR FORCE

Labor Force (16)

Total (million) 3.7

Agriculture 71%

Industry and services 29%

Very few skilled laborers. Agriculture is pastoral nomadism. Data for 1993 est.

Unemployment Rate (17)

Rate not available.

PRODUCTION SECTOR

Electric Energy (18)

Capacity . . . . .144,000 kW

Production . . . . .245 million kWh (1995 est.)

Consumption per capita . . . . .33 kWh (1995 est.)

Capacity shown is prior to the civil war, but now largely shut down due to war damage; some localities operate their own generating plants, providing limited municipal power; UN and relief organizations use their own portable power systems.

Transportation (19)

Highways:

total: 22,100 km

paved: 2,608 km

unpaved: 19,492 km (1996 est.)

Pipelines: crude oil 15 km

Merchant marine: none

Airports: 61 (1997 est.)

Airports—with paved runways:

total: 7

over 3,047 m: 4

2,438 to 3,047 m: 1

1,524 to 2,437 m: 1

914 to 1,523 m: 1 (1997 est.)

MANUFACTURING SECTOR

GDP & Manufacturing Summary (21)

	1980	1985	1990	1992	1993	1994
Gross Domestic Product						
Millions of 1990 dollars	927	1,012	1,062	1,078	1,099	*1,129*
Growth rate in percent	1.79	9.53	−2.70	2.50	2.00	*2.68*
Per capita (in 1990 millions)	138.1	128.5	122.4	121.6	122.8	*124.3*
Manufacturing Value Added						
Millions of 1990 dollars	39	33	41	42	*40*	*42*
Growth rate in percent	9.17	7.55	−6.29	−10.00	*-5.00*	*4.80*
Manufacturing share in percent of current prices	4.7	4.9	3.8	NA	NA	NA

PRODUCTION SECTOR

Airports—with unpaved runways:

total: 54

2,438 to 3,047 m: 3

1,524 to 2,437 m: 14

914 to 1,523 m: 27

under 914 m: 10 (1997 est.)

Top Agricultural Products (20)

Bananas, sorghum, corn, sugarcane, mangoes, sesame seeds, beans; cattle, sheep, goats; fishing potential largely unexploited.

FINANCE, ECONOMICS, & TRADE

Economic Indicators (22)

National product: GDP—purchasing power parity—$8 billion (1996 est.)

National product real growth rate: 4% (1996 est.)

National product per capita: $600 (1996 est.)

Inflation rate—consumer price index: NA%

Exchange Rates (24)

Exchange rates:	
Somali shillings (So. Sh.) per US$1	
November 1997 est.	7,500
January 1996 est.	7,000
1 January 1995	5,000
1 July 1993	2,616
December 1992	4,200

The Republic of Somaliland, a self-declared independent country not recognized by any government, issues its own currency, the Somaliland shilling (Sol. Sh.); estimated exchange rate, Sol. Sh. per US$1—4,000 (November 1997)

Top Import Origins (25)

$269 million (1994 est.) Data are for 1995 est.

Origins	%
Kenya	24
Djibouti	18
Pakistan	6

Top Export Destinations (26)

$130 million (1994 est.) Data are for 1995 est.

Destinations	%
Saudi Arabia	57
Yemen	14
Italy	13
UAE	10
United States (bananas)	NA

NA stands for not available.

Economic Aid (27)

Recipient: ODA, $NA. NA stands for not available.

Import Export Commodities (28)

Import Commodities	Export Commodities
Manufactures	Bananas
Petroleum products	Live animals
Foodstuffs	Fish
Construction materials	Hides

Balance of Payments (23)

	1980	1985
Exports of goods (f.o.b.)	133	91
Imports of goods (f.o.b.)	−401	−331
Trade balance	−268	−240
Services - debits	−139	−123
Services - credits	71	37
Private transfers (net)	143	204
Government transfers (net)	57	19
Overall balance	−136	−103

SOUTH AFRICA

CAPITAL: Pretoria (administrative); Cape Town (legislative); Bloemfontein (judicial).

FLAG: The national flag, adopted in 1994, consists of a blue-black triangle placed vertical to the hoist and bordered in gold-yellow. Bands of red, white, green, white, and blue appear horizontally.

ANTHEM: Two anthems are currently in use: the official anthem, *Die Stem van Suid-Afrika (The Call of South Africa)*, and *Nkosi Sikelel' Afrika (God Bless Africa)*, a hymn adopted by most liberation groups.

MONETARY UNIT: The South African rand (R) is a paper currency of 100 cents. It is used throughout the South African monetary area. There are coins of 1, 2, 5, 10, 20, and 50 cents and 1 rand, and notes of 2, 5, 10, 20, and 50 rand. R1 = $0.21268 (or $1 = R4.702).

WEIGHTS AND MEASURES: The metric system is in use.

HOLIDAYS: New Year's Day, 1 January; Republic Day, 31 May; Kruger Day, 10 October; Day of the Vow, 16 December; Christmas, 25 December; Goodwill Day, 26 December. Movable religious holidays include Good Friday and Ascension; Family Day is a movable secular holiday.

TIME: 2 PM = noon GMT.

LOCATION AND SIZE: The area of South Africa is 1,321,219 square kilometers (510,125 square miles). South Africa also controls two small islands, Prince Edward and Marion, which lie some 1,920 kilometers (1,200 miles) southeast of Cape Town.

South Africa's capital city, Pretoria, is located in the northeastern part of the country.

CLIMATE: South Africa lies almost wholly within the southern temperate zone. In February, the average daily minimum temperature at Durban ranges from 11°C (52°F) in July to 21°C (70°F), while at Port Nolloth, the range is from 7°C (45°F) to 12°C (54°F). At Johannesburg, the average daily minimum is 4°C (39°F) in June and July and 14°C (57°F) in January. While the mean annual rainfall is 46 centimeters (18 inches), 21% of the country receives less than 20 centimeters (8 inches) and 31% gets more than 60 centimeters (24 inches).

Republic of South Africa
Republiek van Suid-Afrika

INTRODUCTORY SURVEY

RECENT HISTORY

A constitution for a united South Africa, which passed the British Parliament as the South Africa Act in 1909, provided for a union of all four territories or provinces, to be known as the Union of South Africa. South Africa sent troops to fight the German Nazis in World War II (1939–45), although many Afrikaaners (as the Boers had come to be called) favored neutrality. In 1948, the National Party (NP) took power, enforcing the policy of apartheid, or racial separation of whites and nonwhites.

South Africa became a republic on 31 May 1961, and the president replaced the British monarch as head of state. There were mounting pressures on the government because of its apartheid policies. On 21 March 1960, a black demonstration had been staged against the "pass laws," laws requiring blacks to carry identification enabling the government to restrict their movement into urban areas. The demonstration resulted in the killing of sixty-nine black protesters by government troops at Sharpeville in Soweto, and provided a focus for local black protests and for widespread international expressions of outrage. During this period, many black leaders were jailed, including Nelson Mandela, the leader of the African National Congress (ANC), a black nationalist group. The ANC was banned as a political party.

In the mid-1970s, the Portuguese colonial empire disbanded and blacks came to power in Mozambique and Angola. The new black-con-

trolled governments in these countries gave aid and political support to the ANC in South Africa. In response, South Africa aided rebel movements in the two former Portuguese territories.

In June 1976, the worst domestic confrontation since Sharpeville took place in Soweto, where blacks violently protested the compulsory use of the Afrikaans language in schools. Suppression of the riots by South African police left at least 174 blacks dead and 1,139 injured.

During the late 1970s, new protest groups and leaders emerged among the young blacks. After one of these leaders, 30-year-old Steven Biko, died while in police custody on 12 September 1977, there were renewed protests. On 4 November, the United Nations Security Council approved a mandatory arms embargo against South Africa—the first ever imposed on a member nation.

In an effort to satisfy nonwhite and international opinion, the government scrapped many aspects of apartheid in the mid-1980s, including the ''pass laws'' and the laws barring interracial sexual relations and marriage. These measures failed to satisfy blacks, however, and as political violence mounted the government imposed states of emergency in July 1985 and again in June 1986.

Further repression in the late 1980s included the banning of the United Democratic Front (UDF) and sixteen other anti-apartheid organizations, suppression of the alternative newspapers *New Nation* and *Weekly Mail*, and assassination of anti-apartheid leaders by secret hit squads identified with the police and military intelligence.

In 1989, President P.W. Botha resigned as head of the NP (National Party) and was replaced by F.W. de Klerk, who was also named acting state president. After the 6 September general election, de Klerk was elected to a five-year term as president. De Klerk launched a series of reforms in September 1989 that led to the release of ANC leader Nelson Mandela and others on 10 February 1990.

The African National Congress (ANC) and other resistance militants, including the Communist Party, were legalized. Mandela had been in prison twenty-seven years and had become a revered symbol of resistance to apartheid. At that point, the ANC began to organize within South Africa and, in August 1990, suspended its armed struggle. Most leaders of the ANC returned from exile. Still, fighting continued, largely between ANC activists and supporters of the Zulu-dominated Inkatha Freedom Party, strongest in Natal province. More than 6,000 people were killed in political violence in 1990 and 1991. In 1991, parliament passed measures to repeal the apartheid laws—the Land Acts (1913 and 1936), the Group Areas Act (1950), and the Population Registration Act (1950).

In July, the ANC convened its first full conference in South Africa in thirty years. Mandela was elected president, and Cyril Ramaphosa was elected secretary general. Meanwhile, negotiations continued through 1991 and 1992 over a transition to majority rule and an end to factional fighting between the ANC and Inkatha, mostly through the Convention for a Democratic South Africa (CODESA), which began in December 1991.

In February 1993, the government and the ANC reached agreement on plans for a transition to democracy. The broad guidelines were agreed upon by the government, the ANC, and other parties in late December 1993. The Conservative Party and Inkatha boycotted the talks on multiparty government, but just a few days before the scheduled elections, Inkatha agreed to participate. The white right was divided on whether to participate in preelection talks, in the election itself, or whether to take up arms as a last resort. The elections proceeded relatively peacefully and with great enthusiasm, and they were pronounced ''free and fair'' by international observers.

The ANC was awarded 252 of the 400 seats in parliament. It became the governing party in all but two of the nine regions. Mandela became president and the ANC's Thabo Mbeki and the NP's de Klerk were made deputy presidents. Zulu leader Mangosuthu Buthelezi was persuaded to take a ministerial post in the cabinet. In May 1994, the Constitutional Assembly met to lay the groundwork for the new constitution. South Africa's new constitution was ratified in February 1997. Local elections were held in November 1995.

The Truth and Reconciliation Commission was established in early 1996 to expose apartheid crimes and abuses committed during the years of white rule.

GOVERNMENT

The terms of the new constitution in February 1997 were determined before the elections of April 1994. There is a 400-seat National Assembly chosen by proportional representation. There is also a Senate of ninety members, 10 from each province or region, who serve as a legislature and also elect the president and deputy presidents. The president names a cabinet, divided proportionally between parties that have gained at least 5% of the vote. The nine provinces have assemblies based on the total number of votes cast in the general election.

Judiciary

The Supreme Court has a supreme appeals division and provincial and local divisions with both original and appeals jurisdictions. The Court of Appeals, with its seat in Bloemfontein, normally consists of the chief justice and a variable number of appeals judges. Judges are appointed by the state president. Trial by jury was abolished in 1969.

Political Parties

Banned in 1960, the African National Congress (ANC) was legalized in 1987 in return for renouncing violence. Headed by Nelson Mandela, it received 62.5% of the vote in the April 1994 elections, making it the ruling party in South Africa.

The National Party (NP), first formed in 1910, was the last party of white rule in South Africa before the 1994 elections, in which it received 20.4% of the vote.

The Inkatha Freedom Party (IFP), headed by Zulu Chief Mangosuthu Buthelezi, captured over 10% of the national vote and won the election for the provincial government in Natal.

Other parties participating in the elections were the Freedom Front (2.2%), the Democratic Party (1.7%), and the Pan-Africanist Congress (1.2%).

DEFENSE

In 1995, South Africa had 137,900 armed personnel; the draft has been abolished. The army had 118,000 troops; the navy, 5,500; and the air force, 8,400. There is also a medical corps of 6,000. In 1995, there were 140,000 active members of the South African Police Service. In 1995, South Africa spent $2.9 billion on defense, or about 2.5% of the gross domestic product (GDP).

ECONOMIC AFFAIRS

The opening of the political process to all South Africans and the election of a new multiracial government in 1994 marked a turning point in South Africa's economic history. With modest agriculture, fabulous mineral wealth, and diverse manufacturing, South Africa's influence extends well beyond its borders.

Real economic growth in the gross domestic product (GDP) was about 2.5%; in 1996. However, the unemployment rate was estimated at 30–40%.

Public Finance

The minister of finance presents the budget to Parliament in March for authorization of expenditures and imposition of the necessary taxes. Government spending in 1992–93 ran 18.2% over budget due to the unexpected addition of R3.4 billion in drought relief efforts. Meanwhile, revenues rose only 4% from the previous year but were still R9.6 billion short of anticipated income. As a result of the recession and tax structure and collection shortcomings, the government borrowing requirement rose to 8.6% of GDP. The CIA estimates that, in 1995, government revenues totaled approximately $30.5 billion and expenditures $38 billion. External debt totaled $22 billion, approximately 5% of which was financed abroad.

Income

In 1998, South Africa's gross national product (GNP) was $119 billion at current prices, or about $2,880 per person. For the period 1985–95, the average annual inflation rate was 13.7%, resulting in a decline in per person GNP of 1.0%.

Industry

Manufacturing is the largest contributor to South Africa's economy. Industry is located mainly in Gauteng, Western Cape, the Durban-Pinetown area of KwaZulu-Natal, and the Port Elizabeth-Uitenhage area of Eastern Cape. The largest industrial area is the metal products and engineering sector. The steel industry supplies a large motor vehicle sector.

Banking and Finance

The South African Reserve Bank (SARB), the central bank of issue, began operations in 1921, and in 1924 assumed liability for the outstanding notes of the commercial banks. It purchases and disposes of the entire gold output. At the end of 1993, the Reserve Bank held 1,008 million troy ounces of gold. Foreign reserves, excluding gold, amounted to $942 million in 1996. Money supply, as measured by M2, was R255,059 million at the end of 1996.

The top four banks—Standard Bank Investment Corp. (Stanbic), Amalgamated Banks of South Africa (ABSA), First National Bank (FNB), and Nedcor—accounted for 80% of total bank assets in the country in 1997.

The Johannesburg Stock Exchange (JSE) ranks 10th in the world in market capitalization. At the end of 1995, its total capitalization was $280 billion.

Economic Development

The recession of 1989 to 1993 was provoked by a drop in investment from 24% to 15%. With the inauguration of multiracial government in 1994, this investment is likely to be restored and is key to the economy's ability to create new jobs and to generate growth. Tremendous changes in the structure of the economy are required as well to relieve the pressures of poverty and inequality that have resulted from apartheid.

SOCIAL WELFARE

South Africa has a comprehensive system of social legislation, which includes unemployment insurance, industrial accident insurance, old-age pensions, disability pensions, war veterans' pensions, pensions for the blind, and maternity grants.

The current African National Congress (ANC) ''government of national unity'' is seeking to provide more social services for its black constituents within the context of the constraints of a weakened economy. Its first priorities are housing, health, education, and the creation of more jobs in the formal economic sector.

Healthcare

As of 1992, the South African government increased its spending in the public and private sectors of health care.

South Africa's governmental policy has been directed toward a more streamlined and equitable public health service to bridge the country's social and ideological divisions.

Hospital care is free for those unable to bear the costs, including nonwhites, but medical treatment is generally conducted on a private basis. In 1990, there were 684 hospitals. In 1992, there was 1 doctor per 1,640 people.

The most prevalent infectious diseases reported in South Africa are tuberculosis, measles, typhoid, malaria, and viral hepatitis. Circulatory disorders are the leading causes of death. By 1990, leprosy had been reduced to less than 1 per 100,000, but malaria and tuberculosis still cause serious problems. In 1994, 3.2% of the adult population was infected with HIV, and there were 10,351 new cases of AIDS reported in 1995. Average life expectancy was 65 years.

Housing

In 1994, the housing backlog was estimated to be 1.2 million homes for the black population, while there is a surplus of white housing units of 83,000. Experts in South Africa forecast that almost 3 million homes will have to be provided by the year 2000 in the urban areas of the country. Recently, there has been an explosive growth of shacks and shantytowns surrounding South Africa's major urban areas. An estimated 66% of the country's population have no access to electricity, and in most black townships there is only one water tap per several thousand people.

EDUCATION

Systems of primary, secondary, and university education are generally provided in separate English-language and Afrikaans-language institutions. Adult literacy was about 82%. In 1994, 20,428 primary schools had a total student enrollment of nearly 8 million. Secondary education lasts an additional seven years.

In 1994, there were 21 universities and 15 other higher education institutions with 617,897 students.

1999 KEY EVENTS TIMELINE

January

- British Prime Minister Tony Blair arrives at the Waterkloof Air Base in Pretoria for his South African visit, amid strong protest from local Muslim groups against continued British and United States air attacks on Iraq.
- Sifiso Nkabinde, the secretary general of the multiracial United Democratic Movement, is assassinated near his home in the western town of Richmond.
- Eleven people are killed in one of two revenge attacks for the death of Nkabinde.
- A bomb explodes in Cape Town outside the city's main police station, wounding 11 people.
- South Africa's murder rate declines over the previous year (1998) compared with trends over the preceding four years, but incidences of robberies with aggravating circumstances surge.

February

- Mandela delivers his final "State of the Nation" address as he prepares to retire after only one term in office.
- The South African Trade and Industry Minister encourages South Africa to be bold in promoting a free trade agreement in the Southern African Development Community as the first step towards creating an economic community similar to the European Union.
- South African scientists develop a low-cost, environmentally friendly water-stabilization system, which could revolutionize water treatment within the nation.
- The U.S.-South African bi-national commission ends its sixth session by establishing two key agreements on the creation of permanent structures to address crime in South Africa and to promote trade and investment.
- Cathy O'Dowd, a South African woman, announces her plans to climb Mount Everest again this year in a historic attempt to reach the summit of Everest from the northern slopes of the mountain.

March

- The discovery of a new fossil species in the Northern Cape province places South Africa as the origin of the most primitive mammals in the world.

- Zimbabwean President Robert Mugabe arrives in South Africa and holds talks with President Nelson Mandela and Foreign Minister Alfred Nzo on the situation in the Democratic Republic of Congo.
- The South African government expresses its anger at the NATO intervention in the former Yugoslavia, saying it is in violation of the United Nations Charter and the accepted norms of international law.
- Two amateur ham radio operators speak to each other through South Africa's first space satellite, Sunsat, which was designed and made at Stellenbosch University near Cape Town and launched from the United States on board a Boeing Delta rocket.

April

- The Truth and Reconciliation Commission (TRC) denies amnesty to 79 members of the governing African National Congress (ANC), including seven cabinet ministers.
- Advance warning that two professional South African hitmen were inside Zambia and plotting with elements in President Chiluba's Movement for Multiparty Democracy (MMD) government to eliminate opposition leader Kenneth Kaunda saves the former president's life in a third attempted hit.
- The demise of Afrikaans as a language of record in South Africa's courts gets a step closer with the news that the influential Judicial Service Commission is to act on the matter.
- President Nelson Mandela, accompanied by his daughter Princess Zenani Mandela-Dlamini and Foreign Minister Alfred Nzo, leave South Africa for Moscow to start Mandela's official and symbolic trip abroad before stepping down from the presidency.

May

- African National Congress Women's League president Winnie Madikizela-Mandela is placed in position number nine of the ANC provisional list despite outrage from opposition parties.
- The South African government expresses pride in its role to ban the use, stockpiling, production, and transfer of anti-personnel mines.
- More than 7,000 would-be immigrants from Mozambique and other African countries have been arrested so far in 1999 while trying to cross the Mozambican border into South Africa.
- Governments of the 21-member COMESA are asked to put political pressure on South Africa to accept goods from some countries of the economic grouping.
- South Africa finally decides to increase its shareholding in the African Development Bank to six percent, making it the second largest single stakeholder in the bank after Nigeria.

June

- The ANC wins 63 percent of the vote in the June 3 parliamentary elections, taking 266 of 400 parliamentary seats, just one short of a two-thirds majority.
- Thabo Mbeki is sworn in as South Africa's second democratically elected president at a glittering inauguration ceremony, which saw Nelson Mandela step down after steering the country away from apartheid rule and oppression.
- Libyan leader Muammar Qadhafi arrives in Johannesburg for the start of a three-day state visit, during which he will attend the inauguration of Thabo Mbeki as the new president of South Africa.
- Cabinet ministers vow to tackle South Africa's crime epidemic head on, warning criminals that the state would use all its might to protect citizens.
- The South African gold mining industry warns that the proposed sale of IMF gold reserves will undermine the current fragile gold market and place further downward pressure on the gold price.

July

- Scientists discover the body of an ancient, well-preserved mummy near Joubertina in the Eastern Cape, the first ancient mummified remains to be discovered in South Africa.
- A nationwide campaign by AIDS organizations to challenge the United States government to show its support for the provision of cheaper drugs to HIV positive people begins in Johannesburg.
- After 90 years of mystery and intrigue surrounding the disappearance of the missing passenger liner the "SS Waratah," the wreckage is discov-

ered lying off the Transkei coast between Durban and East London in the Eastern Cape.

- The combined effects of the International Monetary Fund's proposed sell off of 10 million ounces of gold and the auction by the British government of 25 tonnes of gold spell trouble for South Africa and other gold producing countries in Africa.
- Democratic Republic of Congo President Laurent Kabila vehemently denies report that he is losing the war in his country and seeking asylum in South Africa.

August

- South Africa experiences one of its largest protest actions in years when thousands of workers from twelve unions representing the public sector take to the streets to raise their voices against the government's decision to unilaterally implement a wage offer, which worker unions indicate is unacceptable to them.
- A new book, titled ''The Truth about the Truth Commission'' written by Dr. Anthea Jeffery, criticizes the South African Truth and Reconciliation Commission (TRC).
- The South African government expresses support in efforts to enhance African security and stability in view of promoting the African Renaissance.
- Large parts of Cape Town are declared disaster areas following a freak tornado which left five people dead, 110 injured and 2,000 people homeless.

September

- South African Foreign Minister Nkosazana Zuma announces the appointment of three heads of embassies in the Democratic Republic of Congo, Ghana, and India.
- Three of South Africa's most famous cultural and natural sites, including Robben Island, are expected to be granted World Heritage Site status by UNESCO.
- The South African government expects to raise more than two billion U.S. dollars from the national lottery in the next seven years to fund reconstruction and development programs, sports, and charity organizations.
- Government officials from South Africa and France conclude their first annual consultation on development cooperation to review the flow of French financial support to South Africa since the election of the first democratic government in 1994.
- Presidents Olusegun Obasanjo of Nigeria and Thabo Mbeki of South Africa express their resolve to work together for the good of Africa.

October

- South Africa and Nigeria commit themselves to finalizing trade and investment agreements and collaborating in the petroleum industry and electricity projects.
- Two books titled ''The Boer War'' reveal the crucial role blacks played in the South African Boer War, which began in October 1899.
- South Africa is ranked 32 out of 85 in a list of ''least corrupt'' countries around the world and holds third place in Africa behind Botswana and Namibia.

November

- South Africa condemns European attempts to protect more than 150 ''traditional expressions'' used to describe wines and spirits, saying that it would be betraying other developing countries if it yielded to the European Union's demands.
- Eight people are shot dead and several are injured when rival gangs of taxi drivers fight a gun battle in the town of Empangeni.
- More than 40 people are injured in South Africa when a pipe bomb explodes in a pizza parlor near Cape Town.

December

- Public Enterprises Minister Jeff Rad-Ebe announces South Africa plans to speed reform of its state industries by selling them, listing them, or bringing in foreign partners.

ANALYSIS OF EVENTS: 1999

BUSINESS AND THE ECONOMY

The South African economy is one of the strongest on the continent and it compares favorably with the economies of a number of countries in the West. South Africa is ranked as a middle-income, developing country. The nation has an abundant supply of resources, well-developed financial, legal, communications, energy, and transport sectors, a stock exchange that ranks among the largest in the world, and a modern infrastructure supporting an efficient distribution of goods to major urban centers throughout the southern African region. However, in recent years the economy has grown slowly in part due to structural problems which remain from the apartheid era, especially the problems of poverty and lack of economic empowerment among the disadvantaged groups. The dominance of the South African economy in the southern African region, particularly among the Southern African Development Community (SADC) member states was a contentious issue during 1999.

South African exports to Africa began replacing those of European and U.S. companies in 1999. This has widened the trade imbalance between South Africa and its neighbors since South Africa imports little from other African countries. This has prompted governments of the 21-member Common Market for Eastern and Southern Africa (COMESA) to put political pressure on South Africa to accept goods from some countries of the economic grouping. The South African Trade and Industry Minister was also quick to encourage South Africa to be bold in promoting a free trade agreement in the Southern African Development Community as the first step towards creating an economic community similar to the European Union. This was seen by other countries as an attempt to ease the criticism; but a hopeful start for South Africa to remove structural rigidities, including a complicated and relatively protectionist trade regime.

South Africa also increased its shareholding in the African Development Bank to six percent with the aim of providing investment funds to other cash strapped economies in the region. It is gratifying to note that the new government is committing to open markets, privatization, and a favorable investment climate.

Although ranked third in Africa as the least corrupt country, crime and corruption remain obstacles to direct foreign investment. The 1999 statistics indicated that South Africa's murder rate had declined over the previous year (1998) compared with trends over the preceding four years, but incidence of robberies with aggravating circumstances had surged. The government has vowed to tackle South Africa's crime epidemic head on, warning criminals that the state would use all its might to protect citizens and businesses. The U.S.-South African bi-national commission held its sixth session during the year and concluded two key agreements on the creation of permanent structures to address crime in South Africa and to promote trade and investment.

The year 1999 saw the South African economy experience an external gold price shock. The combined effects of the International Monetary Fund's proposed sell-off of 10 million ounces of gold and the auction by the British government of 25 tonnes of gold spelled trouble for South Africa and other gold producing countries in Africa. South Africa protested strongly the IMF gold sales intended to support debt relief and further concessional lending for highly poor countries, mostly in Africa citing the adverse impact on South Africa and other African gold producers. The steady decline in the price of gold and other precious metals soon after the announcement by Britain, the IMF and other governments to sell off their gold reserves resulted in 5,000 job losses as one of the oldest gold mines shut down. Thousands of migrant laborers from surrounding countries were also sent home to become jobless.

GOVERNMENT AND POLITICS

Early in 1999, Nelson Mandela, president of South Africa since 1994, delivered his final "State of the Nation" address and set a rare example for African leaders by retiring from politics after one term in office as president of South Africa. As the Times of Zambia noted in its editorial of February 8, 1999, "on a continent replete with political crises, and where succession is rarely predictable, President Nelson Mandela, the doyen of the anti-apartheid struggle, has bequeathed to his country, and Africa as a whole, a legacy worthy of praise." Indeed it is not an exaggeration that Mandela's principled nature, stature, integrity and inspiration, despite his brief reign, has led South Africa onto a peaceful political path.

Although 1999 saw some politically motivated violence such as the assassination of Sifiso Nkabinde, the secretary general of the multiracial United Democratic Movement, and the death of 11 people in revenge for his assassination, the year was generally peaceful. The general elections in June were also fairly and peacefully conducted. In the 1994 elections, canvassers who ventured into areas controlled by rival parties risked death. The vote in June 1999 passed without a single political killing, and the outcome of the voting was embraced by all political parties. In the June 3 elections, the ruling African National Congress (ANC) won 266 of 400 parliamentary seats (63 percent), just one seat shy of the two thirds majority required to amend the constitution. Thabo Mbeki was sworn in as South Africa's second democratically elected president at a glittering inauguration ceremony, which saw Nelson Mandela step down after steering the country away from apartheid rule and oppression. However, the one-sided vote in favor of ANC was itself troubling. Critics noted that the dominance of the ANC might lead South Africa to a *defacto* one party state.

It was during 1999 that the Truth and Reconciliation Commission (TRC), under the leadership of Archbishop Tutu, released its report on the atrocities committed during the apartheid era. However, a new book, titled "The Truth about the Truth Commission" written by Dr. Anthea Jeffery was strongly critical of the TRC.

South Africa also played a prominent role in the conflict in the Democratic Republic of the Congo. Numerous meetings between President Mandela, other regional leaders, and the embattled Democratic Republic of Congo President Laurent Kabila were held in South Africa.

CULTURE AND SOCIETY

As South Africa enters the 21st century, it does so with a rich cultural heritage. The year 1999 was full of many cultural events. South African scientists made several discoveries and technological breakthroughs. A team of South African scientists developed a low-cost, environmentally friendly water-stabilization system, which could revolutionize the way water is treated. Scientists at Stellenbosch University near Cape Town designed and produced South Africa's first space satellite, Sunsat, which was later launched in the United States on board a Boeing Delta rocket. The discovery of a new fossil species in the Northern Cape province placed South Africa as the origin of the most primitive mammals in the world. Scientists also discovered the body of an ancient well-preserved mummy near Joubertina in the Eastern Cape, the first ancient mummified remains to be discovered in South Africa.

It was announced that three of South Africa's most famous cultural and natural sites, including Robben Island, would be granted World Heritage Site status by UNESCO. The nation also commemorated the centenary of the Anglo-Boer War. Thousands of Afrikaners across South Africa gathered at war cemeteries to commemorate the 28,000 people who died in concentration camps during the Anglo-Boer War at the turn of the century. Services marking the centenary of the outbreak of the war were held at many of the 130 camps established by the British to intern Boer women and children, as well as hundreds of Africans left homeless during the fighting. Two books entitled "The Boer War" were also published and revealed the crucial role blacks played in the South African Anglo-Boer War. Numerous other books on the war were also published during the course of the year.

DIRECTORY

CENTRAL GOVERNMENT

Head of State

President

Thabo Mvuyelwa Mbeki, Office of the President, Union Buildings, West Wing, 2nd Floor, Government Avenue, Private Bag X955, Pretoria 0001, South Africa
PHONE: +27 (012) 3232502
FAX: +27 (012) 3232573
E-MAIL: mbekit@dpo.pwv.gov.za

Executive Deputy President

Jacob Zuma, Office of the Deputy President, Union Buildings, West Wing, 2nd Floor, Government Avenue, Private Bag X955, Pretoria 0001, South Africa
PHONE: +27 (012) 3232502
FAX: +27 (012) 3232573
E-MAIL: mamoepar@dpo.pwv.gov.za/ Spokesperson

Ministers

Minister of Education

A. K. Asmal, Ministry of Education, Magister Building, Room 910, 123 Schoeman Street, Private Bag X603, Pretoria 0001, South Africa
PHONE: +27 (012) 326 0126
FAX: +27 (012) 323 5989

Minister of Sport and Recreation

N. Balfour, Ministry of Sport and Recreation, Oranje Nassau Building, 3rd Floor, 188 Schoeman Street, Private Bag X869, Pretoria 0001, South Africa
PHONE: +27 (012) 3211781
FAX: +27 (012) 321 8493

Minister of Home Affairs
M. G. Buthelezi, Ministry of Home Affairs, Civitas Building, 10th floor, cor. Andries and Struben Streets, Private Bag X741, Pretoria 0001, South Africa
PHONE: +27 (012) 3268081
FAX: +27 (012) 3216491

Minister of Agriculture and Land Affairs
A. T. Didiza, Ministry of Agriculture and Land Affairs, 184 Jacob Mare Building, Room 328, cor. Jacob Mare and Paul Kruger Streets, Private Bag X844, Pretoria 0001, South Africa
PHONE: +27 (012) 3235212
FAX: +27 (012) 3211244

Minister of Trade and Industry
A. Erwin, Ministry of Trade and Industry, Street House of Trade and Industry, 11th floor, Prinsloo Street, Private Bag X274, Pretoria 0001, South Africa
PHONE: +27 (012) 3227677
FAX: +27 (012) 3227851

Minister of Public Service and Administration
G. J. Fraser-Moleketi, Ministry of Public Service and Administration, Transvaal House, 12th Floor, cor. Vermeulen and van der Walt Streets, Private Bag X884, Pretoria 0001, South Africa
PHONE: +27 (012) 3147911
FAX: +27 (012) 3286565

Minister of Water Affairs and Forestry
R. Kasrils, Ministry of Water Affairs and Forestry, Residensie Building, 185 Schoeman Street, Private Bag X313, Pretoria 0001, South Africa
PHONE: +27 (012) 3381500
FAX: +27 (012) 3284254

Minister of Defense
P. Lekota, Ministry of Defense, Armscor Building Block 5, Level 4, Nossob Street Erasmusrand, Private Bag X427, Pretoria 0001, South Africa
PHONE: +27 (012) 3556119
FAX: +27 (012) 3470118

Minister of Justice
PM Maduna, Ministry of Justice and Constitutional Development, Presidia Building Room 8.4, cor. Paul Kruger and Pretorius Streets, Private Bag X276, Pretoria 0001, South Africa
PHONE: +27 (012) 3238581
FAX: +27 (012) 3211708
E-MAIL: elsa@justisie.wcape.gov

Minister of Finance
T. A. Manuel, Ministry of Finance, 240 Vermeulen Street, 26th Floor, cor. Andries and Vermeulen Streets, Private Bag X115, Pretoria 0001, South Africa
PHONE: +27 (012) 323 8911
FAX: +27 (012) 323 3262

Minister for Posts, Telecommunications and Broadcasting
I. Matsepe-Casaburri, Ministry of Posts, Telecommunications and Broadcasting, Iparioli Office Park, Nkululeko House, 399 Duncan Street, Private Bag X860, Pretoria 0001, South Africa
PHONE: +27 (012) 4278111
FAX: +27 (012) 3626915
E-MAIL: nowjoan@doc.org.za

Minister of Minerals and Energy
P. Mlambo-Ncguka, Ministry of Minerals and Energy, DRC Synodal Center 9th Floor, cor. Andries and Visagie Streets, Private Bag X646, Pretoria 0001, South Africa
PHONE: +27 (012) 3228695
FAX: +27 (012) 3228699
E-MAIL: minsto@mepta.pwv.gov

Minister of Labour
M. M. S. Mdladlana, Ministry of Labour, Laboria House, cor. Schoeman and Paul Kruger Streets, Private Bag X499, Pretoria 0001, South Africa
PHONE: +27 (012) 3226523
FAX: +27 (012) 3201942

Minister of Environmental Affairs and Tourism
M. V. Moosa, Ministry of Environmental Affairs and Tourism, 240 Vermeulen Street, 7th Floor, cor. Andries and Vermeulen Streets, Private Bag X884, Pretoria 0001, South Africa
PHONE: +27 (012) 3103611
FAX: +27 (012) 322082
E-MAIL: pjordan@anc.org.za

Minister of Housing
S. D. Mthembi-Mahanyele, Ministry of Housing, 240 Walker Street, Sunnyside, Private Bag X644, Pretoria 0001, South Africa
PHONE: +27 (012) 4211311
FAX: +27 (012) 3418513

Minister of Provincial Affairs and Constitutional Development
F. S. Mufamadi, Ministry of Provincial Affairs and Constitutional Development, 87 Hamilton

Street, Room 507, Arcadia, Private Bag X802, Pretoria 0001, South Africa
PHONE: +27 (012) 3340705
FAX: +27 (012) 3264478

Minister of Arts, Culture, Science and Technology
B. S. Ngubane, Ministry of Arts, Culture, Science and Technology, Oranje Nassau Building, Room 7060, 188 Schoeman Street, Private Bag X727, Pretoria 0001, South Africa
PHONE: +27 (012) 3378378
FAX: +27 (012) 3242687

Minister of Intelligence
J. M. Nhlanhla, Ministry of Intelligence, PO Box 56450, Arcadia 0007, South Africa
PHONE: +27 (012) 3236738
FAX: +27 (012) 3230718

Minister of Transport
D. Omar, Ministry of Transport, Forum Building, Room 4111, cor. Struben and Bosman Streets, Private Bag X193, Pretoria 0001, South Africa
PHONE: +27 (012) 3093131
FAX: +27 (012) 3283194

Minister of Public Enterprises
J. T. Radebe, Ministry of Public Enterprises, InfoTech Building, Suite 401, 1090 Arcadia Street, Private Bag X15, Hatfield, Pretoria 0028, South Africa
PHONE: +27 (012) 3427111
FAX: +27 (012) 3427226

Minister of Public Works
S. N. Sigcua, Ministry of Public Works, Central Government Building, cor. Bosman and Vermeulen Streets, Private Bag X890, Pretoria 0001, South Africa
PHONE: +27 (012) 3241510
FAX: +27 (012) 3256380

Minister of Correctional Services
B. Skosana, Ministry of Correctional Services, Poyntons Buildings West Block, cor. Schubart and Church Streets, Private Bag X853, Pretoria 0001, South Africa
PHONE: +27 (012) 3238198
FAX: +27 (012) 3234111

Minister of Welfare and Population Development
Z. S. T. Skweyiya, Ministry of Welfare and Population Development, Hallmark Building, Room 501, 226 Vermeulen Street, Private Bag X885, Pretoria 0001, South Africa
PHONE: +27 (012) 3284600
FAX: +27 (012) 3257071
E-MAIL: wels168@welspta.pwv.gov.za

Minister of Health
M. E. Tshabalala-Msimang, Ministry of Health, Civitas Building, Room 2027, cor. Andries and Struben Streets, Private Bag X399, Pretoria 0001, South Africa
PHONE: +27 (012) 3284773
FAX: +27 (012) 3255526
E-MAIL: mnwbd@hltrsa2.pwv.gov.za

Minister of Safety and Security
S. V. Tshwete, Ministry of Safety and Security, Wachthuis, 7th floor, 231 Pretorius Street, Private Bag X463, Pretoria 0001, South Africa
PHONE: +27 (012) 3392800
FAX: +27 (012) 3392819
E-MAIL: mufamadi@saps.org.za

Minister of Foreign Affairs
N. C. Dlamini Zuma, Ministry of Foreign Affairs, Union Buildings, East Wing, Government Avenue, Private Bag X152, Pretoria 0001, South Africa
PHONE: +27 (012) 3510005
FAX: +27 (012) 3510253

POLITICAL ORGANIZATIONS

African National Congress (ANC)

TITLE: President
NAME: Thabo Mbeki

Democratic Party (DP)

TITLE: President
NAME: Tony Leon

Freedom Front (FF)

TITLE: President
NAME: Constand Viljoen

Inkatha Freedom Party (IFP)

TITLE: President
NAME: Mangosuthu Buthelezi

National Party (NP)

TITLE: President
NAME: Marthinus Van Schalkwyk

Pan Africanist Congress

TITLE: President
NAME: Stanley Mogoba

DIPLOMATIC REPRESENTATION

Embassies in South Africa

Belgium
Kanselarij, Chancery, 625 Leyds Street, Muckleneuk, Pretoria 0002, South Africa
PHONE: +27 (012) 443201
FAX: +27 (012) 443216

Canada
1103 Arcadia St., Hatfield 0083, Pretoria, South Africa

China
Taipei Liaison Office, Ground Floor, 12 Baker Street, Rosebank 2196, Johannesburg, South Africa

France
PO Box 1728, Pretoria 0001, South Africa
PHONE: +27 (012) 3484913
FAX: +27 (012) 472980
E-MAIL: pretoria@dree.org

United Kingdom
91 Parliament Street, Cape Town 8001, South Africa
PHONE: +27 (012) 4617220
FAX: +27 (021) 4610017
TITLE: High Commissioner
NAME: Dame Maeve Fort

United States
PO Box 9536, Pretoria 0001, South Africa
PHONE: +27 (012) 3421048
FAX: +27 (012) 3421801

JUDICIAL SYSTEM

Supreme Court

Court of Appeals

FURTHER READING

Articles

Barber, Simon. ''Manuel Rejects Japanese Proposal on Gold.'' *Business Day*, April 28, 1999.

''Four Months in Review: South Africa.'' *Current History* 98 (September 1999): 303.

Hartley, Wyndham. ''US-SA Commission Tackles Crime, Investment.'' *Business Day*, February 19, 1999.

''Mandela and the Anti-Apartheid Struggle.'' *The Times of Zambia*, February 8, 1999.

Mzolo, Bhungani. ''South Africa: AIDS Drugs Protests.'' *The Sowetan*, July 6, 1999.

Books

Ballard, Sebastian. *South Africa Handbook 2000.* 4th ed. Bath: Footprint, 1999.

Brown, Marj. *Land Restitution in South Africa: A Long Way Home.* Cape Town: Institute for Democracy in South Africa, 1999.

Fox, Roddy and Rowntree, Kate. *The Geography of South Africa in a Changing World.* Goodwood, South Africa and Oxford: Oxford University Press, 1999.

Hurt, Karen and Budlender, Debbie. *Money Matters: Women and the Government Budget.* Cape Town: Institute for Democracy in South Africa, 1998.

Priestner, Ian and Reuvid, Jonathan. *Doing Business in South Africa.* 4th ed. London: Kogan Page, 1999.

Internet

South Africa Government Online. Available Online @ http://www.gov.za/ (November 15, 1999).

University of Pennsylvania, African Studies Program. Available Online @ http://www.sas.upenn.edu/African_Studies/Country_Specific/S_Africa.html (November 15, 1999).

SOUTH AFRICA: STATISTICAL DATA

For sources and notes see "Sources of Statistics" in the front of each volume.

GEOGRAPHY

Geography (1)

Area:

Total: 1,219,912 sq km.

Land: 1,219,912 sq km.

Water: 0 sq km.

Note: includes Prince Edward Islands (Marion Island and Prince Edward Island).

Area—comparative: slightly less than twice the size of Texas.

Land boundaries:

Total: 4,750 km.

Border countries: Botswana 1,840 km, Lesotho 909 km, Mozambique 491 km, Namibia 855 km, Swaziland 430 km, Zimbabwe 225 km.

Coastline: 2,798 km.

Climate: mostly semiarid; subtropical along east coast; sunny days, cool nights.

Terrain: vast interior plateau rimmed by rugged hills and narrow coastal plain.

Natural resources: gold, chromium, antimony, coal, iron ore, manganese, nickel, phosphates, tin, uranium, gem diamonds, platinum, copper, vanadium, salt, natural gas.

Land use:

Arable land: 10%

Permanent crops: 1%

Permanent pastures: 67%

Forests and woodland: 7%

Other: 15% (1993 est.)

HUMAN FACTORS

Demographics (2A)

	1990	1995	1998	2000	2010	2020	2030	2040	2050
Population	37,191.5	40,863.9	42,834.5	43,981.8	47,503.0	48,983.2	51,123.5	54,846.9	58,971.7
Net migration rate (per 1,000 population)	NA	NA	NA	NA	NA	NA	NA	NA	NA
Births	NA	NA	NA	NA	NA	NA	NA	NA	NA
Deaths	NA	NA	NA	NA	NA	NA	NA	NA	NA
Life expectancy - males	58.1	56.4	53.6	51.8	46.6	50.4	60.9	70.7	74.2
Life expectancy - females	64.7	60.6	57.8	56.0	49.5	53.4	65.1	76.3	80.0
Birth rate (per 1,000)	29.9	27.8	26.4	25.5	21.4	19.2	17.5	15.6	14.1
Death rate (per 1,000)	9.3	10.8	12.3	13.3	17.8	16.4	11.5	7.9	7.5
Women of reproductive age (15-49 yrs.)	9,378.9	10,472.0	11,066.1	11,405.9	12,476.1	13,083.7	13,739.7	14,609.7	14,860.9
of which are currently married	NA	NA	NA	NA	NA	NA	NA	NA	NA
Fertility rate	3.7	3.3	3.2	3.0	2.4	2.2	2.0	2.0	2.0

Except as noted, values for vital statistics are in thousands; life expectancy is in years.

Health Personnel (3)

Total health expenditure as a percentage of GDP, 1990-1997[a]

Public sector	3.6
Private sector	4.3
Total[b]	7.9

Health expenditure per capita in U.S. dollars, 1990-1997[a]

Purchasing power parity	542
Total	258

Availability of health care facilities per 100,000 people

Hospital beds 1990-1997[a]	NA
Doctors 1993[c]	59
Nurses 1993[c]	175

Health Indicators (4)

Life expectancy at birth

1980	57
1997	65
Daily per capita supply of calories (1996)	2,993
Total fertility rate births per woman (1997)	2.8
Maternal mortality ratio per 100,000 live births (1990-97)	230[c]
Safe water % of population with access (1995)	59
Sanitation % of population with access (1995)	53
Consumption of iodized salt % of households (1992-98)[a]	40

Smoking prevalence

Male % of adults (1985-95)[a]	52
Female % of adults (1985-95)[a]	17
Tuberculosis incidence per 100,000 people (1997)	394
Adult HIV prevalence % of population ages 15-49 (1997)	12.91

Infants and Malnutrition (5)

Under-5 mortality rate (1997)	65
% of infants with low birthweight (1990-97)	NA
Births attended by skilled health staff % of total[a]	82

% fully immunized (1995-97)

TB	95
DPT	73
Polio	73
Measles	76
Prevalence of child malnutrition under age 5 (1992-97)[b]	9

Ethnic Division (6)

Black	75.2%
White	13.6%
Colored	8.6%
Indian	2.6%

Religions (7)

Christian	68%
Muslim	2%
Hindu	1.5%
Traditional and animistic	28.5%

Christian includes most whites and Coloreds, about 60% of blacks, and about 40% of Indians.

Languages (8)

11 official languages, including Afrikaans, English, Ndebele, Pedi, Sotho, Swazi, Tsonga, Tswana, Venda, Xhosa, Zulu.

EDUCATION

Public Education Expenditures (9)

Public expenditure on education (% of GNP)

1980	....
1996	7.9

Expenditure per student

Primary % of GNP per capita

1980	....
1996	15.5

Secondary % of GNP per capita

1980	....
1996	23.7[1]

Tertiary % of GNP per capita

1980	....
1996	64.7[1]

Expenditure on teaching materials

Primary % of total for level (1996)	....
Secondary % of total for level (1996)	4.7
Primary pupil-teacher ratio per teacher (1996)	36
Duration of primary education years (1995)	7

Educational Attainment (10)

Age group (1995)	20+
Total population	22,100,000

Highest level attained (%)

No schooling	13.0

First level

Not completed	17.1
Completed	6.9

Entered second level

S-1	26.7
S-2	25.7
Postsecondary	8.8

Literacy Rates (11A)

In thousands and percent[1]	1990	1995	2000	2010
Illiterate population (15+ yrs.)	4,604	4,731	4,847	5,028
Literacy rate - total adult pop. (%)	79.9	81.8	83.6	86.6
Literacy rate - males (%)	80.3	81.9	83.4	85.9
Literacy rate - females (%)	79.4	81.7	83.8	87.2

GOVERNMENT & LAW

Political Parties (12)

National Assembly	% of seats
African National Congress (ANC)	62.6
National Party (NP)	20.4
Inkatha Freedom Party (IFP)	10.5
Freedom Front (FF)	2.2
Democratic Party (DP)	1.7
Pan-Africanist Congress (PAC)	1.2
African Christian Democratic Party (ACDP)	0.5
Other	0.9

Military Affairs (14B)

	1990	1991	1992	1993	1994	1995
Military expenditures						
Current dollars (mil.)	4,545[e]	3,921[e]	3,281	3,537	3,015	2,895
1995 constant dollars (mil.)	5,223[e]	4,332[e]	3,529	3,708	3,091	2,895
Armed forces (000)	85	80	72	72	102	100
Gross national product (GNP)						
Current dollars (mil.)	107,300	111,200	111,100	116,100	123,700	130,600
1995 constant dollars (mil.)	123,400	122,800	119,500	121,700	126,800	130,600
Central government expenditures (CGE)						
1995 constant dollars (mil.)	41,390	37,730	40,140	40,480	41,690	43,050
People (mil.)	37.2	38.0	38.9	39.6	40.3	41.0
Military expenditure as % of GNP	4.2	3.5	3.0	3.0	2.4	2.2
Military expenditure as % of CGE	12.6	11.5	8.8	9.2	7.4	6.7
Military expenditure per capita (1995 $)	140	114	91	94	77	71
Armed forces per 1,000 people (soldiers)	2.3	2.1	1.9	1.8	2.5	2.4
GNP per capita (1995 $)	3,317	3,232	3,075	3,072	3,148	3,185
Arms imports[6]						
Current dollars (mil.)	260	350	260	260	290	250
1995 constant dollars (mil.)	299	387	280	273	297	250
Arms exports[6]						
Current dollars (mil.)	50	10	90	170	230	100
1995 constant dollars (mil.)	57	11	97	178	236	100
Total imports[7]						
Current dollars (mil.)	18,400	18,830	19,760	20,020	23,390	30,550
1995 constant dollars (mil.)	21,150	20,810	21,260	20,990	23,970	30,550
Total exports[7]						
Current dollars (mil.)	23,550	23,310	23,410	24,260	24,990	27,860
1995 constant dollars (mil.)	27,060	25,750	25,180	25,430	25,610	27,860
Arms as percent of total imports[8]	1.4	1.9	1.3	1.3	1.2	.8
Arms as percent of total exports[8]	.2	0	.4	.7	.9	.4

Government Budget (13B)

Revenues	$30.5 billion
Expenditures	$38 billion
Capital expenditures	$2.6 billion

Data for FY94/95 est.

Crime (15)

Crime rate (for 1997)	
Crimes reported	2,737,500
Total persons convicted	1,488,100
Crimes per 100,000 population	6,300
Persons responsible for offenses	
Total number of suspects	NA
Total number of female suspects	NA
Total number of juvenile suspects	NA

LABOR FORCE

Labor Force (16)

Total (million)	14.2
Services	35%
Agriculture	30%
Industry	20%
Mining	9%
Other	6%

Data for 1996. Total is economically active labor force.

Unemployment Rate (17)

30% (1997 est.); an additional 11% of the workforce is underemployed.

PRODUCTION SECTOR

Electric Energy (18)

Capacity	34.566 million kW (1995)
Production	163.56 billion kWh (1995)
Consumption per capita	3,559 kWh (1995)

Transportation (19)

Highways:

total: 331,265 km

paved: 137,475 km (including 1,142 km of expressways)

unpaved: 193,790 km (1995 est.)

Pipelines: crude oil 931 km; petroleum products 1,748 km; natural gas 322 km

Merchant marine:

total: 9 ships (1,000 GRT or over) totaling 274,797 GRT/270,837 DWT ships by type: container 6, oil tanker 2, roll-on/roll-off cargo 1 (1997 est.)

Airports: 750 (1997 est.)

Airports—with paved runways:

total: 143

over 3,047 m: 10

2,438 to 3,047 m: 4

1,524 to 2,437 m: 46

914 to 1,523 m: 74

under 914 m: 9 (1997 est.)

Airports—with unpaved runways:

total: 607

1,524 to 2,437 m: 35

914 to 1,523 m: 308

under 914 m: 264 (1997 est.)

Top Agricultural Products (20)

Corn, wheat, sugarcane, fruits, vegetables; beef, poultry, mutton, wool, dairy products.

MANUFACTURING SECTOR

GDP & Manufacturing Summary (21)

Detailed value added figures are listed by both International Standard Industry Code (ISIC) and product title.

	1980	1985	1990	1994
GDP ($-1990 mil.)[1]	88,156	94,292	102,167	103,592
Per capita ($-1990)[1]	3,022	2,854	2,756	2,554
Manufacturing share (%) (current prices)[1]	22.6	22.4	24.6	*22.8*
Manufacturing				
Value added ($-1990 mil.)[1]	22,709	21,534	23,181	22,657
Industrial production index	100	103	113	110
Value added ($ mil.)	17,866	12,409	23,181	*25,669*
Gross output ($ mil.)	*53,686*	36,059	68,770	*69,343*
Employment (000)	1,392	1,422	1,525	*1,431*
Profitability (% of gross output)				
Intermediate input (%)	*67*	66	66	*63*
Wages and salaries inc. supplements (%)	*16*	18	*17*	*19*
Gross operating surplus	*17*	17	*17*	*18*

	1980	1985	1990	1994
Productivity ($)				
Gross output per worker	*38,568*	24,982	45,095	*48,462*
Value added per worker	12,835	8,633	15,201	*17,979*
Average wage (inc. supplements)	*6,120*	4,466	*7,708*	*9,348*
Value added ($ mil.)				
311/2 Food products	1,626	1,277	2,220	*2,734*
313 Beverages	458	418	1,055	*1,531*
314 Tobacco products	111	108	83	*113*
321 Textiles	886	408	851	*868*
322 Wearing apparel	477	334	701	*757*
323 Leather and fur products	40	44	75	*99*
324 Footwear	152	113	316	*248*
331 Wood and wood products	213	190	469	*353*
332 Furniture and fixtures	219	138	307	*276*
341 Paper and paper products	591	471	1,208	*1,237*
342 Printing and publishing	549	392	763	*894*
351 Industrial chemicals	1,006	717	932	*1,268*
352 Other chemical products	639	1,047	1,255	*1,272*
353 Petroleum refineries	634	1,038	1,244	*1,281*
354 Miscellaneous petroleum and coal products	111	182	217	*224*
355 Rubber products	297	157	401	*351*
356 Plastic products	355	225	560	*657*
361 Pottery, china and earthenware	28	24	42	*45*
362 Glass and glass products	154	102	292	*337*
369 Other non-metal mineral products	754	481	794	*862*
371 Iron and steel	2,135	986	2,343	*2,303*
372 Non-ferrous metals	555	418	642	*867*
381 Metal products	1,576	860	1,697	*1,588*
382 Non-electrical machinery	1,351	805	1,432	*1,566*
383 Electrical machinery	1,229	607	970	*1,174*
384 Transport equipment	1,258	566	1,705	*2,082*
385 Professional and scientific equipment	49	54	160	*248*
390 Other manufacturing industries	415	246	448	*434*

FINANCE, ECONOMICS, & TRADE

Economic Indicators (22)

National product: GDP—purchasing power parity—$270 billion (1997 est.)

National product real growth rate: 3% (1997 est.)

National product per capita: $6,200 (1997 est.)

Inflation rate—consumer price index: 9.7% (1997 est.)

Balance of Payments (23)

	1992	1993	1994	1995	1996
Exports of goods (f.o.b.)	24,009	24,138	24,947	28,611	29,057
Imports of goods (f.o.b.)	−18,216	−18,287	−21,452	−27,001	−27,027
Trade balance	5,794	5,850	3,494	1,610	2,030
Services - debits	−8,816	−8,550	−8,927	−9,991	−9,100
Services - credits	4,669	4,443	5,064	5,544	5,112
Private transfers (net)	99	101	74	69	−32
Government transfers (net)	−4	29	−23	−53	−42
Overall balance	1,742	1,873	−318	−2,820	−2,032

Exchange Rates (24)

Exchange rates:	
Rand (R) per US$1	
January 1998	4.94193
1997	4.60796
1996	4.29935
1995	3.62709
1994	3.55080
1993	3.26774

Top Import Origins (25)

$28 billion (f.o.b., 1997)

Origins	%
Germany	NA
United States	NA
Japan	NA
United Kingdom	NA
Italy	NA

NA stands for not available.

Top Export Destinations (26)

$31.3 billion (f.o.b., 1997) Data are for 1993.

Destinations	%
France	NA
United Kingdom	NA
China	NA
Germany	NA
Japan	NA

NA stands for not available.

Economic Aid (27)

Recipient: ODA, $NA. Note: current aid pledges include US $600 million over three years, 1994-96; UK $150 million over three years; Australia $21 million over three years; Japan $1.3 billion over two years ending in 1996; EU $833 million over five years. NA stands for not available.

Import Export Commodities (28)

Import Commodities	Export Commodities
Machinery 32%	Gold 20%
Transport equipment 15%	Other minerals and metals 20%-25%
Chemicals 11%	Food 5%
Petroleum products	Chemicals 3%
Textiles	
Scientific instruments	

SPAIN

Kingdom of Spain
España

CAPITAL: Madrid.

FLAG: The national flag, adopted in 1785, consists of three horizontal stripes: a yellow one—equal in size to the other two combined—between two red ones, with the coat of arms on the yellow stripe.

ANTHEM: *Marcha Real Granadera (March of the Royal Grenadier).*

MONETARY UNIT: The peseta (P) is a paper currency of 100 centimos. There are coins of 50 centimos and 1, 5, 25, 50, and 100 pesetas, and notes of 1,000, 2,000, 5,000, and 10,000 pesetas. P1 = $0.0073 (or $1 = P137.38).

WEIGHTS AND MEASURES: The metric system is the legal standard.

HOLIDAYS: New Year's Day, 1 January; St. Joseph's Day, 19 March; Epiphany, 31 March; Day of St. Joseph the Artisan, 1 May; St. James's Day, 25 July; Assumption, 15 August; National Day and Hispanic Day, 12 October; All Saints' Day, 1 November; Immaculate Conception, 8 December; Christmas, 25 December. Movable religious holidays include Holy Thursday, Good Friday, Easter Monday, and Corpus Christi.

TIME: 1 PM = noon GMT.

LOCATION AND SIZE: Occupying the greater part of the Iberian Peninsula, Spain is the third-largest country in Europe, with an area of 504,750 square kilometers (194,885 square miles). Spain has a total boundary length of 7,843 kilometers (4,874 miles). Spain has long claimed Gibraltar, a narrow peninsula on the south coast.

Spain's capital city, Madrid, is located in the center of the country.

CLIMATE: The northern coastal regions are cool and humid, with an average annual temperature of 14°C (57°F). The central plateau is cold in the winter and hot in the summer; Madrid has a winter average of about 8°C (46°F) and a summer average of 23°C (73°F).

The northern coastal regions have an average annual rainfall of 99 centimeters (39 inches); the southern coastal belt has 41–79 centimeters (16–31 inches); and the interior central plain averages no more than 50 centimeters (20 inches) annually.

INTRODUCTORY SURVEY

RECENT HISTORY

Beginning in about 1000 BC, the prehistoric Iberian culture of present-day Spain was changed by Celtic, Phoenician, and Greek invaders. From the sixth to the second century BC, Carthage controlled the Iberian Peninsula up to the Ebro River. From 133 BC until the barbarian invasions of the fifth century AD, Rome ruled Hispania (from which the name Spain is derived). During the Roman period, cities and roads were built, and Christianity and Latin were introduced. The Spanish language grew out of Latin.

In the fifth century, the Visigoths settled in Spain, ruling the country until 711, when it was invaded by the Moors. All of Spain, except for a few northern districts, was under Moorish Muslim rule for periods ranging from 300 to 800 years. A rich civilization arose, characterized by prosperous cities, industries, and agriculture and by brilliant intellectual figures, including Jews as well as Muslims. Throughout this period (711–1492), however, Christian Spain waged periodic wars against the Moors. By the thirteenth century, Muslim rule was restricted to the south of Spain.

In 1492, Spain was unified under Ferdinand II of Aragón and Isabella I of Castile, the "Catholic Sovereigns." Moors and Jews were driven out of Spain. Those who chose to convert to Catholicism and stay in Spain were terrorized by officials of the Inquisition to be sure they were no longer practicing their old religions. Also in 1492, Christopher Columbus sailed to the Americas with Ferdinand

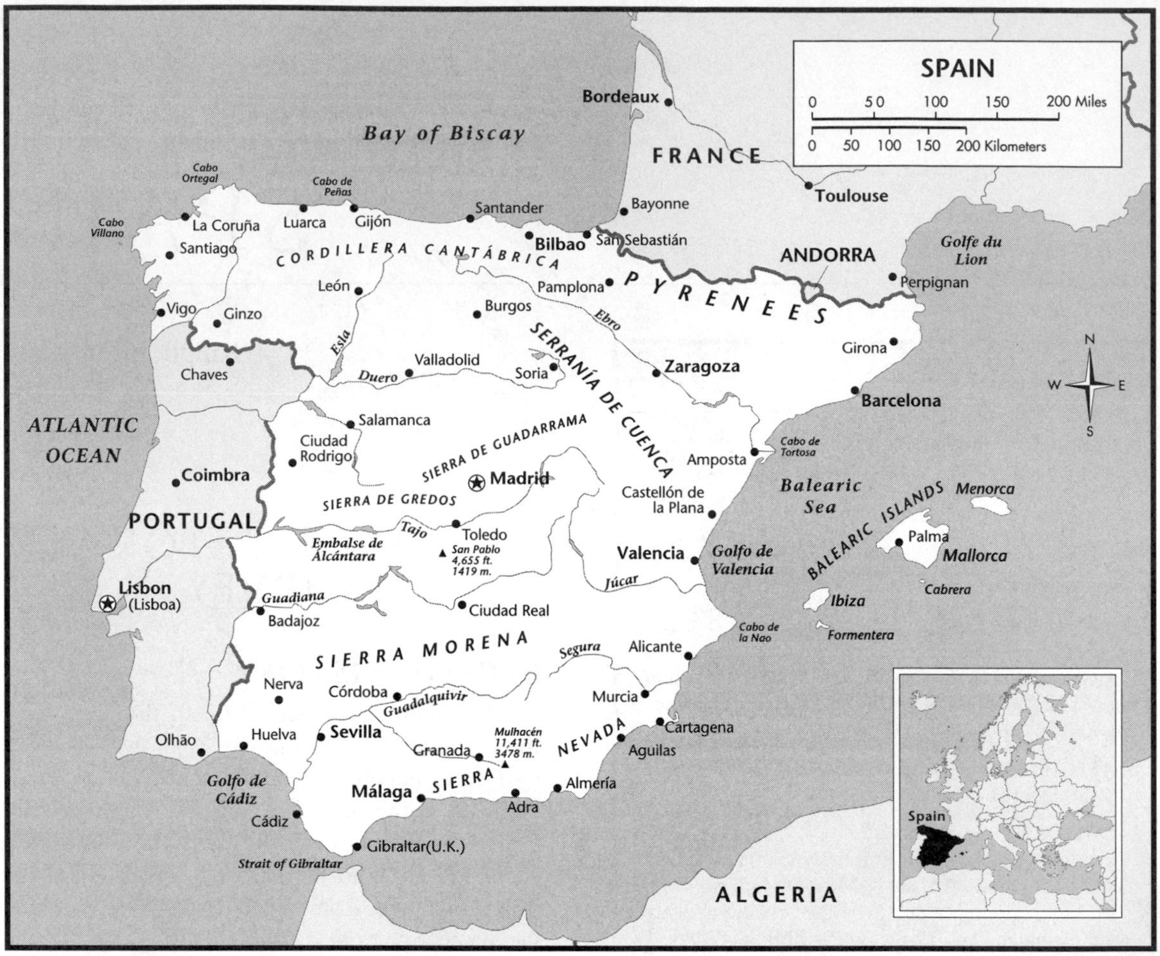

and Isabella's backing. In 1519, Ferdinand Magellan, a Portuguese in the service of Spain, began the first voyage to sail around the world, completed in 1522 by Juan Sebastián Elcano.

The sixteenth century was the golden age of Spain: its empire in the Americas produced vast wealth; its arts flourished; its fleet ruled the high seas; and its armies were the strongest in Europe. By the second half of the sixteenth century, however, religious wars in Europe and the flow of people and resources to the New World had drained the strength of the Spanish nation. In 1588, the ''invincible'' Spanish Armada was defeated by England.

Spain's power on land was ended by wars with England, the Netherlands, and France in the seventeenth century and by the War of the Spanish Succession (1701–14), which established the Bourbon dynasty in Spain.

Much of the nineteenth and early twentieth centuries was spent in passionate struggles between the monarchy and those who wanted Spain to be a republic. Abroad, Spain lost most of its lands in the West to colonial rebellions during the first half of the nineteenth century. Cuba, Puerto Rico, and the Philippines were lost as a result of the Spanish-American War in 1898. Spain remained neutral in World War I (1914–18) but in the postwar period fought to keep its colonial possessions in Morocco.

The Spanish general and dictator Primo de Rivera, who successfully ended the fighting in Morocco in 1927, remained in power under the monarchy until 1930. In 1931, after a public vote, King Alfonso XIII left Spain and a republic was established. Neither right-wing nor left-wing groups had a parliamentary majority, and on the whole the coalition governments were ineffective.

Between 17 July 1936 and 31 March 1939, Spain was ravaged by civil war. The two opposing parties were the Republicans, made up partly of democrats and partly of left-wing groups, and the rebels (Nationalists), who favored the establishment of a right-wing dictatorship. Germany and Italy furnished soldiers and weapons to the Nationalists, while the Soviet Union, Czechoslovakia, and Mexico supported the Republicans. Finally the Republicans were defeated, and General Francisco Franco formed a new government. Under the Franco dictatorship, Spain gave aid to the Axis powers in World War II (1939–45) but was not an active participant.

Spain was admitted to the United Nations in 1955. However, the repressive nature of the Franco regime kept Spain apart from the main social, political, and economic currents of postwar Western Europe. On 20 November 1975, General Franco died at the age of 82, ending a career that had dominated nearly four decades of Spanish history. Two days later, Juan Carlos I was sworn in as king. Labor groups and Catalan and Basque separatists protested the new government. A sharp rise in living costs caused even more political unrest.

On 15 June 1977, the first democratic elections in Spain in 40 years took place. The Union of the Democratic Center (Unión de Centro Democrático—UCD), headed by Adolfo Suárez González won a majority in the new Cortes (parliament). The Cortes prepared a new constitution, which was approved by popular vote and by the king in December 1978.

When Suárez announced his resignation in January 1981, the king named Leopoldo Calvo Sotelo y Bustelo to the premiership. In 1982, 1986, and 1989, the Spanish Socialist Worker's Party (Partido Socialista Obrero Español—PSOE), headed by Felipe González Márquez, won absolute majorities in both houses of parliament. The PSOE failed to win a majority in 1993 but governed with the support of the Basque and Catalan nationalist parties. Political violence, especially murders and kidnappings in the Basque region, has been a continuing problem since the late 1960s. Spain joined the North Atlantic Treaty Organization in 1982, a controversial move that was approved by a majority of Spanish voters in 1986. On 1 January of that year, Spain became a full member of the European Community (EC).

During 1995–97, Basque terrorists continued attacks on civilian, police, and military targets. In August 1995, King Juan Carlos was almost assassinated by terrorists while on vacation.

In 1995, a government conspiracy to locate and kill Basque terrorists living in France was revealed. Before the plot was uncovered, Spain's national police force had cooperated in the killing of 27 suspected Basque terrorists. France and Spain agreed to greater cooperation to legally apprehend suspected terrorists. In 1996, the execution of a university professor by Basque terrorists brought out a half million protesters in Madrid.

GOVERNMENT

A new constitution, approved by the parliament on 31 October 1978, confirmed Spain as a parliamentary monarchy. The king is the head of state. Legislative power rests in the Cortes Generales, consisting of two chambers: the Congress of Deputies (350 members in 1996) and the Senate (254 members). All deputies and 208 senators are popularly elected to four-year terms under universal adult suffrage. The remaining senators (46) are chosen by the assemblies in the 17 autonomous regions. The president, vice-president, and ministers are all appointed by the king and answerable to Congress.

Judiciary

The highest judicial body is the seven-member Supreme Court (Tribunal Supremo). Territorial high courts *(audiencias)* are the courts of last appeal in the 15 regions of the country. Provincial audiencias serve as appeals courts in civil matters and provide the first hearing for criminal cases. The National High Court (Audiencia Nacional), created in 1977, has jurisdiction over criminal cases that cross regional boundaries and over civil cases involving the central state administration. A nine-person jury system was established in 1995, and a new penal code was enacted in 1996.

Political Parties

The Spanish political scene is characterized by changing parties and shifting alliances. The Spanish Socialist Worker's Party (Partido Socialista Obrero Español—PSOE) won absolute majorities in both chambers of the Cortes in 1982, 1986, and 1989. It finished short of a parliamentary majority in 1993 and 1996.

The right-wing element is represented by the People's Alliance (CP), also known as the Popular Party or PP, a coalition made up of the Alianza Popular, the Christian Democratic Partido Demócrata Popular, and the Partido Liberal. In the March 1996 parliamentary elections, the PP won 156 seats, while the PSOE won 141 seats.

The Communist Party (Partido Comunista—PC) was one of Western Europe's most outspoken communist parties in the 1970s. Nationalist parties function in Catalonia, Andalucía, the Basque provinces, and other areas.

DEFENSE

Spain has a 9-month compulsory military service system but is moving toward a smaller, volunteer force. In 1995, the armed forces totaled 206,800, of whom 128,800 were draftees. The 142,200-man army was organized into 5 divisions and 8 special purpose brigades. The navy had 36,100 men. The air force had 28,500 men and about 187 combat aircraft. The Guardia Civil (civil guard) numbered about 75,000. In 1995 Spain spent $8.5 billion on defense. Spain contributes troops and observers to United Nations missions.

ECONOMIC AFFAIRS

Agriculture, livestock, and mining used to be the main supports of the economy. In order to offset the damage suffered during the Civil War, however, the Franco regime concentrated its efforts on industrial expansion. So industry now provides most of the jobs and exports from Spain.

In terms of per person income, Spain still ranks among the lowest in Western Europe. From 1974 through the early 1980s, the Spanish economy was hurt by international factors, especially oil price increases. Consumer prices rose 37% between 1989 and 1995, and unemployment rose from 17.3% to 21.3%.

Public Finance

The public sector deficit in 1996 was equivalent to 4.3% of GDP (compared to 3.8% in 1993 and 4.4% in 1992). Because of Spain's desire to enter the European Monetary Union, it has had to meet stringent limits on its public debt and finances. The current government has trimmed the budget by reducing the civil service payroll and limiting transfers to government-owned companies.

The U.S. Central Intelligence Agency estimates that, in 1994, government revenues totaled approximately $96.8 billion and expenditures $122.5 billion, including capital expenditures of $5.7 billion. External debt totaled $90 billion, approximately 30% of which was financed abroad.

Income

In 1995, Spain's gross national product (GNP) was $532.4 billion, or about $14,350 per person. For the period 1985–95, the average inflation rate was 6.3%, resulting in a real growth rate in per person GNP of 2.6%.

Industry

The chief industries are food and beverages, energy, and transport materials. Chemical production, particularly of superphosphates, sulfuric acid, dyestuffs, and pharmaceutical products, is also significant. Of the heavy industries, iron and steel (13.5 million tons produced in 1994) is the most important. Petroleum refinery production in 1993 totaled 31.9 million tons, and in 1995, 22.3 million tons of cement were produced. Other important products in 1995 included automobiles, 2 million; and commercial vehicles, 298,072.

Banking and Finance

The banking and credit structure centers on the Bank of Spain, the government's national bank of issue since 1874. The bank acts as the government depository as well as a banker's bank for discount and other operations. Other ''official'' but privately owned banks are the Mortgage Bank of Spain, the Local Credit Bank of Spain, the Industrial Credit Bank, the Agricultural Credit Bank, and the External Credit Bank.

In 1997, the private banking system consisted of over 100 banks, comprising national banks, industrial banks, regional banks, local banks, and foreign banks. The liberalization of the banking system and Spain's entry into the EC have raised the number and presence of foreign banks. During the process of financial liberalization required by the EU, the government tried to promote a series of mergers within the banking industry, which it hoped could enable the banks to compete more effectively. As a result, there were two major mergers: Banco de Vizcaya and Banco de Bilbao formed Banco Bilbao Vizcaya (BBV), and Banco Central and Banco Hispanoamericano merged to form Banco Central Hispanoamercano (BCH). The government also brought together all the state-owned banking institutions to form Corporación Bancaria de España, better known by its trade name Argentaria, whose most important component is Banco Exterior (BEX). The government subsequently privatized a 50% stake in Argentaria in 1993 and a further 25% in early 1996. The state's remaining 25% share in Argentaria is expected to be sold in the near future, leaving the banking sector entirely in private hands.

In December 1993 the banking sector was shaken by the decision of the Bank of Spain to take over the running of Banesto, one of the top six banks, and the subsequent discovery of a p 605 billion ($4.75 billion) deficit. Since this takeover, Banesto has staged an impressive recovery, returning to profit in 1995.

Foreign reserves, excluding gold, totaled $45,040 million in April 1996.

Spain has four major stock exchanges, in Madrid, Barcelona, Bilbao, and Valencia. These exchanges are open for a few hours a day, Tuesday through Friday. Since 1961, foreign investment in these exchanges has increased rapidly. The major commercial banks invest in the equity and debt securities of private firms and carry on brokerage businesses as well. In 1995 pre-tax profits of the 380 quoted companies rose by 23.8% to p 2.02 trillion.

Economic Development

After 1939, Spanish economic policy was characterized by the attempt to achieve economic self-sufficiency. This policy, largely imposed by Spain's position during World War II and the isolation to which Spain was subjected in the decade following 1945, was also favored by many Spanish political and business leaders. In 1959, following two decades of little or no overall growth, the Spanish government, acceding to reforms suggested by the IMF, OECD, and IBRD and encouraged by the promise of foreign financial assistance, announced its acceptance of the so-called Stabilization Plan, intended to curb domestic inflation and adverse foreign payment balances.

Long-range planning began with Spain's first four-year development plan (1964–67), providing a total investment of p 355 billion. The second four-year plan (1968–71) called for an investment of p 553 billion, with an average annual growth of 5.5% in GNP. The third plan (1972–75) called for investments of p 871 billion; drastic readjustments had to be made in 1975 to compensate for an economic slump brought on by increased petroleum costs, a tourist slowdown, and a surge in imports. A fifth plan (1976–79) focused on development of energy resources, with investments to increase annually by 9% increments. A stabilization program introduced in 1977 included devaluation of the peseta and tightening of monetary policy. The economic plan of 1979–82 committed Spain to a market economy and rejected protectionism.

Accession to the EU generated increased foreign investment but also turned Spain's former trade surplus with the EU into a growing deficit: the lowering of tariffs boosted imports, but exports did not keep pace. The government responded by pursuing market liberalization and deregulation, in hopes of boosting productivity and efficiency to respond to EU competition. A number of projects, such as the construction of airports, highways, and a high-speed rail line between Madrid and Seville, received EU funding. To prepare Spain for a forthcoming European economic and monetary union, the government in 1992 planned to cut public spending. The currency was devalued three times in 1992–93. Additionally, Spain is a principal beneficiary of the EU's ''harmonization fund.'' This fund provides financial support to poorer EU nations to attempt to reduce the disparities in economic development.

SOCIAL WELFARE

All employed persons aged 14 and older must pay into the national social security program. The program provides for health and maternity insurance, old age and incapacity insurance, a family subsidy, workers' compensation, and job-related disability payments.

One-third more women now work outside the home than a decade ago, but employment levels for women are still relatively low. Roma minorities suffer from housing, education, and employment discrimination. The government provides a legal course of action to help minorities experiencing discrimination.

Human rights abuses have been committed by both the government and Basque separatist groups.

Healthcare

In 1991, there were a total of 175,375 hospital beds. There were 4.1 doctors per 1,000 people.

Average life expectancy for the period 1990–95 was 77 years. Leading causes of death in 1990 were categorized as follows: communicable diseases and maternal/perinatal causes (45 per 100,000 people); noncommunicable diseases (410 per 100,000); and injuries (42 per 100,000).

Housing

Construction has generally fallen short of the 336,000 units needed each year to keep pace with population growth and the deterioration of old buildings. In 1989, 233,063 units were completed and the total number of dwellings in 1991 was 16 million.

EDUCATION

Elementary education is compulsory and free. During 1993, there were 16,540 primary schools with 121,353 teachers and 2.4 million pupils. General secondary schools had 4.7 million pupils with 297,697 teachers. The adult illiteracy rate is estimated at 4%. Students in higher education numbered 1.5 million in 1993 and there were 80,642 instructors.

1999 KEY EVENTS TIMELINE

January

- Spain is one of 11 members of the European Union that adopts the Euro as its currency. The Euro will replace the currency of individual countries by 2002.

March

- Tense relations arise between Spain and England over Gibraltar. The Spanish claim the enclave has become a haven for money laundering and drug traffickers. Spain tightens immigration controls at the border after Gibraltar authorities detain Spanish fishing vessel in January.

June

- Spain offers economic aid to Ecuador, which is going through a deep economic crisis.

July

- Spain's Socialists choose Joaquín Almunia as their presidential candidate. He is chosen after the more favored candidate Josep Borrell withdraws in May.
- Stampeding bulls gore six people, including an American, during the third day of the annual running of the bulls in Pamplona.
- The Ministry of Agriculture proposes a law that would force the owners of dangerous pets to take a psychological test.

August

- Seville plays host to the World Track and Field Championships.
- Niurka Montalvo, 31, wins a gold medal in the long jump at the World Track and Field Championships in Seville. She becomes the first Spanish woman to win a medal in the sport's history. Montalvo, a Cuban by birth, became a Spanish citizen in 1998.
- Spain says it will not accept the use of German as a third language during European Union meetings.
- Unemployment hits a 17-year low, dropping to 15.63 percent. Five years ago, unemployment stood at 25 percent.

September

- Cristina, the youngest daughter in Spain's royal family, gives birth to her first son. He is named Juan.
- The Spanish Supreme Court says Nike can't use its brand name in Spain.

October

- United Kingdom judge says former Chilean dictator Augusto Pinochet can be extradited to Spain.
- Laura Espido Freire, a 25-year-old Basque, wins Spain's highest literary award, the 48th annual Planeta prize. Her previous novels have been widely praised by Spanish critics.
- The Nationalist Party, behind Jordi Pujol, wins 58 seats in Catalonia's parliamentary elections.
- Jordan's King Abdullah II stops in Spain for two-day official visit.
- Spain may have been one of several international sites that housed U.S. nuclear weapons, according to previously classified documents.
- Banco Bilboa Vizcaya announces it will merge with Argentaria to create the nation's second largest bank. BBV is prepared to pay $11.4 billion for Argentaria.

December

- The Basque separatist guerilla group, ETA, announces it will end its 14-month ceasefire on December 3 and resume its campaign of violence in its struggle for Basque independence. Demonstrations in 15 Spanish cities urge ETA not to act on threats of violence.
- Violent winds approaching 130 miles per hour blast the region around Bilbao. Several people are killed in accidents caused by the extreme weather.

ANALYSIS OF EVENTS: 1999

BUSINESS AND THE ECONOMY

As Spain approached the millennium its economy made modest advances in a variety of areas, although the reviews were mixed in September 1999 when the International Monetary Fund praised the country for its economic development but warned that Spain's unemployment, the highest in Europe, was still excessive. In fact unemployment figures had just hit a 17-year low in August 1999, but that figure still stood at nearly 16 percent. This was so even though the participation rate of women in Spain's paid labor force continued at historically low levels. On the other hand, the migration of workers from Spain's rural areas to the cities made the unemployment crisis more obvious and politically undeniable. In addition, Spain's role as a reservoir of Europe's migratory labor force had long since been usurped by Turkey and by other underdeveloped Mediterranean countries. The high youth component of Spain's unemployed workforce (55 percent of Spain's unemployed were under twenty-five years old) contributed to rising criminality, especially in the cities.

One imminent problem was that one of the worst droughts in the last 20 years was beginning to show its effects. Rainfall between September 1998 and August 1999 was the lowest since 1980–81. Through August, the low rainfall cost Spain as much as $3.21 billion in lost production, and that figure was expected to get worse, with a large drop expected in olive oil production. The government was trying to limit inflation to less than 2 percent by the end of 1999, but prices on basic food items were expected to rise by 2 percent because of the drought. Despite the drought, Spain was one of the fastest growing members in the euro zone, with a government forecast of a 3.5 percent rise in gross domestic product for 1999.

GOVERNMENT AND POLITICS

In the geo-political realm Spain pursued an unprecedented quest to extradite and prosecute the Chilean former dictator, Augusto Pinochet for murder and torture during and after the 1973 coup against the government of Salvador Allende. In October 1998, Spanish judge Baltazar Garzon asked British authorities to arrest Pinochet while he recovered from surgery in London. Garzon instantly became both hero and pariah in his own country and in the international community. Human rights groups and leftist parties in Spain and abroad hailed his decision, while conservatives said Spanish courts had no business meddling in Chilean affairs. The case was simply motivated by politics, they said. The United States and other countries assumed a neutral position, and waited while Garzon battled Pinochet's lawyers for most of 1999.

The case brought unwanted attention to Spain, which has had its own shameful chapter of military rule. Offended Chileans, regardless of political tendency, and other critics said that Spain, which emerged from nearly 40 years of military rule in the 1970s, could not give democracy lessons to other countries. Former British Conservative leader Margaret Thatcher criticized Spain for trying to bring Pinochet to justice, and even Cuban authorities said that Pinochet should be tried in Chile and not Spain. Yet Spain was not alone as other countries sought to extradite Pinochet. French President Jacques Chirac's visit on October 3 to pay solemn tribute to Spanish democracy seemed to endorse the Spanish judge's initiative.

But relations were lukewarm at best with Latin America, where many leaders backed Chile. In Chile, where Spain is the top foreign investor, right-wing supporters of Pinochet threatened Spanish citizens, and many left the country. Chilean President Eduardo Frei's government continually pressured Spain and threatened to end diplomatic relations. For most of 1999, Spain withstood the pressure and the many attempts to derail Garzon's case against the General. This also created friction between the judiciary and the government. Spain's conservative government, led by President José Maria Aznar, twice tried to stop Garzon's case through the Spanish courts, arguing that the judge didn't have jurisdiction in the case. Spanish courts came to his rescue on both attempts, and in August, Maria Aznar said he would respect the independence of his country's judicial system. Garzon had accused the government of trying to undermine his case by meeting with Chilean officials, including military officers.

Finally on October 8, a British court ruled that Pinochet could be extradited to Spain to face torture and conspiracy charges. While his opponents danced in the streets singing "he is going to Spain," the fate of the dictator remained unclear.

His lawyers were expected to appeal amidst a new wave of pressure to let him return to Chile because of health reasons.

If anything, many Spaniards privately hope the Pinochet case will be gone before presidential elections in 2000. By some accounts, Aznar remains a favorite, but some supporters worry about his image, and the Socialist threat. The Socialists were ousted in 1996 after fourteen years in power. Aznar rose to power helped in part by corruption scandals in the Socialist government. But in 1999, the Socialists were showing signs of recovery. To blunt the opposition, Aznar seems to be moving away from the right, and claiming a more centrist position. The Economist called Aznar's political shift a Spanish version of Tony Blairs' Third Way. Aznar these days talks about family values and creating jobs. The message has been enough to keep his party in front of the Socialists at the polls.

In May the Socialists had to survive the resignation of their presidential candidate, the popular Jose Borrell. The party crisis was sparked by allegations of wrongdoing by two of Borrell's officials when he was minister of finance under the government of Felipe González. The socialists settled on their secretary-general, Joaquín Almunia to face Aznar.

Spain faced other international challenges, including its role in the bombing of Serbia, a war of words with England over Gibraltar, and frictions with Morocco over the Spanish enclaves of Ceuta and Melilla. Spain claimed Gibraltar had become a haven for money laundering and drug traffickers. It tightened immigration controls at the border, causing major headaches for people traveling in and out of the British enclave. In August, Morocco asked Spain to reconsider the status of its two North African enclaves, but Spanish officials quickly rejected any talks.

CULTURE AND SOCIETY

Spain has been relatively tolerant of foreigners, but several incidents during 1999 showed that the country is not immune to racial troubles. In one incident, two Moroccan men were stabbed by Spanish youths. A march led by skinheads wearing swastikas followed the next day. The march did damage to Moroccan shops and cars. There were other attacks on racial minorities and fear that these problems could get worse as Spain, in spite of its unemployment, begins to depend more on foreign labor. However, cosmopolitan values sometimes won out, as they did in August when Niurka Montalvo, one of the nation's newest immigrants, brought joy to the nation by winning a gold medal in the long jump at the World Track and Field Championships in Seville. A Cuban by birth, she became a Spanish citizen in 1998. Her medal was the first international victory for Spain in women's track and field.

DIRECTORY

CENTRAL GOVERNMENT

Head of State

King
Juan Carlos I de Borbon y Borbon, Monarch, Complejo de la Moncloa, 28071 Madrid, Spain
PHONE: +34 (91) 3353535
FAX: +34 (91) 3900329

Prime Minister
José Maria Aznar

Ministers

Minister of Economy and Finance
Rodrigo Rato Figaredo, Ministry of Economy and Finance

Minister of the Presidency
Francisco Alvarez-Cascos Fernandez, Ministry of the Presidency, Provinicia Numero Uno, 28071 Madrid, Spain
PHONE: +34 (91) 3353535

Minister of Agriculture, Fisheries, and Food
Jesus Maria Posada, Ministry of Agriculture, Fisheries, and Food, Paseo Infanta Isabel, Provincia Numero Uno, 28071 Madrid, Spain

Minister of Defense
Ministry of Defense, Serra Rexach, Eduardo, Baseo de Castellana, 28071 Madrid, Spain
PHONE: +34 (91) 5555000

Minister of Education and Culture
Mariano Rajoy Brey, Ministry of Education and Culture, Alcala, Provincia Numero 34, 28071 Madrid, Spain
PHONE: +34 (91) 7018000

Minister of Environment
Isabel Tocino Biscarolasaga, Ministry of Environment, Placa San Juan de la Cruz, 28071 Madrid, Spain
PHONE: +34 (91) 5977000

Minister of Foreign Affairs
Abel Juan Matutes, Ministry of Foreign Affairs, Place de la Provincia Numero Uno, 28071 Madrid, Spain
PHONE: +34 (91) 3799700

Minister of Health
Jose Manuel Romay Beccaria, Ministry of Health, Paseo del Parado, 18 y 20, 28071 Madrid, Spain
PHONE: +34 (91) 5961000

Minister of Industry
Josep Pique Camps, Ministry of Industry, Paseo de Castellana, 160, 28071 Madrid, Spain
PHONE: +34 (91) 3494000

Minister of Interior
Jaime Mayor Oreja, Ministry of Interior, Amador de los Rios, 7, 28071 Madrid, Spain
PHONE: +34 (91) 5371000

Minister of Justice
Margarita Mariscal de Gante y Miron, Ministry of Justice, San Bernando, 45, 28071 Madrid, Spain
PHONE: +34 (91) 3902000

Minister of Labor and Social Affairs
Javier Arenas Bocanegra, Ministry of Labor and Social Affairs, Agustin de Bedhencourt, 4, 28071 Madrid, Spain
PHONE: +34 (91) 5536000

Minister of Public Administration
Angel Acebes Paniagua, Ministry of Public Administration, Paseo de Castellano, 3, 28071 Madrid, Spain
PHONE: +34 (91) 5861000

Minister of Public Works
Rafael Arias-Salgado Montalvo, Ministry of Public Works, Paseo de la Castellana, 67, 28071 Madrid, Spain
PHONE: +34 (91) 5977000

POLITICAL ORGANIZATIONS

Partido Nacionalista Vasco-PNV (Basque Nationalist Party)

NAME: Xabier Arzallus Antia

Coalición Canaria-CC (Canarian Coalition)

NAME: Lorenzo Ollarte Cullen

Convergència i Unió de Catalunya-CIU (Convergence and Union)

TITLE: Secretary General
NAME: Jordi Pujol

Convergència Democrática de Catalunya-CDC (Democratic Convergence of Catalonia)

NAME: Pere Esteve

Unió Democrática de Catalunya-UDC (Democratic Union of Catalonia)

Partido Popular-PP (Popular Party)

PHONE: +34 (91) 3103265
FAX: +34 (91) 3085587
E-MAIL: oipp@pp.es
NAME: José María Aznar Lopez

Comité Federal del Partido Comunista de España-PCE (Spanish Communist Party)

C/ Toronga 27, 28043 Madrid, Spain
PHONE: +34 (91) 3004969
FAX: +34 (91) 3004744
NAME: Julio Anguita Gonzalez

Partido Socialista Obrero Español-PSOE (Spanish Socialist Workers' Party)

TITLE: Secretary General
NAME: Joaquin Almunia Amann

Izquierda Unida-IU (United Left)

Juventudes Socialistas de España-JSE (Socialist Youth of Spain)

Partido de los Socialistas de Aragón (Party of the Socialists of Aragon)

Partit Socialista de les Illes Balears-PSIB (Socialist Party of the Balearic Islands)

Partido Socialista Canario (Canarian Socialist Party)

Partit dels Socialistes de Catalunya-PSC (Party of the Socialists of Catalonia)

Partido Socialista de Euskadi-Euskadiko Ezkerra-PSE-EE (Socialist Party of the Basque Country)

Partido dos Socialistas de Galicia-PSdeG (Party of the Socialists of Galicia)

Partido Socialista de Navarra-PSN (Socialist Party of Navarre)

Partido Popular-PP (People's Party)

Izquierda Unida-IU (United Left)

Izquierda Unida Asturias (United Left of Asturias)

Izquierda Unida de Castilla y León (United Left of Castile and Leon)

Izquierda Unida de la Región de Murcia (United Left of the Region of Murcia)

Esquerra Unida del País Valencià-EUPV (United Left of the Valencian Country)

Partido Comunista de España-PCE (Communist Party of Spain)

Partido Comunista de Andalucía-PCA (Communist Party of Andalucia)

Partido Comunista de Aragón-PCA-PCE (Communist Party of Aragon)

Partido Comunista de Castilla y León-PCE-PCCL (Communist Party of Castile and Leon)

Partit dels Comunistes de Catalunya-PCC (Party of Catalonian Communists)

Partit Socialista Unificat de Catalunya-PSUC (United Socialist Party of Catalonia)

Euskadiko Partidu Komunista-EPK (Communist Party of the Basque Country)

Partido Comunista de Madrid-PCM (Communist Party of Madrid)

Partido de Acción Socialista-PASOC (Party of Socialist Action)

Izquierda Republicana-IR (Republican Left)

Euzko Alderdi Jeltzalea-EAJ (Basque Nationalist Party)

Esquerra Republicana de Catalunya-ERC (Republican Left of Catalonia)

Bloque Nacionalista Galego-BNG (Galician Nationalist Bloc)

Iniciativa per Catalunya (Initiative for Catalonia)

Unió Valenciana-UV (Valencian Union)

Bloc Nacionalista Valencià-BNV (Nationalist Valencian Bloc)

Partit Valencià Nacionalista-PVN (Valencian Nationalist Party)

Unió Mallorquina-UM (Majorcan Union)

Partido Democratico de la Nueva Izquierda-PDNI (Democratic Party of the New Left)

Los Verdes (The Greens)

Izquierda Andaluza (Andalusian Left)

Partido Humanista de España (Humanist Party of Spain)

Euskadiko Humanistak (Humanist Party of the Basque Country)

Partit per la Independència-PI (Party for the Independence)

Democracia Directa Activa-DDA (Direct Active Democracy)

Partido Revolucionario de los Trabajadores-PRT (Revolutionary Workers' Party)

DIPLOMATIC REPRESENTATION

Embassies in Spain

Argentina
Paseo de la Castellana 53, 28071 Madrid, Spain

Australia
Pza Descubridor Diego de Ordás, 3, 28003 Madrid, Spain
PHONE: +34 (91) 4419300
FAX: +34 (91) 4425362

Austria
Paseo de la Castellana, 91, 28071 Madrid, Spain
PHONE: +34 (91) 5565315

Belgium
Paseo de la Castellana, 18-6°, 28046 Madrid, Spain
PHONE: +34 (1) 5776300
FAX: +34 (1) 4318166

Bolivia
Paseo de la Castellana, 179, 28071 Madrid, Spain
PHONE: +34 (91) 5709858

Brazil
Fernando el Santo, 6, 28071 Madrid, Spain
PHONE: +34 (91) 3080459

Canada
Núñez de Balboa, 35, 28071 Madrid, Spain
PHONE: +34 (91) 4314300
FAX: +34 (91) 4312367

Chile
Lagasca, 88, 28071 Madrid, Spain

PHONE: +34 (91) 4319160

China
Arturo Soria, 113, 28071 Madrid, Spain
PHONE: +34 (91) 5194242

Colombia
Paseo Gral. Martínez Campos, 48, 28071 Madrid, Spain
PHONE: +34 (91) 3103800

Cuba
Paseo de la Habana, 194, 28071 Madrid, Spain
PHONE: +34 (91) 3592500

Czech Republic
Caídos de la División Azul, 28071 Madrid, Spain
PHONE: +34 (91) 3503607

Ecuador
Príncipe de Vergara 73, 28071 Madrid, Spain
PHONE: +34 (91) 5627216

Egypt
Velázquez, 69, 28071 Madrid, Spain
PHONE: +34 (91) 5776308

Finland
Paseo de la Castellana, 15, 28071 Madrid, Spain
PHONE: +34 (91) 3196172

France
Salustiano Olózaga, 9, 28071 Madrid, Spain
PHONE: +34 (91) 4355560

Germany
Fortuny, 8, 28010 Madrid, Spain
PHONE: +34 (91) 3199100
FAX: +34 (91) 3102104

Great Britain
Calle de Fernando el Santo, 16, 28010 Madrid, Spain
PHONE: +34 (91) 3190200
FAX: +34 (91) 3190423

Greece
Avda. Dr. Arce, 24, 28071 Madrid, Spain
PHONE: +34 (91) 5644653

Guatemala
Rafael Salgado, 3, 28071 Madrid, Spain
PHONE: +34 (91) 3440347

India
Avda. Pío XII, 30, 28071 Madrid, Spain
PHONE: +34 (91) 3450406

Indonesia
65 Calle de Agestia, 28043 Madrid, Spain
PHONE: +34 (91) 4130294

Ireland
Claudio Coello, 73, 28071 Madrid, Spain
PHONE: +34 (91) 5763500; 5763508; 5763509
FAX: +34 (91) 4351677

Israel
Calle Velasqeuz, 150-7, 28002 Madrid, Spain
PHONE: +34 (91) 4111357
FAX: +34 (91) 5645974

Italy
Calle Lagasca, 98, 28006 Madrid, Spain
PHONE: +34 (91) 5776529
FAX: +34 (91) 5757776

Japan
Joaquín Costa, 29, 28071 Madrid, Spain
PHONE: +34 (91) 5625546

South Korea
Miguel Angel, 23, 28071 Madrid, Spain
PHONE: +34 (91) 3100053

Libya
Pisuerga, 12, 28071 Madrid, Spain
PHONE: +34 (91) 5635753

Mexico
Paseo de la Castellana, 93, 28071 Madrid, Spain
PHONE: +34 (91) 3692814

Netherlands
Paseo de la Castellana, 178, 28071 Madrid, Spain
PHONE: +34 (91) 5230171
FAX: +34 (91) 5230171

New Zealand
Plaza de la Lealtad, 2, 3rd floor, 28014 Madrid, Spain
PHONE: +34 (91) 5230226; 5315182
FAX: +34 (91) 5230171

Nicaragua
Paseo de la Castellana, 127, 28071 Madrid, Spain
PHONE: +34 (91) 5555510

Nigeria
Segre, 23, 28071 Madrid, Spain
PHONE: +34 (91) 5630911

Norway
Paseo de la Castellana, 31, 28071 Madrid, Spain
PHONE: +34 (91) 3103116

Pakistan
Avda. Pío XII, 2, 28071 Madrid, Spain
PHONE: +34 (91) 3458986

Peru
Príncipe de Vergara, 36, 28071 Madrid, Spain

PHONE: +34 (91) 4314242

Poland
Guisando, 23, 28071 Madrid, Spain
PHONE: +34 (91) 3161365

Portugal
Pinar, 1, 28071 Madrid, Spain
PHONE: +34 (91) 5617800

Romania
Alfonso XIII, 157, 28071 Madrid, Spain
PHONE: +34 (91) 3504436

Russia
Velazquez, 155, E-28002 Madrid, Spain
PHONE: +34 (91) 4110807
FAX: +34 (91) 5629712

Slovakia
Pinar, 29, 28071 Madrid, Spain

South Africa
Claudio Coello, 91, 28071 Madrid, Spain
PHONE: +34 (91) 4356688

Sweden
Calle Caracas, 25, 28010 Madrid, Spain
PHONE: +34 (91) 3081535
FAX: +34 (91) 3081903

Switzerland
Núñez de Balboa, 35, 28071 Madrid, Spain
PHONE: +34 (91) 4313400

Syria
Pza. Platería Martínez 1, 28071 Madrid, Spain
PHONE: +34 (91) 4201602

Thailand
Segre, 29, 28071 Madrid, Spain
PHONE: +34 (91) 5624182

Turkey
Rafael Calvo, 18, 28071 Madrid, Spain
PHONE: +34 (91) 5597114

Venezuela
Capitán Haya, 1, 28071 Madrid, Spain
PHONE: +34 (91) 5558452

JUDICIAL SYSTEM

Supreme Court

Plaza Villa París, s/n, 28071 Madrid, Spain
PHONE: +34 (91) 3971100
FAX: +34 (91) 3194767

FURTHER READING

Articles

''Catalans versus Hollywood,'' *The Economist* (March 6, 1999): 51.

''Catalonia's Election: A New King?'' *The Economist* (October 9, 1999): 59.

''Long Shadows: Conflict Over What to do With Augusto Pinochet Continues.'' *The Economist* (August 7, 1999): 38.

''The Pinochet Case: The Law's Web.'' *The Economist* (October 2, 1999): 65.

''The Spanish Enclaves: The Mayor and the Moroccans.'' *The Economist* (September 4, 1999): 51.

''Spain and the Basques: Tense Times,'' *The Economist* (April 3, 1999): 44.

''Spain and Race: Trouble.'' *The Economist* (July 24, 1999): 47.

''Spain's Socialists: Who next?'' *The Economist* (May 22, 1999): 57.

''Spain: Slinging Mud,'' *The Economist* (May 1, 1999): 50.

SPAIN: STATISTICAL DATA

For sources and notes see "Sources of Statistics" in the front of each volume.

GEOGRAPHY

Geography (1)

Area:

Total: 504,750 sq km.

Land: 499,400 sq km.

Water: 5,350 sq km.

Note: includes Balearic Islands, Canary Islands, and five places of sovereignty (plazas de soberania) on and off the coast of Morocco—Ceuta, Mellila, Islas Chafarinas, Penon de Alhucemas, and Penon de Velez de la Gomera.

Area—comparative: slightly more than twice the size of Oregon.

Land boundaries:

Total: 1,919.1 km.

Border countries: Andorra 65 km, France 623 km, Gibraltar 1.2 km, Portugal 1,214 km, Morocco (Ceuta) 6.3 km, Morocco (Melilla) 9.6 km.

Coastline: 4,964 km.

Climate: temperate; clear, hot summers in interior, more moderate and cloudy along coast; cloudy, cold winters in interior, partly cloudy and cool along coast.

Terrain: large, flat to dissected plateau surrounded by rugged hills; Pyrenees in north.

Natural resources: coal, lignite, iron ore, uranium, mercury, pyrites, fluorspar, gypsum, zinc, lead, tungsten, copper, kaolin, potash, hydropower.

Land use:

Arable land: 30%

Permanent crops: 9%

Permanent pastures: 21%

Forests and woodland: 32%

Other: 8% (1993 est.)

HUMAN FACTORS

Demographics (2A)

	1990	1995	1998	2000	2010	2020	2030	2040	2050
Population	NA	39,072.9	39,134.0	39,208.2	39,179.1	37,849.5	35,715.8	32,988.4	29,404.8
Net migration rate (per 1,000 population)	NA	NA	NA	NA	NA	NA	NA	NA	NA
Births	NA	NA	NA	NA	NA	NA	NA	NA	NA
Deaths	NA	NA	NA	NA	NA	NA	NA	NA	NA
Life expectancy - males	NA	73.2	73.8	74.2	75.8	77.1	78.2	79.1	79.7
Life expectancy - females	NA	81.2	81.6	81.8	82.9	83.9	84.6	85.3	85.8
Birth rate (per 1,000)	NA	9.2	9.7	10.2	8.6	6.6	6.4	6.0	5.2
Death rate (per 1,000)	NA	9.4	9.6	9.8	10.7	11.6	13.1	15.5	18.6
Women of reproductive age (15-49 yrs.)	NA	10,127.5	10,187.9	10,155.3	9,441.5	8,192.6	6,616.9	5,574.8	4,853.2
of which are currently married	NA	NA	NA	NA	NA	NA	NA	NA	NA
Fertility rate	NA	1.2	1.2	1.3	1.3	1.3	1.3	1.3	1.3

Except as noted, values for vital statistics are in thousands; life expectancy is in years.

Health Personnel (3)

Total health expenditure as a percentage of GDP, 1990-1997[a]

- Public sector 5.8
- Private sector 1.6
- Total[b] 7.4

Health expenditure per capita in U.S. dollars, 1990-1997[a]

- Purchasing power parity 1,146
- Total 1,003

Availability of health care facilities per 100,000 people

- Hospital beds 1990-1997[a] 400
- Doctors 1993[c] 400
- Nurses 1993[c] .....NA

Health Indicators (4)

Life expectancy at birth

- 1980 76
- 1997 78

Daily per capita supply of calories (1996) 3,295

Total fertility rate births per woman (1997) 1.1

Maternal mortality ratio per 100,000 live births (1990-97) .....7[c]

Safe water % of population with access (1995) .....

Sanitation % of population with access (1995) 97

Consumption of iodized salt % of households (1992-98)[a] .....

Smoking prevalence

- Male % of adults (1985-95)[a] 48
- Female % of adults (1985-95)[a] 25

Tuberculosis incidence per 100,000 people (1997) 61

Adult HIV prevalence % of population ages 15-49 (1997) 0.57

Infants and Malnutrition (5)

Under-5 mortality rate (1997) 5

% of infants with low birthweight (1990-97) 4

Births attended by skilled health staff % of total[a] .....NA

% fully immunized (1995-97)

- TB .....NA
- DPT 88
- Polio 90
- Measles 90%

Prevalence of child malnutrition under age 5 (1992-97)[b] .....NA

Ethnic Division (6)

Composite of Mediterranean and Nordic types.

Religions (7)

Roman Catholic 99%

Other 1%

Languages (8)

Castilian Spanish 74%, Catalan 17%, Galician 7%, Basque 2%.

EDUCATION

Public Education Expenditures (9)

Public expenditure on education (% of GNP)

- 1980 .....2.3
- 1996 .....4.9[1]

Expenditure per student

- Primary % of GNP per capita
  - 1980 .....
  - 1996 .....15.3[1]
- Secondary % of GNP per capita
  - 1980 .....
  - 1996 .....
- Tertiary % of GNP per capita
  - 1980 .....
  - 1996 .....16.8[1]

Expenditure on teaching materials

- Primary % of total for level (1996) .....
- Secondary % of total for level (1996) .....

Primary pupil-teacher ratio per teacher (1996) .....17[1]

Duration of primary education years (1995) 6

Educational Attainment (10)

Age group (1991) .....25+

Total population 24,667,414

Highest level attained (%)

- No schooling 30.4
- First level
  - Not completed 34.9
  - Completed .....NA
- Entered second level
  - S-1 .....25.5
  - S-2 .....NA
- Postsecondary 8.4

Literacy Rates (11B)

Adult literacy rate

- 1980
  - Male 94%
  - Female 86%
- 1995
  - Male 98
  - Female 96

GOVERNMENT & LAW

Political Parties (12)

Congress of Deputies	% of seats
Popular Party (PP)	38.9
Spanish Socialist Workers Party (PSOE)	37.5
United Left (IU)	10.7
Convergence and Union (CiU)	4.6

Government Budget (13A)

Year: 1995

Total Expenditures: 25,656.7 Billions of Pesetas

Expenditures as a percentage of the total by function:

General public services and public order	5.45
Defense	3.23
Education	4.15
Health	5.54
Social Security and Welfare	39.07
Housing and community amenities	.45
Recreational, cultural, and religious affairs	.69
Fuel and energy	.30
Agriculture, forestry, fishing, and hunting	.52
Mining, manufacturing, and construction	.32
Transportation and communication	2.76
Other economic affairs and services	2.01

Military Affairs (14B)

	1990	1991	1992	1993	1994	1995
Military expenditures						
Current dollars (mil.)	8,382	8,357	7,878	8,806	8,156	8,652
1995 constant dollars (mil.)	9,633	9,234	8,474	9,232	8,361	8,652
Armed forces (000)	263	246	198	204	213	210
Gross national product (GNP)						
Current dollars (mil.)	451,400	479,500	495,000	503,000	524,700	553,800[e]
1995 constant dollars (mil.)	518,700	529,900	532,400	527,300	537,900	553,800[e]
Central government expenditures (CGE)						
1995 constant dollars (mil.)	126,400	129,000	140,200	155,100	150,800	153,700
People (mil.)	38.8	38.8	38.9	39.0	39.1	39.1
Military expenditure as % of GNP	1.9	1.7	1.6	1.8	1.6	1.6
Military expenditure as % of CGE	7.6	7.2	6.0	6.0	5.5	5.6
Military expenditure per capita (1995 $)	248	238	218	237	214	221
Armed forces per 1,000 people (soldiers)	6.8	6.3	5.1	5.2	5.5	5.4
GNP per capita (1995 $)	13,370	13,640	13,680	13,520	13,770	14,160
Arms imports[6]						
Current dollars (mil.)	550	240	250	320	550	675
1995 constant dollars (mil.)	632	265	269	335	564	675
Arms exports[6]						
Current dollars (mil.)	350	90	180	170	360	80
1995 constant dollars (mil.)	402	90	194	178	369	80
Total imports[7]						
Current dollars (mil.)	87,710	93,310	99,760	78,630	92,510	115,000
1995 constant dollars (mil.)	100,800	103,100	107,300	82,430	94,830	115,000
Total exports[7]						
Current dollars (mil.)	55,640	60,180	64,330	59,550	73,290	91,720
1995 constant dollars (mil.)	63,950	66,500	69,200	62,440	75,130	91,720
Arms as percent of total imports[8]	.6	.3	.3	.4	.6	.6
Arms as percent of total exports[8]	.6	.1	.3	.3	.5	.1

Crime (15)

Crime rate (for 1997)

Crimes reported	924,400
Total persons convicted	250,000
Crimes per 100,000 population	2,350

Persons responsible for offenses

Total number of suspects	195,700
Total number of female suspects	18,200
Total number of juvenile suspects	7,650

LABOR FORCE

Labor Force (16)

Total (million)	16.2
Services	64%
Manufacturing, mining, and construction	28%
Agriculture	8%

Data for 1997 est.

Unemployment Rate (17)

21% (1997 est.)

PRODUCTION SECTOR

Electric Energy (18)

Capacity	39.583 million kW (1995)
Production	154.144 billion kWh (1995)
Consumption per capita	4,026 kWh (1995)

Transportation (19)

Highways:

total: 344,847 km

paved: 341,399 km (including 7,747 km of expressways)

unpaved: 3,448 km (1996 est.)

Waterways: 1,045 km, but of minor economic importance

Pipelines: crude oil 265 km; petroleum products 1,794 km; natural gas 1,666 km

Merchant marine:

total: 135 ships (1,000 GRT or over) totaling 1,043,747 GRT/1,651,634 DWT ships by type: bulk 10, cargo 30, chemical tanker 7, combination ore/oil 1, container 8, liquefied gas tanker 3, oil tanker 29, passenger 2, refrigerated cargo 8, roll-on/roll-off cargo 30, short-sea passenger 6, specialized tanker 1 (1997 est.)

Airports: 98 (1997 est.)

Airports—with paved runways:

total: 64

over 3,047 m: 15

2,438 to 3,047 m: 11

1,524 to 2,437 m: 16

914 to 1,523 m: 13

under 914 m: 9 (1997 est.)

Airports—with unpaved runways:

total: 34

1,524 to 2,437 m: 1

914 to 1,523 m: 12

under 914 m: 21 (1997 est.)

Top Agricultural Products (20)

Grain, vegetables, olives, wine grapes, sugar beets, citrus; beef, pork, poultry, dairy products; fish catch of 867,000 metric tons in 1993.

MANUFACTURING SECTOR

GDP & Manufacturing Summary (21)

Detailed value added figures are listed by both International Standard Industry Code (ISIC) and product title.

	1980	1985	1990	1994
GDP ($-1990 mil.)[1]	366,141	395,083	491,957	511,046
Per capita ($-1990)[1]	9,753	10,269	12,527	12,916
Manufacturing share (%) (current prices)[1]	27.0	25.7	22.7	*18.0*
Manufacturing				
Value added ($-1990 mil.)[1]	82,605	84,529	111,315	112,604
Industrial production index	100	102	135	124
Value added ($ mil.)	51,944	33,139	87,679	*81,196*
Gross output ($ mil.)	149,786	104,594	259,945	228,118
Employment (000)	2,383	1,793	1,907	*1,758*
Profitability (% of gross output)				
Intermediate input (%)	65	68	66	*64*
Wages and salaries inc. supplements (%)	20	17	18	*19*
Gross operating surplus	14	15	16	*17*
Productivity ($)				
Gross output per worker	59,041	53,985	127,029	*121,263*

	1980	1985	1990	1994
Value added per worker	20,475	17,112	42,847	*43,162*
Average wage (inc. supplements)	12,852	9,694	24,205	*24,413*
Value added ($ mil.)				
311/2 Food products	5,665	4,193	10,773	*11,072*
313 Beverages	1,932	1,576	4,014	*3,720*
314 Tobacco products	649	471	912	*1,012*
321 Textiles	3,289	1,613	3,314	*2,650*
322 Wearing apparel	1,502	753	2,242	*2,149*
323 Leather and fur products	375	268	614	*465*
324 Footwear	810	415	781	*568*
331 Wood and wood products	1,258	707	2,164	*1,850*
332 Furniture and fixtures	1,262	617	1,534	*1,358*
341 Paper and paper products	1,278	947	2,101	*1,806*
342 Printing and publishing	1,506	1,198	4,403	*4,276*
351 Industrial chemicals	2,006	1,737	3,427	*2,658*
352 Other chemical products	2,506	1,923	5,609	*5,960*
353 Petroleum refineries	1,409	969	1,348	*1,886*
354 Miscellaneous petroleum and coal products	229	191	383	*385*
355 Rubber products	955	597	1,490	*1,429*
356 Plastic products	1,098	814	2,452	*2,562*
361 Pottery, china and earthenware	346	174	432	*312*
362 Glass and glass products	640	442	1,128	*1,029*
369 Other non-metal mineral products	2,522	1,617	4,797	*4,168*
371 Iron and steel	3,255	1,756	3,762	*2,652*
372 Non-ferrous metals	948	616	1,275	*980*
381 Metal products	3,720	2,044	5,437	*4,792*
382 Non-electrical machinery	3,595	2,226	5,745	*5,041*
383 Electrical machinery	3,669	2,064	5,978	*4,736*
384 Transport equipment	4,743	2,776	10,320	*10,545*
385 Professional and scientific equipment	205	122	375	*336*
390 Other manufacturing industries	573	316	870	*801*

FINANCE, ECONOMICS, & TRADE

Economic Indicators (22)

National product: GDP—purchasing power parity—$642.4 billion (1997 est.)

National product real growth rate: 3.3% (1997 est.)

National product per capita: $16,400 (1997 est.)

Inflation rate—consumer price index: 2.1% (1997 est.)

Balance of Payments (23)

	1992	1993	1994	1995	1996
Exports of goods (f.o.b.)	65,826	62,019	73,924	91,003	102,041
Imports of goods (f.o.b.)	−96,247	−76,965	−88,757	−108,664	−116,953
Trade balance	−30,420	−14,946	−14,833	−17,661	−14,912
Services - debits	−41,297	−35,041	−36,306	−39,697	−44,401
Services - credits	48,036	42,575	42,825	53,786	58,484
Private transfers (net)	−84	73	−229	−3,310	−4,786
Government transfers (net)	2,479	1,571	1,727	8,041	7,370
Overall balance	−21,286	−5,767	−6,816	1,159	1,756

Exchange Rates (24)

Exchange rates:	
Pesetas (Ptas) per US$1	
January 1998	153.94
1997	146.41
1996	126.66
1995	124.69
1994	133.96
1993	127.26

Top Import Origins (25)

$118.3 billion (c.i.f., 1995) Data are for 1996.

Origins	%
European Union	65.6
United States	6.6
other developed countries	11.5
Middle East	6.2

Top Export Destinations (26)

$94.5 billion (f.o.b., 1995) Data are for 1996.

Destinations	%
European Union	72.1
United States	4.2
Other developed	7.9

Economic Aid (27)

Donor: ODA, $1.213 billion (1993).

Import Export Commodities (28)

Import Commodities	Export Commodities*
Machinery	Cars and trucks
Transport equipment	Semifinished manufactured goods
Fuels	Foodstuffs
Semifinished goods	Machinery
Foodstuffs	
Consumer goods	
Chemicals	

SRI LANKA

CAPITAL: Colombo.

FLAG: The national flag contains, at the hoist, vertical stripes of green and saffron (orange-yellow) and, to the right, a maroon rectangle with yellow bo leaves in the corners and a yellow lion symbol in the center. The entire flag is bordered in yellow, and a narrow yellow vertical area separates the saffron stripe from the dark maroon rectangle.

ANTHEM: *Sri Lanka Matha (Mother Sri Lanka).*

MONETARY UNIT: The Sri Lanka rupee (R) of 100 cents is a paper currency with one official rate. There are coins of 1, 2, 5, 10, 25, and 50 cents and 1 and 2 rupees, and notes of 10, 20, 50, 100, 500, and 1,000 rupees. R1 = $0.01769 (or $1 = R56.542).

WEIGHTS AND MEASURES: The metric system is the national standard, but British weights and measures and some local units are also used.

HOLIDAYS: Independence Commemoration Day, 4 February; May Day, 1 May; National Heroes Day, 22 May; Bank Holiday, 30 June; Christmas Day, 25 December; Bank Holiday, 31 December. Movable holidays include Maha Sivarathri Day, Milad-an-Nabi, Good Friday, 'Id al-Fitr, Dewali, and 'Id al-'Adha'; in addition, the day of the rise of the full moon of every month of the Buddhist calendar, called a Poya day, is a public holiday.

TIME: 5:30 PM = noon GMT.

LOCATION AND SIZE: Sri Lanka (formerly Ceylon) is an island in the Indian Ocean situated south and slightly east of the southernmost point of India, separated from that country by the Palk Strait which is 23 kilometers (14 miles) wide.

Sri Lanka has a total area of 65,610 square kilometers (25,332 square miles) and a total coastline of 1,340 kilometers (833 miles). Sri Lanka's capital city, Colombo, is located on the Gulf of Mannar coast.

CLIMATE: Sri Lanka has neither summer nor winter. Average rainfall varies from 63 centimeters (25 inches) to 510 centimeters (200 inches), most of the rain coming during the monsoon season. Average temperature is 27°C (80°F).

Democratic Socialist Republic of Sri Lanka

Sri Lanka Prajathanthrika Samajavadi Janarajaya

INTRODUCTORY SURVEY

RECENT HISTORY

With the development of India's nationalist movement in the early twentieth century, nationalists in Ceylon also pressured for greater self-rule. Several democratic political reforms were enacted and in 1948, one year after India became independent, Ceylon became a self-governing dominion within the British Commonwealth.

The period from 1948 through 1970 saw the evolution of Ceylon's multiparty parliamentary system, in which orderly and constitutional elections and changes of government took place. Beginning in 1970, executive power began to be highly centralized under Prime Minister Sirimavo Bandaranaike, who ruled through the use of unpopular emergency powers in support of her socialist, pro-Sinhalese policies. She introduced a new constitution in 1972, changing the dominion of Ceylon to the republic of Sri Lanka.

A constitutional amendment in the fall of 1977 established a presidential form of government. Junius Richard Jayewardene of the more moderate United National Party (UNP) became Sri Lanka's first elected executive president in February 1978. Seven months later, a new, more liberal constitution came into effect, rejecting many of the authoritarian features of the 1972 constitution and introducing proportional representation.

Since 1978, rising tensions and violence between the majority (mostly Buddhist) Sinhalese and minority (mostly Hindu) Tamil communities

have dominated Sri Lankan political life. By the early 1980s, peaceful efforts by moderate Sri Lankan Tamils to make changes in the government to protect their cultural heritage failed. Violent outbreaks resulted in hundreds of deaths and by 1985 Sri Lankan Tamil leadership fell into the hands of extremists advocating violence.

In 1987 Jayewardene and Prime Minister Rajiv Gandhi of India signed an agreement by which the Sri Lankan government reluctantly agreed to give official status to the Tamil language, and to create a separate, independently governed area for the Tamils in the Northern and Eastern provinces. A 100,000-troop Indian peacekeeping force was sent to Sri Lanka to make sure the agreement was followed and to enforce a cease-fire. Yet Tamil separatists led by the extremist Liberation Tigers of Tamil Ealam (LTTE) continued their deadly attacks and by 1990 it was clear that the Indian peacekeeping force was unable to stop the Tamil rebellion. Premadasa's government continued to pursue the possibility of a negotiated settlement, but the LTTE rejected most government terms.

Tamil rebels assassinated President Premadasa in May 1993. The warfare, and the search for a solution, continued under Premadasa's successor with frequently announced cease-fires followed by new outbreaks of fighting. Chandrika Bandaranaike Kumaratunga, daughter of former prime minister Sirimavo Bandaranaike, became prime minister in August 1994. She arranged to partially lift the economic blockade of the rebel-held Jaffna peninsula and offered unconditional talks for a resolution of the dispute. By 1998, however, the death toll of 15 years of civil war was more than 51,000.

GOVERNMENT

The constitution of September 1978 established the Democratic Socialist Republic of Sri Lanka as a free, sovereign, independent state based on universal suffrage at 18 years of age. The president of the republic is directly elected for a six-year term and serves as head of state and as executive head of government, appointing and heading the cabinet of ministers. A prime minister, similarly selected, serves mainly as parliamentary leader. The normal business of legislation is in the hands of a single-chamber parliament consisting of 225 members elected for six-year terms.

Sri Lanka is divided into nine provinces containing a total of 24 districts.

Judiciary

Sri Lanka's judicial system includes district courts, magistrates' courts, courts of request (restricted to civil cases), and rural courts. In criminal cases, the Supreme Court has appeals jurisdiction. Under the 1978 constitution, the other high-level courts are the Court of Appeal, the High Court, and courts of first instance.

Political Parties

The United National Party (UNP) was the main party of the independence movement. Its widely respected leader, D. S. Senanayake, became Ceylon's first prime minister after independence. In 1951, Solomon Bandaranaike left the UNP to form the Sri Lanka Freedom Party (SLFP). Over the years, the SLFP became the island's other major political party. The UNP was friendlier to the West while the SFLP related more to the former Eastern bloc.

After President Premadasa was killed by a Tamil bomber on 1 May 1993, the Parliament unanimously elected Prime Minister Wijetunga as his successor on 7 May 1993. A ''snap'' (unscheduled) election called six months early by President Wijetunga as part of his campaign for re-election backfired on 16 August 1994, when the voters rejected the UNP by a small margin. In its place, they elected to office a seven-party, leftist coalition—now dubbed the People's Alliance—led by the SLFP's Sirimavo Bandaranaike and Chandrika Bandaranaike Kumaratunga, mother and daughter, respectively. More vigorous but less experienced, the younger Kumaratunga promptly became prime minister.

DEFENSE

The defense budget has increased dramatically since 1984 as a result of the Sri Lankan Tamil uprising in the northern half of the island. In 1995, it amounted to $624 million (7% of the gross domestic product, or GDP). Between 1983 and 1995 the army expanded from less than 15,000 to over 90,000. The navy had 10,300 sailors on active duty (with another 8,500 reserves) in 1995. The air force had 10,000, with another 8,500 reserves that year.

The armed forces, traditionally lightly armed, have been re-equipped with United States and European weaponry purchased abroad, at a cost of $160 million since 1986. Sri Lanka has a police force of 80,000, backed up by a volunteer auxiliary force of 20,000 and a home guard of 15,200.

ECONOMIC AFFAIRS

While Sri Lanka remains a primarily agricultural country, expansion of the economy since 1980 has been fueled by strong growth in industry and services.

In the latter half of the 1980s, the national economy faced several grave problems: rising defense costs as a result of the civil war, a series of droughts, and sharply lowered prices for the country's major export crops, tea and coconut-based goods. These conditions led to a rise in inflation, increasing unemployment, and stagnating economic growth.

In 1990–95, annual economic growth averaged 4.8%. In 1996, strong demand for tea and higher rubber prices benefited the economy. That year, however, inflation exceeded 15% and the domestic economy grew by 3.8% as civil war and a drought took their toll.

Public Finance

The U.S. Central Intelligence Agency estimates that, in 1995, government revenues totaled approximately $2.7 billion and expenditures $3.7 billion, including capital expenditures of $851 million. External debt totaled $8.8 billion, approximately 55% of which was financed abroad.

The government projected the 1996 budget deficit to reach 7.8% of GDP, down from 8.5% in 1995, and 10% in 1994. Proceeds from the privatization of government owned enterprises have kept the budget deficits from climbing higher. The civil war, however, continues to drain the economy. The proposed 1997 budget called for a 50% increase in defense spending. The government also proposed a program of low interest loans to help small- and medium-sized businesses escape interest rates three times as high as those in neighboring economies.

Income

In 1998, Sri Lanka's gross national product (GNP) was $15.2 billion, or about $810 per person.

Industry

Since 1977, the government's market-oriented economic policies have encouraged industrial growth, particularly in textiles, wood products, rubber and plastics, food and beverages, and other consumer goods. Manufacturing accounts for 17% of the domestic economy. Sri Lanka's industries produce cement, paper and paperboard, tea, cotton fabric, cotton yarn, plywood, rubber tires, refined petroleum products, and cigarettes.

In 1994, output of textiles, apparel, and leather was valued at $1.58 billion; food, beverages, and tobacco, $911.7 million; chemicals, petroleum, coal, rubber, and plastics, $688.3 million; non-metallic products, $295 million; and fabricated metal products and machinery, $144.1 million.

Banking and Finance

The Central Bank of Sri Lanka, established in 1949, began operations in 1950 with a capital of r 15 million contributed by the government. The sole bank of issue, it administers and regulates the country's monetary and banking systems.

Other statutory banks include the Development Finance Corp. of Ceylon, the National Savings Bank, the National Development Bank of Sri Lanka, and the State Mortgage and Investment Bank (established in 1931 to provide long-term credit primarily for agriculture).

By the end of 1995, the commercial banking system consisted of 26 commercial banks, six local and 17 foreign-owned, such as the Hongkong and Shanghai Banking Corp. According to central bank policy, foreign banks must be capitalized at r 50 million. A percentage of the capital fund is set aside to finance the country's development programs. The two largest Sri Lankan banks, the Ceylon and the People's Bank, are state-owned and account for two-thirds of commercial bank deposits. They are considered to be inefficient, primarily owing to excessive government influence in their lending operations and overstaffing. At the end of 1995, commercial banks operated 901 branches, of which 598 belonged to the two state banks. The foreign banks operated 37 branches. In 1979, the government began to promote Colombo as an international financial center for South Asia by permitting commercial banks to exercise "offshore" banking functions free of strict foreign currency controls.

The Colombo's Brokers Association operates an organized stock market, whose transactions have grown significantly since the 1984 tea export boom increased liquidity in the economy. The Colombo Stock Exchange was established by the Association of Stock Brokers in 1987. The 1990–91 boom continued into the early months of 1994, but then tipped into a bear market over political uncertainties.

Economic Development

Since independence, successive governments have attempted ambitious economic development programs with mixed results. The nationalization in 1962 of three Western oil companies and in 1975 of large rubber and tea plantations was intended to end the nation's economic dependence and neocolonialism, and to create an egalitarian socialist society. The goals of the last five-year plan for 1972–76—to achieve an economic growth rate of 6% annually, to create new jobs, and thereby to ameliorate unemployment—were not met, in part because of drought and unexpected increases in the costs of crude oil, fertilizer, and other imports.

The UNP government elected in 1977 chose as the centerpiece of its development strategy the Mahaweli hydroelectric-irrigation-resettlement program, the largest development project ever undertaken in Sri Lanka. The project involved diverting the Mahaweli Ganga in order to irrigate 364,000 hectares (900,000 acres) and generate 2,037 million kwh of hydroelectricity annually from an installed capacity of 507 Mw. Launched in 1978, construction was largely completed by 1987, at a cost of about $2 billion. Even as the UNP government launched this massive capital program, it sought to encourage private investors, limit the scope of government monopolies, and reduce subsidies on consumer products. Foreign trade, investment, and tourism were all encouraged by the government authorities. In 1986, foreign aid rose 23% in real terms over 1985, largely to finance further massive hydroelectric projects.

While government development policies resulted in moderate growth during the late 1970s and early 1980s, the outbreak of civil war in 1983 led to a rapid rise in defense spending (from 1% of GDP in 1980 to over 4% in 1996), exacerbating structural weaknesses in the Sri Lankan economy. By 1989, rapidly declining economic growth and worsening fiscal and balance of payment problems reached crisis proportions, prompting renewed stabilization and adjustment efforts. Corrective policies involved stimulating savings through new banking regulations and other monetary-tightening measures, reduction of subsidies on wheat and fertilizers, government expenditure reductions, currency devaluation, privatization of many state enterprises, and other incentives for private investment. These measures resulted in greatly improved economic performance in the early 1990s, despite unfavorable weather and lingering insurgency. In 1996, as the market showed signs of weakening, the government reaffirmed its intention to pursue free-market policies as a way to strengthen the economy.

SOCIAL WELFARE

Despite low per person income, Sri Lankans have enjoyed a relatively high standard of living because of generous social welfare programs. In 1978, the government began a vigorous population control program, combining an emphasis on family planning and voluntary sterilization with tax penalties for larger families.

The government gives monthly payments to the aged, sick, and disabled; to poor widows; and to wives of imprisoned or disabled men. To increase private efforts, the government gives grants to volunteer agencies engaged in various welfare activities, particularly orphanages, homes for the aged, and institutions for the mentally and physically handicapped. Although women have equal rights under law, their rights in family matters, including marriage, divorce, child custody, and inheritance, are often dictated by their ethnic or religious group.

Sri Lanka's 1 million ethnic Tamils, who are not entitled to either Indian or Sri Lankan citizenship, experience discrimination. Human rights abuses have been committed by both government and Tamil separatist forces.

Healthcare

The government provides medical service free or at a nominal cost to almost everyone, but its health program is hampered by an increasing shortage of trained personnel and hospital beds. Medical standards, traditionally British, are considered excellent, but in recent years many Sri Lankan physicians and surgeons have moved their practices abroad—particularly to the United States and Britain, where they can make more money.

In 1989, the Department of Health Services had 2,456 physicians, 9,632 nurses, and 5,030 midwives. For the period 1988–93, there was an average of 5,613 people per physician. In 1990, there were 2.8 hospital beds per 1,000 people. About 93% of the population has access to health care services.

Malaria, smallpox, cholera, and plague have been virtually eliminated. Malnutrition, tuberculosis (3,405 cases in 1994), and the gastrointestinal group of infectious diseases are the chief medical problems. In 1990–95, 38% of children under 5 years of age were considered malnourished. Average life expectancy was 73 years.

Housing

Rapid population increase, along with a slowdown in construction during and immediately following World War II (1939–45), led to a serious housing shortage, high rents, high building costs, and many unsanitary and unfit houses in Sri Lanka's first decades after independence. Under the United National Party government's urban development program, 184,860 public-sector housing units were built during the 1978–86 period.

The 1981 census showed a total of 2.8 million housing units, of which 2.1 million were rural, 511,810 urban, and 217,193 situated on farming estates. The average housing unit had 2.5 rooms. Although about 46% of city homes had electricity and 49% had running water, only 8.3% of rural houses were equipped with electricity and 5.1% supplied with piped-in water.

EDUCATION

In 1995, 10% of the adult population was estimated as illiterate with men estimated at 6.6% and women at 12.8%. All education from kindergarten up to and including university training is provided by the government. The estimated expense on education is 10% of the central government's budget. Education is compulsory for 10 years.

Since 1970, the public educational system has consisted of five years of elementary school, six years of secondary, and two years of higher levels. In 1994, there were 9,648 primary schools, with nearly 2 million students and 70,108 teachers. General secondary schools had 2.3 million students the same year. Since 1986, the educational system has been separated into two systems, one which teaches in Sinhala and the other in Tamil.

In 1986 there were nine universities: Colombo, Peradeniya, Moratuwa, Sri Jayawardhanapura, Kelaniya, Jaffna, Ruhuna, Open University, and Batticaloa. Included in the consolidated university system are the former Vidyalankara University, a famous seat of learning for Oriental studies and Buddhist culture; the former Vidyadaya University (established 1959); and the former University of Ceylon (founded 1942). In 1994, universities and other higher-level schools had a total of 39,607 students.

1999 KEY EVENTS TIMELINE

January

- Floods caused by heavy rains displace nearly 8,000 Sri Lankans in the northeastern Trincomalee district and in northern Vavuniya.

April

- Sri Lankan officials announce that investigators found sixteen human skeletons buried in a stadium six miles from a site where soldiers allegedly killed and buried hundreds of ethnic Tamils.

July

- A suicide bomber associated with the Liberation Tigers of Tamil Eelam (LTTE) kills moderate Tamil lawmaker Neelan Thiruchelvan.

August

- A female suicide bomber kills nine policemen and one civilian in Vavuniya.

September

- Air force jets bomb Pudikudiruppu, near the eastern Mullaithivu coast, killing twenty-two Tamil refugees.
- The Sri Lankan government and the Red Cross blame LTTE for the massacre of fifty-six Sinhala civilians in three eastern villages.
- The Tamil Tigers attack a Chinese merchant vessel off the Mullaitivu coast in northeastern Sri Lanka.

October

- Villagers in western Sri Lanka accuse the Voice of America of dumping toxic waste near their water supply.
- At least 125 people are killed in renewed heavy fighting between government troops and separatist Tamil rebels.

December

- Tamil Tiger rebels, attempting to take over a military base in the northern part of the country, are reportedly defeated in two separate incidents by government military forces. The government reports that more than 200 Tamil Tigers and 10 soldiers were killed in the fighting, but the Tamil Tigers report over 100 soldiers were killed. No independent confirmation of the death toll is available.
- President Chandrika Bandaranaike Kumaratunga is wounded and 24 people are killed when a suicide bomber attempts to assassinate her at a political rally. Another suicide bomber kills 12 people at a rally for her main opponent in the presidential elections scheduled to take place in four days.
- President Kumaratunga narrowly wins reelection while she is still in the hospital recovering from the attempt on her life.

ANALYSIS OF EVENTS: 1999

BUSINESS AND THE ECONOMY

Budget minister Lakshman Kadirgamar traveled to Belgium in January to meet with Herman Van Rompuy, Belgium's vice prime minister and minister of budget, to discuss bilateral relations. Sri Lanka's trade with Belgium has grown over the past ten years, with the balance of trade being in favor of Sri Lanka. About thirty Belgian companies have investment projects in Sri Lanka, many in the diamond cutting and polishing sector. Sri Lankan President Chandrika Kumaratunga also signed a free trade agreement with India during a visit to New Delhi.

The economy has been hurt by the ongoing strife within the nation. The billion dollars per year spent financing Sri Lanka's ongoing war has stunted the economy, which grew at a rate of only 4% in 1999. Imports rose 5%, and exports fell 3%. However, Sri Lanka is the world's largest tea exporter, and government officials are hopeful that recent United Nations (UN) reports citing the health benefits of tea will further boost the country's tea exports.

The World Bank approved $29 million in aid to help Sri Lanka finance remediation activities related to the Year 2000 (Y2K) computer problem.

GOVERNMENT AND POLITICS

Election monitors reported widespread violence and ballot box stuffing during a key election in Northwestern Province. The ruling People's Alliance won 30 of the 52 seats at stake on the provincial council.

Ethnic conflict and the war against the Liberation Tigers of Tamil Eelam (LTTE) have heavily militarized Sri Lankan society. Since 1983 over 58,000 have been killed in the fighting. Citizens live with curfews, searches, arrests, detention, and torture on a daily basis. Those who dare to speak out are often killed by the army or by militants. Thousands of Tamil people are arrested each year on suspicion of being members or sympathizers of the LTTE, and many of these detainees are tortured while in custody. According to a recent United Nations study, of all the world's nations, Sri Lanka has the second-highest number of disappeared persons; only Iraq has more.

In December Sri Lankans went to the polls to elect their president. The strongest candidates were expected to be the current president, Chandrika Kumaratunga of the SLFP, and Ranil Wickremesinghe of the UNP. On December 18, four days before the election, a suicide bomber killed 24 people at a political rally and injured Chandrika Kumaratunga in a failed assassination attempt. Another suicide bomber kills 12 people at a UNP rally shortly thereafter. The attacks are blamed on the Tamil rebels. President Kumaratunga goes on to win reelection with 52% of the vote.

CULTURE AND SOCIETY

In 1999 the United States donated $1.3 million to U.N. relief efforts aimed at helping internally displaced persons in Sri Lanka. More than 1 million of Sri Lanka's 18 million citizens have lost their homes in the nineteen years since the LTTE launched its war for independence. Most of the displaced are women and children.

Sri Lanka battles serious social problems. Sexual violence against women is not uncommon. Young unmarried women outnumber available men by as much as five to one. Children in Sri Lanka often face harsh exploitation. Over 15,000 children, some as young as eight, work in the sex trade in Sri Lanka's southern beach resorts. Sri Lanka also has the world's highest suicide rate. More Sri Lankans (7,000) kill themselves each year than are killed in the country's war.

In February 1999 a rally was held in the capital city of Colombo by the National Alliance for Peace and the Interreligious Alliance, which includes Buddhists, Hindus, and Christians.

DIRECTORY

CENTRAL GOVERNMENT

Head of State

President
Chandrika Bandaranaike Kumaratunga, Office of the President, Presidential Secretarial, Secretarial Building, Colombo 1, Sri Lanka
PHONE: +94 324801
FAX: +94 333707

Ministers

Prime Minister
Sirimavo Bandaranaike, Office of the Prime Minister, 58 Sir Ernest de Silva Mawatha, Colombo 7, Sri Lanka
PHONE: +94 575317
FAX: +94 575454

Minister of Foreign Affairs
Lakshman Kadirgamar, Ministry of Foreign Affairs, Republic Building, Colombo 1, Sri Lanka
PHONE: +94 (1) 325371
FAX: +94 (1) 436630

Minister of Defense
Chandrika Bandaranaike Kumaratunga, Ministry of Defense, 15/5 Baladaksha Mawatha, Colombo 3, Sri Lanka
PHONE: +94 (1) 433215
FAX: +94 (1) 541529

Minister of Finance and Planning
Chandrika Bandaranaike Kumaratunga, Ministry of Finance and Planning, Secretariat Building, Colombo 1, Sri Lanka
PHONE: +94 (1) 431028
FAX: +94 (1) 449496

Minister of Public Administration, Home Affairs, and Plantation Industries
Ratnasiri Wickramanayake, Ministry of Public Administration, Home Affairs, and Plantation Industries, Independent Square, Colombo 7, Sri Lanka
PHONE: +94 (1) 698428
FAX: +94 (1) 698427

Minister of Buddha Sasana
Lakshman Jayakody, Ministry of Buddha Sasana, 135 Anagarika Dharmapala Mawatha, Colombo 7, Sri Lanka
PHONE: +94 (1) 437996
FAX: +94 (1) 437997

Minister of Cultural and Religious Affairs
Lakshman Jayakody, Cultural and Religious Affairs, 8th Fl., Sethsiripaya, Battarmulla, Sri Lanka
PHONE: +94 (1) 872330
FAX: +94 (1) 872004

Minister of Science and Technology
Batty Weerakoone, Ministry of Science and Technology, 320 T B Mawatha, Colombo 10, Sri Lanka
PHONE: +94 (1) 695628
FAX: +94 (1) 682057

Minister of Agriculture and Lands
D. M. Jaratyne, Ministry of Agriculture and Lands, Sampathpaya, 82 Rajamalwatte Rd., Battaramulla, Sri Lanka
PHONE: +94 (1) 866636
FAX: +94 (1) 868915

Minister of Livestock Development and Estate Infrastructure
S. Thondaman, Ministry of Livestock Development and Estate Infrastructure, 45 S. Michaels Rd., Colombo 3, Sri Lanka
PHONE: +94 (1) 541365
FAX: +94 (1) 541364

Minister of Fisheries and Aquatic Resources Development
Mahinda Rajapske, Ministry of Fisheries and Aquatic Resources Development, Maligawatte, Colombo 10, Sri Lanka
PHONE: +94 (1) 446187
FAX: +94 (1) 446187

Minister of Education and Higher Education
Richard Pathirana, Ministry of Education and Higher Education
PHONE: +94 (1) 869326
FAX: +94 (1) 865162

Minister of Vocational Training and Rural Industries
Amarasiri Dodangoda, Ministry of Vocational Training and Rural Industries, 1 Alfred House Gardens, Colombo 3, Sri Lanka
PHONE: +94 (1) 597690

Minister of Industrial Development
C. V. Gooneratne, Ministry of Industrial Development, 73/1 Galle Rd., Colombo 3, Sri Lanka
PHONE: +94 (1) 435372
FAX: +94 (1) 421401

Minister of Health and Indigenous Medicine
Nimal Sirpala de Silva, Ministry of Health and Indigenous Medicine, Suwasiripaya, 385 Wimalawansa Mawatha, Colombo 10, Sri Lanka
PHONE: +94 (1) 694132
FAX: +94 (1) 694227

Minister of Samurdhi, Youth Affairs and Sports
D. S. B. Dissanayake, Ministry of Samurdhi, Youth Affairs and Sports, 7A Reid Ave., Colombo 7, Sri Lanka
PHONE: +94 (1) 688947
FAX: +94 (1) 689588

Minister of Internal and International Commerce Affairs and Food
Kingsley T. Wickramaratne, Ministry of Internal and International Commerce Affairs and Food, 21 Vauxhall St., Colombo 2, Sri Lanka
PHONE: +94 (1) 326539
FAX: +94 (1) 447968

Minister of Posts, Telecommunications and the Media
Mangala Samaraweera, Ministry of Posts, Telecommunications and the Media, Levels 7, Floor 17 and 18, West Tower, World Trade Centre, Echelon Square, Colombo 3, Sri Lanka
PHONE: +94 (1) 329567
FAX: +94 (1) 440488

Minister of Tourism and Civil Aviation
Dharmasiri Senanayake, Ministry of Tourism and Civil Aviation, No. 64, 68 Galle Rd., Colombo, Sri Lanka
PHONE: +94 (1) 441464
FAX: +94 (1) 441501

Minister of Shipping, Ports, and Rehabilitation and Reconstruction
M. H. M. Ashraff, Ministry of Shipping, Ports, and Rehabilitation and Reconstruction, 43-89 York St., Colombo 1, Sri Lanka
PHONE: +94 (1) 432249
FAX: +94 (1) 447656

Minister of Irrigation and Power
Anurudhdha Ratwatte, Ministry of Irrigation and Power, 500 T B Jayah Mawatha, Colombo 10, Sri Lanka
PHONE: +94 (1) 687370
FAX: +94 (1) 694968

Minister of Housing and Urban Development
Indika Gunawardena, Ministry of Housing and Urban Development
PHONE: +94 (1) 866444
FAX: +94 (1) 963522

Minister of Transport and Highway
A. H. M. Fowzie, Ministry of Transport and Highway, 1 D R Wijewardena Mawatha, Colombo 10, Sri Lanka
PHONE: +94 (1) 687311
FAX: +94 (1) 694547

Minister of Justice, Constitutional Affairs, Ethnic Affairs and National Integration
G. L. Peiris, Ministry of Justice, Constitutional Affairs, Ethnic Affairs and National Integration, Superior Courts Complex, Colombo 12, Sri Lanka
PHONE: +94 (1) 433192
FAX: +94 (1) 445446

Minister of Co-operative Development
D. P. Wickramasinghe, Ministry of Co-operative Development, 349 Galle Rd, Colombo 3, Sri Lanka
PHONE: +94 (1) 575109
FAX: +94 (1) 573849

Minister of Forestry and Environment
Nandimthra Ekanayake, Ministry of Forestry and Environment, Sixth Floor, Unity Plaza Bldg., Bambalapitiya, Colombo 4, Sri Lanka
PHONE: +94 (1) 594766
FAX: +94 (1) 502566

Minister of Social Services
Berty Premalai Dissanayake, Ministry of Social Services, Fifth Floor, Sethsiripaya Battaramulla, Sri Lanka
PHONE: +94 (1) 877121
FAX: +94 (1) 877126

Minister of Plan Implementation and Parliamentary Affairs
Jeyaraj Fernandopulle, Ministry of Plan Implementation and Parliamentary Affairs, Third Floor, Sethsiripaya Battaramulla, Sri Lanka
PHONE: +94 (1) 862362
FAX: +94 (1) 863019

Minister of Provincial Councils and Local Government
Alavi Moulana, Ministry of Provincial Councils and Local Government, 330 Union Place, Colombo 2, Sri Lanka
PHONE: +94 (1) 326732
FAX: +94 (1) 732599

Minister of Women's Affairs
Hema Ratnayake, Ministry of Women's Affairs, 64-68 Galle Rd, Colombo 3, Sri Lanka
PHONE: +94 (1) 441463
FAX: +94 (1) 441540

Minister of Labour
W. D. J. Seneviratne, Ministry of Labour, Labour Secretariat, Kirula Road, Narahenpita, Colombo 5, Sri Lanka
PHONE: +94 (1) 589267
FAX: +94 (1) 588950

POLITICAL ORGANIZATIONS

Communist Party of Sri Lanka (CPSL)

NAME: K. P. Silva

Democratic People's Liberation Front (DPLF)

Democratic United National Front

NAME: G. M. Premachandra

Ekshat Jathika Pakshaya-UNP (United National Party)

Sirikotha, 400 Kotte Road, Sri Jayawardenepura, Sri Lanka
PHONE: +94 (1) 865375
FAX: +94 (1) 865380
E-MAIL: unp@lanka.net
TITLE: Chairman
NAME: Dingri Banda Wijetunge

Eelam People's Democratic Party (EPDP)

NAME: Douglas Devananda

Lanka Sama Samaja Pakshaya (Sri Lanka Equal Society Party)

Liberal Party

NAME: Chanaka Amaratunga

Sri Lanka Mahajana Pakshaya (Sri Lanka People's Party)

NAME: Ossie Abeygunasekera

Sri Lanka Muslim Congress (SLMC)

NAME: M. H. M. Ahraff

Sri Lanka Nidahas Pakshaya (Sri Lanka Freedom Party)

NAME: Sirimavo Bandaranaike

Sri Lanka Progressive Front

Tamil Vimuktasi Peramuna (Tamil United Liberation Front)

NAME: M. Sivasithambaram

DIPLOMATIC REPRESENTATION

Embassies in Sri Lanka

Australia
3, Cambridge Place, Colombo 7, Sri Lanka
PHONE: +94 (1) 698767

Austria
424, Union Place, Colombo 2, Sri Lanka
PHONE: +94 (1) 691613

Bangladesh
286, Bauddhaloka Mawatha, Colombo 7, Sri Lanka
PHONE: +94 (1) 502198

Belgium
22, Palm Grove, Colombo 3, Sri Lanka
PHONE: +94 (1) 754166; 574453; 576403

Canada
6, Gregory's Road, Colombo 7, Sri Lanka
PHONE: +94 (1) 695841; 698298

China
381, A. Bauddhaloka Mawatha, Colombo 7, Sri Lanka
PHONE: +94 (1) 694494

Cuba
30/58, Malalasekera Mawatha, Colombo 7, Sri Lanka
PHONE: +94 (1) 589778

Cyprus
55, Layards Road, Colombo 5, Sri Lanka
PHONE: +94 (1) 588098

Denmark
264, Grandpass Road, Colombo 14, Sri Lanka
PHONE: +94 (1) 447806

Dominican Republic
108, Barnes Place, Colombo 7, Sri Lanka
PHONE: +94 (1) 697602; 547213

Egypt
39, Dickman's Road, Colombo 5, Sri Lanka
PHONE: +94 (1) 583621

France
89, Rosmead Place, Colombo 7, Sri Lanka
PHONE: +94 (1) 698815; 699750

Finland
81, Barnes Place, Colombo 7, Sri Lanka
PHONE: +94 (1) 698819; 699568

Germany
40, Alfred House Avenue, Colombo 3, Sri Lanka
PHONE: +94 (1) 588325

India
36-38, Galle Road, Colombo 3, Sri Lanka
PHONE: +94 (1) 421605; 422788

Indonesia
1, Police Park Terrace, Colombo 5, Sri Lanka
PHONE: +94 (1) 580113

Iran
17, Bullers Lane, Colombo 7, Sri Lanka
PHONE: +94 (1) 580636; 501137

Iraq
19, Barnes Place, Colombo 7, Sri Lanka
PHONE: +94 (1) 696600

Italy
55, Jawatta Road, Colombo 5, Sri Lanka
PHONE: +94 (1) 588338

Japan
20, Gregory's Road, Colombo 7, Sri Lanka
PHONE: +94 (1) 393831

South Korea
50, Horton Place, Colombo 7, Sri Lanka
PHONE: +94 (1) 699036

JUDICIAL SYSTEM

Supreme Court

FURTHER READING

Articles

Vesilind, Priit J. ''Sri Lanka: A Continuing Ethnic War Tarnishes the Pearl of the Indian Ocean.'' *National Geographic* (January 1997) 110–133.

Welsh, James. ''Sri Lanka: torture continues.'' *The Lancet* 354 (July 31, 1999): 420.

Books

De Silva, Chandra Richard. *Sri Lanka: A History*. New Delhi: Vikas Publishing House, 1997.

Ondaatje, Michael. *Running in the Family*. New York: Penguin Books, 1984.

Sri Lanka: A Country Study. Washington, D.C.: U.S. Government Printing Office, 1990.

Internet

Ceylon Sri Lanka Information. Available Online @ http://www.lankapage.com (November 9, 1999).

Spotlight on Sri Lanka. Available Online @ http://web3.is.lk/spot/ (November 9, 1999).

SRI LANKA: STATISTICAL DATA

For sources and notes see "Sources of Statistics" in the front of each volume.

GEOGRAPHY

Geography (1)

Area:

Total: 65,610 sq km.

Land: 64,740 sq km.

Water: 870 sq km.

Area—comparative: slightly larger than West Virginia.

Land boundaries: 0 km.

Coastline: 1,340 km.

Climate: tropical monsoon; northeast monsoon (December to March); southwest monsoon (June to October).

Terrain: mostly low, flat to rolling plain; mountains in south- central interior.

Natural resources: limestone, graphite, mineral sands, gems, phosphates, clay.

Land use:

Arable land: 14%

Permanent crops: 15%

Permanent pastures: 7%

Forests and woodland: 32%

Other: 32% (1993 est.).

HUMAN FACTORS

Demographics (2A)

	1990	1995	1998	2000	2010	2020	2030	2040	2050
Population	17,192.9	18,289.7	18,933.6	19,355.1	21,481.7	23,338.0	24,731.1	25,679.1	26,145.9
Net migration rate (per 1,000 population)	NA	NA	NA	NA	NA	NA	NA	NA	NA
Births	NA	NA	NA	NA	NA	NA	NA	NA	NA
Deaths	NA	NA	NA	NA	NA	NA	NA	NA	NA
Life expectancy - males	68.6	69.6	69.8	70.0	71.2	72.2	73.2	74.1	74.9
Life expectancy - females	73.5	74.9	75.4	75.8	77.7	79.4	80.8	81.9	82.9
Birth rate (per 1,000)	20.1	19.1	18.4	17.9	16.4	14.3	13.4	12.8	12.3
Death rate (per 1,000)	5.8	5.8	6.0	6.1	6.6	7.4	8.6	10.1	11.2
Women of reproductive age (15-49 yrs.)	4,644.4	5,052.2	5,311.9	5,455.1	5,780.8	5,883.8	5,871.0	5,718.9	5,673.3
of which are currently married	NA	NA	NA	NA	NA	NA	NA	NA	NA
Fertility rate	2.3	2.2	2.1	2.1	2.0	2.0	2.0	2.0	2.0

Except as noted, values for vital statistics are in thousands; life expectancy is in years.

Health Personnel (3)

Total health expenditure as a percentage of GDP, 1990-1997[a]

Public sector	1.4
Private sector	0.4
Total[b]	1.9

Health expenditure per capita in U.S. dollars, 1990-1997[a]

Purchasing power parity	38
Total	11

Availability of health care facilities per 100,000 people

Hospital beds 1990-1997[a]	270
Doctors 1993[c]	23
Nurses 1993[c]	112

Health Indicators (4)

Life expectancy at birth

1980	68
1997	73
Daily per capita supply of calories (1996)	2,263
Total fertility rate births per woman (1997)	2.2
Maternal mortality ratio per 100,000 live births (1990-97)	30[b]
Safe water % of population with access (1995)	70
Sanitation % of population with access (1995)	75
Consumption of iodized salt % of households (1992-98)[a]	47

Smoking prevalence

Male % of adults (1985-95)[a]	55
Female % of adults (1985-95)[a]	1
Tuberculosis incidence per 100,000 people (1997)	48
Adult HIV prevalence % of population ages 15-49 (1997)	0.07

Infants and Malnutrition (5)

Under-5 mortality rate (1997)	19
% of infants with low birthweight (1990-97)	25
Births attended by skilled health staff % of total[a]	94

% fully immunized (1995-97)

TB	96
DPT	97
Polio	98
Measles	94
Prevalence of child malnutrition under age 5 (1992-97)[b]	38

Ethnic Division (6)

Sinhalese	74%
Tamil	18%
Moor	7%
Burgher, Malay, and Vedda	1%

Religions (7)

Buddhist	69%
Hindu	15%
Christian	8%
Muslim	8%

Languages (8)

Sinhala (official and national language) 74%, Tamil (national language) 18%. English is commonly used in government and is spoken by about 10% of the population.

EDUCATION

Public Education Expenditures (9)

Public expenditure on education (% of GNP)

1980	2.7
1996	3.4

Expenditure per student

Primary % of GNP per capita

1980	....
1996	....

Secondary % of GNP per capita

1980	....
1996	14.9[1]

Tertiary % of GNP per capita

1980	65.6[1]
1996	84.8[1]

Expenditure on teaching materials

Primary % of total for level (1996)	....
Secondary % of total for level (1996)	....
Primary pupil-teacher ratio per teacher (1996)	28
Duration of primary education years (1995)	5

Educational Attainment (10)

Age group (1981)	25+
Total population	6,490,502

Highest level attained (%)

No schooling	15.9

First level

Not completed	48.9
Completed	NA

Entered second level

S-1 ....34.1

S-2 ....NA

Postsecondary 1.1

Literacy Rates (11A)

In thousands and percent[1]	1990	1995	2000	2010
Illiterate population (15+ yrs.)	1,307	1,241	1,180	1,005
Literacy rate - total adult pop. (%)	88.7	90.2	91.6	93.9
Literacy rate - males (%)	92.6	93.4	94.1	95.3
Literacy rate - females (%)	84.8	87.2	89.2	92.6

GOVERNMENT & LAW

Political Parties (12)

Parliament	No. of seats
People's Alliance (PA)	105
United National Party (UNP)	94
Eelam People's Democratic Party (EPDP)	9
Sri Lanka Muslim Congress (SLMC)	7
Tamil United Liberation Front (TULF)	5
People's Liberation Organization of Tamil Eelam (PLOTE)	3
Sri Lanka Progressive Front (SLPF)	1
Upcountry People's Front (UPF)	1

Military Affairs (14B)

	1990	1991	1992	1993	1994	1995
Military expenditures						
Current dollars (mil.)	411[e]	451[e]	375	536[e]	541[e]	585[e]
1995 constant dollars (mil.)	472[e]	499[e]	404	562[e]	554[e]	585[e]
Armed forces (000)	22	110	110	110	110	110
Gross national product (GNP)						
Current dollars (mil.)	8,572	9,324	9,983	11,030	11,900	12,820
1995 constant dollars (mil.)	9,852	10,300	10,740	11,560	12,200	12,820
Central government expenditures (CGE)						
1995 constant dollars (mil.)	3,078	3,335	2,960	3,244	3,585	3,738
People (mil.)	17.2	17.5	17.7	17.9	18.1	18.3
Military expenditure as % of GNP	4.8	4.8	3.8	4.9	4.5	4.6
Military expenditure as % of CGE	15.3	15.0	13.6	17.3	15.5	15.7
Military expenditure per capita (1995 $)	27	29	23	31	31	32
Armed forces per 1,000 people (soldiers)	1.3	6.3	6.2	6.1	6.1	6.0
GNP per capita (1995 $)	572	590	607	645	673	699
Arms imports[6]						
Current dollars (mil.)	10	60	5	20	100	160
1995 constant dollars (mil.)	11	66	5	21	103	160
Arms exports[6]						
Current dollars (mil.)	0	0	0	0	0	0
1995 constant dollars (mil.)	0	0	0	0	0	0
Total imports[7]						
Current dollars (mil.)	2,685	3,054	3,445	3,991	4,776	5,185
1995 constant dollars (mil.)	3,086	3,375	3,706	4,184	4,896	5,185
Total exports[7]						
Current dollars (mil.)	1,983	2,039	2,455	2,859	3,208	3,798
1995 constant dollars (mil.)	2,279	2,253	2,641	2,997	3,288	3,798
Arms as percent of total imports[8]	.4	2.0	.1	.5	2.1	3.1
Arms as percent of total exports[8]	0	0	0	0	0	0

Government Budget (13A)

Year: 1997

Total Expenditures: 228,914 Millions of Rupees

Expenditures as a percentage of the total by function:

General public services and public order	11.14[P]
Defense	16.19[P]
Education	9.76[P]
Health	5.30[P]
Social Security and Welfare	13.98[P]
Housing and community amenities	2.10[P]
Recreational, cultural, and religious affairs	-[P]
Fuel and energy	.53[P]
Agriculture, forestry, fishing, and hunting	3.23[P]
Mining, manufacturing, and construction	.89[P]
Transportation and communication	5.36[P]
Other economic affairs and services	4.24[P]

Crime (15)

Crime rate (for 1997)

Crimes reported	71,000
Total persons convicted	NA
Crimes per 100,000 population	400

Persons responsible for offenses

Total number of suspects	NA
Total number of female suspects	NA
Total number of juvenile suspects	NA

LABOR FORCE

Labor Force (16)

Total (million)	6.2
Services	46%
Agriculture	37%
Industry	17%

Data for 1997. Percent distribution for 1997 est.

Unemployment Rate (17)

11% (1997 est.)

PRODUCTION SECTOR

Electric Energy (18)

Capacity	1.557 million kW (1997 est.)
Production	4.86 billion kWh (1997 est.)
Consumption per capita	220 kWh (1997 est.)

Transportation (19)

Highways:

total: 99,200 km

paved: 39,680 km

unpaved: 59,520 km (1996 est.)

Waterways: 430 km; navigable by shallow-draft craft

Pipelines: crude oil and petroleum products 62 km (1987)

Merchant marine:

total: 24 ships (1,000 GRT or over) totaling 204,542 GRT/317,253 DWT ships by type: bulk 2, cargo 13, container 1, oil tanker 2, refrigerated cargo 6 (1997 est.)

Airports: 13 (1997 est.)

Airports—with paved runways:

total: 12

over 3,047 m: 1

1,524 to 2,437 m: 5

914 to 1,523 m: 6 (1997 est.)

Airports—with unpaved runways:

total: 1

1,524 to 2,437 m: 1 (1997 est.)

Top Agricultural Products (20)

Rice, sugarcane, grains, pulses, oilseed, roots, spices, tea, rubber, coconuts; milk, eggs, hides, meat.

MANUFACTURING SECTOR

GDP & Manufacturing Summary (21)

Detailed value added figures are listed by both International Standard Industry Code (ISIC) and product title.

	1980	1985	1990	1994
GDP ($-1990 mil.)[1]	5,242	6,725	7,935	9,829
Per capita ($-1990)[1]	354	417	461	542
Manufacturing share (%) (current prices)[1]	19.0	17.5	18.3	*14.8*
Manufacturing				
Value added ($-1990 mil.)[1]	868	1,081	1,371	1,878
Industrial production index	100	*118*	253	693
Value added ($ mil.)	*376*	*635*	1,112	*1,563*
Gross output ($ mil.)	1,279	1,815	2,519	*2,783*
Employment (000)	*195*	211	283	*266*
Profitability (% of gross output)				
Intermediate input (%)	*71*	*65*	56	*44*

	1980	1985	1990	1994
Wages and salaries inc. supplements (%)	6	6	7	*8*
Gross operating surplus	*23*	*29*	37	*48*
Productivity ($)				
Gross output per worker	*6,572*	8,599	8,910	*10,443*
Value added per worker	*1,931*	*3,001*	3,934	*5,928*
Average wage (inc. supplements)	*407*	529	606	*837*
Value added ($ mil.)				
311/2 Food products	*90*	*181*	240	*282*
313 Beverages	8	*34*	118	*102*
314 Tobacco products	63	*151*	156	*347*
321 Textiles	27	*49*	82	*138*
322 Wearing apparel	12	*39*	142	*251*
323 Leather and fur products	1	*2*	3	*6*
324 Footwear	2	*4*	20	*6*
331 Wood and wood products	5	*8*	9	*12*
332 Furniture and fixtures	2	*2*	1	—
341 Paper and paper products	8	*10*	19	*18*
342 Printing and publishing	*8*	*8*	15	*49*
351 Industrial chemicals	6	*4*	10	*8*
352 Other chemical products	12	*13*	33	*107*
353 Petroleum refineries	55	*30*	*100*	2
354 Miscellaneous petroleum and coal products	*1*	*1*	*6*	—
355 Rubber products	14	*30*	35	*60*
356 Plastic products	4	*3*	9	*11*
361 Pottery, china and earthenware	4	*6*	17	*22*
362 Glass and glass products	2	*2*	4	*7*
369 Other non-metal mineral products	21	*28*	23	*54*
371 Iron and steel	3	*2*	8	*9*
372 Non-ferrous metals	2	*1*	3	*3*
381 Metal products	7	*9*	10	*6*
382 Non-electrical machinery	4	*5*	9	*1*
383 Electrical machinery	10	*4*	7	*7*
384 Transport equipment	4	*2*	25	*18*
385 Professional and scientific equipment	1	—	—	—
390 Other manufacturing industries	1	*5*	10	*34*

FINANCE, ECONOMICS, & TRADE

Economic Indicators (22)

National product: GDP—purchasing power parity—$72.1 billion (1997 est.)

National product real growth rate: 6% (1997 est.)

National product per capita: $3,800 (1997 est.)

Inflation rate—consumer price index: 9.6% (1997)

Balance of Payments (23)

	1992	1993	1994	1995	1996
Exports of goods (f.o.b.)	2,301	2,786	3,208	3,798	4,095
Imports of goods (f.o.b.)	−3,016	−3,528	−4,293	−4,783	−4,872
Trade balance	−715	−742	−1,085	−985	−777
Services - debits	−1,069	−1,109	−1,364	−1,560	−1,580
Services - credits	689	746	898	1,042	941
Private transfers (net)	183	163	167	57	54
Government transfers (net)	462	560	627	675	710
Overall balance	−451	−382	−757	−770	−653

Exchange Rates (24)

Exchange rates:

Sri Lankan rupees (SLRes) per US$1

January 1998	61.479
1997	58.995
1996	55.271
1995	51.252
1994	49.415
1993	48.322

Top Import Origins (25)

$5.4 billion (c.i.f., 1996) Data are for 1996.

Origins	%
India	10.4
Japan	9.1
South Korea	6.5
Hong Kong	6.5
Taiwan	5.3

Top Export Destinations (26)

$4.1 billion (f.o.b., 1996) Data are for 1996.

Destinations	%
United States	34
United Kingdom	9.5
Japan	6.2
Germany	5.8
Belgium-Luxembourg	5.3

Economic Aid (27)

Recipient: ODA, $620 million (1996 est.)

Import Export Commodities (28)

Import Commodities	Export Commodities
Machinery and equipment	Textiles and apparel
Textiles	Tea
Transport equipment	Diamonds and other gems
Petroleum	Rubber products
Building materials	Petroleum products
Sugar	
Wheat	

SUDAN

Republic of the Sudan
Jumhuriyat as-Sudan

CAPITAL: Khartoum.

FLAG: The national flag consists of a tricolor of red, white, and black horizontal stripes, with a green triangle at the hoist.

ANTHEM: *Jundi al-Allah (Soldiers of God).*

MONETARY UNIT: The Sudanese pound (£s) is a paper currency of 100 piasters (qurush) or 1,000 milliemes. There are coins of 1, 2, 5, and 10 milliemes and 2, 5, 10, and 50 piasters, and notes of 25 and 50 piasters and 1, 5, 10, 20, and 50 Sudanese pounds. £s1 = $0.0025 (or $1 = £s400.0).

WEIGHTS AND MEASURES: The metric system is the legal standard, but a highly diverse system based on Egyptian and British standards is in local use.

HOLIDAYS: Independence Day, 1 January; Unity Day, 3 March; Uprising Day, 6 April; Decentralization Day, 1 July; Christmas, 25 December. Movable Muslim religious holidays include the 1st of Muharram (Muslim New Year), 'Id al-Fitr, 'Id al-'Adha', and Milad an-Nabi.

TIME: 2 PM = noon GMT.

LOCATION AND SIZE: Situated in northeast Africa, Sudan is the largest country on the continent. It covers an area of over 2.5 million square kilometers (967,490 square miles), slightly more than one-quarter the size of the United States. Sudan's capital city, Khartoum, is located in the northeastern part of the country.

CLIMATE: Maximum temperatures range from 32°C (90°F) in winter to 42°C (108°F) in summer. Average annual rainfall varies from 120 centimeters (47 inches) in the south to less than 10 centimeters (4 inches) in the north. The most temperate climate occurs in the Red Sea Hills.

INTRODUCTORY SURVEY

RECENT HISTORY

The new Republic of the Sudan was proclaimed on 1 January 1956 under a parliamentary government. In November 1958 a military dictatorship seized power, but the regime was overthrown in October 1964, and civilian politicians ruled for the next five years. A revolutionary council led by Colonel Gaafar Mohammed Nimeiri (Ja'far Muhammad Numayri) overthrew the government in a bloodless coup in May 1969 and established the Democratic Republic of the Sudan. Two years later Nimeiri proclaimed that Sudan would become a one-party state, with the Sudanese Socialist Union the sole political organization. As elected president, Nimeiri brought end to the civil war that had plagued Sudan since independence. In February 1972 the Sudanese government and the South Sudan Liberation Front (the Anyanya rebels) agreed on a cease-fire and on self-rule for the southern provinces.

In 1984 protests and riots over rising prices and government policies broke out. Nimeiri was replaced by a military council headed by General Abdel-Rahman Swar ad-Dhahab in 1985. The country was renamed the Republic of Sudan, the ruling Sudanese Socialist Union was abolished, political and press freedom was restored, and food prices were lowered. Sudan returned to a foreign policy of nonalignment, backing away from its close ties with Egypt and the United States.

General elections held in 1986 resulted in a moderate civilian coalition government, which be-

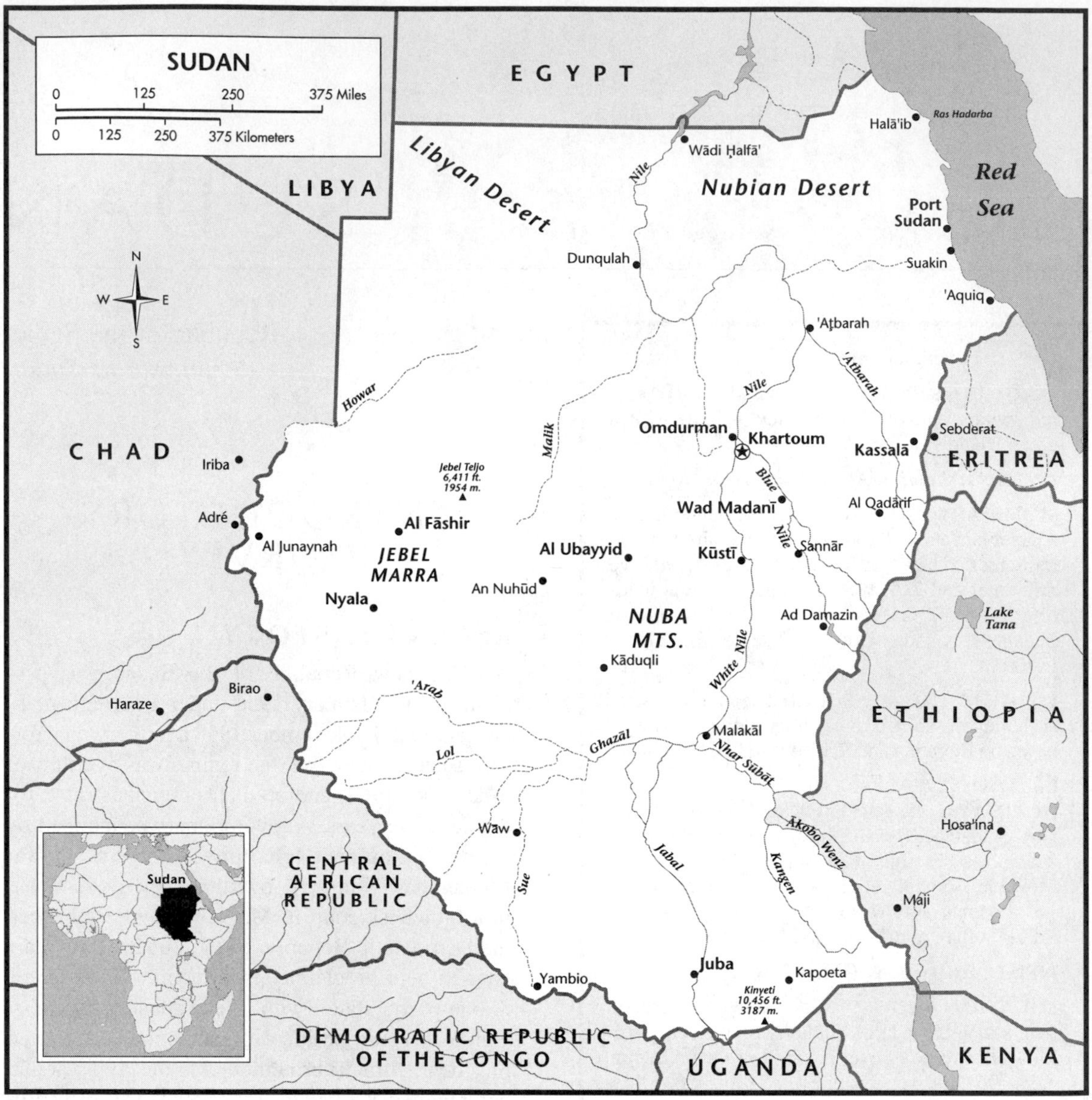

gan searching for a way to unite the country with the Sudanese People's Liberation Army (SPLA). The SPLA controlled much of the south, blocking air traffic (including food relief) to the south and opposing major projects vital to the economy.

In June 1989, a group of army officers led by Brigadier Omar Hassam el-Bashir overthrew the civilian government. The coup makers created a National Salvation Revolutionary Command Council (RCC), a junta (a group of people who run a government, particularly after a coup) composed of 15 military officers assisted by a civilian cabinet. They suspended the 1985 transitional constitution, took away press freedoms, and dissolved all parties and trade unions.

The United Nations estimated that in 1998 some 1.2 million southern Sudanese faced a severe food shortage. Civil war and famine led to the deaths of more than 1.5 million Sudanese since 1983. The United Nations General Assembly condemned Sudan's human rights violations in March 1993.

GOVERNMENT

After the 1989 military coup, the 1985 transitional constitution was set aside. In January 1991, Islamic law was imposed in the six northern provinces. Executive and legislative authority was vested in a 15-member Revolutionary Command Council (RCC). Its chairman, acting as prime minister, appointed a 300-member Transitional National Assembly (TNA). In mid-October 1993, Bashir dissolved the RCC and officially declared himself president of a new civilian government. In 1995, Bashir realigned his cabinet to help the National Islamic Front maintain control over the Ministry of Foreign Affairs. The TNA was replaced by an elected National Assembly in March 1996.

Judiciary

For the Muslim population, justice in personal matters is administered by Muslim law courts, which form the Shari'ah Division of the Sudan judiciary. Civil justice is administered by the Supreme Court, courts of appeal, and lower courts. Criminal justice is administered by major courts, magistrates' courts, and local people's courts, which try civil cases as well.

Political Parties

The Revolutionary Command Council (RCC), which formed the latest military government, dissolved all parties in 1989. The fundamentalist National Islamic Front (NIF), however, continues to function openly and is the strength behind the government. NIF members and supporters hold most key positions and when Bashir dissolved the RCC in October 1993, the NIF further tightened its grip on the state.

The main opposition to the central government is the Sudan's People's Liberation Army (SPLA). In 1997, the SPLA joined forces with the Beja Congress Armed Forces, part of a new alliance of northern rebels known as the National Democratic Alliance. This opposition has been sponsored by Ethiopia and Eritrea, and indirectly by the United States, which holds Sudan responsible for sponsoring terrorism.

DEFENSE

The army has an estimated strength of 85,000, organized into 9 divisions and 29 brigades. The navy, established in 1962, has 1,000 personnel and 7 patrol craft; the air force has 3,000 personnel and 60 combat aircraft, plus air defense missile units. Estimated defense expenditures in 1995 were $389 million. The Sudanese armed forces face an estimated 50,000 rebels of the Sudanese People's Liberation Army.

ECONOMIC AFFAIRS

Sudan has an agricultural economy with great potential for production with the help of irrigation. The livestock sector is sizable as well. However, droughts have led to recent famines, and civil war has led to the virtual collapse of the economy. In 1993, Sudan's failure to pay its international debt, together with its poor human rights record, led to the World Bank's suspension of financing for 15 development projects, and to the International Monetary Fund's (IMF) suspension of Sudan's voting rights in the organization. Sudan is the world's largest debtor to the IMF, owing more than $1.3 billion. The civil war and growing international isolation restrict economic growth.

Public Finance

Sudan's budgets were in deficit throughout the 1960s, 1970s, and 1980s. The budget deficit soared to 22% of GDP in 1991/92, which aggravated inflation. Neither the budget deficit nor inflation shows signs of shrinking as civil war continues in the south, consuming precious budgetary resources. The U.S. Central Intelligence Agency estimates that, in 1995, government revenues totaled approximately $382 million and expenditures $1.06 billion, including capital expenditures of $91 million. External debt totaled $18 billion.

Income

In 1998, Sudan's gross national product (GNP) was $8.2 billion, or about $290 per person.

Industry

Sudan's industrial output has decreased since the early 1980s due to problems with trade and production. Before this difficult period, Sudan's industries supplied many items that had formerly been imported, such as cotton textiles, sugar, household appliances, cement, and tires. Textiles are the largest industry. Factories process cotton seed and groundnuts into oil and cake. The Kenana sugar complex is one of the largest sugar plantations and refining installations in the world.

Banking and Finance

The traditional banking system was inherited from the Anglo-Egyptian condominium (1899–1955). When the National Bank of Egypt opened in Khartoum in 1901, it obtained a privileged position as banker to and for the government, a ''semi-official'' central bank. Other banks followed, but the National Bank of Egypt and Barclays Bank dominated and stabilized banking in Sudan until after World War II. Post-World War II prosperity created a demand for an increasing number of commercial banks.

Before Sudanese independence, there had been no restrictions on the movement of funds between Egypt and Sudan, and the value of the currency used in Sudan was tied to that of Egypt. This situation was unsatisfactory to an independent Sudan, which established the Sudan Currency Board to replace Egyptian and British money. It was not a central bank because it did not accept deposits, lend money, or provide commercial banks with cash and liquidity. In 1959, the Bank of Sudan was established to succeed the Sudan Currency Board and to take over the Sudanese assets of the National Bank of Egypt. In February 1960, the Bank of Sudan began acting as the central bank of Sudan, issuing currency, assisting the development of banks, providing loans, maintaining financial equilibrium, and advising the government.

As of 1994, the money supply, as measured by M2, amounted to £S400.4 million.

In 1996, there were 27 banks in Sudan, of which one, El Nilein Industrial Development Bank, was state-owned. The Bank of Khartoum was privatized at the end of 1995. At the end of 1994, the commercial banks had assets of £S226.3 billion. Banks were nationalized in 1970 but in 1974, foreign banks were allowed to open branches in Sudan.

In December 1990 the government decided to adopt Islamic banking principles. Seven banks in Sudan are based on the principles of Islamic banking that were introduced in September 1984, namely Faisal Islamic Bank of Sudan (FIBS), Islamic Cooperative Development Bank, Tadamun Islamic Bank of Sudan, Sudanese Islamic Bank, Al-Baraka Bank, Islamic Bank of Western Sudan, and Bank of Northern Sudan.

Banks are required to maintain 20% of total deposits as a statutory reserve with the central bank. They must also direct to the agricultural sector 40% of the funds that they have for lending under the new credit ceilings.

Total foreign reserves, excluding gold, amounted to $78.2 million at the end of 1994.

No stock exchange or organized over-the-counter market exists in the Sudan.

Economic Development

The suspension of foreign aid and balance-of-payment support by a growing list of countries has all but stopped economic development. In spite of this, Sudan's government retains food self-sufficiency as a priority goal and seeks to reallocate investment toward agriculture and other productive sectors. Private investment is welcome as the parastatal sector is privatized.

SOCIAL WELFARE

Organized social welfare is administered by the central and local governments, labor unions, and fraternal organizations. Social legislation requires business firms to provide benefits for their employees.

Since the military coup in 1989, the fundamentalist Islamic government has removed many of the basic rights and freedoms of women. Women have been removed from government jobs, have limited educational opportunities, and are arrested for wearing slacks or having an uncovered head. In 1996, the city of Khartoum ordered that women and men must be separated and unable to see each other at public and social events.

Because of the civil war, starvation and malnutrition are widespread. Government and SPLA forces regularly commit human rights abuses such as massacres, kidnapping, enslavement, forced military service, and rape. Sudanese soldiers and Muslim militias armed by the government transport captured southern blacks to the north, where they are used as household slaves. Freedom of speech, press, assembly, association, political choice, and religion are repressed.

Healthcare

In spite of an improvement in medical services and supervision, such diseases as malaria, schistosomiasis, sleeping sickness (trypanosomiasis), tuberculosis (about 211 cases per 100,000 people in 1990), and various forms of dysentery persist. There were 64,608 cases of Guinea worm disease in 1995. Some 34% of children under 5 years old were considered malnourished during 1990–95.

For the period 1990–95, average life expectancy was estimated at about 54 years. Hospital facilities and medical and public health services are free, but only 51% of the population had access to them in 1992. In 1992, there were 9 physicians per 100,000 people. During 1984–92, there were 506,000 civil war-related deaths.

Housing

As of 1983, 60% of housing units were *gottias*, single rooms with round mud walls and a conical straw roof; 36% were *menzils*, multi-room houses with toilet facilities.

A national housing authority provides low-cost housing to government employees, rural schoolteachers, and persons in low-income groups. Khartoum has a number of modern apartment buildings.

EDUCATION

In 1993, 35.6% of the adult male and 62.5% of the adult female population were illiterate.

In 1993, 9,681 primary schools had 2.4 million students and 66,268 teachers. Secondary schools had 718,298 students and 30,642 teachers in 1991.

In 1986, the University of Khartoum's 10 faculties had about 14,000 students. A branch of Cairo University was opened at Khartoum in 1955; by 1986, it had about 20,000 students. Other institutions include the Islamic University of Omdurman and the universities of El-Gezirah (at Wad Madani) and Juba. In 1990, higher level institutions had a total of 59,824 students.

1999 KEY EVENTS TIMELINE

January

- The government announces a six month ceasefire with the Sudanese People's Liberation Army (SPLA) rebels while they negotiate.
- The government is accused of bombing southern hospitals.
- Swiss-based Christian Solidarity International says it has freed 1,050 slaves.

February

- A United States envoy visits Sudan to check on human rights violations.
- Some political prisoners are released.
- *The New York Times* reports there are no traces of chemicals found at a plant bombed by the United States.

March

- The southern rebels attack oil pipelines.
- The Dinka and Nuer tribes of the south form a political alliance.
- Osama bin Laden, accused of the New York World Trade Center bombing, is also accused of buying slaves to work in marijuana fields in Sudan.

April

- The SPLA attacks areas in the Blue Nile and Upper Nile regions.
- Sudan appoints its first woman ambassador, Zeinab Muhammad Mahmoud Abd al-Karim.

May

- The United States will unfreeze the assets of Salah Idris, owner of the bombed pharmaceutical plant in Sudan. The U.S. admits it may have been wrong to assume the plant was making chemical weapons.

June

- As many as seven political parties have registered to present candidates.

August

- Sudan successfully exports its first barrels of oil.
- The United States announces it will ease sanctions against Sudan so that they can receive medical aid and food supplies.

September

- Harry Johnston from the United States is the new presidential envoy appointed to monitor human rights in the Sudan.
- Al-Bashir threatens to close down opposition newspapers.
- Government supporters urge el-Bashir not to allow envoy Harry Johnston in the country.
- The NDA (National Democratic Alliance) blows up a section of pipeline 50 kilometers east of Atbara in northern Sudan.
- The UMMA (opposition party) claims responsibility for the pipeline bombings.

October

- Egypt deports two Sudanese rebels suspected of bombing the pipelines.
- Sudan's ruling party gives speaker of parliament, Hassan Turabi, sweeping powers over el-Bashir.
- Sudan's government accepts a proposal from Kenya for a round of peace talks with the SPLA.
- Al-Bashir declares himself a candidate for elections.
- The ruling National Congress elects el-Bashir as its president.

December

- Sudan and neighboring Uganda sign an agreement not to support rebel factions working to overthrow each other's governments.
- President el-Bashir dissolves the legislature and declares a three-month state of emergency. The speaker of the national assembly, Hassan al-Turabi, accuses el-Bashir of essentially staging a coup. On December 31, el-Bashir announces that the cabinet has resigned.
- Egypt and Sudan announce jointly that they have restored diplomatic relations.

ANALYSIS OF EVENTS: 1999

BUSINESS AND THE ECONOMY

The Sudan, like several other underdeveloped countries that incubated as nations under nineteenth century European imperialism, suffered the results of coming of age. The seeds of civil war and economic stagnation were planted early, in part because of the careless way that the European powers drew the boundaries between their colonies. Culturally, northern and southern Sudan had little in common. The south was Christian, the north was Moslem. The southern black Sudanese had nothing in common with their Arab conquerors. They shared neither race, religion, nor lifestyle.

The northern Arab leadership took it upon itself to determine the identity and direction that the Sudan would take. The el-Bashir government instituted laws forcing the south to convert to Islam. The black Christians living in the south considered the Islamic legal code of *sharia* harsh and unfair. They were forced to wear certain articles of clothing because the Arabs considered it unseemly to be nude or even partially nude. In addition, some enslavement of the Christian Sudanese occurred. This campaign of forced assimilation disrupted civic life and impeded economic development.

The north did not consider the resistance that this policy would evoke in the south. The southern Sudanese fielded an army, the Sudanese People's Liberation Army (SPLA), which waged a ten-year guerrilla and terrorist war against the Islamic government. The northern citizens clearly do not want the war to continue any longer. Many young people have been drafted to fight and the fatalities are high. A northern coalition called the National Democratic Alliance formed to repeal the dictatorship of el-Bashir and to call for a constitutional convention. This was not a cohesive alliance and did not take into account the south's desire for self-government. Many northerners also seemed tired of the strict Islamic government. As of January 1999 opposition political parties could register with the government to compete in elections. This, however, was regarded as sham democracy by the opposition.

The new twist in the war saga was the discovery of oil in the south. This gave life to northern Sudan's war-bankrupted economy. The oil production gave the Sudan almost two million dollars a day. If the country were to succeed economically, it would become clear that the civil war was disadvantageous to all—north and south. But the war continued to obstruct the development of the economy. After ten years of living with continuous civil war, President el-Bashir announced in January 1999 that perhaps the south should be allowed to secede. However, the south is at odds with that solution also. The leader of the SPLA, John Garang, would like to see a united, secular Sudan. Thus, although the possibility exists for a separation of the two parts of the Sudan, as oil production increased it remained to be seen if the north would continue to consider secession a viable option.

GOVERNMENT AND POLITICS

The future looked bleak for Sudan because of a huge incompatibility of two peoples and because the civil war hobbled economic development. Under these conditions it was difficult for the Sudan to move forward with the project of economic development and nation-building when, in addition to the religious and cultural conflict, the rural south was home to tribes that lived in isolation even from one another. The southerners were novices at solidarity with little apprehension of the concept of what a nation was. They had been terrorized and enslaved and wanted to return to a lifestyle in which differences are resolved by elaborate ritual wrestling matches.

Other nearby African states are fighting their own religious and territorial based wars and offer no role models of constructive national development. If President el-Bashir were to allow the southern Sudan to secede, the south would probably soon be swallowed up by a neighboring power with little or no effective resistance from its African tribal residents.

CULTURE AND SOCIETY

The major problems stemming from the long civil war have included slavery and famine. Since its formation as a new country, Sudan has captured and forced the southern Sudanese to fight for the north. The enslavement of the black south has turned many countries against the el-Bashir government. Often young children are stolen from their tribes and forced into slavery and Islamic conversion. The government hopes that this war of attrition will wear down the south.

Churches and other international humanitarian groups have tried without much success to free the slaves. These groups, such as the Swiss-based Christian Solidarity Worldwide, raise money to buy the slaves' freedom. Often this only results in the slaves being re-captured. The United Nations has sent food to aid the starving war victims, but often this too backfires, with the provisions being diverted to sustain the northern army or confiscated by the southern rebel army itself. Sudan's neighbors, Egypt, Libya, and Eritrea, have accused the Sudanese regime of human rights violations. The SPLA has been accused of running a despotic terrorist regime, killing and looting at will. The longer the war continues, the longer it will take for the agrarian south to begin the farming and harvesting of crops to sustain itself.

DIRECTORY

CENTRAL GOVERNMENT

Head of State

President and Prime Minister
Omar Hassan Ahmad el-Bashir, Office of the President, The Palace, Khartoum, Sudan

First Vice President
Ali Osman Mohamed, Office of the First Vice President

Second Vice President
George Kongor Arop, Office of the Second Vice President

Ministers

Minister of Agriculture and Forests
Uthman al-Hadi Ibrahim, Ministry of Agriculture and Forests

Minister of Animal Wealth
Musa al-Muk Kur, Ministry of Animal Wealth

Minister of Aviation
Makki Ali Balayil, Ministry of Aviation

Minister of Cabinet Affairs
Salah Eddin Karar, Ministry of Cabinet Affairs

Minister of Communications and Roads
Alhadi Bushra, Ministry of Communications and Roads

Minister of Defense
Abd al-Rahman Sirr al-Khatim, Ministry of Defense

Minister of Education
Hamid Muhammad Ali Torin, Ministry of Education

Minister of Energy and Mining
Awad al-Jaz, Ministry of Energy and Mining

Minister of External Trade
Adam al-Tahir Hamdun, Ministry of External Trade

Minister of Federal Relations
Ali Alhaj Mohamed, Ministry of Federal Relations

Minister of Finance and National Economy
Abdul Wahab Osman, Ministry of Finance and National Economy

Minister of Foreign Affairs
Mustafa Osman Ismail, Ministry of Foreign Affairs

Minister of Health
Mahdi Babou Nimir, Ministry of Health

Minister of Higher Education and Scientific Research
Ahmed Omer Ibrahim, Ministry of Higher Education and Scientific Research

Minister of Industry
Badr Eddin Mohamed Suleiman, Ministry of Industry

Minister of Information and Culture
Ghazi Salah al-Din, Ministry of Information and Culture

Minister of Interior
Abdul Rahim Mohamed Hussein, Ministry of Interior

Minister of International Cooperation and Investment
Abdalla Hassan Ahmed, Ministry of International Cooperation and Investment

Minister of Irrigation
Kamal Ali Muhammad, Ministry of Irrigation

Minister of Justice
Ali Muhammad Uthman Yasin, Ministry of Justice

Minister of Labor Forces
Agnes Lukudu, Ministry of Labor Forces

Minister of Presidential Affairs
Bakri Hassan Salih, Ministry of Presidential Affairs

Minister of Public Service
Angelo Beda, Ministry of Public Service

Minister of Relations at the National Assembly
Abul Gasim Mohamed Ibrahim, Ministry of Relations at the National Assembly

Minister of Social Planning
al-Tayeb Ibrahim Mohammed Khair, Ministry of Social Planning

Minister of Survey
Galwak Deng, Ministry of Survey

Minister of Tourism and Environment
Mohamed Tahir Eila, Ministry of Tourism and Environment

Minister of Transport and Communications
Lam Akhol Ajawin, Ministry of Transport and Communications

Minister of Information and Culture
Amin Hasan Umar, Ministry of Information and Culture

Minister of State
Isam Siddiq, Ministry of State

POLITICAL ORGANIZATIONS

National Congress

NAME: Umar Hasan Ahmad el-Bashir

DIPLOMATIC REPRESENTATION

Embassies in Sudan

Italy
Street 39, Khartoum 2, Sudan
PHONE: +249 (11) 451614
FAX: +249 (11) 451217

Russia
PO Box 1161, Khartoum, Sudan
PHONE: +249 (11) 41315; 40870

United Kingdom
off Sharia al-Baladia, PO Box 801, Khartoum East, Sudan
PHONE: +249 (11) 777105; 780828; 780856

JUDICIAL SYSTEM

Supreme Court

Special Revolutionary Courts

FURTHER READING

Articles

''Big Fight.'' *Life* 22 (September 1, 1999): 34

Creusvaux, Herve et al. ''Famine in Southern Sudan.'' *The Lancet* 354 (September 4, 1999): 832

''Famine Fallout.'' *U.S. News and World Report* 126 (March 1, 1999): 7.

Gardner, Christine J. ''Slave Redemption.'' *Christianity Today* 43 (Aug. 9, 1999): 28

Goldstein, David. ''Senators Urge End to Sudanese Atrocities.'' *Knight-Ridder Tribune News Service*, 15 June 1999, p. K5563

''International: Rebels Claim Sudan Bombing.'' *The Oil Daily* (September 16, 1999).

Lesch, Ann M. ''Sudan: The Torn Country.'' *Current History* 98 (May 1999): 218

''Nice Words, Pity about the War.'' *The Economist* 350 (February 27, 1999): 42.

''Opposition Sued over Blast.'' *New York Times* 22 September 1999, p. A17

Rosenthal, A.M. ''When is it News?'' *New York Times*, 3 September 1999, p. A19

''Sudan Loses its Chains.'' *The Economist* 351 (June 12, 1999).

''World Watch.'' *Time International* 153 (May 17, 1999): 182.

Books

Anderson, G. Norman. *Sudan in Crisis: the Failure of Democracy.* Gainesville: University Press of Florida, 1999.

Lesch, Ann Mosely. *The Sudan-Contested National Identities.* Bloomington and Indianapolis: Indiana University Press, 1998.

Patterson, Donald. *Inside Sudan: Political Islam, Conflict, and Catastrophe.* Boulder, Colorado: Westview Press, 1999.

SUDAN: STATISTICAL DATA

For sources and notes see "Sources of Statistics" in the front of each volume.

GEOGRAPHY

Geography (1)

Area:

Total: 2,505,810 sq km.

Land: 2.376 million sq km.

Water: 129,810 sq km.

Area—comparative: slightly more than one-quarter the size of the US.

Land boundaries:

Total: 7,687 km.

Border countries: Central African Republic 1,165 km, Chad 1,360 km, Democratic Republic of the Congo 628 km, Egypt 1,273 km, Eritrea 605 km, Ethiopia 1,606 km, Kenya 232 km, Libya 383 km, Uganda 435 km.

Coastline: 853 km.

Climate: tropical in south; arid desert in north; rainy season (April to October).

Terrain: generally flat, featureless plain; mountains in east and west.

Natural resources: petroleum; small reserves of iron ore, copper, chromium ore, zinc, tungsten, mica, silver, gold.

Land use:

Arable land: 5%

Permanent crops: 0%

Permanent pastures: 46%

Forests and woodland: 19%

Other: 30% (1993 est.)

HUMAN FACTORS

Demographics (2A)

	1990	1995	1998	2000	2010	2020	2030	2040	2050
Population	26,627.7	30,556.4	33,550.6	35,530.4	46,572.8	58,621.4	70,858.1	82,696.7	93,624.9
Net migration rate (per 1,000 population)	NA	NA	NA	NA	NA	NA	NA	NA	NA
Births	NA	NA	NA	NA	NA	NA	NA	NA	NA
Deaths	NA	NA	NA	NA	NA	NA	NA	NA	NA
Life expectancy - males	51.8	53.8	55.0	55.8	59.7	63.3	66.6	69.4	71.8
Life expectancy - females	53.4	55.7	57.0	57.9	62.3	66.5	70.3	73.6	76.4
Birth rate (per 1,000)	45.0	41.7	39.9	38.8	33.2	27.7	23.1	19.7	17.2
Death rate (per 1,000)	13.5	11.8	10.9	10.3	8.2	6.8	6.1	5.9	6.3
Women of reproductive age (15-49 yrs.)	6,083.2	6,931.9	7,669.5	8,176.8	11,285.6	14,981.7	18,868.8	22,194.3	24,564.2
of which are currently married	NA	NA	NA	NA	NA	NA	NA	NA	NA
Fertility rate	6.5	6.0	5.7	5.5	4.3	3.4	2.8	2.4	2.2

Except as noted, values for vital statistics are in thousands; life expectancy is in years.

Health Personnel (3)

Total health expenditure as a percentage of GDP, 1990-1997[a]

- Public sector . . . . NA
- Private sector 1.9
- Total[b] 0.2

Health expenditure per capita in U.S. dollars, 1990-1997[a]

- Purchasing power parity . . . . NA
- Total 4

Availability of health care facilities per 100,000 people

- Hospital beds 1990-1997[a] 110
- Doctors 1993[c] 10
- Nurses 1993[c] 70

Health Indicators (4)

Life expectancy at birth

- 1980 48
- 1997 55

Daily per capita supply of calories (1996) 2,391

Total fertility rate births per woman (1997) 4.6

Maternal mortality ratio per 100,000 live births (1990-97) . . . . 370[b]

Safe water % of population with access (1995) 60

Sanitation % of population with access (1995) 22

Consumption of iodized salt % of households (1992-98)[a] 0

Smoking prevalence

- Male % of adults (1985-95)[a] . . . .
- Female % of adults (1985-95)[a] . . . .

Tuberculosis incidence per 100,000 people (1997) 180

Adult HIV prevalence % of population ages 15-49 (1997) 0.99

Infants and Malnutrition (5)

Under-5 mortality rate (1997) 115

% of infants with low birthweight (1990-97) 15

Births attended by skilled health staff % of total[a] . . . 86

% fully immunized (1995-97)

- TB 79
- DPT 75
- Polio 75
- Measles 71

Prevalence of child malnutrition under age 5 (1992-97)[b] . . . . NA

Ethnic Division (6)

- Black 52%
- Arab 39%
- Beja 6%
- Foreigners 2%
- Other 1%

Religions (7)

- Sunni Muslim 70%
- Indigenous beliefs 25%
- Christian 5%

Languages (8)

Arabic (official), Nubian, Ta Bedawie, diverse dialects of Nilotic, Nilo-Hamitic, Sudanic languages, English. Program of Arabization in process.

EDUCATION

Public Education Expenditures (9)

Public expenditure on education (% of GNP)

- 1980 . . . . 4.8
- 1996 . . . .

Expenditure per student

- Primary % of GNP per capita
  - 1980 . . . . 26.9
  - 1996 . . . .
- Secondary % of GNP per capita
  - 1980 . . . .
  - 1996 . . . .
- Tertiary % of GNP per capita
  - 1980 . . . . 589.6
  - 1996 . . . .

Expenditure on teaching materials

- Primary % of total for level (1996) . . . .
- Secondary % of total for level (1996) 0.0

Primary pupil-teacher ratio per teacher (1996) 29

Duration of primary education years (1995) 8

Educational Attainment (10)

Age group (1983)[4] . . . . 25+

Total population 6,492,263

Highest level attained (%)

- No schooling[2] 76.7
- First level
  - Not completed 18.6
  - Completed . . . . NA

Entered second level

S-1 .....1.9

S-2 .....2.0

Postsecondary 0.8

Literacy Rates (11A)

In thousands and percent[1]	1990	1995	2000	2010
Illiterate population (15+ yrs.)	8,088	8,507	8,839	9,154
Literacy rate - total adult pop. (%)	40.3	46.1	51.8	62.6
Literacy rate - males (%)	52.6	57.7	62.4	70.8
Literacy rate - females (%)	28.0	34.6	41.3	54.5

GOVERNMENT & LAW

Political Parties (12)

The legislative branch is a unicameral National Assembly (400 seats; 275 elected by popular vote, 125 elected by a supraassembly of interest groups known as the National Congress). Parties are banned in the new National Assembly following the 1989 coup.

Military Affairs (14B)

	1990	1991	1992	1993	1994	1995
Military expenditures						
Current dollars (mil.)	178	481	NA	NA	426	NA
1995 constant dollars (mil.)	204	531	NA	NA	436	NA
Armed forces (000)	65	65	82	82	82	89
Gross national product (GNP)						
Current dollars (mil.)	4,407	4,659	5,334	5,854[e]	6,406[e]	6,567[e]
1995 constant dollars (mil.)	5,065	5,149	5,738	6,137[e]	6,567[e]	6,567[e]
Central government expenditures (CGE)						
1995 constant dollars (mil.)	NA	NA	NA	NA	1,162	1,325
People (mil.)	26.6	27.4	28.2	28.9	29.6	30.3
Military expenditure as % of GNP	4.0	10.3	NA	NA	6.6	NA
Military expenditure as % of CGE	NA	NA	NA	NA	37.6	NA
Military expenditure per capita (1995 $)	8	19	NA	NA	15	NA
Armed forces per 1,000 people (soldiers)	2.4	2.4	2.9	2.8	2.8	2.9
GNP per capita (1995 $)	190	188	203	212	222	217
Arms imports[6]						
Current dollars (mil.)	100	80	110	90	70	100
1995 constant dollars (mil.)	115	88	118	94	72	100
Arms exports[6]						
Current dollars (mil.)	0	0	0	0	0	0
1995 constant dollars (mil.)	0	0	0	0	0	0
Total imports[7]						
Current dollars (mil.)	619	890	821	945	1,162	1,185
1995 constant dollars (mil.)	711	984	883	991	1,191	1,185
Total exports[7]						
Current dollars (mil.)	374	305	319	417	524	556
1995 constant dollars (mil.)	430	337	343	437	537	556
Arms as percent of total imports[8]	16.2	9.0	13.4	9.5	6.0	8.4
Arms as percent of total exports[8]	0	0	0	0	0	0

Government Budget (13B)

Revenues . . . . $482 million

Expenditures . . . . $1.5 billion

Capital expenditures . . . . $30 million

Data for 1996.

Crime (15)

Crime rate (for 1994)

Crimes reported 529,696

Total persons convicted 20,069

Crimes per 100,000 population 1,830

Persons responsible for offenses

Total number of suspects 407,125

Total number of female suspects . . . . NA

Total number of juvenile suspects 9,975

LABOR FORCE

Labor Force (16)

Total (million) 11

Agriculture 80%

Industry and commerce 10%

Government 6%

Labor shortages for almost all categories of skilled employment. Data for 1996 est.

Unemployment Rate (17)

30% (FY92/93 est.)

PRODUCTION SECTOR

Electric Energy (18)

Capacity . . . . 500,000 kW (1995)

Production . . . . 1.305 billion kWh (1995)

Consumption per capita . . . . 43 kWh (1995)

Transportation (19)

Highways:
total: 11,900 km
paved: 4,320 km
unpaved: 7,580 km (1996 est.)
Waterways: 5,310 km navigable
Pipelines: refined products 815 km
Merchant marine:
total: 4 ships (1,000 GRT or over) totaling 38,093 GRT/49,727 DWT ships by type: cargo 2, roll-on/roll-off cargo 2 (1997 est.)
Airports: 65 (1997 est.)
Airports—with paved runways:
total: 12
over 3,047 m: 1
2,438 to 3,047 m: 8
1,524 to 2,437 m: 3 (1997 est.)
Airports—with unpaved runways:
total: 53
1,524 to 2,437 m: 13
914 to 1,523 m: 29
under 914 m: 11 (1997 est.)

Top Agricultural Products (20)

Cotton, groundnuts, sorghum, millet, wheat, gum arabic, sesame; sheep.

FINANCE, ECONOMICS, & TRADE

Economic Indicators (22)

National product: GDP—purchasing power parity—$26.6 billion (1997 est.)
National product real growth rate: 5% (1997 est.)
National product per capita: $875 (1997 est.)
Inflation rate—consumer price index: 27% (mid-1997 est.)

MANUFACTURING SECTOR

GDP & Manufacturing Summary (21)

	1980	1985	1990	1992	1993	1994
Gross Domestic Product						
Millions of 1990 dollars	5,046	4,969	4,895	5,448	5,448	5,557
Growth rate in percent	−3.41	−2.90	−5.31	11.30	0.00	2.00
Per capita (in 1990 dollars)	270.1	231.5	199.1	210.0	204.5	203.1
Manufacturing Value Added						
Millions of 1990 dollars	519	473	438	445	*446*	*453*
Growth rate in percent	−4.09	−0.26	−11.44	8.37	*0.13*	*1.59*
Manufacturing share in percent of current prices	8.9	8.8	9.2	9.3	*9.1*	NA

Exchange Rates (24)

Exchange rates:	
Sudanese pounds (£Sd) per US$1	
Official rate:	
July 1997	1,602.70
1996	1,250.79
1995	580.87
1994	289.61
1993	159.31
Market rate:	
July 1997	1,612.90
1996	1,250.79
August 1995	571.02
1994	289.61
1993	159.31
1992	97.43

The market rate is a unified exchange rate determined by a committee of local bankers, without official intervention, and is quoted uniformly by all commercial banks.

Top Import Origins (25)

$1.5 billion (1996) Data are for 1996.

Origins	%
Saudi Arabia	10
South Korea	7
Germany	6
Egypt	6

Top Export Destinations (26)

$620 million (f.o.b., 1996) Data are for 1996.

Destinations	%
Saudi Arabia	20
United Kingdom	14
China	11
Italy	8

Economic Aid (27)

Recipient: ODA, $387 million (1993).

Import Export Commodities (28)

Import Commodities	Export Commodities
Foodstuffs	Cotton 23%
Petroleum products	Sesame 22%
Manufactured goods	Livestock/meat 13%
Machinery and equipment	Gum arabic 5%
Medicines and chemicals	
Textiles	

Balance of Payments (23)

	1992	1993	1994	1995	1996
Exports of goods (f.o.b.)	213	306	524	556	620
Imports of goods (f.o.b.)	−810	−533	−1,045	−1,066	−1,339
Trade balance	−597	−227	−521	−510	−719
Services - debits	−298	−131	−240	−177	−202
Services - credits	156	70	78	127	57
Private transfers (net)	109	10	13	—	16
Government transfers (net)	124	75	69	60	21
Overall balance	−506	−202	−602	−500	−827

SURINAME

Republic of Suriname
Republiek Suriname

CAPITAL: Paramaribo.

FLAG: A yellow star is at the center of five stripes: a broad red band in the middle, two white bands, and a green stripe at the top and bottom.

ANTHEM: The *Surinaams Volkslied (National Anthem)* begins ''God zij met ons Suriname'' (''God be with our Suriname'').

MONETARY UNIT: The Suriname guilder (Sf) is a paper currency of 100 cents. There are coins of 1, 5, 10, and 25 cents, and notes of 5, 10, 25, 100, and 500 guilders. Sf1 = $0.00249 (or $1 = Sf401.0).

WEIGHTS AND MEASURES: The metric system is used.

HOLIDAYS: New Year's Day, 1 January; Revolution Day, 25 February; Labor Day, 1 May; National Union Day, 1 July; Independence Day, 25 November; Christmas, 25 December; Boxing Day, 26 December. Movable religious holidays include Holi Phagwah, Good Friday, Easter Monday, and 'Id al-Fitr.

TIME: 8:30 AM = noon GMT.

LOCATION AND SIZE: Situated on the northeast coast of South America, Suriname is the smallest independent country on the continent, with a total area of 163,270 square kilometers (63,039 square miles). The area occupied by Suriname is slightly larger than the state of Georgia. Suriname has a total boundary length of 2,093 kilometers (1,301 miles). Suriname's capital city, Paramaribo, is located on the Atlantic coast.

CLIMATE: The climate is tropical and moist. Daytime temperatures range from 28° to 32°C (82–90°F). At night the temperature drops as low as 21°C (70°F). The annual rainfall in Paramaribo is about 230 centimeters (90 inches).

INTRODUCTORY SURVEY

RECENT HISTORY

Officially under Dutch rule since 1816, a 1954 statute provided full self-rule for Suriname, except in foreign affairs and defense. Suriname became an independent nation on 25 November 1975. For five years, the country was a parliamentary republic under prime minister Henk Arron. In 1980 the government was overthrown in a military coup led by Désiré Bouterse. Parliament was dissolved and the constitution suspended. The following year the new government declared itself a Socialist republic.

The military and Bouterse ruled through a series of supposedly civilian governments, while pressure mounted for a return to genuine civilian rule. When the Surinamese Liberation Army (SLA), a guerrilla movement, began operating in the northeast in July 1986, the government responded by killing civilians suspected of supporting the rebels.

In 1987 the military gave law-making authority to a newly appointed State Council, rather than the elected National Assembly. International pressure mounted, however, and the military soon gave in, scheduling elections in 1991. An anti-military coalition, led by Ronald Venetiaan, won. In 1996, Jules Wijdenbosch of the National Democratic Party was elected to the presidency, marking the first time in independent Suriname's history that one democratically-elected government passed peacefully to another.

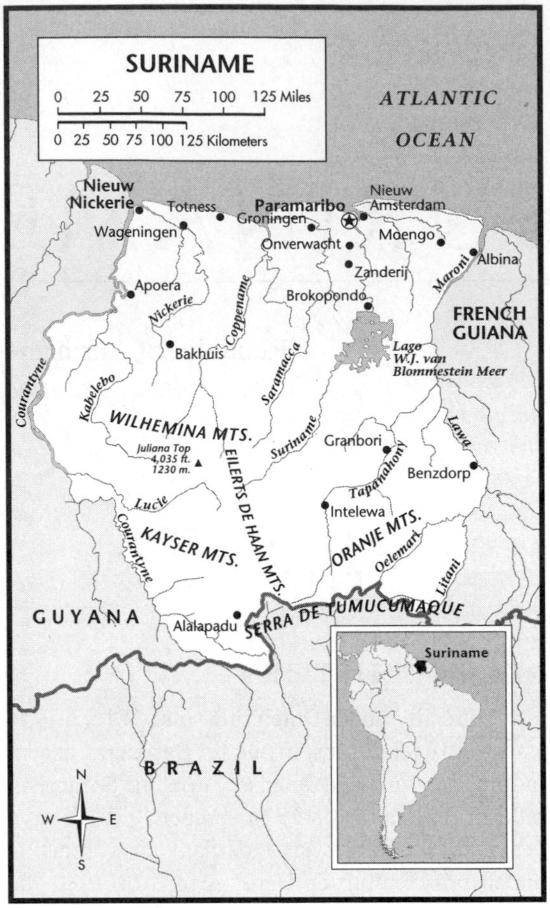

GOVERNMENT

The 1987 constitution provides for a single-chamber, 51-member National Assembly directly elected for a five-year term. The executive branch consists of the president, vice-president, and prime minister, all chosen by the National Assembly. There is also a cabinet and an appointed Council of State. The republic is divided into 10 districts.

Judiciary

The constitution provides the right to a fair public trial before a single judge, the right to legal counsel, and the right to appeal. There is a supreme court.

Political Parties

Suriname's political parties tend to represent particular ethnic groups. The National Party of Suriname (NPS) draws support from the Creole population; the Progressive Reform Party (VHP) is East Indian; and the Indonesian Peasant's Party (KTPI). In 1991, these three parties and the Suriname Labor Party (SPA) formed the New Front (NF) and won a solid victory, gaining 30 of 51 assembly seats.

DEFENSE

The Suriname National Army consists of army, air force, and naval divisions, with a strength of about 1,800 in 1995. Defense spending is below $100 million.

ECONOMIC AFFAIRS

The bauxite industry has traditionally set the pace for Suriname's economy. Next to bauxite, foreign aid is the mainstay of the country's economy. After 1990, the government started a plan to reduce government spending and privatize key sectors of the economy. By 1995, the economy was improving slightly, ending years of decline. In 1997, the inflation rate was officially 50%, but it is believed to be even higher.

Public Finance

For years the Suriname budget operated with a deficit. However, some relief was achieved by absorbing the arrears of principal and interest on Dutch loans into the second five-year plan (1972–76), and government revenues increased following the introduction of a new bauxite levy in 1974. In 1979 and 1980 there were budget surpluses. During the 1980s, however, the military regime in power increased government intervention and participation in the economy, causing public employment and budget deficits to soar. The return to civilian government on 16 September 1991 enticed the Dutch government to resume its development aid program, which could annually amount to $200 million until 1996. Reforms enacted include the reduction of deficit spending, the renunciation of monetary creation as a means of financing deficits, and the deregulation of trade and business licensing systems. According to the 1993 budget, the deficit was about 18% of GDP. The U.S. Central Intelligence Agency estimates that, in 1994, government revenues totaled approximately $300 million and expenditures $700 million, including capital expenditures of $70 million. External debt totaled $180 million.

Income

In 1995, the gross national product (GNP) was $360 million, or about $1,000 per person. For the period 1988–95, the average inflation rate was 48.5%, resulting in a real growth rate in per person GNP of 0.7%.

Industry

The major industries are mining and food processing. The bauxite industry over the years has developed into a complex of factories, workshops, power stations, and laboratories. The bauxite refinery near Paramaribo produced 1.6 million tons of alumina and 32,000 tons of aluminum in 1995.

Banking and Finance

The Central Bank of Suriname has acted as a bank of issue since 1 April 1957. Other banks include the Suriname People's Credit Bank, Post Office Savings Bank, Agricultural Bank, and National Development Bank. There were seven commercial banks in 1997 in Suriname, one of which was Dutch owned.

Money supply, as measured by M2, totaled Sf79,425.7 million at the end of 1995. Foreign reserves, excluding gold, amounted to $96.32 million at the end of 1996.

Economic Development

In wholesale, retail, and foreign trade, the government has been highly interventionist. Quota restrictions or outright bans on many imported items considered nonessential or in competition with local products have been announced. The government forced price rollbacks on domestic items and imposed price controls on essential imports, resulting in some shortages. Although the government was Socialist in principle since 1981, it refrained from nationalizing Suriname's key industries, although it did increase its participation in them. The Action Program announced by the government in May 1982 called for the encouragement of small-scale industry, establishment of industrial parks, development of rural electrification and water supply projects, liberalization of land distribution, and worker participation in management of government enterprises.

In 1975, the Netherlands promised Suriname $110 million annually in grants and loans, for a period of 10–15 years. This aid program and $1.5 million in aid authorized by the U.S. in September 1982 were suspended following the killings of prominent Surinamese in December 1982. In 1983, Brazil and Suriname reached agreement on a trade and aid package, reportedly underwritten by the U.S. By 1986, Suriname had signed trade agreements with several countries, among them the Netherlands.

As of 1994, Suriname was undergoing a comprehensive structural adjustment program (SAP). This program, recommended by the European Community (EC), was designed to establish the conditions for sustained growth of output and employment with relative stability of prices, a viable balance of payments, and protection of the low-income population. However, as of the end of 1996, only minimal progress toward restructuring had been accomplished. Enacting the full SAP in its proper sequence would improve the prospects for the Surinamese economy and living conditions in the late 1990s.

SOCIAL WELFARE

Welfare programs are largely conducted privately, or through ethnic or religious groups. The privately run Home for Women in Crisis Situations is consistently overcrowded, indicating a national problem in the area of violence against women.

Healthcare

Tuberculosis, malaria, and syphilis, once the chief causes of death, have been controlled. Suriname's average life expectancy is 71 years. There is 1 doctor per 1,250 people and an estimated 2.7 hospital beds per 1,000 people.

Housing

Between 1988 and 1990, 82% of the urban and 94% of the rural population had access to a public water supply, while 64% of urban dwellers and 36% of rural dwellers had sanitation services.

EDUCATION

In 1993, there were 87,882 elementary schools pupils; secondary schools had 16,511 pupils. Education is compulsory and free for all children aged 6 through 12. The adult illiteracy rate in 1995 was 7%. Higher education includes four teacher-training colleges, the Technical College, and the University of Suriname. In 1990, higher level institutions reported a total of 495 teaching staff with 4,319 students enrolled.

1999 KEY EVENTS TIMELINE

May

- Shopkeepers close their businesses while angry gangs roam the capital demanding the government's resignation.

June

- Suriname's parliament censures President Jules Wijdenbosch and asks him to call for early elections and step down.

- Frustrated by a collapsing economy, thousands of people march in Suriname's capital to demand President Wijdenbosch's resignation
- President Wijdenbosch calls for early elections in a last-minute bid to avoid being ousted by opponents.

July

- Former military leader Desi Bouterse is convicted of cocaine trafficking by a court in The Hague. A judge sentences him in absentia to 16 years in prison, and fines him $2.18 million.

August

- Interpol seeks a high-ranking official in Suriname's police force on Belgian drug trafficking charges, according to Suriname authorities.

ANALYSIS OF EVENTS: 1999

BUSINESS AND THE ECONOMY

By June 1999, Suriname's currency had lost a third of its value, and inflation was running at about 10 percent a month. The government was quick to blame forces beyond its control. Any disturbances in the economy have profound effects on this small, impoverished nation, where 63 percent of the people live below the poverty line, and frustrated citizens didn't wait to stage protests against the government, asking for the president's resignation. Suriname's finance minister blamed the currency's slide on a major cocaine bust in the Netherlands. According to the government, Dutch importers withheld payment from Surinamese shippers, who scrambled to buy dollars to pay their South American suppliers, according to The Economist. But opposition leaders didn't accept the excuse. In the meantime, the government came up with another idea to save the economy. The President told the nation he had sent a delegation to Phoenix, Arizona, to negotiate for $26 million in aid, but didn't provide any other information. Surinamese were tired of excuses and promises. In October, angry school bus operators joined teachers in a 2-day-old strike to demand raises. Drivers said they had not been paid since May, while operators of 19 ferry boats that carry children to schools in the interior said they were owed money from December.

GOVERNMENT AND POLITICS

National strikes, a sinking economy and protests throughout the country finally convinced President Jules Wijdenbosch to call for early elections in June. The president was trying to win time, and his decision didn't earn any praise from opposition parties, which demanded his immediate resignation. Wijdenbosch, in his third of five years in office, has not been a popular president. He assumed his post with crucial help from the former military leader and dictator Desi Bouterse. His links to the former dictator, who acts as government counselor, have hurt him. Bouterse staged coups in 1980, and 1990, when he held power for five months, and is being pursued by human rights groups over political killings. To make matters worse for the government, Bourterse was convicted of cocaine trafficking by a court in The Hague in July. A judge sentenced him in absentia to 16 years in prison, and fined him $2.18 million. Suriname has protected and defended Bouterse, hurting diplomatic relationships with the Dutch, who keep the tiny nation's economy going.

Yet, it was the economy that turned Suriname against Wijdenbosch. He has been blamed for a major budget deficit that fueled price increases, and devalued the national currency. And despite calling for early elections, Surinamese were not expected to go to the polls for six to nine months, the time it takes to prepare an election for a country of 400,000 people. In the meantime, President Wijdenbosch announced he was firing his entire cabinet, promised emergency measures to improve the economy, and said he would form a new government. Elections are expected to take place in May 2000, and one of the unlikely candidates is Bouterse, whose popularity seems unaffected by the Dutch sentence. A strong opponent may be Stanley Rensch, leader of a human rights group that wants to try Bouterse and his colleagues for political killings and other crimes.

CULTURE AND SOCIETY

The contents of an eighteenth-century Portuguese synagogue in Suriname arrived in Jerusalem in September, after three years of negotiations between Israel and Suriname, an Israel newspaper reported. The Tzedek v'Shalom synagogue was one of two in Paramaribo, where the Jewish community has dwindled to about 200 people. The synagogue was built in 1736, when the Jewish community had become prosperous. In later years, the synagogue fell into disrepair.

Among the items shipped to Jerusalem were the ark of the law, four Torah scrolls, a reader's platform, pews, standing candelabra, brass chandeliers, candle sticks, and six rare silver Torah finials. A Torah crown from 1679 bears the name of Shmuel Cohen Nasi, and may be one of the oldest Jewish ritual objects in the Americas. The items will be used to recreate a synagogue as part of a museum to celebrate the Jews of Suriname.

DIRECTORY

CENTRAL GOVERNMENT

Head of State

President
Jules A. Wijdenbosch, Office of the President, Kl. Combeweg 1, Paramaribo, Suriname
PHONE: +597 472841

Vice President
Pretaapnarian Radhakisun, Office of the Vice President, Dr. Redmondstraataat 118, Paramaribo, Suriname
PHONE: +597 474805

Ministers

Minister of Agriculture
S. Redjosentono, Ministry of Agriculture, Cultuurtuinlaan 10, Paramaribo, Suriname
PHONE: +597 474177

Minister of Defense
R.A. Dwarka-Panday, Ministry of Defense, Kwattawg 29, Paramaribo, Suriname
PHONE: +597 471511

Minister of Education
N. Mahadewsingh, Ministry of Education, Dr. Kafiluddistraat 117, Paramaribo, Suriname
PHONE: +597 498850

Minister of Finance
T. Gobardhan, Ministry of Finance, Onafhankelykheidsplein 2, Paramaribo, Suriname
PHONE: +597 473941

Minister of Foreign Affairs
E.G. Snijders, Ministry of Foreign Affairs, Gravenstraat 8, Paramaribo, Suriname
PHONE: +597 477030

Minister of Health
T. Vishnudath, Ministry of Health, Gravenstraat 64, Paramaribo, Suriname
PHONE: +597 410441

Minister of Interior Affairs
S.W. Kertoidjojo, Ministry of Interior Affairs, Wilhelminastraat 3, Paramaribo, Suriname
PHONE: +597 473141

Minister of Justice and Police
P.R. Sjak-Shie, Ministry of Justice and Police, Gravenstraat 1, Paramaribo, Suriname
PHONE: +597 473841

Minister of Labor
M.A.F. Pierkhan, Ministry of Labor, Wagenwegstraat 22, Paramaribo, Suriname
PHONE: +597 475241

Minister of Transport, Communication and Tourism
D.C. de Bie, Ministry of Transport, Communication and Tourism, Hendrikstraat 2628, Paramaribo, Suriname
PHONE: +597 420422

Minister of Natural Resources
L.A.E. Alibux, Ministry of Natural Resources, Mirandastraat 1115, Paramaribo, Suriname
PHONE: +597 474666

Minister of Planning and Development
W. Nain, Ministry of Planning and Development, Dr. Redmondstraat 118, Paramaribo, Suriname
PHONE: +597 473141

Minister of Public Affairs
B. Mangal, Ministry of Public Affairs, Coppenamestraat 167, Paramaribo, Suriname
PHONE: +597 490666

Minister of Regional Affairs
Y.R.A. Ravales-Resida, Ministry of Regional Affairs, Roseveltkade 471214, Paramaribo, Suriname
PHONE: +597 471269

Minister of Social Affairs and Housing
S. Moestadja, Ministry of Social Affairs and Housing, Waterkant 3031, Paramaribo, Suriname
PHONE: +597 476941

POLITICAL ORGANIZATIONS

Verenigde Hervormings Partij-VHP (United Reformation Party)

PHONE: +597 472065
TITLE: Chairman

NAME: Jagernath Lachmon

Nationale Democratische Partij-NDP (National Democratic Party)

PHONE: +597 499804
E-MAIL: ndpsur@sr.net

DIPLOMATIC REPRESENTATION

Embassies in Suriname

Brazil
Maratakastraat 2, P.O. Box 925, Paramaribo, Suriname
PHONE: +597 400200
FAX: +597 400205

Indonesia
Van Brussellaan 3, P.O. Box 2648, Paramaribo, Suriname
PHONE: +597 497070
FAX: +597 498234

JUDICIAL SYSTEM

Supreme Court

FURTHER READING

Articles

''Arms-for-Drugs Deals on the Rise in Suriname.'' *The Miami Herald*, 12 September 1999.

''Suriname: On Trial.'' *The Economist* (April 3, 1999): 30.

''Suriname Teachers Joined in Strike Action.'' *The Miami Herald*, 6 October 1999.

''Suriname's Wondrous Botch.'' *The Economist* (June 5, 1999): 34.

''Surinamese Police Official Sought by Interpol on Drug Charges.'' Associated Press, 23 August 1999.

SURINAME: STATISTICAL DATA

For sources and notes see "Sources of Statistics" in the front of each volume.

GEOGRAPHY

Geography (1)

Area:

Total: 163,270 sq km.

Land: 161,470 sq km.

Water: 1,800 sq km.

Area—comparative: slightly larger than Georgia.

Land boundaries:

Total: 1,707 km.

Border countries: Brazil 597 km, French Guiana 510 km, Guyana 600 km.

Coastline: 386 km.

Climate: tropical; moderated by trade winds.

Terrain: mostly rolling hills; narrow coastal plain with swamps.

Natural resources: timber, hydropower potential, fish, kaolin, shrimp, bauxite, gold, and small amounts of nickel, copper, platinum, iron ore.

Land use:

Arable land: NA.

Permanent crops: NA.

Permanent pastures: 0%

Forests and woodland: 96%

Other: 4% (1993 est.).

HUMAN FACTORS

Demographics (2A)

	1990	1995	1998	2000	2010	2020	2030	2040	2050
Population	396.2	417.1	428.0	434.1	452.8	460.8	454.1	426.4	379.7
Net migration rate (per 1,000 population)	NA	NA	NA	NA	NA	NA	NA	NA	NA
Births	NA	NA	NA	NA	NA	NA	NA	NA	NA
Deaths	NA	NA	NA	NA	NA	NA	NA	NA	NA
Life expectancy - males	65.8	67.2	68.1	68.6	71.0	73.1	74.7	76.1	77.2
Life expectancy - females	70.8	72.4	73.3	73.9	76.5	78.7	80.5	81.9	83.1
Birth rate (per 1,000)	27.0	24.5	22.5	21.1	17.0	15.3	12.2	10.5	9.3
Death rate (per 1,000)	6.4	6.0	5.8	5.7	5.9	6.4	7.6	10.3	13.7
Women of reproductive age (15-49 yrs.)	102.1	107.4	110.8	113.2	122.4	113.3	102.9	88.9	68.4
of which are currently married	NA	NA	NA	NA	NA	NA	NA	NA	NA
Fertility rate	3.0	2.7	2.6	2.5	2.2	2.0	1.9	1.9	1.8

Except as noted, values for vital statistics are in thousands; life expectancy is in years.

Infants and Malnutrition (5)

Under-5 mortality rate (1997) 30
% of infants with low birthweight (1990-97) 13
Births attended by skilled health staff % of total[a] . . .NA
% fully immunized (1995-97)
TB . . . . .NA
DPT 85
Polio 81
Measles 78
Prevalence of child malnutrition under age 5 (1992-97)[b] . . . . .NA

Ethnic Division (6)

Hindustani (also known locally as "East Indians;" their ancestors emigrated from northern India in the latter part of the 19th century) 37%, Creole (mixed white and black) 31%, Javanese 15.3%, "Maroons" (their African ancestors were brought to the country in the 17th and 18th centuries as slaves and escaped to the interior). 10.3%, Amerindian 2.6%, Chinese 1.7%, white 1%, other 1.1%.

Religions (7)

Hindu 27.4%
Muslim 19.6%
Roman Catholic 22.8%
Protestant 25.2%
Indigenous beliefs 5%

Protestants are predominantly Moravian.

Languages (8)

Dutch (official), English (widely spoken), Sranang Tongo (Surinamese, sometimes called Taki-Taki, is native language of Creoles and much of the younger population and is lingua franca among others), Hindustani (a dialect of Hindi), Javanese.

GOVERNMENT & LAW

Military Affairs (14B)

	1990	1991	1992	1993	1994	1995
Military expenditures						
Current dollars (mil.)	41	43[e]	29[e]	57	39[e]	39[e]
1995 constant dollars (mil.)	47	48[e]	32[e]	60	40[e]	39[e]
Armed forces (000)	4	4	2	2	2	2
Gross national product (GNP)						
Current dollars (mil.)	1,036	1,157	1,348	1,210	1,194	1,302
1995 constant dollars (mil.)	1,191	1,279	1,450	1,269	1,224	1,302
Central government expenditures (CGE)						
1995 constant dollars (mil.)	NA	NA	NA	NA	NA	NA
People (mil.)	.4	.4	.4	.4	.4	.4
Military expenditure as % of GNP	4.0	3.8	2.2	4.7	3.3	3.0
Military expenditure as % of CGE	NA	NA	NA	NA	NA	NA
Military expenditure per capita (1995 $)	119	119	77	144	95	90
Armed forces per 1,000 people (soldiers)	10.0	9.9	4.9	4.8	4.7	4.7
GNP per capita (1995 $)	2,990	3,166	3,536	3,048	2,894	3,030
Arms imports[6]						
Current dollars (mil.)	5	0	0	0	0	0
1995 constant dollars (mil.)	6	0	0	0	0	0
Arms exports[6]						
Current dollars (mil.)	0	0	0	0	0	0
1995 constant dollars (mil.)	0	0	0	0	0	0
Total imports[7]						
Current dollars (mil.)	472	470	468[e]	436[e]	194[e]	NA
1995 constant dollars (mil.)	542	519	503[e]	457[e]	199[e]	NA
Total exports[7]						
Current dollars (mil.)	472	420	391[e]	375[e]	294[e]	NA
1995 constant dollars (mil.)	542	464	421[e]	393[e]	301[e]	NA
Arms as percent of total imports[8]	1.1	0	0	0	0	NA
Arms as percent of total exports[8]	0	0	0	0	0	NA

EDUCATION

Literacy Rates (11A)

In thousands and percent[1]	1990	1995	2000	2010
Illiterate population (15+ yrs.)	22	19	18	15
Literacy rate - total adult pop. (%)	91.7	93.0	94.2	96.0
Literacy rate - males (%)	94.3	95.1	95.9	97.0
Literacy rate - females (%)	89.3	91.0	92.6	95.1

GOVERNMENT & LAW

Political Parties (12)

National Assembly	No. of seats
National Democratic Party (NDP)	16
The New Front (NF)	14
Party for Renewal and Democracy (BVD)	5
Party of National Unity and Solidarity (KTPI)	5
Pendawa Lima	4
Alliance	3
Democratic Alternative '91 (DA '91)	2
Independent Progressive Democratic Alternative (OPDA)	2

Government Budget (13B)

Revenues	$317 million
Expenditures	$333 million
Capital expenditures	$52 million

Data for 1997 est.

LABOR FORCE

Labor Force (16)

Agriculture, industry, services.

Unemployment Rate (17)

20% (1997)

PRODUCTION SECTOR

Electric Energy (18)

Capacity	425,000 kW (1995)
Production	1.601 billion kWh (1995)
Consumption per capita	3,727 kWh (1995)

Transportation (19)

Highways:
total: 4,530 km
paved: 1,178 km
unpaved: 3,352 km (1996 est.)
Waterways: 1,200 km; most important means of transport; oceangoing vessels with drafts ranging up to 7 m can navigate many of the principal waterways.
Merchant marine:
total: 2 ships (1,000 GRT or over) totaling 2,421 GRT/2,990 DWT ships by type: cargo 1, container 1 (1996 est.)
Airports: 45 (1997 est.)
Airports—with paved runways:
total: 5
over 3,047 m: 1
under 914 m: 4 (1997 est.)
Airports—with unpaved runways:
total: 40
914 to 1,523 m: 7
under 914 m: 33 (1997 est.)

MANUFACTURING SECTOR

GDP & Manufacturing Summary (21)

	1980	1985	1990	1992	1993	1994
Gross Domestic Product						
Millions of 1990 dollars	1,659	1,637	1,728	1,879	1,879	1,864
Growth rate in percent	−8.57	2.02	−0.91	5.77	0.00	−0.80
Per capita (in 1990 dollars)	4,672.2	4,340.9	4,321.1	4,594.3	4,538.9	4,459.5
Manufacturing Value Added						
Millions of 1990 dollars	298	239	208	191	180	*181*
Growth rate in percent	−10.52	6.45	−6.85	0.96	−5.66	*0.50*
Manufacturing share in percent of current prices	17.6	12.5	12.3	10.1	14.2	NA

Top Agricultural Products (20)

Paddy rice, bananas, palm kernels, coconuts, plantains, peanuts; beef, chicken; forest products and shrimp of increasing importance.

FINANCE, ECONOMICS, & TRADE

Economic Indicators (22)

National product: GDP—purchasing power parity—$1.44 billion (1997 est.)

National product real growth rate: 4% (1997 est.)

National product per capita: $3,400 (1997 est.)

Inflation rate—consumer price index: 8% (1997 est.)

Exchange Rates (24)

Exchange rates:

Surinamese guilders, gulden, or florins (Sf.) per US$1

central bank midpoint rate:

January 1998	401.00
1997	401.00
1996	401.26
1995	442.23
1994	134.12

parallel rate:

December 1995	412
December 1994	510
January 1994	109

Beginning July 1994, the central bank midpoint exchange rate was unified and became market determined.

Top Import Origins (25)

$490 million (f.o.b., 1997 est.) Data are for 1994.

Origins	%
United States	40
Netherlands	24
Trinidad and Tobago	11
Japan	3

Top Export Destinations (26)

$434.3 million (f.o.b., 1996 est.) Data are for 1994.

Destinations	%
Norway	33
Netherlands	26
United States	13
Japan	6
Brazil	6
United Kingdom	3

Economic Aid (27)

Recipient: the Netherlands provided a 1996 aid package of $224 million to Suriname, Aruba, and the Netherlands Antilles.

Import Export Commodities (28)

Import Commodities	Export Commodities
Capital equipment	Alumina
Petroleum	Aluminum
Foodstuffs	Shrimp and fish
Cotton	Rice
Consumer goods	Bananas

Balance of Payments (23)

	1991	1992	1993	1994	1995
Exports of goods (f.o.b.)	617	609	298	294	416
Imports of goods (f.o.b.)	−619	−486	−214	−194	−293
Trade balance	−2	122	84	99	123
Services - debits	−197	−191	−108	−118	−167
Services - credits	42	42	47	74	106
Private transfers (net)	34	60	24	6	11
Government transfers (net)	−11	−7	−3	−2	−0
Overall balance	−133	25	44	59	73

SWAZILAND

Kingdom of Swaziland

CAPITAL: Mbabane (administrative and judicial); Lobamba (royal and parliamentary).

FLAG: Blue, yellow, crimson, yellow, and blue stripes with the shield and spears of the Emasotsha regiment superimposed on the crimson stripe.

ANTHEM: National Anthem, beginning "O God, bestower of the blessings of the Swazi."

MONETARY UNIT: The lilangeni (pl. emalangeni; E) of 100 cents is a paper currency equal in value to the South African rand, which is also legal tender. There are coins of 1, 2, 5, 10, 20, and 50 cents, 1 lilangeni, and notes of 2, 5, 10, 20, and 50 emalangeni. E1 = $0.21268 (or $1 = E4.702).

WEIGHTS AND MEASURES: The metric system replaced imperial weights and measures in September 1969.

HOLIDAYS: New Year's Day, 1 January; Commonwealth Day, 2nd Monday in March; National Flag Day, 25 April; Birthday of King Sobhuza II, 22 July; Umhlanga (Reed Dance) Day, last Monday in August; Somhlolo (Independence) Day, 6 September; UN Day, 24 October; Christmas Day, 25 December; Boxing Day, 26 December. Movable religious holidays include Good Friday, Holy Saturday, Easter Monday, Ascension, and the Incwala Ceremony.

TIME: 2 PM = noon GMT.

LOCATION AND SIZE: A landlocked country in southern Africa, Swaziland has an area of 17,360 square kilometers (6,703 square miles), slightly smaller than the state of New Jersey. It has a total boundary length of 554 kilometers (344 miles).

Swaziland's capital city, Mbabane, is located in the west central part of the country.

CLIMATE: The high veld has a humid climate with about 140 centimeters (55 inches) of mean annual rainfall. The middle veld and Lubombo are somewhat drier, with about 85 centimeters (33 inches) of rain; the low veld receives about 60 centimeters (24 inches) of rain in an average year. Temperatures range from as low as −3°C (27°F) in winter in the highlands to as high as 42°C (108°F) in summer in the lowlands.

INTRODUCTORY SURVEY

RECENT HISTORY

Under British rule since 1902, Swaziland became an independent nation within the British Commonwealth on 6 September 1968. On 12 April 1973 King Sobhuza II repealed the constitution and assumed supreme executive, legislative, and judicial powers. After Sobhuza died in 1982 a long power struggle took place. Eventually it was decided that 15-year-old Makhosetive, one of Sobhuza's 67 sons, would ascend the throne upon reaching adulthood. He was crowned King Mswati III on 25 April 1986. The new king decreased the power of the Liqoqo, the advisory council, and he has ruled through his prime minister and cabinet.

After signing an agreement with South Africa, Swaziland arrested and deported members of the African National Congress (ANC), the leading South African black nationalist group. In late 1985 and 1986 South African commando squads conducted raids on ANC members and supporters in Swaziland. In September and October 1993 popular elections were held for parliament and a new prime minister, Prince Mbilini, took office. Barnabas Sibusiso Dlamini was appointed prime minister in 1996. That year, the country was disrupted by a general strike supported by the Swaziland Federation of Trade Unions. Strikes have been called as a way to pressure the government for greater democratic control.

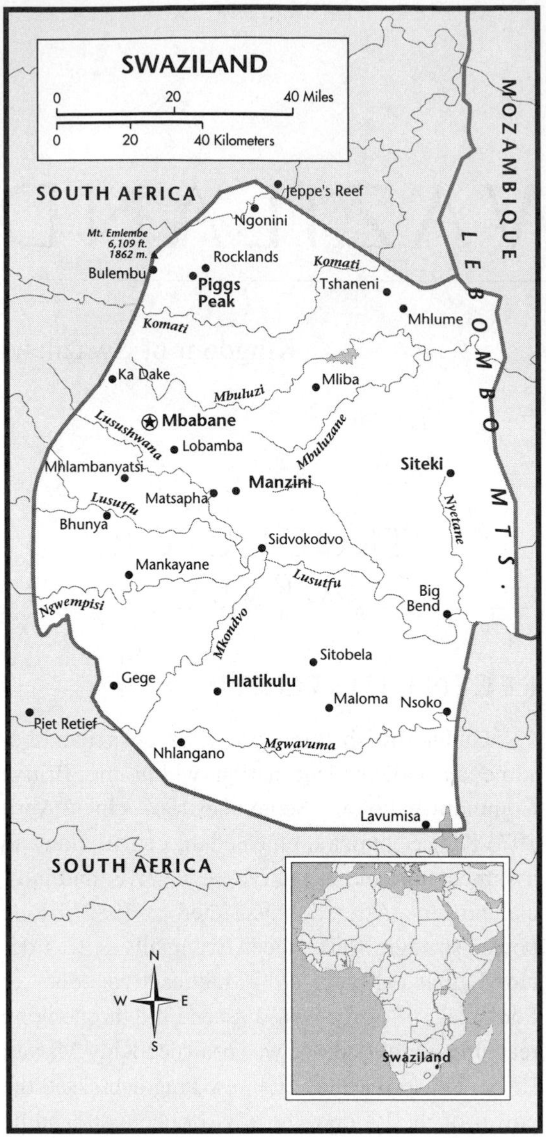

GOVERNMENT

In 1979 a new parliament was created, with a 50-member House of Assembly and a 20-member Senate. To become law, legislation passed by parliament must be approved by the king.

Judiciary

The dual judicial system consists of a set of courts based on a western model and western law, and a set of national courts that follows Swazi law and custom.

Political Parties

Several parties, including the People's United Democratic Movement (Pudemo), the Swaziland United Front, and the Swaziland Progressive Party, operate openly.

DEFENSE

The Umbutfo Swaziland Defense Force has fewer than 3,000 personnel.

ECONOMIC AFFAIRS

The majority of Swazis are engaged in farming. Because of its small size, Swaziland relies on exporting, which is composed primarily of large firms with predominantly foreign ownership. Money sent home from Swazi miners working in South Africa accounts for 20% of national income.

Public Finance

In the past, the government maintained a prudent fiscal policy by avoiding large deficits and restricting public sector growth. From 1987 to 1991 large budgetary surpluses were registered, and the government began making repayments on the external debt as a net creditor to the bank. Budgetary deficits returned in 1992 and 1993.

The U.S. Central Intelligence Agency estimates that, in 1994, government revenues totaled approximately $342 million and expenditures $410 million, including capital expenditures of $130 million. External debt totaled $240 million.

Income

In 1998 Swaziland's gross national product (GNP) was $1.4 billion (U.S.), or about $1,400 per person. The average annual growth in GNP per person for 1998 was −1.3%. Annual average inflation during that period was 8.5%.

Industry

Manufacturing consists mostly of five export-driven industries: wood pulp production, drink processing, fruit canning, refrigerators, and sugar processing. The industrial growth of the 1980s slowed in the early 1990s. Textiles, footwear, gloves, office equipment, candy, furniture, glass, and bricks are also manufactured.

Banking and Finance

The Central Bank of Swaziland is the nation's central bank. The total money supply, as measured by M1, was e 395.4 million at the end of February 1997. The nation's commercial banks are subsidiaries of Barclays Bank and the Standard Chartered Bank. Other financial institutions are the Swaziland Development and Savings Bank and the Union Bank of Swaziland.

Economic Development

The growth experienced in recent years has left unaffected the 66% of Swazis who live on small family farms. While manufacturing employment has risen, about 11% of Swazis are unemployed and actively seeking work. It is hoped that the inauguration of a multiracial government will prove beneficial to ongoing Swaziland-South African economic development.

SOCIAL WELFARE

The government subsidizes workers whose wages fall below specified minimums, and workers' compensation for injuries on the job is also provided.

Women do not have full legal equality with men. A wife may not open a bank account, buy land, or leave the country without the husband's permission.

Healthcare

Major health problems include bilharzia, typhoid, tapeworm, gastroenteritis, malaria, kwashiorkor, and pellagra. In 1995 average life expectancy was 57 years. In 1990 there were 83 doctors, and about 55% of the population had access to health care services. Traditional healers are still consulted by 80% of the population.

Housing

The government hopes to improve housing conditions for low-income groups through self-help schemes and by providing mortgage assistance. In 1985 thirty residential buildings were completed.

EDUCATION

In 1994 there were 521 primary schools with 192,599 pupils and 5,887 teachers. Secondary schools had 52,571 students and 2,872 teachers.

Higher education is provided by the University of Swaziland and the Swaziland College of Technology. Higher level institutions had a total of 515 teachers and 4,183 students in 1993. The adult illiteracy rate was 23% in 1995.

1999 KEY EVENTS TIMELINE

January

- The European Union allocates 50 million Rands to the Government of Swaziland for micro projects in rural areas such as roads, water, clinics, school classrooms or teachers' houses, at the request of local communities.
- The Swaziland Electricity Board intends to join the newly established regional Southern African Power Pool to reduce the cost of electricity, which is imported from South Africa.
- Swaziland's Department of Labour deregisters the country's largest union, the Swaziland Agriculture, Plantation and Allied Workers' Union [SAPAWU], because of its failure to submit annual financial reports for inspection.
- Former Swaziland International Olympics Committee (IOC) member David Sibandze resigns as president of Swaziland's Commonwealth and Olympic Games Association (SOCGA), following his involvement in the Salt Lake City corruption scandal.

February

- A two-day strike by Swaziland's nurses ends following the signing of a recognition agreement between strikers and health authorities.

March

- Swaziland hosts the 20th session of the Interstate Defence and Security Committee, ISDSC, which deliberates on issues of utmost importance to the SADC (Southern African Development Community) member states.
- Swazi bank officials change from a German-based to a British mint firm following a spate of circulated counterfeit bank notes over the past two years.
- Swaziland's Public Service and Information Minister Magwagwa Mdluli is told by Parliament to drastically reduce the size of the country's grossly ''overstaffed and overfunded'' civil service.

April

- Swaziland, with a population of less than one million, now has the third-highest rate of HIV infection in the world.
- Two United States defense officials arrive in the country for a one-day visit to discuss security assistance programs between Swaziland and the United States.

May

- Swaziland mediates for peace between Zambia and Angola in an effort to try and help them resolve their differences following allegations that Zambia is assisting UNITA rebels led by Jonas Savimbi against the Angolan President Eduardo dos Santos.

- The World Health Organization (WHO) welcomes Swaziland's decision to publicly acknowledge that HIV/AIDS is a ''national disaster'' in this tiny mountain kingdom.

June

- Controversy, propaganda, and allegations of irregularity have plagued Swaziland's Constitutional Review Commission since its inception and threaten to shred any credibility of the resulting document.
- Mozambique and Swaziland sign an agreement establishing a legal framework for transport operations between the two countries.

July

- Swaziland reduces corporate tax by 7 percent to 30 percent to encourage local and foreign investment in the kingdom.
- Fears emerge in Swaziland that relations between Swaziland and its powerful neighbor, South Africa, risk being strained due to revelations that Swaziland collaborated with the apartheid regime.
- Water Affairs Minister Ronnie Kasrils travels to Swaziland to meet his Swazi and Mozambican counterparts and to sign agreements on the use of the shared water resources of the countries.

August

- Although the Swazis pride themselves on strong family ties, homelessness is reported to be on the increase in Swaziland's towns.

September

- Bristol-Myers Squibb (BMS) is to provide $100 million over five years to assist South Africa, Botswana, Namibia, Lesotho and Swaziland in finding sustainable solutions for communities to combat AIDS.
- Authorities in Swaziland charge the editor-in-chief of the Times of Swaziland, Bheki Makhubu, with criminal defamation for publishing the profile of a fiancé of King Mswati.
- Swaziland Youth Congress president Bongani calls on the country to ban corporal punishment in school.

October

- More Swazi journalists could be charged with defamation as the country's government prepares new legislation to curb how much newspapers are allowed to publish.
- King Mswati of Swaziland begins an official visit of Belgium and the European Union.
- Swaziland's embattled editor of the Sunday edition of Times of Swaziland has his R3000 bail extended after appearing briefly in the Mbabane magistrate's court on criminal defamation charges.

ANALYSIS OF EVENTS: 1999

BUSINESS AND THE ECONOMY

Swaziland's economy is closely integrated with that of South Africa, with much of its foreign direct investment originating in South Africa and its currency pegged to the South African Rand. Surrounded by South Africa, except for a short border with Mozambique, Swaziland is heavily dependent on South Africa from which it receives nearly all of its imports and to which it sends more than half of its exports. Remittances from Swazi workers in South African mines supplement domestically earned income by as much as 20 percent. Over 60 percent of the population is engaged in subsistence farming. Major problems that the country will need to tackle as it enters into the 21st century include soil overgrazing, soil erosion and the constant threat of drought.

Dependent as it is on South Africa, the country signed a series of agreements with South Africa and neighboring Mozambique concerning the use of the shared water resources of the countries. The government also signed an agreement with Mozambique establishing a legal framework for transport operations between the two countries. It is hoped that Mozambican investment in improved road, rail, and port infrastructure will provide additional trade routes to Swaziland's economy. The three countries have also embarked on a joint venture, the Lubomba Spatial Development Initiative, a multi-million dollar casino resort, game reserve, and lodge development designed to make parts of Swaziland and KwaZulu Natal into a tourist destination.

Swaziland is in the process of lessening its dependence on South Africa. For example, the Swaziland Electricity Board intends to join the newly established regional Southern African Power Pool to reduce the cost of electricity, which is imported from South Africa. Furthermore, Swaziland has reduced corporate tax by 7 percent to 30 percent to encourage local and foreign investment in the kingdom. Aid from the international community has greatly helped to develop rural infrastructure. The European Union allocated 50 million Rands to the Government of Swaziland for micro projects in rural areas such as roads, water, clinics, school classrooms or teachers' houses during the 1999 fiscal year.

GOVERNMENT AND POLITICS

Swaziland is one of the last monarchies in Africa. The country does not have a constitution and King Mswati III rules by decree and appoints anyone he likes to his cabinet. At the time of independence in 1968, Britain left the country with a western style constitution that placed checks on the monarch lessening its power. Mswati's father, King Sobhuza II, abolished that constitution in 1973. Recently a commission has been set up to review the banned constitution and draft a new one. However, controversy and allegations of irregularity have plagued the Constitutional Review Commission since its inception and threaten the credibility of the resulting document. The opposition has voiced its concerns over the make-up of the Constitutional Review Commission, which is full of royalists, traditionalists, and those with the most to lose from the reform.

The media have found it increasingly difficult to function with independence in a country where the monarch's reign is supreme. In September 1999, authorities charged the editor-in-chief of the Times of Swaziland, Bheki Makhubu, with criminal defamation for publishing the profile of a fiancé of King Mswati, the eighth wife to be. It was also revealed that more Swazi journalists could be charged with defamation as the country's government prepares new legislation to curb what newspapers are allowed to publish. However, popular discontent is starting to increase and people are overtly asking for major reforms. Already Mswati has been forced to allow a few reforms and released political prisoners. A group of young political activists has also come to the fore to challenge the authority of Swaziland's Prime Minister and many of King Mswati III's older advisers.

CULTURE AND SOCIETY

A major annual cultural event in Swaziland is the Umhlanga or ''reed dance.'' This is an annual ceremony at which several hundred Swazi maidens dance bare-breasted before the King who then chooses one of them as his bride. The ceremony is attended by many Swazi who are filled with patriotic pride and a sense of national unity. It was at this ceremony that King Mswati chose his eighth bride. Journalist Makhubu published in the Times a true story about the background of the stunningly beautiful 18-year-old Liphovela Senteni, the chosen bride. Makhubu was then arrested for allegedly publishing defamatory articles exposing the difficult background of the king's latest fiancé. The traditional authorities considered it defamatory to call her a ''school dropout'' and a naughty girl who had been kicked out of two high schools.

Apart from the ''reed dance'' other events included the resignation of the former Swaziland International Olympics Committee (IOC) member, David Sibandze, as president of Swaziland's Commonwealth and Olympic Games Association (SOCGA), following his involvement in the Salt Lake City corruption scandal. Swaziland, with a population of less than one million, acknowledged that HIV/AIDS was a ''national disaster'' in this tiny mountain kingdom. It now has the third-highest rate of HIV infection in the world.

DIRECTORY

CENTRAL GOVERNMENT

Head of State

Prime Minister

Barnabas Sibusiso Dlamini, Office of the Prime Minister, PO Box 395, Mbane, Swaziland
PHONE: +268 4042251
FAX: +268 4043943
E-MAIL: bsdlmn@realnet.co.sz

Deputy Prime Minister

Sishayi S. Nxumalo, Office of the Deputy Prime Minister
PHONE: +268 4042723; 4044028
FAX: +268 4040084

Deputy Prime Minister

Arthur Khoza, Office of the Deputy Prime Minister
PHONE: +268 4042723; 4044028
FAX: +268 4040084

Ministers

Minister of Agriculture and Cooperatives
Roy Fanourakis, Ministry of Agriculture and Coopertives, PO Box 162, Mbabane, Swaziland
PHONE: +268 4042731
FAX: +268 4044700
E-MAIL: moac-hg@realnet.co.sz

Minister of Economic Planning and Development
Majozi Sithole, Ministry of Economic Planning and Development, PO Box 602, Mbane, Swaziland
PHONE: +268 4043765
FAX: +268 4042157
E-MAIL: ep_cso@realnet.co.sz

Minister of Education
Abednego Ntshangase, Ministry of Education, PO Box 39, Mbabane, Swaziland
PHONE: +268 4045641
FAX: +268 4043880
E-MAIL: amkhonza@realnet.co.sz

Minister of Enterprise and Employment
Lutfo Dlamini, Ministry of Enterprise and Employment, PO Box 451, Mbabane, Swaziland
PHONE: +268 4046657
FAX: +268 4044711

Minister of Finance
John Carmichael, Ministry of Finance, PO Box 443, Mbabane, Swaziland
PHONE: +268 4042141; 4048145
FAX: +268 4043187
E-MAIL: minfin@realnet.xo.sz

Minister of Foreign Affairs and Trade
Albert Shabangu, Ministry of Foreign Affairs and Trade, PO Box 518, Mbabane, Swaziland
PHONE: +268 4042661
FAX: +268 4042669
E-MAIL: ps_foreign@realnet.xo.sz

Minister of Health and Social Welfare
Phetsile Dlamini, Ministry of Health and Social Welfare, PO Box 5, Mbabane, Swaziland
PHONE: +268 4044016; 4042431; 4048903
FAX: +268 4042092
E-MAIL: minhealth@realnet.xo.sz

Minister of Home Affairs
Prince Sobandla, Ministry of Home Affairs, PO Box 432, Mbabane, Swaziland
PHONE: +268 4042941; 4045941
FAX: +268 4044303

Minister of Housing and Urban Development
Stella Lukhele, Ministry of Housing and Urban Development, PO Box 1832, Mbabane, Swaziland
PHONE: +268 4046035; 4046041
FAX: +268 4045290
E-MAIL: minhouse@realnet.xo.sz

Minister of Justice and Constitutional Affairs
Maweni Simelane, Ministry of Justice and Constitutional Affairs, PO Box 924, Mbabane, Swaziland
PHONE: +268 4046010
FAX: +268 4044796
E-MAIL: legalgpm@realnet.xo.sz

Minister of Natural Resources and Energy
Prince Guduza, Ministry of Natural Resources and Energy, PO Box 57, Mbabane, Swaziland
PHONE: +268 4046245
FAX: +268 4042436
E-MAIL: nergyswa@realnet.xo.sz

Minister of Public Service and Information
Magwagwa Mdluli, Ministry of Public Service and Information, PO Box 338, Mbabane, Swaziland
PHONE: +268 4043251
FAX: +268 4046438
E-MAIL: nhlanhla@realnet.xo.sz

Minister of Public Works and Transport
Peter Dlamini, Ministry of Public Works and Transport, PO Box 58, Mbabane, Swaziland
PHONE: +268 4042321
FAX: +268 4042364
E-MAIL: minister-mopwt@realnet.xo.sz

Minister of Tourism, Environment, and Communication
George Vilakati, Ministry of Tourism, Environment, and Communication, PO Box 2652, Mbabane, Swaziland
PHONE: +268 4042761
FAX: +268 4042774
E-MAIL: mintour@realnet.xo.sz

POLITICAL ORGANIZATIONS

People's United Democratic Movement (PUDEMO)

NAME: Mario Masuku

Swaziland Communist Party (SWACOPA)

NAME: Mphandlana Shongwe

Swaziland Liberation Front (FROLISA)

Convention for Full Democracy in Swaziland (COFUDESWA)

NAME: Sabelo Dlamini

Swaziland National Front (SWANAFRO)

Ngwane Socialist Revolutionary Party (NGWASOREP)

Swaziland Democratic Alliance

NAME: Jerry Nxumalo

Swaziland Federation of Trade Unions (SFTU)

NAME: Jan Sithole

DIPLOMATIC REPRESENTATION

Embassies in Swaziland

China
Warner Street, PO Box 56, Mbabane, Swaziland
TITLE: Ambassador
NAME: Enti Liu

Germany
2nd Floor Dhlanubeka House, PO Box 146, Mbabane, Swaziland
TITLE: Liaison Officer
NAME: Angelina Topfer

Mozambique
Princess Drive, PO Box 1212, Mbabane, Swaziland
TITLE: Ambassador
NAME: Antonio C. F. Sumbana

South Africa
The New Mall, PO Box 2597, Mbabane, Swaziland
TITLE: High Commissioner
NAME: Walter Louw

United Kingdom
Allister Miller Street, Mbabane Private Bag, Mbabane, Swaziland
TITLE: High Commissioner
NAME: John Doble

United States
Central Bank Building, Mbabane, Swaziland
TITLE: Ambassador
NAME: Alan McKee

JUDICIAL SYSTEM

High Court

FURTHER READING

Books

Gillis, D. Hugh. *The Kingdom of Swaziland: Studies in Political History.* Westport, Connecticut: Greenwood Press, 1999.

Hope, Kempe R. *AIDS and Development in Africa: A Social Science Perspective.* New York: Haworth Press, 1999.

Internet

Royal Kingdom of Swaziland. Available Online @ http://members.tripod.com/stewart_marshall/swaziland.htm'' (November 9, 1999).

Swaziland Internet Directory. Available Online @ http://www.directory.sz/internet/ (November 9, 1999).

Swaziland Page. Available Online @ http://www.pitt.edu/~tgsst10/swaziland.html (November 9, 1999).

Swaziland Today. Available Online @ http://www.realnet.co.sz/real/govt/sgt-nl.html (November 9, 1999).

University of Pennsylvania's Swaziland Page. Available Online @ http://www.sas.upenn.edu/African_Studies/Country_Specific/Swaziland.html (November 9, 1999).

SWAZILAND: STATISTICAL DATA

For sources and notes see "Sources of Statistics" in the front of each volume.

GEOGRAPHY

Geography (1)

Area:

Total: 17,360 sq km.

Land: 17,200 sq km.

Water: 160 sq km.

Area—comparative: slightly smaller than New Jersey.

Land boundaries:

Total: 535 km.

Border countries: Mozambique 105 km, South Africa 430 km.

Coastline: 0 km (landlocked).

Climate: varies from tropical to near temperate.

Terrain: mostly mountains and hills; some moderately sloping plains.

Natural resources: asbestos, coal, clay, cassiterite, hydropower, forests, small gold and diamond deposits, quarry stone, and talc.

Land use:

Arable land: 11%

Permanent crops: 0%

Permanent pastures: 62%

Forests and woodland: 7%

Other: 20% (1993 est.)

HUMAN FACTORS

Demographics (2A)

	1990	1995	1998	2000	2010	2020	2030	2040	2050
Population	839.8	909.3	966.5	1,004.1	1,202.2	1,434.0	1,791.7	2,352.9	3,059.3
Net migration rate (per 1,000 population)	NA	NA	NA	NA	NA	NA	NA	NA	NA
Births	NA	NA	NA	NA	NA	NA	NA	NA	NA
Deaths	NA	NA	NA	NA	NA	NA	NA	NA	NA
Life expectancy - males	39.2	38.8	37.3	36.4	36.6	40.9	54.3	67.7	72.0
Life expectancy - females	43.8	41.0	39.8	39.0	37.7	42.5	58.6	74.6	79.4
Birth rate (per 1,000)	43.7	41.6	41.0	40.7	39.4	37.7	35.4	31.9	27.9
Death rate (per 1,000)	20.1	20.6	21.4	22.1	22.6	18.6	9.7	4.2	3.3
Women of reproductive age (15-49 yrs.)	199.2	211.0	224.2	232.8	283.6	343.8	447.1	601.5	783.1
of which are currently married	NA	NA	NA	NA	NA	NA	NA	NA	NA
Fertility rate	6.2	6.1	6.0	5.9	5.4	4.9	4.4	3.9	3.5

Except as noted, values for vital statistics are in thousands; life expectancy is in years.

Infants and Malnutrition (5)

Under-5 mortality rate (1997) 94

% of infants with low birthweight (1990-97) 10

Births attended by skilled health staff % of total[a] . . .NA

% fully immunized (1995-97)

- TB 85
- DPT 82
- Polio 81
- Measles 82

Prevalence of child malnutrition under age 5 (1992-97)[b] . . . . .NA

Ethnic Division (6)

African 97%

European 3%

Religions (7)

Christian 60%

Indigenous beliefs 40%

Languages (8)

English (official, government business conducted in English), siSwati (official).

GOVERNMENT & LAW

Military Affairs (14B)

	1990	1991	1992	1993	1994	1995
Military expenditures						
Current dollars (mil.)	14	14	19	23	25	27
1995 constant dollars (mil.)	17	16	21	24	25	27
Armed forces (000)	3	3	3	3	3	3
Gross national product (GNP)						
Current dollars (mil.)	859	899	941	948	992	1,048
1995 constant dollars (mil.)	987	993	1,012	994	1,017	1,048
Central government expenditures (CGE)						
1995 constant dollars (mil.)	252	274	365	383	416	391
People (mil.)	.9	.9	.9	.9	.9	1.0
Military expenditure as % of GNP	1.7	1.6	2.1	2.5	2.5	2.6
Military expenditure as % of CGE	6.6	5.8	5.7	6.4	6.1	7.0
Military expenditure per capita (1995 $)	19	18	23	27	27	28
Armed forces per 1,000 people (soldiers)	3.5	3.4	3.3	3.3	3.2	3.1
GNP per capita (1995 $)	1,157	1,123	1,130	1,096	1,086	1,084
Arms imports[6]						
Current dollars (mil.)	0	0	0	0	0	0
1995 constant dollars (mil.)	0	0	0	0	0	0
Arms exports[6]						
Current dollars (mil.)	0	0	0	0	0	0
1995 constant dollars (mil.)	0	0	0	0	0	0
Total imports[7]						
Current dollars (mil.)	663	718	866	874	928	1,017
1995 constant dollars (mil.)	762	793	932	916	951	1,017
Total exports[7]						
Current dollars (mil.)	557	595	639	675	746	798
1995 constant dollars (mil.)	640	657	687	708	765	798
Arms as percent of total imports[8]	0	0	0	0	0	0
Arms as percent of total exports[8]	0	0	0	0	0	0

EDUCATION

Educational Attainment (10)

Age group (1986) .... 25+
Total population 221,672
Highest level attained (%)
No schooling 42.0
First level
Not completed 24.0
Completed 10.5
Entered second level
S-1 .... 13.2
S-2 .... 6.3
Postsecondary 3.3

Literacy Rates (11A)

In thousands and percent[1]	1990	1995	2000	2010
Illiterate population (15+ yrs.)	114	114	112	107
Literacy rate - total adult pop. (%)	72.1	76.7	80.4	86.2
Literacy rate - males (%)	73.8	78.0	81.3	86.7
Literacy rate - females (%)	70.6	75.6	79.6	85.8

GOVERNMENT & LAW

Political Parties (12)

The legislative branch is a bicameral Parliament (an advisory body, consists of the Senate (20 seats, 10 appointed by the House of Assembly and 10 appointed by the king; members serve five-year terms) and the House of Assembly (65 seats, 10 appointed by the king and 55 elected by secret, popular vote; members serve five-year terms). Popular parties are banned by the constitution.

Government Budget (13B)

Revenues .... $400 million
Expenditures .... $450 million
Capital expenditures .... $115 million

Data for FY96/97.

Crime (15)

Crime rate (for 1997)
Crimes reported 35,300
Total persons convicted 10,500
Crimes per 100,000 population 3,650
Persons responsible for offenses
Total number of suspects 19,600
Total number of female suspects 2,050
Total number of juvenile suspects 3,450

LABOR FORCE

Labor Force (16)

Total 135,000
Private sector about 70%
Public sector about 30%

Data for 1996.

Unemployment Rate (17)

22% (1995 est.)

PRODUCTION SECTOR

Electric Energy (18)

Capacity .... 130,000 kW (1995)
Production .... 407 million kWh (1995)
Consumption per capita .... 1,062 kWh (1995)

Imports 60% of its electricity from South Africa.

Transportation (19)

Highways:
total: 2,885 km
paved: 814 km
unpaved: 2,071 km (1994 est.)
Airports: 18 (1997 est.)
Airports—with paved runways:
total: 1
2,438 to 3,047 m: 1 (1997 est.)
Airports—with unpaved runways:
total: 17
914 to 1,523 m: 7
under 914 m: 10 (1997 est.)

Top Agricultural Products (20)

Sugarcane, cotton, maize, tobacco, rice, citrus, pineapples, corn, sorghum, peanuts; cattle, goats, sheep.

MANUFACTURING SECTOR

GDP & Manufacturing Summary (21)

Detailed value added figures are listed by both International Standard Industry Code (ISIC) and product title.

	1980	1985	1990	1994
GDP ($-1990 mil.)[1]	611	735	905	963
Per capita ($-1990)[1]	1,091	1,132	1,216	1,158
Manufacturing share (%) (current prices)[1]	19.7	14.1	28.0	29.0
Manufacturing				
Value added ($-1990 mil.)[1]	138	161	254	295
Industrial production index	100	118	131	153
Value added ($ mil.)	104	49	252	344
Gross output ($ mil.)	394	195	615	798
Employment (000)	11	12	20	22
Profitability (% of gross output)				
Intermediate input (%)	74	75	59	57
Wages and salaries inc. supplements (%)	11	12	11	11
Gross operating surplus	16	13	30	33
Productivity ($)				
Gross output per worker	36,595	16,708	30,711	35,882
Value added per worker	9,645	4,165	12,559	15,514
Average wage (inc. supplements)	3,907	2,002	3,409	3,895
Value added ($ mil.)				
311/2 Food products	39	26	74	91
313 Beverages	4	3	101	153
314 Tobacco products	—	—	—	—
321 Textiles	2	1	13	19
322 Wearing apparel	1	—	5	7
323 Leather and fur products	—	—	1	1
324 Footwear	—	—	—	—
331 Wood and wood products	6	1	3	4
332 Furniture and fixtures	2	1	3	4
341 Paper and paper products	24	8	27	35
342 Printing and publishing	7	3	13	18
351 Industrial chemicals	2	—	—	—
352 Other chemical products	4	—	1	1
353 Petroleum refineries	3	—	1	—
354 Miscellaneous petroleum and coal products	—	—	—	—
355 Rubber products	—	—	—	—
356 Plastic products	1	—	—	—
361 Pottery, china and earthenware	—	—	—	—
362 Glass and glass products	—	—	1	1
369 Other non-metal mineral products	1	—	2	3
371 Iron and steel	—	—	—	—
372 Non-ferrous metals	—	—	—	—
381 Metal products	4	3	5	7
382 Non-electrical machinery	1	—	1	1
383 Electrical machinery	1	—	1	1
384 Transport equipment	—	—	—	—
385 Professional and scientific equipment	—	—	—	—
390 Other manufacturing industries	—	—	—	—

FINANCE, ECONOMICS, & TRADE

Economic Indicators (22)

National product: GDP—purchasing power parity—$3.9 billion (1997 est.)

National product real growth rate: 3% (19976 est.)

National product per capita: $3,800 (1997 est.)

Inflation rate—consumer price index: 9.5% (1997)

Exchange Rates (24)

Exchange rates:	
Emalangeni (E) per US$1	
January 1998	4.9417
1997	4.5998
1996	4.2706
1995	3.6266
1994	3.5490
1993	3.2636

The Swazi emalangeni are at par with the South African rand.

Top Import Origins (25)

$1.1 billion (f.o.b., 1996) Data are for FY94/95.

Origins	%
South Africa	88
Japan	NA
United Kingdom	NA
United States	NA

NA stands for not available.

Top Export Destinations (26)

$893 million (f.o.b., 1996) Data are for 1994.

Destinations	%
South Africa	58
European Union	20
Mozambique	6

Economic Aid (27)

Recipient: ODA, $NA. NA stands for not available.

Import Export Commodities (28)

Import Commodities	Export Commodities
Motor vehicles	Soft drink concentrates
Machinery	Sugar
Transport equipment	Wood pulp
Foodstuffs	Cotton yarn
Petroleum products	
Chemicals	

Balance of Payments (23)

	1992	1993	1994	1995	1996
Exports of goods (f.o.b.)	638	685	783	958	887
Imports of goods (f.o.b.)	−780	−789	−831	−989	−964
Trade balance	−141	−104	−48	−31	−77
Services - debits	−281	−361	−308	−318	−299
Services - credits	296	248	230	256	251
Private transfers (net)	127	129	133	131	132
Government transfers (net)	—	—	−1	2	2
Overall balance	—	−88	6	39	9

SWEDEN

CAPITAL: Stockholm.

FLAG: The national flag, dating from 1569 and employing a blue and gold motif used as early as the mid-14th century, consists of a yellow cross with extended right horizontal on a blue field.

ANTHEM: *Du gamla, du fria, du fjallhöga nord (O Glorious Old Mountain-Crowned Land of the North).*

MONETARY UNIT: The krona (Kr) is a paper currency of 100 öre. There are coins of 50 öre and 1, 2, 5, and 10 kronor, and notes of 5, 10, 20, 50, 100, 500, and 1,000 kronor. Kr1 = $0.1277 (or $1 = Kr7.828).

WEIGHTS AND MEASURES: The metric system is the legal standard, but some old local measures are still in use, notably the Swedish mile (10 kilometers).

HOLIDAYS: New Year's Day, 1 January; Epiphany, 6 January; Labor Day, 1 May; Midsummer Day, Saturday nearest 24 June; All Saints' Day, 5 November; Christmas, 25–26 December. Movable religious holidays include Good Friday, Easter Monday, Ascension, Whitmonday.

TIME: 1 PM = noon GMT.

LOCATION AND SIZE: Fourth in size among the countries of Europe, Sweden is the largest of the Scandinavian countries, with about 15% of its total area situated north of the Arctic Circle. Sweden has a total area of 449,964 square kilometers (173,732 square miles), slightly smaller than the state of California. Its total boundary length is 5,423 kilometers (3,370 miles).

Sweden's capital city, Stockholm, is located on the Baltic Sea coast.

CLIMATE: Because of ocean winds, Sweden has higher temperatures than its northerly latitude would suggest. Stockholm averages 3°C (26°F) in February and 18°C (64°F) in July. The climates of northern and southern Sweden differ widely.

Annual rainfall averages 61 centimeters (24 inches) and is heaviest in the southwest and along the frontier between central Sweden and Norway. There is much snowfall, and in the north snow remains on the ground for about half the year.

Kingdom of Sweden
Konungariket Sverige

INTRODUCTORY SURVEY

RECENT HISTORY

Sweden remained neutral in both world wars. During World War II (1939–45), however, Sweden served as a safe place for refugees from the Nazis and let the Danish resistance movement operate on its soil. After the war, Sweden did not join the North Atlantic Treaty Organization, as did its Scandinavian neighbors Norway and Denmark. It did, however, become a member of the United Nations in 1946. In 1953, Sweden joined with Denmark, Norway and Iceland (and later, Finland) to form the Nordic Council.

Carl XVI Gustaf has been king since 1973. In September 1976, a coalition of three non-Socialist parties won a majority in parliamentary elections, ending 44 years of almost uninterrupted Social Democratic rule that had established a modern welfare state. The country's economic situation worsened, however, and the Social Democrats were returned to power in the elections of September 1982. Prime Minister Olof Palme, leader of the Social Democratic Party since 1969, was assassinated for an unknown reason in February 1986.

The environment and nuclear energy were major political issues in the 1980s. Major concerns of the 1990s were conflicts over immigration policies, the economy, and Sweden's relationship to the European Community (EC). In May 1993, the Riksdag changed Sweden's long-standing foreign policy of neutrality. In the future, neutrality would only be followed in time of war. The Riksdag also opened up the possibility of Sweden's participation

in defense alliances. In 1994 Swedes voted to join the European Union and the country officially became a member at the beginning of 1995.

GOVERNMENT

Sweden is a constitutional monarchy. Legislative authority rests in the parliament (*Riksdag*). The monarch performs only ceremonial duties and must belong to the Lutheran Church. In 1980, female descendants were granted the right to the throne. The real chief executive is the prime minister, who is proposed by the speaker of the Riksdag and confirmed by the parliamentary parties. The prime minister appoints a cabinet consisting of 18–20 members.

The Riksdag is a single-chamber body of 349 members serving three-year terms. All members of the Riksdag are directly elected by universal suffrage at age 18. The Riksdag has direct control of the Bank of Sweden and the National Debt Office.

The Riksdag elects one or more ombudsmen (four in 1981) who make sure the courts and public officials follow laws properly. The ombudsmen are concerned especially with protecting the civil rights of individual citizens and of religious and other groups. They may warn or prosecute offenders, although prosecutions are rare.

Sweden is divided into 24 counties and about 278 municipalities, each with an elected council.

Judiciary

Ordinary criminal and civil cases are tried in a local court (*tingsrätt*), consisting of a judge and a panel of citizens appointed by the municipal council. Above these local courts are six courts of appeal (*hovrätter*). The highest court is the Supreme Court (*Högsta Domstolen*), with 24 justices. Special cases are heard by the Supreme Administrative Court and other courts. Capital punishment, last used in 1910, is forbidden by the constitution.

Political Parties

The constitution requires that a party must gain at least 4% of the national popular vote, or 12% in a constituency, to be represented in the Riksdag. Except for a brief period in 1936, the Social Democratic Labor Party was in power almost continuously from 1932 to 1976, either alone or in a coalition with other parties. They lost their parliamentary majority in the elections of September 1976 but returned to power in 1982.

The 1988 election marked a turning point in Sweden with the decline of the Social Democrats and the growth of the Moderates and Liberals. For the first time in 70 years, a new party gained representation in the Riksdag—the Environment Party (MpG) which obtained 20 seats.

In 1994, the Social Democrats held 162 seats; Moderate Party, 80; Center Party, 27; Liberals, 26; Left Party, 22; Greens, 18; and Christian Democrats, 14.

DEFENSE

Because Sweden is neutral in times of war and, therefore, gets no help from any other country, it must have a strong, modern, and independent defense force. The 1996 budget allocated $6.3 billion for defense.

Swedish military defense is based on general draft of all male citizens between the ages of 18 and 47. In wartime a force of 700,000 reservists can be mobilized within 72 hours. The coastal defense is under the command of the Royal Swedish Navy. Regular armed forces in 1995 totaled 62,500 (42,100 draftees). The army had 33,900 (half draftees). The navy had 10,000 men on active duty (4,100 draftees). The air force, with 9,500 regulars (4,100 draftees), had a total of 463 combat aircraft. Reserves in voluntary defense organizations totaled 35,000.

As part of the civil defense program, nuclear-resistant shelters were built over a 10-year period in the large urban areas. The shelters, completed in 1970, are used as garages in peacetime and can hold 6.3 million people in a national emergency.

ECONOMIC AFFAIRS

Sweden is a highly industrialized country. The shift from agriculture to industry began in the 1930s and developed rapidly after World War II (1939–45).

Swedish industry makes high-quality goods and specialized products—ball bearings, high-grade steel, machine tools, glassware—that are in demand worldwide. There is close contact between trade, industry, and finance. Factories are spreading to rural districts. Some natural resources are in good supply, particularly lumber, iron ore, and waterpower. However, Sweden's lack of oil and coal resources makes it dependent on imports for energy production, despite its abundant waterpower.

Swedish living standards and purchasing power are among the highest in the world. However, inflation has been a chronic problem since the early 1970s, with the annual rise in consumer prices peaking at 13.7% in 1980. The rate of price increase declined after that to about 2% each year in the early 1990s. Unemployment shot up from 1.7% in 1990 to 14% in 1994 but has fallen since.

Public Finance

The financial year extends from 1 July to 30 June. Estimates are prepared in the autumn by the Ministry of the Budget and examined by the Riksdag early the following year. The budget contains two sections: an operating budget and a capital budget, the latter generally representing investments in state enterprises. The policy of running a surplus on the budget in boom years and a deficit in depression was used in the period between the two world wars and has been continued as a way of combating inflation. From 1982 to 1989, the budget balance improved from a deficit equivalent to about 13% of GDP to a surplus of nearly 2% of GDP. In 1990, however, a deficit reappeared, equivalent to 1.2% of GDP. In 1991 and 1992 the budget deficits widened to 4.3% and 9.6% of GDP, respectively. The deficit increased to 12.3% of GDP in 1993, before beginning a sharp decline due to austerity measures, put in place by the Social Democrats. By 1994, the deficit had begun to turn around. In 1995, the last year for which the U.S. Central Intelligence Agency has figures, government revenues totaled approximately $109.4 billion and expenditures $146.1 billion. External debt totaled $66.5 billion, approximately 47% of which was financed abroad.

By 1996, the Swedish deficit had been brought largely under control, equaling 2% of GDP. Forecasts were for 1.5% in 1997 and a return to surplus beginning in 1998.

Income

In 1998, Sweden's gross domestic product (GDP) was $175 billion, or about $19,700 per person.

Industry

The basic resources for industrial development are forests, iron ore, and waterpower. Industry accounts for 27% of the gross domestic product. Forest products, machinery, and motor vehicles make up about 60% of the total export value.

Since the end of World War II, emphasis has shifted from production of consumer goods to the manufacture of export items. Swedish-made ships, airplanes, and automobiles are considered outstanding in quality. In 1995, 387,659 automobiles and 91,151 trucks were manufactured. Sweden's motor vehicle producers are Volvo and Saab-Scania.

As exports, transport equipment and iron and steel have become less important, while exports of machinery, chemicals, and paper have been growing in value.

Banking and Finance

The Central Bank of Sweden (Sveriges Riksbank), founded in 1656, is the oldest bank in the world. It is the bank of issue, regulates the value of Swedish currency, controls foreign exchange, and sets the discount rate. The largest commercial bank is the Skandinaviska Enskilda Banken. In the early 1990s, Swedish banks suffered severe losses, the government was forced to intervene and support 2 of the 5 largest commercial banks, Nordbanken and Gota Bank, by taking them over and eventually merging them, and the savings bank Forsta Sparbanken. By the end of 1996, Swedish banks showed improved results, with reduced credit losses and a stricter control of costs since the banking crises set in at the beginning of the 1990s. The smaller banks serve provincial interests. Deposit accounts at various lengths of call are used for short-term credit by industry and trade. At the end of 1994 there were 92 savings banks with 976 offices and 23 commercial banks, including several foreign-owned banks, with 1,711 offices. The deregulation of financial markets has paved the way for foreign banks to open offices in Sweden. In 1997, Sweden's banking sector saw a series of mergers and acquisitions as Svenska Handelsbanken, the nation's largest bank, acquired the country's largest mortgage lender, Stadshypotek. Swedbank and Föreningsbanken merged, creating the second largest bank, and Den Donke Bank, based in Denmark, made the first incursion by a foreign bank into the Swedish retail sector when it purchased Ostgöta Enskilda Bank.

Mortgage banks of various types meet the needs of property owners, home builders, farmers, and shipbuilders. Credit also is extended by some 500 local rural credit societies and by about an equal number of agricultural cooperatives. There are four semi-governmental credit concerns, organized as business companies and created in cooperation with private commercial banks to facilitate long-term lending to agriculture, industry, small industry, and exports. Although the Riksbank's note issue is not tied to its gold reserves, there is an adjustable legal limit.

The money supply, as measured by M1, totaled Kr740.39 billion at the end of the first quarter of 1996. Foreign reserves, excluding gold, totaled $22,373 million at the end of 1995. The krona was allowed to float against foreign currency beginning on 19 November 1992, and soon devalued 20% against the currencies of Western Europe and the U.S. dollar.

The Riksbank lends money to the commercial banks and other credit associations against securities. Traditionally, the Swedish people have preferred to save by placing money in these banks rather than by direct investment, although this seemed likely to change if the Swedish pension reforms of 1997 were fully enacted, allowing workers to decide how to invest a portion of their retirement reserves. In 1992, 118 Swedish companies and 10 foreign companies were listed on the Stockholm Stock Exchange (Stockholms Fondbörs), which was computerized that year. In 1997, the Stockholm Stock Exchange entered into a joint equity trading union with the Danish bourse, creating the first trans-national link of its kind in Europe. The joint equities market became Europe's sixth largest.

Profits from the sale of securities are taxable provided they have been owned for less than five years. The capital gain is wholly taxable for securities held less than two years, but only 40% of the gain is taxable if the shares have been held more than two years. For machinery and equipment a minimum write-off period of three years is prescribed. The 1985 deregulation of the credit market included the removal of ceilings on lending banks, finance houses, and housing credit institutions and had the effect of diminishing part of Sweden's "gray market": direct contact between companies and private individuals with money for loans. Stockbroking is authorized by the Bank Inspection Board.

Economic Development

Between 1946 and 1953, the Swedish economy was dominated by expansion. Thereafter, although production continued to increase (at a lessened rate), inflation was a matter of concern. Domestic investment has remained at about the same level as in 1939, but a larger share has come from public investment. Expansion of output slowed down during the international oil crisis and recession of 1974–75, largely as a result of a weakening of foreign demand for Swedish products, but employment remained high. Thus far, the economy has managed to contain inflationary trends within reasonable limits. Although some industries (the railways, iron-ore mines, etc.) have been nationalized for a long time, private concerns carry on most of Sweden's industry, in terms of both number of workers and value of output.

During periods of unemployment such as the world recession of 1980–81, the central government and the municipalities have expended funds to provide additional employment and to keep the unemployment rate relatively low. The jobless have been put to work building dwellings and highways, extending reforestation work, and constructing water and sewer installations, harbors, lighthouses, railroads, defense projects, and telecommunications facilities. Although the government resorted to stockpiling industrial goods to combat the economic slowdown in the mid-1970s, the cost was considered too high, and the policy was not repeated during the recession of the early 1980s. More recently, the emphasis has been on cutting costs and restraining inflation to make Swedish goods more competitive in the international marketplace.

Regional development has been fostered by the use of investment funds (a tax device permitting enterprises to set aside tax-free reserves during boom years to be used for investment during recessions), relief works, and government lending to small-scale industry. A national program for regional development was introduced in 1972 to develop services and job opportunities in provinces that have lagged behind in industrial development. Projects in northern Sweden benefited most from this program.

In 1991, the government announced a plan to privatize 35 wholly or partially state-owned firms with annual turnovers totaling Kr150 billion. This program was delayed by the economic recession, however. A 10-year, Kr110 billion program of infrastructure investment was announced in 1994. More than 90% of the money would be spent on the road and rail networks, and a bridge that would link Malmö with Copenhagen.

Official development assistance totaled $1.769 billion in 1993.

SOCIAL WELFARE

Sweden has been called the model welfare state; every citizen is guaranteed a minimum subsistence income and medical care. Basic benefits are often increased by cost-of-living supplements.

Old-age pensions are paid to everyone 65 years of age or older, but an earlier retirement is possible, with a reduction in pension benefits. The system also contains provisions for pensions to persons totally disabled before retirement age and for family pensions (widows and orphans). Unemployment insurance is the only kind of voluntary social insurance in Sweden. Administered by the trade unions, it provides benefits according to salary to those who voluntarily enroll. More than half of all employees are covered.

Compulsory health service was introduced in 1955. Hospital care is free for up to two years. Medical services and medicines are provided at very low cost or, in some cases, without charge. Costs of pregnancy and childbirth are covered by health insurance. Since the beginning of 1955, workers' compensation for injuries on the job has been coordinated with the national health service scheme. Public assistance is provided for blind or disabled persons confined to their homes and to people who are in sanitariums, special hospitals, or charitable institutions.

The social services help meet the costs of raising children through monthly family allowances for each child under age 16. A housing allowance is also paid to families with children under 16 to help them have modern, roomy dwellings. For 360 days, a parental benefit is paid upon the birth or adoption of a child to the parent who stays home and takes care of the child. Schoolchildren receive free textbooks, and about 75% of public-school children are given free meals.

Women are less likely to get higher paying jobs, and often receive less pay for equal work. The percentage of women in the work force fell in 1993, for the first time since World War II, to 75.9%.

Healthcare

The national health insurance system, financed by the state and employer contributions, was established in January 1955 and covers all Swedish citizens and alien residents.

In 1990 there were an estimated 28,000 physicians, more than one for every 350 inhabitants. Swedish hospitals, well known for their high standards, had 35,990 short-term beds (hospital) and 85,972 long-term beds (nursing/old-age homes) in 1991.

Many health problems are related to environment and lifestyle (including tobacco smoking, alcohol consumption, and overeating). The most common serious diseases are cardiovascular conditions and cancer. In 1996, there were 1,371 cases of AIDS.

In 1990–95, average life expectancy in Sweden was about 79 years.

Housing

The total number of dwellings was over 4 million in 1990, most of them having two rooms and a kitchen. Construction of new dwellings totaled 66,886 in 1991. Most houses are built by private contractors, but more than half of new housing is designed, planned, and financed by nonprofit organizations and cooperatives.

To ease the housing shortage, the government subsidizes new construction and reconditioning, helps various groups to get better housing, and gives credit at lower interest rates than those in the open market. In 1987, these interest subsidies totaled $1.6 billion; and housing allowances for low-income families were $1 billion.

EDUCATION

Practically the entire adult Swedish population is literate. Education is free and compulsory from age 7 to 17. All students receive the same course of instruction for six years. Beginning in the seventh year, students may choose between a classical and a vocational course. About 80% of all students then enter *gymnasium* (senior high school) or continuation schools.

The senior high school specializes in classical or modern languages or science. After the three-year course, students may take a final graduating examination. The continuation schools offer a two-year curriculum that is more practical and specialized than that of the senior high school and leads more quickly to the practice of a trade. In 1993 there were 600,392 students at the primary level, 607,219 at the secondary level, and 234,466 students enrolled in the universities.

More than 25% of secondary-school graduates attend college or university. Sweden's six universities, all largely financed by the state, are at Uppsala, Lund, Stockholm (1877), Göteborg, Umea, and Linköping. There are also more than two dozen specialized schools and institutions at the university level for such subjects as medicine, dentistry, pharmacology, music, economics, commerce, technology, veterinary science, agriculture, and forestry. Tuition is free, except for some special courses.

Sweden has an active adult education movement in which some 3 million persons participate each year.

1999 KEY EVENTS TIMELINE

January

- American automobile manufacturer Ford Motor Company acquires Volvo car division for $6.45 billion.
- Ostersund officials claim that their town was cheated out of the bid for the 2002 Olympic Games.

February

- Government signs UN treaty to imprison war criminals convicted by Yugoslavia war crimes tribunal.

March

- Government approves compensation to victims of forced sterilization.

April

- Government rejects NATO's request for troops to handle Kosovo refugee crisis in Albania.

May

- UN selects former prime minister Carl Bildt as special envoy to the Balkans.
- Svenska Handelsbanken bids $199 million to purchase Norwegian Bergensbanken.

June

- Film star Greta Garbo's ashes moved to a cemetery outside Stockholm.

July

- Thirteen-hundred people rescued from burning ferry off Swedish coast.

August

- Oresund Fixed Link joins Sweden with Denmark and the rest of Europe.
- Ericsson, the world's third largest mobile phone manufacturer, moves headquarters to London.
- Volvo buys rival truck maker Scania for $7.4 billion.

September

- Sweden and Norway reach final agreement to merge state telecommunications providers Telia and Telecor.

October

- Svenska Handelsbanken makes $3.1 billion bid for Christiania, Norway's second largest bank.
- Volvo partners with Japan's Mitsubishi Motors.

November

- Sweden's prime minister admits his country will inevitably join the euro at some point in the future.
- In a move against "moonshine" production and smuggling, Sweden announces plans to ease its strict alcohol laws, so people can buy wine on Saturdays.

December

- Sweden offers profit-related pay to workers at state-owned companies.
- The government and the Lutheran Church agree to split as of January 1, 2000, ending a relationship that has existed for over 500 years. The agreement to split will change the way the Lutheran Church is funded and the way bishops are elected.
- The Inter-Parliamentary Union in Geneva, Switzerland, reports that Swedish women hold 40 percent of the seats in Parliament, earning Sweden the top spot in gender equity ranking in government.

ANALYSIS OF EVENTS: 1999

BUSINESS AND THE ECONOMY

Merger was the key word in 1999 as several Swedish companies sought to become more competitive in the global marketplace by joining with businesses nationally and abroad.

Described as of the largest corporate events in the history of Sweden and Norway, the countries reached a final agreement in September to merge their state telecommunications providers, Telia and Telecor. The new company, worth an estimated $18.6 billion and employing 51,000, will be headquartered in Stockholm, with international operations controlled in Oslo.

The automobile industry continued its global expansion in January with Ford Motor Company's acquisition of Volvo's car division for $6.45 billion. The merger brought little response from Volvo managers, who, in 1993, had protested a possible merger with Renault of France. Volvo rejected Fiat's bid of approximately $7 billion, fearing the Italian automaker would attempt to exert more control over its manufacturing operations. Ford announced that it would continue to sell the familiar square-shaped cars bearing the Volvo label, and would continue to manufacture the cars in Sweden.

Volvo retained its truck and bus divisions and, in January, acquired a 13 percent share of rival Swedish truck maker Scania, for $666.5 million. Negotiations to purchase Scania stalled in February, but Volvo reached an agreement in August to acquire the company for $7.4 billion.

In October Volvo announced that it would purchase a 5 percent share of Mitsubishi Motors for $270 million. The Japanese motor company, in turn, agreed to buy up to 5 percent of Volvo during the next two years. According to Mitsubishi officials, Mitsubishi, Volvo, and Scania's combined truck and bus production and sales would surpass those of DaimlerChrysler, currently the world's largest truck and bus maker.

Stiffer competition in Sweden and throughout Europe led some Swedish banking and telecommunications businesses to expand beyond the country's borders in 1999. Scandinavia's largest bank, Svenska Handelsbanken, bid $199 million for Bergensbanken, a regional Norwegian bank, in May. Handelsbanken later announced that it would top Finnish-Swedish MeritaNordbanken's $3.1 billion bid for Christiania, Norway's second largest bank. Ericsson, the world's third largest mobile phone manufacturer, moved its headquarters and corporate finance department to London in August to be closer to the European market.

GOVERNMENT AND POLITICS

In 1997 an investigation revealed that the Swedish government had participated in a 40-year program to sterilize citizens considered racially or socially inferior. In June the government attempted to make reparations to the 63,000 victims of the forced sterilization campaign by offering each victim approximately $22,000.

In February Sweden became the fourth country to sign the United Nations treaty to imprison war criminals convicted by The Hague's Yugoslavia war crimes tribunal. The government refused NATO's request for troops to handle the Kosovo refugee crisis in Albania, claiming that it could only send troops for campaigns authorized by the United Nations and the Organization for Security and Cooperation in Europe. The government offered to send logistics staff to help resolve issues, and, in April, agreed to offer asylum to as many as 5,000 Kosovar refugees from Macedonia.

The United Nations selected former Swedish prime minister Carl Bildt as its special envoy to the Balkans in May. Bildt stepped down from his 13-year post as Moderate Party leader in September, and was succeeded by Bo Lundgren.

CULTURE AND SOCIETY

The town of Ostersund received international media attention in January when Swedish officials accused International Olympic Committee delegates of seeking cash and other favors in return for favorable votes in the bid for the 2002 Olympic Games. Ostersund lost the bid to Salt Lake City, Utah, which later became embroiled in a bribe solicitation scandal.

In June family and friends of Swedish film star Greta Garbo, who died in New York City in 1990, gathered for a memorial service at a cemetery outside Stockholm, where her ashes were laid to rest.

Thirteen-hundred people were rescued from a blazing ferry off the west coast of Sweden in July. The fire broke out in the engine room of the Princess Ragnhild, which was en route from Kiel, Germany, to Oslo, Norway. Passengers were transferred from inflatable lifeboats to fishing boats, then to seven helicopters which carried them to safety.

On August 14 Crown Princess Victoria of Sweden and Crown Prince Frederik of Denmark met on the Oresund Bridge to celebrate the completion of a 9.5-mile, $2 billion bridge and tunnel link joining the two countries. Construction of the Oresund Fixed Link, which connects the cities of Malmö and Copenhagen, began in October 1995. The link is scheduled to open to traffic on July 1, 2000.

DIRECTORY

CENTRAL GOVERNMENT

Head of State

King
Carl XVI Gustaf, Monarch, Information and Press Department, Rosenbad 4, 103 33 Stockholm, Sweden
PHONE: +46 (08) 4051000
FAX: +46 (08) 7231171

Prime Minister
Goran Persson

Ministers

Minister of Agriculture, Food, and Fisheries
Margareta Winberg, Ministry of Agriculture, Food, and Fisheries, Vasagatan 8 10, 103 33 Stockholm, Sweden
PHONE: +46 (08) 4051000
FAX: +46 (08) 206496

Minister of Education and Science
Thomas Ostros, Ministry of Education and Science, Drottningg 16, 103 33 Stockholm, Sweden
PHONE: +46 (08) 4051000
FAX: +46 (08) 7231192

Minister of Culture
Marita Ulvskog, Ministry of Culture, Drottningg 16, 103 33 Stockholm, Sweden
PHONE: +46 (08) 4051000

Minister of Defense
Bjorn Von Sydow, Ministry of Defense, Jakobsgatan 9, 103 33 Stockholm, Sweden
PHONE: +46 (08) 4051000
FAX: +46 (08) 7231189

Minister of Environment
Kjell Larsson, Ministry of Environment, Tegelbacken 2, 103 33 Stockholm, Sweden
PHONE: +46 (08) 4051000

Minister of Finance
Bo Ringholm, Ministry of Finance, Drottning 21, 103 33 Stockholm, Sweden
PHONE: +46 (08) 4051000

Minister of Foreign Affairs
Anna Lindh, Ministry of Foreign Affairs, Gustav Adolfs Torg 1, 103 39 Stockholm, Sweden
PHONE: +46 (08) 4051000
FAX: +46 (08) 7231176

Minister of Justice
Laila Freivalds, Ministry of Justice, Rosenbad 4, 103 33 Stockholm
PHONE: +46 (08) 4051000
FAX: +46 (08) 202734

Minister of Health and Social Affairs
Lars Engqvist, Ministry of Health and Social Affairs, Regeringsg 30 32, 103 33 Stockholm, Sweden
PHONE: +46 (08) 4051000
FAX: +46 (08) 7231191

Minister of Industry and Commerce
Bjorn Rosengren, Ministry of Industry and Commerce, Jakobsgatan 26, 103 33 Stockholm, Sweden
PHONE: +46 (08) 4051000
FAX: +46 08 4113616

POLITICAL ORGANIZATIONS

Moderata Samlingspartiet (Moderate Party)

Moderaterna, Box 1243, Schönfeldts grönd 2, 111 82 Stockholm, Sweden
PHONE: +46 (08) 6768000
E-MAIL: info@moderat.se
NAME: Bo Lundgren

Folkpartiet Liberalerna-FP (Liberal Party)

Folkpartiets riksorganisation, Box 6508, Drottninggatan 97, 113 83 Stockholm, Sweden
PHONE: +46 (08) 50911600
NAME: Lars Leijonborg

Centerpartiet (Centre Party)

Centerpartiets riksorganisation, Box 22107, 104 22 Stockholm, Sweden
PHONE: +46 (08) 6173800
E-MAIL: centerpartiet@centerpartiet.se
NAME: Lennart Daléus

Kristdemokratern-KD (Christian Democratic Party)

Kristdemokraterna, Box 451, 101 29 Stockholm, Sweden
PHONE: +46 (08) 243825; 219751
E-MAIL: brev.till@kristdemokrat.se
NAME: Alf Svensson

Miljöpartiet de Grona-MP (Green Party)

Miljöpartiet de gröna, Box 16069, Gamlastan, 103 22 Stockholm, Sweden
PHONE: +46 (08) 54522450
NAME: Lotta Nilsson Hedström; Birger Schlaug

Socialdemokratiska Arbetarepartiet (Social Democratic Party)

Socialdemokratiska partistyrelsen, Sveavägen 68, 105 60 Stockholm, Sweden
PHONE: +46 (08) 7002600
E-MAIL: socialdemokraterna@riksdagen.se
NAME: Gööran Persson

Vänsterpartiet (Left Party)

Vänsterpartiet, Box 12660, Kungsgatan 84, 112 93 Stockholm, Sweden
PHONE: +46 (08) 6540820
FAX: +46 (08) 6532385
NAME: Gudrun Schyman

DIPLOMATIC REPRESENTATION

Embassies in Sweden

Australia
Sergels Torg 12, P.O. Box 7003, 103 86 Stockholm, Sweden
PHONE: +46 (08) 6132900
E-MAIL: reception@austemb.se
TITLE: Ambassador
NAME: Stephen Brady

Austria
Kommendorsgt. 35, 5 tr, 114 58 Stockholm, Sweden
PHONE: +46 (08) 6651770
FAX: +46 (08) 6626928
E-MAIL: austria@algonet.se

Canada
Tegelbacken 4, 7th Floor, P.O. Box 16129, 10323 Stockholm, Sweden
PHONE: +46 (08) 4533000

Ecuador
Engelbrektsgatan 13, P.O. Box 260 95, Stockholm, Sweden
PHONE: +46 (08) 6796043; 6796070
FAX: +46 (08) 6115593
E-MAIL: suecia@embajada ecuador.se

Ethiopian
Erik Dahlbergsallén 15, Box 10148, 100 55 Stockholm, Sweden
PHONE: +46 (08) 6656030
FAX: +46 (08) 6608177
E-MAIL: ethio.embassy@swipnet.se

France
Kommendorsgt. 13, P.O. Box 10241, 100 55 Stockholm, Sweden
PHONE: +46 (08) 4595300
FAX: +46 (08) 4595313

Germany
Skarpogt. 9, P.O. Box 27832, 115 93 Stockholm, Sweden
PHONE: +46 (08) 6701500
FAX: +46 (08) 6615294

Greece
Riddargt. 60, 114 57 Stockholm, Sweden
PHONE: +46 (08) 6608860
FAX: +46 (08) 6605470
TITLE: Nicolaos Ladopoulos

India
Adolf Fredriks Kyrkogata 12, Box 1340, 111 83 Stockholm, Sweden
PHONE: +46 (08) 4113212

Ireland
Ostermalmsgt. 97, P.O. Box 10326, 100 55 Stockholm, Sweden
PHONE: +46 (08) 6618005
FAX: +46 (08) 6601353
E-MAIL: irish.embassy@swipnet.se
TITLE: Ambassador
NAME: M. Burke

Israel
Box 14006, S 104 40 Stockholm, Sweden
PHONE: +46 (08) 6613309
FAX: +46 (08) 6625301

Norway
Strandvagen 113, P.O. Box 27829, 115 93 Stockholm, Sweden
PHONE: +46 (08) 6656340
FAX: +46 (08) 7829899
E-MAIL: info@norgeambassad.se
NAME: Ketil Borde

Panama
Ostermalmsgt. 59, P.O. Box 26146, 100 41 Stockholm, Sweden
PHONE: +46 (08) 6626535
FAX: +46 (08) 6630407

South Africa
Linnégatan 76, 115 23, Stockholm, Sweden
FAX: +46 (08) 6607136
E-MAIL: saemb.swe@telia.com
TITLE: Ambassador
NAME: R.S. Suttner

Spain
Djurgardsvagen 21, 115 21 Stockholm, Sweden
PHONE: +46 (08) 6679430
FAX: +46 (08) 6637965

United Kingdom
Skarpog. 6-8, P.O. Box 27819, 115 93 Stockholm, Sweden
PHONE: +46 (08) 6719000
FAX: +46 (08) 6629989

United States
Strandvagen 101, 115 89, Stockholm, Sweden
PHONE: +46 (08) 7835300
FAX: +46 (08) 6611964

Venezuela
Engelbrektsgatan 35 B, S 114 31 Stockholm, Sweden
PHONE: +46 (08) 4110996; 103176
FAX: +46 (08) 213100
E-MAIL: venezuela.embassy@mbox300.swipnet.se

JUDICIAL SYSTEM

The Supreme Court

Courts of Appeal

District Courts

Parliamentary Ombudsmen

FURTHER READING

Articles

''1,300 Saved as Ferry Burns Off Sweden.'' *New York Times*, 9 July 1999, p. A7.

Andrews, Edmund L. ''Ford-Volvo: A Deal for All Sweden.'' *New York Times*, 30 January 1999, p. C1.

Bradsher, Keith. ''Ford Seen in Deal to Pay $6 Billion for Volvo Car Unit.'' *New York Times*, 28 January 1999, p. A1.

''Carl Bildt, A Good Balkan Swede.'' *The Economist* (June 12, 1999): 50.

Cowell, Alan. ''Swedish Empire, Under Siege, Scrambles to Adapt.'' *New York Times*, 3 April 1999, p. C1.

Deutsch, Claudia H. ''Volvo to Buy Another Swedish Truck Maker for $7.4 Billion.'' *New York Times*, 7 August 1999, p. C2.

Drozdiak, William. ''Swedish Town Says It Was Cheated Out of Olympics: Official Bribes Were Solicited, Rejected.'' *Washington Post*, 16 January 1999, p. A1.

Hoge, Warren. ''Link Finally Anchors Sweden to the Rest of Europe.'' *New York Times*, 5 September 1999.

Neilan, Terence. ''World News Briefs.'' *New York Times*, 24 February 1999, p. A6.

''Pay-Out for Sterilization Victims.'' *The Independent*, 29 June 1999, p. 5.

''Sweden's New Right-Winger.'' *The Economist* (September 11, 1999): 55.

''Swedish Bank Moves to Expand into Norway.'' *New York Times*, 4 May 1999, p. C4.

''Swedish Bank Set to Top Bid for Christiania.'' *Financial Times*, 2 October 1999, p. 8.

SWEDEN: STATISTICAL DATA

For sources and notes see "Sources of Statistics" in the front of each volume.

GEOGRAPHY

Geography (1)

Area:

Total: 449,964 sq km.

Land: 410,928 sq km.

Water: 39,036 sq km.

Area—comparative: slightly larger than California.

Land boundaries:

Total: 2,205 km.

Border countries: Finland 586 km, Norway 1,619 km.

Coastline: 3,218 km.

Climate: temperate in south with cold, cloudy winters and cool, partly cloudy summers; subarctic in north.

Terrain: mostly flat or gently rolling lowlands; mountains in west.

Natural resources: zinc, iron ore, lead, copper, silver, timber, uranium, hydropower potential.

Land use:

Arable land: 7%

Permanent crops: 0%

Permanent pastures: 1%

Forests and woodland: 68%

Other: 24% (1993 est.)

HUMAN FACTORS

Demographics (2A)

	1990	1995	1998	2000	2010	2020	2030	2040	2050
Population	8,558.8	8,827.7	8,886.7	8,938.6	9,114.6	9,197.4	9,019.9	8,588.8	8,051.7
Net migration rate (per 1,000 population)	NA	NA	NA	NA	NA	NA	NA	NA	NA
Births	NA	NA	NA	NA	NA	NA	NA	NA	NA
Deaths	95.2	NA	NA	NA	NA	NA	NA	NA	NA
Life expectancy - males	74.8	76.2	76.5	76.7	77.5	78.3	78.9	79.4	79.8
Life expectancy - females	80.5	81.7	82.0	82.2	83.2	84.0	84.7	85.3	85.8
Birth rate (per 1,000)	14.5	11.7	11.7	12.3	10.1	9.9	8.8	7.8	7.4
Death rate (per 1,000)	11.1	10.8	10.8	10.8	10.7	11.0	12.6	13.5	14.5
Women of reproductive age (15-49 yrs.)	2,048.3	2,044.2	2,002.4	1,984.5	2,018.6	1,960.5	1,869.7	1,747.5	1,553.4
of which are currently married	915.5	NA	NA	NA	NA	NA	NA	NA	NA
Fertility rate	2.1	1.7	1.8	1.9	1.7	1.6	1.5	1.4	1.4

Except as noted, values for vital statistics are in thousands; life expectancy is in years.

Health Personnel (3)

Total health expenditure as a percentage of GDP, 1990-1997[a]

Public sector	7.2
Private sector	1.4
Total[b]	8.6

Health expenditure per capita in U.S. dollars, 1990-1997[a]

Purchasing power parity	1,699
Total	2,222

Availability of health care facilities per 100,000 people

Hospital beds 1990-1997[a]	630
Doctors 1993[c]	299
Nurses 1993[c]	1,048

Health Indicators (4)

Life expectancy at birth	
1980	76
1997	79
Daily per capita supply of calories (1996)	3,160
Total fertility rate births per woman (1997)	1.7
Maternal mortality ratio per 100,000 live births (1990-97)	7[c]
Safe water % of population with access (1995)	....
Sanitation % of population with access (1995)	....
Consumption of iodized salt % of households (1992-98)[a]	....
Smoking prevalence	
Male % of adults (1985-95)[a]	22
Female % of adults (1985-95)[a]	24
Tuberculosis incidence per 100,000 people (1997)	5
Adult HIV prevalence % of population ages 15-49 (1997)	0.07

Infants and Malnutrition (5)

Under-5 mortality rate (1997)	4
% of infants with low birthweight (1990-97)	5
Births attended by skilled health staff % of total[a]	NA
% fully immunized (1995-97)	
TB	12
DPT	99
Polio	99
Measles	96
Prevalence of child malnutrition under age 5 (1992-97)[b]	NA

Ethnic Division (6)

White, Lapp (Sami), foreign-born or first-generation immigrants 12% (Finns, Yugoslavs, Danes, Norwegians, Greeks, Turks).

Religions (7)

Evangelical Lutheran	94%
Roman Catholic	1.5%
Pentecostal	1%
Other	3.5% (1987)

Languages (8)

Swedish. Small Lapp- and Finnish-speaking minorities.

EDUCATION

Public Education Expenditures (9)

Public expenditure on education (% of GNP)	
1980	9.0
1996	8.3
Expenditure per student	
Primary % of GNP per capita	
1980	43.0
1996	27.6[1]
Secondary % of GNP per capita	
1980	15.8[1]
1996	34.8[1]
Tertiary % of GNP per capita	
1980	35.0
1996	75.7[1]
Expenditure on teaching materials	
Primary % of total for level (1996)	3.9
Secondary % of total for level (1996)	....
Primary pupil-teacher ratio per teacher (1996)	11[1]
Duration of primary education years (1995)	6

Educational Attainment (10)

Age group (1995)	16-74
Total population	6,329,913
Highest level attained (%)	
No schooling	NA
First level	
Not completed	18.2
Completed	NA
Entered second level	
S-1	14.7
S-2	44.1
Postsecondary	21.0

GOVERNMENT & LAW

Political Parties (12)

Parliament	% of seats
Social Democrats	45.4
Moderate Party (Conservatives)	22.3
Center Party	7.7
Liberals	7.2
Left Party	6.2
Greens	5.8
Christian Democrats	4.1
New Democracy Party	1.2

Government Budget (13A)

Year: 1997

Total Expenditures: 770.75 Billions of Kronor

Expenditures as a percentage of the total by function:

General public services and public order	8.94
Defense	5.35
Education	5.66
Health	2.38
Social Security and Welfare	47.43
Housing and community amenities	3.33
Recreational, cultural, and religious affairs	.71
Fuel and energy	.08
Agriculture, forestry, fishing, and hunting	1.26
Mining, manufacturing, and construction	.08
Transportation and communication	3.02
Other economic affairs and services	5.60

Military Affairs (14B)

	1990	1991	1992	1993	1994	1995
Military expenditures						
Current dollars (mil.)	5,138	5,476	5,207	5,673	5,832	6,042
1995 constant dollars (mil.)	5,905	6,051	5,601	5,947	5,978	6,042
Armed forces (000)	65	63	45	44	70	51
Gross national product (GNP)						
Current dollars (mil.)	192,400	197,900	198,600	198,200	207,200	218,700[e]
1995 constant dollars (mil.)	221,100	218,700	213,600	207,800	212,400	218,700[e]
Central government expenditures (CGE)						
1995 constant dollars (mil.)	98,540	100,600	105,400	116,400	112,000	103,600
People (mil.)	8.6	8.6	8.7	8.7	8.8	8.8
Military expenditure as % of GNP	2.7	2.8	2.6	2.9	2.8	2.8
Military expenditure as % of CGE	6.0	6.0	5.3	5.1	5.3	5.8
Military expenditure per capita (1995 $)	690	702	646	682	681	683
Armed forces per 1,000 people (soldiers)	7.6	7.3	5.2	5.0	8.0	5.8
GNP per capita (1995 $)	25,830	25,360	24,630	23,830	24,180	24,730
Arms imports[6]						
Current dollars (mil.)	80	160	30	90	300	10
1995 constant dollars (mil.)	92	177	32	94	308	10
Arms exports[6]						
Current dollars (mil.)	575	490	250	60	80	310
1995 constant dollars (mil.)	661	541	269	63	82	310
Total imports[7]						
Current dollars (mil.)	54,260	49,990	50,020	42,680	51,720	64,440
1995 constant dollars (mil.)	62,360	55,240	53,800	44,750	53,020	64,440
Total exports[7]						
Current dollars (mil.)	57,540	55,220	56,120	49,860	61,290	79,910
1995 constant dollars (mil.)	66,130	61,020	60,360	52,270	62,830	79,910
Arms as percent of total imports[8]	.1	.3	.1	.2	.6	0
Arms as percent of total exports[8]	1.0	.9	.4	.1	.1	.4

Crime (15)

Crime rate (for 1997)

Crimes reported	1,196,100
Total persons convicted	299,000
Crimes per 100,000 population	13,500

Persons responsible for offenses

Total number of suspects	85,300
Total number of female suspects	16,200
Total number of juvenile suspects	21,300

LABOR FORCE

Labor Force (16)

Total (million)	4.552
Community, social, and personal services	38.3%
Mining and manufacturing	21.2%
Commerce, hotels, and restaurants	14.1%
Banking, insurance	9.0%
Communications	7.2%
Construction	7.0%
Agriculture, fishing, and forestry	3.2%

Data for 1991. Of total, 84% unionized (1992).

Unemployment Rate (17)

6.6% plus about 5% in training programs (1997 est.)

PRODUCTION SECTOR

Electric Energy (18)

Capacity	35.462 million kW (1995)
Production	142.913 billion kWh (1995)
Consumption per capita	15,996 kWh (1995)

Transportation (19)

Highways:

total: 138,000 km

paved: 105,018 km (including 1,330 km of expressways)

unpaved: 32,982 km (1996 est.)

Waterways: 2,052 km navigable for small steamers and barges.

Pipelines: natural gas 84 km.

Merchant marine:

total: 164 ships (1,000 GRT or over) totaling 2,036,831 GRT/1,919,367 DWT ships by type: bulk 7, cargo 33, chemical tanker 27, combination ore/oil 1, liquefied gas tanker 1, oil tanker 29, railcar carrier 1, refrigerated cargo 1, roll-on/roll-off cargo 41, short-sea passenger 7, specialized tanker 4, vehicle carrier 12 (1997 est.)

Airports: 255 (1997 est.)

Airports—with paved runways:

total: 145

over 3,047 m: 2

2,438 to 3,047 m: 9

1,524 to 2,437 m: 83

914 to 1,523 m: 27

under 914 m: 24 (1997 est.)

Airports—with unpaved runways:

total: 110

914 to 1,523 m: 5

under 914 m: 105 (1997 est.)

Top Agricultural Products (20)

Grains, sugar beets, potatoes; meat, milk.

MANUFACTURING SECTOR

GDP & Manufacturing Summary (21)

Detailed value added figures are listed by both International Standard Industry Code (ISIC) and product title.

	1980	1985	1990	1994
GDP ($-1990 mil.)[1]	189,312	207,212	229,748	222,994
Per capita ($-1990)[1]	22,781	24,816	26,843	25,520
Manufacturing share (%) (current prices)[1]	23.0	23.7	21.4	*22.4*
Manufacturing				
Value added ($-1990 mil.)[1]	37,770	42,805	45,202	48,470
Industrial production index	100	109	119	125
Value added ($ mil.)	30,905	24,486	51,429	35,125
Gross output ($ mil.)	73,194	60,328	115,467	101,975
Employment (000)	853	768	719	594
Profitability (% of gross output)				
Intermediate input (%)	58	59	55	66
Wages and salaries inc. supplements (%)	18	15	16	15
Gross operating surplus	24	26	29	20
Productivity ($)				
Gross output per worker	85,747	78,429	160,549	170,132

	1980	1985	1990	1994
Value added per worker	36,206	31,833	71,509	58,602
Average wage (inc. supplements)	15,835	11,689	24,892	25,235
Value added ($ mil.)				
311/2 Food products	2,719	2,107	4,249	2,541
313 Beverages	338	250	743	424
314 Tobacco products	104	108	257	187
321 Textiles	534	378	620	367
322 Wearing apparel	274	157	199	83
323 Leather and fur products	54	40	52	33
324 Footwear	61	24	27	16
331 Wood and wood products	2,102	1,154	3,046	1,723
332 Furniture and fixtures	452	285	551	373
341 Paper and paper products	2,596	2,230	4,524	3,441
342 Printing and publishing	1,842	1,517	3,158	2,033
351 Industrial chemicals	986	840	1,983	1,369
352 Other chemical products	1,246	1,091	2,544	2,525
353 Petroleum refineries	359	396	1,325	289
354 Miscellaneous petroleum and coal products	137	122	218	119
355 Rubber products	314	225	387	251
356 Plastic products	402	334	786	504
361 Pottery, china and earthenware	87	71	123	72
362 Glass and glass products	175	124	294	160
369 Other non-metal mineral products	801	510	1,129	579
371 Iron and steel	1,650	1,185	2,097	1,702
372 Non-ferrous metals	390	331	640	451
381 Metal products	2,598	2,049	4,448	2,688
382 Non-electrical machinery	3,936	3,185	6,226	4,454
383 Electrical machinery	2,570	2,132	4,021	3,023
384 Transport equipment	3,652	3,153	6,459	4,628
385 Professional and scientific equipment	371	401	1,166	978
390 Other manufacturing industries	154	86	157	111

FINANCE, ECONOMICS, & TRADE

Economic Indicators (22)

National product: GDP—purchasing power parity—$176.2 billion (1997 est.)

National product real growth rate: 2.1% (1997 est.)

National product per capita: $19,700 (1997 est.)

Inflation rate—consumer price index: 2% (1997 est.)

Exchange Rates (24)

Exchange rates:

Swedish kronor (SKr) per US$1

January 1998	8.0085
1997	7.6349
1996	6.7060
1995	7.1333
1994	7.7160
1993	7.7834

Balance of Payments (23)

	1992	1993	1994	1995	1996
Exports of goods (f.o.b.)	55,363	49,348	60,199	79,903	84,690
Imports of goods (f.o.b.)	−48,642	−41,801	−50,641	−63,926	−66,053
Trade balance	6,720	7,548	9,558	15,978	18,636
Services - debits	−37,271	−29,616	−30,220	−38,595	−41,396
Services - credits	24,337	19,716	23,285	30,528	31,268
Private transfers (net)	−2,174	−1,626	−1,674	−1,522	−1,565
Government transfers (net)	−440	−18	−206	−1,448	−1,051
Overall balance	−8,828	−4,159	743	4,941	5,892

Top Import Origins (25)

$66.6 billion (c.i.f., 1996) Data are for 1994.

Origins	%
European Union	62.6
Germany	18.4
United Kingdom	9.5
Denmark	6.6
France	5.5
Finland	6.3
Norway	6.1
United States	8.5

Top Export Destinations (26)

$84.5 billion (f.o.b., 1996) Data are for 1994.

Destinations	%
European Union	59.1
Germany	13.2
United Kingdom	10.2
Denmark	6.9
France	5.1
Norway	8.1
Finland	4.8
United States	8.0

Economic Aid (27)

Donor: ODA, $1.769 billion (1993).

Import Export Commodities (28)

Import Commodities	Export Commodities
Machinery	Machinery
Petroleum and petroleum products	Motor vehicles
Chemicals	Paper products
Motor vehicles	Pulp and wood
Foodstuffs	Iron and steel products
Iron and steel	Chemicals
Clothing 6.3%	Petroleum and petroleum products 8.1%

SWITZERLAND

CAPITAL: Bern.

FLAG: The national flag consists of an equilateral white cross on a red background, each arm of the cross being one-sixth longer than its width.

ANTHEM: The Swiss Hymn begins "Trittst in Morgenrot daher, Seh' ich dich in Strahlenmeer" ("Radiant in the morning sky, Lord, I see that Thou art nigh").

MONETARY UNIT: The Swiss franc (SwFr) of 100 centimes, or rappen, is the national currency. There are coins of 1, 5, 10, 20, and 50 centimes and 1, 2, and 5 francs, and notes of 10, 20, 50, 100, 500, and 1,000 francs. SwFr1 = $0.79618 (or $1 = SwFr1.256).

WEIGHTS AND MEASURES: The metric system is the legal standard.

HOLIDAYS: New Year, 1–2 January; Labor Day, 1 May; Christmas, 25–26 December. Movable religious holidays include Good Friday, Easter Monday, Ascension, and Whitmonday.

TIME: 1 PM = noon GMT.

LOCATION AND SIZE: A landlocked country in west-central Europe, Switzerland has an area of 41,290 square kilometers (15,942 square miles), slightly more than twice the size of New Jersey. Switzerland has a total boundary length of 1,852 kilometers (1,151 miles).

Switzerland's capital city, Bern, is located in the western part of the country.

CLIMATE: The climate of Switzerland north of the Alps is temperate; the average annual temperature is 9°C (48°F). The average rainfall varies from 53 centimeters (21 inches) in the Rhône Valley to 170 centimeters (67 inches) in Lugano. Generally, the areas to the west and north of the Alps have a cool, rainy climate, with winter averages near or below freezing and summer temperatures seldom above 21°C (70°F). The region south of the Alps has a Mediterranean climate, and frost is almost unknown. The climate of the Alps and of the Jura uplands is mostly raw, rainy, or snowy, with frost occurring above 1,830 meters (6,000 feet).

Swiss Confederation
French—*Suisse (Confédération Suisse)*
German—*Schweiz (Schweizerische Eidgenossenschaft)*

INTRODUCTORY SURVEY

RECENT HISTORY

Since the last quarter of the nineteenth century, Switzerland has been concerned primarily with domestic matters, such as social legislation, communications, and industrialization. In foreign affairs, it remained neutral through both World War I (1914–18) and World War II (1939–45), determined to protect its independence with its highly regarded militia. In 1978, Switzerland's 23d canton, Jura, was established by nationwide vote. In 1991, Switzerland celebrated the 700th anniversary of Confederation.

Despite its neutrality, Switzerland has cooperated wholeheartedly in various international organizations, including the League of Nations and the Red Cross. Switzerland has long resisted joining the United Nations, however, partly on the grounds that imposition of sanctions on other countries, as required by various UN resolutions, goes against Switzerland's policy of strict neutrality. Switzerland, however, is a member of most specialized United Nations agencies. In 1992 the Swiss rejected participation in the European Economic Area (EEA) of the European Union (EU).

In early 1998, Swiss banks agreed to set up a plan to compensate Holocaust victims who were looted by the Nazis in World War II. The Nazis deposited large sums of money in Switzerland during the war. The settlement amounted to $1.25 billion.

GOVERNMENT

The Swiss Confederation is a federal union governed under the constitution of 1874, which gave supreme authority to the Federal Assembly (the legislative body) and executive power to the Federal Council.

The Federal Assembly consists of two chambers: the National Council (Nationalrat) of 200 members, elected by direct ballot for four-year terms by all citizens 18 years of age or older; and the Council of States (Ständerat) of 46 members, 2 appointed by each of the 20 cantons and 1 from each of the 6 half-cantons, and paid by the cantons. Legislation must be approved by both houses.

The seven members of the Federal Council, which has no veto power, are the respective heads of the main departments of the federal government.

The Swiss Confederation consists of 23 cantons, 3 of which are subdivided into half-cantons. The cantons have independent rule in all matters not delegated to the federal government by the constitution.

Judiciary

The Federal Court of Justice in Lausanne, composed of 30 permanent members, rules in the majority of cases where a canton or the federal government is involved. It is the highest appeals court for many types of cases. Each canton has its own courts. District courts have three to five members and try lesser criminal and civil cases.

Political Parties

The three strongest parties are the Social Democratic Party, which supports wider state participation in industry and strong social legislation; the Radical Democratic Party, a progressive middle-class party, which favors increased social welfare, strengthening of national defense, and a democratic federally structured government; and the Christian Democrats, which opposes centralization of power.

Other parties include the Center Democratic Union (Swiss People's Party); the League of Independents, a progressive, middle-class consumers' group; the Communist-inclined Workers Party; the Liberal Party; and the Independent and Evangelical Party, which is Protestant and conservative.

The ruling Federal Council is made up of what the Swiss refer to as the ''magic formula'' coalition—the four largest parties always fill the seven positions on the Federal Council. Such a political arrangement is strictly followed, even though it is not an official policy.

DEFENSE

The Swiss army is a well-trained citizen's militia, composed of three field army corps and one alpine field corps. Although in 1995 the standing armed forces had only 3,300 regulars and 18,500 conscript (drafted) trainees, Switzerland can mobilize 396,000 trained militia within 48 hours. All males age 19–20 must serve in the military.

Switzerland's civil defense program, begun in the early 1970s, is able to shelter 90% of the population. Swiss volunteers serve in four overseas peacekeeping operations. Switzerland spent $5.1 billion for defense in 1995.

ECONOMIC AFFAIRS

Because of a lack of minerals and other raw materials and limited agricultural production, Switzerland depends upon imports of food and fodder (livestock feed) and of industrial raw materials. It finances these imports with exports of manufactured goods. Swiss manufacturers focus on quality rather than quantity of output. Other important branches of the economy include international banking, insurance, tourism, and transportation.

Switzerland was less effected than most other nations by the worldwide recession of the early 1980s and experienced a strong recovery beginning in 1983. However, gross national product (GNP) grew by only about 0.7% each year during 1986–92, and by only 1% per year during 1993–95. From 1990 to 1992, the annual inflation rate averaged 5.1%. By 1996, inflation had fallen to 0.8%. Swiss unemployment was 3.8% by 1996.

Public Finance

The Swiss government has been known historically for maintaining a relatively high degree of austerity in comparison to its European neighbors. In 1991, the federal government incurred a budget deficit of over SwFr1.5 billion, the first one in seven years. Cantonal budgets also were in deficit. These deficits continued throughout the 1990s, prompting governments at all levels to take further cost-cutting steps. The overall government deficit is expected to fall to 3% of GDP in 1996. As an international creditor, debt management policies are not relevant to Switzerland, which participates in the Paris Club debt reschedulings and is an active member of the OECD.

The U.S. Central Intelligence Agency estimates that, in 1995, government revenues totaled approximately $31 billion and expenditures $36.9 billion.

In 1998, Switzerland's GDP was nearly $192 billion, or about $26,400 per person.

Industry

Swiss industries are chiefly engaged in the manufacture of highly finished goods. Some industries are concentrated in certain regions: the watch and jewelry industry in the Jura Mountains; machinery in Zürich, Geneva, and Basel; the chemical industry (dyes and pharmaceuticals) in Basel; and the textile industry in northeastern Switzerland.

The machine industry, first among Swiss industries today, produces goods ranging from heavy arms and ammunition to fine precision and optical instruments. Watches and machinery represent about 45% of the total Swiss export value. About 10% of the world's medicines are produced by three companies in Basel. Switzerland has also developed a major food industry, relying in part on the country's capacity for milk production. Condensed milk was first developed in Switzerland, as were two other important processed food products: chocolate and baby food.

Banking and Finance

In 1997, Switzerland had three major banks, 26 cantonal banks, around 130 foreign-owned banks, 210 savings banks, and 209 other banks and finance companies. The bank balance-sheet total per capita in Switzerland is higher than that of any other nation in the world. Moreover, registered banks and bank-like finance companies numbered 494 in 1995, offering the Swiss, on average, the greatest access to banking services of all the world's nations.

The government-supervised Swiss National Bank, incorporated in 1905 and the sole bank of issue, is a semiprivate institution owned by the cantons, by former banks of issue, and by the public. The National Bank acts as a central clearinghouse and participates in many foreign and domestic banking operations. The three big banks, (Union Bank of Switzerland, Swiss Bank Corp., and CS Holdings, which include Credit Suisse and Schweizer Volksbank) dominate the Swiss banking scene and are expanding aggressively overseas. They are universal banks, providing a full range of services to all types of customers.

Regional banks, numbering 127 at the end of 1995, specialize in mortgage lending and credits for small businesses. Since 1994, most of the country's regional banks have been linked in a common holding company providing back-office operations and other services to members in a bid to cut costs.

Foreign banks make up a third of banks active in Switzerland. In contrast to domestic banks, their numbers have risen over the last decade but their business is increasingly focused on asset management, mostly of funds from abroad. Foreign reserves, excluding gold, totaled $36,413 million at the end of 1995. On 1 January 1995 a new banking law came into effect allowing for foreign banks to open subsidiaries, branches, or representative offices in the country without first getting approval of the Federal Banking Commission.

The transactions of private and foreign banks doing business in Switzerland traditionally play a significant role in both Swiss and foreign capital markets; however, precise accounting of assets and liabilities in this sector are not usually made available as public information. Switzerland's strong financial position and its tradition (protected by the penal code since 1934) of preserving the secrecy of individual bank depositors have made it a favorite depository with persons throughout the world. (However, Swiss secrecy provisions are not absolute and have been lifted to provide information in criminal investigations.) The Swiss Office for Compensation executes clearing traffic with foreign countries.

In 1997 Swiss banks came under heavy criticism for losing track of money, gold, and other valuables belonging to Jewish Holocaust victims and held by the banks during World War II. Records also showed the banks had closed thousands of victims' accounts without notice after the war. The banks claimed they had lost the old records, but a group of journalists found the records archived in Lausanne in April of that year.

Also in 1997, an embarrassed Swiss government selected four members to a panel empowered to run a fund for Holocaust victims. Nobel laureate Elie Weisel, a concentration camp survivor, turned down an invitation to serve as one of the three foreign members on the board. The fund, intended to help impoverished Holocaust victims and their families, is supported by funds appropriated by Nazis from Jews sent to concentration camps. Much of the gold, jewels, bonds, and currency taken by the Nazis had been placed in Swiss banks.

Swiss banks were also under fire in 1997 for possibly facilitating money laundering of narcodollars accrued by a former Mexican president's brother and for failing to adequately recover the billions of dollars supposedly plundered by former Zaireian dictator Mobuto Sese Seko, who was overthrown that year. All the negative publicity has caused some to question the usefulness of Swiss banks' much-lauded secrecy.

Stock exchanges operate in Geneva (founded 1850), Basel (1875), and Zürich (1876). The Zürich exchange is the most important in the country. In terms of market capitalization, the Swiss stock exchanges rank seventh in the world, behind New York, Tokyo, Osaka, London, Frankfurt, and Paris, as of 1997. Overall, turnover, including shares, bonds, and options, amounted to SwFr 593.2 billion in 1995, a fivefold increase from 1990. The stock exchanges in Zürich, Geneva, and Basel closed in 1994 when a national electronic stock exchange for all securities trading began operations in Augusta.

Economic Development

Private enterprise is the basis of Swiss economic policy. Although government intervention has traditionally been kept to a minimum (even with monopolistic formations), international monetary crises from late 1974 to mid-1975 led to imposition of various interim control measures; in 1982, with inflation rising, a constitutional amendment mandating permanent government price controls was approved by popular referendum. The Swiss National Bank has followed a general policy of limiting monetary growth. To further raise the standard of living, the government also grants subsidies for educational and research purposes, promotes professional training, and encourages exports. Although certain foreign transactions are regulated, there is free currency exchange and a guarantee to repatriate earnings of foreign corporations.

The cause of the remarkable stability of Switzerland's economy lies in the adaptability of its industries; in the soundness of its convertible currency, which is backed by gold to an extent unmatched in any other country; and in the fact that the particular pattern of Swiss democracy, where every law may be submitted to the popular vote, entails taking into account the wishes of all parties whose interests would be affected by a change in legislation.

Switzerland's development assistance program takes the form of technical cooperation, preferential customs treatment for certain third-world products, and a limited number of bilateral aid arrangements. Official development assistance was $793 million in 1993.

SOCIAL WELFARE

Swiss social legislation has three main components: accident insurance; sickness insurance; and old-age, survivors', and disability insurance. In addition, there is unemployment insurance, military insurance, income insurance, and the farmers' aid organization.

In 1981, a constitutional amendment guaranteeing equal rights to women, particularly in education, work, and the family, was passed by a vote of 797,679 to 525,950. In 1985, 54.7% of voters approved a new law giving women equal rights in marriage.

Healthcare

Health standards and medical care are excellent. The pharmaceuticals industry ranks as one of the major producers of specialized pharmaceutical products. There were 11 hospital beds per 1,000 people in 1990.

There were about 18 cases of tuberculosis per 100,000 people reported in 1990. Life expectancy was averaged at 79 years.

Housing

In 1991, a total of 37,597 new homes were built in communities of 2,000 or more inhabitants. The total housing stock in 1991 stood at 3.2 million. As of 1986, 99.5% of all dwellings had a water supply, 97% had a bath or shower, 78% had central heating, and 100% had a private toilet.

EDUCATION

Primary education is free, and adult illiteracy is nearly nonexistent. Education at all levels is the responsibility of the cantons. Thus, Switzerland has 26 different systems based on differing education laws and varied cultural and linguistic needs. The cantons decide on the types of schools, length of study, teaching materials, and teachers' salaries.

Education is compulsory in most cantons for nine years, and in a few for eight. After primary school, students complete the compulsory portion of their education in various types of secondary Grade I schools, which emphasize vocational or academic subjects to varying degrees. In 1993, 423,399 students were enrolled in primary school, 558,920 were enrolled in secondary school, and 148,664 attended institutions of higher learning.

Switzerland has eight cantonal universities. The largest universities are those of Zürich, Geneva, and Basel; others include those of Lausanne, Bern, Fribourg, and Neuchâtel.

1999 KEY EVENTS TIMELINE

January

- The government reports a 72 percent increase in requests for asylum, largely made up of refugees fleeing civil violence in the Yugoslav province of Kosovo.
- Greenpeace releases a copy on an agreement between Switzerland and Russia indicating the two countries plan to negotiate long-term storage of Russian nuclear waste in Switzerland.
- In response to a request from the Spanish government, Switzerland freezes the accounts of former Argentine general Antonio Bussi. Bussi, now a politician in Argentina, is accused of plundering money from victims of the Argentine military's campaign against leftists and dissidents in the 1970s and 1980s. Over 300 Spanish citizens disappeared in Argentina during this era.
- A Switzerland-based Web site, International Lyric Server (ILS), that publishes song lyrics, is shut down by police after a group of U.S. music publishers brought a criminal copyright infringement suit against it.
- Turkey registers an official complaint with Switzerland after a cable television network replaced

broadcasts of Turkish state-run television TRT with Med TV, a channel suspected of links to Kurdish Workers' Party (PKK) rebels.

February

- Avalanches hit between the villages of Evolene and Les Hauderes, 110 miles form Geneva, crushing about ten chalets. Two people were confirmed dead, and eight were missing and presumed dead.
- A 400-page report published by the government summarizes the advantages and disadvantages of membership in the European Union.

March

- Balloonists Bertrand Piccard of Switzerland and Brian Jones of England begin their quest to circumnavigate the globe in a balloon. The two take off from Chateau d'Oex in western Switzerland on March 1, piloting a balloon weighing nine tons. On March 20, the two become the first to successfully circle the globe nonstop in a balloon.
- The Swiss National Bank releases a report stating that officials ''midjudged the political, legal, and moral aspects'' of its purchases of gold from Hitler. The gold was looted by Nazis and then sold, primarily to the National Bank.
- The Fourteenth Winter World Games for the Deaf are staged from March 6–14.

May

- Days of heavy rain combine with melting snow in the German, Austrian, and Swiss Alps to result in some of the most severe regional flooding of the twentieth century. The flooding causes landslides, a burst dam, and at least five deaths.
- A Congolese man being deported escapes when his fellow passengers on a Swissair flight to Kinshasa, Democratic Republic of the Congo, attacked his three police guards during a stopover in Cameroon. Local authorities in Africa would not allow the plane to continue, and forces the plane with the Swiss officers to return to Switzerland.

June

- Two Swiss banks (United Bank of Switzerland (UBS) and Credit Suisse-First Boston) launch a worldwide effort to contact Holocaust survivors and their heirs. The survivors (and their heirs) may be eligible for a share of a $1.25 billion settlement. The notification process targets 108 countries and will be presented in 29 languages. A class-action lawsuit filed in 1996 resulted in the settlement order in 1997. The suit blamed the banks for indirectly financing the Nazi regime during World War II, including trading gold looted by Germany for hard currency, and alleged that the banks illegally confiscated savings accounts of refugees and Jews who died during the Holocaust.

July

- Twenty-one people die when a July 12 flash flood overcomes their group while canyoning on the Saxeten River in Switzerland. (Canyoning involves jumping or rappelling down waterfalls and mountain streams into river gorges.) Swiss police questioned the five guides who were leading the canyoning party, sponsored by the Swiss company, Adventure World.

September

- An investigation is launched into a canyoning accident that took place in Schliere Gorge in central Switzerland in 1998, leaving an eighteen-year-old Swiss woman paralyzed. This accident occurred a year before the fatal July 1999 accident.
- Switzerland's general prosecutor, Carla Del Ponte, launches an investigation of two U.S.-based companies, Mabetex and Mercata Trading, both of which have links to Russia. The Swiss suspect both companies of paying illegal bribes to government officials. Mabetex, the company that decorated the president's residence, is believed to have paid as much as $10 million in bribes.
- The airline Swissair announces that it will no longer carry passengers who are being deported forcibly from Switzerland because of incidents involving deportees. Some have resisted deportation and disrupted flights, and caused problems for passengers and crew members.
- Swiss graphic artist Herbert Leupin (1916–99) dies. His advertising graphics included the familiar purple cow that advertised Suchard's milk chocolate, and product posters.
- Goma, the first gorilla born in captivity in Europe, celebrates her fortieth birthday at the Basel Zoo.

October

- Lawmakers in the lower house pass legislation to allow the country's central bank, National Bank, to sell half of its 2,600-ton gold reserves starting in spring 2000. The upper house will vote on the bill in December. It will then go to the ruling coalition Cabinet, probably by April 2000.
- For the first time beginning October 15, MFG (Merchant Bank of Geneva), Switzerland will allow depositors to open accounts over the Internet with minimum deposits of $5,000.
- In the October 22 general elections, the Sozialdemokratische Partei der Schweiz (SPS) wins almost 22 percent of the vote, capturing 54 of the 200 seats in the National Council and 6 of the 46 seats in the Council of States. SPS candidate Ruth Dreifuss wins the presidential election.

December

- Legislators in the upper house of the legislature vote on a law to allow the sale of one-half of the National Bank's gold reserves. The law passed in the lower house earlier in the year. Upon passage, the law will be submitted in April 2000 to the cabinet, made up of members of the ruling coalition of the government, for approval.
- Swiss banks are pressured to publicize names of more than 25,000 account holders thought to be victims of Nazi persecution.
- The results of a three-year independent audit states that accounts of an estimated 54,000 people are linked to Nazi persecution, and contends that Swiss banks deliberately issued misleading bank statements to Jews looking for family assets following World War II.

ANALYSIS OF EVENTS: 1999

BUSINESS AND THE ECONOMY

The Swiss economy in 1999 faced accidental set-backs and might also suffer as a result of a downturn in the global economy, but, as a strong, modernized economy, it still managed to have a good year. Tourism, an important component of the Swiss economy, rose 3.2 percent in 1998, but still fell short of the levels of the early 1990s. Fatal canyoning accidents that occurred during 1998 and 1999 was expected to hurt the Swiss tourism industry. As part of an effort to bolster its international image, Switzerland's banks agreed to set up a mechanism for returning funds confiscated during the Nazi era (1930s–1945) to survivors and heirs of survivors. The banks undertook a worldwide campaign to find those who would qualify to file claims. The deadline for submitting claims was October 22, with final court hearings to take place about a month later (November 29). Actual disbursement of the Swiss bank funds was not planned to begin until late December, so it was unlikely that any claimants would receive funds before early 2000.

International Lyric Server (ILS), a Swiss website that published lyrics to almost 100,000 popular songs, was shut down by order of the District Attorney in Basel, in response to a criminal suit filed against ILS under the sponsorship of the U.S. organization, National Music Publishing Association (NMPA). NMPA brought the charges of copyright infringement on behalf of composers and music publishers.

GOVERNMENT AND POLITICS

The environmental activist organization, Greenpeace, released an internal government memo, shedding light on plans being negotiated between Russia and Switzerland for Russian nuclear waste to be exported to Switzerland for processing and storage. Official spokespeople for the two countries confirmed that discussions were taking place, but denied that any agreement had been reached.

Swiss banks have frozen $16.8 million in Russian accounts in response to investigations that U.S. companies have paid illegal bribes to Russian officials. In addition, the Swiss government is providing legal assistance to the United States in an ongoing investigation of a former Ukrainian prime minister, Pavlo Lazarenko. Lazarenko, detained in California, has been accused of stealing $2 million from Ukrainian government and illegally hiding about $4 million in foreign banks.

CULTURE AND SOCIETY

The Swiss government's federal office for refugees report that asylum requests reached 41,300 in 1998, up 72 percent from 1997. In addition, an estimated 12,700 illegal immigrants were arrested at the border, more than twice the number of illegal immigrants apprehended in 1997. More than half of those detained were Yugoslav citizens, almost all of whom were ethnic Albanians fleeing violence in the Serbian province of Kosovo. Over 10,000 were caught on Switzerland's southern border with Italy.

The Swiss balloon pilot Bertrand Piccard, and his partner Brian Jones of England Jones became the focus of an international celebration when they successfully circumnavigated the globe in a balloom. The two were the first to succeed in the challenging expedition.

DIRECTORY

CENTRAL GOVERNMENT

Head of State

President
Ruth Dreifuss, Office of the President

Departments

Chief of Finance
Kasper Villiger, Department of Finance, Bundeshaus West, CH-3003 Bern, Switzerland
PHONE: +41 (31) 3226111
FAX: +41 (31) 3226187

Chief of Foreign Affairs
Joseph Deiss, Department of Foreign Affairs, Bernerhof Bundesgasse 3, CH-3003 Bern, Switzerland
PHONE: +41 (31) 3226111
FAX: +41 (31) 3226187

Chief of Home Affairs
Ruth Dreifuss, Department of Home Affairs, Bundeshaus West, CH-3003 Bern, Switzerland
PHONE: +41 (31) 3229111
FAX: +41 (31) 3227901

Chief of Justice and Police
Ruth Metzler, Department of Justice and Police, Bundeshaus West, CH-3003 Bern, Switzerland
PHONE: +41 (31) 3224111
FAX: +41 (31) 3227832

Chief of Defense, Civil Protection and Sports
Adolf Ogi, Department of Defense, Civil Protection and Sports, Bundeshaus Ost, CH-3003 Bern, Switzerland
PHONE: +41 (31) 3221211
FAX: +41 (31) 3123463

Chief of Economic Affairs
Pascal Couchepin, Department of Economic Affairs, Bundeshaus Ost, CH-3003 Bern, Switzerland
PHONE: +41 (31) 3222111
FAX: +41 (31) 3222056

Chief of Environment, Transport, Energy, and Communications
Moritz Leuenberger, Department of the Environment, Transport, Energy, and Communications, Bundeshaus Nord, CH-2002 Bern, Switzerland
PHONE: +41 (31) 3224111
FAX: +41 (31) 3118392

POLITICAL ORGANIZATIONS

Freisinnig-Demokratische Partei der Schweiz-FDP (Radical Democrats)

Postfach 6136, CH-3001 Bern, Switzerland
PHONE: +41 (31) 3113438; 3120444
FAX: +41 (31) 3121951
TITLE: President
NAME: Franz Steinegger

Christlichdemokratische Volkspartei der Schweiz-CVP (Christian Democrats)

Postfach 5835, CH-3001 Bern, Switzerland
PHONE: +41 (31) 3522364
FAX: +41 (31) 3522430
TITLE: President
NAME: Anton Cottier

Schweiz Sozialdemokratische Partei der Schweiz-SP (Social Democrats)

Postfach, CH-3001 Bern, Switzerland
PHONE: +41 (31) 3110744
FAX: +41 (31) 3115414
TITLE: President
NAME: Peter Bodenmann

Schweizerische Volkspartei-SVP (Swiss People's Party (Agrarians))

Secretariat, Brückfeldstrasste 18, CH-3000 Bern 26, Switzerland
PHONE: +41 (31) 3025858
FAX: +41 (31) 3017585
E-MAIL: gs@svp.ch
TITLE: President
NAME: Ueli Maurer Neugrüt

Liberale Partei der Schweiz-LPS (Liberal Democrats)

Postfach 7107, Spitalgasse 32, CH-3001 Bern, Switzerland
PHONE: +41 (31) 3116404; 3229961
FAX: +41 (31) 3125474; 3329732

TITLE: President
NAME: François Jeanneret

Landesring der Unabhängigen-LdU (Independent Landesring)

Gutenbergstrasse 9, Postfach 7075, CH-3001 Bern, Switzerland
PHONE: +41 (31) 3821636
FAX: +41 (31) 3823695
E-MAIL: ldu@ldu.ch
TITLE: President
NAME: Monika Weber

Evangelische Volkspartei der Schweiz-EVP (Evangelical Party)

Josefstrasse 32, Postfach, CH-8023 Zürich, Switzerland
PHONE: +41 (1) 2727100
FAX: +41 (1) 2721437
TITLE: President
NAME: Otto Zwygart

Partei der Arbeit-PdA (Workers' Party)

25 rue du Vieux-Billard, Case postale 232, CH-1211 Genève 8, Switzerland
PHONE: +41 (22) 3281140
FAX: +41 (22) 3296412
TITLE: President
NAME: Jean Spielmann

Schweizer Demokraten-SD (Swiss Democrats)

Postfach 8116, CH-3001 Bern, Switzerland
PHONE: +41 (31) 3112774
FAX: +41 (31) 3125632
TITLE: President
NAME: Rudolf Keller

Grüne Partei der Schweiz-GPS (Environmentalists)

Waisenhausplatz 21, CH-3011 Bern, Switzerland
PHONE: +41 (31) 3126660
FAX: +41 (31) 3126662
TITLE: President
NAME: Verena Diener

Freiheits-Partei der Schweiz-FP (Freedom Party)

Postfach, CH-4622 Egerkingen, Switzerland
TITLE: President
NAME: Roland F. Borer

Lega dei Ticinesi (Ticino League)

Casella postale 2311, Via Monte Boglia 7, CH-6901 Lugano, Switzerland
TITLE: President
NAME: Giuliano Bignasca

Frauen macht Politik!-FraP! (Women in politics!)

Postfach 9353, CH-8036 Zürich, Switzerland
PHONE: +41 (1) 2424418
TITLE: Secretaries
NAME: Michèle Spieler Marianne Ruegg

Eidgenössisch-Demokratische Union-EDU (Federal Democratic Union)

Postfach, CH-3607 Thun 7, Switzerland
TITLE: President
NAME: Werner Scherrer

Christian Socialists-CSP (Christian Socialists)

Hopfenweg 21, Postfach 5775, CH-3001 Bern, Switzerland
PHONE: +41 (31) 3702102
TITLE: Secretary
NAME: Hedy Jager

DIPLOMATIC REPRESENTATION

Embassies in Switzerland

Australia
2 Chemin des Fins, Case postale 172, CH-1211 Geneva 19, Switzerland
PHONE: +41 (22) 7999100
FAX: +41 (22) 7999178

Belgium
58 rue de Moillebeau, CH-1211 Geneva 19, CP473, Switzerland
PHONE: +41 (22) 7304000
FAX: +41 (22) 7304017
E-MAIL: Geneva@Diplobel.org

China
11 chemin de Surville, 1213 Petit-Lancy, CH-Geneva, Switzerland
PHONE: +41 (22) 7922548
FAX: +41 (22) 7937014
TITLE: Ambassador
NAME: Oiao Zonghuai

Cuba
E-MAIL: mission.cuba@ties.itu.ch
TITLE: Ambassador

NAME: Carlos Amat Fores

Denmark
Thurnstrasse 95, P.O. Box, CH-3006 Bern, Switzerland
PHONE: +41 (31) 3505455
FAX: +41 (31) 3505464
E-MAIL: embassy@denmark.ch

Ecuador
Ensingerstr. 483006, CH-3006 Bern, Switzerland
PHONE: +41 (31) 3511755
FAX: +41 (31) 3512771
E-MAIL: edesuiza@bluewin.ch

France
Schosshaldenstrasse 46, CH-3006 Bern, Switzerland
PHONE: +41 (31) 3592111
FAX: +41 (31) 3592112
E-MAIL: ambassade.fr@iprolink.ch

Georgia
Hotel ''Mon-Repos'' ap.751, 131 Rue de Lausanne, CH-2102 Geneva, Switzerland
PHONE: +41 (22) 7328010
FAX: +41 (22) 7328595
TITLE: Ambassador
NAME: Amiran Kavadze

Hungary
Muristr. 31, CH-3006 Bern, Switzerland

Italy
Tödistrasse 67, CH-8039 Zurich, Switzerland
PHONE: +41 (01) 2866111
FAX: +41 (01) 2011611

Kenya
Bleicherweg 30, P.O. Box 770, CH-8039 Zurich, Switzerland
PHONE: +41 (01) 2022244

Lesotho
Bleicherweg 45, CH-8055 Zurich, Switzerland
PHONE: +41 (01) 4610040

Portugal
220 route de Ferney, 1218 Le Grand-Saconnex (GE), CH-Geneva, Switzerland
PHONE: +41 (22) 7910511; 7910625; 7910663
FAX: +41 (22) 7882503
E-MAIL: consulate.portugal@itu.int
TITLE: Consul General
NAME: Aristides Gonçalves

South Africa
Alpenstrasse 29, CH-3006 Bern, Switzerland
PHONE: +41 (31) 3501313
FAX: +41 (31) 3501325

Swaziland
Talstr. 58, CH-8039 Zurich, Switzerland
PHONE: +41 (01) 2211188

Turkey
Lombachweg 33, CH-3006 Bern, Switzerland
PHONE: +41 (31) 3511691; 3511692; 3511693
FAX: +41 (31) 3528819

United States
Jubilaeumstrasse 93, CH-3005 Bern, Switzerland
PHONE: +41 (31) 3577011
FAX: +41 (31) 3577344
TITLE: Ambassador
NAME: Madeleine May Kunin

JUDICIAL SYSTEM

Federal Supreme Court

Av. du Tribunal fédéral 29, CH-1000 Lausanne 14, Switzerland
PHONE: +41 (21) 3189111
FAX: +41 (21) 3233700

FURTHER READING

Articles

Browne, Malcom W. ''Around-the-World Balloon Pilots Cheered on Return to Switzerland.'' *The New York Times*, 23 March 1999, p. A4.

Olson, Elizabeth. ''45,000 More Wartime Accounts Found by Panel in Swiss Banks.'' *The New York Times*, 17 November 1999, p. A5.

———. ''Swiss Voters, Tilting Right, Unsettle Traditions.'' *New York Times*, 27 October 1999, p. A14.

''U.S. Judge Urged to Approve Settlement Between Swiss Banks, Holocaust Victims.'' *The Wall Street Journal*, 30 November 1999, p. B8.

Books

New, Mitya. *Switzerland Unwrapped: Exposing the Myths.* New York: I.B. Tauris, 1997.

Sandford, John. *Encyclopedia of Contemporary German Culture.* New York: Routledge, 1999.

Steinberg, Jonathan. *Why Switzerland?* 2nd ed. New York: Cambridge University Press, 1996.

Internet

Guide to Switzerland in English Available Online @ http://www.a-switzerland.ch/ (November 30, 1999).

Swiss News Flash Available Online @ http://www.swissnewsflash.com/ (November 30, 1999).

SWITZERLAND: STATISTICAL DATA

For sources and notes see "Sources of Statistics" in the front of each volume.

GEOGRAPHY

Geography (1)

Area:

Total: 41,290 sq km.

Land: 39,770 sq km.

Water: 1,520 sq km.

Area—comparative: slightly less than twice the size of New Jersey.

Land boundaries:

Total: 1,852 km.

Border countries: Austria 164 km, France 573 km, Italy 740 km, Liechtenstein 41 km, Germany 334 km.

Coastline: 0 km (landlocked).

Climate: temperate, but varies with altitude; cold, cloudy, rainy/ snowy winters; cool to warm, cloudy, humid summers with occasional showers.

Terrain: mostly mountains (Alps in south, Jura in northwest) with a central plateau of rolling hills, plains, and large lakes.

Natural resources: hydropower potential, timber, salt.

Land use:

Arable land: 10%

Permanent crops: 2%

Permanent pastures: 28%

Forests and woodland: 32%

Other: 28% (1993 est.).

HUMAN FACTORS

Demographics (2A)

	1990	1995	1998	2000	2010	2020	2030	2040	2050
Population	NA	7,177.8	7,260.4	7,288.7	7,351.7	7,208.6	6,859.8	6,293.6	5,613.8
Net migration rate (per 1,000 population)	NA	NA	NA	NA	NA	NA	NA	NA	NA
Births	NA	NA	NA	NA	NA	NA	NA	NA	NA
Deaths	NA	NA	NA	NA	NA	NA	NA	NA	NA
Life expectancy - males	NA	75.3	75.7	75.9	77.0	77.9	78.7	79.3	79.8
Life expectancy - females	NA	81.9	82.2	82.4	83.3	84.1	84.8	85.3	85.8
Birth rate (per 1,000)	NA	11.6	10.8	10.2	8.0	7.8	6.9	6.1	6.1
Death rate (per 1,000)	NA	9.0	9.0	9.1	9.9	11.4	13.8	16.3	18.4
Women of reproductive age (15-49 yrs.)	NA	1,802.2	1,782.1	1,761.9	1,675.6	1,452.4	1,300.1	1,146.1	961.7
of which are currently married	NA	NA	NA	NA	NA	NA	NA	NA	NA
Fertility rate	NA	1.5	1.5	1.5	1.4	1.3	1.3	1.3	1.3

Except as noted, values for vital statistics are in thousands; life expectancy is in years.

Health Personnel (3)

Total health expenditure as a percentage of GDP, 1990-1997[a]

Public sector	7.1
Private sector	3.1
Total[b]	10.2

Health expenditure per capita in U.S. dollars, 1990-1997[a]

Purchasing power parity	2,527
Total	3,603

Availability of health care facilities per 100,000 people

Hospital beds 1990-1997[a]	2,080
Doctors 1993[c]	301
Nurses 1993[c]	NA

Health Indicators (4)

Life expectancy at birth	
1980	76
1997	79
Daily per capita supply of calories (1996)	3,280
Total fertility rate births per woman (1997)	1.5
Maternal mortality ratio per 100,000 live births (1990-97)	6[c]
Safe water % of population with access (1995)	100
Sanitation % of population with access (1995)	....
Consumption of iodized salt % of households (1992-98)[a]	....
Smoking prevalence	
Male % of adults (1985-95)[a]	36
Female % of adults (1985-95)[a]	26
Tuberculosis incidence per 100,000 people (1997)	11
Adult HIV prevalence % of population ages 15-49 (1997)	0.32

Infants and Malnutrition (5)

Under-5 mortality rate (1997)	5
% of infants with low birthweight (1990-97)	5
Births attended by skilled health staff % of total[a]	NA
% fully immunized (1995-97)	
TB	NA
DPT	NA
Polio	NA
Measles	NA
Prevalence of child malnutrition under age 5 (1992-97)[b]	NA

Ethnic Division (6)

German	65%
French	18%
Italian	10%
Romansch	1%
Other	6%
French	20%
Italian	4%
Romansch	1%
Other	1%

Religions (7)

Roman Catholic	46.1%
Protestant	40%
Other	5%
No religion	8.9% (1990)

Languages (8)

German 63.7%, French 19.2%, Italian 7.6%, Romansch 0.6%, other 8.9%.

EDUCATION

Public Education Expenditures (9)

Public expenditure on education (% of GNP)	
1980	4.8
1996	5.3[1]
Expenditure per student	
Primary % of GNP per capita	
1980	....
1996	18.8[1]
Secondary % of GNP per capita	
1980	30.5
1996	28.4[1]
Tertiary % of GNP per capita	
1980	59.6
1996	44.5[1]
Expenditure on teaching materials	
Primary % of total for level (1996)	....
Secondary % of total for level (1996)	....
Primary pupil-teacher ratio per teacher (1996)	12[1]
Duration of primary education years (1995)	6

Educational Attainment (10)

Age group (1980)	25+
Total population	3,232,206
Highest level attained (%)	
No schooling	NA

Educational Attainment (10)

First level	
Not completed	75.6
Completed	NA
Entered second level	
S-1	8.9
S-2	NA
Postsecondary	11.5

GOVERNMENT & LAW

Political Parties (12)

National Council	No. of seats
Radical Free Democratic Party (FDP)	45
Social Democratic Party (SPS)	54
Christian Democratic People's Party (CVP)	34
Swiss People's Party (SVP)	29
Greens	9
Liberal Party (LPS)	7
Freedom Party (FPS)	7
Alliance of Independents' Party (LdU)	3
Evangelical People's Party (EVP)	2
Swiss Democratic Party (SD)	3
Workers' Party (PdAdS)	3
Others	4

Military Affairs (14B)

	1990	1991	1992	1993	1994	1995
Military expenditures						
Current dollars (mil.)	5,531[e]	4,986	5,028	4,651	4,896	5,034
1995 constant dollars (mil.)	6,357[e]	5,510	5,408	4,876	5,019	5,034
Armed forces (000)	22	22	31	31	39	29[e]
Gross national product (GNP)						
Current dollars (mil.)	267,000	277,700	283,400	288,700	301,500	315,700[e]
1995 constant dollars (mil.)	306,900	306,800	304,800	302,700	309,000	315,700[e]
Central government expenditures (CGE)						
1995 constant dollars (mil.)	NA	71,680	75,030	81,280	NA	NA
People (mil.)	6.8	6.9	7.0	7.1	7.1	7.2
Military expenditure as % of GNP	2.1	1.8	1.8	1.6	1.6	1.6
Military expenditure as % of CGE	NA	7.7	7.2	6.0	NA	NA
Military expenditure per capita (1995 $)	938	796	773	691	705	703
Armed forces per 1,000 people (soldiers)	3.2	3.2	4.4	4.4	5.5	4.0
GNP per capita (1995 $)	45,270	44,350	43,580	42,880	43,430	44,070
Arms imports[6]						
Current dollars (mil.)	625	470	270	230	40	20
1995 constant dollars (mil.)	718	519	290	241	41	20
Arms exports[6]						
Current dollars (mil.)	60	240	450	120	110	100
1995 constant dollars (mil.)	69	265	484	126	113	100
Total imports[7]						
Current dollars (mil.)	69,680	66,480	61,740	56,720	64,070	76,980
1995 constant dollars (mil.)	80,080	73,470	66,410	59,460	65,680	76,980
Total exports[7]						
Current dollars (mil.)	63,780	61,520	61,380	58,690	66,230	77,650
1995 constant dollars (mil.)	73,300	67,980	66,020	61,530	67,890	77,650
Arms as percent of total imports[8]	.9	.7	.4	.4	.1	0

Government Budget (13A)

Year: 1996

Total Expenditures: 96,095 Millions of Francs

Expenditures as a percentage of the total by function:

General public services and public order	4.72
Defense	5.81
Education	2.58
Health	15.58
Social Security and Welfare	51.99
Housing and community amenities	.62
Recreational, cultural, and religious affairs	.40
Fuel and energy	.44
Agriculture, forestry, fishing, and hunting	4.44
Mining, manufacturing, and construction	-
Transportation and communication	6.47
Other economic affairs and services	.39

Crime (15)

Crime rate (for 1997)

Crimes reported	387,300
Total persons convicted	NA
Crimes per 100,000 population	5,450

Persons responsible for offenses

Total number of suspects	103,300
Total number of female suspects	15,400
Total number of juvenile suspects	16,100

LABOR FORCE

Labor Force (16)

Total (million)	3.8
Services	67%
Manufacturing, construction	29%
Agriculture and forestry	4%

Data for 1995. Of total, 850,000 foreign workers, mostly Italian.

Unemployment Rate (17)

5% (1997 est.)

PRODUCTION SECTOR

Electric Energy (18)

Capacity	14.27 million kW (1995)
Production	55 billion kWh (1996)
Consumption per capita	6,850 kWh (1996 est.)

Transportation (19)

Highways:

total: 71,117 km (including 1,594 km of expressways)

paved: NA km

unpaved: NA km (1996 est.)

Waterways: 65 km; Rhine (Basel to Rheinfelden, Schaffhausen to Bodensee); 12 navigable lakes

Pipelines: crude oil 314 km; natural gas 1,506 km

Merchant marine:

total: 22 ships (1,000 GRT or over) totaling 424,261 GRT/733,551 DWT ships by type: bulk 13, cargo 1, chemical tanker 5, oil tanker 2, roll-on/roll-off cargo 1 (1997 est.)

Airports: 67 (1997 est.)

Airports—with paved runways:

total: 42

over 3,047 m: 4

2,438 to 3,047 m: 5

1,524 to 2,437 m: 12

914 to 1,523 m: 6

under 914 m: 15 (1997 est.)

Airports—with unpaved runways:

total: 25

914 to 1,523 m: 1

under 914 m: 24 (1997 est.)

Top Agricultural Products (20)

Grains, fruits, vegetables; meat, eggs.

MANUFACTURING SECTOR

GDP & Manufacturing Summary (21)

Detailed value added figures are listed by both International Standard Industry Code (ISIC) and product title.

	1980	1985	1990	1994
GDP ($-1990 mil.)[1]	183,882	197,004	226,055	226,028
Per capita ($-1990)[1]	29,100	30,141	33,078	31,696
Manufacturing share (%) (current prices)[1]	27.9	25.0	23.7	NA
Manufacturing				
Value added ($-1990 mil.)[1]	49,120	48,482	55,235	59,409
Industrial production index	100	99	112	121
Value added ($ mil.)	27,438	*23,264*	53,557	*60,111*
Gross output ($ mil.)	*73,054*	*60,784*	141,237	174,154

	1980	1985	1990	1994
Employment (000)	927	864	880	776
Profitability (% of gross output)				
Intermediate input (%)	NA	NA	NA	NA
Wages and salaries inc. supplements (%)	NA	NA	NA	NA
Gross operating surplus	NA	NA	NA	NA
Productivity ($)				
Gross output per worker	75,310	68,141	151,677	185,637
Value added per worker	40,009	35,442	60,826	66,570
Average wage (inc. supplements)	NA	NA	NA	NA
Value added ($ mil.)				
311/2 Food products	2,905	1,880	4,004	4,846
313 Beverages	499	323	688	832
314 Tobacco products	292	189	402	487
321 Textiles	972	775	1,410	1,115
322 Wearing apparel	864	299	406	604
323 Leather and fur products	124	52	104	144
324 Footwear	324	114	140	164
331 Wood and wood products	1,078	1,015	2,419	2,747
332 Furniture and fixtures	707	656	1,586	1,826
341 Paper and paper products	624	506	1,081	1,133
342 Printing and publishing	1,471	1,633	3,893	4,422
351 Industrial chemicals	1,529	1,339	3,387	4,451
352 Other chemical products	1,331	1,133	2,766	3,529
353 Petroleum refineries	584	487	700	1,056
354 Miscellaneous petroleum and coal products	96	68	115	126
355 Rubber products	225	170	449	506
356 Plastic products	625	472	1,245	1,404
361 Pottery, china and earthenware	136	135	365	419
362 Glass and glass products	187	185	502	576
369 Other non-metal mineral products	651	596	1,529	1,607
371 Iron and steel	454	359	789	720
372 Non-ferrous metals	583	499	1,179	1,163
381 Metal products	1,921	1,644	3,885	3,837
382 Non-electrical machinery	3,775	3,489	8,248	7,725
383 Electrical machinery	2,859	3,565	8,488	10,153
384 Transport equipment	508	469	1,109	1,038
385 Professional and scientific equipment	1,976	1,111	2,446	3,234
390 Other manufacturing industries	138	103	222	246

FINANCE, ECONOMICS, & TRADE

Economic Indicators (22)

National product: GDP—purchasing power parity—$172.4 billion (1997 est.)

National product real growth rate: 0.4% (1997 est.)

National product per capita: $23,800 (1997 est.)

Inflation rate—consumer price index: -0.1% (1997)

Balance of Payments (23)

	1992	1993	1994	1995	1996
Exports of goods (f.o.b.)	79,870	75,424	82,625	97,139	95,51[illegible]
Imports of goods (f.o.b.)	−80,155	−73,853	−79,295	−93,916	−93,67[illegible]
Trade balance	−285	1,571	3,330	3,223	1,83[illegible]
Services - debits	−29,830	−27,551	−31,702	−34,824	−35,64[illegible]
Services - credits	47,303	46,628	49,365	57,536	58,08[illegible]
Private transfers (net)	−580	−511	−1,013	−1,280	−1,22[illegible]
Government transfers (net)	−2,373	−2,229	−2,411	−2,776	−2,57[illegible]
Overall balance	14,236	17,908	17,569	21,879	20,47[illegible]

Exchange Rates (24)

Exchange rates:	
Swiss francs, franken, or franchi (SFR) per US$1	
January 1998	1.4757
1997	1.4513
1996	1.2360
1995	1.1825
1994	1.3677
1993	1.4776

Top Import Origins (25)

$86.6 billion (c.i.f., 1997) Data are for 1995 est.

Origins	%
European Union	33
Italy	9
Germany	8
France	4
South Korea	5
United States	4
Japan	4

Top Export Destinations (26)

$99.2 billion (f.o.b., 1997) Data are for 1995 est.

Destinations	%
European Union	57
Germany	17
Italy	16
France	11
Lebanon	14
Saudi Arabia	7

Economic Aid (27)

Donor: ODA, $1.034 billion (1995).

Import Export Commodities (28)

Import Commodities	Export Commodities
Machinery 22%	Machinery 29%
Chemicals 20%	Chemicals 26%
Metals 8%	Metals 8%
Agricultural products 9%	Agricultural products 4%

SYRIA

CAPITAL: Damascus (Dimashq).

FLAG: The national flag is a horizontal tricolor of red, white, and black stripes; in the white center stripe are two green five-pointed stars.

ANTHEM: *An-Nashid as-Suri (The Syrian National Anthem)* begins "Protectors of the nation, peace be upon you."

MONETARY UNIT: The Syrian pound (s£) is a paper currency of 100 piasters. There are coins of 25 and 50 piasters and 1 Syrian pound and notes of 1, 5, 10, 25, 50, 100, and 500 Syrian pounds. s£1 = $0.0891 (or $1 = s£11.225).

WEIGHTS AND MEASURES: The metric system is the legal standard, but local units are widely used.

HOLIDAYS: New Year's Day, 1 January; Revolution Day, 8 March; Egypt's Revolution Day, 23 July; Union of Arab Republics Day, 1 September; National Day, 16 November. Muslim religious holidays include 'Id al-Fitr, 'Id al-'Adha', Milad an-Nabi, and Laylat al-Miraj. Christian religious holidays include Easter (Catholic); Easter (Orthodox); and Christmas, 25 December.

TIME: 2 PM = noon GMT.

LOCATION AND SIZE: Situated in southwest Asia, at the eastern end of the Mediterranean Sea, Syria has an area of 185,180 square kilometers (71,498 square miles), slightly larger than the state of North Dakota. Included in this total is the Golan Heights region (1,176 square kilometers/454 square miles), which Israel captured in 1967 and annexed on 14 December 1981.

Syria has a total boundary length of 2,446 kilometers (1,520 miles).

CLIMATE: Average temperatures for Damascus range from about 21° to 43°C (70–109°F) in August and from about −4° to 16°C (25°–61°F) in January. Average rainfall ranges from less than 25 centimeters (10 inches) in the eastern three-fifths of the country to around 125 centimeters (50 inches) in some mountain areas.

Syrian Arab Republic
Al-Jumhuriyah al-'Arabiyah as-Suriyah

INTRODUCTORY SURVEY

RECENT HISTORY

During World War II (1939–45) French and British forces took control of Syria. Later, under pressure from Britain and the United States, the French permitted elections and the formation of a nationalist government. Britain and the United States recognized Syria's independence in 1944.

The Palestine War of 1948–49, which ended with the defeat of the Arab armies and the establishment of an Israeli state, caused Syrians to lose faith in their leadership. Several army factions struggled to gain control of the Syrian state. The 1950s saw growth of pan-Arab (in favor of a united Arab nation) and left-of-center political forces. The strongest of these was the Arab Socialist Ba'th Party, which developed ties with Gamal Abdel Nasser, the pan-Arabist president of Egypt. In late 1957, the Arab Socialist Ba'th Party and Nasser agreed to a union of the two countries, and on 1 February 1958, they proclaimed the union of Syria and Egypt as the United Arab Republic (UAR).

A single-party structure replaced the lively Syrian political tradition; decisions were made in Egypt; land reforms were introduced. Syrians became unhappy with Egyptian rule and in September 1961, Syria seceded from the UAR. After a period of political instability, power was seized in 1963 by the Ba'th Party and a radical socialist government was formed.

During the 1980s, Syria intervened militarily in neighboring Arab states for its own political

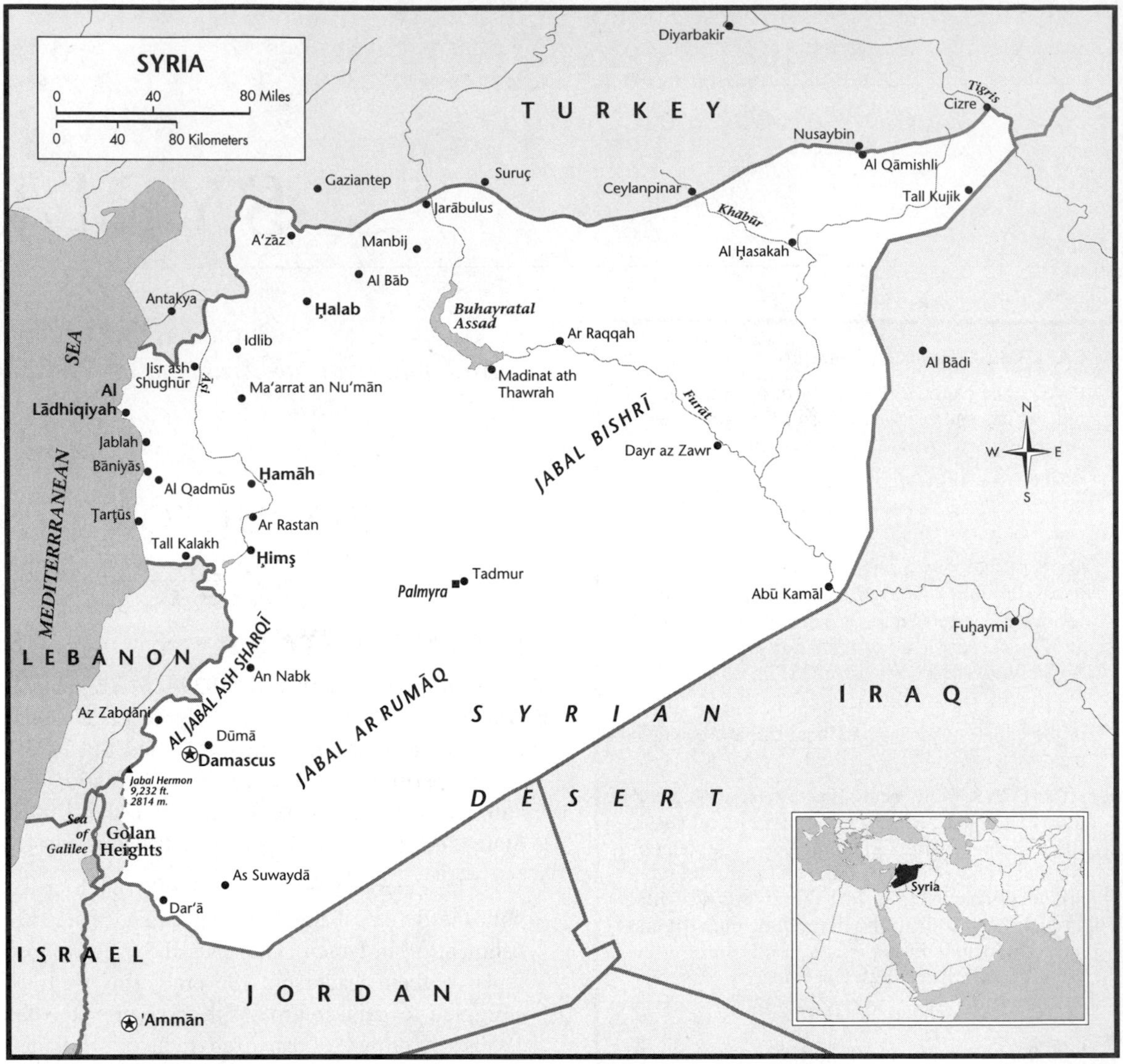

purposes. Syria supported the Palestinians in Jordan's 1970 civil war, aided Christian forces in Lebanon in 1976, and supported Iran in its war against Iraq. After signing a 20-year friendship treaty with the Soviet Union in 1980, Syria placed Soviet antiaircraft missiles in Lebanon's Bekaa Valley. Syria joined the coalition of forces against Iraq in 1990 and agreed to participate in direct peace talks with Israel in 1991. The collapse of the Soviet Union removed Syria's most important outside support.

GOVERNMENT

The constitution of 12 March 1973 gives strong executive power to the president, who is nominated by the Ba'th Party and elected by popular vote to a seven-year term. The single-chamber People's Council has 250 members who are elected every four years, but who have no real power. Suffrage is universal, beginning at age 18. Syria has been under military rule since 1963 (except for 1973–74).

Syria is divided into 13 provinces (*muhafazat*) and Damascus.

Judiciary

There are civil and criminal appeals courts, the highest being the Court of Cassation. Separate State Security Courts rule in cases affecting the security of the government. In addition, Shari'ah courts apply Islamic law in personal cases. The Druze and non-Muslim communities have their own religious courts.

Political Parties

The Arab Socialist Ba'th Party is Syria's strongest political force. It is much larger and more influential than the combined strength of its five partners in the National Progressive Front (NPF), an official political alignment that groups the Communist Party of Syria (SCP) and four small leftist parties—the Syrian Arab Socialist Union (ASU), the Socialist Unionist Party (ASUM), the Democratic Socialist Union Party (DSUP), and the Arab Socialist Party (ASP)—with the Ba'th.

DEFENSE

In 1995, the army had an estimated 315,000 regular troops (250,000 draftees) and 400,000 reserves. The navy had 6,000 men and 8,000 reserves. Naval vessels included 3 submarines, 2 frigates, 27 fast patrol boats, and 1 amphibious landing vessel. The air force had 40,000 men, 579 combat aircraft, and 100 armed helicopters. The air defense command numbered 60,000 with 25 brigades or regiments.

In 1995, budgeted military expenditures totaled $2 billion, or 6% of the gross domestic product (GDP).

ECONOMIC AFFAIRS

Development of the state-owned oil industry and greater use of other mineral resources, particularly phosphates, have helped to expand Syrian industry, which was formerly concentrated in light manufacturing and textiles.

Syria's economy has improved since 1990 due to an increase in oil production, the recovery of the agricultural sector from drought, aid from the Gulf states, and economic reforms that boosted Syrian business. The economy grew by an annual average of 7.5% from 1990 to 1995. However, high government deficits, inflation, and the slow pace of economic reform potentially threaten economic growth over the long term.

Public Finance

Although Syria was able to balance its budget in 1992, large military expenditures and continued subsidization of basic commodities and social services have produced deficits in subsequent years. The U.S. Central Intelligence Agency estimates that, in 1995, government revenues totaled approximately $2.5 billion and expenditures $3.4 billion.

External debt totaled $21.2 billion. The majority of this amount (approximately $12 billion) is military debt owed to the former Soviet Union. It is questionable whether this debt will ever be repaid. Another $3 billion is owed to Western nations.

Income

In 1998, Syria's gross national product (GNP) was $15.6 billion, or about $1,020 per person.

Industry

In 1965, the textile industry was nationalized and reorganized into 13 large state corporations. In 1995, manufacturing and mining accounted for 14% of the gross domestic product (GDP). In the 1970s, the government emphasized the production of iron and steel, fertilizers, chemicals, and household appliances.

Also important are the chemical and engineering industries, the food industry, and oil refining. The largest component of the chemical and engineering sector is the cement industry, which produced 6 million tons in 1995. Some 70,000 tons of crude steel were produced that year, and Syrian refineries produced 61.7 million barrels of residual and distillate fuel oil.

Banking and Finance

Syria's financial services sector is underdeveloped. Besides the Central Bank, there are seven banks and financial institutions in the country, all of which are state-run. The Central Bank, founded in 1956, is the bank of issue for currency, the financial agent of the government, and the cashier for the treasury. The Agricultural Bank makes loans to farmers at low interest; the Industrial Bank (nationalized in 1961), the Popular Credit Bank and the Real Estate Bank (both founded in 1966), and the Commercial Bank of Syria (formed in 1967 by a merger of five nationalized commercial banks) make loans in their defined sectors. So decrepit is the country's financial services sector that most Syrian businessmen and foreigners use banks in either Lebanon or Cyprus. Foreign diplomats in Damascus, for instance, use accounts in the Chtaura, in Lebanon's Beqaa valley, around one hour by car from Damascus.

Private sector groups are calling for reforms such as private participation in banking, the creation of a stock exchange, and separation of the Central Bank of Syria from government.

Economic Development

The transformation of Syria's economy began with the Agrarian Reform Law in 1958, which called for the expropriation of large tracts of land.

During the union with Egypt, laws were passed for the nationalization of banks, insurance companies, and large industrial firms. After the Ba'th Party came to power in 1963, the socialist trend reasserted itself with greater force. A series of laws created a new banking system and instituted public ownership of all large industries. By the early 1970s, however, the government had relaxed many restrictions on trade, foreign investment, and private-sector activity, in an effort to attract private and foreign, especially Arab, contributions to Syria's economic growth.

Since 1961, a series of five-year plans has concentrated on developing the nation's infrastructure and increasing agricultural and industrial production. Investments reached 60% of the target under the first plan (1961–65); the second plan (1966–70) aimed to expand real GDP by 7.2% annually but achieved a yearly growth rate of only 4.7%. The third plan (1971–75) was disrupted by the 1973 Arab-Israeli war, but thanks to aid from other Arab states and large oil price increases, Syria experienced an economic boom with a high annual growth rate of 13%. The fourth plan (1976–80) was hampered by the high cost of Syria's military intervention in Lebanon and a cutoff of aid from Gulf states; economic growth varied widely, from −2.8% in 1977 to 9.2% in 1980.

Under the fifth plan (1981–85), development projects begun during the previous plan were to be continued or completed. Total investment was estimated at s£101 billion, of which 23% was to be provided by the private sector. Real GDP was to grow by 7.7% annually; actual growth rates ranged from 10.2% in 1981 to −3.6% in 1984, averaging 2.3% for the period.

Syria's sixth development plan (1986–90) emphasized increased productivity rather than new projects, with special emphasis on agriculture and agro-industries. Actual investment in agriculture accounted for 18.7% of total spending. The share of the industry and energy sector was at 19.7%, far below the planned 30.9%. Services received the highest share, with 53% of the total.

The seventh five-year plan (1991–95) proposed total investments of s£259 billion, more than double the amount spent under the previous plan. It aims at spending 81.7% of the total on the public sector and 18.3% on the mixed sector/private-sector cooperatives. Officials at the Supreme Planning Commission have stated that agriculture and irrigation continue to receive top priority, with self-sufficiency in cereal production a policy objective. Output in agriculture and manufacturing is planned to expand by 5.6% per annum.

During 1949–86, multilateral assistance to Syria totaled $822.7 million, of which 77% came through the IBRD. U.S. loans and grants during the same period amounted to $581.9 million. Financial aid to Syria from Arab oil-producing states is not made public. Since 1982, Syria has received a million tons of oil annually from Iran, free of charge. Since Syria is in arrears on payments to the World Bank, disbursements were halted in 1988 and projects canceled. Syria has been in violation of the Brooke Amendment since 1985. The improvement in Syria's external payment position in 1989 as well as the resumption of aid flows to Syria in 1990 due to its participation in the coalition against Iraq helped to restore its ability to repay its debt.

SOCIAL WELFARE

The Ministry for Social and Labor Affairs was formed in 1956 to protect the interests of the working population, provide clean housing conditions for workers, and support charities. A system of social insurance, introduced in 1959, provides old age pensions and disability and death benefits. Workers' compensation (for injuries on the job) provides temporary and permanent disability benefits, as well as medical and survivor benefits.

Although the government supports equal pay for equal work and encourages education for women, Islamic beliefs that contradict these policies govern many areas of women's lives.

President Assad's regime has been accused of widespread human rights violations. There are reports that political suspects are imprisoned without trial. Prisoners are subjected to torture and sometimes executed for no particular reason. Public criticism of the Baʿth Party or of government officials is not permitted.

Healthcare

Since World War II, malaria has been virtually eliminated with the aid of the World Health Organization (WHO). Intestinal and respiratory diseases caused by poor living conditions are still common, particularly in rural areas. Average life expectancy was 69 years.

In 1990, there were 1.1 hospital beds per 1,000 people. In 1991, there were 10,114 physicians and 11,957 nurses. In 1993, there was one doctor for every 1,159 people and one nurse for every 1,047 people. That year, about 99% of the population had access to health care services.

Housing

According to the latest available information for 1980–88, total housing units numbered 1.7 million with 6.4 people per dwelling.

EDUCATION

Elementary schooling is free and compulsory for six years. The estimated illiteracy rates in 1995 were 14.3% for men and 44.2% for women. In 1994 there were 2.7 million primary-school pupils with 113,384 teachers in 10,420 schools. At the secondary level, there were 928,882 students and 62,080 teachers that year. Syria has four universities: the University of Damascus; Tishrin University; the University of Aleppo (Halab); and Al-Ba`th University. In 1992, higher level institutions had a total of 194,371 students enrolled.

1999 KEY EVENTS TIMELINE

March

- Hafez al-Assad is sworn in for his fifth term as Syria's president.

July

- Syria purchases fighter aircraft, tanks, and missiles from Russia.

August

- Syria allows the importation of Jordanian newspapers, which had been banned.
- A national holiday is declared in Syria on the occasion of a solar eclipse.

September

- President Clinton meets at the White House to discuss Syrian-Israeli peace talks with the Syrian foreign minister.
- U.S. Secretary of State Madeleine Albright visits Damascus.
- The United Nations World Food Program provides aid after Syria suffers the worst drought to hit the Middle East in decades.

November

- The grooming of Bashar el-Assad, son and apparent successor of President Hafez el-Assad, accelerates with a trip to Paris that analysts say is aimed at raising his profile.

December

- Peace talks with Israel begin in Washington, D.C.

ANALYSIS OF EVENTS: 1999

BUSINESS AND THE ECONOMY

The economy of Syria, where per capita annual income averages about $1,000, has been burdened to the point of near stagnation by huge military expenditures and by a $20 billion debt, much of it owed to Russia. Low oil prices and the worst drought in decades has also had a severely debilitating effect.

In July, a thrice-weekly express train began running along part of the route between Damascus and Amman. The line also includes a weekly express freight run.

In October, electricity ministers from Saudi Arabia, Jordan, Egypt, Kuwait, and Libya met to discuss progress on the linkage of national grids in the Middle East and North Africa. The first stage planned is the connection of Syria, Lebanon, Jordan, Turkey, Iraq, and Egypt. The initial link between Egypt and Jordan was begun in March. Plans are eventually to connect eastern and western Arab countries by 2001, Arab nations with Europe, and the Gulf Cooperation Council nations with one another. The Arab League secretary general urged the creation of an Arab free-trade zone for electronic goods. In May, Syria and Jordan had signed an energy sharing agreement that provided for fossil fuel exploration as well as for the proposed electrical linkage.

GOVERNMENT AND POLITICS

In March, Hafez al-Assad was sworn in for his fifth seven-year term as Syria's president. The dissemination of his inaugural speech through the media rather than by a public reading reignited speculation over his possible successor, a speculation fueled by his limited public presence over the last ten years. Hafez al-Assad seems to be grooming his son Bashar to assume the presidency eventually.

In May, the Iranian president visited Damascus to reinforce ties with Syria, Iran's closest ally in the Middle East. Analysts took the visit as a step toward a regional security arrangement in the Gulf states. On a trip to Moscow in July, the president of Syria signed a $2 billion contract to buy fighter aircraft, tanks, and missiles from Russia to upgrade Syria's military capability.

In September, President Clinton met with the Syrian foreign minister at the White House to discuss Syrian-Israeli peace talks, and in the following month Syria called on the U.S. to persuade Israel to resume peace talks with Syria, which halted during the three-year tenure of Israeli Prime Minister Benjamin Netanyahu. Syria hoped that the recent election of Ehud Barak would mean the resumption of peace talks. Syria insists on what it maintains was a proposal made by Netanyahu's predecessor, Yitzhak Rabin, namely, that Israel would withdraw from the Golan Heights in return for peace with Syria. Barak has said that he agrees to withdrawal in principle, but the extent depends upon Syrian guarantees of security. Through the Norwegian foreign minister, who visited the Middle East in August, Israel conveyed a message to the Syrian president asking for Syrian openness to talks at any time.

Also in August, Barak outlined his concerns in a visit with Clinton in Washington. U.S. Secretary of State Madeleine Albright stated her willingness to mediate talks between Syria and Israel, but the Syrian foreign minister expressed his disappointment with the results of Albright's visit to Damascus in early September. King Abdullah of Jordan announced his support for the position of Syria regarding peace talks with Israel and urged that negotiations resume at the point where they left off in 1996. Further support came from Egypt, which in October refused to participate in multilateral talks with Israel unless Syria and Lebanon were included, a condition Israel was unwilling to meet. In the same month, the foreign minister of Jordan met with the president of Syria to deliver a message from Jordan's King Abdullah and to try to restart peace talks between Israel and Syria.

CULTURE AND SOCIETY

In March, for the first time in ten years, authorities in Syria and Turkey ordered the removal of the wire fencing separating the two countries to allow some hundreds of Syrian citizens to cross the border into Turkey, where they could visit relatives to celebrate the end of Id al-ʿAdhaʾ, an Islamic time of pilgrimage. In a similar move, officials allowed seven Druze brides from Syria to enter Israel to marry husbands in the Golan Heights, while relatives were also permitted to cross the border for the wedding. The brides had to surrender their Syrian passports and citizenship. The Druze are divided by the Syrian-Israeli border in the Golan.

With improved relations between Damascus and Amman since the accession of King Abdullah in Jordan, Syria decided to allow Jordanian newspapers into the country in August. Syria had banned their importation when King Hussein signed a unilateral peace agreement with Israel.

Also in August, a national holiday was declared in Syria on the occasion of a solar eclipse visible across Europe and southern Asia, and a journalists' rights group called Reporters Sans Frontieres (RSF) included Syria in a list of 45 countries denounced by the RSF for interfering in various ways and to varying degrees with access to the Internet.

In the same month, the European Union began to dispense aid to Palestinian and Bedouin refugees in Syria. The EU aid program includes two mobile clinics to dispense health care to the Bedouin, the construction of a new rehabilitation clinic and the improvement of an existing one for Palestinians, and the installation of a medical laboratory at a Palestinian hospital in Damascus. In September, the United Nations World Food Program offered aid in the millions of dollars to Syria to help offset the effects of the worst drought to hit the Middle East in decades. The animals of Bedouin herders in the grazing region of Badia were particularly affected.

DIRECTORY

CENTRAL GOVERNMENT

Head of State

President
Hafez al-Assad

Prime Minister
Mahmoud Zubi, Office of the Prime Minister
PHONE: +963 (11) 2226001
FAX: +963 (11) 2233373

Deputy Prime Minister
Mustafa Tlass, Office of the Deputy Prime Minister

PHONE: +963 (11) 3718306
FAX: +963 (11) 4417546

Ministers

Minister of Agriculture and Agrarian Reform
Assad Mustafa, Ministry of Agriculture and Agrarian Reform
PHONE: +963 (11) 2213613
FAX: +963 (11) 2244078

Minister of Communications
Radwan Martini, Ministry of Communications
PHONE: +963 (11) 2227033
FAX: +963 (11) 2246403

Minister of Defense
Mustafa Tlass, Ministry of Defense
PHONE: +963 (11) 3718306
FAX: +963 (11) 4417546

Minister of Education
Ghassan al-Halabi, Ministry of Education
PHONE: +963 (11) 4444703
FAX: +963 (11) 4420435

Minister of Finance
Khalid al-Mahayni, Ministry of Finance, Al-Sabee Bahrat Sq. Baghdad St., PO Box 13136, Damascus, Syria
PHONE: +963 (11) 2219602
FAX: +963 (11) 2224701

Minister of Foreign Affairs
Farouk al-Sharaa, Ministry of Foreign Affairs
PHONE: +963 (11) 3331200
FAX: +963 (11) 3327620

Minister of Foreign Trade and Economy
Mouhamad Imadi, Ministry of Foreign Trade and Economy, Bawabet el Salheya, Damascus, Syria
PHONE: +963 (11) 2213513
FAX: +963 (11) 2225695
E-MAIL: econ-min@net.sy

Minister of Health
Mouhamad Iyad Shatti, Ministry of Health
PHONE: +963 (11) 2223085
FAX: +963 (11) 3310404

Minister of the Interior
Mohamad Harba, Ministry of the Interior
PHONE: +963 (11) 2211001
FAX: +963 (11) 2223428

Minister of Justice
Hussain Hassoun, Ministry of Justice
PHONE: +963 (11) 2214105
FAX: +963 (11) 2246250

Minister of Transportation
Mufid Abdul al-Karim, Ministry of Transportation
PHONE: +963 (11) 3336801
FAX: +963 (11) 2233317

POLITICAL ORGANIZATIONS

Arab Socialist Union

NAME: Fayiz Ismail

National Progressive Front

TITLE: President
NAME: Hafiz al-Asad

Syrian Arab Socialist Party

NAME: Ghassan `Abd-al-Aziz Uthman

DIPLOMATIC REPRESENTATION

Embassies in Syria

Australia
128/A Farai Street East, Villas Mezzeh Damascus PO Box 3882, Damascus, Syria
PHONE: +963 (11) 6664317
FAX: +963 (11) 6621195

Italy
Avenue al-Mansour 82, Damascus, Syria
PHONE: +963 (11) 3338338
FAX: +963 (11) 3320325

United Kingdom
Kotob Building 11 Mohammad Kurd Ali Street, Malki PO Box 37, Damascus, Syria
PHONE: +963 (11) 3712561
FAX: +963 (11) 3731600

JUDICIAL SYSTEM

Supreme Constitutional Court

High Judicial Council

FURTHER READING

Articles

Contreras, Joseph. ''Cyber-Savvy in Syria.'' *Newsweek International* (April 26, 1999): 54

Klein, Aharon and Scott Macleod. ''Israel's New Syrian View.'' *TIME* 154 (July 5, 1999): 44.

Macleod, Scott, William Dowell, Eric Silver, and Douglas Weller. ''Syria.'' *TIME* (October 18, 1999): 56.

Peterson, Scott. ''A Lyrical Dialogue Between Two Cultures.'' *The Christian Science Monitor*, 22 October 1999, v91 i229, 7.

———. ''Slow Going on the Syrian-Israel Peace Track.'' *The Christian Science Monitor*, 19 October 1999, v91 i226, 7.

Rabinovich, Itamar. ''Israel, Syria and the Road to Peace Not Taken.'' *New York Times*, 1 September 1999, v148 i51632, A23.

Satloff, Robert. ''On Board: Barak Unveils His Plan for Peace.'' *New Republic* (August 16, 1999): 21

''Syria's Nods and Winks.'' *The Economist* (July 24, 1999): 41.

Weymouth, Lally. ''We Want Our Land Back.'' *Newsweek* (October 11, 1999): 55.

Zisser, Eyal. ''Heir Apparent.'' *New Republic* 229 (October 11, 1999): 20

Books

Mannheim, Ivan. *Jordan, Syria and Lebanon Handbook*. Lincolnwood, Ill.: Passport Books, 1998.

Perthes, Volker. *The Political Economy of Syria under Asad*. New York: I.B. Tauris, 1997.

Waldner, David. *State Building and Late Development*. Ithaca, NY: Cornell University Press, 1999.

Internet

ArabNet. ''Syria.'' Available Online @ http://www.arab.net/syria/syria_contents.html (November 2, 1999).

SYRIA: STATISTICAL DATA

For sources and notes see "Sources of Statistics" in the front of each volume.

GEOGRAPHY

Geography (1)

Area:

Total: 185,180 sq km.

Land: 184,050 sq km.

Water: 1,130 sq km.

Note: includes 1,295 sq km of Israeli-occupied territory.

Area—comparative: slightly larger than North Dakota.

Land boundaries:

Total: 2,253 km.

Border countries: Iraq 605 km, Israel 76 km, Jordan 375 km, Lebanon 375 km, Turkey 822 km.

Coastline: 193 km.

Climate: mostly desert; hot, dry, sunny summers (June to August) and mild, rainy winters (December to February) along coast; cold weather with snow or sleet periodically hitting Damascus.

Terrain: primarily semiarid and desert plateau; narrow coastal plain; mountains in west.

Natural resources: petroleum, phosphates, chrome and manganese ores, asphalt, iron ore, rock salt, marble, gypsum.

Land use:

Arable land: 28%

Permanent crops: 4%

Permanent pastures: 43%

Forests and woodland: 3%

Other: 22% (1993 est.).

HUMAN FACTORS

Demographics (2A)

	1990	1995	1998	2000	2010	2020	2030	2040	2050
Population	12,620.2	15,086.6	16,673.3	17,758.9	23,328.9	28,925.7	34,352.1	39,250.9	43,463.3
Net migration rate (per 1,000 population)	NA	NA	NA	NA	NA	NA	NA	NA	NA
Births	NA	NA	NA	NA	NA	NA	NA	NA	NA
Deaths	NA	NA	NA	NA	NA	NA	NA	NA	NA
Life expectancy - males	64.2	65.7	66.5	67.0	69.4	71.5	73.1	74.4	75.5
Life expectancy - females	66.1	68.0	69.1	69.9	73.2	75.9	78.2	80.1	81.5
Birth rate (per 1,000)	42.8	40.4	37.8	36.1	28.3	23.2	19.3	16.6	14.9
Death rate (per 1,000)	6.9	6.0	5.6	5.3	4.3	4.0	4.2	5.0	6.1
Women of reproductive age (15-49 yrs.)	2,660.4	3,269.5	3,704.3	4,014.5	5,814.9	7,737.3	9,371.0	10,467.8	10,917.1
of which are currently married	NA	NA	NA	NA	NA	NA	NA	NA	NA
Fertility rate	6.7	6.1	5.5	5.2	3.6	2.7	2.3	2.1	2.0

Except as noted, values for vital statistics are in thousands; life expectancy is in years.

Health Indicators (4)

Life expectancy at birth

1980 62

1997 69

Daily per capita supply of calories (1996) 3,339

Total fertility rate births per woman (1997) 4.0

Maternal mortality ratio per 100,000 live births (1990-97) . . . . . 180[b]

Safe water % of population with access (1995) 88

Sanitation % of population with access (1995) 71

Consumption of iodized salt % of households (1992-98)[a] 40

Smoking prevalence

Male % of adults (1985-95)[a] . . . .

Female % of adults (1985-95)[a] . . . .

Tuberculosis incidence per 100,000 people (1997) 75

Adult HIV prevalence % of population ages 15-49 (1997) 0.01

Infants and Malnutrition (5)

Under-5 mortality rate (1997) 33

% of infants with low birthweight (1990-97) 7

Births attended by skilled health staff % of total[a] . . . 77

% fully immunized (1995-97)

TB 100

DPT 95

Polio 95

Measles 93

Prevalence of child malnutrition under age 5 (1992-97)[b] 13

Ethnic Division (6)

Arab 90.3%

Kurds, Armenians, and other 9.7%

Religions (7)

Sunni Muslim 74%; Alawite, Druze, and other Muslim sects 16%; Christians (various sects) 10%; Jewish (tiny communities in Damascus, Al Qamishli, and Aleppo).

Languages (8)

Arabic (official); Kurdish, Armenian, Aramaic, Circassian widely understood; French, English somewhat understood.

EDUCATION

Public Education Expenditures (9)

Public expenditure on education (% of GNP)

1980 . . . . . 4.6

1996 . . . . . 4.2

Expenditure per student

Primary % of GNP per capita

1980 . . . . . 8.0

1996 . . . . . 8.2

Secondary % of GNP per capita

1980 . . . . . 15.1

1996 . . . . . 16.4

Tertiary % of GNP per capita

1980 . . . . . 74.7

1996 . . . .

Expenditure on teaching materials

Primary % of total for level (1996) . . . . . 1.9[1]

Secondary % of total for level (1996) 8.2

Primary pupil-teacher ratio per teacher (1996) 23

Duration of primary education years (1995) 6

Literacy Rates (11A)

In thousands and percent[1]	1990	1995	2000	2010
Illiterate population (15+ yrs.)	2,197	2,259	2,295	2,280
Literacy rate - total adult pop. (%)	65.7	70.8	75.6	83.3
Literacy rate - males (%)	82.2	85.7	88.6	92.9
Literacy rate - females (%)	48.9	55.8	62.5	73.7

GOVERNMENT & LAW

Political Parties (12)

People's Council	No. of seats
National Progressive Front	167
Independents	83

Government Budget (13B)

Revenues . . . . . $3.9 billion

Expenditures . . . . . $4.3 billion

Capital expenditures . . . . . $1.9 billion

Data for 1996 est.

Crime (15)

Crime rate (for 1997)

Crimes reported	7,050
Total persons convicted	6,000
Crimes per 100,000 population	42

Persons responsible for offenses

Total number of suspects	1,800
Total number of female suspects	875
Total number of juvenile suspects	900

LABOR FORCE

Labor Force (16)

Total (million)	4.7
Services	40%
Agriculture	40%
Industry	20%

Data for 1995 est. Percent distribution for 1996 est.

Unemployment Rate (17)

12% (1997 est.)

PRODUCTION SECTOR

Electric Energy (18)

Capacity	4.157 million kW (1995)
Production	14.9 billion kWh (1995)
Consumption per capita	970 kWh (1995)

GOVERNMENT & LAW

Military Affairs (14B)

	1990	1991	1992	1993	1994	1995
Military expenditures						
Current dollars (mil.)[3]	4,391	NA	NA	NA	NA	3,563
1995 constant dollars (mil.)[3]	5,045	NA	NA	NA	NA	3,563
Armed forces (000)	408	408	408	408	320	320
Gross national product (GNP)						
Current dollars (mil.)	29,980	33,420	37,970[e]	41,560[e]	45,150[e]	49,520[e]
1995 constant dollars (mil.)	34,460	36,930	40,840[e]	43,570[e]	46,280[e]	49,520[e]
Central government expenditures (CGE)						
1995 constant dollars (mil.)	7,955	10,010	10,180	10,870	13,290	NA
People (mil.)	12.6	13.1	13.6	14.1	14.6	15.1
Military expenditure as % of GNP	14.6	NA	NA	NA	NA	7.2
Military expenditure as % of CGE	63.4	NA	NA	NA	NA	NA
Military expenditure per capita (1995 $)	400	NA	NA	NA	NA	236
Armed forces per 1,000 people (soldiers)	32.3	31.1	30.0	29.0	22.0	21.2
GNP per capita (1995 $)	2,730	2,816	3,005	3,095	3,176	3,283
Arms imports[6]						
Current dollars (mil.)	950	800	380	270	40	70
1995 constant dollars (mil.)	1,092	884	409	283	41	70
Arms exports[6]						
Current dollars (mil.)	0	0	20	0	0	0
1995 constant dollars (mil.)	0	0	22	0	0	0
Total imports[7]						
Current dollars (mil.)	2,400	2,694	3,490	4,140	5,467	4,616
1995 constant dollars (mil.)	2,758	2,977	3,754	4,340	5,604	4,616
Total exports[7]						
Current dollars (mil.)	4,253	3,618	3,093	3,146	3,547	3,970
1995 constant dollars (mil.)	4,888	3,998	3,327	3,299	3,636	3,970
Arms as percent of total imports[8]	39.6	29.7	10.9	6.5	.7	1.5
Arms as percent of total exports[8]	0	0	.6	0	0	0

PRODUCTION SECTOR

Transportation (19)

Highways:

total: 40,480 km

paved: 9,310 km (including 866 km of expressways)

unpaved: 31,170 km (1996 est.)

Waterways: 870 km; minimal economic importance

Pipelines: crude oil 1,304 km; petroleum products 515 km

Merchant marine:

total: 125 ships (1,000 GRT or over) totaling 376,903 GRT/555,679 DWT ships by type: bulk 11, cargo 110, livestock carrier 3, roll-on/roll-off cargo 1 (1997 est.)

Airports: 104 (1997 est.)

Airports—with paved runways:

total: 24

over 3,047 m: 5

2,438 to 3,047 m: 16

914 to 1,523 m: 1

under 914 m: 2 (1997 est.)

Airports—with unpaved runways:

total: 80

1,524 to 2,437 m: 3

914 to 1,523 m: 14

under 914 m: 63 (1997 est.)

Top Agricultural Products (20)

Wheat, barley, cotton, lentils, chickpeas; beef, lamb, eggs, poultry, milk.

MANUFACTURING SECTOR

GDP & Manufacturing Summary (21)

Detailed value added figures are listed by both International Standard Industry Code (ISIC) and product title.

	1980	1985	1990	1994
GDP ($-1990 mil.)[1]	19,254	22,232	23,904	32,636
Per capita ($-1990)[1]	2,212	2,148	1,936	2,303
Manufacturing share (%) (current prices)[1]	3.6	7.7	6.0	NA
Manufacturing				
Value added ($-1990 mil.)[1]	1,343	1,883	1,441	*1,991*
Industrial production index	100	136	113	*159*
Value added ($ mil.)	522	966	1,743	*2,990*
Gross output ($ mil.)	3,259	*6,892*	*10,563*	*15,648*
Employment (000)	195	104	100	*175*
Profitability (% of gross output)				
Intermediate input (%)	84	*86*	*84*	*81*
Wages and salaries inc. supplements (%)	*9*	*6*	*4*	*6*
Gross operating surplus	*6*	*7*	*12*	*13*
Productivity ($)				
Gross output per worker	16,827	*66,400*	*105,942*	*90,096*
Value added per worker	2,691	9,185	17,472	*26,118*
Average wage (inc. supplements)	*1,673*	*4,214*	*4,481*	*5,154*
Value added ($ mil.)				
311/2 Food products	*141*	*-15*	*330*	*393*
313 Beverages	*24*	*-3*	*57*	*70*
314 Tobacco products	*96*	*-10*	*223*	*222*
321 Textiles	*255*	*157*	*461*	*761*
322 Wearing apparel	*12*	*13*	*24*	*45*
323 Leather and fur products	*32*	*14*	*6*	*27*
324 Footwear	*57*	*25*	*11*	*50*
331 Wood and wood products	*27*	*26*	*20*	*59*
332 Furniture and fixtures	*76*	*71*	*56*	*158*
341 Paper and paper products	*3*	*8*	*8*	*11*
342 Printing and publishing	*5*	*16*	*18*	*21*
351 Industrial chemicals	*-9*	*6*	*4*	*9*
352 Other chemical products	*-93*	*56*	*30*	*61*
353 Petroleum refineries	*-179*	*119*	*93*	*203*
354 Miscellaneous petroleum and coal products	*-6*	*4*	*3*	*6*
355 Rubber products	*-34*	*23*	*19*	*40*
356 Plastic products	*-29*	*19*	*14*	*29*
361 Pottery, china and earthenware	*2*	*15*	*10*	*17*
362 Glass and glass products	*5*	*30*	*24*	*47*

	1980	1985	1990	1994
369 Other non-metal mineral products	*22*	*126*	*98*	*203*
371 Iron and steel	—	—	—	—
372 Non-ferrous metals	*13*	*28*	*20*	*57*
381 Metal products	*66*	*131*	*140*	*327*
382 Non-electrical machinery	*13*	*31*	*31*	*68*
383 Electrical machinery	*11*	*49*	*33*	*81*
384 Transport equipment	2	*4*	*5*	*11*
385 Professional and scientific equipment	—	—	—	—
390 Other manufacturing industries	8	23	5	*16*

FINANCE, ECONOMICS, & TRADE

Economic Indicators (22)

National product: GDP—purchasing power parity—$106.1 billion (1997 est.)

National product real growth rate: 4.6% (1997 est.)

National product per capita: $6,600 (1997 est.)

Inflation rate—consumer price index: 15%-20% (1997 est.)

Exchange Rates (24)

Exchange rates: Syrian pounds (£S) per US$1—41.9 (January 1997); official fixed rate 11.225.

Top Import Origins (25)

$5.7 billion (c.i.f., 1997) Data are for 1995 est.

Origins	%
European Union	33
Italy	9
Germany	8
France	4
South Korea	5
United States	4
Japan	4

Top Export Destinations (26)

$4.2 billion (f.o.b., 1997) Data are for 1995 est.

Destinations	%
European Union	57
Germany	17
Italy	16
France	11
Lebanon	14
Saudi Arabia	7

Economic Aid (27)

Recipient: $4.2 billion (1990-92).

Import Export Commodities (28)

Import Commodities	Export Commodities
Machinery and equipment 40%	Petroleum 70%
Foodstuffs/animals 15%	Textiles 12%
Metal and metal products 15%	Food and live animals 10%
Textiles 10%	
Chemicals 10%	
Consumer goods 5%	

Balance of Payments (23)

	1992	1993	1994	1995	1996
Exports of goods (f.o.b.)	3,100	3,203	3,329	3,858	4,298
Imports of goods (f.o.b.)	−2,941	−3,476	−4,604	−4,001	−4,516
Trade balance	159	−273	−1,275	−143	−218
Services - debits	−2,316	−2,329	−2,608	−2,520	−2,488
Services - credits	1,900	1,963	2,501	2,423	2,367
Private transfers (net)	313	60	21	9	6
Government transfers (net)	550	426	570	598	618
Overall balance	605	−153	−791	367	285

TAIWAN

Republic of China
Chung Hwa Min Kuo

CAPITAL: T'aipei.

FLAG: The flag is red with a 12-pointed white sun on the blue upper left quadrant. The 12 points of the sun represent the 12 two-hour periods of the day in Chinese tradition, and symbolize progress.

ANTHEM: *Chung Hwa Min Kuo Kuo Ke* (Chinese National Anthem).

MONETARY UNIT: The new Taiwan dollar (NT$) is a paper currency of 100 cents. There are coins of 50 cents and 1, 5, and 10 dollars, and notes of 50, 100, 500, and 1,000 new Taiwan dollars. NT$1 = US$0.0365 (or US$1 = NT$27.4).

WEIGHTS AND MEASURES: The metric system is employed in government and industrial statistics. Commonly used standards of weights and measures are the catty (1.1 lb or 0.4989 kilograms), the li (0.5 kilometers or 0.31 miles), the ch'ih (0.33 meters or 1.09 feet), and the chia (0.97 hectare or 2.39 acres).

HOLIDAYS: New Year's Day and the Founding of the Republic of China (1912), 1 January; Youth Day (formerly known as Martyrs' Day), 29 March; Tomb-Sweeping Day and Anniversary of the Death of Chiang Kai-shek, 5 April; Birthday of Confucius and Teachers' Day, 28 September; National Day (Double Tenth Day), 10 October; Taiwan Retrocession Day, 25 October; Chiang Kai-shek's Birthday, 31 October; Sun Yat-sen's Birthday, 12 November; Constitution Day, 25 December.

TIME: 8 PM = noon GMT.

LOCATION AND SIZE: Taiwan, the seat of the Republic of China, lies in the western Pacific Ocean, less than 161 kilometers (100 miles) from the southeast coast of mainland China, from which it is separated by the Taiwan Strait. Besides the island proper, Taiwan comprises 21 small islands in the Taiwan group and 64 islands in the P'enghu Ch'üntoo (Pescadores) group, giving it a total area of 35,980 square kilometers.

CLIMATE: Taiwan enjoys an oceanic, subtropical monsoon climate. The average lowland temperature in January is 16°C (61°F) in the north and 20°C (68°F) in the south. The average July temperature is 28°C (82°F) in both the north and south.

INTRODUCTORY SURVEY

RECENT HISTORY

Under Japan's rule since the 1895 conclusion of the First Sino-Japanese War, Taiwan was restored to China following World War II (1939–45). On 8 December 1949, as the Chinese communists were sweeping the nationalist (non-communist) armies off the mainland, the government of the Republic of China (ROC), led by General Chiang Kai-shek (Jiang Jieshi), was officially transferred to Taiwan. Two million mainland Chinese fled to the island with Chiang Kai-Shek. An authoritarian rule began under martial law. Strong government policies contributed to steady economic progress, first in agriculture and then in industry. In the 1950s, with American aid and advice, the ROC undertook a successful program of land redistribution. Japanese investment and the Vietnam War in the 1960s further stimulated economic growth.

The question of which China—the People's Republic of China (PRC), on the Chinese mainland, or the Republic of China (ROC) in Taiwan—should occupy China's seat at the United Nations was debated by the UN General Assembly for more than two decades. Support for Taiwan's representation gradually lessened over the years, and in 1971 the General Assembly voted to remove recognition from the ROC and recognize the PRC. Taiwan successfully avoided worldwide political and economic isolation by maintaining a host of commercial and cultural contacts with other countries.

President Chiang Kai-shek died at age 87 in 1975. While control of the central government had

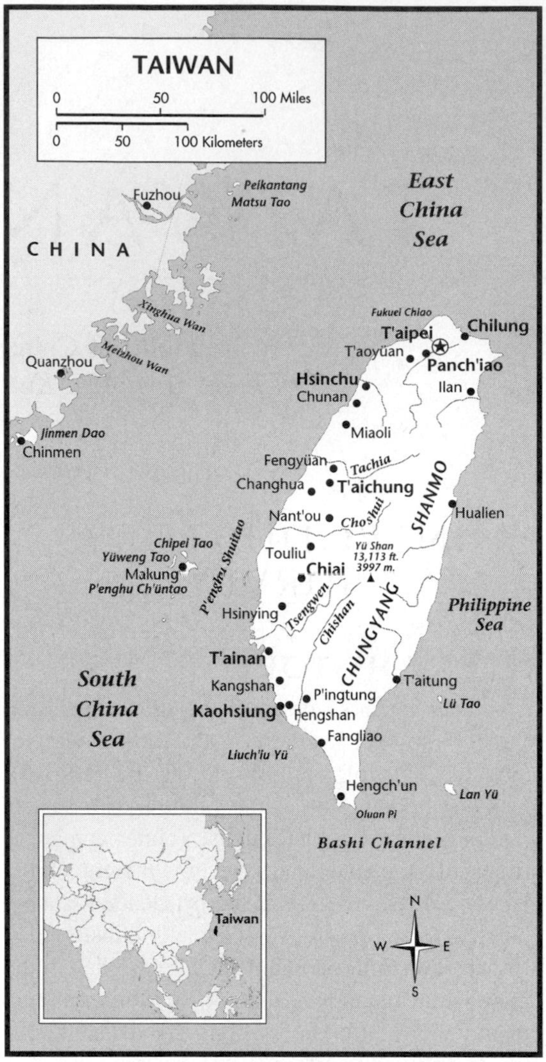

remained in the hands of former mainlanders in the first decades of the nationalists' rule on Taiwan, native-born Taiwanese Chinese increasingly won elections at local levels. Succeeding president Chiang Ching-kuo instituted a policy of bringing more Taiwanese into the Nationalist Party.

By the 1980s, economic development had produced a new middle class. The passage of time, together with intermarriage between former mainlanders and Taiwanese, had brought a new generation for which the distinction between mainlander and Taiwanese held less importance. These factors contributed to popular pressure for a more democratic government. In 1987, following massive protests, martial law was revoked. Press restrictions were eased, citizens were allowed to visit relatives in mainland China, and it became legal to form political parties.

In the early 1990s, Taiwan increasingly began to open up its political system to true democracy. The constitution was rewritten in 1995, requiring a direct election for the president. Since the peaceful return of Hong Kong from British to Chinese control in 1997, however, there has been speculation regarding Taiwan's future relation to the mainland.

GOVERNMENT

The government of the Republic of China in T'aipei claims to be the central government of all of China. The government gets its powers from the National Assembly, which, according to the constitution, exercises political powers on behalf of the people. In 1996 the National Assembly had 334 members. The Assembly has the power to elect and recall the president and the vice-president, amend the constitution, and approve constitutional amendments proposed by the Legislative Yuan (council). In 1994, parliament amended the constitution to provide for the direct election of the president.

The president is the head of state. His term of office is six years, with eligibility for a second term. The constitutional prohibition against third terms was suspended in March 1960 to permit President Chiang Kai-shek to continue in office. He was reelected in 1972 for a fifth six-year term.

Under the president, there are five government branches known as *yuans* (councils or departments): Legislative, Executive, Control, Examination, and Judicial. The Legislative Yuan, elected by popular vote, is the highest lawmaking body. The Executive Yuan is comparable to the cabinet in other countries. The Control Yuan, the highest supervisory branch, exercises audit powers over the government and may impeach public officials. It also supervises the government budget. The Examination Yuan is the equivalent of a civil service commission.

Judiciary

The Judicial Yuan is Taiwan's highest judicial branch. It interprets the constitution and other laws and decrees, rules on administrative suits, and disciplines public officials. The Supreme Court, the highest court of the land, consists of a number of civil and criminal divisions, each of which is formed by a presiding judge and four associate judges.

In 1993 a separate Constitution Court was established. The new court is charged with resolving constitutional disputes and regulating the activities of political parties.

There is no right to a trial by jury, but the right to a fair trial is protected by law and respected in practice.

Political Parties

The Chinese Nationalist Party, better known as the Kuomintang–KMT (or Guomindang–GMD), is the dominant political party in Taiwan. The teachings of Sun Yat-sen (Sun Zhongshan), which stress nationalism, democracy, and people's livelihood, form the ideals of the party.

Under martial law, from 1949 through 1986, the formation of new political parties was illegal. In September 1986, a group of "nonpartisans" formed a new opposition party, the Democratic Progressive Party (DPP), which leaned toward the Taiwanese population and advocated "self-determination." In addition to the DPP, new parties include the Labor Party, the Workers Party, and other minor parties. In 1994, the New Chinese Party (CNP) was formed by former KMT members who favored greater ties with the mainland.

In the 1995 elections for the Legislative Yuan, the KMT won 83 of 164 seats; the DPP, 54; the CNP, 21; and independents, 6. In the National Assembly, the KMT won 183 of 334 seats; the DPP, 99; the CNP, 46; and independents, 6.

DEFENSE

Two years' military service is compulsory for all male citizens. The armed forces totaled 376,000 in 1995. The army had 240,000 members, the navy 38,000, the marines 30,000, and the air force 68,000. In addition, reserves totaled 1.65 million. The navy had 18 destroyers, 4 submarines, 18 frigates, and 98 patrol and coastal combatants. The air force had 392 combat aircraft.

ECONOMIC AFFAIRS

The decade 1971–80 saw the development of the steel, machinery, machine tools, and motor vehicle industries. Such industries, based on imports of raw materials, were encouraged through massive government support for major improvements in roads, railroads, ports, and electricity.

During the 1980s, emphasis was placed on the development of high-technology industries. As a result, between 1981 and 1991, the share of high-technology industries in total manufactures increased from 20% to 29%, making Taiwan the seventh largest producer of computer hardware on the global market. Since the late 1980s the country has undergone a shift towards a services-dominated economy. In 1996, services made up 61% of the domestic economy, compared to less than 50% in the mid-1980s and 44% in the early 1960s. Industry's share of the domestic economy was 36% in 1996. Overall, the economy grew by 5.7% in 1996 and was expected to grow by 6.2% in 1997.

Public Finance

Central government revenues come mostly from taxation, customs and duties, and income from government monopolies on tobacco and wines; other revenues are derived from profits realized by government enterprises. Government accounts continued to show surpluses through the early 1980s, despite the worldwide recession.

Public authorities are anticipating a growing fiscal deficit throughout the late 1990s as Taiwan's six-year development plan requires over $300 billion of investment in public infrastructural construction projects and in upgrading industries. In 1996, the government's deficit was equal to 4% of GDP. Growing demands for social welfare spending and increased defense spending (up 20% in 1996/97, the largest rise in over a decade) continue to put pressure on the budget. Outstanding debt reached 20% of GNP in 1996, up from 7% in 1991, and debt service payments consumed 14% of the central budget in 1997. The government is committed to balancing the budget by 2001. Austerity measures include controlling public sector consumption expenditures, limiting expansion of government expenditures, freezing government employment, limiting public employee pay raises, and encouraging private participation in major public projects. The government is also committed to reducing the public sector's role in the economy. As of 1997, state owned enterprises accounted for about 10% of GDP.

Local governments derive the bulk of their revenues from local taxes on real estate, slaughtering, vehicles, boats, and entertainment. Most revenues are spent on education and police administration.

Income

In 1998 Taiwan's gross domestic product (GDP) was more than US$362 billion, or about US$16,500 per person.

Industry

Industrial production rose spectacularly after the end of World War II (1939–45), especially between 1952 and the early 1980s. Slower economic growth since the mid-1980s and greater investment emphasis on heavy and high-technology industries, as well as services, resulted in declining production figures for some traditional manufactures such as cotton yarn and caustic soda (which fell from 386,505 tons in 1986 to 171,840 tons in 1994). Television set production declined from 5.7 million in 1986 to 1.9 million units in 1994.

Nevertheless, production increases continued to be registered for a number of key products. Cement production grew from 14.8 million tons in 1986 to 22.5 million tons in 1995. Motor vehicle production rose from 170,923 passenger cars in 1986 to 423,318 units in 1994; steel bars from 6.16 million to 18.82 million tons; synthetic fibers from 1.37 million to 2.51 million tons; polyvinyl chloride (PVC) from 795,000 to 1.11 million tons; and shipbuilding from 552,294 to 1.04 million deadweight tons (dwt). Taiwan's computer manufacturing industry was the world's third largest in 1995.

Banking and Finance

Almost all banking institutions are either owned or controlled by the government. The Bank of Taiwan (with 75 branches) issued currency notes, handled foreign exchange, acted as the government's bank, and performed central banking functions in addition to its commercial banking activities before reactivating of the Central Bank of China (CBC) in T'aipei in 1961. The functions of the Central Bank include regulation of the money market, management of foreign exchange, issuance of currency, and service as fiscal agent for the government. The Bank of China is a foreign exchange bank with branch offices in major world capitals. The Bank of Communications is an industrial bank specializing in industrial, mining, and transportation financing. The Export-Import Bank of China, inaugurated 1 February 1979, assists in the financing of Taiwan's export trade. The Central Trust of China acts as a government trading agency and handles most of the procurements of government organizations. The Postal Savings System accepts savings deposits and makes domestic transfers at post offices. Foreign reserves at the end of 1995 totaled $90,310 million. Money supply, as measured by M2, amounted to nt $13,341.6 billion at the end of 1995.

At the end of 1996 there were 32 domestic commercial banks. The government holds majority status in several of the most important, including the Bank of Taiwan, the Cooperative Bank of Taiwan, and the First Commercial Bank. The two largest private banks are the International Commercial Bank of China and the Overseas Chinese Commercial Banking Corp.

In 1990 the government announced the goal of establishing the island as a regional financial center. Its original target of 1996 was far too optimistic, and liberalization will have to be far more thoroughgoing than that to which the authorities are at present committed, but various steps are being taken towards this end. Plans have been announced for the gradual lifting of restrictions on bringing in capital from abroad, raising the limits on capital transfers both in and out of Taiwan by domestic firms and individuals, and giving foreign banks greater freedom of operation. On 18 February 1997, the Finance Ministry set up a 37-member financial reform task fore, headed by the finance minister—finally realizing that reform of Taiwan's hidebound financial sector will take coordinated action. This group was to spend ten months devising proposals in the following four areas: improving the overall efficiency of the banking system; development of capital and derivatives markets, and relaxation of the rules governing the kinds of business banks may conduct; improving market-regulating procedures such as credit evaluation systems, asset management, investor insurance, and insider trading rules; and strengthening banks' internal financial controls. The task force's recommendations were scheduled to be completed by 31 January 1998.

The law stipulates that transactions in stocks and securities can be made only in the registered business places of stock dealers; the first private corporate bond issue was floated in 1958. The first stock exchange in Taiwan opened on 4 February 1962. It is not, as yet, a major source for the supply of capital to local or foreign investors, and the volume of transactions remains relatively small and fluctuates widely. It has functioned far more as a vehicle for speculating with excess liquidity than as a means of raising funds. Its performance has little in common with the fundamentals of the island's real economy. Since the collapse of world stock markets on ''Black Monday'' in October 1987, which took the T'aipei bourse with it, the market has risen to stunning highs and fallen to startling lows. The Taiwan stock market surged in the early months of 1997, with the index smashing through the 8,000-point barrier for the first time since 5 March 1990. This milestone immediately prompted rumblings from the CBC that the market was overheated. Yet by May 1997, the market was flirting with the next resistance level, at 8,500 points.

Economic Development

Since 1950, the government has adopted a series of economic plans to help guide and promote economic growth and industrialization. The first four-year economic development plan (1953–56) emphasized reconstruction and increased production of rice, fertilizers, and hydroelectric power; it resulted in an increase of 36.6% in GNP and 17.4% in income per capita. In the second four-year plan (1957–60), import substitution industries were encouraged. Industry and agriculture both registered significant gains; GNP increased by 31.1%, and national income per capita by 13.3%. The third four-year plan (1961–64) emphasized labor-intensive export industries, basic services, energy development, industries contributing to agricultural growth, and exploration and development of the island's limited natural resources. The results were a 41.5% increase in GNP and a 30.7% increase in per capita income. U.S. loans and grants, totaling U.S.$2.2 billion, and foreign (mostly overseas Chinese) investment financed these early stages of development.

Following the curtailment of AID assistance in 1965, the fourth four-year plan (1965–68) was introduced, followed by the fifth four-year plan (1969–72); increases in GNP for these periods were 45.9% and 55.2%, respectively. By 1971, exports of manufactured goods had registered spectacular increases, and Taiwan's foreign trade pattern changed from one of chronic deficit to consistent trade surpluses. At this point, the government began to redirect its priorities from labor-intensive industries to the development of such capital-intensive sectors as shipbuilding, chemicals, and petrochemicals. The sixth four-year plan (1973–76), adversely affected by the worldwide recession, was terminated in 1975 after producing only an 18.9% increase in GNP. It was replaced by a six-year plan (1976–81) that focused on expansion of basic industries and completion of 10 major infrastructural projects, including rail electrification, construction of the North Link railroad, development of nuclear energy, and construction of the steel mill at Kaohsiung and of the new port of T'aichung.

In 1978, the six-year plan was revised, and 12 new infrastructural projects were added, including completion of the round-the-island railroad, construction of three cross-island highways, expansion of T'aichung Port's harbor, and expansion of steel and nuclear energy facilities. A subsequent four-year plan (1986–89), designed to supplement a longer-range 10-year plan (1980–89), had as a target average annual GNP increase of 6.5%. Among its goals were price stability, annual growth of 7.5% in the service sector, trade liberalization, encouragement of balanced regional development, and redirection of new industrial growth into such high-technology industries as computers, robotics, and bioengineering. In response to flagging export growth and a slowdown in private investment following a stock market collapse in 1990, the government devised a Six-Year Plan for 1991–97 aimed at economic revitalization. The final plan has targeted up to nt $5 trillion for investment mainly in transportation, telecommunications, power generation and pollution control. A ''Statute for Upgrading Industries'' enacted in early 1991 continues the government's efforts to provide incentives for private investment in research and development and high-technology sectors of the economy.

SOCIAL WELFARE

A social insurance system provides medical, disability, old age, survivor, and other benefits. In 1995, a national health insurance program took effect, allowing workers to get medical care from private or public clinics, paid for by the National Health Insurance Bureau. All enterprises and labor organizations must also furnish welfare funds for workers and ''welfare units,'' such as cafeterias, nurseries, clinics, and low-rent housing. Government programs include relief for mainland refugees, disaster-relief assistance, and direct assistance to children in needy families.

In the workplace, women tend to receive lower salaries and less frequent promotion, and are often denied federally mandated maternity leave.

Healthcare

As a result of improved living conditions and mass vaccinations, significant progress has been made in controlling malaria, tuberculosis, venereal disease, leprosy, trachoma, typhoid, diphtheria, and encephalitis. Life expectancy was 77 years. In 1990, there were 22,300 doctors.

Housing

The evacuation of more than 2 million persons from the mainland to an already densely populated island in 1949 made the development of low-cost housing an early priority. Since the 1970s, government housing programs have focused on the cities, with slum clearance and the construction of high-rise apartment dwellings for low-income groups the major goals. The total housing stock stood at 4.7 million units in 1992. The number of people per dwelling was 4.3 as of 1988.

EDUCATION

Taiwan enjoys one of the world's highest literacy rates because of its emphasis on education. About 86% of the people aged fifteen or older are literate. In 1985, 234,674 children attended 2,210 preschools, 2.3 million were in 2,486 primary schools, and 1.1 million attended junior high. After completing nine years of compulsory schooling, 194,757 students chose to attend senior high school, and 421,784 sought vocational training. In 1989–90, there were 116 universities, junior colleges, and independent colleges in Taiwan. Many students attend college in Japan, Europe and the United States.

1999 KEY EVENTS TIMELINE

January

- Taiwan establishes diplomatic ties with Macedonia.

February

- Taiwan leads in number of popular artists winning top awards for Asian music.

March

- China protests inclusion of Taiwan in regional antimissile shield system proposed by U.S.

April

- Taiwan man blamed for internationally spread ''Chernobyl'' computer virus.

July

- Taiwan declares major policy shift in its relations with the People's Republic of China.
- China holds military exercises in response to shift in Taiwan policy.

August

- China protests proposed $550 million U.S. arms sale to Taiwan.

September

- The worst earthquake in Taiwan's history hits central Taiwan, killing over 2,000 persons.

November

- An earthquake measuring 6.9 on the Richter scale strikes off the coast of eastern Taiwan.
- Taiwan tests a defense system capable of shooting down ballistic and cruise missiles.
- Taiwan's defense minister warns the legislature that the island is the target of new missiles being deployed by China.

ANALYSIS OF EVENTS: 1999

BUSINESS AND THE ECONOMY

In May, the U.S. government issued preliminary antidumping penalties against several Taiwanese semiconductor manufacturers whose exports to the United States were considered to be sold below market value, causing injury to competitors in the U.S. Although spokespersons in Taiwan claimed that the impact of the penalties would be minimal, a trade group for Taiwan's semiconductor industry filed countercharges against several U.S. companies exporting similar products to Taiwan.

However, the greatest threat to Taiwan's semiconductor industry, a leader in the world market, came from the major earthquake that shook the country in September, interrupting power and water supplies and causing billions of dollars in property damage. Partial operation in the computer industry's major manufacturing district resumed within a week, but spokesmen predicted that profits for September and the fourth quarter of 1999 could be hurt. However, the industry was expected to rebound in the long term.

Taiwan's overall economy, as well as the semiconductor industry, would need time to recover from the ravages of the quake. Manufacturers in 16 major industrial zones reported losses of

T$1.37 billion in the first week following the disaster. Although economic growth for 1999 was expected to drop at least a quarter of a percent from previous forecasts, in the long run it was expected to rebound in 2000 thanks to the effects of rebuilding efforts, which were expected to take at least two years. One industry in particular that stood to benefit would be the construction sector.

GOVERNMENT AND POLITICS

Taiwan's relationship with mainland China continued to attract international attention to the island nation in 1999. In July, Taiwan's President Lee Teng-Hui unexpectedly announced a major shift in its diplomatic stance toward the People's Republic. Departing from the "one-China" policy in effect since 1991, according to which Taiwan accepted the designation of "renegade province," Lee stated in an interview that relations between the two countries should be conducted on a "special state-to-state" basis. Lee's statement provoked a strong reaction from China, which issued verbal condemnations and held civilian mobilization exercises in Fujian province near Taiwan later in the month.

Tensions between China and Taiwan had risen earlier in the year when the United States proposed to include Taiwan in a proposed antimissile defense system intended to protect Japan and South Korea from possible nuclear attack by North Korea, a measure vigorously opposed by the Chinese, who feared it would strengthen pro-independence forces on the island. Following Lee's July 9 announcement, further events strained relations between the two neighbors: at the end of July, the Chinese seized a Taiwanese freighter bound for the island of Matsu off the Chinese coast, allegedly on smuggling charges, and in August China protested a proposed $550 million U.S. arms sale to Taiwan.

CULTURE AND SOCIETY

On September 20, Taiwan suffered the worst earthquake in its history. The quake, which struck central Taiwan, measured 7.6 on the Richler scale and was the worst the world had seen in over three years. Over 2,000 people were killed, an estimated 100,000 left homeless, and over 6,000 buildings destroyed. Relief agencies throughout the world mobilized, sending rescue workers and aid from over 25 countries. The People's Republic of China offered medical assistance, but it was politely declined.

The earthquake and the thousands of aftershocks that followed it in the next few days trapped thousands of people in the region by damaging main roads and leaving an accumulation of debris. The aftershocks also endangered relief workers struggling to free some 300 people trapped in collapsed buildings. Power and water supplies were interrupted, and thousands of people remained out of doors, fearing the effects of the continuing aftershocks on their homes. The extensive damage to dwellings in central Taiwan was partially blamed on lax enforcement of construction codes. Total property damage was estimated in the billions of dollars.

Earlier in the year, Taiwan had been linked to another type of disaster, one involving high technology. Responsibility for the massively destructive "Chernobyl" computer virus, which affected hundreds of thousands of computers in Asia and the Middle East, was traced to a Taiwanese man who had written the virus program a year earlier, while a student at the Tatung Institute of Technology. The virus, which was programmed to strike on April 26, the anniversary of Russia's Chernobyl nuclear disaster, disabled computers by damaging their hard drives and preventing the machines from being restarted.

In contrast to disasters and tensions in other areas, Taiwan scored a triumph in the realm of popular music in 1999 when it laid claim to the majority of awards for the top 20 musical artists chosen by a leading radio station in Shanghai. Previously, Hong Kong had been considered the capital of Asian music in the region, but the awards showed that Taiwan was becoming a major force in the industry, thanks to its openness to innovation and the sophistication of its audiences.

DIRECTORY

CENTRAL GOVERNMENT

Head of State

President

Lee Teng-hui, Office of the President, Chiehshou Hall, 122 Chungking South Road, Sec. 1, Taipei, Taiwan

PHONE: +886 (2) 23113731

FAX: +886 (2) 23825580

E-MAIL: webmaster@opp.gov.tw

Vice President
Lien Chan, Office of the Vice President, Chiehshou Hall, 122 Chungking South Road, Sec. 1, Taipei, Taiwan
PHONE: +886 (2) 23113731
FAX: +886 (2) 23825580
E-MAIL: webmaster@opp.gov.tw

Premier
Vincent Siew, Office of the Premier, 1 Chung Hsiao East Road, Sec. 1, Taipei, Taiwan
PHONE: +886 (2) 23561500

Vice Premier
Liu Chao-shiuan, Office of the Vice Premier, 1 Chung Hsiao East Road, Sec. 1, Taipei, Taiwan
PHONE: +886 (2) 23561500

Ministers

Minister of the Interior
Huang Chu-Weng, Ministry of the Interior, 5th Floor, 5 Hsu Chou Road, Taipei, Taiwan
PHONE: +886 (2) 23565005
FAX: +886 (2) 23566201
E-MAIL: gethics@mail.moi.gov.tw

Minister of Foreign Affairs
Jason Hu, Ministry of Foreign Affairs, 2 Chiehshou Road, Taipei, Taiwan
PHONE: +886 (2) 23119292
FAX: +886 (2) 23144972

Minister of National Defense
General Chiang Chung-Ling, Ministry of National Defense, P.O. Box 9001, Taipei, Taiwan
PHONE: +886 (2) 23181320
FAX: +886 (2) 2361699

Minister of Finance
Paul Cheng-Hsiung Chiu, Ministry of Finance, 2 Ai Kuo West Road, Taipei, Taiwan
PHONE: +886 (2) 23228000
FAX: +886 (2) 23965829
E-MAIL: root@mof.gov.tw

Minister of Education
Lin Ching-Jiang, Ministry of Education, 5 Chung Shan South Road, Taipei, Taiwan
PHONE: +886 (2) 23566051
FAX: +886 (2) 23976978

Minister of Economic Affairs
Wang Chih-Kang, Ministry of Economic Affairs, 15 Foo Chou Street, Taipei, Taiwan
PHONE: +886 (2) 23212200
FAX: +886 (2) 23919398
E-MAIL: service@moea.gov.tw

Minister of Justice
Liao Cheng-Hao, Ministry of Justice, 130 Chungking South Road, Sec. 1, Taipei, Taiwan
PHONE: +886 (2) 23146871
FAX: +886 (2) 23896759

Minister of Transportation and Communications
Liu Chao Shiuan, Ministry of Transportation and Communications, 2 Chang Sha Street, Sec. 1, Taipei, Taiwan
PHONE: +886 (2) 23492900
FAX: +886 (2) 23118587
E-MAIL: motceyes@motc.gov.tw

POLITICAL ORGANIZATIONS

Chinese Democratic Socialist Party (CDSP)

6 Lane 357, Ho Ping East Road, Sec. 2, Taipei, Taiwan
PHONE: +886 (2) 27072883
TITLE: Chairman
NAME: Wang Peir-Ji

Chinese Republican Party (CRP)

3rd Floor, 26 Lane 90, Jong Shueen Street, Sec. 2, Taipei, Taiwan
PHONE: +886 (2) 29366572
TITLE: Chairman
NAME: Wang Ying-Chyun

China Young Party

2th Floor, 2 Shin Sheng South Road, Sec. 3, Taipei, Taiwan
PHONE: +886 (2) 23626715
TITLE: Chairman
NAME: Jaw Chwen-Shiaw

Liberal Party (DLP)

Kuomintang-KMT (Nationalist Party of China)

53 Jen Ai Road, Sec. 3, Taipei, Taiwan
PHONE: +886 (2) 23121472
TITLE: Chairman
NAME: Lee Teng-Hui

Kungtang-KT (Labour Party)

2nd Floor, 22 Kai Feng Street, Sec. 2, Taipei, Taiwan
PHONE: +886 (2) 23121472
TITLE: Chairman
NAME: Jeng Jau-Ming

Democratic Progressive Party (DPP)

8th Floor, 39 Pei Ping East Road, Taipei, Taiwan
PHONE: +886 (2) 23929989

Chinese New Party (CNP)

4th Floor, 65 Guang Fuh South Road, Taipei, Taiwan
PHONE: +886 (2) 27562222

Taiwan Independence Party (TAIP)

c/o Legislative Yuan, Taipei, Taiwan
TITLE: Chairman
NAME: Hsu Shih-Kai

World United Formosans for Independence (WUFI)

TITLE: Chairman
NAME: George Chang

Green Party

10th Floor, 281 Roosevelt Road, Sec. 3, Taipei, Taiwan
PHONE: +886 (2) 23621362
TITLE: Chairman
NAME: Cau Cherng-Yan

Workers Party

2nd Floor, 181 Fu-hsing South Road, Taipei, Taiwan
PHONE: +886 (2) 27555868

DIPLOMATIC REPRESENTATION

Embassies in Taiwan

Belize
Room 902, 9th Floor 346, Nanking East Road, Sec. 3, Taipei, Taiwan
PHONE: +886 (2) 7783986

Bahamas
18th Floor, 27 Anho Road, Taipei, Taiwan
PHONE: +886 (2) 7218443

Burkina Faso
3rd Floor, 5 Chung Cheng Road, Sec. 2, Taipei, Taiwan
PHONE: +886 (2) 28383776
FAX: +886 (2) 26342701
TITLE: Ambassador
NAME: Jucques Sawadogo

Central African Republic
7th Floor, 59 Yungho Road, Yungho, Sec. 2, Taipei, Taiwan
PHONE: +886 (2) 9225678
FAX: +886 (2) 9202266
TITLE: Ambassador
NAME: Guillaume Mokemat-Kenguemba

Costa Rica
108 Chungcheng Road, Sec. 2, Taipei, Taiwan
PHONE: +886 (2) 8712422
FAX: +886 (2) 22933548
TITLE: Ambassador
NAME: Elena Wachong-Storer

Côte d'Ivoire
110 Yenping South Road, Taipei, Taiwan
PHONE: +886 (2) 3817042

Dominican Republic
6th Floor, 76 Tun Hua South Road, Sec. 2, Taipei, Taiwan
PHONE: +886 (2) 7079006
FAX: +886 (2) 7091429
TITLE: Ambassador
NAME: Victor Manuel Sanchez Pena

El Salvador
15 Lane, 34 Kukung Road, Shihlin, Taipei, Taiwan
PHONE: +886 (2) 8817995
FAX: +886 (2) 8819887
TITLE: Ambassador
NAME: David Ernesto Panamas

Gambia
3rd Floor, 92 Hwang Chi Street, Shih Lin, Taipei, Taiwan
PHONE: +886 (2) 8332434
FAX: +886 (2) 8324336
TITLE: Ambassador
NAME: Antouman Saho

Guatemala
Lane 88, 6 Chien Kuo North Road, Sec. 1, Taipei, Taiwan
PHONE: +886 (2) 5077043
FAX: +886 (2) 5060577
TITLE: Ambassador
NAME: Luis Alberto Noriega Morales

Guinea-Bissau
Lane 77, 6-1 Sung Chiang Road, Taipei, Taiwan
PHONE: +886 (2) 5099052
FAX: +886 (2) 5073111
TITLE: Ambassador
NAME: Inacio Semedo, Jr.

Haiti
3rd Floor, 246 Chungshan North Road, Sec. 6, Taipei, Taiwan
PHONE: +886 (2) 8317086
FAX: +886 (2) 8317086
TITLE: Ambassador
NAME: Sonny Seraphin

Honduras
Room 701, 142 Chunghsiao East Road, Taipei, Taiwan
PHONE: +886 (2) 7518737
FAX: +886 (2) 7120743
TITLE: Ambassador
NAME: Daniel Edgardo Milla Villeda

Liberia
13th Floor, 2 Lane 10, Hsing Yi Road, Peitou, Taipei, Taiwan
PHONE: +886 (2) 8745768
TITLE: Ambassador
NAME: John Cummings

Nicaragua
3rd Floor, 222-6 Jyi Shyan Road, Lu Chow, Taipei, Taiwan
PHONE: +886 (2) 82814512
FAX: +886 (2) 82814515
TITLE: Ambassador
NAME: Salvador Stadthagen

Panama
8th Floor, 46-1 Chungcheng Road, Sec. 2, Taipei, Taiwan
PHONE: +886 (2) 8327872
TITLE: Ambassador
NAME: Carlos Alberto Mendoza

Paraguay
1 Alley, 2 Lane 117 Tienmu East, Taipei, Taiwan
PHONE: +886 (2) 8736310
FAX: +886 (2) 8736312
TITLE: Ambassador
NAME: Ceferino Adrian Valdez Peralta

Swaziland
12th Floor, 127 Jenai Road, Sec. 3, Taipei, Taiwan
PHONE: +886 (2) 7512857

Vatican City
87 Ai Kuo East Road, Taipei, Taiwan
PHONE: +886 (2) 3216847
FAX: +886 (2) 3911926
TITLE: Ambassador
NAME: Joseph Chennoth

JUDICIAL SYSTEM

Judicial Yuan

124 Chungking South Road, Sec. 1, Taipei, Taiwan
PHONE: +886 (2) 23618577
FAX: +886 (2) 23821739

Supreme Court

6 Chang Sha Street, Sec. 1, Taipei, Taiwan
PHONE: +886 (2) 23141160
FAX: +886 (2) 23114246

Taiwan High Court

124 Chungking South Road, Sec. 1, Taipei, Taiwan
PHONE: +886 (2) 23713261

FURTHER READING

Articles

"Taiwan Begins Looking at Timetable and Cost to Rebuild." *The New York Times*, 24 September 1999, p. A12.

"Taiwan Death Toll Rises to 2,000 and Damage to Industry Mounts." *The Wall Street Journal*, 23 September 1999, p. A17.

"Taiwan's Terror." *The Economist* (September 25, 1999): 6.

"Taiwan's Unnerving President Does It Again." *The Economist* (July 17, 1999): 35.

"Taiwan Unseats Hong Kong as a New Music Center." *Billboard* (August 7, 1999): APQ-1.

Books

Copper, John F. *Historical Dictionary of Taiwan (Republic of China).* Lanham, Md.: Scarecrow Press, 1999.

Davison, Gary Marvin, and Barbara E. Reed. *Culture and Customs of Taiwan.* Westport, Conn.: Greenwood Press, 1998.

Kemenade, Willem van. *China, Hong Kong, Taiwan, Inc.: The Dynamics to a New Empire.* New York: Vintage Books, 1998.

TAIWAN: STATISTICAL DATA

For sources and notes see "Sources of Statistics" in the front of each volume.

GEOGRAPHY

Geography (1)

Area:

Total: 35,980 sq km.

Land: 32,260 sq km.

Water: 3,720 sq km.

Note: includes the Pescadores, Matsu, and Quemoy.

Area—comparative: slightly smaller than Maryland and Delaware combined.

Land boundaries: 0 km.

Coastline: 1,448 km.

Climate: tropical; marine; rainy season during southwest monsoon (June to August); cloudiness is persistent and extensive all year.

Terrain: eastern two-thirds mostly rugged mountains; flat to gently rolling plains in west.

Natural resources: small deposits of coal, natural gas, limestone, marble, and asbestos.

Land use:

Arable land: 24%

Permanent crops: 1%

Permanent pastures: 5%

Forests and woodland: 55%

Other: 15%

HUMAN FACTORS

Ethnic Division (6)

Taiwanese (including Hakka)	84%
Mainland Chinese	14%
Aborigine	2%

Demographics (2A)

	1990	1995	1998	2000	2010	2020	2030	2040	2050
Population	20,279.0	21,292.9	21,908.1	22,319.2	24,237.2	25,503.9	26,151.8	26,084.3	25,189.5
Net migration rate (per 1,000 population)	NA	NA	NA	NA	NA	NA	NA	NA	NA
Births	NA	NA	NA	NA	NA	NA	NA	NA	NA
Deaths	NA	NA	NA	NA	NA	NA	NA	NA	NA
Life expectancy - males	70.6	72.2	73.8	75.0	77.2	78.6	79.5	80.1	80.4
Life expectancy - females	76.1	77.9	80.1	81.7	84.1	85.4	86.2	86.6	86.8
Birth rate (per 1,000)	16.6	15.3	14.8	14.5	12.8	10.6	10.1	9.4	8.8
Death rate (per 1,000)	5.4	5.7	5.4	5.2	6.0	7.0	8.8	11.4	13.4
Women of reproductive age (15-49 yrs.)	5,464.2	5,949.5	6,216.6	6,332.9	6,158.7	5,831.0	5,471.9	5,082.9	4,807.3
of which are currently married	NA	NA	NA	NA	NA	NA	NA	NA	NA
Fertility rate	1.8	1.8	1.8	1.8	1.8	1.7	1.7	1.7	1.7

Except as noted, values for vital statistics are in thousands; life expectancy is in years.

Religions (7)

Mixture of Buddhist, Confucian, and Taoist 93%, Christian 4.5%, other 2.5%.

Languages (8)

Mandarin Chinese (official), Taiwanese (Min), Hakka dialects.

GOVERNMENT & LAW

Political Parties (12)

Legislative Yuan	% of seats
Kuomintang (KMT)	46
Democratic Progressive Party (DPP)	33
Chinese New Party (CNP)	13
Independents	8

Government Budget (13B)

Revenues	$40 billion
Expenditures	$55 billion
Capital expenditures	NA

Data for 1998 est. NA stands for not available.

Military Affairs (14B)

	1990	1991	1992	1993	1994	1995
Military expenditures						
Current dollars (mil.)	8,938[e]	9,622	10,300	11,930	11,480	13,140
1995 constant dollars (mil.)	10,270[e]	10,630	11,080	12,500	11,770	13,140
Armed forces (000)	370	370	360	442	425	425
Gross national product (GNP)						
Current dollars (mil.)	169,000	188,900	206,000	224,000	242,900	263,600
1995 constant dollars (mil.)	194,200	208,800	221,600	234,800	248,900	263,600
Central government expenditures (CGE)						
1995 constant dollars (mil.)	NA	32,830[e]	31,680[e]	33,580[e]	35,600[e]	37,700[e]
People (mil.)	20.3	20.5	20.7	20.9	21.1	21.3
Military expenditure as % of GNP	5.3	5.1	5.0	5.3	4.7	5.0
Military expenditure as % of CGE	NA	32.4	35.0	37.2	33.1	34.9
Military expenditure per capita (1995 $)	507	519	536	599	558	618
Armed forces per 1,000 people (soldiers)	18.2	18.1	17.4	21.2	20.2	20.0
GNP per capita (1995 $)	9,575	10,190	10,710	11,240	11,810	12,390
Arms imports[6]						
Current dollars (mil.)	600	800	800	1,000	1,000	1,200
1995 constant dollars (mil.)	690	884	861	1,048	1,025	1,200
Arms exports[6]						
Current dollars (mil.)	10	5	10	10	20	10
1995 constant dollars (mil.)	11	6	11	10	21	10
Total imports[7]						
Current dollars (mil.)	54,830	63,080	72,180	77,100	85,510	103,700
1995 constant dollars (mil.)	63,010	69,700	77,640	80,830	87,650	103,700
Total exports[7]						
Current dollars (mil.)	67,140	76,110	81,390	84,680	92,850	111,600
1995 constant dollars (mil.)	77,160	84,110	87,550	88,770	95,170	111,600
Arms as percent of total imports[8]	1.1	1.3	1.1	1.3	1.2	1.2
Arms as percent of total exports[8]	0	0	0	0	0	0

LABOR FORCE

Labor Force (16)

Total (million)	9.4
Services	52%
Industry	38%
Agriculture	10%

Data for 1997. Percent distribution for 1996 est.

Unemployment Rate (17)

2.7% (1997)

PRODUCTION SECTOR

Electric Energy (18)

Capacity	23.763 million kW (1996)
Production	124.973 billion kWh (1996)
Consumption per capita	5,500 kWh (1995)

Transportation (19)

Highways:

total: 19,701 km

paved: 17,238 km (including 447 km of expressways)

unpaved: 2,463 km (1996 est.)

Pipelines: petroleum products 615 km; natural gas 97 km

Merchant marine:

total: 193 ships (1,000 GRT or over) totaling 5,621,906 GRT/8,583,808 DWT ships by type: bulk 49, cargo 30, combination bulk 2, container 81, oil tanker 18, refrigerated cargo 11, roll-on/roll-off cargo 2 (1997 est.)

Airports: 40 (1997 est.)

Airports—with paved runways:

total: 36

over 3,047 m: 8

2,438 to 3,047 m: 12

1,524 to 2,437 m: 5

914 to 1,523 m: 6

under 914 m: 5 (1997 est.)

Airports—with unpaved runways:

total: 4

1,524 to 2,437 m: 2

under 914 m: 2 (1997 est.)

Top Agricultural Products (20)

Rice, wheat, corn, soybeans, vegetables, fruit, tea; pigs, poultry, beef, milk; fish.

FINANCE, ECONOMICS, & TRADE

Economic Indicators (22)

National product: GDP—purchasing power parity—$308 billion (1997 est.)

National product real growth rate: 6.8% (1997 est.)

National product per capita: $14,200 (1997 est.)

Inflation rate—consumer price index: 0.9% (1997)

Exchange Rates (24)

Exchange rates:

New Taiwan dollars per US$1

Yearend 1997	32.45
1996	27.5
1995	27.4
1994	26.2
1993	26.6
1992	25.4

Balance of Payments (23)

	1985	1990	1991	1992	1993
Exports of goods (f.o.b.)	30,469	66,823	75,535	80,723	84,155
Imports of goods (f.o.b.)	−19,296	−51,895	−59,781	−67,956	−72,750
Trade balance	11,173	14,928	15,754	12,767	11,405
Services - debits	−6,681	−17,673	−19,738	−22,478	−24,430
Services - credits	4,955	14,249	16,250	18,072	20,183
Private transfers (net)	−243	−730	−230	−168	−955
Government transfers (net)	−6	−5	−21	−39	−27
Long-term capital (net)	−1,020	−6,601	−2,827	−3,097	−2,560
Short-term capital (net)	−2,143	−8,549	600	−3,450	−2,056
Errors and omissions	491	463	−129	−240	−19
Overall balance	6,526	−3,918	9,659	1,367	1,541

Top Import Origins (25)

$114.4 billion (c.i.f., 1997) Data are for 1997.

Origins	%
Japan	25.4
United States	20.3
Europe	18.9
Hong Kong	1.7

Top Export Destinations (26)

$122.1 billion (f.o.b., 1997) Data are for 1997.

Destinations	%
United States	24.2
Hong Kong	23.5
Europe	15.1
Japan	9.6

Economic Aid (27)

$NA. NA stands for not available.

Import Export Commodities (28)

Import Commodities	Export Commodities
Machinery and electrical equipment 16.5%	Machinery and electrical Electronic products 14.8%
Electronic products 16.3%	Information/ communications 11.8%
Chemicals 10.0%	Textile products 11.6%
Precision instrument 5.6%	

TAJIKISTAN

Republic of Tajikistan
Jumhurii Tojikistan

CAPITAL: Dushanbe.

FLAG: The flag consists of a broad white horizontal stripe in the center, with a red stripe at the top and a green stripe at the bottom. The national emblem is centered in the white stripe.

MONETARY UNIT: The Tajik ruble (TR) was introduced in May 1995. TR1 = \$0.00352 (or \$1 = TR284).

WEIGHTS AND MEASURES: The metric system is used.

HOLIDAYS: New Year's Day, 1 January; Navruz ("New Day"), 21 March; Independence Day, 9 September.

TIME: 6 PM = noon GMT.

LOCATION AND SIZE: Tajikistan is located in southern Asia, between Uzbekistan and China. Comparatively, it is slightly smaller than the state of Wisconsin, with a total area of 143,100 square kilometers (55,251 square miles). Tajikistan's boundary length totals 3,651 kilometers (2,269 miles).

Its capital city, Dushanbe, is located in the western part of the country.

CLIMATE: The climate ranges from desert-like to polar. The mean temperature is 30°C (86°F) in July and 0°C (32°F) in January. Rainfall in the country averages 12.2 centimeters (4.8 inches).

INTRODUCTORY SURVEY

RECENT HISTORY

Tajikistan became a full Soviet republic in 1929. By the late Soviet period it was the poorest and least developed of the republics. It comprised four separate areas: Khojent, which is geographically and culturally closest to Uzbekistan's Fergana valley, and whose people traditionally hold power; Kuliab, whose families and clans contest those in Kuliab; Badakhshan, home of the poorest and whose land consists mostly of the unfarmable Pamir Mountains; and the Kurgan-Tiube, where the influence of Islam is strong.

With the breakup of the Soviet Union, the republic declared independence on 9 September 1991. Long-time republic leader Rahmon Nabiev was elected president. Civil disorder grew throughout 1992 and by September, when Nabiev was forced from office at gunpoint by a disgruntled democratic faction, full civil war had erupted, with casualties as high as 50,000. By October, Uzbekistan and Russia had joined in the effort to help drive the Tajik government and its supporters out of the country, mostly into neighboring Afghanistan.

In November 1992, the post of president was abolished, and Speaker I. Rakhmonov, of Kŭlob (Kulyab), took power, brutally rooting out all opposition. Although recognized as a legitimate leader by the outside world, Rakhmonov's control of the republic remained weak. Important opposition leaders set up bases in Afghanistan, vowing to return to fight. In 1996, under regional pressure from Russia and Iran, Rakhmonov and Islamic opposition leader Sayid Abdullo Nuri signed a peace treaty in Moscow. The agreement called for a cease-fire and the establishment of a National Reconciliation Commission in 1997.

GOVERNMENT

As of 1996, Tajikistan was considered to be a democracy in name only and was actually a one-party dictatorship. Peace accords signed in 1996 among the main rivals called for creating a National Reconciliation Commission in 1997.

Presidential elections were held in 1994 and were won by Imamali Rakhmonov, who had held the position since he took the position from his predecessor in 1992.

Parliamentary elections were held in 1995, but they were considered fraudulent by international observers. The 80-member Majlis (assembly), replaced the Soviet-era parliament in May 1992.

Judiciary

The judicial system from the Soviet period remains in place. There are courts at the city, district, regional, and republic levels with a separate but parallel system of military courts.

Political Parties

The principal party is still the Communist Party. In the period preceding the Rakhmonov government, there were several opposition parties, now in exile. The most important were the Democratic Party, Rastokhez (Renaissance), a nationalist party primarily of artists and intellectuals, and the Islamic Renaissance Party. Politics in Tajikistan is based more upon regional or clan connections than upon party membership or platform.

DEFENSE

There are an estimated 5,000–7,000 personnel in the armed forces. The army has about 5,200 soldiers with 40 main battle tanks. There are about 16,500 paramilitary border guards. There is also an Islamic movement of 5,000 opposed to the government.

ECONOMIC AFFAIRS

Tajikistan is the poorest of the post-Soviet republics. Agriculture, mainly cotton production, dominates the country's economy. Imports provide practically all of the country's manufactured consumer needs. The civil war severely weakened the economy and left Tajikistan dependent on Russia, Uzbekistan and on international relief. The domestic economy shrank by 12.5% in 1995 and by 4.4% in 1996. However, the government has proposed speeding up the transition to a market-based economy and to enact the fiscal, insurance, and banking reforms needed to stabilize the economy.

Public Finance

Revenues from domestic taxes and resources are limited. Expenditures are largely for grain, the supply of fuel and raw materials for industry, and to maintain the military. Despite proposals to liberalize the economy, the government continues to subsidize inefficient state enterprises. Only 11% of medium and large enterprises have been privatized as of 1997.

Income

In 1998, Tajikistan's gross national product (GNP) was $2.1 billion, or about $350 per person.

Industry

A few state-owned industries dominate Tajikistan's industrial sector, although the state's role is declining through privatization. One of the country's largest enterprises is an aluminum plant near Dushanbe that produced 230,000 tons of aluminum in 1995. Light industry includes cotton cleaning, silk processing, textiles production, knitted goods, footwear, sewing, tanning, and carpet making. The food processing industry uses domestically harvested fruit, wheat, tobacco, and other agricultural products.

Banking and Finance

The National Bank of Tajikistan (NBT) is the country's bank charged with implementing a monetary policy and issuing currency. It was formally established as the central bank in 1991. Commercial and state banks include the Bank for Foreign Investment, three large banks formed from the former Soviet state bank, and three branches from the Russian Commercial Bank. The Law on Banks and Banking Activities, adopted in February 1991, allows banks to compete for resources freely (including the setting of deposit rates) and lifts specialization boundaries. However, competition is very limited. Under IMF pressure, the Tajik government is now seeking to introduce tighter regulation over the banking sector. There is no securities exchange.

Economic Development

Soviet development policy in Tajikistan prioritized the development of the country's agricultural and other primary resources, while capital goods and manufactured consumer goods were imported from elsewhere within the former USSR. Since the late 1970s, greater development of small food processing and consumer plants had been urged by local government officials in order to absorb more of the republic's rural labor force; however, these proposals found little favor with Soviet central planners. After independence, the government targeted the development of hydroelectric power production and a number of other industries (silk, fertilizer, fruit and vegetables, coal, nonferrous metals, and marble production), seen as particularly important for improving the country's export base.

In 1991, a ''Program of Economic Stabilization and Transition to a Market Economy'' was adopted by the newly independent government. In accordance with the program's principles, price liberalization, privatization measures, and fiscal reform were initiated in 1991 and 1992. The government's overthrow in the course of civil war in 1992, however, brought economic development to a virtual standstill and slowed the pace of economic reform. Renewed efforts during 1996–97 to move from a command economy to a market oriented one resulted in proposals to convert medium-sized and large state enterprises to joint-stock companies and to create securities market. Other proposals were aimed at turning land over to private farmers and at privatizing the cotton industry, which continues to dominate agricultural production. As of 1997, the private sector accounted for less than 30% of GDP.

SOCIAL WELFARE

The government's social security systems have been threatened by civil war and economic turmoil. Refugees returning from Afghanistan suffered from malnutrition and many died in the resettlement camps. Resettlement payments were promised by the government but not paid. Women are supposed to be provided with three years' maternity leave and monthly subsidies for each child.

Healthcare

There is a shortage of medical equipment and medicines, creating the potential for an epidemic. In 1995, there were about 10,000 cases of malaria, up from 176 cases in 1990. There were 1,993 reported cases of diphtheria in 1994, mostly near Afghanistan. Life expectancy is 69 years.

Housing

In 1990, Tajikistan had 9.3 square meters (100 square feet) of housing space per person.

EDUCATION

The adult illiteracy rate is estimated at 2%. Education is free and compulsory between the ages of 7 and 17 (for 10 years). There are about 23 primary students per teacher. In 1991, all higher level institutions had a total of 125,100 pupils.

1999 KEY EVENTS TIMELINE

January

- Former Premier, Abdullojonov is implicated in the early November violence in northern Tajikistan.
- The United Tajik Opposition (UTO) leadership warns those opposed to the peace process of severe punishment.
- Gale-force winds and blizzards make 1,500 homeless and cause $1 million worth of damage in Leninabad.
- Supply of gas to the population is limited. Tajikistan's National Reconciliation Commission (NRC) orders the UTO to inventory its stockpiles of weapons and munitions.
- 10% cut at all levels of government does not affect the 30 positions granted to the United Tajik Opposition (UTO).
- Uzbek and Tajik officials agree to resume natural gas supplies from Uzbekistan to Tajikistan.
- Customs officials alter tariffs and import duties in order to facilitate the country's entry into the CIS Customs Union.
- The Security Council reprimands Tajik Security and Interior Ministers for their failure to prevent the November 1998 uprising in Leninabad.
- President Emomali Rakhmonov accuses Uzbekistan of direct involvement in the November insurrection led by Colonel Mahmud Khudoberdiev and Abdumalik Abdullojonov.
- The U.S. embassy, closed since September 1998, reopens.
- Ravshan Gafurov and his supporters are arrested.
- Another eight candidates for government posts, including nominees for the positions of first deputy interior, foreign, and security minister are approved.
- Gafurov confesses to shooting Otakhon Latifi and 25 other murders.

February

- Uzbekistan objects to the continued presence in Tajikistan of the CIS peacekeeping forces of the Commonwealth of Independent States (CIS), formerly members of the Union of Socialist Soviet Republics (USSR). A referendum on changes to the constitution is planned for later this year, as are parliamentary and presidential elections and the 1100th anniversary of the Samanid Dynasty.
- Rakhmonov recognizes Russia as the only reliable partner and guarantor of stability and security in Tajikistan.
- 1999 is declared the year of combating terrorism and organized crime connected with the narcotics business.
- Gafurov is shot and killed by police while trying to escape.
- The Sadirov band dissolves.
- The NRC, at work since 1997, makes its first recommendation: creation of a bicameral parliament. The commission is charged with proposing changes to the country's constitution in order to create conditions for elections to the parliament and the presidency. The commission reached consensus on the formation of the Central Election Commission four of the 15 members of which will be from the UTO.
- Russian and Tajik Muftiat sign an agreement on holding regular conferences on spiritual matters.
- An illness, "unknown to modern medicine," claims the lives of 350 children and elderly at the Tajik–Afghan border.
- Since the start of 1999, 203 cases of typhoid have been reported in the southern Tajik city of Kulab.
- The UN Secretary-General criticizes the slow progress in disarming opposition forces and in the implementation of constitutional reform.
- Rakhmonov warns of a threat by 400 people who undergo sabotage training in Afghanistan with

the goal of "creating chaos" in parts of Tajikistan.

- Russia and three other former Soviet states accept Tajikistan into the customs union.
- The UN Security Council, announcing the opening of two UN offices in Khojand and Qarategin, warns that a loss of confidence between the government and UTO could have dangerous consequences.

March

- Rakhmonov orders a complete integration of soldiers from the UTO into the regular army and amnesties UTO leaders.
- Belarus might help Tajikistan modernize its military hardware, in particular armored vehicles and aircraft.
- UTO fighters are registered and integrated into the regular national army.
- Kazakstan reduces its battalion in Tajikistan from 500 to 300 men.
- Five more UTO members receive government positions in line with the terms of the 1997 Tajik Peace Accord.
- Tajikistan's Supreme Court hands down sentences of 10–14 years in prison to seven supporters of rebel Colonel Makhmud Khudoiberdiev who took part in the abortive attack in October 1997 on presidential guard detachments in Tursunzade, western Tajikistan.
- The UN resumes observer mission in Gorno-Badakhshan, suspended for the murder last summer of three mission members and their driver.
- Tajik Socialist Party leader Safarali Kenjaev is shot dead. Kenjaev was chairman of the Tajik parliament's legislation and human rights committee. He is the third prominent politician assassinated over the past year.

April

- Four Tajik opposition parties warn that Rakhmonov's "regional and ethnic genocide" will lead to a new civil war.
- Narrow political interests are recognized as the cause of delay in implementing both the military and political protocols of the 1997 General Peace Accord.
- UN will resume operations in the Garm region of eastern Tajikistan. Operations there were suspended last summer following the murder of three UN observers and their interpreter.
- A total of 900 former Tajik opposition fighters have been enlisted into three interim army units.
- Rakhmonov fires five senior officials accused of failure to ensure tax collection.
- Tajik and Uzbek authorities discuss cooperation in the spheres of customs, passenger and cargo transit, and the supply and transit of natural gas to Tajikistan.
- While 22 representatives of the UTO have been appointed to government posts, no opposition representative serves on regional and district councils. The UTO also criticized Rakhmonov for his refusal to endorse constitutional amendments agreed on by the CNR, a body composed of both government and opposition representatives.
- Rakhmonov states that only a secular government can guarantee peace in Tajikistan.

May

- Kofi Annan extends the mandate of the UN Observer Mission for another six months.
- The Tajik government representatives agree to drop criminal charges against UTO members and release UTO jailed fighters.
- UN Security Council calls on the Tajik authorities to speed up implementation of the 1997 peace plan by demobilizing fighters, establishing a "broad political dialogue," and creating conditions for holding a referendum and for parliamentary elections.
- 5,500 opposition fighters are amnestied.
- Presidential elections will take place by 6 November 1999, the day Rakhmonov's term expires.
- UTO demands a 30% share in national and local government and 30% representation in banks and foreign embassies.
- Nuri states that those demands—the allocation to the opposition of 30% of posts in national and local governments and the release of 93 imprisoned opposition figures—do not exceed concessions contained in the 1997 peace agreement.

June

- Rakhmonov agrees to a meeting with opposition representatives to discuss their demands.

- The UTO suspends participation in two commissions to protest the leadership's failure to comply with specific conditions of the 1997 peace agreement.
- China invests up to $11 million in Tajikistan to build textile and tobacco factories.
- The UTO creates a working group to discuss conditions for renewing cooperation with the government.
- The World Bank approves a $20 million loan to support the privatization of farms in Tajikistan.
- Tajikistan's trade totals $551.7 million and its foreign deficit amounts to $42.3 million. Rakhmonov names four opposition candidates to government posts.
- The UTO will compile a list of 14 cities and towns in which its nominees will be appointed to head local government bodies.
- Two close associates of rebel Khudoiberdiev are sentenced to death.
- Rakhmonov gives "overall approval" to amendments to the Tajik Constitution. Rakhmonov rejects opposition demand to remove from the constitution an article pledging construction of a secular Tajik state, and the introduction of a bicameral parliament.

July

- Rakhmonov evaluates the damage inflicted by mud slides and coordinates emergency aid.
- Tajikistan's National Bank stabilizes the exchange rate for the Tajik ruble, setting an official rate of 1,400 Tajik rubles to the dollar. The Tajik ruble had fallen to 1,800–2,000. The fall in the value of the national currency leads to steep price rises. Tajik authorities deny rumors that the Tajik government intends to introduce a new national currency—the somon—in September to mark the 1100th anniversary of the Samanid dynasty.
- Haji Amanullo Negmatzoda is re-elected to the chairmanship of the Tajik Muslim Religious Board for another five years.
- More than 70% of the Tajik population currently backs the re-election of Rakhmonov.
- Russia, viewed by Tajikistan as a strategically important partner, plans to establish a military base in Tajikistan.
- Ziyoyev is named minister of emergency situations.
- The U.S. explores ways for helping Tajikistan remove weapons and land mines from its territory.
- Ukraine threatens to stop supplies of alumina to Tajikistan unless Dushanbe pays off its debts with cotton supplies promised earlier. The two countries have signed a deal on partly repaying Tajikistan's $60 million debt to Kiev with cotton supplies totaling 7,000 tons, including 1,300 tons this month. In return, Ukraine pledges to ship 500,000 tons of alumina to Tajikistan in both 1999 and 2000 and increase imports to 600,000 tons in 2001.

August

- The UTO will pull out of the Central Electoral Commission unless a UTO representative is appointed. Once disarmed, the UTO changes from a military into a political force.
- The Tajik Supreme Court lifts a ban on opposition parties imposed for their part in the 1992 civil war in the republic.
- Kofi Annan terms the Tajik Supreme Court action a "significant step" toward peace.
- Rakhmonov and Jiang Zemin sign an agreement on the demarcation of one section of their disputed common border but fail to resolve Chinese territorial claims on parts of Tajikistan's Gorno-Badakhshan Autonomous Oblast.
- Tajikistan protests Uzbekistan's bombing of its Jirgatal district.
- The UTO announces the dissolution of its armed units following Tajik Supreme Court's decision.
- Karimov blames Tajik authorities for allowing armed militants, including ethnic Uzbeks, to cross into neighboring Kyrgyzstan.
- Tajik Security Council Secretary identifies the guerrillas holding hostages in Kyrgyzstan as loyal members of the ethnic Uzbek field commander Juma Namangani.
- Kyrgyzstan and Tajikistan agree to close border crossings between their two countries.

September

- The 8th anniversary of the republic's independence coincides with the celebrations devoted to the 1100th anniversary of the Samanid state.
- The Uzbek official newspaper identifies not the Islamic Movement of Uzbekistan but the UTO as the main support of the hostage-takers.

- Rahmonov's name is put forward by his party, People's Democratic Party, for another term of office.
- Bulgarian diplomat Ivo Petrov is named Kofi Annan's special representative to Tajikistan.
- Tajiks in high numbers turn out to vote on three amendments to the constitution: extending the president's term to seven years, creating a bicameral parliament, and allowing the formation of religious-based parties. The referendum is supported by both the government and the UTO.
- Islamic Renaissance Party nominates Davlat Usmon as its presidential candidate.
- Tajik voters approve constitutional amendments. Turajonzoda argues against nominating a candidate. Tajik opposition party is reregistered.

October

- Unidentified planes bomb remote areas of eastern Tajikistan to flush out Uzbek intruders in southern Kyrgyzstan.
- Opposition candidates call for postponement of presidential vote. Four Japanese hostages are released.
- Tajik Opposition Organization expels Deputy Premier Turajonzoda.

November

- November 6, 1999, is the date for presidential election. 3 million Tajiks are eligible to vote. Opposition party leaders fear it may prove impossible to create conditions for free and fair elections.
- Incumbent president Imamali Rakhmonov is re-elected with 96 percent of the vote in a 98 percent turnout and sworn in as Tajikistan's president for another seven years.
- Russia's prime minister, Vladimir Putin, travels to Dushanbe to congratulate Rakhmonov on his re-election; however, he gives the Tajiks no indication of when Russia might withdraw its troops.

ANALYSIS OF EVENTS: 1999

BUSINESS AND THE ECONOMY

Of all the former Soviet republics on the southern side of the U.S.S.R., Tajikistan was in the worst shape. It had no industrial base and relied on agriculture, primarily cotton, for international trade. But with all the upheaval in the region since the fall of the U.S.S.R. the opportunities for trade had shrunk. In any case, cotton was no substitute for the oil and natural gas that could have aided some of Tajikistan's neighbors out of underdevelopment. Tajikistan had the lowest gross domestic product per capita of any of the former Soviet republics: $990. This meant a hard life for the citizens of Tajikistan and that hard life was reflected in the infant mortality statistics: in 1999 the rate of infant death per thousand live births in Tajikistan was 114 in the first five years of life, compared to 73 per thousand in Turkmenistan.

The second handicap for the Tajik nation is the internally divisive forces which have brought about three changes of government and one civil war since the 1991 dissolution of the U.S.S.R. If Tajikistan was to survive it would have to get over the hatred and suspicion that still characterized the social interaction between the country's warlords.

In 1999 the Tajik government gave the construction of an economic infrastructure top priority. Customs officials altered tariffs and import duties to facilitate Tajikistan's entry into the CIS Customs Union (CIS referred to the Commonwealth of Independent States—the post-1991 confederation of former Soviet republics). Tajikistan was accepted into the customs union. Tajikistan's National Bank stabilized the exchange rate for the Tajik ruble, setting an official rate of 1,400 Tajik rubles to the dollar.

The Tajik government also arranged with Belarus to help modernize Tajik armored vehicles and aircraft. The President of Tajikistan was Emomali Rakhmonov. He generally tried to draw the opposition into the government without giving them any important posts. He also sought to establish friendly relations with the Soviet Union and he stood for a secular state, arguing that the alternative was a never-ending, low-grade civil war, a prospect that would derail the economic development of the nation and might lead to the complete breakdown of order and the death of the state.

The trade agreement between Tajikistan and Uzbekistan illustrates the weakness of the Tajik economy: in the trade agreement struck in 1999 between the two countries Belarus would ship to Tajikistan finished manufactured goods such as motor vehicles, industrial equipment, chemical products, and consumer goods. In return, Tajikistan would sell Uzbekistan extractive or agricultural staples such as aluminum, cotton, and agricultural products. Some of Tajikistan's neighbors expressed interest in building industrial facilities in Tajikistan. China invested $11 million in Tajik textile and tobacco factories. The World Bank approved a $20 million loan to support privatization. Tajikistan's trade amounted to $551.7 million. But its trade deficit grew to $42.3 million. Some of Tajikistan's trading partners carried the debt for awhile. But others had problems of their own. The Ukraine threatens to stop supplies of alumina to Tajikistan unless Tajikistan paid off its debts with cotton. Tajikistan was heavily dependent on its main cash crop, cotton. Its dependence on cotton meant that its economic horizons were set by the size of the harvest and the vagaries of the agricultural futures market.

None of these numbers are staggering in themselves. Any number of western millionaires could have paid off the Tajik trade deficit. The problem was that the Tajik economy was going nowhere. On the one hand, the country was apparently not endowed with rich lodes of useful metals or vast stretches of arable land. Just as important was the issue of violence and anarchy. A society structured by clans and tribes was perhaps not suited for nationhood. This was not just a problem of capitalism. Half a century of ''socialist'' rule had not been able to change this structural fact. Meanwhile, in 1999 Tajikistan's leaders continued to try to buy time and, in spite of everything, build the economy.

GOVERNMENT AND POLITICS

The problems arose from every direction. Trying to stretch the budget, the Tajik leaders cut the number of government jobs by ten percent, making sure not to eliminate the thirty positions that had been granted to the UTO opposition. The civil unrest continued unabated. President Rakhmonov blamed an insurrection that occurred in November on the Republic of Uzbekistan. For its part, Uzbekistan complained about the continued presence of CIS troops in Tajikistan. Rakhmonov recognized Russia as the only reliable guarantor of Tajikistan's stability and security. Russian peacekeeping troops were deployed throughout the country. Rakhmonov continued granting amnesties to UTO leaders and ordered the complete integration of UTO soldiers into the army. Nine hundred former Tajik opposition fighters were inducted into three interim army units and eventually dispersed throughout the army. The government appointed UTO representatives to government posts, but the UTO replied that not enough opposition members were being appointed to regional and district councils.

While the civil war continued on at a reduced level, another part of the government apparatus went about drafting improvements to the nation's constitution. It recommended the creation of a bicameral parliament. Rakhmonov's ''overall approval'' to the constitutional amendments did not include the removal from the constitution of an article pledging construction of a secular Tajik state with a bicameral parliament. This did not appear to bother great numbers of the Tajik population. Seventy percent of them had backed Rakhmonov's reelection. His promise to them was that, as their fighters were disarmed the opposition was becoming transformed from a military into a political force.

While this deliberation went on, Safarali Kenjaev, Tajik Socialist Party leader and Tajik parliament's legislation and human rights committee chair, was shot dead. Given the level of violence, Kofi Annan extended the mandate of the UN Observer Mission for six months. Eventually 5,500 opposition fighters were amnestied. The UTO demanded the allocation of 30 percent of posts in national and local governments and the release of 93 imprisoned opposition figures. Rakhmonov agreed to meet with opposition representatives. He named four UTO candidates to government posts. The UTO compiled a list of 14 cities and towns to which its nominees should be appointed.

As part of the process of creating a peaceful republic the Tajik Supreme Court lifted the 1992 ban on opposition parties. Kofi Annan termed the Court's decision ''a significant step'' toward peace. UTO armed units were being dissolved. Tajikistan protested Uzbekistan's bombing of its Jirgatal district. The Uzbeki spokesman alleged that the Tajik authorities were allowing armed militants to cross into Kyrgyzstan.

The presidential election, which it was hoped would help to restore stability, was scheduled for November 6, 1999. Three million Tajiks were eligible to vote. Rakhmonov's name was put forward by the ruling People's Democratic Party. Tajiks were also to vote on three amendments to the constitution: extending the president's term to seven years, creating a bicameral parliament, and allowing the formation of religious-based parties. One opposition tendency, the IRP, nominated Davlat Usmon as its presidential candidate. The Tajik opposition party reregistered to participate in the election. Opposition candidates called for the postponement of presidential vote.

CULTURE AND SOCIETY

As the political conflict seemed to be winding down, gale-force winds and blizzards made 1,500 Tajik homeless and caused $1 million worth of damage in Leninabad. Russian and Tajik Muftiat agreed on holding conferences on spiritual matters and on opening the madrasahs and mosques. Haji Amanullo Negmatzoda was re-elected to the chairmanship of the Tajik Muslim Religious Board for another five years. The 8th anniversary of the republic's independence coincided with the celebrations devoted to the 1100th anniversary of the Samanid state.

DIRECTORY

CENTRAL GOVERNMENT

Head of State

President
Emomali Rakhmonov, Office of the President, 80 Rudaki Street, Dushanbe 734023, Tajikistan
PHONE: +7 (3772) 212914

Prime Minister
Yakhyo Azimov, Office of the Prime Minister, 80 Rudaki Street, Dushanbe 734023, Tajikistan
PHONE: +7 (3772) 211947

Ministers

Minister of Agriculture
Shodi Kabirov, Ministry of Agriculture, 1 Rudaki Prospekt, Dushanbe, Tajikistan
PHONE: +7 (3772) 211094; 213024

Minister of Culture
Bobokhon Mahmadov, Ministry of Culture

Minister of Defense
Sherali Khayrulloyev, Ministry of Defense

Minister of Economy and External Economic Affairs
Davlat Usmon, Ministry of Economy and External Economic Affairs, 42 Rudaki Street, Dushanbe 734025, Tajikistan
PHONE: +7 (3772) 216400
FAX: +7 (3772) 216914; 213854

Minister of Education
Munira Inoyatova, Ministry of Education

Minister of Emergency Situations
Mirzo Ziyoyev, Ministry of Emergency Situations

Minister of Environmental Protection
Ismail Davlatov, Ministry of Environmental Protection

Minister of Finance
Anvarsho Muzaffarov, Ministry of Finanace, 3 Academic Rajabov Street, Dushanbe, Tajikistan
PHONE: +7 (3772) 273941; 213804
FAX: +7 (3772) 213329

Minister of Foreign Affairs
Talbak Nazarov, Ministry of Foreign Affairs, 40 Rudaki Prospekt, Dushanbe 734025, Tajikistan
PHONE: +7 (3772) 211987; 211808; 213921
FAX: +7 (3772) 210259
E-MAIL: dushanbe@mfaumo.td.silk.org

Minister of Grain Products
Bekmurod Uroqov, Ministry of Grain Products

Minister of Health
Alamkhon Ahmadov, Ministry of Health

Minister of Internal Affairs
Khomiddin Sharipov, Ministry of Internal Affairs

Minister of Justice
Shavrat Ismoilov, Ministry of Justice

Minister of Labor and Employment
Khudoiberdi Kholiknazarov, Ministry of Labor and Employment

Minister of Land Improvement and Water Economy
Davlatbek Makhsudov, Ministry of Land Improvement and Water Economy

Minister of Security
Saidamir Zuhurov, Ministry of Security

Minister of Social Security
Abdussattor Jabborov, Ministry of Social Security

Minister of Transport and Roads
Khayriddin Muhiddinov, Ministry of Transport and Roads

POLITICAL ORGANIZATIONS

Adolatho Party (Justice Party)

TITLE: Chairman
NAME: Abdurahmon Karimov

Citizenship, Patriotism, Unity Party

NAME: Bobokhon Mahmadov

Congress of Popular Unity

NAME: Saifuddin Turayev

Democratic Party (TDP)

TITLE: Chairman
NAME: Jumaboy Niyazov

Islamic Renaissance Party (IRP)

TITLE: Chairman
NAME: Mohammed Sharif Himatzoda

Lali Badakhshan Society

NAME: Atobek Amirbekov

People's Democratic Party of Tajikistan (PPT)

NAME: Emomali Rahmonov

Party of Justice and Development

NAME: Rahmutullo Zainav

Party of Popular Unity and Accord (PPUA)

NAME: Abdumalik Abdullojonov

Rastokhez (Rebirth)

NAME: Takhir Abduzhaborov

Tajik Communist Party (CPT)

NAME: Shodi Shabdolov

Tajikistan Party of Economic and Political Renewal (TPEPR)

DIPLOMATIC REPRESENTATION

Embassies in Tajikistan

India
Hotel Tajikistan, Dushanbe, Tajikstan
PHONE: +7 (3772) 275190; 275369

United States
Oktyabrska Hotel, 105A Prospect Rudaki, Dushanbe 734001, Tajikstan
PHONE: +7 (3772) 210356

JUDICIAL SYSTEM

Supreme Court

FURTHER READING

Government Publications

Central Intelligence Agency. *CIA World Factbook, 1999.* Washington, DC: GPO, 1999.

U.S. Department of State. *Country Reports on Human Rights Practices for 1998: Europe, Canada, and the New Independent States.* Washington, DC: GPO, 1999.

Internet

Radio Free Europe/Radio Liberty. Tajik Service. Available Online @ http://www.rferl.org/bd/ta/index.html (November 2, 1999).

Tajikistan Update. Available Online @ http://www.angelfire.com/sd/tajikistanupdate/ (November 2, 1999).

TAJIKISTAN: STATISTICAL DATA

For sources and notes see "Sources of Statistics" in the front of each volume.

GEOGRAPHY

Geography (1)

Area:

Total: 143,100 sq km.

Land: 142,700 sq km.

Water: 400 sq km.

Area—comparative: slightly smaller than Wisconsin.

Land boundaries:

Total: 3,651 km.

Border countries: Afghanistan 1,206 km, China 414 km, Kyrgyzstan 870 km, Uzbekistan 1,161 km.

Coastline: 0 km (landlocked).

Climate: midlatitude continental, hot summers, mild winters; semiarid to polar in Pamir Mountains.

Terrain: Pamirs and Alay Mountains dominate landscape; western Fergana Valley in north, Kofarnihon and Vakhsh Valleys in southwest.

Natural resources: significant hydropower potential, some petroleum, uranium, mercury, brown coal, lead, zinc, antimony, tungsten.

Land use:

Arable land: 6%

Permanent crops: 0%

Permanent pastures: 25%

Forests and woodland: 4%

Other: 65% (1993 est.).

HUMAN FACTORS

Health Personnel (3)

Total health expenditure as a percentage of GDP, 1990-1997[a]

Public sector	2.4
Private sector	NA
Total[b]	NA

Continued on next page.

Demographics (2A)

	1990	1995	1998	2000	2010	2020	2030	2040	2050
Population	5,332.5	5,813.3	6,020.1	6,194.4	7,368.3	8,889.9	10,366.3	11,854.7	13,261.3
Net migration rate (per 1,000 population)	NA	NA	NA	NA	NA	NA	NA	NA	NA
Births	NA	NA	NA	NA	NA	NA	NA	NA	NA
Deaths	NA	NA	NA	NA	NA	NA	NA	NA	NA
Life expectancy - males	64.5	62.1	61.3	61.0	63.5	66.9	69.8	72.3	74.4
Life expectancy - females	70.1	68.6	67.8	67.4	67.8	73.6	76.9	79.6	81.6
Birth rate (per 1,000)	40.8	30.1	27.7	27.3	29.3	23.7	19.8	18.7	16.6
Death rate (per 1,000)	7.7	7.7	7.8	7.9	8.0	6.2	5.7	6.2	6.8
Women of reproductive age (15-49 yrs.)	1,189.9	1,344.7	1,452.6	1,546.6	2,048.0	2,335.7	2,740.2	2,958.4	3,191.9
of which are currently married	NA	NA	NA	NA	NA	NA	NA	NA	NA
Fertility rate	5.3	3.9	3.5	3.4	3.2	2.9	2.6	2.4	2.3

Except as noted, values for vital statistics are in thousands; life expectancy is in years.

Health Personnel (3)

Health expenditure per capita in U.S. dollars, 1990-1997[a]

Purchasing power parity	NA
Total	NA

Availability of health care facilities per 100,000 people

Hospital beds 1990-1997[a]	880
Doctors 1993[c]	210
Nurses 1993[c]	738

Health Indicators (4)

Life expectancy at birth	
1980	66
1997	68
Daily per capita supply of calories (1996)	2,129
Total fertility rate births per woman (1997)	3.5
Maternal mortality ratio per 100,000 live births (1990-97)	58[b]
Safe water % of population with access (1995)	69
Sanitation % of population with access (1995)	62
Consumption of iodized salt % of households (1992-98)[a]	20
Smoking prevalence	
Male % of adults (1985-95)[a]	....
Female % of adults (1985-95)[a]	....
Tuberculosis incidence per 100,000 people (1997)	87
Adult HIV prevalence % of population ages 15-49 (1997)	<0.005

Infants and Malnutrition (5)

Under-5 mortality rate (1997)	76
% of infants with low birthweight (1990-97)	NA
Births attended by skilled health staff % of total[a]	92
% fully immunized (1995-97)	
TB	99
DPT	95
Polio	92
Measles	95
Prevalence of child malnutrition under age 5 (1992-97)[b]	NA

Ethnic Division (6)

Tajik	64.9%
Uzbek	25%
Russian	3.5%
Other	6.6%

Religions (7)

Sunni Muslim	80%
Shi'a Muslim	5%

Languages (8)

Tajik (official), Russian widely used in government and business.

EDUCATION

Public Education Expenditures (9)

Public expenditure on education (% of GNP)	
1980	....
1996	2.2
Expenditure per student	
Primary % of GNP per capita	
1980	....
1996	....
Secondary % of GNP per capita	
1980	....
1996	....
Tertiary % of GNP per capita	
1980	....
1996	39.5[1]
Expenditure on teaching materials	
Primary % of total for level (1996)	....
Secondary % of total for level (1996)	0.5
Primary pupil-teacher ratio per teacher (1996)	24
Duration of primary education years (1995)	4

Educational Attainment (10)

Age group (1989)	25+
Total population	1,916,494
Highest level attained (%)	
No schooling	9.8
First level	
Not completed	13.0
Completed	NA
Entered second level	
S-1	65.5
S-2	NA
Postsecondary	11.7

Literacy Rates (11B)

Adult literacy rate	
1980	
Male	—
Female	—
1995	
Male	100
Female	100

GOVERNMENT & LAW

Political Parties (12)

Supreme Assembly	No. of seats
Communist Party and affiliates	100
People's Party	10
Party of People's Unity	6
Party of Economic and Political Renewal	1
Other	64

Government Budget (13B)

Revenues	NA
Expenditures	NA
Capital expenditures	NA

NA stands for not available.

Military Affairs (14B)

	1992	1993	1994	1995
Military expenditures				
Current dollars (mil.)	NA	NA	NA	209
1995 constant dollars (mil.)	NA	NA	NA	209
Armed forces (000)	3[e]	3	3	8
Gross national product (GNP)				
Current dollars (mil.)	9,625	8,128	6,882	5,639
1995 constant dollars (mil.)	10,350	8,521	7,055	5,639
Central government expenditures (CGE)				
1995 constant dollars (mil.)	NA	NA	NA	NA
People (mil.)	5.6	5.7	5.8	5.8
Military expenditure as % of GNP	NA	NA	NA	3.7
Military expenditure as % of CGE	NA	NA	NA	NA
Military expenditure per capita (1995 $)	NA	NA	NA	36
Armed forces per 1,000 people (soldiers)	0.5	.5	.5	1.4
GNP per capita (1995 $)	1,849	1,501	1,226	967
Arms imports[6]				
Current dollars (mil.)	0	5	0	0
1995 constant dollars (mil.)	0	5	0	0
Arms exports[6]				
Current dollars (mil.)	0	0	0	10
1995 constant dollars (mil.)	0	0	0	10
Total imports[7]				
Current dollars (mil.)	100[e]	517	900	690[e]
1995 constant dollars (mil.)	108[e]	542	923	690[e]
Total exports[7]				
Current dollars (mil.)	100[e]	314	413	707[e]
1995 constant dollars (mil.)	108[e]	329	423	707[e]
Arms as percent of total imports[8]	0.0	1.0	0	0
Arms as percent of total exports[8]	0.0	0.0	0.0	1.4

LABOR FORCE

Labor Force (16)

Total (million)	1.9
Agriculture and forestry	52%
Manufacturing, mining, and construction	17%
Services	31%

Data for 1996 Percent distribution for 1995.

Unemployment Rate (17)

2.4% includes only officially registered unemployed; also large numbers of underemployed workers and unregistered unemployed people (December 1996).

PRODUCTION SECTOR

Electric Energy (18)

Capacity	4.443 million kW (1995)
Production	14.66 billion kWh (1995)
Consumption per capita	2,302 kWh (1995)

Transportation (19)

Highways:

total: 32,752 km

paved: 21,119 km (note—these roads are said to be hard-surfaced, meaning that some are paved and some are all-weather gravel surfaced)

unpaved: 11,633 km (1992 est.)

Pipelines: natural gas 400 km (1992)

Airports: 59 (1994 est.)

Airports—with paved runways:

total: 14

over 3,047 m: 1

Airports—with paved runways:
2,438 to 3,047 m: 5
1,524 to 2,437 m: 7
914 to 1,523 m: 1 (1994 est.)
Airports—with unpaved runways:
total: 45
914 to 1,523 m: 9
under 914 m: 36 (1994 est.)

Top Agricultural Products (20)

Cotton, grain, fruits, grapes, vegetables; cattle, sheep, goats.

FINANCE, ECONOMICS, & TRADE

Economic Indicators (22)

National product: GDP—purchasing power parity—$4.1 billion (1997 est.)

National product real growth rate: -10% (1997 est.)

National product per capita: $700 (1997 est.)

Inflation rate—consumer price index: 40% (1996 est.)

Exchange Rates (24)

Exchange rates:	
Tajikistani rubles (TJR) per US$1	
January 1997	350
January 1996	284

Top Import Origins (25)

$657 million (1996 est.) Data are for 1997.

Origins	%
Japan	25.6
United States	13.9
Singapore	5
Taiwan	4.6
Germany	4.5
Malaysia	4.1

Top Export Destinations (26)

$768 million (1996 est.) Data are for 1994.

Destinations	%
FSU	78
Netherlands	NA

NA stands for not available.

Economic Aid (27)

Recipient: ODA, $22 million (1993). Note: commitments, $885 million (disbursements $115 million) (1992-95).

Import Export Commodities (28)

Import Commodities	Export Commodities
Fuel	Cotton
Chemicals	Aluminum
Machinery and transport equipment	Fruits
Textiles	Vegetable oil
Foodstuffs	Textiles

MANUFACTURING SECTOR

GDP & Manufacturing Summary (21)

	1980	1985	1990	1992	1993	1994
Gross Domestic Product						
Millions of 1990 dollars	3,227	4,275	4,680	3,388	2,453	2,085
Growth rate in percent	6.68	7.23	4.41	−25.00	−27.60	−15.00
Per capita (in 1990 dollars)	816.2	937.9	885.1	604.5	425.3	351.4
Manufacturing Value Added						
Millions of 1990 dollars	NA	NA	694	507	*468*	*324*
Growth rate in percent	NA	NA	NA	−24.19	*-7.79*	*-30.79*
Manufacturing share in percent of current prices	NA	NA	*14.8*	*15.8*	*17.6*	NA

TANZANIA

CAPITAL: Dar es Salaam.

FLAG: The flag consists of a black diagonal stripe running from the lower left-hand corner to the upper right-hand corner, flanked by yellow stripes. The diagonal stripes separate two triangular areas: green at the upper left and blue at the lower right.

ANTHEM: The Tanzanian National Anthem is a setting to new words of the widely known hymn *Mungu Ibariki Afrika (God Bless Africa).*

MONETARY UNIT: The Tanzanian shilling (Sh) of 100 cents is a paper currency. There are coins of 5, 10, 20, and 50 cents and 1, 5, 10, and 20 shillings, and notes of 10, 20, 50, 100, 200, 500, and 1,000 shillings. Sh1 = $0.0020 (or $1 = Sh494.41).

WEIGHTS AND MEASURES: The metric system is used.

HOLIDAYS: Zanzibar Revolution Day, 12 January; Chama Cha Mapinduzi Day, 5 February; Union Day, 26 April; International Workers' Day, 1 May; Farmers' Day, 7 July; Independence Day, 9 December; Christmas, 25 December. Movable religious holidays include 'Id al-Fitr, 'Id al-'Adha', Milad an-Nabi, Good Friday, and Easter Monday.

TIME: 3 PM = noon GMT.

LOCATION AND SIZE: Situated in East Africa just south of the equator, mainland Tanzania lies between the area of the great lakes—Victoria, Tanganyika, and Malawi (Niassa)—and the Indian Ocean. It contains a total area of 945,090 square kilometers (364,901 square miles). Tanzania has a total boundary length of 5,114 kilometers (3,178 miles). The section of Tanzania known as Zanzibar comprises the islands of Zanzibar and Pemba and all islands within 19 kilometers (12 miles) of their coasts. Tanzania's capital city, Dar es Salaam, is located on the Indian Ocean coast.

CLIMATE: There are four main climatic zones: the coastal area and immediate interior, where conditions are tropical; the central plateau, which is hot and dry; the highland areas; and the high, moist lake regions. The eastern parts of the country average only 75–100 centimeters (30–40 inches of rain, while the western parts receive 200–230 centimeters (80–90 inches).

United Republic of Tanzania
Jamhuri Ya Muungano Wa Tanzania

INTRODUCTORY SURVEY

RECENT HISTORY

Following Germany's World War I defeat, Britain governed Tanganyika in cooperation with the League of Nations. In 1946, Tanganyika became a United Nations trust territory. On 9 December 1961, Tanganyika became an independent nation, and exactly one year later was established as a republic, led by Julius Nyerere. In 1964 Tanganyika merged with Zanzibar and became the United Republic of Tanganyika and Zanzibar, with Nyerere as president. In October, the country's name was changed to Tanzania. Under Nyerere's leadership, Tanzania became steadily more socialist.

Tanzania became one of the strongest supporters of majority rule in southern Africa, backing liberation movements in Mozambique and Rhodesia (now Zimbabwe). In 1978 and 1979 saw military clashes between Tanzania and Uganda. Tensions between the mainland (Tanganyika) and Zanzibar grew in the 1990s, often linked to the ongoing Christian-Muslim conflicts. Also during the 1990s, Tanzania took in 500,000 Rwandan refugees and 200,000 Burundian refugees. The strain on resources, as well as Burundian Tutsis following Hutus, led the government to close its borders in 1995. In 1997, Tanzania implemented a much-criticized plan to return or expel its refugee population.

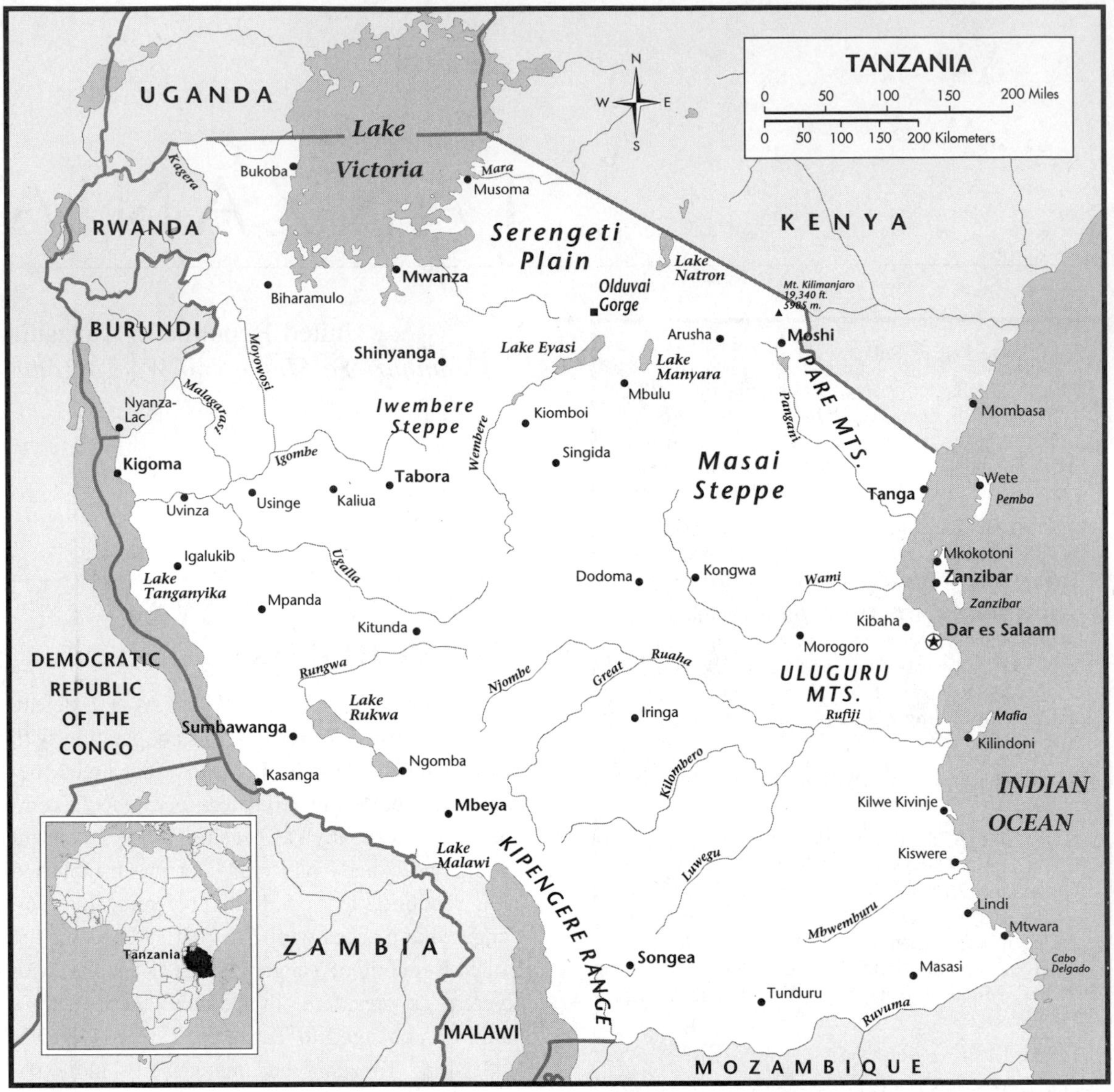

GOVERNMENT

The president, who is both chief of state and head of government, is elected for a five-year term by universal adult voting. Before 1992, the president was nominated by the sole legal party, the Chama Cha Mapinduzi (CCM). He is assisted by a prime minister and cabinet. Formerly there were two vice-presidents. In 1993, however, it was agreed to have just one vice-president, the running mate of the winning presidential nominee. As of 1985, the National Assembly consisted of 216 members elected by universal adult voting for five-year terms and 75 appointed members, many of whom serve by virtue of holding other government posts. The prime minister, who is chosen from the assembly members, heads the assembly.

The government of Zanzibar has exclusive jurisdiction over internal matters, including immigration, finances, and economic policy. In the 1990s, Zanzibar seems to be moving toward even greater self-rule, if not secession.

Judiciary

Mainland Tanzanian law is a combination of British and East African customary law. Local courts are presided over by appointed magistrates. They have limited powers, and there is a right of appeal to district courts, headed by either resident or district magistrates. Appeal can be made to the High Court, which consists of a chief justice and 17 judges appointed by the president. Appeals from the High Court can be made to the five-member

Court of Appeal. Cases concerning the Zanzibar constitution are heard only in Zanzibar courts. All other cases may be appealed to the Court of Appeal of the Republic.

Political Parties

The Tanganyika African National Union (TANU), established in 1954, and the ruling Afro-Shirazi Party of Zanzibar were merged into the Chama Cha Mapinduzi (CCM) Revolutionary Party. It was the only legal political party in Tanzania until 1995. The CCM officially favors nonracism and African socialism. Its basic aims are social equality, self-reliance, economic cooperation with other African states, and ujamaa (familyhood)—the development of forms of economic activity, particularly in rural areas, based on collective efforts. The 172-member National Executive Committee is the main policymaking and directing body of the CCM.

Although Tanzania amended its constitution in 1992 to become a multiparty state, the CCM still controlled the government. National elections were held in 1995. International observers and opposition parties accused the CCM of voter fraud and intimidation of opposition candidates in Zanzibar. Parliamentary election results saw the CCM win 52.9% of the vote and 186 seats; the National Committee for Constitutional Reform, 21.83% and 16 seats; CUF, 5% and 24 seats; and others, 9.5% and 6 seats.

DEFENSE

Tanzania's armed forces totaled 34,600 in 1995. The army had 30,000 members in 5 infantry brigades, 1 tank brigade, and 9 supporting arms battalions. The navy had 1,000 members and 18 craft, and the air force had 3,600 members and 24 combat aircraft. Police field forces, which include naval and air units, number 1,400. The citizens militia totaled 80,000. Defense spending was around $87 million in 1995, or 1.5% of gross domestic product (GDP).

ECONOMIC AFFAIRS

Tanzania has an agricultural economy whose chief commercial crops are sisal, coffee, cotton, tea, tobacco, spices, and cashew nuts. The most important minerals are diamonds and coal. Industry is mainly concerned with the processing of raw materials for export and local consumption. After 25 years of socialism, in 1986 the government moved toward a free-market economy. Since then, the economy has strengthened and donors have pledged funds to help Tanzania renovate its deteriorated infrastructure. In 1996, the economy grew by 3.9%.

Public Finance

The Tanzanian budget covers cash expenditures and receipts for the mainland only, and does not include Zanzibar government revenues and expenditures. Total expenditures include a development budget. The fiscal year ends on 30 June. In the early 1980s, the annual budget deficit went over 10% of GDP, and payment arrears on external debts started to mount. Since 1986, the government has improved its fiscal and monetary policies, with mixed results. The U.S. Central Intelligence Agency estimates that, in 1993, government revenues totaled approximately $495 million and expenditures $631 million, including capital expenditures of $118 million. External debt totaled $6.7 billion.

Income

In 1998 Tanzania's gross national product (GNP) was $6.7 billion, or about $210 per person.

Industry

Tanzanian industry is centered on the processing of local products. Some products are exported to neighboring countries: textiles and clothes, shoes, tires, batteries, transformers and switchgear, electric stoves, bottles, cement, and paper. Other industries include oil refining, fertilizers, rolling and casting mills, metal working, beer and soft drinks, vehicle assembly, bicycles, canning, industrial machine goods, glass and ceramics, agricultural implements, electrical goods, wood products, bricks and tiles, oxygen and carbon dioxide, and pharmaceutical products.

Since the economic reforms of 1986, reform in the industrial sector has been slow. By 1996, 138 of 259 state-owned companies had been privatized.

Banking and Finance

On 5 February 1967, Tanzania nationalized all banks after the adoption of the Arusha Declaration. From then until 1991, banking was a state monopoly. The Bank of Tanzania (BoT), the central bank and bank of issue, provides banking advice to the National Bank of Commerce (NBC), which took over the duties of the former private commercial banks. Other Tanzanian banks include the People's Bank of Zanzibar, the Tanzania Investment Bank, the Tanzania Housing Bank, the Rural Cooperative and Development Bank (CRDB), and the Tanganyika Post Office Savings Bank.

In 1994, the CRDB made a share offer to the public and the government has been urged to take the NBC along the same route. It was revealed in mid-1995 that the NBC had suffered cumulative losses of up to $200 million dollars and its assets quality was described as "unhealthy." NBC is being broken up and privatized.

In 1991, the financial services sector was opened to private and foreign capital. In 1993, the first private banks opened their doors. These were Meridien BIAO and Standard Chartered, the latter being among the UK-owned banks to have been nationalized in 1967. Meridien's Zambian-based African network collapsed in 1995, and Stanbic of South Africa took over the Tanzanian subsidiary after its seizure by the BoT. The Kenyan-owned Trust Bank opened in March 1995, to be followed by Eurafrican Bank (a Belgian-led venture). Also in early 1995, the only private bank to be majority-owned by indigenous Tanzanians, First Adili Bank, began business. Money supply, as measured by M2, totaled Sh613.7 billion at the end of 1995.

Plans are at an advanced stage for the establishment of a local stock market, the Dar es Salaam Stock Exchange (DSE).

Economic Development

The Tanzanian government has focused in recent years on reorganizing and restructuring its economic institutions. Progress has been encouraging and private sector investors are increasingly interested in mining, transport, tourist, and fishing sector opportunities.

The fourth five-year development plan (1981–86) was not fully carried out because of Tanzania's economic crisis. Among the projects implemented were an industrial complex, a pulp and paper project, a machine-tool plant, a phosphate plant, and the development of natural gas deposits. The Economic and Social Action Plan of 1990 scaled back the government's ambitions and sought to continue moderate growth in the economy, improve foreign trade, and alleviate some of the social costs of economic reform. Development planning is now conducted on an annual basis, with recent development priorities set in the areas of transport infrastructure, health, and education.

SOCIAL WELFARE

The government concentrates on community development (including health, labor, and literacy programs) rather than on welfare programs. The elderly, widows, and the physically and mentally handicapped normally are provided for by the traditional tribal system. Orphaned and abandoned children usually are cared for similarly, but missions and voluntary agencies also are active in this field.

The government promotes equal rights and employment for women, but discrimination and violence against women are widespread. Women face considerable discrimination on Zanzibar, which is largely Muslim.

Healthcare

An estimated 80% of the population had access to health care services in 1990–95, and, in 1990, total health care expenditures amounted to 4.7% of the gross domestic product (GDP). Life expectancy is about 51 years.

In 1992, there were 3,000 rural health facilities, 17 regional hospitals, and 3 national medical centers. There is 1 physician for every 28,271 people. Special disease control programs have been carried out with the assistance of the World Health Organization (WHO) and the United Nations Children's Fund (UNICEF) for most major diseases, including malaria, tuberculosis, sleeping sickness, schistosomiasis, poliomyelitis, and yaws.

Tanzania has one of the highest rates of HIV (the virus that causes AIDS) infection in Africa. In 1994, nearly 6.4% of all Tanzanian adults were infected with HIV.

Housing

Tanzania has developed a serious urban housing shortage as a result of the migration of people to the towns. The government in 1951 began a low-cost housing program, which has been continued since that time. As of 1978, 41% of dwellings were constructed with mud and poles, 23% with mud bricks and blocks, and 18% with concrete and stone.

EDUCATION

Education is compulsory for children aged 7 to 14. In 1991, there were 3.5 million students and 98,174 teachers in the 10,437 primary schools. In the same year, there were 183,109 students and 9,904 teachers in the 193 secondary schools. Higher education facilities include the University College in Dar es Salaam, the Sokoine University of Agriculture, the Dar es Salaam Technical College, and the College of African Wildlife Management. Literacy was estimated at 75% for men and 50% for women in 1990. In 1989, all higher level institutions had 5,254 students and 1,206 teaching staff.

1999 KEY EVENTS TIMELINE

January

- Cholera outbreak continues around Lake Rukwa and Lake Tanganyika. Hundreds have been stricken with disease since last year and nearly twenty have died.
- Presidents Benjamin Mkapa of Tanzania, Yoweri Museveni of Uganda, and Daniel Arap Moi of Kenya met in the northern Tanzanian town of Arusha to discuss East African Cooperation. It was originally planned that at this meeting the heads of state would sign the treaty on East African Cooperation. However, it was decided to wait until July 30. Officials are hoping to avoid the problems that resulted in the collapse of the East African Union in 1977. Press reports indicate that disagreements over the nature of East African Institutions had not been fully ironed out.
- Regional leaders lift sanctions on Burundi. Tanzania opens its borders to trade. Sanctions were imposed after the 1996 military coup of Major Pierre Buyoya. While the sanctions were widely flaunted, they did succeed in pushing up the price of imported goods in Burundi.
- Prime Minister Frederick Sumaye announced that the country had a deficit of 600,000 tons of maize and asked for help from the international community in dealing with the situation. Thirteen of Tanzania's twenty regions are expecting poor harvests this year due to unfavorable weather conditions.

March

- Nigeria's president-elect, General Olusegun Obasnjo arrives in Dar es Salaam for a two day state visit aimed at enhancing the fraternal bonds between the two counties.
- The main opposition party NCCR-Mageuzi, suspends its popular populist Chair, former Minister for Home Affairs, Augustine Mrema. In a stormy meeting, Mrema was charged with violating the party's constitution, lack of discipline, and not cooperating with other party officials.

April

- The Bank of Tanzania (the Central Bank) suspends the operations of Greenland Bank.
- The government bans NCCR-Mageuzi's Annual General Conference. Police in Dar es Salaam arrest 200 NCCR-Mageuzi members who protest the government's actions.
- Augustine Mrema, the leading oppostion politician decides to quit the crisis-ridden NCCR-Mageuzi. He joins the relatively obscure TLP (Tanzania Labor Party). The speaker of the National Assembly then declares Mrema's seat in the Parliament vacant and orders a bi-election. The government later prevents Mrema form trying to recontest his seat in Parliament.

June

- Long time politician and the longest serving speaker of the National Assembly, Chief Adam Sapi Mkwawa, dies at age seventy-nine. His death reinforces the notion that national leadership is passing from independence era politicians to a new generation.
- A reconciliation accord signed at the House of Representatives in Zanzibar brings an end to hostilities between the ruling CCM (Chama cha Mapinduzi/Revolutionary Party) and its fierce rival CUF (Civic United Front). The agreement ends a four-year political stalemate over the outcome of the 1995 Zanzibar elections.

July

- The governments of Tanzania, Uganda, and Kenya once again delay the signing of the East African Cooperation Agreement.
- Dar police are mobilized to stop an illegal demonstration planned to protest the retraction of a statement on the part of the Minister for Higher Education who had earlier announced that female students could wear headscarves to school on Fridays for prayers. The government reversed itself again and later announced that female students could wear scarves on Fridays.

August

- The ministry of health announces campaign to eradicate trachoma, an infectious disease which causes blindness as eyelids turn inwards causing damage to the cornea. A New York charity, Edna McConnell Clark Foundation and Pfizer Pharmaceuticals are helping in the campaign which seeks to eliminate the disease by the year 2020. It is estimated that Tanzania has one of the highest infection rates in the world and two million children are currently at risk.

- Idi Simba, the Minister for Industry and Commerce, announces that small businesses, including guest houses and butcheries, would be reserved for Tanzanian nationals. He also stated other activities, such as cargo handling and wholesale distribution of spare parts could only accomodate non-Tanzanians if they owned 50% or less of the business in partnership with nationals.

September

- Traveling in a thick cloud, a plane carrying ten U.S. tourists, a Tanzanian guide, and a pilot crashes into Mt. Meru near the tourist center of Arusha. All aboard are killed.

October

- The first President of Tanzania, Julius Kambarage Nyerere dies from Leukemia in a London hospital. The government announces a period of national mourning to last for one month. Nyerere is buried with full state honors at his home village of Butiama in western Tanzania.
- The long awaited signing of the Treaty for East African cooperation between Uganda, Tanzania, and Kenya was delayed, again.
- A Tanzanian living in South Africa, Khalfan Khamis Mohamed, was arrested in Cape Town in Southe Africa in connection to last year's bombings of the U.S. embassies in Tanzania and Kenya. The suspect was turned over to U.S. authorities and will stand trial in that country.

December

- A committee recommends that Tanzania have a semi-autonomous government, similar to the one that governs the island of Zanzibar. President Banjamin Mkapa attacks the report, noting that a third government would weaken the united government overseeing affairs for both the mainland and the island.
- Two boats capsize on Lake Victoria in separate accidents, and several dozen people are killed.
- Scientists report that the weevils introduced three years ago to eat the overgrowth of water hyacinths on Lake Victoria have been successful.

ANALYSIS OF EVENTS: 1999

BUSINESS AND THE ECONOMY

Tanzania is one of the poorest countries in the world as measured by Gross National Product per capita (less than $250 per year). There is hope that this poverty might be a dark cloud with a silver lining. Tanzania has been seeking and has been promised 100 percent relief from western creditors. This would eliminate the crippling economic burden of debt servicing, which coupled with its strong free market policies might lay the foundation for accelerated economic growth. However, there are conditions that make many within the country anxious concerning whether debt relief will become a reality.

Economic news for the year was encouraging. Finance Minister Daniel Yona said strong performances by the tourism and mining sectors should offset a three-year decline in export cash crop production. While the trade deficit will widen, Yona predicted that the balance of payments debt should stabilize at last year's deficit of $357.7 million. The government announced its expectations that the economy would grow at a healthy real rate of 4.9% in the 1999/2000 (July/August) fiscal year.

Since 1985, Tanzania has gradually been trying to transform a socialist economy into a market one. As part of this process, state owned companies are being privitized. Negotiations regarding one of the most important, NBC (National Bank of Commerce), were close to completion and were being debated in parliament in November. NBC with branches throughout the country, was the largest financial institution in the country. In the past the privitization of state corporations has been a politically sensitive issue and the sale of NBC is no exception.

Various political factions have used privitization to advance their own goals. Those interested in promoting African control over the economy, have insisted that state companies be sold to Tanzanians. Others oppose the sales on principle because they do not favor a private sector economy. Some others are afraid that backroom deals are being made which could hurt the future viability of the companies. Whatever the reasons, opposition to privitization has made the process slow. This is a situation that is not likely to change in the near future.

The slow development of the stock market has been a concern of those interested in strengthening the market economy. The most noteworthy stock market news was the decision of Tanzania Tea Packers Ltd. to become the only fully owned private company to be listed.

Concerning the business environment, Tanzanians were somewhat perplexed by the country's continued poor ranking as one of the most corrupt countries on a study put out by Transparency International. Since this rankings inception, in the mid 1990s, Tanzania has been consistently ranked as one of the most corrupt countries in the world. While even the Tanzanian government admits that corruption is a serious problem, there is debate over whether the country really deserves the dubious distinction as one the most corrupt places on the earth. Most observers agree that there is large qualitative difference between corruption in Tanzania and other countries ranked as most corrupt, like Kenya and Nigeria. Even the head of Transparency International (monitoring election fraud) noted that when he was involved in negotiations with the Tanzanian government on behalf of a Canadian mining company, no government agents acted inappropriately. The head of the observation team also noted that the government of Benjamin Mkapa was making a sincere effort to control this problem.

GOVERNMENT AND POLITICS

The most important political event of the year was the death of ''Baba wa Taifa'' (Father of the Country) Mwalimu (teacher) Julius Kambarage Nyerere. Nyerere, the first President, who had retired from active politics, was seen as kingmaker in Tanzania. It was primarily through Nyerere's efforts that Benjamin Mkapa, Tanzania's current president, securred CCM's nomination, and later on won the General Election in 1995. With the untimely death of Mwalimu Nyerere, Tanzanians are wondering whether the looming election next year will be as peaceful as it was during the patronage of the late father of the nation.

The state of the union between Zanzibar and mainland Tanzania also stands at cross road after Mwalimu's departure. Nyerere was the driving force between the merger of Zanzibar and Tanganika into the United Republic of Tanzania after the 1964 Zanzibar Revolution. He strongly stated that if any changes occur, it should be ''over my dead body.'' A sizable section of population favor a federal system with three government one each for Tanzania mainland, Zanzibar, and the Union.

The Burundi peace talks which Mwalimu Nyerere had been facilitating were also been set back by his death. A new facilitator accepted by both warring Hutu and Tutsi sides needed to be found. The issue of peace in Burundi was of critical regional importance. Although Nyerere made little headway in bringing the waring sides together, it was not for lack of effort. It is unlikely that anyone else could have even matched his modest achievements in this area. The civil war in Burundi showed no signs of letting up. Over 240,000 Burundian refugees are living in camps in western Tanzania. Burundi has accused the Tanzanian government of allowing rebels to operate out of these camps. The United Nations and Tanzania, however, deny that it is providing a safe haven for anti-government guerillas. Early in the year, Tanzania re-opened its border with Burundi, which was closed in 1996 after the military illegally seized power in Burundi. While the fighting in Burundi briefly stopped in October to honor the late Julius Nyerere, there has been an intensification of violence toward the later part of the year. In the refugee camps in Tanzania there is growing tensions between refugees and Tanzanians living near the camps which is causing security concerns for the government.

Other than the death of Julius Nyerere, the second most noteworthy event of the year occurred in the islands of Zanzibar as the ruling party CCM (Chama Cha Mapinduzi/Revolutionary Party) and its rival CUF (Civic United Front) signed a reconciliation agreement. Political tensions in Zanzibar reached a boiling point in the aftermath of the 1995 elections as CCM took control of Zanzibar's government amid accusations by CUF that the elections were rigged. The Zanzibar elections were important because within the union, Zanzibar retains a considerable deal of autonomy over its internal affairs. Many believed that if CUF should win, it might seek to dissolve the union, or at the very least, look to fundamentally revise it. In the aftermath of the elections, CUF launched a boycott of Zanzibar's Parliament and refused to recognize the election results.

The government responded by charging eighteen CUF members with treason, a capital offence. There was also widespread state harassment of CUF supporters. The reconciliation agreement, brokered by the Commonwealth Secretary General's special envoy Moses Anafu, brought the political stalemate on the islands to an end. CUF agreed to end its boycott of Parliament and to recognize the CCM government. For its part, the government agreed to compensate those affected by political persecution and appoint members of CUF to governmental ministers. Despite the agreement, there was apprehension that the deal could fall apart amidst rumors that Zanzibar's president would seek another term in office, contrary to the constitution, and that CUF leaders might be arrested.

CULTURE AND SOCIETY

The one month of national mourning for the "Father of the Nation" Julius Nyerere gave Tanzanians the chance to reflect on the cultural legacy of the late President. No one had done more to shape modern Tanzanian society and culture, than its first President. During the ceremonies to mark his death, hundreds of thousands turned out to pay their last respects to the man who led the country to independence and later forged a national identity out of one hundred and twenty separate ethnic groups. Based on the values of tolerance, equality, and a basic respect for human dignity, Nyerere and Tanzanians crafted a truly multi-cultural society where people of different religions—Muslims, Christians, Hindus, and those who follow traditional African spiritual beliefs—and those of different ethnic and racial groups—Africans, Asians, Arabs, Europeans, Swahili, Nyamwezi, Nyakyusa, Pogoro, Gogo, Ha, Sambaa, and Haya—lived, worked, and studied together.

While Nyerere's vision of an agrarian oriented socialist society was considered an economic failure, Tanzanians note with pride their achievements in building a national culture. The importance of this for Tanzania is illustrated by the ethnic tensions that led to violence in neighboring Rwanda, Burundi, Kenya, and Uganda. It was only fitting that at a church service in London before Mwalimu's body was taken back to Tanzania, it was the country's Mulsim Ambassador to England, Abdulkadir Shareef, who was the main speaker. Most of the political leaders in Africa considered Nyerere honest and committed to improving the life of Africans. Back in Dar es Salaam, ceremonies reflected the strong egalitarian values of the country as the former leader as heads of state mixed with the poor to pay their final respects at the National Stadium in Dar es Salaam. More importantly, Nyerere retained the respect of common Africans because he resisted temptations to build his personal wealth and because he left office voluntarily thereby setting the precident for the peaceful transfer of power.

During colonial rule, when Europeans looked down upon Africa's cultural achievements, Nyerere extolled African values. Nyerere, an intellectual and scholar, translated Shakespeare's Julius Caesar and The Merchant of Venice into Kiswahili. His aim was to prove that Kiswahili was a more suitable language than any European one for the national language of his new country. With a common language and near universal primary school education, all Tanzanians came to share a common bond. The policy of transfering government employees, including teachers, to work outside of their home areas, contributed to building a national culture, as did the shared experience of national youth service. At independence, Nyerere created a ministry of culture charged with building a national identity and by borrowing the best traditions from the country's ethnic groups.

Tanzania's tolerance is reflected by its rich intellectual history. During the 1960s and 1970s the University of Dar es Salaam was a center for radical intellectuals. They did not shy from criticizing Nyerere's policies. In the state and party newspapers, Nyerere encouraged the reporters and editors to speak their minds. While Tanzania's effort to build a multi-cultural society was generally successful, there were also tensions. There had been strains in the relationship between the mainland and Zanzibar, incidents of religious intolerance, and debates over the proper place of non-Africans in society. Another area where the strains of building a multi-cultural nation was evident was the issue of traditional spiritual beliefs and natural healing practices of local African cultures. For many Tanzanians, traditional healing practitioners offer an affordable alternative to expensive western style medicine. For those with diseases that western medicine could not cure, traditional healers offered hope. Others turn to traditional healers to supplement western medical treatment.

While the contribution of natural healing was recognized by the government and most Tanzanians, the practice had its downside. While many traditional healers have spent years learning their trade, others are frauds and con artists engaged in an increasingly lucrative practice. Some people hired witch doctors to cast spells for destructive purposes. In some particularly gruesome incidents near Mbeya, close to the Tanzania border with Malawi and Zambia, a small number of deaths have been attributed to witch doctors from neighboring countries who have reportedly offered large sums of money for human skins. While these incidents have captured the press headlines, perhaps more serious was the hundreds of deaths yearly of suspected witches.

Often poverty and difficult living conditions in the rural areas contributed to the accusations of the elderly engaging in witchcraft. This, in many instances, led to people taking the law into their own hands and killing suspected witches. Around half the murders in Tanzania were thought to be witchcraft related. Some blamed the upsurge on witchcraft in some areas on the disintegration of local communities and their beliefs. This process started with colonialism when local systems of knowledge and spirituality where looked upon as savage and discouraged. In the post-colonial era, grinding poverty and deadly epidemics such as AIDS has torn apart communities making it more difficult for elders to keep in control and channel the use of traditional cultural practices into constructive areas.

DIRECTORY

CENTRAL GOVERNMENT

Head of State

President
Benjamin W. Mkapa, Office of the President, Magogoni Road, PO Box 9120, Dar es Salaam, Tanzania
PHONE: +255 116898

Vice-President
Omar Ali Juma, Office of the Vice-President, Magogoni Road, PO Box 9120, Dar es Salaam, Tanzania
PHONE: +255 116898

Prime Minister
Fredrick Sumaye, Office of the Prime Minister, Magogoni Road, PO Box 9120, Dar es Salaam, Tanzania
PHONE: +255 116898

President of Zanzibar
Salmin Amour, Office of the President of Zanaibar, Magogoni Road, PO Box 9120, Dar es Salaam, Tanzania
PHONE: +255 116898

Ministers

Minister of State
Wilson Masilingi, Ministry of State, Magogoni Road, PO Box 9120, Dar es Salaam, Tanzania
PHONE: +255 116898

Minister of State for Civil Service
Jackson Makwetta, Ministry of State for Civil Service, Magogoni Road, PO Box 9120, Dar es Salaam, Tanzania
PHONE: +255 116898

Minister of State for Planning and Sector Reforms
Nassoro Malocho, Ministry of State for Planning and Sector Reforms, Magogoni Road, PO Box 9120, Dar es Salaam, Tanzania
PHONE: +255 116898

Minister of State for Cabinet Affairs
Mateo Qares, Ministry of State for Cabinet Affairs, Magogoni Road, PO Box 9120, Dar es Salaam, Tanzania
PHONE: +255 116898

Minister of State in the Vice-President's Office for Cabinet Affairs
Edward Lowassa, Ministry of State in the Vice-President's Office for Cabinet Affairs, Magogoni Road, PO Box 9120, Dar es Salaam, Tanzania
PHONE: +255 116898

Minister of State in the Prime Minister's Office
Mohamed Seif Khatib, Ministry of State in the Prime Minister's Office, Magogoni Road, PO Box 9120, Dar es Salaam, Tanzania
PHONE: +255 116898

Government Spokesman in Parliament
Bakari Mbonde, Office of Policies, Magogoni Road, PO Box 9120, Dar es Salaam, Tanzania
PHONE: +255 116898

Minister of Regional and Local Governments
Kingunge Ngombare Mwiru, Ministry of Regional and Local Governments, Magogoni Road, PO Box 9120, Dar es Salaam, Tanzania
PHONE: +255 116898

Minister of Foreign Affairs and International Cooperation
Jakaya Mrisho Kikwete, Ministry of Foreign Affairs and International Cooperation, Kivukoni Road, PO Box 9000, Dar es Salaam, Tanzania
PHONE: +255 111906
FAX: +255 180411

Minister of Home Affairs
Ali Ameir Mohamed, Ministry of Home Affairs, Ohio/Ghana Avenue, PO Box 9223, Dar es Salaam, Tanzania
PHONE: +255 112034
FAX: +255 139675

Minister of Finance
Daniel Yona Ndhiwa, Ministry of Finance, Tancot, PO Box 9111, Dar es Salaam, Tanzania
PHONE: +255 111174
FAX: +255 138573; 117790

Minister of Industry and Trade
Idd Simba, Ministry of Industry and Trade, Co-operative Union Building, Lumumba Road, PO Box 9503, Dar es Salaam, Tanzania
PHONE: +255 181397; 180418
FAX: +255 182481; 112527

Minister of Communications and Transport
Enerst Nyanda, Ministry of Communications and Transport, Tancot House, PO Box 37650, Dar es Salaam, Tanzania
PHONE: +255 668353

Minister of Agriculture and Co-operation
William Kusila, Ministry of Agriculture and Co-operation, Sokoine/Mkwepu Road, PO Box 9192, Dar es Salaam, Tanzania
PHONE: +255 112323; 112324
FAX: +255 113584

Minister of Health
Aaron Chiduo, Ministry of Health, Samora Avenue, PO Box 9083, Dar es Salaam, Tanzania
PHONE: +255 120261

Minister of Education and Culture
Juma Athumani Kapuya, Ministry of Education and Culture, Magogoni Road, PO Box 9121, Dar es Salaam, Tanzania
PHONE: +255 117211; 122005
FAX: +255 113271

Minister of Energy and Mineral Resources
Abdalla Kigoda, Ministry of Energy and Mineral Resources, Sokoine/Mkwepu Street, PO Box 2000/9152, Dar es Salaam, Tanzania
PHONE: +255 117153
FAX: +255 116719

Minister of Water
Mussa Nkhangaa, Ministry of Water, Sokoine/Mkwepu Road, PO Box 9153, Dar es Salaam, Tanzania
PHONE: +255 117153

Minister of Natural Resources and Tourism
Zakia Meghji, Ministry of Natural Resources and Tourism, Samora Avenue, PO Box 9352, Dar es Salaam, Tanzania
PHONE: +255 111062; 112865; 116682
FAX: +255 113082; 114659

Minister of Lands, Housing and Urban Development
Gideon Cheyo, Ministry of Lands, Housing and Urban Development, Samora Avenue/Azikiwe Road, PO Box 1422, Dar es Salaam, Tanzania
PHONE: +255 120419
FAX: +255 113082

Minister of Science, Technology and Higher Education
Pius Ng'wandu, Ministry of Science, Technology and Higher Education, Jamhuri Street, PO Box 2645, Dar es Salaam, Tanzania
PHONE: +255 120419
FAX: +255 113082

Minister of Works
Anna Abdallah, Ministry of Works

Minister of Labour Youth and Development
Paul Kimiti, Ministry of Labour Youth and Development, Hifadhi House, Samora Avenue/Azikiwe Road, PO Box 1422, Dar es Salaam, Tanzania
PHONE: +255 120419
FAX: +255 113082

Minister of Community Development, Women's Affairs and Children
Mary Michael Nagu, Ministry of Community Development, Women's Affairs and Children, Samora Avenue/Azikiwe Road, PO Box 3448, Dar es Salaam, Tanzania
PHONE: +255 115074; 132057; 115635
FAX: +255 132647

Minister of Justice and Constitutional Affairs
Harith Bakari Mwapachu, Ministry of Justice and Constitutional Affairs, Kivukoni Road, PO Box 9050, Dar es Salaam, Tanzania
PHONE: +255 117099

Minister of Defense
Edgar Maokola Majogo, Ministry of Defense, Ismani Road, PO Box 9544, Dar es Salaam, Tanzania
PHONE: +255 117153
FAX: +255 116719

POLITICAL ORGANIZATIONS

Chama cha Demokrasia na-CHADEMA (Party for Democracy and Progress)

Plot No. 922/7,Block No. 186005, Kisutu St., PO Box 5330, Dar es Salaam, Tanzania
TITLE: Chairperson
NAME: Edwin I. Mtei

Chama cha Mapinduzi-CCM (Revolutionary Party of Tanzania)

Kuu St., PO Box 50, Dodoma
PHONE: +255 612282
TITLE: Chairperson
NAME: Benjamin William Mkapa

Civic United Front (CUF)

Mtendeni St., Urban District, PO Box 3637, Zanzibar, Tanzania
PHONE: +255 612282
TITLE: Leader
NAME: Seif Sharrif Hamad

National Convection for Construction and Reform (NCCR-Mageuzi)

Mtendeni St., Urban District, PO Box 5316, Dar es Salaam, Tanzania
TITLE: Chairperson
NAME: Augustine Lyatonga Mrema

National League for Democracy (NLD)

Sinza D/73, PO Box 352, Dar es Salaam, Tanzania
TITLE: Chairperson
NAME: Emmanuel J. E. Makaidi

National Reconstruction Alliance (NRA)

House No 4, Mvita St., PO Box 16542, Dar es Salaam, Tanzania
TITLE: Chairperson
NAME: Ulotu Abubakar Ulotu

Popular National Party (PONA)

Plot No. 104, Songea St., Ilala, PO Box 21561, Dar es Salaam, Tanzania
TITLE: Chairperson
NAME: Wilfrem R. Mwakitwange

Tanzania Democratic Alliance Party (TADEA)

Block 3, Plot No. 37, Buguruni, Malapa, PO Box 63133, Dar es Salaam, Tanzania
TITLE: Chairperson
NAME: Flora M. Kambona

Tanzania People's Party (TPP)

Mbezi Juu, Kawe, PO Box 60847, Dar es Salaam, Tanzania
TITLE: Chairperson
NAME: Alech H. Che

United People's Democratic Party (UPDP)

al-Aziza Restaurant, Kokoni and Narrow Sts., PO Box 3903, Zanibar, Tanzania
TITLE: Chairperson
NAME: Khalfani Ali Abdullah

Union for Multi-Party Democracy of Tanzania (UMD)

77 Tosheka St. Magomeni Mapipa, PO Box 41093, Dar es Salaam, Tanzania
TITLE: Chairperson
NAME: Chief Abdallah Fundikira

DIPLOMATIC REPRESENTATION

Embassies in Tanzania

Albania
93 Msese Rd., PO Box 1034, Kinondoni, Dar es Salaam, Tanzania

Algeria
34 Upanga Rd., PO Box 2963, Dar es Salaam, Tanzania

Angola
PO Box 20793, Dar es Salaam, Tanzania
PHONE: +255 (051) 26689
FAX: +255 (051) 32349

Belgium
NIC Investment House, 7th floor, Samora Machel Ave., PO Box 9210, Dar es Salaam, Tanzania
PHONE: +255 (051) 46047
FAX: +255 (051) 20604

Brazil
IPS Bldg., 9th floor, PO Box 9654, Dar es Salaam, Tanzania

PHONE: +255 (051) 21780
FAX: +255 (051) 20604

Burundi
Plot No. 10007, Lugalo Rd., PO Box 2752, Dar es Salaam, Tanzania
PHONE: +255 (051) 38608

Canada
Mirambo St., PO Box 1022, Dar es Salaam, Tanzania
FAX: +255 (051) 46005

China
Kajificheni Close at Toure Drive, PO Box 1649, Dar es Salaam, Tanzania

Democratic Republic of Congo
Malik Rd., PO Box 975, Dar es Salaam, Tanzania

Cuba
Plot No. 313, Lugalo Rd., PO Box 9282,Upanga, Dar es Salaam, Tanzania

Denmark
Ghana Ave., PO Box 9171, Dar es Salaam, Tanzania
FAX: +255 (051) 46319

Egypt
24 Garden Ave., PO Box 1668, Dar es Salaam, Tanzania
PHONE: +255 (051) 23372

Finland
PO Box 2455, Dar es Salaam, Tanzania
PHONE: +255 (051) 46324
FAX: +255 (051) 41066

France
PO Box 2349, Dar es Salaam, Tanzania
PHONE: +255 (051) 34961
FAX: +255 (051) 41066

Germany
PO Box 9541, Dar es Salaam, Tanzania
PHONE: +255 (051) 463334
FAX: +255 (051) 46292

Guinea
PO Box 2969, Oysterbay, Dar es Salaam, Tanzania
PHONE: +255 (051) 463334
FAX: +255 (051) 46292

Hungary
PO Box 672, Dar es Salaam, Tanzania
PHONE: +255 (051) 34762
FAX: +255 (051) 40193

India
PO Box 2684, Dar es Salaam, Tanzania
PHONE: +255 (051) 46341
FAX: +255 (051) 46747

Indonesia
PO Box 572, Dar es Salaam, Tanzania

Iran
PO Box 3802, Dar es Salaam, Tanzania
PHONE: +255 (051) 34622

Italy
PO Box 2106, Dar es Salaam, Tanzania
PHONE: +255 (051) 46353
FAX: +255 (051) 46354

Japan
PO Box 2577, Dar es Salaam, Tanzania
PHONE: +255 (051) 31215

Kenya
PO Box 5231, Dar es Salaam, Tanzania
PHONE: +255 (051) 31502

South Korea
PO Box 2690, Dar es Salaam, Tanzania

Madagascar
PO Box 5254, Dar es Salaam, Tanzania
PHONE: +255 (051) 41761

Malawi
PO Box 23168, Dar es Salaam, Tanzania
PHONE: +255 (051) 37260

Netherlands
PO Box 9534, Dar es Salaam, Tanzania
PHONE: +255 (051) 46391
FAX: +255 (051) 46189

Nigeria
PO Box 9214, Oysterbay, Dar es Salaam, Tanzania

Norway
PO Box 2646, Dar es Salaam, Tanzania
PHONE: +255 (051) 25195
FAX: +255 (051)46444

Poland
PO Box 2188, Dar es Salaam, Tanzania
PHONE: +255 (051) 46294
FAX: +255 (051) 46294

Romania
PO Box 590, Dar es Salaam, Tanzania

Russia
PO Box 1905, Dar es Salaam, Tanzania
PHONE: +255 (051) 66006
FAX: +255 (051) 66818

Rwanda
PO Box 2918, Dar es Salaam, Tanzania
PHONE: +255 (051) 30119
FAX: +255 (051) 66818

Serbia
PO Box 2838, Dar es Salaam, Tanzania
PHONE: +255 (051) 46377
FAX: +255 (051) 46378

Spain
PO Box 842, Dar es Salaam, Tanzania
PHONE: +255 (051) 66936
FAX: +255 (051) 66818

Sudan
PO Box 2266, Dar es Salaam, Tanzania

Sweden
PO Box 9274, Dar es Salaam, Tanzania
PHONE: +255 (051) 23501
FAX: +255 (051) 66818

Switzerland
PO Box 2454, Dar es Salaam, Tanzania
PHONE: +255 (051) 66008
FAX: +255 (051) 66736

Syria
PO Box 2442, Dar es Salaam, Tanzania
PHONE: +255 (051) 20568

United Kingdom
PO Box 9200, Dar es Salaam, Tanzania
PHONE: +255 (051) 132871; 120880
FAX: +255 (051) 46301

United States
PO Box 9123, Dar es Salaam, Tanzania
PHONE: +255 (051) 66010
FAX: +255 (051) 66701

Vatican City
PO Box 480, Dar es Salaam, Tanzania
PHONE: +255 (051) 68403
FAX: +255 (051) 40193

Yemen
PO Box 349, Dar es Salaam, Tanzania
PHONE: +255 (051) 21722
FAX: +255 (051) 66791

Zambia
PO Box 2525, Dar es Salaam, Tanzania

Zimbabwe
PO Box 20762, Dar es Salaam, Tanzania
PHONE: +255 (051) 30455

JUDICIAL SYSTEM

Court of Appeal of the Republic

Court of Appeal

High Court

FURTHER READING

Articles

''12 Perish in Plane Crash.'' *Daily News*, 8 September 1999, p. 1.

Chege, Wambui. ''African Refugees say World Cares more about Kosovo.'' Reuters, 17 August 1999.

''Dar Police on Guard Against Illegal Demonstration.'' *Daily News*, 27 July 1999, p. 1.

''Isles Reconcialiation Accord Signed'' *Daily News*, 6 June 1999, p. 1.

''The Longest-Serving National Assembly Speaker in the Commonwealth, Mr Sapi Mkwawa Died in Dar Aged 79.'' *East African* (June 21-27, 1999).

''NCCR Meeting 'Suspends' Mrema'' *Daily News*, 19 March 1999, p. 1.

''Statesman Gets Hero's Send-Off.'' *East African* (October 25–31, 1999): 1.

''Tanzania Launches Anti-Blindness Program.'' *Reuters*, 23 August 1999.

Books

Assensoh, A. B. *African Political Leadership: Jomo Kenyatta, Kwame Nkrumah, and Julius K. Nyerere.* Malabar, Fla.: Krieger Pub., 1998.

Maddox, Gregory, James L. Giblin, and Isaria N. Kimambo, eds. *Custodians of the Land: Ecology and Culture in the History of Tanzania.* Athens: Ohio University Press, 1996.

Valk, Peter de. *African Industry in Decline: the Case of Textiles in Tanzania in the 1980s.* New York: St. Martin's Press, 1996.

Internet

Tanzania Tourist Board. Available Online @ http://www.tanzania-web.com/home2.htm (November 15, 1999).

TANZANIA: STATISTICAL DATA

For sources and notes see "Sources of Statistics" in the front of each volume.

GEOGRAPHY

Geography (1)

Area:

Total: 945,090 sq km.

Land: 886,040 sq km.

Water: 59,050 sq km.

Note: includes the islands of Mafia, Pemba, and Zanzibar.

Area—comparative: slightly larger than twice the size of California.

Land boundaries:

Total: 3,402 km.

Border countries: Burundi 451 km, Kenya 769 km, Malawi 475 km, Mozambique 756 km, Rwanda 217 km, Uganda 396 km, Zambia 338 km.

Coastline: 1,424 km.

Climate: varies from tropical along coast to temperate in highlands.

Terrain: plains along coast; central plateau; highlands in north, south.

Natural resources: hydropower potential, tin, phosphates, iron ore, coal, diamonds, gemstones, gold, natural gas, nickel.

Land use:

Arable land: 3%

Permanent crops: 1%

Permanent pastures: 40%

Forests and woodland: 38%

Other: 18% (1993 est.)

HUMAN FACTORS

Demographics (2A)

	1990	1995	1998	2000	2010	2020	2030	2040	2050
Population	24,886.4	28,824.7	30,608.8	31,962.8	39,390.0	46,692.6	55,132.6	65,545.0	76,500.1
Net migration rate (per 1,000 population)	NA	NA	NA	NA	NA	NA	NA	NA	NA
Births	NA	NA	NA	NA	NA	NA	NA	NA	NA
Deaths	NA	NA	NA	NA	NA	NA	NA	NA	NA
Life expectancy - males	46.3	45.4	44.2	43.5	44.7	48.2	57.8	67.4	70.9
Life expectancy - females	49.3	48.7	48.6	48.5	47.6	51.6	63.1	74.7	78.7
Birth rate (per 1,000)	43.6	41.9	40.7	40.0	34.7	30.2	26.4	22.2	19.1
Death rate (per 1,000)	16.9	16.6	16.7	16.8	16.3	14.1	9.1	5.4	5.1
Women of reproductive age (15-49 yrs.)	5,665.4	6,714.3	7,201.7	7,557.5	9,553.8	12,029.6	14,851.0	18,046.1	20,821.3
of which are currently married	NA	NA	NA	NA	NA	NA	NA	NA	NA
Fertility rate	6.2	5.8	5.5	5.3	4.4	3.6	3.0	2.6	2.3

Except as noted, values for vital statistics are in thousands; life expectancy is in years.

Health Personnel (3)

Total health expenditure as a percentage of GDP, 1990-1997[a]

Public sector 1.1

Private sector . . . . .NA

Total[b] . . . . .NA

Health expenditure per capita in U.S. dollars, 1990-1997[a]

Purchasing power parity . . . . .NA

Total . . . . .NA

Availability of health care facilities per 100,000 people

Hospital beds 1990-1997[a] 90

Doctors 1993[c] 4

Nurses 1993[c] 46

Health Indicators (4)

Life expectancy at birth

1980 50

1997 48

Daily per capita supply of calories (1996) 2,028

Total fertility rate births per woman (1997) 5.5

Maternal mortality ratio per 100,000 live births (1990-97) . . . . .530[d]

Safe water % of population with access (1995) 49

Sanitation % of population with access (1995) 86

Consumption of iodized salt % of households (1992-98)[a] 74

Smoking prevalence

Male % of adults (1985-95)[a] . . . .

Female % of adults (1985-95)[a] . . . .

Tuberculosis incidence per 100,000 people (1997) 307

Adult HIV prevalence % of population ages 15-49 (1997) 9.42

Infants and Malnutrition (5)

Under-5 mortality rate (1997) 143

% of infants with low birthweight (1990-97) 14

Births attended by skilled health staff % of total[a] . . .38

% fully immunized (1995-97)

TB 82

DPT 74

Polio 73

Measles 69

Prevalence of child malnutrition under age 5 (1992-97)[b] 31

Ethnic Division (6)

Mainland—native African 99% (of which 95% are Bantu consisting of more than 130 tribes); Other 1% (consisting of Asian, European, and Arab).

Religions (7)

Mainland—Christian 45%

Muslim 35%

Indigenous beliefs 20%

Zanzibar is more than 99% Muslim.

Languages (8)

Kiswahili or Swahili (official), Kiunguju (name for Swahili in Zanzibar), English (official, primary language of commerce, administration, and higher education), Arabic (widely spoken in Zanzibar), many local languages Kiswahili (Swahili) is the mother tongue of Bantu people living in Zanzibar and nearby coastal Tanzania; although Kiswahili is Bantu in structure and origin, its vocabulary draws on a variety of sources, including Arabic and English, and it has become the lingua franca of central and eastern Africa; the first language of most people is one of the local languages.

EDUCATION

Public Education Expenditures (9)

Public expenditure on education (% of GNP)

1980 . . . .

1996 . . . .

Expenditure per student

Primary % of GNP per capita

1980 . . . .

1996 . . . .

Secondary % of GNP per capita

1980 . . . .

1996 . . . .

Tertiary % of GNP per capita

1980 . . . .

1996 . . . .

Expenditure on teaching materials

Primary % of total for level (1996) . . . .

Secondary % of total for level (1996) . . . .

Primary pupil-teacher ratio per teacher (1996) 36

Duration of primary education years (1995) 7

Educational Attainment (10)

Age group (1988)[6] . . . . .25+

Total population 3,598,169

Highest level attained (%)

No schooling	0.0
First level	
Not completed	89.7
Completed	NA
Entered second level	
S-1	7.8
S-2	0.6
Postsecondary	2.0

Literacy Rates (11A)

In thousands and percent[1]	1990	1995	2000	2010
Illiterate population (15+ yrs.)	5,218	5,171	5,011	4,514
Literacy rate - total adult pop. (%)	62.1	67.8	73.3	82.2
Literacy rate - males (%)	75.3	79.4	83.1	88.8
Literacy rate - females (%)	49.5	56.8	63.8	75.9

GOVERNMENT & LAW

Military Affairs (14B)

	1990	1991	1992	1993	1994	1995
Military expenditures						
Current dollars (mil.)	98[e]	NA	106	103	82	69
1995 constant dollars (mil.)	112[e]	NA	114	108	84	69
Armed forces (000)	40	40	46	46	50	35[e]
Gross national product (GNP)						
Current dollars (mil.)	2,829	3,081	3,226	3,399	3,647	3,840[e]
1995 constant dollars (mil.)	3,251	3,405	3,470	3,563	3,739	3,840[e]
Central government expenditures (CGE)						
1995 constant dollars (mil.)	720	876	927	996	1,171	817
People (mil.)	24.8	25.5	26.1	26.7	27.6	28.6
Military expenditure as % of GNP	3.5	NA	3.3	3.0	2.3	1.8
Military expenditure as % of CGE	15.6	NA	12.3	10.9	7.3	8.4
Military expenditure per capita (1995 $)	5	NA	4	4	3	2
Armed forces per 1,000 people (soldiers)	1.6	1.6	1.8	1.7	1.8	1.2
GNP per capita (1995 $)	131	134	133	133	135	134
Arms imports[6]						
Current dollars (mil.)	30	10	5	5	10	0
1995 constant dollars (mil.)	34	11	5	5	10	0
Arms exports[6]						
Current dollars (mil.)	0	0	0	0	0	0
1995 constant dollars (mil.)	0	0	0	0	0	0
Total imports[7]						
Current dollars (mil.)	1,027	1,533	1,510	1,497	1,505	1,619
1995 constant dollars (mil.)	1,180	1,694	1,624	1,569	1,543	1,619
Total exports[7]						
Current dollars (mil.)	415	341	416	450	519	639
1995 constant dollars (mil.)	477	377	447	472	532	639
Arms as percent of total imports[8]	2.9	.7	.3	.3	.7	0
Arms as percent of total exports[8]	0	0	0	0	0	0

GOVERNMENT & LAW

Political Parties (12)

National Assembly	No. of seats
Chama Cha Mapinduzi (CCM)	186
Civic United Front (CUF)	24
National Convention for Construction and Reform (NCCR)	16
Chama Cha Demokrasia na Maendeleo (CHADEMA)	3
United Democratic Party (UDP)	3
Zanzibar House of Representatives	
Chama Cha Mapinduzi (CCM)	26
Civic United Front (CUF)	24

Government Budget (13B)

Revenues	$959 million
Expenditures	$1.1 billion
Capital expenditures	$214 million

Data for FY96/97 est.

Crime (15)

Crime rate (for 1997)	
Crimes reported	528,800
Total persons convicted	NA
Crimes per 100,000 population	1,750
Persons responsible for offenses	
Total number of suspects	46,900
Total number of female suspects	NA
Total number of juvenile suspects	NA

LABOR FORCE

Labor Force (16)

Total (million)	13.495
Agriculture	90%
Industry and commerce	10%

Data for 1995 est.

Unemployment Rate (17)

Rate not available.

PRODUCTION SECTOR

Electric Energy (18)

Capacity	439,000 kW (1995)
Production	895 million kWh (1995)
Consumption per capita	31 kWh (1995)

Transportation (19)

Highways:

total: 88,200 km

paved: 3,704 km

unpaved: 84,496 km (1996 est.)

Waterways: Lake Tanganyika, Lake Victoria, Lake Nyasa

Pipelines: crude oil 982 km

Merchant marine:

total: 8 ships (1,000 GRT or over) totaling 30,371 GRT/41,269 DWT ships by type: cargo 3, oil tanker 2, passenger-cargo 2, roll-on/roll-off cargo 1 (1997 est.)

Airports: 123 (1997 est.)

Airports—with paved runways:

total: 11

over 3,047 m: 2

2,438 to 3,047 m: 2

1,524 to 2,437 m: 5

914 to 1,523 m: 1

under 914 m: 1 (1997 est.)

Airports—with unpaved runways:

total: 112

1,524 to 2,437 m: 17

914 to 1,523 m: 60

under 914 m: 35 (1997 est.)

Top Agricultural Products (20)

Coffee, sisal, tea, cotton, pyrethrum (insecticide made from chrysanthemums), cashews, tobacco, cloves (Zanzibar), corn, wheat, cassava (tapioca), bananas, fruits, vegetables; cattle, sheep, goats.

MANUFACTURING SECTOR

GDP & Manufacturing Summary (21)

Detailed value added figures are listed by both International Standard Industry Code (ISIC) and product title.

	1980	1985	1990	1994
GDP ($-1990 mil.)[1]	2,027	2,102	2,542	2,932
Per capita ($-1990)[1]	109	96	99	102
Manufacturing share (%) (current prices)[1]	10.7	6.1	4.2	*4.8*
Manufacturing				
Value added ($-1990 mil.)[1]	108	83	94	*107*

	1980	1985	1990	1994
Industrial production index	100	81	104	*91*
Value added ($ mil.)	361	278	*99*	*101*
Gross output ($ mil.)	1,266	1,145	*458*	*463*
Employment (000)	101	94	*124*	*149*
Profitability (% of gross output)				
Intermediate input (%)	71	76	*78*	*78*
Wages and salaries inc. supplements (%)	9	9	*5*	*7*
Gross operating surplus	19	16	*16*	*15*
Productivity ($)				
Gross output per worker	12,457	12,141	*3,681*	*3,100*
Value added per worker	3,555	2,952	*797*	*688*
Average wage (inc. supplements)	1,174	1,042	*202*	*205*
Value added ($ mil.)				
311/2 Food products	58	58	*12*	*11*
313 Beverages	14	21	*6*	*6*
314 Tobacco products	12	16	*10*	*11*
321 Textiles	95	43	*15*	*17*
322 Wearing apparel	10	4	*1*	*1*
323 Leather and fur products	7	4	*1*	*1*
324 Footwear	8	6	*1*	*1*
331 Wood and wood products	7	6	2	2
332 Furniture and fixtures	6	3	*1*	*1*
341 Paper and paper products	8	7	*3*	*4*
342 Printing and publishing	14	12	*2*	*3*
351 Industrial chemicals	11	9	*14*	*15*
352 Other chemical products	10	7	*2*	*2*
353 Petroleum refineries	15	10	*3*	*4*
354 Miscellaneous petroleum and coal products	—	—	—	—
355 Rubber products	11	11	*1*	*1*
356 Plastic products	8	2	*1*	*1*
361 Pottery, china and earthenware	—	—	—	—
362 Glass and glass products	—	—	—	—
369 Other non-metal mineral products	11	4	*5*	*6*
371 Iron and steel	*2*	*6*	*2*	*2*
372 Non-ferrous metals	*4*	*4*	*2*	*2*
381 Metal products	20	15	*5*	*4*
382 Non-electrical machinery	3	4	*1*	*1*
383 Electrical machinery	6	6	*1*	*2*
384 Transport equipment	19	19	*6*	*3*
385 Professional and scientific equipment	—	—	—	—
390 Other manufacturing industries	2	2	—	—

FINANCE, ECONOMICS,& TRADE

Balance of Payments (23)

	1992	1993	1994	1995	1996
Exports of goods (f.o.b.)	406	447	519	682	764
Imports of goods (f.o.b.)	−1,335	−1,304	−1,309	−1,340	−1,213
Trade balance	−929	−857	−790	−657	−449
Services - debits	−578	−890	−657	−941	−1,059
Services - credits	178	340	449	614	658
Private transfers (net)	485	522	279	282	324
Government transfers (net)	130	−10	82	112	112
Overall balance	−714	−895	−637	−590	−413

Economic Indicators (22)

National product: GDP—purchasing power parity—$21.1 billion (1997 est.)

National product real growth rate: 4.3% (1997 est.)

National product per capita: $700 (1997 est.)

Inflation rate—consumer price index: 15% (1997 est.)

Exchange Rates (24)

Exchange rates:	
Tanzanian shillings (TSh) per US$1	
January 1998	631.61
1997	612.12
1996	579.98
1995	574.76
1994	509.63
1993	405.27

Top Import Origins (25)

$1.4 billion (c.i.f., 1996) Data are for 1995.

Origins	%
European Union	NA
Kenya	NA
Japan	NA
China	NA
India	NA

NA stands for not available.

Top Export Destinations (26)

$760 million (f.o.b., 1996) Data are for 1995.

Destinations	%
European Union	NA
Japan	NA
India	NA
United States	NA

NA stands for not available.

Economic Aid (27)

Recipient: ODA, $NA. NA stands for not available.

Import Export Commodities (28)

Import Commodities	Export Commodities
Consumer goods	Coffee
Machinery and transportation equipment	Manufactured goods
Crude oil	Cotton
	Cashew nuts
	Minerals
	Tobacco
	Sisal

THAILAND

CAPITAL: Bangkok (Krung Thep).

FLAG: The national flag, adopted in 1917, consists of five horizontal stripes. The outermost are red (symbolizing the Thai people); those adjacent are white (symbolizing Buddhism); the blue center stripe (representing the monarchy) is twice as high as each of the other four.

ANTHEM: There are three national anthems: *Pleng Sansen Phra Barami (Anthem Eulogizing His Majesty); Pleng Chard Thai (Thai National Anthem);* and *Pleng Maha Chati (Anthem of Great Victory),* an instrumental composition.

MONETARY UNIT: The baht (B) is divided into 100 satang. There are coins of 1, 5, 10, 25, and 50 stangs and 1, 5, and 10 baht, and notes of 50 satang and 1, 5, 10, 20, 50, 60, 100, and 500 baht. B1 = $0.03923 (or $1 = B25.49).

WEIGHTS AND MEASURES: The metric system is the legal standard, but some traditional units are also used.

HOLIDAYS: New Year's Day, 1 January; Chakkri Day, 6 April; Songkran Day, mid-April; Coronation Day, 5 May; Queen's Birthday, 12 August; Chulalongkorn Day, 23 October; King's Birthday, 5 December; Constitution Day, 10 December. Movable holidays include Makabuja Day, Plowing Festival, and Visakabuja Day.

TIME: 7 PM = noon GMT.

LOCATION AND SIZE: Comprising an area of 514,000 square kilometers (198,456 square miles) in Southeast Asia, Thailand (formerly known as Siam) extends almost two-thirds down the Malay Peninsula. Comparatively, the area occupied by Thailand is slightly more than twice the size of the state of Wyoming. It has a total boundary length of 8,082 kilometers (5,022 miles).

Thailand's capital city, Bangkok, is located on the Gulf of Thailand coast.

CLIMATE: Thailand has a tropical climate. In most of the country, the temperature rarely falls below 13°C (55°F) or rises above 35°C (95°F), with most places averaging between 24°C and 30°C (75°F and 86°F). The annual rainfall ranges from 102 centimeters (40 inches) in the northeast to over 380 centimeters (150 inches) in the peninsula.

Kingdom of Thailand
Prates Thai

INTRODUCTORY SURVEY

RECENT HISTORY

Although Siam, as Thailand was known until 1939, had established trading contacts with Portugal in the sixteenth century and later with the Dutch, French, and British, the kingdom remained largely isolated from the West until the nineteenth century.

Commercial treaties were signed with Britain (1855) and with the United States and France (1856), and Siam soon became a thriving trading center. The kingdom's economic strength made Siam the only country in Southeast Asia to avoid European colonization in the nineteenth and early twentieth centuries.

In 1932, after a bloodless revolution, the country became a constitutional monarchy. Until 1992 Thailand was ruled by a series of military governments interspersed with brief periods of democracy. Since 1992, a functioning democratic government has prevailed.

At the start of World War II, Thailand, occupied by Japan, declared war on the United States and Britain. After the war, however, Thailand became a U.S. ally and a charter member in the Southeast Asia Treaty Organization (SEATO). During the Vietnam War, United States forces used air bases in Thailand for bombing raids against North Vietnam.

At the end of the twentieth century, Thailand weathered several political upheavals. Contributing to the nation's political instability was the activity

of communist rebels in border areas and the presence of large numbers of refugees from Laos and Cambodia in the 1970s. In 1991 General Chatichai Choonhavan, who became prime minister in July 1988, was ousted in a bloodless coup. Resulting massive civil unrest ended in May 1992 when at least fifty protesters were killed and the prime minister, General Suchinda, was forced to resign. Chuan Leekpai, leader of the Democratic Party, became prime minister in 1992. Coruption charges brought down his governing coalition in 1994, but he took office again in 1997 at the head of another coalition government pledged to economic and political reforms.

GOVERNMENT

The present king, Bhumibol Adulyadej, assumed the throne in 1946. The National Assembly, (Rathasapha) consists of a Senate whose appointed members serve six-year terms and a popularly elected House of Representatives. The voting age is 18. Thailand is divided into 76 administrative provinces, each controlled by an appointed governor.

Judiciary

Judges are appointed by the monarch. Of Thailand's three levels of courts, the highest is the Supreme Court (Sarndika). There is no trial by jury. Islamic courts hear civil cases concerning members of the Muslim minority.

Political Parties

Nine political parties are currently represented in the House of Representatives. Leading parties include the New Aspiration Party (NAP), Democratic Party (DP), Thai Nation Party (TNP), and the Social Action Party (SAP).

DEFENSE

The armed forces are organized as 254,000 active duty members (80,000 draftees) and 200,000 reservists. Thailand has received U.S. military equipment, essential supplies, training and assistance since 1950.

ECONOMIC AFFAIRS

In 1997 after years of impressive economic growth, Thailand's economy suffered a severe setback. In mid-May the stock market plunged, and speculative trading devalued the national currency, the baht. By July, the government was forced to float (offer for sale) the baht, causing its value to drop sharply. By late 1997 the crisis had spread to Singapore, the Philippines, Malaysia, and Indonesia. The Thai economy contracted by 9.4% in 1998 and recovery was still weak in 1999.

Public Finance

Government revenues in 1996/97 were estimated at $24 billion; expenditures at $25 billion, including capital expenditures of $8 billion. The country's external debt totaled $90 billion.

Income

In 1998, Thailand's gross national product (GNP) was US $165.8 billion, or about US $2,740 per person.

Industry

The manufacturing sector was key to Thailand's economic boom of the early 1990s. Computers and electronics, garments and footwear, furniture, wood products, canned food, toys, plastic products, gems and jewelry are expected to flourish again when economic stability is achieved.

Banking and Finance

The Bank of Thailand, established in 1942, operates as an independent body under government supervision; its entire capital is owned by the government. The Bangkok Bank, the largest bank in Thailand, has assets of $40.9 billion in late 1996. Thailand's stock exchange was opened in Bangkok in 1975.

Economic Development

Thailand had a trade surplus of some $12 billion in 1998. Japan is Thailand's largest foreign investor; U.S. investments totaled an estimated $16 billion in 1999. Industry accounts for about 39% of the gross domestic product, while agriculture contributes about 12%. Tourism has recently become a major source of foreign exchange.

Economic growth is centered in Bangkok. As a result there is an increasing income gap between urban rich and rural poor, with 54% of the population still working on farms, cultivating mainly rice. The total labor force is estimated to be some 32.6 million.

SOCIAL WELFARE

A 1990 law established a social security system that includes disability, death, old age, and sickness and maternity benefits. Employers are required to provide workers' compensation coverage, including temporary and permanent disability benefits, and medical and survivor benefits.

Women have equal legal rights in most areas, but inequities remain in domestic areas, including divorce and child support. Many Thai minorities, including many hill tribe members, do not have any form of identification and are denied access to education and health care.

Healthcare

In 1996, life expectancy was 67.3 for men and 71 for women. Infant mortality was six in 1,000 births. Health care facilities are concentrated in the Bangkok metropolitan area. In 1991, there were 51,091 nurses and 10,606 midwives. There was one doctor per 5,000 inhabitants in 1992.

Housing

The Thai government has stimulated housing and community development by providing government mortgages for building, renovation, or purchase of government land and houses. The total number of housing units in 1992 was 11.2 million.

EDUCATION

The adult literacy rate is 93%. Education accounts for 16% of total government expenditures. School attendance is compulsory from age seven through fifteen. In 1992, about 1.1 million students were enrolled in higher education programs.

1999 KEY EVENTS TIMELINE

March

- An Australian accountant is murdered.

May

- An AIDS vaccine is tested on high-risk Thais.

July

- A hospital is accused of selling donated organs.

August

- Military cuts are approved.

September

- The World Wildlife Fund announces a plan to end ivory trade in Thailand.

October

- Myanmar exiles seize embassy.
- Thailand sends 1,500 peacekeeping troops to East Timor.
- The IMF says that the Thai economy is improving.

November

- At an informal meeting in Manila, Philippines, six members of the Association of Southeast Asian Nations (ASEAN)—Brunei, Indonesia, Malaysia, Philippines, Singapore, and Thailand—agree to establish a free-trade zone by eliminating duties on most goods traded in the region by 2010. The remaining four newer and less-developed nation members—Cambodia, Laos, Myanmar (Burma), and Vietnam—will eliminate duties by 2015. Rice will be excluded from trade agreements, however.

December

- The 16-mile Skytrain opens in Bangkok. The Skytrain is an elevated railway, built with private financing at a cost of $1.7 billion, designed to alleviate the city's serious traffic problems.
- Freezing temperatures cause an estimated 30 deaths in record-setting cold weather.

ANALYSIS OF EVENTS: 1999

BUSINESS AND THE ECONOMY

The Thai economy showed improvement in 1999. Blamed for sparking the recent Asian financial crisis after its currency collapsed in 1997, Thailand proved to be one of the first countries to recover from the continent-wide recession.

The Thai government took an active role in sparking economic recovery. In June stock exchange chief Singh Tangtatsawas was appointed president of Krung Thai Bank, a move towards helping the financial sector recover from bad loans; in August a government stimulus package was expanded; and in September Thailand accepted a $1.45 billion loan to help improve its exports and infrastructure.

The economic stimulus package was perhaps the most important of the government's recovery efforts. In April the government announced a modest stimulus package to bolster the economy, and in August the Thai cabinet expanded the package. It focused on business loans, property tax cuts, tariff reductions, and corporate tax relief to improve the business climate. The initial cost of the package was set at $500 million, though it could be expanded to $1 billion.

By most accounts, the expanded stimulus package seemed to work. In September Thailand turned down $3 billion in IMF loans that the country was eligible for, and in October IMF directors released an official statement that cited improvement in the Thai economy. IMF Managing Director Michel Camdessus said, "They [the IMF Board of Directors] noted that the recovery is now underway with growing signs of a rebound in manufacturing, exports, and domestic consumption." Thailand's recovery was much faster than expected, and the Thai government forecasted that year-end economic growth would be between three and four percent.

GOVERNMENT AND POLITICS

On October 1, five student exiles from Myanmar (formerly Burma) stormed the Myanmar embassy in Bangkok. The students demanded democratic reforms in their home of Myanmar, and they held eighty-nine people hostage. The situation was settled peacefully and the hostages were freed when Thai officials met the dissidents' demands and provided helicopter transport to the Myanmar border. Myanmar blamed the hostage situation on lax Thai security and criticized Thailand for providing safe haven to armed Myanmar dissidents.

In January Thailand took steps to distance itself from a grim episode in the history of Southeast Asia. Thailand had been a supporter of the Cambodian Khmer Rouge, whose leaders have been accused of crimes against humanity and genocide, responsible for up to 1.7 million deaths. Thailand has been a shelter for Khmer Rouge exiles, but Thailand said it would cooperate with Cambodia's efforts to bring Khmer Rouge leaders to trial. This offer of cooperation was tested when Khmer Rouge military chief Ta Mok was captured in a Thai-Cambodian border region. Ta Mok sought Thailand's protection, but Thai officials responded by saying that they would not interfere with his trial.

On October 4, responding to the violence in East Timor, Thailand began deploying 1,500 military troops to help the United Nations restore peace and stability in the region.

CULTURE AND SOCIETY

On March 10, an Australian business executive was murdered in Nakhon Sawan. Michael Wansley was driving his van when a motorcycle pulled alongside him. The rider on the back of the motorcycle shot Wansley eight times. Thai investigators linked the murder to Thailand's recent financial troubles, the murder declared part of a cover-up by executives of the debt-ridden sugar mill that Wansley was auditing. The managing director of the sugar mill, Pradit Siriviriyakul, has been charged with orchestrating the murder. Murder charges were also filed against five other men involved in the killing.

In May Thailand became the first developing country to conduct phase III tests of an AIDS vaccine. The AIDSVAX vaccine test will run for four years. Administers of the test are targeting intravenous drug users in Bangkok methadone clinics. Thailand, a country of 60 million people, is estimated to have over one million people infected with HIV, and the Thai government is already credited with slowing the spread of AIDS through an aggressive public education campaign.

The Vachiraprakarn General Hospital in Samut Prakan was accused in July of trading in human organs. A former hospital executive alleged that the hospital pressured the families of terminally-ill patients to sign organ donation documents, and in other cases, the hospital harvested organs without the consent of either the patient or the patient's family. In a few extreme cases, the hospital was even accused of prematurely declaring patients brain-dead in order to have access to their organs. In response to this scandal, the Thai government formed a medical body to oversee organ donations.

DIRECTORY

CENTRAL GOVERNMENT

Head of State

King
Bhumibol Adulyadej, Monarch

Prime Minister and Minister of Defense
Chuan Leekpai, The Secretariat of the Prime Minister, Government House, Phitsanulok Rd., Bangkok 10300, Thailand
PHONE: +66 (2) 2826543
FAX: +66 (2) 2828631
E-MAIL: govspkmn@mozart.inet.co.th

Ministers

Minister of Agriculture and Cooperatives
Pongphon Adireksan, Ministry of Agriculture and Cooperatives, 19 Soi Aree, Phaholyotin Rd., Phayathai Bangkok, Thailand
PHONE: +66 (2) 2797716

Minister of Commerce
Suphachai Phanitchaphak, Ministry of Commerce, 25 Sukumvit Soi 5 Sukumvit Rd., Klongtoey Bangkok, Thailand
PHONE: +66 (2) 2835711
FAX: +66 (2) 2534650

Minister of Communications and Transportation
Suthep Thuaksuban, Ministry of Communications and Transportation, Thanon Ratchadamnoen Nok, Bangkok 10100, Thailand

Minister of Defense
Chuan Leekpai, Ministry of Defense, Government House Phitsanulok Rd., Bangkok 10300, Thailand
PHONE: +66 (2) 2826543
FAX: +66 (2) 2828631
E-MAIL: govspkmn@mozart.inet.co.th

Minister of Education
Somsak Prisananuntagul, Ministry of Education, 319 Wang Chan, Kasem Ratchadamnoen Nok Ave., Bangkok 10300, Thailand

Minister of Finance
Tarrin Nimmanhemin, Ministry of Finance, Rama VI Rd., Bangkok 10400, Thailand

Minister of Foreign Affairs
Surin Phitsuwan, Ministry of Foreign Affairs, 1 Sanamchai Rd., Saranrom Palace, Bangkok 10200, Thailand
E-MAIL: 0100@mserv.mfa.go.th

Minister of Industry
Suwat Liptaphanlop, Ministry of Industry, Thanon Ratchadamnoen Nok, Bangkok 10100, Thailand

Minister of the Interior
Sanan Khachonprasat, Ministry of the Interior, Soi Chawal-preecha, Sanambinnam Rd., Nonthaburi, Thailand

Minister of Justice
Suthat Ngoenmun, Ministry of Justice, Sri Bandhit 3, Bang Greuy Sai noi Rd., Nonthaburi, Thailand

Minister of Labor and Social Welfare
Vudhi Sukosol, Ministry of Labor and Social Welfare, 587 Soi Ratchadanivet, Pracharadbompen Samsennok, Bangkok, Thailand

Minister of Public Health
Kon Thappharangsi, Ministry of Public Health, 111/1 Phaholyothin Soi 5, Phaholyothin Rd., Bangkok 10400, Thailand
PHONE: +66 (2) 2713538

Minister of State University Bureau
Prachuap Chaiyasan, Ministry of State University Bureau, Saranrom Palace, Bangkok 10200, Thailand

POLITICAL ORGANIZATIONS

Chartthai Party

325/74-76 Lukluang Rd., Dusit Bangkok 10300, Thailand
PHONE: +66 (2) 2807054
FAX: +66 (2) 2824003
TITLE: Leader
NAME: Banharn Silpa-Archa

Democrat Party

67 Setsiri Rd., Sansen Ni Phaya, Bangkok 10400, Thailand
PHONE: +66 (2) 2700036
FAX: +66 (2) 2796086
TITLE: Leader
NAME: Chuan Leekpai

Liberal Democratic Party

1133/1617 Nakornchisri Rd,, Dusit Bangkok 10300, Thailand
PHONE: +66 (2) 6682286
TITLE: Leader
NAME: Phinij Jarusombat

Mass Party

630/182 Prapinklao, Bangkok 10700, Thailand
PHONE: +66 (2) 4240851
FAX: +66 (2) 4240851

National Development Party

10 Soi Phahon Yothin, 3 Phahon Yothin Rd., Phayathai Bangkok 10400, Thailand
PHONE: +66 (2) 2793104
FAX: +66 (2) 2794284
TITLE: Leader
NAME: Chatchai Choonhavan

New Aspiration Party

310 Soi. Ruamchit Nakhon Chaisi Rd., Dusit Bangkok 10300, Thailand
PHONE: +66 (2) 2435000
FAX: +66 (2) 2412280
TITLE: Leader
NAME: Chavalit Yongchaiyudh

Palang Dharma Party

445/15 Ramkhamhaeng, 39 Bangkapi, Bangkok 10310, Thailand
PHONE: +66 (2) 7185626
FAX: +66 (2) 7185634
TITLE: Leader
NAME: Ravee Maschamadol

Social Action Party

126 Soi Ongkarak Samsen, 28 Rd., Nakhon Chaisi, Dusit Bangkok 10300, Thailand
PHONE: +66 (2) 2430100
FAX: +66 (2) 2433234
TITLE: Leader
NAME: Montree Pongpanit

Solidarity Party

670/104 Soi Thepnimit, Jaransanitwong Rd., Bangpaid Bangkok 10700, Thailand
PHONE: +66 (2) 4240291
FAX: +66 (2) 4248630
TITLE: Leader
NAME: Chaiyot Sasomsub

Thai Citizen Party

1213/323 Srivara Rd., Bangkok 10310, Thailand
PHONE: +66 (2) 5590008
FAX: +66 (2) 5590016
TITLE: Leader
NAME: Samak Sundaravej

DIPLOMATIC REPRESENTATION

Embassies in Thailand

Argentina
5th Floor, Thaniya Building, 62 Silom Rd., Bangkok, Thailand

Australia
37 South Sathon Rd., Bangkok 10120, Thailand
PHONE: +66 (2) 2872680
FAX: +66 (2) 2872029

Austria
14 Soi Nandha off soi Attakarnprasit Sathorn Tai Rd., Bangkok 10120, Thailand
PHONE: +66 (2) 2873970
FAX: +66 (2) 2873925

Belgium
44 Soi Phya Pipat Silom Rd., Bangkok 10500, Thailand
PHONE: +66 (2) 2360150
FAX: +66 (2) 2367619

Brazil
239 Soi Sarasin Ratchadamri Rd. Lumpini-Pathumwan, Bangkok 10330, Thailand
PHONE: +66 (2) 2526023
FAX: +66 (2) 2542707

Canada
B4 Boonmitr Building, 11th and 12th floor, 138 Silom Rd., PO Box 2090, Bangkok 10500, Thailand
PHONE: +66 (2) 2341561
FAX: +66 (2) 2361941

China
Ratchada Pisek Rd. Dindaeng, Bangkok 10310, Thailand
PHONE: +66 (2) 2457032
FAX: +66 (2) 2468247

France
35 Soi Rong Phasi Kao Charoenkrung 36 Rd., Bangkok 10500, Thailand
PHONE: +66 (2) 2340950
FAX: +66 (2) 2367973

Germany
9 South Sathorn Rd., Bangkok 10120, Thailand
PHONE: +66 (2) 2132331
FAX: +66 (2) 2871776

India
46 Soi Prasammitr, 23 Sukhumvit Rd., Bangkok 10110, Thailand
PHONE: +66 (2) 2580300
FAX: +66 (2) 2584627

Israel
31 Soi Lang Suan Ploenchit Rd., Bangkok 10330, Thailand
PHONE: +66 (2) 2523131
FAX: +66 (2) 2545518

Italy
399 Nang Linchee Rd. Thung Mahamek, Bangkok 10120, Thailand
PHONE: +66 (2) 2854090
FAX: +66 (2) 2854793

Japan
1674 New Phetchaburi Rd., Bangkok 10310, Thailand
PHONE: +66 (2) 2526151
FAX: +66 (2) 2534153

South Korea
23 Thiam-Ruammit Rd. Huaykwang Samsen Nork, Bangkok 10310, Thailand
PHONE: +66 (2) 2477537
FAX: +66 (2) 2477535

Mexico
44/7-8 Convent Rd., Bangkok 10500, Thailand
PHONE: +66 (2) 2340935
FAX: +66 (2) 2368410

Pakistan
31, Soi Nana Nua, 3 Sukhumvit Rd., Bangkok 1010, Thailand
PHONE: +66 (2) 2530288

Russia
108 Sathorn Nua Rd., Bangkok, Thailand
PHONE: +66 (2) 2349824
FAX: +66 (2) 2378488

Serbia
28 Soi 61 Sukhumvit Rd., Bangkok 10110, Thailand
PHONE: +66 (2) 3919090

United Kingdom
Wireless Rd., Bangkok 10330, Thailand
PHONE: +66 (2) 2530191
FAX: +66 (2) 2556051

United States
95 Wireless Rd., Bangkok 10330, Thailand
PHONE: +66 (2) 2525040
FAX: +66 (2) 2542990
TITLE: Ambassador
NAME: Richard Hecklinger

JUDICIAL SYSTEM

Sarndika (Supreme Court)

Court of Appeals

Courts of First Instance

FURTHER READING

Articles

Amorn, Victor. ''Thailand to Announce Bigger-Than-Expected Stimulus.'' Reuters News Service, 9 August 1999.

Berman, Jessica. ''Thailand Attacks AIDS with Two-Pronged Approach.'' *The Lancet* (May 8, 1999): 1600.

Bhatiasevi, Aphaluck. ''Body Parts Trade: Inquiry Finds Allegations Have Grounds.'' *Bangkok Post* (August 7, 1999).

Raedler, John. ''Myanmar Student Gunmen Release Hostages, Leave Thai Capital.'' *CNN Interactive* (October 3, 1999).

''Thailand: the Next Test.'' *The Economist* (July 3, 1999): 32.

THAILAND: STATISTICAL DATA

For sources and notes see "Sources of Statistics" in the front of each volume.

GEOGRAPHY

Geography (1)

Area:

Total: 514,000 sq km.

Land: 511,770 sq km.

Water: 2,230 sq km.

Area—comparative: slightly more than twice the size of Wyoming.

Land boundaries:

Total: 4,863 km.

Border countries: Burma 1,800 km, Cambodia 803 km, Laos 1,754 km, Malaysia 506 km.

Coastline: 3,219 km.

Climate: tropical; rainy, warm, cloudy southwest monsoon (mid- May to September); dry, cool northeast monsoon (November to mid-March); southern isthmus always hot and humid.

Terrain: central plain; Khorat Plateau in the east; mountains elsewhere.

Natural resources: tin, rubber, natural gas, tungsten, tantalum, timber, lead, fish, gypsum, lignite, fluorite.

Land use:

Arable land: 34%

Permanent crops: 6%

Permanent pastures: 2%

Forests and woodland: 26%

Other: 32% (1993 est.)

HUMAN FACTORS

Demographics (2A)

	1990	1995	1998	2000	2010	2020	2030	2040	2050
Population	55,052.3	58,240.9	60,037.4	61,163.8	66,091.7	69,298.2	70,982.2	71,129.7	69,740.7
Net migration rate (per 1,000 population)	NA	NA	NA	NA	NA	NA	NA	NA	NA
Births	NA	NA	NA	NA	NA	NA	NA	NA	NA
Deaths	252.5	NA	NA	NA	NA	NA	NA	NA	NA
Life expectancy - males	65.2	64.7	65.3	65.8	69.4	70.9	73.3	75.7	77.1
Life expectancy - females	71.1	72.3	72.8	73.2	76.5	78.0	80.6	83.1	84.6
Birth rate (per 1,000)	18.4	17.5	16.8	16.1	13.7	12.1	11.1	10.2	9.7
Death rate (per 1,000)	6.6	7.0	7.1	7.2	7.4	8.7	9.7	11.1	12.7
Women of reproductive age (15-49 yrs.)	15,474.4	16,911.6	17,639.6	17,973.6	18,020.9	17,251.1	16,044.1	15,182.1	14,327.8
of which are currently married	9,358.2	NA	NA	NA	NA	NA	NA	NA	NA
Fertility rate	2.0	1.9	1.8	1.8	1.8	1.8	1.7	1.7	1.7

Except as noted, values for vital statistics are in thousands; life expectancy is in years.

Health Personnel (3)

Total health expenditure as a percentage of GDP, 1990-1997[a]

Public sector	2.0
Private sector	1.9
Total[b]	3.9

Health expenditure per capita in U.S. dollars, 1990-1997[a]

Purchasing power parity	230
Total	96

Availability of health care facilities per 100,000 people

Hospital beds 1990-1997[a]	170
Doctors 1993[c]	24
Nurses 1993[c]	99

Health Indicators (4)

Life expectancy at birth

1980	64
1997	69
Daily per capita supply of calories (1996)	2,334
Total fertility rate births per woman (1997)	1.7
Maternal mortality ratio per 100,000 live births (1990-97)	200[c]
Safe water % of population with access (1995)	89
Sanitation % of population with access (1995)	96
Consumption of iodized salt % of households (1992-98)[a]	50

Smoking prevalence

Male % of adults (1985-95)[a]	49
Female % of adults (1985-95)[a]	4
Tuberculosis incidence per 100,000 people (1997)	142
Adult HIV prevalence % of population ages 15-49 (1997)	2.23

Infants and Malnutrition (5)

Under-5 mortality rate (1997)	38
% of infants with low birthweight (1990-97)	6
Births attended by skilled health staff % of total[a]	71

% fully immunized (1995-97)

TB	98
DPT	94
Polio	94
Measles	91
Prevalence of child malnutrition under age 5 (1992-97)[b]	NA

Ethnic Division (6)

Thai	75%
Chinese	14%
Other	11%

Religions (7)

Buddhism	95%
Muslim	3.8%
Christianity	0.5%
Hinduism	0.1%
Other	0.6%

Data are for 1991.

Languages (8)

Thai, English (secondary language of the elite), ethnic and regional dialects.

EDUCATION

Public Education Expenditures (9)

Public expenditure on education (% of GNP)

1980	3.4
1996	4.1[1]

Expenditure per student

Primary % of GNP per capita

1980	8.8
1996	16.0[1]

Secondary % of GNP per capita

1980	....
1996	....

Tertiary % of GNP per capita

1980	60.2
1996	30.7[1]

Expenditure on teaching materials

Primary % of total for level (1996)	1.0
Secondary % of total for level (1996)	4.3[1]
Primary pupil-teacher ratio per teacher (1996)	....
Duration of primary education years (1995)	6

Educational Attainment (10)

Age group (1990)	6+
Total population	49,076,100

Highest level attained (%)

No schooling	10.7

First level

Not completed	69.6
Completed	NA

Entered second level

S-1	13.7
S-2	NA
Postsecondary	5.1

Literacy Rates (11A)

In thousands and percent[1]	1990	1995	2000	2010
Illiterate population (15+ yrs.)	2,572	2,613	2,224	1,607
Literacy rate - total adult pop. (%)	93.3	93.8	95.1	96.8
Literacy rate - males (%)	95.6	96.0	96.8	97.9
Literacy rate - females (%)	91.2	91.6	93.4	95.8

GOVERNMENT & LAW

Political Parties (12)

House of Representatives	No. of seats
New Aspiration Party (NAP)	125
Democratic Party (DP)	123
National Development Party (NDP)	52
Thai Nation Party (TNP)	39
Social Action Party (SAP)	20
Thai Citizen's Party (TCP)	18
Solidarity Party (SP)	8
Liberal Democratic Party (LDP)	4
Mass Party (MP)	2
Other	2

Military Affairs (14B)

	1990	1991	1992	1993	1994	1995
Military expenditures						
Current dollars (mil.)	2,372	2,722	3,154	3,804	3,970	4,014
1995 constant dollars (mil.)	2,726	3,008	3,392	3,988	4,069	4,014
Armed forces (000)	283	283	283	295	290	288
Gross national product (GNP)						
Current dollars (mil.)	96,360	107,900	119,400	132,500	147,100	163,400
1995 constant dollars (mil.)	110,700	119,300	128,400	138,900	150,700	163,400
Central government expenditures (CGE)						
1995 constant dollars (mil.)	15,960	17,730	19,990	22,730	25,870	26,410
People (mil.)	55.1	55.7	56.3	57.0	57.6	58.2
Military expenditure as % of GNP	2.5	2.5	2.6	2.9	2.7	2.5
Military expenditure as % of CGE	17.1	17.0	17.0	17.5	15.7	15.2
Military expenditure per capita (1995 $)	50	54	60	70	71	69
Armed forces per 1,000 people (soldiers)	5.1	5.1	5.0	5.2	5.0	4.9
GNP per capita (1995 $)	2,012	2,141	2,279	2,437	2,616	2,806
Arms imports[6]						
Current dollars (mil.)	290	575	370	140	390	1,100
1995 constant dollars (mil.)	333	635	398	147	400	1,100
Arms exports[6]						
Current dollars (mil.)	0	0	0	0	0	0
1995 constant dollars (mil.)	0	0	0	0	0	0
Total imports[7]						
Current dollars (mil.)	33,380	37,590	40,690	46,210	54,460	70,780
1995 constant dollars (mil.)	38,360	41,540	43,760	48,440	55,820	70,780
Total exports[7]						
Current dollars (mil.)	23,070	28,430	32,470	36,770	45,130	56,460
1995 constant dollars (mil.)	26,510	31,410	34,930	38,550	46,260	56,460
Arms as percent of total imports[8]	.9	1.5	.9	.3	.7	1.6
Arms as percent of total exports[8]	0	0	0	0	0	0

Government Budget (13A)

Year: 1997

Total Expenditures: 901,716 Millions of Baht

Expenditures as a percentage of the total by function:

General public services and public order	10.54P
Defense	11.57P
Education	21.61P
Health	8.58P
Social Security and Welfare	3.74P
Housing and community amenities	5.74P
Recreational, cultural, and religious affairs	1.41P
Fuel and energy	.38P
Agriculture, forestry, fishing, and hunting	9.25P
Mining, manufacturing, and construction	.49P
Transportation and communication	15.86P
Other economic affairs and services	4.29P

LABOR FORCE

Labor Force (16)

Total (million)	32.6
Agriculture	54%
Industry	15%
Services (incl gov't)	31%

Data for 1997 est. Percent distribution for 1996 est.

Unemployment Rate (17)

3.5%

PRODUCTION SECTOR

Electric Energy (18)

Capacity	15.838 million kW (1995)
Production	77.5 billion kWh (1995)
Consumption per capita	1,295 kWh (1995)

Transportation (19)

Highways:

total: 64,600 km

paved: 62,985 km

unpaved: 1,615 km (1996 est.)

Waterways: 3,999 km principal waterways; 3,701 km with navigable depths of 0.9 m or more throughout the year; numerous minor waterways navigable by shallow-draft native craft.

Pipelines: petroleum products 67 km; natural gas 350 km

Merchant marine:

total: 304 ships (1,000 GRT or over) totaling 1,997,060 GRT/3,270,988 DWT ships by type: bulk 48, cargo 145, chemical tanker 7, container 9, liquefied gas tanker 13, multi-function large load carrier 3, oil tanker 62, passenger 1, refrigerated cargo 11, roll-on/roll-off cargo 2, short-sea passenger 1, specialized tanker 2 (1997 est.)

Airports: 106 (1997 est.)

Airports—with paved runways:

total: 55

over 3,047 m: 6

2,438 to 3,047 m: 9

1,524 to 2,437 m: 16

914 to 1,523 m: 20

under 914 m: 4 (1997 est.)

Continued on next page.

FINANCE, ECONOMICS,& TRADE

Balance of Payments (23)

	1992	1993	1994	1995	1996
Exports of goods (f.o.b.)	32,100	36,398	44,478	55,447	54,409
Imports of goods (f.o.b.)	−36,261	−40,695	−48,204	−63,415	−63,897
Trade balance	−4,161	−4,297	−3,726	−7,968	−9,488
Services - debits	−13,608	−16,015	−19,689	−24,719	−26,939
Services - credits	10,820	13,199	14,202	18,646	20,977
Private transfers (net)	42	33	61	42	30
Government transfers (net)	604	717	1,067	445	729
Overall balance	−6,303	−6,363	−8,085	−13,554	−14,692

Transportation (19) cont.

Airports—with unpaved runways:

total: 51

2,438 to 3,047 m: 1

1,524 to 2,437 m: 1

914 to 1,523 m: 15

under 914 m: 34 (1997 est.)

Top Agricultural Products (20)

Rice, cassava (tapioca), rubber, corn, sugarcane, coconuts, soybeans.

FINANCE, ECONOMICS, & TRADE

Economic Indicators (22)

National product: GDP—purchasing power parity—$525 billion (1997 est.)

National product real growth rate: -0.4% (1997 est.)

National product per capita: $8,800 (1997 est.)

Inflation rate—consumer price index: 5.6% (1997 est.)

Exchange Rates (24)

Exchange rates:	
Baht (B) per US$1	
January 1998	53.812
1997	31.364
1996	25.343
1995	24.915
1994	25.150
1993	25.319

Top Import Origins (25)

$73.5 billion (c.i.f., 1996) Data are for 1997.

Origins	%
Japan	25.6
United States	13.9
Singapore	5
Taiwan	4.6
Germany	4.5
Malaysia	4.1

Top Export Destinations (26)

$51.6 billion (f.o.b., 1997) Data are for 1997.

Destinations	%
United States	19.6
Japan	14.9
Singapore	11
Hong Kong	5.7
Malaysia	4.3
United Kingdom	3.7

Economic Aid (27)

Recipient: ODA, $624 million (1993).

Import Export Commodities (28)

Import Commodities	Export Commodities
Capital goods 50%	Manufactures 82%
Consumer goods 10.2%	Agricultural products and fisheries 14%
Fuels 8.7%	

TOGO

Republic of Togo
République Togolaise

CAPITAL: Lomé.

FLAG: The national flag consists of five alternating horizontal stripes of green and yellow. A five-pointed white star is at the center of a red square that spans the height of the top three stripes.

ANTHEM: *Terre de nos aïeux (Land of Our Fathers).*

MONETARY UNIT: The Communauté Financière Africaine franc (CFA Fr) is a paper currency of 100 centimes. There are coins of 1, 2, 5, 10, 25, 50, 100, and 500 CFA francs and notes of 50, 100, 500, 1,000, 5,000, and 10,000 CFA francs. CFA Fr1 = \$0.00196 (or \$1 = CFA Fr510.65).

WEIGHTS AND MEASURES: The metric system is the legal standard.

HOLIDAYS: New Year's Day, 1 January; National Liberation Day, 13 January; Economic Liberation Day, 24 January; Victory Day, 24 April; Independence Day, 27 April; Labor Day, 1 May; Martyrs' Day, 21 June; Assumption, 15 August; All Saints' Day, Anniversary of the failed attack on Lomé, 24 September; 1 November; Christmas, 25 December. Movable religious holidays include Easter Monday, Ascension, Whitmonday, 'Id al-Fitr, and 'Id al-'Adha'.

TIME: GMT.

LOCATION AND SIZE: Situated on the west coast of Africa, Togo has an area of 56,790 square kilometers (21,927 square miles), slightly smaller than the state of West Virginia. It has a total boundary length of 1,703 kilometers (1,058 miles). Togo's capital city, Lomé, is located on the Gulf of Guinea coast.

CLIMATE: Togo has a humid, tropical climate, but receives less rainfall than most of the other countries along the Gulf of Guinea. The heaviest rainfall occurs in the hills of the west, southwest, and center, where the precipitation averages about 150 centimeters (60 inches) a year. The coast gets the least rainfall, about 78 centimeters (31 inches) annually. The average maximum and minimum temperatures are 30°C (86°F) and 23°C (73°F) at Lomé, on the southern coast, and 35°C (95°F) and 15°C (59°F) at Mango, in the north.

INTRODUCTORY SURVEY

RECENT HISTORY

The Republic of Togo became independent from France in 1960. The nation's first president, Sylvanus Olympio, was assassinated in 1963 and succeeded by Nicholas Grunitzky. The Grunitzky government, in which all political parties were represented, was overthrown in a 1967 military coup led by General Gnassingbe Eyadema. Political parties were banned and constitutional freedoms were suspended. Despite several attempts to oust him, General Eyadema has remained in power since 1967.

The country was rocked by political turbulence in the early 1990s resulting in repeated violence and frequent political paralysis. Several armed uprisings by dissidents were put down by displays of lethal and overwhelming military power. President Eyadema was re-elected in 1993 in elections boycotted by the opposition, and retained the presidency in the disputed elections of 1998.

GOVERNMENT

A new constitution calling for multiparty elections was approved on 27 September 1992. Technically, the president is chosen in a direct, popular, multiparty election. There is an 81-seat National Assembly, chosen in multiparty elections.

Judiciary

The Supreme Court sits in Lomé. Other judicial institutions include two Courts of Appeal (one civil, the other criminal); courts for first hearings of

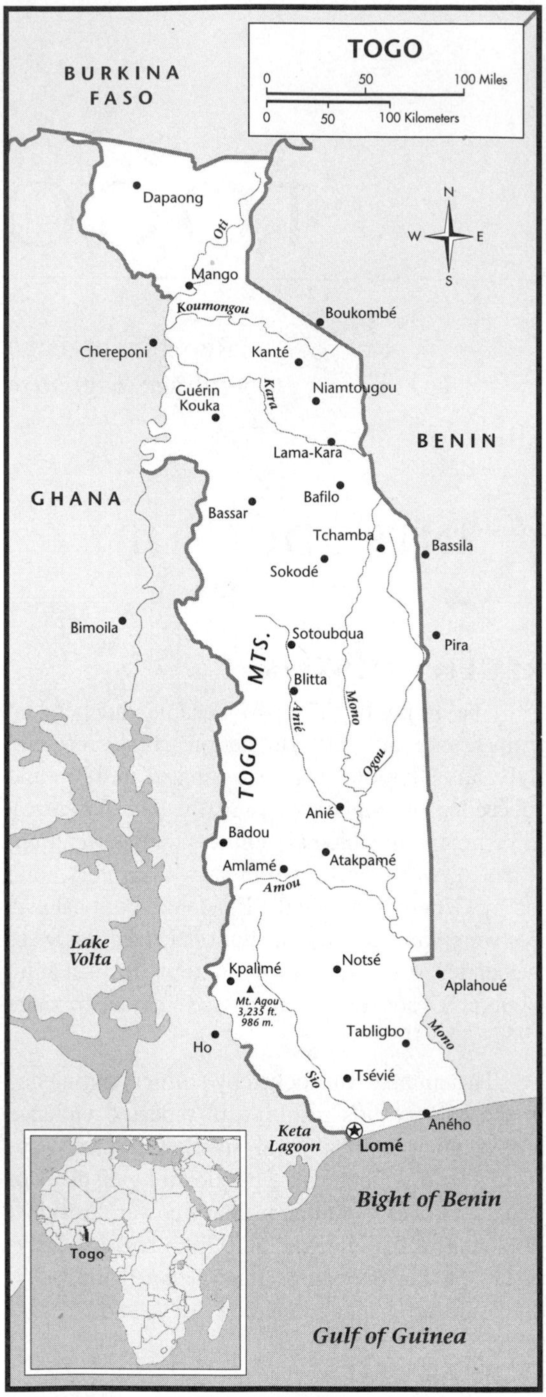

civil, commercial, and criminal cases; labor and children's courts; and the Court of State Security, set up to judge crimes involving foreign or domestic subversion.

Political Parties

Since 1969, the nation has been a one-party state. In that year, the Togolese People's Rally (*Rassemblement du Peuple Togolais*–RPT) was founded as the nation's only legal political party and is headed by President Eyadema. After opposition parties were legalized in 1991, other parties began to function, although threatened by pro-Eyadema forces. Among the new parties are the Togolese Union for Democracy (UTD), the Partí Démocratique Togolais (PDT), and the Action Committee for Renewal (CAR).

DEFENSE

Togo's 6,500-man army consisted of five regiments in 1995. The 250-man air force had 16 combat aircraft, and the 200-member naval unit had two coastal patrol vessels. Defense spending in 1996 was $27 million, or 2% of the gross domestic product (GDP).

ECONOMIC AFFAIRS

Togo's is a subsistence agricultural economy; 80% of its people are engaged in agriculture. The nation also has significant phosphate deposits upon which it depends for foreign trade. Declining prices for Togo's main exports (phosphates, coffee, cocoa, and cotton) continue to have a negative effect on the economy.

In 1994, France devalued the CFA franc, cutting its value in half overnight. The devaluation was designed to encourage investment and discourage the use of cash reserves to buy goods that could be made domestically. However, political instability prevented Togo from taking advantage of the opportunity.

Public Finance

By the late 1970s, public investment expenditures had reached an unsustainable level (exceeding 40% of GDP), touched off by an earlier rise of commodity prices. As a result, large payment arrears on the external debt began to mount. In the mid-1980s, the fiscal deficit was reduced largely through IMP credits and debt reschedulings. The civil unrest of 1991 resulted in decreased revenues and increased expenditures and led to an overall budget deficit of 7.5% of GDP. In 1992 further civil unrest widened the budget deficit to 8.5% of GDP. For 1997 revenues were an estimated US $232 million, but expenditures totaled US $252 million.

Income

In 1998 Togo's gross national product (GNP) was $1.4 billion, or about $330 per person. Nearly one-third of the population lives below the poverty line.

Industry

Industrial production represents 23% part of the economy and includes phosphate mining, agricultural processing, cement, handcrafts, textiles and beverages. Only five percent of the population is involved in industry.

Banking and Finance

The most important commercial and savings banks include the International Bank for Occidental Africa, the Bank of Credit and Commerce International, the Libyan Arab-Togolese Bank of Foreign Commerce, the Togolese Bank for Credit and Industry, and the Union Bank of Togo (the latter two with a state share of 35%).

The banking and credit systems are not well developed, and large sections of the population remain outside the monetary economy. There are no securities exchanges in Togo.

Economic Development

Resumption of international economic aid is linked to democratic reforms that have not yet taken place. In late 1998 the European Union suspended aid and trade preferences for Togo because of concerns over the process of democratization. The World Bank also suspended its disbursements at year end 1998 because Togo was unable to pay its arrears.

SOCIAL WELFARE

The government's social welfare program, established in 1973, includes family allowances and maternity benefits, old age, disability and death benefits, and workers' compensation. Women face discrimination in employment and education.

Healthcare

Medical services include permanent treatment centers and a mobile organization for preventive medicine. In 1992, there was one doctor per 12,500 persons, and about 61% of the population had access to health care services. Yaws, malaria, and leprosy continue to be major medical problems. About 8.5% of the adult population is infected with HIV (the virus that causes AIDS); 6,466 AIDS cases were reported in 1996. Average life expectancy is 59 years.

Housing

The government is attempting to solve the problem of urban overcrowding by promoting housing and establishing sanitation facilities. In 1988, total housing units numbered 470,000 with 6.2 persons per dwelling.

EDUCATION

The illiteracy rate in 1995 was estimated at about 48% (33% for men and 63% for women). Compulsory primary education (ages 6–12) is free. In 1993, there were 633,126 pupils in 2,594 primary schools, and 126,335 students in secondary schools. In 1994, all higher level schools had 10,994 students.

1999 KEY EVENTS TIMELINE

January

- An activist from Togo fights deportation from the United States, claiming she will be tortured and possibly killed if she is sent back.

March

- Togo's opposition parties boycott National Assembly elections, and call on their supporters to stay home. The opposition wants to settle a dispute with the government over results of the 1998 presidential election.

April

- Official election results are announced, with 77 of 81 seats going to the ruling Rally of the Togolese People (RPT). International observers say the elections were fair, but U.S. State Department spokesman James Foley says without the participation of the opposition, the election results did not reflect the will of the Togolese.
- Togo's government warns journalists to respect the national press code or face legal consequences. The Associated Press reports that newspaper reporter Romain Atisso Kudjodji is detained by police for allegedly fabricating a torture story.
- Togo's Interior Minister General Seyi Memene bans independence day celebrations planned by the country's main opposition party, the Union des forces de changement (UFC). The party, led by exiled politician Gilchrist Olympio, tests the country's law that protects freedom of assembly by taking to the streets to observe the 39th anniversary of nationhood.

May

- International mediators travel to Togo to help resolve a political impasse over the June 1998 re-election of President Gnassingbé Eyadéma. The opposition has challenged the election.
- Seven opposition parties call for a stay-at-home protest. They report it is moderately successful.

June

- One year after his election, President Gnassingbé Eyadéma, Africa's Longest-Ruling Leader, continues to tighten control over the nation as opposition parties challenge his 1998 election. However, the government shows some willingness to end political crisis.

July

- Togo's political parties agree to end yearlong political crisis according to President Gnassingbé Eyadéma's promise that he will not seek re-election when his term ends in 2003. The parties also agree to a rerun of the March parliamentary elections.
- French president Jacques Chirac congratulates President Gnassingbé Eyadéma for his government's willingness to negotiate with the opposition.

August

- Chevron Corporation officials say they have not reached a final deal for a West African gas pipeline after signing an agreement with four countries.
- Togo's government said it is considering suing the London-based Amnesty International, a human rights organization that released a 45-page report accusing Togo's armed forces and paramilitary police of killing hundreds of people after the 1998 presidential elections.

October

- Heavy rains affect more than 64,000 people, most of them children, in four regions of Togo. Humanitarian aid agencies ask for $2 million in help.

December

- The Economic Community of West African States (ECOWAS) holds a two-day summit meeting in Lomé, the capital. The main focus of the meeting is the violent unrest in the region.
- Ousted Côte d'Ivoire president Herni Konan Bédié arrives in Togo December 26, and plans to seek asylum in France.

ANALYSIS OF EVENTS: 1999

BUSINESS AND THE ECONOMY

In 1999 the United Nations Development Program announced a $1.6 million program to build schools, drinking water fountains, bathrooms, roads and market stalls. Most of the money will be funneled to Lomé and Tsevie, about 20 miles north of the capital. About 80 percent of Togo's 4.5 million people live in extreme poverty. The poor economy has forced rural villagers to flock to the main cities to find work.

The Chevron and Shell corporations signed a memorandum of understanding with four African nations in August that could bring Nigerian natural gas to Togo, Benin and Ghana within the next three years. Corporate officials said the $400 million offshore project will bring natural gas from the oil fields of Nigeria to the three neighboring countries by 2002. The project is expected to cut greenhouse gas emissions by 120 million cubic feet per day, and at the same time provide energy for power generation, industries and homes in Togo, Benin and Ghana. Domestic use of gas is expected to cut down on deforestation in western Africa, where wood is a common cooking fuel.

GOVERNMENT AND POLITICS

In July, after 10 days of negotiations with opposition parties, President Gnassingbé Eyadéma agreed he would not run for re-election in 2003, ending one of the longest periods of rule by an African leader. Eyadéma seized power in 1967, and was last re-elected in 1998 in a controversial election that challengers said was fraudulent.

Eyadéma had tightened his grip on power in March, when his party, the Rally of the Togolese People (RPT), captured 77 of 81 seats in the National Assembly. The government claimed 66 percent of eligible voters participated in the elections, but opposition parties, which called for a boycott, said only 10 percent voted. The main opposition party, the Union des forces de changement (Union of the Forces of Change or UFC), had called for a boycott to pressure the government to hold new presidential elections. UFC leader, Gilchrist Olympio, who lives in exile in Ghana, claimed he won the 1998 election.

In May, the opposition called for a one-day stay-at-home campaign that was moderately successful. The same month, European mediators arrived in Togo to discuss ways to lift the political impasse. French President Jacques Chirac visited Togo during the negotiations in July, and pressured Eyadéma to find a solution. Eyadéma has been trying to improve relations with the European Union, which suspended cooperation agreements with Togo in 1993. The government also agreed to hold new parliamentary elections in 2000, which will be overseen by a newly independent electoral commission.

During negotiations, the government was embarrassed by a new report published by a Benin-based human rights group, which said Togo's troops executed hundreds of opponents during the 1998 presidential elections. That report confirmed findings by Amnesty International, a London-based human rights group that leveled similar accusations earlier in the year. In its 45-page report, Amnesty International said political opponents were detained without trial and tortured, and that hundreds of people were executed by security forces before and after the elections. The group said bodies of victims, some handcuffed, washed ashore on beaches of Togo and Benin. Eyadéma's government reacted angrily to the report, and in August, it threatened to sue Amnesty International, claiming the group had fabricated a ''pack of lies.'' The government arrested and later released Togo human rights workers who allegedly had provided the information to Amnesty.

CULTURE AND SOCIETY

In July, the government warned parents to keep close track of their children until authorities determined whether Togo was the center for the illegal trade of children. The BBC reported the government made the announcement after several arrests of suspected child traffickers, including two women who were stopped at the Togo-Ghana border with seven children. Other suspected traffickers were stopped with 21 children. The BBC reported that the suspected traffickers claimed they had recruited the children with the consent of their parents to work on cocoa plantations in Côte d'Ivoire.

The CARE organization announced a plan to help residents of the capital city of Lomé to pick up tons of trash from the streets. CARE wants to set up a more efficient trash collection system, and equip collectors with better equipment, including wheelbarrows, forks, shoes and gloves. The group also plans a campaign to teach residents about hygiene and the benefits of proper disposal of trash.

DIRECTORY

CENTRAL GOVERNMENT

Head of State

President
Gnasingbe Eyadema, Office of the President
PHONE: +228 212701

Prime Minister
Eugene Koffi Adoboli, Office of the Prime Minister

Ministers

Minister of Economy and Finance
Abdoul-Hamid Douroudjaye, Ministry of Economy and Finance
PHONE: +228 213554
FAX: +228 210905

Minister of Foreign Affairs and Cooperation
Joseph Kokou Koffigoh, Ministry of Foreign Affairs and Cooperation
PHONE: +228 213601
FAX: +228 213979

Minister of Planning and Territorial Development
Simfe Tchaeou Pre, Ministry of Planning and Territorial Development, B.P. 2748, Lome, Togo
PHONE: +228 213751
FAX: +228 213753

Minister of Health
Kondi Charles Agba, Ministry of Health

Minister of Interior and Security
Seyi Memene, Ministry of Interior and Security

Minister of Labor and Civil Service
Kokou Biossey Tozoun, Ministry of Labor and Civil Service

Minister of National Education and Research
Koffi Sama, Ministry of National Education and Research

Minister of Town Planning and Housing
Hope Agboli, Ministry of Town Planning and Housing

Minister of Transport and Water Resources
Dama Dramani, Ministry of Transport and Water Resources

POLITICAL ORGANIZATIONS

Rally of the Togolese People

TITLE: President
NAME: Gnassingbe Eyadema

Coordination des Forces Nouvelles

NAME: Joseph Koffigoh

Union for Democracy and Solidarity

NAME: Antoine Folly

DIPLOMATIC REPRESENTATION

Embassies in Togo

Great Britain
British School of Lomé, BP 20050 Lome, Togo
PHONE: +228 264606
FAX: +228 214989

JUDICIAL SYSTEM

Supreme Court

Courts of Appeal

Court of State Security

FURTHER READING

Articles

''French President Congratulates Togo on 'Spirit of Openness'.'' Associated Press, 23 July 1999.

''French President Stirs Anger During Visit to West Africa.'' *The New York Times*, 24 July 1999.

''Pipeline Deal Still Pending.'' *The New York Times*, 13 August 1999.

''Ruling Party Sweeps Togo Elections on Opposition Boycott.'' Associated Press, 10 April 1999.

''Togolese Activist Fights Deportation.'' Associated Press, 4 January 1999.

''Togolese Opposition Leader Returns from Exile for Talks.'' Associated Press, 26 July 1999.

''Togolese President Reportedly Plans to Step Down.'' Associated Press, 24 July 1999.

''Togo Police Arrest Newspaper Boss.'' Associated Press, 20 April 1999.

''Togo's President, Africa's Longest-Ruling Leader, Tightens Hold.'' *The New York Times*, 22 June 1999.

TOGO: STATISTICAL DATA

For sources and notes see "Sources of Statistics" in the front of each volume.

GEOGRAPHY

Geography (1)

Area:

Total: 56,790 sq km.

Land: 54,390 sq km.

Water: 2,400 sq km.

Area—comparative: slightly smaller than West Virginia.

Land boundaries:

Total: 1,647 km.

Border countries: Benin 644 km, Burkina Faso 126 km, Ghana 877 km.

Coastline: 56 km.

Climate: tropical; hot, humid in south; semiarid in north.

Terrain: gently rolling savanna in north; central hills; southern plateau; low coastal plain with extensive lagoons and marshes.

Natural resources: phosphates, limestone, marble.

Land use:

Arable land: 38%

Permanent crops: 7%

Permanent pastures: 4%

Forests and woodland: 17%

Other: 34% (1993 est.).

HUMAN FACTORS

Health Personnel (3)

Total health expenditure as a percentage of GDP, 1990-1997[a]

Public sector 1.6

Continued on next page.

Demographics (2A)

	1990	1995	1998	2000	2010	2020	2030	2040	2050
Population	3,680.4	4,410.4	4,905.8	5,262.6	7,401.3	10,145.5	13,386.2	16,970.6	20,725.2
Net migration rate (per 1,000 population)	NA	NA	NA	NA	NA	NA	NA	NA	NA
Births	NA	NA	NA	NA	NA	NA	NA	NA	NA
Deaths	NA	NA	NA	NA	NA	NA	NA	NA	NA
Life expectancy - males	53.2	55.3	56.5	57.4	61.3	64.9	68.1	70.8	73.1
Life expectancy - females	57.0	59.6	61.1	62.2	67.0	71.3	74.9	77.9	80.2
Birth rate (per 1,000)	49.8	46.8	45.2	44.4	40.2	35.2	30.2	26.0	22.4
Death rate (per 1,000)	13.1	11.0	10.0	9.4	7.1	5.5	4.5	4.2	4.2
Women of reproductive age (15-49 yrs.)	831.8	984.1	1,094.4	1,176.9	1,707.6	2,428.0	3,362.2	4,407.2	5,458.8
of which are currently married	NA	NA	NA	NA	NA	NA	NA	NA	NA
Fertility rate	7.2	6.8	6.6	6.4	5.6	4.6	3.8	3.2	2.7

Except as noted, values for vital statistics are in thousands; life expectancy is in years.

Health Personnel (3) cont.

Private sector	2.2
Total[b]	3.4

Health expenditure per capita in U.S. dollars, 1990-1997[a]

Purchasing power parity	59
Total	15

Availability of health care facilities per 100,000 people

Hospital beds 1990-1997[a]	150
Doctors 1993[c]	6
Nurses 1993[c]	31

Health Indicators (4)

Life expectancy at birth	
1980	49
1997	49
Daily per capita supply of calories (1996)	2,155
Total fertility rate births per woman (1997)	6.1
Maternal mortality ratio per 100,000 live births (1990-97)	640[c]
Safe water % of population with access (1995)	55
Sanitation % of population with access (1995)	41
Consumption of iodized salt % of households (1992-98)[a]	1
Smoking prevalence	
Male % of adults (1985-95)[a]	65
Female % of adults (1985-95)[a]	14
Tuberculosis incidence per 100,000 people (1997)	353
Adult HIV prevalence % of population ages 15-49 (1997)	8.52

Infants and Malnutrition (5)

Under-5 mortality rate (1997)	125
% of infants with low birthweight (1990-97)	20
Births attended by skilled health staff % of total[a]	32
% fully immunized (1995-97)	
TB	53
DPT	33
Polio	33
Measles	38
Prevalence of child malnutrition under age 5 (1992-97)[b]	19

Ethnic Division (6)

Native African (37 tribes; largest and most important are Ewe, Mina, and Kabre) 99%, European and Syrian-Lebanese less than 1%.

Religions (7)

Indigenous beliefs	70%
Christian	20%
Muslim	10%

Languages (8)

French (official and the language of commerce), Ewe and Mina (the two major African languages in the south), Kabye (sometimes spelled Kabiye) and Dagomba (the two major African languages in the north).

EDUCATION

Public Education Expenditures (9)

Public expenditure on education (% of GNP)	
1980	5.6
1996	4.7
Expenditure per student	
Primary % of GNP per capita	
1980	8.3
1996	7.8[1]
Secondary % of GNP per capita	
1980	....
1996	....
Tertiary % of GNP per capita	
1980	889.9
1996	501.4[1]
Expenditure on teaching materials	
Primary % of total for level (1996)	0.2
Secondary % of total for level (1996)	8.5
Primary pupil-teacher ratio per teacher (1996)	51[1]
Duration of primary education years (1995)	6

Educational Attainment (10)

Age group (1981)	25+
Total population	1,084,488
Highest level attained (%)	
No schooling	76.5
First level	
Not completed	13.5
Completed	NA
Entered second level	
S-1	8.7
S-2	NA
Postsecondary	1.3

Literacy Rates (11A)

In thousands and percent[1]	1990	1995	2000	2010
Illiterate population (15+ yrs.)	1,055	1,085	1,108	1,119
Literacy rate - total adult pop. (%)	45.4	51.7	57.9	69.5
Literacy rate - males (%)	61.3	67.0	72.0	80.7
Literacy rate - females (%)	30.1	37.0	44.3	58.7

GOVERNMENT & LAW

Political Parties (12)

National Assembly	No. of seats
Action Committee for Renewal (CAR)	36
Rally of the Togolese People (RPT)	35
Togolese Union for Democracy (UTD)	7
Union of Justice and Democracy (UJD)	2
Coordination des Forces Nouvelles (CFN)	1

Military Affairs (14B)

	1990	1991	1992	1993	1994	1995
Military expenditures						
Current dollars (mil.)	34[e]	34[e]	32[e]	37[e]	30[e]	28
1995 constant dollars (mil.)	39[e]	38[e]	34[e]	39[e]	30[e]	28
Armed forces (000)	8	8	6	6	6	12
Gross national product (GNP)						
Current dollars (mil.)	1,104	1,138	1,100	961	1,079	1,227
1995 constant dollars (mil.)	1,269	1,257	1,184	1,008	1,106	1,227
Central government expenditures (CGE)						
1995 constant dollars (mil.)	287[e]	325[e]	NA	NA	NA	274[e]
People (mil.)	3.7	3.8	4.0	4.1	4.3	4.4
Military expenditure as % of GNP	3.1	3.0	2.9	3.9	2.7	2.3
Military expenditure as % of CGE	13.8	11.7	NA	NA	NA	10.2
Military expenditure per capita (1995 $)	11	10	9	10	7	6
Armed forces per 1,000 people (soldiers)	2.2	2.1	1.5	1.5	1.4	2.7
GNP per capita (1995 $)	345	329	299	246	260	278
Arms imports[6]						
Current dollars (mil.)	0	0	0	5	0	0
1995 constant dollars (mil.)	0	0	0	5	0	0
Arms exports[6]						
Current dollars (mil.)	0	0	0	0	0	0
1995 constant dollars (mil.)	0	0	0	0	0	0
Total imports[7]						
Current dollars (mil.)	581	444	395	179	222	386
1995 constant dollars (mil.)	668	491	425	188	228	386
Total exports[7]						
Current dollars (mil.)	268	253	275	136	162	208
1995 constant dollars (mil.)	308	280	296	143	166	208
Arms as percent of total imports[8]	0	0	0	2.8	0	0
Arms as percent of total exports[8]	0	0	0	0	0	0

Government Budget (13B)

Revenues .... $232 million
Expenditures .... $252 million
Capital expenditures .... NA

Data for 1997 est. NA stands for not available.

LABOR FORCE

Labor Force (16)

Total (million) 1.538
Agriculture 65%
Industry 5%
Services 30%

Data for 1993 est. Percent distribution for 1997 est.

Unemployment Rate (17)

Rate not available.

PRODUCTION SECTOR

Electric Energy (18)

Capacity .... 34,000 kW (1995)
Production .... 90 million kWh (1995)
Consumption per capita .... 92 kWh (1995)

Imports electricity from Ghana.

Transportation (19)

Highways:

total: 7,520 km

paved: 2,376 km

unpaved: 5,144 km (1996 est.)

Waterways: 50 km Mono river

Merchant marine: none

Airports: 9 (1997 est.)

Airports—with paved runways:

total: 2

2,438 to 3,047 m: 2 (1997 est.)

Airports—with unpaved runways:

total: 7

914 to 1,523 m: 5

under 914 m: 2 (1997 est.)

Top Agricultural Products (20)

Coffee, cocoa, cotton, yams, cassava (tapioca), corn, beans, rice, millet, sorghum; meat; annual fish catch of 10,000-14,000 tons.

MANUFACTURING SECTOR

GDP & Manufacturing Summary (21)

Detailed value added figures are listed by both International Standard Industry Code (ISIC) and product title.

	1980	1985	1990	1994
GDP ($-1990 mil.)[1]	1,416	1,323	1,636	1,568
Per capita ($-1990)[1]	541	437	463	391
Manufacturing share (%) (current prices)[1]	8.4	6.7	9.9	*6.6*
Manufacturing				
Value added ($-1990 mil.)[1]	156	126	152	*87*
Industrial production index	100	91	112	*64*
Value added ($ mil.)	*51*	*38*	*71*	*47*
Gross output ($ mil.)	*148*	94	226	*178*
Employment (000)	*5*	*5*	*5*	*4*
Profitability (% of gross output)				
Intermediate input (%)	*65*	*75*	*83*	*73*
Wages and salaries inc. supplements (%)	*12*	*10*	*10*	*14*
Gross operating surplus	*23*	*16*	*6*	*13*
Productivity ($)				
Gross output per worker	*27,363*	*21,742*	*53,786*	*39,782*
Value added per worker	*9,458*	*7,695*	*15,402*	*10,872*
Average wage (inc. supplements)	*3,198*	*2,872*	*8,122*	*5,972*
Value added ($ mil.)				
311/2 Food products	4	*11*	*12*	*8*
313 Beverages	16	*14*	*34*	*22*
314 Tobacco products	NA	NA	NA	NA
321 Textiles	*8*	*5*	*10*	*6*
322 Wearing apparel	—	—	—	—
323 Leather and fur products	—	—	—	—

	1980	1985	1990	1994
324 Footwear	6	*2*	*5*	*3*
331 Wood and wood products	1	—	—	—
332 Furniture and fixtures	—	—	—	—
341 Paper and paper products	—	—	—	—
342 Printing and publishing	3	*1*	*1*	*1*
351 Industrial chemicals	3	*1*	*4*	*5*
352 Other chemical products	NA	NA	NA	NA
353 Petroleum refineries	NA	NA	NA	NA
354 Miscellaneous petroleum and coal products	NA	NA	NA	NA
355 Rubber products	NA	NA	NA	NA
356 Plastic products	NA	NA	NA	NA
361 Pottery, china and earthenware	—	—	—	—
362 Glass and glass products	—	—	*1*	—
369 Other non-metal mineral products	6	*2*	*2*	*1*
371 Iron and steel	2	*1*	*1*	—
372 Non-ferrous metals	NA	NA	NA	NA
381 Metal products	1	—	—	—
382 Non-electrical machinery	NA	NA	NA	NA
383 Electrical machinery	NA	NA	NA	NA
384 Transport equipment	NA	NA	NA	NA
385 Professional and scientific equipment	NA	NA	NA	NA
390 Other manufacturing industries	—	—	*1*	*1*

FINANCE, ECONOMICS, & TRADE

Economic Indicators (22)

National product: GDP—purchasing power parity—$6.2 billion (1997 est.)

National product real growth rate: 4.8% (1997 est.)

National product per capita: $1,300 (1997 est.)

Inflation rate—consumer price index: 15.7% (1995)

Exchange Rates (24)

Exchange rates:

CFA francs (CFAF) per US$1

January 1998	608.36
1997	583.67
1996	511.55
1995	499.15
1994	555.20
1993	283.16

Beginning 12 January 1994, the CFA franc was devalued to CFAF 100 per French franc from CFAF 50 at which it had been fixed since 1948.

Balance of Payments (23)

	1990	1991	1992	1993	1994
Exports of goods (f.o.b.)	514	514	420	264	328
Imports of goods (f.o.b.)	−603	−567	−547	−375	−365
Trade balance	−89	−53	−128	−111	−37
Services - debits	−309	−345	−262	−191	−129
Services - credits	182	146	161	88	79
Private transfers (net)	113	91	91	40	25
Government transfers (net)	19	20	−0	1	5
Overall balance	84	−141	−139	−174	−57

Top Import Origins (25)

$404 million (c.i.f., 1996) Data are for 1995 est.

Origins	%
Ghana	17.1
China	13.3
France	12.5
Cameroon	6.0

Top Export Destinations (26)

$196 million (f.o.b., 1996) Data are for 1995 est.

Destinations	%
Canada	9.2
United States	8.1
Taiwan	7.5
Nigeria	6.7

Economic Aid (27)

Recipient: ODA, $NA. NA stands for not available.

Import Export Commodities (28)

Import Commodities	Export Commodities
Machinery and equipment	Cotton
Consumer goods	Phosphates
Petroleum products	Coffee
	Cocoa

TONGA

Kingdom of Tonga
Pule'anga Tonga

CAPITAL: Nuku'alofa.

FLAG: The flag, adopted in 1862, is crimson with a cross of the same color mounted in a white square in the upper left corner.

ANTHEM: *Koe Fasi Oe Tu'i Oe Otu Tonga (Tongan National Anthem)* begins "'E 'Otua Mafimafi Ko homau 'Eiki Koe" ("O Almighty God above, Thou art our Lord and sure defense").

MONETARY UNIT: The Tongan pa'anga (T$) of 100 seniti is a paper currency at par with the Australian dollar. There are coins of 1, 2, 5, 10, 20, and 50 seniti, and 1 and 2 Tongan pa'angas, and notes of 1, 2, 5, 10, 20, and 50 pa'angas. T$1 = US$1.215 (or US$1 = T$0.82305).

WEIGHTS AND MEASURES: The metric system is the legal standard, but some imperial and local weights and measures are also employed.

HOLIDAYS: New Year's Day, 1 January; ANZAC Day, 25 April; Crown Prince's Birthday, 4 May; Independence Day, 4 June; King's Birthday, 4 July; Constitution Day, 4 November; Tupou I Day, 4 December; Christmas, 25–26 December. Movable religious holidays include Good Friday and Easter Monday.

TIME: 1 AM (the following day) = noon GMT.

LOCATION AND SIZE: The Tonga archipelago (island group), also known as the Friendly Islands, lies scattered east of Fiji in the South Pacific Ocean. Consisting of 172 islands of various sizes, Tonga has a total area of 748 square kilometers (289 square miles), slightly more than four times the size of Washington, D.C. Nuku'alofa, the capital, is located on the island of Tongatapu.

CLIMATE: The mean annual temperature is 23°C (73°F), ranging from an average daily minimum of 10°C (50°F) in winter to an average maximum of 32°C (90°F) in summer. Mean annual rainfall ranges from 178 centimeters (70 inches) on Tongatapu to 279 centimeters (110 inches) on Vava'u.

INTRODUCTORY SURVEY

RECENT HISTORY

Prior to European contact, Tonga, whose official name is Pule'anga Fakatu'i 'o Tonga, was inhabited by Austronesian-speaking peoples. The archaeological record indicates the first inhabitants came from Samoa about 3000 years ago. The indigenous society developed a rigidly stratified social system whose ruler held power as far away as the Hawaiian Islands. The first known European contact was Dutch explorer Jakob Le Maire who sighted the island in 1616. However, more substantial contact with Tongans did not occur until British explorer Captain James Cook arrived in 1773 and revisited the islands for the next four years. Cook named the islands Friendly islands. Following an extended period of social turmoil, substantial changes began in 1820 with the rise to power of Taufa'ahau Tupou (George I) on Ha'apai and establishment of a Methodist mission in 1826. By 1845 King Taufa'ahau Tupou ruled over all the Tongan island groups leading to a line of hereditary sacred kings. Toupou instituted a limited monarchy in which Tongans gained ownership of their property from traditional minor chiefs. The king issued a code of laws in 1862 and a constitution in 1875. Tonga's political sovereignty became recognized in a series of treaties with Germany, Great Britain and the United States in the 1870s and 1880s.

Following financial difficulties, Tonga became a British protectorate in 1905 to protect the island from German interest. While remaining part of the British Commonwealth, Tonga again gained politi-

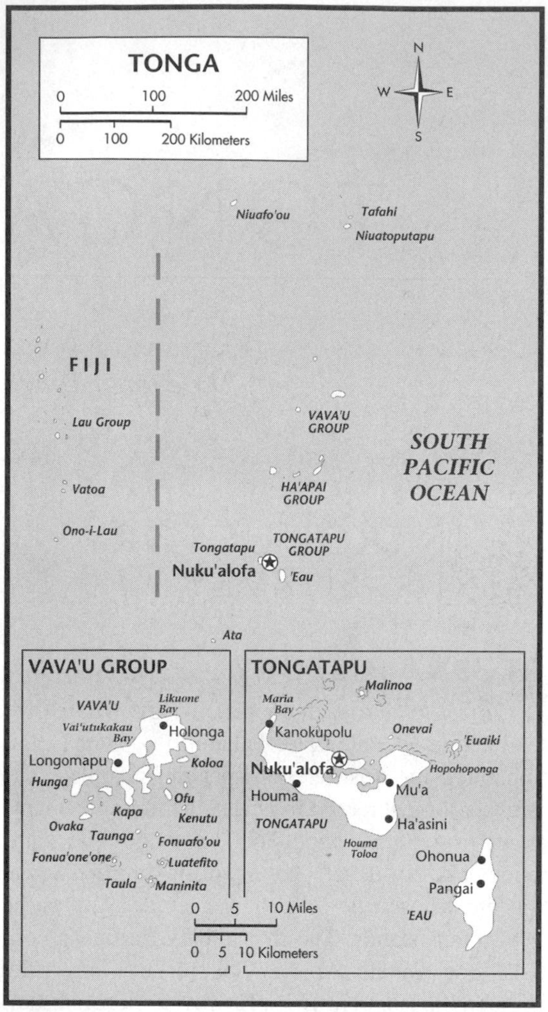

cal independence in 1970. In the 1990s most of Tongans are still Polynesian, living in a rigidly stratified rural society based largely on traditional social patterns with some introduced Western elements. All males of sixteen years of age or older are entitled to a land allotment of approximately 7.5 acres. However, with large estates divided among Tongan nobles a shortage of sufficient land prevents many from exercising the privilege.

GOVERNMENT

Tonga is the only surviving pre-contact Polynesian kingdom. King George Tupou I granted the Tonga constitution in 1875 which remained largely unchanged through the twentieth century. Tonga, a constitutional monarchy, is a member of the British Commonwealth. King Taufa'ahau Toupou IV has been the chief of state since 1965. The Tongan monarch serves as the chief executive and appoints a Privy Council. The council consists of the monarch and a cabinet composed of ten members appointed for life including a prime minister and deputy prime minister. A unicameral legislature is composed of the cabinet, a speaker, nine nobles selected by Tongan nobles and nine representatives elected to three-year terms by literate taxpaying males and literate females over twenty-one years of age. The only forms of local government are elected town and district officials. The primary administrative division of Tonga is the three island groups of Ha'apai, Tongatapu, and Vava'u. Tonga is also part of the Pacific Community and the South Pacific Forum.

Judiciary

Tongan judicial system, based on English law, is composed of the Tongan Supreme Court, the Court of Appeal, the Land Court, and eight Magistrates' courts. Appeals of magistrate rulings go to the Supreme Court, and appeals of the Supreme Court and Land Court decisions proceed to the Court of Appeal which consists of the Privy Council plus the Chief Justice of the Supreme Court. Until 1994 the United Kingdom appointed Tonga justices and subsidized their salaries. Judges are now appointed by the monarch.

Political Parties

Traditionally, no official political parties existed in the monarchy. However, a popular push for democratic reform including increased parliamentary representation for Tongans led to the founding of the Pro-Democracy Movement in 1992. The Movement soon established Tonga's first political party, the People's Party, in 1994.

DEFENSE

The Tonga Defense Services, consisting of only seventeen officers and just over 200 lower rank members in 1992, includes the Royal Tongan Marines, the Tongan Royal Guard, the Maritime Force, and a national police force. Tonga created an Air Wing in 1996 using three aircraft for maritime patrol. In 1989 the percentage of the gross domestic product (GDP) spent on the military was just less than 5%.

ECONOMIC AFFAIRS

Given generally fertile well-drained soils, warm semitropical to tropical climate, and abundant rainfall, agriculture is the primary economic endeavor. Almost a quarter of the land base is arable and approximately half is in permanent crops or pasture. Deforestation has become an environmental concern with expanding agriculture as well as settlement. Agriculture constitutes over 30% of GDP and over 60% of exports. The main exports in 1996 were squash (or pumpkins), coconut oil, vanilla beans, coconuts, and watermelons. Many species are grown as subsistence crops in addition to cattle, pigs, horses, and goats. Fish are abundant with bottom fish primarily exploited. Coconut exports began significantly declining by the 1990s. Efforts to revive the coconut crops expanded as aging coconut plantations had not been effectively replaced.

Tonga had enjoyed a favorable trade balance until 1960 but by the late 1990s Tonga imported four times more than it exported with much of its food imported from New Zealand. To assist with the trade deficit, a regional free trade agreement was established with Australia and New Zealand in the 1980s but market globalization of the 1990s undercut the agreement's effectiveness due to strong competition coming from Asia. To obtain more efficient use of arable lands, reforms to the traditional land tenure system were sought.

Public Finance

Almost half of Tonga's annual revenue is gained through foreign trade taxes with the remainder gained from government services and other direct and indirect taxes. Given Tonga's marked trade deficit, Tonga has received aid from the Asian Development Bank, the European Union, Japan, New Zealand, and Church of Latter-day Saints in the United States. In 1996–97 expenditures were US $120 million while revenues were US $48 million. Tonga received almost $40 million in economic aid in 1995. With increasing dependence on aid, the Tongan government is seeking foreign investment for private economic development.

Income

Despite some economic gains, the per capita income remains relatively low for the Pacific Island region. In 1997–98 the gross domestic product (GDP) was US$232 million at an annual declining rate of 1.5%. Per capita GDP was US$2,100.

Industry

Manufacturing industries, constituting only 10% of the GDP, include crop processing by several cooperative societies, charcoal production, plastic pipe production, leather goods, and metal industries. For example, the Leimatu's Vanilla Society processes vanilla. Most exports go to Japan.

Tourism, based on scenic beauty and mild climate, has further spurred the Tongan economy becoming the primary source of incoming cash. Organizations such as the Tonga Tourist Association and Tonga Visitors' Bureau helped bring in some $10 million (U.S.) annually in tourist trade by the mid-1990s. The Royal Tongan Airlines serves the islands providing connections to Australia and New Zealand among other South Pacific locations. Though still off the beaten track, the tourism industry still brought in earnings exceeding all exports combined. The island of Vava'u was becoming the main tourist center with whale watching and a major South Pacific yacht charter operation.

In 1989 Tonga surprisingly claimed six satellite orbit slots through the International Telecommunications Union. Tonga had launched at least three satellites from Kazakhstan in the 1990s with U.S. assistance. In 1997, Eric Gullichsen began selling Internet domain names from Tonga.

Baking and Finance

The National Reserve Bank of Tonga, established in 1989, is the Tongan bank of issue and manages foreign reserves. Other banks include the Bank of Tonga, the Tonga Development Bank, Mbf Bank Ltd., and Australia and New Zealand Banking Group Ltd.

SOCIAL WELFARE

The monarch's primary emphasis has been health and education programs. A key human rights concern is the traditionally suppressed role of women in Tongan society. Increasing economic opportunities for women and participation in government has been a key reform effort.

Healthcare

Tongans receive free medical and dental care. Tonga has three public hospitals, with one located on each of the three main islands and several health dispensaries on other islands. Life expectancy of Tongans generally is sixty-seven years for males and seventy-two years for females. Generally free of most tropical diseases, the two leading causes of death in the mid-1990s were diseases of the circulatory and nervous systems. Such communicable diseases as influenza, typhoid, and tuberculosis are found in the population. The United Nations and New Zealand provide family planning services. Expenditures on health by Tonga in 1995–96 constituted over 13% of its budget.

Housing

Of the 170 islands in the Kingdom of Tonga, thirty-six are inhabited by over 109,000 residents in 1999 leading to relatively high population densities based on South Pacific standards. Aside from the more European-like urban areas, Tongans live primarily in small villages near their subsistence gardens. Consequently, a government priority has been upgrading rural water supplies and sanitation systems.

EDUCATION

Public education is free for all Tongans. Attendance is compulsory from age six to fourteen at the public and church-operated primary and secondary schools. The government also operates eight vocational training schools including a teacher-training college. Churches also operate some vocational schools. Scholarships are available for continued study abroad. Tongan education expenditures in 1995/96 amounted to over 18% of its annual budget. The literacy rate exceeds 98%.

1999 KEY EVENTS TIMELINE

January

- A regional economic consultant, a radio personality, and a pro-democracy advocate are among 55 candidates who register for Tonga's general election in March. Tongans choose nine people's representatives to the legislative assembly.
- Tonga's Prime Minister, Baron Vaea makes a national appeal for tolerance as concerns grow over the mistreatment of Chinese living in the island.

February

- A member of Tonga's National Millennium Committee criticizes the government and the private sector for failing to come up with any activities for celebrating the new millennium.

March

- Tonga's government collects nearly $40 million by selling passports—primarily to Asians—before it stops the practice in December 1998, government officials said. More than 7,000 passports were sold.

April

- Health authorities ban public feasts in the village of Tatakamotonga after an outbreak of typhoid fever on the main Tongan island of Tongatapu is reported.
- Tonga's former Prime Minister, Prince Fatafehi Tu'ipelehake, dies in Auckland, Royal Palace authorities said. Tu'ipelehake was the younger brother of King Taufa'ahau Tupou IV.

May

- The Asian Development Bank releases a study that concludes Tongan women have a high prevalence for obesity and related problems like diabetes, hypertension and heart disease.
- The Tongan National Museum issues a warning to all Tongans not to sell prized artifacts to overseas museums or individuals.

June

- Tonga's King Taufa'ahau Tupou IV receives a clean bill of health following his latest medical check-up in New Zealand. He turns 81 on July 4.

July

- A United Nations Security Council committee recommends Tonga's admittance to the UN. Tonga established diplomatic relations with China in 1998, ending a 26-year relation with Taiwan and securing the unopposed endorsement of the Security Council, where China has a veto.

August

- Tonga announces plans to introduce an hour of Daylight Saving Time at the end of the 1999 year, making it likely to be the first country to see in the new millennium. Fiji and Tonga will be on the same time, but Tonga is nearer to the International Date Line than Fiji, and can now claim to offer visitors a chance to be among the first to witness the dawn of the new millennium.

September

- Tonga, along with Kiribati and Nauru, becomes a member of the United Nations.
- Foreign and Defense Minister, Prince Lavaka Ata Ulukalala addresses the assembly, along with the presidents of Kiribati and Nauru.
- In World Cup rugby, Siua Taumalolo scores six tries to help Tonga beat Georgia and qualify for the World Cup matches.

October

- Tonga's World Cup campaign ends with a 11-10 loss to England after being reduced to 14 men. Two of Tonga's players catch the attention of British professional teams and are likely to sign contracts.

December

- To celebrate the beginning of the new millennium, 81-year-old King Tafa'ahau Tupou IV leads a prayer and a choir sings Handel's Messiah.

ANALYSIS OF EVENTS: 1999

BUSINESS AND THE ECONOMY

In January, Tonga's cabinet approved budget guidelines to help economy recover from 1998s depression. With gross domestic product and foreign reserve figures continuing to drop and with the budget deficit continuing to grow the government said that Tonga was facing an alarming situation. According to government figures, the Gross Domestic Product (GDP) fell by 4.4 percent in the 1996–97 financial year, and another 1.5 percent in the 1997–98 fiscal year. Droughts and cyclones have contributed to Tonga's depressed economy.

Pro-democracy supporters said the government needed to cut spending to help revive the economy. Some other measures that were proposed included privatizing some government services, including the post office, government stores and the government newspaper, and offering early retirement to government employees.

Despite the growing deficit, the kingdom kept inflation under control. The average inflation rate for 12 months ending in April stood at 4.1 percent. Prices for some items, including food, housing materials, and clothing, saw slight increases, but the cost of transportation was declining slightly. The Tonga Visitors Bureau reported that tourism earnings decreased in 1998, even though there was a slight increase in visitors. In 1998, 27,000 people visited Tonga, an increase of 970 visitors from 1997. But earnings dropped significantly, from $7.65 million in 1997 to only $2.17 million in 1998. Government officials hoped Millenium celebrations would boost tourism in the island.

GOVERNMENT AND POLITICS

Election observers said Tonga's pro-democracy movement lost ground in the March 12 national elections. The Human Rights and Democracy Movement won five of nine seats reserved for commoners in the Parliament. The group, which held six seats, had expected to win seven. The elections had no bearing on the government, however, which is essentially controlled by King Taufa'ahau Tupou IV. The king appoints the Prime Minister and the entire cabinet, all members of Parliament.

The pro-democracy movement endorsed 13 candidates, which attracted about half the national vote. The lackluster performance led a government official to say a referendum on democracy would lose. The pro-democracy group has asked the government to consider whether to hold a referendum asking citizens if they would prefer that all 30 Parliament members were elected instead of just nine.

Another issue agitating at least some portion of the Tonga population was environmental deterioration. In November, Tonga joined seven other Pacific island nations to criticize worldwide lack of action to reduce greenhouse gas emissions. As vulnerable nations they also requested help in adapting to climate changes. The eight nations held a press conference in Bonn, Germany, and described damages already being felt in their nations, and said it was time to stop global warming before problems grew worse.

Tongan spokesman Taniela Tukia said coastal erosion already was affecting the country. Officials from the Ministry of Lands, Survey and Natural Resources, he said, spent a week four years ago searching for survey pegs from a 1927 survey of Tonga. The pegs were finally found under water. In Kiribati, some villages were being forced to move inland because of worsening coastal erosion. And the Cook Islands reported an increase in the frequency and severity of cyclones.

CULTURE AND SOCIETY

Tonga experienced some disturbing instances of racial animosity. In January, Tonga's Prime Minister, Baron Vaea, appealed for tolerance amid growing concerns over the mistreatment of Chinese people living in the island. Chinese representatives in the kingdom brought their concerns to the government, saying crime against members of their community was growing. Armed robbery, burglaries and assault were the most common problems, also some Chinese residents reported being told to return to their homeland.

In a national radio address, Vaea urged Tongans to uphold values of friendship and Christianity, reminding them that more than 200,000 fellow Tongans have migrated to other nations and they are protected by the laws of their adopted nations.

In May, the Asian Development Bank released a study that claimed Tongan women had a high prevalence for obesity and related problems like diabetes, hypertension and heart disease. The study said these lifestyle diseases accounted for the highest number of deaths in the kingdom.

DIRECTORY

CENTRAL GOVERNMENT

Head of State

King
Taufa ahau Tupou IV, Monarch

Prime Minister
Baron Vaea, Office of the Prime Minister

Ministers

Minister of Agriculture
Fatafehi tu Ipelehake, Ministry of Agriculture

Minister of Defense
Lavaka ata Ulukalala, Ministry of Defense

Minister of Education
S. Langi Kavaliku, Ministry of Education

Minister of Finance
Tutoatasi Fakafunua, Ministry of Finance

Minister of Foreign Affairs
Lavaka ata Ulukalala, Ministry of Foreign Affairs

Minister of Health
Vailami Tangi, Ministry of Health

Minister of Justice
Tevita Tupou, Ministry of Justice

POLITICAL ORGANIZATIONS

Tonga People's Party

NAME: Viliami Fukofuka

DIPLOMATIC REPRESENTATION

Embassies in Tonga

Australia
Salote Road, Private Bag 35, Nukualofa, Tonga
PHONE: +676 23244
FAX: +676 23243

New Zealand
P.O. Box 830, Corner Taufa ahau and Salote, Nukualofa, Tonga
PHONE: +676 23122
FAX: +676 23487

United Kingdom
P.O. Box 56, Nukualofa, Tonga
PHONE: +676 21020
FAX: +676 24109

JUDICIAL SYSTEM

Supreme Court

Court of Appeal

Land Court

FURTHER READING

Articles

''Remodel By Royal Neighbor has Neighborhood in Uproar.'' Associated Press, 9 March 1999.

''Rugby World Cup 99: Friendly Tongans Turn Nasty.'' *The Guardian of London*, 16 October 1999.

''Tonga's King Faces Challenges to Power in Elections.'' Associated Press, 9 March 1999.

''Tonga May Use Daylight-Saving Time to Get Jump On Millennium.'' Associated Press, 20 August 1999.

''Tonga: Race Against Time.'' *The New York Times*, 21 August 1999, p. A4.

''UN Admits Three New Members, New General Assembly Session Opens.'' Associated Press, 15 September 1999.

''Warning for the Tongan Dance Routine: The Islanders are Told to Keep Their Distance in Their Pre-Match Ritual.'' *The Guardian of London*, 8 October 1999.

TONGA: STATISTICAL DATA

For sources and notes see "Sources of Statistics" in the front of each volume.

GEOGRAPHY

Geography (1)

Area:

Total: 748 sq km.

Land: 718 sq km.

Water: 30 sq km.

Area—comparative: four times the size of Washington, DC.

Land boundaries: 0 km.

Coastline: 419 km.

Climate: tropical; modified by trade winds; warm season (December to May), cool season (May to December).

Terrain: most islands have limestone base formed from uplifted coral formation; others have limestone overlying volcanic base.

Natural resources: fish, fertile soil.

Land use:

Arable land: 24%

Permanent crops: 43%

Permanent pastures: 6%

Forests and woodland: 11%

Other: 16% (1993 est.)

HUMAN FACTORS

Demographics (2B)

Population (July 1998 est.) 108,207

Age structure:

0-14 years . . . . NA

15-64 years . . . . NA

65 years and over . . . . NA

Population growth rate (1998 est.) 0.81%

Birth rate, 1998 est. (births/1,000 population) 26.43

Death rate, 1998 est. (deaths/1,000 population) . . . 6.07

Net migration rate, 1998 est. (migrant(s)/1,000 population) . . . . -1.23

Infant mortality rate, 1998 est. (deaths/1,000 live births) 38.57

Life expectancy at birth (years):

Total population 69.54

Male 67.51

Female (1998 est.) 71.96

Total fertility rate, 1998 est. (children born/woman) 3.63

Infants and Malnutrition (5)

Under-5 mortality rate (1997) 23

% of infants with low birthweight (1990-97) 2

Births attended by skilled health staff % of total[a] . . . NA

% fully immunized (1995-97)

TB 100

DPT 95

Polio 95

Measles 97

Prevalence of child malnutrition under age 5 (1992-97)[b] . . . . NA

Ethnic Division (6)

Polynesian, Europeans about 300.

Religions (7)

Christian (Free Wesleyan Church claims over 30,000 adherents).

Languages (8)

Tongan, English.

EDUCATION

Educational Attainment (10)

Age group (1986) .25+
Total population .33,911
Highest level attained (%)
No schooling[2] .9.6
First level
Not completed .34.6
Completed .NA
Entered second level
S-1 .51.1
S-2 .NA
Postsecondary .2.8

Literacy Rates (11B)

Adult literacy rate
1980
Male .-
Female .-
1995
Male .-
Female .99%

GOVERNMENT & LAW

Political Parties (12)

Legislative Assembly	No. of seats
Appointed Cabinet ministers	30
Nobles	7
Elected Proreform	7
Traditionalists	2

Government Budget (13B)

Revenues .$49 million
Expenditures .$120 million
Capital expenditures .$75 million

Data for FY96/97 est.

Military Affairs (14A)

Total expenditures .$NA
Expenditures as % of GDP .NA%

NA stands for not available.

LABOR FORCE

Labor Force (16)

Total .36,665
Agriculture .65%

Data for 1994. Percent distribution for 1997 est.

Unemployment Rate (17)

11.8% (FY93/94)

PRODUCTION SECTOR

Electric Energy (18)

Capacity .7,000 kW (1995)
Production .30 million kWh (1995)
Consumption per capita .284 kWh (1995)

MANUFACTURING SECTOR

GDP & Manufacturing Summary (21)

	1980	1985	1990	1992	1993	1994
Gross Domestic Product						
Millions of 1990 dollars	81	120	124	130	129	135
Growth rate in percent	15.81	5.37	−3.95	1.00	−0.20	4.70
Per capita (in 1990 dollars)	875.2	1,323.1	1,289.0	1,336.2	1,333.5	1,381.9
Manufacturing Value Added						
Millions of 1990 dollars	8	8	9	9	*9*	*9*
Growth rate in percent	21.43	6.26	−7.30	2.51	*1.62*	*2.46*
Manufacturing share in percent of current prices	6.1	8.2	NA	NA	NA	NA

PRODUCTION SECTOR

Transportation (19)

Highways:

total: 680 km

paved: 184 km

unpaved: 496 km (1996 est.)

Merchant marine:

total: 4 ships (1,000 GRT or over) totaling 11,278 GRT/16,441 DWT ships by type: bulk 1, liquefied gas tanker 2, roll-on/roll-off cargo 1 (1997 est.)

Airports: 6 (1997 est.)

Airports—with paved runways:

total: 1

2,438 to 3,047 m: 1 (1997 est.)

Airports—with unpaved runways:

total: 5

1,524 to 2,437 m: 1

914 to 1,523 m: 2

under 914 m: 2 (1997 est.)

Top Agricultural Products (20)

Coconuts, copra, bananas, vanilla beans, cocoa, coffee, ginger, black pepper; fish.

FINANCE, ECONOMICS, & TRADE

Economic Indicators (22)

National product: GDP—purchasing power parity—$239 million (1996 est.)

National product real growth rate: 1% (1996 est.)

National product per capita: $2,250 (1996 est.)

Inflation rate—consumer price index: 2% (1997 est.)

Exchange Rates (24)

Exchange rates:	
Pa'anga (T$) per US$1	
November 1997	1.3112
1996	1.2323
1995	1.2709
1994	1.3202
1993	1.3841

Top Import Origins (25)

$82.9 million (f.o.b., 1996) Data are for 1996 est.

Origins	%
NZ	34
Australia	16
United States	10
United Kingdom	8
Japan	6

Top Export Destinations (26)

$15.3 million (f.o.b., 1996) Data are for 1996 est.

Destinations	%
Japan	43
United States	19
Canada	14
NZ	5
Australia	5

Economic Aid (27)

Recipient: ODA, $37 million (1994).

Import Export Commodities (28)

Import Commodities	Export Commodities
Food products	Squash
Live animals	Fish
Machinery and transport equipment	Vanilla
Manufactures	Root crops
Fuels	Coconut oil
Chemicals	

TRINIDAD AND TOBAGO

Republic of Trinidad and Tobago

CAPITAL: Port-of-Spain.

FLAG: On a red field, a black diagonal stripe with a narrow white border on either side extends from top left to bottom right.

ANTHEM: Begins, "Forged from the love of liberty, in the fires of hope and prayer."

MONETARY UNIT: The Trinidad and Tobago dollar (TT$) is a paper currency of 100 cents. There are coins of 1, 5, 10, 25, and 50 cents, and 1 dollar, and notes of 1, 5, 10, 20, and 100 dollars. TT$1 = US$0.16565 (US$1 = TT$6.037).

WEIGHTS AND MEASURES: The metric system is official, but some imperial weights and measures are still used.

HOLIDAYS: New Year's Day, 1 January; Carnival, 14–15 February; Emancipation Day, 1st Monday in August; Independence Day, 31 August; Republic Day, 24 September; Christmas, 25 December; Boxing Day, 26 December. Movable holidays include Carnival, Good Friday, Easter Monday, Whitmonday, Corpus Christi, 'Id al-Fitr, and Dewali.

TIME: 8 AM = noon GMT.

LOCATION AND SIZE: Situated off the northeast coast of South America at the extreme southern end of the Lesser Antilles, the islands of Trinidad and Tobago cover an area of 5,130 square kilometers (1,981 square miles), and are slightly smaller than the state of Delaware. Trinidad, the main island, has an area of 4,828 square kilometers (1,863 square miles). Tobago has an area of 300 square kilometers (116 square miles). In addition, 16 small islands are located off the coasts.

Trinidad and Tobago have a coastline length of 470 kilometers (292 miles). The capital city of Trinidad and Tobago, Port-of-Spain, is located on Trinidad's Gulf of Paria coast.

CLIMATE: There is little variation in temperature conditions through the year. The mean annual temperature for the entire nation is 21°C (70°F). In Port-of-Spain the annual average is 25°C (77°F). In the northern and central hill areas and on Tobago, annual rainfall exceeds 250 centimeters (98.4 inches).

INTRODUCTORY SURVEY

RECENT HISTORY

On 31 August 1962, Trinidad and Tobago became independent from Great Britain but retained membership in the Commonwealth as a British dominion. Eric Williams, the founder of the People's National Movement (PNM), became prime minister and held the office until his death in 1981.

In 1976, Trinidad and Tobago declared itself a republic, and a president replaced the British monarch as chief of state. After losing its majority standing in the 1986 elections, the PNM returned to power in 1991 under Prime Minister Patrick Augustus Mervyn Manning. Since 1995, Basdeo Panday has served as prime minister.

GOVERNMENT

Under its 1961 constitution, as amended in 1976, Trinidad and Tobago is a parliamentary democracy with a two-chamber legislature, and, as head of state, a ceremonial president chosen by Parliament. The 36-member House of Representatives is the more important of the two houses. The Senate consists of 31 members, all appointed by the president. The chief executive officer is the prime minister, who is leader of the majority party. Cabinet ministers are appointed by the president.

In 1980, Tobago attained a degree of self-government when it was granted its own House of Assembly.

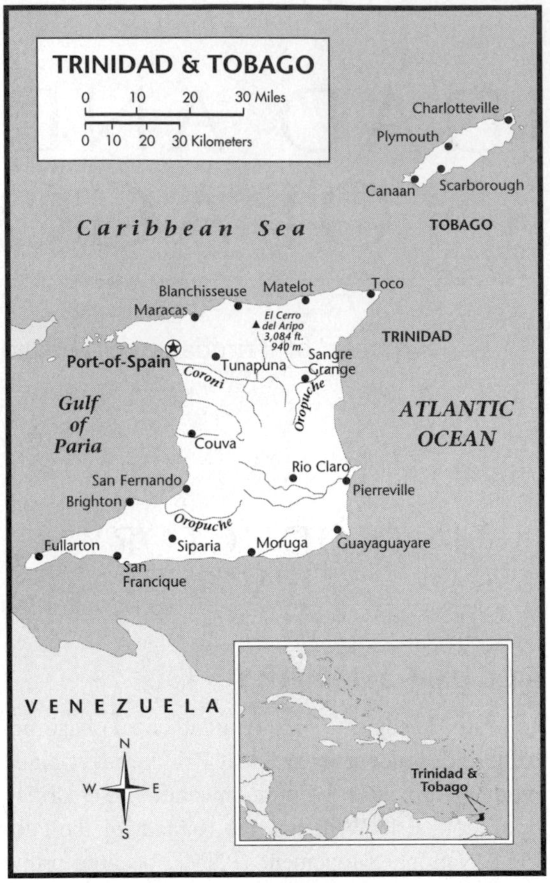

Judiciary

The judicial system is modeled after that of the United Kingdom with some local variations. The Supreme Court of Judicature is made up of the High Court of Justice and the Court of Appeal, both appointed by the president on the advice of the prime minister.

Political Parties

The People's National Movement (PNM), formed in 1956 by Eric Williams, long dominated politics in Trinidad and Tobago. Other parties include the United National Congress (UNC) and the National Alliance for Reconstruction (NAR). Party membership is often based on race and region. The PNM is made up primarily of blacks, and the UNC is composed mostly of Indians.

DEFENSE

The Trinidad and Tobago Defense Force number an estimated 2,100 in one reinforced battalion. There are also a coast guard of 700 and a paramilitary police force numbering 4,800.

ECONOMIC AFFAIRS

Although it is by far the most prosperous of the Caribbean nations, Trinidad and Tobago's high degree of dependence on oil revenues has made it very sensitive to falling oil prices in recent years. Trade and tourism are major segments of the economy. The gross domestic product (GDP) grew by 3.8% in 1994, 2.4% in 1995, and 3.2% in 1996.

Public Finance

In the mid-1980s the government initiated currency devaluation, debt reschedulings, and the adoption of an austere budget that included public service wage reductions and decreased transfers to state enterprises.

The U.S. Central Intelligence Agency estimates that, in 1996, government revenues totaled approximately $1.65 billion and expenditures $1.61 billion.

Income

In 1999, the gross national product (GNP) was US $5.5.billion, or about $4230 per person. For the period 1985–95, the average inflation rate was 6.8%, resulting in a declining rate in per person GNP of 1.6%.

Industry

Long-established industries process sugar, molasses, and rum, followed by fish, lumber, fats and oils, and animal feed. Manufactured products include furniture, matches, angostura bitters, soap, confections and clay products. Newer industries include concrete products, canned citrus, bottled drinks, glass, drugs, chemicals, clothing, building materials, and metal goods.

Banking and Finance

The Central Bank of Trinidad and Tobago (established 1964) is the central regulatory institution and the sole bank of issue. Commercial banking is well established and operated chiefly by Canadian, British, and American interests. Six commercial banks operate in Trinidad and Tobago, five of them locally controlled. The largest is the Republic Bank, formerly Barclays.

Government savings banks as well as credit unions and agricultural credit unions are available for loans and savings throughout the country.

The Trinidad and Tobago Development Finance Co., jointly owned by the government and the private sector, offers medium- and long-term financing to industry.

The Trinidad and Tobago Stock Exchange began operating in 1981.

Economic Development

Government incentives to attract foreign capital include: duty-free imports of equipment and raw materials; income tax holidays; accelerated depreciation allowances; unlimited carryover of losses; and repatriation of capital and profits. To encourage diversification toward non-oil/gas sectors, the government has undertaken comprehensive reforms including major downsizing and refocusing of the public sector with an extensive program of divestment and reduction of public employment.

The country remains largely dependent on oil, even though reserves are being depleted. Key ingredients for future development are: a buildup of reserves consistent with maintaining a competitive exchange rate; monetary policy that contains inflation; and strengthened fiscal balances to reduce government debt and lower interest rates.

SOCIAL WELFARE

A national insurance system provides old age, retirement, and disability pensions, maternity, sickness, and survivors' benefits and funeral grants. A food stamp program was introduced in 1978.

Women constitute 36% of the labor force, and are active in business and the professions.

Healthcare

In 1993, there was 1 doctor for every 1,520 persons. In 1992, 99% of the population had access to health care services. Substantial decreases have been recorded in death rates for malaria, tuberculosis, typhoid, and syphilis.

Housing

There is an acute shortage of adequate housing, and high rents have contributed to inflation. A typical rural home for a large family consists of one to three rooms plus an outside kitchen. Slums and tenements are typical of urban life. Nearly all private dwellings, urban or rural, have toilets and piped-in water.

EDUCATION

About 2% of the population 15 years of age and older is illiterate. In 1993, the islands had 476 primary and intermediate schools with 7,210 teachers and 195,013 students. Secondary schools enrolled 100,278 pupils in 1993. Education is compulsory for six years. In 1990, higher education enrollment was estimated at 4,947 students.

1999 KEY EVENTS TIMELINE

January

- President A.N.R. Robinson declines the request of three independents to be named the official opposition in the Tobago House of Assembly, replacing the one official representative of the opposition party, the People's National Movement (PNM).

February

- Petrochemical companies, hard hit by falling oil and commodity prices attributable to the Asian crisis, low global demand, and OPEC oversupply, unsuccessfully seek financial relief from the government.
- Two former presidents, three former Chief Justices, and the sitting Chief Justice come out in strong opposition to a bill that would give the government more control of the public service, judicial and legal commissions. The bill is seen as threatening to the independence of the judiciary.

May

- Trinidad and Tobago hosts the Miss Universe Pageant. There are widespread complaints about the cost of hosting the pageant, estimated at US$10–13 million (TT$65-80 million).

June

- Having exhausted all legal appeals, a prominent drug lord and eight associates are hanged, three at a time over three days, convicted in the gangland-style slaying of four family members. Amnesty International and other human rights groups protest.

July

- In what is seen by the two main political parties as a prelude to the general elections to be held in 2000, local elections are vigorously contested. Contradicting pollsters' predictions, the opposition PNM party wins ten districts more than the ruling United National Congress (UNC).
- Trinidad and Tobago hosts the twentieth summit of heads of government of the Caribbean Community, the group engaged in promoting regional cooperation. The four larger countries (Jamaica, Trinidad and Tobago, Barbados, and Guyana)

agree to establish a Caribbean Court of Appeal that will replace the U.K. Privy Council.

September

- The leader of the Tobago House of Assembly threatens to sever economic and financial ties with Trinidad.
- The Chief Justice issues a strong accusation claiming that the executive, in collusion with the leadership of the Law Association, is seeking to reduce the independence of the judiciary.

October

- The government reshuffles the cabinet, bringing in a prominent businessman in an unexpected role as ''troubleshooter'' to remove bottlenecks in implementation in government ministries. The move is criticized inasmuch as the new cabinet member's company has millions of dollars in bids for government contracts pending.
- Port of Spain is the site of the third annual World Carnival Conference, sponsored by the Tourism and Industrial Development Company of Trinidad and Tobago and the National Carnival Commission October 19 to 23, 1999.
- Dead fish, including snapper, parrotfish, and angelfish among others, are washing up on beaches around the Caribbean, especially in Barbados, St. Vincent and the Grenadines, Grenada, and Trinidad and Tobago. Health inspectors and scientists are searching for an explanation.
- The four-day World Beat Music Festival takes place. The festival is the sixth-largest contributor to the country's GDP, attracting musicians and audiences from around the world.

ANALYSIS OF EVENTS: 1999

BUSINESS AND THE ECONOMY

Trinidad and Tobago, dependent on oil and related industrial exports, began the year reeling from the effects of the Asian/Russian crisis and resulting low commodity prices. Prices for key products—oil, urea, methanol, ammonia, and, to a lesser extent, gas—plunged drastically. By the third quarter, however, the country was benefiting from a fortuitous upturn in oil and gas prices.

Within the last few years, Trinidad and Tobago have already steered away from oil and toward production of natural gas. In 1999 Phoenix Park Gas Processors, owned by the government and foreign partners, completed a $155 million expansion of its gas processing plant at Point Lisas, making it one of the largest gas processing plants in the Western hemisphere. As of October, bids were being entertained for a $1.5 billion world-scale petrochemical project at Point Lisas. In other areas, the government continued its privatization initiatives, including securing foreign partners for the country's electricity, telephone, and post office operations.

Relations between the government and the business community warmed further during 1999. The chamber of commerce agreed to work with the government on major issues of free trade and specific trade negotiations. Among these issues were the expected establishment of the Single Market and Economy of the Caribbean Community (Caricom), the negotiations for hemispheric free trade (expected by 2005), and discussions in the World Trade Organization (WTO) on various trade arrangements, especially in the area of services. Although Trinidad and Tobago derived less of its Gross Domestic Product from agriculture than most of the other Caribbean islands, continuing WTO negotiations on sugar and bananas stimulated Trinidad's production capacities.

By year's end, tentative guidelines had been negotiated governing discussions of new trade protocols with the European Union. In addition to these trading arrangements, Trinidad and Tobago served as headquarters of the Association of Caribbean States (ACS), a Caribbean Basin-wide economic organization. In 1999 Trinidad hosted the third intersessional meeting of the organization at which the mandate for developing trade, tourism and transportation among member states, and also cooperation arrangements with various Latin American regions, was reviewed.

GOVERNMENT AND POLITICS

In 1999 the ruling United National Congress (UNC) government worked assiduously to promote Trinidad and Tobago as a haven for foreign investors, its image at home was tarnished by a succession of scandals, primarily involving accusations of graft in awarding contracts to local and foreign companies. Questions arose about the bidding process for a proposed airport expansion. Critics accused the government of awarding the contract to a financial backer of the party. Contracts for an electricity co-generation plant went to InnCogen under conditions that seemed too favorable (the overly informal negotiations took place through the offices of a government sympathizer living abroad). In addition, the appointment of a highly paid consultant at the National Petroleum Company proved controversial, leading to the consultant's eventual resignation, and the government was dogged by continuing questions about an earlier poorly implemented rice importation arrangement with the government of India.

The government gained some credit from successfully hosting the Miss Universe contest in May, but popular grumbling about the costs and apparently meager benefits soon eroded any optimism. All these problems hurt the credibility of the government and were seen to be contributing factors in the government's loss of several important districts along with the main urban centers in the local elections held in July. The government's problems were compounded by the perception that the ruling party was trying to control the media, the public service, and the judiciary. During the year, polls showed that the government, the first government with a popular base in the East Indian population, continued to be highly popular with its Indian constituency but that its support in the African population amounted to no more than 25 percent.

Support for the government was also eroded on the sister island of Tobago where secession threats were voiced throughout the year. Tobago had long felt neglected by Trinidad and had been seeking more autonomy in its economic affairs. During the year, relations between the Minister of Tobago Affairs, based in Trinidad, and the Tobago House of Assembly, the smaller island's legislative body, deteriorated considerably.

CULTURE AND SOCIETY

Trinidad and Tobago's population was evenly divided between African and Indian elements. There were also significant groups of Chinese, Syrian, European and mixed persons. Trinidad and Tobago's society exhibited cultural strength, harmoniously blending elements from all the sub-cultures. On the other hand, the existence of large and evenly balanced ethnic divisions in the population also engendered sociopolitical tensions. These divisions worsened over the preceding decade as the supporters of the two main political parties, each reliant on particular ethnic support, became increasingly hostile to each other. The racial character of politics is reflected in polls that show that support for ruling party is very high among the Indian population and very low among Africans. This ethnic division stimulated nationalist sentiment among both Indian and African sub-populations.

As has happened elsewhere, economic liberalization and the reduction in state subsidies have also brought social tensions into play, as increasing numbers of persons, estimated at almost a third of the population, fell into poverty. In 1999, the government tried to ease the plight of the aged by increasing their pensions. It also attempted to appease the general public by postponing new taxes until the year 2000. The trade union movement proved to be especially intransigent. Tensions between labor unions and the government brought a succession of protests, demonstrations and strikes by teachers, oil workers, sugar workers, and civil servants. As the year drew to a close, the arrest late at night of two leading unionists accused of organizing protests in front of a minister of government's home evoked a popular outcry. The government's public relations problem was compounded by its poor relations with the media which charted the Prime Minister with promoting half truths and innuendoes.

Although the country had a highly literate population (above 90 percent literate), the educational system suffered from variable school performance and the lack of free secondary school spots for all. A rigid and allegedly corrupt examination system determined secondary school placement. In 1999, in a move that generated controversy, the government moved to standardize school texts (seen as a cost effective and equalizing measure). The government announced plans to build more schools and phase out the competitive examinations. On the other hand, no notable measures were taken to improve the perennially resource-poor health system, plagued in 1999 by deficiencies that led to a stream of protests by medical personnel and complaints by the general public.

Finally, crime, including drug-related crime, continued to the disrupt and corrupt civil life. Accusations of corruption and misbehavior against the police continued to erode the popular trust. Another major social problem was the high level of domestic abuse. In the first ten months of 1999, 14 women had been murdered, most by men, compared to 13 in all of 1998. Prodded by individual women, concerned citizens in general, and by non-governmental groups, legislature passed a new Domestic Violence Act extending the definition of domestic abuse to sexual, emotional, psychological and financial abuse, and empowering the court to provide financial relief for the victims and their children.

DIRECTORY

CENTRAL GOVERNMENT

Head of State

President
Arthur Robinson, Office of the President
E-MAIL: presoftt@carib-link.net

Prime Minister
Basdeo Panday, Office of the Prime Minister, Level 15, Central Bank Towers, Eric Williams Plaza, Independence Square, Port of Spain, Trinidad and Tobago
PHONE: +(809) 6233653
FAX: +(809) 6273444
E-MAIL: bpanday@trinidad.net

Ministers

Minister of Legal Affairs
Kamla Persad-Bissessar, Ministry of Legal Affairs, Winsure Building, 24-28 Richmond St., Port of Spain, Trinidad and Tobago
PHONE: +(809) 6232010; 6231449
FAX: +(809) 6256530

Minister of Works and Transport
Sadiq Baksh, Ministry of Works and Transport, Level 5, Salvatori Building, Frederick St., Port of Spain, Trinidad and Tobago
PHONE: +(809) 6252643
FAX: +(809) 6254512

Minister of Housing and Settlements
John Humphrey, Ministry of Housing and Settlements, NHA Building, Corner of George St. and South Quay, Port of Spain, Trinidad and Tobago
PHONE: +(809) 6245058; 6247934
FAX: +(809) 6252793

Minister of Public Administration and Information
Wade Mark, Ministry of Public Administration and Information, Central Bank Towers, Independence Square, Port of Spain, Trinidad and Tobago
PHONE: +(809) 6233678
FAX: +(809) 6273444

Minister of Education
Adnesh Nanan, Ministry of Education, Alexandra Street, St. Clair, Port of Spain, Trinidad and Tobago
PHONE: +(809) 6222181
FAX: +(809) 6287818

Minister of Public Utilities
Ganga Singh, Ministry of Public Utilities, Sacred Heart Building, Sackville St., Port of Spain, Trinidad and Tobago
PHONE: +(809) 6274424
FAX: +(809) 6257004

Minister of Energy and Energy Industries
Finbar K. Gangar, Ministry of Energy and Energy Industries, Level 9, Riverside Plaza, Corner of Besson and Picadilly Streets, Port of Spain, Trinidad and Tobago
PHONE: +(809) 6236708
FAX: +(809) 6232726

Minister of Social Development
Manohar Ramsaran, Ministry of Social Development, Level 1 and 4, Salvatori Building, Independence Square, Port of Spain, Trinidad and Tobago
PHONE: +(809) 6238841
FAX: +(809) 6247727

Minister of Sport and Youth Affairs
Pamela Nicholson, Ministry of Sport and Youth Affairs, ISSA Nicholas Building, Corner of Frederick and Duke Streets, Port of Spain, Trinidad and Tobago
PHONE: +(809) 6258875
FAX: +(809) 6240011

Minister of Health
Hamza Rafeeq, Ministry of Health, Corner of Duyncan St. and Independence Square, Trinidad and Tobago
PHONE: +(809) 6270010; 6270012; 6270014
FAX: +(809) 6239628

Minister of Planning and Development
Trevor Sudarna, Ministry of Planning and Development, Level 14, Eric Williams Plaza, Independence Square, Port of Spain, Trinidad and Tobago
PHONE: +(809) 6279700
FAX: +(809) 6238123

Minister of Trade, Industry and Commerce
Mervyn Assam, Ministry of Trade, Industry and Commerce, Level 15, Riverside Plaza, Corner of Besson and Picadilly Streets, Port of Spain, Trinidad and Tobago
PHONE: +(809) 6232931
FAX: +(809) 6238488

Minister of Labor and Cooperatives
Harry Partap, Ministry of Labor and Cooperatives, Level 11, Riverside Plaza, Corner of Besson and Picadilly St., Port of Spain, Trinidad and Tobago
PHONE: +(809) 6234241
FAX: +(809) 6244091

Minister of Foreign Affairs
Ralph Maraj, Ministry of Foreign Affairs, Knowsley Building, 1 Queens Park West, Port of Spain, Trinidad and Tobago
PHONE: +(809) 6234116
FAX: +(809) 6270571

Minister of National Security
Joesph Theodore, Ministry of National Security, Knox St., Port of Spain, Trinidad and Tobago
PHONE: +(809) 6232441
FAX: +(809) 6278044

Minister of Finance and Tourism
Brian Kuei Tung, Ministry of Finance and Tourism, Level 8, Eric Williams Complex, Independence Square, Port of Spain, Trinidad and Tobago
PHONE: +(809) 6279692; 6279693
FAX: +(809) 6276108

Minister of Agriculture, Lands and Marine Resources
Reeza Mohammed, Ministry of Agriculture, Lands and Marine Resources, St. Clair Circle, St. Clair, Port of Spain, Trinidad and Tobago
PHONE: +(809) 6221221
FAX: +(809) 6224246

POLITICAL ORGANIZATIONS

People's National Movement (PNM)

NAME: Patrick Manning

United National Congress (UNC)

NAME: Basdeo Panday

National Alliance for Reconstruction (NAR)

NAME: A.N.R. Robinson

Republican Party

NAME: Nello Mitchell

National Development Party

NAME: Nelson Charles

DIPLOMATIC REPRESENTATION

Embassies in Trinidad and Tobago

Argentina
4th Floor, Tatil Building, 11 Maraval Road, Port of Spain, Trinidad and Tobago, W.I.
PHONE: +(809) 6287557; 6287587
FAX: +(809) 6287544
E-MAIL: embargen-pos@carib-link.net

Brazil
18 Sweet Briar Road, St. Clair, Port of Spain, Trinidad and Tobago, W.I.
PHONE: +(809) 6225771; 6225779
FAX: +(809) 6224323
E-MAIL: brastt@wow.net

Canada
Maple House, 3-3a Sweet Briar Road, St. Clair, Port of Spain, Trinidad and Tobago, W.I.
PHONE: +(809) 6226232
FAX: +(809) 6281830
E-MAIL: chcpspan@opus.co.tt

China
39 Alexandra Street, St. Clair, Port of Spain, Trinidad and Tobago, W.I.
PHONE: +(809) 6226976
FAX: +(809) 6227613
E-MAIL: tian@wow.net

Colombia
16 Queen's Park West, Port of Spain, Trinidad and Tobago, W.I.
PHONE: +(809) 6225904; 6225938
E-MAIL: emtrinidad@trinidad.net

Cuba
2nd Floor, Furness Building, 90 Independence Square, Port of Spain, Trinidad and Tobago, W.I.
PHONE: +(809) 6271306
FAX: +(809) 6273515

France
6th Floor, Tatil Building, 11 Maraval Road, Port of Spain, Trinidad and Tobago, W.I.
PHONE: +(809) 6227446
E-MAIL: francett@wow.net

Germany
7-9 Marli Street, Newtown, Port of Spain, Trinidad and Tobago, W.I.
PHONE: +(809) 6281630

India
6 Victoria Avenue, Port of Spain, Trinidad and Tobago, W.I.
PHONE: +(809) 6277480
FAX: +(809) 6276985
E-MAIL: Hcipos@tstt.net.tt

Jamaica
2 Newbold Street, St. Clair, Port of Spain, Trinidad and Tobago, W.I.
PHONE: +(809) 6224995
FAX: +(809) 6229103

Japan
5 Hayes Street, St. Clair, Port of Spain, Trinidad and Tobago, W.I.
PHONE: +(809) 6285991
FAX: +(809) 6220858
E-MAIL: jpemb@wow.net

Mexico
4th Floor, Algico Plaza, 91-93 St. Vincent Street, Port of Spain, Trinidad and Tobago, W.I.
PHONE: +(809) 6276988; 6276941
FAX: +(809) 6271028

The Netherlands
3rd Floor, Life of Barbados Building, 69-71 Edward Street, Port of Spain, Trinidad and Tobago, W.I.
PHONE: +(809) 6251210; 6251722
FAX: +(809) 6251704
E-MAIL: nethantgov@wow.net

Nigeria
3 Maxwell-Phillip Street, St. Clair, Port of Spain, Trinidad and Tobago, W.I.
PHONE: +(809) 6226834; 6224002
FAX: +(809) 6227162

Panama
Suite 6, 1a Dere Street, Port of Spain, Trinidad and Tobago, W.I.
PHONE: +(809) 6233435
FAX: +(809) 6233440
E-MAIL: embapatt@wow.net

Suriname
PHONE: +(809) 6280704; 6280089
FAX: +(809) 6280086
E-MAIL: AmbSurPDE@opus.co.tt

United Kingdom
19 St. Clair Avenue, St. Clair, P.O. Box 778, Port of Spain, Trinidad and Tobago, W.I.
PHONE: +(809) 6281234; 6282748
FAX: +(809) 6224555
TITLE: High Commissioner
NAME: Gregory Faulkner

United States
15 Queen's Park West, Port of Spain, Trinidad and Tobago, W.I.
PHONE: +(809) 6226371
FAX: +(809) 6255462
E-MAIL: usispos@trinidad.net
TITLE: Ambassador
NAME: Edward Shumaker III

Venezuela
16 Victoria Avenue, Port of Spain, Trinidad and Tobago, W.I.
PHONE: +(809) 6279821; 6279823
FAX: +(809) 6242508
E-MAIL: embaveeneztt@carib-link.net

JUDICIAL SYSTEM

Supreme Court

High Court

Port of Spain, Trinidad and Tobago

FURTHER READING

Articles

Braveboy-Wagner, Jacqueline. ''Trinidad and Tobago,'' in *The Oxford Companion to Politics of the World.* Joel Krieger, ed. New York: Oxford University Press, 1993, p. 920–921.

———. ''Trinidad and Tobago,'' in Charles D. Ameringuer, ed., *Political Parties of the Americas. 1980s to 1990s.* Westport, CT: Greenwood Press, 1992, 573-585.

Books

Ryan, Selwyn. *Pathways to Power: Indians and the Politics of National Unity in Trinidad and Tobago.* Trinidad: Institute for Social and Economic Research, The University of the West Indies, 1996.

Internet

Internet Express: News from Trinidad and Tobago and the Caribbean Region. Available Online @ http://www.trinidadexpress.com (November 24, 1999).

TIDCO (Tourism and Industrial Development Company of Trinidad and Tobago Limited). Available Online @ http://www.tidco.co.tt/ (November 24, 1999).

Trinidad Guardian. Available Online @ http://www.guardian.co.tt/ (November 24, 1999.)

Welcome to Trinidad and Tobago. Available Online @ http://www.visittnt.com/ (November 24, 1999).

TRINIDAD AND TOBAGO: STATISTICAL DATA

For sources and notes see "Sources of Statistics" in the front of each volume.

GEOGRAPHY

Geography (1)

Area:

Total: 5,130 sq km.

Land: 5,130 sq km.

Water: 0 sq km.

Area—comparative: slightly smaller than Delaware.

Land boundaries: 0 km.

Coastline: 362 km.

Climate: tropical; rainy season (June to December).

Terrain: mostly plains with some hills and low mountains.

Natural resources: petroleum, natural gas, asphalt.

Land use:

Arable land: 15%

Permanent crops: 9%

Permanent pastures: 2%

Forests and woodland: 46%

Other: 28% (1993 est.)

HUMAN FACTORS

Health Indicators (4)

Life expectancy at birth	
1980	68
1997	73
Daily per capita supply of calories (1996)	2,751
Total fertility rate births per woman (1997)	1.9
Maternal mortality ratio per 100,000 live births (1990-97)	90[c]
Safe water % of population with access (1995)	96
Sanitation % of population with access (1995)	96

Continued on next page.

Demographics (2A)

	1990	1995	1998	2000	2010	2020	2030	2040	2050
Population	1,198.4	1,155.3	1,116.6	1,086.9	1,031.8	1,075.5	1,084.4	1,076.8	1,057.4
Net migration rate (per 1,000 population)	NA	NA	NA	NA	NA	NA	NA	NA	NA
Births	NA	NA	NA	NA	NA	NA	NA	NA	NA
Deaths	NA	NA	NA	NA	NA	NA	NA	NA	NA
Life expectancy - males	67.3	67.6	68.1	68.3	69.6	70.8	71.9	72.9	73.8
Life expectancy - females	71.9	72.6	73.0	73.3	74.8	76.1	77.3	78.5	79.4
Birth rate (per 1,000)	20.0	16.5	14.9	14.1	14.7	12.5	11.3	11.7	10.8
Death rate (per 1,000)	7.0	7.7	8.0	8.3	9.5	10.1	11.5	12.9	13.0
Women of reproductive age (15-49 yrs.)	301.6	291.5	285.7	281.0	253.5	245.9	261.3	236.1	227.5
of which are currently married	285.6	NA	NA	NA	NA	NA	NA	NA	NA
Fertility rate	2.4	2.2	2.1	2.0	1.9	1.8	1.8	1.8	1.8

Except as noted, values for vital statistics are in thousands; life expectancy is in years.

Health Indicators (4) cont.

Consumption of iodized salt % of households (1992-98)[a] ....
Smoking prevalence
Male % of adults (1985-95)[a] ....
Female % of adults (1985-95)[a] ....
Tuberculosis incidence per 100,000 people (1997) . . .11
Adult HIV prevalence % of population ages 15-49 (1997) 0.94

Infants and Malnutrition (5)

Under-5 mortality rate (1997) 17
% of infants with low birthweight (1990-97) 10
Births attended by skilled health staff % of total[a] . . .98
% fully immunized (1995-97)
TB . . . .NA
DPT 90
Polio 91
Measles 88
Prevalence of child malnutrition under age 5 (1992-97)[b] . . . .NA

Ethnic Division (6)

Black 40%
East Indian 40.3%
Mixed 14%
White 1%
Chinese 1%
Other 3.7%

East Indian is a local term—primarily immigrants from northern India.

Religions (7)

Roman Catholic 32.2%
Hindu 24.3%
Anglican 14.4%
Other Protestant 14%
Muslim 6%
None or unknown 9.1%

Languages (8)

English (official), Hindi, French, Spanish.

EDUCATION

Public Education Expenditures (9)

Public expenditure on education (% of GNP)
1980 . . . .4.0
1996 . . . .3.7
Expenditure per student
Primary % of GNP per capita
1980 . . . .9.2
1996 ....
Secondary % of GNP per capita
1980 . . . .20.1[1]
1996 ....
Tertiary % of GNP per capita
1980 . . . .59.5
1996 ....
Expenditure on teaching materials
Primary % of total for level (1996) . . . .1.0[1]
Secondary % of total for level (1996) 2.8
Primary pupil-teacher ratio per teacher (1996) . . . .25[1]
Duration of primary education years (1995) 7

Educational Attainment (10)

Age group (1990) . . . .25+
Total population 542,425
Highest level attained (%)
No schooling 4.5
First level
Not completed 56.8
Completed . . . .NA
Entered second level
S-1 . . . .32.3
S-2 . . . .NA
Postsecondary 3.4

Literacy Rates (11A)

In thousands and percent[1]	1990	1995	2000	2010
Illiterate population (15+ yrs.)	24	19	14	8
Literacy rate - total adult pop. (%)	97.1	97.9	98.6	99.3
Literacy rate - males (%)	98.3	98.8	99.1	99.4
Literacy rate - females (%)	95.9	97.0	98.1	99.1

GOVERNMENT & LAW

Political Parties (12)

House of Representatives	% of seats
People's National Movement (PNM)	52
United National Congress (UNC)	42.2
National Alliance for Reconstruction (NAR)	5.2

Government Budget (13A)

Year: 1995

Total Expenditures: 8,917.9 Millions of Dollars

Expenditures as a percentage of the total by function:

- General public services and public order 23.28
- Defense 1.92
- Education 14.57
- Health 8.97
- Social Security and Welfare 12.35
- Housing and community amenities 8.97
- Recreational, cultural, and religious affairs 94
- Fuel and energy 17
- Agriculture, forestry, fishing, and hunting 4.78
- Mining, manufacturing, and construction 27
- Transportation and communication 3.38
- Other economic affairs and services 1.57

Crime (15)

Crime rate (for 1997)

- Crimes reported 17,000
- Total persons convicted 2,550
- Crimes per 100,000 population 1,300

Persons responsible for offenses

- Total number of suspects 5,750
- Total number of female suspects 250
- Total number of juvenile suspects 350

Military Affairs (14B)

	1990	1991	1992	1993	1994	1995
Military expenditures						
Current dollars (mil.)	NA	NA	62[e]	63[e]	83	82
1995 constant dollars (mil.)	NA	NA	66[e]	66[e]	85	82
Armed forces (000)	2	2	3	3	3	3
Gross national product (GNP)						
Current dollars (mil.)	3,898	4,128	4,167	4,266	4,461	4,749
1995 constant dollars (mil.)	4,480	4,562	4,482	4,472	4,573	4,749
Central government expenditures (CGE)						
1995 constant dollars (mil.)	NA	1,593	NA	1,647[e]	NA	NA
People (mil.)	1.3	1.3	1.3	1.3	1.3	1.3
Military expenditure as % of GNP	NA	NA	1.5	1.5	1.9	1.7
Military expenditure as % of CGE	NA	NA	NA	4.0	NA	NA
Military expenditure per capita (1995 $)	NA	NA	53	52	67	64
Armed forces per 1,000 people (soldiers)	1.6	1.6	2.4	2.4	2.4	2.4
GNP per capita (1995 $)	3,568	3,618	3,546	3,530	3,602	3,736
Arms imports[6]						
Current dollars (mil.)	0	0	0	0	0	0
1995 constant dollars (mil.)	0	0	0	0	0	0
Arms exports[6]						
Current dollars (mil.)	0	0	0	0	0	0
1995 constant dollars (mil.)	0	0	0	0	0	0
Total imports[7]						
Current dollars (mil.)	1,121	1,667	1,434	1,448	1,131	1,714
1995 constant dollars (mil.)	1,288	1,842	1,542	1,518	1,159	1,714
Total exports[7]						
Current dollars (mil.)	1,718	1,985	1,869	1,612	1,867	2,455
1995 constant dollars (mil.)	1,974	2,193	2,010	1,690	1,914	2,455
Arms as percent of total imports[8]	0	0	0	0	0	0
Arms as percent of total exports[8]	0	0	0	0	0	0

LABOR FORCE

Labor Force (16)

Total 404,500

Construction and utilities 13%

Manufacturing, mining, ...

and quarrying 14%

Agriculture 11%

Services 62%

Data for 1993 est.

Unemployment Rate (17)

16.1% (December 1996)

PRODUCTION SECTOR

Electric Energy (18)

Capacity ....1.15 million kW (1995)

Production ....3.9 billion kWh (1995)

Consumption per capita ....3,068 kWh (1995)

Transportation (19)

Highways:

total: 8,320 km

paved: 4,252 km

unpaved: 4,068 km (1996 est.)

Pipelines: crude oil 1,032 km; petroleum products 19 km; natural gas 904 km

Merchant marine:

total: 1 cargo ship (1,000 GRT or over) totaling 1,336 GRT/2,567 DWT (1997 est.)

Airports: 6 (1997 est.)

Airports—with paved runways:

total: 3

over 3,047 m: 1

2,438 to 3,047 m: 1

1,524 to 2,437 m: 1 (1997 est.)

Airports—with unpaved runways:

total: 3

914 to 1,523 m: 1

under 914 m: 2 (1997 est.)

Top Agricultural Products (20)

Cocoa, sugarcane, rice, citrus, coffee, vegetables; poultry.

MANUFACTURING SECTOR

GDP & Manufacturing Summary (21)

Detailed value added figures are listed by both International Standard Industry Code (ISIC) and product title.

	1980	1985	1990	1994
GDP: ($-1990 mil.)[1]	6,389	5,589	5,068	5,291
Per capita ($-1990)[1]	5,905	4,818	4,100	4,095
Manufacturing share (%) (current prices)[1]	5.2	7.1	8.8	NA
Manufacturing				
Value added ($-1990 mil.)[1]	543	448	438	*492*
Industrial production index	100	86	104	137
Value added ($ mil.)	492	387	683	*605*
Gross output ($ mil.)	*1,518*	1,765	2,344	*2,059*
Employment (000)	44	34	38	*39*
Profitability (% of gross output)				
Intermediate input (%)	*68*	78	71	*71*
Wages and salaries inc. supplements (%)	*18*	*18*	*13*	*13*
Gross operating surplus	*15*	*4*	*16*	*16*
Productivity ($)				
Gross output per worker	*34,268*	52,667	61,801	*52,199*
Value added per worker	11,099	11,715	18,007	*15,358*
Average wage (inc. supplements)	*5,999*	*9,488*	*8,184*	*6,989*
Value added ($ mil.)				
311/2 Food products	67	95	114	*121*
313 Beverages	27	34	58	*53*
314 Tobacco products	14	35	36	*39*
321 Textiles	1	2	5	*4*
322 Wearing apparel	16	13	10	*7*
323 Leather and fur products	—	—	—	—
324 Footwear	4	5	3	2
331 Wood and wood products	6	4	*5*	*4*
332 Furniture and fixtures	9	7	*7*	*5*
341 Paper and paper products	9	14	17	*14*

	1980	1985	1990	1994
342 Printing and publishing	13	19	21	*20*
351 Industrial chemicals	5	6	109	*116*
352 Other chemical products	12	10	23	*19*
353 Petroleum refineries	*190*	*17*	*131*	*68*
354 Miscellaneous petroleum and coal products	*2*	—	*1*	*1*
355 Rubber products	9	10	8	*5*
356 Plastic products	2	8	5	*3*
361 Pottery, china and earthenware	—	—	—	—
362 Glass and glass products	3	4	10	*9*
369 Other non-metal mineral products	23	31	23	*24*
371 Iron and steel	—	—	37	*32*
372 Non-ferrous metals	—	—	—	—
381 Metal products	26	11	14	*12*
382 Non-electrical machinery	13	—	—	—
383 Electrical machinery	3	13	15	*13*
384 Transport equipment	28	43	7	*12*
385 Professional and scientific equipment	—	—	—	—
390 Other manufacturing industries	8	*6*	24	*23*

FINANCE, ECONOMICS, & TRADE

Economic Indicators (22)

National product: GDP—purchasing power parity—$13.2 billion (1996 est.)

National product real growth rate: 3.1% (1996 est.)

National product per capita: $10,400 (1996 est.)

Inflation rate—consumer price index: 3.4% (1996)

Exchange Rates (24)

Exchange rates:

Trinidad and Tobago dollars (TT$) per US$1

January 1998	6.2840
1997	6.2503
1996	6.0051
1995	5.9478
1994	5.9249
1993	5.3511

Top Import Origins (25)

$2.1 billion (c.i.f., 1996) Data are for 1995.

Origins	%
United States	48
Venezuela	10
United Kingdom	8
Germany	NA
Canada	NA

NA stands for not available.

Balance of Payments (23)

	1991	1992	1993	1994	1995
Exports of goods (f.o.b.)	1,774	1,691	1,500	1,778	2,456
Imports of goods (f.o.b.)	−1,210	−996	−953	−1,037	−1,868
Trade balance	564	696	547	741	588
Services - debits	−1,025	−1,040	−832	−907	−709
Services - credits	454	482	394	383	419
Private transfers (net)	2	−1	—	−2	−3
Government transfers (net)	—	1	5	2	−1
Overall balance	−5	139	113	218	294

Top Export Destinations (26)

$2.5 billion (f.o.b., 1996) Data are for 1994.

Destinations	%
United States	48
Caricom countries	15
Latin America	9
European Union	5

Economic Aid (27)

Recipient: ODA, $10 million (1993).

Import Export Commodities (28)

Import Commodities	Export Commodities
Machinery	Petroleum and petroleum products
Transportation equipment	Chemicals
Manufactured goods	Steel products
Food	Fertilizer
Live animals	Sugar
	Cocoa
	Coffee
	Citrus
	Flowers

TUNISIA

CAPITAL: Tunis.

FLAG: Centered on a red ground is a white disk bearing a red crescent and a red five-pointed star.

ANTHEM: *Al-Khaladi (The Glorious).*

MONETARY UNIT: The Tunisian dinar (D) is a paper currency of 1,000 millimes. There are coins of 1, 2, 5, 10, 20, 50, and 100 millimes and of 1, and 5 dinars, and notes of 1, 5, 10, and 20 dinars. D1 = $1.02145 (or $1 = D0.979).

WEIGHTS AND MEASURES: The metric system is the legal standard.

HOLIDAYS: New Year's Day, 1 January; Independence Day, 20 March; Martyrs' Day, 9 April; Labor Day, 1 May; Victory Day, 1 June; Republic Day, 25 July; Women's Day, 13 August; Evacuation Day, 15 October; Accession of President Ben Ali, 7 November. Movable religious holidays include 'Id al-Fitr, 'Id al-'Adha', 1st of Muharram (Muslim New Year), and Milad an-Nabi.

TIME: 1 PM = noon GMT.

LOCATION AND SIZE: Situated on the northern coast of Africa, Tunisia has an area of 163,610 square kilometers (63,170 square miles), slightly larger than the state of Georgia, and has a total boundary length of 2,445 kilometers (1,519 miles). Tunisia's capital city, Tunis, is located on the Mediterranean Sea coast.

CLIMATE: Tunisia consists of two climatic belts, with Mediterranean influences in the north and Saharan in the south. Temperatures are moderate along the coast, with an average annual reading of 18°C (64°F), and hot in the interior south. Temperatures at Tunis range from an average minimum of 6°C (43°F) in January to an average maximum of 33°C (91°F) in August. Precipitation in the north reaches a high of 150 centimeters (59 inches) annually, while rainfall in the extreme south averages less than 20 centimeters (8 inches) a year.

Republic of Tunisia
Al-Jumhuriyah at-Tunisiyah

INTRODUCTORY SURVEY

RECENT HISTORY

On 20 March 1956, France recognized Tunisian independence. In that following April, Habib Bourguiba formed Tunisia's first independent government. Bourguiba was elected president in 1959 and was reelected in 1964, 1969, and 1974, when the National Assembly amended the constitution to make him president for life.

When economic problems and political oppression during the late 1970s led to protests by students and workers, the government began granting more political freedom, but in the late 1980s, Bourguiba cracked down on all forms of dissent. Two legal opposition parties were prevented from taking part in elections. A massive roundup of Islamic fundamentalists took place in 1987. On 27 September 1987, a state security court found 76 defendants guilty of plotting against the government; seven (five in absentia) were sentenced to death.

General Zine al-Abidine Ben Ali, who had conducted the crackdown, was named prime minister in September 1987. Six weeks later, Ben Ali assumed the presidency himself, ousting Bourguiba and promising political liberalization. Almost 2,500 political prisoners were released under his control. The following year, Tunisia's constitution was revised, abolishing the presidency for life. Elections were advanced from 1991 to 1989, and Ben Ali was elected president, running unopposed, and has remained in office ever since.

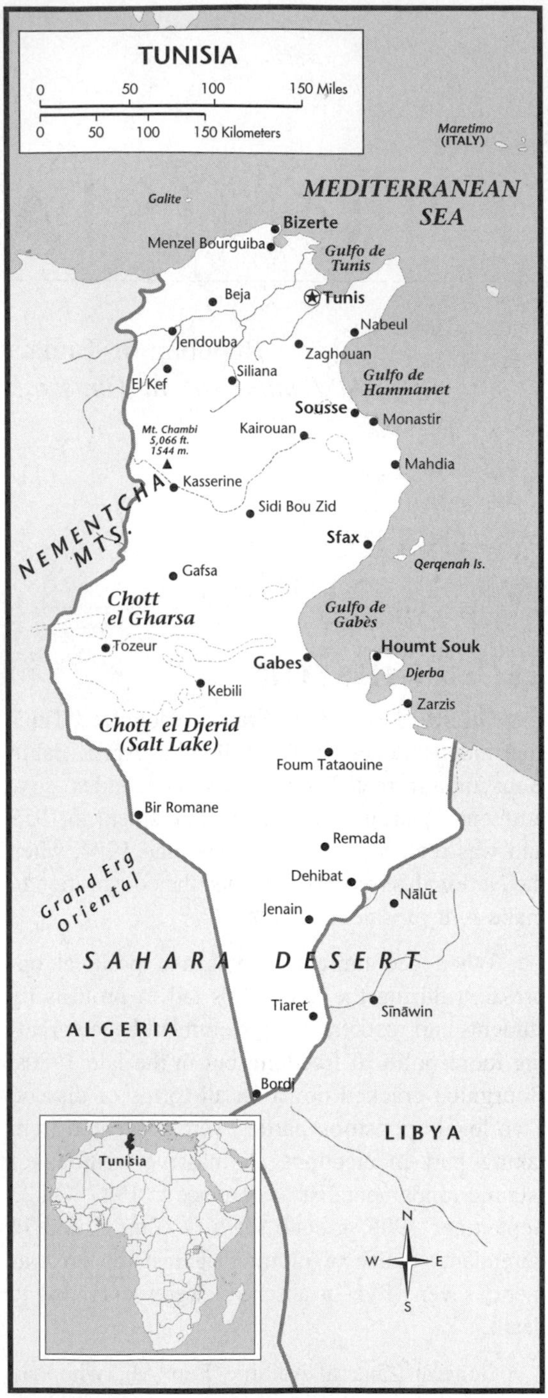

Tunisia has followed a moderate, nonaligned course in foreign relations, complicated by periodic difficulties with its immediate neighbors. Relations with Libya remained tense after ties were resumed in 1987. Algeria signed a border agreement in 1993 and planned a gas pipeline through Tunisia to Italy. Although the United States has provided economic and military aid, Tunisia opposed American support for Kuwait following Iraq's invasion in 1990. In July 1995, Tunisia signed an association agreement with the European Union that in 2007 would make the country part of the free-trade area around the Mediterranean known as the European Economic Area.

GOVERNMENT

The president initiates and directs state policy and appoints judges, provincial governors, the mayor of Tunis, and other high officials. The cabinet, headed by a prime minister, varies in size and is under presidential domination. The single-chamber National Assembly was expanded in 1993 to 163 members, elected by general, free, direct, and secret ballot. All citizens 20 years of age or older may vote. Presidential ratification is required before a bill passed by the legislature can become law, but the assembly may override the president's veto by a two-thirds majority.

Judiciary

Local magistrates are appointed by the president. The ultimate court of appeal is the Court of Cassation. In addition, a high court is set up for the sole purpose of prosecuting a member of the government accused of high treason. A military tribunal, consisting of a presiding civilian judge from the Court of Cassation and four military judges, hears cases involving military personnel as well as national security cases concerning civilians.

Political Parties

The Constitutional Democratic Rally (RCD) holds the majority of assembly seats. The major opposition party is the Movement of Social Democrats. Five other political parties, including the Communist party, are legal. The principal Islamist party, An Nahda, is outlawed.

DEFENSE

As of 1995, Tunisia had an army of 27,000, a navy of 4,500, and an air force of 3,500 personnel. Twelve months of military service are obligatory. The national police number 13,000 men, and the national guard, 10,000.

ECONOMIC AFFAIRS

Agriculture, which engages about one-third of the labor force dominates the Tunisian economy, although minerals (especially crude oil and phosphates) and tourism are the leading sources of income from abroad. Industrial development has increased rapidly since the 1960s. The domestic economy grew by 4.2% annually during 1987–95. A three-year drought ended in 1996 resulting in an economic growth rate of 7%. The private sector accounted for 60% of economic output in 1996.

Public Finance

Oil revenues and levies on imports provide the major sources of current revenue. With the fall of oil prices in 1986, public sector investment and expenditures were cut back. In the late 1980s, a structural adjustment program sponsored by the World Bank and IMF focused on reducing the role of public sector. The U.S. Central Intelligence Agency estimates that, in 1993, government revenues totaled approximately $4.3 billion and expenditures $5.5 billion. External debt totaled $7.7 billion.

Income

In 1998 Tunisia's gross national product (GNP) was $19.4 billion, or about $2,110 per person. For the period 1985–95, the average inflation rate was 6%, resulting in a real growth rate in per person GNP of 1.8%.

Industry

Manufacturing is dominated by textiles and leather operations. In 1996, the textiles industry contributed 6.4% to the gross domestic product (GDP); food processing, 3.3%; chemicals, 2%; and other manufactures, 6.8%. Handicrafts industries produce clothing, rugs, pottery, and copper and leather goods for both local and export markets.

Banking and Finance

The Central Bank of Tunisia is the sole bank of issue. The Tunisian Banking Company, established in 1957, is the leading commercial and investment bank. The state holds 52% of the bank's capital. The total money supply in 1995, as measured by M2, was $7,803 million.

The banking system is a mixture of state-owned and private institutions which offer a variety of financial instruments and services. Of the 12 commercial banks, one is fully state-owned and four others are part-owned by the state. These five banks control 70% of total bank assets.

Activities of the stock exchange in Tunis remain limited to transactions in securities issued by the state and the stocks of a few private or government-owned firms.

Economic Development

A five-year development plan following a foreign exchange crisis in 1986 and the adoption of an IMF sponsored economic rehabilitation scheme, called for expenditure of $10.4 billion in public spending, with another $5.3 billion from private sources and $6 billion from abroad. Services were to receive 39%, agriculture 19%, and manufacturing 16%. Real annual growth of 4% was projected.

The 1992–96 development plan envisaged average annual GDP growth of 6%, based on strong expansion in the manufacturing industry (8.7%) and tourism (22.3%). The plan called for further cuts in consumer subsidies and the privatization of many state assets.

SOCIAL WELFARE

A social security system, to which both employers and workers contribute, includes maternity payments, family allowances, disability and life insurance, and old age insurance. Agricultural workers receive family allowances and old age pensions.

Polygamy was prohibited in 1957; Tunisian women enjoy full civil and political rights under the law. A 1995 law protects children from abandonment and assault.

Healthcare

Health conditions have shown significant improvement in recent years, although diet and sanitation remain deficient. Epidemics have virtually disappeared, and the incidence of contagious diseases has been considerably reduced. Average life expectancy in 1996 was 70 years. Free health services are available to about 70% of the population; about 90% had access to healthcare services in 1993.

Housing

An increase in population and the migration of rural dwellers to urban areas has caused serious housing problems. Squatter communities have sprung up in urban regions. The rate of housing construction lags far behind the need. Housing construction for 1993 was projected at approximately 40,100 units, up from 38,600 units in 1992.

EDUCATION

Primary school enrollment rate is 97% of school-age children. Arabic is the language of instruction in early primary grades but is replaced by French in later grades. The 1995 adult illiteracy rate was 33%.

1999 KEY EVENTS TIMELINE

March

- Tunisian President Zine al-Abidine Ben Ali meets with U.S. First Lady Hillary Rodham Clinton, assuring her that human-rights problems ''will be properly addressed'' in Tunisia.
- Mrs. Clinton delivers an address to a group of professional women in Tunisia to celebrate women's rights, and praises Tunisia for furthering those rights.

April

- Tunisia resumes international flights to Algeria after the United Nations suspends sanctions against Libya.

June

- Three people die and ten are seriously injured as soccer fans of rival teams battle during a Tunisia Cup semifinal match in Beja, northwest of Tunis. President Ben Ali fires the governor of the Beja province after the incident.

July

- President Ben Ali pardons a number of prisoners to mark Republic Day. Some had their terms reduced, while others had the remainder of their sentences commuted.
- Human rights lawyer Radhia Nasraoui, who was defending members of the banned Tunisia Communist Workers Party on trial for threatening public order, receives a suspended sentence of six months for meeting with political activists in her office.

August

- The government announces a record year for tourism earnings, as European arrivals rise sharply during first six months of the year.

October

- The U.N. General Assembly selects Mali, Tunisia, Jamaica, Bangladesh and Ukraine as the newest temporary members of the Security Council, the most powerful U.N. decision-making body.
- As the campaign for Tunisia's first multiparty presidential election officially opens, President Ben Ali promises more telephones, higher incomes, and more freedom for the government controlled press.
- President Zine al-Abidine Ben Ali wins a third term in office, gaining, for the third consecutive time, more than 99 percent of the vote, according to government officials.

November

- Amnesty International reports that Tunisia has released 600 political prisoners, most of whom are members of the Islamic Al-Nahda movement. President Zine al-Abidine Ben Ali also pardons an estimated 2,000 and reduces the sentence of about 1,000 other prisoners to mark his twelfth year in power.
- On November 17, President Ben Ali names Mohamed Ghannouchi to replace Hame Karoui as prime minister.

December

- Tunisian authorities allow a gathering of students and labor activists to demonstrate against the government.

ANALYSIS OF EVENTS: 1999

BUSINESS AND THE ECONOMY

In July, the International Monetary Fund (IMF) reported that Tunisia's economy was expected to grow by 6.5 percent in 1999. Inflation was expected to hold at 3 percent. The report said increased tourism and a good crop year were the main factors in the economy's growth. Tunisia has been trying to live up to the expectations of being a member of the Euro-Mediterranean Partnership with the European Union, which requires the North African nation to dismantle trade tariffs. In October, the government said its trade deficit of about $2.189 billion dinars ($1.87 billion) for the first nine months of the year was slightly higher than the same period in 1998. Tunisia offset its trade deficit with higher food exports, especially with a surge in exports of olive oil. Food exports rose by 64.4 percent over 1998.

In August, the Tunisian government reported that tourism revenues would reach a record by the end of 1999, at about $1.9 billion dinars ($1.6 billion). The previous record was $1.7 billion dinars in 1998. Tourist arrivals during the first half of 1999 were 2.134 million, compared to 1.944 million during the same period in 1998. The number of Tunisian nationals returning to their country for a visit also increased. Tunisian officials said the good economic situation in Europe coupled with political turmoil in the Balkans and the Middle East have boosted tourism to Tunisia. Tourism is Tunisia's largest net foreign currency earner. In August, the government slightly raised the prices of subsidized food, including pasta, couscous, Semolina, and bread. The government also raised the minimum wage.

GOVERNMENT AND POLITICS

In October, Tunisia held its first multi-party presidential elections since independence from France in 1956, moving a little closer to democracy, but not far enough, opponents argued. The outcome was predictable, with President Zine al-Abidine Ben Ali winning a third straight term in office with more than 99 percent of the vote. His two politically moderate opponents gained slightly over a half percent of the vote, with a voter turnout of more than 91 percent, according to government reports.

Ben Ali ran against Behlag Amor, 62, head of the leftist Popular Unity Party, and Abderrahmane Tlili, 56, head of the Unionist Democratic Union, an Arab nationalist party. Both candidates said they didn't expect to win against the powerful Ben Ali. They ran, they said, to help promote pluralism. Parliamentary elections underscored Ben Ali's hold on power. The President's Democratic Constitutional Rally party took 148 of the 182 seats in parliament, while six small opposition parties could not fill a 20 percent quota of seats allotted to them under a recent change in electoral law. But they managed to increase their presence to 34 seats, up from 19. Tunisian Islamic activists held hunger strikes in front of the Tunisian Embassy in London, and some Arab countries to protest against the presidential elections. The strikers protested the exclusion of Islamic candidates from the elections, and called for the release of political prisoners.

During the two-week campaigning period, Ben Ali was portrayed as the ''citizen president'' in the tightly controlled press. He rose to power in 1987, after toppling President-for-life Habib Bourguiba in a bloodless coup. Bourguiba, the nation's founder, built Tunisia into a modern, independent Arab nation respected by the West and its Arab and African neighbors. Ben Ali, a moderate Muslim, continued his legacy, clamping down on religious fundamentalists, and moving the nation toward a market economy.

Laws approved by Ben Ali have given Tunisian women a remarkable level of freedom unseen in other Arab nations. Ben Ali has not been so kind to political opponents. Human rights groups have assailed him for his human rights records, accusing his government of harassing, torturing and imprisoning his opponents. In March, First Lady Hillary Clinton met Ben Ali for an hour to discuss human rights issues. As he had done in the past, Ben Ali acknowledged ''shortcomings'' in his government's record on political freedom, but he promised to address the issue.

CULTURE AND SOCIETY

U.S. First Lady Hillary Clinton visited Tunisia in March to celebrate its record on women's rights. To underscore the importance of what has happened in Tunisia, Mrs. Clinton, speaking to a group of professional Tunisian women, criticized religious fundamentalism that leads to poor treatment of women in Algeria and Afghanistan. She said Islamic countries like Tunisia could serve as an example to other nations where women lack fundamental rights. According to some statistics, women make up 25 percent of the workforce. Former President Bourguiba initiated legislation that helped improve women's rights in Tunisia. Ben Ali continued his efforts, approving legislation that guaranteed women equal rights. By 1999, women held political office, and headed small businesses and large corporations. The National Chamber of Women Heads of Enterprises reported 2,000 members in 1999, but some estimates indicate there are more than 5,000 women business owners.

DIRECTORY

CENTRAL GOVERNMENT

Head of State

President

Zine al-Abidine Ben Ali, Office of the President, Palais de la Presidence, TN-2016 Tunis, Tunisia

Prime Minister

Mohamed Ghannouchi, Office of the Prime Minister, Place du Gouvernement, La Kasbah, TN-1001 Tunis, Tunisia

Ministers

Minister of Agriculture
Sakok Rabah, Ministry of Agriculture

Minister of Commerce
Mondher Zenaidi, Ministry of Commerce

Minister of Communications
Ahmed Friaa, Ministry of Communications

Minister of Culture
Abdelbaki Hermassi, Ministry of Culture

Minister of Development
Abdellatif Saddam, Ministry of Development

Minister of Education
Ridha Ferchiou, Ministry of Education

Minister of Environment and Land Management
Faiza Kefi, Ministry of Environment and Land Management

Minister of Equipment and Housing
Slaheddine Belaid, Ministry of Equipment and Housing

Minister of Finance
Mohamed El Jeri, Ministry of Finance

Minister of Foreign Affairs
Said Ben Mustapha, Ministry of Foreign Affairs

Minister of Higher Education
Dali Jazi, Ministry of Higher Education

Minister of Industry
Moncef Ben Abdallah, Ministry of Industry

Minister of Interior
Ali Chaouch, Ministry of Interior

Minister of International Cooperation and Foreign Investment
Mohamed Ghannouchi, Ministry of International Cooperation and Foreign Investment

Minister of Justice
Abdallah Kallel, Ministry of Justice

Minister of National Defense
Habib Ben Yahia, Ministry of National Defense

Minister of Public Health
Hedi Mhenni, Ministry of Public Health

Minister of Religious Affairs
Ali Chabbi, Ministry of Religious Affairs

Minister of Social Affairs
Chedly Neffati, Ministry of Social Affairs

Minister of State Property and Real Estate Affairs
Mustapha Bouaziz, Ministry of State Property and Real Estate Affairs

Minister of Tourism and Handicrafts
Slaheddine Maaoui, Ministry of Tourism and Handicrafts

Minister of Transportation
Hassine Chouk, Ministry of Transportation

Minister of Vocational Training and Employment
Moncer Rouissi, Ministry of Vocational Training and Employment

Minister of Youth and Childhood
Raouf Najjar, Ministry of Youth and Childhood

POLITICAL ORGANIZATIONS

Mouvement At-Tajdid (At-Tajdid Movement)

Mouvement des Democrates Socialistes (Movement of Socialist Democrats or MDS)

Parti Social Liberal-PSL (Liberal Social Party)

Parti de l'Unite Populaire-PUP (Popular Unity Party)

Rassemblement Constitutionnel Democratique-RCD (Democratic Constitutional Rally)

NAME: Zine al-Abidine Ben Ali

Rassemblement Socialiste Progressiste-RSP (Progressive Socialist Rally)

Renewal Movement

Social Party for Progress

Socialist Progressive Party

Union Democratique Unioniste-UDU (Unionist Democratic Union)

DIPLOMATIC REPRESENTATION

Embassies in Tunisia

Canada
3 Rue de Senegal, Place d'Afrique, Belvedere, Tunis, Tunisia
PHONE: +216 (1) 286577
FAX: +216 (1) 792371

France
Place de l'Independence, Avenue Habib Bourguiba, Tunis, Tunisia
PHONE: +216 (1) 245700

Germany
1 Rue el Hamra, Tunis, Tunisia
PHONE: +216 (1) 786455

United Arab Emirates
9 Ashtrat Street, Tunis, Tunisia
PHONE: +216 782737
FAX: +216 783507

United Kingdom
5 Place do la Victoire, Tunis, Tunisia
PHONE: +216 (1) 2455561
FAX: +216 (1) 345877
TITLE: Ambassador
NAME: Ivor Rawlinson Obe

United States
144 Avenue de la Liberte, TN-1002 Tunis-Belvedere, Tunisia
PHONE: +216 (1) 782566
FAX: +216 (1) 789719
TITLE: Ambassador
NAME: Robin Lynn Raphel

JUDICIAL SYSTEM

Court of Cassation

FURTHER READING

Articles

''Five Nations, Including Tunisia, selected to U.N. Security Council.'' Reuters, 17 October 1999.

''Mrs. Clinton Decries Islamic Violence Against Women.'' Associated Press, 26 March 1999.

''Police Investigate Fatal Riots During Tunisian Soccer Match.'' Associated Press, 16 June 1999.

''Tunisian President Admits Human Rights Shortcomings.'' Associated Press, 27 March 1999.

''Tunisian President Wins Third Term.'' Associated Press, 25 October 1999.

TUNISIA: STATISTICAL DATA

For sources and notes see "Sources of Statistics" in the front of each volume.

GEOGRAPHY

Geography (1)

Area:

Total: 163,610 sq km.

Land: 155,360 sq km.

Water: 8,250 sq km.

Area—comparative: slightly larger than Georgia.

Land boundaries:

Total: 1,424 km.

Border countries: Algeria 965 km, Libya 459 km.

Coastline: 1,148 km.

Climate: temperate in north with mild, rainy winters and hot, dry summers; desert in south.

Terrain: mountains in north; hot, dry central plain; semiarid south merges into the Sahara.

Natural resources: petroleum, phosphates, iron ore, lead, zinc, salt.

Land use:

Arable land: 19%

Permanent crops: 13%

Permanent pastures: 20%

Forests and woodland: 4%

Other: 44% (1993 est.)

HUMAN FACTORS

Health Personnel (3)

Total health expenditure as a percentage of GDP, 1990-1997[a]

Public sector	3.0
Private sector	2.8
Total[b]	5.9

Continued on next page.

Demographics (2A)

	1990	1995	1998	2000	2010	2020	2030	2040	2050
Population	8,206.8	8,971.6	9,380.4	9,645.5	10,959.6	12,215.7	13,235.2	13,972.4	14,398.9
Net migration rate (per 1,000 population)	NA	NA	NA	NA	NA	NA	NA	NA	NA
Births	NA	NA	NA	NA	NA	NA	NA	NA	NA
Deaths	NA	NA	NA	NA	NA	NA	NA	NA	NA
Life expectancy - males	69.8	71.1	71.7	72.2	74.1	75.7	76.9	77.9	78.7
Life expectancy - females	72.3	73.8	74.6	75.1	77.6	79.6	81.2	82.5	83.5
Birth rate (per 1,000)	25.4	20.8	20.1	19.4	17.5	15.4	13.4	12.6	12.1
Death rate (per 1,000)	5.5	5.1	5.1	5.0	5.2	5.5	6.4	8.1	10.0
Women of reproductive age (15-49 yrs.)	2,024.3	2,312.6	2,509.2	2,635.2	3,064.4	3,191.8	3,194.7	3,076.5	3,018.9
of which are currently married	NA	NA	NA	NA	NA	NA	NA	NA	NA
Fertility rate	3.3	2.6	2.4	2.3	2.1	2.0	2.0	2.0	2.0

Except as noted, values for vital statistics are in thousands; life expectancy is in years.

Health Personnel (3) cont.

Health expenditure per capita in U.S. dollars, 1990-1997[a]

Purchasing power parity 260

Total 99

Availability of health care facilities per 100,000 people

Hospital beds 1990-1997[a] 180

Doctors 1993[c] 67

Nurses 1993[c] 283

Health Indicators (4)

Life expectancy at birth

1980 62

1997 70

Daily per capita supply of calories (1996) 3,250

Total fertility rate births per woman (1997) 2.8

Maternal mortality ratio per 100,000 live births (1990-97) . . . . .170[c]

Safe water % of population with access (1995) 90

Sanitation % of population with access (1995) 80

Consumption of iodized salt % of households (1992-98)[a] 98

Smoking prevalence

Male % of adults (1985-95)[a] . . . .

Female % of adults (1985-95)[a] . . . .

Tuberculosis incidence per 100,000 people (1997) . . .40

Adult HIV prevalence % of population ages 15-49 (1997) 0.04

Infants and Malnutrition (5)

Under-5 mortality rate (1997) 33

% of infants with low birthweight (1990-97) 8

Births attended by skilled health staff % of total[a] . . .NA

% fully immunized (1995-97)

TB 93

DPT 96

Polio 96

Measles 92

Prevalence of child malnutrition under age 5 (1992-97)[b] 9

Ethnic Division (6)

Arab 98%

European 1%

Jewish and other 1%

Religions (7)

Muslim 98%

Christian 1%

Jewish and other 1%

Languages (8)

Arabic (official and one of the languages of commerce), French (commerce).

EDUCATION

Public Education Expenditures (9)

Public expenditure on education (% of GNP)

1980 . . . . .5.4

1996 . . . . .6.7

Expenditure per student

Primary % of GNP per capita

1980 . . . . .11.8

1996 . . . . .15.3

Secondary % of GNP per capita

1980 . . . . .37.8

1996 . . . .

Tertiary % of GNP per capita

1980 . . . . .194.7

1996 . . . . .79.3

Expenditure on teaching materials

Primary % of total for level (1996) 2.0

Secondary % of total for level (1996) . . . .

Primary pupil-teacher ratio per teacher (1996) 24

Duration of primary education years (1995) 6

Educational Attainment (10)

Age group (1984) . . . . .25+

Total population 2,714,100

Highest level attained (%)

No schooling 66.3

First level

Not completed 18.9

Completed . . . . .NA

Entered second level

S-1 . . . . .12.0

S-2 . . . . .NA

Postsecondary 2.8

Literacy Rates (11A)

In thousands and percent[1]	1990	1995	2000	2010
Illiterate population (15+ yrs.)	2,005	1,930	1,827	1,502
Literacy rate - total adult pop. (%)	60.2	66.7	72.3	81.6
Literacy rate - males (%)	73.1	78.6	83.2	90.5
Literacy rate - females (%)	47.2	54.6	61.3	72.6

GOVERNMENT & LAW

Political Parties (12)

Chamber of Deputies	% of seats
Constitutional Democratic Rally Party (RCD)	97.7
Movement of Democratic Socialists (MDS)	1.0
Others	1.3

Government Budget (13A)

Year: 1996

Total Expenditures: 6,208.3 Millions of Dinars

Expenditures as a percentage of the total by function:

General public services and public order	14.20
Defense	6.23
Education	18.68
Health	6.87
Social Security and Welfare	16.68
Housing and community amenities	5.15
Recreational, cultural, and religious affairs	2.70
Fuel and energy	-
Agriculture, forestry, fishing, and hunting	8.23
Mining, manufacturing, and construction	.68
Transportation and communication	2.26
Other economic affairs and services	6.17

Military Affairs (14B)

	1990	1991	1992	1993	1994	1995
Military expenditures						
Current dollars (mil.)	342	364	358	384	392	345
1995 constant dollars (mil.)	393	402	386	402	402	345
Armed forces (000)	35	35	35	35	35	35
Gross national product (GNP)						
Current dollars (mil.)	12,460	13,380	14,820	15,410	16,220	17,350
1995 constant dollars (mil.)	14,320	14,780	15,940	16,150	16,630	17,350
Central government expenditures (CGE)						
1995 constant dollars (mil.)	5,454	5,142	5,297	6,393[e]	NA	NA
People (mil.)	8.0	8.2	8.4	8.5	8.7	8.9
Military expenditure as % of GNP	2.7	2.7	2.4	2.5	2.4	2.0
Military expenditure as % of CGE	7.2	7.8	7.3	6.3	NA	NA
Military expenditure per capita (1995 $)	49	49	46	47	46	39
Armed forces per 1,000 people (soldiers)	4.3	4.3	4.2	4.1	4.0	4.0
GNP per capita (1995 $)	1,780	1,801	1,904	1,893	1,913	1,959
Arms imports[6]						
Current dollars (mil.)	40	30	20	20	50	40
1995 constant dollars (mil.)	46	33	22	21	51	40
Arms exports[6]						
Current dollars (mil.)	0	0	0	0	0	0
1995 constant dollars (mil.)	0	0	0	0	0	0
Total imports[7]						
Current dollars (mil.)	5,542	5,190	6,431	6,214	6,581	7,903
1995 constant dollars (mil.)	6,369	5,735	6,917	6,515	6,746	7,903
Total exports[7]						
Current dollars (mil.)	3,526	3,699	4,019	3,802	4,657	5,475
1995 constant dollars (mil.)	4,052	4,087	4,323	3,986	4,774	5,475
Arms as percent of total imports[8]	.7	.6	.3	.3	.8	.5
Arms as percent of total exports[8]	0	0	0	0	0	0

Crime (15)

Crime rate (for 1997)

Crimes reported	128,400
Total persons convicted	108,800
Crimes per 100,000 population	1,450

Persons responsible for offenses

Total number of suspects	128,200
Total number of female suspects	12,800
Total number of juvenile suspects	11,100

LABOR FORCE

Labor Force (16)

Total (million)	2.917
Services	55%
Industry	23%
Agriculture	22%

Shortage of skilled labor. Data for 1993 est. Percent distribution for 1995 est.

Unemployment Rate (17)

15% (1997 est.)

PRODUCTION SECTOR

Electric Energy (18)

Capacity	1.414 million kW (1995)
Production	6.165 billion kWh (1995)
Consumption per capita	696 kWh (1995)

Transportation (19)

Highways:

total: 23,100 km

paved: 18,226 km

unpaved: 4,874 km (1996 est.)

Pipelines: crude oil 797 km; petroleum products 86 km; natural gas 742 km

Merchant marine:

total: 20 ships (1,000 GRT or over) totaling 157,475 GRT/165,922 DWT ships by type: bulk 5, cargo 5, chemical tanker 2, liquefied gas tanker 1, oil tanker 1, roll-on/roll-off cargo 2, short-sea passenger 3, specialized tanker 1 (1997 est.)

Airports: 32 (1997 est.)

Airports—with paved runways:

total: 15

over 3,047 m: 3

2,438 to 3,047 m: 6

1,524 to 2,437 m: 3

914 to 1,523 m: 3 (1997 est.)

Airports—with unpaved runways:

total: 17

1,524 to 2,437 m: 2

914 to 1,523 m: 8

under 914 m: 7 (1997 est.)

Top Agricultural Products (20)

Olives, dates, oranges, almonds, grain, sugar beets, grapes; poultry, beef, dairy products.

MANUFACTURING SECTOR

GDP & Manufacturing Summary (21)

Detailed value added figures are listed by both International Standard Industry Code (ISIC) and product title.

	1980	1985	1990	1994
than half a unit.				
GDP ($-1990 mil.)[1]	8,720	10,709	12,513	14,658
Per capita ($-1990)[1]	1,366	1,475	1,549	1,678
Manufacturing share (%) (current prices)[1]	13.6	13.5	16.9	*16.9*
Manufacturing				
Value added ($-1990 mil.)[1]	1,061	1,486	1,873	2,178
Industrial production index	100	133	157	167
Value added ($ mil.)	939	*1,147*	3,305	*3,818*
Gross output ($ mil.)	3,579	*4,354*	10,611	*12,287*
Employment (000)	125	*166*	*217*	259
Profitability (% of gross output)				
Intermediate input (%)	74	*74*	69	*69*
Wages and salaries inc. supplements (%)	12	*11*	9	*11*
Gross operating surplus	14	*15*	22	*20*
Productivity ($)				
Gross output per worker	28,669	*26,156*	*48,894*	*47,386*
Value added per worker	7,525	*7,106*	*16,463*	*14,723*
Average wage (inc. supplements)	3,499	*2,963*	*4,627*	*5,030*

	1980	1985	1990	1994
Value added ($ mil.)				
311/2 Food products	96	*101*	315	*395*
313 Beverages	49	*55*	92	*129*
314 Tobacco products	22	*40*	207	*257*
321 Textiles	55	*77*	196	*278*
322 Wearing apparel	92	*122*	380	*505*
323 Leather and fur products	6	*9*	*29*	*42*
324 Footwear	21	*24*	*65*	*82*
331 Wood and wood products	12	*18*	*56*	*64*
332 Furniture and fixtures	13	22	*97*	*140*
341 Paper and paper products	24	*21*	*44*	*59*
342 Printing and publishing	17	*15*	*24*	*29*
351 Industrial chemicals	*57*	*55*	39	*104*
352 Other chemical products	*81*	*57*	95	*126*
353 Petroleum refineries	13	*99*	872[d]	*688*
354 Miscellaneous petroleum and coal products	—	NA	[d]	NA
355 Rubber products	8	*11*	28	*36*
356 Plastic products	18	*21*	35	*50*
361 Pottery, china and earthenware	11	*14*	67	*92*
362 Glass and glass products	7	*6*	16	*19*
369 Other non-metal mineral products	156	*161*	263	*236*
371 Iron and steel	45	*60*	69	*66*
372 Non-ferrous metals	8	*5*	*5*	*5*
381 Metal products	53	*69*	*103*	*125*
382 Non-electrical machinery	2	*3*	13	*16*
383 Electrical machinery	35	*39*	101	*150*
384 Transport equipment	30	*33*	60	*81*
385 Professional and scientific equipment	1	*1*	*1*	*1*
390 Other manufacturing industries	5	*8*	31	*43*

FINANCE, ECONOMICS, & TRADE

Economic Indicators (22)

National product: GDP—purchasing power parity—$56.5 billion (1997 est.)

National product real growth rate: 5.6% (1997 est.)

National product per capita: $6,100 (1997 est.)

Inflation rate—consumer price index: 4.6% (1997 est.)

Exchange Rates (24)

Exchange rates:

Tunisian dinars (TD) per US$1

January 1998	1.1612
1997	1.1059
1996	0.9734
1995	0.9458
1994	1.0116
1993	1.0037

Balance of Payments (23)

	1992	1993	1994	1995	1996
Exports of goods (f.o.b.)	4,041	3,746	4,643	5,470	5,519
Imports of goods (f.o.b.)	−6,078	−5,810	−6,210	−7,459	−7,323
Trade balance	−2,037	−2,064	−1,567	−1,989	−1,804
Services - debits	−1,804	−1,985	−2,108	−2,187	−2,289
Services - credits	2,073	2,113	2,338	2,628	2,698
Private transfers (net)	79	100	102	59	22
Government transfers (net)	586	513	697	752	838
Overall balance	−1,103	−1,323	−538	−737	−537

Top Import Origins (25)

$7.4 billion (c.i.f., 1997 est.) Data are for 1996.

Origins	%
European Union	80
North Africa	5.5
Asia	5.5
United States	5

Top Export Destinations (26)

$5.6 billion (f.o.b., 1997 est.).

Destinations	%
Netherlands	75.7
Germany	1.2
Portugal	1.1

Economic Aid (27)

Recipient: ODA, $221 million (1993).

Import Export Commodities (28)

Import Commodities	Export Commodities
Industrial goods and equipment 57%	Hydrocarbons
Hydrocarbons 13%	Textiles
Food 12%	Agricultural products
Consumer goods	Phosphates and chemicals

TURKEY

CAPITAL: Ankara.

FLAG: The national flag consists of a white crescent (open toward the fly) and a white star on a red field.

ANTHEM: *Istiklâl Mar şi (March of Independence).*

MONETARY UNIT: The Turkish lira (TL) is a paper currency of 100 kuruş. There are coins of 1, 5, 10, 20, 25, 50, and 100 liras, and notes of 5,000, 10,000, 20,000, 50,000, 100,000, 250,000, and 500,000 liras. TL1 = $0.000045 (or $1 = TL22,159.8).

WEIGHTS AND MEASURES: The metric system is the legal standard.

HOLIDAYS: New Year's Day, 1 January; National Sovereignty and Children's Day, 23 April; Spring Day, 1 May; Youth and Sports Day, 19 May; Victory Day, 30 August; Independence Day (Anniversary of the Republic), 29 October. Movable religious holidays include Şeker Bayrami (three days) and Kurban Bayrami (four days).

TIME: 3 PM = noon GMT.

LOCATION AND SIZE: The Republic of Turkey consists of Asia Minor, the small area of eastern Turkey in Europe, and a few offshore islands in the Aegean Sea. It has a total area of 780,580 square kilometers (301,384 square miles), which is slightly larger than the state of Texas. Turkey has a total boundary length of 9,827 kilometers (6,106 miles). Its capital city, Ankara, is located in the northwest part of the country.

CLIMATE: The mean temperature range on Turkey's southern and Aegean coasts is 17–20°C (63–68°F), and the annual rainfall ranges from 71 to 109 centimeters (28–43 inches). The Black Sea coast is also relatively mild (14–15°C (57–59°F) and very moist, with 71–249 centimeters (28–98 inches) of rainfall. On the central Anatolian plateau the average annual temperature is 8–12°C (46–54°F), and annual precipitation is 30–75 centimeters (12–30 inches). The eastern third of Turkey is colder (4–9°C (39–48°F), and rainfall averages 41–51 centimeters (16–20 inches).

Republic of Turkey
Türkiye Cumhuriyeti

INTRODUCTORY SURVEY

RECENT HISTORY

Turkey remained neutral during most of World War II (1939–45), but became an early member of the United Nations, the North Atlantic Treaty Organization (NATO), and the Central Treaty Organization, or CENTO (Baghdad Pact).

During the late 1970s, acts of violence by political groups of the extreme left and right, coupled with economic decline, threatened the stability of Turkey's fragile democracy. A five-man military National Security Council (NSC) took power in a bloodless coup in September 1980, imposed martial law, and arrested thousands of suspected terrorists. In a national referendum in 1982, Turkish voters overwhelmingly approved a new constitution. Jurgut Özal, was elected president in 1989. Despite his efforts, Turkey's attempt to join the European Union was deferred. During the Gulf War (1991) Turkey supported the allies against Iraq; in compensation, it received increased Western aid worth $300 million.

When Tansu Ciller became Turkey's first female prime minister in 1993, she faced some major problems: high inflation and unemployment; excessive government regulations; Kurdish rebellions in eastern Turkey; and a challenge to Turkey's secular nationalism from politically militant Islamic groups.

Between 1984 and 1995, fighting between the government and the Kurdistan Workers Party (PKK) resulted in the death of more than 14,000

people. In March 1995, 35,000 Turkish troops, supported by tanks and aircraft, pursued the PKK into Iraq. Later that year, Turkey and Greece narrowly avoided a fight over a group of uninhabited islands in the Aegean Sea.

The Kurdish and Greek problems were complicated by political instability that lasted until the spring of 1996 when Necmettin Erbakan became modern Turkey's first conservative Islamic prime minister. In 1997 Turkey's military leaders threatened to overthrow the government if it did not return to secular policies. Erbakan resigned, and in early 1998, Turkey's highest court ordered that he and four members of parliament be banned from political office for five years.

Following elections held in April 1999, a coalition government was formed including the Democratic Left Party (DSP), the Motherland Party (ANAP), and the Nationalist Action Party (MHP). The coalition has a solid majority in the National Assembly with 352 out of a total of 550 seats.

GOVERNMENT

The constitution, ratified in November 1982, declaring Turkey a democratic and secular republic, vests executive powers in the president of the republic and the Council of Ministers. Legislative functions are delegated to the single-chamber National Assembly, consisting of 550 members elected for five-year terms.

The chief administrative official in each of Turkey's 73 provinces is the provincial governor.

Judiciary

There are four branches of courts: civil, administrative, military, and constitutional. Civil courts are specialized into five sections: civil, enforcement, criminal, commercial, and labor. Decisions of civil courts may be appealed to a High Court of Appeals in Ankara.

The constitution guarantees defendants the right to a public trial. There is no jury system.

Political Parties

Major parties include the Welfare Party; the True Path Party; the Motherland Party: the Democratic Left Party; and Republican People's Party.

Outside the established political system are the Kurdistan Workers Party (PKK) and other smaller separatist parties that have been banned.

DEFENSE

The total armed forces strength in 1993 was 638,000 (including 528,000 draftees), plus 378,700 reserves. The defense budget for 1996 was $5.7 billion. An estimated 30,000 Turkish soldiers were stationed on Cyprus.

ECONOMIC AFFAIRS

Between 1988 and 1993, the annual rate of inflation was between 60–70%. Since 1980, Turkey has deliberately pursued a deflationary policy, allowing the international exchange rate of the lira to fluctuate daily. In 1994, when the lira was devalued, leading to even higher inflation, structural problems in the economy and budget deficits plunged the Turkish economy into its worst recession since World War II. In 1996 the gross national product (GNP) grew by 7% in 1997. Despite the growth, basic structural problems remain, and the economy remains in a boom-bust cycle.

Public Finance

The government responded to the 1994 financial crisis with an austerity program that succeeded in reducing inflation but sent the economy into recession. In 1995 expenditures again exceeded revenues, forcing the government to incur increasing amounts of debt to fund current expenses and support state enterprises. In 1995, government revenues were estimated at approximately $30.2 billion and expenditures $35 billion. External debt totaled $73.8 billion, approximately 58% of which was financed.

Inflation declined to 70% in 1998, down from 99% in 1997, but the public sector fiscal deficit probably remained near 10% of GDP. Interest payments accounted for 42% of central government spending in 1998. An inefficient tax system and huge black market (estimated at 25–50% of the economy) further hampered efforts to increase revenues.

Income

In 1998, Turkey's gross national product (GNP) was $200.5 billion, or about $3160 per person. For the period 1985–95, the average inflation rate was about 64.6%, resulting in a real growth rate in per person GNP of 2.2%.

Industry

Textiles are the largest industrial segment in Turkey after petroleum refineries. Major industrial complexes include a government-owned iron and steel mill. The sugar beet industry ranks first among food-processing enterprises. Manufacturing accounted for 25.9% of the domestic economy in 1995.

Banking and Finance

The Central Bank of the Republic of Turkey has 25 domestic branches and foreign branch offices in New York, London, Frankfurt, and Zurich.

In 1997, Turkey had 69 banks, of which 19 were foreign-owned. Some 50% of total bank assets are concentrated in four banks. The Industrial Development Bank of Turkey, a private bank, stimulates growth of industrial development and channels the flow of capital into the private industrial sector for both short- and long-range development programs.

Turkey's securities exchange is located in Istanbul. With few exceptions, trading is in government bonds.

Economic Development

Long-term economic programs adopted in 1991 and 1994 were designed to reform social security and subsidy programs, implement tax reforms and improve tax administration, and restructure state enterprises, transferring certain inefficient ones to the private sector. The government also delayed several ambitious development proposals, when new foreign credits were not available. By 1996, these plans had reduced the government's role in the economy, but huge budget deficits continued to plague the economy.

In 1999 Turkey was hit by a massive earthquake that led to more than 15,000 deaths and more than 20,000 wounded. Substantial international assistance in the range of $3–4 billion will be needed to help Turkey recover from this devastating blow.

SOCIAL WELFARE

Social security benefits cover industrial accidents and disease, old age, sickness, disability, and maternity insurance. In some localities, the social insurance organization operates its own hospitals and other facilities.

The Civil Code bans sex-based privileges, but stipulates the male as the legal head of the household.

The constitution does not recognize the Kurds as a national, racial, or ethnic minority, but in 1991 the use of the Kurdish language was legalized for non-political purposes.

Healthcare

Free medical treatment, given at state health centers, is provided by the state to any Turkish citizen who demonstrates financial need. Malaria, cholera, and trachoma have been effectively controlled by large-scale public preventive measures. Average life expectancy is 69 years.

Housing

From 1981–85, 305,890 new residential buildings containing 929,104 apartments were completed; virtually all these apartments had electricity, piped water, kitchens, and baths. As of 1985, 71% of all housing units were detached houses, 23%, apartments, and 6%, squatters' houses.

EDUCATION

In 1995 the illiteracy rate was about 8.3% for males and 27.6% for females. Primary, secondary, and much of higher education is free. Education is compulsory for children aged 6 to 14 or until graduation from primary school (grade 5). However, only about 56% of the children of primary school age attend school.

1999 KEY EVENTS TIMELINE

January

- Bulent Ecevit forms a coalition government following the resignation of Mesut Yilmaz because of a corruption scandal.
- Ecevit is sworn in as Prime Minister. Elections are scheduled for April 1999.

February

- Ecevit orders repression of Islamic radicals.
- Abdullah Ocalan (head of Kurdish independence party) is arrested.
- Kurds protest Ocalan's arrest.
- Turkish forces bomb Kurdish camps in northern Iraq.
- Many military clashes with PKK army.

March

- European Union opposes Turkey's entry because of fears of Islamic control of Turkish government.
- Pro-Islamic Party tries to un-seat Ecevit.

April

- Turkey publishes rules for admission to Ocalan's trial.
- Ecevit wins parliamentary election.
- PKK rebels continue fighting the military.

May

- Representatives of Turkey and Greece have a meeting about geological formations in the Aegean. Ownership of these rock formations is contested.

June

- EU criticizes Turkey of human rights violations in its capture of Ocalan.
- Ecevit reiterates his commitment to a secular government. He says Turkey will continue to fight terrorism.
- Turkey states its commitment to the occupation of northern Cyprus.
- Ocalan is sentenced to death. An appeal is to follow.

August

- A huge earthquake (7.4 R) strikes 60 miles east of Istanbul, killing at least 15,000 people.
- Greece is first to come to Turkey's aid.
- A probe into the construction industry reveals many buildings have inferior materials that cannot withstand earthquakes.

September

- Turkey sends aid to Greece after an earthquake kills many in northern Athens.
- Governments of Greece and Turkey agree to meet to discuss Cyprus.

October

- Ecevit comes to Washington, D.C. to meet with President Clinton to resolve Cyprus issue.
- Italy offers amnesty to Ocalan although he is in prison in Turkey.

November

- U.S. president Bill Clinton stops on his nine-day tour of southeastern Europe to visit Turkey, touring Christian sites, comforting earthquake victims, signing pipeline agreements, and meeting 50 other world leaders, including Russia's President Boris Yeltsin, at a summit in Istanbul.

- The Turkish appeals court considering the case of Abdullah Ocalan, leader of the outlawed Kurdistan Workers' Party, decides that the verdict handed down in June—execution for treason—should be upheld.

December

- The European Union reverses its 1997 vote and accepts Turkey as a candidate for membership.
- A Russian oil tanker breaks apart in seas near Turkey, causing an oil spill that experts estimate may take 40 years to clean up.
- Turkey dismisses a final appeal to commute the death sentence of Kurdish rebel leader, Abdullah Ocalan, believed to be responsible for the deaths of 30,000 people during the Kurdish drive for independence. The Turkish Parliament must act on whether or not to carry out the execution, but the European Union is pressuring Turkey to give up the death penalty.
- Hundreds participate in nationwide demonstrations against a government ban on university women wearing headscarves. The secular government of Turkey asserts that the wearing of headscarves is a political Islamic statement, but the protesters contend that the issue is one of freedom of expression. Police detain about 400 protestors.

ANALYSIS OF EVENTS: 1999

GOVERNMENT AND POLITICS

The political leaders of the Republic of Turkey claim to be preparing to enter the modern world with a new economic, political, and humanitarian agenda. Although its most populous city, Istanbul, straddles the European and Asian continents, Turkey has retained, for the most part, an Asian identity. Its almost entirely Islamic population was wrested from the grasp of a religious government by Kemal Mustafa Ataturk in 1923. Since then, in fits and starts, Turkey has struggled to enter the European world. Although Turkey's capital city, Ankara, lies well inside Asia, Istanbul is clearly on the cutting edge of the new Turkey. The year 1999 has been cataclysmic for the Turkish Republic. The country has undergone another tumultuous election, the Kurdish people have seen their leader Abdullah Ocalan arrested, the occupation of Cyprus continues to loom as a threat to war, and a disastrous earthquake killed over 15,000 citizens. The country is still reeling from these events.

As early as January 1999 the government went into a tailspin because of alleged corruption of Prime Minister Mesut Yilmaz and former government official Tansu Ciller. Bulent Ecevit took the reins of government until elections could be held in April. Although a secularist and a left-leaning politician, Ecevit was well respected for his honesty even among his religious opponents. Although Turkey has often given the impression of being on the brink of returning to its religious roots, Ecevit was able to win the April election and used that victory to assure the western world that Turkey has entered modern Europe for good.

Ecevit had three problems looming ahead. With the arrest of Ocalan, the leader of the Kurdish independence party, Ecevit was being criticized with regard to the human rights issue. The Kurds, a long-suffering separatist people, have been ill treated by the Turkish government which has pursued a policy of forced-assimilation. The government forbade the teaching of the Kurdish language in schools and has long been at war with Ocalan, leader of the Kurdish separatist movement. With Ocalan's arrest, conviction, and death sentence, members of the European community condemned Turkey for human rights violations and refused Turkey's bid for entry into the European Union.

Ecevit has maintained that Ocalan is nothing but a common terrorist who is responsible for countless murders and bombings. Ecevit also accused the European Union of discriminating against the first Muslim country requesting membership in the Union and declared that the Ocalan case is just an excuse to deny Turkey's entry. Ecevit has repeatedly stated that he is against capital punishment and is a strong supporter of human rights.

Even though Ocalan told his followers to eschew violence, he has committed numerous acts of terrorism and has caused the deaths of innocent people. The Turkish military was thus able to claim that its repression against Ocalan's Kurdistan Workers' Party (PKK) was a counter-terrorism campaign. The European Union wants Turkey to resolve this problem politically instead of by using military might.

Turkey views this problem as a threat to its political stability which has been based on the outlawing of religiously affiliated parties, thus denying them any voice in government. By assuming the posture of a secular state, Turkey hopes to convince the European Union that Turkey is not a Muslim state and is not associated with the political instability of its neighboring Arab states.

Ironically, the European Union views this forced imposition of a secular state as a human rights violation that excludes Turkey from membership in the EU. Turkey is caught in the contradiction of attempting to conduct a policy of religious toleration when some of its constituent religions are fundamentalist rather than rooted in the tradition of religious and political toleration. The longer Turkey is on the outside looking in, the more likely Islamic and nationalistic groups will find favor with the populace and try to control the government. Then Turkey would stand no chance of being admitted to the EU.

Strangely enough, the arrest of Ocalan, who was turned over to the Turkish government through the Greek embassy, has served to encourage ties of friendship between Turkey and its ancient enemy, Greece. The recent (August 1999) earthquake in Turkey also seems to have benefited relations between Greece and Turkey. Greece was one of the first countries to send economic and medical aid to Turkey following this disaster. One month later, when Greece was hit by a similar quake, Turkey responded by sending medical aid and rescuers. Now the two countries are seemingly on their way to resolving their dispute over their control of the island of Cyprus. Turkish Prime Minister Ecevit went to Washington, D.C. to confer with President Clinton about the resolution of the Cyprus problem.

One current problem in Turkey is the people's lack of confidence in the government because they believe that much of the earthquake damage was the result of corruption in the construction industry and in the governing authorities. The government controlled and gave job-bids to various construction companies that used inferior building material and combined cement with sand to produce large housing units that did not stand up in the earthquake. Turkish citizens are not only homeless, they are angry with the government for causing the needless deaths of so many people.

The housing crisis continues to rankle the people. Before the earthquake, the Turkish government cracked down on homeless people (mostly gypsies) who raised tent cities in Istanbul and other large cities. Now, with many people afraid to re-enter their damaged houses and apartments, the number of tent cities has increased dramatically and the government does not want to seem heartless in its prosecution of earthquake victims. Many more people (about 2000 per day) are migrating to large cities and the cities are hard pressed in finding housing and jobs for them. The Turkish people are at odds with a government that gained their trust by promising them a better, secular world, but now seems to have betrayed them.

An ironical twist of events has made the Ottoman past even more attractive to modern day Turks. Relatives of former sultans, who had formerly been shunned as not politically correct, have begun to emerge as the Turks have re-discovered their once glamorous heritage. Restaurants and hotels have revamped the designs and glamour of the Ottoman Empire as a tourist attraction and as an attempt to capitalize on the return to history. The government that would one time have frowned on this now welcomes these historic themes as tourist attractions.

DIRECTORY

CENTRAL GOVERNMENT

Head of State

President

Suleyman Demirel, Office of the President, Cankaya 06100, Ankara, Turkey
FAX: +90 (312) 4685026
E-MAIL: cankaya@tccb.gov.tr

Prime Minister

Bulent Ecevit, Office of the Prime Minister, Basbakanlik 06573, Ankara, Turkey
FAX: +90 (312) 4170476
E-MAIL: ddlbsl@tccb.gov.tr

Ministers

Minister of Agriculture and Rural Affairs

Husnu Yusuf, Ministry of Agriculture and Rural Affairs, Bakanliklar, Ankara, Turkey
PHONE: +90 (312) 4191677
FAX: +90 (312) 4177168

Minister of Culture

Mustafa Ystemihan Talay, Ministry of Culture, Ataturk Bulvari No.29, 06050 Opera, Ankara, Turkey

PHONE: +90 (312) 3090850
FAX: +90 (312) 3126473
E-MAIL: kultur@kultur.gov.tr

Minister of Energy and Natural Resources
Cumhur Ersumer, Ministry of Energy and Natural Resources, Konya Yolu, Ankara, Turkey
PHONE: +90 (312) 2135330
FAX: +90 (312) 2123816

Minister of the Environment
Fevzi Aytekyn, Ministry of the Environment, Eskipehir Yolu No.8, 06530 Lodmulu/Bilkent, Ankara, Turkey
PHONE: +90 (312) 2879963
FAX: +90 (312) 2852742
E-MAIL: cevre@cevre.gov.tr

Minister of Finance
Sumer Oral, Ministry of Finance, Dikmen Cad., Ankara, Turkey
PHONE: +90 (312) 4250018
FAX: +90 (312) 4250058
E-MAIL: bid@maliye.gov.tr

Minister of Foreign Affairs
Ysmail Cem, Ministry of Foreign Affairs, Balgat 06100, Ankara, Turkey
PHONE: +90 (312) 2872555
E-MAIL: webmaster@mfa.gov.tr

Minister of Forestry
Nami Cadan, Ministry of Forestry, Ataturk Bulvari, Ankara, Turkey
PHONE: +90 (312) 4176000

Minister of Health
Osman Durmup, Ministry of Health, Yenisehir, Ankara, Turkey
PHONE: +90 (312) 4312486
FAX: +90 (312) 4339885

Minister of Industry and Trade
Ahmet Kenan Tanrikulu, Ministry of Industry and Trade, Tandogan, Ankara, Turkey
PHONE: +90 (312) 4314866
FAX: +90 (312) 2304251

Minister of Interior
Sadettin Tantan, Ministry of Interior, Bakanliklar, Ankara, Turkey
PHONE: +90 (312) 4254080
FAX: +90 (312) 4181795

Minister of Justice
Hikmet Sami Turk, Ministry of Justice, Balanliklar, Ankara, Turkey
PHONE: +90 (312) 4191331
FAX: +90 (312) 4173954
E-MAIL: adalet.bakanligi@adelet.gov.tr

Minister of Labor and Social Security
Yapar Okuyan, Ministry of Labor and Social Security, Eskisehir Yolu No.42, Ankara, Turkey
PHONE: +90 (312) 4170727
FAX: +90 (312) 4179765

Minister of National Defense
Sebahattin Cakmakodlu, Ministry of National Defense
PHONE: +90 (312) 4254596
FAX: +90 (312) 4184737

Minister of National Education
Metin Bostanciodlu, Ministry of National Education
PHONE: +90 (312) 4255330
FAX: +90 (312) 4177027

Minister of Public Works and Housing
Koray Aydin, Ministry of Public Works and Housing, Vekaletler Cad. No.1, Ankara, Turkey
PHONE: +90 (312) 4255711
FAX: +90 (312) 4180406

Minister of Transportation
Enis Oksuz, Ministry of Transportation, Sok. No. 5 Emek, Ankara, Turkey
PHONE: +90 (312) 2124416
FAX: +90 (312) 2124930

Minister of Tourism
Erkan Mumcu, Ministry of Tourism, Ismet Inonu Bulvari No. 5, Ankara, Turkey
PHONE: +90 (312) 2128300
FAX: +90 (312) 2136887

POLITICAL ORGANIZATIONS

Anavatan Partisi-ANAP (Motherland Party)

13 Cad. No.3, Balgat, Ankara, Turkey
PHONE: +90 (312) 2865000
FAX: +90 (312) 2865019
TITLE: Chairman
NAME: Mesut Yilmaz

Buyuk Birlik Partisi-BBP (Grand Unity Party)

PHONE: +90 (312) 2548745
E-MAIL: bbpbim@hotmail.com
TITLE: Leader
NAME: Muhsin Yazicioglu

Cumhuriyet Halk Partisi-CHP (Republican People's Party)

Cevre Sok. No.28, Ankara, Turkey
PHONE: +90 (312) 4685969
FAX: +90 (312) 4685969
TITLE: Chairman
NAME: Hikmet Cetin

Demokratik Sol Partisi-DSP (Democratic Left Party)

Fevzi Cakmak Cad. No.17, Ankara, Turkey
PHONE: +90 (312) 2124950
FAX: +90 (312) 2124188
TITLE: Chairman
NAME: Bulent Ecevit

Demokrat Turkiye Partisi-DTP (Democratic Turkey Party)

E-MAIL: mektup@dtp.org.tr
TITLE: Leader
NAME: Husamettin Cindoruk

Dogru Yol Partisi-DYP (True Path Party)

Akay Cad. No.16, Ankara, Turkey
PHONE: +90 (312) 4172239
FAX: +90 (312) 4185657
TITLE: Chairwoman
NAME: Tanu Ciller

Milliyetci Hareket Partisi-MHP (Nationalist Movement Party)

Strazburg Cad. No.36, 06430 Sihhiye, Ankara, Turkey
PHONE: +90 (312) 3218700
FAX: +90 (312) 2311424
E-MAIL: mhp@mhp.org.tr
TITLE: Leader
NAME: Alpasian Turkes

Fazilet Partisi-FP (Virtue Party)

Pehit Danip Tunalygil Cd. No.3, 06570 Maltepe, Ankara, Turkey
PHONE: +90 (312) 2325151
FAX: +90 (312) 2325160
E-MAIL: mata@fp.org.tr
TITLE: Chairman
NAME: Ismail Alptekin

Ozgurluk ve Dayanisma Partisi-ODP (Liberty and Solidarity Party)

E-MAIL: istanil@odp.org.tr
TITLE: Leader
NAME: Ufuk Uras

Refah Partisi-LDP (Welfare Party)

Ziyabey Cad. No.11, Sok. No.24, Balgat, Ankara, Turkey
PHONE: +90 (312) 2873056
FAX: +90 (312) 2877465
TITLE: Chairman
NAME: Necmettin Erbakan

DIPLOMATIC REPRESENTATION

Embassies in Turkey

Afghanistan
Cinnah Cad. No.88, Cankaya, Ankara, Turkey
PHONE: +90 (312) 4381124
FAX: +90 (312) 4387745

Algeria
Pehit Erson Cad. No.42, Cankaya, Ankara, Turkey
PHONE: +90 (312) 4278700
FAX: +90 (312) 4268959

Argentina
Ugur Mumcu Cad. No.60/4, G.O.P., Ankara, Turkey
PHONE: +90 (312) 4462061
FAX: +90 (312) 4462063

Australia
Nenehatun Cad. No.83, G.O.P., Ankara, Turkey
PHONE: +90 (312) 4461180
FAX: +90 (312) 4461188
E-MAIL: ausemank@ibm.net

Austria
Ataturk Bulvari No.189, Kavaklidere, Ankara, Turkey
PHONE: +90 (312) 4190431
FAX: +90 (312) 4189454

Azerbaijan
Cemal Nadir Sok. No.20, Cankaya, Ankara, Turkey
PHONE: +90 (312) 4412620
FAX: +90 (312) 4412600

Bangladesh
Cinnah Cad. No.78/7-10, Cankaya, Ankara, Turkey
PHONE: +90 (312) 4392750
FAX: +90 (312) 4392408

Belarus
Han Sok. No.13/1-2, G.O.P., Ankara, Turkey
PHONE: +90 (312) 4463042

FAX: +90 (312) 4460150

Belgium
Mahatma Gandi Cad. No.55, G.O.P., Ankara, Turkey
PHONE: +90 (312) 4468247
FAX: +90 (312) 4468251
E-MAIL: Ankara@diplobel.org

Bosnia and Herzegovina
Mahatma Gandi Cad. No.91/8-9, G.O.P., Ankara, Turkey
PHONE: +90 (312) 4464090
FAX: +90 (312) 4466228

Brazil
Iran Cad. No.47/1-3, G.O.P., Ankara, Turkey
PHONE: +90 (312) 4685320
FAX: +90 (312) 4685324

Bulgaria
Ataturk Bulvari. No.124, Kavaklidere, Ankara, Turkey
PHONE: +90 (312) 4267455
FAX: +90 (312) 4263178

Finland
Kader Sok. No.44, G.O.P., Ankara, Turkey
PHONE: +90 (312) 4261930
FAX: +90 (312) 4680072

France
Paris Cad. No.70, Kavaklidere, Ankara, Turkey
PHONE: +90 (312) 4681154
FAX: +90 (312) 4679434
TITLE: Ambassador
NAME: Daniel Lequertier

Germany
Ataturk Bulvari No.114, Kavaklidere, Ankara, Turkey
PHONE: +90 (312) 4265465
FAX: +90 (312) 4266959
TITLE: Ambassador
NAME: Hans Joachim Vergau

Italy
Ugur Mumcu Cad. No.100/3, G.O.P., Ankara, Turkey
PHONE: +90 (312) 4461244
FAX: +90 (312) 4461245
TITLE: Ambassador
NAME: Massimiliano Bandini

Malaysia
Kizkulesi Sok. No.44, G.O.P., Ankara, Turkey
PHONE: +90 (312) 4462547
FAX: +90 (312) 4464130

Mexico
Resi Galip Cad. Rabat Sok. No.16, G.O.P., Ankara, Turkey
PHONE: +90 (312) 4460335
FAX: +90 (312) 4462521

Morocco
Rabat Sok. No.11, G.O.P.—Cankaya, Ankara, Turkey
PHONE: +90 (312) 4376020
FAX: +90 (312) 4465733

Russia
Karyagdi Sok. No.5, Cankaya, Ankara, Turkey
PHONE: +90 (312) 4392122
FAX: +90 (312) 4383952
TITLE: Ambassador
NAME: Vadim I. Kuznetsov

Netherlands
Ugur Mumcu Cad. No.77/A, Cankaya, Ankara, Turkey
PHONE: +90 (312) 4460470
FAX: +90 (312) 4463358
TITLE: Ambassador
NAME: Nikolas van Dam

Ukraine
Cemal Nadir Sok. No.9, Cankaya, Ankara, Turkey
PHONE: +90 (312) 4399973
FAX: +90 (312) 4406815

United Arab Emirates
Mahmud Yesari Sok. No.10, Cankaya, Ankara, Turkey
PHONE: +90 (312) 4408410
FAX: +90 (312) 4389854

United Kingdom
Sehit Ersan Cad. No.46/A, Cankaya, Ankara, Turkey
PHONE: +90 (312) 4686230
FAX: +90 (312) 4683214
TITLE: Ambassador
NAME: David Logan

United States
Ataturk Bulvari No.110, Kavaklidere, Ankara, Turkey
PHONE: +90 (312) 4265465
FAX: +90 (312) 4266959
TITLE: Ambassador
NAME: Mark Robert Parris

JUDICIAL SYSTEM

The Constitutional Court

High Court of Appeals

FURTHER READING

Articles

"The Big One." *Newsweek* (August 30, 1999): 20

"Crackdown." *The Economist* (February 13, 1999): 51

"Enemies List." *World Press Review* (September 1999): 10

"Going Wolfish." *The Economist* (June 12, 1999): 49

"It's Rough Going." *The Economist* (March 29, 1999): 56

"A Marriage without Virtue Isn't Easy." *The Economist* (April 10, 1999): 48

Mutuma, Mathiu. "The Kurds Against the World." *World Press Review* (May 1999): 6

"Nation and Tribe the Winners." *The Economist* (April 24, 1999): 53

"The Tragedy of the Kurds." *The Economist* (February 20, 1999): 16

"Turkey, Flattened and in Shock." *The Economist* (August 21, 1999): 39

Books

Fodor's Turkey. New York: Fodor's Travel Publications, 1999.

Kedurie, Sylvia. *Turkey Before and After Ataturk.* London: Frank Cass, 1999.

Pope, Hugh and Nicole Pope. *Turkey Unveiled: A History of Modern Turkey.* Overlook Press, 1999.

TURKEY: STATISTICAL DATA

For sources and notes see "Sources of Statistics" in the front of each volume.

GEOGRAPHY

Geography (1)

Area:

Total: 780,580 sq km.

Land: 770,760 sq km.

Water: 9,820 sq km.

Area—comparative: slightly larger than Texas.

Land boundaries:

Total: 2,627 km.

Border countries: Armenia 268 km, Azerbaijan 9 km, Bulgaria 240 km, Georgia 252 km, Greece 206 km, Iran 499 km, Iraq 331 km, Syria 822 km.

Coastline: 7,200 km.

Climate: temperate; hot, dry summers with mild, wet winters; harsher in interior.

Terrain: mostly mountains; narrow coastal plain; high central plateau (Anatolia).

Natural resources: antimony, coal, chromium, mercury, copper, borate, sulfur, iron ore.

Land use:

Arable land: 32%

Permanent crops: 4%

Permanent pastures: 16%

Forests and woodland: 26%

Other: 22% (1993 est.)

HUMAN FACTORS

Demographics (2A)

	1990	1995	1998	2000	2010	2020	2030	2040	2050
Population	56,124.8	61,438.8	64,568.2	66,620.1	76,574.0	85,650.5	93,485.2	99,648.5	103,656.4
Net migration rate (per 1,000 population)	NA	NA	NA	NA	NA	NA	NA	NA	NA
Births	NA	NA	NA	NA	NA	NA	NA	NA	NA
Deaths	NA	NA	NA	NA	NA	NA	NA	NA	NA
Life expectancy - males	66.7	69.1	70.4	71.3	74.6	76.9	78.4	79.4	80.0
Life expectancy - females	71.2	74.0	75.4	76.4	80.1	82.7	84.3	85.4	86.0
Birth rate (per 1,000)	25.5	22.7	21.4	20.4	17.6	15.3	13.8	12.8	12.1
Death rate (per 1,000)	6.4	5.6	5.3	5.2	5.0	5.4	6.2	7.7	9.3
Women of reproductive age (15-49 yrs.)	13,795.0	15,783.0	16,991.7	17,752.5	20,641.0	22,114.5	22,549.1	22,331.0	22,138.8
of which are currently married	9,709.2	NA	NA	NA	NA	NA	NA	NA	NA
Fertility rate	3.1	2.6	2.5	2.3	2.1	2.0	2.0	2.0	2.0

Except as noted, values for vital statistics are in thousands; life expectancy is in years.

Health Personnel (3)

Total health expenditure as a percentage of GDP, 1990-1997[a]

Public sector	2.7
Private sector	1.1
Total[b]	3.8

Health expenditure per capita in U.S. dollars, 1990-1997[a]

Purchasing power parity	227
Total	113

Availability of health care facilities per 100,000 people

Hospital beds 1990-1997[a]	250
Doctors 1993[c]	103
Nurses 1993[c]	151

Health Indicators (4)

Life expectancy at birth

1980	61
1997	69
Daily per capita supply of calories (1996)	3,568
Total fertility rate births per woman (1997)	2.5
Maternal mortality ratio per 100,000 live births (1990-97)	180[c]
Safe water % of population with access (1995)	....
Sanitation % of population with access (1995)	....
Consumption of iodized salt % of households (1992-98)[a]	18

Smoking prevalence

Male % of adults (1985-95)[a]	63
Female % of adults (1985-95)[a]	24
Tuberculosis incidence per 100,000 people (1997)	41
Adult HIV prevalence % of population ages 15-49 (1997)	0.01

Infants and Malnutrition (5)

Under-5 mortality rate (1997)	45
% of infants with low birthweight (1990-97)	8
Births attended by skilled health staff % of total[a]	76

% fully immunized (1995-97)

TB	73
DPT	79
Polio	79
Measles	76
Prevalence of child malnutrition under age 5 (1992-97)[b]	10

Ethnic Division (6)

Turkish	80%
Kurdish	20%

Religions (7)

Muslim (mostly Sunni)	99.8%
Other (Christian and Jews)	0.2%

Languages (8)

Turkish (official), Kurdish, Arabic.

EDUCATION

Public Education Expenditures (9)

Public expenditure on education (% of GNP)

1980	2.2
1996	2.2[1]

Expenditure per student

Primary % of GNP per capita

1980	6.4
1996	13.4[1]

Secondary % of GNP per capita

1980	....
1996	9.4[1]

Tertiary % of GNP per capita

1980	95.1
1996	51.8[1]

Expenditure on teaching materials

Primary % of total for level (1996)	0.1
Secondary % of total for level (1996)	....
Primary pupil-teacher ratio per teacher (1996)	28[1]
Duration of primary education years (1995)	5

Educational Attainment (10)

Age group (1993)[21]	25+
Total population	NA

Highest level attained (%)

No schooling	30.6

First level

Not completed	6.6
Completed	40.6

Entered second level

S-1	21.9
S-2	NA
Postsecondary	NA

Literacy Rates (11A)

In thousands and percent[1]	1990	1995	2000	2010
Illiterate population (15+ yrs.)	7,616	7,231	6,611	5,112
Literacy rate - total adult pop. (%)	79.2	82.3	85.5	90.9
Literacy rate - males (%)	89.9	91.7	93.7	96.7
Literacy rate - females (%)	68.5	72.4	77.0	85.0

GOVERNMENT & LAW

Political Parties (12)

Grand National Assembly	% of seats
Welfare Party (RP)	21.38
True Path Party (DYP)	19.18
Motherland Party (ANAP)	19.65
Democratic Left Party (DSP)	14.64
Republican People's Party (CHP)	10.71
Independent	0.48

Military Affairs (14B)

	1990	1991	1992	1993	1994	1995
Military expenditures						
Current dollars (mil.)	4,323	4,876	5,437	6,111	6,168	6,606
1995 constant dollars (mil.)	4,968	5,388	5,849	6,406	6,322	6,606
Armed forces (000)	769	804	704	686	811	805
Gross national product (GNP)						
Current dollars (mil.)	123,800	130,800	141,800	157,000	153,000	166,700
1995 constant dollars (mil.)	142,300	144,500	152,500	164,600	156,800	166,700
Central government expenditures (CGE)						
1995 constant dollars (mil.)	24,490	30,160	31,150	40,430	36,410	37,610
People (mil.)	56.1	57.2	58.3	59.3	60.4	61.4
Military expenditure as % of GNP	3.5	3.7	3.8	3.9	4.0	4.0
Military expenditure as % of CGE	20.3	17.9	18.8	15.8	17.4	17.6
Military expenditure per capita (1995 $)	89	94	100	108	105	108
Armed forces per 1,000 people (soldiers)	13.7	14.1	12.1	11.6	13.4	13.1
GNP per capita (1995 $)	2,536	2,526	2,618	2,775	2,597	2,714
Arms imports[6]						
Current dollars (mil.)	1,200	1,200	1,000	1,200	1,100	700
1995 constant dollars (mil.)	1,379	1,326	1,076	1,258	1,128	700
Arms exports[6]						
Current dollars (mil.)	10	30	20	20	30	60
1995 constant dollars (mil.)	11	33	22	21	31	60
Total imports[7]						
Current dollars (mil.)	22,300	21,050	22,870	29,170	23,270	35,710
1995 constant dollars (mil.)	25,630	23,260	24,600	30,590	23,850	35,710
Total exports[7]						
Current dollars (mil.)	12,960	13,590	14,720	15,340	18,110	21,600
1995 constant dollars (mil.)	14,890	15,020	15,830	16,090	18,560	21,600
Arms as percent of total imports[8]	5.4	5.7	4.4	4.1	4.7	2.0
Arms as percent of total exports[8]	.1	.2	.1	.1	.2	.3

Government Budget (13A)

Year: 1996

Total Expenditures: 3,965,948 Billions of Liras

Expenditures as a percentage of the total by function:

General public services and public order	49.17
Defense	8.41
Education	11.16
Health	2.26
Social Security and Welfare	4.45
Housing and community amenities	1.13
Recreational, cultural, and religious affairs	.37
Fuel and energy	2.81
Agriculture, forestry, fishing, and hunting	.85
Mining, manufacturing, and construction	1.21
Transportation and communication	4.11
Other economic affairs and services	1.45

Crime (15)

Crime rate (for 1997)

Crimes reported	331,700
Total persons convicted	221,800
Crimes per 100,000 population	525

Persons responsible for offenses

Total number of suspects	326,900
Total number of female suspects	27,100
Total number of juvenile suspects	30,100

LABOR FORCE

Labor Force (16)

Total (million)	21.6
Agriculture	43.1%
Services	30.1%
Industry	14.4%
Construction	6.0%

About 1.5 million Turks work abroad. Data for 1996. Percent distribution for 1994.

Unemployment Rate (17)

5.9% another 5.1% officially considered underemployed (April 1997)

PRODUCTION SECTOR

Electric Energy (18)

Capacity	21.83 million kW (1997)
Production	103 billion kWh (1997)
Consumption per capita	1,636 kWh (1997)

Transportation (19)

Highways:

total: 381,631 km

paved: 95,408 km (including 1,405 km of expressways)

unpaved: 286,223 km (1996 est.)

Waterways: about 1,200 km

Pipelines: crude oil 1,738 km; petroleum products 2,321 km; natural gas 708 km

Merchant marine:

total: 528 ships (1,000 GRT or over) totaling 6,205,399 GRT/10,400,716 DWT

Airports: 114 (1997 est.)

Airports—with paved runways:

total: 80

over 3,047 m: 17

2,438 to 3,047 m: 21

1,524 to 2,437 m: 18

914 to 1,523 m: 19

under 914 m: 5 (1997 est.)

Airports—with unpaved runways:

total: 34

1,524 to 2,437 m: 1

914 to 1,523 m: 8

under 914 m: 25 (1997 est.)

Top Agricultural Products (20)

Tobacco, cotton, grain, olives, sugar beets, pulses, citrus; livestock.

MANUFACTURING SECTOR

GDP & Manufacturing Summary (21)

Detailed value added figures are listed by both International Standard Industry Code (ISIC) and product title.

	1980	1985	1990	1994
GDP ($-1990 mil.)[1]	88,366	114,593	149,972	163,245
Per capita ($-1990)[1]	1,989	2,276	2,673	2,686
Manufacturing share (%) (current prices)[1]	17.2	18.7	22.2	*21.2*
Manufacturing				
Value added ($-1990 mil.)[1]	16,423	23,735	33,086	36,385
Industrial production index	100	175	243	253
Value added ($ mil.)	10,837	10,448	28,958	*27,459*
Gross output ($ mil.)	29,413	32,470	73,064	64,237
Employment (000)	787	844	975	*913*

	1980	1985	1990	1994
Profitability (% of gross output)				
Intermediate input (%)	63	68	60	*57*
Wages and salaries inc. supplements (%)	16	10	12	*15*
Gross operating surplus	20	23	28	*28*
Productivity ($)				
Gross output per worker	36,960	38,378	74,819	*60,394*
Value added per worker	13,617	12,349	29,656	*25,991*
Average wage (inc. supplements)	6,142	3,717	9,029	*10,227*
Value added ($ mil.)				
311/2 Food products	1,185	973	2,541	*2,824*
313 Beverages	335	330	893	*959*
314 Tobacco products	467	877	1,168	*1,333*
321 Textiles	1,535	1,289	3,222	*3,002*
322 Wearing apparel	60	146	947	*905*
323 Leather and fur products	25	37	60	*49*
324 Footwear	33	22	69	*88*
331 Wood and wood products	118	64	187	*154*
332 Furniture and fixtures	16	55	81	*85*
341 Paper and paper products	205	241	559	*487*
342 Printing and publishing	97	133	434	*400*
351 Industrial chemicals	719	457	1,517	*1,064*
352 Other chemical products	387	394	1,449	*1,294*
353 Petroleum refineries	1,352	1,514	4,525	*4,177*
354 Miscellaneous petroleum and coal products	222	152	458	*340*
355 Rubber products	201	151	452	*436*
356 Plastic products	125	76	328	*290*
361 Pottery, china and earthenware	93	102	466	*427*
362 Glass and glass products	110	167	531	*405*
369 Other non-metal mineral products	535	428	1,365	*1,158*
371 Iron and steel	783	734	1,403	*1,616*
372 Non-ferrous metals	292	181	580	*387*
381 Metal products	395	344	904	*685*
382 Non-electrical machinery	506	456	1,423	*1,242*
383 Electrical machinery	463	531	1,482	*1,661*
384 Transport equipment	541	534	1,743	*1,843*
385 Professional and scientific equipment	8	9	87	*81*
390 Other manufacturing industries	28	48	83	*66*

FINANCE, ECONOMICS, & TRADE

Economic Indicators (22)

National product: GDP—purchasing power parity—$388.3 billion (1997 est.)

National product real growth rate: 7.2% (1997)

National product per capita: $6,100 (1997 est.)

Inflation rate—consumer price index: 99% (1997)

Balance of Payments (23)

	1992	1993	1994	1995	1996
Exports of goods (f.o.b.)	14,891	15,611	18,390	21,975	32,303
Imports of goods (f.o.b.)	−23,081	−29,771	−22,606	−35,187	−41,935
Trade balance	−8,190	−14,160	−4,216	−13,212	−9,632
Services - debits	−7,262	−7,828	−7,936	−9,717	−10,893
Services - credits	10,419	11,787	11,691	16,095	14,628
Private transfers (net)	912	733	383	1,071	555
Government transfers (net)	3,147	3,035	2,709	3,425	3,892
Overall balance	−974	−6,433	2,631	−2,338	−1,450

Exchange Rates (24)

Exchange rates:

Turkish liras (TL) per US$1

January 1998	212,500
1997	151,600
1996	81,405
1995	45,845.1
1994	29,608.7
1993	10,984.6

Top Import Origins (25)

$46.7 billion (f.o.b., 1997) Data are for 1997.

Origins	%
Germany	16
Italy	9
United States	9
France	6
United Kingdom	6

Top Export Destinations (26)

$26 billion (f.o.b., 1997); note—substantial unrecorded exports estimated at $5.8 billion Data are for 1997.

Destinations	%
Germany	20
United States	8
Russia	8
United Kingdom	6
Italy	5

Economic Aid (27)

Recipient: ODA, $195 million (1993).

Import Export Commodities (28)

Import Commodities	Export Commodities
Machinery 26%	Textiles and apparel 37%
Fuels 13%	Iron and steel products 10%
Raw materials 10%	Foodstuffs 17%
Foodstuffs 4%	

TURKMENISTAN

CAPITAL: Ashgabat (Ashkhabad).

FLAG: Green field with claret stripe of five carpet patterns; white crescent and five white stars symbolizing five major regions of Turkmenistan to the right of the stripe. In 1997, two crossed olive branches were added beneath the carpet patterns.

ANTHEM: *Independence Turkmenistan.*

MONETARY UNIT: Manat (MN), the unit of currency, was introduced by the government in November 1993. In 1996, $1 = MN4,000, but exchange rates fluctuate widely.

WEIGHTS AND MEASURES: The metric system is used.

HOLIDAYS: New Year's Day, 1 January; Flag Day, 19 February; International Women's Day, 8 March; Novruz Bairam (first day of spring), 21 March; Victory Day, 9 May; Revival and Unity Day, 18 May; Independence Day, 27 October; Neutrality Day, 12 December.

TIME: 5 PM = noon GMT.

LOCATION AND SIZE: Located in southern Asia, bordering the Caspian Sea between Iran and Uzbekistan, Turkmenistan is slightly larger than the state of California, with a total area of 488,100 square kilometers (188,456 square miles). Its boundary length totals 3,736 kilometers (2,322 miles).

Turkmenistan's capital city, Ashgabat (or Ashkhabad, which means ''city of love''), is located in the central southern part of the country.

CLIMATE: The mean temperature is 28°C (82°F) in July and −4°C (25°F) in January. Daytime temperatures of 122°F in the Kara Kum (Garagum) desert are not unusual. Rainfall averages 25 centimeters (10 inches) a year.

INTRODUCTORY SURVEY

RECENT HISTORY

Present-day Turkmenistan became a Soviet Socialist Republic in 1925. Throughout the Soviet period, Turkmenistan was among the poorest and least assimilated of the republics. Turkmenistan declared independence on 27 October 1991. Saparmuryad Niyazov, appointed during the Gorbachev regime, was elected president by the republic's Supreme Soviet in 1990 and, since he has a large following and great power in the government, has remained in office ever since.

Turkmenistan has played a leading role in attempting to bring stability to Central Asia, especially in neighboring Tajikistan where a civil war lasted from 1992 until a cease-fire was reached in 1997.

GOVERNMENT

The president is both chief of state and head of the government and appoints his own cabinet, the Council of Ministers. The legislative branch comprises two bodies, a 50-member assembly and a 100-member People's Council.

Judiciary

The members of the Supreme Court are appointed by the president. Turkmenistan's court system also includes 61 district and city courts, six provincial courts, military courts and a supreme economic court, which hears cases involving disputes between business enterprises and ministries.

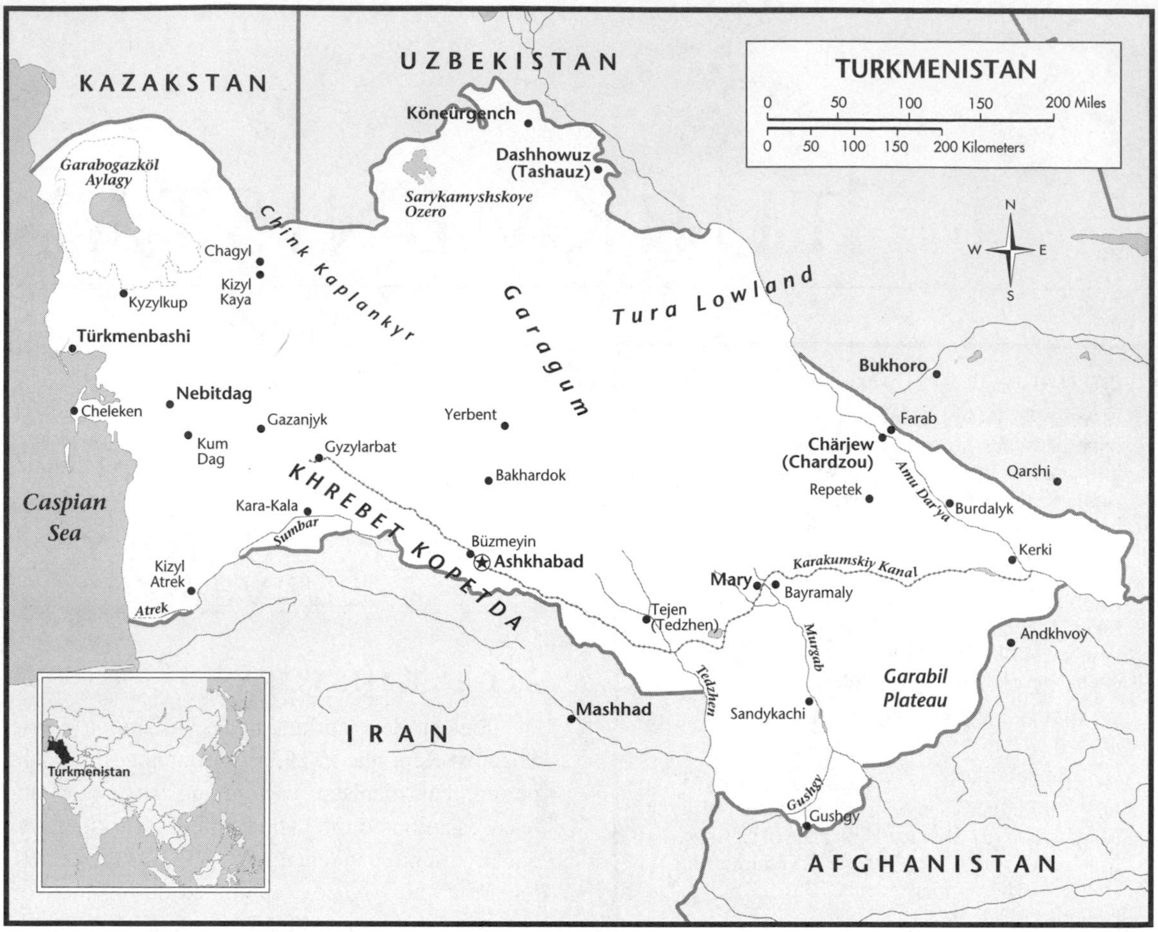

Political Parties

The only legally registered party in the republic is the Democratic Party of Turkmenistan, which is what the Communist Party renamed itself in September 1991.

DEFENSE

The total armed forces consist of 16,000–18,000 personnel (army and air force). The defense budget for 1996 was US $90 million.

ECONOMIC AFFAIRS

Turkmenistan boasts rich mineral deposits, including oil, gas, potassium, sulfur, and salts. It is the world's tenth largest cotton producer and possesses the world's fifth largest reserves of natural gas and substantial oil resources.

In 1994, according to the U.S. Central Intelligence Agency (CIA), Russia's refusal to export Turkmen gas to hard currency markets and mounting debts of its major customers in the former USSR for gas deliveries contributed to a sharp fall in industrial production. The economy bottomed out in 1996, but high inflation continued. The government has recently attempted to export gas to Europe through Iran and Turkey.

Public Finance

Although still a centrally planned economy, Turkmenistan has slowly begun to decrease the size of the public sector's influence. Estimates of 1995 external debt ranged from $400–632.1 million. The CIA estimated the portion of the debt owed to Russia at $275 million.

Income

In 1995, Turkmenistan's gross national product (GNP) was $4.13 billion, or about $940 per person. In 1998, per capita income was estimated by the World Bank at $640. For the period 1985–95 the average annual inflation rate was 381.4%, and the average annual decline in the rate of GNP per person was 9.6%.

Industry

Turkmen carpets are known worldwide for their quality. In 1991, 1.38 million square meters of carpets and tapestry were produced. Fuel-related production (mainly gas and oil) is the second largest component in the industrial sector. In 1995, production of refined petroleum totaled 4.3 million tons. Over 90% of industrial output is produced by state-owned enterprises.

Banking and Finance

The State Central Bank of Turkmenistan (SCBT) issues currency and executes monetary policy. The State Bank for Foreign Economic Activities provides hard currency credits for foreign economic activities.

Sberbank (Savings Bank), holds most household deposits and is still state-owned. The local branch of Vneshekonombank has been incorporated as an independent foreign trade bank and is also state-owned. Investbank, the industrial sector bank, and Agroprombank, the agricultural sector bank, are state-owned via stock distributed to state-owned enterprises.

Economic Development

After gaining independence from the Soviet Union in 1991, the government began efforts to reduce Turkmenistan's dependence on exporting raw materials and importing finished goods and basic foodstuffs. Initiated in time to help avoid some of the most severe economic dislocation experienced in other post-Soviet republics, these efforts included increasing wheat production, developing the country's oil and gas processing capacity, and investing in internal transportation facilities.

The government also initiated a cautious reform agenda aimed at economic liberalization. Guiding principles of this initiative included legislative, fiscal, and monetary measures related to price controls, privatization, and industrial infrastructure development. Only a handful of enterprises had been privatized by 1992, however, and four-fifths of the labor force remained employed in the state-owned sector. As of 1996, the government reported that 1,594 enterprises had been privatized but that over 90% of the value of goods produced by industry could be attributed to state-controlled enterprises. Most industrial enterprises continue to run on the basis of centrally planned state orders and resource allocations.

Twenty-nine percent of the 1992 budget expenditures was allocated to price-differential subsidies paid to retail agencies required to sell food and medicines below wholesale prices. In 1992, a five-year production and investment plan proposed large investments in the development of infrastructure and the energy sector financed by the budget and large inflows of foreign investment.

SOCIAL WELFARE

The government countered the negative impact of post-Soviet economic restructuring on the population by providing allowances for large families, social security payments, and increased pensions. As of 1993 all citizens receive free electricity and water.

Under the constitution, women are protected from discrimination in employment, inheritance, marriage rights, and other areas.

Healthcare

In 1993, there was about 1 doctor for every 306 people, and 1 hospital bed per 93 inhabitants. Average life expectancy is 66 years.

Housing

In 1990, Turkmenistan had 11.1 square feet of housing space per person. In January 1991, 108,000 households (31%) were on waiting lists for urban housing.

EDUCATION

The adult illiteracy rate is estimated at two percent. Education is compulsory from the ages 7 to 17. The government reports 1,764 schools with enrollment of 850,000. In 1991, higher-level institutions had a total of 76,000 pupils enrolled.

1999 KEY EVENTS TIMELINE

January

- Uzbeks and Turkmens discuss rail tariffs, the use of land in border areas, and payment for the transit of Turkmen electricity via Uzbek territory.
- Turkmens celebrate the 118th anniversary of their last major resistance against Tsarist armies at Goektepe.
- Niyazov personally attends peoples' complaints, frees 3,000 prisoners unjustly incarcerated, criti-

cizes the judicial system of Turkmenistan, and advocates new laws for the republic.

February

- In Ashgabat, a meeting of journalists expected to announce the formation of an Independent Journalists' Association is broken up by the Officers of Turkmenistan's National Security Committee.
- Representatives of the Talibans and of the forces under Ahmed Shah Masoud agree to hold talks in Ashgabat.
- Turkmenistan is acceptable to both the Talibans and the opposing forces because the UN officially recognized that republic as a neutral country.
- Aleksandr Petrov from the Moscow office of Human Rights Watch is deported from Turkmenistan for producing ''offending'' materials on Turkmenistan.
- Niyazov predicts that Turkmenistan would export 120 billion cubic meters of natural gas by the year 2,000.
- Projected Trans-Caspian and Trans-Iranian pipelines, along with the existing Trans-Russian pipeline guarantee reliable supplies of gas to Ashgabat.
- Azerbaijan agrees to route the planned Trans-Caspian gas pipeline to Turkey while the U.S. Export-Import Bank contributes approximately $1 billion toward the estimated $3 billion construction costs.
- Two U.S. companies, Bechtel and General Electric Capital, are chosen to head the consortium which is to build the Trans-Caspian pipeline for Turkmenistan.
- Iran and Russia warn Turkmenistan of the consequences of signing the treaty that enables American companies to lead a consortium to build the Trans-Caspian pipeline.

March

- The amount of money in circulation in Turkmenistan increases by 5.4% a month as opposed to the previous 6.9% monthly increase, a good sign.
- The official rate of exchange for the manat remains steady, at 5,200 to $1.
- The European Bank for Reconstruction and Development will participate in financing the construction of the Turkmen segment of Trans-Caspian pipeline project.
- The UN special envoy to Afghanistan, Lakhdar Brahimi, meets with Niyazov in Ashgabat and extends a message of thanks from UN Secretary-General Kofi Annan for hosting the recent round of Afghan peace talks.
- Turkmenistan requires visas from members of the CIS, a requirement that was deemed unnecessary in 1992.
- Niyazov establishes new rules for the transit through Turkmenistan of beer, hard liquor, wine, and tobacco products.
- On the occasion of Nawruz and Kurban Bairami, Niyazov pardons 5,000 prisoners. The total number of people freed this year is 22,000.
- The chairman of the Integration Committee of the CIS Customs Union weighs the pros and cons of Turkmenistan's announcement for visa requirements for CIS members on transit. He feels Turkmenistan not only isolates itself but loses the benefits it receives from international trade.
- Ukraine will ask Turkmenistan to suspend gas shipments beginning April 1999. Kiev currently owes Turkmenistan some $100 million for gas already received.

April

- Turkmenistan continues gas deliveries to Ukraine, despite the latter's growing debt, until all 20 billion cubic meters of gas for 1999 is delivered.
- Niyazov and Nazarbaev agree on exporting Turkmen gas and Kazak oil via China.
- As a payment for Turkmenistan's natural gas, the Ukrainian government undertakes the shipment of barter goods to Turkmenistan.
- At talks in Ashgabat, Niyazov and PSG, the U.S. company that plans to build a gas pipeline across the Caspian, agree to speed up work on the project. The preliminary financial plan and the organization of the multi-company consortium that PSG will head are discussed.

May

- Niyazov transfers some of his oversight powers to the parliament.
- Turkmenistan plans on opening embassies in Azerbaijan, Armenia, Kazakstan, Kyrgyzstan,

Moldova, and Tajikistan to take care of visa requirements imposed on visitors from most CIS states.

- Turkey and Turkmenistan sign an agreement under which Turkey will purchase 750 million kw hours of electricity annually from Turkmenistan between 2000 and 2006.
- The members of the Economic Cooperation Organization (Afghanistan, Azerbaijan, Iran, Kazakstan, Kyrgyzstan, Pakistan, Tajikistan, Turkey, Turkmenistan, and Uzbekistan) discuss the development of a network of export pipelines for oil and gas.
- Turkmenistan plans unilaterally to revoke the open-ended treaty it signed with Russia in 1993, which allows for Turkmen and Russian border troops to jointly guard Turkmenistan's frontiers with Iran and Afghanistan.
- The U.S. announces support for building an underwater Trans-Caspian pipeline to export Turkmenistan's natural gas to Turkey, but opposes the idea of a gas export pipeline via Iran.

June

- The new mandatory visa regime set up by Ashgabat detains fifty-one individuals from flying out of Ashgabat. Acquiring the necessary documentation will take at least a month.
- The prime ministers of Kazakstan, Kyrgyzstan, Tajikistan, and Turkmenistan approve 25 investment projects totaling $50 million. They also agreed to coordinate the operation of their power grids.
- Russia warns Turkmenistan that a ''polarization'' within the CIS may weaken security along the borders of the CIS and create gaps that can be advantageous to adversaries.
- Turkmenistan announces its intention to take over full responsibility for guarding its own frontiers.

July

- Georgia and Turkmenistan conclude a contract whereby the Tbilisi Aviation Plant will repair 45 Turkmen SU-45 fighter aircraft at a cost of $46 million.
- Reconstruction of the existing gas pipeline from Azerbaijan via Georgia will cost between $100–150 million.
- The Shah Deniz reserves are sufficiently large to supply Turkey with enough gas to meet its rapidly growing energy needs.
- During the first six months of 1999, GDP increased by 15%, compared with 1998. Total GDP growth for 1998 was 5%.
- Oil and gas extraction for the first five months of 1999 rose by 10.7%.
- Turkmenistan recently completed its best-ever grain harvest of 1.5 million tons. Such steady increase has made Turkmenistan self-sufficient in grain for two consecutive years.
- To ensure that the names in the country reflect the country's rebirth, Niyazov renames Charjou, Turkmenabad.
- Niyazov approves Turkmenistan's development program for 2001–2010, which entails massive increases in the extraction of oil and natural gas, with production of the former slated to rise from 6.3 million tons to 27–30 million tons. The increase in gas production is predicated on the implementation of the planned Trans-Caspian gas pipeline project.
- More moderate increases are anticipated in cotton and grain production.
- Turkmenistan's population is expected to increase from the present 5 million to 6.5 million in 2005.

August

- The Trans-Caspian gas pipeline pipe will have an annual capacity of 30 billion cubic meters.
- Turkmenistan will not resume shipment of natural gas to Ukraine as long as Ukraine does not pay its debt in full. Forty percent of that debt is to be paid in hard currency and the remainder in barter goods.
- Turkmenistan's foreign trade turnover grew by 57.2 percent to $1.22 billion during the first half of 1999. The trade surplus is $112.1 million. Also during the first six months of 1999, industrial output rose by 19%, primarily as a result of a 160% increase in gas production and a 60% increase in the output of the cotton industry.
- The U.S. urges Turkmenistan, Azerbaijan, Georgia, and Turkey to sign a legal agreement committing their support for the planned Trans-Caspian gas export pipeline.
- Niyazov discusses agriculture and the development of the textile industry.

- Turkmenistan will donate $100,000 to victims of the Turkish earthquake and is prepared to send a team of doctors to the devastated area.
- Moscow is prepared to provide anti-aircraft systems and fighters to Turkmenistan on a ''long-term lease'' basis.

September

- Turkmen government and Iranian energy officials approve a feasibility study and reach agreement on financing construction of a reservoir and dam on the Tedzhen River. The two countries will contribute equally to the estimated $167 million project. The reservoir will have a capacity of 1.2 billion cubic meters, making it possible to irrigate some 20,000 hectares of land on each side of the border.
- Niyazov announces his willingness to pardon and give amnesty to a further 12,000 prisoners before the end of 1999. Twenty-two thousand prisoners, many of them jailed for drug-related offenses, are freed from the country's overcrowded jails.
- China might undertake personnel training for Turkmenistan as well as take over the use and repair of military hardware for that republic.

October

- New program for 2000–2010 envisages increasing oil production to 48 million tons with crude oil exports rising to 33 million tons.

December

- The People's Council is considering legalizing polygamy by allowing a husband to marry again if his first wife gives written permission.
- President Saparmurat Niyazov, who has held the office for 14 years, tells the People's Council not to appoint him president for life. President Niyazov states that he plans to complete his current term, which will end in 2002.
- Turkmenistan becomes the first of the five so-called Central Asian Republics (others are Kazakhstan, Kyrgyzstan, Tajikistan, Uzbekistan) to abolish the death penalty when President Saparmurat Niyazov signs the ban into law.

ANALYSIS OF EVENTS: 1999

BUSINESS AND THE ECONOMY

In 1999 Turkmenistan was a tribal society with a caste of governing officials who had also ruled the country before the dissolution of the USSR. The nation was in a race to develop its oil and natural gas extractive capability before the pressure of its increasing population and the ambitions of acquisitive neighbors disrupted its development plans. Much of the business of governance in 1999 involved setting up mechanisms for the extraction and sale of these petroleum products mostly to other states in the region. Crucial to these goals was the projected Trans-Caspian and Trans-Iranian pipelines, along with the existing Trans-Russian pipeline, which would ensure reliable supplies of gas to Ashgabat.

Inflation was a worry, but in 1999 the amount of money in circulation in Turkmenistan increased by 5.4%. The official rate of exchange for its unit of currency, the *manta,* remained steady at 5,200 to $1. The European Bank for Reconstruction and Development agreed to finance part of the construction of the Turkmen segment of the Trans-Caspian pipeline.

In 1999 Turkmenistan's leadership also had to deal with the modernization of its society in order to facilitate economic development. One example of this was the need to balance the easy flow of commodities across its borders, with the traditional and religious strictures against certain items of trade. One example which received some publicity was the regulation of the transit of beer, hard liquor, wine, and tobacco across its borders. The Islamic faith of ninety percent of its people along with the repressive rule of its leadership inhibited the use of alcohol as well as the cultivation of poppies and the refining of opium-derived drugs. Turkmenistan also had to decide how much to relax its visa requirements for former fellow-republics of the U.S.S.R. Failure to do so might isolate them and diminish its international trade.

As the year 1999 drew to a close the challenges and the time-table of development became clear: Turkmenistan's population was expected to increase to 6.5 million in 2005. The Trans-Caspian gas pipeline was expected to have an annual capacity of 30 billion cubic meters. Bringing these facts into the development of a modernization strategy entailed hard decisions. One was the question of extending credit to satisfy its neighbors' energy needs. During the course of the year Turkmenistan's leaders decided not to resume the shipment of natural gas to the Ukraine until that country paid its debt in full. Forty percent of that debt was to be paid in hard currency, the remainder in barter goods. Such policies helped to bolster Turkmenistan's absolute level of foreign trade by 57.2 percent to $1.22 billion. The trade surplus was $112.1 million, compared with a deficit of $259.4 million the previous year. Industrial output rose by 19 percent, primarily as a result of a 160 percent increase in gas production and a 60 percent increase in the output of the cotton industry. New programs for the years 2000–2010 envisage increasing oil production to 48 million tons with crude oil exports rising to 33 million tons.

GOVERNMENT AND POLITICS

In its ''World Factbook,'' the Central Intelligence Agency of the United States formally categorizes Turkmenistan as a republic, but it was far from the practice of democratic republican governance of the Enlightenment-influenced western European countries. Turkmenistan had a one-party democracy. Formal opposition parties were illegal. On the other hand, President Niazov's exercize of the broad powers at his command was vigorous, decisive, and intelligent.

The favorable prospects of the economy posed questions of how development would transform the other aspects of Turkmen society, possibly in the direction of greater democracy. The results of attempts to establish the institutions of political pluralism, however, are mixed at best. In one incident, security officials disrupted a meeting of journalists who had met to form an Independent Journalists' Association. Aleksandr Petrov from the Moscow office of Human Rights Watch was deported for writing ''offending'' materials on Turkmenistan. Questions of the basic rights of citizens to move about were also at issue. Turkmen leaders adopted a new mandatory visa regime which temporarily detained fifty-one individuals. Acquiring the necessary documentation to travel outside the country's borders was expected to take at least a month. Watching the actions of the nation's leaders as they tried to negotiate economic development, the forces of Islamic fundamentalism were playing a problematic role in Turkmen society. They seemed ready to allow the regime to govern.

Turkmenistan's leaders also had to confront issues involving continuing Cold War reflexes as well as the possibly more disruptive Islamic fundamentalist pressures on the country's developmental policies. Turkmenistan leaders, for example, contracted with the large American Bechtel Company to head the consortium building of Turkmenistan's Trans-Caspian pipeline. Both Iran and Russia warned Turkmenistan of the consequences of signing that agreement. Turkmenistan also decided to require visas for the members of the CIS. Meanwhile, Niyazov and Nazarbaev agreed on exporting Turkmen gas and Kazak oil via China. Turkmenistan planned on opening embassies in Azerbaijan, Armenia, Kazakstan, Kyrgyzstan, Moldova, and Tajikistan. The members of the Economic Cooperation Organization discussed developing a network of export pipelines for oil and gas. America declared its opposition to the export of gas via Iran while Russia warned Turkmenistan not to create ''polarization'' within the CIS. Turkmenistan unilaterally decided to revoke the 1993 treaty that allowed Turkmen and Russian border troops to jointly guard Turkmenistan's frontiers with Iran and Afghanistan. Turkmenistan intended to guard its own frontiers. Reports surfaced that China might be called upon to undertake personnel training for Turkmenistan's military as well to consult on the use and repair of military hardware.

These were all big and complex decisions that interacted on one another in ways that were difficult to predict. Niyazov agreed to transfer some of his oversight powers to the parliament.

CULTURE AND SOCIETY

In 1999 Turkmenistan celebrated the 118th anniversary of its last major resistance against Tsarist armies at Goektepe. The nation was closer in political culture to the society that fought that battle than it was to the more cosmopolitan societies of the Middle East. The evident goals of President Niyazov had more to do with the survival and development of the nation than with the niceties of democratic practice. Given the dangers to his nation, this hard-shelled realism has something to recommend it. Still, his contact with the people of his country resembled the ritual *levés* of an enlightened despot. Even the newspaper headlines bespeak the patriarchal traditions of the land: the president ''personally attends the peoples' complaints, frees 3,000 prisoners unjustly incarcerated, criticizes the judicial system and advocates new laws.''

The UN Secretary-General thanked Niyazov for hosting the recent round of Afghan peace talks. On the national holidays of Nawruz and Kurban Bairami, Niyazov freed 5,000 prisoners. This brought the total number of people freed this year to 22,000. At year's end Niyazov announced his willingness to pardon and give amnesty to a further 12,000 prisoners. As befitting its new role as a developing member of the community of nations, Turkmenistan announced its donation of $100,000 to victims of the Turkish earthquake and prepared to send a team of doctors to the devastated area.

DIRECTORY

CENTRAL GOVERNMENT

Head of State

President
Saparmurad Niyazov, Office of the President, Presidential Palace, 24 Karl Marx Street, Ashgabat 744000, Turkmenistan
PHONE: +7 (12) 354534
FAX: +7 (12) 354388

Chairman of the Supreme Council (Mejlis)
Sakhat Muradov, Office of the Chairman

Ministers

Minister of Agriculture
Serdar Babayev, Ministry of Agriculture, 63 Azadi Street, Ashgabat 744000, Turkmenistan
PHONE: +7 (12) 356691
FAX: +7 (12) 353632

Minister of Autotransportation
Senaguly Rakhmanov, Ministry of Autotransportation, 2 Baba Annanov Street, Ashgabat 744000, Turkmenistan
PHONE: +7 (12) 474992; 474031
FAX: +7 (12) 479480

Minister of Building Materials Industry
Dadebay Annageldiyev, Ministry of Building Materials Industry, 1 Steklozavodskoy Passage, Ashgabat 744000, Turkmenistan
PHONE: +7 (12) 474087
FAX: +7 (12) 475069

Minister of Culture
Orazgeldy Aydogdiyev, Ministry of Culture, 14 Pushkina Street, Ashgabat 744000, Turkmenistan
PHONE: +7 (12) 253560; 253570
FAX: +7 (12) 511991; 256985

Minister of Communications
Rovshan Kerkavov, Ministry of Communications, 40 Neutral Turkmenistan Street, Ashgabat 744000, Turkmenistan
PHONE: +7 (12) 352153
FAX: +7 (12) 390420

Minister of Defense
Batyr Sarjayev, Ministry of Defense

Minister of Economics and Finance
Matkarim Rajapov, Ministry of Economics and Finance

Minister of Education
Abat Rizaeva, Ministry of Education

Minister of Energy and Industry
Saparmurat Nuriyev, Ministry of Energy and Industry, Ministry of Energy and Industry, 6 Nurberdy Pomma Street, Ashgabat 744000, Turkmenistan
PHONE: +7 (12) 510882
FAX: +7 (12) 390682

Minister of Environmental Protection
Pirdjan Kurbanov, Ministry of Environmental Protection

Minister of Foreign Affairs
Boris Shikhmuradov, Ministry of Foreign Affairs

Minister of Health and Medical Industry
Gurganguly Berdimukhamedov, Ministry of Health and Medical Industry, 90 Makhtumkuli Street, Ashgabat 744000, Turkmenistan
PHONE: +7 (12) 351063; 355833; 354521
FAX: +7 (12) 355032

Minister of Internal Affairs
Poran Berdiyev, Ministry of Internal Affairs

Minister of Justice
Gurban Mukhammet Kasimov, Ministry of Justice

Minister of Oil, Gas Industry, and Mineral Resources
Rejepbay Arazov, Ministry of Oil, Gas Industry, and Mineral Resources, 53 Pushkin Street, Ashgabat 744000, Turkmenistan
PHONE: +7 (12) 350230
FAX: +7 (12) 350415

Minister of Textiles
Djamal Geoklenova, Ministry of Textiles, 52 Annadurdyev Street, Ashgabat 744000, Turkmenistan
PHONE: +7 (12) 352588; 355442

FAX: +7 (12) 353228

Minister of Trade and Resources
Dortguly Aidogdyev, Ministry of Trade and Resources

Minister of Transportation
Hudayguly Halykov, Ministry of Transportation, Cabinet of Ministers, 6 Nurberdy Pomma Street, Ashgabat 744000, Turkmenistan
PHONE: +7 (12) 350040
FAX: +7 (12) 351251

POLITICAL ORGANIZATIONS

Democratic Party of Turkmenistan (TDP)

NAME: Saparmurat Niyazov Turkmenbashi

DIPLOMATIC REPRESENTATION

Embassies in Turkmenistan

France
Hotel Regal Alk Altyn Plaza, 141/1 avenue Magtymguly, Ashgabat, Turkmenistan
PHONE: +7 (12) 510623
FAX: +7 (12) 510699

United Kingdom
3rd Floor, Office Building, Ak Altin Plaza Hotel, Ashgabat, Turkmenistan
PHONE: +7 (12) 510616; 510869; 510862
FAX: +7 (12) 510868

United States
9 Pushkin Street, Ashgabat, Turkmenistan
PHONE: +7 (12) 350037; 350042
FAX: +7 (12) 511305
TITLE: Ambassador
NAME: Steven R. Mann

JUDICIAL SYSTEM

Supreme Court

FURTHER READING

Articles

''Shell Joins Pipeline Project.'' *The New York Times*, 7 August 1999, p. B2

Zarakhovich, Yuri. ''Life With Father: In Turkmenistan, a Despotic Ruler Does Things the Old Way.'' *Time International* 153 (March 15, 1999): 22.

Books

Central Intelligence Agency. *CIA World Factbook, 1999.* Washington, DC: GPO, 1999.

Turkmenistan. Washington, D.C.: World Bank, 1994.

U.S. Department of State. *Country Reports on Human Rights Practices for 1998: Europe, Canada, and the New Independent States.* Washington, D.C.: U.S. GPO, 1999.

Internet

Turkmenistan Information Center. Available Online @ http://www.turkmenistan.com/ (November 2, 1999).

TURKMENISTAN: STATISTICAL DATA

For sources and notes see "Sources of Statistics" in the front of each volume.

GEOGRAPHY

Geography (1)

Area:

Total: 488,100 sq km.

Land: 488,100 sq km.

Water: 0 sq km.

Area—comparative: slightly larger than California.

Land boundaries:

Total: 3,736 km.

Border countries: Afghanistan 744 km, Iran 992 km, Kazakhstan 379 km, Uzbekistan 1,621 km.

Coastline: 0 km.

Note: Turkmenistan borders the Caspian Sea (1,768 km).

Climate: subtropical desert.

Terrain: flat-to-rolling sandy desert with dunes rising to mountains in the south; low mountains along border with Iran; borders Caspian Sea in west.

Natural resources: petroleum, natural gas, coal, sulfur, salt.

Land use:

Arable land: 3%

Permanent crops: 0%

Permanent pastures: 63%

Forests and woodland: 8%

Other: 26% (1993 est.)

HUMAN FACTORS

Health Personnel (3)

Total health expenditure as a percentage of GDP, 1990-1997[a]

Public sector	1.2
Private sector	NA
Total[b]	NA

Continued on next page.

Demographics (2A)

	1990	1995	1998	2000	2010	2020	2030	2040	2050
Population	3,667.6	4,086.6	4,297.6	4,435.5	5,187.6	6,083.7	6,918.8	7,698.6	8,422.2
Net migration rate (per 1,000 population)	NA	NA	NA	NA	NA	NA	NA	NA	NA
Births	NA	NA	NA	NA	NA	NA	NA	NA	NA
Deaths	NA	NA	NA	NA	NA	NA	NA	NA	NA
Life expectancy - males	61.2	58.5	57.7	57.3	60.2	64.4	68.1	71.2	73.7
Life expectancy - females	68.4	65.9	65.1	64.7	67.4	71.9	75.8	78.8	81.3
Birth rate (per 1,000)	35.4	29.1	26.2	25.6	24.8	21.6	18.0	17.0	15.5
Death rate (per 1,000)	7.9	8.6	8.7	8.9	7.9	6.8	6.6	7.1	7.6
Women of reproductive age (15-49 yrs.)	872.7	1,014.4	1,102.5	1,165.6	1,457.0	1,628.2	1,818.4	1,919.3	1,999.0
of which are currently married	NA	NA	NA	NA	NA	NA	NA	NA	NA
Fertility rate	4.3	3.6	3.3	3.2	2.8	2.6	2.4	2.3	2.2

Except as noted, values for vital statistics are in thousands; life expectancy is in years.

Health Personnel (3) cont.

Health expenditure per capita in U.S. dollars, 1990-1997[a]

Purchasing power parity	NA
Total	NA

Availability of health care facilities per 100,000 people

Hospital beds 1990-1997[a]	1,150
Doctors 1993[c]	353
Nurses 1993[c]	1,195

Health Indicators (4)

Life expectancy at birth	
1980	64
1997	66
Daily per capita supply of calories (1996)	2,563
Total fertility rate births per woman (1997)	3.0
Maternal mortality ratio per 100,000 live births (1990-97)	44[b]
Safe water % of population with access (1995)	60
Sanitation % of population with access (1995)	60
Consumption of iodized salt % of households (1992-98)[a]	0
Smoking prevalence	
Male % of adults (1985-95)[a]	27
Female % of adults (1985-95)[a]	1
Tuberculosis incidence per 100,000 people (1997)	74
Adult HIV prevalence % of population ages 15-49 (1997)	0.01

Infants and Malnutrition (5)

Under-5 mortality rate (1997)	78
% of infants with low birthweight (1990-97)	5
Births attended by skilled health staff % of total[a]	90
% fully immunized (1995-97)	
TB	97
DPT	98
Polio	99
Measles	100
Prevalence of child malnutrition under age 5 (1992-97)[b]	NA

Ethnic Division (6)

Turkmen	77%
Uzbek	9.2%
Russian	6.7%
Kazak	2%
Other	5.1% (1995)

Religions (7)

Muslim	89%
Eastern Orthodox	9%
Unknown	2%

Languages (8)

Turkmen 72%, Russian 12%, Uzbek 9%, other 7%.

EDUCATION

Public Education Expenditures (9)

Public expenditure on education (% of GNP)	
1980	....
1996	....
Expenditure per student	
Primary % of GNP per capita	
1980	....
1996	....
Secondary % of GNP per capita	
1980	....
1996	....
Tertiary % of GNP per capita	
1980	....
1996	....
Expenditure on teaching materials	
Primary % of total for level (1996)	....
Secondary % of total for level (1996)	....
Primary pupil-teacher ratio per teacher (1996)	....
Duration of primary education years (1995)	4

Literacy Rates (11B)

Adult literacy rate	
1980	
Male	—
Female	—
1995	
Male	99%
Female	97%

GOVERNMENT & LAW

Political Parties (12)

The legislative branch is a under the 1992 constitution, there are two parliamentary bodies, a unicameral People's Council (more than 100 seats, some of which are popularly elected and some are appointed meets infrequently) and a unicameral Assembly (50 seats members are elected by popular vote to serve five-year terms).

Government Budget (13B)

Revenues . . . .$521 million

Expenditures . . . .$548 million

Capital expenditures . . . .$83 million

Data for 1996 est.

Military Affairs (14B)

	1992	1993	1994	1995
Military expenditures				
Current dollars (mil.)	NA	NA	NA	196
1995 constant dollars (mil.)	NA	NA	NA	196
Armed forces (000)	28	28	15	21
Gross national product (GNP)				
Current dollars (mil.)	NA	NA	NA	11,500
1995 constant dollars (mil.)	NA	NA	NA	11,500
Central government expenditures (CGE)				
1995 constant dollars (mil.)	4,227	NA	NA	NA
People (mil.)	3.8	3.9	4.0	4.1
Military expenditure as % of GNP	NA	NA	NA	1.7
Military expenditure as % of CGE	NA	NA	NA	NA
Military expenditure per capita (1995 $)	NA	NA	NA	48
Armed forces per 1,000 people (soldiers)	7.3	7.1	3.7	5.2
GNP per capita (1995 $)	NA	NA	NA	2,822
Arms imports[6]				
Current dollars (mil.)	0	0	0	30
1995 constant dollars (mil.)	0	0	0	30
Arms exports[6]				
Current dollars (mil.)	30	0	0	0
1995 constant dollars (mil.)	32	0	0	0
Total imports[7]				
Current dollars (mil.)	1,008	781	884	720
1995 constant dollars (mil.)	1,084	819	906	720
Total exports[7]				
Current dollars (mil.)	2,146	1,286	1,812	1,737
1995 constant dollars (mil.)	2,308	1,348	1,857	1,737
Arms as percent of total imports[8]	0.0	0	0	4.2
Arms as percent of total exports[8]	1.4	0	0	0

LABOR FORCE

Labor Force (16)

Total (million) 2.34

Agriculture and forestry .44%

Industry and construction 19%

Other 37%

Data for 1996. Percent distribution for 1996.

Unemployment Rate (17)

Rate not available.

PRODUCTION SECTOR

Electric Energy (18)

Capacity . . . . .3.95 million kW (1995)

Production . . . . .9.204 billion kWh (1995)

Consumption per capita . . . . .2,013 kWh (1995)

Transportation (19)

Highways:

total: 24,000 km

paved: 19,488 km (note—these roads are said to be hard-surfaced, meaning that some are paved and some are all-weather gravel surfaced

unpaved: 4,512 km (1996 est.)

Waterways: the Amu Darya is an important inland waterway

Pipelines: crude oil 250 km; natural gas 4,400 km

Merchant marine:

total: 1 oil tanker ship (1,000 GRT or over) totaling 1,896 GRT/3,389 DWT (1997 est.)

Airports: 64 (1994 est.)

Airports—with paved runways:

total: 22

2,438 to 3,047 m: 13

1,524 to 2,437 m: 8

914 to 1,523 m: 1 (1994 est.)

Airports—with unpaved runways:

total: 42

914 to 1,523 m: 7

under 914 m: 35 (1994 est.)

Top Agricultural Products (20)

Cotton, grain; livestock.

FINANCE, ECONOMICS, & TRADE

Economic Indicators (22)

National product: GDP—purchasing power parity—$12.5 billion (1996 est.)

National product real growth rate: -0.3% (1996)

National product per capita: $3,000 (1996 est.)

Inflation rate—consumer price index: 992% (1996 est.)

Exchange Rates (24)

Exchange rates:	
Manats per US$1	
January 1997	4,070
January 1996	2,400

Government established a unified rate in mid-January 1996

Top Import Origins (25)

$1.5 billion from states outside the FSU (1996) Data are for 1996.

Origins	%
FSU	NA
United States	NA
Turkey	NA
Germany	NA
Cyprus	NA

NA stands for not available.

Top Export Destinations (26)

$1.7 billion to states outside the FSU (1996) Data are for 1997.

Destinations	%
United States	24.2
Hong Kong	23.5
Europe	15.1
Japan	9.6

Economic Aid (27)

Recipient: ODA, $10 million (1993). Note: commitments, $1,830 million ($375 million drawn), 1992-95.

Import Export Commodities (28)

Import Commodities	Export Commodities
Machinery and parts	Natural gas
Grain and food	Cotton
Plastics and rubber	Petroleum products
Consumer durables	Textiles
Textiles	Electricity
	Carpets

MANUFACTURING SECTOR

GDP & Manufacturing Summary (21)

	1980	1985	1990	1992	1993	1994
Gross Domestic Product						
Millions of 1990 dollars	3,411	4,453	5,558	5,199	5,069	4,056
Growth rate in percent	−2.31	1.33	5.39	−5.40	−2.50	−20.00
Per capita (in 1990 dollars)	1,190.8	1,380.9	1,519.9	1,356.5	1,292.9	1,011.3
Manufacturing Value Added						
Millions of 1990 dollars	NA	NA	628	765	*795*	*597*
Growth rate in percent	NA	NA	NA	20.30	*4.03*	*-24.98*
Manufacturing share in percent of current prices	NA	NA	*11.3*	*14.7*	NA	NA

TURKS AND CAICOS ISLANDS

CAPITAL: Grand Turk (Cockburn Town).

FLAG: Blue, with the flag of the United Kingdom in the upper hoist side quadrant. The colonial shield is centered on the outer half of the flag. The shield is yellow and contains a conch shell, a lobster, and a cactus. The shield is supported by two flamingos, and as its crest a pelican between two sisal plants.

ANTHEM: *God Save the Queen*

MONETARY UNIT: 1 U.S. dollar (US$) = 100 cents.

WEIGHTS AND MEASURES: The metric system is in force.

HOLIDAYS: Constitution Day, 30 April (1976).

TIME: 7 AM = noon GMT.

LOCATION AND SIZE: The Turks and Caicos Islands are located in the Caribbean Sea. They are composed of two island groups in the North Atlantic Ocean, southeast of the Bahamas. Together, the islands possess a total of 430 sq km, with a coastline of 389 km. The Turks and Caicos Islands are composed of extensive marshes and mangrove swamps.

CLIMATE: The climate is tropical and marine. The Turks and Caicos Islands are moderated by trade winds and weather is generally sunny and relatively dry.

INTRODUCTORY SURVEY

RECENT HISTORY

There is some speculation that Columbus may have made his landfall on Grand Turk or East Caicos on his first voyage of discovery in 1492. The first settlements were by Bermudians, who established solar salt pans in the 1670s. Bahamian, Bermudan, Spanish, French, and British rivalry over the prospering salt trade resulted in numerous invasions and evictions through the first half of the 18th century. In 1787, Loyalists fleeing the American Revolution established settlements and cotton and sisal plantations on several of the larger Caicos Islands. Ten years later, the islands came under the jurisdiction of the Bahamas colonial government. Slavery was abolished in 1834. In 1848, the Turks and Caicos islanders were granted a charter of separation from the Bahamas after they complained of Bahamian taxes on their salt industry.

From 1848 to 1873, the islands were largely self-governing, under the supervision of the governor of Jamaica. Following the decline of the salt industry, the islands became a Jamaican dependency until 1958, when they joined the Federation of the West Indies. When the federation dissolved and Jamaica achieved independence in 1962, Turks and Caicos became a crown colony administered by the British Colonial Office and a local council of elected and appointed members. In 1965, the governor of the Bahamas was also appointed governor of Turks and Caicos, but with the advent of Bahamian independence in 1973, a separate governor was appointed. A new constitution maintaining the status of crown colony and providing for ministerial government was introduced in September 1976. Although independence for Turks and Caicos in 1982 had been agreed upon in principle in 1979, a change in government brought a reversal in policy. The islands are still a crown colony.

The islands were shaken by scandals in the mid-1980s. In March 1985, Chief Minister Norman B. Saunders and two other ministers were arrested in Florida on drug charges and later convicted and sentenced to prison. In July 1986, a commission of inquiry found that Chief Minister Nathaniel Francis and two other ministers had been guilty of ''uncon-

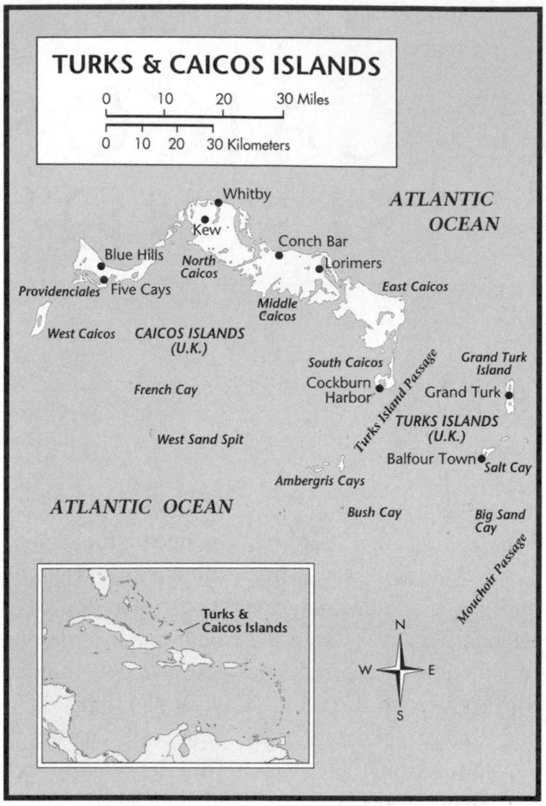

stitutional behavior, political discrimination, and administrative malpractice.'' The governor thereupon ended ministerial government in July 1986 and, with four members of the former Executive Council, formed an Advisory Council to govern until new elections. The islands have returned to their previous form of government, and remain a dependent territory of the UK.

GOVERNMENT

The constitution of 30 August 1976 established a ministerial system in which the governor, representing the crown, retains responsibility for external affairs, defense, and internal security. The governor presides over an Executive Council of seven members: the chief minister, three elected members of the Legislative Council, and three ex officio members (the financial secretary, the chief secretary, and the attorney general). The Legislative Council is unicameral, having 20 seats: a speaker, elected representatives from each of 13 constituencies, 3 nominated members, and the same 3 ex-officio members from the Executive Council. In order to remain in office, the chief minister must command a majority of the Legislative Council. A number of constitutional changes have been suggested and studied, but are still under review by a special commission.

Judiciary

Legislative Council acts, certain laws of the UK Parliament, and a few Jamaican and Bahamian statutes are the law of the land. The administration of justice is in the hands of a magistrate who sits weekly in Grand Turk and may sit in each of the other islands as necessary. The magistrate is also the registrar of deeds and of birth, marriage, death, and company records. Appeals are heard by a nonresident member of the Eastern Caribbean Supreme Court; as of 1985, this judge was from the Bahamas. Appeals in certain cases may also be taken to the UK Judicial Committee of the Privy Council in London.

Political Parties

The Progressive National Party (PNP) and the People's Democratic Movement (PDM) are the two major parties in the system. They differ primarily over the question of independence. The PDM had sought the rapid advance of Turks and Caicos to full internal self-government, and when the UK reportedly insisted on rapid progress toward independence, the PDM appeared to swing toward that view. The PNP, on the other hand, expressed a go-slow attitude, holding that the islands lacked the institutional and financial resources for independence.

With the PNP victory in 1980, Turks and Caicos' independence date, originally scheduled for mid-1982, was postponed indefinitely. During the 1984 elections, neither party raised the issue of independence. In 1987, the PDM won 11 seats and the PNP, 2 seats. In 1991, PNP took 8 of the 13 elected seats in the Legislative Council. In the 1995 elections, PNP took only 4 seats, PDM took 8 seats, and there was 1 independent. The other political parties include the National Democratic Alliance, Turks and Caicos United Party, and the United Democratic Party. The next elections are scheduled for 2000.

ECONOMIC AFFAIRS

Tourism and lobster fishing have replaced salt raking as the main economic activity of the islands, which are very poor. Fishing and subsistence farming are the principal occupations; underemployment and unemployment are estimated at over 40%. Important sources of income include tourism and offshore financial services. The closing of the last U.S. military base in 1983 resulted in the loss of rental payments that accounted for 10% of government revenue. Most of the retail trade on the islands consists of imported goods.

The Turks and Caicos economy is based on tourism, fishing, and offshore financial services. Most food for domestic consumption is imported; there is some subsistence farming mainly corn, cassava, citrus, and beans. The tourism sector expanded in 1995, posting a 10% increase in the first quarter as compared to the same period in 1994. The U.S. was the leading source of tourists in 1995, accounting for upward of 70% of arrivals or about 60,000 visitors. Major sources of government revenue include fees from offshore financial activities and customs receipts as the Islands rely on imports for nearly all consumption and capital goods.

Income

The U.S. Central Intelligence Agency (CIA) maintains statistics gross domestic product (GDP), defined as the value of all final goods and services produced within a nation in a given year. According to CIA estimates, the purchasing power parity of GDP in 1995 was $80.8 million or $6,000 per capita. The CIA estimates that in 1995 the real growth rate of GDP was −1.5%.

Industry

The main industry is the processing of lobster and conch for export. South Caicos and Providenciales have a combined total of four seafood-processing plants. Other industries are related to tourism and offshore services. Most self-employed persons engage in fishing or subsistence farming. Government is the largest employer, followed by the tourist sector and financial services. There is one labor organization, the St. George's Industrial Trade Union, with a membership of approximately 250.

Of the total area of Turks and Caicos, only about 2%, or 1,000 hectares (2,500 acres), is devoted to agriculture. Crop output is small. Rainfall is low, and the cost of transporting produce from island to island is not competitive with imports. Most agricultural production is on North Caicos, although subsistence farming occurs wherever the soil permits. Corn, beans, and other food crops are grown entirely for local or household consumption. Fruit and many other foods must be imported, with Haiti as the principal supplier.

Banking and Finance

Four commercial banks had branches on the principal islands in 1993. In addition to providing the normal financial services, the banks handle securities trading on the principal international exchanges.

Economic Development

As a means of attracting foreign capital, the Encouragement of Development Ordinance grants new investment guarantees against taxes for 10–15 years. In addition, there are no direct taxes on income capital for individuals or companies and no exchange controls. A 400 acre leeward resort project is being developed at the northeast end of Providenciales. This project, estimated at $165 million, was acquired by the Belize based BHI Corporation.

The UK government has historically been the underwriter of economic development planning and investment. Tourism presents the primary economic development opportunity, but growth has so far been small. Since 1979, Turks and Caicos has experienced a flurry of offshore tax haven ventures, but as of 1987 these had little significant on local financial or employment conditions.

The most recent economic development has been on Providenciales, in the form of tourist facilities and retirement home construction. New hotels, resorts, and casinos were inaugurated in the islands in 1993, and the Providenciales airport was expanded at a cost of $8.5 million. In 1991, the PNP government proposed a five-year development program for economic independence called ''Progress through Partnership,'' which aims at achieving over 15% in real growth per year for the next five years.

SOCIAL WELFARE

Healthcare

A modern cottage hospital (30 beds) and an outpatient and dental clinic are located on Grand Turk, and there are 11 outpatient and dental clinics on South, Middle, and North Caicos, Providenciales, and Salt Cay. In 1992, there were 1.25 doctors per 1,000 people. The population in 1994 was only 14,000. The under 5 mortality rate was 24 per 1,000 in 1994, while the infant mortality rate was 25 per 1,000 live births the following year. In 1994, 100% of the population under age 1 was immunized for measles and diptheria, pertussis, and tetanus. In 1993, there were 39 AIDS cases reported.

Housing

Most old wooden shacks—vulnerable to hurricanes and frequently damaged—have been replaced by concrete block structures, especially in the towns. However, some traditional 19th century Bermudian architecture remains. Several old churches and the library are built of limestone.

EDUCATION

Education is free and compulsory between the ages of 7 and 14. Six years of primary school is followed by five years of secondary school. The adult literacy rate is approximately 98%. In 1993 there were 10 primary schools with a total of 1,211 students. Secondary schools had an enrollment of 1,032 students. There are no higher educational institutions on the islands.

1999 KEY EVENTS TIMELINE

January

- Parrot Cay, a 56-room hotel and beach resort, opens in late 1998, and enjoys popularity in early 1999 with tourists.

February

- Twice-weekly air service from Miami, Florida to Grand Turk island is begun by BahamasAir.

September

- The government of Turks and Caicos Islands issues the first license for a casino on the islands on Septmeber 1.
- Charles Palmer, a TCI restaurateur, wins the September 18 Florida Lotto. His lump sum winnings total $3,289,343.
- A tornado strikes Whitby.
- The initial phase is completed of a planned $9-million expansion of the airport on Providenciales, including expansion of both the international arrivals area and the customs and immigration area, the installation of air conditioning and of a baggage carousel.

October

- John Kelly, current governor, will retire from the diplomatic service and his successor, Mervyn Jones is appointed to assume the post in January 2000.
- The first casino on TCI opens October 1.
- ''Empowering the Woman and Girlchild'' is the theme of an event (October 29–31) during which a women's desk in Grand Turk was created to address concerns and issues related to women. National Women in Development cosponsors the event.

ANALYSIS OF EVENTS: 1999

BUSINESS AND THE ECONOMY

Situated one hour (by jet) southeast of Miami, Florida, covering just under 200 square miles, with 230 miles of pristine white-sand beaches and one of the longest coral reefs in the world, the eight islands making up the British overseas territory of Turks and Caicos Islands was mainly sustained by tourism in 1999. Tourism was a key component of the economy although the economy also featured fishing and off-shore financial services. The five percent growth rate of the gross domestic product (GDP) for the previous year was primarily due to increased tourism. Visitor arrivals increased by nearly six percent in 1997, nearly 70 percent of whom were from the United States. The international airport at Providenciales underwent a $9-million upgrade, and twice-weekly air service was initiated between Miami and Grand Turk.

Gambling came to Turks and Caicos Islands in 1999, with the government issuing a gambling license on September 1 and the first licensed casino opening one month later.

In addition to tourism, Turks and Caicos Islands tax laws offer attractive conditions for financial services, making banking and finance another important component of the TCI economy, although the number of finance companies declined by about three percent in 1997.

GOVERNMENT AND POLITICS

Turks and Caicos Islands (TCI) was a territory of the United Kingdom. Mervyn Jones was the appointed governor of TCI, assuming his post at the end of January 2000. Jones was a well-traveled official of the British diplomatic corps. His primary assignment was to assist British exporters.

CULTURE AND SOCIETY

Although a small dependent territory, Turks and Caicos Islands features a healthy and long-lived population of only about 17,000 people. The average life expectancy at birth was over 72 years. The infant mortality rate was 21 deaths per thousand live births. The ethnicity was African and the per capita Gross Domestic Product was $7,700. The popular culture of the islands included an active contribution from women. A government women's desk devoted to women's issues was created during a special event held in Grand Turk October 29–31.

DIRECTORY

CENTRAL GOVERNMENT

Head of State

Monarch
Elizabeth II, Queen of England

Governor
Mervyn Jones, Office of the Governor
PHONE: +649 9462308
FAX: +649 9462903

Chief Minister
Derek H. Taylor, Office of the Chief Minister
PHONE: +649 9462801
FAX: +649 9462777

Ministers

Minister of Finance, Development, and Commerce
Derek Taylor, Ministry of Finance, Development, and Commerce
PHONE: +649 9462801
FAX: +649 9462777

Minister of Tourism, Communications, and Transportation
Oswald Skippings, Ministry of Tourism, Communications, and Transportation

Minister of Home Affairs, Immigration, Labour, and Prison
Hilly Ewing, Ministry of Home Affairs, Immigration, Labour, and Prison, Government Compound Yard, Grand Turk, Turks and Caicos Islands
PHONE: +649 9462801
FAX: +649 9462885

POLITICAL ORGANIZATIONS

People's Democratic Movement (PDM)

Progressive National Party (PNP)

DIPLOMATIC REPRESENTATION

None (Overseas Territory of the United Kingdom)

JUDICIAL SYSTEM

Supreme Court

FURTHER READING

Articles

Latham, Aaron. ''Thar She Blows! Time for a Swim. (Whale-watching in the Caribbean near the Turks and Caicos).'' *The New York Times*, 24 October 1999, p. TR10.

Wasserstein, Wendy. ''Cays to Paradise. (Parrot Cay beach resort, Turks and Caicos Islands).'' *Travel and Leisure* 29 (June 1999): 158.

Books

Baker, Christopher P. *Lonely Planet Turks and Caicos.* Hawthorne, Australia: Lonely Planet, 1998.

Internet

Where, When, How. Available Online @ http://www.WhereWhenHow.com/ (November 19, 1999).

Turks and Caicos Islands Tourist Board. Available Online @ http://www.turksandcaicostourism.com/ (November 19, 1999).

TURKS AND CAICOS ISLANDS: STATISTICAL DATA

For sources and notes see "Sources of Statistics" in the front of each volume.

GEOGRAPHY

Geography (1)

Area:

Total: 430 sq km.

Land: 430 sq km.

Water: 0 sq km.

Area—comparative: 2.5 times the size of Washington, DC.

Land boundaries: 0 km.

Coastline: 389 km.

Climate: tropical; marine; moderated by trade winds; sunny and relatively dry.

Terrain: low, flat limestone; extensive marshes and mangrove swamps.

Natural resources: spiny lobster, conch.

Land use:

Arable land: 2%

Permanent crops: NA%

Permanent pastures: NA%

Forests and woodland: NA%

Other: 98% (1993 est.)

HUMAN FACTORS

Ethnic Division (6)

Black.

Religions (7)

Baptist	41.2%
Methodist	18.9%
Anglican	18.3%
Seventh-Day Adventist	1.7%
Other	19.9%

Data for 1980.

Demographics (2A)

	1990	1995	1998	2000	2010	2020	2030	2040	2050
Population	11.5	14.4	16.2	17.5	23.4	29.3	34.6	38.8	41.8
Net migration rate (per 1,000 population)	NA	NA	NA	NA	NA	NA	NA	NA	NA
Births	NA	NA	NA	NA	NA	NA	NA	NA	NA
Deaths	NA	NA	NA	NA	NA	NA	NA	NA	NA
Life expectancy - males	68.6	69.6	70.2	70.6	72.3	73.8	75.1	76.2	77.1
Life expectancy - females	72.5	73.6	74.2	74.6	76.5	78.1	79.6	80.8	81.8
Birth rate (per 1,000)	33.2	29.5	27.1	25.6	20.7	19.7	17.4	15.6	14.4
Death rate (per 1,000)	6.3	5.4	5.0	4.8	4.6	5.2	6.4	7.7	8.4
Women of reproductive age (15-49 yrs.)	3.2	3.9	4.3	4.6	5.9	7.1	8.2	9.4	9.8
of which are currently married	746.0	NA	NA	NA	NA	NA	NA	NA	NA
Fertility rate	3.6	3.4	3.3	3.3	2.9	2.6	2.4	2.2	2.1

Except as noted, values for vital statistics are in thousands; life expectancy is in years.

Languages (8)

English (official).

EDUCATION

Educational Attainment (10)

Age group (1980)	25+
Total population	2,859
Highest level attained (%)	
No schooling	0.9
First level	
Not completed	74.6
Completed	NA
Entered second level	
S-1	16.9
S-2	NA
Postsecondary	7.7

GOVERNMENT & LAW

Political Parties (12)

Legislative Council	No. of seats
People's Democratic Movement (PDM)	8
Progressive National Party (PNP)	4
Independent	1

Government Budget (13B)

Revenues	$31.9 million
Expenditures	$30.4 million
Capital expenditures	NA

Data for 1995. NA stands for not available.

Military Affairs (14A)

Defense is the responsibility of the UK.

Crime (15)

Crime rate (for 1997)	
Crimes reported	1,050
Total persons convicted	450
Crimes per 100,000 population	7,150
Persons responsible for offenses	
Total number of suspects	500
Total number of female suspects	49
Total number of juvenile suspects	10

LABOR FORCE

Labor Force (16)

Total 4,848. About 33% in government and 20% in agriculture and fishing; large numbers in tourism and financial and other services. Data for 1990 est. Percent distribution for 1997 est.

Unemployment Rate (17)

15% (1996 est.)

PRODUCTION SECTOR

Electric Energy (18)

Capacity	4,000 kW (1995)
Production	5 million kWh (1995)
Consumption per capita	359 kWh (1995)

Transportation (19)

Highways:
total: 121 km
paved: 24 km
unpaved: 97 km
Merchant marine: none
Airports: 7 (1997 est.)
Airports—with paved runways:
total: 4
1,524 to 2,437 m: 3
914 to 1,523 m: 1 (1997 est.)
Airports—with unpaved runways:
total: 3
914 to 1,523 m: 2
under 914 m: 1 (1997 est.)

Top Agricultural Products (20)

Corn, beans, cassava, citrus fruits; fish.

FINANCE, ECONOMICS, & TRADE

Economic Indicators (22)

National product: GDP—purchasing power parity—$110 million (1996 est.)
National product real growth rate: 3.5% (1996 est.)
National product per capita: $7,700 (1996 est.)
Inflation rate—consumer price index: 8% (1994 est.)

Exchange Rates (24)

Exchange rates: US currency is used.

Top Import Origins (25)

$42.8 million (1993). Data are for 1997.

Origins	%
Japan	25.6
United States	13.9
Singapore	.5
Taiwan	4.6
Germany	4.5
Malaysia	4.1

Top Export Destinations (26)

$6.8 million (f.o.b., 1993).

Destinations	%
United States	NA
United Kingdom	NA

NA stands for not available.

Economic Aid (27)

Recipient: ODA, $NA. NA stands for not available.

Import Export Commodities (28)

Import Commodities	Export Commodities
Food and beverages	Lobster
Tobacco	Dried and fresh conch
Clothing	Conch shells
Manufactures	
Construction materials	

TUVALU

CAPITAL: Funafuti.

FLAG: The national flag has the Union Jack in the upper quarter nearest the hoist; nine yellow stars on a light blue field are arranged in the same pattern as Tuvalu's nine islands.

ANTHEM: *Tuvalu mo te Atua (Tuvalu for the Almighty).*

MONETARY UNIT: Both the Australian dollar (A$) and the Tuvaluan dollar (T$) of 100 cents are legal tender. There are coins of 1, 2, 5, 10, 20, and 50 Tuvaluan cents; 1 and 5 Tuvaluan dollars; and notes of 5, 10, 20, 50, and 100 Australian dollars. T$1 = US$0.7008 (or US$1 = T$1.4269).

WEIGHTS AND MEASURES: The metric system is being introduced, but imperial measures are still commonly employed.

HOLIDAYS: New Year's Day, 1 January; National Children's Day, first Monday in August; Tuvalu Day, 1 October; Christmas Day, 25 December; Boxing Day, 26 December. Movable holidays include Commonwealth Day (March), Queen's Official Birthday (June), and Prince of Wales's Birthday (November); movable religious holidays include Good Friday and Easter Monday.

TIME: Midnight = noon GMT.

LOCATION AND SIZE: Tuvalu (formerly the Ellice Islands) is a cluster of nine islands located in the southwestern Pacific Ocean, with a total land area of 26 square kilometers (10 square miles)—about one-tenth the size of Washington, D.C.—and a coastline of 24 kilometers (15 miles). Only three nations are smaller than Tuvalu—Nauru, Monaco, and Vatican City. Tuvalu's capital city, Funafuti, is located on the island of Funafuti.

CLIMATE: The annual mean temperature of 30°C (86°F) is moderated by trade winds from the east. Rainfall averages over 355 centimeters (140 inches).

INTRODUCTORY SURVEY

RECENT HISTORY

The islands were probably settled between the 14th and 17th centuries by Polynesians drifting west with prevailing winds from Samoa and other large islands. The first European to discover Tuvalu is thought to have been the Spanish navigator Alvaro cle Mendaiia de Neyra, who sighted Nui in 1568 and Niulakita in 1595. Further European contact was not made until the end of the 18th century. Between 1850 and 1875, the islands were raided by ships forcibly recruiting plantation workers for South America, Fiji, Hawaii, Tahiti, and Queensland. To help suppress such abuses, the Office of British High Commissioner for the Western Pacific was created in 1877.

In 1892, after ascertaining the inhabitants' wishes, the UK proclaimed the Ellice Islands (as Tuvalu was then known), together with the Gilberts, as a British protectorate. After further consultation, the protectorate became the Gilbert and Ellice Islands Colony in 1916. After the Gilberts were occupied by the Japanese in 1942, U.S. forces occupied the Ellice group in 1943 and drove the Japanese out of the Gilberts. After the war, the ethnic differences between the Micronesians of the Gilberts and the Polynesians of the Ellice Islands led the Ellice Islanders to demand separation. In 1973, a British commissioner appointed to examine the situation recommended administrative separation of the two island groups. The British government agreed, provided that the Ellice Islanders declared their wishes by referendum. The vote,

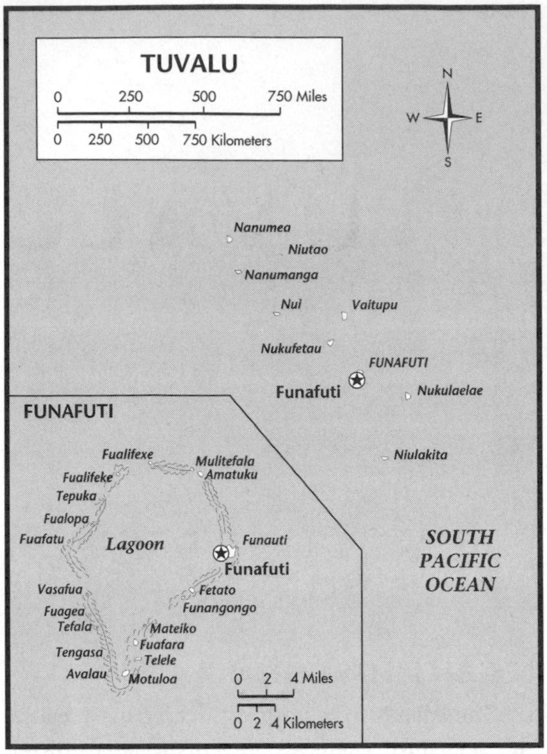

held during August-September 1974 with UN observers in attendance, produced an overwhelming majority of 3,799 to 293 for separation. Accordingly, on 1 October 1975, the Ellice Islands were established as the separate British colony of Tuvalu, and a ministerial system was instituted. Pursuant to a constitutional conference held at London in February 1978, Tuvalu became an independent member of the Commonwealth of Nations on 1 October 1979. Sir Fiatau Penitala Teo became Tuvalu's first governor-general, and Toaripi Lauti, chief minister at the time of independence, took office as Tuvalu's first prime minister. Following new elections in September 1981, Lauti was succeeded in office by Tomasi Puapua, who was reelected in September 1985. In March 1986, Tupua Leupena replaced Sir Fiatau Penitala Teo as governor-general. In a poll held that same year, Tuvaluans rejected the idea that Tuvalu should become a republic. As a result of the 1989 general election the Parliament elected Bikenibeu Paeniu as Prime Minister in September 1989. In the same election, Naama Latasi became the first woman to serve in Tuvalu's Parliament.

In the 1993 legislative elections Paeniu and Puapua, the man who he replaced as Prime Minister, each received six votes from the newly elected 12-member parliament. A second round of votes were held in December that year, from which Puapua withdrew, and Kamuta Latasi was elected Prime Minister.

In 1994, Prime Minister Latasi spearheaded a movement to remove the British Union Jack from the country's flag as a symbolic gesture of independence. In 1995, after conservative French President Jacques Chirac announced his country's intention to conduct above-ground nuclear tests in the South Pacific, Tuvalu emerged as a regional leader in the highly vocal opposition.

In April 1997, the Union Jack was restored as part of Tuvalu's national flag by a vote of 7 to 5 in the Parliament. Newly reelected Prime Minister Bikenibeau Paeniu restored the former flag design, which Latasi had changed without consideration of the views of Tuvalu's citizens.

GOVERNMENT

Tuvalu is an independent constitutional monarchy. The head of state is the British monarch, whose representative on the islands is the governor-general, a Tuvaluan who has the power to convene and dissolve parliament (Manuella Tulaga since June 1994). There is a unicameral legislature, the House of Assembly, with 12 members elected to four-year terms by universal adult suffrage. Four islands—Funafuti, Nanumea, Niutao, and Vaitupu—elect two members each; the other islands elect one each. The cabinet is headed by the prime minister and has four ministers (all House members), plus the attorney-general.

Local administration by elected island councils was established following the creation of the protectorate in 1892. Local governments were established on the eight inhabited islands by a 1966 ordinance that provided the framework for a policy aimed at financing local services at the island level. Funafuti's town council and the other seven island councils (Niulakita has none) each consist of six elected members, including a president.

Judiciary

District magistrates were established with the protectorate in 1892, and a simple code of law based on mission legislation and traditional councils has been observed by native courts. Eight island courts were constituted in 1965 to deal with land disputes, among other local matters. In 1975, a High Court of Justice was set up to hear appeals from district courts. Appeals from the High Court may go to the Court of Appeals in Fiji and ultimately to the UK Privy Council in London.

The right to a fair public trial is respected in practice. Services of the public defender are available to all Tuvaluans free of charge. Defendants have the right to confront witnesses, present evidence, and to appeal. The judiciary is independent and free of governmental interference.

Political Parties

There are no political parties, and political life and elections are dominated by personalities. Small island constituencies with a few hundred kin-related electors judge the leaders by their service to the community.

DEFENSE

Tuvalu has no armed forces except for the local police. For defense the islands rely on Australian-trained volunteers from Fiji and Papua New Guinea.

ECONOMIC AFFAIRS

Prime Minister Toaripi Lauti noted at the time of independence (1979) that all Tuvalu has is sun and a portion of the Pacific. Economic life is simple, but there is no extreme poverty. Subsistence is based on intensive use of limited resources, namely coconuts and fish; copra is the only cash crop. The islands are too small and too remote for development of a tourist industry. The United Nations ranks Tuvalu among the least-developed countries. Tuvalu's GDP was estimated at us$9 million for 1992, equivalent to us$713 per capita. Inflation averaged 3.9% from 1985–93.

Income

Tuvalu's GDP (A$16 million in 1994) is the smallest of any independent state. Wholesale and retail trade account for about one-third of the GDP; government and other services, 9%; agriculture and fishing, 20%; and other sectors for the balance. Major sources of income are British aid and remittances from Tuvaluan workers abroad, which bring in around A$1.6 million a year. The minimum wage in the public sector was $0.40 per hour in 1996. Generally the private sector adopts the same wage.

The U.S. Central Intelligence Agency (CIA) maintains statistics gross domestic product (GDP), defined as the value of all final goods and services produced within a nation in a given year. According to CIA estimates, the purchasing power parity of GDP in 1995 was $7.8 million or $800 per capita. The CIA estimates that in 1995 the inflation rate for consumer prices was 2.9%.

Revenue is obtained principally by means of indirect taxation: stamp sales, the copra export tax, and fishing licenses. Income tax was levied on chargeable income on a sliding scale of 9–50% until 1983, when the rate was changed to a flat 30% of all individual income above A$1,800. A year later, the corporate rate was changed to 40%. Island councils also levy a head tax and a land tax based on territorial extent and soil fertility.

Since a single line tariff was implemented on 1 January 1975, trade preferences are no longer granted to imports from Commonwealth countries. Tariffs, applying mostly to private imports, are levied as a source of revenue. Most duties are ad valorem, with specific duties on alcoholic beverages, tobacco, certain chemicals, petroleum, cinematographic film, and some other goods.

Industry

There is no industry apart from handicrafts, baking, and small-scale construction; the islands lack the population, capital, and resources to make commercial enterprises cost effective. In 1994, manufacturing contributed less than 2% to GDP. Cooperative societies dominate commercial life, controlling almost all retail outlets, the marketing of local handicrafts, and the supply of fish to the capital. Barter remains an important part of the subsistence economy.

Copra, the main cash crop, took many years to recover from the 1972 hurricane and has been affected by fluctuating market prices (although there is a subsidy to producers). Other exports include handicrafts and postage stamps. Most food, fuel, and manufactured goods are imported. Merchandise exports were valued at A$347,000 in 1995, while imports amounted to A $10,327,000 that year. Tuvalu's principal trade partners are Fiji, Australia, and New Zealand.

Although agriculture is the principal occupation, it contributed only 26% to the GDP in 1994. Agriculture is limited because of poor soil quality (sand and rock fragments), uncertain rains, and primitive catchment. Coconuts form the basis of both subsistence and cash cropping; the coconut yield in 1995 was about 3,000 tons. Other food crops are pulaka (taro), pandanus fruit, bananas, and papayas.

The Agricultural Division, based on Vaitupu, has attempted to improve the quality and quantity of livestock to lessen the islands' dependency on imports. Pigs and fowl, which were imported in the 19th century, have been supplanted by goats and rabbits. In 1995, there were some 13,000 pigs on the islands. Honey is also produced.

Sea fishing, especially for tuna and turtle, is excellent. Although fishing is mainly a subsistence occupation, fish is sold in the capital, and beche-de-mer is exported. The fish catch in 1994 was 561 tons, down from 1,460 tons in 1993. The average annual catch during 1990–94 was 713 tons. Japanese aid in 1982 provided a commercial fishing vessel for the islands.

Banking and Finance

The Bank of Tuvalu was founded in Funafuti in 1980 and has branches on all the islands. The bank is jointly owned by the Tuvalu government (75%) and by Barclays Bank, which was responsible for its operation until mid-1985. In 1995, the government bought Westpac's 40% shareholding in the National Bank of Tuvalu and now owns the bank outright. Westpac has managed the bank since it was established in 1980 and is expected to provide an advisory support service.

Economic Development

The cash economy is not sufficiently developed to attract substantial foreign investment. In 1981, the government established the Business Development Advisory Board to promote local and foreign investment in the Tuvalu economy; in 1993, the board became the Development Bank of Tuvalu. As of 1993, a new hotel was slated for construction on Funafuti by Taiwanese interests, and its airstrip was to be paved with aid from the EU.

Development aid, which rose rapidly during the 1960s, peaked at independence in 1979, when the UK undertook to provide £6 million. The Tuvalu Trust Fund was established in 1987 with A$27 million. The Fund receives contributions from Australia, New Zealand, the UK, Japan, Korea, and Tuvalu itself. The net income is paid to the Tuvalu government annually. As of 1994, the Fund amounted to A$36.8 million. Between 1990 and 1992, net total aid from Commonwealth countries and international agencies amounted to us$8.2 million.

SOCIAL WELFARE

Healthcare

There are no serious tropical diseases on the islands except for a dwindling number of leprosy and dysentery cases. In 1990, there were 1.8 hospital beds per 1,000 people. In 1992, there were 31 doctors per 100,000 people. The infant mortality rate was 73.6 per 1,000 live births. By 1991–93, a large portion of Tuvalu's population had access to safe water (100%) and sanitation (85%) in 1993. In 1995, the immunization rates for a child under one were as follows: diphtheria, pertussis, and tetanus (82%); polio (92%); measles (94%); and tuberculosis (88%). About 49% of children under one had been immunized for hepatitis B in 1995. The average life expectancy in 1991–93 was 67.2 years for men and 64 years for women. Malaria was one of the most reported diseases in 1993, with 10,377 cases that year.

Housing

Most islanders live in small villages and provide their own housing from local materials. After the 1972 hurricane, Funafuti was rebuilt with imported permanent materials, but there is still a critical housing shortage on Funafuti and Vaitupu. Government-built housing is largely limited to that provided for civil servants.

EDUCATION

All children receive free primary education from the age of 7. Education is compulsory for nine years. The Tuvaluan school system has seven years of primary and six years of secondary education. Secondary education is provided at Motufoua, a former church school on Vaitupu now jointly administered by the government. In 1994, 1,906 students were enrolled in 11 primary schools while in 1990, secondary schools had 345 students. Tuvalu Marine School was opened in 1979 with Australian aid. In the same year, the University of the South Pacific (Fiji) established an extension center at Funafuti.

1999 KEY EVENTS TIMELINE

April

- Prime Minister Bikenibeu Paeniu is forced out of office by a vote of no confidence.

- The 12-member Tuvalu Parliament elects former Minister of Education Ionatana Ionatana as Prime Minister.

May

- Prime Minister Ionatana holds first official parliamentary meeting.
- Ionatana announces outer island development is one of his government's long-term goals.

July

- James Richard Duckworth, from North England, becomes the new People's Lawyer in Tuvalu.

August

- A severe drought continues to affect Tuvalu; several nations promise help.

September

- Tuvalu officials say small nations are not being heard by world organizations like the United Nations.

ANALYSIS OF EVENTS: 1999

BUSINESS AND THE ECONOMY

With a population of only 10,588 and a natural setting of nine coral atolls in the South Pacific Ocean, halfway between Hawaii and Australia, Tuvalu is a tiny, low-lying Polynesian nation that may entirely disappear if global warming raises the water level high enough.

During part of 1999, Tuvalu suffered through a drought, forcing the small nation to seek economic help. Australia, Japan, New Zealand and Britain said they would help Tuvalu building a desalination plant to help ease water problems. Tuvalu said it needed two desalination plants as water supplies ran low in early August. Japan agreed to provide the plants, while New Zealand said it would help pay to transport them. Australia said it would provide technical assistance and help Tuvalu create water policies.

Government officials also unveiled long-term plans to help develop the outer islands of Tuvalu. They said they would seek funding from the Asian Development Bank to create a trust fund for special projects.

GOVERNMENT AND POLITICS

In April, Prime Minister Bikenibeu Paeniu was forced to give up his post after a vote of no confidence by the 12-member Parliament. Political opponents said growing dissatisfaction with his leadership led to the vote. Reportedly, the parliament members were not happy with Paeniu's private life, and with the slow pace on government projects. Parliament selected Ionatana Ionatana, former Minister of Education, as the new Prime Minister in late April. His new cabinet included Lagitupu Tuilimu as Deputy Prime Minister and Minister of Finance, Samuelu Teo as Minister of Works and Communications, Faimalaga Luka as Minister of Natural Resources and Home Affairs, and Teagai Esekia as Minister of Health and Education.

The government of Tuvalu raises revenue by selling its stamps as well as from proceeds of an international trust fund established in 1987 by Australia, New Zealand, the United Kingdom, supplemented by grants from Japan and South Korea.

CULTURE AND SOCIETY

In June, the Tuvalu Philatelic Bureau issued a special set of 28,000 stamps to help raise funds for Kosovo refugees. Tuvalu officials said 50 percent of the sale price of each stamp set would be donated to humanitarian organizations helping Kosovar refugees.

DIRECTORY

CENTRAL GOVERNMENT

Head of State

Monarch
Elizabeth II, Queen of England

Prime Minister
Ionatana Ionatana, Office of the Prime Minister
PHONE: +688 20100

Governor General
Tomasi Puapua
PHONE: +688 20715

Ministers

Minister of Education, Sports, and Culture
Teagai Esekia, Ministry of Education, Sports, and Culture
PHONE: +688 20407

Minister of Finance and Economic Planning
Lagitupu Tuilimu, Ministry of Finance and Economic Planning
PHONE: +688 20201

Minister of Foreign Affairs
Ionatana Ionatana, Ministry of Foreign Affairs
PHONE: +688 20839

Minister of Health, Women's, and Community Affairs
Teagai Esekia, Ministry of Health, Women's, and Community Affairs
PHONE: +688 20402

Minister of Internal Affairs and Rural and Urban Development
Faimalaga Luka, Ministry of Internal Affairs and Rural and Urban Development
PHONE: +688 20172

Minister of Natural Resources and Environment
Faimalaga Luka, Ministry of Natural Resources and Environment
PHONE: +688 20828

Minister of Tourism, Trade, and Commerce
Lagitupu Tuilimu, Ministry of Tourism, Trade, and Commerce
PHONE: +688 20182

Minister of Works, Energy, and Communications
Samuelu Teo, Ministry of Works, Energy, and Communications
PHONE: +688 20051

POLITICAL ORGANIZATIONS

None

DIPLOMATIC REPRESENTATION

Embassies in Tuvalu

Great Britain
Funafuti, Tuvalu

JUDICIAL SYSTEM

Magistrates Court

PHONE: +688 20837

FURTHER READING

Articles

''Tuvalu Stamps to Help Kosovo Refugees.'' *Pacific Magazine* (September/October 1999): 25.

Internet

Tuvalu Consular Information Sheet. Available Online @ http://travel.state.gov/tuvalu.html (November 2, 1999).

TUVALU: STATISTICAL DATA

For sources and notes see "Sources of Statistics" in the front of each volume.

GEOGRAPHY

Geography (1)

Area:

Total: 26 sq km.

Land: 26 sq km.

Water: 0 sq km.

Area—comparative: 0.1 times the size of Washington, DC.

Land boundaries: 0 km.

Coastline: 24 km.

Climate: tropical; moderated by easterly trade winds (March to November); westerly gales and heavy rain (November to March).

Terrain: very low-lying and narrow coral atolls.

Natural resources: fish.

Land use:

Arable land: 0%

Permanent crops: 0%

Permanent pastures: 0%

Forests and woodland: 0%

Other: 100% (1993 est.)

Note: Tuvalu's nine coral atolls have enough soil to grow coconuts and support subsistence agriculture.

HUMAN FACTORS

Infants and Malnutrition (5)

Under-5 mortality rate (1997)	56
% of infants with low birthweight (1990-97)	3
Births attended by skilled health staff % of total[a]	NA

Continued on next page.

Demographics (2A)

	1990	1995	1998	2000	2010	2020	2030	2040	2050
Population	9.1	10.0	10.4	10.7	12.4	14.5	16.4	18.0	19.6
Net migration rate (per 1,000 population)	NA	NA	NA	NA	NA	NA	NA	NA	NA
Births	NA	NA	NA	NA	NA	NA	NA	NA	NA
Deaths	NA	NA	NA	NA	NA	NA	NA	NA	NA
Life expectancy - males	60.4	61.9	62.7	63.3	65.9	68.2	70.2	71.9	73.3
Life expectancy - females	63.1	64.3	65.1	65.6	68.0	70.2	72.3	74.1	75.8
Birth rate (per 1,000)	29.8	24.8	22.6	21.8	23.7	21.4	18.2	17.8	15.5
Death rate (per 1,000)	9.9	9.1	8.6	8.5	7.6	7.2	7.9	8.5	8.1
Women of reproductive age (15-49 yrs.)	2.5	2.6	2.8	2.9	3.3	3.5	4.2	4.5	4.8
of which are currently married	NA	NA	NA	NA	NA	NA	NA	NA	NA
Fertility rate	3.1	3.1	3.1	3.1	2.9	2.7	2.5	2.3	2.1

Except as noted, values for vital statistics are in thousands; life expectancy is in years.

Infants and Malnutrition (5) cont.

% fully immunized (1995-97)

TB	100
DPT	77
Polio	78
Measles	100
Prevalence of child malnutrition under age 5 (1992-97)[b]	NA

Ethnic Division (6)

Polynesian 96%.

Religions (7)

Church of Tuvalu (Congregationalist)	97%
Seventh-Day Adventist	1.4%
Baha'i	1%
Other	0.6%

Languages (8)

Tuvaluan, English.

EDUCATION

Educational Attainment (10)

Age group (1991)	25+
Total population	4,571
Highest level attained (%)	
No schooling[2]	0.8
First level	
Not completed	71.4
Completed	NA
Entered second level	
S-1	16.2
S-2	NA
Postsecondary	7.0

GOVERNMENT & LAW

Political Parties (12)

The legislative branch is a unicameral Parliament (12 seats — two from each island with more than 1,000 inhabitants, one from all the other inhabited islands members elected by popular vote to serve four-year terms). All seats held by independents.

Government Budget (13B)

Revenues	$4.3 million
Expenditures	$4.3 million
Capital expenditures	NA

Data for 1989 est. NA stands for not available.

Military Affairs (14A)

Total expenditures	$NA
Expenditures as % of GDP	NA%

NA stands for not available.

LABOR FORCE

Labor Force (16)

People make a living mainly through exploitation of the sea, reefs, and atolls and from wages sent home by those working abroad (mostly workers in the phosphate industry and sailors.

Unemployment Rate (17)

Rate not available.

MANUFACTURING SECTOR

GDP & Manufacturing Summary (21)

	1980	1985	1990	1992	1993	1994
Gross Domestic Product						
Millions of 1990 dollars	9	7	8	9	10	10
Growth rate in percent	NA	−1.95	2.49	8.90	8.70	2.60
Per capita (in 1990 dollars)	1,088.0	852.7	897.9	1,024.6	1,002.4	1,028.5
Manufacturing Value Added						
Millions of 1990 dollars	NA	NA	NA	NA	NA	NA
Growth rate in percent	NA	NA	NA	NA	NA	NA
Manufacturing share in percent of current prices	NA	NA	NA	NA	NA	NA

PRODUCTION SECTOR

Electric Energy (18)

Capacity ....2,600 kW (1995)

Production ....3 million kWh (1995)

Consumption per capita ....NA kWh

Transportation (19)

Highways:

total: 8 km (1996 est.)

paved: NA km

unpaved: NA km

Merchant marine:

total: 14 ships (1,000 GRT or over) totaling 53,220 GRT/83,118 DWT ships by type: cargo 8, chemical tanker 4, oil tanker 1, passenger-cargo 1 (1997 est.)

Airports: 1 (1997 est.)

Airports—with unpaved runways:

total: 1

1,524 to 2,437 m: 1 (1997 est.)

Top Agricultural Products (20)

Coconuts; fish.

FINANCE, ECONOMICS, & TRADE

Economic Indicators (22)

National product: GDP—purchasing power parity—$7.8 million (1995 est.)

National product real growth rate: 8.7% (1995)

National product per capita: $800 (1995 est.)

Inflation rate—consumer price index: 3.9% (average 1985-93)

Exchange Rates (24)

Exchange rates:	
Tuvaluan dollars ($T) or Australian dollars ($A) per US$1	
January 1998	1.5281
1997	1.3439
1996	1.2773
1995	1.3486
1994	1.3667
1993	1.4704

Top Import Origins (25)

$4.4 million (c.i.f., 1989)

Origins	%
Fiji	NA
Australia	NA
NZ	NA

NA stands for not available.

Top Export Destinations (26)

$165,000 (f.o.b., 1989).

Destinations	%
Fiji	NA
Australia	NA
NZ	NA

NA stands for not available.

Economic Aid (27)

Recipient: ODA, $1.725 million from Australia (FY96/97 est.); $1.7 million from NZ (FY95/96). Note: substantial annual support from an international trust fund.

Import Export Commodities (28)

Import Commodities	Export Commodities
Food	Copra
Animals	
Mineral fuels	
Machinery	
Manufactured goods	

UGANDA

Republic of Uganda

CAPITAL: Kampala.

FLAG: The national flag consists of six equal horizontal stripes of black, yellow, red, black, yellow, and red (from top to bottom); at the center, within a white circle, is a crested crane, the national bird of Uganda.

ANTHEM: Begins ''O Uganda! May God uphold thee.''

MONETARY UNIT: The new Uganda shilling (NUSh) was introduced in May 1987 with a value equal to 100 old Uganda shillings. NUSh1 = \$0.00092 (or \$1 = NUSh1,088.8). There are coins of 1, 2, and 5 shillings, and notes of 10, 20, 50, 100, 200, 500, and 1,000 shillings.

WEIGHTS AND MEASURES: The metric system is in use.

HOLIDAYS: New Year's Day, 1 January; Labor Day, 1 May; Martyrs' Day, 3 June; Independence Day, 9 October; Christmas Day, 25 December; Boxing Day, 26 December. Movable holidays include Good Friday, Easter Monday, 'Id al-Fitr, and 'Id al-'Adha'.

TIME: 3 PM = noon GMT.

LOCATION AND SIZE: A landlocked country in east-central Africa, situated north and northwest of Lake Victoria, Uganda has a total area of 236,040 square kilometers (91,136 square miles), slightly smaller than the state of Oregon. It has a total boundary length of 2,698 kilometers (1,676 miles).

CLIMATE: Although Uganda is on the equator, its climate is warm rather than hot, and temperatures vary little throughout the year. Most of the territory receives an annual rainfall of at least 100 centimeters (40 inches). On Lake Albert, the mean annual maximum is 29°C (84°F) and the mean annual minimum 22°C (72°F). At Kabale in the southwest, 1,250 meters (4,100 feet) higher, the mean annual maximum is 23°C (73°F), and the mean annual minimum 10°C (50°F).

INTRODUCTORY SURVEY

RECENT HISTORY

Great Britain granted internal self-government to Uganda in 1961. After five years of political infighting Prime Minister Milton Obote in 1966 suspended the constitution, assumed all government powers, and removed the president and vice president. In September 1967 a new constitution made Uganda a republic, granted the president greater powers, and abolished the traditional kingdoms. On January 25, 1971, in a military coup, Idi Amin declared himself president, dissolved the parliament, and amended the constitution to give himself absolute power.

Idi Amin's 8-year rule was disastrous for the country's economy and social structure. In 1978 an international commission estimated that as many as 350,000 Ugandans were murdered during Amin's reign of terror. Amin's removal in 1978 by a combination of Tanzanian and Libyan troops and Ugandan exiles, only ushered in 8 more years of tribal warfare and some of the worst human rights violations reported anywhere.

In 1986, Lt. Gen. Yoweri Museveni seized power, and his National Resistance Movement (NRM), began to put an end to the human rights abuses of earlier governments. Acting on the advice of the International Monetary Fund (IMF), World Bank, and donor governments, Museveni instituted broad economic and political reforms. General elections were held at the end of 1994, and a new constitution was debated and accepted on

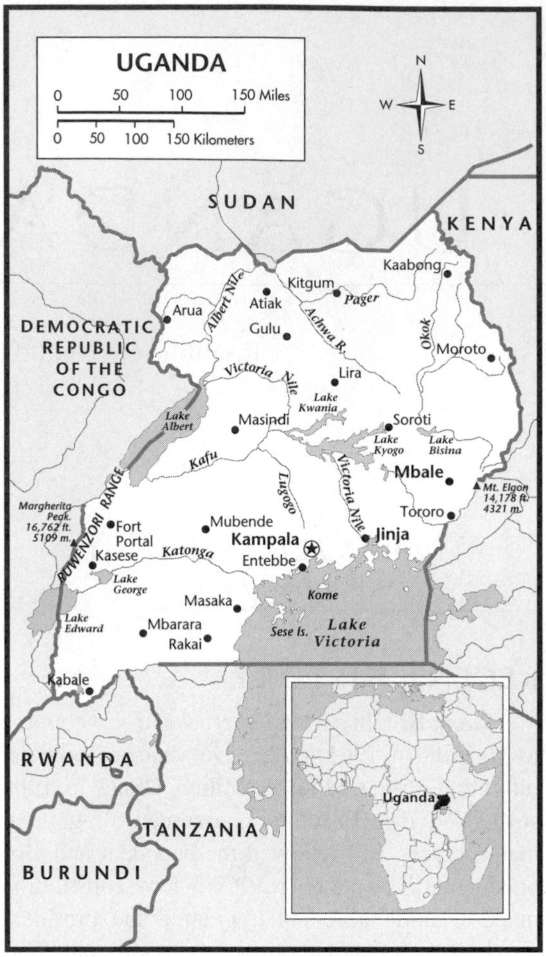

July 12, 1995. However, political party activity was banned for five years.

GOVERNMENT

The president is both chief of state and head of government. The 1995 constitution established a 279-member parliament that included representation for special interest groups (e.g., 10 seats for the army, 39 for women, and 5 for youth). Presidential elections were held under the new constitution in May 1996; parliamentary elections took place a month later. An appointed cabinet advises the president.

Judiciary

The president appoints judges to the two highest courts, the High Court and the Court of Appeal. At the lowest level are three classes of courts presided over by magistrates.

Political Parties

Only one political organization, the National Resistance Movement, is allowed to participate in politics. Political parties were not allowed to participate in the 1996 parliamentary and presidential elections. A referendum on the matter is scheduled in 2000.

DEFENSE

The active armed forces were estimated at 50,000 in 1995, and defense expenditures were estimated at $126 million that year.

ECONOMIC AFFAIRS

Uganda's economy is based on agriculture, which engages about 80% of the population and represents 55% of the total economy. An economic reform process begun in 1986 resulted in some progress; the economic growth rate averaged 6% during 1986–94 and 8% during 1995–97. Even though the size of its economy has doubled in the past ten years, Uganda is still one of the poorest countries in the world and is highly dependent on foreign aid.

Public Finance

The main sources of government revenue are the export duties on coffee and cotton, import duties, income and profit taxes, excise taxes, and sales taxes. Deficits are chronic.

The U.S. Central Intelligence Agency estimates that, in 1995–96, government revenues totaled approximately $869 million and expenditures $985 million. External debt totaled $2.9 billion.

Income

In 1995 Uganda's gross national product (GNP) was $4.67 billion, or about $300 per person. For the period 1985–95 the average inflation rate was 65.5%, resulting in a real growth rate in per person GNP of 2.8%.

Industry

By 1997, industrial production was triple that of 1989. Industries include cotton, coffee, tea, sugar, tobacco, edible oils, dairy products, grain milling, brewing, vehicle assembly, textiles, and steel. Batteries, canned foods, medicines, and salt are other products being produced by Uganda's industrial sector.

Banking and Finance

The Bank of Uganda was established on 16 May 1966 as the bank of issue; the newly created government-owned Uganda Commercial Bank (UCB) was allowed to provide full commercial banking service. In 1997 it had around 50 branches.

The Uganda Development Bank is a government bank that channels long-term loans from foreign sources to Ugandan businesses.

Improvements in the banking industry have been slow to take place. Lack of public confidence in the system was compounded by banking scandals in the 1980s, high inflation, which caused rapid erosion in the value of money, and by the liquidity and insolvency problems of some banks. The UCB's planned privatization and restructuring form a critical element in government's financial reforms.

In 1997 the financial sector consisted of the Bank of Uganda together with 19 commercial banks, 10 credit institutions, 27 insurance companies, 2 development banks, 1 building society, 1 pension fund, and 1 leasing company.

Economic Development

Uganda's economic development policy for the 1990s was outlined in the Economic Recovery Program for 1988–92. State investment was lowered by 42% from the previous plan, and the export sector was to be revived, particularly the nontraditional export sector. The investment budget was divided among the transport and communications (27%), social infrastructure (25.7%), agriculture (23.8%), and industry and tourism sectors (17.2%).

SOCIAL WELFARE

Responsibility for social welfare rests primarily with the Ministry of Culture and Community Development, assisted by voluntary agencies. In 1991 it was estimated that 1.2 million Ugandan children were orphans, as a result of the AIDS epidemic and years of civil war.

Healthcare

AIDS became a severe problem in the 1980s. As of 1995, 1.5 million Ugandans were infected with HIV, the virus that causes AIDS. Containment of other serious diseases, such as cholera, dysentery, tuberculosis, and malaria, is made difficult by poor sanitation and unclean water.

Life expectancy at birth in 1999 was 43 years.

Housing

Most Ugandans live in thatched huts with mud and wattle (interwoven branches and reeds) walls. Even in rural areas, however, corrugated iron is used extensively as a roofing material.

EDUCATION

In 1993, there were 8,431 primary schools with 2.5 million pupils; at the secondary level there were 231,430 students. In 1993, 24,122 students were enrolled in higher education institutions. The adult literacy rate in 1995 was 61.8%.

1999 KEY EVENTS TIMELINE

January

- The government acknowledges problems with the banking sector and vows to restore confidence and boost personal savings rate above its current 3%.
- Inspector General John Odomel resigns amid allegations of fiscal mismanagement and his deputy John Kismebo is sworn in as acting inspector general.

February

- A government report on AIDS and sexual activity shows a moderate decline from 1989 among most ages and categories.
- Ugandan International Olympic Committee (IOC) member Major General Francis Nyangweso angrily protests accusations that he accepted $31,500 as an inducement to sway his vote for the 2000 Olympics Game.
- The government announces the eradication of the Guinea worm disease in 13 of 16 affected districts.
- Eight foreigners and several government officials are killed in Bwindi National Park by rebels from neighboring Congo.

March

- The IMF suspends aid to Uganda, citing uncontrolled defense expenditures and an inability to meet foreign exchange commitments.
- The Uganda People's Defense Force prepares troops for an assignment to Kindu province in eastern Democratic Republic of Congo to support rebels.

- The privatization of Uganda Telecommunications Ltd., with its sale pending to World Tel and Detecom, is suspended by the government.
- President Museveni strongly condemns the killing of eight tourists and Ugandan officials in Bwindi Impenetrable Forest in southwestern Uganda and offers the victims' families condolences.
- A parliamentary censure of Planning and Investment Minister Sam Kuteesa leads to his dismissal.

April

- The government receives a 185 million euro grant for transportation projects for the next six years.
- The government begins a door-to-door voluntary HIV screening campaign.
- Russian and Israeli experts begin training Ugandan Air Force pilots in the eastern part of the country.
- President Museveni reshuffles his cabinet after Prime Minister Kintu Musoke's retirement.
- Bombs explodes in Kampala, killing seven people and injuring twenty. A Zimbabwean national is arrested shortly thereafter.

May

- President Yoweri Museveni offers amnesty to northern rebels and invites former president Milton Obote to return from exile. Later that month, 120 rebels turn in their weapons and surrender.
- The World Bank announces US$13 million in credit to Uganda to revamp its financial system.

June

- Uganda's Chief of Staff Brigadier James Kazini threatens to arrest Congolese rebel military chief Jean Pierre Ondekane for taking advantage of the Ugandan People's Defense Force.

July

- The government announces that the country's population will exceed 22.2 million in 2000, more than triple the 1959 figure. The government urges family planning.

August

- The King of Buganda, Kabaka Mutebi, weds Nabagereka Nagginda in a royal wedding attended by thousands of guests including President Museveni.

September

- Uganda's National Program Manager for Eradication reports that the country achieved 90% eradication of polio through immunization since the program started in 1996.
- State Minister for Environment Kezimbira Miyingo announces the establishment of refrigerant recycling center to combat ozone depletion and expressed a concern over the use of bromide to control pest infestations.

October

- Othieno Akiika, the State Minister for Water, resigns after political pressure resulting from the alleged assault of a police officer in August. He is subsequently arrested.
- Chairman William Kalema of the Ugandan Investment Authority reports that investments into the country plummeted by 25% between 1993 to 1999.
- The Sebutinde Commission on police corruption submits a report to the Minister of Internal Affairs for government action.
- The Ugandan government hires two expert attorneys to defend its policies in a suit at the International Court of Justice (Hague) brought by President Laurent Kabila's government in the Democratic Republic of Congo.
- Parliament prepares to debate a law to regulate party activities for upcoming referendum.

December

- Parliament passes a six-month amnesty for rebels in an effort to stop uprisings against the government by the Lords Resistance Army operating the north, and the Allied Democratic Forces, operating in the west.
- UK Chancellor Gordon Brown reports that the UK is implementing a program to cancel millions of pounds of debt for some of the world's most indebted nations. The first four countries on the list, Uganda, Mozambique, Bolivia, and Mauritania, are expected to qualify for the debt forgiveness program by the end of January 2000.

ANALYSIS OF EVENTS: 1999

BUSINESS AND THE ECONOMY

Uganda experienced slower growth and several areas of dislocation in 1999. Political instability in the neighboring Democratic Republic of Congo affected the government's budget and even tourism. Meanwhile a number of scandals relating to privatization undermined the public sector and resulted in the suspension of International Monetary Fund (IMF) aid. However, by mid-year, the economic picture improved gradually and international investment and aid increased.

The year 1999 began with a crisis in the banking sector. Several government affiliated banks closed their doors. The government in January acknowledged problems with the banking sector and vowed to restore confidence and urged Ugandans to increase their savings rate, which was the lowest in the region at 3 percent. Inflation, which declined to 5 percent last year, was on the upswing, with the government budget ballooning due to commitment of troops in the Congo. In March, the IMF sounded its disapproval of government spending and suspended aid. During the same month, the government announced a suspension of the sale of Uganda Telecommunications Ltd. to World Tel and Detecom, citing increased fraud and irregularities in the privatization process.

In April, the government received a much sought after grant from the European Union totaling 185 million euro over six years for transportation infrastructure improvement in the country. Then in May, the World Bank, citing Uganda's growth over the years, decided to provide a loan of $13 million to stabilize and revamp the country's banking sector. In addition, the World Bank approved close to $50 million for agricultural research and management and the development of land use planning and infrastructure repair capacity of the Kampala City Council. In September, the government announced the establishment of refrigerant recycling centers to combat ozone depletion and also expressed concern over the use of bromide for agricultural pest control.

In March, the killing of eight foreign tourists in Bwindi National Park in southwestern Uganda brought a halt to tourism, shutting down an increasingly important source of foreign currency. The tourists and government officials were allegedly killed by Congolese forces. President Museveni strongly condemned the killing and offered condolences to the victims' families. Meanwhile, the Ugandan military drew fire for failing to protect its border. Questions about military intervention in the Congo increased. In May, however, the government reopened the park.

William Kalema of the Ugandan Investment Authority reported that investment into the country plummeted by 25% between 1993 and 1999, further providing evidence that the economy was slowing. He noted that government corruption and bureaucratic red tape remain the major obstacles to foreign direct investment and urged the government to take action. This came on the heels of a major report from the Sebutinde Commission which exposed major police corruption and evidence tampering. While small enterprises including those owned by Asians did well in 1999, large government owned para-statals saw a halt to privatization because of allegations of fraud.

GOVERNMENT AND POLITICS

The two major themes in Ugandan politics during 1999 were the country's involvement in the Congo's civil war and increased allegations of corruption in the government. A recent report suggested that Uganda was the seventh most corrupt nation in the world. While Museveni remained popular, his legitimacy weakened domestically and internationally as Uganda's involvement in Congo increased and the country's economy slowed and experienced several problems.

In January, the Inspector General of Uganda, John Odomel, resigned amid allegations of fiscal mismanagement. In February, the Ugandan International Olympic Committee (IOC) member Major General Francis Nyangweso angrily protested accusations that he accepted over $30,000 as an inducement to sway his vote for the 2000 Olympics scheduled for Sydney, Australia. This was followed by the Parliamentary censure of Planning and Investment Minister Sam Kuteesa, who subsequently resigned.

The scandals continued into October when Othieno Akiika, the State Minister for Water, resigned and was subsequently arrested for allegedly assaulting a police officer in August. When the Sebutinde Commission submitted a report to the Minister of Internal Affairs detailing numerous cases of police negligence, incompetence, and corruption, the Ugandan justice system was dealt a heavy blow. The report urged Parliament to investigate and correct miscarriages of justice and to stamp out corruption in Uganda's police force.

In addition to corruption and scandal, other causes of change in governmental personnel included the normal shuffling of personalities in the cabinet. When Prime Minister Kintu Musoke retired, Museveni decided to make a number of changes in his cabinet in April. The reconstituted government included eight new ministers or one-third of all cabinet posts. A total of thirteen changes were made, including the designation of Specioza Wandira Kazibwe, a woman, as vice-president.

Meanwhile, the Ugandan government continued committed to the rebel movement in the Democratic Republic of Congo (DRC), with major initiatives underway to destabilize that country's government. The DRC sued Uganda for assisting the rebels and pressed charges at the International Court of Justice (The Hague). The Ugandan government retained two expert attorneys to defend the charges. The Congolese conflict wrecked havoc on the government's budget and led to the IMF cutting off loans in March.

As the country prepared for a referendum next year, the Parliament began to debate laws to control party registration and activities, much to the disappointment of opposition leaders. In a conciliatory move, Museveni invited former president Obote to return from exile and his supporters in northern Uganda to lay down their weapons for amnesty. It was reported that 120 rebels took advantage of the program in May.

CULTURE AND SOCIETY

The government reported in February that Uganda experienced a modest decline in AIDS infection and noted that sexual activity had declined among Ugandans of most ages and categories. The government began a door-to-door AIDS prevention program in April. At the same time, the government urged responsible family planning and reported that the country's population would exceed 22.2 million in 2000, more than triple the 1959 figure. There were several major achievements in the health and disease eradication in 1999. Among them, in February the government announced that the Guinea worm was eradicated in 13 of 16 districts. In September Uganda's National Program Manager for polio eradication announced that the country had achieved a 90 percent decline in the disease since the program started in 1996.

From 1998 to 1999, the government worked to increase the status of women. It was reported by the United Nations that the Sabiny Elders Association received the UN Population Award for promoting women's health issues, including the eradication of female genital mutilation.

In August, the country witnessed a royal wedding of the King of Buganda, Kabaka Mutebi, who wed Nabagereka Nagginda, after many years of batchelorhood. The wedding, attended by President Museveni and thousands of guests created a festive occasion.

A soccer tournament in Uganda took place in October. Dubbed the ''match of the millennium,'' the Ugandan team lost to Zimbabwe. Prior to the match, President Museveni and his cabinet played and won a scrimmage against a team from the Federation of Uganda Football Association. The President contributed one penalty kick goal.

DIRECTORY

CENTRAL GOVERNMENT

Head of State

President
Yoweri Kaguta Museveni, Office of the President, P.O. Box 7168, Kampala, Uganda
PHONE: +256 (41) 254881
FAX: +256 (41) 235462

Vice President
Wandira Specioza Zazibwe, Office of the Vice President, Hannington Road, P.O. Box 7359, Kampala, Uganda
PHONE: +256 (41) 236563; 255299
FAX: +256 (41) 236778

Ministers

Prime Minister
Kintu Musoke, Office of the Prime Minister, Posts and Telecommunications Building, Kitante Road, P.O. Box 341, Kampala, Uganda
PHONE: +256 (41) 25908; 232575
FAX: +256 (41) 242341

First Deputy Prime Minister and Minister of Foreign Affairs
Eriya T. Kategaya, Office of the First Deputy Prime Minister, P.O. Box 7048, Kampala, Uganda

PHONE: +256 (41) 2333922; 244975; 230911
FAX: +256 (41) 258722

Minister of Agriculture, Animal Industry and Fisheries
Specioza Wandira Kazibwe, Ministry of Agriculture, Animal Industry and Fisheries, Lugard Avenue, P.O. Box 102, Entebbe, Uganda
PHONE: +256 (42) 20981
FAX: +256 (42) 21042; 21047

Minister of Defense
Yoweri Kaguta Museveni, Ministry of Defense, Uganda Club, P.O. Box 3798, Kampala, Uganda
PHONE: +256 (41) 270331
FAX: +256 (41) 245911

Minister of Education and Sports
Makubuya Kdhu, Ministry of Education and Sports, Crested Towers, 17/19 Hannington road, P.O. Box 7048, Kampala, Uganda
PHONE: +256 (41) 234440
FAX: +256 (41) 234194

Minister of Finance
Joshua Mayanja-Nkangi, Ministry of Finance, Apollo Kagwa Road, Uganda House, P.O. Box 7048, Kampala, Uganda
PHONE: +256 (41) 235051; 232370
FAX: +256 (41) 234194

Minister of Foreign Affairs
Driya Kategaya, Ministry of Foreign Affairs, Parliamentary Buildings, Parliament Avenue, P.O. Box 7084, Kampala, Uganda

Minister of Gender and Community Development
Hajati Janat B. Mukwaya, Ministry of Gender and Community Development, Udyam House, 4 Jinja Road, P.O. Box 7136, Kampala, Uganda
PHONE: +256 (41) 256375

Minister of Lands, Housing, and Urban Development
Francis Ayume, Ministry of Lands, Housing, and Urban Development, Udyam House, 71 Jinja Road, P.O. Box 7122, Kampala, Uganda
PHONE: +256 (41) 242931

Minister of Local Government
Jaberi Bidandi Ssali, Ministry of Local Government, 10 Kampala Road, P.O. Box 7037, Kampala, Uganda
PHONE: +256 (41) 258435

Minister of Natural Resources
Gerald Sendawula, Ministry of Natural Resources, Amber House, 33 Kampala Road, Kampala, Uganda
PHONE: +256 (41) 230220; 233331
FAX: +256 (41) 243508

Minister of Public Service and Labor
Amanya Musheja, Ministry of Public Service and Labor, Wandegeya, Haji Kasule Road, Kampala, Uganda
PHONE: +256 (41) 251003
FAX: +256 (41) 255467

Minister of Tourism, Wildlife and Antiquities
General Moses Ali, Ministry of Tourism, Wildlife and Antiquities, 1 Parliamentary Avenue, P.O. Box 4241, Kampala, Uganda

Minister of Trade and Industry
Henry Muganwa Kajura, Ministry of Trade and Industry, 3 Parliamentary Avenue, P.O. Box 7000, Kampala, Uganda
PHONE: +256 (41) 243947; 256395
FAX: +256 (41) 245077

Minister of Health
Crispus Kiyonga, Ministry of Health, P.O. Box 8, Entebbe, Uganda
PHONE: +256 (41) 20201
FAX: +256 (41) 20274; 20440

Minister of Information and Broadcasting
Ruhakana Rugunda, Ministry of Information and Broadcasting, Lugard Road, Nakasero Hill, Entebbe, Uganda
PHONE: +256 (41) 254461
FAX: +256 (41) 256888

Minister of Internal Affairs
Ministry of Internal Affairs, 6th/7th Floors, Crested Towers, 17/19 Hannington Road, P.O. Box 7191, Kampala, Uganda
PHONE: +256 (41) 254461
FAX: +256 (41) 256888

Minister of Justice and Constitutional Affairs
Joshua S. Mayanja-Nkangi, Ministry of Justice and Constitutional Affairs, P.O. Box 7183, Kampala, Uganda
PHONE: +256 (41) 233219
FAX: +256 (41) 254828

Minister of Water, Lands, and Environment
Henry Muganwa Kajura, Ministry of Water, Lands, and Environment

Third Deputy Prime Minister; Minister of Disaster Preparedness and Refugees
Tom Butiime, Ministry of Disaster Preparedness and Refugees

Minister of Energy and Mineral Development
Syda Bbumba, Ministry of Energy and Mineral Development

Minister of Public Works, Housing, and Communications
John Nasasira, Ministry of Public Works, Housing, and Communications, P.O. Box 10, Entebbe, Uganda
PHONE: +256 (41) 20101
FAX: +256 (41) 20155

Minister of Finance, Planning, and Economic Development
Gerald Sendawula, Ministry of Finance, Planning, and Economic Development, P.O. Box 8147, Kampala, Uganda
PHONE: +256 (41) 230163
FAX: +256 (41) 232015

Minister of Economic Monitoring
Kweronda Ruhemba, Ministry of Economic Monitoring

Minister of Information
Nsadhu Basoga, Ministry of Information

Minister of Security
Wilson Mukasa Muruli, Ministry of Security

Minister of Parliamentary Affairs
Rebecca Kadaga, Ministry of Parliamentary Affairs

POLITICAL ORGANIZATIONS

National Resistance Movement (NRM)

TITLE: Chairman
NAME: Samson Kisekka

Ugandan People's Congress (UPC)

TITLE: Leader
NAME: Milton Obote

Democratic Party (DP)

TITLE: Leader
NAME: Paul Ssemogerere

Conservative Party (CP)

TITLE: Leader
NAME: Joshua S. Mayanja-Nkangi

DIPLOMATIC REPRESENTATION

Embassies in Uganda

Algeria
P.O. Box 4025, Kampala, Uganda
PHONE: +256 (41) 232928; 232689

Austria
P.O. Box 7457, Kampala, Uganda
PHONE: +256 (41) 235103
FAX: +256 (41) 235160

Belgium
P.O. Box 4379, Kampala, Uganda
PHONE: +256 (41) 221281; 233674
FAX: +256 (41) 250990

Canada
P.O. Box 20115, Kampala, Uganda
PHONE: +256 (41) 258141
FAX: +256 (41) 234518

China
P.O. Box 4106, Kampala, Uganda
PHONE: +256 (41) 236895
FAX: +256 (41) 235087

Democratic Republic of Congo
P.O. Box 4972, Kampala, Uganda
PHONE: +256 (41) 233777

Cuba
P.O. Box 9226, Kampala, Uganda
PHONE: +256 (41) 233742
FAX: +256 (41) 233320

Cyprus
P.O. Box 8717, Kampala, Uganda
PHONE: +256 (41) 235812
FAX: +256 (41) 236053

Denmark
P.O. Box 11234, Kampala, Uganda
PHONE: +256 (41) 256783; 256687
FAX: +256 (41) 254979

Egypt
P.O. Box 4280, Kampala, Uganda
PHONE: +256 (41) 254525; 245152
FAX: +256 (41) 232103

Ethiopia
P.O. Box 7745, Kampala, Uganda
PHONE: +256 (41) 231010

Finland
P.O. Box 7111, Kampala, Uganda
PHONE: +256 (41) 258211
FAX: +256 (41) 231473

France
P.O. Box 7212, Kampala, Uganda
PHONE: +256 (41) 242120; 242176
FAX: +256 (41) 241252

Germany
P.O. Box 7016, Kampala, Uganda
PHONE: +256 (41) 256767
FAX: +256 (41) 343136

Ghana
P.O. Box 1790, Kampala, Uganda
PHONE: +256 (41) 250140

Greece
PHONE: +256 (41) 230953
FAX: +256 (41) 230952

India
P.O. Box 7040, Kampala, Uganda
PHONE: +256 (41) 254943

Indonesia
P.O. Box 6021, Kampala, Uganda
PHONE: +256 (41) 242836
FAX: +256 (41) 230088

Ireland
P.O. Box 7791, Kampala, Uganda
PHONE: +256 (41) 244438
FAX: +256 (41) 244353

Italy
P.O. Box 4646, Kampala, Uganda
PHONE: +256 (41) 341786
FAX: +256 (41) 250448

Japan
P.O. Box 23553, Kampala, Uganda
PHONE: +256 (41) 347317; 347319
FAX: +256 (41) 347348

Kenya
P.O. Box 5220, Kampala, Uganda
PHONE: +256 (41) 258235
FAX: +256 (41) 258239

Libya
P.O. Box 6079, Kampala, Uganda
PHONE: +256 (41) 344924
FAX: +256 (41) 344969

Malta
P.O. Box 2133, Kampala, Uganda
PHONE: +256 (41) 258516

The Netherlands
P.O. Box 7728, Kampala, Uganda
PHONE: +256 (41) 234427
FAX: +256 (41) 231861

Nigeria
P.O. Box 4338, Kampala, Uganda
PHONE: +256 (41) 233691; 233692
FAX: +256 (41) 232543

Norway
P.O. Box 22770, Kampala, Uganda
PHONE: +256 (41) 243621
FAX: +256 (41) 243936

Pakistan
P.O. Box 7022, Kampala, Uganda
PHONE: +256 (41) 231142
FAX: +256 (41) 231142

Poland
PHONE: +256 (41) 245864; 255267

Russia
P.O. Box 7022, Kampala, Uganda
PHONE: +256 (41) 233676; 243808

Rwanda
P.O. Box 2468, Kampala, Uganda
PHONE: +256 (41) 244045; 2345095
FAX: +256 (41) 258547

South Africa
P.O. Box 10194, Kampala, Uganda
PHONE: +256 (41) 231015
FAX: +256 (41) 231007

Spain
P.O. Box 8695, Kampala, Uganda
PHONE: +256 (41) 244331; 245967

Sweden
PHONE: +256 (41) 236636; 236031
FAX: +256 (41) 236852

Switzerland
P.O. Box 4187, Kampala, Uganda
PHONE: +256 (41) 241574; 259894
FAX: +256 (41) 236852

Tanzania
P.O. Box 5750, Kampala, Uganda
PHONE: +256 (41) 257357

Thailand
P.O. Box 5961, Kampala, Uganda
PHONE: +256 (41) 236182
FAX: +256 (41) 236148

United Kingdom
P.O. Box 7070, Kampala, Uganda
PHONE: +256 (41) 257054
FAX: +256 (41) 257304

United States
P.O. Box 7007, Kampala, Uganda
PHONE: +256 (41) 259792

FAX: +256 (41) 235306
TITLE: Ambassador
NAME: Nancy Jo Powell

JUDICIAL SYSTEM

Supreme Court

P.O. Box 7085, Kampala, Uganda

High Court

P.O. Box 7085, Kampala, Uganda

Court of Appeal

Magistrates' Courts

FURTHER READING

Articles

''Horror in the Hills.'' *The Economist* (March 6, 1999).

''Miracle Growth Set to Continue.'' *African Business* (February, 1999).

UGANDA: STATISTICAL DATA

For sources and notes see "Sources of Statistics" in the front of each volume.

GEOGRAPHY

Geography (1)

Area:

Total: 236,040 sq km.

Land: 199,710 sq km.

Water: 36,330 sq km.

Area—comparative: slightly smaller than Oregon.

Land boundaries:

Total: 2,698 km.

Border countries: Democratic Republic of the Congo 765 km, Kenya 933 km, Rwanda 169 km, Sudan 435 km, Tanzania 396 km.

Coastline: 0 km (landlocked).

Climate: tropical; generally rainy with two dry seasons (December to February, June to August); semiarid in northeast.

Terrain: mostly plateau with rim of mountains.

Natural resources: copper, cobalt, limestone, salt.

Land use:

Arable land: 25%

Permanent crops: 9%

Permanent pastures: 9%

Forests and woodland: 28%

Other: 29% (1993 est.).

HUMAN FACTORS

Demographics (2A)

	1990	1995	1998	2000	2010	2020	2030	2040	2050
Population	17,227.4	20,401.2	22,167.2	23,451.7	31,768.0	42,771.6	56,340.3	73,018.3	91,398.0
Net migration rate (per 1,000 population)	NA	NA	NA	NA	NA	NA	NA	NA	NA
Births	NA	NA	NA	NA	NA	NA	NA	NA	NA
Deaths	NA	NA	NA	NA	NA	NA	NA	NA	NA
Life expectancy - males	44.5	40.7	41.8	42.6	46.0	49.4	58.0	66.7	70.0
Life expectancy - females	44.3	41.9	43.4	44.5	49.1	52.8	63.0	73.1	76.8
Birth rate (per 1,000)	52.9	51.9	49.2	48.0	45.3	40.6	34.5	29.0	24.2
Death rate (per 1,000)	19.1	20.6	19.0	17.9	14.4	12.2	7.6	4.5	4.0
Women of reproductive age (15-49 yrs.)	3,755.0	4,324.5	4,615.6	4,874.3	7,017.2	9,870.2	14,004.2	19,212.3	24,633.5
of which are currently married	NA	NA	NA	NA	NA	NA	NA	NA	NA
Fertility rate	7.3	7.3	7.1	7.0	6.1	5.1	4.2	3.4	2.8

Except as noted, values for vital statistics are in thousands; life expectancy is in years.

Health Personnel (3)

Total health expenditure as a percentage of GDP, 1990-1997[a]

Public sector	1.9
Private sector	2.1
Total[b]	3.9

Health expenditure per capita in U.S. dollars, 1990-1997[a]

Purchasing power parity	34
Total	9

Availability of health care facilities per 100,000 people

Hospital beds 1990-1997[a]	90
Doctors 1993[c]	4
Nurses 1993[c]	28

Health Indicators (4)

Life expectancy at birth

1980	48
1997	42
Daily per capita supply of calories (1996)	2,110
Total fertility rate births per woman (1997)	6.6
Maternal mortality ratio per 100,000 live births (1990-97)	550[f]
Safe water % of population with access (1995)	42
Sanitation % of population with access (1995)	67
Consumption of iodized salt % of households (1992-98)[a]	69

Smoking prevalence

Male % of adults (1985-95)[a]	....
Female % of adults (1985-95)[a]	....
Tuberculosis incidence per 100,000 people (1997)	312
Adult HIV prevalence % of population ages 15-49 (1997)	9.51

Infants and Malnutrition (5)

Under-5 mortality rate (1997)	137
% of infants with low birthweight (1990-97)	13
Births attended by skilled health staff % of total[a]	38

% fully immunized (1995-97)

TB	84
DPT	58
Polio	59
Measles	60
Prevalence of child malnutrition under age 5 (1992-97)[b]	26

Ethnic Division (6)

Baganda	17%
Karamojong	12%
Basogo	8%
Iteso	8%
Langi	6%
Rwanda	6%
Bagisu	5%
Acholi	4%
Lugbara	4%
Bunyoro	3%
Batobo	3%
Non-African (European, Asian, Arab)	1%
Other	23%

Religions (7)

Roman Catholic	33%
Protestant	33%
Muslim	16%
Indigenous beliefs	18%

Languages (8)

English (official national language, taught in grade schools, used in courts of law and by most newspapers and some radio broadcasts), Ganda or Luganda (most widely used of the Niger-Congo languages, preferred for native language publications and may be taught in school), other Niger-Congo languages, Nilo-Saharan languages, Swahili, Arabic.

EDUCATION

Public Education Expenditures (9)

Public expenditure on education (% of GNP)

1980	1.2
1996	2.6[1]

Expenditure per student

Primary % of GNP per capita

1980	4.3[1]
1996	....

Secondary % of GNP per capita

1980	....
1996	....

Tertiary % of GNP per capita

1980	1,036.9[1]
1996	....

Expenditure on teaching materials

Primary % of total for level (1996)	....
Secondary % of total for level (1996)	1.3[1]
Primary pupil-teacher ratio per teacher (1996)	35
Duration of primary education years (1995)	7

Educational Attainment (10)

Age group (1991) .25+
Total population 5,455,582
Highest level attained (%)
- No schooling 46.1
- First level
  - Not completed 41.4
  - Completed NA
- Entered second level
  - S-1 8.9
  - S-2 1.3
- Postsecondary 0.5

Literacy Rates (11A)

In thousands and percent[1]	1990	1995	2000	2010
Illiterate population (15+ yrs.)	4,020	4,172	4,142	4,116
Literacy rate - total adult pop. (%)	56.6	61.8	67.0	76.2
Literacy rate - males (%)	69.8	73.7	77.5	83.8
Literacy rate - females (%)	43.9	50.2	56.8	68.8

GOVERNMENT & LAW

Military Affairs (14B)

	1990	1991	1992	1993	1994	1995
Military expenditures						
Current dollars (mil.)	80	119	94	76	71	126
1995 constant dollars (mil.)	92	131	101	80	73	126
Armed forces (000)	60	60	70	70	60	52[e]
Gross national product (GNP)						
Current dollars (mil.)	3,401	3,741	3,933	4,407	4,771	5,421
1995 constant dollars (mil.)	3,909	4,134	4,231	4,621	4,890	5,421
Central government expenditures (CGE)						
1995 constant dollars (mil.)	515	617	868	874	963	946
People (mil.)	17.0	17.6	18.2	18.8	19.3	19.7
Military expenditure as % of GNP	2.4	3.2	2.4	1.7	1.5	2.3
Military expenditure as % of CGE	17.9	21.3	11.7	9.1	7.6	13.3
Military expenditure per capita (1995 $)	5	7	6	4	4	6
Armed forces per 1,000 people (soldiers)	3.5	3.4	3.9	3.7	3.1	2.6
GNP per capita (1995 $)	229	235	233	246	253	275
Arms imports[6]						
Current dollars (mil.)	20	20	10	0	0	0
1995 constant dollars (mil.)	23	22	11	0	0	0
Arms exports[6]						
Current dollars (mil.)	0	0	0	0	0	0
1995 constant dollars (mil.)	0	0	0	0	0	0
Total imports[7]						
Current dollars (mil.)	213	196	439	384	870	1,058
1995 constant dollars (mil.)	245	217	472	403	892	1,058
Total exports[7]						
Current dollars (mil.)	147	200	142	179	424	461
1995 constant dollars (mil.)	169	221	153	188	435	461
Arms as percent of total imports[8]	9.4	10.2	2.3	0	0	0
Arms as percent of total exports[8]	0	0	0	0	0	0

GOVERNMENT & LAW

Political Parties (12)

The legislative branch is a unicameral National Assembly (276 members serve five-year terms; 214 directly elected by universal suffrage, but 62 are nominated by legally established special interest groups and approved by the president—women 39, army 10, disabled 5, youth 5, labor 3). Campaiging by party was not permitted in the last election.

Government Budget (13B)

Revenues	$869 million
Expenditures	$985 million
Capital expenditures	$69 million

Data for FY95/96.

Crime (15)

Crime rate (for 1997)	
Crimes reported	45,000
Total persons convicted	31,000
Crimes per 100,000 population	350
Persons responsible for offenses	
Total number of suspects	42,300
Total number of female suspects	2,950
Total number of juvenile suspects	NA

LABOR FORCE

Labor Force (16)

Total (million)	8.361
Agriculture	86%
Industry	4%
Services	10%

Data for 1993 est. Percent distribution for 1980 est.

Unemployment Rate (17)

Rate not available.

PRODUCTION SECTOR

Electric Energy (18)

Capacity	162,000 kW (1998)
Production	807 million kWh (1995)
Consumption per capita	35 kWh (1995)

Transportation (19)

Highways:

total: 27,000 km

paved: 1,800 km

unpaved: 25,200 km (of which about 4,800 km are all-weather roads) (1990 est.)

Waterways: Lake Victoria, Lake Albert, Lake Kyoga, Lake George, Lake Edward, Victoria Nile, Albert Nile

Merchant marine:

total: 3 roll-on/roll-off cargo ships (1,000 GRT or over) totaling 5,091 GRT/8,229 DWT (1997 est.)

Airports: 29 (1997 est.)

Airports—with paved runways:

total: 5

over 3,047 m: 3

1,524 to 2,437 m: 1

under 914 m: 1 (1997 est.)

Airports—with unpaved runways:

total: 24

2,438 to 3,047 m: 1

1,524 to 2,437 m: 7

914 to 1,523 m: 8

under 914 m: 8 (1997 est.)

MANUFACTURING SECTOR

GDP & Manufacturing Summary (21)

	1980	1985	1990	1992	1993	1994
Gross Domestic Product						
Millions of 1990 dollars	2,393	2,673	3,253	3,489	3,711	4,082
Growth rate in percent	−3.40	1.96	−0.16	3.02	6.35	10.00
Per capita (in 1990 dollars)	182.4	176.9	181.2	181.1	186.1	197.9
Manufacturing Value Added						
Millions of 1990 dollars	98	99	164	221	222	259
Growth rate in percent	6.10	−9.80	4.20	18.08	0.30	16.97
Manufacturing share in percent of current prices	4.2	2.9	5.0	6.4	5.0	5.0

Top Agricultural Products (20)

Coffee, tea, cotton, tobacco, cassava (tapioca), potatoes, corn, millet, pulses; beef, goat meat, milk, poultry.

FINANCE, ECONOMICS, & TRADE

Economic Indicators (22)

National product: GDP—purchasing power parity—$34.6 billion (1997 est.)

National product real growth rate: 5% (1997 est.)

National product per capita: $1,700 (1997 est.)

Inflation rate—consumer price index: 6% (1997)

Exchange Rates (24)

Exchange rates:	
Ugandan shillings (USh) per US$1	
January 1998	1,148.1
1997	1,083.0
1996	1,046.1
1995	968.9
1994	979.4
1993	1,195.0

Top Import Origins (25)

$1.2 billion (c.i.f., 1996) Data are for 1995.

Origins	%
Kenya	26
United Kingdom	12
Japan	8
Germany	8
India	5.5

Top Export Destinations (26)

$604 million (f.o.b., 1996) Data are for 1995.

Destinations	%
Spain	23
France	14
Germany	14
Italy	10
Netherlands	8

Economic Aid (27)

Recipient: ODA, $NA. NA stands for not available.

Import Export Commodities (28)

Import Commodities	Export Commodities
Machinery	Coffee
Chemicals	Gold
Fuel	Cotton
Cotton piece goods	Tea
Transportation equipment	Corn
	Fish
Food	

Balance of Payments (23)

	1992	1993	1994	1995	1996
Exports of goods (f.o.b.)	151	200	463	560	639
Imports of goods (f.o.b.)	−422	−478	−714	−921	−991
Trade balance	−271	−278	−251	−360	−352
Services - debits	−336	−358	−507	−672	−742
Services - credits	39	100	78	122	174
Private transfers (net)	261	281	252	291	278
Government transfers (net)	207	84	255	345	433
Overall balance	−100	−171	−174	−275	−208

UKRAINE

Ukrayina

CAPITAL: Kiev (Kyyiv).

FLAG: Equal horizontal bands of azure blue (top) and yellow.

ANTHEM: *The National Anthem of Ukraine.*

MONETARY UNIT: The official currency, introduced in early 1993, is the hryvnia (HRN), which consists of 100 shahy. $1 = HRN1.76.

WEIGHTS AND MEASURES: The metric system is used.

HOLIDAYS: New Year's Day, 1–2 January; Christmas, 7 January; Women's Day, 8 March; Spring and Labor Day, 1–2 May; Victory Day, 9 May; Ukrainian Independence Day, 24 August.

TIME: 2 PM = noon GMT.

LOCATION AND SIZE: Ukraine, the second largest country in Europe, is located in eastern Europe, bordering the Black Sea, between Poland and Russia. Comparatively, Ukraine is slightly smaller than the state of Texas with a total area of 603,700 square kilometers (233,090 square miles). Its boundary length totals 4,558 kilometers (2,834 miles). Ukraine's capital city, Kiev, is located in the north central part of the country.

CLIMATE: Precipitation is highest in the west and north, least in the east and southeast. Winters vary from cool along the Black Sea to cold farther inland. Summers are warm across the greater part of the country, but hot in the south. The mean temperature is 18°C (66°F) in July and −6°C (21°F) in January. Northern and western Ukraine average 69 centimeters (27 inches) of rainfall a year.

INTRODUCTORY SURVEY

RECENT HISTORY

On 24 August 1991, following the dissolution of the Soviet Union, Ukraine declared itself independent. On 1 December 1991, 90.3% of Ukraine's citizens voted in favor of independence and elected Leonid M. Kravchuk as their first president.

Ukraine joined Russia, Belarus, and other nations of the former Soviet Union in forming the Commonwealth of Independent States (CIS) in December 1991. Since then, Ukrainian-Russian differences have arisen on such issues as the command and control of nuclear weapons, formation of a unified military command, and the character and pace of economic reform.

Since its independence, Ukraine has experienced unrest in some of its predominantly Russian areas, notably in Crimea where demands for secession have complicated Ukrainian-Russian relations.

GOVERNMENT

The constitution adopted in 1996 provides for an elected parliament and president. The Ukrainian parliament consists of a single chamber with 450 seats called the Rada. The prime minister and cabinet are nominated by the president and confirmed by the parliament. In Ukraine's first post-independence presidential elections, held in 1994, the incumbent Leonid Kravchuk was defeated by former prime minister Leonid Kuchma.

Judidiary

The court system has not significantly changed since the former Soviet regime. The three levels of courts include: the people's courts, the provincial or regional courts, and the Supreme Court. All three levels provide first hearings as well as appeals.

Political Parties

The two major parties are the influential Rukh Party and the Communist Party of Ukraine. There

are some 28 other political parties. Independents hold most seats in the Rada.

DEFENSE

Ukraine's armed forces numbered 400,000 in 1995. The defense budget for 1996 was estimated at $749 million.

ECONOMIC AFFAIRS

Ukraine was central to the Soviet agricultural and industrial system. The rich agricultural land of this region (commonly called the ''breadbasket'' of the former Soviet Union) provided 46% of Soviet agricultural output in the 1980s, and 25% of its coal production. Ukraine's economic base is dominated by industry, which accounts for over one-third of GDP. However, agriculture continues to play a major role in the economy, representing about one-fourth of GDP.

Despite resistance from parliament, President Kuchma has introduced economic reforms, including privatization of some segments of the economy, which have slowed hyperinflation. The government also instituted a new currency in 1995. Russia's financial crisis in the mid 1990s resulted in a sharp fall in Ukraine's export revenue and reduced domestic demand. Ukraine's economy remains in crisis since declining industrial and agricultural output and rampant institutional corruption hinders foreign investment, and significant economic restructuring appears unlikely.

Public Finance

Ukraine's public spending levels totaled 44% of GDP in 1996. The government subsidized agriculture, public industrial enterprises, housing, transportation, and telecommunications. Many of these subsidies are targeted for reduction or elimination as privatization progresses. At the end of 1998, Ukraine's outstanding external debt was $10.9 billion.

Income

In 1995, the gross national product (GNP) was $84.1 billion, or about $1,200 per person. By 1998, GNP per capita had fallen to an estimated US $850.

However, according to the World Bank, actual income may be at variance with these figures, since the informal economy has been expanding beyond the reach of Government regulations and taxes. Estimates for the informal economy range as high as 60% of total GDP.

Industry

Ukraine is a major producer of heavy machinery and industrial equipment for industries including mining, steelmaking, and chemicals. Ukraine is also an important supplier of automobiles, clothing, foodstuffs, timber, and paper to other former Soviet republics.

Banking and Finance

The National Bank of Ukraine (NBU), the country's national bank, was established in June 1991. It has since assumed the function of a central bank. The commercial banking sector is dominated by five former state-owned banks (Prominvest Bank, Ukrania, Ukrsotsbank, Eximbank and Oshadbank) in which the state continues to hold a large stake. In addition to these, there are 200 other commercial banks.

The Ukrainian Stock Exchange (USE), established in 1992, coordinates primary and secondary market trading of Ukranian securities. The exchange had 50 broker members in 1997.

Economic Development

In February 1993 Ukraine's parliament tentatively approved a new economic reform plan to stabilize the republic's economy, attract more capital from abroad, and lay the groundwork for a market economy. Proposed reforms included stricter monetary and banking regulation, and break-up of government monopolies. A privatization program is under way in several sectors, including retail trade, services, the food industry, agriculture, and housing.

Since the election of President Kuchma in 1994, the government has implemented a far-reaching economic reform program. Almost all price and trade controls have been abolished in an effort to stabilize the new market economy. Privatization began in earnest in 1995, and a new convertible currency was adopted in 1996.

SOCIAL WELFARE

Ukraine's current welfare system, developed in 1992, provides all employees with old age, disability, and survivor's pensions. Under a dual system of medical benefits, cash benefits are paid to employed persons, while a universal medical care system exists for all residents. The law provides women with the same employment rights as men, but few women rise to high-level positions. Estimates suggest that women may account for as many as 90% of Ukraine's unemployed.

Healthcare

According to United Nations reports, approximately 1 million persons were exposed to unsafe radiation levels following the Chernobyl nuclear accident in 1986.

In 1993 there were 220,000 physicians and a total of 700,000 hospital beds in Ukraine. Average life expectancy at birth in 1999 was 65.9 years. The leading causes of death were cardiovascular and respiratory diseases, cancer, traumas, and accidents.

Housing

In 1991, average housing space per person totaled 18 square meters.

EDUCATION

Ukraine has a 98% literacy rate, with about 70% of adult Ukrainians having a secondary or higher education. Ukraine has about 150 colleges and universities, of which the most important are at Kiev, Lviv, and Kharkiv.

1999 KEY EVENTS TIMELINE

February

- Ukraine's last intercontinental ballistic missile (ICBM) is destroyed in accordance with international disarmament treaties.

May

- An explosion at a coal mine near Donetsk kills 35 miners and injures over 30 others.

July

- U.S. Secretary of Defense William Cohen meets Ukrainian president Leonid Kuchma to discuss Ukrainian military relations with NATO and the U.S.

August

- Ukrainian officials meet with representatives of the International Monetary Fund (IMF) to request continued economic aid.

- German chancellor Gerhard Schroder meets with Ukrainian officials to work out problems with the Chernobyl nuclear reactor.

September

- Nine Ukrainian pilots are detained on suspicion of espionage in Zambia.

November

- Leonid Kuchma is re-elected Ukraine's president for another four-year term against a sworn communist.
- The nuclear power plant at Chernobyl reopens, thirteen years after the world's worst nuclear accident occurred there.

December

- President Leonid Kuchma holds his second term inauguration in the Ukraine Palace instead of in parliament partly due to his strained relationship with the legislative body.
- Parliament rejects President Kuchma's proposed candidate for prime minister, Valery Pustovoitenko. Pustovoitenko has been prime minister since 1996, but was required by law to resign when Kuchma was inaugurated. Kuchma then proposes banker Viktor Yushchenko as prime minister, and Parliament gives its approval.

ANALYSIS OF EVENTS: 1999

BUSINESS AND THE ECONOMY

Two artifacts of the Soviet economy remained as symbols of the difficulty of making the transition to capitalism; one was the collective farm. The intended privatization of Ukraine's famously productive agriculture industry met with little success. Although more than six million Ukrainian farmers had officially become landowning farmers since 1995, the privatized farms, which made up 46 percent of the country's farmland, have remained mismanaged, and are now mostly bankrupt. Out of the total of Ukrainian farms, 2 percent were privately owned by 5 percent of the nation's farmers. But agricultural production had dropped annually since Ukraine's independence, and as much as 70 percent of Ukraine's annual agriculture output is lost before reaching the market.

The other symbol of the Ukraine economy prior to the breakup of the USSR was the Chernobyl nuclear reactor. It remained in operation and the Ukrainian government, while officially committed to the plant's closure, had stated no intention to shut it down in the near future because, they contended, it would not be economically feasible to do so. Ukraine had promised to close Chernobyl in a 1995 agreement with the European Commission and the G-7 countries, but no date was stipulated for the shutdown.

In 1999 the Ukraine was already strapped for energy resources, owing Russia an estimated $1 billion for natural gas, and the nation had no alternatives to the electricity supplied by Chernobyl, which Ukraine expects to last for at least another year. Further strain was put on Ukraine's energy supplies when two nuclear plants, the Rivne and Yuzhnaya power stations, were closed for emergency repairs in July.

Ukraine's proposal to build two new nuclear reactors was blocked by Germany, which, however, was unable to persuade Ukraine to develop other energy sources. The German chancellor met with Ukraine's president over the summer to discuss a number of outstanding issues. Again in a July meeting between the Ukrainian president and representatives of the European Union, Chernobyl and the funding of alternatives to nuclear power plants were major topics. Demonstrators marched in Kiev on the anniversary of the 1986 Chernobyl disaster in April.

The same summit was targeted by the International Federation of the Phonographic Industry (IFPI) as an occasion to confront Ukraine on the issue of music piracy. According to the IFPI, Ukraine has replaced Bulgaria as Europe's center for the illegal reproduction and distribution of compact discs, which deprives the international recording industry of some $120 million annually.

In August Ukrainian officials met with representatives of the International Monetary Fund (IMF) to request the renewal of disbursements of a multi-billion dollar aid package. The IMF officials had halted payment of installments in view of Ukraine's plodding industrial reform and falling tax revenues. Ukraine needed the funds for domestic wages and foreign debt payments. In May, the IMF approved a disbursement totaling over half a billion dollars.

GOVERNMENT AND POLITICS

In 1999 the dissolution of the U.S.S.R. had not entirely done away with the cooperative spirit among some of the surviving member states. In October the president of Georgia, Eduard Shevarnadze, visited Kiev to strengthen Georgian ties with Ukraine. A few days later, the president of Uzbekistan was also in Kiev to sign a ten-year pact of Uzbek-Ukrainian economic cooperation. Kiev was also the site of a September meeting of security officials representing the Commonwealth of Independent States (CIS—the former soviet republics on the southern border of the USSR). The meeting discussed ways of combating terrorism and organized crime.

Not all of Ukraine's foreign relations were so friendly in 1999. Nine Ukrainian pilots aboard an Ilyushin transport were forced to land in Zambia, where authorities accused the Ukrainians of violating restricted air space above a military installation. Zambian officials also accused the pilots of transporting arms to insurgents in Angola. Ukraine paid a bail of $1200 per pilot set by a court in Lukasa, but the pilots were further detained to await a verdict on charges of espionage, an accusation denied by Ukraine.

Another symbol of the post Soviet age was a February 1999 ceremony attended by the U.S. ambassador to Ukraine where the country's last intercontinental ballistic missile (ICBM) was destroyed. Ukraine's nuclear missiles were destroyed under the terms of the 1994 Start One treaty. A U.S. arms reduction program provided over $500 million for Ukraine's nuclear disarmament.

CULTURE AND SOCIETY

The strains of post-soviet life, with its uncertainty and its cut-throat competition, showed in extraordinary actions taken by a number of Ukraine citizens. In October a businessman in Odessa walked into the tax administration office and set himself on fire. The suicide sparked a fire on the premises killing five other people. Later in the month, four women in Dnipropetrovsk doused themselves in gasoline but were prevented by police from committing self-immolation. The women had lost their savings in a pyramid scam.

In spite of these striking demonstrations, the government generally managed to contain effective protest and political opposition. In August the government prevented a small number of people from a group called "Youth Is the Hope of Ukraine" from demonstrating outside the presidential residence in Kiev. The demonstrators had previously received permission to march in order to voice their opposition to capital punishment in Ukraine, where the death penalty is widely supported. Officials cited security reasons for calling off the demonstration. When Ukraine joined the Council of Europe in 1995, its leaders had promised to abolish capital punishment by 1998. The Council voted in June to begin proceedings to suspend Ukraine's membership if it had not taken steps to do away with the death penalty by that time.

Safety conditions in Ukraine's coal industry came under critical public scrutiny as an explosion at a coal mine near Donetsk in eastern Ukraine killed 35 miners and injured over 30 others in May. This amounted to only a third of the total of more than 100 deaths in coal mine accidents in the year to that point. Natural gas was blamed for causing the blast. The president of Ukraine declared a day of national mourning.

Health concerns were prominent in the general population. The United Nations program for AIDS and HIV announced at an August conference in Warsaw, Poland, that Ukraine had the highest incidence of HIV infection among former eastern bloc countries. The Ukrainian ministry of health reported in March that the number of children with tuberculosis had increased by 50% over the past 5 years, and that 8,000 Ukrainians die of the disease each year.

DIRECTORY

CENTRAL GOVERNMENT

Head of State

President
Leonid Kuchma, Office of the President, vul. Bankova, 11, 252220 Kiev, Ukraine
PHONE: +380 (44) 2263265
FAX: +380 (44) 2931001

Prime Minister
Viktor Yushchenko, Office of the Prime Minister
PHONE: +380 (44) 2263263

Ministers

Minister of Agriculture
Borys Supikhanov, Ministry of Agriculture, vul. Khreshchatyk, 24, 252001 Kiev, Ukraine
PHONE: +380 (44) 2263466; 2262504

FAX: +380 (44) 2298756

Minister of the Cabinet of Ministers
Anatoliy Tolstoukhov, Ministry of the Cabinet of Ministers, 12/2 Hrushevskiy Street, 252008 Kiev, Ukraine
PHONE: +380 (44) 2663263; 2935227
FAX: +380 (44) 2932093
E-MAIL: postmaster@rada.kiev.ua

Minister of Coal Industry
Serhiy Tulub, Ministry of Coal Industry, Khmelnytskyi Str., 4, 252001 Kiev, Ukraine
PHONE: +380 (44) 2280372
FAX: +380 (44) 2282131

Minister of Culture and Arts
Dmytro Ostapenko, Ministry of Culture and Arts, vul. Ivan Franka, 19, 252030 Kiev, Ukraine
PHONE: +380 (44) 2244911; 2262645
FAX: +380 (44) 2253257

Minister of Defense
Oleksandr Kuzmuk, Ministry of Defense, Bankivska Street, 6, 252005 Kiev, Ukraine
PHONE: +380 (44) 2262637; 2262656; 2915441
FAX: +380 (44) 2262015

Minister of Economy
Vasyl Rohovyy, Ministry of Economy, vul. Hrushevskoho, 12/2, 252008 Kiev, Ukraine
PHONE: +380 (44) 2930108; 2934465; 2936141
FAX: +380 (44) 2287083

Minister of Education
Mykhaylo Zgurovskyy, Ministry of Education, Peremohy Pr. 10, 252135 Kiev, Ukraine
PHONE: +380 (44) 2263231; 2161049; 2162521
FAX: +380 (44) 2263323

Minister of Emergency Situations
Valeriy Kalchenko, Ministry of Emergency Situations

Minister of Environmental Protection and Nuclear Safety
Vasyl Shevchuk, Ministry of Environmental Protection and Nuclear Safety, Khreshchatyk Street, 5, Kiev, Ukraine
PHONE: +380 (44) 2262428; 2284004
FAX: +380 (44) 2298383; 2280644

Minister of Family and Youth Affairs
Valentyna Dovzhenko, Ministry of Family and Youth Affairs

Minister of Finance
Ihor Mityukov, Ministry of Finance, vul. Hrushevskoho, 12/2, 252008 Kiev, Ukraine
PHONE: +380 (44) 2262044; 2935363
FAX: +380 (44) 2262617; 2932178

Minister of Foreign Affairs
Borys Tarasyuk, Ministry of Foreign Affairs, Mykhaylivska pl. 1, 252018 Kiev, Ukraine
PHONE: +380 (44) 2263379
FAX: +380 (44) 2936950; 2263179; 2263169

Minister of Foreign Economic Relations and Trade
Serhiy Osyka, Ministry of Foreign Economic Relations and Trade, Lvivska Ploscha 8, 252053 Kiev, Ukraine
PHONE: +380 (44) 2265233; 2262733; 2125423
FAX: +380 (44) 2262629

Minister of Health
Andriy Serdyuk, Ministry of Health, vul. Kreshschatik, 5, 252021 Kiev, Ukraine
PHONE: +380 (44) 2263253
FAX: +380 (44) 2287794

Minister of Industrial Policy
Vasyl Hureyev, Ministry of Industrial Policy, vul. Maryny Paskovoi, 15, 252167 Kiev, Ukraine
PHONE: +380 (44) 2262623
FAX: +380 (44) 2274104; 2262004

Minister of Information
Zinoviy Kulyk, Ministry of Information, Prorizna Str., 2, 252003 Kiev, Ukraine
PHONE: +380 (44) 2288769
FAX: +380 (44) 2280991

Minister of Internal Affairs
Yuriy Kravchenko, Ministry of Internal Affairs, Bohomolets Street, 10, 252024 Kiev, Ukraine
PHONE: +380 (44) 2263317
FAX: +380 (44) 2913182

Minister of Justice
Syuzanna Stanik, Ministry of Justice, vul. M. Kotsiubynskoho, 12, 252030 Kiev, Ukraine
PHONE: +380 (44) 2262416
FAX: +380 (44) 2296664

Minister of Labor
Ivan Sakhan, Ministry of Labor, vul. Pushkinska, 28, 252004 Kiev, Ukraine
PHONE: +380 (44) 2246347
FAX: +380 (44) 2245905

Minister of Power and Industry
Oleksiy Shebertsov, Ministry of Power and Industry

POLITICAL ORGANIZATIONS

Ahrarna Partiya Ukrainy (Agrarian Party-AP)

NAME: Kateryna Vashchuk

Vseukrainske Obyednannya "Hromada"-HROMADA (All-Ukrainian Association)

vul. Kutuzova, 18/7, Kiev, Ukraine
PHONE: +380 (44) 2963206; 2951450
NAME: Pavlo Lazarenko

Liberal Democratic Party of Ukraine (LDPU)

vul. Bratyslavska, 40, kv. 110, 252166 Kiev, Ukraine
PHONE: +380 (44) 2907036; 5197970
TITLE: Chairman
NAME: Andriy Koval

Prohresyvna Sotsialistychna Partiya-PSP (People's Democratic Party)

vul. Saksahanskoho, 71, Kiev, Ukraine
PHONE: +380 (44) 2938202; 4124160
TITLE: Chairman
NAME: Valeriy Pustovoytenko

People's Party of Ukraine (PPU)

Dnipropetrovsk, Naberezhna Lenina, 1, Kiev, Ukraine
PHONE: +380 (44) 588032; 587557

Social-Demokratykna Partija Ukrajiny-SDPU (Social Democratic Party of Ukraine)

Tolstoy St., 16/24, Kiev, Ukraine
PHONE: +380 (44) 2290816
NAME: Yuriy Buzduhan

Sotsialistykna Partija Ukrainy-SPU (Socialist Party of Ukraine)

9 Bankova St., 6, Suite 638, Kiev, Ukraine
NAME: Oleksandr Moroz

Selinka Partija Ukrainy-SELPU (Peasants' Party of Ukraine)

vul. Hrinchenka, 1, Room 301, Kiev, Ukraine
PHONE: +380 (44) 2286709
NAME: Serhiy Dovan

Kommunistykna Partija Ukrainy-KPU (Ukrainian Communist Party)

NAME: Petr Symonenk

Narodnyy Rukh Ukrainy-RUKH (Ukrainian People's Movement for Restructuring)

Shevchenko Blvd., 37/122, Kiev, Ukraine
PHONE: +380 (44) 2249151; 2168333; 2742077
TITLE: Chairman
NAME: Vyacheslav Chornovil

DIPLOMATIC REPRESENTATION

Embassies in Ukraine

Algeria
vul. B. Khmelnytskogo, 64, Kiev, Ukraine
PHONE: +380 (44) 2157150

Australia
vul. Kominterna 18, Kiev, Ukraine
PHONE: +380 (44) 2257586

Austrian
vul. I. Franko, 33, Kiev, Ukraine
PHONE: +380 (44) 2205759
FAX: +380 (44) 2275465

Armenia
vul. Instytutskaya, 4, Moskva Hotel, Kiev, Ukraine
PHONE: +380 (44) 2290806
FAX: +380 (44) 2290807

Belarus
vul. Sichnevogo Povstaniya, 6, Kiev, Ukraine
PHONE: +380 (44) 2900201
FAX: +380 (44) 2954667

Belgium
vul. Khmelntyskogo, 58, Kiev, Ukraine
PHONE: +380 (44) 2192910
FAX: +380 (44) 2192717

Brazil
vul. L. Tolstogo, 11, Kiev, Ukraine
PHONE: +380 (44) 2253525
FAX: +380 (44) 2245521

Bulgaria
vul. Gospitalna, 1, Kiev, Ukraine
PHONE: +380 (44) 2245360
FAX: +380 (44) 2349929

Canada
vul. Yaroslaviv Val, 31, Kiev, Ukraine
PHONE: +380 (44) 2123590
FAX: +380 (44) 2120212

China
vul. Grushevskogo, 32, Kiev, Ukraine
PHONE: +380 (44) 2937371

FAX: +380 (44) 2937371

Croatia
vul. Artema. 50/51, Kiev, Ukraine
PHONE: +380 (44) 2165862
FAX: +380 (44) 2246942

Cuba
vul. Bekhterevskiy, 5, Kiev, Ukraine
PHONE: +380 (44) 2162930

Czech Republic
vul. Yaroslaviv Val, 34, Kiev, Ukraine
PHONE: +380 (44) 2120807
FAX: +380 (44) 2246180

Denmark
vul. Volodymyrska, 45, Kiev, Ukraine
PHONE: +380 (44) 2121327
FAX: +380 (44) 2169428

Egypt
vul. Observatorna, 19, Kiev, Ukraine
PHONE: +380 (44) 2121327
FAX: +380 (44) 2169428

Estonia
vul. Kutuzova, 8, Kiev, Ukraine
PHONE: +380 (44) 2962886
FAX: +380 (44) 2948055

Finland
vul. Striletska, 14, Kiev, Ukraine
PHONE: +380 (44) 2284339
FAX: +380 (44) 2282032

France
vul. Chkalova, 84, Kiev, Ukraine
PHONE: +380 (44) 2288728
FAX: +380 (44) 2290870

Georgia
vul. Grushevskogo, 26/1, Kiev, Ukraine
PHONE: +380 (44) 2936957

Germany
vul. Chkalova, 84, Kiev, Ukraine
PHONE: +380 (44) 2161477
FAX: +380 (44) 2169233

Greece
vul. Lypska, 3, National Hotel, Kiev, Ukraine
PHONE: +380 (44) 2918874

Hungary
vul. Reutarska, 33, Kiev, Ukraine
PHONE: +380 (44) 2124004
FAX: +380 (44) 2122090

India
vul. Terehina, 4, Kiev, Ukraine
PHONE: +380 (44) 4356661
FAX: +380 (44) 4356619

Indonesia
vul. Nikolsko-Botanicheska, 14, Kiev, Ukraine
PHONE: +380 (44) 2442762
FAX: +380 (44) 2443804

Iran
vul. Kruglouniversitetska, 12, Kiev, Ukraine
PHONE: +380 (44) 2294463
FAX: +380 (44) 2293255

Israel
Lesi Ukrainki blvd., 34, Kiev, Ukraine
PHONE: +380 (44) 2956925
FAX: +380 (44) 2949736

Italy
vul. Sichnevogo Povstannya, 25, Kiev, Ukraine
PHONE: +380 (44) 2944292
FAX: +380 (44) 2905162

Japan
vul. Museyniy, 4, Kiev, Ukraine
PHONE: +380 (44) 4620020

Kazakhstan
vul. Sichnevogo Povstannya, 6, Kiev, Ukraine
PHONE: +380 (44) 2907721
FAX: +380 (44) 2900610

South Korea
43 Volodymyrska vul., Kiev, Ukraine
PHONE: +380 (44) 2242319
FAX: +380 (44) 2240364

Kyrgyzstan
vul. Kutuzova, 8, Kiev, Ukraine
PHONE: +380 (44) 2955380
FAX: +380 (44) 2959692

Kuwait
vul. Lypska, 3, National Hotel, Kiev, Ukraine
PHONE: +380 (44) 2918734

Latvia
vul. Desyatynna, 4/6, Kiev, Ukraine
PHONE: +380 (44) 2292360
FAX: +380 (44) 2292745

Libya
vul. striletska, 16, Kiev, Ukraine
PHONE: +380 (44) 2446621
FAX: +380 (44) 2446624

Lithuania
vul. Gorkogo, 22, Kiev, Ukraine
PHONE: +380 (44) 2274372

Moldova
vul. Kutuzova, 8, Kiev, Ukraine
PHONE: +380 (44) 2952653

FAX: +380 (44) 2956703

Mongolia
vul. M. Kotsyubinskogo, 3, Kiev, Ukraine
PHONE: +380 (44) 2168891

Netherlands
vul. Turgenevska, 24, Kiev, Ukraine
PHONE: +380 (44) 2161905
FAX: +380 (44) 2168105

Norway
vul. Striletskaya, 15, Kiev, Ukraine
PHONE: +380 (44) 2240066
FAX: +380 (44) 2240655

Peru
vul. Staronavodnitska, 8, Kiev, Ukraine
PHONE: +380 (44) 2957776

Poland
vul. Yaroslaviv Val, 12, Kiev, Ukraine
PHONE: +380 (44) 2246308
FAX: +380 (44) 2293575

Portugal
vul. Chervonoarmiyska, 9/2, Kiev, Ukraine
PHONE: +380 (44) 2272442
FAX: +380 (44) 2272066

Romania
vul. M. Kotsyubinskogo, 8, Kiev, Ukraine
PHONE: +380 (44) 2245261

Russia
vul. Kutuzova, 8, Kiev, Ukraine
PHONE: +380 (44) 2947936

Serbia
vul. Bekhterevskiy 7/11, Kiev, Ukraine
PHONE: +380 (44) 2119734
FAX: +380 (44) 2136969

Slovakia
vul. Yaroslaviv Val, 34, Kiev, Ukraine
PHONE: +380 (44) 2120310
FAX: +380 (44) 2123271

South Africa
vul. Chervonoarmiyska, 9/2, Maculan, Kiev, Ukraine
PHONE: +380 (44) 2277172
FAX: +380 (44) 2207206

Spain
vul. Dekhtyarivskiy, 38–44, Kiev, Ukraine
PHONE: +380 (44) 2131858
FAX: +380 (44) 2130031

Sweden
vul. Lipskaya, 5, National Hotel, Kiev, Ukraine
PHONE: +380 (44) 2918919
FAX: +380 (44) 2916233

Tunisia
vul. Chervonoarmiyska, 44, Kiev, Ukraine
PHONE: +380 (44) 2208655
FAX: +380 (44) 2277220

Turkey
vul. Arsenalna, 18, Kiev, Ukraine
PHONE: +380 (44) 2949964
FAX: +380 (44) 2956423

Turkmenestan
vul. Pushkinska, 6, Kiev, Ukraine
PHONE: +380 (44) 2293034

United Kingdom
vul. Desyatinna, 9, Kiev, Ukraine
PHONE: +380 (44) 4620011
FAX: +380 (44) 4620014

United States
vul. Y. Kotsyubinskogo, 10, Kiev, Ukraine
PHONE: +380 (44) 2447344
FAX: +380 (44) 2447350

Uzbekistan
vul. Mala Zhytomyrska, 20, Kiev, Ukraine
PHONE: +380 (44) 2281246

Vatican City
vul. chervonoarmiyska, 96, Kiev, Ukraine
PHONE: +380 (44) 2699103
FAX: +380 (44) 2692417

Vietnam
vul. Leskova, 5, Kiev, Ukraine
PHONE: +380 (44) 2952837

JUDICIAL SYSTEM

Supreme Court

Constitutional Court

FURTHER READING

Articles

''As 2000 Nears, a Wary Optimism for Ukraine's Nuclear Plants.'' *New York Times,* 19 August 1999, p. A11.

''Could Ukraine revert to Russia?'' *The Economist* 351 (July 10, 1999): 46.

O'Brien, Timothy L. ''A Palace Fit for a Fugitive and Ukraine's Ex-Premier.'' *New York Times,* 1 September 1999, p. A1.

''Ukraine's two minds.'' *The Economist* 351 (June 5): 48.

Internet

Ukrainian Weekly. Available Online @ www.ukrweekly.com/ (November 12, 1999).

UNIAN. Available Online @ www.unian.net:8101/english/index.htm (November 12, 1999).

UKRAINE: STATISTICAL DATA

For sources and notes see "Sources of Statistics" in the front of each volume.

GEOGRAPHY

Geography (1)

Area:

Total: 603,700 sq km.

Land: 603,700 sq km.

Water: 0 sq km.

Area—comparative: slightly smaller than Texas.

Land boundaries:

Total: 4,558 km.

Border countries: Belarus 891 km, Hungary 103 km, Moldova 939 km, Poland 428 km, Romania (south) 169 km, Romania (west) 362 km, Russia 1,576 km, Slovakia 90 km.

Coastline: 2,782 km.

Climate: temperate continental; Mediterranean only on the southern Crimean coast; precipitation disproportionately distributed, highest in west and north, lesser in east and southeast; winters vary from cool along the Black Sea to cold farther inland; summers are warm across the greater part of the country, hot in the south.

Terrain: most of Ukraine consists of fertile plains (steppes) and plateaus, mountains being found only in the west (the Carpathians), and in the Crimean Peninsula in the extreme south.

Natural resources: iron ore, coal, manganese, natural gas, oil, salt, sulfur, graphite, titanium, magnesium, kaolin, nickel, mercury, timber.

Land use:

Arable land: 58%

Permanent crops: 2%

Permanent pastures: 13%

Forests and woodland: 18%

Other: 9% (1993 est.).

HUMAN FACTORS

Health Personnel (3)

Total health expenditure as a percentage of GDP, 1990-1997[a]	
Public sector	3.9
Private sector	NA
Total[b]	NA
Health expenditure per capita in U.S. dollars, 1990-1997[a]	
Purchasing power parity	NA
Total	NA
Availability of health care facilities per 100,000 people	
Hospital beds 1990-1997[a]	990
Doctors 1993[c]	429
Nurses 1993[c]	1,211

Health Indicators (4)

Life expectancy at birth	
1980	69
1997	67
Daily per capita supply of calories (1996)	2,753
Total fertility rate births per woman (1997)	1.3
Maternal mortality ratio per 100,000 live births (1990-97)	30[b]
Safe water % of population with access (1995)	55
Sanitation % of population with access (1995)	49
Consumption of iodized salt % of households (1992-98)[a]	....
Smoking prevalence	
Male % of adults (1985-95)[a]	57
Female % of adults (1985-95)[a]	22
Tuberculosis incidence per 100,000 people (1997)	61
Adult HIV prevalence % of population ages 15-49 (1997)	0.43

Infants and Malnutrition (5)

Under-5 mortality rate (1997) 24
% of infants with low birthweight (1990-97) . . . . . .NA
Births attended by skilled health staff % of total[a] . . .NA
% fully immunized (1995-97)
TB 95
DPT 96
Polio 97
Measles 97
Prevalence of child malnutrition under age 5 (1992-97)[b] . . . . .NA

Ethnic Division (6)

Ukrainian 73%
Russian 22%
Jewish 1%
Other 4%

Religions (7)

Ukrainian Orthodox—Moscow Patriarchate, Ukrainian Orthodox— Kiev Patriarchate, Ukrainian Autocephalous Orthodox, Ukrainian Catholic (Uniate), Protestant, Jewish.

Languages (8)

Ukrainian, Russian, Romanian, Polish, Hungarian.

EDUCATION

Public Education Expenditures (9)

Public expenditure on education (% of GNP)
1980 . . . . .5.6
1996 . . . . .7.2[1]
Expenditure per student
Primary % of GNP per capita
1980 . . . . .21.2
1996 . . . . .43.4[1]
Secondary % of GNP per capita
1980 . . . .
1996 . . . . .7.3[1]
Tertiary % of GNP per capita
1980 . . . . .20.1
1996 . . . . .22.6[1]
Expenditure on teaching materials
Primary % of total for level (1996) . . . .
Secondary % of total for level (1996) . . . .
Primary pupil-teacher ratio per teacher (1996) . . . . .20[1]
Duration of primary education years (1995) 4

HUMAN FACTORS

Demographics (2A)

	1990	1995	1998	2000	2010	2020	2030	2040	2050
Population	51,600.4	51,090.1	50,125.1	49,506.8	47,591.6	46,061.3	44,172.4	42,037.9	39,096.2
Net migration rate (per 1,000 population)	NA	NA	NA	NA	NA	NA	NA	NA	NA
Births	NA	NA	NA	NA	NA	NA	NA	NA	NA
Deaths	NA	NA	NA	NA	NA	NA	NA	NA	NA
Life expectancy - males	65.2	60.9	60.1	60.4	64.4	68.7	72.2	75.0	77.0
Life expectancy - females	74.4	72.0	71.9	71.8	74.2	77.4	80.0	82.1	83.6
Birth rate (per 1,000)	13.2	9.7	9.5	9.6	12.0	8.9	8.6	7.5	6.6
Death rate (per 1,000)	12.3	15.7	16.3	16.4	14.9	13.1	12.7	13.6	14.7
Women of reproductive age (15-49 yrs.)	12,301.3	12,627.2	12,771.6	12,759.0	12,067.2	10,963.2	10,043.7	8,431.0	7,690.0
of which are currently married	NA	NA	NA	NA	NA	NA	NA	NA	NA
Fertility rate	1.9	1.4	1.3	1.3	1.6	1.5	1.5	1.4	1.4

Except as noted, values for vital statistics are in thousands; life expectancy is in years.

Literacy Rates (11B)

Adult literacy rate

1980	
Male	-
Female	-
1995	
Male	98
Female	99

GOVERNMENT & LAW

Political Parties (12)

People's Council	No. of seats
Independents	238
Communist	95
Rukh	22
Agrarians	18
Socialist	15
Republicans	11
Congress of Ukrainian Nationalists	5
Labor	5
Party of Democratic Revival	4
Democratic Party of Ukraine	2
Social Democrats	2
Civil Congress	2
Conservative Republicans	1
Party of Economic Revival of Crimea	1
Christian Democrats	1
Vacant	28

Government Budget (13B)

Revenues	$18 billion
Expenditures	$21 billion
Capital expenditures	NA

Data for 1997 est. NA stands for not available.

Military Affairs (14B)

	1992	1993	1994	1995
Military expenditures				
Current dollars (mil.)	3,905	3,247	4,378	3,588
1995 constant dollars (mil.)	4,200	3,404	4,488	3,588
Armed forces (000)	438[e]	510	495	476
Gross national product (GNP)				
Current dollars (mil.)	250,600	180,400	141,200	123,700
1995 constant dollars (mil.)	221,100	189,100	144,700	123,700
Central government expenditures (CGE)				
1995 constant dollars (mil.)	186,300[e]	NA	NA	46,200
People (mil.)	51.7	51.6	51.3	51.1
Military expenditure as % of GNP	1.9	1.8	3.1	2.9
Military expenditure as % of CGE	2.3	NA	NA	7.8
Military expenditure per capita (1995 $)	81	66	87	70
Armed forces per 1,000 people (soldiers)	8.5	9.9	9.6	9.3
GNP per capita (1995 $)	4,280	3,667	2,819	2,422
Arms imports[6]				
Current dollars (mil.)	0	0	0	0
1995 constant dollars (mil.)	0	0	0	0
Arms exports[6]				
Current dollars (mil.)	0	50	90	160
1995 constant dollars (mil.)	0	52	92	160
Total imports[7]				
Current dollars (mil.)	7,099	9,533	10,750	11,380
1995 constant dollars (mil.)	7,636	9,994	11,020	11,380
Total exports[7]				
Current dollars (mil.)	8,045	7,817	10,300	11,570
1995 constant dollars (mil.)	8,654	8,195	10,560	11,570
Arms as percent of total imports[8]	0.0	0	0	0
Arms as percent of total exports[8]	0.0	.6	.9	1.4

Crime (15)

Crime rate (for 1997)	
Crimes reported	589,200
Total persons convicted	402,400
Crimes per 100,000 population	1,150
Persons responsible for offenses	
Total number of suspects	337,900
Total number of female suspects	58,500
Total number of juvenile suspects	27,700

LABOR FORCE

Labor Force (16)

Total (million)	22.8
Industry and construction	32%
Agriculture and forestry	24%
Health, education, and culture	17%
Trade and distribution	8%
Transport and communication	7%
Other	12%

Data for 1996.

Unemployment Rate (17)

2.6% officially registered; large number of unregistered or underemployed workers (December 1997).

PRODUCTION SECTOR

Electric Energy (18)

Capacity	52 million kW (1997)
Production	177 billion kWh (1997)
Consumption per capita	3,431 kWh (1997)

Transportation (19)

Highways:

total: 172,565 km

paved: 163,937 km (including 1,875 km of expressways); note—these roads are said to be hard-surfaced, meaning that some are paved and some are all-weather gravel surfaced.

unpaved: 8,628 km (1996 est.)

Waterways: 4,400 km navigable waterways, of which 1,672 km were on the Pryp"yat' and Dnipro Rivers (1990)

Pipelines: crude oil 2,010 km; petroleum products 1,920 km; natural gas 7,800 km (1992)

Merchant marine:

total: 202 ships (1,000 GRT or over) totaling 1,498,653 GRT/1,709,393 DWT ships by type: barge carrier 3, bulk 13, cargo 122, chemical tanker 2, combination bulk 1, container 3, multifunction large-load carrier 2, oil tanker 19, passenger 7, passenger-cargo 4, railcar carrier 2, refrigerated cargo 6, roll-on/roll-off cargo 13, short-sea passenger 5 note: Ukraine owns an additional 41 ships (1,000 GRT or over) totaling 515,743 DWT operating under the registries of The Bahamas, Cyprus, Liberia, Malta, Panama, and Saint Vincent and the Grenadines (1997 est.)

Airports: 706 (1994 est.)

Airports—with paved runways:

total: 163

over 3,047 m: 14

2,438 to 3,047 m: 55

1,524 to 2,437 m: 34

914 to 1,523 m: 3

under 914 m: 57 (1994 est.)

Airports—with unpaved runways:

total: 543

over 3,047 m: 7

2,438 to 3,047 m: 7

1,524 to 2,437 m: 16

914 to 1,523 m: 37

under 914 m: 476 (1994 est.)

Top Agricultural Products (20)

Grain, sugar beets, sunflower seeds, vegetables; meat, milk.

MANUFACTURING SECTOR

GDP & Manufacturing Summary (21)

	1980	1985	1990	1992	1993	1994
Gross Domestic Product						
Millions of 1990 dollars	121,758	139,789	155,591	126,021	102,727	77,764
Growth rate in percent	3.81	0.99	1.39	−13.99	−18.48	−24.30
Per capita (in 1990 dollars)	2,437.1	2,745.6	3,013.2	2,441.3	1,992.7	1,511.0
Manufacturing Value Added						
Millions of 1990 dollars	NA	NA	53,966	40,415	33,141	23,861
Growth rate in percent	NA	NA	NA	−13.80	−18.00	−28.00
Manufacturing share in percent of current prices	41.0	38.0	34.7	34.0	32.5	28.9

FINANCE, ECONOMICS, & TRADE

Economic Indicators (22)

National product: GDP—purchasing power parity—$124.9 billion (1997 est.)

National product real growth rate: -3.2% (1997 est.)

National product per capita: $2,500 (1997 est.)

Inflation rate—consumer price index: 10% (yearend 1997 est.)

Balance of Payments (23)

	1994	1995	1996
Exports of goods (f.o.b.)	13,894	14,244	15,547
Imports of goods (f.o.b.)	−16,469	−16,946	−19,843
Trade balance	−2,575	−2,702	−4,296
Services - debits	−1,938	−2,015	−2,300
Services - credits	2,803	3,093	4,901
Private transfers (net)	NA	NA	NA
Government transfers (net)	NA	NA	NA
Overall balance	−1,710	1,624	−1,695

Exchange Rates (24)

Exchange rates:	
Hryvnia per US$1	
February 1998	1.9359
1997	1.8617
1996	1.8295
1995	1.4731
1994	0.3275
1993	0.0453

Top Import Origins (25)

$20.2 billion (1997 est.)

Origins	%
Russia	NA
Turkmenistan	NA
Belarus	NA
Germany	NA
China	NA

NA stands for not available.

Top Export Destinations (26)

$15.2 billion (1997 est.) Data are for 1997.

Destinations	%
Russia	NA
China	NA
Belarus	NA
Turkey	NA
Germany	NA

NA stands for not available.

Economic Aid (27)

Recipient: ODA, $220 million (1993). Note: commitments, 1992-95, $4.5 billion ($4.1 billion drawn).

Import Export Commodities (28)

Import Commodities	Export Commodities
Energy	Ferrous and nonferrous metals
Machinery and parts	Chemicals
Transportation equipment	Machinery and transport equipment
Chemicals	Food products
Plastics and rubber	

UNITED ARAB EMIRATES

Al-Imarat al-`Arabiyah al-Muttahidah

CAPITAL: Abu Dhabi (Abu Zaby).

FLAG: The flag consists of a red vertical stripe at the hoist and three equal horizontal stripes of green, white, and black.

ANTHEM: The National Anthem is an instrumental piece without words.

MONETARY UNIT: The United Arab Emirates dirham (UD), introduced as the currency in May 1973, is divided into 100 fils. There are coins of 1, 5, 10, 25, and 50 fils and 1 and 5 dirham and notes of 5, 10, 50, 100, 200, 500, and 1,000 dirhams. UD1 = $0.2724 (or $1 = UD3.671).

WEIGHTS AND MEASURES: The metric system and imperial and local measures are used.

HOLIDAYS: New Year's Day, 1 January; Accession of the Ruler of Abu Dhabi (Abu Dhabi only), 6 August; National Day, 2 December; Christmas, 25 December. Muslim religious holidays include Lailat al-Miraj, `Id al-Fitr, `Id al-'Adha', Hijra New Year, and Milad an-Nabi.

TIME: 4 PM = noon GMT.

LOCATION AND SIZE: Comprising a total area of approximately 75,581 square kilometers (29,182 square miles), the United Arab Emirates (UAE), is in the eastern Arabian Peninsula. It consists of seven states, or emirates: Abu Dhabi, Dubayy, Ash Shariqah, Ra's al-Khaymah, Al Fujayrah, Umm al-Qaywayn, and `Ajman. Comparatively, the area occupied by UAE is slightly smaller than the state of Maine. UAE has a total boundary length of 2,185 kilometers (1,358 miles). The UAE's capital city, Abu Dhabi, is located on the Persian Gulf.

CLIMATE: The months between May and October are extremely hot, with average temperatures ranging between 28° and 41°C (82–106°F) and high humidity near the coast. Winter temperatures can fall as low as 2°C (36°F) but average between 14° and 28°C (57–82°F). Normal average annual rainfall is from 13 centimeters (5.1 inches).

INTRODUCTORY SURVEY

RECENT HISTORY

When the United Kingdom announced it would withdraw its forces from the Arabian Gulf region, a federation of Arab royal houses (emirates) formed the United Arab Emirates. The UAE was officially proclaimed an independent nation on 2 December 1971.

Land disputes between the new state and its two powerful neighbors, Saudi Arabia and Iran, cropped up during the first two decades of its existence. Border agreements signed with both nations, however, had eased tensions by the end of the century.

In 1981 the UAE helped found the Gulf Cooperation Council (GCC), a political and economic alliance that includes Bahrain, Kuwait, Oman, Qatar, Saudi Arabia, and the UAE. During the Iran-Iraq War (1980–88), the UAE gave aid to Iraq but also maintained diplomatic relations with Iran and sought to mediate the conflict.

In the Persian Gulf War (1990) forces from the UAE fought against Iraq, and the government gave some $4.5 billion to the coalition war effort. The UAE's generous foreign aid to Arab states (over $15 billion through 1991) makes it a significant player in the affairs of the region.

GOVERNMENT

The executive branch of government consists of the Federal Supreme Council, composed of the seven hereditary emirate rulers. The council is re-

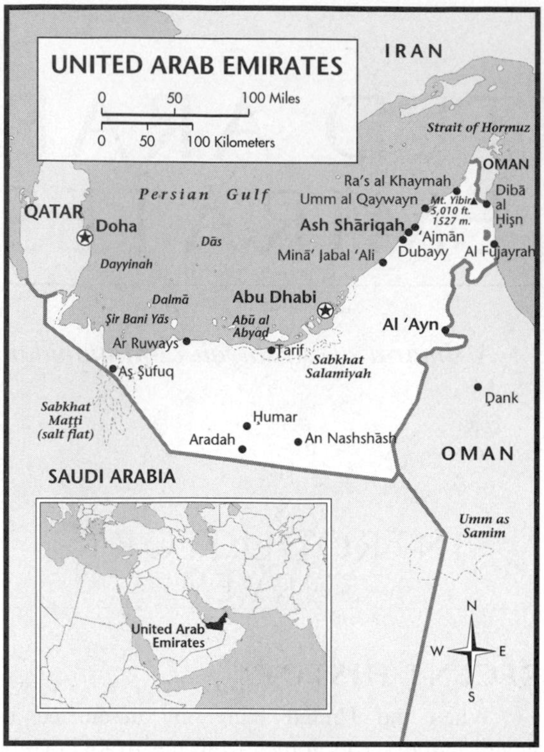

sponsible for establishing general policies, ratifying federal laws, and overseeing the union's budget.

Sheikh Zayid bin Sultan al-Nahayyan, emir of Abu Dhabi, has been president of the council since 1971. The president is assisted by a Council of Ministers, headed by the prime minister. A 40-member legislature, the Federal National Council, can question cabinet ministers and make recommendations to the Federal Supreme Council.

Most of the individual emirates (states) are governed according to tribal traditions.

Judiciary

The 1971 constitution established a Union Supreme Court with judges appointed by the president.

Political Parties

There are no political parties.

DEFENSE

In 1993, the combined UAE forces totaled 54,000 men, all volunteers. One-third were Asian contract soldiers.

ECONOMIC AFFAIRS

The economy of the UAE is based on oil and oil-based industries. However, with its strategic location, modern communications and transportation facilities, and strong banking system, the UAE is also the major trade center in the Gulf region.

In 1995, petroleum accounted for 34% of the gross domestic product (GDP); trade, 12%; government, 11%; construction, 9%; manufacturing, 9%; real estate, 8%; transportation, communications, and storage, 6%; and other sectors, 11%.

Public Finance

Abu Dhabi's oil income accounts for the bulk of federal revenues; under the constitution, each emirate contributes 50% of its net oil income to the federal budget. The U.S. Central Intelligence Agency estimates that, in 1998, government revenues totaled approximately $5.4 billion. External debt was estimated at $14 billion in 1996.

Income

In 1995, United Arab Emirates' gross national product (GNP) was $42.87 billion, or about $17,390 per person. In 1998, the Central Intelligence Agency estimated the nation's GDP at $40 billion; per capita purchasing power parity was $17,400.

Industry

The industrial complex in Abu Dhabi includes an oil refinery with a processing capacity of 120,000 barrels per day; a fertilizer factory, and a gas liquefaction installation. In 1995, the UAE produced 69.3 million barrels of refined petroleum products.

The Dubai industrial port complex includes an aluminum smelter, and other developed industries producing construction-related materials such as cement, asphalt, and concrete blocks.

Banking and Finance

The UAE Central Bank, founded in 1980 and capitalized at $81.7 million, was granted additional capital of $2 billion from the government in 1982, which was to increase by 20% per year until a total deposit of $4 billion had been reached.

In 1996 there were 19 local and 27 foreign banks with a total of 230 and 119 branches respectively.

In 1991 the international Bank of Commerce and Credit International (BCCI), which was based in the UAE and largely owned by the ruling family of Abu Dhabi, was accused of fraud and closed down, causing repercussions all around the world.

Economic Development

The federation used its vast oil wealth during the 1970s to transform the national economy through expansion of roads, ports, airports, communications facilities, electric power plants, and water desalination facilities, and construction of huge oil-processing complexes. With the completion of these projects in the early 1980s, development efforts shifted to diversifying the economy by establishing capital-intensive industries based on oil and gas resources.

Major industrial projects are the Jabal'Ali industrial zone in Dubai and the refinery complex in Abu Dhabi. In mid-1995, 822 companies were operating in the Jabal'Ali Free Zone.

SOCIAL WELFARE

Although there is no social security law in the UAE, many welfare benefits are provided. If the father of a family is unable to work because of illness, disability, or old age, he receives help under the National Assistance Law; should he die or divorce his wife, the woman's future is secure.

Female employment is growing in government service and in occupations such as education and health. A married woman can only work outside the home with her husband's permission. Females are not permitted to leave the country without the permission of a male relative.

Healthcare

Modern hospitals have been built in Abu Dhabi, Dubai, and elsewhere. In 1991, there were 1,526 doctors. Typhoid fever and tuberculosis are rare; malaria remains a problem, however. Average life expectancy is 75 years. In 1991–93, 95% of the population had access to health care services.

Housing

The federal government is attempting to make modern low-cost homes available to poorer families and to supply them with piped water, sewage systems, and electricity.

EDUCATION

Education is compulsory for six years at the primary level. In 1994, there were 262,628 pupils at the primary level and 159,840 pupils in secondary schools. In 1991, all higher level institutions had 10,641 students. The literacy rate is 79.2%.

1999 KEY EVENTS TIMELINE

January

- The Hamriyah Free Zone issues business licenses in the Hamriyah port between Ajman and Umm al-Quwain; the licensing process establishes local and international businesses within the zone.
- While the Water and Power Department has been disbanded, the privatization of eleven new utilities companies begins on January 1 with the creation of the first Independent Water and Power Project (IWPP).
- Distinct differences in policies towards Iraq are exposed as Arab ministers attempt to organize a full summit: Saudi Arabia and Kuwait oppose the summit while Qatar and the UAE support it.
- Under the Islamic law allowing next of kin to have a death sentence set aside, a murdered man's family asks the UAE Supreme Court to grant his murderer, John Aquino, clemency from a 1990 death sentence.
- Sadam Hussein's inflammatory call for Arabs to topple their governments provokes the ministers of Saudi Arabia, Kuwait, Bahrain, Oman, Qatar and the United Arab Emirates.
- During Ramadan, the coast guard works to full capacity as one thousand migrants attempt to enter the UAE illegally; the migrants, mainly Afghan and Pakistani, are arrested and deported back to their home countries.

February

- The Philippine government proposes a plan to the UAE government aimed at protecting Filipino workers in the Gulf. The plan proposes to resolve problems surrounding unscrupulous agencies and employers and the mistreatment of Filipino workers by direct hiring through government labor offices.
- The death of Jordan's King Hussein is acknowledged in the UAE by a declaration of a forty-day state of mourning, with government offices closed for three days.
- UAE signs Y2K declaration as eighteen countries appeal to the World Bank, the Asian Development Bank and other organizations for millions of dollars in aid to help them solve the

potential problems of dealing with the millennium bug.

March

- Low fuel prices threaten the economic stability of the UAE, which derives 60 percent of its income from oil sales.
- The Organization of Oil Exporting Countries (OPEC), including Algeria, Indonesia, Iran, Kuwait, Libya, Nigeria, Qatar, Saudi Arabia, the United Arab Emirates, and Venezuela, agree with non-OPEC members, Mexico and Oman, to cut oil production, increasing the price of crude oil in 1999. Iraq is the only OPEC member to disagree with the cut.

April

- The UAE, Saudi Arabia, Syria, Bahrain, Kuwait, Egypt, and Somalia are accused by Iraq of owing more than $1.6 million in loans and interest to Iraq.
- King Abdullah of Jordan visits three UAE cities, hoping to restore the confidence in Jordan eroded by his father King Hussein's support of Baghdad during the Gulf War.
- Since many sea captains are discharging oil tanks at night in the waters surrounding the UAE, the UAE is considering the death penalty for anyone caught deliberately polluting its environment.
- Rumors circulate that former Pakistani Prime Minister Benazir Bhutto, convicted of corruption charges on April 15, plans to center her anti-Pakistani government campaign in Dubai.

May

- UAE and Oman sign an agreement defining a border between Oman and the emirate of Abu Dhabi; borders between other UAE emirates and Oman await delineation.
- Conservationists worry over the Houbara bustard species as it is hunted close to extinction; however, wildlife experts at the National Avian Research Centre in Abu Dhabi are able to breed birds in captivity through the use of incubators, frozen sperm, and artificial breeding houses, and more than fifty chicks have been hatched.
- The revival of camel racing in the UAE and Saudi Arabia has inspired the sale of children to be trained as camel jockeys; Bangladeshi nationals sell seventeen children but are arrested in West Bengal, India.

June

- Jamil al-Hujailan, head of the GCC, attempts to reassure the United Arab Emirates that improving ties with Iran is a means to help settle the dispute between Iran and the UAE over the three islands.
- Saudi Arabia and the UAE announce that they will work together for the common good of the GCC.

July

- Archaeologists find a mass burial site in the Ras al-Khaimah region that dates to the Uum Al Nar period; the 4,000-year-old site contains at least three hundred bodies, coffins, jewelry and some pottery.
- Despite the need for foreign manual labors, mainly Indian, Pakistani and Bangladeshi, to build and maintain the infrastructure, unskilled foreign labor is no longer desirable in the UAE, and new guidelines to the visa committees that control immigration are in place.

August

- The Congressional Research Service reports that in 1998 the UAE ranked second among the developing world's buyers of weaponry with purchases worth $2.5 billion.
- New visa restrictions bar manual laborers from India and Pakistan, but not those from South Asia; only Indian and Pakistani workers with certificates showing professional qualifications will be given visas.

September

- As a result of the freeze on manual labor from India and Pakistan, Nepalese workers flock to the UAE.
- It is rumored that Barzan Takriti, the half-brother of Iraqi President Sadam Hussein, has left Switzerland and fled to the UAE.
- Arab foreign ministers, including the UAE's Minister of Information and Culture, threaten to boycott Walt Disney millennium celebrations if Jerusalem is presented as the capital of Israel; Arabs want Jerusalem presented as an Arab city.
- Saddam Hussein's brother Barzan al-Tikriti says he is not defecting from Iraq; he is in the UAE to attend to his personal investments, and according to Emirate government sources, he has not applied for political asylum.

- The UAE and 10 other nations participate in Operation Bright Star, the world's largest military training exercise, encompassing 73,000 troops.

October

- Mohammed Makhira al-Dhaheru, UAE Minister for Justice and Islamic Affairs, visits India to sign an extradition treaty and an agreement on assistance in criminal matters.
- The UAE is buying eighty F-16 aircraft from the U.S. in a deal worth at least six billion dollars.

December

- The world's tallest hotel, the 321-meter tall, sail-shaped Burj al-Arab hotel, opens in Dubai.

ANALYSIS OF EVENTS: 1999

BUSINESS AND THE ECONOMY

In 1999 the UAE had one of the world's highest per capita incomes: Gross Domestic Product (GDP) per capita was estimated at $17,400. The UAE's wealth was based on its oil and gas output, about 33 percent of GDP. The UAE's oil and gas reserves were estimated to last for over 100 years at the present rate of production.

The UAE supported the Organization of Oil Exporting Countries (OPEC) efforts to increase oil prices by cutting back crude oil production by 4.3 million barrels per day from March 1999 to March 2000. This strategy more than doubled the price per barrel from $10 per barrel at the end of 1998 to $24 per barrel by September 1999.

Privatization was being encouraged by the government. As power production increased, this sector was being restructured and privatized.

Bankers looked to 1999 as a difficult year for the banking sector. The Asian crisis and the low oil prices were expected to affect economic performance. Big banks with well-capitalized bases were secure, but smaller banks were under pressure to merge, or cooperate with each other.

Estimates in 1996 of the labor force documented these occupational categories: service (60%), industry (32%), and agriculture (8%). In 1997 the labor force was estimated to be 1.3 million. The July 1999 population estimates categorized two-thirds of the UAE population, or 1,576,589 people, as non-nationals, representing this migrant labor force.

There was increasing unemployment among the indigenous Arab population. As the economy tightens the UAE restricted the unskilled foreign labor by placing visa restrictions on Indian and Pakistani manual laborers. These restrictions did not apply to South Asian or Nepalese labor, or to workers with certificates of professional qualification. The UAE remained dependent on foreign labor for menial job categories and for infrastructure maintenance and improvements. Other news issues concerning the transnational migrant labor force included the Philippine government's attempts to protect workers from unscrupulous agencies and employers by direct hiring through UAE labor offices, and the plight of abandoned sailors.

GOVERNMENT AND POLITICS

The government of the United Arab Emirates was formed as a federation of seven sheikhdom (emirates), each with its own ruler. There was no suffrage. The head of the federation in 1999 was President Zaud bin Sultan Al Nuhayan. The UAE had been in a territorial dispute with Iran over three islands in the Persian Gulf, the Great and Lesser Tunb and Abu Musa. Iran seized control of the islands in 1992. Iran exerted unilateral control of the islands: access was restricted. Iran engaged in military exercises in the waters around the islands. The islands were being built up militarily and the UAE was threatened that Iran wanted to develop nuclear weapons. The UAE turned to the Gulf Cooperation Council (GCC) and Arab League for support. Although the gulf states supported the UAE's claim to the islands, as they sought to improve their individual relations with Iran, tension mounts. The UAE wanted a peaceful settlement. It offered to let the International Court at the Hague settle the dispute. Iran refused intervention by any outside body.

Iraq provoked tension in the region that includes the UAE by calling on Arabic populations to topple their governments, claiming unpaid debts in the region, and a strange incident in which it was rumored that the half-brother of Sadam Hussein had defected from Iraq to the UAE. All of the principals involved denied these rumors.

CULTURE AND SOCIETY

Estimates of UAE population as of July 1999 were 2,344,402 people. The country is slightly smaller than the state of Maine. The population is healthy and long-lived. Infant mortality rate in 1999 was estimated at 14.1 deaths per 1000 live births. The total fertility rate estimated for 1999 was 3.5 children born per woman. Life expectancy at birth for the total population was 75.24 years.

In July 1998 it was estimated that 75 percent of the population in the 15–64 age group were non-nationals. Figures for ethnic group composition in 1982 were that 19 percent of the population was Emiri, 23 percent other Arab and Iranian, 50 percent South Asian, and 8 percent were from other expatriates including Westerners and East Asians. The literacy rate for the total population was about 79 percent.

The communication revolution's impact on the UAE was both a boon and a threat. Sheikh Muhammed bin Rashid Al Maktoum, the Crown Prince of Dubai, launched Dubai Internet City, the first free trade zone for business done over the internet. In a poorer Emirate, Ras al-Khaimah, where half the homes have satellite dishes, a concern raised by the police chief is the glut of internet pornography.

Drug trafficking and money laundering had become important issues due to the UAE's proximity to the southwest Asian heroin producing countries and the activity of the Dubai free trade zone. Drug trafficking was made a capital offense in 1998.

Ecological issues also had gained the attention of the people in the UAE. The decline of the Houbara bustard, the favorite prey of falcons, inspired research with some success at the National Avian Research Centre on captive breeding. The Environment Research and Wildlife Development Agency initiated a four year program to track the seacow (also known as the dugong) movements with satellite technology. The UAE had the second highest population of seacows in the world. Oil spills and illegal discharges in the Gulf and along the UAE coast have the UAE considering the death penalty for anyone caught deliberately polluting.

International agencies raised issues regarding the status of women and children: the practice of female genital mutilation, or female circumcision, which is still practiced in the UAE, and the trafficking of children associated with the revival of camel racing as a sport.

DIRECTORY

CENTRAL GOVERNMENT

Head of State

President and Ruler of Abu Dhabi
Zayid ibn Sultan Al Nuhayyan, Office of the President

Vice President, Prime Minister, and Ruler of Dubai
Maktoum bin Rashid al-Maktoum, Office of the Prime Minister, P.O. Box 899, Abu Dhabi, United Arab Emirates

Ruler of Ajman
Hamaid Bin Rashid al Nueimi, Office of the Ruler of Ajman

Ruler of Fujairah
Hamed Bin Mohammed al Sharqi, Office of the Ruler of Fujairah

Ruler of Umm al-Qaiwain
Rashid Bin Ahmed al Mualla, Office of the Ruler of Umm al-Qaiwain

Ruler of Ras al-Khaimah
Saqr Bin Mohammed Al Qassimi, Office of the Ruler of Ras al-Khaimah

Ruler of Sharjah
Sultan Bin Mohammed Al Qassimi, Office of the Ruler of Sharjah

Ministers

Deputy Prime Minister
Sultan Bin Zayed al Nayhan, Office of the Deputy Prime Minister

Minister for Supreme Council Affairs
Mohammed Bin Saqr Bin Mohammed Al Qassimi, Ministry for Supreme Council Affairs, P.O. Box 899, Abu Dubai, United Arab Emirates
PHONE: +971 (2) 273921
FAX: +971 (2) 344137

Minister for Financial and Industrial Affairs
Ahmed Humaid Al Tayer, Ministry for Financial and Industrial Affairs, P.O. Box 433, Abu Dubai, United Arab Emirates
PHONE: +971 (2) 726000
FAX: +971 (2) 773301

Minister for Cabinet Affairs
Saeed Al Ghaith, Ministry for Cabinet Affairs, P.O. Box 899, Abu Dubai, United Arab Emirates
PHONE: +971 (2) 651113
FAX: +971 (2) 652184

Minister of Finance and Industry
Hamdan Bin Rashid Al Maktoum, Ministry of Finance and Industry, P.O. Box 433, Abu Dubai, United Arab Emirates
PHONE: +971 (2) 726000
FAX: +971 (2) 773301

Minister of Youth and Sports
Faisal Bin Dhaled Bin Mohammed Al Qasimi, Ministry of Youth and Sports

Minister of Public Works and Housing
Rakad Bin Salem Bin Rakadh, Ministry of Public Works and Housing, P.O. Box 878, Abu Dubai, United Arab Emirates
PHONE: +971 (2) 651778
FAX: +971 (2) 665598

Minister of Planning
Humaid Bin Ahmed Al Mualla, Ministry of Planning, P.O. Box 904, Abu Dubai, United Arab Emirates
PHONE: +971 (2) 271000
FAX: +971 (2) 269942

Minister of Petroleum and Mineral Resources
Ahmed Saeed Al-Badi, Ministry of Petroleum and Mineral Resources, P.O. Box 59, Abu Dubai, United Arab Emirates
PHONE: +971 (2) 651810
FAX: +971 (2) 663414

Minister of Labor and Social Affairs
Saif Al Jarwan, Ministry of Labor and Social Affairs, P.O. Box 809, Abu Dubai, United Arab Emirates
PHONE: +971 (2) 651890
FAX: +971 (2) 665889

Minister of Justice
Abdulla Bin Omran Taryam, Ministry of Justice, P.O. Box 753, Abu Dubai, United Arab Emirates
PHONE: +971 (2) 652224
FAX: +971 (2) 664944

Minister of Islamic Affairs and Endowments
Mohammed Ahmed Hassan Al-Khazraji, Ministry of Islamic Affairs and Endowments

Minister of Interior
Mohamad Saeed Al Badi, Ministry of Interior, P.O. Box 398, Abu Dubai, United Arab Emirates
PHONE: +971 (2) 414666
FAX: +971 (2) 415780

Minister of Information and Culture
Khalfan Bin Mohammed Al Roumi, Ministry of Information and Culture, P.O. Box 17, Abu Dubai, United Arab Emirates
PHONE: +971 (2) 453000
FAX: +971 (2) 451155

Minister of Health
Ahmed Bin Saeed Al Badi, Ministry of Health, P.O. Box 848, Abu Dubai, United Arab Emirates
PHONE: +971 (2) 330000
FAX: +971 (2) 217722

Minister of Foreign Affairs
Rashid Abdallah Al Noaimi, Ministry of Foreign Affairs, P.O. Box 1, Abu Dubai, United Arab Emirates
PHONE: +971 (2) 652200
FAX: +971 (2) 653849

Minister of Electricity and Water
Humaid Bin Nasser Al Owais, Ministry of Electricity and Water, P.O. Box 629, Abu Dubai, United Arab Emirates
PHONE: +971 (2) 274222
FAX: +971 (2) 269738

Minister of Higher Education and Scientific Research
Nahayan Bin Mubarak Al-Nahayan, Ministry of Higher Education and Scientific Research, P.O. Box 45253, Abu Dubai, United Arab Emirates
PHONE: +971 (2) 766600
FAX: +971 (2) 765262

Minister of Education
Hamad Abdul Rahman Al Madfaa, Ministry of Education, P.O. Box 295, Abu Dubai, United Arab Emirates
PHONE: +971 (2) 213800
FAX: +971 (2) 351164

Minister of Economy and Commerce
Saeed Aimed Ghobash, Ministry of Economy and Commerce, P.O. Box 901, Abu Dubai, United Arab Emirates
PHONE: +971 (2) 268488
FAX: +971 (2) 268339

Minister of Defense
Mohamed Bin Rashid Al Maktoum, Ministry of Defense, P.O. Box 2838, Abu Dubai, United Arab Emirates
PHONE: +971 (2) 532330
FAX: +971 (2) 455033

Minister of Communications
Mohamed Saeed Al Mulla, Ministry of Communications, P.O. Box 900, Abu Dubai, United Arab Emirates
PHONE: +971 (2) 651900
FAX: +971 (2) 667404

Minister of Agriculture and Fisheries
Saeed Mohammed Al Ragabani, Ministry of Agriculture and Fisheries, P.O. Box 213, Abu Dubai, United Arab Emirates
PHONE: +971 (2) 662781
FAX: +971 (2) 654787

POLITICAL ORGANIZATIONS

There are no legal political parties in UAE.

DIPLOMATIC REPRESENTATION

Embassies in United Arab Emirates

Algeria
Box 3070, Abu Dhabi, United Arab Emirates
PHONE: +971 (2) 448949
FAX: +971 (2) 447068

Argentina
Box 3325, Abu Dhabi, United Arab Emirates
PHONE: +971 (2) 436838
FAX: +971 (2) 431392

Australia
Box 9303, Abu Dubai, United Arab Emirates
PHONE: +971 (2) 313444
FAX: +971 (2) 314812

Austria
Box 3095, Abu Dhabi, United Arab Emirates
PHONE: +971 (2) 324103
FAX: +971 (2) 311202

Bahrain
Box 3367, Abu Dhabi, United Arab Emirates
PHONE: +971 (2) 312200
FAX: +971 (2) 311202

Bangladesh
Box 4336, Abu Dubai, United Arab Emirates

Belgium
Box 3686, Abu Dhabi, United Arab Emirates
PHONE: +971 (2) 628966
FAX: +971 (2) 319353

Brazil
Box 3027, Abu Dhabi, United Arab Emirates
PHONE: +971 (2) 665352
FAX: +971 (2) 654559

Canada
Box 52472, Abu Dubai, United Arab Emirates
PHONE: +971 (2) 521717
FAX: +971 (2) 456267

China
Box 2741, Abu Dhabi, United Arab Emirates
PHONE: +971 (2) 434276
FAX: +971 (2) 435440

Croatia
Box 41227, Abu Dhabi, United Arab Emirates
PHONE: +971 (2) 311700
FAX: +971 (2) 338366

Czech Republic
Box 27009, Abu Dhabi, United Arab Emirates
PHONE: +971 (2) 312800
FAX: +971 (2) 316567

Denmark
Box 46666, Abu Dhabi, United Arab Emirates
PHONE: +971 (2) 325900
FAX: +971 (2) 351690

Egypt
Box 4026, Abu Dhabi, United Arab Emirates
PHONE: +971 (2) 445566
FAX: +971 (2) 449878

Finland
Office 1004, Hamdan Street, P.O. Box 3634, Abu Dhabi, United Arab Emirates
E-MAIL: finemb@emirates.net.ae
TITLE: Ambassador
NAME: Matti Pullinen

France
Box 4014, Abu Dhabi, United Arab Emirates
PHONE: +971 (2) 435100
FAX: +971 (2) 434158

Germany
Box 2591, Abu Dhabi, United Arab Emirates
PHONE: +971 (2) 232442
FAX: +971 (2) 270887

Greece
Box 5483, Abu Dhabi, United Arab Emirates
PHONE: +971 (2) 654847
FAX: +971 (2) 656008

Hungary
Box 44450, Abu Dhabi, United Arab Emirates
PHONE: +971 (2) 660107
FAX: +971 (2) 667877

India
P.O. Box 4090, Abu Dhabi, United Arab Emirates
PHONE: +971 (2) 664800
FAX: +971 (2) 651518
E-MAIL: indiauae@emirates.net.ae
TITLE: Ambassador
NAME: K.C. Singh

Indonesia
Box 7256, Abu Dhabi, United Arab Emirates
PHONE: +971 (2) 669233
FAX: +971 (2) 653932

Iran
Box 4080, Abu Dhabi, United Arab Emirates
PHONE: +971 (2) 447618
FAX: +971 (2) 448714

Italy
Box 46752, Abu Dhabi, United Arab Emirates
PHONE: +971 (2) 435622
FAX: +971 (2) 434337

Japan
Box 2430, Abu Dhabi, United Arab Emirates
PHONE: +971 (2) 435696
FAX: +971 (2) 434219

Jordan
Box 4024, Abu Dhabi, United Arab Emirates
PHONE: +971 (2) 447100
FAX: +971 (2) 449157

Kuwait
Box 926, Abu Dhabi, United Arab Emirates
PHONE: +971 (2) 446888
FAX: +971 (2) 444990

Kenya
Box 3854, Abu Dhabi, United Arab Emirates
PHONE: +971 (2) 666300
FAX: +971 (2) 524675

Lebanon
Box 4023, Abu Dhabi, United Arab Emirates
PHONE: +971 (2) 434722
FAX: +971 (2) 435553

Malaysia
Box 3887, Abu Dhabi, United Arab Emirates
PHONE: +971 (2) 656698
FAX: +971 (2) 656697

Mauritania
Box 2714, Abu Dhabi, United Arab Emirates
PHONE: +971 (2) 462724
FAX: +971 (2) 465772

Morocco
Box 4066, Abu Dhabi, United Arab Emirates
PHONE: +971 (2) 433963
FAX: +971 (2) 4433917

Netherlands
Box 46560, Abu Dhabi, United Arab Emirates
PHONE: +971 (2) 321920
FAX: +971 (2) 313158

Oman
Box 2517, Abu Dhabi, United Arab Emirates
PHONE: +971 (2) 463333
FAX: +971 (2) 464633

Pakistan
Box 846, Abu Dhabi, United Arab Emirates
PHONE: +971 (2) 447800
FAX: +971 (2) 447172

Panama
Box 2121, Abu Dubai, United Arab Emirates
PHONE: +971 (2) 263366
FAX: +971 (2) 263315

Philippines
Villa No.1, Jajda Street, P.O. Box 3215, Abu Dhabi, United Arab Emirates
PHONE: +971 (2) 345664; 345665; 334998
FAX: +971 (2) 313559
E-MAIL: info@philembassy.com
TITLE: Ambassador
NAME: Amable R. Aguiluz III

Poland
Box 2334, Abu Dubai, United Arab Emirates
PHONE: +971 (2) 465200
FAX: +971 (2) 462967

Qatar
Box 3503, Abu Dubai, United Arab Emirates
PHONE: +971 (2) 435900
FAX: +971 (2) 434800

Romania
Box 70416, Abu Dubai, United Arab Emirates
PHONE: +971 (2) 666346
FAX: +971 (2) 651598

Russia
Box 8211, Abu Dhabi, United Arab Emirates
PHONE: +971 (2) 721797
FAX: +971(2) 788731

Saudi Arabia
Box 4057, Abu Dhabi, United Arab Emirates
PHONE: +971 (2) 445700
FAX: +971 (2) 446747

Slovakia
Box 3382, Abu Dhabi, United Arab Emirates
PHONE: +971 (2) 321674
FAX: +971 (2) 315839

Somalia
Box 4155, Abu Dhabi, United Arab Emirates
PHONE: +971 (2) 669700
FAX: +971 (2) 651580

Spain
Box 46474, Abu Dhabi, United Arab Emirates
PHONE: +971 (2) 213544
FAX: +971 (2) 342978

Sri Lanka
Box 46534, Abu Dhabi, United Arab Emirates
PHONE: +971 (2) 666688
FAX: +971 (2) 667921

Sudan
Box 4027, Abu Dhabi, United Arab Emirates
PHONE: +971 (2) 666788
FAX: +971 (2) 654231

Sweden
Box 2609, Abu Dhabi, United Arab Emirates
PHONE: +971 (2) 337772
FAX: +971 (2) 332904

Switzerland
Box 46116, Abu Dhabi, United Arab Emirates
PHONE: +971 (2) 343636
FAX: +971 (2) 216127

Syria
Box 4011, Abu Dhabi, United Arab Emirates
PHONE: +971 (2) 448768
FAX: +971 (2) 449387

Thailand
Box 51844, Abu Dubai, United Arab Emirates
PHONE: +971 (2) 492863
FAX: +971 (2) 490932

Tunisia
Box 4166, Abu Dhabi, United Arab Emirates
PHONE: +971 (2) 661331
FAX: +971 (2) 660707

Turkey
Box 3204, Abu Dhabi, United Arab Emirates
PHONE: +971 (2) 655421
FAX: +971 (2) 662691

United States
P.O. Box 4009, Abu Dhabi, United Arab Emirates
PHONE: +971 (2) 436691; 436692
FAX: +971 (2) 434771
E-MAIL: usisamem@emirates.net.ae
TITLE: Ambassador
NAME: Theodore H. Kattouf

Yemen
Box 2095, Abu Dhabi, United Arab Emirates
PHONE: +971 (2) 448454
FAX: +971 (2) 447978

JUDICIAL SYSTEM

Federal Courts of First Instance

Union Supreme Court

FURTHER READING

Books

Ali Rashid, Noor. *The UAE Visions of Change.* Dubai: Motivate Publishing, 1997.

Butti, Obald A. *Imperialism, Tribal Structure, and the Development of Ruling Elites: A Socio-economic History of the Trucial States Between 1892 and 1939.* Ph.D Thesis. Georgetown University: 1992.

Codrai, Ronald. *The Seven Shaikhdoms: Life in the Trucial States Before the Federation of the United Arab Emirates.* London: Stacey International, 1990.

Ghareeb, Edmund and Ibrahim al-Abed, eds. *Perspectives on the United Arab Emirates.* London: Trident, 1997.

Facey, William. *The Emirates by the first photographers.* London: Stacey International, 1996.

Forman, Werner. *Phoenix Rising: the United Arab Emirates, Past, Present and Future.* London: Harvill Press, 1996.

Heard-Bey, Frauke. *From Trucial States to United Arab Emirates: A Society in Transition.* London: Longman, 1996.

Johnson, Julia. *U.A.E.* New York: Chelsea House Publishers, 1987.

Kanafani, Aida S. *Aesthetics and Ritual in the United Arab Emirates: The Anthropology of Food and Personal Adornment Among Arabian Women.* Beirut: American University of Beirut, 1983.

Kay, Shirley. *Emirates Archaeological Heritage.* Dubai, United Arab Emirates: Motivate Publishing, 1988.

Koury, Enver M. *The United Arab Emirates: Its Political System and Politics.* Hyattsville, Maryland: Institute of Middle Eastern and North American Affairs, Inc., 1980.

Mann, Major Clarence. *Abu Dhabi: Birth of an Oil Sheikhdom.* Beirut: Khayats, 1964.

O'Brien, Edna. *Arabian Days.* London: Quartet Books Limited, 1977.

Peck, Malcolm C. *The United Arab Emirates: A Venture in Unity.* Boulder, Colorado: Westview Press, 1986.

Taryam, Abdullah Omran. *The Establishment of the United Arab Emirates 1950–85.* London: Croom-Helm, 1987.

Vine, Peter and Paul Casey. *United Arab Emirates: Profile of a Country's Heritage and Modern Development.* London: Immel, 1992.

Zahlan, Rosemarie Said. *The Making of the Modern Gulf States.* London: Unwin Hyman Ltd., 1989.

Internet

Gulf News. Available Online @ http://www.gulf-news.com/ (November 19, 1999).

Khaleej Times. Available Online @ http://www.khaleejtimes.com/ (November 19, 1999).

UNITED ARAB EMIRATES: STATISTICAL DATA

For sources and notes see "Sources of Statistics" in the front of each volume.

GEOGRAPHY

Geography (1)

Area:

Total: 82,880 sq km.

Land: 82,880 sq km.

Water: 0 sq km.

Area—comparative: slightly smaller than Maine.

Land boundaries:

Total: 867 km.

Border countries: Oman 410 km, Saudi Arabia 457 km.

Coastline: 1,318 km.

Climate: desert; cooler in eastern mountains.

Terrain: flat, barren coastal plain merging into rolling sand dunes of vast desert wasteland; mountains in east.

Natural resources: petroleum, natural gas.

Land use:

Arable land: 0%

Permanent crops: 0%

Permanent pastures: 2%

Forests and woodland: 0%

Other: 98% (1993 est.).

HUMAN FACTORS

Demographics (2A)

	1990	1995	1998	2000	2010	2020	2030	2040	2050
Population	1,952.2	2,181.3	2,303.1	2,386.5	2,852.1	3,286.0	3,576.7	3,818.7	4,057.5
Net migration rate (per 1,000 population)	NA	NA	NA	NA	NA	NA	NA	NA	NA
Births	NA	NA	NA	NA	NA	NA	NA	NA	NA
Deaths	NA	NA	NA	NA	NA	NA	NA	NA	NA
Life expectancy - males	70.6	72.6	73.5	74.2	76.6	78.3	79.3	79.9	80.4
Life expectancy - females	74.0	75.6	76.4	77.0	79.4	81.3	82.8	83.9	84.7
Birth rate (per 1,000)	25.9	18.6	18.6	19.2	21.2	16.7	15.3	15.1	13.3
Death rate (per 1,000)	3.0	2.9	3.1	3.2	4.3	6.0	8.2	8.9	7.6
Women of reproductive age (15-49 yrs.)	350.7	420.9	472.1	506.4	610.0	734.1	872.7	883.1	936.4
of which are currently married	NA	NA	NA	NA	NA	NA	NA	NA	NA
Fertility rate	4.7	3.7	3.6	3.4	2.8	2.4	2.2	2.1	2.0

Except as noted, values for vital statistics are in thousands; life expectancy is in years.

Health Personnel (3)

Total health expenditure as a percentage of GDP, 1990-1997[a]

Public sector	2.0
Private sector	0.5
Total[b]	2.5

Health expenditure per capita in U.S. dollars, 1990-1997[a]

Purchasing power parity	390
Total	379

Availability of health care facilities per 100,000 people

Hospital beds 1990-1997[a]	310
Doctors 1993[c]	168
Nurses 1993[c]	321

Health Indicators (4)

Life expectancy at birth	
1980	68
1997	75
Daily per capita supply of calories (1996)	3,366
Total fertility rate births per woman (1997)	3.5
Maternal mortality ratio per 100,000 live births (1990-97)	26[c]
Safe water % of population with access (1995)	98
Sanitation % of population with access (1995)	95
Consumption of iodized salt % of households (1992-98)[a]	....
Smoking prevalence	
Male % of adults (1985-95)[a]	....
Female % of adults (1985-95)[a]	....
Tuberculosis incidence per 100,000 people (1997)	21
Adult HIV prevalence % of population ages 15-49 (1997)	0.18

Infants and Malnutrition (5)

Under-5 mortality rate (1997)	10
% of infants with low birthweight (1990-97)	6
Births attended by skilled health staff % of total[a]	99
% fully immunized (1995-97)	
TB	98
DPT	94
Polio	94
Measles	35
Prevalence of child malnutrition under age 5 (1992-97)[b]	NA

Ethnic Division (6)

Emiri	19%
Other Arab and Iranian	23%
South Asian	50%
Other expatriates	8% (1982)

Note: less than 20% are UAE citizens (1982). Other expatriates includes Westerners and East Asians.

Religions (7)

Muslim	96%
Shi'a	16%
Christian, Hindu, and other	4%

Languages (8)

Arabic (official), Persian, English, Hindi, Urdu.

EDUCATION

Public Education Expenditures (9)

Public expenditure on education (% of GNP)	
1980	1.3
1996	1.8[1]
Expenditure per student	
Primary % of GNP per capita	
1980	....
1996	....
Secondary % of GNP per capita	
1980	....
1996	....
Tertiary % of GNP per capita	
1980	....
1996	....
Expenditure on teaching materials	
Primary % of total for level (1996)	....
Secondary % of total for level (1996)	...
Primary pupil-teacher ratio per teacher (1996)	16
Duration of primary education years (1995)	6

GOVERNMENT & LAW

Political Parties (12)

The legislative branch is a unicameral Federal National Council (40 seats; members appointed by the rulers of the constituent states to serve two-year terms). There are no political parties.

Government Budget (13A)

Year: 1994

Total Expenditures: 15,693 Millions of Dirhams

Expenditures as a percentage of the total by function:

General public services and public order	17.31[P]
Defense	37.13[P]
Education	17.15[P]
Health	7.35[P]
Social Security and Welfare	3.36[P]
Housing and community amenities	2.01[P]
Recreational, cultural, and religious affairs	3.17[P]
Fuel and energy	3.17[P]
Agriculture, forestry, fishing, and hunting	.71[P]
Mining, manufacturing, and construction	.10[P]
Transportation and communication	.29[P]
Other economic affairs and services	1.01[P]

Crime (15)

Crime rate (for 1997)

Crimes reported	34,200
Total persons convicted	NA
Crimes per 100,000 population	1,400

Persons responsible for offenses

Total number of suspects	40,800
Total number of female suspects	5,300
Total number of juvenile suspects	1,850

Military Affairs (14B)

	1990	1991	1992	1993	1994	1995
Military expenditures						
Current dollars (mil.)	2,590[e]	4,900[e]	2,098	2,125	2,125[e]	1,880[e]
1995 constant dollars (mil.)	2,977[e]	5,415[e]	2,256	2,228	2,178[e]	1,880[e]
Armed forces (000)	66	66	55	55	60	60
Gross national product (GNP)						
Current dollars (mil.)	35,720	35,990	36,700	36,950	38,140	38,950
1995 constant dollars (mil.)	41,050	39,770	39,480	38,740	39,090	38,950
Central government expenditures (CGE)						
1995 constant dollars (mil.)	4,521	4,590	4,562	4,447	4,382	4,900[e]
People (mil.)	2.3	2.4	2.5	2.7	2.8	2.9
Military expenditure as % of GNP	7.3	13.6	5.7	5.7	5.6	4.8
Military expenditure as % of CGE	65.8	118.0	49.5	50.1	49.7	38.4
Military expenditure per capita (1995 $)	1,322	2,268	894	838	780	643
Armed forces per 1,000 people (soldiers)	29.3	27.6	21.8	20.7	21.5	20.5
GNP per capita (1995 $)	18,230	16,660	15,650	14,010	14,010	13,320
Arms imports[6]						
Current dollars (mil.)	1,500	350	340	460	400	875
1995 constant dollars (mil.)	1,724	387	366	482	410	875
Arms exports[6]						
Current dollars (mil.)	5	0	0	0	0	10
1995 constant dollars (mil.)	6	0	0	0	0	10
Total imports[7]						
Current dollars (mil.)	11,200	13,750	17,410	19,520	21,700	NA
1995 constant dollars (mil.)	12,870	15,190	18,730	20,460	22,240	NA
Total exports[7]						
Current dollars (mil.)	23,540	24,440	24,760	23,660[e]	25,300[e]	NA
1995 constant dollars (mil.)	27,060	27,000	26,630	24,810[e]	25,930[e]	NA
Arms as percent of total imports[8]	13.4	2.5	2.0	2.4	1.8	NA
Arms as percent of total exports[8]	0	0	0	0	0	NA

LABOR FORCE

Labor Force (16)

Total (million)	1.05
Services	65%
Industry and commerce	30%
Agriculture	5%

Data for 1996 est. Percent distribution for 1996 est.

Unemployment Rate (17)

Rate not available.

PRODUCTION SECTOR

Electric Energy (18)

Capacity	5.29 million kW (1995)
Production	18 billion kWh (1995)
Consumption per capita	6,155 kWh (1995)

Transportation (19)

Highways:

total: 4,835 km

paved: 4,835 km

unpaved: 0 km (1996 est.)

Pipelines: crude oil 830 km; natural gas, including natural gas liquids, 870 km

Merchant marine:

total: 67 ships (1,000 GRT or over) totaling 945,320 GRT/1,592,164 DWT ships by type: bulk 3, cargo 18, chemical tanker 3, container 7, liquefied gas tanker 1, livestock carrier 1, oil tanker 27, refrigerated cargo 1, roll-on/roll-off cargo 6 (1997 est.)

Airports: 40 (1997 est.)

Airports—with paved runways:

total: 22

over 3,047 m: 9

2,438 to 3,047 m: 3

1,524 to 2,437 m: 3

914 to 1,523 m: 3

under 914 m: 4 (1997 est.)

Airports—with unpaved runways:

total: 18

2,438 to 3,047 m: 1

1,524 to 2,437 m: 4

914 to 1,523 m: 8

under 914 m: 5 (1997 est.)

Top Agricultural Products (20)

Dates, vegetables, watermelons; poultry, eggs, dairy products; fish.

FINANCE, ECONOMICS, & TRADE

Economic Indicators (22)

National product: GDP—purchasing power parity—$54.2 billion (1997 est.)

National product real growth rate: 5% (1997 est.)

National product per capita: $24,000 (1997 est.)

Inflation rate—consumer price index: 3.6% (1997 est.)

Exchange Rates (24)

Exchange rates: Emirian dirhams (Dh) per US$1—3.6710 (fixed rate).

Top Import Origins (25)

$23.5 billion (f.o.b., 1996 est.) Data are for 1995 est.

Origins	%
European Union	33
Italy	9
Germany	8
France	4
South Korea	5
United States	4
Japan	4

Top Export Destinations (26)

$33.2 billion (f.o.b., 1996 est.) Data are for 1996.

Destinations	%
Japan	38
South Korea	7
Singapore	7
India	6
Oman	4
Iran	3

Economic Aid (27)

$NA. NA stands for not available.

Import Export Commodities (28)

Import Commodities	Export Commodities
Manufactured goods	Crude oil 66%
Machinery and transport equipment	Natural gas
Chemicals	Reexports
Food	Dried fish
	Dates

UNITED KINGDOM

CAPITAL: London.

FLAG: The Union Jack, adopted in 1800, is a combination of the banners of England (St. George's flag: a red cross with extended horizontals on a white field), Scotland (St. Andrew's flag: a white saltire cross on a blue field), and Ireland (St. Patrick's flag: a red saltire cross on a white field).

ANTHEM: *God Save the Queen.*

MONETARY UNIT: The pound sterling (£) is a paper currency of 100 pence. Before decimal coinage was introduced on 15 February 1971, the pound had been divided into 20 shillings, each shilling representing 12 pennies (p) or pence; some old-style coins are still in circulation. Under the new system, there are coins of 1, 2, 5, 10, 20, and 50 pence and 1 and 2 pounds, and notes of 5, 10, 20, and 50 pounds. £1 = $1.4822 (or $1 = £0.6747).

WEIGHTS AND MEASURES: Although the traditional imperial system of weights and measures is still in use, a changeover to the metric system is in progress.

HOLIDAYS: New Year's Day, 1 January; Good Friday; Easter Monday (except Scotland); Late Summer Holiday, last Monday in August or 1st in September (except Scotland); Christmas, 25 December; and Boxing Day, 1st weekday after Christmas. Also observed in Scotland are bank holidays on 2 January and on the 1st Monday in August. Northern Ireland observes St. Patrick's Day, 17 March; and Orangeman's Day, 12 July, commemorating the Battle of the Boyne in 1690.

TIME: GMT.

LOCATION AND SIZE: The United Kingdom (U.K.) is situated off the northwest coast of Europe between the Atlantic Ocean and the North Sea. Its total area of 244,820 square kilometers (94,526 square miles) consists of the island of Great Britain—formed by England, Wales, and Scotland—and Northern Ireland, on the island of Ireland.

CLIMATE: The United Kingdom generally has a temperate climate. Mean monthly temperatures range (north to south) from 3°C to 5°C (37°–41°F) in winter and from 12°C to 16°C (54°–61°F) in summer.

United Kingdom of Great Britain and Northern Ireland

INTRODUCTORY SURVEY

RECENT HISTORY

The years following World War II saw the dissolution of the great British Empire that had circled the globe in the late 19th and early 20th centuries. A founding member of the North Atlantic Treaty Organization (NATO) and the European Free Trade Association (EFTA), the United Kingdom joined the European Community (EC) on 1 January 1973. The nation's principal domestic problems in the 1970s were rapid inflation, labor disputes, and the continuing conflict in Northern Ireland.

A Conservative government headed by Prime Minister Margaret Thatcher came to power in 1979 with a program of income tax cuts and reduced government spending. Thatcher, who won reelection in 1983 and 1987, privatized many businesses formerly run by the government.

John Major replaced Thatcher at 10 Downing Street in 1990. Major's government sought to redefine Conservative values with a renewed emphasis on law and order. Labour Party leader Tony Blair was elected prime minister in 1997, ending 18 years of Conservative Party rule. Blair pledged to create regional assemblies for Scotland and Wales and a municipal government for London. He also promised not to exceed Conservative Party spending limits, and pledged no new taxes for five years. Under a settlement approved by voters in both Northern Ireland and the Irish Republic in 1998, Northern Ireland remained a part of the United Kingdom, with an elected assembly, administrative

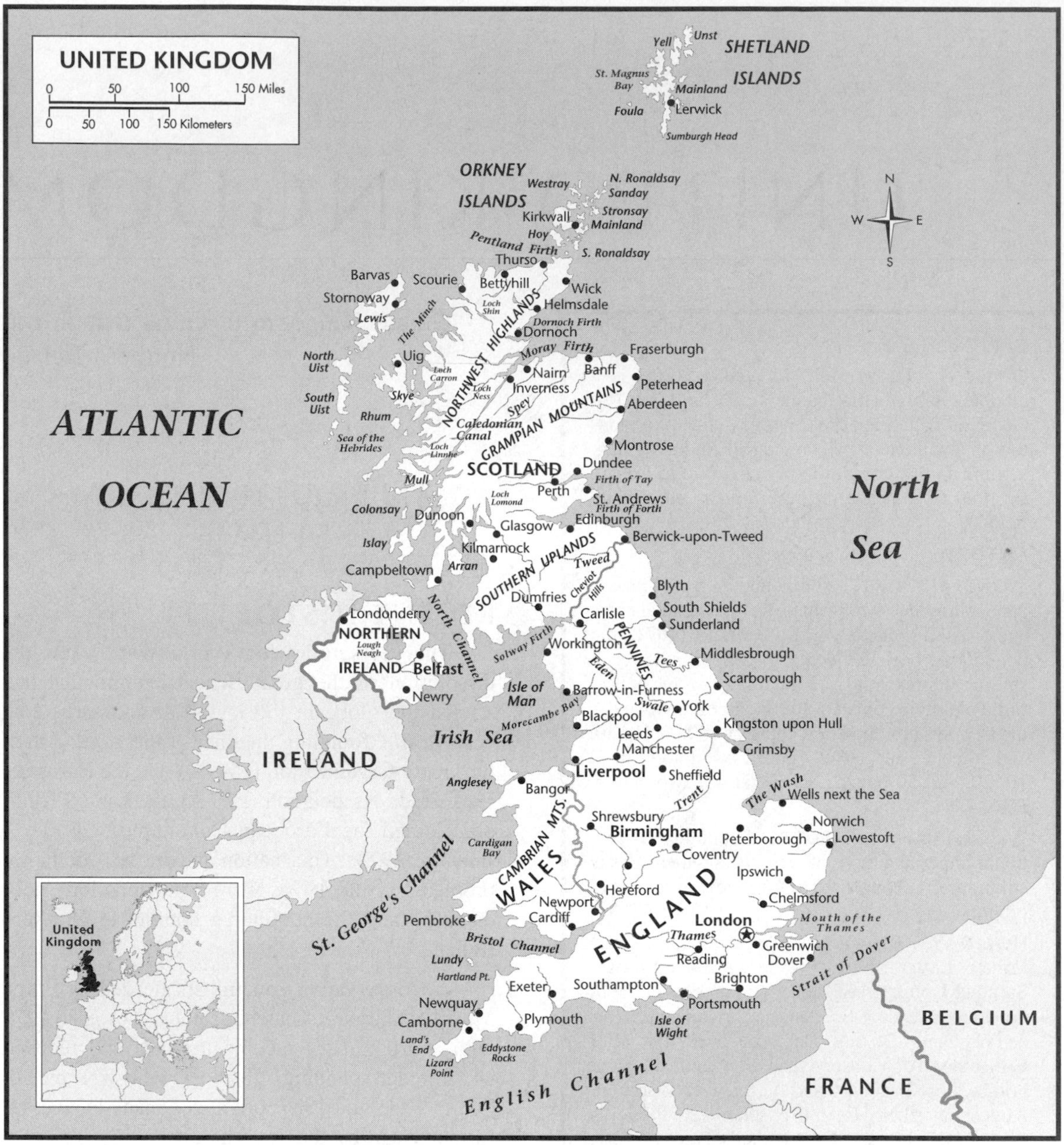

links between the Irish Republic and Northern Ireland, and a council with representatives from the United Kingdom. At the turn of the century, the agreement had not yet been completely implemented.

GOVERNMENT

The United Kingdom is a monarchy in form but a parliamentary democracy in substance. Queen Elizabeth II has reigned since 1952, but rules only by approval of Parliament and acts only on the advice of her ministers.

The United Kingdom is governed, in the name of the sovereign, by Her Majesty's Government—a body of ministers responsible to Parliament. Parliament consists of the sovereign, the House of Lords, and the House of Commons. Over the centuries, the powers of the House of Lords have gradually been reduced; the Blair government drastically reduced the number of hereditary peers in 1999. In 1997 the House of Commons, which is popularly elected, had 659 members, of whom 80% were members from English constituencies; 11% from Scotland; 6% from Wales,; and 3% from Northern Ireland.

All British subjects 18 years and older may vote in national elections.

The chief executive is the prime minister, who, though formally appointed by the sovereign, is traditionally the leader of the majority party in Parliament. He or she is assisted by ministers chosen from the majority party; the most senior ministers compose the cabinet, which decides policy on major issues.

Judiciary

Minor civil cases in England and Wales are heard in some 300 county courts. The High Court is divided into the Chancery court for domestic matters and the Queen's Bench for major civil cases. Appeals from the county courts may be heard in the High Court, the Court of Appeal, and, occasionally, in the House of Lords.

In Scotland, criminal and civil cases are heard at the sheriff court, civil cases may also be heard and in the Outer House of the Court of Session, the supreme civil court in Scotland. Appeals are heard by the Inner House of the Court of Session.

Nearly all (98%) criminal cases are tried in magistrate's courts, consisting of three unpaid magistrates known as justices of the peace. There are about 90 centers of the Crown Court, presided over by a bench of justices or, in the most serious cases, by a High Court judge sitting alone.

All criminal trials in the United Kingdom are held in open court.

Political Parties

Major political parties represented in Parliament include the Conservative Party, the Labour Party, and the Liberal Party. In foreign affairs, there has been little difference between the Labour and Conservative Parties since World War II. They differ mainly on the degree of state control to be applied to industry and commerce.

DEFENSE

Total active army strength in 1995 was 113,000. Naval forces totaled 48,000 and 86,000 men and women served in the Royal Air Force. The defense budget for 1997 was an estimated $33.2 billion, or 3% of gross domestic product (GDP). British troops have been involved in several peace-keeping missions.

ECONOMIC AFFAIRS

The economy of the United Kingdom, one of the world's great trading powers and financial nations, is among the four largest in Western Europe. With few natural resources apart from coal and low-grade iron ore, natural gas and North Sea oil, Britain has long relied on trade. Farming provides 60% of the nation's food. Other food and most raw materials for British industries are imported and paid for largely through exports of manufactured goods.

By the early 1990s, primary energy production accounted for 10% of GDP, one of the highest shares of any industrial nation, and the United Kingdom had far lower dependence on oil imports than in the past. The inflation rate in 1998 was a low 2.6%.

Public Finance

Much of the U.K.'s fiscal deficits have been cyclical, a result of decreased tax receipts and increased benefits during economic downturns, but a significant part of the deficit appears to be structural, a long-term inability of the tax base to support government spending.

The U.S. Central Intelligence Agency estimates that, in 1997, government revenues totaled approximately $487.7 billion and expenditures $492.6 billion.

Income

In 1999 the U.K. gross national product (GNP) was $1.069 trillion, or about $18,410 per person. The purchasing power parity of the gross domestic product (GDP) was estimated in 1998 at $1.252 trillion; per capita GDP purchasing power parity was $21,200.

Industry

The United Kingdom is one of the most highly industrialized countries in the world. Industries such as aerospace, electronics, offshore oil and gas products, and synthetic fibers have overtaken cotton textiles, steel, and shipbuilding in importance to the overall economy. But metals, engineering, and allied industries employ nearly half of all workers in the manufacturing sector.

The U.K. is one of the world's largest exporters of commercial motor vehicles, with an output of 1.3 million automobiles in 1992. And Britain continues to produce high-quality woolen textiles as well as porcelain china, jewelry, and silverware.

Banking and Finance

The United Kingdom is known throughout the world for as a center for banking activities. The City of London has the greatest concentration of banks and the largest insurance market in the world. The Bank of England, established in 1694 as a corporate body and nationalized in 1946, holds the main government accounts, acts as government agent for the issue and registration of government loans and other financial operations, and is the central note-issuing authority. It administers exchange control for the Treasury and is responsible for the application of the government's monetary policy to other banks and financial institutions.

The four major clearing commercial banking groups are Barclays, Lloyds, Midland, and National Westminster. These banks carry out most of the commercial banking in England and Wales. Major Scottish banks include: the Bank of Scotland, the Clydesdale Bank, and the Royal Bank of Scotland.

At the end of 1993, private sector deposits amounted to £233.6 billion. Merchant banks are of great importance in the financing of trade, both domestic and overseas. In addition, 275 overseas banks are directly represented in London.

The Financial Services Act of 1986 established a system of self-regulating organizations (SROs) to oversee operations in various financial markets under the overall control of an umbrella body, the Securities and Investment Board (SIB). Any firm conducting investment business must have authorization to do so from the appropriate SRO.

In 1773 the first stock exchange was opened in London. Some 2,600 companies were listed on the exchange in 1994, with a total market capitalization of £4.8 trillion. The Stock Exchange opened to international competition in October 1986, permitting wider ownership of member firms. Minimum rates of commission on stock sales were abolished. In April 1982, the London Gold Futures Market began operations; it is the only market in Europe making possible worldwide, round-the-clock futures dealings in the metal.

Economic Development

After 1974 the Labour government began to de-emphasize social services and government participation in the economy and to stress increased incentives for private investment. (A notable exception was exploitation of North Sea oil resources.) General investment incentives included tax allowances on new buildings, plants, and machinery. The Conservative government elected in 1979 sought to reduce the role of government in the economy by improving incentives, removing controls, reducing taxes, moderating the money supply, and privatizing several large state-owned companies. Over the past two decades both Labour and Conservative governments have greatly reduced public ownership and contained the growth of social welfare programs.

The U.K. has long been a major source of both bilateral aid (direct loans and grants) and multilateral aid (contributions to international agencies) to developing countries Aid totaled £2,172 million in 1993/94 (compared with £1,463 million in 1988/89). About 70% of the U.K.'s direct, official bilateral development assistance goes to Commonwealth countries.

SOCIAL WELFARE

A system of social security provides national insurance, industrial injuries insurance, and family allowances. Benefits for sickness, unemployment, maternity, and widowhood, as well as guardian's allowances, retirement pensions, and death grants are also provided. Financial assistance for the poor is available through supplementary benefits. For needy families in which the head of the household has full-time employment, a family income supplement is paid.

Women's career progress in most sectors of the economy continues, although employed women earned about 21% less in 1995 than their male counterparts.

Racial discrimination is prohibited by law, but persons of Asian and African origin are sometimes subject to discrimination.

Healthcare

Life expectancy was about 77 years in 1999. Deaths from infectious diseases have been greatly reduced in recent years, although the proportion of deaths from circulatory diseases such as heart attacks and strokes, and cancer has risen.

A comprehensive National Health Service (NHS), established in 1948, provides full medical care to all U.K. residents including general medical, dental, pharmaceutical, and optical services; hospital and specialist services (in patients' homes when necessary) for physical and mental illnesses; and local health authority services (maternity and child welfare, vaccination, prevention of illness, health visiting, home nursing, and other services). The NHS costs Britain's taxpayers more than $73 billion per year. An aging population, more expensive treatments, and a budget crisis have forced some NHS centers to cancel non-emergency treatment. The number of beds available is below the level of demand, causing long waits for treatment.

Housing

In 1991 there were 23 million homes in the United Kingdom. About 50% of families now live in a dwelling built after 1945, typically a two-story house with a small garden.

EDUCATION

Although responsibility for education in the United Kingdom rests with the central government, schools are mainly administered by local education authorities. Nearly the entire adult population is literate. Education is compulsory for all children between the ages of 5 and 16. In 1993, about 5.1 million children attended Britain's 24,000 primary schools.

Britain's 47 universities enrolled 511.123 students in 1993. More than 90% of students in higher education hold awards from public or private funds. In 1997, the government began to change the policy of cost-free tuition by announcing that students would have to start paying for part of the expense.

1999 KEY EVENTS TIMELINE

January

- Prince Edward and Ms. Sophie Rhys-Jones announce plans to marry in June.
- The National Health Service experiences a crisis in hospitals lacking room, funds, and nurses.
- Beachy Head, a large part of the U.K. coastline, falls into the ocean.
- Landmine fund established by the late Princess Diana gives £1 million to landmine charities.
- Ten new provincial government structures are created in Northern Ireland.

February

- British tanks are sent into Kosovo.
- Range Rover closes its Longbridge factory.
- A report on the institutionalized racism of the police force and the murder of Stephen Lawrence lead to the implementation of an ethnic quota system in hiring the police force.
- The debate over genetically modified foods begins trade wars across Europe.

March

- The English Regional Development Agencies (RDAs) creates English regions through which to accomplish its work.
- The Law Lords decide that former Chilean President Augusto Pinochet will stay in the U.K. and then go on trial for crimes against humanity.
- Rosemary Nelson, a lawyer for Irish Catholics, is killed by a car bomb in Northern Ireland set by Protestant loyalist paramilitaries protesting the peace process.
- Four Britons, an American couple, and two New Zealanders, are killed in Uganda by Hutu rebels.

April

- Extradition proceedings continue with former Chilean leader Augusto Pinochet.
- A bomb explodes in a gay bar in London, killing three and wounding seventy. Two other attacks occur, targeting blacks and Asians.
- Truck drivers cause traffic jams in a protest asking the government not to raise taxes that would cause job cuts.
- The police are asked to recruit 8,000 new members from minorities.

May

- David Copeland is charged with three bombings of minority areas in April.
- The Labour Party wins election in the first ever Scottish and Welsh National Assemblies.
- Donald Dewar is elected as the first Minister of Scotland's new regional government.

June

- A Millennium Dome is built in London for New Year's festivities.
- British troops lead the Kosovo liberation.

July

- Ulster Unionists say they will not share power with Sinn Fein as long as the IRA refuses to disarm.

- British cattle ranchers pay for beef advertisements after the mad cow disease scares consumers.
- Clifford Chance, the U.K.'s biggest law firm, creates a global giant with New York's Rogers and Wells and Germany's Pünder, Wolhard, Weber and Aster.
- Michael Asheroft, Tory party treasurer, is accused of mismanagement in Belize banks.

August

- The Scottish National Party (SNP) discusses an independent Scottish arsenal and military.
- The Labour Party in Scotland considers giving workers the right to take their ownership of land from the lairds.

September

- French magistrates clear charges on nine photographers and a press motorcyclist who were implicated in the fatal car crash of Princess Diana.
- The European Court of Human Rights declares Britain's ban on homosexuals in the military to be a violation of the human right to privacy.

October

- British trains collide in London, killing thirty-seven and wounding hundreds.
- Prime Minister Blair replaces Britain's Secretary for Northern Ireland, in a move that pleases Protestants but chagrins Catholics.

November

- Britain's ministers agree to hold more talks aimed at getting France to lift its ban on British beef. Talks will attempt to reassure Paris about the safety of the meat exports.
- The UK government serves notice that it will quit the United Nations Food and Agriculture Organization in two years unless there is immediate reform of the organization's strategy and management.
- Former U.S. senator Ambassador George Mitchell brokers a deal between Ulster Unionists and Sinn Fein, the political wing of the Irish Republican Army (IRA). Sinn Fein makes a statement committing itself to exclusively peaceful means of protest, and the IRA makes its own announcement backing Sinn Fein's statement and promising to align itself with the official body charged with supervising decommissioning.
- Sir Vivian Fuchs, scientist and explorer, dies at age 91.

December

- In preparation for an upcoming European Union summit, Britain's chancellor Gordon Brown, makes clear to the other European finance ministers that Britain will not compromise on the issue of imposing an EU-wide withholding tax on interest from private savings; Britain, almost alone, opposes the tax.
- The Northern Ireland Assembly, a power-sharing parliament, is created in the Northern Ireland province, with ministers from the Democratic Unionist party, led by the Reverend Ian Paisley, and with Martin McGuinness, a former IRA commander, as minister of education.

ANALYSIS OF EVENTS: 1999

BUSINESS AND THE ECONOMY

The spectacle of conservative policies in Great Britain being administered by a Labor government has produced the anomaly of reduced public ownership, yet continued growth of several welfare programs this year, leading to an ever increasing social responsibility for the country's economic health. Like the impact of Ronald Reagan on the post-war liberal consensus in the U.S., the drive during the 1980s of Conservative Prime Minister Margaret Thatcher to privatize or shut down government-owned sectors of the economy has weakened the social democratic ideology of the British Labour Party. Whereas under Thatcher the targets were the shipbuilding industry or the coal mines, the health care system is presently under intense scrutiny. The health of British citizens has come under stress during a National Health Service (NHS) crisis early in the year, in which adequate care, funds, and nurses were not available for patients. The conservative critics charge that although health care may be free for all in this country the quality of care is much below U.S. standards because the lack of competition discourages innovation. Citizens often purchase extra health insurance to make up for the inadequacies of this social welfare program.

Conversion of public sector enterprises to privately owned companies has also been extended to the U.K.'s widely used railway system. In this instance privatization has created problems. Trains and tracks are now owned by a variety of companies, public and private, making communication difficult. The trains that used to run like clockwork are often late and a London train collision in October claimed 37 lives, with many others injured. But the privatization of the British railways may attract investment for the country. With economic growth slowing, and the indicators pointing to a short recession this year, such funds are needed to keep social welfare programs in order. At least one major company, Range Rover, has closed its operation in Longbridge because of high corporate taxes, but many other high-technology companies have found cheap, highly skilled workers in Ireland and Wales. The service sector remains the most profitable area of business in the U.K., while agriculture remains self-sufficient. Within the agricultural sector genetically modified foods have caused quite a controversy, and millions of Pounds have been invested to test the safety of these bio-engineered products. The mad cow disease scare has also put a dent in Britain's agricultural produce earnings, with Europeans boycotting the products and cow herders advertising the safety of their beef.

GOVERNMENT AND POLITICS

As much as the British government would like to remain separate from the European continent, this island depends very much on its relationship with other European Union countries for trade and protection. The issue of joining the single currency euro has been a sticking point for the United Kingdom, where the pound has a higher value than any other currency in Europe. To give up this supremacy and join the faltering Euro strikes many as a step in the wrong direction, although Chancellor of the Exchequer Brown and Prime Minister Tony Blair have supported the move because of the potential value of the euro. Another aspect of the controversy is the £2 million tax rebate that the E.U. returns each year to the U.K. merely for participating in the European Union. Great Britain's military aid during the war in Kosovo this year continued its history of military strength, but it is now no longer the dominant force. It was the presence of peacekeeping NATO troops from several countries that suppressed the Kosovo conflict in June 1999, not just the British.

Thus, although its military presence has waned considerably the former British Empire still influences the world culturally. One example is the importance of British concepts of law. The weight of British Law has grown by accretion, often preserving the intricacies of its ancient jurisprudence. One high-profile legal controversy in 1999 involved the arrest in London of former Chilean military dictator Augusto Pinochet. The former dictator was undergoing medical treatment in England when Spanish judge Baltasar Garzón requested that he be arrested and extradited to Spain. Garzón intended to put Pinochet on trial for crimes against humanity committed against Spanish citizens after the 1973 right-wing coup which toppled the legally elected government of the Chilean socialist, Salvador Allende. The British House of Lords argued for months over whether or not they could detain a former head of state for crimes committed while in office. They also argued whether or not contemporary extradition laws applied to crimes perpetrated in 1973.

The amount of bureaucratic red tape increased dramatically over the year in Britain with the introduction of the English Regional Development Agencies (RDA's) which are meant to create a more equitably based federal system of government. The devolution of power to local government complements the new Scottish Parliament and the Welsh National Assembly elected in May to disperse political power to the regions. These elections returned the ostensibly more liberal Labor Party candidates in both new Assemblies. In Scotland, the nationalists discussed organizing an independent Scottish army and weapons, which falls outside the range of the Scottish Parliament's power vis a vis the British Parliament. A separatist movement is growing in Scotland. Some observers fear that these nationalist tendencies may copy the terrorism and guerrilla tactics of Ireland. In March, a car bombing in Ulster killed a lawyer who supported Irish Catholics. Protestant loyalist paramilitaries claimed responsibility for the attack, which sparked more violence. Peace negotiations this year ended with Ulster Unionists rejecting cooperation unless the Irish Republican Army (IRA) disarm, which was unlikely. Such regional disparities challenge the British government, but perhaps even more complicated is the prevalence of hate crimes and the ordinary everyday discrimination.

CULTURE AND SOCIETY

An eighth month government investigation headed by retired High Court Justice Sir William Macphereson found that British police forces were infested with institutional racism. This investigation exacerbated the popular outrage over the racially induced murder of Mr. Stephen Lawrence by police this year. Police were directed to implement an ethnic quota hiring system and to recruit 8,000 minorities to the police forces. This act also occurred while three bombing attacks in London during April targeted blacks, Asians, and homosexuals. At a gay bar, three people were killed and more than seventy wounded by a bombing. Mr. David Copeland, not affiliated with any hate groups in particular, was charged with all three incidents. In another issue dealing with minorities, the European Court of Human Rights declared that the British military was wrong to discriminate against homosexuals while recruiting because it violated the human right to privacy. Such incidents of abuse of cultural or ethnic minorities are likely to continue considering the wide gap between the poor subclasses and the rich elite. The royal family observed the engagement and marriage of Prince Edward and Ms. Sophie Rhys-Jones this year, but this ceremony did not neutralize the fallout from the death last year of Princess Diana. The French photographers suspected in the car crash which ended her life were cleared from charges in September. The constructive aspects of Princess Diana's legacy continued in Britain and around the world as her landmine fund gave £1 million to charities dedicated to eradicating such devices.

DIRECTORY

CENTRAL GOVERNMENT

Head of State

Queen
Elizabeth II, Monarch, Buckingham Palace, the Visitor Center, London SW1A 1AA, England

Prime Minister, First Lord of the Treasury, Minister for the Civil Service
Tony Blair, Office of the Prime Minister, 10 Downing Street, London SW1A 2AA, England
PHONE: +44 (171) 2703000

Ministers

Deputy Prime Minister and Secretary of the Environment, Transport and the Regions
John Prescott, Department of Environment, Transport and the Regions, Eland House, Bressenden Place, London SW1E 5DU, England
PHONE: +44 (171) 8903000

Chancellor of the Exchequer
Gordon Brown, Office of the Chancellor of the Exchequer

Secretary of Foreign and Commonwealth Affairs
Robin Cook, Department of Foreign and Commonwealth Affairs Office, King Charles Street, Whitehall, London SW1A 2AH, England
PHONE: +44 (171) 2703000

Lord Chancellor
Lord Irvine of Lairg, Lord Chancellor's Department, Selborne House, 54–60 Victoria Street, London SW1E 6QW, England
PHONE: +44 (171) 2108500
E-MAIL: general.queries@lcdhq.gsi.gov.uk

Secretary of the Home Department
Jack Straw, Home Office, 50 Queen Anne's Gate, London SW1H 9AT, England
PHONE: +44 (171) 2734000
FAX: +44 (171) 2732190
E-MAIL: gen.ho@gtnet.gov.uk

Secretary of Education and Employment
David Blunkett, Department of Education and Employment, Sanctuary Buildings, Great Smith Street, London SW1P 3BT, England
PHONE: +44 (171) 9255555
E-MAIL: info@dfee.gov.uk

President of the Council and Leader of the House of Commons
Margaret Beckett, Privy Council Office, 68 Whitehall, London SW1A 2AT, England
PHONE: +44 (171) 2703000

Minister for the Cabinet Office
Marjorie Mo Mowlam, Cabinet Office, 70 Whitehall, London SW1A 2AS, England
PHONE: +44 (171) 2701234

Secretary of Culture, Media and Sport
Chris Smith, Department of Culture, Media and Sport, 2–4 Cockspur Street, London SW1Y 5DH, England
PHONE: +44 (171) 2116200
FAX: +44 (171) 2116210
E-MAIL: enquiries@culture.gov.uk

Secretary of International Development
Clare Short, Department of International Development, 94 Victoria Street, London SW1E 5JL, England
PHONE: +44 (171) 9177000
FAX: +44 (171) 9170019
E-MAIL: enquiry@dfid.gov.uk

Secretary of Social Security
Alistair Darling, Department of Social Security, Richmond House, 79 Whitehall, London SW1A 2NS, England
PHONE: +44 (171) 7122171
FAX: +44 (171) 7122386
E-MAIL: peo@MS41.dss.gsi.gov.uk

Secretary of Health
Alan Milburn, Department of Health, Richmond House, 79 Whitehall, London SW1A 2NS, England
PHONE: +44 (171) 2103000
E-MAIL: dhmail@doh.gsi.gov.uk

Minister of Agriculture, Fisheries and Food
Nick Brown, Ministry of Agriculture, Fisheries and Food, Whitehall Place West, London SW1A 2HH, England
PHONE: +44 (171) 2703000

Leader of the Lords and Minister for Women
Baroness Jay of Paddington, Ministry for Women

Secretary of Trade and Industry
Stephen Byers, Department of Trade and Industry, 1 Victoria Street, London SW1H 0ET, England
PHONE: +44 (171) 2155000

Secretary of Defense
Geoffery Hoon, Department of Defense, Main Building, Whitehall, London SW1A 2HB, England
PHONE: +44 (171) 2189000
E-MAIL: public@ministers.mod.uk

Chief Secretary of the Treasury
Andrew Smith, Department of the Treasury

Minister of the Environment, Transport and the Regions
Lord MacDonald of Tradeston, Ministry of the Environment, Transport and the Regions, Eland House, Bressenden Place, London SW1E 5DU, England
PHONE: +44 (171) 8903000

POLITICAL ORGANIZATIONS

Labour Party

Millbank Tower, London SW1P 4GT, England
E-MAIL: join@labour.org.uk
NAME: Tony Blair

Conservative and Unionist Party

Liberal Democrats

Federal Party Headquarters, 4 Cowley Street, London SW1P 3NB, England
PHONE: +44 (171) 2227999
FAX: +44 (171) 7992170
E-MAIL: englishlibdems@cix.co.uk
TITLE: Leader of the Liberal Democrats
NAME: Charles Kennedy

Scottish National Party

SNP Headquarters, 6 North Charlotte Street, Edinburgh, Scotland
PHONE: +44 (131) 2263661
FAX: +44 (131) 2267373
NAME: Alex Salmond

Party of Wales

Plaid Cymru National Office, 18 Park Grove, Caerdydd CF1 3BN, Wales
PHONE: +44 (1222) 646000
E-MAIL: post@plaidcymru.org
TITLE: President
NAME: Dafydd Wigley

Social Democrat and Liberal Party

121 Ormeau Road, Belfast BT7 1SH, Northern Ireland
PHONE: +44 (1232) 247700
FAX: +44 (1232) 236699
E-MAIL: sdlp@indigo.ie
TITLE: Leader
NAME: John Hume

Democratic Unionist Party

Sinn Féin (We, Ourselves)

51/55 Falls Road, Belfast, Northern Ireland
PHONE: +44 (1232) 624421
FAX: +44 (1232) 622112
TITLE: President
NAME: Gerry Adams

U.K. Unionist

E-MAIL: contactus@ukunionists.freeserve.co.uk

DIPLOMATIC REPRESENTATION

Embassies in the United Kingdom

Algeria
54 Holland Park, London W11 3RS, England
PHONE: +44 (171) 2217800
FAX: +44 (171) 2210448

TITLE: Ambassador Extraordinary and Plenipotentiary
NAME: Ahmed Benyamina

Argentina
65 Brook Street, London W1Y 1YE, England
PHONE: +44 (171) 3181300
FAX: +44 (171) 3181301

Belgium
103 Eaton Square, London SW1W 9AB, England
PHONE: +44 (171) 4703700
E-MAIL: info@belgium-embassy.co.uk

Canada
38 Grosvenor Street, London W1X 0AA, England

Denmark
55 Sloane Street, London SW1X 9SR, England
PHONE: +44 (171) 3330200
FAX: +44 (171) 3330270
E-MAIL: dkembassyuk@compuserve.com
TITLE: Ambassador
NAME: Ole Lønsmann Poulsen

Estonia
16 Hyde Park Gate, London SW7 5DG, England
PHONE: +44 (171) 5893428
FAX: +44 (171) 5893430
E-MAIL: loa@estonia.gov.uk

Finland
38 Chesham Place, London SW1X 8HW, England
PHONE: +44 (171) 8386200
FAX: +44 (171) 2353680
TITLE: Ambassador
NAME: Pertti Salolainen

France
58 Knightsbridge, London SW1X 7JT, England
PHONE: +44 (171) 2011000
FAX: +44 (171) 2011004
E-MAIL: email press@ambafrance.org.uk
NAME: Daniel Bernard

Germany
23 Belgrave Square, London SW1X 8PZ, England
PHONE: +44 (171) 8241300
FAX: +44 (171) 8241435
E-MAIL: mail@german-embassy.org.uk
TITLE: Ambassador
NAME: Hans-Friedrich Von Ploetz

India
India House, Aldwych, London WC2B 4NA, England
TITLE: High Commissioner
NAME: Lalit Mansingh

Israel
2 Palace Green, London W8 4QB, England
PHONE: +44 (171) 9579500
FAX: +44 (171) 9579555
E-MAIL: info@israel-embassy.org.uk

Italy
Three Kings' Yard 14, London W1Y 2EH, England
PHONE: +44 (171) 3122200
FAX: +44 (171) 4992283
E-MAIL: emblondon@embitaly.org.uk
TITLE: Ambassador
NAME: Luigi Amaduzzi

Jamaica
1–2 Prince Consort Rd., London SW7 2BZ, England
PHONE: +44 (171) 8239911
FAX: +44 (171) 5895154
E-MAIL: jamhigh@jhcuk.com
TITLE: High Commissioner
NAME: Derick Heaven

Japan
101–104 Piccadilly, London W1V 9FN, England
PHONE: +44 (171) 4656500
FAX: +44 (171) 4919347
E-MAIL: info@embjapan.org.uk
TITLE: Ambassador
NAME: Sadayuki Hayashi

Jordan
6 Upper Phillimore Gardens, London W8 7HB, England
PHONE: +44 (171) 9373685
FAX: +44 (171) 9378795
NAME: Hussein al-Rifai

Kyrgyzstan
119 Crawford St., London W1H 1AF, England
PHONE: +44 (171) 9351462
FAX: +44 (171) 9357449
E-MAIL: embassy@kyrgyz-embassy.org.uk

Mexico
42 Hertford St., London W1Y 7TF, England
PHONE: +44 (171) 4998586
FAX: +44 (171) 4954035

COLOR FLAGS, SEALS AND REGIONAL MAPS

Color seals for Guadeloupe, Martinique and Serbia are not available at this time.

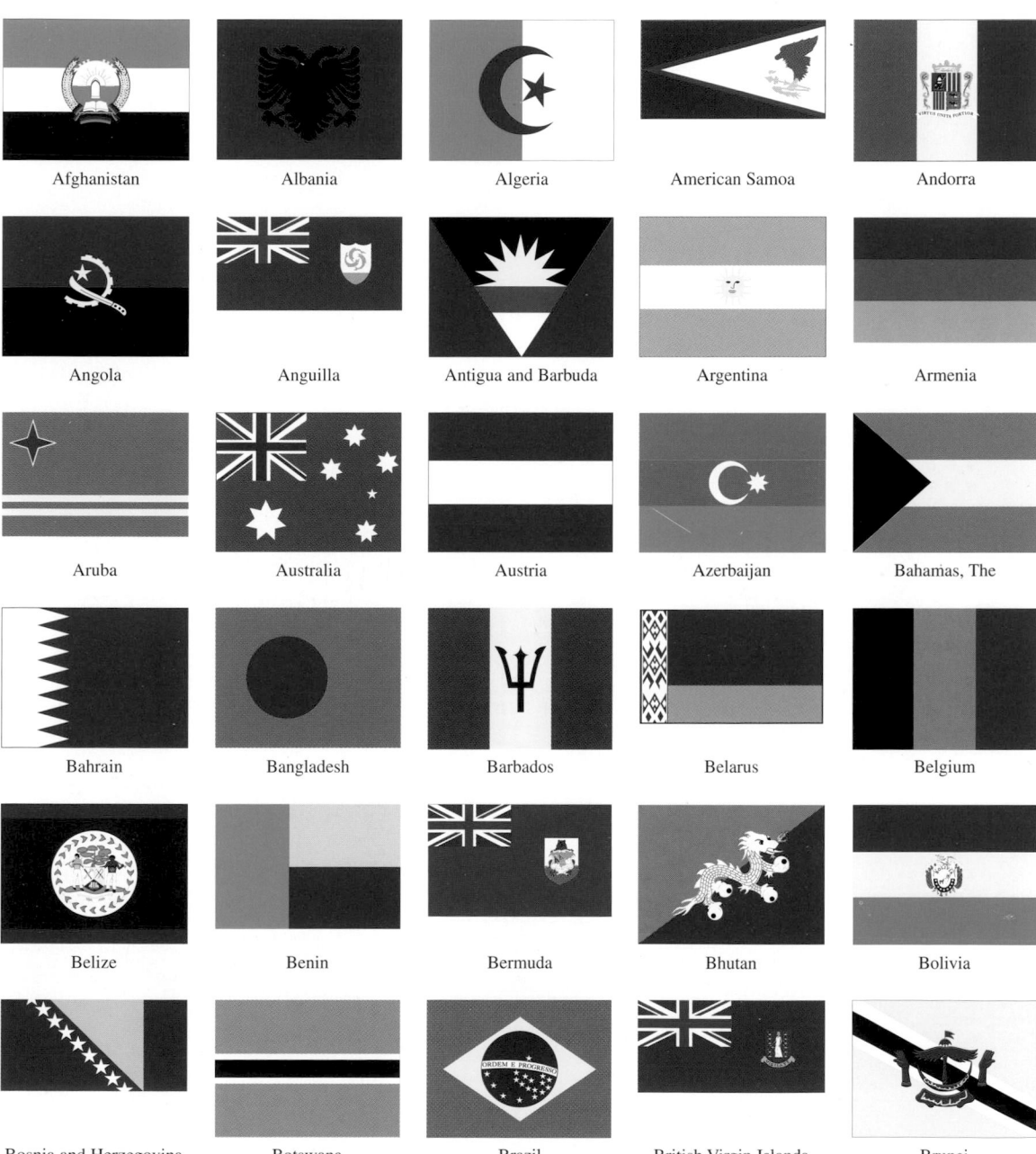

Afghanistan

Albania

Algeria

American Samoa

Andorra

Angola

Anguilla

Antigua and Barbuda

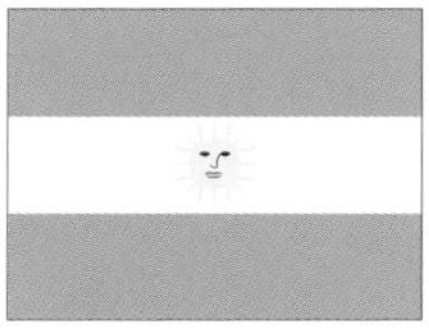

Argentina

Armenia

Aruba

Australia

Austria

Azerbaijan

Bahamas, The

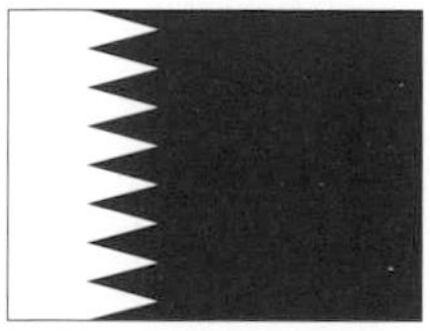

Bahrain

Bangladesh

Barbados

Belarus

Belgium

Belize

Benin

Bermuda

Bhutan

Bolivia

Bosnia and Herzegovina

Botswana

Brazil

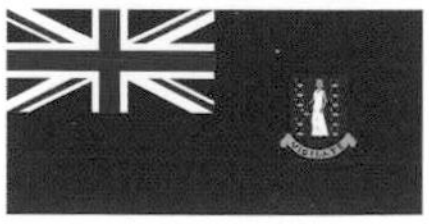

British Virgin Islands

Brunei

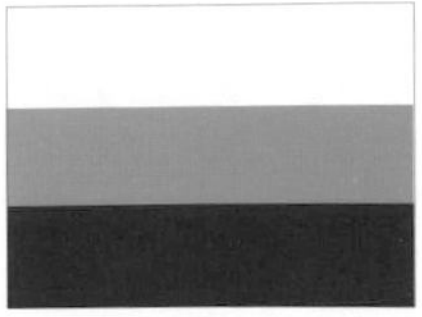
Bulgaria

Burkina Faso

Burundi

Cambodia

Cameroon

Canada

Cape Verde

Cayman Islands

Central African Republic

Chad

Chile

China

Christmas Island

Colombia

Comoros

Congo, Democratic Republic of the

Congo, Republic of the

Cook Islands

Costa Rica

Cote d'Ivoire

Croatia

Cuba

Cyprus

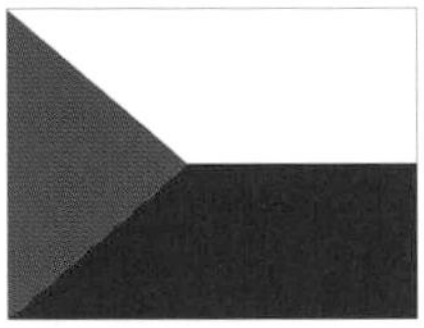
Czech Republic

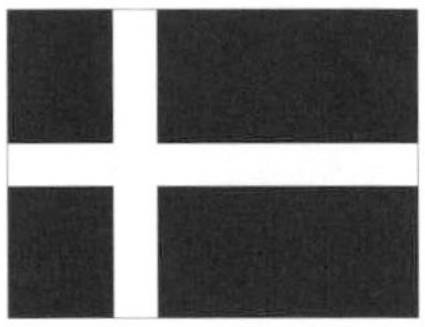
Denmark

Djibouti

Dominica

Dominican Republic

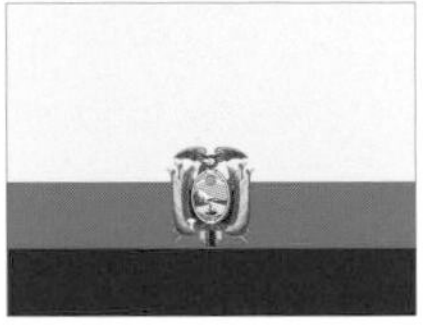
Ecuador

Egypt

El Salvador

Equatorial Guinea

Eritrea

Estonia

Ethiopia

Falkland Islands

Faroe Islands

Fiji

Finland

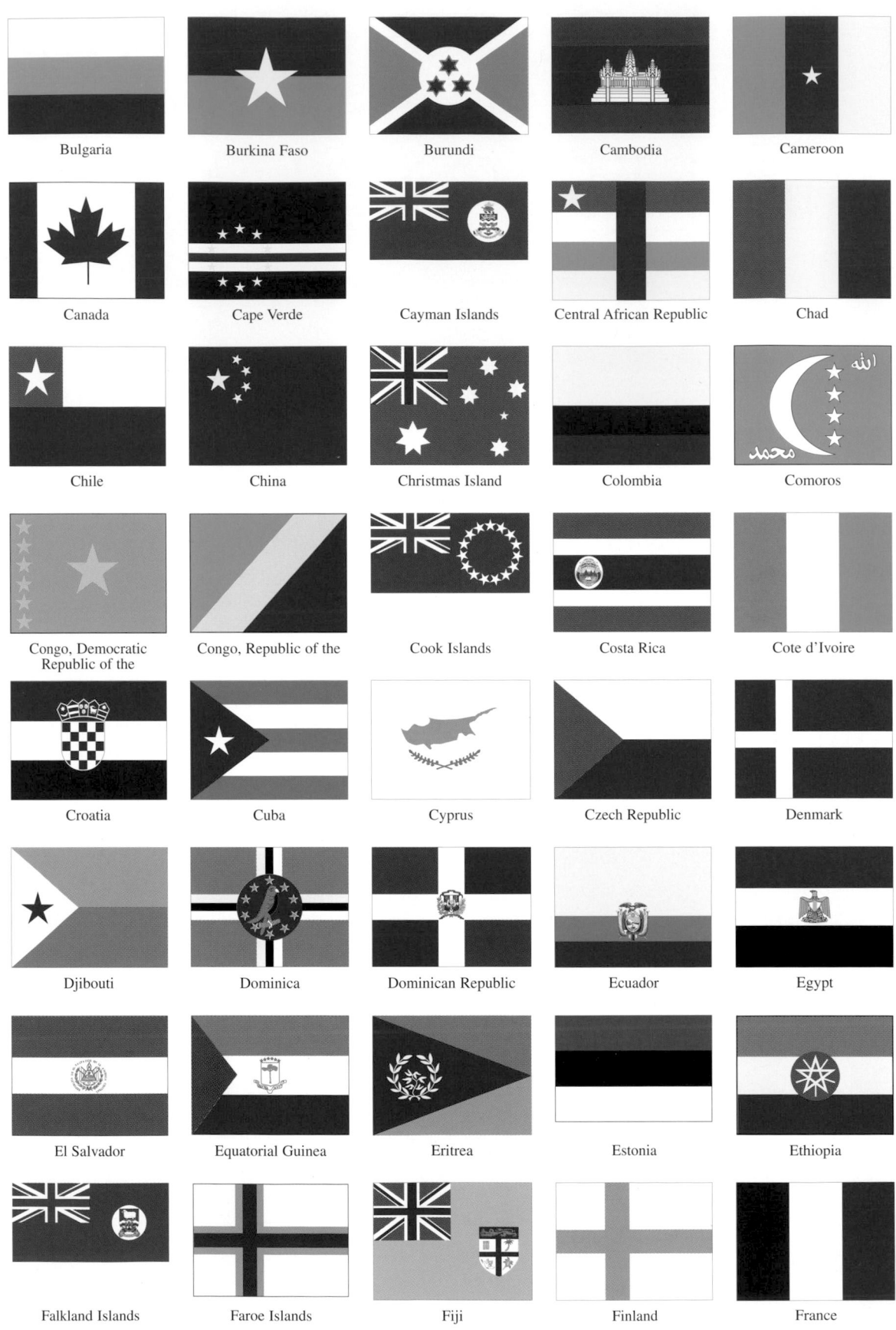

France

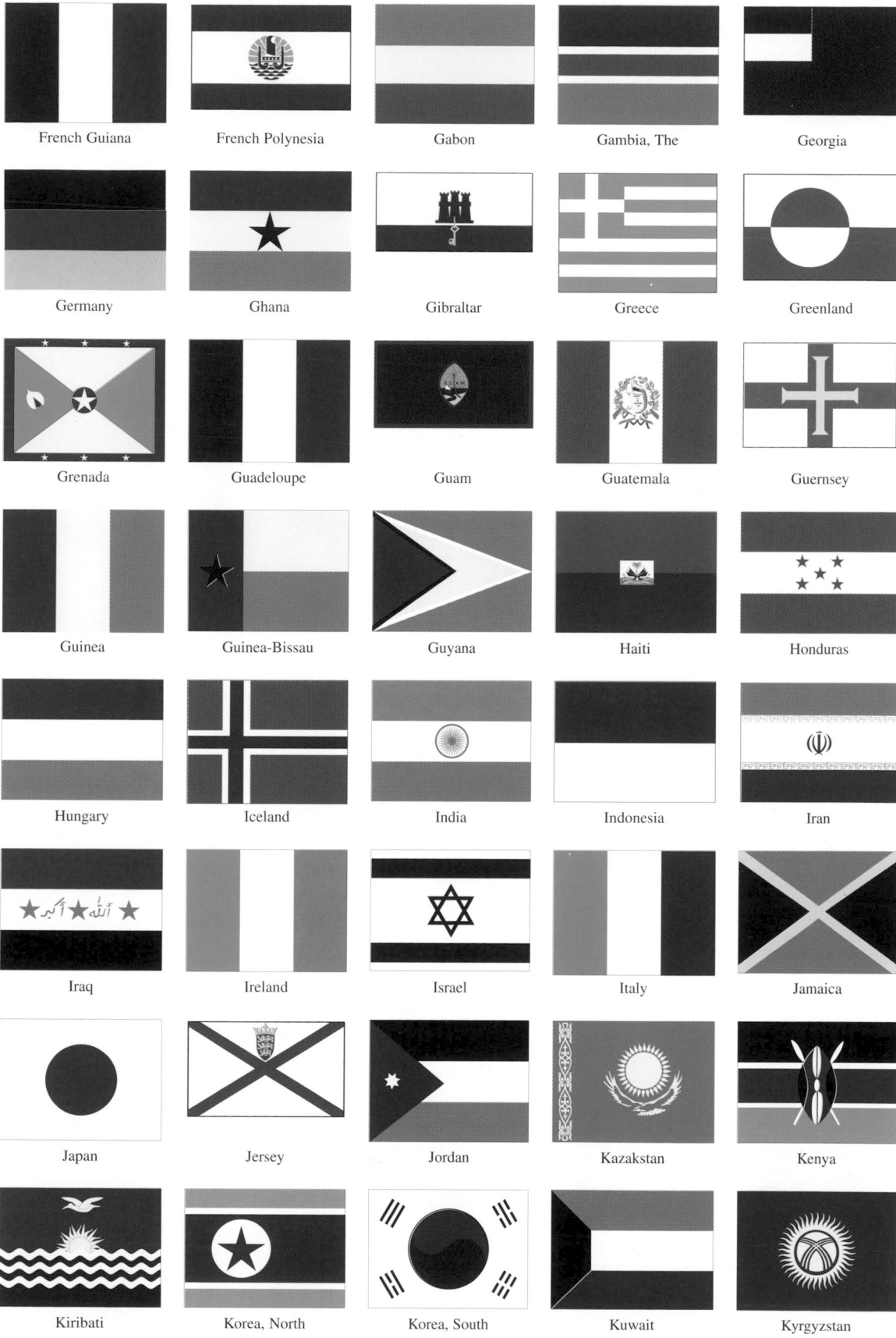

French Guiana

French Polynesia

Gabon

Gambia, The

Georgia

Germany

Ghana

Gibraltar

Greece

Greenland

Grenada

Guadeloupe

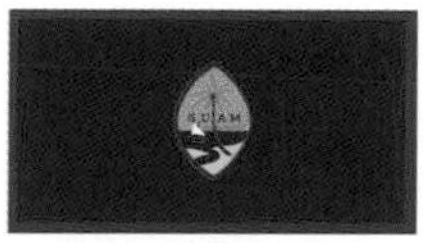
Guam

Guatemala

Guernsey

Guinea

Guinea-Bissau

Guyana

Haiti

Honduras

Hungary

Iceland

India

Indonesia

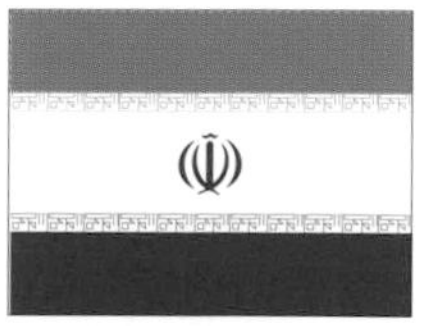
Iran

Iraq

Ireland

Israel

Italy

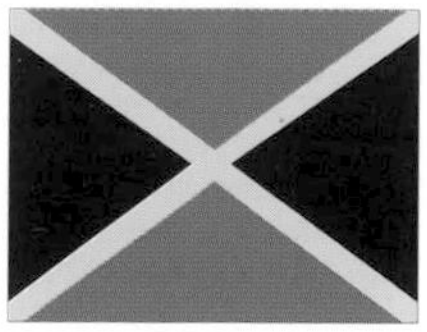
Jamaica

Japan

Jersey

Jordan

Kazakstan

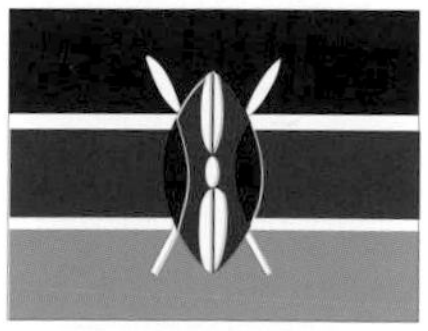
Kenya

Kiribati

Korea, North

Korea, South

Kuwait

Kyrgyzstan

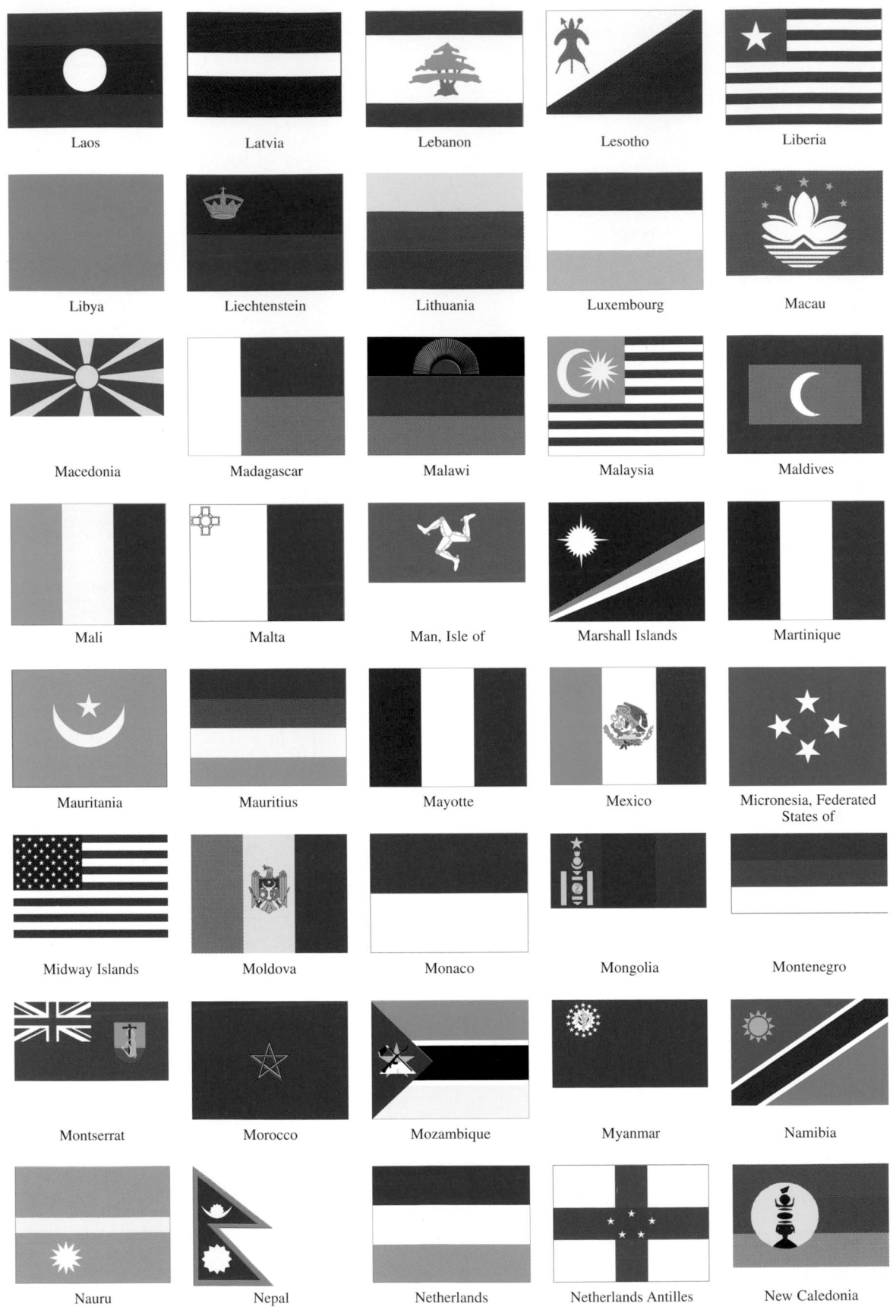

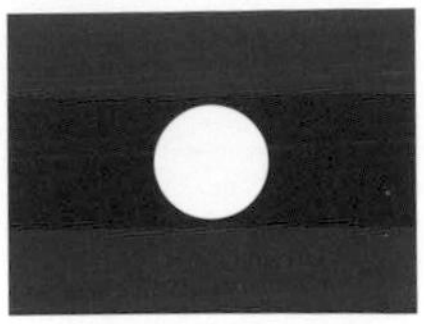
Laos

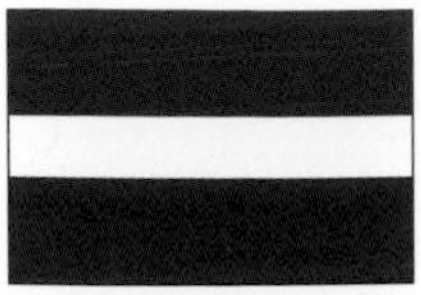
Latvia

Lebanon

Lesotho

Liberia

Libya

Liechtenstein

Lithuania

Luxembourg

Macau

Macedonia

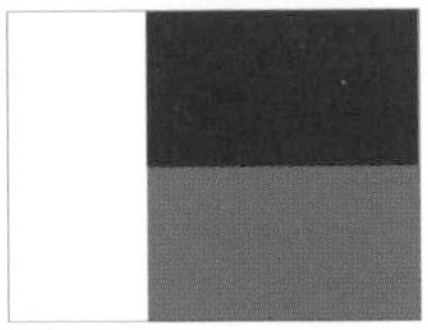
Madagascar

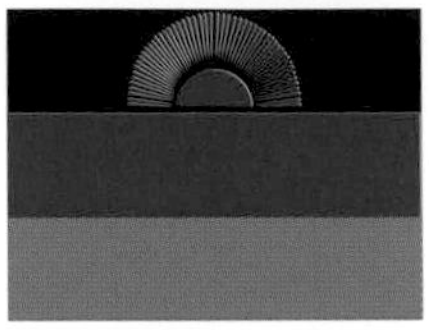
Malawi

Malaysia

Maldives

Mali

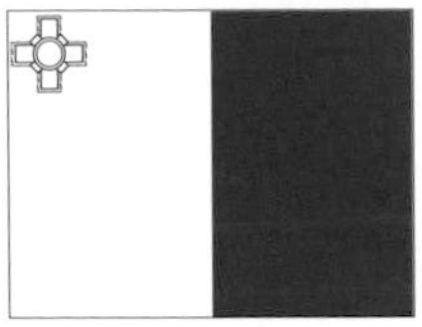
Malta

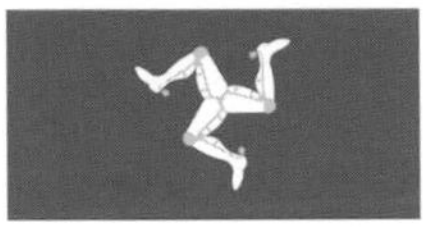
Man, Isle of

Marshall Islands

Martinique

Mauritania

Mauritius

Mayotte

Mexico

Micronesia, Federated States of

Midway Islands

Moldova

Monaco

Mongolia

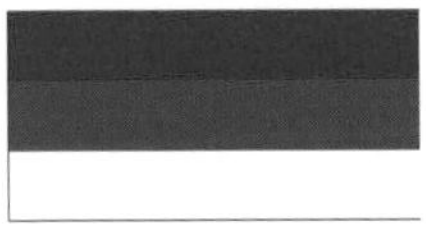
Montenegro

Montserrat

Morocco

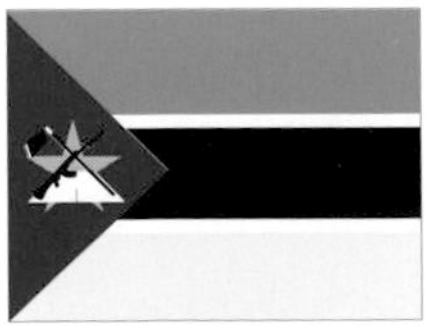
Mozambique

Myanmar

Namibia

Nauru

Nepal

Netherlands

Netherlands Antilles

New Caledonia

New Zealand | Nicaragua | Niger | Nigeria | Niue

Norfolk Island | Northern Mariana Islands | Norway | Oman | Pakistan

Palau | Panama | Papua New Guinea | Paraguay | Peru

Philippines | Poland | Portugal | Puerto Rico | Qatar

Reunion | Romania | Russia | Rwanda | Saint Helena

Saint Kitts and Nevis | Saint Lucia | Saint Pierre and Miquelon | Saint Vincent and the Grenadines | Samoa

San Marino | Sao Tome and Principe | Saudi Arabia | Senegal | Serbia

Seychelles | Sierra Leone | Singapore | Slovakia | Slovenia

Solomon Islands

Somalia

South Africa

Spain

Sri Lanka

Sudan

Suriname

Swaziland

Sweden

Switzerland

Syria

Taiwan

Tajikistan

Tanzania

Thailand

Togo

Tonga

Trinidad and Tobago

Tunisia

Turkey

Turkmenistan

Turks and Caicos Islands

Tuvalu

Uganda

Ukraine

United Arab Emirates

United Kingdom

United States

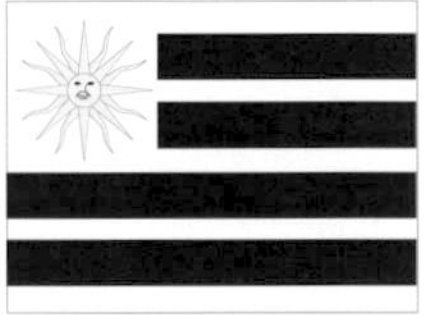
Uruguay

Uzbekistan

Vanuatu

Vatican City

Venezuela

Vietnam

Virgin Islands

Wallis and Futuna

Yemen

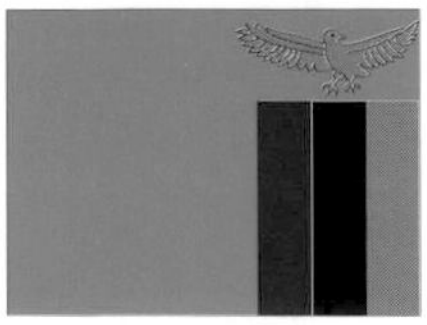
Zambia

Zimbabwe

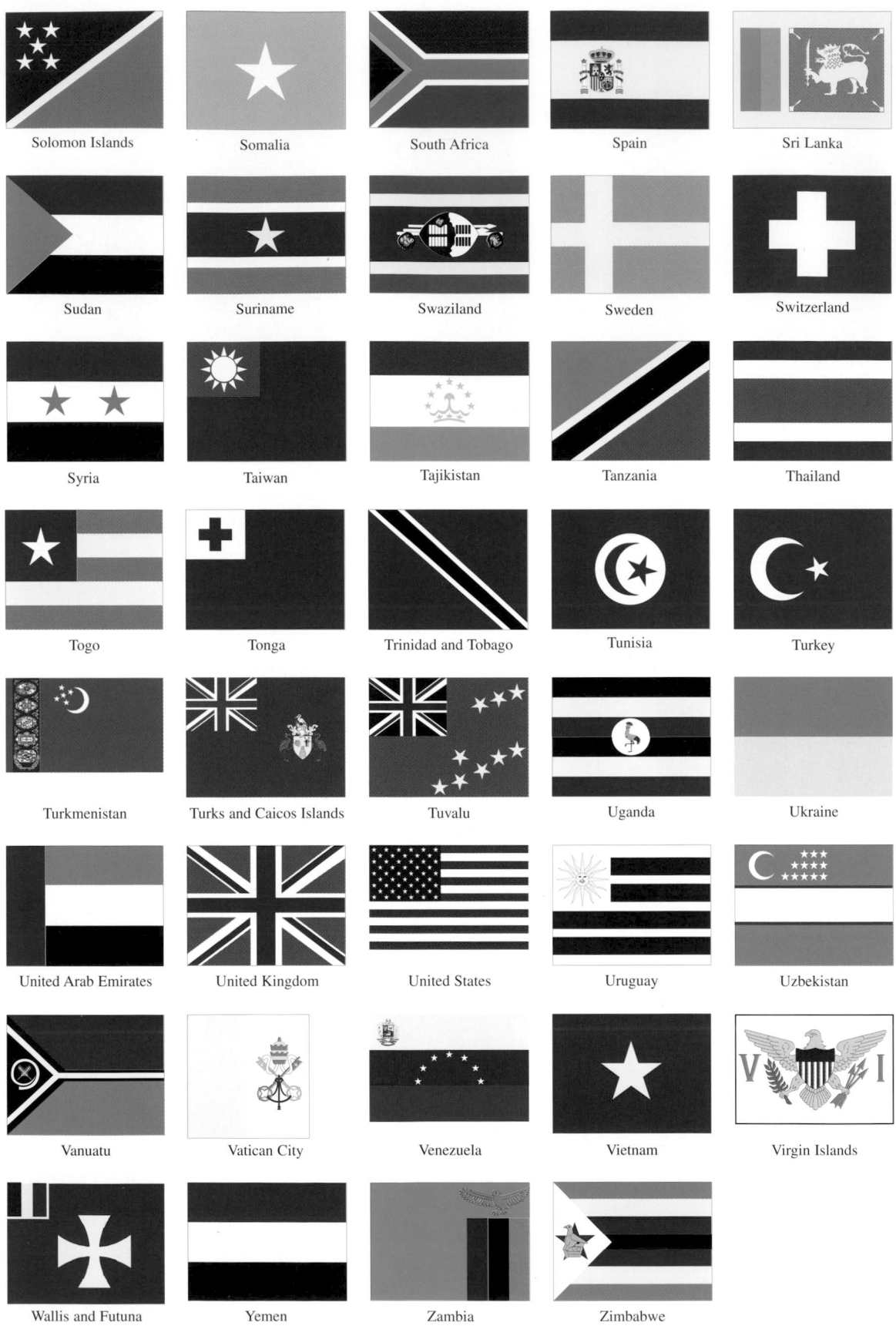

Afghanistan

Albania

Algeria

American Samoa

Andorra

Angola

Anguilla

Antigua and Barbuda

Argentina

Armenia

Aruba

Australia

Austria

Azerbaijan

Bahamas, The

Bahrain

Bangladesh

Barbados

Belarus

Belgium

Belize

Benin

Bermuda

Bhutan

Bolivia

Bosnia and Herzegovina

Botswana

Brazil

British Virgin Islands

Brunei

Bulgaria

Burkina Faso

Burundi

Cambodia

Cameroon

Canada

Cape Verde

Cayman Islands

Central African Republic

Chad

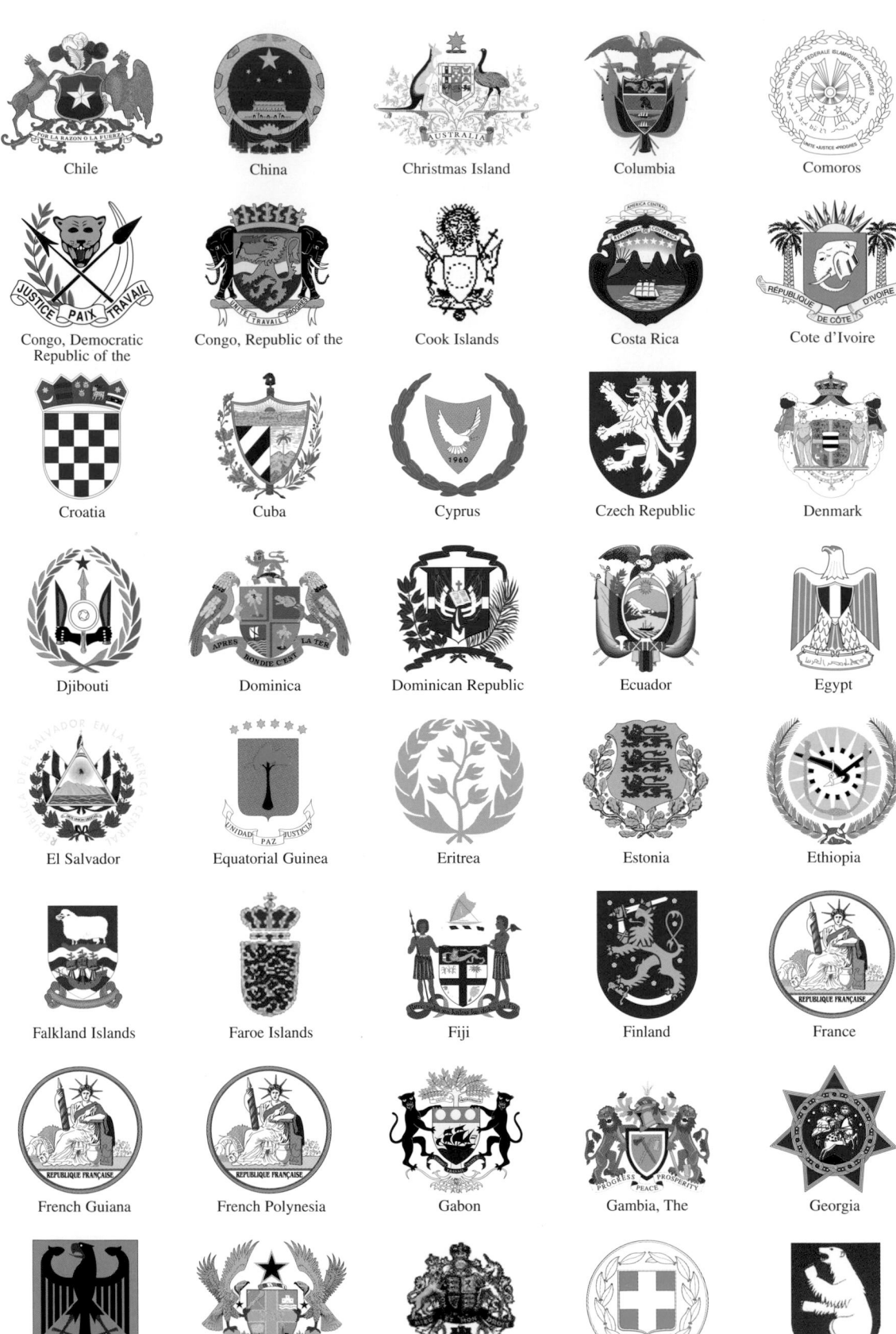

Chile

China

Christmas Island

Columbia

Comoros

Congo, Democratic Republic of the

Congo, Republic of the

Cook Islands

Costa Rica

Cote d'Ivoire

Croatia

Cuba

Cyprus

Czech Republic

Denmark

Djibouti

Dominica

Dominican Republic

Ecuador

Egypt

El Salvador

Equatorial Guinea

Eritrea

Estonia

Ethiopia

Falkland Islands

Faroe Islands

Fiji

Finland

France

French Guiana

French Polynesia

Gabon

Gambia, The

Georgia

Germany

Ghana

Gibraltar

Greece

Greenland

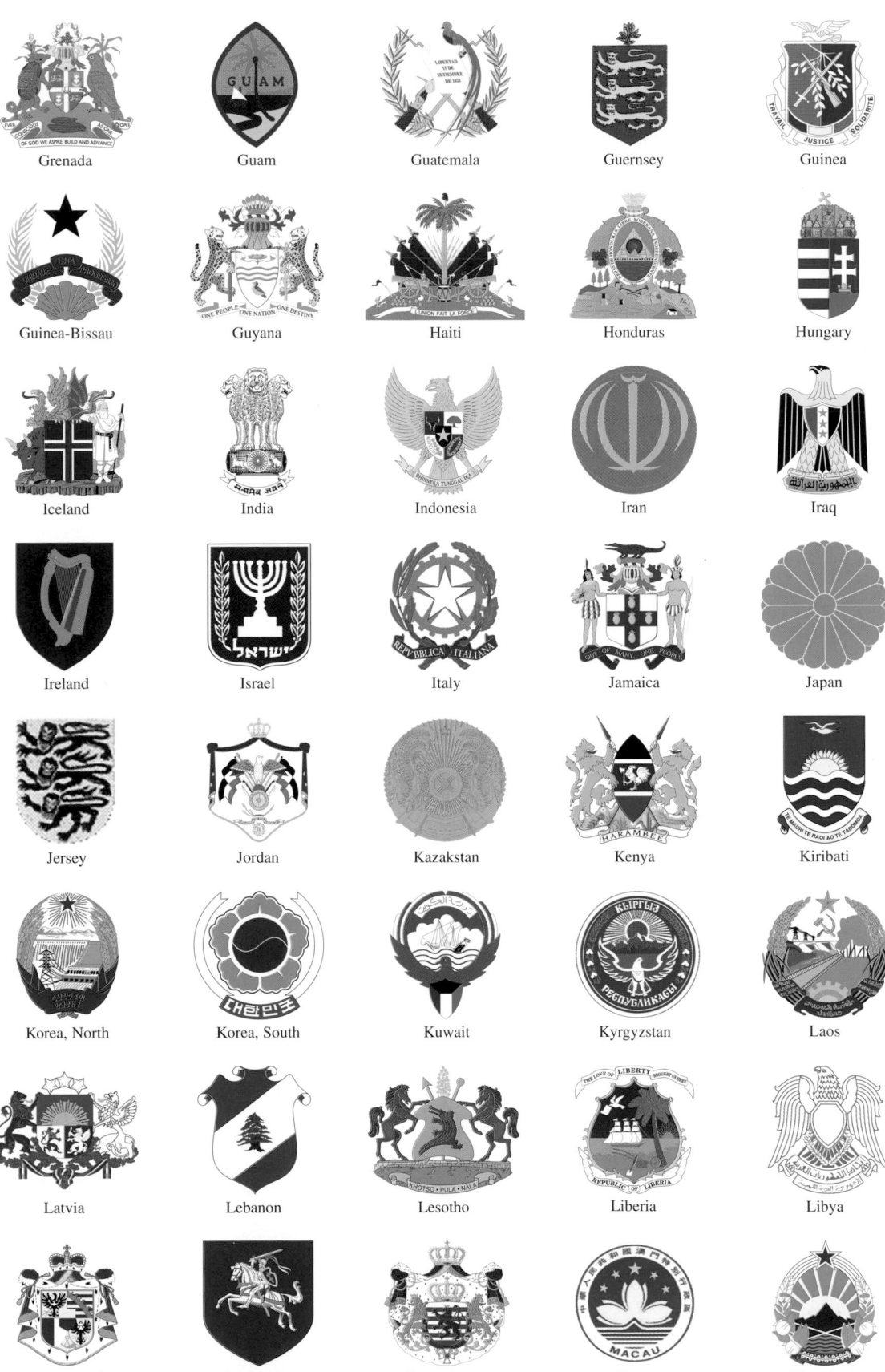

Grenada

Guam

Guatemala

Guernsey

Guinea

Guinea-Bissau

Guyana

Haiti

Honduras

Hungary

Iceland

India

Indonesia

Iran

Iraq

Ireland

Israel

Italy

Jamaica

Japan

Jersey

Jordan

Kazakstan

Kenya

Kiribati

Korea, North

Korea, South

Kuwait

Kyrgyzstan

Laos

Latvia

Lebanon

Lesotho

Liberia

Libya

Liechtenstein

Lithuania

Luxembourg

Macau

Macedonia

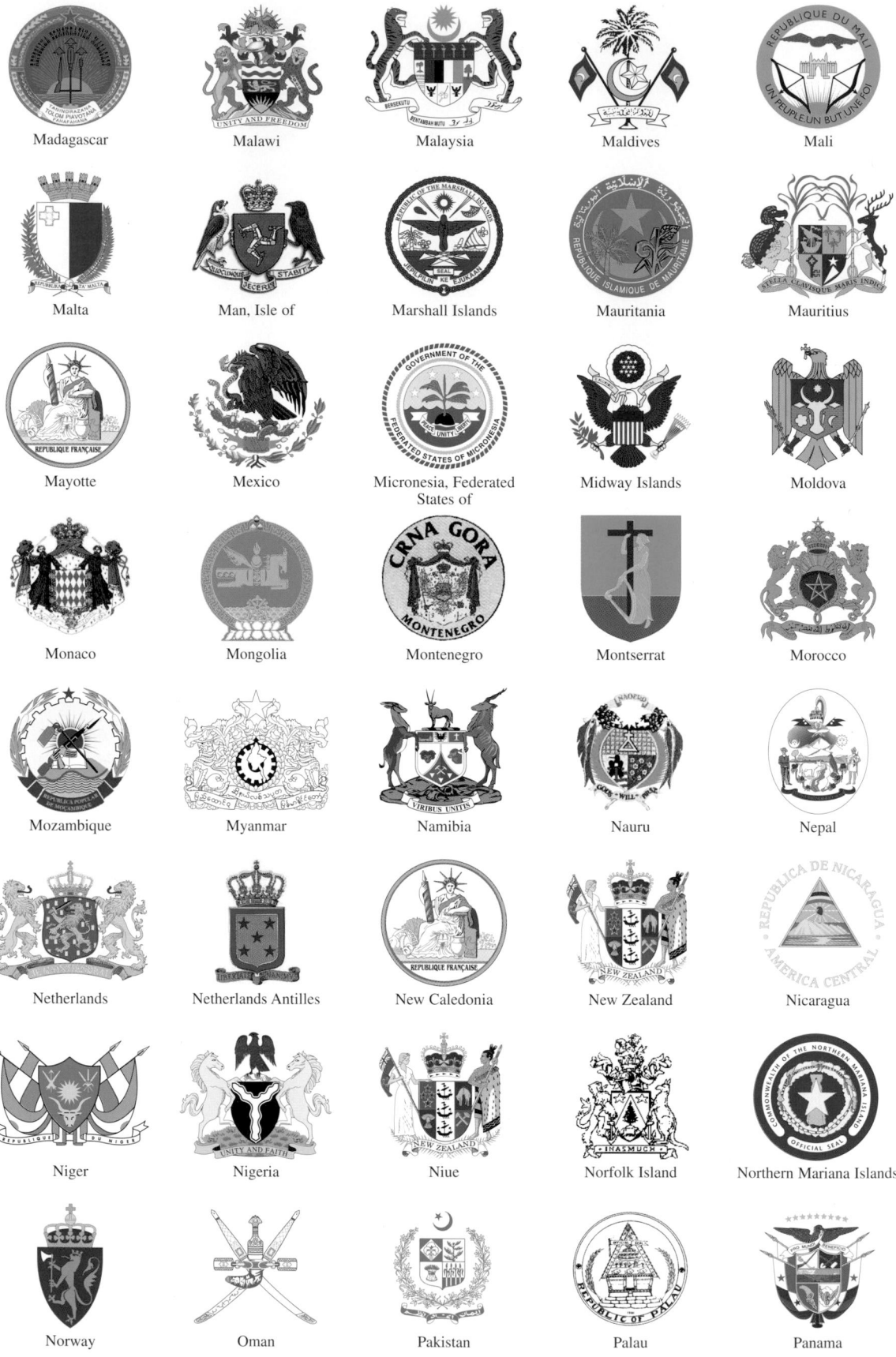

Madagascar

Malawi

Malaysia

Maldives

Mali

Malta

Man, Isle of

Marshall Islands

Mauritania

Mauritius

Mayotte

Mexico

Micronesia, Federated States of

Midway Islands

Moldova

Monaco

Mongolia

Montenegro

Montserrat

Morocco

Mozambique

Myanmar

Namibia

Nauru

Nepal

Netherlands

Netherlands Antilles

New Caledonia

New Zealand

Nicaragua

Niger

Nigeria

Niue

Norfolk Island

Northern Mariana Islands

Norway

Oman

Pakistan

Palau

Panama

Papua New Guinea | Paraguay | Peru | Philippines | Poland

Portugal | Puerto Rico | Qatar | Reunion | Romania

Russia | Rwanda | Saint Helena | Saint Kitts and Nevis | Saint Lucia

Saint Pierre and Miquelon | Saint Vincent and the Grenadines | Samoa | San Marino | Sao Tome and Principe

Saudi Arabia | Senegal | Seychelles | Sierra Leone | Singapore

Slovakia | Slovenia | Solomon Islands | Somalia | South Africa

Spain | Sri Lanka | Sudan | Suriname | Swaziland

Sweden | Switzerland | Syria | Taiwan | Tajikistan

Tanzania

Thailand

Togo

Tonga

Trinidad and Tobago

Tunisia

Turkey

Turkmenistan

Turks and Caicos Islands

Tuvalu

Uganda

Ukraine

United Arab Emirates

United Kingdom

United States

Uruguay

Uzbekistan

Vanuatu

Vatican City

Venezuela

Vietnam

Virgin Islands

Wallis and Futuna

Yemen

Zambia

Zimbabwe

Greenland Sea
ICELAND
Reykjavík
Arctic Circle
Norwegian Sea
Hammerfest
Murmansk
Narvik
Kiruna
Arkhangel'sk
Oulu
Umeå
Gulf of Bothnia
SWEDEN
FINLAND
Tórshavn
Faroe Islands (DENMARK)
Trondheim
Tampere
NORWAY
Shetland Islands
Bergen
Gävle
Helsinki
St. Petersburg
Åland Islands
Tallinn
RUSSIA
Orkney Islands
Oslo
Stockholm
ESTONIA
Outer Hebrides
Gotland
Moscow
North Sea
Göteborg
Riga
UNITED KINGDOM
LATVIA
Öland
Baltic Sea
DENMARK
København (Copenhagen)
LITHUANIA
Vitsyebsk
Belfast
Newcastle upon Tyne
Isle of Man (UK)
Bornholm (DENMARK)
RUSSIA
Vilnius
IRELAND
Dublin
Kaliningrad
Minsk
Liverpool
Gdańsk
Rostock
BELARUS
Irish Sea
Hamburg
POLAND
Homyel'
NETHERLANDS
Amsterdam
Berlin
Cardiff
Hannover
Kharkiv
London
Thames
Warsaw
Poznań
Kiev
GERMANY
Dnipro
Brussels
Leipzig
Wrocław
Rivne
English Channel
BELGIUM
Rhein
Bonn
Kraków
ATLANTIC OCEAN
Guernsey (UK)
Jersey (UK)
Le Havre
Frankfurt
Prague
UKRAINE
Seine
LUXEMBOURG
CZECH REPUBLIC
SLOVAKIA
Paris
Stuttgart
MOLDOVA
Strasbourg
Bratislava
Chişinău
Nantes
Munich
Vienna
AUSTRIA
Budapest
Cluj-Napoca
Odesa
Bay of Biscay
FRANCE
Zürich
Bern
LIECHTENSTEIN
Graz
HUNGARY
SWITZERLAND
SLOVENIA
ROMANIA
Geneva
Lyon
Ljubljana
Pécs
Milan
Zagreb
Bordeaux
CROATIA
Danube
Bucharest
Constanţa
Black Sea
Venice
BOSNIA & HERZEGOVINA
Belgrade
SAN MARINO
MONACO
Genoa
Varna
Bilbao
Adriatic Sea
SERBIA
Florence
Sarajevo
BULGARIA
Oporto
Marseille
ANDORRA
ITALY
MONTENEGRO
Priština
Sofia
Podgorica
Skopje
PORTUGAL
Madrid
Barcelona
Corsica
Rome
ALBANIA
MACEDONIA
VATICAN CITY
Tirana
Salonika
Lisbon
SPAIN
Naples
Sardinia
Tyrrhenian Sea
GREECE
Corfu
Ioánnina
Aegean Sea
Balearic Islands
Sevilla
Athens
Patras
Málaga
Mediterranean Sea
Palermo
Ionian Sea
Gibraltar (UK)
Sicily
Peloponnesus
Rhodes
Crete
MALTA
EUROPE
0 200 400 Miles
0 200 400 Kilometers
ARCTIC OCEAN
UNITED KINGDOM
Bering Strait
Chukchi Sea
Anadyrskiy Zaliv
Bering Sea
East Siberian Sea
Anadyr'
NORWAY
New Siberian Is.
DENMARK
SWEDEN
Barents Sea
Severnaya Zemlya
Laptev Sea
PACIFIC OCEAN
Murmansk
Novaya Zemlya
FINLAND
Kara Sea
RUSSIA
ESTONIA
Tiksi
Verkhoyansk
POLAND
LITHUANIA
LATVIA
St. Petersburg
Arkhangel'sk
Magadan
Noril'sk
BELARUS
Smolensk
Vorkuta
Lena
RUSSIA
Yakutsk
Sea of Okhotsk
Moscow
Kirov
UKRAINE
MOLDOVA
Khanty Mansiysk
Mirnyy
Aldan
Sakhalin
Donetsk
Don
Volga
Kama
Ob'
Yenisey
Angara
Komsomol'sk
Yekaterinburg
Irtysh
Tomsk
Bratsk
Ozero Baykal
Khabarovsk
Amur
Black Sea
Tuapse
Krasnoyarsk
Chita
GEORGIA
T'bilisi
Caspian Sea
Novokuznetsk
Nakhodka
ARMENIA
Yerevan
RUSSIA
0 250 500 Miles
0 250 500 Kilometers

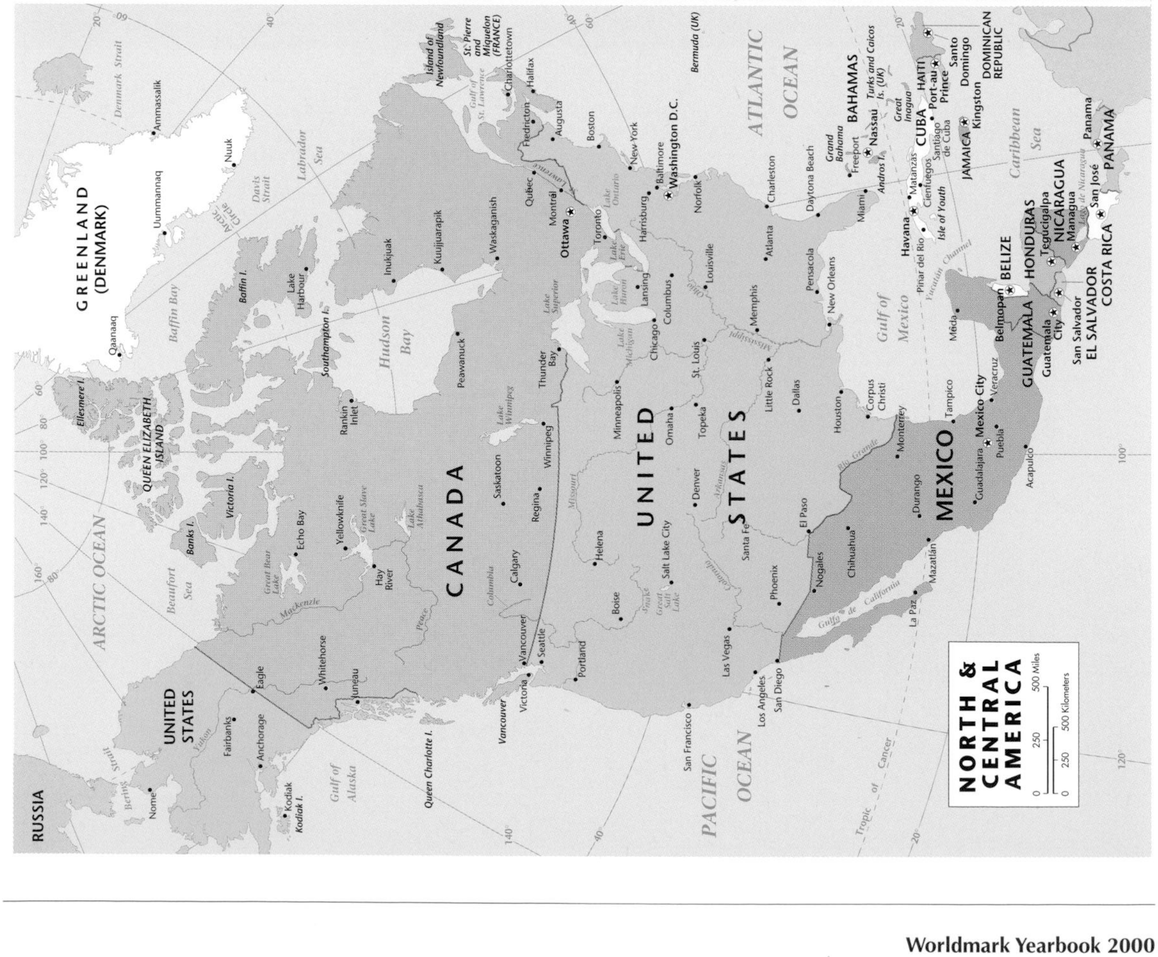
NORTH & CENTRAL AMERICA
0 250 500 Miles
0 250 500 Kilometers
CANADA
UNITED STATES
MEXICO
GREENLAND (DENMARK)
RUSSIA
ARCTIC OCEAN
PACIFIC OCEAN
ATLANTIC OCEAN
Hudson Bay
Gulf of Mexico
Caribbean Sea
GUATEMALA
BELIZE
HONDURAS
EL SALVADOR
NICARAGUA
COSTA RICA
PANAMA
CUBA
JAMAICA
HAITI
DOMINICAN REPUBLIC
BAHAMAS
Ottawa
Washington D.C.
Mexico City

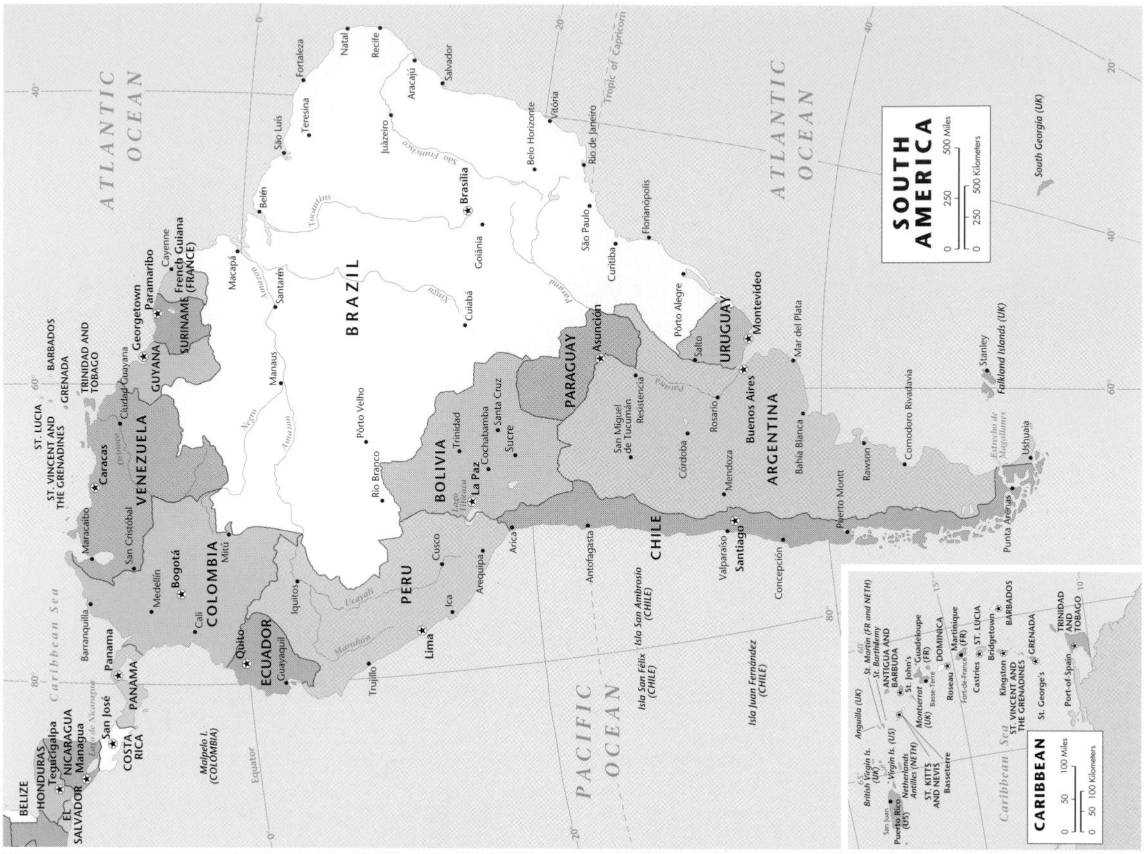
SOUTH AMERICA
0 250 500 Miles
0 250 500 Kilometers
CARIBBEAN
0 50 100 Miles
0 50 100 Kilometers
BRAZIL
ARGENTINA
CHILE
PERU
BOLIVIA
COLOMBIA
VENEZUELA
ECUADOR
PARAGUAY
URUGUAY
GUYANA
SURINAME
French Guiana (FRANCE)
ATLANTIC OCEAN
PACIFIC OCEAN
Caribbean Sea
Brasília
Buenos Aires
Santiago
Lima
Bogotá
Caracas
Quito
La Paz
Asunción
Montevideo
Georgetown
Paramaribo
Falkland Islands (UK)
South Georgia (UK)

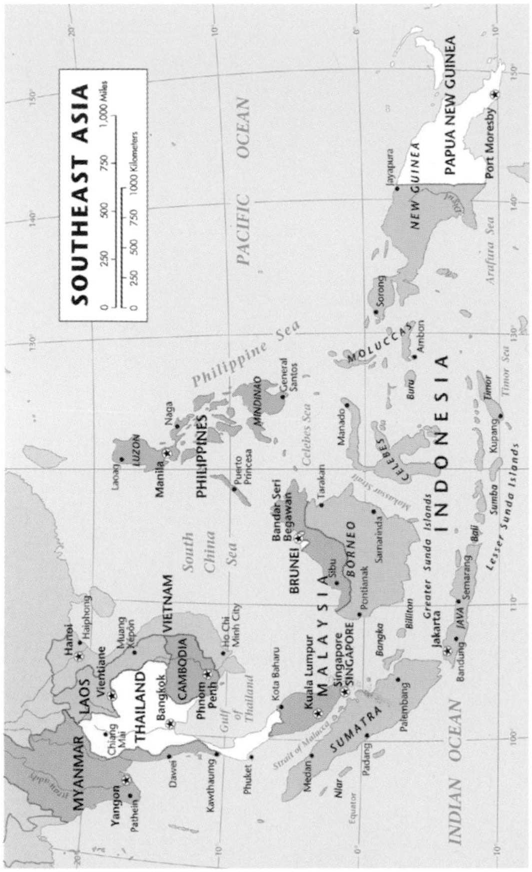
SOUTHEAST ASIA
0 250 500 750 1,000 Miles
0 250 500 750 1000 Kilometers
PACIFIC OCEAN
INDIAN OCEAN
South China Sea
Philippine Sea
MYANMAR
LAOS
THAILAND
CAMBODIA
VIETNAM
PHILIPPINES
MALAYSIA
BRUNEI
SINGAPORE
INDONESIA
PAPUA NEW GUINEA

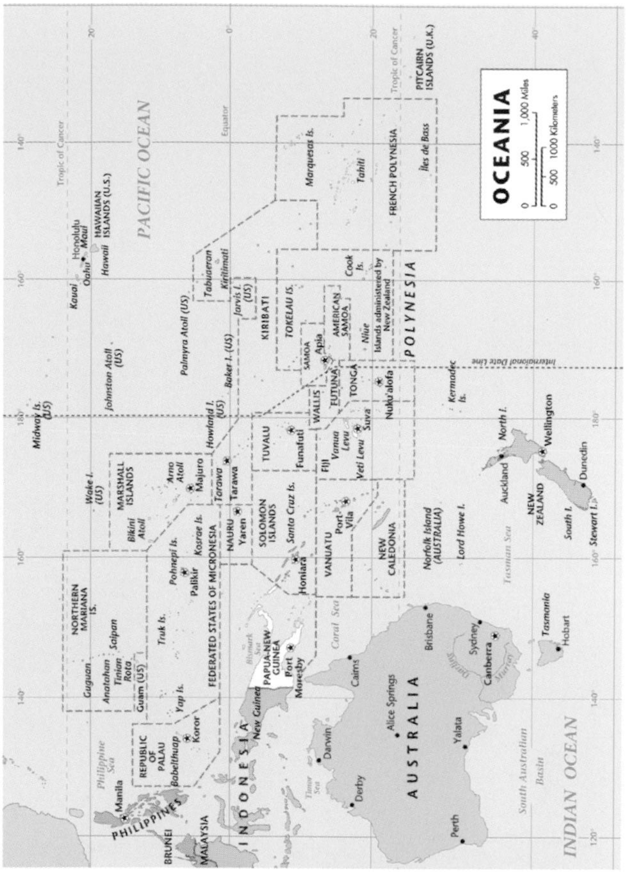
OCEANIA
0 500 1,000 Miles
0 500 1000 Kilometers
PACIFIC OCEAN
INDIAN OCEAN
AUSTRALIA
NEW ZEALAND
POLYNESIA
FEDERATED STATES OF MICRONESIA

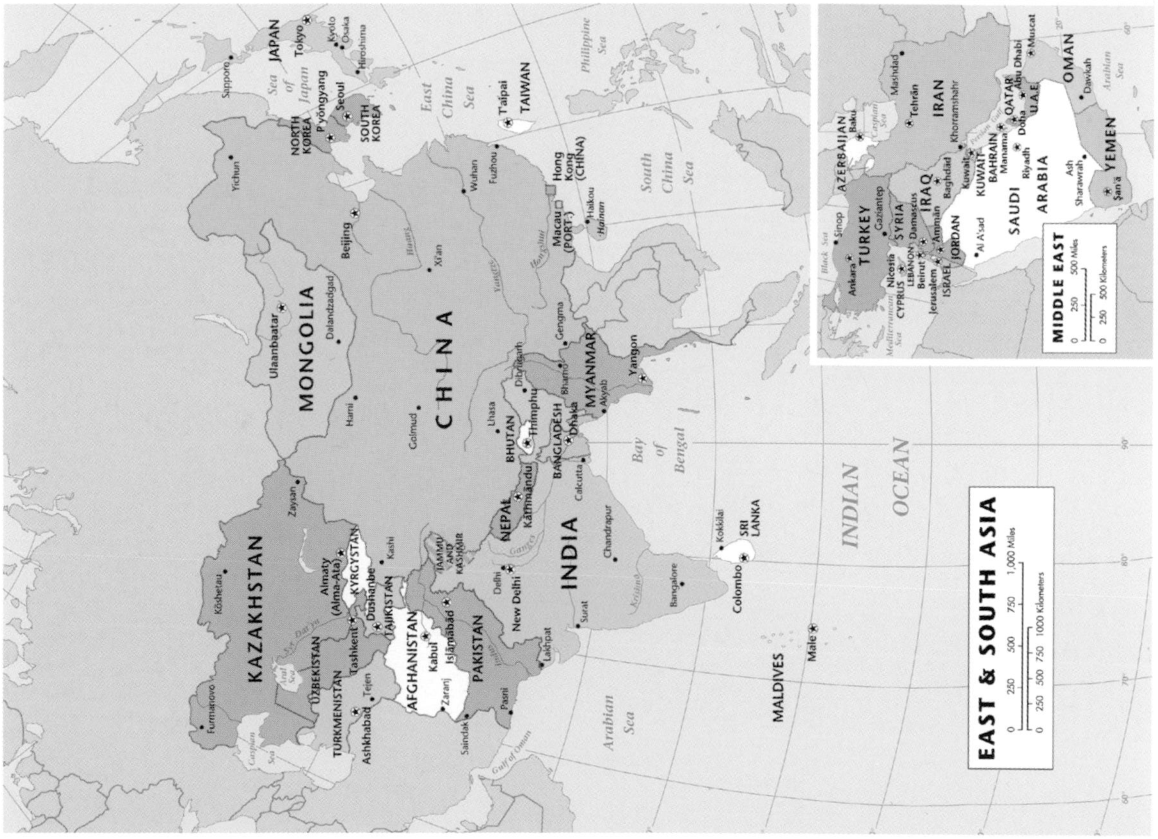
EAST & SOUTH ASIA
0 250 500 750 1,000 Miles
0 250 500 750 1000 Kilometers
KAZAKHSTAN
MONGOLIA
CHINA
INDIA
PAKISTAN
AFGHANISTAN
NEPAL
BHUTAN
BANGLADESH
MYANMAR
SRI LANKA
MALDIVES
JAPAN
NORTH KOREA
SOUTH KOREA
TAIWAN
INDIAN OCEAN
Bay of Bengal
Arabian Sea
MIDDLE EAST
0 250 500 Miles
0 250 500 Kilometers
TURKEY
SYRIA
IRAQ
IRAN
SAUDI ARABIA
YEMEN
OMAN

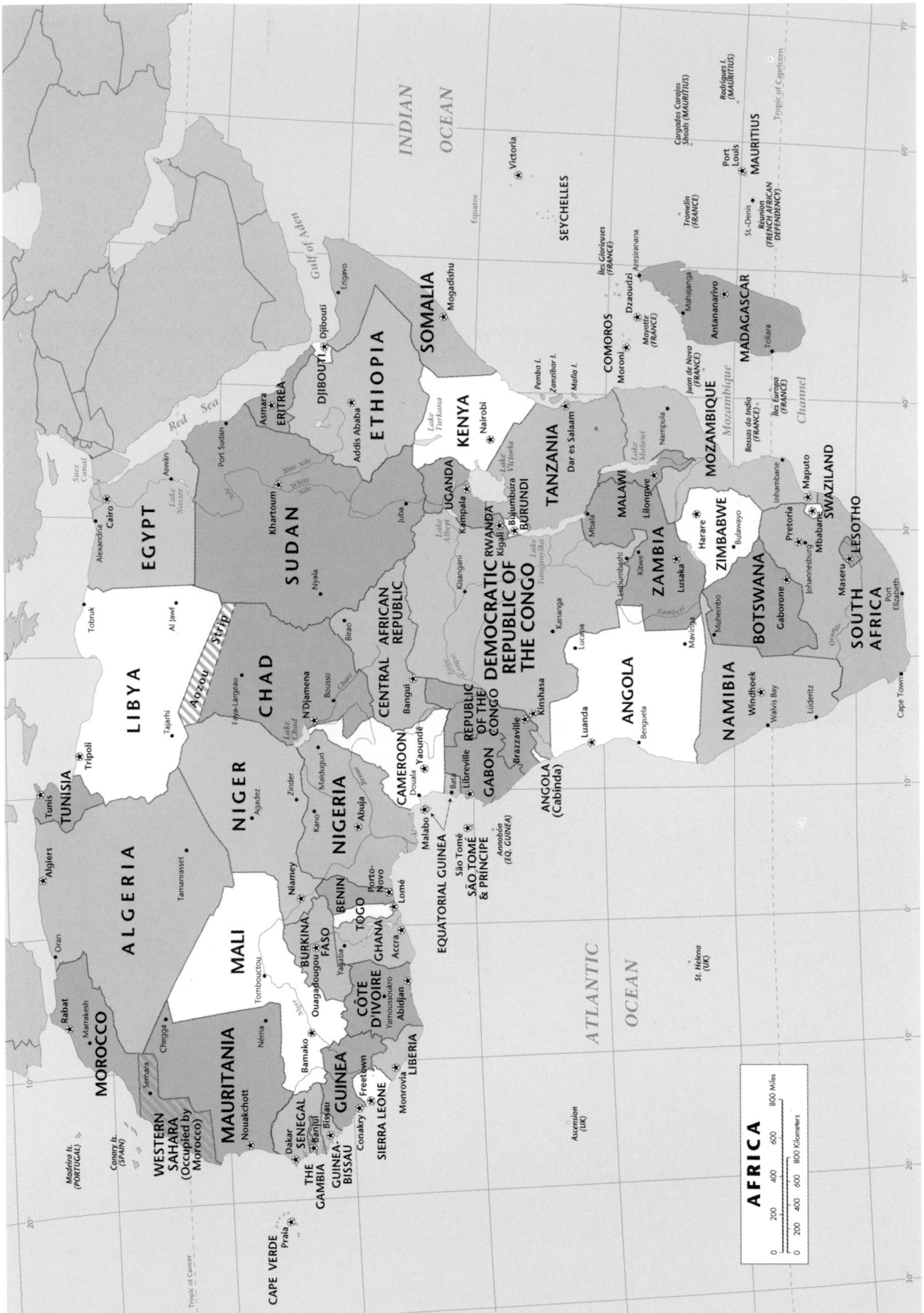
AFRICA
0 200 400 600 800 Miles
0 200 400 600 800 Kilometers
ATLANTIC OCEAN
INDIAN OCEAN
MOROCCO
Rabat
Marrakesh
WESTERN SAHARA (Occupied by Morocco)
Semara
ALGERIA
Algiers
Oran
Tamanrasset
TUNISIA
Tunis
LIBYA
Tripoli
Tobruk
Al Jawf
Aozou Strip
EGYPT
Cairo
Alexandria
Aswân
Lake Nasser
Suez Canal
Red Sea
Gulf of Aden
MAURITANIA
Nouakchott
Chegga
MALI
Tombouctou
Bamako
NIGER
Niamey
Agadez
Zinder
CHAD
N'Djamena
Faya-Largeau
Lake Chad
SUDAN
Khartoum
Port Sudan
Nyala
Juba
ERITREA
Asmara
DJIBOUTI
Djibouti
ETHIOPIA
Addis Ababa
SOMALIA
Mogadishu
Erigavo
SENEGAL
Dakar
THE GAMBIA
Banjul
GUINEA-BISSAU
Bissau
GUINEA
Conakry
SIERRA LEONE
Freetown
LIBERIA
Monrovia
CÔTE D'IVOIRE
Yamoussoukro
Abidjan
BURKINA FASO
Ouagadougou
GHANA
Accra
TOGO
Lomé
BENIN
Porto-Novo
NIGERIA
Abuja
Kano
Maiduguri
CAMEROON
Yaoundé
Douala
EQUATORIAL GUINEA
Malabo
Bata
SÃO TOMÉ & PRINCIPE
São Tomé
GABON
Libreville
REPUBLIC OF THE CONGO
Brazzaville
CENTRAL AFRICAN REPUBLIC
Bangui
DEMOCRATIC REPUBLIC OF THE CONGO
Kinshasa
Kisangani
Kananga
Lubumbashi
ANGOLA
Luanda
Benguela
ANGOLA (Cabinda)
UGANDA
Kampala
KENYA
Nairobi
Lake Turkana
Lake Victoria
RWANDA
Kigali
BURUNDI
Bujumbura
TANZANIA
Dar es Salaam
Lake Tanganyika
ZAMBIA
Lusaka
MALAWI
Lilongwe
MOZAMBIQUE
Maputo
Nampula
Mozambique Channel
ZIMBABWE
Harare
Bulawayo
NAMIBIA
Windhoek
Walvis Bay
BOTSWANA
Gaborone
SOUTH AFRICA
Pretoria
Johannesburg
Cape Town
Port Elizabeth
SWAZILAND
Mbabane
LESOTHO
Maseru
MADAGASCAR
Antananarivo
Toliara
Mahajanga
COMOROS
Moroni
SEYCHELLES
Victoria
MAURITIUS
Port Louis
CAPE VERDE
Praia
Madeira Is. (PORTUGAL)
Canary Is. (SPAIN)
Ascension (UK)
St. Helena (UK)
Tropic of Cancer
Equator
Tropic of Capricorn

Norway
25 Belgrave Square, London SW1X 8QD, England
PHONE: +44 (171) 5915500
FAX: +44 (171) 2456993
E-MAIL: norway.london@mfa.no
TITLE: Ambassador
NAME: Kjell Colding

Poland
47 Portland Place, London W1N 4JH, England
PHONE: +44 (171) 5804324
FAX: +44 (171) 3234018
TITLE: Ambassador
NAME: Stanislaw Komorowski

Russia
5 Kensington Palace Gardens, London W8 4QS, England
PHONE: +44 (171) 2298027
FAX: +44 (171) 2293215

Spain
20 Peel Street, London W8 7PD, England
PHONE: +44 (171) 7272462; 2438535
FAX: +44 (171) 2294965

Sri Lanka
13 Hyde Park Gardens, London W2 2LU, England
PHONE: +44 (171) 2621841
FAX: +44 (171) 2627970
E-MAIL: lancom@easynet.co.uk

Switzerland
16–18 Montagu Place, London W1H 2BQ, England
PHONE: +44 (171) 6166000
FAX: +44 (171) 7247001

Taiwan
50 Grosvenor Gardens, London SW1W OEB, England
PHONE: +44 (171) 3969152; 3969148
FAX: +44 (171) 3969151
E-MAIL: presstro@netcomuk.co.uk

Thailand
29–30 Queen's Gate, London SW7 5JB, England
PHONE: +44 (171) 5892944
FAX: +44 (171) 8239695
E-MAIL: dx42@cityscape.co.uk

Turkey
43 Belgrave Square, London SW1X 8PA, England
PHONE: +44 (171) 3930202
FAX: +44 (171) 3930066
E-MAIL: info@turkishembassy-london.com
TITLE: Ambassador
NAME: Özdem Sanberk

United States
24 Grosvenor Square, London W1A 1AE, England
PHONE: +44 (171) 4999000
TITLE: Ambassador
NAME: Philip Lader

Venezuela
56 Grafton Way, London W1P 5LB, England
PHONE: +44 (171) 3876727
FAX: +44 (171) 3833253
TITLE: Ambassador
NAME: Roy Chaderton-Matos

JUDICIAL SYSTEM

High Court

Lords of Appeal

House of Lords Judicial Office, London SW1A 0PW, England
PHONE: +44 (171) 2193111
FAX: +44 (171) 2192476

Courts of Appeal

Magistrates' Court

Crown Court

The Customer Service Unit, the Court Service, Southside, 105 Victoria Street, London SW1E 6QT, England
E-MAIL: cust.ser.cs@gtnet.gov.uk

FURTHER READING

Articles

Bush, Janet and Philip Webster. "Economy Heading towards Brink of Recession." *The New York Times*, 11 February 1999, p. A1.

Cowell, Alan. "Debut of Euro has the British Underwhelmed." *The New York Times*, 4 January 1999, p. A1, A6.

"Gilt Complex." *The Economist* (August 21 1999): 64.

"Lairds and Land." *The Economist* (January 9, 1999): 53.

Lodge, Robin. "Messages on Bodies Tell of Hate for the British." *The London Times* March 1999, p. A1.

''Not Much to Celebrate.'' *The Economist* (March 27, 1999): 57–58.

''On the Defensive.'' *The Economist* (February 6, 1999) 59.

''Scottish Nationalism: Second Thoughts.'' *The Economist* (September 25, 1999): 67.

''The State of Scotland.'' *The Economist* (May 1, 1999): 53–55.

''Through the Looking Glass.'' *The Economist* (May 1, 1999): 56.

''Towards a Federal Britain.'' *The Economist* (March 27, 1999): 23–25.

''Tragedy of Death Foretold.'' *The Economist* (March 20, 1999): 59–60.

Webster, Philip and Mark Henderson. ''Railtrack Gets Control of Tube Lines.'' *The London Times*, 16 June 1999, p. A1.

Books

Briscoe, Lynden. *Britain's Trade and Economic Structure: The Impact of the European Union*. New York, NY: Routledge, 1999.

Country Finance: United Kingdom. New York, NY: The Economic Intelligence Unit, 1999.

Jones, Bill, ed. *Political Issues in Britain Today*. Manchester, NY: Manchester University Press, 1999.

Milfall, John, ed. *Britain in Europe: Prospects for Change*. Aldershot, Brookfield, VT: Ashgate, 1999.

Northcott, Jim. *Britain's Future: Issues and Choices*. London: Policy Studies Institute, 1999.

Norton, Philip, ed. *Parliaments and Pressure Groups in Western Europe*. London: Frank Cass, 1999.

United Kingdom: The Pinochet Case: Universal Jurisdiction and the Absence of Immunity for Crimes Against Humanity. London: Amnesty International, International Secretariat, 1999.

Wilson, Frank L. *European Politics Today: The Democratic Experience*. Upper Saddle River, NJ: Prentice Hall, 1999.

Internet

British Parliament. Available Online @ http://www.explore.parliament.uk/ (October 18, 1999).

Scottish Parliament. Available Online @ http://www.scottish.parliament.uk/ (October 18, 1999).

UNITED KINGDOM: STATISTICAL DATA

For sources and notes see "Sources of Statistics" in the front of each volume.

GEOGRAPHY

Geography (1)

Area:

Total: 244,820 sq km.

Land: 241,590 sq km.

Water: 3,230 sq km.

Note: includes Rockall and Shetland Islands.

Area—comparative: slightly smaller than Oregon.

Land boundaries:

Total: 360 km.

Border countries: Ireland 360 km.

Coastline: 12,429 km.

Climate: temperate; moderated by prevailing southwest winds over the North Atlantic Current; more than one-half of the days are overcast.

Terrain: mostly rugged hills and low mountains; level to rolling plains in east and southeast.

Natural resources: coal, petroleum, natural gas, tin, limestone, iron ore, salt, clay, chalk, gypsum, lead, silica.

Land use:

Arable land: 25%

Permanent crops: 0%

Permanent pastures: 46%

Forests and woodland: 10%

Other: 19% (1993 est.)

HUMAN FACTORS

Demographics (2A)

	1990	1995	1998	2000	2010	2020	2030	2040	2050
Population	NA	58,476.8	58,970.1	59,247.4	59,955.6	60,177.0	59,430.1	57,250.8	54,115.7
Net migration rate (per 1,000 population)	NA	NA	NA	NA	NA	NA	NA	NA	NA
Births	NA	NA	NA	NA	NA	NA	NA	NA	NA
Deaths	NA	NA	NA	NA	NA	NA	NA	NA	NA
Life expectancy - males	NA	74.1	74.6	74.9	76.3	77.4	78.4	79.1	79.7
Life expectancy - females	NA	79.4	80.0	80.3	81.9	83.2	84.3	85.1	85.7
Birth rate (per 1,000)	NA	12.6	12.0	11.8	10.3	10.0	8.9	8.4	8.1
Death rate (per 1,000)	NA	11.0	10.7	10.6	10.4	10.7	11.8	13.3	14.6
Women of reproductive age (15-49 yrs.)	NA	14,214.2	14,085.7	14,051.1	14,013.6	12,866.9	12,241.8	11,390.6	10,266.2
of which are currently married	NA	NA	NA	NA	NA	NA	NA	NA	NA
Fertility rate	NA	1.7	1.7	1.7	1.6	1.6	1.6	1.5	1.5

Except as noted, values for vital statistics are in thousands; life expectancy is in years.

Health Personnel (3)

Total health expenditure as a percentage of GDP, 1990-1997[a]

Public sector	5.7
Private sector	1.0
Total[b]	6.7

Health expenditure per capita in U.S. dollars, 1990-1997[a]

Purchasing power parity	1,386
Total	1,454

Availability of health care facilities per 100,000 people

Hospital beds 1990-1997[a]	470
Doctors 1993[c]	164
Nurses 1993[c]	NA

Health Indicators (4)

Life expectancy at birth

1980	74
1997	77
Daily per capita supply of calories (1996)	3,237
Total fertility rate births per woman (1997)	1.7
Maternal mortality ratio per 100,000 live births (1990-97)	9[b]
Safe water % of population with access (1995)	100
Sanitation % of population with access (1995)	96
Consumption of iodized salt % of households (1992-98)[a]	....

Smoking prevalence

Male % of adults (1985-95)[a]	28
Female % of adults (1985-95)[a]	26
Tuberculosis incidence per 100,000 people (1997)	18
Adult HIV prevalence % of population ages 15-49 (1997)	0.09

Infants and Malnutrition (5)

Under-5 mortality rate (1997)	7
% of infants with low birthweight (1990-97)	7
Births attended by skilled health staff % of total[a]	NA

% fully immunized (1995-97)

TB	99
DPT	95
Polio	96
Measles	95
Prevalence of child malnutrition under age 5 (1992-97)[b]	NA

Ethnic Division (6)

English	81.5%
Scottish	9.6%
Irish	2.4%
Welsh	1.9%
Ulster	1.8%
West Indian, Indian, Pakistani, and other	2.8%

Religions (7)

Anglican	27,000,000
Roman Catholic	9,000,000
Muslim	1 million
Presbyterian	800,000
Methodist	760,000
Sikh	400,000
Hindu	350,000
Jewish	300,000

UK does not include a question on religion in its census. Data 1991 estimate.

Languages (8)

English, Welsh (about 26% of the population of Wales), Scottish form of Gaelic (about 60,000 in Scotland).

EDUCATION

Public Education Expenditures (9)

Public expenditure on education (% of GNP)

1980	5.6
1996	5.4[1]

Expenditure per student

Primary % of GNP per capita

1980	16.0
1996	18.8[1]

Secondary % of GNP per capita

1980	22.2
1996	20.6[1]

Tertiary % of GNP per capita

1980	79.9
1996	40.9[1]

Expenditure on teaching materials

Primary % of total for level (1996)	2.9
Secondary % of total for level (1996)	....
Primary pupil-teacher ratio per teacher (1996)	18[1]
Duration of primary education years (1995)	6

GOVERNMENT & LAW

Political Parties (12)

House of Commons—	% of seats
Labor	44.5
Conservative	31
Liberal Democratic	17
Other	7.5

Government Budget (13A)

Year: 1995

Total Expenditures: 292,652 Millions of Pounds

Expenditures as a percentage of the total by function:

General public services and public order	7.10
Defense	7.91
Education	4.94
Health	13.96

Continued on next page.

Military Affairs (14B)

	1990	1991	1992	1993	1994	1995
Military expenditures						
Current dollars (mil.)	37,090	39,620	36,580	36,050	35,440	33,400
1995 constant dollars (mil.)	42,630	43,780	39,340	37,790	36,330	33,400
Armed forces (000)	308	301	293	271	257	233
Gross national product (GNP)						
Current dollars (bil.)	902,600	919,900	950,300	996,200	1,058,000	1,110,000[e]
1995 constant dollars (bil.)	1,037,000	1,016,000	1,022,000	1,044,000	1,084,000	1,110,000[e]
Central government expenditures (CGE)						
1995 constant dollars (mil.)	382,100	397,800	423,800	435,000	438,200	462,100
People (mil.)	57.4	57.8	58.0	58.1	58.2	58.4
Military expenditure as % of GNP	4.1	4.3	3.8	3.6	3.4	3.0
Military expenditure as % of CGE	11.2	11.0	9.3	8.7	8.3	7.2
Military expenditure per capita (1995 $)	742	757	679	650	624	572
Armed forces per 1,000 people (soldiers)	5.4	5.2	5.1	4.7	4.4	4.0
GNP per capita (1995 $)	18,070	17,580	17,640	17,980	18,620	19,020
Arms imports[6]						
Current dollars (mil.)	1,100	750	230	320	210	190
1995 constant dollars (mil.)	1,264	829	247	335	215	190
Arms exports[6]						
Current dollars (mil.)	4,600	4,900	4,700	4,600	5,100	5,200
1995 constant dollars (mil.)	5,287	5,415	5,056	4,823	5,228	5,200
Total imports[7]						
Current dollars (mil.)	223,000	209,900	221,600	205,400	227,000	263,700
1995 constant dollars (mil.)	256,300	232,000	238,300	215,300	232,700	263,700
Total exports[7]						
Current dollars (mil.)	185,200	185,000	190,000	180,200	204,900	242,000
1995 constant dollars (mil.)	212,800	204,400	204,400	188,900	210,100	242,000
Arms as percent of total imports[8]	.5	.4	.1	.2	.1	.1
Arms as percent of total exports[8]	2.5	2.6	2.5	2.6	2.5	2.1

GOVERNMENT & LAW

Government Budget (13A) cont.

Social Security and Welfare	31.12
Housing and community amenities	1.77
Recreational, cultural, and religious affairs	.48
Fuel and energy	.76
Agriculture, forestry, fishing, and hunting	1.17
Mining, manufacturing, and construction	.46
Transportation and communication	1.67
Other economic affairs and services	1.87

Crime (15)

Crime rate (for 1997)	
Crimes reported	4,460,600
Total persons convicted	1,249,000
Crimes per 100,000 population	8,600
Persons responsible for offenses	
Total number of suspects	508,100
Total number of female suspects	86,400
Total number of juvenile suspects	121,900

LABOR FORCE

Labor Force (16)

Total (million)	28.2
Services	68.9%
Manufacturing, construction	17.5%
Government	11.3%
Energy	1.2%
Agriculture	1.1%

Data for 1997. Percent distribution for 1996.

Unemployment Rate (17)

5.5% (1997 est.)

PRODUCTION SECTOR

Electric Energy (18)

Capacity	66.149 million kW (1995)
Production	306.62 billion kWh (1995)
Consumption per capita	5,546 kWh (1995)

Transportation (19)

Highways:

total: 372,000 km

paved: 372,000 km (including 3,270 km of expressways)

unpaved: 0 km (1996 est.)

Waterways: 3,200 km under British Waterways Board

Pipelines: crude oil (almost all insignificant) 933 km; petroleum products 2,993 km; natural gas 12,800 km

Merchant marine:

total: 142 ships (1,000 GRT or over) totaling 2,192,956 GRT/2,224,715 DWT ships by type: bulk 5, cargo 26, chemical tanker 5, combination ore/oil 1, container 21, liquefied gas tanker 2, oil tanker 47, passenger 8, passenger-cargo 1, roll-on/roll-off cargo 13, short-sea passenger 12, specialized tanker 1 note: UK owns 337 additional ships (1,000 GRT or over) totaling 13,511,240 DWT that operate under the registries of Bermuda, The Bahamas, Cayman Islands, Cyprus, Hong Kong, Isle of Man, Liberia, Malta, Panama, Singapore, and Saint Vincent and the Grenadines (1997 est.)

Airports: 497 (1997 est.)

Airports—with paved runways:

total: 356

over 3,047 m: 10

2,438 to 3,047 m: 32

1,524 to 2,437 m: 170

914 to 1,523 m: 90

under 914 m: 54 (1997 est.)

Airports—with unpaved runways:

total: 141

1,524 to 2,437 m: 1

914 to 1,523 m: 24

under 914 m: 116 (1997 est.)

Top Agricultural Products (20)

Cereals, oilseed, potatoes, vegetables; cattle, sheep, poultry; fish.

MANUFACTURING SECTOR

GDP & Manufacturing Summary (21)

Detailed value added figures are listed by both International Standard Industry Code (ISIC) and product title.

	1980	1985	1990	1994
GDP				
($-1990 mil.)[1]	760,010	839,360	981,046	1,013,818
Per capita				
($-1990)[1]	13,492	14,825	17,088	17,452
Manufacturing share (%)				
(current prices)[1]	25.9	24.3	22.5	*21.2*
Manufacturing				
Value added				
($-1990 mil.)[1]	190,836	187,499	202,486	206,105

	1980	1985	1990	1994
Industrial production index	100	103	112	111
Value added ($ mil.)	163,790	124,384	254,946	*243,653*
Gross output ($ mil.)	400,930	306,225	579,854	569,891
Employment (000)	6,462	4,935	4,798	*4,169*
Profitability (% of gross output)				
Intermediate input (%)	59	59	56	57
Wages and salaries inc. supplements (%)	23	20	21	*20*
Gross operating surplus	17	20	23	*22*
Productivity ($)				
Gross output per worker	61,483	61,368	119,558	*132,670*
Value added per worker	25,117	24,927	52,653	*57,635*
Average wage (inc. supplements)	14,579	12,520	25,249	*27,894*
Value added ($ mil.)				
311/2 Food products	14,744	12,192	25,143	*25,679*
313 Beverages	5,419	3,554	6,643	*6,682*
314 Tobacco products	1,814	1,479	2,375	*2,459*
321 Textiles	5,419	3,917	7,036	*6,459*
322 Wearing apparel	3,395	2,633	4,679	*4,341*
323 Leather and fur products	558	376	536	*475*
324 Footwear	1,093	752	1,268	*1,086*
331 Wood and wood products	2,349	1,556	3,214	*2,865*
332 Furniture and fixtures	2,558	2,101	4,554	*4,384*
341 Paper and paper products	4,860	3,813	8,036	*8,299*
342 Printing and publishing	9,814	8,807	19,643	*21,385*
351 Industrial chemicals	8,233	7,328	14,179	*13,996*
352 Other chemical products	7,512	6,641	14,893	*17,103*
353 Petroleum refineries	4,512	1,712	4,429	*2,702*
354 Miscellaneous petroleum and coal products	721	428	750	*685*
355 Rubber products	2,349	1,505	3,018	*2,714*
356 Plastic products	3,698	3,087	8,250	*9,449*
361 Pottery, china and earthenware	977	765	1,464	*1,293*
362 Glass and glass products	1,422	960	2,089	*1,988*
369 Other non-metal mineral products	5,698	4,202	9,036	*7,076*
371 Iron and steel	5,860	4,345	8,089	*6,641*
372 Non-ferrous metals	2,581	1,505	2,786	*2,477*
381 Metal products	10,140	7,211	15,018	*13,063*
382 Non-electrical machinery	21,326	15,097	30,071	*27,593*
383 Electrical machinery	15,209	12,387	22,357	*20,454*
384 Transport equipment	17,512	12,931	28,946	*25,401*
385 Professional and scientific equipment	2,209	1,803	3,661	*3,928*
390 Other manufacturing industries	1,791	1,297	2,786	*2,884*

FINANCE, ECONOMICS, & TRADE

Economic Indicators (22)

National product: GDP—purchasing power parity—$1.242 trillion (1997 est.)

National product real growth rate: 3.5% (1997 est.)

National product per capita: $21,200 (1997 est.)

Inflation rate—consumer price index: 3.1% (1997)

Balance of Payments (23)

	1992	1993	1994	1995	1996
Exports of goods (f.o.b.)	188,450	182,060	206,450	241,530	261,870
Imports of goods (f.o.b.)	−211,880	−202,300	−223,410	−259,840	−281,480
Trade balance	−23,430	−20,240	−16,950	−18,310	−19,610
Services - debits	−168,500	−157,970	−161,400	−194,290	−203,730
Services - credits	182,670	170,100	183,550	217,600	227,850
Private transfers (net)	−8,620	−7,360	−7,710	−11,180	−7,870
Government transfers (net)	−470	−60	170	320	470
Overall balance	−18,350	−15,520	−2,340	−5,860	−2,900

Exchange Rates (24)

Exchange rates:	
British pounds (£) per US$1	
January 1998	0.6115
1997	0.6106
1996	0.6403
1995	0.6335
1994	0.6529
1993	0.6658

Top Import Origins (25)

$283.5 billion (f.o.b., 1997) Data are for 1996.

Origins	%
European Union	50.2
Germany	14.2
France	9.0
Netherlands	6.5
United States	13.9

Top Export Destinations (26)

$268 billion (f.o.b., 1997) Data are for 1996.

Destinations	%
European Union	53.2
Germany	12.4
France	9.9
Netherlands	7.8
United States	11.4

Economic Aid (27)

Donor: ODA, $2.908 billion (1993).

Import Export Commodities (28)

Import Commodities	Export Commodities
Manufactured goods	Manufactured goods
Machinery	Machinery
Semifinished goods	Fuels
Foodstuffs	Chemicals
Consumer goods 13.9%	Semifinished goods
	Transport equipment 11.4%

UNITED STATES

United States of America

CAPITAL: Washington, D.C. (District of Columbia).

FLAG: The flag consists of 13 alternate stripes, 7 red and 6 white; these represent the 13 original colonies. Fifty 5-pointed white stars, representing the present number of states in the Union, are placed in 9 horizontal rows.

ANTHEM: *The Star-Spangled Banner.*

MONETARY UNIT: The dollar ($) of 100 cents is a paper currency with a floating rate. There are coins of 1, 5, 10, 25, and 50 cents and 1 dollar, and notes of 1, 2, 5, 10, 20, 50, and 100 dollars.

WEIGHTS AND MEASURES: The imperial system is in common use; however, the use of metrics in industry is increasing, and the metric system is taught in public schools throughout the United States.

HOLIDAYS: New Year's Day, 1 January; Birthday of Martin Luther King, Jr., 3d Monday in January; Presidents' Day, 3d Monday in February; Memorial or Decoration Day, last Monday in May; Independence Day, 4 July; Labor Day, 1st Monday in September; Columbus Day, 2d Monday in October; Election Day, 1st Tuesday after the 1st Monday in November; Veterans or Armistice Day, 11 November; Thanksgiving Day, 4th Thursday in November; Christmas, 25 December.

TIME: Eastern, 7 AM = noon GMT; Central, 6 AM = noon GMT; Mountain, 5 AM = noon GMT; Pacific (includes the Alaska panhandle), 4 AM = noon GMT; Yukon, 3 AM = noon GMT; Alaska and Hawaii, 2 AM = noon GMT; western Alaska, 1 AM = noon GMT.

LOCATION AND SIZE: Located in the Western Hemisphere on the continent of North America, the United States is the fourth-largest country in the world. Its total area, including Alaska and Hawaii, is 9.4 million square kilometers (3.6 million square miles). The continental United States has a total boundary length of 17,563 kilometers (10,913 miles).

CLIMATE: The climate varies from one region of the country to another, from clear contrasts between seasons to moderate differences. Temperatures may average as low as −13°C (9°F) or as high as 43°C (110°F) in different areas of the country.

INTRODUCTORY SURVEY

RECENT HISTORY

During the 20th century, the United States became the world's most powerful nation. In the 1970s President Jimmy Carter negotiated treaties ending U.S. sovereignty over the Panama Canal Zone and in September 1978 mediated the Camp David peace agreement between Israel and Egypt. But an economic recession and a prolonged quarrel with Iran over more than 50 U.S. hostages contributed to his defeat by Ronald Reagan in the 1980 presidential election.

President Reagan used his personal popularity to pass income tax cuts and significantly increase the military budget between 1980 and 1989. With a ballooning national debt, Reagan cut domestic spending. Reagan' decisions to intervene in internal Central American conflicts were controversial, but his administration's foreign policy is often credited with hastening the fall of the former Soviet Union.

Reagan's successor, George Bush, used his personal relationships with foreign leaders to bring about peace talks between Israel and its Arab neighbors, to encourage a peaceful unification of Germany, and to negotiate significant arms reductions with the Russians. Bush sent 400,000 American soldiers to lead a multinational coalition opposing Iraq's invasion of Kuwait in 1990. At home, the collapse of the savings and loan industry in the late 1980s forced the Bush administration to rescue the failed savings and Loans, costing taxpayers over $100 billion.

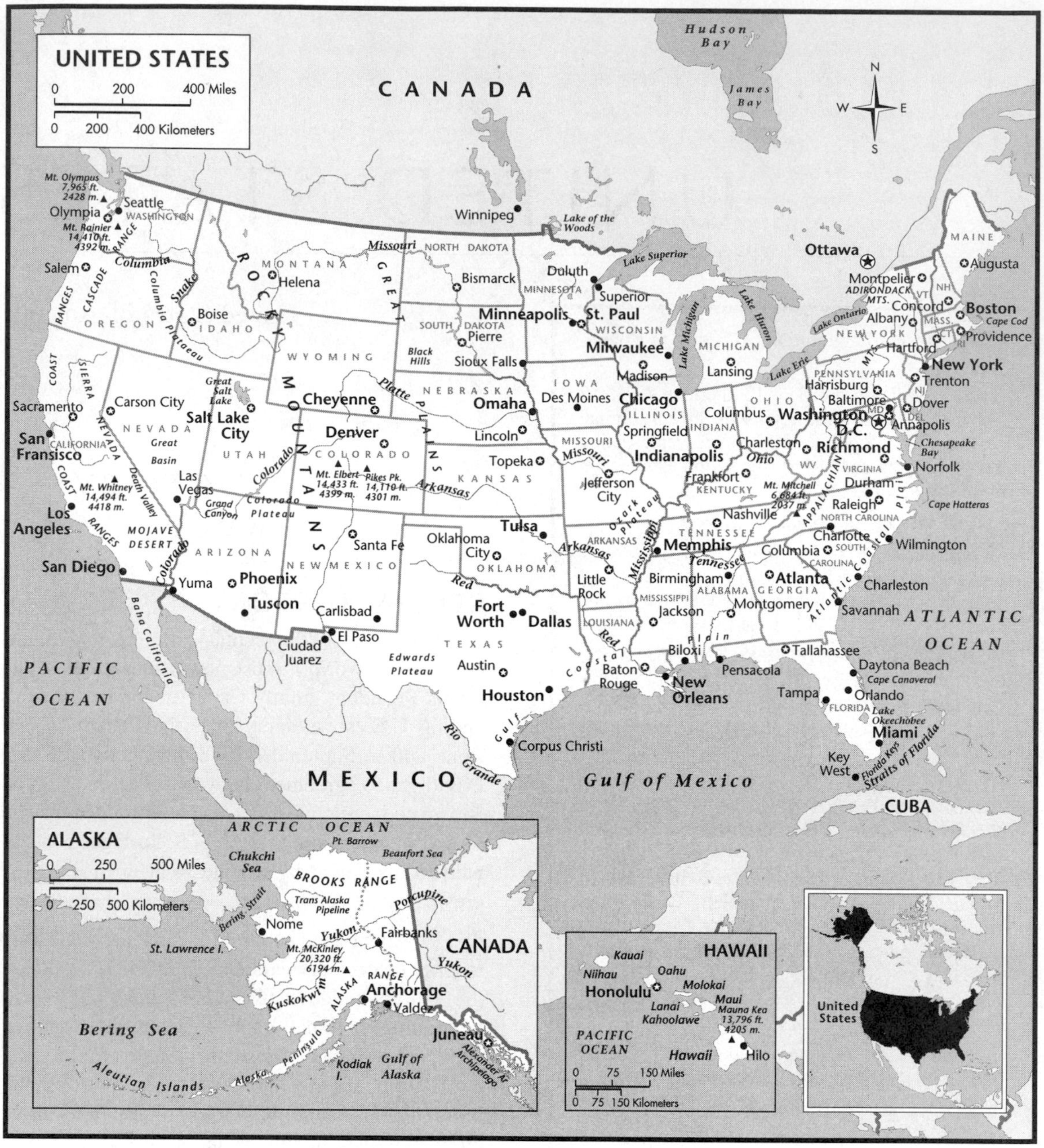

Since being elected president in 1992, Bill Clinton has presided over a period of almost unprecedented economic prosperity. Among his major achievements are the 1993 budget which helped set the economy on its prosperous course; the North American Free Trade Agreement (NAFTA); welfare reform, and deficit reduction legislation.

With the country enjoying prosperity, Clinton was reelected in 1996, becoming the first Democratic president elected to a second term since Franklin D. Roosevelt. Republicans, however, retained control of both houses of Congress.

In 1998, Clinton was impeached after a long investigation by an independent counsel. Although he survived in office, his ability to carry out major policy initiatives was severely compromised.

GOVERNMENT

The Constitution of the United States (1787) is the nation's governing document. The first ten amendments to the Constitution, ratified in 1791 and known as the Bill of Rights, guarantee certain important rights to all American citizens. Later amendments include the 13th Amendment (1865), banning slavery, and the 19th (1920), which gave women the right to vote. Suffrage is universal beginning at age 18.

The United States has a federal form of government, with a constitutionally-defined distribution of powers between federal and state governments. The legislative powers of the federal government rest in Congress, which consists of the 435-member House of Representatives and a 100-member Senate. Each state is apportioned representatives by population. Representatives serve two-year terms. The Senate consists of two senators from each state, elected for six-year terms. One-third of the Senate is elected in every even-numbered year.

The president, as the nation's chief executive, proposes and carries out legislation and with the approval of the Senate, appoints ambassadors, consuls, and all federal judges, including the justices of the Supreme Court. As commander in chief, the president is responsible for the nation's defense, but the power to declare war belongs to Congress. The president conducts foreign relations and makes treaties with the advice and consent of the Senate. The U.S. system of governance is known as "checks and balances."

The president appoints a cabinet, subject to Senate confirmation, consisting of the secretaries who head the departments of the executive branch.

Each of the 50 states is divided into counties and municipalities There are more than 3,000 counties in the United States and more than 19,000 municipalities, including cities, villages, towns, and boroughs.

Judiciary

The Supreme Court, the nation's highest judicial body, consists of the chief justice of the United States and eight associate justices. All justices are appointed for life by the president with the approval of the Senate.

The Supreme Court acts as an appeals court for federal district courts, circuit courts of appeals, and the highest courts in the states. The Supreme Court also exercises the power of judicial review, determining the constitutionality of any state laws, state constitutions, congressional statutes, and federal regulations that are specifically challenged.

Most federal cases are handled by district courts, whose decisions may be carried to the courts of appeals, organized into 13 circuits.

State courts operate independently of the federal judiciary. Most states have a court system that includes courts of general trial jurisdiction and appeals courts. At the highest level of the system is a state supreme or superior court.

Political Parties

Two major parties, Democratic and Republican, have dominated national, state, and local politics since 1860. Independent candidates have won state and local offices, but no candidate has won the presidency without major party backing.

DEFENSE

The United States armed forces in 1996 numbered 1.5 million men and women on active duty and 1.7 million in the reserve. Service in the armed forces has been voluntary since 1973. However, all male citizens must register for the draft at age 18.

By the late 1980s, defense spending was around $300 billion a year and had increased roughly 30% over the decade. By 1997, military spending accounted for $267.2 billion or 3.4% of the nation's GDP.

ECONOMIC AFFAIRS

The United States enjoys a vast variety and quantity of natural resources, but it depends on foreign sources for many raw materials. American dependence on oil imports was dramatically demonstrated during the 1973 OPEC oil embargo, when serious fuel shortages developed in many sections of the country.

The capitalistic American economy in the 1990s is marked by steady growth, low unemployment, low inflation, and rapid advances in technology. The U.S. is the leading industrial power in the world and is highly diversified and technologically advanced. Major industries include: petroleum, steel, motor vehicles, aerospace, telecommunications, chemicals, electronics, food processing, consumer goods, lumber, and mining.

By mid-1997, the recovery that had begun in March 1991 was the third-longest period of economic expansion since World War II.

Public Finance

The President is responsible for preparing the annual federal budget for submission to Congress.

Congress must amend and pass the budget. The fiscal year runs from 1 October to 30 September. Public debt, subject to a statutory debt limit, rose from $43 billion in 1939/40 to more than $3.3 trillion in 1993, but by the end of the decade had been virtually erased.

The U.S. Central Intelligence Agency estimates that, in 1998, U.S. government revenues totaled $1.722 trillion: expenditures were $1.653 trillion.

Income

The gross national product (GNP) in 1998 was about $4.5 trillion. According to the Central Intelligence Agency, the nation's per capita GDP of $31,500 was the largest among major industrial nations.

Industry

Manufacturing accounts for 23% of the nation's GDP. In recent years, high technology industries, led by Microsoft Corp. and other software makers have revolutionized the manufacturing sector.

Large corporations are dominant especially in areas such as steel, automobiles, pharmaceuticals, aircraft, petroleum refining, computers, soaps and detergents, tires, and communications equipment.

Automobile manufacturing struggled in the 1980s, but has rebounded in the 1990s. Passenger car production, which had fallen from 7.1 million in 1987 to 5.4 million in 1991, rose to 6.3 million in 1995.

Banking and Finance

The Federal Reserve Board regulates the money supply and the amount of credit available to the public by adjusting the rediscount rate, buying and selling securities in the open market, setting margin requirements for securities purchases, and altering reserve requirements of member banks in the system. The activities of the board directly affect the interest rate.

Passage of the Financial Services Act of 1999, repealing the Glass-Steagall Act of 1933, which separated commercial and banking activities heralds major changes in the nation's banking system. In 1994, another important piece of legislation removed most barriers to interstate bank acquisitions and interstate banking.

All members of the Federal Reserve System (and other banks that wish to do so) participate in a plan of deposit insurance administered by the Federal Deposit Insurance Corporation (FDIC). Savings and loan associations are insured by the Federal Savings and Loan Insurance Corporation (FSLIC).

By 1996, 51 million individuals and 10,000 institutional investors owned stocks or shares in mutual funds traded on the New York Stock Exchange (NYSE). The two other major U.S. stock markets are the American Stock Exchange (AMEX) and the NASDAQ (National Association of Securities Dealers). From 1995 to 1997, a strong U.S. economy fueled an overall gain in the stock market of 60%.

Economic Development

Recent Republican and Democratic administrations have intervened in various ways in the economy. During the Great Depression the Securities and Exchange Commission and the National Labor Relations Board were established; expansion of regulatory programs accelerated during the 1960s and early 1970s with the creation of the federal Environmental Protection Agency, Equal Employment Opportunity Commission, Occupational Safety and Health Administration, and Consumer Product Safety Commission, among other agencies.

Meanwhile, federal price supports and production subsidies acted to stabilize U.S. agriculture. The federal government stepped in to arrange for guaranteed loans for two large private firms—Lockheed in 1971 and Chrysler in 1980—where thousands of jobs would have been lost in the event of bankruptcy.

The Carter administration deregulated the airline, trucking, and communications industries; subsequently, the Reagan administration relaxed government regulation of bank savings accounts and automobile manufacture as it decontrolled oil and gas prices.

In 1993 Congress approved North American Free Trade Agreement (NAFTA) which extended the Free Trade Agreement between Canada and the United States to include Mexico. NAFTA, by eliminating tariffs and trade barriers, created a free trade zone with a combined market size of $6.5 trillion and 370 million consumers.

SOCIAL WELFARE

Both federal and state governments fund and administer social welfare programs. Old age, survivors, disability, and Medicare (health insurance for seniors) programs are administered by the federal government. A food stamp program, school lunch and breakfast programs, and nutrition programs for the elderly are administered by the Department of Agriculture. Unemployment insurance, dependent child care, and a variety of other public assistance programs are state-administered.

In 1995, nearly 13.7 million Americans received $22 billion under the Aid to Families with Dependent Children (AFDC) program. In 1996, this entitlement program was replaced with a more limited system of assistance funded through grants to states and communities. The new program provides for $16.4 billion in block grants each year through 2002.

In 1995 monthly Social Security benefits totaled $28.1 billion, paid to nearly 43.4 million beneficiaries. Medicare provided $191.2 billion in hospital and medical insurance benefits to some 40 million older Americans in 1996. Medicaid, a program that helps the needy meet the costs of medical, hospital, and nursing home care, disbursed an average $3,311 to 33.3 million recipients in 1995.

Healthcare

The United States health care system is among the most advanced in the world. During the mid-1990s, there was an average of 2.5 doctors per 1,000 people.

Life expectancy in 1998 was 76.1 years. Leading causes of death in 1995 were heart disease (31.9%) and cancer (23.3%).

First identified in 1981, HIV infection and AIDS (acquired immune deficiency syndrome) spread rapidly. By 1995, 513,476 AIDS cases had been reported in the United States. In 1996, 66,885 new cases of AIDS were reported.

Housing

The housing resources of the United States far exceed those of any other country, with 102.2 million housing units as of April 1990, 91.9 million of which were occupied. There were 1.4 million housing starts in 1996. Most dwellings are one-family houses.

EDUCATION

The U.S. literacy rate is estimated at 97%. Education is compulsory in all states and is a responsibility of each state and local government. Primary and secondary school years typically encompass ages 6 to 17.

Government spending (state and federal) on all educational institutions was $529.6 billion in 1995–96.

In 1995 there were 3,688 two-year and four-year colleges and universities in the U.S.

1999 KEY EVENTS TIMELINE

January

- The Senate begins the impeachment trial of President William (Bill) Jefferson Clinton. He is charged with perjury and obstruction of justice.
- Speculation mounts that Hillary Clinton will run for the New York U.S. Senate seat, to be vacated by retiring senator Daniel Patrick Moynihan.
- The U.S. Supreme Court rules that sampling can not be used to apportion seats in the House of Representatives. Proponents of sampling had argued that the current ''head-count'' method undercounts minorities and the urban poor while double-counting people who own second homes.

February

- President Clinton is acquitted on both counts in his impeachment trial.
- The Washington Monument, closed to visitors since August 1998 for renovation, reopens. The monument is encased in a special scaffolding designed by architect Michael Graves. Most of the $9.4 million required to complete the renovations is being supplied by Target Stores and other corporations.

March

- The state of California and the federal government pay Pacific Lumber Company $480 million for about 10,000 acres of old-growth woodlands, thereby saving one of the world's last groves of privately owned old-growth redwood trees. Some of the trees are more than 2000 years old.
- The United States leads NATO air forces against Serbia in the alliance's first war. The campaign is mounted to stop Serbian ''ethnic cleansing'' in the Kosovo province.
- The Bureau of Justice Statistics reports that the nation's prison and jail population reached 1.8 million in 1978, a 4.4 percent increase over the previous year. The number of persons in jail is

nearly double that of 1985, with black men representing almost half of those incarcerated.

- A Michigan jury finds Dr. Jack Kevorkian guilty of murder in the death of a man suffering from Lou Gehrig's disease, whom Kevorkian helped to commit suicide.

April

- President Clinton is found in contempt of court and fined for giving false testimony in a civil suit deposition.
- Two senior male students from Columbine High School in Littleton, Colorado carry out a planed attack on their school, killing one teacher and 12 students, and wounding many others. The two kill themselves before they can be apprehended. The incident reignites the national discussion on gun control and raises questions on the impact of America's culture of violence on its youth.

May

- A Congressional Committee chaired by Rep. Christopher Cox (R-CA) exposes Chinese espionage at American nuclear laboratories.
- Powerful tornadoes ravage Oklahoma, killing 44 and injuring 700.
- In a bombing run over Belgrade, Serbia, U.S. warplanes accidentally strike the Chinese embassy.
- Secretary of the Treasury Robert Rubin announces his resignation.

June

- The federal government reports that legalized gambling now generates more than $50 billion in gross revenues per year. Americans now spend more on gambling than they do on theme parks, video games, spectator sports, and movie tickets combined.
- The Kosovo war comes to a close. NATO peacekeeping forces enter Kosovo.
- 78,972 fans watch the U.S. women's soccer team play Denmark in the opening game of the 1999 Women's World Cup. It is the largest crowed ever to attend a women-only sporting event.

July

- John F. Kennedy, Jr., his wife, and sister-in-law are killed when the private plane goes down in the Atlantic Ocean off the coast of Massachusetts.
- Larry Summers replaces Robert Rubin as Secretary of the Treasury.

August

- Texas governor and presidential candidate George W. Bush wins the Iowa Republican straw poll.
- Attorney General Janet Reno admits that FBI agents used pyrotechnic tear gas in their assault on the Branch Davidian compound in Waco, Texas.

September

- A study by the Center on Budget and Policy Priorities reports that wealth is now more concentrated among the top 1% of American households than at any time since the Depression.
- President Clinton vetoes the budget bill drafted and passed by congressional Republicans.

October

- The Senate rejects the Clinton administration's Comprehensive Test Ban Treaty, dealing a blow to arms-control efforts worldwide.
- An EgyptAir Boeing 767 crashes off the US, killing all passengers and crew aboard.

November

- Federal Judge Thomas Penfield Jackson calls software giant Microsoft a monopoly in his 207-page ''findings of fact'' issued in the U.S. Justice Department's antitrust suit against the company. Although not a final ruling, the findings substantially strengthen the government's position and put pressure on Microsoft to settle the case.
- China and the United States sign a trade deal, paving the way for China's entry into the World Trade Organisation.
- The United Nations strikes a deal with the US, requiring the U.S. to pay at least $350 million of the $600 million it owes in back dues by the end of 1999, or it will lose its seat in the General Assembly.
- Paul Bowles, composer and writer known as ''the only American existentialist,'' dies at age 88.
- First Lady Hillary Rodham-Clinton formally announces that she will run for New York's vacant senate seat next year.
- President Clinton embarks on a nine-day tour through south-eastern Europe.

December

- As many as 1,000 groups and 100,000 demonstrators, including trade unionists, environmentalists, aid lobbyists, consumer-rights campaigners, and human rights activists, rally through the streets of Seattle to protest the World Trade Organization summit talks, focusing on globalization. The opening of the summit is delayed as the protests turn violent; police in full riot gear with armored personnel carriers fire tear gas and rubber bullets at protesters who refuse to move. Seattle is put under curfew as the mayor declares a civil emergency, and the governor calls in the National Guard.
- NASA launches the Mars Polar Lander, a probe scheduled to touch down near the Martian south pole; however, the excursion is a failure, the fourth in a string of recent NASA mishaps, as the probe is lost in space.
- Seattle police chief Norm Stamper resigns after the protests that marred the World Trade Organization conference.

ANALYSIS OF EVENTS: 1999

BUSINESS AND THE ECONOMY

The longest period of American economic expansion continued into 1999 although many economic forecasters saw signs of a pending economic slowdown. The year saw the departure of Treasury Secretary Robert Rubin, the former Goldman-Sachs partner whose policies as treasury secretary are widely credited with helping foster the unprecedented growth. His replacement, academician Larry Summers, has pledged to continue the same policies. For many investors Federal Reserve Chairman Alan Greenspan and his continued vigilance against inflation remains a source of stability. His emphasis on a tight-money policy manifested itself in increases in the prime lending rate designed to prevent inflation.

Perhaps the biggest news was the U.S. budget surplus, the first time the United States budget operated in the black since the days of Lyndon Johnson. During the summer President Clinton announced his intention to eliminate the entire national debt within fifteen years. Most economists and financial experts met the news with hearty approval, although many were also left trying to imagine a future United States Government that did not issue bonds to finance its debts.

Naysayers, however, were quick to point to several adverse signs as the year progressed. Although the Dow Jones Industrial Average soared to record highs over 11,000, by early fall the Dow had slipped to below 10,500 and the markets remained sluggish. The U.S. trade deficit also widened, largely in response to the ''dumping'' (pricing goods below the cost of production) of imports such as steel, primarily from Asia. This continued practice also brought renewed calls by congressional Democrats as well as some Republicans for tighter controls on imports from countries engaging in unfair trade practices.

GOVERNMENT AND POLITICS

The battle between the Republican-controlled Congress and Democratic President Bill Clinton dominated the political scene in the United States throughout most of 1999. Also in the headlines were the U.S.-led NATO air campaign against Serbia, an investigation of Chinese espionage in American nuclear facilities, and struggles between the White House and Congress over the budget and foreign policy. Finally, as the year drew to a close, attention shifted toward the Democratic and Republican presidential hopefuls for 2000.

The year began with the Senate impeachment trial of President Clinton. In December 1998, the House impeached the president on charges of perjury and obstruction of justice relating to his ''inappropriate relationship'' with a former intern, Monica Lewinsky. Following a one-month trial, the Senate acquitted the president on both counts, unable to achieve a two-thirds vote needed to convict and remove him from office. Polls showed the impeachment to be highly unpopular with the American public and Clinton attempted to capitalize on the anti-congressional Republican mood by calling for an overhaul in Social Security, Medicare, and Medicaid in his State of the Union address (given while his Senate trial was ongoing). But the president had spent much-needed political capital on the impeachment fight and he struggled with his legislative agenda throughout the year. As a whole, 1999 proved to be a mixed bag for Clinton in both domestic and foreign policy.

Even though he was acquitted, the president still had trouble mustering the political or moral clout necessary to push his legislative initiatives.

Although subsequently the president got enough moderate House Republicans to join Democrats in support of a Patients' Bill of Rights, the Republicans successfully opposed the president's plan to use the projected budget surplus to fund increased social spending. Instead, the Republican congressional leadership proposed and passed a $700 million tax-cut as opposed to Clinton's proposed $300 million tax cut. When the president vetoed the bill, Congress resorted to passing a continuing resolution that maintained spending limits at the previous year's level until a new budget, mutually acceptable to both congress and the president, could be passed.

Congressional opposition to the White House was even more effective in the field of foreign affairs. Although in March the President found enough backing from moderate Republicans to support NATO air strikes against Serbia, the administration faced relentless and scathing criticism from those who argued that the U.S. was moving too timidly in refusing to consider a ground invasion of Serbia from the outset. Republicans continued to oppose the White House on back payment of UN dues. They demanded that a family planning provision in the organization's budget be eliminated. Failure to pay the dues would cost the United States its vote in the General Assembly beginning in 2000. Finally, in October congress dealt Clinton his worst foreign policy defeat by rejecting the Comprehensive Test Ban Treaty (CTBT). CTBT supporters could not even muster half of the Senate—much less than the two-thirds required—to ratify the treaty. It was the first time the Senate rejected a treaty since the defeat of the Treaty of Versailles in 1920 following World War I.

As the year came to a close, another contentious issue arose at the meeting of the World Trade Organization in Seattle, Washington, in late November and early December. Unexpectedly, the meeting became a focus of a massive demonstration of thousands of critics of "globalization." The protesters, who were steered away from the convention center by the police, ended up confronting the authorities in numerous street clashes. Most of the demonstrators engaged in passive resistance and non-violent obstruction of traffic. Some turned to breaking windows, however, and looting stores. Initially reluctant to crack down on the demonstrators, the Seattle police, the Washington State police, and the National Guard eventually resorted to tear gas, police truncheons and plastic bullets to disperse the protesters. The Seattle mayor imposed a nighttime curfew. Three hundred state police and 200 National Guard troops patrolled the streets.

The demonstrators did manage to make their concerns public as newspaper reporters interviewed the frequently articulate protesters. For the preceding decade the effects of the drive to establish larger free trade zones had been a growing concern to both the advocates of labor and to environmental activists. Earlier in the decade, the passage of the North American Free Trade Agreement (NAFTA) in 1993 had encountered opposition from trade unions. The representatives of labor feared that free trade agreements would lure U.S. companies to shut their domestic manufacturing operations in favor of investing in Mexico or other low wage havens. The environmentalists pointed to the damage to natural habitat that industrialization frequently brings.

The year 1999 seemed destined to go down as a year of scandals, and not just in the streets of Seattle or the presidential boudoir. Other allegations of scandal rocked Washington. The Federal Bureau of Investigation (FBI) and the Department of Justice clashed over the findings of a congressional investigation into the 1993 FBI raid of the Branch Davidian compound in Waco, Texas. The Clinton administration was also shaken by the exposure of Chinese espionage at U.S. nuclear research laboratories. A committee led by Rep. Christopher Cox (R-CA) exposed over twenty years of Chinese spying that provided China with the technology needed to manufacture multiple independently targeted reentry vehicle (MIRV) warheads, thereby allowing a single missile to strike more than one target.

Finally, 1999 marked the official beginning of the 2000 presidential campaigns. On the Democratic side, the early favorite, Vice President Al Gore found an unexpectedly strong challenge from former New Jersey Senator Bill Bradley. The Republican field was more crowded but dominated by a single candidate, Texas Governor George W. Bush (son of former president George H. W. Bush) who staked out an early lead and set fundraising records, amassing over $50 million in his coffers before the year was out. His strongest challenges came from Arizona Senator John McCain and billionaire Steve Forbes. The entry of former Republican commentator Pat Buchanan into the race for the Reform Party nomination promised to add spice to the primary campaign. In addition to the presidential races, another race making the headlines was the New York U.S. Senate seat of retiring Democrat Daniel Patrick Moynihan. First Lady Hillary Rodham Clinton declared she will run as the Democratic nominee against the Republican mayor of New York City, Rudolph Giuliani, in a race the pundits are predicting will be a battle royale.

CULTURE AND SOCIETY

The issue of guns in American society came to the forefront of national discussion in May when two disaffected high school seniors in Littleton, Colorado, embarked on a shooting spree and killed twelve of their fellow classmates before killing themselves. This tragedy, along with the subsequent shootings by a lone gunman in an Atlanta brokerage, elicited renewed public clamor in favor of tougher gun-control laws, particularly increased safety devices that would permit only the owner to fire the weapon. The National Rifle Association (NRA) under the leadership of its president, actor Charlton Heston, prepared to oppose any further restrictions on firearms. For the short term at least, the gun-control advocates appeared to have the upper hand with lawmakers, including some conservatives traditionally opposed to gun-control.

In sports, the Denver Broncos rolled to victory in Super Bowl XXXIII; the San Antonio Spurs took the NBA title; and the New York Yankees won their twenty-fifth World Series. Allegations of unethical lobbying rocked the Olympic world when an investigation revealed members of the Atlanta and Salt Lake City bid groups donated lavish gifts to members of the International Olympic Committee (IOC) in order to sway Olympic site selection votes.

DIRECTORY

CENTRAL GOVERNMENT

Head of State

President
William J. Clinton, Office of the President, 1600 Pennsylvania Ave., Washington, D.C. 20500, United States
PHONE: +(202) 4561414
FAX: +(202) 4562461
E-MAIL: president@whitehouse.gov

Vice President
Albert A. Gore, Jr.

Cabinet

Secretary of Agriculture
Daniel Glickman, Department of Agriculture, 14th and Independence Ave. SW, Washington, D.C. 20250, United States
PHONE: +(202) 7202791

Secretary of Commerce
William M. Daley, Department of Commerce, 14th and Constitution Ave. NW, Room 5854, Washington, D.C. 20230, United States
PHONE: +(202) 4822000
FAX: +(202) 4822741
E-MAIL: WDaley@doc.gov

Secretary of Defense
William S. Cohen, Department of Defense, 1000 Defense Pentagon, Washington, D.C. 20301, United States
PHONE: +(703) 6975737

Secretary of Education
Richard W. Riley, Department of Education, 400 Maryland, SW, Washington, D.C. 20202, United States
PHONE: +(800) 8725327
FAX: +(202) 4010689
E-MAIL: CustomerService@inet.ed.gov

Secretary of Energy
William Richardson, Department of Energy, Forrestal Building, 1000 Independence, SW, Washington, D.C. 20585, United States
PHONE: +(202) 5865000
E-MAIL: the.secretary@hq.doe.gov

Secretary of Health and Human Services
Donna E. Shalala, Department of Health and Human Services, 200 Independence, SW, Washington, D.C. 20201, United States
PHONE: +(202) 6190257
E-MAIL: hhsmail@os.dhhs.gov

Secretary of Housing and Urban Development
Andrew M. Cuomo, Department of Housing and Urban Development, 451 7th St. SW, Washington, D.C. 20410, United States
PHONE: +(202) 4010388

Secretary of the Interior
Bruce Babbitt, Department of the Interior, 1849 C St., NW, Washington, D.C. 20240, United States
PHONE: +(202) 2083100
E-MAIL: Bruce_Babbitt@doi.gov

Attorney General
Janet Reno, Department of Justice, 950 Pennsylvania Ave., NW, Washington, D.C. 205300001, United States
PHONE: +(202) 5142007
FAX: +(202) 5145331
E-MAIL: web@usdoj.gov

Secretary of Labor
Alexis M. Herman, Department of Labor, 200 Constitution Ave., NW, Room S1032, Washington, D.C. 20210, United States
PHONE: +(202) 6934650

Secretary of State
Madeleine K. Albright, Department of State, 2201 C St., NW, Washington, D.C. 20520, United States
PHONE: +(202) 6474000
FAX: +(202) 7367720
E-MAIL: publicaffairs@panet.usstate.gov

Secretary of the Treasury
Lawrence H. Summers, Department of the Treasury, 1500 Pennsylvania Ave., NW, Washington, D.C. 20220, United States
PHONE: +(202) 6222000
FAX: +(202) 6226415
E-MAIL: OPCMail@do.treas.gov

Secretary of Transportation
Rodney E. Slater, Department of Transportation, 400 Seventh St., SW, Washington, D.C. 20590, United States
PHONE: +(202) 3664000

Secretary of Veterans Affairs
Togo D. West, Jr., Department of Veterans Affairs, Consumer Affairs Service (075B), 810 Vermont, NW, Washington, D.C. 20420, United States
PHONE: +(202) 2735771
FAX: +(202) 2735716
E-MAIL: consumeraffairs@mail.va.gov

POLITICAL ORGANIZATIONS

Democratic National Committee

430 S. Capitol St. SE, Washington, D.C. 20003, United States
PHONE: +(202) 8638000
TITLE: DNC National Chair
NAME: Joe Andrew

Republican National Committee

310 First St., SE Washington, D.C. 20003, United States
PHONE: +(202) 8638700
FAX: +(202) 8638820
E-MAIL: chairman@rnc.org
TITLE: Chairman
NAME: Jim Nicholson

Reform Party

P.O. Box 9, Dallas, Texas 75221, United States
PHONE: +(972) 4508800
FAX: +(972) 4508821
E-MAIL: russ.verney@reformparty.org
TITLE: Chairman
NAME: Russell Verney

The Greens/Green Party USA

P.O. Box 1134, Lawrence, MA 01842, United States
PHONE: +(978) 6824353
E-MAIL: gpusa@igc.org

Social Democrats, USA

815 15th St., NW, Suite 921, Washington, D.C. 20005, United States
TITLE: Chairman
NAME: Donald Slaiman

Socialist Party USA

339 Lafayette St., New York, NY 10012, United States
PHONE: +(212) 9824586
E-MAIL: socialistparty@spusa.org
TITLE: National CoChairs
NAME: Susan Dorazio, Don Doumakes

Communist Party USA

235 W. 23rd St., New York, NY 10011, United States
PHONE: +(212) 9894994
FAX: +(212) 2291713
E-MAIL: CPUSA@rednet.org
TITLE: National Chair
NAME: Gus Hall

Socialist Labor Party of America

P.O. Box 218, Mountain View, CA 940420218, United States
PHONE: +(650) 9388359
FAX: +(650) 9388392
E-MAIL: socialists@slp.org
TITLE: National Secretary
NAME: Robert Bills

The Labor Party

P.O. Box 53177, Washington, D.C. 20009, United States
PHONE: +(202) 2345190
FAX: +(202) 2345266
TITLE: Co-Chairpersons
NAME: Robert Clark, Kit Costello, and Baldemar Velasquez

Democratic Socialists of America

180 Varick St., 12th Floor, New York, NY 10014, United States
PHONE: +(212) 7278610
FAX: +(212) 7278616
E-MAIL: dsa@dsausa.org
TITLE: National Director
NAME: Horace Small

DIPLOMATIC REPRESENTATION

Embassies in the United States

Albania
2100 S St., NW, Washington D.C. 20008, United States
PHONE: +(202) 2234942
FAX: +(202) 6287342
TITLE: Ambassador Extraordinary and Plenipotentiary
NAME: Petrit Bushati

Algeria
2118 Kalorama Rd., NW, Washington, D.C. 20008, United States
PHONE: +(202) 2652800
FAX: +(202) 6672174
TITLE: Ambassador Extraordinary and Plenipotentiary
NAME: Ramtane Lamamra

Andorra
Two United Nations Plaza, 25th Floor, New York, NY 10017, United States
PHONE: +(212) 7508064
FAX: +(212) 7506630
TITLE: Ambassador Extraordinary and Plenipotentiary
NAME: Juli Minoves Triquell

Angola
1615 M St., NW, Suite 900, Washington, D.C. 20036, United States
PHONE: +(202) 7851156
FAX: +(202) 7851258
TITLE: Ambassador Extraordinary and Plenipotentiary
NAME: Antonio Dos Santos Franca

Antigua and Barbuda
3216 New Mexico, NW, Washington, D.C. 20016, United States
PHONE: +(202) 3625211
FAX: +(202) 3625225
TITLE: Ambassador Extraordinary and Plenipotentiary
NAME: Lionel A. Hurst

Argentina
1600 New Hampshire, NW, Washington, D.C. 20009, United States
PHONE: +(202) 2386400
FAX: +(202) 2386471
TITLE: Ambassador Extraordinary and Plenipotentiary
NAME: Diego Ramiro Gelar

Armenia
2225 R St., NW, Washington, D.C. 20008, United States
PHONE: +(202) 3191976
FAX: +(202) 3192982
TITLE: Ambassador Extraordinary and Plenipotentiary
NAME: Rouben Robert Shugarian

Australia
1601 Massachusetts Ave., NW, Washington, D.C. 20036, United States
PHONE: +(202) 7973000
FAX: +(202) 7973168
TITLE: Ambassador Extraordinary and Plenipotentiary
NAME: Andrew S. Peacock

Austria
3524 International Court, NW, Washington, D.C. 20008, United States
PHONE: +(202) 8956700
FAX: +(202) 8956750
TITLE: Ambassador Extraordinary and Plenipotentiary
NAME: Peter Moser

Azerbaijan
927 15th St., NW, Suite 700, Washington, D.C. 20005, United States
PHONE: +(202) 8420001
FAX: +(202) 8420004
TITLE: Ambassador Extraordinary and Plenipotentiary
NAME: Hafiz Mir Jalal Pashayev

Bahamas
2220 Massachusetts Ave., NW, Washington, D.C. 20008, United States
PHONE: +(202) 3192660
FAX: +(202) 3192668
TITLE: Ambassador Extraordinary and Plenipotentiary
NAME: Arlington Griffith Butler

Bahrain
3502 International Dr., NW, Washington, D.C. 20008, United States
PHONE: +(202) 3420741
FAX: +(202) 3622192
TITLE: Ambassador Extraordinary and Plenipotentiary
NAME: Muhammad Abdul Ghaffar

Bangladesh
2201 Wisconsin Ave., NW, Washington, D.C. 20007, United States
PHONE: +(202) 3428372
TITLE: Ambassador Extraordinary and Plenipotentiary
NAME: K. M. Shehabuddin

Barbados
2144 Wyoming, NW, Washington, D.C. 20008, United States
PHONE: +(202) 9399200
TITLE: Ambassador Extraordinary and Plenipotentiary
NAME: Courtney Blackman

Belarus
1619 New Hampshire, NW, Washington, D.C. 20009, United States
PHONE: +(202) 9861604
FAX: +(202) 9861805
TITLE: Ambassador Extraordinary and Plenipotentiary
NAME: Valery V. Tsepkalo

Belgium
3330 Garfield St., NW, Washington, D.C. 20008, United States
PHONE: +(202) 3336900
FAX: +(202) 3333079
TITLE: Ambassador Extraordinary and Plenipotentiary
NAME: Alex Reyn

Belize
2535 Massachusetts Ave., NW, Washington, D.C. 20008, United States
PHONE: +(202) 3329636
FAX: +(202) 3326888
TITLE: Ambassador Extraordinary and Plenipotentiary
NAME: James S. Murphy

Benin
2737 Cathedral, NW, Washington, D.C. 20008, United States
PHONE: +(202) 2326656
FAX: +(202) 2651996
TITLE: Ambassador Extraordinary and Plenipotentiary
NAME: Lucien Tonoukouin

Bolivia
3014 Massachusetts Ave., NW, Washington, D.C. 20008, United States
PHONE: +(202) 4834410
FAX: +(202) 3283712
TITLE: Ambassador Extraordinary and Plenipotentiary
NAME: Marcelo Perez Monasterios

Bosnia and Herzegovina
2109 E St., NW, Washington, D.C. 20037, United States
PHONE: +(202) 3371500
FAX: +(202) 3371502
TITLE: Ambassador Extraordinary and Plenipotentiary
NAME: Sven Alkalaj

Botswana
1531 New Hampshire, NW, Washington, D.C. 20036, United States
PHONE: +(202) 2444990
FAX: +(202) 2444164
TITLE: Ambassador Extraordinary and Plenipotentiary
NAME: Archibald Mooketsa Mogwe

Brazil
3006 Massachusetts Ave., NW, Washington, D.C. 20008, United States
PHONE: +(202) 2382700
FAX: +(202) 2382827
TITLE: Ambassador Extraordinary and Plenipotentiary
NAME: Rubens Barbosa

Brunei
3520 International Court, NW, Washington, D.C. 20008, United States
PHONE: +(202) 3420159
FAX: +(202) 3420158
TITLE: Ambassador Extraordinary and Plenipotentiary
NAME: Pengiran Anak Dato Puteh

Bulgaria
1621 22nd St., NW, Washington, D.C. 20008, United States
PHONE: +(202) 3877969
FAX: +(202) 2347973
TITLE: Ambassador Extraordinary and Plenipotentiary
NAME: Philip Dimitrov

Burkina Faso
2340 Massachusetts Ave., NW, Washington, D.C. 20008, United States
PHONE: +(202) 3325577
FAX: +(202) 6671882
TITLE: Ambassador Extraordinary and Plenipotentiary
NAME: Bruno Zidouemba

Burma
2300 S St., NW, Washington, D.C. 20008, United States
PHONE: +(202) 3329044
FAX: +(202) 3329046
TITLE: Ambassador Extraordinary and Plenipotentiary
NAME: Tin Winn

Burundi
2233 Wisconsin Ave., NW, Suite 212, Washington, D.C. 20007, United States
PHONE: +(202) 3422574
FAX: +(202) 3422578
TITLE: Ambassador Extraordinary and Plenipotentiary
NAME: Thomas Ndikumana

Cambodia
4500 16th St., NW, Washington, D.C. 20011, United States
PHONE: +(202) 7267742
FAX: +(202) 7268381
TITLE: Counselor (Charge d'Affaires ad interim)
NAME: Vun Yaung Tan

Cameroon
2349 Massachusetts Ave., NW, Washington, D.C. 20008, United States
PHONE: +(202) 2658790
FAX: +(202) 3873826
TITLE: Ambassador Extraordinary and Plenipotentiary
NAME: Jerome Mendouga

Canada
501 Pennsylvania Ave., NW, Washington, D.C. 20001, United States
PHONE: +(202) 6821740
FAX: +(202) 6827726
TITLE: Ambassador Extraordinary and Plenipotentiary
NAME: Raymond A. J. Chretien

Cape Verde
3415 Massachusetts Ave., NW, Washington, D.C. 20007, United States
PHONE: +(202) 9656820
FAX: +(202) 9651207
TITLE: Ambassador Extraordinary and Plenipotentiary
NAME: Amilcar Spencer Lopes

Central African Republic
1618 22nd St., NW, Washington, D.C. 20008, United States
PHONE: +(202) 4837800
FAX: +(202) 3329893
TITLE: Ambassador Extraordinary and Plenipotentiary
NAME: Henry Koba

Chad
2002 R St., NW, Washington, D.C. 20009, United States
PHONE: +(202) 4624009
FAX: +(202) 2651937
TITLE: Ambassador Extraordinary and Plenipotentiary
NAME: Hassaballah Ahmat Soubiane

Chile
1732 Massachusetts Ave., NW, Washington, D.C. 20036, United States
PHONE: +(202) 7851746
FAX: +(202) 8875579
TITLE: Ambassador Extraordinary and Plenipotentiary
NAME: Genaro Arriagada

China
2300 Connecticut, Washington, D.C. 20008, United States
PHONE: +(202) 3282500
TITLE: Ambassador Extraordinary and Plenipotentiary
NAME: Li Zhao Xing

Colombia
2118 Leroy Place, NW, Washington, D.C. 20008, United States
PHONE: +(202) 3878338
FAX: +(202) 2328643
TITLE: Ambassador Extraordinary and Plenipotentiary
NAME: Luis Alberto Moreno

Comoros
c/o The Permanent Mission of the Federal and Islamic Republic of the Comoros to the United Nations, 420 E. 50th St., New York, NY, 10022 United States
PHONE: +(212) 9728010
FAX: +(212) 9834712
TITLE: Ambassador Extraordinary and Plenipotentiary
NAME: Ahmed Djabir

Democratic Republic of Congo
1800 New Hampshire, NW, Washington, D.C. 20009, United States
PHONE: +(202) 2347690
FAX: +(202) 2370748
TITLE: Minister (Charge d'Affaires ad interim)
NAME: Faida Mitifu

Republic of the Congo
4891 Colorado, NW, Washington, D.C. 20011, United States
PHONE: +(202) 7265500
FAX: +(202) 7261860
TITLE: Minister-Counselor (Charge d'Affaires ad interim)
NAME: Serge Mombouli

Costa Rica
2114 S St., NW, Washington, D.C. 20008, United States
PHONE: +(202) 2342945
FAX: +(202) 2654795
TITLE: Ambassador Extraordinary and Plenipotentiary
NAME: Jaime Daremblum

Côte d'Ivoire
3421 Massachusetts Ave., NW, Washington, D.C. 20007, United States
PHONE: +(202) 7970300
TITLE: Ambassador Extraordinary and Plenipotentiary
NAME: Koffi Moise Koumoue

Croatia
2343 Massachusetts Ave., NW, Washington, D.C. 20008, United States
PHONE: +(202) 5885899
FAX: +(202) 5888936
TITLE: Ambassador Extraordinary and Plenipotentiary
NAME: Miomir Zuzul

Cyprus
2211 R St., NW, Washington, D.C. 20008, United States
PHONE: +(202) 4625772
FAX: +(202) 4836710
TITLE: Ambassador Extraordinary and Plenipotentiary
NAME: Erato Kozakou Marcoullis

Czech Republic
3900 Spring of Freedom St., NW, Washington, D.C. 20008, United States
PHONE: +(202) 3636315
FAX: +(202) 9668540
TITLE: Ambassador Extraordinary and Plenipotentiary
NAME: Alexandr Vondra

Denmark
3200 Whitehaven St., NW, Washington, D.C. 20008, United States
PHONE: +(202) 2344300
FAX: +(202) 3281470
TITLE: Ambassador Extraordinary and Plenipotentiary
NAME: K. Erik Tygesen

Djibouti
1156 15th St., NW, Suite 515, Washington, D.C. 20005, United States
PHONE: +(202) 3310270
FAX: +(202) 3310302
TITLE: Ambassador Extraordinary and Plenipotentiary
NAME: Roble Olhaye

Dominica
3216 New Mexico, NW, Washington, D.C. 20016, United States
PHONE: +(202) 3646781
FAX: +(202) 3646791
TITLE: Ambassador Extraordinary and Plenipotentiary
NAME: Nicholas J. O. Liverpool

Dominican Republic
1715 22nd St., NW, Washington, D.C. 20008, United States
PHONE: +(202) 3326280
FAX: +(202) 2658057
TITLE: Ambassador Extraordinary and Plenipotentiary
NAME: Bernardo Vega

Ecuador
2535 15th St., NW, Washington, D.C. 20009, United States
PHONE: +(202) 2347200
FAX: +(202) 6673482

TITLE: Ambassador Extraordinary and Plenipotentiary
NAME: Ivonne Abaki

Egypt
3521 International Court, NW, Washington, D.C. 20008, United States
PHONE: +(202) 8955400
FAX: +(202) 2444319
TITLE: Ambassador Extraordinary and Plenipotentiary
NAME: Ahmed Maher El Sayed

El Salvador
2308 California St., NW, Washington, D.C. 20008, United States
PHONE: +(202) 2659671
TITLE: Ambassador Extraordinary and Plenipotentiary
NAME: Rene A. Leon

Equatorial Guinea
1712 I St., NW, Suite 410, Washington, D.C. 20006, United States
PHONE: +(202) 2964174
FAX: +(202) 2964195
TITLE: Ambassador Extraordinary and Plenipotentiary
NAME: Pastor Micha Ondo Bile

Eritrea
1708 New Hampshire, NW, Washington, D.C. 20009, United States
PHONE: +(202) 3191991
FAX: +(202) 3191304
TITLE: Ambassador Extraordinary and Plenipotentiary
NAME: Semere Russom

Estonia
2131 Massachusetts Ave., NW, Washington, D.C. 20008, United States
PHONE: +(202) 5880101
FAX: +(202) 5880108
TITLE: Ambassador Extraordinary and Plenipotentiary
NAME: Grigore Kalev Stoicescu

Ethiopia
2134 Kalorama Rd., NW, Washington, D.C. 20008, United States
PHONE: +(202) 2342281
FAX: +(202) 3287950
TITLE: Ambassador Extraordinary and Plenipotentiary
NAME: Berhane Gebrechristos

Fiji
2233 Wisconsin Ave., NW, Suite 240, Washington, D.C. 20007, United States
PHONE: +(202) 3378320
FAX: +(202) 3371996
TITLE: Ambassador Extraordinary and Plenipotentiary
NAME: Napolioni Masirewa

Finland
3301 Massachusetts Ave., NW, Washington, D.C. 20008, United States
PHONE: +(202) 2985800
FAX: +(202) 2986030
TITLE: Ambassador Extraordinary and Plenipotentiary
NAME: Jaakko Laajava

France
4101 Reservoir Rd., NW, Washington, D.C. 20007, United States
PHONE: +(202) 9446000
FAX: +(202) 9446166
TITLE: Ambassador Extraordinary and Plenipotentiary
NAME: Francois V. Bujon

Gabon
2034 20th St., NW, Suite 200, Washington, D.C. 20009, United States
PHONE: +(202) 7971000
FAX: +(202) 3320668
TITLE: Ambassador Extraordinary and Plenipotentiary
NAME: Paul Bundukulatha

Gambia
1155 15th St., NW, Suite 1000, Washington, D.C. 0005, United States
PHONE: +(202) 7851399
FAX: +(202) 7851430
TITLE: Ambassador Extraordinary and Plenipotentiary
NAME: Crispin Greyjohnson

Georgia
1615 New Hampshire, NW, Suite 300, Washington, D.C. 20009, United States
PHONE: +(202) 3872390
FAX: +(202) 3934537
TITLE: Ambassador Extraordinary and Plenipotentiary
NAME: Tedo Japaridze

Germany
4645 Reservoir Rd., NW, Washington, D.C. 20007, United States

PHONE: +(202) 2988141
FAX: +(202) 2984249
TITLE: Ambassador Extraordinary and Plenipotentiary
NAME: Juergen Chrobog

Ghana
3512 International Dr., NW, Washington, D.C. 20008, United States
PHONE: +(202) 6864520
FAX: +(202) 6864527
TITLE: Ambassador Extraordinary and Plenipotentiary
NAME: Kobina Arthur Koomson

Greece
2221 Massachusetts Ave., NW, Washington, D.C. 20008, United States
PHONE: +(202) 9395800
FAX: +(202) 9395824
TITLE: Ambassador Extraordinary and Plenipotentiary
NAME: Alexandre Philon

Grenada
1701 New Hampshire, NW, Washington, D.C. 20009, United States
PHONE: +(202) 2652561
TITLE: Ambassador Extraordinary and Plenipotentiary
NAME: Denis G. Antoine

Guatemala
2220 R St., NW, Washington, D.C. 20008, United States
PHONE: +(202) 7454952
FAX: +(202) 7451908
TITLE: Ambassador Extraordinary and Plenipotentiary
NAME: William Howard Stixrud

Guinea
2112 Leroy Place, NW, Washington, D.C. 20008, United States
PHONE: +(202) 4839420
FAX: +(202) 4838688
TITLE: Ambassador Extraordinary and Plenipotentiary
NAME: Mohamed Aly Thiam

Guinea Bissau
1511 K St., NW, Suite 519, Washington, D.C. 20005, United States
PHONE: +(202) 3473950
FAX: +(202) 3473954
TITLE: Ambassador Extraordinary and Plenipotentiary
NAME: Mario Lopes Da Rosa

Guyana
2490 Tracy Place, NW, Washington, D.C. 20008, United States
PHONE: +(202) 2656900
TITLE: Ambassador Extraordinary and Plenipotentiary
NAME: Mohammed Ali Odeen Ishmael

Haiti
2311 Massachusetts Ave., NW, Washington, D.C. 20008, United States
PHONE: +(202) 3324090
FAX: +(202) 7457215
TITLE: Minister-Counselor (Charge d'Affaires ad interim)
NAME: Louis Harold Joseph

Honduras
3007 Tilden St., NW, Suite 4M, Washington, D.C. 20008, United States
PHONE: +(202) 9667702
FAX: +(202) 9669751
TITLE: Ambassador Extraordinary and Plenipotentiary
NAME: Hugo Noe Pino

Hungary
3910 Shoemaker St., NW, Washington, D.C. 20008, United States
PHONE: +(202) 3626730
FAX: +(202) 9668135
TITLE: Ambassador Extraordinary and Plenipotentiary
NAME: Geza Jeszenszky

Iceland
1156 15th St., NW, Suite 1200, Washington, D.C. 20005, United States
PHONE: +(202) 2656653
FAX: +(202) 2656656
TITLE: Ambassador Extraordinary and Plenipotentiary
NAME: Jon Baldvin Hannibalsson

India
2107 Massachusetts Ave., NW, Washington, D.C. 20008, United States
PHONE: +(202) 9397000
FAX: +(202) 4833972
TITLE: Ambassador Extraordinary and Plenipotentiary
NAME: Naresh Chandra

Indonesia
2020 Massachusetts Ave., NW, Washington, D.C. 20036, United States

PHONE: +(202) 7755200
FAX: +(202) 7755365
TITLE: Ambassador Extraordinary and Plenipotentiary
NAME: Dorodjatun Kuntjoro Jakti

Ireland
2234 Massachusetts Ave., NW, Washington, D.C. 20008, United States
PHONE: +(202) 4623939
FAX: +(202) 2325993
TITLE: Ambassador Extraordinary and Plenipotentiary
NAME: Sean O'Huiginn

Israel
3514 International Dr., NW, Washington, D.C. 20008, United States
PHONE: +(202) 3645500
FAX: +(202) 3645610
TITLE: Ambassador Extraordinary and Plenipotentiary
NAME: Zalman Shoval

Italy
1601 Fuller St., NW, Washington, D.C. 20009, United States
PHONE: +(202) 3285500
FAX: +(202) 4832187
TITLE: Ambassador Extraordinary and Plenipotentiary
NAME: Ferdinando Salleo

Jamaica
1520 New Hampshire, NW, Washington, D.C. 20036, United States
PHONE: +(202) 4520660
FAX: +(202) 4520081
TITLE: Ambassador Extraordinary and Plenipotentiary
NAME: Richard Leighton Bernal

Japan
2520 Massachusetts Ave., NW, Washington, D.C. 20008, United States
PHONE: +(202) 2386700
FAX: +(202) 3282187
TITLE: Ambassador Extraordinary and Plenipotentiary
NAME: Kunihiko Saito

Jordan
3504 International Dr., NW, Washington, D.C. 20008, United States
PHONE: +(202) 9662664
FAX: +(202) 9663110
TITLE: Ambassador Extraordinary and Plenipotentiary
NAME: Marwan Jamil Muasher

Kazakhstan
1401 16th St., NW, Washington, D.C. 20036, United States
PHONE: +(202) 2325488
FAX: +(202) 2325845
TITLE: Ambassador Extraordinary and Plenipotentiary
NAME: Bolat K. Nurgaliyev

Kenya
2249 R St., NW, Washington, D.C. 20008, United States
PHONE: +(202) 3876101
FAX: +(202) 4623829
TITLE: Ambassador Extraordinary and Plenipotentiary
NAME: Samson Kipkoech Chemai

Korea
2450 Massachusetts Ave., NW, Washington, D.C. 20008, United States
PHONE: +(202) 9395600
FAX: +(202) 3870205
TITLE: Ambassador Extraordinary and Plenipotentiary
NAME: Hong Koo Lee

Kuwait
2940 Tilden St., NW, Washington, D.C. 20008, United States
PHONE: +(202) 9660702
FAX: +(202) 9660517
TITLE: Ambassador Extraordinary and Plenipotentiary
NAME: Mohammed Sabah Al-Salim Alsabah

Kyrgyzstan
1732 Wisconsin Ave., NW, Washington, D.C. 20007, United States
PHONE: +(202) 3385141
FAX: +(202) 3385139
TITLE: Ambassador Extraordinary and Plenipotentiary
NAME: Baktybek Abdrissaev

Laos
2222 S St., NW, Washington, D.C. 20008, United States
PHONE: +(202) 3326416
FAX: +(202) 3324923
TITLE: Ambassador Extraordinary and Plenipotentiary
NAME: Vang Rattanavog

Latvia
4325 17th St., NW, Washington, D.C. 20011, United States
PHONE: +(202) 7268213
FAX: +(202) 7266785
TITLE: Ambassador Extraordinary and Plenipotentiary
NAME: Ojars Eriks Kalnins

Lebanon
2560 28th St., NW, Washington, D.C. 20008, United States
PHONE: +(202) 9396300
FAX: +(202) 9396324
TITLE: Ambassador Extraordinary and Plenipotentiary
NAME: Farid Abboud

Lesotho
2511 Massachusetts Ave., NW, Washington, D.C. 20008, United States
PHONE: +(202) 7975533; 7975534; 7975535
FAX: +(202) 2346815
TITLE: Counselor (Charge d'Affaires ad interim)
NAME: Ben T. Nteso

Liberia
5303 Colorado, NW, Washington, D.C. 20011, United States
PHONE: +(202) 7230437
FAX: +(202) 7230436
TITLE: Ambassador Extraordinary and Plenipotentiary
NAME: Rachel Diggs

Lithuania
2622 16th St., NW, Washington, D.C. 20009, United States
PHONE: +(202) 2345860
FAX: +(202) 3280466
TITLE: Ambassador Extraordinary and Plenipotentiary
NAME: Stasys Sakalauskas

Luxembourg
2200 Massachusetts Ave., NW, Washington, D.C. 20008, United States
PHONE: +(202) 2654171
FAX: +(202) 3288270
TITLE: Ambassador Extraordinary and Plenipotentiary
NAME: Arlette Conzemius

Macedonia
3050 K St., NW, Suite 210, Washington, D.C. 20007, United States
PHONE: +(202) 3373063
FAX: +(202) 3373093
TITLE: Ambassador Extraordinary and Plenipotentiary
NAME: Lubica Z. Acevska

Madagascar
2374 Massachusetts Ave., NW, Washington, D.C. 20008, United States
PHONE: +(202) 2655525; 2655526
TITLE: Ambassador Extraordinary and Plenipotentiary
NAME: Zina Andrianarivelo Razafy

Malawi
2408 Massachusetts Ave., NW, Washington, D.C. 20008, United States
PHONE: +(202) 7971007
FAX: +(202) 2650976
TITLE: Ambassador Extraordinary and Plenipotentiary
NAME: Willie Chokani

Malaysia
2401 Massachusetts Ave., NW, Washington, D.C. 20008, United States
PHONE: +(202) 3282700
FAX: +(202) 4837661
TITLE: Ambassador Extraordinary and Plenipotentiary
NAME: Dato Sheikh Abdul Khaled Ghazzali

Mali
2130 R St., NW, Washington, D.C. 20008, United States
PHONE: +(202) 3322249; 9398950
FAX: +(202) 3326603
TITLE: Ambassador Extraordinary and Plenipotentiary
NAME: Cheick Oumar Diarrah

Malta
2017 Connecticut, NW, Washington, D.C. 20008, United States
PHONE: +(202) 4623611; 4623612
FAX: +(202) 3875470
TITLE: Counselor and Consul General (Charge d'Affaires ad interim)
NAME: Alfred A. Farrugia

Marshall Islands
2433 Massachusetts Ave., NW, Washington, D.C. 20008, United States
PHONE: +(202) 2345414
FAX: +(202) 2323236
TITLE: Ambassador Extraordinary and Plenipotentiary
NAME: Banny De Brum

Mauritania
2129 Leroy Place, NW, Washington, D.C. 20008, United States
PHONE: +(202) 2325700
FAX: +(202) 3192623
TITLE: Appointed Ambassador
NAME: Ahmed Ould Khalifa Ould Jiddou

Mauritius
4301 Connecticut, NW, Suite 441, Washington, D.C. 20008, United States
PHONE: +(202) 2441491; 2441492
FAX: +(202) 9660983
TITLE: Ambassador Extraordinary and Plenipotentiary
NAME: Chitmansing Jesseramsing

Mexico
1911 Pennsylvania Ave., NW, Washington, D.C. 20006, United States
PHONE: +(202) 7281600
FAX: +(202) 7281698
TITLE: Ambassador Extraordinary and Plenipotentiary
NAME: Jesus F. Reyes Heroles

Micronesia
1725 N St., NW, Washington, D.C. 20036, United States
PHONE: +(202) 2234383
FAX: +(202) 2234391
TITLE: Ambassador Extraordinary and Plenipotentiary
NAME: Jesse Bibiano Marehalau

Moldova
2101 S St., NW, Washington, D.C. 20008, United States
PHONE: +(202) 6671130
FAX: +(202) 6671204
TITLE: Ambassador Extraordinary and Plenipotentiary
NAME: Ceslav Ciobanu

Mongolia
2833 M St., NW, Washington, D.C. 20007, United States
PHONE: +(202) 3337117
FAX: +(202) 2989227
TITLE: Ambassador Extraordinary and Plenipotentiary
NAME: Jalbuu Choinor

Morocco
1601 21st St., NW, Washington, D.C. 20009, United States
PHONE: +(202) 4627979
FAX: +(202) 2650161
TITLE: Minister (Charge d'Affaires ad interim)
NAME: Mustapha Cherkaoui

Mozambique
1990 M St. NW, Suite 570, Washington, D.C. 20036, United States
PHONE: +(202) 2937146
FAX: +(202) 8350245
TITLE: Ambassador Extraordinary and Plenipotentiary
NAME: Marcos G. Namashulua

Namibia
1605 New Hampshire, NW, Washington, D.C. 20009, United States
PHONE: +(202) 9860540
FAX: +(202) 9860443
TITLE: Ambassador Extraordinary and Plenipotentiary
NAME: Leonard Nangolo Iipumbu

Nepal
2131 Leroy Place, NW, Washington, D.C. 20008, United States
PHONE: +(202) 6674550
FAX: +(202) 6675534
TITLE: Ambassador Extraordinary and Plenipotentiary
NAME: Damodar Prasad Gautam

Netherlands
4200 Linnean, NW, Washington, D.C. 20008, United States
PHONE: +(202) 2445300; 4948594
FAX: +(202) 3623430
TITLE: Ambassador Extraordinary and Plenipotentiary
NAME: Joris M. Vos

New Zealand
37 Observatory Circle, NW, Washington, D.C. 20008, United States
PHONE: +(202) 3284800
FAX: +(202) 6675227
TITLE: Ambassador Extraordinary and Plenipotentiary
NAME: James B. Bolger

Nicaragua
1627 New Hampshire, NW, Washington, D.C. 20009, United States
PHONE: +(202) 9396570
FAX: +(202) 9396542
TITLE: Ambassador Extraordinary and Plenipotentiary
NAME: Francisco Javier Aguirre Sacasa

Niger
2204 R St., NW, Washington, D.C. 20008, United States
PHONE: +(202) 4834224
TITLE: Ambassador Extraordinary and Plenipotentiary
NAME: Joseph Diatta

Nigeria
1333 16th St., NW, Washington, D.C. 20036, United States
PHONE: +(202) 9868400
FAX: +(202) 7751385
TITLE: Ambassador Extraordinary and Plenipotentiary
NAME: Wakili Hassan Adamu

Norway
2720 34th St., NW, Washington, D.C. 20008, United States
PHONE: +(202) 3336000
FAX: +(202) 3370870
TITLE: Ambassador Extraordinary and Plenipotentiary
NAME: Tom Eric Vraalsen

Oman
2535 Belmont Rd., NW, Washington, D.C. 20008, United States
PHONE: +(202) 3871980
FAX: +(202) 7454933
TITLE: Ambassador Extraordinary and Plenipotentiary
NAME: Abdulla Moh'd Aqeel AL Dhahab

Pakistan
2315 Massachusetts Ave., NW, Washington, D.C. 20008, United States
PHONE: +(202) 9396200
FAX: +(202) 3870484
TITLE: Ambassador Extraordinary and Plenipotentiary
NAME: Riaz Khokhar

Palau
1150 18th St. NW, Suite 750, Washington, D.C. 20036, United States
PHONE: +(202) 4526814
FAX: +(202) 4526281
TITLE: Ambassador Extraordinary and Plenipotentiary
NAME: Hersey Kyota

Panama
2862 McGill Terrace, NW, Washington, D.C. 20008, United States
PHONE: +(202) 4831407
TITLE: Ambassador Extraordinary and Plenipotentiary
NAME: Eloy Alfaro

Papua New Guinea
1779 Massachusetts Ave., NW, Suite 805, Washington, D.C. 20036, United States
PHONE: +(202) 7453680
FAX: +(202) 7453679
TITLE: Ambassador Extraordinary and Plenipotentiary
NAME: Nagora Y. Bogan

Paraguay
2400 Massachusetts Ave., NW, Washington, D.C. 20008, United States
PHONE: +(202) 4836960
FAX: +(202) 2344508
TITLE: Counselor (Charge d'Affaires ad interim)
NAME: Elianne Cibils

Peru
1700 Massachusetts Ave., NW, Washington, D.C. 20036, United States
PHONE: +(202) 8339860
FAX: +(202) 6598124
TITLE: Ambassador Extraordinary and Plenipotentiary
NAME: Ricardo V. Luna

Philippines
1600 Massachusetts Ave., NW, Washington, D.C. 20036, United States
PHONE: +(202) 4679300
FAX: +(202) 287614
TITLE: Minister (Charge d'Affaires ad interim)
NAME: Ariel Y. Abadilla

Poland
2640 16th St., NW, Washington, D.C. 20009, United States
PHONE: +(202) 2343800
FAX: +(202) 3286271
TITLE: Ambassador Extraordinary and Plenipotentiary
NAME: Jerzy Kozminski

Portugal
2125 Kalorama Rd., NW, Washington, D.C. 20008, United States
PHONE: +(202) 3288610
FAX: +(202) 4623726
TITLE: Appointed Ambassador
NAME: Joao Rocha Paris

Qatar
4200 Wisconsin Ave., NW, Washington, D.C. 20016, United States

PHONE: +(202) 2741600
TITLE: Ambassador Extraordinary and Plenipotentiary
NAME: Saad Mohamed Al Kobaisi

Romania
1607 23rd St., NW, Washington, D.C. 20008, United States
PHONE: +(202) 3324846; 3324848; 3324851
FAX: +(202) 2324748
TITLE: Ambassador Extraordinary and Plenipotentiary
NAME: Mircea Dan Geoana

Russian Federation
2650 Wisconsin Ave., NW, Washington, D.C. 20007, United States
PHONE: +(202) 2985700
FAX: +(202) 2985735
TITLE: Ambassador Extraordinary and Plenipotentiary
NAME: Yury V. Ushakov

Rwanda
1714 New Hampshire, NW, Washington, D.C. 20009, United States
PHONE: +(202) 2322882
FAX: +(202) 2324544
TITLE: Appointed Ambassador
NAME: Richard Sezibera

Saint Kitts and Nevis
3216 New Mexico, NW, Washington, D.C. 16, United States
PHONE: +(202) 6862636
FAX: +(202) 6865740
TITLE: Ambassador Extraordinary and Plenipotentiary
NAME: Osbert W. Liburd

Saint Lucia
3216 New Mexico, NW, Washington, D.C. 20016, United States
PHONE: +(202) 3646792
FAX: +(202) 3646728
TITLE: Ambassador Extraordinary and Plenipotentiary
NAME: Sonia Merlyn Johnny

Saint Vincent and the Grenadines
3216 New Mexico, NW, Washington, D.C. 20016, United States
PHONE: +(202) 3646730
FAX: +(202) 3646736
TITLE: Ambassador Extraordinary and Plenipotentiary
NAME: Kingsley Cuthbert Augustine Layne

Samoa
800 Second, Suite 400D, New York, NY 10017, United States
PHONE: +(212) 5996196
FAX: +(212) 5990797
TITLE: Ambassador Extraordinary and Plenipotentiary
NAME: Tuiloma Neroni Slade

Saudi Arabia
601 New Hampshire, NW, Washington, D.C. 20037, United States
PHONE: +(202) 3423800
TITLE: Ambassador Extraordinary and Plenipotentiary
NAME: Bandar Bin Sultan

Senegal
2112 Wyoming, NW, Washington, D.C. 20008, United States
PHONE: +(202) 2340540
TITLE: Ambassador Extraordinary and Plenipotentiary
NAME: Mamadou Mansour Seck

Seychelles
800 Second Ave., Suite 400C, New York, NY 10017, United States
PHONE: +(212) 9721785
FAX: +(212) 9721786
TITLE: Ambassador Extraordinary and Plenipotentiary
NAME: Claude Sylvestre Morel

Sierra Leone
1701 19th St., Nw, Washington, D.C. 20009, United States
FAX: +(202) 4831793
TITLE: Ambassador Extraordinary and Plenipotentiary
NAME: John Ernest Leigh

Singapore
3501 International Place, NW, Washington, D.C. 20008, United States
PHONE: +(202) 5373100
FAX: +(202) 5370876
TITLE: Ambassador Extraordinary and Plenipotentiary
NAME: Heng-Chee Chan

Slovakia
2201 Wisconsin Ave., NW, Suite 250, Washington, D.C. 20007, United States
PHONE: +(202) 9655161
FAX: +(202) 9655166

TITLE: Ambassador Extraordinary and Plenipotentiary
NAME: Martin Butora

Slovenia
1525 New Hampshire, NW, Washington, D.C. 20036, United States
PHONE: +(202) 6675363
FAX: +(202) 6674563
TITLE: Ambassador Extraordinary and Plenipotentiary
NAME: Dimitrij Rupel

Solomon Islands
800 Second Ave., Suite 400L, New York, NY 10017, United States
PHONE: +(212) 5996192
FAX: +(212) 6618925
TITLE: Ambassador Extraordinary and Plenipotentiary
NAME: Rex Stephen Horoi

South Africa
3051 Massachusetts Ave., NW, Washington, D.C. 20008, United States
PHONE: +(202) 2324400
FAX: +(202) 2651607
TITLE: Ambassador Extraordinary and Plenipotentiary
NAME: Makate Sheila Sisulu

Spain
2375 Pennsylvania Ave., NW, Washington, D.C. 20037, United States
PHONE: +(202) 4520100; 7282340
FAX: +(202) 8335670
TITLE: Ambassador Extraordinary and Plenipotentiary
NAME: Antonio Oyarzabal

Sri Lanka
2148 Wyoming, NW, Washington, D.C. 20008, United States
PHONE: +(202) 4834025
FAX: +(202) 2327181
TITLE: Ambassador Extraordinary and Plenipotentiary
NAME: Warnasena Rasaputram

Sudan
2210 Massachusetts Ave., NW, Washington, D.C. 20008, United States
PHONE: +(202) 3388565
FAX: +(202) 6672406
TITLE: Ambassador Extraordinary and Plenipotentiary
NAME: Mahdi Ibrahim Mohamed

Suriname
4301 Connecticut, NW, Suite 460, Washington, D.C. 20008, United States
PHONE: +(202) 2447488
FAX: +(202) 2445878
TITLE: Ambassador Extraordinary and Plenipotentiary
NAME: Arnold T. Halfhide

Swaziland
3400 International Dr., NW, Washington, D.C. 20008, United States
PHONE: +(202) 3626683
FAX: +(202) 2448059
TITLE: Ambassador Extraordinary and Plenipotentiary
NAME: Mary M. Kanya

Sweden
1501 M St., NW, Washington, D.C. 20005-1702, United States
PHONE: +(202) 4672600
FAX: +(202) 4672699
TITLE: Ambassador Extraordinary and Plenipotentiary
NAME: Rolf Ekeus

Switzerland
2900 Cathedral, NW, Washington, D.C. 20008, United States
PHONE: +(202) 7457900
FAX: +(202) 3872564
TITLE: Ambassador Extraordinary and Plenipotentiary
NAME: Alfred Defago

Syria
2215 Wyoming, NW, Washington, D.C. 20008, United States
PHONE: +(202) 2326313
FAX: +(202) 2349548
TITLE: Ambassador Extraordinary and Plenipotentiary
NAME: Walid Almoualem

Tanzania
2139 R St., NW, Washington, D.C. 20008, United States
PHONE: +(202) 9396125
FAX: +(202) 7977408
TITLE: Ambassador Extraordinary and Plenipotentiary
NAME: Mustafa Salim Nyang'Anyi

Thailand
1024 Wisconsin Ave., NW, Washington, D.C. 20007, United States

PHONE: +(202) 9443600
FAX: +(202) 9443611
TITLE: Ambassador Extraordinary and Plenipotentiary
NAME: Nitya Pibulsonggram

Togo
2208 Massachusetts Ave., NW, Washington, D.C. 20008, United States
PHONE: +(202) 2344212
FAX: +(202) 2323190
TITLE: Ambassador Extraordinary and Plenipotentiary
NAME: Akoussoulelou Bobjona

Trinidad and Tobago
1708 Massachusetts Ave., NW, Washington, D.C. 20036, United States
PHONE: +(202) 4676490
FAX: +(202) 7853130
TITLE: Ambassador Extraordinary and Plenipotentiary
NAME: Michael Arneaud

Tunisia
1515 Massachusetts Ave., NW, Washington, D.C. 20005, United States
PHONE: +(202) 8621850
TITLE: Ambassador Extraordinary and Plenipotentiary
NAME: Noureddine Mejdoub

Turkey
2525 Massachusetts Ave., NW, Washington, D.C. 20008, United States
PHONE: +(202) 6126700
FAX: +(202) 6126744
TITLE: Ambassador Extraordinary and Plenipotentiary
NAME: Baki Ilkin

Turkmenistan
2207 Massachusetts Ave., NW, Washington, D.C. 20008, United States
PHONE: +(202) 5881500
FAX: +(202) 5880697
TITLE: Ambassador Extraordinary and Plenipotentiary
NAME: Halil Ugur

Uganda
5911 16th St., NW, Washington, D.C. 20011, United States
PHONE: +(202) 7267100; 7260416
FAX: +(202) 7261727
TITLE: Ambassador Extraordinary and Plenipotentiary
NAME: Edith Grace Ssempala

Ukraine
3350 M St., NW, Washington, D.C. 20007, United States
PHONE: +(202) 3330606
FAX: +(202) 3330817
TITLE: Ambassador Extraordinary and Plenipotentiary
NAME: Anton Buteiko

United Arab Emirates
1255 22nd St., NW, Suite 700, Washington, D.C. 20037, United States
PHONE: +(202) 9557999
TITLE: Ambassador Extraordinary and Plenipotentiary
NAME: Mohammad Bin Hussain Alshaali

United Kingdom
3100 Massachusetts Ave., NW, Washington, D.C. 20008, United States
PHONE: +(202) 5886500
FAX: +(202) 5887870
TITLE: Ambassador Extraordinary and Plenipotentiary
NAME: Christopher Meyer

Uruguay
2715 M St., NW, Washington, D.C. 20007, United States
PHONE: +(202) 3311313
FAX: +(202) 3318142
TITLE: Ambassador Extraordinary and Plenipotentiary
NAME: Alvaro Mario Diez de Medina

Uzbekistan
1746 Massachusetts Ave., NW, Washington, D.C. 20036, United States
PHONE: +(202) 8875300
FAX: +(202) 2936804
TITLE: Ambassador Extraordinary and Plenipotentiary
NAME: Sodiq Safaev

Vatican City
Apostolic Nunciature, 3339 Massachusetts Ave., NW, Washington, D.C. 20008, United States
PHONE: +(202) 3337121
TITLE: Apostolic Nuncio
NAME: Gabriele Montalvo

Venezuela
1099 30th St., NW, Washington, D.C. 20007, United States
PHONE: +(202) 3422214
FAX: +(202) 3426820

TITLE: Ambassador Extraordinary and Plenipotentiary
NAME: Alfredo Toro Hardy

Vietnam
1233 20th St., NW, Suite 400, Washington, D.C. 20036, United States
PHONE: +(202) 8610737
FAX: +(202) 8610917
TITLE: Ambassador Extraordinary and Plenipotentiary
NAME: Bang Le

Yemen
2600 Virginia, NW, Suite 705, Washington, D.C. 20037, United States
PHONE: +(202) 9654760
FAX: +(202) 3372017
TITLE: Ambassador Extraordinary and Plenipotentiary
NAME: Abdulwahab A. Alhajjri

Zambia
2419 Massachusetts Ave., NW, Washington, D.C. 20008, United States
PHONE: +(202) 2659717
FAX: +(202) 3320826
TITLE: Ambassador Extraordinary and Plenipotentiary
NAME: Dunstan Weston Kamana

Zimbabwe
1608 New Hampshire, NW, Washington, D.C. 20009, United States
PHONE: +(202) 3327100
FAX: +(202) 4839326
TITLE: Minister Counselor (Charge d'Affaires ad interim)
NAME: Elita Tinoenda Tandi Sakupwanya

JUDICIAL SYSTEM

United States Supreme Court

1 First St., NE, Washington, D.C. 20543, United States
PHONE: +(202) 4793000

United States Court of Appeals

The Administrative Office of the U.S. Courts, 1 Columbus Circle, N.E., Washington, D.C. 20544, United States
PHONE: +(202) 5021504

District Courts

Circuit Courts

FURTHER READING

Articles

''The Accusations and the Defense.'' *Newsweek* (January 25, 1999): 26.

''All Bets Are Off: If There Is a Real Danger in Last Week's Vote Against the Comprehensive Test Ban Treaty, It Is That the Rest of the World May No Longer Trust American Leadership.'' *Newsweek* (October 25, 1999): 28.

Bourge, Christian. ''Quick Dumping Action Urged.'' *American Metal Market* 107 (March 10, 1999): 2.

''Bulls, Bears and Ivory Towers.'' *The Economist* (May 15, 1999): 81.

''Clinton enters WTO as protest arrests rise.'' *USA Today,* 2 December 1999.

Crossette, Barbara. ''U.S. Presses Taliban to Deliver Osama bin Laden.'' *The New York Times,* 19 October 1999, p. A6.

''Four Months in Review, April, May, June, July 1999.'' *Current History* (September 1999): 304.

Huntington, Samuel P. ''The Lonely Superpower.'' *Foreign Affairs* (March 1999): 35.

Melloan, George. ''Summers' Time at Treasury is Starting Badly.'' *The Wall Street Journal,* 17 August 1999, p. A23.

''Portrait Of A Deadly Bond.'' *Time* (May 10, 1999): 26.

Sanger, David E. ''Clinton to See Barak and Arafat in Effort to Speed Peace Moves.'' *The New York Times,* 21 October 1999, p. A10.

Streisand, Betsy. ''What is Known about Columbine.'' *U.S. News and World Report* (October 11, 1999): 57.

Strobeland, Warren P., and Richard J. Newman. ''Score One for Sudan.'' *U.S. News and World Report* (May 17, 1999): 42.

Books

Foreign Trade of the United States. Washington, DC: Bernan Press, 1999.

Grover, William F. and Joseph G. Peschek, eds. *Voices of Dissent: Critical Readings in American Politics.* 3rd ed. New York: Longman, 1999.

Hird, John A. and Michael Reese. *Controversies in American Public Policy.* 2nd ed. New York: St. Martin's/Worth, 1999.

Jones, Jacqueline. *A Social History of the Labouring Classes: from Colonial Times to the Present.* Malden, Mass.: Blackwell Publishers, 1999.

Michel, Sonya et al, eds. *Engendering America: A Documentary History, 1865 to the Present.* Dubuque, IA: McGraw-Hill, 1999.

Perlmutter, Philip. *Legacy of Hate: a Short History of Ethnic, Religious, and Racial Prejudice in America.* Rev. ed. Armonk, N.Y.: M.E. Sharpe, 1999.

Williams, Mary E., ed. *Culture Wars: Opposing Viewpoints.* San Diego, Calif.: Greenhaven Press, 1999.

UNITED STATES: STATISTICAL DATA

For sources and notes see "Sources of Statistics" in the front of each volume.

GEOGRAPHY

Geography (1)

Area:

Total: 9,629,091 sq km.

Land: 9,158,960 sq km.

Water: 470,131 sq km.

Note: includes only the 50 states and District of Columbia.

Area—comparative: about one-half the size of Russia; about three-tenths the size of Africa; about one-half the size of South America (or slightly larger than Brazil); slightly larger than China; about two and one-half times the size of Western Europe.

Continued on next page

HUMAN FACTORS

Demographics (2A)

	1990	1995	1998	2000	2010	2020	2030	2040	2050
Population	249,948.6	263,039.3	270,290.3	274,943.5	298,026.1	323,051.8	347,209.2	370,290.0	394,240.5
Net migration rate (per 1,000 population)	NA	NA	NA	NA	NA	NA	NA	NA	NA
Births	NA	NA	NA	NA	NA	NA	NA	NA	NA
Deaths	2,148.5	NA	NA	NA	NA	NA	NA	NA	NA
Life expectancy - males	NA	72.5	72.9	73.0	74.2	75.7	77.1	78.5	79.7
Life expectancy - females	NA	78.9	79.6	79.7	80.7	81.6	82.5	83.4	84.3
Birth rate (per 1,000)	16.6	14.8	14.4	14.2	14.3	14.2	13.9	14.2	14.4
Death rate (per 1,000)	8.6	8.8	8.8	8.8	8.9	9.0	9.5	10.1	10.1
Women of reproductive age (15-49 yrs.)	65,810.6	68,331.8	69,390.1	69,967.7	71,028.9	72,135.4	77,012.0	81,253.4	85,743.7
of which are currently married	91,671.9	95,315.0	NA	NA	NA	NA	NA	NA	NA
Fertility rate	2.1	2.0	2.1	2.1	2.1	2.1	2.2	2.2	2.2

Except as noted, values for vital statistics are in thousands; life expectancy is in years.

GEOGRAPHY

Geography (1) cont.

Land boundaries:

Total: 12,248 km.

Border countries: Canada 8,893 km (including 2,477 km with Alaska), Cuba 29 km (US Naval Base at Guantanamo Bay), Mexico 3,326 km.

Note: Guantanamo Naval Base is leased by the US and thus remains part of Cuba.

Coastline: 19,924 km.

Climate: mostly temperate, but tropical in Hawaii and Florida and arctic in Alaska, semiarid in the great plains west of the Mississippi River and arid in the Great Basin of the southwest; low winter temperatures in the northwest are ameliorated occasionally in January and February by warm chinook winds from the eastern slopes of the Rocky Mountains.

Terrain: vast central plain, mountains in west, hills and low mountains in east; rugged mountains and broad river valleys in Alaska; rugged, volcanic topography in Hawaii.

Natural resources: coal, copper, lead, molybdenum, phosphates, uranium, bauxite, gold, iron, mercury, nickel, potash, silver, tungsten, zinc, petroleum, natural gas, timber.

Land use:

Arable land: 19%

Permanent crops: 0%

Permanent pastures: 25%

Forests and woodland: 30%

Other: 26% (1993 est.)

HUMAN FACTORS

Health Personnel (3)

Total health expenditure as a percentage of GDP, 1990-1997[a]

Public sector 6.6

Private sector 7.5

Total[b] 14.1

Health expenditure per capita in U.S. dollars, 1990-1997[a]

Purchasing power parity 3,951

Total 4,093

Availability of health care facilities per 100,000 people

Hospital beds 1990-1997[a] 410

Doctors 1993[c] 245

Nurses 1993[c] 878

Health Indicators (4)

Life expectancy at birth

1980 74

1997 76

Daily per capita supply of calories (1996) 3,642

Total fertility rate births per woman (1997) 2.0

Maternal mortality ratio per 100,000 live births (1990-97) . . . . .12[c]

Safe water % of population with access (1995) 73

Sanitation % of population with access (1995) . . . .

Consumption of iodized salt % of households (1992-98)[a] . . . .

Smoking prevalence

Male % of adults (1985-95)[a] 28

Female % of adults (1985-95)[a] 23

Tuberculosis incidence per 100,000 people (1997) . . .7

Adult HIV prevalence % of population ages 15-49 (1997) 0.76

Infants and Malnutrition (5)

Under-5 mortality rate (1997) 8

% of infants with low birthweight (1990-97) 7

Births attended by skilled health staff % of total[a] . . .NA

% fully immunized (1995-97)

TB . . . . .NA

DPT 94%

Polio 84%

Measles 89%

Prevalence of child malnutrition under age 5 (1992-97)[b] 1

Religions (7)

Protestant 56%

Roman Catholic 28%

Jewish 2%

Other 4%

None 10% (1989)

Languages (8)

English, Spanish (spoken by a sizable minority).

EDUCATION

Public Education Expenditures (9)

Public expenditure on education (% of GNP)

1980 . . . . .6.7

1996 . . . . .5.4[1]

Expenditure per student

Primary % of GNP per capita	
1980	27.0
1996	18.5[1]
Secondary % of GNP per capita	
1980	....
1996	23.8[1]
Tertiary % of GNP per capita	
1980	48.2
1996	24.7[1]

Expenditure on teaching materials

Primary % of total for level (1996)	....
Secondary % of total for level (1996)	....
Primary pupil-teacher ratio per teacher (1996)	16[1]
Duration of primary education years (1995)	6

Educational Attainment (10)

Age group (1994)[12]	25+
Total population	164,511,000
Highest level attained (%)	
No schooling	0.6
First level	
Not completed	8.2
Completed	NA
Entered second level	
S-1	44.6
S-2	NA
Postsecondary	46.5

Literacy Rates (11B)

Adult literacy rate

1980	
Male	99%
Female	99%
1995	
Male	-
Female	-

GOVERNMENT & LAW

Political Parties (12)

House of Representatives	No. of seats
Republican Party	227
Democratic Party	205
Independent	1
Vacant	2

Government Budget (13A)

Year: 1997

Total Expenditures: 1,696.61 Billions of Dollars

Expenditures as a percentage of the total by function:

General public services and public order	10.10
Defense	15.71
Education	1.76
Health	20.32
Social Security and Welfare	28.82
Housing and community amenities	2.67
Recreational, cultural, and religious affairs	.31
Fuel and energy	.22
Agriculture, forestry, fishing, and hunting	.95
Mining, manufacturing, and construction	.06
Transportation and communication	2.39
Other economic affairs and services	1.47

Crime (15)

Crime rate (for 1994)

Crimes reported	13,989,500
Total persons convicted	48,978
Crimes per 100,000 population	5,367
Persons responsible for offenses	
Total number of suspects	11,877,188
Total number of female suspects	2,372,426
Total number of juvenile suspects	549,126

LABOR FORCE

Labor Force (16)

Total (million)	136.3
Managerial and professional	29.1%
Technical, sales and admin support	29.6%
Services	13.5%
Manufacturing, mining, transportation, crafts	25.1%
Farming, forestry, fishing	2.7%

Total includes unemployed. Data for 1997.

Unemployment Rate (17)

4.9% (1997)

PRODUCTION SECTOR

Electric Energy (18)

Capacity	741.589 million kW (1995)
Production	3.585 trillion kWh (1995)
Consumption per capita	13,732 kWh (1995)

GOVERNMENT & LAW

Military Affairs (14B)

	1990	1991	1992	1993	1994	1995
Military expenditures						
Current dollars (mil.)	306,200	280,300	305,100	297,600	288,100	277,800
1995 constant dollars (mil.)	351,900	309,700	328,200	312,000	295,300	277,800
Armed forces (000)	2,181	2,115	1,919	1,815	1,715	1,620
Gross national product (GNP)						
Current dollars (bil.)	5,765,000	5,932,000	6,255,000	6,563,000	6,932,000	7,247,000
1995 constant dollars (bil.)	6,625,000	6,555,000	6,729,000	6,881,000	7,106,000	7,247,000
Central government expenditures (CGE)						
1995 constant dollars (bil.)	1,499,000	1,583,000	1,554,000	1,564,000	1,571,000	1,598,000
People (mil.)	249.9	252.6	255.4	258.1	260.7	263.0
Military expenditure as % of GNP	5.3	4.7	4.9	4.5	4.2	3.8
Military expenditure as % of CGE	23.5	19.6	21.1	19.9	18.8	17.4
Military expenditure per capita (1995 $)	1,408	1,226	1,285	1,209	1,133	1,056
Armed forces per 1,000 people (soldiers)	8.7	8.4	7.5	7.0	6.6	6.2
GNP per capita (1995 $)	26,510	25,950	26,340	26,660	27,260	27,550
Arms imports[6]						
Current dollars (mil.)[11]	1,800	1,900	1,600	1,400	1,100	1,000
1995 constant dollars (mil.)[11]	2,069	2,100	1,721	1,468	1,128	1,000
Arms exports[6]						
Current dollars (mil.)	14,200	14,400	13,200	15,200	11,900	15,600
1995 constant dollars (mil.)	16,320	15,910	14,200	15,940	12,200	15,600
Total imports[7]						
Current dollars (mil.)	517,000	508,400	553,900	603,400	689,200	771,000
1995 constant dollars (mil.)	594,100	561,700	595,800	632,600	706,500	771,000
Total exports[7]						
Current dollars (mil.)	393,600	421,700	448,200	464,800	512,600	584,700
1995 constant dollars (mil.)	452,300	466,000	482,100	487,300	525,500	584,700
Arms as percent of total imports[8]	.3	.4	.3	.2	.2	.1
Arms as percent of total exports[8]	3.6	3.4	2.9	3.3	2.3	2.7

PRODUCTION SECTOR

Transportation (19)

Highways:

total: 6.42 million km

paved: 3,903,360 km (including 88,400 km of expressways)

unpaved: 2,516,640 km (1996 est.)

Waterways: 41,009 km of navigable inland channels, exclusive of the Great Lakes

Pipelines: petroleum products 276,000 km; natural gas 331,000 km (1991)

Merchant marine:

total: 286 ships (1,000 GRT or over) totaling 9,627,000 GRT/13,257,000 DWT

Airports: 14,574 (1997 est.)

Airports—with paved runways:

total: 5,167

over 3,047 m: 181

2,438 to 3,047 m: 218

1,524 to 2,437 m: 1,280

914 to 1,523 m: 2,450

under 914 m: 1,038 (1997 est.)

Airports—with unpaved runways:

total: 9,407

over 3,047 m: 1

2,438 to 3,047 m: 6

1,524 to 2,437 m: 164

914 to 1,523 m: 1,686

under 914 m: 7,550 (1997 est.)

Top Agricultural Products (20)

Wheat, other grains, corn, fruits, vegetables, cotton; beef, pork, poultry, dairy products; forest products; fish.

MANUFACTURING SECTOR

GDP & Manufacturing Summary (21)

Detailed value added figures are listed by both International Standard Industry Code (ISIC) and product title.

	1980	1985	1990	1994
GDP ($-1990 mil.)[1]	4,223,246	4,813,597	5,489,600	5,981,621
Per capita ($-1990)[1]	18,543	20,186	21,965	22,951
Manufacturing share (%) (current prices)[1]	21.5	19.4	18.3	17.2
Manufacturing				
Value added ($-1990 mil.)[1]	805,788	900,319	1,032,100	1,162,769
Industrial production index	100	113	128	148
Value added ($ mil.)	769,899	999,439	1,322,110	*1,611,763*
Gross output ($ mil.)	1,857,094	2,266,693	2,861,330	3,400,342
Employment (000)	19,210	17,424	17,502	*17,312*
Profitability (% of gross output)				
Intermediate input (%)	59	56	54	*53*
Wages and salaries inc. supplements (%)	21	21	21	*20*
Gross operating surplus	21	22	26	*27*
Productivity ($)				
Gross output per worker	96,673	130,090	163,486	*194,601*
Value added per worker	40,078	57,188	75,541	*93,199*
Average wage (inc. supplements)	20,044	27,953	33,565	*39,396*
Value added ($ mil.)				
311/2 Food products	63,460	87,970	119,830	*150,961*
313 Beverages	11,810	16,170	21,140	*26,835*
314 Tobacco products	6,160	11,890	22,560	*22,853*
321 Textiles	23,030	26,910	34,960	*43,311*
322 Wearing apparel	19,780	22,150	25,480	*29,373*
323 Leather and fur products	1,850	1,580	2,210	*2,219*
324 Footwear	2,950	2,470	2,320	*2,088*
331 Wood and wood products	12,970	15,390	20,830	*31,320*
332 Furniture and fixtures	9,840	13,250	16,910	*21,079*
341 Paper and paper products	29,790	40,390	57,200	*62,989*
342 Printing and publishing	44,390	73,050	103,180	*128,215*
351 Industrial chemicals	38,920	43,370	73,480	*80,023*
352 Other chemical products	35,530	54,290	81,770	*108,814*
353 Petroleum refineries	23,040	13,890	22,820	*18,910*
354 Miscellaneous petroleum and coal products	2,670	3,450	4,390	*5,961*
355 Rubber products	8,030	10,970	13,430	*16,908*
356 Plastic products	14,540	24,740	37,320	*52,896*
361 Pottery, china and earthenware	1,210	1,300	1,840	*2,254*
362 Glass and glass products	6,470	7,660	10,080	*11,317*
369 Other non-metal mineral products	16,300	19,890	23,980	*26,041*
371 Iron and steel	30,780	24,070	31,780	*37,071*
372 Non-ferrous metals	14,340	11,440	17,510	*19,448*
381 Metal products	53,180	61,810	70,360	*83,136*
382 Non-electrical machinery	102,760	115,550	145,060	*169,614*
383 Electrical machinery	74,850	111,230	112,400	*153,975*
384 Transport equipment	81,280	128,220	154,030	*187,297*
385 Professional and scientific equipment	27,940	40,280	76,520	*94,102*
390 Other manufacturing industries	12,060	13,060	18,720	*22,757*

FINANCE, ECONOMICS, & TRADE

Economic Indicators (22)

National product: GDP—purchasing power parity—$8.083 trillion (1997 est.)

National product real growth rate: 3.8% (1997)

National product per capita: $30,200 (1997 est.)

Inflation rate—consumer price index: 2% (1997)

Exchange Rates (24)

Exchange rates:	
British pounds (£) per US$	
January 1998	0.6115
1997	0.6106
1996	0.6403
1995	0.6335
1994	0.6529
1993	0.6658
Canadian dollars (Can$) per US$	
January 1998	1.4408
1997	1.3846
1996	1.3635
1995	1.3724
1994	1.3656
1993	1.2901
French francs (F) per US$	
January 1998	6.0836
1997	5.8367
1996	5.1155
1995	4.9915
1994	5.5520
1993	5.6632
Italian lire (Lit) per US$	
January 1997	1,787.7
1997	1,703.1
1996	1,542.9
1995	1,628.9
1994	1,612.4
1993	1,573.7
Japanese yen (¥) per US$	
January 1998	129.45
1997	120.99
1996	108.78
1995	94.06
1994	102.21
1993	111.20
German deutsche marks (DM) per US$	
January 1998	1.8167
1997	1.7341
1996	1.5048
1995	1.4331
1994	1.6228
1993	1.6533

Top Import Origins (25)

$822 billion (c.i.f., 1996) Data are for 1995.

Origins	%
Canada	NA
20%	NA
Western Europe	18
Japan	16.5
Mexico	8

NA stands for not available.

Top Export Destinations (26)

$625.1 billion (f.o.b., 1996) Data are for 1995.

Destinations	%
Canada	22
Western Europe	21
Japan	11
Mexico	8

Balance of Payments (23)

	1992	1993	1994	1995	1996
Exports of goods (f.o.b.)	440,350	458,720	504,450	577,740	613,980
Imports of goods (f.o.b.)	−536,460	−589,440	−668,590	−749,430	−803,230
Trade balance	−96,110	−130,720	−164,140	−171,690	−189,250
Services - debits	−228,090	−236,580	−280,260	−337,110	−360,240
Services - credits	303,160	314,670	349,710	413,750	441,260
Private transfers (net)	−19,840	−20,230	−19,130	−14,090	−19,260
Government transfers (net)	−15,780	−17,710	−19,110	−20,060	−21,220
Overall balance	−56,650	−90,570	−132,930	−129,190	−148,720

Economic Aid (27)

Donor: ODA, $9.721 billion (1993).

Import Export Commodities (28)

Import Commodities	Export Commodities
Crude oil and refined petroleum products	Capital goods
Machinery	Automobiles
Automobiles	Industrial supplies and raw materials
Consumer goods	Consumer goods
Industrial raw materials	Agricultural products
Food and beverages	

URUGUAY

Oriental Republic of Uruguay
República Oriental del Uruguay

CAPITAL: Montevideo.

FLAG: The flag, approved in 1830, has four azure blue horizontal stripes on a white background; on a white canton is a golden sun, alternately straight and wavy. This "Sun of May" symbolizes Uruguay's independence.

ANTHEM: *Himno Nacional,* which begins "Orientales, la patria o la tumba" ("Easterners [Uruguayans], our country or death").

MONETARY UNIT: The Uruguayan peso (UP), of 100 centésimos replaced the new peso in 1993 at the rate of UP1 = 1,000 new pesos. There are coins of 10, 20, and 50 centésimos and 1, 2, 5, and 10 new pesos, and notes of 1, 5, 10, 20, and 50 Uruguayan pesos. UP1 = $0.11806 ($1 = UP8.47).

WEIGHTS AND MEASURES: The metric system is the legal standard, but some traditional measures are also used.

HOLIDAYS: New Year's Day, 1 January; Epiphany, 6 January; Landing of the 33, 19 April; Labor Day, 1 May; Battle of Las Piedras, 18 May; Birthday of Artigas, 19 June; Constitution Day, 18 July; Independence Day, 25 August; Columbus Day, 12 October; All Souls' Day, 2 November; Blessing of the Waters, 8 December; Christmas Day, 25 December.

TIME: 9 AM = noon GMT.

LOCATION AND SIZE: The second-smallest South American country, Uruguay is situated in the southeastern part of the continent. It has an area of 176,220 square kilometers (68,039 square miles), slightly smaller than the state of Washington. Uruguay has a total boundary length of 2,063 kilometers (1,282 miles). Uruguay's capital city, Montevideo, is located in the southern part of the country on the Atlantic coast.

CLIMATE: The climate is temperate; the average temperatures are 15°C (59°F) in June and 25°C (77°F) in January. Rainfall is evenly distributed throughout the year; it averages about 109 centimeters (43 inches), varying from 97 centimeters (38 inches) in Montevideo to nearly 127 centimeters (50 inches) farther north. Frost is virtually unknown.

INTRODUCTORY SURVEY

RECENT HISTORY

In the late 1960s, economic and political instability stemming from the decline of wool revenues resulted in the emergence of a well-organized Marxist (communist) guerrilla movement, the Tupamaros, who mounted a campaign of kidnapping, assassination, and bank robbery. These activities, coupled with the worsening economic situation, aggravated Uruguay's political instability.

The military had seized control by 1973, crushing the Tupamaros. Suspending the constitution, military leaders ruled Uruguay until 1981. In 1979, Amnesty International estimated the number of political prisoners at 6,000.

When civilian rule was reestablished in 1985 under Julio María Sanguinetti Cairolo all political prisoners were released, amnesty was declared for former military and police leaders, and talks were initiated between employers and union leaders to reduce social tension. However, slow progress on the economic front led to the 1989 election of the more conservative candidate, Luis Alberto Lacalle, who emphasized deficit reduction, reforms in education, labor, and the civil service, and the return of state enterprises to private ownership. In 1994, Sanguinetti returned to office, and his administration began a program of economic reforms, including a long-range plan for cutting back on generous social security payments.

GOVERNMENT

The president is both head of state and chief executive. The president and vice president are elected on the same ticket by popular vote for five-year terms. A bicameral General Assembly consists of a 30-member Chamber of Senators and a 99-seat Chamber or Representatives, all of whose members are elected by popular vote to serve five-year terms.

Judiciary

The Supreme Court, whose judges are nominated by the president and ratified by the assembly, is the nation's highest court. There are also appeals, civil, and criminal courts. A parallel military court system operates under its own procedure.

Political Parties

Uruguay has Latin America's oldest two-party system. The Colorados are traditional Latin American liberals, representing urban business interests and favoring limitation on the power of the Catholic Church. The Blancos (officially called the National Party) are conservatives, defenders of large landowners and the Church.

DEFENSE

In 1995 the armed forces numbered 25,600, and defense spending totaled $320 million.

ECONOMIC AFFAIRS

Historically, Uruguay's economy has been based on agriculture. Currently, the services sector, especially tourism and financial services, accounts for over 60% of the gross domestic product (GDP). Agriculture, especially livestock, remains important in producing products for export.

Since 1990, the government has lowered tariffs, reduced deficit spending, controlled inflation, reduced the size of government, and entered into the Mercosur free-trade zone with some other South American nations.

Income

In 1998, the gross national product (GNP) was $19.4 billion, or about $6,020 per person. For the period 1985–95, the average inflation rate was 70.5%, resulting in a real growth rate in per person GNP of 3.3%.

Industry

Although foreign trade depends mainly on agriculture, the production of industrial goods for domestic consumption is increasing, primarily in the fields of textiles, rubber, glass, paper, electronics, chemicals, cement, light metallurgical manufactures, ceramics, and beverages. Most industry is concentrated in and around Montevideo. Manufacturing output fell by an average of 1.6% per year during 1990–95.

Banking and Finance

The Bank of the Republic, operates as both the financial agent of the government and as a commercial bank that makes loans and receives deposits. The Central Bank is responsible for currency circulation, thus permitting the Bank of the Republic to concentrate on public and private credit. A third state bank is the Mortgage Bank. There are 22 private commercial banks operating in Uruguay, many of which have been taken over by foreign interests following the introduction in 1982 of new banking regulations authorizing tax-free offshore banking.

A policy of regular mini-devaluations was introduced in mid-1975; currency stability was established in the late 1970s, but in November 1982 the peso, regarded as overvalued, was allowed to float freely.

The Montevideo Stock Exchange is small but active.

Economic Development

After a severe slump in the early 1980s, decisive government action in restructuring the heavy domestic debt burden of the industrial and agricultural sectors, increased confidence in the economy. Domestic investment rose sharply in the decade with some industries at full capacity.

Legislation passed in September 1991 permits partial privatization of certain state-owned enterprises. The Sanguinetti government instituted a new three-stage stabilization program in 1994. The plan increased consumption and payroll taxes, instituted a program designed to downsize government, and planned for long-term reform.

SOCIAL WELFARE

Social legislation provides holidays with pay, minimum wages, annual cash and vacation bonuses, family allowances, compensation for unemployment or dismissal, workers' accident compensation, retirement pensions for rural and domestic workers, old age and disability pensions, and special consideration for working women and minors. The state also provides care for children and mothers, as well as for the blind, deaf, and mute. Free medical attention is available to the poor, as are low-cost living quarters for workers.

In 1995, women made up about one half of the work force but tended to be concentrated in lower paying jobs. Nevertheless, many attend the national university and pursue professional careers.

Healthcare

In 1992 there were 2.9 doctors per 1,000 people. In 1990 there were 4.6 hospital beds per 1,000 people (about 15,000 beds).

For the region, life expectancy is high (74 years); infant mortality is low (19.9 per 1,000 live births in 1990–95). The major causes of death are heart diseases, cancer, and digestive disorders. In 1992, 82% of the population had access to health care services.

Housing

Nearly all housing units are made of durable materials including stone masonry, wood, zinc, or concrete. Of all housing units, 92% had private toilet facilities and 74% had water piped indoors.

EDUCATION

Adult illiteracy in 1995 was approximately 3%, among the lowest in Latin America. Elementary education, which lasts six years, is compulsory. In 1994 there were 2,423 primary schools, with 15,793 teachers and 337,889 students. There were 263,180 students in secondary and technical schools. Enrollment at all institutions of higher learning was 68,227 in 1992.

1999 KEY EVENTS TIMELINE

January

- Former mayor of Montevideo, leftist Tabaré Vázquez is perceived as the early favorite for the October presidential elections.

February

- A record number of Argentines visit the tourist resort of Punta del Este, as a favorable exchange rate has made it the favorite summer spot for Argentines.

March

- After the currency devaluation in Brazil, Uruguay vows to continue supporting the Mercosur economic alliance (composed of Brazil, Argentina, Uruguay, Paraguay and two associate members: Chile and Bolivia).

April

- The construction the Curz del Sur pipeline begins. The pipeline is a 133-mile natural gas pipeline connecting Argentina and Uruguay.

June

- President Sanguinetti attends the Rio Summitt of European, Latin American and Caribbean nations.

July

- Leftist presidential candidate Tabaré Vázquez, facing ruling Colorado Party candidate Jorge Battle and conservative National Party candidate Luis Alberto Lacalle, remains ahead in all polls for the October 30 presidential election.

August

- The IMF reports that Uruguay will suffer a recession in 1999. Uruguay's GNP is expected to fall by 2%, mostly as a result of the negative aftershocks of the January 1999 recession.

- The U.S. announces that Uruguayan citizens will no longer require visas to travel to the United States.
- Uruguay takes a leading role calling on neighboring Argentina and Brazil to coordinate economic policy and strengthen the Mercosur alliance.

November

- With a seven-and-a-half-point victory margin, Jorge Batlle, presidential candidate of the Colorado Party, which (along with the National Party) has dominated Uruguayan politics since the 1850s, defeats his opponent Tabare Vazquez of the left-wing Broad Front.

ANALYSIS OF EVENTS: 1999

BUSINESS AND THE ECONOMY

The negative effects of the Brazilian devaluation and the run against the Argentine Peso hurt Uruguay's economy in 1999. The two largest Mercosur members are Uruguay's main trade partners and the economic crisis in those countries heavily hurt Uruguay's exports. At the same time, lower prices for Brazilian products put pressure on Uruguay's industrial production and service sector. Uruguay's economy is expected have a 2 percent negative growth rate in 1999, but inflation will stay under 5 percent. Because the country is heavily dependent of state sector employment, government spending has increased to offset job losses in the private sector.

An economic recovery in Brazil and Argentina, expected to begin in late 1999, should alleviate Uruguay's negative trade balance and help restore a healthy export-oriented economy. Uruguay's commitment to Mercosur is the only alternative of development for the smallest country in South America. The well-educated population in that country is already employed mostly in the service sector and a healthier Mercosur alliance will open ways for Uruguay to explore markets in the neighboring countries. Infrastructure development and the incorporation of the banking and financial sector into the Mercosur trade pact are the next challenges faced by the Uruguayan business sector as it strives to adapt to the new world economy.

The government has been unable or unwilling to meet the demands of the International Monetary Fund (IMF) and other credit agencies. These large and powerful institutions pose strict demands in return for loans. Some of the demands of the IMF include a drastic reduction in the size of the government's payroll and the adoption of an aggressive privatization program. For the most recent presidential election, all major candidates vowed to defend the welfare state that has made Uruguay one of the most state-controlled economies in the region.

GOVERNMENT AND POLITICS

The presidential elections scheduled for October 30 were poised to alter the traditional political party structure. The centrist Colorado Party and the conservative National Party have alternated in power for most of the century. In 1990, leftist intellectual Tabaré Vázquez was elected mayor of Montevideo, where more than half of Uruguayans reside. As a popular mayor, Vázquez succeeded in presenting himself and his political coalition as a viable national alternative to traditional Colorado and National Party politicians. Vázquez coalition was called Broad Front and it includes a myriad of former leftist guerrillas, ecologists, progressive groups and traditional Colorado politicians and sympathizers dissatisfied with the conservative policies adopted by president Sanguinetti in his second term in office. Vázquez was expected to win the first-round in October, but a likely Colorado-National alliance for the second round is expected to defeat Vázquez in the run-off election to be held in late November. Whatever the end result might be, Vázquez' Broad Front was expected to clinch enough parliamentary seats to become the largest political party in congress. Regardless of who becomes the next president of Uruguay, the Broad Front was expected to consolidate itself as a third major party in a country characterized by bi-polar politics throughout the 20th century.

CULTURE AND SOCIETY

The arrest of General Pinochet, Chile's former dictator in October of 1998 in London sent shockwaves through the countries that had experienced military dictatorships in the 1970s and 1980s. The dictatorships in Argentina and Uruguay were noted for human right violations that included widespread torture and the disappearance of political opponents. In Uruguay, unlike Argentina, no major human right trials were conducted against former military dictatorship officials. In fact, the country had apparently settled the issue when a national referendum on amnesty law was approved by the electorate in 1989. Yet, many of the perpetrators of human rights violations were sued for civil damages and some victims sought justice in other countries. Most recently, an international campaign was launched to find the remains of some of the disappeared. Among the victims, the son of author Juan Gelman became the symbol of the disappeared for many Uruguayans. International personalities like Nobel prize winner José Saramago wrote an open letter to president Sanguinetti asking him to help find the remains of the disappeared. Human rights continue to divide the country, although democracy returned to Uruguay in 1984.

An additional element that is making itself present in Uruguay's society is the growing influence of Argentina in the country. A Mercosur partner, Argentines and Uruguayans have many cultural traditions in common and a similar history. Uruguay is located near Buenos Aires and Uruguay's main trade partner is Argentina. Punta del Este, Uruguay's most important tourist resort, is practically owned by well-to-do Argentines from Buenos Aires who prefer to vacation there than to go to more distant Argentine resorts. Some evidence of nationalist, anti-Argentine sentiments arise particularly during the summer months of January and February, but for the most part Uruguay maintained its reputation as South America's friendliest nation.

DIRECTORY

CENTRAL GOVERNMENT

Head of State

President
Jorge Batlle, Office of the President
E-MAIL: presidente@presidencia.gub.uy

Ministers

Minister of Agriculture and Fishing
Sergio Chiesa, Ministry of Agriculture and Fishing, Millán 4703, Montevideo, Uruguay
FAX: +598 (2) 341935
E-MAIL: dgsa@chasque.apc.org

Minister of Economy and Finance
Luis Mosca, Ministry of Economy and Finance

Minister of Education and Culture
Samuel Lichtensztejn, Ministry of Education and Culture

Minister of Foreign Affairs
Didier Opertti, Ministry of Foreign Affairs

Minister of Foreign Affairs
Roberto Rodriguez Pioli, Ministry of Foreign Affairs

Minister of Housing and Environment
Juan Chiruchi, Ministry of Housing and Environment

Minister of Industry and Energy
Julio Herrera, Ministry of Industry and Energy

Minister of Interior
Luis Jierro Lopez, Ministry of Interior

Minister of Labor and Social Welfare
Ana Lia Pineyrua, Ministry of Labor and Social Welfare

Minister of National Defense
Juan Luis Storace, Ministry of National Defense

Minister of Public Health
Raul Bustos, Ministry of Public Health

Minister of Tourism
Benito Stern, Ministry of Tourism

Minister of Transportation and Public Works
Lucio Caceres, Ministry of Transportation and Public Works

POLITICAL ORGANIZATIONS

Partido Colorado (Colored Party)

NAME: Jorge Batlle

Partido Nacional-Blancos (National Party-Whites)

NAME: Alberto Volonte Berro

Nuevo Espacio (New Sector Coalition)

Eduardo Acevedo 1615, Montevideo, Uruguay
PHONE: +598 (2) 4026989
FAX: +598 (2) 4026991
E-MAIL: larosa@adinet.com.uy
NAME: Hugo Batalla

Encuentro Progresista (Progressive Encounter)

NAME: Tabaré Vázquez

DIPLOMATIC REPRESENTATION

Embassies in Uruguay

Argentina
Cuareim 1470, 11300 Montevideo, Uruguay
NAME: Hernan Maria Patino Mayer

Canada
Edificio Torre Libertad, Plaza Cagancha 1335, Office 1105, 11100 Montevideo, Uruguay
PHONE: +598 (2) 9022030
FAX: +598 (2) 9022029
E-MAIL: mvdeo@dfait-maeci.gc.ca
TITLE: Ambassador
NAME: Brian Northgrave

Italy
Ufficio Commerciale, Montevideo, Uruguay
PHONE: +598 (2) 780542
FAX: +598 (2) 784148
E-MAIL: uffcomit@netgate.com.uy

United States
Lauro Muller 1776, APO AA 34035, Montevideo, Uruguay
PHONE: +598 (2) 236061; 487777
FAX: +598 (2) 488611

JUDICIAL SYSTEM

Supreme Court

FURTHER READING

Articles

Capote, Humberto. "Policymakers Roundtable: the Uruguayan Experience." *Journal of Banking and Finance* (October 1999): 1549.

Eisen, Peter. "Uruguay Gives Go-Ahead For Start of Construction Of Cruz del Sur Pipeline." *The Oil Daily*, 24 March 1999.

"Sour Mercosur." *The Economist* (August 14, 1999).

"Uruguay Stirring." *The Economist* (October 30, 1999).

"Uruguay Turns Left." *Business Week* (November 15, 1999): 69.

Internet

Embassy of Uruguay Washington D.C. Available Online @ http://www.embassy.org/uruguay/ (November 17, 1999).

Falklands-Malvinas. Available Online @ http://www.falkland-malvinas.com/ (November 17, 1999).

URUGUAY: STATISTICAL DATA

For sources and notes see "Sources of Statistics" in the front of each volume.

GEOGRAPHY

Geography (1)

Area:

Total: 176,220 sq km.

Land: 173,620 sq km.

Water: 2,600 sq km.

Area—comparative: slightly smaller than Washington State.

Land boundaries:

Total: 1,564 km.

Border countries: Argentina 579 km, Brazil 985 km.

Coastline: 660 km.

Climate: warm temperate; freezing temperatures almost unknown.

Terrain: mostly rolling plains and low hills; fertile coastal lowland.

Natural resources: fertile soil, hydropower, minor minerals, fisheries.

Land use:

Arable land: 7%

Permanent crops: 0%

Permanent pastures: 77%

Forests and woodland: 6%

Other: 10% (1997 est.)

HUMAN FACTORS

Demographics (2A)

	1990	1995	1998	2000	2010	2020	2030	2040	2050
Population	3,105.7	3,216.4	3,284.8	3,332.8	3,582.5	3,810.6	4,013.5	4,168.1	4,256.0
Net migration rate (per 1,000 population)	NA	NA	NA	NA	NA	NA	NA	NA	NA
Births	NA	NA	NA	NA	NA	NA	NA	NA	NA
Deaths	NA	NA	NA	NA	NA	NA	NA	NA	NA
Life expectancy - males	69.4	71.5	72.4	73.0	75.4	77.1	78.3	79.2	79.8
Life expectancy - females	76.1	78.0	78.8	79.4	81.8	83.5	84.6	85.4	85.9
Birth rate (per 1,000)	18.2	17.1	16.9	16.7	15.3	14.0	13.1	12.2	11.5
Death rate (per 1,000)	9.9	9.1	8.9	8.7	8.4	8.3	8.6	9.3	10.2
Women of reproductive age (15-49 yrs.)	739.6	781.7	802.3	810.7	860.4	903.2	909.0	914.5	907.8
of which are currently married	NA	NA	NA	NA	NA	NA	NA	NA	NA
Fertility rate	2.5	2.3	2.3	2.3	2.2	2.1	2.0	1.9	1.9

Except as noted, values for vital statistics are in thousands; life expectancy is in years.

Health Personnel (3)

Total health expenditure as a percentage of GDP, 1990-1997[a]

Public sector	1.9
Private sector	6.5
Total[b]	8.5

Health expenditure per capita in U.S. dollars, 1990-1997[a]

Purchasing power parity	618
Total	440

Availability of health care facilities per 100,000 people

Hospital beds 1990-1997[a]	450
Doctors 1993[c]	309
Nurses 1993[c]	61

Health Indicators (4)

Life expectancy at birth	
1980	70
1997	74
Daily per capita supply of calories (1996)	2,830
Total fertility rate births per woman (1997)	2.4
Maternal mortality ratio per 100,000 live births (1990-97)	85[c]
Safe water % of population with access (1995)	89
Sanitation % of population with access (1995)	61
Consumption of iodized salt % of households (1992-98)[a]	....
Smoking prevalence	
Male % of adults (1985-95)[a]	41
Female % of adults (1985-95)[a]	27
Tuberculosis incidence per 100,000 people (1997)	31
Adult HIV prevalence % of population ages 15-49 (1997)	0.33

Infants and Malnutrition (5)

Under-5 mortality rate (1997)	21
% of infants with low birthweight (1990-97)	8
Births attended by skilled health staff % of total[a]	96
% fully immunized (1995-97)	
TB	99
DPT	88
Polio	88
Measles	80
Prevalence of child malnutrition under age 5 (1992-97)[b]	4

Ethnic Division (6)

White	88%
Mestizo	8%
Black	4%

Religions (7)

Roman Catholic 66% (less than one-half of the adult population attends Church regularly), Protestant 2%, Jewish 2%, nonprofessing or other 30%.

Languages (8)

Spanish, Portunol, or Brazilero (Portuguese-Spanish mix on the Brazilian frontier).

EDUCATION

Public Education Expenditures (9)

Public expenditure on education (% of GNP)	
1980	2.3
1996	3.3
Expenditure per student	
Primary % of GNP per capita	
1980	11.1
1996	9.3
Secondary % of GNP per capita	
1980	....
1996	....
Tertiary % of GNP per capita	
1980	28.1
1996	24.3
Expenditure on teaching materials	
Primary % of total for level (1996)	5.8
Secondary % of total for level (1996)	....
Primary pupil-teacher ratio per teacher (1996)	20
Duration of primary education years (1995)	6

Educational Attainment (10)

Age group (1996)	25+
Total population	2,242,251
Highest level attained (%)	
No schooling	3.4
First level	
Not completed	53.6
Completed	NA
Entered second level	
S-1	31.7
S-2	NA
Postsecondary	10.1

Literacy Rates (11A)

In thousands and percent[1]	1990	1995	2000	2010
Illiterate population (15+ yrs.)	76	65	55	39
Literacy rate - total adult pop. (%)	96.7	97.3	97.8	98.5
Literacy rate - males (%)	96.2	96.9	97.5	98.3
Literacy rate - females (%)	97.1	97.7	98.1	98.7

GOVERNMENT & LAW

Political Parties (12)

Chamber of Representatives	% of seats
Colorado	.32
National (Blanco) Party	.31
Encuentro Progresista	.31
New Sector Coalition	.5

Military Affairs (14B)

	1990	1991	1992	1993	1994	1995
Military expenditures						
Current dollars (mil.)	301	288[e]	333	290[e]	437	410[e]
1995 constant dollars (mil.)	346	318[e]	359	304[e]	448	410[e]
Armed forces (000)	25	25	25	25	25	25
Gross national product (GNP)						
Current dollars (mil.)	12,160	13,070	14,460	15,020	16,250	17,170
1995 constant dollars (mil.)	13,970	14,440	15,550	15,750	16,660	17,170
Central government expenditures (CGE)						
1995 constant dollars (mil.)	3,806	4,074	4,495	5,444	6,174	5,574
People (mil.)	3.1	3.1	3.1	3.2	3.2	3.2
Military expenditure as % of GNP	2.5	2.2	2.3	1.9	2.7	2.4
Military expenditure as % of CGE	9.1	7.8	8.0	5.6	7.3	7.3
Military expenditure per capita (1995 $)	111	102	114	96	140	127
Armed forces per 1,000 people (soldiers)	8.0	8.0	7.9	7.9	7.8	7.8
GNP per capita (1995 $)	4,499	4,617	4,938	4,966	5,217	5,339
Arms imports[6]						
Current dollars (mil.)	20	30	10	5	0	5
1995 constant dollars (mil.)	23	33	11	5	0	5
Arms exports[6]						
Current dollars (mil.)	0	0	0	0	0	0
1995 constant dollars (mil.)	0	0	0	0	0	0
Total imports[7]						
Current dollars (mil.)	1,343	1,637	2,045	2,324	2,786	2,867
1995 constant dollars (mil.)	1,543	1,809	2,200	2,436	2,856	2,867
Total exports[7]						
Current dollars (mil.)	1,693	1,605	1,703	1,645	1,913	2,106
1995 constant dollars (mil.)	1,946	1,774	1,832	1,725	1,961	2,106
Arms as percent of total imports[8]	1.5	1.8	.5	.2	0	.2
Arms as percent of total exports[8]	0	0	0	0	0	0

Government Budget (13A)

Year: 1997

Total Expenditures: 59,743 Millions of Pesos

Expenditures as a percentage of the total by function:

General public services and public order	10.01
Defense	4.42
Education	7.22
Health	5.88
Social Security and Welfare	60.36
Housing and community amenities	1.43
Recreational, cultural, and religious affairs	.47
Fuel and energy	....
Agriculture, forestry, fishing, and hunting	....
Mining, manufacturing, and construction	....
Transportation and communication	....
Other economic affairs and services	....

Crime (15)

Crime rate (for 1994)

Crimes reported	74,156
Total persons convicted	2,995
Crimes per 100,000 population	2,342

Persons responsible for offenses

Total number of suspects	NA
Total number of female suspects	NA
Total number of juvenile suspects	NA

LABOR FORCE

Labor Force (16)

Total (million)	1.38
Government	25%
Manufacturing	19%
Agriculture	11%
Commerce	12%
Utilities, construction, transport, and communications	12%
Other services	21%

Data for 1997 est. Percentages for 1988 est.

Unemployment Rate (17)

10.3% (December 1997)

PRODUCTION SECTOR

Electric Energy (18)

Capacity	2.055 million kW (1995)
Production	7.6 billion kWh (1995)
Consumption per capita	1,852 kWh (1995)

Transportation (19)

Highways:

total: 8,420 km

paved: 7,578 km

unpaved: 842 km (1996 est.)

Waterways: 1,600 km; used by coastal and shallow-draft river craft

Merchant marine:

total: 2 oil tanker ships (1,000 GRT or over) totaling 44,042 GRT/83,684 DWT (1997 est.)

Airports: 64 (1997 est.)

Airports—with paved runways:

total: 15

2,438 to 3,047 m: 1

1,524 to 2,437 m: 5

914 to 1,523 m: 8

under 914 m: 1 (1997 est.)

Airports—with unpaved runways:

total: 49

1,524 to 2,437 m: 2

914 to 1,523 m: 14

under 914 m: 33 (1997 est.)

Top Agricultural Products (20)

Wheat, rice, corn, sorghum; livestock; fishing.

MANUFACTURING SECTOR

GDP & Manufacturing Summary (21)

Detailed value added figures are listed by both International Standard Industry Code (ISIC) and product title.

	1980	1985	1990	1994
GDP ($-1990 mil.)[1]	7,905	6,897	8,281	9,805
Per capita ($-1990)[1]	2,713	2,293	2,677	3,096
Manufacturing share (%) (current prices)[1]	26.0	27.1	24.7	*16.1*
Manufacturing				
Value added ($-1990 mil.)[1]	2,393	1,815	2,173	*2,076*
Industrial production index	100	76	91	*87*
Value added ($ mil.)	1,286	1,337	2,379	*3,181*
Gross output ($ mil.)	3,302	3,174	5,778	*6,738*
Employment (000)	160	122	168	*136*
Profitability (% of gross output)				
Intermediate input (%)	61	58	59	*53*

	1980	1985	1990	1994
Wages and salaries inc. supplements (%)	*13*	9	10	12
Gross operating surplus	*26*	33	32	*36*
Productivity ($)				
Gross output per worker	20,456	26,023	34,428	*49,487*
Value added per worker	7,966	10,964	14,177	*23,422*
Average wage (inc. supplements)	*2,635*	2,442	3,280	*5,781*
Value added ($ mil.)				
311/2 Food products	165	266	473	*792*
313 Beverages	104	92	181	*393*
314 Tobacco products	90	68	90	*170*
321 Textiles	109	137	235	*253*
322 Wearing apparel	59	43	111	*113*
323 Leather and fur products	31	76	67	*74*
324 Footwear	18	8	19	*14*
331 Wood and wood products	*14*	8	19	*16*
332 Furniture and fixtures	*7*	2	13	*18*
341 Paper and paper products	30	47	51	*72*
342 Printing and publishing	37	27	81	*159*
351 Industrial chemicals	20	26	68	*57*
352 Other chemical products	75	112	162	*256*
353 Petroleum refineries	192	194	239	*99*
354 Miscellaneous petroleum and coal products	2	4	1	*2*
355 Rubber products	40	34	58	*37*
356 Plastic products	24	25	65	*96*
361 Pottery, china and earthenware	13	7	20	*26*
362 Glass and glass products	14	7	21	*22*
369 Other non-metal mineral products	41	24	43	*71*
371 Iron and steel	10	14	31	*27*
372 Non-ferrous metals	3	3	3	*11*
381 Metal products	*53*	32	73	*116*
382 Non-electrical machinery	*16*	12	22	*34*
383 Electrical machinery	33	31	69	*88*
384 Transport equipment	78	32	129	*121*
385 Professional and scientific equipment	1	1	19	*21*
390 Other manufacturing industries	8	6	15	*19*

FINANCE, ECONOMICS, & TRADE

Economic Indicators (22)

National product: GDP—purchasing power parity—$29.1 billion (1997 est.)

National product real growth rate: 5.1% (1997)

National product per capita: $8,900 (1997 est.)

Inflation rate—consumer price index: 15.2% (1997)

Balance of Payments (23)

	1992	1993	1994	1995	1996
Exports of goods (f.o.b.)	1,801	1,732	1,918	2,148	2,440
Imports of goods (f.o.b.)	−1,923	−2,118	−2,624	−2,711	−3,142
Trade balance	−122	−387	−706	−563	−703
Services - debits	−971	−1,189	−1,387	−1,489	−1,487
Services - credits	1,055	1,278	1,613	1,763	1,820
Private transfers (net)	29	22	8	34	31
Government transfers (net)	—	32	33	42	43
Overall balance	−9	−244	−438	−212	−296

Exchange Rates (24)

Exchange rates:

Uruguayan pesos ($Ur) per US$1

January 1998	9.98
1997	9.4448
1996	7.9718
1995	6.3491
1994	5.0529
1993	3.9484

Top Import Origins (25)

$3.7 billion (c.i.f., 1997)

Origins	%
Cote d'Ivoire	NA
France	NA
Togo	NA
Nigeria	NA

NA stands for not available.

Top Export Destinations (26)

$2.7 billion (f.o.b., 1997).

Destinations	%
Brazil	NA
Argentina	NA
United States	NA
Germany	NA
Italy	NA

NA stands for not available.

Economic Aid (27)

Recipient: ODA, $63 million (1994).

Import Export Commodities (28)

Import Commodities	Export Commodities
Machinery and equipment	Wool and textile manufactures
Vehicles	Beef and other animal products
Chemicals	Rice
Minerals	Fish and shellfish
Plastics	Chemicals
Oil	

UZBEKISTAN

Republic of Uzbekistan
Uzbekiston Respublikasi

CAPITAL: Tashkent (Toshkent).

FLAG: Horizontal bands of blue (top), white, and green separated by narrow red bands; white crescent moon and twelve stars on the blue band.

MONETARY UNIT: The som (SOM) is the official currency, introduced when Uzbekistan left the ruble zone in November 1993. SOM1 = $0.02639 (or $1 = SOM37.9).

WEIGHTS AND MEASURES: The metric system is used.

HOLIDAYS: Independence Day, 1 September.

TIME: 5 PM = noon GMT.

LOCATION AND SIZE: Uzbekistan, a republic of the former Soviet Union, is located in central Asia bordering the Aral Sea, between Kazakstan and Turkmenistan. Comparatively, it is slightly larger than the state of California, with a total area of 447,400 square kilometers (172,742 square miles). Uzbekistan's boundary length totals 6,221 kilometers (3,866 miles). Its capital city, Tashkent, is located in the eastern part of the country.

CLIMATE: Temperatures in the desert can reach upwards of 32°C (90°F). In the capital city of Tashkent the mean temperature is 32°C (90°F) in the summer and 3°C (37°F) in January. There is very little rainfall in the country.

INTRODUCTORY SURVEY

RECENT HISTORY

The Uzbek Soviet Socialist Republic was created in 1925. Under the leadership of long-time leader S. Rashidov, Uzbekistan was politically conservative during the 1970s and early 1980s.

In March 1990, Islam Karimov was elected to the newly created post of president by the Uzbek Supreme Soviet. Uzbekistan declared independence from the former Soviet Union on 1 September 1991. Karimov was reelected in December 1991, but he has opposed moving Uzbekistan towards democracy. Opposition parties were banned in 1992 and political reformers were either put in jail or fled the country.

Karimov also opposes the fundamentalist Islamic military and political movements in Afghanistan and Tajikistan. Uzbekistan supplies arms to the secular forces in those countries' civil wars.

GOVERNMENT

The executive branch consists of the president and his appointed prime minister and Cabinet of Ministers. The legislative branch consists of a single-chamber Supreme Soviet of 150 seats. The judicial branch is appointed by the president.

Judiciary

Three levels of courts include: district courts, regional courts, and the Supreme Court.

Political Parties

The ruling National Democratic Party is the renamed Communist Party. In 1992, opposition parties were banned.

DEFENSE

Uzbekistan has an army of 25,000, an air force of 4,000 with 126 combat aircraft, and about 15,300 internal security personnel.

ECONOMIC AFFAIRS

Uzbekistan has one of the lowest per person incomes in Central Asia. The world's third largest cotton producer, the country is evolving from a mainly agricultural economy to include more industry. Agriculture and agricultural processing in-

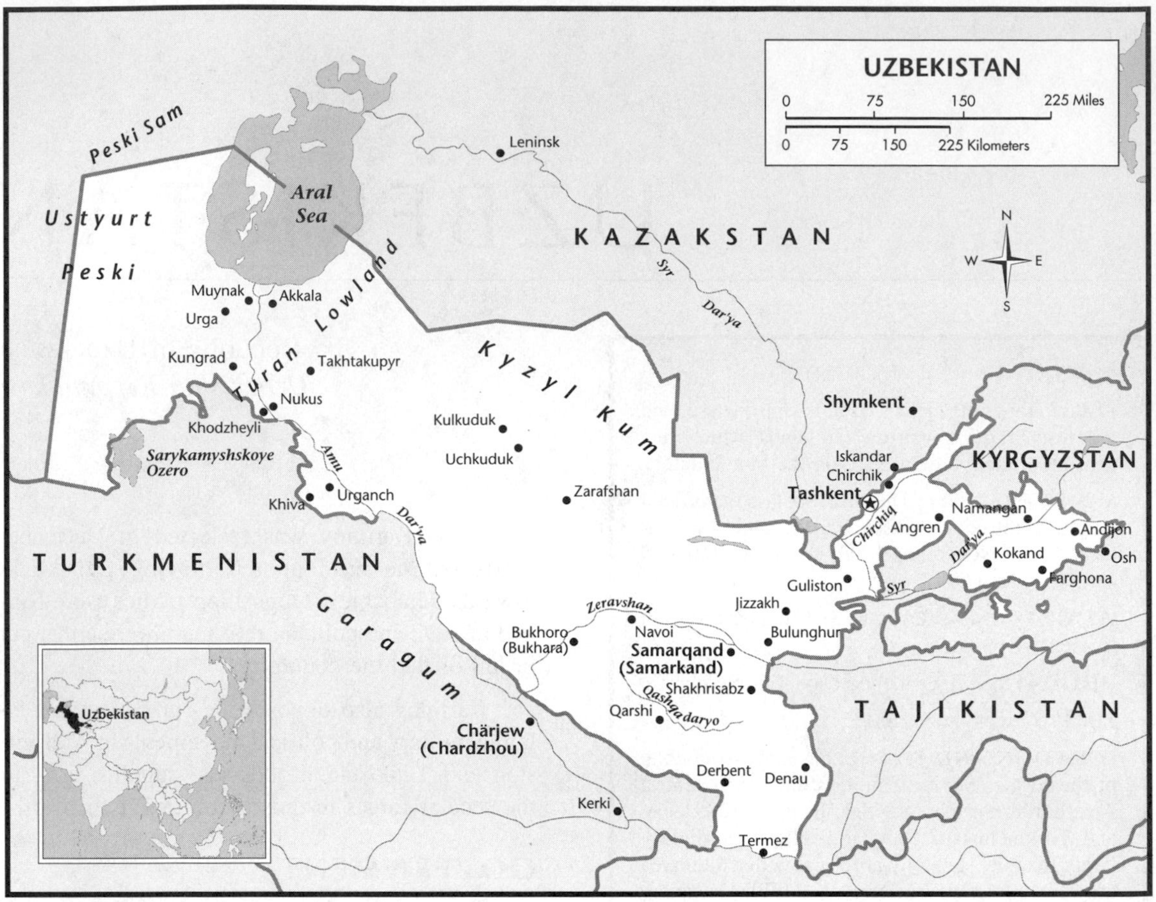

dustries, however, still accounted for half of the gross national product (GNP) in 1996.

Measures toward establishing a market economy were more cautious than in many other post-Soviet countries. The domestic economy shrank by 17% between 1991 and 1994. The government has recently begun to move from a command-driven to a market-driven economy. As a result, economic decline slowed to just 1% in 1996, and the inflation rate fell to 35% (compared to 1,300% in 1994).

Public Finance

Most enterprises, including those in the energy sector, remain in state hands. Privatized enterprises, however, contribute about 16% of government revenues. Subsidies for basic consumer goods (except some food staples and energy products) and subsidized credit to industrial enterprises were substantially reduced during 1994 and 1995. In response, the budget deficit went from a high of 16% in 1993 to 3.3% in 1995. External debt of the end of 1994 was estimated at $1.5 billion. An enterprise profit tax, a value-added tax, and an excise tax on cotton supply the bulk of government revenues.

Income

In 1998, Uzbekistan's gross national product (GNP) was U.S.$20.9 billion, or $810 per person. For the period 1985–95 the average inflation rate was 239%, and the average annual rate of decline in the GNP per person was 3.9%.

Industry

Uzbekistan is the primary producer of machines and heavy equipment in Central Asia. Leading manufactures include: cotton, wool, silk fiber, and processed foods including cottonseed oil, meat, dried fruit, wines, and tobacco.

Banking and Finance

The Central Bank of Uzbekistan manages the country's two-tier banking system and regulates commercial banks by setting reserve requirements and the discount rate. The Uzbek National Bank of Foreign Economic Activities, a state bank, deals exclusively with the foreign exchange rate.

Commercial banks in the country include the Uzbek Commercial Bank and the Uzbek Joint-Stock Innovation Bank. Uzbekistan does not have a security market, but the trading of commodities is widely practiced.

Economic Development

Under centralized Soviet economic planning, extensive stretches of Uzbekistan's land were brought under irrigation for cotton cultivation. Cotton production is expected to continue to play a key role foreign trade. A little over 26% of state investment allocations were directed at the agricultural sector in 1990. Industry received about 22%. Housing and trades and services were other large state investment targets.

In 1995 the government announced a Mass Privatization Program (MPP) with the goal of increasing the private sector's share of the economy from 40% to 60%. Although nearly 60,000 small businesses (96% of the total) and 14,000 farms (accounting for 11% of arable land) had been privatized by 1997, only 20% of Uzbekistan's medium and large-sized enterprises were in private hands.

For the immediate future, developing the country's oil and natural gas fields, bolstering cotton exports through productivity enhancement, and sustaining gold exports are key strategies for procuring some of the necessary financing to support economic development.

In 1992 Uzbekistan agreed to transfer its share of the former Soviet Union's debt to the latter in exchange for relinquishing all claims on Soviet assets. Growing water shortages, severe river and lake pollution caused by heavy use of chemicals in agriculture, the desiccation of the Aral Sea due to massive irrigation, and high levels of both air and water pollution in the country's industrial centers are among the country's most pressing environmental problems.

SOCIAL WELFARE

The social security system, developed in 1994, includes pensions for old age, disability, and survivorship. Sickness, maternity, work injury, and unemployment benefits are also included. Traditional customs decree that women generally marry young and confine their activities to the home.

Healthcare

The average life expectancy is 64 years. Uzbekistan had 3.3 and 12.4 hospital beds per 1,000 persons in 1992.

Services under Uzbekistan's health care system are mostly free of charge. However, the system is inefficient and is under pressure to change its organization and financing.

Housing

In 1990, Uzbekistan had 130 square feet of housing space per person.

EDUCATION

The population is highly educated. The estimated adult illiteracy rate in 1995 was 3%. In recent years, there has been an increased emphasis on teaching Uzbek literature, culture, and history in the schools.

1999 KEY EVENTS TIMELINE

January

- In the first official Tajik and Uzbek meeting since Rakhmonov accused Uzbekistan of harboring Tajik secessionists, Uzbek and Tajik authorities agree on resuming natural gas imports from Uzbekistan.
- Rakhmonov accuses Uzbekistan of direct involvement in the November 1998 insurrection led by Colonel Mahmud Khudoberdiev and former Premier Abdumalik Abdullodjonov.
- The Asian Development Bank plans to extend a $120 million loan to Uzbekistan to upgrade its rail system.

February

- Uzbekistan plans to conduct honest and democratic parliamentary elections in December 1999.
- Uzbekistan will not extend its participation in the CIS Mutual-Defense Treaty which expires in May. Neither does Uzbekistan see a role for Russia on the Tajik/Afghan border after the inclusion of the Mujahedin of the Islamic opposition into the Tajik national army.
- Karimov calls Hezbi Takhriri Islomiya a threat to the administrative boundaries of all Islamic countries. With the substantial financial backing it receives from abroad, Karimov claims, the party intends to remove all political boundaries and install an Islamic caliphate in the region.
- Russia regrets Uzbekistan's decision not to renew its participation in the CIS Mutual-Defense

Treaty, especially in view of NATO's activities near the Russian border.

- Russian authorities praise Uzbekistan as ''the sole country inside the former Soviet Union to maintain its industrial production at the 1990 level. They also praise Uzbekistan for combating Islamic fundamentalism.
- Uzbek customs officials charge $16 for every $100 worth of goods crossing into Uzbekistan from Kyrgyzstan. Uzbekistan's goods for Kazakstan are subject of a 200% tariff.
- Five bombs go off in Tashkent damaging buildings in various parts of the capital. Uzbek authorities blame the explosions on foreign extremist and terrorist organizations, Wahhabis, and Islamic movements.
- February 18 is declared a day of mourning for the victims of the bombings in Tashkent which killed 18 people and injured 150, 12 critically. Karimov threatens to ''cut off the hands of those responsible'' for the bombings.
- Ravshan Salijanov, 27, from the Uzbek city of Namangan, is apprehended in Tashkent. He is one of six men sought by Uzbek authorities in connection with the February 16 bombings in Tashkent.
- Uzbekistan issues new regulations on residency. Foreign citizens, including those from CIS countries, must present an official permit. Foreign visitors staying in Uzbekistan for more than three days must obtain visas.
- Mohammed Solih, the chairman of Erk, Uzbekistan's banned opposition party, is a suspect in the February bombings, training anti-government Uzbek youths in Chechnya, and colluding with the Wahhabis.
- Islam Karimov names Takhir Yuldash and Muhammad Solih as organizers in the February bombing in Tashkent.
- Karimov and Suleyman Demirel attend the opening ceremonies of the joint-venture automotive plant Samkochavto in Samarqand. The plant will produce 5,000 vehicles annually. Turkey's Koc Holding company built the $65 million plant and is co-owner.
- Uzbek citizens who fail to exchange their old Soviet passport for a new Uzbek passport by 1 January 2000 will forfeit their Uzbek citizenship. Since 1995, 12 million people have handed in their old passports and 1.8 million have yet to hand theirs in.
- Japan opens a $107.6 million credit line to finance improvements to Uzbekistan's telephone network.

April

- A Czech government delegation visits Uzbekistan to discuss expanding bilateral economic cooperation and trade. Trade turnover between the two countries totaled $800 million last year.
- Kazakstan bars Uzbek freight trains from transiting its territory until Tashkent pays an $8 million transit debt.
- All perpetrators and the organizers of the 16 February bomb attacks in Tashkent have now been arrested.
- In order to improve cooperation in the airline industry, Uzbekistan and Ukraine form a new ''CIS-Alliance'' air system.

May

- In Tashkent, Masahiko Komura attends the opening of the first Japanese International Cooperation Agency office in Central Asia. The Agency provides technical assistance in the transition to a market economy, environmental protection, and the development of transport, communications, and public health facilities.
- Severe sentences are handed down in the first of a series of trials of persons suspected of involvement in the Tashkent bombings that resulted in 15 deaths.
- Tashkent rejects Kazakstan's request to buy Uzbek natural gas for $30 per thousand cubic meters and the two countries fail to reach an agreement on complex border demarcation issues.

June

- During a visit to Tashkent by Chinese Deputy Premier Qian Qichen, Karimov calls for expanding ties between the two countries as part of an effort to revive ''the ancient Great Silk Road'' linking Europe to Asia.
- Twenty-two individuals are accused of terrorism, attempting to kill President Karimov, drug trafficking, illegal possession of weapons, and robbery. The state prosecutor calls for the death

sentence on 10 of the defendants, and for prison terms of 10–14 years for the remainder.

- Islam Karimov calls on other Central Asian states to work together to promote a rapprochement between the Tajik government and opposition.

July

- A court in Tashkent, in the course of a three hour trial which the Human Rights Watch characterized as "a farce," sentenced 48-year-old Mahbuba Kasymova to five years' imprisonment on charges of fraud and harboring a criminal. Kasymova is a member of the unregistered Independent Human Rights Organization of Uzbekistan and of the banned Birlik opposition party. Mohammad Solih, the banned Erk Democratic Party leader, speaking on Iranian Radio's Uzbek Service, denied Tashkent's charges that he had funded the dissemination of subversive literature in Uzbekistan and that he had been involved in the February bombings in Tashkent.
- Uzbekistan decides to build a special memorial in Tashkent to those who suffered under the Soviet regime in order to foster in young people "respect for their forefathers' heroism and selflessness, belief in the victory of social justice, and devotion to the ideas of independence and patriotism."

August

- The Erk Democratic Party, which has been banned in Uzbekistan, accuses Karimov of either masterminding the bombing himself or recruiting others to undertake it for him.
- Uzbekistan's Foreign Minister confirms earlier reports that some of the militants who had taken four Kyrgyz nationals hostage are Uzbek citizens.
- The Uzbek Foreign Ministry denies any knowledge of the bombings in which four jets approaching from Kyrgyz airspace dropped eight bombs on Tajikistan's Jirgatal district.
- Kyrgyz authorities claim that the ethnic Uzbek guerrilla band headed by Juma Namangani aims to create an Islamic state in the Fergana Valley that would include the Andijan, Fergana, and Namangan oblasts of Uzbekistan and the Leninabad oblast of Tajikistan.
- Karimov criticizes Tajikistan for its inability to control the situation in the eastern part of the country. He notes that Tajik authorities should not have permitted a group of armed militants to cross into neighboring Kyrgyzstan.
- After an unexpected 4.5% gross domestic product increase in 1999, Uzbekistan is heading for 5% economic growth in 2000. Industrial production is forecast to expand 6.1% and agriculture 6% in 1999.
- Elections to a new 250-seat parliament will take place on 5 December, together with elections to city and local councils. The presidential poll will be held on 9 January. Five registered political parties are entitled to field candidates.
- Ricardo Juarez outpoints Uzbekistan's Tulkunbay Turgunov 13–2 to take the gold in featherweight boxing division.
- The hostage situation in neighboring Kyrgyzstan and internal security fears dampen the mood for Uzbekistan's independence day.

September

- A man claiming to be a spokesman for Tohir Yuldashev, military commander of the Uzbek Islamic Movement, told the BBC that the guerrillas want to exchange their hostages for members of the Uzbek Islamic Movement currently imprisoned in Uzbekistan. The Uzbek guerrillas entrenched in southern Kyrgyzstan continue to hold 13 hostages, including four Japanese geologists and a Kyrgyz Interior Ministry general.
- The Boeing company delivers Uzbekistan Airways' first 757–200 to complement its 767 twin-aisle planes. The Airline plans to use the 757–200 fleet for routes connecting Uzbekistan to Europe and the Middle East.
- Zubair ibn Abdurrakim, chairman of the political council of the Islamic Movement of Uzbekistan, announced the beginning of a "holy war" against Uzbekistan with the aim of forcing the release of 50,000 Muslims held in Uzbek prisons.
- Karimov attends the opening ceremony of the Tashkent Islamic University. Established under a presidential decree, the university will teach the history and philosophy of Islam, Islamic law, economy, and natural sciences.
- Uzbekistan expects to harvest 4.1 million tons of raw cotton in 1999.
- Karimov alleges that the hostage-taking in southern Kyrgyzstan has been planned long in ad-

vance and constitutes part of a major conspiracy orchestrated by Islamic terrorists who aim to establish an Islamic state in Central Asia.

- Tashkent tightens control of its borders and places its armed forces on alert.
- Japan is interested in expanding trade with Uzbekistan, which last year stood at $122 million. Japanese companies currently have offices in Tashkent, but a further increase in investment is unlikely because the Uzbek currency is not fully convertible. Japan has invested over $1 billion in Uzbekistan since 1995, of which the Japanese government invested some $334 million.
- The Uzbek official newspaper *Slovo Uzbekistana* alleged that members of the United Tajik Opposition support the hostage-takers in southern Kyrgyzstan, and not the Islamic Movement of Uzbekistan. Leaders of the United Tajik Opposition reject all Uzbek claims. Over the last few weeks around 300 Uzbek prisoners, mainly from the banned Khezbut Takhrir (Liberation Party) religious organization have been released.
- In tennis, the Czech Republic beats Uzbekistan 5–0 in the Davis Cup world group playoff.
- Fidokorlar Congress nominates Karimov for the presidential elections.

October

- Uzbekistan blames Tajikistan's Islamic opposition movement for supporting the 650 Kyrgyzstan intruders who have been holding over a dozen hostages, including four Japanese geologists, for over six weeks.
- Parliament leader Djalilov announces he will contend for presidency in 2000.

November

- Gunmen occupy a tourist area about fifty kilometers from Tashkent, the capital, and six people—three police and three civilians—die in armed clashes with them.
- The Organization for Security and Cooperation in Europe (OSCE) says Uzbek elections laws don't meet international standars, so they won't send observers to monitor the December parliamentary elections.
- At least 24 are dead when an avalanche buries the mountain road on which those killed were traveling.

December

- Police arrest Jafizulloh Nasirov, believed to be the influential leader of the Islamic Party of Liberation or Hezb ut-Tahrir. Bombings in Tashkent linked to Islamic militants have resulted in an increase in arrests of Hezb ut-Tahrir members.
- Parliamentary elections are held December 5, with five parties competing for the 240 seats in the single-chamber Oli Mazhlis (parliament). The Organization for Security and Cooperation in Europe (OSCE) contends that the election does not meet international standards because all candidates running for election have been sanctioned by the government, and that the opposition parties are still banned.
- Authorities burn more than two tons of drugs, mostly heroin and opium, at a scrap metal plant in Tashkent. Narcotics supplies from neighboring Afghanistan offer Uzbek citizens one of the few ways to make quick money.
- Authorities close the airports from 10 p.m. December 31 through noon on January 1, 2000, to avoid any computer problems as the new millennium begins.

ANALYSIS OF EVENTS: 1999

BUSINESS AND THE ECONOMY

Despite the pitfalls and confusion of nation-building in the post-U.S.S.R. era as well as the recurrent challenges of Islamic fundamentalism, Uzbekistan made progress in building its economy during 1999 and appears headed for a more productive year in 2000. International developmental institutions seemed to concur in an optimistic appraisal of Uzbekistan's economic development. The Asian Development Bank extended a $120 million loan to Uzbekistan to upgrade its rail system. Even Russia congratulated Uzbekistan for maintaining its industrial production at the 1990 level.

Part of Uzbekistan's economic resilience is due to its rich endowment of natural resources. It is a major producer of natural gas as well as of gold. Its industrial base includes a well-developed network of chemical and machinery enterprises. It industrialized under socialism and although since

1994 it has allowed some market-based reforms, it continues to rely on a statist approach to fundamental economic decisions. Faced with the failure of socialist economies around it, the government of Uzbekistan has evidently learned how to manipulate its highly centralized economy. It does not appear ready to institute fundamental structural reforms such as the extensive privatization of state-owned enterprises. This fact so displeased the International Monetary Fund (IMF) that it terminated its $185 million credit line to Uzbekistan in 1996.

Uzbekistan continues to search out trading partners that might not be so influenced by IMF or U.S. demands for market-based reforms before engaging in trade. So far the numbers seem to indicate that this refusal to bow to U.S. economic doctrine seems to be working. In 1998, Uzbekistan's gross domestic product (GDP) increased by 4.4%, industrial output by 5.8%, and agricultural production by 4%. Monthly inflation stood at 1.9%, and the Central Bank's monthly refinancing rate stayed below 3%. Retail trade was up by 14% and consumer goods increased in output by 7.2%. Annual foreign investment totaled $1.3 billion, a 22.6% increase over the previous year. The foreign trade surplus reached $240 million.

Commercial relations with neighboring states remain generally good although Kazakstan bars Uzbek freight trains from crossing its territory until Tashkent (an administrative subdivision of eastern Uzbekistan) pays an $8 million transit debt. Since 1991, Turkey has invested $1 billion in 400 Uzbek-Turkish joint ventures. Presidents of Turkey and Uzbekistan attended the opening ceremonies of Samkochavto, the joint-venture automotive plant in Samarqand that is slated to produce 5,000 vehicles annually. The country will also be upgrading its transportation and communication systems. Boeing delivered Uzbekistan Airways' first 757–200 that will connect Uzbekistan to Europe and the Middle East. In addition, Japan plans to open a $107.6 million credit line to finance improvements to Uzbekistan's telephone network.

In 1999 Uzbekistan also attended the opening of the first Japanese International Cooperation Agency. This group offers technical assistance in the transition to a market economy, environmental protection, and in the development of transport, communications, and public health facilities. It is not at all clear that Uzbekistan is ready to break with its state-directed economy. After another 4.5% rise in the gross domestic product in 1999, the country is heading for 5.0% economic growth in 2000. Industrial production is forecast to expand by 6.1% and agriculture by 6.0%. Uzbekistan expects to harvest 4.1 million tons of raw cotton in 1999. This bilateral commercial arrangement has generated over $1 billion since 1995. Last year accounted for $122 million.

GOVERNMENT AND POLITICS

Although economic relations between states appear to be viable, the Uzbek national life continues to suffer internal disruption over the movements of regional autonomy and of Islamic fundamentalism. President Karimov of Tajikistan, Uzbekistan's neighbor to the east, accused Uzbekistan of involvement in an incident in November 1998 in Khujand, a city in the northwest corner of Tajikistan. He claimed that Uzbek insurgents intended to install an Islamic caliphate in Central Asia. Russia, on the other hand, praised Uzbekistan for combating Islamic fundamentalism. In a separate attack in Tashkent, the capital of Uzbekistan, five bombs went off damaging buildings. Foreign terrorist organizations, Wahhabis, and Hezbe Takhriri Islomiya were blamed. Thirty people were arrested. Mohammed Solih, the chairman of Erk, Uzbekistan's banned opposition party, was a suspect in the bombings. Twenty-two individuals allegedly linked to the attack were accused of terrorism, attempting to kill the president, drug trafficking, illegal possession of weapons, and robbery. Ten received the death sentence; the others received 10–14 years each.

Perhaps in reaction to these disturbances, Uzbekistan's government issued tighter regulations on residency, travel, and citizenship. Foreign citizens, including formerly Soviet CIS countries, now have to present an authorized permit to Uzbek authorities. Foreign visitors staying for more than three days must obtain visas. Uzbek citizens who fail to exchange their old Soviet passport for a new Uzbek passport by the last day of 1999 forfeit their Uzbek citizenship.

Uzbekistan confirmed that some of the militants holding four Kyrgyz nationals hostage were Uzbek members of the militant group Juma Namangani which aims to create an Islamic state. The political opposition was blamed for supporting the 650 Kyrgyzstan militants who had been holding over a dozen hostages, including four Japanese geologists. Adding to the unrest, Zubair ibn Abdurrakim, chairman of the political council of the Islamic Movement of Uzbekistan announced a ''holy war'' against Uzbekistan with the aim of forcing the release of 50,000 Muslims held in Uzbek prisons. Karimov alleged that the hostage-taking constituted part of a larger conspiracy orchestrated by Islamic terrorists who aimed to establish an Islamic state in Central Asia. Tashkent tightened control of its borders and placed its armed forces on alert.

Elections to a new 250-seat parliament as well as city and local councils took place on 5 December; a presidential poll is scheduled for 9 January 2000. Five registered political parties were entitled to field candidates. Fidokorlar Congress nominated Karimov for the presidential elections. Parliament leader Djalilov will also contend for presidency in 2000.

CULTURE AND SOCIETY

February 18 was declared a day of mourning for the victims of the Tashkent bombings. Karimov threatened to cut off the hands of those responsible. Karimov called for the revival of the ancient ''Great Silk Road'' linking Europe to Asia. Meanwhile, the political leadership in Uzbekistan built a memorial in Tashkent to those who suffered under the Soviet regime. The Kyrgyzstan hostage situation dampened the mood for Uzbekistan's independence day. President Karimov attended the opening ceremony of the Tashkent Islamic University, established under a presidential decree to teach the history and philosophy of Islam, Islamic law, economy, and natural sciences.

Popular culture continued to turn out heroes and villains. Ricardo Juarez outpointed Uzbekistan's Tulkunbay Turgunov 13–2 to take the gold in the featherweight boxing division. In tennis, the Czech Republic beat Uzbekistan 5–0 in the Davis Cup world group playoff.

DIRECTORY

CENTRAL GOVERNMENT

Head of State

President
Islam Karimov, Office of the President, 43 Uzbekistanskaya Street, 700163, Tashkent, Uzbekistan
PHONE: +7 (371) 1395304
FAX: +7 (371) 1395525; 1395510

Prime Minister
Utkir Sultonov, Office of the Prime Minister, 5 Mustaqillik Sq., Tashkent, Uzbekistan
PHONE: +7 (371) 1398295; 398704
FAX: +7 (371) 1398601; 398463

Ministers

First Deputy Prime Minister
Ismail Jurabekov, 5 Mustaqillik Sq., Tashkent, Uzbekistan
PHONE: +7 (371) 1398281
FAX: +7 (371) 1398601; 398463

Deputy Prime Minister and Minister of Agriculture and Water Management
Bahtiyor Olimjonov, Ministry of Agriculture and Water Management

Deputy Prime Minister
Bakhtiyor Alimjanov, Office of the Deputy Prime Minister

Minister of Agriculture and Water Utilization
Olimjonov Bakhtiyor, Ministry of Agriculture and Water Utilization

Minister of Culture
Hayrulla Juraev, Ministry of Culture, 30 Navoi Street, Tashkent, Uzbekistan
PHONE: +7 (371) 394957

Minister of Defense
Hikmatulla Tursunov, Ministry of Defense

Minister of Public Education
Jura Yoldashev, Ministry of Public Education, 5 Mustaqillik Sq., Tashkent, Uzbekistan
PHONE: +7 (371) 398214

Minister of Finance
Rustam Azimov, Ministry of Finance

Minister of Foreign Affairs
Abdulaziz Komilov, Ministry of Foreign Affairs, 9 Uzbekistan Ave., Tashkent, Uzbekistan
PHONE: +7 (371) 336475

Minister of Foreign Economic Relations
Elyor Ganiyev, Ministry of Foreign Economic Relations, 75 Buyuk Ipak Yuli, Tashkent, Uzbekistan
PHONE: +7 (371) 689256

Minister of Health
Feruz Nazarov, Ministry of Health, 4 Navoi Street, Tashkent, Uzbekistan
PHONE: +7 (371) 411680

Minister of Higher and Secondary Specialized Education
Saidahror Gulomov, Ministry of Higher and Secondary Specialized Education

Minister of Internal Affairs
Zokirjon Almatov, Ministry of Internal Affairs

Minister of Macroeconomics and Statistics
Bakhtiyor Hamidov, Ministry of Macroeconomics and Statistics, 45 a Uzbekistan Ave, Tashkent, Uzbekistan
PHONE: +7 (371) 398216

Minister of Housing and Municipal Economy
Gofurdjon Mukhamedov, Ministry of Housing and Municipal Economy

Minister of Social Security
Oqiljon Obidov, Ministry of Social Security, 20a Avlony Street, Tashkent, Uzbekistan
PHONE: +7 (371) 555371; 532062

Minister of Utilities
Gafur Muhamedov, Ministry of Utilities

Minister of Communications
Fahtullah Abdullaev, Ministry of Communications, 1 Alexey Tolstoy Street, Tashkent, Uzbekistan
PHONE: +7 (371) 336645; 336666

Minister of Justice
Sirojiddin Mirsafoev, Ministry of Justice, 5 Sayilgoh Street, Tashkent, Uzbekistan
PHONE: +7 (371) 331305

Minister of Labor
Anvar Akbarov, Ministry of Labor, 6 Abay Street, Tashkent, Uzbekistan
PHONE: +7 (371) 417706

Minister of Energy
Ergash Shoismatov, Ministry of Energy

Minister of Emergency Situations
Bahodir Kasymov, Ministry of Emergency Situations

Minister of Tourism
Bahtiyor Husanbaev, Ministry of Tourism

POLITICAL ORGANIZATIONS

People's Democratic Party

TITLE: First Secretary
NAME: Abdulkhafiz Jalolov

Adolat (Social Democratic Party)

TITLE: First Secretary
NAME: Turgunpulat Daminov

Vatan Tarakiyoti-VTP (Fatherland Progress Party)

TITLE: First Secretary
NAME: Anwar Yuldashev

Milly Tiklanish-MTP (Adolat Democratic National Rebirth Party)

TITLE: Chairman
NAME: Ibrahim Gafurov

Fidokorlar (Self-sacrificers)

TITLE: Secretary General
NAME: Erkin Norbutaev

DIPLOMATIC REPRESENTATION

Embassies in Uzbekistan

China
79 Gogol Street, Tashkent, Uzbekistan
PHONE: +7 (371) 1338088; 1335375
FAX: +7 (371) 1334735
TITLE: Ambassador
NAME: Li Jinxiang

Egypt
53 Chilanzarskaya Street, Tashkent, Uzbekistan
PHONE: +7 (371) 771328
FAX: +7 (371) 1334735
TITLE: Ambassador
NAME: Mamdouh Shawky

France
25 Akhunbabaev Street, Tashkent, Uzbekistan
PHONE: +7 (371) 71335382; 1337406; 1335384
FAX: +7 (371) 1336210
TITLE: Ambassador
NAME: Jean-Claude Richard

Germany
15 Sharaf Rashidov Street, Tashkent, Uzbekistan
PHONE: +7 (371) 344725; 344763; 346696
FAX: +7 (371) 1206693
TITLE: Ambassador
NAME: Reinhart Bindseil

India
Aleksey Tolstoy Street, 3/5, Tashkent, Uzbekistan
PHONE: +7 (371) 1338357; 1338267; 1330589
FAX: +7 (371) 1333782
TITLE: Ambassador
NAME: Baskar Kumar Mitra

Italy
95 Amir Temur Street, Tashkent, Uzbekistan

PHONE: +7 (371) 352009; 354272; 346649
FAX: +7 (371) 1333782
TITLE: Ambassador
NAME: Jolanda Brunetti Goetz

Japan
52/1 Sadyk Azimov Street, Tashkent, Uzbekistan
PHONE: +7 (371) 1333943; 1334415; 1338921
FAX: +7 (371) 1206514
TITLE: Ambassador
NAME: Kioko Nakayama

Russia
83 Nukus Street, Tashkent, Uzbekistan
PHONE: +7 (371) 557954; 552948; 1526280
FAX: +7 (371) 558774
TITLE: Ambassador
NAME: Alexandr Patsev

Turkey
87 Gogol street, Tashkent, Uzbekistan
PHONE: +7 (371) 1332107; 1332104; 1338037
FAX: +7 (371) 1330833
TITLE: Ambassador
NAME: Umur Abaydin

United Kingdom
67 Gogol street, Tashkent, Uzbekistan
PHONE: +7 (371) 1339847; 1206288; 1206451
FAX: +7 (371) 1206549; 1206575
TITLE: Ambassador
NAME: Barbara Hay

United States
82 Chelanzarskaya, Tashkent, Uzbekistan
PHONE: +7 (371) 1205450
FAX: +7 (371) 1206335
TITLE: Ambassador
NAME: Joseph Presel

JUDICIAL SYSTEM

Supreme Court

FURTHER READING

Articles

''Central Asia: A Matter of Health in Uzbekistan.'' *The Wall Street Journal,* 17 February 1999, p. A18.

Pope, Hugh. ''Six Car Bombs Kill 13 People in Uzbekistan.'' *The New York Times,* 22 September 1999, p. A14.

''Uzebekistan: 300 Freed in Blasts.'' *The Economist* (October 17, 1998): 46.

Books

Kalter, Johannes and Margareta Pavaloi, eds. *Uzbekistan: Heirs to the Silk Road.* New York: Thames and Hudson, 1997.

Karimov, I. A. *Uzbekistan on the Threshold of the Twenty-First Century: Challenges to Stability and Progress.* New York: St. Martin's Press, 1998.

MacLeod, Calum. *Uzbekistan: The Golden Road to Samarkand.* 2nd ed. Lincolnwood, Ill.: Odyssey/Passport, 1997.

Internet

Embassy of the Republic of Uzbekistan, Washington, D.C. Available Online @ http://www.uzbekistan.org/ (November 2, 1999).

Welcome to Uzbekistan. Available Online @ http://www.gov.uz/ (November 2, 1999).

UZBEKISTAN: STATISTICAL DATA

For sources and notes see "Sources of Statistics" in the front of each volume.

GEOGRAPHY

Geography (1)

Area:

Total: 447,400 sq km.

Land: 425,400 sq km.

Water: 22,000 sq km.

Area—comparative: slightly larger than California.

Land boundaries:

Total: 6,221 km.

Border countries: Afghanistan 137 km, Kazakstan 2,203 km, Kyrgyzstan 1,099 km, Tajikistan 1,161 km, Turkmenistan 1,621 km.

Coastline: 0 km.

Note: Uzbekistan borders the Aral Sea (420 km).

Climate: mostly midlatitude desert, long, hot summers, mild winters; semiarid grassland in east.

Terrain: mostly flat-to-rolling sandy desert with dunes; broad, flat intensely irrigated river valleys along course of Amu Darya, Sirdaryo, and Zarafshon; Fergana Valley in east surrounded by mountainous Tajikistan and Kyrgyzstan; shrinking Aral Sea in west.

Natural resources: natural gas, petroleum, coal, gold, uranium, silver, copper, lead and zinc, tungsten, molybdenum.

Land use:

Arable land: 9%

Permanent crops: 1%

Permanent pastures: 46%

Forests and woodland: 3%

Other: 41% (1993 est.)

HUMAN FACTORS

Demographics (2A)

	1990	1995	1998	2000	2010	2020	2030	2040	2050
Population	20,624.1	22,800.0	23,784.3	24,422.5	28,074.7	32,382.7	36,202.0	39,759.6	42,761.6
Net migration rate (per 1,000 population)	NA	NA	NA	NA	NA	NA	NA	NA	NA
Births	NA	NA	NA	NA	NA	NA	NA	NA	NA
Deaths	NA	NA	NA	NA	NA	NA	NA	NA	NA
Life expectancy - males	64.1	61.3	60.5	60.1	62.7	66.3	69.4	72.1	74.3
Life expectancy - females	70.8	68.7	67.9	67.5	69.9	73.7	77.0	79.6	81.7
Birth rate (per 1,000)	35.4	26.4	23.7	23.2	23.5	19.5	16.6	15.9	14.3
Death rate (per 1,000)	7.2	7.6	7.7	7.8	7.4	6.5	6.6	7.4	8.2
Women of reproductive age (15-49 yrs.)	4,790.6	5,557.9	6,058.0	6,426.1	8,036.1	8,756.1	9,606.7	9,708.2	9,910.0
of which are currently married	NA	NA	NA	NA	NA	NA	NA	NA	NA
Fertility rate	4.3	3.2	2.9	2.8	2.5	2.3	2.2	2.2	2.1

Except as noted, values for vital statistics are in thousands; life expectancy is in years.

Health Personnel (3)

Total health expenditure as a percentage of GDP, 1990-1997[a]

Public sector	3.3
Private sector	NA
Total[b]	NA

Health expenditure per capita in U.S. dollars, 1990-1997[a]

Purchasing power parity	NA
Total	NA

Availability of health care facilities per 100,000 people

Hospital beds 1990-1997[a]	830
Doctors 1993[c]	335
Nurses 1993[c]	1,032

Health Indicators (4)

Life expectancy at birth

1980	67
1997	69

Daily per capita supply of calories (1996)	2,550
Total fertility rate births per woman (1997)	3.3
Maternal mortality ratio per 100,000 live births (1990-97)	24[b]
Safe water % of population with access (1995)	57
Sanitation % of population with access (1995)	18
Consumption of iodized salt % of households (1992-98)[a]	0

Smoking prevalence

Male % of adults (1985-95)[a]	40
Female % of adults (1985-95)[a]	1
Tuberculosis incidence per 100,000 people (1997)	81
Adult HIV prevalence % of population ages 15-49 (1997)	<0.005

Infants and Malnutrition (5)

Under-5 mortality rate (1997)	60
% of infants with low birthweight (1990-97)	NA
Births attended by skilled health staff % of total[a]	98

% fully immunized (1995-97)

TB	97
DPT	96
Polio	97
Measles	88
Prevalence of child malnutrition under age 5 (1992-97)[b]	19

Ethnic Division (6)

Uzbek	80%
Russian	5.5%
Tajik	5%
Kazak	3%
Karakalpak	2.5%
Tatar	1.5%
Other	2.5%

Data for 1996 est.

Religions (7)

Muslim (mostly Sunnis)	88%
Eastern Orthodox	9%
Other	3%

Languages (8)

Uzbek 74.3%, Russian 14.2%, Tajik 4.4%, other 7.1%.

EDUCATION

Public Education Expenditures (9)

Public expenditure on education (% of GNP)

1980	....
1996	8.1

Expenditure per student

Primary % of GNP per capita

1980	....
1996	....

Secondary % of GNP per capita

1980	....
1996	....

Tertiary % of GNP per capita

1980	....
1996	....

Expenditure on teaching materials

Primary % of total for level (1996)	....
Secondary % of total for level (1996)	0.3[1]
Primary pupil-teacher ratio per teacher (1996)	21[1]
Duration of primary education years (1995)	4

Literacy Rates (11B)

Adult literacy rate

1980

Male	—
Female	—

1995

Male	100
Female	100

GOVERNMENT & LAW

Political Parties (12)

Supreme Assembly	No. of seats
People's Democratic Party	207
Fatherland Progress Party	12
Other	31

Government Budget (13B)

Revenues	NA
Expenditures	NA
Capital expenditures	NA

NA stands for not available.

Military Affairs (14B)

	1992	1993	1994	1995
Military expenditures				
Current dollars (mil.)	NA	1,751	1,890	2,062
1995 constant dollars (mil.)	NA	1,836	1,937	2,062
Armed forces (000)	40	41	20	21
Gross national product (GNP)				
Current dollars (mil.)	54,290	54,500	53,670	53,970
1995 constant dollars (mil.)	58,390	57,130	55,020	53,970
Central government expenditures (CGE)				
1995 constant dollars (mil.)	26,430	NA	NA	NA
People (mil.)	21.7	22.1	22.3	23.0
Military expenditure as % of GNP	NA	3.2	3.5	3.8
Military expenditure as % of CGE	NA	NA	NA	NA
Military expenditure per capita (1995)	NA	83	86	90
Armed forces per 1,000 people (soldiers)	1.8	1.9	.9	.9
GNP per capita (1995 $)	2,696	2,583	2,440	2,348
Arms imports[6]				
Current dollars (mil.)	0	0	20	0
1995 constant dollars (mil.)	0	0	21	0
Arms exports[6]				
Current dollars (mil.)	0	0	20	10
1995 constant dollars (mil.)	0	0	21	10
Total imports[7]				
Current dollars (mil.)	1,768	2,245	2,844	3,602
1995 constant dollars (mil.)	1,901	2,354	2,915	3,602
Total exports[7]				
Current dollars (mil.)	1,506	2,050	2,626	3,367
1995 constant dollars (mil.)	1,616	2,149	2,692	3,367
Arms as percent of total imports[8]	0.0	0	.7	0
Arms as percent of total exports[8]	0.0	0	.8	.3

Crime (15)

Crime rate (for 1997)	
Crimes reported	67,000
Total persons convicted	NA
Crimes per 100,000 population	300
Persons responsible for offenses	
Total number of suspects	57,800
Total number of female suspects	6,700
Total number of juvenile suspects	3,000

LABOR FORCE

Labor Force (16)

Total (million)	8.6
Agriculture and forestry	44%
Industry and construction	20%
Other	36%

Data for 1996 est. Percent distribution for 1995.

Unemployment Rate (17)

5% plus another 10% underemployed (December 1996 est.)

PRODUCTION SECTOR

Electric Energy (18)

Capacity	11.822 million kW (1995)
Production	45.42 billion kWh (1996 est.)
Consumption per capita	1,916 kWh (1996 est.)

Transportation (19)

Highways:

total: 81,600 km

paved: 71,237 km (note—these roads are said to be hard surfaced, meaning that some are paved and some are all-weather gravel surfaced).

unpaved: 10,363 km dirt (1996 est.)

Waterways: 1,100 (1990)

Pipelines: crude oil 250 km; petroleum products 40 km; natural gas 810 km (1992)

Airports: 3 (1997 est.)

Airports—with paved runways:

total: 3

over 3,047 m: 2

2,438 to 3,047 m: 1 (1997 est.)

Top Agricultural Products (20)

Cotton, vegetables, fruits, grain; livestock.

FINANCE, ECONOMICS, & TRADE

Economic Indicators (22)

National product: GDP—purchasing power parity—$60.7 billion (1997 est.)

National product real growth rate: 2.4% (1997 est.)

National product per capita: $2,500 (1997 est.)

Inflation rate—consumer price index: 55% (1996 est.)

Exchange Rates (24)

Exchange rates:

Uzbekistani soms (UKS) per US$1	
September 1997	75.8
1996	41.1
1995	30.2
1994	11.4
1993	1.0

Top Import Origins (25)

$4.7 billion (1996)

Origins	%
principally other FSU	NA
Czech Republic	NA
Western Europe	NA

NA stands for not available.

Top Export Destinations (26)

$3.8 billion (1996).

Destinations	%
Russia	NA
Ukraine	NA
Eastern Europe	NA
Western Europe	NA

NA stands for not available.

Economic Aid (27)

Recipient: ODA, $71 million (1993). Note: commitments, $2,915 million ($135 million in disbursements) (1992-95).

MANUFACTURING SECTOR

GDP & Manufacturing Summary (21)

	1980	1985	1990	1992	1993	1994
Gross Domestic Product						
Millions of 1990 dollars	15,329	19,717	22,615	18,045	17,612	16,819
Growth rate in percent	7.45	4.05	4.74	−11.13	−2.40	−4.50
Per capita (in 1990 dollars)	961.9	1,088.6	1,107.5	844.2	805.7	752.6
Manufacturing Value Added						
Millions of 1990 dollars	NA	NA	4,623	4,299	*4,454*	*4,499*
Growth rate in percent	NA	NA	NA	−8.20	*3.60*	*1.00*
Manufacturing share in percent of current prices	26.9	29.4	*20.2*	*18.9*	*19.4*	NA

FINANCE, ECONOMICS, & TRADE

Balance of Payments (23)

	1988	1989	1990	1991	1992[1]
Exports of goods (f.o.b.)	NA	NA	NA	NA	869
Imports of goods (f.o.b.)	NA	NA	NA	NA	929
Trade balance	NA	NA	NA	NA	−60
Services - debits	NA	NA	NA	NA	−12
Services - credits	NA	NA	NA	NA	8
Private transfers (net)	NA	NA	NA	NA	2
Government transfers (net)	NA	NA	NA	NA	NA
Long-term capital (net)	NA	NA	NA	NA	129
Short-term capital (net)	NA	NA	NA	NA	NA
Errors and omissions	NA	NA	NA	NA	103
Overall balance	NA	NA	NA	NA	170

Import Export Commodities (28)

Import Commodities	Export Commodities
Grain	Cotton
Machinery and parts	Gold
Consumer durables	Natural gas
Other foods	Mineral fertilizers
	Ferrous metals
	Textiles
	Food products
	Autos

VANUATU

CAPITAL: Port-Vila.

FLAG: Red and green sections are divided horizontally by a gold stripe running within a black border and widening at the hoist into a black triangle on which is a pig's tusk enclosing two crossed yellow mele leaves.

ANTHEM: *Yumi, Yumi, Yumi (We, We, We).*

MONETARY UNIT: As of 1 January 1981, the vatu (VT) replaced at par value the New Hebridean franc as the national currency. There are coins of 100 vatu and notes of 100, 500, 1,000, and 5,000 vatu. VT1 = $0.009 (or $1 = VT111.02).

WEIGHTS AND MEASURES: The metric standard is used.

HOLIDAYS: New Year's Day, 1 January; May Day, 1 May; Independence Day, 30 July; Assumption, 15 August; Constitution Day, 5 October; National Unity Day, 29 November; Christmas Day, 25 December; Family Day, 26 December. Movable religious holidays include Good Friday, Easter Monday, and Ascension.

TIME: 11 PM = noon GMT.

LOCATION AND SIZE: Vanuatu is an irregular Y-shaped chain of some 80 South Pacific islands, with a total land area of about 14,760 square kilometers (5,699 square miles)—slightly larger than the state of Connecticut—and a total coastline of 2,528 kilometers (1,571 miles). The island chain is about 800 kilometers (500 miles) long.

Vanuatu's capital city, Port-Vila, is located on the island of Éfaté.

CLIMATE: Average midday temperatures in Port-Vila range from 25°C (77°F) in winter to 29°C (84°F) in summer. Humidity averages about 74%, and rainfall on Éfaté is about 230 centimeters (90 inches) a year.

Republic of Vanuatu
French—*République de Vanuatu*
Bislama—*Ripablik blong Vanuatu*

INTRODUCTORY SURVEY

RECENT HISTORY

British Captain James Cook discovered, named, and charted most of the southern New Hebrides islands, as Vanuatu was known, in 1774. The next century brought British and French missionaries, planters, and traders to the islands.

A joint Anglo-French naval commission was established in 1887 to protect the lives and interests of the islanders. In 1906, the Anglo-French Condominium (joint rule by nations) was established, largely to settle land claims.

In 1975 a representative assembly replaced the nominated advisory council under which the New Hebrides had been governed. Two years later, the National Party demanded independence, and self-government was agreed on for 1978, to be followed by a 1980 referendum on independence.

In May 1980, however, a dissident French-speaking group attempted to break away and declare an independent government. This attempt was suppressed by the presence of British and French troops sent to Luganville on 24 July, although no shots were fired. The soldiers remained until Vanuatu's formal declaration of independence on 30 July 1980. They were then replaced, at the new government's request, by forces from Papua New Guinea, who were assisted by the local police in putting down the rebellion.

Since independence, Vanuatu has been the only South Pacific nation to follow a nonaligned foreign policy. In January 1987 it signed a contro-

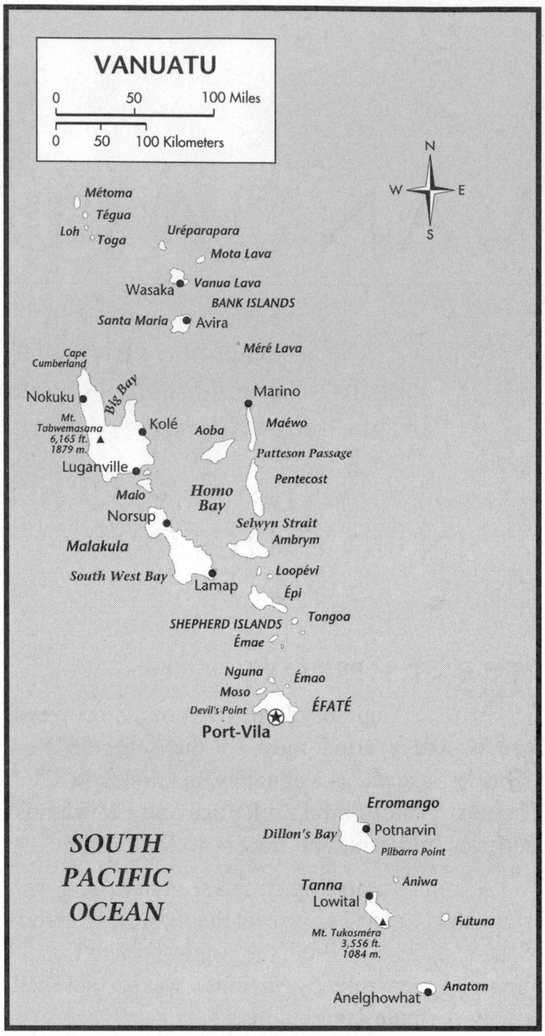

versial fishing agreement with the former Soviet Union.

GOVERNMENT

Vanuatu is an independent republic within Commonwealth of Nations. The head of state is the president; the head of government is the prime minister. The single-chamber legislature consists of 50 members elected by universal adult suffrage to four-year terms. Vanuatu is divided into six administrative districts.

Judiciary

The constitution establishes a Supreme Court, with a chief justice and three other judges, as well as an appeals court. Village and island courts have jurisdiction over customary and other matters.

Political Parties

Political parties include: Union of Moderate Parties (UMP); National United Party (NUP) Vanuatu Party (VP); Melanesian Progressive Party (MPP); Tan Union (TU); Na-Griamel Movement; Friend Melanesian Party: John Frum Movement; and the Vanuatu Republican Party.

DEFENSE

Vanuatuan cadets train in Papua New Guinea for a mobile defense force under the direction of the Australian Ministry of Defense.

ECONOMIC AFFAIRS

Agriculture supports about 75% of the population, but the service industry is playing an increasingly important role in the economy. The absence of personal and corporate income taxes have made Vanuatu an offshore financial center.

Income

In 1998 Vanuatu's gross national product (GNP) was $238 million, or about $1340 per person. For the period 1985–95 the average inflation rate was 5.5%.

Industry

Industries include fish freezing, meat canning, sawmilling, and the production of furniture, soft drinks, and fabricated aluminum goods. Native crafts include basketry, canoe building, and pottery.

SOCIAL WELFARE

The extended family system ensures that no islanders starve, while church missions and the social development section of the Education Ministry concentrate on rural development and youth activities. A provident fund system provides lump-sum benefits for old age, disability, and death. Women are just beginning to emerge from traditional cultural roles, but there are no female leaders in the country's civic, business, or religious institutions.

Healthcare

Malaria is the most serious public health problem, followed by leprosy, tuberculosis, filariasis, and venereal diseases. Life expectancy is 64 years. Medical care is provided by 94 hospitals, health centers, and clinics. The country had 15 physicians in 1991.

Housing

In urban areas only the emerging middle class can afford government-built housing. Other migrants to the towns buy plots of land and build cheap shacks of corrugated iron and waste materials, principally near Port-Vila and Luganville. The vast majority of villagers still build their own homes from local materials.

EDUCATION

The overall literacy rate is 60%. Primary education is available for almost all children except in a few remote tribal areas. In 1992, there were 272 primary schools with 852 teachers and 26,267 students. General secondary schools had 220 teachers and 4,269 students. Education accounted for 20% of the government's budget in 1995.

1999 KEY EVENTS TIMELINE

January

- The South Pacific Forum establishes the Pacific Kava Council to protect regional shares in international profits from kava, a major product of Vanuatu.

February

- A women's rights conference in Papua New Guinea reports that as many as 25% of Vanuatu women live in an abusive relationship.

May

- Four passengers from Vanuatu are among those killed in an airplane crash off the coast of Vanuatu.

July

- Journalists' groups protest when Vanuatu's deputy prime minister threatens violence against one of the country's two independent newspapers.
- The former manager of the Vanuatu Livestock Development Company is found responsible for the company's demise.

June

- Vanuatu joins in a South Pacific Forum call for the establishment of a regional free trade zone.

September

- A United Nations Environment Program report warns that economic obstacles pose a potential threat to Vanuatu's ecology.

November

- A tsunami and a powerful earthquake, centered 90 miles north of Port Vila and measuring 7.1 on the Richter scale, strike Vanuatu, killing at least five people and destroying buildings at 12:21 a.m. local time on November 11. The earthquake is also felt in the Vanuatu islands of Ambrym and Epi, but the island of Pentecost is the hardest hit.

ANALYSIS OF EVENTS: 1999

BUSINESS AND THE ECONOMY

At the South Pacific Forum in June, Vanuatu supported a Pacific Free Trade Area (FTA) that would initially include 14 countries in the region, with the possibility of future expansion. The proposed FTA would offset ramifications of the imminent expiration of the Lome Convention agreement between the European Union (EU) and 71 African, Caribbean, and Pacific (ACP) countries. The Lome Convention agreement is subject to a World Trade Organization (WTO) waiver set to expire in February 2000, though the EU has announced plans to petition for an extension to 2006. The South Pacific FTA would also buffer its members against the expiration in 2006 of the WTO waiver for the U.S. Compact of Free Association, which selectively offers trade preferences to member countries. An official of the South Pacific Forum mentioned Vanuatu's production of kava and beef as an example of the regional diversity of goods that can be traded to the mutual benefit of FTA member nations.

Vanuatu has joined the South Pacific's other major kava producers in moving to avoid being left behind by the current growing interest in kava around the globe. The kava plant is attributed with relaxation properties and is used to make a traditional mild intoxicant in some South Pacific cultures. Kava's reputed benefits as a stress reliever have attracted the attention of producers of herbal medicine, who market kava as an additive and in the form of capsules and tablets and tout the plant's extract as an aphrodisiac and a cure for jet lag. The establishment of kava plantations in Central America has cut into the Pacific Islands' near monopoly on kava, which was until recently all but unknown outside of the region. The South Pacific Forum has set up the Pacific Kava Council to work at establishing and protecting the regional rights to kava and its uses on the bases of chemical and genetic patenting and as intellectual property.

Vanuatu has also begun to profit from further cooperation among small island developing states (SIDS) through the United Nations' SIDSnet, an Internet project designed to overcome the economic hurdles of isolation and small markets faced by SIDS. The Internet program links over 40 island nations in Africa, the Indian Ocean, the Caribbean, the Atlantic, and the South Pacific.

A government-sponsored investigation into the collapse of the once exemplary Vanuatu Livestock Development Company found its former manager to blame. The state-owned concern was left with virtually no assets when the manager, brother-in-law to a cabinet minister, was suspended.

GOVERNMENT AND POLITICS

Various organizations in support of journalistic freedom voiced their concern when Vanuatu's deputy prime minister threatened to wreck the premises of one of the country's two independent newspapers. The deputy prime minister wanted his name struck from an article associating him with a man who had assaulted the paper's publisher. That assault was itself occasioned by an article covering an election dispute. Journalists' groups see these incidents as evidence of increasing attempts at government control of the media in the South Pacific.

The government of Vanuatu, like other nations in the South Pacific, has been hampered by culture and tradition from ratifying the provisions of the Convention on the Elimination of Discrimination Against Women (CEDAW). Religion and culture have been integrated into the legal systems of the nations of the South Pacific, and these often diverge from CEDAW in the realms of the status of women. A conference in Papua New Guinea presented evidence that violence against women is a significant problem in the region, where as many as 25% of Vanuatu women, for example, live in an abusive relationship.

CULTURE AND SOCIETY

A report by the United Nations Environment Program (UNEP) issued in September to coincide with a special General Assembly session on small island developing states (SIDS) listed Vanuatu among Pacific nations whose ecologies are threatened by economic and population pressures. Growing population and increased tourism, a mainstay of island economies, has placed further pressure on the ecologies of Vanuatu and the other countries included in the UNEP study.

In September, the U.N.'s Intergovernmental Panel on Climate Change also announced its predictions that global carbon monoxide emissions, considered the primary factor in global warming, could increase by a factor of five in the next century, with disastrous effects on the world's forests and on ocean levels. Vanuatu was among the oceanic nations mentioned as already affected by rising oceans. Islands like Vanuatu, which are very close to sea level, are suffering from inundation of low-lying areas and coastal erosion.

Four passengers from Vanuatu were among those killed when a small airplane crashed into the ocean off the Vanuatu coast in May. Five people survived by swimming to a nearby island. Two French nationals were also killed. The New Zealand air force helped to locate the wreckage.

DIRECTORY

CENTRAL GOVERNMENT

Head of State

President
John Bani, Office of the President

Prime Minister
Donald Kalpolkas, Office of the Prime Minister

Ministers

Deputy Prime Minister
Willie Jimmy, Office of the Deputy Prime Minister

Minister of Agriculture, Forestry, and Fisheries
John Morrison Willy, Ministry of Agriculture, Forestry, and Fisheries

Minister of Education, Youth, and Sports
Joe Natuman, Ministry of Education, Youth, and Sports

Minister of Finance
Sela Molisa, Ministry of Finance

Minister of Foreign Affairs
Donald Kalpokas, Ministry of Foreign Affairs

Minister of Health
Jean Keasipai, Ministry of Health

Minister of Infrastructure and Public Utilities
Henri Taga, Ministry of Infrastructure and Public Utilities

Minister of Internal Affairs
Vincent Boulekone, Ministry of Internal Affairs

Minister of Lands, Geology and Mines
Silas Hakwa, Ministry of Lands, Geology and Mines

Minister of Trade and Business Development
Willie Jimmy, Ministry of Trade and Business Development

POLITICAL ORGANIZATIONS

Union of Moderate Parties (UMP)

NAME: Serge Vohor

National United Party (NUP)

Vanuatu Party (VP)

NAME: Donald Kalpokas

Melanesian Progressive Party (MPP)

NAME: Barak Sope

Tan Union (TU)

NAME: Vincent Boulkone

Na-Griamel Movement

NAME: Frankie Stevens

Friend Melanesian Party

NAME: Albert Ravutia

John Frum Movement

DIPLOMATIC REPRESENTATION

Embassies in Vanuatu

Australia
KPMG House ORPO Box 111, Port Villa, Vanuatu
PHONE: +678 22777
FAX: +678 23948

Great Britain
KPMG House Rue Pasteur, P.O. Box 567, Port Villa, Vanuatu
PHONE: +678 23100; 25550
FAX: +678 27153

JUDICIAL SYSTEM

Supreme Court

FURTHER READING

Articles

Ballantyne, Tom. ''Islands Out of Stream.'' *Newsweek* (September 7, 1999): 64.

''Islands Fear a Rising Tide.'' *U.S. News and World Report* (September 27, 1999): 43.

Pierce, Charles. ''The Changing Coastline at Mele Beach, Vanuatu.'' *Geography* (April 1999): 149.

''Walter Lini.'' *The Economist* (February 27, 1999): 86.

Internet

National Tourism Office of Vanuatu. Available Online @ http://www.vanuatutourism.com.index1.htm (November 16, 1999).

Vanuatu Online. Available Online @ http:www.vanuatu.net.vu/vol/Vanuatu.html (November 16, 1999).

VANUATU: STATISTICAL DATA

For sources and notes see "Sources of Statistics" in the front of each volume.

GEOGRAPHY

Geography (1)

Area:

Total: 14,760 sq km.

Land: 14,760 sq km.

Water: 0 sq km.

Note: includes more than 80 islands.

Area—comparative: slightly larger than Connecticut.

Land boundaries: 0 km.

Coastline: 2,528 km.

Climate: tropical; moderated by southeast trade winds.

Terrain: mostly mountains of volcanic origin; narrow coastal plains.

Natural resources: manganese, hardwood forests, fish.

Land use:

Arable land: 2%

Permanent crops: 10%

Permanent pastures: 2%

Forests and woodland: 75%

Other: 11% (1993 est.).

HUMAN FACTORS

Infants and Malnutrition (5)

Under-5 mortality rate (1997) 50

% of infants with low birthweight (1990-97) 7

Births attended by skilled health staff % of total[a] . . .NA

% fully immunized (1995-97)

TB 60

DPT 66

Polio 62

Measles 59

Prevalence of child malnutrition under age 5 (1992-97)[b] . . . . .NA

Demographics (2A)

	1990	1995	1998	2000	2010	2020	2030	2040	2050
Population	154.1	173.6	185.2	192.8	230.4	266.0	297.5	324.8	346.8
Net migration rate (per 1,000 population)	NA	NA	NA	NA	NA	NA	NA	NA	NA
Births	NA	NA	NA	NA	NA	NA	NA	NA	NA
Deaths	NA	NA	NA	NA	NA	NA	NA	NA	NA
Life expectancy - males	56.0	57.9	59.0	59.8	63.4	66.6	69.4	71.8	73.8
Life expectancy - females	59.1	61.6	63.1	64.1	68.6	72.7	76.0	78.7	80.8
Birth rate (per 1,000)	36.1	31.3	29.2	27.8	22.9	19.2	16.6	15.0	13.7
Death rate (per 1,000)	10.4	9.1	8.4	8.1	6.9	6.5	6.7	7.3	8.2
Women of reproductive age (15-49 yrs.)	35.5	41.7	45.7	48.4	62.0	72.9	80.1	82.4	82.6
of which are currently married	NA	NA	NA	NA	NA	NA	NA	NA	NA
Fertility rate	5.0	4.1	3.7	3.5	2.6	2.2	2.1	2.0	2.0

Except as noted, values for vital statistics are in thousands; life expectancy is in years.

Ethnic Division (6)

Indigenous Melanesian	94%
French	4%
Vietnamese, Chinese, Pacific Islanders	NA

Religions (7)

Presbyterian	36.7%
Anglican	15%
Catholic	15%
Indigenous beliefs	7.6%
Seventh-Day Adventist	6.2%
Church of Christ	3.8%
Other	15.7%

Languages (8)

English (official), French (official), pidgin (known as Bislama or Bichelama).

Education

Educational Attainment (10)

Age group (1979)	25+
Total population	38,488
Highest level attained (%)	
No schooling	37.2
First level	
Not completed	34.3
Completed	6.5
Entered second level	
S-1	14.7
S-2	7.3
Postsecondary	NA

GOVERNMENT & LAW

Political Parties (12)

Parliament	No. of seats
Vanuatu Party (VP)	18
Union of Moderate Parties (UMP)	12
National United Party (NUP)	11
Other and independent	11

Government Budget (13B)

Revenues	$94.4 million
Expenditures	$99.8 million
Capital expenditures	$30.4 million

Data for 1996 est.

Military Affairs (14A)

Total expenditures	$NA
Expenditures as % of GDP	NA%

NA stands for not available.

LABOR FORCE

Labor Force (16)

Total	NA
Agriculture	65%
Services	32%
Industry	3%

Data for 1995 est. NA stands for not available.

Unemployment Rate (17)

Rate not available.

PRODUCTION SECTOR

Electric Energy (18)

Capacity	11,000 kW (1995)
Production	30 million kWh (1995)
Consumption per capita	173 kWh (1995)

Transportation (19)

Highways:

total: 1,070 km

paved: 256 km

unpaved: 814 km (1996 est.)

Merchant marine:

total: 88 ships (1,000 GRT or over) totaling 1,407,737 GRT/1,761,413 DWT ships by type: bulk 31, cargo 24, chemical tanker 2, combination bulk 1, liquefied gas tanker 4, oil tanker 5, refrigerated cargo 13, vehicle carrier 8 note: a flag of convenience registry; includes ships from 15 countries among which are ships of Japan 30, India 10, US 8, Netherlands 6, Greece 4, Hong Kong 4, Australia 2, Canada 1, China 1, and Poland 1 (1997 est.)

Airports: 31 (1997 est.)

Airports—with paved runways:

total: 2

2,438 to 3,047 m: 1

1,524 to 2,437 m: 1 (1997 est.)

Airports—with unpaved runways:

total: 29

1,524 to 2,437 m: 1

914 to 1,523 m: 10

under 914 m: 18 (1997 est.)

Top Agricultural Products (20)

Copra, coconuts, cocoa, coffee, taro, yams, coconuts, fruits, vegetables; fish, beef.

FINANCE, ECONOMICS, & TRADE

Economic Indicators (22)

National product: GDP—purchasing power parity—$231 million (1996 est.)

National product real growth rate: 3% (1996 est.)

National product per capita: $1,300 (1996 est.)

Inflation rate—consumer price index: 2.2% (1997 est.)

Exchange Rates (24)

Exchange rates:	
Vatu (VT) per US$1	
January 1998	124.56
1997	115.87
1996	111.72
1995	112.11
1994	116.41
1993	121.58

Top Import Origins (25)

$97 million (f.o.b., 1996) Data are for 1996 est.

Origins	%
Japan	47
Australia	23
Singapore	8
New Zealand	6
France	3
Fiji	NA

NA stands for not available.

MANUFACTURING SECTOR

GDP & Manufacturing Summary (21)

	1980	1985	1990	1992	1993	1994
Gross Domestic Product						
Millions of 1990 dollars	94	143	154	163	169	172
Growth rate in percent	−11.46	1.11	4.11	0.77	3.83	2.00
Per capita (in 1990 dollars)	806.8	1,080.2	1,030.5	1,035.8	1,048.8	1,043.8
Manufacturing Value Added						
Millions of 1990 dollars	3	5	9	10	11	13
Growth rate in percent	−11.45	11.21	2.22	8.75	10.80	10.05
Manufacturing share in percent of current prices	4.2	3.8	5.9	NA	NA	NA

FINANCE, ECONOMICS,& TRADE

Balance of Payments (23)

	1991	1992	1993	1994	1995
Exports of goods (f.o.b.)	15	18	17	25	28
Imports of goods (f.o.b.)	−74	−67	−65	−75	−79
Trade balance	−59	−49	−47	−50	−51
Services - debits	−76	−74	−73	−81	−85
Services - credits	91	88	84	88	95
Private transfers (net)	18	17	17	12	13
Government transfers (net)	16	9	9	12	10
Overall balance	−10	−10	12	−19	−18

Top Export Destinations (26)

$30 million (f.o.b., 1996) Data are for 1996 est.

Destinations	%
Japan	28
Spain	21
Germany	14
United Kingdom	7
Cote d'Ivoire	7
Australia	NA
New Caledonia	NA

NA stands for not available.

Economic Aid (27)

Recipient: ODA, $9.6 million from Australia (FY96/97 est.); $3.1 million from NZ (FY95/96).

Import Export Commodities (28)

Import Commodities	Export Commodities
Machines and vehicles	Copra
Food and beverages	Beef
Basic manufactures	Cocoa
Raw materials and fuels	Timber
Chemicals	Coffee

VATICAN CITY

FLAG: The flag consists of two vertical stripes, yellow at the hoist and white at the fly. On the white field, in yellow, are the crossed keys of St. Peter, the first pope, surmounted by the papal tiara (triple crown).

ANTHEM: *Pontifical March.*

MONETARY UNIT: In 1930, after a lapse of 60 years, the Vatican resumed issuance of its own coinage—the lira (L)—but it agreed to issue no more than 300 million lire in any year. There are coins of 10, 20, 50, 100, and 500 lire. Italian notes are also in use. The currencies of Italy and the Vatican are mutually convertible. L1 = $0.0006 (or $1 = L1,611.3).

WEIGHTS AND MEASURES: The metric system is in use.

HOLIDAYS: Roman Catholic religious holidays; the coronation day of the reigning pope; days when public consistory is held.

TIME: 1 PM = noon GMT.

LOCATION AND SIZE: Located within Rome, Vatican City is the smallest state in Europe and in the world. It is a roughly triangular area of 44 hectares (108.7 acres) lying near the west bank of the Tiber River and to the west of the Castel Sant'Angelo. The Vatican area comprises the following: St. Peter's Square; St. Peter's Basilica, the largest Christian church in the world, to which the square serves as an entrance; an area comprised of administrative buildings and Belvedere Park; the pontifical palaces; and the Vatican Gardens, which occupy about half the acreage.

A number of churches and palaces outside Vatican City itself—including the Lateran Basilica and Palace in the Piazza San Giovanni—are under its jurisdiction, as are the papal villa and its environs at Castel Gandolfo, 24 kilometers (15 miles) southeast of Rome, and an area at Santa Maria di Galeria where a Vatican radio station was established in 1957.

CLIMATE: Winters are mild, and although summer temperatures are high during the day, the evenings are cold. Temperatures in January average 7°C (45°F); in July, 24°C (75°F).

The Holy See (State of the Vatican City)
Santa Sede (Stato della Cittá del Vaticano)

INTRODUCTORY SURVEY

RECENT HISTORY

Since the time of St. Peter, regarded by the Church as the first pope, Rome has been the seat of the popes, except in periods of great turbulence, when the pontiffs were forced to take refuge elsewhere, most notably in Avignon, France, from 1309 to 1377. The Roman papal residence before modern times was usually in the Lateran or Quirinal rather than in the Vatican Palace.

The Vatican City State and the places over which the Vatican now exercises jurisdiction are the sole remnants of the States of the Church, or Papal States, which at various times, beginning in 755, included large areas in Italy and, until the French Revolution, even parts of southern France. Most of the papal domain fell into the hands of King Victor Emmanuel II in 1860 in the course of the unification of Italy. By 1870, Pope Pius IX, supported by a garrison of French troops, retained rule over only the besieged city of Rome and a small territory surrounding it. Upon the withdrawal of the French garrison to take part in the Franco-Prussian War, the walls of Rome were breached by the besieging forces on 20 September, and the city fell. On 2 October, following a plebiscite, the city was annexed to the kingdom of Italy and made the national capital.

In May 1871, the Italian government promulgated a Law of Guarantees, which purported to establish the relations between the Italian kingdom and the papacy. The enactment declared the person of the pope to be inviolate, guaranteed him full

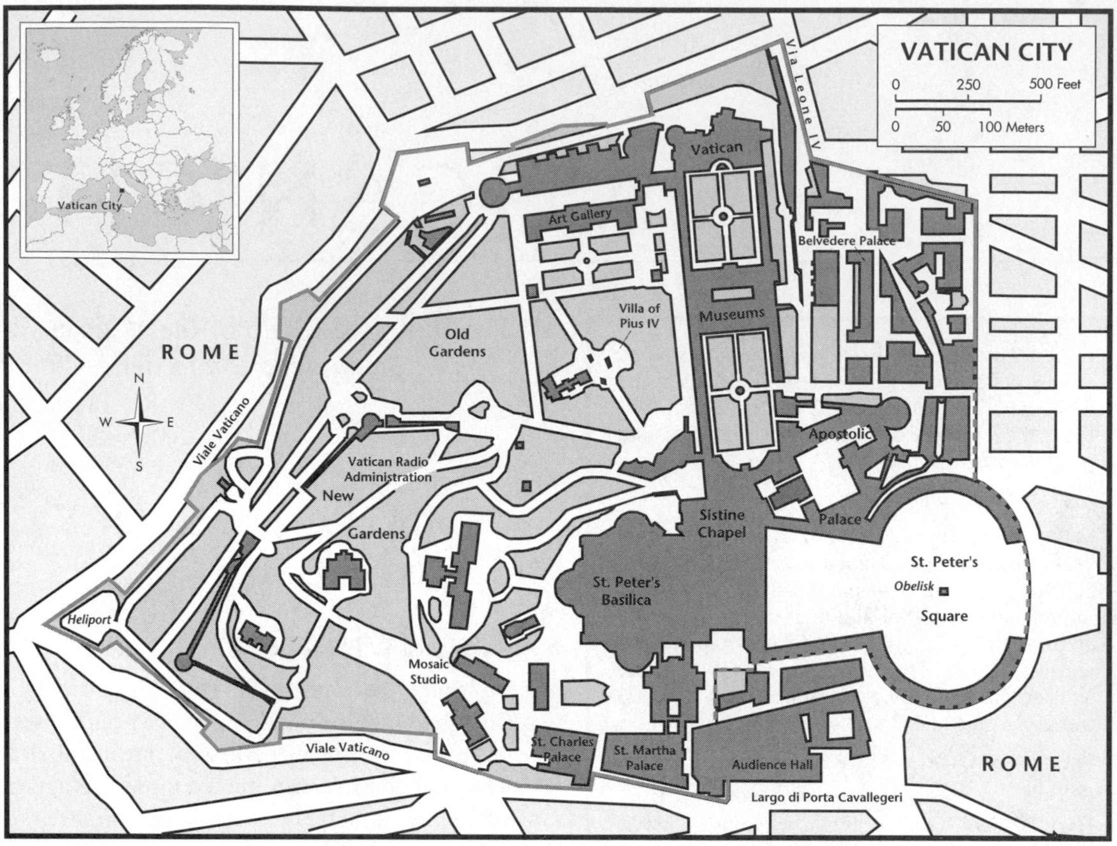

liberty in his religious functions and in the conduct of diplomatic relations, awarded an annual indemnity in lieu of the income lost when the Papal States were annexed, and provided the right of extraterritoriality over the Vatican and the papal palaces. Pius IX refused to accept the law or the money allowance; he and his successors chose to become ''prisoners of the Vatican.'' Until 1919, Roman Catholics were prohibited by the papacy from participating in the Italian government.

The so-called Roman Question was brought to an end by the conclusion on 11 February 1929 of three Lateran treaties between the Vatican and Italy. One treaty recognized the full sovereignty of the Vatican and established its territorial extent. Another treaty was a concordat establishing the Roman Catholic Church as the state church of Italy. The remaining treaty awarded the Vatican 750 million old lire in cash and 1 billion old lire in interest-bearing state bonds in lieu of all financial claims against Italy for annexing the Papal States. The constitution of the Italian Republic, adopted in 1947, substantially embodies the terms of the Lateran treaties. In 1962–65, the Vatican was the site of the Second Vatican Council, the first worldwide council in almost a century. Convened by Pope John XXIII and continued under Paul VI, the Council resulted in modernization of the Church's role in spiritual and social matters.

Ecumenism was the hallmark of the reign (1963–78) of Pope Paul VI. In a move to further Christian unity, he met with Athenagoras, the ecumenical patriarch of the Eastern Orthodox Church, in Jerusalem in 1964. In 1973, Paul VI conferred with the Coptic Orthodox patriarch of Alexandria; later in that same year, he met the exiled Dalai Lama, the first such meeting between a pope and a Buddhist leader. Steps were also taken to improve Roman Catholic-Jewish relations, including a 1965 declaration that Jews are not to be held collectively guilty of the death of Jesus. On doctrinal questions Pope Paul VI was generally conservative, reaffirming papal infallibility, disciplining dissident priests, and reiterating traditional Church opposition to all 'artificial'' methods of contraception, including abortion and sterilization. In September

1972, the concept of an all-male celibate priesthood was upheld. A sign of declining Vatican influence over Italian affairs was the 60% vote by the Italian electorate, in a May 1974 referendum, to retain legislation permitting divorce, which the Church does not sanction.

Pope Paul VI was succeeded by Pope John Paul I, who reigned for only 34 days. John Paul I's sudden death, on 28 September 1978, brought about the election of Polish Cardinal Karol Wojtyla as John Paul II, the first non-Italian pontiff elected in over 450 years. As pope, John Paul II has traveled widely, a practice begun by Paul VI. He has likewise established himself as a conservative in doctrinal matters, as indicated in 1982 by his elevation to the status of personal prelature of Opus Dei, an international organization of 72,000 laity and priests known for its doctrinal fidelity. Pope John Paul II urged Catholic bishops to uphold traditional moral doctrine in a papal encyclical issued in October 1993. The Pope also reaffirmed the male priesthood in 1994. On 13 May 1981, John Paul II was wounded in Vatican Square by a Turkish gunman, who is serving a life sentence. The alleged accomplices, three Bulgarians and three Turks, were acquitted of conspiracy in the assassination attempt on 29 March 1986 because of lack of evidence.

The Lateran treaties of 1929 were superseded in 1984 by a new concordat under which the pope retains temporal authority over Vatican City but Roman Catholicism is no longer Italy's state religion. In December 1993, the Vatican and the Israeli government concluded a mutual recognition agreement. In October 1994 the Vatican established diplomatic relations with the PLO to balance its ties with Israel. In January 1995 Pope John Paul II offered the Chinese government recognition of papal authority over China's Catholics in return for the Vatican's acknowledgment of the country's officially sponsored church, the Chinese Patriotic Catholic Association.

GOVERNMENT

The pope is simultaneously the absolute sovereign of the Vatican City State and the head of the Roman Catholic Church throughout the world. Since 1984, the pope has been represented by the cardinal secretary of state in the civil governance of Vatican City. In administering the government of the Vatican, the pope is assisted by the Pontifical Commission for the Vatican City State. Religious affairs are governed under the pope's direction by a number of ecclesiastical bodies known collectively as the Roman Curia.

The Pontifical Commission consists of seven cardinals and a lay special delegate, assisted since 1968 by a board of 21 lay advisers. Under the commission are the following: a central council (heading various administrative offices); the directorships of museums, technical services, economic services (including the postal and telegraph systems), and medical services; the guard; the Vatican radio system and television center; the Vatican observatory; and the directorship of the villa at Castel Gandolfo, the traditional summer residence of popes.

Much of the work of the Roman Curia is conducted by offices called sacred congregations, each headed by a cardinal appointed for a five-year period. These are the Sacred Congregation for the Doctrine of the Faith (responsible for faith and morals, including the examination and, if necessary, prohibition of books and other writings), the Sacred Congregation for Bishops (diocesan affairs), the Sacred Congregation for the Eastern Churches (relations between Eastern and Latin Rites), the Sacred Congregation for the Sacraments, the Sacred Congregation for Divine Worship, the Sacred Congregation for the Clergy, the Sacred Congregation for Religious Orders and Secular Institutes (monastic and lay communities), the Sacred Congregation for the Evangelization of Peoples (missions), the Sacred Congregation for the Causes of Saints (beatification and canonization), and the Sacred Congregation for Catholic Education (seminaries and religious schools). There are also secretariats for Christian unity, non-Christians, and nonbelievers, and there are permanent and temporary councils and commissions for various other functions.

A pope serves from his election until death. On his decease, the College of Cardinals is called into conclave to choose a successor from their number. The usual method is to vote on the succession; in this case, the cardinal who receives two-thirds plus one of the votes of those present is declared elected. Pending the election, most Vatican business is held in abeyance.

Before the reign of Pope John XXIII, the size of the College of Cardinals was limited to 70. Pope John raised the membership to 88, and his successor, Pope Paul VI, increased the number to 136. Paul VI also decreed that as of 1 January 1971, cardinals would cease to be members of departments of the Curia upon reaching the age of 80 and would lose the right to participate in the election of a pope.

Judiciary

For ordinary legal matters occurring within Vatican territory, there is a tribunal of first instance. Criminal cases are tried in Italian courts. There are three tribunals at the Vatican for religious cases. The Apostolic Penitentiary determines questions of penance and absolution from sin. The Roman Rota deals principally with marital issues but is also competent to handle appeals from any decisions of lower ecclesiastical courts. In exceptional cases, the Supreme Tribunal of the Apostolic Signature hears appeals from the Rota, which ordinarily is the court of last resort.

New codes of canon law for the government of the Latin Rite churches and the administration of the Curia were promulgated in 1918 and 1983. Eastern Rite churches have their own canon law.

Political Parties

There are no political parties.

DEFENSE

The papal patrol force now consists only of the Swiss Guard, who, sometimes armed with such ceremonial weapons as halberds, walk their posts in picturesque striped uniforms supposedly designed by Michelangelo. The force was founded in 1506 and is recruited from several Roman Catholic cantons of Switzerland. It now numbers approximately 100 members. There is also a civilian security force, responsible to the Central Office of Security, which protects Vatican personnel and property and the art treasures owned by the Church. The Vatican maintains its own jail.

ECONOMIC AFFAIRS

The Vatican, being essentially an administrative center, is dependent for its support on the receipt of charitable contributions, the fees charged those able to pay for the services of the congregations and other ecclesiastical bodies, and interest on investments. Funds are also raised from the sale of stamps, religious literature, and mementos and from museum admissions. Vatican City's economy is not commercial in the usual sense.

Public Finance

State income is derived from fees paid by the public for visiting the art galleries and from the sale of Vatican City postage stamps, tourist mementos, and publications. The Vatican also receives income in the form of voluntary contributions (Peter's pence) from all over the world, and from interest on investments. The Prefecture for Economic Affairs coordinates Vatican finances. The U.S. Central Intelligence Agency estimates that, in 1994, government revenues totaled approximately $175.5 million and expenditures $175 million.

Income

Income in 1992 was estimated at $92 million. Residents of Vatican City pay no taxes. Vatican City imposes no customs tariffs.

Industry

A studio in the Vatican produces mosaic work, and a sewing establishment produces uniforms. There is a large printing plant, the Vatican Polyglot Press.

The labor force consists mainly of priests and other ecclesiastics, who serve as consultants or councilors; about 3,000 laborers, who live outside the Vatican; the guards; the nuns, who do the cooking, cleaning, laundering, and tapestry repair; and the cardinals, archbishops, bishops, and other higher dignitaries. Some ecclesiastical officials live outside Vatican City and commute from the secular city. The Association of Vatican Lay Workers, a trade union, has 1,800 members. Lay employees of the Vatican have always had to be Roman Catholics and swear loyalty to the Pope. Under a new set of rules of conduct implemented in October 1995, new employees have to sign a statement binding them to observe the moral doctrines of the Roman Catholic Church.

Banking and Finance

The Vatican bank, known as the Institute for Religious Works (Istituto per le Opere di Religione), was founded in 1942. It carries out fiscal operations and invests and transfers the funds of the Vatican and of Roman Catholic religious communities throughout the world. The Administration of the Patrimony of the Holy See manages the Vatican's capital assets.

Economic Development

The Vatican administers industrial, real estate, and artistic holdings valued in the hundreds of millions of dollars. Investments have been in a wide range of enterprises, with makers of contraceptives and munitions specifically excepted.

SOCIAL WELFARE

Celibacy is required of all Roman Catholic clergy, except permanent deacons. The Church upholds the concept of family planning through such traditional methods as rhythm and abstinence but resolutely opposes such ''artificial methods'' as contraceptive pills and devices, as well as abortion and sterilization. Five important papal encyclicals—Rerum Novarum (1870), Quadragesimo Anno (1931), Mater et Magistra (1961), Pacem in Terris (1963), and Laborem Exercens (1981)—have enunciated the Church position on matters of workers' rights and social and international justice.

Healthcare

The health services directorate, under the Pontifical Commission for the Vatican City State, is responsible for health matters.

Housing

Information is not available.

EDUCATION

The Vatican is a major center for higher Roman Catholic education, especially of the clergy being trained for important positions. Adult literacy is 100%. About 65 papal educational institutions are scattered throughout Rome; some of the more important (all prefixed by the word ''Pontifical'') are the Gregorian University, the Biblical Institute, the Institute of Oriental Studies, the Lateran Athenaeum, the Institute of Christian Archaeology, and the Institute of Sacred Music. There were a total of 11,681 students in 1991 with 1,584 teaching staff in all higher level institutions.

1999 KEY EVENTS TIMELINE

January

- A meeting between John Paul II and a leading Italian leftist signals of improving relations between the Church and the left.
- The Vatican says Christians and Muslims share same beliefs. The message comes during the Islamic holy fasting month of Ramadan.
- The Vatican warns that the devil is still at work, and issues its first updated ritual for exorcism since 1614.
- Pope John Paul visits Mexico and the United States. It is his 85th trip abroad.
- Vatican City owes $23 million in sewer bills. A Vatican spokesman said a 1929 treaty guarantees free water services.

February

- The Vatican intervenes in favor of former Chilean dictator Augusto Pinochet. It sends a letter to the British government concerning Pinochet's extradition trial that would decide whether he could be tried in Spain.

March

- Iranian President Mohammad Khatami shakes hands with Pope John Paul in Vatican City. The pope calls the encounter promising and important.
- The Vatican claims China employs prostitutes to help brainwash Roman Catholic priests and weaken their loyalty to the Church.

June

- Pope John Paul falls and his head hits the ground. Despite slight contusion, he attends Mass in southern Poland.

July

- The Vatican bars the Rev. Robert Nugen and Sister Jeannine Gramick from pastoral work involving homosexual persons. The two Americans have been engaged in joint gay and lesbian ministry since the early 1970s.

August

- A man smuggles gun into St. Peter's Basilica and commits suicide. The death raises security concerns.
- Pope John Paul declines to visit Macao before it reverts to communist China.

September

- Pope John Paul announces the Church will seek pardon for errors of the past. He mentions seeking pardon for ''the failure to respect and defend human rights.''
- A European Union official proposes the Vatican should act as mediator to begin the process of reconciliation in East Timor.
- Pope John Paul and Venezuelan President Hugo Chavez meet to discuss the role of the church in his country.
- The Vatican celebrates the dedication of the restored façade of St. Peter's Basilica. More than

40,000 people join Pope John Paul in the celebrations.

October

- A conservative cleric in Iran blames Muslim disunity in allowing the Vatican to dictate policy in Israel. The Vatican objected to plans to build a mosque near the main Christian shrine in Nazareth.
- Iraq and the Vatican plan a visit by Pope John Paul to Iraq to celebrate the millenium. Later the Vatican announces a papal visit has been put on hold.
- Pope John Paul marks the 21st anniversary of his pontificate with an emotional appeal. He wants Catholics to help him continue his work as leader of the Church.
- Conservative Hindus said they would welcome the Pope in India as long as he recognizes the validity of gods other than the Christian Lord of the Bible.
- The Vatican rejects charges that Pius XII was an anti-Semite; however, the Church decides to review World War II records.
- Pope John Paul marks the role of women in Roman Catholic history by proclaiming three women saints, including a convert from Judaism killed at Auschwitz. They are, he says, spiritual ''co-patronesses of Europe.''

November

- Pope John Paul says the Catholic Church will continue to recruit in India, ignoring Hindu objections.

December

- After a twenty-year restoration effort, the Vatican unveils the frescoes of the Sistine Chapel for viewing.

ANALYSIS OF EVENTS: 1999

BUSINESS AND THE ECONOMY

When the Vatican unveiled the restoration of St. Peter's Basilica façade in a nationally televised ceremony in September, Pope John Paul thanked the state energy company ENI for its ''bountiful generosity which made the restoration possible, employing the most modern techniques.''

The Vatican has come to rely on corporate sponsorships to restore artworks, sculptures, churches and other buildings. ENI spent about $5.4 million to restore St. Peter's façade, and many other companies are lining up to pay for other work. Corporate sponsors also have come from other countries, including Germany and Japan.

GOVERNMENT AND POLITICS

In October, Pope John Paul II celebrated the 21st anniversary of his pontificate. Although 79 years-old and frail, the pope maintained a frenetic pace, and asked Catholics to continue to support his work as leader of the Church. Despite his advancing age, the pope was busy during all of 1999, with visits to Mexico, the United States, and Poland, and planned visits to India, and the former Soviet republic of Georgia. A planned visit to Iraq was delayed under pressures from the United States and Britain. The pope also met with several heads of state, including the new Venezuelan President Hugo Chavez and Iranian President Mohammad Khatani.

John Paul has traveled more than any past popes, with 88 pastoral visits outside of Italy. In 21 years, he has proclaimed 284 saints, created 157 cardinals, written 13 encyclicals, convened 14 synods, and greeted an estimated 14 million faithful in his general audiences.

As the Catholic Church approached 2,000 years of existence, the pope and other church leaders struck a tone of reconciliation. In September, the pope said the Catholic Church would start anew in the millenium, which he declared a Holy Year. More than 30 million visitors are expected to come to Vatican City during the year 2000, more than six times the number in a normal year. During a weekly general audience, the pope said the Church would seek forgiveness for the errors, injustices and human rights violations it committed in the past. While he did not disclose those past errors, it is generally believed, based on previous Vatican documents, that he was talking about the Church's treatment of the Jews, the Inquisition, and human rights abuses, including the forced conversion of native peoples. ''In seeking God's forgiveness at the threshold of the third millenium, the Church wishes to learn from the past,'' the pope said. He said mitigating historical factors were not enough to exonerate the

Church from being ''profoundly sorry for the weaknesses of so many of its sons and daughters which disfigured its image.''

Reconciliation gestures began in January, when the pope met with Italian Prime Minister Massimo D'Alema, a former member of the Communist Party. The historic meeting marked improved relations between the Church and the left, in conflict for more than 50 years in Italy, and throughout the world. That same month, the Vatican made an appeal for understanding between Christians and Muslims, saying both religions share common beliefs about the treatment of fellow human beings. The pope followed with a historic meeting with Iranian President Mohammad Khatami in March. Khatami's nation chairs the 54-nation Islamic Conference, and meeting the pope was considered key in improving relations between two of the world's major religions. Catholics and Muslims claim about 1 billion members each. Khatami said there were ''no quintessential differences'' among faiths. The Vatican was also expected to sign in October a historic statement that would heal a 450-year-old rift with the Protestant churches. The Vatican and the Lutheran World Federation were to sign a ''Common Statement'' on the issue that drove them apart in 1517—the doctrine of justification. That same month, the Vatican hosted representatives of 20 of the world's religions to discuss world problems, including conflicts sparked by religious differences.

The Italian newspaper *La Repubblica* reported in October that Pope John Paul had decided to delay plans for the beatification of Pope Pius XII in 2000. The decision was made to preserve improved relations with Judaism, according to the report. Prominent Jewish organizations have said a joint Catholic-Jewish review of church documents from World War II may not shed much light on the disputed record of Pope Pius XII. Jewish leaders have pressed the Vatican to open its archives for Holocaust research. Some Jews believe Pius didn't do enough to save European Jews from the Nazis. In 1999, Pius was the subject of two books: ''Hitler's Pope,'' which portrays him as an anti-Semite, and the Vatican-sanctioned ''Pius XII and the Second World War.''

Despite the conciliatory tone, Vatican authorities urged Roman Catholic clerics to move toward a new kind of evangelism in Europe, which is ''inextricably bound up with Christianity.'' That message was delivered during a Vatican synod in October, when church leaders mentioned the spread of Islam and other faiths in Europe.

CULTURE AND SOCIETY

In May, Vatican City hosted one of its largest crowds ever, as several hundred thousand people came to witness the beatification of Padre Pio, a Franciscan Capuchin monk who lived most of his life in a monastery in southern Italy. The 20th century monk is said to have borne the wounds of Christ. About 150,000 tickets were distributed for St. Peter's Square alone, and The Via della Conciliazone, a broad and long boulevard that stretches from the Tiber River to the Vatican, was packed with people. Padre Pio died in 1968 at age 81.

In January, the Vatican issued its first updated ritual for exorcism since 1614, warning that the devil was still at work. The pope has denounced the devil as a ''cosmic liar and murderer.'' The new exorcism rite is in Latin. The 84-page book is red, and bounded in leather. The Vatican said the existence of the devil cannot be denied, and anyone who denies its existence would not have the fullness of the Catholic faith. The new book cautions exorcists not to confuse medical problems with actual possession by the devil. The exorcism ritual, which includes sprinkling holy water and ordering the devil to leave the possessed person, did not change.

In June, the Vatican said cloistered nuns would be allowed to surf the Internet. Church authorities will restrict Internet usage so nuns are not ''led away by desire.'' The Vatican also announced it was opening stores to sell replicas of art and relics found in its museums. A store in New York will be the first of 400 around the world that will sell Vatican memorabilia. The profits are expected to go to the poor. The Vatican was also trying to make some contacts among Hollywood movie executives. In September, it sent an ambassador to the Venice Film Festival to promote a spiritual cinema gala. In December, the Vatican was expected to hold its ''Third Millenium Film Festival.'' Church authorities said they consider cinema one of the most important forms of communication today.

DIRECTORY

CENTRAL GOVERNMENT

Head of State

Pope
John Paul II, The Holy See, Apostolic Palace, I-00120 Vatican City, Vatican City
FAX: +39 (6) 69885088

Ministers

Secretary of State
Angelo Cardinal Sodano, Office of the Secretary of State, Apostolic Palace, I-00120 Vatican City, Vatican City
PHONE: +39 (6) 69883913
FAX: +39 (6) 69885255
E-MAIL: vatio26@relstat-segstat.va

Secretary for Relations with States
Archbishop Jean-Louis Tauran, Secretariat for Relations with States, Apostolic Palace, I-00120 Vatican City, Vatican City
PHONE: +39 (6) 69883014
FAX: +39 (6) 69885255
E-MAIL: vatio32@relstat-segstat.va

DIPLOMATIC REPRESENTATION

Embassies in Vatican City

Australia
Via Paola 24, Apt. 10, I-00186 Vatican City, Vatican City
PHONE: +39 (6) 6877688
FAX: +39 (6) 6896255

Austria
Via Reno 9, I-00198 Vatican City, Vatican City
PHONE: +39 (6) 8416262; 8417427; 8555331
FAX: +39 (6) 8543058
E-MAIL: oebvat@rmnet.it

Canada
Via della Conciliazione 4D, I-00193 Vatican City, Vatican City
PHONE: +39 (6) 68307316
FAX: +39 (6) 68806283
E-MAIL: vatcn@dfait.maeci.gc.ca
NAME: Trevor Arnholt

Costa Rica
Vía del Corso 47, I-00186 Vatican City, Vatican City
PHONE: +39 (6) 3215528
TITLE: Ambassador
NAME: Javier Guerra Laspiur

Croatia
Via della Fonte di Fauno 20, I-00153 Vatican City, Vatican City
PHONE: +39 (6) 57300620, 57300640
FAX: +39 (6) 57300650
E-MAIL: velrhvat@tin.it
TITLE: Ambassador
NAME: Marijan Šunjiæ

Egypt
Piazza Della Citta Leonina 9, I-00193 Vatican City, Vatican City
PHONE: +39 (6) 6865878; 6868114
FAX: +39 (6) 6832335
TITLE: Ambassador
NAME: Mohamed Hussein al-Sadr

France
23 via Piave, I-00186 Vatican City, Vatican City
PHONE: +39 (6) 42030900
FAX: +39 (6) 42030968

Germany
Via di Villa Sacchetti 4–6, I-00197 Vatican City, Vatican City
PHONE: +39 (6) 809511
FAX: +39 (6) 80951227

Hungary
Piazza Girolamo Fabrizio 2, I-00161 Vatican City, Vatican City
PHONE: +39 (6) 4402167
TITLE: Ambassador
NAME: Tar Pál

Mexico
Via Ezio 49, 3er. Piso, I-00192 El Vaticano, Vatican City
PHONE: +39 (6) 3230360
FAX: +39 (6) 3230361
E-MAIL: embamex-s.sede@mclink.it
TITLE: Ambassador
NAME: Horacio Sánchez Urzueta

United Kingdom
Via dei Condotti 91, I-00187 Vatican City, Vatican City
PHONE: +39 (6) 69923561
FAX: +39 (6) 69940684

United States
Villa Domiziana, Via Delle Terme Deciane 26, I-00153 Vatican City, Vatican City
PHONE: +39 (6) 46741
FAX: +39 (6) 58300682
E-MAIL: Usinb.holysee@agora.it

TITLE: Ambassador
NAME: Corrine Claiborne Boggs

FURTHER READING

''After a Fall, Pope Keeps Schedule in Poland.'' *The Seattle Times*, 13 June 1999.

''Arts Abroad: A Colorful St. Peter's Raises Eyebrows.'' *The New York Times,* 30 September 1999.

''Iranian President Pays Visit to the Vatican.'' *The Seattle Times*, 11 March 1999.

''Mosque in Nazareth Could Imperil Papal Visit, Vatican Warns.'' *The New York Times,* 15 October 1999.

''New 'Popemobile' is a White Cadillac DeVille.'' *The Seattle Times*, 23 January 1999.

''Nun Opens Vatican Doors to the Net.'' *The New York Times,* 22 July 1999.

''Pontiff Beatifies Mystic Monk.'' *The Seattle Times*, 3 May 1999.

''Pope Calls on Flock to Build 'Continent of Life'.'' *The Seattle Times*, 24 January 1999.

''Pope Sets Forth a Moral Agenda in St. Louis Visit.'' *The New York Times,* 27 January 1999.

''Religious Leaders Meet at Vatican.'' Associated Press, 25 October 1999.

''Rome Journal: 'God's Parking Lot' Runs Into Ancient Grandeur.'' *The New York Times,* 30 September 1999.

''St. Peter's Basilica Cleansed in Ritual After Man's Suicide.'' *The Seattle Times,* 29 August 1999.

''Sign Land Mine Treaty, Pope Urges.'' *The Seattle Times*, 1 March 1999.

''Vatican Panel to Review World War II Record.'' Associated Press, 19 October 1999.

VATICAN CITY: STATISTICAL DATA

For sources and notes see "Sources of Statistics" in the front of each volume.

GEOGRAPHY

Geography (1)

Area:

Total: 0.44 sq km.

Land: 0.44 sq km.

Water: 0 sq km.

Area—comparative: about 0.7 times the size of The Mall in Washington, DC.

Land boundaries:

Total: 3.2 km.

Border countries: Italy 3.2 km.

Coastline: 0 km (landlocked).

Climate: temperate; mild, rainy winters (September to mid-May) with hot, dry summers (May to September).

Terrain: low hill.

Natural resources: none.

Land use:

Arable land: 0%

Permanent crops: 0%

Permanent pastures: 0%

Forests and woodland: 0%

Other: 100% (urban area).

HUMAN FACTORS

Demographics (2B)

Population (July 1998 est.)	860
Population growth rate (1998 est.)	1.15%

Ethnic Division (6)

Italians, Swiss, other.

Religions (7)

Roman Catholic

Languages (8)

Italian, Latin, various other languages.

GOVERNMENT & LAW

Political Parties (12)

The legislative branch is a unicameral Pontifical Commission. There are no parties.

Government Budget (13B)

Revenues	$175.5 million
Expenditures	$175 million
Capital expenditures	NA

Data for 1994. NA stands for not available.

Military Affairs (14A)

Defense is the responsibility of Italy; Swiss Papal Guards are posted at entrances to the Vatican City.

LABOR FORCE

Labor Force (16)

Dignitaries, priests, nuns, guards, and 3,000 lay workers who live outside the Vatican.

PRODUCTION SECTOR

Electric Energy (18)

Capacity 5,000 kW standby. Electricity supplied by Italy.

Transportation (19)

Highways: none; all city streets

Airports: none

Top Agricultural Products (20)

No agricultural products.

FINANCE, ECONOMICS, & TRADE

Exchange Rates (24)

Exchange rates:

Vatican lire (VLit) per US$1

January 1998	1,787.7
1997	1,703.1
1996	1,542.9
1995	1,628.9
1994	1,612.4
1993	1,573.7

The Vatican lira is at par with the Italian lira which circulates freely.

Economic Aid (27)

No data available.

VENEZUELA

CAPITAL: Caracas.

FLAG: The national flag, adopted in 1930, is a tricolor of yellow, blue, and red horizontal stripes. An arc of seven white stars on the blue stripe represents the seven original states. The coat of arms appears on the left side of the yellow stripe.

ANTHEM: *Himno Nacional,* beginning ''Gloria al bravo pueblo'' (''Glory to the brave people'').

MONETARY UNIT: The bolívar (B) is a paper currency of 100 céntimos. There are coins of 5, 25, and 50 céntimos and 1, 2, and 5 bolívars, and notes of 5, 10, 20, 50, 100, 500, and 1,000 bolívars. B1 = \$0.00212 (or \$1 = B472.25).

WEIGHTS AND MEASURES: The metric system is the legal standard.

HOLIDAYS: New Year's Day, 1 January; Declaration of Independence and Day of the Indian, 19 April; Labor Day, 1 May; Army Day and Anniversary of the Battle of Carabobo, 24 June; Independence Day, 5 July; Bolívar's Birthday, 24 July; Civil Servants Day, 4 September; Columbus Day, 12 October; Christmas, 25 December; New Year's Eve, 31 December. Movable holidays are Carnival (Monday and Tuesday before Ash Wednesday), Holy Thursday, Good Friday, and Holy Saturday. Numerous other bank holidays and local festivals are observed.

TIME: 8 AM = noon GMT.

LOCATION AND SIZE: Venezuela, located on the northern coast of South America, covers an area of 912,050 square kilometers (352,144 square miles). It has a total boundary length of 7,609 kilometers (4,729 miles). There are 72 offshore islands. Venezuela's capital city, Caracas, is located in the northern part of the country on the Caribbean Sea coast.

CLIMATE: There are four climatic regions, based mainly on altitude. In the tropical region, mean annual temperatures range from 24° to 35°C (75° to 95°F). In the subtropical region, where Caracas is situated, the means range from 10° to 25°C (50° to 77°F). During the wet season (May to October), the llanos and forest areas are swampy, green, and lush. At the beginning of the dry season, the same areas become dry, brown, and parched.

Republic of Venezuela
República de Venezuela

INTRODUCTORY SURVEY

RECENT HISTORY

For more than a century after Venezuela, led by Simon Bolivar, achieved independence from Spain in 1821, the country was ruled by a series of dictators. Venezuela's first free presidential election took place in 1947. Since the election of Rómulo Betancourt in 1958, Venezuela has had fair and free elections.

The Betancourt government (1959–64) instituted modest financial and agricultural reforms, built schools, and virtually eliminated illiteracy. But, by 1979, when international demand for oil dropped, Venezuela's economic and political systems were seriously threatened. The crisis climaxed with the devaluation of the national currency to one-third of its previous value against the dollar.

Subsequently, the 1980s were years of unrest. In 1988 Carlos Andrés Pérez, became president for the second time. When he removed government subsidies on several consumer goods, including gasoline, prices rose, and Caracas was rocked by rioting. The military was called in, and when the trouble finally died down, thousands had been killed or injured.

In 1992 Venezuela was shocked by two military coup attempts, and Pérez left office under allegations of embezzlement and theft. In December 1993, Venezuelans chose another former president, Rafael Caldera, who ran under a coalition of four parties. The economy fell into recession in 1993, and President Caldera suspended civil rights in

1994 and 1995. Inflation (103% in 1995) affected Venezuela's middle class–doctors, professors, and telephone company workers all went on strike in 1997.

Hugo Chavez Frias was elected president in December 1998.

GOVERNMENT

Venezuela is a republic governed under the constitution of 1961 which stresses social, economic, and political rights. The president is elected by direct popular vote for a five-year term. There is no vice-president.

The bicameral Congress includes a 47-member Senate and 200-seat Chamber of Deputies. Both houses are elected concurrently with the president for five-year terms.

Judiciary

The Supreme Court whose magistrates are elected by Congress for nine-year terms, organizes and directs the lower courts; there is no appeal of its decisions. The Supreme Court can declare a law, or any part of a law or any regulation or act of the president unconstitutional.

The lower branches of the judiciary include courts of appeal and direct courts, whose judges and magistrates are appointed by the Supreme Court.

Political Parties

Since 1958, the dominant force in Venezuelan politics has been the Democratic Action Party. Other major parties include: National Convergence, Social Christian Party; Movement Toward Socialism (MAS); Radical Cause, and Homeland for All (PPT).

DEFENSE

Venezuela's armed forces are professional and well equipped. Total armed strength in 1995 was 79,000, including 23,000 volunteers in an internal security force. Defense expenditures in 1995 were $683 million, or 1.5% of the gross domestic product (GDP).

ECONOMIC AFFAIRS

For half a century, the Venezuelan economy has been dominated by the petroleum industry; in the mid-1980s, oil exports accounted for 90% of all export value. Weakening world oil prices contributed to economic stagnation in the 1980s, when Venezuela had difficulty meeting its payments on short-term loans accumulated by state-owned enterprises.

In 1996 the government devalued its currency, abolished foreign exchange controls, and raised gasoline prices. These moves helped reduce deficits, but resulted in additional inflation.

Income

In 1998 Venezuela's gross national product (GNP) was $81.3 billion, or approximately $3,500 per person.

Industry

Although much of Venezuela's petroleum is exported in crude form, petroleum refining is a major industry. Petroleum products amounted to 368.5 million barrels in 1995. The steel industry, operating at 48% of capacity, produced 3.6 million tons of steel in 1995. Other industries include shipbuilding, automobile production (96,403 vehicles in 1995), and fertilizer manufacture.

Banking and Finance

The state banking system consists of the Central Bank, the Industrial Bank of Venezuela, the Workers' Bank, seven regional and development banks controlled by the Venezuelan Development Corp., and the Agricultural Bank. In the private sector there are commercial banks, investment banks, mortgage banks, and savings and loan associations. The country's first mortgage bank, the Mortgage Bank of Urban Credit, opened in Caracas in 1958.

Caracas's two stock exchanges, the Commercial Exchange and the Commercial Exchange of Miranda State, were merged in 1972. The National Securities Commission, established in 1973, oversees public securities transactions.

Economic Development

Venezuela's economic reform program was initiated in 1989 to move the country from a traditionally state-dominated, oil driven economy, toward a more market-oriented, diversified, and export-oriented economy. In April 1996, with the backing of the IMF, the Caldera administration adopted an additional economic stabilization program to reduce inflation by maintaining a surplus in public sector finances and encouraging real growth in the non-oil economy.

SOCIAL WELFARE

The social security system covers medical care, maternity benefits, disability, retirement and survivors' pensions, burial costs, and a marriage bonus. In July 1982 a reform of the Venezuelan civil code extended women's rights. Women comprise roughly half the student body of most universities, and have advanced in many professions, but are still underrepresented in political and economic life.

Healthcare

Life expectancy averages 73 years. Venezuela is virtually free of malaria, typhoid, and yellow fever. To maintain this status, the Department of Health and Social Welfare continues its drainage and mosquito control programs. It also builds aqueducts and sewers in towns of fewer than 5,000 persons. In 1991–93, 68% of the population had access to safe water, and 55% had adequate sanitation.

Housing

The total number of housing units in 1992 was 3.4 million.

EDUCATION

Venezuela's illiteracy rate was approximately 9% in 1995. Public education from kindergarten through university is free, and education is compulsory for children aged 7 through 13. Approximately 20% of the national budget is assigned to education.

After nine years of elementary school, students attend two to three years of secondary school. In 1993, there were over 4.2 million students enrolled in elementary schools; 311,209 in secondary schools; and 550,873 in colleges and universities.

1999 KEY EVENTS TIMELINE

February

- Hugo Chavez, 44, becomes country's new president. The former army paratrooper who failed to overthrow the government seven years ago promises to carry out a "peaceful revolution."

March

- President Chavez puts soldiers to work cleaning streets. More than half the nation's force is expected to work repairing roads and hospitals, cleaning up shantytowns, and do other work in the next three years.

April

- Venezuelans approve President Chavez's proposal to overhaul the government with a new constitution. Some 88 percent of the voters approved the measure in a referendum; but electoral officials said 60 percent of the 11 million voters abstained.

July

- Chavez's political allies win an overwhelming majority of the seats in a new 131-member assembly that will have the power to rewrite the constitution. The assembly is expected to reinvent the country's democratic system.

August

- The Constitutional Assembly, which earlier limited Congress' power, voted overwhelmingly to shut it down. Critics said Chavez was undermining the continent's oldest democracy.

September

- The president of Petroleos de Venezuela resigns in power struggle with Chavez. The country's vice president and close ally of Chavez is named to head the industry.
- The Constitutional Assembly fires eight judges suspected of corruption and said it would likely fire 50 more. Firings come after two judges throw out charges against two dozen bankers accused in one of Latin America's biggest banking scandals.
- Chavez meets Pope John Paul II in Vatican City. They discuss the role of the church in Venezuela. In the meantime, relations with Colombia sour after Colombian President Andres Pastrana tells Chavez to worry about his own country.

October

- Chavez praises China during a visit, and asks for more Chinese investments in his country. His world tour includes visits to Japan, South Korea, Malaysia, Singapore, the Philippines, and stops in Spain and France.
- Venezuela reiterates claim to territory in Guyana. The claim covers two-thirds of Guyana's territory, an area that is sparsely populated and believed to be rich in minerals.
- Venezuela claims military activities near Guyana were part of an operative against drug traffickers, and not an attempt to enter the Essequibo, a large disputed area that Venezuela claims as its own.
- Venezuela protests the murder of four young Venezuelans and the torture of four others in Colombia. Authorities say the killings and tortures have become a state problem, and will file a complaint with an international tribunal.

November

- Venezuela's business sector lashes out at moves to increase the state's role in the economy.
- Venezuela's constituent assembly approves a new constitution that seeks to strengthen democracy and the judiciary.
- On return from a 19-day tour of nine countries in Asia and Europe, President Hugo Chavez outlines his vision of development for Venezuela; however, his plan is a disappointment, and opponents call for the newly revised constitution to be rejected.

December

- Voters overwhelmingly approve a new Constitution, granting the president greatly increased powers and increasing the president's term from 5 to 6 years, and allowing successive terms. The Constitution also eliminates one chamber of the Congress and increases government control of the military and the economy.
- Hundreds of people are missing and many have died in mudslides and flooding caused by torrential rains. Over 100,000 people have been displaced from their homes.

ANALYSIS OF EVENTS: 1999

BUSINESS AND THE ECONOMY

The driving force in Latin America's fourth largest economy during 1999 was Hugo Chavez. In one of his more visible acts, the president went after the state-run oil industry.

In September, the president of Petroleos de Venezuela resigned in an apparent power struggle for control of the company. Chavez replaced him with vice president Hector Ciavaldini, who is supposed to cut waste and use the company's profits to serve the Venezuelan people. Chavez said he was angered after discovering that some retired executives received pensions of $24,000 a month. The average worker in Venezuela only earns $190 per month, and work is hard to find. About 11 percent of Venezuelans were out of work in 1999, but unofficial figures put the number at 20 percent. Venezuela has one of the world's largest oil reserves outside the Middle East, yet most of the population lives in poverty.

In another one of his symbolic gestures, Chavez said he would sell 23-state owned airplanes to raise $24 million, enough to build 3,000 new houses, he said. His vice president cut his $48,000 per month oil salary in half. These populist measures have raised fears among investors. The power struggle at the oil company, which was considered a model of efficiency, was a warning sign. But Chavez seems to be treading carefully. He has tried to ease fears by raising taxes, keeping a lid on salaries, and courting foreign investment. And he seems intent in cleaning up the system. The Constitutional Assembly in September fired eight judges suspected of corruption and was prepared to fire 50 more. The firings came after two judges threw out charges against two dozen bankers accused in a 1994 scandal that nearly bankrupted the financial system and required a $10 billion government bailout.

Venezuela recorded its lowest inflation rate in nearly 12 years in September, with a figure of 0.9 percent. That marked a sharp decline from a month earlier, when inflation hit 1.5 percent. The country was experiencing one of its worst recessions in more than a decade, and the drop in inflation was welcomed. In the first half of 1999, the economy contracted 9.5 percent. The government wants to cut inflation to 20 percent for 1999, and 15 percent in 2000.

GOVERNMENT AND POLITICS

Forty-four-year-old Hugo Chavez was sworn in as president in February 1999. He had led a failed military coup against the government in 1992, and promised a "peaceful revolution" during his historic inauguration. To some observers, it was one of the most important events in the history of Venezuela, and even of Latin America. As he swore allegiance to the constitution, Chavez said the document was dying and vowed to replace it. Some found his rhetoric alarmingly leftist. His presidential address shook the country's elite, and sent warning signals to western nations that rely on the country's oil.

In no time, Chavez became known as "President Jekyll and Colonel Hyde," in some media circles. "Before, I was the devil. Now I am the President," Chavez admitted to a local television station. If anything, the charismatic Chavez won over the masses. He is a populist—in traditional Latin American style—winning support with promises to destroy the status quo and improve the lives of all Venezuelans. From the pulpit, he resembles Argentina's Peron or Castro on one of his long speeches. He is even married to a beauty queen, certainly a plus in a nation where disappointment sets in if a Venezuelan is not a finalist in the Miss Universe Pageant.

Chavez was not simply talk in 1999. In one of his first actions, he showed an eye for flair and media savvy by sending half the nation's troops into the streets to help clean up the shantytowns and offer medical services to the poor.

He followed small, but highly visible efforts with proposals to overhaul the government with a new constitution. By July, his political allies had won more than 100 seats in a new 131-member Constitutional Assembly. He was reinventing the democratic institutions of his country, he insisted. But opponents saw a *caudillo* in the making. Chavez, they said, was concentrating power in the presidency as he dismantled Congress and the judicial system.

Chavez insisted he was not about to become a dictator. He even called for new presidential elections. To western nations, his message was clear:

Venezuela was open for business. Western developed countries remained cautiously optimistic, even though Chavez has been lashing out against free-market economics. ''The market has gained the status of God,'' he told a Brazilian newspaper. Chavez even traveled to Vatican City to sit with Pope John Paul II to discuss the church's role in Venezuelan society. He is a well-traveled president, with visits to neighboring countries, Europe and Asia. In October, he traveled to China, and praised that nation for offering a buffer to western-style capitalism. In Japan, he asked leaders to buy Venezuelan oil and invest in the country. The Japanese urged him to bring more transparency to Venezuelan law. With his own neighbors Chavez has been less friendly. Venezuela reiterated its claims to two-thirds of Guyana's territory in October, and Venezuelan troops operating near the border spooked the smaller nation. Chavez entangled himself in a war of words with Colombian President Andres Pastrana, who feels the Venezuelan president is meddling in his country's peace process with leftist guerrillas.

The Constitutional Assembly's newly revised and greatly anticipated Constitution, presented in November, was considered a disappointment. In December, however, a second attempt was overwhelmingly approved by voters. The new document grants the president greatly increased powers, increases the president's term from 5 to 6 years, and allows successive terms. The Constitution also eliminates one chamber of the Congress and increases government control of the military and the economy.

CULTURE AND SOCIETY

Chavez's ''peaceful revolution'' aimed to touch every aspect of society, and the president started with the country's judiciary, criminals, and jails. Venezuelan law is considered so inefficient that citizens often take the law into their own hands, with street executions of alleged rapists and other serious criminal offenders. Chavez declared a ''judicial emergency'' and the Constitutional Assembly began firing corrupt judges. Chavez, who spent more than two years in prison for his role in the attempted 1992 coup, vowed to reform the country's prisons, considered among the most violent in the world by human rights groups.

DIRECTORY

CENTRAL GOVERNMENT

Head of State

President
Hugo Chavez Frias, Office of the President

Ministers

Minister of Foreign Affairs
José Vicente Rangel, Ministry of Foreign Affairs

Minister of Defense
Raúl Salazar, Ministry of Defense

Minister of Education
Héctor Navarro, Ministry of Education

Minister of Energy and Mines
Alí Rodriguez Araque, Ministry of Energy and Mines

Minister of Environment and Renewable Natural Resources
Jesús Pérez, Ministry of Environment and Renewable Natural Resources

Ministry of the Family
Lino Martínez, Ministry of the Family

Minister of Finance
Maritza Izaguirre, Ministry of Finance

Minister of Health and Social Assistance
Gilberto Rodríguez Ochoa, Ministry of Health and Social Assistance

Minister of Industry and Commerce
Gustavo Marquez, Ministry of Industry and Commerce

Minister of Interior Relations
Ignacio Arcaya, Ministry of Interior Relations

Minister of Justice
Luís Miquilena, Ministry of Justice

Minister of Labor and Social Development
Lino Martínez, Ministry of Labor and Social Development

Minister of Transportation and Communications
Luís Reyes Reyes, Ministry of Transportation and Communications

Minister for the Secretariat of the Presidency
General Lucas Rincón, Ministry for the Secretariat of the Presidency

Minister of Urban Development
Luís Reyes Reyes, Ministry of Urban Development

POLITICAL ORGANIZATIONS

Acción Democrática-AD (Democratic Action)

Casa Nacional de AD, Calle Los Cedros, La Florida, Caracas 1050, Venezuela
PHONE: +58 (2) 9512655
TITLE: Leader
NAME: Pedro París Montesinos

Convergencia Nacional-CN (National Convergence)

Edif. Tajamar, Mezzanina, Parque Central, Avda. Lecuna, El Conde, Caracas 1010, Venezuela
PHONE: +58 (2) 5769879
FAX: +58 (2) 5768214
TITLE: Leader
NAME: Rafael Caldera Rodríguez

Movimiento Electoral del Pueblo-MEP (People's Electoral Movement)

Edificio Restrepo, Pedrera a Marco Parra, Caracas, Venezuela
PHONE: +58 (2) 410140
TITLE: Leader
NAME: Luís Beltrán Prieto Figueroa

Movimiento de Integración Nacional-MIN (National Integration Movement)

Edif. José María Vargas, 1o, esq. Pajarito, Caracas, Venezuela
PHONE: +58 (2) 5637504
FAX: +58 (2) 5637553
TITLE: Leader
NAME: Gonzalo Pérez Hernández

Movimiento al Socialismo-MAS (Movement Toward Socialism)

Quinta Alemar, Avda. Valencia, Las Palmas, Caracas, Venezuela
PHONE: +58 (2) 7824022
FAX: +58 (2) 7829720
TITLE: Leader
NAME: Argelia Laya. Teodoro Petkoff

Nueva Alternativa-NA (New Alternative)

Edif. José María Vargas, esq. Pajaritos, Apdo 20193, San Martín, Caracas, Venezuela
PHONE: +58 (2) 5637675
TITLE: Leader
NAME: Eduardo Machado

Partido Comunista de Venezuela-PCV (Venezuelan Communist Party)

Edificio Cantaclaro, Esquina San Pedro, San Juan, Caracas, Venezuela
PHONE: +58 (2) 410061
FAX: +58 (2) 4819737
TITLE: Leader
NAME: Trino Meleán

Partido Social Cristiano (Social Christian Party)

esq. San Miguel, Avda Panteón cruce con Fuerzas Armadas, San José, Caracas 1010, Venezuela
PHONE: +58 (2) 516022
FAX: +58 (2) 521876
TITLE: Leader
NAME: Luis Herrera Campíns

Opinión Nacional-OP (National Opinion)

Pájaro a Curamichate 92, 2o, Caracas 101, Venezuela
TITLE: Leader
NAME: Pedro Luis Blanco Peñalver

Movimiento de Izquierda Revolucionaria-MIR (Revolutionary Leftist Movement)

c/o Fracción Parlamentaria MIR, Edif. Tribunales, esq. Pajaritos, Caracas, Venezuela
TITLE: Leader
NAME: Moisés Moleiro

Derecha Emergente de Venezuela-DEV (Venezuelan Emerging Right)

TITLE: Leader
NAME: Vladimir Gessen

Causa Radical (Radical Cause)

TITLE: Leader
NAME: Andrés Velásquez

Apertura, Movimiento de Participacion Nacional (National Participation Movement)

TITLE: Leader
NAME: Carlos Andrés Peréz

Unión Republicana Democrática-URD (Republican-Democrat Union)

TITLE: Leader
NAME: Ismenia Villalba

DIPLOMATIC REPRESENTATION

Embassies in Venezuela

Australia
Quinta Yolanda, Avenida Luis Roche, Between the 6th and 7th transversals, Altamira, Caracas, Venezuela
PHONE: +58 (2) 2634033
FAX: +58 (2) 2613448

Denmark
Torre Centuria, piso 7, Avenida Venezuela/Calle Mohedano, El Rosal, Caracas, Venezuela
PHONE: +58 (2) 9514618
FAX: +58 (2) 9515278
E-MAIL: danmark@cantv.net

France
Calle Madrid con Trinidad Las Mercedes, Apartado 60385, Caracas 1060, Venezuela
PHONE: +58 (2) 9936666
FAX: +58 (2) 9933483
TITLE: Ambassador
NAME: Villemur Patrick

India
Quinta Tagore, San Carlos Nro 12, Urbanización Floresta, Municipio Chacao, Caracas 1060, Venezuela
PHONE: +58 (2) 2857887
FAX: +58 (2) 2865131
E-MAIL: embindia@eldish.net
TITLE: Ambassador
NAME: Niranjan Desai

Italy
Edificio Atrium—P.H. Calle Sorocaima entre venidas Tamanaco y Venezuela El Rosas, Caracas 1060, Venezuela
PHONE: +58 (2) 9527311
FAX: +58 (2) 9524960

Japan
Apartado No.68790, Altamira, Caracas 1062-A, Venezuela
PHONE: +58 (2) 2618333
FAX: +58 (2) 2616780

Russian Federation
Quinta 'Soyuz' Calle Las Lomas, Caracas, Venezuela
PHONE: +58 (2) 7522264
FAX: +58 (2) 7519986

United Kingdom
Av. La Estancia, Torre Las Mercedes, Piso 3, Chuao, Caracas 1060, Venezuela
PHONE: +58 (2) 9934111
FAX: +58 (2) 9939989

United States
Calle F con Calle Suapure, Colinas de Valle Arriba, Caracas 1060, Venezuela
PHONE: +58 (2) 9756411
TITLE: Ambassador
NAME: John F. Maisto

JUDICIAL SYSTEM

Corte Suprema de Justicia

Av. Baralt, Esquina de Dos Pilitas, Foro Libertador, Caracas, Venezuela
PHONE: +58 (2) 8019363
FAX: +58 (2) 8019387
E-MAIL: pcsj@cantv.net

FURTHER READING

Articles

''Chavez Cleans the Slate.'' *The Economist* (July 31, 1999): 29.

''A Rookie's Brave New World.'' *The Seattle Times*, 2 August 1999.

''Venezuela: And Now, The Economy.'' *The Economist* (February 13, 1999): 36.

''Venezuela: Caribbean Jacobinism.'' *The Economist* (August 14, 1999): 29.

''Venezuelan Congressional Leaders May Lose Few Remaining Powers.'' *The Seattle Times*, 30 August 1999.

''Venezuela's Judicial Reform Moves Slowly, Painfully.'' *The Seattle Times*, 4 October 1999.

''Venezuelans Seize Empty Apartments.'' *The Seattle Times*, 28 May 1999.

''Venezuela's President Cultivates Armed Forces.'' *The Seattle Times*, 25 July 1999.

''Venezuela's Reformist President Instills Hope in the Poor.'' *The Seattle Times*, 31 August 1999.

VENEZUELA: STATISTICAL DATA

For sources and notes see "Sources of Statistics" in the front of each volume.

GEOGRAPHY

Geography (1)

Area:

Total: 912,050 sq km.

Land: 882,050 sq km.

Water: 30,000 sq km.

Area—comparative: slightly more than twice the size of California.

Land boundaries:

Total: 4,993 km.

Border countries: Brazil 2,200 km, Colombia 2,050 km, Guyana 743 km.

Coastline: 2,800 km.

Climate: tropical; hot, humid; more moderate in highlands.

Terrain: Andes Mountains and Maracaibo Lowlands in northwest; central plains (llanos); Guiana Highlands in southeast.

Natural resources: petroleum, natural gas, iron ore, gold, bauxite, other minerals, hydropower, diamonds.

Land use:

Arable land: 4%

Permanent crops: 1%

Permanent pastures: 20%

Forests and woodland: 34%

Other: 41% (1993 est.).

HUMAN FACTORS

Demographics (2A)

	1990	1995	1998	2000	2010	2020	2030	2040	2050
Population	19,325.2	21,564.5	22,803.4	23,595.8	27,345.0	30,875.8	33,882.7	36,159.9	37,772.5
Net migration rate (per 1,000 population)	NA	NA	NA	NA	NA	NA	NA	NA	NA
Births	NA	NA	NA	NA	NA	NA	NA	NA	NA
Deaths	89.6	NA	NA	NA	NA	NA	NA	NA	NA
Life expectancy - males	67.9	68.8	69.7	70.3	72.7	74.7	76.3	77.9	78.4
Life expectancy - females	74.1	75.0	75.9	76.5	78.9	80.9	82.5	83.6	84.5
Birth rate (per 1,000)	29.9	25.1	23.0	21.5	18.1	16.3	14.1	13.1	12.4
Death rate (per 1,000)	5.4	5.2	5.0	4.9	4.9	5.4	6.4	7.6	9.1
Women of reproductive age (15-49 yrs.)	4,935.0	5,659.4	6,062.1	6,310.2	7,432.5	8,095.9	8,355.9	8,391.2	8,159.0
of which are currently married	3,251.5	NA	NA	NA	NA	NA	NA	NA	NA
Fertility rate	3.5	3.0	2.7	2.5	2.2	2.0	2.0	2.0	2.0

Except as noted, values for vital statistics are in thousands; life expectancy is in years.

Health Personnel (3)

Total health expenditure as a percentage of GDP, 1990-1997[a]

Public sector	1.0
Private sector	4.5
Total[b]	7.5

Health expenditure per capita in U.S. dollars, 1990-1997[a]

Purchasing power parity	617
Total	213

Availability of health care facilities per 100,000 people

Hospital beds 1990-1997[a]	260
Doctors 1993[c]	194
Nurses 1993[c]	77

Health Indicators (4)

Life expectancy at birth	
1980	68
1997	73
Daily per capita supply of calories (1996)	2,398
Total fertility rate births per woman (1997)	3.0
Maternal mortality ratio per 100,000 live births (1990-97)	120[b]
Safe water % of population with access (1995)	79
Sanitation % of population with access (1995)	72
Consumption of iodized salt % of households (1992-98)[a]	65
Smoking prevalence	
Male % of adults (1985-95)[a]	29
Female % of adults (1985-95)[a]	12
Tuberculosis incidence per 100,000 people (1997)	42
Adult HIV prevalence % of population ages 15-49 (1997)	0.69

Infants and Malnutrition (5)

Under-5 mortality rate (1997)	25
% of infants with low birthweight (1990-97)	9
Births attended by skilled health staff % of total[a]	97
% fully immunized (1995-97)	
TB	89
DPT	60
Polio	76
Measles	68
Prevalence of child malnutrition under age 5 (1992-97)[b]	5

Ethnic Division (6)

Mestizo	67%
White	21%
Black	10%
Amerindian	2%

Religions (7)

Nominally Roman Catholic	96%
Protestant	2%

Languages (8)

Spanish (official), native dialects spoken by about 200,000 Amerindians in the remote interior.

EDUCATION

Public Education Expenditures (9)

Public expenditure on education (% of GNP)	
1980	4.4
1996	5.2[1]
Expenditure per student	
Primary % of GNP per capita	
1980	5.7
1996	2.2[1]
Secondary % of GNP per capita	
1980	23.1[1]
1996	4.8[1]
Tertiary % of GNP per capita	
1980	71.2
1996	....
Expenditure on teaching materials	
Primary % of total for level (1996)	0.7
Secondary % of total for level (1996)	1.7[1]
Primary pupil-teacher ratio per teacher (1996)	....
Duration of primary education years (1995)	9

Educational Attainment (10)

Age group (1993)	10+
Total population	15,628,682
Highest level attained (%)	
No schooling[2]	8.0
First level	
Not completed	43.7
Completed	NA
Entered second level	
S-1	38.3
S-2	NA
Postsecondary	10.1

Literacy Rates (11A)

In thousands and percent[1]	1990	1995	2000	2010
Illiterate population (15+ yrs.)	1,131	1,244	1,154	922
Literacy rate - total adult pop. (%)	90.0	91.1	92.8	95.4
Literacy rate - males (%)	90.9	91.8	93.2	95.4
Literacy rate - females (%)	89.2	90.3	92.3	95.5

GOVERNMENT & LAW

Political Parties (12)

Chamber of Deputies	% of seats
Democratic Action (AD)	25.6
Social Christian Party (COPEI)	24.6
Movement Toward Socialism (MAS)	10.6
National Convergence	8.7
Radical Cause (La Causa R)	19.3

Military Affairs (14B)

	1990	1991	1992	1993	1994	1995
Military expenditures						
Current dollars (mil.)	1,143	2,394	1,809	1,265	1,153	854
1995 constant dollars (mil.)	1,314	2,645	1,946	1,326	1,182	854
Armed forces (000)	75	73	75	75	75	75
Gross national product (GNP)						
Current dollars (mil.)	56,150	64,390	69,160	71,410	70,730	74,550
1995 constant dollars (mil.)	64,530	71,150	74,390	74,860	72,500	74,550
Central government expenditures (CGE)						
1995 constant dollars (mil.)	14,650	16,920	17,210	16,030	17,340	13,650[e]
People (mil.)	19.3	19.8	20.3	20.7	21.1	21.6
Military expenditure as % of GNP	2.0	3.7	2.6	1.8	1.6	1.1
Military expenditure as % of CGE	9.0	15.6	11.3	8.3	6.8	6.3
Military expenditure per capita (1995 $)	68	134	96	64	56	40
Armed forces per 1,000 people (soldiers)	3.9	3.7	3.7	3.6	3.5	3.5
GNP per capita (1995 $)	3,339	3,593	3,671	3,430	3,430	3,457
Arms imports[6]						
Current dollars (mil.)	190	200	80	70	80	90
1995 constant dollars (mil.)	218	221	86	73	82	90
Arms exports[6]						
Current dollars (mil.)	5	0	0	0	0	0
1995 constant dollars (mil.)	6	0	0	0	0	0
Total imports[7]						
Current dollars (mil.)	7,335	11,150	14,070	12,200	8,921	11,970
1995 constant dollars (mil.)	8,430	12,320	15,130	12,790	9,145	11,970
Total exports[7]						
Current dollars (mil.)	17,500	15,150	14,180	14,690	16,090	18,460
1995 constant dollars (mil.)	20,110	16,750	15,260	15,400	16,490	18,460
Arms as percent of total imports[8]	2.6	1.8	.6	.6	.9	.8
Arms as percent of total exports[8]	0	0	0	0	0	0

Government Budget (13B)

Revenues .... $11.99 billion
Expenditures .... $11.48 billion
Capital expenditures .... $3 billion
Data for 1996 est.

Crime (15)

Crime rate (for 1997)
Crimes reported 236,700
Total persons convicted 106,900
Crimes per 100,000 population 1,100
Persons responsible for offenses
Total number of suspects 119,800
Total number of female suspects 7,000
Total number of juvenile suspects 15,100

LABOR FORCE

Labor Force (16)

Total (million) 9.2
Services 64%
Industry 23%
Agriculture 13%
Data for 1997 est.

Unemployment Rate (17)

11.5% (1997 est.)

PRODUCTION SECTOR

Electric Energy (18)

Capacity .... 18.975 million kW (1995)
Production .... 74 billion kWh (1995)
Consumption per capita .... 3,508 kWh (1995)

Transportation (19)

Highways:
total: 84,300 km
paved: 33,214 km
unpaved: 51,086 km (1996 est.)

Waterways: 7,100 km; Rio Orinoco and Lago de Maracaibo accept oceangoing vessels

Pipelines: crude oil 6,370 km; petroleum products 480 km; natural gas 4,010 km

Merchant marine:
total: 28 ships (1,000 GRT or over) totaling 526,832 GRT/933,135 DWT ships by type: bulk 4, cargo 5, combination bulk 1, container 1, liquefied gas tanker 2, oil tanker 9, passenger-cargo 1, roll-on/roll-off cargo 4, short-sea passenger 1 (1997 est.)

Airports: 377 (1997 est.)

Airports—with paved runways:
total: 126
over 3,047 m: 5
2,438 to 3,047 m: 10
1,524 to 2,437 m: 35
914 to 1,523 m: 61
under 914 m: 15 (1997 est.)

Airports—with unpaved runways:
total: 251
1,524 to 2,437 m: 8
914 to 1,523 m: 96
under 914 m: 147 (1997 est.)

Top Agricultural Products (20)

Corn, sorghum, sugarcane, rice, bananas, vegetables, coffee; beef, pork, milk, eggs; fish.

MANUFACTURING SECTOR

GDP & Manufacturing Summary (21)

Detailed value added figures are listed by both International Standard Industry Code (ISIC) and product title.

	1980	1985	1990	1994
GDP ($-1990 mil.)[1]	45,668	42,763	48,598	55,472
Per capita ($-1990)[1]	3,026	2,495	2,492	2,595
Manufacturing share (%) (current prices)[1]	18.6	22.5	20.8	14.6
Manufacturing				
Value added ($-1990 mil.)[1]	8,300	9,087	9,974	11,239
Industrial production index	100	101	114	80
Value added ($ mil.)	14,461	14,071	12,175	10,643
Gross output ($ mil.)	30,213	30,305	24,128	25,881
Employment (000)	426	406	464	446
Profitability (% of gross output)				
Intermediate input (%)	52	54	50	59
Wages and salaries inc. supplements (%)	15	13	9	9

	1980	1985	1990	1994
Gross operating surplus	33	34	42	32
Productivity ($)				
Gross output per worker	67,966	71,154	51,776	57,886
Value added per worker	32,530	33,038	26,127	23,863
Average wage (inc. supplements)	10,358	9,495	4,651	5,367
Value added ($ mil.)				
311/2 Food products	1,425	1,597	1,210	1,282
313 Beverages	953	836	583	760
314 Tobacco products	409	597	273	368
321 Textiles	430	505	291	179
322 Wearing apparel	348	359	160	−40
323 Leather and fur products	57	58	40	27
324 Footwear	197	158	90	194
331 Wood and wood products	106	80	36	33
332 Furniture and fixtures	188	142	65	89
341 Paper and paper products	395	357	277	243
342 Printing and publishing	376	299	182	294
351 Industrial chemicals	325	498	443	542
352 Other chemical products	858	890	662	621
353 Petroleum refineries	4,222	3,634	4,734	2,718
354 Miscellaneous petroleum and coal products	25	30	19	13
355 Rubber products	151	188	139	176
356 Plastic products	394	348	215	190
361 Pottery, china and earthenware	60	39	18	37
362 Glass and glass products	137	132	109	156
369 Other non-metal mineral products	489	378	290	331
371 Iron and steel	651	855	498	458
372 Non-ferrous metals	256	447	788	386
381 Metal products	652	503	336	382
382 Non-electrical machinery	287	241	180	195
383 Electrical machinery	345	307	245	234
384 Transport equipment	605	486	198	691
385 Professional and scientific equipment	38	26	37	42
390 Other manufacturing industries	82	81	66	44

FINANCE, ECONOMICS, & TRADE

Economic Indicators (22)

National product: GDP—purchasing power parity—$185 billion (1997 est.)

National product real growth rate: 5% (1997)

National product per capita: $8,300 (1997 est.)

Inflation rate—consumer price index: 38% (1997)

Balance of Payments (23)

	1992	1993	1994	1995	1996
Exports of goods (f.o.b.)	14,202	14,779	16,110	19,082	23,693
Imports of goods (f.o.b.)	−12,880	−11,504	−8,504	−12,069	−9,937
Trade balance	1,322	3,275	7,606	7,013	13,756
Services - debits	−7,616	−7,839	−8,177	−8,644	−8,191
Services - credits	2,919	2,939	3,195	3,536	3,121
Private transfers (net)	−6	−37	−13	−33	−25
Government transfers (net)	−368	−331	−70	142	163
Overall balance	−3,749	−1,993	2,541	2,014	8,824

Exchange Rates (24)

Exchange rates:	
Bolivares (Bs) per US$1	
January 1998	507.447
1997	488.635
1996	417.333
1995	176.843
1994	148.503
1993	90.826

Top Import Origins (25)

$10.5 billion (f.o.b., 1996).

Origins	%
United States	40
Germany	NA
Japan	NA
Netherlands	NA
Canada	NA

NA stands for not available.

Top Export Destinations (26)

$20.8 billion (f.o.b., 1996).

Destinations	%
United States and Puerto Rico	55
Japan	NA
Netherlands	NA
Italy	NA

NA stands for not available.

Economic Aid (27)

Recipient: ODA, $46 million (1993).

Import Export Commodities (28)

Import Commodities	Export Commodities
Raw materials	Petroleum 78%
Machinery and equipment	Bauxite and aluminum
Transport equipment	Steel
Construction materials	Chemicals
	Agricultural products
	Basic manufactures

VIETNAM

Socialist Republic of Vietnam
Cong Hoa Chu Nghia Viet Nam

CAPITAL: Hanoi.

FLAG: The flag is red with a five-pointed gold star in the center.

ANTHEM: *Tien Quan Ça (Forward, Soldiers!).*

MONETARY UNIT: The dong (D) is a paper currency of 10 hao and 100 xu. There are coins of 1, 2, and 5 xu, and notes of 5 xu, 1, 2, and 5 hao, and 1, 2, 5, and 10 dong. D1 = $0.00091 (or $1 = D 11,000).

WEIGHTS AND MEASURES: The metric system is the legal standard, but some traditional measures are still used.

HOLIDAYS: Liberation of Saigon, 30 April; May Day, 1 May; Independence Day, 2 September. Movable holidays include the Vietnamese New Year (Tet).

TIME: 7 PM = noon GMT.

LOCATION AND SIZE: Situated on the eastern coast of mainland Southeast Asia, the Socialist Republic of Vietnam (SRV) has an area of 329,560 square kilometers (127,244 square miles), slightly larger than the state of New Mexico, and a total boundary length of 7,262 kilometers (4,512 miles).

Vietnam's capital city, Hanoi, is located in the northern part of the country.

CLIMATE: Mean annual rainfall in the north varies from 172 centimeters (68 inches) for the city of Hanoi to over 406 centimeters (160 inches) in the mountains. Daily temperatures in the Red River Delta region can fluctuate from as low as 5°C (41°F) in the dry season to about 30°C (86°F) during the rainy season.

The south is more tropical; temperatures in Ho Chi Minh City vary only between 18° (64°F) and 33°C (91°F) throughout the year. Temperatures in the Central Highlands range from a mean of about 17°C (63°F) in winter to 20°C (68°F) in summer. Annual rainfall averages about 200 centimeters (79 inches) in lowland regions. The typhoon season lasts from July through November, with the most severe storms occurring along the central coast.

INTRODUCTORY SURVEY

RECENT HISTORY

Vietnam, under Ho Chi Minh, achieved independence from France in 1954. There soon followed a long disastrous civil war involving United States troops fighting on the side of South Vietnam (1965–75). The war ended with U.S. withdrawal and capitulation by the South to the communist North. North Vietnam moved quickly to complete reunification of the country. In July 1976, the Socialist Republic of Vietnam (SRV) was established with its capital at Hanoi. Saigon, the former capital of South Vietnam was renamed Ho Chi Minh City.

In January 1989 the first direct talks between Vietnam and China since 1979 resulted in Vietnam's agreement to withdraw its troops from Cambodia and China's pledge to end aid to the Khmer Rouge guerrillas, thereby guaranteeing a measure of peace after a generation of war in Southeast Asia.

On 3 February 1994 President Bill Clinton lifted the American trade embargo against Vietnam.

GOVERNMENT

Vietnam is governed by a president, who is head of state and chair of the National Defense and Security Council, and a prime minister, who heads a cabinet of ministries and commissions. ''People's Committees'' govern in local jurisdictions. There is also a 400-person National Assembly.

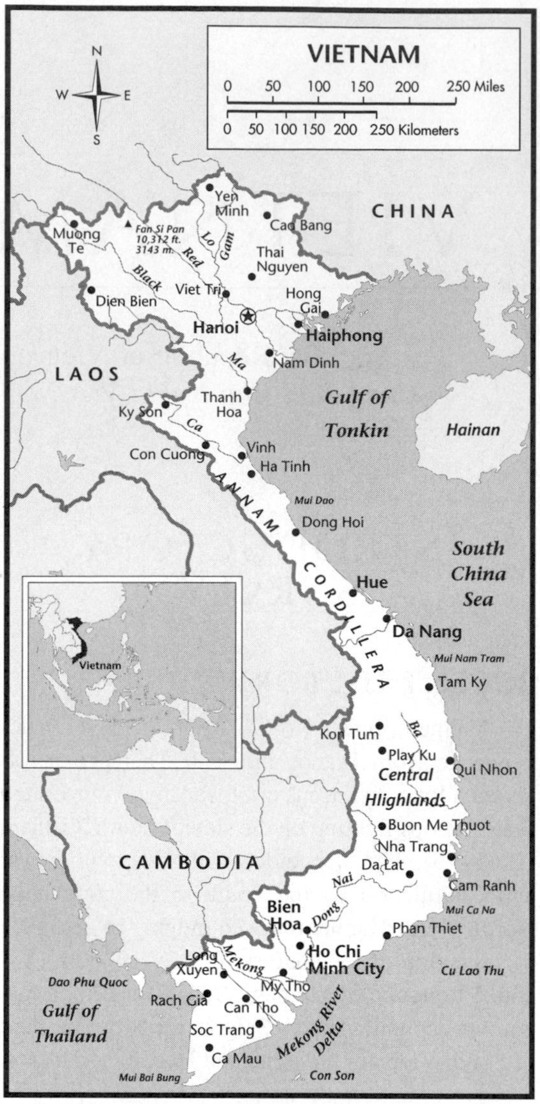

Vietnam is divided into 61 provinces. A special zone, Vung Tau-Con Dao, and three municipalities (Hanoi, Haiphong, and Ho Chi Minh City), are administered by the national government.

Judiciary

The highest court in Vietnam is the Supreme People's Court, whose members are appointed by the National Assembly for five-year terms. There are also local people's courts at each administrative level; military courts; and ''special courts'' established by the National Assembly in certain cases.

Political Parties

Vietnam is a one-party state controlled by the Vietnamese Communist Party (VCP). In theory, two other political parties, the Democratic Party and the Socialist Party, are granted legal existence.

DEFENSE

As of 1995, the People's Army of Vietnam was estimated at 572,000. Military expenditures were estimated to be $975 million of a national income of $19.1 billion. Military service is required for two or three years.

ECONOMIC AFFAIRS

Approximately 73% of the population was engaged in agriculture in 1996. The most diversified area in Southeast Asia in terms of mineral resources, Vietnam is well endowed with coal, tin, tungsten, gold, iron, manganese, chromium, and antimony. Cement, textiles, silk, matches, and paper are the main industrial products.

Vietnam's attempts to change from a state-controlled to a free-market economy have been slowed by the government's fear of changes that promote increased democracy. The labor force, with its low wage base, good skill levels, and high motivation, is an excellent resource for future growth in manufacturing.

Public Finance

Vietnam received $2.2 billion in foreign aid in 1999. Aid from the former Soviet Union, formerly Vietnam's most prominent donor, was greatly reduced after the dissolution of the USSR in 1991.

The U.S. Central Intelligence Agency estimates that, in 1996, government revenues totaled approximately $5.6 billion and expenditures $6 billion.

Income

In 1997 the gross national product (GNP) was 24 billion, or about $310 per person.

Industry

Most heavy and medium industry is concentrated in the north, including the state-owned coal, tin, chrome, and other mining enterprises; an engineering works at Hanoi; power stations; and modern tobacco, tea, and canning factories. Light industry and consumer goods industries, including pharmaceuticals, textiles, and food processing, are more prevalent in the south.

Major industries include: food processing, garments, shoes, machine building, mining, cement, chemical fertilizer, glass, tires, oil, coal, steel, paper. The industrial production growth rate was estimated at 12% in 1998.

State-run industries account for about 45% of the domestic economy.

Banking and Finance

Vietnam has increased its efforts to attract foreign capital and regularize relations with the world financial system. In 1995, the government began to rewrite trade laws and statutes in order to increase international trade opportunities.

Since 1992, Vietnam has moved to a diversified system in which state-owned joint-stock, joint-venture, and foreign banks provide services to a broader customer base. As of December 1995, in addition to four state-owned commercial banks, there were 46 joint-stock banks, 19 foreign bank branches, 4 joint-venture banks and 72 foreign banks with representative offices. The state banks still dominate the system, however, and state enterprises remain the main borrowers.

Much-needed reform of the banking sector is proceeding slowly.

SOCIAL WELFARE

Since 1976, Vietnam's attempts to lay the foundations of an advanced socialist society have been hampered by the relatively poor performance of the economy. There are continuing restrictions on freedom of the press, religion, and assembly. However, free speech and travel restrictions have eased.

Healthcare

Since 1975, some progress in health care has been made. Tuberculosis has been largely controlled, and the incidence of many contagious diseases has been reduced. In 1990–95, 38% of the population had access to safe water, and 21% had adequate sanitation. Life expectancy in 1999 was 68 years.

In the early 1990s, there were 3.5 physicians per 1,000 people, 3.3 hospital beds per 1,000 inhabitants, and about 97% of the population had access to healthcare services.

Housing

Housing is a problem in Hanoi although large flats have been erected in the suburbs to ease the situation. In the countryside, many farm families have built new houses of brick and stone. Similarly, in the rural south, housing is available to meet the requirements of the population because building construction continued at a relatively high level during the war years. In 1989, the majority of housing units were semi-permanent (structures with brick walls and tile roofs lasting about 20 years).

EDUCATION

In 1995 the adult literacy rate was 93.7%. Education is state-controlled at all levels, and five years of primary education is required. In 1994, primary schools had 10 million students; general secondary schools, 4.5 million; and there were some 123,000 students enrolled in higher education institutions.

1999 KEY EVENTS TIMELINE

January

- The Ministry of Finance issues new schedule for import tariffs along with the amended Import-Export Tax Law. Law on Value-Added Tax (VAT) goes into effect.
- Laotian delegation visits Vietnam.
- Twenty seven Vietnamese athletes participate in the 7th Asia-Pacific Olympic Games for the Disabled held in Bangkok. Vietnam wins first gold medal at the games.
- General Doan Khue, former Defense Minister and Politburo member, passes away.
- Secretary General Le Kha Phieu announces agreement on National Program on Tourism 2000.
- Government launches anti-drug campaign in Ho Chi Minh City.

February

- The film ''Lotus'' produced jointly produced by an Algerian and Vietnamese was screened at the Africa Film Festival in Burkina Faso.
- Government budget for 1999 allocates $160 million toward agricultural and rural capital projects.
- Government announces Industrial Zones will export US$1.6 billion in 1999.
- Conference on Vietnam-US Copyright Agreement held in Hanoi and organized by Hanoi's Culture and Information Department.

March

- State Securities Commission and ABN AMRO Bank Corporation holds seminar on stock market operating procedures.
- The Vietnam Chamber of Commerce and Industries, in conjunction with a trade fair and advertisement firm, held a ceremony to promote

Vietnamese products and honor two hundred businesses with a ''quality product stamp'' designation.

- Government announces a new campaign to examine and eradicate corruption in the communist party.
- Government appeals for food aid for central and northern Vietnam, with 1.5 million people affected by drought.

April

- Vietnamese government issues approval to two private firms, Vinh Phat and Thanh Hoa, to begin exporting rice abroad thereby breaking the state monopoly on rice exports.
- Vietnam dissident, Doan Viet Hoat, wants to establish an opposition party in Vietnam, but faces considerable obstacles.
- Prime Minister announces major economic measures to clarify and stabilize foreign business practices in the country.

May

- Ministry of Finance audit reveals $5.8 billion in state assets missing. Government considers need for audits of state-owned enterprises.

June

- Vietnam sends delegation to 10th ASEAN Games held in Pattaya, Thailand
- Government announces that no goods in Vietnam may be sold or have prices quoted in any foreign currency.
- Vietnamese students excel at the 11th Asia-Pacific Mathematics Olympics, placing the country second only to the Republic of Korea.

July

- Rare Javan rhinoceros of Vietnam, long thought to be extinct, found in small numbers by scientists.
- The U.S. and Vietnam announce a tentative agreement on outstanding issues to normalize trading relations between the two nations.

August

- Deputy Head of the Vietnamese Communist Party reiterates strong opposition to multiparty system while corruption appears rampant. Six are sentenced to death and six others to life-in-prison for graft in partially state-owned firm's loss of $280 million.
- Ten new rice strains with improved yield are announced at seminar conducted by the Mekong Delta Rice Institute, the Can Tho College, and the Southern Agricultural Scientific Study Institute.
- Hon Mon Rubber Company receives ISO 9002 designation for bicycle and motorcycle tire quality assurance.
- Government announces a major increase (53.2%) in drug arrests in 1999.
- Vietnamese government blasts foreign firms for using ''illegal'' tactics to dominate local business markets.
- U.S. opens consulate in Ho Chi Minh City.
- The Vietnam Committee on Human Rights expressed concern to the UN Sub-Commission on Human Rights regarding the harassment of political dissidents and Buddhist monks by government authorities.
- Government controlled media launches attack against Vietnam's pop stars as morally decadent and promiscuous.

September

- Vietnam Prime Minister Phan Van Khai and his delegation visit Nordic countries.
- U.S. Secretary of State Albright visits Vietnam for persuasion on final details regarding the trade pact with the US, only to have Prime Minister Khai announce delay later this month.
- Ho Chi Minh Police magazine noted that rape of children increased 25% for the first six months of 1999.

October

- Government allows private firm to mortgage land-use rights with domestic banks.
- Government states that urban unemployment hit 7.4 percent, up from 6.85 percent in 1998.

November

- Starving villagers mob helicopters bringing emergency supplies to central Vietnam as the death toll from floods climbs toward 500.
- At an informal meeting in Manila, Philippines, six members of the Association of Southeast Asian Nations (ASEAN)—Brunei, Indonesia, Malaysia, Philippines, Singapore, and Thailand—agree to establish a free-trade zone by eliminating duties on most goods traded in the

region by 2010. The remaining four newer and less-developed nation members—Cambodia, Laos, Myanmar (Burma), and Vietnam—will eliminate duties by 2015. Rice will be excluded from trade agreements, however.

ANALYSIS OF EVENTS: 1999

BUSINESS AND THE ECONOMY

During 1999, Vietnam's economy continued to show some adverse effects of the Asian Economic Crisis. While the economy grew more rapidly than the previous year, growth remained far lower than the government estimated. Rise in corruption and the government's reluctance to carry forward an economic reform plan weakened the government's credibility in economic matters. The government continued to show mixed signals by attempting to lure foreign investment while failing to significantly alter government policies.

The year began with the Ministry of Finance releasing a new schedule for import tariffs including the reduction, albeit painstakingly slow, of tariffs on imports. At the same time, the government enacted its Value-Added Tax as a way to strengthen the state's fiscal condition after two years of recession. During the year, the government sought to conclude negotiations on a trade pact with the United States and to develop a copyright protection system. While the government was successful in the later, it failed to finalize a trade pact with the United States despite an announcement in July that tentative agreement had been reached. In August, the U.S. opened a consulate in Ho Chi Minh City in anticipation of Secretary of State Madeline Albright's visit in September. During her visit, the U.S. and Vietnam were unable to resolve differences, with the Vietnamese favoring a more complex and gradual relinquishment of controls over domestic tariffs. Finally at the end of September, Prime Minister Khai announced a delay and stated that difficulties would need to be resolved at a later date.

Several successful business enterprises offered prospects for future economic growth, and the trade figures indicated that Vietnam's exports increased over last year. Some progress in invigorating the private sector took place, with the government encouraging tourism and providing the approval of private firms to mortgage land-rights for capital loans. However, the Vietnamese government stated that only domestic banks would be allowed to provide funds. In February, the government noted that the Industrial Zones would export US$1.6 billion in 1999, continuing a trend of increased foreign investment in Vietnam. The government also announced in April that two private firms would be allowed to export rice, a growing source of foreign currency. Other major initiatives included the building of a $5 million meat processing plant in Ninh Binh, and permission for two foreign insurance firms to begin operations this year. The much heralded opening of a stock market was put off until sometime next year. In March, the State Securities Commission and ABN AMRO Bank Corporation jointly sponsored a seminar on equity and market mechanics.

Throughout the year, corruption involving the Vietnamese Communist Party and several businesses underlined the need for creating a more trustworthy accounting system and business environment. After last year's allegation of corruption, in March the government announced a new campaign to examine and eradicate corruption in the Communist Party. In May, however, the Ministry of Finance announced that $5.8 billion in state assets were unaccounted for, prompting the government to suggest that audits of all state enterprises should be conducted.

Also, during May the country's largest corruption trial took place involving 77 businessmen, bankers, and government officials on charges of fraud in the amount of $280 million. In August, six people were sentenced to death and six others to life-in-prison.

In the area of economic reform, the Prime Minister announced a major declaration in April setting the framework for everything from minimum wage rates, to utility charges and land use. The government also decided to make the Vietnamese dong the only currency allowed for commercial activity within the country. The government stated that illegal currency trading at unauthorized rates would lead to prosecution.

GOVERNMENT AND POLITICS

The government continued to contest its image of being out-of-touch and corrupt. Increased crime, drug use, and the influence of western ideology and culture all appeared to cause the Vietnamese Communist Party to lash out against the public with calls of discipline and increased virtue. In January, an important leader, General Doan Khue, former Defense Minister, passed away, signaling the continuing departure of the *ancien régime* from Vietnam's political scene. Also, the government began an anti-drug campaign in Ho Chi Minh City noting increased drug use, smuggling, and arrests in 1999. Two months later, the government announced a campaign to combat corruption in the party and to restore the legitimacy of the party in Vietnam's public life. In August, Dao Duy Quat, head of the Vietnamese Communist Party's Ideology and Culture Commission dismissed any speculation concerning the creation of a multiparty system and stated that no opposition party would be allowed. However, he issued a call for improving the legitimacy of the party. Later that month, the Central Committee of the party met in Hanoi, with discussions centering on administrative reform involving provincial administration and minor leadership changes.

In August, the government reported that drug cases increased over 50 percent over the same period last year and that in 1998 forty-nine people were put to death and 18,000 arrested on drug-related charges. During the month of August, the government attacked foreign firms for ''illegal'' advertising tactics to dominate the domestic market and delivered a scathing criticism of the emerging popular music scene. Government controlled media accused the pop stars of ostentatious lifestyles filled with moral decadence and sexual promiscuity. During the year 1999 the government encouraged the visit of numerous foreign delegations including those of Tajikistan, Laos, the U.S., and Japan. In September Vietnamese Prime Minister Phan Van Khai and his delegation visited Scandanavian countries and developed commercial and cultural links.

CULTURE AND SOCIETY

In 1999, Vietnam witnessed a significant increase in unemployment with the gradual restructuring of state enterprise and the continued effects of the Asian Crisis. The government announced in October that unemployment reached 7.4%, up from 6.85% in 1998. Vietnam experienced drought in March and flooding in August, however the rice harvest escaped major damage.

In February, the film ''Lotus,'' produced by Algerian and Vietnamese nationals in a first ever joint project, was screened at the Africa Film Festival in Burkina Faso. During that month, twenty-seven youths participated in the 7th Asia-Pacific Olympic Games for the Disabled in Bangkok. For the first time, a Vietnamese won a gold medal. In June, Vietnam sent a delegation of athletes to the 10th ASEAN Games in Pattaya, Thailand. Also, during the month of June, Vietnamese students excelled at the 11th Asia-Pacific Mathematics Olympics, placing second only to South Korea.

In August, violence marred a celebration marking the victory of Vietnam's national soccer team over Brunei when a melee involving police and participants of a motorcycle race in Ho Chi Minh resulted in the death of four people and injury to 150. A few days later, the police were placed on high alert for the upcoming game against Myanmar. The police also confiscated over 1,000 motorbikes suspected of being used in the races. In September, the government warned parents to watch over their children more carefully citing a 25 percent increase in child rape in 1999 in Ho Chi Minh City.

DIRECTORY

CENTRAL GOVERNMENT

Head of State

President

Tran Duc Luong, The Governmental Office, 1 Hoang Hoa Tham Street, Hanoi, Vietnam

Ministers

Minister of Agriculture and Rural Development

Le Huy Ngoc, Ministry of Agriculture and Rural Development, Add. 2 Ngoc Ha Street, Ba Dinh Dist., Hanoi, Vietnam
PHONE: +84 (4) 8468160
FAX: +84 (4) 8454319

Minister of Construction

Nguyen Khoa Diem, Ministry of Construction, 37 Le Dai Hanh Street, Hai Ba Trung Dist, Hanoi, Vietnam
PHONE: +84 (4) 8268271; 8254022; 8255497
FAX: +84 (4) 8215591

Minister of Culture and Information

Nguyen Khoa Diem, Ministry of Culture and Information, 51-53 Ngo Quyen Street, Hanoi, Vietnam

PHONE: +84 (4) 8262945; 8262487; 8255349
FAX: +84 (4) 8267101

Minister of Defense
Pham Van Tra, Ministry of Defense, 1A Hoang Dieu Street, Ba Dinh Dist., Hanoi, Vietnam
PHONE: +84 (4) 8468104

Minister or Education and Training
Nguyen Minh Hien, Ministry of Education and Training, 49 Dai Co Viet Street, Hai Ba Trung Dist., Hanoi, Vietnam
PHONE: +84 (4) 8692396; 8694904; 8694795
FAX: +84 (4) 8694085

Minister of Finance
Nguyen Sinh Hung, Ministry of Finance, 8 Ohan Huy Chu Street, Hoan Kiem Dist., Hanoi, Vietnam
PHONE: +84 (4) 8264872; 8262356; 8262357
FAX: +84 (4) 8262266

Minister of Fisheries
Ta Quang Ngoc, Ministry of Fisheries, Ngoc Khank Street, Ba Dinh Dist., Hanoi, Vietnam
PHONE: +84 (4) 8346269
FAX: +84 (4) 8326702

Minister of Foreign Affairs
Nguyen Manh Cam, Ministry of Foreign Affairs, 1 Ton That Dam Street, Ba Dinh Dist., Hanoi, Vietnam
PHONE: +84 (4) 8453973; 8458208; 8458321
FAX: +84 (4) 8445905

Minister of Health
Do Nguyen Phuong, Ministry of Health, 138A Giang Vo Street, Ba Dinh Dist., Hanoi, Vietnam
PHONE: +84 (4) 8464051
FAX: +84 (4) 8464051

Minister of Industry
Dang Vu Chu, Ministry of Industry, 54 Tran Binh Trong Street, Hoan Kiem Dist., Hanoi, Vietnam
PHONE: +84 (4) 8267870
FAX: +84 (4) 8269033

Minister of Interior
Le Minh Huong, Ministry of Interior, 15 Tran Binh Trong Street, Hoan Kiem Dist., Hanoi, Vietnam
PHONE: +84 (4) 8268231
FAX: +84 (4) 8260774; 8260773

Minister of Justice
Nguyen Dinh Loc, Ministry of Justice, 25A Cat Linh Street, Ba Dinh Dist., Hanoi, Vietnam
PHONE: +84 (4) 8231138; 8454765; 8431126
FAX: +84 (4) 8431431

Minister of Labor, War Invalids, and Social Affairs
Nguyen Thi Hang, Ministry of Labor, War Invalids, and Social Affairs, 12 Ngo Quyen Street, Hoan Kiem Dist., Hanoi, Vietnam
PHONE: +84 (4) 8246137; 8269532; 8269536
FAX: +84 (4) 8248036

Minister of Science, Technology and Environment
Chu Tuan Nha, Ministry of Science, Technology and Environment, 39 Tran Hung Dao Street, Hoan Kiem Dist., Hanoi, Vietnam
PHONE: +84 (4) 8252731; 8252732; 8263379
FAX: +84 (4) 8252733

Minister of Trade
Truong Dinh Tuyen, Ministry of Trade, 31 Trang Tien Street, Hoan Kiem Dist., Hanoi, Vietnam
PHONE: +84 (4) 8253881; 8255184; 8264693
FAX: +84 (4) 8264696

Minister of Transport and Communication
Le Ngoc Hoan, Ministry of Transport and Communication, 80 Tran Hung Dao Street, Hoan Kiem Dist., Hanoi, Vietnam
PHONE: +84 (4) 8254012; 8252925; 8252309
FAX: +84 (4) 8267291

Minister of Planning and Investment
Tran Xuan Gia, Ministry of Planning and Investment, 2 Hoang Van Thu Street, Ba Dinh Dist., Hanoi, Vietnam
PHONE: +84 (4) 8455298
FAX: +84 (4) 8232494

Minister of Population and Family Planning
Tran Thi Trung Chien, Ministry of Population and Family Planning

Ministy for Child Care and Protection
Tran Thi Thanh Thanh, Ministry of Child Care and Protection

POLITICAL ORGANIZATIONS

Communist Party of Vietnam (CPV)

TITLE: General Secretary
NAME: Le Kha Phieu

DIPLOMATIC REPRESENTATION

Embassies in Vietnam

Algeria
13 Phan Ch Trinh Street, Hanoi, Vietnam

PHONE: +84 (4) 8253865
FAX: +84 (4) 8260830
TITLE: Ambassador
NAME: Maamar Ahmed

Argentina
8th Floor, Daeha Business Center, 360 Kim Ma Street, Ba Dinh Dist., Hanoi, Vietnam
PHONE: +84 (4) 8315262; 8315578
FAX: +84 (4) 8315577
TITLE: Ambassador
NAME: Armando J.J. Maffei

Australia
66 Ly Thuong Kiet Street, Hanoi, Vietnam
PHONE: +84 (4) 8252763; 8258480
FAX: +84 (4) 8259268
TITLE: Ambassador
NAME: Susan Boyd

Bangladesh
101-108, A1 Van Phuc Quarter, Hanoi, Vietnam
PHONE: +84 (4) 8231625
FAX: +84 (4) 8231628
TITLE: Ambassador
NAME: Kazi Awarul Masud

Belgium
48-50 Nguyen Thai Hoc Street, Hanoi, Vietnam
PHONE: +84 (4) 8452263; 8235006
FAX: +84 (4) 8457165
TITLE: Ambassador
NAME: Benoit Ryelandt

Brazil
E. 16-16 Thuy Khue Street, Hanoi, Vietnam
PHONE: +84 (4) 8430817
FAX: +84 (4) 8432542
TITLE: Ambassador
NAME: Italo Zappa

Brunei Darussalam
5/12 Bac Co Rd., Hanoi, Vietnam
PHONE: +84 (4) 8264816
FAX: +84 (4) 8264821

Bulgaria
Van Phuc Quarter, Street No 358, Hanoi, Vietnam
PHONE: +84 (4) 8452908
TITLE: Ambassador
NAME: Alexander Itov

Cambodia
71A Tran Hung Dao Street, Hanoi, Vietnam
PHONE: +84 (4) 8253788; 8253789
FAX: +84 (4) 8265225
TITLE: Ambassador
NAME: Mease Sip

Canada
31 Hung Vuong Street, Hanoi, Vietnam
PHONE: +84 (4) 8235500
FAX: +84 (4) 8235333
TITLE: Ambassador
NAME: Christine Desloges

China
46 Hoang Dieu Street, Hanoi, Vietnam
PHONE: +84 (4) 8453736; 8453737
FAX: +84 (4) 8232826
TITLE: Ambassador
NAME: Li Jiazhong

Cuba
65 Ky Thuong Kiet Street, Hanoi, Vietnam
PHONE: +84 (4) 8254775
FAX: +84 (4) 8252426
TITLE: Ambassador
NAME: Tania Maceira Delgado

Czech Republic
13 Chu Van An Street, Hanoi, Vietnam
PHONE: +84 (4) 8454131; 8454132
FAX: +84 (4) 8233996
TITLE: Ambassador
NAME: Hana Havlova

Denmark
19 Dien Bien Phu, Hanoi, Vietnam
PHONE: +84 (4) 8231888
FAX: +84 (4) 8231999
TITLE: Ambassador
NAME: Niels Julius Lassen

Egypt
Villa 6, Van Phuc Quarter, Hanoi, Vietnam
PHONE: +84 (4) 8460219
FAX: +84 (4) 8460218
TITLE: Ambassador
NAME: Haney Sharag El Din

Finland
B3B, Giang Vo Quarter, No. 1/2, Hanoi, Vietnam
PHONE: +84 (4) 8456754
FAX: +84 (4) 8232821
TITLE: Ambassador
NAME: Kai Olof Granholm

France
57 Tran Hung Dao Street, Hanoi, Vietnam
PHONE: +84 (4) 8252719; 8254367
TITLE: Ambassador
NAME: Serge Degallaix

Germany
29 Tran Phu Street, PO Box 39, Hanoi, Vietnam
PHONE: +84 (4) 8253836
FAX: +84 (4) 8253838
TITLE: Ambassador
NAME: Klaus Christian Kraemer

Hungary
43-47 Dien Bien Phu Street, Hanoi, Vietnam
PHONE: +84 (4) 8452878
FAX: +84 (4) 8233049

India
25c Lang Ha-Dong Da, Hanoi, Vietnam
PHONE: +84 (4) 8560463; 8560474
FAX: +84 (4) 8560462
TITLE: Ambassador
NAME: Surinder Lal Malik

Indonesia
50 Ngo Quyen Street, Hanoi, Vietnam
PHONE: +84 (4) 8253353; 8253084
FAX: +84 (4) 8259274
TITLE: Ambassador
NAME: Djafar Husin Assegaff

Iran
54 Tran Phu Street, Hanoi, Vietnam
PHONE: +84 (4) 8232068; 8232069
FAX: +84 (4) 8232120
TITLE: Ambassador
NAME: Seyed Kamal Sajadi

Iraq
66 Tran Hung Dao Street, Hanoi, Vietnam
PHONE: +84 (4) 8254141
FAX: +84 (4) 8254055
TITLE: Ambassador
NAME: Mahdi S. Hamoudi Al-Samarrae

Israel
68 Nguyen Tahi Hoc Street, Hanoi, Vietnam
PHONE: +84 (4) 8433140
FAX: +84 (4) 8435760
TITLE: Ambassador
NAME: Uri Halfon

Italy
9 Le Phung Hieu Street, Hanoi, Vietnam
PHONE: +84 (4) 8256246; 8256256
FAX: +84 (4) 8267602
TITLE: Ambassador
NAME: Nario Vittorio Zamboni

Japan
1 Lieu Giai Street, Hanoi, Vietnam
PHONE: +84 (4) 8463000
FAX: +84 (4)8463043
TITLE: Ambassador
NAME: Katsunari Suzuki

South Korea
29 Nguyen Dinh Chieu Street, Hanoi, Vietnam
PHONE: +84 (4) 8226677; 8226678
FAX: +84 (4) 8226328
TITLE: Ambassador
NAME: Bong Kyu Kim

Laos
22 Tran Binh Trong Street, Hanoi, Vietnam
PHONE: +84 (4) 8254576
FAX: +84 (4) 8228414
TITLE: Ambassador
NAME: Khamsing Sayakone

Libya
A3 Van Phuc Quarter, Hanoi, Vietnam
PHONE: +84 (4) 8453379
FAX: +84 (4) 8454977

Malaysia
A3 Van Phuc Quarter, Hanoi, Vietnam
PHONE: +84 (4) 8453371; 8433520
FAX: +84 (4) 8232166
TITLE: Ambassador
NAME: Cheah Sam Kip

Mongolia
39 Tran Phu Street, Hanoi, Vietnam
PHONE: +84 (4) 8453009; 8436256
FAX: +84 (4) 8454954
TITLE: Ambassador
NAME: Ch. Agvaandamdin

Myanmar
A3, Van Phuc Quarter, Hanoi, Vietnam
PHONE: +84 (4) 8253369
FAX: +84 (4) 8252404
TITLE: Ambassador
NAME: Maung Maung Lay

Netherlands
D1 Van Phuc Quarter, Hanoi, Vietnam
PHONE: +84 (4) 8430605
FAX: +84 (4) 8431013
TITLE: Ambassador
NAME: Monique Frank

New Zealand
32 Hang Bai Street, Hanoi, Vietnam
PHONE: +84 (4) 8241481
FAX: +84 (4) 8241480
TITLE: Ambassador
NAME: David Kersey

Norway
Metropole Centre, Suite 701, 56 Ly Thai To Street, Hanoi, Vietnam
PHONE: +84 (4) 8262111
FAX: +84 (4) 8260222
TITLE: Ambassador
NAME: Jens Otterbech

Palestine
E4b, Trung Tu Quarter, Hanoi, Vietnam
PHONE: +84 (4) 8524013; 8522947
FAX: +84 (4) 8263696
TITLE: Ambassador
NAME: Said Khalil Al-Masri

Philippines
27B Tran Hung Dao Street, Hanoi, Vietnam
PHONE: +84 (4) 8257948; 8257873
FAX: +84 (4) 8265760
TITLE: Ambassador
NAME: Rosalinda V. Tirona

Poland
3 Chua Mot Cot Street, Hanoi, Vietnam
PHONE: +84 (4) 8452027; 8453728
FAX: +84 (4) 8236914
TITLE: Ambassador
NAME: Krzysztos Kocel

Romania
5 Le Hong Phong Street, Hanoi, Vietnam
PHONE: +84 (4) 8452014
FAX: +84 (4) 8430922
TITLE: Ambassador
NAME: Valeriu Arteni

Russia
58 Tran Phu Street, Hanoi, Vietnam
PHONE: +84 (4) 8454632
FAX: +84 (4) 8456177
TITLE: Ambassador
NAME: Rachit L. Khamidulin

Serbia
47 Tran Phu Street, Hanoi, Vietnam
PHONE: +84 (4) 8452343
FAX: +84 (4) 8456173

Singapore
41-43 Tran Phu Street, Hanoi, Vietnam
PHONE: +84 (4) 8233965; 8233966
FAX: +84 (4) 8233992
TITLE: Ambassador
NAME: Toh Hock Ghim

Slovakia
6 Le Hong Phong Street, Hanoi, Vietnam
PHONE: +84 (4) 8454335
FAX: +84 (4) 8454145
TITLE: Ambassador
NAME: Jan Gonzor

Spain
104 Tran Hung Dao, 5th Floor, Hanoi, Vietnam
PHONE: +84 (4) 8223438
FAX: +84 (4) 8223441
TITLE: Ambassador
NAME: Ignacio Sagaz

Sweden
2 Nui Truc Street, Van Phuc, Hanoi, Vietnam
PHONE: +84 (4) 8454825; 8232189
FAX: +84 (4) 8232195
TITLE: Ambassador
NAME: Borje Ljunggren

Switzerland
77b Kim Ma Street, Hanoi, Vietnam
PHONE: +84 (4) 8232019; 8232022
FAX: +84 (4) 8232045
TITLE: Ambassador
NAME: Jurg Leutert

Thailand
63 Hoang Dieu Street, Hanoi, Vietnam
PHONE: +84 (4) 8235092
FAX: +84 (4) 8235088
TITLE: Ambassador
NAME: Chalermpon Ake-Uru

United Kingdom
16 Ly Thuong Kiet Street, Hanoi, Vietnam
PHONE: +84 (4) 8252510; 8252349
FAX: +84 (4) 8265762
TITLE: Ambassador
NAME: Peter Keegan Williams

United States
7 Lang Ha Street, Hanoi, Vietnam
PHONE: +84 (4) 8431500
FAX: +84 (4) 8431510

JUDICIAL SYSTEM

The Supreme People's Court

Supreme People's Procuracy

FURTHER READING

Articles

''Disquite Among the Quite: Vietnam's Government is Expert At Silencing Dissent, but Voices of Dissatisfaction are Growing Louder.'' *Time International* (January 18, 1999).

"Will Vietnam's War on Graft Succeed?" *The Asian Wall Street Journal*, 29 March 1999.

"Vietnam Opens to Private Rice exports; in a First Step, Two Firms Get Rights to Sell Small Amounts Overseas." *The Asian Wall Street Journal Weekly*, 19 April 1999.

"Vietnam Puts a Tag on Official Corruption; Finance Ministry Reveals State Agencies Cannot Account for $5.8 billion in Bureaucratic Assets." *The Asian Wall Street Journal Weekly*, 31 May 1999.

"Vietnam Communists Say to Keep Single-Party System." Reuters, 2 August 1999.

"Vietnam Sentences Six to Die Over Graft Scam." Reuters, 4 August 1999.

"Four Killed, 150 Hurt in Vietnam Soccer Violence." Reuters, 6 August 1999.

"Vietnam Communist Party Opens Key Meeting." Reuters, 8 August 1999.

"Vietnam's Drug Crimes, Arrests Soar in 1999." Reuters, 10 August 1999.

Internet

Embassy of the Socialist Republic of Vietnam in the United States. Available Online @ http://www.vietnamembassy-usa.org/ (November 15, 1999).

Vietspace. Available Online @ http://kicon.com (November 15, 1999).

VNN Media. "Vietnam News Network." Available Online @ http://www.vnn-news.com/ (November 15, 1999).

VIETNAM: STATISTICAL DATA

For sources and notes see "Sources of Statistics" in the front of each volume.

GEOGRAPHY

Geography (1)

Area:

Total: 329,560 sq km.

Land: 325,360 sq km.

Water: 4,200 sq km.

Area—comparative: slightly larger than New Mexico.

Land boundaries:

Total: 4,639 km.

Border countries: Cambodia 1,228 km, China 1,281 km, Laos 2,130 km.

Coastline: 3,444 km (excludes islands).

Climate: tropical in south; monsoonal in north with hot, rainy season (mid-May to mid-September) and warm, dry season (mid-October to mid-March).

Terrain: low, flat delta in south and north; central highlands; hilly, mountainous in far north and northwest.

Natural resources: phosphates, coal, manganese, bauxite, chromate, offshore oil and gas deposits, forests.

Land use:

Arable land: 17%

Permanent crops: 4%

Permanent pastures: 1%

Forests and woodland: 30%

Other: 48% (1993 est.).

HUMAN FACTORS

Demographics (2A)

	1990	1995	1998	2000	2010	2020	2030	2040	2050
Population	66,314.5	72,814.6	76,236.3	78,349.5	88,601.7	99,152.9	108,116.9	114,960.3	119,463.5
Net migration rate (per 1,000 population)	NA	NA	NA	NA	NA	NA	NA	NA	NA
Births	NA	NA	NA	NA	NA	NA	NA	NA	NA
Deaths	NA	NA	NA	NA	NA	NA	NA	NA	NA
Life expectancy - males	62.6	64.4	65.4	66.1	69.1	71.7	73.8	75.5	76.8
Life expectancy - females	67.0	69.1	70.3	71.0	74.5	77.3	79.6	81.4	82.8
Birth rate (per 1,000)	30.5	23.8	21.5	20.0	18.2	16.0	13.8	13.1	12.3
Death rate (per 1,000)	8.1	7.1	6.7	6.5	6.0	5.9	6.6	8.1	9.6
Women of reproductive age (15-49 yrs.)	16,519.4	18,947.1	20,599.7	21,738.7	25,502.6	26,668.5	27,194.4	26,336.0	25,631.0
of which are currently married	NA	NA	NA	NA	NA	NA	NA	NA	NA
Fertility rate	3.7	2.8	2.5	2.3	2.0	2.0	2.0	2.0	2.0

Except as noted, values for vital statistics are in thousands; life expectancy is in years.

Health Personnel (3)

Total health expenditure as a percentage of GDP, 1990-1997[a]

Public sector	1.1
Private sector	4.1
Total[b]	5.2

Health expenditure per capita in U.S. dollars, 1990-1997[a]

Purchasing power parity	63
Total	9

Availability of health care facilities per 100,000 people

Hospital beds 1990-1997[a]	380
Doctors 1993[c]	NA
Nurses 1993[c]	NA

Health Indicators (4)

Life expectancy at birth

1980	63
1997	68
Daily per capita supply of calories (1996)	2,502
Total fertility rate births per woman (1997)	2.4
Maternal mortality ratio per 100,000 live births (1990-97)	105[b]
Safe water % of population with access (1995)	47
Sanitation % of population with access (1995)	60
Consumption of iodized salt % of households (1992-98)[a]	65

Smoking prevalence

Male % of adults (1985-95)[a]	73
Female % of adults (1985-95)[a]	4
Tuberculosis incidence per 100,000 people (1997)	189
Adult HIV prevalence % of population ages 15-49 (1997)	0.22

Ethnic Division (6)

Vietnamese	85%-90%
Chinese	3%
Muong, Tai, Meo, Khmer, Man, Cham	NA

Religions (7)

Buddhist, Taoist, Roman Catholic, indigenous beliefs, Islam, Protestant, Cao Dai, Hoa Hao.

Languages (8)

Vietnamese (official), Chinese, English, French, Khmer, tribal languages (Mon-Khmer and Malayo-Polynesian).

EDUCATION

Public Education Expenditures (9)

Public expenditure on education (% of GNP)

1980	....
1996	2.6[1]

Expenditure per student

Primary % of GNP per capita

1980	....
1996	....

Secondary % of GNP per capita

1980	....
1996	....

Tertiary % of GNP per capita

1980	....
1996	....

Expenditure on teaching materials

Primary % of total for level (1996)	....
Secondary % of total for level (1996)	....
Primary pupil-teacher ratio per teacher (1996)	....
Duration of primary education years (1995)	5

Educational Attainment (10)

Age group (1989)	25+
Total population	26,466,214

Highest level attained (%)

No schooling	16.6
First level	
Not completed	69.8
Completed	NA
Entered second level	
S-1	10.6
S-2	NA
Postsecondary	2.6

Literacy Rates (11A)

In thousands and percent[1]	1990	1995	2000	2010
Illiterate population (15+ yrs.)	3,797	2,916	2,332	1,768
Literacy rate - total adult pop. (%)	90.7	93.7	95.6	97.4
Literacy rate - males (%)	94.8	96.5	97.5	98.4
Literacy rate - females (%)	87.0	91.2	93.9	96.4

GOVERNMENT & LAW

Political Parties (12)

National Assembly	% of seats
Communist Party of Vietnam (CPV)	92
Other	8

The 8% other are not CPV members but are approved by the CPV to stand for election.

Government Budget (13B)

Revenues . . . .$5.6 billion
Expenditures . . . .$6 billion
Capital expenditures . . . .$1.7 billion

Data for 1996 est.

Crime (15)

Crime rate (for 1997)
Crimes reported 63,100
Total persons convicted 36,000
Crimes per 100,000 population 84
Persons responsible for offenses
Total number of suspects 52,200
Total number of female suspects . . . . .NA
Total number of juvenile suspects . . . . .NA

Military Affairs (14B)

	1990	1991	1992	1993	1994	1995
Military expenditures						
Current dollars (mil.)	723	720	NA	NA	435	544
1995 constant dollars (mil.)	831	796	NA	NA	446	544
Armed forces (000)	1,052	1,041	550	550	550	550
Gross national product (GNP)						
Current dollars (mil.)	13,900	15,000	16,730	18,490	20,180	21,300
1995 constant dollars (mil.)	15,970	16,580	18,000	19,380	20,680	21,300
Central government expenditures (CGE)						
1995 constant dollars (mil.)	3,131	1,989	2,151	NA	4,613	5,000
People (mil.)	66.3	67.7	69.0	70.3	71.6	72.8
Military expenditure as % of GNP	5.2	4.8	NA	NA	2.2	2.6
Military expenditure as % of CGE	26.5	40.0	NA	NA	9.7	10.9
Military expenditure per capita (1995 $)	13	12	NA	NA	6	7
Armed forces per 1,000 people (soldiers)	15.9	15.4	8.0	7.8	7.7	7.6
GNP per capita (1995 $)	241	245	261	276	289	293
Arms imports[6]						
Current dollars (mil.)	1,100	200	10	10	80	200
1995 constant dollars (mil.)	1,264	221	11	10	82	200
Arms exports[6]						
Current dollars (mil.)	0	0	10	0	0	0
1995 constant dollars (mil.)	0	0	11	0	0	0
Total imports[7]						
Current dollars (mil.)	2,752	2,338	2,541	3,415	4,200[e]	7,500[e]
1995 constant dollars (mil.)	3,163	2,584	2,733	3,580	4,305[e]	7,500[e]
Total exports[7]						
Current dollars (mil.)	2,404	2,087	2,581	2,971	3,600[e]	5,300[e]
1995 constant dollars (mil.)	2,763	2,306	2,776	3,115	3,600[e]	5,300[e]
Arms as percent of total imports[8]	40.0	8.6	.4	.3	1.9	2.7
Arms as percent of total exports[8]	0	0	.4	0	0	0

LABOR FORCE

Labor Force (16)

Total (million)	32.7
Agriculture	65%
Industry and services	35%

Data for 1990 est.

Unemployment Rate (17)

25% (1995 est.)

PRODUCTION SECTOR

Electric Energy (18)

Capacity	5.32 million kW (1995)
Production	12.3 billion kWh (1995)
Consumption per capita	165 kWh (1995)

Transportation (19)

Highways:

total: 93,300 km

paved: 23,418 km

unpaved: 69,882 km (1996 est.)

Waterways: 17,702 km navigable; more than 5,149 km navigable at all times by vessels up to 1.8 m draft

Pipelines: petroleum products 150 km

Merchant marine:

total: 121 ships (1,000 GRT or over) totaling 487,427 GRT/750,000 DWT ships by type: bulk 7, cargo 97, chemical tanker 1, combination bulk 1, oil tanker 9, refrigerated cargo 5, roll-on/roll-off cargo 1 note: Vietnam owns an additional 7 ships (1,000 GRT or over) totaling 97,531 DWT operating under the registries of The Bahamas, Honduras, Liberia, Malta, and Panama (1997 est.)

Airports: 48 (1994 est.)

Airports—with paved runways:

total: 36

over 3,047 m: 8

2,438 to 3,047 m: 3

1,524 to 2,437 m: 5

914 to 1,523 m: 13

under 914 m: 7 (1994 est.)

Airports—with unpaved runways:

total: 12

1,524 to 2,437 m: 2

914 to 1,523 m: 5

under 914 m: 5 (1994 est.)

Top Agricultural Products (20)

Paddy rice, corn, potatoes, rubber, soybeans, coffee, tea, bananas; poultry, pigs; fish.

FINANCE, ECONOMICS, & TRADE

Economic Indicators (22)

National product: GDP—purchasing power parity—$128 billion (1997 est.)

National product real growth rate: 8.5% (1997 est.)

National product per capita: $1,700 (1997 est.)

Inflation rate—consumer price index: 5% (1997)

MANUFACTURING SECTOR

GDP & Manufacturing Summary (21)

	1980	1985	1990	1992	1993	1994
Gross Domestic Product						
Millions of 1990 dollars	3,599	4,980	6,360	7,319	7,912	8,608
Growth rate in percent	−4.81	6.20	5.05	8.60	8.10	8.80
Per capita (in 1990 dollars)	67.0	83.1	95.4	104.9	110.9	118.0
Manufacturing Value Added						
Millions of 1990 dollars	NA	NA	NA	NA	NA	NA
Growth rate in percent	NA	NA	NA	NA	NA	NA
Manufacturing share in percent of current prices	NA	NA	18.5	21.7	21.5	21.5

FINANCE, ECONOMICS, & TRADE

Exchange Rates (24)

Exchange rates:	
New dong (D) per US$1	
January 1998	12,300
December 1996	11,100
1995 average	11,193
October 1994	11,000
November 1993	10,800
July 1991	8,100

Top Import Origins (25)

$11.1 billion (f.o.b., 1996 est.)

Origins	%
Singapore	NA
South Korea	NA
Japan	NA
France	NA
Hong Kong	NA
Taiwan	NA

NA stands for not available.

Top Export Destinations (26)

$7.1 billion (f.o.b., 1996 est.).

Destinations	%
Virgin Islands (US)	NA
Puerto Rico	NA
United States	NA

NA stands for not available.

Economic Aid (27)

Recipient: ODA, $NA. Note: $2.4 billion in credits and grants pledged by international donors for 1997. NA stands for not available.

Import Export Commodities (28)

Import Commodities	Export Commodities
Machinery and equipment	Crude oil
Petroleum products	Marine products
Fertilizer	Rice
Steel products	Coffee
Raw cotton	Rubber
Grain	Tea
Cement	Garments
Motorcycles	Shoes

VIRGIN ISLANDS

Virgin Islands of the United States

CAPITAL: Charlotte Amalie.

FLAG: The flag of the Virgin Islands is white with a modified U.S. coat of arms in the center between the large blue initials V and I; the coat of arms shows a yellow eagle holding an olive branch in one talon and three arrows in the other with a superimposed shield of vertical red and white stripes below a blue panel.

MONETARY UNIT: 1 United States dollar (US$) = 100 cents.

WEIGHTS AND MEASURES: The imperial system is used.

HOLIDAYS: Transfer Day, 31 March.

TIME: 8 AM = noon GMT.

LOCATION AND SIZE: The Virgin Islands of the United States lie about 64 km (40 mi.) N of Puerto Rico and 1,600 km (1,000 mi.) SE of Miami, between 17°40′ and 18°25′N and 64°34′ and 65°3′N. The island group extends 82 km (51 mi.) N°S and 80 km (50 mi.) E°W with a total area of at least 353 sq. km (136 sq. mi.). Only 3 of the more than 50 islands and cays are of significant size: St. Croix, 218 sq. km (84 sq. mi.) in area; St. Thomas, 83 sq. km (32 sq. mi.); and St. John, 52 sq. km (20 sq. mi.).

CLIMATE: The subtropical climate, with temperatures ranging from 21° to 32°C (70 to 90°F) and an average temperature of 25°C (77°F), is moderated by northeast trade winds. Rainfall, the main source of fresh water, varies widely, and severe droughts are frequent. The average yearly rainfall is 114 cm (45 in.), mostly during the summer months.

INTRODUCTORY SURVEY

RECENT HISTORY

Excavations at St. Croix in the 1970s uncovered evidence of a civilization perhaps as ancient as AD 100. Christopher Columbus, who reached the islands in 1493, named them for the martyred virgin St. Ursula. At this time, St. Croix was inhabited by Carib Indians, who were eventually driven from the island by Spanish soldiers in 1555. During the 17th century, the archipelago was divided into two territorial units, one controlled by the British, the other (now the U.S. Virgin Islands) controlled by Denmark. The separate history of the latter unit began with the settlement of St. Thomas by the Danish West India Company in 1672. St. John was claimed by the company in 1683, and St. Croix was purchased from France in 1733. The holdings of the company were taken over as a Danish crown colony in 1754. Sugarcane, cultivated by slave labor, was the backbone of the islands' prosperity in the 18th and early 19th centuries. After brutally suppressing several slave revolts, Denmark abolished slavery in the colony in 1848. A long period of economic decline followed, until Denmark sold the islands to the U.S. in 1917 for $25 million. Congress granted U.S. citizenship to the Virgin Islanders in 1927. In 1931, administration of the islands was transferred from the Department of the Navy to the Department of the Interior, and the first civilian governor was appointed. In the late 1970s, the Virgin Islands government began to consider ways to expand self rule. A UN delegation in 1977 found little interest in independence, however, and a locally drafted constitution was voted down by the electorate in 1979.

GOVERNMENT

The chief executive of the Virgin Islands is the territorial governor, elected by direct popular vote (prior to 1970, territorial governors were appointed by the U.S. president). Constitutionally, the U.S. Congress has plenary authority to legislate for the territory. Enactment of the Revised Organic Act of

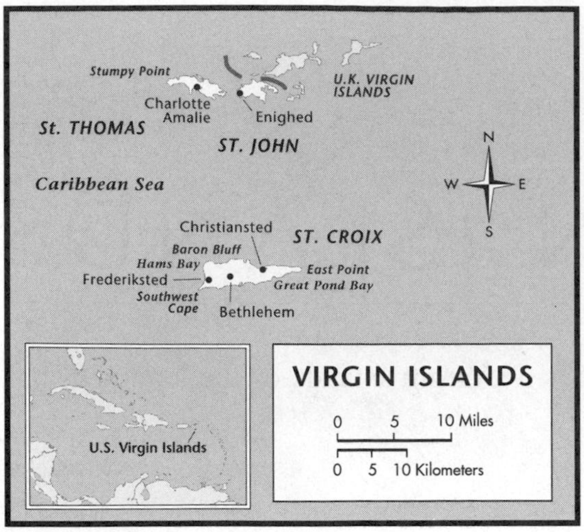

the Virgin Islands on 22 July 1954 vested local legislative power-subject to veto by the governor-in a unicameral legislature. Since 1972, the islands have sent one nonvoting representative to the U.S. House of Representatives.

Judiciary

Courts are under the U.S. federal judiciary; the two federal district court judges are appointed by the U.S. president. Territorial court judges, who preside over misdemeanor and traffic cases, are appointed by the governor and confirmed by the legislature. The district court has appellate jurisdiction over the territorial court.

ECONOMIC AFFAIRS

Tourism, which accounts for approximately 70% of both GDP and employment is the islands' principal economic activity. The number of tourists rose dramatically throughout the late 1960s and early 1970s, from 448,165 in 1964 to 1, 116,127 in 1972/73, but has stagnated since that time.

INCOME

In 1990, median family income was $24,036, and the unemployment rate was 6.7%. Exports in 1985 totaled $3,357.1 million, of which petroleum products constituted 95%; imports of crude petroleum for refining accounted for 82% of the Virgin Islands' $3,740.6 million import bill. The total operating budget of the Virgin Islands government in 1985 was $263.3 million of which taxes on personal and corporate income provided 50%.

Industry

Rum remains an important manufacture, with petroleum refining (on St. Croix) a major addition in the late 1960s.

Economic Development

Economic development is promoted by the US-government-owned Virgin Islands Corp. In 1987 the national product per capita was $11,000. The unemployment rate was 3.7% in 1992. Exports for 1990 totaled $2.8 billion while imports totaled $3.3 billion. The island's primary export is refined petroleum products. Raw crude oil constitutes the Virgin Island's most expensive import. Between 1970 and 1989 economic aid from the nations of the world totaled $42 million. In 1990, median family income was $24,036, and the unemployment rate was 6.7%. Exports in 1985 totaled $3,357.1 million, of which petroleum products constituted 95%; imports of crude petroleum for refining accounted for 82% of the Virgin Islands' $3,740.6 million import bill. The total operating budget of the Virgin Islands government in 1985 was $263.3 million of which taxes on personal and corporate income provided 50%.

SOCIAL WELFARE

Healthcare

The territorial Department of Health provides hospital and medical services, public health services, and veterinary medicine.

EDUCATION

Education is compulsory. The College of the Virgin Islands is the territory's first institution of higher learning.

1999 KEY EVENTS TIMELINE

April

- The governor of the U.S. Virgin Islands announces a financial crisis. The territory claims to have no money to pay its 10,000 workers.
- The former postmaster for the U.S. Virgin Islands is on trial in Tampa, Florida, for twice attacking a Washington lobbyist. He was charged with three counts of sexual battery with slight force in the 1997 attack.

May

- The Hovensa oil refinery postpones several capital and maintenance, and asks contractors to reduce their work force by 200 employees.

June

- Tim Duncan, a native of the U.S. Virgin Islands, leads the San Antonio Spurs to victory in the National Basketball Association championships.
- The U.S. Department of the Interior warns the territory it has 30 days to come up with a financial recovery plan. The U.S. Virgin Islands are more than $1 billion in debt.

July

- Royal Caribbean Cruises agrees to pay an $18 million fine and pleads guilty to 21 felony counts for dumping oil and hazardous chemicals from its cruise ships. Some of the dumping took place in waters near the U.S. Virgin Islands.

August

- The U.S. military proves for any chemical traces at a former chemical weapons storage site in the territory.
- U.S. Immigration officers arrest 21 undocumented immigrants and four men suspected of smuggling them into the territory.

September

- NBA star Tim Duncan signs a contract to promote his native territory in exchange for tax breaks on his planned business enterprises on the islands.
- Almeric L. Christian, the first native of the U.S. Virgin Islands to be named a judge in the territory's District Court, dies at age 79. President Nixon appointed Christian in 1969. He retired in 1988.

October

- Territory officials say a U.S. report that claims the U.S. Virgin Islands are not ready to handle the Year 2000 computer bug is too pessimistic.
- The U.S. Virgin Islands Senate approves controversial land deal to allow a Texas-based rocket company to build its headquarters and assembly plant on a historic site.
- Volunteers save a 17-foot-long pilot whale stranded on the U.S. Virgin Island of St. Croix by pulling the animal back into the Caribbean Sea.
- The 35th Annual Virgin Islands-Puerto Rico Friendship Celebration takes place.

November

- Hurricane Lenny, with winds as high as 145 miles per hour, hits the Virgin Islands and neighboring islands in the Caribbean. High winds and the resulting high tides cause extensive damage to homes, boats, and public buildings along the shore.

ANALYSIS OF EVENTS: 1999

BUSINESS AND THE ECONOMY

During the 1990s the economic activity on the U.S. Virgin Islands provided an object lesson in the perils of running an insufficiently diversified economy. Government officials declared that the islands expected more visitors during the 1999–2000 cruise ship schedule. Part of the reason is that St. Croix was going to be the first U.S. site to greet the new millenium. Officials expected 114 vessels to make port calls on St. Croix between November 1, and April 2000. St. Thomas expected 786 ships, including several vessels making inaugural visits. (The cruise season runs from October 1999 to May 2000.)

The fact that more cruise ships will be making their port of call in the Virgin Islands would seem to be good news for the tourist trade and for the economic health of the islands. Yet, more cruise ships will not necessarily translate into more money for the islands, the Associated Press reported in February. According to the report, more visitors to the island now come in ships, where they eat and sleep, and spend their money on drinks and other items. They are leaving little behind when they disembark.

The U.S. Virgin Islands are concerned by the decrease in the number of tourists who actually stay in hotels, rent cars, shop and eat on the islands. According to government estimates, the number of overnight visitors to the U.S. territory fell from 555,000 in 1988 to 410,000 ten years later, a decline of 26 percent. During the same period, the number of annual arrivals aboard cruise ships increased by half, to 1.6 million. Today, cruise ship passengers account for 80 percent of all visitors, but contribute only 25 percent of the total tourism revenue, according to the government's Development Bank. The result, the AP story says, is that tourism income fell by a third in the past five years, to an annual $600 million.

The Dici Carina Bay resort, which was to open in October in St. Croix, was expected to have the territory's first casino in operation by December. The resort feature 126 oceanfront rooms and 20 villas. The casino will have 275 slot tables and 13 gambling tables.

GOVERNMENT AND POLITICS

During early 1999, the government of the U.S. Virgin Islands found itself in a deep fiscal crisis. In April, the government announced it had no money to pay its 10,000 workers, and had to borrow money to pay them. The U.S. Department of the Interior warned the territory to come up with a financial plan to deal with its $1 billion debt. The Senate established a commercial line of credit with a Puerto Rico bank to avoid future delays in meeting biweekly payrolls, and Governor Charles Turnball was authorized to borrow up to $35 million from commercial banks if necessary to ensure timely payrolls for the remainder of the fiscal year. The territory's biweekly payroll was estimated at $13 million.

In July, the government proposed new fees to generate more money, including mandatory car insurance and a measure to allow cruise ships to open their casinos while docked in St. Thomas. The government also adjusted other fees and fines to bring in more revenues. The fine for littering increased from $250 to $1,000. The U.S. Virgin Islands also received $16 million to fix the Year 2000 computer problem.

''This new package of federal assistance will not only fix our computers but substantially reduce our debt service costs and ease our critical cash flow problems by substituting federal dollars for locally borrowed ones,'' Governor Turnball told the press.

In August, three St. Croix garbage disposal companies stopped service, claiming the government owed them more than $2 million in overdue payments. It was a clear signal that the territory had yet to solve its financial problems.

CULTURE AND SOCIETY

In June, Tim Duncan became one of the most celebrated sons of the U.S. Virgin Islands when he led his San Antonio Spurs to the NBA title. Duncan moved to the U.S. mainland with dreams of becoming an Olympic swimmer. In the NBA title series against the New York Knicks, Duncan was named the most valuable player. In September, Duncan signed a contract to promote his native territory in exchange for tax breaks on his planned business enterprises on the islands.

In August, archaeologists from Michigan Technological University began digging at a former plantation site to collect material on the lives of slaves who worked there. The team found the bottom of a medicine bottle, pieces of china dating back to the 17th century and other relics.

In September, *Travel and Leisure* magazine readers' poll ranked the Ritz-Carlton in St. Thomas among the top hotels in the Caribbean for the second consecutive year. In its ''World's Best Service'' reader's poll, the resort moved from fifth to third place, and ranked 33rd in the Top 100 hotels and resorts in the world.

DIRECTORY

CENTRAL GOVERNMENT

Head of State

Governor
Charles Wesley Turnbull, Office of the Governor

Ministers

Minister of Agriculture
Ministry of Agriculture, Estate Lower Love, Kingshill, St. Croix, U.S. Virgin Islands 00850
PHONE: +(340) 7780997; 7745182
FAX: +(340) 7741823; 7783101

Minister of Education
Ministry of Education, No. 44-46 Kongens Gade, Charlotte Amalie, U.S. Virgin Islands 00802
PHONE: +(340) 7740100
FAX: +(340) 7797153

Minister of Finance
Ministry of Finance, No. 76 Kronprindsens Gade, GERS Building, 2nd Floor, St. Thomas, U.S. Virgin Islands 00802
PHONE: +(340) 7744750
FAX: +(340) 7764028

Minister of Health
Ministry of Health, 48 Sugar Estate, Charlotte Amalie, U.S. Virgin Islands 00802

PHONE: +(340) 7740117; 7736551
FAX: +(340) 7774001; 7731376

Minister of Human Services
Ministry of Human Services, Knud Hansen Complex Building A 1303, Hospital Ground, St. Thomas, U.S. Virgin Islands 00802
PHONE: +(340) 7740930
FAX: +(340) 7743466

Minister of Property and Procurement
Ministry of Property and Procurement, Property and Procurement Bldg. No. 1, Sub Base, 3rd Floor, St. Thomas, U.S. Virgin Islands 00802
PHONE: +(340) 7740828; 7731561
FAX: +(340) 7749704; 7730986

Minister of Public Works
Ministry of Public Works, No. 8 Subbase, Charlotte Amalie, U.S. Virgin Islands 00802
PHONE: +(340) 7731290; 7764844
FAX: +(340) 7745869; 7730670

Minister of Tourism
Ministry of Tourism, P.O. Box 6400, St. Thomas, U.S. Virgin Islands 00804
PHONE: +(340) 7748784
FAX: +(340) 7744390

POLITICAL ORGANIZATIONS

Democratic Party

NAME: James O'Bryon Jr.

Independent Citizens' Movement (ICM)

NAME: Virdin C. Brown

Republican Party

NAME: Charlotte-Poole Davis

DIPLOMATIC REPRESENTATION

None (Territory of the United States)

JUDICIAL SYSTEM

District Court, St. Thomas/St. John Division

5500 Veterans Drive RM310, St. Thomas, U.S. Virgin Islands 00802
PHONE: +(340) 7741800
FAX: +(340) 7778532

District Court, St. Croix Division

3013 Estate Golden Rock Suite 219, St. Croix, U.S. Virgin Islands 00820
PHONE: +(340) 7735021
FAX: +(340) 7732113

FURTHER READING

Articles

''Cruise Line is Fined $18 Million for Dumping at Sea.'' Associated Press, 21 July 1999.

''Duncan, With Title, Now 'Having a Ball'.'' *The Seattle Times*, 17 July 1999.

''Cruises Seen as Mixed Blessing in Virgin Islands.'' *The Seattle Times*, 21 February 1999.

''Ex-Postmaster of Virgin Island Facing Battery Trial in Tampa.'' *The Miami Herald*, 29 April 1999.

''U.S. Virgin Islands Governor Signals Financial Collapse.'' *The Miami Herald*, 7 April 1999.

''U.S. Officials Give Virgin Islands 30 Days for Budget Plan.'' Associated Press, 25 June 1999.

''Tim Duncan to Promote Native U.S. Virgin Islands, Will Get Tax Breaks.'' Associated Press, 24 September 1999.

''Puerto Rico, Virgin Islands Dispute U.S. Findings on Y2K Preparedness.'' Associated Press, 7 October 1999.

''Former Federal Judge Dies in U.S. Virgin Islands.'' Associated Press, 3 September 1999.

''Hills of Trash Polluting Earth, Sea, Fish in U.S. Virgin Islands.'' Associated Press, 17 May 1999.

''Scientists Expect Crowd of Leatherback Turtles in Virgin Islands.'' Associated Press, 19 April 1999.

''Groups Developing Caribbean Slave History into Tourism.'' Associated Press, 30 June 1999.

Internet

Caribbean Week. Available Online @ http://www.cweek.com (November 1, 1999).

VIRGIN ISLANDS: STATISTICAL DATA

For sources and notes see "Sources of Statistics" in the front of each volume.

GEOGRAPHY

Geography (1)

Area:

Total: 352 sq km.

Land: 349 sq km.

Water: 3 sq km.

Area—comparative: twice the size of Washington, DC.

Land boundaries: 0 km.

Coastline: 188 km.

Climate: subtropical, tempered by easterly trade winds, relatively low humidity, little seasonal temperature variation; rainy season May to November.

Terrain: mostly hilly to rugged and mountainous with little level land.

Natural resources: sun, sand, sea, surf.

Land use:

Arable land: 15%

Permanent crops: 6%

Permanent pastures: 26%

Forests and woodland: 6%

Other: 47% (1993 est.).

HUMAN FACTORS

Ethnic Division (6)

Black	80%
White	15%
Other	5%

Demographics (2A)

	1990	1995	1998	2000	2010	2020	2030	2040	2050
Population	104.2	113.9	118.4	121.2	133.5	142.7	147.6	148.8	148.3
Net migration rate (per 1,000 population)	NA	NA	NA	NA	NA	NA	NA	NA	NA
Births	NA	NA	NA	NA	NA	NA	NA	NA	NA
Deaths	NA	NA	NA	NA	NA	NA	NA	NA	NA
Life expectancy - males	71.7	70.8	73.8	74.3	76.3	77.8	78.8	79.5	80.0
Life expectancy - females	79.9	79.3	81.5	81.8	83.2	84.3	85.1	85.6	86.0
Birth rate (per 1,000)	23.0	19.0	17.5	16.6	14.5	13.2	11.5	10.8	10.5
Death rate (per 1,000)	4.9	5.8	5.3	5.4	6.5	8.1	9.8	10.7	11.3
Women of reproductive age (15-49 yrs.)	28.0	29.9	30.6	31.0	33.7	35.1	34.6	33.0	30.5
of which are currently married	26.7	27.4	NA	NA	NA	NA	NA	NA	NA
Fertility rate	3.0	2.7	2.5	2.3	2.0	1.9	1.8	1.8	1.8

Except as noted, values for vital statistics are in thousands; life expectancy is in years.

Religions (7)

Baptist	42%
Roman Catholic	34%
Episcopalian	17%
Other	7%

Languages (8)

English (official), Spanish, Creole.

EDUCATION

Educational Attainment (10)

Age group (1980)[13]	25+
Total population	44,986
Highest level attained (%)	
No schooling	1.5
First level	
Not completed	26.3
Completed	7.8
Entered second level	
S-1	14.4
S-2	25.7
Postsecondary	24.4

GOVERNMENT & LAW

Political Parties (12)

Senate	No. of seats
Independents	6
Democrats	5
Republicans	2
Independent Citizens Movement	2

Government Budget (13B)

Revenues	$364.4 million
Expenditures	$364.4 million
Capital expenditures	NA

Data for 1990 est. NA stands for not available.

Military Affairs (14A)

Defense is the responsibility of the US.

LABOR FORCE

Labor Force (16)

Total	47,443
Agriculture	1%
Industry	20%
Services	62%
Other	17%

Data for 1990 est. Percent distribution for 1990.

Unemployment Rate (17)

6.2% (March 1994)

PRODUCTION SECTOR

Electric Energy (18)

Capacity	316 million kW (1995)
Production	1 billion kWh (1995)
Consumption per capita	10,285 kWh (1995)

Transportation (19)

Highways:

total: 856 km

paved: NA km

unpaved: NA km

Merchant marine: none

Airports: 2 note: international airports on Saint Thomas and Saint Croix; there is an airfield on St. John (1997 est.)

Airports—with paved runways:

total: 2

1,524 to 2,437 m: 2 (1997 est.)

Top AFgricultural Products (20)

Truck garden products, fruit, vegetables, sorghum; Senepol cattle.

FINANCE, ECONOMICS, & TRADE

Economic Indicators (22)

National product: GDP—purchasing power parity—$1.2 billion (1987 est.)

National product real growth rate: NA%

National product per capita: $12,500 (1987 est.)

Inflation rate—consumer price index: NA%

Exchange Rates (24)

Exchange rates: US currency is used

Top Import Origins (25)

$2.2 billion (c.i.f., 1992)

Origins	%
United States	NA
Puerto Rico	NA

NA stands for not available.

Top Export Destinations (26)

$1.8 billion (f.o.b., 1992).

Destinations	%
United States	NA
Puerto Rico	NA

NA stands for not available.

Economic Aid (27)

$NA. NA stands for not available.

Import Export Commodities (28)

Import Commodities	Export Commodities
Crude oil	Refined petroleum products
Foodstuffs	
Consumer goods	
Building materials	

WALLIS AND FUTUNA

Territory of Wallis and Futuna Islands
Territoire des Îles Wallis et Futuna
Wallis et Futuna

CAPITAL: Mata-Utu.

FLAG: The flag of Wallis and Futuna has a large white modified Maltese cross centered on a red background; the flag of France outlined in white on two sides is in the upper hoist quadrant; the flag of France is used for official occassions.

MONETARY UNIT: 1 CFP franc (CFPF) = 100 centimes.

WEIGHTS AND MEASURES: The metric system is used.

HOLIDAYS: National Day, 14 July.

TIME: 12 AM = noon GMT.

LOCATION AND SIZE: Wallis Island and the Futuna, or Hoorn, Islands in the Southwest Pacific constitute a French overseas territory, with the capital at Mata-Utu, on Wallis (also called Uvéa). Lying about 400 km (250 mi.) W of Pago Pago, American Samoa, at 13°22′S and 176°12′W, Wallis, 154 sq. km (59 sq. mi.) in area, is surrounded by a coral reef with a single channel. The Futuna Islands are about 190 km (120 mi.) to the SW at 14°20′S and about 177°30′W. They comprise two volcanic islands, Futuna and Alofi, which, together with a group of small islands, have a total area of about 116 sq. km (45 sq. mi.).

CLIMATE: The islands of Wallis and Futuna have a tropical climate. From November to April it is hot and rainy with a mean temperature between 21–32°C (70–90°F). Their cool, dry season occurs from May to October. The average rainfall is 250–300 cm per year.

INTRODUCTORY SURVEY

RECENT HISTORY

The Futuna Islands was discovered by Dutch sailors in 1616 while Wallis Island (at first called Uvéa) was discovered by the English explorer Samuel Wallis in 1767. A French missionary established a Catholic mission on Wallis in 1837, and missions soon followed on the other islands. In 1842, the French established a protectorate, which was officially confirmed in 1887 for Wallis and in 1888 for Futuna.

GOVERNMENT

A high administrator, representing the French government, is assisted by a Territorial Assembly.

INDUSTRY

Principal commercial activities are the production of copra and fishing for trochus. The total fish catch in 1994 was 193 tons. The cash food crops are yams, taro, bananas, manioc, and arrowroot.

1999 KEY EVENTS TIMELINE

January

- A late December cyclone causes heavy rain and flooding, and the aftermath and clean–up continue.
- The French Ministry of Finance establishes the exchange rate for the Pacific franc and the Euro. One thousand Pacific francs will buy 8.38 Euros.

March

- Celebrations begin on March 19 to commemorate the fortieth anniversary of the reign of King Lavelua Tomasi Kulimoetoke of Wallis, who is eighty-one. His reign began two years before France established territorial control over Wallis and Futuna.

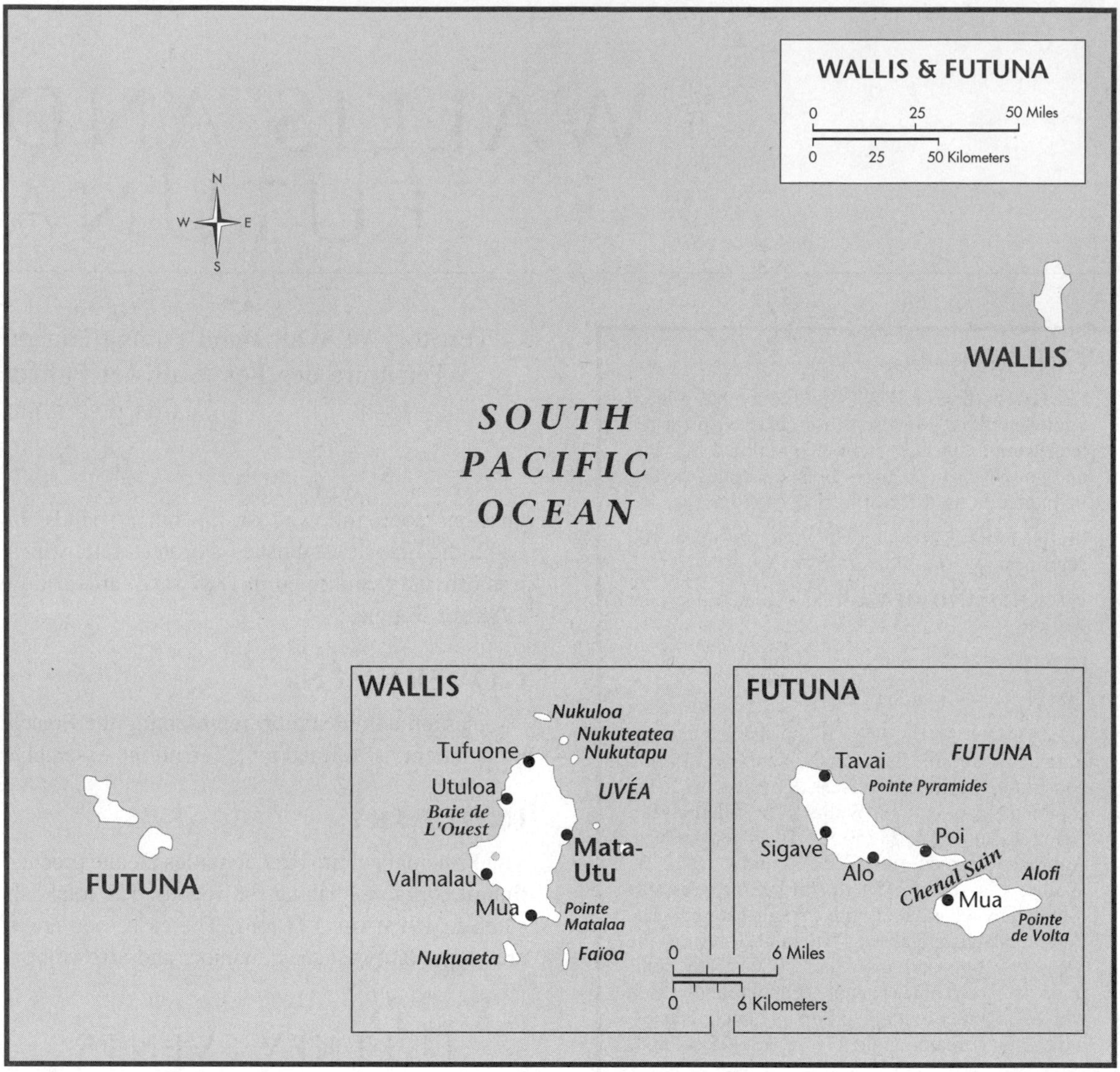

May

- The South Pacific Games open on May 31 in Guam, with the Wallis and Futuna delegation performing dances in front of the main reviewer's stand during the opening ceremonies. The delgates also present artifacts from the islands to the governor of Guam, official host of the Games.
- A delegation from the International Rugby Board (IRB) meets during the South Pacific Games to establish a Oceania Rugby Federation. Twelve nations from the region are competing in the South Pacific Games Rugby Sevens, including Wallis and Futuna.
- The king of Wallis names Atelemo Taofifenua, a 72-year-old who has lived mostly in neighboring New Caledonia since 1952, as the territory's Prime Minister.

June

- Teachers strike on June 30, seeking equal pay. Teachers from France and elsewhere receive higher pay than teachers born on the islands.

August

- Intervention by traditional societal chiefs and the King of Wallis succeeds in convincing teachers to return to work after a seven–week strike that began June 30. The teachers' union, the Force Ouvrière, had been demanding that teachers' salaries be increased to equal those provided in metropolitan France. Some teachers, assisted by parents and other volunteers, had kept the schools open during the strike, and some parents felt the teachers were no longer needed.

- The Prime Minister of Futuna, Niutupea le Taifoi, is admitted into a Noumea hospital for emergency treatment for high blood pressure and an ulcer.

September

- The Wallis and Futuna Territorial Assembly approves the implementation of a water shuttle by investing in a 31-meter (102–foot) high-speed catamaran built in Australia. The water shuttle would replace expensive air service.
- French Assistant Overseas Minister Jean-Jack Queyranne appoints leaders from French Pacific territories to the French Economic and Social Council (Conseil Economique et Social–CES). Camille Ogata, a former member of the Wallis and Futuna Parliament, is appointed to a five-year term on the Council.

ANALYSIS OF EVENTS: 1999

BUSINESS AND THE ECONOMY

Wallis Island and Futuna Island, situated in the South Pacific, two–thirds of the distance from Hawaii and New Zealand, are old French colonies which now carry the designation of ''French Overseas Territories.'' The quality of life in Wallis and Futuna is shaped by the mixture of traditional and modern values. The pace of life is slow, and the growth rate of the Polynesian population of 15,000 is only about 1 percent a year. More people emigrate away from the islands than immigrate to them. The net population movement is a little more than a negative seven per thousand. The two islands are about a hundred miles apart and travel by air between them is expensive, so the Territorial Assembly approved the purchase of a 150-passenger catamaran to provide passenger and cargo shuttle service. Because the purchase was related to the promotion of tourism, it was tax exempt.

Under the Maastricht Treaty outlining the administration of the European Union and its component states, the financial affairs of France's Pacific territories are not included. Therefore, France set the exchange rate, one thousand Pacific francs to 8.38 Euros, when the Euro went into effect as the currency of the European Union.

GOVERNMENT AND POLITICS

When the king of Wallis named 72–year–old New Caledonian Atelemo Taofifenua as Prime Minister, the appointment was regarded as controversial. Taofifenua held an official position in the municipal government of Noumea, the capital of New Caledonia, at the time of his appointment. Taofifenua's first challenge was to deal with the charges of corruption lodged against his predecessor, Make Pilioko, who was living as a refugee from prison with the king. France does not recognize the king's authority in matters of government corruption, and limits his powers to tribal matters.

CULTURE AND SOCIETY

The fortieth anniversary of the reign of King Lavelua Tomasi Kulimoetoke, the fiftieth monarch in the Lavelua line, was celebrated in 1999. The celebrations began with a mass in the cathedral of the capital, Matu Utu. (The native population is overwhelmingly Catholic.) The Lavelua line has ruled Wallis since the fifteenth century. France recognizes the monarch's authority in tribal matters only.

Wallis and Futuna have no military. With the exception of the resort hotels and the Catholic Church (which of course, spans the centuries), the only modern institution (or, one of the main modern institutions) is the teachers' trade union. The islands' teachers struck for seven weeks during the summer, hoping to win salary parity with their counterparts in France and in other French overseas territories. In a stunning gesture mixing ancient folkways with modern institutions, the King and other traditional leaders employed Fakalelei, a traditional forgiveness ceremony, to convince the teachers to return to their classrooms.

Wallis and Futuna participated in the eleventh annual South Pacific Games in Guam. The country's delegates, along with the delegates from Samoa, entertained those on the main reviewing stand with dances during the opening ceremonies, and presented gifts to the governor of Guam just prior to his declaration that the Games could officially begin.

DIRECTORY

CENTRAL GOVERNMENT

Head of State

President of the Territorial Assembly
Victor Brial, Office of the President of the Territorial Assembly, Havelu, BP 31, 98600 Matu-Utu, Wallis and Futuna
PHONE: +681 722504
FAX: +681 722054

Territorial Executive
Claude Pierret, Office of the Territorial Executive, Havelu, BP 16, 98600 Matu-Utu, Wallis and Futuna
PHONE: +681 722727
FAX: +681 722324

Vice Rectorat
Office of the Vice Rectorat, Havelu-Hahake, BP 244, Wallis and Futuna
PHONE: +681 722828
FAX: +681 722040

Ministers

Minister of Territorial Action
Ministry of Territorial Action, BP 131, 98600 Matu-Utu, Wallis and Futuna
PHONE: +681 722667
FAX: +681 722563

Minister of Youth and Sports
Ministry of Youth and Sports, Kafika-Hahake, BP 51, 98600 Matu-Utu, Wallis and Futuna
PHONE: +681 722188
FAX: +681 722322

Minister of Rural Economy and Fishing
Ministry of Rural Economy and Fishing, 98600 Matu-Utu, Wallis and Futuna
PHONE: +681 722823
FAX: +681 722544

Minister of Development
Ministry of Development, 98600 Matu-Utu, Wallis and Futuna
PHONE: +681 722505

Minister of Retirement
Ministry of Retirement, Havelu-Hahake, BP 125, 98600 Matu-Utu, Wallis and Futuna
PHONE: +681 722885
FAX: +681 722282

Minister of Welfare and Families
Ministry of Welfare and Families, Havelu-Hahake, BP 125, 98600 Matu-Utu, Wallis and Futuna
PHONE: +681 722935
FAX: +681 722282

Minister of Work and Social Affairs
Ministry of Work and Social Affairs, Havelu-Hahake, BP 16, 98600 Matu-Utu, Wallis and Futuna
PHONE: +681 722727

POLITICAL ORGANIZATIONS

Rally for the Republic (RPR)

Union populaire locale (UPL)

Union pour la democratie francaise (UDF)

Lua kae tahi

Mouvement des radicaux de gauche (MRG)

Taumu'a Lelei

FURTHER READING

Internet

Wallis Islands. Available Online @ http://wallis–islands.com/index.gb.htm (November 15, 1999).

French Embassy in Suva. Available Online @ http://www.ambafrance.org.fj (November 15, 1999).

WALLIS AND FUTUNA: STATISTICAL DATA

For sources and notes see "Sources of Statistics" in the front of each volume.

GEOGRAPHY

Geography (1)

Area:

Total: 274 sq km.

Land: 274 sq km.

Water: 0 sq km.

Note: includes Ile Uvea (Wallis Island), Ile Futuna (Futuna Island), Ile Alofi, and 20 islets.

Area—comparative: 1.5 times the size of Washington, DC.

Land boundaries: 0 km.

Coastline: 129 km.

Climate: tropical; hot, rainy season (November to April); cool, dry season (May to October); rains 2,500-3,000 mm per year (80% humidity); average temperature 26.6 degrees C.

Terrain: volcanic origin; low hills.

Natural resources: NEGL.

Land use:

Arable land: 5%

Permanent crops: 20%

Permanent pastures: NA%

Forests and woodland: NA%

Other: 75% (1993 est.).

HUMAN FACTORS

Demographics (2B)

Population (July 1998 est.)	14,974
Age structure:	
0-14 years	NA
15-64 years	NA
65 years and over	NA
Population growth rate (1998 est.)	1.06%
Birth rate, 1998 est. (births/1,000 population)	23.02
Death rate, 1998 est. (deaths/1,000 population)	4.78
Net migration rate, 1998 est. (migrant(s)/1,000 population)	−7.61
Infant mortality rate, 1998 est. (deaths/1,000 live births)	20.93
Life expectancy at birth (years):	
Total population	73.82
Male	73.24
Female (1998 est.)	74.4
Total fertility rate, 1998 est. (children born/woman)	2.78

Ethnic Division (6)

Polynesian.

Religions (7)

Roman Catholic 100%.

Languages (8)

French, Wallisian (indigenous Polynesian language).

GOVERNMENT & LAW

Political Parties (12)

The legislative branch is a unicameral Territorial Assembly (20 seats; members are elected by popular vote to serve five-year terms).

Government Budget (13B)

Revenues	$22 million
Expenditures	$22 million
Capital expenditures	NA

Data for 1997 est. NA stands for not available.

Military Affairs (14A)

Defense is the responsibility of France.

LABOR FORCE

Labor Force (16)

Agriculture, livestock, and fishing	80%
Government	4%

Unemployment Rate (17)

Rate not available.

PRODUCTION SECTOR

Electric Energy (18)

No data available.

Transportation (19)

Highways:

total: 120 km (Ile Uvea 100 km, Ile Futuna 20 km)

paved: 16 km (all on Ile Uvea)

unpaved: 104 km (Ile Uvea 84 km, Ile Futuna 20 km)

Waterways: none

Merchant marine:

total: 2 ships (1,000 GRT or over) totaling 44,160 GRT/41,656 DWT ships by type: oil tanker 1, passenger 1 (1997 est.)

Airports: 2 (1997 est.)

Airports—with paved runways:

total: 1

1,524 to 2,437 m: 1 (1997 est.)

Airports—with unpaved runways:

total: 1

914 to 1,523 m: 1 (1997 est.)

Top Agricultural Products (20)

Breadfruit, yams, taro, bananas; pigs, goats.

FINANCE, ECONOMICS, & TRADE

Economic Indicators (22)

National product: GDP—purchasing power parity—$28.7 million (1995 est.)

National product real growth rate: NA%

National product per capita: $2,000 (1995 est.)

Inflation rate—consumer price index: NA%

Exchange Rates (24)

Exchange rates:	
Comptoirs Francais du Pacifique francs (CFPF) per US$1	
January 1998	110.60
1997	106.11
1996	93.00
1995	90.75
1994	100.94
1993	102.96

Linked at the rate of 18.18 to the French franc

Top Import Origins (25)

$13.5 million (c.i.f., 1995 est.)

Origins	%
United States	NA
Puerto Rico	NA

NA stands for not available.

Top Export Destinations (26)

$370,000 (f.o.b., 1995 est.).

Destinations	%
NA	NA

NA stands for not available.

Economic Aid (27)

Recipient: ODA, $NA. NA stands for not available.

Import Export Commodities (28)

Import Commodities	Export Commodities
Foodstuffs	Copra
Manufactured goods	Handicrafts
Transportation equipment	
Fuel	
Clothing	

YEMEN

CAPITAL: Sana (San'a).

FLAG: The national flag is a tricolor of red, white, and black horizontal stripes.

ANTHEM: *Al-Watani (Peace to the Land).*

MONETARY UNIT: The Yemeni riyal (YR) is a paper currency of 100 fils. There are coins of 1, 5, 10, 25, and 50 fils and notes of 1, 5, 10, 20, 50, and 100 riyals. YR1 = $0.0833 (or $1 = YR12.01).

WEIGHTS AND MEASURES: The metric system is being introduced, but local measures remain in common use.

HOLIDAYS: Labor Day, 1 May; Day of National Unity, 22 May; National Day, 14 October; Independence Day, 30 November. Movable Muslim holidays include Laylat al-Miraj, 'Id al-Fitr, 'Id al-'Adha', Milad an-Nabi, and 1st of Muharram.

TIME: 3 PM = noon GMT.

LOCATION AND SIZE: Yemen, located in the southern part of the Arabian Peninsula, is slightly larger than twice the size of the state of Wyoming with a total area of 527,970 square kilometers (203,850 square miles). Yemen has a total boundary length of 3,652 kilometers (2,269 miles).

Sana, the capital, is in the western part of the country.

CLIMATE: Extreme humidity combines with high temperatures—as high as 54°C (129°F) in the shade—to produce a stiflingly hot climate in parts of the country. The climate of the highlands is generally considered the best in Arabia, with temperatures between 22°C (72°F) in June, the hottest month, to 14°C (57°F) in January. Rainfall in the highlands ranges from 41 centimeters (16 inches) to 81 centimeters (32 inches).

Republic of Yemen
Al-Jumhuriyah al-Yamaniyah

INTRODUCTORY SURVEY

RECENT HISTORY

On 22 May 1990, the unified Republic of Yemen was formed with the merger of the Yemen Arab Republic (North Yemen) and the People's Democratic Republic of Yemen (South Yemen). North Yemen had been independent since 1918; South Yemen had attained independence on 30 November 1967 from the United Kingdom.

Since 1990 Lt. Gen. Ali Abdallah Salih, the former president of North Yemen, has served as Yemen's president. Parliamentary elections n 1992 resulted in formation of a three-party coalition government. The coalition started to disintegrate in late 1993, and fighting broke out in May 1994. By July, after thousands of casualties had been suffered, the north had beaten the rebellious south and began restoring order.

GOVERNMENT

The 1990 unity constitution established a political system based on free, multiparty elections. The president, to be elected in the future by popular vote, chooses the vice president and prime minister. A 301-member parliament was elected in 1997 to serve four year terms, but there has not been a direct presidential election. In May 1997, the president appointed a 59-member consultative council to act as an upper house of Parliament.

Judiciary

The separate judicial systems of North and South Yemen were unified at the Supreme Court

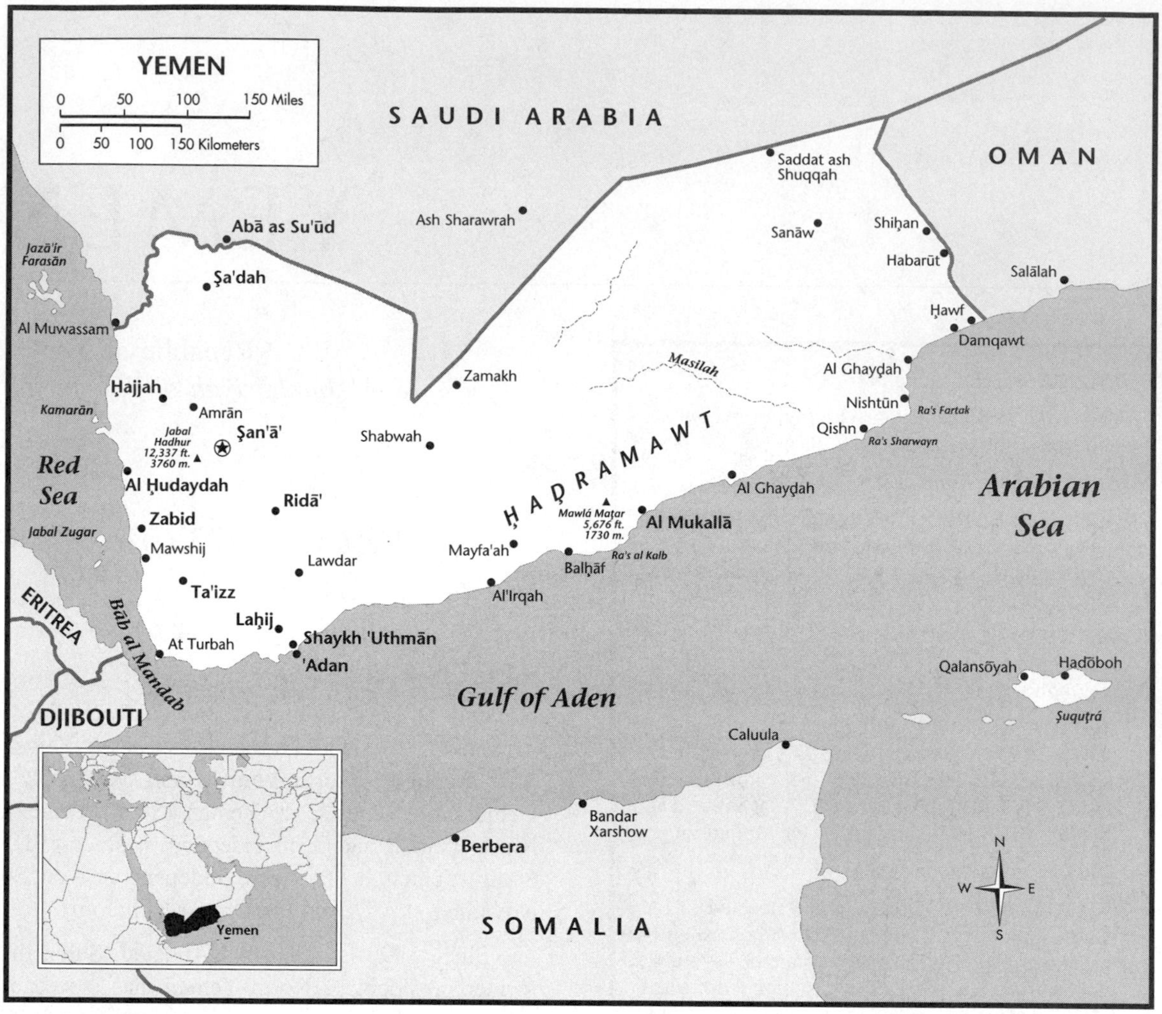

level in 1991. Separate lower court systems continue to function in their respective halves of the country.

Political Parties

After Yemen's unification, General Saleh's General People's Congress (GPC) became the country's largest party. The second largest bloc in the parliament is held by the Islaah Party, a fusion of tribal and Islamic interests which opposed the unity constitution because it did not sufficiently adhere to Islamic principles. At least five smaller parties have been active in the politics of unified Yemen.

DEFENSE

In 1995, the consolidated armed forces numbered 42,000, based on two years of required service. The army of 37,000 has 25,000 draftees. The navy of 1,500 mans 15 patrol and coast combatants and 7 other ships. The air force, with 3,500 members, has 65 combat aircraft. Reserve forces number 50,000.

ECONOMIC AFFAIRS

The 1990 unification of North and South Yemen posed the challenge of merging two different economies. The southern city of Aden, with its port and refinery, is now the economic and commercial center of the country. The Yemeni economy depends on imports of wheat, flour and rice, and other foodstuffs.

Crude oil dominates foreign trade, accounting for 85% of total exports. Following unification, responsibility for development of the oil sector fell to the state-owned general corporation for oil and mineral resources. Yemen also has large reserves of natural gas.

Public Finance

When Yemen aligned itself with Iraq during the Persian Gulf War (1991), Saudi Arabia and the Gulf states, Yemen's main aid donors and hosts to large numbers of Yemeni workers and their families, ended the Yemenis' privileged status. The economic impact of lost remittances was estimated at about $1 billion annually.

After the Gulf crisis, Yemen was confronted with high unemployment, lost income, a sharp cutback in U.S. military and foreign aid programs, and other canceled foreign assistance. Yemen also faced the cost of food imports and social services for the returnees totaling about $500 million. In addition, the civil conflict in 1994 between the north and south increased the need for external aid.

In 1995, the government began a five-year program that removed all controls on the exchange rate. Trade policy reform, privatization, and elimination of price controls also began. The domestic economy grew by 6.2% in 1995, 3.2% in 1996, and an estimated 6.6% in 1997.

Income

In 1999, Yemen's gross national product (GNP) was estimated at $4.4 billion, or about $270 per person.

Industry

Major industries include food processing, clothing, processed food, metal products, soap, and perfumes.

The refinery at Aden processed 100,000 barrels of oil per day in 1995.

Banking and Finance

The Central Bank of Yemen was established in 1971. The state-owned Yemen Bank for Reconstruction and Development (YBRD), founded in 1962, finances development activities, and the International Bank of Yemen, organized in 1980, operates as a commercial bank.

There is no stock exchange.

Economic Development

The government launched a major reform program in 1995 that included tax measures, depreciation of the customs valuation rate, liberalization of cement prices, an increase in petroleum product prices by about 90%, and a 60% rise in electricity tariffs. The government's medium-term goal was to eliminate all subsidies by 1999–2001. Fiscal and monetary measures included: curtailing non-development expenditures, partial reform of the exchange system (including currency depreciation), interest rate reform, and monetary management reforms. Transportation and communication charges were deregulated, health and education fees were increased, and privatization programs were initiated.

In early 1996, the IMF agreed to provide a 15-month standby credit of $191 million, and the World Bank authorized a loan of $80 million to support these reforms. The World Bank also allocated government loans to Yemen worth $365 million during 1996–99, and the EU pledged grants worth $61.7 million in 1996/97, including $30 million in project finance.

SOCIAL WELFARE

Families and tribes typically care for their sick, handicapped, unemployed, and widows and orphans. Those without family or tribal ties beg or have recourse to Islamic pious foundations.

Since1988, a state-run provident fund system has provided old age, disability, survivor, and workers' compensation benefits.

Women have limited access to education, are required to get permission from a male family member before traveling, and are rarely allowed to travel unescorted. Social customs discourage most women from becoming politically active.

Healthcare

Malaria, typhus, tuberculosis, dysentery, whooping cough, measles, hepatitis, schistosomiasis, and typhoid fever are widespread, and inadequate sewage disposal is a general health hazard. Life expectancy in 1999 is an estimated 59.98 years.

In 1993, Yemen averaged 1 doctor per 4,498 and 1 hospital bed per 1,196 persons.

Housing

About one-fourth of urban housing units are huts, tents, or other makeshift structures; much rural housing consists of straw, stone, or baked brick huts. Wealthier Yemenis live in two-to eight-story houses, usually of baked brick with windows outlined in decorative designs.

EDUCATION

In 1990, the literacy rate was estimated to be 38%. The country has about 11,000 primary schools with 2.7 million pupils. At the secondary level, there were 212,129 students in 1993.

1999 KEY EVENTS TIMELINE

March

- Possible link suggested between kidnappers of Western tourists in December 1998 and Saudi terrorist Osama bin Laden.

May

- Three Yemeni Islamic militants are convicted and receive death sentences for the abduction of 16 Western tourists in December 1998.

August

- President Saleh promotes government campaign to discourage use of popular indigenous narcotic, qat, by announcing he will stop using it.
- Explosion at supermarket in Sanaa kills at least three, injures more than 20, and damages surrounding area.
- Ten Islamic militants are convicted of terrorism in connection with bomb plot, and seven are jailed.

September

- President Ali Abdullah Saleh is re-elected in first Yemen's first direct presidential election. Political opposition charges fraud in presidential election.

ANALYSIS OF EVENTS: 1999

BUSINESS AND THE ECONOMY

In 1999 Yemen remained one of the world's twenty poorest nations, with average per capita income totaling only $348. The per capita Gross Domestic Product was only $740 and the unemployment rate was 30 percent. Inflation stood at 11 percent while the 3.3 percent annual growth in population was a ticking time bomb. Infant mortality was high at almost 70 infant deaths per thousand live births but the fertility rate was also high at 7 children born per woman. The life expectancy at birth was just under 60 years.

Like other oil–producing states, Yemen–where oil accounts for more than 80 percent of exports-was hard hit by falling world oil prices in 1999. The government's income was cut in half, necessitating major budget cuts, and unemployment was in the double digits. Low oil prices also endangered another major source of Yemeni income-remittances by Yemenis working abroad, since the neighboring Gulf states that usually hosted these workers were in trouble themselves. Adding to the nation's problems, Yemen's tourist industry, which normally plays an important part in its economy, was endangered by a rash of kidnappings and killings of Westerners that had been brought to public attention by a major kidnapping incident at the end of the previous year.

GOVERNMENT AND POLITICS

In September Yemen carried out the first direct presidential elections ever held on the Arabian peninsula, as voters returned longtime president Ali Abdullah Saleh to office for another five-year term. However, the election was widely seen as having only symbolic significance, for Saleh had no real opposition in the election campaign and a coalition of opposition groups, led by the Socialist Party, boycotted the election. The only other candidate approved by parliament was a minor and largely unknown figure in Saleh's own party.

The political opposition charged the government with a variety of shady political maneuvers including exaggerating the size of voter turnout and outright electoral fraud. Fraud allegations included underage voting, multiple balloting by some individuals, and unauthorized submission of ballots for absent voters. Saleh won 96.3 percent of the vote, to 3.7 percent for his opponent, Najeeb Qahtan al-Shaabi. In spite of its flaws, some observers still felt that the mere act of holding a peaceful and at least nominally democratic election represented progress for Yemen, which had been split in two before 1990 and survived a bitter civil war in 1994.

The kidnapping of 16 Western tourists at the end of 1998, and the killing of four in a botched rescue attempt, continued to have repercussions in 1999. In May three Islamic militants-two Yemenis and an Algerian-were convicted and sentenced to death for the crime, while a fourth received a 20-year prison sentence. Yemeni tribesmen had kidnapped over 100 Westerners in the previous half-dozen years, but these kidnappings had generally been economically rather than ideologically motivated, with the kidnappers demanding economic improvements for their villages and sums of money for themselves. The recent kidnappers, in contrast, had identified themselves as Muslims rather than members of a specific tribe, and they had demanded the release of prisoners held for a kidnapping by another Muslim group. Some thought that the action might be linked to Saudi-born terrorist Osama bin Laden, who had called for attacks on Americans and their allies throughout the world.

In a related development, several Islamic militants had been arrested in December 1998, days before the kidnapping, and charged in a bomb plot possibly related to the kidnapping. In August, ten suspects, including eight British residents born in Arab countries and Pakistan, were convicted of conspiring to commit terrorist acts, and seven were jailed.

CULTURE AND SOCIETY

Starting in the spring, Yemen's government gained attention both at home and abroad when it attempted to combat some of the nation's social and economic problems by embarking on a campaign against one of Yemen's most deeply entrenched cultural mores-the use of a mild, indigenously grown narcotic called qat (or khat). Roughly 75 percent of the population, both male and female, regularly chewed the tender green leaves of the qat plant, which is generally described as inducing relaxation or torpor, although some conversely liken its effects to a ''coffee buzz.''

The use of qat was primarily a social ritual—most users gathered every afternoon to chat while they enjoyed the drug's effects. However, critics claimed that this long-established custom created a serious drain on the nation's productivity, as well as its pocketbooks. Yemenis spent roughly $2.1 billion annually on qat-a figure that is higher than the nation's yearly budget. It was considered commonplace for families-even those living in poverty-to spend between 20 and 50 percent of their income on the substance. In addition, production of the qat plant diverted valuable land and water resources-roughly 40 percent of the nation's irrigated farmland-from other uses. In August, President Saleh, himself a longtime qat user, attempted to set an example for the nation by swearing off the drug and encouraging others to do likewise, and to divert their time from qat-chewing sessions to more productive activities.

Although Saleh was considered secure enough in his position as president. He had held the office he had held for 21 years. Thus most observers believed that he would be able to avoid the fate of Mohsin Al-Aini, a former Prime Minister who had been ousted after attempting to stamp out qat-chewing in 1972. Those same observers, however, discounted the prospects of successfully eradicating this popular national pastime.

DIRECTORY

CENTRAL GOVERNMENT

Head of State

President
Ali Abdullah Saleh, Office of the President
PHONE: +967 (1) 274629

Prime Minister
Abd al-Karim al-Iryani, Office of Prime Minister

Ministers

Minister of Foreign Affairs
Abd al-Qadir al-Ba Jamal, Ministry of Foreign Affairs, Qa'a Ul-Ulofi, Sana'a, Yemen
PHONE: +967 (1) 202555
FAX: +967 (1) 209540

Minister of Immigrant Affairs
Ministry of Immigrant Affairs
E-MAIL: mia@y.net.yee

Minister of Oil and Mineral Resources
Muhammad al-Khadim al-Wajih, Ministry of Oil and Mineral Resources, PO Box 81, Alzubaeri St., Sana'a, Yemen
PHONE: +967 (1) 202340
FAX: +967 (1) 202314
E-MAIL: momrmnstr@y.net.ye

Minister of Supply and Trade
Abd al-Azis al-Jumaim, Ministry of Supply and Trade, PO Box 22210/1804, Sana'a, Yemen
PHONE: +967 (1) 252339
FAX: 967 (1) 252337
E-MAIL: Most@y.net.ye

Minister of Agriculture and Water Resources
Ahmad Salim al-Jabali, Ministry of Agriculture and Water Resources

Minister of Guidance
Nasir Muhammad al-Shaybani, Ministry of Guidance

Minister of Civil Service and Administrative Reform
Muhammad Ahmad al-Junayd, Ministry of Civil Service and Administrative Reform

Minister of Communications
Ahmad Muhammad al-Anisi, Ministry of Communications

Minister of Construction, Housing and Urban Planning
Abdallah Husayn al-Dafai, Ministry of Construction, Housing and Urban Planning

Minister of Culture and Tourism
Abd al-Malik Mansur, Ministry of Culture and Tourism

Minister of Defense
Muhammad Dayfallah Muhammad, Ministry of Defense

Minister of Education
Yahya Muhammad Abdallah al-Shuaybi, Ministry of Education

Minister of Electricity and Water
Ali Hamid al-Sharaf, Ministry of Electricity and Water

Minister of Expatriate Affairs
Ahmad Ali al-Bishari, Ministry of Expatriate Affairs

Minister of Finance
Alawi Salah al-Salami, Ministry of Finance

Minister of Industry
Abd al-Rahman Muhammad Ali Uthman, Ministry of Industry

Minister of Information
Abd al-Rahman al-Akwa, Ministry of Information

Minister of the Interior
Husayn Muhammad al-Arab, Ministry of the Interior

Minister of Justice
Ismail Ahmad al-Wazir, Ministry of Justice

Minister of Labor and Vocational Training
Muhammad Muhammad al-Tayib, Ministry of Labor and Vocational Training

Minister of Local Administration
Sadiq Amin Husayn Abu Ras, Ministry of Local Administration

Minister of Legal and Parliamentary Affairs
Abdallah Ahmad al-Ghanim, Ministry of Legal and Parliamentary Affairs

Minister of Planning and Development
Ahmad Muhammad Abdallah al-Sufan, Ministry of Planning and Development

Minister of Public Health
Abdallah Abd al-Wali Nashir, Ministry of Public Health

Minister of Social Security and Social Affairs
Muhammad Abdallah al-Batani, Ministry of Social Security and Social Affairs

Minister of Transportation
Abd al-Malik Ali al-Sayani, Ministry of Transportation

Minister of Youth and Sports
Abd al-Wahhab al-Rawih, Ministry of Youth and Sports

Minister of State
Faysal Mahmud, Ministry of State

Minister of Cabinet Affairs
Mutahir al-Saidi, Ministry of Cabinet Affairs

POLITICAL ORGANIZATIONS

al-Mu'tammar al-Sha'bi al-'Am (General People's Congress)

E-MAIL: GPC@y.net.ye

al-Tajmu al-Yamani li al-Islah (Yemeni Congregation for Reform)

al-Tantheem al-Wahdawi al-Sha'bi al-Nasseri (Nasserite Unionist People's Organization)

Hizb al-Baath al-'Arabi al-Ishtiraki (Arab Socialist Rebirth Party)

Hizb al-Ishtiraki al-Yaman (Socialist Party of Yemen)

DIPLOMATIC REPRESENTATION

Embassies in Yemen

France
rue 21, PO Box 1286, Sana'a, Yemen
PHONE: +967 (1) 268882; 268888
FAX: +967 (1) 269160
E-MAIL: ambaf@y.net.ye

Italy
Via Safiah Janubia Sana'a, Sana'a, Yemen
PHONE: +967 (1) 269165; 265616
FAX: +967 (1) 266137

United Kingdom
129 Abou al-Hasan al-Hamadani St., Haddah Rd., PO Box 1287, Sana'a, Yemen
PHONE: +967 (1) 264082; 264083; 264084
FAX: +967 (1) 263059

TITLE: Ambassador
NAME: Victor Henders

United States
Dhahr Himyar Zone, Sheraton Hotel District, PO Box 22347, Sana'a, Yemen
PHONE: +967 (1) 238843
FAX: +967 (1) 251563
TITLE: Ambassador
NAME: David G. Newton

JUDICIAL SYSTEM

Supreme Court of the Republic

FURTHER READING

Articles

''Be Happy, Why Worry?'' *The Economist,* September 25, 1999, 52.

Burns, John F. ''Khat-Chewing Yemen Told to Break Ancient Habit.'' *The New York Times (International Pages),* September 19, 1999, 3.

''Vote Fraud is Charged by Opposition in Yemen.'' *The New York Times,* September 26, 1999, A7.

''Yemen's New Terrorists.'' *The Economist,* January 9, 1999, 43.

Books

Kostiner, Joseph. *Yemen: The Tortuous Quest for Unity, 1990–94.* Royal Institute of International Affairs, London: Pinter, 1996.

Internet

Al Thawra. Available Online @ http://www.althawra.com (Accessed November 16, 1999).

YemenNet.Com. Available Online @ http://www.yemennet.com/index.stm (Accessed November 16, 1999).

YEMEN: STATISTICAL DATA

For sources and notes see "Sources of Statistics" in the front of each volume.

GEOGRAPHY

Geography (1)

Area:

Total: 527,970 sq km.

Land: 527,970 sq km.

Water: 0 sq km.

Note: includes Perim, Socotra, the former Yemen Arab Republic (YAR or North Yemen), and the former People's Democratic Republic of Yemen (PDRY or South Yemen).

Area—comparative: slightly larger than twice the size of Wyoming.

Land boundaries:

Total: 1,746 km.

Border countries: Oman 288 km, Saudi Arabia 1,458 km.

Coastline: 1,906 km.

Climate: mostly desert; hot and humid along west coast; temperate in western mountains affected by seasonal monsoon; extraordinarily hot, dry, harsh desert in east.

Terrain: narrow coastal plain backed by flat-topped hills and rugged mountains; dissected upland desert plains in center slope into the desert interior of the Arabian Peninsula.

Natural resources: petroleum, fish, rock salt, marble, small deposits of coal, gold, lead, nickel, and copper, fertile soil in west.

Land use:

Arable land: 3%

Permanent crops: 0%

Permanent pastures: 30%

Forests and woodland: 4%

Other: 63% (1993 est.).

HUMAN FACTORS

Health Indicators (4)

Life expectancy at birth	
1980	49
1997	54
Daily per capita supply of calories (1996)	2,041
Total fertility rate births per woman (1997)	6.4
Maternal mortality ratio per 100,000 live births (1990-97)	1,400[c]
Safe water % of population with access (1995)	39
Sanitation % of population with access (1995)	19
Consumption of iodized salt % of households (1992-98)[a]	21
Smoking prevalence	
Male % of adults (1985-95)[a]	....
Female % of adults (1985-95)[a]	....
Tuberculosis incidence per 100,000 people (1997)	111
Adult HIV prevalence % of population ages 15-49 (1997)	0.01

Infants and Malnutrition (5)

Under-5 mortality rate (1997)	100
% of infants with low birthweight (1990-97)	19
Births attended by skilled health staff % of total[a]	43
% fully immunized (1995-97)	
TB	54
DPT	40
Polio	46
Measles	43
Prevalence of child malnutrition under age 5 (1992-97)[b]	29

Ethnic Division (6)

Predominantly Arab; Afro-Arab concentrations in western coastal locations; South Asians in southern regions; small European communities in major metropolitan areas.

Religions (7)

Muslim including Shaf'i (Sunni) and Zaydi (Shi'a), small numbers of Jewish, Christian, and Hindu.

Languages (8)

Arabic.

EDUCATION

Public Education Expenditures (9)

Public expenditure on education (% of GNP)

1980 ....

1996 ....6.5

Expenditure per student

Primary % of GNP per capita

1980 ....

1996 ....

Secondary % of GNP per capita

1980 ....

1996 ....

Tertiary % of GNP per capita

1980 ....

1996 ....

Expenditure on teaching materials

Primary % of total for level (1996) ....

Secondary % of total for level (1996) ....

Primary pupil-teacher ratio per teacher (1996) 30

Duration of primary education years (1995) 9

Literacy Rates (11B)

Adult literacy rate

1980

Male 14%

Female 3%

1995

Male 53%

Female 26%

HUMAN FACTORS

Demographics (2A)

	1990	1995	1998	2000	2010	2020	2030	2040	2050
Population	NA	14,861.6	16,388.0	17,521.1	24,793.7	34,682.5	46,717.7	60,744.6	76,008.5
Net migration rate (per 1,000 population)	NA	NA	NA	NA	NA	NA	NA	NA	NA
Births	NA	NA	NA	NA	NA	NA	NA	NA	NA
Deaths	NA	NA	NA	NA	NA	NA	NA	NA	NA
Life expectancy - males	NA	56.4	57.7	58.6	62.9	66.7	70.0	72.7	74.8
Life expectancy - females	NA	59.7	61.3	62.5	67.7	72.2	76.0	78.9	81.2
Birth rate (per 1,000)	NA	43.7	43.4	43.3	41.7	36.6	32.0	28.0	24.0
Death rate (per 1,000)	NA	11.6	10.3	9.5	6.7	4.9	4.0	3.6	3.6
Women of reproductive age (15-49 yrs.)	NA	3,059.6	3,460.9	3,762.6	5,504.7	7,963.4	11,332.2	15,163.7	19,443.9
of which are currently married	NA	NA	NA	NA	NA	NA	NA	NA	NA
Fertility rate	NA	7.4	7.1	7.0	6.1	5.2	4.3	3.5	3.0

Except as noted, values for vital statistics are in thousands; life expectancy is in years.

GOVERNMENT & LAW

Political Parties (12)

House of Representatives	No. of seats
General People's Congress (GPC)	189
Islamic Reform Grouping (Islaah)	52
Nasserite Unionist Party	3
National Arab Socialist Baath Party	2
Independents	54
Election pending	1

Government Budget (13B)

Revenues . . . .$2.6 billion

Expenditures . . . .$2.7 billion

Capital expenditures . . . .$1.1 billion

Data for 1998 est.

Crime (15)

Crime rate (for 1997)

Crimes reported 5,100

Total persons convicted 3,150

Crimes per 100,000 population 30

Persons responsible for offenses

Total number of suspects 6,900

Total number of female suspects . . . . .NA

Total number of juvenile suspects . . . . .NA

Military Affairs (14B)

	1990[g]	1991	1992	1993	1994	1995
Military expenditures						
Current dollars (mil.)	848	979[e]	1,116[e]	1,153	NA	NA
1995 constant dollars (mil.)	975	1,082[e]	1,201[e]	1,209	NA	NA
Armed forces (000)	127	127	64	64	69	68
Gross national product (GNP)						
Current dollars (mil.)	7,268	7,142	6,999	7,351	8,373	8,892[e]
1995 constant dollars (mil.)	8,353	7,892	7,529	7,707	8,583	8,892[e]
Central government expenditures (CGE)						
1995 constant dollars (mil.)	3,429	3,712	4,024	4,113	NA	NA
People (mil.)	11.6	11.2	11.7	12.1	12.6	13.0
Military expenditure as % of GNP	11.7	13.7	16.0	15.7	NA	NA
Military expenditure as % of CGE	28.4	29.1	29.8	29.4	NA	NA
Military expenditure per capita (1995 $)	84	96	103	100	NA	NA
Armed forces per 1,000 people (soldiers)	11.0	11.3	5.5	5.3	5.5	5.2
GNP per capita (1995 $)	721	702	645	636	684	683
Arms imports[6]						
Current dollars (mil.)	550	40	5	20	250	140
1995 constant dollars (mil.)	632	44	5	21	256	140
Arms exports[6]						
Current dollars (mil.)	0	0	0	0	0	0
1995 constant dollars (mil.)	0	0	0	0	0	0
Total imports[7]						
Current dollars (mil.)	NA	1,897[e]	2,245[e]	2,156[e]	1,800[e]	NA
1995 constant dollars (mil.)	NA	2,096[e]	2,415[e]	2,260[e]	1,845[e]	NA
Total exports[7]						
Current dollars (mil.)	NA	1,174[e]	999[e]	1,001[e]	1,100[e]	NA
1995 constant dollars (mil.)	NA	1,297[e]	1,075[e]	1,049[e]	1,128[e]	NA
Arms as percent of total imports[8]	NA	2.1	.2	.9	13.9	NA
Arms as percent of total exports[8]	NA	0	0	0	0	NA

LABOR FORCE

Labor Force (16)

Engaged in herding or as expatriate laborers; services, construction, industry, and commerce account for less than one-half of the labor force.

Unemployment Rate (17)

30% (1995 est.)

PRODUCTION SECTOR

Electric Energy (18)

Capacity	810,000 kW (1995)
Production	1.85 billion kWh (1995)
Consumption per capita	126 kWh (1995)

Transportation (19)

Highways:

total: 64,725 km

paved: 5,243 km

unpaved: 59,482 km (1996 est.)

Pipelines: crude oil 644 km; petroleum products 32 km

Merchant marine:

total: 3 ships (1,000 GRT or over) totaling 12,059 GRT/18,563 DWT ships by type: cargo 1, oil tanker 2 (1997 est.)

Airports: 48 (1997 est.)

Airports—with paved runways:

total: 11

over 3,047 m: 2

2,438 to 3,047 m: 6

1,524 to 2,437 m: 1

914 to 1,523 m: 1

under 914 m: 1 (1997 est.)

Airports—with unpaved runways:

total: 37

over 3,047 m: 2

2,438 to 3,047 m: 10

1,524 to 2,437 m: 10

914 to 1,523 m: 12

under 914 m: 3 (1997 est.)

Top Agricultural Products (20)

Grain, fruits, vegetables, qat (mildly narcotic shrub), coffee, cotton; dairy products, poultry, meat; fish.

FINANCE, ECONOMICS, & TRADE

Economic Indicators (22)

National product: GDP—purchasing power parity—$31.8 billion (1997 est.)

National product real growth rate: 5% (1997 est.)

National product per capita: $2,300 (1997 est.)

Inflation rate—consumer price index: 5% (1997 est.)

Exchange Rates (24)

Exchange rates:	
Yemeni rials (YRl) per US$1	
1997	129.158
1996	94.157
1995	40.839
official fixed rate 1991-94	12.010

Top Import Origins (25)

$2.3 billion (f.o.b., 1997 est.) Data are for 1995.

Origins	%
United States	12
France	11
UAE	10
Saudi Arabia	7
United Kingdom	5

MANUFACTURING SECTOR

GDP & Manufacturing Summary (21)

	1980	1985	1990	1991	1992
Gross Domestic Product					
Millions of 1980 dollars	2,779	3,692	5,433	5,868	6,090
Growth rate in percent	6.04	10.31	6.80	8.020	3.79
Manufacturing Value Added					
Millions of 1980 dollars	160	295	363	398	436
Growth rate in percent	7.70	1.46	2.99	9.51	9.67

FINANCE, ECONOMICS, & TRADE

Balance of Payments (23)

	1991	1992	1993	1994	1995
Exports of goods (f.o.b.)	1,197	1,095	1,167	1,824	1,937
Imports of goods (f.o.b.)	−1,897	−1,891	−2,087	−1,522	−1,948
Trade balance	−700	−796	−920	302	−11
Services - debits	−1,384	−1,566	−1,594	−1,223	−1,127
Services - credits	157	200	199	170	217
Private transfers (net)	311	78	54	73	37
Government transfers (net)	953	993	1,014	1,044	1,067
Overall balance	−663	−1,091	−1,248	366	183

Top Export Destinations (26)

$2.3 billion (f.o.b., 1997 est.) Data are for 1995.

Destinations	%
China	23
South Korea	19
Thailand	14
Brazil	13
Japan	12
Thailand	7

Economic Aid (27)

Recipient: ODA, $148 million (1993).

Import Export Commodities (28)

Import Commodities	Export Commodities
Textiles and other manufactured consumer goods	Crude oil
Petroleum products	Cotton
Foodstuffs	Coffee
Cement	Dried and salted fish 7%
Machinery	
Chemicals	

ZAMBIA

Republic of Zambia

CAPITAL: Lusaka.

FLAG: The flag is green, with a tricolor of dark red, black, and orange vertical stripes at the lower corner of the flag, topped by a golden flying eagle.

ANTHEM: *Stand and Sing for Zambia.*

MONETARY UNIT: The kwacha (K) of 100 ngwee replaced the Zambian pound (Z£) on 15 January 1968. There are coins of 1, 2, 5, 10, 20, and 50 ngwee, and notes of 1, 2, 5, 10, 20, 50, 100, and 500 kwacha. K1 = $0.0008 (or $1 = K1,250.0).

WEIGHTS AND MEASURES: The metric system is used.

HOLIDAYS: New Year's Day, 1 January; Youth Day, 11 March; Labor Day, 1 May; African Freedom Day, 24 May; Heroes' Day, 1st Monday after 1st weekend in July; Unity Day, Tuesday after Heroes' Day; Farmers' Day, 5 August; Independence Day, 24 October; Christmas, 25 December. Movable religious holidays include Good Friday and Easter Monday.

TIME: 2 PM = noon GMT.

LOCATION AND SIZE: A landlocked country in south-central Africa, Zambia has an area of 752,610 square kilometers (290,584 square miles), slightly larger than the state of Texas, with a total boundary length of 5,664 kilometers (3,519 miles). Zambia's capital city, Lusaka, is located in the south-central part of the country.

CLIMATE: Although Zambia lies within the tropics, much of it has a pleasant climate because of the altitude. There are wide seasonal variations in temperature and rainfall. The northern and northwestern provinces have an annual rainfall of about 125 centimeters (50 inches), while areas in the far south have as little as 75 centimeters (30 inches). Daytime temperatures may range from 23° to 31°C (73–88°F), dropping at night to as low as 5° C (41°F) in June and July.

INTRODUCTORY SURVEY

RECENT HISTORY

Zambia, formerly Northern Rhodesia, achieved independence from Britain in 1964. Kenneth Kaunda, leader of the ruling United National Independence Party (UNIP), became the nation's first president and proclaimed one-party rule. Reelected in 1969, 1973, 1978, and 1983, Kaunda survived a series of coup attempts. During his regime, continuing civil war in Angola had repercussions in Zambia, bringing disruption of Zambian trade routes and casualties among Zambians along the border.

By 1990, a growing opposition to Kuanda's monopoly of power had coalesced in the Movement for Multiparty Democracy (MMD). In December 1990, after a tumultuous year that included riots in Lusaka and another coup attempt, Kaunda signed legislation ending UNIP's one-party rule. Zambia's new constitution, enacted in August 1991, enlarged the National Assembly, established an electoral commission, and allowed for more than one presidential candidate.

In September 1991. Frederick J. T. Chiluba (MMD) defeated Kaunda, 81% to 15%. Once in power, Chiluba and his party became tyrannical and corrupt, easily winning reelection in November 1996 in voting marred by questionable practices.

GOVERNMENT

Under the 1991 constitution, the president is elected directly by universal suffrage and serves a maximum of two five-year terms. The National

Assembly has 150 directly elected members, also for five-year terms.

Judiciary

The Supreme Court is the highest court in Zambia and serves as the final court of appeal. The chief justice and other judges are appointed by the president. The legal system is based on English common law and customary law.

Political Parties

From 1972 to 1991, Kaunda's United National Independence Party (UNIP) was the only legal party in Zambia. Growing unrest in the late 1980 spawned the Movement for Multiparty Democracy (MMD), which currently holds power. Since 1991, new opposition parties have been formed: the Multi-Racial Party, the National Democratic Alliance, the United Democratic Party, and the United Democratic Congress Party. With the exception of UNIP, over 30 opposition parties operate without government interference.

DEFENSE

In 1995 the strength of the armed forces totaled 21,600; paramilitary forces, consisting of two police battalions, totaled 1,400. The army numbered 20,000; the air force had 1,600 members and 59 combat aircraft. Military service is voluntary.

ECONOMIC AFFAIRS

Copper and other metal exports account for about 75 percent of the country's export income. Continued poor performance of the copper mines, falling copper prices, a series of droughts, and decreasing foreign aid have all taken their toll on the Zambian economy. In 1992 Zambia was classified as a least developed country by the United Nations. The impact of inflation on the poor, the middle class, and business has eroded public support for the government's reform policies.

Public Finance

Heavily dependent on copper, Zambia shows surpluses in public accounts only when the mining industry is prosperous. From 1985 to 1987, Zambia attempted to implement a structural reform program, but in 1987 reverted to deficit spending and monetary creation. Major economic reforms instituted in 1992 have been slow to take effect.

The U.S. Central Intelligence Agency estimated in 1995 that government revenues totaled approximately $835 million and expenditures $888 million, including capital expenditures of $110 million. External debt totaled $7 billion.

Income

In 1998,Zambia's gross national product (GNP) was $3.2 billion, or $330 per capita.

Industry

In 1995 manufacturing output accounted for 30% of the gross domestic product (GDP). The manufacturing sector declined by an average of 1% each year during 1990–95, after growing by and average of 4% per year during the 1980s. Apart from copper refining, the most important industries include construction, foodstuffs, beverages, chemicals, textiles, and fertilizer.

Banking and Finances

The Bank of Zambia (BOZ), a central bank founded in 1964, sets and controls all currency and banking activities in the country. The leading commercial banks are subsidiaries of Barclays, Citibank, Grindlays, and The Standard Bank. There are two development banks: the Development Bank of Zambia and the Lima Bank. In 1985, the first locally and privately owned bank the African Commercial Bank, was formed. Its success led to the establishment of 28 registered commercial banks by the end of December 1994.

Economic Development

Zambia's first priority is to control inflation (43% in 1996), followed by faster implementation of social programs, legal and civil service reform, and privatization. New investment has been slow to form as investors await lower inflation rates.

SOCIAL WELFARE

Social welfare services are provided by the government in association with local authorities and voluntary agencies. Statutory and remedial welfare services include emergency relief, care for the aged, protection of children, adoption, and probation. Employers and employees are required to make contributions toward a worker's retirement at age 55. The Zambia Youth Service operates specially constructed camps that provide vocational training for unemployed and unskilled youth.

Women have full legal rights under the law, but discrimination exists in matters of inheritance, property ownership, marriage, education, and employment.

Healthcare

Life expectancy is only 37 years at birth. Malaria and tuberculosis are major health problems, and hookworm and schistosomiasis afflict a large proportion of the population. In 1994, only 47% of the population had access to safe water, and 42% had adequate sanitation. In 1994, blood surveys indicated that 17% of the adult population was infected with HIV (the virus that causes AIDS).

Housing

Widespread instances of overcrowding and slum growth have for many years focused government attention on urban housing problems. Mining companies have constructed townships for the families of African workers.

EDUCATION

Seven years of primary education is compulsory. Secondary education is for five years. In 1994, 1.5 million pupils were in 3,715 primary schools, and 199,081 students attended secondary schools. All higher level institutions had 15,343 pupils in 1990. Adult illiteracy was estimated at 22%.

1999 KEY EVENTS TIMELINE

January

- Zambian soldiers begin tightening the nation's borders to prevent harmful pests and crop diseases from entering the southern African country.
- Zambia vehemently rejects fresh allegations by the Angolan government that it was providing military and logistical support to Angola's UNITA rebels.

- The World Bank approves $170 million (US) adjustment loan to Zambia to facilitate completion of the privatization of the Zambia Consolidated Copper Mines.
- The Angolan government threatens military action against Zambia, Uganda, Rwanda, Togo and Burkina Faso for allegedly supporting UNITA leader Jonas Savimbi.
- Zambia's finance minister, Edith Nawakwi, presents a 2,227.7 billion kwacha (Zambia's currency) budget which she said had been designed to reverse the economic decline experience in 1998 and stimulate growth (1 U.S. dollar = 2500 Zambia Kwacha).

February

- The Angolan government officially accuses Zambia of providing arms to the UNITA rebels, a charge the Zambian president vehemently denies.
- The Zambian government asks citizens to stop commenting on allegations that some top government officials are engaged in supplying arms to Angola's rebel movement, UNITA.
- Malawian President Muluzi and Zambian President Chiluba discuss the crisis in the Democratic Republic of Congo, DRC, and Angolan accusations against Zambia.
- A member of Zambia's parliament from President Frederick Chiluba's Movement for Multiparty Democracy (MMD) is suspended from the party for six months for allegedly demeaning the president.
- A high-level OAU mission leaves Addis Ababa on a mission aimed at reconciling differences between Angola and Zambia.

March

- Zambia requests assistance from the U.S. to investigate bombings that rocked Lusaka Sunday, February 28, 1999, leaving an Angolan embassy security officer dead and another injured.
- Government pledges to assist women in gaining the right of equal access to land.
- President Laurent-Desire Kabila briefly visits Zambia and conducts discussions with President Frederick Chiluba on issues concerning the DRC and the tension between Angola and Zambia.
- Zambia and the U.S. sign a 20 million dollar ''debt swap for economic development agreement'' to help finance the country's various health care programs.
- The ruling Movement for Multiparty Democracy (MMD) obtains an injunction restraining the newspaper, ''The Post'' from any further publications of articles concerning the security of Zambia and the country's stand-off with Angola.

April

- President Chiluba expresses the hope that the continued cooperation with Britain, whose investment pledges over the last eight years have amounted to $250 million (US), will help strengthen democratic institutions in both countries.
- Zambian president Frederick Chiluba assures his Namibian counterpart, Sam Nujoma, that those calling for the breakaway of Namibia's Caprivi Strip and part of western Zambia have no support that could threaten the ''status and stability of politics in Zambia.''
- Zambia appeals to the international community for immediate assistance to help the government cope with the daily flow of refugees fleeing fighting in the Democratic Republic of Congo (DRC).
- Over 9,000 deaths are recorded since the outbreak of cholera at the beginning of 1999.
- Finance minister, Edith Nawakwi, says Zambia would maintain a tight fiscal and monetary economic programme to foster economic growth and qualify for debt relief under the Highly Indebted Poor Countries Initiative (HIPC) by the end of this year.

May

- Coca-Cola which had increased the price of its product in Zambia by almost 100% in 1998, slashes the cost by about 10 percent.
- Over 800 jobs are to be created through Zambia National Tourism Board's (ZNTB) approval of five new tourism projects worth $57 million (US) in investment pledges.
- According to a recent study conducted by the Lusaka University Teaching Hospital, nearly 40 percent of Zambian women are afflicted by anaemia, mainly due to poverty which robs them of a diet rich in minerals, particularly iron.
- Scores of female children below the age of eleven years are married to elderly men in parts of Zambia who are able to part with wealth in the form of dowry (lobola) with poverty as the major

driving force of early marriages involving female children.

- Zambia secures some $630 million (US) external financing aid for 1999 from the ''Paris Club'' external donors.

June

- Donors (made up of the World Bank, International Monetary Fund and Organization for Economic Cooperation and Development) agree to make available $530 million (US) for Zambia to support the country's economic reforms in 1999.
- Angolan and Zambian representatives sign an agreement that the two countries would forget all past disputes between them.
- President Chiluba and Libyan leader Colonel Mu'ammar al-Qaddafi, after a two day official visit to Zambia, call for peace on the African continent.
- The diesel shortage in Lusaka worsens forcing some motorists to spend nights at filling stations while some inter city bus operators have had to reschedule their departure times.
- Ignoring warnings from the Zambian government, the Lusaka city council razes the stalls of over 1,000 street vendors and orders them to relocate to fifty designated marketplaces.

July

- Envoy says France ready to ''relieve'' Zambia's debt.
- Zambia is said to have one of the highest debt burdens in the world and the highest debt ratios among non-warring nations.
- In Zambia, one in five adults is HIV-positive and 80 percent of families live below the poverty line and 10 percent of the population are orphans.
- The president of Tanzania, Benjamin Mkapa, speaks on investment and trade and praises cooperation with Zambia during a visit to Zambia.

August

- Burdened by a $6 billion (US) foreign debt, Zambia receives a little relief when Britain cancels about 46 million pounds sterling owed it by Zambia and rescheduled the repayment of 65 million pounds debt to 2023.
- No economic growth is to be recorded in Zambia this year.
- Police arrest Mr Mutangelwa Imasiku, the leader of the Barotse Patriotic Front, BPF as he was coming out of the South African High Commissioner's residence where he had sought refuge.
- It is reported that the foreign cargo plane impounded by the Zambia Air Force (ZAF) at the Lusaka International Airport is Ukrainian registered and carrying arms bound for Angola.
- The faction leader of the rebel Rally for Congolese Democracy arrives in Lusaka at the head of a 51-member delegation in preparation for the signing of the ceasefire agreement.

September

- The eleventh International Conference on AIDS and Sexually Transmitted Diseases in Africa is held in Zambia for African researchers to find imaginative, novel and bold solutions to diseases.
- Authorities in Zambia break off negotiations with the Anglo-American Corporation for the sale of the remaining assets of Zambia Consolidated Copper Mines.
- Nine members of a Ukrainian air crew, held in Zambia on suspicion of espionage are released from prison.
- Zambian Non-Governmental Organizations accuse the International Monetary Fund and the World Bank of lacking commitment to the welfare of the poor in Africa by not pressuring rich nations to expand debt forgiveness.
- Some 600 people die from more than 4,000 tuberculosis cases detected in 1998 in Zambia's copper-mining town of Kitwe.

October

- A group of senior executives of leading British companies including the Commonwealth Development Corporation (CDC), arrive in Zambia on a three day trade expedition.
- The number of people awaiting execution at Kabwe Maximum Security Prison shoots up to more than 200 following the recent judgement of the Lusaka High Court which condemned 59 soldiers to death for their role in the 1997 coup attempt.
- The U. S. expresses interest in signing an air service agreement with Zambia to enable both local and American airlines service the routes between the two countries.

- Consumption of practically all foodstuffs, including the staple food maize, per capita in Zambia has drastically fallen over the last nine years.

December

- Members of the Junior Doctors' Association stage an indefinite strike over lack of supplies and support staff.
- Zambia's national football loses to Honduras in a humiliating soccer match in Miami, Florida. The team decides to cut short their tour of America, citing problems with the diet of junk food—primarily burgers and pizza—provided to the competitors by the Miami hosts.
- The nation prepares for privatization of its copper industry, the biggest source of foreign exchange, scheduled to be completed in January 2000.

ANALYSIS OF EVENTS: 1999

BUSINESS AND THE ECONOMY

Zambia entered the year 1999 with a sickly economy and no economic growth in sight. The government of Chiluba has tried to get the economy back on its feet with little success. For the past 8 years Zambia applied the World Bank mandated structural adjustment programs faithfully. The country itself is well endowed with natural resources, its earnings from mineral exports are enormous and, above all, Zambia has enjoyed relative peace during the post-independence era. And yet recessionary condition still continues and the currency is the weakest in the region. In the early 1970s one U.S. dollar was worth 65 Ngwee (100 Ngwee = K1) and now the same U.S. dollar is worth K2,550 (see Bwalya, 1999). A number of reasons have been advanced for the poor performance of Zambia's economy. First the copper industry, the main stay of Zambia's economy has not done well since the early 1970s to the present due to falling copper prices at international markets. Second, the side-effects of the structural adjustment programs have made things far worse than before. The International Monetary Fund and the World Bank have pressured the Zambian Government into embracing de-regulation of exchange control restriction resulting in the flooding of the Zambian economy with imported goods. Third, foreign investors often take profits out to their home countries.

In 1999 Zambia's external debt was estimated at about $7 billion (US), or about $700 (US) per person. With this debt burden, Zambia topped the list of countries with the highest debt burden in the world and the highest debt ratios among non-warring nations. In 1999, several local non-government organizations (NGOS) launched vigorous campaigns to pressure Western nations to cancel the debt. The NGOs accused the IMF and the World Bank of lacking commitment to the welfare of the poor in Africa by not pressuring rich nations to expand debt forgiveness. These campaigns and the faithful implementation of the structural adjustment reforms resulted in a little relief for Zambia. Britain canceled about 46 million pounds sterling owed it by Zambia and rescheduled the repayment of 65 million pounds debt to 2023. France signaled its readiness to ''relieve'' Zambia's debt. The World Bank approved $170 million (US) adjustment loan to Zambia to facilitate completion of the privatization of the Zambia Consolidated Copper Mines. Furthermore, Zambia secured some $630 million (US) external financing aid for 1999 from the ''Paris Club'' donors to support the country's economic reforms in 1999. The U.S. and Zambia signed a 20 million dollar ''debt swap for economic development agreement'' to help finance the country's various health care programs. The government was also hopeful that continued investment from British firms would be forthcoming following a three day visit of a group of senior executives of leading British companies including the Commonwealth Development Corporation. The U. S. also expressed interest in signing an air service agreement with Zambia to enable both local and American airlines service the routes between the two countries. Although burdened by a suffocating debt, 1999 saw some promising gestures from the West to relieve the burden and to invest significantly in the ailing Zambian economy.

GOVERNMENT AND POLITICS

Zambia, along with 8 other countries, was drawn into the Democratic Republic of the Congo (DRC) civil war. Fearing the impact of the war on the already diminishing exports from Zambia, the government of Chiluba, quickly sought to end the fighting through peaceful negotiation. Chiluba was able to bring together the president of the DRC, Laurent-Desire Kabila, the faction leader of the rebel Rally for Congolese Democracy, as well as other regional powers involved in the fighting. This peace initiative resulted in a ceasefire agreement between the warring factions and their respective backers.

Apart from the involvement of Zambia in the DRC conflict, the government of Chiluba was accused by Angola that it was providing military and logistical support to Angola's UNITA rebels. The Angolan government threatened military action against Zambia. Fortunately, a high-level Organization of African Unity delegation managed to reconcile the differences between Angola and Zambia which culminated in the signing of an agreement that the two countries would forget all past disputes between them.

Internally, Chiluba's government faced criticism from different opposition groups for rampant corruption and violation of human rights. Several journalists and leaders of the opposition were harassed and/or arrested. It was also alleged that the government attempted to assassinate former President Kaunda, who was stripped of his citizenship, and is apparently the main opposition to Chiluba's presidency.

CULTURE AND SOCIETY

During the first few months of 1999, Zambia was flooded with a daily flow of thousands of refugees fleeing fighting in DRC prompting the government to appeal to the international community for immediate assistance. The health condition of the Zambian population also worsened due to pervasive poverty and disease. Over 9,000 deaths were attributed to cholera. More than 600 people died as a result of the more than 4,000 tuberculosis cases detected in 1998 in just one town, Kitwe, Zambia's copper-mining town. Another worrisome health hazard is the AIDS epidemic with one in five adults being HIV-positive and 10 percent of the population are orphans, victims of the epidemic. The XIth International Conference on AIDS and Sexually Transmitted Diseases in Africa was this year held in Zambia for African researchers to find imaginative, novel and bold solutions to the epidemic.

Polygamy was said to be well and alive in Zambia with an estimated 20 percent of Zambia marriages being polygamous and an estimated 32 percent of marriages in the southern region falling in this category. Polygamous unions have been blamed for the proliferation of HIV/AIDS and for promoting poverty. Divorces were also on the increase following the breakdown of "old societal values" considered necessary for binding marriages. It was also reported that 80 percent of families live below the poverty line. Consumption of practically all foodstuffs, including the staple food maize, per capita in Zambia has drastically fallen over the last nine years. One of the many consequences of massive poverty has been the marrying off of scores of girl-children below the age of 11 years to elderly men who are able to part with wealth in the form of dowry (lobola). Another consequence, according to a recent study conducted by the Lusaka University Teaching Hospital, is that nearly 40 percent of Zambian women are afflicted by anaemia as poverty robs them of a diet rich in minerals, particularly iron.

DIRECTORY

CENTRAL GOVERNMENT

Head of State

President
Fredrick Chiluba, Office of President

Vice President
Christon Tembo, Office of Vice President

Ministers

Minister of Commerce, Trade, and Industry
David Mpamba, Ministry of Commerce, Trade, and Industry

Minister of Defense
Chitalu Sampa, Ministry of Defense

Minister of Education
Godfrey Miyanda, Ministry of Education

Minister of Energy and Water Development
Benjiman Mwila, Ministry of Energy and Water Development

Minister of Finance
Katele Kalumba, Ministry of Finance

Minister of Foreign Affairs
Sipakeli Walubita, Ministry of Foreign Affairs

Minister of Health
Nkandu Luo, Ministry of Health

Minister of Home Affairs
Peter Machungwa, Ministry of Home Affairs

Minister of Labor and Social Services
Edith Nawakwi, Ministry of Labor and Social Services

Minister of Legal Affairs
Vincent Malambo, Ministry of Legal Affairs

Minister of Mines and Mineral Development
Syamukayumbu Syamujaye, Ministry of Mines and Mineral Development

Minister of Tourism
Anoshi Chipawa, Ministry of Tourism

Minister of Transport and Communications
David Saviye, Ministry of Transport and Communications

POLITICAL ORGANIZATIONS

Agenda for Zambia

NAME: Akashambatwa Lewanika

Liberal Progressive Front

TITLE: President
NAME: Roger Chongwe

Movement for Democratic Process

NAME: Chama Ckakom Boka

National Party

NAME: Daniel Lisulo

DIPLOMATIC REPRESENTATION

Embassies in Zambia

Australia
Memaco House, Cairo Road, Lusaka, Zambia

Canada
Canadian High Commission, PO Box 31312, North End Branch, Barclays Bank Building, Cairo Road, Lusaka, Zambia
PHONE: +260 (1) 228811
FAX: +260 (1) 225160

Italy
Diplomatic Triangle, Embassy Park, Plot 5211, Lusaka, Zambia
PHONE: +260 (1) 250781
FAX: +260 (1) 254929

Mozambique
46 Mulungushi Village, Kundalile Road, Lusaka, Zambia
PHONE: +260 (1) 290451

United Kingdom
British High Commission, PO Box 50050, 5201 Independence Avenue, Lusaka, Zambia
PHONE: +260 (1) 228955
FAX: +260 (1) 253421

United States
PO Box 31617, Independence Avenue, Lusaka, Zambia
PHONE: +260 (1) 228595
FAX: +260 (1) 251578

JUDICIAL SYSTEM

Supreme Court

FURTHER READING

Articles

Atenga, Thomas Hirene. ''Donors Pledge $530 million for Economic Reforms.'' Inter Press Service, June 2, 1999.

Bwalya, Mwimba. ''There is a Hole in Zambia's Economy.'' *The Times of Zambia,* July 6, 1999.

Chipungu, Joel. ''Tuberculosis Kills 600 In Kitwe.'' PanAfrican News Agency, September 29, 1999.

''Congolese Faction Leader in Lusaka to Sign Ceasefire Accord.'' PanAfrican News Agency, August 30, 1999.

Moszynski, Peter. ''A Nation of Orphans Where One in Five Adults has HIV.'' *The Independent*, (London, England), July 27, 1999, 14.

Mulenga, Mildred. ''Zambia Does Not Support Caprivi Strip Secession.'' PanAfrican News Agency, April 14, 1999.

Tembo, Jacqueline. ''Cholera Kills 9,259 Zambians.'' *The Post of Zambia*, April 22, 1999.

''Zambia Appeals for Help as D. R. Congo Refugees Arrive in Large Numbers.'' South African News Agency—SAPA, Johannesburg, BBC Monitoring International Reports (April 17, 1999).

''Zambian Court Bars ' *The Post*' From Reporting Security Issues.'' *'The Post'* web site, Lusaka, BBC Monitoring International Reports, March 25, 1999.

Books

Ferguson, James. *Expectations of Modernity: Myths and Meanings of Urban Life on the*

Zambian Copperbelt. Berkeley: University of California Press,1999.

Hilhorst, Thea and Aarnik, Nettie. *Co-managing the Commons: Setting the Stage in Mali and Zambia,* Amsterdam: Royal Tropical Institute, 1999.

Hope, Kempe R. *AIDS and Development in Africa: A Social Science Perspective.* New York: Haworth Press, 1999.

McIntyre, Chris. *Zambia* Chalfont St. Peter: Bradt,1999.

Internet

The Post (independent Zambia newspaper). Available Online @ http://www.zamnet.zm/zamnet/post/post.html. (Accessed October 14, 1999).

University of Zambia. Available Online @ http://www.unza.zm. (October 14, 1999).

Zambia Daily Mail and *Sunday Mail.* Available Online @ http://www.zamnet.zm/zamnet/zadama/zadama.html. (October 14, 1999).

ZAMBIA: STATISTICAL DATA

For sources and notes see "Sources of Statistics" in the front of each volume.

GEOGRAPHY

Geography (1)

Area:

Total: 752,610 sq km.

Land: 740,720 sq km.

Water: 11,890 sq km.

Area—comparative: slightly larger than Texas.

Land boundaries:

Total: 5,664 km.

Border countries: Angola 1,110 km, Democratic Republic of the Congo 1,930 km, Malawi 837 km, Mozambique 419 km, Namibia 233 km, Tanzania 338 km, Zimbabwe 797 km.

Coastline: 0 km (landlocked).

Climate: tropical; modified by altitude; rainy season (October to April).

Terrain: mostly high plateau with some hills and mountains.

Natural resources: copper, cobalt, zinc, lead, coal, emeralds, gold, silver, uranium, hydropower potential.

Land use:

Arable land: 7%

Permanent crops: 0%

Permanent pastures: 40%

Forests and woodland: 39%

Other: 14% (1993 est.).

HUMAN FACTORS

Demographics (2A)

	1990	1995	1998	2000	2010	2020	2030	2040	2050
Population	7,956.8	8,915.4	9,460.7	9,872.0	12,150.4	14,694.9	17,891.4	22,209.9	26,967.3
Net migration rate (per 1,000 population)	NA	NA	NA	NA	NA	NA	NA	NA	NA
Births	NA	NA	NA	NA	NA	NA	NA	NA	NA
Deaths	NA	NA	NA	NA	NA	NA	NA	NA	NA
Life expectancy - males	38.1	37.1	36.8	36.6	37.8	41.7	53.0	64.3	68.2
Life expectancy - females	38.8	37.7	37.3	37.1	37.8	42.2	55.9	69.6	73.9
Birth rate (per 1,000)	48.0	45.4	44.6	44.4	41.5	36.6	32.0	26.9	22.3
Death rate (per 1,000)	22.7	22.7	22.5	22.6	21.5	18.2	10.7	5.6	4.8
Women of reproductive age (15-49 yrs.)	1,734.3	1,929.5	2,056.1	2,162.2	2,762.5	3,508.3	4,620.5	6,006.4	7,405.9
of which are currently married	NA	NA	NA	NA	NA	NA	NA	NA	NA
Fertility rate	6.9	6.6	6.4	6.3	5.4	4.5	3.7	3.0	2.6

Except as noted, values for vital statistics are in thousands; life expectancy is in years.

Health Personnel (3)

Total health expenditure as a percentage of GDP, 1990-1997[a]

Public sector	2.9
Private sector	0.7
Total[b]	3.3

Health expenditure per capita in U.S. dollars, 1990-1997[a]

Purchasing power parity	31
Total	17

Availability of health care facilities per 100,000 people

Hospital beds 1990-1997[a]	NA
Doctors 1993[c]	NA
Nurses 1993[c]	NA

Health Indicators (4)

Life expectancy at birth

1980	50
1997	43
Daily per capita supply of calories (1996)	1,939
Total fertility rate births per woman (1997)	5.6
Maternal mortality ratio per 100,000 live births (1990-97)	650[d]
Safe water % of population with access (1995)	53
Sanitation % of population with access (1995)	51
Consumption of iodized salt % of households (1992-98)[a]	90

Smoking prevalence

Male % of adults (1985-95)[a]	39
Female % of adults (1985-95)[a]	7
Tuberculosis incidence per 100,000 people (1997)	576
Adult HIV prevalence % of population ages 15-49 (1997)	19.07

Infants and Malnutrition (5)

Under-5 mortality rate (1997)	202
% of infants with low birthweight (1990-97)	13
Births attended by skilled health staff % of total[a]	47

% fully immunized (1995-97)

TB	81
DPT	70
Polio	70
Measles	69
Prevalence of child malnutrition under age 5 (1992-97)[b]	24

Ethnic Division (6)

African	98.7%
European	1.1%
Other	0.2%

Religions (7)

Christian	50%-75%
Muslim and Hindu	24%-49%
Indigenous beliefs	1%

Languages (8)

English (official), major vernaculars—Bemba, Kaonda, Lozi, Lunda, Luvale, Nyanja, Tonga, and about 70 other indigenous languages.

EDUCATION

Public Education Expenditures (9)

Public expenditure on education (% of GNP)

1980	4.2
1996	2.2[1]

Expenditure per student

Primary % of GNP per capita

1980	10.6
1996	5.1[1]

Secondary % of GNP per capita

1980	....
1996	15.4[1]

Tertiary % of GNP per capita

1980	605.6[1]
1996	369.0[1]

Expenditure on teaching materials

Primary % of total for level (1996)	2.8
Secondary % of total for level (1996)	....
Primary pupil-teacher ratio per teacher (1996)	39[1]
Duration of primary education years (1995)	7

Educational Attainment (10)

Age group (1993)[7]	14+
Total population	NA

Highest level attained (%)

No schooling	18.6
First level	
Not completed	54.8
Completed	NA
Entered second level	
S-1	12.9
S-2	12.2
Postsecondary	1.5

Literacy Rates (11A)

In thousands and percent[1]	1990	1995	2000	2010
Illiterate population (15+ yrs.)	1,141	1,082	992	809
Literacy rate - total adult pop. (%)	73.1	78.2	82.9	89.6
Literacy rate - males (%)	82.1	85.6	88.6	92.8
Literacy rate - females (%)	64.8	71.3	77.4	86.6

GOVERNMENT & LAW

Political Parties (12)

National Assembly	No. of seats
Movement for Multiparty Democracy (MMD)	130
National Party (NP)	5
Zambia Democratic Congress (ZADECO)	2
Agenda for Zambia (AZ)	2
Independents	11

Military Affairs (14B)

	1990	1991	1992	1993	1994	1995
Military expenditures						
Current dollars (mil.)	84[e]	NA	51[e]	60[e]	43	102
1995 constant dollars (mil.)	96[e]	NA	54[e]	63[e]	44	102
Armed forces (000)	16	16	16	16	16	22
Gross national product (GNP)						
Current dollars (mil.)	3,151	3,171	3,340	3,597	3,540	3,612
1995 constant dollars (mil.)	3,621	3,504	3,592	3,771	3,628	3,612
Central government expenditures (CGE)						
1995 constant dollars (mil.)	993	1,637	1,111	533	916	810
People (mil.)	8.0	8.2	8.4	8.6	8.8	9.0
Military expenditure as % of GNP	2.7	NA	1.5	1.7	1.2	2.8
Military expenditure as % of CGE	9.7	NA	4.9	11.8	4.8	12.6
Military expenditure per capita (1995 $)	12	NA	6	7	5	11
Armed forces per 1,000 people (soldiers)	2.0	1.9	1.9	1.9	1.8	2.4
GNP per capita (1995 $)	452	425	427	413	413	403
Arms imports[6]						
Current dollars (mil.)	5	20	0	0	5	0
1995 constant dollars (mil.)	6	22	0	0	5	0
Arms exports[6]						
Current dollars (mil.)	0	0	0	0	0	0
1995 constant dollars (mil.)	0	0	0	0	0	0
Total imports[7]						
Current dollars (mil.)	1,220	948	1,107[e]	1,119[e]	845[e]	NA
1995 constant dollars (mil.)	1,402	1,048	1,190[e]	1,173[e]	866[e]	NA
Total exports[7]						
Current dollars (mil.)	1,309	745	756	891	758	760[e]
1995 constant dollars (mil.)	1,504	823	813	934	777	760[e]
Arms as percent of total imports[8]	.4	2.1	0	0	.6	NA
Arms as percent of total exports[8]	0	0	0	0	0	0

Government Budget (13A)

Year: 1996

Total Expenditures: 848.2 Millions of Kwacha

Expenditures as a percentage of the total by function:

General public services and public order	30.26[f]
Defense	4.81[f]
Education	17.74[f]
Health	10.33[f]
Social Security and Welfare	1.44[f]
Housing and community amenities	4.64[f]
Recreational, cultural, and religious affairs	.85[f]
Fuel and energy	.71[f]
Agriculture, forestry, fishing, and hunting	7.58[f]
Mining, manufacturing, and construction	1.08[f]
Transportation and communication	9.00[f]
Other economic affairs and services	4.96[f]

Crime (15)

Crime rate (for 1994)

Crimes reported	71,622
Total persons convicted	40,351
Crimes per 100,000 population	779

Persons responsible for offenses

Total number of suspects	59,381
Total number of female suspects	3,552
Total number of juvenile suspects	NA

LABOR FORCE

Labor Force (16)

Total (million)	3.4
Agriculture	85%
Mining, manufacturing, and construction	6%
Transport and services	9%

Unemployment Rate (17)

22% (1991)

PRODUCTION SECTOR

Electric Energy (18)

Capacity	2.436 million kW (1995)
Production	7.79 billion kWh (1995)
Consumption per capita	668 kWh (1995)

Transportation (19)

Highways:

total: 39,700 km

paved: 7,265 km (including 60 km of expressways)

unpaved: 32,435 km (1996 est.)

Waterways: 2,250 km, including Zambezi and Luapula rivers, Lake Tanganyika

Pipelines: crude oil 1,724 km

Airports: 111 (1997 est.)

Airports—with paved runways:

total: 12

over 3,047 m: 1

2,438 to 3,047 m: 3

1,524 to 2,437 m: 5

914 to 1,523 m: 2

under 914 m: 1 (1997 est.)

Airports—with unpaved runways:

total: 99

2,438 to 3,047 m: 1

1,524 to 2,437 m: 2

914 to 1,523 m: 64

under 914 m: 32 (1997 est.)

Top Agricultural Products (20)

Corn, sorghum, rice, peanuts, sunflower seed, tobacco, cotton, sugarcane, cassava (tapioca); cattle, goats, pigs, poultry, beef, pork, poultry meat, milk, eggs, hides.

MANUFACTURING SECTOR

GDP & Manufacturing Summary (21)

Detailed value added figures are listed by both International Standard Industry Code (ISIC) and product title.

	1980	1985	1990	1994
GDP ($-1990 mil.)[1]	3,438	3,575	3,910	3,771
Per capita ($-1990)[1]	599	521	480	410
Manufacturing share (%) (current prices)[1]	16.5	27.5	24.4	22.0
Manufacturing				
Value added ($-1990 mil.)[1]	588	763	880	856
Industrial production index	100	130	150	146
Value added ($ mil.)	780	555	917	813
Gross output ($ mil.)	1,671	1,249	1,304	1,101
Employment (000)	59	64[1]	67	61
Profitability (% of gross output)				
Intermediate input (%)	53	56	30	26
Wages and salaries inc. supplements (%)	11	12	9	9

	1980	1985	1990	1994
Gross operating surplus	35	33	62	65
Productivity ($)				
Gross output per worker	28,231	19,518	19,439	17,910
Value added per worker	13,173	8,794	13,676	13,352
Average wage (inc. supplements)	3,245	2,324	1,692	1,660
Value added ($ mil.)				
311/2 Food products	92	71	157	134
313 Beverages	193	138	244	219
314 Tobacco products	58	36	71	64
321 Textiles	51	29	64	53
322 Wearing apparel	34	23	39	35
323 Leather and fur products	4	2	—	—
324 Footwear	15	8	1	1
331 Wood and wood products	8	8	8	7
332 Furniture and fixtures	12	9	18	16
341 Paper and paper products	15	7	6	6
342 Printing and publishing	17	13	24	24
351 Industrial chemicals	22	26	53	51
352 Other chemical products	47	42	45	38
353 Petroleum refineries	9	4	4	3
354 Miscellaneous petroleum and coal products	3	2	4	3
355 Rubber products	20	14	16	14
356 Plastic products	7	5	9	9
361 Pottery, china and earthenware	1	1	2	2
362 Glass and glass products	3	3	6	6
369 Other non-metal mineral products	33	35	36	35
371 Iron and steel	10	6	7	6
372 Non-ferrous metals	2	1	2	2
381 Metal products	50	31	45	40
382 Non-electrical machinery	18	8	8	7
383 Electrical machinery	26	17	38	32
384 Transport equipment	28	18	7	6
385 Professional and scientific equipment	—	—	—	—
390 Other manufacturing industries	2	1	2	2

FINANCE, ECONOMICS, & TRADE

Economic Indicators (22)

National product: GDP—purchasing power parity—$8.8 billion (1997 est.)

National product real growth rate: 3.5% (1997 est.)

National product per capita: $950 (1997 est.)

Inflation rate—consumer price index: 43.9% (1996)

Balance of Payments (23)

	1980	1985	1990	1991
Exports of goods (f.o.b.)	1,457	797	1,254	1,172
Imports of goods (f.o.b.)	−1,114	−571	−1,511	−752
Trade balance	343	226	−257	420
Services - debits	−872	−667	−825	−1,059
Services - credits	168	70	109	93
Private transfers (net)	7	6	395	261
Government transfers (net)	−162	−30	−16	−21
Overall balance	−517	−394	−594	−306

Exchange Rates (24)

Exchange rates:

Zambian kwacha (ZK) per US$1

October 1997	1,351.35
1996	1,203.71
1995	857.23
1994	669.37
1993	452.76

Top Import Origins (25)

$990 million (f.o.b., 1996 est.)

Origins	%
South Africa	NA
European Union	NA
Japan	NA
Saudi Arabia	NA
United States	NA
India	NA
Thailand	NA
Malaysia	NA

NA stands for not available.

Top Export Destinations (26)

$975 million (f.o.b., 1996 est.).

Destinations	%
European Union countries	NA
Japan	NA
South Africa	NA
United States	NA
Saudi Arabia	NA

NA stands for not available.

Economic Aid (27)

Recipient: ODA, $2 billion (1995 est.).

Import Export Commodities (28)

Import Commodities	Export Commodities
Machinery	Copper
Transportation equipment	Zinc
Foodstuffs	Cobalt
Fuels	Lead
Petroleum products	Tobacco
Electricity	
Miscellaneous manufactured goods	

ZIMBABWE

CAPITAL: Harare.

FLAG: The flag has seven equal horizontal stripes of green, yellow, red, black, red, yellow, and green. At the hoist is a white triangle, which contains a representation in yellow of the bird of Zimbabwe superimposed on a red star.

ANTHEM: *God Bless Africa.*

MONETARY UNIT: The Zimbabwe dollar (Z$) is a paper currency of 100 cents. There are coins of 1, 5, 10, 20, and 50 cents and 1 dollar, and notes of 2, 5, 10, and 20 dollars. Z$1 = US$0.10039 (or US$1 = Z$9.961).

WEIGHTS AND MEASURES: The metric system is used.

HOLIDAYS: New Year's Day, 1 January; Independence Day, 18 April; Workers' Day, 1 May; Africa Day, 25 May; Heroes' Days, 11–13 August; Christmas Day, 25 December; Boxing Day, 26 December. Movable holidays are Good Friday, Holy Saturday, Easter Monday, and Whitmonday.

TIME: 2 PM = noon GMT.

LOCATION AND SIZE: A landlocked country of south-central Africa, Zimbabwe (formerly Rhodesia) lies between the Zambezi River on the north and the Limpopo River on the south. It has an area of 390,580 square kilometers (150,804 square miles), slightly larger than the state of Montana. Zimbabwe's total boundary length is 3,066 kilometers (1,905 miles). The capital city, Harare, is located in the northeast part of the country.

CLIMATE: Temperatures on the high veld vary from 12–13°C (54–55°F) in winter to 24°C (75°F) in summer. On the low veld the temperatures are usually 6°C (11°F) higher, and summer temperatures in the Zambezi and Limpopo valleys average between 32° and 38°C (90–100°F).

Rainfall decreases from east to west. The eastern mountains receive more than 100 centimeters (40 inches) annually, while Harare has 81 centimeters (32 inches) and Bulawayo 61 centimeters (24 inches). The south and southwest receive little rainfall.

Republic of Zimbabwe

INTRODUCTORY SURVEY

RECENT HISTORY

Zimbabwe became independent in April 1980, when agreements with Britain brought to an end 15 years of unilaterally declared independence by the former white-minority government of Rhodesia led by Ian Smith. Robert Mugabe became prime minister and formed a coalition government that included his rival, Joshua Nkomo; the new parliament opened on 14 May 1980.

At first Zimbabwe made significant economic and social progress, but internal dissent was a growing problem. A long-simmering rivalry erupted between Mugabe's dominant Zimbabwe African National Union (ZANU)-Patriotic Front Party, which represented the majority Shona tribes, and Nkomo's Zimbabwe African People's Union (ZAPU), which had the support of the minority Ndebele. Mugabe ousted Nkomo from the cabinet in February 1982.

By the time that ZANU and ZAPU, agreed to merge in 1987, the political instability of Zimbabwe's neighbors to the south and east was threatening the country. In 1986 10,000 Zimbabwean troops were deployed in Mozambique to keep antigovernment forces in that country from severing Zimbabwe's rail, road, and oil-pipeline to coastal ports.

In the meantime, dissatisfaction with the economic situation at home led to growing popular opposition to Mugabe. The government responded by restricting human and political rights, placing

checks on the formerly independent judiciary, and tampering with the election process. In March 1996, Mugabe was reelected by a 92.7 majority in the electoral college.

Price increases on basic foods and utilities have led to recent demonstrations by students, union workers, and farmers. Land redistribution remains a volatile issue. Rural poverty, and urban unemployment problems complicate any economic reform efforts.

GOVERNMENT

Under the 1980 constitution, independent Zimbabwe had a two-chamber Parliament. The lower house, the House of Assembly, had 100 members, and the upper house, or Senate, had 40 members. The two houses of Parliament were merged after the 1990 elections into a chamber of 150 members. Another constitutional change created an executive presidency and abolished the office of prime minister. The president is directly elected for a 6-year term and may be reelected. There is universal suffrage beginning at age 18.

Judiciary

A four-member Supreme Court, headed by the chief justice, hears cases involving violations of fundamental rights guaranteed in the constitution and hears appeals of other cases. There is a High Court consisting of general and appeals divisions. Below the High Court are regional courts with civil jurisdiction and courts with both civil and criminal jurisdiction.

Political Parties

The Zimbabwe African National Union-Patriotic Front (ZANU-PF), in power since 1980, grew out of the Zimbabwe African People's Union (ZAPU) led by Joshua Nkomo and the Zimbabwe African National Union (ZANU), led by the Reverend Ndabaningi Sithole, and later by Robert Mugabe.

ZAPU and ZANU merged in late 1987 under the name of ZANU. In January, 1995 long-time Mugabe rival Sithole returned from exile and created his own party, also using the ZANU name as ZANU-Ndonga. Other parties include: Zimbabwe Unity Movement (ZUM); the Democratic Party (DP); the Forum Party of Zimbabwe; and United Parties.

The ZANU-PF has undisputed control over the executive and legislative branches of government. The party controls most of the large newspapers, radio stations, and television stations, and gets its funding from the government.

DEFENSE

Regular armed forces numbered 43,000 in 1995 The national police force has 19,500 members, with 2,300 support personnel. Estimated defense expenditures for 1995 were $233 million, or 4% of the gross domestic product (GDP).

ECONOMIC AFFAIRS

Zimbabwe has abundant agricultural and mineral resources and well-developed industry, but. the unemployment rate was almost 45% in 1994 and thousands remain chronically dependent on food support.

Public Finance

Zimbabwe derives its principal revenues from income taxes, sales tax, customs and excise duties, and interest, dividends, and profits. Principal categories of expenditure are education, defense, debt service, and agriculture. An extensive reform program implemented in 1991, was set back by a severe drought in 1992; the government's deficit rose to more than 10% of GDP in 1993.

The U.S. Central Intelligence Agency estimates that, in 1997, government revenues totaled approximately $2.5 billion and expenditures $2.9 billion, including capital expenditures of $279 million. External debt totaled $5 billion.

Income

In 1999, Zimbabwe's gross national product (GNP) was $6.9 billion or $610 per person. For the period 1985–95 the average inflation rate was 20.9%, resulting in a decline in per person GNP of 0.6%.

Industry

Zimbabwe has a substantial cotton and textile industry and a sizeable metal and engineering sector. Manufacturing accounted for 30% of the domestic economy in 1995. Industries include mining (coal, clay, numerous metallic and nonmetallic ores), copper, steel, nickel, tin, wood products, cement, chemicals, fertilizer, footwear, foodstuffs, and beverages.

Banking and Finance

The Reserve Bank of Zimbabwe (RBZ) administers all monetary and exchange controls and is the sole bank of issue. The Zimbabwe Development Bank was established in 1983 as a development finance institution.

Commercial banks include Barclays Bank International, Standard Bank, Zimbank, Bank of Credit and Commerce Zimbabwe (43% government owned), and Grindlays Bank. The Post Office Savings Bank is an important savings institution.

The Zimbabwe Stock Exchange in Harare deals in government securities and the securities of many privately owned companies.

Economic Development

Since independence, Zimbabwe has received large amounts of foreign aid from the World Bank, the United States, the European Community, the former Soviet Union, and China. In 1999 the World Bank had 10 active projects in Zimbabwe with commitments totaling US $512 million. These funds support projects in the financial, public sector management, agriculture, environment, infrastructure and social sectors.

SOCIAL WELFARE

Government programs are available for child welfare, delinquency, adoption, family problems, refugees, the aged, and public assistance. Voluntary welfare organizations providing facilities for the aged, the handicapped, and care of children receive some government assistance.

Healthcare

About 85% of the population had access to health care services in 1992. For the period 1988 to 1993, there was 1 doctor per 5,994 persons. Average life expectancy is 38 years.

Tuberculosis, malaria, measles and schistosomiasis are commonly reported. In 1994, 17.4% of all adults in Zimbabwe were infected with HIV, the virus that leads to AIDS.

Housing

In rural areas housing is mainly of brick or mud and stick construction with thatch or metal roofs. Urban housing is generally of brick.

EDUCATION

The adult illiteracy rate in 1995 is about 15%. In 1993 there were 2.4 million students in 4,578 primary schools; 639,559 in general secondary schools; and 61,553 students in all higher level institutions.

1999 KEY EVENTS TIMELINE

January

- The Department of Civil Aviation is replaced by the Civil Aviation Authority of Zimbabwe under a commercialization program.
- The flood waters of the Pungue river in the central Mozambican province of Sofala were threatening to cut off the highway between Zimbabwe and the Indian Ocean port of Beira.
- Military police detain Mark Chavunduka, editor of the weekly *Standard* for reporting that soldiers had tried to overthrow President Robert Mugabe's government in December of 1998.

February

- The legal fraternity lends its unequivocal support to the Supreme Court and High Court Judges for petitioning President Mugabe to reaffirm the rule of law in Zimbabwe.
- The South African Human Rights Commission (SAHRC) calls on the South African government to play an active role in ending Zimbabwe's human rights abuses.

March

- Congo-Rebels fighting in the Democratic Republic of the Congo claim shooting down a Zimbabwean MiG jet fighter.
- Zimbabwe, once ranked as one of the best performing economies with potential in sub-Sahara Africa, is fast losing that status.
- Persecuted and harassed, Zimbabwe's independent media vows to continue challenging Robert Mugabe's government on the issue of public accountability.
- Three American men arrested at Harare's International Airport with a large cache of weapons are charged with terrorism and international espionage against two African nations.

April

- Gold producer, Ashanti Goldfields, records a 32 percent increase in 1998 from 1.2 million ounces in 1997 to 1.5 million ounces in 1998.
- Zimbabwe ratifies Common Market for East and Southern Africa (COMESA) Treaty.
- President Mugabe confirms that the government was considering severing ties with the International Monetary Fund (IMF) and this revelation creates another crisis for the already troubled economy.
- Debate rages within the local women's organizations following a Supreme Court Judgement which recently ruled that couples married under the customary law cannot successfully sue for damages when their partners commit adultery.

May

- Tension over land reform between Zimbabwe's commercial farmers and the government is set to ease following recent cabinet approval of a framework plan which the farmers' union has described as "significant."
- Iranian and Zimbabwean presidents hold first round of talks in Tehran.
- Many of Zimbabwe's middle and low income earners struggle to afford basic commodities such as bread, sugar, beef, chicken, pork and even the staple food, mealie meal. Government, food manufacturers, and the business community fail to curb recent devastating price rises.
- A serious fuel shortage looms in Zimbabwe because the state-owned oil company has run out of credit to buy imports; supplies may have to be rationed.

June

- The widely publicized Magaya vs Magaya ruling of the Supreme Court of Zimbabwe that denied women the right to inherit under customary law fuels a review of the interface between customary and general law as regards women.
- Despite the cabinet's approval to award the outstanding 5 percent backdated to January, civil

servants go on strike demanding a 20-percent cost-of-living adjustment.

- For the first time in 20 years, Zimbabwe fails to service its external debt due to a shortage of foreign currency.
- Over four thousand people gather in Chitungwiza for the National Constitutional Assembly (NCA) and call for an all inclusive and participatory constitution making process.
- Zimbabwe's major milling firms close down amid a stand-off with government over its pricing policy on the staple maize meal, prompting fears of a new food crisis.

July

- Zimbabweans are plunged into mourning following the announcement of the death of Vice-President Joshua Nkomo, one of Zimbabwe's founding fathers.
- Maize meal, Zimbabwe's national staple, is readily available again after three weeks of severe shortages that ended after the government was forced to back down to demands by millers for a 42 percent price increase.
- The International Monetary Fund (IMF) agrees to lend 200 million U.S. dollars to Zimbabwe over a 14-month period, after clarification of the country's war spending.
- Mozambique expresses its dissatisfaction and indignation over the forced repatriation of Mozambicans from Zimbabwe.
- The Harare High Court trial of three Americans accused of gun-running comes to an abrupt halt after President Robert Mugabe issues a proclamation overriding a series of court orders to improve the conditions under which the three are being held in prison.

August

- President Robert Mugabe says Zimbabwean Troops in the Democratic Republic of Congo (DRC) are there to defend the government of President Laurent Kabila against rebels backed by invasion forces from Rwanda and Uganda.
- The Harare city council plans a campaign to convince people to break with cultural traditions and accept cremation in an attempt to save space as authorities face soaring AIDS deaths and overflowing cemeteries.

September

- Citing evidence that the three American defendants were tortured, a judge gives them unusually light sentences of six months each for possessing weapons of war and trying to smuggle guns aboard a plane.
- Zimbabwe's national labour union leaders end two years of shadowy political maneuvering by launching an opposition party to challenge President Robert Mugabe's 19 year rule.
- Zimbabwean president Robert Mugabe is to visit Cuba after he ends his visit to Brazil.
- A week long strike for better pay and working conditions by junior doctors shakes Zimbabwe's already troubled health delivery system.
- The Netherlands suspends its $15 million (US) dollar annual development aid to Zimbabwe, accusing the government of President Robert Mugabe of extravagance, lack of accountability, and abuse of human rights.

October

- A Zimbabwean human rights organization, the Catholic Commission for Justice and Peace, begins exhuming remains of victims of atrocities committed by government troops in the 1980s.
- The African Development Bank grants Zimbabwe a loan of $130 million (US).
- Zimbabwe denies it misled the IMF about its military expenditure in the Democratic Republic of Congo to win a crucial loan from the fund two months ago.
- Zimbabwean journalists Mark Chavunduka and Ray Choto are this year's winners of the world renowned International Press Freedom award.

November

- The Zimbabwe dollar plunges, due to a combination of economic factors, including the close of tobacco auctions in September, the worldwide drop in prices resulting in a loss of income from mining, and government spending on activities such as keeping approximately 11,000 troops in Congo.
- Contrary to IMF recommendations to free the exchange rate, the government of President Mugabe is determined to keep the exchange rate at its current level until after the parliamentary elections in Apri, 2000.

ANALYSIS OF EVENTS: 1999

BUSINESS AND THE ECONOMY

Zimbabwe, once ranked as one of the best performing economies with great potential in sub-Sahara saw its fortunes and status wane during the 1999 period. The economy was set back by labor unrest, a severe food crisis and the AIDS epidemic with an estimated HIV infection rate of one in four people and a massive increase in deaths of the economically active age-group.

The government blamed the economic crisis sweeping the country on low commodity prices on the world market and the devastating effects of the structural adjustment programs. It also blamed the political problems on domestic and Western opponents attempting to oust Mugabe's regime. During 1999, many of Zimbabwe's middle and low income earners struggled to afford basic commodities, such as bread, sugar, and the staple food, mealie meal. The government, the food manufacturers, and the business community failed to curb price rises. Inflation rose at an annual rate of forty-six percent, causing a fifty percent devaluation of the Zimbabwe dollar.

The country also faced a serious fuel shortage as the state-owned oil company ran out of credit to buy imported oil. The donor community, particularly the International Monetary Fund (IMF), was not amused by President Mugabe's continued commitment to the costly war in the Democratic Republic of the Congo. Mugabe deployed thousands of troops in support of President Laurent Kabila. The IMF suspended its loan program to Zimbabwe pending Mugabe's explanation of the source of funds used to support his involvement in the Congo. The Netherlands also suspended its annual development aid to Zimbabwe, accusing the government of extravagance, lack of accountability and abuse of human rights.

These difficulties not withstanding, Mugabe was positive that the economy would grow by 1.5 percent in 1999 from 1.6 percent the previous year and against a target of at least 5 percent. To increase exports within the region, Mugabe's government quickly ratified the Common Market for East and Southern Africa (COMESA) Treaty. The African Development Bank eased the pressure on Zimbabwe by granting it a loan of 130 million U.S. dollars. Gold producer, Ashanti Goldfields, also announced a 32 percent increase in gold production from 1.2 million ounces in 1997 to 1.5 million ounces in 1998. Later in the year the IMF ended its squabble with Mugabe and reinstated the loan program.

GOVERNMENT AND POLITICS

The political arena was dominated by issues concerning human rights, the harassment of the media, wide-spread corruption in government circles, and the plunging of the country into the costly Congo war. The opposition has alleged that the main motive for Mugabe's involvement in the Congo war is to make economic gains (using state machinery) from investment and trade deals made with the natural resource rich Congo before and during the war. Another headache for the government of Zimbabwe was the wide-spread media coverage of the arrest of a trio of journalists over an article that said a restive military had planned to topple Mugabe. Three American men were also arrested at Harare's International Airport with a large cache of weapons and were charged with terrorism and international espionage in Zimbabwe and the Congo. While in custody the journalists and the American men were tortured raising concerns over human rights and the rule of law in Zimbabwe. The deepening social crisis rooted in government repression, corruption, mismanagement and stifling of reforms has called forth a massive outcry for a new constitution from opposition groups. The public raised such an outcry over these scandals that Mugabe had to appoint a commission to draft a new constitution. However the opposition, a coalition of various social groups called the National Constitutional Assembly (NCA), has launched its own parallel commission to draft an alternative constitution, a move which threatens a clash with Mugabe. These were big problems for a nation with an already sickly economy.

CULTURE AND SOCIETY

Gender issues, in particular the status of women in Zimbabwean culture, took center stage during 1999. The widely publicized Magaya vs Magaya ruling of the Supreme Court of Zimbabwe that denied women the right to inherit under customary law fueled a review of the interface between customary and general law as regards women. In a landmark five-to-zero decision, the court said the "nature of African society" dictates that women are not equal to men, especially in family relationships. The court said centuries-old African cultural norms, which are not written down, say women should never be considered adults within the family, but only as a ''junior male,'' or teenager. Angry women took to the streets to protest the court decision and to deliver a toughly-worded petition.

Other important happenings during the year included the death of Vice-President Joshua Nkomo, one of Zimbabwe's founding fathers. Tension over land reform between Zimbabwe's commercial farmers and the government was lessened following the approval by the government of a framework plan which the farmers' union have described as ''significant.'' The out of hand AIDS epidemic prompted the Harare city council to launch a campaign to convince people to break with cultural traditions and accept cremation in an attempt to save space as mortality due to AIDS soared and cemeteries filled up quickly. The media scored a hit with Zimbabwean journalists Mark Chavunduka and Ray Choto being declared as the 1999 joint winners of the world renowned International Press Freedom award.

DIRECTORY

CENTRAL GOVERNMENT

Head of State

President
Robert Mugabe, Office of the President, Munhumutapa Buildinga, Samora Machel Avenue, Harare, Zimbabwe
PHONE: +263 (4) 726666

Vice President
Simon Vengai Muzenda, Office of the Vice President

Vice President
Joshua Nkomo, Office of the Vice President

Ministers

Minister of Defense
Moven Mahachi, Ministry of Defense, Munhumutapa Building, Samora Machel Avenue, Private Bag 7713, Causeway, Harare, Zimbabwe
PHONE: +263 (4) 700155; 728271

Minister of Education, Sports, and Culture
Gabriel Machinga, Ministry of Education, Sports, and Culture, Ambassador House, Union Avenue, PO Box CY 121, Causeway, Harare, Zimbabwe
PHONE: +263 (4) 734051; 734067

Minister of Finance
Herbert Murerwa, Ministry of Finance, Munhumutapa Building, Samora Machel, Private Bag 7705, Causeway, Harare, Zimbabwe
PHONE: +263 (4) 794571

Minister of Foreign Affairs
Stanislaus Mudenge, Ministry of Foreign Affairs, Munhumutapa Building, Samora Machel, Zimbabwe
PHONE: +263 (4) 794681; 727005; 704704

Minister of Health and Child Welfare
Timothy Stamps, Ministry of Health and Child Welfare, Kaguvi Building, PO Box CY 1122, 4th Street/Central Avenue, Causeway, Harare, Zimbabwe
PHONE: +263 (4) 730011

Minister of Higher Education and Technology
Ignatius Chombo, Ministry of Higher Education and Technology, PO Box UA 275, Union Avenue, Harare, Zimbabwe
PHONE: +263 (4) 7022361

Minister of Home Affairs
Dumiso Dabengwa, Ministry of Home Affairs, 11th Floor, Mukwati Building, 4th Street/Selous, Private Bag 505D, Harare, Zimbabwe
PHONE: +263 (4) 723653

Minister of Industry and Commerce
Nathan Shamuyarira, Ministry of Industry and Commerce, 13th Floor, Mukwati Building, 4th Street/Livingstone Avenue, Private Bag 7708, Causeway, Harare, Zimbabwe
PHONE: +263 (4) 702731; 729801

Minister of Information, Posts, and Telecommunications
Chen Chimutengwende, Ministry of Information, Posts, and Telecommunications, Linquenda House, PO Box CY 1276, N. Mandela Ave., Causeway, Harare, Zimbabwe
PHONE: +263 (4) 706891

Minister of Justice, Legal, and Parliamentary Affairs
Emmerson Mnangagwa, Ministry of Justice, Legal, and Parliamentary Affairs, 4th Floor, Corner House, Samora Machel Avenue, Private Bag 7704, Causeway, Zimbabwe
PHONE: +263 (4) 790902

Minister of Lands and Agriculture
Kumbirai Kangai, Ministry of Lands and Agriculture, 1 Borrowdale Road, Private Bag 7701, Causeway, Harare, Zimbabwe
PHONE: +263 (4) 7006081

Minister of Local Government, Rural and Urban Development
John Nkomo, Ministry of Local Government, Rural and Urban Development, Mukwati Building, Private Bag 7706, Causeway, Harare, Zimbabwe
PHONE: +263 (4) 790601

Minister of Mines and Environment
Simon Moyo, Ministry of Mines and Environment, Zimre Centre, L. Takawira Avenue, Private Bag 7709, Causeway Harare, Zimbabwe
PHONE: +263 (4) 732881

Minister of National Affairs, Employment Creation, and Cooperatives
Tenjiwe Lesaba, Ministry of National Affairs, Employment Creation, and Cooperatives, New Zanu, PF HQ, Rotten Row, Private Bag 7762, Causeway, Harare, Zimbabwe
PHONE: +263 (4) 792353

Minister of Public Service, Labor, and Social Welfare
Florence Chitauro, Ministry of Public Service, Labor, and Social Welfare, 12th Floor, Compensation House, Central Avenue/4th Street, Private Bag 7707, Causeway, Harare, Zimbabwe
PHONE: +263 (4) 790871

Minister of Rural Resources and Water Development
Joyce Mujuru, Ministry of Rural Resources and Water Development

Minister of Transport and Energy
Enos Chikowore, Ministry of Transport and Energy, 4th Street/Central Avenue, PO Box CY595, Causeway, Harare, Zimbabwe
PHONE: +263 (4) 700991; 700693; 707121
FAX: +263 (4) 708225

Minister of State for National Security
Sydney Sekeramayi, Ministry of State for National Security

Minister of State
Tsungirai Hungwe, Ministry of State

Minister of State
Sithembiso Nyoni, Ministry of State

Minister of State
Oppah Rushesha, Ministry of State

POLITICAL ORGANIZATIONS

Zimbabwe African National Union-Patriotic Front (ZANU-PF)

NAME: Robert Mugabe

Zimbabwe African National Union-NDONGA (ZANU-NDONGA)

NAME: Ndabaningi Sithole

Zimbabwe Unity Movement (ZUM)

NAME: Edgar Tekere

Democratic Party (DP)

NAME: Emmanuel Magoche

Forum Party of Zimbabwe

NAME: Enock Dumbutshena

DIPLOMATIC REPRESENTATION

Embassies in Zimbabwe

Australia
4th Floor, Karigamombe Centre, 53 Samora Machel Avenue, PO Box 4541, Harare, Zimbabwe
PHONE: +263 (4) 757774
FAX: +263 (4) 757770

Austria
30 Samora Machel Avenue, Room 216, New Shell House, PO Box 4120, Harare, Zimbabwe
PHONE: +263 (4) 702921
FAX: +263 (4) 705877

Brazil
Old Mutual Centre, cor. 3rd Street/Jason Moyo Avenue, PO Box 2530, Harare, Zimbabwe
PHONE: +263 (4) 702921
FAX: +263 (4) 737782

Canada
45 Baines Avenue, PO Box 1430, Harare, Zimbabwe
PHONE: +263 (4) 733881
FAX: +263 (4) 732917

China
Oakwood Building, 30 Baines Avenue, PO Box 4749, Harare, Zimbabwe
PHONE: +263 (4) 724572
FAX: +263 (4) 794959

France
Ranelage Road, off Orange Grove Drive, Highlands, PO Box 1378, Harare, Zimbabwe
PHONE: +263 (4) 498096
FAX: +263 (4) 495657

Germany
14 Samora Machel Avenue, PO Box 2168, Harare, Zimbabwe
PHONE: +263 (4) 731955
FAX: +263 (4) 790680

India
12 Natal Road, Belgravia, PO Box 4620, Harare, Zimbabwe
PHONE: +263 (4) 795955
FAX: +263 (4) 722324

Italy
7 Bartholomew Close, Greendale, PO Box 1062, Harare, Zimbabwe
PHONE: +263 (4) 497373
FAX: +263 (4) 498199

Japan
18th Floor, Karigamombe Centre, 53 Samora Machel Avenue, PO Box 2710, Harare, Zimbabwe
PHONE: +263 (4) 757861
FAX: +263 (4) 757864

Kenya
95 Park Lane, PO Box 4069, Harare, Zimbabwe
PHONE: +263 (4) 790847
FAX: +263 (4) 723042

Nigeria
36 Samora Machel Avenue, PO Box 4742, Harare, Zimbabwe
PHONE: +263 (4) 790765
FAX: +263 (4) 725004

Russia
70 Fife Avenue, PO Box 4250, Harare, Zimbabwe
PHONE: +263 (4) 720358
FAX: +263 (4) 700534

South Africa
Temple Bar House, Baker Avenue, PO Box 121, Harare, Zimbabwe
PHONE: +263 (4) 753147
FAX: +263 (4) 753185

Spain
16 Phillips Avenue, Belgravia, PO Box 3300, Harare, Zimbabwe
PHONE: +263 (4) 738681
FAX: +263 (4) 795440

United Kingdom
7th Floor, Corner House, L. Takawira Street/S. Machel Avenue, PO Box 4490, Harare, Zimbabwe
PHONE: +263 (4) 772990
FAX: +263 (4) 774617

United States
Arax House, 172 Herbert Chitepo Avenue, PO Box 3340, Harare, Zimbabwe
PHONE: +263 (4) 794521
FAX: +263 (4) 796488

JUDICIAL SYSTEM

Supreme Court

FURTHER READING

Articles

''Law, What Law?'' *The Economist.* January 30, 1999, 43.

Ndlovu, Khumbulani. ''No End to Spiral of Price Rises.'' Africa Information Afrique. May 26, 1999.

''New Party to Stimulate Zimbabwean Politics.'' PanAfrican News Agency. September 13, 1999.

Shoko, Rangarirai. ''Zimbabwe Secures 130 Million Dollars From ADB.'' PanAfrican News Agency. October 7, 1999.

''Zimbabwe Faces Critical Shortage of Maize Meal as Millers Shut.'' Agence France-Presse. June 21, 1999.

''Zimbabwe: Mugabe Reportedly Overrides Judges in Trial of Three Americans.'' South African News Agency—SAPA, Johannesburg, BBC Monitoring International Reports. July 22, 1999.

''Zimbabwe Rights Organisation Starts Exhuming Massacre Victims.'' Agence France-Presse. October 1, 1999.

''Zimbabweans Mourn Vice President Joshua Nkomo.'' *PanAfrican News Agency.* July 1, 1999.

Books

Munro, William Andrew. *The Moral Economy of the State: Conservation, Community Development, and State-making in Zimbabwe.* Athens: Ohio University Center for International Studies, 1998.

Skelnes, Tor. *The Politics of Economic Reform in Zimbabwe: Continuity and Change in*

Development. New York: St. Martin's Press, 1995.

Summers, Carol. *From Civilization to Segregation.* Athens: Ohio University Press, 1994.

Internet

Africa Online. Available Online@ http://www.africaonline.com/AfricaOnline/travel/zimbabwe/zimbabwe.html. (October 15, 1999).

African Journals. Available Online @ http://www.oneworld.org/inasp/ajol/index.html. (October 15, 1999).

The Herald (Harare). Available Online @http://www.zimsurf.co.zw/theherald/muzimbabwe/nhasi/. (October 15, 1999).

ZIMBABWE: STATISTICAL DATA

For sources and notes see "Sources of Statistics" in the front of each volume.

GEOGRAPHY

Geography (1)

Area:

Total: 390,580 sq km.

Land: 386,670 sq km.

Water: 3,910 sq km.

Area—comparative: slightly larger than Montana.

Land boundaries:

Total: 3,066 km.

Border countries: Botswana 813 km, Mozambique 1,231 km, South Africa 225 km, Zambia 797 km.

Coastline: 0 km (landlocked).

Climate: tropical; moderated by altitude; rainy season (November to March).

Terrain: mostly high plateau with higher central plateau (high veld); mountains in east.

Natural resources: coal, chromium ore, asbestos, gold, nickel, copper, iron ore, vanadium, lithium, tin, platinum group metals.

Land use:

Arable land: 7%

Permanent crops: 0%

Permanent pastures: 13%

Forests and woodland: 23%

Other: 57% (1993 est.).

HUMAN FACTORS

Demographics (2B)

Population (July 1998 est.) 11,044,147

Age structure:

0-14 years (male 2,439,907; female 2,397,761) 44%

15-64 years (male 2,914,336; female 3,000,442) 54%

65 years and over, July 1998 est. (male 133,232; female 158,469) 2%

Population growth rate (1998 est.) 1.12%

Birth rate, 1998 est. (births/1,000 population) 31.32

Death rate, 1998 est. (deaths/1,000 population) 20.09

Net migration rate (migrant(s)/1,000 population) . . .NA

Sex ratio (male(s)/female):

At birth 1.03

Under 15 years 1.01

15-64 years 0.97

65 years and over (1998 est.) 0.84

Infant mortality rate, 1998 est. (deaths/1,000 live births) 61.75

Life expectancy at birth (years):

Total population 39.16

Male 39.12

Female (1998 est.) 39.19

Total fertility rate, 1998 est. (children born/woman) .3.86

There is a small but steady flow of Zimbabweans into South Africa in search of better paid employment.

Health Personnel (3)

Total health expenditure as a percentage of GDP, 1990-1997[a]

Public sector 1.7

Private sector 3.1

Total[b] 4.7

Health expenditure per capita in U.S. dollars, 1990-1997[a]

Purchasing power parity 133

Total 41

Availability of health care facilities per 100,000 people

Hospital beds 1990-1997[a] 50

Doctors 1993[c] 14

Nurses 1993[c] 164

Health Indicators (4)

Life expectancy at birth

1980 55

1997 52

Daily per capita supply of calories (1996) 2,083

Total fertility rate births per woman (1997) 3.8

Maternal mortality ratio per 100,000 live births (1990-97) . . . . .280[d]

Safe water % of population with access (1995) 77

Sanitation % of population with access (1995) 66

Consumption of iodized salt % of households (1992-98)[a] 80

Smoking prevalence

Male % of adults (1985-95)[a] 36

Female % of adults (1985-95)[a] 15

Tuberculosis incidence per 100,000 people (1997) 543

Adult HIV prevalence % of population ages 15-49 (1997) 25.84

Infants and Malnutrition (5)

Under-5 mortality rate (1997) 80

% of infants with low birthweight (1990-97) 14

Births attended by skilled health staff % of total[a] . . .69

% fully immunized (1995-97)

TB 82

DPT 78

Polio 79

Measles 73

Prevalence of child malnutrition under age 5 (1992-97)[b] 16

Ethnic Division (6)

African 98%

Shona 71%

Ndebele 16%

Other 11%

White 1%

Mixed and Asian 1%

Religions (7)

Syncretic (part Christian, part indigenous beliefs) 50%, Christian 25%, Indigenous beliefs 24%; Muslim and other 1%.

Languages (8)

English (official), Shona, Sindebele (the language of the Ndebele, sometimes called Ndebele), numerous but minor tribal dialects.

EDUCATION

Public Education Expenditures (9)

Public expenditure on education (% of GNP)

1980 . . . . .6.6

1996 . . . . .8.3

Expenditure per student

Primary % of GNP per capita

1980 . . . . .24.3

1996 . . . . .202[1]

Secondary % of GNP per capita

1980 . . . .

1996 . . . . .38.2[1]

Tertiary % of GNP per capita

1980 . . . . .407.1

1996 . . . . .406.5[1]

Expenditure on teaching materials

Primary % of total for level (1996) . . . . .9.8[1]

Secondary % of total for level (1996) . . . .

Primary pupil-teacher ratio per teacher (1996) 39

Duration of primary education years (1995) 7

Educational Attainment (10)

Age group (1992) . . . . .25+

Total population 3,445,195

Highest level attained (%)

No schooling 22.3

First level

Not completed 53.2

Completed . . . . .NA

Entered second level

S-1 . . . . .19.4

S-2 . . . . .NA

Postsecondary 4.9

Literacy Rates (11A)

In thousands and percent[1]	1990	1995	2000	2010
Illiterate population (15+ yrs.)	972	940	881	780
Literacy rate - total adult pop. (%)	82.3	85.1	87.6	91.7
Literacy rate - males (%)	88.3	90.4	92.3	95.2
Literacy rate - females (%)	76.5	79.9	83.1	88.3

GOVERNMENT & LAW

Political Parties (12)

House of Assembly	No. of seats
Zimbabwe African National Union-Patriotic Front (ZANU-PF)	117
Zimbabwe African National Union-NDONGA (ZANU-NDONGA)	2
Independent	1

Government Budget (13B)

Revenues . . . .$2.5 billion

Expenditures . . . .$2.9 billion

Capital expenditures . . . .$279 million

Data for FY96/97 est.

Crime (15)

Crime rate (for 1997)

Crimes reported 134,000

Total persons convicted 906,200

Crimes per 100,000 population 1,050

Persons responsible for offenses

Total number of suspects 98,600

Total number of female suspects . . . . .NA

Total number of juvenile suspects . . . . .NA

Military Affairs (14B)

	1990	1991	1992	1993	1994	1995
Military expenditures						
Current dollars (mil.)	261	292[e]	277[e]	235[e]	220	231
1995 constant dollars (mil.)	300	323[e]	298[e]	246[e]	226	231
Armed forces (000)	45	45	48	48	43	40
Gross national product (GNP)						
Current dollars (mil.)	4,923	5,220	5,047	5,470	5,709	5,754
1995 constant dollars (mil.)	5,658	5,768	5,429	5,734	5,852	5,754
Central government expenditures (CGE)						
1995 constant dollars (mil.)	2,305	2,311	2,511[e]	2,340[e]	NA	NA
People (mil.)	10.1	10.4	10.6	10.9	11.1	11.2
Military expenditure as % of GNP	5.3	5.6	5.5	4.3	3.9	4.0
Military expenditure as % of CGE	13.0	14.0	11.9	10.5	NA	NA
Military expenditure per capita (1995 $)	30	31	28	23	20	21
Armed forces per 1,000 people (soldiers)	4.4	4.3	4.5	4.4	3.9	3.6
GNP per capita (1995 $)	559	556	511	526	528	516
Arms imports[6]						
Current dollars (mil.)	60	50	90	10	10	0
1995 constant dollars (mil.)	69	55	97	10	10	0
Arms exports[6]						
Current dollars (mil.)	0	0	5	0	10	0
1995 constant dollars (mil.)	0	0	5	0	10	0
Total imports[7]						
Current dollars (mil.)	1,847	2,055	2,203	1,820	2,241	2,315
1995 constant dollars (mil.)	2,123	2,271	2,370	1,908	2,297	2,315
Total exports[7]						
Current dollars (mil.)	1,726	1,532	1,445	1,568	1,885	2,119
1995 constant dollars (mil.)	1,984	1,693	1,554	1,644	1,932	2,119
Arms as percent of total imports[8]	3.2	2.4	4.1	.5	.4	0
Arms as percent of total exports[8]	0	0	.3	0	.5	0

LABOR FORCE

Labor Force (16)

Total (million) 4.228
Agriculture 27%
Transport and services 46%
Industry 27%

Data for 1993 est.

Unemployment Rate (17)

At least 45% (1994 est.).

Electric Energy (18)

Capacity . . . . .2.148 million kW (1995)
Production . . . . .7.1 billion kWh (1995)
Consumption per capita . . . . .792 kWh (1995)

Transportation (19)

Highways:
total: 18,338 km
paved: 8,692 km
unpaved: 9,646 km (1996 est.)
Waterways: the Mazoe and Zambezi rivers are used for transporting chrome ore from Harare to Mozambique
Pipelines: petroleum products 212 km
Airports: 468 (1997 est.)
Airports—with paved runways:
total: 20
over 3,047 m: 3
2,438 to 3,047 m: 2
1,524 to 2,437 m: 5
914 to 1,523 m: 10 (1997 est.)
Airports—with unpaved runways:
total: 448
1,524 to 2,437 m: 3
914 to 1,523 m: 221
under 914 m: 224 (1997 est.)

Top Agricultural Products (20)

Corn, cotton, tobacco, wheat, coffee, sugarcane, peanuts; cattle, sheep, goats, pigs.

MANUFACTURING SECTOR

GDP & Manufacturing Summary (21)

Detailed value added figures are listed by both International Standard Industry Code (ISIC) and product title.

	1980	1985	1990	1994
GDP: ($-1990 mil.)[1]	4,596	5,656	6,811	7,001
Per capita ($-1990)[1]	645	674	688	636
Manufacturing share (%) (current prices)[1]	24.1	22.2	25.8	30.4
Manufacturing				
Value added ($-1990 mil.)[1]	1,006	1,131	1,404	1,313
Industrial production index	100	112	140	131
Value added ($ mil.)	1,480	1,278	2,232	1,867
Gross output ($ mil.)	3,579	3,020	4,749	3,855
Employment (000)	161	163	186	224
Profitability (% of gross output)				
Intermediate input (%)	59	58	53	52
Wages and salaries inc. supplements (%)	17	18	15	13
Gross operating surplus	24	25	32	35
Productivity ($)				
Gross output per worker	22,265	18,452	25,557	17,154
Value added per worker	9,205	7,816	12,014	8,363
Average wage (inc. supplements)	3,848	3,241	3,939	2,239
Value added ($ mil.)				
311/2 Food products	193	130	241	306
313 Beverages	92	189	302	215
314 Tobacco products	55	72	76	88
321 Textiles	147	114	255	157
322 Wearing apparel	70	55	102	52
323 Leather and fur products	4	4	7	10
324 Footwear	34	42	66	43
331 Wood and wood products	38	17	43	34
332 Furniture and fixtures	26	15	32	18
341 Paper and paper products	30	37	64	52
342 Printing and publishing	59	45	94	49
351 Industrial chemicals	58	67	98	68
352 Other chemical products	80	78	127	85
353 Petroleum refineries	—	1	1	—
354 Miscellaneous petroleum and coal products	7	8	16	13

	1980	1985	1990	1994
355 Rubber products	30	24	37	34
356 Plastic products	25	37	47	28
361 Pottery, china and earthenware	3	2	3	3
362 Glass and glass products	9	5	9	7
369 Other non-metal mineral products	44	28	54	55
371 Iron and steel	194	105	184	291
372 Non-ferrous metals	10	9	13	14
381 Metal products	132	78	135	93
382 Non-electrical machinery	39	22	43	20
383 Electrical machinery	44	36	88	61
384 Transport equipment	38	48	81	63
385 Professional and scientific equipment	2	1	2	2
390 Other manufacturing industries	17	9	13	8

FINANCE, ECONOMICS, & TRADE

Economic Indicators (22)

National product: GDP—purchasing power parity—$24.9 billion (1996 est.)

National product real growth rate: 8.1% (1996 est.)

National product per capita: $2,200 (1996 est.)

Inflation rate—consumer price index: 21.4% (1996)

Exchange Rates (24)

Exchange rates:

Zimbabwean dollars (Z$) per US$1

January 1998	18.7970
1997	11.8906
1996	9.9206
1995	8.6580
1994	8.1500
1993	6.4725

Top Import Origins (25)

$2.2 billion (f.o.b., 1996 est.) Data are for 1996 est.

Origins	%
South Africa	38
United Kingdom	9
United States	5
Japan	5

Top Export Destinations (26)

$2.5 billion (f.o.b., 1996 est.) Data are for 1996 est.

Destinations	%
South Africa	12
United Kingdom	12
Germany	6
Japan	6

Economic Aid (27)

Recipient: ODA, $362 million (1993).

Balance of Payments (23)

	1990	1991	1992	1993	1994
Exports of goods (f.o.b.)	1,748	1,694	1,528	1,609	1,961
Imports of goods (f.o.b.)	−1,505	−1,646	−1,782	−1,487	−1,803
Trade balance	243	48	−255	122	158
Services - debits	−782	−905	−963	−851	−1,033
Services - credits	287	300	331	407	411
Private transfers (net)	108	95	242	179	—
Government transfers (net)	4	6	41	27	40
Overall balance	−140	−457	−604	−116	−425

Import Export Commodities (28)

Import Commodities	Export Commodities
Machinery and transportation equipment 41%	Agricultural 38% (tobacco 28%)
Other manufactures 24%	Manufactures 34%
Chemicals 13%	Gold 12%
Fuels 10%	Textiles 4%
	Ferrochrome 7%

INTERNATIONAL ORGANIZATIONS

ACADEMIA EUROPAEA (AE)

31 Old Burlington St., London W1X 1LB, England **PHONE:** 44 171 7345402 **FX:** 44 171 2875115 **E-MAIL:** acadeuro@compuserve.com **FOUNDED:** 1988. **OFFICERS:** Peter J. Colyer, Exec.Sec. **MEMBERS:** 1,800. **STAFF:** 4. **BUDGET:** US$400,000. **LANGUAGES:** English.

DESCRIPTION: Members are academics and other individuals with an interest in European thought. Promotes increased understanding and appreciation of European culture and scholarship. Facilitates communication and cooperation among members; sponsors research and educational programs. **PUBLICATIONS:** *European Review* (in English), quarterly.

ACADEMY FOR EDUCATIONAL DEVELOPMENT (AED)

1825 Connecticut Ave. NW, Washington, DC 20009-5721 **PHONE:** (202) 884-8000 **FX:** (202) 884-8400 **E-MAIL:** admindc@aed.org **WEBSITES:** http://www.aed.org **FOUNDED:** 1961. **OFFICERS:** Stephen F. Moseley, Pres. & CEO. **STAFF:** 600. **BUDGET:** US$124,000,000. **LANGUAGES:** English.

DESCRIPTION: Seeks to address human resource and economic development needs through education, communication, and the dissemination of information. Works with educational institutions, community organizations, foundations, corporations, policy leaders, non-governmental and community-based organizations, international multilateral and bilateral fundraisers, and governmental and international agencies to increase access to learning, transfer skills and technology, and support the development of educational institutions. Projects and activities are organized in three program areas: Education and Exchange Services, including basic education and training programs, the Center for Youth Development and Policy Research, the National Institute for Work and Learning, and international exchange programs; Social Development, including health, nutrition, family planning, environment, telecommunications, development information services; School and Community Services; and Human Resources and Institutional Development, including higher education management services, management development services, computer and systems services, and vocational and technical training programs. **PUBLICATIONS:** *Academy News*, semiannual. Includes program updates.; *Investing in Change*; *Linking Progress to People*; *New Directions for Parent Involvement*; *Results and Realities: A Decade of Experience in Communication for Child Survival*; *School Reform and Youth Transition*; *Testing to Learn, Learning to Test: A Policymaker's Guide to Better Educational Testing*.

ACCION INTERNATIONAL (ACCION)

120 Beacon St., Somerville, MA 02143 **PHONE:** (617) 492-4930 **FX:** (617) 876-9509 **E-MAIL:** info@accion.org **WEBSITES:** http://www.accion.org **FOUNDED:** 1961. **OFFICERS:** Robin Ratciffe, VP & Dept. Mgr. **STAFF:** 70. **BUDGET:** US$6,300,000. **LANGUAGES:** English, Portuguese, Spanish.

DESCRIPTION: Provides financial services to the self-employed poor in 15 Latin American countries and 9 U.S. cities. Works to reduce poverty and unemployment by providing loans and business support to microentrepreneurs in low-income communities through a network of affiliated institutions. **PUBLICATIONS:** *ACCION International Publications*, annual; *Ventures*, triennial; Annual Report; Books; Brochures; Catalog; Monographs; Papers.

ACP-EC AGRICULTURAL COMMODITIES COMMITTEE

c/o ACP, Avenue Georges Henri 451, B-1200 Brussels, Belgium **PHONE:** 32 2 7430600 **FX:** 32 2 7355573 **TX:** 26558 ACP-B **OFFICERS:** Ng'andu Peter Magande. **LANGUAGES:** English, French.

DESCRIPTION: A joint project of the African, Caribbean and Pacific Group of States and the European Community. Government agencies involved in the regulation of the production and trade of commodities. Promotes increased international trade in agricultural commodities. Serves as a forum for the discussion of international trade issues arising from the transfer of commodities.

ACP-EC COMMITTEE ON INDUSTRIAL COOPERATION

c/o ACP, Avenue Georges Henri 451, B-1200 Brussels, Belgium **PHONE:** 32 2 7430600 **FX:** 32 2 7355573 **TX:** 26558 ACP-B **OFFICERS:** Ng'andu Peter Magande. **LANGUAGES:** English, French.

DESCRIPTION: A joint project of the African, Caribbean and Pacific Group of States and the European Community. Members are government agencies involved in the regulation of industries. Promotes increased international trade in industrial products. Serves as a forum for the discussion of international trade issues arising from the transfer of manufactured goods.

ACP-EC COUNCIL OF MINISTERS

c/o Council of the European Union, Batiment Justusu Lipsius, Rue de la Loi 175, B-1048 Brussels, Belgium **PHONE:** 32 2 2856111 **FX:** 32 2 2858476 **TX:** 21711 Consil B **FOUNDED:** 1975. **OFFICERS:** Flemming Bjornekaer. **LANGUAGES:** English, French, Portuguese.

DESCRIPTION: A joint project of the African, Caribbean and Pacific Group of States and the European Community. Elected government officials. Promotes communication among members and economic and political cooperation among members' countries. Serves as a forum for the discussion of international economic and social issues. **PUBLICATIONS:** Annual Report (in French).

ACP-EC CUSTOMS COOPERATION COMMITTEE

c/o Council of the European Union, Rue de la Loi 175, B-1048 Brussels, Belgium **PHONE:** 32 2 2856111 **TX:** 21711 Consil B **LANGUAGES:** English, French.

DESCRIPTION: A joint project of the African, Caribbean and Pacific Group of States and the European Community. Members are customs officials and agencies. Promotes coordination of international customs protocols to simplify and expedite international trade. Serves as a forum for the discussion of international trade and customs issues.

ACP-EC DEVELOPMENT FINANCE COOPERATION COMMITTEE

c/o ACP, Avenue Georges Henri 451, B-1200 Brussels, Belgium **PHONE:** 32 2 7430600 **FX:** 32 2 7355573 **TX:** 26558 ACP-B **OFFICERS:** Ng'andu Peter Magande. **LANGUAGES:** English, French.

DESCRIPTION: A joint project of the African, Caribbean and Pacific Group of States and the European Community. Membership consists of international development authorities and government agencies responsible for the regulation development. Promotes coordination of international development projects; seeks to ensure that developing countries avoid economically damaging foreign debt. Serves as a forum for the discussion of international development and finance issues.

ACP-EC JOINT ASSEMBLY

c/o Secretariat General, European Parliament, Plateau du Kirchberg, L-2929 Luxembourg, Luxembourg **PHONE:** 352 43004996 **FX:** 352 43004995 **TX:** 3494 EUPARL LU **LANGUAGES:** English, French.

DESCRIPTION: A joint project of the African, Caribbean and Pacific Group of States (ACP) and the European Community (EU). Representatives of the member states of the ACP and EC. Seeks to ensure peaceful relations between members; encourages free international trade. Serves as a forum for the discussion of international trade, political, and social issues.

ACTION AGAINST HUNGER (AAH) (Action Contre la Faim)

9, rue Dareau, F-75014 Paris, France **PHONE:** 33

1 43358888 **FX:** 33 1 43358800 **TX:** 202539 F **E-MAIL:** acf@acf.imaginet.fr **FOUNDED:** 1979. **OFFICERS:** Jean Luc Bodin, Exec.Dir. **MEMBERS:** 250. **STAFF:** 60. **BUDGET:** 250,000,000 Fr. **LANGUAGES:** English, French, Spanish.

DESCRIPTION: Individuals concerned with the effects of world hunger. Objectives are: to provide immediate aid to persons throughout the world who suffer daily from poverty and hunger; to sensitize the public to the problem of world hunger. Has organized local committees to be responsible for a community in the Third World, undertaking a program of specific aid. Fosters self-sufficiency in underdeveloped countries. Programs include: sending food and sanitary brigades to Third World countries; obtaining municipal budget funds for projects; and participating with existing organizations to provide a vehicle for immediate aid. Conducts surveys. Holds training programs in health care, agriculture, management, and trades. **PUBLICATIONS:** *Interventions*, periodic.

ACTION COMMITTEE FOR LATIN AMERICAN COOPERATION AND COORDINATION ON IRON AND STEEL (Comote de Accion para la Cooperacion y Concentracion Latinoamericana en el Sector Siderurgico)

Avenida La Estancia, Edificio Maraven -P-10 Chuao, Caracas, Venezuela **PHONE:** 58 2 2080979, 58 2 2080980 **OFFICERS:** Rafeal Angel Carrasquel, Sec.

DESCRIPTION: Promotes the iron and steel industries throughout Latin America. Fosters cooperation and information exchange between producers in Latin American countries.

ACTION IN EUROPE FOR EDUCATION, INVENTION AND INNOVATION (AEEII)

c/o Espace Entreprise, 27 rue du Champ de Mars, F-57200 Sarreguemines, France **PHONE:** 33 3 87987575 **FX:** 33 3 87982727 **LANGUAGES:** English, French.

DESCRIPTION: Members are educators and inventors. Seeks to improve the conditions of employment of members; promotes increased protection of the intellectual property rights of inventors and innovators. Serves as a forum for the exchange of information among members; conducts educational programs.

ACTION FOR GREENING SAHEL

Nihon-son-kyoiku-kenkyusho Bldg., 3F, 227 Minamisuna, Koto-ku, Tokyo 136, Japan **PHONE:** 81 3 56323029 **FX:** 81 3 56323070 **FOUNDED:** 1991. **OFFICERS:** Kazuma Takahashi. **MEMBERS:** 232. **STAFF:** 31. **BUDGET:** 57,646,282¥. **LANGUAGES:** French, Japanese.

DESCRIPTION: Works to combat desertification in the Sahel region of Africa. Promotes local agricultural self-sufficiency through introduction of improved erosion control and introduction of more productive crop strains. Supports introduction of appropriate technologies, including more efficient cooking stoves, to reduce dependence on wood as a fuel. Conducts tree planting programs; makes available technical assistance to local cooperatives in Chad and Burkina Faso. **PUBLICATIONS:** *La Foret, C'est la vie* (in Japanese), quarterly. Activities of A.G.S.

ACTION IN INTERNATIONAL MEDICINE (AIM)

125 High Holborn, London WC1V 6QA, England **PHONE:** 44 0171 4053090 **FX:** 44 0171 4053093 **E-MAIL:** actintmed@aol.com **FOUNDED:** 1988. **OFFICERS:** Dr. Christopher Rose **MEMBERS:** 100. **LANGUAGES:** English.

DESCRIPTION: Members are health care institutions and health services. Promotes adoption of of "CCI," a self-help health improvement program which empowers local communities to address their health and poverty issues. Works to establish health care infrastructure in developing areas; provides support and assistance to grass roots health services. **PUBLICATIONS:** *AIM Bulletin* (in English), quarterly.

ACTION FOR SOUTHERN AFRICA (ASA)

28 Penton St., London N1 9SA, England **PHONE:** 44 0171 8333133 **FX:** 44 0171 8373001 **E-MAIL:** actsa@geo2.poptel.org.uk **OFFICERS:** Ben Jackson, Dir. **LANGUAGES:** English.

DESCRIPTION: Development organizations with operations in southern Africa. Promotes sustainable and appropriate development in the region.

Coordinates members' activities; provides support and assistance to locally administered agribusinesses and microenterprises; sponsors educational programs.

ACTIONAID - FREE POST

Leach Road, Chard, Somerset TA20 1FR, England **PHONE:** 44 1460 238000 **E-MAIL:** mail@actionaid.org.uk **WEBSITES:** http://www.oneworld.org/actionaid **FOUNDED:** 1972. **OFFICERS:** John Batten, Chief Exec. **MEMBERS:** 110,000. **STAFF:** 2,500. **BUDGET:** £37,500,000. **LANGUAGES:** English.

DESCRIPTION: A British-founded international development charity working with over three million of the world's poorest people in 20 countries in Africa, Asia and Latin America. Projects are undertaken in partnership with local communities and organizations. The focus is on integrated, long-term development in health, education, natural resource management, income generation and savings and credit schemes. Also works to limit the impact of conflict and natural disasters through its emergencies unit. **PUBLICATIONS:** *Annual Report* (in English); *Common Cause* (in English), semiannual.

ADVENTIST DEVELOPMENT AND RELIEF AGENCY INTERNATIONAL (ADRA)

12501 Old Columbia Pike, Silver Spring, MD 20904 **TF:** (800) 424-2372 **FX:** (301) 680-6370 **TX:** 440186, SDA-UI **WEBSITES:** http://www.adra.org **FOUNDED:** 1956. **OFFICERS:** Ralph S. Watts Jr., Pres. **STAFF:** 92. **BUDGET:** US$85,000,000. **LANGUAGES:** English.

DESCRIPTION: A nonsectarian relief and development agency of the Seventh-day Adventist Church serving developing countries. Sponsors institutional and community development, preventive health care, agricultural, and nutrition programs in Africa, Asia, the Pacific region, and Latin America. Teaches AIDS awareness and prevention. Conducts adult education and training classes; participates in refugee programs. Helps construct earthquake-resistant housing; builds schools, roads, and water and sewage systems. Provides food, clothing, and other forms of aid to disaster-stricken areas; assists food processing factories. Supports hospitals, dispensaries, clinics, sanitariums, retirement homes, and orphanages. Operates speakers' bureau. **PUBLICATIONS:** *ADRA Works*, quarterly; Newsletter, quarterly.

ADVISORY COMMITTEE ON EQUAL OPPORTUNITIES FOR WOMEN AND MEN (ACEOWM)

European Commn. DG V, J-27, 1/27 Rue de al Loi 200, B-1049 Brussels, Belgium **OFFICERS:** E. Van Winckel. **LANGUAGES:** English, French.

DESCRIPTION: Representatives of women's and human rights organizations in Europe. Promotes equality of opportunity and protection under the law regardless of gender. Develops and implements social and legal programs advancing the equality of women.

ADVISORY COMMITTEE ON PROTECTION OF THE SEA (ACOPS)

11 Dartmouth St., London SW1H 9BN, England **PHONE:** 44 171 7993033 **FX:** 44 171 7992933 **E-MAIL:** enquiries@acops.org **FOUNDED:** 1952. **OFFICERS:** Dr. Viktor Sebek, Exec.Dir. **MEMBERS:** 70. **STAFF:** 4. **LANGUAGES:** Croatian, English, French, Greek, Italian, Portuguese, Russian, Serbian, Spanish.

DESCRIPTION: Members are local authority associations, port and fishery associations, industry and tourist organizations, trade unions, and wildlife groups in 18 countries. Promotes preservation and protection of the world's seas from pollution by human activities. Compiles statistics. Sponsors seminars for local authority representatives. Organizes international conferences on sustainable tourism and national and global plans of action to reduce marine pollution from landbased activities. **PUBLICATIONS:** *ACOPS Annual Oil Pollution Survey of two UK Coasts*; *ACOPS Newsletter*, semiannual; *ACOPS Yearbook*, biennial.

AEROSPACE MEDICAL ASSOCIATION (AsMA)

320 S. Henry St., Alexandria, VA 22314 **PHONE:** (703) 739-2240 **FX:** (703) 739-9652 **WEBSITES:** http://www.asma.org **FOUNDED:** 1929. **OFFICERS:** Russell B. Rayman M.D., Exec.Dir. **MEMBERS:** 4,200. **BUDGET:** US$700,000. **LANGUAGES:** English.

DESCRIPTION: Members are medical and scientific personnel engaged in clinical, operational, and research activities in aviation, space, and environmental medicine. Sponsors continuing professional education programs. **PUBLICATIONS:** *Aviation, Space, and Environmental Medicine*, monthly. Covers the medical aspects of survival in aviation, space, and undersea exploration. Includes annual index; book reviews; meetings calendar.; *Aviation, Space and Environmental Medicine*, annual; *Scientific Papers*, annual.

AFRICA 2000 NETWORK

Selous Ave. 60, PO Box 4775, Harare, Zimbabwe **PHONE:** 263 4 700926, 263 4 700937 **FX:** 263 4 700946 **E-MAIL:** khethiwe@afan.icon.co.zw **WEBSITES:** http://www.africa2000.com/ **FOUNDED:** 1989. **OFFICERS:** Khethiwe Moyo Mhlanga. **MEMBERS:** 150. **STAFF:** 6. **LANGUAGES:** English.

DESCRIPTION: Membership consists of development organizations with operations in Africa; individuals interested in African economic and social development. Promotes sustainable development appropriate for local needs and capacities. Monitors development projects; formulates model development strategies; sponsors educational programs. **PUBLICATIONS:** Newsletter, bimonthly.

AFRICA EXPORT-IMPORT BANK

World Trade Centre Bldg., PO Box 404 Gezira, Corniche el-Nil, Cairo 11451, Egypt **FX:** 20 2 5780277 **OFFICERS:** Christopher Edordu, Pres. **LANGUAGES:** English.

DESCRIPTION: Promotes growth of international trade involving Africa. Provides financial and technical assistance to national banks throughout Africa; makes available banking and financial services to African companies engaged in international trade.

THE AFRICA FUND (AF)

50 Broad St., Ste. 711, New York, NY 10004 **PHONE:** (212) 785-1024 **FX:** (212) 785-1078 **E-MAIL:** africafund@igc.org **WEBSITES:** http://www.prairienet.org/acas/afund.html **FOUNDED:** 1966. **OFFICERS:** Jennifer Davis, Exec.Dir. **MEMBERS:** 10,000. **STAFF:** 6. **BUDGET:** US$450,000. **LANGUAGES:** English.

DESCRIPTION: Established by the American Committee on Africa. Works to support the struggle for democracy and economic justice. Defends human rights, including working for the release of political prisoners. Provides assistance to Africans, educates Americans on African issues and supports a U.S. policy that supports democracy and economic development. **PUBLICATIONS:** *Africa Fund News*, semiannual; *Africa Fund Perspectives*, periodic; Annual Report.

AFRICA AND MIDDLE EASTERN COMMITTEE OF PARLIAMENTARIANS ON POPULATION AND DEVELOPMENT

c/o Parliament of Zimbabwe, PO Box C4298, Causeway, Harare, Zimbabwe **OFFICERS:** M. Chinamasa. **LANGUAGES:** English.

DESCRIPTION: Members are legislators representing African and Middle Eastern nations. Promotes increased understanding of population and development issues. Coordinates members' activities; encourages implementation of population control programs.

AFRICA WATER NETWORK (AWN)

c/o Environmental Liason Centre International, PO Box 10538, Nairobi, Kenya **PHONE:** 254 2 556943, 254 2 555579 **FX:** 254 2 555513 **TX:** 23240 ELC KE **E-MAIL:** awn@elei.gn.apc.org **FOUNDED:** 1990. **OFFICERS:** Dr. Maurice Ndege, Exec.Dir. **MEMBERS:** 273. **STAFF:** 12. **BUDGET:** US$100,000. **LANGUAGES:** English, French, Swahili.

DESCRIPTION: Works to conserve natural water sources and maintain pure drinking supplies. Promotes awareness of ecological problems dealing with water sources. **PUBLICATIONS:** *Bank Watch Africa*, quarterly; *Droplets* (in English and French), bimonthly.

AFRICAN AIRLINES ASSOCIATION (AFRAA)

Box 20116, Nairobi, Kenya **PHONE:** 254 2 502418 **FX:** 254 2 601173 **TX:** NBOXA8X **FOUNDED:** 1968. **OFFICERS:** Capt. Mohammed

Ahmed, Sec.Gen. **MEMBERS:** 34. **STAFF:** 19. **LANGUAGES:** English, French.

DESCRIPTION: Members are chief executives of African airlines in 44 countries. Seeks to: promote and develop African air transport services that are safe, reliable, economic, and efficient; strengthen economic, commercial, and technical cooperation among members, particularly in the areas of ground handling, sales promotions, and interline services; promote commerce and tourism in Africa. Encourages policy coordination among members in selecting facilities and equipment with a preference for using, where possible, the facilities and equipment of members. Fosters the pooling of resources among members; seeks to reduce members' operating costs. Serves as a forum for discussion of members' common interests. Conducts activities for the benefit of members as well as African peoples and governments. Offers management seminars for executives and training courses for employees. **PUBLICATIONS:** *Annual Reports*; *Newsletter/ Bulletin* (in English and French), quarterly; *Proceedings AGA* (in English and French), annual.

AFRICAN ASSOCIATION

5 Ahmet Hishmat St., 11561 Zamalik, Cairo, Egypt **PHONE:** 20 2 3407658 **OFFICERS:** Mohamed Fouad el-Bedewy, Sec.Gen. **LANGUAGES:** Arabic, English.

DESCRIPTION: Individuals with an interest in Africa. Promotes increased understanding and appreciation of African history, culture, politics, and development. Conducts research and educational programs.

AFRICAN ASSOCIATION FOR PUBLIC ADMINISTRATION AND MANAGEMENT

PO Box 48677, Nairobi, Kenya **PHONE:** 254 2 521944 **FX:** 254 2 521845 **OFFICERS:** A.D. Yahaya, Sec.Gen. **LANGUAGES:** English.

DESCRIPTION: Membership consists of administrators and managers of public sector programs and institutions. Promotes more efficient operations within the public sector. Conducts continuing professional development programs; facilitates communication and exchange of information among members.

AFRICAN BIOSCIENCES NETWORK (ABN)

UNESCO Office, 12 avenue de Rome, Boite Postale 3311, Dakar, Senegal **PHONE:** 221 235082, 221 219669 **FX:** 221 238393 **TX:** 21735-51410 SG **E-MAIL:** uhdak@unesco.org **LANGUAGES:** English, French.

DESCRIPTION: Members are biologists and other scientists with an interest in the biological sciences. Seeks to advance the study, teaching, and practice of the biological sciences. Promotes practical commercial and social application of biosciences. Serves as a forum for the exchange of information among members; provides consulting services to development programs and government agencies; sponsors research and educational programs.

AFRICAN BUSINESS ROUND TABLE (ABRT)

Boite Postale, Abidjan 01, Cote d'Ivoire **PHONE:** 225 216138 **FX:** 225 216272 **TX:** 23717 **OFFICERS:** Dr. Karamo Sonko. **LANGUAGES:** English, French.

DESCRIPTION: Membership consists of businesses and trade organizations. Promotes growth and development of African businesses; seeks to increase international trade involving Africa. Serves as a clearinghouse on African business and trade; facilitates communication and cooperation among members; sponsors research; conducts promotional activities.

AFRICAN CAPACITY BUILDING FOUNDATION (ACBF)

Southampton Life Centre, 7th Fl., Corner Jason Moyo Ave. and 2nd St., PO Box 1562, Harare, Zimbabwe **PHONE:** 263 4 738520, 263 4 790398 **FX:** 263 4 702915 **E-MAIL:** root@acbf.samara.co.zw **OFFICERS:** Abel L. Thoahlane. **LANGUAGES:** English.

DESCRIPTION: Members are economic policy and development professionals. Seeks to ensure sustainable and appropriate economic development throughout Africa. Cooperates with government economic and development authorities to devise economic policies conducive to economic growth.

AFRICAN, CARIBBEAN AND PACIFIC GROUP OF STATES (Groupo de Estados de Africa, del Caribe y del Pacifico)

Avenue Georges Henri 451, B-1200 Brussels, Belgium **PHONE:** 32 2 7430600 **FX:** 32 2 7355573

DESCRIPTION: Promotes improved economic and social relations in regards to international development between member states and third world countries. **PUBLICATIONS:** *ACP-EEC Courier*, semimonthly; *Directory of ACP Technical Institutions*; *Directory of ACP Universities*.

AFRICAN CENTRE FOR DEMOCRACY AND HUMAN RIGHTS STUDIES (ACDHRS)

Karaiba Ave. and Kombo St., Mary Division, Banjul, Gambia **PHONE:** 220 394961, 220 370005 **FX:** 220 394962 **E-MAIL:** acdhrs@commit.gm **FOUNDED:** 1989. **OFFICERS:** Zoe Tembo, Exec.Dir. **LANGUAGES:** English, French.

DESCRIPTION: Promotes observance of human and peoples' rights and democratic principles throughout Africa. Works to increase public awareness of human rights issues; maintains information and documentation center. Conducts research and educational programs; provides training courses for human rights personnel. **PUBLICATIONS:** *African Human Rights Newsletter*, quarterly; Brochure, periodic.

AFRICAN CHRISTIAN RELIEF (ACR)

7941 E Lakeside Pkwy., Ste. 109, Tucson, AZ 85730 **PHONE:** (520) 722-8447 **FX:** (520) 298-1404 **OFFICERS:** Bryan Gregory, Pres. **LANGUAGES:** English.

DESCRIPTION: Christian relief organizations with operations in Africa. Promotes increased economic prosperity and self-sufficiency among developing countries in Africa. Conducts agricultural training programs; provides technical assistance to rural development projects; makes available food, medical supplies, and educational materials to people living in economically disadvantaged areas in Africa.

AFRICAN COMMISSION ON AGRICULTURAL STATISTICS (ACAS)

c/o FAO Regional Office for Africa, United Nations Agency Bldg., PO Box 1628, Accra, Ghana **PHONE:** 233 21 666851, 233 21 666852 **FX:** 233 21 668427 **TX:** 2139 **E-MAIL:** fao-raf@field.fao.org **LANGUAGES:** English.

DESCRIPTION: Organization of national agencies responsible for compiling agricultural statistics and other information. Seeks to increase the accuracy and availability of African agricultural data. Fosters collaboration and standardization of techniques among members; develops more accurate methods of quantifying agricultural production.

AFRICAN COMMISSION OF HEALTH AND HUMAN RIGHTS PROMOTERS (ACHHRP)

Rabito Clinic, PO Box 7286, North, Accra, Ghana **PHONE:** 233 21 774526 **FX:** 233 21 777465 **LANGUAGES:** English.

DESCRIPTION: Members are health and human rights organizations. Promotes respect for the rights of the individual and the rule of law; seeks to increase availability of health care services among previously underserved populations. Publicizes human rights abuses; makes available health care programs.

AFRICAN COMMISSION ON HUMAN AND PEOPLES' RIGHTS (ACHPR)

Kariba Ave., PO Box 673, Banjul, Gambia **PHONE:** 220 392962 **FX:** 220 390764 **TX:** 2346 OAU BJL GV **OFFICERS:** Germain Baricako. **LANGUAGES:** English.

DESCRIPTION: Members are human rights and indigenous peoples' organizations. Promotes respect for the rights of the individual and the rule of law in developing areas of Africa. Monitors the human rights situation in developing areas and publicizes abuses; makes available legal services.

AFRICAN COMMITTEE FOR DEVELOPMENT (ACD)

38 avenue Fernand Lefebvre, F-78300 Poissy, France **OFFICERS:** Ernest Mihami. **LANGUAGES:** French.

DESCRIPTION: Members are development organizations. Promotes implementation of sustainable economic and social development programs appropriate for local needs and administrative capacities. Serves as a clearinghouse on African development organizations and projects; provides consulting services.

AFRICAN COUNCIL OF AIDS SERVICE ORGANIZATIONS

c/o ENDA Tiers Monde, 6 rue Calmette, Boite Postale 3370, Dakar, Senegal **PHONE:** 221 229695, 221 236617 **FX:** 221 236615 **E-MAIL:** aficaso@enda.sn **OFFICERS:** Abdelkader Bacha. **LANGUAGES:** English, French.

DESCRIPTION: Organizations providing services to people with AIDS. Seeks to improve the quality of life of people with aids; promotes increased availability of AIDS services; works to stop the spread of AIDS. Serves as a forum for communication among members; coordinates members' activities; functions as a clearinghouse on AIDS and AIDS services.

AFRICAN DEVELOPMENT BANK (ADB)

Boite Postale 1387, Abidjan 01, Cote d'Ivoire **PHONE:** 225 204444 **FX:** 225 217753 **E-MAIL:** adbabjacos@gn.apc.org **LANGUAGES:** English, French.

DESCRIPTION: Members are representatives of national financial institutions. Seeks to further economic development of member states. Serves as a clearinghouse on national and regional economic development. Provides financial support and consulting services to development programs and agencies. Sponsors research and educational programs.

AFRICAN DEVELOPMENT FUND (ADF)

Boite Postale 1387, Abidjan 01, Cote d'Ivoire **PHONE:** 225 204444 **FX:** 225 217753 **LANGUAGES:** English, French.

DESCRIPTION: Membership consists of development and financial organizations. Seeks to increase the effectiveness of African economic and community development programs. Provides financial assistance and consulting services to development projects.

AFRICAN ECONOMIC COMMUNITY (AEC)

c/o OAU, PO Box 3243, Addis Ababa, Ethiopia **PHONE:** 251 1 517700 **FX:** 251 1 512622 **TX:** 21046 **LANGUAGES:** English.

DESCRIPTION: Organization of national economic agencies. Promotes economic growth and increased trade among African countries. Formulates national and international economic policies and trade agreements; serves as a clearinghouse on African economic activity.

AFRICAN ECONOMIC RESEARCH CONSORTIUM (AERC)

International House, 8th Fl., PO Box 62882, Nairobi, Kenya **PHONE:** 254 2 225234 **FX:** 254 2 219308 **E-MAIL:** aerc@elci.gn.apc.org **OFFICERS:** Benno J. Ndulu. **LANGUAGES:** English.

DESCRIPTION: Members are economists. Seeks to advance economic research and scholarship. Facilitates exchange of information among members; sponsors research programs.

AFRICAN ENERGY POLICY RESEARCH NETWORK (AFREPREN)

PO Box 30979, Nairobi, Kenya **PHONE:** 254 2 566032, 254 2 571467 **FX:** 254 2 561464 **E-MAIL:** skarekezi@form-net.com **FOUNDED:** 1987. **OFFICERS:** Stephen Karekezi. **MEMBERS:** 98. **STAFF:** 9. **LANGUAGES:** French, Spanish.

DESCRIPTION: Promotes the study of energy sources in Africa. Objective is to strengthen local research capacity and to utilize it in the service of energy policy making and planning. Distributes important energy documents to network members and interested energy agencies. **PUBLICATIONS:** *AFREPREN News* (in English), quarterly. Over 100 books, working papers and newsletters published annually.

AFRICAN ENVIRONMENTAL RESEARCH AND CONSULTING GROUP (AERCG)

14912 Walmer St., Overland Park, KS 66223-1161 **PHONE:** (913) 897-6132 **FX:** (913) 897-6132 **E-MAIL:** aercgc31@freewwweb.com **WEBSITES:** http://www.africaenviro.org **FOUNDED:** 1990. **OFFICERS:** Peter A. Sam,

Chair. **MEMBERS:** 1,300. **STAFF:** 10. **LANGUAGES:** English.

DESCRIPTION: Members are African and African-American professionals. Promotes mitigation of environmental hazards in Africa. Provides voluntary technical assistance and financial support to environmental protection and mitigation programs; promotes collaboration between African and American scientists working in pollution control and related fields; seeks to raise public awareness in the United States regarding environmental problems in Africa. Conducts preliminary assessments to identify environmental hazards; facilitates technology transfers. **PUBLICATIONS:** *International Environmental Consulting Practice* (in English).

AFRICAN EXPORT IMPORT BANK (AEIB)

World Trade Center, 1191 Corniche El-Nil St., PO Box 404, Cairo 11568, Egypt **PHONE:** 20 2 5780281 **FX:** 20 2 570276 **TX:** 20003 AFRXM-UN **OFFICERS:** John Washington T. Otieno. **LANGUAGES:** Arabic, English.

DESCRIPTION: Members are banks and other financial institutions with an interest in international trade. Seeks to remove barriers to international trade; promotes growth and development of African national economies. Functions as a clearinghouse on African business and trade; provides financial support and consulting services to businesses engaged in the importation or exportation of goods.

AFRICAN FARMERS ASSOCIATION (AFA)

PO Box 14, Giza 12211, Egypt **PHONE:** 20 2 3604679, 20 2 3607403 **FX:** 20 2 3609134 **TX:** 21381 AFRIC UN **OFFICERS:** Mohamed Idris. **LANGUAGES:** Arabic, English.

DESCRIPTION: Made up of farmers' organizations. Promotes agricultural advancement. Serves as a forum for the exchange of information among members.

AFRICAN FEDERATION OF WOMEN ENTREPRENEURS

c/o African Centre for Women, PO Box 3001, Addis Ababa, Ethiopia **PHONE:** 251 1 517200, 251 1 517301 **FX:** 251 1 512785 **FOUNDED:** 1993. **LANGUAGES:** English.

DESCRIPTION: A program of the United Nations Economic Commission for Africa. Seeks to play an active role in the formulation and implementation of strategies for women entrepreneurs, and to develop mechanisms to promote increased interaction among women entrepreneurs throughout Africa. Emphasizes development of rural businesses to create more diverse economies. Makes available credit and other financial assistance and entrepreneurial training to women wishing to enter business. **PUBLICATIONS:** *Directory of African Women Entrepreneurs* (in English), periodic.

AFRICAN FEED RESOURCES RESEARCH NETWORK (AFRNET)

c/o International Livestock Centre for Africa, PO Box 46847, Nairobi, Kenya **PHONE:** 254 2 632013 **FX:** 254 2 631481 **OFFICERS:** Dr. John Ndikumama, Coor. **LANGUAGES:** English.

DESCRIPTION: Made up of agricultural research organizations. Promotes improved feeding of livestock. Conducts feed research; educates farmers about new feeds and feeding techniques.

AFRICAN FORESTRY AND WILDLIFE COMMISSION (AFWC)

c/o FAO Regional Office for Africa, United Nations Agency Bldg., PO Box 1628, Accra, Ghana **PHONE:** 233 21 666851, 233 21 666852 **FX:** 233 21 668427 **TX:** 2139 **E-MAIL:** FAO-RAF@field.fao.org **LANGUAGES:** English.

DESCRIPTION: Members are national forestry and conservation organizations. Promotes environmental protection and sustainable use of natural resources. Facilitates communication among members and coordinates members' activities; sponsors research; serves as a clearinghouse on environmental protection.

AFRICAN GROUNDNUT COUNCIL (AGC) (Conseil Africain de l'Arachide - CAA)

Trade Fair Complex, PO Box 3025, Marina, Lagos, Nigeria **PHONE:** 234 1 880982, 234 1 886121 **FX:** 234 1 880982 **TX:** 21366 AFNUCO NG **FOUNDED:** 1964. **OFFICERS:** Mour

Mamadou Samb, Exec.Sec. **MEMBERS:** 6. **STAFF:** 10. **BUDGET:** US$270,000. **LANGUAGES:** English, French.

DESCRIPTION: Representatives of Gambia, Mali, Niger, Nigeria, Senegal, and Sudan. Promotes consumption of the groundnut (also known as the peanut) and its by-products. Attempts to resolve problems related to production, handling, and export of groundnuts; calls for the exchange of technical information. Encourages solidarity among members. Works to protect the price of the groundnut worldwide. **PUBLICATIONS:** *Groundnut Newsletter* (in English and French), periodic; *Groundnut Review* (in English and French), periodic; *Groundnut Statistical.*

AFRICAN HIGHLAND INITIATIVE ECO-REGIONAL PROGRAM (AHIERP)

c/o ASARECA, PO Box 765, Entebbe, Uganda **PHONE:** 256 42 20556, 256 42 20212 **FX:** 256 42 21126 **TX:** 61287 NATURE UG **E-MAIL:** asareca@imul.com **LANGUAGES:** English.

DESCRIPTION: Membership consists of environmental organizations with an interest in African highland ecosystems. Promotes preservation of habitats and sustainable use of natural resources. Facilitates communication and cooperation among members; sponsors environmental protection and habitat rehabilitation projects.

AFRICAN INSTITUTE (Institut Africain)

c/o Africa Museum, Centre d'Etude et de Documentation Africaines, Leuvenssesteenweg 13, B-3080 Tervuren, Belgium **PHONE:** 32 2 7681993 **FX:** 32 2 7681995 **E-MAIL:** institu.fricain@euronet.be **FOUNDED:** 1988. **OFFICERS:** Gauthier de Villers, Dir. **STAFF:** 4. **BUDGET:** 8,300,000 BFr. **LANGUAGES:** Dutch, English, French.

DESCRIPTION: Individuals and research organizations with an interest in Africa. Promotes improved understanding of development and related issues facing Africa today. Conducts research; provides support and assistance to development and research projects. **PUBLICATIONS:** *Cahiers Africains* (in French), bimonthly.

AFRICAN INSTITUTE FOR ECONOMIC AND SOCIAL DEVELOPMENT (AIESD) (Institut Africain pour le Developpement Economique et Sociale - INADES)

15 avenue Jean-Mermoz, Cocody, BP 2088, Abidjan 08, Cote d'Ivoire **PHONE:** 225 441594, 225 392186 **FX:** 225 448438 **E-MAIL:** guerym@ci.refer.org **FOUNDED:** 1962. **OFFICERS:** Michel Guery, Dir. **LANGUAGES:** English, French.

DESCRIPTION: Works to develop African communities in the areas of agriculture, management of development projects, and leadership skills; strives to prepare rural dwellers for urban life in areas under development. Makes available educational and training programs and correspondence courses for people affected by development programs and development workers. Conducts economic and social research; maintains documentation and information center. **PUBLICATIONS:** *Le Fichiez Afrique*, bimonthly.

AFRICAN IRON AND STEEL ASSOCIATION (AISA)

Block No. 2, Flat No. 1, Tamale St., Bamenda, Wuse, Zone 3, PMB 268 Garki, Abuji, Nigeria **PHONE:** 234 9 5233170 **FX:** 234 9 5233275 **OFFICERS:** Dr. Sanusi A. Mohammed. **LANGUAGES:** English.

DESCRIPTION: Producers of iron and steel. Seeks to advance the steel industry on the African continent. Represents members' interests before labor and trade organizations, government agencies, and the public.

AFRICAN JURISTS ASSOCIATION (AJA)

37 boulevard Ornan, F-75018 Paris, France **PHONE:** 33 1 42230830 **FX:** 33 1 42580376 **TX:** 660141 Cordija **OFFICERS:** M. Fethsahlis. **LANGUAGES:** English, French.

DESCRIPTION: Members are judicial officials in African countries. Promotes respect for the rights of the individual and the rule of law. Monitors legal developments publicizes human rights abuses; advocates against the use of cruel and unusual punishments in criminal justice.

AFRICAN MEDICAL AND RESEARCH FOUNDATION (AMREF)

PO Box 30125, Nairobi, Kenya **PHONE:** 254 2 501301, 254 2 500508 **FX:** 254 2 609518 **TX:** 23254 AMREF **E-MAIL:** amrefinf@africaonline.co.ke **WEBSITES:** http://www.amref.org **FOUNDED:** 1957. **OFFICERS:** Dr. John Batten, Dir.Gen. **STAFF:** 450. **BUDGET:** US$14,000,000. **LANGUAGES:** English, Kiswahili.

DESCRIPTION: Purpose is to improve the health of people living in 5 east and south African countries. Focuses efforts on developing low-cost healthcare for people in rural areas. Program includes: training of community health workers; planning and evaluation of health projects; consulting services; primary healthcare education; operation of a medical radio communication network within the region; ground mobile health services for nomadic and pastoral peoples; the Flying Doctor Service (funded by the Flying Doctors' Society of Africa), which provides airborne support for remote medical, surgical, and public health facilities. Conducts medical research, particularly into the control of malaria, sleeping sickness, and hydatid disease; applies behavioral and social sciences in health improvement. Offers computerized services. **PUBLICATIONS:** *Afya* (in English), quarterly. Covers developments in medical education.; *AMREF Annual Report* (in English); *AMREF News* (in English), quarterly; *Cobasheca* (in English), quarterly; *Defender* (in English), quarterly; *Research Papers*, periodic; *Training Manuals*, periodic; Reports.

AFRICAN MOUNTAIN ASSOCIATION (AMA)

PO Box 12760, Addis Ababa, Ethiopia **FX:** 251 1 552350 **E-MAIL:** epa@padis.gn.apc.org **OFFICERS:** Dr. Tewolde Berhan Gebre Egziabher. **LANGUAGES:** English.

DESCRIPTION: Members are individuals and organizations with an interest in the ecology of mountainous regions of Africa. Promotes protection of mountain ecosystems. Serves as a clearinghouse on mountain ecology; conducts environmental protection programs.

AFRICAN NETWORK ON AGRI-FOOD, TRADE, ENVIRONMENT AND DEVELOPMENT

Boite Postale 8230, Lome, Togo **OFFICERS:** Kokougan Apaloo. **LANGUAGES:** English, French.

DESCRIPTION: Membership consists of international development, agricultural, and environmental organizations. Promotes appropriate and sustainable development in rural areas of Africa. Coordinates members' activities; provides support and assistance to development projects.

AFRICAN NETWORK ON DEBT AND DEVELOPMENT (ANDD)

c/o MWENGO, PO Box H6 817, Highlands, Harare, Zimbabwe **PHONE:** 263 4 364664 **FX:** 263 4 722363 **E-MAIL:** afrodad@zimtap.tool.ne **OFFICERS:** Opa Kapisimpanga. **LANGUAGES:** English.

DESCRIPTION: Membership consists of businesses, financial institutions, international trade organizations, and government agencies. Promotes increased awareness of the impact of international debt on the economies of developing African nations. Serves as a clearinghouse on debt and development; works to devise solutions to economic problems caused by international debt.

AFRICAN NETWORK FOR DEVELOPMENT (AND)

PO Box 60233, Addis Ababa, Ethiopia **PHONE:** 251 1 514963 **FX:** 251 1 515833 **OFFICERS:** Troare Gaoussou. **LANGUAGES:** English.

DESCRIPTION: Members are individuals and organizations with an interest in development. Promotes appropriate and sustainable economic and community development in Africa. Provides advice and assistance to development programs.

AFRICAN NETWORK ON HIV/AIDS - EUROPE

Focus Consultancy, 32B Warwick Sq., London SW1 V2AQ, England **PHONE:** 44 171 9320072 **FX:** 44 171 9320074 **E-MAIL:** london@fcwsq.demon.co.uk **OFFICERS:** Dawn Hill. **LANGUAGES:** English.

DESCRIPTION: Members are national European AIDS organizations with an interest in the AIDS epidemic in Africa. Seeks to slow the spread of AIDS and increase the availability of AIDS treatment services in African countries. Makes available educational and consulting services.

AFRICAN NETWORK FOR PREVENTION AND PROTECTION AGAINST CHILD ABUSE AND NEGLECT

PO Box 71420, Nairobi, Kenya **PHONE:** 254 2 722496, 254 2 726794 **FX:** 254 2 721 **E-MAIL:** anppcan@arcc.or.ke **WEBSITES:** http://www.africaonline.co.ke/anppcan/ **OFFICERS:** Dr. Philista Onyango. **LANGUAGES:** English.

DESCRIPTION: Membership includes individuals and organizations. Seeks to prevent child abuse and neglect. Conducts research into the underlying causes of child abuse; makes available support and services to abused and neglected children.

AFRICAN NETWORK OF SCIENTIFIC AND TECHNOLOGICAL INSTITUTIONS (ANSTI) (Reseau Africain d'Institutions Scientifiques et Technologiques - RAIST)

UNESCO-ROSTA, PO Box 30592, Nairobi, Kenya **PHONE:** 254 2 622620, 254 2 622619 **FX:** 254 2 215991 **TX:** 22275 NAIROBI **E-MAIL:** j.massaquoi@unesco.org **FOUNDED:** 1980. **OFFICERS:** J.G.M. Massaquoi, Coordinator. **MEMBERS:** 87. **STAFF:** 4. **BUDGET:** US$300,000. **LANGUAGES:** English, French.

DESCRIPTION: Memberhip consists of schools of engineering, science, and technology, and research centers and institutes in 33 African countries. Encourages cooperation between such facilities in the areas of research and training. Also encourages study in Africa. Sponsors seminars and workshops. **PUBLICATIONS:** *African Journal of Science and Technology* (in English and French), semiannual. Series include technology and science.; *ANSTI Newsletter* (in English), periodic; *Directory of ANSTI Institutions* (in English), periodic. Also publishes textbooks and occasional reports.

AFRICAN OIL PALM DEVELOPMENT ASSOCIATION (AFOPDA)

15 Boite Postale 341, Abidjan 15, Cote d'Ivoire **PHONE:** 225 251518 **OFFICERS:** Baude-Laire Sourou, Exec.Sec. **LANGUAGES:** English, French.

DESCRIPTION: Members are oil palm producers. Promotes growth and development of the oil palm industries. Represents members' interests.

AFRICAN PETROLEUM PRODUCERS' ASSOCIATION (APPA)

Boite Postale 4540, Pointe-Noire, Congo **PHONE:** 242 949710 **FX:** 242 949717 **FOUNDED:** 1987. **OFFICERS:** Maxime Obiang-nze, Exec.Sec. **MEMBERS:** 12. **STAFF:** 11. **LANGUAGES:** Arabic, English, French, Portuguese, Spanish.

DESCRIPTION: Organization of petroleum producing countries in Africa. Promotes growth and development in the African petroleum industries. Facilitates communication and cooperation among petroleum producers. **PUBLICATIONS:** *Bulletin APPA* (in English and French), quarterly.

AFRICAN PRESERVATION PROGRAMME (APP)

PO Box 534, Kraaifontein 7569, Republic of South Africa **OFFICERS:** John Spence. **LANGUAGES:** Afrikaans, English.

DESCRIPTION: Membership includes individuals and organizations. Promotes preservation of endangered species of African wildlife and their habitats. Gathers and disseminates information on environmental protection and wildlife conservation.

AFRICAN PROTESTANT CHURCH (Eglise Protestante Africaine)

Boite Postale 26, Lolodorf, Cameroon **OFFICERS:** Rev. Marnia Woungly-Massaga, Dir.-Gen. **LANGUAGES:** English, French.

DESCRIPTION: Organization of Protestant denominations. Promotes the growth of Protestantism. Facilitates communication and cooperation among members; participates in religious, social welfare, and educational activities.

AFRICAN PUBLISHERS NETWORK

PO Box 3773, Harare, Zimbabwe **PHONE:** 263 4 726405, 263 4 705105 **FX:** 263 4 705106 **E-MAIL:** apnet@mango.zw **FOUNDED:** 1992. **OFFICERS:** Richard Crabbe, Chm. **MEMBERS:** 27. **STAFF:** 5. **LANGUAGES:** English, French, Portuguese.

DESCRIPTION: Book publishers, nonprofit organizations, and educational institutions interested in publishing and the dissemination of information using the printed word. Promotes increased activity among African publishing institutions and authors. **PUBLICATIONS:** *African Publishing Review* (in English, French, and Portuguese), periodic. Contains country reports and letters from members. Includes reviews of international publishers' meetings worldwide.; *Seminar Papers*; *Thematic Catalogues*.

AFRICAN REGIONAL CENTRE FOR TECHNOLOGY (ARCT) (Centre Regional Africain de Technologie - CRAT)

BP 2435, Dakar, Senegal **PHONE:** 221 237712 **FX:** 221 237713 **TX:** 61282 CRATEC SG **E-MAIL:** arct@endadak.gn.apc.org **FOUNDED:** 1977. **OFFICERS:** Dr. Ousmane Kane, Deputy Dir. **MEMBERS:** 31. **STAFF:** 30. **BUDGET:** US$1,300,000. **LANGUAGES:** English, French.

DESCRIPTION: Member countries of the Economic Commission for Africa and the Organization of African Unity united to strengthen African technological capabilities. Advocates the development of indigenous technological capabilities, human resources and rural development, and the use of technology to stimulate socioeconomic growth. Seeks to contribute to the development and use of technology in member states; works to heighten awareness of technological development. Promotes the use of technology in development projects; assists in the formulation of technology policies as an integral part of scientific, technological, and socioeconomic development planning. Encourages research and training in the methodologies of technology planning; seeks improvement of member states' technology importation terms and conditions; promotes the management and dissemination of technology and technological information. Assesses the social implications of the development, transfer, and adaptation of technology. Assists member states in the establishment of national institutions to examine such implications. Provides advice and identification on technological options; promotes, through national institutions, effective links between technology producers and users; promotes the exchange of technical, managerial, and research personnel among member states. Assists in the training of technical and managerial personnel. Maintains a technology register of African research institutions' programs and achievements, institutions outside Africa concerned with technologies relevant to the needs of member states, and available specialists in various technological fields. **PUBLICATIONS:** *African Technodevelopment* (in English and French), semiannual. Contains articles on food and energy technologies, policies, case studies, and technology transfer.; *Alert Africa* (in English and French), quarterly; *Bibliography on Energy in Africa*, annual; *Bibliography on Post-Harvest Technologies in Africa*, annual; *Biogas Manual: Construction Techniques and Comparison of Four Biodigester Types*, periodic; *Directory of Institutions and Experts in Science and Technology in Africa*, periodic; *Directory of Scientific and Engineering Societies in Sub-Saharan Africa*, periodic; *Directory of Technician Training Institutions in Africa*, periodic; *Guide to Directories*, periodic; *Guide to New and Renewable Energy Projects in Africa*, periodic; *Infonet* (in English), semiannual; Annual Report (in English and French); Manuals; Reports, periodic. Containing information on food and energy.

AFRICAN REGIONAL ORGANIZATION FOR STANDARDIZATION (ARSO) (Organisation Regionale Africaine de Normalisation - ORAN)

PO Box 57363, Nairobi, Kenya **PHONE:** 254 2 330882, 254 2 224561 **FX:** 254 2 218792 **TX:** 22097 ARSO **E-MAIL:** Sec-gen@arso.sasa.unon.org **FOUNDED:** 1977. **OFFICERS:** Dr. Adebayo Oyejola. **MEMBERS:** 24. **STAFF:** 15. **LANGUAGES:** English, French.

DESCRIPTION: Organization of the member states of the Economic Commission for Africa and the Organization of African Unity. Seeks to enhance national standards and measurement capabilities of members. Encourages the effective application of science and technology to advance industrial and socioeconomic development of member states. Formulates standards to provide for consumer protection and human safety. Supports the elaboration of regional standards in order to contribute toward the development and expansion of intra-African trade. Provides technical information and consultation services; operates a regional standards documentation and information service. Conducts quality control and metrology activities, certification marking operations, and laboratory tests. Offers training programs; organizes expert groups.

PUBLICATIONS: *ARSO Annual Report*; *ARSO-DISNET Constitution and Guide*; *ARSO Newsletter*, semiannual; *Catalogue of African Regionalal Standards*, annual; *Experts in Standardization in Africa*, triennial; *Workshop Proceedings*, periodic; Papers. Containing technical information.

AFRICAN RURAL AND AGRICULTURAL CREDIT ASSOCIATION (AFRACA) (Association Africaine de Credit Rural et Agricole - AFRACA)

PO Box 41378, Nairobi, Kenya **PHONE:** 254 2 717911, 254 2 715991 **FX:** 254 2 710082 **E-MAIL:** afraca@afracaonline.co.ke **FOUNDED:** 1977. **OFFICERS:** Mr. S.I. Ijioma, Sec.Gen. **MEMBERS:** 41. **STAFF:** 7. **BUDGET:** US$400,000. **LANGUAGES:** English, French.

DESCRIPTION: Members are national banks and other financial institutions, government departments, and training institutions dealing with agricultural credit in 27 countries. Encourages cooperation among agricultural credit institutions. Works to improve the planning and management of financial services for rural development; advocates the extension of banking services to small-scale farmers and women. Advises member institutions in their efforts to enlist the help of international financial organizations and to design viable agricultural assistance programs. Operates West African Centre for Agricultural Credit Training in Freetown, Sierra Leone; East African Centre for Agricultural Credit Training in Dar Es Salaam, Tanzania; and the Southern African Centre for Agricultural Credit Training in Swaziland designed to train employees of member institutions. Organizes in-service training exchange programs; sponsors national and regional seminars, courses, and workshops in agricultural credit and banking. **PUBLICATIONS:** *AFRACA Brochure* (in English and French), biennial; *AFRACA News* (in English and French), quarterly; *General Assembly Report* (in English and French), biennial. Also publishes seminar reports, promotional information, and case studies.

AFRICAN SOCIETY OF INTERNATIONAL AND COMPARATIVE LAW

Kairaba Ave., Private Bag 52, Banjul, Gambia **PHONE:** 220 224968 **FX:** 220 224969 **OFFICERS:** Emile Yakpo, Sec. **LANGUAGES:** English.

DESCRIPTION: Members are attorneys and legal scholars. Promotes the study and practice of international and comparative law. Conducts continuing professional development programs.

AFRICAN STUDIES ASSOCIATION (ASA)

c/o Dr. Allen J. Green, Rutgers the State University of New Jersey, 132 George St.-Douglass Campus, New Brunswick, NJ 08901-1400 **PHONE:** (732) 932-8173, (732) 932-3391 **E-MAIL:** callasa@rcl.rutgers.edu **FOUNDED:** 1957. **OFFICERS:** Dr. Allen J. Green, Acting Exec.Dir. **MEMBERS:** 3,700. **STAFF:** 7. **BUDGET:** US$500,000. **LANGUAGES:** English.

DESCRIPTION: Members are persons specializing in teaching, writing, or research on Africa including political scientists, historians, geographers, anthropologists, economists, librarians, linguists, and government officials; persons who are studying African subjects; institutional members are universities, libraries, government agencies, and others interested in receiving information about Africa. Seeks to foster communication and to stimulate research among scholars on Africa. Sponsors placement service; conducts panels and discussion groups; presents exhibits and films. **PUBLICATIONS:** *African Studies Review*, 3/year; *ASA News*, quarterly; *Directory of African and Afro-American Studies in the U.S.*, periodic; *History in Africa*, annual; *Issue: A Journal of Opinion*, semiannual. Also publishes scholarly and bibliographical material.

AFRICAN TIMBER ORGANIZATION (ATO)

Boite Postale 1077, Libreville, Gabon **PHONE:** 241 732928 **FX:** 241 734030 **TX:** OAB 5620 GO **E-MAIL:** oab.gabon@internetgabon.com **FOUNDED:** 1976. **OFFICERS:** Paul Ngatse Obala. **MEMBERS:** 13. **STAFF:** 22. **BUDGET:** 2,250,000 Fr. **LANGUAGES:** English, French.

DESCRIPTION: Organization of African lumber producing countries. Promotes sustainable use of African timber resources. Facilitates communication and cooperation among member countries; conducts promotional activities. **PUBLICATIONS:** *ATO Bulletin of Information* (in English and French), annual.

AFRICAN TRAINING AND RESEARCH CENTRE IN ADMINISTRATION FOR DEVELOPMENT (ATRCAD) (Centre Africain de Formation et de Recherche Administratives pour le Developpement - CAFRAD)

avenue Mohamed V, Boite Postale 310, Tangiers, Morocco **PHONE:** 212 9 942632 **FX:** 212 9 941415 **TX:** 33664 **OFFICERS:** Mamadou Thiam, Dir.Gen. **LANGUAGES:** English, French.

DESCRIPTION: Researchers and other individuals with an interest in the administration and operation of development programs. Seeks to improve the performance and efficiency of development projects and organizations through more effective management. Conducts continuing professional development courses for development organization administrators; provides support and assistance to development projects.

AFRICAN UNION OF DEVELOPMENT BANKS (Union Africaine des Banques pour le Developpement)

Boite Postale 2045, Cotonou, Benin **PHONE:** 229 301500 **FX:** 229 300284 **TX:** 5024 **OFFICERS:** Kouanvi Tigoue, Exec.Sec. **LANGUAGES:** English, French.

DESCRIPTION: African development banks. Seeks to insure adequate financing for sustainable economic and social development. Makes available banking and financial advice and services to development projects.

AFRICAN WOMEN DEVELOPMENT AND COMMUICATION NETWORK (AWDCN)

PO Box 54562, Nairobi, Kenya **PHONE:** 254 2 741301 **FX:** 254 2 741320 **E-MAIL:** femnet@elci.gn.apc.org **OFFICERS:** Safiatu K. Singhateh. **LANGUAGES:** English.

DESCRIPTION: Members are individuals and organizations. Seeks to advance the economic, legal, and social status of African women. Works with development organizations to ensure increased participation by women in development; sponsors educational and training programs.

AFRICARE

440 R St. NW, Washington, DC 20001 **PHONE:** (202) 462-3614 **FX:** (202) 387-1034 **FOUNDED:** 1971. **OFFICERS:** C. Payne Lucas, Pres. **MEMBERS:** 2,300. **STAFF:** 100. **BUDGET:** US$26,502,640. **LANGUAGES:** English, French, Portuguese.

DESCRIPTION: Seeks to improve the quality of life in rural Africa. Provides health and environmental protection services in rural areas of Africa; works to improve African water and agricultural resources; conducts public education programs in the United States on African development. Operates speakers' bureau. **PUBLICATIONS:** *Development Education Series*; Annual Report (in English, French, and Portuguese); Newsletter, periodic.

AFRICATRACK

8, rue Juliette Lamber, F-75017 Paris, France **PHONE:** 33 1 47312727 **FX:** 33 1 47312757 **FOUNDED:** 1987. **OFFICERS:** Henri-Jean Vittecoq. **MEMBERS:** 50. **BUDGET:** US$954,386. **LANGUAGES:** English, French.

DESCRIPTION: Members are private companies, farmers, mechanics, doctors, nurses, and others in 5 countries. Raises funds to buy agricultural equipment, seeds, fertilizers, and medical supplies for villages in Africa. Sponsors Africatrack, whereby agricultural experts, medical staff, and volunteers drive in a convoy of tractors and trailers from various locations in Europe to selected African villages and distribute tractors, trailers, and agricultural supplies. Believes that aid should be relevant and should promote selfhelp and long-term development. Organizes training programs.

AFRO-ASIAN PEOPLE'S SOLIDARITY ORGANIZATION (AAPSO)

89 Abdel Aziz al Seoud St., Manial, Cairo, Egypt **PHONE:** 20 2 3636081, 20 2 3622946 **FX:** 20 2 3637361 **E-MAIL:** aapso@idsc.gov.eg **FOUNDED:** 1958. **OFFICERS:** E.A. Vidyasekera, Sec.Coor. **MEMBERS:** 88. **STAFF:** 35. **BUDGET:** £E 300,000. **LANGUAGES:** Arabic, English, French.

DESCRIPTION: African and Asian solidarity committees, movements, and parties. Purpose is to coordinate and accelerate what the group sees as the liberation struggle of Afro-Asian peoples against colonialism, fascism, imperialism, neocolonialism, racism, and Zionism. Seeks to ensure complete national independence and cultural, economic, and social development. Supports decolonization and the right of all people to self-determination and freedom. Encourages unity among African and Asian democratic, patriotic, and progressive forces based on ''militant solidarity'' and in support of the principles of the Charter of the United Nations and U.N. declarations on the New International Economic Order, disarmament, and other matters. Sponsors conferences, research projects and symposia. **PUBLICATIONS:** *Bulletin* (in Arabic, English, and French), monthly; *Development and Socio-Economic Progress*, quarterly; *Proceedings of Conferences and Meetings*, periodic.

AFRO-ASIAN RURAL RECONSTRUCTION ORGANISATION (AARRO) (Organisation Afro-Asiatique pour la Reconstruction Rurale)

2 State Guest Houses Complex, Chana Kyapuri, New Delhi 100 021, Delhi, India **PHONE:** 91 11 600474, 91 11 6877783 **FX:** 91 11 6115937 **E-MAIL:** aarohq@hub.nic.in **FOUNDED:** 1962. **OFFICERS:** Dr. S.M. Ilyas. **MEMBERS:** 24. **STAFF:** 31. **BUDGET:** US$840,920. **LANGUAGES:** Arabic, English, French.

DESCRIPTION: Inter-country governmental organization uniting African and Asian governments to develop a better understanding of the economy among their rural populations. Purpose is to explore opportunities for coordinating efforts that improve the general welfare of rural people and eliminate poverty. Promotes cooperative activities; provides for the exchange of farmers and experts. Compiles statistics; conducts educational, research, and training programs. Sponsors workshops, seminars, and pilot projects; offers project formulation and consulting services. **PUBLICATIONS:** *AARRO Newsletter* (in Arabic and English), quarterly. Provides information on organization activities.; *Annual Report* (in Arabic and English); *Rural Reconstruction* (in English), semiannual. Contains research and review articles on agriculture and rural development. Also publishes workshop and seminar reports.

AFS INTERCULTURAL PROGRAMS (AFS)

71 W. 23rd St., 17th Fl., New York, NY 10010 **PHONE:** (212) 807-8686 **TF:** (800) AFS-INFO **FX:** (212) 807-1001 **TX:** AFSIUI666379 UW(WUI) **E-MAIL:** info@afs.org **WEBSITES:** http://www.afs.org **FOUNDED:** 1914. **OFFICERS:** Richard Spencer, Pres. **MEMBERS:** 100,000. **BUDGET:** US$32,000,000. **LANGUAGES:** English.

DESCRIPTION: Former American Field Service ambulance drivers, students, and host families in 80 countries who have participated in AFS programs. Promotes international understanding, primarily through exchange of secondary school students, 16 to 18 years of age. Conducts a variety of exchange programs providing family living experiences to fit the needs of the participants. Also conducts investigations of the nature and impact of intercultural experiences, program quality, cultural training, specific cultures, and global education. **PUBLICATIONS:** *AFS Orientation Handbook*, annual; *AFS Research Reports*, periodic; *AFS World*, quarterly. For alumni and host families.; *Directions*, bimonthly; *Papers in Intercultural Learning*, semiannual.

AGENCY FOR COOPERATION AND RESEARCH IN DEVELOPMENT (ACORD)

Dean Bradley House, 52 Horseferry Rd., London SW1P 2AF, England **PHONE:** 44 0171 2278600 **FX:** 44 0171 7991868 **E-MAIL:** acord@gn.apc.org **OFFICERS:** Idriss Jazairy. **LANGUAGES:** English.

DESCRIPTION: Membership consists of economic and community development organizations. Promotes sustainable development; seeks to ensure cooperation among members and between members and local government authorities and indigenous peoples in developing areas. Coordinates members' activities; provides technical assistance and administrative support to development projects.

AGENCY FOR THE PROHIBITION OF NUCLEAR WEAPONS IN LATIN AMERICA AND THE CARIBBEAN (OPANAL) (Organismo para la Proscripcion de las Armas Nucleares en la America Latina y el Caribe - OPANAL)

Temistocles 78, Colonia Polanco, 11560 Mexico City, DF, Mexico **PHONE:** 52 5 2804923, 52 5 2805064 **FX:** 52 5 2802965 **FOUNDED:** 1969. **OFFICERS:** Enrique Roman Morey, Sec.Gen. **MEMBERS:** 31. **STAFF:** 7. **BUDGET:** US$507,000. **LANGUAGES:** English, Spanish.

DESCRIPTION: Members are nations that have signed and ratified the Treaty of Tlatelolco for the Prohibition of Nuclear Weapons in Latin America and whose territories are entirely free of nuclear weapons. Purposes are to: administer the treaty and promote its goal toward prohibition of the production and use of nuclear weapons; establish procedures for an international control system ensuring observance of the treaty and proper operation of the control system; encourage exclusively peaceful use of nuclear materials and facilities within their jurisdictions; generally promote total disarmament worldwide. **PUBLICATIONS:** Report. Includes texts of treaties, studies, and conference resolutions.

AGRICULTURAL FOOD MARKETING ASSOCIATION FOR THE NEAR EAST AND NORTH AFRICA

PO Box 910725, Al Shimiesani, Amman, Jordan **PHONE:** 962 9 690418, 962 9 661192 **FX:** 962 9 690418 **TX:** 21654 UNDP JO **OFFICERS:** Magmout Al Hiyari. **LANGUAGES:** Arabic, English.

DESCRIPTION: Members are food and agricultural organizations. Promotes sustainable rural and agricultural development. Provides support and assistance to approved development projects; serves as a clearinghouse on food production.

AGUDATH ISRAEL WORLD ORGANIZATION (AIWO) (Organisation Mondiale Agudath Israel - OMAI)

Hacherat Sq., PO Box 326, IL-91002 Jerusalem, Israel **PHONE:** 972 2 5384357 **FX:** 972 2 5383634 **FOUNDED:** 1912. **OFFICERS:** Rabbi Yeudah Meir Abramowitz, Chmn. **MEMBERS:** 500,600. **STAFF:** 486. **BUDGET:** US$5,284,000. **LANGUAGES:** English, French, German, Hebrew, Spanish.

DESCRIPTION: Individuals united to solve, in the spirit of traditional Judaism, issues confronting Jewish people in Israel and worldwide. Coordinates Orthodox Jewish religious, educational, and philanthropic efforts. Organizes Beth Jacob program of educational seminaries for girls, with a special section on youth and adult education and school financing. Sponsors COPE Vocational Institute and Jewish Education Program. Operates Project RISE offering aid and education to Russian emigrants and provides aid to needy individuals in Eastern Bloc countries. Sponsors children's services. **PUBLICATIONS:** *Coalition (New York)*, periodic; *Hamodia (Israel)* (in Hebrew), daily; *Jedion (Belgique)* (in French and Hebrew), monthly; *Jewish Observer (New York)*, monthly; *Jewish Tribune (London)* (in English and Yiddish), weekly; *Judische Stimme (Zurich)* (in German), monthly; *La Voz Judia (Buenois Aires)* (in Spanish), monthly; *Perspectives (Toronto)*, 5/year; *World Congress Book* (in Hebrew), quinquennial; *Yiddishe Worte (New York)* (in Yiddish), monthly. Also publishes *Fifty Years of Agudath Israel* and other books on educational, historical, and ideological issues.

AID TO THE CHURCH IN NEED (ACN) (Kirche in Not Ostpriesterhilfe)

Bischof-Kaller-Strasse 3, Postfach 12 09, D-61452 Konigstein, Germany **PHONE:** 49 6174 2910 **FX:** 49 6174 3423 **E-MAIL:** kinop@compuserve.com **FOUNDED:** 1947. **OFFICERS:** Jose Correa. **BUDGET:** US$70,000,000. **LANGUAGES:** Dutch, English, French, German, Italian, Portuguese, Spanish.

DESCRIPTION: Participants are men and women in 15 countries. Works to promote Catholicism worldwide. Offers financial, material, and spiritual support to refugees and Christians in need in communist and former communist countries, and the oppressed in the Third World. Maintains information department. **PUBLICATIONS:** *Echo der Liebe* (in Dutch, English, French, German, Italian, Portuguese, and Spanish), 8/year; Books.

AID INTERNATIONAL (AIDI)

1776 Peachtree St. NW, Ste. 450, Atlanta, GA 30309 **PHONE:** (404) 875-3891 **FX:** (404) 872-6253 **OFFICERS:** O. J. Brown, Pres. **LANGUAGES:** English.

DESCRIPTION: Members are health care personnel. Seeks to improve the quality and availability of health care and social services in developing regions worldwide. Provides emergency relief and medical services; conducts training programs for medical technicians and paramedics. Maintains emergency disaster management team.

AIDS CARE EDUCATION AND TRAINING (ACET)

PO Box 3693, London SW15 2BQ, England **PHONE:** 44 0181 7800400 **FX:** 44 0181 7800450 **E-MAIL:** acet@acetuk.org **WEBSITES:** http://www.acetuk.org **FOUNDED:** 1988. **OFFICERS:** Richard Evans, Development Dir. **STAFF:** 30. **BUDGET:** £1,300,000. **LANGUAGES:** English.

DESCRIPTION: Promotes effective and unconditional care for people with HIV and AIDS worldwide. Conducts educational and training programs for AIDS care providers; works with national and international AIDS care organizations to improve the quality and availability of AIDS care. **PUBLICATIONS:** *ACET Newsletter*, 3/year; *HIV - Facts for Life*.

AIR SERV INTERNATIONAL

615 Brookside Ave., Ste. B, PO Box 3041, Redlands, CA 92373-0993 **PHONE:** (909) 793-2627 **FX:** (909) 793-0226 **TX:** 4720427 AIRSERV **E-MAIL:** asi@xc.org **WEBSITES:** http://www.airserv.org **FOUNDED:** 1984. **OFFICERS:** Kenneth W. Frizzell, Pres. **STAFF:** 85. **BUDGET:** US$4,000,000. **LANGUAGES:** English.

DESCRIPTION: Provides air transport of relief and development personnel, medicines, and cargo to countries devastated by war or natural disasters. Utilizes small, light aircraft to deliver supplies and personnel to restricted areas. Conducts charitable programs.

AIRPORTS COUNCIL INTERNATIONAL - AFRICA (ACIA)

Cairo Intl. Airport, Terminal 2, Rm. 2118, Cairo, Egypt **PHONE:** 20 2 3046955 **FX:** 20 2 3046954 **E-MAIL:** aci2000afr10@hotmail.com **LANGUAGES:** Arabic, English.

DESCRIPTION: International airports and airport authorities. Fosters cooperation among members and between members and other partners in aviation. Works to maintain an air transport system that is safe, secure, efficient, and environmentally sustainable. Represents members before the International Civil Aviation Organization.

AIRPORTS COUNCIL INTERNATIONAL - ASIA (ACIA)

c/o Institute of Aviation Management, Intl. Airports Authority of India, Patterson Farm, Old Gurgeon Rd., New Delhi 110037, Delhi, India **PHONE:** 91 11 5652307 **FX:** 91 11 5652674 **LANGUAGES:** English.

DESCRIPTION: International airports and airport authorities. Fosters cooperation among members and between members and other partners in aviation. Works to maintain an air transport system that is safe, secure, efficient, and environmentally sustainable. Represents members before the International Civil Aviation Organization.

AIRPORTS COUNCIL INTERNATIONAL - EUROPE (ACI-Europe)

Square de Meeus 6, B-1000 Brussels, Belgium **PHONE:** 32 2 5220978 **FX:** 32 2 5132606 **E-MAIL:** info@aci-europe.org **WEBSITES:** http://www.aci-europe.org **FOUNDED:** 1991. **OFFICERS:** Danielle Michel, Mem.Ser.Man. **MEMBERS:** 400. **STAFF:** 13. **LANGUAGES:** English, French.

DESCRIPTION: International airports and airport authorities. Fosters cooperation among members and between members and other partners in aviation. Works to maintain an air transport system that is safe, secure, efficient, and environmentally sustainable. Represents members before the International Civil Aviation Organization. **PUBLICATIONS:** *Communique* (in English and French), monthly.

AIRPORTS COUNCIL INTERNATIONAL - LATIN AMERICA-CARIBBEAN (ACILAC)

c/o IAAIM, Apartado 146, La Guaira, Caracas, Venezuela **PHONE:** 58 2 3551232, 58 31 551235 **FX:** 58 2 3551654 **E-MAIL:** acilac@telcel.net.ve **WEBSITES:** http://www.acilac.com.ve **FOUNDED:** 1991. **OFFICERS:** Fernando Oliveira, Reg.Sec. **MEMBERS:** 60. **STAFF:** 3. **LANGUAGES:** English, Spanish.

DESCRIPTION: International airports and airport authorities. Fosters cooperation among members and between members and other partners in aviation. Works to maintain an air transport system that is safe, secure, efficient, and environmentally sustainable. Represents members before the International Civil Aviation Organization. **PUBLICATIONS:** *Boletin de ACI/LAC* (in Spanish), bimonthly.

AIRPORTS COUNCIL INTERNATIONAL - PACIFIC (ACIP)

PO Box 23750, Airport Postal Outlet, Richmond, BC, Canada V7B 1Y7 **PHONE:** (604) 276-6773, (604) 276-6753 **FX:** (604) 276-6070 **E-MAIL:** idaley@aci-pacific.org **WEBSITES:** http://www.aci-pacific.org **FOUNDED:** 1991. **OFFICERS:** Laura Daley, Admin. **STAFF:** 3. **LANGUAGES:** English, French.

DESCRIPTION: International airports and airport authorities. Fosters cooperation among members and between members and other partners in aviation. Works to maintain an air transport system that is safe, secure, efficient, and environmentally sustainable. Represents members before the International Civil Aviation Organization. **PUBLICATIONS:** *Dateline* (in English), 3/year.

ALBERT EINSTEIN INTERNATIONAL ACADEMY FOUNDATION (AEIAF)

U.S. Federal Bldg., Rms. 201-201A, 301 W. Lexington, Independence, MO 64050 **PHONE:** (816) 224-1233 **FX:** (501) 821-2398 **E-MAIL:** uidba@usa.net **FOUNDED:** 1965. **OFFICERS:** Dr. Marcel Dingli-Attard-Inguanez, Pres. **LANGUAGES:** English.

DESCRIPTION: Participants are scholars dedicated to promoting world peace through the implementation of a world government based on terms advocated by Albert Einstein (1879-1955). Proposes that the United Nations charter be revised to: transform the U.N. from a ''league of sovereign states into a government deriving its specific powers from the peoples of the world'' through direct vote; establish the U.N. Security Council as the executive branch, the U.N. General Assembly as the legislative branch, and independent tribunals as the judicial branch of a world government; confer upon the U.N. some of the functions and responsibilities of a state, including establishment and maintenance of laws governing the production and use of ''weapons of mass destruction'' and the enforcement of a Bill of Rights designed to protect individual freedom worldwide. Promotes world peace, literature, science, the fine arts and other fields of knowledge. Conducts research on world peace, climate, ocean affairs, and foreign affairs. Maintains information on important educational institutions worldwide. **PUBLICATIONS:** *Einstein on Peace*, periodic.

ALBERT SCHWEITZER ECOLOGICAL CENTRE (ASEC) (Centre Ecologique Albert Schweitzer - CEAS)

rue de la Cote 2, CH-2000 Neuchatel, Switzerland **PHONE:** 41 32 7250836 **FX:** 41 32 7251507 **E-MAIL:** ceas.ne@bluewin.ch **FOUNDED:** 1980. **OFFICERS:** Daniel Schneider, Dir. **MEMBERS:** 200. **STAFF:** 5. **BUDGET:** 1,500,000 SFr. **LANGUAGES:** English, French.

DESCRIPTION: Members are technical workers and researchers interested in ecological solutions to agricultural problems. Encourages the application of appropriate technology, solar energy, and other ecological methods to combat desertification and increase agricultural production. Coordinates such projects as the production and distribution of water pumps, solar heaters, and dryers with African organizations. Conducts vocational training in appropriate technology. Compiles documentation. **PUBLICATIONS:** *L'Avenir est entre vos mains* (in French and German), periodic; *Tropical Fruits Processing Manual* (in English and French). Contains rules and procedures for the management and operation of small fruit processing enterprises, and recipes.

ALL AFRICA CONFERENCE OF CHURCHES (AACC) (Conference des Eglises de Toute l'Afrique - CETA)

The Gen. Sec., Waiyaki Way, Westlands, PO Box 14205, Nairobi, Kenya **PHONE:** 254 2 441338, 254 2 441339 **FX:** 254 2 443241 **TX:** 22175 AACC **E-MAIL:** aacc-selfood@maf.org **FOUNDED:** 1963. **MEMBERS:** 137. **LANGUAGES:** English, French.

DESCRIPTION: Christian churches (130) and councils (22) of several denominations from 33 African countries. Associate members made up of Theological Colleges. Examines international affairs and the church's role in allaying contemporary crises. Conducts development activities for the benefit of women, young people, and refugees. Encourages the sharing of experience and information among members; provides placement and training services to African Christian churches. Sponsors programs focusing on human rights abuses, war, and the needs of refugees and exiles; undertakes research in areas such as family life education, women's leadership programs, and the preservation of important historical documents. Offers research, documentation, language, consultation, communications, and technical services. Works in collaboration with the World Council of Churches. Sponsors charitable program. Compiles statistics. **PUBLICATIONS:** *Baobab* (in English and French), quarterly; *TAM-TAM* (in English and French), bimonthly.

ALL AFRICA PRESS SERVICE (APS)

PO Box 14205, Nairobi, Kenya **PHONE:** 254 2 442215, 254 2 440224 **FX:** 254 2542 443241 **TX:** 22175 NAIROBI **FOUNDED:** 1979. **STAFF:** 15. **BUDGET:** US$100,000. **LANGUAGES:** English, French.

DESCRIPTION: Disseminates news and information in Africa and abroad from an African perspective. Highlights development alternatives; features the concerns of indigenous peoples. Works to minimize African dependence on foreign media agencies for news and information. Fosters improved communication between African countries and their peoples. Is operated in conjunction with the All Africa Conference of Churches, the Lutheran World Federation, the World Council of Churches, and the World Association for Christian Communication. **PUBLICATIONS:** *AANA News* (in English and French), weekly; *APS Bulletin* (in English), weekly.

ALLIANCE OF NGOS ON CRIME PREVENTION AND CRIMINAL JUSTICE (ANCPCJ)

PO Box 81826, Lincoln, NE 68501 **PHONE:** (402) 464-0602 **FX:** (402) 464-5931 **E-MAIL:** garyhill@cega.com **FOUNDED:** 1972. **OFFICERS:** Gary Hill. **MEMBERS:** 26. **STAFF:** 1. **LANGUAGES:** English.

DESCRIPTION: Membership consists of: International nongovernmental organizations having consultative status with the United Nations and focusing on crime prevention, criminal justice administration, or treatment of the offender (16); observers (10) not having consultative status or not having a major portion of their organizational activity in those areas. Provides for regularized communication with the UN Crime Prevention Branch. Facilitates and provides a structure for NGOs working together on issues of common concern. A major area of activity has been on the transfer of imprisoned aliens to their home countries for serving of their sentences.

ALLIANCE OF SMALL ISLAND STATES (ASIS)

Trinidad and Tobago Mission to the United Nations, 820 2nd Ave., 5th Fl., New York, NY 10017 **PHONE:** (212) 697-7620 **FX:** (212) 682-3580 **OFFICERS:** Annette des Iles. **LANGUAGES:** English, French.

DESCRIPTION: Representatives of small island and coastal states. Promotes appropriate and sustainable economic and social development in member countries. Serves as a clearinghouse on island and coastal states; represents members' interests before the United Nations and international trade and development organizations.

ALLIANCE FOR SOUTH ASIAN AIDS PREVENTION (ASAAP)

399 Church St., 3rd Fl., Toronto, ON, Canada M5B 2J6 **PHONE:** (416) 599-2727 **FX:** (416) 351-8994 **OFFICERS:** Anthony Mohammed, Exec.Dir. **LANGUAGES:** English, French.

DESCRIPTION: Membership includes individuals and organizations. Seeks to stop the spread of HIV and AIDS in Asia. Conducts AIDS and HIV prevention education programs. **PUBLICATIONS:** *WASPNews*, periodic.

ALTERNATIVE TRADE ORGANIZATION OF NONTRADITIONAL PRODUCTS AND DEVELOPMENT FOR LATIN AMERICA

PO Box 14-0233, Parquee Industrial Villa El Salvador, Lote 9, MZDF, Lima 14, Peru **PHONE:** 51 1 2875995, 51 1 2873703 **FX:** 51 1 2873703 **FOUNDED:** 1989. **OFFICERS:** Gaston Vizcarra, Pres. **LANGUAGES:** Spanish.

DESCRIPTION: Promotes economic development through formation of local cooperative organizations and the establishment of microenterprises. Works to ensure that indigenous institutions retain control of economic development programs, and to enhance the economic and social potential of women in traditional societies. Provides technical and financial assistance to local business enterprises and cooperatives.

AMAZONIAN COOPERATION COUNCIL (ACC)

Tratado de Cooperacion Amazonica, Avenida Prolongacion Primavera 654, Chacarilla, Surco, Lima 33, Peru **PHONE:** 51 14 499084, 51 14 389662 **FX:** 51 14 498718 **LANGUAGES:** English, Spanish.

DESCRIPTION: Representatives of countries controlling territory in the Amazon Basin. Promotes environmentally sustainable development of the Amazon region. Monitors development activities in the Amazon; facilitates communication and cooperation among members.

AMER MEDICAL DIVISION, AMERICAN NEAR EAST REFUGEE AID (AMER)

1522 K St. NW, Ste. 202, Washington, DC 20005 **PHONE:** (202) 347-2558 **FX:** (202) 682-1637 **E-MAIL:** anera@anera.org **WEBSITES:** http://www.anera.org **FOUNDED:** 1948. **OFFICERS:** Nina Dodge, VP. **STAFF:** 1. **BUDGET:** US$500,000. **LANGUAGES:** English.

DESCRIPTION: Has operated since 1971 as a division of American Near East Refugee Aid. Solicits and ships medical supplies and pharmaceuticals for use by Palestinians, Lebanese and Jordanians in the Middle East and helps with various other medical projects in the region.

AMERICAN AND COMMON MARKET CLUB (ACMC)

17, rue Emile Claus, bte. 3, B-1050 Brussels, Belgium **PHONE:** 32 2 6475801 **FX:** 32 2 6475801 **FOUNDED:** 1961. **OFFICERS:** Ambassador Alfred Cahen, Pres. **MEMBERS:** 400. **STAFF:** 1. **BUDGET:** 600,000 BFr. **LANGUAGES:** English, French.

DESCRIPTION: Corporate chief executive officers, EEC officials, and others. Brings together professionals who are nationals of the United States or the European Communities and who are interested in promoting the development of the European Communities. **PUBLICATIONS:** *ACMC Directory* (in English and French), annual. Contains listings of membership and speakers.

AMERICAN FRIENDS SERVICE COMMITTEE (AFSC)

1501 Cherry St., Philadelphia, PA 19102 **PHONE:** (215) 241-7000 **TF:** (800) 558-AFSC **FX:** (215) 241-7247 **E-MAIL:** afscinfo@afsc.org **WEBSITES:** http://www.afsc.org **FOUNDED:** 1917. **OFFICERS:** Kara Newell, Exec.Dir. **STAFF:** 356. **BUDGET:** US$27,661,703. **LANGUAGES:** English.

DESCRIPTION: Founded by and related to the Religious Society of Friends (Quakers); supported and staffed by individuals sharing basic values regardless of religious affiliation. Attempts to relieve human suffering and find new approaches to world peace and social justice through nonviolence. Work in 22 countries and 43 areas of the United States includes development and refugee relief, peace education, and community organizing. Conducts programs with U.S. communities on the problems of minority groups such as housing, employment, and denial of legal rights. Maintains Washington, D.C., office to present AFSC experience and perspectives to policymakers. Seeks to build informed public resistance to militarism and the military-industrial complex. A co-recipient of the 1947 Nobel Peace Prize. Programs are multiracial, nondenominational, and international. **PUBLICATIONS:** *Literature Resource Catalogue*; *Quaker Service Bulletin*, semiannual; Annual Report (in English), annual. Contains program activities and financial summary. Also publishes program literature; prepares video materials on current world problems.

AMERICAN INSTITUTE FOR INTERNATIONAL STEEL (AIIS)

1325 G St. NW Ste. 980, Washington, DC 20005-3104 **PHONE:** (202) 628-3878 **FX:** (202) 737-3134 **E-MAIL:** aiis@msn.com **FOUNDED:** 1990. **OFFICERS:** David H. Phelps, Exec.Dir. **MEMBERS:** 200. **LANGUAGES:** English.

DESCRIPTION: Importers and exporters of steel produced worldwide.

AMERICAN JEWISH JOINT DISTRIBUTION COMMITTEE (AJJDC)

711 3rd Ave., New York, NY 10017-4014 **PHONE:** (212) 687-6200 **FX:** (212) 682-7262 **E-MAIL:** info@jdcny.org **WEBSITES:** http://www.jdc.org **FOUNDED:** 1914. **OFFICERS:** Michael Schneider, Exec.VP. **STAFF:** 286. **BUDGET:** US$65,000,000. **LANGUAGES:** English.

DESCRIPTION: Maintains health, welfare, assistance and social programs for needy Jews in nearly 60 countries in Asia, Africa, Europe, the former

Soviet Union, and Latin America. Provides funds for secular and religious education, feeding and medical programs, economic aid, summer camps, community development, manpower training, and aid to the aged and handicapped. The JDC program in Israel provides a broad range of services for the aged, the disabled and children youth at risk, and participates with local agencies in developing health, welfare, and rehabilitation services, vocational training and placement, social integration, and community center programs. Financially supported by Jewish federations and welfare funds through the United Jewish Appeal. **PUBLICATIONS:** *American Jewish Joint Distribution Committee—Annual Report*. Provides information on the JDC, its programs, and countries where the organization operates. Includes budget information.; Brochures.

AMERICAN JEWISH WORLD SERVICE (AJWS)

989 Ave. of the Americas, Fl. 10, New York, NY 10018-5410 **PHONE:** (212) 736-2597 **FX:** (212) 736-3463 **E-MAIL:** jws@jws.org **WEBSITES:** http://www.ajws.org **FOUNDED:** 1985. **OFFICERS:** Stepanie Fingeroth, Asst. to Pres. **MEMBERS:** 30,000. **STAFF:** 12. **BUDGET:** US$2,600,000. **LANGUAGES:** English.

DESCRIPTION: A Jewish-sponsored humanitarian and relief organization working on an exclusively non-sectarian basis in the developing world. AJWS' aim is to provide the means for disadvantaged people particularly women and children and threatened minorities, to move toward self-sufficiency. The organization's Jewish Volunteer Corps sends professionals to work with non-governmental organizations for periods of one to nine months. **PUBLICATIONS:** *AJWS Reports*, 3/year. Includes update of programming and volunteer activities; Annual Report.

AMERICAN REFUGEE COMMITTEE (ARC)

2344 Nicollet Ave. S., Ste. 350, Minneapolis, MN 55404-3305 **PHONE:** (612) 872-7060 **FX:** (612) 607-6499 **E-MAIL:** archq@archq.org **WEBSITES:** http://www.archq.org **FOUNDED:** 1979. **OFFICERS:** Judy Marcoviller, Recruitment. **STAFF:** 18. **BUDGET:** US$13,467,687. **LANGUAGES:** English.

DESCRIPTION: Works for the survival, health and well-being of refugees, displaced persons and those at risk. ARC staff and volunteers, provide primary health care, training and other health related services to more than one million people in Asia, Africa, Europe, and Central America. **PUBLICATIONS:** *Bridges Newsletter*, 4/year; Annual Report.

AMERICAS SOCIETY (AS)

680 Park Ave., New York, NY 10021 **PHONE:** (212) 249-8950 **FX:** (212) 249-5868 **WEBSITES:** http://www.americas-society.org; http://www.as-coa.org **FOUNDED:** 1982. **OFFICERS:** Ambassador Thomas E. McNamara, Pres. **MEMBERS:** 800. **STAFF:** 50. **BUDGET:** US$3,000,000. **LANGUAGES:** English, Portuguese, Spanish.

DESCRIPTION: Works to educate U.S. citizens about the other nations of the Americas. Fosters mutual understanding of the economic, political, and cultural issues facing Latin America, the Caribbean, and Canada today. Promotes appreciation of the cultures of the Western Hemisphere. **PUBLICATIONS:** *Review*, biennial. Focuses on Latin American literature and arts.; Books; Brochure; Catalogs.

AMIDEAST

1730 M St. NW, Ste. 1100, Washington, DC 20036-4505 **PHONE:** (202) 776-9600 **FX:** (202) 776-7000 **E-MAIL:** inquiries@amideast.org **WEBSITES:** http://www.amideast.org **FOUNDED:** 1951. **OFFICERS:** Hon. William A. Rugh, Pres. and CEO. **MEMBERS:** 165. **STAFF:** 202. **BUDGET:** US$36,400,000. **LANGUAGES:** Arabic, French.

DESCRIPTION: Dedicated to improving understanding between Americans and the people of the Middle East and North Africa through education, information and development programs. Conducts educational advising and testing services for Arab students and institutions interested in U.S. educational opportunities. Administers educational and training programs for a variety of government, corporate, and institutional sponsors of Arab students. Maintains English-language programs for the general public and corporate and government agency clients in Bahrain, Egypt, Kuwait, Lebanon, Tunisia, and Yemen. Provides public outreach services in the form of publications and videotapes to support educational exchanges and materials to improve teaching about the Arab world in American secondary schools and Colleges. Directs technical assistance programs to support institution-building in the Arab world, with an emphasis on educational institutions, public administration, judiciary bodies, legislatures, and nongovernmental organizations. Maintains a network of field offices in Bahrain, Egypt, Jordan, Kuwait, Lebanon, Morocco, Syria, Tunisia, the West Bank/Gaza and the Yemen Arab Republic. **PUBLICATIONS:** *The Advising Quarterly.* Newsletter serving as a periodic reference for international education exchange professions. Contains book, software, and video reviews.; *AMIDEAST News*, quarterly. Contains project activities updates.; *Education in the Arab World.* Describes the educational systems of Algeria, Bahrain, Egypt, Jordan, Kuwait, Lebanon, and Morocco.; *Introduction to the Arab World.* Lays the foundation for exploring the diversity and unity, cultural traditions, and contemporary concerns of the Arab world.; *Planning Your Future: Resources on Careers and Higher Education.* Lists over 1000 field-of-study resources providing the most current information on career and educational advising books, videos and software.; Annual Report. Includes program review, membership list, project descriptions, and financial statement.; Various educational resources on the Arab world.

AMNESTY INTERNATIONAL (AI) (Amnistia Internacional - AI)

Intl. Secretariat, 1 Easton St., London WC1X 8DJ, England **PHONE:** 44 171 4135500 **FX:** 44 171 9561157 **TX:** 28502 AMNSTY G **E-MAIL:** amnestyis@amnesty.org **WEBSITES:** http://www.amnesty.org **FOUNDED:** 1961. **OFFICERS:** Pierre Sane, Sec.Gen. **MEMBERS:** 1,100,000. **STAFF:** 270. **LANGUAGES:** Arabic, English, French, Spanish.

DESCRIPTION: A human rights movement working to secure the immediate and unconditional release of all prisoners of conscience (individuals who have not used or advocated violence and have been imprisoned due to their beliefs, color, sex, religion, ethnic origin, or language); urges immediate, fair trials of political prisoners. Seeks an end to torture, executions and ''disappearances,'' abuses by opposition groups, hostage taking, arbitrary killing, and other inhuman, cruel, or degrading treatment or punishment. Conducts research and organizes worldwide action by local members on behalf of prisoners held in countries other than their own; initiates national and international campaigns to publicize patterns of human rights abuses; sends delegates to observe trials and visit countries to interview prisoners and meet with government officials. Has declared an Annual Amnesty International Week. Chooses 3 Worldwide Appeal Cases to be printed in its newsletter, so that concerned individuals may write letters to the prisoners, organize petitions on their behalf, and publicize each prisoner's situation. **PUBLICATIONS:** *Amnesty International Report* (in Arabic, English, French, and Spanish), annual; *Newsletter* (in Arabic, English, French, and Spanish), monthly; Handbook, periodic.

AMNESTY FOR WOMEN

Grosse Berg Str. 231, D-22767 Hamburg, Germany **PHONE:** 49 40 384753, 49 40 38613654 **FX:** 49 40 385758 **E-MAIL:** amnesty4women@t-online.de **FOUNDED:** 1986. **OFFICERS:** Pat Mix. **MEMBERS:** 20. **STAFF:** 10. **BUDGET:** DM 200,000. **LANGUAGES:** Bulgarian, Croatian, English, German, Polish, Portuguese, Russian, Spanish, Tagalog, Thai, Turkish.

DESCRIPTION: Strives to eliminate the exploitation, oppression, and mistreatment of women. Organizes programs and activities to raise public consciousness. Lobbies governmental agencies and organizations for recognition and protection of women's rights; supports development programs which improve the status of women. **PUBLICATIONS:** *Deutschland - Ein Paradies fur Frauen?* (in English, German, Polish, Portuguese, Russian, Spanish, Tagalog, and Thai), periodic. Information brochure for women immigrating to Germany.; *Rundbrief Amnesty for Women* (in German), quarterly; Reports (in English), periodic.

ANCIENT FOREST INTERNATIONAL

PO Box 1850, Redway, CA 95560 **PHONE:** (707) 923-3015, (707) 923-3001 **FX:** (707) 923-3015 **E-MAIL:** afi@igc.org **WEBSITES:** http://www.ancientforests.org **FOUNDED:** 1988. **OFFICERS:** Esteban Millard, Off.Mgr. **STAFF:** 5. **BUDGET:** US$100,000. **LANGUAGES:** English, Spanish.

DESCRIPTION: Individuals interested in the conservation and wise use of natural resources, particularly native forests. Promotes public awareness of environmental issues; gathers and disseminates information on ecological problems and their solutions; upholds the human rights of indigenous peoples. Principle focus is the acquisition of old-growth forests in rainforests of America, south & north. **PUBLICATIONS:** *Chile's Native Forests.*

ANDEAN AIRLINE ASSOCIATION (AAA) (Asociacion Andina de Lineas Aereas - AALA)

c/o JUNAC, Paseo de Republica 3895, Casilla 18-1177, Lima 27, Peru **PHONE:** 51 14 2212222 **FX:** 51 14 2213329 **TX:** JUNAC 20104 PU **LANGUAGES:** English, Spanish.

DESCRIPTION: Members are airlines serving the Andean countries. Promotes growth and development of the air travel and shipping industries. Represents members before trade and labor organizations, government agencies, and the public.

ANDEAN ASSOCIATION OF SHIPOWNERS (AAO) (Asociacion Andina de Armadores - AAA)

c/o JUNAC, Paseo de la Republica 3895, Casilla 18-1177, Lima 27, Peru **PHONE:** 51 14 2212222 **FX:** 51 14 2213329 **TX:** JUNAC 20104 PU **LANGUAGES:** English, Spanish.

DESCRIPTION: Members are ship-owners. Promotes growth and development of the maritime shipping industries of the Andean countries. Represents members' commercial and regulatory interests; conducts promotional activities.

ANDEAN COMMISSION OF JURISTS (ACJ) (Comision Andina de Juristas - CAJ)

Los Sauces 285, San Isidro, Lima 27, Peru **PHONE:** 51 14 407907, 51 14 428094 **FX:** 51 14 426468 **E-MAIL:** dgs@cajpe.org.pe **FOUNDED:** 1982. **OFFICERS:** Diego Garcia-Sayan, Exec.Dir. **MEMBERS:** 16. **STAFF:** 32. **LANGUAGES:** English, Spanish.

DESCRIPTION: Jurists from Bolivia, Chile, Colombia, Ecuador, Peru, and Venezuela working to protect human rights and promote democracy. Offers counseling and education regarding legal and human rights issues to peoples of the region. Conducts research; holds seminars. Maintains consultative status at the Economic and Social Council of the UN. **PUBLICATIONS:** *Andean Newsletter* (in English and Spanish), monthly.

ANDEAN COMMUNITY OF NATIONS (Grupo Andino - Junta del Acuerdo de Cartagena)

Avenida Paseo de la Republica 3895, Casilla Postal 18-1177, Lima 27, Peru **PHONE:** 51 221 2222 **FX:** 51 221 3329 **TX:** 20104 PE **FOUNDED:** 1969. **OFFICERS:** Sebastian Alegrett Ruiz, Gen.Sec. **MEMBERS:** 5. **STAFF:** 250. **LANGUAGES:** Spanish.

DESCRIPTION: Members are Andean countries including Bolivia, Colombia, Ecuador, Peru, and Venezuela. Promotes the balance and harmonious economic development of member countries; encourages their growth by economic integration; facilitates their Latin American integration with the aim of raising the standard of living in the area. Compiles statistics. **PUBLICATIONS:** *Gaceta Oficial* (in Spanish), periodic; *Gaceta Oficial del Acuerdo de Cartagena* (in Spanish), periodic.

ANDEAN INFORMATION NETWORK (AIN)

Casilla 4817, Cochabamba, Bolivia **PHONE:** 591 42 24383 **FX:** 591 42 52401 **E-MAIL:** paz@albatross.cnb.net **WEBSITES:** http://www.scbbs-bo.com/ain **OFFICERS:** Kathryn Ledebur, Coor. **LANGUAGES:** English, Spanish.

DESCRIPTION: Membership includes individuals and organizations. Promotes respect for human rights and the rule of law in the Andean region. Publicizes human rights abuses; monitors legislative and judicial development in the region.

ANDEAN PATAGONIAN NATURALIST SOCIETY (APNS)

Villegas 369, 2e piso A, Carlos de Bariloch 84, Rio Negro, Argentina **PHONE:** 54 94 426800 **OFFICERS:** Miguel Christie. **LANGUAGES:** English, Spanish.

DESCRIPTION: Environmentalists with an interest in the ecosystems of Patagonia. Promotes preservation of endangered Patagonian wildlife and habitats. Serves as a clearinghouse on Patagonian ecology; sponsors research and educational programs.

ANDEAN SOCIAL MANAGEMENT CORP. (Corporacion Andina de Gerencia Social - COANDES)

Casilla Postal 17-11-6618, Quito, Ecuador **PHONE:** 593 2 464852, 593 2 466496 **FX:** 593 2 464852 **FOUNDED:** 1989. **OFFICERS:** Alfredo Bastidas Torres, Exec.Pres. **MEMBERS:** 130. **STAFF:** 32. **BUDGET:** US$28,000. **LANGUAGES:** English, Spanish.

DESCRIPTION: Promotes integrated, sustainable economic growth throughout the Andes Mountains. Conducts market studies and analyzes development programs to assess their appropriateness for the region. Consults with development agencies planning programs for the region; monitors results and impacts of existing projects to determine their ongoing fitness. **PUBLICATIONS:** *Social Management* (in Spanish), semiannual. Covers social development topics.

ANDEAN TRIBUNAL OF JUSTICE (ATJ)

Calle Rocha 450, Apartado Postal 17079054, Quito, Ecuador **PHONE:** 593 2 237264, 593 2 528 **FX:** 593 2 554533 **TX:** 21263 ED **LANGUAGES:** English, Spanish.

DESCRIPTION: Members are attorneys, jurists, and legal scholars. Seeks to advance the legal systems of the countries of the Andes. Sponsors legal research; makes available legal services in international disputes.

ANGELCARE (CAI)

PO Box 600370, San Diego, CA 92160-0370 **PHONE:** (619) 562-0631 **TF:** (800) 842-2810 **FX:** (619) 258-8671 **E-MAIL:** angcare@aol.com **WEBSITES:** http://www.angelcare.org **FOUNDED:** 1977. **OFFICERS:** Dr. T. J. Grosser, CEO & Pres. **STAFF:** 60. **BUDGET:** US$4,500,000. **LANGUAGES:** English.

DESCRIPTION: Nondenominational charitable program that provides nutritional, medical, and educational assistance to needy children in Southeast Asia, Africa, Latin America, Eastern Europe, and U.S. Provides resources to meet health needs, including nutrition supplements (''Nutri-Paks''), medicine, and the services of physicians, nurses, nutritionists, and paramedics. Operates primary health care clinics in Nairobi, Kenya, and Malaysia. Participates in foreign grant-matching programs; operates mobile nutrition, blood collection, and distribution vans; establishes and operates ''health posts'' to examine and care for malnourished children and mothers. Conducts ''feed and teach programs,'' which train mothers in nutrition and in identifying, cultivating, and preparing nutritious foods available from local sources. Educates community members in preventive health care, proper sanitation, and growing nutritious foods. Angelcare is funded through contributions and the operation of a child sponsorship program in Indonesia, Thailand, Guatemala, and Kenya in which individuals financially support and correspond with a needy child. Provides the opportunity for individuals or groups to donate gift ''Nutri-Paks,'' Medical Gift-Paks, equipment, and training materials. Sponsors projects to provide a training ground for medical students in treatment of malnourished, ill, and destitute children. Operates speakers' bureau. **PUBLICATIONS:** *Briefs*, MON. Includes member news and information on programs.

ANGLO-JAPANESE ECONOMIC INSTITUTE (AJEI)

6 Hugh St., London SW1V 1RP, England **PHONE:** 44 171 8217980 **FX:** 44 171 8217981 **E-MAIL:** ajei@dial.pipex.com **FOUNDED:** 1961. **OFFICERS:** David Morris, Deputy Dir. **STAFF:** 4.

DESCRIPTION: Monitors and reports on economic and policy matters in the United Kingdom and Japan. Promotes economic relations between the two countries. **PUBLICATIONS:** *Euro-Japanese Journal*, 3/year; *Japanese Addresses in the United Kingdom* (in English), semiannual. Lists branch offices of Japanese companies operating in the United Kingdom. Includes company name index.

ANTI-DISCRIMINATION ALLIANCE (ADA)

Rue de la Concorde 53, B-1050 Brussels, Belgium **PHONE:** 32 2 5480491 **FX:** 32 2 5480499 **E-MAIL:** d.maes@ecas.org **OFFICERS:** Tony Venables. **LANGUAGES:** English, French.

DESCRIPTION: Individuals and organizations. Promotes equal protection under the law of all persons. Seeks to eradicate discrimination based on race, gender, age, or belief. Monitors legal, political, and judicial activities; publicizes human rights abuses and cases of discrimination.

ANTI-DRUGS DIVISION EUROPEAN UNION (ADDEU)

Square Charles Maurice Wiser 13/1, B-1040 Brussels, Belgium **PHONE:** 32 2 2310597 **FX:** 32 2 2310597 **OFFICERS:** G. Seronveau. **LANGUAGES:** English, French.

DESCRIPTION: Members are national drug abuse and trafficking prevention organizations and agencies. Seeks to end the trade in illegal drugs; promotes rehabilitation of substance abusers. Facilitates communication and cooperation among members.

ANTI-SLAVERY INTERNATIONAL (ASI) (Socieded Contra la Esclavitud)

The Stableyard, Broomgrove Rd., London SW9 9TL, England **PHONE:** 44 171 9249555 **FX:** 44 171 7384110 **E-MAIL:** antislavery@gn.apc.org **FOUNDED:** 1839. **OFFICERS:** Mike Dottridge, Dir. **MEMBERS:** 2,000. **STAFF:** 17. **BUDGET:** £400,000. **LANGUAGES:** English, French, German, Portuguese, Spanish.

DESCRIPTION: Works to: eradicate slavery in all its forms; abolish labor systems analogous to slavery; promote the well-being of indigenous peoples. Works to make conventions of the United Nations on human rights more effective. Initiates research into slavery and exploitation of indigenous peoples. **PUBLICATIONS:** *Annual Report* (in English); *Reporter*, quarterly; Reports, periodic. On human rights, child labor, and indigenous people.

ANZUS COUNCIL

Dept. of Foreign Affairs, Canberra, ACT 2600, Australia **FX:** 61 6 2612151 **FOUNDED:** 1951. **OFFICERS:** A.J. Tyler **MEMBERS:** 3. **LANGUAGES:** English.

DESCRIPTION: Australian and New Zealand ministers for foreign affairs and U.S. Secretary of State. Administers the ANZUS security treaty drafted by members whose purpose is to provide for their collective defense and to preserve peace and security in the Pacific area. The United States has accused New Zealand of violating the treaty by refusing entry of U.S. ships to New Zealand ports. New Zealand has a policy of refusing to let nuclear-powered or nuclear-armed ships enter its ports and has enforced that policy against the United States. The U.S. will neither confirm nor deny the presence of nuclear arms on its warships. Because of this disagreement, the U.S. in 1986 formally suspended its security commitment to New Zealand. While Australia disagrees with New Zealand's policies, it has retained defense ties with both the U.S. and New Zealand. **PUBLICATIONS:** *Communique* (in English), annual.

AOAC INTERNATIONAL

481 N. Frederick Ave., No. 500, Gaithersburg, MD 20877-2504 **PHONE:** (301) 924-7077 **FX:** (301) 924-7089 **E-MAIL:** aoac@aoac.org **WEBSITES:** http://www.aoac.org **FOUNDED:** 1884. **OFFICERS:** Raymond Matulis, Exec.Dir. **MEMBERS:** 3,800. **STAFF:** 30. **BUDGET:** US$3,000,000. **LANGUAGES:** English.

DESCRIPTION: Government, academic, and industry analytical scientists who develop, test, and collaboratively study methods for analyzing fertilizers, foods, feeds, pesticides, drugs, cosmetics, and other products related to agriculture and public health. Offers short courses for analytical laboratory personnel in chemical and microbiological quality assurance, lab waste management, statistics, giving expert testimony, and technical writing. **PUBLICATIONS:** *Journal of AOAC International*, bimonthly. Includes papers on original research and new techniques and applications, new product information, collaborative studies, and meeting symposia.; *Official Methods of Analysis of the AOAC*, quinquennial. Includes annual supplements.; *Referee*, 12/year; Manuals; Membership Directory, annual; Monographs; Proceedings.

ARAB AIR CARRIERS ORGANIZATION (AACO) (Al-Itihad al-Arabi Linakl al-Jawi)

PO Box 13-5468, Beirut, Lebanon **PHONE:** 961 1 861297, 961 1 861298 **FX:** 961 1 603140 **TX:** STTA BEYXAXB **E-MAIL:** info@aaco.org **WEBSITES:** http://www.aaco.org **FOUNDED:** 1965. **OFFICERS:** Adli Dajani, Sec.Gen. **MEMBERS:** 17. **STAFF:** 12. **BUDGET:** US$542,500. **LANGUAGES:** Arabic, English.

DESCRIPTION: Air carriers registered in member nations of the League of Arab States. Purpose is to promote economic and technical cooperation among members working in aviation and tourism, especially in areas of commercial training, maintenance, and ground services. Studies air tariffs within the Arab world as well as those imposed on travel to and from other countries. Publishes tariffs manuals and tables and annual statistical bulletins of Arab airport operations and freight, passenger, and mail traffic. Compiles statistics.

ARAB AUTHORITY FOR AGRICULTURAL INVESTMENT AND DEVELOPMENT (AAAID)

PO Box 2102, Khartoum, Sudan **PHONE:** 249 773752, 249 773753 **FX:** 249 772600 **TX:** 23017 AAAID SD **OFFICERS:** Yousif Abdal Latif Alserkal. **LANGUAGES:** Arabic, English.

DESCRIPTION: Government agencies responsible for agricultural investment and development. Promotes sustainable agricultural development throughout the Arab World. Provides financial and techical assistance to development projects; formulates model development strategies.

ARAB BIOSCIENCES NETWORK (ABN)

c/o IUBS Secretariat, 51 boulevard de Montmorency, F-75016 Paris, France **PHONE:** 33 1 45250009 **FX:** 33 1 45252029 **E-MAIL:** lubs@paris7.jussieu.fr **LANGUAGES:** Arabic, English.

DESCRIPTION: Biologists and other bioscientists. Seeks to advance the study, teaching, and practice of the biosciences. Serves as a forum for the exchange of information among members; sponsors research and educational programs.

ARAB BUREAU OF EDUCATION FOR THE GULF STATES (ABEGS)

PO Box 3908, Riyadh, Saudi Arabia **PHONE:** 966 1 4800555 **FX:** 966 1 4802839 **TX:** 401441 Tarbia SJ **OFFICERS:** Dr. Ali M. Al-Towagry. **LANGUAGES:** Arabic, English.

DESCRIPTION: Organization of educators and educational organizations and agencies. Seeks to improve the quality and availability of educational opportunity in the Persian Gulf states. Serves as a clearinghouse on education; conducts training programs for educators and other school personnel.

ARAB CENTER FOR THE STUDIES OF ARID ZONES AND DRY LANDS (ACSAD)

PO Box 2440, Damascus, Syrian Arab Republic **PHONE:** 963 11 5323087, 963 11 5323039 **FX:** 963 11 5323063 **TX:** ACSAD 412697 SY **E-MAIL:** rudcsad@rusys.eg.net **FOUNDED:** 1971. **OFFICERS:** Dr. Hassan Seoud, Dir.Gen. **STAFF:** 150. **BUDGET:** US$5,400,000.

DESCRIPTION: Intergovernmental body established by the League of Arab States. Aims to conserve and develop arid land resources, raise the standard of living through better crop production, and curb desertification. Sponsors agronomic, water and soil resource, and livestock studies. Provides training for Arab scientists and technicians in research, ecology, rangeland management, soil and water use and conservation, and hydrological technology. Collaborates with national researchers and international groups. **PUBLICATIONS:** *Agriculture and Water in Arid Zones of the Arab World* (in Arabic), biennial; *Camel Newsletter* (in English), biennial. Includes summary in Arabic and French.; Papers, periodic. Containing research information.; Reports, periodic. Containing technical information and information on projects and seminars.

ARAB ECONOMIC AND SOCIAL COUNCIL (AESC)

c/o LAS, Midan el Tahrir, PO Box 11642, Cairo 11642, Egypt **PHONE:** 20 2 750511 **FX:** 20 2 740331 **TX:** 92111 ALSUN **OFFICERS:** Dr. Abdul-Hassan Zalzala. **LANGUAGES:** Arabic, English.

DESCRIPTION: Members are economic development and trade organizations. Promotes international trade and economic cooperation to raise the standard of living in Arab countries. Serves as a forum for the resolution of trade disputes and the removal of barriers to international economic cooperation.

ARAB FEDERATION FOR FOOD INDUSTRIES (AFFI)

PO Box 13025, Baghdad, Iraq **PHONE:** 964 1 7175990, 964 1 7199383 **TX:** 213556 AFFIB-IK **FOUNDED:** 1976. **OFFICERS:** Dr. Falah Said Jabar. **MEMBERS:** 650. **STAFF:** 35. **LANGUAGES:** Arabic, English.

DESCRIPTION: Members are food producers, processors, distributors, and retailers. Promotes growth and development in the food industries;

seeks to strengthen economic relations within the Arab World. Serves as a forum for communication and cooperation among members; conducts promotional activities. **PUBLICATIONS:** *Arab Food Industries* (in Arabic), quarterly; *Food and Future* (in Arabic), quarterly.

ARAB FEDERATION OF SOCIAL WORKERS (AFSW)

Union of Social Professors of Egypt, 32 Gaber Ebn Halan St., Dokki, Cairo, Egypt **OFFICERS:** Abd El Monseff. **LANGUAGES:** Arabic, English.

DESCRIPTION: Organization of social workers. Seeks to improve the standard of living in the Arab World. Represents members' professional and economic interests; serves as a forum for exchange of information among members.

ARAB HISTORIANS ASSOCIATION (ARABHA)

PO Box 4085, Baghdad, Iraq **PHONE:** 964 1 5433928 **FX:** 964 1 5373895 **TX:** 2606 UOAH **FOUNDED:** 1974. **OFFICERS:** Prof. Mustafa A.K. Al-Najjar, Pres. **MEMBERS:** 1,200. **STAFF:** 50. **LANGUAGES:** Arabic, English.

DESCRIPTION: Arab historians with master's or doctoral degrees in history. Works for greater communication among Arab historians and historians of other countries in an effort to foster better relations between Arab and non-Arab nations. Conducts courses at U.S. universities, symposia, and seminars. **PUBLICATIONS:** *Arab Historian* (in Arabic and English), quarterly.

ARAB INDUSTRIAL DEVELOPMENT AND MINING ORGANIZATION (AIDMO)

Khatauat Junction on France St., PO Box 8019, Rabat, Morocco **PHONE:** 212 7 772600, 212 7 772601 **FX:** 212 7 772188 **TX:** 36763 M **OFFICERS:** Mohamed Karbid. **LANGUAGES:** Arabic, English.

DESCRIPTION: Members are national industrial development and mining agencies. Promotes growth of the industrial capacities of Arab countries. Formulates and disseminates model national and international industrial development strategies. Coordinates international industrial development policies and activities; facilitates communication and cooperation among members.

ARAB INSTITUTE FOR HUMAN RIGHTS (AIHR)

10 rue Ibn Massoud, Par la rue Elmoz, 1004 Tunis, Tunisia **PHONE:** 216 1 767003, 216 1 767889 **FX:** 216 1 750911 **E-MAIL:** aihr@apc.org **OFFICERS:** Frej Fenniche. **LANGUAGES:** Arabic, English.

DESCRIPTION: Human rights organizations. Promotes respect for the rights of the individual and the rule of law in the Arab World. Monitors social, judicial, and legislative developments impacting upon human rights and publicizes human rights abuses. Makes available legal services.

ARAB INVESTMENT COMPANY (AIC)

PO Box 4009, Riyadh 11491, Saudi Arabia **PHONE:** 966 1 4760601 **FX:** 966 1 4760514 **LANGUAGES:** Arabic, English.

DESCRIPTION: Members are development finance institutions and agencies. Promotes investment in economic and community development programs operating in the Arab World. Serves as a clearinghouse on investment and development; provides financial support and other assistance to development projects.

ARAB IRON AND STEEL UNION (AISU)

Route de Cheraga, W Tipaza, Boite Postale 04 Cherava, Algiers, Algeria **PHONE:** 213 781579, 213 781580 **TX:** SOLBARA 63 158 ALGER **OFFICERS:** Mohamed Laid Lachgar. **LANGUAGES:** Arabic, English, French.

DESCRIPTION: Organization of iron and steel producers. Promotes growth and development of the iron and steel industries in the Arab World. Facilitates communication and cooperation among members; represents the commercial and regulatory interests of the iron and steel industries; conducts promotional activities.

ARAB LABOUR ORGANIZATION (ALO)

PO box 814, Al-Doqi, Cairo, Egypt **LANGUAGES:** Arabic, English.

DESCRIPTION: Organization of trade unions. Seeks to advance the trade union movement in the Arab World. Provides support and services to union organizing initiatives; facilitates communication and cooperation among unions representing workers in different trades and countries.

ARAB LAWYERS UNION (ALU) (Union des Avocats Arabes - UAA)

13 Arab Lawyers' Union St., Garden City, Cairo, Egypt **PHONE:** 20 2 3557132, 20 2 3352486 **FX:** 20 2 3547719 **TX:** 22266 ALU UN **E-MAIL:** alu@starnet.com.eg **FOUNDED:** 1944. **OFFICERS:** Farouk Abu Eissa, Sec.Gen. **MEMBERS:** 26. **STAFF:** 15. **LANGUAGES:** Arabic, English, French.

DESCRIPTION: A pan Arab confederation of all the independent Bar Associations and Law Societies in the Arab countries, in consultative status with UN ECOSOC and UNESCO. Established in 1944 to defend the freedom of the Arab World and to promote and protect, the rule of law, the independence of the legal profession, civil society, human rights and democratization and modernization of the legal system in the Arab Countries. **PUBLICATIONS:** *Le Droit*, semiannual.

ARAB LEAGUE EDUCATIONAL, CULTURAL AND SCIENTIFIC ORGANIZATION (ALECSO) (Organisation Arabe pour l'Education, la Culture et la Science - ALECSO)

BP 1120, Tunis, Tunisia **PHONE:** 216 1 784466 **FX:** 216 1 784965 **TX:** 18825 TN **FOUNDED:** 1970. **OFFICERS:** Mr. Mohamed El-Mili Brahimi, Dir.Gen. **MEMBERS:** 21. **STAFF:** 187. **BUDGET:** US$17,000,000. **LANGUAGES:** English, French.

DESCRIPTION: Fosters cultural unity among Arab countries so that these nations may play a positive role in world progress. Conducts educational, cultural, and scientific activities; sponsors seminars and professional training courses. Organizes competitions. **PUBLICATIONS:** *Arab Journal of Culture*, semiannual; *Arab Journal of Education*, semiannual; *Arab Journal of Information*, semiannual; *Arab Journal of Science*, semiannual; *Arab Newsletter of Publication*, semiannual; *Journal of the Arabization Bureau*, semiannual; *Journal of the Institute of the Arab Manuscript*, semiannual; Books, periodic.

ARAB MAHGREB UNION (AMU)

27 avenue Okba Agdal, Rabat, Morocco **PHONE:** 212 7 772682, 212 7 772676 **FX:** 212 7 772693 **OFFICERS:** Mohamed Amamou. **LANGUAGES:** Arabic, English, French.

DESCRIPTION: Representative of the states of the Maghreb Region. Promotes appropriate and sustainable development of the Maghreb. Coordinates the efforts of development projects operating in the Maghreb; facilitates international trade and cultural exchange.

ARAB MEDICAL UNION (AMU)

Union des Medecins Arabes, Mahrajene, PO Box 290, 1082 Tunis, Tunisia **PHONE:** 216 1 886800 **FX:** 216 1 889293 **OFFICERS:** Aziz El Matri. **LANGUAGES:** Arabic, English.

DESCRIPTION: Physicians and other health care professionals. Seeks to increase availability and quality of health care services in the Arab World; promotes professional advancement of members. Makes available health services; sponsors research; conducts continuing professional development courses.

ARAB NETWORK FOR ENVIRONMENT AND DEVELOPMENT (ANED)

PO Box 2, Magles El Shaab, Cairo, Egypt **PHONE:** 20 2 3041634 **FX:** 20 2 3041635 **E-MAIL:** aoye@ritsec1.com.eg **FOUNDED:** 1990. **MEMBERS:** 200. **LANGUAGES:** Arabic, English.

DESCRIPTION: Members are NGOs. Coordinates the flow and exchange of skills, experiences, and information between regional community organizations. Creates new grassroots activities to be implemented by RAED NGOs members. Encourages the inclusion of community participation projects in government programs to achieve sustainable development. Gathers and disseminates regional and international data on different environmental and developmental problems.

ARAB OFFICE FOR CRIMINAL POLICE (AOCP)

Maysaloun St., Shawkat Alaaidi Ave., behind Dar-es-Salaam School, Damascus, Syrian Arab Republic **LANGUAGES:** Arabic, English.

DESCRIPTION: Members are law enforcement agencies. Promotes excellence in the practice of criminal law enforcement. Facilitates communication and cooperation among members; sponsors exchange and training programs.

ARAB OFFICE FOR NARCOTICS AFFAIRS (AONA)

PO Box 17225, Amman, Jordan **LANGUAGES:** Arabic, English.

DESCRIPTION: Members are law enforcement agencies specializing in the control of the drug trade. Seeks to improve the effectiveness of narcotics trade control efforts. Facilitates communication and cooperation among members; sponsors exchange and training programs.

ARAB ORGANIZATION FOR HUMAN RIGHTS (AOHR)

91 Al-Marghany St., Heliopolis, Cairo, Egypt **PHONE:** 20 4181396, 20 4188378 **FX:** 20 4185346 **E-MAIL:** aohr@link.com.eg **WEBSITES:** http://www.gf.net/AOHR **FOUNDED:** 1983. **OFFICERS:** Mohammed Fayek, Sec.Gen. **MEMBERS:** 2,020. **STAFF:** 15. **LANGUAGES:** Arabic, English.

DESCRIPTION: Individuals with an interest in human rights in the Arab World. Promotes and seeks to protect human rights in the Arab World in accordance with international standards. Serves as a clearinghouse on human rights in Arab countries; conducts research and educational programs; provides information and assistance to other organizations pursuing similar goals. **PUBLICATIONS:** *Human Rights in the Arab World* (in Arabic and English), annual; *Nadawat Fikria*, periodic; Newsletter, monthly; Annual Report.

ARAB STATES NEWS EXCHANGE (ASNE)

El Menzah, PO Box 65, 1004 Tunis, Tunisia **PHONE:** 216 1 766155 **FX:** 216 1 767411 **LANGUAGES:** Arabic, English.

DESCRIPTION: News agencies and other press and media outlets. Promotes free flow of information on current events within the Arab World. Gathers and disseminates news stories and other information; sponsors current events research programs.

ARAB TOWNS ORGANIZATION (ATO)

PO Box 68160, Kaifar 71962, Kuwait **PHONE:** 965 484970568, 965 4849603 **FX:** 965 4849319 **TX:** MODON 46390 KT **FOUNDED:** 1967. **OFFICERS:** Abdul-Aziz Al Adasani, Sec.Gen. **MEMBERS:** 400. **STAFF:** 22. **LANGUAGES:** Arabic.

DESCRIPTION: Arab towns united to encourage cooperation and exchange of expertise. Works toward: raising the standards of municipal services and utilities; preserving the character and cultural heritage of Arab towns; developing and modernizing municipal and local government institutions and legislation. Provides loans to members for development projects through the Arab Towns Development Fund. Offers training courses for municipal officials; conducts conferences, seminars, and symposia. Sponsors annual competitions and ''The Day of the Arab Town'' event. **PUBLICATIONS:** *Al-Madinah Al-Arabiyah* (in Arabic), bimonthly; *Proceedings of Congress* (in Arabic), triennial. Also publishes brochures, pamphlets, and books (in Arabic and English).

ARAB UNION OF RAILWAYS (AUR)

PO Box 6599, Aleppo, Syrian Arab Republic **PHONE:** 963 21 220302 **TX:** 331009 CFS SY **OFFICERS:** Mourhaf Sabouni. **LANGUAGES:** Arabic, English.

DESCRIPTION: Organization of railroads. Promotes international rail transportation within the Arab World; seeks to expand the rail infrastructure of Arab countries. Represents members' commercial and regulatory interests; facilitates communication and cooperation among members.

ARABIAN GULF STATES JOINT PROGRAMME PRODUCTION INSTITUTE (AGSJPPI)

PO Box 24391, Al-Safah, Kuwait **PHONE:** 965 5327164 **FX:** 965 5327162 **WEBSITES:** http://www.imarabe.org/ **LANGUAGES:** Arabic, English.

DESCRIPTION: National information agencies of the Persian Gulf States. Promotes free access to accurate information on public projects and policies in the region. Facilitates discussion of issues of mutual interest to members; serves as a clearinghouse on public information.

ARABO-AFRICAN ARBITRATORS ASSOCIATION (AAAA)

3 Abo El-Feda St., Zamalik, Cairo, Egypt **PHONE:** 20 2 3401333, 20 2 3401335 **FX:** 20 2 3401336 **TX:** 22261 RCIAC UN **OFFICERS:** Dr. M. I. M. Aboul-Enein. **LANGUAGES:** Arabic, English.

DESCRIPTION: Members are professional arbitrators and mediators. Promotes equitable resolution of legal, political, and economic disputes. Makes available arbitration and mediation services; conducts continuing professional development programs; facilitates exchange of information among members.

ARCTIC NETWORK (AN)

PO Box 102252, Anchorage, AK 99510 **PHONE:** (907) 272-2452 **FX:** (907) 272-2453 **E-MAIL:** arnet@alaska.net **OFFICERS:** Margie Gibson, Exec.Dir. **LANGUAGES:** English.

DESCRIPTION: Environmentalists, scientists, local communities and organizations with an interest in the Arctic. Promotes preservation of Arctic wildlife habitat and traditional native cultures. Monitors commercial and scientific activities in the Arctic; sponsors research and educational programs. **PUBLICATIONS:** *Leads*, quarterly.

ARTS CENTRE GROUP (ACG)

The Courtyard, 59A Portobello Rd., London W11 3DB, England **PHONE:** 44 171 2434550 **FX:** 44 171 2217689 **E-MAIL:** acg@dial.pipex.com **WEBSITES:** http://www.dspace.dial.pipex.com/acg **FOUNDED:** 1971. **OFFICERS:** Carla Moss, Admin. **MEMBERS:** 700. **STAFF:** 1. **LANGUAGES:** English.

DESCRIPTION: Christians in 12 countries working professionally in the arts, the media, and entertainment. Promotes an understanding of the link between members' Christian identities and their artistic expression. Encourages Christian artists to develop their professional abilities and assists them in merging their faith and careers. Evangelizes other artists by encouraging them to express Christian ideals through their work. Organizes events ranging from debates, discussions and performance to an annual lecture, an annual poetry competition, prayer and bible studies and social events. **PUBLICATIONS:** *Articulate* (in English), annual; *Artyfact* (in English), bimonthly; *Telling Images*. On the role of Christians in the arts and media.

ASEAN BANKERS ASSOCIATION (ABA)

180 Cecil St., 17-00, Singapore 069546, Singapore **PHONE:** 65 2247155 **FX:** 65 2250727 **TX:** RS 28021 **E-MAIL:** aseanbc@pacific.net.sg **FOUNDED:** 1976. **OFFICERS:** Mr. Tay Kah Chye. **MEMBERS:** 7. **LANGUAGES:** English.

DESCRIPTION: Made up of national bank associations in Brunei, Indonesia, Malaysia, Philippines, Singapore, Thailand and Vietnam. Promotes the development of banking and financial system professions in member countries of the Association of South East Asian Nations. Seeks to accelerate economic growth and to identify potential growth opportunities for banks in ASEAN member countries. Fosters cooperation among members and encourages collaboration on projects. Conducts educational programs on topics including mergers and acquisitions, bank selling skills, and bank management. **PUBLICATIONS:** *Directory of Banks in ASEAN* (in English), biennial.

ASEAN BUSINESS FORUM

Wisma Bakrie, 5th Fl., Jalan H.R. Rasuna Said, Kav. B-1, 12920 Jakarta, Indonesia **PHONE:** 62 21 5257855 **FX:** 62 21 5205488 **OFFICERS:** Ranjit Gill. **LANGUAGES:** English, Indonesian.

DESCRIPTION: Organization of businesses and business organizations in southeastern Asia. Promotes increased international trade among the ASEAN countries. Serves as a clearinghouse on business and trade; sponsors promotional activities.

ASEAN CHEMICAL INDUSTRIES CLUB

c/o ASEAN-CCI, ASEAN Secretariat Bldg., 3rd Fl., 70A Jalan Sisingamangaraja, 12110 Jakarta, Indonesia **PHONE:** 62 21 7267325, 62 21 7262991 **FX:** 62 21 7267326 **E-MAIL:** acci@asean.or.id **LANGUAGES:** English, Indonesian.

DESCRIPTION: Members are chemical manufacturers in southeastern Asia. Promotes growth and development of the chemical industries in Southeast Asia. Represents members' commercial and regulatory interests; sponsors promotional activities.

ASEAN CONFEDERATION OF WOMEN'S ORGANIZATIONS

c/o SCWO, Entrepreneurship Development Centre, NTU, Nanyang Ave., Nanyang Technological Univ., Singapore 639798, Singapore **PHONE:** 65 7994760 **FX:** 65 7920415 **OFFICERS:** Tan Seok Buay. **LANGUAGES:** Chinese, English.

DESCRIPTION: Membership consists of women's organizations in Southeast Asia. Seeks to increase the legal and social status of women; promotes full participation by women in economic, political, and social life. Facilitates communication and cooperation among members; conducts educational and advocacy programs.

ASEAN COUNCIL ON PETROLEUM (ASCOPE)

c/o PERTAMINA Head Office, Grqanadi Bldg., 8th Fl., Jalan H.R. Rasuna Said Kav. 8-9, Block 10/1, Kuningan, 12950 Jakarta, Indonesia **PHONE:** 62 21 2523377 **FX:** 62 21 2520551 **TX:** 44441 **FOUNDED:** 1975. **OFFICERS:** Mr. Njoman Sudibia. **LANGUAGES:** English.

DESCRIPTION: A council of the Association of South East Asian Nations. Member oil companies of ASEAN from 6 countries working to develop the petroleum industry. Sponsors training programs and studies in the technical field. **PUBLICATIONS:** *ASCOPE Directory* (in English), biennial; *Proceedings of Conferences* (in English), quadrennial; *Technical Papers* (in English), semiannual.

ASEAN FEDERATION OF ELECTRICAL, ELECTRONICS AND ALLIED INDUSTRIES

c/o ASEAN-CCI, ASEAN Secretariat Bldg., 3rd Fl., Jalan Sisingamangaraja, 12110 Jakarta, Indonesia **PHONE:** 62 21 7267325, 62 21 7262991 **FX:** 62 21 7267326 **E-MAIL:** acci@asean.or.id **LANGUAGES:** English, Indonesian.

DESCRIPTION: Members are manufacturers of electrical and electronic devices and components. Promotes growth and development of the electrical and electronics industries in Southeast Asia. Facilitates communication and cooperation among members; represents the commercial and regulatory interests of the electrical and electronics industries.

ASEAN FEDERATION OF FOOD PROCESSING INDUSTRIES

31 Jalan Pulo Mas Raya, PO Box 3005/jkt, Jakarta, Indonesia **LANGUAGES:** English, Indonesian.

DESCRIPTION: Organization of food processors. Promotes growth and development of the food processing industries in Southeast Asia. Facilitates communication and cooperation among members; represents the commercial and regulatory interests of the food processing industries; conducts promotional activities.

ASEAN FREE TRADE AREA

c/o ASEAN Secretariat, 70A Jalan Sisingamangaraja, Kebayoran Baru, PO Box 2072, 12110 Jakarta, Indonesia **PHONE:** 62 21 712272, 62 21 711988 **FX:** 62 21 7398234 **TX:** 47213 ASEAN JKT **LANGUAGES:** English, Indonesian.

DESCRIPTION: Members are government trade agencies, international trade organizations, and businesses. Promotes increased international trade involving the countries of Southeast Asia. Serves as a clearinghouse on Southeast Asian trade; advocates for removal of barriers to trade; sponsors promotional activities.

ASEAN INDUSTRY CLUB

c/o ACCI, 4711 Jalan Sungai Pandan, Kuala Belait 6004, PO Box 416, Begara, Brunei Darussalam **PHONE:** 673 3 331619, 673 3 331670 **FX:** 673 3 334799 **LANGUAGES:** Chinese, English.

DESCRIPTION: Organization of industrial concerns. Promotes industrial growth and development in Southeast Asia. Facilitates communication and cooperation among members; represents the interests of industry before labor and trade organizations, government agencies, and the public.

ASEAN INSTITUTE FOR HEALTH DEVELOPMENT

25/5 Phutthamonthon 4 Rd., Salaya, Mahido Univ., Nakhon Pathom 73170, Thailand **PHONE:** 66 2 4419040, 66 2 4419870 **FX:** 66 2 4419044 **E-MAIL:** directad@mahidol.ac.th **OFFICERS:** Dr.

Boongium Tragoolvonse. **LANGUAGES:** English, Thai.

DESCRIPTION: Organization of national health agencies. Seeks to improve the availability and quality of primary health care services in Southeast Asia. Facilitates establishment of local health services and facilities; serves as a clearinghouse on public health policies and programs.

ASEAN LAW ASSOCIATION

OCBC Centre, 65 Julia St., Ste. 27-01, Singapore 049513, Singapore **OFFICERS:** Charles Lim. **LANGUAGES:** Chinese, English.

DESCRIPTION: Members are attorneys and law firms and organizations. Promotes understanding of law and legal issues affecting Southeast Asia. Facilitates communication and exchange among members; sponsors research and educational programs.

ASEAN NETWORK ON ENERGY CONSERVATION AND MANAGEMENT

ASEAN-SCNCER Secretariat, Res. and Development Centre for Applied Physics, Indonesian Inst. of Sciences, P3FT-LIPI, Jalan Cisitu, 40135 Bandung, Indonesia **PHONE:** 62 22 2503052, 62 22 2504833 **FX:** 62 22 2503050 **LANGUAGES:** English, Indonesian.

DESCRIPTION: Scientific organizations and government agencies with an interest in energy conservation. Promotes application of science to problems surrounding energy use and conservation. Serves as a clearinghouse on energy conservation; sponsors research and educational programs; provides consulting services to energy producers and consumers.

ASEAN PORT AUTHORITIES ASSOCIATION

c/o Philippine Ports Authority, Marsman Bldg., 22 Muelle de San Francisco St., South Harbor, Ports Area, Manila 1018, Philippines **PHONE:** 63 2 5274746, 63 2 5276432 **FX:** 63 2 5274749 **OFFICERS:** Alda P. Dizon. **LANGUAGES:** English, Filipino.

DESCRIPTION: Organization of port authorities. Seeks to improve the efficiency of maritime transportation and port handling in southeastern Asia. Facilitates exchange of information among members; works to remove barriers to international maritime trade; conducts research to improve the performance of port facilities.

ASEAN REGIONAL FORUM

c/o ASEAN, 70A Jalan Sisingamangaraja, Kebayoran Baru, PO Box 2072, 12110 Jakarta, Indonesia **PHONE:** 62 21 7262410, 62 21 7262991 **FX:** 62 21 7398234 **TX:** 47214 ASEAN JKT **WEBSITES:** http://www.dfat.gov.au/arf/arfhome.html **LANGUAGES:** English, Indonesian.

DESCRIPTION: Organization of Southeast Asian governments. Promotes international trade involving southeastern Asia; seeks to maintain friendly relations among members and between members and the rest of the world. Serves as a forum for discussion of international trade and diplomatic issues affecting southeastern Asia.

ASEAN REGIONAL NETWORK ON ENVIRONMENTAL EDUCATION

c/o IESAM, College Los Banos, Laguna 4031, Philippines **LANGUAGES:** Filipino.

DESCRIPTION: Environmental educators and educational institutions. Seeks to advance environmental education; promotes increased public awareness of environmental issues. Serves as a clearinghouse on environmental education; sponsors research and training programs.

ASEAN-EC JOINT COMMITTEE

c/o ASEAN Secretariat, 70A Jalan Sisingamangaraja, Jakarta, Indonesia **PHONE:** 62 21 712272, 62 21 712991 **FX:** 62 21 7398234 **TX:** 47213 ASEAN JKT **OFFICERS:** M. C. Abad. **LANGUAGES:** English, Indonesian.

DESCRIPTION: Representatives of the countries comprising the Association of South East Asian Nations (ASEAN) and the European Community (EC). Promotes increased trade and improved relations between the countries of the ASEAN and the EC. Serves as a forum for the discussion of issues of mutual interest to members; facilitates removal of barriers to international trade.

ASIA CRIME PREVENTION FOUNDATION (ACPF) (Asia Keisei Zaidan)

Sunshine 60 Bldg., 16-11, 3-1-1, Higashi-Ikebukuro, Toshima-ku, Tokyo 170, Japan **PHONE:** 81 3 39871444 **FX:** 81 3 59512505 **FOUNDED:** 1982. **OFFICERS:** Tetsuro Takizawa, Sec. **MEMBERS:** 3,400. **BUDGET:** 36,000,000¥. **LANGUAGES:** English, Japanese.

DESCRIPTION: Attorneys, judges, and individuals working in criminal justice or involved in probation supervision. Promotes regional cooperation in the prevention of crime. Encourages improvement in criminal justice systems and increased cooperation on the part of the public sector. Supports research and field work in the area of crime prevention; collects and disseminates information; organizes symposiums, workshops, and seminars. Conducts training in conjunction with the United Nations Asia and Far East Institute for the Prevention of Crime and the Treatment of Offenders. **PUBLICATIONS:** *ACPF Today* (in English), annual; *Newsletter* (in Japanese), semiannual.

ASIA AND FAR EAST INSTITUTE FOR THE PREVENTION OF CRIME AND THE TREATMENT OF OFFENDERS

1-26 Harumi-cho, Fuchu-shi, Tokyo 183, Japan **PHONE:** 81 423 337021, 81 423 337023 **FX:** 81 423 337024 **E-MAIL:** ldj00272@niftyserve.or.jp **OFFICERS:** Toichi Fujwara. **LANGUAGES:** English, Japanese.

DESCRIPTION: Law enforcement and criminal justice organizations. Seeks to prevent crime and ensure the humane incarceration and effective rehabilitation of convicted offenders. Serves as a clearinghouse on crime and criminal justice; sponsors research and educational programs.

THE ASIA FOUNDATION (TAF)

465 California St., 14th Fl., San Francisco, CA 94104 **PHONE:** (415) 982-4640 **FX:** (415) 392-8863 **E-MAIL:** webmaster@asiafound.org **WEBSITES:** http://www.asiafoundation.org **FOUNDED:** 1954. **OFFICERS:** Carolyn Iyoya Irving, Communications & Outreach Mgr. **MEMBERS:** 300. **STAFF:** 250. **LANGUAGES:** English.

DESCRIPTION: Supported by U.S. government grants and private contributions. Works to: strengthen government and institutions in Asia and the Pacific Islands; promote Asian-Pacific cooperation; and develop closer Asian-American relations. Supports projects that strengthen representative government, build effective legal systems, foster market economies, increase accountability in the public and private sectors, develop independent and responsible media, and encourage the development of human resources. Maintains the following programs: Books for Asia distributes over one million books and journals each year to Asian schools, universities, libraries, and research centers; and Asian-American Exchange offers in-country opportunities to Asians and Pacific Islanders to share professional training and academic study with Americans. **PUBLICATIONS:** Annual Report. Working Paper Series; Asian Perspectives Series; Projects List; Occasional Reports.

ASIA LABOURERS SOLIDARITY (ALS)

4-20-11 Sakae Maruzen Biru 403, Naka-ku, Nagoya 460, Japan **PHONE:** 81 52 9913519 **FX:** 81 52 9913519 **OFFICERS:** Yoi Shigeau, Rep. **LANGUAGES:** English, Japanese.

DESCRIPTION: Labor unions and other workers' organizations. Promotes solidarity among laborers in Asia. Facilitates communication and cooperation among members.

ASIA MONITOR RESOURCE CENTER (AMRC)

444-446 Nathan Rd., 8th Fl., Flat B, Kowloon, Hong Kong **PHONE:** 852 23321346 **FX:** 852 23855319 **E-MAIL:** amrc@hk.super.net **WEBSITES:** http://www.hk.super.net/-amrc **FOUNDED:** 1979. **OFFICERS:** Leung Po Lam **MEMBERS:** 16. **STAFF:** 7. **BUDGET:** HK$250,000. **LANGUAGES:** Chinese, English.

DESCRIPTION: Promotes development of democratic and independent labor movements throughout Asia; seeks to facilitate the empowerment of workers. Encourages and supports development of grass roots labor and human rights organizations; provides assistance to labor organizations providing education, training, health, safety, and human rights services to workers. Assists international aid organizations in identifying local collaborators for development programs. Gathers and disseminates information on Asian labor organizations and issues including wages, employment levels, occupational safety and health, and the activities of transnational corporations; conducts research **PUBLICATIONS:** *Asian Labour Update* (in English), quarterly. Features news bulletin on labor issues in the Asia-Pacific region. Additional publications available upon request.

ASIA PACIFIC ACADEMY OF OPHTHALMOLOGY (APAO)

Gleneagles Hospital, Annexe Block 02-38, 6A Napier Rd., Singapore 258500, Singapore **PHONE:** 65 4666666 **FX:** 65 7333360 **E-MAIL:** wll@pacific.net.sg **OFFICERS:** Arthur S. M. Lim. **LANGUAGES:** Chinese, English.

DESCRIPTION: Organization of ophthalmologists. Seeks to advance ophthalmological study, teaching, and practice. Facilitates exchange of information among members; conducts continuing professional development courses.

ASIA-PACIFIC BROADCASTING UNION (ABU)

PO Box 1164, 59700 Kuala Lumpur, Malaysia **PHONE:** 60 3 2823592, 60 3 2823108 **FX:** 60 3 2825292 **TX:** ABU MA 32227 **E-MAIL:** sg@abu.org.my **WEBSITES:** http://www.abu.org.my **FOUNDED:** 1964. **OFFICERS:** Hugh Terence Leonard, Sec.Gen. **MEMBERS:** 100. **STAFF:** 24. **BUDGET:** US$700,000. **LANGUAGES:** English.

DESCRIPTION: Broadcasting associations in 50 countries and areas united to coordinate and promote the study of problems related to broadcasting and to ensure communication among broadcasting services. Works to facilitate all aspects of broadcasting for the purpose of education and national development. Contributes to the advancement of international friendship and goodwill through broadcasting. Offers training courses on television news exchange. Conducts competitions. **PUBLICATIONS:** *ABU News* (in English), bimonthly; *ABU Technical Review* (in English), bimonthly.

ASIA AND PACIFIC COMMISSION ON AGRICULTURAL STATISTICS (APCAS)

FAO Regional Office for Asia and the Pacific, Maliwan Mansion, 39 Phra Atit Rd., Bangkok 10200, Thailand **PHONE:** 66 2 2817844 **FX:** 66 2 2800445 **TX:** 82815 FOODAG TH **OFFICERS:** H. Som. **LANGUAGES:** English, Thai.

DESCRIPTION: Agencies responsible for the compilation of agricultural statistics. Promotes increased understanding of the agricultural production of Asia and the Pacific. Serves as a forum for the exchange of information among members; conducts statistical research; functions as a clearinghouse on Asian and Pacific agriculture.

ASIA/PACIFIC CULTURAL CENTRE FOR UNESCO (ACCU)

6 Fukuromachi, Shinjuku-ku, Tokyo 162, Japan **PHONE:** 81 3 32694435 **FX:** 81 3 32694510 **E-MAIL:** general@accu.or.jp **WEBSITES:** http://www.accu.or.jp **FOUNDED:** 1971. **OFFICERS:** Muneharu Kusaba, Dir.Gen. **MEMBERS:** 795. **STAFF:** 30. **BUDGET:** US$346,000,000. **LANGUAGES:** English, Japanese.

DESCRIPTION: Promotes mutual understanding and cultural cooperation, according to the principles of UNESCO, among peoples of Asia and the Pacific by preserving and promoting culture and by developing books, book publishing, and the design of literacy programs and materials within the region. Coordinates joint regional programs such as co-publication of children's books, co-production of audio-visual materials on music and culture, and joint production of literacy materials for neoliterates in rural areas. Organizes training programs in book development and literacy material development; organizes seminars for cultural promoters. Sponsors photo and children's picture book illustrations contests. **PUBLICATIONS:** *Asian Cultural Centre News* (in Japanese), monthly; *Asian/Pacific Book Development* (in English), quarterly.

ASIA PACIFIC FINANCE ASSOCIATION (APFA)

Dept. of Economics and Finance, 83 Tat Chee Ave., Univ. Coll. of Hong Kong, Kowloon Tong, Kowloon, Hong Kong **PHONE:** 852 7888800 **FX:** 852 7888806 **OFFICERS:** Richard Ho. **LANGUAGES:** Chinese, English.

DESCRIPTION: Financial institutions and academics with an interest in finance. Promotes increased understanding of finance and economics; seeks to ensure availability of capital for business expansion. Serves as a clearinghouse on finance; sponsors research and educational programs.

ASIA-PACIFIC FORUM FOR CHILD WELFARE (APFCW)

Jalan Teuku Umar 10, 10350 Jakarta, Indonesia **PHONE:** 62 21 3905746 **FX:** 62 21 3905746 **E-MAIL:** apfcw95@rad.net.id **WEBSITES:** http://www.apfcw.or.id **FOUNDED:** 1994. **OFFICERS:** Palupi Widjajanti Ph.D., Sec.Gen. **MEMBERS:** 24. **STAFF:** 2. **BUDGET:** US$39,772. **LANGUAGES:** English.

DESCRIPTION: Organizations working for the welfare of children. Seeks to improve the performance of child welfare agencies and organizations through staff development and training. Promotes improved quality of life for children in Asia and the Pacific. Facilitates communication and cooperation among members and between members and other organizations with similar goals and activities. Sponsors research; organizes staff exchanges; gathers and disseminates information; undertakes public education campaigns to raise awareness of child welfare issues. **PUBLICATIONS:** Position papers.

ASIA PACIFIC FORUM ON WOMEN, LAW AND DEVELOPMENT (APWLD)

PO Box 12224, 50770 Kuala Lumpur, Malaysia **PHONE:** 60 3 6510648, 60 3 6511649 **FX:** 60 3 6511371 **TX:** MA 31655 MPS **FOUNDED:** 1988. **OFFICERS:** Joy Oraa, Regional Coord. **MEMBERS:** 55. **STAFF:** 4. **BUDGET:** US$500,000. **LANGUAGES:** English.

DESCRIPTION: Women's organizations comprised of women's activists, lawyers, human rights activists, and interested individuals. Seeks to enable women in the Asia Pacific region to utilize the law as an effective instrument to empower them in struggles for justice and equality. Promotes women's development in family, society, economics, politics, and national development. Fosters exchange of information among members. Seeks to establish a Women's Rights Charter. Lobbies Asian Pacific governments to ratify the UN Convention on the Elimination of All Forms of Discrimination Against Women. Advocates the basic concept of human rights. **PUBLICATIONS:** *Forum News* (in English), quarterly; Reports.

ASIA PACIFIC MARKETING FEDERATION (APMF)

APMF Secretariat, Level 2, 464 St. Kilda Rd., Melbourne, VIC 3004, Australia **PHONE:** 61 3 98208788 **FX:** 61 3 98208650 **E-MAIL:** r.cameron@uws.edu.au **OFFICERS:** Dr. Ross C. Cameron. **LANGUAGES:** English.

DESCRIPTION: Marketing organizations. Promotes increased cooperation within the marketing industries of Asia and the Pacific. Serves as a forum for the discussion of issues of mutual interest to members; represents the commercial and regulatory interests of the marketing industry.

ASIA PACIFIC MOUNTAIN NETWORK (APMN)

c/o ICIMOD, PO Box 3226, Jawalakhel, Lalitpur, Kathmandu, Nepal **PHONE:** 977 1 525313 **FX:** 977 1 524409 **E-MAIL:** shahid@icimod.org.np **WEBSITES:** http://www.mtnforum.org **FOUNDED:** 1995. **OFFICERS:** Dr. Shahid Akhtar. **MEMBERS:** 260. **STAFF:** 1. **LANGUAGES:** English, Nepali.

DESCRIPTION: Membership consists of individuals and organizations. Promotes sustainable development of mountainous regions in Asia and the Pacific. Serves as a clearinghouse on mountain development; monitors development projects; formulates model development strategies. **PUBLICATIONS:** *Asia-Pacific Mountain News Bulletin* (in English), semiannual.

ASIA AND PACIFIC PLANT PROTECTION COMMISSION (APPPC)

FAO Regional office for Asia and the Pacific, Phra Atit Rd., Bangkok 10200, Thailand **PHONE:** 66 2 2817844 **FX:** 66 2 2800445 **TX:** 82815 FOODAG TH **E-MAIL:** chongyao.shen@fao.org **FOUNDED:** 1956. **OFFICERS:** Prof. Chong-yao Shen, Exec.Sec. **MEMBERS:** 25. **LANGUAGES:** English, French.

DESCRIPTION: Ministers of agriculture from member nations working to disseminate information on regional plant protection. Seeks to prevent the introduction of destructive plant pests and diseases.

ASIA-PACIFIC TELECOMMUNITY (APT)

12/49 Soi 5, Chaengwattana Rd., Bangkok

10210, Thailand **PHONE:** 66 2 5736893 **FX:** 66 2 5744226 **TX:** 84198 APTELBK TH **E-MAIL:** apthq@mozart.inet.co.th **WEBSITES:** http://www.aptsec.org **FOUNDED:** 1979. **OFFICERS:** Mr. Jong-oon Lee, Exec.Dir. **MEMBERS:** 70. **STAFF:** 21. **BUDGET:** US$1,785,470. **LANGUAGES:** English.

DESCRIPTION: Membership is open to any state within the region. Works in conjunction with the United Nations Economic and Social Commission for Asia and the Pacific and the International Telecommunications Union to promote development of telecommunications and coordinate planning, programming, and management of telecommunications networks and services. Acts as a consultative body for the settlement of telecommunications matters. Provides exchange program for engineers and managerial personnel; sends consultants to developing counties; offers training fellowships. Conducts seminars and workshops. Collects and disseminates information on developments in telecommunication management. Conducts technical studies; compiles statistics. Cooperates with United Nations agencies and other international and regional organizations. **PUBLICATIONS:** *Asia Pacific Telecommunication Journal*, semiannual. Also publishes reports of study groups and seminars.

ASIA SOCIETY

725 Park Ave., New York, NY 10021 **PHONE:** (212) 288-6400 **FX:** (212) 517-8555 **WEBSITES:** http://www.asiasociety.org **FOUNDED:** 1956. **OFFICERS:** Ambassador Nicholas Platt, Pres. **MEMBERS:** 6,500. **STAFF:** 100. **BUDGET:** US$14,000,000. **LANGUAGES:** English.

DESCRIPTION: Dedicated to fostering understanding of Asia and communication between Americans and the peoples of Asia and the Pacific. Presents a wide range of programs including high-level conferences, symposia, international study missions, press briefings, publications, major art exhibitions and performing arts productions. **PUBLICATIONS:** *Archives of Asian Art*, annual; *Asia Newsletter*, 3/year. Contains chapter news and calendar of events.; *Asian Update*; *China Briefing*, annual; *India Briefing*, annual; *Korea Briefing*, annual; Annual Report, annual; Books; Brochures; Catalogs; Catalogs; Monographs; Videos.

ASIA SOIL CONSERVATION NETWORK FOR THE HUMID TROPICS (ASCNHT)

Blok VII, Fl. 6, Jalan Gatot Subroto, PO Box 7632 JKP, 10076 Jakarta, Indonesia **PHONE:** 62 21 5720206 **FX:** 62 21 5700263 **E-MAIL:** cuasoc76@centrin.net.id **OFFICERS:** Manggala Wanabakti. **LANGUAGES:** English, Indonesian.

DESCRIPTION: Soil and other agricultural scientists. Promotes soil conservation through adoption of improve farming techniques. Develops soil-conserving farming methods and trains farmers in their use; sponsors research and continuing professional development programs.

ASIAN ASSOCIATION FOR AGRICULTURAL ENGINEERING (AAAE)

c/o Program Agricultural and Food Engineering, AIT, PO Box 4, Klong Luang, Pathumthani 12120, Thailand **PHONE:** 66 2 5245479, 66 2 5245450 **FX:** 66 2 5246200 **TX:** 84276 TH **E-MAIL:** aaae@ait.ac.th **FOUNDED:** 1990. **OFFICERS:** Prof. V. M. Salokhe. **MEMBERS:** 300. **STAFF:** 1. **LANGUAGES:** English.

DESCRIPTION: Agricultural engineers and scientists. Promotes introduction of improved farming techniques in Asia; seeks to advance agricultural scholarship. Serves as a clearinghouse on agricultural engineering; sponsors research and educational programs. **PUBLICATIONS:** *AAAE Newsletter* (in English), quarterly; *International Agricultural Journal* (in English), quarterly.

ASIAN BANKERS ASSOCIATION (ABA)

c/o CACCI Secretariat, 7/F No. 3 Sung-shou Rd., Taipei 110, Taiwan **PHONE:** 886 2 7255663, 886 2 7255664 **FX:** 886 2 7255665 **FOUNDED:** 1981. **OFFICERS:** Lawrence T. Liu, Sec.-Treas. **MEMBERS:** 150. **STAFF:** 4. **LANGUAGES:** Chinese, English.

DESCRIPTION: Banks and other financial institutions. Promotes growth and development of the Asian banking industries; seeks to ensure financial stability among Asian economies. Serves as a forum for the exchange of information among members; represents the collective interests of the banking industries. **PUBLICATIONS:** *ABA Newsletter* (in English), monthly; *Journal of Banking and Finance*, semiannual.

ASIAN CENTRE FOR ORGANIZATION, RESEARCH AND DEVELOPMENT (ACORD)

C-126 Greater Kailash-I, New Delhi 110 048, Delhi, India **PHONE:** 91 11 6410616, 91 11 6435993 **FX:** 91 11 6479397 **TX:** 73119 **E-MAIL:** acord@del2.vsnl.net.in **FOUNDED:** 1981. **OFFICERS:** Ms. Kiron Wadhera, Pres. **MEMBERS:** 9. **STAFF:** 50. **LANGUAGES:** English, Hindi.

DESCRIPTION: Development and research organizations. Promotes effective implementation of development projects. Serves as a clearinghouse on organization, research, and development; makes available consulting services to development organizations and projects; conducts educational and training programs. **PUBLICATIONS:** *Acord News - Corporate* (in English), semiannual; *Acord News - Development* (in English), semiannual.

ASIAN CLEARING UNION (ACU)

c/o Central Bank of the Islamic Republic of Iran, PO Box 11365/8531, Tehran, Iran **PHONE:** 98 21 2842076 **FX:** 98 21 2847677 **TX:** 88-216868 **E-MAIL:** acusecret@neda.net **FOUNDED:** 1974. **OFFICERS:** Mohammad Firouzdor, Sec.Gen. **MEMBERS:** 7. **LANGUAGES:** English.

DESCRIPTION: Central banks, monetary authorities, and treasuries of Asian countries. Works to economize on the use of exchange reserves; promotes shifting of national banking services to domestic banks; seeks to enhance economic, financial, and commercial cooperation among Asian nations. Provides short-term credit facilities. **PUBLICATIONS:** *ACU Annual Report* (in English); *ACU Newsletter*, monthly.

ASIAN COALITION ON DEBT AND ADJUSTMENT (ACDA)

c/o FDC, 34 Matiaga St., Central District, Quezon City 1101, Philippines **PHONE:** 63 2 9211985 **FX:** 63 2 9214381 **E-MAIL:** fdc@phil.gn.apc.org **LANGUAGES:** English, Filipino.

DESCRIPTION: Development organizations and financial institutions. Seeks to reduce the foreign debt of developing countries in Asia. Formulates and implements debt-reduction strategies; serves as a clearinghouse on international debt.

ASIAN COALITION FOR HOUSING FINANCE (ACHF)

c/o Housing Development Finance Corp., Ramon House, 4th Fl., 169 Backbay Reclamation, Bombay 400 020, Maharashtra, India **PHONE:** 91 22 2850000 **FX:** 91 22 2852336 **OFFICERS:** Nasser Munjel **LANGUAGES:** English, Hindi.

DESCRIPTION: Organization of housing finance corporations. Promotes growth and development of the housing finance industry in Asia. Represents members' commercial and regulatory interests; conducts promotional activities.

ASIAN COALITION OF HUMAN RIGHTS ORGANIZATIONS (ACHRO)

Intl. Center for Law in Development, 777 United Nations Plaza, New York, NY 10017 **OFFICERS:** Dr. Clarence Dias. **LANGUAGES:** English.

DESCRIPTION: Members are human rights organizations. Promotes respect for the rights of the individual and the rule of law in Asia. Facilitates communication and cooperation among members; monitors human rights issues and publicizes abuses.

ASIAN CONFEDERATION OF TEACHERS (ACT)

Wisma DTC, 2nd Fl., 3455B Jalan Sultansh Zainab, Kelantan, 15050 Kota Bharu, Malaysia **OFFICERS:** Hj Muhammad Mustapha. **LANGUAGES:** English, Malay.

DESCRIPTION: Teachers and teachers' unions. Seeks to obtain optimal conditions of employment for teachers; promotes the trade union movement within the educational professions. Coordinates members' activities; represents teachers in negotiations with educational authorities and institutions.

ASIAN COORDINATING GROUP FOR CHEMISTRY (ACGC)

UNESCO, 8 Poorvi Marg, Vasant Vihar, New Delhi 110 057, Delhi, India **PHONE:** 91 11 676308, 91 11 677310 **FX:** 91 11 6873351 **LANGUAGES:** English, Hindi.

DESCRIPTION: Chemists, chemistry educators and students, and chemistry organizations. Seeks to advance scholarship in the field of chemistry. Facilitates exchange of information among members; sponsors research and educational programs.

ASIAN CULTURAL FORUM ON DEVELOPMENT (ACFD)

494 Ladprao 101 Rd., Sol 11, Bangkapi, Bangkok 10240, Thailand **PHONE:** 66 2 3779357 **FX:** 66 2 3740464 **E-MAIL:** acfod@ksc15.th.com **OFFICERS:** Bantorn Ondam. **LANGUAGES:** English, Thai.

DESCRIPTION: Cultural and development organizations. Promotes preservation of indigenous cultures in areas undergoing economic development in Asia. Documents traditional cultures; consults with development agencies and projects to minimize their impact on indigenous cultures.

ASIAN DEVELOPMENT BANK (ADB)

PO Box 789, Manila 0980, Philippines **FX:** 63 2 6362444 **TX:** 63587 ADB PN (ETPI) **WEBSITES:** http://www.asiandevbank.org **FOUNDED:** 1966. **OFFICERS:** Mitsuo Sato, Pres. **MEMBERS:** 55. **STAFF:** 1,940. **BUDGET:** US$188,266,000. **LANGUAGES:** English.

DESCRIPTION: Development finance institution engaged in promoting the economic and social progress of developing member countries in the Asian and Pacific region. Provides loans and technical assistance for development projects and programs. Promotes investment of public and private capital for development. Assists in coordinating development policies and plans for member countries. Supports regional economic cooperation. **PUBLICATIONS:** *ADB Business Opportunities*, monthly; *Asian Development Outlook*, annual; *Asian Development Review*, semiannual; Annual Report.

ASIAN ECOLOGICAL SOCIETY (AES)

Tunghai Univ., Biology Dept., Box 843, Taichung 400, Taiwan **PHONE:** 886 4 3595622, 886 4 3590991 **FX:** 886 4 3595622 **E-MAIL:** edgarlin@ms5.hinet.net **FOUNDED:** 1978. **OFFICERS:** Lin Jun-Yi, Pres. **MEMBERS:** 100. **STAFF:** 2. **BUDGET:** US$10,000. **LANGUAGES:** English, Mandarin.

DESCRIPTION: Conducts research on theoretical and applied ecological policies. Fosters cooperation among ecological groups. **PUBLICATIONS:** *Nuclear Report from Taiwan* (in English), bimonthly.

ASIAN ECOTECHNOLOGY NETWORK (AEN)

M.S. Swaminathan Res. Found., Centre for Res. on Sustainable Agriculture and Rural Development, Third Cross Rd., Taramani Institutional Area, Madras 600 113, Tamil Nadu, India **PHONE:** 91 44 2351319 **E-MAIL:** mssrf.madras@sm8.springtrpg.sprint.com **LANGUAGES:** English, Hindi.

DESCRIPTION: Members are individuals and organizations. Promotes environmentally sustainable economic development. Identifies sustainable development strategies and technologies; provides consulting services to development projects.

ASIAN ENERGY INSTITUTE (AEI)

Tata Energy Res. Inst., Darbon Seth Block, Habitat Pl., Lodhi Rd., New Delhi 110 003, Delhi, India **OFFICERS:** Preety Bhandari. **LANGUAGES:** English, Hindi.

DESCRIPTION: Scientists with an interest in energy and related issues. Promotes development of environmentally sustainable energy sources; encourages more efficient use of nonrenewable energy resources. Serves as a clearinghouse on energy; sponsors research and educational programs; provides assistance to corporations and organizations wishing to make use of alternative energy sources.

ASIAN FEDERATION OF CATHOLIC MEDICAL ASSOCIATIONS (AFCMA)

Catholic Univ. Medical Coll., 505 Banpo-dong, Sochu-ku, Seoul 137 701, Republic of Korea **PHONE:** 82 2 5935141 **FX:** 82 2 5323112 **LANGUAGES:** English, Korean.

DESCRIPTION: Catholic medical associations. Seeks to increase the availability of quality health care services in previously underserved areas of Asia; promotes adherence to high standards of ethics and practice among Catholic health care services and institutions. Facilitates communication and cooperation among members; makes available health services; conducts exchange and continuing professional development programs.

ASIAN FEDERATION FOR MEDICAL CHEMISTRY (AFMC)

c/o Korean Chemical Society, Science and Technology Bldg., Rm. 703, 635-4 Yoksam-dong, Kangnam-gu, Seoul 135 703, Republic of Korea **PHONE:** 82 2 34533781 **FX:** 82 2 34533785 **E-MAIL:** kcschem@neon.kcsnet.or.kr **FOUNDED:** 1946. **OFFICERS:** Dai Woon Lee, Pres. **MEMBERS:** 5,000. **STAFF:** 7. **BUDGET:** US$670,000. **LANGUAGES:** English, Korean.

DESCRIPTION: Chemists, physicians, and other scientists and health care personnel with an interest in medical chemistry. Promotes medical chemistry research and scholarship. Serves as a clearinghouse on medical chemistry; conducts research and educational programs. **PUBLICATIONS:** *Bulletin of the Korean Chemical Society*, monthly; *Chemical Education* (in Korean), quarterly; *ChemWorld* (in Korean), monthly; *Journal of the Korean Chemical Society* (in English, French, German, and Korean), bimonthly; *Korean Journal of Medicinal Chemistry* (in English), semiannual.

ASIAN FOOD COUNCIL (AFC)

Central Food Technological Res. Inst., Suryakanti, Hinkel 570 017, Mysore, India **PHONE:** 91 821 42335 **FX:** 91 821 27697 **OFFICERS:** Dr. H. A. B. Parpia. **LANGUAGES:** English, Hindi.

DESCRIPTION: Scientists with an interest in the processing of food products. Seeks to advance the technology applied to food processing; promotes increased availability of affordable food products. Serves as a clearinghouse on food sciences; sponsors research and educational programs.

ASIAN FORUM FOR HUMAN RIGHTS AND DEVELOPMENT (Forum-Asia)

c/o Union for Civil Liberty, 109 Suthisarnwinichai Rd., Samsennok, Huaykwang, Bangkok 10310, Thailand **PHONE:** 66 2 2769846, 66 2 2769847 **FX:** 66 2 6934939 **E-MAIL:** forumasiabkk@mozart.inet.co.th **WEBSITES:** http://www.forumasia.org **FOUNDED:** 1991. **OFFICERS:** Somchai Homlaor. **MEMBERS:** 25. **STAFF:** 15. **BUDGET:** US$400,000. **LANGUAGES:** English, Thai.

DESCRIPTION: Human rights and development organizations. Promotes and protects the civil and human rights, coordination and collaboration with regional human rights organizations to achieve Thai civil society and sustainable development. Monitors civil rights policies of development projects and publicizes abuses.

ASIAN FORUM OF PARLIAMENTARIANS ON POPULATION AND DEVELOPMENT (AFPPD)

c/o Neurological Research Foundation, Sala Pamnak, 312 Rajavithi Rd., Phyathal, Bangkok 10400, Thailand **PHONE:** 66 2 2486726, 66 2 2486727 **FX:** 66 2 2468827 **E-MAIL:** skhare@mozart.inet.co.th **OFFICERS:** Shiv Khare. **LANGUAGES:** English, Thai.

DESCRIPTION: Organization of Asian legislators. Promotes population control as an integral aspect of sustainable economic development. Serves as a clearinghouse on population issues; formulates development and population control strategies; conducts motivational and public education and programs.

ASIAN HEALTH INSTITUTE (AHI)

987-30 Minamiyama, Komenogi, Nisshin-Shi, Aichi-gun, Aichi 470-01, Japan **PHONE:** 81 5617 31950 **FX:** 81 5617 31990 **FOUNDED:** 1980. **OFFICERS:** Sato Hikaru M.D., Gen.Sec. **MEMBERS:** 7,700. **STAFF:** 10. **BUDGET:** 88,000,000¥. **LANGUAGES:** English, Japanese.

DESCRIPTION: Promotes improved public health, particularly in Asia. Conducts public educational campaigns to create indigenous leadership for public health initiatives in the region. **PUBLICATIONS:** *Ajia no Kenko* (in Japanese), bimonthly. Member Newsletter; *Aju no Kodomo* (in English and Japanese), annual. Children's magazine on the live of children in Asian countries; *Asian Health Institute* (in English), quarterly. Forum for training course alumni, grassroots health workers. Occasional reports on study tours, special research projects, retrospectives, and so on.

ASIAN HUMAN RIGHTS COMMISSION (AHRC)

Mongkok Commercial Centre, Unit D, 7th Fl., 16-16B Argyle St., Kowloon, Hong Kong **PHONE:** 852 26986339 **FX:** 852 26986367

E-MAIL: alegrcen@hk.super.net **WEBSITES:** http://www.hk.super.net/-ahrchk/ **LANGUAGES:** Chinese, English.

DESCRIPTION: Human rights organizations and agencies. Promotes respect for the rights of the individual and the rule of law in Asia. Monitors human rights issues and publicizes abuses; makes available legal services.

ASIAN MEDIA INFORMATION & COMMUNICATION CENTRE (AMIC)

SCS Bldg., NTU, PO Box 360, Singapore 916412, Singapore **PHONE:** 65 7927570 **FX:** 65 7927129 **E-MAIL:** amicline@singnet.com.sg **WEBSITES:** http://www.amic.org.sg **FOUNDED:** 1971. **OFFICERS:** Mr. Vijay Menon, Sec.Gen. **MEMBERS:** 594. **STAFF:** 22. **LANGUAGES:** English.

DESCRIPTION: Mass communications institutions and individuals in 33 countries. Assists in the socioeconomic development of communications media in Asia. Promotes, coordinates, and conducts applied research in communications. Organizes conferences, seminars, workshops, and training programs for professionals in the field. Provides consulting services; serves as liaison among organizations and governments conducting programs in the field. Acts as resource center and clearinghouse, providing microfiche, microfilm, and photocopies. **PUBLICATIONS:** *Asian Journal of Communication* (in English), semiannual. Features book reviews, editorials, and research notes.; *Asian Mass Communication Bulletin* (in English), bimonthly; *Mass Communication Periodical Literature Index* (in English), biennial; *Media Asia* (in English), quarterly; *Studies on Mass Communications in Asia* (in English), periodic. Also publishes bibliographies, monographs, brochures, papers, and reports; has published *AMIC Directory of Mass Communication Institutions in Asia* and *Broadcasting Glossary*.

ASIAN NETWORK ON PROBLEM SOILS (ANPS)

c/o RAPA, Maliwan Mansion, Phra Atit Rd. 39, Bangkok 10200, Thailand **PHONE:** 66 2 2817844 **FX:** 66 2 2800445 **TX:** 828 15 FOODAG TH **OFFICERS:** Dr. F. J. Dent. **LANGUAGES:** English, Thai.

DESCRIPTION: Soil scientists and soil conservation agencies. Promotes improved erosion control; seeks to advance scientific understanding of problem soils. Gathers and disseminates information on problem soils and erosion control; sponsors research and educational programs.

ASIAN AND PACIFIC COCONUT COMMUNITY (APCC)

3rd Fl., Linà Bldg. KAV-B7, JL H.R. Rasuna Said, Kuningan, 12920 Jakarta, Indonesia **PHONE:** 62 21 5221712, 62 21 5221713 **FX:** 62 21 5221714 **TX:** 62209 APCC IA **E-MAIL:** apcc@indo.net.id **WEBSITES:** http://www.apcc.org.sg **FOUNDED:** 1969. **OFFICERS:** P.G. Punchihewa, Exec.Dir. **STAFF:** 5. **LANGUAGES:** English.

DESCRIPTION: Governments of 14 coconut producing countries in Asia and the Pacific. Encourages technical, economic, and other forms of international cooperation within the coconut industry; fosters development, promotion, and protection of the coconut industry. Provides information, studies, training, and coordination efforts. Compiles statistics. **PUBLICATIONS:** *COCOINFO International* (in English), semiannual; *Cocomunity Newsletter* (in English), semimonthly; *Coconut Production and Productivity Series*; *Cocunut Processing Technology Information Documents Series*; *Cord-Coconut Research and Development*, semiannual; *Directory of Coconut Traders and Equipment Manufacturers*, periodic; *Proceedings of Technical Meeting-Cocotech*; *Processing of Coconut Products Series*; *Statistical Yearbook*.

ASIAN PACIFIC DENTAL FEDERATION (APDF/APRO)

242 Tanjong Katong Rd., Singapore 437030, Singapore **PHONE:** 65 3453125 **FX:** 65 3442116 **FOUNDED:** 1955. **OFFICERS:** Dr. Oliver Hennedige, Sec.Gen. **MEMBERS:** 19. **LANGUAGES:** English.

DESCRIPTION: National dental associations in Australia, Bangladesh, Guam, Hong Kong, India, Indonesia, Japan, Malaysia, Mongolia, Myanmar, Nepal, New Zealand, Pakistan, Philippines, Republic of Korea, Singapore, Sri Lanka, Taiwan, and Thailand. Works to improve dental and general health in the Asia Pacific region. Encourages education and research links between national dental associations. **PUBLICATIONS:** *APDF/APRO Technical Report*, periodic; *Dentistry in the Asian Pacific Region*, periodic.

ASIAN AND PACIFIC DEVELOPMENT CENTRE (APDC)

Pesiaran Duta, PO Box 12224, 50770 Kuala Lumpur, Malaysia **PHONE:** 60 3 6511088 **FX:** 60 3 6510316 **TX:** MA 30676 APDEC **E-MAIL:** info@apdc.po.my **WEBSITES:** http://www.apdc.com.my/apdc **FOUNDED:** 1980. **OFFICERS:** Dr. Mohd. Noor Hj. Harun, Dir. **MEMBERS:** 21. **STAFF:** 59. **LANGUAGES:** English.

DESCRIPTION: Representatives of Asian and Pacific countries. Objectives are: to enhance the socioeconomic growth of member countries by suggesting viable development policies, strategies, and programs, with special emphasis on innovative planning geared to specific regional needs. Furthers the research and educational capabilities of members by working to improve national educational institutions and by creating research networks. Analyzes national development problems from a regional point of view. Fosters programs to optimally utilize human resources in the region. Sponsors Asian and Pacific Energy Planning Network. Organizes training projects on topics such as agricultural development, international trade and regional cooperation, new technologies and public management, energy planning and management, and women working in development. **PUBLICATIONS:** *Creating the Vision: Microfinancing the Poor in Asia-Pacific (Issues, Constraints, Capacity Building)*; *Executive Summary - Regional Consultation on Refugee Women and Women in Situations of Armed Conflict*; *Gender Training Resources in the Asian and Pacific Region: A Select Annotated Bibliography*; *Microfinance Capacity Assessment - Bangladesh, India, Indonesia, Philippines, South Pacific*; *Proceedings of Consultative ANWIM (Asian Network on Women & International Migration)*; *Proceedings of the Bank Poor Regional Workshop*.

ASIAN AND PACIFIC FEDERATION OF ORGANIZATIONS FOR CANCER RESEARCH AND CONTROL (APFOCRC)

Seoul Natl. Univ. Hospital, 28 Yunkun-dong, Conggno-ku, Seoul 110-744, Republic of Korea **PHONE:** 82 2 7602314, 82 2 7635110 **FX:** 82 2 7448307 **OFFICERS:** Dr. J. Kim. **LANGUAGES:** English, Korean.

DESCRIPTION: Cancer researchers and research institutions; organizations providing care to people with cancer. Seeks to discover more effective methods for the treatment and eventual cure of cancer; works to improve the quality of life of people with cancer. Serves as a clearinghouse on cancer and cancer research; facilitates exchange of information among cancer researchers; provides support and services to people with cancer.

ASIAN AND PACIFIC REGIONAL AGRICULTURAL CREDIT ASSOCIATION (APRACA)

Maliwan Mansion, Phra Atit Rd., Bangkok 10200, Thailand **PHONE:** 66 2 2817844 **FX:** 66 2 2800445 **TX:** 82815 FOODAG TH **OFFICERS:** Zulkifli M. Noor. **LANGUAGES:** English, Thai.

DESCRIPTION: Financial institutions providing credit to farmers and agricultural enterprises. Promotes agricultural development; seeks to ensure adequate financing of agricultural enterprises. Functions as a clearinghouse on agricultural finance; provides financial support to rural development initiatives; facilitates communication and cooperation among members.

ASIAN PACIFIC REGIONAL COORDINATING COMMITTEE (APRCC)

c/o USI-USO House, USO Rd., New Delhi 110 067, Delhi, India **PHONE:** 91 11 6561103, 91 11 6525146 **FX:** 91 11 6856283 **FOUNDED:** 1994. **MEMBERS:** 24. **LANGUAGES:** English, Hindi.

DESCRIPTION: Representatives of Asian and Pacific countries. Promotes international understanding and cooperation within the region. Functions as a forum for the resolution of disputes and the discussion of issues of mutual interest to members. **PUBLICATIONS:** *Asia Pacific Link* (in English), triennial.

ASIAN PACIFIC RURAL AND AGRICULTURAL CREDIT ASSOCIATION (APRACA)

39 Maliwan Mansion, Phra Atit Rd., Bangkok 10200, Thailand **PHONE:** 66 2 2800195 **FX:** 66 2 2801524 **E-MAIL:** apraca@ksc15.th.com

WEBSITES: http://www.apraca.th.com **FOUNDED:** 1977. **OFFICERS:** Mr. Yong-Jin Kim, Sec.Gen. **MEMBERS:** 54. **STAFF:** 10. **BUDGET:** US$309,000. **LANGUAGES:** English.

DESCRIPTION: Banking institutions and agricultural research training institutions in 19 countries in the Asia-Pacific region. Objectives are to: stimulate improved financial planning for agricultural development in the region; coordinate training programs on agricultural credit, rural finance, and other common problems; act as information clearinghouse on agricultural credit in the region. Acts in cooperation with the Food and Agriculture Organization of the United Nations and other international and regional organizations. Provides technical assistance to member countries. Conducts training programs, cooperative programs, staff exchange programs, and study tours. Maintains living banks and rural youth and marketing finance facilities for rural entrepreneurs. Provides consultancy services on rural finance and development and instruction for trainers working in agricultural banking institutions; conducts quarterly conferences, seminars, and workshops. **PUBLICATIONS:** *Asia-Pacific Rural Finance*, quarterly; *News Digest*, quarterly. Also publishes reports, books, brochure, and case studies.

ASIAN PRODUCTIVITY ORGANIZATION (APO)

4-14, Akasaka 8-Chome, Minato-ku, Tokyo 107-0052, Japan **PHONE:** 81 3 34087221 **FX:** 81 3 4087220 **E-MAIL:** apo@apo-tokyo.com **FOUNDED:** 1961. **OFFICERS:** Takahashi Tajima, Secy. Gen. **MEMBERS:** 18. **STAFF:** 45. **BUDGET:** US$18,828,800. **LANGUAGES:** English.

DESCRIPTION: Asian and Pacific governments of 18 countries that are members of the United Nations Economic and Social Commission for Asia and the Pacific; governments outside the region can apply for associate members. Works to enhance quality of life by increasing productivity and accelerating economic development and environmental improvement in the Asian and Pacific region. Develops human resources and asseminates information on techniques and experiences in modern productivity to agriculture, industry, and service sectors. Studies productivity issues; identifies members' needs; plans and designs programs based on findings. Conducts research, surveys, study missions, and training courses. Provides technical experts services. Develops training manuals and audiovisual training materials. Acts as think-tank, institution builder, regional adviser, clearinghouse, and catalyst for productivity management. **PUBLICATIONS:** *APO Annual Report* (in English); *APO News*, monthly; *Catalog of Publications*; *Catalogue of Audio-Visuals*; *Directory of National Productivity Organizations in APO Member Countries*, biennial; *Proceedings of Governing Body Meeting*, annual; *Productivity Journal*, semiannual. Contains technical information.

ASIAN REINSURANCE CORP. (ARC)

Chamman Phenjati Center, Tower B, 17th Fl., 65 Rama IX Rd., Huaykwang, Bangkok 10320, Thailand **PHONE:** 66 2 2452169, 66 2 2452197 **FX:** 66 2 2488011 **TX:** 87231 ASRE TH **E-MAIL:** asianre@loxinfo.co.th **FOUNDED:** 1979. **OFFICERS:** Armando S. Malabanan. **STAFF:** 25. **BUDGET:** US$770,000. **LANGUAGES:** English, Thai.

DESCRIPTION: Insurance and reinsurance corporations. Promotes growth and development of the Asian insurance industries. Represents the commercial and regulatory interests of members; facilitates communication and cooperation among Asian insurance companies.

ASIAN RURAL INSTITUTE (ARI)

442-1 Tsukinokizawa, Nishinasuno-machi, Nasu-gun, Tochigi 329-2703, Japan **PHONE:** 81 287 363111 **FX:** 81 287 375833 **E-MAIL:** ari@nasu-net.or.jp **WEBSITES:** http://www.nasu-net.or.jp/-ari/; http://www.ari.edu **FOUNDED:** 1973. **OFFICERS:** Rev. Jintaro Ueda, Dir. **STAFF:** 12. **BUDGET:** US$1,200,000. **LANGUAGES:** English, Japanese.

DESCRIPTION: Training institute for rural leaders from Asia, Africa, Latin America, and the Pacific Islands. Conducts training programs designed to increase local administrative and entrepreneurial capabilities in rural areas, and to develop skill in organic and sustainable farming methods. Seeks to educate the public regarding environmental and development issues faced by rural areas. Sponsors surveys of agricultural techniques and economic development. Operates work camps. **PUBLICATIONS:** *Ajia No Tsuchi* (in Japanese), periodic; *Take My Hand* (in English), semiannual.

ASIAN SOCIETY OF AGRICULTURAL ECONOMISTS (ASAE)

c/o Korea Rural Economists Institute, 4-102 Hoegi-dong, Dongdaemun-ku, Seoul 130-050, Republic of Korea **PHONE:** 82 2 9627311, 82 2 9571593 **FX:** 82 2 9656950 **E-MAIL:** jsupchoi@kreisun.krei.re.kr **FOUNDED:** 1991. **OFFICERS:** Yang-Boo Choe, Sec.Treas. **MEMBERS:** 180. **LANGUAGES:** English.

DESCRIPTION: Professors and researchers in agricultural economics; agricultural economic institutes; government officials. Serves as a forum for discussion of agricultural and rural development issues; seeks to advance scientific knowledge in the field of agricultural economics and the structure of rural societies. Plans to conduct international conferences linking agricultural economists worldwide.

ASIAN STUDENTS' ASSOCIATION (ASA)

353 Shanghai St. 4/F, Kowloon, Hong Kong **PHONE:** 852 3880515 **FX:** 852 7825535 **FOUNDED:** 1969. **OFFICERS:** Norman Uy Camay, Sec. **MEMBERS:** 52. **STAFF:** 3. **LANGUAGES:** English.

DESCRIPTION: Student unions and organizations. Supports academic freedom and students' rights; promotes active student involvement in the educational process. Encourages development of student organizations and activist movements that will advocate issues including human rights for indigenous populations, improved working conditions for Asian peasants, and world peace. Fosters cooperation and understanding among members. Organizes workshops on topics including economics and development, peace and disarmament, and human rights. Sponsors training projects and student exchange programs. **PUBLICATIONS:** *Asian Student News* (in English), quarterly; *Movement News Roundup*, monthly; *Workshop Reports*.

ASIAN VEGETABLE RESEARCH AND DEVELOPMENT CENTER (AVRDC)

PO Box 42, Shanhua 741, Taiwan **PHONE:** 886 6 5837801 **FX:** 886 6 5830009 **TX:** 73560 AVRDC **E-MAIL:** avrdcbox@netra.avrdc.org.tw **WEBSITES:** http://www.avrdc.org.tw **FOUNDED:** 1971. **OFFICERS:** Dr. Samson Tsou, Dir.Gen. **MEMBERS:** 14. **STAFF:** 300. **BUDGET:** US$9,000,000. **LANGUAGES:** English.

DESCRIPTION: Representatives of Australia, Germany, France, Japan, Republic of China, the Philippines, Republic of Korea, Thailand, and the United States. Conducts research to advance the development of vegetable crops in humid and subhumid tropical areas; collects and analyzes germ plasm to develop improved vegetable varieties. Serves as a network for the sharing of results of research and development. Maintains regional programs and networks aimed at improving research capacity of national partners. Sponsors vegetable research skills training courses and advanced training programs for researchers and graduate students. **PUBLICATIONS:** *Annual Progress Report* (in English); *Annual Report* (in English); *Centerpoint* (in English), semiannual; *Seminar Proceedings* (in English), periodic; *Technical Bulletins* (in English), periodic; *TVIS Newsletter*, semiannual; *Working Papers* (in English), periodic; Monographs.

ASIAN WOMEN WORKERS' CENTER (Azia Joshi Rodosha Koryu Center)

Nishi-waseda 2-3-18-34, Shinjuku-ku, Tokyo 169-0051, Japan **PHONE:** 81 3 32024993 **FX:** 81 3 32024993 **E-MAIL:** awwc@mail.webnik.ne.jp **FOUNDED:** 1983. **OFFICERS:** Miyoko Shiozawa, Dir. **MEMBERS:** 700. **STAFF:** 5. **BUDGET:** 11,000,000¥.

DESCRIPTION: Seeks to improve the status of women working in Asia. Works to deepen mutual understanding and strengthen solidarity with Asian people and to provide accurate information and resources to Japanese on Asian issues. Provides information on women's labor in Japan, the problems of Asian women workers, and labor laws concerning women workers. Participates in exchange programs and exposure trips. Organizes study tours to one of the Asian countries every summer. **PUBLICATIONS:** *Asian Friends* (in Japanese), bimonthly; *Discrimination Against Women Workers in Japan*; *Resource Materials on Women's Labor in Japan* (in English), semiannual; *Stories of Japanese Women Workers*.

ASIAN WOMEN'S HUMAN RIGHTS COUNCIL (AWHRC)

2124 1st A Cross, 16 B Main, H.A.L. II Stage,

Bangalore 560 008, Karnataka, India **PHONE:** 63 2 9246406 **FX:** 63 2 9246381 **E-MAIL:** awhrc@phil.gn.apc **FOUNDED:** 1986. **OFFICERS:** Ms. Corinne Kumar, Coord. **MEMBERS:** 25. **STAFF:** 4. **LANGUAGES:** English, Filipino.

DESCRIPTION: Women lawyers and feminist activists involved in the promotion and defense of basic human rights and women's rights. Encourages study of human rights. Fosters cooperation and solidarity between women's groups and individuals advocating human rights recognition. Compiles information on national policies throughout Asia which affect human rights. **PUBLICATIONS:** *Asian Womenews* (in English), semiannual; *The Quilt Journal.*

ASIAN-JAPAN WOMEN'S RESOURCE CENTER (Asia-Josei-Siryo-Center)

Shibuya Co-op 311, 14-10, Sakuragaoka, Shibuya-ku, Tokyo 150, Japan **PHONE:** 81 3 37805245 **FX:** 81 3 34639752 **E-MAIL:** ajwrc@jca.or.jp **FOUNDED:** 1995. **OFFICERS:** Yayori Matsai. **MEMBERS:** 650. **STAFF:** 4. **BUDGET:** 119,000¥. **LANGUAGES:** Japanese.

DESCRIPTION: Works to advance the status of Asian women by providing a variety of educational, community, and legal services. **PUBLICATIONS:** *Women's Asia 21* (in Japanese), quarterly; *Women's Asia 21 - Voices from Japan* (in English), annual.

ASIAN-PACIFIC NEWS NETWORK (APNN)

c/o Xinhua News Agency, 57 Xuanwumen Xidajie, Beijing, People's Republic of China **PHONE:** 86 10 63072707 **FX:** 86 10 672707 **OFFICERS:** Yu Jiafu. **LANGUAGES:** Chinese, English.

DESCRIPTION: News agencies and press and media outlets. Promotes free flow of accurate current events information in Asia. Gathers and disseminates news and current information reports; provides news gathering and research services to members.

ASIA-PACIFIC 2000

PO Box 12544, 50782 Kuala Lumpur, Malaysia **PHONE:** 60 3 2559122 **FX:** 60 3 2552870 **OFFICERS:** Anwar Fazal. **LANGUAGES:** English, Malay.

DESCRIPTION: Membership made up of development organizations. Promotes appropriate and environmentally sustainable economic and community development in Asia and the Pacific. Serves as a clearinghouse on development and the environment; sponsors research, educational, and advocacy programs.

ASIA-PACIFIC ASSOCIATION OF AGRICULTURAL RESEARCH INSTITUTIONS (APAARI)

FAO Office in India, PO Box 3088, New Delhi 110 003, Delhi, India **PHONE:** 91 11 4628877, 91 11 4693060 **E-MAIL:** fao-ind@cgnet.com **OFFICERS:** Dr. R. S. Paroda. **LANGUAGES:** English, Hindi.

DESCRIPTION: Agricultural research institutions, educational institutions, and agricultural scientists. Seeks to advance the study, teaching, and practice of the agricultural sciences. Serves as a clearinghouse on agricultural science; sponsors research and educational programs.

ASIA-PACIFIC BUSINESS ASSOCIATION (APBA)

Champs Elysees B/D, b-dong, 2nd Fl., Taechi 1-dong, Kangnam-gu, Seoul 135-281, Republic of Korea **PHONE:** 82 2 5638513 **FX:** 82 2 5549927 **OFFICERS:** Jae-Hong Chun. **LANGUAGES:** English, Korean.

DESCRIPTION: Businesses and business and trade organizations. Promotes a social and political climate conducive to profitable business in Asia and the Pacific. Facilitates exchange of information among members; serves as a clearinghouse on Asian and Pacific business; represents the commercial and regulatory interests of members.

ASIA-PACIFIC CENTRE OF EDUCATIONAL INNOVATION FOR DEVELOPMENT (APCEID)

c/o UNESCO PROAP, 920 Sukumvit Rd., PO Box 967, Prakanong Post Office, Bangkok 10110, Thailand **PHONE:** 66 2 3910879, 66 2 3910291 **FX:** 66 2 3910866 **TX:** 20591 TH

E-MAIL: uhbgk@unesco.org **OFFICERS:** Rupert MacLean. **LANGUAGES:** English, Thai.

DESCRIPTION: Educators and educational institutions. Promotes increased availability of educational opportunity in developing regions of Asia and the Pacific. Provides support and educational services to development organizations and projects; sponsors training programs.

ASIA-PACIFIC COUNCIL OF AIDS SERVICES ORGANIZATIONS (APCASO)

Yayasan Citra Usadha Indonesia, Jalan Belimbing Gang Y No. 4, 80231 Denpasar, Indonesia **PHONE:** 62 361 246757 **FX:** 62 361 246757 **E-MAIL:** apcaso@idola.net.id **OFFICERS:** T. Merati. **LANGUAGES:** English, Indonesian.

DESCRIPTION: Organizations providing services to people with AIDS and their families. Promotes increased availability of AIDS services in Asia and the Pacific. Facilitates communication and cooperation among members; coordinates provision of AIDS services by multiple agencies.

ASIA-PACIFIC ECONOMIC COOPERATION (APEC)

Alexandra Point Bldg., 14th Fl. 01/04, 438 Alexandra Rd., Singapore 119958, Singapore **PHONE:** 65 2761880 **FX:** 65 2761775 **E-MAIL:** info@mail.apecsec.org.sg **WEBSITES:** http://www.apecsec.org.sg/ **OFFICERS:** Dato'Noor Adlan. **LANGUAGES:** Chinese, English.

DESCRIPTION: Trade, business, and economic organizations; corporations; government agencies. Promotes increased trade and economic growth among the countries of Asia and the Pacific. Represents the interests of trade and business before international trade organizations; facilitates establishment of economic joint ventures; lobbies for removal of barriers to international trade.

ASIA-PACIFIC PEOPLE'S ENVIRONMENT NETWORK (APPEN)

c/o Sahabat Alam Malaysia, 27, Lorong Maktab, 10250 Penang, Malaysia **PHONE:** 60 4 2275705, 60 4 2276930 **FX:** 60 4 2275705 **TX:** CAPPG MA 40989 **FOUNDED:** 1982. **OFFICERS:** S.M. Mohd Idris, Coor. **LANGUAGES:** English.

DESCRIPTION: Organizations in 25 countries in the Asia-Pacific region concerned with the problems of depleting natural resources and degradation of the environment. Collects and disseminates information; investigates and reports on environmental issues in the region. Sponsors educational and research programs. **PUBLICATIONS:** *The Bhopal Tragedy - One Year After*; *Damming the Narmada*; *Decimation of World Wildlife - Japan as Number One*; *Environment, Development and Natural Resource Crisis in Asia and Pacific*; *Environmental Crisis in Asia-Pacific, Forest Resources Crisis in the Third World*; *Global Anti-Golf Movement Updates* (in English), semiannual; *Global Development and Environmental Crisis - Has Humankind a Future?*.

ASIA-PACIFIC RELIEF ORGANIZATIONS NETWORK (APRON)

AMDA Intl., 310-1 Narazu, Okayama 701 12, Japan **PHONE:** 81 86 2847730 **FX:** 81 86 2848959 **E-MAIL:** koikc@amda.or.jp **LANGUAGES:** English, Japanese.

DESCRIPTION: National and international relief organizations. Promotes cooperation among members to ensure effectiveness of relief efforts in Asia and the Pacific. Coordinates relief activities in the region; provides support and assistance to relief projects; facilitates exchange of information among members.

ASIA-PACIFIC STRATEGIC CENTRE FOR DEVELOPMENT FINANCE (APSCDF)

c/o ADFIAP, Skyland Plaza, Senator Gil J. Puyat Ave., Makati, Metro Manila 1200, Philippines **PHONE:** 63 2 8161672, 63 2 7926116 **FX:** 63 2 8176498 **LANGUAGES:** English, Filipino.

DESCRIPTION: Development and finance organizations. Seeks to ensure adequate financial backing for development programs without creating unbearable debt within the area to be developed. Formulates model development finance strategies; makes available consulting services; serves as a clearinghouse on development finance.

ASIA-PACIFIC TELEVISION NEWS EXCHANGE (APTNE)

PO Box 1164, 59700 Kuala Lumpur, Malaysia

PHONE: 60 3 2823592 **FX:** 60 3 2825292 **TX:** 32227 (ABUMA) Kuala **E-MAIL:** sg@abu.org.my **WEBSITES:** http://www.abu.org.my/abu/ **LANGUAGES:** English, Malay.

DESCRIPTION: Television news agencies and media outlets. Promotes free flow of current events information in Asia and the Pacific. Gathers and disseminates current events information for commercial use by members; sponsors research and news gathering activities.

ASSEMBLY OF THE BLOCK OF NATIONS IN EUROPE AND ASIA FOR FREEDOM AND DEMOCRACY

Zeppelinstrasse 67, D-81669 Munich, Germany **PHONE:** 49 89 482532 **FX:** 49 89 486519 **E-MAIL:** 100114.335@compuserve.com **OFFICERS:** Ms. Slava Stetsko. **LANGUAGES:** English, German.

DESCRIPTION: National organizations supporting political freedom and democratic political systems. Promotes establishment and strengthening of democratic institutions worldwide. Serves as a clearinghouse on political freedom and democracy; publicizes the excesses of authoritarian regimes; sponsors educational programs.

ASSEMBLY OF CARIBBEAN COMMUNITY PARLIAMENTARIANS (ACCP)

c/o CARICOM, Bank of Guyana Bldg., 3rd Fl., Avenue of the Republic, PO Box 10827, Georgetown, Guyana **PHONE:** 592 2 69281, 592 2 69289 **FX:** 592 2 66091 **TX:** 2263 CARISEC GY **LANGUAGES:** English.

DESCRIPTION: Representatives of parliamentary governments in the Caribbean Community. Promotes cooperation among the members of the Caribbean Community to facilitate sustainable regional economic development. Serves as a forum for the deliberation of issues of mutual interest to members; formulates reports on regional development issues.

ASSEMBLY OF EUROPEAN REGIONS (AER) (Assemblee des Regions d'Europe - ARE)

Immeuble Europe, 20, place des Halles, F-67054 Strasbourg Cedex, France **PHONE:** 33 3 88220707 **FX:** 33 3 88756719 **E-MAIL:** aerpress@sdv.fr **FOUNDED:** 1985. **OFFICERS:** Luc Van den Brande, Pres. **MEMBERS:** 286. **STAFF:** 17. **LANGUAGES:** English, French, German, Italian, Spanish.

DESCRIPTION: Regional authorities and interregional organizations in 26 countries. Works to facilitate members' participation in European development by increasing their participation and representation in European institutions. Encourages cooperation, communication, and information exchange among the regions from western, central, and eastern Europe. Promotes regionalism and federalism in Europe. Examines such issues as employment, tourism, culture, education, international relations, and environmental problems. Cooperates with related international organizations. **PUBLICATIONS:** *AER Online - Are en Direct* (in English, French, and German), quarterly; *Cartes de Visite*, semiannual.

ASSIST INTERNATIONAL (AI)

Scotts Valley, CA 95067-6396 **PHONE:** (831) 438-4582 **FX:** (831) 439-9602 **E-MAIL:** assistintl@aol.com **WEBSITES:** http://www.assistintl.org **OFFICERS:** Robert J. Pagett, Pres. **LANGUAGES:** English.

DESCRIPTION: Health care organizations. Procures used cardiac care monitoring systems and other medical equipment in the United States for distribution among underserved populations in the developing world. Distributes medical equipment, food, pharmaceuticals, and other supplies in needy areas worldwide.

ASSISTANCE AUX CREATEURS D'ENTREPRISES DU NORD-OUEST EUROPEEN (ACENOE)

c/o Patrick C. Locufier, Residence Daniel Balavoine, 85, ave. de la Liberte, F-59831 Lambersart Cedex, France **FOUNDED:** 1982. **OFFICERS:** Patrick C. Locufier, President. **BUDGET:** 11,000,000 Fr. **LANGUAGES:** Dutch, English, Esperanto, French, German, Hebrew, Spanish, Yiddish.

DESCRIPTION: Promotes economic development in Europe by assisting the creation of new businesses and jobs. Sponsors unemployment programs to assist governments in improving the economic climate of the European community. Provides aid to refugees throughout Europe in order to establish a more economically stable environment.

ASSISTANCE FOR ECONOMIC RESTRUCTURING IN THE COUNTRIES OF CENTRAL AND EASTERN EUROPE

c/o European Commission, Coordination Unit DG I-G24, Rue de la Loi 200, B-1049 Brussels, Belgium **PHONE:** 32 2 2992244 **FX:** 32 2 2358341 **TX:** COMEU B 21877 **OFFICERS:** Herman de Lange. **LANGUAGES:** English, French.

DESCRIPTION: Governments of the European Community. Promotes sustainable economic development among the countries of Central and Eastern Europe. Coordinates development and financial activities undertaken by members in Central and Eastern Europe.

ASSOCIATED COUNTRY WOMEN OF THE WORLD (ACWW)

Vincent House, 6th Fl., Vincent Sq., London SW1P 2NB, England **PHONE:** 44 171 8348635 **FX:** 44 171 2336205 **FOUNDED:** 1930. **OFFICERS:** Anna Frost, Gen.Sec. **STAFF:** 11. **LANGUAGES:** English.

DESCRIPTION: Rural women's societies. Promotes friendly and helpful relations among women's organizations in over 60 countries and assists in the economic, social, and cultural development of their members and countries. Maintains consultative status with the United Nations Economic and Social Council. Offers leadership training course, nutrition education, and functional literacy programs. Serves as information clearinghouse. **PUBLICATIONS:** *The Countrywoman* (in English), quarterly; Reports, periodic.

ASSOCIATION FOR THE ADVANCEMENT OF POLICY, RESEARCH AND DEVELOPMENT IN THE THIRD WORLD (AAPRDTW)

PO Box 70257, Washington, DC 20024-0257 **PHONE:** (202) 508-1441, (202) 785-0048 **FX:** (202) 331-3759 **FOUNDED:** 1981. **OFFICERS:** Dr. Mekki Mtewa, Exec.Dir. **MEMBERS:** 3,450. **STAFF:** 10. **LANGUAGES:** English.

DESCRIPTION: Scholars; private industries; international, national, and local government personnel; interested individuals. Promotes science, technology, and development through exchange and generation of practical solutions to problems facing governments in developing countries. Encourages the effective and improved utilization of resources, including human development and planning institutions. Advocates ethical standards and supports interest in government research and development policies and programs. Facilitates the exchange of experience and ideas between professionals in developing countries. Disseminates information on policy management practices and current research in the field. Compiles statistics; offers consultations; organizes research symposia. Provides placement service, charitable program, and speakers' bureau. Maintains 31 committees and 7 regional sections. **PUBLICATIONS:** *Basic Principles for the Equal Treatment and Protection of Overseas Scholars*, periodic; *Comparative Development and African Administration*; *Credential: Journal of Science, Technology and International Development*; *Credentials: International Journal of Science, Research and Development*, annual. Contains reviews of contemporary development issues and information on research practices and funding for professional institutions.; *Development Futures: The Coming Challenges*; *International Science and Technology*; *Perspectives in International Development, and Science, Technology, and Development*; Membership Directory, periodic.

ASSOCIATION FOR THE ADVANCEMENT OF WOMEN IN AFRICA

PO Box 20378, Kitwe, Zambia **PHONE:** 243 229634, 243 217537 **FOUNDED:** 1985. **OFFICERS:** Rosemary Mumbi. **MEMBERS:** 2,000. **STAFF:** 27. **BUDGET:** US$200,000. **LANGUAGES:** English.

DESCRIPTION: Works to enhance the social, political, and economic status of women in Africa. Provides training in organic farming, counseling groups, and one-on-one counseling. Conducts research. **PUBLICATIONS:** *Women's Exclusive Africa* (in English), monthly. Covers social, political, policies, and economic issues.

ASSOCIATION OF AFRICAN DEVELOPMENT FINANCE INSTITUTIONS (AADFI)

c/o ADB, 01, Boite Postale 1387, Abidjan 01, Cote d'Ivoire **PHONE:** 225 204444 **FX:** 225 227344 **TX:** 23717 **OFFICERS:** J.A. Hammond, Sec.Gen. **LANGUAGES:** English, French.

DESCRIPTION: Development banks and related financial institutions in Africa. Seeks to coordinate development activities and financing throughout the continent. Provides financial services to development projects and organizations.

ASSOCIATION OF AFRICAN MARITIME TRAINING INSTITUTES (AAMTI)

Gamal Abdel Nasser St., PO Box 1029, Alexandria, Egypt **PHONE:** 20 3 5481163, 20 3 5611816 **FX:** 20 3 5602144 **TX:** 54160 ACAD UN **OFFICERS:** Dr. R. M. Rashad. **LANGUAGES:** Arabic, English.

DESCRIPTION: Educational institutions offering training programs for maritime personnel. Seeks to ensure the availability of trained personnel for merchant marine service; promotes excellence in maritime training. Maintains placement service; facilitates communication and exchange among members.

ASSOCIATION OF AFRICAN TRADE PROMOTION ORGANIZATIONS (AATPO)

Boite Postale 23, Tangiers, Morocco **PHONE:** 212 9 41536, 212 9 41687 **FX:** 212 9 41536 **TX:** AOAPC 33695 **E-MAIL:** aoapc@mtds.com **OFFICERS:** Adeyinka W. Orimalade. **LANGUAGES:** English, French.

DESCRIPTION: National trade organizations and authorities. Promotes increased international trade involving African countries. Serves as a forum for the discussion of international trade and related issues; facilitates negotiations involving the removal of national barriers to international trade.

ASSOCIATION OF AFRICAN UNIVERSITIES (AAU) (Association des Universites Africaines - AUA)

PO Box 5744, Accra-North, Accra, Ghana **PHONE:** 233 21 774495, 233 21 761588 **FX:** 233 21 774821 **TX:** 2284 ADUA GH **E-MAIL:** secgen@aau.org **WEBSITES:** http://www.aau.org **FOUNDED:** 1967. **OFFICERS:** Prof. Narcisso Matos. **MEMBERS:** 136. **STAFF:** 23. **LANGUAGES:** Arabic, English, French.

DESCRIPTION: Representatives of universities in 42 countries. Promotes intellectual exchange and cooperation among African universities, particularly in developing curricula and establishing the equivalence of degrees; seeks to identify, study, and publicize the educational needs of African universities and to coordinate the efforts aimed at meeting these needs; stimulates contacts with the international academic community. Sponsors the exchange of professors in an effort to satisfy the particular research and administrative needs of member universities. Offers undergraduate and graduate fellowships and scholarships; provides academic placement. Promotes the increased use of African languages. Operates documentation and information centre which gathers and disseminates information on member universities. Sponsors research and educational programs and seminars for university teachers, administrators, and other professionals concerned with higher education in Africa. Maintains liaison with the United Nations Educational, Scientific and Cultural Organization, the Organization of African Unity, and other international groups. Provides reader services. **PUBLICATIONS:** *AAU Newsletter* (in English and French), 3/year; *The African Experience with Higher Education*; *Creating The African University Emerging Issues of the 1970s*; *Directory of African Universities* (in English and French), biennial; *Reports of Seminars* (in English and French), periodic; *Study on Cost Effectiveness and Efficiency in African Universities*; *University Productive Sector Linkages in Ghana: Universities and Small and Medium Scale Enterprises*; *University Productive Sector Linkages Review of the State of the Art in Africa*. Also publishes library acquisition list.

ASSOCIATION OF AFRICAN WOMEN FOR RESEARCH AND DEVELOPMENT (AAWORD) (Association des Femmes Africaines pour la Recherche sur le Developpement - AFARD)

BP 15367, Sicop Amitie II Villa N 4050, Dakar, Senegal **PHONE:** 221 8242053, 221 259822 **FX:** 221 8242056 **FOUNDED:** 1977. **OFFICERS:** Ms.

Yassine Fall, Exec.Sec. **MEMBERS:** 700. **STAFF:** 4. **BUDGET:** 500,000 Fr CFA. **LANGUAGES:** English, French.

DESCRIPTION: African women in 42 countries working in the field of social science; national research groups; researchers and research groups outside of Africa. Aims to increase the participation of African women in their society; to this end, conducts research into issues affecting African women, identifies resources to assist members in such research, and provides networking among African women researchers. Addresses labor, demographic, educational, political, professional, and ideological issues. Studies the situation of women in developing countries. Sponsors seminars. Organizes training programs and specialized research workshops, such as the Working Group on Women and Reproduction in Africa, in an effort to identify research sources and develop methodologies of study. **PUBLICATIONS:** *ECHO* (in English and French), quarterly. Also publishes occasional papers and bibliographical series.

ASSOCIATION OF ARAB INSTITUTES AND CENTRES FOR ECONOMIC AND SOCIAL RESEARCH (AAICESR)

c/o Institut d'Economie Quantitive, 27 rue de Liban, 1000 Tunis, Tunisia **PHONE:** 216 1 283633, 216 1 280904 **FX:** 216 1 787034 **OFFICERS:** Sami Cherif. **LANGUAGES:** Arabic, English, French.

DESCRIPTION: Economics and social sciences research institutes. Promotes excellence in economic and social scholarship. Provides consulting services to public policy agencies and development programs; sponsors research and educational programs.

ASSOCIATION OF ARAB UNIVERSITIES (AARU)

PO Box 401 Jebeyha, Amman, Jordan **PHONE:** 962 6 845131, 962 6 840135 **FX:** 962 6 832994 **TX:** 23855 AARU JO **FOUNDED:** 1969. **OFFICERS:** Dr. Ehab Ismail, Sec.Gen. **MEMBERS:** 119. **STAFF:** 21. **LANGUAGES:** Arabic.

DESCRIPTION: Arab universities united to promote cooperation; works to raise the standard of university education in the Arab world. Considers academic programs and student-staff affairs in a national and international capacity. Sponsors book exhibitions, seminars, symposia, and sports festivals. **PUBLICATIONS:** *Guide to Arab Universities* (in Arabic); *Journal of the AARU* (in Arabic), annual; *Palestinian Question*. Two volumes.; *Statistical Guide for Arab Universities*; *Studies in Arab Societies*; Newsletter (in Arabic), semiannual; Membership Directory; Proceedings.

ASSOCIATION OF ASIA PACIFIC AIRLINES (AAPA)

Corporate Business Centre, 151 Paseo de Roxas, Makati, Metro Manila, Philippines **PHONE:** 63 2 8403191 **FX:** 63 2 8103518 **TX:** SITA MNLOAXD **OFFICERS:** Richard T. Stirland. **LANGUAGES:** English, Filipino.

DESCRIPTION: Airlines based in Asia and the Pacific. Promotes growth and development among the Asian and Pacific air freight and passenger carriers. Represents members' interests before labor and aviation organizations, government agencies, and the public; conducts promotional programs.

ASSOCIATION OF ASIAN CONFEDERATION OF CREDIT UNIONS (ACCU)

PO Box 24-171, Bangkok 10240, Thailand **PHONE:** 66 2 3743170, 66 2 3745101 **FX:** 66 2 3745321 **E-MAIL:** accuran@ksc.th.com **FOUNDED:** 1971. **OFFICERS:** Ranjith Hettiarachchi, Gen.Mgr. **MEMBERS:** 17. **STAFF:** 8. **BUDGET:** 12,000,000 Bht. **LANGUAGES:** English.

DESCRIPTION: Credit union leagues, federations, and promotion centers in 15 countries. Promotes the development of Asian credit unions that will better serve the socioeconomic needs of the region. Seeks to attain a unified regional credit union system by pooling financial and human resources, encouraging legal recognition of credit unions, and soliciting national and international support for the Asian credit union movement. Provides technical, financial, and educational assistance to member credit unions. Offers courses for trainers and managers. Conducts research on cooperatives and credit unions. **PUBLICATIONS:** *ACCU News* (in English), bimonthly, always January, March, May, July, September, and November. Covers activities of ACCU and member organizations.; *ACCU Report and Directory* (in English), annual; Reports.

ASSOCIATION OF BANKING SUPERVISORY AUTHORITIES OF LATIN AMERICA AND THE CARIBBEAN

c/o CEMA, Durango 54, Cuahtemoc, Colonia Roma, 06700 Mexico City 7, DF, Mexico **PHONE:** 52 5 5330300, 52 5 2073325 **FX:** 52 5 5146554 **TX:** 1771229 CEMLME **OFFICERS:** Edgardo Mimica. **LANGUAGES:** English, Spanish.

DESCRIPTION: Banking and financial authorities. Promotes effective supervision of financial and banking activities; seeks to ensure financial stability in Latin America and the Caribbean. Formulates and enforces standards of banking supervisory practice and ethics; serves as a forum for the discussion of international financial issues.

ASSOCIATION OF BORDERLANDS SCHOLARS

c/o Anthony Popp, New Mexico State Univ., Dept. of Economics, Box 30001, Dept. 3CQ, Las Cruces, NM 88003 **PHONE:** (505) 646-5198 **FX:** (505) 646-1915 **FOUNDED:** 1976. **OFFICERS:** Stephen P. Mumme, Pres. **MEMBERS:** 3,004. **LANGUAGES:** English.

DESCRIPTION: Scholars at academic and governmental institutions representing the Americas, Asia, Africa, and Europe. Dedicated to the systematic interchange of ideas and information relating to international border areas, including the United States-Mexico borderlands region, Asia, Africa, and Europe. Studies contemporary issues such as regional economic integration, the emergence of new post-communist nation-states, and the need to institutionalize management of transboundary problems ranging from immigration to shared natural resources. **PUBLICATIONS:** *Journal of Borderlands Studies*, semiannual. Contains research; *La Frontera*, semiannual. Reports on meetings and professional news.

ASSOCIATION OF CARIBBEAN STATES (ACS)

ACS Secretariat, 11-13 Victoria Ave., PO Box 660, Port of Spain, Trinidad and Tobago **PHONE:** (868) 623-2783 **FX:** (868) 623-2679 **E-MAIL:** ACS-AEC@Trinidad.net **WEBSITES:** http://www.ACS-AEC.org/ **OFFICERS:** Simon Molina Duarte. **LANGUAGES:** English, French, Spanish.

DESCRIPTION: Representatives of the Caribbean states. Promotes regional cooperation in economic, political, and social matters. Serves as a forum for the discussion of issues of mutual concern to members; arbitrates economic and political disputes involving Caribbean states; formulates and enforces international trade protocols.

ASSOCIATION OF CARIBBEAN UNIVERSITY, RESEARCH AND INSTITUTIONAL LIBRARIES

Apartado Postal 23317, San Juan, Puerto Rico 00931 **PHONE:** 787 7630000 **FX:** 787 7635685 **E-MAIL:** vortes@upracd.upr.clu.edu **OFFICERS:** Oneida R. Ortiz. **LANGUAGES:** English, Spanish.

DESCRIPTION: Organization of special libraries serving universities and research institutions. Promotes growth and development of libraries and archives; encourages high standards of ethics and practice among library personnel. Encourages communication and cooperation among members.

ASSOCIATION OF CATHOLIC INSTITUTES OF EDUCATION (ACIE)

Universite Catholique de l'Ouest, Boite Postale 808, F-49008 Angers Cedex, France **OFFICERS:** Michel Soetard. **LANGUAGES:** English, French.

DESCRIPTION: Catholic educational institutions. Promotes excellence among Catholic educational systems and schools. Conducts training programs for Catholic educators and administrators.

ASSOCIATION OF THE CENTRAL ALPS (ACA)

ARGE ALP, Amt der Tiroler Landestregierung, A-6010 Innsbruck, Austria **PHONE:** 43 512 5082340 **FX:** 43 512 5082345 **WEBSITES:** http://www.argealp.at/ **LANGUAGES:** English, French, German, Italian.

DESCRIPTION: Governments with territory in the central Alps. Promotes international cooperation in the development of the central Alps. Serves as a forum for the discussion of interests of mutual interest to members; arbitrates disputes involving members. Formulates model development strategies for the central Alps.

ASSOCIATION FOR CHILDHOOD EDUCATION INTERNATIONAL (ACEI)

17904 Georgia Ave., Ste. 215, Olney, MD 20832 **PHONE:** (301) 570-2111 **TF:** (800) 423-3563 **FX:** (301) 570-2212 **E-MAIL:** aceihq@aol.com **WEBSITES:** http://www.udel.edu/bateman/acei **FOUNDED:** 1892. **OFFICERS:** Gerald Odland, Exec.Dir. **MEMBERS:** 11,000. **STAFF:** 14. **BUDGET:** US$1,000,000. **LANGUAGES:** English.

DESCRIPTION: Teachers, parents, and other caregivers in 31 countries interested in promoting good educational practices for children from infancy through early adolescence. Seeks to: promote the inherent rights, education, and well-being of all children in home, school, and community; promote desirable conditions, programs, and practices for children; raise the standard of preparation for teachers and others who are involved with the care and development of children; encourage continuous professional growth of educators; promote cooperation among all individuals and groups concerned with children; inform the public of the needs of children and the ways in which various programs must be adjusted to fit those needs and rights. Conducts workshops and travel/study tours abroad. Conducts research and educational programs; maintains Hall of Fame and speakers bureau. Maintains liaison with government agencies, cooperating organizations, teaching institutions, and manufacturers and designers of materials and equipment for children. **PUBLICATIONS:** *ACEI Exchange*, bimonthly; *Childhood Education*, bimonthly, 6/year; *Focus on Elementary*; *Focus on Infancy*; *Focus on Middle School*; *Focus on Prek-k*; *Journal of Research in Childhood Education*, 2/year; Booklets; Books.

ASSOCIATION OF CHRISTIAN INSTITUTES FOR SOCIAL CONCERN IN ASIA

Ecumenical Christian Centre, Whitefield, Bangalore 560 066, Karnataka, India **PHONE:** 91 80 8452270, 91 80 8452653 **FX:** 91 80 8452653 **FOUNDED:** 1963. **OFFICERS:** Rev. Dr. Prof. M.J. Joseph. **MEMBERS:** 22. **STAFF:** 65. **BUDGET:** US$6,000,000. **LANGUAGES:** English, Hindi.

DESCRIPTION: Christian denominations operating social welfare programs in Asia. Promotes sustainable economic development throughout Asia; seeks to ensure the spiritual advancement of people in developing areas. Provides support and assistance to development projects; conducts social welfare and service programs; sponsors religious and educational programs. **PUBLICATIONS:** *Theology for our Times* (in English).

ASSOCIATION OF COFFEE PRODUCING COUNTRIES (ACPC)

c/o Brazilian Embassy, 32 Green St., London W1Y 4AT, England **PHONE:** 44 171 4990877 **FX:** 44 171 4934790 **OFFICERS:** Roberio Oliveira Silva, Sec.Gen. **LANGUAGES:** English.

DESCRIPTION: Promotes increased demand for coffee and related products; works to remove barriers to the international coffee trade. Represents members' interests; conducts promotional activities; serves as a clearinghouse on coffee.

ASSOCIATION OF COMMITTEES OF SIMPLIFIED PROCEDURES FOR INTERNATIONAL TRADE WITHIN THE EUROPEAN COMMUNITY AND THE EUROPEAN FREE TRADE ASSOCIATION

SIPRICOM, Office Belge du Commerce Exterieur, Boulevard Emile Jacqmain 162, B-1000 Brussels, Belgium **OFFICERS:** E. Goffin. **LANGUAGES:** English, French.

DESCRIPTION: National business and trade committees working with the European Community (EC) and the European Free Trade Association (EFTA). Seeks to simplify EC and EFTA regulations as a means of promoting international trade. Drafts model trade policies; conducts advocacy activities.

ASSOCIATION FOR THE COOPERATION OF CHURCHES IN THE ENVIRONMENT AND DEVELOPMENT OF CENTRAL AFRICA (Association pour la Cooperation des Eglises, l'Environnement et le Developpement de l'Afrique Centrale)

Boite Postale 1199, Brazzaville, Congo **PHONE:** 242 813383 **FX:** 242 814751 **OFFICERS:** Daniel Kilem-Mbila. **LANGUAGES:** English, French.

DESCRIPTION: Central African churches. Promotes increased involvement of religious organizations in environmental protection and economic development initiatives. Assists churches wishing to establish social service programs; conducts training programs for church social service personnel; provides consulting services to development and environmental protection organizations with operations in central Africa.

ASSOCIATION FOR COOPERATION IN DEVELOPMENT (ACD) (Associacao de Cooperacao para o Desenvolvimento - SUL)

Travessa Maria da Fonte, 5-Esgueira, C.P. 263, Caixa Postal 3803, Aveiro, Portugal **PHONE:** 351 34 23676, 351 34 315292 **FX:** 351 34 22787 **E-MAIL:** bgrui@student.ua.pt **FOUNDED:** 1990. **OFFICERS:** Rui Correja, Exec.Comm.Pres. **STAFF:** 6. **BUDGET:** US$90,000. **LANGUAGES:** English, Portuguese, Spanish.

DESCRIPTION: Development organizations with programs in Portugal and its former territories of Cape Verde, Sao Tome and Principe, Guinea-Bissau, Mozambique, Angola and Brazil. Promotes communication and cooperation among members to ensure locally appropriate and sustainable economic development. Gathers and disseminates information to increase public awareness of, and support for, development programs and issues. Provides support and assistance to development projects operating in areas including environmental protection, human rights, population control, tourism, water supply and sanitation, the rights of indigenous people, and education. Specific groups working with the issue of East Timor, A Solidarity Group Tibet and with Western Sahara. **PUBLICATIONS:** *Meridialis* (in Portuguese), quarterly.

ASSOCIATION COOPERATIVE UNESCO

Boite Postal 493, Brazzaville, Congo **PHONE:** 242 811303, 242 831986 **FOUNDED:** 1984. **OFFICERS:** Bahoumina Gabin Blaise. **MEMBERS:** 60. **STAFF:** 10. **BUDGET:** US$404,000. **LANGUAGES:** English, French.

DESCRIPTION: Promotes and supports the activities of the United Nations Educational, Scientific and Cultural Organization. Facilitates mutual cooperation between development agencies and local communities, and among communities undergoing development. Conducts rural development programs in areas including watershed management, water supply and sanitation, road building and maintenance, and agricultural and vocational education. **PUBLICATIONS:** *Nous Les Volontaires de Developement* (in French), quarterly.

ASSOCIATION FOR CULTURAL ECONOMICS INTERNATIONAL (ACEI)

Dept. of Economics, 301 Lake Hall, Northeastern Univ., Boston, MA 02115 **PHONE:** (617) 373-2839 **FX:** (617) 373-3640 **E-MAIL:** acei@neu.edu **WEBSITES:** http://www.acei.neu.edu **OFFICERS:** Neil O. Alper. **LANGUAGES:** English.

DESCRIPTION: Economists and economics educators and students. Seeks to advance cultural economics scholarship and practice. Serves as a clearinghouse on cultural economics; sponsors research and educational programs. **PUBLICATIONS:** *Journal of Cultural Economics* (in English), quarterly.

ASSOCIATION FOR THE DEFENCE OF HUMAN RIGHTS AND DEMOCRATIC LIBERTIES IN THE ARAB WORLD

46 rue de Vaugirard, F-75006 Paris, France **PHONE:** 33 1 47812861 **FX:** 33 1 42741698 **LANGUAGES:** Arabic, English, French.

DESCRIPTION: Members are human rights organizations. Promotes respect for the rights of the individual and the rule of law in the Arab world; seeks to develop and strengthen democratic political systems and institutions in Arab countries. Monitors human rights issues in the Arab world and publicizes alleged abuses; sponsors educational programs.

ASSOCIATION FOR THE DEVELOPMENT OF EDUCATION IN AFRICA (ADEA)

c/o IIEP, 7-9 rue Eugene Delacroix, F-75116 Paris, France **PHONE:** 33 1 45033796 **FX:** 33 1 45033965 **E-MAIL:** adea@iiep.unesco.org **WEBSITES:** http://www.bellanet.org/partners/adea/ **OFFICERS:** Richard Sack **LANGUAGES:** English, French.

DESCRIPTION: Educational organizations with operations in Africa. Promotes development of African educational institutions and systems; seeks to ensure availability of educational opportunity throughout Africa. Provides support and assistance to members; makes available consulting services to African educational institutions and agencies.

ASSOCIATION FOR THE DEVELOPMENT OF EVALUATION METHODS IN EDUCATION (Association pour le Developpement des Methodologies d'Evaluation en Education)

UCL, Dept. des Sciences de l'Education, Place de l'Universie 1, B-1348 Louvain-la-Neuve, Belgium **PHONE:** 32 10 474467 **OFFICERS:** L. Paquay. **LANGUAGES:** English, French.

DESCRIPTION: Membership consists of educational organizations. Promotes development of effective methods for the evaluation of educational programs and personnel. Develops evaluation schemes; sponsors research and training programs.

ASSOCIATION OF DEVELOPMENT FINANCING INSTITUTIONS IN ASIA AND THE PACIFIC (ADFIAP)

2F Skyland Plaza, Sen. Gil Puyat Ave., Makati City, Metro Manila 1200, Philippines **PHONE:** 63 2 8161672, 63 2 8430932 **FX:** 63 2 8176498 **E-MAIL:** inquire@adfiap.org **WEBSITES:** http://www.adfiap.org **FOUNDED:** 1976. **OFFICERS:** Orlando P. Pena, Sec.Gen. **MEMBERS:** 81. **STAFF:** 20. **BUDGET:** US$319,200. **LANGUAGES:** English.

DESCRIPTION: Institutions from 33 countries whose main activity is the financing of development in Asia and the Pacific in order to foster economic development and growth in these areas. Aims to: develop financing institutions in Asia and the Pacific; promote regional and international cooperation among member institutions in the interest of regional economic development; facilitate the exchange of information; coordinate investment promotion policies; raise the level of expertise among member institutions. Conducts surveys and research; organizes seminars, symposia, and training programs. **PUBLICATIONS:** *ADFIAP Annual Report* (in English); *ADFIAP Newsletter* (in English), bimonthly; *Asian Banking Digest* (in English), bimonthly; *Development Notes* (in English), monthly; *Directory of DFIs in Asia and the Pacific* (in English), quinquennial; *Journal of Development Finance* (in English), quarterly; *Summary of Proceedings* (in English), annual.

ASSOCIATION OF THE EASTERN ALPS (AEA)

Via Gazzoletti 2, I-38100 Trent, Italy **PHONE:** 39 461 201410 **FX:** 39 461 201409 **OFFICERS:** Silvano Longo. **LANGUAGES:** English, Italian.

DESCRIPTION: Governments with territory in the eastern Alps. Promotes international cooperation in the development of the eastern Alps. Serves as a forum for the discussion of interests of mutual interest to members; arbitrates disputes involving members. Formulates model development strategies for the eastern Alps.

ASSOCIATION OF EASTERN CARIBBEAN MANUFACTURERS (AECM)

c/o ECSEDA, PO Box 961, Roseau, Dominica **PHONE:** (809) 448-6555 **FX:** (809) 448-6735 **LANGUAGES:** English.

DESCRIPTION: Membership consists of manufacturing firms. Promotes growth and development of the manufacturing industries of the eastern Caribbean. Represents members' interests before labor and international trade organizations, government agencies, and the public; sponsors promotional programs.

ASSOCIATION OF EUROPEAN AIRLINES (AEA)

350, ave. Louise, bte. 4, B-1050 Brussels, Belgium **PHONE:** 32 2 6270600 **FX:** 32 2 6484017 **TX:** 22918 **FOUNDED:** 1954. **OFFICERS:** Karl-Heinz Neumeister, Sec.Gen. **MEMBERS:** 26. **STAFF:** 23. **LANGUAGES:** English.

DESCRIPTION: Representatives of 26 major European airlines in 26 countries. Aims to contribute to the improvement and development of European commercial air transport. Promotes infrastructure capacity (airports, air traffic control) to match growing air traffic demand. Coordinates the exchange of data among members and addresses commercial, environmental, technical and political issues. Liaises with the European Community Institutions, international bodies, and national public agencies and ministries. **PUBLICATIONS:** *AEA Yearbook*, annual. Reviews trends and achievements of AEA airlines in preceding calendar year and also includes information on current issues facing the industry.; *Medium-Term Forecast of European Scheduled Passenger Traffic*, annual.

Contains forecasts developed using a basic regression model form.; *Statistical Appendices*. Contains extensive statistics on revenue, traffic capacity and production data for each airline.

ASSOCIATION OF EUROPEAN BORDER REGIONS (AEBR)

Staatsminister im Bundeskanzleramt, Adenauerallee 139, D-5300 Bonn, Germany **LANGUAGES:** English, German.

DESCRIPTION: Representatives of European countries. Promotes peaceful resolution of border disputes involving members. Serves as a forum for the discussion of European border issues; arbitrates disputes among members.

ASSOCIATION OF EUROPEAN CHAMBERS OF COMMERCE AND INDUSTRY (AECCI)

Rue Archimede 5, Boite 4, B-1000 Brussels, Belgium **PHONE:** 32 2 2820850 **FX:** 32 2 2300038 **E-MAIL:** eurocham@mail.interpac.be **WEBSITES:** http://www.eurochambres.be/ **OFFICERS:** Frank Friedrich. **LANGUAGES:** English, French.

DESCRIPTION: European chambers of commerce and industry. Promotes advancement of business and industry in Europe. Represents the interests of business before labor organizations, government and international trade organizations, and the public; sponsors promotional campaigns.

ASSOCIATION OF EUROPEAN CITIES AND REGIONS FOR CULTURE (AECRC)

25 rue Deparcieux, F-75014 Paris, France **PHONE:** 33 1 45387013 **FX:** 33 1 45387013 **E-MAIL:** rencontr@club-internet.fr **OFFICERS:** Roger Tropeano. **LANGUAGES:** English, French.

DESCRIPTION: Municipal and regional cultural organizations and agencies. Seeks to ensure access of individuals throughout Europe to cultural exhibitions and performances. Coordinates members' activities; sponsors educational and cultural programs.

ASSOCIATION OF EUROPEAN CONJUNCTURE INSTITUTES (AIECE) (Association d'Instituts Europeens de Conjoncture Economique - AIECE)

c/o Institut de Recherches Economiques, 3, place Montesquieu, BP 4, B-1348 Louvain, Belgium **PHONE:** 32 10 474152 **FX:** 32 10 473965 **E-MAIL:** olbrecht@ires.ucl.ac.be **WEBSITES:** http://www.econ.ucl.ac.be/conj/aiece.fr.html; http://www.econ.ucl.ac.be/conj/aiece.en.html **FOUNDED:** 1957. **OFFICERS:** P. Olbrechts, Adm.Sec. **MEMBERS:** 40. **LANGUAGES:** English, French.

DESCRIPTION: Economic research institutes in 21 countries. Studies analysis methods of economic activity, conjunctural evolution in Europe, and forms of European integration.

ASSOCIATION OF EUROPEAN FRUIT AND VEGETABLE PROCESSING INDUSTRIES (AEFVPI)

Avenue de Roodebeek 30, B-1030 Brussels, Belgium **PHONE:** 32 2 7438730 **FX:** 32 2 7368175 **E-MAIL:** sia01@sia-dvi.be **OFFICERS:** Ms. P. Keppenne. **LANGUAGES:** English, French.

DESCRIPTION: Organization of fruit and vegetable processors. Promotes advancement of the European processed foods industries. Represents members' interests before labor and agricultural organizations, government and international trade agencies, and the public.

ASSOCIATION OF EUROPEAN JOURNALISTS (AEJ)

Cedaceros 11, 3F, E-28014 Madrid, Spain **PHONE:** 34 1 4296869, 34 1 4291754 **FX:** 34 1 4292754 **OFFICERS:** Miguel Angel Anguilar. **LANGUAGES:** English, Spanish.

DESCRIPTION: Professional journalists. Promotes freedom of the press throughout Europe; seeks to ensure adherence to high standards of ethics and practice by members. Serves as a forum for the exchange of information among members; sponsors continuing professional development programs.

ASSOCIATION OF EUROPEAN REGIONAL FINANCIAL CENTRES (AERFC)

Associacao Barbelona Centre Financier Europeu, Avenida Diagonal 537, E-08029 Barcelona, Spain **PHONE:** 34 3 4029105 **FX:** 34 3 4029101 **OFFICERS:** Carles Tusquets. **LANGUAGES:** English, Spanish.

DESCRIPTION: European financial centers and business and economic organizations. Promotes unification of European financial and business systems. Facilitates communication and cooperation among members; arbitrates international disputes involving members.

ASSOCIATION OF EUROPEAN UNIVERSITIES (Association des Universites Europeennes)

10, rue du Conseil General, CH-1211 Geneva 4, Switzerland **PHONE:** 41 22 3292644, 41 22 3292251 **FX:** 41 22 3292821 **E-MAIL:** cre@uni2a.unige.ch **WEBSITES:** http://www.unige.ch/cre/ **FOUNDED:** 1959. **OFFICERS:** Dr. Andris Barblan, Sec.Gen. **MEMBERS:** 528. **STAFF:** 12. **BUDGET:** US$2,000,000. **LANGUAGES:** English, French.

DESCRIPTION: Universities in 40 European countries. Promotes cooperation among European universities. Acts as a forum for inter-European discussions and informal meetings; informs members and other interested parties about developments in university policy in Europe; represents members before governmental and nongovernmental bodies concerned with higher education in Europe; provides services in the fields of institutional management and development. Conducts programs in areas including institutional evaluation, emerging technologies, history of education, and international academic communication. **PUBLICATIONS:** *CRE action Directory* (in English and French), annual. Contains members and organizations concerned with higher education; *CRE Info* (in English and French), quarterly; *CREaction* (in English and French), semiannual. Contains one CRE directory and one thematic issue.; *History of the University in Europe, Vol. I: Universities in the Middle Ages*; *Implementing Sustainable Development at University Level.* Additional publications available on request.

ASSOCIATION OF EXPORT CREDIT INSURERS AND OF EXPORT PROMOTION ORGANIZATIONS

56 avenue Faidherbe, PO Box 3939, Dakar, Senegal **PHONE:** 221 224234 **FX:** 221 213611 **OFFICERS:** Amadou Saloum. **LANGUAGES:** English, French.

DESCRIPTION: Export credit insurance companies and export promotion organizations. Promotes increased international trade through simplification of export credit and tariff regulations worldwide. Develops model international trade and export credit protocols; serves as a forum for the discussion of issues of mutual interest to members; provides consulting services to businesses engaged in international trade.

ASSOCIATION OF FOOD AND AGRICULTURAL MARKETING AGENCIES IN ASIA AND THE PACIFIC (AFAMA)

c/o FAO Regional Office for Asia and the Pacific, Phra Atit Rd., Bangkok 10200, Thailand **PHONE:** 66 2 2817844, 66 2 6291203 **FX:** 66 2 6291203 **TX:** 82815 FOODAG TH **FOUNDED:** 1983. **OFFICERS:** M. R. Satyal, Exec.Dir. **MEMBERS:** 25. **STAFF:** 3. **BUDGET:** US$226,000. **LANGUAGES:** English.

DESCRIPTION: Government institutions, apex cooperatives, and private sector food and agricultural marketing concerns from 13 Asian countries involved in the marketing of grains, livestock and livestock products, fruits, and vegetables. Encourages economic and technical cooperation among regional food marketing institutions; facilitates the exchange of experience and information among members. Sponsors managerial and technical training and exchange programs. Carries out joint projects with the Food and Agriculture Organization of the United Nations and member governments. Conducts research on policy issues affecting food marketing systems. **PUBLICATIONS:** *Food Marketing News* (in English), quarterly. Contains regional news, legislative and regulatory information, statistics, food and marketing policies, and marketing strategies.; *Food Marketing System Improvement Policies and Programmes in Asian and Pacific Countries*, periodic; *Food Marketing Training Facilities in Asian and Pacific Countries*, periodic; *Improvement of Post-Harvest Fresh Fruits and Vegetables: A Manual*, periodic; Brochures, periodic; Reports, periodic. Information on various training workshops, seminars, and expert consultations on aspects of food and agricultural marketing.

ASSOCIATION OF FREE TRADE ZONES OF LATIN AMERICA AND THE CARIBBEAN

Cento Comercial El Pueblito, Oficina 204, Bocagrande, Avenida San Martin, Cartagena de Indias, Colombia **PHONE:** 57 5 6550575, 57 5 6550579 **FX:** 57 5 6550576 **OFFICERS:** Hector Trujillo Velez. **LANGUAGES:** English, Spanish.

DESCRIPTION: International free trade zones located in Latin America and the Caribbean. Promotes creation of free trade zones to encourage international exchange of goods and services. Works to ensure efficient operation of existing free trade zones; advocates for the creation of new free trade zones and the removal of barriers to international trade.

ASSOCIATION OF GEOSCIENTISTS FOR INTERNATIONAL DEVELOPMENT - LATIN AMERICA OFFICE (AGIDLAO)

c/o Institute of Geosciences, Univ. of Sao Paulo, CP 11 348, 05422-970 Sao Paulo, Sao Paulo, Brazil **PHONE:** 55 11 8184231 **FX:** 55 11 8184207 **E-MAIL:** abmacedo@usp.br **WEBSITES:** http://agid.igc.usp.br **FOUNDED:** 1974. **OFFICERS:** Dr. Ariel B. Macedo **MEMBERS:** 2,030. **STAFF:** 1. **BUDGET:** US$10,000. **LANGUAGES:** English, Portuguese, Spanish.

DESCRIPTION: Earth scientists and other individuals interested in international development; governmental agencies, societies, and foundations. Promotes activities in the geosciences related to the needs of individuals living in regions undergoing economic development; seeks to advance application of the geosciences to international development issues. Provides support and assistance to development programs; sponsors research; conducts training courses. **PUBLICATIONS:** *AGID Update* (in English), 1-3/year; *Geoscience and Development Journal*, annual.

ASSOCIATION TO HELP CHERNOBYL

1-10 Rakuen-apart 137, Rakuencho, Showa-ku, Nagoya-shi, Aichi 4660822, Japan **PHONE:** 81 52 8361073 **FX:** 81 52 8361073 **FOUNDED:** 1990. **OFFICERS:** Masaharu Kawata CMR. **MEMBERS:** 2,800. **STAFF:** 13. **BUDGET:** US$250,000. **LANGUAGES:** English, Japanese, Russian, Ukrainian.

DESCRIPTION: Provides medical supplies and supplements to victims of the Chernobyl disaster. Invites Ukrainians to Japan for study of medical techniques for treating people exposed to radiation. Works to educate the public regarding the dangers of nuclear power. Conducts charitable activities benefiting Chernobyl victims; sponsors children's services. **PUBLICATIONS:** *Poleshe* (in Japanese), bimonthly.

ASSOCIATION OF THE INTERNATIONAL CHRISTIAN YOUTH EXCHANGE IN EUROPE (AICYEE)

Naamsesteenweg 164, B-3001 Louvain, Belgium **PHONE:** 32 16 233762 **FX:** 32 16 233925 **FOUNDED:** 1977. **OFFICERS:** Jan Bal, Sec.Gen. **MEMBERS:** 15. **STAFF:** 2. **LANGUAGES:** Dutch, English, French, Spanish.

DESCRIPTION: European national committees of the International Christian Youth Exchange, which organizes 1-year exchange programs for youth in 31 countries. Coordinates European activities, candidate selection, family recruitment, social placements, information weekends, co-workers' training, language courses, and contacts with other European youth organizations. **PUBLICATIONS:** *Conference and Seminar Reports* (in English), after each activity; *Euronews* (in English), quarterly; *ICYE Europe's Yearly Report*; *ICYE Handbook on European Youth Exchanges*; *ICYE Songbook*.

ASSOCIATION OF INTERNATIONAL CHURCHES IN EUROPE AND THE MIDDLE EAST

Gjorlingsvei 10, DK-2900 Hellerup, Denmark **PHONE:** 45 39624785 **FX:** 45 39624785 **E-MAIL:** 100545.2077@compuserve.com **WEBSITES:** http://www.sic.no/aiceme/home.html **OFFICERS:** Rev. Richard Andersen. **LANGUAGES:** Danish, English.

DESCRIPTION: Churches with international programs. Promotes religious observance and seeks to advance the role of churches in social, economic, and political life. Facilitates communication and cooperation among members; sponsors religious, community service, and educational programs.

ASSOCIATION OF INTERNATIONAL COLLEGES AND UNIVERSITIES (AICU)

1301 S. Noland Rd., Independence, MO 64055 **PHONE:** (816) 461-3633 **FX:** (816) 461-3634 **E-MAIL:** tiuf@aol.com **FOUNDED:** 1973. **OFFICERS:** Dr. John Wayne Johnston, Pres. **MEMBERS:** 7,940. **STAFF:** 5. **LANGUAGES:** English.

DESCRIPTION: Membership consists of individuals, organizations, and institutions. Promotes international cooperation and understanding; assists in achieving a coordinated program among institutions of higher education to help eliminate ignorance, superstition, disease, and poor communication; develops alternative solutions and institutions of higher education. Maintains a network of scholars and learning centers. Compiles statistics. **PUBLICATIONS:** *Association of International Colleges and Universities—Directory*, annual; *Association of International Colleges and Universities—Newsletter*, quarterly.

ASSOCIATION OF INTERNATIONAL HEALTH RESEARCHERS (AIHR)

2665 Pleasant Valley Rd., Mobile, AL 36606 **PHONE:** (334) 473-3946 **FOUNDED:** 1982. **OFFICERS:** Dr. Roy E. Kadel, Pres. **MEMBERS:** 123. **LANGUAGES:** English.

DESCRIPTION: Members are individuals interested in quality health research. Works to: promote a better understanding of scientifically effective research techniques and methodologies; encourage interaction among individuals in international health research. Compiles statistics.

ASSOCIATION OF INTERNATIONAL MARKETING (AIM)

Balaam St., PO Box 70, London E13 8BQ, England **PHONE:** 44 181 9867539 **FX:** 44 181 9867539 **FOUNDED:** 1983. **OFFICERS:** C. Oham, Dir. & Gen.Sec. **MEMBERS:** 600. **STAFF:** 2. **LANGUAGES:** English.

DESCRIPTION: Executives in 10 countries who have at least 3 years experience in international marketing or have completed a specified course offered by the organization. Promotes the advancement and exchange of information and ideas in international marketing. Offers certification courses in international marketing. **PUBLICATIONS:** *A guide to Marketing in Europe* (in English). Marketing book.; *AIM Membership Handbook*, periodic; *Aspects of International Marketing and Career Opportunities*; *Dictionary of International Marketing*; *Directory of International Marketing*; *Employment Found in International Marketing*; *European Marketing Tips and Terms for the Single Market*; *Guide on How to Start Own International Marketing Agency*; *International Marketing Agencies Consultants' Registration*; *International Marketing News*, bimonthly; *Introduction to the International Marketing Agency*; *Journal of International Marketing*, annual.

ASSOCIATION FOR INTERNATIONAL PRACTICAL TRAINING (AIPT)

10400 Little Patuxent Pky. Ste. 250, Columbia, MD 21044 **PHONE:** (410) 997-2200 **FX:** (410) 992-3924 **E-MAIL:** aipt@aipt.org **WEBSITES:** http://www.aipt.org **FOUNDED:** 1950. **OFFICERS:** Elizabeth Chazottes, CEO. **STAFF:** 35. **BUDGET:** US$2,900,000. **LANGUAGES:** English.

DESCRIPTION: Sponsors and facilitates on-the-job training and in-the-laboratory practical training exchanges for students and professionals between the Untied States and more than 60 other countries. Also sponsors training visits to the U.S. **PUBLICATIONS:** *AIPT Annual Report*. Includes information on each year's exchanges, participating employers, financial statements and list of officers/directors.; *International IAESTE Annual Report*. Includes the results of each year's exchange of students among the 65 IAESTE member countries. Also includes national reports and statistics.; *Practically Speaking*, quarterly. Articles on events, trainee exchanges, industry news.

ASSOCIATION INTERNATIONALE DE LA MUTUALITE (AIM)

Rue d'Arlon 50, B-1000 Brussels, Belgium **PHONE:** 32 2 2345700, 32 2 2345700 **FX:** 32 2 2345708 **WEBSITES:** http://www.hurisc.org/aim **FOUNDED:** 1950. **OFFICERS:** Willy Palm, Dir. **MEMBERS:** 42. **STAFF:** 4. **BUDGET:** 700,000 SFr. **LANGUAGES:** English, French, German.

DESCRIPTION: National and regional health and social insurance federations, unions, and institutions in 23 countries. Works to promote sound management of mutual benefit funds in all countries. Engages in political representation of members' interest, at national and international levels. Coordinates and guides efforts and activities of affiliated groups; facilitates exchange and comparison of information and experience concerning the financing of health care in various countries. Publishes documentation and information of general interest in the field. Maintains working bodies to study specific issues and problems; compiles statistics. Organizes international meetings. **PUBLICATIONS:** *Aims* (in English, French, and German), quarterly; *Committee and Export Group Reports* (in English, French, and German).

ASSOCIATION OF LATIN AMERICAN AND CARIBBEAN ECONOMISTS (ALACE)

Asociacion Nacional de Economistas de Cuba, Call 22 No. 901, Esquina a 9na, Playa, Havana, Cuba **PHONE:** 53 7 249261, 53 7 293303 **FX:** 53 7 223456 **LANGUAGES:** English, Spanish.

DESCRIPTION: Economists in Latin America and the Caribbean. Seeks to advance economics scholarship and practice. Facilitates communication and cooperation among members; sponsors research, exchange, and educational programs.

ASSOCIATION FOR THE MONETARY UNION OF EUROPE (AMUE) (Association pour l'Union Monetaire de l'Europe - AUME)

26, rue de la Pepiniere, F-75008 Paris, France **PHONE:** 33 1 44706030 **FX:** 33 1 45223377 **E-MAIL:** info@amue.org **WEBSITES:** http://www.amue.org **FOUNDED:** 1987. **OFFICERS:** E. Davignon, Pres. **MEMBERS:** 400. **STAFF:** 13. **LANGUAGES:** English, French, German.

DESCRIPTION: Promotes the economic and monetary union of the European Union. Fosters cooperative economic, business, and political policies within the European Community. Conducts research programs on economic and monetary union. **PUBLICATIONS:** *European Monetary Union for Business* (in English), monthly. Provides information on institutional and market developments in relation to the European Monetary Union.; Reports. On economic studies.

ASSOCIATION MONTESSORI INTERNATIONALE (AMI)

Koninginneweg 161, NL-1075 CN Amsterdam, Netherlands **PHONE:** 31 20 6798932 **FX:** 31 20 6767341 **E-MAIL:** ami@xsyall.nl **WEBSITES:** http://www.montessori-ami.org **FOUNDED:** 1929. **OFFICERS:** G.J. Portielje, Pres. **MEMBERS:** 3,000. **STAFF:** 5. **LANGUAGES:** English.

DESCRIPTION: Teacher training centres, Montessori societies and individuals interested in furthering and learning of the ideas and principles of Dr. Maria Montessori (1870-1952), Italian physician and educator. The Montessori method, stresses the development of the child's initiative and sensory-motor training through individual work with developmental materials; it also emphasizes the freedom of the child with the teacher as a supervisor and guide rather than as formal instructor. The aims of the AMI are to: ensure the Montessori method is implemented with integrity; spread knowledge of the physical, intellectual, moral, social and mental development of the child, from conception to maturity, at home and in society; promote recognition of the child's fundamental rights as envisaged by Montessori, irrespective of racial, religious, political, or social environment; cooperate with other bodies and organizations that promote the development of education, human rights, and peace. Offers teacher training centres affiliated with the AMI to prepare adults to work in the classroom with children in various age groups (0-3, 3-6, 6-12); the AMI conducts a four-stage Training of Trainers' programme which prepares Montessori teachers to become qualified AMI-trainers; provides pedagogical guidance to AMI-approved manufacturers of Montessori materials; provides for the publication of Montessori's books in different languages; conducts study conferences, seminars, and lectures. **PUBLICATIONS:** *Communications* (in English), quarterly.

ASSOCIATION OF NATIONAL, EUROPEAN AND MEDITERRANEAN SOCIETIES OF GASTROENTEROLOGY (Association des Societes Nationales, Europeennes et Mediterraneennes de Gastroenterologie - ASNEMGE)

University Hospital, Dept. of Gastroenterology, P.O. Box 85500, NL-3508 GA Utrecht, Netherlands **PHONE:** 31 30 2507004, 31 30 2506275 **FX:** 31 30 2507371 **FOUNDED:** 1947. **OFFICERS:** Prof. G.P. van Berge Henegouwen. **MEMBERS:** 37. **LANGUAGES:** English, French, German.

DESCRIPTION: A branch of the World Organization of Gastroenterology. Members are national societies for gastroenterology in 37 European and Mediterranean countries. Promotes the exchange of scientific information in the field of gastroenterology and co-organizer of yearly United European Gastroenterology week. **PUBLICATIONS:** Proceedings, quadrennial.

ASSOCIATION OF OFFICIAL SEED ANALYSTS (AOSA)

PO Box 81152, Lincoln, NE 68501-1152 **PHONE:** (402) 476-3852 **FX:** (402) 476-6547 **E-MAIL:** assoc@navix.net **WEBSITES:** http://www.zianet.com/aosa **FOUNDED:** 1908. **OFFICERS:** Dick Lawson, Pres. **MEMBERS:** 75. **STAFF:** 2. **BUDGET:** US$100,000. **LANGUAGES:** English.

DESCRIPTION: Members are officials of 60 federal, state, and provincial seed testing and research laboratories. Seeks to: develop uniform rules for testing field, vegetable, flower, and tree seeds; encourage the use of high quality seed; promote research; foster the training of seed analysts. **PUBLICATIONS:** *Cultivar Purity Testing Handbook*; *Directory of Members*, quadrennial; *Rules for Testing Seed.* Includes rules for testing seeds, seedling evaluation handbook, and uniform classification of weed and crop seeds.; *Seed Analysts Training Manual*; *Seed Technologist Newsletter*, 3/year. Reports on research on seed technology, offers information regarding purity and germination testing of seeds, includes annual meeting proceedings.; *Seed Technology*, semiannual; *Uniform Classification of Crop Weed & Seed*; *Vigor Testing Handbook.*

ASSOCIATION FOR THE PREVENTION OF TORTURE (APT) (Association pour la Prevention de la Torture)

Case Postale 2267, CH-1211 Geneva 2, Switzerland **PHONE:** 41 22 7342088 **FX:** 41 22 7345649 **E-MAIL:** apt@apt.ch **WEBSITES:** http://www.apt.ch **FOUNDED:** 1977. **OFFICERS:** Claudine Haenni, Sec.Gen. **MEMBERS:** 500. **STAFF:** 10. **BUDGET:** 1,100,000 SFr. **LANGUAGES:** English, French, Spanish.

DESCRIPTION: Mandate is to prevent torture and treatment contrary to human dignity. Seeks to ensure the implementation of international laws and principles forbidding torture and to reinforce mechanisms for the prevention of ill-treatment. Proposed first draft of the European Convention for the Prevention of Torture and Inhuman or Degrading Treatment or Punishment. Continues to work towards the effective implementation of the convention today. Also strives for the adoption of the draft Optional Protocol to the United Nations Convention against Torture and other Cruel, Inhuman or Degrading Treatment or Punishment. Cooperates with various bodies concerned with the prevention of torture or active in related fields. Has five programmes covering activities in Asia, Europe, Africa, Latin America, and the United Nations. **PUBLICATIONS:** *European Convention for the Prevention of Torture* (in English and French); *How to Combat Torture*; *Reports on Detention in Various European Countries*; Newsletter (in English, French, and Spanish), quarterly. Gives information on prevention of torture related subjects and APF work. Comparative studies.

ASSOCIATION OF PRIVATE EUROPEAN CABLE OPERATORS (APEC)

1, blvd. Anspach, bte. 25, B-1000 Brussels, Belgium **PHONE:** 32 2 2232591 **FX:** 32 2 2230696 **FOUNDED:** 1992. **OFFICERS:** Marcel de Sutter, Pres. **MEMBERS:** 22. **STAFF:** 5. **BUDGET:** 3,000,000 BFr. **LANGUAGES:** English, French, German.

DESCRIPTION: Professional organizations of national associations dealing with cable TV, telecommunications and value added services. Encourages the development of distribution by cable; ensures exchange of documentation; conducts research on relevant technical, legal, and programming questions. **PUBLICATIONS:** *Membership List*, periodic; Annual Report; Newsletter.

ASSOCIATION FOR THE PROMOTION OF AFRICAN COMMUNITY INITIATIVES (APACI) (Association pour la Promotion des Initiatives Communautaires Africaines - APICA)

c/o Dusnane Kornio, BP 5946, Douala, Cameroon **PHONE:** 237 370405, 237 370404 **FX:** 237 370402 **E-MAIL:** apica@camnet.cm **FOUNDED:** 1980. **OFFICERS:** Ousmane Kornio, Exec. Officer. **MEMBERS:** 24. **STAFF:** 42. **BUDGET:** 600,000,000 Fr CFA. **LANGUAGES:** English, French.

DESCRIPTION: Individuals in 16 countries working in the field of community development. Provides technical, organizational, and administrative support to local organizations, private voluntary organizations, ad hoc groups, and non-governmental organizations engaged in development projects in areas such as public health, sanitation, agriculture, and food manufacture and processing. Studies appropriate technologies with a view toward improving traditional methods and developing innovative alternatives. Disseminates research findings and information in an effort to assist development programs by proposing, evaluating, and testing solutions to practical problems. Maintains contacts and channels for information exchange with organizations working in development. Offers training in management and organizational skills. Operates documentation center. **PUBLICATIONS:** *Annual Report of Activities* (in French). Includes summaries in English; *Communantes Africaines* (in English and French), quarterly; Brochures.

ASSOCIATION OF RURAL COOPERATION IN AFRICA AND LATIN AMERICA (ACRA)

Plaza Adela Zarudio, 70, Sopacachi, La Paz, Bolivia **PHONE:** 591 2 415932, 591 2 410708 **LANGUAGES:** English, Spanish.

DESCRIPTION: Promotes cooperation and coordination of programs among international organizations and national agencies conducting rural development programs. Conducts projects in areas including agricultural and animal husbandry instruction, the status of women, primary education, environmental protection, and transportation.

ASSOCIATION OF SECRETARIES GENERAL OF PARLIAMENTS (ASGP) (Association des Secretaires Generaux des Parlements)

Comm. Office, House of Commons, London SW1A 0AA, England **PHONE:** 44 171 2193259 **FX:** 44 171 2196864 **FOUNDED:** 1938. **OFFICERS:** Yusef Azad, Sec. **MEMBERS:** 225. **LANGUAGES:** English, French.

DESCRIPTION: Members are secretaries general from parliamentary assemblies in 95 countries, and/or their deputies. Facilitates contacts between secretaries general for the purposes of studying the law, procedure, practice and working methods of different parliaments; proposes measures for improving those methods and for securing cooperation between the services of different parliaments. **PUBLICATIONS:** *Constitutional and Parliamentary Information* (in English and French), semiannual.

ASSOCIATION OF SOUTH PACIFIC AIRLINES (ASPA)

Box 9817, Nadi Airport, Nadi, Fiji **PHONE:** 679 723526 **FX:** 679 720196 **FOUNDED:** 1979. **OFFICERS:** George E. Faktaufon, Sec.Gen. **MEMBERS:** 16. **STAFF:** 2. **LANGUAGES:** English.

DESCRIPTION: Airlines headquartered in or providing service to the South Pacific region. Seeks to develop commercial aviation in the Pacific region. Advises governments on problems affecting Pacific civil aviation. Coordinates planning and administrative policies of member airlines. Standardizes equipment, scheduling, and marketing campaigns. Establishes regional employee training centers. **PUBLICATIONS:** *South Pacific Aviation* (in English), quarterly.

ASSOCIATION OF SOUTHEAST ASIAN INSTITUTIONS OF HIGHER LEARNING (ASAIHL)

c/o Dr. Ninnat Olanvoravuth, Ratasastra Bldg. 2, Chulalongkorn Univ., Henri Dunant Rd., Bangkok 10330, Thailand **PHONE:** 66 2 2516966 **FX:** 66 2 2537909 **FOUNDED:** 1956. **OFFICERS:** Dr. Ninnat Olanvoravuth, Sec.Gen. **MEMBERS:** 160. **STAFF:** 3. **BUDGET:** US$200,000. **LANGUAGES:** English.

DESCRIPTION: Institutions in 14 countries. Assists members in achieving international recognition in teaching, research, and public service through mutual assistance. Offers numerous fellowships. Maintains speakers' bureau and placement service. Compiles statistics; sponsors competitions. **PUBLICATIONS:** *ASAIHL Newsletter*, semiannual; *Handbook of Southeast Asian Institutes of Higher*

Learning (in English), triennial. Also publishes seminar reports and papers.

ASSOCIATION OF SOUTH EAST ASIAN NATIONS (ASEAN)

Jalan Sisingamangaraja 70 A, Jakarta, Indonesia **PHONE:** 62 21 712272 **FX:** 62 21 7398234 **TX:** 47213 ASEAN JKT **FOUNDED:** 1967. **OFFICERS:** Latit Tuah. **MEMBERS:** 6. **STAFF:** 60. **LANGUAGES:** English.

DESCRIPTION: Representatives of Brunei, Indonesia, Malaysia, Philippines, Singapore, and Thailand. Goals are to: accelerate economic growth among member nations; develop intraregional economic cooperation; encourage social and cultural progress; promote regional peace and stability. Fosters collaboration among members for mutual benefit; promotes Southeast Asian studies; seeks to cooperate with international and regional organizations with similar aims. **PUBLICATIONS:** *Annual Report* (in English); *Annual Report of the ASEAN Standing Committee* (in English); *ASEAN* (in English), bimonthly; *ASEAN Documents Series* (in English), periodic; *Statements of Foreign Ministers and Dialogue Partners* (in English), annual.

ASSOCIATION FOR THE TAXONOMIC STUDY OF TROPICAL AFRICAN FLORA

c/o National Botanic Garden of Belgium, Domein Van Bouchout, 3-1860, B-1860 Meise, Belgium **PHONE:** 322 2693905 **FX:** 322 2701567 **E-MAIL:** rammetoo@br.fgov.be **FOUNDED:** 1950. **OFFICERS:** J. Rammeloo, Sec.Gen. **LANGUAGES:** English.

DESCRIPTION: Botanists, taxonomists, and other individuals from 58 countries with an interest of the plants of tropical Africa. Seeks to advance the study and practice of botany. Facilitates cooperation between botanists engaged in the study of flora of tropical Africa. **PUBLICATIONS:** *AETFAT*, annual.

ASSOCIATION FOR TEACHER EDUCATION IN EUROPE (ATEE) (Association pour la Formation des Enseignants en Europe)

60, rue de la Concorde, B-1050 Brussels, Belgium **PHONE:** 32 2 5127405, 32 2 5128425 **FX:** 32 2 5128425 **E-MAIL:** info@atee.org **WEBSITES:** http://www.atee.org **FOUNDED:** 1976. **OFFICERS:** Mara Garofalo. **MEMBERS:** 650. **STAFF:** 2. **BUDGET:** 4,500,000 BFr. **LANGUAGES:** English, French, German.

DESCRIPTION: Institutions and individuals in 26 countries active in the field of teacher education. Objectives are: to enhance basic and continued education of teachers in Europe; to promote and coordinate research and innovation in the field of teacher education, with special emphasis on a comparative analysis of different approaches to teacher education; to accumulate, exchange, and disseminate information on teacher education; to encourage cooperation among professionals and institutions throughout Europe. Offers consultancy services to the Commission of the European Communities, European Cultural Foundation, and other international groups. Organizes seminars on education for teachers. **PUBLICATIONS:** *ATEE - Guide to Institutions of Teacher Education in Europe* (in English); *ATEE News*, quarterly; *Cahiers*, periodic. Provides information on special themes and projects; *Conference Proceedings*, annual; *European Journal of Teacher Education*, 3/year.

ASSOCIATION OF TIN PRODUCING COUNTRIES (ATPC)

c/o SNTEE, Rua da Quitanda, no. 62 sala 802-Centro, Jalan Sultan Hishamuddin, 20011-030 Rio de Janeiro, Brazil **E-MAIL:** snicc@infolink.com.br **FOUNDED:** 1983. **OFFICERS:** Gonzalo Martinez, Exec.Sec. **MEMBERS:** 6. **STAFF:** 5. **BUDGET:** US$323,000. **LANGUAGES:** English.

DESCRIPTION: Representatives of government and industry. Aims to increase tin consumption and obtain remuneration for producers. Activities include research and development and market studies. Compiles statistics. **PUBLICATIONS:** *Annual Report* (in English).

ATLANTIC BRIDGE (AB) (Atlantik-Bruecke - AB)

c/o Dr. Beate Lindemann, Adenauerallee 131, Postfach 1147, D-53113 Bonn, Germany **PHONE:** 49 228 214160, 49 228 214260 **FX:** 49 228 214659 **FOUNDED:** 1952. **OFFICERS:** Dr. Beate Lindemann, Exec.V.Chm. **MEMBERS:** 350. **STAFF:** 5. **LANGUAGES:** English, German.

DESCRIPTION: Leaders from the fields of academia, business, government, journalism, trade unions, and law. Fosters German-American friendship. Organizes: travel programs for U.S. journalists; seminars for American military officers, editors, and teachers stationed in Germany; German-American conferences and workshops. **PUBLICATIONS:** *East-West Issues* (in English), biennial; *Meet United Germany*; *These Strange German Ways*; *The United States of America and the United Germany: Pillars of a Post Wall World*; *Off the Wall: A Wacky History of Germany Since 1989*; *Speaking Out: Jewish Voices from United Germany America Within Us.*

ATLANTIC TREATY ASSOCIATION (ATA) (Association du Traite Atlantique - ATA)

10 rue Crevaux, F-75116 Paris, France **PHONE:** 33 1 45532880 **FX:** 33 1 47554963 **E-MAIL:** ata.sg@wanadoo.fr **FOUNDED:** 1954. **OFFICERS:** Alfred Cahen, Sec.Gen. **MEMBERS:** 32. **STAFF:** 2. **LANGUAGES:** English, French.

DESCRIPTION: National associations belonging to the Atlantic Alliance; signatories of NATO PFP agreements. Promotes the ideas and objectives of the North Atlantic Treaty Organization.

AVIATION SANS FRONTIERES

Brussels Natl. Airport, Brucargo 706, PO Box 7513/4, B-1931 Brucargo, Belgium **PHONE:** 32 2 7532470 **FX:** 32 2 7532471 **E-MAIL:** asf-belgium@online.be **FOUNDED:** 1983. **OFFICERS:** Alain Peeters, Gen.Mgr. **MEMBERS:** 200. **STAFF:** 1. **BUDGET:** 30,000 BFr.

DESCRIPTION: Pilots who volunteer their services to deliver relief supplies to needy or disaster-stricken areas of Africa. Undertakes charitable programs.

BALTIC AND INTERNATIONAL MARITIME COUNCIL (BIMCO)

Bagsvaerdvej 161, DK-2880 Bagsvaerd, Denmark **PHONE:** 45 44444500 **FX:** 45 44444450 **TX:** 19086 **E-MAIL:** mailbox@bimco.dk **WEBSITES:** http://www.bimco.dk **FOUNDED:** 1905. **OFFICERS:** Finn Frandsen, Sec.Gen. **MEMBERS:** 2,658. **STAFF:** 47. **LANGUAGES:** English.

DESCRIPTION: Membership consists of: Shipowners (971); shipbrokers (1592); clubs (52); associate members (43). Purpose is to keep members informed of current conditions in the maritime industry including unfair charges and claims, objectionable charter parties, and other matters. Represents members in international shipping negotiations with trade organizations, governments, and port and canal authorities; takes action on matters affecting the industry; offers advice in disputes. Prepares and issues shipping documents, standard charter parties, and bills of lading for: bare-boat and time chartering; combined transports; dry cargo, liner, and tanker trades; vegetable and animal fats and oils. Interprets charter terms and other documents. Conducts research and educational programs on topics including seafarers shortages, tax reform, drug abuse, and decision making. Provides information on: actual port experiences, customs, and expenses; disbursement accounts; ice and navigation; ports and trade routes; rates of loading and discharging; stowage coefficients; taxation. **PUBLICATIONS:** *BIMCO Bulletin* (in English), bimonthly. Features new shipping documents, legal decisions, and arbitration awards as well as information on current developments in the shipping industry; *BIMCO Holiday Calendar*, annual; *BIMCO Weekly Ice Report*, (in season); *BIMCO Weekly News*; *Check Before Fixing*, annual; *Costs, Sanctions and Security*, annual. Agency Tariffs-Annual.

BANKING FEDERATION OF THE EUROPEAN UNION (BFEU) (Federation Bancaire de l'Union Europeenne - FBUE)

c/o Nikolas Bomcke, 10, rue Montoyer, B-1040 Brussels, Belgium **PHONE:** 32 2 5083711 **FX:** 32 2 5112328 **FOUNDED:** 1960. **OFFICERS:** Nikolaus Bomcke, Sec.Gen. **MEMBERS:** 18. **LANGUAGES:** English, French.

DESCRIPTION: Banking associations in member countries of the European Economic Community and in European Free Trade Association countries. Represents members before European organizations.

BAPTIST WORLD ALLIANCE (BWA)

6733 Curran St., Mc Lean, VA 22101-6005 **PHONE:** (703) 790-8980 **FX:** (703) 893-5160 **E-MAIL:** bwa@bwanet.org **WEBSITES:** http://

www.bwanet.org **FOUNDED:** 1905. **OFFICERS:** Denton Lotz, Gen.Sec.-Treas. **MEMBERS:** 191. **STAFF:** 20. **BUDGET:** US$2,000,000. **LANGUAGES:** English.

DESCRIPTION: Members are international Baptist bodies representing more than 42,000,000 individuals. ''To exist as an expression of the essential oneness of Baptist people in the Lord Jesus Christ; to impart inspiration to the fellowship; and to provide channels for sharing concerns and skills in witness and ministry.'' Focuses major activities in the areas of: communications; evangelism and education; world aid; study and research; promotion and development. Conducts relief programs, such as sending aid to distressed peoples and assisting in the rehabilitation of refugees from political and other oppressions. Encourages observation of annual Baptist World Alliance Day on the first Sunday of February. Maintains speakers' bureau and biographical archives; compiles statistics. **PUBLICATIONS:** *The Baptist World*, quarterly; *Baptist World Alliance Information Service*, monthly; *Baptist World Alliance News*, monthly; *World Congress Reports*; *World Youth Conference Reports*; Booklets; Yearbook.

BENELUX ECONOMIC UNION (BEU) (Benelux Economische Unie - BEU)

39, rue de la Regence, B-1000 Brussels, Belgium **PHONE:** 32 2 5193811 **FX:** 32 2 5134206 **FOUNDED:** 1960. **OFFICERS:** B.M.J. Hennekam, Sec.Gen. **MEMBERS:** 3. **STAFF:** 78. **BUDGET:** 200,000,000 BFr. **LANGUAGES:** Dutch, French.

DESCRIPTION: Representatives of Belgium, Luxembourg, and Netherlands. Works to establish legislation with respect to intellectual property. Strives for the dovetailing of policies in the framework of cross-border cooperation, especially in the following fields: physical planning, environment issues, transportation and infrastructure, public health, social and fiscal problems of transfrontier workers and cooperation between local governments or local governmental bodies. Sponsors Court of Justice; maintains College of Arbitrators to settle disputes regarding application of the present treaty of conventions related to the aims of the treaty. **PUBLICATIONS:** *Textes de Base Benelux*; Brochure (in Dutch and French), periodic; Newsletter, monthly.

BIBLE LITERATURE INTERNATIONAL (BLI)

625 E. North Broadway, Columbus, OH 43214 **PHONE:** (614) 267-3116 **TF:** (800) 326-WORLD **FX:** (614) 267-3116 **E-MAIL:** atg@atgi.com **WEBSITES:** http://www.bli.org **FOUNDED:** 1923. **OFFICERS:** James R. Falkenberg, Pres. **STAFF:** 23. **LANGUAGES:** English.

DESCRIPTION: Provides foreign language religious literature and funds for producing such literature to missionaries in 170 nations representing more than 400 different mission boards. **PUBLICATIONS:** *Quiet Miracle*, quarterly.

BINATIONAL CENTER FOR HUMAN RIGHTS (BCHR) (Centro Binacional de Derechos Humanos - CBDH)

Av. Paseo del Centenario 3-B, Depto 11, De Sarrollo Urbano Rio Tijuana, Apartado Postal 848, 22000 Tijuana, Baja California Norte, Mexico **PHONE:** 52 66 828550 **FX:** 52 66 828550 **FOUNDED:** 1987. **OFFICERS:** Victor Clark-Alfaro, Dir. **MEMBERS:** 5. **STAFF:** 11. **BUDGET:** US$30,000.

DESCRIPTION: Individuals interested in preventing violence against juveniles and migrants in the U.S. and Mexico. Gathers information on alleged human rights violations. Undertakes investigations; maintains migrant defense and border monitoring programs; sponsors sex workers defense project; works to insure respect for the human rights of indigenous peoples. Conducts research and educational programs. **PUBLICATIONS:** *Informe Anual de Derechos Humanos*, annual.

BIRD LIFE INTERNATIONAL

Wellbrook Ct., Girton Rd., Cambridge, Cambs. CB3 0NA, England **PHONE:** 44 1223 277318 **FX:** 44 1223 277200 **TX:** 818794 ICBP G **E-MAIL:** birdlife@birdlife.org.uk **FOUNDED:** 1922. **OFFICERS:** Dr. Michael Rands, Dir. & Chief Exec. **MEMBERS:** 61. **STAFF:** 38. **BUDGET:** £1,050,000. **LANGUAGES:** Dutch, English, French, German, Spanish.

DESCRIPTION: A partnership of natural conservation organisations. The mission is to conserve all bird species on earth and their habitats, and through this, to work for the world's biological diversity and the sustainability of human use of natural resources. Advises governments and private bodies on significant national and international bird conservation issues. **PUBLICATIONS:** *Birdlife International Annual Review*; *World Birdwatch* (in English), quarterly; Monograph, periodic. Also publishes a series of bird conservation books.

BLESSINGS INTERNATIONAL (BI)

5881 S Garnett Rd., Tulsa, OK 74146-6812 **PHONE:** (918) 250-8101 **FX:** (918) 250-1281 **E-MAIL:** blessingsint@compuserve.com **WEBSITES:** http://www.blessing.org **OFFICERS:** Harold Harder, Pres. **LANGUAGES:** English.

DESCRIPTION: Members are health care professionals and other individuals. Seeks to increase the availability and quality of health care services among previously underserved communities worldwide. Procures and distributes pharmaceuticals and medical and surgical supplies; sends teams of volunteers to assist in the operation of local health services in developing and economically disadvantaged areas.

B'NAI B'RITH INTERNATIONAL (BBI)

1640 Rhode Island Ave. NW, Washington, DC 20036 **PHONE:** (202) 857-6600 **FX:** (202) 857-1099 **FOUNDED:** 1843. **OFFICERS:** Dr. Sidney M. Clearfield, Exec.VP. **MEMBERS:** 500,000. **STAFF:** 280. **BUDGET:** US$15,000,000. **LANGUAGES:** English.

DESCRIPTION: Members are Jewish men, women, and youth ''of good moral character.'' Offers religious, cultural, civic, and social programs for teenagers and Jewish students and faculty of some 400 college campuses in 12 countries. Maintains museum, and speakers' bureau. Conducts programs on important Jewish issues. **PUBLICATIONS:** *ADL Bulletin*, monthly; *B'nai B'rith Jewish Book News*, periodic; *Bottom Line*, monthly; *International Jewish Monthly*; *Judaism Pamphlet Series*, periodic; *The Shofar*, monthly.

BOTH ENDS

Damrak 28-30, NL-1012 LJ Amsterdam, Netherlands **PHONE:** 31 20 6230823 **FX:** 31 20 6208049 **E-MAIL:** info@bothends.org **WEBSITES:** http://www.bothends.org **FOUNDED:** 1986. **OFFICERS:** Theo van Koolwijk, Exec.Dir. **MEMBERS:** 25. **STAFF:** 15. **BUDGET:** 2,000,000 f. **LANGUAGES:** Dutch, English, French, Spanish.

DESCRIPTION: Aims to strengthen NGOs that work on environmental and social justice issues, especially in the South. Supports NGOs in information sharing and fundraising. Promotes collaboration with academic institutions and the private sector. Assists in capacity building programmes and facilitates the participation of NGOs in policy formulation, implementation, and evaluation. **PUBLICATIONS:** *Green and Grey below Sea Level* (in English), periodic.

BRITISH INSTITUTE OF HUMAN RIGHTS (BIHR)

King's Coll. London, 8th Fl., 75-79 York Rd., London SE1 7AW, England **PHONE:** 44 171 4012712, 44 171 4012722 **FX:** 44 171 4012695 **E-MAIL:** bihr@kcl.ac.uk **WEBSITES:** http://www.kcl.ac.uk/bihr; http://www.kcl.ac.uk/depsta/rel/bihr **FOUNDED:** 1970. **OFFICERS:** Sarah Cooke, Dir. **LANGUAGES:** English.

DESCRIPTION: Strives to increase awareness of human rights issues. Concerns include the effects of international human rights law on prisoners' rights, religious freedom, and torture. Promotes the teaching of human rights topics at all levels of education. Communicates with government representatives. Conducts research. Sponsors seminars, symposia, conferences, and lectures. Works in Romania and the former Yugoslavia. **PUBLICATIONS:** *Conference and Lecture Proceedings*, periodic; *Human Rights Case Digest*, monthly; Brochure, periodic; Reports, periodic.

BROTHER TO BROTHER INTERNATIONAL (BBI)

PO Box 27634, Tempe, AZ 85285-7634 **PHONE:** (602) 345-9200 **FX:** (602) 345-2747 **E-MAIL:** bbi@worldvision.org **WEBSITES:** http://www.bbi.org **FOUNDED:** 1982. **OFFICERS:** Mike Veitenhans, Program Mgr. **STAFF:** 9. **LANGUAGES:** English.

DESCRIPTION: Works to procure excess inventory, such as pharmaceuticals, medical supplies, personal care products, clothing, educational material, seed and food, from corporations throughout the United States. Donated excess inventory or technical information is then provided to national and international non-profits across the United States and around the world. Provides a network for U.S. charities during domestic and international emergencies. **PUBLICATIONS:** *For the Children.*

THE BROTHER'S BROTHER FOUNDATION (BBF)

1501 Reedsdale St., Ste. 3005, Pittsburgh, PA 15233-2341 **PHONE:** (412) 321-3160 **FX:** (412) 321-3325 **E-MAIL:** bbfound@aol.com **WEBSITES:** http://www.brothersbrother.com **FOUNDED:** 1958. **OFFICERS:** Luke L. Hingson, Pres. **STAFF:** 11. **BUDGET:** US$54,240,411. **LANGUAGES:** English.

DESCRIPTION: Delivers surplus goods including pharmaceuticals and other medical supplies, food, books, seeds, tools, intraocular lenses, used eyeglasses, and other gifts-in-kind to 35 developing nations. Maintains speakers' bureau. **PUBLICATIONS:** *Field News*, semiannual; Annual Report, annual; Brochure.

BROTHERS OF CHARITY (BC) (Freres de la Charite - FC)

c/o Fr. Oscar Duym, Via G.B. Pagano 35, Casella Postale 9082, I-00167 Rome, Italy **PHONE:** 39 6 6604901 **FX:** 39 6 6631466 **E-MAIL:** fcduymsg@pcn.net **FOUNDED:** 1807. **OFFICERS:** Fr. Oscar Duym, Sec.Gen. **MEMBERS:** 600. **LANGUAGES:** Dutch, English, French.

DESCRIPTION: Catholic order of lay brothers in 22 countries. Conducts mission work; teaches underprivileged youth and physically and mentally handicapped people. Offers programs on education, nursing, and psychiatry. **PUBLICATIONS:** *De tout Coeur* (in French), monthly; *Psychiatrie & Verpleegkunde* (in Dutch), bimonthly; *Van Harte* (in Dutch), monthly.

BUDDHIST PEACE FELLOWSHIP (BPF)

Dept. E, PO Box 4650, Berkeley, CA 94704 **PHONE:** (510) 655-6169 **FX:** (510) 655-1369 **E-MAIL:** bpf@bpf.org **WEBSITES:** http://www.bpf.org/ **FOUNDED:** 1978. **OFFICERS:** Alan Senauke, Dir. **MEMBERS:** 4,500. **STAFF:** 5. **BUDGET:** US$280,000. **LANGUAGES:** English.

DESCRIPTION: Devoted to the cultivation of world wide peace, social justice, nonviolence, and environmental activism. Objectives include witnessing to the Buddhist commitment to nonviolence as a means of social change, and promoting national and international Buddhist peace projects. BPF members work in disarmament and environmental campaigns, and provide support for homeless people and Buddhist prisoners. Supports peace and relief efforts worldwide. **PUBLICATIONS:** *Turning Wheel*, quarterly. *Journal of the Buddhist Peace Fellowship*; Also publishes texts on the historical and scriptural roots of Buddhist activism.

BUSINESS WOMAN - C.I.S.

Merzlyakovsky ln. 8-5, Rm. 7, 121814 Moscow, Russia **PHONE:** 7 95 2906326, 7 95 2906326 **FX:** 7 95 2001207 **TX:** 411089 OMNIA SU **FOUNDED:** 1992. **OFFICERS:** Lyudmila Konareva, Pres. **MEMBERS:** 50. **LANGUAGES:** English.

DESCRIPTION: Encourages the involvement of women in international business. Promotes women's economic independence. Offers legal and financial assistance to women in business ventures. Fosters networking and communication between members.

CANADA ASIA ACCORD ASSOCIATION (CAAA)

Rm 24, 6th Fl., 7 Winston Churchhill Sq., Edmonton, AB, Canada T5J 2V5 **PHONE:** (780) 474-7891 **E-MAIL:** apac@freenet.edmonton.ab.ca **FOUNDED:** 1988. **OFFICERS:** C.Z. Shaw, Pres. **MEMBERS:** 50. **LANGUAGES:** Chinese, English, French, Japanese.

DESCRIPTION: Individuals and organizations with an interest in international relations involving Canada and Asia. Promotes improved Canadian awareness and understanding of Asian cultures as a stimulus to enhanced business and social links between Canada and the nations of Asia. Conducts research and educational programs; facilitates international exchanges; gathers and disseminates information on Asian culture, economy, and history. **PUBLICATIONS:** *Asia Pacific Bulletin*, quarterly.

CANADA-INDONESIA BUSINESS COUNCIL (CIBC)

110-260 Adelaide St. E, Toronto, ON, Canada

M5A 1N1 **PHONE:** (416) 366-8490 **FX:** (416) 947-1534 **FOUNDED:** 1973. **OFFICERS:** John Walker, Sec. **MEMBERS:** 480. **STAFF:** 6. **BUDGET:** US$250,000. **LANGUAGES:** Bahasa Indonesia, English, French.

DESCRIPTION: Canadian businesses trading with Indonesia. Promotes increased trade between Canada and Indonesia. Represents members' interests before business and trade organizations and government agencies. Facilitates exchange of information among members.

CANADIAN COUNCIL FOR INTERNATIONAL COOPERATION (CCIC) (Conseil Canadien pour la Cooperation Internationale)

1 Nicholas St., Ste. 300, Ottawa, ON, Canada K1N 7B7 **PHONE:** (613) 241-7007 **FX:** (613) 241-5302 **E-MAIL:** gauris@ccic.ca **WEBSITES:** http://fly.web.net/ccic **OFFICERS:** Denise Fournier, Communications Officer. **MEMBERS:** 116. **STAFF:** 30. **BUDGET:** US$2,851,000. **LANGUAGES:** English, French.

DESCRIPTION: International development organizations. Promotes ''global development in a peaceful and healthy environment, with social justice, human dignity, and participation for all.'' Facilitates communication and cooperative action among members. Represents members' interests before government agencies concerned with the operation of nonprofit organizations and international aid programs.

CANADIAN LUTHERAN WORLD RELIEF (CLWR)

1080 Kingsbury Ave., Winnipeg, MB, Canada R2P 1W5 **PHONE:** (204) 694-5602 **TF:** (800) 661-2597 **FX:** (204) 694-5460 **OFFICERS:** Ruth E. Jensen, Exec.Dir. **LANGUAGES:** English, French.

DESCRIPTION: Relief operations conducted by the Lutheran churches of Canada. Seeks to improve the quality of life of victims of natural disasters and human conflicts; promotes spread of the Lutheran faith. Provides support and assistance to Lutheran relief projects; sponsors educational and religious programs.

CANADIAN NATIONAL COMMITTEE FOR PACIFIC ECONOMIC COOPERATION (CANCPEC) (Comite National pour la Cooperation Economique avec la Region du Pacifique)

c/o Asia Pacific Foundation of Canada, 999 Canada Pl., Ste. 666, Vancouver, BC, Canada V6C 3E1 **PHONE:** (604) 684-5986 **FX:** (604) 681-1370 **TX:** 04-53332 **E-MAIL:** paul@apfc.apfnet.org **WEBSITES:** http://www.apfc.ca **FOUNDED:** 1985. **OFFICERS:** Paul Irwin. **MEMBERS:** 28. **STAFF:** 2. **LANGUAGES:** Chinese, English, French, Japanese.

DESCRIPTION: Representatives of Canadian businesses, government agencies, and trade organizations. Promotes increased and more profitable participation by Canada in Asian and Pacific trade. Represents Canada's interests in the Pacific Economic Cooperation Council.

CANADIAN-AMERICAN BUSINESS COUNCIL (CABC)

1629 K St. NW, Ste. 1100, Washington, DC 20006 **PHONE:** (202) 785-6717 **FX:** (202) 331-4212 **E-MAIL:** canambusco@aol.com **FOUNDED:** 1987. **OFFICERS:** Paul S. Weller Jr., Exec.Dir. **MEMBERS:** 100. **STAFF:** 3. **LANGUAGES:** English, French.

DESCRIPTION: Individuals, corporations, institutions, and organizations with an interest in trade between the United States and Canada. Promotes free trade. Gathers and disseminates information; maintains speakers' bureau. **PUBLICATIONS:** *CABCommunique* (in English and French), periodic.

CARE

151 Ellis St. NE, Atlanta, GA 30303-2439 **PHONE:** (404) 681-2552, (800) 422-7385 **TF:** (800) 521-CARE **FX:** (404) 589-2658 **WEBSITES:** http://www.care.org **FOUNDED:** 1945. **OFFICERS:** Peter D. Bell, CEO/Pres. **STAFF:** 9,000. **BUDGET:** US$454,000,000. **LANGUAGES:** English.

DESCRIPTION: International aid and development organization providing disaster aid and self help development programs overseas. Maintains volunteer committees in many U.S. cities and 12 regional and district offices. Work is supported by contributions from individuals, governments, and organizations; U.S. government provides Food-for-Peace agricultural commodities and financial grants for emergency relief and development programming. Host governments share internal operating costs and may contribute material support and personnel. CARE stresses shared-cost, selfhelp partnership development programs with long-term sustainability in mind. Programs are in the areas of renewable resources development, small enterprise development, construction of potable water systems, primary health care and nutrition, agricultural development, education, and community development. Originally founded to send aid to World War II victims in Europe, CARE now operates in 66 countries in Asia, Africa, Eastern Europe, and Latin America. **PUBLICATIONS:** *CARE Briefs*; *CARE World Report*, quarterly; *What and Where of CARE*; Annual Report.

CARIBBEAN ASSOCIATION OF INDUSTRY AND COMMERCE (CAIC)

Trinidad Hilton, PO Box 442, Lady Young Rd., Ste. 351, St. Anns, Trinidad and Tobago **E-MAIL:** caic@trinidad.net **WEBSITES:** http://www.trinidad.net/caic/ **FOUNDED:** 1955. **OFFICERS:** Felipe M. Noguera, CEO. **MEMBERS:** 122. **STAFF:** 7. **BUDGET:** BD$1,500,000. **LANGUAGES:** English.

DESCRIPTION: Firms, banks, and companies belonging to national chambers of commerce or manufacturers' associations. Coordinates private sector ventures within the Caribbean Community. Encourages agricultural development and expansion of industries based on indigenous agricultural products. Provides technical assistance to business and industry. Monitors legislative developments and lobbies in the interests of members. Sponsors training and leadership programs; offers scholarships and bursaries in support of studies connected with the aims of the association. Sponsors seminars and exhibitions. **PUBLICATIONS:** *CAIC News* (in English), monthly; *CAIC Times* (in English, French, and Spanish), quarterly. Also publishes research papers.

CARIBBEAN COMMUNITY AND COMMON MARKET (CARICOM)

Bank of Guyana Bldg., PO Box 10827, Georgetown, Guyana **PHONE:** 592 2 69281, 592 2 69289 **FX:** 592 2 67816 **TX:** 2263 CARISEC GY **E-MAIL:** carisec1@caricom.org **WEBSITES:** http://www.caricom.org **FOUNDED:** 1973. **OFFICERS:** Edwin W. Carrington, Sec.Gen. **MEMBERS:** 14. **STAFF:** 230. **LANGUAGES:** English.

DESCRIPTION: Members are the governments of Antigua and Barbuda, Bahamas, Barbados, Belize, Dominica, Grenada, Guyana, Jamaica, Montserrat, Saint Christopher-Nevis, Saint Lucia, Saint Vincent and the Grenadines, and Trinidad and Tobago, and Suriname. Objectives are to: promote cooperation and understanding among member states; integrate the economies of member states through the Caribbean Common Market; coordinate the foreign policies of member states; provide common services and cooperation in functional matters such as health, education and culture and industrial relations. Compiles statistics. **PUBLICATIONS:** *CARICOM Perspective* (in English), annual; *Report of the Secretary-General of the Caribbean Community*, annual. Also publishes news releases and abstracts, articles, directories, bibliographies, statistical reports, and monographs.

CARIBBEAN CONFERENCE OF CHURCHES (Conferencia de Iglesias del Caribe)

PO Box 616, Bridgetown, Barbados **PHONE:** (246) 427-2681 **FX:** (246) 429-2075 **E-MAIL:** cccbdos@ndl.net **FOUNDED:** 1973. **MEMBERS:** 33. **STAFF:** 30. **LANGUAGES:** English.

DESCRIPTION: Christian churches in the Caribbean. Promotes unity, renewal, and joint action among members; seeks social justice and human dignity worldwide. Assists Christian councils in the region in fostering consultations; provides and stimulates programs in experimentation, research, and study in order to aid churches in understanding what the group believes is the ''decisive action of God. In terms of their cultural experience and needs.'' Provides a forum for members and local and national Christian councils. Offers disaster and emergency relief. Conducts surveys. Organizes communication and training courses; sponsors development and scholarship programs; provides loans. Holds conferences, seminars, and workshops. Maintains the Caribbean Appropriate Technology Centre to promote the development and application of appropriate technologies and a related network to serve the needs of the poor in the Caribbean. **PUBLICATIONS:** *Christian Action/Ho-*

rizons (in English and Spanish), quarterly; *Directory of Organisations and Individuals Active in the Promotion of Appropriate Technologies in the Eastern Caribbean*, periodic; *Directory of Projects*, annual. Includes updates.; *Handbook of Churches in the Caribbean* (in English), quadrennial; Annual Report (in English). Also publishes *Fashion Me a People* religious educational text for use in Sunday Schools, *Mission to Haiti* (in English and French), monograph, and brochures.

CARIBBEAN CONSERVATION ASSOCIATION (CCA) (Association Caraibe pour l'Environnement)

Savannah Lodge, The Garrison, St. Michael, Barbados **PHONE:** (246) 426-6933, (246) 426-9635 **E-MAIL:** cca@carbisurf.com **WEBSITES:** http://www.tidco.co.tt/local/seduweb/cca/cca.html **FOUNDED:** 1967. **OFFICERS:** Glenda Medina, Dir. **MEMBERS:** 350. **STAFF:** 6. **BUDGET:** US$300,000. **LANGUAGES:** English, French, Spanish.

DESCRIPTION: Caribbean governments, business organizations, nongovernmental organizations, individuals, and students united to promote the preservation, development, and conservation of the natural and cultural environments of the wider Caribbean. Objectives include: collecting information on completed, active, and planned environmental projects; coordinating activities that satisfy the area's conservation needs; fostering greater awareness of cultural and natural resources of the Caribbean. Offers technical advice to members. Acts as a forum for the exchange of information. Organizes training courses, seminars, and workshops. **PUBLICATIONS:** Annual Report (in English, French, and Spanish), annual; Monographs (in English, French, and Spanish), periodic.

CARIBBEAN DEVELOPMENT BANK (CDB)

PO Box 408, Wildey, St. Michael, Barbados **PHONE:** (246) 431-1600 **FX:** (246) 426-7269 **TX:** WB 2287 **FOUNDED:** 1970. **OFFICERS:** Sir Neville Vernon Nicholls, Pres. **MEMBERS:** 25. **STAFF:** 188. **BUDGET:** US$14,042,000. **LANGUAGES:** English.

DESCRIPTION: Caribbean states and territories that belong to the United Nations or International Atomic Energy Agency. Promotes the harmonious development of Caribbean member countries in an effort to build an integrated Caribbean economy, with particular regard to less developed member countries. Provides financial and technical assistance for projects in member countries. Compiles statistics; organizes research and educational programs. **PUBLICATIONS:** *Annual Report* (in English); *CDB News* (in English), quarterly; *CTCS TEU Update* (in English), quarterly; *Statement by the President* (in English), annual; *Summary of Proceedings* (in English), annual. Also publishes study reports, policy papers, and information documents.

CARIBBEAN FOOD CORPORATION (CFC)

30 Queen's Park West, PO Bag 264B, Port of Spain, Trinidad and Tobago **PHONE:** (868) 622-5827, (868) 622-5211 **FX:** (868) 622-4430 **E-MAIL:** cfc@trinidad.net **FOUNDED:** 1976. **OFFICERS:** E.C. Clyde Parris, Mng.Dir. **MEMBERS:** 12. **STAFF:** 8. **BUDGET:** US$1,000,000. **LANGUAGES:** English.

DESCRIPTION: Governmental representatives of member states of the Caribbean Community. Works to: facilitate increased self-sufficiency in the food systems of member countries; improve nutritional levels; organize bulk purchases of agricultural inputs and commodities; mobilize funds and technical and managerial skills within and outside the Community; identify, plan, and implement projects in agricultural production, processing, storage, transportation, distribution, and marketing. Arranges and finances technical, managerial, and marketing services for projects. Serves as regional trade agent or commercial representative. **PUBLICATIONS:** *Annual Report* (in English); *News from Caribbean Food Corp.* (in English), annual.

CARIBBEAN FOOD CROPS SOCIETY (CFCS) (Sociedad Caribena de Cultivos Alimenticios)

Secretariat, Cooperative Extension Service, RR02, Box 10000, St. Croix, Virgin Islands of the United States 00850 **PHONE:** (809) 778-0246, (809) 692-4080 **FX:** (809) 692-4085 **E-MAIL:** kboaten@uvi.edu **FOUNDED:** 1962. **OFFICERS:** Mr. Kofi Boateng, Sec. **MEMBERS:** 450. **LANGUAGES:** Dutch, English, French, Spanish.

DESCRIPTION: Firms and individuals from 30 countries interested in agricultural sciences and crop production. Promotes and fosters Caribbean food production, processing, and distribution in order to improve levels of nutrition and standards of living in the Caribbean region. **PUBLICATIONS:** *News Letter* (in English and Spanish), 3/year; *Proceedings of Annual Meeting*. Also publishes *Professional Register*.

CARIBBEAN HUMAN RIGHTS NETWORK

Third Ave., No. 5, Belleville, St. Michael, Barbados **PHONE:** (246) 436-9456 **FX:** (246) 436-0645 **E-MAIL:** crights@caribsurf.com **FOUNDED:** 1986. **OFFICERS:** Sheila Stuart, Coordinator. **MEMBERS:** 9. **STAFF:** 3. **LANGUAGES:** English, French, Spanish.

DESCRIPTION: Members include: Ecumenical Centre for Human Rights, Grand Bahama Human Rights Association, Guyana Human Rights Association, Independent Jamaica Council for Human Rights, Puerto Rican Institute for Civil Liberties, St. Vincent and the Grenadines Human Rights Association, Centre for Ecumenical Planning and Action, and MOIWANA '86. Promotes the advancement of human rights in the Caribbean through coordination of members' efforts, strengthening of local organizations advocating similar goals, and public educational programs. Cooperates with regional and international human rights organizations; publicizes human rights abuses in regional media, focusing on issues of police brutality, fair elections, and freedom of expression.

CARIBBEAN/LATIN AMERICAN ACTION (C/CAA)

1818 N St. NW, Ste. 500, Washington, DC 20036 **PHONE:** (202) 466-7464 **FX:** (202) 822-0075 **E-MAIL:** info@claa.org **WEBSITES:** http://www.claa.org **FOUNDED:** 1980. **OFFICERS:** Peter Johnson, Exec.Dir. **MEMBERS:** 98. **STAFF:** 14. **BUDGET:** US$1,600,000. **LANGUAGES:** English.

DESCRIPTION: Caribbean Basin and United States leaders from corporate, international, government, and nonprofit organizations. Aims to: utilize U.S. private sector resources to meet Caribbean development needs; promote awareness and responsive policies by the U.S. public and government; strengthen the leadership role of private sector institutions within Caribbean societies; improve relations and maintain communication and cooperation between the people of the U.S. and the region. Introduces U.S. companies to the Caribbean for investment and joint venture projects; stimulates information network to encourage trade, investments, and tourism, and to provide technical assistance. Assists in the creation and organization of ''business for development'' organizations. Works with U.S. development agencies to develop a private sector aid strategy in the Caribbean Basin. Conducts seminars, press conferences, and meetings with business and government leaders. **PUBLICATIONS:** *Caribbean Basin Commercial Profile*, annual.

CARIBBEAN METEOROLOGICAL ORGANISATION (CMO)

St. Ann's Ave., PO Box 461, Port of Spain, Trinidad and Tobago **PHONE:** (868) 624-4481 **FX:** (868) 623-3634 **E-MAIL:** cebcmo@carib-link.net **FOUNDED:** 1951. **OFFICERS:** C.E. Berridge, Coordinating Dir. **MEMBERS:** 16. **STAFF:** 60. **BUDGET:** US$2,000,000. **LANGUAGES:** English.

DESCRIPTION: Members are Caribbean nations that maintain a meteorological service. Seeks to contribute to the economic growth of member nations through the development and use of meteorology. Provides training in meteorology, operational hydrology, and specialized studies in subjects such as wind and solar energy; sponsors seminars. Maintains Caribbean Meteorological and Hydrological Institute (CMHI) located in Bridgetown, Barbados. Sponsors cooperative programs. Conducts research; compiles statistics. **PUBLICATIONS:** *Climatological Summaries* (in English), quarterly. Meterological Hydrological parameters.

CARIBBEAN TOURISM ORGANIZATION (CTO)

Sir Frank Walcott Bldg., 2nd Fl., Culloden Farm Complex, St. Michael, Barbados **PHONE:** (246) 427-5242 **FX:** (246) 429-3065 **FOUNDED:** 1989. **OFFICERS:** Mr. Jean S. Holder, Sec.Gen. **MEMBERS:** 32. **STAFF:** 39. **BUDGET:** US$2,500,000. **LANGUAGES:** English.

DESCRIPTION: Members include governments of

32 countries and the Caribbean Hotel Association. Collects and disseminates information; coordinates projects; offers courses and technical assistance; serves as resource for the tourism industry. Studies the social, ecological, and economic aspects of increased tourism in the Caribbean. Works to develop the market for uniquely Caribbean tourism and to preserve and protect the indigenous arts and crafts of the region. Motivates Caribbean people to use their resources for promotion of tourism to increase economic independence, employment, and equal distribution of wealth. Provides marketing and consulting services; conducts training. Bestows awards; compiles statistics. **PUBLICATIONS:** *Caribbean Tourism Statistical Report*, annual; *Statistical News* (in English), quarterly; Books (in English), periodic; Reports, periodic.

CARMELITE MISSIONARIES (CM) (Carmelitane Missionarie)

Via del Casaletto 115, I-00151 Rome, Italy **PHONE:** 39 6 535472, 39 6 5827216 **FX:** 39 6 58232279 **FOUNDED:** 1860. **OFFICERS:** Sr. Maria Pilar Miguel Garcia, Sec.Gen. **MEMBERS:** 1,847. **LANGUAGES:** English, Italian, Spanish.

DESCRIPTION: Members are Roman Catholic women in 36 countries. Provides primary and secondary education; maintains health centers, hospitals, missions, mobile clinics, and welfare centers, pastoral work, and spirituality centers. **PUBLICATIONS:** *Shelahani* (in English, French, Italian, Korean, Polish, and Spanish), annual. Contains the activities of the congregation in the different countries where the Carmelite Missionaries are present.

CATHOLIC INSTITUTE FOR INTERNATIONAL RELATIONS (CIIR) (Institut Catholique pour les Relations Internationales)

Unit 3 Canonbury Yard, 190A New North Rd., Islington, London N1 7BJ, England **PHONE:** 44 171 3540883 **FX:** 44 171 3590017 **E-MAIL:** ciirlon@gn.apc.org **FOUNDED:** 1940. **OFFICERS:** Dr. Ian Linden, Exec.Dir. **MEMBERS:** 3,000. **STAFF:** 39. **BUDGET:** £2,500,000. **LANGUAGES:** English, French, Spanish.

DESCRIPTION: Catholic and non-Catholic development agencies, justice and peace groups, governments, politicians, and media and church organizations. Acts as a center for research, information, and education on human rights, economic, justice, and development issues (including agriculture and food security, drugs trade, development, gender, rapid economic change, and transnational corporations) in Latin America, Southern Africa, and Asia, including East Timor; promotes a better understanding of international justice issues and a greater awareness of the church's social teachings. Supports development projects in the Dominican Republic, Ecuador, Haiti, El Salvador, Guatemala, Nicaragua, Yemen, Zimbabwe, Namibia, and Peru, providing skilled workers in agriculture, health, formal and non-formal education, and environment projects. Arranges forums providing resource material for members of British and European parliaments, policy-makers, and the media. **PUBLICATIONS:** *Briefing Papers, Comment Series*, periodic. Consists publications on development issues.; *CIIR Newsletter*, quarterly; *Comment*, 3/year; *Timor Link*, quarterly. Contains information on East Timor.

CATHOLIC INTERNATIONAL EDUCATION OFFICE (CIEO) (Office International de l'Enseignement Catholique - OIEC)

60, rue des Eburons, B-1000 Brussels, Belgium **PHONE:** 32 2 2307252 **FX:** 32 2 2309745 **E-MAIL:** maria@ciateq.mx **WEBSITES:** http://sparc.ciateq.conacyt.mx/-maria/oiec-0c.htm **FOUNDED:** 1952. **OFFICERS:** Andres Delgado-Hernandez, Sec.Gen. **MEMBERS:** 127. **STAFF:** 8. **LANGUAGES:** English, French, Spanish.

DESCRIPTION: Countries with one national secretariat of Catholic education; religious congregations and associations; interested individuals. Represents the Catholic Church throughout the world, particularly in the field of education. Provides a forum for the examination and discussion of problems in the adaptation of Catholic education to modern society. Collects and disseminates information on and coordinates national activities in Catholic education. Organizes study groups and programs in international relations. **PUBLICATIONS:** *Acts of the 1994 World Congress* (in English, French, and Spanish); *Bulletin OIEC* (in English, French, and Spanish), quarterly; *Educating the Inner Self*; *Eduquer a la Liberte et a l'Amour*; *The Riches of Interiority*.

CATHOLIC INTERNATIONAL FEDERATION FOR PHYSICAL AND SPORTS EDUCATION (CIFPSE) (Federation Internationale Catholique d'Education Physique et Sportive - FICEP)

22, rue Oberkampf, F-75011 Paris, France **PHONE:** 33 1 43385057 **FX:** 33 1 43140665 **FOUNDED:** 1911. **OFFICERS:** Jacques Gautheron. **MEMBERS:** 10. **LANGUAGES:** French, German.

DESCRIPTION: National Catholic gymnastic and sporting associations. Seeks to develop and promote worldwide physical and sports education emphasizing Christian attitudes and principles. Encourages and regulates international meetings, competitions, and championships; sponsors sporting events and seminars.

CATHOLIC MEDIA COUNCIL (CAMECO) (Medienplanung fur Entwicklungslander, Mittel und Osteuropa)

Anton-Kurze-Allee 2, Postfach 1912, D-52074 Aachen, Germany **PHONE:** 49 241 73081 **FX:** 49 241 73462 **TX:** 832719 MIRA D **E-MAIL:** cameco@compuserve.com **FOUNDED:** 1969. **OFFICERS:** Hans Peter Gohla, Exec.Dir. **MEMBERS:** 12. **STAFF:** 11. **LANGUAGES:** Dutch, English, French, German, Italian, Polish, Portuguese, Russian, Spanish.

DESCRIPTION: International Catholic media organization. Evaluates and coordinates communication projects in and for developing countries of church-funding agencies. Advises individual communicators on questions concerning projects and their execution in various media; cooperates and coordinates activities with similar organizations on an ecumenical basis and with international secular media organizations and funding agencies for developing countries. Conducts studies on the possibilities and needs of Christian media work in developing countries; collates information on communication media and related topics. Sponsors projects on press and publishing, radio and television, film, and audiovisual and group media. **PUBLICATIONS:** *Media Forum* (in English, French, and Spanish), quarterly; *Service Paper*, periodic.

CATHOLIC MEDICAL MISSION BOARD (CMMB)

10 W. 17th St., New York, NY 10011-5765 **PHONE:** (212) 242-7757 **TF:** (800) 678-5659 **FX:** (212) 807-9161 **E-MAIL:** info@cmmb.org **WEBSITES:** http://www.cmmb.org **FOUNDED:** 1928. **OFFICERS:** Terry Kirch, Pres./Dir. **STAFF:** 30. **BUDGET:** US$3,100,000. **LANGUAGES:** English.

DESCRIPTION: Provides health care assistance to clinical facilities in developing and transitional countries. Financial aid granted to students matriculated in accredited health care education programs in their own mission countries. Placement program assists health providers interested in volunteering for short- and long-term tours of service at selected clinical sites around the world. Conducts charitable programs. **PUBLICATIONS:** *Medical Mission News*, quarterly. Provides information on CMMB's programs. Recipients of shipments and health care placement opportunities are features.; Annual Report.

CATHOLIC NEAR EAST WELFARE ASSOCIATION (CNEWA)

1011 1st Ave., New York, NY 10022-4195 **PHONE:** (212) 826-1480 **TF:** (800) 442-6392 **FX:** (212) 838-1344 **TX:** 910250 144ONYKU0 **E-MAIL:** bad@cnewa.org **FOUNDED:** 1926. **OFFICERS:** Msgr. Robert L. Stern, Sec.Gen. **STAFF:** 60. **BUDGET:** US$19,000,000. **LANGUAGES:** Arabic, Armenian, English, French, Italian.

DESCRIPTION: Catholic organization of individuals and unrelated groups. Raises funds to assist humanitarian projects in 28 countries, primarily in the Near and Middle East; pays costs of education for native priests and sisters and provides money for chapels and rectories, orphanages, convents, and schools. Sponsors health care programs and maintains clinics. Promotes interest in the Eastern Rites and issues related to church unity. Participates in interfaith dialogue with non-Christian religions. Maintains speakers' bureau. Compiles statistics. **PUBLICATIONS:** *Catholic Near East*, bimonthly. Includes coverage of the events and activities of projects throughout the Near and Middle East.; Brochures. Also publishes resource materials guides and instructional work sheets.

CELTIC LEAGUE (CL)

33 Ceide na Griandige, Rath Cvil, Coiatha

Cliath, Eire, Ireland **PHONE:** 353 14589795 **FX:** 353 14589795 **FOUNDED:** 1961. **OFFICERS:** J.B. Moffatt, Gen.Sec. **MEMBERS:** 3,000. **LANGUAGES:** Breton, Gaelic, Welsh.

DESCRIPTION: Interested individuals and members of political-nationalist and cultural organizations in Celtic regions (Wales, Scotland, Brittany, Ireland, Cornwall, Isle of Man) and throughout the world. Fosters cooperation between national movements in the Celtic regions; seeks to make known the struggle and achievements of the Celtic national movements. **PUBLICATIONS:** *CARN*, quarterly; *For a Celtic Future.*

CENTER FOR COMMUNITY ACTION OF B'NAI B'RITH INTERNATIONAL

1640 Rhode Island Ave. NW, Washington, DC 20036 **PHONE:** (202) 857-6582 **FX:** (202) 857-1099 **E-MAIL:** cca@bnaibrith.org **WEBSITES:** http://www.bnaibrith.org/ **FOUNDED:** 1843. **OFFICERS:** Rhonda Love, Exec.Dir. **MEMBERS:** 250,000. **LANGUAGES:** English.

DESCRIPTION: A department of B'nai B'rith International. Works to resolve social problems such as hunger and poverty through community service; encourages and trains members to become community volunteers. Sponsors food drives, voter registration campaigns, and nursing home visitation services. Provides assistance to veterans, handicapped persons, the elderly, and incarcerated persons. Offers children's services. Provides ''hands on'' assistance & financial assistance for disasters nationally and internationally. **PUBLICATIONS:** *International Jewish Monthly*. Also publishes how-to guides on public service programs.

CENTER OF CONCERN (CC)

1225 Otis St., NE, Washington, DC 20017 **PHONE:** (202) 635-2757 **FX:** (202) 832-9494 **E-MAIL:** coc@coc.org **FOUNDED:** 1971. **OFFICERS:** James E. Hug, Exec.Dir. **STAFF:** 14. **LANGUAGES:** English.

DESCRIPTION: People working with grass roots and international networks to show the connections between global and local justice. Promotes social analysis, theological reflection, policy advocacy, research, and public education on global and local issues such as poverty, underemployment, unemployment, hunger, women's rights, economic justice, and social development. **PUBLICATIONS:** *Catholic Social Teaching: Our Best Kept Secret*; *Center Focus*, bimonthly. Contains articles with a theological approach on international justice, peace, Latin America, Horn of Africa, and competing economic systems.; *Hope and Crisis in Africa*; *Opting for the Poor: The Task for North Americans*; *Social Analysis: Linking Faith and Justice*; *Trouble and Beauty: Women Encounter Catholic Social Teaching*; *Women Connecting*; Books.

CENTER FOR THE DEVELOPMENT OF HUMAN RESOURCES IN RURAL ASIA (CENDHRRA)

PO Box 458, Greenhills, Metro Manila 1502, Philippines **PHONE:** 63 2 8232912, 63 2 8232928 **FX:** 63 2 8237707 **E-MAIL:** cendhrra@mnl.sequel.net **FOUNDED:** 1975. **OFFICERS:** Dr. Antonio L. Ledesma, Dir. **STAFF:** 7. **BUDGET:** US$100,000. **LANGUAGES:** English.

DESCRIPTION: Nongovernmental organizations in Asian and Pacific countries involved in integrated rural development activities. Works to overcome ''moral, political, technical, and structural obstacles to full human development.'' Encourages dialogue and cooperation among individuals, institutions, and organizations in the Asian and Pacific regions that are concerned with human development. Provides information sharing, evaluative studies, consulting services, and assistance in research, management, and training to Asian individuals, agencies, and groups serving marginal, low-income, and rural communities. Promotes establishment of networks of individuals of various religions and cultures who live and work with the rural poor to set up self-reliant, participatory living structures within their communities. Prepares country reports on the participation of non-governmental organizations in integral rural development projects; conducts analyses and comparative studies on ongoing projects. Compiles statistics on rural project profiles. **PUBLICATIONS:** *CENDHRRA Partnership Newsletter*, quarterly; *Dialogue With Asia's Rural Man*, periodic; *Sollicitudo*, triennial; Monographs. Includes updates on village development work; Monographs.

CENTER FOR DEVELOPMENT SERVICES, MIDEAST

Citibank Bldg., 6th Fl., 4 Ahmed Pasha St.,

Garden City, Cairo, Egypt **PHONE:** 20 2 3546599, 20 2 3557558 **FX:** 20 2 3547278 **FOUNDED:** 1990. **OFFICERS:** Dr. Alaa Saber, Mng.Dir. **STAFF:** 30. **BUDGET:** US$800,000. **LANGUAGES:** Arabic, English.

DESCRIPTION: Community development professionals and development educators. Promotes creation and implementation effective community development projects throughout the Middle East. Conducts training programs for development agency personnel; provides technical support to development projects.

CENTER FOR DOCUMENTATION, RESEARCH AND TEACHING ON THE INDIAN OCEAN REGION (CEDREFI)

Boite Postale 91, Rose Hill, Mauritius **PHONE:** 230 4655036 **FX:** 230 4651422 **E-MAIL:** pynee@syfed.mu.refer.org **FOUNDED:** 1981. **OFFICERS:** Pynee A. Chellaperhal, Dir. **MEMBERS:** 15. **STAFF:** 2. **LANGUAGES:** English, French.

DESCRIPTION: Representatives of governments in the Indian Ocean region. Promotes human-centered and environmentally sustainable economic and social development in the Indian Ocean basin. Identifies, collects, and disseminates information on the region and its development; conducts research and feasibility studies; makes available training for development organization staff; initiates and supports grass roots development projects. Publicizes environmental and development issues. Plans to produce audiovisual educational materials. **PUBLICATIONS:** *Bulletin IBION* (in French), quarterly; *Mauritius: Research Environment*; Brochure; Directory, periodic. Lists organizations and individuals engaged in geopolitical, developmental, environmental, and regional cooperation issues.

CENTER FOR INTERNATIONAL ENVIRONMENTAL LAW (CIEL)

1367 Connecticut Ave. NW, Ste. 300, Washington, DC 20036 **PHONE:** (202) 785-8700 **FX:** (202) 785-8701 **E-MAIL:** cielus@igc.org **WEBSITES:** http://www.econet.apc.org/ciel/ **FOUNDED:** 1989. **OFFICERS:** Dana Clark **STAFF:** 24. **BUDGET:** US$1,000,000. **LANGUAGES:** English, French, German, Italian, Portuguese, Russian.

DESCRIPTION: Environmental attorneys. Seeks to strenghten national, international, and comparative environmental law and public policy worldwide. Conducts environmental and legal research and provides recommendations to policy makers.

CENTER FOR INTERNATIONAL POLICY (CIP)

1755 Massachusetts Ave. NW, Ste. 312, Washington, DC 20036 **PHONE:** (202) 232-3317 **FX:** (202) 232-3440 **E-MAIL:** cip@ciponline.org **WEBSITES:** http://www.ciponline.org **FOUNDED:** 1975. **OFFICERS:** William C. Goodfellow, Dir. **STAFF:** 10. **BUDGET:** US$550,000. **LANGUAGES:** English.

DESCRIPTION: Monitors peace efforts in Central America and the Caribbean and examines the impact of U.S. foreign policy on human rights in the Third World. Studies and reports on the implications of U.S. foreign assistance. **PUBLICATIONS:** *International Policy Reports*, bimonthly. Each issue is devoted to a specific topic; Monographs; Also publishes regional updates.

CENTER FOR RESEARCH-INFORMATION-ACTION FOR DEVELOPMENT IN AFRICA (CRIAA)

PO Box 140, Auckland Park, Republic of South Africa **PHONE:** 27 11 6465748 **FX:** 27 11 646514 **OFFICERS:** Benoit Allanic **LANGUAGES:** Afrikaans, English, French.

DESCRIPTION: National development organizations in France, Mozambique, Namibia, South Africa, and Zimbabwe. Promotes appropriate and sustainable economic and community development in southern Africa. Places development professionals with projects requiring assistance; makes available administrative support to development programs; serves as a clearinghouse on regional development initiatives.

CENTER FOR WOMEN WAR VICTIMS (CWWW)

Radnicki dol 20, CT-10000 Zagreb, Croatia **PHONE:** 385 1 4823258, 385 1 4823188 **FX:** 385 1 4823258 **E-MAIL:** cenzena@zamir.net **FOUNDED:** 1992. **MEMBERS:** 25. **STAFF:** 15.

BUDGET: DM 410,000. **LANGUAGES:** Croatian, English.

DESCRIPTION: Social workers and other individuals concerned about the plight of people displaced by conflict in the Balkans. Seeks to ensure that refugees and other displaced persons contact relief agencies and organizations. Provides psychological and social counseling; makes available education, health care, and immigration referral service. **PUBLICATIONS:** *Annual Reports Collection* (in Croatian and English). Contains information about centers and activities in the context of fo Yugoslavia.

CENTRAL AMERICAN BANK OF ECONOMIC INTEGRATION (Banco Centroamericano de Integracion Economica)

Apartado Postal 772, Tegucigalpa, Honduras **PHONE:** 372230 **FX:** 370793 **FOUNDED:** 1961. **OFFICERS:** Jose Manuel Pacas, Pres. **MEMBERS:** 6. **STAFF:** 458. **BUDGET:** US$20,100,000. **LANGUAGES:** English.

DESCRIPTION: Promotes economic integration of the six member countries of Costa Rica, El Salvador, Guatemala, Honduras, Nicaragua, and Panama. Interested in the balanced economic growth and development of members. **PUBLICATIONS:** *Annual Report (Memoria Anual)*; *Boletin Estadistico Historico*, annual; *Boletin Estadistico Mensual* (in Spanish), monthly; *Revista de la Integracion y el Desarrollo de Centroamerica*, periodic.

CENTRAL AMERICAN MONETARY COUNCIL (Consejo Monetario Centroamericano)

Apartado Postal 5438, San Jose 1000, Costa Rica **PHONE:** 506 2336044, 506 2234890 **FX:** 506 2215643 **E-MAIL:** secmsa@sol.racsa.co.cr **WEBSITES:** http://www.cmca.or.cr **FOUNDED:** 1974. **OFFICERS:** Lic. Jose Paiz Moreira, Exec.Off. **MEMBERS:** 5. **STAFF:** 12. **BUDGET:** US$1,200,000. **LANGUAGES:** Spanish.

DESCRIPTION: Promotes the coordination of money, credit, and exchange policies to form throughout Central America the basis of the Central American Monetary Union. **PUBLICATIONS:** *Boletin Estadistico* (in Spanish), annual. Examines the macroeconomic evolution of Central American countries; *Informe Economico* (in Spanish), annual.

CENTRAL COMMISSION FOR THE NAVIGATION OF THE RHINE (CCNR)

2, place de la Republique, Palais du Rhin, F-67000 Strasbourg Cedex, France **PHONE:** 33 88 3522010 **FX:** 33 88 3321072 **E-MAIL:** ccmn@wanadoo.fr **FOUNDED:** 1815. **OFFICERS:** M. Fulda, Pres. **LANGUAGES:** Dutch, English, French, German, Italian.

DESCRIPTION: Members are Belgium, Germany, France, the Netherlands, and Switzerland. Promotes unrestricted navigation on the river Rhine. Activities include: establishment of rules regarding safety and environmental protection for inland navigation including manning regulations and labour conditions; measures to improve the economy; standardization of river law and transport law; customs regulations; arbitration of disputes involving river traffic; and river maintenance. **PUBLICATIONS:** Annual Report (in Dutch, French, and German).

CENTRE FOR APPLIED STUDIES IN INTERNATIONAL NEGOTIATIONS (CASIN) (Centre d'Etudes Pratiques de la Negociation Internationale)

7bis Ave. de la Paix, CH-1202 Geneva, Switzerland **PHONE:** 41 22 9061695 **FX:** 41 22 7310233 **E-MAIL:** Freymond@hei.unige.ch **FOUNDED:** 1979. **OFFICERS:** Jean F. Freymond, Dir. **STAFF:** 10. **BUDGET:** US$1,500,000. **LANGUAGES:** English, French, Spanish.

DESCRIPTION: Purpose is to provide a forum for education and dialogue in areas related to selected international issues, negotiations, and conflict resolution. Offers professional training and education in negotiations, conflict prevention and resolutions, diplomacy, development, international trade and the environment. Organizes policy dialogues on issues related to the environment, consumer affairs, development, and African agriculture. Conducts research; maintains two documentation centers, one on NGOs. Offers professional training programs tailored to the needs of requesting institutions. **PUBLICATIONS:** *Developing African Agriculture: New Initiatives for Institutional Cooperation* (in English), annual; *Environmental NGO's in Central and Eastern Europe*; *Management of*

Interdependence, periodic; *NGO's and the GATT: A Report*; *NGO's Information Digest*, monthly; *Report of Activities*, annual; *Seminar Proceedings*, periodic; Monographs, periodic. Contains information on international negotiations; Papers, periodic. Presented at roundtables.

CENTRE FOR ECONOMIC POLICY RESEARCH

90-98 Goswell Rd., London EC1V 7DB, England **OFFICERS:** Anitra Hume-Wright, Personnel Officer. **MEMBERS:** 200. **LANGUAGES:** English, French, German.

DESCRIPTION: Research fellows throughout Europe. Enables members to collaborate in economic research; disseminates results to the public. Areas of interest include open economy macroeconomics, trade policy, and European economic integration.

CENTRE FOR THE INDEPENDENCE OF JUDGES AND LAWYERS (CIJL) (Centre pour l'Independance des Magistrats et des Avocats - CIMA)

26, chemin de Joinville, BP 160, CH-1216 Geneva Cointrin, Switzerland **PHONE:** 41 22 7884792 **FX:** 41 22 7884880 **TX:** 418531 ICJ CH **FOUNDED:** 1978. **OFFICERS:** Mona Rishmawi, Dir. **STAFF:** 2. **LANGUAGES:** English, French, Spanish.

DESCRIPTION: Established by the International Commission of Jurists to promote and protect the independence of the legal profession and of the judiciary. Collects and disseminates information worldwide on legal guarantees for the independence of legal and judiciary professionals, violations of their rights, and facts in cases of harassment, repression, or victimization of individual judges and lawyers. Encourages individual organizations to assist their colleagues by representing their concerns to the authorities of the country concerned and by taking action in appropriate cases. **PUBLICATIONS:** *Attacks on Justice: The Harassment and Persecution of Judges and Lawyers* (in English), annual; *CIJL Yearbook* (in English, French, and Spanish), annual.

CENTRE ON INTEGRATED RURAL DEVELOPMENT FOR AFRICA (CIRDA)

PO Box 6115, Arusha, United Republic of Tanzania **PHONE:** 255 57 2576 **FX:** 255 57 8532 **TX:** 42053 **OFFICERS:** Dr. Abdel-Moneim M. Elsheikh, Dir. **LANGUAGES:** English.

DESCRIPTION: Members are rural development organizations. Promotes appropriate and sustainable development of rural areas in Africa. Conducts research; provides consulting services and other support to rural development projects.

CENTRE FOR LATIN AMERICAN MONETARY STUDIES (Centro de Estudios Monetarios Latinoamericanos - CEMLA)

Durango 54, Delegacion Cuauhtemoc, 06700 Mexico City, DF, Mexico **PHONE:** 52 5 5330309, 52 5 5330300 **FX:** 52 5 5146554 **FOUNDED:** 1952. **OFFICERS:** Lic. Sergio Ghigliazza Garcia. **MEMBERS:** 65. **STAFF:** 50. **BUDGET:** US$2,900,000. **LANGUAGES:** English, French, Portuguese, Spanish.

DESCRIPTION: Central banks of Latin America and the Caribbean (associates); central banks of Canada, France, Italy, Japan, Philippines, Portugal, Spain, and the United States (collaborators); government and regional organizations and banks of Latin America. Conducts research in monetary and financial policies and procedures. Facilitates cooperation among members in financial matters. Sponsors training programs for Latin American central bank personnel. **PUBLICATIONS:** *Boletin* (in Spanish), bimonthly; *Monetaria* (in Spanish), quarterly; *Money Affairs* (in English), semiannual; *Seminarios*, periodic. Includes seminar reports. Also publishes books, studies, and essays.

CERCLE INTERNATIONAL POUR LA PROMOTION DE LA CREATION

BP 1256, Bafoussam, Cameroon **PHONE:** 237 446267, 237 446668 **FX:** 237 446669 **E-MAIL:** cipcre@geod.geonet.de **FOUNDED:** 1990. **OFFICERS:** Pasteur Jean-Blaise Kenmogne. **MEMBERS:** 35. **STAFF:** 30. **BUDGET:** 100,000,000 Fr CFA. **LANGUAGES:** English, French.

DESCRIPTION: Works to raise awareness of environmental issues. Promotes environmental protection and the responsible use of natural resources. Disseminates information. **PUBLICATIONS:** *Al'affut SU*, bimonthly; *Ecovoy* (in French), quarterly. Contains information on sustainable development.

CHEMICAL SOCIETIES OF THE NORDIC COUNTRIES (CSNC) (Nordiska Kemistradet - NK)

c/o Svenska Kemistsamfundet, Wallingatan 24, 3tr, S-111 24 Stockholm, Sweden **PHONE:** 46 8 4115260 **FX:** 46 8 106678 **E-MAIL:** agneta@chemsoc.se **WEBSITES:** http://www.chemsoc.se **FOUNDED:** 1959. **OFFICERS:** Dr. Agneta Sjogren, Acting Sec. **MEMBERS:** 4. **LANGUAGES:** Danish, Norwegian, Swedish.

DESCRIPTION: Membership consists of national chemical societies. Purpose is to promote chemistry in Nordic countries by coordinating the activities of member societies. Maintains contacts with the Federation of European Chemical Societies and other international groups.

CHILDREN OF THE AMERICAS (COA)

c/o W.O. Mills, III, PO Box 140165, Dallas, TX 75214-0165 **PHONE:** (214) 823-7000 **FX:** (214) 823-7991 **FOUNDED:** 1979. **OFFICERS:** W. O. Mills III, Pres. **MEMBERS:** 600. **STAFF:** 2. **LANGUAGES:** English, Spanish.

DESCRIPTION: Individuals and organizations committed to aiding orphaned or abandoned children ages 6 through 16 who live on the streets. Provides first aid and emergency medical care, counseling, and social services. Conducts student/faculty research programs; sponsors student and volunteer street healthcare workers. Efforts are currently concentrated in Santo Domingo, Dominican Republic, and Bogota, Colombia. Maintains speakers' bureau; offers photo service. **PUBLICATIONS:** *A Global Perspective on Gamins*; *COA Update*, quarterly; *Health Care in the Street.*

CHILDREN AND DEVELOPMENT (Enfants et Developpement)

13 rue Jules Simon, F-75015 Paris, France **PHONE:** 33 1 48422303 **FX:** 33 1 48424175 **LANGUAGES:** English, French.

DESCRIPTION: Social welfare and development programs. Seeks to improve the quality of life of children living in economically disadvantaged areas worldwide. Collaborates with government agencies to ensure full community participation in development projects. Sponsors programs in areas including maternal and child health, education, vocational training, nutrition, and water supply and sanitation.

CHILDREN'S WISH FOUNDATION INTERNATIONAL (CWFI)

8615 Roswell Rd., Atlanta, GA 30350 **PHONE:** (770) 393-9474 **TF:** (800) 323-WISH **FX:** (770) 393-0683 **E-MAIL:** childrenswish@mindspring.com **WEBSITES:** http://www.childrenswish.org **FOUNDED:** 1978. **STAFF:** 19. **BUDGET:** US$25,000,000. **LANGUAGES:** English.

DESCRIPTION: Seeks to fulfill the wishes of terminally ill children under 18 years old. Maintains speakers' bureau; compiles statistics. Provides children's services throughout North America, Europe, and Asia. **PUBLICATIONS:** *Wishing Well*, quarterly; Brochures, quarterly.

CHOL CHOL FOUNDATION FOR HUMAN DEVELOPMENT (CCFHD)

4431 Garrison St. NW, Washington, DC 20016 **PHONE:** (301) 259-4518 **FX:** (301) 259-4518 **E-MAIL:** cholchol@chilepac.net **FOUNDED:** 1971. **OFFICERS:** Catherine T. Jensen, Chair. **STAFF:** 25. **BUDGET:** US$200,000. **LANGUAGES:** English.

DESCRIPTION: Participants are individuals seeking an end to world hunger. Objective is to assist poor Third World citizens in developing economic self-reliance through programs and instruction in agriculture, nutrition, forestry, livestock management, animal husbandry, and other related skills. Sponsors teams of technicians offering technical assistance; provides a revolving loan fund for agricultural development; conducts research. **PUBLICATIONS:** *Newsnotes*, monthly. Provides information about CCFHD's activities. Includes member profiles and statistics; Brochure (in English and Spanish), periodic.

CHRISTIAN AID

35 Lower Marsh, Waterloo, London SE1 7RT, England **PHONE:** 44 171 6204444 **FX:** 44 171 6200719 **TX:** 916504 CHRAID **E-MAIL:**

info@christian-aid.org **WEBSITES:** http://www.christian-aid.org.uk **FOUNDED:** 1945. **OFFICERS:** Judith Knight, Admin. **STAFF:** 279. **BUDGET:** £40,000,000. **LANGUAGES:** English.

DESCRIPTION: Aid and development arm of the Council of Churches for Britain and Ireland. Supports community development programs designed to help some of the world's poorest people. Works in partnership with other organizations in 70 countries throughout the world. Seeks to combat the causes of poverty through advocacy, campaigns, and public education. **PUBLICATIONS:** *Christian Aid News* (in English), 3/year; Reports. Produces reports on topical issues such as debt, fair trade, and child labor.

CHRISTIAN CHILDREN'S FUND (CCF)

2821 Emerywood Pky., Box 26484, Richmond, VA 23261-5066 **PHONE:** (804) 756-2700 **TF:** (800) 776-6767 **FX:** (804) 756-2718 **TX:** 6844270 CCF-UW **WEBSITES:** http://www.christianchildrensfund.org **FOUNDED:** 1938. **OFFICERS:** Dr. Margaret C. McCullough, Pres. **STAFF:** 587. **BUDGET:** US$117,000,000. **LANGUAGES:** English.

DESCRIPTION: International, nonsectarian child development organization providing food, clothing, medical care and educational opportunities to children of all races and creeds. Works through the donations of sponsors who maintain personal contact with the child they help support. Provides assistance to needy children and their families in 32 countries including the U.S.A. Member of American Council for Voluntary International Action, NGO Committee on UNICEF, and International Council of Voluntary Agencies. **PUBLICATIONS:** *Children's Circle News*, quarterly. Contains planned giving news and information.; *Childworld*, 3/year. Includes articles on third world problems of children, families, and communities.; *State of CCF's Children*, annual; Annual Report.

CHRISTIAN CONFERENCE OF ASIA (CCA)

Pan Tin Village, Mei Tin Rd., Shatin, Hong Kong **PHONE:** 852 2 6911068 **FX:** 852 2 6924378 **E-MAIL:** cca@hk.super.net **FOUNDED:** 1957. **OFFICERS:** Dr. Feliciano Carino, Gen.Sec. **MEMBERS:** 108. **STAFF:** 20. **BUDGET:** HK$2,350,000. **LANGUAGES:** English.

DESCRIPTION: A forum of cooperation among the churches and national Christian bodies in Asia within the framework of the wider ecumenical movement. Believes the purpose of God for the church in Asia is life together in a common obedience of witness to the mission of God in the world. Encourages Asian contribution to Christian thought, wisdom and action throughout the world. Promotes the protection of human dignity and the promotion of carrying for the creation. **PUBLICATIONS:** *CCA News* (in English), quarterly; *CTC Bulletin* (in English), 3/year; *Voices* (in English), quarterly.

CHRISTIAN ENGINEERS IN DEVELOPMENT

The Byre, 15 Catrail Rd., Galasheils, Selkirkshire TD1 1NW, England **PHONE:** 44 1896 755498 **FX:** 44 1896 757891 **FOUNDED:** 1985. **OFFICERS:** Gareth R. Cozens, Dir. & Sec. **MEMBERS:** 50. **STAFF:** 6. **BUDGET:** US$120,000. **LANGUAGES:** English.

DESCRIPTION: Christian engineers and other professionals interested in equitable and sustainable development of economically underdeveloped areas worldwide. Seeks to apply engineering principles to the design and implementation of economic development programs; makes available technical assistance to local programs sharing similar goals. Promotes environmental protection and the development of indigenous leadership to guide local development. **PUBLICATIONS:** *CEO Newsletter* (in English), quarterly. Contains news of projects and activities.

CHRISTIAN FOUNDATION FOR CHILDREN AND AGING (CFCA)

1 Elmwood Ave., Kansas City, KS 66103-3719 **PHONE:** (913) 384-6500 **TF:** (800) 875-6564 **FX:** (913) 384-2211 **E-MAIL:** cfca@sky.net **WEBSITES:** http://www.cfcausa.org **FOUNDED:** 1981. **OFFICERS:** Louis Finocchario, Exec.Dir. **MEMBERS:** 131,102. **STAFF:** 76. **BUDGET:** US$30,470,000. **LANGUAGES:** English, Spanish.

DESCRIPTION: Seeks to advance the physical, mental, spiritual, and social welfare of the economically disadvantaged, especially children and aging persons in developing countries. U.S. sponsors provide financial support and correspond with individuals in need; volunteers help provide social services, including medical, educational, and nutritional programs. Provides Christian education and guidance Conducts orientation program for volunteers and Mission Awareness trips to Mexico and Central America. **PUBLICATIONS:** *Country Report*, periodic; *Project Update*, monthly; *Sacred Ground* (in English and Spanish), 3/yr. News of sponsorship and CFCA mission activity.

CHRISTIAN LITERACY ASSOCIATES (CLA)

541 Perry Hwy., Pittsburgh, PA 15229 **PHONE:** (412) 364-3777 **E-MAIL:** drliteracy@aol.com **FOUNDED:** 1977. **OFFICERS:** Dr. William E. Kofmehl Jr., Pres. **MEMBERS:** 4,500. **STAFF:** 35. **BUDGET:** US$95,000. **LANGUAGES:** English.

DESCRIPTION: Christian professionals and paraprofessionals who prepare basic adult literacy materials for use through churches; volunteers also train tutors and plan literacy campaigns. In the United States, works with local congregations, councils of churches, denominations, jails and jail ministries, retarded and disabled groups, refugee help groups, and school districts. Outside the United States, works with local councils of churches, denominations, seminaries, missionary groups, and international ministries. Produces adult basic literacy primers in several languages in cooperation with native speakers. Maintains volunteer staff of professional educators to help design and prepare supplementary reading materials and to train potential tutors. Assists local churches and groups in organizing regional and national basic literacy programs. Conducts training workshops for literacy tutors. Model programs include Christian Literacy Associates of Haiti and the Allegheny County Literacy Council. **PUBLICATIONS:** *Christian Literacy Outreach*, quarterly. Contributions newsletter; *Christian Literacy Tutor Workshop*. Literacy training workshops; *40 Bible Stories for New Readers*; *The Four R's*; *Gameplans and Guidelines*; *Guide to Establishing a Christian Literacy Outreach*; *Light is Coming*. Basic reading Series, in nineteen languages; *Managing Methods and Materials*.

CHRISTIAN PEACE CONFERENCE (CPC) (Conference Chretienne pour la Paix - CCP)

Prokopova 4, CZ-130 00 Prague 3, Czech Republic **PHONE:** 420 2 22781800 **FX:** 420 2 22781801 **FOUNDED:** 1958. **OFFICERS:** Sylvia Weber, Dir. **STAFF:** 2. **LANGUAGES:** English, French, German, Russian, Spanish.

DESCRIPTION: Churches, Christian associations, and individuals in 80 countries. Acts as a forum for Christians to meet and deal with political, social, and economic issues. Works for world peace through disarmament, human rights, international cooperation, and peaceful coexistence efforts. Promotes social and economic structures aimed toward eliminating oppression, exploitation, racial discrimination, hunger, and illiteracy; opposes imperialism. Conducts research. **PUBLICATIONS:** *CPC Information* (in English and German), bimonthly; Booklets (in English and German), periodic. Also publishes study series.

CHRISTIAN SOLIDARITY INTERNATIONAL (CSI)

c/o Rev. Hansjurg Stuckelberger, Zelglistr. 64, Postfach 70, CH-8122 Binz, Switzerland **PHONE:** 41 1 9804700 **FX:** 41 1 9804715 **E-MAIL:** csi-int@csi-int.ch **WEBSITES:** http://www.csi-int.ch **FOUNDED:** 1977. **OFFICERS:** Rev. Hansjurg Stuckelberger, Pres. **MEMBERS:** 13. **STAFF:** 35. **BUDGET:** US$5,000,000. **LANGUAGES:** English, French, German.

DESCRIPTION: Members are national committees with more than 40,000 supporters worldwide. Works for religious freedom, as defined in the Universal Declaration of Human Rights, Charter of the United Nations, and other international documents. Assists persecuted Christians and others in need by providing legal and financial aid and personal contact. Conducts educational programs on persecuted churches in the East and West. **PUBLICATIONS:** *Appell* (in German), bimonthly. Contains news for CSI groups; *CSI Family News*, monthly. Containing news from individual committees; *CSI Magazine* (in Czech, English, French, German, Hungarian, and Italian), monthly; *Information Statement*, bimonthly.

CHURCH WORLD SERVICE AND WITNESS (CWSW)

475 Riverside Dr., Rm. 678, New York, NY 10115-0050 **PHONE:** (212) 870-2061, (212) 870-2257 **TF:** (800) 456-1310 **FX:** (212) 870-3523 **E-MAIL:** rondah@cws.ncccusa.org **FOUNDED:**

1946. **OFFICERS:** Rodney I. Page, Exec.Dir. **MEMBERS:** 33. **STAFF:** 170. **BUDGET:** US$50,000,000. **LANGUAGES:** English.

DESCRIPTION: Division of the National Council of Churches of Christ in the U.S.A. providing worldwide development and emergency aid to the poor in east and southern Asia, Africa, Latin America, Middle East, and Eastern Europe. Engages in works of Christian compassion, relief, technical assistance, reconstruction, interchurch aid, and ministering to the victims of war and other emergencies such as famines and floods. There are no restrictions regarding race, religion or ethnicity. Sends food, clothing, medicines, hospital supplies, blankets and self-help equipment overseas annually. Additional food and funds are raised through CROP, the community hunger appeal of CWSW; medical and hospital supplies are solicited by Interchurch Medical Assistance, which allocates donations to its member agencies and medical missionary boards. Promotes global education; monitors human rights and international economic and political issues; cooperates with and seeks to respond to the leadership and insights provided by World Council of Churches, and partner councils, churches, and ecumenical groups abroad. Sponsors Immigration and Refugee Program. Supports or is directly responsible for self-help development projects in more than 70 countries. Has resettled over 374,273 refugees in the U.S. in the last 43 years. **PUBLICATIONS:** *CWS Up-Date*, biweekly; *Notebook*, monthly; *One Great Hour of Sharing Materials*, annual; *Regional Newsletter*, periodic; Annual Report, annual. Also prepares promotional materials, films, posters, folders, and fact sheets.

CHURCHES' COMMISSION FOR MIGRANTS IN EUROPE (CCME) (Kommission der Kirchen fur Migranten in Europa)

174, rue Joseph II, B-1040 Brussels, Belgium **PHONE:** 32 2 2312011 **FX:** 32 2 2311413 **FOUNDED:** 1964. **OFFICERS:** Dr. Jan Niessen, Gen.Sec. **MEMBERS:** 14. **STAFF:** 2. **BUDGET:** 7,000,000 BFr. **LANGUAGES:** English, French, German.

DESCRIPTION: Membership consists of Protestant, Orthodox, and Anglican churches and church agencies from 14 countries. Informs churches and the public of the problems of immigrants and ethnic minorities in Europe; encourages churches and the public to take responsibility for solving such problems. Issues of concern include: racism in Europe; political rights of migrants; migrant women. Works to safeguard the rights of immigrants and ethnic minorities. Monitors pertinent legislation. **PUBLICATIONS:** *Ad-Hoc Position Papers*, periodic; *Briefing Papers* (in English, French, and German), quarterly. Provides information on current events and activities; *Migration Newssheet* (in English and French), monthly. Details legal issues pertaining to migrant rights and incidents of discrimination.; *Report of Assembly*, biennial.

CLIMATE NETWORK EUROPE

44 rue du Taciturne, B-1000 Brussels, Belgium **PHONE:** 32 2 2310180 **FX:** 32 2 2305713 **E-MAIL:** canron@gn.apc.org **WEBSITES:** http://www.climatenetwork.org **FOUNDED:** 1989. **OFFICERS:** Lynne Clark. **MEMBERS:** 75. **STAFF:** 3. **LANGUAGES:** English, French, German.

DESCRIPTION: European nongovernmental organizations with an interest in global climate and climatic change. Advocates a commitment to a 20% decrease in carbon dioxide emissions by the industrialized countries of the world over the next ten years. Gathers and disseminates information on greenhouse gases and global warming; works to increase public awareness of the implications of global climatic change. **PUBLICATIONS:** *Hotspot* (in English), 5/year.

CLUB DU SAHEL

c/o OECD, 2 rue Andre Pascal, F-75775 Paris, France **PHONE:** 33 1 45248200, 33 1 45248787 **FX:** 33 1 45249031 **TX:** 640048 **E-MAIL:** jacqueline.damon@oecd.org **LANGUAGES:** Arabic, English, French.

DESCRIPTION: Countries of the Sahel, a sub-Saharan region of north central Africa. Promotes sustainable economic and social development in the region. Facilitates cooperation among development organizations operating in the Sahel; conducts research; provides support and assistance to development projects.

COADY INTERNATIONAL INSTITUTE (CII)

St. Francis Xavier Univ., PO Box 5000,

Anitgonish, NS, Canada B2G 2W5 **PHONE:** (902) 867-3961 **FX:** (902) 867-3907 **E-MAIL:** coady@stfx.ca **WEBSITES:** http://www.stfx.ca/pinstitutes **FOUNDED:** 1959. **OFFICERS:** Mary Coyle, Dir. **STAFF:** 17. **BUDGET:** US$1,800,000. **LANGUAGES:** English, French.

DESCRIPTION: Works to insure a ''full and abundant life for all in a just, inclusive, participatory, and sustainable society'' through training of development organization personnel and indigenous peoples affected by development programs. Conducts extension education programs enabling youth, artisans, the unemployed, women, and traditional farmers and fisherman to participate more fully in the development of their local regions; sponsors on-site educational programs for development agency staff. Makes available consulting services; creates model development programs and methods; facilitates linkage between Canadian and overseas development organizations. **PUBLICATIONS:** *Coady Connection*, semiannual. Development news and issues for graduates and partners; *Creating a Balance: Developing New Relationships Between Men and Women*; *Gender and Development: A Resource Directory*.

COALITION FOR WOMEN IN INTERNATIONAL DEVELOPMENT (CWID)

1611 Kent St., Ste. 600, Arlington, VA 22209 **PHONE:** (703) 525-9430 **FOUNDED:** 1976. **OFFICERS:** Elise Smith, Chmn. **MEMBERS:** 140. **LANGUAGES:** English.

DESCRIPTION: Membership consists of organizations (50) and individuals (90). Promotes the participation of American women in U.S. foreign policy and of women throughout the world in the economic, social, and political development of their respective countries. Strives to inform and influence U.S. policymakers about issues affecting women and their families in developing countries. Conducts research programs. Lobbies and advises on issues concerning U.S. or foreign government policies. **PUBLICATIONS:** *Fact Sheet*, periodic; *Minutes of the Quarterly Meetings*.

COCOA PRODUCERS' ALLIANCE (CPA) (Alliance des Pays Producteurs de Cacao)

8/10 Broad St., 11th Fl., PO Box 1718, Lagos, Nigeria **PHONE:** 234 1 2635574, 234 1 2635506 **FX:** 234 1 2635684 **TX:** 28288 **FOUNDED:** 1962. **OFFICERS:** D.S. Kamga, Sec.Gen. **MEMBERS:** 13. **STAFF:** 26. **BUDGET:** US$1,000,000. **LANGUAGES:** English, French, Portuguese, Spanish.

DESCRIPTION: Government representatives of cocoa-producing countries. Purposes are to exchange technical and scientific information, to discuss problems of mutual interest, and to advance social and economic relations between producers. Seeks to ensure adequate supplies to the market at remunerative prices. Offers technical consultation on agricultural extension. Promotes increased consumption of cocoa. Compiles statistics regarding cocoa production forecasts by member states. Conducts soils workshops. **PUBLICATIONS:** *Proceedings of International Cocoa Research Conferences* (in English, French, Portuguese, and Spanish), triennial.

COLOMBO PLAN BUREAU (CPB)

12 Melbourne Ave., PO Box 596, Colombo 4, Sri Lanka **PHONE:** 94 1 581853, 94 1 581813 **FX:** 94 1 58174 **TX:** 21537 METALIX CE **FOUNDED:** 1951. **OFFICERS:** Dr. Hak Sui Kim, Dir. **MEMBERS:** 24. **STAFF:** 21. **BUDGET:** US$232,188. **LANGUAGES:** English.

DESCRIPTION: Members are independent countries. Permanent administrative organ of the Colombo Plan for Cooperative Economic Development in Asia and the Pacific, a British Commonwealth initiative previously known as the Colombo Plan for Cooperative Economic Development in South and South-East Asia, until expansion in 1977. The Colombo Plan (as it is more simply known) is an arrangement of principles and set policies whereby a developing member country may, on a one-to-one basis, negotiate for assistance from a developed member country. Such assistance may be technical or in the form of capital or commodity aid. The plan provides for the existence not only of its bureau but also of: the Consultative Committee, a deliberative body of government ministers; the Colombo Plan Council, comprising representatives from member countries, which promotes and monitors all assistance under the plan; the Colombo Plan Staff College for Technician Education. The bureau monitors assistance and disseminates information on plan activities. Conducts research on development problems; compiles statistics. **PUBLICATIONS:** *Annual Report of the Co-*

lombo Plan Council (in English); Brochures; Papers.

COLUMBIAN SQUIRES (CS)

1 Columbus Plz., New Haven, CT 06510-3326 **PHONE:** (203) 772-2130 **FX:** (203) 772-1923 **WEBSITES:** http://www.kofc-supreme-council.org **FOUNDED:** 1925. **OFFICERS:** Ronald J. Tracz, VP. **MEMBERS:** 24,355. **STAFF:** 6. **LANGUAGES:** English, French, Spanish.

DESCRIPTION: International fraternity of young Catholic men (12-18 years old) throughout the United States, Canada, Mexico, the Philippines, Puerto Rico, the Bahamas, the Virgin Islands and Guam. Promotes leadership building activities for youth. Sponsored by Knights of Columbus. **PUBLICATIONS:** *Squires Newsletter*, monthly. Includes information for members, counselors, and others.

COMITE EUROPEEN DES ASSURANCES (CEA)

3 bis, rue de la Chaussee d'Antin, F-75009 Paris, France **PHONE:** 33 1 44831183 **FX:** 33 1 47700375 **WEBSITES:** http://www.cea.assur.org **FOUNDED:** 1953. **OFFICERS:** Francis Loheac, Sec.Gen. **MEMBERS:** 29. **STAFF:** 31. **LANGUAGES:** English, French, German.

DESCRIPTION: Members are insurance associations representing 6000 companies in western Europe. Monitors developments in insurance law and practice and disseminates information to members; facilitates exchange of information among members. Represents insurers before government and international agencies concerned with insurance of reinsurance. **PUBLICATIONS:** *CEA Codification of European Insurance Directives* (in English, French, and German), periodic; *CEA, ECO European Insurance Data* (in English, French, and German), periodic; *CEA INFO, European Insurance Topics* (in English, French, and German), bimonthly; *CEA Info, Special Issues* (in English, French, and German); *European Insurance in Figures* (in English), periodic; *Health Technical Compendium* (in English, French, and German), periodic; *Indirect Taxation on Insurance Contracts in Europe* (in English, French, and German), periodic; *Pensions in Europe* (in English, French, and German), bimonthly. Latest developments in European insurance.; *Vandemecum on Non-Life Insurance Freedom of Services* (in English, French, and German), periodic. Additional publications available on request.

COMITE EUROPEEN DE DROIT RURAL (CEDR)

P/a Vlaamse Landmaatschappij, 72, ave. de la Toison d'Or, B-1060 Brussels, Belgium **PHONE:** 32 2 5437208 **FX:** 32 2 5437399 **E-MAIL:** marc.heyerick@vlm.be **FOUNDED:** 1957. **OFFICERS:** Dr. Marc Heyerick, Sec.Gen. **MEMBERS:** 14. **LANGUAGES:** English, French, German, Italian, Spanish.

DESCRIPTION: National associations representing 3750 specialists in agrarian law and related social sciences. Promotes and provides for the exchange of views on questions of agrarian law and related issues. Publishes the opinions of members on the integration of agrarian law and other social sciences; conducts research and disseminates findings on agrarian law, particularly as enforced in the European Economic Community. Presents proposals and opinions before international organizations and governments; maintains consultative status with the Council of Europe and the Food and Agriculture Organization of the United Nations. Establishes educational facilities and collaborates with those already established. **PUBLICATIONS:** *Bulletin of the European Council for Agricultural Law*, periodic; *Colloquium Report*, biennial; *Symposium Report*, periodic.

COMMISSION FOR THE CONSERVATION OF ANTARCTIC MARINE LIVING RESOURCES (CCAMLR) (Comision para la Conservacion de los Recursos Vivos Marinos Antarticos)

23 Old Wharf, Hobart, TAS 7000, Australia **PHONE:** 61 3 62310366 **FX:** 61 3 62349965 **E-MAIL:** ccamlr@ccamlr.org **FOUNDED:** 1982. **OFFICERS:** Mr. Esteban de Salas, Exec.Sec. **MEMBERS:** 23. **STAFF:** 20. **BUDGET:** $A 1,970,000. **LANGUAGES:** English, French, Russian, Spanish.

DESCRIPTION: Works to conserve the Antarctic marine ecosystem, while allowing for rational use of resources. Members represent 22 countries and the European Community. **PUBLICATIONS:** *Basic Documents* (in English, French, Russian, and Spanish), periodic. Includes text of conventions, rules of procedures, etc.; *CCAMLR Inspectors Manual* (in

English, French, Russian, and Spanish); *CCAMLR Newsletter*, annual; *CCAMLR Science*; *CCAMLR Scientific Observers Manual* (in English, French, Russian, and Spanish); *CCAMLR Statistical Bulletin* (in English, French, Russian, and Spanish), annual; *Conservation Measures in Force*; *Reports of Commission and Scientific Committee Meetings*, annual; *Scientific Abstracts*.

COMMISSION ON CRIME PREVENTION AND CRIMINAL JUSTICE (CCPCJ)

c/o Centre for International Crime Prevention, United Nations Office for Drug Control and Crime Prevention, Postfach 500, A-1400 Vienna, Austria **PHONE:** 43 1 260604269 **FX:** 43 1 260605898 **E-MAIL:** bdoering@unov.or.at **WEBSITES:** http://www.ifs.univie.ac.at/-uncjin/uncjin.html **FOUNDED:** 1992. **OFFICERS:** B. Doering, Sec. **MEMBERS:** 40. **LANGUAGES:** Arabic, Chinese, English, French, Russian, Spanish.

DESCRIPTION: Created by resolution of the United Nations, membership consists of member states elected by the council. Promotes international cooperation in the field of crime prevention and criminal justice; coordinates activities in this area for the Economic and Social Council. **PUBLICATIONS:** *Crime Prevention and Criminal Justice Newsletter* (in English, French, and Spanish), biennial; *International Review of Criminal Policy* (in English, French, Russian, and Spanish), annual.

COMMISSION FOR THE DEFENSE OF HUMAN RIGHTS IN CENTRAL AMERICA (CODEHUCA) (Comision para la Defensa de los Derechos Humanos en Centroamerica - CODEHUCA)

Paseo de los Estudiantes, Apartado 189-1002, San Jose 1006-1002, Costa Rica **PHONE:** 506 2342935, 506 2250270 **FX:** 506 2342935 **TX:** 3286 CODEHUCA CR **FOUNDED:** 1978. **OFFICERS:** Mirna Perla de Anaya. **MEMBERS:** 12. **STAFF:** 24. **LANGUAGES:** English, Spanish.

DESCRIPTION: National Central American human rights organizations and Central Americans involved in the protection of human rights. Investigates reported human rights violations; serves as information network on human rights. Conducts research on legal aspects of human rights in Central America. Develops educational materials; conducts seminars and workshops on the status of human rights in Central America, courses on documentation and information techniques, and training for human rights promoters. Compiles statistics; sponsors competitions. **PUBLICATIONS:** *Annual Report*; *BRECHA* (in Spanish), monthly; *Brecha* (in English), bimonthly; *Documentacion Sobre Derechos Humanos* (in English and Spanish), monthly. Also publishes papers.

COMMISSION FOR RACIAL EQUALITY (CRE)

Elliot House, 10/12 Allington St., London SW1E 5EH, England **PHONE:** 44 171 8287022 **FX:** 44 171 6307605 **E-MAIL:** info@cre.gov.uk **WEBSITES:** http://www.cre.gov.uk **FOUNDED:** 1977. **OFFICERS:** Sir Herman Ouseley. **MEMBERS:** 15. **STAFF:** 200. **BUDGET:** £15,000,000. **LANGUAGES:** English.

DESCRIPTION: Works to eliminate discrimination based on race. Promotes equal opportunity among persons of different races. **PUBLICATIONS:** *Connections* (in English), quarterly. Includes news and features on race relations.

COMMISSION DES SCIENCES ARTS ET BELLES-LETTRES

Maison de la Fondation Europeenne, 19, rue de la Reine, B-4500 Huy, Belgium **PHONE:** 32 85 212671 **FX:** 32 85 211071 **FOUNDED:** 1965. **OFFICERS:** H.E. the Rt. Hon. Earl Pierre Pasleau de Charnay, Hon. Ambassador. **LANGUAGES:** English, French, Italian, Spanish.

DESCRIPTION: Members are scholars, artists, writers, composers, instructors, and artisans in European countries. Works for world peace through educational and cultural collaboration. Promotes freedom of speech and protests repressive actions including censorship, arbitrary arrest and imprisonment, torture, and execution. Denounces the destruction of cultural heritage such as books and works of art. Feels that artists, writers, and scholars are often the victims of persecution when their philosophy diverges from that of the political system of their society. Believes that the freedom to criticize the government and institutions is indispensable to improved economic and political organization; places crucial importance on combating distortion of the truth by the media for political gain. Organizes seminars, exhibitions, and receptions. Petitions governments to uphold the Universal Declaration of Human Rights. Provides information, documentation, and technical services.

COMMISSION ON THE STATUS OF WOMEN (CSW)

United Nations, Rm. DC2-1220, PO Box 20, New York, NY 10017 **PHONE:** (212) 963-5086 **FX:** (212) 963-3463 **FOUNDED:** 1946. **OFFICERS:** Angela King, Asst.Sec.Gen. **MEMBERS:** 45. **STAFF:** 15. **BUDGET:** US$202,400. **LANGUAGES:** Arabic, Chinese, English, French, Russian, Spanish.

DESCRIPTION: Representatives from United Nations member countries. Promotes women's rights in political, economic, civil, social, and educational fields. Encourages cooperation between organizations seeking to advance the status of women, and advises the UN and member bodies on situations requiring immediate attention. Acts as a preparatory body for the world conference on women.

COMMITTEE OF AGRICULTURAL ORGANISATIONS IN THE EU (Comite des Organisations Professionnelles Agricoles de l'EU - COPA)

23-25, rue de la Science, bte. 3, B-1040 Brussels, Belgium **PHONE:** 32 2 2872711 **FX:** 32 2 2872700 **E-MAIL:** mail@copa-cogeca.be **FOUNDED:** 1958. **OFFICERS:** Risto Volanen, Sec.Gen. **MEMBERS:** 30.

DESCRIPTION: Federation of national agricultural organizations in the European Union. Represents the interests of EU farmers and works to secure fair income and living and working conditions for them and their families. Joint Secretariat with the General Committee of Agricultural Cooperation in the EU. **PUBLICATIONS:** *Annual Report*, annual. Includes information on the agricultural situation in the EC.; Papers.

COMMITTEE OF EEC SHIPBUILDERS' ASSOCIATIONS (CESA)

c/o Fabrimetal, 21, rue des Drapiers, B-1050 Brussels, Belgium **PHONE:** 32 2 5102311 **FX:** 32 2 5102301 **TX:** 21078 **FOUNDED:** 1968. **OFFICERS:** J.E. Perez, Sec.Gen. **MEMBERS:** 11. **LANGUAGES:** English.

DESCRIPTION: National associations of shipbuilders and shipbuilding companies in member countries of the European Economic Community. Purposes are to: represent shipyards before EEC institutions; increase cooperation among shipyards in member countries; study and seek solutions to problems.

COMMITTEE ON THE ELIMINATION OF RACIAL DISCRIMINATION (CERD)

United Nations High Commissioner for Human Rights, Palais des Nations, CH-1211 Geneva 10, Switzerland **PHONE:** 41 22 9179290 **FX:** 41 22 9179022 **TX:** 412962 **E-MAIL:** rhusbands.hchr@unog.ch **FOUNDED:** 1969. **OFFICERS:** Robert Husbands, Sec. **MEMBERS:** 18. **LANGUAGES:** Chinese, English, French, Russian, Spanish.

DESCRIPTION: Created by the International Convention on the Elimination of All Forms of Racial Discrimination. Members are eighteen human rights experts elected by the convention's 153 state parties. Established to monitor the implementation of the convention's provisions concerning the elimination of all forms of racial discrimination in member states. Examines petitions from individuals or groups who are victims of racial discrimination; examines reports submitted by states and parties to the convention and addresses violations of convention provisions by individual states. **PUBLICATIONS:** *Annual Report to the General Assembly of the U.N.*; *Reports by State Parties to the Convention*, periodic; *Summary Records of Proceedings*, periodic.

COMMITTEE FOR EUROPEAN CONSTRUCTION EQUIPMENT (CECE) (Europaisches Baumachinen Komite)

c/o VDMA, 101, Rue De Stassart, B-1050 Brussels, Belgium **PHONE:** 32 2 5127202 **FX:** 32 2 5025442 **FOUNDED:** 1959. **OFFICERS:** Mr. P. Juliens. **LANGUAGES:** English.

DESCRIPTION: Membership consists of national committees representing more than 1000 construction equipment manufacturers. Facilitates contact between members in an effort to improve market conditions and increase productivity. Represents members' views and exchanges data with national and international organizations including the European Economic Community and the International Organization for Standardization. Encourages international and uniform standards for design, safety regulations, terminology, test procedures, and nomenclature. Works for technical progress through research aimed at improving working conditions. Develops product nomenclature; compiles statistics. Offers computerized services. Conducts periodic market trend seminars and press conferences.

COMMITTEE FOR INTERNATIONAL SELF-RELIANCE (CIS)

Rubaga Rd., PO Box 2923, Kampala, Uganda **PHONE:** 256 41 272158 **FOUNDED:** 1985. **OFFICERS:** Jesse Kiisa. **MEMBERS:** 101. **STAFF:** 12. **BUDGET:** US$375,000. **LANGUAGES:** English.

DESCRIPTION: Promotes local autonomy in economic development, with emphasis on the development of rural areas. Creates and implements locally administered public and dental health programs; conducts social service projects including antilitter, pollution control, waste management operations. Maintains women's self-help program. Runs water and sanitation, maternal and child health, community health and nutrition as well as general social mobilization through community drama. **PUBLICATIONS:** *Community Dentistry* (in English), quarterly. Provides information on dental health issues.; *Food Hygiene*; *Health and You* (in English), quarterly.

COMMITTEE FOR THE PROMOTION AND ADVANCEMENT OF COOPERATIVES (COPAC)

15 route des Morillons, 1218 Grand Saconnex, CH-1218 Geneva, Switzerland **PHONE:** 41 22 9298825 **FX:** 41 22 7984122 **TX:** 610181 FOODAGRI ROME **E-MAIL:** copac@coop.org **WEBSITES:** http://www.copacgva.org **FOUNDED:** 1971. **OFFICERS:** Ms. MariaElena Chavez, Coord. **MEMBERS:** 7. **STAFF:** 1. **BUDGET:** US$112,500. **LANGUAGES:** English.

DESCRIPTION: Representatives of non-governmental organizations of the cooperative movement together with farmers' and workers' organizations, and the United Nations and its agencies. Promotes and coordinates the development of sustainable cooperative enterprises through policy dialogue, technical cooperation and information, and concrete collaborative activities. Organizes symposia and consultations for cooperative leaders, officials from development agencies, civil servants, and researchers; and publishes materials on cooperatives. **PUBLICATIONS:** *COPAC Directory of Agencies Assisting Developing Countries* (in English), biennial; Reports.

COMMITTEE TO PROTECT JOURNALISTS (CPJ)

330 7th Ave., 12th Fl., New York, NY 10001 **PHONE:** (212) 465-1004, (212) 465-9344 **FX:** (212) 465-9568 **TX:** 910 2504794 **E-MAIL:** info@cpj.org **WEBSITES:** http://www.cpj.org **FOUNDED:** 1981. **OFFICERS:** Ann K. Cooper, Exec.Dir. **MEMBERS:** 468. **STAFF:** 20. **BUDGET:** US$1,800,000. **LANGUAGES:** Arabic, French, German, Hebrew, Russian, Spanish.

DESCRIPTION: Membership consists of media organizations, human rights groups, and journalists. Supports journalists around the world who have been subjected to human rights violations. Is concerned about efforts by governments to limit the ability of foreign correspondents and local journalists to practice their profession. Keeps U.S. and foreign journalists informed about such practices and organizes protests on behalf of those whose rights have been violated. Compiles and publicizes human rights abuses. Brings exiled journalists to the U.S. for interviews and press conferences; sends delegations of journalists to areas of the world where there are especially serious and continuous problems. Maintains liaison with press groups worldwide to exchange information and launch campaigns on urgent human rights abuse cases. Maintains information files and speakers' bureau; compiles statistics. Conducts research programs. **PUBLICATIONS:** *Attacks on the Press*, annual. Documents incidents of abuse worldwide.; *Bouch Pe: The Crackdown on Haiti's Media Since the Overthrow of Aristide*; *Dangerous Assignments*, quarterly. Covers recent cases of press abuse.; *Don't Force Us to Lie (The Struggle of Chinese Journalists in the Reform Era)*; *Journalism Under Occupation*; *Journalists Advisory on Yugoslavia*.

COMMITTEE ON SCIENCES FOR FOOD SECURITY

c/o International Council of Scientific Unions, 51 Boulevard de Montmorency, F-75016 Paris, France **OFFICERS:** W.E.H. Blum, Chm. **LANGUAGES:** English, French.

DESCRIPTION: A committee of the International Council of Scientific Unions. Members are scientists, researchers, and other individuals and organizations and institutions with an interest in food science. Promotes advancement of food and related sciences; seeks to increase global food output. Facilitates exchange of information among food scientists and researchers worldwide; serves as a clearinghouse on food and related sciences; conducts research and educational programs.

COMMITTEE ON WORLD FOOD SECURITY (CFS)

c/o Barbara Huddleston, Food and Agriculture Orgn. of the U.N., ESAF, Rm. C-336, Via delle Terme di Caracalla, I-00100 Rome, Italy **PHONE:** 39 6 57053052 **FX:** 39 6 57055522 **TX:** 625852 FAO I **E-MAIL:** barbara.huddleston@fao.org **WEBSITES:** http://www.fao.org **FOUNDED:** 1975. **OFFICERS:** Barbara Huddleston, Sec. **MEMBERS:** 109. **LANGUAGES:** Arabic, Chinese, English, French, Spanish.

DESCRIPTION: An intergovernmental unit of the Food and Agriculture Organization of the United Nations. Reviews and disseminates information on the food security situation, in particular in low-income food-deficit countries. Periodically evaluates available supplies of basic foodstuffs to assess the needs of domestic and world markets and the need for food aid in the event of crop shortage or failure. Recommends policies necessary for maintaining minimum world food security, increasing food production, and improving access to food by urban and rural poor. **PUBLICATIONS:** *Committee Documents* (in Arabic, Chinese, English, French, and Spanish), annual; *Report of Annual Session* (in Arabic, Chinese, English, French, and Spanish).

COMMON MARKET FOR EASTERN AND SOUTHERN AFRICA (COMESA)

Comesa Centre, Ben Bella Rd., PO Box 30051, Lusaka, Zambia **PHONE:** 260 1 229726, 260 1 229732 **FX:** 260 1 225107 **E-MAIL:** comesa@comesa.int **WEBSITES:** http://www.comesa.int **FOUNDED:** 1981. **OFFICERS:** Dr. Binguwa Mutharika, Sec.Gen. **MEMBERS:** 21. **STAFF:** 111. **BUDGET:** US$5,000,000.

DESCRIPTION: Members are the governments of Angola, Burundi, Comoros, Democratic Republic of the Congo, Djibouti Egypt, Eritrea, Ethiopia, Kenya, Lesotho, Madagascar, Malawi, Mauritius, Mozambique, Namibia, Rwanda, Sudan, Swaziland, Tanzania, Uganda, Zaire, Zambia, and Zimbabwe. Promotes political and economic cooperation among members to facilitate sustainable economic growth in southern and eastern Africa. Maintains free trade area, customs union, and common visa arrangement among members; plans to offer establishment of payments union. Operates Court of Justice to mediate disputes among members. Conducts studies and develops plans for regional economic and infrastructural development; undertakes agricultural research to augment and supplement traditional farming methods and crop selections; monitors and regulates exploitation of natural resources in the region. **PUBLICATIONS:** *COMESA in Brief* (in English); *Traders Directory*, periodic; Journal, quarterly. Newsletters, bulletins, and surveys.

THE COMMONWEALTH

c/o Commonwealth Secretariat, Marlborough House, Pall Mall, London SW1Y 5HX, England **PHONE:** 44 171 8393411 **FX:** 44 171 9300827 **TX:** 27678 **E-MAIL:** info@commonwealth.int **WEBSITES:** http://www.thecommonwealth.org **FOUNDED:** 1947. **OFFICERS:** Chief Emeka Anyaoku, Sec.Gen. **MEMBERS:** 54. **BUDGET:** £10,500,000. **LANGUAGES:** English.

DESCRIPTION: A voluntary association of 54 independent sovereign states, including the United Kingdom and most of its former dependencies. The Commonwealth is governed by consensus among its member governments, it is not a federation and has no formal constitution or binding obligations. The Commonwealth Secretariat (COMSEC) was organized in 1965 to manage the organization's daily activities. Conducts cooperative technical projects, programs, services committees, and international conferences; acts as an information clearinghouse. Focuses on training, technical assistance, economics, industry, agriculture, food production, education, law, science, youth, health, human rights, and women and development. Areas of interest include: post-apartheid South Africa; recognition of democratically elected leaders; environmental protection and sustainable development; measures to counter international drug trafficking; alleviation of poverty; combating AIDS; international economic issues. Administers the Commonwealth Fund for Technical Co-Operation, the Commonwealth Science Council, and the Commonwealth Youth Programme. **PUBLICATIONS:** *Commonwealth Currents*, quarterly; *Commonwealth Law Bulletin*, quarterly; *Secretary-General's Report*, biennial.

COMMONWEALTH FOUNDATION (CF)

Marlborough House, Pall Mall, London SW1Y 5HY, England **PHONE:** 44 171 9303783 **FX:** 44 171 8398157 **TX:** 27678 COMSEC **E-MAIL:** geninfo@commonwealth.int **WEBSITES:** http://www.oneworld.org/com_fnd/ **FOUNDED:** 1966. **OFFICERS:** Dr. Humayun Khan, Dir. **MEMBERS:** 47. **STAFF:** 13. **BUDGET:** £1,840,000. **LANGUAGES:** English.

DESCRIPTION: An intergovernmental organisation with a mandate to support the work of the non-governmental sector in the Commonwealth. Funding is provided by Commonwealth governments. Through the provision of travel grants and specific awards, supports professional development, training opportunities and the sharing of skills, experience and information among the people of the 54 Commonwealth member countries. Foundation programmes and grants benefit non-governmental organisations (NGOs), professional associations and cultural bodies. Encourages activities which facilitate co-operation between developing countries in priority areas including the eradication of poverty, rural development, health, non-formal education, community enterprise, gender and development, disability and the arts and culture. **PUBLICATIONS:** *Common Path* (in English), quarterly; *Non-Governmental Organizations: Guidelines for Good Policy and Practice*; *Workshop Reports.*

COMMONWEALTH FUND FOR TECHNICAL CO-OPERATION (CFTC)

Commonwealth Secretariat, Marlborough House, Pall Mall, London SW1Y 5HX, England **PHONE:** 44 171 8393411 **FX:** 44 171 9300827 **BUDGET:** £25,000,000.

DESCRIPTION: A fund of the Commonwealth Secretariat. Provides technical assistance by offering experts, training, and consultancy services to developing Commonwealth countries in administration. **PUBLICATIONS:** *Skills for Development*, biennial.

COMMONWEALTH HUMAN ECOLOGY COUNCIL (CHEC) (Conseil du Commonwealth pour l'Ecologie Humaine)

146 Buckingham Palace Rd., London SW1W 9TR, England **PHONE:** 44 171 7308668 **FX:** 44 171 7308688 **E-MAIL:** chec@unl.ac.uk **FOUNDED:** 1969. **OFFICERS:** Mrs. Zena Daysh, Exec.V.Chm. **MEMBERS:** 500. **STAFF:** 3. **LANGUAGES:** English.

DESCRIPTION: Members are governmental departments, academicians and universities, ecological associations, and interested individuals in 41 countries. Seeks to make the human ecology field a major component in national planning and policy and to increase awareness and improve programs concerned with the interrelationship of humans in the total environment; works for creation of human ecology institutions in Commonwealth countries. Conducts courses, seminars, and workshops. **PUBLICATIONS:** *CHEC Journal*, periodic; *CHEC Points*, periodic; *Report Series*, 1-10/year; Annual Report; Bibliographies, periodic.

COMMONWEALTH LAWYERS' ASSOCIATION (CLA)

c/o Law Society, 114 Chancery Ln., London WI2A 1PL, England **PHONE:** 44 171 2421222, 44 171 2421222 **FX:** 44 171 8310057 **TX:** 261203 **E-MAIL:** helen.ramsey@lawsociety.org.uk **FOUNDED:** 1986. **OFFICERS:** Helen Potts, Exec.Sec. **MEMBERS:** 850. **STAFF:** 2. **LANGUAGES:** English.

DESCRIPTION: Lawyers, bar associations, and law societies in Commonwealth countries. Works to raise and maintain the standard of legal services in the Commonwealth. Protects members' interests; conducts workshops and professional development programs. Seeks to stimulate a reappraisal of values, institutions, and methodologies among members of the legal profession. Facilitates the exchange of information among members concerning developments relevant to the organization and useful to legal professionals. **PUBLICATIONS:** *Clarion* (in English), quarterly; Papers, periodic.

COMMONWEALTH TRADE UNION COUNCIL (CTUC)

Congress House, 23-28 Great Russell St., London WC1B 3LS, England **PHONE:** 44 171 6310728 **FX:** 44 171 4360301 **E-MAIL:** ctuc-london@geo2.poptel.org.uk **FOUNDED:** 1979. **OFFICERS:** Annie Watson, Dir. **STAFF:** 6. **LANGUAGES:** English.

DESCRIPTION: Trade union organizations in British Commonwealth countries. Represents trade union views to Commonwealth governments and institutions; stimulates consultations, information sharing, and consensus. Promotes a better understanding of global interdependence; encourages efforts to improve the North-South dialogue and to create a fair international economic order. (The North-South dialogue represents efforts to create economic equality between the poverty-stricken countries of the South and the developed countries of the North.) Provides practical assistance to trade unions in developing countries through worker education and training programs. Compiles a register of members' needs and seeks aid-giving organizations for sponsorship of specific projects. Maintains liaison with the Commonwealth Secretary-General and the Commonwealth Secretariat. **PUBLICATIONS:** Annual Report (in English), annual.

COMMUNICATIONS INTERNATIONAL (CI) (Internationale de Communications)

38, ave. du Lignon, CH-1219 Geneva, Switzerland **PHONE:** 41 22 9792020 **FX:** 41 22 7963975 **E-MAIL:** ci@union-network.org **WEBSITES:** http://194.209.82.11 **FOUNDED:** 1911. **OFFICERS:** Philip Bowyer, Gen.Sec. **MEMBERS:** 208. **LANGUAGES:** English, French, German, Italian, Japanese, Spanish, Swedish.

DESCRIPTION: Trade union organizations in 117 countries engaged in work relating to the transmission and processing of messages representing 4,200,000 postal, telegraph, telephone, and radio workers. Encourages solidarity among workers of free trade unions; maintains friendly relations among international free trade union organizations. Prevents or halts developments that threaten peace, freedom, or democracy; safeguards workers' economic or moral interests; Supports affiliated unions and promotes the formation, growth, and strengthening of free trade unions. Conducts research; sponsors training courses for members, particularly in South America, Asia, and Africa. **PUBLICATIONS:** *PTT Euro-Info* (in English, French, German, Spanish, and Swedish), quarterly; *PTTI African News* (in English and French), bimonthly; *PTTI Asia News*, monthly; *PTTI News* (in Arabic, English, French, German, Japanese, Spanish, and Swedish), monthly; *PTTI Studies*, quarterly; *Unity* (in English and Spanish), monthly. On trade union rights, impact of technology, and women workers.

COMMUNITY DEVELOPMENT SOCIETY (CDS)

1123 N. Water St., Milwaukee, WI 53202 **PHONE:** (414) 276-7106 **FX:** (414) 276-7704 **E-MAIL:** cole@svinicki.com **WEBSITES:** http://www.comm-dev.org **FOUNDED:** 1969. **OFFICERS:** Jane Svinicki, Admin. **MEMBERS:** 1,100. **BUDGET:** US$75,000. **LANGUAGES:** English.

DESCRIPTION: Professionals and practitioners in community development; international, national, state, and local groups interested in community development efforts. Provides a forum for exchange of ideas and experiences; disseminates information to the public; advocates excellence in community programs, scholarship, and research; promotes citizen participation as essential to effective community development. Sponsors educational programs. **PUBLICATIONS:** *CDS Membership Directory*, annual; *Journal of the Community Development Society*, semiannual; *Vanguard*, periodic; Books; Brochures. Community Development Practice (quarterly - 608 pages).

COMPARATIVE EDUCATION SOCIETY IN EUROPE (CESE)

Univ. of Copenhagen, Dept. of Education, Philosophy and Rhetoric, Njalsgade 80, DK-2300 Copenhagen, Denmark **PHONE:** 45 35328861 **FX:** 45 35328850 **FOUNDED:** 1961. **OFFICERS:** Dr. Thyge Winter-Jensen. **MEMBERS:** 300. **LANGUAGES:** English, French, German.

DESCRIPTION: University teachers, research workers, and others in 42 countries concerned with studies in comparative and international education. Promotes comparative and international studies in education. **PUBLICATIONS:** *Conference Proceedings*, biennial; Newsletter, 3/year.

COMPASSION INTERNATIONAL (CI)

3955 Cragwood Dr., PO Box 7000, Colorado Springs, CO 80918 **PHONE:** (719) 594-9900 **TF:** (800) 336-7676 **FX:** (719) 594-6271 **E-MAIL:** ciinfo@us.ci.org **WEBSITES:** http://www.ci.org **FOUNDED:** 1952. **OFFICERS:** Wesley K. Stafford, Pres. **STAFF:** 510. **BUDGET:** US$70,000,000. **LANGUAGES:** English, Spanish.

DESCRIPTION: Offers ministry to children in underdeveloped and developing countries; provides for educational, physical, material, and spiritual needs. Programs include: schools (financial support sent to schools); family helper plans (care within family unit); children's homes (financial support); hostels (financial support sent to hostels); special care centers (physical and mental health care facilities); meal sponsorship (hot lunch five days/week for school children); scholarship (enables children to attend schools away from their homes and villages); student centers (assists children with their studies); commodities (provides pharmaceuticals and other materials to medical facilities around the world); domestic (assists needy American children in inner cities and Indian reservations). Maintains speakers' bureau. **PUBLICATIONS:** *ChildLink*, semiannual. Focuses on sponsored children and the impact sponsors make on their lives; Audiotapes; Brochures; Videos. Also publishes curriculum.

COMPATIBLE TECHNOLOGY (CT)

1536 Hewitt Ave., St. Paul, MN 55104-1284 **PHONE:** (612) 659-3183 **FX:** (612) 659-3184 **E-MAIL:** ctimpla@aol.com **FOUNDED:** 1981. **OFFICERS:** Dr. Rolfe Leary. **BUDGET:** US$79,099. **LANGUAGES:** English.

DESCRIPTION: Individuals and organizations with an interest in farming. Seeks to increase the food production capabilities of small farms in the developing world through the introduction of more effective farming techniques and technologies. Develops, introduces, and trains indigenous people to make use of appropriate technologies and improved productive techniques. Encourages establishment of agricultural microenterprises to increase the economic viability of rural areas in the developing world.

COMPOSITES MANUFACTURING ASSOCIATION OF THE SOCIETY OF MANUFACTURING ENGINEERS (CMA/SME)

1 SME Dr., PO Box 930, Dearborn, MI 48121-0930 **PHONE:** (313) 271-1500 **TF:** (800) 743-4SME **FX:** (313) 271-2861 **TX:** 297742 SME UR **WEBSITES:** http://www.sme.org **FOUNDED:** 1990. **OFFICERS:** Cheri Skomra, Assoc. Mgr. **MEMBERS:** 1,500. **LANGUAGES:** English.

DESCRIPTION: A division of the Society of Manufacturing Engineers. Members are composites manufacturing professionals and students in 21 countries. Addresses design, tooling, assembly, producibility, supportability, and future trends of composites materials and hardware; promotes advanced composites technology. Analyzes industry trends; evaluates composites usage. Conducts educational programs; facilitates exchange of information among members; operates placement service. **PUBLICATIONS:** *Bibliography of Technical Resources on Composites*; *Composites Applications: The Future Is Now*; *Composites in Manufacturing*, quarterly; *Directory of Composites Manufacturers, Suppliers, Consultants, and Research Organizations*; *Fundamentals of Composites Manufacturing*; *Introduction to Composites Technology*; *Manufacturing Engineering*, 10/year; Books; Papers; Reprints.

COMPUTER AND AUTOMATED SYSTEMS ASSOCIATION OF SOCIETY OF MANUFACTURING ENGINEERS (CASA/SME)

One SME Dr., PO Box 930, Dearborn, MI 48121-0930 **PHONE:** (313) 271-1500 **FX:** (313) 271-2861 **TX:** 297742 SME UR **WEBSITES:** http://www.sme.org/casa **FOUNDED:** 1975. **OFFICERS:** Sandra B. Marshall, Mgr. **MEMBERS:** 7,000. **STAFF:** 2. **LANGUAGES:** English.

DESCRIPTION: Sponsored by Society of Manufacturing Engineers. Members are automation implementation professionals, manufacturers, consultants, vendors, academics, and students in 35 countries. Promotes computer automation and enterprise integration for the advancement of research, design, installation, operation, maintenance, and communication in manufacturing. Acts as liaison between industry, government, and academia to identify areas that need further technological development; encourages companies to develop completely integrated manufacturing facilities. Conducts seminars and workshops on automated manufacturing. **PUBLICATIONS:** *CIM Implementation Guide*; *How to Implement Concurrent Engineering and Improve Your Time to Market*; *Management Guide to CIM*; *The New Manufacturing Enterprise Wheel*; *SME Blue Book Series*.

CONCERN WORLDWIDE (CW)

104 E 40th St., Rm. 903, New York, NY 10016 **PHONE:** (212) 557-8000 **FX:** (212) 557-8004 **E-MAIL:** info@concern-ny.org **WEBSITES:** http://www.concernusa.org **OFFICERS:** Siobhan Walsh, Exec.Dir. **LANGUAGES:** English.

DESCRIPTION: Members are individuals in the United States who provide volunteer support to relief and development organizations worldwide. Promotes more effective provision of relief to victims of disaster; seeks to ensure appropriate and sustainable economic development. Conducts programs in areas including water supply and sanitation, women's empowerment, health and nutrition, and education and training.

CONFEDERATION OF ASIAN-PACIFIC CHAMBERS OF COMMERCE AND INDUSTRY (CACCI)

7F, 3 Sungshou Rd., Taipei 110, Taiwan **PHONE:** 886 2 7255663, 886 2 7255664 **FX:** 886 2 7255665 **TX:** 11144 CTCOM **E-MAIL:** cacci@ms1.showtower.com.tw **FOUNDED:** 1966. **OFFICERS:** Lawrence T. Liu, Dir.Gen. **MEMBERS:** 20. **STAFF:** 10. **BUDGET:** US$350,000. **LANGUAGES:** English.

DESCRIPTION: Members are national and local chambers of commerce and individual businesses in 18 countries. Promotes economic cooperation and development. Encourages cooperation between chambers of commerce, businesses, and business people by promoting joint ventures, collaboration with other organizations, and by offering reciprocal benefits to business representatives visiting other member countries. Conducts research; provides training to member chambers of commerce; promotes laws and regulations to aid economic cooperation; compiles statistics. **PUBLICATIONS:** *CACCI Profile* (in English), monthly; *Journal of Commerce and Industry* (in English), semiannual.

CONFEDERATION OF THE FOOD AND DRINK INDUSTRIES OF THE EU (CIAA) (Vereinigung der Ernahrungsindustriens der EWG)

74, rue de la Loi, B-1040 Brussels, Belgium **PHONE:** 32 2 5141111 **FX:** 32 2 5112905 **FOUNDED:** 1959. **OFFICERS:** Raymond Destin, Gen.Dir. **MEMBERS:** 15. **STAFF:** 12. **LANGUAGES:** English, French.

DESCRIPTION: Members are food and drink federations in 15 West European countries comprising 14,000 enterprises and 2,500,000 individuals. Represents members' interests before the European Parliament, Council of Europe, and other institutions of the European Communities. Serves to coordinate food and drink industries when affected by European Union actions concerning food legislation, free movement of goods, external trade, and other issues of agriculture and trade. **PUBLICATIONS:** *CIAA Status Report: Food Law in the European Union* (in English), 3/year, always December, April, and August.

CONFEDERATION OF INTERNATIONAL TRADING HOUSES ASSOCIATIONS (CITHA) (Confederation des Associations d'Entreprises de Commerce International)

Adriaan Goekooplaan 5, NL-2517 JX The Hague, Netherlands **PHONE:** 31 70 3546811 **FX:** 31 70 3512777 **E-MAIL:** citha@verbondgroothandel.nl **FOUNDED:** 1955. **OFFICERS:** Dr. L. Antonini, Gen.Sec. **MEMBERS:** 9. **STAFF:** 2. **LANGUAGES:** English.

DESCRIPTION: Members are international trading house associations; enterprises directly engaged in international trade. Represents member interests internationally; examines problems affecting international trade. Fosters exchange of information in order to solve international problems and liberalize trade; serves as a coordinating body for national associations representing companies and individuals engaged in international trade. Works to justify the existence of free trade and its development; conducts symposia.

CONFEDERATION OF NORDIC BANK, FINANCE AND INSURANCE EMPLOYEES' UNIONS (NFU)

Birger Jarlsgatan 31, Box 7375, S-103 91 Stockholm, Sweden **PHONE:** 46 8 6140300 **FX:** 46 8 6113898 **E-MAIL:** lidstrom@nfufinance.org **WEBSITES:** http://www.nfufinance.org **FOUNDED:** 1999. **OFFICERS:** Jan-Erik Lidstrom, Gen.Sec. **MEMBERS:** 10. **STAFF:** 3. **BUDGET:** 3,900,000 SKr. **LANGUAGES:** Danish, Finnish, Icelandic, Norwegian, Swedish.

DESCRIPTION: Members are bank and insurance employees' trade unions of Denmark, Finland, Iceland, Norway, and Sweden representing 143,000 individuals. Promotes the interests of bank and insurance employees in Scandinavia. Conducts biennial training course for shop stewards. **PUBLICATIONS:** *Annual Report* (in Swedish); *Annual Statistical Report*; *Press Cuttings*, 4-5/year.

THE CONFERENCE BOARD (TCB)

845 3rd Ave., New York, NY 10022 **PHONE:** (212) 759-0900 **FX:** (212) 980-7014 **E-MAIL:** info@conference-board.org **WEBSITES:** http://www.conference-board.org **FOUNDED:** 1916. **OFFICERS:** Richard E. Cavanagh, Pres. & CEO. **MEMBERS:** 3,000. **STAFF:** 250. **LANGUAGES:** English.

DESCRIPTION: Members are corporations, government agencies, libraries, colleges, and universities. Fact-finding institution that conducts research and publishes studies on business economics and management experience. Holds more than 100 conferences, council meetings, and seminars per year in the U.S., Latin America, and Europe where members exchange ideas and keep abreast of business trends and developments. Makes research available to secondary schools, colleges, and universities at minimum cost. Disseminates research data to the public. **PUBLICATIONS:** *Across the Board*, 10/year; *Business Cycle Indicators*, monthly; *The Conference Board—Cumulative Index*, annual; *Consumer Confidence Survey*, monthly; *International Economic Scoreboard*, quarterly; *Regional Economies and Markets*, quarterly; *StraightTalk*, 10/year; *Work-Family Roundtable*, quarterly.

CONFERENCE OF EUROPEAN CHURCHES (CEC) (Konferenz Europaischer Kirchen - KEK)

150, rte. de Ferney, BP 2100, CH-1211 Geneva 2, Switzerland **PHONE:** 41 22 7916111 **FX:** 41 22 7916227 **TX:** 415730 OIK CH **E-MAIL:** reg@wcc_coe.org **FOUNDED:** 1959. **OFFICERS:** Rev. Dr. Keith W. Clements, Gen.Sec. **MEMBERS:** 125. **STAFF:** 21. **BUDGET:** 2,600,000 SFr. **LANGUAGES:** English, French, German, Russian.

DESCRIPTION: Ecumenical fellowship of Christian churches in Europe. Seeks mutual understanding among European churches; encourages cooperation among members to facilitate their service, as churches, to modern Europe. Activities include ecumenical studies, peace work, and defense of human rights. Conducts consultations on theology, human rights, justice and peace, and Christian service. **PUBLICATIONS:** *KEK Communique* (in English, French, and German), periodic; *List of Member Churches* (in English), annual; *Monitor* (in English, French, and German), quarterly; Monographs (in English, French, and German), periodic.

CONFERENCE OF EUROPEAN MINISTERS OF LABOUR (Conference des Ministres Europeens du Travail)

Coun. of Europe, BP 431 R6, F-67075 Strasbourg Cedex, France **PHONE:** 33 88 412331, 33 88 412162 **FX:** 33 88 412718 **TX:** 870943 F **FOUNDED:** 1972. **OFFICERS:** Sinon Tonelli. **MEMBERS:** 40. **LANGUAGES:** English, French.

DESCRIPTION: Members are labor ministers of Council of Europe member states. Maintains liaison with EEC, European Free Trade Association, International Labor Organization, and Organization for Economic Cooperation and Development.

CONGREGATION OF THE HOLY SPIRIT (CSSp)

Clivo di Cinna 195, I-00136 Rome, Italy **PHONE:** 39 0635 40461, 39 0635 404632 **FX:** 39 0635 450676 **E-MAIL:** cssp@rm.nettuno.it **FOUNDED:** 1703. **OFFICERS:** James Hurley CSSp, Gen.Sec. **MEMBERS:** 3,097. **LANGUAGES:** English, French, Portuguese.

DESCRIPTION: Members are catholic priests, brothers, and students in 54 countries. Carries out evangelical work, development activities, and missionary education among underprivileged individuals in Africa and Latin America. Operates charitable and training programs. **PUBLICATIONS:** *Directory of Personnel* (in English and French), triennial; *Esprit Saint* (in French), quarterly; *General Bulletin* (in English, French, and Portuguese), triennial; *Missionary Outlook* (in English), bimonthly; *Pentecote Sur le Monde* (in French), bimonthly; *Regional* (in English, German, and Portuguese); *Spiritan Missionary News* (in English), quarterly; *Spiritus* (in French), quarterly.

CONGREGATION OF THE MISSIONARIES OF MARIANHILL (CMM)

c/o Fr. Yves La Fontaine, Via S. Giovanni Eudes 91, I-00163 Rome, Italy **PHONE:** 39 6 66411909, 39 6 66419023 **FX:** 39 6 66414128 **E-MAIL:** mariannhill@alt.it **FOUNDED:** 1909. **OFFICERS:** Fr. Yves La Fontaine. **MEMBERS:** 400. **LANGUAGES:** English, German.

DESCRIPTION: Roman Catholic order of priests and lay brothers in 13 countries involved in evangelization. Conducts educational programs in Third World countries. **PUBLICATIONS:** *CMM News* (in English and German), bimonthly; *Family News* (in English and German), semiannual; *Leaves* (in English), bimonthly; *Mariannhill Missions Magazine* (in German), 8/year; *Revue Missionnaire Mariannhill* (in French), 5/year; *Status Membrorum CMM* (in English and German), triennial.

CONSERVATION FOUNDATION

1 Kensington Gore, London SW7 2AR, England **PHONE:** 44 171 5913111 **FX:** 44 171 5913110 **E-MAIL:** conservef@gn.apc.org **WEBSITES:** http://www.newsnet-21.org.uk **FOUNDED:** 1982. **OFFICERS:** David Shreeve. **STAFF:** 4. **LANGUAGES:** English.

DESCRIPTION: Creates and manages environmental programmes many of which are sponsored by commercial organizations. Covers all environmental interests in the United Kingdom, Europe, and internationally. Operates as a registered charity housed at the headquarters of the Royal Geographical Society. Produces monthly Information Diary for the United Kingdom's environmental journalists and programme makers. **PUBLICATIONS:** *Network 21* (in English), semiannual; Also produces videos and films.

CONSERVATION INTERNATIONAL (CI)

2501 M St., NW Ste. 200, Washington, DC 20037 **PHONE:** (202) 429-5660 **FX:** (202) 887-5188 **E-MAIL:** newmember@conservation.org **WEBSITES:** http://www.conservation.org **FOUNDED:** 1987. **OFFICERS:** Peter Seligmann, CEO & Chm. **MEMBERS:** 50,000. **STAFF:** 400. **BUDGET:** US$12,000,000. **LANGUAGES:** English, French, German, Italian, Portuguese, Spanish.

DESCRIPTION: Members are corporations and individuals in 15 countries interested in environmental protection and conservation. Cooperates with governments and other organizations to help all nations develop the ability to sustain biological diversity and the ecosystems that support life on earth while addressing basic economic and social needs. Has sponsored an agreement whereby a portion of Bolivia's debt to United States banks was forgiven in exchange for Bolivia's promise to protect a part of the Amazon rainforest; other project sites include Costa Rica and Mexico. Promotes scientific understanding through research programs; encourages and facilitates ecosystem management and conservation-based development. Conducts public outreach programs; assists in formulation and implementation of policy. Maintains small library.

CONSULTATIVE COUNCIL OF JEWISH ORGANIZATIONS (CCJO)

420 Lexington Ave., Ste. 1733, New York, NY 10170 **PHONE:** (212) 808-5437 **FX:** (212) 983-0094 **FOUNDED:** 1946. **OFFICERS:** Warren Green, Deputy Sec.Gen. **MEMBERS:** 3. **LANGUAGES:** English, French.

DESCRIPTION: Representatives from the Alliance Israelite Universelle in Paris, France, the Anglo-Jewish Association in London, England, and the Canadian Friends of the Alliance Israelite Universelle in Montreal, PQ, Canada. Promotes and advances human rights and freedom accomplished through cooperation with the United Nations and its specialized agencies. **PUBLICATIONS:** *Alliance Review*, annual. Covers activities included in the annual journal of the American Friends of the Alliance.

CONSULTATIVE GROUP ON INTERNATIONAL AGRICULTURAL RESEARCH (CGIAR)

1818 H St. NW, Washington, DC 20433 **PHONE:** (202) 473-8918 **FX:** (202) 473-8110 **E-MAIL:** cgiar@cgnet.com **WEBSITES:** http://www.cgiar.org **FOUNDED:** 1971. **OFFICERS:** Alexander Von Der Osten, Exec.Sec. **MEMBERS:** 40. **STAFF:** 12. **BUDGET:** US$300,000,000. **LANGUAGES:** English.

DESCRIPTION: Members are countries, multilateral development agencies, regional development banks, and private foundations. Supports network of 18 agricultural research centers purporting to improve the quality and quantity of food production and protect natural resources in developing countries. Supports research on critical aspects of food production in these countries that are not covered by other research facilities and that are of wide usefulness regionally or globally. Emphasizes development of new technologies suited to the resource-poor farmer; promotes collection and storage of germplasm in gene banks accessible to developing countries. Maintains a technical advisory committee comprising agricultural and social scientists which reviews scientific and technical aspects of all center programs and advises on emergent needs, priorities, and opportunities for research. **PUBLICATIONS:** *CGIAR Highlights*, quarterly; Annual Report. Also publishes policy booklets and informational brochures.

CONSUMERS INTERNATIONAL

24 Highbury Crescent, London N5 1RX, England **PHONE:** 44 171 2266663 **FX:** 44 171 3540607 **E-MAIL:** consint@consint.org.uk **WEBSITES:** http://www.consumersinternational.org **FOUNDED:** 1960. **OFFICERS:** Julian Edwards, Dir.Gen. **MEMBERS:** 260. **STAFF:** 70. **BUDGET:** 2,000,000 f. **LANGUAGES:** English, French, Spanish.

DESCRIPTION: Global association for consumers organizations from more than 100 countries. Objectives are to support the formation and development of strong and effective consumer organizations; foster international cooperation amongst consumer organizations and other supporting bodies; pursue an agenda for international action to protect consumers, especially the poor, the marginalised and the disadvantaged; and establish an authoritative and influential presence in global and regional policy making bodies. Produces publications including consumer-related policy papers, book, handbooks, and kits. Works on such issues as sustainable consumption, food and technical standards, health concerns and international trade. Works closely with the United Nations agencies and other international decision-making bodies. Helped found and works closely with Health Action International, Pesticides Action Network, and International Baby Food Action Network. **PUBLICATIONS:** *Consumer Currents* (in English); *Regional*; *World Consumer*, quarterly; *World Consumer Rights Day Kit*, annual, usually January. Published to celebrate World Consumer Rights Day on March 15th. Also publishes seminar and conference proceedings and position papers.

CONVENTION FOR BIODIVERSITY

15 chemin des Anemones, CP 356, CH-1219 Geneva, Switzerland **PHONE:** 41 22 9799111 **FX:** 41 22 7972512 **OFFICERS:** C. Juma, Exec.Sec. **LANGUAGES:** English, French, German, Italian.

DESCRIPTION: Scientists, researchers, and scientific and environmental organizations and institutions. Promotes maintenance of global biodiversity through protection of endangered habitats and species. Provides support and assistance to conservation and environmental protection programs.

CONVENTION ON CLIMATE CHANGE (UNFCCC)

PO Box 124, D-53153 Bonn, Germany **PHONE:** 49 228 8151000 **FX:** 49 228 8151999 **E-MAIL:** secretariat@unfccc.de **WEBSITES:** http://www.unfccc.de **FOUNDED:** 1996. **OFFICERS:** Nardos Assefa. **STAFF:** 80. **BUDGET:** US$12,000,000. **LANGUAGES:** Arabic, Chinese, English, French, German, Russian.

DESCRIPTION: Researchers, scientists, and organizations and agencies with an interest in global climatic change. Seeks to advance understanding of climatology and fluctuations in the earth's air temperature. Facilitates communication and cooperation among members; gathers and disseminates information; conducts research and educational programs. **PUBLICATIONS:** *Climate Change Bulletin* (in English, French, and Spanish), periodic.

COOPERATION COUNCIL FOR THE ARAB STATES ON THE GULF

PO Box 7153, Riyadh 11462, Saudi Arabia

PHONE: 1 4827777 **FX:** 403635 **FOUNDED:** 1981. **OFFICERS:** Sheikh Jamil Ibrahim Alhejailan, Sec. Gen. **LANGUAGES:** English.

DESCRIPTION: Members are Arab states on the Persian Gulf. Aims to promote coordination and cooperation among members in economic, social, and cultural affairs. Formulates regulations in fields including economic and financial affairs, commerce, customs, communications, education and culture, social and health affairs, tourism, and legislative and administrative affairs. Stimulates scientific and technological progress in industry, mineralogy, agriculture, and water and animal resources; encourages establishment of scientific research centers and projects. Seeks to consolidate the efforts and goals of member states in legal, economic, political, and environmental issues.

CORPORACION DE INVESTIGACIONES ECONOMICAS PARA LATINOAMERICA

Mac-Igr 125, 5° y 17 piso, Casilla 16496, Correo 9, Santiago, Chile **PHONE:** 56 2 6333836 **FX:** 56 2 6334411 **FOUNDED:** 1977. **OFFICERS:** Dagmarn Raczynski, Dir.

DESCRIPTION: Evaluates and investigates the economic stability of Latin American businesses and corporations. Works to forecast the productivity of the industrial and financial sectors of Latin American society. **PUBLICATIONS:** *Coleccion Estudios Cieplan* (in Spanish), 3/year.

COUNCIL OF ARAB ECONOMIC UNITY (CAEU)

1191 Corniche el-Nil, 12th Floor, POB 1, Mohammed Fareed, Cairo, Egypt **PHONE:** 2 5755321 **FX:** 2 5754090 **FOUNDED:** 1957. **MEMBERS:** 12. **LANGUAGES:** Arabic, English.

DESCRIPTION: Members are Arab countries. Aims to bring about the economic unification of the Arab world in the interest of the region's socioeconomic well-being and prosperity. Has created organizations, such as joint Arab companies and Arab specialized federations, and has initiated economic agreements between Arab nations. Coordinates national plans for economic development in an effort to design regional economic strategies. Supports the development of member countries' economic infrastructures, with particular emphasis on sectors, such as transportation, that play a role in regional integration. Encourages regional integration in agriculture and industry. Offers technical assistance to members. Conducts empirical research on economic unification. Compiles statistics intended to aid in planning and integration activities. Maintains Arab Centre for Statistics. Publishes books, foreign trade bulletins, demographic and statistical studies, and technical dictionaries. Cooperates with the League of Arab States and other regional and international organizations. **PUBLICATIONS:** *Annual Bibliography of the Library of the Council*; *Annual Bulletin* (in Arabic and English); *Journal of Arab Economic Unity* (in Arabic), semiannual; *Statistical Yearbook for Arab Countries* (in Arabic and English).

COUNCIL OF THE BARS AND LAW SOCIETIES OF THE EUROPEAN UNION (Conseil des Barreaux de l'Union - CCBE)

40, rue Washington, B-1050 Brussels, Belgium **PHONE:** 32 2 6404274, 32 2 6400931 **FX:** 32 2 6477941 **E-MAIL:** ccbe@ccbe.org **WEBSITES:** http://www.ccbe.org **FOUNDED:** 1960. **OFFICERS:** Valerie Bauer, Sec.Gen. **MEMBERS:** 18. **STAFF:** 8. **LANGUAGES:** English, French.

DESCRIPTION: Members are national bars of the countries of the European Union and eight observer states. Addresses questions affecting the legal profession and coordinates and harmonizes professional practice. Acts as liaison between members and between EU institutions and members. Addresses questions pertaining to consumer protection, education, lawyers' rights of establishment, professional conduct, new technology, and the rights of defense. Maintains Permanent Delegations to the European Court of Justice and to the Court of Human Rights. **PUBLICATIONS:** *Cross Border Handbook* (in English). Updates statements of practice in all member countries.

COUNCIL FOR THE DEVELOPMENT OF SOCIAL SCIENCE RESEARCH IN AFRICA

avenue Cheikh Anta Diop, Angle Canal IV, Boite Postale 3304, Dakar, Senegal **PHONE:** 221 259822 **FX:** 221 241289 **TX:** 61339 **LANGUAGES:** English, French.

DESCRIPTION: Members are social scientists and researchers in the social sciences. Seeks to advance the study, teaching, and practice of the social sciences. Conducts research and educational programs.

COUNCIL FOR EDUCATION IN THE COMMONWEALTH (CEC)

c/o College of Preceptors, Coppice Row, Theydon Bois, Essex CM16 7DN, England **PHONE:** 44 1992 812727 **FX:** 44 1992 814690 **E-MAIL:** preceptor@mailbox.ulcc.ac.uk **FOUNDED:** 1959. **OFFICERS:** Trevor Bottomley, Hon.Sec. **MEMBERS:** 300. **BUDGET:** £5,000. **LANGUAGES:** English.

DESCRIPTION: An organization based in the British parliament with academic institutions, corporations, nongovernmental organizations and interested individuals as supporters in 60 countries. Encourages discussion on problems in education and training in the Commonwealth, particularly in developing countries, and the role that Great Britain should play in the solution of these problems. Promotes educational cooperation; works to influence public opinion. Makes recommendations to the Commonwealth Secretariat, the Commission of the European Communities, the European Parliament, and other bodies. Initiates studies. **PUBLICATIONS:** *CEC Newsletter* (in English), annual.

COUNCIL OF EUROPE (CE) (Conseil de l'Europe)

c/o Dr. Caroline LE Tarnec, European Dept. for the Quality of Medicines, F-67075 Strasbourg Cedex, France **PHONE:** 33 3 88412815 **FX:** 33 3 88412771 **TX:** STRASBOURG 870943 F **E-MAIL:** pressunit@coe.fr **WEBSITES:** http://www.pheur.org **FOUNDED:** 1949. **OFFICERS:** David Tarschys, Sec.Gen. **MEMBERS:** 40. **STAFF:** 1,200. **BUDGET:** 1,100,000 Fr. **LANGUAGES:** English, French.

DESCRIPTION: Members are European states. Purpose is to foster greater unity and cooperation among the peoples and countries of Europe through their governments and parliaments. Works to safeguard and promote parliamentary democracy and human dignity throughout Europe. Seeks to create a society in which technological progress does not interfere with individual values and freedoms. Works to harmonize laws of member states through a series of agreements and conventions. Organizes activities concerning human rights, media, legal cooperation, local and regional authorities, environment, migrations, social affairs, public health, sports, youth, and culture. Functions through 2 main bodies: the Council of Ministers comprised of foreign ministers of member countries, and the Parliamentary Assembly consisting of representatives appointed by national parliaments. Maintains the European Commission of Human Rights and the Court of Human Rights to review complaints of alleged violations of the European Convention on Human Rights and Fundamental Freedoms and the European Social Charter, which provides for the harmonization of social policies. Administers the Council of Europe Social Development Fund which provides finances for social programs such as aid for refugees and migrant workers. Maintains Standing Conference of Local and Regional Authorities of Europe to familiarize locally elected representatives with the work of European construction; sponsors European Cultural Convention to encourage states to add a cultural dimension to the process of economic and social development in education, culture, and sports. Maintains the European Youth Centre, an international educational establishment and meeting place for young people in Strasbourg, France and the European Youth Foundation, which provides financial support for youth organizations' European activities. Acts to end drug use and trafficking within Europe; works to eliminate national bias from school textbooks. Sponsors European Schools Day competition. **PUBLICATIONS:** *Catalogue of Publications*, annual; *Crafts and Heritage*; *Cultural Policy* (in French), 3/year; *FORUM* (in English, French, German, and Italian), quarterly; *NATUROPA* (in English, French, German, and Italian), 3/year; Reports, periodic. On human rights, mass media, crime, population, the environment, law, equality between women and men, social policy, and youth.

COUNCIL OF EUROPEAN AND JAPANESE NATIONAL SHIPOWNERS' ASSOCIATIONS (CENSA)

12 Carthusian St., London EC1M 6EB, England **PHONE:** 44 171 6005405 **FX:** 44 171 6005398 **E-MAIL:** censa@marisec.org **FOUNDED:** 1956. **OFFICERS:** Robert T. Bishop, Sec.Gen. **MEMBERS:** 12. **STAFF:** 3. **LANGUAGES:** English.

DESCRIPTION: Members are national shipowners' associations in Denmark, United Kingdom, Finland, France, Germany, Greece, Italy, Japan, Netherlands, Norway, and Sweden. Purpose is to promote, develop, and protect sound policies in all sectors of shipping. Coordinates, represents, and facilitates the exchange of members' views among other shipowner groups and before European liner conferences and shippers' councils. Keeps members informed of legislative and policy developments around the world that may affect them. Serves as adviser and corresponding commercial body to member countries' governments. **PUBLICATIONS:** *CENSA Information Bulletin*, semiannual.

COUNCIL ON INTERNATIONAL EDUCATIONAL EXCHANGE (CIEE)

205 E. 42nd St., New York, NY 10017 **PHONE:** (212) 822-2600 **FX:** (212) 822-2699 **TX:** 423 227/ 6730395 **FOUNDED:** 1947. **OFFICERS:** Stevan Trooboff, Pres/CEO. **MEMBERS:** 326. **STAFF:** 700. **BUDGET:** US$275,000,000. **LANGUAGES:** English, French, Spanish.

DESCRIPTION: Dedicated to helping people gain understanding, acquire knowledge, and develop skills for living in a globally interdependent and culturally diverse world. Council Study Centers provide credit-bearing study abroad programs at 36 universities on six continents. Sponsors exchange programs between secondary school, operat programs for teachers of English, as well as homestay, language and touring programs, and campus-based ESL programs. Administer a series of bilateral student work exchange programs with 12 countries. Places U.S. participants in short-term international volunteer projects worldwide. Operates a retail travel network of over 50 offices worldwide that focus on student/youth market. **PUBLICATIONS:** *Student Travels*, semiannual. Contains information on foreign travel and study programs.; *Update*, semiannual. Issued to institutions and interested educators and administrators.; *Work Abroad*, annual. For students; Brochures, semiannual. Contain information on foreign travel and study programs. Travel, work, study abroad.; Directories. Contain information on foreign travel and study programs.

COUNCIL FOR INTERNATIONAL ORGANIZATIONS OF MEDICAL SCIENCES (CIOMS) (Conseil des Organisations Internationales des Sciences Medicales)

World Health Org., Ave. Appia, CH-1211 Geneva 27, Switzerland **PHONE:** 41 22 7913406, 41 22 7913467 **FX:** 41 22 7910746 **TX:** 415416 **FOUNDED:** 1949. **OFFICERS:** Dr. Zbigniew Bankowski, Sec. Gen. **MEMBERS:** 103. **STAFF:** 3. **LANGUAGES:** English, French.

DESCRIPTION: International organizations of medical sciences. Promotes and coordinates medical and scientific activities of member associations and national institutions affiliated with the council. Maintains collaborative relations with the World Health Organization and United Nations Educational, Scientific and Cultural Organization. Serves the scientific interests of the international biomedical community. **PUBLICATIONS:** *Calendar of International and Regional Congresses of Medical Sciences*, annual; *International Nomenclature of Diseases*; *Proceedings of Round Table Conferences*; Directory, periodic.

COUNCIL OF NORDIC TEACHERS' UNIONS (CNTU) (Nordiske Laererorganisationers Samrad - NLS)

c/o Lillemor Darinder, Vandkunsten 3, stuen, DK-1467 Copenhagen K, Denmark **PHONE:** 45 33141114 **FX:** 45 33142205 **FOUNDED:** 1968. **OFFICERS:** Lillemor Darinder, Sec.Gen. **MEMBERS:** 20. **STAFF:** 2. **BUDGET:** 3,000,000 SKr. **LANGUAGES:** Danish, English, Finnish, French, Norwegian, Swedish.

DESCRIPTION: Associations representing all 600,000 teachers in all the Nordic countries. Promotes cooperation between teachers in Nordic countries and internationally. Negotiates with Nordic governmental bodies. Assists education development in Africa.

COUNCIL OF NORDIC TRADE UNIONS (NFS) (Nordens Fackliga Samorganisation - NFS)

Barnhusgatan 16, S-111 23 Stockholm, Sweden **PHONE:** 46 8 209880 **FX:** 46 8 7898868 **FOUNDED:** 1972. **OFFICERS:** Bjorgulv Froyn, Gen.Sec. **MEMBERS:** 8,000,000. **STAFF:** 5. **LANGUAGES:** Danish, Finnish, Icelandic, Norwegian, Swedish.

DESCRIPTION: Members are trade unions in Nordic countries. Purpose is to defend the interests of members before employers and government agencies.

THE COUSTEAU SOCIETY (TCS)

870 Greenbrier Cir., Ste. 402, Chesapeake, VA 23320 **PHONE:** (757) 523-9335 **TF:** (800) 441-4395 **FX:** (757) 523-2747 **TX:** 6974570 **E-MAIL:** cousteau@infi.net **WEBSITES:** http://www.cousteau.org; http://www.dolphinlog.org **FOUNDED:** 1973. **OFFICERS:** Francine Cousteau, Pres. **MEMBERS:** 125,000. **STAFF:** 40. **BUDGET:** US$10,000,000. **LANGUAGES:** English, French.

DESCRIPTION: Environmental education organization dedicated to the protection and improvement of the quality of life for present and future generations. Objectives are education and evaluation of relationships between humans and nature. **PUBLICATIONS:** *Calypso Log* (in English and French), bimonthly. Covers the world's natural environment. Includes news of TCS activities and ecological issues.; *Dolphin Log*, bimonthly. For children ages seven to 15 covering all areas of science, history, and arts related to the world's water system; Books; Films.

DANISH REFUGEE COUNCIL (DRC) (Dansk Flygtningehjaelp)

Bledgamsvej 104. C, 2., DK-2100 Copenhagen, Denmark **PHONE:** 45 31 421811 **FX:** 45 31 423185 **TX:** 19581 REFUCO DK **E-MAIL:** drc@drc.dk **WEBSITES:** http://www.drc.dk **FOUNDED:** 1956. **OFFICERS:** Mr. Arne Piel Christensen, Gen.Sec. **MEMBERS:** 22. **STAFF:** 700. **LANGUAGES:** Danish, English.

DESCRIPTION: Humanitarian and community organizations in Denmark. Aims to increase awareness of the refugee problem worldwide. Raises funds for international projects; provides assistance to refugees. Conducts fact-finding missions; offers courses. **PUBLICATIONS:** *Arsberetning* (in Dutch), annual; *Flygtninge*, 5/year. Contains organization updates; *Flygtninge Tal*, periodic. Contains statistics. Also publishes *A Guide to DRC*; *An Asylum Seekers's Way Through the System*; *Current Asylum Policy and Humanitarian Principles*; *Role of Airline Companies in the Asylum Procedure*; and books, magazines, and reports.

DANUBE TOURIST COMMISSION (Die Donau)

Margaretenstrasse 1, A-1040 Vienna, Austria **PHONE:** 43 1 58866 **FX:** 43 1 5886620 **E-MAIL:** deutsch@oewwien.via.at **FOUNDED:** 1972. **OFFICERS:** Ursula Deutsch, Gen.Sec. **MEMBERS:** 7. **STAFF:** 3. **LANGUAGES:** English, German.

DESCRIPTION: Works to increase awareness of the Danube Region's tourism opportunities. Conducts tours of the area for journalists and a press service. **PUBLICATIONS:** *Press Service of the Danube Tourist Commission* (in English, French, German, and Italian), quarterly.

DEFENCE FOR CHILDREN INTERNATIONAL (DCI) (Defense des Enfants - International - DEI)

1, rue Varembe, Case Postale 88, CH-1211 Geneva 20, Switzerland **PHONE:** 41 22 7340558 **FX:** 41 22 7401145 **E-MAIL:** dci-hq@ping.ch **FOUNDED:** 1979. **OFFICERS:** Ricardo Dominice, Sec.Gen. **STAFF:** 10. **BUDGET:** US$1,000,000. **LANGUAGES:** English, French, Spanish.

DESCRIPTION: Individuals and groups in over 70 countries united to ensure ongoing international action directed toward promoting and protecting the rights of the child. Monitors and evaluates children's rights. Seeks to establish an international standard for child protection. Maintains consultative status with the Council of Europe, the United Nations Economic and Social Council, the United Nations Children's Fund UNESCO. Serves as referral center on comparative legislation. Responds to information requests and provides technical assistance. **PUBLICATIONS:** *International Children's Rights Monitor* (in English, French, and Spanish), quarterly.; Newsletter, bimonthly.; Reports, periodic.

DEMOCRAT YOUTH COMMUNITY OF EUROPE (DEMYC) (Union Juventudes Democratas Europeas)

c/o Arthur Winkler-Hermsden, Kaerntnerstrasse 51, A-1010 Vienna, Austria **PHONE:** 43 1 40126630 **FX:** 43 1 40126639 **TX:** 111771 UPPRA **FOUNDED:** 1964. **OFFICERS:** Matthias Peterlik, Sec.Gen. **MEMBERS:** 34. **STAFF:** 4. **LANGUAGES:** English, French, German.

DESCRIPTION: International political youth organization composed of youth factions from Christian-Democratic, liberal, and conservative parties in 31 European countries and Israel. Aims to strengthen ties and cooperation between members in working toward a united Europe. Encourages interest in and understanding of international affairs among young people. Sponsors training programs intended to familiarize members with international policy. Offers Annual summer school. **PUBLICATIONS:** *DEMYC Directory* (in English), biennial; *DEMYC-Info* (in English), monthly; *The European Democrat* (in English), quarterly. Also publishes policy papers and booklets.

DESERT LOCUST CONTROL ORGANIZATION FOR EASTERN AFRICA (DLCO-EA)

PO Box 4255, Addis Ababa, Ethiopia **PHONE:** 251 1 611475, 251 1 611464 **FX:** 251 1 611648 **TX:** 21510 DLCO ET **FOUNDED:** 1962. **OFFICERS:** Dr. A.H.M. Karrar, Dir. **MEMBERS:** 8. **STAFF:** 260. **BUDGET:** US$2,100,000. **LANGUAGES:** English, French.

DESCRIPTION: Members are ministries of agriculture of Djibouti, Eritrea, Ethiopia, Kenya, Somalia, Sudan, Tanzania, and Uganda. Promotes effective control of the desert locust in East Africa. Works to coordinate and reinforce national action against crop-damaging and disease-carrying pests including locusts, the tsetse fly, the Quelea bird, and armyworms. Conducts training in pest control spray techniques and pesticide analysis. **PUBLICATIONS:** *Newsletter DLCO-EA* (in English), quarterly; *Technical Report* (in English), periodic; Report (in English), annual.

DEUTSCH-ARMENISCHE GESELLSCHAFT (DAG)

c/o Josef Jaschke, Schilstr. 15, D-88718 Daisendorf, Germany **PHONE:** 49 75325442, 49 7531882412 **FX:** 49 7531883739 **E-MAIL:** josef.jaschke@uni-konstanz.de **WEBSITES:** http://www.deutsch-armenische-gesellschaft.de **FOUNDED:** 1914. **OFFICERS:** Elvira Kiendl, Sec. **MEMBERS:** 220. **STAFF:** 4. **BUDGET:** DM 25,000. **LANGUAGES:** Armenian, German.

DESCRIPTION: Promotes friendship and cultural exchange between Germany and the Armenian people. Works to improve mutual understanding between Germans and Armenians; conducts cultural programs, exhibits, and conferences. **PUBLICATIONS:** *Armenisch-Deutsche Korrespendenz* (in Armenian and German), quarterly. Contains news and information on politics, economy, religion, history, human rights, and book reviews.

DEVELOPING COUNTRIES FARM RADIO NETWORK (DCFRN) (Red de Radio Rural de los Paises en Desarrollo)

366 Adelaide St. W Ste. 706, Toronto, ON, Canada M5V 1R9 **PHONE:** (416) 971-6333 **FX:** (416) 971-5299 **E-MAIL:** dcfrn@web.net **FOUNDED:** 1979. **OFFICERS:** Nancy Bennett, Exec.Dir. **MEMBERS:** 1,300. **STAFF:** 7. **BUDGET:** C$400,000. **LANGUAGES:** English, French, Spanish.

DESCRIPTION: Rural broadcasters, writers, agricultural extension and health workers, missionaries, and teachers in 121 developing countries. Provides free broadcast material on simple, practical, low-cost techniques that any rural family can use to increase food supplies and improve their nutrition and health. **PUBLICATIONS:** *Farm Radio Network Broadcast Package* (in English, French, and Spanish), quarterly; *Voices/Voces/Echos* (in English, French, and Spanish), quarterly.

DEVELOPMENT BANK OF SOUTHERN AFRICA (DBSA)

PO Box 8604, Halfway House 1685, Republic of South Africa **LANGUAGES:** Afrikaans, English.

DESCRIPTION: Seeks to ensure availability of capital for the financing of economic and community development programs. Promotes use of local funds to finance development. Makes available financial and other assistance to development projects; serves as a clearinghouse on development finance in southern Africa.

DEVELOPMENT INNOVATIONS AND NETWORKS (Innovations et Reseaux pour le Developpement - IRED)

3, rue de Varembe, Case Postale 116, CH-1211 Geneva 20, Switzerland **PHONE:** 41 22 7341716 **FX:** 41 22 7400011 **E-MAIL:** ired@worldcom.ch **FOUNDED:** 1980. **OFFICERS:** Rudo Chitiga, Sec.Gen. **MEMBERS:** 225. **BUDGET:** US$2,000,000. **LANGUAGES:** English, French, Spanish.

DESCRIPTION: Members are nongovernmental development organizations. Seeks to establish local networks; encourages and fosters local development programs and activities; facilitates mutual technological support. Sponsors study tours and projects in self-management. **PUBLICATIONS:** *Alternative Financing* (in English and French), Manual intended for development agents and NGO leaders to help them better achieve economical and financial self sufficiency; *IRED Forum* (in English, French, and Spanish), quarterly; *Manual of Practical Management, Vol I & II*, Contains informtion for those with grass roots managers and people responsible for conducting, managing, and inspiring UGO Associations in the 3rd World.

DIRECT RELIEF INTERNATIONAL (DRI)

27 S. La Patera Ln., Santa Barbara, CA 93117 **PHONE:** (805) 964-4767 **FX:** (805) 681-4838 **E-MAIL:** info@directrelief.org **WEBSITES:** http://www.directrelief.org **FOUNDED:** 1948. **OFFICERS:** Max Goff, Exec.Dir. **STAFF:** 23. **BUDGET:** US$28,500,000. **LANGUAGES:** Spanish.

DESCRIPTION: Donates contributed pharmaceuticals, medical supplies, and equipment to health facilities and locally coordinated health projects in medically underdeveloped areas of the world. Provides emergency assistance to refugees and other victims of natural disaster and civil strife. **PUBLICATIONS:** *Presidents' Report*, annual, contains a Summary of year's accomplishments with financial report included; *Program Reports*, periodic; *Response*, 3/year.

DISCALCED BROTHERS OF THE MOST BLESSED VIRGIN MARY OF MOUNT CARMEL (Ordo Fratrum Excalceatorum Beatissimae Mariae Virginis de Monte Carmelo)

Corso d'Italia 38, I-00198 Rome, Italy **PHONE:** 39 6 854431 **FX:** 39 6 85350206 **FOUNDED:** 1568. **OFFICERS:** Fr. Camilo Maccise OCD, Superior Gen. **MEMBERS:** 3,765. **LANGUAGES:** English, French, German, Italian, Latin, Spanish.

DESCRIPTION: Members are catholic men in 58 countries. Promotes religious living. Activities include: parish and mission work; teaching; ministerial programs. Operates the Teresianum, an educational institution granting degrees in theology, spirituality, and theological anthropology. Conducts symposia. Maintains separate branches for women, the Discalced Carmelite Nuns, and for lay individuals, the Secular Order of Discalced Carmelites. **PUBLICATIONS:** *Acta Ordinis Carmelitarum Discalceatorum*, annual. Includes collection of official documents of the Order in the orginal language; *Comminicationes OCD*, bimonthly; *Conspectus OCD* (in Latin), periodic; *Servitium Informatium Carmelitanum (SIC)*, semiannual.

DIVINE WORD MISSIONARIES (DWM) (Societa del Verbo Divino - SVD)

Via dei Verbiti 1, I-00154 Rome, Italy **PHONE:** 39 6 5754021 **FX:** 39 6 5783031 **TX:** 5042014 CI VA **E-MAIL:** dwm@mwci.net **WEBSITES:** http://www.vais.net/-svd **FOUNDED:** 1875. **OFFICERS:** Very Rev. Heinrich Barlage SVD, Superior Gen. **MEMBERS:** 5,780. **LANGUAGES:** English, German, Spanish.

DESCRIPTION: Members are Roman Catholic priests and brothers in 60 countries. Promotes the evangelization of all peoples. Provides high school and university education. Maintains seminaries. **PUBLICATIONS:** *Anthropos* (in English and German), quarterly; *Catalogus*, annual; *Missionschronik* (in English, German, and Italian), annual.

DOCARE INTERNATIONAL (DI)

1750 NE 168th St., North Miami Beach, FL 33162-3021 **FOUNDED:** 1961. **OFFICERS:** Anslie M. Stark, Exec.Sec. **MEMBERS:** 150. **LANGUAGES:** English.

DESCRIPTION: Volunteer organization of medical doctors, osteopathic physicians, nurses, dentists, veterinarians, pharmacists, optometrists, podiatrists, and laypersons with special skills. Serves as a medical outreach program providing health care services to people in remote areas of Mexico, Central America, and the Caribbean. Is concerned with those deprived of medical care due to terrain, language, and cultural barriers. Conducts two to three one-week medical missions per year to areas in need until physicians or health care specialists are provided by the host country government. Has provided care to the Tarahumara Indians of northern Mexico, the Tepehuan Indians of Central Mexico, Mayan Indians in the Yucatan jungle, and an orphanage in Honduras. **PUBLICATIONS:** *DOCARE Flyer*, quarterly. Includes stories about members, meetings, planned missions, and election of officers.

DOCTORS WITHOUT BORDERS

6 East 39th St., 8th Floor, New York, NY 10016 **PHONE:** (212) 679-6800 **TF:** (888) 392-0392 **FX:** (212) 679-7016 **E-MAIL:** doctors@newyork.msf.org **WEBSITES:** http://www.dwb.org **FOUNDED:** 1971. **OFFICERS:** Joelle Tanguy, Exec.Dir. **STAFF:** 10. **BUDGET:** US$6,000,000. **LANGUAGES:** English.

DESCRIPTION: Members are medical and non-medical professionals. Provides assistance to victims of war, natural and man-made disasters, and epidemics, and to others who lack access to health care. Each year more than 2,000 volunteers provide relief in more than 80 countries. **PUBLICATIONS:** *Alert*, 3/year. Contains articles on the fields where Doctors Without Borders are active; *Populations in Dangers*, annual. Contains a look at 5 people in crisis and the humanitarian response.

DOCTORS OF THE WORLD (DW)

375 W Broadway, 4th Fl., New York, NY 10012-4324 **PHONE:** (212) 529-1556 **FX:** (212) 529-1571 **E-MAIL:** dow@interserv.com **OFFICERS:** Derick G. Wong, Exec.Dir. **LANGUAGES:** English.

DESCRIPTION: Members are physicians and other health care professionals. Seeks to increase the availability and quality of health services in developing regions worldwide. Recruits and trains volunteers from the United States to provide professional assistance to local health care services in developing areas.

DOUBLE HARVEST (DH)

556 Jeffries Rd., Fletcher, NC 28732 **PHONE:** (704) 891-4116 **FX:** (704) 891-8581 **E-MAIL:** doubleharvst@aol.com **OFFICERS:** H. J. Lemkes, Admin. **LANGUAGES:** English.

DESCRIPTION: Promotes increased agricultural production and reforestation in developing regions worldwide. Provides land, equipment, and technical assistance to farms and nurseries; conducts training in water resources management and soil conservation.

EARTHACTION INTERNATIONAL (EI)

30 Cottage St., Amherst, MA 01002 **PHONE:** (413) 549-8118 **FX:** (413) 549-0544 **E-MAIL:** amherst@earthaction.org **WEBSITES:** http://www.earthaction.org **FOUNDED:** 1988. **OFFICERS:** Lois Barber, Intl. Coord. **MEMBERS:** 1,700. **STAFF:** 6. **BUDGET:** US$800,000. **LANGUAGES:** English, French, Spanish.

DESCRIPTION: Members are individuals and organizations in 151 countries. Seeks to mobilize international public pressure on key decisionmakers when important global decisions are being made. **PUBLICATIONS:** *EarthAction Alert* (in English, French, and Spanish), monthly. Each Action Kit focuses on one global environment, development, peace or human rights issue; *Editorial Advisories*; *Parliamentary Alerts*.

EARTHSAVE INTERNATIONAL

600 Distillery Commons, Ste. 200, Louisville, KY 40206-1922 **E-MAIL:** earthsave@aol.com **WEBSITES:** http://www.earthsave.org **FOUNDED:** 1988. **OFFICERS:** Patricia Carney, Exec.Dir. **MEMBERS:** 7,000. **STAFF:** 20. **BUDGET:** US$1,500,000. **LANGUAGES:** English.

DESCRIPTION: EarthSave promotes food choices that are healthy for the planet. We educate, inspire and empower people to take positive action for all life on Earth. **PUBLICATIONS:** *EarthSave*, quarterly. 16 page full color; *EarthSave Educational Series*, periodic; *Healthy School Lunch Action Guide*; *Our Food Our World: The Realities of an Animal-Based Diet*; *Youth for Environmental Sanity: Youth Action Guide*.

EAST ASIA TRAVEL ASSOCIATION (EATA) (Higashi Asia Kanko Kyokai)

Japan Natl. Tourist Orgn., 2-10-1, Yurakucho, Chiyoda-ku, Tokyo 100, Japan **PHONE:** 81 3 32162910 **FX:** 81 3 32147680 **FOUNDED:** 1966. **OFFICERS:** Mr. Ichiro Tanaka, Sec.Gen. **MEMBERS:** 7. **STAFF:** 2. **BUDGET:** US$648,000. **LANGUAGES:** English.

DESCRIPTION: National government tourist offices, airlines, and travel agent and hotel associations in 6 countries. Promotes tourism in East Asia; facilitates tourist travel to the region. Holds chapters in 9 regions.

EASTERN AFRICA ASSOCIATION (EAA)

2 Vincent St., London SW1P 4LD, England
PHONE: 44 171 8285511 **FX:** 44 171 8285251
E-MAIL: cmb@westafricacom.demon.co.uk
FOUNDED: 1964. **OFFICERS:** J.A. Wood, Chief Exec. **MEMBERS:** 160. **STAFF:** 4. **BUDGET:** £130,000. **LANGUAGES:** English.

DESCRIPTION: Members are companies from Australia, Belgium, Denmark, France, Germany, India, Italy, Netherlands, South Africa, Switzerland, Sweden, Hong Kong, the United Kingdom, and the United States with business interests in Eritrea, Ethiopia, Kenya, Madagascar, Mauritius, Seychelles, Tanzania and Uganda. Interests include mining, manufacturing, agriculture, shipping, transportation, engineering, petroleum, publishing, insurance, and banking. Encourages foreign participation in the economic development of Eastern Africa. Works for the mutual benefit of Eastern African countries and foreign investors. Represents members interests before governments; maintains a network of contacts; disseminates information. information. **PUBLICATIONS:** *Eastern Africa Association Newsletter* (in English), periodic.

EASTERN REGIONAL ORGANIZATION FOR PUBLIC ADMINISTRATION (EROPA) (Organizacion Regional del Oriente para la Administracion Publica - OROAP)

Univ. Of the Philippines, Coll. of Public Administration, Diliman, Quezon City, Metro Manila, Philippines **PHONE:** 63 2 9285411, 63 2 9279085 **FX:** 63 2 9283861 **FOUNDED:** 1960. **OFFICERS:** Patricia A. Sto. Tomas, Sec.Gen. **MEMBERS:** 485. **STAFF:** 8. **BUDGET:** US$45,000. **LANGUAGES:** English.

DESCRIPTION: Organizations, individuals, and nations in the Asia-Pacific region united to improve public administration in order to advance the economic and social development of the region. Seeks to develop public appreciation of the importance of public administration. Promotes adoption of more effective administrative systems. Fosters managerial talent at the executive and middle management levels; cooperates with international public administration organizations. Operates the EROPA Development Management Centre in Seoul, Republic of Korea, the EROPA Local Government Centre in Tokyo, Japan, and the EROPA Training Centre in New Delhi, India. Offers training programs including the Asian Foreign Service Program, the Executive Policy Development Course, and local government courses. Administers the EROPA Educational/Development Fund, which finances scholarships, research projects, and related activities. **PUBLICATIONS:** *Administrative Reform towards Promoting Productivity in Bureaucratic Performance (1991)* (in English); *Asian Review of Public Administration* (in English), semiannual; *EROPA Bulletin* (in English), quarterly; *New Trends in Public Administration for the Asia Pacific Region: Decentralization*; *Public Administration and Sustainable Development (1994)*; *16th General Assembly and 42nd Executive Council Meeting, Tokyo, Japan*; Monographs; Reports.

EASTERN AND SOUTHERN AFRICA TRADE AND DEVELOPMENT BANK

Boite Postale 1750, Bujumbura, Burundi
PHONE: 257 22 5432 **FX:** 257 22 4983 **TX:** 5142
OFFICERS: Martin Ogang, Pres. **LANGUAGES:** English, French.

DESCRIPTION: Seeks to insure sound financial backing for trade and development organizations and projects in eastern and southern Africa. Makes available financial assistance and banking services to development and trade organizations.

EASTERN AND SOUTHERN AFRICAN MINERAL RESOURCES DEVELOPMENT CENTRE

PO Box 9573, Dar es Salaam, United Republic of Tanzania **PHONE:** 255 51 650321, 255 51 650347 **FX:** 255 51 650319 **TX:** 41401 **E-MAIL:** geodesa@cats-net.com **FOUNDED:** 1977. **OFFICERS:** Antonio A. Pedro, Dir.Gen. **MEMBERS:** 6. **STAFF:** 26. **BUDGET:** US$543,000. **LANGUAGES:** English.

DESCRIPTION: Geologists, geophysicists, and other individuals, businesses, and organizations with an interest in mining and minerals. Promotes profitable and sustainable extraction of the mineral resources of eastern and southern Africa. Seeks to develop more efficient methods of identifying and extracting minerals. Conducts research; facilitates communication and cooperation among mineral interests in the region. **PUBLICATIONS:** *Technical Reports* (in English), semiannual; Annual Report.

ECMA

114 Rue du Rhone, CH-1204 Geneva, Switzerland **PHONE:** 41 22 8496000 **FX:** 41 22 8496001 **E-MAIL:** webmaster@ecma.ch **WEBSITES:** http://www.ecma.ch **FOUNDED:** 1961. **OFFICERS:** Jan Van Den Beld, Sec.Gen. **MEMBERS:** 46. **STAFF:** 6. **BUDGET:** 3,000,000 SFr. **LANGUAGES:** Dutch, French, German, Italian.

DESCRIPTION: Members are manufacturers of computers and telecommunications products and services. Facilitates communication and cooperation among members; monitors international developments in the computer and telecommunications industry and trade regulations; gathers and disseminates information. **PUBLICATIONS:** *ECMA Brochure*; *ECMA Momento*; *Standards and Technical Reports* (in English), biennial.

ECOBUSINESS CENTRE (EBC)

1 Bulgaria Sq. Conference Centre, National Palace of Culture, Administration Bldg., 9th Fl., Rm. 10, BG-1414 Sofia, Bulgaria **PHONE:** 359 2 91662774, 359 2 91666688 **FX:** 359 2 657053 **E-MAIL:** drumev@sf.icn.bg **FOUNDED:** 1991. **OFFICERS:** Dobrin Pamukov, Chair. **MEMBERS:** 24. **STAFF:** 3. **BUDGET:** US$12,600. **LANGUAGES:** Bulgarian, English, French, German, Italian, Russian.

DESCRIPTION: Businesses with an interest in environmental protection. Promotes development of environmentally friendly businesses. Conducts public relations campaigns to raise public awareness of business and environmental issues; publicizes environmentally friendly products. Provides services to members including provision of contacts and consulting; gathers and disseminates information on ecobusiness and related issues; coordinates members' activities. Sponsors research programs. **PUBLICATIONS:** *Ecobusiness*, bimonthly.

ECOLES INTERNATIONAL NGO (EINGO)

Darkhan Village, 702017 Tashkent, Uzbekistan **PHONE:** 7 3712 246039, 7 3712 246063 **FX:** 7 3712 214722 **FOUNDED:** 1996. **OFFICERS:** A. Kayimov **MEMBERS:** 11. **LANGUAGES:** English, Uzbekistani.

DESCRIPTION: Members are industries, scientific and research institutes, social welfare organizations, and individuals. Promotes rehabilitation of ecosystems negatively impacted by human activities. Establishes green areas in degraded ecosystems; sponsors research and educational programs.

ECOLOGIC DEVELOPMENT FUND

c/o Shaun Paul, 1692 Massachusetts Ave., Cambridge, MA 02138-1803 **PHONE:** (617) 441-6300 **FX:** (617) 441-6307 **E-MAIL:** enews@ecologic.org **WEBSITES:** http://www.ecologic.org **FOUNDED:** 1993. **OFFICERS:** Shaun Paul, Dir. **STAFF:** 7. **LANGUAGES:** English.

DESCRIPTION: Works to preserve the biodiversity of tropical ecosystems and promote the well-being of threatened local habitats through small-scale, community-based development. Supports the productive use of local resources to meet local needs. Encourages communities to identify and solve their ecological and economic problems in ways that respect both their cultural integrity and the natural limits of their ecosystem. Works with communities to design programs that allow the communities to retain or reclaim traditional knowledge while providing access to resources and information that will enable them to improve their living conditions. Conducts research. Maintains speakers' bureau **PUBLICATIONS:** Newsletter, quarterly.

ECONOMIC COMMISSION FOR LATIN AMERICA AND THE CARIBBEAN (ECLAC) (Comision Economica para America Latina y el Caribe - CEPAL)

Casilla 179-D, Santiago, Chile **PHONE:** 56 2 2102000 **FX:** 56 2 2080252 **TX:** 441054 UNSGO CZ **FOUNDED:** 1948. **OFFICERS:** Gert Rosenthal, Exec.Sec. **MEMBERS:** 47. **STAFF:** 600. **LANGUAGES:** English, French, Spanish.

DESCRIPTION: A regional agency of the United Nations. Contributes to the economic and social development of the Latin American and Caribbean regions. Conducts studies and programs, in collaboration with governments of the region, in the fields of foreign trade, agricultural production, industrial development, transport and communications, regional integration, statistics, natural resources, and the environment. Sponsors social development activities such as population censuses, development of information media, child and youth programs, and urbanization efforts. Provides countries with technical assistance, human resources, training, and services enabling them to use sources of compiled and processed data on economic and social development. Sponsors the Latin American Institute for Economic and Social Planning and the Latin American Demographic Center. **PUBLICATIONS:** *CEPAL Review* (in English and Spanish), 3/year; *CEPALINDEX* (in Spanish), annual; *Cooperation and Development* (in English and Spanish), quarterly; *Demographic Bulletin* (in English and Spanish), semiannual; *Economic Survey of Latin America* (in English and Spanish), annual; *FAL Bulletin* (in English and Spanish), bimonthly; *Latin American Population Abstracts* (in English and Spanish), semiannual; *Micronoticias* (in Spanish), weekly; *Notas Sobre la Economia y el Desarrollo de America Latina* (in English and Spanish), biweekly; *PLANINDEX* (in Spanish), semiannual; *Population Notes* (in English and Spanish), 3/year; *Statistical Yearbook for Latin America* (in English and Spanish), annual; *Terminology Newsletter* (in English and Spanish), 3/year.

ECONOMIC COMMUNITY OF CENTRAL AFRICAN STATES (ECCAS) (Communaute Economique des Etats de l'Afrique Centrale - CEEAC)

BP 2112, Libreville, Gabon **PHONE:** 241 733547, 241 733677 **TX:** 5780 GO **FOUNDED:** 1983. **OFFICERS:** Mr. Kasasa Cinyanta Mutati. **MEMBERS:** 10. **STAFF:** 80. **BUDGET:** US$3,500,000. **LANGUAGES:** English, French, Portuguese, Spanish.

DESCRIPTION: Promotes economic cooperation among Burundi, Cameroon, Central African Republic, Chad, Equatorial Guinea, Gabon, Republic of Congo, Rwanda, Sao Tome and Principe, and Zaire. Works to ease movement of people and goods within the region. Seeks to reduce trade restrictions and encourages establishment of common customs tariffs.

ECONOMIC COMMUNITY OF WEST AFRICAN STATES (ECOWAS) (Communaute Economique des Etats de l'Afrique de l'Ouest - CEDEAO)

Secretariat Bldg., No. 60, Yakubu Gowon Crescent, Asokoro District, Abuja, Nigeria **PHONE:** 234 9 5231858 **FX:** 234 9 2347648 **E-MAIL:** ecosummit@hotmail.com **WEBSITES:** http://www.cedeao.org; http://www.ecowas.net **FOUNDED:** 1975. **OFFICERS:** Mr. Lansana Kouyate, Exec.Sec. **MEMBERS:** 16. **BUDGET:** US$7,521,000. **LANGUAGES:** English, French.

DESCRIPTION: Representatives of the governments of 16 West African states. Works toward economic integration and cooperation among West African states. Compiles statistics; conducts workshops on telecommunications and colloquia on economic integration. Sponsors competitions. **PUBLICATIONS:** *ECOWAS Economic and Social Indicators* (in English and French), periodic; *ECOWAS Economic Profile Series* (in English and French), periodic; *ECOWAS News* (in English and French), periodic; *ECOWAS Policies and Programme Series* (in English and French), periodic; *Official Journal of the Economic Community of West African States* (in English and French), semiannual; *Statistics of Intra-ECOWAS Trade* (in English and French), periodic; *Yearbook of International Trade Statistics of ECOWAS* (in English and French).

ECONOMIC AND SOCIAL COMMISSION FOR WESTERN ASIA (ESCWA)

Riad el-Solh Square, POB 11-8575, Beirut, Lebanon **PHONE:** 1 981301 **FX:** 1 981510 **TX:** 216917 **FOUNDED:** 1974. **OFFICERS:** Hazem el-Belwabi, Exec. Sec. **MEMBERS:** 13. **BUDGET:** US$24,750,000.

DESCRIPTION: A regional council of the United Nations (UN), ESCWA's members are UN member states from the Middle East. Palestine is a member of ESCWA, where it is represented by the Palestinian Liberation Organization, but Israel is not. ESCWA's purpose is to foster economic and social development in its member states. It conducts research into issues of concern to its members and develops programs to address these concerns. ESCWA often works with other United Nations organizations and with international and regional organizations in Western Asia. **PUBLICATIONS:** *External Trade Bulletin*, annual; *National Accounts Studies*, annual; *Population Bulletin*, biennial; *Review of Monetary and Banking Policy in ESCWA Region*, annual; *Statistical Abstract of ESCWA Region*, annual; *Survey and Assessment of Energy-*

Related Activities and Development in the ESCWA Region, annual; *Survey of Economic and Social Developments in the ESCWA Region*, annual.

ECONOMIC AND SOCIAL RESEARCH FOUNDATION (ESRF)

PO Box 31226, Dar es Salaam, United Republic of Tanzania **PHONE:** 255 51 75115 **FX:** 255 51 34723 **OFFICERS:** Sam Wangwe, Exec.Dir. **LANGUAGES:** English.

DESCRIPTION: Economists and other social scientists; educational institutions. Seeks to advance the study, teaching, and practice of the social sciences. Facilitates communication and cooperation among members. Conducts research and educational programs.

ECUMENICAL DEVELOPMENT COOPERATIVE SOCIETY (EDCS) (Societe Cooperative Ecumenique de Developpement - SCOD)

P.C. Hooftlaan 3, NL-3818 HG Amersfoort, Netherlands **PHONE:** 31 33 4224040 **FX:** 31 33 4650336 **E-MAIL:** office@edcs.nl **FOUNDED:** 1975. **OFFICERS:** Gert van Maanen, Gen.Mgr. **MEMBERS:** 392. **STAFF:** 30. **LANGUAGES:** Dutch, English, French, German, Spanish.

DESCRIPTION: National church, church-related, and voluntary support associations in 80 countries. Provides loans and capital to underprivileged people for enterprise development. Seeks to create jobs and income, particularly in developing countries, through self-supporting and viable economic activity. **PUBLICATIONS:** *Annual Report* (in Dutch, English, French, German, and Spanish).; Newsletter (in Dutch, English, French, German, and Spanish), biennial.

EDUCATION INTERNATIONAL (Internationale de l'Education)

155, Blvd. Emile-Jacqmain, (8), B-1210 Brussels, Belgium **PHONE:** 32 2 2240611 **FX:** 32 2 2240606 **E-MAIL:** educint@ei-ie.org **WEBSITES:** http://www.ei-ie.org **FOUNDED:** 1993. **OFFICERS:** Fred van Leeuwen, Gen.Sec. **MEMBERS:** 284. **STAFF:** 30. **LANGUAGES:** English, French, Spanish.

DESCRIPTION: National teachers' organizations representing 23,000,000 individuals in 149 countries. Promotes international understanding and goodwill through education. Works to improve teaching methods and professional training of teachers; defends the rights of the teaching profession and encourages closer relationships between teachers in different countries. Conducts leadership training programs. **PUBLICATIONS:** *The Education International Quarterly Magazine* (in English, French, and Spanish). Contains articles and dossiers on issues of interest to members; *EI Monthly Monitor* (in English, French, and Spanish).

END CHILD PROSTITUTION, PORNOGRAPHY AND TRAFFICKING (ECPAT 'UK') (Thomas Clarkson House)

Stableyard, Broomgrove Rd., London SW9 9TL, England **PHONE:** 44 71 9249555 **FX:** 44 71 7384110 **E-MAIL:** antislavery@gn.apc.org **FOUNDED:** 1994. **OFFICERS:** Helen Veitch, Campaign Coor. **MEMBERS:** 400. **STAFF:** 2.

DESCRIPTION: Individuals and organizations. Seeks to erradicate the commercial sexual exploitation of children worldwide, and to end "child sex tourism." Coordinates activities of British organizations working to end child prostitution; conducts educational programs to raise public awareness of sexual tourism; lobbies government agencies to insure British compliance with Article 34 of the United Nations Convention on the Rights of the Child. Works with governments worldwide to improve enforcement of laws banning the sexual exploitation of children. **PUBLICATIONS:** *ECPAT News*, bimonthly; Brochure.; Student Pack.; Resource Pack.; Youth Work Pack.

ENGLISH SPEAKING UNION OF THE COMMONWEALTH (ESU)

37 Charles St., Dartmouth House, London W1X 8AB, England **PHONE:** 44 171 4933328 **FX:** 44 171 4956108 **E-MAIL:** esu@esu.org **WEBSITES:** http://www.esu.org **FOUNDED:** 1918. **OFFICERS:** Mrs. Valerie Mitchell, Dir.Gen. **MEMBERS:** 6,000. **STAFF:** 29. **BUDGET:** £1,770,000. **LANGUAGES:** English.

DESCRIPTION: Companies, schools, and individuals. Fosters international friendship and understanding through the use of English as a global language. Offers scholarships for the study of English literature, music, and business. Sponsors research projects and study programs for teachers and individuals in business. Organizes foreign exchange programs, including an Intern Exchange, in which participants work in the British House of Commons or the United States Congress. Sponsors English language public speaking and debating competitions. Hold educational and literary events. literary events. **PUBLICATIONS:** *Ambassador Booklist*, annual; *Concord* (in English), semiannual. Offers forum for ideas and information on the world of the ESU; *ESU Newsletter* (in English), monthly; *International Directory* (in English), annual.

ENTENTE COUNCIL (EC) (Conseil de l'Entente - CE)

BP 3734, Abidjan 01, Cote d'Ivoire **PHONE:** 225 331001, 225 332835 **FX:** 225 331149 **TX:** 23558 **FOUNDED:** 1959. **OFFICERS:** Paul Kouame, Sec.Gen. **MEMBERS:** 5. **LANGUAGES:** French.

DESCRIPTION: Heads of state of Benin, Burkina Faso, Cote d'Ivoire, Niger, and Togo. Administers the Mutual Aid and Loan Guaranty Fund to insure and grant loans for projects involving agriculture, commerce, industry, infrastructure, and tourism intended to foster economic development in the member states.

ENTRAIDE MISSIONAIRE

15, De Castelnau Ouest, Montreal, PQ, Canada H2R 2W3 **PHONE:** (514) 270-6089 **FX:** (514) 270-6156 **E-MAIL:** emi@web.net **WEBSITES:** http://www.web.net/-emi **FOUNDED:** 1957. **OFFICERS:** Gerardo Aiquel. **MEMBERS:** 120. **STAFF:** 5. **BUDGET:** C$350,000. **LANGUAGES:** English, French.

DESCRIPTION: Members are religious communities and organizations. Promotes respect for human rights and the rule of law in Africa and Latin America. Conducts educational programs to raise public awareness of global human rights issues. **PUBLICATIONS:** *EMI en Bref*, periodic; *Info-Zaire*, periodic.

ENVIRONMENT LIAISON CENTRE INTERNATIONAL (ELCI)

Ndemi Rd., off Ngong Rd., PO Box 72461, Nairobi, Kenya **PHONE:** 254 2 576114, 254 2 562022 **FX:** 254 2 562175 **TX:** 23240 ELCKE **E-MAIL:** library@elci.sasa.unon.org **FOUNDED:** 1974. **OFFICERS:** Rob Sinclair, Exec. Dir. **MEMBERS:** 800. **STAFF:** 25. **BUDGET:** 1,500,000 KSh. **LANGUAGES:** English, French, Spanish.

DESCRIPTION: An information and communication network of more than 850 non-governmental and community-based organisations (NGOs and CBOs) in over 100 countries. Strives for a more sustainable world; members share their information and ideas, learn from each other's experiences and contribute to a growing environmental movement worldwide. **PUBLICATIONS:** *Development Classified*; *ECOFORUM*, periodic; *Grassroots Reflection on Agenda 21*; *Where There is no Librarian*.

ENVIRONMENTAL LAW ALLIANCE WORLDWIDE (ELAW)

1877 Garden Ave., Eugene, OR 97403 **PHONE:** (541) 687-8454 **FX:** (541) 687-0535 **E-MAIL:** elawus@elaw.org **WEBSITES:** http://www.elaw.org **FOUNDED:** 1989. **OFFICERS:** Bern Johnson, Exec.Dir. **STAFF:** 9. **BUDGET:** US$622,000. **LANGUAGES:** English, French, Russian, Spanish.

DESCRIPTION: Global alliance of public interest attorneys and scientists in 50 countries who defend the environment through law. Gives advocates around the world access to the lessons and resources of U.S. efforts to protect the environment including regulatory models, case precedents from U.S. courts, U.S. legal scholarship and jurisprudence, and information about multinational corporations. **PUBLICATIONS:** *E-LAW Update*, quarterly.

EURAIL COMMUNITY (EC) (Eurail-Gemeinschaft)

N.V. Nederlandse Spoorwegen, Postbus 2025, NL-3500 HA Utrecht, Netherlands **PHONE:** 31 30 357077 **FX:** 31 30 319621 **E-MAIL:** eurail@ns.nl **FOUNDED:** 1959. **MEMBERS:** 30. **STAFF:** 4. **LANGUAGES:** Dutch, English, French, German.

DESCRIPTION: Members are rail and shipping companies in Europe that participate in the sale of Eurail passes and Eurail ticket. (A Eurailpass or Eurailticket allows the purchaser to travel on European trains and ships within a specified period of time, for a pre-paid set price.) Goals are to: promote the sale of Eurailpasses and Eurailtickets outside Europe; create transportation specially adapted to the needs of non-Europeans. Compiles statistics. **PUBLICATIONS:** *Eurail Time Table* (in English, French, Japanese, Portuguese, and Spanish), annual; *Eurail Traveler's Guide/Map* (in Chinese, English, French, Hebrew, Indonesian, Japanese, Korean, Portuguese, Spanish, and Thai), annual.

EUREKA

rue Neerveld, 107, B-1200 Brussels, Belgium **PHONE:** 32 2 7770950 **FX:** 32 2 7707495 **E-MAIL:** eureka.secretariat@es.eureka.be **WEBSITES:** http://www.eureka.be **OFFICERS:** L.J.A.M. Van Den Bergen, Head of the Eureka Secretariat. **MEMBERS:** 25. **LANGUAGES:** English, French, German, Italian, Spanish.

DESCRIPTION: Members are European businesses, institutes, and organizations with an interest in market-oriented research and development. Facilitates collaborative and market-driven European research and development inititatives. Coordinates members' activities. **PUBLICATIONS:** *Eureka News* (in English, French, German, Italian, and Spanish), quarterly.

EURO INSTITUTE

8, rue President Carnot, F-69002 Lyon, France **PHONE:** 33 72 564232 **FX:** 33 72 418491 **TX:** 33049 **E-MAIL:** euroinstitut@asi.fr **WEBSITES:** http://www.euro-institut.org **FOUNDED:** 1992. **OFFICERS:** Frans Andriessen, Pres. **STAFF:** 15. **LANGUAGES:** English, French.

DESCRIPTION: Promotes the study, analysis, and evaluation of the economic and legal ramifications of adopting a single monetary unit in the European Community. Fosters communication between European international and national institutions and international experts. Compiles statistics; disseminates information. **PUBLICATIONS:** *Banking Supervision in the European Community*; *Competition Between Major International Currencies and the Future Role of the Single Currency*; *Euro: La Revue de la Monnaie Unique* (in French), bimonthly; *The Legal and Practical Guide to the Use of the ECU*, semiannual. Contain study and survey results.; Reports. Contain study and survey results.

EURONAID

Houtweg 60, Postbus 12, NL-2501 CA The Hague, Netherlands **PHONE:** 31 70 3305757 **FX:** 31 70 3641701 **TX:** 30960 **E-MAIL:** euronaid@euronaid.nl **FOUNDED:** 1980. **OFFICERS:** Bernd Dreesmann, Sec.Gen. **MEMBERS:** 29. **STAFF:** 22. **LANGUAGES:** Dutch, English, French, German, Spanish.

DESCRIPTION: European association of nongovernmental organisations for food aid and food security. Facilitates access of NGOs to institutional donors, (mainly the European Commission). Strives to ensure that NGOs of different sizes and countries of the European Union may benefit from services, thereby increasing opportunities for support of food aid and food security programmes. **PUBLICATIONS:** *Activity Review* (in English, French, German, and Spanish).

EUROPEAN ACADEMY OF ARTS, SCIENCES AND HUMANITIES (EAASH) (Academie Europeenne des Sciences, des Arts et des Lettres - AESAL)

60, rue Monsieur le Prince, F-75006 Paris, France **PHONE:** 33 1 46330531 **FX:** 33 1 46342367 **FOUNDED:** 1979. **OFFICERS:** Nicole Lemaire d'Agaggio, Gen.Sec. **MEMBERS:** 600. **BUDGET:** 200,000 Fr. **LANGUAGES:** Dutch, English, French, German, Italian, Portuguese, Spanish.

DESCRIPTION: Members are heads of state, ministers, and presidents of intergovernmental organizations, individuals belonging to national academies, and organizations representing 60 countries. Seeks to foster peace through international collaboration in education, the arts, sciences, and humanities. Advocates a multidisciplinary approach rather than academic specialization in addressing contemporary issues. Coordinates AIDS research in cooperation with UNESCO. Conducts exhibitions and meetings of experts. **PUBLICATIONS:** *Aspects Ethiques et Juridiques de la Sauvegarde des Especes Vivantes*; *Innovation et Societe*; *Life Sciences and Society*; *Science, Culture, et Sante du Monde*; Books; Reports; Reports, periodic. Scientific reports from members of the European Network.

EUROPEAN ALLIANCE OF PRESS AGENCIES

Rue Pelletier 8B, B-1030 Brussels, Belgium **PHONE:** 32 2 7431311 **FX:** 32 2 7351874 **E-MAIL:** dir@belganews.be **FOUNDED:** 1957. **OFFICERS:** R.V. De Ceuster, Sec.Gen. **MEMBERS:** 30. **LANGUAGES:** Dutch, English, French, German.

DESCRIPTION: National press agencies. Promotes improved professional and technical cooperation among members; represents members' interests.

EUROPEAN ALUMINIUM ASSOCIATION (EAA)

ave. de Broqueville 12, B-1150 Brussels, Belgium **PHONE:** 32 2 7756311 **FX:** 32 2 7790531 **E-MAIL:** eaa@eaa.be **FOUNDED:** 1981. **OFFICERS:** Dick Dermer, Sec.Gen. **LANGUAGES:** English, French, German.

DESCRIPTION: Members are producers of primary, secondary, and wrought aluminum and aluminum foil; national associations in 15 countries. Promotes technical cooperation and information exchange within the aluminum industry and represents members' interests before international organizations and national governments. **PUBLICATIONS:** *Aluminium Abstracts*, monthly; *Aluminum Quarterly Report*; *European Aluminum Statistics*, annual.

EUROPEAN ASSOCIATION FOR ANIMAL PRODUCTION (EAAP) (Federation Europeenne de Zootechnie - FEZ)

Via A. Torlonia, 15 A, I-00161 Rome, Italy **PHONE:** 39 6 8840785, 39 6 44238013 **FX:** 39 6 44241466 **E-MAIL:** zoorec@rmnet.it **FOUNDED:** 1949. **MEMBERS:** 36. **STAFF:** 3. **LANGUAGES:** English, French, German, Russian.

DESCRIPTION: Members are national professional, academic, scientific, and technical organizations in 36 countries in Europe and the Mediterranean basin. Works to improve technical and economic aspects of animal production through organization and coordination of scientific research, experimentation, and application. Sponsors seminars, symposia, and workshops. Compiles statistics; bestows awards. **PUBLICATIONS:** *Livestock Production Science*, 16/year.

EUROPEAN ASSOCIATION FOR BANKING AND FINANCIAL LAW (EABFL) (Association Europeene pour le Droit Bancaire et Financier - AEDBF)

c/o AEDBF Belgium, Boulevard A. Reyers 103, Boite 30, B-1030 Brussels, Belgium **LANGUAGES:** English, French.

DESCRIPTION: Members are attorneys and legal organizations with an interest in banking and financial law. Promotes effective statutory regulation of the banking and financial industries. Serves as a clearinghouse on banking and financial law; provides consulting services to legislative and judicial bodies responsible for the formation and application of banking and financial law; conducts research and educational programs.

EUROPEAN ASSOCIATION OF COOPERATIVE BANKS (EACB) (Groupement Europeen des Banques Cooperatives)

23-25, rue de la Science, bte. 9, B-1040 Brussels, Belgium **PHONE:** 32 2 2301124, 32 2 2301419 **FX:** 32 2 2300649 **FOUNDED:** 1970. **OFFICERS:** Oliver Rohlfs, Asst. to the Sec.Gen. **MEMBERS:** 31. **STAFF:** 8. **LANGUAGES:** English, French, German.

DESCRIPTION: Members are national associations of cooperative banks and individual cooperative banks operating in Europe. Promotes cooperative banking; represents members' interests. Coordinates activities with other European banking and cooperative associations. Compiles statistics. **PUBLICATIONS:** *Activity Report* (in English, French, and German), biennial; *Information Brochures*, periodic.

EUROPEAN ASSOCIATION OF EXPLORATION OF GEOSCIENTISTS AND ENGINEERS (EAGE)

PO Box 59, NL-3990 DB Hauten, Netherlands **PHONE:** 31 30 6354055, 31 30 6354066 **FX:** 31 30 6343524 **E-MAIL:** eage@eage.nl **FOUNDED:** 1951. **MEMBERS:** 5,000. **STAFF:** 10. **LANGUAGES:** English.

DESCRIPTION: Geophysientist, engineers, and others in 81 countries actively engaged in the application of geosciences and related engineering subjects. Promotes the science of geophysics as it applies to exploration and related sciences and seeks to enhance fellowship and cooperation among persons interested in the field. Conducts educational courses on seismic interpretation, demonstrating how geological structures are expressed on seismic sections and how maps of the interpreted structures can be drawn. **PUBLICATIONS:** *Basin Research* (in English), quarterly; *First Break* (in English), monthly; *Geophysical Prospecting* (in English), bimonthly; *Petroleum Geoscience* (in English), quarterly; *Practical Aspects of Seismic Inversion*. Covers proceedings of the 1989 Workshop on stratigraphic inversion.; *Practical Aspects of Semismic Data Inversion, The Marmousi Experience*. Contains proceedings of the 1990 EAEG Workshop dealing with structual inversion and migration.; *Seismic Atlas of Structural and Stratigraphic Features*; *Shear Waves, Anisotropy and Polarization*. Contains proceedings of the 1987 Workshop; *Yearbook* (in English).

EUROPEAN ASSOCIATION FOR HEALTH INFORMATION AND LIBRARIES (EAHIL)

c/o ICP-NTI, PO Box 23213, NL-1100 DS Amsterdam, Netherlands **PHONE:** 31 20 5662095 **FX:** 31 20 6963228 **E-MAIL:** eahil@amc.uva.nl **WEBSITES:** http://www.eahil.org **FOUNDED:** 1987. **OFFICERS:** Elisabeth Husem, Pres. **MEMBERS:** 480. **LANGUAGES:** English, French.

DESCRIPTION: Members are individuals, institutions, and collectives in 28 European countries. Seeks to promote the interests of health libraries worldwide; disseminates health information to members. **PUBLICATIONS:** *Newsletter to European Health Librarians* (in English and French), quarterly; *Proceedings of the Biennial Conferences*.

EUROPEAN ASSOCIATION OF MANUFACTURERS OF RADIATORS (EURORAD) (Europaische Vereinigung der Hersteller von Heizkoerpern - EURORAD)

Konradstrasse 9, Postfach 7190, CH-8023 Zurich, Switzerland **PHONE:** 41 1 2719090 **FX:** 41 1 2719292 **E-MAIL:** gerster@jgp.ch **WEBSITES:** http://www.jgp.ch **FOUNDED:** 1960. **OFFICERS:** Kurt Egli, Exec. Officer. **MEMBERS:** 12. **STAFF:** 1. **LANGUAGES:** English, French, German.

DESCRIPTION: Members are national associations of radiator manufacturers. Encourages cooperation among members in technical and economic matters.

EUROPEAN ASSOCIATION FOR POPULATION STUDIES (EAPS) (Association Europeenne pour l'Etude de la Population - AEEP)

Postbus 11676, NL-2502 AR The Hague, Netherlands **PHONE:** 31 70 3565200, 31 70 3565229 **FX:** 31 70 3647187 **E-MAIL:** eaps@nidi.nl **WEBSITES:** http://www.nidi.nl/eaps/ **FOUNDED:** 1983. **OFFICERS:** Gys Beets, Exec.Sec. **MEMBERS:** 525. **LANGUAGES:** English, French, German.

DESCRIPTION: Members are individuals from 40 countries interested or engaged in European population studies. Promotes European population studies and seeks to stimulate a greater public interest in population issues. Fosters cooperation among researchers and interested parties. Disseminates information. Maintains liaison with population research centers and national demographic societies throughout Europe. **PUBLICATIONS:** *EAPS Newsletter* (in English), semiannual; *European Journal of Population* (in English and French), quarterly.

EUROPEAN ASSOCIATION FOR RESEARCH ON PLANT BREEDING (Europaische Gesellschaft fur Zuchtungsforschung)

PO Box 315, NL-6700 AH Wageningen, Netherlands **FX:** 31 317 483457 **E-MAIL:** marjo.dejeu@users.pr.wau **FOUNDED:** 1956. **OFFICERS:** Mrs.Dr.Ir. M.J. de Jeu, Sec.Gen. **MEMBERS:** 1,100. **LANGUAGES:** English, French, German.

DESCRIPTION: Members are geneticists and plant breeders; institutes and plant breeding firms. Aim is to promote international scientific and technical cooperation in the field of plant breeding, in order to contribute to its further progress. Excludes activities connected with commercial interest. **PUBLICATIONS:** *Eucarpia Bulletin* (in English), annual; Proceedings, triennial. Contains congresses and section meetings.

EUROPEAN ASSOCIATION FOR THE SCIENCE OF AIR POLLUTION (EURASAP)

c/o Prof. P. Builtjes, TNO, PO Box 342, NL-7300 AH Delft, Netherlands **PHONE:** 31 555793035, 31 555293591 **FX:** 31 55493252 **E-MAIL:** builtjes@mep.tno.nl **FOUNDED:** 1986. **OFFICERS:** Prof. Peter Builtjes, Chm. **MEMBERS:** 350. **BUDGET:** £2,000. **LANGUAGES:** English.

DESCRIPTION: Scientists from 20 countries working in the field of air pollution and atmospheric chemistry. Objectives are to facilitate communication and information exchange among scientists and coordinate research and practical application of air pollution studies. Conducts educational and research programs.

EUROPEAN ASSOCIATION OF SCIENCE EDITORS (EASE)

c/o Mrs. Jennifer Gretton, PO Box 426, Guildford GU4 7ZH, England **PHONE:** 44 1483 211056 **FX:** 44 1483 211056 **E-MAIL:** secretary@ease.org.uk **WEBSITES:** http://www.ease.org.uk/ **FOUNDED:** 1982. **OFFICERS:** Jennifer Gretton, Sec.-Treas. **MEMBERS:** 960. **BUDGET:** £30,000. **LANGUAGES:** English.

DESCRIPTION: Members are editors of serial and other scientific publications in 49 countries; others responsible for editing or managing such a publication; individuals representing scientific publications or publishing bodies. Aims are to: promote improved communication in science by encouraging cooperation among editors in all disciplines of science; assist in the efficient operation of publications in the field. Encourages discussion on topics including: finding and keeping authors, editors, readers, publishers, and printers; producing publications quickly and economically; keeping up with modern technology in editing and printing; intellectual and practical problems in the transfer of scientific information. **PUBLICATIONS:** *European Science Editing* (in English), 3/year. Contains articles, meeting reports, bibliography, and news of interest to editors.; *Science Editors' Handbook.*

EUROPEAN ASSOCIATION FOR THE STUDY OF DIABETES (EASD)

Merowingerstr, 29, D-40223 Dusseldorf, Germany **PHONE:** 49 211 316738 **FX:** 49 211 3190987 **E-MAIL:** easd@uni-duesseldorf.de **FOUNDED:** 1964. **OFFICERS:** Viktor Joergens M.D., Exec.Dir. **MEMBERS:** 5,500. **STAFF:** 2. **LANGUAGES:** English.

DESCRIPTION: Individuals and firms in 55 countries. Promotes research into the disease of diabetes. Sponsors postgraduate education courses. **PUBLICATIONS:** *Diabetologia* (in English), periodic; *Membership List*, triennial.

EUROPEAN ASSOCIATION FOR THE TRADE IN JUTE AND RELATED PRODUCTS (EATJRP)

Adriaan Goekooplaan 5, NL-2517 JX The Hague, Netherlands **PHONE:** 31 70 3546811 **FX:** 31 70 3512777 **FOUNDED:** 1970. **OFFICERS:** H.J.J. Kruiper. **MEMBERS:** 28. **STAFF:** 2. **LANGUAGES:** English.

DESCRIPTION: Members are European companies in 8 countries directly involved in the trading of jute products. (Jute is a glossy fiber of certain East Indian plants used chiefly for sacking, burlap, and twine.) Works to improve relations among individual companies by promoting contact between them. Defends the interests of members at the international level; acts as representative for trade in jute products.

EUROPEAN BAPTIST FEDERATION (EBF) (Europaische Baptistische Foderation)

Postfach 61 03 40, D-22423 Hamburg, Germany **PHONE:** 49 40 5509723 **FX:** 49 40 5509725 **E-MAIL:** 100422.2213@compuserve.com **WEBSITES:** http://www.ebf.org **FOUNDED:** 1949. **OFFICERS:** Karl-Heinz Walter, Exec. Officer. **MEMBERS:** 50. **STAFF:** 3. **BUDGET:** DM 400,000. **LANGUAGES:** English, German.

DESCRIPTION: European Baptist unions representing over 2.5 million individuals in 46 countries. Aims to draw together and inspire Baptists in carrying out European mission work. Represents European Baptists before the Baptist World Alliance; promotes understanding among Baptists and other Christians. Calls for peace, reconciliation, social justice, and human rights. Sponsors evangelical, educational, relief, and church aid programs. Sponsors International Baptist Theological Seminary; maintains the Institute for Baptist and Anabaptist Studies, the Institute for Mission and Evangelism, and the Seminary Institute of Theological Education, and International Lay Academy. **PUBLICATIONS:** *European Baptist Press Service News Bulletin*, biweekly; Directory.

EUROPEAN BRAIN AND BEHAVIOUR SOCIETY (EBBS)

c/o Dr. Susan J. Sara, Institut des Neurosciences, 9 quai St. Bernard, F-75005 Paris, France **PHONE:** 33 1 44273252 **FX:** 33 1 44273251 **E-MAIL:** ebbs@snv.jussieu.fr **FOUNDED:** 1966. **OFFICERS:** Susan J. Sara, Sec.Gen. **MEMBERS:** 550. **STAFF:** 1. **LANGUAGES:** Dutch, English, French.

DESCRIPTION: Members are scientists with an interest in the study of the brain and behavior. Promotes and facilitates exchange of information among members and between members and others working in related fields. Conducts research and educational programs. **PUBLICATIONS:** Books (in English), biennial; Newsletter, periodic.

EUROPEAN BREWERY CONVENTION (EBC)

Postbus 510, NL-2380 BB Zoeterwoude, Netherlands **PHONE:** 31 71 5456047, 31 71 5456614 **FX:** 31 71 5410013 **E-MAIL:** secretariat@ebc-nl.com **FOUNDED:** 1946. **OFFICERS:** Mrs. M. van Wijngaarden, Sec.Gen. **MEMBERS:** 21. **STAFF:** 3. **LANGUAGES:** English, French, German.

DESCRIPTION: Members are national brewery organizations. Encourages scientific cooperation within the malting and brewing industries of Europe. Initiates and coordinates research; serves as a forum for exchanging experiences, information, and research results among European and non-European brewery experts. Cooperates with the BRI. **PUBLICATIONS:** *Advances in Malting Barley*, annual; *Analytical Methods*; *EBC Thesaurus*; *Elsevier's Dictionary of Brewing*; *Monographs of Symposia*, annual; *Proceedings of Congress*, biennial.

EUROPEAN BUSINESS AVIATION ASSOCIATION (EBAA)

Brusselssteenweg 2a, B-3080 Tervuren, Belgium **PHONE:** 32 2 7660070 **FX:** 32 2 7681325 **E-MAIL:** ebaa@compuserve.com **FOUNDED:** 1977. **OFFICERS:** Fernand M. Francois. **MEMBERS:** 100. **STAFF:** 10. **LANGUAGES:** English.

DESCRIPTION: Represents, promotes, and protects the interests of business aviation. Areas of concern include air traffic congestion, route charges, airport slots, crew licenses and hours, and noise. Disseminates information and advice; conducts research; facilitates contacts among members. **PUBLICATIONS:** *EBAA Bulletin*, 10/year; *Who's Who* (in English), annual. Also publishes position papers and guidelines.

EUROPEAN CENTRE FOR CULTURE (ECC) (Centre Europeen de la Culture - CEC)

European Centre for Culture, Villa Moynier, 120b rue de Lausanne, CH-1202 Geneva, Switzerland **PHONE:** 41 22 7322803 **FX:** 41 22 7384012 **E-MAIL:** cecge@vtx.ch **WEBSITES:** http://www.europeans.ch **FOUNDED:** 1950. **OFFICERS:** Jean-Fred Bourquin, Exec.Pres. **MEMBERS:** 500. **STAFF:** 11. **BUDGET:** 1,400,000 SFr. **LANGUAGES:** English, French, German.

DESCRIPTION: Non-governmental organization dealing with important issues facing Europeans. Strives to underline through European projects, the essential role played by culture in our society to ease tension between the different areas in Europe where people of different cultures live side by side. Encourages the participation of Europeans in the unification and democratic processes. Conducts studies and research. **PUBLICATIONS:** *Newsletter* (in French), periodic; *Temps Europeens* (in French), quarterly. Also publishes books.

EUROPEAN CENTRE FOR ECOTOXICOLOGY AND TOXICOLOGY OF CHEMICALS (ECETOC)

Michelangelo Bldg., 4, ave. Van Nieuwenhuyse, bte. 6, B-1160 Brussels, Belgium **PHONE:** 32 2 6753600 **FX:** 32 2 6753625 **E-MAIL:** info@ecetoc.org **FOUNDED:** 1978. **OFFICERS:**

Dr. F.M. Carpanini. **MEMBERS:** 52. **STAFF:** 8. **BUDGET:** US$1,600,000. **LANGUAGES:** English.

DESCRIPTION: Members are individuals and companies in 13 countries active in the chemical industry. Collects information pertinent to the health of individuals who work with chemicals; seeks to reduce the ecological impact of the manufacture, processing, and use of chemicals. Disseminates scientific, ecological, and toxicological information. Develops methods for the handling of chemicals from ecological and toxicological standpoints. Cooperates with governmental, health, and public bodies concerned with ecological and toxicological problems relating to chemicals. Organizes literature studies and experimental testing programs; investigates epidemiological studies. Maintains task forces; conducts periodic technical meetings. **PUBLICATIONS:** *Information Sheet*, monthly; *Technical Reports*.

EUROPEAN CHEMICAL INDUSTRY COUNCIL (CEFIC)

Avenue Van Nieuwenhuyse, 4 box 1, B-1160 Brussels, Belgium **PHONE:** 32 2 6767211 **FX:** 32 2 6767300 **TX:** 62444 **E-MAIL:** mail@cefic.be **WEBSITES:** http://www.cefic.be **FOUNDED:** 1972. **OFFICERS:** Dr. Hugo Lever, Dir.Gen. **MEMBERS:** 64. **STAFF:** 80. **LANGUAGES:** English.

DESCRIPTION: Chemical federations and manufacturers. Represents the chemical industry in Europe. Coordinates positions and policies; addresses issues of importance to the chemical industry, such as international trade, environment, health and safety, transport and distribution of chemicals, energy and raw materials, and supplies. Compiles statistics. Cooperates with international bodies; maintains liaison with plastics and petrochemical associations. **PUBLICATIONS:** *Annual Report*; *Spotlight* (in English), 10/year. Also publishes seminar proceedings and descriptive pamphlet.

EUROPEAN CIVIL AVIATION CONFERENCE (ECAC) (Conference Europeenne de l'Aviation Civile - CEAC)

3 bis, villa Emile Bergerat, F-92522 Neuilly-sur-Seine, France **PHONE:** 33 1 46418545 **FX:** 33 1 47381367 **E-MAIL:** 101575.1313@compuserve.com **WEBSITES:** http://www.ecac-ceac.org **FOUNDED:** 1955. **OFFICERS:** R. Benjamin, Exec.Sec. **MEMBERS:** 37. **STAFF:** 16. **BUDGET:** US$2,400,000. **LANGUAGES:** English, French.

DESCRIPTION: Members are national, governmental civil aviation departments. Reviews developments in European aviation and promotes the continued development of a safe, efficient, and sustainable European air transport system. Adopts recommendations and measures to assist members in the preparation of national regulations and to lend guidance to the practical work of aeronautical authorities. Promotes harmonization of member states' policies, particularly with regard to cooperation among airlines, charter operations, tariff policies, air traffic control, and facilitation and security matters. Reviews policy affecting economic regulation of scheduled air transport. Develops recommendations on practical measures for improvement of the security system. Examines problems in the fields of acoustic and environmental nuisances and aircraft accidents. **PUBLICATIONS:** *Catalogue of Publications* (in English and French), annual; *Reports of Sessions* (in English and French), triennial; *Statistical Digest* (in English and French), periodic.

EUROPEAN COMMISSION OF HUMAN RIGHTS (ECHR) (Commission Europeenne des Droits de l'Homme - CEDH)

Conseil de l'Europe, F-67075 Strasbourg Cedex, France **PHONE:** 33 88 412018 **FX:** 33 88 412792 **TX:** 870943F **WEBSITES:** http://www.dhcommhr.coe.fr **FOUNDED:** 1953. **OFFICERS:** M. de Salvia, Deputy Sec. **MEMBERS:** 33. **STAFF:** 100. **LANGUAGES:** English, French.

DESCRIPTION: Members are individuals elected by the Committee of Ministers of the Council of Europe. Reviews individual and interstate petitions of human rights violations of the European Convention on Human Rights and rules on the admissibility of such petitions. Acts as an intermediary between parties in order to reach a friendly settlement; failing a settlement, issues opinions of alleged convention breach; may send cases to the Court of Human Rights. **PUBLICATIONS:** *Decisions and Reports* (in English and French), 4-6/year; *Stock-Taking on the European Convention on Human Rights* (in English and French), periodic; *Survey and Statistics* (in English and French), annual.

EUROPEAN COMMISSION FOR INDUSTRIAL MARKETING (CEMI) (Commission Europeenne de Marketing Industriel - CEMI)

18 St. Peters Steps, Brixham, Devon, England **FOUNDED:** 1969. **OFFICERS:** Dr. A.J. Williamson, Pres. **MEMBERS:** 38. **STAFF:** 5. **LANGUAGES:** English, French, Spanish.

DESCRIPTION: Members are associations of industrial marketing executives in 16 countries. Represents members before the European and international industrial products market; facilitates the exchange of information among members. Develops training programs in industrial marketing and research through the European College of Marketing and Marketing Research in conjunction with the European Marketing Association. Compiles statistics. **PUBLICATIONS:** *European Marketing Newsletter* (in English), bimonthly; *Journal of International Marketing and Marketing Research* (in English), 3/year.

EUROPEAN COMMITTEE OF ASSOCIATIONS OF MANUFACTURERS OF AGRICULTURAL MACHINERY (CEMA) (Comite Europeen des Groupements de Constructeurs du Machinisme Agricole - CEMA)

19, rue Jacques Bingen, F-75017 Paris, France **PHONE:** 33 1 42128590 **FX:** 33 1 40549560 **TX:** SGCTMA 640362 F **FOUNDED:** 1959. **OFFICERS:** Jacques Dehollain, Sec.Gen. **MEMBERS:** 16. **LANGUAGES:** English, French, German.

DESCRIPTION: Members are national tractor and agricultural machinery manufacturers' associations. Promotes the economic and technical interests of the European agricultural machinery industry. Represents members before international, governmental, and private organizations. **PUBLICATIONS:** *Terminologie Illustree du Machinisme Agricole*, periodic.

EUROPEAN COMMITTEE FOR STANDARDIZATION (ECS) (Comite Europeen de Normalisation - CEN)

36, rue de Stassart, B-1050 Brussels, Belgium **PHONE:** 32 2 5500811 **FX:** 32 2 5500819 **E-MAIL:** cen@cenclc.be **WEBSITES:** http://www.cenorm.be **FOUNDED:** 1961. **OFFICERS:** Georg Hongler, Sec.Gen. **MEMBERS:** 19. **STAFF:** 100. **BUDGET:** 5,275,000 BFr. **LANGUAGES:** English, French, German.

DESCRIPTION: Member countries of the European Economic Community and the European Free Trade Association. Strives to eliminate technical barriers to trade through the preparation of European standards and harmonization documents to assure that goods produced in accordance with standards developed by CEN may be sold on the markets of other member countries. Implements procedures for the mutual recognition of test and inspection results and certification systems within Europe. Encourages uniform implementation of international standards issued by the International Organization for Standardization. Maintains 184 technical committees, including CENCER Standing Committee which operates a registered conformity marking system; goods produced in accordance with CEN European standards and harmonization documents bear such a mark. CENCER also maintains a system for mutual recognition of respective national test and inspection results without mandatory uniformity. **PUBLICATIONS:** *CEN Memento*. Offers a survey of the structure of CEN.; *CEN Work Programme*. Lists the title of all working items.; *The New Approach*. Discusses the legislation and standards on the free movement of goods in Europe; *Standards for Access to the European Market*. Gives an overview of present activities in European standardization. Additional publication information available upon request.

EUROPEAN COMMODITIES EXCHANGE (ECE) (Bourse de Commerce Europeenne)

10, place Gutenberg, F-67081 Strasbourg Cedex, France **PHONE:** 3 88 752576 **FX:** 3 88 752579 **FOUNDED:** 1960. **OFFICERS:** Sonia Kleiss-Stark, Gen.Sec. **MEMBERS:** 46. **LANGUAGES:** English, French, German.

DESCRIPTION: Members are commodities exchanges of European Economic Community countries involved in the trade of corn, cattle feed, and other ground products. Works to facilitate the exchange of agricultural products.

EUROPEAN COMMUNITIES CHEMISTRY COUNCIL (ECCC)

c/o Royal Soc. of Chemistry, Burlington House,

Piccadilly, London W1V 0BN, England **PHONE:** 44 171 4403303 **FX:** 44 171 4378883 **E-MAIL:** mcewane@rsc.org **FOUNDED:** 1973. **OFFICERS:** Evelyn K. McEwan, Sec. **MEMBERS:** 25. **LANGUAGES:** English.

DESCRIPTION: Members are Western European national organizations whose interests include the theoretical and practical aspects of chemistry. Collaborates with the Commission of the European Communities to advise and consult in matters involving the science and practice of chemistry in member nations of the European Union and EEA. Awarding body for the designation of European Chemist.

EUROPEAN COMMUNITY BANKING FEDERATION (ECBF) (Federation Bancaire de l'Union Europeenne - FBCE)

10, rue Montoyer, B-1040 Brussels, Belgium **PHONE:** 32 2 5083711 **FX:** 32 2 5112328 **FOUNDED:** 1957. **OFFICERS:** Nikolaus Bomcke, Sec.Gen. **MEMBERS:** 18.

DESCRIPTION: Promotes and represents the interests of the banking industry in the European Community and EFTA. Establishes industry standards. Compiles statistics. **PUBLICATIONS:** *Annual Report*; Papers, periodic.

EUROPEAN CONFEDERATION OF AGRICULTURE (CEA) (Confederation europeene de l'Agriculture)

Rue de la Science 23-25, Boite 23, B-1040 Brussels, Belgium **PHONE:** 32 2 2304380 **FX:** 32 2 2304677 **E-MAIL:** cea@pophost.eunet.be **FOUNDED:** 1889. **OFFICERS:** Christophe Hemard. **MEMBERS:** 300. **STAFF:** 3. **BUDGET:** 700,000 SFr. **LANGUAGES:** English, French, German.

DESCRIPTION: Members are agricultural associations and other organizations indirectly involved with some aspect of agriculture representing 30 countries. (Defines agriculture as including forestry, vine-growing, horticulture, market gardening, fishing, and production and processing of agricultural commodities.) Represents and protects the interests of European agriculture in economic and social matters; works to strengthen its family base. Fosters technical, scientific, economic, and social development of the agricultural profession; promotes recognition of agriculture's roles in food supply and new material production, landscape conservation, and environmental protection. Protects and encourages existence of independent family farms and of mutual aid and cooperation among farms. Opposes state control of agriculture and collective ownership of land. Sponsors symposia; organizes competitions. **PUBLICATIONS:** *CEA Dialog* (in English, French, and German), annual; *Index of the Organs and Members of CEA* (in French and German), biennial; *Reports of Annual Meeting*. Also publishes reports and proceedings.

EUROPEAN CONFEDERATION OF INDEPENDENTS

Hauptgeschaftsstelle, Bur. Principal, Oberbexbacher Str., D-66450 Bexbach, Germany **PHONE:** 49 6826 1470, 49 6826 2188 **FX:** 49 6826 50904 **E-MAIL:** info@bvd-cedi.de **FOUNDED:** 1973. **OFFICERS:** Andre Vonner, Pres. **MEMBERS:** 1,400,000. **STAFF:** 15. **LANGUAGES:** French, German.

DESCRIPTION: Promotes the development and importance of small and medium sized enterprises in the European Community. Upholds the ideals of a unified European Community. Compiles statistics. **PUBLICATIONS:** *Gewerbe-Report*, monthly; Annual Report; Brochures; Surveys.

EUROPEAN CONFEDERATION OF PAINT, PRINTING INK AND ARTISTS' COLOURS MANUFACTURERS ASSOCIATIONS (CEPE) (Europaische Vereinigung der Verbande der Lack-, Druckfarben-, und Kunstlerfarbenfabrikanten)

4, ave. E. Van Nieuwenhuyse, B-1160 Brussels, Belgium **PHONE:** 32 2 6767480 **FX:** 32 2 6767490 **E-MAIL:** cepe@mail.interpac.be **FOUNDED:** 1951. **OFFICERS:** Jean Schoder, Sec.Gen. **MEMBERS:** 39. **STAFF:** 6. **LANGUAGES:** English, French, German.

DESCRIPTION: Members are Western European national associations representing 1400 manufacturers in 17 countries. Objectives are to: promote the paint, varnish, printing, ink, and artists' colors industry in Europe; collect and disseminate information; study problems connected with the industry, particularly those affecting the environment and human health, in an effort to find comprehensive solutions; work with international authorities and groups to establish regulations and guidelines for the industry; represent members before the Council of Europe, European Free Trade Association, World Health Organization, and other international groups. **PUBLICATIONS:** *CEPE Annual Review* (in English); *CEPE News* (in English, French, German, Italian, and Spanish), quarterly.

EUROPEAN CONFEDERATION OF WOODWORKING INDUSTRIES (Confederation Europeenne des Industries du Bois - CEI-BOIS)

Hof-Ter-Vleestdreef 5, Bus 4, B-1070 Brussels, Belgium **PHONE:** 32 2 5562585 **FX:** 32 2 5562595 **E-MAIL:** euro.wood.fed@skynet.be **FOUNDED:** 1952. **OFFICERS:** Dr. Guy Van Steertegem, Sec.Gen. **MEMBERS:** 20. **STAFF:** 3. **LANGUAGES:** English, French, German.

DESCRIPTION: Members are national European woodworking federations in Austria, Belgium, Denmark, England, France, Germany, Hungary, Italy, Netherlands, Portugal, Republic of Ireland, Slovakia, Sweden, and Switzerland; the European Federation of Brushware Manufacturer, European Federation of Fibreboard Manufacturer, European Federation of MDF Manufacturer, European Federation of Manufacturer of Wooden Building Components, European Federation of Particleboard Manufacturer, European Federation of Plywood Manufacturer, and Western European Institute for Wood Preservation. Objectives are to: act as a permanent liaison among national and regional federations; represent the interests of European woodworking industries before European and non-European governments; engage in research into the trade, technical, economical, social, and fiscal problems and issues of the industry. Disseminates information. **PUBLICATIONS:** *Circular Letters* (in English), weekly; Papers.

EUROPEAN CONFEDERATION OF YOUTH CLUB ORGANIZATIONS (ECYC)

ornevej 45, DK-2400 Copenhagen NV, Denmark **PHONE:** 45 38108038 **FX:** 45 38104655 **E-MAIL:** ecycdk@centrum.dk **FOUNDED:** 1976. **OFFICERS:** Iram Ahmed, Sec.Gen. **MEMBERS:** 3,700,000. **STAFF:** 15. **BUDGET:** 1,750,000 DKr. **LANGUAGES:** English, French.

DESCRIPTION: Members are young people ages 16 to 22. Conducts exchange programs and youth camps. Organizes research and development programs; conducts seminars. **PUBLICATIONS:** *Course-Directors' Manual* (in English and French); *Directory* (in Danish, English, French, and German), periodic; *ECYC News Bulletin* (in Danish, English, and French), monthly; *European Youth Club Manual* (in English and French); *Get Built: Survey on Youth Club Premises* (in English and French); *Violence Package* (in English and French); *Youth Clubs in Europe* (in Danish, English, and French), 3/year.

EUROPEAN CONFERENCE OF MINISTERS OF TRANSPORT (ECMT) (Conference Europeenne des Ministres des Transports)

2, rue Andre Pascal, F-75775 Paris Cedex 16, France **PHONE:** 33 1 45249711 **FX:** 33 1 45249742 **E-MAIL:** ecmt.contact@oecd.org **WEBSITES:** http://www.oecd.org/cem/ **FOUNDED:** 1953. **OFFICERS:** Gerhard Aurbach, Sec.Gen. **MEMBERS:** 36. **STAFF:** 20. **LANGUAGES:** English, French.

DESCRIPTION: Members are European ministers of transport. Promotes the development and efficiency of Europe's inland transport by coordinating the activities of European transport authorities. Issues of interest include transport policy, liberalization and harmonization of rules, international railway transport, urban traffic, transport for handicapped persons, inland waterway and combined transports, transport market organization, new technologies, and coordination of European road traffic rules, signs, signals, and traffic safety. **PUBLICATIONS:** *Annual Report* (in English and French); *ECMT Newsletter* (in English), semi-annual; *Research Bulletin*, annual; *Round Table Proceedings*, quarterly; *Statistical Series*, annual; *Symposia Proceedings*, triennial; Reports. With studies and technical information.

EUROPEAN CONSORTIUM FOR POLITICAL RESEARCH (ECPR)

Univ. of Essex, Wivenhoe Park, Colchester, Essex CO4 3SQ, England **PHONE:** 44 1206 872501 **FX:** 44 1206 872500 **E-MAIL:** ecpr@essex.ac.uk **WEBSITES:** http://www.essex.ac.uk/ecpr **FOUNDED:** 1970.

OFFICERS: Ken Newton, Exec.Dir. **MEMBERS:** 230. **STAFF:** 3. **LANGUAGES:** English.

DESCRIPTION: European universities and institutes in 20 countries engaged in teaching and research in political science. Aids scholars in their research and promotes the development of political science in Europe. Fosters collaboration and increases the levels of training. Enhances the dissemination of research information by facilitating exchanges and visits of European political science scholars and teachers. Holds 2 summer schools on quantitative research methods, taught in English and French, with workshops and research sessions. Research projects include the Political Economy of Interest Group Power and the Future of Party Government. **PUBLICATIONS:** *Directory of European Political Scientists*, periodic. Contains biographical and professional information, credentials, affiliations, publications, and fields of interest and research.; *ECPR News* (in English), 3/year. Contains current information on members, activities, conferences, fellowships, and vacant university posts.; *European Journal of Political Research* (in English), 8/year. Contains scholarly articles of a broadly theoretical or comparative nature.; *Exchange and Mobility*; *I Sistemi de Partito*; *Index of Workshop Papers*; *LOGOPOL Newsletter*, periodic. Contains information on matters of interest to persons involved in local government or politics.; *Political Science in Europe*, periodic. Contains information on member institutions including courses offered, research interests, and staff member lists.; *Sage Modern Politics Series*.

EUROPEAN CONVENTION FOR CONSTRUCTIONAL STEELWORK (ECCS)

32/36 avenue des Ombrages, Ste. 20, B-1200 Brussels, Belgium **PHONE:** 32 2 7620429 **FX:** 32 2 7620935 **E-MAIL:** eccs@steelconstruct.com **WEBSITES:** http://www.steelconstruct.com **FOUNDED:** 1955. **OFFICERS:** R.V. Salkin, Sec.Gen. **MEMBERS:** 24. **STAFF:** 2. **LANGUAGES:** English, French, German.

DESCRIPTION: Members are national associations representing manufacturers of structural steel and construction companies engaged in the building of steel structures. Facilitates communication and cooperation among members. Represents members' interests. Conducts research and educational programs; compiles statistics. **PUBLICATIONS:** *ECCS Review* (in English), quarterly.

EUROPEAN COPPER INSTITUTE

Avenue de Tervueren 168, B-1150 Brussels, Belgium **PHONE:** 32 2 7777070 **FOUNDED:** 1989. **OFFICERS:** Robert A. Zehnder, CEO. **MEMBERS:** 14. **STAFF:** 4. **BUDGET:** £15,000,000. **LANGUAGES:** Dutch, English, French.

DESCRIPTION: Membership restricted to ICA and IWCC members as well as the national copper centres in Europe. Concerned with the detailed planning, coordination, and funding of European copper promotion under the direction of its Executive Board.

EUROPEAN COUNCIL OF CONSCRIPTS ORGANISATIONS (ECCO)

Postbus 2384, NL-3500 GJ Utrecht, Netherlands **PHONE:** 31 30 2443425 **FX:** 31 30 2422195 **FOUNDED:** 1979. **OFFICERS:** Marc Hulst, Sec.Gen. **MEMBERS:** 12. **STAFF:** 1. **BUDGET:** 275,000 Fr. **LANGUAGES:** English.

DESCRIPTION: Members are national organizations of military conscripts in Europe. Seeks to draw international attention to the poor working and living conditions of conscripts; cites examples of human rights violations of conscripts as well as news of army reform in various countries. Works to ensure for military personnel the same democratic and human rights that are afforded civilians; provides for the exchange of information and experience among member organizations. Cooperates with European institutions such as the European Union, the Council of Europe, and the OSCE-Process. Has consultative status with the Council of Europe. **PUBLICATIONS:** *ECCO - Echo* (in English), 3/year. Contains information on current developments of conscription in Europe; *European Charter on the Rights of Conscripts*.; *Everything You Always Wanted to Know About ECCO*; *Guide Book for Creating a Representation System for Conscripts*.

EUROPEAN COUNCIL OF INTERNATIONAL SCHOOLS (ECIS)

21 Lavant St., Petersfield, Hants. GU32 3EL, England **PHONE:** 44 1730 268244 **FX:** 44 1730 267914 **E-MAIL:** ecis@ecis.org **WEBSITES:** http://www.ecis.org **FOUNDED:** 1965. **OFFICERS:** Michael Maybury, Exec.Sec. **MEMBERS:** 1,800.

STAFF: 25. **BUDGET:** £1,000,000. **LANGUAGES:** English.

DESCRIPTION: Members include international schools, colleges, and institutions involved in the field of education worldwide; manufacturers of educational equipment and publishers; individuals interested in international education. Represents members' interests and serves as a liaison between member-schools and institutions of higher learning throughout the world. Evaluates and accredits schools according to standards set by international educational agencies. Provides guidance to schools regarding educational programs, teaching methods, and organizational questions. Assists schools in the recruitment of directors and teaching staff. Conducts research and gathers information of interest to members. Holds training courses on education in an international setting. Compiles statistics; operates placement service. **PUBLICATIONS:** *ECIS International Schools Directory* (in English), annual. Lists information about international schools worldwide.; *Higher Education Directory* (in English), annual; *International Schools Journal* (in English), semiannual; *IS* (in English), 3/year; *Resource Books.*

EUROPEAN CULTURAL FOUNDATION (ECF) (Fondation Europeenne de la Culture - FEC)

Jan van Goyenkade 5, NL-1075 HN Amsterdam, Netherlands **PHONE:** 31 20 676022 **FX:** 31 20 6752231 **TX:** 18710 FEC NL **E-MAIL:** eurocult@eurocult.org **WEBSITES:** http://www.eurocult.org **FOUNDED:** 1954. **OFFICERS:** Stephan Ruediger, Sec.Gen. **STAFF:** 14. **BUDGET:** 3,600,000 f. **LANGUAGES:** Dutch, English, French, German.

DESCRIPTION: Aims to promote cultural cooperation in Europe by developing new projects and programmer, and by providing subsidies through a grants programme. Priority areas are: Central & Eastern Europe. The Mediterranean Region and projects dealing with cultural pluralism: policies and practices. **PUBLICATIONS:** *Annual Facts and Figures* (in Dutch, English, French, and German); *Annual List of Grants Awarded* (in English); *Guidelines for Grants* (in English and French); *Profile of the ECF* (in English and French); Newsletter (in English), semiannual. Also publishes occasional papers.

EUROPEAN DEMOCRAT UNION (EDU) (Union Democratique Europeene)

Reichsratsstrasse 11, A-1010 Vienna, Austria **PHONE:** 43 1 4051686 **FX:** 43 1 4051680 **E-MAIL:** eduwein@via.at **WEBSITES:** http://members.magnet.at/edu/theedu.htm **FOUNDED:** 1978. **OFFICERS:** Alexis Wintoniak, Exec.Sec. **MEMBERS:** 40. **STAFF:** 4. **LANGUAGES:** English, French, German, Spanish.

DESCRIPTION: Christian Democrat, conservative, and similar political parties in 29 countries. Supports the growth of democracy throughout the world. Encourages co-operation between Christian Democrat, Conservative, and like-minded parties in Europe. Advocates uniform national political programs and foreign and European policy; coordinates activities of European political organizations. Offers support to member parties; encourages co-operation among members. Examines the political climate in nations worldwide. **PUBLICATIONS:** *EDU-Bulletin* (in English, French, and German), 3/year; *EDU-Yearbook* (in English); *Names and Addresses* (in English, French, and German), annual; *Press Release* (in English, French, and German), 10/year. Also publishes reports.

EUROPEAN DIRECT MARKETING ASSOCIATION (EDMA)

439 Avenue de Tervueren, B-1150 Brussels, Belgium **PHONE:** 32 2 7794268, 32 2 7789920 **FX:** 32 2 7794269 **E-MAIL:** edma@skynet.be **FOUNDED:** 1976. **OFFICERS:** Carol R. Rog, Chm. **MEMBERS:** 650. **STAFF:** 9. **BUDGET:** 24,000,000 BFr. **LANGUAGES:** English, French, German.

DESCRIPTION: Companies, associations, and institutions engaged in direct marketing. Promotes direct marketing as a communications method for creating mutually profitable relationships between sellers and buyers; facilitates contacts and exchange of ideas and techniques among countries and members; supplies information on services, products, companies, and people within the industry. Conducts research and provides statistics. **PUBLICATIONS:** *EDMA Membership Directory*, annual; *EDMAFlash* (in English), quarterly; *EDMAGram* (in English), quarterly.

EUROPEAN ENVIRONMENTAL BUREAU (EEB) (Bureau Europeen de l'Environnement - BEE)

34, Bol de Waterloo, B-1000 Brussels, Belgium **PHONE:** 32 2 2891090 **FX:** 32 2 2891099 **FOUNDED:** 1974. **OFFICERS:** John Hontelez, Sec.Gen. **MEMBERS:** 128. **STAFF:** 10. **BUDGET:** 7,250,000 BFr. **LANGUAGES:** English, French.

DESCRIPTION: Members are environmental organizations within the European Economic Community. Promotes environmental conservation and the restoration and efficient use of human and natural resources. Advocates an equitable and sustainable lifestyle; seeks to heighten public awareness of environmental problems; submits proposals and recommendations to authorities responsible for environmental quality. Acts as liaison between European institutions and non governmental environmental organizations; coordinates lobbying activities of such groups. Holds conferences. Maintains Regular Information Systems on Environment and Development. **PUBLICATIONS:** *Annual Report* (in English and French); *Metamorphosis*, 5/year. This publication is listed upon request.

EUROPEAN EVANGELICAL ALLIANCE (EEA) (Europaische Evangelische Allianz - EEA)

Postfach 23, A-1037 Vienna, Austria **PHONE:** 43 1 7149151 **FX:** 43 1 7138382 **FOUNDED:** 1954. **OFFICERS:** Stuart McAllister, Gen.Sec. **MEMBERS:** 22. **STAFF:** 1. **BUDGET:** US$40,000. **LANGUAGES:** English, German.

DESCRIPTION: European national Evangelical alliances. Promotes cooperation between Western and Eastern European evangelical groups on issues such as human rights and evangelism. Organizes international prayer weeks and conferences of faith.

EUROPEAN FEDERATION OF AGRICULTURAL WORKERS' UNIONS

38, rue Fosse-aux-Loups, bte. 8, B-1000 Brussels, Belgium **PHONE:** 32 02 2185308 **FX:** 32 02 2199926 **E-MAIL:** efa.weipert@skynet.be **FOUNDED:** 1958. **OFFICERS:** Wolfgang Weipert, Sec.Gen. **MEMBERS:** 2,000,000. **STAFF:** 3. **BUDGET:** 14,000,000 BFr. **LANGUAGES:** English, French, German, Italian.

DESCRIPTION: Promotes the freedom of association and collective self-management, in a spirit of solidarity, of active, unemployed, seasonal farm workers or small farmers as well as environmental protection and safeguard. **PUBLICATIONS:** *EFA-Express* (in English, French, and Italian). Includes topics in agriculture and the environment.

EUROPEAN FEDERATION OF ASSOCIATIONS OF INSULATION CONTRACTORS (FESI) (Federation Europeenne des Syndicats d'Entreprises d'Isolation - FESI)

c/o J. Schmoldt, Hamptverband der Deutschen Banindustine e.v., Karl-Liebknecht-Strasse 33, D-10178 Berlin, Germany **PHONE:** 49 30 2426863 **FX:** 49 30 2425597 **FOUNDED:** 1970. **OFFICERS:** J. Schmoldt, Gen.Sec. **MEMBERS:** 15. **LANGUAGES:** English, German.

DESCRIPTION: Members are groups representing firms involved in thermal, sound-proof, and fire-proof insulation. Facilitates contacts among members; addresses questions concerning the industry; defends members' interests before international bodies.

EUROPEAN FEDERATION OF BIOTECHNOLOGY (EFB)

c/o DECHEMA, Postfach 15 01 04, D-60061 Frankfurt, Germany **PHONE:** 49 69 7564279 **FX:** 49 69 7564201 **TX:** 412490 DCHA D **E-MAIL:** efb@dechema.de **WEBSITES:** http://www.dechema.de/efb.htm **FOUNDED:** 1978. **OFFICERS:** Prof. Gerhard Kreysa, Gen.Sec. **MEMBERS:** 81. **LANGUAGES:** English, French, German.

DESCRIPTION: Technical and scientific associations representing 26 European and non-European countries. Works to further biotechnology (the study of micro-organisms and cultured tissue cells) as a multidisciplinary field of scientific inquiry combining the methodologies and accomplishments of biology, microbiology, biochemistry, and engineering. Collects and disseminates information on biotechnology. Cooperates with other international organizations. **PUBLICATIONS:** *EFB Newsletter* (in English), quarterly; *Proceedings of European Congress of Biotechnology*, periodic.

EUROPEAN FEDERATION OF BUILDING SOCIETIES (EFBS) (Europaische Bausparkassenvereinigung)

Dottendorfer Str. 82, D-53129 Bonn, Germany **PHONE:** 49 228 239041 **FX:** 49 228 239046 **FOUNDED:** 1962. **OFFICERS:** Andreas J. Zehnder, Mng.Dir. **MEMBERS:** 68. **LANGUAGES:** English, French, German.

DESCRIPTION: Organizations in 19 countries providing financing for housing. Purpose is to represent members' interests before European Union bodies. Objectives are to: enhance the development of member organizations and to coordinate their activities; study the correlation of competition among members and the flow of capital; cooperate with building societies, or groups with similar goals, in European countries not belonging to the European Union. Conducts comparative studies. **PUBLICATIONS:** *Annual Report* (in English, French, and German); *Directory of Members* (in English, French, and German), periodic; Newsletter (in English, French, and German), bimonthly.

EUROPEAN FEDERATION OF CHEMICAL ENGINEERING (EFCE) (Europaische Foderation fur Chemie-Ingenieur-Wesen - EFCIW)

c/o DECHEMA eV, Theodor-Heuss-Allee 25, Postfach 15 01 04, D-60486 Frankfurt, Germany **PHONE:** 49 69 7564143 **FX:** 49 69 7564201 **FOUNDED:** 1953. **MEMBERS:** 50. **LANGUAGES:** English, French, German.

DESCRIPTION: Scientific and technical societies in the field of chemical engineering. Encourages progress in the area of chemical engineering and fosters the scientific and economic development and application of manufacturing processes. Collects and disseminates information and ideas pertaining to chemical engineering. Sponsors training programs, symposia, and workshops on chemical engineering and its technical applications. **PUBLICATIONS:** *EFCE Newsletter* (in English), periodic; Proceedings.

EUROPEAN FEDERATION OF CORROSION (EFC)

c/o DECHEMA, Postfach 15 01 04, D-60061 Frankfurt, Germany **PHONE:** 49 69 7564209 **FX:** 49 69 7564201 **FOUNDED:** 1955. **MEMBERS:** 30. **LANGUAGES:** English, French, German.

DESCRIPTION: Works to advance interest in and study of corrosion by promoting cooperation between scientific and technical societies in Europe and collaboration with similar groups throughout the world. Promotes research and strives to avoid duplication of experimental work. Acts as a forum for the exchange of information on corrosion and its control; provides sources of information and education; promotes the documenting of information. **PUBLICATIONS:** *EFC Newsletter*, periodic; *European Federation of Corrosion Publication Series.*

EUROPEAN FEDERATION OF FOOD SCIENCE AND TECHNOLOGY (EFFoST)

c/o ATO-DLO, PO Box 17, NL-6700 AA Wageningen, Netherlands **PHONE:** 31 317 475000 **FX:** 31 317 475347 **E-MAIL:** effost@ato.dlo.nl **WEBSITES:** http://www.ato.dlo.nl/effost/ **FOUNDED:** 1986. **OFFICERS:** Daniella Stijnen. **STAFF:** 2. **LANGUAGES:** English.

DESCRIPTION: European affiliates of the International Union of Food Science and Technology. Objectives are: to develop closer contact between food producers and distributors, universities and research institutions; enhance rapid technology transfer from ideas/research into industrial applications to improve European competitiveness. Promotes continuing professional development and educational excellence within food science and technology. Maintains a collaborative network of research organizations with the European food industry aimed at cooperation and knowledge sharing. **PUBLICATIONS:** *Calendar of Food Science, Technology, and Engineering Meetings in Europe and Major International Events* (in English), quarterly; *Cereals in a European Context*; *Directory of Research and Education in Food Science, Technology, and Engineering* (in English); *Education and Training in Food Science*; *Guide to Educational Developments in Europe* (in English); *Minimal Processing of Food - Conference Proceedings* (in English); Pamphlet. Contains description of EFFoST; Pamphlet. Contains description of EFFoST.

EUROPEAN FEDERATION OF GREENS PARTIES (EG)

European Parliament - Leo 2C85, Rue Wiertz, B-1047 Brussels, Belgium **PHONE:** 32 2 2845135, 32 2 2847135 **FX:** 32 2 2849135 **E-MAIL:** rmonroe@europarl.eu.int **WEBSITES:** http://www.europeangreens.org **FOUNDED:** 1984. **OFFICERS:** Ralph Monoe, Sec.Gen. **MEMBERS:** 30. **STAFF:** 2. **BUDGET:** 5,000,000 BFr. **LANGUAGES:** English, French.

DESCRIPTION: Green parties in 28 European countries. Coordinates green party activities in Europe. **PUBLICATIONS:** *Update* (in English), monthly.

EUROPEAN FEDERATION OF HANDLING INDUSTRIES (FEM) (Federation Europeenne de la Manutention - FEM)

Kirchenweg 4, CH-8032 Zurich, Switzerland **PHONE:** 41 1 3844844 **FX:** 41 1 3844848 **FOUNDED:** 1953. **OFFICERS:** Dr. K. Meier, Sec.Gen. **MEMBERS:** 16. **LANGUAGES:** English, French, German.

DESCRIPTION: Members are national committees representing European associations of materials handling manufacturers. Promotes cooperation among manufacturers of handling equipment; works to safeguard the economic and technical interests of the industry. Encourages legislation and standardization intended to facilitate import-export activities. Encourages technical progress and safety in the workplace.

EUROPEAN FEDERATION FOR INTERCULTURAL LEARNING (EFIL)

18-24, rue des Colonies, B-1000 Brussels, Belgium **PHONE:** 32 2 5145250 **FX:** 32 2 5142929 **E-MAIL:** info@efil.be **WEBSITES:** http://www.afs.org/efil **FOUNDED:** 1971. **OFFICERS:** Elisabeth Hardt, Sec.Gen. **MEMBERS:** 19. **STAFF:** 3. **BUDGET:** US$500,000. **LANGUAGES:** English.

DESCRIPTION: Members are national volunteer organizations concerned with intercultural learning for young people. Promotes peace by stimulating an awareness of common humanity both between and within nations. Encourages a wider understanding of the world's social, cultural, and physical environments with the belief that peace can be threatened as much by social injustices between and within nations as by international tensions. Fosters exchange of ideas and experience among researchers, practitioners, and experts in the field of intercultural learning. Activities include: 3-month and 12-month exchanges for secondary school students; short- and long-term exchanges for young workers; short and long-term exchanges for volunteers in community services; exchange programs for teachers in Africa and those involved in multicultural education. Conducts conferences, research meetings, and seminars. **PUBLICATIONS:** *Latest Edition* (in English), monthly; *Seminar Report*, periodic; Report, biennial.

EUROPEAN FEDERATION OF MANAGERIAL STAFF OF AGRICULTURAL AND ALIMENTARY INDUSTRY AND COMMERCE (Federation Europeenne du Personnel d'Encadrement des Industries et Commerces Agricoles et Alimentaires - FEPEDICA)

c/o Pierre Broquet, 59/63, rue du Rocher, F-75008 Paris, France **PHONE:** 33 1 55301330 **FX:** 33 1 55301331 **FOUNDED:** 1990. **OFFICERS:** Pierre Broquet, Pres. **MEMBERS:** 12,000. **STAFF:** 5. **LANGUAGES:** Dutch, English, French, German, Italian.

DESCRIPTION: Members are individuals holding managerial positions in the agricultural and food industries in Europe and the Canadian province of Quebec. Objectives are: to promote the well-being and the professional interests of individual members; to contribute to general economic progress; to represent members' interests before regional groups and the European Union. **PUBLICATIONS:** *INFORCADRE* (in Dutch and French), quarterly; *Vouloir Magazine* (in French), quarterly.

EUROPEAN FEDERATION OF NATIONAL ENGINEERING ASSOCIATIONS (FEANI) (Federation Europeenne D'Associations Nationales d'Ingenieurs - FEANI)

Rue Du Beau Site 21, B-1000 Brussels, Belgium **PHONE:** 32 2 6390390 **FX:** 32 2 6390399 **E-MAIL:** barbel.hakini@feani.com **FOUNDED:** 1951. **OFFICERS:** Ms. Barbel Hakini, Sec. **MEMBERS:** 27. **STAFF:** 5. **BUDGET:** 3,000,000 Fr. **LANGUAGES:** English, French, German.

DESCRIPTION: National engineering associations representing 1,500,000 engineers. Conducts seminars. **PUBLICATIONS:** *Engineering Development International* (in English), annual, always in the fall. Deals with the latest technology at the forefront of engineering development; *Feani Handbook*, annual. Provides a convenient reference on national engineering institutions, education and training, and major engineering achievements across Europe; *FEANI Index*, periodic. Authoritative reference material providing details of the FEANI Register, the FEANI Code of Ethics, educational system info, and other info; *FEANI News* (in English), quarterly.

EUROPEAN FEDERATION OF THE PLYWOOD INDUSTRY (EFPI) (Federation Europeenne de l'Industrie du Contreplaque - FEIC)

33 rue de Naples, F-75008 Paris, France **PHONE:** 33 1 47201732 **FX:** 33 1 47207631 **FOUNDED:** 1957. **OFFICERS:** P. Lapeyre, Conseiller du President. **MEMBERS:** 15. **STAFF:** 1. **LANGUAGES:** English, French, German.

DESCRIPTION: Members are national European plywood organizations. Facilitate contact between members and organizes joint research on problems facing the industry. Supports efforts to promote the use of plywood. Represents its members vis-a-vis governments and international organizations. **PUBLICATIONS:** Annual Report.

EUROPEAN FEDERATION OF THE TRADE IN DRIED FRUIT, EDIBLE NUTS, PRESERVED FOOD, SPICES, HONEY, AND SIMILAR FOODSTUFFS (FRUCOM)

Grosse Baeckerstr. 4, D-20095 Hamburg, Germany **PHONE:** 49 40 3747190 **FX:** 49 40 37471926 **FOUNDED:** 1959. **OFFICERS:** Dr. Klaus Hanebuth, Sec. **MEMBERS:** 12. **LANGUAGES:** English.

DESCRIPTION: Works to protect the interests of the import trade in dried fruit, almonds and other nuts, preserved food, spices, honey, and similar foods. Promotes these products within the European Community, particularly among authorities and institutions in the EC.

EUROPEAN FESTIVALS ASSOCIATION (EFA) (Association Europeenne des Festivals - AEF)

120 B, rue de Lausanne, CH-1202 Geneva, Switzerland **PHONE:** 41 22 7386873 **FX:** 41 22 7384012 **E-MAIL:** geneva@eurofestivals.efa.ch **WEBSITES:** http://www.eurofestivals.efa.ch **FOUNDED:** 1952. **OFFICERS:** Tamas Klenjanszky, Gen.Sec. **MEMBERS:** 75. **STAFF:** 2. **LANGUAGES:** English, French, German.

DESCRIPTION: Works to develop and maintain high artistic standards of music, theatre, and dance festivals. Promotes the festivals through dissemination of information and publicity highlighting the artistic significance of music festivals to European culture. **PUBLICATIONS:** *Festivals*, annual. Festival guidebook.

EUROPEAN FINANCE ASSOCIATION (EFA)

c/o European Inst. for Advanced Studies in Management, 13, rue d'Egmont, B-1000 Brussels, Belgium **PHONE:** 32 2 5119116 **FX:** 32 2 5121929 **E-MAIL:** efa@ciasm.be **WEBSITES:** http://www.eiasm.be/efa.html **FOUNDED:** 1974. **OFFICERS:** Gerry Van Dyck, Exec.Sec. **MEMBERS:** 450. **LANGUAGES:** English.

DESCRIPTION: Members are academics and practitioners interested in financial management and theory and its application. Fosters dissemination and exchange of information; provides forum for presentation of research results in the areas of company finance, investment, financial markets, and banking. **PUBLICATIONS:** Newsletter, quarterly.

EUROPEAN FINANCIAL MANAGEMENT AND MARKETING ASSOCIATION (EFMA) (Association Europeene de Management et Marketing Financiers - EFMA)

16, rue d'Aguesseau, F-75008 Paris, France **PHONE:** 33 1 47425272 **FX:** 33 1 47425676 **TX:** 280288 F **WEBSITES:** http://www.efma.com **FOUNDED:** 1971. **OFFICERS:** Michel Barnich, Sec.Gen. **MEMBERS:** 140. **STAFF:** 18. **LANGUAGES:** English, French.

DESCRIPTION: Members are European financial organizations in 17 countries. Goals are to: establish communication among individuals who work with European financial organizations and who support the concept of marketing; encourage innovation in the field; foster initiation of financial marketing research projects; represent the interests of European financial marketing. Sponsors seminars and professional training sessions. Maintains documentation center. Compiles data on credit card systems. **PUBLICATIONS:** *EFMA Newsletter* (in English and French), bimonthly; *Press Review* (in English and French), weekly; *Special Issues* (in English and French), periodic. Offers transcripts of speeches.

EUROPEAN FOUNDATION FOR MANAGEMENT DEVELOPMENT (EFMD) (Fondation Europeenne pour le Developpement du Management)

88, rue Gechard, B-1050 Brussels, Belgium **PHONE:** 32 2 6480385 **FX:** 32 2 6460768 **E-MAIL:** info@efmd.be **WEBSITES:** http://www.efmd.be **FOUNDED:** 1971. **OFFICERS:** Bernadette Conraths, Dir.Gen. **MEMBERS:** 400. **STAFF:** 20. **BUDGET:** US$1,600,000. **LANGUAGES:** English, French.

DESCRIPTION: Members are corporations, educational institutions, employers associations, management consultants, and individuals in 45 countries with an interest in management development, training, and education. Seeks to identify, research, and address leading management development issues. Fosters development of professional competence of those responsible for management development within companies and educational institutions; promotes education, development, and research in the field through working groups, seminars and conferences. Strives to organize effective interaction among all those involved in the management development process. Administers the Euro-China Association for Management, the European Business Ethics Network. Coordinates development projects on behalf of organizations such as the European Communities, including Integrating Environmental Issues into Management Development Project and the Management Education for Small and Medium-sized Enterprises in Europe Project. **PUBLICATIONS:** *Annual Report*; *Documentation on Books, Cases, and Other Teaching Materials in Management*, bimonthly; *EFMD Journal and Bulletin*, 3/year; *Material in Management*, bimonthly; *Membership List*, annual. Also publishes reports on topics such as training needs of European managers, internal and external training, and management education.

EUROPEAN FREE TRADE ASSOCIATION (EFTA) (Association Europeenne de Libre-Echange - AELE)

9-11, rue de Varembe, CH-1211 Geneva 20, Switzerland **PHONE:** 41 22 7491111 **FX:** 41 22 7339291 **TX:** 414102 EFTA CH **E-MAIL:** efta-mailbox@secrbru.efta.be **WEBSITES:** http://www.efta.int **FOUNDED:** 1960. **OFFICERS:** Kjartan Johannsson, Sec.Gen. **MEMBERS:** 4. **STAFF:** 60. **BUDGET:** 16,045,000 SFr. **LANGUAGES:** English, French.

DESCRIPTION: Members are representatives of Iceland, Liechtenstein, Norway, and Switzerland. Objectives are to: eliminate obstacles to trade in Western Europe; promote economic prosperity in member states; ensure fair trade competition. **PUBLICATIONS:** Annual Report (in English), annual. Also publishes periodic factsheets.

EUROPEAN FURNITURE MANUFACTURERS FEDERATION (UEA) (Union Europeenne de l'Ameublement - UEA)

Chaussee de Haecht, 35, B-121 Brussels, B-1000 Brussels, Belgium **PHONE:** 32 2 2181889, 32 2 2233964 **FX:** 32 2 2192701 **E-MAIL:** secretariat@uea.be **WEBSITES:** http://www.u-e-a.com **FOUNDED:** 1950. **OFFICERS:** Mr. B. De Turck, Sec.Gen. **MEMBERS:** 15. **STAFF:** 4. **BUDGET:** 10,000,000 BFr. **LANGUAGES:** English, French, German.

DESCRIPTION: Members are national furniture manufacturers' federations. Supports the interests of the European furniture industry and serves as liaison among professional associations on a European level. Presents and defends federation positions before appropriate government, national, and international bodies. Conducts studies on professional problems. Compiles statistics. **PUBLICATIONS:** *The Bedroom Furniture Industry in the European Countries* (in English), annual. Provides industry statistics; *Exports & Imports of All Types of Furniture of the EU Countries*; *Exports & Imports of Bedroom Furniture of the EU Countries*; *Exports & Imports of Dining Room Furniture of the EU Countries*; *Exports & Imports of Kitchen Furniture of the EU Countries*; *Exports & Imports of*

Mattresses Furniture of the EU Countries; *Exports & Imports of Office Furniture of the EU Countries*; *Exports & Imports of Other Furniture of the EU Countries*; *Exports & Imports of Parts of Furniture of the EU Countries*; *Exports & Imports of Plastic Furniture of the EU Countries*; *Exports & Imports of Rattan Furniture of the EU Countries*; *Office Furniture Industry in the European Countries*; *UFA Newsletter* (in Dutch, English, French, German, and Italian), bimonthly; *The Upholstered Furniture Industry in the European Countries.*

EUROPEAN GLASS CONTAINER MANUFACTURERS COMMITTEE (EGM)

Northumberland Rd., Sheffield, S. Yorkshire S10 2UA, England **PHONE:** 44 114 2686201 **FX:** 44 114 2681073 **FOUNDED:** 1951. **OFFICERS:** Dr. W.G.A. Cook, Sec. **MEMBERS:** 24. **LANGUAGES:** English.

DESCRIPTION: Members are national organizations in 15 countries that manufacture glass containers. To promote communication among members.

EUROPEAN HEALTHCARE MANAGEMENT ASSOCIATION (EHMA)

Vergemount Hall, Clonskeagh, Dublin 6, Ireland **PHONE:** 353 1 2839299 **FX:** 353 1 2838653 **E-MAIL:** ehma@iol.ie **WEBSITES:** http://www.iol.ie/-EHMA/ **FOUNDED:** 1966. **OFFICERS:** Philip C. Berman, Dir. **MEMBERS:** 200. **STAFF:** 4. **LANGUAGES:** English, French.

DESCRIPTION: Members are policy makers, senior managers, personnel directors, academic institutions, and research organizations in the healthcare sector. Seeks to improve healthcare in Europe by raising standards of managerial performance in the health sector. Fosters cooperation between health service organizations and institutions in the field of healthcare management education and training. Promotes the continuing education and development of healthcare managers. Offers advice and support to national governments in Europe; evaluates members' management development programs. **PUBLICATIONS:** *Conference Proceedings* (in English), annual; *Health Services Administration Education and Research* (in English), annual; *Management Development for Health Care: An International Perspective* (in English); *Management Education and Training in the Health Sector: A Perspective for Italy* (in English); Newsletter (in English), bimonthly.

EUROPEAN HEALTH POLICY FORUM (EHPF) (Forum Europeen de Politique de Sante - EHPF)

School of Public Health, Leuven Univ., PO Box 214, B-3000 Louvain, Belgium **PHONE:** 32 16 336978 **FX:** 32 50 220541 **E-MAIL:** 106053.2765@compuserve.com **FOUNDED:** 1981. **OFFICERS:** Prof. Mia Defever, Dir. **STAFF:** 3. **LANGUAGES:** English.

DESCRIPTION: Members are leading policymakers in a broad range of health care fields and industries representing 25 countries. Purposes are: to inform members of new developments in health care policy and research; to facilitate communication between policymakers in different fields; to draw attention to new research and its implications for health policy. Sponsors the development and transfer of health information systems. **PUBLICATIONS:** *Health Policy*, monthly; Papers, periodic; Proceedings, periodic.

EUROPEAN INDEPENDENT STEELWORKS ASSOCIATION (EISA)

205, rue Belliard, B-1040 Brussels, Belgium **PHONE:** 32 2 2307962 **FX:** 32 2 2300136 **E-MAIL:** malois@eisa.org **WEBSITES:** http://www.eisa.org/ **FOUNDED:** 1981. **OFFICERS:** Maria Alois, Dir.Gen. **MEMBERS:** 15. **STAFF:** 2. **LANGUAGES:** English, French, German, Greek, Italian, Spanish, Swedish.

DESCRIPTION: Privately-owned steel companies. Aims to promote the interests of steelworks companies before the European Union.

EUROPEAN INSTITUTE FOR CRIME PREVENTION AND CONTROL AFFILIATED WITH THE UNITED NATIONS (Institut d'Europe pour la Prevention du Crime et la Lutte Contre la Delinquance Affile a l'Organisation des Nations Unies)

PO Box 161, FIN-00131 Helsinki, Finland **PHONE:** 358 9 18257880 **FX:** 358 9 18257890 **E-MAIL:** heuni@om.vn.fi **WEBSITES:** http://www.vn.fi/om/suomi/heuni **FOUNDED:** 1981.

OFFICERS: Matti Joutsen, Dir. **STAFF:** 5. **BUDGET:** FM 2,400,000. **LANGUAGES:** English, French, Russian.

DESCRIPTION: Functions as the European branch of the United Nations Crime and Criminal Justice Programme. Assists European countries in responding to the challenges presented by crime. Gathers and disseminates information on crime and criminal justice; conducts seminars and research projects; makes available scholarships to Europeans. Participates in planning and developing the United Nations' information system on crime and criminal justice. **PUBLICATIONS:** *Crime and Criminal Justice in Europe and North America* (in English, French, and Russian); *Crime Prevention Strategies in Europe and North America* (in English, French, and Russian); *Criminal Law and the Environment* (in English, French, and Russian); *Directory of Computerized Criminal Justice Information Systems* (in English, French, and Russian); *Foreigners in Prison* (in English, French, and Russian); *Prison Health* (in English, French, and Russian); *Profiles of Criminal Justice Systems in Europe and North America* (in English, French, and Russian); *Selected Issues in Criminal Justice* (in English, French, and Russian); *Towards a Victim Policy in Europe* (in English, French, and Russian).

EUROPEAN INSTITUTE FOR THE MEDIA (EIM) (Europaisches Medieninstitut - IEC)

Kaistrasse 13, D-40221 Dusseldorf, Germany **PHONE:** 49 211 901040 **FX:** 49 211 9010456 **E-MAIL:** info@eim.org **WEBSITES:** http://www.eim.org **FOUNDED:** 1983. **OFFICERS:** Prof. Dr. Bernd-Peter Lange, Dir.Gen. **MEMBERS:** 20. **STAFF:** 50. **BUDGET:** DM 5,000,000. **LANGUAGES:** English, French, German, Russian.

DESCRIPTION: Acts as a forum for the discussion of media aims and policies by representing European public and professional media interests. Undertakes research on the developing roles and influence of the media and proposes appropriate media policies in Europe. Encourages European aid and technical assistance to Eastern European countries for media development. Conducts research on the interdependence of European countries in the communications field. **PUBLICATIONS:** *The Bulletin* (in English, French, and German), quarterly. Overview of European media developments; *Macedonian Media Bulletin* (in English and Macedonian); *Media Facts* (in English, French, and German), periodic; *Media Monitoring Reports of Elections in Central and Eastern Europe*; *Media Monographs*, 3-4/year; *SREDA European-Russian Media Magazine* (in English and Russian); *Television and Film Forum*, annual; *Ukrainian Media Bulletin* (in English, Russian, and Ukrainian).

EUROPEAN INVESTMENT BANK (EIB) (Banque Europeenne d'Investissement - BEI)

100, blvd. Konrad Adenauer, L-2950 Luxembourg, Luxembourg **PHONE:** 352 43791 **FX:** 352 437704 **TX:** 3530 BNKEU LU **E-MAIL:** info@cib.org **WEBSITES:** http://www.eib.org **FOUNDED:** 1958. **OFFICERS:** Henry Marty-Gauquie, Dir. of Communications. **MEMBERS:** 15. **STAFF:** 970.

DESCRIPTION: Purpose is to finance capital investment projects that further the European Union's internal and external economic objectives. Focuses on regional development, communications infrastructure, environmental protection, industrial competitiveness, and energy. Supports growth and employment within Europe. Provides loans primarily within the European Union, but does finance investment in some 120 countries outside the EU within the framework of the Union's external cooperation policies. **PUBLICATIONS:** *Annual Report and Annual Brochure* (in Danish, Dutch, English, Finnish, French, German, Greek, Italian, Portuguese, Spanish, and Swedish). Also publishes European Investment Bank: Statutes, and briefing series.

EUROPEAN LEAGUE AGAINST RHEUMATISM (EULAR) (Ligue Europeenne Contre le Rhumatisme)

Witikonerstrasse 15, CH-8032 Zurich, Switzerland **PHONE:** 41 1 3839690 **FX:** 41 1 3839810 **E-MAIL:** eular@bluewin.ch **WEBSITES:** http://www.eular.org **FOUNDED:** 1947. **OFFICERS:** Fred K. Wyss, Exec.Sec. **MEMBERS:** 12,000. **STAFF:** 2. **LANGUAGES:** English, French, German.

DESCRIPTION: Members of social and scientific organizations and pharmaceutical firms in 40 countries. Maintains small collection of rheumatology journals. **PUBLICATIONS:** *Annals of the Rheumatic Diseases* (in English), monthly.

EUROPEAN LEAGUE OF INSTITUTES OF THE ARTS (ELIA)

Waterloo Plein 219, NL-1001 PG Amsterdam, Netherlands **PHONE:** 31 20 6203936 **FX:** 31 20 6205616 **E-MAIL:** elia@elia.ahk.nl **WEBSITES:** http://www.elia.ahk.nl **FOUNDED:** 1990. **OFFICERS:** Carla Delfos, Exec.Dir. **MEMBERS:** 330. **STAFF:** 7. **LANGUAGES:** English.

DESCRIPTION: Institutes of higher education in the arts are members; individuals with an interest in artistic education are associates. Promotes effective education in all artistic disciplines; facilitates communication and exchange of information among members and between members and other international arts organizations. Lobbies government agencies and international organizations on matters pertaining to artistic education and arts funding. **PUBLICATIONS:** *ELIA News Letter* (in English), every 5 weeks. Includes reports on international conferences, EC funding programmes, and announcements of arts activities; *European Journal of Arts Education*, semiannual.

EUROPEAN LIBERAL, DEMOCRAT, AND REFORM PARTY (Parti Europeen des Liberaux, Democrates, et Reformateurs)

Bureau Leo 5 1/2C 54, 97, rue Belliard, B-1047 Brussels, Belgium **PHONE:** 32 2 2843242, 32 2 2843169 **FX:** 32 2 2311907 **FOUNDED:** 1976. **OFFICERS:** Christian Ehlers, Sec.Gen. **MEMBERS:** 33. **STAFF:** 5. **BUDGET:** 12,000,000 BFr. **LANGUAGES:** English, French.

DESCRIPTION: Political parties in 12 member countries of the European Union, and affiliates from non-EU member countries in Europe. Works to unite political parties in the European Community within the framework of liberal and democratic ideas. **PUBLICATIONS:** *ELDR Newsletter* (in English and French), quarterly. Contains information on the activities of the Party and its members.; *Electoral Programme*; *Nutshell*; *Short History*; *Vade-mecum*, periodic.

EUROPEAN LIVESTOCK AND MEAT TRADING UNION (UECBV) (Union Europeenne du Commerce du Betail et de la Viande - UECBV)

c/o Jean Luc Meriaux, 81 A, rue de la Loi, bte. 9, B-1040 Brussels, Belgium **PHONE:** 32 2 2304603 **FX:** 32 2 2309400 **E-MAIL:** uecbv@pophost.eunet.be **FOUNDED:** 1952. **OFFICERS:** Jean Luc Meriaux, Sec.Gen. **MEMBERS:** 40. **STAFF:** 4. **LANGUAGES:** English, French, German.

DESCRIPTION: National livestock and meat federations; members of the European Association of Livestock Markets. Promotes the international livestock and meat trade. Represents and defends members' interests before the European Economic Commission and internationally.

EUROPEAN MARKETING ASSOCIATION (EMA)

18 St. Peters Hill, Brixham, Devon, England **FOUNDED:** 1969. **OFFICERS:** Philip Allen, Gen.Sec. **STAFF:** 10. **LANGUAGES:** English.

DESCRIPTION: Members are marketing executives and researchers; manufacturing companies. Promotes the marketing profession in Europe. Facilitates the exchange of information and experience among members; informs members of current developments in marketing techniques. Offers training course in marketing and industrial marketing research through the European College of Marketing and Marketing Research sponsors seminars. **PUBLICATIONS:** *European Industrial Marketing Research Personal Contact Directory* (in English), periodic; *European Marketing Newsletter*, quarterly; *Journal of International Marketing and Marketing Research*, 3/year; *Journal of International Selling & Sales Management*, semiannual.

EUROPEAN AND MEDITERRANEAN PLANT PROTECTION ORGANIZATION (EPPO) (Organisation Europeenne et Mediterraneenne pour la Protection des Plantes - OEPP)

1, rue Le Notre, F-75016 Paris, France **PHONE:** 33 1 45207794 **FX:** 33 1 42248943 **E-MAIL:** hq@eppo.fr **WEBSITES:** http://www.eppo.org **FOUNDED:** 1951. **OFFICERS:** Dr. I.M. Smith, Dir.Gen. **MEMBERS:** 41. **STAFF:** 12. **BUDGET:** 5,000,000 Fr. **LANGUAGES:** English, French.

DESCRIPTION: Membership consists of national governments. Promotes cooperation among European governments in the area of crop and forest protection, with an emphasis on preventing the introduction and spread of pests. **PUBLICATIONS:**

EPPO Bulletin/Bulletin OEPP (in English and French), quarterly. Includes conference proceedings; *EPPO Technical Documents*.

EUROPEAN METAL UNION (EMU) (Europaische Metall Union - EMU)

c/o Metaalunie, Einsteinbaan 1, PO Box 2600, NL-3430 GA Nieuwegein, Netherlands **PHONE:** 31 30 6053344 **FX:** 31 30 6053115 **E-MAIL:** info@metaalunie.nl **FOUNDED:** 1954. **OFFICERS:** Harm-Jan Keijer, Sec.Gen. **MEMBERS:** 6. **LANGUAGES:** English, French, German.

DESCRIPTION: Members are national unions of employers in the metal trade and associations with similar interests. Works to coordinate and exchange information about professional and vocational training in the field. Compiles statistics. Member of European Organisation for Crafts and Small and Medium Sized Enterprises (UEAPME).

EUROPEAN MINE, CHEMICAL, AND ENERGY WORKERS' FEDERATION (EMCEF) (Europaische Foderation der Bergbeu-, Chemie-, und Energie Gewerkschaften)

109, ave. Emile de Beco, B-1050 Brussels, Belgium **PHONE:** 32 2 6262180 **FX:** 32 2 6460685 **E-MAIL:** s.eannon@emcef.be **FOUNDED:** 1996. **OFFICERS:** Franco Bisegna, Gen.Sec. **MEMBERS:** 3,000,000. **STAFF:** 8. **LANGUAGES:** English, French, German.

DESCRIPTION: Members are representatives of mining, cardboard, cellulose, cement, ceramics, chemical, glass, oil refinery, energy, paper, plastic, and rubber industry unions from 27 European countries. Promotes the interests of workers.

EUROPEAN MOLECULAR BIOLOGY ORGANIZATION (EMBO)

Postfach 10 22 40, D-69012 Heidelberg, Germany **PHONE:** 49 6221 383031 **FX:** 49 6221 384879 **TX:** 461613 EMBL D **E-MAIL:** embo@embl-heidelberg.de **WEBSITES:** http://www.embo.org **FOUNDED:** 1963. **OFFICERS:** Dr. Frank Gannon, Exec.Dir. **MEMBERS:** 900. **STAFF:** 3. **BUDGET:** DM 20,000,000. **LANGUAGES:** English, French, German.

DESCRIPTION: Promotes the advancement of molecular biology in Europe and neighboring countries. Administers program, funded by the European Molecular Biology Conference, consisting of fellowships and courses. Holds courses and workshops. **PUBLICATIONS:** *EMBO Journal* (in English), semimonthly.

EUROPEAN MORTGAGE FEDERATION (EMF) (Federation Hypothecaire Europeenne)

14, ave. de la Joyeuse Entree, bte. 2, B-1040 Brussels, Belgium **PHONE:** 32 2 2854030 **FX:** 32 2 2854031 **E-MAIL:** emf.be@skynet.be **FOUNDED:** 1967. **OFFICERS:** Judith Hardt, Sec.Gen. **MEMBERS:** 37. **STAFF:** 8. **BUDGET:** 35,000,000 BFr. **LANGUAGES:** English, French, German.

DESCRIPTION: Members are national associations of mortgage-lending institutions and individual mortgage institutions in 16 countries. Promotes measures within the European Union in the interest of mortgage credit and lending institutions; provides European authorities with advice concerning mortgage credit and lending practices. Publishes information about mortgage credit and property markets. **PUBLICATIONS:** Annual Reports, annual; Books, periodic; Directories, periodic; Proceedings, periodic; Reports, periodic; Surveys, periodic.

EUROPEAN NETWORK OF POLICEWOMEN

Postbus 1102, NL-3800 BC Amersfoort, Netherlands **PHONE:** 31 33 4654019 **FX:** 31 33 4654083 **E-MAIL:** info@enp.nl **WEBSITES:** http://www.enp.nl **FOUNDED:** 1989. **OFFICERS:** Trudy Manders, Dir. **MEMBERS:** 1,500. **STAFF:** 3. **LANGUAGES:** Dutch, English, French, German, Italian, Spanish.

DESCRIPTION: Seeks to enhance the status of female police officers throughout Europe. Fosters mutual support and exchange of experiences among female police officers. Supports international research on topics of interest to policewomen. Encourages formation of national networks of policewomen. **PUBLICATIONS:** *ENP, Five Years European Network of Policewomen* (in English and French); *Equal Treatment of Policewomen in the European Community* (in Dutch and English); *Everything You Always Wanted to Know about Policewomen but Were Afraid to Ask*;

Facts, Figures and General Information about Policewomen in Europe (in English); *How to Combat Sexual Harassment within the European Police Services* (in English); *Women in European Policing, 'What's It All About!'* (in English). Provides country by country description of nine European police forces, including information of legislation, equality, history, and development of women.; Annual Report (in English), annual; Newsletter, quarterly.

EUROPEAN NEWSPAPER PUBLISHERS' ASSOCIATION (ENPA) (Association Europeenne des Editeurs de Journaux)

rue des Pierres 29/8, B-1000 Brussels, Belgium **PHONE:** 32 2 5510190 **FX:** 32 2 5510199 **E-MAIL:** vanderstraaten@enpa.be **WEBSITES:** http://www.enpa.be **FOUNDED:** 1961. **OFFICERS:** Michel Vander Straeten, Dir.Gen. **MEMBERS:** 20. **STAFF:** 3. **BUDGET:** 20,000,000 BFr. **LANGUAGES:** Dutch, English, French, German.

DESCRIPTION: Members are national associations of daily newspaper publishers in member countries of the European Union. Represents members' interests before the Union the European Parliament, and other European bodies. **PUBLICATIONS:** Newsletter (in English, French, and German), monthly.

EUROPEAN ORGANISATION FOR CIVIL AVIATION EQUIPMENT (EUROCAE) (Organisation Europeenne pour l'Equipement de l'Aviation Civile - EUROCAE)

17, rue Hamelin, F-75783 Paris Cedex 16, France **PHONE:** 33 1 45057188, 33 1 45057235 **FX:** 33 1 45057230 **TX:** SYCELEC 611045 F **E-MAIL:** eurocae@compuserve.com **WEBSITES:** http://www.eurocae.org **FOUNDED:** 1963. **OFFICERS:** Francis Grimal, Sec.Gen. **MEMBERS:** 76. **STAFF:** 4. **BUDGET:** 2,400,000 Fr. **LANGUAGES:** English, French.

DESCRIPTION: Members are manufacturers of electronic equipment, aircraft and engine builders, and national aviation administrations in 10 countries. Objectives are to: advance the application and technologies of electronics to civil aviation; study and resolve technical problems facing users and manufacturers of aviation equipment; advise and assist international bodies in the establishment of international standards.

EUROPEAN ORGANISATION FOR THE EXPLOITATION OF METEOROLOGICAL SATELLITES (EUMETSAT) (Europaische Organisation fur die Nutzung von Wettersatelliten)

Am Kavalleriesand 31, D-64295 Darmstadt, Germany **PHONE:** 49 6151 8077 **FX:** 49 6151 807555 **TX:** 419 320 metsatd **E-MAIL:** ops@eumetsat.de **WEBSITES:** http://www.eumetsat.de **FOUNDED:** 1986. **OFFICERS:** Dr. Tillmann Mohr, Dir. **MEMBERS:** 17. **STAFF:** 136. **LANGUAGES:** English, French.

DESCRIPTION: Seeks to establish and maintain the long-term continuity of European systems of operational meteorological satellites. Contributes to a global meteorological satellite observing system coordinated with other space-faring nations. Sponsors students attending the International Space University. **PUBLICATIONS:** *Image* (in English and French), semiannual; Brochures; Annual Report. Additional publications available on request.

EUROPEAN ORGANIZATION FOR NUCLEAR RESEARCH (CERN) (Organisation Europeenne pour la Recherche Nucleaire - CERN)

CH-1211 Geneva 23, Switzerland **PHONE:** 41 22 7676111 **FX:** 41 22 7676555 **TX:** 419000 CER CH **WEBSITES:** http://www.cern.ch **FOUNDED:** 1954. **OFFICERS:** Prof. Luciano Maiani, Dir.Gen. **MEMBERS:** 19. **STAFF:** 2,700. **BUDGET:** 913,000,000 SFr. **LANGUAGES:** English, French, German.

DESCRIPTION: Members are countries participating in a collaborative program of subnuclear research. Goals are: development of new technologies; provision of particle physics research facilities unavailable to physicists in member countries. (Particle physics, also known as subnuclear or high energy physics, involves the study of the behavior of the smallest particles of matter.) Organization conducts research primarily for scientific goals and is not concerned with the development of nuclear power systems or weapons. Research is carried out at the European Laboratory for Particle Physics located in Geneva, Switzerland. **PUBLICATIONS:** *Annual Report* (in English and French); *CERN Courier* (in English and French), 10/year.

EUROPEAN ORGANIZATION FOR QUALITY (EOQ) (Organisation Europeenne pour la Qualite - EOQ)

Postfach 5032, CH-3001 Bern, Switzerland **PHONE:** 41 31 3206166, 41 31 3206720 **FX:** 41 31 3206828 **E-MAIL:** eoq@aey.ch **FOUNDED:** 1956. **OFFICERS:** Mr. Martin Hodlei, Sec.Gen. **MEMBERS:** 31. **STAFF:** 3. **LANGUAGES:** English, French.

DESCRIPTION: Members are European national quality management organizations and institutions; institutions and individuals from non-European countries. Objectives are to: enhance European competitiveness; facilitate the exchange of information and experience; develop knolwdge in the area of quality theory and practice. **PUBLICATIONS:** *EOQ Glossary of Terms Used in the Management of Quality*; *European Quality* (in English), bimonthly. Includes editorials, reprints of papers, information on activities and conferences, and book and periodical reviews.; *Proceedings of Annual Congress*; *Seminar Proceedings*; Annual Report (in English); Handbook, annual; Proceedings.

EUROPEAN ORGANIZATION FOR RESEARCH AND TREATMENT OF CANCER (EORTC) (Organisation Europeenne pour la Recherche et le Traitement du Cancer)

83, ave. Mounier, bte. 11, B-1200 Brussels, Belgium **PHONE:** 32 2 7741630, 32 2 7741611 **FX:** 32 2 7723545 **E-MAIL:** eortc@eortc.be **WEBSITES:** http://www.eortc.be **FOUNDED:** 1962. **OFFICERS:** F. Meunier M.D., Dir.Gen. **MEMBERS:** 2,500. **STAFF:** 70. **BUDGET:** 5,937,000 BFr. **LANGUAGES:** English.

DESCRIPTION: Members are doctors, pharmacologists, clinicians, statisticians, computer analysts, and others in 25 countries involved in the development of anticancer therapies. Aims to develop cancer research in Europe through the coordination of joint research projects by hospitals and laboratories. Maintains screening program of potential anticancer agents and clinical research groups formed to carry out trials with new therapeutic agents. Maintains EORTC New Drug Development Program to coordinate the development of new anticancer medications; operates EORTC Central Office and Data Center to coordinate and conduct cancer clinical trials performed by a network of 2000 doctors in over 300 hospitals in Europe; conducts educational and research programs as well as health, economics, and quality of life studies. Compiles statistics. **PUBLICATIONS:** *EORTC Organisation, Activities and Current Research* (in English), annual; *European Journal of Cancer* (in English), monthly.

EUROPEAN ORGANIZATION OF THE WORLD CONFEDERATION OF TEACHERS (EOWCT)

33 Rue De Treves, B-1040 Brussels, Belgium **PHONE:** 32 2 2854729 **FX:** 32 2 2308722 **E-MAIL:** wct@cmt-wcl.org **FOUNDED:** 1963. **OFFICERS:** Gaston De La Haye, Sec.Gen. **MEMBERS:** 6,000,000. **STAFF:** 2. **BUDGET:** 8,000,000 BFr. **LANGUAGES:** Dutch, English, French.

DESCRIPTION: Members are national labor unions representing teachers in Europe. Seeks to advance the trade union movement; promotes economic and professional advancement of teachers. Facilitates establishment of teachers' unions on the local, state, national, and regional levels; serves as a forum for the exchange of information among members; sponsors continuing professional development and vocational education programs.

EUROPEAN ORTHODONTIC SOCIETY (EOS)

49 Hallam St., Flat 31, London W1N 5LL, England **PHONE:** 44 171 9352795 **FX:** 44 171 9352795 **E-MAIL:** eoslondon@compuserve.com **FOUNDED:** 1907. **OFFICERS:** Prof. J.P. Moss, Hon.Sec. **MEMBERS:** 2,500. **STAFF:** 1. **LANGUAGES:** English.

DESCRIPTION: Members are orthodontists in 78 countries promoting the science of orthodontics. **PUBLICATIONS:** *European Journal of Orthodontics* (in English), bimonthly.

EUROPEAN PATENT OFFICE (EPO) (Europaisches Patentamt - EPA)

Erhardtstrasse 27, D-80331 Munich, Germany **PHONE:** 49 89 23990 **FX:** 49 89 23994465 **TX:** 523656 EPMU D **WEBSITES:** http://www.european-patent-office.org **FOUNDED:** 1977. **OFFICERS:** Ingo Kober, Pres. **MEMBERS:**

19. **STAFF:** 4,000. **BUDGET:** DM 1,173,400,000. **LANGUAGES:** English, French, German.

DESCRIPTION: Aims to make the process of acquiring patents easier, less expensive, and more reliable by establishing a single European legal procedure for granting patents in the 19 states contracted to the European Patent Convention. Grants patents that are enforceable, in contracting states, as nationally granted patents. **PUBLICATIONS:** *Annual Report* (in English, French, and German); *Directory of Professional Representatives*, annual; *European Patent Bulletin*, weekly; *How to Get a European Patent*; *List of Publications* (in English, French, and German); *Official Journal of the European Patent Office* (in English, French, and German), monthly; Brochures.

EUROPEAN PEOPLES' PARTY/ CHRISTIAN DEMOCRATS (EPP) (Parti Populaire Europeen - PPE)

67, rue d'Arlon, B-1040 Brussels, Belgium **PHONE:** 32 2 2854144 **FX:** 32 2 2854141 **FOUNDED:** 1976. **OFFICERS:** Klaus Welle, Sec.Gen. **MEMBERS:** 20. **STAFF:** 12. **BUDGET:** US$800,000. **LANGUAGES:** Dutch, English, French, German, Greek, Italian, Spanish.

DESCRIPTION: Seeks the unification of European states within a federal political structure, rooted in Christian standards, that protects the well-being of all citizens, allows individuals to enjoy their freedom in solidarity, favors the development of social relationships, protects the environment, and acts for peace and security. **PUBLICATIONS:** *CD-News* (in Dutch, English, French, German, Italian, and Spanish), weekly; *EPP-News* (in Dutch, English, French, German, Italian, and Spanish), weekly.

EUROPEAN RAILWAY WAGON POOL (ERWP) (Communaute des Wagons EUROP)

SNCB/NMBS-Reseau Gerant, Departement T - Service 10.3, 85, rue de France, B-1060 Brussels, Belgium **PHONE:** 32 2 5254130, 32 2 5253143 **FX:** 32 2 5254453 **FOUNDED:** 1953. **OFFICERS:** J. Dekempeneer, Exec. officer. **MEMBERS:** 9. **LANGUAGES:** French, German, Italian.

DESCRIPTION: Members are European railway administrations. Promotes members' interests in dealing within and among nations. Disseminates information to members.

EUROPEAN SALT PRODUCERS ASSOCIATION (ESPA)

17, rue Daru, F-75008 Paris, France **PHONE:** 33 1 47665290 **FX:** 33 1 47665266 **E-MAIL:** bmoinier@eu-salt.com **WEBSITES:** http://www.eu-salt.com **FOUNDED:** 1958. **OFFICERS:** Bernard Moinier, Sec.Gen. **MEMBERS:** 13. **LANGUAGES:** English, French, German.

DESCRIPTION: Members are salt producers from 13 countries. Defends members' interests; provides representation for producers before European and international bodies. Maintains a documentation center to facilitate technical advancement in the salt industry. **PUBLICATIONS:** *Espa*, periodic; *Regulations* (in English), semiannual; *Salt Echo* (in English), quarterly. For members only; *Salt Throughout the World*, annual. Contains information on technical matters, medical matters, and patents.

EUROPEAN SOCIETY OF CULTURE (ESC) (Societe Europeenne de Culture - SEC)

Giudecca S4P, I-30133 Venice, Italy **PHONE:** 39 41 5230210 **FX:** 39 41 5231033 **FOUNDED:** 1950. **OFFICERS:** Michelle Campagnolo-Bouvier, Gen.Sec. **MEMBERS:** 1,900. **STAFF:** 3. **LANGUAGES:** English, French, German, Italian.

DESCRIPTION: Members are scientists, writers, and interested others in 60 countries. Serves to protect and affirm culture by uniting individuals of different origins, discussing solutions to world crises, and debating other philosophical and political issues. Organizes exhibition of documents. **PUBLICATIONS:** *Comprendre: Revue de Politique de la Culture et Organe de la SEC* (in English and French), periodic.

EUROPEAN SOCIETY FOR ENVIRONMENT AND DEVELOPMENT (IPRE)

c/o W.J. Cairns, blvd. Lambermont, 432, BP 1, B-1030 Brussels, Belgium **PHONE:** 32 2 2450032 **FX:** 32 2 2458523 **E-MAIL:** 100432.2650@compuserve.com **FOUNDED:**

1976. **OFFICERS:** Mark Dubrulle, Pres. **MEMBERS:** 200. **LANGUAGES:** Dutch, English, French, German, Italian, Spanish.

DESCRIPTION: Members are individuals in 20 countries working professionally in environmental affairs and involved in public administration, industry, or employed at an international organization, university, research institute, or public interest group. Goals are to provide a forum for the discussion of environmental problems and to promote an awareness of the complexity of these problems. Conducts workshops and symposia. **PUBLICATIONS:** *Future European Environmental Policy and Subsidiary*; *Symposia Proceedings* (in English), semiannual.

EUROPEAN SOLIDARITY TOWARD EQUAL PARTICIPATION OF PEOPLE (EUROSTEP)

115 Rue Stevin, B-1000 Brussels, Belgium **PHONE:** 32 2 2311659, 32 2 2311709 **FX:** 32 2 2303780 **E-MAIL:** admin@eurostep.org **WEBSITES:** http://www.oneworld.org/eurostep **FOUNDED:** 1990. **OFFICERS:** Simon Stocker, Dir. **MEMBERS:** 21. **STAFF:** 3. **BUDGET:** US$285,714. **LANGUAGES:** Dutch, English, French, German.

DESCRIPTION: Members are European nongovernmental development organizations. Promotes justice and equal opportunity for all people worldwide. Works to influence official development policies of national governments, the European Union and multilateral institutions and to advocate development models based on the perspective of NGO's. Facilitates communication and exchange of information among members. Conducts research and educational programs on topics including sustainable development, gender, lome treaty, and social watch. **PUBLICATIONS:** *The Reality of Aid* (in English), annual. An independent review of development cooperation.

EUROPEAN SPACE AGENCY (ESA) (Agence Spatiale Europeene - ASE)

8-10, rue Mario Nikis, F-75738 Paris Cedex 15, France **PHONE:** 33 1 53697654 **FX:** 33 1 53697560 **TX:** ESA 202746 **WEBSITES:** http://www.esa.int **FOUNDED:** 1975. **OFFICERS:** M. Antonio Rodota, Dir.Gen. **MEMBERS:** 14. **STAFF:** 1,707. **BUDGET:** 2,600,000 Fr. **LANGUAGES:** English, French, German.

DESCRIPTION: Members are representatives of Austria, Belgium, Denmark, Finland, France, Germany, Italy, Netherlands, Norway, Republic of Ireland, Spain, Sweden, Switzerland, and United Kingdom. Provides the means whereby West European nations may cooperatively advance research into space and space technology; applies such research in the development of exclusively peaceful and primarily scientific operational space systems. Works for a long-term European space policy that will culminate in a fully integrated European space program; seeks an industrial policy, agreed upon by member nations, that will meet the agency's industrial needs while strengthening European industry. Has launched and operates several scientific and applications satelites. Conducts development programs in the fields of space, science, applications, and transportation. Maintains the European Space Research and Technology Centre (ESTEC), the European Space Operations Centre (ESOC), and the European Space Research Institute (ESRIN). Cooperates closely with the National Aeronautics and Space Administration in the United States. **PUBLICATIONS:** *Conference Proceedings*, periodic; *Earth Observation Quarterly* (in English and French); *ECSL News*, quarterly; *ESA Bulletin*, quarterly; *Microgravity News* (in English and French), 3/year; *Procedures, Standards, and Specifications*, periodic; *Reaching for the Skies*, quarterly. Contains information on space transportation systems; Brochures.

EUROPEAN TELECOMMUNICATIONS SATELLITE ORGANIZATION (EUTELSAT) (Organisation Europeenne de Telecommunications par Satellite)

70, rue Balard, Cedex 15, F-75502 Paris Cedex 15, France **PHONE:** 33 1 53984747 **FX:** 33 1 53984679 **TX:** 203823 **WEBSITES:** http://www.eutelsat.org **FOUNDED:** 1977. **OFFICERS:** Jean Grenier, Dir.Gen. **MEMBERS:** 45. **STAFF:** 270. **LANGUAGES:** English, French.

DESCRIPTION: Members are public and private telecommunications operators in 45 countries. Designs, develops, constructs, operates, and maintains the space segment of the European telecommunications satellite system. Provides space segment capacity for telephone, television, business communications, transponder lease services, and land mobile services. **PUBLICATIONS:** Annual Report; Brochures; Newsletter, quarterly.

EUROPEAN TRAVEL COMMISSION (ETC) (Commission Europeenne de Tourisme - CET)

Rue Du Marche aux Herbes 61, B-1000 Brussels, Belgium **PHONE:** 32 2 5040303 **FX:** 32 2 5141843 **E-MAIL:** etc@planetinternet.be **WEBSITES:** http://www.goeurope.com **FOUNDED:** 1948. **OFFICERS:** Walter Leu, Exec.Dir. **MEMBERS:** 29. **STAFF:** 3. **BUDGET:** US$1,200,000. **LANGUAGES:** English, French.

DESCRIPTION: Members are national tourist organizations from 29 countries working to promote Europe as a tourist destination overseas. Conducts research; sponsors seminars. **PUBLICATIONS:** Annual Report (in English and French), annual. Also publishes pamphlets on trends of tourism in Europe.

EUROPEAN UNION (EU)

c/o The European Commission, 200 rue de la Loi, 1049 Brussels, Belgium **PHONE:** 2 299 11 11 **FX:** 2 295 01 38 **TX:** 21877 **WEBSITE:** http://europa.eu.int **FOUNDED:** 1952. **MEMBERS:** 15. **BUDGET:** ECU 89,136,900,000. **LANGUAGES:** Danish, Dutch, English, Finnish, French, German, Greek, Italian, Portuguese, Spanish, Swedish.

DESCRIPTION: Members are European states that are signatories of the Treaty on European Union. According to this treaty, the ultimate goal of the EU is to create "an ever closer union among the peoples of Europe, in which decisions are taken as closely as possible to the citizen." To reach this goal the EU works to eliminate physical and economic boundaries between its members, strengthen economic and social cohesion, and establish an economic and monetary union and single European currency. The EU coordinates the economic and foreign policies of its members, adjudicates disputes between its members on these matters, and represents the interests of its member states to other nations and international organizations. The EU also seeks to represent the interests and uphold the rights of citizens of its member states, all of whom are also considered to be citizens of the European Union. Member states retain their sovereignty and independence of action within the limitations established by treaty. In what is called the principle of subsidiarity, the actions of the EU are restricted to those that cannot be sufficiently achieved by its member states. The EU divides its legislative and executive functions between the European Commission, the European Council, the Council of the European Union (often called the Council of Ministers) and the European Parliament. The Court of Justice of the European Communities is the judicial body of the EU and ensures that the EU and its individual members obey their treaty requirements. In addition to these governing bodies, the EU maintains a large number of subsidiary organizations that advise the governing bodies and assist them in carrying out policy. The EU has its origins in a series of treaties between European states which established international organizations. First among these organizations was the European Coal and Steel Community, established in 1952. The Treaty of Rome (effective 1958) established the European Economic Community and the European Atomic Energy Community. In 1965 another treaty established a single council and a single commission to govern all three of these Communities. Many additional treaties expanded or amended these communities, culminating in the Treaty on European Union in 1993 that formally established the European Union. The original Communities continue to exist within the framework of the EU. **PUBLICATIONS:** *Bulletin of the European Union*, 10/year; *European Economy*, semiannual; *European Voice*, weekly; *General Report on the Activities of the European Union*, annual. Many other reports and statistical documents.

EUROPEAN UNION OF COACHBUILDERS (EUC) (Union Europeenne de la Carrosserie - UEC)

46, blvd. de la Woluwe, bte. 14, B-1200 Brussels, Belgium **PHONE:** 32 2 7786200 **FX:** 32 2 7786222 **E-MAIL:** mail@federauto.be **FOUNDED:** 1938. **OFFICERS:** Hilde Vander Stichele, Gen.Sec. **MEMBERS:** 7. **STAFF:** 3. **LANGUAGES:** Dutch, English, French, German.

DESCRIPTION: Members are national associations of coachbuilders. Works to harmonize vehicle construction regulations. Compiles statistics. Offers professional training.

EUROPEAN UNION OF THE FRUIT AND VEGETABLE WHOLESALE, IMPORT AND EXPORT TRADE (EUCOFEL) (Union Europeenne du Commerce de Gros, d'Expedition, d'Importation, et d'Exportation en Fruits et Legumes)

29, Rue Jenneval, B-1000 Brussels, Belgium **PHONE:** 32 2 7361584, 32 2 7361654 **FX:** 32 2 7321747 **E-MAIL:** eucofel.fruittrade.org@skynet.be **FOUNDED:** 1958. **OFFICERS:** Mr. V.A. van Dijk, Sec.Gen. **MEMBERS:** 29. **STAFF:** 3. **LANGUAGES:** English, French.

DESCRIPTION: Members are fruit and vegetable importers, exporters, and wholesalers. Promotes the interests of the produce wholesale, import, and export trade; represents the trade within the European Economic Communities and before other international bodies. Conducts research into problems peculiar to the fruit and vegetable trade. **PUBLICATIONS:** *Activity Report* (in English and French), annual; *Compendium Eu Quality Standards*; *Information Bulletin* (in English and French), monthly.

EUROPEAN UNION OF JEWISH STUDENTS (EUJS) (Europaische Union Judischer Studenten)

Avenue Antoine Depage 3, B-1050 Brussels, Belgium **PHONE:** 32 2 6477279 **FX:** 32 2 6482431 **E-MAIL:** eujsoffice@compuserve.com **FOUNDED:** 1978. **OFFICERS:** Joelle Fiss, Chair. **MEMBERS:** 170,000. **STAFF:** 5. **BUDGET:** US$1,400,000. **LANGUAGES:** English, French, Hebrew.

DESCRIPTION: Encourages Jewish educational activities on European campuses and seeks to heighten awareness and appreciation of the Hebrew language and Jewish history and religion. Fosters greater understanding and cooperation among religious and national student organizations. Promotes positive identification among European Jews with Israel through publications, debates, and visits; makes members aware of educational opportunities in Israel. Campaigns for kosher facilities on European campuses and for improvement of facilities and subsidies for European students studying in Israel. Represents Jewish students before European forums dealing with cultural and political issues. Works to fight all forms of fascism, racism, and anti-Semitism. Maintains charitable program, speakers' bureau, and placement service. Conducts research; sponsors seminars. **PUBLICATIONS:** *EUJS News* (in English and French), bimonthly; *Newsletter* (in English), 10/year.

EUROPEAN UNION OF MEDICAL SPECIALISTS (UEMS)

20, ave. de la Couronne, B-1050 Brussels, Belgium **PHONE:** 32 2 6495164 **FX:** 32 2 6403730 **E-MAIL:** uems@optinet.be **FOUNDED:** 1958. **OFFICERS:** Dr. R. Peiffer, Gen.Sec. **MEMBERS:** 24. **LANGUAGES:** English, French.

DESCRIPTION: Members are medical specialists in 18 countries. Works to ensure high quality care for patients. Fosters communication among members; promotes members' interests.

EUROPEAN UNION OF THE NATURAL GAS INDUSTRY (EUROGAS) (Europaische Vereinigung der Erdgaswirtschaft)

4, ave. Palmerston, B-1000 Brussels, Belgium **PHONE:** 32 2 2371126, 32 2 2371129 **FX:** 32 2 2306291 **E-MAIL:** eurogas@arcadis.be **FOUNDED:** 1990. **OFFICERS:** Peter Claus, Sec.Gen. **MEMBERS:** 13. **STAFF:** 5. **BUDGET:** 40,000,000 BFr. **LANGUAGES:** English, French, German.

DESCRIPTION: Members are national gas industries and trade associations of the gas industry. Fosters improved relations between members and international government organizations. Conducts economic studies. **PUBLICATIONS:** Brochure (in English), annual. Contains statistics.

EUROPEAN UNION OF WOMEN (EUW) (Europaische Union of Women - EFU)

Auklands, Gloucester Rd., Thornsby, Bristol BS12 1JM, England **PHONE:** 44 1454 413865 **FX:** 44 1454 412490 **E-MAIL:** euwintnat@aol.com **WEBSITES:** http://www.pfi.co.uk/euw **FOUNDED:** 1953. **OFFICERS:** Mrs. Pam Richards, Gen.Sec. **LANGUAGES:** English, French, German.

DESCRIPTION: Members are women, in 21 countries, who are active in public life or are members of parliaments and local authorities. Seeks to strengthen the influence of women on political and civic life in European nations and safeguard the social and economic rights of all individuals. Furthers cooperation among members in order to foster the conservation of European cultural heritage and support peace efforts worldwide.

EUROPEAN YOUTH FORUM (YFEU) (Forum Europeen de la Jeunesse - FEJ)

120, rue Joseph II, B-1000 Brussels, Belgium
PHONE: 32 2 2306490 **FX:** 32 2 2302123
E-MAIL: youthforum@brussels.blackbox.at
FOUNDED: 1996. **OFFICERS:** Hronn Petursdotbr, Sec.Gen. **MEMBERS:** 93. **STAFF:** 17.
LANGUAGES: English, French.

DESCRIPTION: Members are national youth committees and international youth organizations. Serves as a political platform for youth organizations; represents members' interests before institutions of the European Union Council of Europe, and UN. Promotes greater involvement of youth in the activities and development of the European institutions; encourages mutual understanding and regard for the rights of all European citizens. Addresses such issues as employment and labor market policy, poverty, youth rights, youth mobility and exchanges, North/South issues, equal rights and fair treatment of women, racism, peace and international cooperation, environment, and education and vocational training. Maintains documentation center. **PUBLICATIONS:** *Annual Report* (in English and French); *Charter of Youth Rights*; *Guide to Member Organisations* (in English and French). Lists Youth Forum member organizations.; *Is There Anybody Out There* (in English and French). Comes with accompanying report.; *Lives on Hold?* (in English and French); *Youth Opinion* (in English, French, German, and Spanish), quarterly.

EUROPEAN YOUTH FOUNDATION (EYF)

c/o Council of Europe, 30 rue Pierre-de-Coubertin, F-67000 Strasbourg Cedex, France
PHONE: 33 3 88412019 **FX:** 33 3 90214964
E-MAIL: eyf@coe.fr **FOUNDED:** 1973. **OFFICERS:** Andre-Jacques Dodin, Principal Admin. **STAFF:** 4. **BUDGET:** 16,000,000 Fr. **LANGUAGES:** English, French.

DESCRIPTION: Representatives from 40 European countries. A department established by the Council of Europe in order to involve youth organizations in the process of European cooperation and integration. Supports international youth activities that promote peace, understanding, and cooperation; encourages respect for human rights and fundamental freedoms. Provides funding to international nongovernmental youth groups for seminars, conferences, and study camps; supports projects including the production of publications, posters, and records. Conducts educational and research programs; compiles statistics. The European Youth Foundation is part of the Council of Europe's Youth Directorate. **PUBLICATIONS:** *Annual Report.*

EUROTRANSPLANT INTERNATIONAL FOUNDATION (EIF)

PO Box 2304, NL-2301 CH Leiden, Netherlands
PHONE: 31 71 5795795 **FX:** 31 71 5790057
WEBSITES: http://www.eurotransplant.nl
FOUNDED: 1966. **OFFICERS:** Dr. Bernard Cohen, Dir. **LANGUAGES:** English.

DESCRIPTION: International non-profit organ exchange organization. **PUBLICATIONS:** *Eurotransplant Newsletter* (in English), monthly.

FAUNA AND FLORA INTERNATIONAL (FFI)

Great Eastern House, Tenison Rd., Cambridge CB1 2DT, England **PHONE:** 44 1223 571000 **FX:** 44 1223 461481 **E-MAIL:** info@fauna.flora.org
WEBSITES: http://www.ffi.org.uk/ **FOUNDED:** 1903. **OFFICERS:** Mark Rose, Dir. **MEMBERS:** 4,000. **STAFF:** 20. **BUDGET:** US$1,350,000.

DESCRIPTION: Members are individuals, libraries, universities, natural history societies, other wildlife conservation organizations, and governmental departments responsible for wildlife, national parks, and tourism in 80 countries. Purpose is to prevent the extinction of species of wild animals and plants by promoting the conservation of wildlife, the establishment of new national parks, the enactment and enforcement of laws to protect wildlife, and the education of governments and individuals in the value of world wildlife as a non-renewable natural resource. Conducts research relating to endangered species. **PUBLICATIONS:** *Fauna and Flora News*, semiannual; *Good Bulb Guide*, annual; *Mountain Gorilla Update*, semiannual; *Oryx* (in English), quarterly; *Sound Wood*, semiannual.

FDI WORLD DENTAL FEDERATION

7 Carlisle St., London W1V 5RG, England **PHONE:** 44 171 9357852 **FX:** 44 171 4860183 **E-MAIL:** worldental@fdi.org.uk **WEBSITES:** http://www.fdi.org.uk **FOUNDED:** 1900. **OFFICERS:** Dr. Per Ake Zillen, Exec.Dir. **MEMBERS:** 650,000. **STAFF:** 19. **LANGUAGES:** French, German, Spanish.

DESCRIPTION: National Dental Associates and individual dentists worldwide. A federation of National Dental Association and individual member dentists with the aim of helping to promote oral health worldwide. **PUBLICATIONS:** *Community Dental Health* (in English), quarterly; *The European Journal of Prosthodontics and Restorative Dentistry* (in English), quarterly; *FDI World* (in English, French, German, and Spanish), bimonthly; *International Dental Journal* (in English, French, German, and Spanish), bimonthly.

FEDERAL UNION OF EUROPEAN NATIONALITIES (FUEN) (Foderalistische Union Europaischer Volksgruppen - FUEV)

Schiffbrucke 41, D-24939 Flensburg, Germany **PHONE:** 49 461 12855 **FX:** 49 461 180709 **E-MAIL:** info@fuen.org **WEBSITES:** http://www.fuen.org **FOUNDED:** 1949. **OFFICERS:** Armin Nickelsen, Sec.Gen. **MEMBERS:** 101. **STAFF:** 10. **BUDGET:** DM 400,000. **LANGUAGES:** English, French, German, Russian.

DESCRIPTION: Members are national groups in 29 countries united to preserve the cultures and rights of European nationals. Supports the United Nations and the Council of Europe in securing human rights and freedoms. Operates information service for scientists, politicians, and the press. Negotiates with international institutions and governments concerning nationality problems. **PUBLICATIONS:** *Information - FUEN Actuel* (in English, French, German, and Russian), semiannual.

FEDERATION OF ARAB SCIENTIFIC RESEARCH COUNCILS (FASRC)

PO Box 13027, Baghdad, Iraq **PHONE:** 964 1 8881709 **TX:** 212466ACARS1K **FOUNDED:** 1976. **OFFICERS:** Prof. Taha T. Al-Naimi, Secretary General. **STAFF:** 10. **BUDGET:** US$50,000. **LANGUAGES:** Arabic, English.

DESCRIPTION: Works to improve collaboration between scientific institutions in the Arab World, especially in the areas of food, agriculture, environment, energy, and medicine. Conducts educational and research programs. **PUBLICATIONS:** *FASRC Newsletter* (in Arabic), bimonthly.

FEDERATION OF ASIAN SCIENTIFIC ACADEMIES AND SOCIETIES (FASAS)

c/o Indian National Science Academy, Bahadur Shah Zafar Marg, New Delhi 110002, Delhi, India **PHONE:** 91 11 3232066, 91 11 3232075 **FX:** 91 11 3235648 **TX:** 3161835 INSA IN **FOUNDED:** 1984. **OFFICERS:** S.K. Sahni, Exec.Sec. **MEMBERS:** 14. **STAFF:** 3. **BUDGET:** US$25,000. **LANGUAGES:** English, Hindi.

DESCRIPTION: Members are national scientific academies and societies in Asia. Seeks to increase interest in the sciences and to advance scientific teaching, study, and practice. Facilitates communication and cooperation among members; encourages application of the sciences in national economic, social, and agricultural development programs. **PUBLICATIONS:** *Conservation of Biodiversity for Sustainable Development*; *FASAS Newsletter* (in English), annual.

FEDERATION OF COMMODITY ASSOCIATIONS (FCA) (Federation Europeene des Associations des Professionnels en Produits de Base)

Gafta House, 6 Chapel Pl., Rivington St., London EC2A 3SH, England **PHONE:** 44 171 8149666 **FX:** 44 171 8148383 **TX:** 886984 **E-MAIL:** fca@gafta.demon.co.uk **FOUNDED:** 1943. **OFFICERS:** Pamela Kirby Johnson, Sec. **MEMBERS:** 58. **STAFF:** 16. **LANGUAGES:** English, French, Italian, Spanish.

DESCRIPTION: Members are commodity trade associations; companies, firms, or individual traders in commodities. Promotes and protects the commercial interests of members. Seeks to establish an effective liaison between members and the European Commission through the exchange of information and the coordination of problem-solving procedures. Offers information service on current and projected legislation. Maintains close links with educational establishments. **PUBLICATIONS:** *Book of Rules*.

FEDERATION OF EUROPEAN BIOCHEMICAL SOCIETIES (FEBS)

c/o Prof. Vito Turk, Dept. of Biochemistry and Mol. Biology, Jozef Stefan Inst., Jamova 39, SLO-1000 Ljubljana, Slovenia **PHONE:** 386 61 1257080, 386 61 1773900 **FX:** 386 61 273594 **TX:** 31296 JOSTIN **E-MAIL:** vito.turk@ijs.si **FOUNDED:** 1964. **OFFICERS:** Prof. Vito Turk, Gen.Sec. **MEMBERS:** 39,000. **BUDGET:** 7,000,000 Din. **LANGUAGES:** English, French, German.

DESCRIPTION: Purpose is to further research and education in the field of biochemistry and to disseminate research findings. Holds advanced courses; offers fellowships. **PUBLICATIONS:** *European Journal of Biochemistry*, biweekly; *FEBS Letters* (in English), weekly.

FEDERATION OF EUROPEAN CHEMICAL SOCIETIES (FECS) (Federation des Societes Chimiques Europeennes - FECS)

Royal Soc. of Chemistry, Burlington House, Piccadilly, London W1V 0BN, England **PHONE:** 44 171 4403303 **FX:** 44 171 4378883 **E-MAIL:** mcewane@rsc.org **WEBSITES:** http://www.chemsoc.org/gateway/fecs.htm **FOUNDED:** 1970. **OFFICERS:** Evelyn K. McEwan, Sec.Gen. **MEMBERS:** 40. **LANGUAGES:** English, French, German, Russian.

DESCRIPTION: Members are national societies throughout Europe representing 200,000 chemists. Fosters cooperation among member societies. Encourages discussion in all fields of chemistry.

FEDERATION OF THE EUROPEAN DENTAL INDUSTRY (Federation de l'Industrie Dentaire en Europe - FIDE)

Kirchweg 2, D-50858 Cologne, Germany **PHONE:** 49 221 9486280 **FX:** 49 221 483428 **FOUNDED:** 1957. **OFFICERS:** Harald Russegger, Sec.Gen. **MEMBERS:** 11. **LANGUAGES:** English.

DESCRIPTION: Members are national associations of companies engaged in the manufacture of dental instruments and supplies. Functions as a platform for coordination of development and works to harmonize international standards within the industry; represents members' interests before government and European Community agencies. Promotes environmental protection, bar coding (HIBC), market statistics (project). **PUBLICATIONS:** *FIDE - European Dental Industry* (in English), biennial. Membership brochure; *FIDE News* (in English), 1-2/year.

FEDERATION OF FRENCH-LANGUAGE GYNECOLOGISTS AND OBSTETRICIANS (Federation des Gynecologues et Obstetriciens de Langue Francaise - FGOLF)

c/o Prof. Jean Rene Zorn, Clinique Universitaire Baudelocque, 123, blvd. de Port-Royal, F-75674 Paris Cedex 14, France **PHONE:** 33 1 42341143 **FX:** 33 1 42341231 **FOUNDED:** 1950. **OFFICERS:** Prof. Jean-Rene Zorn **MEMBERS:** 700. **LANGUAGES:** French.

DESCRIPTION: Members are French-speaking gynecologists and obstetricians. Purpose is to promote scientific study in the French language of all aspects of the biology of human reproduction. Conducts training sessions and travel and exchange programs. Maintains permanent committees to deal with special topics. **PUBLICATIONS:** *Journal de Gynecologie Obstetrique et Biologie de la Reproduction* (in French), bimonthly.

FEDERATION OF INTERNATIONAL CORPS AND ASSOCIATIONS CONSULAIRES (FICAC) (Federation Internationale des Corps et Associations Consulaires - FICAC)

PO Box 4679, Nicosia 1302, Cyprus **PHONE:** 35 72 442483 **FX:** 35 72 455785 **FOUNDED:** 1982. **OFFICERS:** Vagn Jespersen, Hon.Pres. **MEMBERS:** 44. **BUDGET:** US$10,000. **LANGUAGES:** English.

DESCRIPTION: Members are national consular associations representing 16,000 individuals in 44 countries. Promotes contact with international organizations and improved relations between members. Encourages study of trade, culture, and social activities; works to improve members' immunities, privileges, and security in receiving states; represents members at meetings dealing with the status of consuls; strives for uniform treatment of individual consuls by receiving states. Organizes periodic seminars. **PUBLICATIONS:** *Introduction to International Consular Cooperation*; *Membership List* (in English), periodic; *Priveleges and Obligations for Honorary Consuls Under the Vienna Convention on Consular Relations*; Newsletter, periodic.

FEDERATION OF ISLAMIC ASSOCIATIONS IN THE U.S. AND CANADA (FIA)

25351 5 Mile & Aubery Rd., Redford, MI 48239 **E-MAIL:** michaol2@jumo.com **FOUNDED:** 1951. **MEMBERS:** 300,000. **STAFF:** 7. **BUDGET:** US$125,000. **LANGUAGES:** Arabic, English.

DESCRIPTION: Members are religious, political, social, and educational organization that acts as an umbrella for Muslim groups in the U.S. and Canada. Objectives are to: defend the human rights of Muslims and all oppressed people through democratic, political means; promote the spirit, ethics, philosophy, and culture of Muslim heritage; answer questions and correct misconceptions about Islam; promote friendly relations between Muslims and non-Muslims of North America. Conducts charitable program, specialized education, and youth programs; distributes literature to schools, universities, and libraries. Offers placement and computerized services. Sponsors weekly radio program, Muslim Star: Voice of American-Canadian Muslims. **PUBLICATIONS:** Books. Also issues press releases.

FELLOWSHIP OF COUNCILS OF CHURCHES IN EASTERN AND SOUTHERN AFRICA (FCCESA)

PO Box 30315, Lusaka, Zambia **PHONE:** 260 1 229551 **FX:** 260 1 224308 **OFFICERS:** Rev. Violet Sampa-Bredt. **LANGUAGES:** English.

DESCRIPTION: Members are national councils of churches in eastern and southern Africa. Promotes religious observance; seeks to ensure adherence to Christian principles among development projects operating in the region. Facilitates communication and cooperation among members. Conducts community service, development, religious, and educational programs.

FEMCONSULT

Koninginnegracht 53, NL-2514 AE The Hague, Netherlands **PHONE:** 31 70 3655744 **FX:** 31 70 3623100 **E-MAIL:** gender@femconsult.nl **FOUNDED:** 1985. **OFFICERS:** Catharina M.G. van Heel, Mgr.Dir. **MEMBERS:** 600. **STAFF:** 25. **LANGUAGES:** Dutch, English, French, German, Spanish.

DESCRIPTION: Members are women consultants in 30 countries with experience in Africa, Asia, and Latin America. Promotes the participation of women in Third World development programs. Strives to: acquire assignments for women in mainstream development programs; implement development projects in Third World nations focusing on women; increase awareness of the role played by women in these development projects; maintain high standards of practice. Undertakes gender-based assignments in areas such as agriculture, irrigation, animal husbandry, environment, small scale enterprises, credit, socioeconomics, nutrition, and public health. Conducts training courses and education programs. **PUBLICATIONS:** *FEMCONSULT Newsletter* (in English), periodic; *Project Report*, periodic; *Skills Bank* (in Dutch and English), semiannual; *Symposium Report*, periodic.

FIND YOUR FEET (FYF)

37/39 Great Guildford St., London SE1 0ES, England **PHONE:** 44 20 74018794 **FX:** 44 20 77717226 **E-MAIL:** fyf@gn.apc.org **FOUNDED:** 1967. **OFFICERS:** Dan Taylor, Dir.

DESCRIPTION: Works in partnership with NGOs in Southern Africa, South Asia, and Latin America in longterm development programs promoting sustainable livelihoods. Helping to break the cycle of poverty by providing investment to enable people to increase their income and achieve longterm food security. **PUBLICATIONS:** *Seeds of Change* (in English), bimonthly.

FLYING DOCTORS OF AMERICA (FDoA)

4015 Holcomb Bridge Rd., Ste. 350922, Norcross, GA 30092 **PHONE:** (770) 209-9277 **FX:** (770) 446-9634 **E-MAIL:** fdoamerica@aol.com **FOUNDED:** 1990. **OFFICERS:** Allan Gathercoal, Founder/Director. **MEMBERS:** 7,500. **STAFF:** 7. **LANGUAGES:** English.

DESCRIPTION: Members are volunteer physicians, dentists, pharmacists, and support members. Flies volunteer health care professionals and other individuals to Third World countries to provide medical care to impoverished people. Maintains speakers' bureau. **PUBLICATIONS:** *Touch and Go*, monthly; Brochure.

FOOD AND AGRICULTURE ORGANIZATION OF THE UNITED NATIONS (FAO) (Organizacion de la Naciones Unidas para la Agricultura y la Alimentacion)

c/o Peter Lowrey, Viale delle Terme di Caracalla, I-00100 Rome, Italy **PHONE:** 39 6 57051 **FX:** 39 6 57053152 **TX:** 610181 FAO I **E-MAIL:** telex-room@fao.org **WEBSITES:** http://www.fao.org **FOUNDED:** 1945. **OFFICERS:** Jacques Diouf, Dir.Gen. **MEMBERS:** 175. **STAFF:** 4,400. **BUDGET:** US$650,000,000. **LANGUAGES:** Arabic, Chinese, English, French, Spanish.

DESCRIPTION: Objectives are to: raise the nutritional levels and living standards of people worldwide; improve food and agricultural production and distribution; increase the quality of life of rural people. Works to achieve world food security wherein all people at all times have both physical and economic access to the food they need. Implements advisory and technical assistance programs for the world agricultural community on behalf of governments and development funding agencies and advises governments on all aspects of agrarian policy and planning. Acts as forum for member nations to meet and discuss topical problems and collects, analyzes, and disseminates food and agriculture information. Promotes integrated rural development through programs involving food and fodder cultivation, animal husbandry, small-scale fisheries, and the sustainable development of forests. Conducts field programs such as reforestation, grain processing, and crop species introduction. Provides relief, emergency assistance, investment project preparation, and specialized advice to requesting countries; maintains Associate Professional Officer Scheme, whereby developed countries assign and pay technicians to work on field projects; assists developing countries in assessing needs and obtaining external capital to build agricultural systems; cooperates with regional development and financial institutions. Maintains food shortage early warning system. Sends missions to countries to analyze aspects of agricultural development strategies. Contributes to countries' policies in areas of nutrition assessment, planning, and surveillance. Recommends international policy concerning world trade in food and agricultural commodities and legislation on genetic resources, food standards and the environment. Fosters cooperation with other international organizations in specific subjects and geographic areas through FAO's commissions and other statutory bodies. Administers World Food Programme in conjunction with the United Nations; cooperates with U.N. agencies including the United Nations Educational, Scientific and Cultural Organization, International Atomic Energy Agency, and the World Health Organization. Hosts expert consultations. **PUBLICATIONS:** *Animal Health Yearbook*; *Ceres*, bimonthly; *Commodity Review and Outlook*, annual; *FAO Annual Review*; *FAO Bulletin of Statistics*, quarterly; *FAO Fertilizer Yearbook*; *FAO Plant Protection Bulletin*, quarterly; *FAO Production Yearbook*; *FAO Trade Yearbook*; *Food and Agricultural Legislation*, annual; *Food and Nutrition*, semiannual; *Food Outlook*, monthly; *Rural Development*, annual; *The State of Food and Agriculture*, annual; *Unasylva*, quarterly; *Yearbook of Fisheries Statistics*; *Yearbook of Forest Products*.

FOOD AND AGRICULTURE ORGANIZATION OF THE UNITED NATIONS - REGIONAL OFFICE FOR AFRICA

United Nations Agency Bldg., N. Maxwell Rd., PO Box 1628, Accra, Ghana **TX:** 2139 **OFFICERS:** B.F. Dada. **LANGUAGES:** English.

DESCRIPTION: Regional office of the Food and Agriculture Organization of the United Nations. Works to raise the standard of living of people in Africa. Supports and provides technical assistance to government agencies and nongovernmental organizations working to raise African nutritional and living standards by improving agricultural, forestry, and fisheries production.

FOOD AID COMMITTEE (FAC)

c/o International Grains Council, One Canada Sq., Canary Wharf, London E14 5AE, England **PHONE:** 44 171 5131122 **FX:** 44 171 5130630 **TX:** 8813241 **E-MAIL:** igc-fac@igc.org.uk **WEBSITES:** http://www.igc.org.uk **FOUNDED:** 1967. **OFFICERS:** Mr. G.A. Denis, Exec.Dir. **MEMBERS:** 23. **LANGUAGES:** English, French, Spanish.

DESCRIPTION: Implements the Food Aid Convention (FAC), which is part of the International Grains Agreement, 1995. The FAC is an intergovernmental agreement whose parties commit minimum annual amounts of grains as aid to developing countries. **PUBLICATIONS:** *Food Aid Shipments* (in English), annual. Includes statistics.

FOOD CORPS, U.S.A.

Brandeis Univ., 415 South St., Waltham, MA 02254-9110 **PHONE:** (617) 736-2764 **OFFICERS:** Ruth Morgenthau, Pres. **LANGUAGES:** English.

DESCRIPTION: Members are national liaison committees in Guinea, Mali, Mexico, Peru, Sri Lanka, Tanzania, Togo, and Zimbabwe. Seeks to increase the productive capacity of small farms. Conducts research to develop sustainable variations of traditional agricultural techniques; facilitates international exchanges involving agricultural and extension personnel; forms local organizations to advocate on behalf of small farmers.

FOOD FOR THE HUNGRY INTERNATIONAL

58 Beulah Rd., Tunbridge Wells TN1 2NR, England **PHONE:** 44 1892 534410 **FX:** 44 1892 534410 **E-MAIL:** fh.uk@btinternet.com **FOUNDED:** 1971. **OFFICERS:** Niall M. Watson. **STAFF:** 1. **BUDGET:** £600,000. **LANGUAGES:** English.

DESCRIPTION: Christian organization seeking to meet the nutritional needs of people in economically developing countries and areas beset by natural or man-made disasters. Develops and implements programs to help local communities in economically developing areas improve their agricultural techniques; provides emergency services including food, shelter, and medical care to victims of disasters.

FOOD FOR THE POOR

c/o James J. Cavnar, 550 SW 12th Ave., Bldg. 4, Deerfield Beach, FL 33442 **PHONE:** (954) 427-2222 **FX:** (954) 570-7654 **WEBSITES:** http://www.foodforthepoor.com **FOUNDED:** 1982. **OFFICERS:** Ferdinand Mahfood, Founder/Dir. **MEMBERS:** 520,000. **STAFF:** 145. **LANGUAGES:** English.

DESCRIPTION: Works to improve the health, economic, spiritual, and social conditions of the poor, primarily in the Caribbean and Latin America. Provides educational and charitable programs, speakers' bureau, and Children's services. **PUBLICATIONS:** *Food for the Poor*, quarterly.

FOREST PEOPLES PROGRAMME

Fosseway Business Center, Ste. 1C, Stratford Rd., Moreton-in-Marsh GL56 9NQ, England **PHONE:** 44 1608 652893 **FX:** 44 1608 652878 **E-MAIL:** info@fppwrm.gn.apc.org **OFFICERS:** Marcus Colchester, Dir. **STAFF:** 5. **LANGUAGES:** English, French, Spanish.

DESCRIPTION: Citizen's groups. Promotes protection of rainforests and the people and wildlife that inhabit them worldwide. Campaigns for the defense of the civil and human rights of people indigenous to the rainforest; provides support and assistance to environmental protection organizations operating in rainforests.

FORUM INTERNATIONAL: INTERNATIONAL ECOSYSTEMS UNIVERSITY (IEU)

91 Gregory Ln., No. 21, Pleasant Hill, CA 94523 **PHONE:** (925) 671-2900 **FX:** (925) 946-1500 **E-MAIL:** forum@ix.netcom.com **FOUNDED:** 1965. **OFFICERS:** Dr. Nicolas D. Hetzer, Dir. **MEMBERS:** 42,000. **STAFF:** 26. **BUDGET:** US$9,872,005. **LANGUAGES:** English, French, German, Greek, Italian, Portuguese, Romanian, Russian, Spanish, Swedish.

DESCRIPTION: Works to create a forum for education, research, and action to deal with problems such as environmental deterioration, socioeconomic change, poverty, overpopulation, and lack of educational opportunity. Sponsors ecosystems field studies in Africa, Europe, Latin America, and North America. **PUBLICATIONS:** *Ecosphere*, bimonthly. Describes theory and practice of ecosystemic, whole-world-oriented, transdisciplinary, value-based education, and research.

FORUM TRAIN EUROPE (FTE)

c/o Swiss Federal Railways, Hochschulstrasse 6, CH-3030 Bern, Switzerland **PHONE:** 41 512 201111 **FX:** 41 512 202302 **TX:** 991121 **WEBSITES:** http://www.sbb.ch/ **FOUNDED:** 1923. **MEMBERS:** 55. **LANGUAGES:** English, French, German, Italian, Russian.

DESCRIPTION: European railway and shipping organizations whose trains and ships provide connections between 2 or more European countries. Determines international passenger rail and maritime connections. **PUBLICATIONS:** *Membership List*, periodic.

FOUNDATION FOR ECOLOGICAL DEVELOPMENT ALTERNATIVES (FEDA) (Stichting Mondiaal Alternatief - SMA)

PO Box 151, NL-2130 AD Hoofddorp, Netherlands **PHONE:** 31 23 5632305 **FX:** 31 23 5641359 **E-MAIL:** ronaldg@camels.nld.toolnet.org **FOUNDED:** 1974. **OFFICERS:** Mr. R. Gerrits. **MEMBERS:** 1,500. **STAFF:** 10. **LANGUAGES:** Dutch, English.

DESCRIPTION: Membership consists of scientists and interested individuals seeking to: increase awareness of the dangers of pollution and deforestation; emphasize the dependence of mankind on the ecosytem; encourage respect for all life. Offers alternative solutions to ecological problems. Promotes the application of ecologically safe pest control techniques that also foster self-sufficiency. Opposes the exportation of pesticides to Third World countries; communicates with producers of pesticides; works to modify the policies of pesticide companies in order to limit the production of the most harmful pesticides. Strives to slow deforestation; has created fuel alternatives for small communities in developing nations. Lobbies government authorities on conservation issues. Conducts surveys; compiles statistics. **PUBLICATIONS:** *Ecoscripts* (in English), periodic. Research reports; Brochures, periodic.

FOUNDATION EUROPALIA INTERNATIONAL (FEI)

10, rue Royale, B-1000 Brussels, Belgium **PHONE:** 32 2 5078550 **FX:** 32 2 5135488 **WEBSITES:** http://www.europalia.com **FOUNDED:** 1969. **OFFICERS:** Luc Stainier, Gen. Mgr. **STAFF:** 11. **LANGUAGES:** Dutch, English, French, Spanish.

DESCRIPTION: Promotes diversity and expansion of European culture by sponsoring and organizing annual festival highlighting a specific country. The three month festival includes exhibits, demonstrations, debates, workshops, and seminars that encompass all artistic and cultural disciplines of the featured country. Maintains museum.

FOUNDATION EUROPEENNE

19, rue de la Reine, B-4500 Huy, Belgium **PHONE:** 32 85 212671 **FX:** 32 85 211071 **FOUNDED:** 1965. **OFFICERS:** H.E. the Rt. Hon. Earl Pierre Pasleau de Charnay, Hon. Ambassador. **LANGUAGES:** English, French, Italian, Spanish.

DESCRIPTION: Operates under the eminent patronage of H.R.H. the Prince Henrik of Denmark. Created by a group of men of letters and journalists to honor men and women of standing and enterprise who have become renowned through their particular business success, for adding value to the culture of the 15 countries of the European Union. **PUBLICATIONS:** *Dictionnaire des Distinctions Honorifiques.*

FOUNDATION FOR INTERNATIONAL SCIENTIFIC CO-ORDINATION (FISC) (Fondation pour la Science - Centre International de Synthese)

12, rue Colbert, F-75002 Paris, France **PHONE:** 33 1 42975068 **FX:** 33 1 42974646 **E-MAIL:** synthese@filnet.fr **FOUNDED:** 1924. **OFFICERS:** Mm. Eric Brian. **STAFF:** 2. **LANGUAGES:** French.

DESCRIPTION: Scientists and philosophers interested in uniting diverse disciplines in the utilization of scientific research. Seeks to develop and coordinate pure scientific research and communication among scholars, researchers, and professors in the field. Discourages what the foundation sees as ''too narrow specialization and exclusively utilitarian preoccupations.'' Facilitates contact among members in rural areas and foreign countries and those in Paris. **PUBLICATIONS:** *L'Evolution de l'Humanite*; *Revue de Synthese* (in English, French, and German), quarterly; *Revue d'Histoire des Sciences* (in English and French), quarterly; *Semaines de Synthese, Cahiers de Synthese* (in French), periodic.

FOUNDATION OF THE PEOPLES OF THE SOUTH PACIFIC - INTERNATIONAL (FSPI)

PO Box 951, Port Vila, Vanuatu **PHONE:** 678 22915 **FX:** 678 24510 **E-MAIL:** fspi@wantok.org.uv **FOUNDED:** 1992. **OFFICERS:** Kathy Fry, Regional Mgr. **MEMBERS:**

10. **STAFF:** 3. **BUDGET:** 500,000 V. **LANGUAGES:** English.

DESCRIPTION: Devises and manages regional development programs through a network of National affiliates. Programs emphasize economic and environmental sustainability, and facilitate an increased role for women in local economic life. Conducts health and nutrition activities focusing on improving nutrition for infants and young children, community forestry, small enterprise development, conflict resolution and much more.

FOUNDATION FOR THE SUPPORT OF INTERNATIONAL MEDICAL TRAINING (FSIMT)

417 Center St., Lewiston, NY 14092 **PHONE:** (716) 754-4883 **FX:** (519) 836-3412 **E-MAIL:** iamat@sentex.net **WEBSITES:** http://www.sentex.net/-iamat **FOUNDED:** 1960. **OFFICERS:** Mrs. M. A. Uffer, Pres. **MEMBERS:** 6,000,000. **STAFF:** 25. **LANGUAGES:** English.

DESCRIPTION: Individuals and corporations organized to provide information regarding the availability of competent medical care overseas and information concerning sanitary conditions, health hazards, and climatic conditions in various parts of the world. Offers detailed guidance on vaccination and immunization requirements and tropical diseases. **PUBLICATIONS:** *Be Aware of Schistosomiasis*; *How to Protect Yourself Against Malaria*, annual; *Immunization Chart*, annual; *Set of 24 World Climate Charts*, annual; *Traveller Clinical Chart*, annual; *When Hiking Through Latin America, Be Alert to Chagas' Disease*, annual; *World Malaria Risk Chart*, annual; *World Schistosomiasis Risk Chart*, annual; Brochure, annual; Directory, annual.

FREELANCE PROFESSIONALS NETWORK (Reseau des Pigistes Professionels)

933, Station H, Montreal, PQ, Canada H3G 2M9 **E-MAIL:** info@f-pro.ca **FOUNDED:** 1992. **OFFICERS:** Philip McMaster, Founder. **MEMBERS:** 100,000. **LANGUAGES:** English, French.

DESCRIPTION: Freelance professionals serving the communications, film, broadcasting, and publishing industries. Assists members in selling their services outside their native countries. Maintains Publisher Information Service to protect the copyrights of freelance writers; develops and implements international standards for freelance work. **PUBLICATIONS:** *FREEWorld*, quarterly. Downloadable internet publication.

FRENCH POPULAR RELIEF (FPR) (Secours Populaire Francais - SPF)

9-11 rue Froissart, F-75140 Paris Cedex 03, France **PHONE:** 33 1 42785048 **FX:** 33 1 42740101 **LANGUAGES:** English, French.

DESCRIPTION: French relief and development organizations with operations in Vietnam. Promotes appropriate and sustainable development; seeks to improve the quality of life in developing regions. Sponsors public health, general education, and nutrition programs; makes available emergency relief services.

FRENCH-AMERICAN FOUNDATION (FAF)

509 Madison Ave. Ste. 310, New York, NY 10022-5501 **PHONE:** (212) 288-4400 **FX:** (212) 288-4769 **E-MAIL:** french_amerfon@msn.com **FOUNDED:** 1976. **OFFICERS:** Mr. Michael Iovenko, Pres. **STAFF:** 6. **BUDGET:** US$1,500,000. **LANGUAGES:** English, French.

DESCRIPTION: Works to strengthen relations between the United States and France by creating opportunities for French and American professionals to discuss and address problems of major concern to both societies and to stimulate change through cooperation. Projects include exchange of specialists, internships, study tours, conferences, fellowships, surveys, and special studies. Sponsors bicentennial fellowships for U.S. doctoral candidates, a continuing Chair in American Civilization at a university in Paris, and a two month professional exchange for American and French journalists. **PUBLICATIONS:** *End of Year Report*, annual, always December; *Project Reports*, periodic.

FRIENDS OF THE EARTH INTERNATIONAL (FOEI)

Postbus 19199, NL-1000 GD Amsterdam, Netherlands **PHONE:** 31 20 6221369 **FX:** 31 20 6392181 **E-MAIL:** foeint@antenna.nl **WEBSITES:** http://www.xs4all.nl/-foeint **FOUNDED:** 1971. **OFFICERS:** Ann Doherty, Infomation Officer.

MEMBERS: 59. **STAFF:** 6. **BUDGET:** US$7,000,000. **LANGUAGES:** English, French, Spanish.

DESCRIPTION: National groups committed to the preservation, restoration, and responsible use of the environment. **PUBLICATIONS:** *Annual Report* (in English); *Link* (in English), bimonthly; Pamphlets.

FRIENDS OF TEMPERANCE (FT) (Raittiuden Ystavat - RY)

Annankatu 29 A 9, FIN-00100 Helsinki, Finland **PHONE:** 358 9 6944177 **FX:** 358 9 6944407 **E-MAIL:** tom.anthoni-lcoivaluhta@raitis.fi **WEBSITES:** http://www.raitis.fi **FOUNDED:** 1853. **OFFICERS:** Tom Anthnoi, Mgr.Dir. **MEMBERS:** 10,000. **STAFF:** 17. **BUDGET:** US$400,000. **LANGUAGES:** English, Finnish, Swedish.

DESCRIPTION: Temperance and substance abuse prevention and treatment organizations. Promotes increased involvement in temperance issues by the public and by development organizations worldwide. Conducts temperance education courses; facilitates cooperation among temperance organizations. **PUBLICATIONS:** *Ystavien Kesken* (in Finnish), QRT.

FRIENDS OF THE THIRD WORLD (FTW)

611 W. Wayne St., Fort Wayne, IN 46802-2167 **PHONE:** (219) 422-6821 **FX:** (219) 422-1650 **E-MAIL:** fotw@apc.org **WEBSITES:** http://www.parlorcity.com/secop/fotw **FOUNDED:** 1972. **OFFICERS:** James F. Goetsch, Exec. Officer. **MEMBERS:** 600. **STAFF:** 6. **BUDGET:** US$500,000. **LANGUAGES:** English, French, Spanish.

DESCRIPTION: Individuals in 40 countries concerned with voluntary action against poverty; organizations of low-income persons including handicraft cooperatives, neighborhood associations, and agencies working on the poverty issue. Objective is to: demonstrate the existence and viability of an alternative system of trade as an effective means to deal with poverty and unemployment; educate the public regarding poverty-related problems and possible solutions. Provides information, consultation, and training skills for low-income persons in graphic arts/retail sales, building maintenance, bookkeeping, and management; also provides marketing assistance for handicrafts produced by low-income persons and groups. Distributes books on poverty-related topics and organizes educational activities. Operates technical assistance and resource center; conducts workshops. **PUBLICATIONS:** *Alternative Trading News*, quarterly. Covers organization's activities. Includes produce profile, marketing reports, calendar of events, and resource list; *Book Catalog*, periodic; Pamphlets.

FRIENDS WORLD COMMITTEE FOR CONSULTATION (FWCC) (Comite Mundial de Consulta de los Amigos - CMCA)

4 Byng Pl., London WC1E 7JH, England **PHONE:** 44 20 73880497 **FX:** 44 20 73834644 **E-MAIL:** fwccworldofficelondon@compuserve.com **FOUNDED:** 1937. **OFFICERS:** Elizabeth Duke, Gen.Sec. **MEMBERS:** 285,000. **STAFF:** 16. **BUDGET:** US$700,000. **LANGUAGES:** English, French, Kiswahili, Spanish.

DESCRIPTION: Works to encourage and strengthen the worldwide character and spiritual life of the Religious Society of Friends (Quakers). Promotes understanding between Friends and members of other religious demonations throughout the world. Sponsors intervisitation among Friends of differing nationalities and cultures. Conducts studies. Represents concerns of Quakers at the United Nations on peace, disarmament, economic justice, human rights, and sustainable development. Disseminates information concerning United Nations affairs. **PUBLICATIONS:** *Calendar of Yearly Meetings* (in English), annual; *Faith in Action: Encounters with Friends*; *Friends World News*, semiannual. Includes annual report in first issue of the year; *FWCC Friends Directory - Section of the Americas*, periodic; *Quakers Around the World* (in English), every 6 years. Lists all Quaker organizations and self statements of their history and purpose; *Quakers World Wide*. Contains history of the FWCC.

FRONTIERS FOUNDATION OPERATION BEAVER (FFOB)

2615 Danforth Ave., Ste. 203, Toronto, ON, Canada M4C 1L6 **PHONE:** (416) 690-3930 **FX:** (416) 690-3934 **E-MAIL:** frontier@globalserve.net **WEBSITES:** http://www.amtak.com/frontiers

FOUNDED: 1968. **OFFICERS:** Marco A. Guzman, Exec.Dir. **MEMBERS:** 100. **STAFF:** 7. **BUDGET:** C$1,500,000. **LANGUAGES:** English, French, Spanish.

DESCRIPTION: Volunteers united to advance the social and economic standards of communities in Canada and Third World countries. Volunteers work with host communities in Canada on community-based development projects such as building and/or renovating homes and organizing youth activities. Strives to develop construction skills and abilities, leadership experience, and cross-cultural experience as well as educational programs. **PUBLICATIONS:** *Beaver Tales* (in English and French), semiannual; *Special 25th Anniversary Edition*; Brochures, periodic; Videos, periodic; Videos, periodic.

FULBRIGHT ASSOCIATION (FA)

1130 17th St. NW, Ste. 310, Washington, DC 20036 **PHONE:** (202) 331-1590 **FX:** (202) 331-1979 **E-MAIL:** fulbright@fulbright.org **WEBSITES:** http://www.fulbright.org **FOUNDED:** 1977. **OFFICERS:** Jane L. Anderson, Exec.Dir. **MEMBERS:** 6,200. **STAFF:** 4. **BUDGET:** US$600,000. **LANGUAGES:** English.

DESCRIPTION: Past participants in the Fulbright program of international exchange; interested others. Coordinates membership support of international educational and cultural exchange programs and public service projects. Local chapters offer hospitality, enrichment, and mentor programs to visiting Fulbright scholars, teachers, and students. **PUBLICATIONS:** *Fulbright Association Newsletter*, quarterly. Covers the Fulbright Program, international exchange, international education, chapter news, available scholarships, alumni achievements and activities; *Report of Nations*, periodic.

FUNDACION BARILOCHE (FB)

Av. 12 de Octubre, 8400 Rio Negro, Argentina **PHONE:** 54 944 22050 **FX:** 54 944 22050 **E-MAIL:** fb@bariloche.com.ar **FOUNDED:** 1963. **OFFICERS:** Carlos E. Suarez, Exec.Pres. **STAFF:** 40. **BUDGET:** US$1,100,000. **LANGUAGES:** Spanish.

DESCRIPTION: Seeks to improve the quality of life in Latin America. Works in the following areas: rural development; energy; environmental conservation; municipal and urban development. Provides service in: project supervision; evaluation; technical assistance; computer systems; publications; investigation; participation in the identification and planning of government programs. **PUBLICATIONS:** *Boletin Bibliografico* (in Spanish), semiannual, always March and October; *Dessarrollo y Energia* (in Spanish), semiannual.

GAMBIA RIVER BASIN DEVELOPMENT ORGANISATION (Organisation pour la Mise en Valeur du Fleuve Gambie - OMVG)

13, passage Leblanc, Ave. Nelson Mandela, Dakar, Senegal **PHONE:** 221 8223159 **FX:** 221 8225926 **E-MAIL:** omvg@telecomplus.sn **FOUNDED:** 1978. **OFFICERS:** Mamadou Nassirou Diallo, Exec. Sec. **MEMBERS:** 4. **STAFF:** 15. **BUDGET:** 116,304,907 Fr CFA. **LANGUAGES:** English, French.

DESCRIPTION: Governments of Gambia, Guinea, Guinea-Bissau, and Senegal. Objectives are to: promote and coordinate the development of the Gambia and its tributaries' basins within members' national boundaries; realize member-initiated projects related to the technical and economic development of the area. Maintains documentation center. **PUBLICATIONS:** *Annual Activity Report* (in English and French); *Semi-Annual Activity Report* (in English and French); *Technical Report* (in English and French), periodic; Proceedings.

GANDHI MEMORIAL INTERNATIONAL FOUNDATION (GMIF)

4 Bates Blvd., Orinda, CA 94563-2804 **FOUNDED:** 1983. **OFFICERS:** Yogesh K. Gandhi, Pres. **MEMBERS:** 9,000. **STAFF:** 9. **LANGUAGES:** English.

DESCRIPTION: Seeks to eliminate spiritual and material poverty throughout the world. The foundation believes that only through individual change can humanity begin to deal with and solve problems facing us today. Promotes the universal principle of nonviolence. Aims to develop a school curriculum on nonviolence and to implement it in the U.S., the U.K., Canada, and India through pilot programs. Plans to build low cost houses for the underprivileged and to increase support to organizations that help the poor. Offers children's services and charitable, educational, and research programs; maintains speaker's bureau. **PUBLICATIONS:** *Gandhi Today*, quarterly.

GEMS OF HOPE

8 King St. E., No. 1105, Toronto, ON, Canada M5C 1B5 **PHONE:** (416) 362-4367 **FX:** (416) 362-4170 **E-MAIL:** gems@web.net **WEBSITES:** http://www.argun.com/gemsofhope **FOUNDED:** 1982. **OFFICERS:** John Paterson, Exec.Dir. **STAFF:** 3. **BUDGET:** US$500,000. **LANGUAGES:** English, French, Spanish.

DESCRIPTION: Individuals, corporations, and organizations with an interest in responsible global development. Promotes economic self-sufficiency among families and communities in the developing world. Supports development initiatives focusing on health care, the empowerment of women, basic education, and local enterprise. **PUBLICATIONS:** *Gems*, semiannual.

GENERAL ARAB WOMEN'S FEDERATION (GAWF)

Hay-Al-Maghreb, 33 Zugaq 5, Mahala 304, Baghdad, Iraq **PHONE:** 964 1 4227117 **FX:** 964 1 4252372 **FOUNDED:** 1944. **OFFICERS:** Dr. Manal Younis Abdul-Razzaq, Sec.Gen. **MEMBERS:** 23. **STAFF:** 15. **LANGUAGES:** English, French.

DESCRIPTION: Works to enhance the social and economic status of Arab women. Gathers and disseminates information; conducts educational programs. **PUBLICATIONS:** *Arab Women* (in Arabic), semiannual.

GENERAL ASSOCIATION OF MUNICIPAL HEALTH AND TECHNICAL EXPERTS (GAMHTE) (Association Generale des Hygienistes et Techniciens Municipaux - AGHTM)

83 Av. Foch, BP 3916, F-75761 Paris Cedex 16, France **FOUNDED:** 1905. **OFFICERS:** Alain Lasalmonie, Sec.Gen. **MEMBERS:** 1,200. **STAFF:** 5. **LANGUAGES:** French.

DESCRIPTION: Members are engineers, engineering consultants, technicians, architects, scientists, administrators, and municipal and private services in 32 countries. Objectives are to: stimulate fundamental and applied research on urban and rural public hygiene; disseminate findings of such research; encourage the exchange of information and ideas among members. Areas of interest include city planning, energy management, traffic problems, and pollution prevention. Operates information service; organizes field tours. **PUBLICATIONS:** *Techniques Sciences-Methodes*, monthly. Also publishes monographs.

GERMAN AGRO ACTION (GAA) (Deutsche Welthungerhilfe - DWHH)

Adenauerallee 134, Box 120509, D-53113 Bonn, Germany **PHONE:** 49 228 22880, 49 228 2288212 **FX:** 49 228 220710 **TX:** 8869697 DWHH **E-MAIL:** 100073.432@compuserve.com **FOUNDED:** 1962. **OFFICERS:** Dr. Volker Hausmann, Sec.Gen. **STAFF:** 90. **BUDGET:** US$86,000,000. **LANGUAGES:** English, French, German, Spanish.

DESCRIPTION: Promotes creation of self-help programs to address nutrition and hunger concerns in developing countries. Seeks to ensure respect for human rights and the rule of law in developing regions. Supports programs in areas including agricultural and rural development, nutrition education, water and sanitation, desertification and reforestation, food security, and microenterprise formation. Conducts educational programs to raise awareness in Europe of issues faced by developing countries. Makes available emergency assistance to victims of natural disasters.

GERMAN MEDICAL WELFARE ORGANISATION, ACTION MEDEOR (GMWOAM) (Deutsches Medikamenten-Hilfswerk, Action Medeor - DMH)

St.-Toeniser-Strasse 21, D-47918 Toenisvorst, Germany **PHONE:** 49 2156 97880 **FX:** 49 2156 80632 **TX:** 8531064 AMED **E-MAIL:** action.medeor@t-online.de **WEBSITES:** http://www.medeor.org **FOUNDED:** 1964. **OFFICERS:** Bernd Pastors, Gen.Mgr. **MEMBERS:** 45. **STAFF:** 35. **BUDGET:** DM 1,700,000. **LANGUAGES:** English, French, German, Spanish.

DESCRIPTION: Relief organization working to combat poverty and illness in underdeveloped nations through the distribution of prepared and packaged pharmaceutical products, medical equipment, and information. Provides supplies to over 10,500 medical stations in 128 countries. Works closely with contacts in recipient countries to ensure equitable distribution of supplies. Liaises with embassies, forwarding companies, and customs authorities. Sponsors lectures and exhibits. Organization is named for the list of 184 drugs compiled by the World Health Organization which WHO maintains are sufficient to treat all diseases and illnesses. **PUBLICATIONS:** *Action Medeor* (in English, German, and Spanish). Contains summaries of important medical publications; *Health Promotes Development*. Documentation and reports from Krefeld Symposium and Organisation reports; *Medeor Manual*; *Medeor Report*, semiannual.

GLOBAL COALITION FOR AFRICA (GCA)

1750 Pennsylvania Ave. NW, Ste. 1204, Washington, DC 20433 **PHONE:** (202) 676-0845, (202) 458-4111 **FX:** (202) 522-3259 **TX:** RCA 248423 **E-MAIL:** smcleague@worldbank.org **OFFICERS:** Ahmedou Ould-Abdallah. **LANGUAGES:** English.

DESCRIPTION: Government agencies and nongovernmental organizations with an interest in the economic and community development of Africa. Promotes environmentally sustainable development appropriate to local needs and capacities throughout Africa. Monitors development projects to ensure their viability; provides technical support and other assistance to selected development programs.

GLOBAL CROP PROTECTION FEDERATION (GCPF)

143, ave. Louise, B-1050 Brussels, Belgium **PHONE:** 32 2 5420410 **FX:** 32 2 5420419 **TX:** 62120 **E-MAIL:** gcpf@pophost.eunet.be **WEBSITES:** http://www.gcpf.org **FOUNDED:** 1960. **OFFICERS:** Mr. K.P. Vlahodimos, Dir.Gen. **MEMBERS:** 6. **STAFF:** 3. **BUDGET:** 90,000,000 BFr. **LANGUAGES:** English, French.

DESCRIPTION: Federation of national and regional crop protection associations. Represents the interests of the crop protection product industry. Seeks to promote understanding of the crop protection industry's contributions to society. Promotes protection of crops through proper use of agrochemicals; works to ensure that such products conform to the needs of agriculture and society. Advocates safe and sensible production, handling, packing, shipment, and application of agrochemicals by establishing and recommending uniform standards and regulations in compliance with international rules. Encourages cooperation among national and international lawmakers concerning control, testing, and approval of agrochemicals. Provides forum for exchange of information, discussion, and problem solving. Seeks to educate the public on the nature, purpose, and benefits of agrochemicals. Acts as liaison between member associations and interstate/international organizations regarding matters such as regulation, toxicology, ecology, residue, and legislation. Organizes working groups and committees to study current problems on an international scale. **PUBLICATIONS:** Publishes brochures, guidelines for the safe use and handling of pesticides, technical monographs, and educational and training material for retailers and agriculture instructors.

GLOBAL FUND FOR WOMEN

425 Sherman Ave., Ste. 300, Palo Alto, CA 94306 **PHONE:** (650) 853-8305 **FX:** (650) 328-0384 **E-MAIL:** gfw@globalfundforwomen.org **WEBSITES:** http://www.globalfundforwomen.org **FOUNDED:** 1987. **OFFICERS:** Kavita N. Ramdas, Pres. **STAFF:** 14. **BUDGET:** US$2,300,000. **LANGUAGES:** English.

DESCRIPTION: Grantmaking organization providing support to women's groups working on emerging, controversial, or difficult issues. Supports overseas groups working on projects related to female human rights. Provides funds to seed, strengthen, and link groups that are committed to women's well being and that work for their full participation in society. **PUBLICATIONS:** *Network News* (in English and Spanish), semiannual; Annual Report, annual; Brochure.

GLOBAL HEALTH COUNCIL

1701 K St. NW, Ste. 600, Washington, DC 20006 **PHONE:** (202) 833-5900 **FX:** (202) 833-0075 **E-MAIL:** gnc@globalhealthcouncil.org **WEBSITES:** http://www.globalhealthcouncil.org **FOUNDED:** 1971. **OFFICERS:** Nils Daulaire, Pres. **MEMBERS:** 1,600. **STAFF:** 10. **BUDGET:** US$1,200,000. **LANGUAGES:** English.

DESCRIPTION: Membership organization made up of private voluntary organizations, health and medical associations, universities, government agencies, foundations, corporations, consulting firms, and individuals interested in promoting greater and more effective U.S. participation in practical international health and development programs. Seeks to strengthen U.S. public and private sector participation in international health activities. Areas of concern include: HIV/AIDS; women's health; improving primary health care worldwide; environmental health; population and family planning; tropical and preventive medicine; appropriate health technology. Supports improved health and development legislation. Conducts career service in conjunction with annual conference. **PUBLICATIONS:** *AIDS Link*, bimonthly. Contains the latest global information on HIV/AIDS issues.; *Career Network*, monthly. Contains listings of employment opportunities in International Health.; *Directory of U.S.-Based Agencies Involved in International Health Assistance*, periodic. Lists geographical areas served and types of workers sought by U.S. health agencies.; *Global Learning for Health*; *Healthlink*, 10/year. Contains information on international health policy issues and calendar of events.; *NCIH Membership Directory*, periodic. Contains individual and organizational members with phone numbers and key contacts.

GLOBAL WARMING INTERNATIONAL CENTER (GWIC)

22W381 75th St., Naperville, IL 60565-9245 **PHONE:** (630) 910-1551 **FX:** (630) 910-1561 **E-MAIL:** syshen@megsinet.net **WEBSITES:** http://www.globalwarming.net/ **FOUNDED:** 1986. **OFFICERS:** Dr. Sinyan Shen, Dir. **MEMBERS:** 12,381. **STAFF:** 17. **BUDGET:** US$75,000,000. **LANGUAGES:** English.

DESCRIPTION: Ministerial agencies and industrial corporations. Concerned with impacts and effects of global warming. Provides a focus for governments, the private sector, and academia to share information on global warming internationally. Coordinates training for personnel dealing with environmental issues, energy planning, and natural resource management through the Institute for World Resource Research. Establishes the Global Warming Index (GWI) and the Extreme Event Index (EEI) for international standardization. Maintains speaker's bureau; compiles statistics; operates placement service; conducts research and educational programs. **PUBLICATIONS:** *World Resource Review*, quarterly.

GRASSROOTS INTERNATIONAL (GRI)

179 Boylston St., 4th Fl., Boston, MA 02130 **PHONE:** (617) 524-1400 **FX:** (617) 524-5525 **E-MAIL:** grassroots@igc.org **WEBSITES:** http://www.grassroots.org **FOUNDED:** 1983. **OFFICERS:** Kevin Murray, Exec.Dir. **STAFF:** 7. **BUDGET:** US$1,330,000. **LANGUAGES:** English.

DESCRIPTION: Purpose is to fund community-based relief and development projects in Africa, Asia, Latin America, and the Middle East, and to provide educational and information programs in the U.S. on peace and justice issues. Assists local social change organizations in Eritrea, South Africa, the West Bank and Gaza, Haiti, Mexico, and the Philippines. Provides humanitarian assistance to people in Eritrea and Lebanon. Supports rehabilitation and development projects examining the causes of famine. Organizes forums and media outreach programs to schools; operates speakers' bureau. **PUBLICATIONS:** *Grassroots International—Insights*, 3/year. Provides information on GRI's activities; Videos. Also publishes program profiles, fact sheets, and news releases; disseminates slideshow and photo exhibit.

GREENPEACE EUROPEAN UNIT

rue de la Tourelle, 37, Torekenstraat 37, B-1040 Brussels, Belgium **PHONE:** 32 2 2801400 **FX:** 32 2 2308413 **E-MAIL:** eec.unit@diala.greenpeace.org **OFFICERS:** Jackie Lilly, Admin.

DESCRIPTION: Regional unit of Greenpeace International. Serves as liaison between GI and its national affiliates and the industrial and environmental protection agencies of the European Community. Seeks to develop public policy solutions to environmental problems; conducts lobbying activities.

GREENPEACE INTERNATIONAL (GPI)

Keizersgracht 176, NL-1016 DW Amsterdam, Netherlands **PHONE:** 31 20 5236222 **FX:** 31 20 5236200 **TX:** 18775 GPINT NL **E-MAIL:** sdesk@ams.grenpeace.org **WEBSITES:** http://www.greenpeace.org **FOUNDED:** 1971. **OFFICERS:** Thilo Bode, Exec.Dir. **MEMBERS:** 4,500,000. **STAFF:** 1,000. **BUDGET:** US$25,000,000. **LANGUAGES:** English.

DESCRIPTION: Works to stop and reverse destruction of the atmosphere and biosphere. Promotes disarmament and an end to nuclear testing; supports the use of alternative and renewable energy sources. Areas of interest include protection of marine animals and their habitats and prevention of land, air, and water pollution, especially dumping of toxic waste. Engages in non-violent direct action in defense of the environment. Has coordinated campaigns on selected issues of environmental abuse, incorporating lobbying, scientific reporting, public education, and direct action.

GREENPEACE MEDITERRANEAN (GM)

6 Manol Mansions, De Paule Ave., Balzan BZN 02, Malta **PHONE:** 356 49078415 **FX:** 356 490782 **E-MAIL:** gpmedite@diala.greenpeace.org **WEBSITES:** http://www.greenpeace.org **OFFICERS:** Mario Damato. **LANGUAGES:** English, Greek, Italian, Spanish, Turkish.

DESCRIPTION: Regional branch of Greenpeace International. Works to develop solutions to ecological problems through political action. Seeks to change government policies regarding pollution discharge and control.

GROUP FOR ENVIRONMENTAL MONITORING (GEM)

PO Box 30684, Braamfontein 2017, Republic of South Africa **PHONE:** 27 11 4037666, 27 11 8385449 **FX:** 27 11 4037563 **E-MAIL:** ewilliam@gem.org.za **WEBSITES:** http://www.gem.org.za **OFFICERS:** Eric Williams. **LANGUAGES:** Afrikaans, English.

DESCRIPTION: Promotes environmental protection and sustainable use of natural resources in southern Africa. Monitors the activities of industrial and development organizations and publicizes instances of environmental degradation.

GROUP OF THE PARTY OF EUROPEAN SOCIALISTS (PES) (Groupe du Parti des Socialistes Europeens)

c/o Joan Cornet Prat, 79, rue Belliard, B-1047 Brussels, Belgium **PHONE:** 32 2 2842111 **FX:** 32 2 2306664 **TX:** 63988 SOCEP **E-MAIL:** pesnet@europarl.eu.int **WEBSITES:** http://www.europarl.eu.int/pes; http://www.eurosocialist.org **FOUNDED:** 1954. **OFFICERS:** Joan Cornet Prat, Sec.Gen. **MEMBERS:** 214. **STAFF:** 169. **LANGUAGES:** English, French, German, Spanish.

DESCRIPTION: An associated member of the Socialist International. Socialist members of the European Parliament. Prepares and conducts parliamentary debates focusing on social and economic prosperity, universal respect for human rights, social justice as a basis for global socioeconomic organization, world security and peace, technological and scientific development based on social responsibility, the political and economic unification of Europe, and other political goals promulgated in the manifesto of the Party of European Socialists, formerly Confederation of Socialist Parties and Social Democratic Parties of the European Community. **PUBLICATIONS:** *EuroMP, Le Carnet, SXB Express* (in English, French, and German), monthly; *Secretariat du Groupe du Parti des Socialistes Europeens* (in English and French), periodic; Directory, periodic.

GULF ORGANIZATION FOR INDUSTRIAL CONSULTING (GOIC)

PO Box 5114, Doha, Qatar **PHONE:** 974 858888 **FX:** 974 831465 **TX:** 4619 GOIC DH **E-MAIL:** goic@goic.org.qa **WEBSITES:** http://www.goic.org.qa; http://www.goic.org.qa **FOUNDED:** 1976. **OFFICERS:** Dr. Abdul Rahman Ahmed Al-Ja'fary, Sec.Gen. **MEMBERS:** 6. **BUDGET:** 24,150,000 QRl. **LANGUAGES:** Arabic, English.

DESCRIPTION: States of the Arabian Gulf working for mutual cooperation and coordination. Collects and disseminates information on industrial development projects, studies, and policies; proposes establishment of common industrial projects in member states. Promotes technical and economic cooperation among existing or planned industrial companies and establishments. Provides technical assistance in the preparation and evaluation of industrial projects. Sponsors training courses and seminars. Maintains technical advisory committees. **PUBLICATIONS:** *Annual Report* (in Arabic and English); *Industrial Cooperation in the Arabian Gulf* (in Arabic), quarterly; *Industrial Focus* (in Arabic and English), bimonthly; Bulletin (in Arabic and English), monthly. Also publishes books, booklets, industrial studies, and press releases.

HAGUE CONFERENCE ON PRIVATE INTERNATIONAL LAW (Conference de la Haye de Droit International Prive)

Scheveningseweg 6, NL-2517 KT The Hague, Netherlands **PHONE:** 31 70 3633303 **FX:** 31 70 3604867 **E-MAIL:** secretariat@hcch.net **WEBSITES:** http://www.hcch.net **FOUNDED:** 1893. **OFFICERS:** J.H.A. van Loon, Sec.Gen. **MEMBERS:** 46. **STAFF:** 12. **BUDGET:** 3,119,705 f. **LANGUAGES:** English, French.

DESCRIPTION: Intergovernmental organization working for the progressive unification of private international law. **PUBLICATIONS:** *Actes et Documents des Sessions de la Conference* (in English and French), quadrennial; *Collection of Conventions*, quadrennial; *Practical Handbooks*, periodic.

HANSARD SOCIETY FOR PARLIAMENTARY GOVERNMENT

St. Philips Bldg. N, Sheffield St., London WC2A 2EX, England **PHONE:** 44 171 9557478 **FX:** 44 171 9557492 **E-MAIL:** harvard-society@ue.ac.uk **FOUNDED:** 1944. **OFFICERS:** Stephen Coleman, Dir., Scholars Prog. **MEMBERS:** 400. **STAFF:** 7. **LANGUAGES:** English.

DESCRIPTION: Promotes knowledge of and interest in parliamentary governments and procedures. Conducts research, educational programs, and training courses for companies and teachers; sponsors international student exchange programs. Organizes academic competitions and lobbying seminars. **PUBLICATIONS:** *Agenda for Change*; *Making the Law: The Legislative Process*; *Parliament and Government Pocket Book*; *Parliament at Work*; *Parliamentary Affairs*, monthly; *Report of the Hansard Commission on Women at the Top*; *Report on Regulation of Privatised Utilities*; *What Price Hansard?*

HOLT INTERNATIONAL CHILDREN'S SERVICES (HICS)

PO Box 2880, Eugene, OR 97402 **PHONE:** (541) 687-2202 **FX:** (541) 683-6175 **E-MAIL:** info@holtintl.org **WEBSITES:** http://www.holtintl.org **FOUNDED:** 1956. **OFFICERS:** John L. Williams, Exec.Dir. **STAFF:** 100. **BUDGET:** US$11,000,000. **LANGUAGES:** English.

DESCRIPTION: To deinstitutionalize children in developing countries by rehabilitating biological families, encouraging adoption within the developing country, and arranging inter-country adoption when in the best interest of the child. Offers assistance to children in Korea, India, the Philippines, Thailand, Vietnam, Hong Kong, China, Romania, and Latin America, as well as the U.S. Provides funds for food, clothing, housing, and medical care until an adoptive home can be found. Offers nutrition education, foster homes, well-baby clinics, physical, occupational, and speech therapy, and special care and education for the retarded. All services are supported through donations, sponsorships, and adoption fees. **PUBLICATIONS:** *Holt International Children's Services Annual Report*; *Holt International Families!*, bimonthly. For Holt supporters and families who have adopted children from foreign countries. Includes list of children ready for adoption, letter, and photos.

HOPE UNLIMITED (Espoir Sans Frontieres - ESF)

BP 12241, Sicap Sacre Coeur I No. 8250, Dakar, Senegal **PHONE:** 221 8256699 **FX:** 221 8256699 **E-MAIL:** esfsiege@telecomplus.sn **FOUNDED:** 1988. **OFFICERS:** Etienne Sokeng, Pres. **MEMBERS:** 4,500. **STAFF:** 25. **BUDGET:** US$215,237. **LANGUAGES:** French.

DESCRIPTION: Promotes the rights and improved welfare of children in Africa and South America. Maintains a network of groups and individuals supporting activities assisting children. Offers information and counseling services for young mothers on child rearing and health and nutrition. Provides financial, medical, and personal assistance to fatherless and orphaned children. **PUBLICATIONS:** *Demain l'Enfant* (in French), monthly. Contains information from articles and studies done on children's problems; *Feminin Pluriel* (in French), periodic; *La Lettre d'Espoir*; *L'Ecoute Comme Therapie Sociale 8 Ans D'Experience*; *Les Enfants Martyrs* (in French), periodic; *To Be a Young Mother in Africa*. Contains information for young mothers on the problems of early marriage and domestic violence.

HUMAN RIGHTS ADVOCATES INTERNATIONAL (HRAI)

14 E. 48th St., No. 5, New York, NY 10017-1008 **PHONE:** (212) 715-0176 **FX:** (212) 715-0183 **FOUNDED:** 1979. **OFFICERS:** Sanford

Mevorah, Exec.Dir. **MEMBERS:** 300. **LANGUAGES:** English.

DESCRIPTION: Law firms, attorneys, academicians, and individuals worldwide interested in constitutional law and human rights. Protects and promotes human rights by coordinating and providing legal services, constitutional law resources, counseling, and investigations of alleged human rights violations. Represents exiles, individual citizens, and member states of the United Nations. Acts as sole representative for Amerasian children in Vietnam. Offers seminars on human rights law and international organization affairs; operates speakers' bureau. Conducts research on constitutionalism. **PUBLICATIONS:** *Constitutions of Dependencies and Special Sovereignties*; *Constitutions of the Countries of the World*. Also publishes research results.

HUMAN RIGHTS INFORMATION AND DOCUMENTATION SYSTEMS INTERNATIONAL (HURIDOCS)

48 Chemin du Grand Montfleury, CH-1290 Versoix, Switzerland **PHONE:** 41 22 7555252 **FX:** 41 22 7411768 **E-MAIL:** huridocs@comlink.org **FOUNDED:** 1982. **OFFICERS:** Manuel Guzman, Exec.Dir. **STAFF:** 4. **BUDGET:** US$1,100,000. **LANGUAGES:** English, French, Spanish.

DESCRIPTION: Network of human rights organizations with contacts in 156 countries. Aims to improve access to and dissemination of information on human rights through more effective, appropriate, and compatible methods and techniques of information handling. Coordinates the development of documentation systems and promotes the use of standard formats. Studies problems of information handling, particularly the capabilities and needs of nongovernmental organizations. Develops and disseminates formats and software for use in documenting human rights: bibliographic documentation, documentation of human rights violations. Maintains task forces; offers training courses and workshops for human rights documentalists. **PUBLICATIONS:** *Human Rights, Refugees, Migrants & Development: Directory of NGOS in OECD Countries (with OECD and UNHCR)*; *HURIDOCS News* (in English, French, and Spanish), 2-3/year; *HURIDOCS Standard Formats for the Recording and Exchange of Bibliographic Information on Human Rights*; *Information for Human Rights: A HURIDOCS Reader for Information Workers*; *Standard Formats: A Tool in the Documentation of Human Rights Violations*.

HUMAN RIGHTS NETWORK IN ASIA (HRNA)

4-5-16-301 Iidabashi, Chiyoda-ku, Tokyo 102-0072, Japan **PHONE:** 81 3 32370217 **FX:** 81 3 32370287 **E-MAIL:** cfrtyo@aol.com **FOUNDED:** 1993. **OFFICERS:** Ken Arimitsu, Coordinator. **MEMBERS:** 220. **STAFF:** 2. **BUDGET:** 1,200,000¥. **LANGUAGES:** English, Japanese.

DESCRIPTION: Volunteers working for human rights organizations. Promotes respect for the rights of the individual and the rule of law in Asia. Provides support and services to relief, development, and human rights projects; monitors the status human rights in Asia and publicizes abuses. **PUBLICATIONS:** *Fax Information* (in English and Japanese), weekly.

HUMAN RIGHTS RESEARCH AND EDUCATION CENTRE (HREC)

Univ. of Ottawa, 57 Louis Pasteur, PO Box 450, Station A, Ottawa, ON, Canada K1N 6N5 **PHONE:** (613) 562-5775 **FX:** (613) 562-5125 **E-MAIL:** hrrec@uottawa.ca **WEBSITES:** http://www.uottawa.ca/hrrec/ **FOUNDED:** 1981. **OFFICERS:** Errol Mendes, Dir. **LANGUAGES:** English, French.

DESCRIPTION: Seeks to further ''discussion of the linkages between human rights, governance, legal reform and development.'' Conducts research and educational programs; provides support and services to human rights organizations worldwide. Evaluates Canadian social justice institutions and suggests reforms. **PUBLICATIONS:** *Human Rights: Chinese and Canadian Perspectives*; *Human Rights Research and Education Bulletin* (in English and French), semiannual; Bulletin, periodic; Reports.

HUMANIST INSTITUTE FOR CO-OPERATION WITH DEVELOPING COUNTRIES (HIVOS) (Instituto Humanista para la Cooperacion con los Paises en Desarrollo)

Raamweg 16, NL-2596 HL The Hague,

Netherlands **PHONE:** 31 70 3765500 **FX:** 31 70 3624600 **E-MAIL:** hivos@hivos.nl **WEBSITES:** http://www.hivos.nl **FOUNDED:** 1968. **OFFICERS:** J.J. Dijkstra, Exec.Off. **STAFF:** 86. **BUDGET:** US$49,000,000. **LANGUAGES:** Bahasa Indonesia, Dutch, English, Portuguese, Russian, Spanish.

DESCRIPTION: Promotes self-determination in a pluralistic society; seeks just distribution of knowledge and economic power. Supports 300 development programs in the Third World, with respect to local social structures and resources. Promotes the activities of local organizations to facilitate the creation of job opportunities. **PUBLICATIONS:** *Hivos Magazine* (in Dutch), quarterly.

HUMANISTIC INSTITUTE FOR COOPERATION WITH DEVELOPING COUNTRIES (HICDC)

PO Box 2227, Harare, Zimbabwe **PHONE:** 263 4 706704 **FX:** 263 4 791781 **FOUNDED:** 1968. **OFFICERS:** Jan Vossen. **LANGUAGES:** Dutch, English.

DESCRIPTION: International development organizations. Seeks to increase the economic opportunities available to people living in developing regions; promotes sustainable development suited to local needs and capacities. Provides financial support to approved development projects; conducts educational programs to raise awareness of development issues.

ICA COMMITTEE ON CO-OPERATIVE COMMUNICATIONS

c/o ICA, 15, rte. des Morillons, Grand-Saconnex, CH-1218 Geneva, Switzerland **PHONE:** 41 22 9298888 **FX:** 41 22 7984122 **E-MAIL:** treacy@coop.org **WEBSITES:** http://www.coop.org; gopher://wiscinfo.wisc.edu:70/11/.info-source/.coop **FOUNDED:** 1968. **OFFICERS:** Mary Treacy, Dir. Communications. **MEMBERS:** 45. **STAFF:** 1. **LANGUAGES:** English.

DESCRIPTION: A member of International Co-operative Alliance. Editors, journalists, publishers, and other communicators employed by cooperative organizations. Serves as a forum for press and other communications issues. Works to improve members' professional competence, effectiveness, and understanding in cooperative information media responsibilities; contributes to the development of the cooperative press. Fosters contacts among members; encourages exchange of experiences and information. Provides consultation and assistance to members.

ICC COMMERCIAL CRIME SERVICES

Maritime House, 1 Linton Rd., Barking, London IG11 8HG, England **PHONE:** 44 181 5913000 **FX:** 44 181 5942833 **TX:** 8956492 **E-MAIL:** ccs@dial.pipex.com **FOUNDED:** 1981. **OFFICERS:** P. Mukundan, Dir. **LANGUAGES:** English.

DESCRIPTION: Works to combat commercial fraud. Maintains ICC Counterfeiting Intelligence Bureau, ICC International Maritime Bureau, and ICC Commercial Crime Bureau. **PUBLICATIONS:** Brochure; Bulletin; Newsletter, periodic.

ICC INTERNATIONAL MARITIME BUREAU

Maritime House, 1 Linton Rd., Barking, London IG11 8HG, England **PHONE:** 44 181 5913000 **FX:** 44 181 5942833 **TX:** 8956492 **E-MAIL:** ccs@dial.pipex.com **FOUNDED:** 1981. **OFFICERS:** P. Mukundan, Dir. **LANGUAGES:** English.

DESCRIPTION: Works to prevent, investigate, and combat maritime and trading crime.

IFRA

Washingtonplatz 1, D-64287 Darmstadt, Germany **PHONE:** 49 6151 7336 **FX:** 49 6151 733800 **E-MAIL:** expressions@ifra.com **WEBSITES:** http://www.ifra.com **FOUNDED:** 1961. **OFFICERS:** Guenther W. Boettcher, Mng.Dir. **MEMBERS:** 1,520. **STAFF:** 65. **BUDGET:** DM 13,000,000. **LANGUAGES:** English, French, German, Spanish.

DESCRIPTION: Newspaper publishers in more than 50 countries; manufacturers of newspaper materials and equipment are associate members. Promotes modernization of the newspaper industry by: developing, testing, and evaluating machines and techniques; evaluating standard specifications for raw materials; investigating economy and quality improvements. Maintains research institute. **PUBLICATIONS:** *Newspaper Techniques* (in English, French, and German), monthly. Contains information about technical development in the newspaper industry; *Special Report.*

IIEC-ASIA

8 Sukhumvit Soi 49/9, Bangkok 10110, Thailand **PHONE:** 66 2 3810814, 66 2 7126057 **FX:** 66 2 3810815 **E-MAIL:** iiecasia@loxinfo.co.th **FOUNDED:** 1984. **OFFICERS:** Terry Kraft-Oliver. **STAFF:** 12. **BUDGET:** US$600,000. **LANGUAGES:** English, Thai.

DESCRIPTION: Promotes efficient use of energy in economically developing countries. Creates local businesses to develop an indigenous stake in the preservation of community energy sources; facilitates communication and cooperation between local businesses and international businesses and development agencies. Conducts educational programs and practical demonstrations of energy-efficient technologies.

INDIAN MUSLIM RELIEF COMMITTEE (IMRC)

c/o Manzoor Ghori, 800 San Antonio, Suite 1, Palo Alto, CA 94303 **PHONE:** (650) 856-0440 **FX:** (650) 856-0444 **FOUNDED:** 1981. **OFFICERS:** Manzoor Ghori, Chm. **MEMBERS:** 4,000. **STAFF:** 2. **BUDGET:** US$750,000. **LANGUAGES:** English.

DESCRIPTION: Aids India's Muslims in achieving security, freedom, and equality—their rights as citizens of India. Provides economic and educational assistance to Yateem/Miskeen children, immediate relief to those affected by natural disasters or to victims of communal violence, and to disseminate information about the Muslims of India. Provides educational charitable programs. Offers legal aid services. **PUBLICATIONS:** Brochure; Bulletin; Newsletter; Reports.

INDIAN OCEAN COMMISSION (IOC)

Ave. Sir Guy Forget, BP 7, Quatre Bornes, Mauritius **PHONE:** 230 4259564, 230 4251652 **FX:** 230 4252709 **E-MAIL:** coi7@bow.intnet.mu **FOUNDED:** 1982. **OFFICERS:** Caabi Elvachrdutu Mohamed, Sec.Gen. **MEMBERS:** 5. **STAFF:** 50. **BUDGET:** MRs 7,000,000. **LANGUAGES:** French.

DESCRIPTION: Representatives from Comoros Islands, Madagascar, Mauritius, Reunion, and Seychelles. Promotes cooperation among member countries. **PUBLICATIONS:** *Korail Ocean Indien* (in French), monthly. Contains regional news; Annual Report.

INDONESIA HUMAN RIGHTS CAMPAIGN (TAPOL)

111 Northwood Rd., Thornton Heath, Surrey CR7 8HW, England **PHONE:** 44 181 7712904 **FX:** 44 171 4975355 **E-MAIL:** hops@gn.apc.org **WEBSITES:** http://www.gn.apc.org/tapol **FOUNDED:** 1973. **OFFICERS:** Carmel Budiardjo, Dir. **MEMBERS:** 750. **STAFF:** 5. **LANGUAGES:** English, Indonesian.

DESCRIPTION: Objectives are to: inform individuals worldwide about human rights violations in Indonesia, East Timor, and West Papua; obtain the assistance of British and other governments to pressure the Indonesian military government to change; stop all British arms sales to Indonesia; cooperate with British and other peace organizations in order to further human rights concerns. Indonesia has been under military rule since 1965, when the state military, supplied with British and U.S. weapons, took control. Since then millions of Indonesians have been denied jobs, been forced to resettle, been executed or imprisoned, and suffered other human rights abuses for holding beliefs differing from those of the military government. The government, still in possession of weapons supplied by the United Kingdom and the United States, invaded East Timor and retains control of West Papua. Both these island nations now suffer human rights abuses similar to those suffered by Indonesians, in addition to being denied the right to self-determination. **PUBLICATIONS:** *Reports*, periodic; *Tapol Bulletin* (in English), bimonthly. Contains information, news and analysis of current human rights issues in Indonesia.

INDUSTRIAL WORKERS OF THE WORLD (IWW)

103 W. Michigan Ave., Ypsilanti, MI 48197 **PHONE:** (734) 483-3548 **FX:** (734) 483-4050 **E-MAIL:** ghq@iww.org **WEBSITES:** http://www.iww.org **FOUNDED:** 1905. **OFFICERS:** Fred Chase, Gen.Sec.-Treas. **MEMBERS:** 926. **STAFF:** 2. **BUDGET:** US$50,000. **LANGUAGES:** English, Finnish, German, Italian, Russian, Spanish, Swedish.

DESCRIPTION: Individuals of all nationalities, religions, or political affiliations, who work for wages or salary. Works for the abolition of the wage system, improvement of conditions through militant unionism, and ultimately "the elimination of social and economic problems at their root, through the establishment of a cooperative commonwealth to replace exploitation of this planet and its people for power or profit." Conducts periodic lectures. Maintains historical archives at Wayne State University, Detroit, Michigan. **PUBLICATIONS:** *Industrial Worker*, monthly. Covers world labor issues, economic developments, and organization news; includes book reviews; *Little Red Songbook*; *One Big Union*; Monographs.

INFORMATION CENTER FOR WOMEN FROM ASIA, AFRICA, AND LATIN AMERICA (Fraueninformationszentrum fur Frauen aus Afrika, Asien, und Latein Amerika - FIZ)

Quellenstrasse 25, CH-8005 Zurich, Switzerland
PHONE: 41 1 2718282 **FX:** 41 1 2725074
E-MAIL: fiz-mail@access.ch **FOUNDED:** 1985.
MEMBERS: 1,000. **STAFF:** 4. **BUDGET:** 420,000 SFr. **LANGUAGES:** English, French, German, Portuguese, Spanish, Thai.

DESCRIPTION: Members are development policy institutes; religious and social welfare organizations; women's associations; individuals. Seeks to increase awareness of the problems facing Third World women who emigrate to countries outside the Third World. Offers counseling and information programs to women from Asia, Africa, and Latin America currently living in Switzerland or contemplating a move there. Disseminates information on Swiss working and living conditions and studies the extent and consequences of trafficking of women from Asia, Africa, and Latin America into Switzerland. Serves as a liaison between women and institutions designed to aid them. Works to educate the public about the migration of women and the structural violence against women. **PUBLICATIONS:** *Annual Report* (in German); *Documentation* (in German); *Gekauftes Ungluck* (in French and German); *Migration von Frauen aus Mittel- und Osteuropa in die Schweiz* (in German); *Rundbrief FIZ* (in German), semiannual, always spring and autumn; *Swiss News* (in English, French, and Spanish), semiannual; Pamphlets (in English, French, Portuguese, Spanish, and Thai).

INFOSHARE INTERNATIONAL (II)

743 Addison St., Ste. A, Berkeley, CA 94710
PHONE: (510) 204-9099 **FX:** (510) 843-4066
E-MAIL: Infoshare@aol.com **FOUNDED:** 1993.
OFFICERS: Julie Stachowiak, Exec.Dir. **STAFF:** 3. **BUDGET:** US$230,000. **LANGUAGES:** English, Russian.

DESCRIPTION: AIDS organizations in Russia and the United States. Promotes increased public awareness of AIDS and HIV and their prevention and treatment. Serves as a clearinghouse on AIDS and HIV; provides clothing, food, and support services to people with HIV and their families; makes available technical and administrative assistance to grass roots AIDS organizations; conducts educational programs.

INSOL INTERNATIONAL (II)

2/3 Philpot Ln., London EC3M 8AQ, England
PHONE: 44 171 9296679 **FX:** 44 171 9296678
E-MAIL: claireb@insol.ndirect.co.uk **WEBSITES:** http://www.insol.org **FOUNDED:** 1982.
OFFICERS: Claire Broughton, Exec.Dir.
MEMBERS: 26. **STAFF:** 2. **BUDGET:** US$200,000.
LANGUAGES: English.

DESCRIPTION: Organizations representing 6500 insolvency practitioners from 62 countries worldwide. Seeks international professional recognition of insolvency practitioners. Works to improve communication and cooperation among members; facilitates the exchange of information; establishes working committees to examine international issues of insolvency practice. Is developing a bibliographic data base of insolvency publications. **PUBLICATIONS:** *INSOL International Membership* (in English), annual; *INSOL World*; *International Insolvency Review* (in English), 3/year. Contains academic commentary on key insolvency issues worldwide.

INSTITUTE OF CULTURAL AFFAIRS (ICAI)

8, rue Amedee Lynen, B-1210 Brussels, Belgium
PHONE: 32 2 2190087 **FX:** 32 2 2190406
E-MAIL: icai@linkline.be **FOUNDED:** 1977.
OFFICERS: Richard Alton, Sec.Gen. **MEMBERS:** 30. **STAFF:** 4. **BUDGET:** US$88,000.
LANGUAGES: Dutch, English, French, German, Spanish.

DESCRIPTION: Individuals in 30 countries promoting world development. Supports work in lifelong education, sustainable development, organizational transformation, and planetary ecology. Works with the United Nations and related international organizations. Provides hostel and conference facilities; develops educational programs. Offers placement service and speakers' bureau. **PUBLICATIONS:** *Between Two Worlds*; *Directory of ICA Locations and Activities*, annual; *Methods for Active Participation*; *Network Exchange*, 10/year. Documents activities of ICAs; *Participation Works*; *Winning Through Participation*. Book: Beyond Prince and Merchant: Citzem.

INSTITUTE FOR DEMOCRACY IN EASTERN EUROPE (IDEE) (IDEE Instytut na rzecz Demokracji w Europie Wschodniej)

PO Box 311, 00-950 Warsaw, Poland **PHONE:** 48 22 6271845 **FX:** 48 22 6271846 **E-MAIL:** idee@idee.ngo.pl **WEBSITES:** http://sunsite.icm.edu.pl/home/idee/index_a.htm **FOUNDED:** 1991. **OFFICERS:** Irena Lasota, Chm. **STAFF:** 5. **BUDGET:** US$205,000.

DESCRIPTION: Upholds the principles of democracy and pluralism. Offers assistance to independent newspapers and publishers in eastern Europe. **PUBLICATIONS:** *The Centers for Pluralism Newsletter* (in English), quarterly. Reports on the activities of the CEE by NGOs working in the field of civic education; *Informacionyi Biuletin* (in Russian). Offers additional manuals for NGOs, journalists, and publishers.

INSTITUTE OF ENERGY (IE)

18 Devonshire St., London W1N 2AU, England **PHONE:** 44 171 5807124 **FX:** 44 171 5804420 **E-MAIL:** info@ioe.org.uk **WEBSITES:** http://www.instenergy.org.uk **FOUNDED:** 1927. **OFFICERS:** Ms. D.P. Davy, Sec. and Chief Exec. **MEMBERS:** 4,900. **STAFF:** 7. **BUDGET:** £600,000. **LANGUAGES:** English.

DESCRIPTION: Individuals interested in energy issues. Maintains charitable program; bestows awards. **PUBLICATIONS:** *Energy World* (in English), 10/year; *Energy World Yearbook* (in English); *Fuel and Energy Abstracts* (in English), bimonthly; *Journal of the Institute of Energy* (in English), quarterly; Proceedings, periodic.

INSTITUTE OF EUROPEAN RESEARCH AND STUDIES (Institut de Recherches et d'Etudes Europeennes - IREE)

51, ave. du Globe, bte. 4, B-1190 Brussels, Belgium **PHONE:** 32 2 3455751 **FOUNDED:** 1976. **OFFICERS:** Paul van Oye, Pres. **MEMBERS:** 50. **STAFF:** 13. **LANGUAGES:** French.

DESCRIPTION: University administrators, professors, and researchers studying the problems of European unification. Compiles statistics. **PUBLICATIONS:** *Studia Europa* (in French), monthly; Reports, periodic.

INSTITUTE FOR THE INTEGRATION OF LATIN AMERICA AND THE CARIBBEAN (INTAL) (Instituto para la Integracion de America Latina y el Caribe)

c/o Mr. Uziel Nogueira, Casilla de Correo 39, Sucursal 1, 1401 Buenos Aires, Argentina **PHONE:** 54 11 43201850 **FX:** 54 11 43201865 **E-MAIL:** uzieln@iadb.org **WEBSITES:** http://www.iadb.org/intal **FOUNDED:** 1964. **OFFICERS:** Mr. Juan Jose Taccone, Dir. **MEMBERS:** 30. **STAFF:** 12. **LANGUAGES:** English, Spanish.

DESCRIPTION: International organization created by an agreement between the Inter-American Development Bank (IDB) and the government of Argentina. Promotes regional integration through studies, training of public officials, disseminating information, and by providing technical assistance to countries and various sub-regional integration schemes. Assists the region's governments in bringing the private sector, including representatives of the business community, labor,and non-governmental organizations (NGOs) into its integration policy deliberations. **PUBLICATIONS:** *Carta Mensual INTAL* (in English, Portuguese, and Spanish), monthly. Provides information on regional events, meetings, agreements, and publications. Also includes bibliography section.; *Integration and Trade* (in English and Spanish), 3/year. Provides information on integration processes and trade relations within Latin America. Also publishes monographs.

INSTITUTE OF INTERNATIONAL EDUCATION (IIE)

809 United Nations Plz., New York, NY 10017-3580 **PHONE:** (212) 883-8200 **FX:** (212) 984-

5452 **WEBSITES:** http://www.iie.org **FOUNDED:** 1919. **OFFICERS:** Dr. Allan E. Goodman, Pres./CEO. **STAFF:** 400. **BUDGET:** US$105,000,000. **LANGUAGES:** English.

DESCRIPTION: Seeks to develop better understanding between the people of the U.S. and the peoples of other countries through higher educational exchange and training programs for students, scholars, artists, leaders, and specialists; assists in developing educational programs to serve the economic and social needs of emerging nations. Serves as an information clearinghouse and provides consultative services on all phases of educational and cultural exchange. Provides information and advice on higher education in the U.S. and abroad. Conducts research. Sponsors Arts International Program. Administers higher education programs for undergraduates, graduate students, and professionals. **PUBLICATIONS:** *Academic Year Abroad* (in English), annual. Guide to over 2,600 overseas study programs offered by U.S. colleges and universities during the academic year.; *English Language and Orientation Programs in the U.S.*, biennial. Lists intensive English programs for foreign students admitted to U.S. postsecondary schools.; *Financial Resources for International Study*. Lists sources of financial aid for U.S. students studying abroad.; *Funding for U.S. Study*. Lists sources of Financial aid for International students studying in the U.S.; *Institute of International Education—Annual Report*; *Open Doors*, annual. Includes reports of annual statistics on foreign students enrolled at U.S. colleges and universities.; *Vacation Study Abroad*, annual. Guide to over 2,200 overseas study programs offered to U.S. nationals during the summer.

INSTITUTE OF INTERNATIONAL FINANCE (IIF)

2000 Pennsylvania Ave. NW, Ste. 8500, Washington, DC 20006-1812 **PHONE:** (202) 857-3600 **FX:** (202) 775-1430 **WEBSITES:** http://www.iif.com **FOUNDED:** 1983. **OFFICERS:** Charles H. Dallara, Mng.Dir. **MEMBERS:** 230. **STAFF:** 60. **BUDGET:** US$15,000,000. **LANGUAGES:** English.

DESCRIPTION: Financial institutions. Seeks to improve upon the process of sovereign lending, trade and project financing, and the long-term efficiency of international credit markets. Functions are to gather accurate economic information on individual countries; to discuss economic plans, assumptions, and financing needs with borrower countries; to serve as a focal point for dialogue between the international banking community and multilateral institutions, central banks, and bank supervisory authorities; to support members' risk management, asset allocation, and business development in emerging markets. **PUBLICATIONS:** *Economic Review*, monthly; Annual Report, annual; Brochure.

INSTITUTE OF INTERNATIONAL LAW (IIL) (Institut de Droit International - IDI)

c/o M. Ch. Dominice, La Vague, 33, route de Suisse, CH-1297 Founex, Switzerland **PHONE:** 41 22 7760646, 41 22 7311730 **FX:** 41 22 7384306 **FOUNDED:** 1873. **OFFICERS:** Christian Dominice, Sec.Gen. **MEMBERS:** 132. **LANGUAGES:** English, French.

DESCRIPTION: Individuals in 44 countries who have made outstanding contributions to international law in theory and practice. Promotes the progress of international law. Disseminates information to members. **PUBLICATIONS:** *Annuaire* (in English and French), biennial.

INSTITUTE OF MANUFACTURING (IMANF)

58 Clarendon Ave., Leamington Spa, Warwickshire CV32 4SA, England **PHONE:** 44 01926 855498 **FX:** 44 01926 513100 **FOUNDED:** 1978. **OFFICERS:** Mr. H.J. Hammons, Pres. and Founder. **MEMBERS:** 400. **STAFF:** 4. **LANGUAGES:** English.

DESCRIPTION: Individuals involved in manufacturing in 40 countries. Provides information to members on new developments, techniques, processes, and methods in the field of manufacturing. Conducts seminars and programs in management training and education. **PUBLICATIONS:** *Manufacturing Management and Manufacturers* (in English), annual.

INSTITUTE FOR THE STUDY OF GENOCIDE (ISG)

899 10th Ave., Rm. 325, New York, NY 10019 **PHONE:** (212) 582-2537 **FX:** (212) 582-9127 **FOUNDED:** 1982. **OFFICERS:** Dr. Helen Fein, Exec.Dir. **MEMBERS:** 50. **LANGUAGES:** English.

DESCRIPTION: Professors, scholars, and other individuals. United to further research and scholarship on the causes, prevention, and consequences of genocide. Goals are to: sponsor historical, contemporary, and predictive research on the causes of genocide, mass political killing, and other violations of human rights of people; research and evaluate responses to genocide; monitor contemporary reports of human rights violations; considers deterrents to genocide and strategies to impede genocide and to assist the potential victims. Establishes liaison with related groups internationally and with archives and libraries specializing in the history of indigenous peoples. Reports on collections of scholarly material and related lectures and seminars. **PUBLICATIONS:** *ISG Newsletter*, semiannual.

INSTITUTION OF CHEMICAL ENGINEERS (IChemE)

Davis Bldg., 165-189 Railway Terr., Rugby, Warwickshire CV21 3HQ, England **PHONE:** 44 1788 578214 **FX:** 44 1788 560833 **E-MAIL:** library@icheme.org.uk **WEBSITES:** http://www.icheme.org **FOUNDED:** 1922. **OFFICERS:** Dr. T.J. Evans, Chief Exec. **MEMBERS:** 24,000. **STAFF:** 85. **LANGUAGES:** English.

DESCRIPTION: Chemical engineers united to promote and develop the science of chemical engineering and to further scientific and economic development and application of manufacturing processes in which chemical and physical changes of materials are involved. Promotes research into chemical engineering and communicates with governments on behalf of industry; compiles statistics. Holds symposia; offers continuing education courses. **PUBLICATIONS:** *The Chemical Engineer* (in English), biweekly; *Chemical Engineering Research and Design*, 8/year; *Environmental Protection Bulletin*; *Food and Bioproducts Processing*, quarterly; *Loss Prevention Bulletin*; *Process Safety and Environmental Protection*, quarterly.

INTER PRESS SERVICE - THIRD WORLD NEWS AGENCY (IPS) (Inter Press Service Tercer Mundo)

c/o Dr. Roberto Savio, Via Panisperna 207, I-00184 Rome, Italy **PHONE:** 39 6 4742497, 39 6 4751918 **FX:** 39 6 4817877 **TX:** 610574 IPSROMI **E-MAIL:** romaser@ips.org **WEBSITES:** http://www.ips.org **FOUNDED:** 1964. **OFFICERS:** Dr. Roberto Savio, Dir.Gen. **MEMBERS:** 127. **BUDGET:** US$13,000,000. **LANGUAGES:** English, Spanish.

DESCRIPTION: Cooperative nongovernmental news agency of professional journalists in 91 countries. Reports issues and developments in the Third World for the media and other uses in the Third World and in developed countries. Works to facilitate and improve the exchange of information within and the quality of information about the Third World. Produces daily news broadcasts in Dutch, English, Finnish, French, German, Hungarian, Nepalese, Norwegian, Portuguese, Spanish, and Swahili. Provides professional in-house training, and in-field placements for Third World journalists. Conducts research in the field of information exchange; organizes annual training seminar. Compiles statistics. **PUBLICATIONS:** *Central America*, weekly; *Church Bulletin*, weekly; *Columnist Service*, periodic; *Communication*, weekly; *Cultural Bulletin*, weekly; *Development Round-Up*, weekly; *Education*, weekly; *Environment*, weekly; *G77 Bulletin*, monthly; *Human Rights*, weekly; *Investment and Development in the Third World*, weekly; *IPS TWN - South Commission*, daily; *Metropolis 2000*, weekly; *Mineral and Siderurpical*, weekly; *Petrol*, weekly; *Population*, weekly; *South-North Development Monitor*, periodic; *Voice of the OPEC Countries*, weekly; *Women's Feature Services*, weekly. Contains information on Africa, Asia, and Latin America.

INTER-AFRICAN COFFEE ORGANIZATION (IACO) (Organisation Interafricaine du Cafe - OIAC)

BP V210, Abidjan, Cote d'Ivoire **PHONE:** 225 216131 **FX:** 225 216212 **TX:** 22406 OICAFE ABIDJAN **FOUNDED:** 1960. **OFFICERS:** Arega Worku, Sec.Gen. **MEMBERS:** 25. **STAFF:** 15. **LANGUAGES:** English, French.

DESCRIPTION: Coffee-producing countries on the African continent and its bordering islands. Works to establish close contact among national, regional, and international organizations concerned with problems related to the production, processing, and consumption of coffee in the world. Seeks uniform policies for the marketing of coffee. Interests include: marketing; research; and development of coffee production. **PUBLICATIONS:** *Exporter Directory of African Coffee*, biennial.

INTERAFRICAN NETWORK FOR HUMAN RIGHTS AND DEVELOPMENT (INHRD)

PO Box 31145, Lusaka, Zambia **PHONE:** 260 1 226544, 260 1 238913 **FX:** 260 1 238911 **E-MAIL:** afronet@zamnet.zm **OFFICERS:** Ngande Mwanajiti, Exec.Dir. **LANGUAGES:** English.

DESCRIPTION: Human rights and development organizations. Seeks to ensure respect for human rights and the rule of law in developing regions. Monitors the human rights situation in areas undergoing development and publicizes abuses; makes available legal services.

INTER-AFRICAN UNION OF LAWYERS

12 rue du Prince Moulay Abdullah, Casablanca, Morocco **PHONE:** 212 2 271017 **FX:** 212 2 204686 **OFFICERS:** Francois Xavier Agonjo-Okawe, Sec.Gen. **LANGUAGES:** Arabic, English.

DESCRIPTION: Attorneys. Promotes professional advancement of members. Facilitates communication and cooperation among members; represents members' interests; conducts continuing professional development programs.

INTER-AMERICAN BAR ASSOCIATION (IABA)

1211 Connecticut Ave., Ste. 202 NW, Washington, DC 20036 **PHONE:** (202) 393-1217 **FX:** (202) 393-1241 **E-MAIL:** laba@iaba.org **FOUNDED:** 1940. **OFFICERS:** Louis G. Ferrand, Sec.Gen. **MEMBERS:** 3,000. **STAFF:** 3. **LANGUAGES:** English, French, Portuguese, Spanish.

DESCRIPTION: An organization of national, regional, and special associations of attorneys (50); individual lawyers (3000). Purposes are to: advance the science of jurisprudence, and in particular, the study of comparative law; promote uniformity in commercial legislation; further the knowledge of laws of Western Hemisphere countries; propagate justificative administration through the creation and maintenance of independent judicial systems; protect and defend civil, human, and political rights of individuals; uphold the honor of the legal profession; encourage geniality and brotherhood among members. **PUBLICATIONS:** *Inter-American Bar Association—Conference Proceedings* (in English and Spanish). Collection of papers presented at IABA conferences; *Inter-American Bar Association—Letter to Members*, quarterly. Includes awards and calendar of events; *Inter-American Bar Association—Membership Directory*, annual.

INTER-AMERICAN COMMERCIAL ARBITRATION COMMISSION (I-ACAC)

OAS Administration Bldg., 19th & Constitution Ave. NW, Rm. 211, Washington, DC 20006 **PHONE:** (202) 458-3249 **FX:** (202) 458-3293 **E-MAIL:** sice@oas.org **FOUNDED:** 1934. **OFFICERS:** Charles R. Norberg, Dir.Gen. **STAFF:** 3. **LANGUAGES:** English, Spanish.

DESCRIPTION: Companies, businessmen, bankers, lawyers, investors, and interested others in 32 North and South American countries. Arbitrates or negotiates adjustment of international trade controversies. Promotes effective national arbitration laws in member countries of the Organization of American States. Conducts promotional and educational programs.

INTER-AMERICAN COMMISSION ON HUMAN RIGHTS (IACHR)

1889 F St. NW, 8th Fl., Washington, DC 20006 **PHONE:** (202) 458-6002 **FX:** (202) 458-3992 **E-MAIL:** cidhoea@oas.org **WEBSITES:** http://www.lachr.org; http://www.lachr.org **FOUNDED:** 1960. **OFFICERS:** Dr. Jorge Taiana, Exec.Sec. **MEMBERS:** 7. **STAFF:** 23. **BUDGET:** US$2,800,000. **LANGUAGES:** English, French, Portuguese, Spanish.

DESCRIPTION: Citizens of member nations of the Organization of American States. Promotes and protects human rights in the Caribbean, and North, Central, and South America. Maintains 5000 volume library specializing in human rights law. **PUBLICATIONS:** *Special County Reports*; Annual Report (in English, French, and Spanish); Reports. Covers all member states of the Organzation of American States (OAS).

INTER-AMERICAN COMMISSION OF WOMEN (CIM)

c/o Org. of Amer. States, 1889 F St. NW, Rm. 880, Washington, DC 20006 **PHONE:** (202) 458-6084 **FX:** (202) 458-6094 **FOUNDED:** 1928.

OFFICERS: Carmen Lomellin, Exec.Sec. **MEMBERS:** 34. **STAFF:** 7. **BUDGET:** US$1,500,000. **LANGUAGES:** English, French, Portuguese, Spanish.

DESCRIPTION: Specialized agency of the Organization of American States dealing with issues concerning women. Commission is composed of one delegate for each member country of OAS. Mobilizes, trains, and organizes women ''so that they may fully participate in all fields of human endeavor, on a par with men, as two beings of equal value, coresponsible for the destiny of humanity.'' Informs the OAS general assembly and member governments on: civil, political, social, economic, and cultural status of women in the Americas; progress achieved in the field as well as problems to be considered; development of a plan of action following the Decade of Women (1976-85) of strategies for full and equal participation by women by the year 2000. Serves as liaison for women's groups throughout the hemisphere and conducts research on laws affecting women. Operates a regional information center in Santiago, Chile; finances development projects in Latin America and the Caribbean. **PUBLICATIONS:** *Final Report-Assembly of Delegates*, biennial; *Series: Studies*, periodic; Reports. Proceedings of technical meetings.

INTER-AMERICAN CONFEDERATION FOR CATHOLIC EDUCATION (CIEC) (Confederacion Interamericana de Educacion Catolica)

Calle 78, Numero 12-16, Apartado Aereo 90036, Bogota, Colombia **PHONE:** 57 1 2550513 **FX:** 57 1 2553676 **E-MAIL:** ciec@latino.net.co **FOUNDED:** 1945. **OFFICERS:** Sister M. Constanza Arangra FMS. **MEMBERS:** 26. **LANGUAGES:** English, Portuguese, Spanish.

DESCRIPTION: Purposes are to promote the principles and rules of Catholic education and to maintain and improve the quality of Catholic schools and teachers. Sponsors peace education program and instruction in social values and ethics. Compiles statistics. **PUBLICATIONS:** *Central* (in Spanish), periodic; *Educacion Hoy*, quarterly; *Radiar Collection*, periodic; *Textos*, periodic.

INTER-AMERICAN CONFERENCE ON SOCIAL SECURITY (IACSS) (Conferencia Interamericana de Seguridad Social - CISS)

Calle San Ramon s/n Esq. Ave. San Jeronimo, Colonia San Jeronimo Lidice, Delegacion Magdalena Contreras, 10100 Mexico City, DF, Mexico **PHONE:** 52 5 5950011 **FX:** 52 5 6838524 **TX:** 1772223210 CETEME **E-MAIL:** ciss@data.net.mx **WEBSITES:** http://www.ciss.org.mx **FOUNDED:** 1942. **OFFICERS:** Maria Eluira Contreras, Sec.Gen. **MEMBERS:** 60. **STAFF:** 30. **LANGUAGES:** English, French, Spanish.

DESCRIPTION: Social security institutions in 37 North and South American countries. Seeks to develop and facilitate cooperation between social security programs in the Americas. Permanent Inter-American Committee on Social Security acts as the group's executive body. Operates Inter-American Center for Social Security Studies and American Commissions on Social Security. Conducts courses, seminars, and Annual programs; compiles statistics. **PUBLICATIONS:** *Bulletin CISS* (in English and Spanish), bimonthly. Contains social security news, information about meetings, courses; *Social Security Journal* (in English and Spanish), bimonthly. Also publishes directory of members, series of social security studies, and series of social security monographs.

INTER-AMERICAN COUNCIL FOR EDUCATION, SCIENCE AND CULTURE (CIECC)

c/o Organization of American States, 1889 F St. NW, 2nd FL., Washington, DC 20006 **PHONE:** (202) 458-3783 **FX:** (202) 458-3526 **FOUNDED:** 1948. **OFFICERS:** L. Zuniga, Exec.Sec. **MEMBERS:** 32. **BUDGET:** US$27,085,000. **LANGUAGES:** English.

DESCRIPTION: Representatives of member countries of the Organization of American States; affiliated members are representative from 22 countries with permanent observer status. Fosters the development of education, science, technology, and culture through its Regional Program of Educational Development, Regional Program of Scientific and Technological Development, Regional Program of Cultural Development, and Fellowship and Training Program. **PUBLICATIONS:** *Ciencia Interamericana*.

INTER-AMERICAN DEVELOPMENT BANK (IDB)

1300 New York Ave., NW, Washington, DC 20577 **PHONE:** (202) 623-1000 **FX:** (202) 623-3096 **WEBSITES:** http://www.iadb.org **FOUNDED:** 1959. **LANGUAGES:** English, French, Portuguese, Spanish.

DESCRIPTION: Members are financial institutions in North, Central, and South America. Promotes social and economic development of countries in these regions; makes available technical assistance. **PUBLICATIONS:** *Economic and Social Progress in Latin America* (in English, French, Portuguese, and Spanish), annual; *The IDB* (in English, French, Portuguese, and Spanish), monthly; Report (in English, French, Portuguese, and Spanish), annual.

INTER-AMERICAN INSTITUTE OF HUMAN RIGHTS (IIHR) (Instituto Interamericano de Direitos Humanos)

Apartado Postal 10081, San Jose 1000, Costa Rica **PHONE:** 506 2340404 **FX:** 506 2340955 **TX:** 2233 CORTE CR **E-MAIL:** instituto@iidh.ed.cr **FOUNDED:** 1980. **OFFICERS:** Juan E. Mendez, Exec.Dir. **STAFF:** 75. **LANGUAGES:** English, Spanish.

DESCRIPTION: Created by an agreement between the Inter-American Court of Human Rights and the Republic of Costa Rica. Seeks to promote and strengthen democracy and human rights throughout the Americas by means of specialized training, research, education, political mediation, and technical assistance for governmental and civil society bodies and international organizations. Conducts an interdisciplinary course on human rights annually. Maintains programs on refugees, indigenous populations, gender, prevention of torture, and human rights training for nongovernmental organization (NGO) staff. Also conducts programs on the administration of justice, security forces, and ombudsmen. Operates the Center for Electoral Promotion and Assistance. **PUBLICATIONS:** *Annual Report*; *Boletin Informativo*, quarterly; *Revista IIDH*, semiannual. Also publishes research and project reports, monographs, articles, and books.

INTER AMERICAN PRESS ASSOCIATION (IAPA)

2911 NW 39th St., Miami, FL 33142 **PHONE:** (305) 634-2465 **TF:** (800) 542-3732 **FX:** (305) 635-2272 **E-MAIL:** sipiapa@aol.com **WEBSITES:** http://www.sipiapa.org **FOUNDED:** 1942. **OFFICERS:** Julio Munoz, Exec.Dir. **MEMBERS:** 1,383. **STAFF:** 10. **BUDGET:** US$900,000. **LANGUAGES:** English, Portuguese, Spanish.

DESCRIPTION: Newspapers, magazines, educators, and other individuals in allied fields. Promotes and protects freedom of the press in the Americas. **PUBLICATIONS:** *Hora de Cierre*, quarterly; *IAPA News*, monthly; *Noticiero SIP* (in Spanish), monthly.

INTER-AMERICAN TROPICAL TUNA COMMISSION (IATTC)

c/o Scripps Institution of Oceanography, 8604 La Jolla Shores Dr., La Jolla, CA 92037-1508 **PHONE:** (858) 546-7100 **FX:** (858) 546-7133 **E-MAIL:** rallen@iattc.ucsd.edu **FOUNDED:** 1950. **OFFICERS:** Dr. Robin L. Allen, Dir. **STAFF:** 50. **BUDGET:** US$3,227,000. **LANGUAGES:** English, Spanish.

DESCRIPTION: Appointed commissioners representing Japan, France, Nicaragua, Mexico, Ecuador, El Salvador, Costa Rica, Panama, Vanuatu, Venezuela and the United States. Conducts studies on Pacific Ocean tunas and dolphins associated with tunas. Recommends conservation measures to member governments in order to maintain optimum levels of tuna stock and maximum level of dolphin stock. (Dolphins are frequently caught and inadvertently killed in tuna nets.) Monitors population and mortality levels; conducts research. **PUBLICATIONS:** *Special Reports* (in English and Spanish), periodic; Annual Report (in English and Spanish), annual; Bulletin (in English and Spanish), periodic.

INTER-CHURCH COMMITTEE ON HUMAN RIGHTS IN LATIN AMERICA (ICCHRLA) (Comite Inter-Eglises des Droits Humains en Amerique Latine - CIEDHAL)

129 St. Clair Ave. W, Toronto, ON, Canada M4V 1N5 **PHONE:** (416) 921-0801 **FX:** (416) 921-3843 **E-MAIL:** icchrla@web.apc.org **WEBSITES:** http://www.web.net/-icchrla **FOUNDED:** 1977. **OFFICERS:** Rusa Jeremic, Admin.Coor. **MEMBERS:** 17. **STAFF:** 4. **LANGUAGES:** English, French, Spanish.

DESCRIPTION: Christian churches and Catholic religious orders in Canada. Intervenes on behalf of victims of human rights abuses; disseminates information and analyses of human rights situations; makes recommendations to relevant organizations on appropriate responses to human rights violations; monitors Canadian foreign policy toward Latin American countries. Organizes fact-finding missions. Maintains close links with churches and human rights groups in Latin America. **PUBLICATIONS:** *ICCHRLA ALERTA* (in English), quarterly; *Peace What Peace: Confronting Central Americas New Economic War*; *Special Reports* (in English and Spanish), periodic.

INTERCHURCH MEDICAL ASSISTANCE (IMA)

College Ave. at Blue Ridge, Box 429, New Windsor, MD 21776 **PHONE:** (410) 635-8720 **FX:** (410) 635-8726 **E-MAIL:** ima@brethren.org **WEBSITES:** http://www.interchurch.org **FOUNDED:** 1961. **OFFICERS:** Paul Derstine, Pres. **MEMBERS:** 12. **STAFF:** 7. **BUDGET:** US$15,500,000. **LANGUAGES:** English.

DESCRIPTION: Denominational-founded autonomous organization for the solicitation, collection, and distribution of pharmaceutical, medical, dental, and hospital supplies for use in the overseas charity medical programs of American Protestant churches, relief agencies, and other American charitable organizations. **PUBLICATIONS:** *Interchurch Medical Newsletter*, quarterly; Annual Report, annual.

INTERGOVERNMENTAL AUTHORITY ON DEVELOPMENT (IGAD)

Boite Postale 2653, Djibouti, Djibouti **PHONE:** 253 354050 **FX:** 253 356994 **TX:** 5978 **OFFICERS:** Dr. David Muduuli, Sec.Gen. **LANGUAGES:** English, French.

DESCRIPTION: African government agencies responsible for the oversight and regulation of development programs. Promotes appropriate and sustainable development. Conducts research; drafts model legislation for the regulation of development projects.

INTERGOVERNMENTAL COPYRIGHT COMMITTEE (ICC) (Comite Intergouvernemental du Droit d'Auteur - CIDA)

c/o UNESCO, Maison de l'UNESCO, 7, place de Fontenoy, F-75700 Paris, France **PHONE:** 33 1 45684705, 33 1 45684702 **FX:** 33 1 45685589 **TX:** 204461 **E-MAIL:** d.bax@unesco.org **FOUNDED:** 1952. **OFFICERS:** Mr. S. Abada, Chief, SDCCIC. **MEMBERS:** 18. **LANGUAGES:** Arabic, English, French, Russian, Spanish.

DESCRIPTION: National government representatives who study problems concerning the application and operation of the Universal Copyright Convention, which was adopted in 1952 and revised in 1971. Prepares for periodic revision of the convention; keeps states party to the convention updated on its activities; studies problems regarding international protection of copyrights. **PUBLICATIONS:** *Copyright Bulletin* (in Chinese, English, French, Russian, and Spanish), quarterly. Includes doctrin articles, news from states, information on activities in Ch. field of copyrights.

INTERGOVERNMENTAL GROUP ON RICE (IGRC)

Via delle Terme di Caracalla, I-00100 Rome, Italy **PHONE:** 39 6 57054136 **FX:** 39 6 57054495 **E-MAIL:** concepcion.calpe@fao.org **FOUNDED:** 1956. **OFFICERS:** Concepcion Calpe. **LANGUAGES:** English, Italian.

DESCRIPTION: Representatives of governments of rice-producing nations that are members of the Food and Agriculture Organization of the United Nations. Promotes environmentally sustainable production of rice; seeks to ease restrictions to international rice trade. Facilitates consultation and communication among members; works to secure international trade arrangements. Gathers and disseminates information on rice production, environmental protection, and trade. **PUBLICATIONS:** Pamphlets; Reports.

INTERGOVERNMENTAL OCEANOGRAPHIC COMMISSION (IOC) (Commission Oceanographique Intergouvernementale - COI)

c/o UNESCO, Maison de l'UNESCO, 7, place de Fontenoy, F-75700 Paris, France **PHONE:** 33

1 45683983, 33 1 45685810 **TX:** 204461 PARIS **E-MAIL:** g.kullunberg@unesco.org **WEBSITES:** http://www.unesco.org:80/ioc **FOUNDED:** 1960. **OFFICERS:** Dr. P. Bernal, Exec.Sec. **MEMBERS:** 125. **STAFF:** 60. **LANGUAGES:** English, French, Russian, Spanish.

DESCRIPTION: National government representatives united to promote scientific investigations to the nature of oceans and ocean resources. Operates Marine Environmental Data Information Referral System providing referrals concerning availability, location, and characteristics of marine environmental data. Operates ocean observations, sea level, currents, contamination, coastal studies: develops the Gloval Ocean Observing System. Deals with subjects such as: bathymetry (measurement of ocean depths); physical, chemical, biological, and other aspects of oceanography; hydrography; remotely sensed observations of the sea; marine environment, biology, geology, geophysics, and meteorology; marine organic, inorganic, and radioactive pollution. Supports international, ocean education and study groups. Operates in association with: World Meteorological Organization; International Atomic Energy Agency; International Council for the Exploration of the Sea; International Maritime Organization; Scientific Committee on Oceanic Research; United Nations; United Nations Environment Programme; Food and Agriculture Organization of the United Nations. **PUBLICATIONS:** *Annual Report of Activities* (in English, French, Russian, and Spanish); *Guides*, periodic; *Reports of Groups of Experts*, periodic; *Summary Reports of Governing and Major Subsidiary Bodies*, periodic; *Technical Series*, periodic; *Training Course Reports*, periodic; *Workshop Reports*, periodic; Manuals, periodic.

INTERGOVERNMENTAL ORGANIZATION FOR INTERNATIONAL CARRIAGE BY RAIL (OTIF) (Organisation Intergouvernementale pour les Transports Internationaux Ferrovaires - OTIF)

Gryphenhubeliweg 30, CH-3006 Bern, Switzerland **PHONE:** 41 31 3591010 **FX:** 41 31 3591011 **E-MAIL:** otif@otif.ch **FOUNDED:** 1890. **OFFICERS:** Michel Burgmann, Dir.Gen. **MEMBERS:** 39. **STAFF:** 13. **BUDGET:** 3,000,000 SFr. **LANGUAGES:** French, German.

DESCRIPTION: Representatives of countries adhering to the Convention on International Carriage by Rail regulating the transport of goods, passengers, and luggage by rail. Aims to establish a uniform system of law applicable to the carriage of passengers, luggage, and goods in international rail traffic between member states. Establishes rules for the conveyance of dangerous goods. Advises, arbitrates, and conciliates on disputes between railways; examines requests for amendment to the Convention. Maintains Central Office for International Carriage by Rail. Disseminates information on behalf of international transport services. Organizes training courses. **PUBLICATIONS:** *Bulletin des Transports Internationaux Ferroviaires* (in French and German), bimonthly; *Rapport de gestion/Geschaftsbericht* (in French and German), annual.

INTERNATIONAL ACADEMIC UNION (IAU) (Union Academique Internationale - UAI)

Palais des Academies, 1, rue Ducale, B-1000 Brussels, Belgium **PHONE:** 32 2 5502200, 32 2 5502211 **FX:** 32 2 5502205 **FOUNDED:** 1919. **OFFICERS:** Philippe Roberts-Jones, Adm.Sec. **MEMBERS:** 47. **STAFF:** 1. **BUDGET:** 4,000,000 BFr. **LANGUAGES:** English, French.

DESCRIPTION: Affiliated academies and bodies of similar scientific standing in 40 countries in the fields of philology, archaeology, history, and social sciences. Promotes cooperation among members through 60 collective projects and coordination of activities. **PUBLICATIONS:** *Proceedings of General Assembly* (in French), annual; Book, periodic.

INTERNATIONAL ACADEMY OF ASTRONAUTICS (IAA) (Academie Internationale d'Astronautique - AIA)

c/o Dr. Jean Michel Contant, 6, rue Galilee, PO Box 1268-16, F-75766 Paris Cedex 16, France **PHONE:** 33 1 47238215 **FX:** 33 1 47238216 **TX:** 651767 F **E-MAIL:** sgeneral@iaanet.org **WEBSITES:** http://www.iaanet.org **FOUNDED:** 1960. **OFFICERS:** Dr. Jean-Michel Contant, Sec.Gen. **MEMBERS:** 1,136. **BUDGET:** US$200,000. **LANGUAGES:** English, French, German, Russian, Spanish.

DESCRIPTION: Individuals in 60 countries who have distinguished themselves in the basic, engineering, life, and social sciences in areas connected with astronautics. Promotes the advancement of astronautics by conducting studies and holding symposia and scientific meetings. Cooperates with national academies. **PUBLICATIONS:** *Acta Astronautica*, monthly. Also publishes dictionaries, reports, and newsletters.

INTERNATIONAL ACCOUNTING STANDARDS COMMITTEE (IASC) (Comite International des Normes Comptables)

166 Fleet St., London EC4A 2DY, England **PHONE:** 44 171 3530565 **FX:** 44 171 3530562 **E-MAIL:** iasc@netcomuk.co.uk **FOUNDED:** 1973. **OFFICERS:** Sir Bryan Carsberg, Sec.Gen. **MEMBERS:** 119. **STAFF:** 12. **LANGUAGES:** English.

DESCRIPTION: Professional accountancy bodies representing more than 1,900,000 accountants in 88 countries. Aim is to formulate and publish, in the public interest, International Accounting Standards to be observed in the presentation of audited financial statements and to promote their worldwide acceptance and observance. (International Accounting Standards are statements regulating the representation and disclosure of figures within financial statements; such standards do not nullify local laws and may not be in accordance with local legislature.) Works for the improvement and harmonization of regulations, accounting standards, and procedures relating to the presentation of financial statements. Provides standards developed in response to the needs and problems of both developed and developing nations. Provides liaison between members and stock exchanges, representative bodies of accountants, industry groups, and other international organizations. Operates several technical committees to research current programs; maintains consultative group of representatives from banks, business, labor, law, securities commissions, stock exchanges, and international governmental organizations who offer advice on the process of setting International Accounting Standards. **PUBLICATIONS:** *IASC Insight*, quarterly. Available only through an annual subscription service.; *IASC Update*, 3/year; *International Accounting Standards*.

INTERNATIONAL ACTION FOR THE RIGHTS OF THE CHILD (IARC) (Action Internationale pour les Droits de l'Enfant - AIDE)

BP 427, F-75870 Paris Cedex 18, France **FX:** 33 1 42055552 **FOUNDED:** 1986. **OFFICERS:** Jean Dallais, Gen.Mgr. **MEMBERS:** 1,100. **STAFF:** 6. **BUDGET:** 1,000,000 Fr. **LANGUAGES:** English, French, Portuguese, Spanish, Tagalog.

DESCRIPTION: Provides food, lodging, and medical and psychological support for exploited and incarcerated children, without regard for politics, race, or religious affiliation. Promotes the ideal that no child ever deserves to be imprisoned. Conducts research to locate families. Provides remedial education and attempts to mainstream children into public and private educational systems. Operates a residential aid center, La Porte Ouverte, and plans to open a second facility. **PUBLICATIONS:** Newsletter, periodic.

INTERNATIONAL ADVERTISING ASSOCIATION (IAA)

521 5th Ave., Ste. 1807, New York, NY 10175 **PHONE:** (212) 557-1133 **FX:** (212) 983-0455 **E-MAIL:** iaa@iaaglobal.org **WEBSITES:** http://www.iaaglobal.org **FOUNDED:** 1938. **OFFICERS:** Norman Vale, Dir.Gen. **MEMBERS:** 5,000. **STAFF:** 9. **LANGUAGES:** English.

DESCRIPTION: Global network of advertisers, advertising agencies, the media and related services, spanning 95 countries. Demonstrates to governments and consumers the benefits of advertising as the foundation of diverse, independent media. Protects and advances freedom of commercial speech and consumer choice, encourages greater practice and acceptance of advertising self-regulation, provides a forum to debate emerging professional marketing communications issues and their consequences in the fast-changing world environment, and takes the lead in state-of-the-art professional development through education and training for the marketing communications industry of tomorrow. Conducts research on such topics as restrictions and taxes on advertising, advertising trade practices and related information, and advertising expenditures around the world. Sponsors IAA Education Program. Has compiled recommendations for international advertising standards and practices. **PUBLICATIONS:** *The Case for Advertising Self-Regulation*; *IAA Annual Report and Mem-*

bership Directory. Lists names, business affiliations, addresses, telephone, fax, and cable numbers of all members; includes index of members by name and organization; *IAA World News*, quarterly. Covers the advertising industry, with emphasis on the value of advertising, freedom of commercial speech, and consumer choice.; *Monographs on Severely Restricted or Forbidden Advertising Practices*; Pamphlets.

INTERNATIONAL AFRICAN INSTITUTE (IAI) (Institut African International)

SOAS, Thornhaugh St., Russell Sq., London WC1H 0XG, England **PHONE:** 44 171 3236035, 44 171 3236108 **FX:** 44 171 3236118 **E-MAIL:** iai@soas.ac.uk **WEBSITES:** http://www.oneworld.org/iai/ **FOUNDED:** 1926. **OFFICERS:** Prof. Paul Spencer, Hon.Dir. **MEMBERS:** 40. **STAFF:** 4. **LANGUAGES:** English, French.

DESCRIPTION: Individuals or corporate bodies interested in African studies or concerned with the social development of African peoples. Promotes the serious study of African peoples, including their languages, cultures, social life, traditional patterns of ethnic organization and associated beliefs and values, and emerging social forms and cultural developments together with publications on these topics. Facilitates communication between African scholars and Africanists. **PUBLICATIONS:** *Africa*, quarterly; *Africa Bibliography*, annual; *Africa Issues Development Series*, 2-3/year; *African Languages* (in English and French), periodic; *Classics in Africa Anthropology*; *International African Library*; *International African Seminars*; *International Guide to African Studies Research*, periodic; *Thesaurus of African Languages*.

INTERNATIONAL AGENCY FOR THE PREVENTION OF BLINDNESS (IAPB)

c/o Mrs. M.J. Haws, Admin., Grosvenor Hall, Bolnore Rd., Haywards Heath, W. Sussex RH16 4BX, England **PHONE:** 44 1444 458810 **FX:** 44 1444 458810 **E-MAIL:** mhaws@dircon.co.uk **FOUNDED:** 1975. **OFFICERS:** Dr. R. Pararajasegaram, IAPB Pres. **STAFF:** 1. **LANGUAGES:** English.

DESCRIPTION: Ophthalmic societies and societies for the prevention of blindness whose members include ophthalmologists, public health officers, nutritionists, geneticists, and other health workers. Coordinates international research into the causes of impaired vision or blindness; promotes measures calculated to eliminate such causes; disseminates knowledge worldwide on preventing blindness and on matters pertaining to care of the eyes. Cooperates with the World Health Organization, United Nations Children's Fund, and other international agencies. **PUBLICATIONS:** *IAPB Fifth General Assembly*. Proceedings document; *IAPB News* (in English); *World Blindness and Its Prevention Vols. 1-4*. Based on proceedings of respective General Assemblies.

INTERNATIONAL AGENCY FOR RESEARCH ON CANCER (IARC) (Centre International de Recherche sur le Cancer - CIRC)

150, cours Albert Thomas, F-69372 Lyon Cedex 08, France **PHONE:** 33 472738485 **FX:** 33 472738575 **TX:** 380023 **E-MAIL:** lastname@iarc.fr **WEBSITES:** http://www.iarc.fr **FOUNDED:** 1965. **OFFICERS:** Dr. Paul Kleihues, Dir. **MEMBERS:** 18. **STAFF:** 180. **BUDGET:** US$18,000,000. **LANGUAGES:** English, French.

DESCRIPTION: Cancer research arm of the World Health Organization. Representatives of nations involved in international collaboration in cancer research. Generates and disseminates information on the causes and prevention of cancer; conducts research in the field of cancer epidemiology, biostatistics, and environmental carcinogenesis. Evaluates and examines populations with unusually high or low frequencies of cancer and identifies the role of environmental factors including cultural and dietary habits and chemicals. Assists governments in cancer control programs. Maintains laboratories and collaborates with scientists working in national laboratories. Organizes training courses; compiles statistics. **PUBLICATIONS:** *Biennial Report* (in English and French); *Directory of On-Going Research in Cancer Epidemiology* (in English), biennial; *IAR Cancer Disc CD-ROM*, annual; *Scientific Publications Series* (in English), periodic; *Technical Report Series* (in English and French), periodic; Monographs (in English), 3/year.

INTERNATIONAL AIDS SOCIETY (IAS)

PO Box 5619, S-114 86 Stockholm, Sweden

PHONE: 46 8 4596621 **FX:** 46 8 6626095 **E-MAIL:** ias@congrex.se **WEBSITES:** http://www.ias.se/ **FOUNDED:** 1988. **OFFICERS:** Lars Olof Kallings MD,PhD, Hec.Gen. **MEMBERS:** 5,700.

DESCRIPTION: Promotes the study of AIDS internationally. **PUBLICATIONS:** *IAS Membership Directory*; *IAS Newsletter* (in English), 3/year.

INTERNATIONAL AIR CARRIER ASSOCIATION (IACA) (Association Internationale de Charter Aerien)

Minervastraat 4, Keiberg 2, B-1930 Zaventem, Belgium **PHONE:** 32 2 7205303 **FX:** 32 2 7205137 **TX:** 20074 IACA B **FOUNDED:** 1967. **OFFICERS:** Marcel Pisters, Dir.Gen. **MEMBERS:** 36. **LANGUAGES:** English.

DESCRIPTION: Independent charter air carriers and specialized airlines in Europe, the United States, and the Middle East. Objectives are to: increase understanding and recognition of benefits derived from international air charter operations; broaden the base of air travel by encouraging government agencies to permit the free flow of air charter operations; develop communications and cooperation among international charter companies; improve quality of international air charter services.

THE INTERNATIONAL ALLIANCE, AN ASSOCIATION OF EXECUTIVE AND PROFESSIONAL WOMEN (TIA)

3510 Derry St., Harrisburg, PA 17111 **PHONE:** (410) 472-4221, (717) 901-3596 **FX:** (410) 472-2920 **E-MAIL:** MAXX28b@prodigy.com **WEBSITES:** http://www.t-i-a.com **FOUNDED:** 1980. **OFFICERS:** Rosemary McAvoy, Exec.Dir. **MEMBERS:** 10,000. **STAFF:** 2. **BUDGET:** US$125,000. **LANGUAGES:** English.

DESCRIPTION: Local networks (32) comprising 10,000 professional and executive women in 12 countries; individual businesswomen without a network affiliation (225) are alliance associates. Seeks to: promote recognition of the achievements of women in business; encourage placement of women in senior executive positions; maintain high standards of professional competence among members. Facilitates communication on an international scale among professional women's networks and their members. Represents members' interests before policymake business and government. Sponsors programs that support equal opportunity and enhance members' business and professional skills Operates appointments and directors service. Maintains speakers' bureau. **PUBLICATIONS:** *The Alliance*, bimonthly. Includes calendar of events and research updates; Membership Directory, annual.

INTERNATIONAL ALLIANCE OF FOOD PRODUCTS (IAFPA)

c/o Grocery Manufacturers of America, 1010 Wisconsin Ave. NW, 9th fl., Washington, DC 20007 **PHONE:** (202) 337-9400 **FX:** (202) 337-4508 **E-MAIL:** info@gmabrands.com **WEBSITES:** http://www.gmabrands.com **FOUNDED:** 1982. **OFFICERS:** Dr. Steven Liller. **MEMBERS:** 20. **LANGUAGES:** English.

DESCRIPTION: Grocery manufacturers' associations of Australia, Austria, Belgium, Brazil, Canada, Denmark, France, Germany, Italy, Japan, Mexico, New Zealand, South Africa, Switzerland, the United Kingdom, the United States, and Venezuela. Promotes high standards in the manufacture and distribution of grocery products. Coordinates activities of grocery manufacturers worldwide.

INTERNATIONAL ALLIANCE OF WOMEN (IAW) (Alliance Internationale des Femmes - AIF)

10 Queen St., Melbourne 3000, Australia **PHONE:** 61 3 96293653, 61 8 93285194 **FX:** 61 3 96292904 **FOUNDED:** 1902. **OFFICERS:** Priscilla Todd, Hon.Sec. **STAFF:** 2. **LANGUAGES:** English, French.

DESCRIPTION: Women's organizations and individuals in 70 countries. Objectives are: to secure reforms necessary to establish a real equality of liberties, status, and opportunities between men and women; to urge women to use their rights and influence in public life to ensure that the status of every individual shall be based on respect for human personality and not sex, race, or creed; to work for understanding between nations. Promotes exchange of views and experiences. **PUBLICATIONS:** *Action Programme* (in English and French), triennial; *Congress Reports*, triennial; *Family Planning for All*; *International Women's News Journal* (in English and French), quarterly.

INTERNATIONAL AMATEUR ATHLETIC FEDERATION (IAAF)

Stade Louis II, Avenue Prince Hereditaire Albert, MC-98000 Monaco Cedex, Monaco **PHONE:** 377 92 057068 **FX:** 377 92 057069 **WEBSITES:** http://www.iaaf.org **FOUNDED:** 1912. **OFFICERS:** Istvan Gyulai, Gen.Sec. **MEMBERS:** 209. **STAFF:** 48. **LANGUAGES:** English, French.

DESCRIPTION: National athletics (track and field) federations. World governing body of athletics. Facilitates the participation of any country or individual in international athletic competitions. Compiles rules and regulations for international men's and women's amateur competitions; establishes regulations for world and Olympic amateur athletic records. Organizes international athletic competitions and works to ensure that competitions are held in accordance with international regulations. Has established the Development Aid Programme to organize courses on athletics for coaches and officials worldwide. **PUBLICATIONS:** *Handbook* (in English), biennial. Features the Constitutional and Technical Rules of the IAAF; *IAAF Directory and Calendar*, annual; *IAAF Magazine* (in English and French), quarterly; *IAAF Newsletter* (in English), bimonthly; *New Studies in Athletics* (in English), quarterly.

INTERNATIONAL AMATEUR RADIO UNION (IARU)

PO Box 310905, Newington, CT 06131-0905 **PHONE:** (860) 594-0200 **FX:** (860) 594-0259 **E-MAIL:** iaru@iaru.org **WEBSITES:** http://www.iaru.org **FOUNDED:** 1925. **OFFICERS:** Larry E. Price, Sec. **MEMBERS:** 148. **LANGUAGES:** English.

DESCRIPTION: Federation of national amateur radio societies. Purpose is to represent the interests of radio amateurs, encourage effective agreements among national amateur radio societies, enhance and promote amateur radio, and further international goodwill. **PUBLICATIONS:** *Calendar*, semiannual.

INTERNATIONAL ANGEL ASSOCIATION (IAA)

85 Aza-onzuka, Goganzuka, Itami-shi, Hyogo 664, Japan **PHONE:** 81 727 847504 **FX:** 81 727 844608 **FOUNDED:** 1982. **OFFICERS:** Yuriko Kawamura. **MEMBERS:** 400. **STAFF:** 15. **BUDGET:** US$998,870. **LANGUAGES:** English, Japanese.

DESCRIPTION: Works to ''protect the dignity of life and spirit of love for humanity'' worldwide. Provides support and assistance to international development and human services programs adhering to these ideals. Maintains an orphanage and medical clinic in Bangladesh; constructs schools in Bangladesh, Cambodia, and the Philippines. Conducts literacy, agricultural training, and technical education courses.

INTERNATIONAL ASSOCIATION FOR ACCIDENT AND TRAFFIC MEDICINE (IAATM)

IAATM Office, Kizilirmak Cad 53/5, Kocatepe, TR-06640 Ankara, Turkey **PHONE:** 90 312 2850202, 90 312 2850303 **FX:** 90 312 2872390 **E-MAIL:** trafrak@dialup.ankara.edu.tr **FOUNDED:** 1960. **OFFICERS:** Erdal Cila M.D., Exec.Dir. **MEMBERS:** 370. **STAFF:** 3. **BUDGET:** US$15,000. **LANGUAGES:** English.

DESCRIPTION: Physicians and other professionals in 56 countries who are interested in motor vehicle and traffic related accidents; national associations for accident and traffic medicine. Fosters communication among governments and other organizations concerned with traffic problems. Conducts research; disseminates information. **PUBLICATIONS:** *Congress Proceedings*; *Journal of Traffic Medicine* (in English), quarterly.

INTERNATIONAL ASSOCIATION AGAINST NOISE (Association Internationale Contre le Bruit - AICB)

Hirschenplatz 7, CH-6004 Lucerne, Switzerland **PHONE:** 41 41 4103013 **FX:** 41 41 4109093 **FOUNDED:** 1959. **OFFICERS:** Dr. Willy Aecherli, Sec.Gen. **STAFF:** 1. **BUDGET:** 4,000 SFr. **LANGUAGES:** English, French, German, Swedish.

DESCRIPTION: National and regional organizations in 12 countries seeking noise control. Works to promote noise control at the international level. Encourages cooperation and the exchange of information among members. Prepares supranational measures and internationally recognized standards and regulations. Conducts research. **PUBLICA-**

TIONS: Directory (in English and German), periodic. Also publishes information bulletins.

INTERNATIONAL ASSOCIATION OF AGRICULTURAL ECONOMISTS (IAAE)

1211 W. 22nd St., Ste. 216, Oak Brook, IL 60523 **PHONE:** (630) 571-9393 **FX:** (630) 571-9580 **E-MAIL:** iaae@farmfoundation.org **WEBSITES:** http://www.ag.iastate.edu/journals/agecon **FOUNDED:** 1929. **OFFICERS:** Dr. Walter J. Armbruster, Sec.-Treas. **MEMBERS:** 1,600. **LANGUAGES:** English.

DESCRIPTION: Agricultural economists in 97 countries engaged in research, teaching, or administrative work. Fosters development of the science of agricultural economics and furthers application of the results of economic investigation into agricultural processes to improve economic and social conditions. **PUBLICATIONS:** *International Association of Agriculture Economists—Proceedings of Conferences*, triennial; Bulletin; Papers.

INTERNATIONAL ASSOCIATION OF AGRICULTURAL INFORMATION SPECIALISTS (IAALD)

c/o Ms. Margot Bellamy, CAB Intl., Wallingford, Oxon. OX10 8DE, England **PHONE:** 44 1491 832111 **FX:** 44 1491 833508 **E-MAIL:** m.bellamy@cabi.org **FOUNDED:** 1955. **OFFICERS:** Ms. Margot Bellamy, Sec.Treas. **MEMBERS:** 800. **LANGUAGES:** English, French, Spanish.

DESCRIPTION: Agriculturists, librarians, documentalists, and related associations and institutions in over 80 countries. Promotes library science and documentation and the professional interests of agricultural librarians and documentalists, including forestry, agricultural engineering, veterinary science, fisheries, food and nutrition, and agricultural and food industries. **PUBLICATIONS:** *Quarterly Bulletin of the IAALD* (in English); *World Congress Proceedings*, quinquennial; *World Directory of Agricultural Information Resource Centers.*

INTERNATIONAL ASSOCIATION OF AGRICULTURAL MEDICINE AND RURAL HEALTH (IAAMRH) (Association Internationale de Medecine Agricole et de Sante Rurale)

Saku Central Hospital, 197, Usuda, Minamisaku-gun, Nagano 384-03, Japan **PHONE:** 81 267 823131 **FX:** 81 267 829602 **FOUNDED:** 1961. **OFFICERS:** Shosui Matsushima M.D., Acting Sec.Gen. **MEMBERS:** 450. **LANGUAGES:** English, French, German, Russian.

DESCRIPTION: Physicians, nurses, paramedics, and other health care professionals in 40 countries. Studies problems of agricultural medicine and rural health around the world. Seeks means of averting the detrimental effects of certain agricultural and rural work conditions. **PUBLICATIONS:** *Agricultural Medicine and Rural Health* (in English), semiannual; *Journal*, quarterly.

INTERNATIONAL ASSOCIATION OF ALLERGOLOGY AND CLINICAL IMMUNOLOGY (IAACI)

611 E. Wells St., Milwaukee, WI 53202 **PHONE:** (414) 276-6445 **FX:** (414) 272-6070 **E-MAIL:** info@iaaci.org **FOUNDED:** 1951. **OFFICERS:** Rick Iber, Exec.Sec. **MEMBERS:** 25,000. **BUDGET:** US$1,500,000. **LANGUAGES:** English.

DESCRIPTION: Medical practitioners in allergology (the study of allergies and their treatment) and immunology, from 49 national allergology societies in 45 countries. Conducts research; increases awareness of completed and in-progress research programs; develops world standardization of categorization of allergies; establishes criteria for and monitors training of allergists; assesses world environmental conditions. **PUBLICATIONS:** *Allergy and clinical immunology International*, bimonthly. Includes research updates; *Progress in Allergy and Clinical Immunology*, triennial.

INTERNATIONAL ASSOCIATION OF APPLIED LINGUISTICS (AILA)

c/o David Singleton, Centre for Language and Communications Studies, Arts Bldg., Trinity College, Dublin 55455, Ireland **PHONE:** 353 1771 626 **FOUNDED:** 1964. **OFFICERS:** Jacques Girard. **MEMBERS:** 38. **LANGUAGES:** English.

DESCRIPTION: National associations of applied linguistics, representing 4500 individuals, united to coordinate and encourage research in the field of applied linguistics. Holds seminars, symposia, and workshops. **PUBLICATIONS:** *AILA News*, 3/year; *AILA Review*, annual. Series.

INTERNATIONAL ASSOCIATION OF ART CRITICS (AICA) (Association Internationale des Critiques d'Art - AICA)

11, rue Berryer, F-75008 Paris, France **PHONE:** 33 1 42561753 **FX:** 33 1 42560842 **FOUNDED:** 1948. **OFFICERS:** Kim Levin, Pres. **MEMBERS:** 3,700. **LANGUAGES:** English, French, Spanish.

DESCRIPTION: Membership consists of individuals in 72 countries. Fosters international cooperation through the creation and diffusion of the culture of art. Promotes critical discipline in the artistic domain. Seeks to protect and represent the moral and professional interests of members at international gatherings. Encourages collaboration, harmony, and the exchange of information among members and furthers the formation of national and international art and cultural organizations. **PUBLICATIONS:** *AICARC Bulletin*, semiannual; Directory, periodic.

INTERNATIONAL ASSOCIATION OF ASTHMOLOGY (INTERASMA) (Asociacion Internacional de Asmologia - INTERASMA)

c/o P. Godard, M.D., Hopital Arnaud de Villeneuve, Maladies Respiratoires, F-34295 Montpellier Cedex 5, France **PHONE:** 33 67336117 **FX:** 33 67521848 **E-MAIL:** phgodard.asmanet@paris.net **WEBSITES:** http://www.asmanet.com/interasma.html **FOUNDED:** 1954. **OFFICERS:** Hugo Neffen, Sec.Gen. **MEMBERS:** 1,200. **LANGUAGES:** English, French, German, Russian, Spanish.

DESCRIPTION: Doctors and other individuals interested in asthmology. Promotes a broadening of knowledge of bronchial asthma and related diseases, especially with regard to their etiology, pathological mechanisms, patho-physiology, therapy, epidemiology, rehabilitation, and prevention. Sponsors specialized training in allergology. **PUBLICATIONS:** *INTERASMA Membership Directory*, triennial; *Journal of Investigational Allergology and Clinical Immunology*, bimonthly; *Newsletter of Interasma* (in English and French), quarterly.

INTERNATIONAL ASSOCIATION AUTO THEFT INVESTIGATORS (IAATI)

PO Box 1176, Cross City, FL 32628 **PHONE:** (352) 498-3346 **FX:** (352) 493-0021 **E-MAIL:** gmbrown@iaati.org **WEBSITES:** http://www.iaati.org **FOUNDED:** 1951. **OFFICERS:** H. Lee Ballard, Exec. Dir. **MEMBERS:** 2,000. **LANGUAGES:** English.

DESCRIPTION: Law enforcement officers in 28 countries responsible for investigating vehicle thefts; vehicle manufacturers, insurance investigators, claims handlers, U.S. and Canadian auto theft bureau agents, members of motor vehicle and rental vehicle agencies. Objectives are to: encourage high professional standards among auto theft investigators; promote exchange of information and cooperation between law enforcement agencies; facilitate dissemination of technical information among members. **PUBLICATIONS:** *The APB*, quarterly; *Training and Education*, quarterly; *Vehicle Theft Investigators Manual*; Membership Directory, annual.

INTERNATIONAL ASSOCIATION OF BIBLIOPHILES (IAB) (Association Internationale de Bibliophilie - AIB)

Bibliotheque Nationale de France, Reserve des livres vares, Quai Francois-Mauriac, F-75706 Paris Cedex 13, France **PHONE:** 33 1 53795481 **FX:** 33 1 53795460 **FOUNDED:** 1963. **OFFICERS:** Jean-Marc Chatelain, Gen.Sec. **MEMBERS:** 500. **LANGUAGES:** English, French.

DESCRIPTION: Libraries; museums; book collectors, and clubs. Seeks to stimulate the development of bibliophily and establish a permanent link between bibliophiles of different countries. Organizes exhibitions and visits to public and private collections. **PUBLICATIONS:** *Bulletin du Bibliophile* (in English, French, German, and Italian), biennial.

INTERNATIONAL ASSOCIATION FOR BRIDGE AND STRUCTURAL ENGINEERING (IABSE) (Internationale Vereinigung fur Bruckenbau und Hochbau)

ETH-Honggerberg, CH-8093 Zurich, Switzerland **PHONE:** 41 1 6332647 **FX:** 41 1 6331241 **E-MAIL:** secretariat@iabse.ethz.ch **WEBSITES:** http://www.iabse.ethz.ch **FOUNDED:** 1929. **OFFICERS:** Alain Golay, Exec.Dir. **MEMBERS:** 4,200. **STAFF:** 6. **BUDGET:** US$700,000. **LANGUAGES:** English, French, German.

DESCRIPTION: Individuals in 90 countries; collective members are universities, institutes of technology, libraries, consulting engineers' firms, contractors, associations, syndicated firms, and public authorities. Deals with all problems in structural engineering concerning planning, design, construction, and maintenance. Promotes international collaboration between engineers and researchers, particularly representatives of science, industry, and public authorities. Initiates and supports research and testing. **PUBLICATIONS:** *Congress Report*; *Membership List*, biennial; *Structural Engineering Documents*; *Structural Engineering International*, quarterly, always February, May, August, and November; Reports.

INTERNATIONAL ASSOCIATION OF BROADCASTING (IAB) (Asociacion Internacional de Radiodifusion - AIR)

Brandzen 1961/402, 11200 Montevideo, Uruguay **PHONE:** 598 2 4088121, 598 2 4088129 **FOUNDED:** 1946. **OFFICERS:** Dr. Hector Oscar Amengual, Dir.Gen. **MEMBERS:** 168. **LANGUAGES:** English, French, Spanish.

DESCRIPTION: Private radio and television companies and associations of radio and/or television broadcasters operating in 40 European and North, Central, and South American countries. Principal aim is the defense of freedom of expression, especially as it pertains to the communications media. Promotes cooperation among broadcasters and public interest projects. **PUBLICATIONS:** *La Gaceta de AIR* (in English and Spanish), bimonthly. Also publishes *Basic Rules*, *Communication Media: Instrument of Freedom and Culture*, *Doctrine of Private Broadcasting of the Americas*, *The Role of Broadcasting in Education and Culture*, and monographs.

INTERNATIONAL ASSOCIATION OF BUDDHIST STUDIES (IABS)

c/o Inst. of Buddhist Studies, Dept. of Philosophy & Religion, Univ. of N. Carolina of Wilmington, Wilmington, NC 28403 **PHONE:** (910) 642-3547 **FOUNDED:** 1976. **OFFICERS:** Lou Lancaster, Treas. **MEMBERS:** 500. **STAFF:** 1. **LANGUAGES:** English.

DESCRIPTION: Scholars of philosophy, religion, history, sociology, anthropology, art, archaeology, and psychology, committed to the nonpolitical, intellectual treatment of Buddhism. Fosters the development of interdisciplinary Buddhist studies and stimulates interest in obscure and lesser-known aspects of Buddhism. Promotes the Buddhist principles of peace, understanding, friendliness, and tolerance. Encourages greater academic organization and coordinates members' activities. Plans to establish a library. **PUBLICATIONS:** *Journal of the International Association of Buddhist Studies*, semiannual.

INTERNATIONAL ASSOCIATION OF BUYING AND MARKETING GROUPS (IABG) (Internationale Vereinigung von Einkaufs- und Marketingverbaenden - IVE)

c/o Dr. Guenter Olesch, Vorgebirgsstrasse 43, D-53119 Bonn, Germany **PHONE:** 49 228 985840 **FX:** 49 228 985841 **E-MAIL:** igiorgini.zgv@t-fonline.de **FOUNDED:** 1951. **OFFICERS:** Prof.Dr. Guenter Olesch, Gen.Sec. **MEMBERS:** 600. **STAFF:** 5. **LANGUAGES:** English, French, German, Italian, Spanish.

DESCRIPTION: Conducts research on work methods and internal organization; compiles statistics. Organizes study sessions and meetings of experts. **PUBLICATIONS:** Newsletter.

INTERNATIONAL ASSOCIATION OF CHIEFS OF POLICE (IACP)

515 N. Washington St., Alexandria, VA 22314 **PHONE:** (703) 836-6767 **TF:** (800) THE-IACP **FX:** (703) 836-4543 **WEBSITES:** http://www.theiacp.org **FOUNDED:** 1893. **OFFICERS:** Dan Rosenblatt, Exec.Dir. **MEMBERS:** 16,000. **STAFF:** 50. **BUDGET:** US$12,000,000. **LANGUAGES:** English.

DESCRIPTION: Police executives who are commissioners, superintendents, chiefs, and directors of national, state, provincial, and municipal departments; assistant and deputy chiefs; division or district heads. Provides consultation and research services in all phases of police activity. Conducts educational programs. **PUBLICATIONS:** *Directory of IACP Members*, annual; *Model Policies*, periodic; *Police Buyers' Guide*, annual; *Police Chief: The Professional Voice of Law Enforcement*, monthly. Covers all aspects of law enforcement duties, from improved administrative techniques to operational practices, legislative issues, and technology.; *Training Keys*, periodic.

INTERNATIONAL ASSOCIATION FOR CHILD AND ADOLESCENT PSYCHIATRY AND ALLIED PROFESSIONS (IACAPAP)

PO Box 207900, New Haven, CT 06520-7900 **PHONE:** (203) 785-5759 **FX:** (203) 785-7402 **E-MAIL:** donald.cohen@yale.edu **FOUNDED:** 1948. **OFFICERS:** Donald J. Cohen M.D., Pres. **MEMBERS:** 50. **LANGUAGES:** English, French.

DESCRIPTION: National societies; others in the field of child and adolescent psychiatry and allied professions. Promotes collaboration among related professions including pediatrics, psychology, public health, social work, education, nursing, and others involved in research and practice in the field of child and adolescent psychiatry. **PUBLICATIONS:** *IACAPAP Newsletter*, quarterly; Monograph, semiannual.

INTERNATIONAL ASSOCIATION FOR COMPUTER SYSTEMS SECURITY (IACSS)

6 Swathmore Ln., Dix Hills, NY 11746 **PHONE:** (516) 499-1616 **FX:** (516) 462-9178 **FOUNDED:** 1981. **OFFICERS:** Robert J. Wilk, Founder & Pres. **MEMBERS:** 800. **STAFF:** 14. **BUDGET:** US$400,000. **LANGUAGES:** English.

DESCRIPTION: Organizations and individuals in 32 countries interested in the security of their computerized information systems. Offers a testing program to certify individuals as Computer Systems Security Professionals; upholds a code of professional ethics. Supports continuing education through workshops; furthers awareness of security issues both within the industry and the government. Conducts in-house management security awareness programs and monthly seminars and workshops; distributes information on state-of-the-art methods of protecting computer and communication resources. Maintains speakers' bureau, sponsors lectures, and compiles statistics. Presents Distinguished Service Award. **PUBLICATIONS:** *Computer Systems Security Guide*; *Computer Systems Security Newsletter*, quarterly; *International Directory of Computer Systems Security Products and Services*, periodic; *Proceedings Regional 1 IACSS Conference*, annual; *Proceedings Regional 2 IACSS Conference*, annual; Books; Brochures; Papers.

INTERNATIONAL ASSOCIATION OF CRAFTS AND SMALL AND MEDIUM SIZED ENTERPRISES (IACME) (Schweizerischer Gewerbeverband - IGU)

Centre patronal, Avenue Agassiz 2, CH-1001 Lausanne, Switzerland **PHONE:** 41 21 3197111 **FX:** 41 21 3197910 **E-MAIL:** cenrtrepatron@tx.ch **FOUNDED:** 1947. **OFFICERS:** Pierre Triponez, Dir. **MEMBERS:** 26. **STAFF:** 2. **LANGUAGES:** English, French, German.

DESCRIPTION: National organizations representing 6 million individuals. Seeks to maintain and develop individual work and services of high quality worldwide and to insure freedom for individual initiative in the economic system. Activities include the exchange of documentation, studies of economic and professional problems of members, and promotion of international trainee exchange programs.

INTERNATIONAL ASSOCIATION FOR CROSS CULTURAL COMMUNICATION

3840, rue Marcil, Montreal, PQ, Canada H4A 2Z4 **PHONE:** (514) 485-8114 **FX:** (514) 485-7968 **E-MAIL:** girardij@videotron.ca **FOUNDED:** 1965. **OFFICERS:** Jacques Girard. **MEMBERS:** 850. **LANGUAGES:** English, French.

DESCRIPTION: Linguists, semioticians, and audiovisual communication experts from 75 countries. Goals are to conduct research relating to theoretical and practical aspects of communication in its widest sense and to disseminate findings. Fields of operation are: audiovisual and structuro-global methods in relation to the communication sciences, education, sociology, and psychology; promotion of audiovisual techniques; interdisciplinary studies. Activities include: organization of workgroups consisting of specialists residing in various countries; creation and direction of research centers in the existing institutions; establishment and direction of research and recycling centers; promotion and stimulation of personal research; arranging of seminars and colloquia on clearly defined topics; publication of work by researchers, work-groups, and research centers. Conducts language and literacy programs. **PUBLICATIONS:** *Degres*, quarterly; *Etudes Linguistiques*, 3/year; *Langues et Culture*, 3-4/year.

INTERNATIONAL ASSOCIATION FOR DENTAL RESEARCH (IADR)

1619 Duke St., Alexandria, VA 22314 **PHONE:** (703) 548-0066 **FX:** (703) 548-1883 **E-MAIL:** research@iadr.com **WEBSITES:** http://www.iadr.com **FOUNDED:** 1920. **OFFICERS:** Eli Schwarz DDS, MPH, PHD, Exec.Dir. **MEMBERS:** 12,000. **STAFF:** 14. **BUDGET:** US$2,500,000. **LANGUAGES:** English.

DESCRIPTION: Individuals engaged or interested in advancing research in the various aspects of dental and related sciences. **PUBLICATIONS:** *Abstracts, A Special Issue of Journal of Dental Research*, quarterly; *Advances in Dental Research*, periodic. Covers developments in dental research and the chemistry, biology, and function of the oral cavity. Also includes conference proceedings; *Critical Reviews in Oral Biology and Medicine*, quarterly; *IADReports*, 5 times a year. Includes calendar of events; *Journal of Dental Research*, monthly. Disseminates new information and knowledge on all sciences relevant to dentistry, the oral cavity, and associated structures in health and disease; *Program and Abstracts*, annual; *Special Care in Dentistry*, bimonthly.

INTERNATIONAL ASSOCIATION OF DENTAL STUDENTS (IADS)

c/o FDI World Dental Federation, 7 Carlisle St., London W1V 5RG, England **PHONE:** 44 171 9357852 **FX:** 44 171 4860183 **E-MAIL:** liz.stockell@fdi.org.uk **WEBSITES:** http://www.unite.co.uk/customers/iads/ **FOUNDED:** 1951. **MEMBERS:** 52. **LANGUAGES:** English.

DESCRIPTION: Coordinating body between National Association of Dental students. Association members in 52 countries. Promotes international contact among dental students; facilitates exchange of students between member countries; develops international programs. Organises annual concerts. Conducts competitions. **PUBLICATIONS:** *ADS Newsletter*, semiannual; *IADS Exchange Guide* (in English).

INTERNATIONAL ASSOCIATION FOR THE DEVELOPMENT OF DOCUMENTATION, LIBRARIES AND ARCHIVES IN AFRICA

Boite Postale 375, Dakar, Senegal **PHONE:** 221 240954 **OFFICERS:** Emmanuel K.W. Dadzie, Sec. **LANGUAGES:** English, French.

DESCRIPTION: Libraries, archives, and organizations and individuals with an interest in information management. Promotes improved access to information throughout Africa. Facilitates establishment of libraries, archives, and other clearinghouses.

INTERNATIONAL ASSOCIATION FOR THE DEVELOPMENT OF RESEARCH AND INSTRUCTION IN FRENCH-SPEAKING COUNTRIES (Association Internationale pour le Developpement de la Recherche en Didactique du Francais Langue Maternelle)

Ecole Normale Superieure, Parc de Saint-Cloud, Grille d'Honneur, F-92211 Saint-Cloud Cedex, France **PHONE:** 33 1 47719111, 33 1 20370158 **FX:** 33 1 46023911 **OFFICERS:** Yves Reuter, Pres. **LANGUAGES:** English, French.

DESCRIPTION: Researchers, educators, and educational institutions. Seeks to advance research and education in French-speaking countries. Serves as a clearinghouse on research and education; sponsors training programs for educators; facilitates exchange among educational institutions.

INTERNATIONAL ASSOCIATION OF DREDGING COMPANIES (IADC)

Duinweg 21, NL-2585 JV The Hague, Netherlands **PHONE:** 31 70 3523334 **FX:** 31 70 3512654 **E-MAIL:** iadc@compuserve.com **WEBSITES:** http://www.gasandoil.com/iadc; http://www.iadc-dredging.com/ **FOUNDED:** 1965. **OFFICERS:** P.J.A. Hamburger, Sec.Gen. **MEMBERS:** 125. **STAFF:** 9.

DESCRIPTION: International dredging companies representing 34 countries. Purpose is to defend members' international interests. Activities include improvement of the standards of conduct, development of research possibilities, organization of courses and symposia, and cooperation with organizations and authorities to improve contract procedures. **PUBLICATIONS:** *Terra et Aqua: International Journal on Public Works, Ports and Waterways Development* (in English), quarterly. Also publishes articles in the business press and information on dredging developments. Special publications on environmental aspects.

INTERNATIONAL ASSOCIATION FOR EARTHQUAKE ENGINEERING (IAEE) (Kokusai Jishin Kogaku-kai)

Kenchiku Kaikan, 3rd Fl., 5-26-20, Shiba, Minato-ku, Tokyo 108, Japan **PHONE:** 81 3 34531281 **FX:** 81 3 34530428 **FOUNDED:** 1963. **OFFICERS:** Tsuneo Katayama, Sec.Gen. **MEMBERS:** 47. **BUDGET:** US$10,000. **LANGUAGES:** English.

DESCRIPTION: Promotes international cooperation among scientists and engineers in the field of earthquake engineering through interchange of information, ideas, and results of research and practical experience. Disseminates information and provides technical assistance. **PUBLICATIONS:** *Basic Concepts of Seismic Codes*; *Earthquake Resistant Regulations - A World List*, quadrennial; *International Directory of Earthquake Engineering Research*, quadrennial; *International Journal of Earthquake Engineering and Structural Dynamics*, quarterly; *Proceedings of World Conference*, quadrennial.

INTERNATIONAL ASSOCIATION FOR ECOLOGY

Drawer E, Aiken, SC 29802 **PHONE:** (803) 725-2472 **FX:** (803) 725-3309 **FOUNDED:** 1967. **OFFICERS:** Dr. Rebecca R. Sharitz, Sec.Gen. **MEMBERS:** 1,300. **STAFF:** 2. **LANGUAGES:** English.

DESCRIPTION: Libraries and institutions; national and international ecological associations; and students and other individuals. Promotes and communicates the science of ecology and the application of ecological principles to global needs. Coordinates application of ecological principles and encourages public awareness of the economic and social importance of ecology. Collects and evaluates information; acts as clearinghouse and center for coordination and dissemination of information and materials related to ecology and the global environment; reports on programs and legislative events affecting the environment. Represents ecologists in international forums. Conducts a global census of ecologists; determines the status of environmental research in developing countries. Sponsors lectures series; fosters research; compiles statistics. **PUBLICATIONS:** *Ecology International*, annual; *INTECOL Newsletter*, bimonthly; Books; Monographs.

INTERNATIONAL ASSOCIATION FOR EDUCATIONAL ASSESSMENT (IAEA)

PO Box 6665, Princeton, NJ 08541 **FOUNDED:** 1975. **OFFICERS:** Frances M. Ottobre, Exec.Sec. **LANGUAGES:** English.

DESCRIPTION: Educational agencies worldwide involved in the application of assessment techniques in order to improve educational processes. Maintains that ''assessment is essential if education throughout the world is to be improved and if its benefits are to be extended to increasing numbers of people.'' Addresses topics of international interest in educational assessment. Provides a network for collaboration on research projects. Has conducted a feasibility study to develop a scholastic ability test for Arabic, Chinese, English, and Portuguese. Provides referral service of organizations specializing in educational assessment. Sponsors projects. **PUBLICATIONS:** *Teacher's Guide to Assessment*.

INTERNATIONAL ASSOCIATION FOR EDUCATIONAL AND VOCATIONAL GUIDANCE (IAEVG) (Internationale Vereinigung fur Schul-und Berufsberatung)

c/o Lyn Barham, Acting Sec.Gen., The Mount, 41 Rowden Hill, Chippenham, Wilts. SN15 2AQ, England **PHONE:** 44 1249 460659 **FX:** 44 1249 460659 **E-MAIL:** lynbarham@aol.com **FOUNDED:** 1951. **OFFICERS:** Ms. Lyn Barham, Acting Sec.Gen. **MEMBERS:** 300. **STAFF:** 1. **BUDGET:** US$35,000. **LANGUAGES:** English, French, German, Spanish.

DESCRIPTION: Individuals, institutions, and national associations in 54 countries concerned with educational and vocational guidance. Collects and distributes information pertaining to educational and vocational guidance; promotes professional training and encourages research by granting scholarships for study and travel. Collaborates with international organizations, governmental and nongovernmental, and individuals involved in educational and vocational guidance and related matters; participates in activities relating to educational and vocational guidance in research as well as in practical application. Organizes seminars, colloquia, conferences, workshops, and study tours in conjunction with related organizations. Advises government and national and international organizations on the development of guidance systems. **PUBLICATIONS:** *Educational and Vocational Guidance Bulletin*, SAN; *Glossary of Educational and Vocational Guidance Terms in English, French, German, and Spanish*; *Newsletter*, 3/year; Bibliographies, periodic; Monographs.

INTERNATIONAL ASSOCIATION OF EDUCATORS FOR WORLD PEACE (IAEWP)

PO Box 3282, Mastin Lake Sta., Huntsville, AL 35810 **PHONE:** (256) 534-5501 **FX:** (256) 536-1018 **TX:** 9102405482 **E-MAIL:** mercieca@hiwaay.net **FOUNDED:** 1969. **OFFICERS:** Charles Mercieca Ph.D., Exec.VP. **MEMBERS:** 20,500. **STAFF:** 30. **LANGUAGES:** English, French, Russian, Spanish.

DESCRIPTION: Teachers, students, attorneys, medical doctors, social workers, clergy, businesspeople, and other individuals united to achieve world peace and better international relations through education. Promotes improved curriculum and methods of instruction in schools and seeks to implement the United Nations Universal Declaration of Human Rights. Cooperates with programs organized by the United Nations Educational, Scientific, and Cultural Organization. Maintains speakers' bureau and children's services. **PUBLICATIONS:** *Education for Peace: What it Entails*; *Il Bung*, monthly. Published in Korea; *International Association of Educators for World Peace—Circulation Newsletter*, bimonthly. Provides updates on the association's worldwide chapter events, (presently active in 94 countries), including professional meetings and seminars; *International Association of Educators for World Peace—Directory of International Officers*, biennial. Includes biographical data.; *Mismanagement in Higher Education*; *Peace Education*, semiannual. Includes book reviews.; *Peace Progress*, annual. Covers national and international educational and social issues.; Newsletters. Published in various countries.

INTERNATIONAL ASSOCIATION OF ELECTRICAL CONTRACTORS (AIE) (Association Internationale des Entreprises d'Equipement Electrique - AIE)

5, rue Hamelin, F-75116 Paris, France **PHONE:** 33 1 44058420 **FX:** 33 1 44058405 **FOUNDED:** 1954. **OFFICERS:** Mr. D. Hannotin, Gen.Sec. **MEMBERS:** 19. **STAFF:** 3. **BUDGET:** US$100,000. **LANGUAGES:** English, French, German.

DESCRIPTION: National electrical contractors associations. Provides liaison between members and promote discussion among them. Collects and disseminates technical and commercial information. **PUBLICATIONS:** *International Vocabulary* (in English, French, and German).

INTERNATIONAL ASSOCIATION OF EUROPEAN UNION COUNTRIES BORDERING THE MEDITERRANEAN (Association Internationale des Pays du Pourtour de la Mediterranee et de l'Union Europeene)

Universidad Complutense de Madrid, Hermanos San Roman 5, Pozuelo, E-28224 Madrid, Spain **OFFICERS:** Antonio Marquina. **LANGUAGES:** English, Spanish.

DESCRIPTION: Representatives of Mediterranean countries in the European Union. Promotes understanding and cooperation among members. Serves as a forum for the discussion of issues of mutual interest to members; represents members before the European Union; arbitrates disputes involving members.

INTERNATIONAL ASSOCIATION FOR THE EXCHANGE OF STUDENTS FOR TECHNICAL EXPERIENCE (IAESTE)

PO Box 6104, Swords, Dublin, Ireland **PHONE:** 353 1 8402786, 353 1 8402786 **FX:** 353 1 8402055 **E-MAIL:** jinereidgsiaeste@tinet.ie **FOUNDED:** 1948. **OFFICERS:** James E. Reid, Gen.Sec. **MEMBERS:** 65. **BUDGET:** 300,000 SFr. **LANGUAGES:** English.

DESCRIPTION: National committees (56) and cooperating institutions (9). Represents academic, industrial, and student interests in the organization of technical exchange programs supplementing university and college education. Promotes international understanding by providing students with technical experience abroad. Maintains placement service. **PUBLICATIONS:** *Activity Report*, quarterly; *Annual Report* (in English).

INTERNATIONAL ASSOCIATION FOR FINANCIAL PLANNING (IAFP)

5775 Glenridge Dr. NE, Ste. B-300, Atlanta, GA 30328-5364 **PHONE:** (404) 845-0011 **TF:** (800) 945-IAFP **FX:** (404) 845-3660 **E-MAIL:** info@iafp.org **WEBSITES:** http://www.iafp.org **FOUNDED:** 1969. **OFFICERS:** Janet McCallen, Exec.Dir. **MEMBERS:** 17,000. **STAFF:** 30. **BUDGET:** US$7,000,000. **LANGUAGES:** English.

DESCRIPTION: Individuals in 22 countries involved in the financial planning aspects of the financial services industry. Seeks to increase members' expertise and professional standing in financial planning by assisting them in obtaining ideas and educational opportunities that would otherwise not be available. Promotes standards of business ethics, personal integrity, and professional conduct in the financial planning field. Conducts specialized education programs; maintains speakers' data bank; compiles statistics. **PUBLICATIONS:** *Planning Matters*, monthly.

INTERNATIONAL ASSOCIATION OF FISH AND WILDLIFE AGENCIES (IAFWA)

444 N. Capitol St. NW, Ste. 544, Washington, DC 20001 **PHONE:** (202) 624-7890 **FX:** (202) 624-7891 **E-MAIL:** iafwa@sso.org **WEBSITES:** http://www.sso.org/iafwa **FOUNDED:** 1902. **OFFICERS:** R. Max Peterson, Exec.VP. **MEMBERS:** 450. **BUDGET:** US$300,000. **LANGUAGES:** English.

DESCRIPTION: State and provincial fish and wildlife agencies (68) and officials (382). Educates the public about the economic importance of conserving natural resources and managing wildlife property as a source of recreation and a food supply; supports better conservation legislation, administration, and enforcement. **PUBLICATIONS:** *International Association Convention Proceedings*, annual; Newsletter, bimonthly.

INTERNATIONAL ASSOCIATION OF GEODESY (IAG) (Association Internationale de Geodesie)

c/o Dept. of Geophysics, Juliane Maries Vej 30, DK-2100 Copenhagen 0, Denmark **PHONE:** 45 35 320600 **FX:** 45 35 365357 **E-MAIL:** iag@gfy.ku.dk **WEBSITES:** http://www.gfy.Ku.dk/-iag/ **FOUNDED:** 1864. **OFFICERS:** C.C. Tscherning, Gen.Sec. **MEMBERS:** 80. **BUDGET:** US$25,000. **LANGUAGES:** English, French.

DESCRIPTION: A member of the International Union of Geodesy and Geophysics. Objectives are: to promote geodetic research and encourage study in geodesy; to collect and disseminate scientific information; to provide a forum for geodesists worldwide to present recent discoveries, learn new methods and results, and build working relationships with other scientists. Organizes special study groups. **PUBLICATIONS:** *Geodesist's Handbook*, quadrennial; *Journal of Geodesy*, monthly; *Proceedings*, periodic. Also publishes special publications.

INTERNATIONAL ASSOCIATION OF GEOMAGNETISM AND AERONOMY (IAGA)

c/o J.A. Joselyn, NOAA R/E/SE, 325 Broadway, Boulder, CO 80303-3328 **PHONE:** (303) 497-5147, (303) 494-4258 **FX:** (303) 497-3645 **E-MAIL:** jjoselyn@sec.noaa.gov **FOUNDED:** 1919. **OFFICERS:** Jo Ann Joselyn, Sec.Gen. **MEMBERS:** 3,000. **BUDGET:** £50,000. **LANGUAGES:** English, French.

DESCRIPTION: Representatives in 72 countries united to promote research into and understanding of the subjects of geomagnetism and aeronomy. **PUBLICATIONS:** *IAGA News* (in English), periodic; *News*, annual.

INTERNATIONAL ASSOCIATION OF GROUP PSYCHOTHERAPY (IAGP)

c/o SANDAHL, Bondegatan 21, S-116 33 Stockholm, Sweden **PHONE:** 46 8 50608800 **FX:** 46 8 50608840 **E-MAIL:** christer.sandahl@sandahls.se **WEBSITES:** http://

www.psych.mcgill.ca/labs/iagp/IAGP.html **FOUNDED:** 1954. **OFFICERS:** Christer Sandahl PhD, Sec. **MEMBERS:** 800.

DESCRIPTION: Physicians, psychologists, social workers, nurses, and mental health workers in 32 countries interested in group psychotherapy. Seeks to further communication and education between professionals in the practice and study of group psychotherapy. **PUBLICATIONS:** *Forum* (in English), quarterly. Articles and member information; Directory, triennial.

INTERNATIONAL ASSOCIATION FOR THE HISTORY OF RELIGIONS (IAHR) (Association Internationale pour l'Histoire des Religions)

c/o Prof. Armin Geertz, Sec. Gen., Dept. of the Study of Religion, Univ. of Arhus, Main Bldg., DK-8000 Arhus, Denmark **PHONE:** 45 89422306 **FX:** 45 86130490 **E-MAIL:** geertz@teologi.aau.dk **FOUNDED:** 1950. **OFFICERS:** Prof. Armin W. Gertz, Gen.Sec. **MEMBERS:** 23. **LANGUAGES:** English, French, German, Italian.

DESCRIPTION: National organizations promoting the scholarly study of the history of religions. **PUBLICATIONS:** *NUMEN*, semiannual; *Science of Religion* (in English), quarterly; Monograph, periodic.

INTERNATIONAL ASSOCIATION OF HORTICULTURAL PRODUCERS (AIPH) (Internationaler Verband des Erwerbsgartenbaues)

Bezuidenhoutseweg 153, NL-2509 AB The Hague, Netherlands **PHONE:** 31 70 3041234 **FX:** 31 70 3470956 **TX:** 31406 PGES/NL **FOUNDED:** 1948. **OFFICERS:** Dr. J. Rotteveel, Gen.Sec. **MEMBERS:** 25. **STAFF:** 1. **LANGUAGES:** English, French, German.

DESCRIPTION: National associations of horticultural producers. Promotes the general interests of horticulturists in the international field. Studies economic problems; compiles international statistics on production, trade, consumption, and markets. Seeks uniform application of Plant Breeders' Rights. Publishes a calendar of national and international horticultural exhibitions. **PUBLICATIONS:** *Directives for International Horticultural Exhibitions*; *Yearbook of International Horticultural Statistics.*

INTERNATIONAL ASSOCIATION FOR HUMAN RIGHTS OF THE KURDS (IMK) (Internationaler Verein Fuer Menschenrechte der Kurden)

Postfach 200738, D-53137 Bonn, Germany **PHONE:** 49 228 362802, 49 228 3680712 **FX:** 49 228 363297 **E-MAIL:** imk-bonn@t-online.de **FOUNDED:** 1991. **OFFICERS:** Abubekir Saydam, Dir. **LANGUAGES:** English, German, Kurdish, Turkish.

DESCRIPTION: Promotes primacy of the rule of law and international standards of human rights in Kurdistan, a region populated by the Kurdish minorities of Turkey, Iraq, Iran, and Syria, as well as Kurds living in exile. **PUBLICATIONS:** *IMK Weekly Information Service* (in English and German). News bulletin on human rights situation in Kurdistan; Annual Report, annual. Covering various aspects of human rights in Kurdistan, e.g. freedom of expression.

INTERNATIONAL ASSOCIATION FOR HYDRAULIC RESEARCH (IAHR)

Postbus 177, NL-2600 MH Delft, Netherlands **PHONE:** 31 15 2858557 **FX:** 31 15 2858582 **E-MAIL:** jolien.mans@iahr.nl **FOUNDED:** 1935. **OFFICERS:** H.J. Overbeek, Sec.Gen. **MEMBERS:** 2,520. **STAFF:** 5. **LANGUAGES:** English, French.

DESCRIPTION: Members are engineering graduates, scientists, and others engaged in hydraulic research (2300); corporate members (220) are hydraulic laboratories, organizations, and institutions carrying out hydraulic research and engineering works. Maintains 3 divisions and 16 sections including Hydroinformatics, Fluid Phenomena in Energy Exchanges, Hydraulic Machinery and Cavitation, and Water Resources Management; ecohydraulics. **PUBLICATIONS:** *Hydraulic Structures Design Manual Series* (in English); *IAHR*, bimonthly; *Journal of Hydraulic Research* (in English), bimonthly; *Proceedings of Congress*, biennial; Monographs, periodic.

INTERNATIONAL ASSOCIATION FOR HYDROGEN ENERGY (IAHE)

PO Box 248266, Coral Gables, FL 33124 **PHONE:** (305) 284-4666 **FX:** (305) 284-4792 **E-MAIL:** verirogl@eng.miam.edu **FOUNDED:** 1974. **OFFICERS:** T. Nejat Veziroglu, Pres.

MEMBERS: 2,500. **STAFF:** 12. **BUDGET:** US$150,000. **LANGUAGES:** English.

DESCRIPTION: Scientists and engineers professionally involved with the use of hydrogen; students, educational institutions, professional societies, and corporate groups. Promotes discussion and publication of ideas furthering realization of a clean, inexhaustible energy system based on hydrogen. Organizes independent and joint research projects, short courses, and conferences on various aspects of hydrogen energy. **PUBLICATIONS:** *International Journal of Hydrogen Energy*, monthly.

INTERNATIONAL ASSOCIATION OF HYDROLOGICAL SCIENCES (IAHS) (Association Internationale des Sciences Hydrologiques - AISH)

c/o Gordon J. Young, Gen. Sec., IAHS SG, Dept. of Geography, Wilfrid Laurier Univ., Waterloo, ON, Canada N2L 3C5 **PHONE:** (519) 884-0710 **FX:** (519) 725-1342 **E-MAIL:** 44iahs@wlu.ca **WEBSITES:** http://www.wlu.ca/-wwwiahs/index.html **FOUNDED:** 1922. **OFFICERS:** Dr. Gordon J. Young, Sec.Gen. **MEMBERS:** 2,000. **STAFF:** 3. **BUDGET:** US$100,000. **LANGUAGES:** English, French.

DESCRIPTION: Individuals from 70 member states of the International Union of Geodesy and Geophysics. Promotes the study of hydrology as an aspect of the earth sciences and of water resources; studies the hydrological cycle in the earth and waters of the continents. Initiates, facilitates, and coordinates research into those hydrological problems that require international cooperation. Provides for discussion, comparison, and publication of research results. **PUBLICATIONS:** *Catalogue of IAHS Publications*, biennial; *Hydrological Sciences Journal*, bimonthly; *IAHS Handbook*, quadrennial; *IAHS Newsletter*, 3/year; *Proceedings of Symposia*; Reports, 6-12/year.

INTERNATIONAL ASSOCIATION OF ISLAMIC BANKS

Queen's Bldg., 23rd Fl., Al-Balad District, PO Box 9707, Jeddah 21423, Saudi Arabia **PHONE:** 966 2 6431276 **FX:** 966 2 6447239 **OFFICERS:** Samir A. Sheikh, Sec.Gen. **LANGUAGES:** Arabic, English.

DESCRIPTION: An organization of banks and other financial institutions. Promotes investment and economic development in Islamic countries. Provides banking and financial services to development organizations and programs. Represents members' interests.

INTERNATIONAL ASSOCIATION OF JEWISH LAWYERS AND JURISTS (IAJLJ)

10 Daniel Frish St., IL-64731 Tel Aviv, Israel **PHONE:** 972 3 6910673 **FX:** 972 3 6953855 **FOUNDED:** 1969. **OFFICERS:** Ophra Kidron, Exec.Dir. **MEMBERS:** 15,000. **STAFF:** 2. **LANGUAGES:** English, French, Hebrew, Spanish.

DESCRIPTION: Judges, lawyers, and law professors in 52 countries. Objectives are to: contribute toward the establishment of an international legal order based on the Rule of Law; promote human rights, the principles of equality, and the right of all states and peoples to live in peace; enhance the study of, and research into, the source of Jewish law with reference to the legal concepts of other nations. Seeks information on the legal and personal status of Jews in various countries. Fosters the study of legal problems of special interest to Jewish communities within the framework of international and domestic law; provides legal assistance in certain cases. **PUBLICATIONS:** *Justice* (in English), quarterly.

INTERNATIONAL ASSOCIATION OF JUDGES (IAJ)

Palazzo di Giustizia, Piazza Cavour, I-00193 Rome, Italy **PHONE:** 39 6 68837051 **FX:** 39 6 6883420 **E-MAIL:** iajuim@tin.it **WEBSITES:** http://space.tin.it/edicola/masbonom **FOUNDED:** 1953. **OFFICERS:** Massimo Bonomo, Sec.Gen. **MEMBERS:** 52. **STAFF:** 6. **BUDGET:** US$50,000. **LANGUAGES:** English, French, German, Italian, Spanish.

DESCRIPTION: National associations of judges in 40 countries. Promotes protection of human rights and freedom by safeguarding the independence of judiciaries throughout the world; promotes and encourages cordial relations between judges of different countries. Conducts research program on comparative law. Organizes seminars and exchange visits between judges. **PUBLICATIONS:** *Bulletin* (in English and French), annual.

INTERNATIONAL ASSOCIATION OF LAWYERS (Union Internationale des Avocats - UIA)

25 rue du Jour, F-75001 Paris, France **PHONE:** 33 1 45088234 **FX:** 33 1 45088231 **E-MAIL:** 10771.2060@compuserve.com **WEBSITES:** http://www.uianet.org **FOUNDED:** 1927. **OFFICERS:** Nathalie Alabert-Brouard, Admn.Mgr. **MEMBERS:** 3,300. **STAFF:** 6. **LANGUAGES:** English, French, German, Spanish.

DESCRIPTION: Membership consists of attorneys (3000) and legal firms and organizations (300). Promotes development of jurisprudence in all fields of law. Seeks to establish permanent relations and exchanges involving bar associations worldwide. Represents and defends the interests of attorneys; conducts research and studies to improve the organization of the legal profession. Provides consulting services to international agencies including the United Nations, Council of Europe, and the International Criminal Court. **PUBLICATIONS:** *Juriste International* (in English, French, and German), monthly; *Nouvelles News Noticias* (in English, French, and Spanish), monthly.

INTERNATIONAL ASSOCIATION OF LAWYERS AGAINST NUCLEAR ARMS

A. Paulownastraat 103, NL-2518 BC The Hague, Netherlands **PHONE:** 31 70 3634484 **FX:** 31 70 3455951 **E-MAIL:** ialana@antenna.nl **WEBSITES:** http://www.ddh.nl/org/ialana **FOUNDED:** 1989. **OFFICERS:** Peter Weiss, Co-president. **LANGUAGES:** Dutch, English.

DESCRIPTION: National lawyers' organizations. Coordinates national disarmament campaigns of members, with emphasis on nuclear disarmament and safe disposal of obsolete nuclear weapons. **PUBLICATIONS:** Newsletter, quarterly.

INTERNATIONAL ASSOCIATION OF LEGAL SCIENCE (IALS) (Association Internationale des Sciences Juridiques - AIJS)

c/o CISS-UNESCO, 1, rue Miollis, F-75015 Paris, France **PHONE:** 33 1 45682558, 33 1 45671410 **FX:** 33 1 43068798 **E-MAIL:** ipsa.secretary@dial.pipex.com **FOUNDED:** 1950. **OFFICERS:** Mr. M. Leker, Sec.-Gen. **MEMBERS:** 58. **STAFF:** 3. **BUDGET:** US$25,000. **LANGUAGES:** English, French.

DESCRIPTION: National committees from 58 countries devoted to comparative legal studies. Encourages worldwide development of legal science; works to increase mutual understanding among member nations through the comparative study of national legal systems. Promotes international exchanges by arranging meetings between jurists and providing access to legal sources, publications, and documents. **PUBLICATIONS:** *National Legal Bibliographies*, periodic; *News*, periodic, 2-3/year; *Proceedings of the Annual Colloquia* (in English and French), 2-3/year; Booklet, biennial.

INTERNATIONAL ASSOCIATION OF LIGHTHOUSE AUTHORITIES (IALA) (Association Internationale de Signalisation Maritime - AISM)

20 ter, rue Schnapper, F-78100 St. Germain-en-Laye, France **PHONE:** 33 1 34517001 **FX:** 33 1 34518205 **TX:** 695499 IALAISM F **E-MAIL:** aismiala@easynet.fr **WEBSITES:** http://www.iala-aism.org **FOUNDED:** 1957. **OFFICERS:** T. Kruuse, Sec.Gen. **MEMBERS:** 205. **STAFF:** 7. **BUDGET:** 5,500,000 Fr. **LANGUAGES:** English, French.

DESCRIPTION: Members are principal lighthouse authorities; organizations providing aids to navigation or scientific agencies; manufacturers of aids to navigation. Goal is to encourage the continued improvement of aids to navigation through any appropriate technical means for the safe and expeditious movement of vessels. Special tasks are: discussing matters relevant to the association's field of activity and technical questions of general interest; collecting and circulating information about the activities of lighthouse authorities as well as encouraging, supporting, and making known recent developments in the aids to navigation fields; establishing technical committees or working groups to study special problems and formulate appropriate reommendations; promoting assistance to services or organizations requesting help in connection with their aids to navigation problems; maintaining liaison with intergovernmental, international, and other organizations relevant to maritime safety, especially regarding hydrography, oceanography, communications, and meteorology; collecting statistics concerning aids to navigation. **PUBLICATIONS:** *Bulletin* (in English and French), quarterly;

Conference Reports, quadrennial; *List of IALA Members*, annual. Also publishes dictionaries, manuals, and recommendations.

INTERNATIONAL ASSOCIATION OF LITERARY CRITICS (Association Internationale des Critiques Litteraires - AICL)

38, rue du Fauborg-Saint-Jacques, F-75014 Paris, France **PHONE:** 33 1 40513300 **FX:** 33 1 43549299 **FOUNDED:** 1970. **OFFICERS:** Robert Andre, Pres. **MEMBERS:** 720. **STAFF:** 5. **BUDGET:** 70,000 Fr. **LANGUAGES:** English, French.

DESCRIPTION: Scholars, academics, literary critics, unions of critics, and writers in 37 countries. Examines problems related to criticism in literature. Strives to: raise cultural levels throughout the world; foster mutual understanding among cultures; improve relationships among critics. **PUBLICATIONS:** *Revue* (in English and French), biennial; *Revue de l'Association Internationale des Critiques Litteraires*, semiannual; Proceedings.

INTERNATIONAL ASSOCIATION OF LOGOPEDICS AND PHONIATRICS (IALP)

c/o Prof. Dolores Battle, 141 S Ln., Orchard Park, NY 14127 **PHONE:** (716) 878-6210 **FX:** (716) 878-6234 **E-MAIL:** battle@buffalostate.edu **WEBSITES:** http://www.ldc.lu.se/logopedi/ialp **FOUNDED:** 1924. **OFFICERS:** Prof. Dolores Battle, Exec. Officer. **MEMBERS:** 450. **LANGUAGES:** English, French, German.

DESCRIPTION: Phoniatricians (voice disorder specialists), logopedists (speech-language defect therapists), and audiologists (hearing specialists) from 59 affiliated societies in 38 countries. Fosters and conducts scientific study of disorders in human communication. **PUBLICATIONS:** *Aspects of Bilingual Aphasia*; *Folia Phoniatrica et Logopaedica* (in English, French, and German), bimonthly; *IALP Directory*, biennial.

INTERNATIONAL ASSOCIATION FOR MATHEMATICAL GEOLOGY (IAMG)

c/o Dr. Thomas A. Jones, 5211 Braeburn Dr., Bellaire, TX 77401-4814 **PHONE:** (713) 431-6546 **FX:** (713) 431-6336 **WEBSITES:** http://www.iamg.org **FOUNDED:** 1968. **OFFICERS:** Dr. Ricardo Olea, Pres. **MEMBERS:** 525. **LANGUAGES:** English.

DESCRIPTION: Membership consists of professional geologists, mathematicians, statisticians, and interested individuals. Promotes cooperation in the application and use of mathematics and statistics in geological research and technology. **PUBLICATIONS:** *Computers and Geosciences*, 10/year; *IAMG Newsletter*, semiannual; *Mathematical Geology*, 8/year; *Natural Research*, quarterly; *Studies in Mathematical Geology*, periodic.

INTERNATIONAL ASSOCIATION FOR MEDIA AND COMMUNICATION RESEARCH

Universidad Autonoma de Barcelona, Faculdad de Ciencas de la Comunicacion, 01893 Bellaterra, Barcelona, Spain **PHONE:** 34 3 5811545 **FX:** 34 3 5811545 **E-MAIL:** iamcr@selene.uab.es **FOUNDED:** 1957. **OFFICERS:** Prof. Hamid Mowlana, Pres. **MEMBERS:** 2,500. **LANGUAGES:** English, French, Spanish.

DESCRIPTION: Individuals, institutes, universities, research groups, and other organizations in the field of media and communication.

INTERNATIONAL ASSOCIATION OF MEDICINE AND BIOLOGY OF ENVIRONMENT (IAMBE) (Association Internationale de Medecine et de Biologie de l'Environnement - AIMBE)

115, rue de la Pompe, F-75116 Paris, France **PHONE:** 33 1 45534504, 33 1 53651082 **FX:** 33 1 45534175 **FOUNDED:** 1971. **OFFICERS:** Dr. Richard Abbou, Pres. **BUDGET:** US$2,000,000. **LANGUAGES:** English, French, Spanish.

DESCRIPTION: Individuals, associations, and firms in 72 countries concerned with ecological medicine and biology. Purpose is to study the adaptation of mankind to the environment and to study and treat sicknesses resulting from this adaptation. Examines natural cycles and balances; promotes research in ecological medicine and biology and corollary sciences; collects and disseminates information concerning the protection of mankind and the environment. Facilitates contacts with persons who deal professionally with problems related to the protection of humans and their surroundings. Organizes symposia and congresses; conducts courses and seminars.

INTERNATIONAL ASSOCIATION OF METEOROLOGY AND ATMOSPHERIC SCIENCES (IAMAS)

Dept. of Physics, Univ. of Toronto, Toronto, ON, Canada M5S 1A7 **PHONE:** (416) 978-2982, (416) 445-2947 **FX:** (416) 978-8905 **E-MAIL:** list@atmosp.physics.utoronto.ca **WEBSITES:** http://iamas.org **FOUNDED:** 1919. **OFFICERS:** R. List, Sec.Gen. **LANGUAGES:** English.

DESCRIPTION: Member countries of the International Union of Geodesy and Geophysics. Promotes research in all aspects of atmospheric sciences; initiates, facilitates, and coordinates research in the field; stimulates discussion and provides for the publication of research results. **PUBLICATIONS:** *Proceedings of Assemblies*, biennial.

INTERNATIONAL ASSOCIATION OF MUSEUMS OF ARMS AND MILITARY HISTORY (IAMAM) (Association Internationale des Musees D'Armes et D'Histoire Militaire)

c/o Dr. Claude Gaier, Musee d'Armes de Liege, Quai de Maestricht 8, B-4000 Liege, Belgium **PHONE:** 32 4 2219417, 32 4 2219416 **FX:** 32 4 2219401 **FOUNDED:** 1957. **OFFICERS:** Dr. Claude Gaier, Pres. **MEMBERS:** 270. **LANGUAGES:** English, French.

DESCRIPTION: Recognized public museums and similar public institutions in 70 countries which exclusively collect weapons, uniforms, and other military equipment or have important collections of this kind. Establishes contact between museums and similar institutions in the same field; promotes the study of the objects that make up these museums; furthers the aims of the International Council of Museums. **PUBLICATIONS:** *Mohonk Courier* (in English and French), semiannual, Featuring desktop; *Triennial Congress Report* (in English and French). Contains congress lectures.

INTERNATIONAL ASSOCIATION OF OFFICIAL HUMAN RIGHTS AGENCIES (IAOHRA)

444 N. Capitol St., Ste. 536, Washington, DC 20001 **PHONE:** (202) 624-5410 **FX:** (202) 624-8185 **FOUNDED:** 1949. **OFFICERS:** Melanie Campbel-Hill, Pres. **MEMBERS:** 187. **STAFF:** 1. **BUDGET:** US$100,000. **LANGUAGES:** English.

DESCRIPTION: Governmental human rights agencies with legal enforcement powers. Objectives are to foster better human relations and to enhance human rights procedures under the law. Conducts training services that include: administration and management training; technical assistance in civil rights compliance and curriculum development for colleges and universities, business, industry, and other organizations; training for administrators and commissioners to promote awareness and capability in current literature, theory, and philosophy relative to equal opportunity. Maintains ongoing liaison with federal agencies involved with civil rights enforcement in order to coordinate development of state legislation. Has developed and conducted training and technical assistance workshops for regional planning units and state planning agencies. Plans to establish a human rights training institute. Sponsors workshops; compiles statistics. **PUBLICATIONS:** *IAOHRA Newsletter*, quarterly; Membership Directory, annual. Also publishes technical notes.

INTERNATIONAL ASSOCIATION OF PAPYROLOGISTS (Association Internationale de Papyrologues - AIP)

c/o Fondation Egyptologique Reine Elisabeth, parc du Cinquantenaire, 10, B-1000 Brussels, Belgium **PHONE:** 32 2 7417364 **FOUNDED:** 1946. **OFFICERS:** Prof. A. Martin, Sec. **MEMBERS:** 400. **LANGUAGES:** Dutch, English, French, German, Italian.

DESCRIPTION: Papyrologists, sociologists, and historians of ancient law and antiquity. (Papyrology is the study of papyrus documents). Promotes international cooperation; supports and assists members in papyrological research. **PUBLICATIONS:** *Bibliographie Papyrologique*, quarterly. Reference books and specialized bibliographies.

INTERNATIONAL ASSOCIATION OF PENAL LAW (IAPL) (Association Internationale de Droit Penal)

c/o Dr. Helmut Epp, Austrian Parliament, A-1017 Vienna, Austria **PHONE:** 43 1 401104421

FX: 43 1 401104822 **E-MAIL:** h.epp@magnet.at **FOUNDED:** 1924. **OFFICERS:** Dr. Helmut Epp, Sec.Gen. **MEMBERS:** 3,500. **STAFF:** 2. **LANGUAGES:** English, French.

DESCRIPTION: Jurists, criminologists, judges, lawyers, administrators, and others with an interest in criminal justice. Serves as an international forum for discussion of multi- and interdisciplinary aspects of the criminal sciences, with emphasis on criminal policy and codification of national criminal law, comparative criminal justice, and international law. Works for the modernization of national legislation and development of the criminal sciences. Cooperates with agencies of the United Nations in addressing matters of international criminal justice and crime prevention. Operates Arab human rights program, which promotes extension and protection of the rights of individuals within the Arab World; plans to conduct African program. Maintains committees of criminal justice experts. Makes available postgraduate-level courses in human rights and criminal law. **PUBLICATIONS:** *Letter d'Information*, annual; *Nouvelles Etudes Penales*, periodic; *Revue Internationale de Droit Penal* (in English and French), semiannual.

INTERNATIONAL ASSOCIATION FOR THE PHYSICAL SCIENCES OF THE OCEAN (IAPSO)

c/o Dr. Fred E. Camfield, PO Box 820440, Vicksburg, MS 39182-0440 **PHONE:** (601) 636-1363 **FX:** (601) 629-9640 **E-MAIL:** camfield@vicksburg.com **WEBSITES:** http://www.olympus.net/IAPSO **FOUNDED:** 1919. **OFFICERS:** Dr. Fred E. Camfield, Sec.Gen. **BUDGET:** US$30,000. **LANGUAGES:** English.

DESCRIPTION: Member country of the International Union of Geodesy and Geophysics. Promotes the mathematical, physical, and chemical study of problems relating to the ocean and interactions taking place at its boundaries. Coordinates research and encourages discussion, comparison, and publication of results. **PUBLICATIONS:** *Proces-Verbaux*, quadrennial. Proceedings of the general assembly. Also publishes scientific papers.

INTERNATIONAL ASSOCIATION FOR PLANT PHYSIOLOGY (IAPP) (Association Internationale pour la Physiologie des Plantes)

CSIRO Div. of Food Science and Technology, PO Box 52, North Ryde, NSW 2113, Australia **PHONE:** 61 2 98878333 **FX:** 61 2 98873107 **TX:** AA 23407 **FOUNDED:** 1964. **OFFICERS:** Dr. Douglas Graham, Gen.Sec.-Treas. **MEMBERS:** 40. **BUDGET:** US$1,080. **LANGUAGES:** English.

DESCRIPTION: National and regional societies of plant physiologists representing 10,000 individual members in 40 countries. Promotes plant physiology internationally and encourages the formation of national societies, especially in developing countries. Sponsors exchange among plant physiologists on topics of regional significance in an effort to encourage cooperation between scientists in developed countries and their colleagues in developing countries. Conducts regional seminars or conferences. **PUBLICATIONS:** Newsletter, annual.

INTERNATIONAL ASSOCIATION FOR PLANT TAXONOMY (IAPT) (Association Internationale pour la Taxonomie Vegetale)

Botanischer Garten und Botanisches Museum Berlin-Dahlem, Konigin-Luise-Strasse 6-8, D-14191 Berlin, Germany **PHONE:** 49 30 8316010 **FX:** 49 30 83006218 **E-MAIL:** iapt@zedat.fu-berlin.de **WEBSITES:** http://www.bgbm.fu-berlin.de/iapt/default.htm **FOUNDED:** 1950. **OFFICERS:** Prof. Werner Greuter, Exec.Sec. **MEMBERS:** 2,800. **STAFF:** 1. **LANGUAGES:** English, French, Spanish.

DESCRIPTION: Coordinates work related to plant taxonomy and international codification of plant names. Bestows awards. **PUBLICATIONS:** *Regnum Vegetabile* (in English), periodic; *Taxon*, quarterly; Handbook (in English), periodic.

INTERNATIONAL ASSOCIATION OF PORTS AND HARBORS (IAPH) (Kokusai Kowan Kyokai)

Kono Bldg., 1-23-9 Nishi-Shimbashi, Minato-ku, Tokyo 105, Japan **PHONE:** 81 3 35914261 **FX:** 81 3 35800364 **TX:** 2222516 IAPH J **E-MAIL:** iaph@msn.com **WEBSITES:** http://www.iaph.or.jp **FOUNDED:** 1955. **OFFICERS:** Hiroshi Kusaka, Sec.Gen. **MEMBERS:** 331. **STAFF:** 6. **BUDGET:** 370,000,000¥. **LANGUAGES:** English, Japanese.

DESCRIPTION: Public, private, and governmental authorities, board commissioners, agencies, and organizations (in 90 countries) which have a direct and responsible connection with the operation, management, and development of ports and their related facilities. Aim is to increase the efficiency of ports and harbors by furthering knowledge in port organization, management, administration, operation, development, and promotion, thereby advancing international friendship and understanding and the growth of waterborne commerce. Monitors and conducts activities in areas including: planning, development, and operation of cargo handling facilities and systems; construction, maintenace, and safe marine operation of ports and harbors and port environment protection; procedures facilitating trade through ports and harbors; training, education, and technical assistance to developing ports; international law affecting port interests; public affairs as they relate to port development. Maintains consultative status with International Maritime Organization, United Nations Conference on Trade and Development, and United Nations Economic and Social Council. **PUBLICATIONS:** *Conference Proceedings* (in English), biennial; *Membership Directory* (in English), annual; *Ports and Harbors* (in English), 10/year; *Proceedings of Biennial Conferences.*

INTERNATIONAL ASSOCIATION FOR RELIGIOUS FREEDOM (IARF)

c/o General Secretariat, 2 Market St., Oxford, England OX1 3EF, England **PHONE:** 44 1865 202744 **FX:** 44 1865 202746 **E-MAIL:** iarf@interfaith-center.org **WEBSITES:** http://www.iarf-religiousfreedom.net **FOUNDED:** 1900. **OFFICERS:** Rev.Dr. Robert Traer, Gen.Sec. **MEMBERS:** 3,000. **STAFF:** 4. **LANGUAGES:** Bangla, Dutch, English, French, German, Japanese.

DESCRIPTION: Religious groups in 30 countries subscribing to the principles of openness. Collaborates in worldwide efforts to liberate religion from exclusionary tendencies. Conducts interreligious dialogues and intercultural encounters; sponsors social service network on development programs. Operates charitable program. Maintains computerized services. **PUBLICATIONS:** *Congress Proceedings* (in English), triennial; *IARF World* (in English), semiannual. Also publishes occasional papers.

INTERNATIONAL ASSOCIATION FOR RESEARCH IN INCOME AND WEALTH (IARIW)

c/o New York University, Dept. of Economics, Rm. 700, 269 Mercer St., New York, NY 10003 **PHONE:** (212) 924-4386 **FX:** (212) 366-5067 **E-MAIL:** iariw@fasecon.econ.nyu.edu **WEBSITES:** http://www.econ.nyu.edu/dept/iariw **FOUNDED:** 1947. **OFFICERS:** Jane Forman, Exec.Sec. **MEMBERS:** 400. **STAFF:** 2. **LANGUAGES:** English.

DESCRIPTION: Specialists in the field of national income accounting. Facilitates communication among members. Disseminates information pertaining to the definition and measurement of national income and wealth and social accounting and their use in economic budgeting, international comparisons, aggregations of national income and wealth, problems of statistical methodology, and related matters. **PUBLICATIONS:** *Bibliography on Income and Wealth.* Covers 1937 to 1960; *The Review of Income and Wealth*, quarterly; Membership Directory, periodic.

INTERNATIONAL ASSOCIATION FOR THE RHINE VESSELS REGISTER

Vasteland 12E, NL-3011 BL Rotterdam, Netherlands **PHONE:** 31 10 4116070 **FX:** 31 10 4129091 **WEBSITES:** http://www.ivr.nl **FOUNDED:** 1948. **OFFICERS:** Mrs. T. Hacksteiner. **MEMBERS:** 1,000. **LANGUAGES:** Dutch, English, French, German.

DESCRIPTION: Inland navigation companies, transportation experts, and government transportation and navigation authorities with an interest in shipping on the Rhine River. Promotes safe transportation of goods along the Rhine. Compiles statistics. **PUBLICATIONS:** *IVR Register*, annual.

INTERNATIONAL ASSOCIATION OF SCHOLARLY PUBLISHERS (IASP)

c/o Michael Huter, WUV Universitatsverlag, Berggasse 5, A-1090 Vienna, Austria **PHONE:** 45 86197033 **FX:** 45 86198433 **FOUNDED:** 1972. **OFFICERS:** Michael Huter, Pres. **MEMBERS:** 100. **LANGUAGES:** English.

DESCRIPTION: Individuals representing 51 countries organized for mutual advice and assistance on academic publishing and the development of new scholarly publishing facilities. **PUBLICATIONS:**

IASP Newsletter (in English), bimonthly; *Proceedings of Meeting*, periodic.

INTERNATIONAL ASSOCIATION OF SCHOOLS OF SOCIAL WORK (IASSW) (Association Internationale des Ecoles de Service Social)

c/o Prof. Lena Dominelli, Dept. of Social Work Studies, Southampton Univ., Southampton SO17 1BJ, England **PHONE:** 44 1703 593054 **FX:** 44 1703 51156 **E-MAIL:** ld@socsci.soton.ac.uk **FOUNDED:** 1928. **OFFICERS:** Lena Dominelli, Pres. **MEMBERS:** 600. **BUDGET:** US$125,000. **LANGUAGES:** English, French, Spanish.

DESCRIPTION: Schools of social work in 80 countries; national and regional associations of schools of social work. Provides international leadership and encourages high standards in social work education; represents the interests of social work education in connection with activities of other public and private international bodies. Sponsors study groups on social work education; offers consultative to educational institutions; undertakes projects of concern to social work educators; maintains an updated information system on social work education worldwide. **PUBLICATIONS:** *Directory of Members* (in English, French, and Spanish), biennial; *Human Rights Teaching and Learning About Human Rights*; *International Social Work*, bimonthly; *Newsletter*, 3/year; *Proceedings of International Congress*, biennial; *Social Work Field: Instruction in Post-Communist Societies.*

INTERNATIONAL ASSOCIATION OF SCIENTIFIC EXPERTS IN TOURISM (AIEST) (Asociacion Internacional de Expertos Cientificos del Turismo)

Varnbuelstrasse 19, CH-9000 St. Gallen, Switzerland **PHONE:** 41 71 2242530 **FX:** 41 71 2242536 **E-MAIL:** aiest@unisg.ch **FOUNDED:** 1949. **OFFICERS:** Prof. Dr. Thomas Bieges, Sec.Gen. **MEMBERS:** 380. **STAFF:** 1. **LANGUAGES:** English, French, German, Italian, Spanish.

DESCRIPTION: Individuals and organizations in 45 countries working in the tourism field. Fosters scientific work in tourism and supports efforts to create or to enlarge education and training possibilities at university level for managerial positions in tourism. Facilitates the exchange of information between members, provides documentation, and supports research and training centers specializing in tourism. **PUBLICATIONS:** *Booklet*, annual; *Tourist Review* (in English, French, German, and Italian), quarterly. Also publishes congress reports and lectures.

INTERNATIONAL ASSOCIATION OF SEDIMENTOLOGISTS (IAS) (Association Internationale de Sedimentologistes)

c/o Prof. Andre Strasser, Institut de Geologie, Perolles, CH-1700 Fribourg, Switzerland **PHONE:** 41 26 3008978 **FX:** 41 26 3009792 **E-MAIL:** andreas.strasser@unifr.ch **WEBSITES:** http://www.blacksci.co.uk/uk/society/ias **FOUNDED:** 1952. **OFFICERS:** Prof. Andre Strasser, Gen.Sec. **MEMBERS:** 2,200. **LANGUAGES:** English.

DESCRIPTION: Academic and industrial geologists in 98 countries. Promotes the study of sedimentology and the interchange of research, particularly where international cooperation is desirable. **PUBLICATIONS:** *Newsletter*, bimonthly; *Reprint Volumes*, periodic; *Sedimentology* (in English), bimonthly; *Special Publications*, periodic. Also publishes abstracts and guide books.

INTERNATIONAL ASSOCIATION OF THE SOAP, DETERGENT AND MAINTENANCE PRODUCTS INDUSTRY (Association Internationale de la Savonnerie, de la Detergence et des Produits d'Entretien - AISE)

Sq. Marie-Louise 49, B-1000 Brussels, Belgium **PHONE:** 32 2 2308371 **FX:** 32 2 2308288 **TX:** 23167 FECHIM B **E-MAIL:** a.i.s.e@infoboard.be **FOUNDED:** 1952. **OFFICERS:** Pierre V. Costa, Sec.Gen. **MEMBERS:** 19. **STAFF:** 8. **BUDGET:** 40,000,000 BFr. **LANGUAGES:** English.

DESCRIPTION: National associations in 23 countries representing the soap, detergent, cleaning, and maintenance products industry in Europe. Represents the industry in contacts with international official and professional bodies. Cooperates with international authorities in the preparation of legislation affecting the activities of the industry. Studies issues relevant to the industry, and seeks collective solutions to concerns about human and environmental health. **PUBLICATIONS:** *Information Bulletin*, quarterly.

INTERNATIONAL ASSOCIATION OF SOUND AND AUDIOVISUAL ARCHIVES (IASA)

c/o Albrecht Haefner, Suedwestfunk, Postfach 820, D-76522 Baden-Baden, Germany **PHONE:** 49 7221 9293487 **FX:** 49 7221 9292094 **E-MAIL:** haefner@swf.de **WEBSITES:** http://www.llgc.org.uk/iasa/index.htm **FOUNDED:** 1969. **OFFICERS:** Albrecht Haefner, Sec.Gen. **MEMBERS:** 380. **LANGUAGES:** English, French, German.

DESCRIPTION: Organizations and individuals with an interest in sound and audiovisual archives; institutions which preserve sound recordings. Seeks to facilitate international cooperation among members representing archives of music, history, literature, drama, and folklore recordings; collections of natural history, bio-acoustic, and medical sounds; recorded linguistic and dialect surveys; and radio and television sound archives. **PUBLICATIONS:** *Directory of Members*, biennial; *IASA Information Bulletin* (in English), quarterly; *IASA Journal* (in English), semiannual; Manuals, periodic.

INTERNATIONAL ASSOCIATION OF SOUTH-EAST EUROPEAN STUDIES (IASEES) (Association Internationale d'Etudes du Sud-Est Europeen - AIESEE)

Soseaua Kiseleff 47, Sect. 1, R-71268 Bucharest 2, Romania **PHONE:** 40 1 2225409, 40 1 2233063 **FX:** 40 1 2233063 **FOUNDED:** 1963. **OFFICERS:** Razvan Theodorescu, Sec.Gen. **MEMBERS:** 25. **STAFF:** 9. **LANGUAGES:** English, French.

DESCRIPTION: National and international research committees from 24 European and non-European countries; the International Committee of Thracology. Purpose is to foster a climate of mutual respect among members and to facilitate research in the human sciences. Promotes studies of Balkan and South-East European civilizations. Organizes and collaborates on foreign study exchange programs, roundtables, congresses, symposia, and practical courses designed to aid in the development of young specialists in the field of South-East European studies. **PUBLICATIONS:** *Bulletin de l'AIESEE* (in English, French, and German), quarterly. Contains scientific studies. Also publishes brochures, documents, studies, and minutes of international symposia including: *Etudes et Documents Concernant le Sud-Esteuropeen*, *Bibliotheque d'Etudes du Sud-Est Europeen*, and *Actes des Colloques*.

INTERNATIONAL ASSOCIATION OF STUDENTS IN ECONOMICS AND MANAGEMENT (IASEM)

AIESEC Netherlands, Erasmus Univ., Burgemeester Oudiaan 50, NL-3062 PA Rotterdam, Netherlands **PHONE:** 31 10 4081793 **FX:** 31 10 4523514 **E-MAIL:** nl@aiesec.nl **WEBSITES:** http://www.aiesec.nl **FOUNDED:** 1948. **OFFICERS:** Thomas van Bock. **MEMBERS:** 200. **LANGUAGES:** Dutch, English, French.

DESCRIPTION: Members are students of economics and management. Promotes excellence in economics and business scholarship. Facilitates communication and cooperation among members, with emphasis on international exchange. **PUBLICATIONS:** *A4* (in Dutch), monthly.

INTERNATIONAL ASSOCIATION FOR THE STUDY OF THE LIVER (IASL) (Association Internationale pour l'Etude du Foie)

Q.I.M.R., Liver Unit, The Bancroft Centre, 300 Herston Rd., Brisbane, QLD 4029, Australia **PHONE:** 61 7 33712056, 61 7 33620170 **FX:** 61 7 33620191 **E-MAIL:** jhallid@tpgi.com.au **WEBSITES:** http://enterprise.powerup.com.au/-iasl/ **FOUNDED:** 1958. **OFFICERS:** Prof. June W. Halliday, Sec.-Treas. **MEMBERS:** 879. **LANGUAGES:** English.

DESCRIPTION: Members of regional societies for the study of the liver representing 67 countries. Supports the training of experts in hepatology; encourages basic and clinical research on the liver and its diseases; acts to facilitate prevention, recognition, and treatment of diseases of the liver. **PUBLICATIONS:** Newsletter, biennial.

INTERNATIONAL ASSOCIATION FOR THE STUDY OF ORGANIZED CRIME (IASOC)

c/o University of Illinois at Chicago, Sam Houston St. Univ., Criminal Justice Ctr., Huntsville, TX 77341-2296 **PHONE:** (409) 294-1632 **FX:** (409) 294-1653 **WEBSITES:** http://www.acsp.uic.edu/iasoc/crim_org/vol12/

index.cfm **FOUNDED:** 1984. **OFFICERS:** Richard Ward, Dir., Operations. **MEMBERS:** 500. **STAFF:** 2. **BUDGET:** US$10,000. **LANGUAGES:** English.

DESCRIPTION: Researchers, investigators, educators, and students representing 22 countries interested in the study of organized crime. Provides a network for those involved in the study of organized crime. Operates a dispository for the collection and dissemination of information on organized crime and related subjects to provide research assistance to criminal justice agencies. Maintains speakers' bureau; operates expertise pool. **PUBLICATIONS:** *Criminal Organizations*, semiannual. Includes book and movie reviews; Membership Directory, biennial.

INTERNATIONAL ASSOCIATION OF TECHNOLOGICAL UNIVERSITY LIBRARIES (IATUL)

c/o Mr. M. Breaks, Heriot-Watt Univ. Library, Edinburgh EH14 4AS, Scotland **PHONE:** 44 131 4513570 **FX:** 44 131 4513164 **E-MAIL:** john@lib.chalmers.se **FOUNDED:** 1955. **OFFICERS:** Mr. M. Breaks, Sec. **MEMBERS:** 198. **LANGUAGES:** English.

DESCRIPTION: Libraries of academic institutions in 41 countries that offer courses in engineering or technology at the doctoral level; nonvoting associate members are libraries of academic institutions that offer such courses to the Master's or equivalent level; observer members are libraries housed in national patent offices and science museums with technological collections that meet research standards. Works to facilitate international cooperation among member libraries to stimulate and develop library projects of international and regional importance. **PUBLICATIONS:** *IATUL Conference Proceedings* (in English), annual; *IATUL Newsletter*, quarterly.

INTERNATIONAL ASSOCIATION OF THEORETICAL AND APPLIED LIMNOLOGY (IATAL)

c/o Dr. Robert G. Wetzel, Univ. of Alabama, Dept. of Biology, Tuscaloosa, AL 35487-0206 **PHONE:** (205) 348-1793, (205) 348-1787 **FX:** (205) 348-1403 **E-MAIL:** rwetzel@biology.as.us.edu **WEBSITES:** http://www.limnology.org **FOUNDED:** 1922. **OFFICERS:** Dr. Robert G. Wetzel, Gen.Sec.-Treas. **MEMBERS:** 3,250. **STAFF:** 1. **BUDGET:** US$450,000. **LANGUAGES:** English.

DESCRIPTION: Promotes the study of all aspects of limnology (freshwater research). Encourages scientific discourse among those pursuing academic research and those concerned with practical fishery, pollution, and water-supply problems. Organizes lectures, excursions, and symposia. **PUBLICATIONS:** *Mitteilungen*. Series consisting of accounts of methods, reviews, and the proceedings of symposia; *Verhandlungen*, triennial. Provides a summary of current research in the field of fresh water biology. Includes membership directory.

INTERNATIONAL ASSOCIATION OF TRADING ORGANIZATIONS FOR A DEVELOPING WORLD (ASTRO) (Association Internationale des Organismes de Commerce pour un Mond en Developpement)

Dunajska Cesta 104, PO Box 2592, SLO-1001 Ljubljana, Slovenia **PHONE:** 386 61 344771 **FX:** 386 61 1681451 **TX:** 39425 ASTRO SI **FOUNDED:** 1984. **OFFICERS:** Prof. K.L.K. Rao, Exec.Dir. **MEMBERS:** 55. **STAFF:** 7. **BUDGET:** US$500,000. **LANGUAGES:** English, French, Spanish.

DESCRIPTION: International non-governmental association comprising trading organizations (TOs) of 34 developing countries. Aims to strengthen the organizational and professional skills of TOs and their managers. Provides for exchange of trade information, consulting services, and trade contact among TOs. Promotes economic and technical cooperation among developing countries and countries in transition. Sponsors on-the-job training programs, seminars, and workshops on trade expansion stategies, TO management, commodities and futures, and other matters involving international trade. Participates in international and regional symposia, conferences, and round tables. **PUBLICATIONS:** *ASTRO Update* (in English), quarterly; *Countertrade Survey*, bimonthly; *Handbook of STOs in the Developing Countries*. Also published *Buy-Back Arrangements and Trade Expansion*, *Comprehensive Reference Service on Countertrade*, *Manual of Special Transactions*, and reports.

INTERNATIONAL ASSOCIATION OF UNIVERSITIES (IAU) (Association Internationale des Universites - AIU)

1, rue Miollis, F-75732 Paris Cedex 15, France **PHONE:** 33 01 45682545, 33 01 45682626 **FX:** 33 01 47347605 **E-MAIL:** centre.iau@unesco.org **WEBSITES:** http://www.unesco.org/iau **FOUNDED:** 1950. **OFFICERS:** C. Langlois, Dir. **MEMBERS:** 610. **STAFF:** 14. **LANGUAGES:** English, French.

DESCRIPTION: Members are universities and other institutions of higher education in over 120 countries; associate members are regional and national associations of universities. Promotes cooperation at the international level among universities and institutions of similar rank as well as among other bodies concerned with higher education and research. Acts as clearinghouse for information about university institutions and the organization of higher education worldwide. Conducts research into higher education policy, investigating areas such as: higher education administration and financial planning, academic freedom and university autonomy, and information technology in higher education. Provides consulting services in matters concerning the recognition and equivalence of degrees and diplomas and other university problems. **PUBLICATIONS:** *Higher Education Policy*, quarterly; *IAU Newsletter/nouvelles de l'Aiu*, bimonthly; *International Handbook of Universities*, biennial. In conjunction with Elsevier Pergamon; *Issues in Higher Education*, periodic. In conjunction with Elsevier Pergamon.; *Quarterly Journal of the International Association of Universities* (in English); *Seminar Report*, periodic; *World Academic Database CD-ROM*, annual.

INTERNATIONAL ASSOCIATION FOR VEGETATION SCIENCE (IAVS) (Internationale Vereinigung fur Vegetationskunde - IVV)

JBN-DLO, PO Box 23, NL-6700 Wageningen, Germany **PHONE:** 31 317 477914 **FX:** 31 317 424988 **E-MAIL:** j.h.j.schaminee@ibn.dlo.nl **FOUNDED:** 1937. **OFFICERS:** Dr. J.H.J. Schaminee, Sec.Gen. **MEMBERS:** 1,400. **LANGUAGES:** English, French, German, Spanish.

DESCRIPTION: Vegetation scientists in 67 countries including botanists and ecologists. Fosters contacts among vegetation scientists worldwide; promotes development of vegetation science. Works to define and clarify problems regarding taxonomy, nomenclature, and ecology of plant communities. Cooperates with other societies for the protection of nature. **PUBLICATIONS:** *Journal of Vegetation Science* (in English), bimonthly; *Report of the International Symposium*, annual.

INTERNATIONAL ASSOCIATION OF VOLCANOLOGY AND CHEMISTRY OF THE EARTH'S INTERIOR (IAVCEI) (Association Internationale de Volcanologie et de Chimie de l'Interieur de la Terre)

Australian Geological Survey Org., GPO Box 378, Canberra, ACT 2601, Australia **E-MAIL:** wjohnson@agso.gov.au **WEBSITES:** http://geont1.lanl.gov/HEIKEN/one/iavcei_home_page.htm **FOUNDED:** 1919. **OFFICERS:** Dr. R.W. Johnson, Sec.Gen. **MEMBERS:** 720. **LANGUAGES:** English.

DESCRIPTION: Representatives of 47 countries participating in the International Union of Geodesy and Geophysics. Fosters geological, geophysical, and geochemical studies of volcanoes and their eruptions. Initiates and coordinates research and promotes cooperation in these studies. Bestows awards; arranges conferences and symposia. **PUBLICATIONS:** *Bulletin of Volcanology*, 8/year;. Also publishes *Catalogue of Active Volcanoes of the World* and research reports.

INTERNATIONAL ASSOCIATION ON WATER QUALITY (IAWQ)

Duchess House, 20 Masons Yard, Duke St., St. James, London SW1Y 6BU, England **PHONE:** 44 171 8398390 **FX:** 44 171 8398299 **E-MAIL:** iawq@compuserve.com **WEBSITES:** http://www.iawq.org.uk **FOUNDED:** 1965. **OFFICERS:** Anthony Milburn, Exec.Dir. **MEMBERS:** 6,850. **STAFF:** 14. **BUDGET:** US$1,600,000. **LANGUAGES:** English.

DESCRIPTION: An organization of individuals, national organizations, municipal authorities, pollution control agencies, government departments, research institutes, and commercial concerns in 120 countries. Objectives are to: contribute to the advancement of research, development, and applications in drinking water treatment and supply, wastewater treatment, water pollution control and water quality management; encourage communication and a better understanding among those engaged in the solution of water quality problems and water quality management. Promotes the exchange of information on drinking water treatment and supply, wastewater treatment, water pollution control, water quality management, and its practical application. **PUBLICATIONS:** *IAWQ Yearbook* (in English), annual; *Water Quality International*, bimonthly; *Water Research*, monthly; *Water Science and Technology*, bimonthly; Reports, periodic. Contain scientific and technical information.

INTERNATIONAL ASTRONAUTICAL FEDERATION (IAF) (Federation Internationale d'Astronautique)

3-5, rue Mario Nikis, F-75015 Paris, France **PHONE:** 33 1 45674260 **FX:** 33 1 42732120 **E-MAIL:** iafam@iplus.fr **WEBSITES:** http://www.iafastro.com **FOUNDED:** 1950. **OFFICERS:** Mr. Claude Gourdet, Exec.Sec. **MEMBERS:** 129. **LANGUAGES:** English, French, German, Russian, Spanish.

DESCRIPTION: National and international space agencies, government laboratories and research centers, main space industrial companies and their trade associations as well as many learned societies. Plays an important role in disseminating information, in pulling together world experts in space development and utilization, and in providing a significant world-wide network aimed at ensuring that space activity benefits humanity. **PUBLICATIONS:** *Proceedings of Congress*, annual. Also publishes working group reports.

INTERNATIONAL ATLANTIC ECONOMIC SOCIETY (IAES)

4949 West Pine Blvd., Second Floor, St. Louis, MO 63108-1431 **PHONE:** (314) 454-0100 **FX:** (314) 454-9109 **E-MAIL:** iaes@iaes.org **WEBSITES:** http://www.iaes.org **FOUNDED:** 1973. **OFFICERS:** Dr. John M. Virgo, Exec.VP. **MEMBERS:** 1,500. **STAFF:** 6. **BUDGET:** US$400,000. **LANGUAGES:** English.

DESCRIPTION: Economists and business specialists from academic, government, and private organizations. Furthers economics and related subjects on both the theoretical and applied levels and increases the exchange of new ideas and research worldwide. **PUBLICATIONS:** *Atlantic Economic Journal*, quarterly. Includes book reviews and anthologies; *International Advances in Economic Research*, quarterly. Includes the top 15% of papers presented at the North American and European conferences of the IAES and conference abstracts.

INTERNATIONAL ATOMIC ENERGY AGENCY (IAEA) (Agence Internationale de l'Energie Atomique - Inatom Vienna)

Vienna Intl. Centre, Wagramerstrasse 5, Postfach 100, A-1400 Vienna, Austria **PHONE:** 43 1 26000 **FX:** 43 1 26007 **TX:** 112645 ATOMA **E-MAIL:** offical.mail@iaea.org **WEBSITES:** http://www.iaea.or.at/ **FOUNDED:** 1957. **OFFICERS:** Dr. Mohamed El Baradei, Dir.Gen. **MEMBERS:** 128. **STAFF:** 2,216. **BUDGET:** US$224,300,000. **LANGUAGES:** Arabic, Chinese, English, French, Russian, Spanish.

DESCRIPTION: Objectives are to accelerate and to enlarge the contribution of atomic energy to peace, health, and prosperity throughout the world. Ensures that any assistance given by the agency is not used for military purposes. Works as verification organization for the Treaty on the Non-Proliferation of Nuclear Weapons as well as treaties on nuclear weapon free zones in Africa, Latin America and the Pacific. Encourages research and development of the practical applications of atomic energy for peaceful uses throughout the world and fosters the exchange of scientific and technical information among nations. Establishes health and safety standards; advises member states in the safe operation of nuclear installations; applies safeguards in accordance with agreements established between member states on non-proliferation objectives. Operates laboratories in Seibersdorf, Austria and Monaco and a physics research institute in Trieste, Italy. **PUBLICATIONS:** *CINDA* (in English), annual. Index to literature and computer files on microscopic neutron data.; *INIS Atomindex* (in English, French, Russian, and Spanish), semimonthly. Contains bibliographic descriptions and abstracts.; *International Atomic Energy Agency Bulletin*, quarterly; *Meetings on Atomic Energy*, quarterly; *Newsbriefs*, bimonthly; *Nuclear Fusion*, monthly; Directories; Proceedings; Reports.

INTERNATIONAL AUTOMOBILE FEDERATION (FIA) (Federation Internationale de l'Automobile)

2, Chemin de Blandonnet, Case Postale 196, CH-1215 Geneva 15, Switzerland **PHONE:** 41 22

5444400 **FX:** 44 22 5444450 **E-MAIL:** merindol@fia.imaginet.fr **WEBSITES:** http://www.fia.com **FOUNDED:** 1904. **OFFICERS:** Max Mosley, Pres. **MEMBERS:** 149. **LANGUAGES:** English, French, Spanish.

DESCRIPTION: Federation of automobile clubs or associations in 116 countries. Aims to: develop and organize motor touring in all countries; assist motorists with day-to-day traffic problems; organize, promote, and regulate world motor sports; study traffic, touring, and technical problems related to the motor vehicle; protect the interests of motor vehicle users. Acts as intermediary between motorists and international institutions. Participates in international governmental meetings and in the drafting of important conventions and agreements; maintains consultative status with the United Nations and the Council of Europe. Endeavors to provide motorists at home and abroad with information and assistance. Works to protect the environment and national heritage and to open up new touring areas. Deals with matters of highway legislation, road and vehicle safety, motorists in society, clean air standards, fuel economy, technical progress in car design, and simplification of customs formalities. Organizes subcommissions and working groups to study problems of current interest. Maintains International Court of Appeals to give definitive rulings on disputes between members and appeals against decisions of national sporting authorities when they affect drivers, organizers, or manufacturers from different countries. **PUBLICATIONS:** *Bulletin Officiel du Sport Automobile Federation Internationale de l'Automobile* (in English and French), monthly; *FIA Yearbook* (in English, French, and Spanish), annual; *Yearbook of Automobile Sport*, annual.

INTERNATIONAL BACCALAUREATE ORGANISATION (IBO) (Organisation du Baccalaureat International - OBI)

15, rte. des Morillons, CH-1218 Grand-Saconnex, Switzerland **PHONE:** 41 22 797140 **FX:** 41 22 7910277 **E-MAIL:** ibhq@ibo.org **WEBSITES:** http://www.ibo.org **FOUNDED:** 1965. **OFFICERS:** Prof. Dvek Blackma, Dir.Gen. **MEMBERS:** 770. **STAFF:** 120. **BUDGET:** US$15,000,000. **LANGUAGES:** English, French, Spanish.

DESCRIPTION: Schools in 95 countries that prepare and enter students for the International Baccalaureate curriculum and examinations. (The IB diploma program is a 2-year, pre-university curriculum leading to the IB diploma or to a certificate recording separate subject examinations. IB students are required to become proficient in language, mathematics, social science, experimental sciences, and to develop aesthetic and moral values.) Seeks to provide an internationally acceptable curriculum for secondary schools and to promote the internationally recognized diploma, which permits admission to universities worldwide. Encourages research into methods and techniques of evaluation and assessment. Holds seminars and workshops for administrators and teachers. Conducts Middle Years Programme for ages 11-16 and Primary Years Progamme for ages 3-12. **PUBLICATIONS:** *IB World*, 3/year; *Publications Catalogue*, periodic; *Schools Directory*, annual; Annual Report (in English), annual.

INTERNATIONAL BADMINTON FEDERATION (IBF)

Manor Park Place, Rutherford Way, Cheltenham GL5L 9TU, England **PHONE:** 44 1242 234904 **FX:** 44 1242 221030 **E-MAIL:** info@intbadfed.org **WEBSITES:** http://www.intbadfed.org **FOUNDED:** 1934. **OFFICERS:** Neil Cameron, Exec.Dir. **MEMBERS:** 138. **STAFF:** 15. **LANGUAGES:** English.

DESCRIPTION: Organization of national badminton associations in 138 countries. Promotes the sport of badminton; manages and controls the game internationally. Organizes major events. Upholds the established Laws of Badminton and promotes their uniform enforcement. Acts to settle disputes among national organizations. **PUBLICATIONS:** *Badminton a Simple Way*. Coaching book.; *Event Organisation Manual*. Instructions for organizing a Badminton Event; *Physical Training for Badminton*. Training manual; *Statute Book*, annual; *World Badminton*, quarterly; Videos, periodic.

INTERNATIONAL BANK FOR ECONOMIC CO-OPERATION (IBEC)

c/o V.S. Khokhlov, 11 Masha Poryvaeva St., 107815 Moscow, Russia **PHONE:** 7 95 9753861, 7 95 9753851 **FX:** 7 95 9752202 **TX:** MOSCOW 411391 MZBKRU **FOUNDED:** 1963. **OFFICERS:** V.S. Khokhlov, Board Chm. **MEMBERS:** 9. **STAFF:** 210. **LANGUAGES:** Russian.

DESCRIPTION: Members are bankers and financial officials representing the governments of Bulgaria, Cuba, Czech Republic, Mongolia, Poland, Romania, Slovakia, and Vietnam. Promotes the development of economic ties among member countries. Seeks to facilitate the transition of member countries to a market economy. Offers credit and settlement facilities, and other banking services to interested organizations. Cooperates with the United Nations Conference on Trade and Development and the Economic Commission for Europe. **PUBLICATIONS:** *Information Bulletin* (in Russian), monthly; Annual Report (in English and Russian).

INTERNATIONAL BANK FOR RECONSTRUCTION AND DEVELOPMENT (IBRD)

1818 H St. NW, Washington, DC 20433 **PHONE:** (202) 477-1234 **FX:** (202) 477-6391 **TX:** 248423 **E-MAIL:** pic@worldbank.org **WEBSITES:** http://www.worldbank.org **FOUNDED:** 1945. **OFFICERS:** James D. Wolfensohn, Pres. **LANGUAGES:** English.

DESCRIPTION: Member governments of the International Monetary Fund. Assists the reconstruction and productive growth of Third World nations by facilitating the investment of capital in development projects and activities. Encourages foreign investment and the balanced growth of international trade. Provides financial aid for specified programs. Works in conjunction with the International Development Association and the International Finance Corporation to carry out policies and strategies of the World Bank. **PUBLICATIONS:** *Global Development Finance*, annual; *Global Economic Prospects and Developing Countries*, annual; *World Bank Annual Report*; *World Bank Atlas*, annual; *World Development Indicators*, annual. Many additional reports, CD-ROMS.

INTERNATIONAL BAR ASSOCIATION (IBA) (Association Internationale du Barreau)

271 Regent St., London W1R 7PA, England **PHONE:** 44 171 6291206 **FX:** 44 171 4090456 **E-MAIL:** sbl@int-bar.org **WEBSITES:** http://www.ibanet.org **FOUNDED:** 1947. **OFFICERS:** Paul Hoddinott, Exec.Dir. **MEMBERS:** 18,000. **STAFF:** 42. **BUDGET:** £4,100,000. **LANGUAGES:** English, French, German, Spanish.

DESCRIPTION: A federation of national bar associations and individual members of the legal profession in 183 countries. Works to advance the science of jurisprudence; promotes uniformity in related legal fields and administration of justice under law. Seeks to establish and maintain friendly relations among members of the legal profession worldwide. Supports the legal principles and aims of the United Nations. **PUBLICATIONS:** *Directory of Members*, annual; *International Bar News*, quarterly; *International Business Lawyer*, 11/year; *International Legal Practitioner*, quarterly; *Journal of Energy & Natural Resources Law*, quarterly; Books.

INTERNATIONAL BASKETBALL FEDERATION (FIBA) (Federation Internationale de Basketball - FIBA)

Boschetsrieder Str. 67, Postfach 700607, D-81306 Munich, Germany **PHONE:** 49 89 7481580 **FX:** 49 89 74815833 **TX:** 5213054 FIBA D **E-MAIL:** secretariat@office.fiba.com **WEBSITES:** http://www.fiba.com **FOUNDED:** 1932. **OFFICERS:** Borislav Stankovic, Sec.Gen. **MEMBERS:** 206. **STAFF:** 22. **LANGUAGES:** English, French, German, Russian, Spanish.

DESCRIPTION: National basketball federations united to promote, supervise and direct the sport of basketball throughout the world. Establishes game rules; specifies equipment, facilities, and all internal executive and sport regulations including the appointment of international referees, and the transfer of players from one country to another. **PUBLICATIONS:** *Basketball Bulletin* (in English and French), semiannual; *European Directory*, annual; *FIBA Basketball Monthly*; *FIBA Directory*, annual; *FIBA Media Guide*, annual.

INTERNATIONAL BEE RESEARCH ASSOCIATION (IBRA) (Association Internationale de Recherche Apicole)

18 North Rd., Cardiff, S. Glam CF1 3DY, Wales **PHONE:** 44 1222 372409 **FX:** 44 1222 665522 **E-MAIL:** ibra@cardiff.ac.uk **WEBSITES:** http://www.cardiff.ac.uk/ibra/index.html **FOUNDED:** 1949. **OFFICERS:** Dr. Pamela Munn, Deputy Dir. **MEMBERS:** 800. **STAFF:** 4. **BUDGET:** £150,000. **LANGUAGES:** English, French, German, Welsh.

DESCRIPTION: Members are individuals, beekeeping societies, and research organizations in 130 countries. Promotes and coordinates bee research work and research on pollination. Provides worldwide information service through publications, correspondence, and journals. Aids beekeepers and promotes beekeeping as a sustainable activity in developing countries. **PUBLICATIONS:** *Apicultural Abstracts* (in English), quarterly. Reviews of world literature on bees, beekeeping, and related subjects.; *Bee World*, quarterly; *Journal of Apicultural Research*, quarterly.

INTERNATIONAL BIOMETRIC SOCIETY (IBS)

1444 I St. NW, Ste. 700, Washington, DC 20005-2210 **PHONE:** (202) 712-9049, (202) 216-9623 **FX:** (202) 216-9646 **E-MAIL:** ibs@bostromdc.com **WEBSITES:** http://www.tibs.org **FOUNDED:** 1947. **OFFICERS:** Charles McGrath CAE, Exec.Dir. **MEMBERS:** 6,500. **STAFF:** 3. **BUDGET:** US$473,000. **LANGUAGES:** English.

DESCRIPTION: Members are biologists, statisticians, and others interested in applying statistical techniques to research data. Works to advance subject-matter sciences through the development of quantitative theories and the development, application, and dissemination of effective mathematical and statistical techniques. Compiles statistics. **PUBLICATIONS:** *Biometric Bulletin* (in English and French), quarterly. Includes membership information, announcements, abstracts, and calendar of events; *Biometrics*, quarterly. Emphasizes the role of statistics and mathematics in the biological sciences. Includes book reviews and annual index.; *The International Biometric Society—Membership Directory*, biennial.

INTERNATIONAL BOARD ON BOOKS FOR YOUNG PEOPLE (IBBY) (Union Internationale pour les livres de jeunesse)

Nonnenweg 12, Postfach, CH-4003 Basel, Switzerland **PHONE:** 41 61 2722917 **FX:** 41 61 2722757 **E-MAIL:** ibby@eye.ch **FOUNDED:** 1953. **OFFICERS:** Leena Maissen, Exec.Dir. **MEMBERS:** 64. **STAFF:** 2. **BUDGET:** US$200,000. **LANGUAGES:** English, French, German, Spanish.

DESCRIPTION: Members consist of institutes, publishing houses, individuals and groups in 64 countries interested in literature for children and young people. Recognizes that books play an important role in the development of children. Aims to increase international understanding and goodwill among all people through children's literature; addresses concerns specific to literature for young people; encourages the promotion and distribution of high quality books for children. Maintains the IBBY Documentation Center of Books for Disabled Young People. Compiles biennial IBBY Honour List of outstanding children's books published in member countries. Sponsors International Children's Book Day in April. **PUBLICATIONS:** *Bookbird* (in English), quarterly. Covers many facets of international children's literature including news from IBBY and the IBBY National Sections; *Congress Proceedings*, biennial; *IBBY Honour List*, biennial; *What is IBBY?*. Regional publications and national newsletters.

INTERNATIONAL BOOK BANK (IBB)

2201 Eagle St., Unit D., Baltimore, MD 21222 **PHONE:** (410) 362-0334 **E-MAIL:** ibbusa@worldnet.att.net **FOUNDED:** 1987. **OFFICERS:** James A. W. Rogerson, Exec.Dir. **MEMBERS:** 1. **STAFF:** 7. **BUDGET:** US$815,000. **LANGUAGES:** English.

DESCRIPTION: Collects new and used books from schools, libraries, and publishers for distribution to schools in developing countries. Offers procurement and shipping services to government agencies, foundations, educational institutions, service clubs, book bank agencies, and the general public. Collaborates with the National Association of College Stores, the Peace Corps, and the United States Information Agency. Operates speakers' bureau. **PUBLICATIONS:** *Year in Review*.

INTERNATIONAL BOOK PROJECT (IBP)

1440 Delaware Ave., Lexington, KY 40505 **PHONE:** (606) 254-6771 **FX:** (606) 367-0242 **E-MAIL:** ibookp@iglou.com **WEBSITES:** http://www.ibook.org **FOUNDED:** 1966. **OFFICERS:** Ken DeGilio, Exec.Dir. **MEMBERS:** 2,500. **STAFF:** 4. **BUDGET:** US$150,000. **LANGUAGES:** English.

DESCRIPTION: Individuals, civic groups, and churches working to collect funds for the distribution of books to needy institutions such as schools, hospitals, clinics, seminaries, churches, and urban and village libraries in 100 developing countries, including Bahamas, Belize, Ghana, India, Indonesia, Kenya, Liberia, Nigeria, and Philippines. Seeks to provide information to people who do not have access to books and journals and to encourage Americans to become involved with the people and conditions in developing countries through correspondence with people of these countries. Encourages members to send books and journals to these countries; provides mailing tips. Compiles statistics. **PUBLICATIONS:** *Books Abroad*, quarterly.

INTERNATIONAL BOOKSELLERS FEDERATION (IBF) (Internationale Buchhaendler-Vereinigung - IBV)

rue du Grand Hoslice 34A, B-1000 Brussels, Belgium **PHONE:** 32 2 2234940 **FX:** 32 2 2234941 **E-MAIL:** eurobooks@skynet.be **FOUNDED:** 1956. **OFFICERS:** Christiane Vuidar, Gen.Sec. **MEMBERS:** 220. **STAFF:** 1. **LANGUAGES:** English, French, German.

DESCRIPTION: Booksellers' associations (20) are members and booksellers (200) are associate members. Promotes international cooperation among booksellers and associations of booksellers. Encourages exchange of ideas and experiences and discussion of common problems. **PUBLICATIONS:** *Booksellers International* (in English), annual; *Bulletin*, 2-3/year; *Membership List*, periodic; Directory.

INTERNATIONAL BRAIN RESEARCH ORGANIZATION (IBRO) (Organisation Internationale de Recherche sur le Cerveau)

c/o Dr. Carlos Belmonte, 51, Blvd. de Montmorency, F-75016 Paris, France **PHONE:** 33 1 46479292 **FX:** 33 1 45206006 **E-MAIL:** ibro@wanadoo.fr **WEBSITES:** http://www.ibro.org **FOUNDED:** 1960. **OFFICERS:** Dr. Carlos Belmonte, Sec.Gen. **MEMBERS:** 51,000. **STAFF:** 2. **LANGUAGES:** English, French.

DESCRIPTION: Scientists working in neuroanatomy, neuroendocrinology, the behavioral sciences, neurocommunications and biophysics, brain pathology, and clinical and health-related sciences. Works to promote international cooperation in research on the nervous system. Sponsors fellowships, exchange of scientific workers, and traveling teams of instructors to supplement local teachings. Organizes international neuroscience symposia and workshops. **PUBLICATIONS:** *Directory of Members*, periodic; *Neuroscience*, bimonthly; *News*, annual. Also publishes handbook series.

INTERNATIONAL BRIDGE, TUNNEL AND TURNPIKE ASSOCIATION (IBTTA)

2120 L St. NW, Ste. 305, Washington, DC 20037-1527 **PHONE:** (202) 659-4620 **FX:** (202) 659-0500 **E-MAIL:** ibtta@ibtta.org **WEBSITES:** http://www.IBTTA.org **FOUNDED:** 1932. **OFFICERS:** Neil D. Schuster, Exec.Dir. **MEMBERS:** 235. **STAFF:** 8. **BUDGET:** US$2,000,000. **LANGUAGES:** English.

DESCRIPTION: Public and private agencies operating toll bridges, tunnels, and turnpikes and companies providing support services and equipment. Monitors and reports on events and legislative action affecting transportation systems worldwide. Conducts research programs; compiles statistics. **PUBLICATIONS:** *ETTM System Survey*; *Tollways* (in English), monthly. Member newsletter focuses on trends, developments and news about the worldwide toll industry.; *Tollways*, monthly. Member newsletter focuses on trends, developments and news about the worldwide toll industry.

INTERNATIONAL BUREAU OF EDUCATION (IBE) (Bureau International d'Education - BIE)

Case Postale 199, CH-1211 Geneva 20, Switzerland **PHONE:** 41 22 9177800 **FX:** 41 22 9177801 **TX:** 415771 BIE **E-MAIL:** j.hallak@ibe.unesco.org **WEBSITES:** http://www.ibe.unesco.org **FOUNDED:** 1925. **OFFICERS:** Jacques Hallak, Dir. **MEMBERS:** 186. **STAFF:** 19. **BUDGET:** US$7,500,000. **LANGUAGES:** English, French.

DESCRIPTION: A center of comparative education within the United Nations Educational, Scientific and Cultural Organization. Monitors developments in educational content, teaching and learning strategies, and educational trends and innovations. Works to develop and promote education information systems and services at national, regional, and international levels. Prepares educational documentation; conducts comparative studies and publishes results. **PUBLICATIONS:** *Educational Innovation and Information* (in English, French, and Spanish), quarterly; *Prospects* (in Arabic, Chinese, English, French, Russian, and Spanish), quarterly; *Studies in Comparative Education Series*, periodic; *Study Abroad* (in English, French, and Spanish), biennial. Guide to study offers by universities, specialized schools, and international organizations.

INTERNATIONAL BUREAU FOR EPILEPSY (IBE) (Bureau International pour l'Epilepsie)

Postbus 21, NL-2100 AA Heemstede, Netherlands **PHONE:** 311 23 5291019, 31 23 5237411 **FX:** 31 23 5470119 **E-MAIL:** ibe@xs4all.nl **FOUNDED:** 1961. **OFFICERS:** Mr. Richard Holmes, Pres. **MEMBERS:** 51. **STAFF:** 2. **BUDGET:** US$100,000. **LANGUAGES:** English.

DESCRIPTION: National organizations and individuals interested in the medical, social, and scientific aspects of epilepsy. Focuses on aspects of daily life with epilepsy. Facilitates exchange of information and experience regarding the care of persons with epilepsy. Provides material on how to organize and finance non-medical societies. Organizes training sessions. Works to build an international film library on epilepsy. **PUBLICATIONS:** *A Manual for Epilepsy Self-Help Groups*; *Employing People with Epilepsy: Principles for Good Practice* (in English, Italian, Portuguese, and Spanish); *Epilepsy Education Manual*; *Epilepsy in Focus*, biennial. Catalogue of audiovisual materials; *Epilepsy Passport*. Guidelines for travel by people with epilepsy; *International Epilepsy News* (in English and Japanese), quarterly. Contains information on epilepsy.

INTERNATIONAL BUREAU OF FISCAL DOCUMENTATION (IBFD) (Internationaal Belasting Documentatie Bureau - IBDB)

PO Box 20237, NL-1000 HE Amsterdam, Netherlands **PHONE:** 31 20 5540100 **FX:** 31 20 6228658 **TX:** 13217 INTAX NL **WEBSITES:** http://www.ibfd.nl **FOUNDED:** 1938. **OFFICERS:** Prof. H.M.A.L. Hamaekers, Chief Exec. **STAFF:** 55. **BUDGET:** US$4,000,000. **LANGUAGES:** Chinese, Dutch, English, Finnish, French, German, Icelandic, Italian, Norwegian, Spanish, Swedish.

DESCRIPTION: Independent, nonprofit research institute specialized in international tax and investment legislation. Publishes information in a wide range of journals, loose-leaf services, books, and databases. Also organizes both in-company and open courses on this subject on a postgraduate level. **PUBLICATIONS:** *African Tax Systems* (in English and French), quarterly; *Annual Report*; *Bulletin for International Fiscal Documentation*, monthly; *European Taxation*, monthly; *International VAT Monitor*, monthly; *Supplementary Service to European Taxation*, monthly; *Tax News Service*, biweekly; *Tax Treatment of Transfer Pricing*, semiannual; *Taxation and Investment in Canada and the United States*, biennial; *Taxation in European Socialist Countries*, semiannual; *Taxation in Latin America*, quarterly; *Taxation of Companies in Europe*, quarterly; *Taxation of Patent Royalties, Dividends and Interest in Europe*, semiannual; *Taxation of Private Investment Income*, semiannual; *Taxes and Investment in Asia and the Pacific*, quarterly; *Taxes and Investment in the Middle East*, quarterly; *Value Added Taxation in Europe*, quarterly. Also publishes monographs, studies, and a handbook on the U.S.-German Tax Convention.

INTERNATIONAL BUREAU OF THE PERMANENT COURT OF ARBITRATION (PCA) (Bureau International de la Cour Permanente d'Arbitrage)

Peace Palace, Carnegieplein 2, NL-2517 KJ The Hague, Netherlands **PHONE:** 31 70 3024242, 31 70 3024165 **FX:** 31 70 3024167 **E-MAIL:** pca@euronet.nl **WEBSITES:** http://www.lawschool.cornell.edu/library/pca **FOUNDED:** 1899. **OFFICERS:** P.J.H. Jonkman, Sec.Gen. **MEMBERS:** 88. **STAFF:** 9. **LANGUAGES:** English, French.

DESCRIPTION: International organization administering a range of mechanisms for the resolution of disputes involving at least one State party or international organization. **PUBLICATIONS:** *Annual Report of the Administrative Council* (in English and French); *Convention for the Pacific Settlement of International Disputes (1899)*; *Convention for the Pacific Settlement of International Disputes (1907)*; *The Permanent Court of Arbitration - New Directions, 1991*; *Permanent Court of Arbitration Optional Conciliation Rules, 1996*; *The Permanent Court of Arbitration - Optional Rules for Arbitrating Disputes Between Two Parties of which only One is a State, 1993*; *Permanent Court of Arbitration Optional Rules for Arbitrating Disputes Between Two States, 1992*; *Permanent Court of Arbitration Optional Rules for Arbitration Between International Organizations and Private Parties, 1996*; *Permanent Court of Arbitration Optional Rules for Arbitration Involving International Organizations and States, 1996*; *The Permanent Court of Arbitration - What It is and What It Can Do.*

INTERNATIONAL BUREAU FOR THE STANDARDISATION OF MANMADE FIBRES (BISFA) (Bureau International pour la Standardisation des Fibres Artificielles - BISFA)

4, ave. Van Nieuwenhuyse, B-1160 Brussels, Belgium **PHONE:** 32 2 6767455 **FX:** 32 2 6767454 **E-MAIL:** van@cirfs.org **WEBSITES:** http://www.bisfa.org **FOUNDED:** 1928. **OFFICERS:** Alex Krieger, Sec.Gen. **MEMBERS:** 67. **STAFF:** 2. **LANGUAGES:** English, French, German, Italian.

DESCRIPTION: Members are man-made fiber producers in 20 countries. Works to establish technical rules relating to specifications and characteristics of different types of man-made fibers and technical standards for testing purposes. Conducts studies of all types relating to designation, classification, and standardization of fibers. **PUBLICATIONS:** *Rules for Standardization of Various Categories of Man-Made Fibres.*

INTERNATIONAL BUREAU OF WEIGHTS AND MEASURES (BIPM) (Bureau International des Poids et Mesures)

Pavillon de Breteuil, F-92312 Sevres Cedex, France **PHONE:** 33 1 45077070 **FX:** 33 1 45342021 **E-MAIL:** info@bipm.fr **WEBSITES:** http://www.bipm.fr **FOUNDED:** 1875. **OFFICERS:** T.J. Quinn, Dir. **MEMBERS:** 48. **STAFF:** 70. **LANGUAGES:** French.

DESCRIPTION: Responsible for establishing the fundamental standards and scales for measurement of the principal physical quantities and maintaining the international prototypes. **PUBLICATIONS:** *Annual Report* (in English); *Circular*, monthly; *Metrologia*, periodic; *Reports of General Conference of Weights and Measures*, quadrennial; Reports, periodic. From nine specialized committees.

INTERNATIONAL BUTCHERS CONFEDERATION (Internationaler Metzgermeisterverband)

Rue Jacques de Lalaing 4-box 10, B-1040 Brussels, Belgium **PHONE:** 32 2 2303876 **FX:** 32 2 2303451 **E-MAIL:** info@cibc.be **FOUNDED:** 1907. **OFFICERS:** Mrs. Kirsten Diessner, Dir. of the Sec. Gen. **MEMBERS:** 18. **BUDGET:** US$29,830. **LANGUAGES:** English, French, German.

DESCRIPTION: Comprised of 18 national butchers' associations and meat traders' federations representing more than 150,000 butcher and meat-cutting companies which together employ nearly 1 million people. Defends the common interests of the meat trading and catering industry through social and economic policy, agricultural policy, knowledge and legislation on foodstuffs and meat products, and vocational training. Represents the affairs of the meat-cutting and catering industry before the European Council, the European Commission, the European Parliament, the Economic and Social Committee, as well as other associations and organizations in Brussels.

INTERNATIONAL CADMIUM ASSOCIATION - EUROPEAN OFFICE (ICdA)

ave. de Tervuren, 168/4, B-1150 Brussels, Belgium **PHONE:** 32 2 7770560 **FX:** 32 2 7770565 **TX:** 261286 **E-MAIL:** icda@village.uunet.be **FOUNDED:** 1976. **OFFICERS:** Dr. Raymond Sempels, Dir. **MEMBERS:** 48. **STAFF:** 2. **BUDGET:** US$400,000. **LANGUAGES:** English, French.

DESCRIPTION: Cadmium companies, producers, and users; mining companies. Purpose is to provide current, authoritative information and advice to users and potential users of cadmium as well as governments and other official bodies concerned with codes, standards, and regulations regarding cadmium processing and cadmium products. Collects and disseminates technical data on the properties and uses of cadmium and its alloys and compounds. Provides a forum for the discussion of problems arising from the production, processing, and use of cadmium. Sponsors seminars on the uses of cadmium in alloys, batteries, coatings, and in pigments and stabilizers for plastics and coatings. **PUBLICATIONS:** *Conference Proceedings*, triennial; *Industry Surveys*, periodic; *NiCd Battery Update*, biennial; *Seminar Proceedings*, periodic; *Technical Notes on Cadmium*, periodic.

INTERNATIONAL CARGO HANDLING COORDINATION ASSOCIATION (ICHCA)

71 Bondway, London SW8 1SH, England

PHONE: 44 171 7931022 **FX:** 44 171 8201703 **E-MAIL:** postmaster@ichca.org.uk **WEBSITES:** http://www.ichca.org.uk **FOUNDED:** 1952. **OFFICERS:** Gerry Askham, CEO. **MEMBERS:** 2,000. **STAFF:** 8. **BUDGET:** £500,000. **LANGUAGES:** English.

DESCRIPTION: Members are individuals and organizations in 84 countries with interests in international handling and transport of goods. Works to increase efficiency and economy in the handling and movement of goods from origin to destination by all modes and at all phases of transportation. Maintains information service and expert committees; sponsors research programs, and intergovernmental representation (NGO). **PUBLICATIONS:** *Cargo Today* (in English), quarterly. Includes cargo handling database abstracts; *ICHCA Buyer's Guide to Manufacturers*, annual; *Who's Who in Cargo Handling*, annual; *World of Cargo Handling - ICHCA Annual Review*; Report, biennial.

INTERNATIONAL CARTOGRAPHIC ASSOCIATION (ICA) (Association Cartographique Internationale)

c/o M. Pierre PLANQUES, Secretaire General, 138, bis rue de Grenelle, F-75700 Paris, France **PHONE:** 33 1 43988295 **FX:** 33 1 439884100 **E-MAIL:** jean-philippe.grelot@ign.fr **FOUNDED:** 1959. **OFFICERS:** Jean Philippe Grelot **MEMBERS:** 80. **BUDGET:** US$35,000. **LANGUAGES:** English, French.

DESCRIPTION: National societies and committees for cartography. Works to advance the study of cartographic source material and the design, construction, and reproduction of maps; coordinates cartographic research involving international cooperation. **PUBLICATIONS:** *Basic Cartography*; *Cartography: Past, Present, and Future*; *Compendium of Cartographic Techniques*; *Elements of Spatial Data Quality*; *Inventory of World Topographic Mapping*; *Satellite Imagery*; Newsletter, semiannual.

INTERNATIONAL CATHOLIC CHILD BUREAU (BICE) (Bureau International Catholique de l'Enfance - BICE)

63, rue de Lausanne, CH-1202 Geneva, Switzerland **PHONE:** 41 22 7313248 **FX:** 41 22 7317793 **E-MAIL:** bice@dial.eunet.ch **FOUNDED:** 1948. **OFFICERS:** Dr. Francois Ruegg, Sec.Gen. **MEMBERS:** 242. **STAFF:** 15. **LANGUAGES:** English, French, Spanish.

DESCRIPTION: Organization founded in 1948 involved with children on a national and international level. Serves the holistic growth of all children in a Christian perspective. Develops short, medium and long-range projects to promote spiritual growth, intercultural awareness, and the rights of the child. **PUBLICATIONS:** *BICE Info* (in English), semester; *Children and Prostitution: Don't Give Up On Me?* (in English and French); *Children Worldwide* (in English and French), annual. Includes annual report; *Enfants de Partout* (in French), quarterly.

INTERNATIONAL CATHOLIC MIGRATION COMMISSION (ICMC) (Comision Catolica Internacional de Migracion)

37-39, rue de Vermont, Case Postale 96, CH-1211 Geneva 20, Switzerland **PHONE:** 41 22 9191020 **FX:** 41 22 9191048 **TX:** 414122 **E-MAIL:** admin@icmc.dpn.ch **FOUNDED:** 1951. **OFFICERS:** William Canny, Sec.Gen. **LANGUAGES:** English, French, Spanish.

DESCRIPTION: Operational arm of the Catholic Church, specifically mandated to foster and facilitate the coordination of Catholic assistance to refugees, migrants, and displaced persons. Guided by the belief in the sanctity of the individual and the family to promote humane treatment of all people. **PUBLICATIONS:** *Annual Report* (in English, French, and Spanish).

INTERNATIONAL CATHOLIC UNION OF THE PRESS (UCIP) (Union Catholique Internationale de la Presse - UCIP)

37-39, rue de Vermont, Case Postale 197, CH-1211 Geneva 20, Switzerland **PHONE:** 41 22 7340017, 41 22 7347416 **FX:** 41 22 7340053 **FOUNDED:** 1927. **OFFICERS:** Joseph Chittilappilly, Sec.Gen. **MEMBERS:** 6,900. **STAFF:** 4. **LANGUAGES:** English, French, German, Spanish.

DESCRIPTION: Confederation of journalists, news agencies, church press associations, federations of dailies and periodicals, and the International Catholic Association of Teachers and Research Fellows in the Sciences and Techniques of Information. Purposes are to: link Catholics who influence public opinion through the press and mass media; inspire a high standard of professional conscience; represent the interests of the Catholic press before international organizations; develop Catholic information in the Third World; promote the professional training of Catholic journalists. Sponsors International Network of Young Jounalists program. **PUBLICATIONS:** *Communication, Culture, Religion*; *Freedom of Journalist*; *Journalism for World Peace and Development*; *UCIP Information* (in English, French, German, and Spanish), quarterly; Books.

INTERNATIONAL CELL RESEARCH ORGANIZATION (ICRO) (Organisation Internationale de Recherche sur la Cellule)

c/o UNESCO SC/BSC, 1, rue Miollis, F-75732 Paris, France **FX:** 33 1 45685818 **E-MAIL:** icro@unesco.org **FOUNDED:** 1962. **OFFICERS:** Dr. G. Cohen, Exec.Sec. **MEMBERS:** 400. **LANGUAGES:** English, French.

DESCRIPTION: Members are scientists, researchers, and laboratories in 75 countries. Encourages and facilitates the exchange of information on basic cell biology research. Organizes international training courses in microbiology, biotechnology, and cell and molecular biology. Compiles statistics.

INTERNATIONAL CENTER FOR THE HEALTH SCIENCES (ICHS)

Barracks Hill, PO Box 4744, Charlottesville, VA 22905-4744 **PHONE:** (804) 971-6921 **FX:** (804) 971-7605 **FOUNDED:** 1991. **OFFICERS:** Warren E. Grupe MD, Medical Dir. **STAFF:** 5. **LANGUAGES:** English.

DESCRIPTION: Works to improve the health of all people through creative health education programs designed to meet the unique needs and capabilities of communities around the world. Sponsors collaborative research; develops and strengthens current preventative and curative care; manages and implements health science education programs in cooperation with US institutions in Central America, Africa and the former Soviet Union; advocates for the health care needs of all people.

INTERNATIONAL CENTER FOR NOT-FOR-PROFIT LAW (ICNL)

733 15th St. NW, Ste. 420, Washington, DC 20005-2112 **PHONE:** (202) 624-0766 **FX:** (202) 624-0767 **E-MAIL:** dcincnl@aol.com **WEBSITES:** http://www.icnl.org **FOUNDED:** 1992. **OFFICERS:** Karla Simon, Exec.VP. **STAFF:** 20. **BUDGET:** US$1,500,000. **LANGUAGES:** English, French, Mandarin dialects, Polish, Russian, Spanish.

DESCRIPTION: Seeks to assist in the formation of laws and regulatory systems that stimulate activities of nonprofit organizations worldwide. Provides technical assistance to agencies entrusted with the writing of laws and regulations governing nonprofit organizations; assists in the privatization of social, educational, and cultural assets and programs in formerly socialist countries.

INTERNATIONAL CENTER FOR THE SOLUTION OF ENVIRONMENTAL PROBLEMS (ICSEP)

3355 Navigation, Ste. 310, Houston, TX 77003 **PHONE:** (713) 527-8711 **FX:** (713) 527-8025 **E-MAIL:** icsep@neosoft.com **WEBSITES:** http://www.neosoft.com/-icsep/ **FOUNDED:** 1976. **OFFICERS:** Joseph L. Goldman Ph.D., Tech.Dir. **STAFF:** 5. **LANGUAGES:** English.

DESCRIPTION: Institutes, corporations, and individuals engaged in scientific, engineering, management, economic, and offshore environmental activities. Uses a multi-disciplinary approach to anticipate upcoming environmental problems and provide solutions for avoiding or reducing them. Is involved in numerous civic activities related to the environment. Conducts research projects on: anticipating the effects of storms and reducing their causes and occurrences off- and onshore; effects of storms related to city design; changes in surface vegetation caused by spreading urbanization; fiood control; reduction of land erosion; remote sensing applications; land subsidence and flooding by overpumping; decline in land value resulting from these changes; the effect of weather on soil and agriculture; the effect of herbicide sprays on users and groundwater. **PUBLICATIONS:** Brochure; Newsletter, annual.

INTERNATIONAL CENTRE FOR CHILDHOOD AND THE FAMILY (Centre International de l'Enfance et de la Famille)

c/o Dr. Olivier Brasseur, Carrefour de

Longchamp, Bois de Boulogne, F-75016 Paris, France **PHONE:** 33 1 44302000 **FX:** 33 1 45257367 **E-MAIL:** cidef@compuserve.com **FOUNDED:** 1949. **OFFICERS:** Dr. Olivier Brasseur, Dir.Gen. **STAFF:** 81. **BUDGET:** 43,000,000 Fr. **LANGUAGES:** English, French, Spanish.

DESCRIPTION: Seeks to bring about the highest attainable standards of health and welfare among children and their families in France and around the world. Acts as an interface between researchers, decision makers, professionals, associations, and institutions. Works to identify and analyze the legal, economic, cultural, educational, and institutional dynamics that have an impact on the well-being of children and families. Disseminates information needed by those involved in child and family protection. Advises decision makers on the implementation and assessment of programs and activities to assist children and their families. **PUBLICATIONS:** *Children in the Tropics* (in English, French, and Spanish), bimonthly. Focuses on children's health with contributions from all disciplines relevant to children's health services and activities; one double issue per year. Additional publication information available on request.

INTERNATIONAL CENTRE FOR CRIMINAL LAW REFORM AND CRIMINAL JUSTICE POLICY

1822 East Mall, Vancouver, BC, Canada V6T 1Z1 **PHONE:** (604) 822-9875 **FX:** (604) 822-9317 **E-MAIL:** icclr@law.ubc.ca **WEBSITES:** http://www.icclr.law.ubc.ca **FOUNDED:** 1991. **OFFICERS:** Daniel C. Prefontaine Q.C., Dir. and CEO. **LANGUAGES:** English, French.

DESCRIPTION: As part of the United Nations Crime Prevention and Criminal Justice program Network of Institutes, the International Centre contributes to policy development and law reform nationally and internationally and provides implementation assistance. Its objective is the promotion of democratic principles, the rule of law, and respect of human rights in criminal law and the administration of criminal justice. Focuses its activities on technical cooperation, research training and advisory services. It is currently involved in projects in China, South-East Asia and Latin America. **PUBLICATIONS:** Newsletter, quarterly.

INTERNATIONAL CENTRE OF INSECT PHYSIOLOGY AND ECOLOGY (ICIPE)

PO Box 30772, Nairobi, Kenya **PHONE:** 254 2 861680, 254 2 802501 **FX:** 254 2 803360 **TX:** 22053 ICIPE **E-MAIL:** directorgeneral@icipe.org **WEBSITES:** http://www.icipe.org **FOUNDED:** 1970. **OFFICERS:** Dr. Hans R. Herren, Dir.Gen. **STAFF:** 310. **BUDGET:** US$12,000,000. **LANGUAGES:** English, French, Kiswahili.

DESCRIPTION: Seeks to contribute to increased food production by investigating pests of major crops, major livestock diseases, and insect carriers of human diseases in the tropics. Goals are to: promote advanced insect biology research and knowledge for environmentally friendly management, conservation, and utilization of arthropods; provide research training to doctoral and postdoctoral fellows; act as a forum for the exchange of knowledge among scientists and increase development-oriented research such as integrated and sustainable pest management in developing countries. Maintains research programs on cereal stemborers, podborers, tsetse flies, locusts, livestock ticks, and vectors of tropical diseases such as malarial mosquitos. Offers technical support services in areas of insect and animal breeding and field-station, outreach, workshop, and laboratory management; encourages development of nonchemical pest control technologies. Promotes local economic self-sufficiency through implementation of apicultural and sericultural projects. Compiles biostatistics. **PUBLICATIONS:** *Annual Report* (in English). Describes research, capacity building, technologies and outreach activities.; *Current Themes in Tropical Science*. Features a series of books.; *ICIDE Update* (in English), monthly; *ICIPE Profile* (in English and French), periodic. Contains original research reports, review articles, and book reviews.; *ICIPE Vision and Strategic Framework* (in English), periodic; *Insect Science and Its Application* (in English and French), quarterly. Contains original research reports, review articles, and book reviews.; *Proceedings of Workshops* (in English and French), periodic; *Research Highlights* (in English), periodic; *Review Papers*, periodic. Insect science teaching manuals and textbooks; scholarly works in the fields of medicine, social science, biology, agriculture, and economics.

INTERNATIONAL CENTRE FOR THE LEGAL PROTECTION OF HUMAN RIGHTS (ICLPHR)

33 Islington High St., London N1 9LH, England **PHONE:** 44 0171 2783230 **FX:** 44 0171 2784334 **E-MAIL:** ir@interights.org **FOUNDED:** 1982. **OFFICERS:** Emma Playfair, Exec.Dir. **STAFF:** 13. **BUDGET:** US$512,000. **LANGUAGES:** English.

DESCRIPTION: Members are attorneys and other individuals with an interest in human rights. Seeks to protect and expand the rights of the individual by strengthening their statutory underpinning worldwide. Drafts model human rights legislation and statutes; provides legal representation for individuals and groups appearing before international human rights tribunals; sponsors public education programs. **PUBLICATIONS:** *Interights Bulletin* (in English), quarterly.

INTERNATIONAL CENTRE FOR LOCAL CREDIT (ICLC)

Koninginnegracht 2, NL-2514 AA The Hague, Netherlands **PHONE:** 31 70 3750850 **FX:** 31 70 3750406 **E-MAIL:** centre@bng.nl **WEBSITES:** http://www.bng.nl **FOUNDED:** 1958. **OFFICERS:** Mr. N.M. Heijstek. **MEMBERS:** 21. **LANGUAGES:** English.

DESCRIPTION: Aim is to promote local authority credit. Endeavors to achieve this aim by: gathering, exchanging, and distributing information and advice as to the organization, management, and activities of member institutions; gathering, exchanging, and distributing information on local authority credit and subjects connected with or affecting the same; studying, jointly if necessary, important subjects in the field of local authority credit; and other activities. **PUBLICATIONS:** *Informational Bulletin* (in English), periodic; Newsletter (in English and French), periodic; Reports.

INTERNATIONAL CENTRE OF RESEARCH AND INFORMATION ON PUBLIC AND CO-OPERATIVE ECONOMY (CIRIEC)

Univesite de Liege au Sart-Tilman, Batiment B33, bte. 6, B-4000 Liege, Belgium **PHONE:** 32 41 43662746 **FX:** 32 41 43662958 **E-MAIL:** ciriec@ulg.ac.be **FOUNDED:** 1947. **OFFICERS:** Prof. Bernard Thiry, Dir. **STAFF:** 6. **LANGUAGES:** English, French, German.

DESCRIPTION: Agencies and organizations concerned with public and cooperative economy; scientific and cultural institutions with special interest in the problems of collective economy; individuals. Purposes are to conduct research and promote the dissemination of information on public and cooperative economy. According to CIRIEC, public and cooperative economy has a direct aim of service, not profit, and includes state intervention in economic affairs, public enterprise, activities of municipal authorities, cooperation, and the role of trade unions in economics. **PUBLICATIONS:** *Annals of Public and Cooperative Economics* (in English and French), quarterly. An international scientific journal; *Working Papers*; Books.

INTERNATIONAL CENTRE FOR SETTLEMENT OF INVESTMENT DISPUTES (ICSID)

1818 H St. NW, Washington, DC 20433 **PHONE:** (202) 477-1234 **FX:** (202) 522-2615 **FOUNDED:** 1966. **OFFICERS:** Ibrahim F. I. Shihata, Sec.Gen. **MEMBERS:** 110. **LANGUAGES:** English.

DESCRIPTION: Members are countries that are members of the International Bank for Reconstruction and Development, or, by invitation, parties to the Statute of the International Court of Justice. Provides facilities for the conciliation and arbitration of investment disputes between members and nationals of other members in accordance with the provisions of the Convention on the Settlement of Investment Disputes between States and Nationals of Other States. Collects and disseminates information relating to legislation, international agreements, and other investment matters. **PUBLICATIONS:** *ICSID Review: Foreign Investment Law Journal*, semiannual. Contains information on laws and practices relating to foreign investment. Covers domestic legislation, investment treaties, and contractual trends.; *International Centre for Settlement of Investment Disputes—Annual Report*; *Investment Laws of the World*, periodic. Texts of basic investment laws of over 90 countries.; *Investment Treaties*, annual. Provides texts of selected bilateral promotion and protection treaties entered into since 1960 by developed and developing countries.; *News From ICSID*, semiannual. Covers disputes before the centre and centre activities; *Rules and Regulations*.

INTERNATIONAL CENTRE FOR THE STUDY OF THE PRESERVATION AND RESTORATION OF CULTURAL PROPERTY (ICCROM) (Centre International d'Etudes pour la Conservation et la Restauration des Biens Culturels - ICCROM)

c/o Marc Laenen, Via di San Michele 13, I-00153 Rome, Italy **PHONE:** 39 6 585531 **FX:** 39 6 58553349 **E-MAIL:** iccrom@iccrom.org **WEBSITES:** http://www.iccrom.org **FOUNDED:** 1959. **OFFICERS:** Marc Laenen, Dir.Gen. **MEMBERS:** 94. **STAFF:** 36. **LANGUAGES:** English, French.

DESCRIPTION: Members are state governments; associate members are nonprofit public or private institutions of a scientific or cultural nature. Promotes heritage awareness and the preservation, restoration, and preventive conservation of historic buildings, archaeological sites, museum collections, and library and archival material. Collects, examines, and disseminates documentation concerning scientific and technical problems of preserving and restoring cultural property. Offers advice and recommendations involving cultural property. Conducts seminars and courses on restoration work. **PUBLICATIONS:** *ICCROM Newsletter* (in English and French), annual; *International Directory on Training in Conservation of Cultural Heritage* (in English), periodic; *Library List of Acquisitions* (in English and French), 3/year; *Stop Press* (in English and French), semiannual.

INTERNATIONAL CHAMBER OF COMMERCE (ICC) (Chambre de Commerce Internationale - CCI)

c/o M-C Psimenos de Metz-Noblat, 9, Rue d'Anjou, F-75008 Paris, France **PHONE:** 33 1 42651266 **FX:** 33 1 49240639 **E-MAIL:** cnfcci@dial.oleane.com **WEBSITES:** http://www.iccwbo.org **FOUNDED:** 1920. **OFFICERS:** Maria Livanos Cattaui, Sec.Gen. **MEMBERS:** 7,058. **STAFF:** 80. **LANGUAGES:** English, French.

DESCRIPTION: Business associations and individual corporations in 130 countries. Represents and offers services to international business. Seeks to evaluate and express the consensus of those businesses involved in trade and international investment; represents members before the United Nations and governmental agencies. Works to secure effective and consistent action in the development and improvement of business conditions worldwide and to harmonize and standardize trading practices. Facilitates contact, cooperation, and exchange among business people and business organizations throughout the world. Conducts research and educational programs on topics such as banking practices, commercial arbitration, and international contracts. Operates ICC Court of Arbitration, ICC Institute of International Business Law and Practice, ICC International Bureau of Chambers of Commerce, ICC Corporate Security Service, ICC International Maritime Bureau, ICC Centre for Maritime Cooperation, ICC Counterfeiting Intelligence Bureau and numerous commissions, standing committees, and working parties. **PUBLICATIONS:** *ICC Handbook*, annual; Report, annual.

INTERNATIONAL CHAMBER OF SHIPPING (ICS) (Chmabre Internationale de la Marine Marchande)

c/o Carthusian Court, 12 Carthusian St., London EC1M 6EB, England **PHONE:** 44 171 4178844 **FX:** 44 171 4178877 **FOUNDED:** 1921. **OFFICERS:** Mr. J.C.S. Horrocks, Sec.Gen. **MEMBERS:** 40. **LANGUAGES:** English.

DESCRIPTION: Members are national associations of shipowners and individual shipping companies. Promotes interests of members worldwide in shipping matters such as documentation, insurance, marine safety, maritime law, navigation, and pollution control. Facilitates exchange of information and ideas. Represents members' concerns before governments and intergovernmental organizations throughout the world in an effort to formulate policies for national and international application. Works with the Oil Companies International Marine Forum to disseminate information on facilitation, pollution, and safety. Maintains consultative status with the International Maritime Organization and other agencies of the United Nations; participates in the work of international organizations with common interests.

INTERNATIONAL CHILD CARE U.S.A. (ICCUSA)

PO Box 14485, Columbus, OH 43214-0485 **TF:** (800) 722-4453 **FX:** (614) 447-1123 **WEBSITES:** http://www.intlchildcare.org **FOUNDED:** 1965. **OFFICERS:** Dr. John Yates, CEO. **BUDGET:** US$499,000. **LANGUAGES:** English.

DESCRIPTION: Nondenominational, church- and privately-funded organization that seeks to: create a better life for children in Haiti through nutritional care, education, and medical aid; help the peoples of developing countries become self-sufficient and assume responsibility for their own programs. Conducts Crusade Against Tuberculosis, a TB prevention and vaccination program for children and young adults up to the age of 20; also provides primary health care for all Haitians. Works in rural clinics and missions to train government-certified TB agents in order to allow native people to diagnose and treat TB and provide preventive health care education such as training in basic sanitation and hygiene. Sponsors program whereby TB drugs are dispensed at cost to Haitian clinics and hospitals. Trains local leaders in the Dominican Republic in community-based health promotion programs. Maintains Grace Children's Hospital in Port-au-Prince, Haiti. Operates speakers' bureau. **PUBLICATIONS:** *Grace*, quarterly; Brochure.

INTERNATIONAL CHILD RESOURCE INSTITUTE (ICRI)

1581 Le Roy Ave., Berkeley, CA 94708-1941 **E-MAIL:** icrichild@aol.com **FOUNDED:** 1981. **OFFICERS:** Ken Jaffe, Exec.Dir. **STAFF:** 30. **BUDGET:** US$3,000,000. **LANGUAGES:** English, Portuguese, Spanish, Swedish.

DESCRIPTION: Individuals interested in issues regarding day care for children, including maternal and child health, child abuse prevention, neglect, and other children's issues. Organizations and companies that furnish or are engaged in child care. Implements model projects to gather information on techniques and practices involved in innovative forms of child care and child health. Provides technical assistance to individuals, corporations, and government agencies that wish to establish and maintain child care centers and other children's programs. Serves as a clearinghouse for information on children's issues. Maintains offices in Brazil, Ethiopia, Kenya, and Malaysia. Conducts speakers' bureau. **PUBLICATIONS:** *ICRI's World Child Report*, periodic. Reviews issues facing families and children worldwide.

INTERNATIONAL CHILD WELFARE ORGANIZATION (ICWO)

Soroti Rd., Lots 22-30, PO Box 124, Kumi, Uganda **FX:** 256 41 254576 **FOUNDED:** 1989. **OFFICERS:** Rev. James Arikosi, Exec.Dir. **MEMBERS:** 186. **STAFF:** 19. **BUDGET:** US$26,346. **LANGUAGES:** English, Kiswahili.

DESCRIPTION: Individuals and organizations working to improve the quality of life of children made homeless and orphaned by political upheaval. Provides food, shelter, health care, and other assistance to displaced children. **PUBLICATIONS:** *ICWO Fact File* (in English), annual; *Voice of the Children* (in English), annual.

INTERNATIONAL CHIROPRACTORS ASSOCIATION (ICA)

1110 N. Glebe Rd., Ste. 1000, Arlington, VA 22201 **PHONE:** (703) 528-5000 **TF:** (800) 423-4690 **FX:** (703) 528-5023 **E-MAIL:** chiro@chiropractic.org **WEBSITES:** http://www.chiropractic.org/ **FOUNDED:** 1926. **OFFICERS:** Ronald Hendrickson, VP. **MEMBERS:** 6,000. **STAFF:** 15. **LANGUAGES:** English.

DESCRIPTION: Professional society of chiropractors, chiropractic educators, students, and laypersons. Sponsors professional development programs and practice management seminars. **PUBLICATIONS:** *Congressional Directory*, annual; *ICA Today*, bimonthly. Covers membership and association activities, includes legislative information and research updates.; *International Chiropractors Association Membership Directory*, annual; *International Review of Chiropractic*, bimonthly. Also publishes materials on patient education.

INTERNATIONAL CHRISTIAN SERVICE FOR PEACE (Internationaler Christlicher Friedensdienst)

Postfach 1322, D-56503 Neuwied, Germany **PHONE:** 49 2631 83790 **FX:** 49 2631 31160 **E-MAIL:** eirene-int@oln.comlink.apc.org **WEBSITES:** http://www.eirene.org **FOUNDED:** 1957. **OFFICERS:** Eckehard Fricke, Gen.Sec. **MEMBERS:** 6. **STAFF:** 15. **BUDGET:** DM 4,500,000. **LANGUAGES:** English, French, German, Spanish.

DESCRIPTION: Peace churches and international peace organizations that provide volunteers for long-term service in both the northern and southern hemispheres in an effort to promote peace and nonviolence worldwide. Facilitates communication and collaboration among groups practicing the equitable distribution and use of material resources, and those seeking to free themselves from cultural, economic, and political isolation and repression. Stimulates the planning and execution of projects in these areas provided they do not necessitate the replacement of native workers, do not require substantial financial investment, and can be continued independently in the future. Seeks to raise public awareness of the close connection between development and peace service. Examines problems of injustice and suggests credible, nonviolent alternatives. Supports the work of conscientious objectors, offers alternatives to military service, and cooperates in peace activities. **PUBLICATIONS:** *Circular Letter* (in German), quarterly; Annual Report (in Dutch, French, and German).

INTERNATIONAL CHRISTIAN YOUTH EXCHANGE (ICYE)

International Office, Grosse Hamburger Str. 30, D-10115 Berlin, Germany **PHONE:** 49 30 28390550 **FX:** 49 30 28390552 **E-MAIL:** icyeio@ipn-b.de **WEBSITES:** http://www.icye.org **FOUNDED:** 1949. **OFFICERS:** Salvatore Romagna, Prog.Off. **MEMBERS:** 950. **STAFF:** 3. **BUDGET:** DM 1,200,000. **LANGUAGES:** English, Spanish.

DESCRIPTION: Operates youth exchange programs in 35 countries for young people between the ages of 16-30. Organizes and arranges for participants to undertake voluntary community service and educational programs. **PUBLICATIONS:** Report, annual.

INTERNATIONAL CIVIL AVIATION ORGANIZATION (ICAO) (Organisation de l'Aviation Civile Internationale)

999 University St., Montreal, PQ, Canada H3C 5H7 **PHONE:** (514) 954-8219 **FX:** (514) 954-6077 **TX:** 05-24513 **E-MAIL:** icaohq@icao.org **WEBSITES:** http://www.icao.int; http://www.cam.org/-icao **FOUNDED:** 1944. **OFFICERS:** Assad Kotaite, Pres. **MEMBERS:** 185. **BUDGET:** US$52,191,000. **LANGUAGES:** Arabic, Chinese, English, French, Russian, Spanish.

DESCRIPTION: Organization of national government representatives. Seeks to develop the standards and procedures in international air navigation and to foster the planning and development of international air transport so as to insure safe and orderly growth of international civil aviation. Carries out activities in air navigation, air transport, and legal matters. Provides advice, assistance, and training. Compiles statistics. **PUBLICATIONS:** *ICAO Journal* (in English, French, and Spanish), 10/year.

INTERNATIONAL CIVIL DEFENCE ORGANIZATION (ICDO) (Organization International de Protection Civile)

10-12, chemin de Surville, PO Box 172, CH-1213 Geneva 2, Switzerland **PHONE:** 41 22 7934433 **FX:** 41 22 7934428 **TX:** 423786 CH **E-MAIL:** icdo@icdo.org **WEBSITES:** http://www.icdo.org **FOUNDED:** 1972. **OFFICERS:** Sadok Znaidi, Sec.Gen. **MEMBERS:** 61. **LANGUAGES:** Arabic, English, French, Spanish.

DESCRIPTION: Members are the governments of 51 countries. Conducts organizational, training, and equipment research in the field of civil defense. Maintains liaison among governmental, professional, and volunteer organizations concerned with the protection and safety of civilian populations. Advocates public education against home, industrial, and natural hazards. Conducts research in accident prevention and response; sponsors training of field personnel and disaster managers. **PUBLICATIONS:** *International Civil Defence Journal* (in Arabic, English, French, Russian, and Spanish), quarterly.

INTERNATIONAL COALITION FOR DEVELOPMENT ACTION (ICDA) (Coalition Internationale pour l'Action au Developpement - CIAD)

115, rue Stevin, B-1000 Brussels, Belgium **PHONE:** 32 2 2300430 **FX:** 32 2 2305237 **E-MAIL:** icda@skynet.be **WEBSITES:** http://www.icda.be **FOUNDED:** 1975. **OFFICERS:** Janice Goodson Foerde, Chair. **MEMBERS:** 500. **STAFF:** 4. **LANGUAGES:** English, French.

DESCRIPTION: Development and environment and workers' organizations in 22 CECO countries that address issues concerning developing countries working closely with southern NGOs and NGO networks. Serves as a network committed to building a just and equitable international order focussing on international trade and related issues by: undertaking campaigns to raise public awareness of development issues and their underlying causes; exchanging ideas and experiences on action models and campaign strategies; creating channels of communication between members to develop an understanding of the problems these groups face; initiating and maintaining active links between members and other groups such as trade unions, women's organizations, and peace and environmental, consumer action groups; keeping members informed about issues and events affecting relations between developed and developing countries. The ICDA serves as a vehicle through which national development groups are able to strengthen the international aspect of their work; it further believes that cooperation increases political pressure and improves the effectiveness of lobbying and action. Coordinates member participation at international negotiations, conferences, and meetings; sponsors policy and lobbying coordination conferences. Concerns include: role of transnational corporations in developing countries; industrial restructuring to develop stronger links with the trade union movement. **PUBLICATIONS:** *ICDA Update* (in English), quarterly. Additional publication information available on request.

INTERNATIONAL COCOA ORGANIZATION (ICCO) (Organisation Internationale du Cacao)

22 Berners St., London W1P 3DB, England **PHONE:** 44 171 6373211 **FX:** 44 171 6310114 **TX:** 28173 ICOCOA G **E-MAIL:** info@icco.org **WEBSITES:** http://www.icco.org **FOUNDED:** 1973. **OFFICERS:** Dr. Edouard Kouame, Exec.Dir. **MEMBERS:** 40. **STAFF:** 22. **BUDGET:** £31,000,000. **LANGUAGES:** English, French, Russian, Spanish.

DESCRIPTION: International organization of cocoa-exporting and importing countries established to implement the International Cocoa Agreement. (The ICCA was first negotiated in 1972 to bring stability to the world cocoa market by preventing excessive fluctuation in the price of cocoa, which adversely affects the long-term interests of producers and consumers.) Acts as a center for the collection and distribution of information on all aspects of the world cocoa economy. Compiles statistics. Is governed by the International Cocoa Council. **PUBLICATIONS:** *Cocoa Newsletter* (in English, French, Russian, and Spanish), semiannual. Features statistical, technical, and scientific articles on cocoa.; *ICCO Annual Report* (in English, French, Russian, and Spanish). Contains information on ICCO activities during the preceding cocoa year.; *Quarterly Bulletin of Cocoa Statistics*; *World Cocoa Directory* (in English, French, and Spanish), annual. Lists over 3000 associations and companies involved in cocoa; *The World Cocoa Market to the Year 2000*.

INTERNATIONAL COLLEGE OF SURGEONS (ICS)

1516 N. Lake Shore Dr., Chicago, IL 60610 **PHONE:** (312) 642-3555 **FX:** (312) 787-1624 **E-MAIL:** info@icsglobal.org **WEBSITES:** http://www.icsglobal.org **FOUNDED:** 1935. **OFFICERS:** Max Downham, Exec.Dir. **MEMBERS:** 14,000. **STAFF:** 12. **BUDGET:** US$2,000,000. **LANGUAGES:** English.

DESCRIPTION: General surgeons and surgical specialists in 110 countries maintaining official relations with the World Health Organization. Promotes the universal teaching and advancement of surgery and its allied sciences. Maintains International Museum of Surgical Sciences containing specialty rooms showing the growth and perfection of many surgical specialties. Maintains library open to researchers, individuals working in the profession, and the public. Organizes postgraduate clinics around the world; conducts lecture series and periodic congresses; offers grants, scholarships, and loans for residencies, research, and advanced study in surgery. Sends surgical teaching teams to developing countries. Bestows honorary fellowship. **PUBLICATIONS:** *International College of Surgeons Newsletter*, quarterly. Covers college news and business; *International Surgery*, quarterly. Presents papers on clinical, experimental, cultural, and historical topics pertinent to surgery and related fields. Contains book reviews.

INTERNATIONAL COMMISSION OF AGRICULTURAL ENGINEERING (CIGR) (Commission Internationale du Genie Rural - CIGR)

Nussalice 5, D-53115 Bonn, Germany **PHONE:** 49 2 28732389 **FX:** 49 2 28739644 **E-MAIL:** cigr@uni-bonn.de **WEBSITES:** http://wwworg.nlh.no/CIGR/ **FOUNDED:** 1930.

OFFICERS: Dr. Peter Schulze Lammers, Sec.Gen. **MEMBERS:** 3,500. **LANGUAGES:** English, French.

DESCRIPTION: Membership consists of: national associations (27) representing agricultural architects, engineers, professors, research workers, technicians, and other agricultural equipment specialists; interested others (6). Purpose is to provide assistance to projects of common interest to members in the field of agricultural engineering. Promotes the art, science, and technique of agricultural engineering in agricultural fields such as district planning and environmental improvement. Fosters and coordinates scientific and technical research on topics such as: soil protection and conservation; irrigation; land improvement and reclamation; use of electricity in rural areas; agricultural machinery and buildings; food processing; management and ergonomics; and engineering and plant production. Advocates and seeks to improve the education and training of agricultural engineers, architects, and technicians. Works to establish a documentation service for the collection, verification, and dissemination of data. Maintains 6 technical sections. Operates educational programs. **PUBLICATIONS:** *CIGR Newsletter* (in English and French), quarterly; *Proceedings of International CIGR Congress of Agricultural Engineering*, quadrennial; *Proceedings of Symposia, Seminars, and Workshops*; *Reports of Working Groups*.

INTERNATIONAL COMMISSION ON CIVIL STATUS (ICCS) (Commission Internationale de l'Etat Civil - CIEC)

3, pl. Arnold, F-67000 Strasbourg, France
PHONE: 33 388 611862 **FX:** 33 388 605879
FOUNDED: 1950. **OFFICERS:** Jacques Massip, Sec.Gen. **MEMBERS:** 13. **STAFF:** 2. **LANGUAGES:** French.

DESCRIPTION: National governmental representatives from 13 countries united to improve and internationally coordinate member countries' municipal and family laws. Acts as forum for international exchange of information on each country's municipal, personal, and family laws; exchanges legal texts and documentation. Coordinates efforts with the Hague Conference on Private International Law, the Council of Europe, the European Community, and the United Nations High Commission for Refugees. **PUBLICATIONS:** *Convention Proceedings*; *Guide Pratique International de l'Etat Civil* (in French), annual; *Proces-Verbaux*, semiannual.

INTERNATIONAL COMMISSION FOR FOOD INDUSTRIES (Commission Internationale des Industries Agricoles et Alimentaires - CIIA)

16, rue Claude Bernard, F-70005 Paris, France
PHONE: 33 01 433136 **FX:** 33 01 43313202
FOUNDED: 1939. **OFFICERS:** Guy Dardenne, Sec.Gen. **LANGUAGES:** English, French, German, Spanish.

DESCRIPTION: Representatives of governments; universities, training establishments, research laboratories, and professional associations. Promotes the agricultural, fishery, and food industries by developing international cooperation among them. The commission is responsible for activities concerning processing industries for agricultural and fishery products. Contributes to developments in areas such as agricultural production, distribution and consumption of products, and production of industrial equipment. Guides research in the field and studies means of adapting such work; advocates the constant application of research results to industrial practice. Seeks to protect public health by ensuring the quality of finished products; protects the rights of inventors and innovators in the food industries. Facilitates interchange among members and industrialized and developing countries. Offers training courses. **PUBLICATIONS:** *Calendar*, semiannual; *Food and Agriculture Industries Revue* (in French), 10/year; *Proceedings of Congresses and Symposia*, annual.

INTERNATIONAL COMMISSION ON GLASS (ICG) (Commission Internationale du Verre - CIV)

c/o Dr. F. Nicoletti, Stazione Sperimentale del Vetro, Via Briati 10, I-30141 Murano, Italy
PHONE: 39 41 739422 **FX:** 39 41 739420
E-MAIL: spevetro@ve.nettuno.it **FOUNDED:** 1932. **OFFICERS:** Dr. F. Nicoletti, Hon.Sec. **MEMBERS:** 27. **LANGUAGES:** English, French, German.

DESCRIPTION: National organizations and institutes interested in the science and technology of glass. Coordinates related activities and disseminates technical information. Maintains 17 technical committees.

INTERNATIONAL COMMISSION ON ILLUMINATION (CIE) (Commission Internationale de l'Eclairage - CIE)

Kegelgasse 27, A-1030 Vienna, Austria **PHONE:** 43 1 71431870 **FX:** 43 1 713083818 **E-MAIL:** ciecb@ping.at **WEBSITES:** http://www.cie.co.at/cie/home.html **FOUNDED:** 1913. **OFFICERS:** Mag. Christine Hermann, Gen.Sec. **MEMBERS:** 41. **STAFF:** 3. **LANGUAGES:** English, French, German.

DESCRIPTION: National committees united for international cooperation and exchange of information on all matters relating to the art, science, and technology of light and lighting. Objectives are to: provide an international forum for all matters relating to lighting; promote the study of lighting; maintain liaison and technical interaction with other international organizations concerned with light and lighting matters. Conducts studies and educational programs. **PUBLICATIONS:** *CIE-News* (in English), quarterly; *CIE-Roster*, annual; *Proceedings of Sessions and Symposia*; *Technical Reports*, periodic.

INTERNATIONAL COMMISSION OF JURISTS (ICJ) (Commission Internationale de Juristes - CIJ)

PO Box 216, 81 a, ave. de Chatelaine, CH-1219 Geneva, Switzerland **PHONE:** 41 22 9793800 **FX:** 41 22 9793801 **TX:** 418531 ICJCH **E-MAIL:** icjch@gn.apc.org **FOUNDED:** 1952. **OFFICERS:** Adama Dieng, Sec.Gen. **MEMBERS:** 45. **STAFF:** 17. **LANGUAGES:** English, French, Spanish.

DESCRIPTION: Jurists representing the main legal systems of the world; interested others. Works to promote and protect human rights through the rule of law worldwide. Conducts studies of factual and legal aspects of situations that impinge on human rights. Recent studies have included South Africa, Sri Lanka, the West Bank, Indonesia, Peru, Uruguay, Iraq, Nigeria, and the United States of America. **PUBLICATIONS:** *ICJ Newsletter* (in English), quarterly; *Review* (in English, French, and Spanish), semiannual; Reports; Reports.

INTERNATIONAL COMMISSION ON LARGE DAMS (ICOLD) (Commission Internationale des Grands Barrages - CIGB)

151, blvd. Haussmann, F-75008 Paris, France **PHONE:** 33 1 40426824, 33 1 40427733 **FX:** 33 1 40426071 **E-MAIL:** secretaire.general@icold-cigb.org **WEBSITES:** http://genepi.louis-jean.com/cigb/index.html **FOUNDED:** 1928. **OFFICERS:** Jacques Lecornu, Sec.Gen. **MEMBERS:** 81. **STAFF:** 4. **LANGUAGES:** English, French.

DESCRIPTION: Organization of national committees on large dams. Objectives are to promote improvement in the design, construction, maintenance, and operation of large dams and to act as a clearinghouse. Conducts study tours of large dams and organizes studies and experiments. **PUBLICATIONS:** *Abstracts of ICOLD Publications*, periodic; *Bulletin*, periodic; *Congress Proceedings and Transactions*, triennial; *ICOLD Directory*, triennial.

INTERNATIONAL COMMISSION ON MICROBIOLOGICAL SPECIFICATIONS FOR FOODS (ICMSF) (Committee Internationale pour la Definition des Caracteristiques Microbiologiques des Aliments - CIDCMA)

c/o Dr. M. van Schothorst, Ave. Nestle 55, CH-1800 Vevey, Switzerland **PHONE:** 41 21 9244241 **FX:** 41 21 9211885 **E-MAIL:** michiel.van-schothorst@ch02.nestle.com **WEBSITES:** http://www.dfst.csiro.au/icmsf.htm **FOUNDED:** 1962. **OFFICERS:** Dr. M. van Schothorst. **MEMBERS:** 17. **LANGUAGES:** English.

DESCRIPTION: Food microbiologists employed by governments, industries, universities, research institutes, and hospitals. Studies, recommends, and provides information on the microbiology of foods and food safety. Seeks to: establish comparable standards of microbiological judgment among countries; foster safe movement of foods in international commerce; overcome difficulties caused by differing microbiological standards. Maintains Latin American, Balkan-Danubian, and Middle East-North African subcommissions to handle problems peculiar to these regions. **PUBLICATIONS:** *International Commission on Microbiological Specifications for Foods—Proceedings of the General Conference*, annual. Includes annual report and membership list. Also publishes books, research papers, and documents on microbiological methods, sampling, microbial ecology of foods, and food safety control methods.

INTERNATIONAL COMMISSION FOR OPTICS (ICO) (Commission Internationale d'Optique - CIO)

Institut d'Optique - CNRS, Universite de Paris Sud, Bte. Postale 147, F-91403 Orsay Cedex, France **PHONE:** 33 1 69358741 **FX:** 33 1 69358700 **E-MAIL:** pierre.chavel@iota.u-psud.fr **WEBSITES:** http://www.ico-optics.org **FOUNDED:** 1947. **OFFICERS:** Dr. P. Chavel, Sec.Gen. **MEMBERS:** 45. **BUDGET:** US$25,500. **LANGUAGES:** English, French.

DESCRIPTION: Organization of national committees interested in the field of optics. Promotes progress in theoretical, applied, and physiological optics; facilitates the rapid exchange of knowledge, works to standardize optics nomenclature, measurement units, and symbols. Compiles information on optics, including holography, optical computing, lens design, and spectroscopy. Encourages shipment of optics publications to Third World optics groups; participates in optics education courses in Third World countries; operates summer school program. **PUBLICATIONS:** *ICO General Meeting Proceedings* (in English), triennial; *ICO Newsletter* (in English), quarterly; *ICO Report*, triennial.

INTERNATIONAL COMMISSION ON PHYSICS EDUCATION (ICPE)

c/o Prof. Paul J. Black, School of Education, Kings Coll. London, Cornwell House, Waterloo Rd., London SE1 8WA, England **PHONE:** 44 171 8723166, 44 171 8723189 **FX:** 44 171 8723182 **E-MAIL:** paul.black@kcl.ac.uk **WEBSITES:** http://www.physics.umd.edu/ripe/icpe; http://www.physics.ohio-state.edu/-jossem/icpe/books.html **FOUNDED:** 1960. **OFFICERS:** Prof. Paul J. Black, Chm. **MEMBERS:** 13. **LANGUAGES:** English, French.

DESCRIPTION: Commission of the International Union of Pure and Applied Physics. Aim is to improve the teaching of physics at all levels of education throughout the world mainly by coordination and distribution of information and by sponsorship of meetings. **PUBLICATIONS:** *Connecting Research on Physics Education with Teacher Education*; *ICPE Newsletter*, semiannual; *Physics Examination for University Entrance an International Study*; Reports; Surveys.

INTERNATIONAL COMMISSION FOR THE PRESERVATION OF ISLAMIC CULTURAL HERITAGE (ICPICH) (Commission Internationale pour la Sauvegarde du Patrimoine Culturel Islamique)

Barbaros Bulvari, Yildiz Sarayi, Seyir Kosku, Besiktas, TR-80700 Istanbul, Turkey **PHONE:** 90 212 2591742 **FX:** 90 212 2584365 **TX:** 26484 ISAM TR **E-MAIL:** ircica@ihlasnet.tr **FOUNDED:** 1983. **OFFICERS:** Prof. Ekmeleddin Ihsanoglu, Sec. **MEMBERS:** 17. **STAFF:** 10. **BUDGET:** US$650,000. **LANGUAGES:** Arabic, English, French.

DESCRIPTION: Subsidiary of the Organization of the Islamic Conference. Representatives of Islamic states; scholars and groups of experts. Works for the conservation, restoration, and reconstruction of monuments representing Islamic cultural heritage. Sponsors calligraphy competition; conducts research and educational programs; compiles statistics; maintains collections of Islamic art. **PUBLICATIONS:** Newsletter (in Arabic, English, and French), 3/year.

INTERNATIONAL COMMISSION FOR THE PREVENTION OF ALCOHOLISM AND DRUG DEPENDENCY (ICPA)

12501 Old Columbia Pike, Silver Spring, MD 20904 **PHONE:** (301) 680-6719 **FX:** (301) 680-6707 **E-MAIL:** 74617.2242@compuserve.com **WEBSITES:** http://www.icpa-dd.org **FOUNDED:** 1952. **OFFICERS:** Thomas R. Neslund, Exec.Dir. **LANGUAGES:** English.

DESCRIPTION: Representatives of national public health committees and other individuals interested in the physical and social effects of alcoholism and drug dependency. Fosters the scientific study of alcohol and drugs, their effects on the physical, mental, and moral powers of the individual, and their effects on social, economic, political, and religious life. Encourages preventive education; disseminates information on drug and alcohol abuse. Serves as a liaison with similar groups around the world. Sponsors exchange and research programs. Conducts film shows, forums, and radio and television events. **PUBLICATIONS:** *ICPA Alert*, periodic; *ICPA Dispatch*, periodic; *ICPA Reporter*, quarterly.

INTERNATIONAL COMMISSION FOR THE PROTECTION OF ALPINE REGIONS (ICPAR)

Im Bretscha 22, FL-9494 Schaan, Liechtenstein **PHONE:** 423 75 2374030 **FX:** 423 75 2374031 **E-MAIL:** cipra@cipra.org **FOUNDED:** 1952. **OFFICERS:** Dr. Andreas Goetz. **MEMBERS:** 7. **STAFF:** 4. **BUDGET:** US$200,000. **LANGUAGES:** French, German, Italian, Slovene.

DESCRIPTION: Representatives from Austria, France, Germany, Italy, Liechtenstein, Switerland, and Slovenia. Promotes conservation and preservation of alpine regions in Europe. Consults with interested parties involved environmental questions impacting alpine habitats; makes available advise regarding the preservation of alpine flora and fauna.

INTERNATIONAL COMMISSION FOR THE PROTECTION OF THE RHINE AGAINST POLLUTION (ICPR) (Internationale Kommission zum Schutze des Rheins Gegen Verunreinigung - IKSR)

Hohenzollernstrasse 18, Postfach 3 09, D-56003 Koblenz, Germany **PHONE:** 49 261 12495 **FX:** 49 261 36572 **E-MAIL:** iksr@rz-online.de **FOUNDED:** 1963. **OFFICERS:** Koos Wieriks, Gen. Sec. **MEMBERS:** 6. **STAFF:** 9. **BUDGET:** DM 1,700,000. **LANGUAGES:** English, French, German.

DESCRIPTION: Governmental representatives of France, Germany, Luxembourg, Netherlands, and Switzerland; a representative of the European Economic Community. Objectives are: to determine the origin, nature, and extent of pollution of the river Rhine; to devise viable measures for protection against pollution of the Rhine; to prepare antipollution agreements and treaties between members. Sponsors Action Program Rhine which aims to: improve the ecosystem of the Rhine; guarantee the production of drinking water in future years; reduce the sediment toxicity; and improve the ecosystem of the North Sea. Maintains a monitoring program. Development of programmes or the protection against flooding. **PUBLICATIONS:** *Activity Report* (in French and German), annual; *Statistical Report on Physico-Chemical Analysis* (in French and German), annual; *Topic Rhine* (in English, French, and German), 3/year.

INTERNATIONAL COMMISSION ON RADIATION UNITS AND MEASUREMENTS (ICRU)

7910 Woodmont Ave., Ste. 800, Bethesda, MD 20814 **PHONE:** (301) 657-2652 **TF:** (800) 229-2652 **FX:** (301) 907-8768 **E-MAIL:** ncp@ncp.com **WEBSITES:** http://www.ncp.com **FOUNDED:** 1925. **OFFICERS:** William M. Beckner, Exec.Dir. **MEMBERS:** 101. **STAFF:** 3. **LANGUAGES:** English.

DESCRIPTION: Commission members, senior advisors, consultants, and representatives of report committees in 12 countries. Develops internationally acceptable recommendations regarding: quantities and units of radiation and radionuclides; procedures suitable for the measurement and application of these quantities in clinical radiology and radiobiology; physical data needed in the application of these procedures, the use of which tends to assure uniformity in reporting. The commission has divided its field of interest into 15 technical areas. **PUBLICATIONS:** *ICRU News*, periodic. Contains scientific papers and information of interest; *ICRU Reports*, periodic.

INTERNATIONAL COMMISSION ON RADIOLOGICAL PROTECTION (ICRP) (Commission Internationale de Protection Contre les Radiations)

S-171 16 Stockholm, Sweden **PHONE:** 46 8 7297275 **FX:** 46 8 7297298 **E-MAIL:** jack.valentin@ssi.se **WEBSITES:** http://www.icrp.org **FOUNDED:** 1928. **OFFICERS:** Jack Valentin, Scientific Sec. **MEMBERS:** 70. **STAFF:** 1. **BUDGET:** US$300,000. **LANGUAGES:** English.

DESCRIPTION: Qualified individuals from 20 countries organized to consider the fundamental principles upon which appropriate radiation protection measures can be based, while leaving to national protection bodies the responsibility of formulating the specific advice and codes of practice or regulations that are best suited to the needs of individual countries. **PUBLICATIONS:** *Annals of ICRP* (in English), periodic; *EML* (in English), periodic; *ICRP Database of Dose Coefficients.*

INTERNATIONAL COMMISSION OF SUGAR TECHNOLOGY (CITS) (Commission Internationale Technique de Sucrerie - CITS)

Masktbreiter Str. 74, D-97155 Ochsenfurt, Germany **PHONE:** 49 9331 91450 **FX:** 49 9331 91462 **FOUNDED:** 1948. **OFFICERS:** Dr. Guiseppe Vaccari, VP. **MEMBERS:** 45. **LANGUAGES:** English, French, German.

DESCRIPTION: Individuals affiliated with sugar companies or sugar institutes. Promotes technical research within the sugar industry. Facilitates contact and cooperation among members and research and technical workers in the industry. **PUBLICATIONS:** *Proceedings of General Assembly*, quadrennial.

INTERNATIONAL COMMISSION ON ZOOLOGICAL NOMENCLATURE (ICZN) (Commission Internationale de Nomenclature Zoologique - CINZ)

Natural History Museum, Cromwell Rd., London SW7 5BD, England **PHONE:** 44 171 9389387 **E-MAIL:** iczn@nhm.ac.uk **WEBSITES:** http://www.iczn.org **FOUNDED:** 1895. **OFFICERS:** Dr. P.K. Tubbs, Exec.Sec. **MEMBERS:** 26. **STAFF:** 3. **BUDGET:** US$108,000. **LANGUAGES:** English.

DESCRIPTION: Zoologists, paleozoologists, scientists, and others interested in zoological nomenclature in 19 countries. Serves as a bureau to assist zoologists worldwide with problems in zoological nomenclature. **PUBLICATIONS:** *Bulletin of Zoological Nomenclature* (in English), quarterly; *International Code of Zoological Nomenclature*; *Official Lists and Indexes of Names and Works in Zoology*; *Towards Stability in the Names of Animals*.

INTERNATIONAL COMMITTEE ON AERONAUTICAL FATIGUE (ICAF) (Comite International de la Fatigue du Structures Aeronautique)

c/o Prof. O. Buxbaum, Fraunhofer Institut fuer Betriebsfestigkeit LBF, Bartningstrasse 47, D-64289 Darmstadt, Germany **PHONE:** 49 6151 7051 **FX:** 49 6151 705214 **FOUNDED:** 1951. **OFFICERS:** Prof. O. Buxbaum, Gen.Sec. **MEMBERS:** 13. **LANGUAGES:** English.

DESCRIPTION: Fosters exchange of research reports and information concerning aeronautical fatigue. (Fatigue is the process whereby metals fail when subjected to cyclical stress.) **PUBLICATIONS:** *Minutes of Conference*, biennial; *Proceedings of Symposia*, biennial.

INTERNATIONAL COMMITTEE FOR ANIMAL RECORDING (ICAR)

Via A. Torlonia 15 A, I-00161 Rome, Italy **PHONE:** 39 6 44238013 **FX:** 39 6 44241466 **E-MAIL:** zoorec@rmnet.it **FOUNDED:** 1951. **MEMBERS:** 39. **STAFF:** 3. **LANGUAGES:** English, French, German.

DESCRIPTION: Animal breeder associations, organizations involved in recording information on milk and meat production, boards of agriculture, and herd book associations in 36 countries. Works to promote and improve methods of recording and testing the productivity of farm animals. **PUBLICATIONS:** *Annual Statistics*; *Conference Proceedings*, biennial; *Interbull Bulletin*, annual.

INTERNATIONAL COMMITTEE OF CATHOLIC NURSES AND MEDIO-SOCIAL ASSISTANTS (ICCN) (Comite International Catholique des Infirmieres et Assistantes Medico Sociales - CICIAMS)

43 Sq. Vergote, B-1030 Brussels, Belgium **PHONE:** 32 2 7321050 **FX:** 32 2 7348460 **FOUNDED:** 1928. **OFFICERS:** An Verlinde, Gen.Sec. **MEMBERS:** 79. **STAFF:** 1. **LANGUAGES:** English, French, German.

DESCRIPTION: Professional Catholic nursing associations, Catholic nursing and medico-social work schools, and other Catholic groups representing the nursing profession in 57 countries. Works to encourage the development of members and ensure their technical ability in accordance with Christian moral principles. Promotes development of the nursing profession in general; fosters health and social welfare measures consistent with Christian principles and scientific progress while respecting individual religious convictions. Provides assistance to nursing schools and associations in developing countries; facilitates exchange of statistics between hospital establishments and medico-social organizations. **PUBLICATIONS:** *CICIAMS News* (in English, French, and German), quarterly.

INTERNATIONAL COMMITTEE OF FOUNDRY TECHNICAL ASSOCIATIONS (Comite International des Associations Techniques de Fonderie - CIATF)

Konradstr. 9, Postfach 7190, CH-8023 Zurich, Switzerland **PHONE:** 41 1 2719090 **FX:** 41 1 2719292 **E-MAIL:** gerster@jgp.ch **WEBSITES:** http://www.jgp.ch **FOUNDED:** 1927. **OFFICERS:** Dr. J. Gerster, Gen.Sec. **MEMBERS:** 36. **LANGUAGES:** English, French, German.

DESCRIPTION: National foundry technical associations. Represents the interests of the foundry and related industries.

INTERNATIONAL COMMITTEE FOR HISTORICAL SCIENCES (CISH)

44, rue de l'Aminal Mouchez, F-75014 Paris, France **PHONE:** 33 1 45809046 **FX:** 33 1 45654350 **FOUNDED:** 1926. **OFFICERS:** Francois Bedarida, Sec.Gen. **MEMBERS:** 93. **LANGUAGES:** English, French.

DESCRIPTION: National committees united to promote the historical sciences. Conducts historical research. Cooperates with international groups and commissions. **PUBLICATIONS:** *Bulletin d'information du CISH*, annual.

INTERNATIONAL COMMITTEE FOR HUMAN RIGHTS IN TAIWAN (ICHRT)

PO Box 15182, Chevy Chase, MD 20825 **PHONE:** (301) 468-5932 **E-MAIL:** taiwandc@globescope.com **WEBSITES:** http://www.taiwandc.org **FOUNDED:** 1976. **OFFICERS:** Dr. G. van der Wees, Chm. **MEMBERS:** 2,080. **BUDGET:** US$20,000.

DESCRIPTION: Individuals (2000) and organizations (80) in 25 countries interested in human rights and democracy in Taiwan. Collects and disseminates information; campaigns for the release of political prisoners. Supports establishment of a free and democratic political system in Taiwan and membership the United Nations. Organizes lectures on political developments. **PUBLICATIONS:** *Taiwan Communique*, bimonthly.

INTERNATIONAL COMMITTEE OF MILITARY MEDICINE (ICMM) (Comite International de Medecine Militaire - CIMM)

Quartier Saint-Laurent, 79 rue Saint-Laurent, B-4000 Liege, Belgium **PHONE:** 32 4 2222183 **FX:** 32 4 2222150 **FOUNDED:** 1921. **OFFICERS:** Colonel Dr. J. Sanabria, Sec.Gen. **MEMBERS:** 89. **STAFF:** 2. **LANGUAGES:** English, French.

DESCRIPTION: Medical representatives of the military forces of nations worldwide. Works to encourage and maintain professional cooperation among those in various countries whose mission is to care for the sick and wounded of the armed forces. Aims to improve such care by standardization of related disciplines and techniques, complete documentation and dissemination of military-medical information, and development of an international medical law. Conducts courses. **PUBLICATIONS:** *International Review of the Armed Forces Medical Services* (in English and French), quarterly.

INTERNATIONAL COMMITTEE OF THE RED CROSS (ICRC) (Comite International de la Croix-Rouge - CICR)

19, ave. de la Paix, CH-1202 Geneva, Switzerland **PHONE:** 41 22 7346001 **FX:** 41 22 7332057 **TX:** 414226 **E-MAIL:** pres.gva.@gwn.icrc.org **WEBSITES:** http://www.icrc.org **FOUNDED:** 1863. **OFFICERS:** Cornelio Sommaruga, Pres. **MEMBERS:** 20. **STAFF:** 8,000. **BUDGET:** 800,000,000 SFr.

DESCRIPTION: Founding institution of the Red Cross and Red Crescent Movement. Serves as a neutral institution that operates mainly during times of war, civil war, or internal strife, maintains the principles of the Red Cross, and ensures application of the Geneva Conventions and protocols. Aids in the protection and care of all wounded and is concerned with their spiritual well-being, dignity, and family rights. Seeks to ban torture and cruel treatment, summary executions and mass killings, deportations, taking of hostages, and pillage and the merciless destruction of civilian property. Participates in relief programs in aid of refugees and displaced people who are the victims of armed conflicts. Works to ensure that political prisoners' detention conditions conform to internationally accepted standards; delegates are allowed to that political prisoners' detention conditions conform to the internationally accepted standards; delegates are allowed to visit prisoners of war and civilian internees and to speak with detainees without witnesses. Provides a radio communications network service comprising 53 stations directly linked to Geneva, Switzerland. Maintains medical division, which assesses medical needs in crisis areas, and provides hospital services during hostilities. Operates relief division whose duties include assessing the needs in the field and possibilities of local purchases and storage, collecting cash and basic goods for aid programs, and supervising distribution of relief supplies. **PUBLICATIONS:** *Annual Report* (in Arabic, English, French, German, and Spanish); *The Geneva Conventions*; *ICRC Bulletin*, monthly; *International Review of the Red Cross* (in Arabic, English, French, German, and Spanish), bimonthly; *The Protocols Additional*. Also publishes *The Geneva Conventions*, *The Protocols Additional*, and other publications on humanitartian law.

INTERNATIONAL COMMITTEE FOR SOCIAL SCIENCE INFORMATION AND DOCUMENTATION (ICSSD)

Barnsteen Straat ig, Alphen A/D, NL-2403 BW Ryn, Netherlands **PHONE:** 31 172 442848 **FX:** 31 172 442848 **FOUNDED:** 1950. **OFFICERS:** Dr. A.F. Marks, Sec.-Gen. **MEMBERS:** 30. **STAFF:** 1. **BUDGET:** US$90,000. **LANGUAGES:** Dutch, English.

DESCRIPTION: Gathers, preserves, and arranges information in the social sciences. Compiles social science bibliographies; promotes research and study in the social sciences. Conducts educational programs. **PUBLICATIONS:** Newsletter.

INTERNATIONAL COMMUNITY CORRECTIONS ASSOCIATION (ICCA)

PO Box 1987, La Crosse, WI 54602 **PHONE:** (608) 785-0200 **FX:** (608) 784-5335 **E-MAIL:** icca@execpe.com **WEBSITES:** http://www.iccaweb.org **FOUNDED:** 1964. **OFFICERS:** Peter Kinziger, Exec.Dir. **MEMBERS:** 1,300. **STAFF:** 3. **BUDGET:** US$450,000. **LANGUAGES:** English.

DESCRIPTION: Agencies and individuals from North America and other countries working in community based correctional programs. Purposes are to assist members in functioning more effectively through the exchange of information regarding management and treatment; promote the development of community-based correctional programs; develop and implement a program of public information and education in the field of community based treatment; assist social institutions within communities to accept the responsibility for coping with crime, substance abuse, mental health, delinquency, and related social problems. Sponsors programs in the areas of corrections. Conducts research programs and Margaret Mead Lecture Series. **PUBLICATIONS:** *ICCA Journal*, quarterly. Provides the latest information on activities, articles on community corrections, and book reviews; *ICCA Membership Directory*, every 3-4 yrs. Listing of current programs in the U.S. and Canada; Magazines.

INTERNATIONAL COMMUNITY FOR THE RELIEF OF STARVATION AND SUFFERING (ICROSS)

PO Box 15619, Nyoonyorrie Clinic, Mbagathi, Kenya **PHONE:** 2542 560494 **FX:** 2542 566811 **FOUNDED:** 1981. **OFFICERS:** Dr. Michael Elmore Meegan, Intl.Dir. **STAFF:** 68. **BUDGET:** 700,000 KSh. **LANGUAGES:** French, Swahili.

DESCRIPTION: Promotes the eradication of hunger among rural nomads in the drought stricken semiarid areas of Africa. Serves individuals suffering from dehydration, diarrhea, chronic infection, disease, and other forms of pain and illness. Offers family planning, immunization, and AIDS education services. Conducts research and training on epidemics.

INTERNATIONAL CONFEDERATION OF AGRICULTURAL CREDIT (Confederation Internationale du Credit Agricole - CICA)

Birmensdorferstr. 67, CH-8004 Zurich, Switzerland **PHONE:** 41 1 2910575 **FX:** 41 1 2910766 **TX:** 817675 CICA CH **FOUNDED:** 1950. **OFFICERS:** Yves Barsalou, Pres. **MEMBERS:** 250. **STAFF:** 2. **LANGUAGES:** English, French, German, Italian, Spanish.

DESCRIPTION: Agricultural credit banks and organizations involved in the development or study of agricultural credit. Acts as a liaison and clearinghouse for documentation and information concerning agricultural credit. Represents agricultural credit institutions and organizations on an international level. Maintains contacts with governments and international organizations involved with agricultural credit. Undertakes pertinent studies and publications of documents. **PUBLICATIONS:** *Bulletin* (in English and French), 2-3/year; *CICA-Information* (in English, French, German, and Italian), monthly.

INTERNATIONAL CONFEDERATION OF ART DEALERS (Confederation Internationale des Negociants en Oeuvres d'Art - CINOA)

32, rue Ernest Allard, B-1000 Brussels, Belgium **PHONE:** 32 2 5134831, 32 2 5020686 **E-MAIL:** rudolf.otto@usa.net **WEBSITES:** http://www.cinoa.org **FOUNDED:** 1935. **OFFICERS:** Rudolf Otto, Pres. **MEMBERS:** 29. **STAFF:** 1. **LANGUAGES:** English, French, German.

DESCRIPTION: National trade associations and federations in 24 countries representing 5,000 individuals. Coordinates the art works of chambers, unions, associations, and federations of dealers. Contributes, by legal means, to artistic and economic expansion. Organizes exhibitions. Maintains inquiry and research bureau. **PUBLICATIONS:** *List of Members* (in English and French), periodic; Directory, periodic.

INTERNATIONAL CONFEDERATION OF EUROPEAN SUGAR-BEET GROWERS (CIBE) (Confederation Internationale des Betteraviers Europeens - CIBE)

29, rue du General Foy, F-75008 Paris, France **PHONE:** 33 1 44693900 **FX:** 33 1 42932893 **FOUNDED:** 1925. **OFFICERS:** Hubert Chavanes, Sec.Gen. **MEMBERS:** 24. **STAFF:** 7. **LANGUAGES:** English, French, German, Italian, Spanish.

DESCRIPTION: National and regional beet growers' organizations and federations of beet processing cooperatives representing 18 countries. Defends the rights of beet growers on an international level, particularly within the European Economic Community. Facilitates the exchange of economic and technical information concerning sugar beet farming and the sugar industry. Compiles statistics. **PUBLICATIONS:** *Congress Report* (in English, French, German, Italian, and Spanish), biennial.

INTERNATIONAL CONFEDERATION OF FREE TRADE UNIONS (ICFTU) (Confederation Internationale des Syndicats Libres)

Boulevard Emile Jacqmain, 155, B-1210 Brussels, Belgium **PHONE:** 32 2 2240211 **FX:** 32 2 2015815 **E-MAIL:** internetpo@icftu.org **WEBSITES:** http://www.icftu.org **FOUNDED:** 1949. **OFFICERS:** Bill Jordan, Gen.Sec. **MEMBERS:** 174. **STAFF:** 70. **LANGUAGES:** English, French, German, Spanish.

DESCRIPTION: Trade unions in 143 countries and territories representing 124,000,000 individuals. Promotes the interests of workers worldwide; works to improve standards of living, full employment, and social security. Seeks to reduce the disparity between wealth and poverty within nations and on the international level. Assists in the organization of workers worldwide to ensure recognition of individual organizations as free bargaining agents. Offers educational courses and vocational training. Maintains International Solidarity Fund to provide assistance to workers who are victims of oppression and to help in building-up trade unions; supports trade union-sponsored social and economic development projects such as cooperatives, vocational training schemes, and health services. Maintains consultative status with the United Nations. **PUBLICATIONS:** *Annual Survey of Violations of Trade Union Rights* (in English, French, and Spanish); *Conference Reports* (in English, French, German, and Spanish); *Congress Reports* (in English); *Features*, periodic; *Trade Union World* (in English, French, and Spanish), monthly; Monographs (in English, French, German, and Spanish); Pamphlets (in English, French, German, and Spanish); Videos.

INTERNATIONAL CONFEDERATION OF POPULAR BANKS (CIBP) (Confederation Internationale des Banques Populaires - CIBP)

Le Ponant, 5, rue Leblanc, F-75511 Paris Cedex 15, France **PHONE:** 33 1 40396619 **FX:** 33 1 40396060 **TX:** 270483 CSBPCP **E-MAIL:** cicp@cyber.banquepopulaire.fr **FOUNDED:** 1950. **OFFICERS:** Pierre Klein, Sec.Gen. **MEMBERS:** 13. **STAFF:** 6. **LANGUAGES:** English, French, German, Italian, Spanish.

DESCRIPTION: Acts as a liaison between financial and banking institutions to promote growth of small and medium-sized companies. Member groups provide easier access to a wide range of loans through extensive networks and fundamental knowledge of local and regional markets. **PUBLICATIONS:** *Congress Proceedings* (in English, French, German, Italian, and Spanish), triennial; Papers, periodic. Contains studies and statements of working groups and commissions.

INTERNATIONAL CONFERENCE OF AGRICULTURAL CHAMBERS (Verband der Landwirtschaftskammern)

Godesberger Allee 142-148, D-53175 Bonn, Germany **PHONE:** 49 228 308010 **FX:** 49 228 374431 **FOUNDED:** 1979. **OFFICERS:** Dr. Ortwin Wagner **MEMBERS:** 29.

DESCRIPTION: Exchanges information on modern methods in agriculture.

INTERNATIONAL CONFERENCE ON LARGE HIGH VOLTAGE ELECTRIC SYSTEMS (Conference Internationale des Grands Reseaux Electriques a Haute Tension - CIGRE)

21 Rue D'Artois, F-75008 Paris, France **PHONE:** 33 1 53891290, 33 1 53891292 **FX:** 33 1 53891299 **E-MAIL:** secretary-general@cigre.org **WEBSITES:** http://www.cigre.org **FOUNDED:** 1921. **OFFICERS:** Marc Herouard, Sec.Gen. **MEMBERS:** 7,500. **STAFF:** 8. **BUDGET:** 10 Fr. **LANGUAGES:** English, French.

DESCRIPTION: Individuals and institutions active in the area of high voltage power systems. Facilitates and promotes the exchange of technical information among countries in the field of electricity generation and transmission at high voltages. Areas of concern include: technical aspects of electricity generation; construction and operation of substations and transformer stations and their associated equipment; construction, insulation, and operation of high voltage electrical lines; interconnection of systems and the operation and protection of interconnected systems. Maintains 15 study committees. **PUBLICATIONS:** *Electra* (in English and French), bimonthly, always February, April, June, August, October, and December; *Proceedings of the Session*, biennial; *Symposia Papers*, periodic.

INTERNATIONAL CONGRESS OF AFRICAN STUDIES

c/o School of Oriental and African Studies, Thornbaugh St., London WC1H 0XG, England **PHONE:** 44 171 3236035 **FX:** 44 171 3236118 **LANGUAGES:** English.

DESCRIPTION: Institutions with African studies programs; educators and students in the field. Promotes advancement in the teaching of African studies. Facilitates communication and cooperation among members.

INTERNATIONAL COOPERATION FOR DEVELOPMENT AND SOLIDARITY (CIDSE) (Cooperation Internationale pour le Developpement et la Solidarite)

16, rue Stevin, B-1000 Brussels, Belgium **PHONE:** 32 2 2307722 **FX:** 32 2 2307082 **E-MAIL:** postmaster@cidse.be **WEBSITES:** http://www.cidse.be **FOUNDED:** 1966. **OFFICERS:** Jef Felix, Sec.Gen. **MEMBERS:** 16. **STAFF:** 8. **BUDGET:** 35,000,000 BFr. **LANGUAGES:** English, French, German, Spanish.

DESCRIPTION: 16 Catholic development organizations in Europe, North America and New Zealand. Aims to promote common strategy on development projects in developing countries. **PUBLICATIONS:** *CIDSE Advocacy Newsletter* (in English), quarterly; *CIDSE Progress Report* (in English and French), annual.

INTERNATIONAL CO-OPERATIVE ALLIANCE (ICA) (Alliance Cooperative Internationale - ACI)

15, rte. des Morillons, Grand-Soconnex, CH-1218 Geneva, Switzerland **PHONE:** 41 22 9298888 **FX:** 41 22 7984122 **TX:** 415620 ICA CH **E-MAIL:** ica@coop.org **WEBSITES:** http://www.coop.org **FOUNDED:** 1895. **OFFICERS:** Bruce Thordarson, Dir. General. **MEMBERS:** 224. **LANGUAGES:** English, French, German, Russian, Spanish.

DESCRIPTION: Cooperative organizations representing over 750 million individuals in 93 countries. Advocates cooperative principles and methods. Works to: protect the economic interests of cooperatives; strengthen the cooperative movement worldwide; assist cooperatives in influencing local policy and legislation; mobilize financial resources for the development of cooperatives. Encourages the development of cooperatives in Third World countries. Conducts research on cooperatives. Sponsors educational seminars. **PUBLICATIONS:** *Annual Report*; *ICA News*, quarterly; *Membership Directory*, annual; *Review of International Cooperation*, quarterly. Also publishes pamphlets and directories of educational establishments, research institutes, and other cooperative groups.

INTERNATIONAL COPYRIGHT SOCIETY (Internationale Gesellschaft fur Urheberrecht - INTERGU)

Rosenheimer Strasse 11, D-81667 Munich, Germany **PHONE:** 49 89 4800300 **FX:** 49 89 48003408 **TX:** 522306 **FOUNDED:** 1954. **OFFICERS:** Prof. Dr. Reinhold Kreile, Pres. **MEMBERS:** 397. **LANGUAGES:** English, French, German, Italian, Portuguese, Russian, Spanish.

DESCRIPTION: Scholars, lawyers, and other persons in 48 countries active in the legal protection of intellectual creation. Engages in scientific inquiry concerning the natural rights of authors and seeks to apply this knowledge in a practical manner worldwide, particularly in the area of legislation. Objective is to lay the foundations for modern copyright while serving the interests of the public. Cooperates with the Commission of the European Communities, the Council of Europe, the United Nations Economic and Social Council, the United Nations Educational, Scientific and Cultural Organization, and the World Intellectual Property Organization. **PUBLICATIONS:** *Author's Right or Copyright*; *Book Series* (in English, French, German, Russian, and Spanish), periodic; *Gemeinschaftsrecht und Einzelstaatliches Recht bei der Schaffungeiner Europaischen wirtschaftlichen Interessenvereinigung* (in German); *Publication Series*, periodic; *Yearbook*, periodic.

INTERNATIONAL COTTON ADVISORY COMMITTEE (ICAC)

1629 K St., Ste. 702, Washington, DC 20006 **PHONE:** (202) 463-6660 **FX:** (202) 463-6950 **TX:** 40827289 **E-MAIL:** secretariat@icac.org **WEBSITES:** http://www.icac.org **FOUNDED:** 1939. **OFFICERS:** L. H. Shaw, Exec.Dir. **MEMBERS:** 42. **STAFF:** 10. **BUDGET:** US$1,200,000. **LANGUAGES:** Arabic, English, French, Russian, Spanish.

DESCRIPTION: Observes developments affecting the cotton industry; collects and disseminates statistics on cotton production, trade, consumption, stocks, and prices. Offers recommendations to members on measures leading to development of a sound cotton economy. Forum for international discussion of cotton matters. **PUBLICATIONS:** *Cotton: Monthly Update of the World Situation*; *COTTON: Review of the World Situation* (in English, French, and Spanish), bimonthly; *Cotton: World Statistics*. Provides world cotton supply/demand statistics since 1980, by country on a crop year basis.; *Country Statements at the 53rd Plenary Meeting*; *Current Research Projects in Cotton*; *Fiber Characteristics and the Spinners Perspective: A Look Into the Future*; *ICAC documents on CD-ROM*; *The ICAC Recorder* (in English, French, and Spanish), quarterly; *Outlook for Cotton Supply*. Provides an overview of factors affecting world cotton prices; provides statistics on aggregate world cotton supply and use, with price forecasts.; *Proceedings of the 54th Plenary Meetings* (in English, French, and Spanish), annual; *Survey of Cotton Production Practices*. Provides information on how cotton is grown in 28 countries; *Survey of the Cost of Production of Raw Cotton*. Compares the cost of production of cotton grown under a wide range of growing conditions; *The World Cotton Market: Prospects for the Nineties*. Results of a joint ICAC-FAO econometric study to forecast developments in cotton supply and demand to the year 2000; *World Cotton Trade*. Discusses trade developments in raw cotton since 1980; analyzes world trade by region, and provides import/export projections by country.; *World Textile Demand*. Analyses and projections of world end-use consumption of textiles, mill use, and production of cotton and chemical yarn and fabric for 100 countries.

INTERNATIONAL COUNCIL FOR ADULT EDUCATION (ICAE) (Conseil International d'Education des Adultes)

720 Bathurst St., Ste. 500, Toronto, ON, Canada M5S 2R4 **PHONE:** (416) 588-1211 **FX:** (416) 588-5725 **E-MAIL:** icae@web.net **WEBSITES:** http://www.web.net/icae **FOUNDED:** 1973. **OFFICERS:** Lalita Ramdas, Pres. **MEMBERS:** 101.

STAFF: 6. **BUDGET:** C$1,600,000. **LANGUAGES:** Arabic, English, French, Spanish.

DESCRIPTION: National and regional associations of adult educators in 92 countries. Seeks to advance the knowledge, skills, and competencies of individuals and groups worldwide through education and participation in the determination and achievement of their economic, social, and cultural development; strives to enhance international understanding and world peace. Fosters the activities of member organizations and aids in the establishment of such organizations; works in cooperation with other related nongovernmental organizations, universities, and research institutes. Acts as liaison with agencies of the United Nations and various intergovernmental organizations. Maintains resource center that disseminates information on education, peace and human rights, environmental issues, participatory research, gender issues, developmental, and health issues. **PUBLICATIONS:** *Convergence* (in English), quarterly. International journal of adult education.; *ICAE News* (in English and French), quarterly; *Pachamama*, periodic. Publishes news of the ICAE Learning for Environmental Action Programme(LEAP); Bibliographies, periodic; Books, periodic; Papers, periodic.

INTERNATIONAL COUNCIL OF AIRCRAFT OWNER AND PILOT ASSOCIATIONS (IAOPA)

421 Aviation Way, Frederick, MD 21701 **PHONE:** (301) 695-2220 **FX:** (301) 695-2375 **E-MAIL:** ruth.moser@aopa.org **WEBSITES:** http://www.iaopa.org **FOUNDED:** 1962. **OFFICERS:** Phil Boyer, Pres. **MEMBERS:** 51. **LANGUAGES:** English.

DESCRIPTION: Pilots (400,000) represented by member groups who fly general aviation airplanes for business and recreational purposes. Facilitates the movement of general aviation aircraft for peaceful purposes; develops airports, air routes, communications, navigation facilities, and services designed and operated to fill the needs of general aviation. Works to eliminate barriers that impede the utilization of general aviation aircraft for international flights. **PUBLICATIONS:** *IAOPA Bulletin* (in English), quarterly.

INTERNATIONAL COUNCIL ON ALCOHOL AND ADDICTIONS (ICAA) (Conseil International sur les Problemes de l'Alcoolisme et des Toxicomanies - CIPAT)

Case Postale 189, CH-1001 Lausanne, Switzerland **PHONE:** 41 21 3209865, 41 21 3209866 **FX:** 41 21 3209817 **E-MAIL:** icaa@pingnet.ch **WEBSITES:** http://www.icaa.ch; http://www.icaa.ch **FOUNDED:** 1907. **OFFICERS:** Eva Tongue LL.D., Dir. **MEMBERS:** 500. **STAFF:** 5. **BUDGET:** 750,000 SFr. **LANGUAGES:** English, French, German, Hungarian, Italian, Spanish.

DESCRIPTION: Persons and organizations interested in or working with problems of alcoholism or drug addiction in 180 countries. Encourages interdisciplinary exchange of information and experience in research, prevention, treatment, and rehabilitation for alcoholism and drug addiction. Organizes international conferences on the prevention and treatment of alcohol and drug dependence. Also focuses on issues related to tobacco dependence and gambling. **PUBLICATIONS:** *Alcoholism* (in English), semiannual; *Conference Proceedings*, triennial; *ICAA News*, quarterly.

INTERNATIONAL COUNCIL ON ARCHIVES (ICA) (Conseil International des Archives)

60, rue des Francs-Bourgeois, F-75003 Paris, France **FX:** 33 1 42722065 **E-MAIL:** 100640.54@compuserve.com **WEBSITES:** http://data1.archives.ca/ica/ **FOUNDED:** 1948. **OFFICERS:** M. Charles Kecskemeti, Sec.Gen. **MEMBERS:** 1,528. **STAFF:** 5. **LANGUAGES:** English, French.

DESCRIPTION: International organisation of national and international archival associations, central archival administrations, archival institutions, and individuals in more than 160 countries. Encourages preservation of archives and works to advance their administration. Strives to simplify the use of existing archives by making their contents more widely known and by encouraging greater freedom of access and making reproductions more readily available. Promotes archival development in the Third World. Conducts research on descriptive standards and the impact of information technology on archives; Created under the auspices of the United Nations Educational, Scientific and Cultural Organization. **PUBLICATIONS:**

Archivum (in English, French, German, Italian, and Spanish), annual. Deals with a single theme per issue; *Guide to the Sources for the History of Nations*, periodic; *Janus*, semiannual; *List of Publications*, annual; Bulletin, semiannual; Directory, annual.

INTERNATIONAL COUNCIL FOR BUILDING RESEARCH, STUDIES, AND DOCUMENTATION (CIB) (Conseil International du Batiment pour la Recherche, l'Etude et la Documentation - CIB)

Kruisplein 25-G, Postbus 1837, NL-3000 BV Rotterdam, Netherlands **PHONE:** 31 10 4110240 **FX:** 31 10 4334372 **E-MAIL:** secretariat@cibwold.nl **WEBSITES:** http://www.cibworld.nl **FOUNDED:** 1953. **OFFICERS:** Dr. W.J.P. Bakens, Sec.Gen. **MEMBERS:** 500. **STAFF:** 5. **BUDGET:** 1,200,000 f. **LANGUAGES:** Dutch, English.

DESCRIPTION: Individuals and institutions in 70 countries seeking to develop international cooperation in building, housing, and planning research and documentation. Studies technical, economic, social, and environmental aspects of the building, housing, and planning fields. Promotes progress in these fields by improving quality, reducing costs, and increasing productivity. Coordinates international research through 60 working commissions and task groups. Studies subjects such as building economics, performance concept in building, low cost housing, water supply and drainage, housing sociology, heating and cooling, international procurement, and post construction liability and insurance. Sponsors conferences, symposia, and workshops. **PUBLICATIONS:** *Information Bulletin* (in English), bimonthly; *International Directory of Building Research, Information and Development Organizations*, triennial; *Journal: Building Research and Practice*, bimonthly; Reports, periodic. Also publishes reports, proceedings, and analyses.

INTERNATIONAL COUNCIL OF CHRISTIANS AND JEWS (ICCJ) (Internationaler Rat der Christen und Juden)

Martin Buber House, Werlestrasse 2, Postfach 1129, D-64629 Heppenheim, Germany **PHONE:** 49 6252 5041 **FX:** 49 6252 68331 **E-MAIL:** iccj-buberhouse@t-online.de **FOUNDED:** 1962. **OFFICERS:** Rev. Friedhelm Pieper, Gen.Sec. **MEMBERS:** 29. **STAFF:** 4. **BUDGET:** DM 775,000. **LANGUAGES:** English, French, German.

DESCRIPTION: Members are national organizations striving to make Christians and Jews aware of their unique historical and religious ties. Sponsors educational efforts to combat misunderstanding and prejudice, especially in the fields of interfaith, interracial, and international relations. Deals with the problems facing human rights, particularly those related to the well-being of minority groups. Encourages research and development in the field of interreligious and intergroup dialogue. Has studied issues such as the Holocaust, aspects of violence and terrorism, neo-Nazism, anti-Semitism, and ecological and peace related matters. Sponsors study groups and exhibitions. Conducts research on revision of prejudicial material in religious and historical textbooks. Engages also in Jewish-Christian-Muslim dialogue. **PUBLICATIONS:** *From the Martin Buber House* (in English), periodic; *ICCJ Nachrichten* (in German), periodic; *ICCJ News* (in English), periodic.

INTERNATIONAL COUNCIL FOR COMMERCIAL ARBITRATION (ICCA) (Conseil International pour l'Arbitrage Commercial)

PO Box 16050, S-103 21 Stockholm, Sweden **PHONE:** 46 8 6131825 **FX:** 46 8 7230176 **TX:** 15638 **E-MAIL:** ulf.franke@chamber.se **FOUNDED:** 1972. **OFFICERS:** Mr. Ulf Franke, Sec.Gen. **MEMBERS:** 36. **LANGUAGES:** English.

DESCRIPTION: Individuals active in arbitration organizations; specialists in the field of international commercial arbitration. Encourages the development of international commercial arbitration and cooperation with organizations of the United Nations and other related groups. Maintains commissions. **PUBLICATIONS:** *Congress Series*, periodic; *International Handbook Commercial Arbitration*, periodic; *Yearbook Commercial Arbitration.*

INTERNATIONAL COUNCIL OF COMMUNITY CHURCHES (ICCC)

21116 Washington Pkwy., Frankfort, IL 60423-3112 **PHONE:** (815) 464-5690 **FX:** (815) 464-5692 **WEBSITES:** http://www.icc.i.go.2

FOUNDED: 1950. **OFFICERS:** Rev. Michael E. Livingston, Exec.Dir. **MEMBERS:** 800. **STAFF:** 6. **LANGUAGES:** English.

DESCRIPTION: Promotes the fellowship of community churches internationally and provides an instrument through which community-minded and freedom-loving churches can cooperate in making a contribution toward a united church. Maintains placement bureau for ministers. **PUBLICATIONS:** *The Christian Community*, 8/year; *Community Yearbook*; *Pastor's Journal*, 3/year.

INTERNATIONAL COUNCIL ON EDUCATION FOR TEACHING (ICET)

c/o National Louis University, 1000 Capitol Dr., Wheeling, IL 60090-7201 **PHONE:** (703) 525-5253 **FX:** (703) 351-9381 **FOUNDED:** 1953. **OFFICERS:** Mrs. Sandra Klassen, Exec.Dir. **STAFF:** 4. **BUDGET:** US$500,000. **LANGUAGES:** English.

DESCRIPTION: Individuals, universities, and organizations concerned with the professional preparation of teachers, school administrators, and specialists. Serves as the voice of teacher education; provides a forum for examination and discussion of issues, trends, problems, and innovations in teacher education. Encourages cooperation in the preparation of educational specialists; conducts cooperative research projects. Training division operates administrative fellowship programs, selects and oversees technical assistance personnel, develops and manages specialized training programs including individually-tailored programs for university administrators. Maintains consultative status with United Nations Educational, Scientific and Cultural Organization. **PUBLICATIONS:** *International Yearbook on Teacher Education*; *W. Clement Stone Lecture*, annual; Newsletter, quarterly; Pamphlets; Proceedings, annual.

INTERNATIONAL COUNCIL OF ENVIRONMENTAL LAW (ICEL) (Conseil International du Droit de l'Environnement)

Godesberger Ailee 108-112, D-53175 Bonn, Germany **PHONE:** 49 228 2692240 **FOUNDED:** 1969. **OFFICERS:** W.E. Burhenne, Exec.Gov. **MEMBERS:** 334. **STAFF:** 2. **LANGUAGES:** English, French.

DESCRIPTION: Members are individuals (314) and organizations (20) in the fields of environmental law, policy, and administration united to facilitate professional contact between members and to foster the exchange and dissemination of information on the legal and policy aspects of environmental conservation. **PUBLICATIONS:** *Environmental Policy and Law* (in English and French), bimonthly; *References to Environmental Policy and Law Literature*, quarterly; Directory, semiannual.

INTERNATIONAL COUNCIL FOR THE EXPLORATION OF THE SEA (ICES) (Conseil International pour l'Exploration de la Mer - CIEM)

Palaegade 2-4, DK-1261 Copenhagen K, Denmark **PHONE:** 45 33154225 **FX:** 45 33934215 **E-MAIL:** ices.info@ices.dk **WEBSITES:** http://www.ices.dk **FOUNDED:** 1902. **OFFICERS:** Prof. Christopher C.E. Hopkins, Gen.Sec. **MEMBERS:** 19. **STAFF:** 38. **BUDGET:** 20,000,000 DKr. **LANGUAGES:** English, French.

DESCRIPTION: Members are governments of 19 countries. Encourages and facilitates research and investigation of the sea and its living resources; develops programs and coordinates activities of participating governments. Maintains databanks for oceanography, fisheries, and marine contaminants for the North Atlantic; collects and publishes fishery statistics from member countries. Publishes marine and fishery science periodicals and monographs. Acts as scientific advisory body to the International Baltic Sea Fishery Commission, North-East Atlantic Fisheries Commission, North Atlantic Salmon Conservation Organization, Oslo and Paris Commissions, Commission of the European Union, and others. **PUBLICATIONS:** *ICES Cooperative Research Reports* (in English), periodic; *ICES Fisheries Statistics*, annual; *ICES Identification Leaflets for Diseases and Parasites of Fish and Shellfish*; *ICES Identification Leaflets for Plankton*; *ICES Journal of Marine Science*, bimonthly; *ICES Marine Science Symposia*, periodic; *ICES Techniques in Marine Environmental Sciences*, periodic; *Oceanographic Data Lists and Inventories.*

INTERNATIONAL COUNCIL OF FRENCH-SPEAKING RADIO AND TELEVISION (CIRTEF) (Conseil Internationale des Radio-Televisions d'Expression Francaise - CIRTEF)

c/o RTBF Local 9M58, 52, blvd. Auguste-Reyers, B-1044 Brussels, Belgium **PHONE:** 32 2 7324585, 32 2 7368958 **FX:** 32 2 7326240 **E-MAIL:** cirtef@rtbf.be **FOUNDED:** 1978. **OFFICERS:** Abdelkader Marzouki, Sec.Gen. **MEMBERS:** 42. **STAFF:** 5. **BUDGET:** US$1,400,000. **LANGUAGES:** French.

DESCRIPTION: Radio and television stations and educational broadcast authorities in 32 countries that broadcast wholly or partly in the French language. Purposes are to maintain a permanent dialogue among radio and television stations and to promote the role of radio and television in community development. Provides professional training. Collaborates with agencies of the United Nations. Organizes discussion groups. **PUBLICATIONS:** *Bulletin CIRTEF en Bref* (in French), monthly. Contains information for members and supporters; *DOC/CIRTEF*, periodic; *Guide CIRTEF*, periodic. Lists members' names and addresses.

INTERNATIONAL COUNCIL OF GRAPHIC DESIGN ASSOCIATIONS (ICOGRADA) (Conseil International des Associations Graphique)

PO Box 398, London W11 4UG, England **PHONE:** 44 171 6038494 **FX:** 44 171 3716040 **E-MAIL:** 106065.2235@compuserve.com **FOUNDED:** 1963. **OFFICERS:** Mary V. Mullin, Sec.Gen. **MEMBERS:** 58. **STAFF:** 3. **LANGUAGES:** English, French.

DESCRIPTION: Members are 58 national associations in 38 countries. Seeks to raise the standards of graphic design and professional practice internationally; assists in improving the professional status of graphic designers. Promotes training and interchange of designers, teachers, and students among countries; conducts research; sponsors annual students' seminar. Compiles professional codes and regulations, surveys, and reports. Friends of ICOGRADA, individuals actively supporting the objectives of ICOGRADA; and ICOGRADA Foundation, established to promote the international development of graphic design education. Maintains speakers' bureau. **PUBLICATIONS:** *Code of Conduct*. Includes regulations for international design competition; *Fax News*, monthly; *Message Board* (in English), quarterly.

INTERNATIONAL COUNCIL FOR HEALTH, PHYSICAL EDUCATION, RECREATION, SPORT, AND DANCE (ICHPERSD)

1900 Association Dr., Reston, VA 20191 **PHONE:** (703) 476-3486, (703) 476-3462 **TF:** (800) 213-7193 **FX:** (703) 476-9527 **E-MAIL:** ichper@aahperd.org **FOUNDED:** 1958. **OFFICERS:** Dr. Dong Ja Yang, Sec.Gen. **MEMBERS:** 1,500. **LANGUAGES:** English.

DESCRIPTION: National groups and professional organizations concerned with programs, policies, and the educational aspects of health, physical education, sports, recreation, and dance; institutions of higher education, libraries, and HPERD departments and professional individuals in 114 countries. Serves as a clearinghouse for exchange of information and ideas; represents members' interests in the field. Sponsors consultations and seminars. Prepares exhibits of books, photographs, films, pictures, and other materials. Compiles statistics; conducts study and research in cooperation with national groups and governmental organizations such as UNESCO. **PUBLICATIONS:** *Congress Proceedings*, biennial; *Journal of the International Council for Health, Physical Education, Recreation, Sport, and Dance*, quarterly. Contains association news, coverage of developments in international athletic competition, and information about new techniques in physical education.; Report. Covers physical education and games in curriculum. Published in conjunction with UNESCO.; Report. Covers teacher preparation for physical education. Published in conjunction with UNESCO.; Report. Covers the status of teachers in physical education. Published in conjunction with UNESCO.

INTERNATIONAL COUNCIL OF JEWISH WOMEN (ICJW)

24-32 Stephenson Way, London NW1 2JW, England **PHONE:** 44 171 3888311 **FX:** 44 171 3872110 **E-MAIL:** hq@icjw.demon.co.uk **WEBSITES:** http://www.icjw.org.uk **FOUNDED:** 1912. **OFFICERS:** June Jacobs, Pres. **MEMBERS:** 1,750,000. **LANGUAGES:** English, French, Spanish.

DESCRIPTION: Members are national organizations linking nearly 1,500,000 Jewish women. Objectives are to: promote friendly relations, understanding, and mutual support among Jewish women; uphold and strengthen the bonds of Judaism; show solidarity with Israel and support the efforts of Israel to secure a just and lasting peace; economic security and social, educational, and cultural development in Israel; further the highest interests of humanity in the fields of international relations, government, social welfare and education; cooperate with national and international organizations working for goodwill among all peoples and for equal rights for humanity; supports the Universal Declaration of Human Rights of the United Nations, and encourages work for the improvement of the social, economic, and legal status of all women under Jewish and civil law; encourage and assist in the education, training and use of volunteers. **PUBLICATIONS:** *Cooking Time Around the World*; *Directory of ICJW Affiliates* (in English), triennial; *ICJW Newsletter* (in English and Spanish), semiannual; *Links Around the World - Community Services* (in English and Spanish), 3/year.

INTERNATIONAL COUNCIL FOR LABORATORY ANIMAL SCIENCE (ICLAS)

c/o Steven P.Pakes, DVM, 5323 Harry Hines Blvd., Dallas, TX 75235-9073 **PHONE:** (214) 648-3218 **FX:** (214) 648-2659 **E-MAIL:** spakes@mednet.swmed.edu **WEBSITES:** http://www.iclas.org **FOUNDED:** 1956. **OFFICERS:** S.P. Pakes, Sec.Gen. **LANGUAGES:** English.

DESCRIPTION: Laboratories, research institutions, and other facilities making use of laboratory animals for experimental purposes. Promotes humane and appropriate use of live animal experimentation. Facilitates exchange of information among laboratory animal experimentation programs worldwide.

INTERNATIONAL COUNCIL ON METALS AND THE ENVIRONMENT (ICME)

506-294 Albert St., Ottawa, ON, Canada K1P 6E6 **PHONE:** (613) 235-4263 **FX:** (613) 235-2865 **E-MAIL:** info@icme.com **WEBSITES:** http://www.icme.com **FOUNDED:** 1991. **OFFICERS:** Gary Nash, Sec.Gen. **MEMBERS:** 27. **STAFF:** 11. **LANGUAGES:** English.

DESCRIPTION: Producers of nonferrous and precious metals. Promotes environmentally responsible production and disposal of metals and increased use of recycling in the metal-producing industries. Disseminates information on environmentally sustainable mining and metal production. **PUBLICATIONS:** *ICME Newsletter*, quarterly. Features 8 pages long newsletter presenting views on key issues and provides information on developments related to the environment and health.

INTERNATIONAL COUNCIL OF MUSEUMS (ICOM) (Conseil International des Musees)

Maison de l'Unesco, 1, rue Miollis, F-75732 Paris Cedex 15, France **PHONE:** 33 1 47340500 **FX:** 33 1 43067862 **TX:** UNESCO 270602 **E-MAIL:** secretariat@icom.org **WEBSITES:** http://www.icom.org/ **FOUNDED:** 1946. **OFFICERS:** Manus Brinkman, Sec.Gen. **MEMBERS:** 15,000. **STAFF:** 10. **BUDGET:** 9,000,000 Fr. **LANGUAGES:** English, French.

DESCRIPTION: Institutions and individuals involved in the museum profession in 148 countries. Fosters international cooperation among museums and serves as the coordinating and representative international body furthering museum interests. Conducts educational and research programs. Maintains 256 international committees. **PUBLICATIONS:** *ICOM News* (in English, French, and Spanish), quarterly; *Nouvelles de l'ICOM* (in English, French, and Spanish).

INTERNATIONAL COUNCIL OF NURSES (ICN) (Conseil International des Infirmieres - CII)

3, place Jean-Marteau, CH-1201 Geneva, Switzerland **PHONE:** 41 22 9080100 **FX:** 41 22 9080101 **E-MAIL:** icn@uni2a.unige.ch **WEBSITES:** http://www.icn.ch **FOUNDED:** 1899. **OFFICERS:** Judith A. Oulton, Exec.Dir. **MEMBERS:** 112. **STAFF:** 15. **LANGUAGES:** English, French, Spanish.

DESCRIPTION: ICN works in collaboration with its national nurses' association. Provides a medium through which members can work together to promote the health of people and the care of the sick.

Objectives are to: improve the standards and status of nursing; promote the development of strong national nurses' associations; serve as the authoritative voice for nurses and the nursing profession worldwide. **PUBLICATIONS:** *International Nursing Review* (in English), bimonthly. Contains information on nursing and health issues; Brochures, periodic.

INTERNATIONAL COUNCIL FOR OPEN AND DISTANCE EDUCATION (ICDE) (Conseil International de l'Enseignement a Distance)

Gjerdrumsvei 12, N-0486 Oslo 4, Norway **PHONE:** 47 22 950630 **FX:** 47 22 950719 **E-MAIL:** icde@icde.no **WEBSITES:** http://www.icde.org **FOUNDED:** 1938. **OFFICERS:** Reidar Roll, Sec.Gen. **MEMBERS:** 9,000. **STAFF:** 4. **LANGUAGES:** English, Spanish.

DESCRIPTION: Individuals and institutions in 100 countries interested in distance education programs whereby the teacher and the student carry out their essential tasks while apart, replacing classroom contact with communication through other media. Distance education includes participation in programs such as proprietary and government correspondence schools, rural development projects, educational broadcasting, extension or external studies departments, teleconferencing networks, open universities, consortia of traditional institutions, community education projects, publishers, instructional packages for computers, and satellite linkages. Acts as a clearinghouse and provides advisory services. Monitors and takes part in United Nations Educational, Scientific and Cultural Organization activities. Is developing data bank on distance education institutions. **PUBLICATIONS:** *Conference Book*, periodic; *Open Praxis*, semiannual; *Proceedings*, periodic.

INTERNATIONAL COUNCIL OF PSYCHOLOGISTS (ICP)

Psych Dept., Southwest Texas State Univ., San Marcos, TX 78666-4601 **PHONE:** (512) 245-7605, (512) 245-2526 **FX:** (512) 245-3153 **E-MAIL:** jd04@academia.swt.edu **FOUNDED:** 1942. **OFFICERS:** John M. Davis Ph.D., Sec.General. **MEMBERS:** 1,300. **STAFF:** 2. **LANGUAGES:** English.

DESCRIPTION: Psychologists and individuals professionally active in fields allied to psychology. Seeks to advance psychology and further the application of its scientific findings. Conducts continuing education workshops and educational programs. **PUBLICATIONS:** *International Psychologist*, quarterly. Includes book reviews, calendar of events, and information on members; *World Psychology*, quarterly. Developments in international psychology, broad research reviews; cross-cultural research articles; teaching, education and training in psychology around the world.; Directory, biennial.

INTERNATIONAL COUNCIL FOR SCIENTIFIC AND TECHNICAL INFORMATION (ICSTI)

51, blvd. de Montmorency, F-75016 Paris, France **PHONE:** 33 1 45256592 **FX:** 33 1 42151262 **E-MAIL:** icsti@dial.oleane.com **WEBSITES:** http://www.cisti.nrc.ca/icsti/icsti.html **FOUNDED:** 1952. **OFFICERS:** Marthe Orfus, Exec.Sec. **MEMBERS:** 54. **STAFF:** 2. **LANGUAGES:** English.

DESCRIPTION: International scientific unions; abstracting and indexing services in the field of natural and physical sciences; information centers; academic societies; libraries; online vendors. Seeks to increase users' awareness of technical and scientific information by facilitating access to and transfer of information on an interdisciplinary and international level. Studies specific economic, legal, and technological constraints affecting the flow of scientific information. Analyzes methods of collecting, storing, organizing, and disseminating information in an effort to satisfy the research needs of the international scientific community. Arranges for the exchange of information and ideas among members. **PUBLICATIONS:** *Biotechnology Information Sources: North and South America*; *Forum* (in English), quarterly; *International Classification Scheme for Physics*; *Multilingual Thesaurus of Geosciences 2nd Ed.*; *Numeric Databases: A Directory*; *Squaring the Information Circle*.

INTERNATIONAL COUNCIL OF SCIENTIFIC UNIONS (ICSU)

51, blvd. de Montmorency, F-75016 Paris, France **PHONE:** 33 1 45250329 **FX:** 33 1 42889431 **TX:** 645554 F **E-MAIL:**

secretariat@icsu.org **WEBSITES:** http://www.icsu.org **FOUNDED:** 1931. **OFFICERS:** J.F. Stuyck-Taillandier, Exec.Dir. **MEMBERS:** 117. **STAFF:** 9. **BUDGET:** US$3,000,000. **LANGUAGES:** English, French.

DESCRIPTION: Members are national research councils or academies (95) and scientific unions of various scientific disciplines (25). Facilitates and coordinates the activities of international scientific unions in the exact and natural sciences. Acts as the coordinating center for national organizations adhering to the council. Through the intermediary of member organizations, enters into relations with governments of their respective countries to promote scientific research. Collaborates with the United Nations and its agencies. **PUBLICATIONS:** *Annual Report*; *Science International*, quarterly; *Year Book.*

INTERNATIONAL COUNCIL OF SHOPPING CENTERS (ICSC)

665 5th Ave., New York, NY 10022 **PHONE:** (212) 421-8181 **FX:** (212) 486-0849 **TX:** 128285 **E-MAIL:** icsc@icsc.org **WEBSITES:** http://www.icsc.org/ **FOUNDED:** 1957. **OFFICERS:** John Riordan, Pres. **MEMBERS:** 35,000. **STAFF:** 100. **BUDGET:** US$25,000,000. **LANGUAGES:** English.

DESCRIPTION: Owners, developers, retailers, and managers of shopping centers; architects, engineers, contractors, leasing brokers, promotion agencies, and others who provide services and products for shopping center owners, shopping center merchant associations, retailers, and public and academic organizations. Promotes professional standards of performance in the development, construction, financing, leasing, management, and operation of shopping centers throughout the world. Engages in research and data gathering on all aspects of shopping centers; compiles statistics. Sponsors school for professional development annually, offering courses in all areas of the industry; leading to the designations CSM (Certified Shopping Center Manager) and CMD (Certified Marketing Director), and CLS (Certified Leasing Specialist). Holds 200 seminars and conferences annually. **PUBLICATIONS:** *Directory of Products and Services*, annual. Lists suppliers of products and securities to the shopping center industry; *Government Affairs Report*, quarterly; *ICSC Membership Directory*, annual; *ICSC Research Quarterly*; *Journal of Shopping Center Research*, semiannual; *Legal Update*, triennial; *Monthly Mall Merchandise Indy*; *Retail Challenge*, quarterly; *Shopping Centers Today*, monthly.

INTERNATIONAL COUNCIL FOR SMALL BUSINESS (ICSB)

c/o Jefferson Smurfit Center for Entrepreneurial Studies, St. Louis Univ., 3674 Lindell Blvd., St. Louis, MO 63108 **PHONE:** (314) 977-3628 **FX:** (314) 977-3627 **E-MAIL:** icsb@slu.edu **WEBSITES:** http://www.icsb.org **FOUNDED:** 1957. **OFFICERS:** Sharon Bower, Sec. **MEMBERS:** 1,800. **STAFF:** 1. **LANGUAGES:** English.

DESCRIPTION: Management educators, researchers, government officials, and professionals in 80 countries. Fosters discussion of topics pertaining to the development and improvement of small business management. **PUBLICATIONS:** *ICSB Bulletin*, quarterly; *Journal of Small Business Management*, quarterly; *List of Members*, annual; Proceedings, annual.

INTERNATIONAL COUNCIL ON SOCIAL WELFARE (ICSW) (Conseil International de l'Action Sociale - CIAS)

380 St. Antoine St. W, Ste. 3200, Montreal, PQ, Canada H2Y 3X7 **PHONE:** (514) 287-3280 **FX:** (514) 287-9702 **E-MAIL:** icswintl@colba.net **WEBSITES:** http://www.icsw.org/ **FOUNDED:** 1928. **OFFICERS:** Dirk Jarre, Pres. **MEMBERS:** 95. **STAFF:** 5. **LANGUAGES:** English, French, Spanish.

DESCRIPTION: Members are national committees (62) and international organizations (20) representing persons and groups interested in social welfare, social justice, and social development. Promotes international cooperation in the field of social welfare; promotes and conducts research on matters affecting social welfare. Provides information and referral services to members. **PUBLICATIONS:** *Biennial Report, 1992-94* (in English, French, and Spanish); *ICSW Information* (in English), quarterly; *Social Development Review* (in English, French, and Spanish), quarterly.

INTERNATIONAL COUNCIL ON SOCIAL WELFARE - LATIN AMERICA AND CARIBBEAN (ICSW/Brazil) (Consejo Internacional de Bienestar Social - CIBES)

c/o Ana Luzia, 685-2 Andar, 20030-0040 Rio de Janeiro, Rio de Janeiro, Brazil **PHONE:** 55 21 2208174, 55 21 2208274 **FX:** 55 21 2208274 **TX:** 513140 **FOUNDED:** 1946. **OFFICERS:** Therezinha Arnaut, V.Pres. **MEMBERS:** 380. **STAFF:** 16. **BUDGET:** US$100,000. **LANGUAGES:** English, French, Spanish.

DESCRIPTION: Development organization concerned with all aspects of social welfare, social development policy, and its practice. Provides a bridge between voluntary and governmental sectors from the grassroots to the international level. Promotes a generalist approach to social problems and collaborative social action. Organizes training activities, workshops, and regional and international conferences. Offers expertise in the following areas: strategic planning; project supervision and evaluation; technical assistance; publications. **PUBLICATIONS:** *Debates Sociais*, semiannual.

INTERNATIONAL COUNCIL OF SOCIETIES OF INDUSTRIAL DESIGN (ICSID)

Yrjonkatu 11 E, FIN-00120 Helsinki, Finland **PHONE:** 358 9 607611 **FX:** 358 9 607875 **E-MAIL:** icsidsec@icsid.org **WEBSITES:** http://www.icsid.org **FOUNDED:** 1957. **OFFICERS:** Kaarina Pohto, Sec.Gen. **MEMBERS:** 150. **STAFF:** 3. **BUDGET:** US$200,000. **LANGUAGES:** English.

DESCRIPTION: National societies in 54 countries; includes councils, institutes, and other groups of professionals and nonprofessionals interested in promoting industrial design. Offers advisory service for industry, governments, and designers. Is concerned with problems of professional practice and conduct. Organizes competitions and exhibitions. **PUBLICATIONS:** *Guides* (in English); *ICSID News*, bimonthly; *World Directory of Design Schools*, periodic.

INTERNATIONAL COUNCIL OF TANNERS (ICT) (Conseil International des Tanneurs - CIT)

Leather Trade House, Kings Park Rd., Moulton Park, Northampton, Northhants. NN3 6JD, England **PHONE:** 44 1604 679917 **FX:** 44 1604 679998 **E-MAIL:** ict@blcleathertech.com **FOUNDED:** 1926. **OFFICERS:** Mr. R. Paul Pearson, Sec. **MEMBERS:** 36. **STAFF:** 2. **BUDGET:** £30,000. **LANGUAGES:** English, French, German.

DESCRIPTION: Leather trade associations united to promote the leather industry throughout the world. Fosters research and development in the industry. **PUBLICATIONS:** *ICT Update*, periodic; *International Glossary of Leather Terms.*

INTERNATIONAL COUNCIL FOR TRADITIONAL MUSIC (ICTM)

c/o Center for Ethnomusicology, Columbia Univ., New York, NY 10027 **E-MAIL:** ictmqwoof@music.columbia.edu **FOUNDED:** 1947. **OFFICERS:** Dieter Christensen, Sec.Gen. **MEMBERS:** 1,300. **STAFF:** 1. **BUDGET:** US$55,000. **LANGUAGES:** French, German.

DESCRIPTION: Individuals and institutions. Promotes the preservation, study, dissemination, and practice of all forms of traditional music including folk, dance, popular, classical, and urban. Exchanges recorded materials with radio and television organizations. **PUBLICATIONS:** *Directory of Traditional Music*, biennial; *International Council of Traditional Music*, semiannual; *Yearbook for Traditional Music* (in English, French, and German).

INTERNATIONAL COUNCIL OF WOMEN (ICW) (Conseil International des Femmes - CIF)

13, rue Caumartin, F-75009 Paris, France **PHONE:** 33 1 47421940 **FX:** 33 1 42662623 **E-MAIL:** icw-cif@wanadoo.fr **FOUNDED:** 1888. **OFFICERS:** Marie-Christine Lafargue, Gen.Sec. **MEMBERS:** 76. **STAFF:** 1. **LANGUAGES:** English, French.

DESCRIPTION: Members are national councils of women comprising national and local women's organizations. Serves as a medium for consultation among women on those actions necessary to promote the welfare of humankind, the family, children, and the individual. Advises women of their rights and their civic, social, and political responsibilities; works for the equal legal status of women and for the removal of all that restricts women from full participation in life. Supports international peace and arbitration. Areas of interest include advancement of women; education; human rights; literacy; role of women in economic and social development. Maintains consultative status with the Economic and Social Council of the United Nations, United Nations Educational, Scientific and Cultural Organization, United Nations Children's Fund, World Health Organization, and the Council of Europe. **PUBLICATIONS:** *Children's Stories from Many Lands* (in English and French); *Side by Side*; *Triennial Report*; Newsletter (in English and French), periodic.

INTERNATIONAL COUNCIL FOR WOMEN IN THE ARTS (ICWA)

PO Box 226, Lafayette, CA 94549 **PHONE:** (925) 256-0808 **FX:** (925) 944-9479 **E-MAIL:** icwacvar@aol.com **FOUNDED:** 1990. **OFFICERS:** Salwa Mikdadi Nashashibi, Pres. **LANGUAGES:** English.

DESCRIPTION: Seeks to support and encourage artists from Asia, Africa, Latin America, and the Middle East in order to open new avenues for intercultural communication within the United States and to provide American audiences with better understanding of contemporary non-western cultures. Offers educational programs, art exhibits. **PUBLICATIONS:** *Forces of Change: Artists of the Arab World*. Exhibition Catalog.

INTERNATIONAL CRIMINAL POLICE ORGANIZATION (ICPO) (Organisation Internationale de Police Criminelle - OIPC)

c/o Mr. Serge Sabourin, 200, quai Charles de Gaulle, F-69006 Lyon, France **PHONE:** 33 4 72447000 **FX:** 33 4 72447163 **TX:** OIPC 301987 F **WEBSITES:** http://www.kenpubs.co.uk/interpol-pr/ **FOUNDED:** 1923. **OFFICERS:** R.E. Kendall, Sec.Gen. **MEMBERS:** 177. **STAFF:** 320. **BUDGET:** US$26,000,000. **LANGUAGES:** Arabic, English, French, Spanish.

DESCRIPTION: Intergovernmental organization responsible for promoting the widest possible mutual assistance between law enforcement agencies around the world. Through its computer network and databases the organization serves as a central point of coordination for investigators of a transnational nature. **PUBLICATIONS:** *International Crime Statistics* (in Arabic, English, French, and Spanish), annual; *International Criminal Police Review* (in Arabic, English, French, and Spanish), bimonthly.

INTERNATIONAL CROPS RESEARCH INSTITUTE FOR THE SEMI-ARID TROPICS (ICRISAT)

PO Box 776, Bulawayo, Zimbabwe **PHONE:** 263 9 838311, 263 9 838312 **FX:** 263 9 838253 **E-MAIL:** icrisatzw@sgiar.org **FOUNDED:** 1983. **OFFICERS:** Dr. G.M. Heinrich. **STAFF:** 32. **LANGUAGES:** English.

DESCRIPTION: Works to improve the quality and reliability of food production in semi-arid tropical climates. Conducts research on improved crop strains, agricultural techniques, and more effective watershed management. Serves as a clearinghouse on agriculture and crops; facilitates development and transfer of technologies benefiting farmers in semiarid tropical regions.

INTERNATIONAL CYSTIC FIBROSIS (MUCOVISCIDOSIS) ASSOCIATION (ICFMA)

Avda Campanar, 106-3o-6.a, E-46015 Valencia, Spain **PHONE:** 34 6 1414, 34 6 9464 **FX:** 34 6 4047 **FOUNDED:** 1964. **OFFICERS:** Aisha Ramos, Sec. **MEMBERS:** 51. **BUDGET:** US$20,000.

DESCRIPTION: National medical/lay volunteer organizations concerned with cystic fibrosis. Works to: further the interests of persons with cystic fibrosis; improve medical care available to them and the psychological and social care available to them and their families. Stimulates, supports, and advances research in the nature, cause, prevention, treatment, alleviation, and care of cystic fibrosis. Coordinates information services and the interchange of knowledge of all phases of cystic fibrosis. **PUBLICATIONS:** *Directory of International Treatment Locations for Cystic Fibrosis Patients*, periodic.

INTERNATIONAL DEMOCRAT UNION (IDU)

32 Smith Sq., London SW1P 3HH, England **PHONE:** 44 171 2220847 **FX:** 44 171 2221459 **E-MAIL:** idu@compuserve.com **FOUNDED:** 1983.

OFFICERS: Graham Wynn, Exec.Sec. **MEMBERS:** 35. **STAFF:** 3. **LANGUAGES:** English.

DESCRIPTION: Conservative and similar central right political parties in over 40 countries. Seeks to foster among member parties the common philosophy of a free, open, and democratic society emphasizing the rule of law, social justice, the role of the family, and a free competitive market economy. Encourages closer cooperation among member parties and its regional groups, the European Democrat Union, the Pacific Democrat Union, and the Americas Democrat Union. Provides a forum for the exchange of ideas and information.

INTERNATIONAL DEVELOPMENT ASSOCIATION (IDA)

1818 H St., NW, Washington, DC 20433
PHONE: (202) 477-1234 **FX:** (202) 477-6391 **TX:** 248423 **E-MAIL:** pic@worldbank.org **WEBSITES:** http://www.worldbank.org **FOUNDED:** 1960. **OFFICERS:** James D. Wolfensohn, Pres. **MEMBERS:** 160. **LANGUAGES:** English.

DESCRIPTION: Functional member of the World Bank Group, closely affiliated with the International Bank for Reconstruction and Development. Membership is open to countries of the World Bank. Promotes the economic development of the World Bank's poorer member countries by extending credits on easier terms than are normally available. Makes loans for projects aimed at strengthening the economies of developing countries in Asia, the Middle East, Africa, and the Western Hemisphere. Provides economic advice. **PUBLICATIONS:** *Annual Report.*

INTERNATIONAL DEVELOPMENT FOUNDATION (IDF)

PO Box 70257, Washington, DC 20024-0257
PHONE: (202) 723-7010, (202) 508-1441 **FX:** (202) 723-7010 **FOUNDED:** 1984. **OFFICERS:** Dr. Mekki Mtewa, Chm. **MEMBERS:** 100. **STAFF:** 4. **LANGUAGES:** English.

DESCRIPTION: Policymakers, business executives, educators, community leaders, and international organizations. Fosters recognition of contributions toward international understanding; seeks to provide a network of institutions interested in international development. Disseminates advice; encourages the funding and execution of research. Administers academic exchange programs for leadership training and intellectual growth. Maintains speakers' bureau. **PUBLICATIONS:** *Conference Proceedings*; *Directory of Inter-Regional Development Organizations*, annual; *Inter-Regional Connexions*, quarterly.

INTERNATIONAL DIABETES FEDERATION (IDF) (Federation Internationale du Diabete - FID)

1, rue Defacqz, B-1000 Brussels, Belgium
PHONE: 32 2 5385511 **FX:** 32 2 5385114 **E-MAIL:** idf@idf.org **WEBSITES:** http://www.idf.org **FOUNDED:** 1949. **OFFICERS:** H. Williams, Exec.Dir. **MEMBERS:** 1,100. **STAFF:** 9. **LANGUAGES:** English, French, Spanish.

DESCRIPTION: Organization consists of: National diabetes associations; diabetes sections of national academies; endocrinology, metabolic, and diabetes societies; diabetes supplies companies are supporting members; association represents over one million individuals through its national associations. Objectives are: to improve the quality of life in the global diabetic community; to promote the exchange of information; to improve standards of treatment; to develop educational methods designed to give patients a better understanding of their disease; to educate the public in the early recognition of the disease and the importance of its medically supervised treatment; to encourage medical, scientific, and socioeconomic research. Maintains liaison with the World Health Organization. Compiles statistics. Provides educational grants through the IDF Educational Foundation. **PUBLICATIONS:** *Diabetes Voice* (in English), quarterly; *The Economics of Diabetes & Diabetes Care: Costing Diabetes, the Cause for Prevention* (in English); *IDF Bulletin* (in English), quarterly; *IDF Diabetes Voice*, 3/year; *Triennial Report.*

INTERNATIONAL ECONOMIC ASSOCIATION (IEA) (Association Internationale des Sciences Economiques - AISE)

23, rue Campagne Premiere, F-75014 Paris, France **PHONE:** 33 1 43279144 **FX:** 33 1 42799216 **TX:** 264918 TRACE F **FOUNDED:** 1950. **OFFICERS:** Prof. Jean-Paul Fitoussi, Sec.Gen. **MEMBERS:** 60. **STAFF:** 1. **LANGUAGES:** English.

DESCRIPTION: Members are national academic, professional, and scientific associations and committees representing economists. Promotes and facilitates the advancement of economic knowledge by initiating and coordinating measures of international cooperation. Provides for global dissemination of economic information and ideas including abstracts, bibliographies, dictionaries, and translations. Facilitates the development of personal contacts among members. **PUBLICATIONS:** *Conference and Congress Proceedings*, triennial; *List of Members and Officers*, periodic; *Newsletter*, biennial; Brochures.

INTERNATIONAL ELECTRICAL TESTING ASSOCIATION (NETA)

PO Box 687, Morrison, CO 80465 **PHONE:** (303) 697-8441 **FX:** (303) 697-8431 **E-MAIL:** neta@netaworld.org **WEBSITES:** http://www.netaworld.org **FOUNDED:** 1972. **OFFICERS:** Dr. Mary R. Jordan, Exec.Dir. **MEMBERS:** 1,500. **STAFF:** 4. **BUDGET:** US$1,000,000. **LANGUAGES:** Spanish.

DESCRIPTION: Independent firms involved in testing, analysis, and maintenance of electrical power systems; firms supplying construction, maintenance, engineering, or similar services to the power systems industry; interested individuals. Seeks to represent, promote, and advance the interests of the electrical testing industry through safety and technical advancements, sound competition, establishment of standards, and dissemination of related data. Offers training programs and technical certification. Provides technical support in procedures and specifications. **PUBLICATIONS:** *Electrical Acceptance Testing Specifications*, quadrennial. Specifications for electrical acceptance testing.; *NETA Maintenance Specifications*. Specifications for electrical maintenance testing.; *NETA World*, quarterly. Includes safety and technical tips and calendar of events.; *Technical Conference Papers*. Covers annual technical conference.

INTERNATIONAL ELECTROTECHNICAL COMMISSION

3, rue de Varembe, Case Postale 131, CH-1211 Geneva 20, Switzerland **PHONE:** 41 22 9190211 **FX:** 41 22 9190300 **E-MAIL:** info@iec.ch **WEBSITES:** http://www.iec.ch **FOUNDED:** 1906. **OFFICERS:** Mr. A. Amit, Gen.Sec. **MEMBERS:** 60. **STAFF:** 114. **BUDGET:** 18,000,000 SFr. **LANGUAGES:** English, French, Russian.

DESCRIPTION: Members are national committees representing their country's electrical and electronic interests and composed of manufacturers, users, trade associations, government bodies, scholars, and engineers. Develops international electrotechnical standards. Operates 200 technical committees and subcommittees. **PUBLICATIONS:** *Catalogue of Publications* (in English and French), annual. With bimonthly updates; *IEC Multilingual Dictionary of Electricity, Electronics, and Telecommunications*; *Standards* (in English and French), periodic; Bulletin (in English and French), bimonthly; Annual Report (in English and French), annual.

INTERNATIONAL ENERGY AGENCY (IEA) (Agence Internationale de l'Energie)

9, rue de la Federation, F-75739 Paris Cedex 16, France **PHONE:** 33 1 40576554 **FX:** 33 1 40576559 **E-MAIL:** info@ie.org **WEBSITES:** http://www.iea.org **FOUNDED:** 1974. **OFFICERS:** Robert Priddle, Exec.Dir. **MEMBERS:** 23. **STAFF:** 120. **LANGUAGES:** English, French.

DESCRIPTION: Industrialized oil-consuming countries that carry out an international energy program designed to build and sustain strong energy economies. Objectives are: to improve energy supply and demand balance; to develop oil-alternative energy sources; to coordinate international oil market information; to maintain an emergency oil sharing system. Compiles statistics. **PUBLICATIONS:** *Energy Policies of IEA Countries*, annual; *Energy Statistics*; *IEA Oil Market Report* (in English), monthly; *Reviews of Coal and Other Energy Sources*.

INTERNATIONAL EPIDEMIOLOGICAL ASSOCIATION (IEA)

c/o Dr. Harotune Armenian, 111 Market Pl., Ste. 840, Baltimore, MD 21202 **PHONE:** (410) 223-1625 **FX:** (410) 223-1620 **E-MAIL:** htelljoh@jhsph.edu **FOUNDED:** 1950. **OFFICERS:** Kunio Aoki, Pres. **MEMBERS:** 2,250. **STAFF:** 2. **LANGUAGES:** English.

DESCRIPTION: Individuals interested in epidemiology (science dealing with incidence, distribution, and control of disease in populations). Studies methods and applications of disease control, clinical medicine, and health services. Conducts seminars and workshops. Encourages epidemiologic research. **PUBLICATIONS:** *Dictionary of Epidemiology*; *International Journal of Epidemiology*, bimonthly; Manuals; Membership Directory, triennial; Monographs.

INTERNATIONAL EQUESTRIAN FEDERATION (IEF) (Federation Equestre Internationale - FEI)

24, ave. Mon-Repos, BP 157, CH-1000 Lausanne 5, Switzerland **PHONE:** 41 21 3125656 **FX:** 41 21 3128677 **TX:** 454802 FEICH **FOUNDED:** 1921. **OFFICERS:** Dr. Bo Helander, Sec.Gen. **MEMBERS:** 113. **STAFF:** 24. **LANGUAGES:** English, French.

DESCRIPTION: International equestrian federations. Serves as the sole international authority for international dressage, jumping, 3-day events, and other equestrian events. Promotes the organization of international equestrian activities. Seeks to develop, standardize, coordinate, and publish rules and regulations for equestrian competitions and supervise their technical organization. Provides a means for discussion and understanding between national federations and offers support to strengthen their authority. Encourages instruction in riding, driving, and horsemanship for recreational purposes; sponsors courses for judges, organizers, and course designers. Compiles statistics. **PUBLICATIONS:** *Directory*, annual; *FEI Bulletin* (in English and French), 10/year; Annual Report (in English and French), annual.

INTERNATIONAL EROSION CONTROL ASSOCIATION (IECA)

Box 774904, Steamboat Springs, CO 80477-4904 **PHONE:** (970) 879-3010 **TF:** (800) 455-4322 **FX:** (970) 879-8563 **E-MAIL:** ecinfo@ieca.org **WEBSITES:** http://www.ieca.org **FOUNDED:** 1972. **OFFICERS:** Ben Northcutt, Exec.Dir. **MEMBERS:** 1,200. **STAFF:** 4. **BUDGET:** US$100,000. **LANGUAGES:** English.

DESCRIPTION: Landscape contractors, government officials, landscape architects, engineers, manufacturers and suppliers, and others in 40 countries. Encourages the exchange of information and ideas concerning effective and economical methods of erosion control. Recognizes the need for an organized discipline in soil erosion and sediment control so that laws, specifications, procedures, and restrictions concerning land disturbances may be written by qualified professionals. Offers short courses. **PUBLICATIONS:** *Erosion Control Journal*, bimonthly; *Proceedings of Conferences*, annual; *Products and Services Directory*, biennial; Membership Directory, annual. Also publishes material on erosion and sediment control technology, revegetation, and related subjects.

INTERNATIONAL EUROPE-BASED INDUSTRY ASSOCIATION FOR STANDARDIZING INFORMATION AND COMMUNICATION SYSTEMS

114, rue du Rhone, CH-1204 Geneva, Switzerland **PHONE:** 41 22 8496000 **FX:** 41 22 8496001 **E-MAIL:** helpdesk@ecma.ch **WEBSITES:** http://www.ecma.ch **FOUNDED:** 1961. **OFFICERS:** Mr. J. van den Beld, Sec.Gen. **MEMBERS:** 51. **STAFF:** 6. **BUDGET:** 2,400,000 SFr. **LANGUAGES:** English.

DESCRIPTION: Industrial companies that develop, produce, and market hardware or software products or services used to process digital information for business, scientific, and control purposes. Aims to scientifically study the methods and procedures necessary to facilitate and standardize the use of information processing and telecommunication systems. Promulgates standards on the functional design and use of information processing and telecommunication systems. **PUBLICATIONS:** *ECMA Memento, ECMA Standards, ECMA Technical Reports* (in English), semiannual; Reports, periodic.

INTERNATIONAL FARM MANAGEMENT ASSOCIATION (IFMA)

Farm Management Unit, Univ. of Reading, PO Box 236, Reading, Berks. RG6 6AT, England **PHONE:** 44 118 9351458 **FX:** 44 118 9756467 **FOUNDED:** 1974. **OFFICERS:** P.J. James, Hon.Sec./Treas. **MEMBERS:** 1,000. **STAFF:** 1. **LANGUAGES:** English.

DESCRIPTION: Farmers, extension workers, academics, resource use planners, and managers in 68 countries concerned with the planning, production, and marketing in agriculture. Furthers the knowledge and understanding of farm business management and fosters the exchange of ideas and information about farm management theory and practice worldwide. **PUBLICATIONS:** *Congress Proceedings* (in English), biennial; *Journal of International Farm Management* (in English), semiannual.

INTERNATIONAL FEDERATION OF ACCOUNTANTS (IFAC)

535 5th Ave., 26th Fl., New York, NY 10017 **PHONE:** (212) 286-9344 **FX:** (212) 286-9570 **E-MAIL:** mariahermann@ifac.org **WEBSITES:** http://www.ifac.org **FOUNDED:** 1977. **MEMBERS:** 125. **STAFF:** 10. **BUDGET:** US$1,200,000. **LANGUAGES:** English, French, Spanish.

DESCRIPTION: Membership consists of accounting bodies recognized by law or general consensus representing over 1,000,000 individuals in 78 countries. Seeks to achieve international technical, ethical, and educational guidelines and standards for the accountancy profession. Fosters cooperation among members and encourages development of regional groups with similar goals. **PUBLICATIONS:** *International Federation of Accountants—Annual Report*. Reviews the federation's activities in the past year; provides a description of IFAC and financial statements.; *International Federation of Accountants—Newsletter*, quarterly. Covers worldwide meetings and activities of interest to accountants. Includes calendar of events.

INTERNATIONAL FEDERATION OF ACTION OF CHRISTIANS FOR THE ABOLITION OF TORTURE (FIACAT) (Accion de los Cristianos para la Abolocion de la Tortura)

27, rue de Maubeuge, F-75009 Paris, France **PHONE:** 33 1 42800160 **FX:** 33 1 42802089 **E-MAIL:** fi.acat@wanadoo.fr **FOUNDED:** 1987. **OFFICERS:** Guy Aurenche, Pres. **MEMBERS:** 30,000. **STAFF:** 1. **BUDGET:** 684,000 Fr. **LANGUAGES:** English, French.

DESCRIPTION: Organization of national associations of Christian Action for the Abolition of Torture. Aims to create a unified international Christian front that campaigns against torture in all countries. Urges international action on behalf of victims of atrocities and seeks to influence policies toward countries with a poor human rights record; coordinates the activities of member groups; maintains close relations with Christian groups in all countries to gather first-hand accounts of the practice of torture. Organizes colloquia. **PUBLICATIONS:** *FIACAT*; *FIACAT News* (in English, French, and Spanish), quarterly; *Torturers, Tortured, and Christian Hope* (in English, French, and German); Books (in French).

INTERNATIONAL FEDERATION OF AGRICULTURAL PRODUCERS (IFAP) (Federation Internationale des Producteurs Agricoles)

60, rue St. Lazare, F-75009 Paris, France **PHONE:** 33 1 45260553 **FX:** 33 1 48747212 **E-MAIL:** info@ifap.org **WEBSITES:** http://www.ifap.org/influence.html **FOUNDED:** 1946. **OFFICERS:** D.L.J. King, Sec.Gen. **MEMBERS:** 80. **STAFF:** 10. **BUDGET:** US$900,000. **LANGUAGES:** English, French.

DESCRIPTION: Members are national farm organizations in 58 countries. Promotes the well-being of all who obtain their livelihood from the land, and works to assure them adequate and stable remuneration. Encourages exchange of information and ideas and fosters sustainability in production, and marketing of agricultural commodities. Provides advice and assistance to appropriate individual bodies, especially the United Nations and its organizations, on matters affecting the interests and welfare of agricultural primary producers. **PUBLICATIONS:** *Conference Proceedings*, biennial; *Farming for Development*, periodic; *IFAP Newsletter* (in English and French), periodic; Booklet.

INTERNATIONAL FEDERATION OF AIR LINE PILOTS ASSOCIATIONS (IFALPA) (Federation Internationale des Associations de Pilotes de Ligne)

Interpilot House, Gogmore Ln., Chertsey, Surrey KT16 9AP, England **PHONE:** 44 1932 571711 **FX:** 44 1932 570920 **E-MAIL:** admin@ifalpa.org **WEBSITES:** http://www.ifalpa.org **FOUNDED:** 1948. **OFFICERS:** Cathy Bill, Exec.Dir. **MEMBERS:** 94. **STAFF:** 19. **BUDGET:** £1,250,000. **LANGUAGES:** English.

DESCRIPTION: Organization of national pilot associations representing more than 100,000 pilots. Promotes the development of a safe and orderly system of air transportation and the protection of the interests of airline pilots. Activities include: the regular exchange of information and ideas; examination of common problems; coordination of policies. Maintains team of accident investigation experts to assist pilots involved in accidents. Works to standardize legislation concerning hijacking. Conducts surveys. Maintains liaison with the International Civil Aviation Organization and the International Air Transport Association. **PUBLICATIONS:** *IFALPA Quarterly Review*; *''Introducing IFALPA,''* periodic.

INTERNATIONAL FEDERATION OF AIR TRAFFIC CONTROLLERS' ASSOCIATIONS (IFATCA) (Federation Internationale des Associations de Controleurs du Trafic Aerien - FIACTA)

1255 University St., Ste. 408, Montreal, PQ, Canada **PHONE:** (514) 866-7040 **FX:** (514) 866-7612 **E-MAIL:** ifatca@sympatico.ca **FOUNDED:** 1961. **OFFICERS:** Ms. Maura Estrada, Office Mgr. **MEMBERS:** 114. **STAFF:** 2. **BUDGET:** £200,000. **LANGUAGES:** English.

DESCRIPTION: Air traffic controllers' associations representing 114 countries. Promotes safety and efficiency in international air navigation and a high standard of knowledge and professional efficiency among air traffic controllers. Assists and advises in the development of safe air traffic control systems; collects and distributes information on professional problems and developments; sponsors and supports legislation and regulations that will improve working conditions for air traffic controllers. Works for a worldwide federation of air traffic controllers' associations. Operates professional subject and technical libraries. **PUBLICATIONS:** *The Controller*, quarterly.

INTERNATIONAL FEDERATION OF AIRWORTHINESS (IFA) (Federation Internationale de Navigabilite Aerospatiale)

UK Secretariat, 58 Whiteheath Ave., Ruislip, Middlesex HA4 7PW, England **PHONE:** 44 1895 672504 **FX:** 44 1895 676656 **E-MAIL:** dksmith@rmplc.co.uk **FOUNDED:** 1975. **OFFICERS:** D.K. Smith, Dir. **MEMBERS:** 122. **STAFF:** 6. **LANGUAGES:** English.

DESCRIPTION: Aerospace manufacturers, airlines, aircraft engineering and service facilities, international flight safety associations, professional aeronautical societies, and airworthiness authorities in 47 countries. Provides a forum for the exchange of experience and ideas on all areas of airworthiness including maintenance, design, and operations. Encourages a mutual understanding between airlines and airworthiness authorities. Organizes working parties to investigate specific problems in the aerospace industry. **PUBLICATIONS:** *Annual Report of Accounts*; *Conference Proceedings*, annual; *IFA News* (in English), quarterly; *Memos*, periodic.

INTERNATIONAL FEDERATION OF ASSOCIATED WRESTLING STYLES (IFAWS) (Federation Internationale des Luttes Associees - FILA)

Ave. Juste-Olivier 17, CH-1006 Lausanne, Switzerland **PHONE:** 41 21 3128426 **FX:** 41 21 3236073 **TX:** 455958 FILA CH **E-MAIL:** filalausanne@bluewin.ch **WEBSITES:** http://www.fila-wrestling.org **FOUNDED:** 1912. **OFFICERS:** Jane Fruttiger, Sec. **MEMBERS:** 141. **STAFF:** 3. **LANGUAGES:** English, French.

DESCRIPTION: Organization of national wrestling federations. Participates in activities including international wrestling events, world championships, and the Olympics. Organizes clinics for referees and coaches. **PUBLICATIONS:** *Wrestling Review* (in English and French), semiannual.

INTERNATIONAL FEDERATION OF ASSOCIATION FOOTBALL (IFAF) (Federation Internationale de Football Association - FIFA)

c/o Fifa House, 11 Hitzigweg, CH-8030 Zurich, Switzerland **PHONE:** 41 1 3849595 **FX:** 41 1 3849696 **TX:** 817240 **E-MAIL:** webmaster@www.en-linea.com **WEBSITES:** http://www.fifa.com **FOUNDED:** 1904. **OFFICERS:** Joseph S. Blatter, Gen.Sec. **MEMBERS:** 198. **STAFF:** 50. **LANGUAGES:** English, French, German, Spanish.

DESCRIPTION: National Football (soccer) associations from 198 countries. Promotes the game of association football; encourages football matches at all levels—amateur, non-amateur, and professional; controls football by preventing infringements of the rules and by preventing introduction of other improper methods or practices in the game; works to prevent racial, religious, or political discrimination among players and associations; provides a means for resolving differences that may arise among the national associations. Organizes courses of instruction for coaches, referees, team doctors, and administrators; organizes the FIFA World Cup and World Championships for youths, women, and indoor football. **PUBLICATIONS:** *Directory*, annual; *FIFA Magazine* (in English, French, German, and Spanish), bimonthly; *FIFA News* (in English, French, German, and Spanish), monthly; *Handbook*; *Technical Report of World Cup*, quadrennial.

INTERNATIONAL FEDERATION OF ASSOCIATIONS OF TEXTILE CHEMISTS AND COLOURISTS (Internationale Foderation der Vereine der Textilchemiker und Coloristen - IFATCC)

c/o Markus Krayer, Postfach 403, CH-4153 Reinach, Switzerland **PHONE:** 41 61 6365221 **FX:** 41 61 6373063 **E-MAIL:** markus.krayer@cibasc.com **FOUNDED:** 1931. **OFFICERS:** Markus Krayer, Sec. **MEMBERS:** 21. **LANGUAGES:** English, French, German.

DESCRIPTION: National associations united to promote scientific and technical cooperation in the field. Fosters permanent personal and professional relations among members; maintains liaison with similar groups abroad.

INTERNATIONAL FEDERATION OF BEEKEEPERS' ASSOCIATIONS (APIMONDIA) (Federation Internationale des Associations d'Apiculture)

Corso Vittorio Emanuele 101, I-00186 Rome, Italy **PHONE:** 39 6 6852286 **FX:** 39 6 6852286 **TX:** 612533 **E-MAIL:** apimondia@mclink.it **FOUNDED:** 1949. **OFFICERS:** Riccardo Jannoni-Sebastianini, Sec.Gen. **MEMBERS:** 55. **STAFF:** 4. **BUDGET:** US$100,000. **LANGUAGES:** English, French.

DESCRIPTION: National beekeepers associations in 49 countries united to encourage the dissemination of information regarding new techniques, the results of scientific research, and economic developments in beekeeping. Seeks to ensure efficient coordination of the activities of member associations; compiles statistics relating to all aspects of beekeeping. Conducts symposia; sponsors competitions. **PUBLICATIONS:** *Apiacta* (in English, French, German, and Spanish), quarterly. International technical magazine of apicultural and economic information; *Congress Proceedings*, biennial; Books; Membership Directory.

INTERNATIONAL FEDERATION OF THE BLUE CROSS (IFBC) (Federation Internationale de la Croix-Bleue - FICB)

Lindenrain 5A, PO Box 6813, CH-3001 Bern, Switzerland **PHONE:** 41 31 3005860 **FX:** 41 31 3005869 **E-MAIL:** ifbc.bern@bluewin.ch **FOUNDED:** 1877. **OFFICERS:** Hans Ruttimann, Gen.Sec. **MEMBERS:** 72,946. **STAFF:** 1,180. **LANGUAGES:** English, French, German, Portuguese.

DESCRIPTION: National societies and local groups representing individuals in 43 countries. Works to help alcoholics and drug addicts. Disseminates information concerning alcoholism. Operates cure homes and clinics. **PUBLICATIONS:** *Editorial*, bimonthly; *INFO*, semiannual.

INTERNATIONAL FEDERATION OF BUSINESS AND PROFESSIONAL WOMEN (IFBPW)

Studio 16, Cloisters Business Centre, 8 Battersea Park Road, London SW8 4BG, England **PHONE:** 44 171 7388323 **FX:** 44 171 6228528 **E-MAIL:** bpwi_hq@compuserve.com **FOUNDED:** 1930. **OFFICERS:** Sylvia G. Perry, Pres. **MEMBERS:** 100,000. **STAFF:** 3. **BUDGET:** US$245,000. **LANGUAGES:** English, French, Spanish.

DESCRIPTION: Promotes the status of women worldwide. Seeks higher business and professional standards. **PUBLICATIONS:** *A Measure Filled* (in English and Spanish). Biography of founder Dr. Lena Madesin Phillips; *BPW News International* (in English, French, and Spanish), monthly; *IFBPW Trade Directory*, periodic; *In Pride and with Promise* (in English); *Roster*, semiannual; *UN Bulletin*, monthly.

INTERNATIONAL FEDERATION OF CATHOLIC PAROCHIAL YOUTH COMMUNITIES (Federation Internationale des Mouvements Catholiques d'Action Paroissiales - FIMCAP)

St. Karliquai 12, CH-6000 Lucerne, Switzerland **PHONE:** 41 41 4194747 **FX:** 41 41 4194711

FOUNDED: 1961. **OFFICERS:** Annette Leiner, Gen.Sec. **MEMBERS:** 35. **STAFF:** 5. **BUDGET:** 3,500,000 BFr. **LANGUAGES:** English, French.

DESCRIPTION: National Catholic youth organizations in 24 countries representing 2,000,000 individuals. Seeks to provide young people with a basic Christian parish life and to prepare them for active roles in contributing to a better world and a living church. Aims to unite members into one Christian community and to create an understanding of needs throughout the world. Sponsors religious activities, courses, group discussions, and instruction; studies religious, social, and human problems. Conducts World Camp summer programs. Promotes activities of national groups; represents members' interests before other international organizations. Provides exchange programme for local groups (mainly in Europe). **PUBLICATIONS:** *Link* (in English, French, and Spanish), bimonthly. Newsletter for national responsive and interested readers.

INTERNATIONAL FEDERATION OF CELL BIOLOGY (IFCB)

c/o Dept. of Cellular & Structural Biology, Univ. of Texas Health Science Center, 7703 Floyd Curl Dr., San Antonio, TX 78229 **PHONE:** (210) 567-3817 **FX:** (210) 567-3803 **FOUNDED:** 1972. **OFFICERS:** Dr. Ivan L. Cameron, Sec.Gen. **MEMBERS:** 21. **LANGUAGES:** English, French.

DESCRIPTION: Members are national (7) and regional (14) associations of cell biologists. Works to: promote international cooperation among scientists working in cell biology and related fields; contribute to the advancement of cell biology in all of its branches. Acts as coordinating body that initiates special studies and encourages research in subjects outside the normal scope of national societies, such as the problem of scientific communication. Conducts seminars. **PUBLICATIONS:** *Cell Biology International Reports*, monthly; *International Cell Biology - Review of International Congresses.*

INTERNATIONAL FEDERATION OF CLINICAL CHEMISTRY AND LABORATORY MEDICINE (IFCC) (Federation Internationale de Chimie Clinique)

30, Rue Lionnois, F-54000 Nancy, France **PHONE:** 33 385352616 **FX:** 33 383321322 **WEBSITES:** http://www.ifcc.org **FOUNDED:** 1952. **OFFICERS:** Chantal Thirion, Office Mgr. **MEMBERS:** 75. **BUDGET:** $A 500,000. **LANGUAGES:** English, French.

DESCRIPTION: Members are national societies of clinical chemistry representing 25,000 individuals. Purposes are to advance the science and practice of clinical chemistry and to enhance its service to health and medicine. Authorizes and sponsors international congresses of clinical chemistry; encourages joint participation of members in related congresses. Conducts studies, reviews, and reports on clinical chemistry problems of international concern. Offers advice, consultation, and recommendations on clinical chemistry problems. Provides the basis for closer liaison and the free exchange of professional information and data among clinical chemists worldwide. Promotes international cooperation and coordination of clinical chemistry in matters of research, reference methods and materials, uniform regulations and statutes, and related subjects. Assists in the improvement of professional standards and codes of ethics wherever possible. Contributes in other ways wherever feasible to the improvement of clinical chemistry and its services to humanity. Maintains expert committees for investigation into such subjects as instrumentation, theory of reference values, nomenclature and principles of quality control, assessment of nutritional status, assessment of drugs of abuse, assessment of molecular biological techniques, proteins, immunoassay, enzymes, quantities, and units. Organizes training courses. **PUBLICATIONS:** *Journal of the International Federation of Clinical Chemistry*, bimonthly; *Recommendations on Clinical Chemistry*, periodic; Directory, periodic.

INTERNATIONAL FEDERATION OF CONSULTING ENGINEERS (FIDIC) (Federation Internationale des Ingenieurs Conseils - FIDIC)

BP 86, CH-1000 Lausanne 12, Switzerland **PHONE:** 41 21 6544411 **FX:** 41 21 6535432 **E-MAIL:** fidic@pobox.com **WEBSITES:** http://www.fidic.org **FOUNDED:** 1913. **OFFICERS:** Marshall Gysi, Mng.Dir. **MEMBERS:** 67. **STAFF:** 7. **BUDGET:** 1,500,000 SFr. **LANGUAGES:** English, French, German, Spanish.

DESCRIPTION: Members are national associations of independent consulting engineers. Promotes the interests of independent consulting engineers and represents electrical, civil, environmental, geotechnical, metallurgical, agricultural, and other engineering fields. Standardizes forms of agreement, contracts, and guidelines for consulting engineers; acts as clearinghouse. **PUBLICATIONS:** *FIDIC International Directory*, biennial; *Guide to Quality Management in the Consulting Engineering Industry*; *Independent Consulting Engineer*, 3/year; *INFO*, annual; *The White Book Guide*. Provides insights into the rationale for the White Book's provisions; Papers; Reports.

INTERNATIONAL FEDERATION OF DISABLED WORKERS AND CIVILIAN HANDICAPPED (FIMITIC) (Federation Internationale des Mutiles, des Invalides du Travail et des Invalides Civils - FIMITIC)

Beethovenallee 56-58, D-53173 Bonn, Germany **PHONE:** 49 228 95640 **FX:** 49 228 9564132 **E-MAIL:** fimitic@t-online.de **FOUNDED:** 1953. **OFFICERS:** Marija Stiglic, Sec.Gen. **MEMBERS:** 5,000,000. **STAFF:** 3. **LANGUAGES:** English, French, German.

DESCRIPTION: National organizations in 28 countries united to promote physical and vocational rehabilitation and full employment for the disabled. Seeks to foster international cooperation in the development of rehabilitation services; sponsors the International Day of the Disabled. Holds seminars. Acts as clearinghouse. **PUBLICATIONS:** *Congress Reports*; *Nouvelles* (in English and German), quarterly.

INTERNATIONAL FEDERATION OF EDUCATIVE COMMUNITIES (FICE) (Federation Internationale des Communautes Educatives - FICE)

Piazza SS. Annunziata 12, I-50122 Florence, Italy **PHONE:** 39 55 2469162 **FX:** 39 55 2347041 **E-MAIL:** fice@lycosmail.com **WEBSITES:** http://www.freeweb.org/associazioni/fice **FOUNDED:** 1948. **OFFICERS:** Gianluca Babanotti, Sec.Gen. **LANGUAGES:** English, French, German.

DESCRIPTION: Founded under the auspices of UNESCO and Social Council, United Nations Educational, Scientific and Cultural Organization, and United Nations Children's Fund. Members are professionals in 27 countries working in the field of child education. Promotes improvement in residential care and alternative education, in accordance with the U.N. Charter of the Child. Addresses problems involving the education of children deprived of a normal home environment. **PUBLICATIONS:** *FICE Bulletin* (in English, French, and German), semiannual; *Policymaking, Research and Staff in Residential Care*; *Recent Changes and New Trends in Extrafamilial Child: An International Perspective*. Containing information on the situation and development of extrafamilial child care; *Training of Residential Child and Youth Care Staff*. Additional publication information available upon request.

INTERNATIONAL FEDERATION OF EUROPE HOUSES (Federation Internationale des Maisons de l'Europe - FIME)

Pestelstrasse 2, D-66119 Saarbrucken, Germany **PHONE:** 49 681 954520, 49 681 9545222 **FX:** 49 681 9545250 **E-MAIL:** fime-gs@t-online.de **FOUNDED:** 1962. **OFFICERS:** Arno Krause, Pres. **MEMBERS:** 104. **LANGUAGES:** English, French, German, Italian.

DESCRIPTION: Institutions in 23 countries known as Europe-Houses that conduct activities aimed at fostering European unity and overcoming problems related to European integration. Member institutions work as independent bodies within the systems of each country. Promotes cooperation among members in the preparation and presentation of information; organizes professional training programs. **PUBLICATIONS:** *FIME* (in English, French, German, and Italian), annual; *FIME-INFO* (in English, French, German, and Italian), quarterly; *FIME-Program* (in English, French, German, and Italian), annual.

INTERNATIONAL FEDERATION OF FREE JOURNALISTS (IFFJ)

4 Overton Rd., London N14 4SY, England **PHONE:** 44 181 3602991 **FOUNDED:** 1942. **OFFICERS:** Krystyna Asipowicz, Gen.Sec. **MEMBERS:** 94. **LANGUAGES:** English, French, German, Polish.

DESCRIPTION: Journalists of Central and Eastern European origin in 7 countries. Promotes and defends the principles of a free and honest press; works in the defense of freedom of information and of human and national rights. Strives to raise standards and ethics of journalism and to foster cooperation among journalistic organizations. Provides articles to the British and East European press; sponsors periodic seminars on political, social, cultural, and economic problems in Eastern and Central Europe.

INTERNATIONAL FEDERATION OF FREIGHT FORWARDERS ASSOCIATIONS (FIATA) (Federation Internationale des Associations de Transitaires et Assimilies - FIATA)

Baumackerstrasse 24, Postfach 8493, CH-8050 Zurich, Switzerland **PHONE:** 41 1 3116511 **FX:** 41 1 3119044 **E-MAIL:** info@fiata.com **WEBSITES:** http://www.fiata.com **FOUNDED:** 1926. **OFFICERS:** Marco A. Sangaletti, Dir. **MEMBERS:** 2,500. **STAFF:** 7. **LANGUAGES:** English, French, German.

DESCRIPTION: Forwarding firms in 95 national organizations and more than 2,500 associated members in 150 countries united to protect the interests of the forwarding trade at the international level. Organizes FIATA Standard Training Programme and forums; maintains consultative status with the United Nations. **PUBLICATIONS:** *FIATA*, periodic; *FIATA Review*, 5/year; *Membership List*, annual; *News*, 10/year; Brochures, periodic.

INTERNATIONAL FEDERATION OF GYNECOLOGY AND OBSTETRICS (FIGO) (Federation Internationale de Gynecologie et d'Obstetrique - FIGO)

27 Sussex Pl., Regent's Park, London NW1 4RG, England **PHONE:** 44 171 7232951 **FX:** 44 171 2580737 **E-MAIL:** secret@figo.win-uk.net **WEBSITES:** http://www.figo.org **FOUNDED:** 1954. **OFFICERS:** Dr. Giuseppe Benagiano, Sec.Gen. **MEMBERS:** 100. **STAFF:** 4. **LANGUAGES:** English, French, Spanish.

DESCRIPTION: Objectives are to: promote and assist in the development of scientific and research work relating to all facets of gynecology and obstetrics; improve the physical and mental health of women, mothers, and their children; provide an exchange of information and ideas; improve teaching standards; promote international cooperation among medical bodies. Acts as liaison with World Health Organization and other international organizations. **PUBLICATIONS:** *FIGO Annual Report on the Results of Treatment of Gynecologic Cancer.* Provides results of treatment in gynecological cancer; *FIGO Newsletter*, 3/year; *International Journal of Gynecology and Obstetrics*, monthly.

INTERNATIONAL FEDERATION FOR HOUSING AND PLANNING (IFHP) (Federation Internationale pour l'Habitation, l'Urbanisme et l'Amenagement des Territoires - FIHUAT)

Wassenaarseweg 43, NL-2596 CG The Hague, Netherlands **PHONE:** 31 70 3244557 **FX:** 31 70 3282085 **E-MAIL:** ifhp.nl@inter.nl.net **FOUNDED:** 1913. **OFFICERS:** Elsbeth van Hylckama Vlieg, Sec.Gen. **MEMBERS:** 500. **STAFF:** 5. **LANGUAGES:** English, French, German.

DESCRIPTION: Planners, architects, municipal officials, housing experts, sociologists, economists, public health experts, politicians, interested laymen, governmental departments, local authority agencies, national associations, project developers, consultancy agencies, financing institutions, university departments, and research institutes. Studies and promotes the improvement of housing, town planning, regional planning, and the preservation and improvement of the human environment. Maintains international contacts and information exchange. Establishes working groups. **PUBLICATIONS:** *IFHP Membership List and Directory*, biennial; *IFHP Newsletter* (in English, French, and German), quarterly. Contains reports on IFHP activities and events. Provides articles about new developments.; *Latest Developments in the Field, News from the IFHP Bureau*, annual. Also publishes monographs and reports on congresses, conferences and other IFHP activities. Edited a Trilingual Dictionary of Housing and Planning.

INTERNATIONAL FEDERATION OF INDUSTRIAL ENERGY CONSUMERS (IFIEC) (Federation Internationale des Industries Consommatrices d'Energie)

7, chemin des Tattes, Vesenaz, CH-1222 Geneva, Switzerland **PHONE:** 41 22 7522364 **FX:** 41 22 7525624 **FOUNDED:** 1983. **OFFICERS:** Dr. Werner K. Veith, Sec.Gen. **MEMBERS:** 19. **LANGUAGES:** English.

DESCRIPTION: Executive representatives of national organizations and companies from 13 countries. Concerns itself with industrial energy consumption and its technological, economic, and governmental policy implications. Promotes the interests of consumers of industrial energy and supports research on energy, technology, and economics. Monitors and seeks to influence related governmental policies. Works to assure adequate energy supply at reasonable cost; encourages related public debate and publicizes accomplishments.

INTERNATIONAL FEDERATION FOR INFORMATION AND DOCUMENTATION (FID) (Federation Internationale d'Information et de Documentation - FID)

Prins Willem-Alexanderhof 5, Postbus 90402, NL-2509 LK The Hague, Netherlands **PHONE:** 31 70 3140671 **FX:** 31 70 3140667 **E-MAIL:** fid@python.konbib.nl **WEBSITES:** http://fid.conicyt:8000.cl **FOUNDED:** 1895. **OFFICERS:** Stephen Parker, Exec.Dir. **MEMBERS:** 400. **STAFF:** 5. **LANGUAGES:** English.

DESCRIPTION: Individuals and organizations who develop, produce, research and use information products and who are involved in information management. Promotes the development of information professionals and users. Focuses on new trends in: information management and service; information in corporate environments; business, finance and industrial information; information policy research; marketing information systems and services; application of information technology to information services. **PUBLICATIONS:** *Education and Training Newsletter*, quarterly; *FID Directory*, biennial; *International Forum on Information and Documentation*, quarterly.

INTERNATIONAL FEDERATION FOR INFORMATION PROCESSING (IFIP) (Federation Internationale pour le Traitement de l'Information)

IFIP Secretariat, Hofstrasse 3, A-2361 Laxenburg, Austria **PHONE:** 43 2236 73616 **FX:** 43 2236 736169 **E-MAIL:** ifip@ifip.or.at **WEBSITES:** http://www.ifip.or.at **FOUNDED:** 1960. **OFFICERS:** K. Bauknecht, Pres. **MEMBERS:** 44. **STAFF:** 2. **LANGUAGES:** English.

DESCRIPTION: Professional-technical information processing societies representing 63 countries. Fosters cooperation and exchange between information technology organizations. Organizes symposia and workshops. **PUBLICATIONS:** *Congress and Conference Proceedings*; *Information Bulletin*, annual; *Newsletter*, quarterly; *What is IFIP*, periodic; Journals.

INTERNATIONAL FEDERATION OF JOURNALISTS (IFJ) (Federation Internationale des Journalistes - FIJ)

c/o Aidan White, Sec. Gen., Rue Royale, 266, B-1210 Brussels, Belgium **PHONE:** 32 2 2232265 **FX:** 32 2 2192976 **TX:** 61275 IPC **E-MAIL:** ifj@pophost.eunet.be **WEBSITES:** http://www.ifj.org **FOUNDED:** 1926. **OFFICERS:** Aidan White, Gen.Sec. **MEMBERS:** 94. **STAFF:** 11. **LANGUAGES:** English, French, German, Spanish.

DESCRIPTION: National trade unions and professional organizations of journalists in the press, radio, and television fields in 75 countries representing over 320,000 individuals. Works to: safeguard freedom of the press and freedom of journalists according to Article 19 of the Universal Declaration of Human Rights; uphold the IFJ Declaration of Principles, a code of conduct for journalists. Aims to defend and advance the moral and material interests of journalists, especially by providing factual information and model contracts for use and guidance in collective bargaining. Issues protests against acts threatening freedom of expression and trade union freedom including the harassment, persecution, and detention without trial of journalists, and censorship and suppression of mass media. Organizes fact-finding missions of IFJ representatives in countries where press freedom is under pressure or is being suppressed and offers courses in unionism to journalists worldwide, especially in developing countries and Eastern Europe. Conducts surveys and publishes results; makes recommendations on activities common to journalists and not determined by law or practice. Promotes goodwill, assistance, and protection for members traveling and working in other member countries. Assists in professional and trade union training programs. Issues International Press Card certifying the holder as a professional journalist. Participates in activities of the United Nations, the United Nations Educational, Scientific and Cultural Organization, the Council of Europe, World Intellectual Property Organization, and other international organizations concerned with journalists and the press. **PUBLICATIONS:** *Direct Line* (in English,

French, German, and Spanish), monthly; *IFJ Information* (in English, French, and Spanish), annual. Special reports.

INTERNATIONAL FEDERATION OF LIBERAL AND RADICAL YOUTH (IFLRY) (Federacion International de Juventudes Liberales y Radicales)

BP 781, B-1000 Brussels, Belgium **PHONE:** 32 2 5124457 **FX:** 32 2 5024122 **E-MAIL:** iflry@unicall.be **WEBSITES:** http://www.worldlib.org/iflry **FOUNDED:** 1947. **MEMBERS:** 80. **STAFF:** 1. **BUDGET:** 200,000 BFr. **LANGUAGES:** English, French, Spanish.

DESCRIPTION: Members are international youth organizations representing 1,500,000 individuals in 52 countries. Strives to foster peace, freedom, justice, and international unity. Has staged a political pressure campaign to denounce racism and xenophobia worldwide. Supports continued democratization of Eastern Europe. Conducts bimonthly seminar. **PUBLICATIONS:** *ILFLRY Newsletter* (in English), monthly; *Libel* (in English and Spanish), bimonthly. Also publishes seminar and congress reports.

INTERNATIONAL FEDERATION OF LIBRARY ASSOCIATIONS AND INSTITUTIONS (IFLA)

c/o Mr. Leo Voogt, Gen. Sec., Postbus 95312, NL-2509 CH The Hague, Netherlands **PHONE:** 31 70 3140884, 31 70 3140755 **FX:** 31 70 3834827 **E-MAIL:** ifla.hq@ifla.nl **WEBSITES:** http://www.ifla.org **FOUNDED:** 1927. **OFFICERS:** Leo Voogt, Sec.Gen. **MEMBERS:** 1,650. **STAFF:** 9. **BUDGET:** 1,000,000 f. **LANGUAGES:** English, French, German, Russian, Spanish.

DESCRIPTION: Library associations and institutions in 153 countries. Promotes international understanding, cooperation, discussion, research, and development in all fields of library activity including bibliography, information services, and the education of personnel. Represents members in matters of international interest and communicates on behalf of the library profession worldwide. Studies practical questions of international librarianship. Aims to standardize the field of libraries and books. Maintains 45 committees. **PUBLICATIONS:** *IFLA Council Report* (in English), annual; *IFLA Directory* (in English), biennial; *IFLA Journal* (in English), bimonthly. Includes abstracts in French, German, Spanish, and Russian.; *IFLA Professional Reports* (in English), bimonthly. French and Spanish translations.; *IFLA/Saur Publications Series* (in English), quarterly; *IFLA's Medium Term Programme 1992-1997, 1998-2003*; *International Cataloguing and Bibliographic Control*, quarterly.

INTERNATIONAL FEDERATION OF MEDICAL STUDENTS ASSOCIATIONS (IFMSA) (Federacion Internacional de Asociaciones de Estudiantes de Medicina)

c/o WMA, Boite Postale 63, R-01212 Ferney-Voltaire Cedex, France **PHONE:** 33 450404759 **FX:** 33 450405937 **FOUNDED:** 1951. **OFFICERS:** Mrs. Mia Hilhorst. **MEMBERS:** 50. **STAFF:** 1. **LANGUAGES:** English, French, Spanish.

DESCRIPTION: National medical student associations in 50 countries. Objectives are: to provide an international forum for medical students to discuss topics of interest and to formulate policies from such discussions; to provide a professional exchange for medical students; to facilitate contacts with other worldwide organizations; to enhance members' influence in fundraising activities for IFMSA-recognized projects. Trains medical personnel to meet the health needs of their societies. Organizes clerkships and study tours. Places students in hospitals worldwide through its Professional Exchange Committee. Promotes practical teaching in primary health care. Has established relations with the European Economic Community, and organizations of the United Nations and the World Health Organization. **PUBLICATIONS:** *IFMSA Annual Report*; *IFMSA Newsletter*, quarterly; *Medical Student International*; *The SCOPE Handbook*.

INTERNATIONAL FEDERATION FOR MODERN LANGUAGES AND LITERATURES (Federation Internationale des Langues et Litteratures Modernes - FILLM)

c/o Prof. David A. Wells, Birkbeck Coll., Malet St., London WC1E 7HX, England **PHONE:** 44 171 6316103 **FX:** 44 171 3833729 **FOUNDED:** 1928. **OFFICERS:** Prof. David A. Wells, Sec.Gen. **MEMBERS:** 18. **LANGUAGES:** English, French, German, Italian, Russian, Spanish.

DESCRIPTION: Federation of modern language and literature societies. Objectives are to establish permanent contact between literary scholars, to develop facilities for their work, and encourage their academic disciplines under the aegis of United Nations Educational, Scientific and Cultural Organization. Promotes the study of the history of medieval and modern languages and literatures. **PUBLICATIONS:** *Acta*, triennial.

INTERNATIONAL FEDERATION OF MULTIPLE SCLEROSIS SOCIETIES (IFMSS)

c/o MS Society of Great Britain and Northern Ireland, 25 Effie Rd., London SW6 1EE, England **PHONE:** 44 171 6107171 **FX:** 44 171 7369861 **E-MAIL:** info@ifmss.org.uk **WEBSITES:** http://www.ifmss.org.uk **FOUNDED:** 1967. **OFFICERS:** Richard Hamilton, Sec. General. **MEMBERS:** 36. **STAFF:** 6. **LANGUAGES:** English, French.

DESCRIPTION: Key aims are to stimulate scientific research at a global scale, disseminate information internationally, assist the development of national MS societies, and encourage full integration and participation of all people affected by MS. **PUBLICATIONS:** *Annual Report* (in English); *Federation Updates*, biennial; *MS Management*; *MS Research in Progress*; *MS Therapeutic Claims.*

INTERNATIONAL FEDERATION OF MUSICIANS (Federation Internationale des Musiciens - FIM)

21 bis rue Victor Masse, F-75009 Paris, France **PHONE:** 33 01 45263123 **FX:** 33 01 45263157 **E-MAIL:** 106340.1224@compuserve.com **FOUNDED:** 1948. **OFFICERS:** M. Jean Vincent, Gen.Sec. **MEMBERS:** 50. **STAFF:** 2. **BUDGET:** 475,000 SFr. **LANGUAGES:** English, French, German, Spanish.

DESCRIPTION: National unions of musicians representing 102,000 individuals. Works to protect and further the economic, legal, social, and artistic interests of musicians organized in member unions. Protects members from unauthorized broadcasting, recording, and public performance of their work worldwide. Serves as liaison between musicians and other international organizations. Offers moral and material support to member unions in the interests of the profession. Seeks to improve working conditions in the field. **PUBLICATIONS:** *FIM-Bulletin* (in English, French, German, and Spanish), quarterly.

INTERNATIONAL FEDERATION OF OPERATIONAL RESEARCH SOCIETIES (IFORS) (Federation Internationale des Societes de Recherche Operationnelle - FISRO)

c/o Helle R. Welling, Technical University of Denmark, IMSOR, Bldg. 321, Rm. 209, DK-2800 Lyngby, Denmark **PHONE:** 45 45253410, 45 45881433 **FX:** 45 45881397 **E-MAIL:** hw@imm.dtu.dk **WEBSITES:** http://www.ifors.org **FOUNDED:** 1959. **OFFICERS:** Loretta Pelegrina, Sec. **MEMBERS:** 48. **STAFF:** 1. **LANGUAGES:** English, French.

DESCRIPTION: Members are national operational research societies (44) and other organizations interested in activities of the federation (4) representing 30,000 individuals in 44 countries. (Operational research is a scientific approach to the solution of problems in the management of complex systems.) Primary objective is the advancement of operational research as a unified science worldwide. Encourages the teaching of operational research, the establishment of national operational research societies, and the exchange of information among national groups. **PUBLICATIONS:** *International Abstracts of Operational Research*, 8/year; *International Transactions of Operational Research*, 4-5/year; Bulletin, 6-8/year.

INTERNATIONAL FEDERATION OF OPHTHALMOLOGICAL SOCIETIES (IFOS)

c/o Bruce E. Spivey. MD, Sec.-General, Columbia-Cornell Care, L.L.C., 900 Third Ave., Ste. 500, New York, NY 10022 **PHONE:** (212) 588-7301 **FX:** (212) 588-7307 **E-MAIL:** bspivey@cccare.com **WEBSITES:** http://www.icoph.org **FOUNDED:** 1857. **OFFICERS:** Bruce E. Spivey MD, Sec.-Gen.

DESCRIPTION: The ICO represents opthalmologic organizations throughout the world. Dedicated to international exchange in ophthalmology. Encourages the study and improvement of ophthalmologic education; formulates international standards. Advocates the prevention and treatment of preventable blindness in developing nations, particularly

Africa. Supports the International Agency for the Prevention of Blindness.

INTERNATIONAL FEDERATION OF OTO-RHINO-LARYNGOLOGICAL SOCIETIES (IFOS) (Federation Internationale des Societes Oto-Rhino-Laryngologiques)

7-219 Eaton N, 200 Elisabeth St., Toronto, ON, Canada M5G 2C4 **PHONE:** (416) 340-4190 **FX:** (416) 340-4209 **E-MAIL:** peter.alberti@utoronto.ca **WEBSITES:** http://www.ifosworld.org **FOUNDED:** 1965. **OFFICERS:** Dr. P.W. Alberti, Sec.Gen. **MEMBERS:** 95. **LANGUAGES:** English, French, German, Japanese, Spanish.

DESCRIPTION: Members are national (83) and international otorhinolaryngological societies (12) in 72 countries. Promotes the advancement of otorhinolaryngology (ORL), the prevention and control of diseases and disorders of the ear, nose, and throat, and the training of otorhinolaryngologists. Participates, as an affiliate to the World Health Organization, in the World-wide Prevention Action on Hearing Impairment. Encourages international cooperation between otorhinolaryngologists and members. Plans to establish a museum. Sponsors film and video competitions. **PUBLICATIONS:** *IFOS Newsletter* (in English), quarterly; *International ORL Directory*, periodic; *Year Book*; Films, periodic; Reports, periodic; Videos, periodic.

INTERNATIONAL FEDERATION FOR PARENT EDUCATION (IFPE) (Federation Internationale pour l'Education des Parents - FIEP)

1, ave. Leon-Journault, F-92311 Sevres Cedex, France **PHONE:** 33 1 45072164 **FX:** 33 1 46266927 **FOUNDED:** 1964. **OFFICERS:** Moncef Guitouni, Pres. **MEMBERS:** 130. **LANGUAGES:** English, French.

DESCRIPTION: Members are nongovernmental associations or institutions and individuals interested in encouraging the establishment of family education worldwide. Studies the problems confronting various cultures arising out of the need experienced by families throughout the world to adjust to changing ideas and ways of life. Provides parents with access to the educational work carried out in different countries. Holds Annual colloquium; conducts international seminars and research programs sponsored by United Nations Educational, Scientific and Cultural Organization.

INTERNATIONAL FEDERATION OF PARK AND RECREATION ADMINISTRATION (IFPRA) (Federation Internationale des Services des Espaces Verts et de la Recreation)

The Grotto, Lower Basildon, Reading, Berks. RG8 9NE, England **PHONE:** 44 1491 874800 **FX:** 44 1491 874801 **E-MAIL:** ifpra@ilam.co.uk **FOUNDED:** 1957. **OFFICERS:** Alan Smith, Gen.Sec. **MEMBERS:** 456. **LANGUAGES:** English, French, German.

DESCRIPTION: Members are individuals from the fields of parks, recreation, amenity, leisure, and related services; government departments, municipal and public authorities, universities, and scientific and educational institutions; national allied professional associations are affiliates. Objectives are to: establish a world coordinating center; collect and disseminate general and statistical information to members; promote the establishment of national associations for affiliation with the federation. Attempts to facilitate the exchange of students and professionals between countries and to adopt internationally acceptable training and qualification standards. Establishes special committees to study matters of professional interest; promotes research. **PUBLICATIONS:** *IFPRA Bulletin* (in English), quarterly; Proceedings, periodic.

INTERNATIONAL FEDERATION OF THE PERIODICAL PRESS (FIPP) (Federation Internationale de la Presse Periodique)

Queens House, 55/56 Lincoln's Inn Fields, London WC2A 3LJ, England **PHONE:** 44 0171 4044169 **FX:** 44 0171 4044170 **E-MAIL:** info@fipp.com **WEBSITES:** http://www.fipp.com **FOUNDED:** 1925. **OFFICERS:** Mr. Per R. Mortensen, Pres. and COO. **MEMBERS:** 114. **STAFF:** 3. **BUDGET:** £150,000. **LANGUAGES:** English, French, Norwegian, Spanish.

DESCRIPTION: Members are national associations of periodical publishers (34) and individual periodical publishing companies (100) in 31 countries. Organized to: support press freedom; protect and represent the interests of its members before international bodies; compile and distribute information internationally; promote the image of the periodical press as an advertising medium; encourage the adoption of uniform standards. Maintains liaison with United Nations Educational, Scientific and Cultural Organization, International Chamber of Commerce, and Universal Postal Union. **PUBLICATIONS:** *FIPP/FAEP Environment Position Paper*; *FIPP Membership Directory*, annual; *FIPP Specification for European Gravure Printing*; *FIPP/Zenith World Magazine Trends* (in English); *Magazine World*, bimonthly; books and papers.

INTERNATIONAL FEDERATION OF PHARMACEUTICAL MANUFACTURERS ASSOCIATIONS (IFPMA) (Federation Internationale de l'Industrie du Medicament - FIIM)

30, rue St. Jean, PO Box 9, CH-1211 Geneva 18, Switzerland **PHONE:** 41 22 3401200 **FX:** 41 22 3401380 **E-MAIL:** mcone@ifpma.org **WEBSITES:** http://www.ifpma.org **FOUNDED:** 1968. **OFFICERS:** Dr. Harvey E. Bale Jr., Dir.Gen. **MEMBERS:** 56. **STAFF:** 8. **LANGUAGES:** English, French, Spanish.

DESCRIPTION: Members are pharmaceutical manufacturers' associations. Objectives are to: facilitate communication among members; answer questions in the field of health legislation, science, and research; contribute to the advancement of worldwide health and welfare; promote development of ethical principles and practices throughout the pharmaceutical industry. Operates the IFPMA Code of Marketing. Examines scientific and policy issues relating to the international production, use, and supply of pharmaceuticals. Offers expertise and cooperation to governmental and nongovernmental international organizations. Operates joint scheme with the World Health Organization for training government quality control laboratory personnel in developing countries. **PUBLICATIONS:** *Code of Pharmaceutical Marketing Practices* (in English, French, and Spanish); *GATT Trips and the Pharmaceutical Industry: A Review* (in English); *Health Horizons* (in English, French, and Spanish), 3/year; *IFPMA Structure and Activities* (in English), periodic; *International Pharmaceutical Issues* (in English); *Symposia Proceedings*, periodic; Booklets, periodic.

INTERNATIONAL FEDERATION OF PHILOSOPHICAL SOCIETIES (IFPS) (Federation Internationale des Societes de Philosophie - FISP)

Seminaire de Philosophie de l'Universite, CH-1700 Fribourg, Switzerland **PHONE:** 41 26 4242669 **FX:** 41 26 3009786 **E-MAIL:** philosophie@unifr.ch **WEBSITES:** http://www.bu.edu/wcp/ **FOUNDED:** 1948. **OFFICERS:** Prof. Ioanna Kucuradi. **MEMBERS:** 102. **BUDGET:** US$12,000. **LANGUAGES:** English, French, German, Russian, Spanish.

DESCRIPTION: Members are national and international philosophical societies, academies, and unions. Promotes international cooperation in the field of philosophy. Conducts research programs. **PUBLICATIONS:** *Congress Proceedings*, quinquennial; *FISP Newsletter*, annual; *Ideas Underlying World Problems*; *Philosophes Critiques d'Eux-Memes*, periodic; *Philosophy and Cultural Development*.

INTERNATIONAL FEDERATION OF THE PHONOGRAPHIC INDUSTRY (IFPI)

IFPI Secretariat, 54 Regent St., London W1R 5PJ, England **PHONE:** 44 171 8787900 **FX:** 44 171 8787950 **TX:** 919044 IFPI G **E-MAIL:** info@ifpi.org **WEBSITES:** http://www.ifpi.org **FOUNDED:** 1933. **OFFICERS:** Nicholas Garnett, Dir.Gen. **MEMBERS:** 1,319. **LANGUAGES:** English, French, German, Italian, Spanish.

DESCRIPTION: Members are producers and distributors of phonograms and videograms in 72 countries. Promotes the interests of members through the use of statutes, case law, contracts, and agreements. Represents members on national and international copyright issues; coordinates anti-piracy actions. Compiles statistical information on the international recording industry. Provides industry contacts and advisory services. **PUBLICATIONS:** *Annual Review*; *Newsletter*, quarterly; Reports, periodic.

INTERNATIONAL FEDERATION FOR PHYSICAL EDUCATION (Federation Internationale d'Education Physique - FIEP)

4 Cleevecroft Ave., Bishops Cleeve, Cheltenham, Glos. GL52 4JZ, England **PHONE:**

44 1242 673674 **FX:** 44 1242 673674 **FOUNDED:** 1923. **OFFICERS:** John C. Andrews, Pres. **LANGUAGES:** English, French, Spanish.

DESCRIPTION: Members are colleges, universities, and physical education and sports institutes; administrators; interested individuals; libraries. Works to develop physical education programs and activities in all countries and to promote international exchange and cooperation. Deals with scientific, technical, pedagogical, and social aspects of physical education and sports; does not take part in political, religious, or racial discussions or discrimination. Offers courses. **PUBLICATIONS:** *FIEP Bulletin* (in English, French, and Spanish), 3/year.

INTERNATIONAL FEDERATION OF PRESS CUTTING AGENCIES (IFPCA) (Federation Internationale des Bureaux d'Extraits de Presse - FIBEP)

Streulistrasse 19, CH-8030 Zurich, Switzerland **PHONE:** 41 1 3888200 **FX:** 41 1 3888201 **FOUNDED:** 1953. **OFFICERS:** Dr. Dieter Henne, Sec.Gen. **MEMBERS:** 67. **LANGUAGES:** English, French, German.

DESCRIPTION: Members are press cutting agencies (organizations supplying newspaper clippings to their subscribers). Seeks to: improve the standing of the profession; prevent unfair and illegal practices such as copyright infringement; develop friendly business relations among press cutting bureaus worldwide.

INTERNATIONAL FEDERATION OF RED CROSS AND RED CRESCENT SOCIETIES

P.O. Box 372, CH-1211 Geneva 19, Switzerland **PHONE:** 41 22 7304222 **FX:** 41 22 7331727 **TX:** 4121333 **FOUNDED:** 1919. **OFFICERS:** George Weber, Sec.Gen. **MEMBERS:** 169. **BUDGET:** 175,000,000 SFr. **LANGUAGES:** Arabic, English, French, Spanish.

DESCRIPTION: Members are national Red Cross and Red Crescent Societies representing over 250 million individuals. Objectives are: to conduct and coordinate international disaster relief; to be a permanent organ of liaison, coordination, and study among members; to assist members societies in organizing and carrying out their work on both national and international levels; to provide care for refugees outside conflict areas. Supports establishment and development of national societies; sponsors Red Cross and Red Crescent Youth. Offers advisory services on community health and blood tranfusion. Research and educational programs include: regional institutes; training courses; health education; youth programs; information network; community services; blood programs; development programs; pre-disaster planning. Operates documentation centre; compiles statistics. Administers charitable program; offers children's services. Operates speakers' bureau; maintains museum and hall of fame. **PUBLICATIONS:** *Annual Review*; *Red Cross, Red Crescent*, 3/year; *Weekly News*. Also publishes *Red Cross, Red Crescent Directory*.

INTERNATIONAL FEDERATION OF SCIENCE EDITORS (IFSE)

School for Scientific Communication, Mario Negri Sud Inst., I-66030 Imbaro, Italy **PHONE:** 39 872 570316, 39 872 570303 **FX:** 39 872 570317 **E-MAIL:** balaban@cmns.mnegri.it **WEBSITES:** http://www.cmns.mnegri.it/ifse **FOUNDED:** 1978. **LANGUAGES:** English.

DESCRIPTION: Members are editors of scientific periodicals. Promotes free exchange of scientific information worldwide. Conducts continuing professional development courses for science periodical staff.

INTERNATIONAL FEDERATION OF SENIOR POLICE OFFICERS (FIFSP) (Federation Internationale des Fonctionnaires Superieurs de Police)

26, rue de Cambaceres, F-75008 Paris, France **PHONE:** 33 1 49274067 **FX:** 33 1 49240113 **FOUNDED:** 1950. **OFFICERS:** J.P. Havrin, Sec.Gen. **MEMBERS:** 10,000. **LANGUAGES:** English, French, German.

DESCRIPTION: Members are senior police officers and police associations in 45 countries. Focuses its concerns on: police and human rights; creation of an international code of police ethics; social and preventive police actions; and police relations with the public. Conducts research programs on intellectual property (counterfeiting) cybercrime. **PUBLICATIONS:** *Congress Proceedings*, triennial; *International Police Information*, 2-4/year; Directory, periodic. Also publishes monographs.

INTERNATIONAL FEDERATION OF SHIPMASTERS' ASSOCIATIONS (IFSMA) (Federation Internationale des Associations de Patrons de Navires - FIAPN)

202 Lambeth Rd., London SE1 7JY, England **PHONE:** 44 171 2610450 **FX:** 44 171 9289030 **TX:** 934089 MARSOC G **E-MAIL:** hq@ifsma.org **FOUNDED:** 1974. **OFFICERS:** Capt. R. Clipsham, Gen.Sec. **MEMBERS:** 8,000. **STAFF:** 3. **LANGUAGES:** English.

DESCRIPTION: Members are national associations of qualified seagoing master mariners and shipmasters in 40 countries. Promotes safety at sea and serves as professional representative at meetings of the International Maritime Organization. **PUBLICATIONS:** *Conference Papers* (in English); *IFSMA Newsletter-The International Shipmasters Link*, quarterly; Annual Report (in English), annual. Includes directory; Membership Directory.

INTERNATIONAL FEDERATION OF SOCIAL SCIENCE ORGANIZATIONS (IFSSO) (Federation Internationale des Organisations de Sciences Sociales - FIOSS)

c/o Secretariat of the International Federation of Social Science Organizations, 14 Via dei Laghi, I-00198 Rome, Italy **PHONE:** 39 6 8848943 **FX:** 39 6 8848943 **FOUNDED:** 1979. **OFFICERS:** Dr. Giuseppe Ciccarone, Exec.Sec. **MEMBERS:** 21. **STAFF:** 2. **LANGUAGES:** English, French.

DESCRIPTION: Members are national and regional social science organizations and research agencies. Aims to support and develop the social sciences and encourage international cooperation in the field. Contributes to the development of the social sciences, especially in developing countries, and works to further active participation of the social sciences in discussions on problems of modern society and in efforts to find solutions. Cooperates with member organizations in sponsoring special projects. Provides international forum for communication and exchange of information. **PUBLICATIONS:** *International Directory of Social Science Organizations* (in English, French, and Spanish), biennial; *Newsletter* (in English), semiannual; Papers, periodic.

INTERNATIONAL FEDERATION OF THE SOCIETIES OF CLASSICAL STUDIES (FIEC) (Federation Internationale des Associations d'Etudes Classiques - FIEC)

6, chemin Aux-Folies, CH-1293 Bellevue, Switzerland **PHONE:** 41 22 7742656 **FX:** 41 22 7742656 **FOUNDED:** 1948. **OFFICERS:** Francois Paschoud, Sec.Gen. **MEMBERS:** 79. **LANGUAGES:** English, French.

DESCRIPTION: Members are societies of classical studies representing 44 nationalities on 5 continents; includes 14 international groups. Promotes the study, in all relevant disciplines, of Greco-Latin antiquity. Areas of interest include philology, linguistics, literature, paleography, papyrology, archaeology, history, epigraphy, numismatics, and history of religion, law, and sciences. Makes recommendations regarding subsidy grants to classical studies publications of the United Nations Educational, Scientific and Cultural Organization.

INTERNATIONAL FEDERATION OF SOCIETIES OF ELECTRON MICROSCOPY

c/o Australian Key Centre for Microscopy, FO9, Univ. of Sydney, Sydney, NSW 2006, Australia **PHONE:** 61 2 93512351 **FX:** 61 2 93517682 **E-MAIL:** djhc@emu.su.oz.au **OFFICERS:** Prof. D. Cockayne.

DESCRIPTION: Members are national and international societies of electron microscopists. Seeks to advance standards of training and practice in the field. Facilitates communication and cooperation among electron microscopists worldwide.

INTERNATIONAL FEDERATION OF STOCK EXCHANGES (Federation Internationale des Bourses de Valeurs - FIBV)

22, blvd. de Courcelles, F-75017 Paris, France **PHONE:** 33 1 44010545 **FX:** 33 1 47549422 **E-MAIL:** secretarial@fibv.com **WEBSITES:** http://www.fibv.com **FOUNDED:** 1961. **OFFICERS:** Mr. Gerrit de Marez Oyens, Sec.Gen. **MEMBERS:** 51. **STAFF:** 7. **LANGUAGES:** English, French.

DESCRIPTION: Members are stock exchanges and stock exchange associations in 40 countries. Objectives are: to encourage collaboration among individual stock exchanges and associations; to increase public awareness of the role securities markets play in national economies; to promote securities trading. Exchanges information on recent developments concerning financial markets. Operates documentation center; compiles statistics. Maintains working committee to research specific topics such as increasing internationalization of capital markets and tax incentives to stock market investment in various countries. Organizes workshops to study development in communications, data processing technology, and automation as they relate to stock market procedures, public affairs, clearing and settlements, and derivative products. **PUBLICATIONS:** *Focus* (in English), monthly; *Statistics Report*, monthly; *Statistics Yearbook*, annual; Annual Report; Brochure.

INTERNATIONAL FEDERATION OF SURGICAL COLLEGES (IFSC)

c/o S.W.A. Gunn, M.D., M.S., FRCSC, Hon. Sec.-Gen., CH-1279 Bogis-Bossey, Switzerland **PHONE:** 41 22 7762161 **FX:** 41 22 7766417 **E-MAIL:** ifsc.us-muldoon@mail.med.upenn.edu **FOUNDED:** 1958. **OFFICERS:** Prof. John Tetblanche, Pres. **MEMBERS:** 795. **STAFF:** 1. **BUDGET:** US$89,000. **LANGUAGES:** English.

DESCRIPTION: Members are national colleges, academies, and associations of surgery in 64 countries; interested individuals. Works to improve standards of surgery throughout the world. Promotes cooperation and exchange of medical and surgical information among surgical institutions. Encourages high standards of education, training, and research in surgery and its allied sciences. Supports clinical and scientific congresses in the surgical community. Fosters cooperation in developing the best possible standards of surgical facilities and treatment, and in providing appropriate surgical training in countries requesting aid. Collaborates with the World Health Organization in attempts to strengthen rural surgical health services in developing countries; maintains distribution facility for journals to Third World nations. **PUBLICATIONS:** *IFSC News* (in English), biennial; *World Journal of Surgery* (in English, French, and German), monthly. Additional publications include collected papers.

INTERNATIONAL FEDERATION OF TEACHERS OF MODERN LANGUAGES (FIPLV) (Federation Internationale des Professeurs de Langues Vivantes)

Seestrasse 247, CH-8038 Zurich, Switzerland **PHONE:** 41 1 4855251 **FX:** 41 1 4825054 **TX:** 815250 **FOUNDED:** 1931. **OFFICERS:** Eric Steenbergen, Pres. **MEMBERS:** 38. **LANGUAGES:** English, French, German, Italian, Russian, Spanish.

DESCRIPTION: Members are multilingual (32) and unilingual (6) associations of modern language teachers. Promotes cooperation among language teachers of various countries; coordinates efforts and research work to improve teaching methods; stimulates exchanges including teachers, books, teaching materials, and audiovisual aids; advises national and international organizations on foreign language teaching reforms. Seeks to improve initial and in-service training of teachers. Organizes working groups to study particular problems. Sponsors courses and research work for modern language teachers. Acts as liaison with governmental organizations and the United Nations Educational, Scientific and Cultural Organization. **PUBLICATIONS:** *FIPLV Newsletter*, quarterly. Also publishes books, papers, and symposia proceedings.

INTERNATIONAL FEDERATION FOR THEATRE RESEARCH (IFTR) (Federation Internationale pour la Recherche Theatrale - FIRT)

Samuel Beckett Centre, Trinity Coll., Dublin 2, Ireland **PHONE:** 353 1 6081550 **FX:** 353 1 6793488 **E-MAIL:** bsnglton@tcd.ie **FOUNDED:** 1957. **OFFICERS:** Dr. Brian Singleton. **MEMBERS:** 400. **LANGUAGES:** English, French.

DESCRIPTION: Members are public and private organizations in over 40 countries devoting all or part of their activity to theatre research; individual researchers and other interested persons. Promotes and coordinates the study of theatre history; disseminates scholarly, technical, and other important works on theatre research; assists in the preservation of historic theatre buildings and theatre material. Aids theatrical researchers in obtaining facilities in libraries and museums; helps members obtain research grants. Promotes relations with other historical and artistic studies departments. **PUBLICATIONS:** *FIRT/IFTR - Sibmas Bulletin*, periodic; *News Bulletin*, semiannual; *Theatre Research International* (in English), 3/year.

INTERNATIONAL FEDERATION FOR THE THEORY OF MACHINES AND MECHANISMS (IFToMM) (Federation Internationale pour la Theorie des Machines et des Mechanismes)

Univ. of Oulu, Dept. of Mechanical Engg., PO Box 4800, SF-9041 Oulu, Finland **PHONE:** 358 8 5532050 **FX:** 358 8 **TX:** 32375 oylin sf **E-MAIL:** tatu@me.oulu.fi **WEBSITES:** http://www.cim.mcgill.ca/-iftomm **FOUNDED:** 1969. **OFFICERS:** Prof. Tatu Leinonen, Sec.Gen. **MEMBERS:** 45. **STAFF:** 1. **BUDGET:** US$16,000. **LANGUAGES:** English, French, German, Russian.

DESCRIPTION: Members are technical university professors, engineers, and other specialists in the theory of machines and mechanisms. Promotes development in machines and mechanisms by theoretical and experimental research and its practical applications. Fosters exchange of scientific and engineering information; encourages international visits by specialists and students; assists developing nations in their work on the theory of machines and mechanisms by providing personnel, organizing courses, and coordinating other educational activities. Acts as liaison between members and other international associations and unions. Honors eminent scientists, engineers, and organizations in the field. **PUBLICATIONS:** *IFToMM*, annual; *Mechanisms and Machine Theory*, bimonthly; *Terminology of the Theory of Machines and Mechanisms*; *World Congress Proceedings*, quadrennial.

INTERNATIONAL FEDERATION OF TOURIST CENTRES (FICT) (Federation Internationale de Centres Touristiques - FICT)

c/o Walter Rosli, Direktor Bern Tourismus, Hauptbahnhof, Postfach, CH-3001 Bern, Switzerland **PHONE:** 41 31 3111212 **FX:** 41 31 3121233 **FOUNDED:** 1949. **MEMBERS:** 67. **BUDGET:** 14,000 SFr. **LANGUAGES:** French, German.

DESCRIPTION: Members are managers of tourist offices, information bureaus, and similar organizations. Disseminates technical information; develops cooperative activities.

INTERNATIONAL FEDERATION OF UNIVERSITY WOMEN (IFUW) (Federation Internationale des Femmes Diplomees des Universites - FIFDU)

8, rue de l'Ancien-Port, CH-1201 Geneva, Switzerland **PHONE:** 41 22 7312380 **FX:** 41 22 7380440 **E-MAIL:** info@ifuw.org **WEBSITES:** http://www.ifuw.org **FOUNDED:** 1919. **OFFICERS:** Murielle Joye, Sec.Gen. **MEMBERS:** 180,000. **LANGUAGES:** English, French, Spanish.

DESCRIPTION: Members are national associations representing 180,000 women possessing degrees from institutions recognized by IFUW. Promotes understanding and friendship among university women of the world, irrespective of race, religion, or political opinions. Encourages international cooperation and the full application of members' knowledge and skills to national, regional, or worldwide problems. Works to further the development of education. Represents university women in international organizations. Provides assistance for displaced university women. Undertakes studies and compiles reports dealing with the legal, social, political, economic and educational status of women. **PUBLICATIONS:** *IFUW News* (in English), bimonthly; *Triennial Report*; Papers, periodic. Contains study information.

INTERNATIONAL FEDERATION OF VEXILLOLOGICAL ASSOCIATIONS (FIAV)

504 Branard St., Houston, TX 77006-5018 **PHONE:** (713) 529-2545, (713) 655-2742 **FX:** (713) 752-2304 **E-MAIL:** sec.gen@fiav.org **FOUNDED:** 1969. **OFFICERS:** Charles Spain, Sec.-Gen. **MEMBERS:** 42. **STAFF:** 3. **LANGUAGES:** English, French, German, Spanish.

DESCRIPTION: Members are associations and institutions throughout the world whose object is the pursuit of vexillology. (Vexillology is the scientific and scholarly study of flag history and symbolism.) FIAV promotes the organization of biennial International Congresses of Vexillology, encourages the creation in all countries of associations and institutions dedicated to vexillology, and sanctions international standards for vexillology. **PUBLICATIONS:** *Flag Bulletin* (in English), bimonthly. Provides information on vexillology.; *Info-FIAV*, semiannual. Includes summaries of the work of the general assembly and board.; *Report of the International Congress of Vexillology*, biennial.

INTERNATIONAL FEDERATION OF WORKERS' EDUCATIONAL ASSOCIATIONS (IFWEA) (Federation International des Associations pour l'Education des Travailleurs)

PO Box 8705, N-0028 Oslo, Norway **PHONE:** 47 23061288 **FX:** 47 23061270 **E-MAIL:** jmehlum@online.no **WEBSITES:** http://www.ifwea.org **FOUNDED:** 1947. **OFFICERS:** Jan Mehlum, Gen.Sec. **MEMBERS:** 90. **STAFF:** 2. **BUDGET:** US$200,000. **LANGUAGES:** English, French, German, Spanish.

DESCRIPTION: Members are organizations in 41 countries concerned with adult and worker education. Aims to increase opportunities for workers to improve themselves intellectually, culturally, socially, and economically. Promotes new organizations for worker education where appropriate movements do not already exist. Represents the interests of worker education to United Nations Educational, Scientific and Cultural Organization and other governmental and nongovernmental international organizations. Acts as an information and literature clearinghouse. **PUBLICATIONS:** *Workers' Education* (in English), quarterly; Newsletter, quarterly; Pamphlets; Proceedings; Reports.

INTERNATIONAL FELLOWSHIP OF RECONCILIATION (IFOR) (Mouvement International de la Reconciliation)

Spoorstraat 38, NL-1815 BK Alkmaar, Netherlands **PHONE:** 31 72 5123014 **FX:** 31 72 5151102 **E-MAIL:** office@ifor.org **WEBSITES:** http://www.gn.apc.org/ifor **FOUNDED:** 1919. **OFFICERS:** Johanna S.M. Kooke, Gen.Sec. **MEMBERS:** 125,000. **STAFF:** 12. **BUDGET:** 600,000 f. **LANGUAGES:** English, French, German, Spanish.

DESCRIPTION: International spiritually-based movement, with branches and groups in 50 countries. IFOR members believe in the power of love and truth to create justice and restore community. They promote active nonviolence both as a way of life and as a means of personal, social and political change. They extend the boundaries of community and affirm its diversity of cultures and religious traditions; identify with those in every nation, race, religion and sex who are victims of injustice and exploitation; show respect for all people and reverence for all creation. The International Secretariat and IFOR branches and groups engage in activities including: nonviolence education and training; peace education; human rights support; justice struggles; disarmament campaigns and development of social defense alternatives; engagement with churches and other religious bodies on issues of peace, justice and the integrity of creation; bridge building between people through international delegations. Provides seminars and workshops on active nonviolence; is engaged in human rights solidarity and campaign work; promotes women's development and leadership. **PUBLICATIONS:** *Nonviolence Training in Africa* (in English and French), 3/year; *Nonviolent Possibilities for the Palestinian-Israeli Conflict*; *Patterns in Reconciliation* (in English), semiannual; *RI* (in English), bimonthly; Pamphlets, periodic.

INTERNATIONAL FERTILIZER INDUSTRY ASSOCIATION (IFA) (Association Internationale de l'Industrie des Engrais - IFA)

28, rue Marbeuf, F-75008 Paris, France **PHONE:** 33 1 53930500 **FX:** 33 1 53930547 **TX:** 640481 F **E-MAIL:** ifa@fertilizer.org **WEBSITES:** http://www.fertilizer.org **FOUNDED:** 1926. **OFFICERS:** L.M. Maene, Dir.Gen. **MEMBERS:** 500. **STAFF:** 24. **LANGUAGES:** English, French.

DESCRIPTION: Members are fertilizer and raw materials producers in 80 countries; national and regional fertilizer associations; fertilizer trading, engineering, shipping, and consultancy companies; governmental bodies. Objectives are: to promote the discussion of all matters affecting the fertilizer industry; to facilitate the exchange of information among members concerning the production, marketing, and use of fertilizers and their raw materials; to prepare and publish statistics and reports which promote a better international understanding of the industry's problems and projects. Holds technical and marketing meetings. Maintains liaison with United Nations agencies. **PUBLICATIONS:** *Fertilizers and Agriculture* (in English), periodic.

INTERNATIONAL FINANCE CORPORATION (IFC)

2121 Pennsylvania Ave., NW, Washington, DC 20433 **PHONE:** (202) 477-1234 **FX:** (202) 947-4384 **E-MAIL:** webmaster@ifc.org **WEBSITES:** http://www.ifc.org **FOUNDED:** 1956. **OFFICERS:**

Peter Woicke, Exec.VP. **MEMBERS:** 174. **STAFF:** 1,528. **LANGUAGES:** English.

DESCRIPTION: Member of the World Bank Group. Works to improve the economies of developing countries by promoting private sector growth. Provides venture capital for enterprises that develop local markets. Offers financial and technical assistance to development finance companies to stimulate the flow of private capital. IFC works in conjunction with the International Development Association and the International Bank for Reconstruction and Development to carry out policies and strategies of the World Bank.

INTERNATIONAL FINANCIAL SOCIETY FOR INVESTMENT AND DEVELOPMENT IN AFRICA (IFSIDA) (Societe Internationale Financiere pour les Investissements et le Developpement en Afrique - SIFIDA)

22 rue Francois-Perreard, Boite Postale 310, Chene-Bourg, CH-1225 Geneva, Switzerland **PHONE:** 41 22 3486000 **FX:** 41 22 3482161 **TX:** 418647 **OFFICERS:** Philippe Sechaud, Mng.Dir. **LANGUAGES:** English, French.

DESCRIPTION: Members are financial institutions providing funds or services to development organizations working in Africa. Promotes international investment in African development. Provides support and services to members. Gathers and disseminates business, trade, and financial information.

INTERNATIONAL FISCAL ASSOCIATION (IFA) (Association Fiscale Internationale)

World Trade Centre, Beursplein 37, PO Box 30215, NL-3001 DE Rotterdam, Netherlands **PHONE:** 31 10 4052990 **FX:** 31 10 4055031 **E-MAIL:** n.gensecr@ifa.nl **WEBSITES:** http://www.ifa.nl **FOUNDED:** 1938. **OFFICERS:** J. Frans Spierdijk, Sec.Gen. **MEMBERS:** 9,500. **STAFF:** 3. **LANGUAGES:** English, French, German.

DESCRIPTION: Members are corporations and individuals including tax lawyers and advisers, accountants, professors, and government officials representing 90 countries. Promotes the study and advancement of international and comparative law regarding public finance, international and comparative fiscal law, and the financial and economic aspects of taxation. Conducts research. Sponsors seminars. **PUBLICATIONS:** *Cahiers de Droit Fiscal International* (in English, French, German, and Spanish), annual. Scientific books; *IFA Congress Seminar Series*, annual; *Yearbook* (in English, French, German, and Spanish).

INTERNATIONAL FOOD INFORMATION SERVICE PUBLISHING (IFISP)

Lane End House, Shinfield, Reading RG2 9BB, England **PHONE:** 44 1189 883895 **FX:** 44 1189 885065 **TX:** 847204 IFIS G **E-MAIL:** ifis@ifis.org **WEBSITES:** http://www.ifis.org **FOUNDED:** 1968. **OFFICERS:** Dr. John Metcalfe. **STAFF:** 30. **LANGUAGES:** English.

DESCRIPTION: Gathers and disseminates information on food science and technology. **PUBLICATIONS:** *Food Science and Technology Abstracts*, monthly.

INTERNATIONAL FOODSERVICE MANUFACTURERS ASSOCIATION (IFMA)

180 N. Stetson, Ste. 4400, Chicago, IL 60601 **PHONE:** (312) 540-4400 **FX:** (312) 540-4401 **E-MAIL:** ifma@prodigy.com **WEBSITES:** http://www.ifmaworld.com **FOUNDED:** 1952. **OFFICERS:** Michael J Licata, Pres. **MEMBERS:** 630. **STAFF:** 17. **LANGUAGES:** English.

DESCRIPTION: Members are national and international manufacturers and processors of food, food equipment, and related products for the away-from-home food market. Associate and allied members provide support services to the industry through marketing, publishing, distribution, consulting, promotion, research, advertising, public relations, and brokering. Activities are aimed at marketing, merchandising, sales training, and market research. Compiles statistics. **PUBLICATIONS:** *IFMA World*, 9/year. Includes membership activity information.; Membership Directory, annual. Also publishes forecast data, chain restaurant market data; casual dining data; multiple concept data.; Canadian and Mexican information, international markets report.; compensation survey and top 100 speciality chain data.

INTERNATIONAL FOUNDATION FOR THE CONSERVATION OF WILDLIFE (Fondation Internationale pour la Sauvegarde de la Faune - IGF)

15, rue de Teheran, F-75008 Paris, France **PHONE:** 33 1 45635133, 33 1 45635133 **FX:** 33 1 45633294 **E-MAIL:** igf@cirad.fr **FOUNDED:** 1977. **OFFICERS:** Bertrand des Clers, Dir. **STAFF:** 4. **BUDGET:** US$300,000. **LANGUAGES:** English, French.

DESCRIPTION: Works to raise international awareness of conservation issues. Conducts educational programs; promotes conservation of biodiversity, wise use of renewable natural resources, and sustainable development. **PUBLICATIONS:** *Africaine*; *Faune Sauvage: La Ressource Oubliee*; Articles; Proceedings.

INTERNATIONAL FOUNDATION OF THE HIGH-ALTITUDE RESEARCH STATIONS JUNGFRAUJOCH AND GORNERGRAT (HFSJG)

Sidlerstrasse 5, CH-3012 Bern, Switzerland **PHONE:** 41 31 6314051, 41 31 6314052 **FX:** 41 31 6314405 **FOUNDED:** 1931. **OFFICERS:** Prof. H. Debrunner **MEMBERS:** 6. **STAFF:** 6. **BUDGET:** 900,000 SFr. **LANGUAGES:** French, German, Italian.

DESCRIPTION: Members are national associations of scientists interested in high alpine ecosystems. Promotes and supports the operation of a research station and an astronomical observatory in the Swiss Alps. Conducts research programs.

INTERNATIONAL FRAGRANCE ASSOCIATION (IFRA)

8, rue Charles-Humbert, CH-1205 Geneva, Switzerland **PHONE:** 41 22 3213548 **FX:** 41 22 7811860 **E-MAIL:** ifra@dial.eunet.ch **WEBSITES:** http://www.ifraorg.org **FOUNDED:** 1973. **OFFICERS:** F. Grundschober, Gen.Sec. **MEMBERS:** 14. **STAFF:** 3. **LANGUAGES:** English, French, German.

DESCRIPTION: Members are national associations of fragrance manufacturers. Promotes safety evaluation and regulations for fragrance ingredients. Issues guidelines and code of practice for the use of fragrance ingredients. **PUBLICATIONS:** *Code of Practice*, periodic. Features guidelines with regular updates; *Information Letters*, periodic.

INTERNATIONAL FRIENDS OF NATURE (IFN) (Naturfreunde Internationale)

Diefenbachgasse 36, A-1150 Vienna, Austria **PHONE:** 43 1 8923877 **FX:** 43 1 8129789 **E-MAIL:** nfi@nfi.at **WEBSITES:** http://www.nfi.at; http://www.nfhouse.org **FOUNDED:** 1895. **OFFICERS:** Manfred Pils, Sec.Gen. **MEMBERS:** 600,000. **STAFF:** 5. **LANGUAGES:** English, French, German.

DESCRIPTION: Objectives are to: improve international understanding and friendship; contribute to the protection and growth of nature and the countryside; promote sustainable development; promote sustainable development and ecological tourism; promote transborder protection of the environment; promote alpine and sporting activities including mountain climbing, winter and water sports, camping, and exploring; encourage participation in cultural events including art, literature, theatre, photography, music, and folk dancing. Organizes special groups for children and youth. Maintains Nature Friend Houses, holiday homes, and camping facilities. Conducts seminars. **PUBLICATIONS:** *Bulletin* (in English, French, and German), quarterly. Also publishes booklet.

INTERNATIONAL FROZEN FOOD ASSOCIATION (IFFA)

2000 Corporate Ridge, Ste. 1000, Mc Lean, VA 22102 **PHONE:** (703) 821-0770 **FX:** (703) 821-1350 **FOUNDED:** 1973. **OFFICERS:** Leslie G. Sarasin, Sec. **LANGUAGES:** English.

DESCRIPTION: Members are companies and associations engaged in some aspect of the production, distribution, or marketing of frozen food, and firms supplying goods and services in support of those activities. Objectives are to: encourage sound development of the frozen food industry and the production of high quality frozen food in sanitary plants; encourage practical trade in frozen foods through development of appropriate standards of identity, quality, sanitation, labeling, and other trade factors; provide a clearinghouse of industry information, news, and technical reports to implement the above. Is developing library. **PUBLICATIONS:** *IFFA World Review*, monthly. Includes new product information, research updates, and statistics.

INTERNATIONAL FUND FOR AGRICULTURAL DEVELOPMENT (IFAD)

Via del Serafico 107, I-00142 Rome, Italy **PHONE:** 39 6 54591 **FX:** 39 6 5043463 **TX:** 620330 IFAD **E-MAIL:** ifad@ifad.org **WEBSITES:** http://www.ifad.org **FOUNDED:** 1977. **OFFICERS:** Fawzi Hamad Al-Sultan, Pres. **MEMBERS:** 161. **STAFF:** 239. **BUDGET:** US$54,135,000. **LANGUAGES:** Arabic, English, French, Spanish.

DESCRIPTION: Members are representatives of developed and developing nations, including oil-exporting developing nations. Provides financial assistance for agricultural development programs intended to introduce, expand, or improve food production systems and to strengthen agricultural policies and institutions in member nations. Works to improve the nutritional level and living conditions of poor populations in developing countries. Organizes programming missions charged with advising member countries on methods of maximizing the effectiveness of assistance programs. **PUBLICATIONS:** *Governing Council Report* (in Arabic, English, French, and Spanish), annual; *IFAD Annual Report* (in Arabic, English, French, and Spanish); *Africa: Sowing the Seeds of Self-Sufficiency*; *Breaking the Pattern of Hunger and Poverty*; *From Destitution to the Entrepreneurship*; *Moving Along the Participatory Way*; *Providing Food Security for All*; *Report on Rural Women Living in Poverty*; *Rural Women in Agricultural Investment Projects 1977-84*; *The Poor Are Bankable: Rural Credit the IFAD Way*; *Voices from the Field*.

INTERNATIONAL FUR TRADE FEDERATION (IFTF)

PO Box 318, Walton, Surrey KT12 2WH, England **PHONE:** 44 1932 232866 **FX:** 44 1932 232656 **WEBSITES:** http://www.igu.org **FOUNDED:** 1949. **OFFICERS:** Mrs. J. Bailey, Exec.Dir. **MEMBERS:** 33. **LANGUAGES:** English.

DESCRIPTION: Members are associations that trade, manufacture, or process fur skins. Seeks to: promote and organize joint action; collect statistics and other information; promote conservation of species scientifically proven to be threatened; develop and protect the fur trade. Seeks to establish commercial trade uniformity. Appoints or acts as arbitrator in disputes. **PUBLICATIONS:** *Newsletter*, periodic.

INTERNATIONAL GAS UNION (IGU) (Union Internationale de l'Industrie du Gaz - UIIG)

PO Box 19, NL-9700 Groningen, Netherlands **PHONE:** 31 505 212999 **FX:** 31 505 255951 **E-MAIL:** secr.igu@gasunie.nl **WEBSITES:** http://www.igu.org **FOUNDED:** 1931. **OFFICERS:** J.F. Meeder, Gen.Sec. **MEMBERS:** 57. **STAFF:** 1. **LANGUAGES:** English, French.

DESCRIPTION: Members are national technical gas associations in 57 countries. Purpose is to study the gas industry with the aim of advancing its economic and technical progress. Facilitates contact, cooperation, and exchange of information and experience among members and other gas engineers; acts as liaison among international organizations concerned with energy. Encourages research and training programs at the national level. Compiles statistics. **PUBLICATIONS:** *Dictionaries of the Gas Industry* (in Arabic, English, French, German, Italian, Portuguese, Russian, and Spanish); *Proceedings of Gas Research Conference* (in English), triennial; *Proceedings of LNG Conference* (in English and French), triennial; *Proceedings of the World Gas Conference* (in English and French), triennial.

INTERNATIONAL GEOGRAPHICAL UNION (IGU) (Union Geographique Internationale - UGI)

c/o Prof. Dr. E. Ehlers, Dept. of Geography, Univ. of Bonn, Meckenheimer Allee 166, D-53115 Bonn, Germany **PHONE:** 49 228 739287 **FX:** 49 228 739272 **E-MAIL:** secretariat@igu.bn.uunet.de **WEBSITES:** http://www.helsinki.fi/science/igu **FOUNDED:** 1922. **OFFICERS:** Prof. Dr. E. Ehlers, Sec.Gen. **MEMBERS:** 83. **STAFF:** 1. **BUDGET:** US$110,000. **LANGUAGES:** English, French.

DESCRIPTION: Members are academies of science, research councils, or similar bodies in 83 countries. Promotes the study of geographical problems; initiates and coordinates geographical research requiring international cooperation and promotes its scientific discussion and publication. Provides for the participaton of geographers in the work of similar international organizations. Facilitates the collection and diffusion of geographical data and documentation in and between member countries. Participates in various forms of international cooperation with the goal of advancing the study and application of geography. Maintains 23 commissions and 8 study groups concerned with specific areas of geographic investigations. **PUBLICATIONS:** *IGU Bulletin* (in English and French), semiannual; Proceedings, periodic; Proceedings, periodic.

INTERNATIONAL GEOSPHERE-BIOSPHERE PROGRAMME

IGBP Secretariat, Royal Swedish Acad. of Science, Box 50005, Lilla Frescativagen 4, S-104 05 Stockholm, Sweden **PHONE:** 46 8 166448 **FX:** 46 8 166405 **E-MAIL:** sec@igbp.kva.se **WEBSITES:** http://www.igbp.kva.se **FOUNDED:** 1986. **OFFICERS:** Dr. W.S. Steffen, Exec.Dir. **STAFF:** 8. **LANGUAGES:** English, Swedish.

DESCRIPTION: A committee of the International Council of Scientific Unions. Ecologists, biophysicists, and other earth scientists and researchers with an interest in bio- and geospheres. Promotes improved understanding of the ''interactive physical, chemical and biological processes that regulate the total earth system, the unique environment that it provides for life, the changes that are occurring in this system, and the manner in which they are influenced by human actions.'' Facilitates exchange of information among earth and environmental scientists and researchers worldwide; conducts research and educational programs. **PUBLICATIONS:** *Global Change Newsletter* (in English), quarterly. Contains features on IGBP projects.

INTERNATIONAL GLACIOLOGICAL SOCIETY (IGS)

Lensfield Rd., Cambridge CB2 1ER, England **PHONE:** 44 1223 355974 **FX:** 44 1223 336543 **E-MAIL:** 100751.1667@compuserve.com **FOUNDED:** 1936. **OFFICERS:** Simon Ommanney, Sec.Gen. **MEMBERS:** 800. **STAFF:** 2. **LANGUAGES:** English.

DESCRIPTION: Persons with a scientific, practical, or general interest in any aspect of ice and snow study; membership comprises scientists and others from 33 countries in such fields as physics, meteorology, oceanography, geology, geography, engineering, and chemistry. The society also has 500 libraries as subscribers. Aim is to stimulate research and interest in the practical and scientific problems of snow and ice. Conducts symposia, discussions, and meetings. **PUBLICATIONS:** *Annals of Glaciology* (in English), 1-2/year. Contains conference proceedings; *Ice*, 3/year; *Journal of Glaciology*, 3/year; Books. Contains information on snow and ice studies.

INTERNATIONAL GRAINS COUNCIL (IGC)

1 Canada Sq., Canary Wharf, London E14 5AE, England **PHONE:** 44 171 5131122 **FX:** 44 171 5130630 **E-MAIL:** igc-fac@igc.org.uk **WEBSITES:** http://www.igc.org.uk **FOUNDED:** 1949. **OFFICERS:** Mr. G.A. Denis, Exec.Dir. **MEMBERS:** 33. **STAFF:** 18. **LANGUAGES:** English, French, Russian, Spanish.

DESCRIPTION: Administers the Grains Trade Convention (1995), which is part of the International Grains Agreement (1995), an intergovernmental agreement whose main objectives are international cooperation in all aspects of grains, the expansion of grains trade and its freest possible flow, stability of international grains markets and enhanced food security. Provides a forum for the exchange of information and discussion of members' concerns regarding grains. **PUBLICATIONS:** *Grain Market Report* (in English, French, Russian, and Spanish), monthly; *Report of the Council* (in English, French, Russian, and Spanish), annual; *Wheat and Coarse Grain Shipments*, annual; *World Grain Statistics*, annual.

INTERNATIONAL GYMNASTIC FEDERATION (FIG) (Federation Internationale de Gymnastique)

10, rue des Oeuches, PO Box 359, CH-2740 Moutier 1, Switzerland **PHONE:** 41 32 4946410 **FX:** 41 32 4946419 **E-MAIL:** fig.gymnastics@worldsport.org **FOUNDED:** 1881. **OFFICERS:** Norbert Bueche, Sec.Gen. **MEMBERS:** 126. **LANGUAGES:** English, French, German, Russian, Spanish.

DESCRIPTION: Members are national gymnastics federations representing 29 million individuals. Works to develop the sport of gymnastics and to increase participation worldwide. Sponsors courses for judges and trainers. **PUBLICATIONS:** *Bulletin FIG* (in English and French), quarterly; *Codes*; Manuals; Reports.

INTERNATIONAL HEADQUARTERS OF THE SALVATION ARMY (IHSA)

c/o Gen. Paul A. Rader, 101 Queen Victoria St., London EC4P 4EP, England **PHONE:** 44 171 2365222 **FX:** 44 171 2364981 **FOUNDED:** 1865. **OFFICERS:** Gen. Paul A. Rader. **MEMBERS:** 1,500,000. **STAFF:** 74,000.

DESCRIPTION: Members are ordained ministers, volunteers, and others donating time to religious and social welfare activities in 103 countries and colonies. Embraces Christian ideals and high moral standards; seeks to minister to the physical, spiritual, and emotional needs of mankind. Serves to propagate Christianity, provide education, relieve poverty, and establish charitable projects. Works for the betterment of the poor through evangelistic and social enterprises including alcohol and drug rehabilitation programs, hostels for the homeless, children's homes, schools, hospitals, clinics, and institutes for the blind and handicapped. Preaches the Gospel and publishes in over 4 languages. Cooperates with international relief agencies and governments. **PUBLICATIONS:** *All the World*, quarterly; *Salvation Army YearBook*, annual; *Young Soldier*, weekly.

INTERNATIONAL HEALTH EXCHANGE (IHE)

8-10 Dryden St., London WC2E 9NA, England **PHONE:** 44 0171 8365833 **FX:** 44 0171 3791239 **E-MAIL:** info@ihe.org.uk **WEBSITES:** http://www.ihe.org.uk **FOUNDED:** 1980. **OFFICERS:** Alice Tligui, Dir. **MEMBERS:** 1,200. **STAFF:** 3. **BUDGET:** £200,000. **LANGUAGES:** English.

DESCRIPTION: Members are health care professionals; organizations operating health programs in developing areas worldwide. Promotes availability of effective health services in previously underserved areas. Recruits health staff and volunteers and places them with health services and aid agencies in developing areas that require assistance. Conducts research and educational programs; maintains speakers' bureau. **PUBLICATIONS:** *Health Exchange Magazine* (in English), monthly; Annual Report. Calendar and guides.

INTERNATIONAL HEALTH POLICY AND MANAGEMENT INSTITUTE (IHPMI)

c/o Paul Detrick, Christian Health Services Dev. Corp., 10133 Dunn Rd., Ste. 400, St. Louis, MO 63136 **FOUNDED:** 1983. **OFFICERS:** Paul Detrick, Treas. **MEMBERS:** 100. **LANGUAGES:** English.

DESCRIPTION: Members are health policymakers, hospital presidents, physicians, business leaders, and educators. Dedicated to improving and expanding knowledge of health care economics and management systems. Goals are to: facilitate discussion for the exchange of health care techniques, strategies, and ideas at the theoretical and applied levels to improve the financing and delivery of health care; encourage research and understanding of health policy issues; examine and compare health care systems of industrialized and developing countries; promote the need to maintain access to quality health care while limiting costs within a free market framework. Conducts research and cross-cultural comparative studies on multihospital systems, government health regulations, health systems analysis and planning, health care finance and investment, alternative health care delivery systems, economics of aging, health care cost effectiveness within free market systems, and medical ethics. Disseminates research results; monitors and reports on events affecting health care; conducts demonstration projects. **PUBLICATIONS:** *International Perspectives*, quarterly. Contains excerpts from other publications. Also publishes books, monographs, and proceedings.

INTERNATIONAL HELSINKI FEDERATION FOR HUMAN RIGHTS (IHF)

Rummelhardt Gasse 2/18, A-1090 Vienna, Austria **PHONE:** 43 1 4027387, 43 1 4088822 **FX:** 43 1 4087444 **E-MAIL:** helsinki@ping.at **FOUNDED:** 1982. **OFFICERS:** Aaron Rhodes, Exec.Dir. **MEMBERS:** 31. **STAFF:** 7. **BUDGET:** US$850,000. **LANGUAGES:** English.

DESCRIPTION: Members are national organizations monitoring compliance with the human rights provisions of the 1975 Helsinki Final Act. Defends the right of organizations and individuals to monitor human rights and compliance with the Final Act within the 52 signatory countries. Promotes human rights and the continuation of the Helsinki process. Hopes to establish Helsinki monitoring committees in all participating countries. Conducts research, does advocacy with intergovernmental organizations and organizes projects with NGO's in Eastern Europe. **PUBLICATIONS:** *Annual Report*; *Conference Proceedings*, periodic; *Final Report: Strategies for the Strengthening of Human Rights*; *Human Rights and Civil Society*, quarterly; *Monitoring Human Rights in Europe*; Reports, periodic.

INTERNATIONAL HOSPITAL FEDERATION (IHF) (Federacion Internacional de Hospitales)

46 Grosvernor Gardens, London SW1W 0EB, England **PHONE:** 44 171 8819222 **FX:** 44 171 8819223 **E-MAIL:** 101662,1262@compuserve.com **WEBSITES:** http://www.ihf.co.uk **FOUNDED:** 1929. **OFFICERS:** Dr. Errol Pickering, Dir.Gen. **MEMBERS:** 1,694. **STAFF:** 8. **LANGUAGES:** English, French, Spanish.

DESCRIPTION: Members are individuals (1117); organizations (388); professional firms (114); government representatives (75). Promotes improvement in the planning and management of hospitals and health services through study tours, information services, and research and development projects. Serves as an advocate for hospitals and related health service organizations in world health affairs. Sponsors regional conferences and courses for senior hospital and health services managers from developing countries. Compiles statistics. **PUBLICATIONS:** *Global Healthcare* (in English), quarterly; *Health Service International* (in French and Spanish), quarterly; *Hospital Management International*, annual; *World Hospitals*, 3/year. Includes summaries in French and Spanish; Books; Reports, periodic.

INTERNATIONAL HOTEL AND RESTAURANT ASSOCIATION (IH&RA)

251, rue du Faubourg, St. Martin, F-75010 Paris, France **PHONE:** 33 1 44893400 **FX:** 33 1 40367330 **E-MAIL:** infos@ih-ra.com **WEBSITES:** http://www.ih-ra.com **FOUNDED:** 1946. **OFFICERS:** Christine Clech, Sec.Gen. **MEMBERS:** 2,500. **STAFF:** 20. **LANGUAGES:** English, French, Spanish.

DESCRIPTION: Members are national and regional hotel associations in more than 150 countries and territories; hotels and restaurants with an international market; persons engaged in the hotel or restaurant trade; national and international hotel chains; national hotel and restaurant associations. Allied members are hotel training centers, and students, reservation centers, tourism organizations, international organizations, travel and trade publications; suppliers and service companies. Seeks to raise the professionalism and profile of the international hotel industry. Promotes advancement of the hospitality industries. Works with government and intergovernmental organizations in all activities and pending legislation affecting the international hotel industry; distributes information likely to assist members on questions affecting hotels and restaurants. Encourages the employment of qualified hotel staff and the exchange of students; provides information on professional and international training. **PUBLICATIONS:** *Competitive Advantage*; *Environmental Action Pack for Hotels*; *Hotels: The International Magazine of the Hotel and Restaurant Industry* (in English), monthly; *IH&RA Handbook* (in English and French), annual; *IH&RA White Paper on the Global Hospitality Industry*; *Info* (in English, French, and Spanish), quarterly, every 3 weeks; *International Careers and Choices*; *Scanning the Business Environment*; *Trends, Challenges, and Opportunities*; Proceedings, annual. Proceedings of the annual conference. Case studies, conference reports, codes of practice and conduct, white papers.

INTERNATIONAL HUMAN RIGHTS LAW GROUP (IHRLG)

1200 18th St. NW, Ste. 602, Washington, DC 20036 **PHONE:** (202) 822-4600 **FX:** (202) 822-4606 **E-MAIL:** humamrights@hrlawgroup.org **FOUNDED:** 1978. **OFFICERS:** Gay J. McDougall, Exec.Dir. **STAFF:** 100. **BUDGET:** US$4,000,000. **LANGUAGES:** English, French, Spanish.

DESCRIPTION: Members are human rights and legal professionals engaged in human rights advocacy, litigation and training around the world. Supports and helps empower advocates to expand the scope of human rights protection for men and women and promotes broad participation in building human rights standards and procedures at the national, regional and international levels.

INTERNATIONAL HYDROGRAPHIC ORGANIZATION (IHO) (Bureau Hydrographique International - BHI)

4, quai Antoine 1er, BP 445, MC-98011 Monte Carlo Cedex, Monaco **PHONE:** 377 93 108100 **FX:** 377 93 108140 **TX:** 479164MC-INHORG **E-MAIL:** info@ihb.mc **WEBSITES:** http://www.iho.shom.fr **FOUNDED:** 1921. **OFFICERS:** R.Adm. Giuseppe Angrisano, Pres. **MEMBERS:** 67. **STAFF:** 22. **BUDGET:** US$1,600,000. **LANGUAGES:** English, French.

DESCRIPTION: Members are representatives of national hydrographic offices. Seeks uniformity in nautical charts and documents and adoption of reliable and efficient methods of carrying out hydrographic surveys and development projects. Serves as a consultative and technical body. Maintains IHO Commission on Promulgation of Radio Navigational Warnings, which monitors the WorldWide Navigational Warning Service, a program for the transmission of messages to mariners warning of potential hazards; these hazards include shipwrecks resting in shipping lanes, changes or failures of navigational aids, underwater cable-laying or hydrocarbon exploration activities that may prevent passage of ships. **PUBLICATIONS:** *IHO Yearbook*; *International Hydrographic Bulletin*, monthly; *International Hydrographic Review*, semiannual; *Report of Proceedings of Quinquennial Conference*. Also publishes *Chart Specifications of the IHO*, *Repertory of Technical Resolutions*, *IHO Transfer Standard for Digital Hydrographic Data*, and *Precise Positioning Systems for Hydrographic Surveying*, special publications on technical subjects, and catalog of publications.

INTERNATIONAL INFORMATION CENTRE AND ARCHIVES FOR THE WOMEN'S MOVEMENT (International Informatiecentrum en Archief voor de Vrouwenbeweging)

Obiplein 4, NL-1094 RB Amsterdam, Netherlands **PHONE:** 31 20 6654552, 31 20 6650820 **FX:** 31 20 6655812 **E-MAIL:** info@iiav.nl **WEBSITES:** http://www.iiav.nl **FOUNDED:** 1935. **OFFICERS:** Nicolette van der Post. **LANGUAGES:** Dutch, English.

DESCRIPTION: Promotes the advancement of women worldwide. Fosters the exchange of information related to international women's movements. Conducts research programs. **PUBLICATIONS:** *LOVER, Tydschrift Over Feminisme, Cultuur en Wetenschap* (in Dutch), quarterly. Contains English surveys.

INTERNATIONAL INFORMATION MANAGEMENT CONGRESS (IMC)

1650 38th St., No. 205W, Boulder, CO 80301 **PHONE:** (303) 440-7085 **FX:** (303) 440-7234 **E-MAIL:** info@iimc.org **WEBSITES:** http://www.iimc.org **FOUNDED:** 1962. **OFFICERS:** John A. Lacy, Pres. & CEO. **MEMBERS:** 800. **STAFF:** 10. **BUDGET:** US$1,000,000. **LANGUAGES:** English.

DESCRIPTION: Trade association for the document imaging/management industry. Seeks to communicate document-based technologies and applications to an international audience through conferences, exhibitions, publications, and various membership interactions. Promotes understanding and cooperation among organizations engaged in furthering the progress and application of document-based information systems. **PUBLICATIONS:** *IMC Journal*, bimonthly.

INTERNATIONAL INSTITUTE OF ADMINISTRATIVE SCIENCES (IIAS) (Institut International des Sciences Administratives)

1, rue Defacqz, B-1000 Brussels, Belgium **PHONE:** 32 2 5389165 **FX:** 32 2 5379702 **E-MAIL:** iias@agoranet.be **FOUNDED:** 1930. **OFFICERS:** Giancarlo Vilella, Dir.Gen. **MEMBERS:** 178. **STAFF:** 11. **BUDGET:** US$1,213,000. **LANGUAGES:** English, French.

DESCRIPTION: Members in 97 countries. Examines administrative experience in different countries; developments of rational administrative methods and general principles; research on and plans for improving administrative law and practice and for technical assistance therein. Specific areas of interest include: data processing and administration; productivity; inservice training; institutional development; and innovation. Works in conjunction with the Council of Europe, the United Nations, and the United Nations Educational, Scientific and Cultural Organization. **PUBLICATIONS:** *IIAS* (in English and French), 3/year; *International Review of Administrative Sciences* (in Arabic, English, and French), quarterly; Books; Proceedings.

INTERNATIONAL INSTITUTE FOR CHILDREN'S LITERATURE AND READING RESEARCH (IICLRR) (Internationales Institut fur Jugendliteratur und Leseforschung)

Mayerhofgasse 6, A-1040 Vienna, Austria **PHONE:** 43 1 5050359, 43 1 5052831 **FX:** 43 1 505035917 **E-MAIL:** kidlit@netway.at **WEBSITES:** http://www.netway.at/kidlit **FOUNDED:** 1965. **OFFICERS:** Mag. Karin Sollat, Exec. Officer. **MEMBERS:** 500. **STAFF:** 5. **LANGUAGES:** English, French, German.

DESCRIPTION: Members are individuals, publishing houses, institutions, libraries, schools, and universities in 28 countries. Promotes and evaluates international research in children's literature; sponsors conferences and seminars on problems of juvenile literature. Undertakes studies on juvenile literature and founds study groups. Serves as advisory and documentation center for children's books. Creates special activities for promoting good children's books. **PUBLICATIONS:** *1001 1 Buch* (in German), quarterly. Also publishes a series of booklets on problems dealt with in children's literature, an international bibliography of technical literature, and lists of recommendations and suggestions for translations.

INTERNATIONAL INSTITUTE OF COMMUNICATIONS (IIC) (Institut International des Communications)

Tavistock House South, Tavistock Sq., London WC1H 9LF, England **PHONE:** 44 171 3880671 **FX:** 44 171 3800623 **E-MAIL:** iic@mailbox.ulcc.ac.uk **FOUNDED:** 1968. **OFFICERS:** Vicki MacLeod. **MEMBERS:** 1,500. **STAFF:** 8. **LANGUAGES:** English.

DESCRIPTION: Members are broadcasters, academics, professionals, technologists, journalists and others; broadcasting organizations, telecommunications, computer and electronic firms, news agencies, corporations, foundations, and universities. Objectives are the provision of an independent forum for those concerned with present use and future development of communication and the promotion and dissemination of research and policy studies concerned with the impact of electronic media on society and with the economic, social, legal, and political implications of contemporary communication technology. Examines issues such as ownership, control of the media, cultural ecology, international telecommunications structures, communications, and development. **PUBLICATIONS:** *InterMedia* (in English), bimonthly; *Reports of Seminars*; Books; Papers.

INTERNATIONAL INSTITUTE FOR ETHNIC GROUP RIGHTS AND REGIONALISM (INTEREG) (Internationales Institut fur Nationalitatenrecht und Regionalismus)

c/o Dr. Rudolf Hilf, Hessstrasse 26, Postfach 34 01 61, D-80799 Munich 34, Germany **PHONE:** 49 89 2721498 **FX:** 49 89 2716475 **FOUNDED:** 1977. **OFFICERS:** Dr. Rudolf Hilf. **MEMBERS:** 107. **STAFF:** 1. **LANGUAGES:** Arabic, English, French, German, Italian, Spanish.

DESCRIPTION: Members are scholars and politicians in 40 countries working to promote nationality rights, regionalist independence, and self-determination in order to reduce armed conflicts and maintain peace. Conducts educational program on group rights issues; holds seminars. Sponsors research projects and field studies. **PUBLICATIONS:** *Directory*, periodic; *Europa Ethnica*, quarterly; *Regional Contact*, quarterly. Also publishes books.

INTERNATIONAL INSTITUTE OF HUMAN RIGHTS (IIHR) (Institut International des Droits de l'Homme - IIDH)

2, Allee Rene Cassin, F-67000 Strasbourg, France **PHONE:** 33 3 88 458445 **FX:** 33 3 88 458450 **E-MAIL:** iidhiihr@mail.sdv.fr **FOUNDED:** 1969. **OFFICERS:** Jean-Bernard Marie, Sec.Gen. **STAFF:** 8. **LANGUAGES:** Arabic, English, French, Spanish.

DESCRIPTION: Promotes the growth and protection of human rights; fosters the belief that such rights are essential to maintaining world peace. Encourages the study and research of human rights and promotes the training of teachers in the field. Primary activity is a 4-week study/teaching session focusing on various issues including human rights protection, international humanitarian law, racial discrimination, refugee protection, and women's rights. Sponsors annual study sessions at the International Training Center for University Human Rights Teaching. Collects and disseminates documentation. **PUBLICATIONS:** *Collection of Lectures* (in English and French), annual; *Glossary of Human Rights*; *Human Rights Law Journal*, quarterly, Published until 1979; *Journal of the International Institute of Human Rights* (in English and French), periodic.

INTERNATIONAL INSTITUTE OF HUMANITARIAN LAW (IIHL) (Institut International de Droit Humanitaire - IIDH)

Villa Ormond, Corso Cavallotti 115, I-18038 San Remo, Italy **PHONE:** 39 184 541848 **FX:** 39 184 541600 **E-MAIL:** iihl@sistel.it **WEBSITES:** http://www.tdm.dmw.it/iihl **FOUNDED:** 1970. **OFFICERS:** Judge Ugo Genesio, Sec.Gen. **MEMBERS:** 140. **STAFF:** 6. **BUDGET:** US$700,000. **LANGUAGES:** English, French, Italian, Spanish.

DESCRIPTION: Members are institutions, associations, universities, and individuals in 45 countries distinguished in the field of humanitarian law. Promotes the development and dissemination of international humanitarian law. Conducts courses on the law of war for officers and courses on international refugee law for government officials. Provides a center for the study and research of all areas of humanitarian law; sponsors seminars. Works in cooperation with the International Red Cross and organizations of the United Nations including the United Nations High Commissioner for Refugees. Maintains collection of literature and proceedings. **PUBLICATIONS:** *Humanitarian Dialogue*, 3/year; *Proceedings of Conferences and Courses*, periodic; *San Remo Manual on International Law Applicable to Armed Conflicts at Sea*; *Yearbook*; Booklets; Monographs.

INTERNATIONAL INSTITUTE FOR INFRASTRUCTURAL, HYDRAULIC AND ENVIRONMENTAL ENGINEERING

PO Box 3015, NL-2601 DA Delft, Netherlands **PHONE:** 31 15 2151715 **FX:** 31 15 2122921 **E-MAIL:** ihe@ihe.nl **FOUNDED:** 1957. **OFFICERS:** A.S. Ramsundersingh. **STAFF:** 122. **BUDGET:** US$22,500,000. **LANGUAGES:** Dutch, English.

DESCRIPTION: Members are engineers, scientists, development organization personnel, and other individuals and organizations with an interest in global economic and social development. Promotes and facilitates transfer of engineering expertise and technology to ensure effective execution of development programs. Conducts research and training courses for engineers and other professionals involved in infrastructure development; creates international networks of development professionals to encourage communication and cooperation among programs.

INTERNATIONAL INSTITUTE FOR LABOUR STUDIES (IILS) (Institut International d'Etudes Sociales)

Case Postale 6, CH-1211 Geneva 22, Switzerland **PHONE:** 41 22 7997628 **FX:** 41 22 7998542 **TX:** 415647 IIO CH **E-MAIL:** inst@ilo.org **FOUNDED:** 1960. **OFFICERS:** Padmanabha Gopinath, Dir. **STAFF:** 22. **LANGUAGES:** English, French, Spanish.

DESCRIPTION: Institute established by the International Labor Organization. Promotes a better understanding of and encourages research into labor problems on topics such as labor institutions, social exclusion and labour, business and society. Provides leadership training in social policy for potential leaders in government service and employers' and workers' organizations; Collects and disseminates information. **PUBLICATIONS:** *Bibliography Series*, periodic; *Discussion Papers Series*, periodic; *Research Series*, periodic; Books, periodic.

INTERNATIONAL INSTITUTE FOR LIGURIAN STUDIES (IILS) (Institut International d'Etudes Ligures - IIEL)

Via Romana 39, I-18012 Bordighera, Italy **PHONE:** 39 184 263601 **FX:** 39 184 266421 **E-MAIL:** iisl@masterweb.it **FOUNDED:** 1947. **OFFICERS:** Cosimo Costa, Pres. **MEMBERS:** 1,721. **STAFF:** 21. **LANGUAGES:** English, French, Italian, Spanish.

DESCRIPTION: Members are individuals in 7 countries. Sponsors research into ancient Ligurian history and archaeology. (Ancient Liguria was a region in southwestern Europe inhabited by Ligurians during the Paleolithic Period.) Offers educational programs. Maintains several museums. **PUBLICATIONS:** *Cahiers Ligures de Prehistoire et de Protohistoire* (in French and Italian), annual. Contains articles on prehistory and protohistory in ancient Liguria.; *Collana Storica della Liguria Orientale* (in Italian); *Collana Storica dell'Oltregiogo Ligure* (in Italian); *Collana Storica dell'Oltremare*

Ligure (in Italian); *Collana Storico-Archeologica della Liguria Occidentale* (in Italian); *Collezione Di Monografie Preistoriche E Archeologiche* (in French and Italian); *Giornale Storico della Lunigiana* (in Italian), annual; *Itinerari Liguri* (in French and Italian); *Itinerari Liguri - Musei e Monumenti* (in Italian); *Rivista di Studi Liguri* (in English, French, Italian, and Spanish), annual; *Rivista Ingauna e Intemelia* (in Italian), annual; *Studi Genuensi* (in Italian), annual.

INTERNATIONAL INSTITUTE FOR MANAGEMENT DEVELOPMENT

Chemin de Bellerive 23, Po Box 915, CH-1001 Lausanne, Switzerland **PHONE:** 41 21 6180111 **FX:** 41 21 6180707 **E-MAIL:** info@imd.ch **WEBSITES:** http://www.imd.ch/ **OFFICERS:** Dr. Derek Abell. **LANGUAGES:** French, German, Italian.

DESCRIPTION: Works to develop local business management and entrepreneurial skills in economically underdeveloped areas. Promotes sustainable economic growth under local management; assists in the establishment of microenterprises. Conducts educational programs.

INTERNATIONAL INSTITUTE FOR PEACE (IIP) (Internationales Institut fur den Frieden)

Moellwaldplatz 5, A-1040 Vienna, Austria **PHONE:** 43 1 50464370 **FX:** 43 1 5053236 **E-MAIL:** iip@aon.at **FOUNDED:** 1957. **OFFICERS:** Erwin Lanc. **MEMBERS:** 150. **BUDGET:** 500,000 AS. **LANGUAGES:** English, German.

DESCRIPTION: Members are individuals and corporate bodies in 25 countries. Carries out activities to preserve, safeguard and consolidate peace efforts throughout the world. Cooperates with individuals and organizations interested in peace. Organizes colloquia and symposia. Strives for a confrontation of opinions and the resulting dialogue, as well as the elaboration of joint conclusions and practical recommendations, in order to increase the impact on the public and responsible circles in its endeavors to promote peace. **PUBLICATIONS:** *Occasional Paper* (in English), periodic; *Peace and Security - the IIP research quarterly* (in English). Also publishes monographs and books.

INTERNATIONAL INSTITUTE OF PHILOSOPHY (IIP) (Institut International de Philosophie)

8, rue Jean-Calvin, F-75005 Paris, France **PHONE:** 33 1 43363911 **FX:** 33 1 47077794 **FOUNDED:** 1937. **OFFICERS:** Catherine Champniers, Chef du Secretariat. **MEMBERS:** 97. **STAFF:** 4. **LANGUAGES:** English, French, German, Italian, Spanish.

DESCRIPTION: Members are individuals in 40 countries interested in the study of philosophy. Promotes mutual understanding among philosophers from different traditions and cultural backgrounds. **PUBLICATIONS:** *Bibliography of Philosophy* (in English, French, German, Italian, and Spanish), quarterly; *Controverses Philosophiques*; *Philosophers on Their Own Work*; *Philosophical Problems Today*; *Philosophy and World Community*. Series; Proceedings, annual.

INTERNATIONAL INSTITUTE OF PUBLIC FINANCE (IIPF) (Institut International de Finances Publiques - IIFP)

Univ. of Saarland, PO Box 151150, D-66041 Saarbrucken, Germany **PHONE:** 49 681 3023653 **FX:** 49 681 3024369 **E-MAIL:** iipf@rz.uni-sb.de **WEBSITES:** http://www.wiwi.uni-sb.de/iipf **FOUNDED:** 1937. **OFFICERS:** Prof. Robert Haveman, Pres. **MEMBERS:** 970. **STAFF:** 1. **BUDGET:** US$50,000. **LANGUAGES:** English, French, German.

DESCRIPTION: Members are individuals and associations in 65 countries active in the field of public finance and economics. Purpose is to research public finance and public economics. Facilitates the international exchange of information and experience and the establishment of scientific contacts among members and others in the field. Provides a forum for the discussion and resolution of problems of public economics and issues of implementation and principle. Sponsors seminars, training programs, and other public interest activities. **PUBLICATIONS:** *International Tax and Public Finance* (in English), quarterly.

INTERNATIONAL INSTITUTE OF REFRIGERATION (IIR) (Institut International du Froid - IIF)

177, blvd. Malesherbes, F-75017 Paris, France **PHONE:** 33 1 42273235 **FX:** 33 1 47631798 **E-MAIL:** iifiir@ibm.net **WEBSITES:** http://www.iifiir.org **FOUNDED:** 1908. **OFFICERS:** L. Lucas, Exec.Dir. **MEMBERS:** 1,500. **STAFF:** 15. **BUDGET:** US$1,400,000. **LANGUAGES:** English, French.

DESCRIPTION: In addition to the 60 IIR member countries, IIR associate members are companies, organizations, institutions, associations, scientists, and professionals in more than 90 countries. Works for the popularization of cryogenics and refrigeration science and technology at the international level as applied to food, agriculture, industry, transportation, cryology, air conditioning, human hygiene, health, heat pumps, and energy conservation. Encourages research. Conducts training courses and seminars. **PUBLICATIONS:** *Bulletin of the International Institute of Refrigeration*, bimonthly. Provides information on abstracts.; *International Journal of Refrigeration* (in English and French), 8/year; *International Refrigeration Dictionary*; *Series: Refrigeration Science and Technology*, periodic.

INTERNATIONAL INSTITUTE OF SOCIAL ECONOMICS (IISE) (Institut International d'Economie Sociale)

Enholmes Hall, Patrington, Hull, Humberside HU12 0PR, England **PHONE:** 44 1964 630033 **FX:** 44 1964 631716 **FOUNDED:** 1972. **OFFICERS:** Prof. Barrie O. Pettman, Dir. **MEMBERS:** 550. **STAFF:** 2. **LANGUAGES:** English.

DESCRIPTION: Members are persons experienced in, studying, or interested in social economics. Aims are to: assist in the development of social economics as a recognized discipline with a scientific foundation and accepted standards of qualifications and ethics; enhance communication and the exchange of ideas; help social economists to understand and apply newly developed ideas and techniques and translate them into practical terms; aid and motivate colleges and universities to develop and maintain sound and adequate social economics teaching. Seeks to broaden the discipline of social economics through research, publication, and meetings. Conducts comparative studies concerning: the definition of social economics; economic systems; population growth and its ramifications; changes in the workforce; income distribution and policy. Develops relationships with international bodies associated with specific facets of social economics; makes presentations to government departments and related institutions to improve the quality and extent of statistical sources. **PUBLICATIONS:** *International Journal of Social Economics* (in English), monthly.

INTERNATIONAL INSTITUTE OF SPACE LAW (IISL) (Institut International de Droit Spatial)

3-5 rue Mario Nikis, F-75015 Paris, France **PHONE:** 33 1 45674260 **FX:** 33 1 42732120 **E-MAIL:** jtmasson@cyberway.com.sq **WEBSITES:** http://www.iafastro.com **FOUNDED:** 1959. **OFFICERS:** Dr. N. Jasentuliyana, Pres. **MEMBERS:** 350. **LANGUAGES:** English.

DESCRIPTION: Members are individuals in 35 countries contributing to or interested in developments in space law. Fosters the social science aspects of astronautics, exploration, and space travel. **PUBLICATIONS:** *History of the International Institute of Space Law*; *Newsletter* (in English), semiannual. Provides current information on institution's activities; *Proceedings of the Colloquium on the Law of Outer Space* (in English), annual; Brochure (in English), periodic.

INTERNATIONAL INSTITUTE FOR STRATEGIC STUDIES (IISS) (Institut International d'Etudes Strategiques)

23 Tavistock St., London WC2E 7NQ, England **PHONE:** 44 171 3797676, 44 171 8720770 **FX:** 44 171 8363108 **TX:** 9312102499 G **E-MAIL:** iiss@iiss.org.uk **WEBSITES:** http://www.isn.ethz.ch/iiss; http://www.fsk.ethz.ch/iiss **FOUNDED:** 1958. **OFFICERS:** Cdr. J.A.A. McCoy, Adm.Dir. **MEMBERS:** 2,600. **STAFF:** 36. **BUDGET:** £2,500,000. **LANGUAGES:** English.

DESCRIPTION: Members are journalists, politicians, businesspersons, academic personnel, retired service officers, economists, and interested others in 80 countries; associate members are active service officers and government officials; corporate members are newspapers, universities, television stations, embassies and government ministries, service colleges, and other corporate entities. Prepares studies on topical strategic subjects. **PUBLICATIONS:** *Adelphi Papers* (in English), 8/year; *Military Balance*, annual; *Strategic Comments*, monthly; *Strategic Survey*, annual; *Survival*, quarterly.

INTERNATIONAL INSTITUTE FOR TRADITIONAL MUSIC (IITM)

Winklerstrasse 20, D-14193 Berlin, Germany **E-MAIL:** iitm@netmbx.netmbx.de **FOUNDED:** 1963. **OFFICERS:** Prof. Max Peter Baumann, Dir. **STAFF:** 16. **LANGUAGES:** English, French, German, Portuguese, Spanish.

DESCRIPTION: Promotes traditional non-European art music as well as traditional European music. Arranges exchange of internationally recognized scholars and artists; sponsors lectures and musical performances for experts and the public. Conducts monthly colloquium of ethnomusicologists. Organizes concerts and festivals. **PUBLICATIONS:** *Intercultural Music Studies*; *Musikbogen: Wege zum Berstandnis fremder Musikkulturen*; *The World of Music* (in English), 3/year. Also publishes brochures on special forms of traditional music and edits the CD-series *Traditional Music of the World.*

INTERNATIONAL INSTITUTE OF TROPICAL AGRICULTURE (IITA) (Institut International d'Agriculture Tropicale - IITA)

Oyo Rd., Private Mail Bag 5320, Ibadan, Oyo, Nigeria **PHONE:** 234 2 2412626 **FX:** 234 2 2412221 **TX:** 31159 **E-MAIL:** iita@cgnet.com **WEBSITES:** http://www.cgiar.org/iita **FOUNDED:** 1967. **OFFICERS:** Dr. Lukas Brader, Dir.Gen. **STAFF:** 1,100. **LANGUAGES:** English, French.

DESCRIPTION: Works to: improve the quantity and quality of such tropical food crops as cassava, cowpeas, maize, plantain, soybeans, and yams; develop sustainable agricultural systems to replace bush fallow and slash-and-burn cultivation methods in humid and subhumid tropical regions. Conducts research and international cooperation programs which include training and germplasm exchange activities. Disseminates information. **PUBLICATIONS:** *Annual Report* (in English and French); *Research Bulletin* (in English and French), semiannual; Monographs, periodic; Proceedings, periodic.

INTERNATIONAL INSTITUTE FOR THE UNIFICATION OF PRIVATE LAW (Institut International pour l'Unification du Droit Prive - UNIDROIT)

Via Panisperna 28, I-00184 Rome, Italy **PHONE:** 39 6 69941372 **FX:** 39 6 69941394 **FOUNDED:** 1926. **OFFICERS:** Luigi Ferrari Bravo. **MEMBERS:** 56. **LANGUAGES:** English, French, German, Italian, Spanish.

DESCRIPTION: Members are governments interested in examining ways of harmonizing and coordinating private law of states and groups of states. Prepares drafts of laws and conventions with the aim of establishing uniform law and improving international relations in the field; works to prepare various states for the adoption of uniform rules of private law. Organizes conferences, seminars, study groups, symposia, and work programs for academics and practitioners interested in international trade law and the legal and technical aspects of unification; sponsors studies in comparative private law. Collaborates with independent experts and organizations in the field. Originally established as an auxillary organ of the League of Nations, UNIDROIT was reorganized as an independent intergovernmental organization in 1940. **PUBLICATIONS:** *Digest of Legal Activities of International Organizations and Other Institutions*, periodic; *News Bulletin*, semiannual; *Uniform Law Review*, quarterly.

INTERNATIONAL INSTITUTE OF WELDING (IIW) (Institut International de la Soudure - IIS)

90, rue des Vanesses, F-93420 Villepinte, France **PHONE:** 33 1 49903608, 33 1 49903679 **FX:** 33 1 49903680 **E-MAIL:** iiw@wanadoo.fr **FOUNDED:** 1948. **OFFICERS:** M. Bramat, Chief Exec. **MEMBERS:** 40. **STAFF:** 5. **LANGUAGES:** English.

DESCRIPTION: Members are associations involved in scientific and technical aspects of welding and related processes. Promotes welding and its application; provides for the exchange of scientific and technical information concerning welding research and education. Formulates international standards for welding. Encourages the creation of national welding associations in countries where there are none. Organizes research and educational programs; maintains study groups. **PUBLICATIONS:**

Welding in the World (in English and French), bimonthly; Monograph, periodic; Papers, periodic.

INTERNATIONAL INVESTMENT BANK (IIB) (Mezhdunarodnyi Investitsionnyi Bank - MIB)

Masha Poryvayeva St. 7, 107078 Moscow, Russia **PHONE:** 7 95 9753829, 7 95 9754008 **FX:** 7 95 9752070 **TX:** 411358 IIB RU **FOUNDED:** 1970. **OFFICERS:** Igor M. Novikov, Chm. **MEMBERS:** 10. **STAFF:** 240. **BUDGET:** 1,300,000 Rb. **LANGUAGES:** Russian.

DESCRIPTION: Members are representatives of Bulgaria, Russian Federation, Cuba, Hungary, Mongolia, Poland Romania, Vietnam, Slovak Republic and Czech Republic. Promotes the switching the national economies of member countries to a market basis for more complete integration of the countries in the world economy. Grants credits for construction, expansion, reconstruction, and technical renovation projects of interest to its member states. Concentrates its efforts in the areas of production acceleration and scientific and technical progress. To date (1971–1961), the IIB has extended loans totalling around ECU 7 billion for financing 159 projects in its member-countries and other states. **PUBLICATIONS:** *Annual Report* (in English and Russian). Outlines the yearly acitivities of the IIB.

INTERNATIONAL IRON AND STEEL INSTITUTE (IISI)

120, rue Colonel Bourg, B-1140 Brussels, Belgium **PHONE:** 32 2 7028900 **FX:** 32 2 7028899 **TX:** 22639 **E-MAIL:** steel@iisi.be **WEBSITES:** http://www.worldsteel.org **FOUNDED:** 1967. **OFFICERS:** Lenhard J. Holschuh, Sec.Gen. **MEMBERS:** 200. **STAFF:** 20. **LANGUAGES:** English.

DESCRIPTION: Members are steel companies, national and international steel federations and steel research associations in more than 50 countries. Total membership (200) produces over 70 percent of total world steel production. Serves as a world forum on various aspects of the international steel industry. Conducts research into economic, financial, technological, environmental, and promotional aspects of world steel. Collects, evaluates, and disseminates world steel statistics. **PUBLICATIONS:** *Conference Papers* (in English), annual. Contains proceedings from annual conference.; *Iron and Crude Steel Production* (in English), monthly. Pig iron and crude steel production statistics.; *Steel Statistical Yearbook* (in English), annual; *World Steel in Figures* (in English), annual; *World Steel Statistics Monthly* (in English). Contains comprehensive world statistics on crude steel and steel products, trade, and raw materials; Reports, periodic. Contains information on the environment, economics, raw materials, technology, market promotion, human resources, and statistics.

INTERNATIONAL ISLAMIC FEDERATION OF STUDENT ORGANIZATIONS (IIFSO) (Federacion Islamica Internacional de Organizaciones Estudiantes)

PO Box 8631, Salimiyah 22057, Kuwait **PHONE:** 965 2443548 **FX:** 965 2443549 **E-MAIL:** sou.tahan@usa.net **FOUNDED:** 1969. **OFFICERS:** Mustafa Mohammed Tahan, Exec.Dir. **MEMBERS:** 44. **LANGUAGES:** Arabic, English.

DESCRIPTION: Members are Muslim student organizations. Objectives are: to unite Muslim students worldwide and strengthen Islamic fraternal bonds; to encourage study of the Islamic faith and the Arabic language; to spread the principles of Islam around the world. Conducts seminars and youth camps. Maintains information center. **PUBLICATIONS:** *AL-AKHBAR* (in Arabic and English), biweekly; *Connection* (in English), monthly. Also publishes Islamic books in 69 languages.

INTERNATIONAL ISLAMIC NEWS AGENCY (IINA)

PO Box 5054, Jeddah 21422, Saudi Arabia **PHONE:** 966 2 6652056, 966 2 6658561 **FX:** 966 2 6659358 **TX:** 601090 INPRES S **E-MAIL:** iina@mail.gcc.com.bh **WEBSITES:** http://www.islamicnews.org **FOUNDED:** 1971. **OFFICERS:** Abdul Wahab Kashif, Acting Dir.Gen. **MEMBERS:** 52. **STAFF:** 8. **BUDGET:** US$1,000,000. **LANGUAGES:** Arabic, English, French.

DESCRIPTION: Members are national news agencies of Islamic states. Primary aim is to support the causes of Islamic states by disseminating information to the public concerning their progress and achievements. **PUBLICATIONS:** *Events of Islamic World: A Yearbook* (in Arabic); *News Bulletin* (in Arabic, English, and French), daily.

INTERNATIONAL JOINT COMMISSION (IJC)

1250 23rd St. NW, Ste. 100, Washington, DC 20440 **PHONE:** (202) 736-9000 **FX:** (202) 736-9015 **E-MAIL:** commission@washington.ijc.org **WEBSITES:** http://www.ijc.org **FOUNDED:** 1911. **MEMBERS:** 6. **STAFF:** 22. **BUDGET:** US$3,700,000. **LANGUAGES:** English, French.

DESCRIPTION: Joint U.S.-Canada quasi-judicial and advisory tribunal on boundary and transboundary water problems. Established from Boundary Waters Treaty of 1909 to prevent disputes on the use of boundary and transboundary waters and investigate questions arising from transboundary issues. Approves and disapproves applications from governments, companies, and individuals for obstructions, uses, and diversions of water that affect the natural level and flow of water on the other side of the international boundary; investigates particular questions, reports findings to the U.S. and Canadian governments, and offers recommendations; monitors compliance with IJC orders of approval. Maintains advisory boards of scientists, engineers, and other experts to supply IJC with technical studies and field work; answers student inquiries regarding the environmental quality of the Great Lakes. Conducts public hearings. **PUBLICATIONS:** *Biennial Report on Great Lakes Water Quality*. Contains information on water quality in the Great Lakes system.; *Focus*, 3/year. Contains information on current activities.; Reports. On governmental efforts to reduce pollution in the Great Lakes.

INTERNATIONAL JUDO FEDERATION (IJF) (Federation Internationale de Judo)

Doosan Tower 33rd Fl., 18-12, Ulchi-Ro-6-Ka, Chung-Ku, Seoul 100-730, Republic of Korea **PHONE:** 82 2 33981017 **FX:** 82 2 33981020 **E-MAIL:** yspark@ifj.org **WEBSITES:** http://www.ijf.org **FOUNDED:** 1951. **OFFICERS:** Heinz Kempa, Gen.Sec. **LANGUAGES:** English, French.

DESCRIPTION: Members are judo federations in 154 countries united to: promote cordial and friendly relations among members; coordinate and supervise judo activities worldwide; promote the spread and development of the sport. Compiles standards and statistics. Conducts courses for coaches and referees; organizes seminars and world championships. **PUBLICATIONS:** *Judo Magazine* (in English, French, and Spanish), quarterly.

INTERNATIONAL JUTE ORGANIZATION (IJO)

145 Monipuripara, Old Airport Rd., Dhaka 1215, Bangladesh **FX:** 880 2 9125248 **TX:** 642792 IJO BJ **FOUNDED:** 1984. **OFFICERS:** Mr. K.M. Rabbani, Exec.Dir. **MEMBERS:** 27. **STAFF:** 22. **BUDGET:** US$1,000,000. **LANGUAGES:** Arabic, Chinese, English, French.

DESCRIPTION: Members are jute importing and exporting countries. (Jute is a plant fiber, native to East India, that is commonly used for sacking, carpet backing cloth, burlap, and mats.) Works to: improve the structure of the jute market; increase promotion of jute and jute products; conduct research; maintain and enlarge present markets and develop new markets. **PUBLICATIONS:** *Annual Report*; *Jute* (in English), quarterly.

INTERNATIONAL LABOUR ORGANIZATION (ILO) (Organisation Internationale du Travail - OIT)

4, route des Morillons, CH-1211 Geneva 22, Switzerland **PHONE:** 41 22 7996111 **FX:** 41 22 7988685 **TX:** 415647 ILO CH **E-MAIL:** webinfo@ilo.org **WEBSITES:** http://www.ilo.org **FOUNDED:** 1919. **OFFICERS:** Mr. Juan Somavia, Dir.Gen. **MEMBERS:** 174. **STAFF:** 2,258. **BUDGET:** US$569,080,000. **LANGUAGES:** English, French, Spanish.

DESCRIPTION: A specialized agency associated with the United Nations. National delegations of government, employer, and worker representatives attending conferences and meetings of the ILO. Exists to promote the voluntary cooperation of nations to improve labor conditions and raise living standards, thereby improving prospects of peace by fostering economic and social stability throughout the world. Sets standards covering all aspects of work life, including human rights, through the adoption of conventions and recommendations. Aids Third World development through a technical cooperation program. Works for human rights, employment promotion, and managerial and vocational training for improvement of working conditions and environment. Received Nobel Peace Prize in 1969. Provides permanent secretariat for the International Social Security Association; has established the International Occupational Safety and Health Information Centre, the International Institute for Labor Studies. **PUBLICATIONS:** *International Labour Documentation* (in English,

French, and Spanish); *International Labour Review* (in English, French, and Spanish), quarterly; *Labour Education*, quarterly; *Official Bulletin*, 2-3/year; *World Employment Report*; *World Labour Report*; *World of Work* (in English, French, and Spanish), periodic.

INTERNATIONAL LAW ASSOCIATION (ILA) (Association de Droit International - ADI)

Charles Clore House, 17 Russell Sq., London WC1B 5DR, England **PHONE:** 44 171 3232978 **FX:** 44 171 3233580 **FOUNDED:** 1873. **OFFICERS:** Barbara Osorio, Sec. **MEMBERS:** 4,500. **STAFF:** 1. **LANGUAGES:** English, French.

DESCRIPTION: Members are lawyers and representatives in 85 countries active in the shipping, commercial, and banking industries. Fosters interest in the study, advancement and unification of international public and private law and comparative law, and in resolving legal conflicts. Conducts seminars. **PUBLICATIONS:** *Conference Reports*, biennial; *The Effect of Independence on Treaties*; *The Present State of International Law*; Newsletter (in English and French), semiannual. Additional publications available upon request.

INTERNATIONAL LAW STUDENTS ASSOCIATION (ILSA)

2223 Massachusetts Ave. NW, Washington, DC 20008-2864 **PHONE:** (202) 939-6030 **FX:** (202) 265-0386 **E-MAIL:** ilsa@access.digex.net **WEBSITES:** http://www.kentlaw.edu/ilsa **FOUNDED:** 1961. **OFFICERS:** Yvette Roozenbeek, Exec.Dir. **MEMBERS:** 10,000. **STAFF:** 2. **LANGUAGES:** English, French, Spanish.

DESCRIPTION: Members are law societies at law schools worldwide; interns and associate members. Seeks to promote interest in international legal problems through cooperative development of programs. Provides support to local groups for on-campus programming and coordinates regional, national, and international events. Conducts annual Philip C. Jessup International Law Moot Court Competition, open to all international law societies. Administers the U.S. portion of the Student Trainee Exchange Program, a student administered internship exchange between U.S. and European law firms and legal organizations, and the Transamerica Student Exchange Program between the U.S. and Mexico. Provides research opportunities to undergraduate, graduate, and law students in the field of international law. Maintains speakers list; coordinates a book donation program. **PUBLICATIONS:** *Ad Rem*, 4/year; *Handbook for Student International Law Societies*; *ILSA Guide to Education and Career Development in International Law*; *ILSA Journal of International and Corporate Law*, annual; *Jessup Competition Materials, 1987-1993*; Videos.

INTERNATIONAL LEAD ZINC RESEARCH ORGANIZATION (ILZRO)

2525 Meridian Pky., PO Box 12036, Research Triangle Park, NC 27709 **PHONE:** (919) 361-4647 **FX:** (919) 361-1957 **E-MAIL:** jcole@ilzro.org **WEBSITES:** http://www.ilzro.org **FOUNDED:** 1958. **OFFICERS:** Jerome F. Cole, Pres. **MEMBERS:** 77. **STAFF:** 16. **BUDGET:** US$5,000,000. **LANGUAGES:** English.

DESCRIPTION: Research organization sponsored by major producers, smelters, and refiners of lead and/or zinc from 15 countries. Seeks to develop new applications for lead and zinc. Seeks to improve current uses of lead and zinc; compiles technical information on these metals. Directs approximately 150 research programs through its contracts with universities, governments, independent laboratories, industrial companies, and member companies. Research and development projects deal with die castings, wrought zinc, alloys, galvanized steel, plating, welding, lead and zinc chemistry, environmental studies, batteries, lead for architectural uses, and other subjects. **PUBLICATIONS:** *R&D Focus and Environmental Update*, periodic; Articles; Manuals. Also publishes data sheets and contractors' reports.

INTERNATIONAL LEAGUE AGAINST EPILEPSY (ILAE)

c/o Dr. Peter Wolf, Klinik Mara I, Maraweg 21, D-33617 Bielefeld, Germany **PHONE:** 49 521 1444897, 49 521 1443686 **FX:** 49 521 1444637 **E-MAIL:** iku@mara.de **FOUNDED:** 1909. **OFFICERS:** Dr. Peter Wolf. **MEMBERS:** 62.

DESCRIPTION: National organizations united to encourage scientific research on epilepsy, and to promote optimal treatment and rehabilitation of epileptic patients. Fosters development of and cooperation among associations with common interests. **PUBLICATIONS:** *Epigraph*; *Epilepsia*, monthly. Contains scientific papers and meeting abstracts for professional researchers in the field of epilepsy. Includes book reviews and research reports.

INTERNATIONAL LEAGUE FOR HUMAN RIGHTS (ILHR)

432 Park Ave. S., Ste. 1103, New York, NY 10016 **FX:** (212) 684-1696 **E-MAIL:** ilhr@ilhr.org **WEBSITES:** http://www.ilhr.org **FOUNDED:** 1942. **OFFICERS:** Catherine Q. Fitzpatrick, Exec.Dir. **MEMBERS:** 2,700. **STAFF:** 3. **BUDGET:** US$350,000. **LANGUAGES:** English.

DESCRIPTION: Individuals and national affiliates promoting human rights, including political and civil rights, racial and religious freedom, and the implementation of the Universal Declaration of Human Rights. Serves as nongovernmental agency accredited by the United Nations, International Labor Organization, United Nations Educational, Scientific and Cultural Organization, and Council of Europe. Participates in studies and programs on human rights. Advocates effective procedures to protect human rights, including protection of minorities; deals with issues of torture, political imprisonment, due process of law, racial discrimination, genocide, apartheid, treatment of prisoners, status of women, and religious freedom; promotes ability of local human rights groups to exist and work unimpeded by government. Intervenes directly with governments concerning violations of human rights. Sends special investigators to areas where human rights violations exist and sends observers to political trials. **PUBLICATIONS:** *Crime and Servitude*. An expose of the traffic in women for prostitution from the newly independent states.; *In Brief*, 10-15/year.; *Petitions Before the UN Trusteeship Council*; *Report of a Medical Fact-Finding Mission to El Salvador*; Booklets; Books; Pamphlets; Reports. Covers special reports on worldwide human rights.

INTERNATIONAL LESBIAN AND GAY ASSOCIATION (ILGA)

Kolenmarkt, 81, B-1000 Brussels, Belgium **PHONE:** 32 2 5022471 **FX:** 32 2 5022471 **E-MAIL:** ilga@ilga.org **WEBSITES:** http://www.ilga.org **FOUNDED:** 1978. **OFFICERS:** G. Thomas Hoemig. **MEMBERS:** 550. **STAFF:** 1. **BUDGET:** 2,000,000 BFr. **LANGUAGES:** English, French, Spanish.

DESCRIPTION: Gay, lesbian, bisexual and transgender groups in 70 countries. Fights discrimination against homosexuals in all sectors and promotes the recognition of lesbian and gay rights by applying pressure on governments, international groups, and the media. Serves as information clearinghouse on gay oppression and liberation issues. **PUBLICATIONS:** *Conference Report*, annual; *ILGA Bulletin*, 5/year; *ILGA Pink Book*. Also publishes press releases.

INTERNATIONAL LIAISON CENTRE FOR FILM AND TELEVISION SCHOOLS (Centre International de Liaison des Ecoles de Cinema et de Television - CILECT)

8, rue Theresienne, B-1000 Brussels, Belgium **PHONE:** 32 2 5119839 **E-MAIL:** hverh.cilect@skynet.be **WEBSITES:** http://www.cilect.org **FOUNDED:** 1955. **OFFICERS:** Henry Verhasselt, Exec.Sec. **MEMBERS:** 128. **LANGUAGES:** English, French.

DESCRIPTION: Members are film and television schools having the status of an institution of higher education; film and/or television research institutes; individuals who have contributed significantly to the field of television and film research or teaching; membership represents 49 countries. Objectives are: to promote cooperation among film and television higher teaching and research institutions, staffs, and students; to improve teaching and educational standards to benefit future creative film and television program makers and scholars worldwide; to provide a forum for the exchange of educational, cultural, artistic, and technical developments among members. Encourages research and scholarly work in the history, practice, and theory of film and television. Is concerned with audiovisual developments and offers assistance to individuals contributing information, documentation, and research to the communications media. Sponsors student film festivals; offers consulting services; organizes the exchange of student-produced films. **PUBLICATIONS:** *CILECT News* (in English), quarterly; *Conference Proceedings*, periodic; *Membership List*, periodic.

INTERNATIONAL LIFELINE

PO Box 32714, Oklahoma City, OK 73123 **E-MAIL:** negpa@aol.com **FOUNDED:** 1978. **OFFICERS:** Robert E. Watkins, Pres. **STAFF:** 3. **LANGUAGES:** Creole, French.

DESCRIPTION: Members are volunteer medical personnel. Members donate time and services in emerging nations in the Caribbean for short-term assignments. Conducts charitable programs; offers children's services. **PUBLICATIONS:** none. **Convention/Meeting:** none.

INTERNATIONAL LIVESTOCK RESEARCH INSTITUTE (ILRI)

PO Box 30709, Nairobi, Kenya **PHONE:** 254 2 630743 **FX:** 254 2 631499 **TX:** 22040 **OFFICERS:** Dr. Hank Fitzhugh, Dir. **LANGUAGES:** English.

DESCRIPTION: Members are agricultural researchers and livestock organizations. Seeks to advance the study and practice of animal husbandry. Works to improve the production and profitability of livestock operations. Conducts research and educational programs; gathers and disseminates information.

INTERNATIONAL MAIZE AND WHEAT IMPROVEMENT CENTER (Centro Internacional de Mejoramiento de Maiz y Trigo - CIMMYT)

c/o Corinne de Gracia, Head Librarian, Lisboa 27, Apartado Postal 6-641, 06600 Mexico City, DF, Mexico **PHONE:** 52 5 8042004, 52 5 8047558 **FX:** 52 5 8047559 **TX:** 1772023 CIMTME **E-MAIL:** cimmyt@cgiar.org **WEBSITES:** http://www.cimmyt.mx; http://www.cgiar.org **FOUNDED:** 1966. **OFFICERS:** Prof. Timothy Reeves, Dir. Gen. **STAFF:** 796. **BUDGET:** US$35,000,000. **LANGUAGES:** English, Spanish.

DESCRIPTION: An internationally funded, nonprofit, scientific research and training organization. Participates in a worldwide research program for sustainable maize and wheat systems, with emphasis on helping the poor while protecting natural resources in developing countries. **PUBLICATIONS:** *CIMMYT Annual Report* (in English and Spanish). Explains how research by CIMMYT and collaborating institutions continues to bolster agricultural productivity w/o doing damage to the environment; *CIMMYT Research Briefs*, periodic; *CIMMYT Research Report Series*, periodic; *CIMMYT Summaries*, periodic; *CIMMYT World Maize Facts and Trends*, biennial; *CIMMYT World Wheat Facts and Trends*, biennial; *Literature Update on Maize*, bimonthly; *Literature Update on Wheat, Barley, and Triticale*, periodic.

INTERNATIONAL MARINE CONTRACTORS ASSOCIATION

Carlyle House, 235 Vauxhall Bridge Rd., London SW1 1EJ, England **PHONE:** 44 171 9318171 **FX:** 44 171 9318935 **E-MAIL:** imca@imca-int.com **WEBSITES:** http://www.imca-int.com **FOUNDED:** 1995. **OFFICERS:** Tony Read, Chief Exec. **MEMBERS:** 130. **STAFF:** 6.

DESCRIPTION: Members are companies active in the offshore marine contracting industry, including vessel owners/operators, and marine and underwater contractors. Represents members on an international basis with particular reference to improvements in safety standards. **PUBLICATIONS:** Newsletter, 3/year.

INTERNATIONAL MARINE TRANSIT ASSOCIATION (IMTA)

34 Otis Hill Rd., Hingham, MA 02043 **PHONE:** (781) 749-0078 **FX:** (781) 749-0078 **FOUNDED:** 1976. **OFFICERS:** Martha A. Reardon, Sec.-Treas. **MEMBERS:** 500. **LANGUAGES:** English.

DESCRIPTION: Members are marine transit operators, shipyard operators and suppliers, manufacturers, government and regulatory authorities, naval architects, marine engineering and planning consultants, academics, and specialists in maritime training. Objectives are to: promote and support waterborne transit operations, ferries, hovercraft, and related craft; conduct research and compile information on developments affecting the ferry service industry; exchange information and technical data; stimulate industry cooperation and advancement. **PUBLICATIONS:** Membership Directory, annual; Proceedings, periodic.

INTERNATIONAL MARINELIFE ALLIANCE

2800 4th St. N, Ste. 123, St. Petersburg, FL 33704 **PHONE:** (813) 896-8626 **E-MAIL:**

Prubec@compuserve.com **WEBSITES:** http://www.imamarinelife.org/ **LANGUAGES:** English.

DESCRIPTION: Members are individuals dedicated to preservation of marine species. Promotes captive breeding and keeping of marine organisms in danger of becoming extinct in the wild. Works to increase public awareness of environmental crises facing marine flora and fauna; conducts research and educational programs.

INTERNATIONAL MARITIME COMMITTEE (IMC) (Comite Maritime International - CMI)

Markgravestr. 9, B-2000 Antwerp, Belgium **PHONE:** 32 3 2273526 **FX:** 32 3 2273528 **TX:** 31653 VOET B **E-MAIL:** admini@cmi.inc.org **FOUNDED:** 1897. **OFFICERS:** Baron Leo Delwaide, Admin. **MEMBERS:** 54. **LANGUAGES:** English, French.

DESCRIPTION: Members are national maritime law associations. Strives to promote and contribute to the unification of maritime and commercial law, maritime customs, usages, and practices. Cooperates with other worldwide organizations. **PUBLICATIONS:** *Annuaire CMI Yearbook* (in English and French), annual; *CMI News Letter* (in English and French), quarterly; *Proceedings of International Conferences of CMI*, quadrennial.

INTERNATIONAL MARITIME INDUSTRIES FORUM (IMIF)

15 A Hanover St., London W1R 9HG, England **PHONE:** 44 171 4934559 **FX:** 44 171 4910736 **FOUNDED:** 1975. **OFFICERS:** J.G. Davis CBE, Chm. **MEMBERS:** 130. **BUDGET:** £80,000. **LANGUAGES:** English.

DESCRIPTION: Members are shipowners and builders, shipbreakers, oil companies, insurance companies, classification societies, and bankers in 25 countries. Seeks to: maintain a healthy commercial and financial climate for all sectors of shipping, including ownership, operation, construction, and international trade; encourage discussions of mutual interest; foster change and stimulate action to benefit the maritime industry. Strives to upgrade the standards of ships, port state control and to establish shipbreaking plants in the Third World to promote its large market for rerolled and recycled ship scrap. **PUBLICATIONS:** *All About IMIF* (in English). Contains history and objectives of IMIF; *Newsletter*, 2-3/year.

INTERNATIONAL MARITIME ORGANIZATION (IMO) (Organisation Maritime Internationale - OMI)

4 Albert Embankment, London SE1 7SR, England **PHONE:** 44 171 7357611 **FX:** 44 171 5873210 **TX:** 23588 **E-MAIL:** info@imo.org **WEBSITES:** http://www.imo.org **FOUNDED:** 1958. **OFFICERS:** W.A. O'Neil, Sec.Gen. **MEMBERS:** 155. **STAFF:** 300. **BUDGET:** £36,612,200. **LANGUAGES:** English, French, Spanish.

DESCRIPTION: Members are governments involved in promoting the safety of international merchant shipping and preventing pollution at sea caused by ships. **PUBLICATIONS:** *IMO News*, quarterly; Catalog, periodic.

INTERNATIONAL MARITIME PILOTS ASSOCIATION (IMPA)

c/o N.F. Matthews, HQS Wellington, Temple Stairs, Victoria Embankment, London WC2R 2PN, England **PHONE:** 44 171 2403973 **FX:** 44 171 2403518 **FOUNDED:** 1970. **OFFICERS:** N.F. Matthews, Sec.Gen. **LANGUAGES:** English.

DESCRIPTION: Members are associations of maritime pilots in 35 countries. Disseminates information on matters of mutual interest to members. **PUBLICATIONS:** *IMPA Newsletter*, periodic; *Required Boarding Arrangements for Pilots*; *Shipmaster Guide to Pilot Transfer by Helicopter*, periodic.

INTERNATIONAL MATHEMATICAL UNION (IMU) (Union Mathematique Internationale - UMI)

c/o J. Palis, IMU, IMPA, Instituto de Matematica Pura e Aplicada, Estrada Dona Castorina 110, 22460 Rio de Janeiro, Rio de Janeiro, Brazil **PHONE:** 55 21 5111749 **FX:** 55 21 5124112 **E-MAIL:** dalitz@zib.de **WEBSITES:** http://elib.zib.de/IMU/ **FOUNDED:** 1952. **OFFICERS:** Prof. David Mumford, Pres. **MEMBERS:** 62. **STAFF:** 1. **LANGUAGES:** English, French, Russian.

DESCRIPTION: Members are national academies, mathematical societies, and research councils. Promotes international cooperation in the field of mathematics; encourages international mathematical activities. **PUBLICATIONS:** *IMU Bulletin* (in English), 1-2/year; *World Directory of Mathematicians* (in English), quadrennial.

INTERNATIONAL MEASUREMENT CONFEDERATION (IMEKO) (Internationale Messtechnische Konfoderation - IMEKO)

Postafiok 457, H-1371 Budapest V, Hungary **PHONE:** 36 1 1531562 **FX:** 36 1 1531406 **TX:** 225792 MTESZ H **E-MAIL:** imeko.ime@mtesz.hu **FOUNDED:** 1958. **OFFICERS:** Dr. Tamas Kemeny, Sec.Gen. **MEMBERS:** 33. **BUDGET:** US$45,000. **LANGUAGES:** English.

DESCRIPTION: Members are national scientific and technical societies concerned with measurement science, technology, and instrumentation. Promotes the international exchange of scientific and technical information relating to developments in measuring techniques, instrument design and manufacture, and to the application of instrumentation in scientific research and in industry. Facilitates cooperation among scientists and engineers in studying problems in this field. Conducts seminars, symposia, and conferences. Maintains consultative status with United Nations Educational, Scientific, and Cultural Organization and the United Nations Industrial Development Organization. **PUBLICATIONS:** *ACTA IMEKO* (in English), triennial. Includes Congress papers.; *IMEKO Bulletin*, semiannual; *Measurement*, quarterly; *Technical Committee Conference*, semiannual.

INTERNATIONAL MEAT SECRETARIAT

64 rue Taitbout, F-75009 Paris, France **PHONE:** 33 1 42800472 **FX:** 33 1 42806745 **E-MAIL:** meat.ims@wanadoo.fr **WEBSITES:** http://www.meat-ims.org **FOUNDED:** 1974. **OFFICERS:** Laurence Wrixon, Sec.Gen. **MEMBERS:** 95. **LANGUAGES:** English, French, Spanish.

DESCRIPTION: Members are organizations, corporations, and individuals involved in the trading of meat and meat products; organizations representing livestock breeders, meat packers, and related trades; government bodies concerned with the meat trade. Works to improve the economic potential of the global meat industry; facilitates communication and cooperation among members; maintains contact with national government and international agencies involved in the meat and related trades. Conducts research and educational programs. **PUBLICATIONS:** *IMS Newsletter* (in English), bimonthly. Contains economic, commercial, technical information on the meat sector.; *Meat Processing International Edition*, bimonthly.

INTERNATIONAL MEDICAL SERVICES FOR HEALTH (INMED)

45449 Severn Way, Ste. 161, Sterling, VA 20166 **PHONE:** (703) 444-4477 **TF:** (800) 521-1175 **FX:** (703) 444-4471 **E-MAIL:** inmed@ix.netcom.com **WEBSITES:** http://www.inmed.org **FOUNDED:** 1986. **OFFICERS:** Linda Pfeiffer Ph.D., Pres. **STAFF:** 14. **BUDGET:** US$2,560,000. **LANGUAGES:** Portuguese, Spanish.

DESCRIPTION: Works to help disadvantaged people worldwide to improve the health of their families and communities. Assists community based programs in 98 countries to achieve lasting changes for better health. Operates MotherNet America, a lay home visiting program to reduce infant mortality, low birthweight child abuse and neglect and other risks facing disadvantaged families; conducts the Children as Agents of Change treatment and education program; provides access to medicine for non-profit organizations; offers technical assistance and training. **PUBLICATIONS:** *InScope*, quarterly; *MotherNet America Bulletin*, periodic; *PEC Bulletin*, periodic; *TB Explorer*, quarterly.

INTERNATIONAL METALWORKERS FEDERATION (IMF) (Federation Internationale des Organisations de Travailleurs de la Metallurgie - FIOM)

54 bis, rte. des Acacias, Case Postale 1516, CH-1227 Geneva, Switzerland **PHONE:** 41 22 3085050 **FX:** 41 22 3085055 **TX:** 423298 METAL CH **E-MAIL:** imf@mfmetal.org **WEBSITES:** http://www.imfmetal.org **FOUNDED:** 1893. **OFFICERS:** Marcello Malentacchi, Gen.Sec. **MEMBERS:** 18,000,000. **STAFF:** 30. **LANGUAGES:** English, French, German, Italian, Japanese, Portuguese, Spanish, Swedish.

DESCRIPTION: Members are trade unionists. Conducts conferences for industrial departments. Provides regional assistance and specialized education programs. **PUBLICATIONS:** *IMF News*, periodic.

INTERNATIONAL MOBILE SATELLITE ORGANIZATION

99 City Rd., London EC1Y 1AX, England **PHONE:** 44 171 7281000 **FX:** 44 171 7281044 **TX:** 297201 **WEBSITES:** http://www.inmarsat.org **FOUNDED:** 1979. **OFFICERS:** Mr. W. Grace, Dir.Gen. **MEMBERS:** 79. **LANGUAGES:** English, French, Russian, Spanish.

DESCRIPTION: Members are satellite telecommunications investors. Operates a satellite system of mobile telecommunications services for commercial, distress, and safety applications at sea, in the air, and on land. Supports direct-dial telephone, telex, fax, electronic mail, and data connections for maritime applications; flight-deck voice and data, automatic position and status reporting, and direct-dial passenger telephone for aircraft; and two-way data communications, position reporting, and fleet management for land transport. **PUBLICATIONS:** *Aeronautical Satellite News*, quarterly; *Ocean Voice*, quarterly; *Transat*, quarterly.

INTERNATIONAL MONETARY FUND (IMF)

700 19th St. NW, Washington, DC 20431 **PHONE:** (202) 623-7000 **FX:** (202) 623-4661 **TX:** (RCA) 248331 IMF UR **FOUNDED:** 1945. **OFFICERS:** Michel Camdessus, Mng.Dir. **MEMBERS:** 175. **STAFF:** 2,400. **LANGUAGES:** English.

DESCRIPTION: Members are national governments. Works to: facilitate monetary cooperation through consultation and collaboration among member nations; assist in the balanced expansion of trade and thus contribute to the internal development and prosperity of member nations; maintain stability in monetary exchange arrangements, particularly to avoid exchange depreciations; participate in establishing a multilateral system of payments between member nations and in eliminating exchange restrictions that hamper trade; make available the resources of the fund to provide member nations with a means of assuaging economic difficulties. Maintains the IMF Institute, which conducts training courses and seminars and provides lecturers on subjects such as compilation of statistics and formulation and execution of balance of payment policies. Offers technical assistance on monetary matters to member nations and their dependencies and to multinational institutions. Acts as a depository of information and statistical data regarding the economic affairs of member nations. Operates library, in conjunction with the World Bank, on finance and economic development. **PUBLICATIONS:** *Annual Report of Executive Board*; *Annual Report on Exchange Arrangements and Exchange Restrictions*; *Balance of Payments Statistics*, annual; *Direction of Trade Statistics*, annual; *Finance and Development*, quarterly; *Government Finance Statistics Yearbook*; *IMF Survey*, 23/year; *International Financial Statistics*, annual; *Staff Papers*, quarterly; *Summary Proceedings*, annual; *World Economic Outlook*, semiannual. Also publishes pamphlet and working paper series and books.

INTERNATIONAL MOVEMENT AGAINST ALL FORMS OF DISCRIMINATION AND RACISM (IMADR)

c/o Buraku Liberation Research Institute, 1-6-12 Kuboyoshi, Naniwa-ku, Osaka-shi, Osaka 556, Japan **PHONE:** 81 6 5611093, 81 6 5680905 **FX:** 81 6 5680714 **FOUNDED:** 1990. **OFFICERS:** Kinhide Mushakoji. **MEMBERS:** 21. **STAFF:** 8. **BUDGET:** US$587,147. **LANGUAGES:** English, Japanese.

DESCRIPTION: National branch of the international organization. Works in conjunction with United Nations agencies to ensure respect for human rights and eliminate racism and discrimination in Japan. Organizes campaigns promoting international human rights declarations; conducts research on racism, both in Japan and abroad. **PUBLICATIONS:** *Human Rights and the World Today* (in Japanese), annual; *Imadar-JC Tsushin* (in Japanese), monthly. Focuses on specific national and international human rights issues.

INTERNATIONAL MOVEMENT ATD FOURTH WORLD (IMATDFW) (Mouvement International ATD Quart Monde)

107, ave. du General Leclerc, F-95480 Pierrelaye, France **PHONE:** 33 1 34304622 **FX:** 33 1 30376572 **E-MAIL:** qtdint@qtd_quartmonde.org **FOUNDED:** 1957. **OFFICERS:** Eugen Brand, Sec.Gen. **LANGUAGES:** Dutch, English, French, German, Italian, Spanish.

DESCRIPTION: Members are volunteers working in ghettos, barrios, and public housing in 106 countries. Seeks to: reappraise poverty policies and programs; encourage and enable Fourth World families to work to improve their positions; foster public recognition and representation of Fourth World families; mobilize national and international efforts in support of Fourth World activities. (The term Fourth World refers to families at the bottom of the social scale who are not asked to participate in societal progression.) Conducts petition drives, training programs, field work, and research. Offers internships. Organizes preschools, street libraries, clubs of knowledge, cultural and drop-in centers, and literacy programs. **PUBLICATIONS:** *Cahiers du Quart Monde* (in Dutch, English, and German), annual; *Feuille de Route* (in French), monthly; *Letter to Friends Around the World* (in English, French, and Spanish), 3/year. Also publishes *Pere Joseph Wresinski's writings* (book) and newsletters, handbooks, leaflets, posters, and brochures.

INTERNATIONAL MOVEMENT OF CATHOLIC JURISTS (IMCJ) (Mouvement International des Juristes Catholiques)

4, sq. la Bruyere, F-75009 Paris, France **PHONE:** 33 1 42804954 **FX:** 33 1 48741500 **FOUNDED:** 1946. **OFFICERS:** Louis Pettitti, Pres. **MEMBERS:** 500. **STAFF:** 20. **LANGUAGES:** English, French, Italian, Spanish.

DESCRIPTION: Members are Catholic lawyers, judges, and university professors in 30 countries. Areas of concern include human rights, international and family law, and ethical and cultural issues. Arranges missions to countries where violations of human rights systematically occur. Conducts research on legal matters; organizes ecumenical action to work for international freedom of religion. Sponsors seminars on various aspects of human rights including ethics, genetics, and family law. **PUBLICATIONS:** *Convergences* (in French), quarterly.

INTERNATIONAL MUSIC CENTRE (IMZ - IMZ)

c/o Franz Patay, Speisinger Strasse 121-127, A-1230 Vienna, Austria **PHONE:** 43 1 8890315 **FX:** 43 1 889031577 **E-MAIL:** office@imz.at **WEBSITES:** http://www.imz.magnet.at/imz/ **FOUNDED:** 1961. **OFFICERS:** Dr. Franz Patay, Sec.Gen. **MEMBERS:** 200. **STAFF:** 8. **LANGUAGES:** English, French, German.

DESCRIPTION: Members are public and private television and radio stations, record and video companies, music academies, opera houses, and editors in 30 countries. Works to promote music productions through television, video, film, radio, and records. Maintains index of international television-music broadcasts. Organizes seminars. Conducts annual Avant Premiere screening of new music programs; organizes annual Dance on Video competition. **PUBLICATIONS:** *Dance Screen*, periodic; *IMC Directory*, periodic; *IMZ Bulletin/Music in the Media* (in English, French, and German), bimonthly. New productions of member organizations; News about Music/Dance in the audiovisual media; *Opera Screen*, periodic.

INTERNATIONAL MUSIC COUNCIL (IMC) (Conseil International de la Musique)

1, rue Miollis, F-75732 Paris Cedex 15, France **PHONE:** 33 1 45682550 **FX:** 33 1 43068798 **E-MAIL:** imc_cim@compuserve.com **FOUNDED:** 1949. **OFFICERS:** Guy Huot, Gen.Sec. **MEMBERS:** 110. **STAFF:** 3. **LANGUAGES:** English, French.

DESCRIPTION: Members are national music committees; international music organizations; individuals; members of honor. Created by the United Nations Educational, Scientific and Cultural Organization to implement the policies of UNESCO in the field of music. Acts through national music committees and international organizations in 70 countries which provide representation of and effective liaison between the different elements constituting the musical life of the country; through its 30 international member organizations it touches most fields of musical endeavour. Promotes music creation, music performance, music education, and music research in the musical cultures of the world. Acts as a link between composers and interpreters, cultural authorities, the mass media, and the trade and industries of music. **PUBLICATIONS:** *Resonance* (in English and French), 3/year; *The Universe of Music: A History*; Newsletter, monthly.

INTERNATIONAL MUSICOLOGICAL SOCIETY (IMS) (Internationale Gesellschaft fur Musikwissenschaft)

Postfach 1561, CH-4001 Basel, Switzerland **FX:** 41 1 9231027 **E-MAIL:** imsba@swissonline.ch **WEBSITES:** http://www.ims-online.ch **FOUNDED:**

1927. **OFFICERS:** Dr. Dorothea Baumann, Sec.Gen. **MEMBERS:** 1,200. **STAFF:** 1. **LANGUAGES:** English, French, German, Italian, Spanish.

DESCRIPTION: Members are musicological societies, university libraries, and individuals in 46 countries. Works to further musicological research through international cooperation. **PUBLICATIONS:** *Acta Musicologica* (in English, French, German, Italian, and Spanish), semiannual; *Communiques*, 1-2/year; *Congress Reports*, quinquennial; Bibliographies, periodic.

INTERNATIONAL NARCOTICS CONTROL BOARD (INCB) (Junta Internacional de Fiscalizacion de Estupefacientes - JIFE)

United Nations Vienna Intl. Centre, Postfach 500, A-1400 Vienna, Austria **PHONE:** 43 1 26060 4277 **FX:** 43 1 26060 5867 **TX:** 135612unoa **E-MAIL:** incb@undcp.org **WEBSITES:** http://www.incb.org **FOUNDED:** 1968. **OFFICERS:** Herbert Schaepe, Sec. **MEMBERS:** 13. **STAFF:** 28. **LANGUAGES:** Arabic, Chinese, English, French, Russian, Spanish.

DESCRIPTION: Established by the 1961 Convention on Narcotic Drugs, this treaty organization is composed of 13 members elected by the Economic and Social Council of the United Nations to serve in their personal capacity as independent experts. Work of the board consists of continuous supervision and overall coordination of the national implementation throughout the world of the provisions of the various existing drug control treaties. **PUBLICATIONS:** *Annual Report of the Board* (in Arabic, Chinese, English, French, Russian, and Spanish); *Psychotropic Substances*, annual. Assessments of medical and scientific requirements for substances in Schedules II, III, and IV.

INTERNATIONAL NATURAL RUBBER ORGANIZATION (INRO) (Organisation Internationale du Caoutchouc Naturel - INRO)

Wisma SPK, 4th Fl., Jalan Sultan Ismail, 50712 Kuala Lumpur, Malaysia **PHONE:** 60 3 2486466 **FX:** 60 3 2486485 **TX:** MA 31570 **E-MAIL:** inro@po.jaring.my **FOUNDED:** 1980. **OFFICERS:** Mr. Ahmad Zubeir Haji Nordin, Exec.Dir. **MEMBERS:** 24. **STAFF:** 29. **BUDGET:** US$1,725,000. **LANGUAGES:** Chinese, English, French.

DESCRIPTION: Members are rubber exporting and importing countries. Purpose is to achieve balanced growth between the supply of and demand for natural rubber. Promotes stable conditions in the trade of natural rubber by avoiding excessive price fluctuations and by stabilizing these prices without distorting long-term trends in the market; provides incentives for increased production and the resources for heightened economic and social development. Fosters research and development and facilitates improvements in the processing, market competitiveness, and distribution of raw natural rubber. Seeks to expand rubber trade and to facilitate market access for natural rubber and its products. Encourages international cooperation in and consultation within the industry on matters affecting supply and demand. INRO is governed by the International Natural Rubber Council. **PUBLICATIONS:** *INRO Annual Report* (in English); *INRO Newsletter* (in English), quarterly.

INTERNATIONAL NAVIGATION ASSOCIATION (INA) (Association Internationale de Navigation)

Graaf de Ferraris - 11eme etage - Bte. 3, 156, Blvd. Emile Jacqmain, B-1000 Brussels, Belgium **PHONE:** 32 2 5537160 **FX:** 32 2 5537155 **FOUNDED:** 1885. **OFFICERS:** C. van Begin, Sec.Gen. **MEMBERS:** 2,471. **STAFF:** 3. **LANGUAGES:** English, French.

DESCRIPTION: Members are government delegates; civil engineers, contractors, and other individuals; dredging companies, laboratories, and port authorities. Promotes the maintenance and operation of inland and ocean navigation through the design, construction, improvement, maintenance, and operation of inland and maritime waterways, inland and maritime ports, and coastal areas. Compiles and publishes information about issues in the field; encourages studies on particular problems; organizes international and national committees for the exchange of experience and research. **PUBLICATIONS:** *Account of Proceedings* (in English and French), quadrennial. Congress proceedings; *Bulletin of the PIANC*, quarterly; *Dictionary* (in Dutch, English, French, German, Italian, and Spanish), periodic; *Reports of Study Commissions and Committees*.

INTERNATIONAL NETWORK FOR CHEMICAL EDUCATION (INCE)

Intl. Centre for Chemical Studies, Vegova 4, PO Box 18/1, SLO-61001 Ljubljana, Slovenia **PHONE:** 386 61 214326, 386 61 214374 **FX:** 386 61 226170 **TX:** 31572 CHEMIN YU **E-MAIL:** mtf_iccs@uni-lf.si **FOUNDED:** 1980. **OFFICERS:** Prof. Aleksandra Kornhauser. **STAFF:** 28. **LANGUAGES:** English.

DESCRIPTION: Operates under the auspices of the United Nations Educational, Scientific and Cultural Organization. Chemists and institutions in 65 countries involved in chemical research and education. Purpose is to organize international cooperation among chemists and scientific institutions, particularly in the area of university-industry cooperation, in an effort to improve university chemistry programs worldwide. Offers support to science faculties in developing countries in their efforts to develop programs for teaching chemistry and cooperation with industry. Provides seminars and workshops. Sponsors joint projects in university-industry cooperation, and dissemination of scientific information. Promotes the free flow of scientific information, with particular emphasis on the transfer of scientific information from research centers to industry. Encourages members to exchange and donate expertise, equipment, publications, and services in the interest of international scientific progress. **PUBLICATIONS:** *INCE Directory of Cooperating Institutions and Individuals* (in English), periodic. Also publishes reports on projects, seminars, and workshops.

INTERNATIONAL NETWORK OF CHILDREN OF JEWISH HOLOCAUST SURVIVORS (INCJHS)

c/o Rosita Kenigsberg, Florida Intl. Univ., North Miami Campus, 3000 NE 151st St., North Miami Beach, FL 33181 **PHONE:** (305) 919-5690 **FX:** (305) 919-5691 **FOUNDED:** 1981. **OFFICERS:** Rosita Kenigsberg, Exec. VP. **MEMBERS:** 250,000. **LANGUAGES:** English.

DESCRIPTION: Seeks to: provide a liaison among organizations of children of Holocaust survivors and coordinate their activities; provide these groups with a unified voice on issues including the rise of neo-Nazism, anti-Semitism, and desecration of the Holocaust. **PUBLICATIONS:** *Bimonthly.*

INTERNATIONAL NETWORK FOR SUSTAINABLE ENERGY (INFORSE)

PO Box 2059, DK-1013 Copenhagen K, Denmark **PHONE:** 45 33 121307 **FX:** 45 33 121308 **E-MAIL:** inforse@inforse.dk **WEBSITES:** http://www.inforse.dk **FOUNDED:** 1992. **OFFICERS:** Michael Kvetny, Sec. **MEMBERS:** 170. **STAFF:** 6. **BUDGET:** US$150,000. **LANGUAGES:** Danish, English.

DESCRIPTION: Network of organizations working in the field of sustainable and renewable energy sources. Promotes increased communication and cooperation among members. Conducts research to develop sustainable energy strategies. Sponsors educational programs to raise public and political awareness of energy issues.

INTERNATIONAL NON-GOVERNMENTAL ORGANIZATION CORP.

6 Volgina St., 117485 Moscow, Russia **PHONE:** 7 95 3308492, 7 95 3308647 **FX:** 7 95 3308492 **E-MAIL:** incorvuz@interset.ru **FOUNDED:** 1989. **OFFICERS:** Gennady Kaliouzhny, Dir.Gen. **MEMBERS:** 133. **STAFF:** 15. **BUDGET:** US$120,000. **LANGUAGES:** English, Russian.

DESCRIPTION: Members are graduates of educational institutions located in the former Soviet Union; educational institutions at all levels of study. Seeks to advance the educational systems of the countries of the former Soviet Union; promotes continuing professional development among educators and educational administrators. Conducts evaluations of educational programs and makes recommendations for their improvement; makes available fellowships; sponsors educational research. UNESCO/INCORVUZ chair and network for the development of NGOs in countries in transition: promotion of NGOs activities and their networking, evaluation studies, exchange of information and experience among NGOs, NGOs consultations, seminars, workshops, research on civil society issues. **PUBLICATIONS:** *INCORVUZ* (in English and Russian), biennial; Report, periodic.

INTERNATIONAL NUCLEAR LAW ASSOCIATION (INLA) (Association Internationale du Droit Nucleaire - AIDN)

c/o V. Verbraeken, 29, sq. de Meeus, B-1000 Brussels, Belgium **PHONE:** 32 2 5475841 **FX:** 32 2 5030440 **E-MAIL:** aidn.inla@skynet.be **FOUNDED:** 1970. **OFFICERS:** B. Van Cauter. **MEMBERS:** 500. **LANGUAGES:** English, French.

DESCRIPTION: Members are individuals in 31 countries. Arranges for and promotes the study of legal problems related to the peaceful utilization of nuclear energy. Advocates the protection of humankind and the environment. Cooperates on a scientific basis with associations and institutions with similar interests. Sponsors discussions.

INTERNATIONAL NUMISMATIC COMMISSION (Commission Internationale de Numismatique - CIN)

c/o Dr. A.M. Burnett, Dept. of Coins and Medals, British Museum, London WC1B 3DG, England **PHONE:** 44 171 3238227 **FX:** 44 171 3238171 **E-MAIL:** aburnette@british-museum.ac.uk **FOUNDED:** 1937. **OFFICERS:** Dr. A.M. Burnett, Sec. **MEMBERS:** 137. **BUDGET:** 50,000 SFr. **LANGUAGES:** French, German.

DESCRIPTION: Members are national numismatic organizations and societies in 35 countries; mints representing member nations. Acts as a center of information; ensures that numismatics is represented at congresses of related sciences; organizes international numismatic congresses. **PUBLICATIONS:** *Coins and Compoters*, semiannual; *Compte-Rendu* (in English, French, and German), annual; *International Numismatic Newsletter*, semiannual.

INTERNATIONAL OIL POLLUTION COMPENSATION FUND (IOPC Fund) (Fonds International d'Indemnisation pour les Dommages dus a la Pollution par les Hydrocarbures - FIPOL)

4 Albert Embankment, London SE1 7SR, England **PHONE:** 44 171 5822606 **FX:** 44 171 7350326 **TX:** 23588 **E-MAIL:** iopcfund@dircon.co.uk **FOUNDED:** 1978. **OFFICERS:** Mans Jacobsson, Dir. **MEMBERS:** 82. **STAFF:** 20. **LANGUAGES:** English, French.

DESCRIPTION: Members are states which adhere to the International Convention on the Establishment of an International Fund for Compensation for Oil Pollution Damage elaborated by the International Maritime Organization. Provides supplementary compensation to victims of oil spills when compensation obtained from shipowners or their insurers under the 1969 International Convention on Civil Liability for Oil Pollution Damage is insufficient. **PUBLICATIONS:** *Claims Manual* (in English, French, and Spanish); *Explanatory Note Prepared by the Secretriat* (in English, French, and Spanish); *Statistics* (in English), annual; Report (in English and French), annual.

INTERNATIONAL OLIVE OIL COUNCIL (IOOC) (Consejo Oleicola Internacional)

Principe de Vergara, 154, E-28002 Madrid, Spain **PHONE:** 34 1 5630071, 34 1 5903638 **FX:** 34 1 5631263 **E-MAIL:** iooc@mad.servicom.es **FOUNDED:** 1959. **OFFICERS:** Fausto Luchetti, Exec.Dir. **MEMBERS:** 10. **STAFF:** 52. **LANGUAGES:** Arabic, English, French, Italian, Spanish.

DESCRIPTION: Intergovernmental organization that administers the International Agreement on Olive Oil and Table Olives. Fosters international cooperation for the integrated development of the world, economy for olive products. Designs and promotes programs of technical cooperation in olive cultivation, olive oil extraction, and table olive processing. Seeks to balance production and consumption of olive products. Works to eliminate unfair competition through the preparation of a code of standard fair trade practices, the adoption of standard international contracts for transactions involving olive products, and the development of an international conciliation and arbitration office to deal with disputes. Promotes the establishment of uniform standards for physical and chemical characteristics for olive oil and its analysis. Conducts research programs and undertakes any activities and measures that could highlight the biological value of olive oil and table olives with particular reference to their national qualities and therapeutic properties. Initiates and conducts generic promotional campaigns (educational and informational) to expand world consumption of olive products. Fosters the coordination of production, industrialization and marketing policies of these products. **PUBLICATIONS:** *IOOC Information Sheet* (in French and Spanish), biweekly; *National Policies for Olive Products* (in English, French, Italian, and Spanish), annual; *OLIVAE* (in English, French, Italian, and Spanish), 5/year; *Proceedings* (in Arabic, English, French, Italian, and Spanish), periodic; Reports, periodic. Containing information on the olive and olive oil sector.

INTERNATIONAL OLYMPIC COMMITTEE (IOC) (Comite International Olympique - CIO)

Chateau de Vidy, Case Postale 356, CH-1007 Lausanne, Switzerland **PHONE:** 41 21 6216111 **FX:** 41 21 6216216 **FOUNDED:** 1894. **OFFICERS:** Juan Antonio Samaranch. **MEMBERS:** 111. **STAFF:** 80. **BUDGET:** 36,985 SFr. **LANGUAGES:** English, French.

DESCRIPTION: Contributes to building a peaceful and better world by educating youth through sport practised without discrimination of any kind and in the Olympic spirit, which requires mutual understanding with a spirit of friendship, solidarity and fair play. Aims to encourage the coordination, organization and development of sport and sports competitions; collaborate with the competent public or private organizations and authorities in the endeavour to place sport at the service of humanity; ensure the regular celebration of the Olympic Games; fight against any form of discrimination affecting the Olympic Movement; support and encourage the promotion of sports ethics; dedicate efforts to ensuring that in sports the spirit of fair play prevails and violence is banned; lead the fight against the use of drugs in sports; take measures to ensure the prevention of endangering the health of athletes, oppose any political or commercial abuse of sports and athletes; ensure that Olympic Games are held in conditions which demonstrate a responsible concern for environmental issues; support the International Olympic Academy (IOA), and support other institutions which devote themselves to Olympic education. **PUBLICATIONS:** *Olympic Charter*, annual; *Olympic Directory*, annual; *Olympic Review*, bimonthly. Also publishes books and pamphlets.

INTERNATIONAL OMBUDSMAN INSTITUTE (IOI) (Institut International du Mediateur)

Univ. of Alberta, I.O.I. Faculty of Law, Weir Library, Rm. 205D, Edmonton, AB, Canada T6G 2H5 **PHONE:** (780) 492-3196 **FX:** (780) 492-4924 **E-MAIL:** dcallan@law.ualberta.ab.ca **FOUNDED:** 1978. **OFFICERS:** Mrs. D. Callan, Office Mgr. **MEMBERS:** 150. **STAFF:** 2. **LANGUAGES:** English, French, German, Spanish.

DESCRIPTION: Ombudsman offices, complaint handling organizations, institutions, libraries, and individuals in 74 countries. Promotes concept of ombudsmanship and supports research and educational efforts in the field. Disseminates information about ombudsmanship; participates in seminars concerning the ombudsman concept. **PUBLICATIONS:** *Court Cases of Interest to the Ombudsman Institution*. Includes supplements; *Occasional Paper Series*, 5/year; *Ombudsman Profiles*; Directory, semiannual; Journal, annual; Newsletter, bimonthly.

INTERNATIONAL OPTICIANS ASSOCIATION

6 Hurlingham Business Park, Sulivan Rd., London SW6 3DU, England **PHONE:** 44 171 7360088 **FX:** 44 171 7315531 **E-MAIL:** general@abdo.demon.co.uk **FOUNDED:** 1985. **OFFICERS:** Malcolm Hunt. **MEMBERS:** 7. **STAFF:** 2. **BUDGET:** £1,000.

DESCRIPTION: Members are companies engaged in the business of retail opticians which are optical employers. Represents the interests of optical employers, in legislation, relating with the Department of Health, with the General Optical Council and other optical bodies. Issues advice and guidance to members and is involved in European matters through the Joint Optical Committee on the EC.

INTERNATIONAL ORGANISATION FOR THE ELIMINATION OF ALL FORMS OF RACIAL DISCRIMINATION (Organisation Internationale pour l'Elimination de Toutes les Formes de Discrimination Raciale)

5, rte. des Morillons, Bureau No. 475, CP 2100, CH-1211 Geneva 2, Switzerland **PHONE:** 41 22 7886233 **FX:** 41 22 7886245 **FOUNDED:** 1976. **OFFICERS:** Abdullah Sharfelddin, Pres. **BUDGET:** US$250,000. **LANGUAGES:** English.

DESCRIPTION: Members are individuals and groups in 17 countries involved in human rights issues and who oppose racism and racial discrimination. Promotes equality, dignity, rights for all peoples and individuals. Provides support to non-governmental organizations with similar aims. Cooperates with universities in different countries and United Nations bodies on the issue of racism and racial discrimination; disseminates information. Plans to establish a library and research centers. **PUBLICATIONS:** *Without Prejudice* (in English), quarterly; Bulletin, periodic; Monographs, periodic; Papers, periodic.

INTERNATIONAL ORGANIZATION FOR BIOLOGICAL CONTROL OF NOXIOUS ANIMALS AND PLANTS (IOBC) (Organisation Internationale de Lutte Biologique Contre les Animaux et les Plantes Nuisibles - OILB)

c/o Dr. E. Wajnberg, INRA-Ecologie des Parasitoides, 37, Blvd. du Cap, F-06600 Antibes, France **PHONE:** 33 4 93678892 **FX:** 33 4 93678897 **E-MAIL:** wajnberg@antibes.inra.fr **FOUNDED:** 1956. **OFFICERS:** Dr. E. Wajnberg, Sec.Gen. **MEMBERS:** 2,000. **BUDGET:** US$600,000. **LANGUAGES:** English, French, German, Spanish.

DESCRIPTION: A section of the International Union of Biological Sciences. Public and private institutions including government departments, universities, and academies; interested individuals. Aims to promote, coordinate, and intensify research on biological and integrated control of animals and plants harmful to agriculture in member countries. Operates identification service; organizes commissions; conducts symposia and workshops. **PUBLICATIONS:** *BioControl* (in English, French, German, and Spanish), quarterly; *IOBC Newsletter*, 02Y; *Section Bulletin*, periodic. Also publishes periodic working group newsletters, books, and information carriers.

INTERNATIONAL ORGANIZATION OF CITRUS VIROLOGISTS (IOCV)

c/o C.N. Roistacher, Sec., Dept. Plant Pathology, Univ. of California, Riverside, Riverside, CA 92521 **PHONE:** (909) 684-0934 **FX:** (909) 684-4324 **E-MAIL:** Chester.R@worldnet.att.net **FOUNDED:** 1957. **OFFICERS:** Patricia Barkley, Chrm. **MEMBERS:** 220. **LANGUAGES:** English.

DESCRIPTION: Members are individuals engaged in research in citriculture, plant virus diseases, or plant protection; persons involved in nursery operations, fruit production, processing, and marketing related to citrus. Promotes cooperative study of citrus fruit disease and the exchange of information regarding their identity, relationships, effects, importance, means of spread, control, and/or prevention. Encourages personal contacts and the preparation and distribution of materials relevant to the study of citrus virus disease. Seeks to facilitate the development of mutual understanding among individuals, institutions, and agencies concerned with the production of citrus fruits. **PUBLICATIONS:** *International Organization of Citrus Virologists—Proceedings*, triennial. Provides research reports on viral diseases affecting citrus trees; Newsletter, every two or three months.

INTERNATIONAL ORGANIZATION OF THE FLAVOR INDUSTRY (IOFI)

8, rue Charles-Humbert, CH-1205 Geneva, Switzerland **PHONE:** 41 22 3213548 **FX:** 41 22 7811860 **E-MAIL:** iofi@dial.eunet.ch **FOUNDED:** 1969. **OFFICERS:** F. Grundschober, Gen.Sec. **MEMBERS:** 21. **STAFF:** 3. **LANGUAGES:** English, French, German.

DESCRIPTION: Members are national associations of flavor manufacturers. Active in areas of safety evaluation and regulation of flavoring substances. Compiles and studies control measures; issues a code of practice for the flavor industry. Provides current information to members, authorities, and international organizations. Sponsors working groups and committees. **PUBLICATIONS:** *Code of Practice* (in English, French, and German); *Documentation Bulletin*, periodic; *Information Letters*, periodic.

INTERNATIONAL ORGANIZATION OF JOURNALISTS (IOJ) (Organizacion Internacional de Periodistas)

Celetna 2, CZ-110 01 Prague, Czech Republic **PHONE:** 42 2 2365916 **FX:** 42 2 2368804 **TX:** 122631 JOUR C **FOUNDED:** 1946. **OFFICERS:** Gerard Gatinot, Sec.Gen. **MEMBERS:** 265,000. **LANGUAGES:** Arabic, English, French, Portuguese, Russian, Spanish.

DESCRIPTION: Members are national journalists' unions, groups, and committees and journalists in 139 countries. Objectives are: to promote free, honest, and accurate journalism as a means of enhancing international understanding and friendship among peoples; to protect the well-being and rights of journalists, particularly the right to freely practice their profession; to work against journalistic misinformation and dishonest reporting. Provides assistance to journalists through the International Journalists' Solidarity Lottery. Offers moral and material support to journalists suffering persecution for their dedication to progressive issues. Organizes seminars, symposia, training courses, and exhibitions. Bestows International Journalism Prize and the Julius Fucik Medal of Honour for significant contributions to peace and understanding among nations. **PUBLICATIONS:** *Democratic Journalist* (in English, French, Russian, and Spanish), monthly; *Interpressgraphic* (in English), quarterly; *IOJ Newsletter* (in Arabic, English, French, German, Russian, and Spanish), biweekly.

INTERNATIONAL ORGANIZATION OF LEGAL METROLOGY (OIML) (Organisation Internationale de Metrologie Legale - OIML)

11, rue Turgot, F-75009 Paris, France **PHONE:** 33 1 48781282, 33 1 48781282 **FX:** 33 1 42821727 **E-MAIL:** biml@oiml.org **WEBSITES:** http://www.oiml.org **FOUNDED:** 1955. **OFFICERS:** B. Athane, Dir. **MEMBERS:** 97. **STAFF:** 10. **LANGUAGES:** English, French.

DESCRIPTION: Purpose is to harmonize and coordinate worldwide administrative and technical regulations in the field of metrology. (Metrology is the science of measurements; legal metrology subjected to regulations by law or governmental decree.) Seeks to facilitate free commerce between countries with regard to measuring instruments and all commodities and services related to measurement. Provides guidance for manufacturers and users; ensures high quality instruments; sponsors training programs in the field. Fosters close cooperation among members. Maintains 70 technical committees, and subcommittees, dealing with all facets of legal metrology, and a special council with technical secretariats for addressing metrology problems in developing nations. Maintains documentation center for members. **PUBLICATIONS:** *Bulletin of International Organization of Legal Metrology* (in English), quarterly; *International Recommendations and Documents* (in English and French), periodic.

INTERNATIONAL ORGANIZATION FOR MEDICAL PHYSICS (IOMP)

c/o Prof. Hans Svensson, Radiation Physics Dept., Univ. Hospital, 90185 UMEA, Stockholm, Sweden **FOUNDED:** 1963. **OFFICERS:** Prof. Gary D. Fullerton. **BUDGET:** US$50,000.

DESCRIPTION: A member of the International Union of Physical and Engineering Sciences in Medicine. National organizations of medical physics representing 10,000 individuals. Fosters international cooperation in medical physics; promotes communication between various branches of medical physics and allied subjects. Conducts training programs. Has established 43 libraries in developing countries. **PUBLICATIONS:** *Clinical Physics and Physiological Measurement*, bimonthly; *Medical Physics World*, semiannual; *Physics in Medicine and Biology*, monthly.

INTERNATIONAL ORGANIZATION FOR MIGRATION (IOM) (Organizacion Internacional para las Migraciones)

17, rte. des Morillons, Case Postale 71, CH-1211 Geneva 19, Switzerland **PHONE:** 41 22 7179111 **FX:** 41 22 7986150 **TX:** 415722 **E-MAIL:** info@iom.int **WEBSITES:** http://www.iom.int **FOUNDED:** 1951. **OFFICERS:** James N. Purcell Jr., Dir.Gen. **MEMBERS:** 59. **STAFF:** 1,600. **BUDGET:** US$238,113,010. **LANGUAGES:** English, French, Spanish.

DESCRIPTION: Members are representatives of governments of 59 nations and 42 observer states. Fosters orderly and planned migration of refugees, displaced persons, and other individuals to countries offering resettlement opportunities. Aids transfer of specialized human resources to promote economic, educational and social advancement of developing countries. Maintains Centre for Information on Migration in Latin America. Promotes cooperation and coordination on migration issues. Provides advisory services to governments and other institutions on migration administration, legislation, and policy as well as a forum for discussions and exchanges. **PUBLICATIONS:** *International Migration* (in English), quarterly. Includes summaries in French and Spanish.; *IOM is . . .* (in English, French, and Spanish). Information brochure providing information on programmes and activities.; *IOM Latin American Migration Journal* (in English and Spanish), 3/year; *IOM News* (in English, French, and Spanish); *Trafficking in Migrants* (in English), quarterly.

INTERNATIONAL ORGANIZATION FOR MOTOR TRADES AND REPAIRS (IOMTR) (Organisation Internationale du Commerce et de la Reparation Automobile)

Kosterijland 15, NL-3981 AJ Bunnik, Netherlands **PHONE:** 31 30 6595301 **FX:** 31 30 6564982 **E-MAIL:** iomtr@rdc.nl **WEBSITES:** http://www.rdc.nl/iomtr **FOUNDED:** 1947. **OFFICERS:** Henk W.G. Van Dijk, Exec.Dir. **MEMBERS:** 32. **STAFF:** 2. **LANGUAGES:** English, French, German.

DESCRIPTION: Members are associations of employers in 24 countries involved in the sale, repair, and servicing of motor vehicles. Objectives are to: collect and disseminate information affecting employers in any aspect of the motor trade; stimulate contact and discussion; encourage and promote the international adoption of ethical codes of trading; advise, support, and counsel members to take joint action in the international field. Conducts conferences and various programs and activities. **PUBLICATIONS:** *Thirty Years of IOMTR*; Newsletter (in English, French, and German), quarterly.

INTERNATIONAL ORGANIZATION OF MOTOR VEHICLE MANUFACTURERS (IOMVM) (Organisation Internationale des Constructeurs d'Automobiles - OICA)

4, rue de Berri, F-75008 Paris, France **PHONE:** 33 1 43590013 **FX:** 33 1 45638441 **E-MAIL:** otca@club.internet.fr **FOUNDED:** 1919. **OFFICERS:** Mr. Jean M. Muller, Sec.Gen. **MEMBERS:** 35. **STAFF:** 7. **LANGUAGES:** English, French.

DESCRIPTION: Members are national automobile manufacturers' associations; associate members are importers' associations. Objectives are to act as a clearinghouse for and further the interests of the automobile industry and to promote the study of economic and recreational matters affecting automobile construction. Represents members' interests involving technical, industrial and economic policy, and exhibitions. **PUBLICATIONS:** *Yearbook of the World's Motor Industry* (in English and French), annual.

INTERNATIONAL ORGANIZATION OF SECURITIES COMMISSIONS (IOSCO) (Organisation Internationale des Commissions de Valeurs - OICV)

PO Box 171, Stock Exchange Tower, 800 Square Victoria, Ste. 4210, Montreal, PQ, Canada H4Z 1C8 **PHONE:** (514) 875-8278 **FX:** (514) 875-2669 **E-MAIL:** mail@oicv.iosco.org **WEBSITES:** http://www.iosco.org **FOUNDED:** 1983. **OFFICERS:** Peter Clark, Sec.Gen. **MEMBERS:** 158. **STAFF:** 6. **LANGUAGES:** English, French, Portuguese, Spanish.

DESCRIPTION: Members are government agencies regulating securities trading. Facilitates cooperation and exchange of information among members. Conducts research programs. **PUBLICATIONS:** Annual Report, annual.

INTERNATIONAL ORGANIZATION FOR STANDARDIZATION (ISO) (Organisation Internationale de Normalisation)

1, rue de Varembe, Case Postale 56, CH-1211 Geneva 20, Switzerland **PHONE:** 41 22 7490111 **FX:** 41 22 7333430 **E-MAIL:** central@iso.ch **WEBSITES:** http://www.iso.ch/ **FOUNDED:** 1947. **OFFICERS:** Lawrence D. Eicher, Sec.Gen. **MEMBERS:** 133. **STAFF:** 165. **LANGUAGES:** English, French, Russian.

DESCRIPTION: Worldwide Federation of national standards bodies which promotes standardization worldwide. Develops and publishes International Standards to facilitate exchange of goods and services and foster mutual cooperation in intellectual, scientific, technological and economic spheres of endeavor. Maintains 186 technical committees and 2,673 subcommittees and working groups; sponsors advisory committees relating to conformity assessment, consumer policy, information systems and services, and the needs of developing countries. **PUBLICATIONS:** *ISO Annual Report*; *ISO Bulletin* (in English and French), monthly. Contains news on standardization work, publications, and events; *ISO Catalogue* (in English and French), annual. Contains information on ISO standards and other publications; *ISO International Standards*, periodic; *ISO Memento* (in English and French), annual. Contains information on ISO organization and structure; *ISO 9000 and ISO 14000 News* (in English and French), bimonthly. Contains information on the ISO 9000 quality management and ISO

14000 environmental management standards and activities; *ISO Standards Handbooks*, periodic.

INTERNATIONAL ORGANIZATION FOR THE STUDY OF THE OLD TESTAMENT (IOSOT) (Organisation Internationale pour l'Etude de l'Ancien Testament)

c/o Prof. A. van der Kooij, Faculteit der Godgeleerdheid, Postbus 9515, NL-2300 RA Leiden, Netherlands **PHONE:** 31 71 5272577 **FX:** 31 71 5272571 **FOUNDED:** 1950. **OFFICERS:** Prof. A. van der Kooij, Gen.Sec. **LANGUAGES:** English, French, German.

DESCRIPTION: Promotes international cooperation in Old Testament study. Encourages the exchange of information. **PUBLICATIONS:** *Vetus Testamentum* (in English, French, and German), quarterly. Includes Supplements.

INTERNATIONAL ORGANIZATION OF SUPREME AUDIT INSTITUTIONS (INTOSAI) (Organisation Internationale des Institutions Superieures de Control de Finances Publiques)

Dampfschiffstrasse 2, Postfach 240, A-1033 Vienna, Austria **PHONE:** 43 1 711718350 **FX:** 43 1 7180969 **E-MAIL:** intosai@rechnungshof.gv.at **WEBSITES:** http://www.intosai.org **FOUNDED:** 1953. **OFFICERS:** Dr. Franz Fiedler, Sec.Gen. **MEMBERS:** 179. **LANGUAGES:** Arabic, English, French, German, Spanish.

DESCRIPTION: Members are government organizations that perform the auditing function for their states. Promotes the exchange of ideas and experiences between supreme audit institutions in the field of public auditing. Works to standardize finance and auditing terminology. Sponsors UN/ INTOSAI seminars for auditing staffs of supreme audit institutions in developing countries. **PUBLICATIONS:** *Congress Reports*, triennial; *International Journal of Government Auditing*, quarterly; *INTOSAI Circulars*, periodic; *INTOSAI Documents*, periodic; *INTOSAI Informations*, periodic.

INTERNATIONAL PEACE ACADEMY (IPA)

777 United Nations Plz., 4th Fl., New York, NY 10017 **PHONE:** (212) 687-4300 **FX:** (212) 983-8246 **E-MAIL:** ipa@ipacademy.org **WEBSITES:** http://www.ipacademy.org **FOUNDED:** 1970. **OFFICERS:** David Malone, Pres. **STAFF:** 20. **BUDGET:** US$1,600,000. **LANGUAGES:** English.

DESCRIPTION: Private postgraduate educational institute for professional training in skills of conflict management. All meetings and programs of the academy are by invitation only and not open to the public. Conducts international training seminars in the Middle East, Europe, Africa, Latin and North America, and Asia. Conducts research on practical problems in the areas of peacekeeping, mediation, negotiation, ocean dispute settlement mechanisms, and disarmament skills. Holds off-the-record meetings for assessing the negotiating process during specific violent conflicts. **PUBLICATIONS:** *International Peacekeeping*, quarterly. Examines theory and practice of peacekeeping.; *IPA Initiatives*, semiannual; *IPA Policy Briefing Series*; *Keeping the Peace: Multidimensional UN Operations in Cambodia and El Salvador*; *Occasional Papers*, periodic; *Peacemaking and Peacekeeping for the New Century*; *Rights & Reconciliation: UN Strategies in El Salvador*.

INTERNATIONAL PEACE BUREAU (IPB) (Bureau International de la Paix)

41, rue de Zurich, CH-1201 Geneva, Switzerland **PHONE:** 41 22 7316429 **FX:** 41 22 7389419 **E-MAIL:** info@ipb.org **WEBSITES:** http://www.ipb.org **FOUNDED:** 1892. **OFFICERS:** Colin Archer, Gen.Sec. **MEMBERS:** 160. **STAFF:** 2. **BUDGET:** 150,000 SFr. **LANGUAGES:** English, French.

DESCRIPTION: Members are international and national peace organizations in over 40 countries. Serves the cause of peace by promoting disarmament, international cooperation, and the peaceful solution of international conflicts. Facilitates communication among members. Organizes and coordinates peace events and activities. Represents members before governments and United Nations agencies. **PUBLICATIONS:** *Facing Tomorrow*. Documents the 1985 World Conference on Women in Nairobi.; *From Hiroshima to the Hague*. Discusses the campaign to obtain an Advisory Opinion from the International Court of Justice on the legal status of nuclear weapons.; *IPB Centenary Exhibition Catalogue*. Illustrated survey of the peace work as published by the UN's League of Nations

Archives.; *IPB News* (in English), quarterly; *Mass Media in Times of War*. Collection of articles which examines wartime media bias.; *100 Years of Peacemaking* (in English, Finnish, German, and Swedish). Provides a broad history of the peace movement by former IPB Sec.Gen. Rainer Santi.; *The Right to Refuse Military Orders*. Provides legal, historical, and personal perspectives on the right to refuse military order from various countries.; *Tackling the Flow of Arms*. Summarizes government-level proposals for regulation.; *Women and the Military System*. Covers women's roles in relation to militarism in many societies.; *World Military and Social Expenditure*; *Youth and Conscription*. Features international studies and information on the history of conscription and conscientious objection.

INTERNATIONAL PEACE RESEARCH ASSOCIATION (IPRA)

Fredericiagade 18, DK-1310 Copenhagen K, Denmark **PHONE:** 45 3345 5050 **FX:** 45 3345 5060 **E-MAIL:** bm@vip.cybercity.dk **WEBSITES:** http://www.copri.dk/ipra/ipra.html **FOUNDED:** 1965. **OFFICERS:** Biorn Moller, Sec.Gen. **MEMBERS:** 1,800. **STAFF:** 1. **LANGUAGES:** English, French, German, Spanish.

DESCRIPTION: Members are scientific associations, research institutes, and individuals in 76 countries engaged in research on the problems of war and peace, development, and peace education. Works to advance interdisciplinary research into the conditions of peace and the causes of war and other forms of violence. Fosters cooperation and contact between scholars and educators; encourages national studies and educational courses on the pursuit of world peace; promotes the establishment of new research centers and institutes, particularly in the Third World; encourages the worldwide dissemination of results of peace research. Maintains observer status with United Nations Educational, Scientific and Cultural Organization and nongovernmental organization status with the United Nations and the United Nations Conference on Trade and Development. **PUBLICATIONS:** *International Peace Research Newsletter* (in English), quarterly.

INTERNATIONAL PEAT SOCIETY (IPS) (Internationale Moor und Torf-Gesellschaft - IMTG)

Kuokkalantie 4, FIN-40520 Jyvaskyla, Finland **PHONE:** 358 14 674042 **FX:** 358 14 677405 **E-MAIL:** ips@peatsociety.fi **WEBSITES:** http://www.peatsociety.fi **FOUNDED:** 1968. **OFFICERS:** Mr. Raimo Sopo, Sec.Gen. **MEMBERS:** 1,300. **STAFF:** 2. **BUDGET:** US$50,000. **LANGUAGES:** English, German, Russian.

DESCRIPTION: Members are national committees, institutions, companies, and individuals engaged in peat and peatland utilization in 31 countries. Promotes understanding and cooperation concerning the study and sustainable utilization of peatlands, bogs, peat, and related materials, such as sapropel, for scientific, technical, and economic progress. Conducts research programs and periodic symposia. Maintains 7 scientific and technical commissions; compiles statistics. **PUBLICATIONS:** *Congress Proceedings*, quadrennial; *Global Peat Resources*; *IPS Bulletin*, annual; *IPS Journal*, annual; *List of Members*, periodic; *Peat Dictionary* (in English, Finnish, German, Russian, and Swedish); *Proceedings of Symposia*, semiannual.

INTERNATIONAL P.E.N. (P.E.N.)

9/10 Charterhouse Bldgs., Goswell Rd., London EC1M 7AT, England **PHONE:** 44 171 2534308 **FX:** 44 171 2535711 **E-MAIL:** intpen@dircon.co.uk **WEBSITES:** http://www.oneworld.org/internatpen **FOUNDED:** 1921. **OFFICERS:** Terry Carlbom, Sec. **MEMBERS:** 14,000. **STAFF:** 6. **BUDGET:** US$103,600. **LANGUAGES:** English, French, Spanish.

DESCRIPTION: Members are poets, playwrights, essayists, novelists, editors, translators, radio and television scriptwriters, historians, and other types of writers from 94 countries. (P.E.N. stands for poets, playwrights, essayists, editors, and novelists.) Objectives are to: act as a forum to promote intellectual exchange, friendship, and goodwill among writers internationally; support freedom of expression and promote freedom for the exchange of literature among all countries regardless of political situations; and inform publishers, editors, librarians, and university departments about literature in languages of lesser currency. Encourages translation of contemporary literature in minority languages. Works to defend writers suffering from governmental harassment, imprisonment, or other forms of oppression; operates P.E.N. Foundation Emergency Fund headquartered in the Netherlands, which offers aid to the dependents of harassed or imprisoned writers. Holds special subject conferences and literary sessions. **PUBLICATIONS:** *Pen International* (in English and French), semiannual. Reviews of books in small languages, pen news, papers from pen conferences, poems, and stories; Pamphlets; Papers.

INTERNATIONAL PENAL AND PENITENTIARY FOUNDATION (IPPF) (Fondation Internationale Penale et Penitentiaire - FIPP)

Bundesministerium der Justiz, Aussenstelle Berlin, D-10104 Berlin, Germany **PHONE:** 49 30 20259226 **FX:** 49 30 20259525 **FOUNDED:** 1951. **OFFICERS:** Dr. Konrad Hobe, Sec.Gen. **MEMBERS:** 60. **LANGUAGES:** English, French.

DESCRIPTION: Members are specialists in the fields of criminal and penitentiary law and criminal policy in 28 countries. Promotes research and development in methods of crime prevention and advances in the treatment of offenders. **PUBLICATIONS:** *Proceedings of Colloquia and Meetings* (in English and French), periodic.

INTERNATIONAL PEPPER COMMUNITY (IPC)

LINA Bldg., 4th Fl., Jl. H.R. Rasuna Said, Kav. B.7, 12920 Jakarta, Indonesia **PHONE:** 62 21 5224902, 62 21 5224903 **FX:** 62 21 5224905 **TX:** 60739 IPC IA **E-MAIL:** ipc@indo.net.id **FOUNDED:** 1972. **OFFICERS:** K.P.G. Menon, Exec.Dir., IPC. **MEMBERS:** 8. **STAFF:** 10. **LANGUAGES:** English.

DESCRIPTION: Intergovernmental organization of pepper-producing countries of Brazil, India, Indonesia, Malaysia, Thailand, Sri Lanka, and Micronesia Papua New Guinea. Coordinates, harmonizes, and promotes all activities relative to the pepper economy. **PUBLICATIONS:** *Directory of Pepper Importers/Grinders/End-Users/Food Processors*; *Directory of Pepper Researchers*; *International Pepper News Bulletin* (in English), quarterly. Contains pepper prices, market reviews, and articles relating to the industry.; *Pepper Exporters Directory*; *Pepper Statistical Yearbook* (in English).

INTERNATIONAL PHARMACEUTICAL FEDERATION (IPF) (Federation Internationale Pharmaceutique - FIP)

PO Box 84200, NL-2508 AE The Hague, Netherlands **PHONE:** 31 70 3021976 **FX:** 31 70 3021999 **TX:** 32781 FIP NL **E-MAIL:** pauline@fip.nl **WEBSITES:** http://www.pharmweb.net/pwmirror/pw9/fip/pharmweb92.html **FOUNDED:** 1912. **OFFICERS:** A.W. Davidson, Gen.Sec. **MEMBERS:** 4,000. **STAFF:** 6. **BUDGET:** 4,000,000 f. **LANGUAGES:** English, French, German, Spanish.

DESCRIPTION: Members are pharmaceutical organizations, pharmacists, and other interested organizations in 85 countries. Objectives are to: develop pharmacy at the international level in professional and scientific fields; extend the role of the pharmacist in the health care field; foster communication among members; act as a clearinghouse; collaborate with efforts to improve pharmaceutical structures in various countries; advocate and support measures to ensure distribution, dispensation, and proper use of medicines. Exchanges opinions on professional and ethical issues; develops research and study programs; fosters cooperation among pharmacists, teachers, research workers, and the practitioners in the field of drug information; improves methods for assembling, selecting, summarizing, indexing, storing, classifying, and analyzing clinical and social surveys. Subjects studied include biopharmaceutics, pharmacokinetics and drug metabolism, administrative and social pharmacy, pharmacognosy, and the history of pharmacy. Collaborates with World Health Organization. **PUBLICATIONS:** *International Pharmacy Journal* (in English, French, German, and Spanish), bimonthly; *Scientific Congress Proceedings*, biennial. Also publishes abstracts.

INTERNATIONAL PHARMACEUTICAL STUDENTS' FEDERATION (IPSF)

c/o IPSF Secretariat, Andries Bickerweg 5, NL-2517 JP The Hague, Netherlands **PHONE:** 31 70 3631925, 31 70 3632771 **FX:** 31 70 3633914 **E-MAIL:** ipsf@fip.nl **WEBSITES:** http://www.pharmweb.net; http://www.pharmweb.net/pharmweb/fip.html **FOUNDED:** 1949. **OFFICERS:** Alison Sutherland, Pres. **MEMBERS:** 300,000. **STAFF:** 7. **BUDGET:** 100,000 f. **LANGUAGES:** English, French, German, Spanish.

DESCRIPTION: Members are national organizations of pharmacy students who represent 50% of all pharmacy students in their countries; or local pharmacy student organizations; pharmacy students and graduates who have been graduated less than 4 years. Objectives are to study and promote interests of pharmacy students and encourage international cooperation. Student exchange program; development fund; Activities include collection of documentation and information on student issues; book appeal. **PUBLICATIONS:** *IPSF News Bulletin*, 3-4/year. Also publishes reports on meetings and symposia papers.

INTERNATIONAL PHILATELIC FEDERATION (FIP) (Federation Internationale de Philatelie)

Zollikerstrase 128, CH-8008 Zurich, Switzerland **PHONE:** 41 1 4223839 **FX:** 41 1 3831446 **FOUNDED:** 1926. **OFFICERS:** Marie-Louise Heiri, Sec.Gen. **MEMBERS:** 78. **STAFF:** 1. **BUDGET:** US$200,000. **LANGUAGES:** English, French, German, Russian, Spanish.

DESCRIPTION: Seeks to further interest in stamp collecting. Represents members' interests. Sponsors exhibitions; disseminates information. **PUBLICATIONS:** *FLASH* (in English, French, German, and Spanish), quarterly. Contains information about worldwide philatelic activities.

INTERNATIONAL PHONETIC ASSOCIATION (IPA) (Association Phonetique Internationale)

CLCS, Arts Building, Trinity College, Dublin, Ireland **PHONE:** 353 1 6081348 **FX:** 353 1 6772694 **E-MAIL:** esling@uvic.ca **WEBSITES:** http://www.arts.gla.ac.uk/IPA/ipa.html **FOUNDED:** 1886. **OFFICERS:** J.H. Esling, Sec. **MEMBERS:** 800. **LANGUAGES:** English.

DESCRIPTION: Members are students, teachers, and university and college libraries in 50 countries. Promotes the scientific study of phonetics and its applications. Has devised the International Phonetic Alphabet, which is widely used in the phonetic transcription of languages. Conducts examinations in English phonetics and grants certificates to qualified persons. **PUBLICATIONS:** Journal, semiannual.

INTERNATIONAL PHOTOBIOLOGY ASSOCIATION (AIP) (Association Internationale de Photobiologie)

c/o Dr. M.H.V. Van Regenmortel, Centre National De La Recherche Scientifique, IBMC, I.B.M.C. 15, Rou DesCartes, F-67084 Strasbourg, France **PHONE:** 33 88 417022 **FX:** 33 88 610680 **E-MAIL:** vanregen@ibmc.u-strasbg.fr **WEBSITES:** http://pol.newi.ac.uk/aip_info.html **FOUNDED:** 1928. **OFFICERS:** T.M.A.R. Dubbelman, Sec.Gen. **MEMBERS:** 2,000. **LANGUAGES:** English, French.

DESCRIPTION: Members are associated national and international photobiology groups representing 14 countries. Conducts research in physics, chemistry, and climatology of non-ionising radiations (ultra-violet, visible, and infra-red) in relation to their biological effects and the effects of the application of these radiations in biology and medicine. Maintains Niels Finsen Foundation. **PUBLICATIONS:** *Congress Proceedings*, quadrennial.

INTERNATIONAL PHYCOLOGICAL SOCIETY (IPS)

c/o Prof. M.D. Guiry, The Martin Ryan Science Institute, University College Galway, Galway, Ireland **PHONE:** 353 91 750410, 353 87 2519917 **FX:** 353 91 750702 **E-MAIL:** mike.guiry@seaweed.ucg.ie **FOUNDED:** 1960. **OFFICERS:** Prof. M.D. Guiry, Pres. **MEMBERS:** 800. **STAFF:** 15. **LANGUAGES:** English.

DESCRIPTION: Members are scientists working to develop phycology (a division of botany concerned with algae and seaweed). Promotes international cooperation among individual phycologists and phycological organizations. Disseminates information. Sponsors competitions and bestows awards. **PUBLICATIONS:** *Phycologia* (in English), bimonthly. Contains articles dealing with biology, ecology, cytology, physiology, and biochemistry of algae. Includes book reviews and obituaries.

INTERNATIONAL PHYSICIANS FOR THE PREVENTION OF NUCLEAR WAR (IPPNW)

727 Massachusetts Ave., 2nd Fl., Cambridge, MA 02139-3323 **PHONE:** (617) 868-5050 **FX:** (617) 868-2560 **E-MAIL:** ippnwbos@igc.apc.org **WEBSITES:** http://www.healthnet.org/ippnw/

FOUNDED: 1980. **OFFICERS:** Michael Christ, Exec.Dir. **MEMBERS:** 200,000. **STAFF:** 12. **BUDGET:** US$1,600,000. **LANGUAGES:** English.

DESCRIPTION: Federation of national physicians' organizations representing 200,000 physicians in 60 countries dedicated to mobilizing the influence of the medical profession against the threat of war and its weapons. Seeks to focus international attention on the medical consequences of nuclear or conventional war. Works to delegitimize, and therefore abolish, all forms of nuclear weaponry. Promotes alternatives to violence and armed conflict. Educates the public, world leaders, and the medical profession about the link between militarism, underdevelopment, and the destruction of the environment. Works with the United Nations and World Health Organization on projects related to world health and the arms race. Sponsors educational campaigns; prepares curricula on nuclear war at leading medical schools; conducts continued research on the medical, psychological, and biospheric effects of nuclear war. Coordinates regional symposia. Currently campaigns against the use of land mines and nuclear weapons, and in favor of gun control. **PUBLICATIONS:** *Abolition 2000*; *Affiliate Directory*, semiannual; *Crude Nuclear Weapons: Proliferation and the Terrorist Threat Report*; *Nuclear Wastelands. A global guide to nuclear weapon production and its health & environmental effects.*; *Opportunities for International Control of Weapons-Usable Fissile Materials*; *Plutonium: Deadly Gold of the Nuclear Age*; *Radioactive Heaven and Earth: The Health and Environmental Effects of Nuclear Weapons Testing In, On, and Above the Earth*; *Vital Signs*, quarterly; *The War in Nicaragua: The Effects of Low-Intensity Conflict on an Underdeveloped Country*; Annual Report.

INTERNATIONAL PLANNED PARENTHOOD FEDERATION, WESTERN HEMISPHERE REGION (IPPF/WHR)

120 Wallstreet, 9th Fl., New York, NY 10005 **PHONE:** (212) 248-6400 **FX:** (212) 248-4221 **TX:** 620661 **E-MAIL:** info@ippfwhr.org **WEBSITES:** http://www.ippfwhr.org **FOUNDED:** 1952. **OFFICERS:** Hernan Sanhueza, Regional Dir. **MEMBERS:** 46. **STAFF:** 65. **BUDGET:** US$21,779,220. **LANGUAGES:** English, Spanish.

DESCRIPTION: A division of the International Planned Parenthood Federation. Independent family planning organizations in Canada, Latin America, Caribbean Islands, and the United States. Views family planning as "the expression of the human right of couples to have only the children they want and to have them when they want them." Works to extend the practice of voluntary family planning by providing information, education, and services to couples. Seeks to persuade governments to establish national family planning programs. Conducts research programs; maintains speakers' bureau; sponsors specialized education programs. **PUBLICATIONS:** *Forum* (in English and Spanish), quarterly. Includes calendar of events. Distributed primarily to affiliated family planning programs.; Annual Report (in English and Spanish). Also publishes occasional monographs, studies, and position papers.

INTERNATIONAL POLICE ASSOCIATION (IPA)

Intl. Administration Centre, 1 Fox Rd., West Bridgford, Notts. NG2 6AJ, England **PHONE:** 44 115 9455985 **FX:** 44 115 9822578 **E-MAIL:** wendy@ipa-iac.org **WEBSITES:** http://www.ipa-iac.org **FOUNDED:** 1950. **OFFICERS:** Alan F. Carter, Int'l.Sec.Gen. **MEMBERS:** 269,205. **LANGUAGES:** English, French, German.

DESCRIPTION: Members are police officers, either on active service or retired. Seeks to unite all active and retired members of the police service to establish ties of friendship and mutual aid among them. Organizes exchange holidays, pen-friendships, and group visits; encourages and stimulates public service to promote respect for law and the maintenance of order among members of the police service in all countries. Engages in social and cultural activities; Works to establish a correspondence service. NGO (consultative status) with UN and council of Europe. **PUBLICATIONS:** *Information Guide-World Edition*, annual; *International Executive Council Proceedings*, annual; *IPA Newsletter*, periodic; *IPA Section Handbooks*, annual; *Scholarship Papers*, periodic.

INTERNATIONAL POLITICAL SCIENCE ASSOCIATION (IPSA) (Association Internationale de Science Politique - AISP)

Univ. College Dublin, Dept. of Politics, Belfield, Dublin 4, Ireland **PHONE:** 353 1 7068182 **FX:**

353 1 7061171 **E-MAIL:** ipsa@ucd.ie **WEBSITES:** http://www.ucd.ie/-ipsa/index.html **FOUNDED:** 1949. **OFFICERS:** Dr. John Coakley, Sec.Gen. **MEMBERS:** 1,500. **STAFF:** 2. **BUDGET:** US$130,000. **LANGUAGES:** English, French.

DESCRIPTION: Members are professional political scientists (1350); associations, libraries, and institutions (100); national associations (41). Promotes the advancement of political science throughout the world and fosters contact between the political scientists of various regions and political systems of the world. Collaborates on interdisciplinary work with other international associations. Sponsors 12 study groups and 38 research committees; conducts research seminars and roundtables. **PUBLICATIONS:** *Advances in Political Science: An International Series*. Eight volumes.; *International Political Science Abstracts*, bimonthly; *International Political Science Review*, quarterly; *Participation*, 3/year. Also publishes microfiche of papers presented at congresses.

INTERNATIONAL PRESS INSTITUTE (IPI)

Spiegelgasse 2/29, A-1010 Vienna, Austria **PHONE:** 43 1 5129011 **FX:** 43 1 5129014 **TX:** 112344 IPI **E-MAIL:** ipi.vienna@xpoint.at **WEBSITES:** http://www.freemedia.at **FOUNDED:** 1950. **OFFICERS:** Johann P. Fritz, Dir. **MEMBERS:** 2,000. **STAFF:** 9. **BUDGET:** 1,700,000 AS. **LANGUAGES:** English, French, German, Spanish.

DESCRIPTION: Journalists. Defend freedoms of expression and of the press worldwide. Organizes media campaigns publicizing violations of press freedom; facilitates cooperation among press organizations and journalists worldwide; works to insure the free flow of news. Sponsors investigative efforts where press freedoms appear to be endangered, and documents abuses of rights; provides protection to journalists whose rights are threatened. Conducts research and educational programs. **PUBLICATIONS:** *Global Network of Editors and Media Executives*; *IPI Report*, quarterly; *World Press Freedom Review*, annual.

INTERNATIONAL PRIMARY ALUMINIUM INSTITUTE (IPAI) (Institut International d'Aluminium Primaire - IIAP)

New Zealand House, Haymarket, London SW1Y 4TE, England **PHONE:** 44 171 9300528 **FX:** 44 171 3210183 **FOUNDED:** 1972. **OFFICERS:** Alan Payne, Sec.Gen. **MEMBERS:** 35. **STAFF:** 8. **LANGUAGES:** English.

DESCRIPTION: Producers of primary aluminum from 23 countries. Seeks to increase world understanding of the aluminum industry and to develop additional uses of primary aluminum. Provides a forum for the exchange of information and the discussion of developments affecting the industry; conducts studies on problem areas affecting the industry including energy, the environment, and health and safety; collects and disseminates information. Bestows honorary membership to individuals who have contributed significantly to the IPAI. **PUBLICATIONS:** *Alumina Annual Production Capacity*, semiannual; *Alumina Production*, quarterly; *Aluminium Recovered from Purchased or Tolled Scrap*, annual; *Directory*, biennial; *Electrical Power Utilization*, annual; *Energy for Production of Metallurgical Alumina*, annual; *Movement of Primary Aluminium*, quarterly; *Primary Aluminium Annual Production Capacity*, semiannual; *Primary Aluminium Production*, monthly; *Unwrought and Total Aluminium Inventories*, monthly.

INTERNATIONAL PRIMATOLOGICAL SOCIETY (IPS)

Psychology Department, Univ. of Georgia, Athens, GA 30602 **PHONE:** (706) 542-3036 **FX:** (706) 542-3275 **E-MAIL:** doree@arches.uga.edu **WEBSITES:** http://www.primate.wisc.edu/pin/ips.html **FOUNDED:** 1966. **OFFICERS:** Dr. D.M. Fragaszy, Sec.Gen. **MEMBERS:** 1,500. **LANGUAGES:** English.

DESCRIPTION: Members are individuals interested in primate research. Facilitates cooperation among primatologists and fosters conservation and judicious use of primates in research. **PUBLICATIONS:** *International Journal of Primatology*, quarterly; *IPS Bulletin*, biennial; *Members' Handbook*, annual; *Proceedings of the International Congress*, biennial.

INTERNATIONAL PRISONERS AID ASSOCIATION (IPAA)

Dept. of Sociology, Louisville, KY 40292 **PHONE:** (502) 852-6836 **FX:** (502) 852-0099 **OFFICERS:** Dr. Badr-El-Din Ali, Exec.Dir. **MEMBERS:** 92. **LANGUAGES:** English.

DESCRIPTION: Members are agencies and individuals in 45 countries concerned with aid programs for prisoners. Assists nongovernmental organizations in their efforts to prevent crime, rehabilitate offenders, stimulate social action and legislation, and disseminate information concerning sound methods of crime control. Maintains consultative status with the United Nations and the Council of Europe. **PUBLICATIONS:** *International Directory of Prisoners Aid Agencies*, periodic; Newsletter, 3/year.

INTERNATIONAL PROGRAMME ON CHEMICAL SAFETY (IPCS) (Programme International sur la Securite des Substances Chimiques - PISSC)

c/o World Health Orgn., Avenue Appia, CH-1211 Geneva 27, Switzerland **PHONE:** 41 22 7913588 **FX:** 41 22 7914848 **TX:** 415416 OMS **FOUNDED:** 1980. **OFFICERS:** Dr. Michel Mercier, Dir. **MEMBERS:** 33. **STAFF:** 25. **BUDGET:** US$4,500,000. **LANGUAGES:** English, French.

DESCRIPTION: Members are states belonging to the World Health Organization, the International Labour Organization, and the United Nations Environment Programme. Evaluates the health risks posed to humans and the environment by exposure to chemicals. Proposes methods and guidelines for measuring chemical exposure and for assessing health risks. Encourages the use and improvement of methods for laboratory testing and epidemiological studies. Provides guidelines on safe levels of chemical exposure through daily intake of food additives and pesticide and veterinary drug residues. Disseminates information regarding diagnosis and treatment of chemical poisoning; promotes international cooperation in dealing with chemical emergencies. Seeks to enhance the scientific basis for health risk assessment for determination of chemical hazards controls. Conducts training courses on subjects such as ecotoxicology, clinical toxicology, the detection of mutagenesis, and occupational hazards and human reproduction; coordinates international research programs; maintains expert advisory groups and offers advisory services for technical cooperation in regulation and control of chemicals. Organizes symposia and workshops on chemical safety. **PUBLICATIONS:** *Environmental Health Criteria Document* (in English, French, and Spanish), 15/year; *Health and Safety Guides* (in English), 20/year; *IPCS Newsletter* (in English), 2-3/year; *Poison Information Monographs* (in English, French, and Spanish), periodic; *Technical Document on Food Additives and on Pesticide Residues* (in English), periodic. Also publishes monographs, research reports, guideline documents, and a glossary; makes available *International Chemical Safety Cards*.

INTERNATIONAL PROGRAMMERS GUILD (IPG)

6535 Millcreek Dr., Unit No. 44, Mississauga, ON, Canada L5N 2M2 **PHONE:** (905) 812-8500 **FX:** (905) 812-1953 **E-MAIL:** info@ipgnet.com **FOUNDED:** 1984. **OFFICERS:** Warren Schmidt, Pres. **MEMBERS:** 10,000. **STAFF:** 10. **LANGUAGES:** English, French, German.

DESCRIPTION: Members are professional, student, and gifted computer programmers. Furthers careers, contact and exchange of programming data among members. Conducts professional certification program. Represents members' interests. IPG supplies employment and education to interested programmers as part of its career development program. **PUBLICATIONS:** *Guild News* (in English), quarterly.

INTERNATIONAL PROGRESS ORGANIZATION (IPO) (Organisation Internationale pour le Progres - OIP)

Kohlmarkt 4, A-1010 Vienna, Austria **PHONE:** 43 1 5332877 **FX:** 43 1 533296221 **E-MAIL:** info@i-p-o.org **WEBSITES:** http://i-p-o.org **FOUNDED:** 1972. **OFFICERS:** Dr. Hans Koechler, Pres. **LANGUAGES:** English, French, German.

DESCRIPTION: Promotes peaceful coexistence among nations through cultural and academic exchange and fosters tolerance toward and greater understanding of other cultures. Seeks to further the cause of human rights for all cultures and to overcome racial discrimination worldwide; supports all cultural entities in their development toward self-realization. Adheres to the principles of internationalism, universalism, and pacifism; encourages individual reflection on and critique of social and political values. Facilitates cultural dialogue and the exchange of experts, scientific information, and experiences among members; collaborates with other international organizations. Maintains consultative status with the United

Nations and United Nations Educational, Scientific and Cultural Organization. Coordinates research programs in the field of international relations and conflict resolution. **PUBLICATIONS:** *Studies in International Relations* (in English and French), periodic; Brochures, periodic; Monographs, periodic.

INTERNATIONAL PSYCHOANALYTICAL ASSOCIATION TRUST (IPA) (Association Psychanalytique Internationale)

Broomhills, Woodside Ln., London N12 8UD, England **PHONE:** 44 181 4468324 **FX:** 44 181 4454729 **FOUNDED:** 1910. **OFFICERS:** Valerie Tufnell, Adm.Dir. **MEMBERS:** 9,200. **STAFF:** 7. **LANGUAGES:** English, French, German, Spanish.

DESCRIPTION: Members are organizations in 30 countries involved in psychoanalysis. Encourages communication among members and promotes high educational standards. Organizes training programs. **PUBLICATIONS:** *Bulletin*, annual; *Information Booklet*, periodic; *Newsletter*, semiannual; *Roster*, annual; Monographs, annual.

INTERNATIONAL PUBLISHERS ASSOCIATION (IPA) (Union Internationale des Editeurs - UIE)

3, ave. de Miremont, CH-1206 Geneva, Switzerland **PHONE:** 41 22 3463018 **FX:** 41 22 3475717 **E-MAIL:** secretariat@upa-uie.org **WEBSITES:** http://www.ipa-uie.org **FOUNDED:** 1896. **OFFICERS:** J.A. Koutchoumow, Sec.Gen. **MEMBERS:** 74. **STAFF:** 5. **LANGUAGES:** English, French, German, Spanish.

DESCRIPTION: Members are national publishers' associations, in 65 countries. Objectives are: to uphold and defend the complete freedom of publishers to publish and distribute works of thought; to secure international cooperation among members; to maintain the international flow of books free from tariffs and other obstructions; to help overcome illiteracy and the lack of books and other educational materials; to take part in the revisions and drafting of new copyright conventions.

INTERNATIONAL RADIATION PROTECTION ASSOCIATION (IRPA)

Postbus 662, NL-5600 AR Eindhoven, Netherlands **PHONE:** 31 40 2473355 **FX:** 31 40 2435020 **TX:** 51163 TUEHV NL **E-MAIL:** irpa.exof@sbd.tue.nl **WEBSITES:** http://www.tue.nl/sbd/irpa/ **FOUNDED:** 1966. **OFFICERS:** Chris J. Huyskens, Exec. Officer. **MEMBERS:** 16,000. **LANGUAGES:** English.

DESCRIPTION: Members are scientists in radiation protection fields united to provide an international forum for interaction and exchange of information among those involved in radiation protection activities. Seeks to protect human beings and the environment from the hazards of ionizing radiation. Works to promote the use of radiation and atomic energy for the good of humankind. **PUBLICATIONS:** *IRPA Bulletin*, quarterly; Brochure, quarterly.

INTERNATIONAL RAIL TRANSPORT COMMITTEE (CIT) (Comite International des Transports Ferroviaires - CIT)

Legal Div., Swiss Federal Railways, Hoch Schulstrasse 6, CH-3030 Bern, Switzerland **PHONE:** 41 512202234, 41 512202806 **FX:** 41 512203457 **TX:** 991212 BERN/SWITZERL **FOUNDED:** 1902. **OFFICERS:** Thomas Leimgruber, Sec. **MEMBERS:** 300. **STAFF:** 6. **LANGUAGES:** English, French, German, Italian.

DESCRIPTION: Members are transport establishments including railways, road transport companies, and shipping firms in 38 countries. Purpose is to improve international rail transport law based on the Convention Concerning International Carriage by Rail (COTIF) and its appendices A, Uniform Rules Concerning the Contract for International Carriage of Passengers and Luggage by Rail (CIV), and B, Uniform Rules Concerning the Contract for International Carriage of Goods by Rail (CIM). Represents members' interests at conferences, conventions, and other meetings concerning the problems of transport law. Collaborates with other world organizations in ensuring the uniformity and improvement of railway operating and business conditions. Facilitates the exchange of information and ideas among members. **PUBLICATIONS:** *CIT INFOS* (in English, French, and German), periodic, 2-4/year.

INTERNATIONAL RAILWAY CONGRESS ASSOCIATION (IRCA) (Association Internationale du Congres des Chemins de Fer - AICCF)

85, rue de France, B-1060 Brussels, Belgium **PHONE:** 32 2 5207831 **FX:** 32 2 5254084 **TX:** 4620424 BERAILB **FOUNDED:** 1885. **OFFICERS:** A. Martens, Sec.Gen. **MEMBERS:** 124. **STAFF:** 15. **BUDGET:** US$400,000. **LANGUAGES:** English, French, German.

DESCRIPTION: Members are railway administrations, government agencies, and organizations. Promotes exchange of railway experience and knowledge among members. Supplies members with information on specific problems. **PUBLICATIONS:** *Rail International* (in English, French, and German), monthly. Contains leading papers by top railway managers and scientific papers.

INTERNATIONAL RAYON AND SYNTHETIC FIBRES COMMITTEE (CIRFS) (Comite International de la Rayonne et des Fibres Synthetiques)

4, ave. E. Van Nieuwenhuyse, B-1160 Brussels, Belgium **PHONE:** 32 2 6767455 **FX:** 32 2 6767454 **E-MAIL:** pur@cirfs.org **FOUNDED:** 1950. **OFFICERS:** F.B. Blaisse, Pres. **MEMBERS:** 30. **STAFF:** 12. **LANGUAGES:** English, French.

DESCRIPTION: Members are national rayon and manmade fiber companies. Purpose is to develop, improve, and increase the use of rayon and artificial and synthetic fibers and products made from them. Maintains technical advisory committees. **PUBLICATIONS:** *Information on Man-Made Fibres* (in English, French, and German), annual.

INTERNATIONAL READING ASSOCIATION (IRA)

800 Barksdale Rd., PO Box 8139, Newark, DE 19714-8139 **PHONE:** (302) 731-1600 **FX:** (302) 731-1057 **E-MAIL:** jbutler@reading.org **WEBSITES:** http://www.reading.org **FOUNDED:** 1956. **OFFICERS:** Alan Farstrup, Exec.Dir. **MEMBERS:** 94,000. **STAFF:** 90. **BUDGET:** US$8,000,000. **LANGUAGES:** English, French, Spanish.

DESCRIPTION: Members are teachers, reading specialists, consultants, administrators, supervisors, researchers, psychologists, librarians, and parents interested in promoting literacy. Seeks to improve the quality of reading instruction and promote literacy worldwide. Disseminates information pertaining to research on reading, including information on adult literacy, early childhood and literacy development, international education, literature for children and adolescents, and teacher education and professional development. Maintains over 40 special interest groups and over 70 committees. **PUBLICATIONS:** *Journal of Adolescent & Adult Literacy*, 8/year. Offers information on the theory and practice of teaching reading and study skills to adolescents and adults.; *Lectura y Vida* (in Spanish), quarterly. Covers reading theory research and practice in Latin America.; *Reading Research Quarterly*. Includes original research reports and articles on theory in teaching reading and learning to read.; *Reading Teacher*, 8/year. Contains articles on the theory and practice of teaching reading skills to preschool and elementary school children.; *Reading Today*. Provides news on reading education; includes Washington D.C. updates and news for parents.; Books; Catalog; Monographs; Monographs.

INTERNATIONAL RED LOCUST CONTROL ORGANIZATION FOR CENTRAL AND SOUTHERN AFRICA (IRLCO-CSA)

PO Box 240252, Ndola, Zambia **PHONE:** 260 2 614284, 260 2 615684 **FX:** 260 2 614285 **TX:** LOCUST ZA 30072 **E-MAIL:** locust@zamnet.zm **FOUNDED:** 1949. **OFFICERS:** Mr. E.K. Byaruhanga, Dir. **MEMBERS:** 8. **STAFF:** 27. **BUDGET:** US$1,000,000. **LANGUAGES:** English.

DESCRIPTION: Representatives of Kenya, Malawi, Mozambique, Swaziland, Tanzania, Uganda, Zambia, and Zimbabwe. Surveys and controls populations of red locusts in identified outbreak areas; seeks also to control other migratory pests including locusts, grain-eating birds, armyworms, and tsetse flies in member countries. Conducts operational research regarding locust control. Holds seminars and training courses. **PUBLICATIONS:** *Annual Report* (in English); *Quarterly Report* (in English); *Scientific Papers* (in English), periodic.

INTERNATIONAL REHABILITATION MEDICINE ASSOCIATION (IRMA)

1333 Moursund Ave., Rm. A-221, Houston, TX 77030 **PHONE:** (713) 799-5086 **FX:** (713) 799-5058 **FOUNDED:** 1966. **OFFICERS:** Donna Jones, Exec.Dir. **MEMBERS:** 2,000. **LANGUAGES:** English.

DESCRIPTION: Members are physicians from all specialties of medicine and surgery interested in promoting improvement of health through understanding and utilization of rehabilitation medicine. Goal is to convince governments and society that rehabilitation medicine services should be provided to aged, chronically diseased, and disabled persons who need help in entering the vocational and social mainstream of their communities. Provides a forum for continuous graduate medical education; provides speakers on rehabilitation medicine. Maintains liaisons with the International Federation of Physical Medicine and Rehabilitation and the World Health Organization and is a member of the Council for International Organizations of Medical Sciences and Organizations. **PUBLICATIONS:** *Book of Abstracts*, quadrennial; *IRMA*, biennial; *News and Views*, quarterly; *Scientific Monographs*, semiannual.

INTERNATIONAL RELIEF FRIENDSHIP FOUNDATION (IRFF)

177 White Plains Rd., 50F, Tarrytown, NY 10591 **PHONE:** (914) 366-0558 **FX:** (914) 366-0558 **E-MAIL:** irffint@aol.com **FOUNDED:** 1976. **OFFICERS:** Kathy Winings, Exec.Dir. **LANGUAGES:** English.

DESCRIPTION: Seeks to address the problems of sickness, poverty, illiteracy, and other forms of human suffering. Supports programs that encourage the active participation of the intended beneficiaries. Operates programs in agricultural and technical training, children's education, health education, direct relief aid, and interreligious youth service programs. Operates Religious Youth Service, which provides opportunities for young people of all faiths to work together in needy communities. Projects running in Bulgaria, Croatia, Moldova, Albania, Russia, Ghana, Ivory Coast, Rwanda, Cameroon, Nigeria, Peru, Bangladesh, Sri Lanka, Uganda, Brazil, Mexico, Haiti. **Convention/Meeting:** none. **PUBLICATIONS:** *East-West Perspective* (in German); *Frontiers in Development*, periodic; *Frontiers in Development* (in Italian).

INTERNATIONAL REPUBLICAN INSTITUTE (IRI)

1212 New York Ave. NW, Ste. 900, Washington, DC 20005-6107 **PHONE:** (202) 408-9450 **FX:** (202) 408-9462 **TX:** 5106000161 **E-MAIL:** web@iri.org **WEBSITES:** http://www.iri.org **FOUNDED:** 1983. **OFFICERS:** Lorne Craner, Pres. **STAFF:** 60. **BUDGET:** US$13,000,000. **LANGUAGES:** English.

DESCRIPTION: Conducts programs outside the United States to promote democracy and strengthen free-markets, and the rule-of-law. Conducts programs in Albania, Angola, Azerbaijan, Belarus, Bulgaria, Burma, Cambodia, China, Cuba, El Salvador, Georgia, Guatemala, Haiti, Indonesia, Liberia, Macedonia, Mexico, Moldove, Mongolia, Morocco, Nicaragua, Poland, Romania, Russia, Serbia, Slovakia, South Africa, Turkey, Thailand, Ukraine, Venezuela, Vietnam, West Bank and Gaza Strip, Western Sahara, and Zimbabwe. **PUBLICATIONS:** *Country Specific Reports*, periodic; *IRI Newsletter*, quarterly.

INTERNATIONAL RESCUE COMMITTEE (IRC)

122 E. 42nd St., New York, NY 10168 **PHONE:** (212) 551-3000 **FX:** (212) 551-3180 **E-MAIL:** kburch@intrescom.org **WEBSITES:** http://www.intrescom.org **FOUNDED:** 1933. **OFFICERS:** Renold Levy, Pres. **STAFF:** 500. **BUDGET:** US$7,690,536. **LANGUAGES:** English.

DESCRIPTION: Nonsectarian, nonpartisan, voluntary agency founded by Albert Einstein and supported by individuals, foundations, corporations, unions, and civic, educational, human rights, and community groups. Assists refugee victims of religious, political, and racial persecution, civil strife, famine, and war. Current programs are located in Africa, Asia, Central America, Europe, North America, and the Middle East. **PUBLICATIONS:** *Children in Flight*; *Field Report*, 3/year; *Flight*; *International Rescue Committee—Annual Report.* Review of agency's refugee activities and the worldwide refugee situation.

INTERNATIONAL RESEARCH GROUP ON WOOD PRESERVATION (IRG) (Groupe International de Recherches sur la Preservation du Bois - GIRPB)

Brinellvagen 34, S-100 44 Stockholm, Sweden **PHONE:** 46 8 101453 **FX:** 46 8 108081 **E-MAIL:** irg@sp.se **WEBSITES:** http://irg-wp.com **FOUNDED:** 1969. **OFFICERS:** Joran Jermer, Sec.Gen. **MEMBERS:** 325. **STAFF:** 1. **BUDGET:**

1,200,000 SKr. **LANGUAGES:** English, French, German, Spanish.

DESCRIPTION: Members are wood preservation scientists, specialists, and other interested individuals from 52 countries. Promotes worldwide research and cooperation among researchers; coordinates collaborative projects. Encourages exchange of technical information; conducts research projects for other international bodies such as standards and approval organizations. **PUBLICATIONS:** *Documents* (in English), 150/year; *International Directory of Members and Sponsors*, annual; Annual Report (in English); Newsletter (in English), 2-5/year.

INTERNATIONAL RESEARCH INSTITUTE FOR MEDIA, COMMUNICATION AND CULTURAL DEVELOPMENT (MEDIACULT)

Schoenburgstr. 27/4, A-1040 Vienna, Austria **PHONE:** 43 1 5041316, 43 1 5041317 **FX:** 43 1 50413164 **E-MAIL:** mediacult@eunet.at **FOUNDED:** 1969. **OFFICERS:** Alfred Smudits, Sec.Gen. **MEMBERS:** 63. **STAFF:** 4. **LANGUAGES:** English, French, German.

DESCRIPTION: Members are committees, federations, institutes, schools, and other organizations (27) and individuals (36) involved in communications, performing arts, and cultural research. Purpose is to undertake research in the field of cultural development, particularly in the music sector and the role of new communication technologies in the arts sector. Promotes coordination of research. Sponsors studies on topics such as changes in cultural communication as a result of electronic media, economic difficulties within the performing arts, and democratic cultural representation at the community level. **PUBLICATIONS:** *Mediacult.doc*, periodic. Containing information on research projects and seminars of Mediacult and its members.; *MEDIACULT News* (in English, French, and German), semiannual. Research news; *Studies and Research Projects*, periodic.

INTERNATIONAL RICE COMMISSION (IRC) (Commission Internationale du Riz - CIR)

Food and Agriculture Orgn. of the U.N., Via delle Terme di Caracalla, I-00100 Rome, Italy **PHONE:** 39 6 57055769 **FX:** 39 6 57056347 **TX:** 610181 FAO I **E-MAIL:** dat.tran@fao.org **WEBSITES:** http://www.fao.org/waicent/faoinfo **FOUNDED:** 1949. **OFFICERS:** Dat Van Tran, Exec.Sec. **MEMBERS:** 61. **STAFF:** 3. **BUDGET:** US$100,000. **LANGUAGES:** English, French, Spanish.

DESCRIPTION: A program of the Food and Agriculture Organization of the United Nations. Promotes national and international action relating to the production, conservation, distribution, and consumption of rice, without entering into international commerce. **PUBLICATIONS:** *International Rice Commission* (in English, French, and Spanish), annual. Contains information on rice research, development, and the world rice situation and outlook.; *IRC Session Proceedings*, quadrennial.

INTERNATIONAL RICE RESEARCH INSTITUTE (IRRI)

PO Box 933, Manila 1099, Philippines **PHONE:** 63 2 8181926 **FX:** 63 2 8911292 **TX:** 40890 RICE PM **FOUNDED:** 1960. **OFFICERS:** Dr. Ronald P. Cantrell, Dir.Gen. **STAFF:** 854. **BUDGET:** US$37,470,571. **LANGUAGES:** English, Spanish.

DESCRIPTION: Conducts basic research on the rice plant and its cultural management, with the goal of improving the well-being of present and future generations of rice farmers and consumers, particularly those with low incomes; disseminates results of research and plant materials; operates a training program for rice scientists. Operates Riceworld, a museum and learning center about rice and rice-related culture and research. **PUBLICATIONS:** *Corporate Report*, annual; *International Rice Research Notes*, 3/year; *Program Report*, annual; *Rice Literature Update*, 3/year. Also publishes scientific books, literature, serials, journals, and brochures.

INTERNATIONAL ROAD FEDERATION (IRF)

1010 Massachusetts Ave. NW, Ste. 410, Washington, DC 20001 **PHONE:** (202) 371-5544 **FX:** (202) 371-5565 **E-MAIL:** info@irfnet.org **FOUNDED:** 1948. **OFFICERS:** Gerald P. Shea, Dir.Gen. **MEMBERS:** 750. **STAFF:** 7. **LANGUAGES:** English, French, German, Spanish, Vietnamese.

DESCRIPTION: Members are road associations, private sector firms, and public sector firms in 70 countries. Encourages the development and improvement of highways and highway transportation and the exchange of technologies. Provides educational grants to select countries for graduate training through the International Road Educational Foundation. **PUBLICATIONS:** *World Highways*, 8/year; *World Road Statistics*, annual. Also produces videotape training aids and reports.

INTERNATIONAL ROAD SAFETY ORGANIZATION (PRI) (Prevention Routiere Internationale - PRI)

75, rue de Mamer, BP 40, L-8005 Bertrange, Luxembourg **PHONE:** 352 318341 **FX:** 352 311460 **E-MAIL:** int.road.safety@pri.lu **WEBSITES:** http://www.pri.lu **FOUNDED:** 1959. **OFFICERS:** Leon Nilles, Pres. **MEMBERS:** 70. **STAFF:** 4. **LANGUAGES:** English, French, German.

DESCRIPTION: Members are national road safety associations working to promote methods of improving road safety. Researches accident prevention. Maintains committees dealing with specific road safety issues including road safety in the armed forces, contents of road safety communications, evaluation of road safety work, traffic safety education, and codes of good practice. **PUBLICATIONS:** *Congress Proceedings* (in English, French, and German), biennial; *International Road Safety*, triennial; Annual Report.

INTERNATIONAL ROAD TRANSPORT UNION (IRU) (Union Internationale des Transports Routiers)

BP 44, CH-1211 Geneva 20, Switzerland **PHONE:** 41 22 9182700 **FX:** 41 22 9182741 **E-MAIL:** iru@iru.org **WEBSITES:** http://www.iru.org **FOUNDED:** 1948. **OFFICERS:** Martin Marmy, Sec.Gen. **MEMBERS:** 151. **STAFF:** 125. **LANGUAGES:** English, French.

DESCRIPTION: Members are national associations of road transport operators (of goods and passengers), vehicle manufacturers, individual companies, and international organizations in 63 countries. Seeks to standardize international regulations affecting road transport. Encourages the exchange of information and disseminates information on topics such as regulations, traffic conditions, and international meetings. Provides promotional materials. **PUBLICATIONS:** *Conference Reports*; *Handbook of International Road Transport* (in English, French, and German), biennial. Contains information about national regulations applying to road transport.; *Selection of International Road Transport*, semiannual; Papers.

INTERNATIONAL ROWING FEDERATION (Federation Internationale des Societes d'Aviron - FISA)

CH-3653 Oberhofen, Switzerland **PHONE:** 41 33 435053 **FX:** 41 33 435073 **FOUNDED:** 1892. **OFFICERS:** John Boultbee, Sec.Gen. **MEMBERS:** 94. **STAFF:** 6. **BUDGET:** 2,250,000 SFr. **LANGUAGES:** English, French.

DESCRIPTION: Members are national rowing federations. Encourages organization of international and Olympic regattas governed by rules of racing adapted to the development of the sport; maintains the principles of amateurism in all competitions according to the definition of an amateur oarsman and federation rules. Works to develop the sport of rowing. Facilitates formation of national federations in countries where none exist. **PUBLICATIONS:** *Coaches Conference Report*, annual; *FISA Coach* (in English), quarterly; *FISA Directory*, annual; *FISA Information* (in English, French, and German), bimonthly.

INTERNATIONAL RUBBER RESEARCH AND DEVELOPMENT BOARD (IRRDB) (Organisation Internationale de La Recherches et du Developpement Sur Le Caoutchouc)

Chapel Bldg., Brickendonbury, Hertford, Herts. SG13 8NP, England **PHONE:** 44 1992 584829, 44 1992 584966 **FX:** 44 1992 504267 **E-MAIL:** irrdb@aol.com **WEBSITES:** http://www.irrdb.org **FOUNDED:** 1937. **OFFICERS:** Mr. K.P. Jones, Sec. **MEMBERS:** 15. **STAFF:** 2. **LANGUAGES:** English.

DESCRIPTION: Members are national institutes for rubber research. Purpose is to coordinate research and facilitate the exchange of information among members; conducts seminars. Sponsors cooperative projects in research and development. **PUBLICATIONS:** *Newsletter*, quarterly; Annual Report (in English and French).

INTERNATIONAL RUBBER STUDY GROUP (IRSG) (Groupe International d'Etudes du Caoutchouc)

Heron House, 1st Fl., Wembley Hill Rd., Wembley HA9 8OA, England **PHONE:** 44 181 9037727 **FX:** 44 181 9032848 **E-MAIL:** lrsg@compuserve.com **WEBSITES:** http://www.rubberstudy.com **FOUNDED:** 1944. **OFFICERS:** Mr. M.E. Cain, Sec.Gen. **MEMBERS:** 19. **STAFF:** 9. **LANGUAGES:** English, French.

DESCRIPTION: Members are 19 governments of countries producing and consuming natural and synthetic rubber and rubber products. Provides a forum to discuss the problems affecting the rubber industry worldwide including natural and synthetic raw materials and manufacturing. Conducts economic studies; makes available statistics on the rubber industry. **PUBLICATIONS:** *International Rubber Digest* (in English), monthly. Contains rubber market analysis and information.; *International Rubber Forum*, annual. Contains meeting proceedings.; *Outlook for Elastomers*, annual; *Rubber Statistical Bulletin* (in English), monthly; *Rubber Statistics Yearbook*, annual; *Secretariat Papers*, periodic. Containing information on rubber economics and statistics.; *World Rubber Economy Challenges and Changes*, periodic.

INTERNATIONAL SACERDOTAL SOCIETY SAINT PIUS X (FSSPX) (Fraternite Sacerdotale Saint Pie X - FSSP)

Schwandegg, CH-6313 Menzingen, Switzerland **PHONE:** 41 41 7553636 **FX:** 41 41 7551444 **WEBSITES:** http://www.fsspx.org **FOUNDED:** 1970. **OFFICERS:** Bishop Bernard Fellay, Sup.Gen. **MEMBERS:** 528. **STAFF:** 9. **LANGUAGES:** English, French, German, Italian, Spanish.

DESCRIPTION: Members are Roman Catholic priests, seminarians, oblates, and brothers in 27 countries. Seeks to preserve and strengthen the priesthood and its ministry and to uphold the Holy Mass and other Roman Catholic traditions. Maintains 6 seminaries; operates worldwide program of Catholic missions. **PUBLICATIONS:** *The Angelus* (in English), monthly; *Fideliter* (in French), bimonthly; *Letter to Friends and Benefactors* (in English, French, German, Portuguese, and Spanish), semiannual; *Mitteilungsblatt der Priestbruderschaft* (in German), monthly; *St. Mary's Magazine*, bimonthly; Books; Pamphlets.

INTERNATIONAL SAVE THE CHILDREN ALLIANCE (SCA)

Rantzausgade 60, DK-2200 Copenhagen, Denmark **PHONE:** 45 70206120 **FX:** 45 70206220 **E-MAIL:** redbarnet@redbarnet.dk **FOUNDED:** 1979. **OFFICERS:** Red Barnet. **MEMBERS:** 22. **STAFF:** 2. **BUDGET:** US$286,000. **LANGUAGES:** English.

DESCRIPTION: Volunteer organizations working for the improvement of living conditions for disadvantaged children. Work is carried out in 90 countries and focuses on assistance in development programs and defense of children's rights. **PUBLICATIONS:** *Annual Report* (in English, French, and Spanish); *Newsletter* (in English), semiannual.

INTERNATIONAL SCHOOLS ASSOCIATION (ISA) (Association des Ecoles Internationales)

CIC Case 20, CH-1211 Geneva 20, Switzerland **PHONE:** 41 22 7336717 **FX:** 41 22 7347082 **FOUNDED:** 1951. **OFFICERS:** Cyril Ritchie. **MEMBERS:** 80. **STAFF:** 2. **LANGUAGES:** English, French.

DESCRIPTION: Members are international schools and other educational institutions. Seeks to establish an interchangeable curriculum and graduation standards for international schools and to advance the recognition of equivalent educational qualifications worldwide. Provides advisory and consultative services; promotes and carries out studies on educational or administrative questions; encourages establishment of new schools; fosters an interest in international affairs in regular schools. Facilitates teacher exchanges; initiated the International Baccalaureate and the International Schools Middle Years Programme, and is promoting an international primary education curriculum. **PUBLICATIONS:** *Bulletin* (in English), semiannual. Contains pedagogical information.; *Curriculum for the Middle School*, periodic; *Curriculum for the Young Child*, periodic; Reports, periodic.

INTERNATIONAL SECURITIES MARKET ASSOCIATION

7 Limeharbour, London E14 9NQ, England **PHONE:** 44 171 5385656 **FX:** 44 171 5384902 **E-MAIL:** info@isma.co.uk **FOUNDED:** 1969. **OFFICERS:** Tim Dickerson, Head of Corporate Communications. **MEMBERS:** 770. **STAFF:** 105.

DESCRIPTION: Members are all major European banks, investment houses and financial institutions active in the international securities markets as well as a number of firms in N. America, the Middle East and Australasia. Establishment of uniform market practices which now govern all transactions in international securities; the provision of market information and statistical services by daily electronic data and printed publications and the operation of TRAX (an industry wide trade matching and reporting computer system); educational seminars.

INTERNATIONAL SEED TESTING ASSOCIATION (ISTA) (Association Internationale d'Essais de Semences)

ISTA Secretariat, Reckenholzstr. 191, Postfach 412, CH-8046 Zurich, Switzerland **PHONE:** 41 1 3776000 **FX:** 41 1 3776001 **E-MAIL:** istach@iprolink.ch **WEBSITES:** http://www.seedtest.org **FOUNDED:** 1924. **OFFICERS:** Heinz Schmid, Exec. Officer. **MEMBERS:** 146. **STAFF:** 5. **BUDGET:** 700,000 SFr. **LANGUAGES:** English, French, German.

DESCRIPTION: Members are seed testing stations in 66 countries. Objectives are to: develop, adopt, and publish standard procedures for sampling and testing seeds, and to promote uniform application of these procedures for evaluation of seeds moving in international trade; promote research in all areas of seed science and technology, including sampling, testing, storing, processing, and distributing seeds; encourage variety (cultivar) certification; participate in congresses and training courses aimed at furthering these objectives; establish and maintain liaison with other organizations having common or related interests in seed. Conducts research and educational programs; maintains 17 technical committees. **PUBLICATIONS:** *International Rules for Seed Testing* (in English, French, and German), triennial; *Multilingual Glossary of Common Plant-Names*; *News Bulletin*, 3/year; *Seed Science and Technology* (in English), 3/year; Proceedings, periodic.

INTERNATIONAL SERICULTURAL COMMISSION (ISC) (Commission Sericicole Internationale - CSI)

25, quai Jean-Jacques Rousseau, F-69350 La Mulatiere, France **PHONE:** 33 478 504198 **FX:** 33 478 860957 **FOUNDED:** 1962. **OFFICERS:** Dr. G. Chavancy, Sec.Gen. **MEMBERS:** 12. **STAFF:** 1. **BUDGET:** US$100,000. **LANGUAGES:** English, French.

DESCRIPTION: An intergovernmental organization that encourages the development and improvement of sericulture. (Sericulture is the production of raw silk by raising silk worms.) Compiles statistics; operates documentation service. **PUBLICATIONS:** *Directory of Sericulture and Bacology* (in English and French); *Sericologia* (in English and French), quarterly.

INTERNATIONAL SHIP SUPPLIERS ASSOCIATION (ISSA) (Association Internationale des Approvisionneurs de Navires - AIAN)

Woodcock House, Gibbard Mews, 37/38 High St., Wimbledon SW19 5BY, England **PHONE:** 44 181 9710010 **FX:** 44 181 9710000 **TX:** 935741 SGISSA G **E-MAIL:** issa@dial.pipex.com **WEBSITES:** http://www.shipsupply.org **FOUNDED:** 1955. **OFFICERS:** J. S. Eade, Sec. **MEMBERS:** 1,620. **STAFF:** 3. **LANGUAGES:** English.

DESCRIPTION: Members are companies in 80 countries engaged in the supplying of merchant shipping with ship stores; associated groups. Aims to unite the industry for the improvement, efficiency, and modernization of the system and method of supplying merchant vessels. Conducts studies, public relations program, and merchant marine school programs. **PUBLICATIONS:** *Code of Ethics*; *International Ship Suppliers Register*; *Ship Stores Catalogue*; *Ship Supplier* (in English), quarterly.

INTERNATIONAL SHIPPING FEDERATION (ISF) (Federation Internationale des Armateurs)

c/o Carthusian Court, 12 Carthusian Court, London EC1M 6EB, England **PHONE:** 44 171 4178844 **FX:** 44 171 4178877 **FOUNDED:** 1909.

OFFICERS: Mr. J.C.S. Horrocks, Sec.Gen. **MEMBERS:** 32. **STAFF:** 4. **LANGUAGES:** English.

DESCRIPTION: Members are national shipowners' associations and maritime employers' federations in 30 countries. Monitors and disseminates information on all aspects of maritime employment and social affairs; proposes and coordinates international shipowners' positions concerning employment and other public issues; represents members before governments and unions, particularly before the international bodies concerned with these issues. Promotes the exchange of views and policies among members on questions pertaining to the employment of seamen, such as pay conditions and benefits of officers and ratings at the national level, relations with unions, developments in manning organization on board, the role of management in ship operations, safe working practices, and maritime training. **PUBLICATIONS:** *ISF Guide to International Ship Registers*; *ISF Maritime Labour Supply Guide*; *ISF Training Record Book for Deck Cadets*; *ISF Training Record Book for Engine Cadets*; *Piracy and Armed Robbery at Sea: A Master's Guide*; *The STCW Convention: A Guide for the Shipping Industry*.

INTERNATIONAL SHOOTING SPORT FEDERATION (ISSF)

c/o Horst G. Schreiber, Bavariaring 21, D-80336 Munich, Germany **PHONE:** 49 89 5443550 **FX:** 49 89 54435544 **E-MAIL:** issfmunich@compuserve.com **WEBSITES:** http://www.worldsport.com/worldsport/sports/shooting/home.html; http://www.issf-shooting.org **FOUNDED:** 1907. **OFFICERS:** Horst G. Schreiber, Sec.Gen. **MEMBERS:** 149. **STAFF:** 5. **LANGUAGES:** English, French, German, Spanish.

DESCRIPTION: Members are national shooting federations governing all amateur shooting sports in their respective countries. Acts as international governing body for all amateur shooting sport disciplines. Seeks to unite national federations and to foster permanent relations among them. Organizes world and continental championships and supervises shooting events in the Olympic games. Endeavors to establish technical regulations for all branches of the sport. Offers educational programs for trainers, referees, and judges. **PUBLICATIONS:** *ISSF News* (in English, French, German, and Spanish), bimonthly. Contains technical, medical, and activities articles concerning shooting sports.; *Official Statistics Rules and Regulations* (in English and German), annual; *Official UIT Rule Book 1997-2000*; *Provisional Special Technical Rules for Rapid Air Pistol*; *VIT History Book*. Additional publication information available upon request.

INTERNATIONAL SILK ASSOCIATION (ISA) (Association Internationale de la Soie - AIS)

34, rue de la Charite, F-69002 Lyon, France **PHONE:** 33 4 78421079 **FX:** 33 4 78375672 **TX:** 330949 F CODE 115 **FOUNDED:** 1948. **OFFICERS:** Ronald Currie, Sec.Gen. **MEMBERS:** 350. **STAFF:** 2. **BUDGET:** 1,000,000 Fr. **LANGUAGES:** English, French.

DESCRIPTION: Representatives of professional or governmental bodies, and firms involved in silk processing in 39 countries. Addresses issues pertinent to the manufacture and trade of silk. Facilitates the exchange of silk materials. Works for the standardization of testing and grading methods and uniformity of sales techniques. Disseminates information on sericulture and the silk trade. Promotes and finances research intended to discover new uses for silk. Compiles statistics. **PUBLICATIONS:** *Congress Report* (in English and French), biennial; *ISA Directory* (in English and French), biennial; Newsletter (in English and French), monthly.

INTERNATIONAL SKATING UNION (ISU)

Chemin de Primerose 2, CH-1007 Lausanne, Switzerland **PHONE:** 41 21 6126666 **FX:** 41 21 6126677 **E-MAIL:** info@isu.ch **WEBSITES:** http://www.isu.org **FOUNDED:** 1892. **OFFICERS:** Mr. Fredi Schmid, Gen.Sec. **MEMBERS:** 73. **STAFF:** 8. **LANGUAGES:** English.

DESCRIPTION: Works to administer the Figure Skating and Speed Skating Sports at international levels and encompassing the national federations and special clubs administering these sports at national levels. The national skating associations are located in 54 countries. Guides and promotes amateur ice skating and its organized development on the basis of friendship and mutual understanding among those involved in the sport. Sponsors courses for judges, referees, and starters. **PUBLICATIONS:** *ISU Constitution and General Regulations* (in English), biennial; *ISU Constitution and Minutes of the Congress* (in English), biennial; *ISU*

Special Regulations Figure Skating (in English), biennial; *ISU Special Regulations Ice Dancing* (in English), biennial; *ISU Special Regulations Speed Skating and Short Track Skating* (in English), biennial; *ISU Special Regulations Synchronized Skating* (in English), biennial.

INTERNATIONAL SKI FEDERATION (FIS) (Federation Internationale de Ski)

Blochstr. 2, CH-3653 Oberhofen, Switzerland **PHONE:** 41 33 2446161 **FX:** 41 33 2435353 **E-MAIL:** lewis@fisski.ch **FOUNDED:** 1924. **OFFICERS:** Gian Franco Kasper, Pres. **MEMBERS:** 100. **LANGUAGES:** English, French, German, Russian.

DESCRIPTION: Members are national associations concerned with skiing. Establishes rules and serves as court of appeal in case of dispute. Organizes world championships and skiing events. **PUBLICATIONS:** *Bulletin*, quarterly; *Calendar*, annual; *Congress Minutes*, biennial.

INTERNATIONAL SOCIAL SCIENCE COUNCIL (ISSC) (Conseil International des Sciences Sociales - CISS)

c/o UNESCO, 1, rue Miollis, F-75732 Paris Cedex 15, France **PHONE:** 33 1 45682558, 33 1 45682829 **FX:** 33 1 45667603 **TX:** 204461 **E-MAIL:** issclak@unesco.org **WEBSITES:** http://www.uta.fi/laitokset/hallinto/issc.htm **FOUNDED:** 1952. **OFFICERS:** Leszek A. Kosinski, Sec.Gen. **MEMBERS:** 34. **STAFF:** 2. **BUDGET:** US$500,000. **LANGUAGES:** English, French.

DESCRIPTION: International non-governmental and non-profit making scientific organization of global, international, regional, and national social and behavioral science organisations throughout the world. Promotes the social sciences and their application to major contemporary problems and fosters international cooperation among specialists in the social sciences. Facilitates and coordinates interdisciplinary activities of autonomous member associations. Encourages and contributes to the organization and development of research in the field. Renders necessary support to developing countries seeking to establish or reinforce national or regional structures in social science research. Collaborates with the United Nations and its specialized agencies, particularly UNESCO. **PUBLICATIONS:** *Dynamics of Societal Learning About Global Environmental Change*. Presents available information on studies designed to understand the behavior, attitudes, values, and learning processes of people worldwide.; *Ecological Disorder in Amazonia: Social Aspects*. Contains results from the Second International Seminar held in Rio de Janeiro in 1989 as part of an ISCC project on ecological disorder in Amazonia.; *Families: Celebration and Hope in a World of Change*. Presents a unique, international perspective on today's family as an enduring, living, and evolving institution faced with recent social issues.; *ISSC Newsletter*, quarterly. Disseminates information on the activities of the Council.; *Le Devenir de la Famille. Dynamique Familiale dans les Differentes Aires Culturelles*. Attempts to give a general view of different family structures and forms and explains how modernization has effected them.; *Population Data and Global Environmental Change*. Provides information from a survey of global population data sources and evaluates the adequacy of these data for Global Environmental Change studies.; *Post-War Development Theories and Practice*. In conjunction with the UNESCO Division of Philosophy; *Poverty in the 1990's: The Responses of Urban Women.*. In conjunction with the UNESCO Division of Human Rights; Proceedings; Reports. Additional publicaton information available on request.

INTERNATIONAL SOCIAL SECURITY ASSOCIATION (ISSA) (Association Internationale de la Securite Sociale)

4 routes des Morillons, Case Postale 1, CH-1211 Geneva 22, Switzerland **PHONE:** 41 22 7996617 **FX:** 41 22 7998509 **E-MAIL:** aissweb@ilo.org **WEBSITES:** http://www.issa.int/ **FOUNDED:** 1927. **OFFICERS:** Dalmer D. Hoskins, Sec.Gen. **MEMBERS:** 340. **STAFF:** 50. **LANGUAGES:** English, French, German, Spanish.

DESCRIPTION: Members are social security administrations in 130 countries. Encourages mutual technical assistance and information exchange among members. Conducts research into social security questions and collaborates with other international organizations in the field of social security. Organizes discussions. **PUBLICATIONS:** *African News Sheet* (in English and French), periodic; *Asian and Pacific News Sheet*, periodic; *Contacto* (in Spanish), periodic; *Documentacion de Seguridad Social: Serie Americana* (in Spanish), periodic; *EDP Bulletin* (in English, French, German, and Spanish), semiannual; *Estudios de la Seguridad Social* (in Spanish), periodic; *International Social Security Review* (in English, French, German, and Spanish), quarterly; *ISSA Bulletin* (in English, French, German, and Spanish), quarterly; *Safety Worldwide* (in English, French, German, and Spanish), semiannual; *Social Security Documentation: African Series* (in English and French), periodic; *Social Security Documentation: Asian and Pacific Series*, periodic; *Social Security Documentation: Caribbean Series*, periodic; *Social Security Documentation: Data Processing Series* (in English, French, German, and Spanish), periodic; *Social Security Documentation: European Series* (in English and French), periodic; *Social Security Documentation: Pacific Series*, periodic; *Social Security Documentation: Provident Fund Series*, periodic; *Studies and Research Series* (in English and French), periodic; *Trends in Social Security* (in English, French, German, and Spanish), quarterly.

INTERNATIONAL SOCIAL SERVICE (ISS) (Service Social International - SSI)

32, quai du Seujet, CH-1201 Geneva, Switzerland **PHONE:** 41 22 9067700 **FX:** 41 22 9067701 **E-MAIL:** iss.gs@span.ch **WEBSITES:** http://www.childhub.ch/iss **FOUNDED:** 1924. **OFFICERS:** Damien Ngabonziza, Sec.Gen. **STAFF:** 9. **LANGUAGES:** English, French, Spanish.

DESCRIPTION: Members are national independent organizations that form part of the existing social service structure in their respective countries. Assists individuals who, as a consequence of voluntary or forced migration or other social problems of an international character, have to overcome personal or family difficulties requiring coordinated efforts in several countries for solution; studies the conditions and consequences of migration in relation to individual and family life. Often assists individuals through correspondents, branches, and local programs in 110 countries. Collaborates with organizations having common interests. Maintains specialized documentation center on migration, refugees, children's rights, and family law. **PUBLICATIONS:** *Guidelines on Procedures for Intercountry Adoption* (in English and Spanish); *ISS Newsletter*, biennial; *Unaccompanied Children in Emergencies: A Field Guide for Their Care and Protection* (in English and Spanish). Internal and Intercountry Adoption Laws.

INTERNATIONAL SOCIETY OF AFRICAN SCIENTISTS (ISAS)

PO Box 9209, Wilmington, DE 19809 **E-MAIL:** isas@universal.dca.net **WEBSITES:** http://www.dca.net/isas **FOUNDED:** 1982. **OFFICERS:** Dr. Senyo Opong, Pres. **MEMBERS:** 100. **LANGUAGES:** English.

DESCRIPTION: Members are scientists and engineers of African descent. Provides a means by which members can use their professional expertise in solving the technical problems facing African and Caribbean nations; updates members on technical developments throughout the world, specifically in Africa and the Caribbean; fosters better communication and cooperation among scientists, engineers, and scientific organizations. Compiles and maintains a list of experts; provides a database of technical information. **PUBLICATIONS:** *ISAS Newsletter*, quarterly; *ISAS Transactions of Annual Technical Conference*.

INTERNATIONAL SOCIETY OF ART AND PSYCHOPATHOLOGY (ISAP) (Societe Internationale de Psychopathologie de l'Expression)

Centre Hospitalier de Ste. Anne, Clinique de la Faculte, Centre d'Etude de l'Expression, 100, rue de la Sante, F-75674 Paris Cedex 14, France **PHONE:** 33 1 45895521, 33 1 59276974 **FX:** 33 1 59216974 **FOUNDED:** 1959. **OFFICERS:** Dr. Guy Roux, Pres. **MEMBERS:** 600. **LANGUAGES:** English, French, German.

DESCRIPTION: Members are physicians and specialists including aestheticians, critics, artists, writers, art historians, psychologists, sociologists, and criminologists. Brings together specialists interested in the problems of expression and artistic activities in connection with psychiatric, sociological, and psychological research, as well as in the use of methods applied in fields other than that of mental and neurological illness. Organizes international discussion panels and symposia. **PUBLICATIONS:** Newsletter (in English and French), quarterly.

INTERNATIONAL SOCIETY FOR BIOMEDICAL RESEARCH ON ALCOHOLISM (ISBRA)

Department of Pharmacology, Univ. of Colorado Health Sciences Center, 4200 E. Ninth Ave., C-

236, Denver, CO 80262 **PHONE:** (303) 315-5690 **FX:** (303) 315-7097 **FOUNDED:** 1980. **OFFICERS:** Dr. Paula L. Hoffman, Treas. **MEMBERS:** 730. **LANGUAGES:** English.

DESCRIPTION: Members are physicians, psychologists, biologists, and other scientists in over 20 countries who conduct research on biological factors in the etiology and treatment of alcoholism and its medical complications. Official collaborating agency of the World Health Organization. Conducts international collaborative research projects and training courses. **PUBLICATIONS:** *Advances in Biomedical Alcohol Research* (in English), biennial. Contains proceedings of the biennial congresses; *Alcoholism: Clinical and Experimental Research* (in English), bimonthly.

INTERNATIONAL SOCIETY OF BIOMETEOROLOGY (ISB) (Societe Internationale de Biometeorologie - SIB)

c/o Dr. Paul J. Beggs, Department of Physical Geography, Macquarie Univ., **PHONE:** 61 2 98508399 **FX:** 61 2 98508420 **E-MAIL:** pbeggs@ocs1.ocs.mq.edu.au **WEBSITES:** http://www.es.mq.edu.au/ISB/ **FOUNDED:** 1956. **OFFICERS:** Dr. Paul J. Beggs, Sec. **MEMBERS:** 262. **LANGUAGES:** English, French, German.

DESCRIPTION: Members are scientists from 37 countries working in the field of biometeorology. Works to unite into one international body all biometeorologists working in the fields of agriculture, botany, entomology, forestry, as well as human, veterinary, and zoological biometeorology. Conducts studies on the influence of weather and climate on all living organisms, and on physiochemical processes both on Earth and in extraterrestrial environments. Maintains study groups. **PUBLICATIONS:** *Biometeorology Bulletin* (in English), semiannual; *International Journal of Biometeorology*, quarterly; *Proceedings of the International Congress of Biometeorology* (in English); *Progress in Biometeorology*, periodic.

INTERNATIONAL SOCIETY OF BLOOD TRANSFUSION (ISBT) (Societe Internationale de Transfusion Sanguine - SITS)

PO Box 111, Royal Lancaster Infirmary, Ashton Rd., Lancaster LA1 4GT, England **PHONE:** 44 1524 306272 **FX:** 44 1524 306273 **TX:** 603218 CNTS **E-MAIL:** isbt_nbs@compuserve.com **FOUNDED:** 1937. **OFFICERS:** Dr. H.H. Gunser, Sec.Gen. **MEMBERS:** 1,300. **LANGUAGES:** English, French.

DESCRIPTION: Members of national blood bank societies in 105 countries. Works toward solving the scientific, technical, social, and ethical problems related to the transfusion of blood. Encourages closer relations among individuals dealing with such problems; standardizes methods and equipment. Facilitates the exchange of information among members. **PUBLICATIONS:** *Transfusion Today* (in English), quarterly; *Vox Sanguinis*, quarterly. Also publishes congress proceedings and technical guides.

INTERNATIONAL SOCIETY FOR BUSINESS EDUCATION (Societe Internationale pour l'Enseignement Commercial - SIEC)

3550 Anderson St., Madison, WI 53704-2599 **PHONE:** (608) 837-7518 **FX:** (608) 834-1301 **E-MAIL:** gkantin@madison.tech.wi **FOUNDED:** 1901. **OFFICERS:** G. Leekantin, Gen.Sec. **MEMBERS:** 2,000. **STAFF:** 1. **BUDGET:** 60 SFr. **LANGUAGES:** English, French, German, Italian, Spanish.

DESCRIPTION: Members are educators involved in business education; heads of in-company training institutions; firms; schools and universities at various levels. Aims to promote the international exchange of ideas and experiences in business education and to further the education of teachers in business fields. Organizes courses in economic and business education. **PUBLICATIONS:** *International Review for Business Education* (in English, French, German, Italian, and Spanish), semiannual.

INTERNATIONAL SOCIETY FOR CARDIOVASCULAR SURGERY (ISCVS)

13 Elm St., Manchester, MA 01944-1314 **PHONE:** (978) 526-8330 **FX:** (978) 526-4018 **E-MAIL:** iscvs@prri.com **FOUNDED:** 1951. **OFFICERS:** William T. Maloney, Exec.Dir. **MEMBERS:** 2,500. **STAFF:** 12. **LANGUAGES:** English.

DESCRIPTION: Encourages exchange and cooperation between cardiovascular specialists. Promotes discussion of ideas pertinent to the cardiovascular disease field and stimulates investigation and study of cardiovascular diseases. **PUBLICATIONS:** *Cardiovascular Surgery*, bimonthly.

INTERNATIONAL SOCIETY OF CERTIFIED ELECTRONICS TECHNICIANS (ISCET)

2708 W. Berry, Ste. 3, Fort Worth, TX 76109 **PHONE:** (817) 921-9101 **FX:** (817) 921-3741 **E-MAIL:** iscetfw@aol.com **WEBSITES:** http://www.iscet.org **FOUNDED:** 1970. **OFFICERS:** Clyde Nabors, Exec.Dir. **MEMBERS:** 2,000. **STAFF:** 8. **BUDGET:** US$300,000. **LANGUAGES:** English, Spanish.

DESCRIPTION: Members are technicians in 43 countries who have been certified by the society. Seeks to provide a fraternal bond among certified electronics technicians, raise their public image, and improve the effectiveness of industry education programs for technicians. Offers training programs in new electronics information. Maintains library of service literature for consumer electronic equipment, including manuals and schematics for out-of-date equipment. Offers all FCC licenses. Sponsors testing program for certification of electronics technicians in the fields of audio, communications, computer, consumer, industrial, medical electronics, radar, radio-television, and video. **PUBLICATIONS:** *ISCET Update*, quarterly. Includes job listings.; *ProService Magazine*, bimonthly; *Technical Log*, quarterly.

INTERNATIONAL SOCIETY OF CITY AND REGIONAL PLANNERS (AIU IGSRP)

Mauritskade 23, NL-2514 HD The Hague, Netherlands **PHONE:** 31 70 3462654 **FX:** 31 70 3617909 **E-MAIL:** isocarp@bart.nl **WEBSITES:** http://www.soc.titech.ac.jp/isocarp/ **FOUNDED:** 1965. **OFFICERS:** Judy A.L.Y. van Hemert, Exec.Sec. **MEMBERS:** 450. **STAFF:** 3. **LANGUAGES:** English, French, German.

DESCRIPTION: Members are planners who have presented original works of merit during the 5 years preceding their candidacy; non-planners who have made notable contributions in the field serve as advisory members. Seeks to promote the profession of city and regional planning and encourage exchange of information and experience between planners. Works to improve planning education. Provides planning information and advice. Sponsors research projects and comparative studies on topics such as policies, methods, and planning legislation. Compiles current documentation material worldwide. Participates in major national and international planning events and discussions. **PUBLICATIONS:** *Case Studies* (in English), annual; *Congress*; *Habitat II - Forum Proceedings 1996*; *International Manual of Planning Practice, Vols. I, II, and III* (in English). Contains comparative studies on national planning methods, policies, and legislation; *Membership List*, periodic; *Network*, quarterly; *Special Bulletin* (in English), semiannual.

INTERNATIONAL SOCIETY FOR CONTEMPORARY MUSIC

c/o Gaudeamus, Swammerdamstraat 38, NL-1091 RV Amsterdam, Netherlands **PHONE:** 31 20 6947349 **FX:** 31 20 6947258 **E-MAIL:** iscm@xs4all.nl **WEBSITES:** http://www.xs4all.nl/-iscm **FOUNDED:** 1923. **OFFICERS:** H. Heuvelmans, Sec.Gen. **MEMBERS:** 52. **BUDGET:** 60,000 f. **LANGUAGES:** Dutch, English, French, German.

DESCRIPTION: Promotes the creation of contemporary music. Facilitates communication and cooperation among composers worldwide. **PUBLICATIONS:** *World New Music Magazine* (in English), annual.

INTERNATIONAL SOCIETY OF CRIME PREVENTION PRACTITIONERS (ISCPP)

266 Sandy Point Rd., Emlenton, PA 16373-2524 **FOUNDED:** 1978. **OFFICERS:** Jim Howell, Exec.Dir. **MEMBERS:** 1,300. **STAFF:** 2. **BUDGET:** US$150,000. **LANGUAGES:** English.

DESCRIPTION: Members are crime prevention practitioners, private security officers, and interested individuals from 14 countries. Works to facilitate exchange of ideas on crime prevention practices and programs. Sponsors basic crime prevention training. **PUBLICATIONS:** *Basic Crime Prevention Curriculum*, annual. A guide for basic crime prevention ideas and techniques.; *Practitioner*, quarterly. Contains current crime prevention information.; Membership Directory, annual.

INTERNATIONAL SOCIETY OF DEVELOPMENTAL BIOLOGISTS (ISDB)

Netherlands Inst. for Developmental Biology, Hubrecht Laboratory, Uppsalalaan 8, NL-3584 CT Utrecht, Netherlands **PHONE:** 31 30 2510211 **FX:** 31 30 2516464 **E-MAIL:** postmaster@niob.knaw.nl **WEBSITES:** http://www.niob.knaw.nl **FOUNDED:** 1911. **OFFICERS:** Prof. Dr. S.W. de Laat, Sec.-Treas. **MEMBERS:** 900. **BUDGET:** US$10,000.

DESCRIPTION: Members are scientists from 31 countries. Promotes the study of developmental biology by encouraging research and communication in the field. **PUBLICATIONS:** *Mechanisms of Development*, monthly.

INTERNATIONAL SOCIETY OF DISASTER MEDICINE (ISDM) (Societe Internationale de Medecine de Catastrophe - SIMC)

Case Postale 133, Jussy, CH-1254 Geneva, Switzerland **PHONE:** 41 22 7591312, 41 22 7590510 **FX:** 41 22 7590550 **E-MAIL:** hdca@bluewin.ch **FOUNDED:** 1975. **OFFICERS:** Dr. Marcel R. Dubouloz, Gen.Sec. **MEMBERS:** 500. **LANGUAGES:** Arabic, English, French, Spanish.

DESCRIPTION: Physicians in 42 countries who promote study and advancement in the field of disaster medicine. Conducts symposia and research programs; organizes scientific commissions. **PUBLICATIONS:** *Proceedings* (in English and French), 5/year.

INTERNATIONAL SOCIETY FOR EDUCATION THROUGH ART (INSEA) (Internationaler Kunsterzieher Verband)

c/o Diederik Schonau, PO Box 1109, NL-6801 Arnhem, Netherlands **FX:** 31 26 3521202 **E-MAIL:** insea@cito.nl **FOUNDED:** 1951. **OFFICERS:** Diederik Schonau, Sec.-Treas. **MEMBERS:** 1,000. **LANGUAGES:** English, French.

DESCRIPTION: Individuals in 62 countries engaged in creative education through art; regional and national art societies. Promotes the exchange of information and experience in the field of art and design education through publications, papers, and study groups. Seeks to encourage and advance creative education through art. Holds regional conferences. Sponsors research programs; bestows awards. **PUBLICATIONS:** *Arts in Cultural Diversity*; *Newsletter*, 3/year; *Report of World Congress*, triennial.

INTERNATIONAL SOCIETY FOR ENVIRONMENTAL TOXICOLOGY AND CANCER (ISETC)

PO Box 134, Park Forest, IL 60466 **PHONE:** (708) 758-3242 **FX:** (708) 758-3276 **E-MAIL:** jir@interaccess.com **FOUNDED:** 1983. **OFFICERS:** Dr. George Scherr, Sec.Treas. **MEMBERS:** 700. **LANGUAGES:** English.

DESCRIPTION: Clinicians and researchers working in the fields of environmental toxicology and oncology. Promotes research and information exchange. Founded the World Institute of Ecology and Cancer jointly with the European Institute of Ecology and Cancer and the Panafrican Institute of Ecology and Cancer. **PUBLICATIONS:** *Journal of Environmental Pathology, Toxicology and Oncology*, bimonthly.

INTERNATIONAL SOCIETY OF FINANCIERS (ISF)

PO Box 18508, Asheville, NC 28814 **PHONE:** (828) 252-5907 **FX:** (828) 251-5061 **E-MAIL:** insofin@bellsouth.net **WEBSITES:** http://www.insofin.com **FOUNDED:** 1979. **OFFICERS:** Ronald I. Gershen, Pres. **MEMBERS:** 250. **BUDGET:** US$70,000. **LANGUAGES:** English.

DESCRIPTION: Membership in more than 50 countries includes: real estate, minerals, commodities, and import-export brokers; corporate, industrial, and private lenders; and other financial professionals. Provides information and referrals on major domestic and international financial projects and transactions, and fosters integrity and professionalism among members. **PUBLICATIONS:** *International Financier: Information and Referrals on International Projects Requiring the Expertise of Members Throughout the Free World*, monthly.

INTERNATIONAL SOCIETY FOR GENERAL SEMANTICS (ISGS)

PO Box 728, Concord, CA 94522 **PHONE:** (925) 798-0311 **FX:** (925) 798-0312 **E-MAIL:** isgs@a.crl.com **WEBSITES:** http://www.crl.com/-isgs/isgshome.html **FOUNDED:** 1943. **OFFICERS:**

Paul Dennithorne Johnston, Exec.Dir. **MEMBERS:** 2,000. **STAFF:** 2. **BUDGET:** US$150,000. **LANGUAGES:** English.

DESCRIPTION: Members are educators, business and professional people, scientists, and others interested in general semantics and improving communication. Fosters knowledge of and inquiry into general semantics and non-Aristotelian systems through publications and lectures. **PUBLICATIONS:** *Et Cetera: A Review of General Semantics*, quarterly.

INTERNATIONAL SOCIETY FOR HORTICULTURAL SCIENCE (ISHS) (Societe Internationale de la Science Horticole - SISH)

Kardinaal Mercierlaan 92, B-3001 Louvain, Belgium **PHONE:** 32 16 229427 **FX:** 32 16 229450 **E-MAIL:** info@ishs.org **WEBSITES:** http://www.ishs.org **FOUNDED:** 1959. **OFFICERS:** Ir. J. Van Assche, Exec.Dir. **MEMBERS:** 2,950. **STAFF:** 5. **BUDGET:** 1,400,000 f. **LANGUAGES:** English, French.

DESCRIPTION: Horticultural scientists, universities, institutes, research stations, and other organizations in 86 countries. Seeks to further all branches of horticultural science by improving international contacts and cooperation in the study of related scientific and technical problems. **PUBLICATIONS:** *Acta Horticulturae*, 25-30/year. Contains symposia proceedings; *Chronica Horticulturae*, quarterly; *Congress Proceedings*, quadrennial; *Horticultural Research International*, quinquennial; *Scientia Horticulturae*, monthly.

INTERNATIONAL SOCIETY FOR HUMAN RIGHTS (ISHR) (Internationale Gesellschaft fur Menschenrechte - IGFM)

Borsigallee 16, D-60388 Frankfurt am Main, Germany **PHONE:** 49 69 4201080 **FX:** 49 69 4201083 **FOUNDED:** 1972. **OFFICERS:** Hans Born. **MEMBERS:** 4,000. **STAFF:** 28. **BUDGET:** DM 4,200,000. **LANGUAGES:** Croatian, English, French, German, Italian, Russian, Spanish.

DESCRIPTION: Students, teachers, lawyers, doctors, and other interested individuals in 42 countries united to promote fundamental human rights and provide practical help without violence for people and organizations in totalitarian countries. Promotes the Universal Declaration of Human Rights advocating: the right to freedom of opinion and expression; the right to leave one's country and to return to it; the right of religious freedom; the right to found organizations, and to receive and disseminate information; the right to be respected. Organizes work groups whose activities include assisting in individual cases, writing and circulating appeals and documents, and organizing and participating in demonstrations and press conferences. Encourages public disclosure and denunciation of all violations of human rights; assists politicians and state institutions. Seeks to improve international tolerance and understanding; collaborates with international organizations. Provides information service. **PUBLICATIONS:** *Fur die Menschenrechte*, bimonthly; *Human Rights Worldwide* (in English), bimonthly; *Press Releases* (in English, German, Russian, and Spanish); *Zeitschrift Menschenrechte* (in German), bimonthly; Newsletter (in Spanish), quarterly; Pamphlets.

INTERNATIONAL SOCIETY OF INTERNAL MEDICINE (ISIM) (Societe Internationale de Medecine Interne)

c/o Dr. Rolf A. Streuli, Regional Hospital, CH-4900 Langenthal, Switzerland **PHONE:** 41 62 9163102 **FX:** 41 62 9164155 **E-MAIL:** r.streuli@slo.ch **WEBSITES:** http://www.acponline.org/isim/ **FOUNDED:** 1948. **OFFICERS:** Dr. Rolf A. Streuli, Sec.Gen. **MEMBERS:** 2,000. **STAFF:** 2. **LANGUAGES:** English.

DESCRIPTION: Members are internal medicine specialists. Promotes scientific knowledge in internal medicine; furthers the education of young internists; encourages friendship among physicians in all countries. Federation of Societies of Internal Medicine from 43 countries.

INTERNATIONAL SOCIETY FOR LABOR LAW AND SOCIAL SECURITY (ISLLSS) (Societe Internationale de Droit du Travail et de la Securite Sociale)

Case Postale 500, CH-1211 Geneva 22, Switzerland **PHONE:** 41 22 7996343 **FX:** 41 22 7996260 **E-MAIL:** servais@ilo.ch **FOUNDED:** 1958. **OFFICERS:** Jean-Michel Servais, Sec.Gen.

MEMBERS: 1,500. **LANGUAGES:** English, French, German, Spanish.

DESCRIPTION: Lawyers interested in the scientific study of labor law and social security law at the national and international levels. Promotes the exchange of ideas and information; encourages close collaboration between jurists and other experts in labor law and social security. **PUBLICATIONS:** *Congress Proceedings*, triennial.

INTERNATIONAL SOCIETY OF LYMPHOLOGY (ISL)

c/o Dr. Marlys H. Witte, Univ. of Arizona, Coll. of Medicine/Surgery, 1501 N. Campbell Ave., No. 4406, Tucson, AZ 85724 **PHONE:** (520) 626-6118 **FX:** (520) 626-0822 **E-MAIL:** lymph@cu.it.arizona.edu **FOUNDED:** 1966. **OFFICERS:** Marlys H. Witte M.D., Sec.Gen. **MEMBERS:** 385. **LANGUAGES:** English.

DESCRIPTION: Professionals active in the medical, biological, and technical sciences. Promotes the study of lymphology and seeks to advance and disseminate knowledge in the field. Activities include: stimulating and strengthening experimentation and clinical investigation in lymphology; establishing relations between researchers and clinicians in the different fields of lymphology; encouraging contact and exchange of ideas among members and national and international organizations. Organizes postgraduate courses. **PUBLICATIONS:** *Lymphology*, quarterly; *Progress in Lymphology*, biennial.

INTERNATIONAL SOCIETY FOR MILITARY LAW AND LAW OF WAR (Societe Internationale de Droit Militaire et de Droit de la Guerre)

c/o Auditorat General pres la Cour Militaire, Palais de Justice, B-1000 Brussels, Belgium **PHONE:** 32 2 5086025 **FX:** 32 2 5086087 **E-MAIL:** soc-mil-law@skynet.be **WEBSITES:** http://www.soc-mil-law.org **FOUNDED:** 1955. **OFFICERS:** M. Fobe, Sec.Gen. **MEMBERS:** 1,000. **STAFF:** 1. **BUDGET:** 1,500,000 BFr. **LANGUAGES:** English, French.

DESCRIPTION: Membership consists of lawyers, jurists, judge advocates, and officers in over 50 countries. Organized for the study of military law and the law of war. Conducts research on harmonization of internal law and international conventions; promotes recognition of war law heedful of human rights. **PUBLICATIONS:** *Military Law and Law of War Review* (in Dutch, English, French, German, Italian, and Spanish), quarterly; *Recueils of the International Society for Military Law and Law of War* (in Dutch, English, French, German, Italian, and Spanish), triennial; Articles (in Dutch, English, French, German, Italian, and Spanish).

INTERNATIONAL SOCIETY FOR MUSIC EDUCATION (ISME) (Societe Internationale pour l'Education Musicale)

Univ. of Reading, ICRME, Bulmershe Ct., Reading, Berks. RG6 1HY, England **PHONE:** 44 118 9318846 **FX:** 44 118 9352080 **E-MAIL:** e.smith@reading.ac.uk **WEBSITES:** http://www.isme.org; http://www.isme.org **FOUNDED:** 1953. **OFFICERS:** Elizabeth Smith, Admin. **MEMBERS:** 1,500. **STAFF:** 1. **LANGUAGES:** English.

DESCRIPTION: Associations, educational institutions, and individual educators in 60 countries interested in music education. Promotes music education throughout the world as both a profession and an integral part of general education and community life. Cooperates with international government groups such as United Nations Educational, Scientific and Cultural Organization. **PUBLICATIONS:** *International Journal of Music Education* (in English), semiannual. Includes abstracts in French, German, and Spanish; *ISME Edition*, periodic.

INTERNATIONAL SOCIETY OF NEUROPATHOLOGY (Societe Internationale de Neuropathologie)

c/o Dr. Janice R. Anderson, Dept. of Histopathology, Addenbrooke's Hospital, Cambridge CB2 2QQ, England **PHONE:** 44 1 223217170 **FX:** 44 1 223216980 **E-MAIL:** jra20@cam.ac.uk **WEBSITES:** http://www.medschl.cam.ac.uk/npsoc/; http://www.upmc.edu/divisions/neuropath/bpath.html; http://www.neuropathology2000.co.uk **FOUNDED:** 1972. **OFFICERS:** Dr. Janice R. Anderson, Sec.Gen. **MEMBERS:** 2,500. **LANGUAGES:** English.

DESCRIPTION: Members of national societies of neuropathology in 30 countries. Works to foster the formation of national and regional societies of neuropathology and to promote cooperation among these societies. Maintains liaison with international organizations in various fields of neurological sciences. Encourages the exchange of information and persons engaged in neuropathology. Initiates research projects. **PUBLICATIONS:** *Brain Pathology*, quarterly; *Membership Directory*, periodic.

INTERNATIONAL SOCIETY FOR ONEIRIC MENTAL IMAGERY TECHNIQUES (ISMIT) (Societa Internazionale per le Techniche d'Imagerie Mentale Onirique)

Chez Dr. Andre Virel, 69, rue Vasco de Gama, F-75015 Paris, France **PHONE:** 33 1 01484681 **FOUNDED:** 1966. **OFFICERS:** Andre Virel, Pres. **MEMBERS:** 80. **STAFF:** 7. **BUDGET:** 23,000 Fr.

DESCRIPTION: Fosters inquiry and interaction among researchers and practitioners of mental imagery techniques in psychotherapy. Conducts research on mental imagery psychotherapy; disseminates research results. Organizes seminars on psychotherapeutic practice. **PUBLICATIONS:** *Decentration*, periodic; *Mental Imagery*; *Vocabulary of Psychotherapies*. Additional publications available on request.

INTERNATIONAL SOCIETY OF ORTHOPAEDIC SURGERY AND TRAUMATOLOGY (Societe Internationale de Chirurgie Orthopedique et de Traumatologie - SICOT)

40, rue Washington, bte. 9, B-1050 Brussels, Belgium **PHONE:** 32 2 6486823 **FX:** 32 2 6498601 **E-MAIL:** hq@sicot.org **FOUNDED:** 1929. **OFFICERS:** Mr. A.J. Hall, Sec.Gen. **MEMBERS:** 3,000. **STAFF:** 2. **LANGUAGES:** English, French.

DESCRIPTION: Orthopaedic surgeons in 103 countries united to contribute to the progress of science by conducting studies related to orthopedic surgery and traumatology. **PUBLICATIONS:** *International Orthopedics* (in English and French), bimonthly. Includes Newsletter.

INTERNATIONAL SOCIETY FOR THE PERFORMING ARTS (ISPA)

PO Box 909, Rye, NY 10580-0909 **PHONE:** (914) 921-1550 **FX:** (914) 921-1593 **E-MAIL:** info@ispa.org **WEBSITES:** http://www.ispa.org **FOUNDED:** 1948. **OFFICERS:** David Watson, Exec.Dir. **MEMBERS:** 700. **STAFF:** 3. **BUDGET:** US$800,000. **LANGUAGES:** English.

DESCRIPTION: Executives and directors of concert and performance halls, festivals, performing companies, and artist competitions; government cultural officials; artists' managers; and other interested parties with a professional involvement in the performing arts from more than 50 countries in every region of the world, and in every arts discipline. **PUBLICATIONS:** *Performing Arts Forum*, bimonthly. Contains information on employment opportunities, member news, and statistics.; Membership Directory, annual.

INTERNATIONAL SOCIETY FOR THE PREVENTION OF WATER POLLUTION (ISPWP) (Comitato Internazionale Contra l'Inquinemento dell'Acqua)

Little Orchard, Bentworth, Alton, Hants. GU34 5RB, England **PHONE:** 44 1420 562225 **FOUNDED:** 1980. **OFFICERS:** Earl Maitland, Chm. **MEMBERS:** 750. **STAFF:** 6. **BUDGET:** £30,000. **LANGUAGES:** English, French, Greek, Italian.

DESCRIPTION: Experts in water problems from 14 countries united to prevent the pollution or contamination of water throughout the world. Raises and provides funds for research into the causes and effects of, and solutions to, water pollution problems. Seeks to attract the attention of governments, states, and municipalities affected by water pollution; recommends improvements. Offers children's services. Conducts research and charitable programs. Maintains speakers' bureau. **PUBLICATIONS:** *ISFTPOWP Newsletter* (in English, French, Greek, and Italian), periodic. Includes information on water pollution.

INTERNATIONAL SOCIETY FOR ROCK MECHANICS (ISRM) (Societe Internationale de Mecanique des Roches - SIMR)

c/o Laboratorio Nacional de Engenharia Civil, Avenida do Brasil 101, P-1799 Lisbon Codex, Portugal **PHONE:** 351 1 8482131 **FX:** 351 1 8478187 **E-MAIL:** delgado@lnec.pt **FOUNDED:**

1962. **OFFICERS:** Jose Delgado Rodrigues, Sec.Gen. **MEMBERS:** 6,000. **STAFF:** 2. **BUDGET:** US$70,000. **LANGUAGES:** English, French, German.

DESCRIPTION: Individuals and collective bodies in 62 countries working to stimulate the development of rock mechanics. Purposes are to: initiate and coordinate international cooperation in the science of rock mechanics; promote the international publication of the results of scientific investigations in rock mechanics; encourage individuals and local organizations to form national groups primarily or solely concerned with the science of rock mechanics; maintain liaisons with other bodies representing fields of science related to rock mechanics including geology, geophysics, soil mechanics, mining engineering, petroleum engineering, and civil engineering; encourage the teaching of rock mechanics. **PUBLICATIONS:** *Membership List* (in English), quadrennial; *News/Journal*, quarterly; *Reports of the Commissions*.

INTERNATIONAL SOCIETY FOR SOIL MECHANICS AND GEOTECHNICAL ENGINEERING (Societe Internationale de Mecanique des Sols et de la Geotechnique)

Univ. Engineering Dept., Trumpington St., Cambridge CB2 1PZ, England **PHONE:** 44 1223 355020 **FX:** 44 1223 359675 **FOUNDED:** 1936. **OFFICERS:** Dr. R.H.G. Parry, Sec.Gen. **MEMBERS:** 70. **STAFF:** 2. **BUDGET:** US$100,000. **LANGUAGES:** English, French.

DESCRIPTION: Organization national societies of engineers in soil mechanics and related areas representing 17,000 individuals. Fosters international fellowship and cooperation among engineers and scientists for the advancement of knowledge in the field of geotechnics and its engineering applications. Facilitates exchange of information among member societies. Maintains 29 technical committees. **PUBLICATIONS:** *Conference Proceedings*, biennial; *ISSMFE News*, quarterly; *Lexicon of Soil Mechanics Terms*. Published in 8 different languages.

INTERNATIONAL SOCIETY FOR SOILLESS CULTURE (ISOSC)

Postbus 52, NL-6700 AB Wageningen, Netherlands **PHONE:** 31 317 413809 **FX:** 31 317 423457 **FOUNDED:** 1955. **OFFICERS:** Abram A. Steiner, Sec.Gen. & Treas. **MEMBERS:** 490. **BUDGET:** US$13,000. **LANGUAGES:** English.

DESCRIPTION: Persons in 66 countries engaged in research, advisory work, or practical application of soilless culture (hydroponics). Purposes are to: promote the study of problems of soilless culture; organize or promote courses and excursions; prepare or support works on scientific and practical aspects of soilless culture; promote international cooperation in research and advisory service. Disseminates information. **PUBLICATIONS:** *Bibliography* (in English), biennial. Literature on soilless culture; *Proceedings*, quadrennial.

INTERNATIONAL SOCIETY FOR STEREOLOGY (ISS) (Societe Internationale pour la Stereologie)

c/o Dr. J.R. Nyengaard, Stereol Research Lab, Bartholin Bldg., Aarhus Univ., DK-8000 Arhus, Denmark **PHONE:** 45 8949 3654 **FX:** 45 8949 3650 **E-MAIL:** stereotsj@srfcd.aau.dk **FOUNDED:** 1961. **OFFICERS:** Dr. Jens R. Nyengaard. **MEMBERS:** 560. **LANGUAGES:** English.

DESCRIPTION: Researchers in life sciences and materials science in 37 countries. Objectives are to: promote the application of stereology to the description of microstructures in the life and materials sciences; provide a quantitative, scientific basis concerning the properties of structure in space. (Stereology is a branch of science concerned with the development and testing of inferences about the 3-dimensional properties of objects or matter ordinarily observed from a 2-dimensional point of view.) **PUBLICATIONS:** *Acta Sterologica*, semiannual; *Journal of Microscopy*, monthly; *Membership Directory*, annual; *Newsletter*, annual.

INTERNATIONAL SOCIETY FOR THE STUDY OF MEDIEVAL PHILOSOPHY (SIEPM) (Societe Internationale pour l'Etude de la Philosophie Medievale - SIEPM)

Institut Superieur de Philosophie, College Mercier, 14, place du Cardinal Mercier, B-1348 Louvain, Belgium **PHONE:** 32 10 474807 **FX:** 32 10 478285 **E-MAIL:** danhier@sofi.ucl.ac.be **FOUNDED:** 1958. **OFFICERS:** Prof. Jacqueline Hamesse, Sec. **MEMBERS:** 590. **LANGUAGES:** English, French, German, Italian, Spanish.

DESCRIPTION: Membership consists of: individuals pursuing scientific research in medieval philosophy and related fields; national and local groups with the same goal. Promotes international cooperation in the field. **PUBLICATIONS:** *Bulletin de philosophie medievale*, annual.

INTERNATIONAL SOCIETY OF SURGERY (ISS) (Societe Internationale de Chirurgie - SIC)

Netzibodenstr. 34, PO Box 1527, CH-4133 Pratteln, Switzerland **PHONE:** 41 61 8159666, 41 61 8159667 **FX:** 41 61 8114775 **E-MAIL:** surgery@nbs.ch **WEBSITES:** http://www.surgery.nbs.ch **FOUNDED:** 1902. **OFFICERS:** Dr. Thomas Ruedi, Sec.Gen. **MEMBERS:** 4,000. **STAFF:** 3. **BUDGET:** US$320,000. **LANGUAGES:** English.

DESCRIPTION: Surgeons from 90 countries wishing to contribute to the progress of science by researching and discussing surgical problems at congresses and general assemblies. **PUBLICATIONS:** *ISS/SIC* (in English), 2-3/year; *Membership Booklet*, biennial; *World Journal of Surgery* (in English), monthly. Scientific publication on general surgery.

INTERNATIONAL SOCIETY FOR TECHNOLOGY IN EDUCATION (ISTE)

1787 Agate St., Eugene, OR 97403-1923 **PHONE:** (541) 346-4414 **TF:** (800) 336-5191 **FX:** (541) 346-5890 **E-MAIL:** iste@oregon.uoregon.edu **WEBSITES:** http://www.iste.org **FOUNDED:** 1979. **OFFICERS:** Maia S. Howes, Exec.Sec. **MEMBERS:** 10,000. **STAFF:** 32. **LANGUAGES:** English.

DESCRIPTION: Teachers, administrators, computer and curriculum coordinators, and others interested in improving the quality of education through the innovative use of technology. Facilitates exchange of information and resources between international policy makers and professional organizations; encourages research and evaluation relating to the use of technology in education. Maintains the Private Sector Council to promote cooperation among private sector organizations to identify needs and establish standards for hardware, software, and other technology-based educational systems, products, and services. **PUBLICATIONS:** *Journal of Research on Computing in Education*, quarterly. Includes articles on research, system or project descriptions and evaluations, and assessments and theoretical positions on educational computing.; *Learning and Leading with Technology*, 8/year. Includes information on language arts, Logo, science, mathematics, telecommunications, equity, and international connections.; *Update*, 7/year. Includes activities information, news on events in Washington, and conference calendar.

INTERNATIONAL SOCIOLOGICAL ASSOCIATION (ISA) (Association Internationale de Sociologie - AIS)

Faculty Political Sciences and Sociology, Univ. Complutense, E-28223 Madrid, Spain **PHONE:** 34 913527650 **FX:** 34 913524945 **E-MAIL:** isa@sis.ucm.es **WEBSITES:** http://www.ucm.es/info/isa **FOUNDED:** 1949. **OFFICERS:** Izabela Barlinska, Exec.Sec. **MEMBERS:** 3,000. **STAFF:** 3. **BUDGET:** US$200,000. **LANGUAGES:** English, French, Spanish.

DESCRIPTION: National sociological societies, international and multinational regional associations of sociologists, research institutions, university departments, scholars, and supporting institutions in 167 countries. Encourages personal contacts among sociologists worldwide; promotes international dissemination and exchange of information; facilitates research. Maintains 50 research committees and 5 working groups. **PUBLICATIONS:** *Current Sociology*, quarterly; *Directory of Individual and Collective Members*, annual; *International Sociology*, quarterly; *Sage Studes in International Sociology*; Bulletin, 3/year.

INTERNATIONAL SOLAR ENERGY SOCIETY (ISES)

Villa Tannheim, Wiesentalstrasse 50, D-79115 Freiburg, Germany **PHONE:** 49 761 4590650 **FX:** 49 761 4590699 **E-MAIL:** hq@ises.org **WEBSITES:** http://www.ises.org/ **FOUNDED:** 1954. **OFFICERS:** Burkhard Holder, Exec.Dir. **MEMBERS:** 4,000. **STAFF:** 8. **LANGUAGES:** English.

DESCRIPTION: Members are solar energy researchers, scientists, engineers, and organizations; interested individuals. Principal activity is the publication of various materials. **PUBLICATIONS:** *International Congress Proceedings*, biennial; *Solar Energy Journal*, monthly; *Sunworld*, quarterly.

INTERNATIONAL SOLID WASTES ASSOCIATION (ISWA)

Overgaden Oven Vandet 48 E, DK-1415 Copenhagen K, Denmark **PHONE:** 45 32 961588 **FX:** 45 32 961584 **E-MAIL:** iswa@inet.uni2.dk **WEBSITES:** http://www.iswa.org **OFFICERS:** Suzanne Arup Veltze. **MEMBERS:** 1,200. **LANGUAGES:** Danish, English.

DESCRIPTION: Members are national solid waste treatment and public sanitation committees. Promotes creation and implementation of environmentally sustainable solid waste management systems. Conducts research on new solid waste treatment technologies and practices. **PUBLICATIONS:** *ISWA Times*, quarterly; *Waste Management and Research*, bimonthly.

INTERNATIONAL SOLIDARITY FOUNDATION (Kansainvalinen Solidaarisuussaatio)

Agricolankatu 4, FIN-00530 Helsinki, Finland **PHONE:** 358 9 7011200 **FX:** 358 9 7731702 **E-MAIL:** finsolid@tsl.fi **WEBSITES:** http://www.finsolid.fi **FOUNDED:** 1970. **OFFICERS:** Helena Laukko, Exec.Dir. **MEMBERS:** 15,000. **STAFF:** 6. **BUDGET:** US$1,710,088. **LANGUAGES:** English, Finnish, Spanish.

DESCRIPTION: Promotes public awareness of living conditions in developing countries. Conducts long-term integrated community-based development programs. Programs include: support to women's groups, adult education, support for street children, and income-generating activities. Works to improve food security. **PUBLICATIONS:** *Apuraportti* (in Finnish), semiannual. Contains articles about development cooperation.

INTERNATIONAL SPECIAL TOOLING AND MACHINING ASSOCIATION (ISTMA) (Association Internationale de l'Outillage Special)

Postfach 71 08 64, D-60498 Frankfurt, Germany **PHONE:** 49 69 66031470 **FX:** 49 69 66032251 **E-MAIL:** brodtman_pwa@vdma.org **FOUNDED:** 1973. **OFFICERS:** Thilo K. Brodtmann. **MEMBERS:** 17. **STAFF:** 1. **LANGUAGES:** English, French, German.

DESCRIPTION: National associations of tool, die, and mold manufacturers. Works to facilitate the exchange of business and technical information on a permanent basis. Compiles statistics. **PUBLICATIONS:** *Abstract Services*, periodic; *Special Bulletins*, periodic. Also publishes study papers; is compiling a glossary of technical terms.

INTERNATIONAL SPORT PRESS ASSOCIATION (ISPA) (Association Internationale de la Presse Sportive - AIPS)

c/o Togay Bayatli, Hold u. 1, H-1054 Budapest, Hungary **PHONE:** 36 1 3112689, 36 1 3119080 **FX:** 36 1 3533807 **FOUNDED:** 1924. **OFFICERS:** Togay Bayalti, Sec.Gen. **MEMBERS:** 130. **STAFF:** 3. **LANGUAGES:** English, French, Russian, Spanish.

DESCRIPTION: Organization of national sport press associations representing 10,000 sports journalists in press, photography, television, and radio. Works for members' ethical and professional interests. Encourages mutual assistance and solidarity among sports journalists; obtains work facilities for journalists. Conducts seminars; compiles statistics; operates speakers' bureau. **PUBLICATIONS:** *International Sports Magazine* (in English, French, and Spanish), annual; *Newsletter*, 5/year.

INTERNATIONAL STATISTICAL INSTITUTE (ISI) (Institut International de Statistique)

428 Prinses Beatrixlaan, PO Box 950, NL-2270 AZ Voorburg, Netherlands **PHONE:** 31 70 3375737 **FX:** 31 70 3860025 **E-MAIL:** isi@cbs.nl **WEBSITES:** http://www.cbs.nl/isi **FOUNDED:** 1885. **OFFICERS:** Dr. M.P.R. VandenBroecke, Sec. **MEMBERS:** 1,997. **STAFF:** 8. **LANGUAGES:** Arabic, Chinese, English, French, German, Italian, Russian, Spanish.

DESCRIPTION: Persons from more than 80 countries who have contributed to the development or application of statistical methods or to the administration of statistical services. Works toward the development and improvement of statistical methods and their application worldwide. Sponsors statistics course at the ISEC in Calcutta, India. Compiles and publishes information pertaining to international statistics; has established an abstracting service of statistical publications. **PUBLICATIONS:** *Bulletin of ISI* (in English and French), biennial. Proceedings from the biennial session;

Directory of National Statistical Agencies, biennial; *Directory of Statistical Societies*, annual; *International Statistical Review*, 3/year.

INTERNATIONAL STUDIES ASSOCIATION (ISA)

324 Social Sciences Bldg., Univ. of Arizona, Tucson, AZ 85721 **PHONE:** (520) 621-7715 **FX:** (520) 621-5780 **E-MAIL:** isa@u.arizona.edu **WEBSITES:** http://www.isanet.org **FOUNDED:** 1959. **OFFICERS:** Thomas J. Volgy, Exec.Dir. **MEMBERS:** 2,900. **BUDGET:** US$400,000. **LANGUAGES:** English.

DESCRIPTION: Social scientists and other scholars from a wide variety of disciplines who are specialists in international affairs and cross-cultural studies; academicians; government officials; officials in international organizations; business executives; students. Promotes research, improved teaching, and the orderly growth of knowledge in the field of international studies; emphasizes a multidisciplinary approach to problems. Conducts workshops and discussion groups. **PUBLICATIONS:** *International Studies Newsletter*, 10/year. Contains information on awards, grants, and fellowships; *International Studies Notes*, 3/year. Contains research, curricular, and program reports on international affairs; *International Studies Quarterly*.

INTERNATIONAL SUGAR ORGANIZATION (ISO) (Organizacion Internacional del Azucar - OIA)

1 Canada Sq., Canary Wharf, Docklands, London E14 5AA, England **PHONE:** 44 171 5131144 **FX:** 44 171 5131146 **E-MAIL:** isolondon@compuserve.com **WEBSITES:** http://www.sugarinfo.co.uk/iso.html **FOUNDED:** 1969. **OFFICERS:** Dr. P. Baron, Exec.Dir. **MEMBERS:** 52. **STAFF:** 12. **BUDGET:** £800,000. **LANGUAGES:** English, French, Russian, Spanish.

DESCRIPTION: Governments of sugar exporting and importing countries. Is currently administering the 1992 International Sugar Agreement. Promotes the sugar industry. **PUBLICATIONS:** *Far East Sugar Economics*; *Latin American and Caribbean Sugar Economics*; *Market Report - Seminar Proceedings on Eastern Europe*; *Middle East Sugar Economics and North Africa*; *Statistical Bulletin* (in English), monthly; *Sugar Yearbook* (in English). Additional publication information available upon request.

INTERNATIONAL TABLE TENNIS FEDERATION (ITTF) (Federation Internationale de Tennis de Table)

53 London Rd., St. Leonards-on-Sea, E. Sussex TN37 6AY, England **PHONE:** 44 1424 721414 **FX:** 44 1424 431871 **E-MAIL:** hq@ittf.cablenet.co.uk **WEBSITES:** http://www.ittf.com **FOUNDED:** 1926. **OFFICERS:** Mr. Adam Sharara, Deputy Pres. CEO. **MEMBERS:** 180. **STAFF:** 5. **LANGUAGES:** English, French, German, Spanish.

DESCRIPTION: Organization of national associations of table tennis players. Purpose is to: uphold the laws of table tennis and revise them as necessary; help the spread of table tennis throughout the world and raise its technical level. Sponsors biennial World Championships; presents the World Cup annually. **PUBLICATIONS:** *Bulletin* (in English, French, and Spanish), bimonthly; *Handbook*, biennial; *ITTF Directory*, annual; *Match Officials Handbook*, biennial; *Register of International Umpires* (in English), biennial; *Rules Booklet* (in English), biennial; *Table Tennis Digest* (in English), quarterly; *Table Tennis Illustrated*, bimonthly; *Tournament Referees*, biennial.

INTERNATIONAL TANKER OWNERS POLLUTION FEDERATION (ITOPF)

Staple Hall, Stone House Ct., 87-90 Houndsditch, London EC3A 7AX, England **PHONE:** 44 171 6211255 **FX:** 44 171 6211783 **E-MAIL:** central@itopf.com **WEBSITES:** http://www.itopf.com **FOUNDED:** 1968. **OFFICERS:** Dr. Ian C. White, Mng.Dir. **MEMBERS:** 3,900. **STAFF:** 18. **BUDGET:** £1,400,000. **LANGUAGES:** English, French, Spanish.

DESCRIPTION: Membership consists of independent and oil company tanker owners and government-owned companies in over 100 countries. Responds to spills from tankers; offers advice on clean-up, investigates the impact of spills, and assesses the technical merits of subsequent claims for compensation for clean-up costs and damage. Undertakes contingency planning and training assignments and maintains various databases, including are an oil spills from tankers. Produces various technical publications. **PUBLICATIONS:** *Ocean Or-*

bit (in English), annual; *Response to Marine Oil Spills* (in English, French, and Spanish); *Technical Information Papers*, periodic. Annual Review.

INTERNATIONAL TEA COMMITTEE (ITC) (Comite International du The)

Sir John Lyon House, 5 High Timber St., London EC4V 3NH, England **PHONE:** 44 171 2484672 **FX:** 44 171 3296955 **E-MAIL:** inteacom@globalnet.co.uk **FOUNDED:** 1933. **OFFICERS:** Peter Abel, Consultant/Sec. **MEMBERS:** 16. **STAFF:** 3. **LANGUAGES:** English.

DESCRIPTION: Tea-producing and consuming governments and associations. Collects and disseminates information and statistics on tea. **PUBLICATIONS:** *Annual Bulletin of Statistics*; *Monthly Statistical Summary*; *World Tea Statistics 1910-90*.

INTERNATIONAL TELECOMMUNICATION UNION (ITU) (Union Internationale des Telecommunications)

c/o Mrs. F. Lambert, Palais des Nations, CH-1211 Geneva 20, Switzerland **PHONE:** 41 22 7305111 **FX:** 41 22 7337256 **TX:** 421000 CHUIT **E-MAIL:** itumail@itu.int **WEBSITES:** http://www.itu.int **FOUNDED:** 1865. **OFFICERS:** Mrs. F. Lambert. **MEMBERS:** 188. **STAFF:** 717. **BUDGET:** 327,655,000 SFr. **LANGUAGES:** Arabic, Chinese, English, French, Russian, Spanish.

DESCRIPTION: Adopts international regulations and treaties governing all terrestrial and space uses of the frequency spectrum and the use of the geostationary-satellite orbit. Develops standards to ensure the interconnection of telecommuncation systems on a worldwide scale regardless of the type of technology used. Fosters development of telecommunications in developing countries. Establishes medium-term development policies and strategies in consultation with other partners. Provides specialized technical assistance in telecommunication policies, choice and transfer of technologies, management, financing of investment projects and mobilization of resources, installation and maintenance of networks, management of human resources, and research and development. **PUBLICATIONS:** *ITU News* (in English, French, and Spanish), 10/year.

INTERNATIONAL TELECOMMUNICATIONS SATELLITE ORGANIZATION (INTELSAT)

3400 International Dr. NW, Washington, DC 20008-3098 **PHONE:** (202) 944-6800 **FX:** (202) 944-7898 **WEBSITES:** http://www.intelsat.int **FOUNDED:** 1964. **OFFICERS:** Conny Kullman, Dir.Gen. & CEO. **MEMBERS:** 143. **LANGUAGES:** English, French, Spanish.

DESCRIPTION: Members are governments that adhere to 2 international telecommunications agreements. Each government designates a telecommunications entity, either public or private, as its signatory to the INTELSAT Operating Agreement. Seeks to unify the design, development, construction, establishment, maintenance, and operation of the space segment of the global communications satellite system. The space segment provides overseas telecommunications services and live television, enables a number of domestic communications systems, and includes communication satellites and the telemetry, control, command, monitoring, and related facilities and equipment required to support satellite operations. As of January 1998 there were 19 satellites in geosynchronous orbit. **PUBLICATIONS:** Annual Report; Brochures. Also publishes services materials.

INTERNATIONAL TENNIS FEDERATION (ITF) (Federation Internationale de Tennis)

Bank Ln., Roehampton, London SW15 SX2, England **PHONE:** 44 181 8786464 **FX:** 44 181 8787799 **E-MAIL:** recephon@itftennis.com **WEBSITES:** http://www.itftennis.com **FOUNDED:** 1913. **OFFICERS:** Brian Tobin, Pres. **MEMBERS:** 201. **STAFF:** 85. **LANGUAGES:** English, French, Spanish.

DESCRIPTION: Organization of national tennis associations. Works to promote and develop the game and teaching of tennis at all levels throughout the world. Organizes the Davis Cup, the Fed Cup, the World Youth Cup, and other international competitions including the tennis event at the Olympic Games. **PUBLICATIONS:** *ITF World* (in English), quarterly; *Junior World Ranking Rules, Regulations, Code of Conduct and Calendar* (in English, French, and Spanish), annual; *Junior World Ranking Tournament Guide*, annual; *Regulations for the Davis Cup Competition* (in English, French, and

Spanish), annual; *Regulations for the Women's International Team Competition - Fed Cup* (in English, French, and Spanish), annual; *Rules and Standing Orders of the ITF* (in English, French, and Spanish), annual; *Rules of Tennis* (in English, French, and Spanish), annual; *This is the ITF* (in English), annual; *Veteran Calendar*, annual; *Veteran Handbook*, annual; *World of Tennis* (in English), annual; *World Youth Cup Regulations* (in English), annual.

INTERNATIONAL TEXTILE MANUFACTURERS FEDERATION (ITMF) (Federation Internationale des Industries Textiles)

Am Schanzengraben 29, Postfach, CH-8039 Zurich, Switzerland **PHONE:** 41 1 2017080 **FX:** 41 1 2017134 **E-MAIL:** secretariat@itmf.org **FOUNDED:** 1904. **OFFICERS:** Dr. Herwig Strolz, Dir.Gen. **STAFF:** 4. **LANGUAGES:** English, French, German.

DESCRIPTION: Textile manufacturers and allied organizations in 50 countries. Provides neutral forum for worldwide textile industries; acts as clearinghouse and spokesman for industry in matters relating to raw materials such as cotton and synthetic fibers. Maintains official liaison with intergovernmental organizations and permanent liaison with international nongovernmental organizations interested in the textile industry. Compiles statistics; conducts research. **PUBLICATIONS:** *Country Statements*, annual; *Directory*, biennial; *International Cotton Industry Statistics*, annual; *International Production Cost Comparison - Spinning/Weaving/Knitting*, biennial; *International Textile Machinery Shipment Statistics*, annual; *International Textile Manufacturing*, annual; *State of Trade Report*, quarterly; *Survey on Cotton Contamination, Foreign Matter, and Stickiness*, biennial.

INTERNATIONAL THEATRE INSTITUTE (ITI) (Institut Internationale du Theatre)

UNESCO, 1, rue Miollis, F-75732 Paris Cedex 15, France **PHONE:** 33 1 45682650 **FX:** 33 1 45665040 **E-MAIL:** iti@unesco.org **WEBSITES:** http://www.iti-worldwide.org **FOUNDED:** 1948. **OFFICERS:** Andre-Louis Perinetti, Sec.Gen. **MEMBERS:** 95. **STAFF:** 2. **LANGUAGES:** English, French.

DESCRIPTION: Organization of national theatre institute centers and theatre professionals. Encourages communication worldwide and facilitates the exchange of information and theatre experience. Seeks to establish a creative atmosphere for those involved in theatre; promotes public appreciation of artistic creation. Maintains formal associate relations with UNESCO. **PUBLICATIONS:** *NEWS from the Secretariat* (in English and French), 3/year; *World of Theatre* (in English and French), biennial; *World Theatre Directory*, biennial; Newsletter (in English and French), quarterly.

INTERNATIONAL TOBACCO GROWERS ASSOCIATION

Apartado 5, 6001 Castelo Branco, Portugal **PHONE:** 351 72 325901 **FX:** 352 72 325906 **E-MAIL:** tobaccoleaf@dial.pipex.com **WEBSITES:** http://www.tobaccoleaf.org **FOUNDED:** 1984. **OFFICERS:** David J. Walder, Chief Exec. **MEMBERS:** 18. **LANGUAGES:** French, Spanish.

DESCRIPTION: Members are national organizations representing tobacco growers worldwide. Serves as a forum for discussion and exchange of information among members. **PUBLICATIONS:** *Tobacco Briefing*, periodic; *Tobacco Courier*, periodic; *Tobacco in the Developing World Vol. II*; *The Use of Woodfuel for Curing Tobacco*.

INTERNATIONAL TUNGSTEN INDUSTRY ASSOCIATION (ITIA)

Unit 7 Hackford Walk, Hackford Rd., No. 119-123, London SW9 0QT, England **PHONE:** 44 171 5822777 **FX:** 44 171 5820556 **E-MAIL:** itia_imoa@compuserve.com **WEBSITES:** http://www.itia.org.uk/ **FOUNDED:** 1988. **OFFICERS:** Michael Maby, Sec.Gen. **MEMBERS:** 55. **STAFF:** 2. **BUDGET:** £55,000. **LANGUAGES:** English, French.

DESCRIPTION: Organizations and companies engaged in the production, processing, and consumption of tungsten; trading companies; assayers. Promotes cooperation among members in research, production, processing, and use of tungsten. Protects the common interests of members in technical, environmental, and health matters. Collects and disseminates information; compiles statistics. **PUBLICATIONS:** *Newsletter*, semiannual; *''Tungsten''*.

INTERNATIONAL UNION AGAINST CANCER (UICC) (Union Internationale Contre le Cancer)

3, rue du Conseil General, CH-1205 Geneva, Switzerland **PHONE:** 41 22 8091811 **FX:** 41 22 8091810 **E-MAIL:** info@uicc.org **WEBSITES:** http://www.uicc.ch/; http://www.uicc.org **FOUNDED:** 1933. **OFFICERS:** A.J. Turnbull, Exec.Dir. **MEMBERS:** 296. **STAFF:** 19. **BUDGET:** US$4,000,000. **LANGUAGES:** English.

DESCRIPTION: Voluntary cancer leagues and societies, national organizations, private or public cancer research institutions, and ministries of health in 86 countries. Promotes a comprehensive international campaign against cancer. Directs activities in fields of prevention, research, and treatment; sponsors special projects in fields including cervical cancer, head and neck cancer, and exchange of information on unproven methods in cancer treatment. Facilitates training courses for researchers and health professionals; makes available advisory visits by cancer experts. Bestows numerous fellowships and awards. **PUBLICATIONS:** *International Calendar of Meetings on Cancer* (in English), semiannual; *International Journal of Cancer and Predictive Oncology*, 30/year; *UICC International Directory of Cancer Institutes and Organizations*, quadrennial; *UICC News*, quarterly. Also publishes monographs, reports, publication lists, and new book announcements; produces cancer education packs and workshop guidelines.

INTERNATIONAL UNION AGAINST TUBERCULOSIS AND LUNG DISEASE (IUATLD) (Union Internationale Contre la Tuberculose et les Maladies Respiratoires - UICTMR)

68, blvd. St. Michel, F-75006 Paris, France **PHONE:** 33 1 44320360 **FX:** 33 1 43299087 **WEBSITES:** http://www.iuatld.org **FOUNDED:** 1920. **OFFICERS:** Dr. Nils Billo, Exec.Dir. **MEMBERS:** 4,500. **STAFF:** 12. **LANGUAGES:** English, French, Spanish.

DESCRIPTION: National associations, physicians, and other individuals worldwide dedicated to the fight against tuberculosis and other respiratory diseases. Promotes community health and, within its framework, the delivery of preventive, diagnostic, and therapeutic measures against tuberculosis. Aids governmental efforts to improve health services; helps countries formulate, prepare, and execute national tuberculosis control programs. Provides training, advice, antituberculosis drugs and vaccines, medical equipment, and transportation. Encourages community participation. Activities include: dissemination of scientific knowledge; exchange of experience and conferences; operational and applied research; field projects under the Mutual Assistance Programme. Acts as liaison with the World Health Organization and other organizations. **PUBLICATIONS:** *The International Journal of Tuberculosis and Lung Disease* (in English), monthly. With summaries in French and Spanish; *IUATLD Newsletter* (in English, French, and Spanish), 3/year. Contains news and reports on the activities of the IUATLD and information on conferences and courses; *Management of Asthma in Adults* (in English and French). A guide for low-income countries; *Management of the Child with Cough or Difficult Breathing* (in English); *Tuberculosis Guide for Low Income Countries* (in English, French, and Spanish).

INTERNATIONAL UNION OF AIR POLLUTION PREVENTION AND ENVIRONMENTAL PROTECTION ASSOCIATIONS (IUAPPA) (Union Internationale des Associations pour la Prevention de la Pollution de l'Air et la Protection de l'Environnement - UIAPPA)

136 North St., Brighton, E. Sussex BN1 1RG, England **PHONE:** 44 1273 326313 **FX:** 44 1273 735802 **E-MAIL:** cleanair@mistral.co.uk **FOUNDED:** 1965. **OFFICERS:** R. Mills, Dir.Gen. **MEMBERS:** 34. **STAFF:** 3. **BUDGET:** US$50,000. **LANGUAGES:** English, French.

DESCRIPTION: National air pollution prevention and environmental protection associations in 36 countries. Promotes global public education relating to the importance of clean air and methods and results of pollution control. Facilitates the exchange of information and publications; encourages the use of uniform scientific and technical terminology; promotes uniform methods of measurement and monitoring. Acts as liaison with other international and national scientific and technical organizations. Maintains information service. **PUBLICATIONS:** *Clean Air Around the World* (in English), triennial; *IUAPPA Members Handbook*, triennial; *IUAPPA Newsletter* (in English and French), quarterly; *IUAPPA Regional Conference*

Proceedings; *IUAPPA World Congress Proceedings*, triennial.

INTERNATIONAL UNION OF ANTHROPOLOGICAL AND ETHNOLOGICAL SCIENCES (IUAES) (Union Internationale des Sciences Anthropologiques et Ethnologiques - UISAE)

c/o Office of the Sec.Gen., Institute of Cultural and Social Studies, Univ. of Leiden, PO Box 9555, NL-2300 Leiden, Netherlands **PHONE:** 31 71 5273992 **FX:** 31 71 5273619 **E-MAIL:** nas@rulfsw.leidenuniv.nl **FOUNDED:** 1948. **MEMBERS:** 10,000. **STAFF:** 2. **BUDGET:** US$40,000. **LANGUAGES:** English.

DESCRIPTION: National associations, institutes, academies, research councils, professional anthropologists, ethnologists, and interested individuals in 60 countries involved in related fields such as ethnomusicology, folklore, human genetics, sociology, psychology, linguistics, and palaeontology. Purposes are to further anthropological and ethnological studies and to stimulate cooperation among international organizations in the field of human sciences. **PUBLICATIONS:** *IUAES Newsletter* (in English), 3/year.

INTERNATIONAL UNION OF BIOCHEMISTRY AND MOLECULAR BIOLOGY (IUBMB)

18 Leyden Crescent, Saskatoon, Canada S7J 2S4 **PHONE:** (306) 374-1304 **FX:** (306) 955-1314 **E-MAIL:** f.vella@sk.sympatico.ca **WEBSITES:** http://www.iubmb.unibe.ch **FOUNDED:** 1955. **OFFICERS:** Dr. F. Vella, Gen.Sec. **MEMBERS:** 65. **BUDGET:** C$450,000. **LANGUAGES:** English.

DESCRIPTION: National academies, research councils, or biochemical societies; associated bodies represent national biochemical and molecular biology societies; special members are organizations representing industrial and other groups. Promotes international cooperation in the research, discussion, and publication of matters relating to biochemistry and molecular biology. Seeks to: standardize methods, nomenclature, and symbols used in biochemistry and molecular biology; contribute to the advancement of biochemistry and molecular biology; promote high standards; aid biochemists and molecular biologists in developing countries. Supports interest group meetings. **PUBLICATIONS:** *Biochemical Education*, quarterly; *Biochemical Nomenclature and Related Documents*, periodic. Standardizes and codifies the nomenclature of natural products; *Biochemistry and Molecular Biology International*, monthly; *BioFactors*, quarterly; *Biotechnology and Applied Biochemistry*, bimonthly. Covers applications of biochemical research with emphasis on biotechnology; *Trends in Biochemical Sciences*, monthly.

INTERNATIONAL UNION OF BIOLOGICAL SCIENCES (IUBS) (Union Internationale des Sciences Biologiques)

c/o Dr. Talal Younes, 51, blvd. de Montmorency, F-75016 Paris, France **PHONE:** 33 1 45250009 **FX:** 33 1 45252029 **E-MAIL:** iubs@paris7.jussieu.fr **FOUNDED:** 1919. **OFFICERS:** Dr. Talal Younes, Exec.Dir. **MEMBERS:** 117. **STAFF:** 2. **BUDGET:** US$600,000. **LANGUAGES:** French.

DESCRIPTION: National adhering organizations (42) and international scientific associations and commissions (75) engaged in the study of biological sciences. Promotes the study of biological sciences; initiates, facilitates, and coordinates research and other scientific activities that demand international cooperation. Works to ensure the discussion and dissemination of the results of cooperative research; assists in the publication of their reports. **PUBLICATIONS:** *Biology International* (in English), semiannual; *Biology International: Special Edition*, periodic; *Proceedings of IUBS General Assembly*, triennial. Includes directory. Also publishes monograph and methodology series.

INTERNATIONAL UNION FOR CONSERVATION OF NATURE AND NATURAL RESOURCES (IUCN) (Union Internationale pour la Conservation de la Nature et de ses Resources - UICN)

Rue Mauverney 28, CH-1196 Gland, Switzerland **TX:** 419624 IUCN CH **E-MAIL:** mail@hq.iucn.org **WEBSITES:** http://www.iucn.org **FOUNDED:** 1948. **OFFICERS:** David McDowell, Dir.Gen. **MEMBERS:** 913. **STAFF:** 130. **BUDGET:** 65,000,000 SFr. **LANGUAGES:** English, French, Spanish.

DESCRIPTION: National governments, agencies, and nongovernmental organizations in 138 countries. Initiates and promotes actions ensuring the ''perpetuation of man's natural environment.'' Seeks to maintain genetic diversity, ensure sustainable use of renewable resources and maintain essential ecological processes. Continuously reviews and assesses world environmental problems and management policies; promotes research contributing to their solution and effectiveness. Maintains a network of over 8000 volunteer experts; promotes effective management; focuses public attention on protected area issues; provides advisory services. Maintains close working relations with the United Nations system; cooperates with Council of Europe, Organization of African Unity, Organization of American States, and other intergovernmental bodies; collaborates with International Council for Bird Preservation, International Council of Scientific Unions, U.S. World Wildlife Fund, and other nongovernmental groups. Disseminates information; cosponsors World Conservation Monitoring Centre in Kew and Cambridge, England (with United Nations Environment Programme and World Wide Fund for Nature,); sponsors IUCN Environmental Law Centre in Bonn, Germany. **PUBLICATIONS:** *Caring for the Earth*; *Interact*, 3/year; *IUCN Bulletin*, quarterly; *Nature Herald*, periodic; *Protected Areas Systems Reviews*, periodic; *Red Data Books*; *U.N. List of National Parks and Protected Areas*; *World Conservation Strategy*. Also publishes reports, books, and monographs.

INTERNATIONAL UNION OF CRYSTALLOGRAPHY (IUCr) (Union Internationale de Cristallographie - UIC)

2 Abbey Sq., Chester, Cheshire CH1 2HU, England **PHONE:** 44 1244 345431 **FX:** 44 1244 344843 **E-MAIL:** execsec@iucr.org **WEBSITES:** http://www.iucr.org **FOUNDED:** 1948. **OFFICERS:** Mr. M.H. Dacombe, Exec.Sec. **MEMBERS:** 40. **STAFF:** 21. **BUDGET:** US$3,000,000. **LANGUAGES:** English, French, German, Russian.

DESCRIPTION: National academies, national research councils, scientific societies, and similar bodies in 40 countries. Objectives are to: promote international cooperation in crystallography (the study of crystal form, structure, and modes of aggregation); contribute to all aspects of its advancement including related topics concerning the noncrystalline states; facilitate international standardization of methods used such as units, nomenclature, and symbols; form a focus for the relations of crystallography to other sciences. Organizes training schools on various aspects of crystallography. Maintains 18 commissions. **PUBLICATIONS:** *Acta Crystallographica Section A* (in English, French, German, and Russian), bimonthly. Fundamentals of crystallography; *Acta Crystallographica Section B*, bimonthly; *Acta Crystallographica Section C*, monthly; *Acta Crystallographica Section D*, monthly; *International Tables for Crystallography*; *Journal of Applied Crystallography*, bimonthly; *Synchrotron Radiation*, bimonthly; *World Directory of Crystallographers*.

INTERNATIONAL UNION FOR ELECTRICITY APPLICATIONS (UIE) (Union Internationale pour la application d l'indtricate - UIE)

c/o G. Vanderschueren, CNIT - Espace ELEC, BP10 2, place de la Defense, F-92053 Paris Cedex 06, France **PHONE:** 33 1 41265648 **FX:** 33 1 41265649 **E-MAIL:** uie@uie.org **WEBSITES:** http://www.uie.org **FOUNDED:** 1953. **OFFICERS:** Ronnie Belmans, Pres. **MEMBERS:** 22. **STAFF:** 1. **LANGUAGES:** English, French.

DESCRIPTION: National committees (12); organizations, companies, and individuals (10) in the field of the electricity applications. Studies problems related to electricity used in large and small industries in the production, transformation, and treatment of all types of products, and in space heating and air conditioning. Strives to increase user interest in electric heating; provides information on new techniques. Works to adapt the technical and financial conditions of energy supply to industrial situations. **PUBLICATIONS:** *Activity Report* (in English and French), annual. Electronic information; *Congress Proceedings*, quadrennial; *News Flash*, quarterly. Also publishes technical reports.

INTERNATIONAL UNION OF FAMILY ORGANIZATIONS (IUFO) (Union Internationale des Organismes Familiaux - UIOF)

28, place St.-Geroges, F-75009 Paris, France **PHONE:** 33 1 48780759 **FX:** 33 1 42829524 **FOUNDED:** 1947. **OFFICERS:** Dr. Deisi Noel Weber Kusztra, Pres. **MEMBERS:** 375. **STAFF:** 24. **BUDGET:** 14,000,000 Fr. **LANGUAGES:** Arabic, English, French.

DESCRIPTION: Members are: family associations;

child welfare, parent-teacher, and women's organizations; marriage counseling and guidance organizations; social work councils and associations; research centers of demographic social and family problems; social security offices; national insurance departments; institutes concerned with housing problems; public institutes for population and family problems; ministries and ministerial departments. Purposes are to: bring together regardless of race or creed all organizations, both public and private, that are working for the well-being of the family; cooperate in national and international research undertaken on family problems; gather and disseminate information on family problems; make known to the global public and international organizations the fundamental needs, rights, and requirements of the family; develop and establish among all families in the world an atmosphere favorable to the establishment of lasting peace. Conducts research and educational programs on family housing, marriage, parent-teacher relations, family allowances and benefits, and family policies. **PUBLICATIONS:** *Families of the Work.* Information from international and regional conferences and current events in family policies; *IUFO Members' Handbook* (in English and French), annual. Guide to worldwide family movements; *World's Families*; Monographs. Different documents published by specialised commissions.

INTERNATIONAL UNION OF FOOD, AGRICULTURAL, HOTEL, RESTAURANT, CATERING, TOBACCO, AND ALLIED WORKERS' ASSOCIATION (IUF) (Internationale Union der Lebensmittel-, Landwirschafts-, Hotel-, Restaurant-, Cafe- und Genussmittelarbeiter-Gewerkschaften)

8, rampe du Pont-Rouge, Petit-Lancy, CH-1213 Geneva, Switzerland **PHONE:** 41 22 7932233 **FX:** 41 22 7932238 **E-MAIL:** iuf@iuf.org **FOUNDED:** 1920. **OFFICERS:** Ron Oswald, Gen.Sec. **MEMBERS:** 2,600,000. **STAFF:** 18. **BUDGET:** 6,000,000 SFr. **LANGUAGES:** English, French, German, Spanish, Swedish.

DESCRIPTION: Unions in 118 countries representing 2,600,000 workers in agriculture, plantation, food, and tobacco processing, beverage, hotel, and catering industries. Works to advance members' interests through international coordination. Maintains forum for exchange of ideas and experiences between member unions. Monitors activities of companies in the food and allied industries. Sponsors educational programs; disseminates information on collective bargaining, occupational health and safety, and employment, trade and technology matters. **PUBLICATIONS:** *IUF News Bulletin* (in English, French, German, Spanish, and Swedish), bimonthly; *Labour Issues*, periodic.

INTERNATIONAL UNION OF FORESTRY RESEARCH ORGANIZATIONS (IUFRO) (Internationaler Verband Forstlicher Forschungsanstalten)

Seckendorff-Gudentweg 8, Tirolergarten, A-1131 Vienna, Austria **PHONE:** 43 1 8770151 **FX:** 43 1 8779355 **E-MAIL:** iufro@forvie.ac.at **WEBSITES:** http://gis.umn.edu/iufro/ **FOUNDED:** 1892. **OFFICERS:** H. Schmutzenhofer, Sec. **MEMBERS:** 793. **STAFF:** 10. **LANGUAGES:** English, French, German, Spanish.

DESCRIPTION: Universities and research institutions representing 15,000 scientists in 115 countries involved in the study of forestry, forest products, and related disciplines. Works to standardize techniques of research and systems of measurement in studies regarding all facets of forestry and forest products. **PUBLICATIONS:** *Annual Report* (in English, French, German, and Spanish); *Informaciones de IUFRO* (in Spanish), periodic; *IUFRO News* (in English), quarterly; *IUFRO World Series*, annual; *Proceedings of Congresses and Meetings*, periodic. Also publishes an occasional paper.

INTERNATIONAL UNION OF FRENCH-LANGUAGE JOURNALISTS AND PRESS (IUFLJP) (Union Internationale des Journalistes et de la Presse de Langue Francaise - UIJPLF)

3, cite Bergere, F-75009 Paris, France **PHONE:** 33 1 47700280 **FX:** 33 1 48242632 **TX:** 250303 PARIS **FOUNDED:** 1950. **OFFICERS:** Georges Gros, Sec.Gen. **MEMBERS:** 2,000. **STAFF:** 5. **LANGUAGES:** French.

DESCRIPTION: Professionals involved in written as well as audiovisual journalism and publishing in the French language. Supports the conservation of the French language; facilitates international communication between editors, directors, and journalists. Works to develop and coordinate liaisons between publications and radio and television stations throughout the world. Fosters universal coordination in technical and informational matters and professional training. Conducts surveys on written and audiovisual French-language press. Maintains Pressotheque de Langue Francaise, a center that regularly receives 1500 French-language periodicals from 70 countries with the objective of recognizing French-language publications published outside of France. **PUBLICATIONS:** *La Gazette de la Presse Francophone* (in French), bimonthly.

INTERNATIONAL UNION OF GEODESY AND GEOPHYSICS (IUGG)

c/o Dr. Georges Balmino, Bur. Gravimetrique Intl., 18, ave. Edouard Belin, F-31401 Toulouse Cedex 4, France **PHONE:** 33 561 332980, 33 561 332889 **FX:** 33 561 253098 **E-MAIL:** balmino@pontos.cst.cnes.fr **WEBSITES:** http://www.obs-mip.fr/uggi/; http://www.obs-mip.fr/uggi/ **FOUNDED:** 1919. **OFFICERS:** Dr. Georges Balmino, Sec.Gen. **MEMBERS:** 75. **LANGUAGES:** English, French.

DESCRIPTION: One of 20 scientific unions in the International Council of Scientific Unions. Members are countries represented in the Union's 7 semi-autonomous constituent Associations: International Association of Geodesy; International Association of Geomagnetism and Aeronomy; International Association of Hydrological Sciences; International Association of Meterology and Atmospheric Sciences; International Association for the Physical Sciences of the Ocean; International Association of Seismology and Physics of the Earth's Interior; International Association of Volcanology and Chemistry of the Earth's Interior. Promotes and coordinates studies carried out by member Associations, in an effort to utilize scientific knowledge in the service of society. Data gathered during such efforts is available through the World Data Centre system and the Federation of Astronomical and Geophysical Data Analysis Service, established in conjunction with ICSU. Offers financial aid to younger scientists so that they are able to attend symposia. Participates via ICSU in Inter-Union Commissions on such subjects as Antarctic research and space research. Maintains Inter-Union Committee for Developing Countries. Cooperates with United Nations Educational, Scientific and Cultural Organization in the study of natural catastrophes. **PUBLICATIONS:** *Compte Rendu of the General Assembly* (in English), quadrennial; *IUGG Yearbook* (in English and French), annual.

INTERNATIONAL UNION OF GEOLOGICAL SCIENCES (IUGS) (Union Internationale des Sciences Geologiques)

c/o Geological Survey of Norway, Leiv Erikssons Vei 39, PO Box 3006 - Lade, N-7002 Trondheim, Norway **PHONE:** 47 73 904040 **FX:** 47 73 502230 **E-MAIL:** iugs.secretariat@ngu.no **WEBSITES:** http://www.iugs.org **FOUNDED:** 1961. **OFFICERS:** Prof. A. Boriani, Sec.Gen. **MEMBERS:** 114. **STAFF:** 2. **LANGUAGES:** English.

DESCRIPTION: Representatives of national committees for geological science. Seeks to: facilitate and encourage the study of problems pertaining to geological sciences; enhance international cooperation in geology and related sciences; provide continuity in international cooperation. **PUBLICATIONS:** *Episodes* (in English), quarterly; *International Union of Geological Sciences Directory*, periodic.

INTERNATIONAL UNION OF GOSPEL MISSIONS (IUGM)

1045 Swift, North Kansas City, MO 64116 **PHONE:** (816) 471-8020 **FX:** (816) 471-3718 **E-MAIL:** iugm@iugm.org **WEBSITES:** http://www.iugm.org **FOUNDED:** 1913. **OFFICERS:** Rev. Stephen E. Burger, Exec.Dir. **MEMBERS:** 1,000. **STAFF:** 10. **BUDGET:** US$1,200,000. **LANGUAGES:** English.

DESCRIPTION: Rescue ministry executives and staff, and concerned individuals in 6 countries. Promotes rescue mission work for all persons experiencing crisis. Missions sponsor: coffee-houses for youths; emergency shelters for men and women, women with children, and families; day camps, resident camps, and wilderness camps for inner-city children; cafeterias for low-income persons. Missions also serve meals and provide sleeping space to individuals in need. Operates Disaster Aid Association, which provides assistance and services to members in times of disaster or crisis. Maintains speakers' bureau; offers placement service; compiles statistics. **PUBLICATIONS:** *How to Have a Better Board of Directors*; *IUGM Directory*, biennial; *Membership and Resource Directory*; *RESCUE*, bimonthly. Includes instructional articles for rescue ministry personnel; *Rescue Happenings*, monthly. Covers IUGM activities, includes calendar of events.; *Rescue Mission Salary Survey*; *Sample Staff Policy Manual*; Surveys.

INTERNATIONAL UNION FOR HEALTH PROMOTION AND EDUCATION (IUHPE) (Union Internationale de Promotion de la Sante et d'Education pour la Sante - UIPES)

2, rue Auguste Comte, F-92170 Vanves, France **PHONE:** 33 1 46450059 **FX:** 33 1 46450045 **E-MAIL:** iuhpemcl@worldnet.fr **FOUNDED:** 1951. **OFFICERS:** Marie-Claude Lamarre, Exec.Dir. **MEMBERS:** 2,000. **STAFF:** 5. **LANGUAGES:** English, French, Spanish.

DESCRIPTION: Global association of people and organizations working in the fields of health promotion and health education, dedicated to the promotion of world health through education, community action and the development of healthy public policies. Unites people from many sectors to address policy, program and practice issues. Provides an interdisciplinary forum for health promotion and health education professionals from around the world to share knowledge, experience and views. **PUBLICATIONS:** *Conference Proceedings*, triennial; *Promotion and Education* (in English, French, and Spanish), quarterly. Contains articles on the theory and practice of health promotion and health education as well as international news about major events; Brochure, periodic; Monographs, periodic. Focuses on research and health promotion and education; Newsletters, periodic.

INTERNATIONAL UNION OF THE HISTORY AND PHILOSOPHY OF SCIENCE (IUHPS) (Union Internationale d'Histoire et de Philosophie des Sciences)

c/o Sec.Gen., Dept. of Philosophy, Helen C. White Hall, Rm. 5185, Univ. of Wisconsin, Madison, WI 53706 **PHONE:** (608) 263-3700 **FX:** (608) 265-3701 **FOUNDED:** 1956. **OFFICERS:** E. Sober, Sec.Gen. **LANGUAGES:** English, French.

DESCRIPTION: Members are scholars from 36 countries. Promotes research and international congresses on logic and the history and philosophy of science. **PUBLICATIONS:** *Bulletin*, annual; *Newsletter*, 3/year.

INTERNATIONAL UNION FOR HOUSING FINANCE

111 E. Wacker Dr., Ste. 900, Chicago, IL 60601-4389 **PHONE:** (312) 946-8201, (312) 946-8200 **FX:** (312) 946-8202 **TX:** 24538 BSA G **WEBSITES:** http://www.housingfinance.org **FOUNDED:** 1914. **OFFICERS:** Donald R. Holton, Sec.Gen. **MEMBERS:** 350. **STAFF:** 2. **LANGUAGES:** English, French, German, Spanish.

DESCRIPTION: Private and publicly owned thrift- and home-financing institutions in 60 countries; building societies and savings and loan associations at all levels; government agencies concerned with housing and home finance; interested individuals. Disseminates information on housing finance policies and techniques worldwide. Encourages comparative study of methods and practices and the establishment of suitable educational bodies; organizes 2-week international training courses; conducts seminars, conferences, and study groups. **PUBLICATIONS:** *Housing Finance International* (in English), quarterly; *Secondary Mortgage Markets International Perspectives*; Newsletter, 2-3/year.

INTERNATIONAL UNION OF IMMUNOLOGICAL SOCIETIES (IUIS)

Lister Research Laboratories, Univ. Department of Surgery, Royal Infirmary, Laurston Pl., Edinburgh EH3 9YW, Scotland **PHONE:** 44 131 5363831 **FX:** 44 131 6676190 **FOUNDED:** 1969. **OFFICERS:** Keith James, Sec.Gen. **MEMBERS:** 51. **BUDGET:** US$100,000. **LANGUAGES:** English.

DESCRIPTION: National professional societies of basic and applied immunologists. Encourages the orderly development and utilization of the science of immunology. Promotes the application of new developments to clinical and veterinary problems and standardizes reagents and nomenclature. Conducts educational symposia and scientific meetings. **PUBLICATIONS:** *The Immunologist*, bimonthly.

INTERNATIONAL UNION FOR INLAND NAVIGATION (IUIN) (Union Internationale de la Navigation Fluviale - UINF)

7, quai du General Koenig, F-67085 Strasbourg Cedex, France **PHONE:** 33 88 362844 **FX:** 33 88

370482 **FOUNDED:** 1952. **OFFICERS:** M. Ruscher. **MEMBERS:** 9. **BUDGET:** 65,000 SFr. **LANGUAGES:** French, German.

DESCRIPTION: Members are national organizations of inland waterways carriers. Furthers research activities on inland navigation. Represents industry interests. **PUBLICATIONS:** *Yearly Report* (in French and German).

INTERNATIONAL UNION OF LOCAL AUTHORITIES (IULA) (Union Internationale des Villes et Pouvoirs Locaux)

Postbus 90646, NL-2509 LP The Hague, Netherlands **PHONE:** 31 70 3066066 **FX:** 31 70 3500496 **E-MAIL:** iula@iula-hq.nl **WEBSITES:** http://www.iula.org **FOUNDED:** 1913. **OFFICERS:** Jacques Jobin, Dir. **MEMBERS:** 400. **STAFF:** 7. **BUDGET:** 1,000,000 f. **LANGUAGES:** English, French, German, Spanish.

DESCRIPTION: Organization consists of: national unions of cities, counties, and other local authorities; ministries of local and regional governments; institutes of public administration; municipal credit institutions; individuals in more than 100 countries of Africa, the Americas, Asia, Europe, the Mediterranean, and the Middle East, the Arab World and the Pacific. Works to promote the cause of local development, raise standards of local administration, and foster individual participation in civic affairs. Supplies information on local affairs; stimulates town affiliations and sister-city arrangements; has official consultative status with organizations of the United Nations. **PUBLICATIONS:** Newsletter (in English).

INTERNATIONAL UNION OF MICROBIOLOGICAL SOCIETIES (IUMS) (Union Internationale des Societes de Microbiologie)

c/o Dr. Marc H. Van Regenmortel, IBMC, IUMS, IBMC/CNRS, Laboratoire d'immunochimie, 15 Rue Descartes, F-67000 Strasbourg, France **PHONE:** 33 3 88417022 **FX:** 33 3 88610680 **E-MAIL:** vanregen@ibmcu-strasbg.fr **FOUNDED:** 1930. **OFFICERS:** Dr. Marc H. van Regenmortel, Sec.Gen. **MEMBERS:** 106. **BUDGET:** US$150,000.

DESCRIPTION: National microbiological societies in 62 countries representing 100,000 microbiologists united to maintain contact with microbiological societies throughout the world. **PUBLICATIONS:** *Archives of Virology*, monthly; *International Journal of Food Microbiology*, bimonthly; *International Journal of Systematic Bacteriology*, quarterly; *IUMS Directory*, quadrennial; *Journal of Biological Standardization*, quarterly; *World Journal of Microbiology and Biotechnology*, bimonthly.

INTERNATIONAL UNION OF NUTRITIONAL SCIENCES (IUNS)

c/o Prof. Osman M. Galal, UCLA School of Public Health, Intl. Health Program, 10833 Le Conte Ave., Los Angeles, CA 95172 **PHONE:** (310) 206-9639, (310) 206-8444 **FX:** (310) 794-1805 **E-MAIL:** ogalal@ucla.edu **WEBSITES:** http://www.monash.edu.au/IUNS/ **FOUNDED:** 1946. **OFFICERS:** Prof. O.M. Galal. **MEMBERS:** 65. **STAFF:** 14. **LANGUAGES:** English.

DESCRIPTION: Organization of national nutritional societies. Promotes international cooperation in the scientific study of nutrition and its applications. Encourages research and the exchange of scientific information. Cooperates with the Food and Agriculture Organization of the United Nations, the United Nations Educational, Scientific and Cultural Organization, and the World Health Organization. Maintains 24 committees. **PUBLICATIONS:** *Annual Report*; *Directory*, quadrennial; *Newsletter*, 1-2/year.

INTERNATIONAL UNION OF PHARMACOLOGY (IUPHAR)

c/o Prof. W.C. Bowman, University of Strathclyde, Dept. of Physiology & Pharmacology, 204 George St., Glasgow G1 1XW, Scotland **PHONE:** 44 141 5524400, 44 1357 300622 **FX:** 44 141 5522562 **E-MAIL:** w.c.bowman@strath.ac.uk **FOUNDED:** 1966. **MEMBERS:** 52. **BUDGET:** US$60,000. **LANGUAGES:** English.

DESCRIPTION: National and international societies in pharmacology and related disciplines representing approximately 30,000 individuals. Purpose is to promote cooperation between pharmacological societies and encourage free international exchange of ideas and research. Acts as a forum for participation between related scientific bodies.

Works to standardize the use of drugs worldwide and rationally define the receptors and ion channels on which they act. **PUBLICATIONS:** *Congress Proceedings*, quadrennial; *Directory of IUPHAR*, annual; *IUPHAR Newsletter*, semiannual.

INTERNATIONAL UNION OF PHYSIOLOGICAL SCIENCES (IUPS)

c/o Secretariat, Batiment Cervi, Hopital de la Pitie-Salpetriere, 83, blvd. de l'Hopital, F-75651 Paris Cedex 13, France **PHONE:** 33 1 42177537 **FX:** 33 1 42177575 **E-MAIL:** suorsoni@infobiogen.fr **FOUNDED:** 1953. **OFFICERS:** Susan Orsoni, Exec.Sec. **MEMBERS:** 52. **STAFF:** 1. **LANGUAGES:** English.

DESCRIPTION: Physiological societies united to exchange scientific information. Coordinates research and educational programs. **PUBLICATIONS:** *News in Physiological Sciences* (in English), quarterly; *World Directory of Physiologists* (in English), periodic.

INTERNATIONAL UNION OF POLICE SYNDICATES (Union Internationale des Syndicats de Police - UISP)

617 rue de Neudorf, L-2220 Luxembourg, Luxembourg **PHONE:** 352 434961 **FX:** 352 435182 **FOUNDED:** 1927. **OFFICERS:** Hermann Lutz, Sec.Gen. **MEMBERS:** 18. **STAFF:** 2. **LANGUAGES:** English, French, German.

DESCRIPTION: National and regional organizations representing nearly 500,000 individuals. Acts as clearinghouse and seeks communal resolutions to common police-related problems. Promotes association among police officials; advocates crime prevention and international cooperation regarding economic and environmental crimes; seeks to include women in police service. Opposes human rights violations, racial discrimination, and apartheid. Collects and disseminates information regarding international humanitarian law and other matters of police service. **PUBLICATIONS:** *Report*, triennial.

INTERNATIONAL UNION OF PREHISTORIC AND PROTOHISTORIC SCIENCES (IUPPS) (Union Internationale des Sciences Pre- et Protohistoriques)

Dept. of Archaeology, Univ. of Ghent, Blandijnberg 2, B-9000 Gent, Belgium **PHONE:** 32 92 644111, 32 92 2223692 **FX:** 32 92 2644173 **E-MAIL:** uispp@ping.be **FOUNDED:** 1931. **OFFICERS:** Prof. J. Bourgeois, Sec.Gen. **MEMBERS:** 560. **LANGUAGES:** English, French.

DESCRIPTION: Professional archaeologists from 140 countries. Purpose is to promote research in prehistoric and protohistoric archaeology. Maintains working commissions and 31 committees to study all periods ranging from the earliest to the early medieval era. **PUBLICATIONS:** *Atlas of African Prehistory*; *Corpus of Prehistoric Amber*; *Encyclopedia Berbere*; *Inventaria Archaeologica*; *Prahistorische Bronzcfunde*.

INTERNATIONAL UNION OF PSYCHOLOGICAL SCIENCE (IUPsyS)

School of Psychology, Univ. of Ottawa, Ottawa, ON, Canada K1N 6N5 **PHONE:** (613) 562-5800 **FX:** (613) 562-5169 **E-MAIL:** pritchie@uottawa.ca **WEBSITES:** http://www.faseb.org/iups **FOUNDED:** 1951. **OFFICERS:** Piare Ritchie, Sec.Gen. **MEMBERS:** 51. **LANGUAGES:** English, French.

DESCRIPTION: National psychological societies and committees united for: the development of psychological science through exchange of ideas and scientific information; the exchange of scholars and students; the organization of international congresses of psychology; publication and documentation; cooperation with related scientific groups interested in the development of psychology as a discipline and a profession. Conducts educational programs and international research network project. **PUBLICATIONS:** *Congress Proceedings*, quadrennial; *International Directory of Psychologists*, quinquennial; *International Journal of Psychology*, bimonthly; *Monograph*, periodic.

INTERNATIONAL UNION FOR PURE AND APPLIED BIOPHYSICS (IUPAB) (Organisation Internationale de Biophysique Pure et Appliquee)

Dept. of Biochemistry and Molecular Biology, Univ. of Leeds, Leeds LS2 9JT, England **PHONE:** 44 113 2333023 **FX:** 44 113 2333167 **E-MAIL:** actn@viovax.leeds.ac.uk **WEBSITES:** http://bmbsgi11.leeds.ac.uk/iupab/ **FOUNDED:** 1966. **OFFICERS:** Prof. A.C.T. North, Sec.Gen.

MEMBERS: 43. **STAFF:** 2. **BUDGET:** US$100,000. **LANGUAGES:** English.

DESCRIPTION: National committees appointed by academies and research councils representing 45 countries. Purposes are to: organize international cooperation in biophysics and promote communication between branches of biophysics and allied disciplines; encourage cooperation between the societies that represent the interests of biophysics; contribute to the advancement of biophysics. **PUBLICATIONS:** *IUPAB Biophysics Series*, periodic; *IUPAB News*, periodic; *Quarterly Review of Biophysics.*

INTERNATIONAL UNION OF PURE AND APPLIED CHEMISTRY (IUPAC)

c/o Dr. John W. Jost, PO Box 13757, Research Triangle Park, NC 27709-3757 **PHONE:** (919) 485-8700 **FX:** (919) 485-8706 **E-MAIL:** secretariat@iupac.org **WEBSITES:** http://www.iupac.org **FOUNDED:** 1919. **OFFICERS:** Dr. J.W. Jost, Exec.Dir. **MEMBERS:** 45. **STAFF:** 5. **BUDGET:** US$934,000. **LANGUAGES:** English.

DESCRIPTION: National organizations united to investigate and make recommendations for action on chemical matters of international importance that need regulation, standardization, or codification. Cooperates with other international organizations that deal with topics of a chemical nature; promotes continuing cooperation among the chemists of the member countries. Contributes to the advancement of chemistry in all aspects. **PUBLICATIONS:** *Chemistry International*, bimonthly; *Pure and Applied Chemistry*, monthly.

INTERNATIONAL UNION FOR QUATERNARY RESEARCH (INQUA) (Union Internationale pour l'Etude du Quarternaire)

c/o Prof. Sylvi Haldorsen, PO Box 5028, N-1432 Aas, Norway **PHONE:** 47 64948252 **FX:** 47 64947485 **E-MAIL:** sylvi.haldorsen@ijvf.nlh.no **WEBSITES:** http://inqua.nlh.no **FOUNDED:** 1928. **OFFICERS:** Prof. Edward Derbyshire, Sec. **MEMBERS:** 37. **LANGUAGES:** English, French, German, Russian, Spanish.

DESCRIPTION: National academies. Encourages interdisciplinary study of problems dealing with the Quaternary Period as defined in geologic stratigraphy. Maintains 16 commissions. Conducts research programs. **PUBLICATIONS:** *Congress Proceedings*, quadrennial; *Quaternary International*, 8/year; *Quaternary Perspective*, semiannual; Handbook, periodic; Reports, periodic.

INTERNATIONAL UNION OF RADIO SCIENCE (IURS) (Union Radio Scientifique Internationale - URSI)

Sint-Pietersnieuwstraat 41, B-9000 Gent, Belgium **PHONE:** 32 9 2643320 **FX:** 32 9 2644288 **E-MAIL:** heleu@intec.rug.ac.be **FOUNDED:** 1919. **MEMBERS:** 49. **LANGUAGES:** English, French, German.

DESCRIPTION: National committees of scientists and research and scientific organizations with an interest in radiotelegraphy, telecommunications, and electronics. Promotes research, study, and advancement in the fields of telecommunications and electronics. Facilitates exchange of information among researchers and scientists working in telecommunications and related fields; conducts research programs; encourages international standardization of electronics and communications equipment and terminology. **PUBLICATIONS:** *Modern Radio Science*, triennial; *Radio Science*, bimonthly; *Radio Science Bulletin*, quarterly; *Review of Radio Science*, triennial.

INTERNATIONAL UNION OF RADIOECOLOGY (Union Internationale de Radioecologie - UIR)

Rue Defacqz, 1, Bte 1, B-1050 Brussels, Belgium **PHONE:** 32 14 300318 **FX:** 32 14 301278 **E-MAIL:** 76403.3162@compuserve.com **FOUNDED:** 1979. **OFFICERS:** Peter Coughtrey, Gen.Sec. **MEMBERS:** 500. **STAFF:** 15. **BUDGET:** 1,500,000 BFr. **LANGUAGES:** English, French.

DESCRIPTION: Radioecologists in 35 countries studying the environmental implications of the peaceful use of nuclear energy. Radioecology is the scientific discipline analyzing the effects of radioactivity in the environment on living organisms. Promotes cooperation and information exchange among members; defends the independence and ethical orientation of radioecology as a profession before authorities. Conducts workshops and seminars. Offers professional training to young scientists. **PUBLICATIONS:** *Newsletter* (in English and

French), quarterly; *Proceedings of Workshops* (in English), periodic.

INTERNATIONAL UNION OF RAILWAYS (UIC) (Union Internationale des Chemins de Fer - UIC)

c/o Adam Wieladek, Chm., 16, rue Jean Rey, F-75015 Paris, France **PHONE:** 33 1 44492020 **FX:** 33 1 44492029 **TX:** 270835 UNINFER F **E-MAIL:** communication@uic.asso.fr **WEBSITES:** http://www.uic.asso.fr **FOUNDED:** 1922. **OFFICERS:** Paul Veron, Communications Dir. **MEMBERS:** 137. **STAFF:** 130. **LANGUAGES:** English, French, German.

DESCRIPTION: Organization consists of: railways situated in Europe, Africa, Asia, North and South America, and Australia; metro companies, shipping operators, and like entities. Seeks to develop international railway traffic through technical interoperability. Studies problems common to both railway and related industries. Promotes standardization of technology and conducts technical research and development programs for member railways. Organizes colloquia and congresses. Compiles statistics. **PUBLICATIONS:** *International Railway Statistics* (in English, French, and German), annual; *Rail International* (in English, French, and German), monthly; *Rail Lexic*. Contains international railway terminology; *Railway Terminology* (in Dutch, English, Esperanto, French, German, Hungarian, Italian, Polish, Portuguese, Spanish, and Swedish), periodic; *UIC Activities* (in English, French, and German); *UIC Panorama*, quarterly; *Vademecum Guide to Members*, annual; Reports; Reports.

INTERNATIONAL UNION FOR THE SCIENTIFIC STUDY OF POPULATION (IUSSP) (Union Internationale pour l'Etude Scientifique de la Population)

34, rue des Augustins, B-4000 Liege, Belgium **PHONE:** 32 42 224080 **FX:** 32 42 223847 **TX:** 42648 POP UN **E-MAIL:** fdevpop1@um1.ulg.ac.be **FOUNDED:** 1928. **OFFICERS:** Jane Verrall, Exec.Sec. **MEMBERS:** 1,815. **STAFF:** 11. **BUDGET:** US$2,000,000. **LANGUAGES:** English, French.

DESCRIPTION: Demographers, economists, sociologists, physicians, public health officials, public administrators, statisticians, and other individuals in 127 countries engaged in the statistical study of populations, the effect of births, marriages, mortality, and health upon populations, and the relationship between these population phenomena and other economic, social, and biological phenomena. Stimulates interest in demography among governments, national and international organizations, scientific bodies, and the public. Organizes conferences and seminars. Sponsors training workshop for African demographers. **PUBLICATIONS:** *Conference Proceedings* (in English and French), quadrennial; *Directory of Members*, periodic; *International Bibliography of Historical Demography*, annual; *IUSSP Newsletter* (in English and French), 3/year; *IUSSP Papers* (in English and French), periodic. Also publishes *Multilingual Demographic Dictionary* and other books.

INTERNATIONAL UNION OF SOIL SCIENCE (IUSS) (Union Internationale de la Science du Sol - UISS)

Universitaet fuer Bodenkultur, Institut fuer Bodenforschung, Gregor-Mendel-Str. 33, A-1180 Vienna, Austria **PHONE:** 43 1 4789107 **FX:** 43 1 4789110 **TX:** 111010 TZSTA **E-MAIL:** iuss@edv1.boku.ac.at **WEBSITES:** http://www.cirad.fr//isss.html **FOUNDED:** 1924. **OFFICERS:** Dr. W.E.H. Blum, Sec.Gen. **MEMBERS:** 45,000. **STAFF:** 1. **BUDGET:** US$50,000. **LANGUAGES:** English, French, German.

DESCRIPTION: Non-profit, non-governmental, scientific society with individuals from 163 countries engaged in the study and application of soil science. Promotes contacts among scientists; stimulates scientific research to further the application of soil research. **PUBLICATIONS:** *Bulletin of the ISSS* (in English, French, German, and Spanish), semiannual. Contains news about the society and its activities, reports of meetings, news from National and Regionaal Societies, and upcoming events.; *Membership List*, quadrennial; *Proceedings of the World Soil Conference*, quadrennial.

INTERNATIONAL UNION OF SUPERIORS GENERAL (IUSG) (Union Internationale des Superieures Generales - UISG)

Piazza di Ponte San Angelo 28, I-00186 Rome, Italy **PHONE:** 39 6 6840020 **FX:** 39 6 68400239 **FOUNDED:** 1965. **OFFICERS:** Sr. Marguerite

Letourneau, Sec.Gen. **MEMBERS:** 2,000. **STAFF:** 10. **LANGUAGES:** Dutch, English, French, German, Italian, Spanish.

DESCRIPTION: Women in 105 countries who are superiors general of apostolic religious institutes or societies that are pontifical or diocesan. Objectives are to: foster, at the international level, the life and mission of religious women in the church; develop communications with other church organizations and arrange representation before international bodies. Conducts research programs; sponsors seminars, workshops, and special subject conferences. Maintains task forces and working groups. **PUBLICATIONS:** *Bulletin UISG* (in Dutch, English, French, German, Italian, and Spanish), 3/year.

INTERNATIONAL UNION OF TECHNICAL ASSOCIATIONS AND ORGANIZATIONS (UATI) (Union Internationale des Associations et Organismes Techniques)

1, rue Miollis, Maison de l'UNESCO, F-75732 Paris Cedex 15, France **PHONE:** 33 1 45682770 **FX:** 33 1 43062927 **TX:** 204461 PARIS **E-MAIL:** uati@unesco.org **WEBSITES:** http://www.unesco.org/uati **FOUNDED:** 1950. **OFFICERS:** Noel Murati, Sec.Gen. Associate. **MEMBERS:** 30. **STAFF:** 5. **LANGUAGES:** English, French.

DESCRIPTION: Large, international, nongovernmental engineering organizations working in such areas as energy, transportation, water management, rural development, civil engineering, industrial technology, and the environment. Works to identify common interests of members and to promote, coordinate, and facilitate member action within those interests. Maintains consultative status to ECOSOC, UNESCO, and UNIDO. **PUBLICATIONS:** *Directory of the National Committees and Representatives of the Union's Member Associations*, quadrennial; *UATI Bulletin* (in English and French), semiannual; *Proceedings*, biennial; *Reports*, periodic.

INTERNATIONAL UNION OF TENANTS (IUT)

Box 7514, S-103 92 Stockholm, Sweden **PHONE:** 46 8 7910250 **FX:** 46 8 204344 **E-MAIL:** nic.nilsson@hyresgasterna.se **FOUNDED:** 1955. **OFFICERS:** Nic Nilsson, Gen.Sec. **MEMBERS:** 40. **STAFF:** 3. **BUDGET:** US$150,000. **LANGUAGES:** English.

DESCRIPTION: Members are national tenants' associations. Seeks social housing policies and legal protection of occupation rights for all tenants. Acts as medium for cooperation among members. Maintains speakers' bureau; compiles statistics. **PUBLICATIONS:** *Global Tenant* (in English), quarterly; *Tenants Charter*; *Tenants Democracy*; *Tenants Forum*.

INTERNATIONAL UNION OF TESTING AND RESEARCH LABORATORIES FOR MATERIALS AND STRUCTURES (RILEM) (Reunion Internationale des Laboratoires d'Essais et de Recherches sur les Materiaux et les Constructions)

Rilem Secretariat General, ENS-Pavillon des Jardins, 61 avenue de President Wilson, F-94235 Cachan Cedex, France **PHONE:** 33 1 47402397 **FX:** 33 1 47400113 **E-MAIL:** Pascale.callec@rilem.ens-cachan.fr **WEBSITES:** http://www.rilem.ens-cachan.fr **FOUNDED:** 1947. **OFFICERS:** Mr. M. Brusin, Sec.Gen. **MEMBERS:** 900. **STAFF:** 5. **LANGUAGES:** English, French.

DESCRIPTION: Individuals and organizations from 80 countries directly involved or concerned with testing and research in the field of building materials and structures. Purposes are: to research the properties and performance of materials and structures in laboratory and service conditions; to unify test methods on an international scale and develop new experimental procedures; to apply research results to the structural use of materials in building and civil engineering fields. Maintains 54 technical committees; conducts symposia. Publishes technical recommendations, proceedings of RILEM Symposia and RILEM Reports **PUBLICATIONS:** *Materials and Structures/Materiaux et Constructions* (in English and French), 10/year; *Studies in Terminology*. Provides a trilingual dictionary of materials and structures.; *Directory*, periodic. Also publishes newsletters, memoranda of technical committees, proceedings of symposia, and RILEM technical recommendations for testing methods.

INTERNATIONAL UNION OF THEORETICAL AND APPLIED MECHANICS (IUTAM)

c/o Prof. Michael A. Hayes, Dept. of

Mathematical Physics, Univ. College Dublin, F-202 Arts Bldg., Belfield, Dublin 4, Ireland **PHONE:** 353 1 7068370, 353 1 2692229 **FX:** 353 1 7061172 **E-MAIL:** michaelhayes@ucd.ie **WEBSITES:** http://www.iutam.org **FOUNDED:** 1946. **OFFICERS:** Prof. Werner Schiehlen, Pres. **MEMBERS:** 49. **LANGUAGES:** English, French, German.

DESCRIPTION: National organizations involved in the field of mechanics. Objectives are to: promote mechanics as a branch of science; bring together individuals and organizations involved in theoretical and experimental aspects of mechanics and related disciplines. **PUBLICATIONS:** *Annual Report* (in English). Includes directory; *Proceedings of Congress*, quadrennial; *Proceedings of Symposia*, 9/year; Newsletter, semiannual.

INTERNATIONAL UNION OF THERAPEUTICS (Union Therapeutique Internationale)

c/o Prof. A. Pradalier, Hopital Universitaire Louis Mourier, 178, rue des Renouillers, F-92701 Colombes Cedex, France **PHONE:** 33 47 606705 **FX:** 33 47 606072 **FOUNDED:** 1934. **OFFICERS:** Prof. Andre Pradalier, Pres. **MEMBERS:** 350. **LANGUAGES:** English, French.

DESCRIPTION: International organizations in 22 countries interested in the development of therapeutics. Conducts training, research, and exchange programs. **PUBLICATIONS:** *Abstracts or Communications' Book of the Congress* (in English and French), biennial.

INTERNATIONAL UNION FOR VACUUM SCIENCE, TECHNIQUE AND APPLICATIONS (IUVSTA)

c/o William D. Westwood, 7 Mohawk Cr., Nepean, ON, Canada K2H 7G7 **PHONE:** (613) 829-5790 **FX:** (613) 829-3061 **E-MAIL:** westwood@istar.ca **WEBSITES:** http://www.vacuum.org/IUVSTA **FOUNDED:** 1961. **OFFICERS:** Dr. William D. Westwood, Sec.Gen. **MEMBERS:** 30. **LANGUAGES:** English, French, German.

DESCRIPTION: Members are national vacuum science and engineering organizations. Purposes are to: promote and develop vacuum science, technique, and applications in all countries; coordinate activities of members; establish international working groups to study areas such as standardization and training of specialists; develop and instigate exchanges, meetings, and communications in vacuum science in cooperation with members and other associations. **PUBLICATIONS:** *News Bulletin*, quarterly.

INTERNATIONAL UNIVERSITY FOUNDATION (IUF)

1301 S. Noland Rd., Independence, MO 64055 **PHONE:** (816) 461-3633 **FOUNDED:** 1973. **OFFICERS:** Dr. John Wayne Johnston, Pres. **MEMBERS:** 62,337. **STAFF:** 4. **LANGUAGES:** English.

DESCRIPTION: Membership consists of individuals (60824) and institutions (30) united to advance the free flow of information and personnel between universities worldwide. Encourages international cooperation in higher education as a means of resolving global dilemmas in a rational, informed manner. Attempts to integrate and coordinate the educational efforts of member institutions via communication networks. Sponsors training programs. Maintains museum; compiles statistics. **PUBLICATIONS:** *International University Foundation—Directory*, annual; *International University Foundation—Report*, annual; Books.

THE INTERNATIONAL URBAN DEVELOPMENT ASSOCIATION (INTA) (Association Internationale du Developpement Urbain - AIVN)

Nassau Dillenburgstraat 44, NL-2596 AE The Hague, Netherlands **PHONE:** 31 70 3244526 **FX:** 31 70 3280727 **E-MAIL:** intainfo@inta-aivn.org **FOUNDED:** 1975. **OFFICERS:** Michel Sudarskis, Sec.Gen. **MEMBERS:** 700. **STAFF:** 5. **BUDGET:** US$500,000. **LANGUAGES:** English, French.

DESCRIPTION: National municipal and administrative bodies and agencies; corporations involved in consulting, construction, planning, and development; and consultants, specialists, university professors, and individuals responsible for planning, building, and management. Purpose is to provide a forum for the exchange of information among those involved in the planning, construction, development, administration, and financing of existing and new towns or other large-scale community development projects. Encourages world studies in the field and the circulation of findings. Sponsors conferences, seminars, and study tours. Collects and disseminates information on the planning, building, and management of large-scale development and on the control of urban growth. **PUBLICATIONS:** *Conference/Seminar Proceedings* (in English and French); Newsletter (in English and French), quarterly; Reports.

INTERNATIONAL VINE AND WINE OFFICE (OIV) (Office International de la Vigne et du Vin - OIV)

18, rue d'Aguesseau, F-75008 Paris, France **PHONE:** 33 1 44948081 **FX:** 33 1 42669063 **E-MAIL:** oivmail@oiv.wine.int **FOUNDED:** 1924. **OFFICERS:** Georges Dutruc-Rosset, Dir.Gen. **MEMBERS:** 45. **STAFF:** 5. **LANGUAGES:** French.

DESCRIPTION: Representatives of governments of vine-growing countries. Addresses scientific, technical, economic, and legal issues concerning viticulture and vine-derived products such as wine, grape juice, table grapes, and raisins. Determines standards regarding vine products and advises member governments on accepted norms. Works to create international research programs and to encourage information exchange among scholars and research institutions. Conducts technological research with a view toward rationalizing the production process and reducing production costs; compiles statistics. Strives to develop a general viticultural policy based on the resources and specific needs of members. Cooperates with the Food and Agriculture Organization of the United Nations and other international groups. Offers courses on marketing and management of wine and spirits and on viticulture and enology in hot climates. **PUBLICATIONS:** *Bulletin de l'OIV* (in English and French), bimonthly; *La Lettre de l'OIV*, monthly; *Lexique de la Vigne et du Vin* (in English, French, German, Italian, Portuguese, Russian, and Spanish); *Monographs*, periodic.

INTERNATIONAL VOLLEYBALL FEDERATION (FIVB) (Federation Internationale de Volleyball)

12, ave. de la Gare, CH-1000 Lausanne, Switzerland **PHONE:** 41 21 3453535 **FX:** 41 21 3453545 **TX:** 455234 FIVB CH **E-MAIL:** info@mail.fivb.ch **WEBSITES:** http://www.fivb.ch **FOUNDED:** 1947. **OFFICERS:** Mr. Alain Coupat, Mgmt.Dir. **MEMBERS:** 214. **STAFF:** 22. **BUDGET:** 15,000,000 SFr. **LANGUAGES:** Arabic, English, French, Russian, Spanish.

DESCRIPTION: National volleyball federations united to promote worldwide expansion of all forms of volleyball, including beach volleyball, to establish international rules, and to develop friendly relations among national federations. Maintains high technical standards in the sport. Organizes international, world, and Olympic tournaments. Conducts administrators, coaches, and referees courses. Sponsors educational programs; compiles statistics. **PUBLICATIONS:** *Coaches Manual* (in English and French); *FIVB X-Press* (in English and French), monthly. Includes competition highlights and special events; *Volley Tech* (in English and German), bimonthly. Technical analysis; *Volley World* (in Arabic, English, French, and Spanish), bimonthly. Includes competition highlights; *Volleyball Rules* (in Arabic, English, French, and Spanish). Also publishes clips, guidelines, manuals, and videotapes.

INTERNATIONAL WATER RESOURCES ASSOCIATION (IWRA)

c/o Southern Illinois University, 4535 Fanen Hall, Albuquerque, NM 87131 **PHONE:** (618) 453-5138 **FX:** (618) 453-2671 **WEBSITES:** http://www.iwra.siu.edu **FOUNDED:** 1972. **OFFICERS:** Ben Dzieyielewski, Exec.Dir. **MEMBERS:** 2,000. **STAFF:** 7. **LANGUAGES:** English.

DESCRIPTION: Individuals and organizations interested in water resources development. Seeks to: advance the planning, development, management, education, and research of water resources; establish an international forum for those concerned with water resources; encourage international programs in the field of water resources; cooperate with other organizations of common interest. **PUBLICATIONS:** *IWRA Update*, quarterly; *Water International*, quarterly. Covers association news, IWRA activities, and events in the international water resources field. Includes research reports, and calendar of events.

INTERNATIONAL WEIGHTLIFTING FEDERATION (IWF)

Postafiok 614, H-1374 Budapest, Hungary **PHONE:** 36 1 1530530 **FX:** 36 1 1530199 **TX:** 227553 AISHK H **FOUNDED:** 1905. **OFFICERS:**

Dr. Tamas Ajan, Gen.Sec. **MEMBERS:** 160. **STAFF:** 2. **LANGUAGES:** Arabic, English, French, German, Russian, Spanish.

DESCRIPTION: National weightlifting federations and associations. Objectives are to: organize, control, and develop weightlifting and weightlifting events on an international scale; promote cooperation among federations and lifters; resolve disputes that may arise between federations; set rules; verify world records. Conducts several local and regional clinics. **PUBLICATIONS:** *Constitution and Technical Rules* (in Arabic, English, French, and Spanish), quadrennial; *World Weightlifting* (in English and Spanish), quarterly; Books, periodic.

INTERNATIONAL WHALING COMMISSION (IWC)

The Red House, 135 Station Rd., Impington, Cambridge CB4 9NP, England **PHONE:** 44 1223 233971 **FX:** 44 1223 232876 **E-MAIL:** iwcoffice@compuserve.com **WEBSITES:** http://ourworld.compuserve.com/homepages/iwcoffice **FOUNDED:** 1946. **OFFICERS:** R. Gambell, Sec. **MEMBERS:** 39. **STAFF:** 15. **BUDGET:** £1,170,000. **LANGUAGES:** English.

DESCRIPTION: Commissioners representing member countries. Regulates whaling in order to promote the conservation of whale stocks. Encourages and helps organize studies on whales and whaling; compiles statistics on current condition of whale stocks; disseminates information on methods of estimating the size of whale stocks. Collects and publishes whaling statistics previously compiled by Norway's Bureau of International Whaling Statistics. **PUBLICATIONS:** *Journal of Cetacean Research and Management* (in English), 3/year. Includes papers and reports of scientific committee and summary of annual meeting; *Scientific Papers*, periodic; Books, periodic. Special issues on different aspects of cetaceans.

INTERNATIONAL WOMEN'S HEALTH COALITION (IWHC)

24 E. 21st St., 5th Fl., New York, NY 10010 **PHONE:** (212) 979-8500 **FX:** (212) 979-9009 **E-MAIL:** info@iwhc.org **WEBSITES:** http://www.iwhc.org **FOUNDED:** 1980. **OFFICERS:** Joan B. Dunlop, Pres. **STAFF:** 22. **BUDGET:** US$3,500,000. **LANGUAGES:** English.

DESCRIPTION: Seeks to promote and provide high quality reproductive health care for women in the Southern countries. Provides technical assistance, supports innovative health care projects and policy-oriented field research in Africa, Asia and Latin America. Produces public education materials. **PUBLICATIONS:** *The Sexuality Connection in Reproductive Health*. Explores the importance of the study of sexuality and gender-based power relations to reproductive health policies and programs; *Special Challenges in Third World Women's Health: Reproductive Tract Infections, Cervical Cancer, and Contraceptive Safety*; *Women's Political Action and Population Policy in Three Developing Countries*; Reports.

INTERNATIONAL WOMEN'S RIGHTS ACTION WATCH (IWRAW)

Univ. of Minnesota, Humphrey Inst. of Public Affairs, 301 19th Ave. S, Minneapolis, MN 55455 **PHONE:** (612) 625-5093 **FX:** (612) 624-0068 **E-MAIL:** iwraw@hhh.umn.edu **WEBSITES:** http://www.igc.org/iwraw **OFFICERS:** Marsha Freeman, Dir. **LANGUAGES:** English.

DESCRIPTION: Resource and communication center for an international network of over 5,300 individuals and groups concerned with implementation of the Convention on the Elimination of Discrimination Against Women (CEDAW). Works closely with Non Governmental Organizations (NGOs) in developing countries and with the CEDAW secretariat and the members of the CEDAW Committee. Attends the CEDAW session and reports on the Committees work. Conducts on the status of women in countries under review each session and provide the Committee with additional information to use in its review. Provides a great deal of information to NGO activist around the world, not only on the Convention but on what others are doing and developments in law and policy concerning women. **PUBLICATIONS:** *Assessing the Status of Women: A Guide to Reporting Under the Convention on the Elimination of All Forms of Discrimination Against Women*; *From the Fourth World Conference on Women: The Beijing Declaration and Platform for Action*; *Implementing the International Right to Sexual Nondiscrimination*; *Women's Human Rights and Reproductive Rights: Capacity and Choice* (in English, Portuguese, and Spanish); *Women's Watch* (in English and Spanish), quarterly. Also publishes books, videos, and bulletins about women's rights.

INTERNATIONAL WOOL TEXTILE ORGANIZATION (IWTO) (Federation Lainiere Internationale - FLI)

19-21, rue du Luxembourg, bte. 14, B-1040 Brussels, Belgium **PHONE:** 32 2 5130620 **FX:** 32 2 5140665 **E-MAIL:** iwto@skynet.be **FOUNDED:** 1928. **OFFICERS:** W.H. Lakin, Sec.Gen. **MEMBERS:** 28. **STAFF:** 3. **BUDGET:** 15,000 BFr. **LANGUAGES:** English, French.

DESCRIPTION: Organizations in 23 countries interested in the merchandising, processing, spinning, and weaving of wool and allied fibers. Aims to promote and take positions on measures affecting trade and the industry. Encourages the study and solution of economic and commercial questions; collects and disseminates information. Works to ensure the functioning of the International Arbitration Agreement. **PUBLICATIONS:** *International Wool - Textile Overview*; *IWTO Brochure*, annual; *Specifications*, annual; *Statistics*, annual; *Wool Statistics*. Also publishes *IWTO Blue Book*, *White Papers*, and statistical updates.

INTERNATIONAL WROUGHT COPPER COUNCIL (IWCC)

6 Bathurst St., Sussex Sq., London W2 2SD, England **PHONE:** 44 171 7237465 **FX:** 44 171 7240308 **FOUNDED:** 1953. **OFFICERS:** Simon Payton, Sec.Gen. **MEMBERS:** 19. **LANGUAGES:** English.

DESCRIPTION: National trade associations of the copper and copper alloy fabricating industries. Promotes information exchange and cooperation within the industry; conducts promotional activities. **PUBLICATIONS:** *Survey of Capacities of Copper Mines, Smelters, Refineries and Copper Wire Rod Plants*, semiannual; *Directory*, periodic.

INTERNATIONAL YOUNG CHRISTIAN WORKERS (IYCW) (Jeunesse Ouvriere Chretienne Internationale - JOCI)

4, avenue G. Rodenbach, B-1030 Brussels, Belgium **PHONE:** 32 2 2421811 **FX:** 32 2 2424800 **E-MAIL:** international.secretariat@jociycw.net **WEBSITES:** http://www.jociycw.net **FOUNDED:** 1945. **OFFICERS:** Helio Antonio Alves, Pres. **MEMBERS:** 70. **STAFF:** 5. **LANGUAGES:** English, French, Spanish.

DESCRIPTION: Aims to educate young workers to accept present and future responsibilities in their commitment to the working world and to help them overcome all obstacles preventing professional fulfillment. Maintains working commissions. **PUBLICATIONS:** *Africa INFO* (in English, French, and Spanish), semiannual; *American INFO* (in English, French, and Spanish), semiannual; *Asia Pacific* (in English), quarterly; *International INFO* (in English, French, and Spanish), 3/year; *IYCW Bulletin* (in English, French, and Spanish), 3/year; Books.

INTERNATIONAL YOUTH HOSTEL FEDERATION (IYHF) (Federation Internationale des Auberges de Jeunesse - FIAJ)

9 Guessens Rd., Welwyn Garden City, Herts. AL8 6QW, England **PHONE:** 44 1707 324170 **FX:** 44 1707 323980 **E-MAIL:** iyhf@iyhf.demon.co.uk **WEBSITES:** http://www.iyhf.org **FOUNDED:** 1932. **OFFICERS:** Rawdon Lau, Sec.Gen. **MEMBERS:** 60. **STAFF:** 20. **LANGUAGES:** Arabic, English, French, German, Spanish.

DESCRIPTION: National youth hostel associations in 60 countries representing 4 million members. Encourages cooperation among national youth hostel associations and seeks to educate young people of all nations by promoting youth tourism and an appreciation of travel. Provides low-cost accommodation and programs on outdoor education, recreation, and touring; holds periodic travel seminars; sponsors annual international youth travel forum and manager training program. Carries out development projects in Asia and Latin America. Conducts research and compiles statistics. Maintains numerous committees. **PUBLICATIONS:** *Annual Report* (in English). Includes statistics; *Guide to Budget Accommodation*, annual; *International Bulletin*, bimonthly.

INTERNATIONAL YOUTH LIBRARY (IYL) (Internationale Jugendbibliothek - IJB)

Schloss Blutenburg, D-81247 Munich, Germany **PHONE:** 49 89 8912110 **FX:** 49 89 8117553 **E-MAIL:** bib@ijb.de **FOUNDED:** 1948. **OFFICERS:**

Dr. Barbara Scharioth, Dir. **STAFF:** 35. **BUDGET:** DM 2,500,000. **LANGUAGES:** English, French, German, Italian, Japanese, Russian, Spanish, Swedish.

DESCRIPTION: A project of the United Nations Educational, Scientific and Cultural Organization. Experts on children's literature working to foster knowledge and international understanding. Serves as a liaison between institutions, publishers, and interested individuals. Disseminates information and advice. Organizes travelling book exhibits and professional seminars for teachers and librarians. **PUBLICATIONS:** *Catalogs of Children's Literature* (in German); *IJB Report* (in English and German), semiannual, always July and December; *The White Ravens* (in English), annual, always April.

INTERNATIONALE ALPENSCHUTZKOMISSION (Comissione Internationale per la Protezione delle Alpi - CIPRA)

Im Bretscha 22, FL-9494 Schaan, Liechtenstein **PHONE:** 423 75 2374030 **FX:** 423 75 2374031 **E-MAIL:** cipra@cipra.org **WEBSITES:** http://www.cipra.org **FOUNDED:** 1952. **OFFICERS:** Andreas Gotz, Dir. **MEMBERS:** 7. **STAFF:** 4. **BUDGET:** US$200,000. **LANGUAGES:** French, German, Italian, Slovene.

DESCRIPTION: Representatives from Austria, France, Germany, Italy, Liechtenstein, Switzerland, and Slovenia involved in nature management, protection, natural history, regional development, and environmental protection of alpine regions. Consults interested parties and resolves problems related to protecting alpine regions and biotopes, which support plants and animals susceptible to dangers caused by man's recreation. **PUBLICATIONS:** *CIPRA Info* (in French, German, Italian, and Slovene), quarterly. Includes problems, developments and important items in the Alps; *CIPRA-Little Serie*, periodic; *Conference Proceedings*, annual.

INTERNATIONALER SOCIALDIENST (ISD)

1 Deutscher Zweig e.v., Stockborn 5-7, D-60439 Frankfurt, Germany **PHONE:** 49 69 9580702 **FX:** 49 69 95807465 **E-MAIL:** issger@t-online.de **FOUNDED:** 1924. **OFFICERS:** Ingrid Baer, Dir. **STAFF:** 40. **BUDGET:** DM 3,000,000. **LANGUAGES:** English, French, German.

DESCRIPTION: Works to ensure availability of effective social services and related programs worldwide. Gathers and disseminates information.

INTERNET SOCIETY

11150 Sunset Hills Rd., Ste. 100, Reston, VA 20190-5321 **PHONE:** (703) 326-9880 **FX:** (703) 326-9881 **E-MAIL:** membership@isoc.org **WEBSITES:** http://www.isoc.org **FOUNDED:** 1992. **OFFICERS:** Donald M. Heath, Pres./CEO. **MEMBERS:** 9,000. **STAFF:** 14. **BUDGET:** US$5,000,000. **LANGUAGES:** English.

DESCRIPTION: Members are technologists, developers, educators, researchers, government representatives, and business people. Seeks to ensure global cooperation and coordination for the Internet and related internetworking technologies and applications. Supports the development and dissemination of standards for the Internet. Promotes the growth of Internet architecture and Internet-related education and research. Encourages assistance to technologically developing countries in implementing local Internet infrastructures. **PUBLICATIONS:** *ISOC Forum*, monthly. Contains electronic information; *On the Internet*, bimonthly. Includes articles about various aspects of the Internet and its impact.

INTER-PARLIAMENTARY UNION (IPU) (Union Interparlementaire - UIP)

Place du Petit-Saconnex, Case Postale 438, CH-1211 Geneva 19, Switzerland **PHONE:** 41 22 9194150 **FX:** 41 22 7333141 **E-MAIL:** postbox@mail.ipu.org **WEBSITES:** http://www.ipu.org **FOUNDED:** 1889. **OFFICERS:** Anders B. Johnsson, Sec.Gen. **MEMBERS:** 136. **STAFF:** 30. **BUDGET:** 99,000,000 SFr. **LANGUAGES:** English, French.

DESCRIPTION: Members are national groups formed in legislative assemblies established by sovereign states. Objectives are to: work for peace and international cooperation; strengthen and develop representative democracy; improve the working methods of parliamentary institutions; promotes the status of women and women's participation in politics; and defends the human rights of MPs. Conducts research and disseminates information on various aspects of the parliamentary process and fosters the objective study of economic, social, and cultural problems of international significance. Debates and adopts resolutions on main issues before the international community. **PUBLICATIONS:** *Chronicle of Parliamentary Elections* (in English and French), annual. Provides comprehensive information on national legislative elections throughout the world; *Codes of Conduct for Elections*; *Conference Proceedings*, semiannual; *Democracy, Principals and Realisation*; *Free and Fair Elections: International Law and Practice*; *Inter-Parliamentary Bulletin*, semiannual; *Men and Women in Politics, Democracy Still in the Making*; *Presiding Officers of National Parliamentary Assemblies*; *Women in Parliaments - 1945-1995: A World Statistical Survey*; *Women, What the IPU is Doing?*; *World Directory of Parliaments*, annual.

INTER-UNIVERSITY EUROPEAN INSTITUTE ON SOCIAL WELFARE (IEISW) (Institut Europeen Interuniversitaire de l'Action Sociale - IEIAS)

179, rue du Debarcadere, B-6001 Marcinelle, Belgium **PHONE:** 32 71 447267, 32 71 447264 **FX:** 32 71 471104 **FOUNDED:** 1970. **OFFICERS:** S. Mayence, Dir.Gen. **MEMBERS:** 20. **STAFF:** 6. **BUDGET:** US$1,500,000. **LANGUAGES:** English, French.

DESCRIPTION: Representatives of European universities from 12 countries interested in the study of social welfare. Promotes interdisciplinary research in the field of social welfare; studies viable methods of satisfying individual material, cultural, and psychological needs, thereby improving the relationship between the individual and society. Analyzes the efficacy of social policy in establishing social services. Seeks uniform social welfare policy throughout Europe. Collects and disseminates information on social welfare in Europe. Maintains liaison with the Council of Europe, the European Economic Community, the United Nations, and the European Centre of Social Welfare. Operates documentation center, translation bureau, and publications service; organizes day sessions, and graduate courses.

ISIS INTERNATIONAL

PO Box 1837 Main, Quezon City, Metro Manila 1100, Philippines **PHONE:** 63 2 9267297, 63 2 9241065 **FX:** 63 2 9241065 **E-MAIL:** isis@mnl.sequel.net **FOUNDED:** 1974. **OFFICERS:** Marianita C. Villariba, Dir. **MEMBERS:** 8. **STAFF:** 17. **BUDGET:** US$550,000. **LANGUAGES:** English, Spanish.

DESCRIPTION: Women's information and communication service. Facilitates global communication among women and disseminates information and materials produced by women and women's groups in an effort to foster solidarity among women. Examines issues such as women's role in development, health, education, food and nutrition, media, prostitution and violence against women, employment, and theories of feminism. Provides technical assistance and training in communication skills and information management. Seeks to mobilize support and solidarity among women on an international scale. (The group is named for Isis, the Egyptian goddess of knowledge and fertility). Isis still maintains its global character while having a special focus on the Asia and Pacific regions. **PUBLICATIONS:** *Against All Odds: Essays on Women, Religion and Development from India and Pakistan* (in English); *Asia Pacific Women and Health Directory*; *Clip Art Book*, annual; *Directory of Third World Women's Publications*; *Health Beyond Borders: The Asia-Pacific Women's Health Network Report*; *Information Pack*, quarterly; *Isis and Other Guides to Health*; *ISIS Women in Development Resource Guide*; *Let Our Silenced Voices be Heard: The Traffic on Asian Women*; *Organizing Strategies in Women's Health: An Information and Action Handbook*; *Powerful Images: A Woman's Guide to Audiovisual Resources*, periodic; *Teenage Pregnancy in the South: Charting Our Destiny*; *Women Empowering Communication: A Resource Book on Women and Globalization of Media*; *Women Envision*, monthly. Information on preparatory activities for the World Conference and NGO forum on women; *Women in Action*, 3/year; *Women's Actions for the Environment Information Pack*; *Women's Perspectives on Population Issues*.

ISLAMIC CENTRE FOR DEVELOPMENT OF TRADE (ICDT) (Centre Islamique pour le Developpement du Commerce - CIDC)

Ave. des FAR, Tours des Habous, 11 eme etage, Casablanca 20000, Morocco **PHONE:** 212 2314974 **FX:** 212 2310110 **TX:** 46296 M **E-MAIL:**

icdt@icdt.org **WEBSITES:** http://www.icdt.org **FOUNDED:** 1981. **OFFICERS:** Dr. Badre-Eddine Allali, Dir.Gen. **MEMBERS:** 54. **STAFF:** 23. **BUDGET:** US$1,500,000. **LANGUAGES:** Arabic, English, French.

DESCRIPTION: A subsidiary of the Organization of the Islamic Conference. Purpose is to establish, maintain, and strengthen trade exchange among Islamic countries. Assists member states in their efforts to create and maintain national trade organizations; offers advice concerning trade matters. Collects and disseminates information; compiles foreign trade statistics for member countries. Conducts research aimed at furthering regional trade in the context of regional and global economic development. Organizes specialized training sessions with a view toward developing a body of trade experts. **PUBLICATIONS:** *Directory of Foreign Trade Operators in the Islamic Countries*, periodic; *Directory of Trade Training Organisations*, periodic; *Inter-Islamic Trade* (in Arabic, English, and French), annual; *L'Uruguay Round: Enjeux et Implications pour les Pays Islamiques* (in English and French); *Tijaris: Magazine of International and Inter-Islamic Trade* (in Arabic, English, and French), quarterly; Papers, periodic. Contains notes on Islamic summits and ministerial conferences.

ISLAMIC DEVELOPMENT BANK (IDB)

PO Box 5925, Jeddah 21432, Saudi Arabia **PHONE:** 966 2 6361400 **FX:** 966 2 6366871 **TX:** 601137/ISDB SJ **E-MAIL:** archives@sairti00.bitnet **FOUNDED:** 1975. **OFFICERS:** Dr. Ahmed Mohamed Ali, Pres. **MEMBERS:** 52. **STAFF:** 700. **BUDGET:** US$132,549,000. **LANGUAGES:** Arabic, English, French.

DESCRIPTION: Members are: African and Asian Muslim states; Turkey and the Palestine Liberation Organization. Purpose is to enhance the economic and social development of member countries and Muslim communities worldwide, in accordance with Islamic law. Provides financial and technical assistance to members; offers equity capital and interest-free loans for projects in member countries; operates special funds for aid to Muslim communities in nonmember countries; establishes trust funds; promotes foreign trade between members, particularly in capital goods; conducts research in an effort to make all financial transactions conform to Islamic law. Provides training for individuals working in the area of social and economic development. Maintains Islamic Research and Training Institute. **PUBLICATIONS:** *Annual Report* (in Arabic, English, and French); *Directory of Research Institutes*; *Information Bulletin* (in Arabic, English, and French), weekly; *IRTI Magazine* (in Arabic, English, and French), monthly.

ISLAMIC EDUCATIONAL, SCIENTIFIC AND CULTURAL ORGANIZATION (IESCO)

Hay Rd., Boite Postale 2775, Rabat 10104, Morocco **PHONE:** 212 7 772433 **FX:** 212 7 777459 **TX:** 32645 **OFFICERS:** Dr. Abdulaziz bin Othman Al-Twaijri, Dir.Gen. **LANGUAGES:** Arabic, English.

DESCRIPTION: Organization of Islamic art, science, and cultural organizations. Promotes appreciation of Islamic arts and culture; encourages participation in artistic, scientific, and cultural activities. Conducts research and educational programs; sponsors cultural events.

ISLAMIC SOLIDARITY FUND

c/o OIC Secretariat, PO Box 178, Jeddah 21411, Saudi Arabia **PHONE:** 966 2 6800800 **FX:** 966 2 6873568 **OFFICERS:** Sheikh Nasir Abdullah bin Hamdan, Chm. **LANGUAGES:** Arabic, English.

DESCRIPTION: Members are Islamic states and communities. Promotes Islamic worship. Provides support and assistance to Muslim organizations and congregations.

ISLAMIC STATES BROADCASTING ORGANIZATION (ISBO) (Organisation des Radiodiffusions des Etats Islamiques)

c/o Hussein M. Al-Askeri, PO Box 6351, Jeddah 21442, Saudi Arabia **PHONE:** 966 2 6721121, 966 2 6722269 **FX:** 966 2 6722600 **TX:** 601442 ISBOSJ **FOUNDED:** 1975. **OFFICERS:** Hussein M. Al-Askari, Sec.-Gen. **MEMBERS:** 51. **STAFF:** 18. **LANGUAGES:** Arabic, English, French.

DESCRIPTION: Members are national radio and television corporations of Islamic states. Objectives are to: propagate Islamic ideals through broadcasting services; produce Islamic radio and television programs; coordinate program exchanges among member states. Maintains cooperative relations with international organizations. Organizes competitions for production of high-standard Islamic programs; conducts symposia and audiovisual Arabic language instruction. **PUBLICATIONS:** *Programmes' Directory* (in Arabic), periodic; *Report*, periodic.

ISTITUTO DELLE SUORE MAESTRE DI SANTA DOROTEA (ISMSD)

Casa Generalizia, Via Raffaele Conforti 25, I-00166 Rome, Italy **PHONE:** 39 6 6224041, 39 6 6630040 **FX:** 39 6 6628925 **E-MAIL:** seggen.msd@pcn.net **FOUNDED:** 1838. **OFFICERS:** Suor Attilia Almici, Sec.Gen. **MEMBERS:** 725.

DESCRIPTION: Members are Roman Catholic women in Bolivia, Burundi, Colombia, Italy, Brazil, Cameroon, Madagascar and Congo who prepare young people, in particular women, for leadership. Teaching in rural parochial and scholastic settings, they provide spiritual and educational assistance, striving to develop leaders who have been ''risen'' in the Christian community. **PUBLICATIONS:** *Ardere per Accendere*, quarterly.

JAPAN INTERNATIONAL RESCUE ACTION COMMITTEE (JIRAC)

Kyoei Bldg., 401-2-14-5 Shibuya, Shibuya-ku, Tokyo 150, Japan **PHONE:** 81 3 54676530 **FX:** 81 3 54676531 **WEBSITES:** http://www.volunteer-post.mpt.go.jp/index_e.html **OFFICERS:** Kentaro Hayashi **MEMBERS:** 400. **STAFF:** 5. **BUDGET:** US$643,996. **LANGUAGES:** English, Japanese.

DESCRIPTION: Promotes world peace and the non-violent solution of conflicts. Provides emergency assistance to refugees and other victims of conflict worldwide. Current programs include provision of aid-in-kind to Kurdish refugees, and to the aged, infants, and the poor in the Russian Far East.

JEWISH DOCUMENTATION CENTER

c/o Simon Wiesenthal, Salztorgasse 6/IV/5, A-1010 Vienna, Austria **PHONE:** 43 1 5339805, 43 1 5339131 **FX:** 43 1 5350397 **FOUNDED:** 1947. **OFFICERS:** Simon Wiesenthal, Exec. Officer. **STAFF:** 3. **LANGUAGES:** English, German, Polish, Russian.

DESCRIPTION: Objectives are to: search and document war crimes and war criminals; search out witnesses for trials and contacts with courts, police offices and various archives, the Office of Special Investigation of the United States Department of Justice, and the Central Prosecutor's Office in Ludwigsburg, Germany; fight against anti-Semitic and neo-Nazi propaganda worldwide and bring charges against publishers and distributors of such literature. **PUBLICATIONS:** *Annual Report*; Reports, periodic.

JEWISH WOMEN INTERNATIONAL (JWI)

1828 L St. NW, Ste. 250, Washington, DC 20036 **PHONE:** (202) 857-1300 **TF:** (800) 343-2823 **FX:** (202) 857-1380 **E-MAIL:** jwi@jwi.org **WEBSITES:** http://www.jewishwomen.org **FOUNDED:** 1897. **OFFICERS:** Gail Rubinson, Exec.Dir. **MEMBERS:** 75,000. **STAFF:** 30. **LANGUAGES:** English.

DESCRIPTION: Jewish Women International strengthens the lives of women, children, and families through education, advocacy and action. Focusing on family violence and the emotional health of children, JWI serves as an agent for change. Community activities include domestic violence awareness, self-esteem projects, interfaith forums, holocaust awareness, hospital humor carts, and other youth projects. Founded and maintains a residential treatment center for emotionally disturbed children in Jerusalem, Israel. **PUBLICATIONS:** *Jewish Woman*, quarterly. Includes book reviews and features.

JOINT FAO/WHO/OAU REGIONAL FOOD AND NUTRITION COMMISSION FOR AFRICA (Commission Regional Mixte FAO/OMS/OUA de l'Alimentation et de la Nutrition en Afrique)

c/o E. Ndiaye, c/o FAO Regional Office for Africa, PO Box 1628, Accra, Ghana **PHONE:** 233 21 666851, 233 21 666854 **FX:** 233 21 668427 **TX:** 2139 **E-MAIL:** fao-rafr@cgnet.com **FOUNDED:** 1963. **MEMBERS:** 44. **STAFF:** 10. **BUDGET:** US$500,000. **LANGUAGES:** English, French.

DESCRIPTION: Representatives of the Food and Agriculture Organization of the United Nations, World Health Organization, and Organization of African Unity. Purpose is to gather, analyze, and disseminate information on food and nutrition. **PUBLICATIONS:** *Food and Nutrition Bulletin* (in English and French), semiannual; *Special Papers* (in English and French), periodic.

JOINT ORGANIZATION FOR CONTROL OF DESERT LOCUSTS AND BIRD PESTS (JOCDLBP) (Organisation Commune de Lutte Antiacridienne et de Lutte Antiaviarie - OCLALAV)

Route des Peres Maristes, Boite Postale 1066, Dakar, Senegal **PHONE:** 221 323280 **FX:** 221 320487 **OFFICERS:** Abdullah Ould Soueid Ahmed, Dir.Gen. **LANGUAGES:** English, French.

DESCRIPTION: Members are organizations and individuals with an interest in the control of desert locusts and avian pests. Seeks to advance the study and practice of agricultural pest control. Conducts research and educational programs.

JUNIOR CHAMBER INTERNATIONAL (JCI)

400 University Dr., Coral Gables, FL 33134 **PHONE:** (305) 446-7608 **FX:** (305) 442-0041 **WEBSITES:** http://www.juniorchamber.org **FOUNDED:** 1944. **OFFICERS:** Benny L. Ellerbe, Sec.Gen. **MEMBERS:** 400,000. **STAFF:** 43. **BUDGET:** US$3,700,000. **LANGUAGES:** Chinese, English, French, Japanese, Korean, Spanish.

DESCRIPTION: National Jaycee organizations representing 400,000 individuals between the ages of 18 and 40 dedicated to the principles of leadership training through community development. Sponsors a course on individual development; conducts charitable programs; sponsors competitions. Maintains program information library. **PUBLICATIONS:** *JCI News*, quarterly; Directory, annual; Proceedings, annual. Also publishes pamphlets.

KATALYSIS PARTNERSHIP

1331 N. Commerce St., Stockton, CA 95202 **PHONE:** (209) 943-6165 **FX:** (209) 943-7046 **E-MAIL:** Katalysis2@aol.com **FOUNDED:** 1984. **OFFICERS:** Gerald Hildebrand, Pres. & CEO. **STAFF:** 21. **BUDGET:** US$2,100,000. **LANGUAGES:** English, Spanish.

DESCRIPTION: Supports low income people to gain self-reliance by helping them to improve their economic and social conditions. Works through multilateral partnerships with community based organizations using participatory processes to provide training and technical assistance in microenterprise development, women's community banking, and institutional strengthening. Practices a partnership model of international development which allows all participants to relate as equals, relaxing the hierarchical mindset and replacing power with process as the means of effecting organizational goals. **PUBLICATIONS:** *Beyond the Annual Campaign: A Handbook for Sustainable Fundraising*; *Choosing Partnership: The Evolution of the Katalysis Model*; *Fieldnotes*, semiannual; *Katalysis*; *Perfecting the Alliance: Viable Fundraising for International Partnerships.*

KING BAUDOUIN FOUNDATION (KBF)

rue Brederodestraat 21, B-1000 Brussels, Belgium **PHONE:** 32 2 5111840, 32 2 5115221 **FX:** 32 2 5115221 **FOUNDED:** 1975. **STAFF:** 80. **BUDGET:** 800,000,000 BFr. **LANGUAGES:** English, French.

DESCRIPTION: Seeks to improve the quality of life and increase the scope of economic opportunity available to people in developing regions worldwide. Works with indigenous government agencies, businesses, and associations to stimulate educational, community development, and social welfare programs. Encourages creation of small businesses and microenterprises to reduce unemployment in developing regions.

LAKE CHADE BASIN COMMISSION (LCBC)

Boite Postale 727, N'Djamena, Chad **PHONE:** 235 514145 **FX:** 235 514137 **TX:** 5251 **OFFICERS:** Abubakar B. Jauro, Exec.Sec. **LANGUAGES:** English, French.

DESCRIPTION: Members are government agencies responsible for water resources management and economic development in the Lake Chad Basin. Promotes appropriate and sustainable development along Lake Chad. Provides assistance and support to water resources management and development projects.

LALMBA ASSOCIATION

7685 Quartz St., Arvada, CO 80007 **PHONE:** (303) 420-1810 **FX:** (303) 467-1232 **E-MAIL:** lalmba@aol.com **WEBSITES:** http://www.lalmba.org **FOUNDED:** 1963. **OFFICERS:** Hugh Downey, Pres. **MEMBERS:** 35. **LANGUAGES:** English.

DESCRIPTION: International relief organization working primarily in health related areas in Kenya, Mexico, Ethiopia, and Eritrea. Coordinates and supports community programs that enable local people to assume responsibility for their own development. Places 5-15 volunteers in overseas assignments each year. **PUBLICATIONS:** *Hugh's News*, quarterly.

LATIN AMERICAN ASSOCIATION FOR ASIA AND AFRICA STUDIES (LAAAAS) (Asociacion Latinoamericana de Estudios de Asia y Africa Afroasiaticos - ALADAA)

c/o CEAA, El Colegio de Mexico, Camino al Ajusco 20, 10740 Mexico City, DF, Mexico **PHONE:** 52 5 6455955, 52 5 6454954 **FX:** 52 5 6450464 **TX:** 1777585 COLME **FOUNDED:** 1976. **OFFICERS:** Prof.Dr. Machiko Tanaka, Sec.Gen. **MEMBERS:** 550. **LANGUAGES:** English, French, Spanish.

DESCRIPTION: Members are scholars interested or engaged in Afro-Asian studies in 15 Latin American countries. Promotes Asian and African studies in Latin America. **PUBLICATIONS:** *ALADAA - Members*, periodic.

LATIN AMERICAN ASSOCIATION OF DEVELOPMENT FINANCING INSTITUTIONS (Asociacion Latinoamericana de Instituciones Financieras de Desarrollo - ALIDE)

Apartado Postal 3988, Lima 100, Peru **PHONE:** 511 4 422400 **FX:** 511 4 428105 **E-MAIL:** postmaster@alide.org.pe **WEBSITES:** http://www.alide.org.pe **FOUNDED:** 1968. **OFFICERS:** Dr. Carlos Garatea Yori, Sec.Gen. **MEMBERS:** 90. **STAFF:** 15. **BUDGET:** US$1,000,000. **LANGUAGES:** English, French, Portuguese, Spanish.

DESCRIPTION: Public and private development financing institutions; institutions outside of Latin America that contribute to the region's economic development; institutions with goals similar to those of ALIDE. Aims to mobilize resources for the financing of Latin American economic and technological development; encourages members to cooperate and contribute to the integration of Latin American economies. Works to strengthen member banking institutions; promotes the exchange of specialized personnel. Disseminates information on national and international facilities for financing, development, technical cooperation, and personnel training; reports on Latin American investment projects and coordinates the collaboration of members in such projects. Offers training programs and courses to develop theoretical and practical skills in financing personnel; provides consulting services. Operates ALIDE Information Network and ALIDE Documentation Center. Administers Financial and Technological Information Network on Projects and Investments. Maintains consultative status with the Organization of American States. **PUBLICATIONS:** *Abstract Services* (in English and Spanish), monthly; *ALIDE Bulletin* (in English and Spanish), bimonthly; *ALIDE-News* (in Spanish), monthly; *Annual Report*; *Economic and Legal Studies*; *Education and Training Program*, periodic; *Latin American Directory of Development Financing Institutions* (in Spanish), biennial; *Work Program*, annual.

LATIN AMERICAN ASSOCIATION OF DEVELOPMENT ORGANIZATIONS (Asociacion Latinoamericana de Organizaciones de Promocion Costa Rica - ALOP)

Bo Escalante, De La Iglesia, Sta. Teresita 300 Norte, Y 275 Este, Casa No. 3144, San Jose 1350, Costa Rica **PHONE:** 506 2833018, 506 2832122 **FX:** 506 2835898 **E-MAIL:** alopse@sol.racsa.co.cr **WEBSITES:** http://www.alop.or.cr **FOUNDED:** 1979. **OFFICERS:** Manuel O. Chiriboga, Exec.Sec. **MEMBERS:** 43. **STAFF:** 8. **BUDGET:** US$1,800,000. **LANGUAGES:** English, Portuguese, Spanish.

DESCRIPTION: Members are nongovernmental organizations in 20 countries. Promotes communication and cooperation among members; works to improve cooperation between all nations in the Western Hemisphere. Advocates on behalf of indigenous peoples living in areas affected by development projects; supports development plans relying on technologies appropriate for sustained local use. Makes available medical, technical, legal, and emergency assistance; gathers and disseminates information on international development projects. Lobbies for sustainable development before international agencies, multinational institutions, and government agencies. **PUBLICATIONS:** *Latin America's Report on the Reality of AIDS*; *Tianguis* (in Spanish), quarterly.

LATIN AMERICAN BANKING FEDERATION (LABF) (Federacion Latinoamericana de Bancos - FELABAN)

Carrera 11A, No. 93-67, Of. 202, Bogota, Colombia **PHONE:** 57 1 6218617, 57 1 6218490 **FX:** 57 1 6218021 **E-MAIL:** felaban@latinbanking.com **WEBSITES:** http://www.latinbanking.com **FOUNDED:** 1965. **OFFICERS:** Maricielo Glen de Tobon, Gen.Sec. **MEMBERS:** 19. **STAFF:** 10. **BUDGET:** US$800,000. **LANGUAGES:** Spanish.

DESCRIPTION: Members are banking associations in Latin American countries. Works toward an integrated Latin American banking system.

LATIN AMERICAN BUREAU OF SOCIETY OF JESUS (Secretariado Latinoamericano de la Compania de Jesus - SELASI)

Almagro 6, E-28010 Madrid, Spain **PHONE:** 34 1 3197581, 34 1 3080254 **FX:** 34 1 7598908 **E-MAIL:** secretariado@intelred.es **FOUNDED:** 1968. **OFFICERS:** Cesareo Garcia del Cerro, Dir. **MEMBERS:** 5,200. **STAFF:** 12. **BUDGET:** US$1,500,000. **LANGUAGES:** English, French, Spanish.

DESCRIPTION: Pastoral and religious organization engaging in fundraising campaigns in 126 countries. Provides aid to programs developed by Jesuits in Latin American countries. Sends teachers and volunteers to poverty-stricken areas in Latin America. Works to improve health, education, and agricultural techniques in Latin America. Operates speakers' bureau, children's services, and charitable programs. Compiles statistics. **PUBLICATIONS:** Newsletter (in Spanish), quarterly.

LATIN AMERICAN AND CARIBBEAN COMMUNICATION AGENCY (Agence Latinoamericana y Caribe de Communication)

Apartado 14-0225, Lima 14, Peru **PHONE:** 51 1 2211488 **FX:** 51 1 2212877 **E-MAIL:** info@alc.org.pe **LANGUAGES:** English, Spanish.

DESCRIPTION: Members are communications organizations. Promotes increased flow of information among the countries of Latin America and the Caribbean. Facilitates exchange of information among members; works to establish communications networks.

LATIN AMERICAN CATHOLIC PRESS UNION (UCLAP)

Apartado Postal 1/-21-1/8, Quito, Ecuador **PHONE:** 593 2 548046, 593 2 501654 **FX:** 593 2 501654 **OFFICERS:** Adalid Contreras, Exec.Sec. **MEMBERS:** 26. **STAFF:** 10. **BUDGET:** US$150,000. **LANGUAGES:** English, Spanish.

DESCRIPTION: Members are national Catholic press organizations; Catholic media outlets. Facilitates coordination of activities among members; conducts training programs. **PUBLICATIONS:** *Arandu* (in Spanish), periodic.

LATIN AMERICAN CENTER SOCIAL ECOLOGY (CLAES) (Centro Latino American Ecologia Social)

Casilla Correo 13125, 11700 Montevideo, Uruguay **PHONE:** 598 2 922362, 598 2 922363 **FX:** 598 2 201908 **E-MAIL:** claes@adinet.com.uy **FOUNDED:** 1982. **OFFICERS:** E. Gudynas, Dir. **STAFF:** 12. **LANGUAGES:** English, Portuguese.

DESCRIPTION: Members are environmental nongovernmental organizations in Latin America. Monitors the programs and advancements of NGOs throughout Latin America concerned with the environment and development. Analyzes information provided by member organizations. Conducts research on alternative development strategies, environmental ethics, natural resources management, and local government policies. Coordinates the Latin American Network on Social Ecology. **PUBLICATIONS:** *SICA-AL*; *Studies Social Ecology* (in Spanish), periodic; *Teko-HA* (in Spanish), periodic; Booklets.

LATIN AMERICAN CENTRAL OF WORKERS (CLAT) (Central Latinoamericana de Trabajadores - CLAT)

Apartado 6681, Caracas 1010-A, Venezuela **PHONE:** 58 32 720878, 58 32 720794 **FX:** 58 32 720463 **FOUNDED:** 1954. **OFFICERS:** Emilio Maspero, Gen.Sec. **MEMBERS:** 51. **STAFF:** 120. **LANGUAGES:** English, French, Portuguese, Spanish.

DESCRIPTION: National federations, trade federations, union organizations, cooperatives, and rural, professional women's, and young workers' associations representing 21,000,000 individuals in 35 countries. Seeks to liberate workers socially, culturally, economically, and politically. Works for democracy and to defend workers' human, trade union, and political rights. Organizes and trains trade union leaders; conducts seminars. Compiles statistics. Offers placement service. **PUBLICATIONS:** *CLAT Boletin*, monthly; *CLAT Directory*, annual; *Noticias CLAT* (in Spanish), monthly.

LATIN AMERICAN CENTRE OF SOCIAL ECOLOGY (Centro Latinamericano de Ecologia Social)

Casilla Correo 13000, 11700 Montevideo, Uruguay **PHONE:** 598 2 922362, 598 2 922363 **FX:** 598 2 793132 **FOUNDED:** 1989. **OFFICERS:** Eduardo Gudynas, Exec. Dir. **STAFF:** 6. **LANGUAGES:** English, Portuguese.

DESCRIPTION: Studies the relationship of humans to the environment. Coordinates the Latin American Network on Social Ecology. Conducts research on alternative development strategies, environmental ethics, natural resource management, and local governmental environmental policies. **PUBLICATIONS:** *Ecology, Market and Development* (in Spanish); *Estudios en Ecologia Social* (in Spanish); *Introduction to the Methodologies of Social Ecology* (in Spanish); *Teko-Ha* (in English and Spanish), annual. Lists feature articles, news, bibliography on social ecology, book reviews, and announcements.; *Temas Clave* (in Spanish), Irregular; Reprints. Papers published by staff in other journals.

LATIN AMERICAN CIVIL AVIATION COMMISSION (LACAC) (Comision Latinoamericana de Aviacion Civil - CLAC)

Aeropuerto Internacional, Zona Comercial, Edificio CORPAC, Piso 2, Apartado Postal 4127, Lima 100, Peru **PHONE:** 51 15 753664, 51 15 751646 **FX:** 51 15 751743 **E-MAIL:** clacsec@chavin.rcp.net.pe **FOUNDED:** 1973. **OFFICERS:** Marco Ospina Yepez, Sec. **MEMBERS:** 21. **STAFF:** 5. **BUDGET:** US$138,250. **LANGUAGES:** Spanish.

DESCRIPTION: Organization of Latin American countries, as represented by civil aviation directors. Purpose is to harmonize policies and rules concerning civil aviation in an effort to regulate air transport in the region. Compiles origin and destination statistics of passengers, freight, and mail in Latin America. **PUBLICATIONS:** *Digest of Statistics* (in Spanish), annual.

LATIN AMERICAN COUNCIL OF CHURCHES (CLAI) (Consejo Latinoamericano de Iglesias - CLAI)

Calle Inglaterra 943 y Mariana de Jesus, Casilla 17-08-8522, Quito, Ecuador **PHONE:** 593 2 553996, 593 2 529933 **FX:** 593 2 553996 **E-MAIL:** manuel@clai.ecuanex.net.ec **WEBSITES:** http://www.ecuanex.apc.org/clai **FOUNDED:** 1982. **OFFICERS:** Rev. Ysrael Batista. **MEMBERS:** 152. **STAFF:** 40. **BUDGET:** US$1,000,000. **LANGUAGES:** English, Portuguese, Spanish.

DESCRIPTION: Churches and associated organizations in 19 countries representing 11 Christian denominations. Promotes unity among churches in Latin America. Sponsors charitable programs; offers placement and children's services. Conducts educational programs. Holds workshops, seminars, and training sessions. Operates speakers' bureau. **PUBLICATIONS:** *CLAI News* (in English), quarterly; *Rapidas* (in Spanish), monthly; *Signos de Vida* (in Spanish); Books.

LATIN AMERICAN ECONOMIC SYSTEM (SELA) (Systeme Economique Latino-Americain)

Avenida Francisco de Miranda, Edificio Centro Empresarical Pargee de Este, Apartado de Correos 17035, Urbanizacion La Carlota, Caracas 1010-A, Venezuela **PHONE:** 58 2 9055111, 58 2 9055208 **FX:** 58 2 2388923 **TX:** 23294 **E-MAIL:** difusion@sela.org **WEBSITES:** http://www.sela.org **FOUNDED:** 1975. **OFFICERS:** Carlos Moneta, Permanent Sec. **MEMBERS:** 27. **STAFF:** 72. **BUDGET:** US$3,900,000. **LANGUAGES:** English, French, Portuguese, Spanish.

DESCRIPTION: Organization of Latin American and Caribbean countries. Promotes the principles of equality, sovereignty and independence of states, solidarity, non-intervention in internal affairs, and respect for differences in political, economic, and social systems. Encourages intraregional cooperation in order to accelerate the economic and social development of member states. Provides a permanent system of consultation and coordination for the adoption of common positions and strategies on economic and social issues in international forums. Areas of interest include: regional integration and cooperation; international trade; industrialization; economic relations between member states and other contries or groups of countries; participation of member states in international economic and environmental organizations. Activities have included: coordination of the Latin American and Caribbean position at the Uruguay Round; formulation of regional industrialization policy; evaluation of problems facing the Latin American and Caribbean regions. **PUBLICATIONS:** *Capitulos del Sela* (in Spanish), quarterly, semiannually in English; *Meeting Documents*, annual; *SELA Antenna in the United States* (in English and Spanish), monthly; *Strategig Issues* (in English and Spanish), monthly. Also publishes *Papeles del Sela*.

LATIN AMERICAN ENERGY ORGANIZATION (OLADE) (Organizacion Latinoamericana de Energia - OLADE)

Av. Mariscal Antonio Jose de Sucre N58-63, Sector San Carlos, PO Box 17-11-6413, Quito, Ecuador **PHONE:** 593 2 598122, 593 2 597995 **FX:** 593 2 539684 **TX:** 2728 OLADE ED **E-MAIL:** olade1@olade.org.ec **WEBSITES:** http://www.olade.org.ec **FOUNDED:** 1973. **OFFICERS:** Mr. Luiz A.M. da Fonseca, Exec.Sec. **MEMBERS:** 25. **STAFF:** 48. **LANGUAGES:** English, French, Portuguese, Spanish.

DESCRIPTION: Latin American and Caribbean nations united to develop and protect their shared and individual energy resources. Serves to collectively defend against external sanction and coercion applied in reaction to the energy measures of member states; seeks to create a Latin American energy market. Coordinates negotiations between members to ensure an adequate supply of energy for the national development of each state; encourages, within and among the member states, the industrialization of energy resources, the expansion of industries, and the implementation of energy projects. Promotes the creation of institutions to finance energy and energy-related projects. Works to develop fair energy pricing policies and to ensure adequate means of transporting energy resources. Fosters technical cooperation, the exchange of scientific, juridical, and contractual information, and the development and sharing of energy technology. Advocates measures to prevent environmental pollution due to energy activities. Offers courses; compiles statistics. **PUBLICATIONS:** *Energy - Economic Statistics and Indicators of Latin America and the Caribbean* (in English and Spanish), annual; *Energy Magazine* (in English and Spanish), 3/year; *Latin-American Energy Update* (in English and Spanish), quarterly; *Publications List* (in English and Spanish), annual.

LATIN AMERICAN EPISCOPAL COUNCIL (Consejo Epicopal Latinoamericano)

Carrera 5 A, No. 118-31, Usaquen, Apartado Aereo 5278, Bogota 51086, Colombia **PHONE:** 57 1 6121620, 57 1 6128990 **FX:** 57 1 6121929 **E-MAIL:** celam@celam.org **WEBSITES:** http://www.celam.org **FOUNDED:** 1955. **OFFICERS:** Mons. Jorge E. Jimenez, Pres. **LANGUAGES:** English, French, Portuguese, Spanish.

DESCRIPTION: Organization of cardinals, archbishops, and bishops of the Catholic church in Latin American and Caribbean countries. **PUBLICATIONS:** Books (in Portuguese and Spanish), quarterly; Magazine (in Portuguese and Spanish), quarterly.

LATIN AMERICAN IRON AND STEEL INSTITUTE (ILAFA) (Instituto Latinoamericano del Fierro y el Acero - ILAFA)

Casilla 16065, Correo 9, Las Condes, Santiago 9, Chile **PHONE:** 56 2 2330545, 56 2 2337742 **FX:** 56 2 2330768 **E-MAIL:** ilafa@entelchile.net **WEBSITES:** http://www.ilafa.org **FOUNDED:** 1959. **OFFICERS:** Anibal Gomez, Sec.Gen. **MEMBERS:** 131. **STAFF:** 13. **LANGUAGES:** Portuguese, Spanish.

DESCRIPTION: Latin American iron and steel enterprises; producers of iron ore, ferroalloys, and other raw material; research and national iron and steel institutes; engineering and equipment manufacturers. Affiliated members are nonregional enterprises and institutions. Purposes are to: promote the study of scientific, technical, industrial, and economic problems dealing with production, marketing and use of steel, steelmaking, raw materials, and other inputs and equipment; further the creation and dissemination of technical knowledge; foster standardization. Organizes training courses, technical meetings, and roundtables; arranges personal contacts; offers information services; cooperates with regional and international organizations; compiles statistics. Conducts seminars. **PUBLICATIONS:** *ACERO Latinoamericano* (in Spanish), bimonthly; *Proceedings of Technical Meeting* (in Portuguese and Spanish), periodic; *Repertorio Siderurgico Latinoamericano 1995* (in Spanish); *Statistical Yearbook* (in English and Spanish).

LATIN AMERICAN RAILWAYS ASSOCIATION (Asociacion Latinoamericana de Estradas de Ferro)

Av. Belgrano 863, piso 1, 1092 Buenos Aires, Argentina **PHONE:** 54 1 3311298, 54 1 3430593 **FX:** 54 1 3312747 **E-MAIL:** alaf@alaf.int.ar **FOUNDED:** 1964. **OFFICERS:** Agustin Rafael, Gen.Sec. **MEMBERS:** 83. **STAFF:** 19. **BUDGET:** US$300,000. **LANGUAGES:** Portuguese, Spanish.

DESCRIPTION: Members include: railways (22), metropolitan railways (6), and industrial railway enterprises (55) in 18 countries. Purposes are to: promote economical, efficient, and reliable railway transport; encourage trade by railway; coordinate the development of Latin American railways and related industries, thereby enhancing the prosperity of Latin American peoples. Encourages cooperation among developing countries for the advancement of the railway industry. Administers scholarships offered by members. Compiles statistics. Coordinates professional training programs. **PUBLICATIONS:** *Anuario Estadistico Ferrocarriles Latinoamericano* (in Spanish), annual; *Revista ALAF*, 3/year.

LATIN AMERICAN STUDIES ASSOCIATION (LASA)

946 William Pitt Union, Univ. of Pittsburgh, Pittsburgh, PA 15260 **PHONE:** (412) 648-7929 **FX:** (412) 624-7145 **WEBSITES:** http://www.pitt.edu/-lasa/ **FOUNDED:** 1966. **OFFICERS:** Reid Reading, Exec.Dir. **MEMBERS:** 4,900. **STAFF:** 4. **BUDGET:** US$200,000. **LANGUAGES:** Portuguese, Spanish.

DESCRIPTION: Organization of persons and institutions with scholarly interests in Latin America, inclusive of non-U.S. nationals. Encourages more effective training, teaching, and research in Latin American studies; aids in interchange of professional personnel; works to improve library resource materials in this field; provides centralized information services; fosters research and publications. **PUBLICATIONS:** *Brochure* (in English, Portuguese, and Spanish), annual; *LASA Forum*, quarterly. Reports on anthropology, economics, geography, political science, human rights, literature, the arts, and Spanish and Portugese languages; *Latin American Research Review*, 3/year. Provides articles, research reports and notes, and review essays. Also issues special reports, reprints, and the publications of the Consortium of Latin American Studies.

LATVIAN NATIONAL FOUNDATION (LNF)

Box 108, S-101 22 Stockholm 1, Sweden **PHONE:** 46 8 6414821, 46 8 208222 **FOUNDED:** 1947. **OFFICERS:** Mara Strautmane, Chm. **MEMBERS:** 6,000. **BUDGET:** 1,000,000 SKr. **LANGUAGES:** English, French, German, Latvian, Spanish, Swedish.

DESCRIPTION: Members are individuals, Latvian organizations, and church groups in 9 countries. Encourages support for political, economic, and social reforms taking place in Latvia. Maintains records on individuals that were deported under Soviet rule. Provides economic assistance to families in Latvia. Supports research and Latvian studies; promotes exchange programs between Latvia and other countries. Distributes political and educational materials. Compiles demographic statistics. **PUBLICATIONS:** *LNF Informacija*, semiannual; Books.

LAUBACH LITERACY INTERNATIONAL (LLI)

Box 131, 1320 Jamesville Ave., Syracuse, NY 13210 **PHONE:** (315) 422-9121 **TF:** (888) 528-2224 **FX:** (315) 422-6369 **E-MAIL:** info@laubach.org **WEBSITES:** http://www.laubach.org **FOUNDED:** 1955. **OFFICERS:**

Robert F. Caswell, Pres. & CEO. **MEMBERS:** 80,000. **STAFF:** 109. **BUDGET:** US$11,000,000. **LANGUAGES:** English.

DESCRIPTION: Seeks to reduce adult illiteracy worldwide. Motivates and supports teaching of the world's estimated 965 million illiterate adults and older youths to a level of listening, speaking, reading, writing, and basic computational skills enabling them to solve their daily problems. Laubach sponsors more than 1100 literacy programs in the US and 71 partner programs in 36 developing countries worldwide. Provides training, technical assistance, and resources for local adult literacy programs using trained volunteers. Operates New Readers Press (NRP), a division that publishes educational materials for adults with limited reading skills and resources for teachers and tutors in the U.S. Signal Hill, an imprint of Laubach Literacy, publishes informational and pleasure reading materials by and for adults with limited reading skills. **PUBLICATIONS:** *Laubach Literacy Action Directory*, annual; *Laubach Literacy International—Annual Report*. Highlights LLI overseas programs and activities of Laubach Literacy Action and New Readers Press. Includes financial statements.; *Laubach LitScape*, quarterly. Newsletter for tutors and members of LLA, covers literacy issues and LLA programs and information.; *Literacy Advocate*, periodic. Covers U.S. and international programs and membership activities.; *New Readers Press—Catalog*, periodic. Lists LLI's publications and materials for teenagers and low-level readers.; *News for You*, weekly. For older teens and adults with special reading needs. Includes 4th to 6th grade reading level material and international and national news.; *Signal Hill Publications*, periodic. Lists informational and pleasure reading materials for adults and older teens with limited reading skills.

LAW ASSOCIATION FOR ASIA AND THE PACIFIC (LAWASIA)

11th Floor, NT House, 22 Mitchell St., Darwin, NT 0800, Australia **PHONE:** 61 8 89469500 **FX:** 61 8 89469505 **E-MAIL:** lawasia@lawasia.asn.au **WEBSITES:** http://www.lawasia.asn.au **FOUNDED:** 1966. **OFFICERS:** Roslyn West, Sec.Gen. **MEMBERS:** 2,500. **STAFF:** 4. **LANGUAGES:** English.

DESCRIPTION: Lawyers, organizations, firms, and corporations in 24 countries. Facilitates an international and professional network for lawyers to update, reform and develop the law in the region. Provides members the opportunity to be updated in the laws of the region by offering specialty conferences throughout the year and a biennial copnference held in every odd year. **PUBLICATIONS:** *LAWASIA* (in English), quarterly; *LAWASIA Directory of Members* (in English), annual; *LAWASIA Human Rights Newsletter*, bimonthly; *LAWASIA Newsletter*, quarterly. Also publishes reports.

LAWYERS COMMITTEE FOR HUMAN RIGHTS (LCHR)

333 7th Ave., 13th Fl., New York, NY 10001 **PHONE:** (212) 845-5200 **FX:** (212) 845-5299 **TX:** 5106005783 **FOUNDED:** 1978. **OFFICERS:** Michael Posner, Exec.Dir. **MEMBERS:** 800. **STAFF:** 28. **BUDGET:** US$2,200,000. **LANGUAGES:** English.

DESCRIPTION: Public interest law center that works to promote international human rights and refugee law and legal procedures. Focuses on cases where volunteer lawyers may help promote international human rights standards. Is involved in the pro bono representation of indigent political asylum applicants in the U.S. Investigates human rights issues in the justice system for follow-ups with local lawyers, U.S. foreign policy issues in markers, and intergovernmental organizations. Conducts training sessions and educational workshops for attorneys on international human rights and refugee law. Bestows Roger Baldwin Medal of Liberty every two years to individuals and organizations making a significant contribution to human rights in any part of the world. **PUBLICATIONS:** *Lawyers Committee for Human Rights—Newsbriefs*, quarterly. Reports on committee activities and developments in asylum law. Includes obituaries. Also publishes periodic reports on human rights abuses, refugee law, and related topics.

LAY MISSION-HELPERS ASSOCIATION (LMHA)

3424 Wilshire Blvd., Los Angeles, CA 90010-2241 **PHONE:** (213) 637-7222 **FX:** (213) 637-6223 **E-MAIL:** propla@relaypoint.net **WEBSITES:** http://www.relaypoint.net/-propla **FOUNDED:** 1955. **OFFICERS:** Msgr. Michael Meyers, Dir. **MEMBERS:** 700. **STAFF:** 2. **LANGUAGES:** English.

DESCRIPTION: Roman Catholic laymen and laywomen serving as missionaries overseas. Objectives are to develop the spiritual life of members and provide assistance in missionary areas. Trains and assigns missionaries to serve in apostolates for a period of two to three years in areas of Africa, Papua New Guinea, Micronesia, and Thailand; conducts eight-month training course with classes in theology, spirituality, scripture, liturgy, missiology, anthropology, and other subjects; bases assignments on applicants' aptitude and training. Mission activities are in areas such as education, academic instruction at all grade levels, paramedical treatment, nursing, participation in medical and hospital projects, building and construction, mechanics, bookkeeping, secretarial and clerical work, and other crafts and trades. Provides missionaries with room and board, health care, and money for incidental expenses and necessities. Works in cooperation with Mission Doctors Association.

LEAD DEVELOPMENT ASSOCIATION INTERNATIONAL (LDA)

42 Weymouth St., London W1N 3LQ, England **PHONE:** 44 171 4998422 **FX:** 44 171 4931555 **E-MAIL:** eng@ldaint.org **WEBSITES:** http://www.ldaint.org/default.htm **FOUNDED:** 1956. **OFFICERS:** Dr. D.N. Wilson, Dir. **MEMBERS:** 20. **STAFF:** 5. **BUDGET:** £300,000. **LANGUAGES:** English.

DESCRIPTION: Members are mining companies, metal producers, semi-fabricators, associations, and user groups in 10 countries. Promotes use of lead in all forms; represents the lead industry at national and international levels. Provides promotional and technical information concerning lead to developing countries. Organizes seminars on technical and environmental aspects of the production and use of lead. Conducts market surveys. **PUBLICATIONS:** *Proceedings*, biennial.

LEAGUE OF ARAB STATES (LAS) (Jamiat Adduwal Alarabia - JAA)

Tahrir Sq., PO Box 11642, Cairo, Egypt **PHONE:** 2 5750511 **TX:** 92111 **FX:** 5775626 **FOUNDED:** 1945. **OFFICERS:** Dr. Ahmed Esmat Abd Al-Meguid, Sec.Gen. **MEMBERS:** 21. **LANGUAGES:** Arabic.

DESCRIPTION: Representatives of Algeria, Bahrain, Djibouti, Egypt, Iraq, Jordan, Kuwait, Lebanon, Libya, Mauritania, Morocco, Oman, Palestine, Qatar, Saudi Arabia, Somalia, Sudan, Syria, Tunisia, United Arab Emirates, and Yemen. Works toward peace and security in the Arab region. Settles regional disputes; has dealt with the Arab-Israeli conflict by formulating a common strategy toward Palestine and a common economic policy toward Israel. Promotes cooperation among member states in the areas of economic affairs, military cooperation, communication, cultural affairs, and passport, extradition, and health matters. Assists national liberation movements and aids in combating colonialism. Maintains observer status with the United Nations. Holds semiannual committee and council meetings. Has established Academy of Arab Music, Arab Authority for Exhibitions, Institute for Forestry, and Special Bureau for Boycotting Israel. **PUBLICATIONS:** *Abstracts from the Zionist Media* (in Arabic), weekly; *Analytical Bulletin of Arab Countries in Foreign Trade with EEC* (in Arabic and English), periodic; *Analytical Bulletin of Arab Trade with U.S.A.* (in Arabic and English), periodic; *Housing* (in Arabic), quarterly; *LAS Information Bulletin* (in Arabic); *Military Affairs* (in Arabic), quarterly; *The Palestine Question* (in Arabic), monthly; *Population Bulletin* (in Arabic and English), periodic; *Sh'oun Arabiyya* (in Arabic), quarterly.

LEATHERHEAD FOOD RESEARCH ASSOCIATION (LFRA)

Randalls Rd., Leatherhead, Surrey KT22 7RY, England **PHONE:** 44 1372 376761 **FX:** 44 1372 386228 **E-MAIL:** jtomlinson@lfra.co.uk **WEBSITES:** http://www.lfra.co.uk **FOUNDED:** 1919. **OFFICERS:** Dr. Mark Kierstan, Dir. **MEMBERS:** 800. **STAFF:** 220. **LANGUAGES:** Dutch, English, French, German, Italian, Spanish.

DESCRIPTION: Members are manufacturers and suppliers of food products. Seeks to advance members' commercial interests. Provides national and international support and services to food products manufacturers and suppliers. **PUBLICATIONS:** *European Legislation Manual* (in English), annual; *Market Report*, periodic; *UK Legislation Manual* (in English), annual; *Worldwide Legislation Manual* (in English), annual. Contains scientific and technical reports; Journal (in English), monthly.

LIAISON GROUP OF THE EUROPEAN MECHANICAL, ELECTRICAL, ELECTRONIC, AND METALWORKING INDUSTRIES (Organisme de Liaison des Industries Mecaniques, Electriques, Electroniques, et de la Transoformation des Metaux en Europe - ORGALIME)

Diamant Building, Fifth floor, Boulevard A. Reyers 80, B-1030 Brussels, Belgium **PHONE:** 32 2 7068235 **FX:** 32 2 27068250 **E-MAIL:** secretariat@orgalime.be **FOUNDED:** 1954. **OFFICERS:** Patrick Knox-Peebles, Sec.Gen. **MEMBERS:** 25. **STAFF:** 10.

DESCRIPTION: Members are national trade associations representing 40,000 mechanical, electrical/ electronic engineering, and metalworking firms in 16 western European countries. Represents members' interests to the European Commission, Council, and Parliament. Seeks to influence European policies affecting the industry. Encourages cooperation and information exchange between members. Prepares model contracts and guides for use by members. Arranges seminars. **PUBLICATIONS:** *ORGALIME News* (in English).

LIBERAL INTERNATIONAL (LI) (Internationale Liberale)

1 Whitehall Pl., London SW1A 2HD, England **PHONE:** 44 171 8395905 **FX:** 44 171 9252685 **E-MAIL:** li@worldlib.org **WEBSITES:** http://www.worldlib.org/ **FOUNDED:** 1947. **OFFICERS:** Julius Maaten, Sec.Gen. **MEMBERS:** 80. **STAFF:** 5. **LANGUAGES:** English, French, German, Spanish.

DESCRIPTION: Organization of national liberal political parties and groups. Promotes: the welfare and protection of economic, racial, political, and ethnic minorities; the respect of human rights; decentralization of political and economic power; individual liberty and tolerance; conflict resolution by peaceful means. Monitors national elections. Coordinates international activities of liberal parties; conducts study visits. Cooperates with liberal forces in the developing world and cultural and research institutions. **PUBLICATIONS:** *A Changing World*. Features the history of liberal international.; *A Sense of Liberty*. Contains basic documents of the Liberal International.; *Congress Resolutions*, annual; *Facts and Figures about the Liberal International.*, semiannual; *London Aerogramme* (in English), bimonthly; *Strengthening of the United Nations*. Report of Vayrynen Committee.

LIGA INTERNATIONAL (LI)

19531 Campus Dr., Ste. 20, Santa Ana, CA 92707 **PHONE:** (714) 852-8611 **FX:** (714) 852-8739 **E-MAIL:** liga@earthlink.net **WEBSITES:** http://www.sdaworld.com/liga// **FOUNDED:** 1948. **OFFICERS:** Jacquelyn Hanson, Pres. **MEMBERS:** 700. **STAFF:** 1. **LANGUAGES:** English, Spanish.

DESCRIPTION: Members are physicians, dentists, nurses, pilots, technicians, assistants, educators, and laypeople interested in providing medical and educational assistance to impoverished people of rural Mexico. Liga (Spanish word for ''league'') seeks to stimulate interest and support for establishing and maintaining educational, charitable, and medical programs among underprivileged inhabitants of Mexico; exchange scientific information between medical and educational groups. Sponsors monthly trips to clinics in Ocoroni, El Fuerte, San Blas, in Sinaloa Mexico. Operates speakers' bureau. **PUBLICATIONS:** *Liga High Flying Times*, quarterly; Brochure.

LIGHTHOUSE INTERNATIONAL

Minami-fukuin Shinryojo, 161-4 Kitamotojuku, Kitamoto-shi 364, Saitama, Japan **PHONE:** 81 485 917191 **FX:** 81 485 919688 **E-MAIL:** todai87@jca.ax.apc.org **FOUNDED:** 1987. **OFFICERS:** Akira Hattori **MEMBERS:** 400. **STAFF:** 15. **BUDGET:** US$469,046. **LANGUAGES:** English, Japanese.

DESCRIPTION: Promotes the welfare of Afghan refugees in Pakistan. Works for improved educational opportunities and increased access to health care for Afghan refugees. Operates medical centers in Afghanistan and Pakistan; maintains schools in refugee camps. Conducts fundraising activities. **PUBLICATIONS:** *Todai* (in Japanese), quarterly.

LIONS CLUBS INTERNATIONAL (LCI)

300 22nd St., Oak Brook, IL 60523 **PHONE:** (630) 571-5466 **FX:** (630) 571-8890 **E-MAIL:** lions@lionsclubs.org **WEBSITES:** http://www.lionsclubs.org **FOUNDED:** 1917. **OFFICERS:** Win Hamilton, Exec.Admin. **MEMBERS:** 1,425,000. **STAFF:** 325. **BUDGET:** US$46,000,000. **LANGUAGES:** Chinese, English, Finnish, French, German, Italian, Japanese, Korean, Portuguese, Spanish, Swedish.

DESCRIPTION: Organization of local clubs representing business and professional men and women in 185 countries and geographic areas. Provides community service in order to increase international understanding and cooperation. Fosters awareness of environmental, social, and health related problems. Activities include: blindness prevention; work with the deaf; drug awareness programs; international youth camp programs; youth exchange. Maintains Leo Clubs (for young adults), and the Lions Clubs International Foundation for sight conservation and work with the blind. **PUBLICATIONS:** *The Lion Magazine*, 10/year. Includes award and convention news, and obituaries. In 20 languages.

LIPTAKO-GOURMA INTEGRATED DEVELOPMENT AUTHORITY (Autorite de Developpement Integre de la Region Liptako-Gourma)

avenue M. Thevenond, Boite Postale 619, Ouagadougou, Burkina Faso **PHONE:** 226 306148 **TX:** 5247 **OFFICERS:** Isagne Dembele, Sec.-Gen. **LANGUAGES:** English, French.

DESCRIPTION: Organization of countries encompassing the Liptako-Gourma region. Promotes appropriate and sustainable development in Liptako-Gourma. Coordinates development activities in the region; gathers and disseminates information.

LUTHERAN WORLD FEDERATION (LWF) (Federation Lutherienne Mondiale - FLM)

150, rte. de Ferney, BP 2100, CH-1211 Geneva 2, Switzerland **PHONE:** 41 22 7916111 **FX:** 41 22 7988616 **TX:** 415 546 LWS CH **FOUNDED:** 1947. **OFFICERS:** Rev.Dr. Ishmael Noko, Gen.Sec. **MEMBERS:** 122. **STAFF:** 94. **BUDGET:** US$131,000,000. **LANGUAGES:** English, French, German, Spanish, Swedish.

DESCRIPTION: Organization of Lutheran churches representing 57,340,000 individuals in 60 countries. Acts to coordinate relief and development work, conduct ecumenical dialogues, and facilitate joint action among Lutherans. Conducts bilateral dialogue worldwide; assists Third World churches in communications, development, and missions projects; aids in the development of theological education in Africa; coordinates mission activities worldwide; aids member churches in coping with global issues including human rights, peace and justice, and developing church ties between the East and West. **PUBLICATIONS:** *Lutheran World Federation Directory* (in English), annual. Lists LWF member churches, recognized congregations, and other contacts.; *Lutheran World Information*, bimonthly; *LWF Documentation*, 2-3/year; *LWF Today*, quarterly. Contains description of LWF aims and activities.

MAGNUS HIRSCHFELD CENTER FOR HUMAN RIGHTS (MHCHR)

c/o Crosswicks House, 551 Valley Rd., Ste. 169, Upper Montclair, NJ 07043-1832 **PHONE:** (201) 237-3406 **FX:** (973) 746-3147 **E-MAIL:** crosswix@hotmail.com **WEBSITES:** http://www.envirolink.org/orgs/-magnus; http://www.angelfire.com/nj/hirschfeldcentre/index.html **FOUNDED:** 1995. **OFFICERS:** William A. M. Courson, Exec.Dir. **MEMBERS:** 6. **STAFF:** 3. **LANGUAGES:** English, French, Spanish.

DESCRIPTION: Promotes respect for the civil and human rights of the individual worldwide, with particular emphasis on protecting the rights of gay, lesbian, bisexual and transgendered people. Documents cases involving abuse of human rights; conducts legal research and educational programs and works with international organizations to address human rights abuses. Makes available legal services, with respect to legal representation of victims of human rights abuses before international and regional fora. **PUBLICATIONS:** *Occasional Papers*, quarterly. Copies of Legal Pleadings.

MAN AND THE BIOSPHERE PROGRAMME (MAB) (Programme sur l'Homme et la Biosphere - MAB)

Division of Ecological Sciences, UNESCO, 1, rue Miollis, F-75725 Paris Cedex 15, France **PHONE:** 33 1 45684067 **FX:** 33 1 45685804 **TX:** 204461 **E-MAIL:** mab@unesco.org **WEBSITES:** http://www.unesco.org/mab **FOUNDED:** 1971. **OFFICERS:** Dr. Pierre Lasserre, Dir. **MEMBERS:** 136. **STAFF:** 30. **BUDGET:** US$4,000,000. **LANGUAGES:** Arabic, Chinese, English, French, Russian, Spanish.

DESCRIPTION: Carries out, through its national committees, projects focusing on problems of resource management involving both natural and social scientists. Studies human impact on tropical forests, grazing lands, arid zones, and other ecosystems; also examines the effect of environmental modifications on humans. Oversees the coordination of efforts and the exchange of information among scientists working on related problems in similar ecological zones. Has established a World Network of Biosphere Reserves composed of 352 sites in 87 countries. Works in cooperation with the Food and Agriculture Organization of the United Nations, the International Council of Scientific Unions, the International Conservation Union, the World Conservation Union (IUCN), the WWF, United Nations Environment Programme, the World Health Organization, and the World Meteorological Organization. **PUBLICATIONS:** *InfoMAB* (in English and French), semiannual; *International Bulletin on Biosphere Reserves* (in English and French), periodic. Also publishes brochures, digests, reports, the MAB book series, posters on biodiversity, and folding maps on biosphere reserves (in several languages).

MAP INTERNACIONAL - OFICINA REGIONAL PARA AMERICA LATINA (MAP)

Isla Espanola 222 y Rio Coca, Casilla 17-08-8184, Quito, Ecuador **PHONE:** 593 2 469384, 593 2 250859 **FX:** 593 2 250859 **E-MAIL:** map@map.ecx.ec **FOUNDED:** 1976. **OFFICERS:** Mauricio Solis Paz **STAFF:** 12. **LANGUAGES:** English, Spanish.

DESCRIPTION: Promotes and supports awareness of environmental concerns throughout Latin America. Provides a forum for improved relations among church leaders. Facilitates health programs. Encourages the church to fulfill its biblical mandate to be a healing community engaged in transforming individuals, families, and institutions. **PUBLICATIONS:** *Boletin SIDA* (in Spanish), quarterly. Provides information on AIDS.

MAP INTERNATIONAL (MAP)

PO Box 215000, Brunswick, GA 31521-5000 **PHONE:** (912) 265-6010 **TF:** (800) 225-8550 **FX:** (912) 265-6170 **E-MAIL:** mapus@map.org **WEBSITES:** http://www.map.org **FOUNDED:** 1954. **OFFICERS:** Paul B. Thompson, Pres./CEO. **STAFF:** 100. **BUDGET:** US$111,000,000. **LANGUAGES:** English, French, Spanish.

DESCRIPTION: Non-profit Christian relief and development organization that promotes the health of people living in the world's poorest communities. Works with partners in the areas of community health development, disease prevention and eradication, relief and rehabilitation and global health advocacy. Promotes access to health services and essential medicines in more than 100 countries each year. **PUBLICATIONS:** *MAP International Report*, bimonthly; *Our Health*, semiannual; Annual Report.

MARANGOPOULOS FOUNDATION FOR HUMAN RIGHTS

1 Lycavittou St., GR-106 72 Athens, Greece **PHONE:** 30 1 3637455, 30 1 3613527 **FX:** 30 1 3622454 **FOUNDED:** 1977. **OFFICERS:** L.A. Sicilianos, Dir. **STAFF:** 9. **BUDGET:** 75,000,000 Dr. **LANGUAGES:** English, French, German, Greek.

DESCRIPTION: Promotes and works to protect fundamental human rights and freedoms, with primary emphasis on the advancement of human rights education and training. Works to increase public interest in issues of human rights, peace, and the development of democratic institutions. Conducts and supports research programs; holds colloquys and international conferences; disseminates information on human rights violations to government bodies; cooperates with organizations holding similar goals. Drafts conventions governing the establishment and maintenance of human rights policies. Makes available free legal aid to victims of human rights abuses. Organization derives its name from, and was founded as the legacy of, George N. Marangopoulos, Chief Justice of the Supreme Administrative Court of Greece and defender of human rights under the Greek military junta of 1967-73. **PUBLICATIONS:** *Aspects of the Protection of Individual and Social Rights* (in English, French, German, and Greek); *The Human Dimension of the CSCE* (in English and French). Contains the proceedings from the colloquy of the CSCE.; *The International Protection of Human Rights* (in Greek); *New Forms of Discrimination: Migrants, Refugees, Minorities* (in English and French); *Protection of Human Rights in the European Framework* (in English, French, German, and Greek); *Protection of the Rights of Migrant Workers and Their Families - The International and the National Dimension* (in English, French, and Greek); *Women's*

Rights - Human Rights (in English and Greek); Proceedings, periodic.

MARYKNOLL MISSION FAMILY (MMF)

PO Box 307, Maryknoll, NY 10545-0307 **PHONE:** (914) 941-7590, (914) 941-7636 **FX:** (914) 941-4705 **E-MAIL:** mkweb@maryknoll.org **WEBSITES:** http://www.maryknoll.org **FOUNDED:** 1911. **OFFICERS:** Raymond Finch, Superior Gen. **MEMBERS:** 631. **LANGUAGES:** English.

DESCRIPTION: Members are priests, religious brothers, associate priests and seminarians. Purposes are to: bring the Gospel of Jesus Christ to poor and oppressed people in 27 foreign countries; assist the Maryknoll Society to become more aware of human rights and religious freedom issues within those countries; stimulate greater concern for peace in the U.S. and throughout the world; serve as a forum for communicating and disseminating information on topics and issues of justice and peace to members, other individuals, and organizations; establish links with and collaborate in various programs and activites with churches and secular groups sharing mutual concerns. Encourages scholarship relating to missionary topics. **PUBLICATIONS:** *Maryknoll Magazine*, monthly. Mission magazine works of Maryknoll missioners overseas; *Newsnotes*, bimonthly; *Orbis Books*; *Revista Maryknoll* (in English and Spanish), monthly. Bi-Lingual Mission magazine; *World Parish*, monthly.

MEDIC ALERT FOUNDATION INTERNATIONAL (MAFI)

2323 Colorado Ave., Turlock, CA 95382 **PHONE:** (209) 668-3333, (800) 228-6222 **TF:** (800) 432-5378 **FX:** (209) 669-2495 **E-MAIL:** inquiries@medicalert.org **WEBSITES:** http://www.medicalert.org **FOUNDED:** 1956. **OFFICERS:** Tanya J. Glazebrook, Pres./CEO. **MEMBERS:** 4,700,000. **STAFF:** 137. **LANGUAGES:** English, Spanish.

DESCRIPTION: Medical information service open to all. Computerized, confidential medical file, Medic Alert emblem on pendant or bracelet engraved with primary medical condition and can call collect number to access 24-hour Emergency Response Center. Information is used to speed diagnosis, accurate treatment and save lives. Conducts continuous public and professional education program. **PUBLICATIONS:** *Medic Alert Foundation International*, quarterly.

MEDICAL REHABILITATION CENTER FOR TORTURE VICTIMS (MRCTV)

9 Lycabettous St., GR-106 72 Athens, Greece **PHONE:** 30 1 3604967, 30 1 3646807 **FX:** 30 1 3612273 **E-MAIL:** mrct@compulink.gr **FOUNDED:** 1989. **OFFICERS:** Maria Piniou-Kalli, Dir. **MEMBERS:** 100. **STAFF:** 10. **BUDGET:** US$180,000. **LANGUAGES:** English, Greek.

DESCRIPTION: Members are physicians and other health care professionals. Promotes ratification of the International Conventions for Human Rights. Advocates for the worldwide abolition of torture. Provides medical and psychological and legal assistance to victims of torture. Conducts research on medical aspects of torture; sponsors public educational programs. **PUBLICATIONS:** *Anthropina* (in English and Greek), quarterly.

MEDICAL WOMEN'S INTERNATIONAL ASSOCIATION (MWIA)

Herbert-Lewin-Strasse 1, D-50931 Cologne, Germany **PHONE:** 49 221 4004558, 49 221 4004559 **FX:** 49 221 4004557 **TX:** 08882161 BAEK **E-MAIL:** mwia@aol.com **WEBSITES:** http://members.aol.com/mwia/index.htm **FOUNDED:** 1919. **OFFICERS:** Dr. Waltraud Diekhaus, Sec.Gen. **MEMBERS:** 20,000. **STAFF:** 2. **LANGUAGES:** English.

DESCRIPTION: Members are women involved in or interested in medicine in 70 countries. Provides women with an opportunity to exchange information about medical problems with worldwide implications; promotes friendship and understanding between women; secures members' cooperation in matters relating to international health. Seeks to encourage women to enter the field of medicine and allied sciences and to overcome discrimination against female physicians. Aids women in developing countries in obtaining fellowships and grants for research and travel; offers information and advice to members visiting other countries. **PUBLICATIONS:** *Congress Report* (in English), triennial; *MWIA Update* (in English), quarterly; *Women Physicians of the World*.

MEKONG RIVER COMMISSION (MRC)

PO Box 1112, 364 M.V. Preah Monivong, Sangkat Phsar Doerm Thkouv, Khan Chamkar Mon, Phnom Penh, Cambodia **PHONE:** 855 23 720979 **FX:** 855 23 720972 **E-MAIL:** mrcs@bigpond.com.kh **FOUNDED:** 1995. **OFFICERS:** Mr. Yasunobu Matoba, CEO. **MEMBERS:** 4. **STAFF:** 100. **BUDGET:** 10,000,000 Bht. **LANGUAGES:** English.

DESCRIPTION: Established by the Agreement on the Cooperation for the Sustainable Development of the Mekong River Basin, replacing the former Mekong Committee and the subsequent Interim Mekong Committee. Aims to cooperate and promote in a constructive and mutually beneficial manner in the sustainable development, utilization, conservation, and management of the Mekong River Basin water and related resources for navigational and non-navigational purposes, as well as for socio and economic development. Strives to ensure that all efforts are consistent with the needs to protect, preserve, enhance, and manage the environmental and aquatic conditions and maintenance of the ecological balance to this river basin. **PUBLICATIONS:** *Lower Mekong Hydrologic Yearbook* (in English), annual; *Mekong Annual Report* (in English); *Mekong News* (in English), semiannual; *Work Programme* (in English), annual.

MENNONITE ECONOMIC DEVELOPMENT ASSOCIATES (MEDA)

302-280 Smith St., Winnipeg, MB, Canada R3C 1K2 **PHONE:** (204) 956-6430 **FX:** (204) 942-4001 **E-MAIL:** meda@meda.org **WEBSITES:** http://www.meda.org **FOUNDED:** 1953. **OFFICERS:** Ben Sprunger **MEMBERS:** 3,358. **BUDGET:** US$2,900,000. **LANGUAGES:** English.

DESCRIPTION: Promotes business and community development in Bolivia, Canada, Haiti, Honduras, Nicaragua, Russia, the United States, Tanzania, and Zimbabwe. Seeks to create economic self-sufficiency by providing credit, technical assistance, training, consulting, and appropriate technology to developing small businesses. Facilitates international trading partnerships. **PUBLICATIONS:** *The Marketplace* (in English), bimonthly.

MENSA INTERNATIONAL (MIL)

15 The Ivories, 6-8 Northampton St., London N1 2HY, England **PHONE:** 44 171 2266891 **FX:** 44 171 2267059 **FOUNDED:** 1946. **OFFICERS:** Edward J. Vincent, Exec.Dir. **MEMBERS:** 100,000. **STAFF:** 3. **LANGUAGES:** English.

DESCRIPTION: Individuals from 100 countries whose intelligence, as measured by standardized tests, is within the top 2 percent of the general population. Aims to promote: social contact among intelligent people; research in psychology and the social sciences; the identification and development of human intelligence. Mensa's policy is to include intelligent people of every opinion and background. Maintains no religious or political affiliations and is open to all who meet the intelligence criterion. Conducts research on public opinions and sociological potential of the highly intelligent. Provides volunteers for research workers requiring a high I.Q. group. Sponsors projects to foster intelligence and provides educational facilities. Sponsors periodic competitions; maintains speakers' bureau. Offers children's services and charitable program. **PUBLICATIONS:** *International Journal* (in English), monthly; *Registers*, periodic; Bulletin, periodic.

MERCY CORPS INTERNATIONAL

3030 SW 1st Ave., Portland, OR 97201-4796 **PHONE:** (503) 796-6800 **TF:** (800) 292-3355 **FX:** (503) 796-6844 **E-MAIL:** programs@mercycorps.org **WEBSITES:** http://www.mercycorps.org **FOUNDED:** 1979. **OFFICERS:** Neal Keny-Guyer, CEO. **STAFF:** 500. **BUDGET:** US$41,000,000. **LANGUAGES:** English.

DESCRIPTION: Assists poverty-striken communities throughout the world to achieve self-reliance, productivity, and human dignity. Motivates and educates the public about the plight of the poor and to work for peace and justice. Provides emergency relief, self-help development projects, agriculture projects, awareness programs, and development education. Maintains speakers' bureau. **PUBLICATIONS:** *Mercy Corps Report*, quarterly.

MEXICAN COMMISSION FOR THE DEFENSE AND PROMOTION OF HUMAN RIGHTS (Comision Mexicana de)

Tehuantepec 155, Colonia Roma, 06760 Mexico City, DF, Mexico **PHONE:** 52 5849116, 52 5642592 **FX:** 52 5842731 **E-MAIL:** cmdpdh@laneta.apc.org **WEBSITES:** http://www.laneta.apc.org/cmdpdh **FOUNDED:** 1989.

OFFICERS: Mariclaire Acosta, Pres. **MEMBERS:** 65. **STAFF:** 20. **BUDGET:** US$360,000. **LANGUAGES:** English, Spanish.

DESCRIPTION: Individuals and organizations with an interest in the protection of human rights worldwide. Promotes creation of public opinion favorable to the defense of human rights, leading to the establishment of a political and diplomatic resolution to protect human rights globally. Carries out investigations to identify human rights abuses worldwide; gathers and disseminates information; conducts public awareness and education campaigns. Organizes public campaigns to apply political pressure to governments found to be violating the human rights of their citizens. Monitors human rights policies and activities of the Mexican government. Conducts research; sponsors charitable activities; produces educational materials. **PUBLICATIONS:** *Guion. Derechos Humanos en Mexico* (in Spanish), bimonthly; Books; Report, periodic.

MICRONESIA INSTITUTE

PO Box 58064, Washington, DC 20037-8064 **PHONE:** (202) 342-9272 **FX:** (202) 342-9351 **E-MAIL:** actinc@erols.com **FOUNDED:** 1983. **OFFICERS:** Patricia Luce Chapman, CEO. **BUDGET:** US$200,000. **LANGUAGES:** English.

DESCRIPTION: Promotes and encourages efforts to advance the standard of living and quality of life in Micronesia. (Micronesia refers to the group of islands in the western Pacific which includes the Marshall Islands, Palau, FSM, Kiribati, Nauru, Northern Mariana Islands, and Guam.) Works to strengthen the friendly relations between Micronesians and Americans, and inform Americans about Micronesia. Serves as an independent link between Micronesian communities and public and private resources in the U.S. Encourages tourism and business growth, supports education and research, and initiates cooperative exchange programs. Provides programs to bring modern business practices to Micronesia. Develops counseling systems to help Micronesian students choose colleges in the U.S. Conducts hospitality and other events to recognize visiting Micronesians, their cultures, and arts. Provides technical assistance and support networks. **PUBLICATIONS:** *The Pacific Navigator*, quarterly. Provides current information on Micronesia and the institute.

MIDDLE EAST DEVELOPMENT AND SCIENCE INSTITUTE (MEDSI)

2657G Old Annapolis Rd., Ste. 144, Hanover, MD 21076 **PHONE:** (202) 310-1758 **FX:** (301) 912-1136 **E-MAIL:** medsi@aol.com **OFFICERS:** A. Y. Zohny, CEO & Pres. **STAFF:** 6. **BUDGET:** US$150,000. **LANGUAGES:** Arabic, English.

DESCRIPTION: Members are scientists and development professionals. Promotes effective, sustainable, and appropriate economic, political, and community development in the Middle East. Conducts educational and advocacy activities; provides advice and support to development organizations operating in the Middle East.

MIDDLE EAST NEUROSURGICAL SOCIETY (Societe de Neurochirurgie du Moyen Orient)

Amer. Univ. Medical Center, PO Box 113-6044, Beirut, Lebanon **PHONE:** 961 1 353486 **FX:** 961 1 373 1450231 **TX:** LE 20801 AMUNOB **FOUNDED:** 1958. **OFFICERS:** Dr. Fuad Sami Haddad, Pres. **MEMBERS:** 545. **LANGUAGES:** English.

DESCRIPTION: Members are individuals in 12 countries. Disseminates information to the medical profession and to the public on clinical advancements and scientific research in neurosurgery and similar disciplines.

MINEWATCH

218 Liverpool Rd., London N1 1LE, England **PHONE:** 44 0171 6091852 **FX:** 44 0171 7006189 **OFFICERS:** Christine Lancaster, Coordinator. **LANGUAGES:** English.

DESCRIPTION: Individuals and organizations concerned about the human and environmental danger posed by land mines. Promotes efforts to remove land mines left behind following military conflicts; seeks to secure an international ban on future land mine use. Serves as a clearinghouse on land mines; sponsors advocacy activities.

MINORITY RIGHTS GROUP INTERNATIONAL (MRGI)

379 Brixton Rd., London SW9 7DE, England **PHONE:** 44 171 9789498 **FX:** 44 171 7386265 **E-MAIL:** minority.rights@mrgmail.org **WEBSITES:**

http://www.minorityrights.org **FOUNDED:** 1969. **OFFICERS:** Alan Phillips, Exec.Dir. **STAFF:** 20. **LANGUAGES:** English, French.

DESCRIPTION: An information and research organization dedicated to obtaining justice for minority and non-dominant majority groups. Aims to enhance awareness of groups suffering discrimination. Conducts educational projects in schools and attends meetings of the United Nations. Maintains nongovernmental organization and consultative status with the Economic and Social Council. **PUBLICATIONS:** *MRG Reports* (in English), bimonthly; *Outsider* (in English), semiannual; *World Directory of Minorities*.

MISSION D'AIDE AU DEVELOPPEMENT DES ECONOMIES RURALES (MADERA)

3, rue Roubo, F-75011 Paris, France **PHONE:** 33 1 43705007 **FX:** 33 1 43706007 **E-MAIL:** madera@globenet.org **FOUNDED:** 1988. **OFFICERS:** Michel Verron, Chm. **MEMBERS:** 30. **STAFF:** 11. **BUDGET:** 20,000,000 Fr. **LANGUAGES:** French.

DESCRIPTION: Promotes rural and social development in developing countries. Fosters effective use of agricultural resources. Encourages communication among industrial and developing nations. **PUBLICATIONS:** *Barg e Sabz* (in Persian), bimonthly.

MISSION DOCTORS ASSOCIATION (MDA)

3424 Wilshire Blvd., Los Angeles, CA 90010 **PHONE:** (626) 285-8868 **FX:** (626) 309-1716 **E-MAIL:** missiondrs@earthlink.net **WEBSITES:** http://www.missiondoctors.org **FOUNDED:** 1957. **OFFICERS:** Timothy Lefevre M.D., Pres. **STAFF:** 1. **LANGUAGES:** English.

DESCRIPTION: Recruits, trains, and supports volunteer Catholic physicians and sends them to serve in Third World hospitals or clinics for a period of 2-3 years. Mission doctors accepted in the program must display a genuine spirit of sacrifice and must participate in a 9-month course on theology, missiology, scripture, and Third World culture. Provides for complete support of doctors and their families while overseas; also provides small monthly stipend. Screens and trains physicians' spouses to be part-time lay missionaries.

MISSIONARIES OF THE COMPANY OF MARY (SMM)

Casa Generalizia Monfortani, Viale dei Monfortani 65, I-00135 Rome, Italy **PHONE:** 39 6 3052332 **FX:** 39 6 35505742 **E-MAIL:** smm@pcn.net **FOUNDED:** 1705. **OFFICERS:** Rev.Fr. William Considine, Superior Gen. **MEMBERS:** 1,040. **LANGUAGES:** Dutch, English, French, Italian, Spanish.

DESCRIPTION: Priests and brothers in 37 countries. Activities include missionary work in developing countries. Promotes the teachings of St. Louis Marie de Montfort (1673-1716). Conducts educational, charitable, and research programs. **PUBLICATIONS:** Publishes newsletters, magazines, and directories (in Dutch, English, French, German, and Italian).

MISSIONARIES OF THE HOLY SPIRIT (MSpS) (Misioneros del Espiritu Santo - MSpS)

Avenida Universidad 1702, 04010 Mexico City, DF, Mexico **PHONE:** 52 5 6587433 **FOUNDED:** 1914. **OFFICERS:** Eugenio Sanchez Sierra MSpS, Sec.Gen. **MEMBERS:** 450. **LANGUAGES:** English, French, German, Italian, Spanish.

DESCRIPTION: Catholic priests, lay brothers, and clerical students in Germany, Italy, Mexico, Peru, Spain, and the United States. Provides religious guidance. **PUBLICATIONS:** *Cor Unum*, monthly; *Directorio de los Misioneros del Espiritu Santo* (in Spanish), biennial; *La Cruz* (in Spanish), monthly; *Pentecostes* (in Spanish), monthly; *Temas. Espiritualidad de la Cruz*; Books, periodic.

MISSIONARY SISTERS OF OUR LADY OF THE HOLY ROSARY (MSHR)

c/o Sr. Monica Devine, 23 Cross Ave., Blackrock, Dublin, Ireland **PHONE:** 353 1 2881708, 353 1 2881709 **FX:** 353 1 2836308 **E-MAIL:** mshrgen@indigo.ie **FOUNDED:** 1924. **OFFICERS:** Sr. Monica Devine, Superior Gen. **MEMBERS:** 385. **LANGUAGES:** English.

DESCRIPTION: Catholic sisters in Africa, Europe, and North and South America. Provides educational, medical, pastoral, and social services. **PUBLICATIONS:** *Information Bulletin* (in English), bimonthly; *The Second Burial of Bishop Shanahan* (in English); *Vincula* (in English), bimonthly.

MORAL RE-ARMAMENT (MRA)

Mountain House, CH-1824 Caux, Switzerland **PHONE:** 41 22 7330920 **FX:** 41 22 7330267 **E-MAIL:** mrageneva@compuserve.com **WEBSITES:** http://www.caux.ch **FOUNDED:** 1938. **LANGUAGES:** English, French, German, Italian.

DESCRIPTION: Individuals and organizations dedicated to ''global transformation carried forward by people of different faiths who seek God's inspiration for individual and common action.'' Encourages individuals to deepen their own religious faith and thereby obtain the strength to take action in favor of peace and social justice. Organizes grass roots associations to undertake activities including helping individuals and families in need, rebuilding hope and a sense of community in cities, ''healing the wounds of history,'' and ''strengthening the moral and spiritual dimensions of democracy.'' **PUBLICATIONS:** *Healing the Past - Forging the Future*, periodic.

MORALOGY INTERNATIONAL RELIEF COMMITTEE (MIRC)

c/o Institute of Moralogy, 2-1-1 Hikarigaoka, Kashiwa-shi, Chiba 277, Japan **PHONE:** 81 471 733216 **FX:** 81 471 761177 **FOUNDED:** 1980. **OFFICERS:** Tsunehisa Miwa. **BUDGET:** US$147,044. **LANGUAGES:** English, Japanese.

DESCRIPTION: Provides aid to victims of natural disasters, famines, and other emergencies worldwide. Supports humanitarian and educational programs of the Japan Volunteer Center in Cambodia and Ethiopia; cooperates with international development and relief agencies operating in Bangaldesh. Conducts educational programs to raise public awareness of development issues; produces educational materials.

MOVEMENT AGAINST RACISM AND FOR FRIENDSHIP BETWEEN PEOPLES (Mouvement Contre le Racisme et pour l'Amitie Entre les Peuples - MRAP)

43, boulevard de Magenta, F-75010 Paris, France **PHONE:** 33 1 53389999 **FX:** 33 1 40409098 **E-MAIL:** mrap@ras.eu.org **FOUNDED:** 1949. **OFFICERS:** Tony St. Clair, CEO. **MEMBERS:** 10,000. **STAFF:** 10. **BUDGET:** US$600,000. **LANGUAGES:** English, French.

DESCRIPTION: Aim is to eliminate racism in all its forms. Believes that all people should be assured of their rights, liberties, and respect for their dignity. Promotes mutual knowledge, understanding, and friendship among all people as a contribution to world peace. Achieves objectives by: waging campaigns and appeals to public opinion against racist acts; encouraging agencies having jurisdiction to punish those responsible for racist acts; providing financial, legal, and moral aid to victims of racism; petitioning and proposing laws to government authorities; demonstrating and holding public ceremonies; supporting teachers who work toward instilling respect for humanity and civic anti-racism in young people. Operates charitable program; provides legal services. Organizes summer schools. **PUBLICATIONS:** *Differences* (in French), monthly; Newsletter (in French), monthly. Also publishes books, brochures, postcards, and posters.

MULTILATERAL INVESTMENT GUARANTEE AGENCY (MIGA)

1818 H Street, NW, Washington DC 20433,USA **PHONE:** (202) 477-1234 **FX:** (202) 477-6391 **TX:** 248423 **WEBSITE:** http://www.ipanet.com **FOUNDED:** 1988. **OFFICERS:** James D. Wolfensohn, Pres. **MEMBERS:** 142.

DESCRIPTION: An organization within the World Bank Group, MIGA's members are states belonging to that institution. Its purpose is to guarantee investments in developing nations against losses from non-commercial risks such as war, civil instability, misappropriation of resources by the host government, and inconvertibility of local currency.

NATIONAL COUNCIL OF NEGRO WOMEN - INTERNATIONAL DIVISION

PO Box 850, Harare, Zimbabwe **PHONE:** 263 4 702480, 263 4 702480 **FX:** 263 4 704546 **E-MAIL:** ncnwsaro@africaonline.co.zw **OFFICERS:** Dorothy Height, CEO. **MEMBERS:** 500,000. **LANGUAGES:** English, Shona.

DESCRIPTION: Supports the cultural and educational development of women throughout southern Africa. Encourages women's economic and political participation in society. Fosters communication among members. **PUBLICATIONS:** *Sisters* (in English), periodic. Contains journals and newsletters.

NATIONAL LATINA HEALTH ORGANIZATION (NLHO)

PO Box 7567, Oakland, CA 94601 **PHONE:** (510) 534-1362 **FX:** (510) 534-1364 **E-MAIL:** latinahlth@aol.com **FOUNDED:** 1986. **OFFICERS:** Luz Alvarez Martinez, Exec.Dir. **LANGUAGES:** English, Spanish.

DESCRIPTION: Members are Puerto Rican, Chicana, Mexican, Cuban, Caribbean and South and Central American women. Works to increase awareness of health issues among Latin American women. Works to achieve bilingual access to quality health care and self-empowerment of Latinas through culturally sensitive educational programs, health advocacy, outreach, research, and the development of public policy. Cooperates with other organizations to defend reproductive rights, affordable birth control, sex education, prenatal care, and freedom from sterilization abuse. Offers technical training for community health facilitators. Organizes forums. **PUBLICATIONS:** Newsletter, periodic. Includes Latina health issues, legislation affecting Latinas, and calendar of events and activities on health and reproductive rights.

NATIONAL ORGANIZATION FOR RELIEF (NOFR)

15th St. at South Kasr Ave., PO Box 2504, Khartoum, Sudan **PHONE:** 11 613827 **TX:** 22043 MECCA **FOUNDED:** 1984. **OFFICERS:** E.A. Elfaki. **MEMBERS:** 200. **STAFF:** 10. **BUDGET:** £S 100,000,000. **LANGUAGES:** Arabic, English.

DESCRIPTION: Provides first aid and assistance to people affected by natural disasters. Encourages conservation and protection of the natural environment. Offers technical assistance to small farmers. Assists the disadvantaged.

NAUTICAL INSTITUTE (NI)

202 Lambeth Rd., London SE1 7LQ, England **PHONE:** 44 171 9281351 **FX:** 44 171 4012537 **E-MAIL:** sec@nautinst.org **FOUNDED:** 1972. **OFFICERS:** C.J. Parker, Sec. **MEMBERS:** 6,000. **STAFF:** 8. **BUDGET:** £350,000. **LANGUAGES:** English.

DESCRIPTION: Members are qualified mariners in navies and the merchant marine; membership represents 70 countries. Promotes high standards of knowledge and competence among those operating sea-going vessels. Conducts research and educational programs. Sponsors competitions. **PUBLICATIONS:** *Seaways*, monthly; Proceedings, periodic.

NEED

12601 N. Cave Creek Rd., PO Box 54541, Phoenix, AZ 85078 **PHONE:** (623) 879-9676 **FX:** (623) 971-4399 **E-MAIL:** need123@aol.com **FOUNDED:** 1985. **OFFICERS:** Dulal Borpujari, Pres. **STAFF:** 2. **BUDGET:** US$380,000. **LANGUAGES:** English.

DESCRIPTION: Provides disaster relief, food distribution, and medical and building supplies to needy areas of the world. Supports a school and health clinic in Calcutta, India; maintains charitable program; offers children's services. Provides development assistance; conducts educational programs. Activities conducted primarily in Southeast Asia.

NEIGHBOR TO NEIGHBOR (N2N)

1611 Telegraph Ave. Ste. 1111, Oakland, CA 94612 **PHONE:** (510) 419-0101 **TF:** (800) 366-8289 **FX:** (510) 419-0202 **FOUNDED:** 1985. **OFFICERS:** Jeff Robinson, Exec.Dir. **MEMBERS:** 52,000. **STAFF:** 15. **BUDGET:** US$1,500,000. **LANGUAGES:** Spanish.

DESCRIPTION: Began as a project of the Institute for Food and Development Policy. Publicized, organized, and lobbied to change U.S. policy in Central America. Now works for comprehensive, secure healthcare in the U.S. (single payer health care) as a social justice issue. Seeks to refocus U.S military aid to reconstruction aid. Maintains funds: Action Fund for lobbying and legislative activities; Education Fund for public education through grass roots organization and the media. Encourages participation in grass roots organizing to affect legislation. Distributes information packets. Conducts public opinion polls. **PUBLICATIONS:** *Grounds for Action*, quarterly; Brochures.

NETWORK OF WOMEN IN DEVELOPMENT EUROPE

Rue du Commerce 70, B-1040 Brussels, Belgium **PHONE:** 32 2 5459070 **FX:** 32 2 5127342 **E-MAIL:** wide@gn.apc.org **FOUNDED:** 1985. **OFFICERS:** Maija Sala, Interim Coor. **MEMBERS:** 15. **STAFF:** 5.

DESCRIPTION: European network of women, national women's groups, NGOs, and development platforms from 14 European countries concerned with gender and development issues. Works to influence European and International policies and to raise awareness on gender and development issues among important sectors of opinion in Europe, with the objective of empowering women worldwide. Enhances the objectives of the ''Foward Looking Strategies'' (FLS) of the 1985 United Nations World Conference on Women in Nairobi, which call for wide-reaching changes in the policies and development programs of governments, multilateral institutions, and NGOs in the North and South pursuing increased awareness of women's importance in development. Promotes a process of joint reflection and interaction between Northern and Southern women, aiming for joint action on international and national levels without denying the diversity among women based on their economic and political environments, class, race, and religion. Responds to major, global challenges, contributing with specific actions, networking, and lobbying on concrete issues related to development, debt, trade, environment, population, and human rights, with attention to the indivisible linkages between them and the interdependence between North and South in the search for solutions. Develops strategies along the following lines: exchanging information and experiences concerning national policies on gender issues; gender blindness in Northern NGDOs and useful instruments on gender training and evaluation; strengthening national WIDE platforms in order to be more effective in bringing a gender perspective into national NGDOs and into lobbying national governments; promoting dialogue and joint action among Southern women, European women, and Southern women living in Europe; lobbying European and International institutions on gender related development issues in close cooperation with other European networks. **PUBLICATIONS:** Newsletter, quarterly; Bulletin, semiannual; Paper; Reports.

NEW YORK CITY COMMISSION FOR THE UNITED NATIONS, CONSULAR CORPS, AND INTERNATIONAL BUSINESS (NYCC)

2 United Nations Plz., 27th Fl., New York, NY 10017 **FOUNDED:** 1962. **STAFF:** 15. **BUDGET:** US$4,000,000. **LANGUAGES:** English.

DESCRIPTION: The mayor's official liaison between the city of New York and the international diplomatic and business communities. Responsible for promoting the role of the city as headquarters of the United Nations and as one of the main centers of foreign consulates and foreign businesses. Assists in the total adjustment of the international diplomatic and business communities to the life and culture of the city; provides advice and counseling on landlord-tenant problems, school placement, and general casework. Serves as liaison among the UN community, the Consular Corps, the police and traffic departments, and related agencies. Sponsors orientation program for newly arrived diplomats and international business personnel and their spouses as well as numerous forums and meetings. Seeks ways to improve housing, educational and child care arrangements, and relocation problems for international families. Administers program of Sister Cities International with: Budapest, Hungary; San Juan, Puerto Rico; Rome, Italy; Jerusalem, Israel; Tokyo, Japan; Beijing, China; Madrid, Spain; Cairo, Egypt; Santo Domingo, Dominican Republic. Maintains speakers' bureau; conducts educational and research programs. Compiles statistics. **PUBLICATIONS:** *Diplomatic World Bulletin*, monthly; *Diplomatically Speaking*, annual; *Economic Impact of the Diplomatic Community on the City of New York*, periodic; *Guide for Consular Officers*, annual; *Guidebook to New York City* (in English, French, and Spanish), periodic; *NYC Foreign Accents*, quarterly; *Register of Foreign Consulates*, annual. Lists consulates located in New York City; *Sister City International Newsletter*, quarterly; *Sister City Update Reports*, periodic.

NGO COMMITTEE ON DISARMAMENT (NGOCD)

777 United Nations Plz., Rm. 3-B, New York, NY 10017 **PHONE:** (212) 687-5340 **FX:** (212) 687-1643 **FOUNDED:** 1973. **OFFICERS:** Vermon C. Nichols, Pres. **MEMBERS:** 150. **STAFF:** 1. **BUDGET:** US$30,000. **LANGUAGES:** English.

DESCRIPTION: Representatives of nongovernmental organizations at United Nations headquarters concerned with disarmament. Seeks to exert influence on UN disarmament activities and reflect such activities to members' constituencies. Arranges seminars; undertakes educational projects. Sponsors forums. **PUBLICATIONS:** *Disarmament*

Times, 6/year. Newspaper reporting on UN activities pertaining to arms control and disarmament. Also publishes forum proceedings.

NGO COMMITTEE ON HUMAN RIGHTS

866 United Nations Plz., Ste. 120, New York, NY 10017-1822 **OFFICERS:** Techeste Ahderom, Exec. Officer. **MEMBERS:** 120. **BUDGET:** US$1,000. **LANGUAGES:** English.

DESCRIPTION: Representatives of non-governmental organizations (NGOs) at the United Nations. Works to make human rights a priority for the United Nations. Serves as a forum for discussion and education on human rights issues, especially those relating to the work of the United Nations.

NICKEL DEVELOPMENT INSTITUTE

c/o European Technical Information Centre, The Holloway, Alvechurch, Birmingham B48 7QB, England **PHONE:** 44 1527 584777 **FX:** 44 1527 585562 **E-MAIL:** pcutler@nidi.org **FOUNDED:** 1984. **OFFICERS:** Dr. Peter Cutler, Tech. Dir. **MEMBERS:** 15. **STAFF:** 3.

DESCRIPTION: Membership is open to primary nickel producers. Associate membership is available to other industries. Head office in Toronto can advise on this. The market development and applications research organisation of the primary nickel industry. Its objective is sustained growth in the consumption of nickel. It carries out market development and technical research, and provides a technical service of information and advice to nickel and nickel alloy users worldwide. **PUBLICATIONS:** *Communique* (in English), quarterly; *Nickel* (in English), quarterly.

NIGER BASIN AUTHORITY (Autorite du Bassin du Niger)

Boite Postale 729, Niamey, Niger **PHONE:** 227 723102 **OFFICERS:** Othman Mustapha, Exec.Sec. **LANGUAGES:** English, French.

DESCRIPTION: Countries encompassing the Niger River Basin. Promotes appropriate and sustainable development in the region. Provides support and assistance to development projects and coordinates development activities in the Niger Basin; gathers and disseminates information.

NIPPON INTERNATIONAL COOPERATION FOR COMMUNITY DEVELOPMENT (NICCO) (Nippon Kokusai Minkan Kyoryoku Kikan)

Takada-cho 500, Ainomachi-dori, Oike-agaru, Nakagyo-ku, Kyoto 604, Japan **PHONE:** 81 75 2410681 **FX:** 81 75 2410682 **E-MAIL:** kyoto-nicco@msn.com **FOUNDED:** 1979. **OFFICERS:** Satoyo Ono, Pres. **MEMBERS:** 1,045. **STAFF:** 5. **BUDGET:** 71,313,735¥. **LANGUAGES:** English, Japanese.

DESCRIPTION: Offers assistance in rural communities of developing nations. **PUBLICATIONS:** *Relief Action* (in Japanese), biennial. Provides report on on-going projects.

NOBEL FOUNDATION (NF) (Nobelstiftelsen - NS)

Sturegatan 14, Box 5232, S-102 45 Stockholm, Sweden **PHONE:** 46 8 6630920 **FX:** 46 8 6603847 **WEBSITES:** http://www.nobel.se **FOUNDED:** 1900. **OFFICERS:** Mr. Michael Sohlman, Exec.Dir. **MEMBERS:** 9. **LANGUAGES:** English, French, German, Swedish.

DESCRIPTION: Administers trust fund established by Swedish scientist and inventor Alfred Nobel (1833-96). Confers Nobel Prizes annually to individuals who have made significant contributions in physics, chemistry, physiology or medicine, literature, and peace. Also confers the Bank of Sweden's Prize in Economic Science in memory of Alfred Nobel.

NORDIC BUILDING RESEARCH COOPERATION GROUP (NBS) (Nordiska Byggforskningsorgans Samarbetsgrupp - NBS)

c/o Building Research Institute, Statensbyggeforstningsinstitutt, Postboks 119, DK-2970 Horsholm, Denmark **PHONE:** 45 42865533 **FX:** 45 42867535 **FOUNDED:** 1957. **OFFICERS:** Hans Jorgen Larsen, Dir. **MEMBERS:** 8. **STAFF:** 1. **LANGUAGES:** Danish, Norwegian, Swedish.

DESCRIPTION: Building research institutes in Nordic countries. Encourages cooperation and efficient use of resources in member countries; promotes exchange of ideas among Nordic building researchers.

NORDIC COUNCIL

PO Box 3043, DK-1021 Copenhagen, Denmark **PHONE:** 45 33 960400 **FX:** 45 33 111870 **WEBSITES:** http://www.norden.org; http://www.baltchild.org **FOUNDED:** 1952. **OFFICERS:** Ms. Berglind Asgeirsdottir, Gen.Sec. **MEMBERS:** 87. **STAFF:** 18. **LANGUAGES:** Danish, English, Norwegian, Swedish.

DESCRIPTION: Representatives from Denmark (including Faroe Islands and Greenland), Finland (including Aland Islands), Iceland, Norway, and Sweden. Seeks to arrange cooperative efforts among the Nordic countries on economic, social, cultural, legal, labor, and environmental issues. **PUBLICATIONS:** *Norden i Veckan* (in Danish, English, Finnish, Icelandic, Norwegian, and Swedish), weekly; *Politik i Norden* (in Danish, Finnish, Norwegian, and Swedish), 8/year; *Yearbook of Nordic Statistics* (in English and Swedish).

NORDIC COUNCIL FOR ADULT STUDIES IN CHURCH (NCASC) (Nordiska Kyrkliga Studieradet - NKS)

Stifstradet i Borga Stift, Kajsaniemig 10, FIN-00100 Helsinki, Finland **PHONE:** 358 9 22886634 **FX:** 358 9 22886601 **FOUNDED:** 1965. **OFFICERS:** Maj-Britt Palmgren, Exec. Officer. **MEMBERS:** 11. **BUDGET:** US$30,000. **LANGUAGES:** Danish, Finnish, Icelandic, Norwegian, Swedish.

DESCRIPTION: Members are education boards of the national churches of Denmark, Finland, Iceland, Norway, and Sweden. Promotes joint study projects among Nordic churches, with particular emphasis on biblical studies. Encourages information exchange among members; organizes adult church educational programs. **PUBLICATIONS:** *STUDIE-bulletin* (in Danish, Finnish, Icelandic, Norwegian, and Swedish), semiannual.

NORDIC COUNCIL OF MINISTERS (NCM) (Nordisk Ministerrad)

Store Strandstraede 18, DK-1255 Copenhagen K, Denmark **PHONE:** 45 33 960200 **FX:** 45 33 960202 **WEBSITES:** http://www.norden.org **FOUNDED:** 1971. **OFFICERS:** Soren Christensen, Sec.Gen. **MEMBERS:** 8. **STAFF:** 85. **BUDGET:** 702,600,000 DKr. **LANGUAGES:** Danish, Finnish, Icelandic, Norwegian, Swedish.

DESCRIPTION: Members are ministers of cooperation, representing the governments of Denmark, Finland, Iceland, Norway, and Sweden. Works for cooperation between Nordic countries in all fields except foreign policy, national security, and defense. Executes the recommendations of the Nordic Council. Promotes economic growth. **PUBLICATIONS:** *Newsletter* (in English, Estonian, French, German, Latvian, Lithuanian, and Russian), bimonthly.

NORDIC INDUSTRY WORKERS FEDERATION (Nordiska Industriarbetarefederationen - NIF)

c/o Svenska Pappersindustriarbetareforbundet, Box 1127, S-111 81 Stockholm, Sweden **PHONE:** 46 8 7966100 **FX:** 46 8 114179 **FOUNDED:** 1901. **OFFICERS:** Sune Ekbage, Pres. **MEMBERS:** 16. **LANGUAGES:** Danish, Finnish, Icelandic, Norwegian, Swedish.

DESCRIPTION: Members are nordic trade union organizations representing 440,000 individuals from 5 countries. Encourages cooperation among members.

NORDIC PROJECT FUND (NOPEF) (Nordiska Projektexportfonden)

PO Box 241, FIN-00171 Helsinki, Finland **PHONE:** 358 9 1800350 **FX:** 358 9 650113 **FOUNDED:** 1982. **OFFICERS:** M. Backstrom **STAFF:** 6. **LANGUAGES:** Danish, English, Finnish, French, German, Norwegian, Swedish.

DESCRIPTION: Owned and financed by the governments of the Nordic countries. Supports Nordic companies' international competitiveness and the strengthening of industrial collaboration among the Nordic countries. Grants financial assistance to Nordic entities conducting feasibility studies for specific projects outside Europe.

NORTH ATLANTIC ASSEMBLY (Assemblee de l'Atlantique Nord - AAN)

3, place du Petit Sablon, B-1000 Brussels, Belgium **PHONE:** 32 2 5132865 **FX:** 32 2 5141847 **TX:** 24809 AANAA B **E-MAIL:** secretariat@naa.be **WEBSITES:** http://www.nato.int/related/naa **FOUNDED:** 1955.

OFFICERS: Simon Lunn, Sec.Gen. **MEMBERS:** 188. **STAFF:** 30. **LANGUAGES:** English, French.

DESCRIPTION: Members are parliamentarians from the 16 member countries of the Atlantic Alliance; representatives from 15 newly-democratic states in central and eastern Europe are associates. Objectives are to: review issues relating to the Atlantic Alliance; strengthen cooperation and understanding and develop a feeling of solidarity among member countries of the Atlantic Alliance; encourage governments and parliaments of member countries to take into account Atlantic Alliance viewpoints when drafting national legislation; act as a link between Alliance parliaments and the North Atlantic Treaty Organization authorities. Sponsors project to assist in the establishment of parliamentary democracy in Eastern and Central European nations. Conducts study tours and seminars. Sponsors subcommittees and working groups on subjects such as: civilian security and cooperation; the future of the armed forces; defence and security cooperation between Europe and North America; northern security issues: east-west economic cooperation and convergence; NATO enlargement and the new democracies; transatlantic and European relations; the proliferation of military technology; Balkan stability; Mediterranean issues. **PUBLICATIONS:** *Annual Policy Recommendations* (in English and French); *Committee and Subcommittee Reports*, annual; *List of Documents*, annual; *NATO Enlargement Watch*, quarterly.

NORTH ATLANTIC TREATY ORGANIZATION (NATO) (Organisation du Traite de l'Atlantique Nord - OTAN)

B-1110 Brussels, Belgium **PHONE:** 32 2 7074111 **FX:** 32 2 7074579 **TX:** 23867 **E-MAIL:** natodoc@hq.nato.int **WEBSITES:** http://www.NATO.INT **FOUNDED:** 1949. **OFFICERS:** Javier Solana, Sec.Gen. **MEMBERS:** 19. **LANGUAGES:** English, French.

DESCRIPTION: Alliance of 19 independent nations dedicated to collective defense through political cooperation and joint defense planning. (The organization was established through the signing of the North Atlantic Treaty in Washington, DC on Apr. 4, 1949, and links Canada and the United States with 17 European nations.) Reaffirms the purpose and principles of the United Nations charter with regard to the security of Europe and of the Euro-Atlantic area. Since the Brussels Summit in 1994, the Alliance has transformed its policies and structures in response to the changes in Europe, including by establishing close cooperation and security links with the countries of central and Eastern Europe. Other changes include increased coordination and cooperation with other international institutions, a reduced and more flexible force structure, and active involvement in international crisis management and peacekeeping. Consultation and cooperation take place on a variety of political, military, economic, and scientific matters. **PUBLICATIONS:** *NATO Review*, quarterly; *Science and Society Newsletter*, triennial; Books, periodic; Brochures, periodic; Pamphlets, periodic.

NORTH CALOTTE COUNCIL (NCC) (Nordkalottradet - NKR)

c/o Jouko Jama, Box 8056, FIN-96101 Rovaniemi, Finland **PHONE:** 358 16 3301231, 358 16 3301232 **FX:** 358 16 346658 **E-MAIL:** jouko.jama@lapponia.reg.fi **FOUNDED:** 1976. **OFFICERS:** Jouko Jama, Gen.Sec. **MEMBERS:** 13. **STAFF:** 2. **BUDGET:** FM 6,500,000. **LANGUAGES:** Finnish, Norwegian, Swedish.

DESCRIPTION: A committee of the Nordic Council of Ministers. Representatives of labor and planning agencies in Finland, Norway, and Sweden. Promotes cooperation between member agencies as it relates to employment and development in the North Calotte area (a region encompassing Lappland in northern Finland; Nordland, Troms, and Finnmark in central and northern Norway; and Norrbotten in northern Sweden). **PUBLICATIONS:** *Nordkalottkommittens Promemorior*; *Nordkalottkommittens Publikationsserie*, 2-4/year.

NORTH PACIFIC ANADROMOUS FISH COMMISSION (NPAFC)

889 W. Pender, Ste. 502, Vancouver, BC, Canada V6C 3B2 **PHONE:** (604) 775-5550 **FX:** (604) 775-5577 **E-MAIL:** wmorris@unixg.ubc.ca **WEBSITES:** http://www.npafc.org **FOUNDED:** 1993. **OFFICERS:** Dr. Irina Shestakova, Exec.Dir. **MEMBERS:** 4. **STAFF:** 4. **BUDGET:** C$540,000.

DESCRIPTION: Intergovernmental organization of association representatives, state administrators, scientists, lawyers, and others interested in development of fisheries in the North Pacific Ocean. Established by convention between the United States, Canada, Japan, and Russian Federation.

Works to prohibit salmon fisheries on the high seas of the North Pacific. **PUBLICATIONS:** *Annual Report* (in English, Japanese, and Russian); *Statistical Yearbook*, annual.

NORTHERN SHIPOWNERS' DEFENCE CLUB (NSDC) (Nordisk Skibsrederforening - NORDISK)

Kristinelundveien 22, Postboks 3033 El., N-0207 Oslo 9, Norway **PHONE:** 47 2 2135600 **FX:** 47 2 2430035 **TX:** 76825 NORTH N **E-MAIL:** post@nordisk-skibsrderforening.no **FOUNDED:** 1889. **OFFICERS:** Nicholas Hambro, Gen.Mgr. **STAFF:** 26. **BUDGET:** 65,000,000 NKr. **LANGUAGES:** English, Norwegian.

DESCRIPTION: Members are shipowners and others in 10 countries owning or chartering vessels and rigs used in drilling for gas and oil. Provides members with advice concerning maritime law on matters resulting from business activities. Settles disputes among members; undertakes litigation on members' behalf. **PUBLICATIONS:** *List of Agents*; *Medlemsblad* (in English), semiannual; *Members Circular*; Annual Report (in English), annual.

NORTHWEST ATLANTIC FISHERIES ORGANIZATION (NAFO) (Organisation des Peches de l'Atlantique Nord-Ouest)

PO Box 638, Dartmouth, NS, Canada B2Y 3Y9 **PHONE:** (902) 468-5590 **FX:** (902) 468-5538 **E-MAIL:** nafo@fox.nstn.ca **WEBSITES:** http://www.nafo.ca **FOUNDED:** 1979. **OFFICERS:** Dr. L. Chepel, Exec. Sec. **MEMBERS:** 17. **STAFF:** 12. **BUDGET:** C$750,000. **LANGUAGES:** English.

DESCRIPTION: Members are contracting parties working to investigate, protect, and conserve the fishery resources of the Northwest Atlantic Ocean. Provides resource management. Compiles statistics; maintains research programs. **PUBLICATIONS:** *Annual Report*; *Journal of Northwest Atlantic Fishery Science*, annual; *List of Fishing Vessels*, triennial; *NAFO Index of Meeting Documents*, quinquennial; *Sampling Yearbook*; *Scientific Council Reports*, annual; *Scientific Council Studies*, annual; *Statistical Bulletin*, annual.

OBLATES OF MARY IMMACULATE (OMI) (Oblats de Marie Immaculee)

Casella Postale 9061, I-00100 Rome, Aurelio, Italy **PHONE:** 39 6 398771 **FX:** 39 6 39375322 **E-MAIL:** omigen@networld.it **FOUNDED:** 1816. **OFFICERS:** Alexandre Tache' OMI, Sec.Gen. **MEMBERS:** 4,850. **LANGUAGES:** English, French, German, Italian, Polish, Spanish.

DESCRIPTION: Members are Roman Catholic missionaries in 62 countries. Activities include missionary and pastoral work. Provides religious training. **PUBLICATIONS:** *OMI Documentation* (in English, French, German, Italian, Polish, and Spanish), 7/year; *OMI Information* (in English, French, German, Italian, Polish, and Spanish), monthly.

OECD NUCLEAR ENERGY AGENCY (NEA) (Agence de l'OCDE pour l'Energie Nucleaire - AEN)

Le Seine St. Germain, 12, blvd. des Iles, F-92130 Issy-les-Moulineaux, France **PHONE:** 33 1 45241000 **FX:** 33 1 45241110 **TX:** 640048 AENNEA **FOUNDED:** 1958. **OFFICERS:** Mr. Luis E. Echavarri, Dir.Gen. **MEMBERS:** 27. **STAFF:** 80. **LANGUAGES:** English, French.

DESCRIPTION: Government representatives from member countries of the Organisation for Economic Cooperation and Development. Seeks to promote cooperation in the safety and regulation of nuclear power and to develop nuclear energy as a contributor to economic progress. Encourages the exchange of scientific and technical information and the harmonization of government regulatory policies and practices. Coordinates and supports research and development programs. Reviews technical and economic aspects of the nuclear fuel cycle; assesses supply and demand of nuclear power and its effect on energy demand. Areas of work include safety, radiation protection, waste management, development, science, and legal affairs. Sponsors European Nulear Energy Tribunal. **PUBLICATIONS:** *NEA Activity Report*, annual; *NEA Newsletter*, semiannual; *Nuclear Energy Data*, annual; *Nuclear Law Bulletin*, semiannual; *Red Book: Uranium Resources, Production and Demand*, biennial; Reports, periodic.

OFFICE OF THE HIGH COMMISSIONER FOR HUMAN RIGHTS (OHCHR) (Office Du Haut Commissaire Aux Droits de l'Homme - UNOG)

Geneva 10, CH-1211 Geneva 10, Switzerland

PHONE: 41 22 917900 **FX:** 41 22 9170213 **TX:** 412962 **E-MAIL:** webadmn.hchr@unog.ch **WEBSITES:** http://www.unhchr.ch **OFFICERS:** Ms. Mary Robinson, Under-Sec.Gen. **STAFF:** 400. **LANGUAGES:** Arabic, Chinese, English, French, Russian, Spanish.

DESCRIPTION: The OHCHR has the overall responsibility for the coordination of UN activities. In the promotion and protection of the human rights and fundamental freedoms put forth in the Universal Declaration of Human Rights and other human rights resolutions of the U.N. General Assembly. Carries out research; follows up and prepares reports on the implementation of human rights programs. Investigates confidential communications concerning allegations of human rights violations; provides special working groups to carry out onsite missions. Administers a program of advisory services and technical assistance in the field of human rights. Supervises training courses and an annual program of fellowships for government officials who deal with human rights issues. Coordinates activities with nongovernmental organizations, human rights institutions, and the media. Collects and disseminates information. **PUBLICATIONS:** *Fact Sheet* (in Arabic, Chinese, English, French, Russian, and Spanish), periodic; *Human Rights and Elections: A Handbook on the Legal Technical and Human Rights Aspects of Elections*; *Human Rights and Social Work: A Manual for Schools of Social Work and the Social Work Profession*; *Proceedings* (in English and French), periodic. Additional publication information available upon request.

OIL COMPANIES INTERNATIONAL MARINE FORUM (OCIMF) (Forum Maritime International des Compagnies Petrolieres)

96 Victoria St., 15th Fl., London SW1E 5JW, England **PHONE:** 44 171 8287696, 44 171 8286283 **FX:** 44 171 2452921 **TX:** 24942 **FOUNDED:** 1970. **OFFICERS:** R.C. Oldham, Dir. **MEMBERS:** 40. **STAFF:** 8.

DESCRIPTION: Members are oil companies worldwide united to promote safety in the transportation and storage of crude oil and its products, including gas and petrochemicals, and to prevent pollution from tankers and at terminals. Works to establish guidelines for equipment at terminals and offshore moorings, and for tanker and gas carrier manifolds. Conducts studies and research projects on such projects as effects of wind and currents, safe terminal moorings of large ships, tanker salvage, and the handling of disabled ships. Is involved with other organizations in the consideration of governmental and industrial contingency plans to handle spills. Participates in work of the International Maritime Organization in areas of carriage of bulk chemicals and gases, equipment, fire protection, navigation safety, ship design, standards of training, and watchkeeping. Works with governments represented by IMO and provides information and assistance. Cooperates with other industry organizations, including the International Chamber of Shipping and the Society of International Gas Tanker and Terminal Operators, on the operation of tankers and gas carriers such as the ship to ship transfer for oil tankers and gas carriers, and planned passage in the Malacca/Singapore Straits. **PUBLICATIONS:** *Clean Seas Guide for Oil Tankers*; *Disable Tankers - Report Studies on Ship Drift and Towage*; *Inspection Guidelines for Bulk Oil Carriers and Gas Carriers*, periodic; *International Safety Guide for Oil Tankers and Terminals*, periodic. Also publishes books and pamphlets on marine transport and ships.

ONE PERCENT FOR DEVELOPMENT FUND (OPDF) (Fonds un pour Cent pour le Developpement - FUPCD)

Palais des Nations, Rm. A-825, CH-1211 Geneva 10, Switzerland **PHONE:** 41 22 7997705 **FOUNDED:** 1976. **OFFICERS:** Ms. Sylvie Pichelin, Coord. **MEMBERS:** 350. **BUDGET:** 270,000 SFr. **LANGUAGES:** English, French, Spanish.

DESCRIPTION: Members are United Nations staff members in Austria, Italy, Switzerland, the United States, and other countries who contribute 1% of their salaries to finance community development projects worldwide. Aims to foster concern for development issues and encourage self-reliance at the community level while increasing available funds for grassroots projects. Projects include: constructing schools, libraries, and hospital wings; developing animal husbandry, mixed farming, and progressive agriculture; installing wells. Also organizes activities promoting community development; sponsors film shows and exhibitions.

OPEC FUND FOR INTERNATIONAL DEVELOPMENT

Parkring 8, A-1011 Vienna, Austria **PHONE:** 43 1 515640 **FX:** 43 1 5139238 **TX:** 131734 FUND A **FOUNDED:** 1976. **OFFICERS:** Dr. Y. Seyyid Abdulai, Dir.Gen. **MEMBERS:** 13. **STAFF:** 80. **BUDGET:** US$6,260,000. **LANGUAGES:** English.

DESCRIPTION: A multilateral finance institution which seeks to promote cooperation between member states of the organization of the Petroleum Exporting Countries (OPEC) and other developing countries by providing financial support to the latter to assist in their economic and social advancement. Provides concessional loans for development purposes. Makes contributions to other international development agencies whose work benefits developing countries. Provides grants to finance technical assistance and research. Contributes to international emergency relief and food aid operations. **PUBLICATIONS:** *Annual Report* (in Arabic, English, French, and Spanish). Contains summary of annual and cumulative operations.; *OPEC Aid Institutions: A Profile* (in English), annual; *OPEC Fund Newsletter*, 3/year; Books; Brochures; Pamphlets; Papers.

OPERATION SMILE INTERNATIONAL (OSI)

6435 Tidewater Dr., Norfolk, VA 23509-1600 **PHONE:** (757) 321-7645, (757) 625-0375 **FX:** (757) 321-7660 **E-MAIL:** rfishkin@operationsmile.org **WEBSITES:** http://www.operationsmile.org **FOUNDED:** 1982. **OFFICERS:** Thomas Fox Ph.D, CEO. **STAFF:** 80. **LANGUAGES:** English.

DESCRIPTION: Operation Smile is the not-for-profit, volunteer medical services organization providing free reconstructive surgery to children suffering from facial deformities. Based in Norfolk, Virginia, since its founding in 1982, Operation Smile's volunteers have cared for more than 45,000 children in 16 developing countries and across the United States. Operation Smile, emphasizes medical training and educational exchanges during international missions to build long-term self-sufficiency. **PUBLICATIONS:** *Operation Smile Newsletter*, 3/year. Contains highlights of Operation Smile activities.

ORDER OF FRIARS MINOR (OFM)

Via Santa Maria Mediatrice 25, I-00165 Rome, Italy **PHONE:** 39 6 68491365 **FX:** 39 6 68491364 **E-MAIL:** comgen@ofm.org **WEBSITES:** http://www.ofm.org **FOUNDED:** 1209. **OFFICERS:** Fr. Giacomo Bini OFM, Minister Gen. **MEMBERS:** 19,000. **LANGUAGES:** English, French, German, Italian, Spanish.

DESCRIPTION: Members are Catholic priests and brothers in 97 countries. Activities include: providing religious guidance; maintaining chaplaincies, parishes, and schools; conducting missionary evangelization. **PUBLICATIONS:** *Acta Ordinis Minorum* (in English, French, German, Italian, Latin, Portuguese, and Spanish), 3/year; *Fraternitas* (in English, French, German, Italian, Polish, Portuguese, and Spanish), 10/year.

ORDER OF ST. CAMILLUS

Piazza della Maddalena 53, I-00186 Rome, Italy **PHONE:** 39 6 6797796, 39 6 6797797 **FX:** 39 6 6789418 **FOUNDED:** 1582. **OFFICERS:** Fr. Angelo Brusco, Superior Gen. **MEMBERS:** 1,070. **STAFF:** 10. **LANGUAGES:** Chinese, English, French, German, Italian, Polish, Portuguese, Spanish, Thai.

DESCRIPTION: Members are priests, brothers, and individuals in training or formation in 32 countries. Provides physical, mental, and spiritual assistance to the sick in public and private institutions as well as in private homes. Operates the International Institute for the Pastoral Theology of Health Care known as the CAMILLIANUM. **PUBLICATIONS:** *Camiilianum* (in English, French, German, Italian, Polish, Portuguese, and Spanish). Healthcare theology.; *Camilliani* (in English, German, Italian, Portuguese, and Spanish), monthly; *Catalogo delle Case* (in Italian), annual; *Edizioni Camilliane*; *Ethos Today*; *Humanizar*; *Missione Salute*; *Missioni Camilliani*; *S. Camillo e la Sua Opera*.

ORDRE DE L'ETOILE DE L'EUROPE (OEE)

Maison de la Fondation Europeenne, 19, rue de la Reine, B-4500 Huy, Belgium **PHONE:** 32 85 212671 **FX:** 32 85 211071 **FOUNDED:** 1965. **OFFICERS:** H.E. the Rt. Hon. Earl Pierre Pasleau de Charnay, Hon. Ambassador. **LANGUAGES:** English, French, Italian, Spanish.

DESCRIPTION: Operates under the eminent patronage of H.R.H. the Prince Henrik of Denmark. Individuals who have made notable contributions to European culture. Promotes a prosperous, unified Europe through expansion of liberal economic policies. Group maintains that economic liberalism is directly linked to the social progress and cultural elevation of the individual. Seeks to cultivate an intellectual corps capable of responding to issues on the European level. Bestows 5 honorary titles of distinction to individuals displaying merits and talents dedicated to the ''European ideal.'' Serves as a liaison between members. Organizes literary and artistic expositions. Sponsors radio and television broadcasts. Arranges debates. **PUBLICATIONS:** *Newsletter*, periodic.

ORGANISATION OF EASTERN CARIBBEAN STATES, ECONOMIC AFFAIRS SECRETARIAT (OECS/EAS)

PO Box 822, High St., St. Johns, Antigua-Barbuda **PHONE:** (268) 462-3500 **FX:** (268) 462-1537 **TX:** 2157 ECONSEC AK **E-MAIL:** economico@candw.ag **FOUNDED:** 1981. **OFFICERS:** Dr. J. Bernard Yankey, Dir. **MEMBERS:** 9. **STAFF:** 36. **LANGUAGES:** English.

DESCRIPTION: A division of the Organisation of Eastern Caribbean States. Representatives of Anguilla, Antigua and Barbuda, British Virgin Islands, Dominica, Grenada, Montserrat, St. Christopher and Nevis, St. Lucia, and St. Vincent and the Grenadines. Promotes economic integration in the areas of trade, tourism, energy, agriculture, and industry according to the provisions of the agreement that established the East Caribbean Common Market. Encourages economic planning and manpower development training. Compiles statistics. **PUBLICATIONS:** *Annual Digest of Statistics* (in English); *Documentation Centre Current Awareness Bulletin* (in English), quarterly; *Energy Bulletin* (in English), periodic; *National Accounts Digest* (in English), annual; *OECS - Occasional Paper Series* (in English); *Statistical Pocket Digest* (in English), periodic; *Trade Digest* (in English), annual.

ORGANISATION FOR ECONOMIC CO-OPERATION AND DEVELOPMENT (OECD) (Organisation Cooperation et Developpment Economique)

2, rue Andre Pascal, F-75775 Paris Cedex 16, France **PHONE:** 33 1 45248200 **FX:** 33 1 45248500 **TX:** 640048 PARIS **E-MAIL:** news.contact@oecd.org **WEBSITES:** http://www.oecd.org/ **FOUNDED:** 1961. **OFFICERS:** Donald J. Johnston, Sec.Gen. **MEMBERS:** 29. **STAFF:** 1,500. **LANGUAGES:** English, French.

DESCRIPTION: Representatives of European governments, Australia, Canada, Korea, Japan, Mexico, New Zealand, and the United States. Seeks to stimulate economic cooperation between member countries, expand world trade, and coordinate aid for less developed areas of the world. Compiles statistics. **PUBLICATIONS:** *Energy Prices and Taxes*, quarterly; *Financial Market Trends*, 3/year; *Financial Statistics*, biweekly; *Indicators of Industrial Activity*, quarterly; *Main Economic Indicators*, monthly; *Monthly Statistics of Foreign Trade*; *OECD Economic Outlook*, biennial; *OECD Economic Studies*, biennial; *OECD Economic Survey*, annual. Provides information for each country.

ORGANISATION OF EUROPEAN ALUMINIUM REFINERS AND REMELTERS

OEA Technical Office, Broadway House, Calthorpe Rd., Five Ways, Birmingham B15 1TN, England **PHONE:** 44 121 4561103 **FX:** 44 121 4562274 **E-MAIL:** oeato@alfed.org.uk **WEBSITES:** http://www.metalnet.co.uk **FOUNDED:** 1960. **OFFICERS:** Mark Askew. **MEMBERS:** 50. **STAFF:** 1. **LANGUAGES:** English.

DESCRIPTION: Members are companies in the secondary aluminium industry producing foundry ingot, deoxidant for the steel industry, master alloys, rolling slab and extrusion billet predominantly from recycled aluminium. Represents the common interests of smelters in Europe. It helps members to operate in an environmentally acceptable way and assists in the search for new markets. Represents members interests to the public, to legislative organisations and provides a forum for members cooperation on technical and economic areas. **PUBLICATIONS:** *Comparison of National Standards for Aluminium Casting Alloys* (in English, French, and German), annual; *OEA*, annual.

ORGANIZATION OF AFRICAN TRADE UNION UNITY (OATUU) (Organisation de l'Unite Syndicale Africaine - OUSA)

PO Box M 386, Accra, Ghana **PHONE:** 233 21

774531, 233 21 772574 **FX:** 233 21 772621 **TX:** 2673 OATUU GH **FOUNDED:** 1973. **OFFICERS:** Hassan A. Sunmonu, Sec.Gen. **MEMBERS:** 72. **STAFF:** 30. **BUDGET:** Cd 600,000. **LANGUAGES:** Arabic, English, French.

DESCRIPTION: Members are African national trade union organizations in 52 African countries. Seeks to build trade union unity on national and continental levels and to coordinate and guide the actions of national trade unions in Africa. Promotes the material, cultural, and moral interests of the African working class; supports a policy of full employment and equitable distribution of national income. Works for independence as well as standardization of legislation on labor and collective bargaining in member countries. Provides assistance to trade unions, especially those under colonial and foreign domination; opposes imperialism, colonialism, neocolonialism, feudalism, and any other form of oppression or exploitation in Africa. Represents the interests of African workers before appropriate organizations and institutions. Conducts educational and research programs. Plans to establish library, conference hall, and labor college. **PUBLICATIONS:** *African Trade Union and Foreign Debt* (in English and French). Provides a voice for African workers.; *Economic Integration in Africa: Strengthening the Role of Workers and Trade Unions*; *The Role and Place of the Public Sector in Development*; *The Trade Union, Rural Population and Food Self-Sufficiency in Africa*; *Voice of African Workers*; *Women Participation in Development and Trade Union Activities.*

ORGANIZATION OF AFRICAN UNITY (OAU) (Organisation de l'Unite Africaine - OUA)

PO Box 3243, Addis Ababa, Ethiopia **PHONE:** 251 1 517700 **FX:** 251 1 513063 **TX:** 21046 **FOUNDED:** 1963. **OFFICERS:** H.E. Salim A. Salim, Sec.Gen. **MEMBERS:** 51. **BUDGET:** US$29,000,000. **LANGUAGES:** Arabic, English, French, Portuguese.

DESCRIPTION: Members are independent African states. Objectives are to: promote unity and solidarity among members and defend their sovereignty, territorial integrity, and independence; eliminate all forms of colonialism in Africa; encourage and coordinate members' efforts to improve the quality of life in Africa; further cooperation between members and non-African states, in accordance with the U.N. Charter and the Universal Declaration of Human Rights. **PUBLICATIONS:** *Brief History*; *Cultural Charter*; *Economic Declaration*; *Fundamental Changes Taking Place in the World*; *OAU in a Month* (in English and French); *The Week in the OAU*; *Weekly Press Brief.*

ORGANIZATION OF AMERICAN STATES (OAS)

17th St. & Constitution Ave. NW, Washington, DC 20006 **PHONE:** (202) 458-3000, (202) 458-6046 **FX:** (202) 458-6421 **E-MAIL:** pi@oas.org **WEBSITES:** http://www.oas.org **FOUNDED:** 1890. **OFFICERS:** Cesar Gaviria Trujillo, Sec.Gen. **MEMBERS:** 35. **STAFF:** 688. **LANGUAGES:** English, French, Portuguese, Spanish.

DESCRIPTION: International organization created to achieve an order of peace and justice among the American nations, to promote their solidarity, to strengthen their collaboration, and to defend their sovereignty, territorial integrity, and independence. Operates through agencies and institutions throughout the hemisphere. Maintains museum of modern Latin American art. Provides speakers' bureau. **PUBLICATIONS:** *Americas* (in English and Spanish), bimonthly. Discusses culture, economy, history, and lifestyles of the Americas.; *Secretary General Annual Report*; Catalog; Journals; Papers; Proceedings; Reports.

ORGANIZATION OF ARAB PETROLEUM EXPORTING COUNTRIES (OAPEC) (Organization de Paises Arabes Exportadores de Petroleo)

PO Box 20501, Kuwait City 13066, Kuwait **PHONE:** 20 965 4844500 **FX:** 20 965 4815747 **TX:** 22166 NAFARAB **E-MAIL:** oapec@kuwait.net **FOUNDED:** 1968. **OFFICERS:** Abdul-Aziz Al-Turki, Sec.Gen. **MEMBERS:** 10. **STAFF:** 50. **BUDGET:** £E 1,388,000. **LANGUAGES:** Arabic.

DESCRIPTION: Members are Arab nations producing and exporting petroleum. Seeks to: promote cooperation among member countries in economic activities of the petroleum industry; determine methods to safeguard members' interests; assist member countries in exchanging information and expertise. Maintains Arab Center for Energy Studies. Studies petroleum industry developments. Compiles statistics and projections on energy production and consumption, petroleum reserves, and production of petroleum products. Conducts technical and international cooperation seminars. **PUBLICATIONS:** *OAPEC Monthly Bulletin* (in Arabic and English); *Oil and Arab Cooperation* (in Arabic and English), quarterly; *Secretary General's Annual Report* (in Arabic and English). Also publishes *Energy Terminology: A Multilingual Glossary*, legal documents, monographs, proceedings, and technical studies.

ORGANIZATION FOR COORDINATION AND COOPERATION IN THE STRUGGLE AGAINST ENDEMIC DISEASES (OCCSED) (Organisation de Coordination et de Cooperation pour la Lutte Contre les Grandes Endemies - OCCGE)

01 Boite Postale 153, Bobo-Dioulasso 01, Burkina Faso **PHONE:** 226 970101 **FX:** 226 970099 **OFFICERS:** Abdoulaye Rhaly, Sec.Gen. **LANGUAGES:** English, French.

DESCRIPTION: Members are health care professionals, researchers, and services. Seeks to eradicate selected contagious diseases. Conducts medical research; delivers medical services to populations prone to endemic diseases.

ORGANIZATION FOR THE DEVELOPMENT OF THE SENEGAL RIVER (ODSR) (Organisation pour la Mise en Valeur du Fleuve Senegal - OMVS)

46 rue Carnot, Boite Postale 3152, Dakar, Senegal **PHONE:** 221 223679 **FX:** 221 234762 **TX:** 51670 **OFFICERS:** Alpha Oumar Konare, Chm. **LANGUAGES:** English, French.

DESCRIPTION: Members are countries encompassing the Senegal River Basin. Promotes appropriate and sustainable development in the region. Provides support and assistance to development programs and coordinates development activities in the Senegal Basin; gathers and disseminates information.

ORGANIZATION OF IBERO-AMERICAN STATES FOR EDUCATION, SCIENCE AND CULTURE (OEI) (Organizacion de Estados Iberoamericanos para la Educacion, la Ciencia y la Cultura - OEI)

c/o Bravo Murillo 38, E-28015 Madrid, Spain **PHONE:** 34 91 5944382, 34 91 5944442 **FX:** 34 91 5943286 **E-MAIL:** oeimad@oei.es **WEBSITES:** http://www.oei.es **FOUNDED:** 1949. **OFFICERS:** Jose Torreblanca Prieto, Sec.Gen. **LANGUAGES:** Portuguese, Spanish.

DESCRIPTION: Members are government representatives. Encourages educational, cultural, and scientific cooperation and technical assistance among Spanish- and Portuguese-speaking countries. Conducts training courses, postgraduate courses, and basic and adult education projects. **PUBLICATIONS:** *Guias y Repertorios*, periodic; Monographs; Pamphlets.

ORGANIZATION FOR INDUSTRIAL, SPIRITUAL AND CULTURAL ADVANCEMENT INTERNATIONAL (OISCA)

6-12 Izumi 3-chome, Suginami-ku, Tokyo 168-0063, Japan **PHONE:** 81 3 33225164, 81 3 33225161 **FX:** 81 3 33247111 **E-MAIL:** oisca@oisca.org **WEBSITES:** http://www.oisca.org **FOUNDED:** 1961. **OFFICERS:** Yuichi Kobayashi, Sec.Gen. **MEMBERS:** 15,000. **STAFF:** 100. **BUDGET:** US$10,000,000. **LANGUAGES:** English, Japanese.

DESCRIPTION: Members are individuals, corporations, and associations representing 80 countries. Promotes international cooperation and understanding in an effort to ease tension and conflict throughout the world. Assists in agricultural and rural development in Asian-Pacific countries by sending specialists to render technical aid. Organizes missions to study local needs in Asian Pacific nations; offers scholarships to students and trainees from both Asia and Africa for study in Japan; conducts youth training programs, Asian Pacific Youth Forums for Community Development, and Love Green, a tree-planting program. Organizes periodic seminar on education and development; sponsors youth forum and North-South Forum. **PUBLICATIONS:** *Friendship Through Training*, annual; *OISCA Bulletin Board* (in English), bimonthly; *OISCA Journal* (in Japanese), monthly.

ORGANIZATION FOR INTERNATIONAL ECONOMIC RELATIONS (IER) (Organisation pour les Relations Economiques Internationales)

Rudolfsplatz 5, A-1010 Vienna, Austria **PHONE:** 43 1 5338105, 43 1 5333810 **FX:** 43 1 533356817 **TX:** 116195 IERA **FOUNDED:** 1947. **OFFICERS:** Dr. W. Hendricks. **MEMBERS:** 1,000. **STAFF:** 7. **LANGUAGES:** English, German.

DESCRIPTION: Members are enterprises in Western Europe interested in economic cooperation with the Southeast-, Central-, and East-European countries. Works to establish and foster economic relations among countries, companies, and organizations. **PUBLICATIONS:** *Donaueuropaischer Informationsdienst* (in English and German), periodic; *Internationale Wirtschaftszeitschrift ''West-Ost-Journal,''* periodic.

ORGANIZATION OF THE ISLAMIC CONFERENCE (OIC) (Organisation de la Conference Islamique - OCI)

Kilo 6, Mecca Rd., PO Box 178, Jeddah 21411, Saudi Arabia **PHONE:** 966 2 6800880 **FX:** 2 6873568 **TX:** 601366 **FOUNDED:** 1969. **OFFICERS:** Azeddine Laraki, Sec.Gen. **MEMBERS:** 46. **STAFF:** 122. **LANGUAGES:** Arabic, English, French.

DESCRIPTION: Members are Islamic states. Promotes solidarity, peace, and strengthened cooperation among members in the areas of politics, economics, culture, science, and technology. Works for the achievement of welfare and harmony among Muslims worldwide. Seeks: elimination of racial segregation and discrimination; eradication of colonialism in all its forms; the support of international peace and security founded on justice; protection of sites of importance to Islamic worship; support for the struggle of Palestinians to regain the land they believe belongs to them. Maintains Islamic Jurisprudence Academy.

ORGANIZATION OF THE ISLAMIC SHIPOWNERS' ASSOCIATION

PO Box 14900, Jeddah 21434, Saudi Arabia **PHONE:** 966 2 6653379 **FX:** 966 2 6604920 **TX:** 607303 **OFFICERS:** Dr. Abdullatif A. Sultan, Sec.Gen. **LANGUAGES:** Arabic, English.

DESCRIPTION: Members are shipowners' associations. Promotes growth and development of the shipping industries of Islamic countries. Represents members' interests before government agencies and international trade and labor organizations.

ORGANIZATION OF THE PETROLEUM EXPORTING COUNTRIES (OPEC) (Organisation des Pays Exportateurs de Petrole - OPEP)

Obere-Donaustrasse 93, A-1020 Vienna, Austria **PHONE:** 43 1 211120 **FX:** 43 1 2149827 **TX:** 134474 **E-MAIL:** info@opec.org **WEBSITES:** http://www.opec.org **FOUNDED:** 1960. **OFFICERS:** Dr. Rilwanu Lukman, Sec.Gen. **MEMBERS:** 11. **LANGUAGES:** English.

DESCRIPTION: Representatives from Algeria, Indonesia, Iran, Iraq, Kuwait, Libya, Nigeria, Qatar, Saudi Arabia, United Arab Emirates, and Venezuela. Principal aims are the coordination and unification of the petroleum policies of member countries and the determination of the best means for safeguarding their individual and collective interests. Devises means of ensuring: the stabilization of prices in international oil markets with a view toward eliminating harmful and unnecessary fluctuations, due regard being given at all times to the interests of the producing nations; a steady income to the producing countries; an efficient, economic, and regular supply of petroleum to consuming nations; a fair return on capital of those investing in the petroleum industry. Conducts research; compiles statistics. **PUBLICATIONS:** *OPEC Annual Statistical Bulletin*; *OPEC Bulletin* (in English), monthly; *OPEC Review*, quarterly. Also publishes books and various booklets.

ORIENT AIRLINES ASSOCIATION (OAA)

5/F Corporate Business Centre, 151 Paseo de Roxas, Makati, Metro Manila 1226, Philippines **PHONE:** 63 2 8190151, 63 2 8403191 **FX:** 63 2 8103518 **TX:** MNLOAXD **FOUNDED:** 1966. **OFFICERS:** Richard Thomas Stirland, Dir.Gen. **MEMBERS:** 16. **STAFF:** 21. **BUDGET:** US$625,000. **LANGUAGES:** English.

DESCRIPTION: Members are airline companies of Australia, Brunei-Darussalam, Hong Kong, Indonesia, Japan, Malaysia, New Zealand, Papua New Guinea, Philippines, Republic of Korea, Singapore, Taiwan, and Thailand. Promotes the development of air commerce in the region, with particular emphasis on passengers' interests and safety; encourages collaboration among members in order to eliminate harmful competition. Works to create conditions favorable to the travel and tourism industry and to strengthen the role of member airlines in regional economic, cultural, and social cooperation. Provides for the exchange of opinions on subjects of common interest. Conducts seminars on topics including fraud prevention, flight safety, and automation; sponsors air transport course. **PUBLICATIONS:** *OAA Annual and Statistical Report* (in English); *Orient Aviation*, bimonthly.

OSPAR COMMISSION

New Ct., 48 Carey St., London WC2A 2JQ, England **PHONE:** 44 171 2429927 **FX:** 44 171 8317427 **E-MAIL:** secretariat@ospar.org **WEBSITES:** http://www.ospar.org **FOUNDED:** 1992. **OFFICERS:** Ben van de Wetering, Sec. **MEMBERS:** 16. **STAFF:** 12. **BUDGET:** £849,940. **LANGUAGES:** English, French.

DESCRIPTION: Members are governments of Belgium, Denmark, Germany, Finland, France, Iceland, Luxembourg, Netherlands, Norway, Portugal, Republic of Ireland, Spain, Sweden, Switzerland, the United Kingdom, and the Commission of European Communities. **PUBLICATIONS:** Annual Report (in English and French).

OXFAM

274 Banbury Rd., Oxford OX2 7DZ, England **PHONE:** 44 1865 311311 **FX:** 44 1865 312600 **TX:** 83610 OXFAM G **E-MAIL:** oxfam@oxfam.org.uk **WEBSITES:** http://www.oxfam.org.uk/; http://www.oneworld.org/oxfam **FOUNDED:** 1942. **OFFICERS:** David Bryer, Dir. **STAFF:** 1,343. **BUDGET:** £91,800,000. **LANGUAGES:** English.

DESCRIPTION: Supported by 180,000 financial donors and assisted by 30,000 volunteers. Provides food and shelter to people in emergency situations. Assists people in their efforts to gain economic self-sufficiency. Feels that all people have a right to good living conditions; believes in the fundamental dignity of people and their inherent ability to overcome obstacles imposed by geopolitical and socioeconomic hardships. Contends that the world's material resources can, if equitably distributed, satisfy the basic needs of all people. Administers charitable program; maintains 50 field offices and operates over 2900 relief and development projects in more than 70 countries. Conducts educational programs and campaigns in the United Kingdom and the Republic of Ireland. **PUBLICATIONS:** *Development in Practice* (in English), quarterly; *Gender and Development* (in English), quarterly; Catalog.

OXFAM INTERNATIONAL ADVOCACY OFFICE

1511 K St. NW, Ste. 1044, Washington, DC 20005 **PHONE:** (202) 393-5332 **FX:** (202) 783-8739 **E-MAIL:** oxfamintdc@igc.apc.org **OFFICERS:** Justin Forsyth, Dir. **MEMBERS:** 9. **LANGUAGES:** English.

DESCRIPTION: Members are national OXFAM organizations in the United States, Belgium, Canada, Hong Kong, Ireland, New Zealand, and the United Kingdom Community Aid Abroad of Australia, and the Netherlands Organization for International Development. Seeks to relieve "poverty, distress, and suffering in every part of the world without regard to political and religious beliefs." Encourages participation by indigenous populations in the operation and administration of development and relief projects. Conducts projects in areas including health, agriculture, small business development, and education. Coordinates the relief and development efforts of public and private sector organizations worldwide. Makes available emergency relief services.

PACIFIC AREA NEWSPAPER PUBLISHERS ASSOCIATION (PANPA)

PO Box 1087, Gosford 2250, Australia **PHONE:** 612 4322 1066 **FX:** 612 4322 1141 **E-MAIL:** panpa@panpa.org.au **WEBSITES:** http://www.aap.com.au/panpa/service.nsf **FOUNDED:** 1969. **OFFICERS:** Frank Kelett, Exec.Dir. **MEMBERS:** 320. **STAFF:** 5. **LANGUAGES:** English.

DESCRIPTION: Members are chief executive officers, production managers, marketing/advertising managers, and editors in 18 countries involved in newspaper publication. Acts as a forum for the exchange of ideas. Conducts seminars and workshops on newspaper advertising, sales, management, production, and editorial topics. **PUBLICATIONS:** *PANPA Bulletin* (in English), monthly; Journal.

PACIFIC COMMUNITY (Communaute du Pacifique)

BP D5, Noumea 98848, New Caledonia **PHONE:**

687 263818 **FX:** 687 263818 **E-MAIL:** spc@spc.org.nc **WEBSITES:** http:// www.spc.org.nc/ **FOUNDED:** 1947. **MEMBERS:** 27. **STAFF:** 170. **BUDGET:** 2,044,409,400 CFP Fr. **LANGUAGES:** English, French.

DESCRIPTION: Members are Pacific Island countries and territories, Australia, France, New Zealand, the United Kingdom and the United States. Purpose is to provide technical advice, training, research and assistance in areas of: agriculture and plant protection; fisheries; health; socio-economic and statistical services; rural development and technology; women's affairs; media; youth; culture. Organizes training courses and technical meetings. **PUBLICATIONS:** *Report of the South Pacific Conference*, annual. Also issues technical papers, handbooks, information circulars, posters, and leaflets.

PACIFIC ECONOMIC COOPERATION COUNCIL (PECC)

4 Nassim Rd., Singapore 258372, Singapore **PHONE:** 65 7379822, 65 7379823 **FX:** 65 7379824 **TX:** 04-53332 **E-MAIL:** peccsec@pacific.net.sg **WEBSITES:** http:// www.pecc.net **FOUNDED:** 1980. **OFFICERS:** Paul Irwin. **MEMBERS:** 23. **STAFF:** 6. **LANGUAGES:** Chinese, English, French, Japanese.

DESCRIPTION: Members are representatives of businesses, governments, and economic and trade research organizations from the Asia-Pacific region. Promotes increased understanding of factors influencing international trade and economic development; promotes increased trade among Asian and Pacific countries. Conducts research and disseminates information in areas including trade policy, economic forecasts, financial and capital markets, human resource development, science and technology, telecommunications, transportation, tourism, fisheries, and agriculture. **PUBLICATIONS:** *Pacific Economic Outlook*, annual; *PECC Link*, quarterly; Books; Reports.

PACIFIC RIM CONSORTIUM (PRC)

750 State St., Ste. 204, San Diego, CA 92101 **PHONE:** (619) 238-0088 **FX:** (619) 238-7371 **E-MAIL:** patrickon@mindspring.com **FOUNDED:** 1991. **OFFICERS:** Patricia Rickon, Coordinator. **STAFF:** 55. **LANGUAGES:** English.

DESCRIPTION: Members are women's and development organizations. Promotes increased participation by women in economic development programs. Trains women to establish and administer income-generating enterprises; collaborates with development programs to ensure inclusion of women in development activities; provides technical, financial, and administrative assistance to selected development projects and microenterprises. Maintains foundry, dance studio, and music conservatory in the former Soviet Union.

PACIFIC SCIENCE ASSOCIATION (PSA)

1525 Bernice St., Honolulu, HI 96817 **PHONE:** (808) 848-4139 **FX:** (808) 847-8252 **E-MAIL:** psa@bishop.bishop.hawaii.org **FOUNDED:** 1920. **OFFICERS:** L.G. Eldredge, Exec.Sec. **MEMBERS:** 100. **STAFF:** 1. **LANGUAGES:** English.

DESCRIPTION: Members are countries within or bordering the Pacific Ocean; scientists, scientific societies, institutions, and corporations. Initiates and promotes the study of scientific problems relating to the Pacific region; encourages cooperation among scientists of all Pacific countries. **PUBLICATIONS:** *Coral Reef Newsletter*, semiannual; *The Evolution and Organization of Prehistoric Society in Polynesia*; *Information Bulletin*, quarterly; *International Fax Directory for Biologists*; *Pacific Research Titles*, periodic; Proceedings, periodic.

PACIFIC TELECOMMUNICATIONS COUNCIL (PTC)

2454 S. Beretania St., Ste. 302, Honolulu, HI 96826 **PHONE:** (808) 941-3789 **FX:** (808) 944-4874 **E-MAIL:** info@ptc.org **WEBSITES:** http:// www.ptc.org/ **FOUNDED:** 1980. **OFFICERS:** Richard J. Barber, Exec.Dir. **MEMBERS:** 630. **STAFF:** 15. **BUDGET:** US$1,700,000. **LANGUAGES:** English.

DESCRIPTION: Members are organizations and professionals involved as providers, users, policymakers, and analysts in telecommunications development in the Americas, Oceania, and Asia. Provides a forum for discussion between governments, academia, and telecommunications users, planners, and providers of equipment and services. Organizes conferences and seminars to exchange views and information on telecommunication services and systems in the Pacific region. Addresses immediate and future-oriented telecommunications, information services/systems and broadcasting/multimedia issues. Conducts workshops on telecommunications skills; conducts research programs. **PUBLICATIONS:** *Member's Bulletin* (in English), 6/yr. Membership activities newsletter.; *Pacific Telecommunications Council—Membership Directory*, annual. Listing of members of PTC and public relations representatives of associated firms.; *Pacific Telecommunications Review (PTR)*, quarterly. Reports on council activities and provides academic articles about telecommunications in the Pacific. Covers international policy issues.; *PTC Proceedings*, annual.

PAN AFRICAN INSTITUTE FOR DEVELOPMENT (PAID)

3, rue de Varembe, CH-1211 Geneva 20, Switzerland **PHONE:** 41 22 7336016, 41 22 7336017 **FX:** 41 22 7330975 **E-MAIL:** ipdge@iproliuk.ch **FOUNDED:** 1964. **OFFICERS:** Faya Kondiano. **MEMBERS:** 250. **STAFF:** 120. **BUDGET:** 3,000,000 SFr. **LANGUAGES:** French.

DESCRIPTION: Members are individuals concerned with development activities, particularly rural development. Promotes economic, social, and cultural development, preferably through grass roots community structures, with priority given to activities that aid the neediest people in rural areas. Sponsors research programs; offers consulting services.

PAN AFRICAN INSTITUTE FOR DEVELOPMENT - EASTERN AND SOUTHERN AFRICA (PAIDESA)

PO Box 80448, Kabwe, Zambia **PHONE:** 260 5 223651, 260 5 222341 **FX:** 260 5 223451 **TX:** ZA 81290 **E-MAIL:** paidesa@zamnet.zm **FOUNDED:** 1964. **OFFICERS:** Luther Banga, Dir. **LANGUAGES:** English, French.

DESCRIPTION: Members are regional development and training institutes. Promotes sustainable social and economic development in eastern and southern Africa. Provides consulting services and technical support to economic and community development organizations; conducts research and educational programs.

PAN-AFRICAN UNION OF SCIENCE AND TECHNOLOGY

Boite Postale 2339, Brazzaville, Congo **PHONE:** 242 832265 **FX:** 242 832185 **TX:** 5511 **OFFICERS:** Levy Makany, Sec.Gen. **LANGUAGES:** English, French.

DESCRIPTION: Members are scientists and other individuals with an interest in technology and the sciences. Seeks to advance the study and teaching of science; encourages technology transfer. Conducts research and educational programs.

PAN-AFRICAN WRITERS' ASSOCIATION (PAWA)

Cantonments, PO Box C450, Accra, Ghana **OFFICERS:** Atukwei Okai, Sec.Gen. **LANGUAGES:** English.

DESCRIPTION: Professional writers. Promotes recognition and advancement of members. Represents members' interests.

PAN-AMERICAN ASSOCIATION OF OPHTHALMOLOGY (PAAO)

1301 S. Bowen Rd., Ste. 365, Arlington, TX 76013 **PHONE:** (817) 265-2831 **FX:** (817) 275-3961 **E-MAIL:** paao@flash.net **WEBSITES:** http://www.flash.net/-paao **FOUNDED:** 1939. **OFFICERS:** Teresa J. Bradshaw, Admin. **MEMBERS:** 10,000. **STAFF:** 2. **BUDGET:** US$200,000. **LANGUAGES:** English, Portuguese, Spanish.

DESCRIPTION: Members are ophthalmologists throughout the Western Hemisphere. Seeks to improve the treatment of eye diseases and prevention of blindness in the Americas through the exchange of ideas and treatments. **PUBLICATIONS:** *Noticiero* (in Spanish), quarterly; *Pan-American Association of Ophthalmology-The Pan American*, semiannual. Includes obituaries and conference calendar.

PANAMERICAN CULTURAL CIRCLE (PCC)

16 Malvern Pl., Verona, NJ 07044 **PHONE:** (973) 239-3125 **FX:** (973) 239-3125 **FOUNDED:** 1963. **OFFICERS:** Elio Alba-Buffill Ph.D., Exec.Sec. **MEMBERS:** 800. **LANGUAGES:** English, Spanish.

DESCRIPTION: Members are professors, scholars, writers, poets, artists, and others interested in the culture and democratic ideals of Panamerica. Seeks to: promote the ideals of Panamericanism; disseminate information on Latin American culture; stimulate literary activity and the study of Spanish and Latin American literature. **PUBLICATIONS:** *Circulo Poetico* (in English and Spanish), annual. Contains poetry.; *Circulo: Revista de Cultura* (in English and Spanish), annual. Contains scholarly works on Latin American and Spanish literature and culture.; *Estudios Literarios Sobre Hispanoamerica*; *Jose Marti en el Centenario de su Muerte*; *Marti Ante la Critica Actual*; *Marti Ante la Critica Actual.*

PAN AMERICAN DEVELOPMENT FOUNDATION (PADF)

2600 16th St., NW, Washington, DC 20009-4202 **PHONE:** (202) 458-3969 **FX:** (202) 458-6316 **FOUNDED:** 1962. **OFFICERS:** Sarah Horsey-Barr, Exec.Dir. **STAFF:** 25. **BUDGET:** US$23,000,000. **LANGUAGES:** Portuguese, Spanish.

DESCRIPTION: Utilizes grants from the public and private sectors to improve the quality of life in Latin America and the Caribbean. Fulfills its mission through municipal training and democracy strengthening, environmental preservation, microenterprise development, and in-kind medical and vocational tools donations. **PUBLICATIONS:** *Pan American Development Foundation—Newsletter*, semiannual. Covers foundation activities and programs.; Annual Report (in English and Spanish), annual.

PAN AMERICAN HEALTH AND EDUCATION FOUNDATION (PAHEF)

525 23rd St. NW, Washington, DC 20037 **PHONE:** (202) 974-3416 **FX:** (202) 974-3658 **E-MAIL:** pahef@paho.org **FOUNDED:** 1968. **OFFICERS:** Jess Gersky, Exec.Sec. **STAFF:** 5. **BUDGET:** US$3,000,000. **LANGUAGES:** English.

DESCRIPTION: Seeks to mobilize financial and human resources for the improvement of health and education, particularly in Latin America; to advance the objectives of the Pan American Health Organization and World Health Organization. Cosponsors Program for Textbooks and Instructional Materials, which makes needed items available for the training of health personnel at all levels, including professional, technical, and auxiliary. Works cooperatively with organizations and governmental bodies which share the same objectives. **PUBLICATIONS:** *Boletin de Medicamentos y Terapeutica* (in Spanish), quarterly; Manuals. Provides information for primary health care workers.

PAN AMERICAN HEALTH ORGANIZATION (PAHO)

525 23rd St. NW, Washington, DC 20037 **PHONE:** (202) 974-3000 **FX:** (202) 338-0869 **WEBSITES:** http://www.paho.org **FOUNDED:** 1902. **OFFICERS:** Dr. Bryna Brennan, Dir., Public Information. **MEMBERS:** 41. **STAFF:** 1,100. **LANGUAGES:** English, Spanish.

DESCRIPTION: Members are governments of Western Hemisphere nations united to improve physical and mental health in the Americas. Coordinates regional activities combating disease including exchange of statistical and epidemiological information, development of local health services, and organization of disease control and eradication programs. Encourages development in health systems and technology; provides consulting services; conducts educational courses on public health topics including environmental health, food and nutrition, and tropical diseases. Has established Emergency Preparedness and Disaster Relief Coordination Program in order to increase the ability of health institutions to effectively handle emergencies. Operates the Natural Disaster Relief Voluntary Fund to support disaster relief activities. Maintains the Pan American Sanitary Bureau, the regional office for the Americas of the World Health Organization. Develops health documentaries and coordinates teleconferences. **PUBLICATIONS:** *Disaster Preparedness in the Americas*, quarterly. PAHO Emergency Preparedness and Disaster Relief Coordination Program. Covers disaster preparedness, mitigation, and management.; *EPI Newsletter*, bimonthly. Provides information on immunization programs in the Americas. Covers new technologies available for the execution of programs.; *Epidemiological Bulletin* (in English and Spanish), bimonthly. Disseminates epidemiological information regarding communicable and noncommunicable diseases of public health importance.; *Health in the Americas*, quadrennial. Each edition, different price.; *Revista Panamericana de Salud Publica/Pan American Journal of Public Health* (in English, Portuguese, and Spanish), monthly. Serves as a reference source regarding health problems in the Americas and progress made toward solutions. Includes book reviews and results.; Manuals; Monographs; Reports. Also publishes scientific books and technical paper.

PANEUROPEAN UNION (PEU) (Paneuropa-Union - PEU)

Karlstrasse 57, D-80333 Munich, Germany **PHONE:** 49 89 554683 **FX:** 49 89 594768 **E-MAIL:** karsten.braeuker@paneuropa.org **WEBSITES:** http://www.paneuropa.org **FOUNDED:** 1923. **OFFICERS:** Otto Von Habsburg, Pres. **MEMBERS:** 32. **LANGUAGES:** English, French, German, Italian.

DESCRIPTION: Members are national organizations. Seeks to unite all European peoples into a nonpartisan political and economic union based on principles of liberty, self-determination, and Christian values. **PUBLICATIONS:** *Paneuropa* (in Croatian, Czech, French, and Slovak), monthly; *Paneuropa Deutschland* (in German), quarterly; *Paneuropa-Intern* (in German), 17/year; *Paneuropa Osterreich* (in German), monthly. Also publishes documentation on Paneuropean topics.

PAN-PACIFIC AND SOUTH EAST ASIA WOMEN'S ASSOCIATION

PO Box 362, Apia, Western Samoa **PHONE:** 685 21823 **FOUNDED:** 1958. **OFFICERS:** Aliitasi Taua'a. **MEMBERS:** 500.

DESCRIPTION: Promotes cooperation among women of the Asia-Pacific region. Works to improve the social, economic, and cultural conditions.

PAN-PACIFIC SURGICAL ASSOCIATION (PPSA)

1360 S Beretania St., Ste. 304, Honolulu, HI 96814-1520 **PHONE:** (808) 528-1180 **FX:** (808) 528-1188 **E-MAIL:** congress@ppsa.org **FOUNDED:** 1929. **OFFICERS:** Gayle Yoshida, Exec.Admin. **MEMBERS:** 2,716. **STAFF:** 1. **LANGUAGES:** English.

DESCRIPTION: Members are professional international association of surgeons. Coordinates exchange programs and educational projects. Conducts volunteer pogram in which surgeons go to Micronesia and the Pacific for two or more weeks. Sponsors educational programs; operates speakers' bureau. **PUBLICATIONS:** *Pan-Pacific Surgical Association Bulletin*, quarterly. Also publishes program.

PARLIAMENTARIANS FOR GLOBAL ACTION (PGA)

211 E. 43rd St., Ste. 1604, New York, NY 10017 **PHONE:** (212) 687-7755 **FX:** (212) 687-8409 **E-MAIL:** parglobal@aol.com **FOUNDED:** 1979. **OFFICERS:** Shazia Rafi. **MEMBERS:** 1,200. **STAFF:** 13. **BUDGET:** US$1,120,000. **LANGUAGES:** English, Spanish.

DESCRIPTION: Members are parliamentarians from 95 nations seeking solutions to problems of the environment, uncontrolled population growth, poverty, and war which require the action of the world community. **PUBLICATIONS:** *Parliamentarians for Global Action* (in English), annual; *Parliamentarians Global Action Update*, quarterly.

PARTNERS OF THE AMERICAS

1424 K St. NW, Ste. 700, Washington, DC 20005 **PHONE:** (202) 628-3300 **FX:** (202) 628-3306 **TX:** 6421 NAPAR **E-MAIL:** info@partners.poa.com **WEBSITES:** http://www.partners.net **FOUNDED:** 1964. **OFFICERS:** Norman Brown, Pres. **MEMBERS:** 20,000. **STAFF:** 31. **BUDGET:** US$8,400,000. **LANGUAGES:** English.

DESCRIPTION: Members are volunteer private citizens organized in 60 partnerships linking U.S. states with Latin American and Caribbean countries. Goals are to encourage innovative, community-based projects and joint planning among local partnerships; to provide a stimulus to local partnerships in generating additional resources for their technical and cultural exchange projects; to assist in the implementation of effective and ongoing development projects in the Latin American, Mexican, and Caribbean partner areas. Partnerships form the basis for exchange projects in agriculture, public health, culture, rehabilitation, community education, sports, and other areas of hemispheric development. **PUBLICATIONS:** *A Good Idea That Works*; *Partners*, quarterly. Reports on technical assistance projects and exchanges between the United States and Latin America.; Annual Report; Booklets; Brochures.

PARTY OF EUROPEAN SOCIALISTS (PES)

European Parliament, rue Wiertz, B-1047 Brussels, Belgium **PHONE:** 32 2 2842976, 32 2 2842978 **FX:** 32 2 2301766 **E-MAIL:** pes@pes.org **WEBSITES:** http://www.pes.org **FOUNDED:** 1974. **OFFICERS:** Jean-Francois Vallin, Gen.Sec. **STAFF:** 17. **LANGUAGES:** French, German, Spanish.

DESCRIPTION: Members are socialist, democratic socialists, and labor parties in European Economic Community countries. **PUBLICATIONS:** *Fax Info* (in English, French, German, and Spanish), monthly.

PAX ROMANA, INTERNATIONAL CATHOLIC MOVEMENT FOR INTELLECTUAL AND CULTURAL AFFAIRS (ICMICA) (Pax Romana, Movimento Internacional de Intelectuales Catolicos)

c/o Anselmo Lee, 15, rue du Grand-Bureau, CP 85, CH-1211 Geneva 24, Switzerland **PHONE:** 41 22 8230707 **FX:** 41 22 8230708 **E-MAIL:** miicmica@paxromana.int.ch **FOUNDED:** 1947. **OFFICERS:** Anselmo Lee, Sec.Gen. **MEMBERS:** 30,000. **STAFF:** 3. **LANGUAGES:** English, French, Spanish.

DESCRIPTION: An autonomous arm of Pax Romana. National federations of Catholic intellectuals, professionals, and others. Promotes a living faith and openness of mind in the cultural, economic, political, social, and technical spheres; encourages Christian thinking on contemporary problems; participates in the apostolic action of the church. Holds symposia and seminars on topics such as: human rights; new international economic order; North-South dialogue; the status of women; poverty and peace; development and technology; theological reflections. **PUBLICATIONS:** *Convergence*, quarterly; *Newsletter* (in English, French, and Spanish), 3/year. Also publishes *Guide to ICMICA*, special reports, and monographs.

PENAL REFORM INTERNATIONAL (PRI)

169 Clapham Rd., London SW9 0PU, England **PHONE:** 44 171 5826500 **FX:** 44 171 7354666 **OFFICERS:** Vivien Stern. **LANGUAGES:** English.

DESCRIPTION: Promotes humane operation of criminal justice systems worldwide. Meets with national criminal justice officials to discuss issues and propose solutions to problems; advocates alternative sentencing and treatment of offenders.

PERMANENT INTERNATIONAL COMMITTEE OF LINGUISTS (CIPL) (Comite International Permanent des Linguistes - CIPL)

c/o Prof. van Sterkenburg, Instituut voor Nederlandse Lexicologie, PO Box 9515, NL-2300 Leiden, Netherlands **PHONE:** 31 71 5141648 **FX:** 31 71 5272115 **E-MAIL:** secretariaat@rulxha.leidenuniv.nl **FOUNDED:** 1928. **OFFICERS:** Prof. Sterkenburg, Sec.Gen. **MEMBERS:** 50. **LANGUAGES:** English, French, German.

DESCRIPTION: Members are national and international linguistic organizations united to assist in the development of linguistic science. Attempts to further linguistic research and to coordinate activities for the advancement of linguistics. Promotes contact among linguists and disseminates research results. Sponsors special linguistic projects. **PUBLICATIONS:** *Linguistic Bibliography*, annual. Also publishes congress proceedings locally.

PERMANENT SECRETARIAT OF THE GENERAL TREATY ON CENTRAL AMERICAN ECONOMIC INTEGRATION (Secretaria Permanente del Tratado General de Integracion Economica Centroamericana - SIECA/CACM)

4A av. 10-25, zona 14, Codigo Postal 01901, Guatemala City, Guatemala **PHONE:** 502 2 3682151, 502 2 3682152 **FX:** 502 2 3373750 **TX:** 6203 ANSTEL GU **E-MAIL:** info@sieca.org.gt **WEBSITES:** http://www.sieca.org.gt **FOUNDED:** 1961. **OFFICERS:** Hersson Rodriguez Sierra, Dir. **MEMBERS:** 5. **LANGUAGES:** Spanish.

DESCRIPTION: Representatives of Costa Rica, El Salvador, Guatemala, Honduras, and Nicaragua. Purpose is to enforce the terms of over 30 treaties, agreements, and protocols signed by Central American countries. Develops studies intended to contribute toward the economic integration of Central America. Offers computerized services. Compiles statistics; operates speakers' bureau. **PUBLICATIONS:** *Series Estadisticas Seleccionadas* (in Spanish), annual; *Series Estadisticas Seleccionadas de Centroamerica* (in Spanish), annual;

Sistema Aracelario Centoamericano (in Spanish), periodic.

PERMANENT TRIPARTITE COMMISSION FOR EAST AFRICAN COOPERATION

Intl. Conf. Centre, Arusha, United Republic of Tanzania **OFFICERS:** Francis Kirimi Muthaura, Exec.Sec. **LANGUAGES:** English.

DESCRIPTION: Members are East African governments. Seeks to insure appropriate and sustainable development in the region. Works to coordinate development activities in East Africa.

PHYSICIANS FOR SOCIAL RESPONSIBILITY (PSR) (Laarkarin Sosiaalinen Vastuu - LSV)

c/o Lena Hyppouen, KTL Mannerheimintie 166, FIN-00300 Helsinki, Finland **PHONE:** 358 9 4744230 **OFFICERS:** Prof. Pirjo Makela, Chm. **BUDGET:** US$200,000. **LANGUAGES:** English, Finnish.

DESCRIPTION: Members are physicians, veterinarians, dentists, and students. Seeks to ensure availability of primary health care services; promotes effective and appropriate economic development worldwide. Conducts public educational programs to raise awareness of global issues including health care, peace, and environmental protection. Operates health care and vaccination campaigns in Bangladesh and Peru. **PUBLICATIONS:** *LSV-Tiedote* (in Finnish), bimonthly; Books.

PLANETARY ASSOCIATION FOR CLEAN ENERGY (PACE) (Societe Planetaire pour l'Assainissement de l'Energie - SPAE)

100 Bronson, No. 1001, Ottawa, ON, Canada K1R 6G8 **PHONE:** (613) 236-6265 **FX:** (613) 235-5876 **FOUNDED:** 1976. **OFFICERS:** Dr. A. Michrowski, Pres. **MEMBERS:** 3,550. **STAFF:** 3. **BUDGET:** C$200,000. **LANGUAGES:** English, French, Italian, Spanish.

DESCRIPTION: Members are researchers, individuals, corporations, and institutions in 60 countries seeking to facilitate research, development, demonstration, and evaluation of clean energy systems. The association defines clean energy systems as those which utilize natural sources, and are inexpensive, non-polluting, and universally applicable. Concerns include the bioeffects of low level electromagnetics, bioenergetics, and new energy technology. Tests and recommends products that facilitate the implementation of clean energy systems. Serves as a consultant to governments and other agencies. Maintains speakers' bureau. **PUBLICATIONS:** *Newsletter*, periodic; Monograph, periodic; Proceedings, periodic.

PLENTY INTERNATIONAL

PO Box 394, Summertown, TN 38483 **PHONE:** (931) 964-4864, (931) 964-4391 **FX:** (931) 964-4864 **E-MAIL:** plentyi@usit.net **WEBSITES:** http://www.plenty.org **FOUNDED:** 1974. **OFFICERS:** Peter Schweitzer, Exec.Dir. **MEMBERS:** 3,500. **STAFF:** 6. **BUDGET:** US$300,000. **LANGUAGES:** English.

DESCRIPTION: Members are volunteers committed to assisting indigenous people around the world, while allowing them to maintain their cultural identity. Charitable relief and development organization based on the concept that ''there's plenty for everyone - if we all share.'' Helps the needy provide themselves with adequate food, housing, clean water, and medical care; focuses on projects that help create self-sufficiency through appropriate training and development of life-supporting technologies. Current programs include refugee aid and development, soybean propagation and utilization, alternative energy sources, soy dairies, reforestation, human rights litigation, Caribbean basin small business development, and community development on U.S. Indian reservations. Operates Natural Rights Center and De Colores Trading Company. Maintains collection of documentaries and speakers' bureau. **PUBLICATIONS:** *Climate in Crisis: The Greenhouse Effect and What We Can Do.* Provides information on growing and using soybeans for food.; *The Global Kitchen: A Collection of Vegeterian Recipes*; *Plenty Bulletin*, quarterly.

PLUNKETT FOUNDATION

23 Hanborough Business Park, Long Hanborough, Oxford OX8 8LH, England **PHONE:** 44 1993 883636 **FX:** 44 1993 883576 **E-MAIL:** plunkett@gn.apc.org **WEBSITES:** http://www.plunkett.co.uk **FOUNDED:** 1919. **OFFICERS:** Simon Rawlinson, C.E.O. **MEMBERS:** 300. **STAFF:** 10.

DESCRIPTION: Training and consultancy for the

agricultural and horticultural cooperative sector and other farmer-controlled businesses in the UK and internationally. Provides human resource development and associated consultancy. **PUBLICATIONS:** *The World of Co-operative Enterprise* (in English), annual. Compilation of international articles on co-operatives topics. Directory and Statistics of UK Agricultural Co-operatives.

PONTIFICAL MISSION FOR PALESTINE (PMP)

1011 1st Ave., New York, NY 10022-4195 **PHONE:** (212) 826-1480 **TF:** (800) 442-6392 **FX:** (212) 838-1344 **E-MAIL:** bad@cnewa.org **WEBSITES:** http://www.cnewa.org **FOUNDED:** 1949. **OFFICERS:** Msgr. Robert L. Stern, Sec.Gen. **STAFF:** 40. **LANGUAGES:** English.

DESCRIPTION: Papal agency for humanitarian and charitable assistance. Provides aid to victims of conflict in the Middle East without regard to their national or religious affiliation. Encourages and supports programs providing relief, rehabilitation, and development in the Middle East. **PUBLICATIONS:** *Catholic Near East Magazine*, bimonthly. Includes trip information; *Resource Guide*.

POPULATION COMMUNICATIONS INTERNATIONAL (PCI)

777 United Nations Plz., 5th Fl., New York, NY 10017-3521 **PHONE:** (212) 687-3366 **FX:** (212) 661-4188 **TX:** MCI 66560 UN **E-MAIL:** pciny@population.org **WEBSITES:** http://www.population.org **FOUNDED:** 1985. **OFFICERS:** David J. Andrews, Pres. **MEMBERS:** 7,000. **STAFF:** 15. **BUDGET:** US$3,000,000. **LANGUAGES:** English.

DESCRIPTION: Works worldwide in the fields of population, environment and development. Develops motivational communication campaigns on family planning and other population-related topics for use in countries with the highest population growth rates. Promotes increased individual understanding of the relationships between family size, the environment, and the health, happiness, and prosperity of individuals, families, and communities. Uses a broad, multi-media approach, including family planning soap operas, to promote family health, family planning, and the small family norm. **PUBLICATIONS:** *Beyond Cairo*, quarterly. Documents NGO events and activities.; *International Dateline/Teletipo Poblacion* (in English and Spanish), monthly. Provides a population news service.; *Member News*, quarterly. Provides updates for members on PCI activities.

POPULATION CONCERN

178-202 Great Portland St., London W1N 5T8, England **PHONE:** 44 171 6379582, 44 171 6311546 **FX:** 44 171 4362143 **E-MAIL:** info@populationconcern.org.uk **WEBSITES:** http://www.populationconcern.org.uk **FOUNDED:** 1974. **OFFICERS:** Mrs. Wendy Thomas, Chief Exec. **STAFF:** 22. **BUDGET:** £2,500,000. **LANGUAGES:** English.

DESCRIPTION: Seeks to heighten public awareness of global population and related issues. Raises funds to support and deliver, in partnership with local organizations, sexual and reproductive health programs worldwide including family planning with an education and advocacy program in the U.K. Sees free and informed reproductive choice and access to confidential sexual and reproductive health services as a basic human right, integral to the improvement of the quality of life. **PUBLICATIONS:** *Annual Review* (in English); *Update* (in English), annual; Books, periodic; Pamphlets, periodic.

POPULATION COUNCIL (PC)

1 Dag Hammarskjold Plz., New York, NY 10017 **PHONE:** (212) 339-0500 **FX:** (212) 755-6052 **TX:** 9102900660 POPCO **E-MAIL:** pubinfo@popcouncil.org **WEBSITES:** http://www.popcouncil.org **FOUNDED:** 1952. **OFFICERS:** Linda G. Martin, Pres. **STAFF:** 430. **BUDGET:** US$50,000,000. **LANGUAGES:** Arabic, English, French, Spanish.

DESCRIPTION: Aims to improve reproductive health and achieve a balance between people and resources. Analyzes demographic trends; conducts biomedical research to develop new contraceptives; works with public and private agencies to improve the quality of family planning and reproductive health services. Helps governments design and implement just and sustainable population and development policies; communicates the results of research in the population field to a broad audience; and helps build research capacities in developing countries. Areas of study include: contraceptive development; reproductive physiology; family planning and fertility; reproductive health (safe abortion, HIV/AIDS and other sexually transmitted diseases, postpartum care, safe motherhood, emergency contraception), gender, family, and development (girls' opportunities, sexuality and sexual violence, men as partners); population policy; transition to adulthood; population growth. **PUBLICATIONS:** *Population and Development Review* (in English), quarterly. Covers the interrelationships between population and soci-economic change and related public policy issues.; *Population Briefs*, quarterly. Results of research by Council scientists in biomedical sciences, public health, and social sciences.; *Population Council Annual Report*; *Studies in Family Planning*, quarterly. Covers family planning and related health and development issues, including such aspects as fertility regulation and contraceptive technology.

POPULATION INSTITUTE (PI)

107 2nd St. NE, Washington, DC 20002 **PHONE:** (202) 544-3300 **TF:** (800) 787-0038 **FX:** (202) 544-0068 **E-MAIL:** web@populationinstitute.org **WEBSITES:** http://www.populationinstitute.org **FOUNDED:** 1969. **OFFICERS:** Werner Fornos, Pres. **MEMBERS:** 81,000. **STAFF:** 12. **BUDGET:** US$1,350,000. **LANGUAGES:** English.

DESCRIPTION: U.S. members (25,000) are doctors, lawyers, businessmen, educators, religious leaders, and other concerned individuals. Overseas members (7000) are those influential in their country's leadership structure and those working directly in the population/family planning field. Seeks to marshall public opinion on global overpopulation problems. Keeps members abreast of initiatives and developments to curb world overpopulation. Members convey their views on global population issues to their elected representatives on matters of major legislation. Sponsors educational programs. Maintains an accredited intern program that sponsors 20-30 interns per year and speakers' bureau. **PUBLICATIONS:** *Popline* (in English, French, and Spanish), bimonthly. Focuses on policy, research, and demographic developments related to world population problems and issues.; *Towards the 21st Century*, quarterly; *World Population News Service*.

POSTAL UNION OF THE AMERICAS, SPAIN, AND PORTUGAL (PUASP) (Union Postal de las Americas, Espana, y Portugal - UPAEP)

Calle Cebollati 1470, Casilla de Correos 20042, Montevideo, Uruguay **PHONE:** 598 2 400114, 598 2 4000070 **FX:** 598 2 4005046 **E-MAIL:** secretaria@upaep.com.uy **WEBSITES:** http://www.upaep.com.uy **FOUNDED:** 1911. **OFFICERS:** Mario Felmer Klenner, Sec.Gen. **MEMBERS:** 27. **STAFF:** 20. **BUDGET:** US$2,300,000. **LANGUAGES:** Spanish.

DESCRIPTION: Members are postal administrations united to: extend and improve the postal services through increased cooperation among the postal administrations of member countries; conduct studies on topics such as the introduction of new services; upgrade the level of professional training of postal officials and improve the management of postal services and working systems; establish an effective means for representing the interests of postal unions. Conducts postal training classes. **PUBLICATIONS:** *Annual Report* (in Spanish). Also publishes council and congress documents, postal studies, and training manuals.

PRESIDING BISHOP'S FUND FOR WORLD RELIEF (PBF/WR)

PO Box 12043, Newark, NJ 07101 **TF:** (800) 334-7626 **FX:** (212) 983-6377 **E-MAIL:** pbfwr@dfms.org **WEBSITES:** http://www.pbfwr.org **FOUNDED:** 1940. **OFFICERS:** Mary Becchi, Dir. of Grants. **STAFF:** 8. **BUDGET:** US$6,000,000. **LANGUAGES:** English.

DESCRIPTION: Relief and development arm of the Episcopal church. Purpose is to provide relief of worldwide human suffering. Responds to natural disasters and other emergencies; communicates appeals for financial aid at times of emergencies; makes grants for relief, rehabilitation, and development. Provides follow-up response to crisis situations through rehabilitation grants. Bestows development grants that focus on such areas as: agricultural education; purchase of seeds and farming equipment; distribution of tools; well-digging projects; upgrading of livestock; funding for doctors, nurses, and nutritionists; elimination of the root causes of famine and disease. Works to alleviate drought conditions. Work is accomplished in cooperation with Anglican as well as ecumenical and private voluntary agencies. **PUBLICATIONS:**

Annual Report; Brochures; Newsletter, periodic; Pamphlets. Also publishes posters and press releases; produces films and slide presentations.

PRIVACY INTERNATIONAL

Morgan Towers, Bromley BR1 3QE, England **PHONE:** 44 181 4020737 **FX:** 44 181 3133726 **FOUNDED:** 1990. **OFFICERS:** Simon Davies, Dir.Gen.

DESCRIPTION: Members are human rights advocates, journalists, information technology experts, academics, and data protection experts. Works to protect personal privacy and monitor surveillance activities by governments and other organizations worldwide. Establishes guidelines for police files in emerging democracies.

PROGRAMA DE LAS NACIONES UNIDAS PARA EL MEDIO AMBIENTE (PNUMA)

Oficina Regional para America Latina y Caribe, Blvd. de los Virreyes No. 155, Col. Lomas Virreyes, 11000 Mexico City, DF, Mexico **PHONE:** 52 5 2024841, 52 5 2026943 **FX:** 52 5 2020950 **E-MAIL:** uneprolac@igc.apc.org **WEBSITES:** http://www.unep.mx; http://www.rolac.unep.mx **FOUNDED:** 1972. **OFFICERS:** Arsenio Rodriguez, Dir. **STAFF:** 12. **BUDGET:** US$1,000,000. **LANGUAGES:** English, French, Spanish.

DESCRIPTION: A division of the United Nations Environment Programme. Assists governments in environmental protection activities. Serves as secretariat for the Intergovernmental Regional Forum on the Environment in Latin America and the Caribbean.

PROJECT CONCERN INTERNATIONAL (PCI)

3550 Afton Rd., San Diego, CA 92123 **PHONE:** (619) 279-9690 **FX:** (619) 694-0294 **E-MAIL:** postmaster@projcon.cts.com **WEBSITES:** http://www.serve.com/pci **FOUNDED:** 1961. **OFFICERS:** Heather Girard, Event/Media Coord. **STAFF:** 35. **BUDGET:** US$13,000,000. **LANGUAGES:** English.

DESCRIPTION: Works with communities worldwide to ensure low-cost, basic health care for those most in need, particularly mothers and children. Provides education, training, and medical assistance to safeguard the world's impoverished children. Works with volunteers to prepare local communities to care for their own children with long-term, self-sustaining projects. Maintains programs in Bolivia, Guatemala, El Salvador, Mexico, Nicaragua, India, Indonesia, Romania, and the United States. **PUBLICATIONS:** *Concern News*, quarterly. Lists volunteer opportunities available to members.; *Project Concern International—Annual Report*. Contains financial statements.

PUBLIC HEALTH COMMITTEE OF THE COUNCIL OF EUROPE

Coun. of Europe, F-67075 Strasbourg Cedex, France **PHONE:** 33 388 412000 **E-MAIL:** peter.baum@coe.org **OFFICERS:** Dr. Peter Baum, Adm. Officer.

DESCRIPTION: A committee of the Council of Europe (CE). Representatives of national ministries of health. Coordinates activities of the committees of experts assembled by CE under the Partial Agreement in the Social and Pubic Health Field. The goal of the agreement is to assist in the development of uniform legislation in participating countries in the areas of public health and consumer protection.

QUE ME (VCHR) (Uy ban Bao ve Quyen lam Nguoi Viet Nam)

25, rue Jaffeux, F-92230 Gennevilliers, France **PHONE:** 33 1 47931081 **FX:** 33 1 47914138 **FOUNDED:** 1976. **OFFICERS:** Vo Van Ai, Pres. **STAFF:** 6. **LANGUAGES:** English, French, Vietnamese.

DESCRIPTION: Individuals. Strives to: eliminate oppression in Vietnam; increase awareness of human rights violations in Vietnam; assist Vietnamese refugees in adapting to Western culture; promote positive relations between the Vietnamese and Western societies. Communicates concerns to national and international authorities; submits petitions to international agencies such as the United Nations. Has sponsored a rescue mission for the boat people in the South China Sea. Organizes fundraising campaigns. **PUBLICATIONS:** *Dictionaries*; *Human Rights Violations in Vietnam* (in English and French), bimonthly; *Letter from Vietnam* (in English and French), periodic; *Que Me* (in Vietnamese), monthly; *Vietnam Today* (in English,

French, German, Italian, and Swedish), annual; Books; Bulletins.

RAINFOREST ALLIANCE (RA)

65 Bleecker St., 6th Fl., New York, NY 10012
PHONE: (212) 677-1900 **TF:** (888) MY-EARTH
FX: (212) 677-2187 **E-MAIL:** canopy@ra.org
WEBSITES: http://www.rainforest-alliance.org
FOUNDED: 1986. **OFFICERS:** Daniel Katz, Exec. Dir. **MEMBERS:** 14,000. **STAFF:** 40. **BUDGET:** US$2,900,000. **LANGUAGES:** English.

DESCRIPTION: Works for the conservation of tropical forests for the benefit of the global community. Develops and promotes economically viable and socially desirable alternatives to the destruction of tropical forests, an endangered, biologically diverse natural resource. Educates and researches the social and natural sciences; develops cooperative partnerships with businesses, governments and local peoples. **PUBLICATIONS:** *The Canopy*, bimonthly. Covers RA special events and breaking news.; *Eco-Exchange* (in English and Spanish), bimonthly. Covers breaking environmental news in the tropics.

RAINFOREST INFORMATION CENTRE (RIC)

PO Box 368, Lismore, NSW 2480, Australia
PHONE: 61 266 218505 **FX:** 61 266 222339
E-MAIL: rainfaus@mullum.com.au **WEBSITES:** http://www.forests.org/ric/ **FOUNDED:** 1980.
OFFICERS: John Revington. **MEMBERS:** 100.
STAFF: 30. **BUDGET:** $A 100,000. **LANGUAGES:** English.

DESCRIPTION: Members are environmental activists in 12 countries interested in the protection of dwindling rainforest resources. Generates public awareness of dangers to tropical forests through information campaigns. Provides aid and sponsors development programs. **PUBLICATIONS:** *Australian Rainforest Memorandum.*

RAV TOV INTERNATIONAL JEWISH RESCUE ORGANIZATION (RTIJRO)

500 Bedford Ave., Brooklyn, NY 11211 **PHONE:** (718) 963-1991 **FX:** (718) 935-9772 **FOUNDED:** 1973. **OFFICERS:** Divid Niederman, Exec.Dir. **STAFF:** 40. **LANGUAGES:** English.

DESCRIPTION: International network of offices that aids refugees who wish to immigrate to the U.S. or any other Western country. Provides financial assistance to immigrants until they become self-supporting; sponsors domestic assistance program. Assists with visa applications and housing once destination is reached; provides moral and religious support for those in transit and those who have reached destination; maintains employment service with direct job placement and on-the-job training; offers preschool, kindergarten, elementary school, and adult religious instruction; conducts English as a second language training program. **PUBLICATIONS:** *Achievement Bulletin*, quarterly.

RED BANCOS (RB)

Jackson 1136, 11200 Montevideo, Uruguay
PHONE: 598 2 4096192 **FX:** 598 2 4019222
E-MAIL: redbanco@chasque.apc.org **WEBSITES:** http://www.item.org.uy **FOUNDED:** 1984.
OFFICERS: Roberto Bissio. **MEMBERS:** 25. **STAFF:** 2. **LANGUAGES:** English, Spanish.

DESCRIPTION: Members are organizations working with refugees and other displaced people in Latin America and the Caribbean. Seeks to improve the quality of life of displaced people; encourages integration of displaced people in their new homelands. Makes available emergency relief to refugees; sponsor cultural and educational programs.

REFUGEE RELIEF INTERNATIONAL (RRI)

5735 Arapahoe Ave., Ste. A-5, Boulder, CO 80303-1340 **PHONE:** (303) 449-3750 **FX:** (303) 444-5617 **FOUNDED:** 1982. **OFFICERS:** Alexander M. S. McColl, Pres. **LANGUAGES:** English.

DESCRIPTION: Members are physicians, paramedics, and nurses with prior military experience. Provides medical and other help to refugees and other victims of war and oppression throughout the world. Major efforts have been in Central America, although significant contributions to multi-national, multi-agency relief efforts have been made in Afghanistan, Azerbaijan, and in support of the Karens in Burma. Transports and distributes medical supplies and equipment. Conducts classes on first aid, hygiene, public health and sanitation for indigenous paramedics, refugees and others.

REFUGEE WOMEN IN DEVELOPMENT (RefWID)

5225 Wisconsin Ave. NW, Ste. 502, Washington, DC 20015-2034 **PHONE:** (703) 931-6442 **TF:** (703) 931-5906 **E-MAIL:** refwind@erals.com **FOUNDED:** 1981. **OFFICERS:** Sima Wali, Pres. **STAFF:** 2. **BUDGET:** US$200,000. **LANGUAGES:** English.

DESCRIPTION: Members are refugee women who have resettled in the U.S. Seeks to enable refugee women worldwide to attain social and economic independence and security through acculturation, economic security, ethnic preservation, and emotional support. Focuses on low-income working-age refugee women with limited skills and those suffering escape trauma. Sponsors education and research programs; advocates improvements in programs and services for refugee women. Develops program models, training curricula, and community involvement approaches. Programs included leadership development and capacity-building, human rights/protection, domestic violence prevention, and international representation. Priorities include conducting training in domestic violence prevention and intervention; developing practical programs models for leadership development and coalition building; carrying out education and advocacy with refugee and mainstream organizations, policymakers, and the general public. **PUBLICATIONS:** *Leadership Development Model for Refugee Women: A Preliminary Report*; *Leadership Development Model for Refugee Women: A Replication Guide*; *The Production and Marketing of Ethnic Handcrafts in the U.S.*; *Understanding Family Violence Within U.S. Refugee Communities: A Training Manual.*

REFUGEES INTERNATIONAL (RI)

1705 N St. NW, Washington, DC 20036 **PHONE:** (202) 828-0110 **TF:** (800) REF-UGEE **FX:** (202) 828-0819 **E-MAIL:** ri@refintl.org **WEBSITES:** http://www.refintl.org **FOUNDED:** 1979. **OFFICERS:** Lionel A. Rosenblatt, Pres. **STAFF:** 12. **BUDGET:** US$1,141,500. **LANGUAGES:** English, French, Italian, Korean, Spanish, Thai.

DESCRIPTION: Provides advocacy, information, public education, and community support to refugees and displaced persons worldwide. Through voluntary action, seeks alternative means of handling refugee migration and permanent resettlement. Operates emergency need assessment program. Works to assist and support existing refugee relief and resettlement programs. Monitors and reports on events pertaining to refugees. Fosters voluntary support in the form of funds, sponsorship of refugee families, letters urging governmental support, medical services, and relief for refugees around the world. **PUBLICATIONS:** *Bosnia Relief Watch*, periodic; *Refugee & Relief Alert*, quarterly; *RI Bulletin*, periodic. Also publishes crisis generated press releases.

REGIONAL ASSOCIATION OF OIL AND NATURAL GAS COMPANIES IN LATIN AMERICA ANDTHE CARIBBEAN (Asociacion Regional de Empresas de Petroleo, Gas Natural en Latino America, el Caribe - ARPEL)

Calle Javier de Viana 2345, Casilla de Correo 1006, 11200 Montevideo, Uruguay **PHONE:** 598 2 4006993, 598 2 4007454 **FX:** 598 2 4009207 **E-MAIL:** arpel@adinet.com.uy **WEBSITES:** http://www.arpel.org **FOUNDED:** 1965. **OFFICERS:** Andres Tierno Abreu. **MEMBERS:** 20. **STAFF:** 8. **LANGUAGES:** English, Portuguese, Spanish.

DESCRIPTION: Members are oil companies in 18 Latin American countries. Conducts research projects benefiting sectors of the petroleum industry such as equipment and materials, conservation of hydrocarbon resources, development programs, commercial transactions, and international cooperation. Examines such industry issues to formulate recommendations leading to mutually beneficial collaboration among members that may in turn contribute to the economic and technological integration of Latin America. Conducts lectures.

REGIONAL CENTRE FOR SERVICES IN SURVEYING, MAPPING AND REMOTE SENSING

PO Box 18118, Nairobi, Kenya **PHONE:** 254 2 803320 **FX:** 254 2 802767 **TX:** 25285 **OFFICERS:** Simon Ndyetabula, Dir.Gen. **LANGUAGES:** English.

DESCRIPTION: Members are cartographers and surveyors. Seeks to advance the science and practice of cartography, surveying, and remote sensing. Conducts research and educational programs.

REGIONAL COORDINATED PROGRAM OF POPULAR EDUCATION (Programa Regional Coordinado de Educaion Popular)

Apartado 1272-2050, San Jose 1272, Costa Rica **PHONE:** 506 2531015, 506 2806540 **FX:** 506 2537023 **E-MAIL:** asistencia@alforja.or.cr **WEBSITES:** http://www.alforja.or.cr **FOUNDED:** 1981. **OFFICERS:** Oscar Jara Holliday, Reg.Coor.

DESCRIPTION: Works to improve educational standards and programs in Central America.

REGIONAL ISLAMIC DA'WAH COUNCIL OF SOUTHEAST ASIA AND THE PACIFIC (RISEAP) (Majlis Da'wah Islamiah Serantau Asia Tenggara dan Pasifik)

Perkim Bldg., 5th Fl., Jalan Ipoh, 51200 Kuala Lumpur, Malaysia **PHONE:** 60 3 4428166, 60 3 4412417 **FX:** 60 3 4410655 **TX:** RISEAP **E-MAIL:** riscap@asiapac.net **WEBSITES:** http://ngo.asiapac.net/riseap/index.html **FOUNDED:** 1980. **OFFICERS:** Dato Ahmad Nordin, Hon.Sec.Gen. **MEMBERS:** 57. **STAFF:** 18. **BUDGET:** US$600,000. **LANGUAGES:** Arabic, Chinese, English, Malay.

DESCRIPTION: Members are Muslim voluntary organizations active in welfare work and evangelism in 20 countries. Objectives are to: encourage unity of the Islamic movement and to unite Islamic organizations to work for the betterment of Muslims in the fields of Islamic education and social welfare; establish and support mosques and Islamic centers; disseminate Islamic teachings; promote, organize, and subsidize relief and welfare activities for the sick, poor, and needy. Supports regional libraries, information and documentation centers, research bureaus, and schools. Provides financial assistance to students. Maintains speakers' bureau. Conducts research on religious, educational, social, and other problems of Muslim minorities and on discrimination and repression of Muslims. Provides training for Muslim workers and youth. Conducts dialogues, seminars, conferences, symposium, and training course on Islam in various countries. **PUBLICATIONS:** *Al-Nahdah* (in English), quarterly. Includes information on Muslim issues.; *Report*, biennial; *Riseap Newsletter* (in English), monthly. Also publishes *Islam for Children*, *A Glimpse of Islam*, *Muhammad in the Bible*, *Arabic Lessons in Thai*, and *Marvellous Stories from the Life of Muhammad*, available in English, Chinese, Khmer, Vietnamese, and Pidgin English.

REGIONAL ORGANIZATION FOR THE PROTECTION OF THE MARINE ENVIRONMENT (ROPME)

PO Box 26388, Safat 13124, Kuwait **PHONE:** 965 5312140, 965 5312142 **FX:** 965 5324172 **E-MAIL:** ropmek@kuwait.net **WEBSITES:** http://www.kuwait.net/-ropmek **FOUNDED:** 1979. **OFFICERS:** Dr. Abdul Rahman Al-Awadi, Exec.Sec. **MEMBERS:** 8. **BUDGET:** 1,000,000 KD. **LANGUAGES:** Arabic, English, Persian.

DESCRIPTION: Members are government representatives of Bahrain, Iran, Iraq, Kuwait, Oman, Qatar, Saudi Arabia, and the United Arab Emirates working to implement the Kuwait Regional Convention for Cooperation on the Protection of the Marine Environment from Pollution ratified in 1979. Organizes scientific development programs and projects outlined in the Kuwait Action Plan adopted with the convention. Operates the Marine Emergency Mutual Aid Centre which facilitates cooperation and coordination of activities in the event of marine emergencies such as oil spills or dumping of waste products. Works in conjunction with United Nations agencies and organizations; conducts training programs in marine science and related fields. Conducts seminars and specialized training workshops. **PUBLICATIONS:** *ROPME Newsletter* (in Arabic and Persian), quarterly. Also publishes Annual children's booklets, technical reports, manuals and directories of oil combating services.

REGIONAL STUDIES ASSOCIATION (RSA)

Wharfdale Projects, 15 Micawber St., London N1 7TB, England **PHONE:** 44 171 4901128 **FX:** 44 171 2530095 **E-MAIL:** rsa@mailbox.ulcc.ac.uk **WEBSITES:** http://www.regional-studies-assoc.ac.uk **FOUNDED:** 1965. **OFFICERS:** Sally Hardy, Dir. **MEMBERS:** 800. **STAFF:** 4. **LANGUAGES:** English, French, German, Spanish.

DESCRIPTION: Members are individuals (550) and corporate groups (250) such as government departments, ministries, county councils, local authorities, research bodies, educational institutions, consultants' offices, geographers, economists, town planners, architects, and engineers interested in regional planning. Promotes education and research in the field of regional planning and development internationally; provides a forum for exchange of ideas and information; disseminates research results. Maintains European Regional Research Network which produces reports, directories, surveys, and lists of contact addresses in Europe; promotes regional studies and research at European rather than national levels; develops contact among teachers and researchers throughout Europe. **PUBLICATIONS:** *Annual Report* (in English); *European Urban and Regional Research Directory*, annual; *Regional Policy and Development Book Series*; *Regional Studies*, 9/year; *Regions: The Newsletter of the Regional Studies Association*, bimonthly.

REHABILITATION INTERNATIONAL (RI)

25 E. 21st St., New York, NY 10010 **PHONE:** (212) 420-1500 **FX:** (212) 505-0871 **FOUNDED:** 1922. **OFFICERS:** Susan Parker, Sec.Gen. **MEMBERS:** 150. **LANGUAGES:** English.

DESCRIPTION: Members are organizations in 90 countries conducting programs for the rehabilitation of people with physical and mental disabilities. Disseminates information on every phase of disability prevention and rehabilitation. Assists experts in planning work or study programs outside their own countries; aids in practical development of national rehabilitation programs. **PUBLICATIONS:** *International Journal of Rehabilitation Research*, quarterly; *International Rehabilitation Review*, triennial. Provides coverage of United Nations agencies, discusses the elimination of architectural and environmental barriers for the disabled.; *One-In-Ten* (in Arabic, English, French, and Spanish), quarterly; *Proceedings of World Congress*, quadrennial; *Rehabilitation* (in Spanish), semiannual; Directory, annual. Also publishes booklets and papers.

REHABILITATION RESEARCH FOUNDATION (RRF)

Box 1425, Columbia, MO 65205-1425 **PHONE:** (573) 874-8791 **FX:** (314) 874-0814 **FOUNDED:** 1958. **OFFICERS:** Dr. Earl-Clayton Grandstaff Sr., Exec. Officer. **MEMBERS:** 12,000. **STAFF:** 8. **BUDGET:** US$50,000,000.

DESCRIPTION: Plans, funds, and evaluates innovative programs in the areas of prison reform, offender re-socialization, and mental health. Acts as consultant to governments (national, state, local, and foreign) and private agencies and societies. Holds retreats. Sponsors charitable program; operates hall of fame and speakers' bureau. Compiles statistics. **PUBLICATIONS:** Newsletter, quarterly.

RESEARCH CENTRE FOR ISLAMIC HISTORY, ART AND CULTURE (IRCICA) (Centre de Recherche sur l'Histoire, l'Art et la Culture Islamiques)

PO Box 24, Besiktas, TR-80692 Istanbul, Turkey **PHONE:** 90 212 2591742, 90 212 2605988 **FX:** 90 212 2584365 **TX:** 26484 ISAM TR **E-MAIL:** ircica@superonline.com **FOUNDED:** 1979. **OFFICERS:** Prof.Dr. Ekmeleddin Ihsanoglu, Dir.Gen. **STAFF:** 60. **BUDGET:** US$1,910,000. **LANGUAGES:** Arabic, English, French.

DESCRIPTION: A subsidiary of the Organization of the Islamic Conference. Scholars and university students in 55 countries; member states of the OIC. Studies the common cultural legacy of Islamic countries to foster understanding among Muslims; urges Moslem writers and historians to join forces in opposing foreign misconceptions and prejudices against Islam. Strives to identify and rectify distortions of Islam found in textbooks; encourages the exchange of books on history, art, and culture among universities and research institutions throughout the Islamic world. Acts as Secretariat of the International Commission for the Preservation of Islamic Cultural Heritage. Holds exhibitions, seminars, and lectures; conducts research program. **PUBLICATIONS:** *International Directory of Islamic Cultural Institutions* (in English), periodic; *Library Accession List* (in English), quarterly; Newsletter (in Arabic, English, and French), quarterly; Report (in Arabic, English, and French), quarterly.

RESEARCH GROUP FOR AN ALTERNATIVE ECONOMIC STRATEGY (RGAES) (Groupe de Recherche pour une Strategie Economique Alternative - GRESA)

Rue Royale 11, B-1000 Brussels, Belgium **PHONE:** 32 2 2197076 **FX:** 32 2 2196486 **E-MAIL:** gresea@village.uunet.be **FOUNDED:** 1978. **OFFICERS:** Anne Peeters, Sec.Gen. **STAFF:** 9. **BUDGET:** US$145,000.

DESCRIPTION: Members are economists, researchers, and other individuals with an interest in global economic development. Works to improve the economic understanding of organizations involved in international development. Serves as a forum for the discussion of global economic development issues. Conducts research and gathers and disseminates information; holds courses for development organization personnel in topics including global economics and the creation of economic strategies. Maintains speakers' bureau. **PUBLICATIONS:** *Gresea Echos* (in French), quarterly. Comprises of free information of issues and publication of the association.

REUTER FOUNDATION

85 Fleet St., London EC4P 4AJ, England **PHONE:** 44 171 5427015 **FX:** 44 171 5428599 **E-MAIL:** foundation@reuters.com **WEBSITES:** http://www.foundation.reuters.com **FOUNDED:** 1982. **OFFICERS:** Stephen Somerville, Dir. **STAFF:** 5. **BUDGET:** £3,000,000. **LANGUAGES:** English.

DESCRIPTION: Promotes professional development of journalists worldwide. Conducts study and training programs for journalists from the developing world and from central and eastern Europe; sponsors research and educational programs and schools and universities; provides financial and technical support for charitable and cultural organizations. New service launched in 1997 AlertNet - online rapid news and communications service for international disaster relief organizations. **PUBLICATIONS:** *Reuterlink* (in English), 3/year; *Reuterlink Extra* (in English), 3/year. Also publishes various educational books.

RHEMA INTERNATIONAL (RI)

414 East St., Rochester, MI 48307 **PHONE:** (248) 652-2450 **FX:** (248) 608-9249 **FOUNDED:** 1977. **OFFICERS:** Patricia Gruits, Pres. **BUDGET:** US$200,000. **LANGUAGES:** English.

DESCRIPTION: Participants are individuals concerned about the quality of life in Haiti. Seeks to afford to Haitians the opportunity to live better lives by providing educational and medical services and establishing self-help programs. Operates medical and dental clinics; conducts paramedical training courses in areas including prenatal care and physical hygiene. Maintains Hope Academy International, which trains Haitians to organize medical, health education, and spiritual enlightenment programs for the benefit of rural villages. Accepts donated funds, equipment, and medicine; cooperates with other non-profit charities and organizations to distribute aid to needy Haitians. Conducts job training courses to teach young Haitians marketable skills. Offers seminars for pastors, ministers, and Christian educators. **PUBLICATIONS:** *Rhema Newsletter*, monthly; *Understanding God and His Covenants* (in English, French, Russian, and Spanish); *Understanding the Master's Voice.*

RODALE INSTITUTE (RI)

611 Siegfriedale Rd., Kutztown, PA 19530 **PHONE:** (610) 683-6009, (610) 683-1400 **FX:** (215) 683-8548 **E-MAIL:** info@rodaleins.org **WEBSITES:** http://www.rodaleinstitute.org **FOUNDED:** 1985. **OFFICERS:** Anthony Rodale. **STAFF:** 40. **LANGUAGES:** English, French, Russian, Spanish.

DESCRIPTION: Serves as an international network for sharing information and technologies in the fields of regenerative agriculture, gardening, and community development. Works in partnership with small-scale producers and entrepreneurs in Africa, Asia, and Latin America. Works to establish information networks, services, and products that advance regenerative farming principles and enhance indigenous resources. Provides training in regenerative farming techniques. Conducts research in resource-efficient farming methods.

ROTARY INTERNATIONAL (RI)

1 Rotary Center, 1560 Sherman Ave., Evanston, IL 60201-3698 **PHONE:** (847) 866-3000 **FX:** (847) 328-8554 **TX:** 724465 **WEBSITES:** http://www.rotary.org/ **FOUNDED:** 1905. **OFFICERS:** Aaron Hyatt, Gen.Sec. **MEMBERS:** 1,197,308. **STAFF:** 490. **LANGUAGES:** English.

DESCRIPTION: Members are business and professional executives in 150 countries and additional 34 geographical regions. Undertakes community development programs; promotes high ethical standards in business and professions; fosters ''international understanding, goodwill, and peace.'' Supports polio immunization campaigns. Maintains Rotary Foundation which offers scholarships to outstanding men and women, enabling them to study or teach in other countries. The foundation also bestows grants for international development projects and organizes exchange programs for business and professional people. Administers Interact and Rotaract programs for youths. **PUBLICATIONS:** *Convention Proceedings*, annual; *The Rotarian*, monthly; *The Rotary Basic Library.*

ROYAL SOCIETY OF CHEMISTRY (RSC)

c/o Stanley Langer, Burlington House, Piccadilly, London W1V 0BN, England **PHONE:** 44 171 4403325 **FX:** 44 171 7341227 **E-MAIL:** langers@rsc.org **FOUNDED:** 1841. **OFFICERS:** Dr. T.D. Inch, Sec.Gen. **MEMBERS:** 45,000. **STAFF:** 330. **BUDGET:** US$30,000,000. **LANGUAGES:** English.

DESCRIPTION: Members are chemists and interested persons in most countries of the world. Fosters the growth and application of chemistry by disseminating information. Establishes standards of qualification and conduct for the profession. Seeks to serve the public in an advisory and consultative capacity in matters relating to science and chemistry; monitors developments in chemistry. Encourages communication and cooperation between higher education and industry; offers career guidance and continuing education courses; funds small research projects. Maintains the Benevolent Fund, which helps members, former members, and their dependents who are in need of aid and the Corday-Morgan Memorial Fund, which enables members to travel to developing countries to lecture and exchange information. Organizes symposia, conferences, scientific meetings, and exhibitions. Provides Schools Publications Service, offering subscriptions for schools in Canada, England, Ireland, and the United States. Compiles statistics; sponsors competitions. **PUBLICATIONS:** *Analyst*, monthly; *Analytical Abstracts*, monthly; *Analytical Atomic Spectroscopy*; *Analytical Proceedings*, monthly; *Chemical Engineering Abstracts*, monthly; *Chemical Hazards in Industry*, monthly; *Chemical Society Reviews*, quarterly; *Chemistry in Britain*, monthly; *Chemistry in Europe*, quarterly; *Current Biotechnology Abstracts*, monthly; *Dalton Transactions*, monthly; *Education in Chemistry*, monthly; *Faraday Discussions*, semiannual; *Faraday Transactions*, monthly; *Journal of Analytical Atomic Spectrometry*, bimonthly; *Journal of Chemical Research*, monthly; *Journal of Materials Chemistry*, monthly; *Journal of the Chemical Society Chemical Communications*, semimonthly; *Laboratory Hazards Bulletin*, periodic; *Mass Spectrometry Bulletin*, monthly; *Mendeleev Communications*, periodic; *Methods in Organic Synthesis*, monthly; *Natural Product Reports*, bimonthly; *Perkin Transactions I*, monthly; *Perkin Transactions II*, monthly; *Professional Bulletin*, monthly; *Royal Society of Chemistry Annual Report.*

ROYAL SOCIETY FOR THE ENCOURAGEMENT OF ARTS, MANUFACTURES, AND COMMERCE (RSA)

8 John Adam St., London WC2N 6EZ, England **PHONE:** 44 171 9305115 **FX:** 44 171 8395805 **WEBSITES:** http://www.rsa.org.uk **FOUNDED:** 1754. **OFFICERS:** Penny Egan, Dir. **MEMBERS:** 21,500. **STAFF:** 65. **BUDGET:** £2,000,000. **LANGUAGES:** English.

DESCRIPTION: Members are individuals in 70 countries. Works to create a "civilised society based on a sustainable economy." Stimulates discussion as a means to developing ideas and encouraging action. Fields of interest include: business and industry; design and technology; education and the arts; the environment. **PUBLICATIONS:** *RSA Journal* (in English), monthly.

SAINT PATRICK'S MISSIONARY SOCIETY (SPSFM)

St. Patrick's, Kiltegan, Wicklow, Ireland **PHONE:** 353 508 73600 **FX:** 353 508 73644 **E-MAIL:** spsgen@iol.ie **FOUNDED:** 1932. **OFFICERS:** Very Rev. Kieran Birmingham, Superior Gen. **MEMBERS:** 355. **STAFF:** 100. **LANGUAGES:** English.

DESCRIPTION: Members are missionary priests and deacons working in Brazil, and 15 African and Caribbean countries. **PUBLICATIONS:** *Africa* (in English), 9/year.

SALESIAN YOUTH MOVEMENT (SYM) (Movimiento Juvenil Salesiano - MJS)

Via della Pisana 1111, I-00163 Rome, Italy **PHONE:** 39 6 6561121 **FX:** 39 6 65612556 **E-MAIL:** adomenech@sdb.org **WEBSITES:** http://www.sdb.org **FOUNDED:** 1854. **OFFICERS:** Antonio Domenech, Counselor. **MEMBERS:** 450,000. **STAFF:** 4,800. **LANGUAGES:** English, French, Italian, Spanish.

DESCRIPTION: A movement with free adherence,

composed of youth from Salesian environments, present in 118 countries. Aims to provide a Unitarian Christian education to Catholic youth; promotes education as an advancement of the human-Christian experience. Recognizes the necessity of mediation and creativity within the church as an integral part of free education for young people. Involves youth in sports and tourism activities; guides youth through the stages toward full communion in the church with study in church doctrine and missionary and other volunteer or apostolic works. **PUBLICATIONS:** *Dossier PG: Esperienze a Confronto* (in English, French, Italian, and Spanish), periodic; *Mision Joven* (in Spanish), monthly; *Note di Pastorale Giovanile* (in Italian), monthly; *Radar ADS* (in Italian), monthly; *Youth Pastoral Documents* (in English, Italian, and Spanish), periodic.

THE SALVATION ARMY (SA)

615 Slaters Ln., PO Box 269, Alexandria, VA 22313 **PHONE:** (703) 684-5500 **TF:** (800) SALARMY **FX:** (703) 684-3478 **WEBSITES:** http://www.salvationarmyusa.org **FOUNDED:** 1865. **OFFICERS:** Commissioner Robert A. Watson, Natl.Cmdr. **MEMBERS:** 450,312. **STAFF:** 30,930. **BUDGET:** US$1,354,000. **LANGUAGES:** English.

DESCRIPTION: Commissioned officers are ordained ministers devoting full time to religious and social welfare activities; members of local church or corps community centers are known as soldiers. An international Christian religious and charitable movement, organized on a paramilitary pattern, dedicated to meeting the physical, spiritual, and emotional needs of mankind. Work is carried out through local centers of operation which include adult rehabilitation centers, clinics, outpatient programs for unwed mothers, recreation centers, camping programs for children and adults, senior and children's day care, senior housing and activity centers, and emergency feeding and shelter stations; and through service extension units located in communities not supporting a full Salvation Army program, which extend aid in emergencies. Maintains speakers' bureau and 38 divisions; compiles statistics. Offers placement and referral services at local level. Provides officers' training schools. **PUBLICATIONS:** *Marching to Glory*; *ProgramAids*, quarterly. Contains weekly in-house programs for women, including ideas for decorations, games, refreshments, and devotionals.; *War Cry*, biweekly. Features articles on Christian topics, includes association news and Bible studies.; *What Is The Salvation Army?*; *Young Salvationist*, 10/year. For high-school age volunteers of The Salvation Army; covers issues confronting teenagers from a Christian perspective. Also includes fiction,; Annual Report.

SALVATION ARMY HOME LEAGUE (SAHL) (Ligue de Foyer - LF)

c/o Gen. Paul Rader, 101 Queen Victoria St., London EC4 4EP, England **PHONE:** 44 171 2365222 **FX:** 44 171 2366272 **FOUNDED:** 1907. **OFFICERS:** Commissioner Kay F. Rader, Pres. **MEMBERS:** 418,411. **LANGUAGES:** English.

DESCRIPTION: Members are women over the age of 16 from 94 countries. Provides for education, fellowship, service, and worship.

SARAWAK CAMPAIGN COMMITTEE (SCC)

2F, 3-17-24 Mejiro, Toshima-ku, Tokyo 171-0031, Japan **PHONE:** 81 3 39543510 **FX:** 81 3 39511084 **E-MAIL:** scc@kiwi.ne.jp **WEBSITES:** http://www.vcom.or.jp/-scc/ **FOUNDED:** 1990. **OFFICERS:** Mihoko Uramoto, Gen.Sec. **MEMBERS:** 550. **STAFF:** 10. **BUDGET:** US$12,370,560. **LANGUAGES:** English, Japanese.

DESCRIPTION: Members are individuals (500) and organizations (50) united to protect the tropical rain forests and the rights of indigenous peoples on the Island of Sarawak, Malaysia. Organizes communities in Japan to reduce the use of tropical hardwoods; conducts public education campaigns to raise awareness of environmental protection and human rights issues. **PUBLICATIONS:** *Muri no kee*.

SCANDINAVIAN RESEARCH COUNCIL FOR CRIMINOLOGY (SRCC) (Pohjoismaiden Kriminologinen Yhteistyoneuvosto)

Det retsvidenskabelige institut D, Sankt Peders Streede 19, DK-1453 Copenhagen, Denmark **PHONE:** 45 35 323339 **FX:** 45 35 323334 **FOUNDED:** 1962. **OFFICERS:** Nina Krarup, Sec.Gen. **MEMBERS:** 15. **STAFF:** 3. **LANGUAGES:** Danish, Norwegian, Swedish.

DESCRIPTION: Members are government representatives from the 5 Scandinavian countries. Combines national resources to form one research unit which assists and advises Scandinavian authorities and the Nordic Council. Initiates and conducts criminological research, including research on hidden criminality, police, and victimology. Collects and disseminates information to Scandinavian criminologists. Organizes international seminars to which non-Scandinavian criminologists are invited. **PUBLICATIONS:** *Nordisk Kriminologi* (in Danish, Norwegian, and Swedish), 3/year; *Scandinavian Studies in Criminology* (in English); Proceedings (in English), periodic.

SCHOOL SISTERS OF NOTRE DAME (SSND) (Armen Schulschwestern von Unserer Lieben Frau)

c/o Sister. Rosemary Howarth, SSND, Via della Stazione Aurelia 95, I-00165 Rome, Italy **PHONE:** 39 6 66418065, 39 6 66418075 **FX:** 39 6 66411212 **WEBSITES:** http://www.ssnd.org/ssnd/ **FOUNDED:** 1833. **OFFICERS:** Sister Rosemary Howarth SSND, Gen. Superior. **MEMBERS:** 5,018. **LANGUAGES:** Chamorro, Czech, English, German, Hungarian, Italian, Japanese, Polish, Portuguese, Romanian, Slovak, Spanish.

DESCRIPTION: Members are Catholic women in 35 countries. Promotes religious life, justice, and peace. Provides: staffing for women's colleges and residences; information to mothers in developing nations; courses on prayer and theology; basic elementary and secondary education; an alternative high school for girls; services to the elderly and education for the handicapped; ministry to the sick. **PUBLICATIONS:** *Generalate News* (in English and German), periodic. Internal communication for school Sisters of Notre Dame.

SCIENTIFIC COMMITTEE ON PROBLEMS OF THE ENVIRONMENT (SCOPE) (Comite Scientifique sur les Problemes de l'Environnement)

51, blvd. de Montmorency, F-75016 Paris, France **PHONE:** 33 1 45250498 **FX:** 33 1 42881466 **E-MAIL:** scope@paris7.jussieu.fr **FOUNDED:** 1969. **OFFICERS:** Philippe Bourdeau, Pres. **MEMBERS:** 221. **LANGUAGES:** English, French.

DESCRIPTION: Members are international scientific unions and 39 national committees working to ascertain the influence of humans on their environment and the effects of environmental changes on people and their health and welfare. Serves as an interdisciplinary consultant to government and to intergovernmental and nongovernmental bodies with respect to environmental problems. Works on projects in the areas of: practices and policies; ecosystem processes and biodiversity; health and environment; with an overall focus on sustainability. **PUBLICATIONS:** *SCOPE Newsletter* (in English), 3/year; *SCOPE Series*, 3/year. Contains reports on environmental issues.

SECRETARIAT OF THE CONVENTION ON INTERNATIONAL TRADE IN ENDANGERED SPECIES OF WILD FAUNA AND FLORA (CITES) (Convention sur le Commerce International des Espices de le Faune et de Flore Sauvages Menacees d'Extinction - CITES)

c/o Secretariat, Geneva Executive Centre, 15 Chemin des Anemones, CH-1219 Chatelaine, Switzerland **PHONE:** 41 22 9799149 **FX:** 41 22 7973420 **TX:** 454584 CTES CH **FOUNDED:** 1973. **OFFICERS:** Izgred Topkov, Sec.Gen. **MEMBERS:** 112. **STAFF:** 21. **BUDGET:** US$6,000,000. **LANGUAGES:** English, French, Spanish.

DESCRIPTION: Monitors and coordinates information on international trade of wild flora and fauna and quota programs. Collects scientific information on various species; initiates and formulates field projects for conservation studies and for the development of wildlife management methods. Conducts regional seminars on the enforcement of import regulations. **PUBLICATIONS:** *Notifications to the Parties* (in English, French, and Spanish), periodic; *Proceedings of Conference of the Parties* (in English, French, and Spanish), biennial; *Species Identification Manual* (in English and French). Also publishes directory.

SECRETARIAT OF THE U.N. FRAMEWORK CONVENTION ON CLIMATE CHANGE (UNFCCC)

Haus Carstanjen, Martin Luther King Strasse 8, PO Box 260124, D-53153 Bonn, Germany

PHONE: 49 228 8151000 **FX:** 49 228 8151999 **E-MAIL:** scretariat@unfccc.de **WEBSITES:** http://www.unfccc.de **OFFICERS:** Micheal Zammit Cutajar, Exec.Sec. **MEMBERS:** 165. **BUDGET:** 75,000,000 SFr.

DESCRIPTION: Member countries (165) united to stabilize greenhouse gas concentrations in the atmosphere at a level that would prevent dangerous anthropogenic interference with the climate system. Works to achieve this goal within a timeframe sufficient to allow ecosystems to adapt naturally to climate change, to ensure that food production is not threatened, and to enable economic development to proceed in a sustainable manner.

SEVA FOUNDATION (SF)

1786 5th St., Berkeley, CA 94710-1716 **PHONE:** (510) 845-7382 **TF:** (800) 223-7382 **FX:** (415) 845-7410 **E-MAIL:** admin@seva.org **WEBSITES:** http://www.seva.org **FOUNDED:** 1978. **OFFICERS:** James O'Den, Exec.Dir. **MEMBERS:** 27,000. **STAFF:** 10. **BUDGET:** US$1,300,000. **LANGUAGES:** English.

DESCRIPTION: Seva (a Sanskrit word for service) works to prevent blindness in India, Nepal, and Tibet; community development in Mexico and Guatemala; and provides diabetes treatment for Native Americans on reservations. In addition, Seva provides small grants for local projects that serve homelessness, youth at risk, and persons with HIV/AIDS. **PUBLICATIONS:** *Epidemiology of Blindness in Nepal: Report of the 1981 Nepal Blindness Survey*; *Gift of Service Catalog*, annual. Describes gifts that individuals can give to support needy communities, including traditional gifts and gifts of training.; *Seva Foundation—Progress Report*, annual. Includes achievements, goals, and financial statement.; *Special Project Reports*, periodic; *Spirit of Service*, annual. Reports on national and international humanitarian activities. Includes project updates and annual gift catalog.

SISTEMA DE LA INTEGRACION CENTROAMERICANA (SICA)

Boulevard de la Orden de Malta No. 470, Santa Elena, El Salvador **PHONE:** 503 2896131 **FX:** 503 2896124 **E-MAIL:** sgsica@sicanet.org.sv **WEBSITES:** http://www.sicanet.org.sv **FOUNDED:** 1951. **OFFICERS:** Ernesto Leal Sanchez, Gen.Sec. **MEMBERS:** 6.

DESCRIPTION: Composed of the Summit of Central American Presidents, the Council of Ministers, the Executive Committee, the General Secretariat, the Consultative Committee, four Technical Secretariats (economic-SIECA, social-SISCA, ecological-CCAD, cultural-CECC) and more than 15 specialized agencies (bank for development, health and nutrition, industrial research, public administration, telecommunications, air navigation, etc.) Other bodies of SICA are the central American Parliament (PARLACEN) and the central American Court of Justice (CCJ). Main objective is to follow-up on the decisions of the Summit of Presidents and to coordinate Secretariats and specialized agencies. **PUBLICATIONS:** *Centroamerica Mensual* (in Spanish), monthly.

SISTERS OF CHARITY OF SAINT VINCENT DE PAUL (SC)

150 Bedford Hwy., Halifax, NS, Canada B3M 3J5 **PHONE:** (902) 457-3500 **FX:** (902) 457-3506 **FOUNDED:** 1849. **OFFICERS:** Sr. Joan Verner, Cong.Sec. **MEMBERS:** 840. **STAFF:** 190. **LANGUAGES:** English.

DESCRIPTION: Members are Catholic sisters in 6 countries involved in education, health care, social services, and pastoral ministry. **PUBLICATIONS:** *Changing Times* (in English), 3/year.

SISTERS OF THE GOOD SHEPHERD (SGS) (Soeurs du Bon Pasteur)

Via Raffaello Sardiello 20, I-00165 Rome, Italy **PHONE:** 39 6 66418545, 39 6 66418546 **FX:** 39 6 66418864 **FOUNDED:** 1835. **OFFICERS:** Sr. Digna Maria Rivas, Gen.Sec. **MEMBERS:** 5,465. **LANGUAGES:** English, French, German, Italian, Portuguese, Spanish.

DESCRIPTION: Members are Catholic women in 64 countries. Dedicated to promoting and addressing the social needs of women and children. Conducts charitable programs. **PUBLICATIONS:** *Newsletter* (in Dutch, English, French, German, Italian, Japanese, Portuguese, Spanish, and Tamil), quarterly.

SISTERS OF SAINT JOSEPH OF THE SACRED HEART (RSJ) (Hermanas de San Jose del Sagrado Corazon)

11 Mount St., North Sydney, NSW 2059, Australia **PHONE:** 61 2 9297344 **FX:** 61 2 9297994 **E-MAIL:** sosjclt@internet-australia.com **FOUNDED:** 1866. **OFFICERS:** Sr. Giovanni Farquer RSJ, Leader. **MEMBERS:** 1,195. **LANGUAGES:** English, Spanish.

DESCRIPTION: Members are sisters and volunteers in Australia, Republic of Ireland, New Zealand, and Peru. Has established 270 centers, mostly in rural areas, that offer educational, medical, and social services. Group is also actively engaged in pastoral work. Offers children's services and charitable programs. Sponsors ministry conferences for members of the congregation. **PUBLICATIONS:** *Soundings* (in English), semiannual.

SISTERS OF SAINT LOUIS (SSL) (Soeurs de Saint Louis)

3 Beech Ct., Ballinclea Rd., Killiney 6, Dublin, Ireland **PHONE:** 353 1 2350304 **FX:** 353 1 2350345 **E-MAIL:** sslgen@tinet.ie **FOUNDED:** 1842. **OFFICERS:** Sr. Margaret Healy SSL, Gen.Sec. **MEMBERS:** 534. **LANGUAGES:** English, French, Portuguese.

DESCRIPTION: Members are religious women ministering in Brazil, England, France, Ghana, Liberia, Nigeria, Republic of Ireland, and the United States. Promotes the Christian values of justice, freedom, peace, and dignity. Serves the poor through education, medical assistance and primary health care; provides counseling. Performs ecumenical work, especially for the good of the poor and the handicapped. Works with other groups in different kinds of co-operative endeavors. **PUBLICATIONS:** *Newsletter* (in English, French, and Portuguese), biennial; Newsletter (in English, French, and Portuguese), semiannual.

SMALL RIVERS INTERNATIONAL ASSOCIATION (SRIA)

ul. Slavinskogo, d. 1/2, Minsk, Belarus **PHONE:** 375 172 635333, 375 172 635906 **FX:** 375 172 642734 **E-MAIL:** gerard@wri.belpak.minsk.by **FOUNDED:** 1995. **OFFICERS:** Tamara M. Pushkareva. **LANGUAGES:** English, Russian.

DESCRIPTION: Members are individuals and organizations interested in the health of small rivers. Promotes conservation and environmental protection; seeks to raise public awareness of environmental issues affecting small rivers. Conducts educational programs; facilitates networking and cooperation among nongovernmental organizations working in conservation and sustainable natural resource management.

SOCIALIST INTERNATIONAL WOMEN (SIW) (Internationale Socialiste des Femmes)

Maritime House, Old Town, Clapham, London SW4 OJW, England **PHONE:** 44 171 6274449 **FX:** 44 171 7204448 **TX:** 261735 SISEC G **E-MAIL:** socintwomen@gn.apc.org **FOUNDED:** 1907. **OFFICERS:** Marlene Haas, Sec.Gen. **MEMBERS:** 130. **LANGUAGES:** English, French, Spanish.

DESCRIPTION: Promotes action programs to combat sex discrimination. Works for human rights, development, and peace. **PUBLICATIONS:** *Women and Politics* (in English), quarterly.

SOCIETE DE BIOGEOGRAPHIE

57, rue Cuvier, F-75005 Paris, France **FOUNDED:** 1924. **OFFICERS:** Marc Salomon, Sec.Gen. **MEMBERS:** 300. **STAFF:** 1. **BUDGET:** 50,000,000 Fr. **LANGUAGES:** English, French, Spanish.

DESCRIPTION: Members are botanists, zoologists, ethnologists, geologists, and geographers in 15 countries. Promotes conservation efforts in France. Focuses attention on the plight of endangered species. Conducts demographic studies. **PUBLICATIONS:** *Biogeographica* (in English, French, and Spanish), quarterly; *Memoires de la Societe de Biogeographie*.

SOCIETE FRANCAIS DE CHIMIE (SFC)

250, rue St. Jacques, F-75005 Paris, France **PHONE:** 33 1 40467160 **FX:** 33 1 40467161 **E-MAIL:** sfc@sfc.fr **WEBSITES:** http://www.sfc.fr **FOUNDED:** 1983. **OFFICERS:** Jean-Claude Brunie, Sec.Gen. **MEMBERS:** 5,000. **STAFF:** 7. **BUDGET:** 5,000,000 Fr. **LANGUAGES:** English, French.

DESCRIPTION: An international scientific society of chemists working in academic institutions and industry; libraries; industrial firms. Purpose is to bring together industrial chemists and their academic colleagues in an effort to further chemistry as a theoretical and applied scientific discipline. Objectives are: to enhance the development of all areas of chemistry; to enable members and other interested persons to exchange information and ideas; to ensure the publication of members' research findings; to disseminate members' opinions on topics concerning the profession, with particular emphasis on industrial, economic, educational, and research questions; to establish and develop research contacts with chemical societies within the Federation of European Chemical Societies; to represent chemists and their concerns before professional and public institutions. Organizes educational programs and research meetings. Offers children's services. **PUBLICATIONS:** *Analusis* (in French), 10/year; *Annuaire de la Societe Francaise de Chimie* (in French); *Bulletin de la Societe Chimique de France* (in English and French), bimonthly; *European Journal of Inorganic Chemistry*; *Journal de Chimie Physique* (in English and French), monthly; *Journal of Chemical Research* (in English, French, and German), periodic; *L'Actualite Chimique* (in French), monthly.

SOCIETE UNIVERSITAIRE EUROPEENNE DE RECHERCHES FINANCIERES (SUERF)

Herengracht 205, NL-1016 BE Amsterdam, Netherlands **PHONE:** 31 20 5208565 **FX:** 31 20 5208606 **E-MAIL:** suerf@wxs.nl **FOUNDED:** 1963. **OFFICERS:** Ingrid Haye, Exec. Officer. **MEMBERS:** 500. **STAFF:** 1. **LANGUAGES:** English, French.

DESCRIPTION: Members are academics, bank economists, and interested individuals (300); financial institutions (100) representing 25 countries. Develops contacts among members in order to discuss monetary and financial questions. Sponsors research in monetary, economic, and financial areas. **PUBLICATIONS:** *SUERF* (in English and French), 7/year; *SUERF Newsletter*; *SUERF Papers* (in English and French); *SUERF Reprints* (in English and French); *SUERF Series* (in English and French); *Suerf Studies* (in English and French); *SUERF Translations* (in English and French).

SOCIETY OF AFRICAN MISSIONS

23 Bliss Ave., Tenafly, NJ 07670 **PHONE:** (201) 567-0450 **FX:** (201) 567-7156 **E-MAIL:** smausa-c@smafathers.org **WEBSITES:** http://www.smafathers.org **FOUNDED:** 1856. **OFFICERS:** Rev. Ulick Bourke SMA. **MEMBERS:** 1,238. **LANGUAGES:** English.

DESCRIPTION: Members are Roman Catholic missionary priests, brothers, and lay associates from 10 countries in North America and Europe working in Africa. Seeks to establish the Catholic church throughout the missionary world and to aid in the creation of a native clergy. Helps to fulfill people's needs in clinics, schools, leprosariums, and hospitals. Promotes socio-economic development, peace and justice projects, and agricultural programs. Sponsors educational and charitable programs. Maintains museum of African arts and crafts, mainly from West Africa. **PUBLICATIONS:** *Society of African Missions—Frontline Report*, bimonthly. Highlights the activities of Roman Catholic missionaries.

SOCIETY OF AUTOMOTIVE ENGINEERS (SAE)

400 Commonwealth Dr., Warrendale, PA 15096-0001 **PHONE:** (412) 776-4841 **FX:** (412) 776-5944 **E-MAIL:** sae@sae.org **WEBSITES:** http://www.sae.org **FOUNDED:** 1905. **OFFICERS:** Max E. Rumbaugh Jr. **MEMBERS:** 74,000. **STAFF:** 305. **BUDGET:** US$45,000,000.

DESCRIPTION: Members are engineers, business executives, educators, and students from more than 80 countries who come together to share information and exchange ideas for advancing the engineering of mobility systems. SAE is the major source of technical information and expertise used in designing, building, maintaining and operating self-propelled vehicles, whether land-, sea-, air-, or space-based. Collection, organization, storage and dissemination of information on cars, trucks, aircraft, space vehicles, off-highway vehicles, marine equipment, and engines of all types. Produces technical publications, conducts numerous meetings, seminars and educational activities, and fosters information exchange among the worldwide automotive and aerospace communities. **PUBLICATIONS:** *Aerospace Engineering*, 11/year; *Automotive Engineering*, monthly; *Off-Highway Engineering*, quarterly; *SAE Roster*, annual; *SAE Show Daily*; *SAE Update*, monthly; *Technical Literature*; *Truck Engineering*, semiannual.

SOCIETY OF BUSINESS ECONOMISTS (SBE)

11 Bay Tree Walk, Watford, Herts. WD1 3RX, England **PHONE:** 44 1923 237287 **E-MAIL:** admin@sbe.co.uk **WEBSITES:** http://www.sbe.co.uk **FOUNDED:** 1953. **OFFICERS:** A.W. Sentance, Chm. **MEMBERS:** 650. **LANGUAGES:** English.

DESCRIPTION: Members are economists in 14 countries working in commerce, finance, and industry. **PUBLICATIONS:** *Business Economist* (in English), 3/year.

SOCIETY OF COMPARATIVE LEGISLATION (Societe de Legislation Comparee)

28, rue St.-Guillaume, F-75007 Paris, France **PHONE:** 33 1 44398623 **FX:** 33 1 44398628 **E-MAIL:** slc@sky.fr **FOUNDED:** 1869. **OFFICERS:** Madame Marie-Anne Gallot-Le-Lorier, Gen.Sec. **MEMBERS:** 600. **STAFF:** 4. **LANGUAGES:** French.

DESCRIPTION: Members are university professors, scholars, judges, attorneys, and barristers in 46 countries. Promotes the study and comparison of the laws and legal systems of various countries as well as the means to improve the different branches of law. Collects documents and translations of foreign laws; studies current problems in the fields of private, public, and criminal law. **PUBLICATIONS:** *Actualite de la propriete dans les pays d'Europe*; *Droit du Travail - Hier et Demain* (in French); *Etude de Droit Japonais* (in French); *Jalons - Dits et Ecrits d'Andre Tunc* (in French); *La Responsabilite des Prestataires de Services* (in English and French); *Le droit compare: aujourd'hui et demain* (in French); *Tables Analytiques de la RIDC 1949-1973* (in French).

SOCIETY FOR COMPUTERS AND LAW (SCL)

10 Hurle Crescent, Clifton, Bristol, Avon BS8 2TA, England **PHONE:** 44 1179 237393 **FX:** 44 1179 239305 **E-MAIL:** ruth.baker@su.org **WEBSITES:** http://www.scl.org **FOUNDED:** 1973. **OFFICERS:** Ruth Baker, Gen.Mgr. **MEMBERS:** 2,700. **STAFF:** 2. **LANGUAGES:** English.

DESCRIPTION: Members are academics, computer scientists, lawyers, librarians, and individuals in local government (2700) in 34 countries. Promotes the use of high technology in law for the benefit of lawyers as well as the public. Objectives are to: study the use of computers in legal research and practice; advance education in the implications of computers applied to the law; promote the development of legal computer systems and legal information retrieval systems, monitoring their performance on behalf of the public and the legal profession. Collaborates with other organizations holding similar objectives. **PUBLICATIONS:** *Computers and Law* (in English), bimonthly; *Conference Proceedings* (in English), periodic.

SOCIETY FOR GEOLOGY APPLIED TO MINERAL DEPOSITS (SGA) (Societe de Geologie Appliquee aux Gites Mineraux - SGA)

c/o Czech Geological Survey, Klarov 131, CZ-118 00 Prague 1, Czech Republic **PHONE:** 420 2 5817390 **FX:** 420 2 5818748 **E-MAIL:** pasava@cgu.cz **WEBSITES:** http://www.immr.tu-clausthal.de/sga.html **FOUNDED:** 1965. **OFFICERS:** Jan Pasava, Sec. **MEMBERS:** 750. **LANGUAGES:** English, French, German.

DESCRIPTION: Members are geologists in 68 countries concerned with economic geology, mineral resources, industrial minerals and environmental aspects related to mineral deposits who work in industry, research, and education. Supports and promotes the application of scientific knowledge to the study and development of mineral resources and the profession of geology in science and industry. Seeks to improve professional and ethical standards in the field. Organizes field excursions. **PUBLICATIONS:** *Mineralium Deposita and SGA News* (in English), bimonthly; Membership Directory, biennial; Membership Directory, biennial.

SOCIETY FOR HUMAN ECOLOGY (SHE)

c/o Melville Cote, Coll. of the Atlantic, 105 Eden St., Bar Harbor, ME 04609 **PHONE:** (207) 288-5015 **FX:** (207) 288-4126 **WEBSITES:** http://www.coa.edu **FOUNDED:** 1981. **OFFICERS:** Melville P. Cote, Exec.Dir. **MEMBERS:** 150. **LANGUAGES:** English.

DESCRIPTION: Members are educators, health practitioners, scientists, and other professionals in 30 countries studying human ecology and its applications. Focuses attention on the consequences of human action on manufactured, natural, and social environments. Promotes interdisciplinary collaboration; facilitates the exchange of information; identifies problems and recommends solutions from an ecological perspective. Conducts workshops and symposia. Participates in human ecology consortium. Organization is distinct from the International Organization for Human Ecology, formerly known as the Society for Human Ecology. **PUBLICATIONS:** *Human Ecology: A Gathering of Perspectives*. Proceedings from SHE's First International Conference, held in 1985; *Human Ecology and Decision Making: An International and Interdisciplinary Collaboration*; *Human Ecology Bulletin*, semiannual; *Human Ecology—Coming of Age: An International Overview*. Proceedings of the symposium organized at the International Congress of Ecology in 1990; *Human Ecology: Crossing Boundaries*. Selected papers from the 6th conference of SHE; *Human Ecology: Research and Applications*. Proceedings from SHE's Second International Conference, held in 1988; *Human Ecology: Steps to the Future*. Proceedings from SHE's Third International Conference; *Human Ecology: Strategies for the Future*. Selected papers from the 4th conference of the SHE, held in 1990; *International Directory of Human Ecologists*, periodic. Lists over 700 human ecologists worldwide, with descriptions of their work, research, and activities; includes addresses and phone numbers; Brochure.

SOCIETY FOR INTERNATIONAL DEVELOPMENT (SID) (Societe Internationale pour le Developpement - SID)

207 via Panisperna, I-00184 Rome, Italy **PHONE:** 39 6 4872172 **FX:** 39 6 4872170 **E-MAIL:** info@sidint.org **FOUNDED:** 1957. **OFFICERS:** Boutros Boutros-Ghali, Pres. **MEMBERS:** 6,000. **STAFF:** 17. **BUDGET:** US$1,600,000. **LANGUAGES:** English, French, Spanish.

DESCRIPTION: Global forum of individuals and institutions concerned with participative, dynamic development. Promotes aspirations, new ideas, and policy action designed to transform development priorities towards a world which is people-centered, sustainable, just, democratic, and incisive. **PUBLICATIONS:** *Bridges*, bimonthly; *Development* (in English), quarterly. Reports of all projects and events.

SOCIETY OF INTERNATIONAL TREASURERS (SIT)

2 Tereslake Green, Westbury - On-Trym, Bristol BS10 GLT, England **PHONE:** 44 117 9508019 **FX:** 44 117 9508019 **E-MAIL:** mail@socintrs.com **FOUNDED:** 1977. **OFFICERS:** Geoffrey Jones, Dir.Gen. **MEMBERS:** 350. **LANGUAGES:** English.

DESCRIPTION: Members are individuals in 15 countries involved in corporate treasury management within major multinational corporations, excluding financial service corps.

SOCIETY OF SAINT VINCENT DE PAUL - INTERNATIONAL (SSVP) (Societe de Saint Vincent de Paul)

5, rue du Pre-aux-Clercs, F-75007 Paris, France **PHONE:** 33 1 44553655 **FX:** 33 1 42617256 **FOUNDED:** 1833. **OFFICERS:** Amin de Tarrazi, Intl. V. Pres. **MEMBERS:** 875,000. **LANGUAGES:** English, French, Spanish.

DESCRIPTION: Members are Catholic lay persons engaged in a spirit of justice and charity and united by a personal commitment to serve those who suffer. Seeks, through personal contact, to relieve suffering and promote the dignity and integrity of the individual. Conducts research. **PUBLICATIONS:** *Vincenpaul* (in English, French, and Spanish), quarterly.

SOCIETY FOR THREATENED PEOPLES (Gesellschaft fuer Bedrohte Voelker - GfbV)

Postfach 20 24, D-37010 Goettingen, Germany **PHONE:** 49 551 499060 **FX:** 49 551 58028 **E-MAIL:** info@gfbv.de **WEBSITES:** http://www.gfbv.de **FOUNDED:** 1970. **OFFICERS:** Tilman Zuelch, Chm. **MEMBERS:** 7,300. **STAFF:** 18. **BUDGET:** US$1,562,000. **LANGUAGES:** English, French, German, Spanish.

DESCRIPTION: Members are individuals in 20 countries working to protect ethnic and religious groups worldwide that are victims of human rights violations. Publicizes and condemns violations of human rights of threatened peoples. **PUBLICATIONS:** *Pogrom* (in German), bimonthly. Academic, scholarly, publication.

SOCIO-ECOLOGICAL UNION (SEU) (Sotsialno-Ekologicheskiy Soyuz - SEU)

PO Box 211, 121019 Moscow, Russia **PHONE:** 7 95 1247934, 7 95 1247934 **FX:** 7 95 921716 **E-MAIL:** soceco@glas.apc.org **WEBSITES:** http://www.ecoline.ru/seu **FOUNDED:** 1988. **OFFICERS:** Vladimir Zakhazov, Dir. of Center for Coord. & Info. **MEMBERS:** 25,000. **STAFF:** 30. **BUDGET:** US$1,000,000. **LANGUAGES:** English, Russian.

DESCRIPTION: Members are organizations representing 30,000 individuals in 18 countries. Seeks improvement of public health and community mental health through increased environmental protection and preservation of biodiversity. Conducts demonstrations and rallies, circulates petitions, and holds mass meetings. Works for international cooperation to help solve problems. Conducts research and educational programs. **PUBLICATIONS:** *Bereginya* (in Russian), monthly; *SEU Times* (in English and Russian), monthly.

SOLIDARITY WITH WOMEN IN DISTRESS (SOLWODI)

Propstei - Str. 2, 5407 Boppard, D-3 Hirzennach, Germany **PHONE:** 49 6741 2232 **FX:** 49 6741 2310 **E-MAIL:** solwodi@t-online.de **WEBSITES:** http://www.solwodi.de **FOUNDED:** 1987. **OFFICERS:** Lea Ackerman. **MEMBERS:** 7. **STAFF:** 28. **LANGUAGES:** English, French, German, Polish, Russian, Spanish.

DESCRIPTION: Seeks to help women from Asia, Africa, Latin America, and central and eastern Europe adjust to German society. Acts as an intermediary between the women and authorities to defend the social and legal rights of women. Offers counseling and advice. Encourages women's self-reliance. Conducts research and educational programs. **PUBLICATIONS:** *Die Frau nach Katalog* (in German); *SOLWODI* (in English, French, and German), quarterly.

SOROPTIMIST INTERNATIONAL OF THE AMERICAS (SIA)

2 Penn Center Plaza, Ste. 1000, Philadelphia, PA 19102 **PHONE:** (215) 557-9300 **TF:** (800) 942-4629 **FX:** (215) 568-5200 **E-MAIL:** siang@soroptimist.org **WEBSITES:** http://www.siahg.com **FOUNDED:** 1921. **OFFICERS:** Leigh Wintz, Exec.Dir. **MEMBERS:** 50,000. **STAFF:** 22. **BUDGET:** US$2,500,000. **LANGUAGES:** English.

DESCRIPTION: Members are executive and professional women in 21 countries in North and South America and Asia. Areas of activity include economic and social development, education, environment, health, human rights, the status of women, and international understanding. Sponsors ''S'' Clubs (high school service clubs), Sigma Societies and Venture Clubs of the Americas. Conducts Youth Forum. **PUBLICATIONS:** *The Soroptimist of the Americas*, 7/year. Includes program and technical notes and listings of new clubs and Training Awards Program winners.

SOUTH ASIA ASSOCIATION OF NONGOVERNMENTAL ORGANIZATIONS (SANGO)

PO Box 1170, 29 Mauve Area, G-9/1, Islamabad 44000, Pakistan **PHONE:** 92 51 858972, 92 51 260373 **FX:** 92 51 261386 **TX:** 5945 CIOIB PK **OFFICERS:** Dr. M. Sadiq Malik. **STAFF:** 10. **BUDGET:** PRs 850,000. **LANGUAGES:** English.

DESCRIPTION: Serves as a forum for communication among nongovernmental development organizations operating programs in southern Asia. Promotes the growth of nongovernmental organizations fostering economic development. Functions as a liaison between members and government agencies at all levels; facilitates networking and cooperation among members. Gathers and disseminates information; advocates on behalf of members before government agencies and the public. **PUBLICATIONS:** *RDF Newsletter* (in English), monthly; *SANGO News*, monthly.

SOUTH ASIAN ASSOCIATION FOR REGIONAL COOPERATION (SAARC)

c/o SAARC Secretariat, PO Box 4222, Kathmandu, Nepal **PHONE:** 977 1 221794, 977 1 221785 **FX:** 977 1 227033 **TX:** 2561 SAARC NP **E-MAIL:** saarc@mos.com.np **FOUNDED:** 1985. **OFFICERS:** Naeem U. Hassan, Sec.Gen. **MEMBERS:** 7. **LANGUAGES:** English.

DESCRIPTION: Representatives of Bangladesh, Bhutan, India, the Maldives, Nepal, Pakistan, and Sri Lanka. Promotes economic development and cooperation among members. Seeks to improve the standard of living of South Asians. Convenes meetings of members to discuss issues including regional economics, terrorism, women and development, children's health and rights, disabled people, and youth. Facilitates adoption of commercial, trade, immigration, human rights, and development agreements among members and between members and related international organizations. Maintains food security reserve; operates SAARC Agricultural Information Centre, SAARC Tuberculosis Centre, SAARC Documentation Centre, SAARC Meteoroligical Research Centre. **PUBLICATIONS:** *SAARC in Brief*, periodic; *SAARC Newsletter*, periodic; Directory, periodic; Books; Reports.

SOUTH CENTRE

Case Postale 228, CH-1211 Geneva 19, Switzerland **PHONE:** 41 22 7918050 **FX:** 41 22 7988531 **E-MAIL:** south@southcentre.org **WEBSITES:** http://www.southcentre.org **FOUNDED:** 1995. **OFFICERS:** Branislav Gosovic, Principal Off. **MEMBERS:** 46. **STAFF:** 8. **LANGUAGES:** English, French, Spanish.

DESCRIPTION: Intergovernmental body of developing countries. Works for the benefit of the South making efforts to ensure that all developing countries and interested groups have access to its publications and the results of its work irrespective of membership. Works to assist in developing points of view of the South on major policy issues and to generate ideas and action-oriented proposals for consideration by the collectively of South governments, institutions of co-operation, intergovernmental organizations of the South and non-governmental organizations and the community at large. **PUBLICATIONS:** *Environment and Development Towards a Common Strategy of the South in the UNCED Negotiations and Beyond, Geneva, November 1991*; *Press Release*, periodic; *South Centre: Laying New Foundations*; *South Letter*, quarterly; Reports. Additional publication information available upon request.

SOUTH PACIFIC ACTION COMMITTEE FOR HUMAN ECOLOGY AND ENVIRONMENT (SPACHEE)

c/o USP, PO Box 1168, Suva, Fiji **PHONE:** 679 312371 **FX:** 679 303053 **FOUNDED:** 1982. **OFFICERS:** Asenaca Ravuvu, Education Officer. **MEMBERS:** 200. **STAFF:** 2. **LANGUAGES:** English.

DESCRIPTION: Promotes conservation and environmental protection in the South Pacific. Works to ensure that economic development programs in the region have minimal environmental impact. **PUBLICATIONS:** *Envirowatch* (in English), quarterly; *Spachtee Newsletter* (in English), quarterly.

SOUTH PACIFIC APPLIED GEOSCIENCE COMMISSION (SOPAC)

Private Mail Bag, GPO, Suva, Fiji **PHONE:** 679 381377, 679 381139 **FX:** 679 370040 **E-MAIL:** alf@sopac.org.fj **WEBSITES:** http://www.sopac.org.fj/ **FOUNDED:** 1972. **OFFICERS:** Alf Simpson, Dir. **MEMBERS:** 16. **STAFF:** 40. **BUDGET:** $F 6,000,000. **LANGUAGES:** English.

DESCRIPTION: Members are Pacific and South Pacific island states. Investigates the mineral and energy resources of member states. Assists members in the development and implementation of mineral resources extraction programs. Serves as a clearinghouse on the environmental and social effects of mineral resources development; works to insure maintenance of water and environmental quality in developing areas; assists members in the extraction of minerals in coastal zones; supports national disaster preparedness authorities in the South Pacific region; conducts educational and training programs. **PUBLICATIONS:** *Proceedings of Annual Session* (in English); *SOPAC News*, quarterly; *SOPAC Projects*, periodic; *Technical Bulletin*, periodic. Also publishes technical and cruise reports and charts.

SOUTH PACIFIC FORUM SECRETARIAT (SPFS)

Private Mail Bag, Suva, Fiji **PHONE:** 679 312600 **FX:** 301102 **TX:** 2229 **E-MAIL:** info@forumsec.org **WEBSITES:** http://www.forumsec.org.fj **FOUNDED:** 1972. **OFFICERS:** Noel Levi, Sec.Gen. **MEMBERS:** 15. **STAFF:** 40. **BUDGET:** US$9,160,000. **LANGUAGES:** English.

DESCRIPTION: Works to promote regional cooperation on aid, trade, economic development, and environmental issues among member nations of the Pacific region. Administers assistance programs provided by the European Community and other donor organizations. Compiles statistics. **PUBLICATIONS:** *Annual Report* (in English); *Directory of Aid Agencies* (in English), periodic; *Forum News* (in English), quarterly; *National Handcraft Book* (in English), periodic; *Profiles of Forum Island Countries* (in English), periodic; *South Pacific Organizations Coordinating Committee* (in English), periodic. Also publishes brochures.

SOUTH PACIFIC REGIONAL ENVIRONMENT PROGRAMME (SPREP) (Programme Regional Oceanien de l'Environnement - PROE)

PO Box 240, Apia, Western Samoa **PHONE:** 685 21929 **FX:** 685 20231 **E-MAIL:** sprep@pctok.peg.apc.org **FOUNDED:** 1982. **OFFICERS:** Mr. Tamari'i Tutangata, Dir. **MEMBERS:** 26. **STAFF:** 56. **BUDGET:** US$8,000,000. **LANGUAGES:** English, French.

DESCRIPTION: Founded as a regional program of the United Nations Environment Programme and the South Pacific Commission. Now operates autonomously to protect the natural resources and environment of the region. Provides environmental education; monitors and conducts research on pollution and protected area management; assists in the development of national environmental policy. Works with Association of South Pacific Environmental Institutions and other regional, national, and international nongovernment organizations. **PUBLICATIONS:** *CASO Link* (in English), quarterly; *Environmental Newsletter* (in English and French), quarterly; *South Pacific Biodiversity Conservation Programme*; *South Pacific Climate Change Project*; *South Pacific Sea Level and Climate Change* (in English), quarterly; *SREP Brochure.*

SOUTH EAST ASIA IRON AND STEEL INSTITUTE (SEAISI)

PO Box 7094, 40702 Shah Alam, Selangor, Malaysia **PHONE:** 60 3 5591102 **FX:** 60 3 5591159 **E-MAIL:** seaisi@pop1.jaring.my **WEBSITES:** http://www.jaring.my/seaisi **FOUNDED:** 1971. **OFFICERS:** Takashi Kitamura, Sec.Gen. **MEMBERS:** 700. **STAFF:** 13. **LANGUAGES:** English.

DESCRIPTION: Steel-producing enterprises and ministerial representatives from Australia, Indonesia, Japan, Malaysia, Philippines, Singapore, Republic of Korea, Taiwan, Thailand, and Vietnam. Promotes and represents scientific, technological, and economic development of the iron and steel industry in member countries. Fosters regional cooperation among governments, industries, and organizations; advocates steel product standardization. Provides a forum for the exchange of information. Encourages the establishment and extension of training programs for personnel employed in the iron and steel industry. Compiles statistics on iron and steel production, consumption, and trade in Southeast Asia. Offers advisory services; conducts Annual seminar. **PUBLICATIONS:** *SEAISI Country Report* (in English), annual; *SEAISI Directory* (in English), annual; *SEAISI Newsletter* (in English), monthly; *SEAISI Quarterly Journal* (in English); *SEAISI Steel Statistical Yearbook* (in English), annual. Also publishes technical papers presented at conference.

SOUTHEAST ASIAN MINISTERS OF EDUCATION ORGANIZATION (SEAMEO)

Darakarn Bldg., 920 Sukhumvit Rd., Bangkok 10110, Thailand **PHONE:** 66 2 3910144 **FX:** 66 2 3812587 **TX:** 22683 SEAMES TH **E-MAIL:** secretariat@seameo.org **FOUNDED:** 1965. **OFFICERS:** Dr. Suparak Racha-Intra, Dir. **MEMBERS:** 16. **STAFF:** 36. **LANGUAGES:** English.

DESCRIPTION: Members are representatives of Brunei-Darussalam, Cambodia, Indonesia, Laos, Malaysia, Myanmar Philippines, Singapore, Thailand, and Vietnam; associate members are representatives of Australia, Canada, France, Germany, New Zealand, and the Netherlands. Facilitates cooperation among members through educational, scientific, and cultural programs and projects in order to promote peace, prosperity, security, respect for justice, the rule of law, human rights, and fundamental freedoms for all people. Collaborates in efforts to advance mutual knowledge and understanding of people throughout the world. Assists members in the development of educational activities. Promotes preservation of the independence, integrity, and diversity of the various cultures and educational systems of member nations. **PUBLICATIONS:** *SEAMEO Calendar of Activities* (in English), annual. Lists training seminars, workshops, and other activities of SEAMEO.; *SEAMEO Canada Chronicler*, semiannual; *SEAMEO Directory* (in English), annual; *SEAMEO Forum*, semiannual; *SEAMEO Publications*, annual; *SEAMEO Update* (in English), bimonthly; Reports, periodic.

SOUTHERN AFRICA FEDERATION FOR AIDS

Morkel House, 17 Beveridge, Avondale, Harare, Zimbabwe **PHONE:** 263 4 336193 **LANGUAGES:** English.

DESCRIPTION: Social welfare and health care organizations. Promotes increased awareness of AIDS and HIV and their prevention; seeks to improve the quality of life of people with AIDS and their families. Serves as a clearinghouse on AIDS and HIV.

SOUTHERN AFRICAN CATHOLIC BISHOPS' CONFERENCE - JUSTICE AND PEACE (SACBC-JP)

PO Box 941, Pretoria 0001, Republic of South Africa **PHONE:** 27 12 3236458 **FX:** 27 12 3266218 **E-MAIL:** sacbcjp@wn.apc.org **FOUNDED:** 1967. **OFFICERS:** Mr. Ashley Green-Thompson, Dir. **MEMBERS:** 1,000. **STAFF:** 5. **BUDGET:** R 100,000. **LANGUAGES:** English.

DESCRIPTION: Members are bishops in Botswana, South Africa, and Swaziland. Objectives are to: enhance awareness of the social teachings of the Catholic Church as contained in various Vatican documents; increase understanding of the Southern African situation in order that the teachings may be effectively applied; heighten Catholic compassion for the oppressed; promote reconciliation among different social classes. Coordinates the work of diocesan commissions and other human rights' groups. Communicates with the Pontifical Commission for Justice and Peace in Rome, Italy and with other national commissions and groups worldwide. Works with other ecumenical groups. Conducts research into social problems. Promotes economic justice and monitors government regarding macro-economic strategy. **PUBLICATIONS:** Annual Report (in English).

SOUTHERN AFRICAN DEVELOPMENT COMMUNITY (SADC)

SADC Building, Private Bag 0095, Gaborone, Botswana **PHONE:** 3151863 **FX:** 372848 **WEBSITES:** http://www.sadcreview.com **FOUNDED:** 1980. **OFFICERS:** Dr. S.H.S. Makoni, Exec.Sec. **MEMBERS:** 10. **STAFF:** 30. **LANGUAGES:** English, Portuguese.

DESCRIPTION: Members are majority-ruled states of Southern Africa: Angola, Botswana, Lesotho, Malawi, Mozambique, Namibia, Swaziland, Tanzania, Zambia, and Zimbabwe. Works to accelerate economic growth to improve the living conditions of the peoples of southern Africa. Aims to achieve equitable regional integration through coordination of member states' development strategies. **PUBLICATIONS:** *Conference Proceedings* (in English), annual; *SADC Handbook.* Also publishes survey reports.

SOUTHERN AFRICAN REGIONAL DEMOCRACY FUND (SARDF)

PO Box 2427, Gaborone, Botswana **PHONE:** 267 324449 **FX:** 267 324404 **FOUNDED:** 1994. **OFFICERS:** Peter Olson. **LANGUAGES:** English.

DESCRIPTION: Members are individuals and organizations working for community and economic programs in southern Africa. Seeks to establish and strengthen democratic political systems and institutions in the region. Promotes increased participation by indigenous peoples and women in international development projects operating in southern Africa. Facilitates creation and delivery of civic education courses; provides financial and other assistance to democratic and development organizations; gathers and disseminates information.

SPECIAL COMMITTEE ON THE SITUATION WITH REGARD TO THE IMPLEMENTATION OF THE DECLARATION ON THE GRANTING OF INDEPENDENCE TO COLONIAL COUNTRIES AND PEOPLES

United Nations, Rm. S-2977, New York, NY 10017 **PHONE:** (212) 963-4272, (212) 963-3651 **FX:** (212) 963-5305 **TX:** 420544 **E-MAIL:** cherniavsky@un.org **WEBSITES:** http://www.un.org **FOUNDED:** 1961. **MEMBERS:** 24. **STAFF:** 4. **LANGUAGES:** Arabic, Chinese, English, French, Russian, Spanish.

DESCRIPTION: United Nations committee comprising representatives of 24 nations concerned with the progress of people under colonial rule toward self-determination and independence. Considers situations in 17 territories based on information received from administering powers or local governments, nongovernmental organizations, published sources, and observations of the committee's visiting missions. Reviews military activities and activities of foreign economic interests in colonial territories; enlists support of United Nations specialized agencies and international institutions to assist decolonization efforts, especially through aid to colonial people; seeks to mobilize public opinion in support of decolonization by disseminating information. **PUBLICATIONS:** *Decolonization Bulletins*, quarterly.

SRI LANKA PROJECT (SLP)

c/o British Refugee Council, 3 Bondway, London SW8 1SJ, England **PHONE:** 44 71 8203000 **FX:** 44 71 5829929 **E-MAIL:** refcounciluk@gn.apc.org **WEBSITES:** http://www.gn.apc.org/refugeecounciluk **FOUNDED:** 1987. **OFFICERS:** Nick Hardwick, Chief Exec. **LANGUAGES:** English, Tamil.

DESCRIPTION: Promotes and facilitates provision of humanitarian assistance to displaced and dispossessed people in Sri Lanka. Maintains liaison and seeks to coordinate efforts of a network of nongovernmental relief and development organizations with programs in Sri Lanka. Gathers and disseminates to Sri Lankan refugees in Europe and North America information about political events in their home areas. **PUBLICATIONS:** *Sri Lanka Monitor*, monthly.

STOCKHOLM INTERNATIONAL PEACE RESEARCH INSTITUTE (SIPRI) (Stockholms Internationella Fredsforskningsinstitut)

Sign alistgatan 9, Frosunda, S-169 70 Solna, Sweden **PHONE:** 46 8 6559700 **FX:** 46 8 6559733 **E-MAIL:** sipri@sipri.se **WEBSITES:** http://www.sipri.se **FOUNDED:** 1966. **OFFICERS:** Dr. Adam Daniel Rotfeld, Dir. **STAFF:** 55. **LANGUAGES:** English.

DESCRIPTION: Independent, international research institute established to conduct and publish studies relating to arms control and disarmament issues. Topics include international security, peacekeeping and regional security, chemical and biological warfare, military expenditure, arms transfers, arms production, military technology and arms control. Compiles statistics. **PUBLICATIONS:** *Nuclear Weapons and Arms Control in South Asia*; *Peace, Security and Conflict Prevention*. SIPRT-UNESCO Handbook; *SIPRI Yearbook*, annual. Covers armaments, disarmaments and international security; Monographs, bimonthly; Reports.

SUGAR INDUSTRY TECHNOLOGISTS (SIT)

164 N. Hall Dr., Sugar Land, TX 77478 **PHONE:** (281) 494-2046 **FX:** (281) 494-2304 **E-MAIL:** exedirsit@aol.com **FOUNDED:** 1941. **OFFICERS:** L. Anhaiser, Exec.Dir. **MEMBERS:** 600. **STAFF:** 2. **BUDGET:** US$80,000. **LANGUAGES:** English.

DESCRIPTION: Members are cane sugar refinery technical engineers, bacteriologists, chemists, processors, production workers, and administrators. Purpose is to serve the professional interests of members by disseminating information on scientific and other technical aspects of sugar refining and by encouraging original research. **PUBLICATIONS:** *Roster*, annual; *Sugar Industry Technologists—Proceedings of Annual Meeting*.

SULABH INTERNATIONAL SOCIAL SERVICE ORGANISATION

Sulabh Bhawan, Mahavir Enclave, Palam Dabri Rd., New Delhi 110 045, India **PHONE:** 91 11 5032617, 91 11 5032631 **FX:** 91 11 5036122 **E-MAIL:** sulabh1@nde.vsnl.net.in **FOUNDED:** 1970. **OFFICERS:** B.B. Sahay, Hony. Chm. **MEMBERS:** 11. **STAFF:** 50,000. **BUDGET:** US$30,000,000. **LANGUAGES:** English.

DESCRIPTION: Strives to promote human rights, environmental sanitation, health and hygiene, nonconventional sources of energy, waste management and social reforms through education, training and campaigns. It has developed a scavenging-free two-pit pourflush, safe and hygienic on-site human waste disposal technology; a new concept of construction and maintenance of pay-and-use public toilets, and generation of biogas and biofertiliser produced from excreta-based plants. Runs English medium school for wards of liberated scavengers and vocational training courses for relieved scavenger families. Its cost effective and appropriate sanitation system has been recognised as a global urban practice by U.N. Centre for Human Settlements at the Habitat II Conference, Istanbul in June 1996. It has also been given a special consultative status with the Economic and Social Council of the U.N. **PUBLICATIONS:** *Human Resources Development in India*; *Man and His Mission*; *Power Generation from Human Excreta*; *Road to Freedom*; *Sulabh*

India (in English and Hindi), monthly; *Sulabh Sansar* (in English and Hindi), monthly; *Sulabh Shauchalaya: A Simple Idea that Worked*; *Sulabh Shauchalaya: A Study of Directed Change.*

SYNDESMOS

c/o Orthodox Press Service (OPS), BP 44, F-92333 Sceaux, France **PHONE:** 33 1 46601774 **FX:** 33 1 46604554 **E-MAIL:** 10041.1333@compuserve.com **FOUNDED:** 1953. **OFFICERS:** Mr. Alexander Belopopsky **MEMBERS:** 74. **STAFF:** 4. **BUDGET:** US$500,000. **LANGUAGES:** English, French, Greek, Russian.

DESCRIPTION: Eastern Orthodox youth organizations and movements, theological academies, schools, and faculties in 32 countries. Promotes Orthodox Christian youth movements and works to create a bond of unity among them. Assists Orthodox youth in their relations with other Christian groups and peoples of other faiths. Fosters a deeper understanding of the Orthodox faith and the responsibilities of the church in the modern world. Furthers cooperation with Oriental Orthodox churches through common youth activities. Sponsors summer camps and youth festivals on Orthodox theological education; provides consultation on religious education. Conducts conferences and study seminars. (Syndesmos is Greek for ''bound together'' and comes from the New Testament.) **PUBLICATIONS:** *Syndesmos Directory*, quadrennial; *Syndesmos Journal* (in English), semiannual; *Syndesmos News* (in English, Greek, and Russian), quarterly; Handbooks; Reports.

SYNERGOS INSTITUTE

9 East 69th St., New York, NY 10021 **PHONE:** (212) 517-4900 **FX:** (212) 517-4815 **WEBSITES:** http://www.synergos.org **FOUNDED:** 1986. **OFFICERS:** S. Bruce Schearer, Pres. **STAFF:** 14. **BUDGET:** US$1,700,000. **LANGUAGES:** English.

DESCRIPTION: A private, non-governemntal, non-profit organization funded by grants from foundations, corporations, international agencies, and individual contributors. Works with voluntary organizations and other groups in supporting local communities to develop effective, sustainable solutions to poverty problems. Also works closely with Associates and a dozen partner organizations in 18 countries. Seeks to create new models and avenues of action to address poverty. Conducts programs in Mexico; works to conserve the Chimalapas rain forest and achieve improved and secured livelihood for it residents. **PUBLICATIONS:** *A Survey of Endowed Grantmaking Foundations in Africa, Asia, Eastern Europe, Latin America and the Caribbean*; *Building Development Projects in Partnership with NGOs and Communities: An Action Agenda for Government Policymakers and Donors*; *Community-Based Development Experience: Can It Be Replicated from South to North, and If So, How?*; *Endowed National Community Development Foundations: A New Approach to Financing the Non-Profit Sector in African Countries*; *Establishing Endowed National Foundation in Countries of the South*; *Holding Together: Collaborations and Partnerships in the Real World*; *How Community Development Foundations Can Help Strengthen Civil Society*; *Informal Survey of Activities Being Conducted in Japan by U.S. Non-Profit Organizations Working on Environment and Conservation Issues*; *Multiparty Cooperation for Development in Asia*; *Proposal to Promote Democracy and Community Self-Reliance in Twenty Countries by Establishing Endowed Private Development Foundations*; *Sustaining Collaborative Problem Solving: Stratagies from a Study in Six Asian Countries.*

TAIGA RESCUE NETWORK

Ajtte, PO Box 116, S-962 23 Jokkmokk, Sweden **PHONE:** 46 97117039, 46 97112547 **FX:** 46 97112057 **E-MAIL:** taiga@jokkmokk.se **FOUNDED:** 1992. **OFFICERS:** Anne Janssen, Int.Coor. **MEMBERS:** 130. **STAFF:** 2.

DESCRIPTION: Members are individuals interested in the conservation and wise use of natural resources, particularly native forests. Works to support local struggles and strengthen the cooperation between individuals, NGO's and indigenous peoples and nations concerned with the protection, restoration and sustainable use of the boreal forests by means that ensure the intergerity of natural processes and dynamics. **PUBLICATIONS:** *Taigas News on Boreal Forest* (in English), quarterly; *The Talga Trade: A Report on the Production, Consumption and Trade of Boreal Wood Products.*

TECHNOSERVE

49 Day St., Norwalk, CT 06854 **PHONE:** (203) 852-0377 **TF:** (800) 99-WORKS **FX:** (203) 838-

6717 **E-MAIL:** technoserve@tns.org **WEBSITES:** http://www.technoserve.org **FOUNDED:** 1968. **OFFICERS:** Peter A. Reiling, Pres. & CEO. **MEMBERS:** 95. **STAFF:** 350. **BUDGET:** US$14,000,000. **LANGUAGES:** English, Spanish.

DESCRIPTION: Works to improve the economic and social well-being of low-income people in Latin America, Africa, and Eastern Europe. Provides agriculture and business training to help the poor build their own self-sustaining enterprises. Works with more than 175 enterprises and institutions. Also provides feasibility assessment; design and implementation of management, production and financial systems and controls; monitoring and evaluating enterprise performance and impact. **PUBLICATIONS:** *Working Solution*, semiannual; Annual Report. Includes membership listing. Also publishes research findings, case histories, and sector studies.

TELEWORK, TELECOTTAGE, AND TELECENTRE ASSOCIATION (TCA)

Shortwood, Nailsworth GL6 0SH, England **PHONE:** 44 1203 696986, 44 1453 834874 **FX:** 44 1453 836174 **E-MAIL:** 100272.3137@compuserve.com **WEBSITES:** http://ourworld.compuserve.com/homepages/teleworker **OFFICERS:** Alan Denbigh **MEMBERS:** 2,300. **STAFF:** 2. **BUDGET:** £150,000.

DESCRIPTION: Members largely fall into one of four categories: individuals (teleworkers) working from home supported by technology, people running centres supporting home based teleworkers, regional development agencies, companies introducing teleworking schemes. Concerned with the improvement in employment, training and services for people living in rural areas and the development of local economies through the use of IT and telecommunications including share facilities in local centres. **PUBLICATIONS:** *Teleworker Magazine* (in English), bimonthly; *Teleworking Handbook.*

TERRE DES HOMMES

Case Postale 388, En Budron C Le Mont, CH-1000 Lausanne 9, Switzerland **PHONE:** 41 21 654 6666 **FX:** 41 21 654 6677 **TX:** 454 042 TDH CH **E-MAIL:** tdh@tdh.ch **FOUNDED:** 1960. **OFFICERS:** Francois Droz, Dir. **BUDGET:** 20,000,000 SFr. **LANGUAGES:** English, French, German, Italian.

DESCRIPTION: Promotes sustainable economic growth in Africa, with particular emphasis on issues of drought and the plight of abandoned children. Makes available financial and technical assistance to communities affected by economic development programs. Conducts educational campaigns to enable rural communities to operate basic health care services and make better use of water resources.

THAI NATIONAL COMMITTEE OF THE INTERNATIONAL ASSOCIATION ON WATER QUALITY (TNCIAWQ)

c/o Environmental Engineers' Assn. of Thailand, Faculty of Engineering, Chulalongkorn Univ., Bangkok 10330, Thailand **PHONE:** 66 2 2527511, 66 2 2186669 **FX:** 66 2 2527510 **FOUNDED:** 1984. **OFFICERS:** Prof. T. Panswad, Sec.Gen. **MEMBERS:** 1,000. **STAFF:** 15. **BUDGET:** US$70,000. **LANGUAGES:** Thai.

DESCRIPTION: Members are engineers involved in pollution control and waste minimization in 5 countries. Operates under the auspices of the Environmental Engineers Association of Thailand. Works to develop new methods of combatting pollution. Conducts research and charitable activities; performs equipment testing and certification. Organizes training sessions and seminars; offers placement services; disseminates information. **PUBLICATIONS:** *EEAT Newsletter* (in English and Thai), bimonthly; Manuals; Proceedings.

THEOSOPHICAL SOCIETY (TS)

Adyar, Chennai 600 020, Tamil Nadu, India **PHONE:** 91 44 4912815, 91 44 4912474 **FX:** 91 44 4902706 **E-MAIL:** para.vidya@gems.vsnl.net.in **FOUNDED:** 1875. **OFFICERS:** Mrs. Radha Burnier, International. **MEMBERS:** 30,000. **STAFF:** 150. **LANGUAGES:** English.

DESCRIPTION: Works for the ''universal brotherhood of humanity'' without distinctions of race, creed, color, sex, or social class. Encourages the study of comparative religion, philosophy, and science. Investigates unexplained laws of nature and the ''powers latent in man.'' Conducts study courses; sponsors research into oriental culture, philosophy, and religion. Operates Theosophical Order of Service, comprising individuals who make and distribute clothes for poor children, maintains school, hostel, and welfare center for the poor. Operates Camping Centre for scouts, youths, and children. Sponsors study classes and training camps. **PUBLICATIONS:** *Adyar Newsletter* (in English), quarterly; *Brahma-vidya, the Adyar Library Bulletin* (in English), annual; *The Theosophist* (in English), monthly; *Wake Up India* (in English), quarterly.

THIRD WORLD ACADEMY OF SCIENCES (TWAS)

Intl. Centre for Theoretical Physics, Enrico Fermi Building, Via Beirut 6, I-34014 Trieste, Italy **PHONE:** 39 40 2240327 **FX:** 39 40 224559 **TX:** 460392 ICTPI **E-MAIL:** twas@ictp.trieste.it **WEBSITES:** http://www.ictp.trieste.it/-twas/TWAS.html **FOUNDED:** 1983. **OFFICERS:** Prof. Mohamed Hag Ali Hassan, Exec.Dir. **MEMBERS:** 480. **STAFF:** 10. **LANGUAGES:** English.

DESCRIPTION: Members are scientists in 75 developing countries. Assists developing countries by promoting scientific research and development. Promotes information exchange between researchers in developing countries; provides scientists with needed facilities; assists in developing scientific manpower by supporting young scientists. **PUBLICATIONS:** *Conference Proceedings* (in English), biennial; *Newsletter* (in English), quarterly; *Yearbook* (in English). Also publishes brochures, annual report, and program information.

THIRD WORLD FORUM (TWF) (Forum du Tiers Monde)

39 Dokki St., PO Box 43, Orman, Cairo, Egypt **PHONE:** 20 2 3488092 **FX:** 20 2 3480668 **TX:** 93919 CADSA UN **FOUNDED:** 1978. **OFFICERS:** Dr. Ismail-Sabri Abdalla, Chm. **MEMBERS:** 350. **STAFF:** 1. **LANGUAGES:** Arabic, English, French, Spanish.

DESCRIPTION: Members are social scientists and intellectuals. Promotes the exchange of ideas for improving the quality of life in Third World countries. Conducts research, seminars, and conferences. **PUBLICATIONS:** *Arab Alternative Futures AAF Dossier* (in Arabic and English), quarterly; *Collective Self Reliance* (in Arabic); *Images of the Arab Futures* (in English); *Third World Forum Newsletter*, quarterly.

TIBETAN AID PROJECT (TAP)

2910 San Pablo Ave., Berkeley, CA 94702 **PHONE:** (510) 848-4238 **TF:** (800) 33TIBET **FX:** (510) 548-2230 **E-MAIL:** tap@dnai.com **WEBSITES:** http://www.nyingma.org **FOUNDED:** 1974. **OFFICERS:** Tarthang Tulku, Pres. **MEMBERS:** 10,000. **STAFF:** 5. **BUDGET:** US$300,000. **LANGUAGES:** English, German, Japanese, Portuguese.

DESCRIPTION: A project of the Tibetan Nyingma Relief Foundation. Offers assistance to individuals in Tibet and Tibetan refugees in India, Nepal, Bhutan, and Sikkim. Conducts Tibetan Student Support Program; sponsors relief distribution to monasteries and schools for support of religious and community activities. Supports shedeas (college-level philosophy schools). **PUBLICATIONS:** *Bodh Gaya World Peace Cermony 1994*. Covers Bodh Gaya Stupa history.; *From the Roof of the World: Refugees of Tibet*; Brochures, annual.

TIMORESE INTERNATIONAL SECTION FOR HUMAN RIGHTS

PO Box 10, NL-2280 ZH Aarijswijk, Netherlands **OFFICERS:** Paulino Gama, Pres. **LANGUAGES:** Bahasa Indonesia, Dutch, Indian Dialects.

DESCRIPTION: Members are individuals and organizations with an interest in Timor. Seeks to publicize the Timorese nationalist movement. Promotes observance of human rights by Indonesian authorities in Timor. Gathers and disseminates information on the Timorese independence movement and human rights abuses in Timor.

TRADE UNION ADVISORY COMMITTEE TO THE OECD (TUAC OECD) (Commission Syndicale Consultative aupres de l'OCDE)

26, ave. de la Grande Armee, F-75017 Paris, France **PHONE:** 33 1 47634263 **FX:** 33 1 47549828 **E-MAIL:** tuac@oecd.org **WEBSITES:** http://www.tuac.org **FOUNDED:** 1948. **OFFICERS:** John Evans, Gen.Sec. **MEMBERS:** 50. **STAFF:** 7. **BUDGET:** 5,000,000 Fr. **LANGUAGES:** English, French.

DESCRIPTION: Members are national trade union centers in 29 countries representing 70 million workers. Coordinates and represents the views of trade unions before governments of industrialized democracies. Areas of focus include: economic and social policies; employment and globalization policies; multinational enterprises; environment; education. Maintains consultative relations with the Organization for Economic Co-Operation and Development. **PUBLICATIONS:** *Adaptability versus Flexibility* (in English and French); *Fighting Unemployment and Managing Structural Changes* (in English); *Industrial Restructuring and Employment in Central and Eastern Europe* (in English); *Korea and the OECD* (in English); *Problems and Opportunities of Restructuring Industrial Regions* (in English). Jobs First-Trade Unions and the Modernisation of the Labour Market; Language(s): English.

TRANSFRIGOROUTE INTERNATIONAL (TI)

c/o ASTAG, Weissenbuhlweg 3, B.P. 581, CH-3007 Bern, Switzerland **PHONE:** 41 31 3708585 **FX:** 41 31 3708589 **E-MAIL:** transfrigo@bluewin.ch **WEBSITES:** http://www.transfrigo.com **FOUNDED:** 1955. **OFFICERS:** Hanspeter Tanner, Gen.Sec. **MEMBERS:** 1,700. **LANGUAGES:** English, French, German.

DESCRIPTION: Members are national firms specializing in refrigerated and food tanker transportation in 28 countries. Promotes the increase of road transport of goods under controlled temperatures. Works for continual development of technical, economic, and structural cooperation in the industry. Cooperates with intergovernmental organizations, including the United Nations. **PUBLICATIONS:** *Directory* (in English, French, and German), annual. Contains the addresses of all affiliates and factual information on the international refrigerated road transport industry.; *Driver's Guide*; *Frigoriscope* (in English, French, and German), quarterly; *Technical and Economic Rules*.

TRANSTEC

N.V. Transtec S.A. Researchparc, 75, ave. de Tyras, B-1120 Brussels, Belgium **PHONE:** 32 2 2664950 **FX:** 32 2 2664965 **E-MAIL:** transtec@transtec.be **FOUNDED:** 1983. **OFFICERS:** Olivier Bribosia, Project Mgr. **STAFF:** 90. **BUDGET:** US$14,500,000. **LANGUAGES:** Arabic, Dutch, English, French, German, Italian, Portuguese, Russian, Spanish.

DESCRIPTION: Provides consulting services and technical assistance in NSO coordination, constitutional strengthening, export promotion, and tourism development. Maintains such programs as refugee aid, assistance to native peoples to resettle former land, promotion of health and education programs, development of basic communication infrastructures, increased agricultural production, and establishment of small businesses based on local crafts and agriculture.

TRILATERAL COMMISSION (TC)

345 E. 46th St., New York, NY 10017 **PHONE:** (212) 661-1180 **FX:** (212) 949-7268 **E-MAIL:** trilat@panix.com **WEBSITES:** http://www.trilateral.org **FOUNDED:** 1973. **OFFICERS:** Charles B. Heck, N.Amer.Dir. **MEMBERS:** 330. **STAFF:** 11. **LANGUAGES:** English.

DESCRIPTION: Members are distinguished private citizens from North America, Western Europe, and Japan, including academic, business, labor, and media professionals. Encourages closer cooperation among these three democratic industrialized regions. Meets annually to analyze major issues confronting the ''trilateral'' area; seeks to improve public understanding of these issues. Works to develop proposals for joint action and to nurture habits of working together. **PUBLICATIONS:** *Trialogue*, annual; *Triangle Papers*, semiannual. Reports of TC task forces containing topics such as monetary matters, trade, the energy crisis, and institution reform. Also publishes brochure.

TROPICAL FOREST RESOURCE GROUP (TFRG)

c/o Oxford Forestry Institute, Dept. of Plant Science, South Parks Rd., Oxford, Oxon. OX4 4LL, England **PHONE:** 44 1865 271035 **FX:** 44 1865 271036 **E-MAIL:** pbacon@ermine.on.ac.uk **FOUNDED:** 1992. **OFFICERS:** Dr. Philip Bacon, Sec. **MEMBERS:** 8. **STAFF:** 4. **BUDGET:** £10,000. **LANGUAGES:** English.

DESCRIPTION: Organizations, universities, research institutes, private companies, and government agencies seeking to inform development project and policy developers in areas of forest management and conservation. Gathers and disseminates information on subjects including national and regional forest planning, land and soil evaluation, economic and social system assessments, and biomass energy resource management. Conducts educational and training programs. Initiates and coordinates responses to charitable public and private organizations worldwide.

TROPICAL GROWERS' ASSOCIATION (TGA)

c/o Assn. of the Intl. Rubber Trade, 606 The Chandlery, 50 Westminster Bridge Rd., London SE1 7QY, England **PHONE:** 44 171 7217440 **FX:** 44 171 7217459 **FOUNDED:** 1907. **OFFICERS:** Mr. P.D. Gatland, Dir. **MEMBERS:** 80. **STAFF:** 1. **LANGUAGES:** English.

DESCRIPTION: Individuals and companies in 35 countries interested in cultivating tropical trees or plants including rubber, oil palm, cocoa, and coconut. Organizes seminars; compiles statistics. **PUBLICATIONS:** Bulletin (in English), bimonthly.

UKRAINIAN PATRIARCHAL WORLD FEDERATION (UPWF)

4437 McKinley, Warren, MI 48091 **PHONE:** (810) 755-1575 **FX:** (810) 755-3706 **FOUNDED:** 1974. **OFFICERS:** Vasyl Kolodchin, Pres. **LANGUAGES:** English.

DESCRIPTION: Members are national societies of Ukrainian Catholics from 8 countries. Seeks organizational autonomy of the Ukrainian Catholic Church and recognition of the Ukrainian Patriarchate by the Holy See of Rome. Works to enrich the religious lives of laymen and engage them in ecumenical dialogue. Coordinates activities of Ukrainian Patriarchal societies. Organizes periodical symposia on theological topics. **PUBLICATIONS:** *The Patriarchate*, monthly.

UNAIDS

20, avenue Appia, CH-1221 Geneva, Switzerland **LANGUAGES:** English, French, German, Italian.

DESCRIPTION: Members are health care personnel and other individuals and organizations interested in stopping the global AIDS epidemic. Seeks to reduce the suffering of people with HIV and AIDS; works to ''counter the impact of the pandemic on individuals, communities, and societies.'' Conducts educational programs in areas including AIDS prevention and treatment; makes available health and social services to people with HIV and AIDS.

UNESCO INTERNATIONAL INSTITUTE FOR EDUCATIONAL PLANNING (IIEP) (Institut International de Planification de l'Education)

7-9, rue Eugene Delacroix, F-75116 Paris, France **PHONE:** 33 1 45037700 **FX:** 33 1 40728366 **E-MAIL:** information@iiep.unesco.org **WEBSITES:** http://www.unesco.org/iiep **FOUNDED:** 1963. **OFFICERS:** Jacques Hallak, Dir. **STAFF:** 62.

DESCRIPTION: Members are states belonging to the United Nations Educational, Scientific and Cultural Organization united to form an international center for advanced training and research in educational planning. Aim is to contribute to the development of education worldwide by increasing both knowledge and the number of competent professionals in educational planning. Conducts research on issues such as improving the quality of basic education, secondary education policies and strategies, meeting the needs of underprivileged groups, promoting efficient educational organization, management, and financing. **PUBLICATIONS:** *IIEP Newsletter* (in English, French, Portuguese, Russian, and Spanish), quarterly.

UNESCO'S CO-ACTION PROGRAMME

UNESCO, 7, place de Fontenoy, SP 07, F-75352 Paris, France **PHONE:** 33 1 45681000, 33 1 45681472 **FX:** 33 1 45671690 **WEBSITES:** http://www.unesco.org **FOUNDED:** 1950. **OFFICERS:** Ms. Astrid Gillet, Asst.Prog.Spec. **LANGUAGES:** English, French.

DESCRIPTION: Operates under the auspices of the United Nations Educational, Scientific and Cultural Organization. Selects small scale projects presented by NGOs, local community groups and institutions and brings them to the attention of potential donors, mainly individuals, foundations, charitable organizations and educational institutions, through the publishing of a six-monthly project catalogue. Aims to promote development through self-reliance. **PUBLICATIONS:** *Co-Action Catalogue* (in English and French), semiannual.

Also publishes brochure (English-French); presentation of the Co-Action programme.

UNION OF ARAB JURISTS (UAJ) (Union des Juristes Arabes - UJA)

PO Box 6026, Baghdad, Iraq **PHONE:** 964 1 8850132, 964 1 8840051 **FX:** 964 1 8849973 **TX:** 212661 HUQUQIYN IK **FOUNDED:** 1975. **OFFICERS:** Shibib Al-Maliki, Sec.Gen. **MEMBERS:** 16. **LANGUAGES:** Arabic, English.

DESCRIPTION: Members are Arab jurists' associations. Objectives are: to politically and economically liberate what the union sees as the Arab homeland, and to work toward unity; to establish and unify Arab laws; to formulate the constitutional and legal framework for political and social principles in the Arab homeland; to defend human rights and develop a strong relationship between Arab jurists and similar organizations. **PUBLICATIONS:** *Arab Jurist*, quarterly.

UNION FOR THE COORDINATION OF THE PRODUCTION AND TRANSMISSION OF ELECTRICITY (UCPTE) (Union pour la Coordination de la Production et du Transport de l'Electricite)

2 rue Pierre d'Aspelt, Bp 37, L-2010 Luxembourg, Luxembourg **PHONE:** 352 44902400 **FX:** 352 44902401 **E-MAIL:** secretariat.ucpte@edinfor.pt **WEBSITES:** http://www.ucpte.org **FOUNDED:** 1951. **OFFICERS:** J.A. Allen Lima, Pres. **MEMBERS:** 27. **LANGUAGES:** French, German.

DESCRIPTION: Members are companies representing electrical utilities and electrical energy authorities of Austria, Belgium, Bosnia-Hercegovina, Croatia, France, Germany, Greece, Italy, Luxembourg, Netherlands, Portugal, Slovenia, Spain, Fyrom, Republic of Yugoslavia, and Switzerland. Works toward the coordination of the operation of interfaces with power systems, particularly with regard to security of operations. Establish the technical rules of the games for the reliability of system interconnection. **PUBLICATIONS:** *UCPTE-Annual Report* (in English, French, and German).

UNION OF EUROPEAN FOOTBALL ASSOCIATIONS (UEFA) (Union des Associations Europeennes de Football)

Chemin de la Redoute 54, Case Postale 2, CH-1260 Nyon, Switzerland **PHONE:** 41 22 9944444 **FX:** 41 22 9944488 **WEBSITES:** http://www.uefa.com **FOUNDED:** 1954. **OFFICERS:** Gerhard Aigner, Gen.Sec. **MEMBERS:** 51. **STAFF:** 96. **LANGUAGES:** English, French, German.

DESCRIPTION: Members are national football associations in 51 countries. Promotes the sport of football (soccer) in Europe. Fosters agreement on common policy and good relations between member associations. Organizes competitions and tournaments and offers biennial referees' and trainers' courses. Compiles statistics. **PUBLICATIONS:** *Annual Agenda*; *First Division Clubs in Europe*, annual; *UEFA Bulletin* (in English, French, and German), quarterly; *UEFA Flash* (in English, French, and German), monthly.

UNION OF EUROPEAN RAILWAY INDUSTRIES (UNIFE) (Union des Industries Ferroviaires Europeenes - UNIFE)

221 ave. Louise, bve 11, B-1050 Brussels, Belgium **PHONE:** 32 2 6261260 **FX:** 32 2 6261261 **E-MAIL:** unife@v.verso.pophost.eunet **FOUNDED:** 1975. **OFFICERS:** Mr. Drewin Nieuwenhuis, Gen.Mgr. **MEMBERS:** 100. **STAFF:** 7. **LANGUAGES:** English, French, German.

DESCRIPTION: Members are companies belonging to the European Community or the European Free Trade Association which develop, produce, repair or assemble material for guided land transport including locomotives, powered rail vehicles, rollling stock, embarked and signalling equipment, and fixed installations. Promotes guided land transport industry. Represents members' interests. **PUBLICATIONS:** *UNIFE Newsletter* (in English), bimonthly; Brochure.

UNION OF INDUSTRIAL AND EMPLOYERS' CONFEDERATIONS OF EUROPE (UNICE) (Union des Confederations de l'Industrie et des Employeurs D'Europe - UNICE)

40, rue Joseph II, bte. 4, B-1000 Brussels, Belgium **PHONE:** 32 2 2376511 **FX:** 32 2

2311445 **E-MAIL:** main@unice.be **FOUNDED:** 1958. **OFFICERS:** Peter Kettlewell. **MEMBERS:** 33. **STAFF:** 40. **LANGUAGES:** English, French.

DESCRIPTION: Members are federations of industries and employers' associations in 25 European countries. Works to promote the professional interests of member federations and ensure their participation in European industrial advancement. Acts as a spokesperson for European business and industry. Seeks to influence and coordinate the formation of policies affecting European business and industry. **PUBLICATIONS:** *UNICE Information* (in English and French), bimonthly. Contains information on the immediate past work and future trends of the five operating departments.; *UNICE Position Papers* (in English and French), semiannual.

UNION OF INTERNATIONAL ASSOCIATIONS (UIA) (Union des Associations Internationales - UAI)

40, rue Washington, B-1050 Brussels, Belgium **PHONE:** 32 2 6401808, 32 2 6404109 **FX:** 32 2 6436199 **E-MAIL:** uiaweb@uia.be **WEBSITES:** http://www.uia.org **FOUNDED:** 1907. **OFFICERS:** Jacques Raeymaeckers, Sec.Gen. **MEMBERS:** 136. **STAFF:** 19. **LANGUAGES:** English, French.

DESCRIPTION: Members are individuals and international associations, foundations, corporations, and other institutions in 37 countries. Promotes the development of nonprofit organizations and the understanding of how international bodies represent human interests. Collects, analyzes, and publishes data on international governmental and nongovernmental organizations. Conducts research on legal and administrative concerns of international organizations such as possibilities of interorganizational networking, international conference organization, and principles governing associations' actions under pressure from national initiatives. **PUBLICATIONS:** *Encyclopedia of World Problems and Human Potential*, periodic; *International Association Statutes*; *International Congress Calendar*, quarterly; *Transnational Associations* (in English and French), bimonthly; *Who's Who in International Organizations*; *Yearbook of International Organizations* (in English and French).

UNION OF INTERNATIONAL FAIRS (Union des Foires Internationales - UFI)

35 bis, rue Jouffroy-d'Abbans, F-75017 Paris, France **PHONE:** 33 1 42679912 **FX:** 33 1 42271929 **TX:** UFI 644097 **E-MAIL:** info@ufinet.org **WEBSITES:** http://www.ufinet.org **FOUNDED:** 1925. **OFFICERS:** Gerda Marquardt, Sec.Gen. **MEMBERS:** 202. **STAFF:** 7. **LANGUAGES:** English, French, German.

DESCRIPTION: Members are exhibition centers, national fair exhibition associations, statistical control organizations, and international organizers of trade fair exhibitions representing 67 countries. Promotes international economic exchange by making international trade fairs more effective in their operation and services rendered. Encourages the exchange of ideas and experiences among members; represents members' interests before other international and intergovernmental organizations. Sanctions only those international fairs and exhibitions that have been held for at least 3 years and satisfy market needs of the country in which they are held. Promotes standardization of the industry and aids developing countries in organizing events. Discourages fairs with no apparent economic value. Conducts research; sponsors technical and training seminars. Presents awards; compiles statistics. **PUBLICATIONS:** *Associate Members List*; *Calendar of International Fairs and Exhibitions*, annual; *Seminar Proceedings*; *This is UFI*, periodic; *UFI Information*, semiannual; *Who's Who*, annual. Provides membership information.

UNION OF LATIN AMERICAN UNIVERSITIES (ALAU) (Union de Universidades de America Latin - UDUAL)

Apartado Postal 70232, Ciudad Universitaria, Coyoacan, 04510 Mexico City, DF, Mexico **PHONE:** 52 5 6161414 **FX:** 52 5 6162383 **TX:** 1764112 UUALME **E-MAIL:** udual@servidor.unam.mx **WEBSITES:** http://www.unam.mx/udual **FOUNDED:** 1949. **OFFICERS:** Dr. Abelardo Villegas Maldonado, Gen.Sec. **MEMBERS:** 160. **STAFF:** 38. **LANGUAGES:** Spanish.

DESCRIPTION: Members are Latin American universities, schools, and cultural institutions in 22 countries. Encourages cultural and economic integration of Latin American nations by providing a link between institutions of culture and higher education. Strives to contribute to social, economic, and cultural change both locally and regionally and to assist in creating a free, peaceful, and humanistic society. Provides information on Latin American and other universities. Serves as a regional center for understanding and cooperation through its relations with international cultural organizations. Maintains university information and statistics center. **PUBLICATIONS:** *Directorio UDUAL* (in Spanish), annual. Contains information on 1000 Latin American universities.; *Gaceta UDUAL* (in Spanish), quarterly; *Gaceta Udual*, quarterly; *Sistema de Informacion de Educacion Superior en America Latina v el Caribe* (in Spanish), biennial; *Universidades* (in Spanish), semiannual. Covers Latin American higher education analysis and problems.; *Ventana*, bimonthly; Bulletin, monthly.

UNION OF LIBERAL TRADE UNION ORGANISATIONS (ULTUO) (Bund Freier Gewerkschaften)

Badenerstr. 41, CH-8004 Zurich, Switzerland **PHONE:** 41 1 2410757 **FX:** 41 1 2410914 **FOUNDED:** 1948. **OFFICERS:** Dr. Andreas Hubli. **STAFF:** 1. **LANGUAGES:** French, German.

DESCRIPTION: Members are trade unions in 5 countries. Objectives are: to improve the material and moral status of manual and intellectual workers on the basis of a free and democratic state; to secure the liberty and dignity of man; to recognize the rights of minorities; to oppose class struggle and dictatorship; to encourage all population groups to collaborate in a liberal economy.

UNION OF NATIONAL RADIO AND TELEVISIONS ORGANIZATIONS OF AFRICA (UNRTA) (Union des Radiodiffusions et Televisions Nationales d'Afrique - URTNA)

101 rue Carnot, Boite Postale 3237, Dakar, Senegal **PHONE:** 221 215970 **FX:** 221 225113 **TX:** 650 **OFFICERS:** Efoe Adodo Mensah, Sec.Gen. **LANGUAGES:** English, French.

DESCRIPTION: Members are national organizations representing broadcasters. Promotes high technical, professional, and ethical standards in broadcasting. Represents members' interests before government agencies, international broadcasting and labor organizations, and the public.

UNITAS MALACOLOGICA (UM)

c/o Dr. Peter Morgan, The Natural History Museum, Cromwell Rd., London SW7 5BD, England **PHONE:** 31 71 5687614, 31 71 5275146 **FX:** 31 71 5627666 **FOUNDED:** 1962. **OFFICERS:** Dr. Peter Morgan, Sec. **MEMBERS:** 400. **LANGUAGES:** English.

DESCRIPTION: Members are malacologists (zoologists specializing in the study of mollusks) in 47 countries who are members of museums and scientific institutions; interested individuals and institutions. Furthers the worldwide study of mollusks. **PUBLICATIONS:** *UM Newsletter*, biennial.

UNITED BIBLE SOCIETIES (UBS)

Reading Bridge House, Reading, Berks. RG1 8PJ, England **PHONE:** 44 118 9500200 **FX:** 44 118 9500857 **WEBSITES:** http://www.biblesociety.org **FOUNDED:** 1946. **OFFICERS:** Rev. Fergus Macdonald, Sec.Gen. **MEMBERS:** 135. **LANGUAGES:** English, French, Spanish.

DESCRIPTION: Members are Bible societies rooted in the life of their own peoples and churches and committed to the widest possible effective distribution of the Holy Scriptures. Maintains team of technical consultants. Coordinates activities and advises on Scripture translation, production, and distribution. Also coordinates arrangements to provide Scriptures in countries without Bible societies and in emergency situations. Conducts training institutes for tranlators, distributors, Scripture producers, and administrative personnel; produces Scriptures for new literates. Organizes periodic workshops. Compiles statistics. **PUBLICATIONS:** *The Bible Translator*, quarterly; *Bulletin*, semiannual; *Prayer Booklet*, annual; *World Report* (in English, French, and Spanish), monthly.

UNITED NATIONS (UN)

United Nations Headquarters, New York, NY 10017 **PHONE:** (212) 963-1234 **FX:** (212) 963-4879 **E-MAIL:** inquiries@un.org **WEBSITES:** http://www.un.org **FOUNDED:** 1945. **OFFICERS:** Kofi Annan, Sec.Gen. **MEMBERS:** 185. **STAFF:** 4,700. **BUDGET:** US$1,300,000,000. **LANGUAGES:** Arabic, Chinese, English, French, Russian, Spanish.

DESCRIPTION: Members are countries around the world representing 98% of the world's population. Established following World War II to: identify and solve international disputes that threaten world peace and security; advocate respect for human rights, the "dignity and worth of the human person, and the equal rights of men and women and of nations large and small;" work to create conditions where justice and respect for treaties and international law can be maintained. Members pledge to live peacefully beside their neighbors, and not to use armed force except in the common interest. Attempts to ensure that member and nonmember countries act to maintain peace and security; promotes social progress, a better standard of living, and freedom for all people. Provides a forum for members to consider major international issues, establish broad standards, and create bodies to study problems, formulate policy, and administer programs. Encourages the discussion of international issues, including nuclear arms, the arms race, population growth, human rights, environmental changes, water and food supply, raw materials, and economic development. Initiates studies and makes recommendations to further international cooperation and develop and codify international law. Maintains several branches that formulate and implement programs: the Secretariat serves as an international civil servant and works to resolve international disputes, administer peace-keeping operations, sponsor studies and surveys, and assure that policies are properly implemented; the Security Council works to establish and maintain world peace by adopting resolutions and enforcing them through negotiations, arbitrations, and sanctions; the Economic and Social Council coordinates and fosters economic and social progress by initiating activities relating to development and industrialization, world trade, population, social welfare, human rights, science and technology, and other issues; the Trusteeship Council acts as an advocate for the "Trust Territories," and promotes their progressive development toward self-government or independence; the International Court of Justice acts as the highest international judicial authority, by pronouncing judgements and clarifying issues of international judicial concern. Maintains the following programs: United Nations Development Programme; International Maritime Organization; United Nations Centre for Human Settlements; United Nations Conference on Trade and Development; United Nations High Commission for Refugees; United Nations Industrial Development Organization; United Nations Institute for Training and Demographic Research; United Nations International Research and Training Institute for the Advancement of Women; United Nations Relief and Works Agency for Palestinian Refugees in the Near East; Universal Postal Union; World Food Council; World Food Programme; International Fund for Agricultural Development; United Nations Children's Fund (UNICEF); United Nations Environment Programme; International Finance Corporation; United Nations Centre for Science and Technology; United Nations Centre for Transnational Corporations; United Nations Disaster Relief Organization; United Nations Fund for Population Activities; United Nations Institute for Disarmament and Research; United Nations University; World Census Programme. Compiles statistics; collects, analyzes, and publishes data. **PUBLICATIONS:** *Bulletin on Narcotics*, semiannual; *CEPAL Review* (in English and Spanish), 3/year. Contains essays and studies focusing on economic trends and implementation of reforms, industrialization, income distribution, and monetary systems.; *Commodity Trade Statistics*, 28/year; *CTC ReportER*, semiannual. With supplement. Reports on questions about transnational corporations in a governmental and nongovernmental context.; *Current Bibliographic Information* (in English and French), monthly; *IRPTC Bulletin*, semiannual. Contains information on hazardous chemicals.; *Monthly Bibliography, Part I* (in English and French), bimonthly. Subject compilation of newly acquired books, official documents, and periodicals.; *Monthly Bibliography, Part II* (in English and French). Lists selected articles on political, legal, economic, financial, and other questions of the day.; *Monthly Bulletin of Statistics* (in English and French), quarterly. Covers population, food, trade, production, finance, and national income. Includes quarterly data on regions.; *Objective Justice*, semiannual. Contains articles on the crucial implications of apartheid, racial discrimination, and colonialism.; *Permanent Missions to the United Nations*, semiannual; *Population and Vital Statistics Report*, quarterly; *Statistical Indicators of Short-Term Economic Changes in ECE Countries*, monthly; *U.N. Chronicle* (in Arabic, English, French, and Spanish), quarterly. Reports on problems related to food, health, nuclear disarmament, world economy, and other issues. Documents each session of the General Assembly.; *UNDOC: Current Index*, quarterly. UN documents index, with annual cumulative index on microfiche.; *Yearbook* *of the United Nations* (in English and French); Books; Reports. Also issues daily news releases; sells souvenir items. Committees, councils, departments, institutes, and programs also publish materials.

UNITED NATIONS CENTRE FOR HUMAN SETTLEMENTS (UNCHS)

Information, Audio Visual and Documentation Division, United Nations Ave., Gigiri, PO Box 30030, Nairobi, Kenya **PHONE:** 254 2 621234 **FX:** 254 2 624266 **TX:** 22996 UNHAB KE **E-MAIL:** habitat@unchs.org **WEBSITES:** http://www.bestpractices.org; http://www.habitat.org; http://www.habitat.unchs.org/home.htm **FOUNDED:** 1978. **OFFICERS:** Mr. Klaus Toepfes, Acting Exec.Dir. **MEMBERS:** 58. **BUDGET:** US$36,000,000. **LANGUAGES:** Arabic, Chinese, English, French, Russian, Spanish.

DESCRIPTION: Operates under the auspices of the U.N. Commission on Human Settlements. Purposes are to: coordinate, monitor, and assess United Nations projects to improve human settlements, particularly in developing countries; provide for the exchange of information on human settlements; promote, monitor and evaluate the implementation of the Habitat Agenda at the local, national and regional levels and among partners, including local authorities, non-governmental organizations and the private sector. Coordinates U.N. human settlement projects and evaluates their effectiveness. Conducts research on technical, practical, and socioeconomic aspects of constructing settlements and distributes related printed and audiovisual materials; ensures the implementation of urban and rural settlement projects. Conducts training for projects. Compiles statistics. **PUBLICATIONS:** *Habitat Debate* (in English), quarterly; *UNCHS Habitat News* (in Arabic, English, French, and Spanish), 3/year; *UNCHS Shelter Bulletin* (in English and French), 3/year. Also publishes abstracts, audiovisual materials, bibliographies, and technical monographs.

UNITED NATIONS CHILDREN'S FUND (UNICEF)

633 3rd Ave., 23rd Fl., New York, NY 10017 **PHONE:** (212) 824-6275, (800) 252-KIDS **TF:** (800) 553-1200 **TX:** 7607848 **E-MAIL:** netmaster@unicef.org **WEBSITES:** http://www.unicefusa.org **FOUNDED:** 1946. **OFFICERS:** Carol Bellamy, Exec.Dir. **STAFF:** 6,755. **BUDGET:** US$804,000,000. **LANGUAGES:** English, French, Spanish.

DESCRIPTION: Semi-autonomous U.N. agency working for sustainable human development to ensure the survival, protection, and development of children. Cooperates with governments in the developing world to develop and implement low-cost community-based programs in social service, health, nutrition, education, water and sanitation, environment, and women in development. Works for universal ratification and implementation of the Convention on the Rights of the Child and achievement of the objectives and the goals of the 1990 World Summit for Children. Provides universal immunization against six childhood diseases— diphtheria, measles, poliomyelitis, whooping cough, tetanus, and tuberculosis. Promotes the use of oral rehydration therapy to treat diarreheal dehydration, which is one of the leading causes of death in children in developing countries. Works to eliminate poliomyelitis in selected countries and regions, neonatal tetanus, Vitamin A deficiency, iodine deficiency disorders, and guinea worn diseases. In cooperation with the World Health Organization, has launched a ''baby-friendly hospital initiative'' to advance breastfeeding. Provides training for health workers and traditional birth attendants; furnishes technical supplies and equipment for health centers. Advocates with and helps governments in strengthening basic education for all, particularly in increasing schooling opportunities for girls. Supports efforts to improve the status of women by enhancing educational and vocational opportunities and supporting small-scale income generating projects and credit schemes to increase the earning power of women. Assists Urban Basic Services programs to address the health and sanitation needs of the urban poor and supports local safe water supply projects. Seeks to help children in especially difficult circumstances, including children in armed conflicts. Provides relief and rehabilitation assistance in response to the needs of children and women affected by emergencies. Works with 37 national committees, mostly in industrialized countries, providing fund-raising and advocacy support. More than 190 non-governmental organizations (NGOs) maintain consultative relationships with the organization. **PUBLICATIONS:** *Facts and Figures* (in English, French, and Spanish), annual. Lists facts and statistical data on children and women.; *First Call for Children* (in

English, French, and Spanish), quarterly; *The Progress of Nations* (in English, French, and Spanish), annual; *State of the World's Children* (in Arabic, English, French, and Spanish), annual. Offered in 20 other national languages.; *UNICEF Annual Report* (in English, French, and Spanish). Summarizes UNICEF policies and programs.; *UNICEF at a Glance* (in English, French, and Spanish).

UNITED NATIONS CONFERENCE ON TRADE AND DEVELOPMENT (UNCTAD) (Conference des Nations Unies sur le Commerce et le Developpement - CNUCED)

Palais des Nations, CH-1211 Geneva 10, Switzerland **PHONE:** 41 22 9071234 **FX:** 41 22 9070043 **TX:** 412962 **E-MAIL:** ers@unctad.org **FOUNDED:** 1964. **OFFICERS:** Mr. Rubens Ricupero, Sec.Gen. **MEMBERS:** 188. **STAFF:** 450. **BUDGET:** US$55,000,000. **LANGUAGES:** Arabic, Chinese, English, French, Russian, Spanish.

DESCRIPTION: Serves as the focal point within the United Nations for the integrated treatment of development and interrelated issues in the areas of trade, finance, technology, investment, and sustainable development. Facilitates the integration of developing countries into the international trading system. Provides a forum for discussions aimed at considering development strategies and policies in a globalized world economy. Analyzes and exchanges successful developmental experiences. Also examines issues related to: trade preferences and the role of the Generalized System of Preferences (GSP); trade and environment issues from a development perspective; the role of foreign direct investment in the development process and issues relevant to a possible multilateral framework on investment. **PUBLICATIONS:** *Advanced Technology Assessment Systems Bulletin*, periodic; *Commodity Price Bulletin* (in English, French, and Spanish), monthly; *Guide to UNCTAD Publications*, annual; *Handbook of International Trade and Development Statistics* (in English and French), annual; *Least Developed Countries Report*, annual; *Quarterly Bulletin of the UNCTAD Committee on Tungsten* (in English); *Review of Maritime Transport*, annual; *Science and Technology Update*, quarterly; *Trade and Development Report* (in Arabic, English, French, and Spanish), annual; *Transnational Corporations*, 3/year; *UNCTAD Commodity Yearbook*, annual; *World Investment Report*, annual.

UNITED NATIONS DEVELOPMENT FUND FOR WOMEN (UNIFEM)

304 E. 45th St., 6th Fl., New York, NY 10017 **PHONE:** (212) 906-6400, (212) 906-6962 **FX:** (212) 906-6705 **FOUNDED:** 1976. **OFFICERS:** Noeleen Heyzer, Dir. **STAFF:** 18. **BUDGET:** US$13,000,000. **LANGUAGES:** English, French, Spanish.

DESCRIPTION: Autonomous fund operating in association with the United Nations Development Programme and created by the UN General Assembly following the International Women's Year, 1975. The Fund's purpose is to support the efforts of women in the developing world to achieve their objectives for economic and social development and for equality, and by doing so, to improve the quality of life for all. The Fund works in three key program areas of strategic importance to women: Agriculture, Trade and Industry, and Macro Policy and National Planning. Initiatives here are complemented by technical support for credit, technology transfer, small business development, and training. The approach stresses capacity-building, empowerment and collaboration with appropriate partners. The Fund also addresses issues which are on the international agenda, and which critically affect women as beneficiaries and contributors to the development process. **PUBLICATIONS:** *A Guide to Community Revolving Loan Funds* (in English, French, and Spanish); *Development Review* (in English, French, and Spanish), semiannual; *Information Booklets* (in English, French, and Spanish), periodic; *UNIFEM News* (in English, French, and Spanish), quarterly; Papers (in English, French, and Spanish).

UNITED NATIONS DEVELOPMENT PROGRAMME (UNDP)

1 United Nations Plz., New York, NY 10017 **PHONE:** (212) 906-5315 **FX:** (212) 906-5364 **WEBSITES:** http://www.undp.org **FOUNDED:** 1965. **OFFICERS:** James Gustave Speth, Admin. **BUDGET:** US$1,000,000,000. **LANGUAGES:** English, French, Spanish.

DESCRIPTION: Formed as a result of United Nations General Assembly vote to merge two existing development operations, the Expanded Programme of Technical Assistance and the Special Fund. Currently funds 6000 projects in more than 170 developing countries and territories. Through a network of 132 offices in developing countries it cooperates with governments, at their request, and with 40 specialized and technical UN agencies and non-governmental organizations. Goals are to enhance self-reliance and promote sustainable human development. Priority fields are: poverty alleviation and grassroots development; environment and natural resources; management development; technical coopertion among developing countries; transfer and adaptation of technology; women in development. Experts, consultants, equipment and fellowships are provided in agriculture, industry, education, energy, transport, communications, public administration, health, housing, trade, and related areas to help governments make progress in these areas. Operates the Global Environment Facility in cooperation with the World Bank and the United Nations Environment Programme, which provides grants to enable developing countries to address the global dimensions of their environmental problems. **PUBLICATIONS:** *Choices* (in English, French, Japanese, and Spanish), 3/year; *Compendium of Approved Projects*, annual; *Cooperation South*, periodic; *Flash* (in Arabic, English, French, Russian, and Spanish), weekly; *UNDP Annual Report of the Administrator* (in Arabic, English, French, Japanese, and Spanish); *UNDP Human Development Report*, annual.

UNITED NATIONS DIVISION FOR THE ADVANCEMENT OF WOMEN

2 United Nations Plaza, DC2, 12th Fl., New York, NY 10017 **E-MAIL:** daw@un.org **WEBSITES:** http://www.un.org/womenwatch/daw/ **LANGUAGES:** English.

DESCRIPTION: Members are individuals and organizations with an interest in the economic, political, and social status of women. Promotes increased participation by women in society; seeks to protect and expand the legal rights of women. Conducts research; gathers and disseminates information.

UNITED NATIONS ECONOMIC COMMISSION FOR AFRICA (ECA) (Commission Economique pour l'Afrique - CEA)

Africa Hall, PO Box 3001, Addis Ababa, Ethiopia **PHONE:** 251 1 517200, 251 1 180065 **FX:** 251 1 514416 **TX:** 21029 **E-MAIL:** mailroom.uneca@un.org **FOUNDED:** 1958. **OFFICERS:** K.Y. Amoako, Exec.Sec. **MEMBERS:** 56. **STAFF:** 720. **BUDGET:** US$146,720,000. **LANGUAGES:** Arabic, English, French.

DESCRIPTION: A subsidiary of the United Nations. Independent African states. Seeks to facilitate socioeconomic development in Africa and to strengthen economic relations among African nations and with other countries of the world; assists in the formulation of policy favoring African economic and technical growth. Operates field activities through a system known as Multinational Programming and Operational Centres. Studies economic and technical problems and disseminates results; collects and evaluates statistical, economic, and technical information. Holds seminars, workshops, and symposia. Provides training and advisory services. Sponsors research and educational programs. **PUBLICATIONS:** *African Target*, quarterly; *Direction of Trade*, semiannual; *Survey of Economic and Social Conditions*, periodic; *Trade By Commodity*, semiannual; Report, annual.

UNITED NATIONS ECONOMIC COMMISSION FOR EUROPE (UN/ECE) (Commission Economique pour L'Europe - CEE-ONU)

Palais des Nations, CH-1211 Geneva 10, Switzerland **FX:** 41 22 9170505 **E-MAIL:** info.ece@unece.org **WEBSITES:** http://www.unece.org **FOUNDED:** 1947. **OFFICERS:** Yves Berthelot, Exec.Sec. **MEMBERS:** 55. **STAFF:** 200. **BUDGET:** US$49,310,800. **LANGUAGES:** English, French, Russian.

DESCRIPTION: A regional commission of the United Nations. Assists countries in transition to market economies; formulates interdisciplinary approaches to economic cooperation in industry; protects the environment; and solves transportation and trade problems. Compiles statistics; provides economic analysis. **PUBLICATIONS:** *Annual Report* (in English, French, and Russian); *ECE Highlights*, quarterly; *Economic Bulletin for Europe*, annual; *Economic Survey of Europe*, annual; *Statistical Journal*, quarterly. Additional publication information available upon request.

UNITED NATIONS ECONOMIC AND SOCIAL COMMISSION FOR ASIA AND THE PACIFIC (ESCAP)

United Nations Bldg., Rajadamnern Ave., Bangkok 10200, Thailand **PHONE:** 66 2 881234 **FX:** 66 2 881000 **TX:** 82315 ESCAPTH **E-MAIL:** mooy.unescap@un.org **WEBSITES:** http://www.unescap.org **FOUNDED:** 1947. **OFFICERS:** Adrianus Mooy, Exec.Sec. **MEMBERS:** 51. **STAFF:** 760. **BUDGET:** US$62,875,000. **LANGUAGES:** Chinese, English, French, Russian.

DESCRIPTION: Serves as a regional intergovernmental forum of the United Nations to help member countries in their economic and social development. Conducts research projects and provides technical assistance and training programs for member countries. **PUBLICATIONS:** *Asia Pacific Development Journal*, periodic; *Asia-Pacific Population Journal*, quarterly; *Atlas of Mineral Resources of the ESCAP Region - Monographic Series*, periodic; *Economic and Social Survey of Asia and the Pacific*, annual; *Energy Resources Development Series*, periodic; *Foreign Trade Statistics of Asia and Pacific*, annual; *Quarterly Bulletin of Statistics for Asia and the Pacific*; *Small Industry Bulletin for Asia and the Pacific*, periodic; *Statistical Yearbook for Asia and the Pacific*, annual; *Studies in Trade and Investment* (in English), quarterly; *TISNET Trade and Investment Bulletin*, 23/year; *Water Resources Journal*, quarterly.

UNITED NATIONS EDUCATIONAL, SCIENTIFIC AND CULTURAL ORGANIZATION (UNESCO) (Organisation des Nations Unies pour l'Education, la Science et la Culture)

7, place de Fontenoy, F-75352 Paris, France **PHONE:** 33 1 45681000 **FX:** 33 1 45671690 **TX:** 204461 PARIS **WEBSITES:** http://www.unesco.org **FOUNDED:** 1946. **OFFICERS:** Federico Mayor, Dir.Gen. **MEMBERS:** 186. **STAFF:** 3,086. **BUDGET:** US$455,000,000. **LANGUAGES:** Arabic, Chinese, English, French, Russian, Spanish.

DESCRIPTION: Advocates collaboration among nations in the areas of education, science, and culture as a means of realizing peace, prosperity, and a world in which justice, law, and human rights are respected. Strives to protect and preserve mankind's cultural heritage and to instill in member nations a sense of cultural and historic identity. Promotes equal educational opportunity for all by advising members on viable educational methods and policies and by providing structures for international cooperation. Seeks to further collaboration among members in all fields of intellectual endeavor and to maximize the accessibility of information. Works toward the abolition of iliteracy and the worldwide availability of elementary educations. Promotes culture of Paris. Strives to facilitate developing nations' access to advanced technology and encourages cultural interaction; coordinates the activities of the World Decade of Cultural Development. Attempts to define the role of education, science, culture, and communication in social development. Encourages nations to participate in international cultural and scientific programs. Sponsors global programs to further education; organizes international research programs in the field of science; offers advisory services to members; conducts programs to study and preserve cultural traditions and monuments of archaeological importance. Conducts literacy campaigns; offers teacher training, education for refugees, and instruction for jounalists. Advocates the establishment of national news agencies and social science institutions. Sponsors the Intergovernmental Oceanographic Commission, the International Geological Correlation Programme, the International Hydrological Programme, the Man and the Biosphere Programme, the Intergovernmental Programme in Informatics, and the International Programme for Development of Communication. Compiles statistics. **PUBLICATIONS:** *Copyright Bulletin* (in English, French, and Spanish), quarterly; *Museum* (in English, French, Russian, and Spanish), quarterly; *Nature and Resources* (in English, French, Russian, and Spanish), quarterly; *Prospects* (in Arabic, English, French, and Spanish), quarterly; *Sources* (in English and French), periodic; *UNESCO Courier* (in Arabic, Basque, Bulgarian, Catalan, Chinese, Czech, Dutch, English, Finnish, French, German, and Greek), monthly. Also publishes reference books, bibliographies, manuals, statistical reports, and monographs on science, social science, education, culture, and communication.

UNITED NATIONS ENVIRONMENT PROGRAMME (UNEP) (Programme des Nations Unies pour l'Environnement - PNUE)

PO Box 30552, Nairobi, Kenya **PHONE:** 254 2 621234, 254 2 623292 **FX:** 254 2 623927 **TX:**

22068 KNEPKE **WEBSITES:** http://www.unep.org **FOUNDED:** 1972. **OFFICERS:** Ms. Elizabeth Dowdeswell. **MEMBERS:** 185. **STAFF:** 658. **BUDGET:** 9,900,000 KSh. **LANGUAGES:** Arabic, Chinese, English, French, Russian, Spanish.

DESCRIPTION: Representatives of 185 member nations of the United Nations. Advocates the protection of natural resources and encourages the integration of environmental considerations in development activities. Works in conjunction with U.N. agencies including the World Health Organization and the Food and Agriculture Organization of the United Nations. Monitors changes in the environment through Earthwatch, a worldwide surveillance system that collects data, offers information, studies threats to the environment such as chemical pollution, and warns against impending environmental crises. Works with national governments on such trans-boundary problems as ozone depletion, climatic change and global warming, desertification, species migration, oil spill management, acid rain and hazardous wastes, and biological diversity; sponsors Oceans and Coastal Areas Program for nations sharing common coastal waters. Sponsors conferences; offers guidelines on the environmental considerations important to different industries. Sponsors training programs; conducts workshops. **PUBLICATIONS:** Also publishes regional newsletters, books, and materials on environmental and natural resources issues.

UNITED NATIONS ENVIRONMENT PROGRAMME - REGIONAL OFFICE FOR ASIA AND THE PACIFIC (Programme des Nations Unies pour L'Environnement - PNUE)

The United Nations Bldg., Rajadamnern Ave., Bangkok 10200, Thailand **PHONE:** 66 2 2829161, 66 2 28299381 **FX:** 66 2 2803829 **TX:** 82392 BANGKOK **E-MAIL:** listproc@pan.cedar.univie.ac.at **FOUNDED:** 1975. **OFFICERS:** Dr. Suvit Yodmani, Dir./Reg.Rep. **STAFF:** 24. **BUDGET:** US$2,000,000. **LANGUAGES:** English, Thai.

DESCRIPTION: Advocates the protection of natural resources. Raises global awareness of environmental issues by providing technical and public information to mobilize nations and people to take environmental action. Monitors changes in the environment; disseminates information. **PUBLICATIONS:** *Environment Asia-Pacific* (in English), quarterly; *Environment is Your Business Too*; *NETTLAP* (in English), quarterly; *UNEP/ROAP Profile: Toward Environmental Citizenship*; Catalog.

UNITED NATIONS HIGH COMMISSIONER FOR REFUGEES (UNHCR) (Haut Commissariat des Nations Unies pour les Refugies - HCR)

c/o Public Information Dept., 94 rue Ruppard, Momtbrillant, CH-1202 Geneva GE, Switzerland **PHONE:** 41 22 7398111 **FX:** 41 22 7397377 **FOUNDED:** 1951. **LANGUAGES:** Arabic, Chinese, English, French, Spanish.

DESCRIPTION: Subsidiary of the United Nations General Assembly. Representatives of 43 countries. Promotes and defends the rights of refugees and offers them international protection. Seeks lasting solutions to refugee problems through voluntary repatriation, local integration in the country providing first asylum, or resettlement in other countries. Supplies material and legal aid to refugees. Seeks the establishment of international standards for the treatment of refugees in situations involving employment, education, residence, and freedom of movement. Operates emergency relief and material assistance program. Maintains Centre for Documentation on Refugees. **PUBLICATIONS:** *Magazine Refugees* (in English, French, and Spanish), monthly. Includes periodical supplements in Arabic, German, and Italian; *Refugee Abstracts* (in English, French, and Spanish), quarterly.

UNITED NATIONS INDUSTRIAL DEVELOPMENT ORGANIZATION (UNIDO) (Organisation des Nations Unies pour le Developpement Industriel - ONUDI)

Vienna Intl. Centre, Postfach 300, A-1400 Vienna, Austria **PHONE:** 43 1 260260 **FX:** 43 1 2692669 **TX:** 135612 UNO A **E-MAIL:** unido-pinfo@unido.org **WEBSITES:** http://www.unido.org; http://www.unido.org/plweb-cgi/ida_ixacctpl **FOUNDED:** 1966. **OFFICERS:** Mr. Carlos Alfredo Magarinos, Dir. **MEMBERS:** 168. **STAFF:** 730. **BUDGET:** US$129,500,000. **LANGUAGES:** Arabic, Chinese, English, French, Russian, Spanish.

DESCRIPTION: Specialized agency of the United Nations dedicated to the promotion of sustainable industrial development in countries with developing and transition economies by harnessing the joint forces of governments and the private sector to foster competitive industrial production, developing international industrial partnerships, and promoting socially equitable and environmentally friendly industrial development. Main priority areas are; strategies, policies and institution-building for global economic integration; environment and energy; small- and medium-scale enterprises; policies, networking and basic technical support; innovation, productivity and quality for international competitiveness; industrial information, investment and technology promotion; rural industrial development; Africa and the least developed countries; linking industry and agriculture. Offers feasibility studies; planning assistance; installation of pilot plants; provision of industrial facilities for research and development; dissemination of industrial, technological and marketing information; industrial training and in-plant group training programs; study tours and seminars; compiles statistics. **PUBLICATIONS:** *Annual Report* (in English, French, and Spanish); *Country Industrial Development Reviews* (in English), periodic; *Global Report* (in English and French), annual; *Industry Africa* (in English and French), annual; *International Yearbook of Industrial Statistics*, annual.

UNITED NATIONS OFFICE FOR DRUG CONTROL AND CRIME PREVENTION (UNODCCP)

Vienna International Centre, POB 500, 1400 Vienna, Austria **PHONE:** 1 213450 **FX:** 1 21345-5866 **TX:** 135612 **E-MAIL:** undcp-hq@undcp.un.or.at **WEBSITE:** http://undcp.org **FOUNDED:** 1998. **OFFICERS:** Pino Arlachi, Exec. Dir.

DESCRIPTION: The UNODCCP is the primary drug control and law enforcement agency of the United Nations (UN). Its declared purpose is "to address the interrelated issues of drug control, crime prevention and international terrorism in all its forms." The UNODCCP coordinates all of the crime and drug control efforts of the UN and works with national and international crime and drug control organizations. It was formed in 1998 by the union of the United Nations International Drug Control Programme and the United Nations Centre for International Crime Prevention. **PUBLICATIONS:** *Bulletin on Narcotics*, quarterly; Reports, periodic.

UNITED NATIONS POPULATION FUND (UNFPA)

220 E. 42nd Street, New York, NY, 10017, USA, **PHONE:** (212) 297-5020 **FX:** (212) 557-6416 **WEBSITE:** http://www.unfpa.org **FOUNDED:** 1967. **OFFICERS:** Dr. Nadis Safik, Exec. Dir. **BUDGET:** US$305,000,000.

DESCRIPTION: The UNFPA is the body of the United Nations (UN) charged with the study of population and related issues, and with promoting awareness of population problems. It is also mandated to assist nations that request its aid to cope with their population pressures and family planning needs. **PUBLICATIONS:** *Annual Report*; *AIDS Update*, annual; *Populi* (English, French, Spanish), 10/year; *The State of World Population*, annual.

UNITED NATIONS RELIEF AND WORKS AGENCY FOR PALESTINE REFUGEES IN THE NEAREAST (UNRWA) (Office de Secours et de Travaux des Nations Unies)

Vienna International Centre, Public Information Office, PO Box 700, A-1400 Vienna, Austria **PHONE:** 972 7 6777333, 972 7 6777526 **FX:** 972 7 6777526 **TX:** 135310 UNRA A **E-MAIL:** unrwapio@unrwa.org **WEBSITES:** http://www.unrwa.org **FOUNDED:** 1949. **OFFICERS:** Peter Hansen, Commissioner Gen. **STAFF:** 21,302. **BUDGET:** US$666,600,000. **LANGUAGES:** Arabic, English, French.

DESCRIPTION: Purpose is to provide education, health, relief and social services to 3.6 million registered Palestinian refugees living in five fields of its area of operations i.e. Jordan, Lebanon, Syria, and the West Bank and Gaza Strip. Operates elementary and preparatory schools, vocational and teacher training centers, and health centers. Provides special assistance programs for refugee families living in hardship, and supports community rehabilitation, youth activities and women's program centers. **PUBLICATIONS:** *Annual Report* (in Arabic, Chinese, English, French, Russian, and Spanish); *Palestine Refugees Today* (in Arabic and English), semiannual; *UNRWA News* (in Arabic and English), monthly; Booklets; Videos.

UNITED NATIONS RESEARCH INSTITUTE FOR SOCIAL DEVELOPMENT (UNRISD) (Institut de Recherche des Nations Unies pour le Developpement Social - UNRISD)

Palais des Nations, CH-1211 Geneva 10, Switzerland **PHONE:** 41 22 9173020 **FX:** 41 22 9170650 **TX:** 412962 UNO CH **E-MAIL:** info@unrisd.org **WEBSITES:** http://www.unrisd.org **FOUNDED:** 1963. **OFFICERS:** Thandika Mkandawire, Dir. **STAFF:** 40. **BUDGET:** US$5,000,000. **LANGUAGES:** English, French, Spanish.

DESCRIPTION: Believes that for effective development policies to be formulated an understanding of the social and political context is crucial, as is an accurate assessment of how such policies affect different social groups. Examines the causes of poverty worldwide and the factors perpetuating it; seeks to clarify contributing factors to determine appropriate courses of action. Conducts in-depth research, primarily in developing countries. **PUBLICATIONS:** *UNRISD News* (in English, French, and Spanish), biennial; Books (in English, French, and Spanish); Monographs (in English, French, and Spanish); Papers, periodic.

UNITED NATIONS VOLUNTEERS (UNV) (Voluntarios de las Naciones Unidas)

Palais des Nations, CH-1211 Geneva 10, Switzerland **PHONE:** 49 228 8152000 **FX:** 49 228 8152001 **E-MAIL:** hq@unv.org **FOUNDED:** 1970. **OFFICERS:** Bill Jackson. **STAFF:** 120. **BUDGET:** 6,500,000 SFr. **LANGUAGES:** English, French, Spanish.

DESCRIPTION: Participants are adult volunteer-specialists from 126 countries working for economic and social progress in 134 developing nations. Host countries request and approve candidates who work within a community for 2 years as middle- and upper-level specialists in fields including agriculture, forestry, education, and economic development planning and administration. Works with other volunteer organizations such as the Peace Corps, British Voluntary Service Overseas, and World University Service of Canada in identifying and funding qualified volunteers. Today, volunteer specialists are also serving in humanitarian projects, which may be of shorter duration. **PUBLICATIONS:** *International Volunteer Day News* (in English and French), annual; *Serving Abroad with UNV: Vacant Posts* (in English, French, and Spanish), quarterly; *UNV at a Glance*, annual. Statistics; *UNV News* (in English, French, and Spanish), quarterly; *UNV Spectrum*, semiannual. Roster update. Also publishes booklets, flyers, monographs, and reports.

UNITED PRESS INTERNATIONAL (UPI)

1510 H. St., Ste. 600, Washington, DC 20005 **PHONE:** (202) 898-8000 **TF:** (800) 796-4874 **FX:** (202) 898-8057 **FOUNDED:** 1958. **OFFICERS:** Paul Heigel, CEO. **LANGUAGES:** English.

DESCRIPTION: Press association wire service that gathers news and photographs of current events to distribute to newspapers, periodicals, cable systems, and radio and television stations throughout the world; maintains 204 local news bureaus in 79 countries. Provides speakers.

UNITED TOWNS ORGANISATION (UTO) (Federation Mondiale des Cites Unies et Villes Jumelees - FMCU-VJ)

22, rue d'Alsace, F-92522 Levallois-Perret Cedex, France **PHONE:** 33 1 47393686 **FX:** 33 1 47393685 **FOUNDED:** 1957. **OFFICERS:** Michel Bescond, Dir.Gen. **MEMBERS:** 1,800. **STAFF:** 9. **LANGUAGES:** Arabic, English, French, German, Italian, Portuguese, Russian, Spanish.

DESCRIPTION: Members are towns, local and regional authorities, and various organizations. Purpose is to improve international understanding through the development of cultural, economic, and technical exchanges through town twinning and other forms of international cooperation. Programs include: integration of immigrant populations and the disabled; access of women to public office; environment; peace; youth; sanitation; transport. Compiles statistics. **PUBLICATIONS:** Publishes research booklets.

UNITY-AND-DIVERSITY WORLD COUNCIL (UDC)

5521 Grosvenor Blvd., Los Angeles, CA 90066-6915 **PHONE:** (310) 577-1968, (310) 390-2192 **FX:** (310) 578-1028 **E-MAIL:** udcworld@gte.net **WEBSITES:** http://www.globalvisions.org/cl/udcworld **FOUNDED:** 1965. **OFFICERS:** Leland P. Stewart, Founder/Central Coord. **MEMBERS:** 200.

STAFF: 3. **BUDGET:** US$40,000. **LANGUAGES:** English.

DESCRIPTION: Worldwide coordinating body for cultural, scientific, educational, and religious nonprofit organizations and related businesses, as well as individuals. Fosters the emergence of a new universal person and civilization based on unity-and-diversity among all peoples and all life. Seeks to aid in establishing a new, worldwide civilization based upon the reality of the whole person by applying the methods and discoveries of modern science coupled with the insights of religion, philosophy, and the arts. Interfaith Celebration. **PUBLICATIONS:** *General Assembly Bulletin*, monthly; *Interfaith Celebration*, monthly; *Spectrum Magazine*, quarterly; *Spectrum Update*, quarterly; *Unity and Diversity Celebration*, weekly. Announces weekly celebrations and other meetings; *Unity-and-Diversity Spiritual Celebration Guide*; *Worldview Exploration Publications*.

UNIVERSAL ALLIANCE OF DIAMOND WORKERS (UADW) (Alliance Universelle des Ouvriers Diamantaires - AUDOD)

c/o Jef Hoymans, Lange Kievitstraat 57/3, B-2018 Antwerp, Belgium **PHONE:** 32 3 2321557 **FX:** 32 3 2264009 **E-MAIL:** uadw@planetinternet.be **FOUNDED:** 1905. **OFFICERS:** Jef Hoymans, Gen.Sec. **MEMBERS:** 300,000. **STAFF:** 3. **LANGUAGES:** Dutch, English, French, German.

DESCRIPTION: International diamond workers' union concerned with improving labor conditions.

UNIVERSAL ESPERANTO ASSOCIATION (UEA) (Universala Esperanto-Asocio - UEA)

Nieuwe Binnenweg 176, NL-3015 BJ Rotterdam, Netherlands **PHONE:** 31 10 4361044 **FX:** 31 10 4361751 **E-MAIL:** uea@inter.nl.net **WEBSITES:** http://www.uea.org **FOUNDED:** 1908. **OFFICERS:** Mr. Osmo Buller, Dir.-Gen. **MEMBERS:** 26,000. **STAFF:** 12. **BUDGET:** 1,000,000 f. **LANGUAGES:** Esperanto.

DESCRIPTION: Members are individuals from 120 countries who use and support the international language Esperanto; national and other organizations of Esperantists worldwide. Objectives are to: promote and spread the use of Esperanto throughout the world; facilitate international communication by transcending linguistic boundaries; further good international relations and the respect for human rights, in accordance with the Universal Declaration of Human Rights; stimulate cooperation in all areas of human endeavor and foster friendly relations among all people; cultivate a feeling of universal human solidarity among members; encourage members to develop an understanding and appreciative attitude toward all cultures and traditions. Coordinates the activities of national Esperanto associations, including efforts to increase the number of Esperanto courses taught via radio and television and in schools and universities. Engages in publicity work on an international scale in an effort to acquaint non-Esperantists with the Esperanto movement; supports cultural events dedicated to the cause of Esperanto. Sponsors competitions; compiles statistics; conducts research and maintains documentation on language problems and language planning throughout the world. Maintains operational status with the United Nations Educational, Scientific and Cultural Organization, special consultative status with the Economic and Social Council of the United Nations, and consultative status with the United Nations Children's Fund, the Council of Europe, and the Organization of American States. **PUBLICATIONS:** *Esperanto* (in Esperanto), monthly. Includes information on the promotion and use of Esperanto, includes running bibliography.; *Esperanto Documents* (in English, Esperanto, and French), periodic; *Jarlibro* (in Esperanto); *Libroservo de UEA* (in Esperanto), periodic. Contains a publications catalog.; Audiotapes; Bibliographies; Books; Brochures.

UNIVERSAL POSTAL UNION (UPU) (Union Postale Universelle)

Case Postale, CH-3000 Bern 15, Switzerland **PHONE:** 41 31 3503111 **FX:** 41 31 3503110 **TX:** 912761 UPU CH **E-MAIL:** ib.info@ib.upu.org **WEBSITES:** http://www.upu.int **FOUNDED:** 1874. **OFFICERS:** T.E. Leavey, Dir.Gen. **MEMBERS:** 189. **STAFF:** 167. **BUDGET:** 35,700,000 SFr. **LANGUAGES:** Arabic, Chinese, English, French, German, Portuguese, Russian, Spanish.

DESCRIPTION: A specialized agency of the United Nations. Countries united for the exchange of international postal services, the organization and improvement of the postal services, and the development of international collaboration in the postal sphere. Conducts studies and gives opinions on technical operation and economic questions; sponsors seminars. Disseminates information; compiles statistics. **PUBLICATIONS:** *Directory of Chiefs of Postal Administrations* (in English, French, and Spanish), periodic. Includes monthly update.; *Post*Code 1997*. Contains alphabetical list of 500,000 localities in 189 countries.; *Union Postale* (in Arabic, Chinese, English, French, German, Russian, and Spanish), quarterly. Contains inserts in: French, English with Arabic, Chinese, German, Russian and Spanish.

UNREPRESENTED NATIONS AND PEOPLES ORGANIZATION (UNPO)

444 N. Capital St., Ste. 846, Washington, DC 20001-1570 **PHONE:** (202) 637-0475 **FX:** (202) 637-0585 **E-MAIL:** unposf@igc.apc.org **WEBSITES:** http://www.unpo.org **FOUNDED:** 1991. **OFFICERS:** Karen E. Onthank, Exec.Dir. **LANGUAGES:** English.

DESCRIPTION: Seeks to provide a voice to those nations not represented in the United Nations and other established international forums. Upholds the principles of self-determination, democracy, nonviolence, and environmental protection. Works to protect citizens from the human rights abuses of government. Conducts fact-finding missions and election monitoring. Provides education and training in diplomacy, conflict management, and democratic processes. **PUBLICATIONS:** *A UNPO Report on the National and Human Rights Situation of the Albanians in Kosova*; *Human Rights Dimensions of Population Transfer*. Conference report on UNPO International Conference held in Tallinn, Estonia.; *The Position of UNPO in the International Legal Order*; *The Reemergence of Self-Determination*; *Report of a UNPO Mission to Abkhazia, Georgia, and the Northern Caucasus*; *Self-Determination in Relation to Individual Human Rights, Democracy, and the Protection of the Envrionment*. UNPO Conference Report; *Summary Report of the UNPOI Mission to Kosova to Monitor Parliamentary and Presidential Elections*; *UNPO Members Fact Book*. Contains geographic data, and information on religion, human rights violations, language, and political systems; *UNPO News*, 3/year.

VOLUNTEER OPTOMETRIC SERVICES TO HUMANITY/INTERNATIONAL (VOSH)

132 S. First Ave., Iowa City, IA 52245 **PHONE:** (319) 338-3267, (319) 338-4370 **FX:** (319) 338-2499 **E-MAIL:** vosher@aol.com **FOUNDED:** 1973. **OFFICERS:** Phillip E. Hottel O.D., Pres. **MEMBERS:** 900. **LANGUAGES:** English.

DESCRIPTION: Members are optometrists, optometric students, opticians, and other interested individuals. Conducts missions to disadvantaged countries, mostly in the Caribbean, Central and South America, Africa, Asia, and Eastern Europe to provide free visual care to the needy. Members pay their own travel, food, and lodging expenses, establish temporary clinics, and examine and distribute eyeglasses to an average of 2500 patients during each one-week mission. (Glasses are donated by civic clubs, churches, private citizens, and optical companies.) **PUBLICATIONS:** Brochure; Newsletter, quarterly.

WAR RESISTERS' INTERNATIONAL (WRI) (Internationale des Resistants a la Guerre - IRG)

5 Caledonian Rd., London N1 9DX, England **PHONE:** 44 171 2784040 **FX:** 44 171 2780444 **E-MAIL:** warresisters@gn.apc.org **FOUNDED:** 1921. **OFFICERS:** Dominique Saillard, Coord. **MEMBERS:** 150,000. **STAFF:** 2. **BUDGET:** £60,000. **LANGUAGES:** English, French, German, Spanish.

DESCRIPTION: Pacifist organization of individuals and movements participating in war resistance and peace activities in 35 countries. Disseminates information on training for nonviolent action and the problems of violent and nonviolent revolutionary change. Acts as information and coordinating center for conscientious objectors. **PUBLICATIONS:** *Broken Rifle* (in English, French, German, and Spanish), 5/year; *Peace News* (in English), 11/year. Incorporates former *WRI Newsletter*.; *WRI Women* (in English), 3/year; Pamphlets, periodic.

WAR ON WANT (WOW)

Fenner Brockway House, 37-39 Great Guildford St., London SE1 0ES, England **PHONE:** 44 171 6201111 **FX:** 44 171 2619291 **E-MAIL:** wow@gn.apc.org **FOUNDED:** 1952. **OFFICERS:** Catherine Matheson, Exec.Dir. **MEMBERS:** 3,500.

STAFF: 10. **BUDGET:** £1,800,000. **LANGUAGES:** English, French, Spanish.

DESCRIPTION: Members are individuals and groups in the United Kingdom. Campaigns against the root causes of world poverty; provides funding for projects in developing countries. **PUBLICATIONS:** *Annual Report*; *Upfront* (in English), quarterly.

WEST AFRICA COMMITTEE (WAC) (Comite pour l'Afrique Occidentale)

Hope House, 45 Great Peter St., London SW1P 3LT, England **PHONE:** 44 171 2221077 **FX:** 44 171 2221079 **FOUNDED:** 1956. **OFFICERS:** Brigadier G.G. Blakely, Advisor. **MEMBERS:** 180. **STAFF:** 5. **LANGUAGES:** English.

DESCRIPTION: Members are firms and companies located outside West Africa with substantial commercial interests in the area's countries. Seeks to aid and stimulate the economic development of West African countries. Represents overseas private sector operators considering operations in West Africa that are mutually beneficial to the operator and country involved. Represents members' interests before governmental bodies. Provides information to members on economic and political developments in West African countries. **PUBLICATIONS:** *Country Reports* (in English), periodic; *Nigerian Bulletin* (in English), monthly; Newsletter (in English), monthly.

WEST AFRICA RICE DEVELOPMENT ASSOCIATION (WARDA) (Association pour le Developpement de la Riziculture en Afrique de l'Ouest - ADRAO)

c/o Dr. Kanayo F. Nwanze, BP 2551, Bouake 01, Cote d'Ivoire **PHONE:** 225 634514 **FX:** 225 634714 **TX:** 69138 ADRAO CI **E-MAIL:** warda@cgnet.com **FOUNDED:** 1971. **OFFICERS:** Dr. Kanayo Nwanze, Dir.Gen. **MEMBERS:** 17. **STAFF:** 347. **BUDGET:** US$11,900,000. **LANGUAGES:** English, French.

DESCRIPTION: Autonomous intergovernmental research association comprised of 17 member states in West Africa. Belongs to the Consultative Group on International Agricultural Research (CGIAR), a network of 16 international research centers supported by 50 public and private sector donors. Works in partnership with the national agriculture research systems of its member states, with advanced research institutes, NGOs, the private sector, and with its target group: the resource poor smallholder farm families of West Africa. Works to strengthen sub-Saharan Africa's capability for technology generation, technology transfer, and policy formulation regarding rice-based crooping systems. Aims to increase the sustainable productivity while conserving the natural resource base of these rice systems, and to contribute to the food security of poor rural and urban households. **PUBLICATIONS:** *Annual Report* (in English and French); *Current Contents at WARDA* (in English and French), monthly; *Trainerlink* (in English and French), periodic; *WARDA News* (in English and French), monthly. Also publishes proceedings, training manuals, and occasional papers.

WEST AFRICAN BANKERS' ASSOCIATION

22 Wilberforce St., Private Mail Bag 1012, Freetown, Sierra Leone **FX:** 232 22 229024 **LANGUAGES:** English.

DESCRIPTION: Members are banks and other financial institutions. Promotes growth and development of banking and financial capabilities in western Africa. Facilitates communication and cooperation among members; represents members' interests.

WEST AFRICAN CLEARINGHOUSE (WACH) (Chambre de Compensation de l'Afrique de l'Ouest - CCAO)

Kissy House, 54 Siaka Stevens St., Private Mail Bag 218, Freetown, Sierra Leone **PHONE:** 232 22 224485, 232 22 224486 **FX:** 232 22 026501 **TX:** 3368 WACH SL **FOUNDED:** 1975. **OFFICERS:** E.O. Akinnifesi, Exec.Sec. **MEMBERS:** 9. **STAFF:** 32. **BUDGET:** US$500,000. **LANGUAGES:** English, French.

DESCRIPTION: Members are central banks serving 15 West African countries. Aims to help members save their foreign reserves through promoting the use of domestic currencies in subregional trade; advocates liberalized trade policies and monetary cooperation. Encourages consultation among members. Offers seminars for commercial bankers and economists. **PUBLICATIONS:** *Annual Report* (in English and French). Also publishes brochures.

WEST AFRICAN DEVELOPMENT BANK (Banque Ouest Africaine de Developpement - BOAD)

PO Box 1172, Lome, Togo **PHONE:** 228 214244, 228 215906 **FX:** 228 215267 **TX:** 5289 BOAD TG **FOUNDED:** 1973. **OFFICERS:** Abou Bakar Baba-Moussa, Pres. **MEMBERS:** 13. **STAFF:** 165. **BUDGET:** 3,530,700 Fr CFA. **LANGUAGES:** English, French.

DESCRIPTION: Representatives of Benin, Burkina Faso, Cote d'Ivoire, Mali, Niger, Senegal, Togo, and the Central Bank of West African States; associate members are representatives of the European Economic Community, France, and Germany. Aims to promote the balanced economic development of member states and to provide for the economic integration of West Africa; emphasis is placed on facilitating the most efficient means of development of disadvantaged states while fostering a unified regional economy. Provides financial assistance to members. Cooperates with similar regional and international organizations. Provides professional training including a recruitment and training program, specialized training, and personnel retraining. Maintains 16 divisions and 5 departments. **PUBLICATIONS:** *BOAD INFO* (in French), 3/year. Contains information on international economic problems.; *Rapport d'Activite* (in English and French), annual. Also publishes project reports and manuals.

WEST AFRICAN DISCUSSION AGREEMENT (WADA)

180 Minerva Dr., Colonial Heights, NY 10710-2107 **PHONE:** (914) 779-1622 **FX:** (914) 779-3841 **FOUNDED:** 1945. **OFFICERS:** Harold J. Moran, Exec.Admin. **MEMBERS:** 7. **STAFF:** 1. **BUDGET:** US$20,000. **LANGUAGES:** English.

DESCRIPTION: Members are steamship companies flying the flags of 6 countries and serving West African ports, U.S. North Atlantic and Gulf. Promotes trade and freight rates, charges and practices relating to the transportation of cargo by sea to and from the U.S. and 18 nations of West Africa. The Discussion Agreement is unusual among the agreements approved by the Federal Maritime Commission in that its scope is both eastbound and westbound, serving U.S. and West African ports. **PUBLICATIONS:** Publishes ocean freight rate tariffs.

WEST AFRICAN RICE DEVELOPMENT ASSOCIATION (WARDA) (Association pour le'Developpement de la Riziculture en Afrique de l'ouest - ADRAO)

01 BP 2551, Bouake 01, Cote d'Ivoire **PHONE:** 225 634515 **FX:** 225 634714 **E-MAIL:** warda@cgnet.com **FOUNDED:** 1971. **OFFICERS:** Dr. Kanayo F. Nwanze, Dir.Gen. **MEMBERS:** 17. **STAFF:** 360. **BUDGET:** L$9,800,000. **LANGUAGES:** English, French.

DESCRIPTION: Members are individual member states (17). Promotes the agricultural development of rice crops and yields throughout Western Africa. Objectives include: increasing the quality and quantity of rice produced in West Africa; encouraging production and use of varieties suited to conditions of the member countries; promoting storage processing as well as marketing; and contributing to the extension of rational production methods. Research and activities focus on establishing coordinated plant introduction and protection trials, phyto-sanitary services, and integrated projects for irrigated rice. Maintains a development department which researches economic and sociological factors associated with rice production. **PUBLICATIONS:** Annual Report; Brochure.

WESTERN EUROPEAN UNION (WEU) (Union de l'Europe Occidentale - UEO)

4, rue de la Regence, B-1000 Brussels, Belgium **PHONE:** 32 2 5004411, 32 2 5004455 **FX:** 32 2 5113915 **E-MAIL:** ueo.presse@skynet.be **FOUNDED:** 1955. **OFFICERS:** Myriam Sochacki, Chief. **MEMBERS:** 10. **LANGUAGES:** English, French.

DESCRIPTION: Full members are Belgium, France, Germany, Greece, Italy, Luxembourg, Netherlands, Portugal, Spain, and the United Kingdom. Associate members are Iceland, Norway and Turkey. Observers are Austria, Denmark, Finland, Ireland, and Sweden. Associate Partners are Bulgaria, the Czech Republic, Hungary, Poland, Romania, Slovakia, Estonia, Latvia, Lithuania and Slovenia. Based on the Brussels Treaty of 1948, modified in 1954. ''Plays a pivotal role'' between the European Union (EU) and the North Atlantic Treaty Organization (NATO) in the development of a European Security and Defence Identity (ESDI). Purpose is to conduct European military operations in the humanitarian, peacekeeping, and crisis management fields. Currently assists in the Albanian police reorganization. **PUBLICATIONS:** *Annual Report of the Council to the Assembly* (in English and French); *Chaillot Papers*; *Proceedings of the Assembly of WEU*, semiannual.

WINGS OF HOPE (WOH)

18590 Edison Ave., Chesterfield, MO 63005 **PHONE:** (314) 537-1302 **FX:** (314) 537-3139 **E-MAIL:** woh206@earthlink.net **WEBSITES:** http://www.wings-of-hope.org **FOUNDED:** 1962. **OFFICERS:** Donald Malvern, Chairman. **MEMBERS:** 1,365. **STAFF:** 2. **BUDGET:** US$715,000. **LANGUAGES:** English.

DESCRIPTION: Members are humanitarian aviation and radio communication service helping to bring equipment, training, and the services of hospitals, educators, missionaries, health, world-aid, and development groups to people in need in remote areas of the world. Enables volunteer professional pilots to fly in the bush; offers bush pilot training. Provides speakers' bureau. **PUBLICATIONS:** *Wings of Hope*, quarterly. Reports on organization field operations.

WOMANKIND WORLDWIDE (WW)

3/4 Albion Pl., Galena Rd., London W6 0LT, England **PHONE:** 44 181 5638607, 44 181 5638608 **FX:** 44 181 5638611 **E-MAIL:** info@womankind.org.uk **WEBSITES:** http://www.oneworld.org/womankind **FOUNDED:** 1989. **OFFICERS:** Maggie Baxter, Exec.Dir. **STAFF:** 10. **BUDGET:** £1,000,000. **LANGUAGES:** English, French, Italian, Spanish.

DESCRIPTION: Promotes, supports, and funds women's initiatives in developing countries. Seeks to create a more peaceful society through the equal participation of women in determining values, directions and governance of society at all levels. **PUBLICATIONS:** *Annual Review*; Newsletter (in English), semiannual.

WOMEN IN DEVELOPMENT EUROPE (WIDE)

Rue Du Commerce 70, B-1040 Brussels, Belgium **PHONE:** 32 2 5459070 **FX:** 32 2 5127342 **E-MAIL:** wide@gn.apc.org **WEBSITES:** http://www.eurosur.org/wide/porteng.htm **FOUNDED:** 1985. **OFFICERS:** Mieke Van der Veken, Coor. **MEMBERS:** 12. **STAFF:** 5. **BUDGET:** IR£21,000,000. **LANGUAGES:** English, French, German, Spanish.

DESCRIPTION: All women working in NGOs in Europe and who support WIDE. Provides a forum for the exchange of information and builds up more expertise on issues of importance to women's development needs in developing countries. Encourages and strengthens the national networks, development policies, priorities, especially in southern Europe. Promotes purposeful contacts with women in partner countries so that their development priorities will become guiding principles of the organizations' activities. Lobbies European and international institutions. **PUBLICATIONS:** *WIDE Bulletin* (in English and Spanish), quarterly. Covers information on a different topic in each edition.; *WIDE Newsletter* (in English), bimonthly. Contains articles on WIDE activities and current events. Also publishes WIDE position papers and lobby briefings.

WOMEN, LAW AND DEVELOPMENT INTERNATIONAL

1350 Connecticut Ave. NW, Ste. 407, Washington, DC 20036 **PHONE:** (202) 463-7477 **FX:** (202) 463-7480 **E-MAIL:** wld@wld.org **WEBSITES:** http://www.wld.org **FOUNDED:** 1993. **OFFICERS:** Margaret A. Schuler Ph.D., Exec.Dir. **STAFF:** 4. **BUDGET:** US$350,000. **LANGUAGES:** Arabic, English, French, Hindi Suriname Hindustanti, Spanish, Urdu.

DESCRIPTION: Committed to the defense and promotion of women's rights globally. Links individuals from activist groups, research institutions, and advocacy and human rights organizations throughout the world in order to: discern a global consensus on fundamental gender rights; develop the legal and political skills women need to transform this consensus into concrete action for change; advocate favorable UN and governmental policies affecting women's rights; and expand and strengthen the women's rights network. Conducts educational and training programs. **PUBLICATIONS:** Annual Report; Brochures; Handbooks; Reports.

WOMEN AND LAW IN SOUTHERN AFRICA PROJECT (WLSA)

PO Box UA 171, Union Ave., Harare, Zimbabwe **PHONE:** 263 4 772943, 263 4 773071

FX: 263 4 772935 E-MAIL: kilsa@samara.co.zw FOUNDED: 1988. OFFICERS: Sara Mvududu, Reg.Coor. MEMBERS: 90. STAFF: 16. BUDGET: Z$962,000. LANGUAGES: English.

DESCRIPTION: Works to improve the quality of life for women living in Botswana, Lesotho, Mozambique, Swaziland, Zambia, and Zimbabwe through improving their legal rights. Sponsors programs to educate women about their legal rights. Offers legal advice. Organizes campaigns to change laws that are biased against women. Offers a forum for exchange of information for international women's organizations. Conducts research on issues relevant to women and law, such as family and inheritance laws. Disseminates information; compiles statistics. PUBLICATIONS: *Uncovering Reality*; *Women and the Law in Southern Africa* (in English), quarterly; *Your Child Too*; Reports.

WOMEN'S CAMPAIGN FOR SOVIET JEWRY

One to One, Pannell House, 779/781 Pinchley Road, London NW11 8DN, England PHONE: 44 181 4587147, 44 181 4587148 FX: 44 181 4589971 FOUNDED: 1970. OFFICERS: Margaret Rigal, Co-Chwm. MEMBERS: 5,000. LANGUAGES: English, Hebrew, Russian.

DESCRIPTION: Members are human rights groups, members of the media and parliament, trade unionists, religious bodies, and concerned individuals in 16 countries. Seeks to call attention to the treatment accorded to Jews in the Commonwealth of Independent States who are imprisoned or persecuted for having tried to obtain permission to emigrate or for having observed religious or cultural conventions. Furthers public awareness of such injustices; exerts pressure on members of parliament and influential organizations; stages demonstrations and petition drives. Offers assistance to Jews immigrating to Israel. Provides speakers; holds seminars. Offers children's services; maintains charitable program. Compiles statistics. Organises treks in Israel, Egypt and Jordan to raise funds for the disadvantaged among the new immigrant families. PUBLICATIONS: *Newsletter*, biweekly; *35's Circular*, periodic; Pamphlets, periodic.

WOMEN'S ENVIRONMENT AND DEVELOPMENT ORGANIZATION (WEDO)

355 Lexington Ave., 3rd Fl., New York, NY 10017 PHONE: (212) 973-0325 FX: (212) 973-0335 E-MAIL: wedo@igc.org WEBSITES: http://www.wedo.org FOUNDED: 1990. OFFICERS: Susan Davis, Exec.Dir. LANGUAGES: English.

DESCRIPTION: International network of women activists and leaders concerned with environment, development, and social justice from the community to the global level. Works to make women more visible as participants, policy-makers, and leaders in fate-of-the-earth decisions. Campaigns for gender balance by the year 2001 in all governmental and non-governmental bodies, commissions, committees, and boards that deliberate on environment, development, and women's rights issues. Monitors implementation of governments' Earth Summit ICPD, Social Summit and Women's Conference commitments. Participates in United Nations conferences and facilitates the women's caucus for NGO advocacy. Undertakes Women, Health and the Environment: Action for Cancer Prevention Campaign, public hearings, activitist conferences, and public education. PUBLICATIONS: *News and Views*; Brochures; Reports. Promotes action through action alert mailings, primers, press and political briefings, advertising, flyers and postcards. Also produces a video and radio documentaries.

WOMEN'S SERVICE WORLDWIDE (WSW)

c/o Dr. Kate Alexander, PO Box 16727, Encino, CA 91416 PHONE: (818) 464-3522 FX: (818) 464-3522 E-MAIL: williamwallaceworld@worldnet.att.net FOUNDED: 1955. OFFICERS: Dr. Kate Alexander, CEO & Pres. LANGUAGES: Chinese, English, French, Italian, Japanese.

DESCRIPTION: Members are individuals and organizations. Seeks to improve the quality of life of women worldwide. Provides assistance to women, including legal services in child support and stalking cases, secretarial and information services, credit, and political advocacy and action campaigns. Conducts environmental protection activities; sponsors research and educational programs; makes available children's services; maintains museum and hall of fame; participates in charitable initiatives; compiles statistics. **PUBLICATIONS:** *William Wallace Worldwide*, monthly; Journal, periodic; Directory, periodic; Brochure; Bulletin.

WOMEN'S WORLD BANKING

8 W 40th St., New York, NY 10018 **PHONE:** (212) 768-8513 **FX:** (212) 768-8519 **E-MAIL:** swwb@swwb.org **FOUNDED:** 1979. **OFFICERS:** Nicola C. Armacoast. **MEMBERS:** 45. **STAFF:** 30. **LANGUAGES:** English, French, Spanish.

DESCRIPTION: Mission is to expand low income women's economic assets, participation and power by opening access to finance, information and markets. Provides and organizes support to affiliates who in turn offer direct services to low income women. Builds learning and change networks comprised of leading microfinance instituions and banks. Works with policy makers to build financials systems that work for the ''poor majority.'' Seeks to create an environment that will help a low income woman build her business, improve her living conditions, keep her family well-fed and healthy, educate her children, develop respect at home and in her community, and secure a political voice. **PUBLICATIONS:** *What Works*, periodic; Manuals.

WORLD ACADEMY OF ART AND SCIENCE (World Academy)

46891 Grisson St., Sterling, VA 20165 **PHONE:** (703) 450-0428 **FX:** (703) 450-0429 **E-MAIL:** ananda@igc.apc.org **FOUNDED:** 1960. **OFFICERS:** Harlan Cleveland, Pres. **MEMBERS:** 480. **STAFF:** 2. **BUDGET:** US$200,000. **LANGUAGES:** English.

DESCRIPTION: Members are individuals in 55 countries who are eminent in art, the natural and social sciences, and in humanistic studies and whose knowledge and understanding transcend the boundaries of their specific competencies. Endeavors to support and to become an effective instrument for the realization of a world order in which human dignity is honored in deed as in word. Expresses the concern of scientists, professionals, and the public for the social consequences and policy implications of knowledge. Encourages development of new ideas. Sponsors projects, symposia, conferences, and meetings in various countries. **PUBLICATIONS:** *Directory of Fellows*; *World Academy of Art and Science News*, 2-3/year; Books; Monographs.

WORLD AIRLINES CLUBS ASSOCIATION (WACA)

c/o International Air Transport Assn., 800 Place Victoria, PO Box 113, Montreal, PQ, Canada H4Z 1M1 **PHONE:** (514) 874-0202 **FX:** (514) 874-1753 **E-MAIL:** winterbota@iata.org **WEBSITES:** http://www.waca.org; http://www.waca.org **FOUNDED:** 1966. **OFFICERS:** Aubrey Winterbotham, Mgr. **MEMBERS:** 3,000. **STAFF:** 1. **LANGUAGES:** English.

DESCRIPTION: Members are airline/interline clubs in 50 countries representing 3,000 employees of the civil and commercial airline industry. Promotes airline transportation as a mode of travel and works to provide better service to the traveling public. Seeks to increase public awareness of the contribution of the airlines to world understanding and to extend, promote, and publicize the airline/interline clubs movement; advises on and coordinates the activities of member clubs worldwide. Promotes economic and social programs for the betterment of airlines and the community and country they serve. Fosters discussion on civil and commercial aviation; organizes annual and ongoing projects in the field. Arbitrates disputes among members. Offers children's services; sponsors charitable programs; compiles statistics. Conducts sports tournaments and special Olympic competitions for mentally handicapped individuals. **PUBLICATIONS:** *Airline/Interline Club Newsletter* (in English), 3/year. Contains world news from WACA.; *WACA Contact* (in English), semiannual; *WACA News* (in English), 3/year.

WORLD ALLIANCE OF REFORMED CHURCHES (WARC) (Alliance Reformee Mondiale - ARM)

c/o Prof. Milan Opocensky, 150, rte. de Ferney, CH-1211 Geneva 2, Switzerland **PHONE:** 41 22 7916238 **FX:** 41 22 7916505 **TX:** 415730 OIK CH **E-MAIL:** mil@warc.org **WEBSITES:** http://www.warc.ch **FOUNDED:** 1875. **OFFICERS:** Prof. Milan Opocensky, Gen.Sec. **MEMBERS:** 214. **STAFF:** 10. **BUDGET:** 2,000,000 SFr. **LANGUAGES:** English, French, German, Spanish.

DESCRIPTION: Members are Reformed, Presbyterian, and Congregational churches in 105 countries, with a total estimated membership of 75 million people. Facilitates the contribution of members' experiences and insights to the ecumenical movement; works toward one Church of Jesus Christ for the sake of the Gospel. Provides forum for members to test their religious convictions and ecumenical commitment; offers opportunities for exploration of other Reformed churches and churches of other traditions. Encourages exchange of resources, experiences, and information; fosters members' participation in interchurch projects of the World Council of Churches designed to assist churches and individuals in need. Advocates both individual and collective human rights throughout the world including religious and civil liberties; sponsors departments to study and publish findings on such topics as the theological basis of human rights. Advocates a fuller representation of Christian women in all leading bodies of member churches and in the alliance itself; participates in world-level dialogue with most Christian traditions; monitors doctrinal developments in member churches, and shares programs of interest. **PUBLICATIONS:** *Handbook of Member Churches*, periodic; *Reformed World*, quarterly; *Update*, quarterly. Also publishes studies (series), monographs, and papers.

WORLD ALLIANCE OF YOUNG MEN'S CHRISTIAN ASSOCIATIONS (WAYMCA) (Alliance Universelle des Unions Chretiennes de Jeunes Gens)

12 Clos Belmont, CH-1208 Geneva, Switzerland **PHONE:** 41 22 8495100 **FX:** 41 22 8495110 **E-MAIL:** office@ymca.int **WEBSITES:** http://www.ymca.int **FOUNDED:** 1855. **OFFICERS:** John W. Casey, Sec.Gen. **MEMBERS:** 90. **STAFF:** 15. **LANGUAGES:** English, French, German, Spanish.

DESCRIPTION: Members are national YMCA movements consisting of interdependent local associations in 130 countries. Purpose is to ''unite those young men who, regarding Jesus Christ as their God and Saviour, according to the Holy Scripture, desire to be His disciples in their faith and in their life, and to associate their efforts for the extension of His Kingdom amongst young men.'' **PUBLICATIONS:** *YMCA Directory* (in English, French, German, and Spanish), biennial. Includes addresses of YMCAs.; *YMCA World* (in English), quarterly; Reports, periodic. Contains information on meetings, consultations, World Councils, etc.

WORLD APOSTOLATE OF FATIMA (THE BLUE ARMY) (WAF)

Avenida Beato Nuno, Caixa Postal 38, P-2496 Fatima, Portugal **PHONE:** 351 49 532931 **FX:** 351 49 532931 **FOUNDED:** 1947. **OFFICERS:** Bishop Constantino Luna, Pres. **MEMBERS:** 120,000,000. **LANGUAGES:** English, French, German, Italian, Spanish, Swahili.

DESCRIPTION: Members are religious lay persons, and others in 145 countries. Aims to spread the message of world peace through prayer which is believed to have been conveyed to 3 children in Fatima, Portugal, by the apparition of the Virgin Mary in 1917. Encourages the praying of the rosary, acts of penance, adoration of the Blessed Sacrament, and the wearing of the scapular of Our Lady of Mount Carmel. **PUBLICATIONS:** *Dear Bishop* (in French); *Heartbeat* (in English, French, German, and Spanish), quarterly; *Hearts Aflame*, bimonthly. Provides information for interested youths.; *International Directory* (in English), annual; *Sol di Fatima* (in Spanish), monthly; *Soul* (in English and Spanish), bimonthly; Books, periodic; Pamphlets, periodic.

WORLD ASSEMBLY OF MUSLIM YOUTH (WAMY)

Prince Abdulaziz Bin Mosaid Bin, Jelewi Str., Sulaimaniya, PO Box 10845, Riyadh 11443, Saudi Arabia **PHONE:** 966 1 4641669, 966 1 4641663 **FX:** 966 1 4641710 **TX:** 400413 ISLAMI SJ **E-MAIL:** wamy@kacst.edu.sa **FOUNDED:** 1972. **OFFICERS:** Dr. Maneh Hammad Al-Johani, Sec.Gen. **MEMBERS:** 400. **STAFF:** 65. **BUDGET:** US$1,500,000. **LANGUAGES:** Arabic, English, French.

DESCRIPTION: Members are Muslim youth organizations and Muslim organizations that include activities for youths among their programs. Promotes better understanding of the Islamic faith among Muslims and others. Conducts regional and international youth camps. Translates books into regional languages. Compiles statistics. **PUBLICATIONS:** *Conference Proceedings* (in Arabic), triennial; *WAMY Newsletter - Islamic Future* (in Arabic, English, and French), monthly.

WORLD ASSEMBLY OF SMALL AND MEDIUM ENTERPRISES (WASME) (Assemblee Mondiale des Petites et Moyennes Entreprises - AMPME)

301-302 Saraswati House, 27 Nehru Pl., New Delhi 110 019, Delhi, India **PHONE:** 91 11 6414058, 91 11 6411417 **FX:** 91 11 6852170 **E-MAIL:** cdawasme@gemsvsnl.net.in **FOUNDED:** 1980. **OFFICERS:** Dr. Chakradhari Agrawal Ph.D., Sec.Gen. **MEMBERS:** 84. **STAFF:** 36. **BUDGET:** Rs 5,000,000. **LANGUAGES:** English, French, Russian, Spanish.

DESCRIPTION: Members are banks, governmental and nongovernmental organizations, and related organizations in 107 countries. Seeks economic growth through the creation of new employment opportunities within a framework of stability and social justice. Promotes the creation of small and medium enterprises, an environment of economic freedom, self-employment, and a broadened industrial base; seeks greater technical cooperation among small and medium-sized enterprises. Encourages closer relations between governmental and nongovernmental organizations; conducts critical examination of policies for economic growth in developing countries. Compiles statistics; disseminates information and documentation. Provides placement services. Operates Technology Promotion and Exchange Center to provide advice and practical help to small and medium businesses. Maintains a Roster of Experts for assignments with the United Nations Industrial Development Organization. Works in cooperation with the International Labor Organization, International Trade Centre, and United Nations Conference on Trade and Development. **PUBLICATIONS:** *India - An Economic Overview*, monthly; *List of Members, Associates, and Observers*, annual; *Role of NGOs in Africa in Restructuring of SMEs*; *SMEs - A Global Overview*; *WASME Newsletter* (in English), monthly; Journal, annual.

WORLD ASSEMBLY OF YOUTH (WAY) (Assemble Mondiale de la Jeunesse - AMJ)

Ved Bellahoj 4, Bronshoj, DK-2700 Copenhagen, Denmark **PHONE:** 45 38607770 **FX:** 45 38605797 **E-MAIL:** wayouth@centrum.dk **WEBSITES:** http://www.jaring.my/way **FOUNDED:** 1949. **OFFICERS:** Heikki Pakarinen, Sec.Gen. **MEMBERS:** 96. **STAFF:** 2. **BUDGET:** US$300,000. **LANGUAGES:** English, French, Spanish.

DESCRIPTION: Members are national youth committees composed of principal voluntary democratic youth organizations in 59 countries. Promotes youth programs, primarily in the fields of development and human rights, and fosters understanding among young people of each other's racial, religious, and national backgrounds. Supports local youth leaders in promoting the Universal Declaration of Human Rights, the economic and social advancement of all peoples, and the idea that people can help themselves through their own efforts. Seeks to provide constructive leadership and to assist young people in facing the challenges of changing economic, social, and political realities in the world and in contributing fully to their nations' development. Fosters interracial respect and world understanding. Works toward solutions to agrarian problems and questions of international development assistance and the sharing of technology. Facilitates the collection and dissemination of information on the needs and problems of youth. Sponsors programs for young people on topics such as crime prevention, drug abuse, the environment, family planning, leadership training, nonviolence, population, rural village development, services for children and the elderly, and social defense. Maintains International Youth Press Service. Solicits funding; provides technical assistance and the basis for generating publicity and for establishing nationwide youth programs. Conducts development management and youth leadership training. **PUBLICATIONS:** *WAY Forum* (in English), quarterly; *WAY Information* (in English and Spanish), bimonthly; *Youth Roundup* (in English), bimonthly. Also publishes program reports.

WORLD ASSOCIATION FOR ANIMAL PRODUCTION (WAAP)

Via A. Torlonia, 15 A, I-00161 Rome, Italy **PHONE:** 39 6 44238013 **FX:** 39 6 44241466 **E-MAIL:** zoorec@rmnet.it **FOUNDED:** 1965. **MEMBERS:** 17. **LANGUAGES:** English.

DESCRIPTION: Members are national and regional societies of animal production, animal science, and dairy science. Promotes international cooperation and participation in world conferences on animal production for the purpose of reviewing scientific, technical, and educational problems in the field.

PUBLICATIONS: *News Items*, semiannual. Also publishes ad-hoc issues.

WORLD ASSOCIATION OF CHILDREN'S FRIENDS (WACF) (Association Mondiale des Amis de l'Enfance - AMADE)

16, blvd. de Suisse, MC-98000 Monaco Cedex, Monaco **PHONE:** 52 377 93154241 **FX:** 52 377 93154245 **FOUNDED:** 1964. **OFFICERS:** M. Jacques Dandis, Gen.Sec. **MEMBERS:** 20. **STAFF:** 2. **LANGUAGES:** English, French, German, Italian, Portuguese, Spanish.

DESCRIPTION: Protects children and adolescents from dangers that may jeopardize their development or destroy their happiness. Works on behalf of young people for the acquisition of financial aid to promote the development of charitable institutions, including funds for buildings and equipment and the hiring of technical personnel. Supervises charitable activities and seeks to ensure the correct use of funds. Through group studies, determines what is good for young people and stimulates appropriate action. Lobbies against television violence and the liberalization of what the group refers to as ''so-called soft drugs.'' Advocates ratification of and strict adherence by all countries to the International Labor Office Convention 138 and the European Charter of Human Rights, which govern child and adolescent labor and the physical and moral health of young people. Campaigns on such issues as stricter regulations on the use of children in pictorial advertisements, further legislative acceptance of a minor being consulted discreetly by the judge before custody is awarded to 1 of the divorced parents, and the reintroduction of good citizenship courses in schools. Protests against child prostitution and defends the rights of the unborn child. Coordinates, at the international level, the efforts of national groups; offers guidance and advice in regard to national projects and programs. **PUBLICATIONS:** *Children First* (in English, French, and Spanish), biennial.

WORLD ASSOCIATION FOR CHRISTIAN COMMUNICATION

357 Kennington Ln., London SE11 5QY, England **PHONE:** 44 171 5829139 **FX:** 44 171 7350340 **E-MAIL:** wacc@wacc.org.uk **WEBSITES:** http://www.wacc.org.uk **STAFF:** 20. **LANGUAGES:** English.

DESCRIPTION: Promotes communication among development programs worldwide, with particular emphasis on communication among Christian ministries and programs. Conducts research and educational programs. **PUBLICATIONS:** *Action*; *Media Development* (in English and Spanish), quarterly.

WORLD ASSOCIATION FOR EDUCATIONAL RESEARCH (WAER) (Association Mondiale des Sciences de l'Education - AMSE)

Universiteit Gent, Hertstraat 66, B-9000 Gent, Belgium **PHONE:** 32 9 2255526 **FX:** 32 9 2646490 **E-MAIL:** vibence.henderick@rug.ac.be **FOUNDED:** 1953. **OFFICERS:** Dr. M.L. van Herreweghe, Gen.Sec. **MEMBERS:** 680. **LANGUAGES:** Dutch, English, French.

DESCRIPTION: Members are individuals in 50 countries conducting educational research at the university level. Fosters the international development of educational research. Provides for the exchange of information among members. **PUBLICATIONS:** *Communications* (in English, French, and German), semiannual; *Proceedings*, quadrennial. Also publishes scientific texts.

WORLD ASSOCIATION OF GIRL GUIDES AND GIRL SCOUTS (WAGGGS) (Asociacion Mundial de las Guias Scouts)

World Bur., Olave Centre, 12 C Lyndhurst Rd., London NW3 5PQ, England **PHONE:** 44 171 7941181 **FX:** 44 171 4313764 **E-MAIL:** wagggs@wagggsworld.org **WEBSITES:** http://www.wagggsworld.org **FOUNDED:** 1928. **OFFICERS:** Lesley Bulman, Dir. World Bureau. **MEMBERS:** 136. **STAFF:** 80. **LANGUAGES:** English, French, Spanish.

DESCRIPTION: Members are national organizations representing in excess of 10,000,000 individuals. Promotes unity of purpose and common understanding in the fundamental principles of the Girl Guide and Girl Scout movement throughout the world; encourages friendship and mutual understanding among girls and young women of all nations. Holds training sessions at world centers. Conducts charitable programs through community development projects. At the cutting edge of issues affecting girls and young women. **PUBLICATIONS:** *Challenging Movement* (in English, French, and Spanish), triennial; *Our World News* (in English,

French, and Spanish), bimonthly; *Trefoil Round the World* (in English), triennial. Contains statistics and information from member organizations; *World Issues* (in English, French, and Spanish), periodic. Covers issues relevant to girls and young women such as homelessness, AIDS, and nutrition.; Books, periodic.

WORLD ASSOCIATION OF INDUSTRIAL AND TECHNOLOGICAL RESEARCH ORGANIZATIONS (WAITRO) (Association Mondiale des Organisations de Recherche Industrielle et Technologique)

Danish Technological Inst., PO Box 141, DK-2630 Taastrup, Denmark **PHONE:** 45 43504350, 45 43507045 **FX:** 45 43507250 **TX:** 33146 TEKTAA DK **E-MAIL:** waitro@teknologisk.dk **WEBSITES:** http://waitro.dti.dk **FOUNDED:** 1970. **OFFICERS:** M.D. Mengu, Dep.Sec.Gen. **MEMBERS:** 180. **STAFF:** 7. **BUDGET:** US$700,000. **LANGUAGES:** English, French, Spanish.

DESCRIPTION: Organizations in 80 countries actively engaged in technological research and interested in encouraging, promoting, and supporting industrial research. Aims are: to assist the development and improve the capabilities of members, especially in developing countries; to identify fields of research needing cooperation or assistance by outside agencies; to promote coordination and cooperation among members; to encourage the transfer of research results and technical expertise between members; and to promote exchange of experience in research management among members. Finds finances for improving research facilities, training of personnel, and information exchange. Implements exchange of research workers and research facilities; sponsors seminars. Serves as a link between members and international associations and federations. Represents members on matters of common interest. **PUBLICATIONS:** *Directory of Sources of Funds for International R&D Cooperation* (in English), biennial. Lists and profiles organizations supporting technological R&D in developing countries.; *WAITRO News* (in English), quarterly. Various seminars proceedings and technical reports.

WORLD ASSOCIATION OF JUDGES (WAJ)

1000 Connecticut Ave. NW, Ste. 202, Washington, DC 20036 **PHONE:** (202) 466-5428 **FX:** (202) 452-8540 **E-MAIL:** wja@wja-wptlc.org **FOUNDED:** 1966. **OFFICERS:** Margaret M. Henneberry, Exec. Officer. **LANGUAGES:** English, French, German, Spanish.

DESCRIPTION: Members of courts in countries throughout the world. Promotes the expansion of the rule of law in the world community and the advancement and improvement of the administration of justice for all people.

WORLD ASSOCIATION OF LAW PROFESSORS (WALP)

1000 Connecticut Ave. NW, Ste. 202, Washington, DC 20036 **PHONE:** (202) 466-5428 **FX:** (202) 452-8540 **E-MAIL:** wja@wja-wptlc.org **FOUNDED:** 1975. **OFFICERS:** Margaret M. Henneberry, Exec. Officer. **LANGUAGES:** English, French, German, Spanish.

DESCRIPTION: Members are law professors united to further international cooperation through the development of international law. Promotes the objectives and programs of World Jurist Association.

WORLD ASSOCIATION OF LAWYERS (WAL)

1000 Connecticut Ave. NW, Ste. 202, Washington, DC 20036 **PHONE:** (202) 466-5428 **FX:** (202) 452-8540 **E-MAIL:** wja@wja-wptlc.org **FOUNDED:** 1975. **OFFICERS:** Margaret M. Henneberry, Exec. Officer. **LANGUAGES:** English, French, Spanish.

DESCRIPTION: Seeks the development of international law as a basis for the peaceful resolution of conflict. Promotes the goals and activities of World Jurist Association. **PUBLICATIONS:** *Law/Technology*, quarterly. Reports on the relation of technology to law.; *World Jurist*, bimonthly.

WORLD ASSOCIATION OF NEWSPAPERS (WAN)

25, rue d'Astorg, F-75008 Paris, France **PHONE:** 33 1 47428500 **FX:** 33 1 47424948 **TX:** 290513 F **E-MAIL:** contact_us@wan.asso.fr **WEBSITES:** http://www.fiej.org **FOUNDED:** 1948. **OFFICERS:**

Timothy Balding, Dir.Gen. **MEMBERS:** 57. **STAFF:** 21. **LANGUAGES:** English, French, German, Spanish.

DESCRIPTION: Members are national associations of newspaper publishers; individual news executives; news agencies; and regional press organizations. Promotes and defends press freedom including the economic independence of newspapers; fosters communications and contacts between newspaper executives. Sponsors research and supports the press in developing countries. **PUBLICATIONS:** *WAN Newsletter* (in English, French, German, and Spanish), quarterly; *World Press Trends Survey* (in English), annual. Books.

WORLD ASSOCIATION FOR PUBLIC OPINION RESEARCH (WAPOR)

Univ. of North Carolina, School of Journalism and Mass Communication, CB-3365, Howell Hall, Chapel Hill, NC 27599-3365 **PHONE:** (919) 962-6396 **FX:** (919) 962-4079 **E-MAIL:** kcole@email.unc.edu **WEBSITES:** http://www.wapor.org **FOUNDED:** 1946. **OFFICERS:** Ms. Katherine Cole, Office Mgr. **MEMBERS:** 433. **LANGUAGES:** English.

DESCRIPTION: Members are academic survey research scholars and commercial survey researchers. Purposes are to establish and promote contacts between survey researchers on opinions, attitudes, and behaviors of people in various countries; further the use of objective scientific survey research in international affairs. Activities include improving methods and professional standards, training personnel, and coordinating and integrating international polls. **PUBLICATIONS:** *International Journal of Public Opinion Research*, quarterly; Directory, annual; Newsletter, quarterly.

WORLD ASSOCIATION OF RESEARCH PROFESSIONALS (ESOMAR)

J.J. Viottastraat 29, NL-1071 JP Amsterdam, Netherlands **PHONE:** 31 20 6642141 **FX:** 31 20 6642922 **E-MAIL:** email@esomar.nl **WEBSITES:** http://www.esomar.nl **FOUNDED:** 1948. **OFFICERS:** Guergen Schwoerer, Dir.Gen. **MEMBERS:** 3,500. **STAFF:** 20. **LANGUAGES:** English, French, German, Spanish.

DESCRIPTION: Members are users and providers of market, opinion, or social research. Promotes and seeks to develop the highest professional and technical standards in marketing and market research to enhance effective decision-making in the private and public sectors. Aims to further the professional interests of members. Facilitates exchange of information among members; conducts educational programs. **PUBLICATIONS:** *Congress Papers* (in English, French, and German), annual; *Consumer Market Research Handbook*; *Directory of ESOMAR members*, annual; *ESOMAR Directory of Marketing Research Organisations* (in English), annual; *Multilingual Glossaries of Technical Terms*; *News Brief* (in English, French, and German), monthly; *Report of ESOMAR Annual Industry Study*; *Seminar and Conference Papers* (in English), 8-10/year; *What is Market Research?*.

WORLD ASSOCIATION OF SOCIETIES OF PATHOLOGY - ANATOMIC AND CLINICAL (WASP) (Association Mondiale des Societes de Pathologie - Anatomique et Clinique)

c/o Professor Mikio Mori, Koshigaya Hospital, Dept. of Clinical Pathology, Dokkyo University School of Medicine, 2-1-50, Minamikoshigaya, Koshigaya 343-8555, Japan **PHONE:** 81 48 9656798 **FX:** 81 48 9654853 **E-MAIL:** mickey@dokkyomed.ac.jp **WEBSITES:** http://www.wasp.or.jp/ **FOUNDED:** 1947. **OFFICERS:** Mikio Mori M.D. **MEMBERS:** 54. **LANGUAGES:** English.

DESCRIPTION: Members are national societies of anatomic and clinical pathology in 40 countries united to foster cooperation between members and improve standards in anatomic and clinical pathology. Maintains a variety of committees and educational programs. **PUBLICATIONS:** *Directory* (in English), biennial; *News Bulletin of the World Association of Societies of Pathology/Commission on World Standards*, quarterly.

WORLD ASSOCIATION OF VETERINARY FOOD-HYGIENISTS (WAVFH) (Asociacion Mundial de Veterinarios Higienistas de los Alimentos - AMVHA)

c/o Federal Institute for Health Protection of Consumers and Veterinary Medicine (BGVV), Diedersdorferweg 1, D-12277 Berlin, Germany **PHONE:** 49 30 84122101 **FX:** 49 30 84122121

E-MAIL: p.teufel@bgvv.de **FOUNDED:** 1956. **OFFICERS:** Dr. L. Ellerbroek, Pres. **BUDGET:** US$2,000. **LANGUAGES:** English, French, German.

DESCRIPTION: Members are national veterinary societies in 37 countries. Facilitates international exchange of research results that involve food microbiology, food hygiene, food-related zoonotic diseases, and environmental pollution affecting either animal or human food. Organizes roundtables and meetings. **PUBLICATIONS:** *Symposia Proceedings* (in English), quadrennial. Also publishes conference reports.

WORLD ASSOCIATION OF VETERINARY MICROBIOLOGISTS, IMMUNOLOGISTS, AND SPECIALISTS IN INFECTIOUS DISEASES (WAVMI) (Association Mondiale des Veterinaires Microbiologistes, Immunologistes et Specialistes des Maladies Infectieuses - AMVMI)

7, ave. du General de Gaulle, F-94704 Maisons-Alfort Cedex, France **PHONE:** 33 1 43967021 **FX:** 33 1 43967022 **FOUNDED:** 1968. **OFFICERS:** Charles Pilet, Pres. **LANGUAGES:** English, French.

DESCRIPTION: Facilitates international exchange among veterinary microbiologists, immunologists, and specialists in infectious diseases. Sponsors international course in animal immunology; conducts symposia. **PUBLICATIONS:** *Comparative Immunology, Microbiology and Infectious Diseases*, quarterly.

WORLD BANK (WB)

1818 H St. NW, Washington, DC 20433 **PHONE:** (202) 477-1234 **FX:** (202) 477-6391 **WEBSITES:** http://www.worldbank.org **OFFICERS:** James D. Wolfensohn, Pres. **LANGUAGES:** English.

DESCRIPTION: Comprises the International Bank for Reconstruction and Development, the International Development Association, International Finance Corporation and the Multilateral Investment Guarantee Agency. Established by the United Nations to assist in raising the standards of living in developing countries by channelling financial resources from developed countries. Emphasis is placed on investments which foster active participation in the development process. Programs concentrate on rural and urban development, agriculture, and education. Activities include improving water and sewage facilities, building low-cost housing, and increasing the productivity of small industries. Assists organizations with identifying, designing, and executing development projects; offers financial aid to national development institutions. Encourages discussion on common development problems such as income distribution, rural poverty, unemployment, excessive population growth, and rapid urbanization. Conducts research programs on topics including economic planning and public utilities. Works in association with the United Nations Development Programme and executes many UNDP projects.

WORLD BLIND UNION (WBU) (Union Mondiale des Aveugles - UMA)

La Coruna, 18, E-28020 Madrid, Spain **PHONE:** 34 91 5713685 **FX:** 34 91 5715777 **E-MAIL:** umc@once.es **WEBSITES:** http://www.once.es/wbu **FOUNDED:** 1984. **OFFICERS:** Pedro Zurita, Sec.Gen. **MEMBERS:** 155. **LANGUAGES:** English, French, Spanish.

DESCRIPTION: Members are national associations of the blind and international agencies for the blind. Fosters international cooperation among organizations working for the welfare of the blind and the prevention of blindness in 155 countries. Encourages creation and development of national organizations of and for the blind; seeks to provide necessary technical and material aid to such groups. Provides guidance in fields of education, rehabilitation, vocational training, and employment. Works toward the introduction and improvement of minimum standards for the welfare of the blind worldwide. Encourages and conducts studies in the field of service to the blind and the prevention of blindness. Collects and disseminates information on the conditions of the blind in different countries; informs members of social and legislative matters relating to blindness and its prevention. Maintains relations with numerous United Nations agencies including the World Health Organization, the International Council on Disability, and the International Agency for the Prevention of Blindness; sponsors the International Council for the Education of the Visually Impaired. **PUBLICATIONS:** *General Assembly Proceedings*, periodic; *Review* *of the European Blind* (in English, French, German, and Russian), semiannual; *The World Blind* (in English, French, and Spanish), biennial. Also available in braille and on casette.

WORLD BOXING ORGANIZATION (WBO)

c/o Nick P. Kerasiotis, 412 Colorado Ave., Aurora, IL 60506 **PHONE:** (630) 897-4765, (312) 814-2719 **FX:** (630) 897-1134 **FOUNDED:** 1920. **OFFICERS:** Nick P. Kerasiotis, Treas. **LANGUAGES:** English.

DESCRIPTION: Membership composed of the athletic commission or any duly authorized body legally organized to regulate, control, or supervise boxing in any country, territorial or political subdivision, province, or city. Purpose is to regulate professional boxing. **PUBLICATIONS:** *Ratings*, bimonthly; Bulletin, monthly.

WORLD BUREAU OF METAL STATISTICS (WBMS)

27-A High St., Ware, Herts. SG12 9BA, England **PHONE:** 44 1920 461274 **FX:** 44 1920 464258 **TX:** 817746 **E-MAIL:** wbms@compuserve.com **FOUNDED:** 1946. **OFFICERS:** S.M. Eales, Gen.Mgr. **MEMBERS:** 20. **STAFF:** 7. **LANGUAGES:** English.

DESCRIPTION: Collects and disseminates current, accurate statistics on metal production, consumption, stocks, and international trade. **PUBLICATIONS:** *Surveys on Minor Metals*; *World Copper Statistics Since 1950*, annual; *World Flows of Unwrought Copper, Lead, Zinc, Nickel, Tin, and Aluminum*, annual; *World Metal Statistics*, monthly; *World Metal Statistics Yearbook*; *World Stainless Steel Statistics*, annual.

WORLD CHESS FEDERATION (Federation Internationale des Echecs - FIDE)

PO Box 166, CH-1000 Lausanne, Switzerland **FOUNDED:** 1924. **OFFICERS:** Casto P. Abundo, Sec. **MEMBERS:** 151. **STAFF:** 5. **BUDGET:** 725,000 SFr. **LANGUAGES:** Arabic, English, French, German, Portuguese, Russian, Spanish.

DESCRIPTION: Members are national chess federations. Promotes the diffusion and development of chess among all nations. Hopes to raise the level of chess culture and knowledge on a scientific, creative, and cultural basis. Supports close international cooperation among chess devotees in all fields of chess activity. Sponsors World Chess Championship and European, Asian, African, and American championships. Awards international titles of Grandmaster, Master, FIDE Master, and Arbiter. **PUBLICATIONS:** *Address List*; *FIDE Forum* (in English), bimonthly; *Minutes of Congress With Annexes*, annual; *President's Circular Letters and Chess Calendar*, 5/year; *Rating List*, biennial; *Regulations Handbook*, annual.

WORLD CHRISTIAN LIFE COMMUNITY (WCLC) (Weltgemeinschaft Christliches Lebens)

CP 6139, Borgo Santo Spirito 8, I-00195 Rome, Italy **PHONE:** 39 6 6868079 **FX:** 39 6 6868079 **E-MAIL:** mcvx.wici@agora.stm.it **WEBSITES:** http://maple.lemoyne.edu/jesuit/clc **FOUNDED:** 1953. **OFFICERS:** Giles Michaud, Exec.Sec. **MEMBERS:** 150,000. **STAFF:** 4. **BUDGET:** 200,000,000 Lr. **LANGUAGES:** English, French, Spanish.

DESCRIPTION: Members are Catholic adults and youth in 60 countries wholly committed to serving the church and the world in every area of life. Encourages greater collaboration and mutual assistance among members in order to promote true Christian life communities. **PUBLICATIONS:** *Progressio* (in English, French, and Spanish), triennial. With two supplements.

WORLD CONCERN (WC)

PO Box 33000, 19303 Fremont Ave. N., Shoreline, WA 98133 **PHONE:** (206) 546-7201 **FX:** (206) 546-7269 **E-MAIL:** wconcern@crista.org **WEBSITES:** http://www.worldconcern.org **FOUNDED:** 1975. **OFFICERS:** Paul Kennel, Exec.Dir. **STAFF:** 45. **BUDGET:** US$21,000,000. **LANGUAGES:** English.

DESCRIPTION: Members are Christians working to empower refugees and poor people through relief and self-help development strategies. Seeks to bring hope to the poor through programs which restore health, enable families to attain self-sufficiency, ensure basic education, prevent diseases, protect livestock, and improve the environment. Programs include: refugee resettlement; primary health care; community planning; housing; small business development; veterinary training; reforestation; farming assistance. **PUBLICATIONS:** *Loveline Newsletter*, monthly.

WORLD CONFEDERATION OF JEWISH COMMUNITY CENTERS (WCJCC) (Confederacion Mundial de Centros Comunitarios Judios)

12 Hess St., IL-94185 Jerusalem, Israel **PHONE:** 972 26 251265 **FX:** 972 26 247767 **FOUNDED:** 1977. **OFFICERS:** Sara Bogen, Exec.Dir. **STAFF:** 2. **BUDGET:** 220,000 IS. **LANGUAGES:** English, French, Hebrew, Russian, Spanish.

DESCRIPTION: Serves as a council of Jewish Community Center movements in various countries. Promotes communication and cooperation among constituents; organizes exchange visits of lay leaders and professionals; encourages the sharing of publications. Offers program information, training seminars in Israel. **PUBLICATIONS:** *Lexicon of Terms*; *Newsletter*, periodic; *World Directory of JCCs*, biennial.

WORLD CONFEDERATION OF LABOUR (WCL) (Confederation Mondiale du Travail - CMT)

33, rue de Treves, B-1040 Brussels, Belgium **PHONE:** 32 2 2854700 **FX:** 32 2 2308722 **E-MAIL:** info@cmt-wcl.org **WEBSITES:** http://www.cmt-wcl.org **FOUNDED:** 1920. **OFFICERS:** W. Thys, Sec.Gen. **MEMBERS:** 26,000,000. **STAFF:** 30. **LANGUAGES:** Dutch, English, French, German, Spanish.

DESCRIPTION: Members are trade union organizations representing 26,000,000 individuals in 114 countries. Works to create economic, social, cultural, and political institutions consistent with the developmental needs of individuals and society. Organizes leadership training for African, Asian, Latin American, and middle and western European leaders; provides technical assistance to confederations in developing countries. Forms committees for document preparation. **PUBLICATIONS:** *Events* (in Dutch, English, French, German, and Spanish), semiannual. Review on trade union information and training; *Flash* (in Dutch, English, French, German, and Spanish), semimonthly; *Labor* (in Dutch, English, French, German, and Spanish), bimonthly; Directory.

WORLD CONFEDERATION FOR PHYSICAL THERAPY (WCPT) (Confederation Mondiale pour la Therapie Physique)

46-48 Grosvenor Gardens, London SW1W 0EB, United Kingdom **PHONE:** 44 171 8819234 **FX:** 44 171 8819239 **E-MAIL:** wcpt@dial.pipex.com **FOUNDED:** 1951. **OFFICERS:** Ms. B.J. Myers, Sec.Gen. **MEMBERS:** 65. **STAFF:** 2.

DESCRIPTION: Members are national physical therapy associations. Aims to improve and strengthen contacts among physical therapists, associations representing physical therapists, and others involved in the physical therapy and rehabilitation fields. Sponsors educational programs. **PUBLICATIONS:** *Annual Report* (in English); *WCPT News* (in English), semiannual; Books, periodic; Reports, periodic.

WORLD CONFEDERATION OF TEACHERS (WCT) (Confederation Syndicale Mondiale des Enseignants - CSME)

c/o Gaston De La Haye, 33, rue de Treves, B-1040 Brussels, Belgium **PHONE:** 32 2 2854752 **FX:** 32 2 2308722 **TX:** 26966 CMTWCL B **E-MAIL:** wct@cmt-wcl.org **FOUNDED:** 1963. **OFFICERS:** Gaston De La Haye, Gen.Sec. **STAFF:** 5. **LANGUAGES:** Dutch, English, French, German, Spanish.

DESCRIPTION: Members are national teachers' trade unions with a total membership of 1,200,000 individuals in 42 countries. Works to foster international cooperation among teachers' trade unions, especially through the United Nations Educational, Scientific and Cultural Organization and the International Labor Organization. Studies questions relating to employment, employment stability, work duration, class sizes, factors of nervous tension, safety and hygiene in schools, teacher's health care, and situations in rural and remote areas. Studies and drafts proposals designed to improve the general and legal status of teachers. Seeks to further improve both teachers' training and in-service training. Encourages formation and growth of new teachers' union in countries where such groups do not exist. **PUBLICATIONS:** *Labor*, monthly.

WORLD CONFEDERATION OF UNITED ZIONISTS (WCUZ)

130 E. 59th St., 12th Flr., New York, NY 10022 **PHONE:** (212) 371-1452 **FX:** (212) 371-3265 **FOUNDED:** 1946. **OFFICERS:** Kalman Sultanik, Exec.Co-Pres. **MEMBERS:** 600,000. **LANGUAGES:** English, French, Hebrew, Spanish.

DESCRIPTION: Unaffiliated with any Israeli political party, group promotes Zionist education, information, and welfare activities on behalf of Israel. Encourages private and collective agriculture and industry in Israel; strives for an ''Israel-centered creative Jewish survival in Diaspora.'' **PUBLICATIONS:** *Zionist Information Views* (in English and Spanish), monthly. Reports on membership activities.

WORLD CONFERENCE ON RELIGION AND PEACE (WCRP/USA)

777 United Nations Plz., New York, NY 10017 **PHONE:** (212) 687-2163 **FX:** (212) 983-0566 **FOUNDED:** 1966. **OFFICERS:** Dr. William F. Vendley, Sec.Gen. **MEMBERS:** 10,000. **STAFF:** 17. **BUDGET:** US$1,500,000. **LANGUAGES:** English.

DESCRIPTION: Members are leadership and institutions from major world religions. Develops effective international, multireligious cooperation in peace efforts. Works for a just international economic order, nuclear and conventional arms limitation, human rights, conservation of natural resources, conflict resolution, child issues and education for peace. Has cosponsored world conferences on religion and peace. **PUBLICATIONS:** *Religion for Peace: A Newsletter on Inter-Religious Dialogue and Action for Peace Issued by the World Conference on Religion and Peace*, 3/year. Focuses on the activities of the conference. Publicizes problems facing the developed and the developing world such as poverty and malnutrition; *WCRP/USA Occasional Papers*, periodic; Brochures; Monographs.

WORLD CONGRESS OF FAITHS (WCF) (Congres Mondial des Religions)

2 Market St., Oxford OX1 3EF, England **PHONE:** 44 1865 202751 **FX:** 44 1865 202746 **WEBSITES:** http://www.interfaith-center.org/oxford/wcf **FOUNDED:** 1936. **OFFICERS:** David Storey, Hon.Treas. **MEMBERS:** 500. **BUDGET:** £35,000. **LANGUAGES:** English.

DESCRIPTION: Members are individuals in 7 countries concerned with interfaith fellowship and understanding. Encourages the study of world religions. Seeks to: enhance international understanding; end prejudice; encourage dialogue, understanding, and cooperation among the various faith communities of the world. Sponsors study tours and retreats. **PUBLICATIONS:** *Conference Proceedings*; *One Family* (in English), 3/year. Contains U.K. events; *World Faiths Encounter* (in English), 3/year.

WORLD CONSTITUTION AND PARLIAMENT ASSOCIATION (WCPA)

8800 W 14th Ave., Lakewood, CO 80215-4817 **PHONE:** (303) 233-3548 **FX:** (303) 526-2185 **TX:** 3712957 EARTH RESCUE **FOUNDED:** 1959. **OFFICERS:** Philip Isely, Sec.Gen. **MEMBERS:** 2,000. **LANGUAGES:** English.

DESCRIPTION: Members are individuals and organizations in 50 countries interested in implementing a world constitution establishing a federal world government that would ''achieve world peace and solve problems for the good of humanity.'' Promotes national ratification of the Constitution for the Federation of Earth; hopes to sponsor a popular referenda on this issue. **PUBLICATIONS:** *Across Frontiers*, bimonthly; *Constitution for the Federation of Earth*; *Design and Action for a New World*; *Quantum Leap for Survival*; Pamphlets.

WORLD COUNCIL OF CHURCHES (WCC) (Conseil Oecumenique des Eglises)

150 route de Ferney, POB 2100, CH-1211 Geneva 2, Switzerland **PHONE:** 41 22 7916111 **FX:** 41 22 7910361 **TX:** 415730 **FOUNDED:** 1948. **OFFICERS:** Rev. Dr. Konrad Raiser, Gen.Sec. **MEMBERS:** 330. **STAFF:** 270. **BUDGET:** 300,000,000 SFr. **LANGUAGES:** English, French, German, Spanish.

DESCRIPTION: Ecumenical fellowship of Eastern Orthodox, Oriental Orthodox, Lutheran, Reformed, Methodist, Anglican, Old Catholic, Pentecostal, Baptist, United, and Independent denominations in 105 countries. Conducts wide range of programs of witness, service, unity, and renewal throughout the world. Among its concerns are: faith and order; mission and evangelism; church and society; dialogue with people of living faiths and ideologies; interchurch and refugee service; international affairs; racism; Christian medical assistance to developing countries; youth work; renewal and congregational life; women in the church and society; biblical studies. Maintains Ecumenical Institute, an educational center with a graduate seminar, near Geneva, Switzerland. **PUBLICATIONS:** *Ecumenical News International* (in English and French), bimonthly; *Ecumenical Press Service*, weekly; *Ecumenical Review*, quarterly; *International Review of Mission*, quarterly; *Service Oecumenique de Presse et d'Information*, weekly. Also publishes 25-30 books annually.

WORLD COUNCIL OF CREDIT UNIONS (WOCCU)

5710 Mineral Point Rd., PO Box 2982, Madison, WI 53701 **PHONE:** (608) 231-7130 **FX:** (608) 238-8020 **E-MAIL:** mail@woccu.org **WEBSITES:** http://www.woccu.org **FOUNDED:** 1970. **OFFICERS:** Chris Baker, CEO. **MEMBERS:** 21. **STAFF:** 50. **LANGUAGES:** English, French, Spanish.

DESCRIPTION: Members are regional confederations and national credit union leagues representing 84 national credit union movements and 95 million credit union members. Serves as a liaison and representative voice for the international credit union movement. Fosters communication between affiliates, development and donor groups, and governmental and nongovernmental agencies. Provides technical assistance to developing credit union movements. Conducts research on credit union development, savings mobilization, and microfinance. Organizes conferences, seminars, and symposia on credit union and cooperative development issues concerning the international credit union movement. Compiles statistics; assists visitation of member nations by credit union travelers. **PUBLICATIONS:** *Credit Union World*, quarterly. Source for International credit union information.; *International Credit Union Statistics* (in English, French, and Spanish), annual. Compilation of country-by-country, regional and international credit union statistics of member nations.; *International Credit Union Statistics* (in English, French, and Spanish), annual. Compilation of country-by-country, regional and international credit union statistics of member nations.; *Perspectives*, bimonthly. Contains information on WOCCU activities and development projects.

WORLD COUNCIL OF OPTOMETRY (WCO)

c/o Anthony F Di Stefano OD, 1200 W. Godfrey Ave., Fitch Hall, Philadelphia, PA 19141 **PHONE:** (215) 276-6220 **FX:** (215) 276-6081 **E-MAIL:** 6932932@mcimail.com **FOUNDED:** 1927. **OFFICERS:** Anthony F. Di Stefano, Exec.Dir. **MEMBERS:** 65. **STAFF:** 2. **LANGUAGES:** English.

DESCRIPTION: Members are national optical and optometric organizations in 60 countries. Promotes optometry and vision care worldwide. Works to develop optometry through education, legal and legislative advice, discussion with authorities, and support of similar associations. Is concerned with protecting optometry from what the league sees as proposed restrictive legislation, and apathy toward the profession. Establishes educational standards; offers evaluation program for schools and colleges of optometry; provides consulting services. **PUBLICATIONS:** *International Country Contact Register*, periodic; *Interoptics* (in English), quarterly; *Members Journals*, periodic; *Optometric Mailing Directory* (in English), periodic. Lists information on optometric organizations.; *Publications List*, periodic; *Register of Schools and Colleges of Optometry*, periodic; Brochure, periodic.

WORLD CRAFTS COUNCIL (WCC) (Conseil Mondial de l'Artisanat)

Miyako-Messe, 2F, 9-1, Seishoji-cho, Okazaki, Sakyo-ku, Kyoto 606, Japan **PHONE:** 81 75 7618550 **FX:** 81 75 7618610 **E-MAIL:** wcckyoto@mbox.kyoto-inet.or.jp **FOUNDED:** 1964. **MEMBERS:** 91. **STAFF:** 2. **BUDGET:** US$500,000. **LANGUAGES:** English, French, German, Spanish.

DESCRIPTION: Members are international crafts organizations dedicated to strengthening and maintaining the status of crafts as a vital part of cultural life. Serves as a clearinghouse for information on crafts. Promotes respect for and recognition of the work of crafts; offers encouragement, assistance, and advice to craftspersons. Organizes an international exchange of craftspersons among member countries for study, work, and travel. Sponsors research program on the role of crafts in development. Sponsors meetings, conferences, workshops, and exhibitions. Maintains information service. Establishes world craft data base and image data bank of crafts. **PUBLICATIONS:** *Annual Report*; *WCC*

International Newsletter (in English and French), semiannual. Also publishes regional craft directories, survey of programs offered by schools, universities, and academies, and other materials.

WORLD CUSTOMS ORGANIZATION (WCO) (Organisation mondiale des douanes - OMD)

Rue du Marche 30, B-1210 Brussels, Belgium **PHONE:** 32 2 2099211 **FX:** 32 2 2099292 **E-MAIL:** information@wcoomd.org **WEBSITES:** http://www.wcoomd.org **FOUNDED:** 1952. **OFFICERS:** M. Danet, Sec.Gen. **MEMBERS:** 150. **STAFF:** 130. **LANGUAGES:** English, French.

DESCRIPTION: Enhances the effectiveness and efficiency of customs administrations in the areas of compliance with trade regulations, protection of society and revenue collection, thereby contributing to the economic and social well-being of nations. **PUBLICATIONS:** *WCO News* (in English and French), semiannual; Brochures, periodic; Handbooks, periodic.

WORLD ECONOMIC FORUM (WEF)

53, chemin des Hauts-Crets, Cologny, CH-1223 Geneva, Switzerland **PHONE:** 41 22 8691212 **FX:** 41 22 7862744 **TX:** 413280 **FOUNDED:** 1971. **OFFICERS:** Klaus Schwab, Pres. **MEMBERS:** 1,000. **STAFF:** 70. **LANGUAGES:** English.

DESCRIPTION: Members are business persons, politicians, and academicians. Integrates leaders from business, government, and academia into a partnership committed to improving the state of the world. Operates under the legal supervision of the Swiss Federal Council and has consultative status with the United Nations. **PUBLICATIONS:** *World Competitiveness Report* (in English), annual; *World Link* (in English), bimonthly; Proceedings.

WORLD EDUCATION FELLOWSHIP (WEF)

58 Dickens Rise, Chigwell, E. Sussex IG7 6NY, England **PHONE:** 44 181 2817122 **FX:** 44 181 2817122 **FOUNDED:** 1921. **OFFICERS:** George John, Gen.Sec. **MEMBERS:** 14,000.

DESCRIPTION: Educational charitable organization promoting international understanding and cooperation through education. Fosters improvements in education that encourage individual contributions to society. Promotes the continuation of education into adult life. Provides a forum for teachers, parents, and students to discuss educational issues. Has examined concerns such as the role of fine arts in education, cross-cultural understanding, environmental education, and peace studies. Maintains consultative status with the United Nations Educational, Scientific and Cultural Organization. **PUBLICATIONS:** *The New Era in Education*, 3/year.

WORLD ENERGY COUNCIL (WEC) (Conseil Mondial de l'Energie)

34 St. James's St., London SW1A 1HD, England **PHONE:** 44 171 9303966 **FX:** 44 171 9250452 **TX:** 264707 WECIHQ G **E-MAIL:** info@wec.co.uk **WEBSITES:** http://www.worldenergy.org/wec-geis **FOUNDED:** 1924. **OFFICERS:** S. Doucet, Sec.Gen. **MEMBERS:** 91. **STAFF:** 15. **LANGUAGES:** English, French.

DESCRIPTION: Members are energy ministries, fuel and power corporations, engineering industries, universities, research organizations, and manufacturers in 91 countries involved in the production, supply, and study of energy resources. Promotes the development and peaceful use of energy resources by considering: potential resources and all means of production, transportation, and utilization; energy consumption in its overall relationship to the growth of economic activity in the area; and the social and environmental aspects of energy supply and utilization. Specific topics include worldwide survey of energy resources, national energy data profiles, and energy terminology. Maintains 8 technical and topical study committees. **PUBLICATIONS:** *Energy Dictionary* (in English, French, German, and Spanish), every 6 years; *National Energy Data* (in English), triennial; *World Energy Council Journal* (in English and French), biennial; *World Energy Resources* (in English and French), triennial; Proceedings, periodic; Reports, periodic.

WORLD ENVIRONMENT CENTER (WEC)

419 Park Ave. S., Ste. 1800, New York, NY 10016 **PHONE:** (212) 683-4700 **FX:** (212) 683-5053 **E-MAIL:** webmaster@wec.org **WEBSITES:** http://www.wec.org **FOUNDED:** 1974. **OFFICERS:** Antony Marcil, Pres. & CEO. **STAFF:** 46.

BUDGET: US$1,000,000. **LANGUAGES:** Bengali, Bulgarian, Estonian, Latvian, Lithuanian, Polish, Romanian, Russian, Spanish, Tamil, Thai, Ukrainian.

DESCRIPTION: Works to strengthen industrial and urban environmental, health, and safety policies by establishing and promoting partnerships among industry, government, and nongovernmental organizations. Facilitates information and expertise. Encourages corporate environmental leadership and responsibility. Provides training and technical cooperation programs utilizing volunteer experts. **PUBLICATIONS:** *Country Profiles*. Summary of environmental, health, and safety regulations and statutes for 30 national and subnational jurisdictions.

WORLD EVANGELICAL FELLOWSHIP (WEF)

PO Box WEF, Wheaton, IL 60189-8004 **PHONE:** (630) 668-0440 **FX:** (630) 668-0498 **E-MAIL:** wef-na@xc.org **WEBSITES:** http://www.worldevangelical.org **FOUNDED:** 1951. **OFFICERS:** Agustin B. Vencer Jr., Int'l. Dir. **MEMBERS:** 150,000,000. **BUDGET:** US$1,900,000. **LANGUAGES:** English, French, Portuguese, Spanish.

DESCRIPTION: Works to establish and help regional and national evangelical alliances empower and mobilize local churches and Christian organizations to ''disciple the nations for Christ.'' Seeks to: strengthen evangelical leadership; develop a ''world evangelical identity''; enhance the interdependence of churches through sharing of information and resources. **PUBLICATIONS:** *Directory of Evangelical Associations*, biennial; *Evangelical Review of Theology*, quarterly; *Evangelical World*, bimonthly; *WEF Religious Liberty Commission Update on Christian Persecution*, 6/year.

WORLD EXPORT PROCESSING ZONES ASSOCIATION (WEPZA)

PO Box 986, Flagstaff, AZ 86002 **PHONE:** (520) 779-0052 **FX:** (520) 774-8589 **E-MAIL:** wepza@aol.com **WEBSITES:** http://www.wepza.org/world/ **FOUNDED:** 1978. **OFFICERS:** Richard L. Bolin, Dir. **MEMBERS:** 50. **STAFF:** 6. **BUDGET:** US$250,000. **LANGUAGES:** English.

DESCRIPTION: Members are export processing zones and free economic zones, public and private. (EPZs are where materials are brought in to be assembled or processed. These zones are customs-controlled, allowing for inexpensive production, and later export, of goods. EPZs are found throughout the world.) Fosters information exchange and represents members' interests before international bodies. Assists in training staff of EPZs; conducts management program for EPZ-user managers; sponsors conferences of managers and export zone users and periodic training workshops. Maintains professional code of conduct; compiles statistics. **PUBLICATIONS:** *Journal of The Flagstaff Institute*, semiannual. Discusses free trade zones, free economic zones, development of a global market, and investment attraction in developing countries.; *Juarez/2000*. Includes data summaries from interviews of managers of maquila plants and workers.; *Linking the Export Processing Zone to Local Industry*; *Production Sharing: A Conference With Peter Drucker*. Provides a global perspective on production sharing which is economic integration by stages of the productive process across international borders.; *Sonora/2010*. Examines the accelerating march to the Mexican interior of the maquila industry.; *WEPZA International Directory of Export Processing Zones and Free Trade Zones*; *WEPZA Newsletter*, monthly. Publishes various books.

WORLD FEDERALIST MOVEMENT

777 U.N. Plaza, 12th Fl., New York, NY 10017 **PHONE:** (212) 599-1320, (212) 687-2176 **FX:** (212) 599-1332 **E-MAIL:** wfm@igc.apc.org **WEBSITES:** http://www.igc.apc.org/icc/; http://www.worldfederalist.org **FOUNDED:** 1947. **OFFICERS:** Sir Peter Ustinov, Pres. **MEMBERS:** 30,000. **STAFF:** 6. **LANGUAGES:** English.

DESCRIPTION: Strives to achieve a just world order by strengthening the United Nations. Advocates extending the power of the U.N. to include the authority to make and enforce laws for the peaceful settlement of disputes, to govern the high seas and outer space, and to raise revenue under limited powers of taxation. Calls for improved international cooperation in the areas of the environment, development, and disarmament. Organizes seminars. Sponsors educational and research program. **PUBLICATIONS:** *WFNews*, semiannual.

WORLD FEDERATION OF ADVERTISERS (WFA) (Federation Mondiale des Annonceurs - FMA)

18-24, rue des Colonies, bte. 6, B-1000 Brussels, Belgium **PHONE:** 32 2 5025740 **FX:** 32 2 5025666 **E-MAIL:** info@wfa.be **WEBSITES:** http://www.wfa.be **FOUNDED:** 1953. **OFFICERS:** Bernhard Adriaensens, Mng.Dir. **MEMBERS:** 66. **STAFF:** 4. **LANGUAGES:** English, French.

DESCRIPTION: Members are international advertisers' associations in 40 countries. Represents members' interests before international and supranational authorities and trade and consumer associations; maintains close relationships with decision-makers within the European Community Commission, the Council of Europe, and United Nations agencies. **PUBLICATIONS:** *Agency Contract Guidelines*; *Charter on Media Issues*; *Harmonization of TV Audience Research in Europe*; *Management of Agency Services*; *Media Buying Groups*; *Portrayal of Women in Advertising*; *Report and Review*, annual; *The 10 Golden Rules for Lobbying*; *WFA E-Monitor* (in English), monthly; Papers, periodic.

WORLD FEDERATION OF THE DEAF (WFD)

Magnus Ladulasgatan 63 4tr, 11827 Stockholm, Sweden **FX:** 46 8 4421499 **E-MAIL:** carol.lee.aquiline@wfdnews.org **WEBSITES:** http://www.wfdnews.org **FOUNDED:** 1951. **OFFICERS:** Carol-Lee Aquiline, Gen.Sec. **MEMBERS:** 120. **STAFF:** 1. **LANGUAGES:** English.

DESCRIPTION: Ordinary members are National associations of the deaf that are legally constituted and represent in the widest sense the deaf in their country; other members are national or international associations, societies, and bodies for and of the deaf; health, educational, social, and similar establishments that accept the aims of the federation; professionals interested in deafness and related subjects; persons performing special tasks connected with the aims of the federation; parents and friends of the deaf. Represents the worldwide deaf community in international forums such as the United Nations. Makes policy statements and recommendations. Coordinates research and other projects with the deaf community; conducts surveys. Has established a network of experts on deafness and related issues. Promotes the use of national sign language as the first language of deaf people. **PUBLICATIONS:** *Proceedings of International Congresses*, quarterly; *WFD Manual on How to Establish and Run an Organization of the Deaf*; *WFD News* (in English), quarterly; *WFD Report on the Status of Sign Language*; *WFD Survey of Deaf People in the Developing World.*

WORLD FEDERATION OF ENGINEERING ORGANIZATIONS (WFEO) (Federation Mondiale des Organisations d'Ingenieurs - FMOI)

Maison de l'UNESCO, 1, rue Miollis, F-75015 Paris, France **PHONE:** 33 1 45684892 **FX:** 33 1 45684914 **E-MAIL:** pdeboigne@fmoi.org **FOUNDED:** 1968. **OFFICERS:** Mr. M. Pierre Edouard de Boigne, Exec.Dir. **MEMBERS:** 85. **STAFF:** 3. **BUDGET:** US$200,000. **LANGUAGES:** English, French.

DESCRIPTION: Members are national and international organizations representing the engineering profession. Purposes are: to encourage the formation and activities of national and international associations of engineers; to facilitate the exchange of engineering information; and to foster engineering education and training. Promotes cooperation among engineering organizations and between engineering and other organizations. Encourages the freedom of movement of engineers between different countries. Sponsors scientific and engineering meetings, symposia, and congresses. Undertakes special projects relating to the professional interests of members of the Federation. Cooperates with governmental and nongovernmental organizations. **PUBLICATIONS:** Newsletter (in English), semiannual.

WORLD FEDERATION OF HUNGARIAN JEWS (WFHJ)

136 E. 39th St., New York, NY 10016 **PHONE:** (212) 683-5377 **FX:** (212) 684-6327 **E-MAIL:** figyelo@hotmail.com **FOUNDED:** 1951. **OFFICERS:** Ervin Farkas, Pres. **LANGUAGES:** English, German, Hungarian.

DESCRIPTION: Members are Hungarian Jewish associations and individuals in 15 countries. United to represent Hungarian Jews worldwide in seeking restitution by the German government for World War II war crimes, as well as punishment of Hungarian war-criminals. Promotes the cultural values of Hungarian Jewry; carries on charitable work in

Israel and in Hungary. **Convention/Meeting:** none. **PUBLICATIONS:** *Newyorki Figyelo: Voice of American Hungarian Jewry* (in English and Hungarian), monthly. Contains articles on Hungarian-Jewish history and cultural heritage; also includes book reviews and literary and poetic pieces by Hungarian writers.

WORLD FEDERATION FOR MEDICAL EDUCATION (WFME) (Federation Mondiale pour l'Enseignement Medical - FMEM)

c/o University of Copenhagen, Panum Institute, Blegdamsvej 3, DK-2200 Copenhagen, Denmark **PHONE:** 45 35 32790009 **FX:** 45 35 327070 **FOUNDED:** 1972. **OFFICERS:** Dr. Hans Karle, Pres. **STAFF:** 4. **BUDGET:** £6. **LANGUAGES:** English.

DESCRIPTION: Members are regional medical associations and associations of medical schools. Promotes the integrated study of medical education worldwide. Evaluates the effectiveness of medical education in meeting the needs of contemporary society. Acts as international representative of medical education before the World Health Organization, UNICEF, UNESCO, United Nations Development Programme, and the World Bank. **PUBLICATIONS:** *Report of the World Conference on Medical Education, Edinburgh, 1988/Proceedings of the World Summit on Medical Education, Edinburgh, 1994*; *Report of the World Summit on Medical Education, Edinburgh, 1994.*

WORLD FEDERATION FOR MENTAL HEALTH (WFMH)

1021 Prince St., Alexandria, VA 22314 **PHONE:** (703) 838-7543 **FX:** (703) 519-7648 **E-MAIL:** wfmh@erols.com **WEBSITES:** http://www.wfmh.org **FOUNDED:** 1948. **OFFICERS:** Richard Hunter, Dep.Sec.Gen. **MEMBERS:** 3,500. **STAFF:** 6. **BUDGET:** US$432,000. **LANGUAGES:** English.

DESCRIPTION: Members are associations and individuals dedicated to achieving the highest level of public mental health. Objectives include charitable, scientific, literary, and educational activities in the field of mental health. Organizes training programs. Sponsors World Mental Health Day. **PUBLICATIONS:** Newsletter, quarterly.

WORLD FEDERATION OF NEUROSURGICAL SOCIETIES (WFNS)

c/o Dr. Edward Laws, Univ. of Virginia Health Services, Department of Neuro-Societies, Box 212, Charlottesville, VA 22908 **PHONE:** (804) 924-2650 **FX:** (804) 924-5894 **FOUNDED:** 1957. **OFFICERS:** Dr. Edward Laws, Sec. **MEMBERS:** 64. **STAFF:** 1. **BUDGET:** US$65,000. **LANGUAGES:** English.

DESCRIPTION: Members are national neurosurgical societies representing approximately 17,400 neurosurgeons. Works for the advancement of neurological surgery. **PUBLICATIONS:** *Critical Reviews in Neurosurgery*; *Federation News*, quarterly; *Proceedings of International Congress*, quadrennial; *World Directory*, periodic.

WORLD FEDERATION OF OCCUPATIONAL THERAPISTS (WFOT) (Federation Mondiale des Ergotherapeutes)

c/o Mrs. Carolyn Webster, 53 Ord St., PO Box 441, West Perth 6872, Australia **PHONE:** 61 894269200 **FX:** 61 894269200 **E-MAIL:** wfot@multiline.com.au **FOUNDED:** 1952. **OFFICERS:** Mrs. Caroline Webster, Pres. **MEMBERS:** 50. **LANGUAGES:** English.

DESCRIPTION: Membership is comprised of 50 countries which have Occupational Therapy Associations with approved constitutions and educational programs; associate members, which are countries with approved constitutions; and individual members, who are qualified occupational therapists. Objectives are to: promote the occupational therapy profession; provide for exchange of information and publications; encourage international cooperation; establish standards of education for occupational therapists; maintain professional ethics; facilitate international exchange and placement of therapists and students. Promotes research; assists developing countries in formulating education programs for occupational therapists. **PUBLICATIONS:** *Conference Proceedings*, quadrennial; *WFOT Bulletin*, semiannual; Bibliographies, periodic. Contains information on standards and requirements; Manuals, periodic. Contains information on study courses; Pamphlets, periodic. Contains information on standards and requirements.

WORLD FEDERATION OF PUBLIC HEALTH ASSOCIATIONS (WFPHA)

c/o Amer. Public Health Assn., 1015 15th St. NW, Ste. 300, Washington, DC 20005 **PHONE:** (202) 789-5696, (202) 789-5651 **FX:** (202) 789-5661 **FOUNDED:** 1967. **OFFICERS:** Dr. Richard Levinson, Associte Exec.Dir. **MEMBERS:** 48. **LANGUAGES:** English.

DESCRIPTION: Members are national public health associations united ''to strengthen the public health profession and to improve community health throughout the world.'' Encourages formation of national public health associations. Organizes field projects; conducts special studies; offers lectures; compiles reports. **PUBLICATIONS:** *Reports of Triennial International Congresses*; *WFPHA Report*, quarterly.

WORLD FEDERATION OF TRADE UNIONS (WFTU) (Federation Syndicale Mondiale - FSM)

Branicka 112, CZ-14701 Prague 4, Czech Republic **PHONE:** 420 2 44462140, 420 2 44462085 **FX:** 420 2 44461378 **E-MAIL:** wftu@login.cz **FOUNDED:** 1945. **OFFICERS:** Alexander Zharikov, Gen.Sec. **MEMBERS:** 86. **STAFF:** 20. **LANGUAGES:** English, French, Spanish.

DESCRIPTION: Members are trade unions in 120 countries with a combined membership of 130,000,000 individuals. Seeks to develop and coordinate joint action of trade unions at the international level on questions of common interest. Aims are to: defend the right to work and the right to decent living conditions; defend and extend trade union rights and democratic freedoms; limit the power of transnational corporations; establish a more just international economic order. Opposes apartheid and all forms of racial discrimination, the nuclear arms race, and aggressive imperialist movements. Organizes solidarity campaigns with struggling trade unions and gives material and moral support. Is particularly concerned with the problems of women, children, and migrant workers. Cooperates with international democratic organizations such as the World Federation of Democratic Youth and the Women's International Democratic Federation. **PUBLICATIONS:** *Flashes from the Trade Unions* (in English, French, and Spanish), biweekly. Contains news and features on world labor. Also publishes bulletins, booklets, posters, and leaflets.

WORLD FEDERATION OF UNITED NATIONS ASSOCIATIONS (WFUNA) (Federation Mondiale des Associations pour les Nations Unies - FMANU)

c/o Palais des Nations, CH-1211 Geneva 10, Switzerland **PHONE:** 41 22 7330730 **FX:** 41 22 7334838 **TX:** 412962 UNO CH **FOUNDED:** 1946. **MEMBERS:** 80. **STAFF:** 4. **LANGUAGES:** English, French.

DESCRIPTION: Members are national associations supporting the United Nations. Promotes education and understanding of the role of the U.N. and its agencies and advocates its development. **PUBLICATIONS:** *Membership List*, semiannual; *WFUNA News* (in English and French); Reports, periodic.

WORLD FELLOWSHIP OF BUDDHISTS (WFB) (Fraternite Mondiale des Bouddhistes)

616 Benjasiri Park, Soi Medhinivet, Off.Soi.Sukhumvit 24, Sukhumvit Rd., Bangkok 10110, Thailand **PHONE:** 66 1 1284 **FX:** 66 1 0555 **E-MAIL:** wfb_hq@asianet.co.th **FOUNDED:** 1950. **OFFICERS:** Mr. Phan Wannamethee, Pres. **MEMBERS:** 123. **STAFF:** 14. **BUDGET:** 5,000,000 Bht. **LANGUAGES:** English.

DESCRIPTION: Members are Buddhist organizations representing 37 countries. Promotes strict observance and practice of teachings of Buddha. Strives to secure solidarity among Buddhists and to propagate the doctrine of the Buddha. Organizes and participates in activities including social, educational, cultural, and other humanitarian services. Works with other organizations toward securing peace and happiness among all individuals. Makes available exchange program for books and magazines particularly in Dhamma field. **PUBLICATIONS:** *Book Series* (in English and Thai); *WFB Journal* (in Thai), bimonthly; *WFB Newsletter* (in English), monthly; *WFB Review* (in English), quarterly.

WORLD FOOD PROGRAMME (WFP) (Programme Alimentaire Mondiale - PAM)

Via Cesare Giulio Vioke, 68, Paro dei Medici, I-00148 Rome, Italy **PHONE:** 39 6 65131 **FX:** 39 6 6590632 **TX:** 626675 WFP I **E-MAIL:** webadministrator@wfp.org **WEBSITES:** http://www.wfp.org **FOUNDED:** 1961. **OFFICERS:** Ms. Catherine Bertini, Exec.Dir. **STAFF:** 4,000. **BUDGET:** US$1,400,000,000. **LANGUAGES:** Arabic, English, French, German.

DESCRIPTION: Frontline United Nations agency mandated to combat hunger and to encourage longer-term food security in the poorest regions of the globe. Operational since 1963, the WFP works to: support the most vulnerable when food needs are critical so that they are better able to attain their human potential; help the hungry poor become self-reliant and build assets such as, schools, and irrigation systems in their communities; and help those caught up in humanitarian crises. **PUBLICATIONS:** *WFP Food Aid Review* (in English, French, and Spanish), annual; *WFP Journal* (in English, French, and Spanish), quarterly. Also publishes *Food Aid Works for the Environment*, *WFP in Africa*, *When Food is Hope*, *More Than Food Aid*, *Food Aid and the Well-Being of Children in Developing Countries*, *Women and Food Aid*, *Emergency Handbook: Food Aid in Emergencies*, *WFP Food Aid Review*, brochures, research studies, and policy papers.

WORLD HEALTH ORGANIZATION (WHO) (Organisation Mondiale de la Sante - OMS)

20, ave. Appia, CH-1211 Geneva 27, Switzerland **PHONE:** 41 22 7912111 **FX:** 41 22 7910746 **TX:** 415416 UNISANTE-GENE **FOUNDED:** 1948. **OFFICERS:** Hiroshi Nakajima M.D., Dir.Gen. **MEMBERS:** 192. **STAFF:** 4,500. **BUDGET:** US$842,654,000. **LANGUAGES:** Arabic, Chinese, English, French, Russian, Spanish.

DESCRIPTION: International health agency of the United Nations consisting of countries working toward the goal of ''health for all,'' seeking to obtain the highest level of health care for all people. Believes health is a fundamental right of every human being without distinction of race, religion, political belief, economic situation, or social conditions and holds that all people deserve equal access to health services to enable them to lead socially and economically productive lives. Objectives are to: act as directing and coordinating authority on international health work; ensure valid and productive technical cooperation; promote research; prevent and combat diseases; generate and transfer information. Strives to eliminate poverty. Emphasizes the health needs of developing countries lacking resources and funds for modern medical technologies; works toward developing new techniques that will fulfill these needs by utilizing available resources, integrating educational, agricultural, town planning, and sanitation programs with health programs, combining peripheral health services and existing health systems, and applying appropriate technologies at reasonable costs. Establishes standards for food, biological, and pharmaceutical needs, develops standardized diagnostic procedures, and determines environmental health criteria. Promotes 8 elements of primary health care including: health education on prevention and cures; proper food supply and nutrition; adequate supply of safe water and sanitation; maternal and child health care; immunization; control of endemic diseases; and provisions of essential drugs. Acts as clearinghouse; coordinates activities with the United Nations on health and socioeconomic development; works with international nongovernmental organizations in the health sector. Supports International Agency for Research on Cancer. Maintains expert and scientific committees. **PUBLICATIONS:** *Bulletin of WHO* (in English and French), bimonthly; *International Digest of Health Legislation* (in English and French), quarterly; *Weekly Epidemiological Record* (in English and French); *WHO Drug Information* (in English, French, and Spanish), quarterly; *World Health* (in English, French, German, Portuguese, Russian, and Spanish), 10/year; *World Health* (in Arabic and Farsi); *World Health Forum* (in Arabic, Chinese, English, French, Russian, and Spanish), quarterly; *World Health Statistics Annual* (in English, French, and Russian); *World Health Statistics Quarterly* (in English and French). Includes summaries in Arabic, Chinese, Russian, and Spanish. Also publishes *WHO Technical Report Series*, *WHO Environmental Health Criteria Series*, *WHO AIDS Series*, *WHO Health and Safety Guidelines*, directories, and non-serial publications.

WORLD INTELLECTUAL PROPERTY ORGANIZATION (WIPO) (Organisation Mondiale de la Propriete Intelectuelle - OMPI)

34, chemin des Colombettes, CH-1211 Geneva

20, Switzerland **PHONE:** 41 22 3389111, 41 22 3389246 **FX:** 41 22 3388810 **TX:** 412912 OMPICH **WEBSITES:** http://www.wipo.int **FOUNDED:** 1970. **OFFICERS:** Dr. Kamil Idris, Dir.Gen. **STAFF:** 650. **BUDGET:** 400,000,000 SFr. **LANGUAGES:** Arabic, Chinese, English, French, Russian, Spanish.

DESCRIPTION: Advocates the protection of intellectual property worldwide. (Intellectual property is described as having 2 major categories: industrial property, namely patents and other rights involving technological inventions as well as trademarks and industrial designs; and copyrights, involving literary, musical, and artistic works.) Encourages creative intellectual activity; facilitates the transfer of technology to and among developing nations. Administers international intellectual property treaties including the Paris Convention for the Protection of Industrial Property and the Berne Convention for the Protection of Literary and Artistic Works. Undertakes preparation and adaptations of treaties to meet with changes in international industrial, trade, and cultural relations. Provides assistance to governments in modernizing their legislation and institutions. Conducts training programs for developing nations. Administers treaties for the registration of industrial property rights and an arbitration and mediation center for intellectual property disputes between private parties. **PUBLICATIONS:** *Gazette OMPI des Marques Internationales/WIPO Gazette of International Marks* (in English and French), bimonthly; *Industrial Property and Copyright* (in English, French, and Spanish), monthly; *Intellectual Property in Asia and the Pacific* (in English), quarterly; *International Designs Bulletin* (in English and French), monthly; *Les Appelations d'Origine* (in French), periodic; *PCT Gazette* (in English and French), weekly; *PCT Newsletter* (in English), bimonthly; *WIPO General Information* (in Arabic, Chinese, English, French, German, Japanese, Portuguese, Russian, and Spanish), periodic; Papers, periodic. Containing patent laws, treaties, and convention proceedings.

WORLD JEWISH CONGRESS (WJC)

501 Madison Ave., 17th Fl., New York, NY 10022 **PHONE:** (212) 755-5770 **TF:** (800) 755-5883 **FX:** (212) 755-5883 **TX:** 236129 WJC UR **FOUNDED:** 1936. **OFFICERS:** Elan Steinberg, Exec.Dir. **MEMBERS:** 33. **STAFF:** 66. **LANGUAGES:** English.

DESCRIPTION: Members are national Jewish organizations representing 3,000,000 individuals in 68 countries. Seeks to secure and safeguard the rights, status, and interests of Jews and Jewish communities throughout the world. Is committed to supporting national and international protection of human rights, without distinction on grounds of race or religion. Maintains 10,000 volume library. **PUBLICATIONS:** *Batefutsot* (in Hebrew), monthly; *Boletin Informative OJI* (in Spanish), biweekly; *Christian Jewish Relations*, quarterly; *Coloquio* (in Spanish), quarterly; *Gesher* (in Hebrew), quarterly; *News and Views From the WJC*, 6-8/year; *Patterns of Prejudice*, quarterly; *Soviet Jewish Affairs*, annual. Also publishes books and research reports.

WORLD JURIST ASSOCIATION (WJA)

1000 Connecticut Ave. NW, Ste. 202, Washington, DC 20036 **PHONE:** (202) 466-5428 **FX:** (202) 452-8540 **TX:** 440456 **E-MAIL:** wja@wja-wptlc.org **WEBSITES:** http://www.wja-wptlc.org **FOUNDED:** 1963. **OFFICERS:** Margeret M. Henneberry, Exec.VP. **MEMBERS:** 8,000. **STAFF:** 4. **LANGUAGES:** English, French, German, Spanish.

DESCRIPTION: Lawyers, judges, law professors, jurists, law students, and nonlegal professionals in 140 countries and territories. Seeks to build laws and legal institutions for international cooperation. Conducts Global Work Program to recommend research and voluntary action for development of international law as a basis for promoting the rule of law and the resolution of disputes by peaceful means. Sponsors biennial World Law Day. Maintains biographical archives. WJA contains 21 Sections, including Constitutional Law, Foreign Trade and Investment, Human Rights, and Litigation. **PUBLICATIONS:** *Law and Judicial Systems of Nations* (in English), periodic. Offers descriptions of different legal systems around the world.; *Law Technology*, quarterly. Contains articles about the relation of technology to law. Includes bibliography of computers and law and book reviews.; *Publications Directory*, periodic; *The World Jurist*, bimonthly. For those seeking to promote the rule of law and improvement of administration of justice in the world community.; Pamphlets; Papers.

WORLD MEDICAL ASSOCIATION (WMA) (Association Medicale Mondiale - AMM)

28, ave. des Alpes, BP 63, F-01212 Ferney-Voltaire, France **PHONE:** 33 450 407575 **FX:** 33 450 405937 **E-MAIL:** wma@iprolink.fr **WEBSITES:** http://www.wma.net **FOUNDED:** 1947. **OFFICERS:** Dr. Delon Human, Sec.Gen. **MEMBERS:** 71. **STAFF:** 8. **BUDGET:** US$1,000,000. **LANGUAGES:** English, French, Spanish.

DESCRIPTION: Federation of national medical associations throughout the world. Goal is to achieve the highest international standards in medical education, medical science, medical ethics, and health care for people worldwide. Promotes closer ties and better communication among medical organizations and doctors of the world; studies professional problems in different countries. Represents and protects the rights and interests of physicians and people internationally. Encourages proper nutrition in developing countries; urges the teaching of human values in the practice of medicine. Seeks to improve maternal and child health care. Issues declarations on ethical topics including: abortion; AIDS; abuse of children and the elderly; euthanasia; genetic engineering; organ transplantation; the use and misuse of psychotropic drugs; torture; biomedical research on human subjects; medical care in rural areas; an international code of medical ethics; and rights of the patient. **PUBLICATIONS:** *World Medical Journal* (in English), bimonthly. Contains medical and medico-political articles.

WORLD METEOROLOGICAL ORGANIZATION (WMO) (Organisation Meteorologique Mondiale - OMM)

7 bis avenue de la Paix, CP 2300, CH-1211 Geneva 2, Switzerland **PHONE:** 41 22 7308111 **FX:** 41 22 7308181 **TX:** 414199 OMMCH **E-MAIL:** ipa@www.wmo.ch **WEBSITES:** http://www.wmo.ch **FOUNDED:** 1950. **OFFICERS:** Prof. G.O.P. Obasi, Sec.Gen. **MEMBERS:** 185. **STAFF:** 260. **BUDGET:** US$255,000,000. **LANGUAGES:** Arabic, Chinese, English, French, Russian, Spanish.

DESCRIPTION: Members are states, territories, and groups of territories. Supports international cooperation in establishing networks for meteorological observations and hydrological and geophysical observations related to meteorology. Promotes the creation of centers offering meteorological services and systems to facilitate exchange of information. Encourages uniform publication of observations and statistics and the application of meteorology as it involves aviation, shipping, water problems, and agriculture. Fosters activities in operational hydrology and cooperation between meterological and hydrological services. Organizes courses in meteorology and operational hydrology. **PUBLICATIONS:** *Annual Report* (in English, French, Russian, and Spanish); *Composition of WMO* (in English and French), quarterly; *Reports of Constituent Bodies* (in English, French, Russian, and Spanish), periodic; *WMO* (in English, French, Russian, and Spanish), quarterly.

WORLD METHODIST COUNCIL (WMC)

575 Lakeshore Dr., PO Box 518, Lake Junaluska, NC 28745 **PHONE:** (828) 456-9432 **FX:** (828) 456-9433 **FOUNDED:** 1881. **OFFICERS:** Joe Hale, Gen.Sec. **MEMBERS:** 29,069,376. **BUDGET:** US$500,000. **LANGUAGES:** English.

DESCRIPTION: Fraternal and cooperative association of Methodist, Wesleyan, and other related united churches. Seeks to ''draw the branches of the Wesleyan Movement closer together in fellowship and devotion to their mutual heritage and to promote among them evangelistic, educational, historical, and other cooperative movements.'' Exercises no legislative power over separate denomination members but gives focus and leadership to mutually agreed upon goals and programs. Maintains World Methodist Museum of rare books, records, paintings, sculptures, and other materials pertaining to John Wesley and the beginnings of early Methodism. **PUBLICATIONS:** *Encyclopedia of World Methodism*; *Proceedings of World Methodist Conferences*; *World Methodist Council—Handbook of Information*, quinquennial. Includes community and membership statistics of World Methodism.; *World Parish: International Organ of the World Methodist Council*, bimonthly.

WORLD MOVEMENT OF CHRISTIAN WORKERS (WMCW) (Mouvement Mondial des Travailleurs Chretiens - MMTC)

Bld. du Jubile 124, B-1080 Brussels, Belgium **PHONE:** 32 2 4215840, 32 2 4215841 **FX:** 32 2

4215849 **E-MAIL:** mmtc@skynet.be **FOUNDED:** 1966. **OFFICERS:** Norbert Klein, Gen.Sec. **MEMBERS:** 55. **LANGUAGES:** English, French, German, Spanish.

DESCRIPTION: Federation of Christian workers' organizations. Objectives are to serve and unite people committed to the liberation of the working class and to aid in the spiritual fulfillment of workers and their families. Acts as forum for discussion among key leaders of movements in different countries. Promotes coordination of efforts in order to maximize continental and regional benefits. Participates in international labor events. **PUBLICATIONS:** *Executive Council Directory*, annual; *INFOR-WMCW* (in English, French, German, Portuguese, and Spanish), bimonthly. Also publishes *The Evolving WMCW* (booklet); *The World Economy (cartoon)* in 5 languages.

WORLD NEIGHBORS (WN)

4127 NW 122 St., Oklahoma City, OK 73120-9933 **PHONE:** (405) 752-9700 **TF:** (800) 242-6387 **FX:** (405) 752-9393 **E-MAIL:** info@wn.org **WEBSITES:** http://www.wn.org **FOUNDED:** 1951. **OFFICERS:** Ron Burkard, Exec.Dir. **MEMBERS:** 10,000. **STAFF:** 33. **BUDGET:** US$3,200,000. **LANGUAGES:** English.

DESCRIPTION: Seeks to eliminate hunger, disease, and poverty in Asia, Africa, and Latin America. Helps people to analyze and solve their own problems by developing and testing simple technologies at the community level and training local leaders to spread successful methods. Programs focus on food production, community-based health, family planning, water and sanitation, environmental conservation, and small business. Provides speakers, videocassettes, films, filmstrips, literature, and other training materials on long-term economic development and causes of poverty and hunger to schools, religious and study groups, and other organizations. **PUBLICATIONS:** *Neighbors*, quarterly; *Training Materials Catalog*; *Two Ears of Corn - A Guide to People-Centered Agricultural Improvement*; Annual Report.

WORLD NEIGHBORS - SOUTHEAST ASIA

PO Box 71, Ubud, 80571 Bali, Indonesia **PHONE:** 62 361 975707 **FX:** 62 361 976487 **E-MAIL:** wnisea@denpasar.wasantara.net.id **FOUNDED:** 1951. **OFFICERS:** Stefan Wodicka, Area Rep.

DESCRIPTION: Works to coordinate the activities of organizations working to eliminate hunger, disease, and poverty. Conducts programs to strengthen the human resources of economically developing areas, focusing on indigenous leadership development, family planning, food production, water and sanitation, small business development, and environmental protection. Sponsors educational programs.

WORLD ORGANISATION OF SYSTEMS AND CYBERNETICS (WOSC) (Organisation Mondiale pour la Systemique et la Cybernetique - OMSC)

c/o Prof. Robert Vallee, 2, rue de Vouille, F-75015 Paris, France **PHONE:** 33 1 45336246 **FX:** 33 1 45336276 **FOUNDED:** 1969. **OFFICERS:** Prof. Robert Vallee, Dir.Gen. **MEMBERS:** 33. **STAFF:** 12. **LANGUAGES:** English, French.

DESCRIPTION: Members are national societies interested in cybernetics, systems, robotics, computer science, artificial intelligence, and related areas. Objectives are to: sponsor national and international activities in the fields of cybernetics and systems, and generate interest in related disciplines; weed out pseudo-cybernetic claims and base cybernetics on sound scientific foundations. Topics of interest include cybernetics modeling, computer simulation, biocybernetics, economic and social systems, ecosystems, adaptive systems, and philosophy of cybernetics. Acts as clearinghouse on robotics and promotes development in aspects such as: sensory perception; control devices; design of effectors. Organizes conventions and exhibitions. Maintains hall of fame. **PUBLICATIONS:** *Kybernetes* (in English), 9/year; Proceedings.

WORLD ORGANIZATION AGAINST TORTURE (Organisation Mondiale Contre la Torture/SOS-Torture)

37-39 Rue de Vermont, PO Box 119, CH-1211 Geneva 20, Switzerland **PHONE:** 41 22 7333140 **FX:** 41 22 7331051 **E-MAIL:** omct@omct.org **WEBSITES:** http://www.omct.org **FOUNDED:** 1985. **OFFICERS:** Eric Sottas, Dir. **MEMBERS:** 200. **STAFF:** 8. **BUDGET:** 1,200,000 SFr. **LANGUAGES:** English, French, Spanish.

DESCRIPTION: Members are nongovernmental organizations in 95 countries. Participates in the struggle against disappearances, summary executions, physical and mental torture, and psychiatric internment for political reasons. Offers rehabilitative services to any victim of torture in any country, regardless of that state's political affiliations. Maintains consultative status with the Economic and Social Council and the International Labor Organization and African Commission for Human and People's Rights. **PUBLICATIONS:** *Africa: A New Lease on Life* (in English); *Development and Human Rights, The Least Developed Countries* (in English and French); *El Terrorismo de Estado en Colombia* (in Spanish); *Exactions & Enfants* (in French); *Into the Lions Den. Gross Violations Against Human Rights Defenders 1992-1996*; *La Promotion et la Protection des droits de l'homme a l'heure des ajustements structurels* (in French); *Practical Guide to the International Procedure* (in English, French, and Spanish); *Report on Investigative Mission in Chile* (in English, French, and Spanish); *Soins Medicaux Dans Les Prisons du Maroc* (in French); *SOS-Torture* (in English), quarterly.

WORLD ORT UNION (Union Mondiale ORT)

ORT House, 126 Albert St., London NW1 7NE, England **PHONE:** 44 171 4468500 **FX:** 44 171 4468653 **E-MAIL:** help.desk@ort.org **WEBSITES:** http://www.ort.org **FOUNDED:** 1880. **OFFICERS:** Dr. Ellen Isler, Dir.Gen. **MEMBERS:** 120,300. **STAFF:** 600. **BUDGET:** US$300,000,000. **LANGUAGES:** English, French, German, Hebrew, Portuguese, Russian, Spanish.

DESCRIPTION: Members are volunteer committees, women's groups, and professional groups in 31 countries. Aims to develop industrial and agricultural computer skills among Jews and others in an effort to help individuals become economically self-sufficient. Promotes the highest standards of production and improvement of the economy in affiliated countries. Provides vocational and technical training in schools in 24 countries; collaborates with various governments in sponsoring programs of technical assistance in developing nations; offers apprenticeship opportunities and placement services. Organizes apprenticeship programs. (The acronym ORT stands for Organization for Rehabilitation Through Training). **PUBLICATIONS:** *Front Line News* (in English, French, Hebrew, and Russian), bimonthly; *ORTdata*, quarterly; *What in the World is ORT?* (in English, French, Hebrew, and Russian), annual.

WORLD PEACE FOUNDATION (WPF)

1 Eliot Sq., 3rd Fl., Cambridge, MA 02138-4952 **PHONE:** (617) 491-5085, (617) 491-8543 **FX:** (617) 491-8588 **E-MAIL:** mmcnally@hiid.harvard.edu **FOUNDED:** 1910. **OFFICERS:** Robert I. Rotberg, Pres. **STAFF:** 4. **LANGUAGES:** English.

DESCRIPTION: Seeks to advance the cause of world peace through study, analysis, and the advocacy of wise action. Interests include: international relations and foreign affairs, regional security, international economic issues, preventive diplomacy, and the creation of early warning systems. **PUBLICATIONS:** *The Caribbean Prepares for the Twenty-First Century*; *Collective Security in an Changing World*; *Combatting Cocaine in the Supplying Countries: Challenges and Strategies*; *Democratic Transitions in Central American and Panama*; *International Organization*, quarterly. Scholarly journal on international organization and world order studies.; *Third World Security in the Post-Cold War Era*; *The United States and Latin American Democracy: Lessons from History*. Also publishes a variety of reports and books available from their individual publishers.

WORLD PETROLEUM CONGRESSES - A FORUM FOR PETROLEUM SCIENCE, TECHNOLOGY, ECONOMICS, AND MANAGEMENT (WPC) (Congres Mondiaux du Petrole)

61 New Cavendish St., London W1M 8AR, England **PHONE:** 44 171 4677137, 44 171 4677100 **FX:** 44 171 5802230 **E-MAIL:** secretariat@world-petroleum.org **WEBSITES:** http://www.world-petroleum.org **FOUNDED:** 1933. **OFFICERS:** Paul Tempest, Dir.Gen. **MEMBERS:** 54. **STAFF:** 15. **BUDGET:** US$20,000,000. **LANGUAGES:** English.

DESCRIPTION: Members are national committees. Advances petroleum science and technology and the study of economic, financial, and management issues in the petroleum industry. **PUBLICATIONS:** *Newsletter* (in English), quarterly; *Proceedings of the World Petroleum Congress*, triennial; *WPC Di-*

rectory, periodic; Brochures, periodic; Papers, periodic.

WORLD PLOUGHING ORGANISATION (WPO) (Organisation Mondiale de Labourage)

26 Gable Ave., Cockermouth, Cumbria CA13 9BU, England **PHONE:** 44 1900 603096 **FOUNDED:** 1952. **OFFICERS:** Don Richardson, Treas. **MEMBERS:** 28. **LANGUAGES:** English.

DESCRIPTION: Members are nationally representative plowing competition organizations. Objectives are: to preserve, foster, and develop the art of plowing; to promote world plowing championship; to provide for demonstrations and trade displays; to encourage the development and adoption of improved techniques and aids to those in all branches of agriculture; to promote cooperation and enterprise in producing food for an increasing world population; to support fellowship and understanding among people worldwide. Conducts training programs; offers travel opportunities; sponsor periodic seminars. **PUBLICATIONS:** *World Ploughing Contest Handbook*, annual; *WPO Bulletin of News and Information*, 1-2/year.

WORLD POLICY INSTITUTE (WPI)

65 5th Ave., Ste. 413, New York, NY 10003 **PHONE:** (212) 229-5808 **FX:** (212) 229-5579 **E-MAIL:** robertsa@newschool.edu **WEBSITES:** http://www.worldpolicy.org **FOUNDED:** 1948. **OFFICERS:** Steven Schlesinger, Dir. **STAFF:** 7. **BUDGET:** US$600,000. **LANGUAGES:** English.

DESCRIPTION: Formulates and promotes practical policy recommendations on U.S. and world economic and security issues; develops positive initiatives that reflect the shared needs and interests of all nations. Conducts ongoing research program on U.S. international policy. Offers seminars, briefings, and lectures. **PUBLICATIONS:** *World Policy Journal*, quarterly. Examines the complex cultural, political, and historical issues that are coming to shape our lives.; Pamphlets.

WORLD PRESS FREEDOM COMMITTEE (WPFC)

11690 Sunrise Valley Dr., Reston, VA 20191-1409 **PHONE:** (703) 715-9811 **FX:** (703) 620-6790 **E-MAIL:** freepress@wpfc.org **FOUNDED:** 1976. **OFFICERS:** Marilyn J. Greene, Exec.Dir. **STAFF:** 4. **LANGUAGES:** English.

DESCRIPTION: Members are journalistic organizations united to support freedom of the press, especially in Third World and Eastern European countries. Speaks out against ''those who seek to deny truth in news and those who abuse newsmen.'' Encourages high professional standards and performance of the news media. Offers professional and technical print and broadcast assistance to Third World journalists; conducts seminars. Has sponsored over 150 projects in Africa, Asia, Latin America, Eastern Europe, and the Caribbean including organization of training programs and preparation of manuals for practitioners. **PUBLICATIONS:** Newsletter, 4-8/year. Also publishes papers, manuals, declarations, and reports.

WORLD PSYCHIATRIC ASSOCIATION (WPA)

c/o Mt. Sinai School of Medicine-CUNY, Univ. of NY, Fifth Ave. & 100th St., Box 1093, New York, NY 10029-6574 **PHONE:** (212) 241-6133 **FX:** (212) 426-0437 **E-MAIL:** wpa@dti.net **WEBSITES:** http://www.wpanet.org **FOUNDED:** 1961. **OFFICERS:** Prof. Juan E. Mezzich Jr., Sec.Gen. **MEMBERS:** 465. **STAFF:** 18. **LANGUAGES:** English, French, German, Spanish.

DESCRIPTION: Members are psychiatric societies and individuals in 90 countries. Objectives are: to promote international cooperation in the field of psychiatry; to advance inquiry into the etiology, pathology, and treatment of mental illness; to strengthen relations among psychiatrists working in various fields. Encourages the exchange of information concerning the medical problems of mental diseases; sponsors educational and research programs. Comprises 40 sections representing different specialties in psychiatry. **PUBLICATIONS:** *WPA Bulletin* (in English, French, German, and Spanish), quarterly; *WPA News*, quarterly.

WORLD RESOURCES INSTITUTE (WRI)

1709 New York Ave. NW, Ste. 700, Washington, DC 20006 **PHONE:** (202) 638-6300, (202) 662-2589 **FX:** (202) 638-0036 **TX:** 64414 WRI WASH **E-MAIL:** wri@igc.apc.org **FOUNDED:** 1982. **OFFICERS:** Deanna Madvin

Wolfire. **STAFF:** 120. **BUDGET:** US$10,000,000. **LANGUAGES:** English.

DESCRIPTION: Members are scientists and other academics with an interest in the environment and development. Promotes environmentally sustainable economic and community development globally; seeks to identify and introduce alternative and renewable energy sources. Assists government agencies, development organizations, and private sector organizations address the needs of human beings without causing environmental harm. Produces and distributes educational materials; sponsors research; identifies emerging energy sources and appropriate technologies and conducts training courses in their use; conducts natural resources management assessments. Maintains Center for International Development and Environment. **PUBLICATIONS:** Books; Papers; Reports.

WORLD ROAD ASSOCIATION (WRA) (Association Mondiale de la Route - AIPCR)

La Grande Arche, Paroi Nord, F-92055 Paris Ccdcx, Francc **PHONE:** 33 1 47968121 **FX:** 33 1 49000202 **E-MAIL:** piarc@wanadoo.fr **WEBSITES:** http://www.piarc.lcpc.fr **FOUNDED:** 1909. **OFFICERS:** Jean Francois Coste, Sec.Gen. **MEMBERS:** 2,900. **STAFF:** 10. **BUDGET:** 2,500,000 Fr. **LANGUAGES:** English, French.

DESCRIPTION: Members are national governments, regional authorities, public bodies, associations, commercial organizations, and individuals concerned with roads and road traffic. Fosters the construction, improvement, maintenance, use, and economic development of roads; encourages the growth of road systems throughout the world. Facilitates discussion of questions and research concerning roads and road traffic; collects and disseminates statistical and general information. Encourages: improvement and standardization of methods of administration, finance, design, construction, and maintenance of roads and engineering structures; international unification of highway traffic codes; research on roads, road traffic, and engineering structures. **PUBLICATIONS:** *PIARC Lexicon* (in English and French), periodic. Containing information on policy issues, technical issues, and recommendations.; *Report of Congress Proceedings*, quadrennial; *Reports of Committees*, quadrennial; *Routes/Roads* (in English and French), quarterly; *Terminology: Technical Dictionary of Road Terms*. Published in 11 languages; Book.

WORLD SOCIETY FOR THE PROTECTION OF ANIMALS (WSPA) (Societe Mondiale pour la Protection des Animaux)

2 Langley Lane, Vauxhall, London SW8 1TJ, England **PHONE:** 44 171 7930540 **FX:** 44 171 7930208 **E-MAIL:** wspa@wspa.org.uk **WEBSITES:** http://www.wspa.org.uk **FOUNDED:** 1981. **OFFICERS:** Andrew Dickson, Chief Exec. **MEMBERS:** 100,000. **STAFF:** 50. **BUDGET:** US$5,000,000. **LANGUAGES:** English, French, German, Spanish.

DESCRIPTION: Members are national animal welfare societies in 70 countries whose membership comprises over 100,000 individuals. Purposes are: to undertake and promote the conservation and protection of animals worldwide; to study international and national legislation relating to animal welfare; to promote efforts for the protection of animals and the conservation of their environment; to prevent cruelty to animals and relieve their suffering. Operates Emergency Rescue Service. **PUBLICATIONS:** *Animals International* (in English), semiannual. Contains information on animal welfare.

WORLD SQUASH FEDERATION LTD. (WSF)

6 Havelock Rd., Hastings, E. Sussex TN34 1BP, England **PHONE:** 44 1424 429245, 44 1424 429246 **FX:** 44 1424 429250 **E-MAIL:** squash@wsf.cablenet.co.uk **WEBSITES:** http://www.squash.org **FOUNDED:** 1967. **OFFICERS:** Edward J. Wallbutton. **MEMBERS:** 115. **STAFF:** 3. **BUDGET:** US$250,000.

DESCRIPTION: Members are national squash rackets associations united to promote, administer, and regulate the game of squash worldwide. Produces rules of the game and specifications for equipment. **PUBLICATIONS:** *Executive Committee Report* (in English), annual; *World Squash Address Book* (in English), annual.

WORLD STUDENT CHRISTIAN FEDERATION (WSCF) (Federation Universelle des Associations Chretiennes d'Etudiants)

5, rte. des Morillons, CH-1218 Geneva, Switzerland **PHONE:** 41 22 7988953 **FX:** 41 22 7982370 **FOUNDED:** 1895. **OFFICERS:** Clarissa Balan-Sycip, Co-Sec.Gen. **MEMBERS:** 300,000. **STAFF:** 6. **BUDGET:** US$1,000,000. **LANGUAGES:** English, French, Spanish.

DESCRIPTION: Members are national student Christian movements in 90 countries united for social justice, ecumenism, and educational purposes. Fosters leadership exchange and development through conferences, courses, and publications. Examines ecumenical and political questions. **PUBLICATIONS:** *Federation News*, quarterly; *WSCF Directory*, quadrennial; *WSCF Journal*, quarterly.

WORLD SUSTAINABLE AGRICULTURE ASSOCIATION (WSAA)

2025 I St. NW, Ste. 512, Washington, DC 20006 **PHONE:** (202) 293-2155 **FX:** (202) 293-2209 **E-MAIL:** wsaa@igc.apc.org **WEBSITES:** http://ourworld.compuserve.com/homepages/wsaa **FOUNDED:** 1991. **OFFICERS:** Linda Elswick, Dir. of Policy Directorate. **BUDGET:** US$408,000. **LANGUAGES:** English.

DESCRIPTION: Promotes adoption of sustainable agriculture systems and policies worldwide.

WORLD TRADE CENTERS ASSOCIATION (WTCA)

1 World Trade Center, Ste. 7701, New York, NY 10048 **PHONE:** (212) 435-7168 **FX:** (212) 435-2810 **WEBSITES:** http://www.wtca.org **FOUNDED:** 1970. **OFFICERS:** Guy F. Tozzoli, Pres. **MEMBERS:** 303. **STAFF:** 24. **BUDGET:** US$4,300,000. **LANGUAGES:** English.

DESCRIPTION: Regular members are organizations substantially involved in the development or operation of a World Trade Center. Affiliate members are chambers of commerce and organizations sponsoring Executive Business clubs, libraries, exhibit facilities, and other trade center-related activities. Seeks to encourage expansion of world trade, promote international business relationships, and increase Third World participation in world trade. Maintains the WTC Network, providing low-cost message service, identification of qualified business opportunities worldwide, and promotion of subscribers' businesses in a select international community. Developing global video-conferencing network and an international electronic financial instrument. **PUBLICATIONS:** *World Trade Centers Association—Proceedings of the General Assembly*, annual; *World Traders Magazine*, quarterly; *WTCA/GALE World Business Directory*, annual; *WTCA Membership Directory*, semiannual; *WTCA News*, monthly. Covers membership activities online; *WTCA Services Directory*, annual; Brochures.

WORLD TRADE ORGANIZATION (Organizacion Mundial del Comercio)

Centre William Rappard, 154, rue de Lausanne, CH-1211 Geneva 21, Switzerland **PHONE:** 41 22 7395111, 41 22 7395007 **FX:** 41 22 7595458 **TX:** 412324 OMCWTOCH **E-MAIL:** enquiries@wto.org **WEBSITES:** http://www.wto.org **FOUNDED:** 1995. **OFFICERS:** Renato Ruggiero, Dir.Gen. **MEMBERS:** 132. **STAFF:** 450. **BUDGET:** 115,000,000 SFr. **LANGUAGES:** English, French, Spanish.

DESCRIPTION: Legal and institutional foundation of the multilateral trading system. Provides the principal contractual obligations detemining how governments frame and implement domestic trade legislation and regulations. Provides a forum for collective debate, negotiation and adjudication regarding trade relations. **PUBLICATIONS:** *Basic Instruments and Selected Documents*, annual; *International Market for Meat*, annual; *Studies in International Trade*, periodic; *Trade and Environment Bulletin*, monthly; *Trade Policy Reviews*, periodic; *World Market for Dairy Products*, annual; *WTO Annual Report (2 Volumes)*; *WTO Focus*, monthly.

WORLD UNDERWATER FEDERATION (Confederation Mondiale des Activites Subaquatiques - CMAS)

Viale Iziano 74, I-00196 Rome, Italy **PHONE:** 39 6 36858480 **FX:** 39 6 36858490 **TX:** 621610 CONIES I **WEBSITES:** http://www.cmas.org **FOUNDED:** 1959. **OFFICERS:** Pierre Dernier, Sec.Gen. **MEMBERS:** 102. **STAFF:** 7. **BUDGET:** US$580,000. **LANGUAGES:** English, French, Spanish.

DESCRIPTION: Members are federations and national associations in 90 countries. Objectives are to: develop and promote the sport of diving and other underwater sports and activities; train individuals to teach the techniques of underwater activities; promote quality standards of diver training worldwide; improve current underwater equipment and make suggestions on additional equipment needs. Conducts technical, scientific, sports, medical, and environmental protection activities. Conducts educational and research programs. **PUBLICATIONS:** *CMAS One Star Diver Training Programme*; *CMAS Standards and Requirements*; *CMAS 2 Star Diver Training Programme*; *Diving Standards*; *Finswimming Manual*; *Scientific Diving: A General Code of Practice* (in Chinese, English, French, and Russian), semiannual.

WORLD UNION (WU) (Union Mondiale)

Sri Aurobindo Ashram, Pondicherry 605 002, Tamil Nadu, India **PHONE:** 91 413 34834 **FOUNDED:** 1958. **OFFICERS:** Samar Basu, Gen.Sec.-Treas. **MEMBERS:** 1,500.

DESCRIPTION: Members are persons in 30 countries interested in the ideal of human unity, peace, and progress based on a spiritual foundation. Seeks to ''foster awakening to the fact that all life is one, that unity is already present, that we must wake up to it and live in our highest consciousness.'' Is presently participating in the ratification campaign of the Constitution for the Federation of Earth adopted by the World Constituent Assembly in June 1977 at Innsbruck, Austria. World Union Centers have been established in India and other countries; each is autonomous and sponsors its own study groups and programs. Has established Forum for World Peace at Sardar Patel University in India and at other Indian universities; coordinates Forum for Human Unity. Holds seminars. Conducts research programs. **PUBLICATIONS:** *World Union* (in English), quarterly. Comes with accompanying newsletter.; Bulletins; Brochures; Books; Booklets.

WORLD UNION OF CATHOLIC WOMEN'S ORGANIZATIONS (WUCWO) (Union Mondiale des Organisations Feminines Catholiques - UMOFC)

18, Notre Dame des Champs, F-75006 Paris, France **PHONE:** 33 1 45442765, 44 1303 261003 **FX:** 33 1 42840480 **E-MAIL:** wucwoparis@wanadoo.fr **WEBSITES:** http://home.wxs.nl/-wucwo **FOUNDED:** 1910. **OFFICERS:** Mrs. Gillian Badcock, Sec.Gen. **MEMBERS:** 25,000,000. **STAFF:** 3. **LANGUAGES:** English, French, German, Spanish.

DESCRIPTION: Coordinates activities of organizations of Catholic women in 57 countries and represents them on an international level. **PUBLICATIONS:** *WUCWO Newsletter* (in English, French, German, and Spanish), 3/year.

WORLD UNION OF JEWISH STUDENTS (WUJS)

PO Box 7914, IL-91077 Jerusalem, Israel **PHONE:** 972 2 610133 **FX:** 972 2 610741 **E-MAIL:** wujs@jerl.co.il **FOUNDED:** 1924. **OFFICERS:** Claude Kaliyoti, Chmn. **MEMBERS:** 200,000. **STAFF:** 12. **BUDGET:** US$600,000. **LANGUAGES:** English, Hebrew, Spanish.

DESCRIPTION: Association of Jewish student organizations and unions in 40 countries. Purposes are to unite national associations of Jewish students in all countries; to further and protect Jewish student interests and to ensure adequate representation of Jewish student opinion at meetings of internatinal and other organizations; to cooperate with interested organizations in promoting interests of students in general and to strengthen ties of Jewish students with Israel; to combat anti-Semitism and racism; to assist Jewish communities. Sponsors International Graduate Center for Hebrew and Jewish Studies in Arad, Israel. Conducts charitable, educational, research, and leadership programs. Maintains regional offices in Jerusalem, Israel; New York City, United States; Brussels, Belgium; Mexico City, Mexico; Sydney, NSW, Australia; and Johannesburg, South Africa. **PUBLICATIONS:** *History and Heritage*; *Jewish Student Activist Handbook* (in English), semiannual; *Shalom Elanu.* Peace activities; *WUJS Leads* (in English), quarterly; *WUJS Report*, quarterly; Monographs, periodic.

WORLD UNION OF MAPAM (WUM) (Union Mondiale du Mapam - UMM)

13 Leonardo Da Vinci St., IL-61400 Tel Aviv, Israel **PHONE:** 972 3 6925352, 972 3 6925331 **FX:** 972 3 6925335 **FOUNDED:** 1948. **OFFICERS:** Hanan Cohen, Sec.Gen. **MEMBERS:** 50,000. **STAFF:** 10. **LANGUAGES:** English, French, Hebrew, Spanish.

DESCRIPTION: Members are individuals from 18 countries supporting humanistic policies within a socialist/Zionist political framework. Sponsors Hashomer Hatzair youth movement. Maintains regional committees. **PUBLICATIONS:** *Encuentro*, monthly; *Viewpoint*, monthly.

WORLD UNION FOR PROGRESSIVE JUDAISM (WUPJ)

633 3rd Ave., New York, NY 10017-6778 **PHONE:** (212) 650-4090 **FX:** (212) 650-4289 **E-MAIL:** 5448032@mcimail.com **WEBSITES:** http://www.rj.org/wupj/index.html **FOUNDED:** 1926. **OFFICERS:** Rabbi Imiel Hirsh, North American Dir. **MEMBERS:** 1,500,000. **STAFF:** 6. **LANGUAGES:** English.

DESCRIPTION: Members are congregations of Reform, Liberal, and Progressive Jews in 29 countries. Promotes the cause of Progressive Judaism; establishes new congregations and national constituencies; arranges for the training of rabbis and teachers; coordinates activity between constituencies. Represents progressive Jewry at the United Nations and United Nations Educational, Scientific and Cultural Organization. **PUBLICATIONS:** *International Conference Reports*, biennial.

WORLD UNIVERSITY SERVICE

14 Dufferin St., London EC1Y 8PD, England **PHONE:** 44 171 4265800, 44 171 4265820 **FX:** 44 171 2511314 **E-MAIL:** overseas@wusuu.org **FOUNDED:** 1920. **OFFICERS:** Caroline Nursey, Dir. **MEMBERS:** 400. **STAFF:** 23. **BUDGET:** US$2,500,000. **LANGUAGES:** English.

DESCRIPTION: Works with refugees in the UK to provide advice on educational issues. Cooperates with partner organizations in Africa, Latin America, and the Middle East to educate refugees and other marginalized groups.

WORLD VETERANS FEDERATION (WVF) (Federation Mondiale des Anciens Combattants - FMAC)

17, rue Nicolo, F-75116 Paris, France **PHONE:** 33 1 40726100 **FX:** 33 1 40728058 **TX:** FMACWVF 643253 F **E-MAIL:** fmacwvf@cybercable.fr **FOUNDED:** 1950. **OFFICERS:** Marek Hagmajer, Sec.Gen. **MEMBERS:** 145. **STAFF:** 6. **LANGUAGES:** English, French.

DESCRIPTION: Members are 145 national associations of war veteran, victim of war, prisoner-of-war, deportee, former resistant and former personnel of peace-keeping operations organizations in 77 countries representing more than 27 million individuals. Seeks to maintain international peace and security through the application of the Charter of the United Nations. and international instruments for human rights. Defends the spiritual and material interests of war veterans and victims. Encourages cooperartion and direct relations among national organizations of war veterans and victims. Supports measures for the peaceful settlement of international conflicts and sponsors research on rehabilitation of the disabled, legislation concerning war veterans, protection of human rights, disarmament, peace-keeping, and the development of international humanitarian law. Addresses psychomedical problems relating to war and other severe stress experiences. Elaborates programs of rehabilitation and social reinsertion of war veterans and victims of war. Consultative status general category with the U.N. **PUBLICATIONS:** *WVF in Brief* (in English and French), quarterly.

WORLD VETERINARY ASSOCIATION (WVA) (Welt-Tierarztegesellschaft)

Rosenlunds Alle 8, DK-2720 Vanlose, Denmark **PHONE:** 45 38710156 **FX:** 45 38710322 **E-MAIL:** wva@ddd.dk **WEBSITES:** http://www.worldvet.org **FOUNDED:** 1959. **OFFICERS:** Dr. Lars Holsaae. **MEMBERS:** 96. **STAFF:** 2. **BUDGET:** US$130,000. **LANGUAGES:** English, French, German, Spanish.

DESCRIPTION: Association of national (70), international (20), and commercial (4) organizations. Purposes are to: unify and promote the veterinary profession; encourages high standards of health and welfare in all species of animals; improve veterinary science through the exchange of information and education; establish uniform nomenclature. Sends volunteers to developing countries to assist in food production, processing and distribution, to teach basic techniques in disease control and eradication and to develop national veterinary services. **PUBLICATIONS:** *Report on Resolutions*, periodic; *World Veterinary Directory* (in English), periodic; Bulletin (in English), semiannual; Membership Directory, annual.

WORLD WATCH INSTITUTE (WWI)

1776 Massachusetts Ave. NW, Washington, DC 20036-1904 **FX:** (202) 296-7365 **FOUNDED:** 1974. **OFFICERS:** Hilary French **STAFF:** 31. **BUDGET:** US$3,100,000. **LANGUAGES:** English.

DESCRIPTION: Members are environmental and development researchers. Seeks to raise public understanding of global environmental issues. Sponsors lobbying and advocacy campaigns in support of environmental protection. Conducts research and educational programs.

WORLD WIDE FUND FOR NATURE - WWF INTERNATIONAL (WWF) (Fonds Mondial pour la Nature)

Avenue du Mont-Blanc, CH-1196 Gland, Switzerland **PHONE:** 41 22 3649111 **FX:** 41 22 3645358 **WEBSITES:** http://www.panda.org **FOUNDED:** 1961. **OFFICERS:** Dr. Claude Martin, Dir.Gen. **MEMBERS:** 4,700,000. **STAFF:** 130. **BUDGET:** US$37,100,000. **LANGUAGES:** English, French.

DESCRIPTION: Worlds largest working organization in the conservation of nature and natural resources. Purpose is to promote the conservation of the natural environment and ecological processes essential to life on earth. Seeks to generate the strongest possible moral and financial support for the safeguarding of the living environment; encourages action based on scientific priorities. Seeks to heighten public awareness of threats to the environment. Works in conjunction with the International Union for Conservation of Nature and Natural Resources.

WORLD YOUNG WOMEN'S CHRISTIAN ASSOCIATION (WORLD YWCA) (Alliance Mondiale des Unions Chretiennes Feminines)

16 Ancienne Route, Grand-Saconnex, CH-1218 Geneva, Switzerland **PHONE:** 41 22 9296040 **FX:** 41 22 9296044 **E-MAIL:** worldoffice@worldywca.ch **WEBSITES:** http://www.worldywca.org **FOUNDED:** 1894. **OFFICERS:** Dr. Mumsimbi Kanyoro, Sec.Gen. **LANGUAGES:** English, French, Spanish.

DESCRIPTION: Members are associations in 92 countries. Seeks to build a worldwide fellowship for young women to learn of and express the love of God. Works for the elimination of racism and racial discrimination, for economic and social justice, and for the building of a world community. Sponsors consultations on world issues and study programs for members. Attempts to make the public aware of world problems. Conducts human rights and rehabilitation projects and services for refugees and immigrants. Current emphasis is on studying the problems of women in a changing world, peace, the environment, housing, health, and political and family life education. Conducts seminars. **PUBLICATIONS:** *Annual Report*; *Common Concern*, quarterly; *Directory*, annual.

WORLDVIEW INTERNATIONAL FOUNDATION

8 Kinross Ave., Colombo 4, Sri Lanka **PHONE:** 94 1 589648, 94 1 592057 **FX:** 94 1 589225 **TX:** 21494 GLOBAL CE **FOUNDED:** 1979. **OFFICERS:** Fjortoft Arne, Sec.Gen. **MEMBERS:** 300. **BUDGET:** US$3,500,000. **LANGUAGES:** Arabic, Bangla, English, French, Nepalese Dialects, Norwegian, Sinhalese, Spanish, Thai.

DESCRIPTION: Members are individuals involved in media work in 50 countries. Works to improve communication strategies assisting the development efforts of underprivileged people in Third World countries. Distributes information on issues such as safe drinking water, nutrition, hygiene, primary health care, population activities, food production, and other basic needs. Conducts training and consultancy programs to help students and media workers in developing countries make better use of television, film, radio, audiocassette, and print media for the purpose of development communication. Organizes seminars, conferences, and workshops on issues related to development communication and alternative information strategies. Utilizes the mass media to discuss and promote rural communication. Sponsors competitions. Produces video programs to help development workers present visual messages to rural audiences; also produces documentaries, journalistic feature reports, and dramatized stories. **PUBLICATIONS:** *WIF Newsletter*, quarterly.

YACHT BROKERS, DESIGNERS AND SURVEYORS ASSOCIATION (YBDSA)

Wheel House, Petersfield Rd., Whitehill, Bordon, Hants. GU35 9BU, England **PHONE:** 44 1420

473862 **FX:** 44 1420 488328 **FOUNDED:** 1987. **OFFICERS:** Ms. Rae Boxall, CEO. **MEMBERS:** 250. **STAFF:** 4. **LANGUAGES:** English.

DESCRIPTION: Members are yacht and small craft brokers, designers, and surveyors in 10 countries. Objectives are to: conduct and maintain standards of professional competence; agree on common forms of contract; arbitrate disputes; provide for discussion and exchange of information. Conducts teaching programs and workshops. **PUBLICATIONS:** *Membership Directory* (in English), annual; Brochure, periodic; Papers, periodic.

YAD VASHEM, THE HOLOCAUST MARTYRS' AND HEROES' REMEMBRANCE AUTHORITY (YV)

Har Hazikaron, PO Box 3477, IL-91034 Jerusalem, Israel **PHONE:** 972 26 751611 **FX:** 972 26 433511 **E-MAIL:** info@yad-vashem.org.il **WEBSITES:** http://www.yadvashem.org.il **FOUNDED:** 1953. **OFFICERS:** Mr. Avner Shalev, Chm. **STAFF:** 150. **LANGUAGES:** English, French, German, Greek, Hebrew, Hungarian, Polish, Portuguese, Russian, Spanish, Yiddish.

DESCRIPTION: Conducts activities in 12 countries. Seeks to: preserve the memory of the 6,000,000 victims of the Holocaust; study the historical roots of anti-Semitism; transmit the lessons to future generations. Collects and publishes testimony of events surrounding the Holocaust; establishes memorial projects. Maintains historical museum of photographs, documents, and other objects concerning Jews in the ghettos and as victims of the Nazis; operates museum of art. Maintains Children's Memorial commemorating the 1,500,000 Jewish children who died in the Holocaust; has established Avenue of the Righteous and Garden of the Righteous and awards title Righteous Among the Nations to honor gentiles who risked their lives to hide, and otherwise save Jews. Operates Hall of Remembrance and Hall of Names to collect and list names of victims. Has established Valley of the Communities in commemoration of 5000 Jewish communities destroyed or damaged in the Holocaust. Maintains Memorial Cave of stones engraved with names of victims. Has established memorial to the deportees; an original cattle car, in memory of Jews deported to the extermination camps. Maintains International School for Holocaust Studies and International Institute for Holocaust Research, to train teachers and prepare curricula for use in schools; conducts lectures. Operates charitable program and speakers' bureau; conducts seminars. Compiles statistics. Maintains and expends archives of primary source material, currently holding over 50 million pages. **PUBLICATIONS:** *Proceedings of International Conferences* (in English and Hebrew), periodic; *Yad Vashem Magazine* (in English and Hebrew), quarterly; *Yad Vashem Studies* (in English and Hebrew), annual; Books (in English and Hebrew). Includes diaries and historical texts.

YOUTH FOR CHRIST INTERNATIONAL (YFCI) (Jeunesse pour Christ - JPC)

Raffles City, PO Box 214, Singapore 911708, Singapore **PHONE:** 65 3387944 **FX:** 65 3368776 **E-MAIL:** 76613.1333@compuserve.com **WEBSITES:** http://www.gospelcom.net/yfc/yfci **FOUNDED:** 1944. **OFFICERS:** Rev. Jean-Jacques Weiler, Int. Pres./CEO. **STAFF:** 2,600. **BUDGET:** US$50,000,000. **LANGUAGES:** English.

DESCRIPTION: Autonomous national Youth for Christ programs in 76 nations. Purpose is to: coordinate the world-wide work of Youth For Christ; establish a youth-to-youth philosophy of evangelism on a scriptural basis best suiting the cultures involved; guide the development of the ministry in new countries; promote and protect the public image world-wide; encourage youth to live a balanced Christian life and maintain a responsibility to church and community and world-wide involvement in evangelism and social concerns. Sponsors Youth Leadership Training Schools. Compiles statistics; maintains speakers' bureau. **PUBLICATIONS:** *World Perspective*, quarterly; *YFCI World Directory*, periodic.

YOUTH FOR DEVELOPMENT AND COOPERATION (YDC)

Rijswijkstr 141 G, NL-1062 ES Amsterdam, Netherlands **PHONE:** 31 20 6142510 **FX:** 31 20 6175545 **E-MAIL:** ydc@geo2.poptel.org.uk **FOUNDED:** 1947. **OFFICERS:** Bas Auer, Sec.Gen. **MEMBERS:** 53. **STAFF:** 2. **BUDGET:** US$200,000. **LANGUAGES:** English.

DESCRIPTION: Members are national organizations of young people concerned with international order, social justice, and issues of international development including disarmament and the responsible use and fair distribution of the world's resources. Focuses on issues involving the Third World such as: disarmament; poverty; Third World debt; technology transfer; the position of women in society; political, social, cultural, and environmental development. Conducts research and seminars; holds conferences. Organizes campaigns focusing on underdevelopment. Maintains a network of student organizations and others involved in development issues. Uses simulation games. **PUBLICATIONS:** *FLASH* (in English), monthly; *IMPACT* (in English), biennial.

YUGNTRUF - YOUTH FOR YIDDISH (YYY)

200 W. 72nd St., Rm. 40, New York, NY 10023
PHONE: (212) 787-6675 **FX:** (718) 231-7905
E-MAIL: ruvn@aol.com **FOUNDED:** 1964.
OFFICERS: Brukhe Caplan, New York Coordinator. **MEMBERS:** 1,000. **STAFF:** 3.
LANGUAGES: Yiddish.

DESCRIPTION: Members are college and high school students and young adults with an interest in Yiddish. (Yugntruf means ''call to youth.'') Objectives are: to perpetuate Yiddish cultural heritage; to develop contemporary Yiddish literature and culture; to strengthen Yiddish as a spoken language in the members' respective countries, especially among young people; to oppose cultural and linguistic assimilation among Jews. Sponsors picnics, concerts, lectures, and discussions. Makes available ''Vaserl'' record of Yiddish songs, an anthology, Vidervuks, of young Yiddish writers, Dos Kleyne Vokerl, children's T-shirts and buttons in Hebrew lettering. Sponsors Pripetshik conducted Sunday school entirely in Yiddish, for preschoolers and elementary school children as well as a Yiddish retreat, ''Yiddish Vokh.'' **PUBLICATIONS:** *Vidervuks: An Anthology of 20 Young Yiddish Writers* (in Yiddish); *Yiddish Source Finder*. Lists Yiddish schools, publishing houses, publications, records, radio programs, clubs, and reading circles; *Yugntruf*, periodic. Contains Yiddish stories, poems, and articles. Includes Yugntruf news; Brochures.

ZINC DEVELOPMENT ASSOCIATION (ZDA) (Association pour le Developpement du Marche du Zinc)

42 Weymouth St., London W1N 3LQ, England
PHONE: 44 171 4996636 **FX:** 44 171 4931555
E-MAIL: enq@zda.org **FOUNDED:** 1938.
OFFICERS: D.N. Wilson, Dir. **MEMBERS:** 8.
STAFF: 3. **LANGUAGES:** English.

DESCRIPTION: Members are mine and metal producers, zinc users, and others interested in the zinc industry. Objectives are: to promote the use of zinc in all forms; to represent the interests of the zinc industry before national and international groups concerned with standardization, economic questions, and technical developments within the industry. Provides a zinc use advisory service on topics including suitable applications, market developments, and aspects of manufacturing technology which improve efficiency and profitability. **PUBLICATIONS:** Brochures; Manuals; Newsletter, periodic.

ZONTA INTERNATIONAL (ZI)

557 W. Randolph St., Chicago, IL 60661-2206
PHONE: (312) 930-5848 **FX:** (312) 930-0951
WEBSITES: http://www.zonta.org **FOUNDED:** 1919. **OFFICERS:** Janet Halstead, Exec.Dir.
MEMBERS: 35,000. **STAFF:** 18. **BUDGET:** US$2,500,000. **LANGUAGES:** English.

DESCRIPTION: Members are business and professional executives worldwide ''working together to advance the status of women.'' Coordinates service projects that provide health, and income-generation, including cooperative projects with U.N. agencies. Conducts program to assist young women in the fields of public affairs and policy-making. Has consultative status with Council of Europe, International Labor Organization, United Nations Educational, Scientific and Cultural Organization, and United Nations Economic and Social Council, UNICEF, and UNIFEM. Local club members conduct community projects to assist women. **PUBLICATIONS:** *The Zontian* (in English), quarterly; Directory, annual.

INDEX

This index includes personal names, international organizations, geographic locators, as well as general subject indexing. Country entries including statistics are in bold. "See" and "See also" references are italicized.

B

C

F

G

H

I

J

K

M

S

V

W